Encyclopedia of
Image Processing

Encyclopedias from the Taylor & Francis Group

Print	Online

Agriculture

Encyclopedia of Agricultural, Food, and Biological Engineering, 2nd Ed., 2 Vols.   Pub'd. 10/21/10
K10554 (978-1-4398-1111-5)   K11382 (978-1-4398-2806-9)

Encyclopedia of Animal Science, 2nd Ed., 2 Vols.   Pub'd. 2/1/11
K10463 (978-1-4398-0932-7)   K10528 (978-0-415-80286-4)

Encyclopedia of Biotechnology in Agriculture and Food   Pub'd. 7/16/10
DK271X (978-0-8493-5027-6)   DKE5044 (978-0-8493-5044-3)

Business and Computer Science

Encyclopedia of Computer Science & Technology, 2nd Ed., 2 Vols.   Pub'd 12/21/2016
K21573 (978-1-4822-0819-1)   K21578 (978-1-4822-0822-1)

Encyclopedia of Image Processing   Publishing 2018
K23510 (978-1-4822-4490-8)   K23510 (978-1-3510-3274-2)

Encyclopedia of Information Assurance, 4 Vols.   Pub'd. 12/21/10
AU6620 (978-1-4200-6620-3)   AUE6620 (978-1-4200-6622-7)

Encyclopedia of Information Systems and Technology, 2 Vols.   Pub'd. 12/29/15
K15911 (978-1-4665-6077-2)   K21745 (978-1-4822-1432-1)

Encyclopedia of Library and Information Sciences, 4th Ed.   Pub'd. 11/13/17
K15223 (978-1-4665-5259-3)   K15224 (978-1-4665-5260-9)

Encyclopedia of Software Engineering, 2 Vols.   Pub'd. 11/24/10
AU5977 (978-1-4200-5977-9)   AUE5977 (978-1-4200-5978-6)

Encyclopedia of Supply Chain Management, 2 Vols.   Pub'd. 12/21/11
K12842 (978-1-4398-6148-6)   K12843 (978-1-4398-6152-3)

Encyclopedia of U.S. Intelligence, 2 Vols.   Pub'd. 12/19/14
AU8957 (978-1-4200-8957-8)   AUE8957 (978-1-4200-8958-5)

Encyclopedia of Wireless and Mobile Communications, 2nd Ed., 3 Vols.   Pub'd. 12/18/12
K14731 (978-1-4665-0956-6)   KE16352 (978-1-4665-0969-6)

Chemistry, Materials and Chemical Engineering

Encyclopedia of Aluminum and Its Alloys   Publishing 2018
(978-1-4665-1080-7)   K14799 (978-1-3510-3274-2)

Encyclopedia of Chemical Processing, 5 Vols.   Pub'd. 11/1/05
DK2243 (978-0-8247-5563-8)   DKE499X (978-0-8247-5499-0)

Encyclopedia of Chromatography, 3rd Ed.   Pub'd. 10/12/09
84593 (978-1-4200-8459-7)   84836 (978-1-4200-8483-2)

Encyclopedia of Iron, Steel, and Their Alloys, 5 Vols.   Pub'd. 1/6/16
K14814 (978-1-4665-1104-0)   K14815 (978-1-4665-1105-7)

Encyclopedia of Plasma Technology, 2 Vols.   Pub'd 12/12/2016
K14378 (978-1-4665-0059-4)   K21744 (978-1-4822-1431-4)

Encyclopedia of Polymer Applications   Publishing 2018
(978-1-4987-2993-2)   K26102 (978-1-3510-1942-9)

Encyclopedia of Supramolecular Chemistry, 2 Vols.   Pub'd. 5/5/04
DK056X (978-0-8247-5056-5)   DKE7259 (978-0-8247-4725-1)

Encyclopedia of Surface & Colloid Science, 3rd Ed., 10 Vols.   Pub'd. 8/27/15
K20465 (978-1-4665-9045-8)   K20478 (978-1-4665-9061-8)

Engineering

Dekker Encyclopedia of Nanoscience and Nanotechnology, 3rd Ed., 7 Vols.   Pub'd. 3/20/14
K14119 (978-1-4398-9134-6)   K14120 (978-1-4398-9135-3)

Encyclopedia of Energy Engineering and Technology, 2nd Ed., 4 Vols.   Pub'd. 12/1/14
K14633 (978-1-4665-0673-2)   KE16142 (978-1-4665-0674-9)

Encyclopedia of Optical and Photonic Engineering, 2nd Ed., 5 Vols.   Pub'd. 9/22/15
K12323 (978-1-4398-5097-8)   K12325 (978-1-4398-5099-2)

Environment

Encyclopedia of Environmental Management, 4 Vols.   Pub'd. 12/13/12
K11434 (978-1-4398-2927-1)   K11440 (978-1-4398-2933-2)

Encyclopedia of Environmental Science and Engineering, 6th Ed., 2 Vols.   Pub'd. 6/25/12
K10243 (978-1-4398-0442-1)   KE0278 (978-1-4398-0517-6)

Encyclopedia of Natural Resources, 2 Vols.   Pub'd. 7/23/14
K12418 (978-1-4398-5258-3)   K12420 (978-1-4398-5260-6)

Medicine

Encyclopedia of Biomaterials and Biomedical Engineering, 2nd Ed.   Pub'd. 5/28/08
H7802 (978-1-4200-7802-2)   HE7803 (978-1-4200-7803-9)

Encyclopedia of Biomedical Polymers and Polymeric Biomaterials, 11 Vols.   Pub'd. 4/2/15
K14324 (978-1-4398-9879-6)   K14404 (978-1-4665-0179-9)

Concise Encyclopedia of Biomedical Polymers and Polymeric Biomaterials, 2 Vols.   Pub'd. 8/14/17
K14313 (978-1-4398-9855-0)   KE42253 (978-1-315-11644-0)

Encyclopedia of Biopharmaceutical Statistics, 4th Ed.   Publishing 2018
H100102 (978-1-4398-2245-6)   HE10326 (978-1-4398-2246-3)

Encyclopedia of Clinical Pharmacy   Pub'd. 11/14/02
DK7524 (978-0-8247-0752-1)   DKE6080 (978-0-8247-0608-1)

Encyclopedia of Dietary Supplements, 2nd Ed.   Pub'd. 6/25/10
H100094 (978-1-4398-1928-9)   HE10315 (978-1-4398-1929-6)

Encyclopedia of Medical Genomics and Proteomics, 2 Vols.   Pub'd. 12/29/04
DK2208 (978-0-8247-5564-5)   DK501X (978-0-8247-5501-0)

Encyclopedia of Pharmaceutical Science and Technology, 4th Ed., 6 Vols.   Pub'd. 7/1/13
H100233 (978-1-84184-819-8)   HE10420 (978-1-84184-820-4)

Routledge Encyclopedias

Encyclopedia of Public Administration and Public Policy, 3rd Ed., 5 Vols.   Pub'd. 11/6/15
K16418 (978-1-4665-6909-6)   K16434 (978-1-4665-6936-2)

Routledge Encyclopedia of Modernism   Pub'd 5/11/16
    Y137844 (978-1-135-00035-6)

Routledge Encyclopedia of Philosophy Online   Pub'd. 11/1/00
    RU22334 (978-0-415-24909-6)

Routledge Performance Archive   Pub'd. 11/12/12
    Y148405 (978-0-203-77466-3)

Encyclopedia titles are available in print and online.
To order, visit https://www.crcpress.com
Telephone: 1-800-272-7737
Email: orders@taylorandfrancis.com

Encyclopedia of Image Processing

Edited by
Dr. Phillip A. Laplante

CRC Press
Taylor & Francis Group
Boca Raton London New York

CRC Press is an imprint of the
Taylor & Francis Group, an **informa** business

CRC Press
Taylor & Francis Group
6000 Broken Sound Parkway NW, Suite 300
Boca Raton, FL 33487-2742

© 2019 by Taylor & Francis Group, LLC
CRC Press is an imprint of Taylor & Francis Group, an Informa business

No claim to original U.S. Government works

Printed in Canada on acid-free paper

International Standard Book Number-13: 978-1-4822-4490-8

This book contains information obtained from authentic and highly regarded sources. Reasonable efforts have been made to publish reliable data and information, but the author and publisher cannot assume responsibility for the validity of all materials or the consequences of their use. The authors and publishers have attempted to trace the copyright holders of all material reproduced in this publication and apologize to copyright holders if permission to publish in this form has not been obtained. If any copyright material has not been acknowledged please write and let us know so we may rectify in any future reprint.

Except as permitted under U.S. Copyright Law, no part of this book may be reprinted, reproduced, transmitted, or utilized in any form by any electronic, mechanical, or other means, now known or hereafter invented, including photocopying, microfilming, and recording, or in any information storage or retrieval system, without written permission from the publishers.

For permission to photocopy or use material electronically from this work, please access www.copyright.com (http://www.copyright.com/) or contact the Copyright Clearance Center, Inc. (CCC), 222 Rosewood Drive, Danvers, MA 01923, 978-750-8400. CCC is a not-for-profit organization that provides licenses and registration for a variety of users. For organizations that have been granted a photocopy license by the CCC, a separate system of payment has been arranged.

Trademark Notice: Product or corporate names may be trademarks or registered trademarks, and are used only for identification and explanation without intent to infringe.

Visit the Taylor & Francis Web site at
http://www.taylorandfrancis.com

and the CRC Press Web site at
http://www.crcpress.com

Dedication

To the beloved dogs that I have had in my life: Ginger, Francis, Teddy, Maggie, and Henry, and those yet to come.

Brief Content

3D Medical Imaging	1
Artificial Neural Networks	11
Biomedical Image Registration	24
Boolean Algebras	40
Cantor, Fuzzy, Near, and Rough Sets	46
Color	54
Color Image Denoising	72
Color Image Segmentation Metrics	95
Depth Statistics	114
Digital Halftoning: Computational Intelligence-Based Approach	123
Dimensionality Reduction	135
Discrete Fourier Transforms	148
Facial Animation	154
First-Order Statistics	165
Floodplain Mapping	177
Forgery Detection in Digital Images	183
Fourier Analysis	194
Fuzzy Set Theory	216
Gradients, Edges, and Contrast	234
Grayscale Morphological Image Enhancement	254
Image and Scene Analysis	270
Image Enhancement	281
Image Inpainting Evolution: A Survey	293
Image Interpolation	308
Image Processing and Measurement	316
Image Retrieval	335
Image Secret Sharing	354
Image Super Resolution	365
Information Sets for Edge Detection	378
JPEG: Symmetric Exponential Quantization	395
JPEG: YCgCb Color Space vs. YCbCr A Comparison	400
Large-Scale Remote Sensing Image Processing	404
Markov Random Fields	411
Mathematical Topics from 3-D Graphics	421
Missing Depth Data In-Painting	431
Modern Cryptography	439
Modern Hardware Design Practices	460
Multibiometric Fusion and Template Protection Strategies	473
Optical Character Recognition (OCR)	486
Optical Coding Schemes: Applications	494
Optical Coding Schemes: Basics	496
Optical Coding Schemes: Hardware	512
Perceptual Image and Video Quality	517
Perceptual Image Enhancement	531
Photographic Image Characteristics	548
Raindrop Imaging	557
Real-Time Object Detection and Tracking	566
Reflectance Modeling	583
Remote Sensing Image Fusion	588
Remotely Sensed Images: Lossy Compression	601
Scattering Features of Complex Mediums	615
Semantic Processing	649
Sparse Modeling	664
The Human Visual System	672
Thermal Infrared Remote Sensing	688
Three-Dimensional Intensity-Curvature Measurement	697
Time and Motion	714
Topology in Geographic Information System	726
Traffic Analysis	734
UAS Imaging	746
UAV Imaging for Disaster Management	759
Unusual Trajectory Detection	766
Video Analytics in the Compressed Domain	782
Virtual Worlds	799
Visual Cryptography Applications	809
Wavelet Analysis	821

Encyclopedia of Image Processing

Editor-in-Chief

Phillip A. Laplante
Professor of Software Engineering, Great Valley School of Graduate Professional Studies, Pennsylvania State University, Malvern, Pennsylvania, U.S.A.

Editorial Advisory Board

Luciano da F. Costa
University of São Paolo, São Paolo, Brazil

Gang Hua
Stevens Institute of Technology, Hoboken, New Jersey, U.S.A.

Nasser Kehtarnavaz
University of Texas at Dallas, Richardson, Texas, U.S.A.

Robert Loce
Palo Alto Research Center, Xerox Corporation, Webster, New York, U.S.A.

Rastislav Lukac
Foveon, Inc., San Jose, California, U.S.A.

Paisarn Muneesawang
Naresuan University in Thailand, Tha Pho, Thailand

Stuart William Perry
University of Technology Sydney, Ultimo, New South Wales, Australia

Konstantinos N. Plataniotis
University of Toronto, Toronto, Canada

Wenping Wang
University of Hong Kong, Pok Fu Lam, Hong Kong

Topical Editors/Reviewers

Reviewing an article for the Encyclopedia is an important and substantial task. Therefore, gratitude is extended to the following individuals who reviewed one or more articles:

Tayfun Aytac
TÜBİTAK BİLGEM, Ankara, Turkey

Cigdem Beyan
Istituto Italiano di Tecnologia, Genoa, Italy

Antony J. Bourdillon
UHRL, San Jose, California, U.S.A.

Alan C. Bovik
University of Texas, Austin, Texas, U.S.A.

Matthias F. Carlsohn
Engineering & Consultancy Dr. Carlsohn, Bremen, Germany

Arpitam Chatterjee
Jadavpur University, Kolkata, India

Carlo Ciulla
University of Information Science and Technology, Ohrid, Macedonia

Yuri A. Gadzhiev
Dagestan State Technical University, Makhachkala, Russia

Hamid Abdullah Jalab
Faculty of Computer Science and Information Technology, University of Malaya, Kuala Lumpur, Malaysia

Fabrizio Lamberti
Dipartimento di Energia, Politecnico di Torino, Torino, Italy

Rubisley Lemes
Federal University of Bahia, Salvador, Brazil

John M. Libert
National Institute of Standards and Technology, Gaithersburg, Maryland, U.S.A.

Robert Loce
PARC, Webster, New York, U.S.A.

Jorge D. Mendiola-Santibañez
Universidad Autonoma de Queretaro, Queretaro, Mexico

Philippos Mordohai
Stevens Institute of Technology, Hoboken, New Jersey, U.S.A.

Stuart William Perry
University of Technology Sydney, Ultimo, New South Wales, Australia

Maurício Segundo
Federal University of Bahia, Salvador, Brazil

Ju Shen
University of Dayton, Dayton, Ohio, U.S.A.

Matthew Shreve
Xerox, Webster, New York, U.S.A.

Contributors

Gabriel Ambrósio Archanjo / *Marvin Project, Itatiba, Brazil*
Shaveta Arora / *The NorthCap University, Gurgaon, India*
A. R. Bednarek / *University of Florida, Gainesville, Florida, U.S.A.*
Paul A. Benkeser / *Wallace H. Coulter Department of Biomedical Engineering, Georgia Institute of Technology, Atlanta, Georgia, U.S.A.; Emory University, Atlanta, Georgia, U.S.A.*
Edgar A. Bernal / *Palo Alto Research Center, Xerox Corporation, Webster, New York, U.S.A.*
Cigdem Beyan / *Pattern Analysis and Computer Vision (PAVIS) Department, Istituto Italiano di Tecnologia (IIT), Genoa, Italy*
Bernd Borchert / *Universität Tübingen, Tübingen, Germany*
Gloria Bordogna / *Italian National Research Council, Institute for the Dynamics of Environmental Processes, Dalmine, Italy*
Alan C. Bovik / *Department of Electrical and Computer Engineering, The University of Texas at Austin, Austin, Texas, U.S.A.; Laboratory of Image and Video Engineering, The University of Texas at Austin, Austin, Texas, U.S.A.*
Yi Cao / *Department of Computer Science and Information Engineering, Nanjing University of Information Science and Technology, Nanjing, China*
Vittorio Castelli / *T. J. Watson Research Center, IBM, Yorktown Heights, New York, U.S.A.*
Rajat Subhra Chakraborty / *Indian Institute of Technology Kharagpur, West Bengal, India*
Arpitam Chatterjee / *Department of Printing Engineering, Jadavpur University, Kolkata, India*
Sen-ching Samson Cheung / *Department of Electrical and Computer Engineering, University of Kentucky, Lexington, Kentucky, U.S.A.*
Lark Kwon Choi / *Department of Electrical and Computer Engineering, The University of Texas at Austin, Austin, Texas, U.S.A.; Laboratory of Image and Video Engineering, The University of Texas at Austin, Austin, Texas, U.S.A.*
P. K. Choudhury / *Institute of Microengineering and Nanoelectronics, Universiti Kebangsaan Malaysia, Bangi, Malaysia*
Carlo Ciulla / *University of Information Science and Technology, Ohrid, Macedonia*
Douglas W. Cunningham / *Brandenburg University of Technology, Cottbus, Germany*
Fletcher Dunn / *Valve Software, Bellevue, Washington, U.S.A.; Computer Science and Engineering, University of North Texas, Denton, Texas, U.S.A.*
Yuri A. Gadzhiev / *Dagestan State Technical University (DSTU), Makhachkala, Russia*
Deepti Ghadiyaram / *Laboratory of Image and Video Engineering, The University of Texas at Austin, Austin, Texas, U.S.A.; Department of Computer Sciences, The University of Texas at Austin, Austin, Texas, U.S.A.*
Ali Gholipour / *Department of Radiology, Boston Children's Hospital, Harvard Medical School, Boston, Massachusetts, U.S.A.*
Jose Gomis-Cebolla / *Global Change Unit, Image Processing Laboratory, University of Valencia, Valencia, Spain*

Todd Goodall / Department of Electrical and Computer Engineering, Laboratory of Image and Video Engineering, The University of Texas at Austin, Austin, Texas, U.S.A.
Genady Ya. Grabarnik / St. John's University, Queens, New York, U.S.A.
Gaurav Gupta / The NorthCap University, Gurgaon, India
Madasu Hanmandlu / Indian Institute of Technology Delhi, Delhi, India
Majid Harouni / Computer Science, Dolatabad Branch, Islamic Azad University, Isfahan, Iran
Fang Huang / School of Resources and Environment, University of Electronic Science and Technology of China, Chengdu, China; Institute of Remote Sensing Big Data, Big Data Research Center, University of Electronic Science and Technology of China, Chengdu, China
Qunying Huang / Geography, University of Wisconsin-Madison, Madison, Wisconsin, U.S.A.
Yuxia (Lucy) Huang / Computing Sciences, Texas A&M University Corpus Christi, Corpus Christi, Texas, U.S.A.
N. Iqbal / Institute of Microengineering and Nanoelectronics, Universiti Kebangsaan Malaysia, Bangi, Malaysia; Department of Physics, University of Gujrat, Hafiz Hayat, Pakistan
Juan Carlos Jimenez / Global Change Unit, Image Processing Laboratory, University of Valencia, Valencia, Spain
Yves Julien / Global Change Unit, Image Processing Laboratory, University of Valencia, Valencia, Spain
Mohan S. Kankanhalli / National University of Singapore, Singapore
Ariyo Kanno / Graduate School of Science and Engineering, Yamaguchi University, Ube, Japan
Nasser Kehtarnavaz / Department of Electrical Engineering, University of Texas at Dallas, Richardson, Texas, U.S.A.
Vignesh Kothapalli / Electronics and Electrical Engineering, Indian Institute of Technology Guwahati, Guwahati, India
Donald Kraft / Department of Computer Science, U.S. Air Force Academy, Colorado Springs, Colorado, U.S.A.
Prasanna Kumar / Mathematics, Birla Institute of Technology and Science – Goa Campus, Zuarinagar, India
Wing C. Kwong / Hofstra University, Hempstead, New York, U.S.A.
Dave K. Kythe / Security Leader & Advisor, Los Angeles, California, U.S.A.
Prem K. Kythe / University of New Orleans, New Orleans, Louisiana, U.S.A.
Ann Latham Cudworth / Ann Cudworth Projects/Alchemy Sims, New York University, New York, New York, U.S.A.
Xin Li / Lane Department of Computer Science and Electrical Engineering, West Virginia University, Morgantown, West Virginia, U.S.A.
Robert P. Loce / Palo Alto Research Center, Xerox Corporation, Webster, New York, U.S.A.
Matthew C. Mariner / George A. Smathers Libraries, University of Florida, Gainesville, Florida, U.S.A.
Mile Matijević / Faculty of Graphic Arts, University of Zagreb, Zagreb, Croatia
Barry McCullagh / Dublin, Ireland
Bombaywala Md.Salman R. / Uka Tarsadia University, Surat, India

Jorge Domingo Mendiola Santibañez / *Center of Research and Technological Development in Electrochemistry (CIDETEQ), Querétaro, Mexico; Facultad de Ingeniería, Autonomous University of Querétaro (UAQ), Querétaro, Mexico*

Evangelia Micheli-Tzanakou / *Department of Biomedical Engineering, Rutgers University, Piscataway, New Jersey, U.S.A.*

Miroslav Mikota / *Faculty of Graphic Arts, University of Zagreb, Zagreb, Croatia*

Debdeep Mukhopadhyay / *Indian Institute of Technology Kharagpur, West Bengal, India*

Phivos Mylonas / *Department of Informatics, Ionian University, Corfu, Greece*

Priya Narayanan / *Department of Geography, School of Earth Science, Central University of Karnataka, Kalaburagi, India*

Sankar K. Pal / *Machine Intelligence Unit, Indian Statistical Institute, Kolkata, India*

S. Palanikumar / *Information Technology, Noorul Islam Centre for Higher Education, Thuckalay, India*

Frederic I. Parke / *Department of Visualization, College of Architecture, Texas A&M University, College Station, Texas, U.S.A.*

Gabriella Pasi / *Department of Informatics, Systems and Communication, University of Studies of Milano Bicocca, Milan, Italy*

Kanai Chandra Paul / *Department of Printing Engineering, Jadavpur University, Kolkata, India*

Chirag N. Paunwala / *Electronics and Communication Department, Sarvajanik College of Engineering and Technolgy, Surat, India*

Mita C. Paunwala / *Electronics and Communication Department, C. K. Pithawala College of Engineering and Technology, Surat, India*

Lluís Pesquer / *Grumets Research Group, CREAF, Universitat Autònoma de Barcelona, Catalonia, Spain*

James F. Peters / *Computational Intelligence Laboratory, Electrical and Computer Engineering, University of Manitoba, Winnipeg, Manitoba, Canada*

Christine Pohl / *Institute of Computer Science, University of Osnabrueck, Osnabrueck, Germany*

Tania Pouli / *Technicolor, Rennes, France*

N.N. Ramaprasad / *Department of Geography, School of Earth Science, Central University of Karnataka, Kalaburagi, India*

Erik Reinhard / *Technicolor Research and Innovation, Rennes, France*

Klaus Reinhardt / *Universität Tübingen, Tübingen, Germany*

Irina Rish / *IBM, Yorktown Heights, New York, U.S.A.*

Palungbam RojiChanu / *Department of Electronics and Communication Engineering, National Institute of Technology Nagaland, Dimapur, India*

John C. Russ / *Department of Materials Science and Engineering, College of Engineering, North Carolina State University, Raleigh, North Carolina, U.S.A.*

J. R. Saylor / *Department of Mechanical Engineering, Clemson University, Clemson, South Carolina, U.S.A.*

Amisha J. Shah / *Electronics and Communication Department, C. K. Pithawala College of Engineering and Technology, Surat, India*

Ju Shen / *Department of Computer Science, University of Dayton, Dayton, Ohio, U.S.A.*

Kh. Manglem Singh / *Department of Computer Science and Engineering, National Institute of Technology Manipur, Imphal, India*

Drazen Skokovic / *Global Change Unit, Image Processing Laboratory, University of Valencia, Valencia, Spain*

Oskar Skrinjar / *Wallace H. Coulter Department of Biomedical Engineering, Georgia Institute of Technology, Atlanta, Georgia, U.S.A.; Emory University, Atlanta, Georgia, U.S.A.*

Jose Antonio Sobrino / *Global Change Unit, Image Processing Laboratory, University of Valencia, Valencia, Spain*

Guillem Soria / *Global Change Unit, Image Processing Laboratory, University of Valencia, Valencia, Spain*

Lihong Su / *Harte Research Institute for Gulf of Mexico Studies, Texas A&M University-Corpus Christi, Corpus Christi, Texas, U.S.A.*

Xingming Sun / *Department of Computer Science and Information Engineering, Nanjing University of Information Science and Technology, Nanjing, China*

Jian Tao / *Texas A&M Engineering Experiment Station and High Performance Research Computing, Texas A&M University, College Station, Texas, U.S.A.*

Iván Terol Villalobos / *Center of Research and Technological Development in Electrochemistry (CIDETEQ), Querétaro, Mexico*

Bipan Tudu / *Department of Electronics and Instrumentation Engineering, Jadavpur University, Kolkata, India*

Bhaumik Vaidya / *Electronics and Communication Department, Sarvajanik College of Engineering and Technolgy, Surat, India*

J.L. van Genderen / *Department of Earth Observation Science, Faculty of Geoinformation Science and Earth Observation (ITC), University of Twente, Enschede, The Netherlands*

Aarohi Vora / *Electronics and Communication Department, Sarvajanik College of Engineering and Technology, Surat, India*

Caixia Wang / *Geomatics, University of Alaska Anchorage, Anchorage, Alaska, U.S.A.*

Keith Waters / *Kwaters Consulting, Boston, Massachusetts, U.S.A.*

Jonathan Weir / *Queen's University, Belfast, U.K.*

WeiQi Yan / *Queen's University, Belfast, U.K.*

Ching-Nung Yang / *Jiangsu Engineering Center of Network Monitoring and School of Computer and Software, National Dong Hwa University, Hualien, Taiwan*

Guu-Chang Yang / *National Chung Hsing University, Taichung City, Taiwan*

Hadi Yazdani Baghmaleki / *Computer Science, Dolatabad Branch, Islamic Azad University, Isfahan, Iran*

Alaitz Zabala / *Grumets Research Group. Dep Geografia, Universitat Autònoma de Barcelona, Catalonia, Spain*

Zhili Zhou / *Department of Computer Science and Information Engineering, Nanjing University of Information Science and Technology, Nanjing, China*

Contents

Contributors xiii
Preface xix
Acknowledgments xxi
About the Editor-in-Chief xxiii

3D Medical Imaging / *Oskar Skrinjar and Paul J. Benkeser* 1
Artificial Neural Networks / *Evangelia Micheli-Tzanakou* 11
Biomedical Image Registration / *Ali Gholipour and Nasser Kehtarnavaz* 24
Boolean Algebras / *A. R. Bednarek* 40
Cantor, Fuzzy, Near, and Rough Sets / *James F. Peters and Sankar K. Pal* 46
Color / *Tania Pouli, Erik Reinhard, and Douglas W. Cunningham* 54
Color Image Denoising / *Palungbam RojiChanu and Kh. Manglem Singh* 72
Color Image Segmentation Metrics / *Majid Harouni and Hadi Yazdani Baghmaleki* . 95
Depth Statistics / *Tania Pouli, Erik Reinhard, and Douglas W. Cunningham* 114
Digital Halftoning: Computational Intelligence-Based Approach / *Arpitam Chatterjee, Kanai Chandra Paul, and Bipan Tudu* 123
Dimensionality Reduction / *Tania Pouli, Erik Reinhard, and Douglas W. Cunningham* ... 135
Discrete Fourier Transforms / *Dave K. Kythe and Prem K. Kythe* 148
Facial Animation / *Frederic I. Parke and Keith Waters* 154
First-Order Statistics / *Tania Pouli, Erik Reinhard, and Douglas W. Cunningham* ... 165
Floodplain Mapping / *N. N. Ramaprasad and Priya Narayanan* 177
Forgery Detection in Digital Images / *Zhili Zhou, Yi Cao, Xingming Sun, and Ching-Nung Yang* ... 183
Fourier Analysis / *Tania Pouli, Erik Reinhard, and Douglas W. Cunningham* 194
Fuzzy Set Theory / *Donald Kraft, Gloria Bordogna, and Gabriella Pasi* 216
Gradients, Edges, and Contrast / *Tania Pouli, Erik Reinhard, and Douglas W. Cunningham* ... 234
Grayscale Morphological Image Enhancement / *Jorge Domingo Mendiola Santibañez and Iván Ramón Terol Villalobos* 254
Image and Scene Analysis / *James F. Peters* 270
Image Enhancement / *S. Palanikumar* 281
Image Inpainting Evolution: A Survey / *Bombaywala Md.Salman R. and Chirag N. Paunwala* ... 293
Image Interpolation / *Xin Li* 308
Image Processing and Measurement / *John C. Russ* 316
Image Retrieval / *Vittorio Castelli* 335
Image Secret Sharing / *WeiQi Yan, Jonathan Weir, and Mohan S. Kankanhalli* ... 354
Image Super Resolution / *Mita C. Paunwala and Amisha J. Shah* 365
Information Sets for Edge Detection / *Shaveta Arora, Vignesh Kothapalli, Madasu Hanmandlu, and Gaurav Gupta* 378
JPEG: Symmetric Exponential Quantization / *Yuri A. Gadzhiev* 395
JPEG: YCgCb Color Space vs. YCbCr—A Comparison / *Yuri A. Gadzhiev* 400
Large-Scale Remote Sensing Image Processing / *Fang Huang and Jian Taoc* 404
Markov Random Fields / *Tania Pouli, Erik Reinhard, and Douglas W. Cunningham* . 411
Mathematical Topics from 3-D Graphics / *Fletcher Dunn and Ian Parberry* 421

Missing Depth Data In-Painting / *Ju Shen, Sen-ching Samson Cheung, Chen Chen, and Ruixu Liu*	431
Modern Cryptography / *Debdeep Mukhopadhyay and Rajat Subhra Chakraborty*	439
Modern Hardware Design Practices / *Debdeep Mukhopadhyay and Rajat Subhra Chakraborty*	460
Multibiometric Fusion and Template Protection Strategies / *Aarohi Vora, Chirag N. Paunwala, and Mita Paunwala*	473
Optical Character Recognition (OCR) / *Matthew C. Mariner*	486
Optical Coding Schemes: Applications / *Wing C. Kwong and Guu-Chang Yang*	494
Optical Coding Schemes: Basics / *Wing C. Kwong and Guu-Chang Yang*	496
Optical Coding Schemes: Hardware / *Wing C. Kwong and Guu-Chang Yang*	512
Perceptual Image and Video Quality / *Deepti Ghadiyaram, Todd Goodall, Lark Kwon Choi, and Alan C. Bovik*	517
Perceptual Image Enhancement / *Lark Kwon Choi, Alan C. Bovik, Todd Goodall, and Deepti Ghadiyaram*	531
Photographic Image Characteristics / *Miroslav Mikota and Mile Matijević*	548
Raindrop Imaging / *J. R. Saylor*	557
Real-Time Object Detection and Tracking / *Bhaumik Vaidya and Chirag N. Paunwala*	566
Reflectance Modeling / *Ariyo Kanno*	583
Remote Sensing Image Fusion / *Christine Pohl and J. L. van Genderen*	588
Remotely Sensed Images: Lossy Compression / *Lluís Pesquer and Alaitz Zabala*	601
Scattering Features of Complex Mediums / *P. K. Choudhury and N. Iqbal*	615
Semantic Processing / *Phivos Mylonas*	649
Sparse Modeling / *Irina Rish and Genady Ya. Grabarnik*	664
The Human Visual System / *Tania Pouli, Erik Reinhard, and Douglas W. Cunningham*	672
Thermal Infrared Remote Sensing / *Juan Carlos Jimenez, Jose Antonio Sobrino, Guillem Soria, Yves Julien, Drazen Skokovic, and Jose Gomis-Cebolla*	688
Three-Dimensional Intensity-Curvature Measurement / *Carlo Ciulla*	697
Time and Motion / *Tania Pouli, Erik Reinhard, and Douglas W. Cunningham*	714
Topology in Geographic Information System / *Prasanna Kumar*	726
Traffic Analysis / *Gabriel Ambrósio Archanjo and Barry McCullagh*	734
UAS Imaging / *Yuxia (Lucy) Huang and Lihong Su*	746
UAV Imaging for Disaster Management / *Caixia Wang and Qunying Huang*	759
Unusual Trajectory Detection / *Cigdem Beyan*	766
Video Analytics in the Compressed Domain / *Edgar A. Bernal and Robert P. Loce*	782
Virtual Worlds / *Ann Latham Cudworth*	799
Visual Cryptography Applications / *Bernd Borchert and Klaus Reinhardt*	809
Wavelet Analysis / *Tania Pouli, Erik Reinhard, and Douglas W. Cunningham*	821

Preface

Image processing is important in so many disciplines that it would be very difficult to list them all. But clearly image processing advances have enabled important applications in health care, avionics, robotics, natural resource discovery, defense, and so on. It is my hope, then, that this encyclopedia will be of interest to specialists in these and other disciplines including electrical engineers, computer scientists, computer engineers, computing professionals, managers, software professionals, and other technology professionals. Compiling this encyclopedia was quite challenging. The most significant challenge was finding willing and capable authors as experts are always busy. There was a constant search for new authors, and committed authors often needed extra encouragement. As the articles begin to be delivered by the authors, peer reviews for the articles are needed to be organized. Finding expert peer reviewers, who were also busy, was not always easy. The articles and review reports were then returned to the authors for revision, and in many cases, another round of reviews. The process was not dissimilar to editing a special issue of a scholarly journal. The final articles then needed to be edited by expert copy editors, then returned to the authors for another check. The EIC conducted a final check.

In some cases, in order to round out coverage, we mined content from other books published by Taylor & Francis (T&F) and restructured that content to encyclopedia articles. This step was also time consuming as it required an analysis of hundreds of books in the T&F library. It should be no surprise, then, that the process of building this encyclopedia, from start to finish, took 4 years.

I hope you are pleased with the result. This encyclopedia is the result of the work of many contributors and reviewers from industry and academia across the globe. We tried to be as correct and comprehensive as possible, but of course, in a work of this grand scope, there are bound to be holes in coverage, as well as typographical, possibly even factual errors. I take full responsibility for these errors, and hope that you will contact me at plaplante@psu.edu to notify me of any errors.

Phillip A. Laplante, Editor

Acknowledgments

Compiling an encyclopedia is a significant undertaking, and the role of editor-in-chief is similar to that of the captain of an aircraft carrier—the captain merely articulates the mission of the ship and its destination and provides general guidance along the way—hundreds of others do the real work. This encyclopedia really did involve hundreds of people: contributors, reviewers, editors, production staff, and more, so I cannot thank everyone personally. But some special kudos are required.

Collectively, I thank the authors of the articles and the reviewers—without them, of course, there would be no encyclopedia. Members of the Editorial Advisory Board also provided a great deal of advice, encouragement, and hard work, and I am grateful to them for those. And there are many staff members at Taylor & Francis in the acquisitions, editing, production, marketing, and sales departments that deserve credit. But I would like to single out Molly Pohlig and Stephanie DeRosa, who served as my eyes, ears, and hands throughout the encyclopedia development process. I am grateful for their enthusiasm and counsel. I also want to thank my family for indulging me when I worked on this project in our family room over many days and evenings.

About the Editor-in-Chief

Dr. Phillip A. Laplante is professor of software and systems engineering at Penn State University's Great Valley School of Graduate Professional Studies. Previously, he was a professor and academic administrator at several colleges and universities. Prior to his academic experiences, Dr. Laplante worked as a professional software engineer for almost 8 years. He was involved in requirements engineering and software delivery for such projects as the Space Shuttle Inertial Measurement Unit, commercial CAD software, and major projects for Bell Laboratories.

Laplante's research, teaching, and consulting currently focus on the areas of requirements engineering, software testing, project management, and cyberphysical systems. He serves on a number of corporate and professional boards, and is a widely sought speaker and consultant.

Dr. Laplante has published more than 250 technical papers and 34 books, including three dictionaries and three other encyclopedias (the Encyclopedia of Software Engineering, the Encyclopedia of Information Systems and Technology, and the Encyclopedia of Computer Science and Technology, Second Edition) all published by CRC Press/Taylor & Francis. He also edits the following Taylor & Francis book series: Applied Software Engineering, Image Processing, and What Every Engineer Should Know About.

He holds a BS degree in systems planning and management, a master's degree in electrical engineering, and a PhD degree in computer science, all from Stevens Institute of Technology. He also holds an MBA degree from the University of Colorado. He is a fellow of the Institute of Electrical and Electronics Engineers and the International Society for Optics and Photonics, and is a licensed professional engineer in the Commonwealth of Pennsylvania.

Encyclopedia of Image Processing

- 3D Medical Imaging–Color Image
- Depth Statistics–Fuzzy Set Theory
- Gradients–Image Super Resolution
- Information Sets–Multi-Biometric Fusion
- Optical–Reflectance Modeling
- Remote Sensing–Time and Motion
- Topology–Wavelet Analysis

3D Medical Imaging

Oskar Skrinjar and Paul J. Benkeser
Wallace H. Coulter Department of Biomedical Engineering, Georgia Institute of Technology and Emory University, Atlanta, Georgia, U.S.A.

Abstract
This entry describes the instrumentation of a number of three-dimensional (3-D) imaging technologies commonly used in the practice of medicine, together with a description of the image processing and analysis techniques that aid in the interpretation of the imagery produced by these technologies.

INTRODUCTION

Medical imaging technology has come a long way since its beginnings in 1895 with Röentgen's discovery of X rays and their ability to image structures in two dimensions within the body. Many modern imaging technologies are capable of acquiring, displaying, and manipulating imagery in three dimensions to aid in the diagnosis and treatment of disease and injuries. This article describes the instrumentation of a number of three-dimensional (3-D) imaging technologies commonly used in the practice of medicine, together with a description of the image processing and analysis techniques that aid in the interpretation of the imagery produced by these technologies.

MAGNETIC RESONANCE IMAGING

Magnetic resonance imaging (MRI) is an imaging technique used to produce images of structures inside the body. It is based on the principles of nuclear magnetic resonance (NMR), a standard spectroscopic technique used to obtain chemical and physical information about molecules.[1] The nucleus of atoms with an odd atomic number or an odd number of neutrons can be detected using NMR techniques. There are many such atoms in the body, but for the purposes of MRI, the hydrogen atom (which consists of a single proton) is of primary interest due to its large magnetic moment. Largely inspired by Lauterbur's development in 1972[2] of the spatial encoding principles used to form images from NMR data, Kumar, Welti, and Ernst devised a two-dimensional (2-D) Fourier transform approach using field gradient pulses and phase encoding to form the modern foundation for MRI instrumentation.[3] Such instrumentation is used to acquire multidimensional image data sets in which the contrast is a function of the density, distribution, and relaxational properties of hydrogen protons within the region of the body being imaged. MRI is used to image both structures and function within the body.

Magnetic resonance imaging systems in clinical use today consist of the following components: 1) a magnet and shim coils to generate a static homogeneous magnetic field with a field strength typically in the range of 0.2 to 3 T; 2) a coil to transmit radiofrequency (RF) signals into the body; 3) a receiver coil to detect the RF signals transmitted by the hydrogen nuclei; 4) gradient coils to provide spatial localization of the RF signals; 5) a computer to control the system timing and reconstruct the RF signals into the image; and 6) a video display system. An example of such a system is illustrated in Fig. 1.

Magnetic resonance imaging systems have a number of attractive features for medical imaging applications. They can generate 2-D sectional images at a variety of orientations, 3-D volumetric images, or four-dimensional (4-D) images representing spatial-spectral distributions, all without any mechanical movement of the instrumentation.[4] Unlike single photon emission computed tomography (SPECT) and positron emission tomography (PET), MRI does not require injection of radioisotopes. In addition, the imaging process involves the use of RF energy that lies in the nonionizing region of the electromagnetic spectrum.

X-RAY COMPUTED TOMOGRAPHY

The introduction of X-ray computed tomography (CT) in the early 1970s was one of the most profound developments in the field of medical imaging since the time of Röentgen.[5,6] X-ray CT involves generating X-ray photons, exposing the body to them, detecting and digitizing those photons that are transmitted through the body, and tomographically reconstructing the digitized data to form an image. The function being imaged is the distribution

Encyclopedia of Image Processing, First Edition
DOI: 10.1081/E-EBBE2-120013956
Copyright © 2018 by Taylor & Francis. All rights reserved.

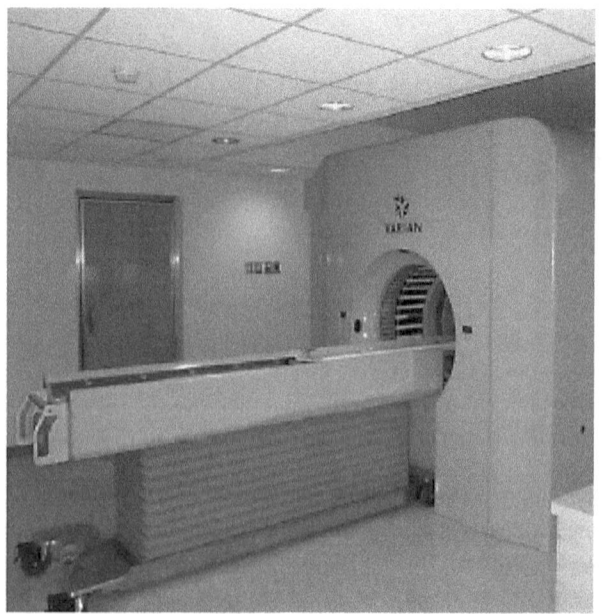

Fig. 1 Varian 3T magnetic resonance imaging system

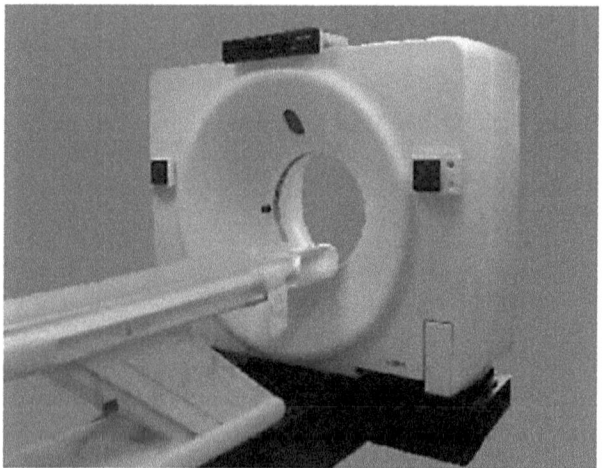

Fig. 2 GE CT/i™ x-ray computed tomography system

of the linear attenuation coefficient within the body. The logarithm of the ratio of intensities of the generated and transmitted photons is proportional to the integral of the attenuation coefficient along the path. Contrast in X-ray CT is thus dependent on the spatial differences in the attenuation coefficient, which is strongly correlated to tissue density. X-ray CT is primarily used to image structures within the body. However, with high-speed systems and the introduction of contrast agents, it may also be used to image functional changes in organs, such as blood flow rate.

X-ray CT systems in clinical use today typically consist of the following components: 1) a gantry containing a fan-beam X-ray source and a circular array of 600 to 4800 (depending on the manufacturer) independent X-ray detectors that surrounds the patient; 2) a table on which the patient lies that may be translated through the hole in the center of the gantry; 3) a computer for controlling the image acquisition parameters and for reconstructing the image; and 4) a video display system. In fourth-generation systems, the X-ray source is an X-ray tube that is rotated 360 degrees around the patient to acquire multiple projections needed to reconstruct the image. In fifth-generation systems, a high-energy electron beam is electronically scanned along a tungsten strip anode to produce the fan beam of X-rays without mechanical rotation of any component. With these fifth-generation systems, projection data can be acquired fast enough to image the beating heart without significant motion artifacts. Spiral scanning systems, consisting of fourth-generation systems that acquire projection data while the patient is translated through the gantry in a continuous motion, also can be used to image moving structures. An example of an X-ray CT system is shown in Fig. 2.

NUCLEAR TOMOGRAPHIC IMAGING

This is a class of nuclear medicine imaging techniques that utilize tomographic reconstruction methods to image the nuclear radiation emitted from radiopharmaceuticals that have been ingested, inhaled, or injected into the body. This radiation is in the form of either high-energy gamma rays or positrons, which travel very short distances before they annihilate to produce two 511 keV photons. This differs from X-ray CT in that the emission source, not the attenuation, is measured. There are two different types of nuclear tomographic imaging techniques: single photon emission computed tomography (SPECT) and positron emission tomography (PET). Both of these techniques are used to image function, not structure. For example, in cardiology applications, with the appropriate radiopharmaceutical (e.g., 99 m-Tc and FDG) these techniques can be used to measure perfusion in the myocardium.

Single Photon Emission Computed Tomography (SPECT)

The SPECT method was initially investigated by Kuhl and Edwards in 1963 prior to the development of X-ray CT, PET, or MRI.[7] In a manner similar to X-ray CT, SPECT requires the rotation of a gamma-ray photon detector array around the body to acquire projection data from multiple angles. SPECT systems in clinical use today typically consist of the following components: 1) single or dual photon detector arrays (aka gamma cameras), each with its own collimator; 2) gantry for rotating the camera(s); 3) table on which the patient lies that is positioned in the center of the gantry; 4) computer for controlling the image acquisition parameters and for reconstructing the image; and 5) video display system.

The image reconstruction task for SPECT is far more challenging than for X-ray CT because the emission sources are at unknown positions and densities inside the body. The problem is compounded by the need to use a collimator to acquire projection data, which rejects all photons that are not at a particular trajectory angle, thus resulting in a low detection efficiency. Losses due to attenuation of the photons in the body further reduce the obtainable resolution. The latter losses can be partially compensated for during the reconstruction process.[8] The typical resolution achievable by SPECT systems is on the order of 7 to 15 mm. An example of a SPECT system and typical images are shown in Fig. 3.

Positron Emission Tomography (PET)

The development of computed tomography concepts in the early 1970s spurred the development of the early PET systems.[9] PET has several inherent advantages over SPECT. First, no collimator is required because the pair of detected 511 keV gamma-ray photons produced when the positron is annihilated travel along the same line and thus can be traced back to their origin. Second, higher energy photons experience less attenuation, and the attenuation that does occur can be more easily corrected for compared to SPECT. Third, PET does not require the mechanical rotation of gamma cameras around the patient. Instead, modern PET systems encircle the patient with photon detectors that are electronically coupled in such a manner as to discriminate the coincident detection of the two photons from individual annihilation events. The resolution achievable by this detector system is limited primarily by the uncertainty associated with the path length of the positron prior to annihilation. These advantages are the primary reasons why PET is capable of achieving resolutions three to four times better than SPECT. Although SPECT imaging resolution is inferior to that of PET, the cost of a typical SPECT imaging procedure is currently a factor of two to three times lower than the cost of similar PET procedures. In addition to its higher cost, PET requires the use of radiopharmaceuticals with extremely short half-lives, requiring a close proximity to a particle accelerator. An example of a PET system and typical images are shown in Fig. 4.

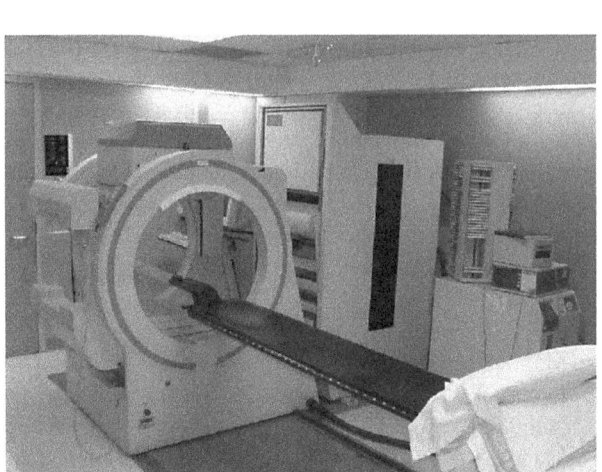

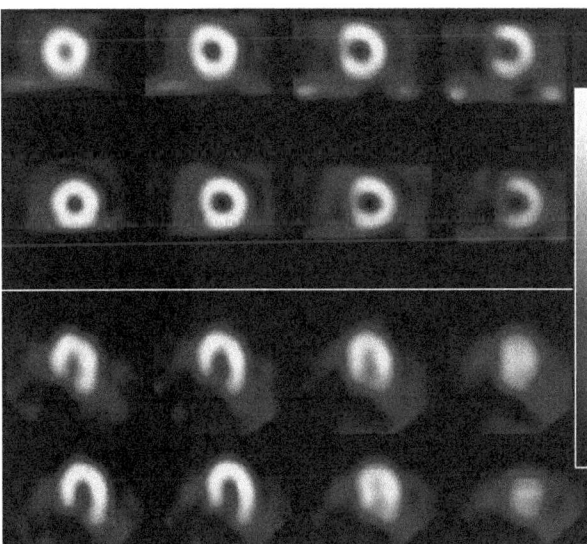

Fig. 3 (top) ADAC Vertex™ dual-head SPECT imaging system; (bottomb) normal SPECT myocardial perfusion images

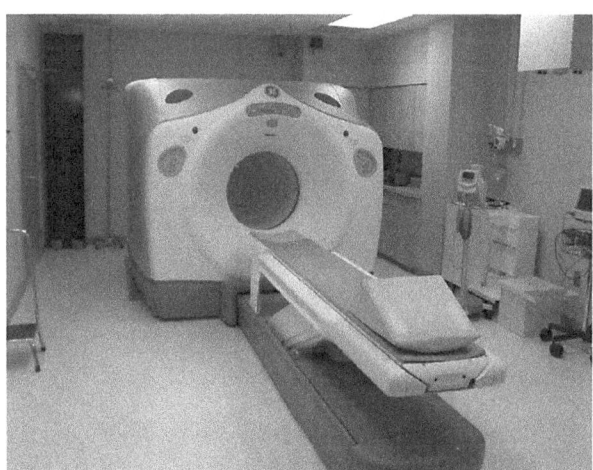

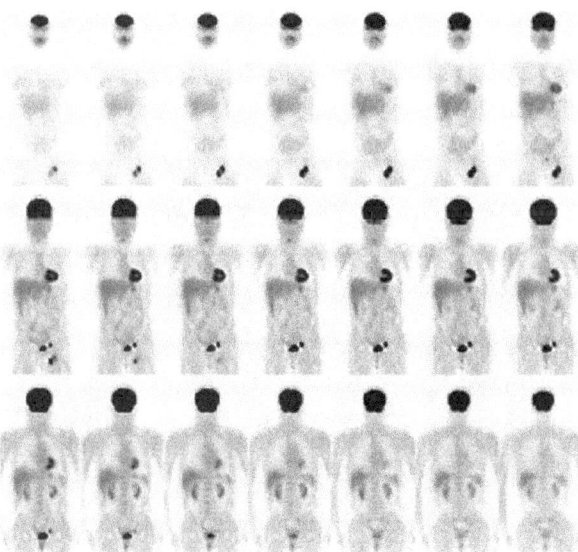

Fig. 4 (top) GE Medical Systems Discovery LS™ with PET and CT imaging capabilities; (bottom) PET whole-body images

OTHER CLINICAL 3-D IMAGING TECHNOLOGIES

Although MRI, PET, SPECT, and X-ray CT are the most pervasive 3-D imaging technologies used in clinical settings today, a number of others are worth noting. Three-dimensional imaging can be performed with ultrasound using either mechanical scanning of one-dimensional (1-D) array transducers or through electronic scanning of 2-D array transducers.[10] Electrical impedance tomography has received considerable attention in the research community, but has found only limited clinical utility to date.[11] Angiography systems have been developed with solid-state digital X ray detectors mounted on rotating C-arms, as shown in Fig. 5, to provide 3-D insight into vascular pathologies.[12]

MEDICAL IMAGE PROCESSING AND ANALYSIS

The development of medical image acquisition techniques has provided some of the most effective diagnostic tools in medicine. These systems are also used for surgical planning and navigation, as well as for imaging in biology. Two-, three-, and higher-dimensional images contain vast amounts of data that need to be interpreted in a timely, accurate, and meaningful manner to benefit clinical and research applications. To assist in interpretation of biomedical images, a number of automated and semiautomated image processing and analysis techniques have been developed.[13] This overview presents the most standard methods used in interpretation of 3-D medical images: segmentation, registration, and visualization. In addition, a brief summary of other methods is given at the end.

SEGMENTATION

Image segmentation is the partitioning of an image into regions that are homogeneous with respect to one or more characteristics or features.[13,14] In medical applications, image segmentation is important for feature extraction, shape analysis, image measurements, and image display. There is no one standard segmentation method that can be successfully used in all imaging applications. Rather, a variety of segmentation techniques have been proposed, many of which are application-specific.[17,18,19] A recent overview of medical image segmentation methods has been published.[33] An example of 3-D medical image segmentation is illustrated in Fig. 6.

Image segmentation methods can be classified in several ways:[13]

- Manual, semiautomatic, and automatic.
- Pixel-based (local methods) and region-based (global methods).
- Manual delineation, low-level segmentation, and model-based segmentation.
- Classical (thresholding, edge-based, and region-based techniques), statistical, fuzzy, and neural network techniques.

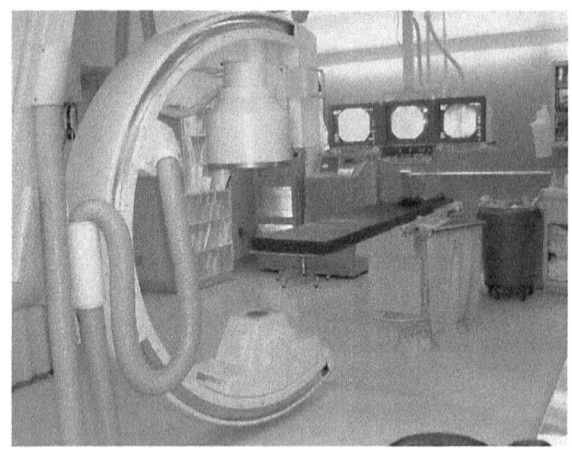

Fig. 5 Phillips Integris™ 3-D rotational angiography system

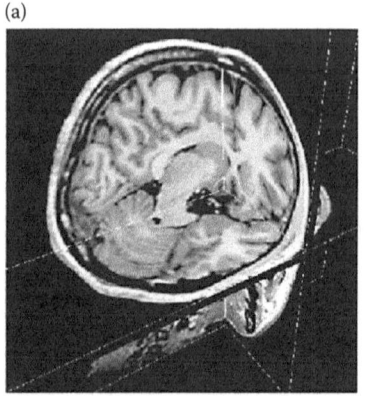 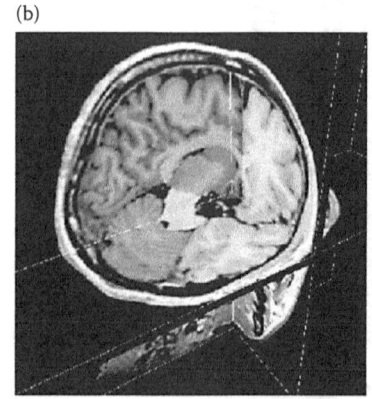 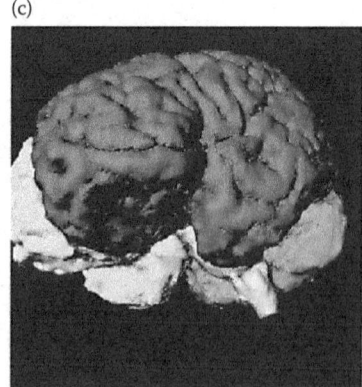

Fig. 6 Example of 3-D medical image segmentation showing (a) three orthogonal sections through a 3-D MR image of the human head; (b) the same image segmented into the left hemisphere (red), right hemisphere (green), cerebellum (blue), and brain stem (yellow); and (c) the 3-D surface models generated for each of the four segmented regions

Furthermore, most segmentation techniques can be classified as region-based (techniques that look for regions satisfying a given homogeneity criterion) or boundary-based (techniques that look for edges between regions with different characteristics). Most image segmentation methods use one image modality (MRI, CT, ultrasound, PET, SPECT, etc.). Their performance can be enhanced by combining images of different modalities (multispectral segmentation[26,27]) or images acquired over time (dynamic, motion, or temporal segmentation[20,28,29]).

Among region-based segmentation methods, the most common ones are thresholding, clustering, region splitting and merging, and region growing.[13,20] Thresholding is the simplest region-based segmentation method.[19,20,21] In this technique a threshold is selected and an image is divided into two groups of pixels: one having values less than the threshold and the other with values greater or equal to the threshold. The threshold can be set globally for the entire image, or can be locally adjusted in an adaptive manner. In addition, methods have been proposed for automatic setting of an optimal threshold value.[20] Segmentation approaches based on thresholding can produce acceptable results only in simple situations. One of the rare cases in which thresholding is successfully used in medical applications is the segmentation of bony structures from CT images (see Fig. 7 for an example). Thresholding approaches can be enhanced with mathematical morphology.[20] Such algorithms, although usually simple, can provide effective and computationally efficient ways to segment complex medical structures. An example of the output of such an algorithm[32] is illustrated in Fig. 8. Clustering segmentation methods partition the image into clusters of pixels that have strong similarity in the feature space.[13,22,23,24] Each pixel is examined and assigned to the cluster that best represents the value of its characteristic vector of features of interest. Region splitting and merging is another region-based segmentation method.[20] The image is subdivided initially into a set of arbitrary disjointed regions that are

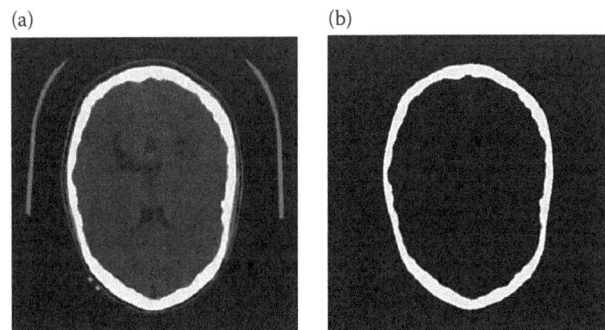

Fig. 7 Skull segmentation (b) from a CT image (a) of the head by global thresholding (an axial slice through the 3-D volume is shown)

then merged and split in an attempt to satisfy certain conditions. Region growing starts from a set of "seed" points and assigns adjacent pixels or regions to the same group if their image values are similar enough, according to some predefined criterion.[13,20] A typical representative of region growing methods is watershed segmentation.[20] The idea behind boundary-based segmentation methods is to extract object boundaries and to segment regions enclosed by the boundaries.[13,20,25] These algorithms typically rely on an edge detector, such as Sobel, Prewitt, Roberts, or Canny.[20] One of the most successful classes of boundary-based segmentation strategies are methods based on active boundaries. This line of work was inspired by the pioneer work on active contours (also known as snakes),[34] which is a 2-D version of the approach. The idea is that the object contours are modeled as deformable curves that automatically deform and attempt to align with image edges, at the same time preserving certain predefined properties (e.g., smoothness, proximity to certain landmarks, sharp corners, etc). The work was extended to 3-D and applied to medical images.[35,36,37]

Often, the complexity of medical structures requires the design of application-specific segmentation methods

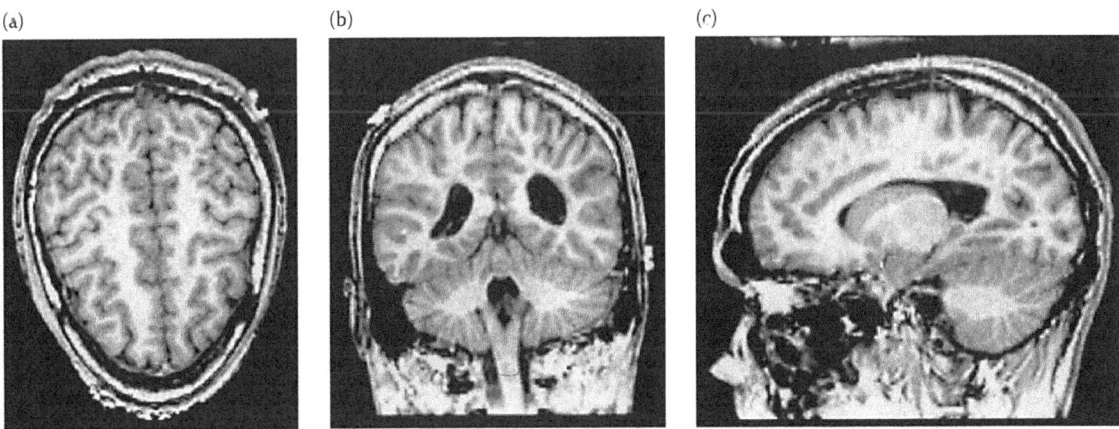

Fig. 8 3-D brain segmentation from MR images using thresholding and mathematical morphology. Contours of the segmented brain are shown in axial (a), coronal (b), and sagittal (c) sections through the 3-D MR image of the head

that embody concepts from both region-based and boundary-based approaches. For example, an algorithm for the segmentation of the cortex is described[38] that, in addition to using image information, restricts the cortical thickness to be within a predefined range, which greatly improves its performance. A more general way to incorporate prior knowledge into segmentation algorithms is by statistical approaches. The idea is to segment a number of objects from a certain class, learn the statistics of the shape variability of each object, and then use each to constrain the segmentation of a new object from the same class. Such an approach has been used to segment the corpus callosum from 3-D MR images of the head.[39]

REGISTRATION

Image registration is the process of aligning images so that corresponding features can be easily related.[15] The inputs to an image registration algorithm are the two images to be registered; the output is the geometric transformation that maps points in one image to points in the other.[14] Many image registration methods have been proposed, and they may be classified in a number of ways.[30,31] Medical image registration has a wide range of applications that include, but are not limited to: combining information from multiple imaging modalities (e.g., combining images that show function and images that show anatomical structures); monitoring changes in size, shape, location, or image intensity over time; relating preoperative images to the data acquired intraoperatively or during therapy; and relating an individual's anatomy to a standardized atlas.[15]

Medical image registration methods can be classified based on the following classification criteria: image dimensionality, registration basis, geometric transformation, degree of interaction, optimization procedure, modality, subject, and object.[14] Each of the criteria is briefly discussed in the remainder of the section.

Medical image registration typically involves 3-D images (3-D to 3-D), but sometimes 2-D images (2-D to 2-D) or images of different dimensions (3-D to 2-D) need to be registered. Figures 9 and 10 illustrate the 3-D to 3-D case, in which functional and anatomical 3-D images are registered and fused.[40] An example of the 2-D to 2-D case is the registration of temporal mammograms, which are 2-D images.[41] Registration of 3-D computed tomography images and 2-D portal images,[42] which is an example of the 3-D to 2-D case, is often needed in prostate radiation therapy. Portal images are the X-ray images obtained with a radiation treatment beam during the prostate radiation therapy.

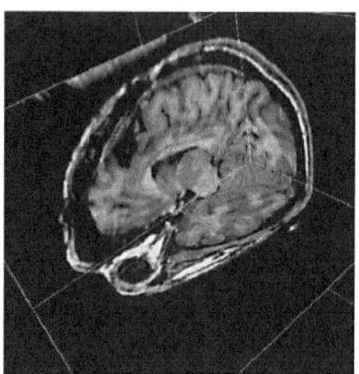
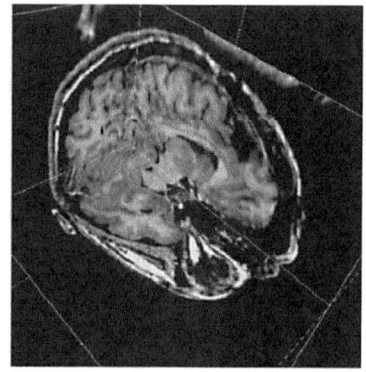
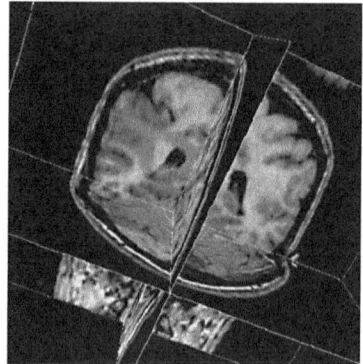

Fig. 9 Rigidly registered and fused SPECT and anatomical MR 3-D images

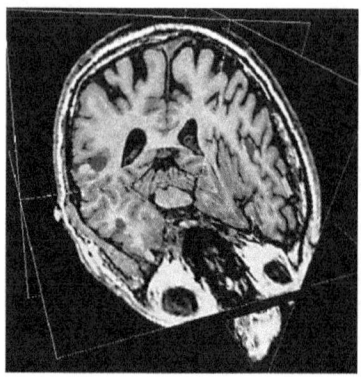
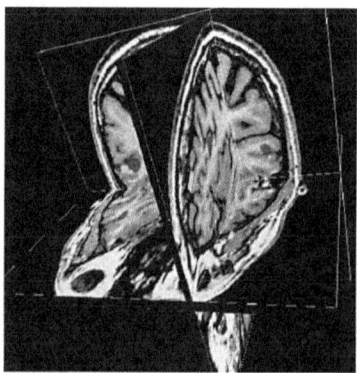
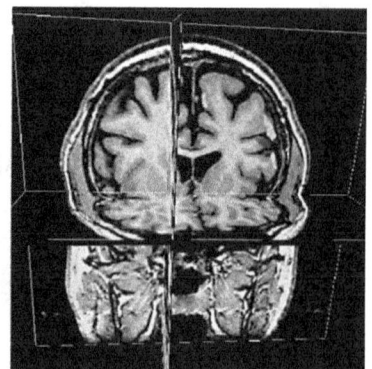

Fig. 10 Rigidly registered and fused functional MR and anatomical MR 3-D images

Registration basis refers to the underlying mechanics of the registration. By this criterion, image registration algorithms are classified as feature-based, intensity-based, or hybrid. Feature-based algorithms use various features (usually points, curves, and surfaces that are manually or automatically extracted from the two images) to determine the geometric transformation necessary to register the two images.[43,44,45] Image-based algorithms use image intensity as the only information to register the images,[46] and hybrid methods combine both approaches.[47]

Geometric transformation is the characteristic of image registration algorithms that determines the mathematical form of the geometric mapping used to align the two images. Geometric transformations used in practice range from rigid[48] to affine[49] to nonrigid.[46] Rigid transformation models (composed of translation and rotation) are used in applications where there is no deformation between the two images except for a change in the location and orientation. For example, in the case of registration of 3-D images of the head of the same patient, the skull, falx, and tentorium provide almost perfectly rigid support for the soft brain tissues, and one can assume a rigid transformation between the two (see Figs. 9 and 10). Affine transformation, in addition to translation and rotation, can account for global scale and shear between the two images. Nonrigid transformation models are needed in cases where the deformation between two images is more complex than affine, e.g., in cases with soft-tissue deformation (cardiac images, pre- and postoperative images, etc.) or in aligning a patient's anatomy to that in an atlas.

Degree of interaction refers to the amount of manual interaction needed by a human operator in performing image registration. The manual interaction may range from no interaction in a fully automated algorithm, which is the ideal situation, to simple initialization of certain parameters or to manual adjustments needed throughout the registration process.

Optimization procedure refers to the approach employed to determine the geometric transformation, i.e., its parameters. Usually, a function of the two images and the mapping between them is used to measure the quality of the registration, and the goal is to optimize (minimize or maximize) the function with respect to the parameters of the transformation. Ideally, the algorithm is based on a closed-form solution that is guaranteed to produce the global extremum of the function. The global extremum can also be determined by an exhaustive search, which in most cases is computationally too expensive, and in some cases it can be determined by means of dynamic programming. In cases in which it is computationally prohibitive to compute the global extremum, iterative optimization schemes are used. Their main disadvantage is that they might find a local extremum that is not the global one, but they are computationally less expensive and for this reason widely used in practice.

Modality refers to the type of images to be registered. When the two images are of the same modality (e.g., MRI–MRI or CT–CT), the registration is called intramodality or single-modality or monomodality. When the two images are of different modalities (e.g., MRI–SPECT as in Fig. 9 or MRI–fMRI as in Fig. 10), the registration is called intermodality or multimodality.

Medical image registration algorithms based on the subject imaged are classified as: intrapatient (the two images are of the same patient), interpatient (the two images are of different patients), and patient-to-atlas (an image of a patient needs to be registered to an atlas image, which itself is typically based on one or more images of different patients).

VISUALIZATION

Three-dimensional visualization generally refers to transformation and display of 3-D objects so as to effectively represent the 3-D nature of the objects. Although an extensive discussion of various visualization techniques for 3-D medical imaging has been published,[16] here we briefly present the most commonly used ways to display data from 3-D medical images.

Sections through 3-D image volume can be displayed as individual 2-D images (e.g., Figs. 7a and 8a–c). These 2-D image sections can be displayed together in a 3-D coordinate system as texture-mapped rectangular plane sections (e.g., Figs. 6a–b, 9, and 10). Such displays provide a better insight about spatial relationships between image structures compared to the case of individually displayed 2-D sections. Usually, the displayed 2-D sections are orthogonal to the three main axes, but sometimes oblique 2-D sections are used.

In order to provide even better appreciation for the spatial relationships between objects in the 3-D image, the objects of interest are segmented and a surface model is generated around each object, which tessellates the surface with polygons allowing for rapid computer manipulation. These models are then surface-rendered, often combined with other visualization techniques. One can control the color, transparency, visibility, smoothness, complexity, and other properties of these models. Examples of surface-rendered objects are given in Figs. 6c, 11, and 12b–c.

Another way to display 3-D images is volume rendering, during which the entire image is displayed and transparency values are assigned to each voxel in the data set. Ray tracing techniques then give the sense of depth perception without the need for a priori segmentation. Typically only a part of the image volume is shown by using clip planes. Examples are shown in Fig. 12, where Fig. 12b–c illustrates a way to combine surface and volume rendering techniques for visualization of medical 3-D image data.

Note that more complex displays usually require a number of image processing methods to prepare the data for

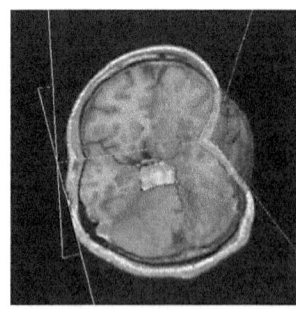 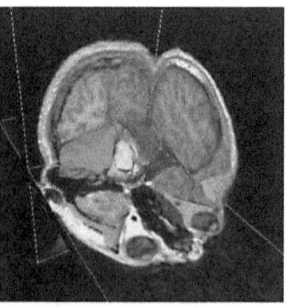

Fig. 11 Examples of surface-rendered models displayed together with orthogonal sections through an MR 3-D image of the human head. Models of the right (green) and left (red) cerebral hemisphere are displayed as transparent surfaces, and models of the cerebellum (blue) and brain stem (yellow) are displayed as opaque surfaces

visualization. For example, in order to generate the displays in Fig. 12b–c, the head was segmented, its surface model (which represents the skin) was generated, and the anatomical MR and function MR images were (rigidly) registered.

OTHER TECHNIQUES AND VALIDATION

Although segmentation, registration, and visualization are the most widely used techniques for interpretation of 3-D medical image data, a number of other methods have been developed. These methods include image enhancement and restoration, feature extraction, statistical analysis, biomechanical modeling, and quantification.[13,14,16]

A critical component of every medical image interpretation technique is its validation, because its clinical application affects the health and sometimes the life of the patient. Abnormalities, pathological cases, variability of normal human structures, limited image quality, and image artifacts are among the factors that make automated medical image interpretation difficult. On the other hand, even small errors in interpretation of medical images can negatively and sometimes fatally affect the patient's health. For this reason, any medical image interpretation technique that is intended to be used clinically needs to be validated on a large number of subjects under various conditions. An ideal validation study involves comparison of the output of the medical image interpretation method against ground truth. However, ground truth is rarely available and other nonideal validation approaches need to be used. These include simulations, phantom studies, and qualitative and quantitative comparisons of expert-generated output to the output produced by the medical image interpretation method.

CONCLUSION

Modern imaging technologies are capable of acquiring, displaying, and manipulating imagery in three dimensions to aid in the diagnosis and treatment of disease and injuries. There is no single imaging technology available to satisfy the needs of the entire spectrum of clinical diagnostic procedures employed. Rather, a collection of technologies is available for clinicians to select from in order to achieve the best possible contrast and resolution for a particular imaging procedure.

Image processing and analysis techniques, ranging from simple thresholdbased segmentation to more sophisticated methods, including deformable models, statistical methods, and multimodality image registration, are available for use with the imagery produced with these technologies. The goal of these techniques is to enhance medical procedures by improving their accuracy and reducing the length and cost. The complexity of medical problems often requires application-specific solutions, which need to be extensively validated in order to be acceptable for clinical use. Finally, the 3-D nature of these problems requires

(a) (b) (c)

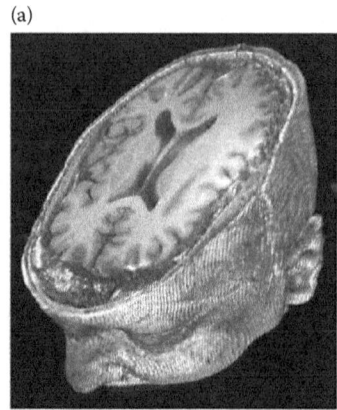 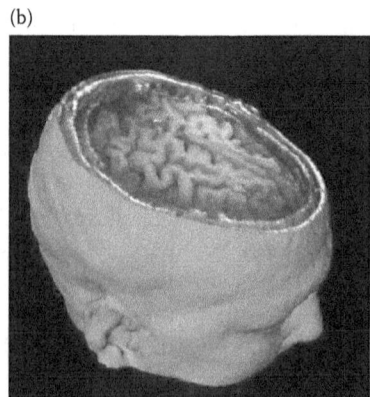 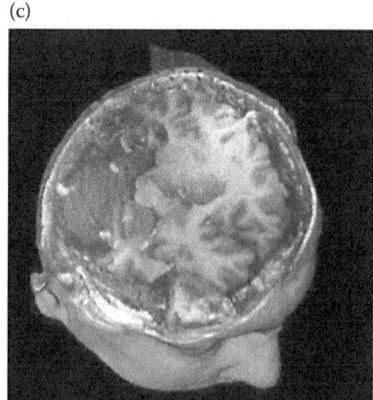

Fig. 12 Examples of visualization of 3-D medical image data. In (a), a purely volume-rendered MR 3-D image of a human head is shown cut with an oblique clip plane; in (b) and (c), volume-rendered fused anatomical MR and functional MR images are displayed together with a surface-rendered skin model

advanced visualization techniques in order to properly display the data and allow for an easier interpretation.

ARTICLES OF FURTHER INTEREST

Electrical Impedance Imaging; Functional MRI: Applications; Magnetic Resonance Imaging in Temperature Measurement; Microcomputed Tomography and Its Applications; Optical Coherence Tomography; Positron Emission Tomography (PET); Real Time Tomographic Reflection; Ultrasound Doppler

REFERENCES

1. Röentgen, W.K. On a new kind of rays. Nature **1896**, *53*, 274–276.
2. Hornak, J.P. *The Basics of MRI*. Available at http://www.cis.rit.edu/htbooks/mri/, (accessed in June 2003).
3. Lauterbur, P. C. Image formation by induced local interactions: Examples employing nuclear magnetic resonance. Nature **1973**, *242*, 190–191.
4. Kumar, A.; Welti, D.; Ernst, R. NMR Fourier zeugmatography. J. Magn. Reson **1975**, *18*, 1–69.
5. Liang, Z.P.; Lauterbur, P.C. Introduction. In *Principles of Magnetic Resonance Imaging*; Akay, M.; Ed.; IEEE Press: New York, 2000, 1–12.
6. Cho, Z.H.; Jones, J.P.; Singh, M. X-ray computerized tomography. In *Foundations of Medical Imaging*; John Wiley & Sons: New York, 1993, 148–164.
7. Kuhl, D.E.; Edwards, R.Q. Image separation radioisotope scanning. Radiology. **1963**, *80*, 653–661.
8. Zaidi, H.; Hasegawa, B. Determination of the attenuation map in emission tomography. J. Nucl. Med. **2003**, *44* (2), 291–315.
9. Cho, Z.H.; Chan, J.K.; Eriksson, L. Circular ring transverse axial position camera for 3-D reconstruction of radionuclide distribution. IEEE Trans. Nucl. Sci. **1976**, *23*, 613–622.
10. Fenster, A.; Downey, D.B.; Cardinal, H.N. Three-dimensional ultrasound imaging. Phys. Med. Biol. **2001**, *46* (5), R67–R99.
11. Zou, Y.; Guo, Z. A review of electrical impedance techniques for breast cancer detection. Med. Eng. Phys. **2003**, *25* (2), 79–90.
12. Hagen, G.; Wadstrom, J.; Magnusson, A. 3D rotational angiography of transplanted kidneys. Acta Radiol. **2003**, *44* (2), 193–198.
13. Bankman, I.N.; Ed.; *Handbook of Medical Imaging. Processing and Analysis*; Academic Press: San Diego, 2000.
14. Medical image processing and analysis. In *Handbook of Medical Imaging, Vol. 2*; Sonka, M.; Fitzpatrick, J.M.; Eds.; Spie Press: Bellingham, WA, 2000.
15. Hajnal, J.V.; Hill, D.L.G.; Hawkes, D.J.; Eds.; *Medical Image Registration*; Biomedical Engineering Series, CRC Press: Boca Raton, FL, 2001.
16. Robb, R.A. *Biomedical Imaging, Visualization, and Analysis*; Wiley-Liss, John Wiley & Sons: New York, 2000.
17. Bezdek, J.C.; Hall, L.O.; Clarke, L.P. Review of MR image segmentation techniques using pattern recognition. Med. Phys. **1993**, *20* (4), 1033–1048.
18. Clarke, L.P.; Velthuizen, R.P.; Camacho, M.A.; Heine, J.J.; Vaidyanathan, M.; Hall, L.O.; Thatcher, R.W.; Silbiger, M. L. MRI segmentation: Methods and applications. Magn. Reson. Imag. **1995**, *13* (3), 343–368.
19. Pal, N.R.; Pal, S.K. A review of image segmentation techniques. Pattern Recogn. **1993**, *26* (9), 1227–1249.
20. Gonzalez, R.C.; Woods, R.E. *Digital Image Processing*, 2nd Ed; Prentice Hall: Upper Saddle River, NJ, 2002.
21. Sahoo, P.K.; Soltani, S.; Wond, A.K.; Chen, Y.C. A survey of thresholding techniques. Comput. Vis. Graph. Image Process **1988**, *41*, 233–260.
22. Bramdt, M.E.; Bohan, T.P.; Kramer, L.A.; Fletcher, J. M. Estimation of CSF, white and gray matter volumes in hydrosephalic children using fuzzy clustering of MR imaging. Comput. Med. Imag. Graph **1994**, *18* (1), 25–34.
23. Duda, R.O.; Hart, P.E. *Pattern Recognition and Scene Analysis*; Wiley: New York, 1973.
24. Jain, A.K.; Flynn, P.J. Image segmentation using clustering. In *Advances in Image Understanding*; Bowyer, K.; Ahuja, N.; Eds.; IEEE Computer Society Press: Los Alamitas, CA, 1996.
25. Haralick, R.M.; Shapiro, L.G. Survey: Image segmentation techniques. Comput. Vis. Graph. Image Process **1985**, *29*, 100–132.
26. Fletcher, L.M.; Marsotti, J.B.; Hornak, J.P. A multispectral analysis of brain tissues. Magn. Reson. Med. **1993**, *29*, 623–630.
27. Reddick, W.E.; Glass, J.O.; Cook, E.N.; Elkin, T.D.; Deaton, R.J. Automated segmentation and classification of multispectral magnetic resonance images of brain using artificial neural networks. IEEE Trans. Med. Imag. **1997**, *16* (6), 911–918.
28. Lucas-Quesada, F.A.; Sinha, U.; Sinha, S. Segmentation strategies for breast tumors from dynamic MR images. J. Magn. Reson. Imag. **1996**, *6*, 753–763.
29. Rogowska, J.; Preston, Jr., K.; Hunter, G.J.; Hamberg, L. M.; Kwong, K.K.; Salonen, O.; Wolf, G.L. Applications of similarity mapping in dynamic MRI. IEEE Trans. Med. Imag. **1995**, *14* (3), 480–486.
30. Maintz, J.B.A.; Viergever, M.A. A survey of medical image registration. Med. Image Anal **1998**, *2*, 1–36.
31. Maurer, C.D.; Fitzpatrick, J.M. A review of medical image registration. In *Interactive Image-Guided Neurosurgery*; Maciunas, R.J.; Ed.; American Association of Neurological Surgeons: Park Ridge, IL, 1993, 17–44.
32. Stokking, R. *Integrated Visualization of functional and Anatomical Brain Images*; PhD thesis; Utrecht University, 1998.
33. Duncan, J.S.; Ayache, N. Medical image analysis: Progress over two decades and the challenges ahead. IEEE Trans. Pattern Anal. Mach. Intell. **2000**, *22* (1), 85–106.
34. Kass, M.; Witkin, A.; Terzopoulos, D. Snakes: Active contour models. Int. J. Comput. Vis. **1988**, *1* (4), 321–331.
35. Miller, J.V.; Breen, D.E.; Lorensen, W.E.; O'Bara, R.M.; Wozny, M.J. Geometrically deformed models: A method for extracting closed geometric models from volume data. Comput. Graph **1991**, *25* (4), 217–226.
36. McInerney, T.; Terzopoulos, D. A dynamic finite element surface model for segmentation and tracking in

multidimensional medical images with application to cardiac 4D image analysis. Comput. Med. Imag. Graph **1995**, *19* (1), 69–83.
37. Cohen, I.; Cohen, L.D.; Ayache, N. Using deformable surfaces to segment 3D images and infer differential structures. CVGIP Image Underst. **1992**, *56* (2), 242–263.
38. Zeng, X.; Staib, L.H.; Schultz, R.T.; Duncan, J.S. Segmentation and measurement of the cortex from 3-D MR images using coupled-surfaces propagation. IEEE Trans. Med. Imag. **1999**, *18* (10), 927–937.
39. Leventon, M.E.; Grimson, W.E.L.; Faugeras, O. Statistical shape influence in geodesic active contours. IEEE Conf. Comput. Vis. Pattern Recognit. **2000**, *1*, 316–323.
40. Studholme, C.; Hill, D.L.G.; Hawkes, D.J. An overlap invariant entropy measure for 3d medical image alignment. Pattern Recogn. **1999**, *32* (1), 71–86.
41. Richard, F.J.P.; Cohen, L.D. A new image registration technique with free boundary constraints: Application to mammography. Comput. Vis. Image Underst. **2003**, *89*, 166–196.
42. Bansal, R.; Staib, J.H.; Zhe, C.; Rangarajan, A.; Knisely, J.; Nath, R.; Duncan, J.S. Entropy-based dual-portal-to-3-DCT registration incorporating pixel correlation. IEEE Trans. Med. Imag. **2003**, *22* (1), 29–49.
43. Haili, C.; Rangarajan, A. A feature registration framework using mixture models. In *IEEE Workshop on Mathematical Methods in Biomedical Image Analysis, Proceedings*; IEEE Computer Society: Los Alamitos, CA, 2000, 190–197.
44. Pelizzari, C.A.; Chen, G.T.Y.; Spelbring, D.R.; Weichselbaum, R.R.; Chen, C.T. Accurate three-dimensional registration of CT, PET, and/or MR images of brain. J. Comput. Assist. Tomogr. **1989**, *13*, 20–26.
45. Cuchet, E.; Knoplioch, J.; Dormont, D.; Marsault, C. Registration in neurosurgery and neuroradiotherapy applications. J. Image Guid. Surg. **1995**, *1*, 198–207.
46. Rueckert, D.; Sonoda, L.I.; Haynes, C.; Hill, D.L.G.; Leach, M.O.; Hawkes, D.J. Non-rigid registration using free-form deformations: Application to breast MR images. IEEE Trans. Med. Imag. **1999**, *18* (8), 712–721.
47. Johnson, H.J.; Christensen, G.E. Consistent landmark and intensity-based image registration. IEEE Trans. Med. Imag. **2002**, *21* (5), 450–461.
48. Holden, M.; Hill, D.L.G.; Denton, E.R.E.; Jarosz, J. M.; Cox, T.C.S.; Rohlfing, T.; Goodey, J.; Hawkes, D. J. Voxel similarity measures for 3D serial MR brain image registration. IEEE Trans. Med. Imag. **2002**, *19* (2), 94–102.
49. Maes, F.; Collignon, A.; Vandermeulen, D.; Marchal, G.; Suetens, P. Multimodality image registration by maximization of mutual information. IEEE Trans. Med. Imag. **1997**, *16* (2), 187–198.

Artificial Neural Networks

Evangelia Micheli-Tzanakou
Department of Biomedical Engineering, Rutgers University, Piscataway, New Jersey, U.S.A.

Abstract
Neural networks have been a much publicized topic of research in recent years and are now beginning to be used in a wide range of subject areas. One of the strands of interest in neural networks is to explore possible models of biological computation.

INTRODUCTION

Neural networks have been a much publicized topic of research in recent years and are now beginning to be used in a wide range of subject areas. One of the strands of interest in neural networks is to explore possible models of biological computation. Human brains contain about 1.5×10^{11} neurons of various types, with each receiving signals through 10 to 10^4 synapses. The response of a neuron happens in about 1 to 10 msec. Yet, one can recognize an old friend's face and call him in about 0.1 sec. This is a complex pattern recognition task that must be performed in a highly parallel way, because the recognition occurs in about 100 to 1000 steps. This indicates that highly parallel systems can perform pattern recognition tasks more rapidly than current conventional sequential computers. As yet, very-large-scale integration (VLSI) technology, which is essential planar implementation with at most two or three layer cross-connections, is far from achieving these parallel connections that require 3-D interconnections.

ARTIFICIAL NEURAL NETWORKS

Although originally the neural networks were intended to mimic a task-specific subsystem of a mammalian or human brain, recent research has mostly concentrated on the ANNs, which are only vaguely related to the biological system. Neural networks are specified by 1) the net topology; 2) node characteristics; and 3) training or learning rules.

Topological consideration of the Artificial Neural Networks (ANNs) for different purposes is provided in review papers.[1,2] Because our interests in the neural networks are in classification only, the feed-forward multilayer perceptron topology is considered, leaving the feedback connections to the references.

The topology describes the connection with the number of the layers and the units in each layer for feed-forward networks. Node functions are usually non-linear in the middle layers but can be linear or non-linear for output layer nodes. However, all of the units in the input layer are linear and have fan-out connections from the input to the next layer.

Each output y_j is weighted by w_{ij} and summed at the linear combiner, represented by a small circle in Fig. 1. The linear combiner thresholds its inputs before it sends them to the node function φ_j. The unit functions are non-linear, monotonically increasing, and bounded functions as shown at the right in Fig. 1.

USAGE OF NEURAL NETWORKS

One use of a neural network is classification. For this purpose, each input pattern is forced adaptively to output the pattern indicators that are part of the training data; the training set consists of the input covariate x and the corresponding class labels. Feed forward networks, sometimes called multilayer perceptrons (MLP), are trained adaptively to transform a set of input signals, X, into a set of output signals, G. Feedback networks begin with the initial activity state of a feedback system. After state transitions have taken place the asymptotic final state is identified as the outcome of the computation. One use of the feedback networks is in the case of associative memories: upon being presented with a pattern near a prototype X it should output pattern X', and as autoassociative memory or contents addressable memory by which the desired output is completed to become X.

In all cases the network learns or is trained by the repeated presentation of patterns with known required outputs (or pattern indicators). Supervised neural networks find a mapping $f: X \rightarrow G$ for a given set of input and output pairs.

OTHER NEURAL NETWORKS

The other dichotomy of the neural networks family is unsupervised learning, that is clustering. The class

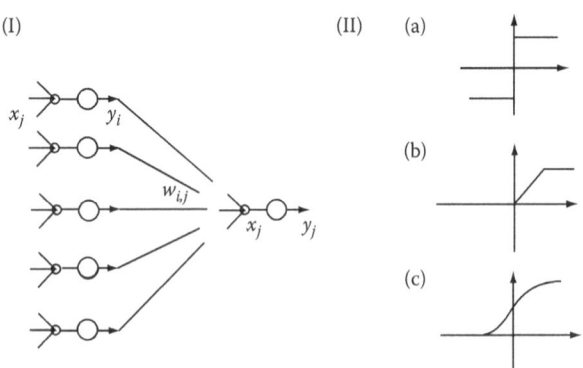

Fig. 1 (I) The linear combiner output $x_j = \sum_{i=1}^{n} y_i w_{ik}$. (II) Possible node functions: hard limiter (a) threshold (b) and sigmoid (c) non-linear function

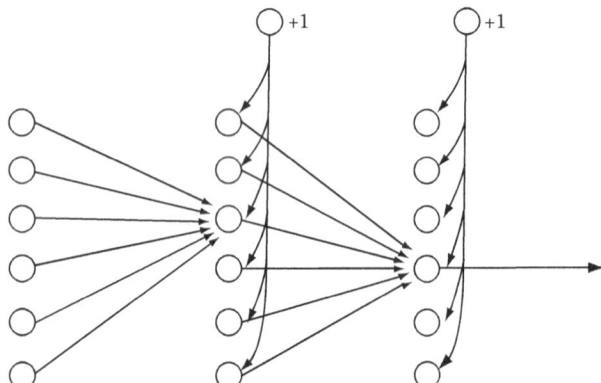

Fig. 2 A generic feed-forward network with a single hidden layer. For bias terms the constant inputs with 1 are shown and the weights of the constant inputs are the bias values, which will be learned as training proceeds

information is not known or is irrelevant. The networks find groups of similar input patterns.

The neighboring code vectors in a neural network compete in their activities by means of mutual lateral interactions and develop adaptively into specific detectors of different signal patterns. Examples are the Self Organizing Map[3] and the Adaptive Resonance Theory (ART) networks.[4] The ART is different from other unsupervised learning networks in that it develops new clusters by itself. The network develops a new code vector if there exist sufficiently different patterns. Thus, the ART is truly adaptive, whereas other methods require the number of clusters to be specified in advance.

FEED-FORWARD NETWORKS

In feed-forward networks the signal flows only in the forward direction; no feedback exists for any node. This is perhaps best illustrated graphically in Fig. 2. This is the simplest topology and is good enough for most practical classification problems.[5]

The general definition allows more than one hidden layer and also allows "skip layer" connections from input to output. With this skip layer, one can write a general expression for a network output y_k with one hidden layer,

$$y_k = \varphi_k \left\{ b_k + \sum_{i \to k} w_{ik} x_i + \sum_{j \to k} w_{jk} \varphi_j \left\{ b_j + \sum_{i \to j} w_{ij} x_i \right\} \right\}, \qquad (1)$$

where the b_j and b_k represent the thresholds for each unit in the jth hidden layer and the output layer, which is the kth layer. Because the threshold values b_j, b_k are to be adaptive, it is useful to have a threshold for the weights for constant input value of 1, as in Fig. 2. The function $\varphi(\ )$ is almost inevitably taken to be a linear, sigmoidal ($\varphi(x) = e^x/(1+e^x)$) or threshold function ($\varphi(x) = I(x>0)$).

Rumelhart et al.[6] demonstrated that the feed-forward multilayer perceptron networks can learn using gradient values obtained by an algorithm called error backpropagation.[1]

This contribution is a remarkable advance since 1969, when Minsky and Papert[7] claimed that the non-linear boundary required for the XOR problem can be obtained using a multilayer perceptron. The learning method was unknown at the time.

Rosenblatt[8] introduced the one-layer single perceptron learning method called the *perceptron convergence* procedure and research on the single perceptron was active until the counterexample of the XOR problem, which the single perceptron could not solve, was introduced.

In multilayer network learning the usual objective or error function to be minimized has the form of a squared error

$$E(w) = \sum_{p=1}^{p} \| t^p - f(x^p; w) \|^2 \qquad (2)$$

that is to be minimized with respect to **w**, the weights in the network. Here p represents the pattern index, $p = 1,2,\ldots,P$ and t^p is the target or desired value when x^p is the input to the network. Clearly, this minimization can be obtained by any number of unconstrained optimization algorithms; gradient methods or stochastic optimization are possible candidates.

The updating of weights has a form of the steepest descent method,

$$w_{ij} = w_{ij} - \eta \frac{\partial E}{\partial w_{ij}}, \qquad (3)$$

---

[1]A comment on the terminology backpropagation is given in the next section. There, backpropagation is interpreted as a method of finding the gradient values of a feed-forward multilayer perceptron network rather than as a learning method. A pseudosteepest descent method is the learning mechanism used in the network.

where the gradient value $\partial E/\partial w_{ij}$ is calculated for each pattern being present; the error term $E(w)$ in the on-line learning is not the summation of the squared error for all P patterns.

Note that the gradient points are in the direction of maximum increasing error. In order to minimize the error it is necessary to multiply the gradient vector by -1 and by a learning rate η.

The updating method (Eq. 3) has a constant learning rate η for all weights and is independent of time. The original method of steepest descent has the time-dependent parameter η_k; hence, η_k must be calculated as iterations progress.

ERROR BACKPROPAGATION

Backpropagation was first discussed by Bryson and Ho,[9] later researched by Werbos[10] and Parker,[11] and rediscovered and popularized by Rumelhart et al.[6,12] Each pattern is presented to the network and the input x_j and output y_j are calculated as in Fig. 3. The partial derivative of the error function with respect to weights is

$$\Delta E(t) = \left(\frac{\partial E(t)}{\partial w(t)} \cdots \frac{\partial E(t)}{\partial w_n(t)} \right)^T \quad (4)$$

where, n is the number of weights and t is the time index representing the instance of the input pattern presented to the network.

The former indexing is for the on-line learning in which the gradient term of each weight does not accumulate. This is the simplified version of the gradient method that makes use of the gradient information of all training data. In other words, there are two ways to update the weights using Eq. 4.

$$w_{ij}^{(p)} \leftarrow w_{ij}^{(p)} - \eta \left(\frac{\partial E}{\partial w_{ij}} \right)^{(p)} \text{ temporal learning} \quad (5)$$

$$w_{ij} \leftarrow w_{ij} - \eta \sum \left(\frac{\partial E}{\partial w_{ij}} \right)^{(p)} \text{ epoch learning} \quad (6)$$

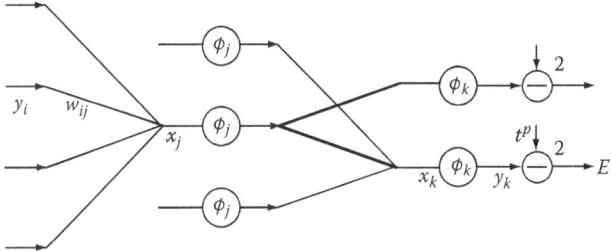

Fig. 3 Error backpropagation. The δ_j for weight w_{ij} is obtained. δ_k's are then back propagated via thicker weight lines, w_{jk}'s

One way is to sum all the P patterns to get the sum of the derivatives in Eq. 6 and the other way (Eq. 5) is to update the weights for each input and output pair temporally without summation of the derivatives. The temporal learning, also called on-line learning (Eq. 5), is simple to implement in a VLSI chip, because it does not require the summation logic and storing of each weight, whereas the epoch learning in Eq. 6 does. However, the temporal learning is an asymptotic approximation version of the epoch learning, which is based on minimizing an objective function (Eq. 2).

With the help of Fig. 3, the first derivatives of E with respect to a specific weight w_{jk} can be expanded by the chain rule:

$$\frac{\partial E}{\partial w_{jk}} = \frac{\partial E}{\partial x_k} \frac{\partial x_k}{\partial w_{jk}} = \frac{\partial E}{\partial x_k} y_j = \varphi'_k(x_k) \frac{\partial E}{\partial y_k} y_j \quad (7)$$

Or

$$\frac{\partial E}{\partial w_{jk}} = \frac{\partial \varphi_k(x)}{\partial x_k} \frac{\partial E}{\partial y_k} y_j = \delta_k y_j. \quad (8)$$

For output units, $\partial E/\partial y_k$ is readily available, i.e., $2(y_k - t^p)$, where y_k and t^p are the network output and the desired target value for input pattern x^p. The $\varphi'_k x_k$ is straightforward for the linear and logistic non-linear node functions; the hard limiter, on the other hand, is not differentiable.

For the linear node function,

$$\varphi'(x) = 1 \quad \text{with} \quad y = \varphi_x = x,$$

whereas for the logistic unit the first-order derivative becomes

$$\varphi'(x) = y(1-y) \quad (9)$$

when $\quad \varphi(x) = \dfrac{e^x}{1+e^x} \quad (10)$

The derivative can be written in the form

$$\frac{\partial E}{\partial w_{ij}} = \sum_P y_i^p \delta_j^p, \quad (11)$$

which has become known as the *generalized delta rule*.

The δ's in the generalized delta rule (Eq. 11) for output nodes, therefore, become

$$\delta_k = 2y_k(1-y_k)(y_k - t^p) \text{ for a logistic output unit} \quad (12a)$$

and

$$\delta_k = 2(y_p - t^p) \text{ for a linear output unit.} \quad (12b)$$

The interesting point in the backpropagation algorithm is that the δ's can be computed from output to input through hidden layers across the network; δ's for the units in earlier layers can be obtained by summing the δ's in the higher layers. As illustrated in Fig. 3, the δj are obtained as

$$\begin{aligned}\delta_j &= \phi'_j(x_j)\frac{\partial E}{\partial y_j} \\ &= \phi'_j(x_j)\sum_{j\to k} w_{jk}\frac{\partial E}{\partial x_k} \\ &= \phi'_j(x_j)\sum_{j\to k} w_{jk}\delta_k\end{aligned} \quad (13)$$

The δ_k's are available from the output nodes. As the updating or learning progresses backward, the previous (or higher) δ_k are weighted by the weights w_{jk}'s and summed to give the δ_j's. Because Eq. 13 for δ_j only contains terms at higher layer units, it can be calculated backward from the output to the input of the network; hence, the name backpropagation.

Madaline Rule III for multilayer network with sigmoid function

Widrow took an independent path in learning as early as the 1960s.[13,14] After some 30 years of research in adaptive filtering, Widrow and colleagues returned to neural network research[13] and extended Madaline I with the goal of developing a new technique that could adapt multiple layers of adaptive elements using the simpler hard-limiting quantizer. The result was Madaline Rule II (or simply MRII), a multilayer linear combiner with a hard-limiting quantizer.

Andes (unpublished, 1988) modified the MRII by replacing the hard-limiting quantizer, resulting in MRIII by a sigmoid function in the Madaline, i.e., a single-layer linear combiner with a hard-limiting quantizer. It was proven later that MRIII is, in essence, equivalent to backpropagation. The important difference from the gradient-based backpropagation method is that the derivative of the sigmoid function is not required in this realization; thus, the analog implementation becomes feasible with this MRIII multilayer learning rule.

A comment on the terminology backpropagation

The terminology backpropagation has been used differently from what it should mean. To obtain the partial derivatives of the error function (at the system output node) with respect to the weights of the units lower than the output unit, the terms in the output unit are propagated backward, as in Eq. 13. However, the network (actually the weights) learns (or weights are updated) using the pseudo steepest descent method (Eq. 3); it is pseudo because a constant term is used, whereas the steepest descent method requires an optimal learning rate for each weight and time instance, i.e., $\eta_{ij}(k)$.

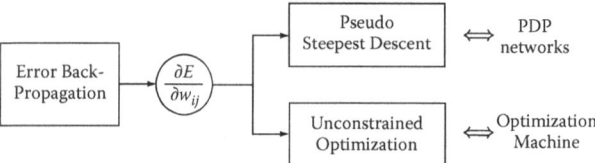

Fig. 4 Functional diagram for an optimization machine

The error backpropagation is indeed used to find the necessary gradient values in the updating rule. Thus, it is not a good idea to call the backpropagation a learning method: the learning method is a simple version of the steepest descent method, which is one of the classical minimizer finding algorithms. Backpropagation is an algorithm used to find the gradient of E in a feed-forward multilayer perceptron network.

Optimization machines with feed forward multilayer perceptrons

Optimization in multilayer perceptron structures can be easily realized by gradient-based optimization methods with the help of backpropagation. In the multilayer perceptron structure, the functions can be minimized/maximized via any gradient-based unconstrained optimization algorithm, such as Newton's method or the steepest descent method.

The optimization machine is functionally described as depicted in Fig. 4 and consists of two parts: gradient calculation and weight (or parameter) updating.

The gradient of E of the multilayer perceptron network is obtained by error backpropagation. If this gradient is used in an on-line fashion with the constant learning rate, as in Eq. 3, then this structure is the neural network used earlier.[12] This on-line learning structure possesses a desirable feature in VLSI implementation of the algorithm because it is temporal: no summation over all the patterns is required but the weights are updated as the individual pattern is presented to the network. It requires little memory but sometimes the convergence is too slow.

The other branch in Fig. 4 illustrates unconstrained optimization of the non-linear function. The optimization machine obtains the gradient information as before, but various and well-developed unconstrained optimizations can be used to find the optimizer. The unconstrained non-linear minimization is divided basically into two categories: gradient methods and stochastic optimization. The gradient methods are deterministic and use the gradient information to find the direction for the minimizer. Stochastic optimization methods such as the ALgorithm Of Pattern Extraction (ALOPEX) are discussed in another section of this chapter and in other references.[15–17]

Comparisons of ALOPEX with backpropagation are reported by Zahner and Micheli-Tzanakou.[15]

Justification for gradient methods for non-linear function approximation

Getting stuck in local minima is a well-known problem for gradient methods. However, the weights (or the dimensionality of the weight space in neural networks) are usually much larger than the dimensionality of the input space $X \subset R^p$ that is preferred when searching for optimization. The redundant degrees of freedom used to find a better minimizer justify the gradient methods used in neural networks.

Another justification for use of the gradient method in optimization may result from the approximation of highly nonlinear functions using Taylor expansion,[5] where the first- and second-order approximations, i.e., a quadratic approximation to the non-linear function, are used. The quadratic function in a covariate x has a unique minimum or maximum.

TRAINING METHODS FOR FEED-FORWARD NETWORKS

Two basic methods exist to train the feed-forward networks: gradient-based learning and stochastic learning. Training or learning is essentially an unconstrained optimization problem. A number of algorithm optimizations can be applied to the function approximated by the network in a structured way defined by the network topology.

In gradient-based methods, the most popular learning involves the steepest descent/ascent method with the error backpropagation algorithm to obtain the required gradient of the minimizing/maximizing error function with respect to the weights in the network.[6,12] Another method using the gradient information is Newton's method, which is basically used for zero finding of a non-linear function. The function optimization problem is the same as the zero finding of the first derivative of a function; hence, Newton's method is valid.

All the deterministic (as opposed to stochastic) minimization techniques are based on one or both of the steepest descent and Newton's methods. The objective function to be optimized is usually limited to a certain class in network optimization. The square of the error $\|t - \hat{y}\|^2$ and the information theoretic measuring the Kullback–Leibler distance are objective functions used in feed-forward networks. This is because of the limitation in calculating the gradient values of the network utilized by the error backpropagation algorithm.

The recommended method of optimization according to Broyden, Fletcher, Goldfarb, and Shannon (BFGS) is the well-known Hessian matrix update in Newton's method of unconstrained optimization,[18] which requires gradient values. For the optimization machine in Fig. 4, the feed-forward network with backpropagation provides the gradients and the Hessian approximation is obtained using the BFGS method.

The other dichotomy of the minimization of an unconstrained non-linear multivariate function is grouped into the so-called stochastic optimization. The representative algorithms are simulated annealing,[19] Boltzmann machine learning,[20] and ALOPEX.[21,22] Simulated annealing[19] has been successfully used in combinatoric optimization problems such as the traveling salesman problem, VLSI wiring, and VLSI placement problems. An application of feed-forward network learning was reported,[23] with the weights being constrained to be integers or discrete values rather than a continuum of the weight space.

Boltzmann machine learning, by Hinton and Sejnowski,[20] is similar to simulated annealing except that the acceptance of randomly chosen weights is possible even when the energy state is decreased. In simulated annealing, the weights yielding the decreased energy state are always accepted, but in the Boltzmann machine probability is used in accepting the increased energy states.

The simulated annealing and the Boltzmann machine learning (a general form of Hopfield network[24] for the associative memory application) algorithms are mainly for combinatoric optimization problems with binary states of the units and the weights. Extension from binary to M-ary in the states of the weights has been reported for classification problems[24] in simulated annealing training of the feed-forward perceptrons.

The ALOPEX was originally used for construction of visual receptive fields,[21,22,25,26] but with some modifications it was later applied to the learning of any type of network not restricted to multilayer perceptrons. The ALOPEX is a random walk process in each parameter in which the direction of the constant jump is decided by the correlation between the weight changes and the energy changes.

THE ALOPEX

The ALOPEX process is an optimization procedure that has been successfully demonstrated in a wide variety of applications. Originally developed for receptive field mapping in the visual pathway of frogs, ALOPEX's usefulness, combined with its flexible form, increased the scope of its applications to a wide range of optimization problems. Since its development by Tzanakou and Harth in 1973,[21,22] ALOPEX has been applied to real-time noise reduction,[27] and pattern recognition,[28] in multilayer neural network training, among other applications.

Optimization procedures in general attempt to maximize or minimize a function $F(\ )$. The function $F(\ )$ is called the cost function and its value depends on many parameters or variables. When the number of parameters is large, finding the set that corresponds to the optimal solution is exceedingly difficult. If N were small, then one could perform an exhaustive search of the entire parameter space to find the "best" solution. As N increases, intelligent algorithms are needed to quickly locate the solution. Only an exhaustive search can guarantee that a global optimum is found; however, near-optimal solutions are acceptable because of the tremendous speed improvement over exhaustive search methods.

Backpropagation, described earlier, being a gradient descent method, often gets stuck in local extrema of the cost function. The local stopping points often represent unsatisfactory convergence points. Techniques have been developed to avoid the problem of local extrema, with simulated annealing[19] being the most common. Simulated annealing incorporates random noise, which acts to dislodge the process from local extremes. It is crucial to the convergence of the process that the random noise be reduced as the system approaches the global optimum. If the noise is too large, the system will never converge and can be mistakenly dislodged from the global solution.

The ALOPEX is another process that incorporates a stochastic element to avoid local extremes in search of the global optimum of the cost function. The cost function or response is problem dependent and is generally a function of a large number of parameters. The ALOPEX iteratively updates all parameters simultaneously based on the cross-correlation of local changes, ΔX_i, and the global response change ΔR, plus additive noise. The cross-correlation term $\Delta X_i \Delta R$ helps the process move in a direction that improves the response (Eq. 14). Table 1 illustrates how this can be used to find a global maximum of R.

All parameters X_i are changed simultaneously in each iteration according to

$$X_i(n) = X_i(n-1) + \gamma \Delta X_i(n) \Delta R(n) + r_i(n). \quad (14)$$

The basic concept is that this cross-correlation provides a direction of movement for the next iteration. For example, take the case where $X_i\downarrow$ and $R\uparrow$. This means that the parameter X_i decreased in the previous iteration, and the response increased for that iteration. The product $\Delta X_i \Delta R$ is a negative number, and thus X_i would be decreased again in the next iteration. This makes perfect sense because a decrease in X_i produced a higher response; if one is looking for the global maximum then X_i should be decreased again. Once X_i is decreased and R also decreases, then $\Delta X_i \Delta R$ is now positive and X_i increases.

These movements are only tendencies because the process includes a random component $r_i(n)$, which will act to move the weights unpredictably, avoiding local extrema of the response. The stochastic element of the algorithm helps avoid local extrema at the expense of slightly longer convergence or learning periods.

The general ALOPEX updating equation (Eq. 14) is explained as follows. $X_i(n)$ are the parameters to be updated, n is the iteration number, and $R(\ )$ is the cost function, of which the best solution in terms of X_i is sought. Gamma, γ, is a scaling constant, $r_i(n)$ is a random number from a gaussian distribution whose mean and standard deviation are varied, and $\Delta X_i(n)$ and $\Delta R(n)$ are found using

Table 1 Combinations of parameter changes and their contribution to the ALOPEX convergence

	ΔX_i		ΔR	$\Delta X_i \Delta R$
$X_i\uparrow$	+	$R\uparrow$	+	+
$X_i\uparrow$	+	$R\downarrow$	−	−
$X_i\downarrow$	−	$R\uparrow$	+	−
$X_i\downarrow$	−	$R\downarrow$	−	+

$$\Delta X_i(n) = X_i(n-1) - X_i(n-2) \quad (15)$$

$$\Delta R(n) = R(n-1) - R(n-2). \quad (16)$$

The calculation of $R(\ )$ is problem dependent and can be easily modified to fit many applications. This flexibility was demonstrated in the early studies of Harth and Tzanakou.[22] In mapping receptive fields, no a priori knowledge or assumptions were made about the calculation of the cost function, instead a response was measured with microelectrodes inserted in the brain. Using action potentials as a measure of the response,[22,25,26] receptive fields could be determined using the ALOPEX process to iteratively modify the stimulus pattern until it produced the largest response. This response was obtained from the stimulated neurons. (This was the first computer-to-brain interfacing ever used with electrodes inserted in an animal brain to measure neuronal activity and use it as feedback to the computer.)

Because of its stochastic nature, efficient convergence depends on the proper control of both the additive noise and the gain factor γ. Initially, all parameters X_i are random and the additive noise has a gaussian distribution with mean 0 and standard deviation σ, initially large. The standard deviation decreases as the process converges to ensure a stable stopping point (Fig. 5). Conversely, γ increases with iterations (Fig. 6). As the process converges, ΔR becomes smaller and smaller, requiring an increase in γ to compensate.

Additional constraints include a maximal change permitted for X_i, for a single iteration. This bounded step size prevents the algorithm from drastic changes from one iteration to the next.

These drastic changes often lead to long periods of oscillation, during which the algorithm fails to converge.

MULTILAYER PERCEPTRON (MLP) NETWORK TRAINING WITH ALOPEX

An MLP can also be trained for pattern recognition using ALOPEX. A response is calculated for the jth input pattern based on the observed and desired output,

$$R_j(n) = O_k^{des} - (O_k^{obs}(n) - O_k^{des})^2, \quad (17)$$

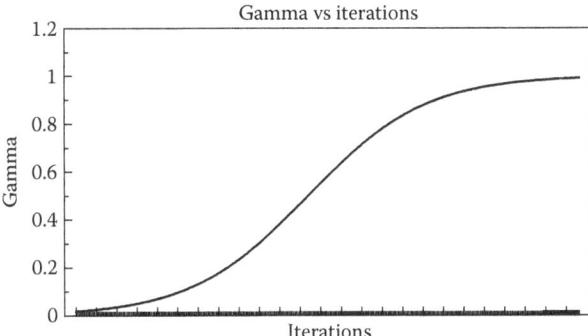

Fig. 5 Gamma vs. Iterations

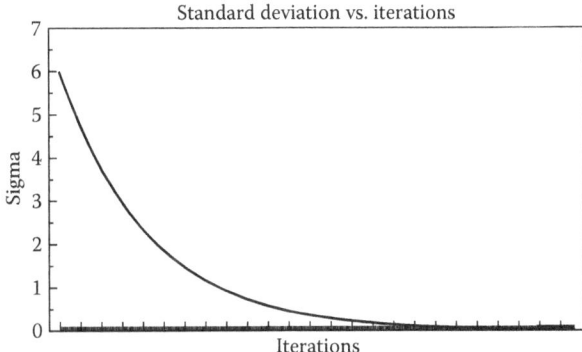

Fig. 6 Gaussian noise standard deviation vs. Iterations

where O_k^{obs} and O_k^{des} (the observed and desired outputs, respectively) are vectors corresponding to O_k for all k. The total response for iteration n is the sum of all individual template responses, $R_j(n)$.

$$R(n) = \sum_{j=1}^{m} R_j(n) \qquad (18)$$

In Eq. 18, m is the number of templates used as inputs. The ALOPEX iteratively updates the weights using both the global response information and the local weight histories, according to the following equations,

$$W_{ij}(n) = r_i(n) + \gamma \Delta W_{ij}(n) \Delta R(n) + W_{ij}(n-1) \qquad (19)$$

$$W_{jk}(n) = r_i(n) + \gamma \Delta W_{jk}(n) \Delta R(n) + W_{ik}(n-1), \qquad (20)$$

where γ is an arbitrary scaling factor, $r_i(n)$ is the additive gaussian noise, ΔW represents the local weight change, and ΔR represents the global response information. These values are calculated by

$$\Delta W_{ij}(n) = W_{ij}(n-1) - W_{ij}(n-2) \qquad (21)$$

$$\Delta W_{jk}(n) = W_{jk}(n-1) - W_{jk}(n-2) \qquad (22)$$

$$\Delta R(n) = R(n-1) - R(n-2). \qquad (23)$$

After the network was trained, it was tested for correct recognition using incomplete or noisy input patterns. The results demonstrate the robustness of the system to noise-corrupted data. Regardless of which training procedure was used, backpropagation or ALOPEX, the recognition ability of the system was the same. The only difference was in how the response grew with iterations.

The neural network's robustness is derived from its parallel architecture and depends on the network topology rather than the learning scheme used to train. The network used was a three-layer feed-forward network with 35 input nodes, 20 hidden nodes, and 10 output nodes. The network's recognition ability was tested with noisy input patterns. Each 5×7 digit of the training set was subjected to noise of varying gaussian distribution and tested for correct recognition. The single-value response feedback, as opposed to the extensive error propagation schemes of other neural network training algorithms, makes ALOPEX suitable for fast VLSI implementation.

In our laboratory, an analog VLSI chip was designed to implement the ALOPEX algorithm, as described in another section. Making full use of the algorithm's tolerance to noise, an analog design was chosen. Analog designs offer larger and faster implementations than digital designs.

ISSUES IN NEURAL NETWORKS

Universal approximation

More than 30 years ago, Minsky and Papert[7] reported that simple two-layer (no hidden layers) networks cannot approximate the non-linearly separating functions (e.g., XOR problems), but a multilayer neural network could do the job. Many results on the capability of multilayer perceptron have been reported. Some theoretical analyses for the network capability of the universal approximator are listed below and are extensively discussed by Hornik et al.[29]

Kolmogorov[30] tried to answer the question of Hilbert's 13th problem, i.e., the multivariate function approximation by a superposition of the functions of one variable. The superposition theory sets the upper limit of the number of hidden units to $(2n+1)$ units, where n is the dimensionality of the multivariate function to be approximated. However, the functional units in the network are different for the different functions to be approximated, while one would like to find an adaptive method to approximate the function from the given training data at hand. Thus, Kolmogorov's superposition theory says nothing about the capability of a multilayer network or which method is best.

More general views have been reported. Le Cun[31] and Lapedes and Farber[32] demonstrated that monotone squashing functions can be used in the two hidden layers to approximate the functions. Fourier series expansion of a function was realized by a single-layer network by Gallant and White[33] with cosine functions in the units.

Further related results using the sigmoidal or logistic units are demonstrated by Hecht-Nielsen.[34] Hornik et al.[29] presented a general approximation theory of one hidden layer network using arbitrary squashing functions such as cosine, logistic, hyperbolic tangent, etc., provided that sufficiently many hidden units are available. However, the number of hidden units is not considered to attain any given degree of approximation.[29]

The number of hidden units obviously depends on the characteristics of the training data set, i.e., the underlying function to be estimated. It is intuitive to say that the more complicated functions to be trained, the more hidden units are required.

For the number of the hidden units Baum and Haussler[35] limit the size of general networks (not necessarily the feed-forward multilayer perceptrons) by relating it to the size of the training sample. The authors analytically demonstrated that if the size of the sample is N, and we want to correctly classify future observations with at least a fraction $(1-\varepsilon/2)$, then the size of the sample has a lower bound given by

$$N \geq 0\left(\frac{W}{\varepsilon}\log\frac{N}{\varepsilon}\right),$$

where W is the number of the weights and N is the number of the nodes in a network. This, however, does not apply to the interesting feed-forward neural networks and the given bound is not useful for most applications.

There seems to be no rule of thumb for the number of hidden units.[5] Finding the size of the hidden units can usually be performed by cross-validation or any other resampling methods. The usual starting value for the size is suggested to be about the average of the number of the input and output nodes.[15] Failure in learning can be attributed to three main reasons:[29]

- Inadequate learning.
- Inadequate number of hidden units.
- The presence of a stochastic rather than a deterministic relation between input and target in the training data, i.e., noisy training data.

THE VLSI IMPLEMENTATION OF ALOPEX

Artificial neural networks have existed for many years, yet because of recent advances in technology they are again receiving much attention. Major obstacles in ANN, such as a lack of effective learning algorithms, have been overcome in recent years. Training algorithms have advanced considerably, and now VLSI technology may provide the means for building superior networks. In hardware, the networks have much greater speed, allowing for much larger architectures.

The tremendous advancement in technology during the past decades, particularly in VLSI technology, renewed interest in ANNs. Hardware implementations of neural networks are motivated by a dramatic increase in speed over software models. The emergence of VLSI technology has and will continue to lead neural network research in new directions. The past few years saw a considerable advancement in VLSI. Chips are now smaller, faster, contain larger memories, and are becoming cheaper and more reliable to fabricate.

Neural network architectures are varied, with over 50 different types explored in the research.[34] Hardware implementations can be electronic, optical, or electro-optical in design. A major problem in hardware realizations is often not a result of the network architecture, but the physical realities of the hardware design. Optical computers, while they may eventually become commercially available, suffer far greater problems than do VLSI circuits. Thus, for the immediate and near future, neural network hardware designs will be dominated by VLSI.

Much debate exists as to whether digital or analog VLSI design is better suited for neural network applications. In general, digital designs are easier to implement and better understood. In digital designs, computational accuracy is only limited by the chosen word length. Although analog VLSI circuits are less accurate, they are smaller, faster, and consume less power than digital circuits.[36] For these reasons, applications that do not require great computational accuracy are dominated by analog designs.

Learning algorithms, especially backpropagation, require high precision and accuracy in modifying the weights of the network, which led some to believe that analog circuits are not well suited for implementing learning algorithms.[37] Analog circuits can achieve high precision at the cost of increasing the circuit size. Analog circuits with high precision (8 bits) tend to be equally large compared with their digital counterparts.[38] Thus, high-precision analog circuits lose their size advantage over digital circuits. Analog circuits are of greater interest in applications requiring only moderate precision.

Early studies demonstrate that analog circuits can realize learning algorithms, provided that the algorithm is tolerant to hardware imperfections such as low precision and inherent noise. Macq et al. present a fully analog implementation of a Kohonen map, one type of neural network, with on-chip learning.[39] Because analog circuits have been proven capable of the computational accuracy necessary for weight modification, they should continue to be the choice of neural network research.

Analog circuits are far superior to digital circuits in size, speed, and power consumption, and these areas constrain most neural network applications. To achieve greater network performance, the size of the network must be increased. The ability to implement larger, faster networks is the major motivation for hardware implementation, and analog circuits are superior in these areas. Power consumption is also a major concern as networks become larger.[40] As the number of transistors per chip increases, power consumption becomes a major limitation. Analog circuits dissipate less power than digital circuits, thus permitting larger implementations.

Beside its universality to a wide variety of optimization procedures, the nature of the ALOPEX algorithm makes it suitable for VLSI implementation. The ALOPEX is a biologically influenced optimization procedure that uses a single-value global response feedback to guide weight movements toward their optimum. This single-value feedback, as opposed to the extensive error propagation schemes of other neural network training algorithms, makes ALOPEX suitable for fast VLSI implementation.

Recently, a digital VLSI approach to implementing the ALOPEX algorithm was undertaken by Pandya et al.[41] Results of their study indicated that ALOPEX could be implemented using a single-instruction multiple-data (SIMD) architecture. A simulation of the design was carried out in software and good convergence for a 4×4 processor array was demonstrated.

The importance of VLSI to neural networks has been demonstrated. For neural networks to achieve greater abilities, larger and faster networks must be built. In addition to size and speed advantages, other factors, including cost and reliability, make VLSI implementations the current trend in neural network research. The design of a fast analog optimization algorithm, ALOPEX, is discussed below.

Let us assume that we have an array of 64 pixels that we call $I_i(n)$, where n represents the iteration. The additive gaussian white noise is denoted by $r_i(n)$ and $R_j(n)$ is the response (or cost function) of the jth template at iteration n.

Let us assume that there are 16 templates to choose from, each with 64 pixels. The ALOPEX process is run on each of them with the objective being to recognize (converge to) an input pattern. As a result of the iterative behavior, if allowed to run long enough ALOPEX will eventually converge to each of the templates. However a "match" can be found by choosing the template that took the least amount of time to converge.

By convergence we mean finding either the global maximum or the minimum of the response function. This response function can be calculated many different ways, depending on the application. To allow this chip to be general enough to handle many applications, the response is computed off the chip. A Programmable Read Only Memory (PROM) is used to compute the response based on the error between the input, $I_i(n)$, and the template. The PROM enables the response function to be changed to meet the needs of the application.

Although the chip design is limited to only 64 ALOPEX subunits, the parallel nature of ALOPEX enables many chips to be wired together for larger applications. Parallel implementations are made easy because each subunit receives a single global response feedback, which governs its behavior. Backpropagation, on the other hand, requires dense interconnections and communication between each node. This flexibility is a tremendous advantage when it comes to hardwired implementations.

Originally, the ALOPEX chip was designed using digital VLSI techniques. Digital circuitry was chosen over analog because it is easier to test and design. Floating point arithmetic was used to insure a high degree of accuracy. The digital design consisted of shift registers, floating point adders, and floating point multipliers. However, after having done much work toward the digital design it was abandoned in favor of an analog design. The performance of the digital design was estimated and was much slower than that of an analog design. The chip area of the digital design was much larger than that of an analog design would be. Also, the ALOPEX algorithm would be tolerant of analog imperfections because of its stochastic nature. For these reasons, it seemed clear that a larger, faster network could be designed with analog circuitry.

The analog design required components similar to the digital design to implement the algorithm. The main requirements were an adder, a multiplier, a difference amplifier, a sample and hold mechanism, and a multiplexing scheme. These cells each perform a specific function and are wired together in a way that implements the ALOPEX process.

The chip is organized into 64 ALOPEX subunits, one for each pixel in the input image. They are stacked vertically, wiring by abutment. Each subunit is composed of smaller components that are wired together horizontally and contains the following cells: a group selector, demultiplexor, follower aggregator, multiplier, transconductance amplifier, multiplexor, and another group selector.

The gaussian white noise required for the ALOPEX process is added to the input before it reaches the chip. This will allow precise control of the noise, which is very important in controlling the stability of the algorithm. If there is too much noise, the system will not converge. If there is too little noise, the system will get stuck in local minima of the cost function. By controlling the noise during execution, using a method similar to simulated annealing[19] where the noise decays with time, the convergence time can be improved.[28] By having direct control of the added noise, the component and functional testing can be performed with no noise added, greatly simplifying the testing.

The addition, multiplication, and subtraction required by the ALOPEX algorithm are performed by the follower aggregator, the Gilbert multiplier,[36] and the transconductance amplifier, respectively. The follower aggregation circuit computes the weighted average of its inputs. By weighing the inputs equally, the circuit computes the average of the two inputs. The average was chosen instead of the sum because the circuit is more robust, in that the output never has to exceed the supply voltage. A straight summer is more difficult to design because voltages greater than the supply voltage could be required. The output of the follower aggregator is sent to the multiplier, where a

C-switch acts as a sample and hold, to store the value of the previous iteration. The difference between these signals is one input to the multiplier. The previous two responses, calculated off chip, are the other two inputs. The output of the multiplier is controlled by the control signal γ. The error signal is computed by a trans-conductance amplifier.

In designing the chip, much effort was made in making it controllable and testable, while making the chip general enough that it could be used in a wide variety of applications. This is why the gaussian white noise is added off chip and also why the error signal is taken off chip for the computation of the response. This not only allows the response function to be changed to meet the requirements of the specific application, but also provides the operator with accessible test points.

Although backpropagation is the most widely used software tool for training neural networks, it is less suitable for VLSI hardware implementation than ALOPEX for many reasons. Whereas backpropagation converges quickly as a result of its gradient descent method, it can often get stuck in local extrema. The ALOPEX tends to avoid local extrema by incorporating the random noise component at the expense of slightly longer convergence times.

The major differences arise when hardware implementation is discussed. Backpropagation is computationally taxing because of the error computation required for each node in the network. Each error is a function of many parameters (i.e., all the weights of the following layer). In hardware, very complex interconnections between all nodes are required to compute this error.

The ALOPEX is ideal for VLSI implementation for a couple of reasons. First, the algorithm is tolerant to small amounts of noise; in fact, noise is incorporated to help convergence. Second, all parameters change based on their local history and a single-value global response feedback. This single-value feedback is much simpler to implement than the error propagation used in backpropagation.

NEUROMORPHIC MODELS

Since the era of the vacuum tube, a multitude of neuronal models composed of discrete components and off-the-shelf Integrated Circuits (ICs) have been published. Similar efforts in custom VLSI, however, are far fewer in number. A good introduction to a number of neuronal attributes was presented by Linares-Barranco et al.[42] Complementary Metal Oxide Semiconductor (CMOS)-compatible circuits for approximating a number of mathematical models of cell behavior are described. In its simplest form, this model represents the cell membrane potential in the axon hillock as nothing more than a linear combination of an arbitrary number, n, of dendritic inputs, X, each of which is weighted by a unique multiplier, W, summed, and processed by a non-linear range-limiting operator, f. The mathematical equation for this relationship is

$$Y_k = f\left\{\sum_{i=1}^{n} W_i X_i\right\} = f\{S_k\}, \qquad (24)$$

This relationship is realized in a CMOS circuit. This circuit is totally static and makes no provision for time courses of changes in input or output signals or intracellular relationships. An operational transconductance amplifier (OTA), as described by Wolpert and Micheli-Tzanakou,[43] is used in lieu of operational amplifiers for this and most other VLSI neural network applications. Highly compatible with CMOS circuit technology, it is structurally simple and compact, realizable with only nine transistors, and provides reasonable performance. The only consideration it warrants is that its transfer function is a transconductance. As such, operations performed on its output signals must be oriented to its current, rather than its voltage. When driving high load impedances, as is usually the case with CMOS circuitry, this is only a minor inconvenience, necessitating buffering for lower load impedances. In fact, under some circumstances, such as when algebraic summation is being performed, a current output may actually be an advantage, allowing output nodes to be simply tied together.

The non-linear range-limiting operator, f, mentioned earlier, is necessitated by the observation that, for a given biological neuron, there are limits on the strength of the electrochemical gradients the cell's ionic pumps can generate. This imposes limits on how positive and negative cell membrane potential may go. Because a neuron may receive inputs from many other neurons, there is no such limit on the aggregate input voltage applied. As a result, an *activation function*, a non-linearity of the relationship between aggregate input potential and output potential of a neuron, must be imposed. This is typically done in one of three ways: the binary hard-limiter, which assumes one of only two possible states, active or inactive; the linear-graded threshold, which assumes a linear continuum of active states between its minimal and maximal values; and the sigmoid, which assumes a sigmoidal distribution of values between its negative minimal and positive maximal output values.

Which type of activation function is employed depends on the type of artificial neuron and network in which it is implemented. In networks where cell outputs are all-or-none, such as the McCullouch and Pitts models,[44] the binary threshold model is used. In networks where neurons are theorized to have variable output levels applied to distinctly designated excitatory and inhibitory inputs, such as Hopfield networks,[24] the linear threshold model is used. In networks where a synaptic connection must be both excitatory and inhibitory, depending on the level of activity, the sigmoid threshold is used. In either of the latter two activation functions, the slope of the overall characteristic can be varied to suit the sensitivity of the cell in question.

Another well-developed implementation of individual artificial nerve cells is that by Wolpert and Micheli-Tzanakou.[43,45] Whereas most neuromorphic models are based on the Hodgkin–Huxley equations, this one uses a sequencer to synthesize the action potential in three distinct phases. It also employs a different formulation for cell membrane and threshold potentials known as an integrate-and-fire model, presented and implemented in discrete components by French and Stein in 1970.[46] It makes use of the aforementioned leaky integrator and provides off-chip control over the response and persistence of stimuli assimilated into the membrane potential. The model affords similar controls over the resting level and time constant of the cell threshold potential and allows refraction to be recreated. This organization also affords control over the shape, resting level, and duration of the action potential and produces a Transistor-Transistor Logic (TTL)-compatible pulse in parallel with the action potential. These controls, all of which are continuously and precisely adjustable, make this model ideal for replicating the behavior of a wide variety of individual nerve cells, and it has been successfully applied as such.

The Wolpert and Micheli-Tzanakou model is organized around three critical nodes, the somatic potential, the axonal potential, and the threshold potential. Each of these nodes is biased off-chip with an R-C network so that its resting level and time constant are independently and continuously controllable. Stimuli to the cell are buffered and standardized by truncation into 10-μsec impulses. Synaptic weight inputs on the excitatory and inhibitory pathways allow this value to be increased or decreased from off-chip. The impulses are then applied to a somatic potential by a push–pull Metal-Oxide-Semiconductor Field-Effect Transistor (MOSFET) stage and compared with threshold potential by an OTA acting as a conventional voltage comparator. When threshold is exceeded, an action potential is synthesized and outputted. This waveform is then binarized and buffered to form a binary-compatible output pulse. At the same time, threshold is elevated to form the refractory period. The circuit consists of approximately 130 transistors plus a few on-chip and discrete resistors and capacitors and was implemented in a conventional CMOS technology, requiring a single-ended DC supply of 4–10 V DC and occupying 0.6 mm² of chip area.

With its critical nodes bonded out off-chip, Wolpert and Micheli-Tzanakou's neuromime rate of operation may be accelerated from a biologically compatible time frame over several orders of magnitude. This model was first implemented in 1986[43] and was intended as a flexible and accurate esthetic, rather than a mathematical model of cell behavior. Since then, it has been used to successfully recreate a number of networks from well-documented biological sources. Waveforms obtained in these recreations have demonstrated a striking similarity to intracellular recordings taken from their biological counterparts. It has also been successfully applied to problems in robotics and rehabilitation.

Neurological Process Modeling

Lateral inhibition is the process in which a cell containing some level of information encoded as its output level acts to inhibit and is inhibited by a similar adjoining cell. For many years, this process has been observed with striking regularity in both 1- and 2-D arrays of sensory receptors in a variety of systems and a variety of organisms. In numerous morphological, mathematical, and circuit studies, it has been identified as a key image preprocessing step that optimizes a sensory image to facilitate fast and accurate recognition in subsequent operations. Lateral inhibition accomplishes this by amplifying differences, enhancing image contrast, lending definition to its outward shape, and isolating the image from its background. Whereas a digital computer would accomplish this process one pixel at a time, biological systems manage it in a manner that is both immediate and simultaneous.

Laterally inhibited behavior has been observed in pairs of cells implemented in hardware and software models by many researchers, but in dedicated VLSI by only a few. Among them, Nabet and Pinter of Drexel University and Darling of the University of Washington have extensively studied the stability and effectiveness of both pairs and linear strings of mutually inhibiting cells in CMOS VLSI and obtained results that correlate well with biological data.[47] This line of work has been explored in two dimensions in another series of VLSI-based models by Wolpert and Micheli-Tzanakou.[48] Arrays of mutually inhibiting cells that inhibit via continuously active connections and cells that inhibit by dynamic or strobed controls both offered stable and variable control over the degree of inhibition. Arrays of hexagonally interconnected cells were more stable than the square array, which tended to "checkerboard" when significant levels of inhibition were attempted. Feedback inhibition, where one array is used to store both the initial and the inhibited images, was as effective, but less convenient to access than feed-forward inhibition, where separate input and inhibited images are maintained.

Characterization of lateral inhibition in the context of a more specific biological model has been pursued in another noteworthy effort by Andreou et al.[49] Multiple facets of cell–cell interactions, including both mutual inhibition and leakage of information between adjoining cells, were implemented in VLSI as a model of early visual processing in the mammalian retina.[49] There, adjacent cells on the photoreceptor layer intercommunicate through gap junctions, where their cell membrane potentials couple through a resistive path. Simultaneously, optical information from the photoreceptor cells is downloaded to corresponding cells of the horizontal layer, which have mutually inhibitory connections. One-dimensional arrays, and, subsequently, 2-D models of these relationships were implemented in analog VLSI and tested. Although little numerical data were published from these arrays, the 2-D array produced a number of optical effects associated with the human visual system, including Mach bands, simultaneous contrast enhancement, and the Herman–Herring illusion, all of which are indicative of the real-time image processing known to occur in the mammalian retina.

Finally, the definitive VLSI implementation of a 2-D array is the well-known silicon retina devised by Mead and described in his text, "Analog VLSI and Neural Systems."[50] Mead presents a comprehensive treasury of analog VLSI circuits for a variety of mathematical operations necessary to implement neural networks in VLSI. The book then goes on to discuss several applications of analog ANNs, culminating in an auditory model of the cochlea and a visual model of the retina.

ARTICLES OF FURTHER INTEREST

Biopotential Amplifiers; Biosensors; Compression of Digital Biomedical Signals; Control of Movement; Digital Biomedical Signal Acquisition and Processing: Basic Topics; Digital Biomedical Signal Processing: Advanced Topics; Excitable Tissue, Electrical Stimulation Of; Modeling, Biomedical

REFERENCES

1. Lippmann, R.P. An introduction to computing with neural nets. IEEE ASSP Mag. **1987**, April, 4–22.
2. Hush, D.; Horne, B. Progress in supervised neural networks. IEEE Signal Process. Mag. **1993**, January, 8–39.
3. Kohonen, T. The self-organizing map. Proc. IEEE **1990**, *78* (9), 1464–1480.
4. Carpenter, G.A.; Grossberg, S. Art2 Self-organization of stable category recognition codes for analog input patterns. Appl. Opt. **1987**, *26*, 4919–4930.
5. Ripley, B.D. Statistical aspects of neural networks. In *Chaos and Networks: Statistical and Probabilistic Aspects*; Barndorff-Neilsen, O.E.; Cox, D.R.; Jensen, J.L.; Kendall, S.S.; Eds.; Chapman & Hall: London, 1993.
6. Rumelhart, D.E.; Hinton, G.E.; Williams, R.J. Learning internal representations by error back propagation. In *Parallel Distributed Processing Explorations in the Microstructure of Cognition, I: Foundations*; Rumelhart, D.E.; McClelland, J.L.; the PDP Research Group, Eds.; MIT Press: Cambridge, MA, 1986, Chap. 8.
7. Minsky, M.; Papert, S. *Perceptrons: An Introduction to Computational Geometry*; MIT Press: Cambridge, MA, 1969.
8. Rosenblatt, F. *Principles of Neurodynamics*; Spartan Books: New York, 1959.
9. Bryson, A.E.; Ho, Y.C. *Applied Optimal Control*; Bleisdell: New York, 1969.
10. Werbos, P.J. *Beyond Regression: New Tools for Prediction and Analysis in the Behavioral Sciences*; PhD thesis; Harvard University, Cambridge, MA, 1974.
11. Parker, D.B. *Learning Logic*; Technical Report; Center for Computational Research in Economics and Management Science, MIT: Cambridge, MA, 1985.
12. Rumelhart, D.E.; Hinton, G.E.; Williams, R.J. Learning representations by back-propagating errors. Nature **1986**, *323*, 533–536.
13. Widrow, B.; Lehr, M. 30 years of adaptive neural networks: Perceptron, madaline and backpropagation. Proc. IEEE **1990**, *78* (9), 1415–1442.
14. Widrow, B. Generalization and information storage in networks of adaline "Neurons." In *Self-Organizing Systems*; Yovitz, M.; Jacobi, G.; Goldstein, G.; Eds.; Spartan Books: Washington, DC, 1962, 435–461.
15. Zahner, D.; Micheli-Tzanakou, E. Alopex and backpropagation. In *Supervised and Unsupervised Pattern Recognition*, Feature Extraction in Computational Intelligence; Micheli-Tzanakou, E.; Ed.; CRC Press: Boca Raton, FL, 2000, Chap. 2, 61–78.
16. Micheli-Tzanakou, E.; Uyeda, E.; Sharma, A.; Ramanujan, K.S.; Dong, J. Face recognition: Comparison of neural networks algorithms. Simulation **1995**, *64*, 37–51.
17. Micheli-Tzanakou, E. Neural networks in biomedical signal processing. In *The Biomedical Eng. Handbook*, 2nd Ed.; Bronzino, J.; Ed.; CRC Press: Boca Raton, FL, 1995, Chap. 60, 917–931.
18. Peressini, A.L.; Sullivan, F.E.; Uhl, J.J., Jr. *The Mathematics of Nonlinear Programming*; Springer-Verlag: New York, 1988.
19. Kirkpatrick, S.; Gelatt, C.D. Jr.; Vecchi, M.P. Optimization by simulated annealing. Science **1983**, *220* (4598), 671–680.
20. Hinton, G.E.; Sejnowski, T.J. Learning and relearning in Boltzmann machines. In *Parallel Distributed Processing: Explorations in the Microstructure of Cognition I: Foundations*; Rumelhart, D.E.; McClelland, J.L.; and the PDP Research Group; Eds.; MIT Press: Cambridge, MA, 1986, Chap. 7.
21. Tzanakou, E.; Harth, E. Determination of visual receptive fields by stochastic methods. Biophys. J. **1973**, *15*, 42a.
22. Harth, E.; Tzanakou, E. Alopex: A stochastic method for determining visual receptive fields. Vis. Res. **1974**, *14*, 1475–1482.
23. Engel, J. Teaching feed-forward neural networks by simulated annealing. Complex Syst. **1988**, *2*, 641–648.
24. Hopfield, J.J.; Tank, D.W. Neural computation of decisions in optimization problems. Biol. Cybernet. **1985**, *52*, 141–152.
25. Tzanakou, E.; Michalak, R.; Harth, E. The ALOPEX process: Visual receptive fields by response feedback. Biol. Cybernet. **1979**, *35*, 161–174.
26. Micheli-Tzanakou, E. Non-linear characteristics in the frog's visual system. Biol. Cybern. **1984**, *51*, 53–63.
27. Ciaccio, E.; Tzanakou, E. *The ALOPEX process: Application to real-time reduction of motion artifact. In. Annual International Conference of IEEE-EMBS*; Philadelphia, **1990**, *12* (3), 1417–1418.
28. Dasey, T.J.; Micheli-Tzanakou, E. A pattern recognition application of the Alopex process with hexagonal arrays. In *International Joint Conference on Neural Networks*, Vol. II; Washington, DC, 1989, 119–125.
29. Hornik, K.; Stichcombe, M.; White, H. Multilayer feedforward networks are universal approximators. Neural Netw. **1989**, *2*, 359–366.

30. Kolmogorov, A.N. On the representation of continuous functions of many variables by superposition of continuous functions of one variable and addition. Doklady Akad. Nauk SSSR **1957**, *114*, 953–956.
31. Le Cun, Y. *Modeles Connexionistes de L'apprentissage*; PhD thesis; Universite' Pierre et Marie Curie, Paris, 1987.
32. Lapedes, A.; Farber, R. *How Neural Networks Work*. Technical Report; Los Alamos National Laboratory: Los Alamos, NM, 1988.
33. Gallant, A.R.; White, J. There exists a neural network that does not make avoidable mistakes. In *IEEE Second International Conference on Neural Networks*, Vol. I; San Diego, CA, 1988, 657–664.
34. Hecht-Nielsen, R. Theory of the back propagation neural network. In *Proceedings of the International Joint Conference on Neural Networks*, Vol. I; San Diego, CA, 1989, 593–608.
35. Baum, E.; Haussler, D. What size net gives valid generalization? Neural Comput. **1989**, *1*, 151–160.
36. Mead, C.; Ismail, M.; Eds.; *Analog VLSI Implementation of Neural Systems*; Kluwer Academic: Boston, 1989.
37. Ramacher, U.; Ruckert, U.; Eds.; *VLSI Design of Neural Networks*; Kluwer Academic: Boston, MA, 1991.
38. Graf, H.P.; Jackel, L.D. Analog electronic neural network circuits. In *IEEE Circuits and Devices Magazine*; 1989, July, 44–55.
39. Macq, D.; Verlcysen, M.; Jespers, P.; Legat, J. Analog implementation of a Kohonen map with on-chip learning. IEEE Trans. Neural Netw. **1993**, *4* (3), 456–461.
40. Andreou, A.; et al. VLSI Neural Systems. IEEE Trans. Neural Netw. **1991**, *2* (2), 205–213.
41. Pandya, A.S.; Shandar, R.; Freytag, L. An SIMD architecture for the Alopex neural network. In *Parallel Architectures for Image Processing*, Vol. 1246; SPIE: Portland, OR, 1990, 275–287.
42. Linares-Barranco, B.; Sanchez-Sinencio, E.; Rodriguez-Vazquez, A.; Huertas, J.L. A CMOS implementation of Fitzhugh–Nagumo model. IEEE. J. Solid-State Circuits **1991**, *26*, 956–965.
43. Wolpert, S.; Micheli-Tzanakou, E. An integrated circuit realization of a neuronal model. In *Proceedings of the IEEE Northeast Bioengeering Conference*, New Haven, CT, 1986, March 13–14.
44. McCullouch, W.S.; Pitts, W. A logical calculus of the ideas imminenet in nervous activity. Bull. Math. Biophys. **1943**, *5*, 115–133.
45. Wolpert, S.; Micheli-Tzanakou, E. A neuromime in VLSI. IEEE Trans. Neural Netw. **1995**, *6* (6), 1560–1561.
46. French, A.S.; Stein, R.B. A flexible neuronal analog using integrated circuits. IEEE Trans. Biomed. Eng. **1970**, *17*, 248–253.
47. Nabet, B.; Pinter, R.B. *Sensory Neural Networks: Lateral Inhibition*; CRC Press: Boca Raton, FL, 1991.
48. Wolpert, S.; Micheli-Tzanakou, E. Silicon models of lateral inhibition. IEEE Trans. Neural Netw. **1993**, *4* (6), 955–961.
49. Andreou, A.G.; Boahen, K.A.; Pouliquen, P.O.; Pavasovic, A.; Jenkins, R.E.; Strohbehn, K. Current-mode subthreshold MOS circuits for analog VLSI neural systems. IEEE Trans. Neural Syst. **1991**, *2* (2).
50. Mead, C.A. *Analog VLSI and Neural Systems*; Addison-Wesley: Boston, MA, 1989.

Biomedical Image Registration

Ali Gholipour
Department of Radiology, Boston Children's Hospital, Harvard Medical School, Boston, Massachusetts, U.S.A.

Nasser Kehtarnavaz
Department of Electrical Engineering, University of Texas at Dallas, Richardson, Texas, U.S.A.

Abstract

Image registration, which aims at transforming images to match or align their correspondences, has many applications in biomedical image processing. These applications include, but are not limited to, image and structure alignment for motion correction, compensation of spatial and/or temporal variations in anatomy and variations in position or field of view, as well as image reconstruction and analysis, atlas construction, and automatic image segmentation. A typical registration algorithm has four main components: a transformation model, a correspondence basis, an interpolation method, and an optimization technique. Among these, the choices of the transformation model and the correspondence basis are particularly important due to the wide range of applications, their implications, and the nature of geometric and photometric differences between images. This chapter provides an algorithmic overview of biomedical image registration with a focus on its components and a brief discussion of its applications and validation. Some key references are provided to guide further reading.

Keywords: Alignment; Correspondence; Deformation; Image registration; Similarity measure; Spatial normalization; Transformation.

INTRODUCTION

Image registration is one of the most fundamental and influential tools in biomedical image processing. The domain of its influence ranges from computer-aided diagnosis to image-guided surgery. Many of the biomedical image processing tools and resources, such as atlases of the anatomy, automated image segmentation, and intramodality or multimodality image reconstruction and fusion techniques, are based on image registration. Due to the wide range of applications and indications, an image registration technique should be very carefully designed and validated. The goal of this chapter is to provide a framework to understand the structure and components of an image registration algorithm. The choice of the components that are discussed in detail are as follows: (1) a transformation model, (2) a correspondence basis, (3) an interpolation method, and (4) an optimization technique. Example applications of biomedical image registration will also be discussed to provide a better understanding of the methods in use.

Among the main components of image registration, the choice of the transformation model and the correspondence basis are particularly important due to their use in different applications. The choice of a transformation model depends on the nature of geometric differences between images. The transformation model may change from simple rigid transformation to correct for rigid motion to large and dense deformations to account for nonrigid motion or changes in anatomy or complex anatomic variability between subjects or samples. The choice of a correspondence basis depends on the source and target images and may be based on point correspondences or similarity metrics such as correlation coefficient or information-theoretic measures. The correspondences established through biomedical image registration are the bases of many high-level biomedical image analysis tasks. These tasks include multispectral and multimodality image analysis, spatial normalization for group and longitudinal analysis, atlas construction, automatic atlas-based image segmentation, and image reconstruction. The key to achieve accurate and reliable registration is careful analysis of the nature of the differences between the source and target images.

Finally, all components of an image registration algorithm, including the interpolation and optimization methods, are designed based on the type of the application, the complexity of the transformation and the correspondences, and the computation time/cost vs. accuracy requirements.

An algorithmic overview of image registration with its formulation in vector spaces is provided in Section "Components of Image Registration." This section continues with a detailed description of the main components of an image registration algorithm. Applications of image registration and validation are discussed in the sections that follow and the chapter then concludes with a brief summary and some references for further reading.

COMPONENTS OF IMAGE REGISTRATION

Image registration is normally formulated as an optimization problem to find the best global or local spatial transformation or deformation model that aligns a source image to a target image. The source and target images can be represented by $I_s : \mathfrak{R}^n \to \mathfrak{R}^m$ and $I_t : \mathfrak{R}^n \to \mathfrak{R}^m$, which associate a vector of scalar photometric (intensity) values in the m-dimensional vector space $\mathfrak{R}^m$ with points with coordinates $\mathbf{x}$ in the n-dimensional vector space $\mathfrak{R}^n$. For typical grayscale medical images in the three-dimensional (3D) physical space, $n = 3$, $\mathbf{x} = (x, y, z)$, and $m = 1$. For two-dimensional (2D) images, $n = 2$; for RGB images, $m = 3$; and for tensor images, $m = 6$.

Figure 1 depicts an algorithmic overview of a typical image registration algorithm. The target image is fixed, and the source image undergoes a spatial transformation of the form $T(\mathbf{x}) : \mathfrak{R}^n \to \mathfrak{R}^n$ with parameters $\boldsymbol{\theta}$. Registration is set up as an optimization problem that finds the appropriate parameters of the transformation model by minimizing a cost function, maximizing the overlap of correspondences, or the similarity of the source and target images. We will show the similarity of the two images by $S(I_t, I_s \circ T)$, but the formulations can be expressed through other types of differences, distances, or cost functions based on the source and target images. In some formulations such as point matching, the registration problem may be solved in one step through a closed-form solution; however, oftentimes numerical iterative optimization is required.

The target image is fixed and the source image is transformed and resampled in each iteration of the registration algorithm by a fast and accurate interpolation method. The problem is formulated as follows:

$$(a) \quad T^* = \underset{T \in \mathrm{T}}{\mathrm{argmax}} \left(S(I_t, I_s \circ T) \right); \quad \text{or}$$
$$(b) \quad T^* = \underset{T \in \mathrm{T}}{\mathrm{argmin}} \left(D(I_t, I_s \circ T) \right) \quad (1)$$

where T is the set of all plausible transformations under the transformation model and $D(.)$ in (b) is a cost function that may include photometric image distance or differences and regularization terms. Note that image registration is defined as maximization of similarity between images, shown in (a), or minimization of a distance or cost function between images, shown in (b). The problem may be solved hierarchically in multiple stages based on multi-resolution or multi-scale representations of the source and target images or based on increased complexity of the transformation models. Appropriate

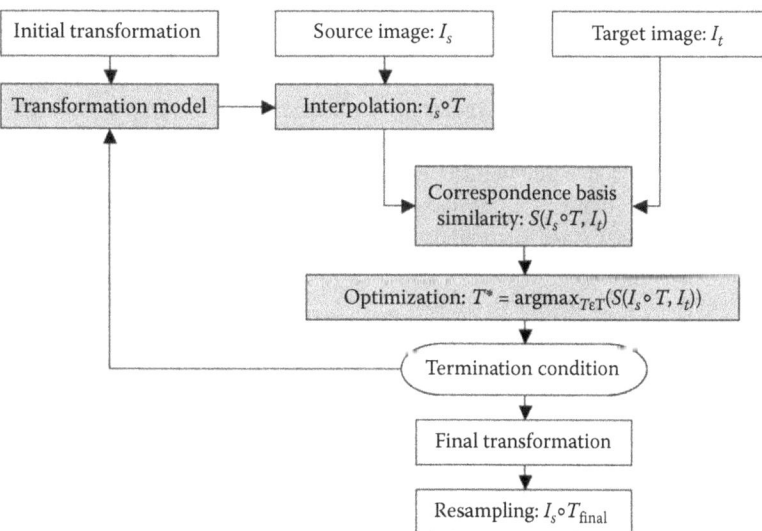

Fig. 1 An overview of a typical image registration algorithm, which consists of four main components: a transformation model, a correspondence basis, an interpolation method, and an optimization technique. The source image is transformed and interpolated in each iteration of registration, and its similarity to the target image is computed and maximized for registration. The algorithm may be implemented in a multistage, multiscale, or multi-resolution framework. The initialization of the transformation may be critical depending on the application. The initial transformation may be set to identity when the geometric differences between images are small or local. In the case of difference coordinates or positions, an initial global rigid transformation may be computed through matching the center of gravity and the principal axes of the images

initialization of the transformation is also important. The transformation can be initialized as identity transform when the geometric differences between images are small or computed through center of gravity and principal axes (i.e., matching image moments) when images are in different coordinate systems. A termination condition is defined for iterative registration by a threshold on the variation of optimization function. In what follows, the four components (transformation model, correspondence basis, interpolation method, and optimization technique) of an image registration algorithm are considered. The section that follows discusses the use of these components for various applications.

Transformation Model

The transformation T is defined as a function or operation that acts on a vector $\mathbf{x}$ in the real n-dimensional vector space $\mathfrak{R}^n$ and generates a transformed vector $\mathbf{x}' = T(\mathbf{x})$. The complexity of the transformation model is defined by its degrees of freedom (DOF) or the number of its parameters, that is, the elements of $\boldsymbol{\theta}$. Next, we review the most widely used transformation models in biomedical image registration. These range from rigid and affine transformation models to high-dimensional small or large deformations based on biophysical/biomechanical, elastic material, or viscous fluid models. The choice of the transformation model depends on the nature of the application and the geometric differences between images.

Rigid Transformation

A rigid transformation in the Euclidean vector space preserves distances between every pair of points, thus preserves the shape and size of objects. A rigid transformation involves translations and rotations: $T(\mathbf{x}) = R\mathbf{x} + \mathbf{t}$, where $\mathbf{t}$ is a vector defining the translation of the origin, and R is an orthogonal rotation matrix, that is, $R^{\mathrm{T}} = R^{-1}$, and $\det R = 1$. A 2D rigid transformation has 3 DOF, that is, two translations $\mathbf{t} = (t_x, t_y)$ and a rotation angle α. In 3D, a rigid transformation has 6 DOF, that is, three translation and three rotation parameters. A 3D rigid transformation is shown as follows:

$$T_{\text{rigid}} = \begin{pmatrix} x' \\ y' \\ z' \\ 1 \end{pmatrix} = \begin{pmatrix} & & & t_x \\ & R & & t_y \\ & & & t_z \\ 0 & 0 & 0 & 1 \end{pmatrix} \begin{pmatrix} x \\ y \\ z \\ 1 \end{pmatrix} \quad (2)$$

A rotation in 3D can be represented by different formalisms with a fixed point called center of rotation. Some of these formalisms are as follows: (1) a rotation matrix (R), (2) orthonormal basis (three unit vectors that make an orthonormal basis), (3) Euler axis and angle, (4) Euler angles (precession, nutation, and intrinsic rotations), and (5) quaternions (versors in 4D vector space). Simple mathematical formula exists to convert between these representations. Among these, the quaternions are popular due to computational benefits.

It is worth noting that the same formalisms can be used to represent orientation, which is an important geometric characteristic of a biomedical image. Orientation and origin of an image have to be carefully taken into account whenever geometric transformations are applied to an image.

Rigid transformation is an appropriate model for many biomedical image registration applications. For example, it is appropriate for registering in vitro images of tissue, assuming they are fixed as well as head and brain images of a subject assuming that the skull and thus the brain may only move rigidly in 3D, their anatomy and shape do not change, and there is no nonrigid geometric distortion in the imaging process. In the applications section that follows, it is discussed how these assumptions may alter under different conditions thus requiring the use of more complex transformation models.

Affine Transformation

An affine transformation preserves points, straight lines, and planes. Parallel lines and planes remain parallel after affine transformation, but angles between lines and planes as well as distances between points are not preserved. An affine transformation is a composition of a translation and a linear map, therefore, can be represented by an augmented matrix. In 3D, a general affine transformation has 12 DOF: three rotations, three translations, three scale, and three shear parameters. This can be shown by an augmented 4×4 matrix with $[0\,0\,0\,1]$ in the last row and 12 elements in the upper rows. The affine transformation can be obtained as follows:

$$T_{\text{affine}}(\mathbf{x}) = T_{\text{shear}} \cdot T_{\text{scale}} \cdot T_{\text{rigid}} \cdot \begin{bmatrix} \mathbf{x} \\ 1 \end{bmatrix} \quad (3)$$

where

$$T_{\text{scale}} = \begin{bmatrix} s_x & 0 & 0 & 0 \\ 0 & s_y & 0 & 0 \\ 0 & 0 & s_z & 0 \\ 0 & 0 & 0 & 1 \end{bmatrix} \quad (4)$$

and a shear in the x–y plane can be written as

$$T^{xy}_{\text{shear}} = \begin{bmatrix} 1 & 0 & h_x & 0 \\ 0 & 1 & h_y & 0 \\ 0 & 0 & 1 & 0 \\ 0 & 0 & 1 & 1 \end{bmatrix} \quad (5)$$

With this representation, any number of affine transformations can be combined by matrix multiplication.

Projective Transformation

Another set of transformations that can be represented by an augmented matrix—similar to the affine transformation—are projective transformations. Projective transformation may be used to project or register 3D images such as CT or magnetic resonance imaging (MRI) to 2D images such as X-ray. In practice, this transformation may be combined with a rigid transformation to account for the differences in the coordinate systems. Perspective projections are another type of transformations in this category. While parallel projection transformations project points onto the image plane along parallel lines, perspective projection transforms points onto the image plane along lines that originate from a point called center of projection.

Nonrigid Transformation by Harmonic Basis Functions

One approach to describe a nonrigid transformation is by a combination of harmonic basis functions ψ_i:

$$T_{\text{HBF}}(\mathbf{x}) = \begin{bmatrix} \mathbf{x} \\ 1 \end{bmatrix} = \begin{bmatrix} a_{10} & \cdots & a_{1l} \\ \vdots & \cdots & \vdots \\ a_{n0} & \cdots & a_{nl} \\ 0 & \cdots & 0 \end{bmatrix} \begin{bmatrix} \psi_1(\mathbf{x}) \\ \vdots \\ \psi_l(\mathbf{x}) \\ 1 \end{bmatrix} \quad (6)$$

A common choice for the ψ_i is Fourier or trigonometric basis functions which provide a spectral representation of the deformation field. Each ψ_i represents a specific frequency of the deformation, thus the choice of l defines the frequency spectrum of the deformation model.[1] A small l defines a low-dimensional (and low-frequency) transformation.

Nonrigid Transformation by Thin Plate Splines

Many times, it is possible or desired to describe a nonrigid transformation based on a set of control points or landmarks on the source and target images. Scattered data interpolation by splines is a useful approach to interpolate between control points and approximate a smooth deformation field that maps the control points from the source image to the target image. Such a nonrigid transformation can be modeled by thin plate splines (TPSs). Given a set of P control points $\varphi_p \in \mathfrak{R}^n$, a TPS function $t_i(\mathbf{x}): \mathfrak{R}^n \to \mathfrak{R}$ in the ith dimension ($i = 1 \ldots n$) can be defined as a linear combination of radial basis functions ψ_i:

$$t_i(\mathbf{x}) = \begin{bmatrix} a_{i0} & \cdots & a_{in} \end{bmatrix} \begin{bmatrix} 1 \\ \mathbf{x} \end{bmatrix} + \sum_{p=1}^{P} b_{ip} \psi_i\left(\left|\varphi_p - \mathbf{x}\right|\right) \quad (7)$$

The transformation is described by n separable TPS functions as $T = \begin{bmatrix} t_1 & \cdots & t_n \end{bmatrix}^T$ with parameters a and b that define the affine and non-affine parts of the transformation, respectively. This model has $n(P + n + 1)$ parameters. The control point interpolation conditions provide nP equations, and a set of another $n(n + 1)$ equations are obtained from model assumptions as follows:

$$\begin{bmatrix} \Psi & \Phi \\ \Phi^T & 0 \end{bmatrix} \begin{bmatrix} \mathbf{b} \\ \mathbf{a} \end{bmatrix} = \begin{bmatrix} \Phi' \\ 0 \end{bmatrix} \quad (8)$$

where $\mathbf{a}$ is an $n(n + 1)$ vector of affine parameters, $\mathbf{b}$ is an nP vector of non-affine parameters, and Φ is a matrix of control point locations as follows:

$$\Phi = \begin{bmatrix} 1 & \varphi_1 \\ \vdots & \vdots \\ 1 & \varphi_P \end{bmatrix} \quad (9)$$

Ψ is the kernel matrix with elements defined as $\Psi_{ij} = \psi\left(\left|\varphi_i - \varphi_j\right|\right)$, and Φ' is the matrix of transformed control point locations. The radial basis function of TPS is $\psi(r) = r^2 \log r^2$ in 2D and $\psi(r) = r$ in 3D.

The augmented matrix on the right-hand side of Eq. 8 is symmetric and can be inverted or solved through LU (lower–upper) decomposition to find the transformation parameters $\mathbf{a}$ and $\mathbf{b}$.

B-Spline Free-Form Deformation Model

In free-form deformations (FFDs), the deformation field is modeled by displacements of a grid of control points. Although different functions may be used to generate a smooth and continuous deformation field that interpolates between control points, B-splines are a common choice due to their excellent smoothing and locality properties.[2] In 3D, assuming a $n_x \times n_y \times n_z$ grid of control points $\varphi_{i,j,k}$ with uniform control point spacing δ, the B-spline FFD can be written as the 3D tensor product of one-dimensional (1D) cubic B-splines:

$$T_{\text{FFD}}(\mathbf{x}) = \sum_{l=0}^{3} \sum_{l'=0}^{3} \sum_{l''=0}^{3} B_l(u) B_{l'}(v) B_{l''}(w) \varphi_{i+l, j+l', k+l''} \quad (10)$$

where $i = \frac{x}{\delta} - 1$, $j = \frac{y}{\delta} - 1$, $k = \frac{z}{\delta} - 1$, $u = \frac{x}{\delta} - \frac{x}{\delta}$, $v = \frac{y}{\delta} - \frac{y}{\delta}$, $w = \frac{z}{\delta} - \frac{z}{\delta}$, and B_l is the lth B-spline basis function:

$$\begin{aligned} B_0(u) &= (1-u)^3/6 \\ B_1(u) &= (3u^3 - 6u^2 + 4)/6 \\ B_2(u) &= (-3u^3 + 3u^2 + 3u + 1)/6 \\ B_3(u) &= u^3/6 \end{aligned} \quad (11)$$

Cubic B-spline basis functions have a local support, so they generate a locally controlled FFD model. This transformation is computationally efficient when compared to TPS deformations.

Elastic Models for Small Deformations

One approach to the deformation of shapes in images is to model them as elastic material. In continuum mechanics, elasticity of an object is governed by two forces: an external force that acts to deform the elastic material and an internal force that counteracts its deformation. The two forces counteract each other until the elastic material reaches an equilibrium position. The behavior of an elastic material under the external force $\mathbf{f}$ is described by the Navier linear elastic partial differential equations (PDEs):

$$\mu \nabla^2 \mathbf{u}(\mathbf{x}) + (\lambda + \mu) \nabla(\nabla . \mathbf{u}(\mathbf{x})) + \mathbf{f}(\mathbf{x}) = 0 \quad (12)$$

where $\mathbf{u}$ is the deformation field, ∇ and ∇^2 are the gradient and Laplace operators, respectively, and λ and μ are the Lame's elasticity constants of the elastic material. In deformable image registration, the PDE is solved through finite differences as well as a deformation field is obtained at any pixel or voxel of the image. In 3D, $\mathbf{u}$ has three elements at each voxel of the source image that define the displacement of that voxel in x, y, and z directions. Registration is derived by the external force $\mathbf{f}$, which is typically the gradient of a cost function or a measure of correspondence between the source and target images. The elastic model is naturally smooth and does not allow highly localized deformations, thus is used for small deformations.

Viscous Fluid Models for Large Deformations

In contrast to the elastic model that is appropriate for small deformations, a viscous fluid model seems to be an appropriate model of large, highly localized deformations. The deformation $\mathbf{u}(\mathbf{x})$ in the viscous fluid model is described by the integral of a time-dependent velocity field $\mathbf{v}(\mathbf{x})$:

$$\mathbf{u}(\mathbf{x},1) = \int_0^1 \mathbf{v}(\mathbf{u}(\mathbf{x},t)) dt \quad (13)$$

The deformation model is then defined on the velocity field rather than on the deformation field, by the Navier–Stokes PDE:

$$\mu \nabla^2 \mathbf{v}(\mathbf{x}) + (\lambda + \mu) \nabla(\nabla . \mathbf{v}(\mathbf{x})) + \mathbf{f}(\mathbf{x}) = 0 \quad (14)$$

which is solved for each time point. Here λ and μ are the viscosity coefficients. The PDEs can be solved using finite differences or successive over-relaxation. Variants of this model may be solved efficiently by a convolution filter in scale space assuming viscosity is constant. Such a filter can be designed as the impulse response of a linear operator defined in its eigenfunction basis as follows:

$$L = \mu \Delta \mathbf{u}(\mathbf{x}) + (\lambda + \mu) \nabla(\nabla . \mathbf{v}(\mathbf{x})) \quad (15)$$

The deformations obtained from this model are differentiable and can also be constrained to carry other desired properties. Several properties that are frequently sought after in deformable image registration are symmetry, inverse consistency, topology preservation, and diffeomorphism. Symmetric registration is possible through cost functions that are symmetric to estimate the transformation between source and target images. Inverse consistency can be achieved through constraining forward and backward transformations to be inverse of one another. Topology preserving or homeomorphic deformations are continuous and locally one to one, and have a continuous inverse. The Jacobian determinant of a homeomorphism is greater than zero and is computed as follows:

$$|J(x,y,z)| = \det \begin{bmatrix} \frac{\partial T_x}{\partial x} & \frac{\partial T_x}{\partial y} & \frac{\partial T_x}{\partial z} \\ \frac{\partial T_y}{\partial x} & \frac{\partial T_y}{\partial y} & \frac{\partial T_y}{\partial z} \\ \frac{\partial T_z}{\partial x} & \frac{\partial T_z}{\partial y} & \frac{\partial T_z}{\partial z} \end{bmatrix} \quad (16)$$

Finally, a diffeomorphic transformation is invertible and both the deformation and its inverse are differentiable. A diffeomorphism is topology preserving and maps a differentiable manifold to another.

Diffusion Model for Deformation

In this case, the deformation is modeled by the diffusion equation, which is a PDE that describes the dynamics of a material undergoing diffusion:

$$\Delta \mathbf{u} + f(\mathbf{x}) = 0 \quad (17)$$

The use of this model for image registration involves an iterative procedure that is built upon the concept of Maxwell's demons. In this algorithm, each image element is considered as a demon. Demon forces are calculated by considering optical flow constraints. The calculated deformation field is regularized through convolution with a Gaussian kernel (the Green's function of the diffusion equation). The image is then deformed and interpolated using the calculated deformation field.[3] In this approach, diffeomorphism can be obtained by applying an exponential map to the deformation field.[4] With diffeomorphic demons, symmetry is achieved through averaging the forward and backward driving forces.

Biomechanical/Biophysical Models of Deformation

When prior information about physical or anatomical properties of deformations is available, specific biomechanical or biophysical models may be designed and implemented through finite element methods. Such models may be used for the registration of breast images, tumor growth, brain shift in craniotomy, as well as cardiac and respiratory motion.

Correspondence Basis

The second component of a registration algorithm is the correspondence basis. The correspondence can be established based on a set of paired points, features, or landmarks detected on the source and target images, or based on similarity or distances that measure the alignment or misalignment of the source and target images. Correspondences can be established extrinsically or intrinsically. Extrinsic correspondences may be established by markers attached to the imaged structures that are visible on the images. Image registration using extrinsic markers is usually fast and accurate but requires preparation prior to imaging and is limited to specific applications where such markers are available or can be attached. Intrinsic correspondences are used in a wider range of applications as they do not require prospective planning and are established based on image contents. The disadvantage of intrinsic correspondences is that they may not be easily identified or may not be precise.

Point-Based Correspondences

Correspondences may be established between the source and target images in the form of paired point sets. The points may be extrinsic or intrinsic and detected manually or automatically. Registration can be solved through the alignment of the center of the two point sets followed by a rotation that minimizes the sum of squared distances between the point pairs.[5] This is achieved through singular value decomposition. For a 6-DOF rigid registration, in 3D, at least three noncollinear point pairs are required. The point pairs may be used as control points to calculate a nonrigid transformation based on TPSs.

If two point sets, $\mathbf{p}$ and $\mathbf{q}$ in $\Re^n$, are available on the source and target images, but are not paired, that is, the correspondences between the point sets are not known a priori, an iterative algorithm is needed to simultaneously find the point correspondences and solve the registration problem. The iterative closest point (ICP) algorithm,[6] designed for this purpose, starts by an estimate of correspondences between point sets using

$$\mathbf{y}_i = \min_{\mathbf{q}_j \in \mathbf{q}} \left\| \mathbf{q}_j - T(\mathbf{p}_i) \right\|^2 \tag{18}$$

and minimizes the Euclidean distances between the matched point sets through the following equation:

$$T^* = \min_{T \in \mathrm{T}} \sum_i \left\| \mathbf{y}_i - T(\mathbf{p}_i) \right\|^2 \tag{19}$$

The computed transformation T^* is then used in Eq. 18 to update the point correspondence estimates, and the iterations between Eqs. 18 and 19 continue until a solution is converged. Various computational techniques may be used to enhance the efficiency of this algorithm, especially in estimating point correspondences. Methods such as Kalman filter or unscented Kalman filter have also been used to solve the problem of point matching in point-based image registration.

Feature-Based Correspondences

Image features, such as surfaces, may be extracted and used to establish point correspondences that are then solved through an algorithm such as ICP. On the other hand, local feature correspondences may be established based on image blocks or patches. Feature correspondences may be established through scale invariant feature transform descriptors leading to key points or through photometric similarity of image blocks or patches. In block matching, a brute-force search may be used to search for matching blocks on the source and target images, but more efficient search methods may also be used. There are four sets of parameters in a block-matching algorithm: the size of the blocks on both images, the grid step size for the blocks on the source image and on the target image, and the size of the search window on the target image. A multi-resolution approach is often helpful. Measures such as correlation coefficient or other measures discussed later in this section may be used to calculate the similarity of blocks. Blocks should be large enough for reliable calculation of similarity measures. For local deformations, a combination of photometric and spatial distances may be used to measure similarity of blocks.

Statistical Similarity or Distance Measures

As compared to feature matching techniques, which require extracting features from source and target images, image intensity information may be directly used to solve a registration problem. When images are of the same modality and obtained under similar imaging conditions, the differences in the intensity values of two images should be minimal (i.e., resembling the imaging noise) when they are in alignment. Under these conditions, a distance measure for registration involves a norm function of photometric distances (i.e., intensity differences) between the source and target images:

$$D_{SSD} = \sum_N \left(I_t - I_s \circ T\right)^2 \qquad (20)$$

This is the sum of squared differences (SSDs) of intensity values. In the case of non-scalar intensity values, a norm function over the vector space $\mathfrak{R}^m$ may be used. If the imaging noise is Gaussian distributed, the SSD is the optimal maximum likelihood solution of the registration problem. Under heavy-tailed noise, like Laplacian noise, the optimal solution can be obtained from l_1 norm optimization. Statistical methods can be used to verify the distribution of noise and validate the registration process.

The assumption of equality for the intensity values of the source and target images is rather strong. Cross-correlation (CC) of the intensity values of the two images measures the linear relationship of image intensity values:

$$S_{CC} = \sum_N \left(I_t - \mu_t\right)\left(I_s \circ T - \mu_s\right) \qquad (21)$$

where μ_s and μ_t are the mean intensity values of the source and target images, respectively. CC is maximum when the images are in alignment, so is maximized in an image registration process. In practice, normalized CC (NCC), also referred to as correlation coefficient, is commonly used as it accounts for variations in brightness and contrast between images:

$$S_{NCC} = \frac{\sum_N \left(I_t - \mu_t\right)\left(I_s \circ T - \mu_s\right)}{\sqrt{\left(\sum_N I_t - \mu_t\right)^2 \left(\sum_N I_s \circ T - \mu_s\right)^2}} \qquad (22)$$

Another statistical measure used in image registration is variance of intensity ratios (VIR), which is defined based on a ratio image calculated on the overlapping region (Ω) of the two images by $R(\mathbf{x}) = \frac{I_t(\mathbf{x})}{I_s \circ T(\mathbf{x})}$. With $N = \text{Card}(\Omega)$ and $\bar{R}(\mathbf{x}) = \frac{1}{N}\sum_{\mathbf{x}\in\Omega} R(\mathbf{x})$, the VIR is defined as

$$D_{VIR} = \frac{\sqrt{\frac{1}{N}\sum_{\mathbf{x}\in\Omega}\left(R(\mathbf{x}) - \bar{R}(\mathbf{x})\right)^2}}{\bar{R}(\mathbf{x})} \qquad (23)$$

Note that VIR is the inverse of a similarity measure and has to be minimized in registration. This corresponds to reducing the nonuniformity of the ratio image. By defining $\Omega_i = \{\mathbf{x} \in \Omega, I_s \circ T(\mathbf{x}) = i\}$, and its cardinals $N_i = \text{Card}(\Omega_i)$, a similarity measure called correlation ratio (CR) is obtained as follows:

$$S_{CR} = 1 - \frac{1}{N\sigma^2}\sum_i N_i \sigma_i^2 \qquad (24)$$

where $\sigma^2 = \frac{1}{N}\sum_{\mathbf{x}\in\Omega} I_t(\mathbf{x}) - \left(\frac{1}{N}\sum_{\mathbf{x}\in\Omega} I_t(\mathbf{x})\right)^2$ and $\sigma_i^2 = \frac{1}{N_i}\sum_{\mathbf{x}\in_i} I_t(\mathbf{x}) - \left(\frac{1}{N_i}\sum_{\mathbf{x}\in_i} I_t(\mathbf{x})\right)^2$. CR takes on values between 0 and 1 and is maximized in image registration.

Information-Theoretic Measures

The assumption of the existence of a linear relationship between the intensity values of two images may be restrictive in comparing images of different modalities. The conceptual use of NCC in measuring image similarity can be generalized to more complex, probably nonlinear, intensity relationships through analyzing the joint histogram of two images. The joint histogram of two images is more dispersed when the images are not in alignment. This leads to quantifying the information content of the joint histogram of the source and target images, hence the use of information-theoretic measures in image registration.

The first intuitive measure to consider is the joint entropy of the two images, which quantifies the amount of information in the joint histogram of the target and transformed source images:

$$S_{JE} = H\left(I_t, I_s \circ T\right) = -\sum_v \sum_\kappa p(v,\kappa)\log p(v,\kappa) \qquad (25)$$

where $p(v,\kappa)$ is the joint probability density function (PDF) of the target and transformed source images; $0 \leq v \leq L_t$ and $0 \leq \kappa \leq L_s$ denote the indices of the joint histogram bins in computing the PDF, and L_t and L_s are the number of histogram bins over the intensity values of the target and source images, respectively.

A more widely used information-theoretic measure is the mutual information (MI), which is derived from the joint and marginal entropies of the two images:

$$S_{MI} = H\left(I_t\right) + H\left(I_s \circ T\right) - H\left(I_t, I_s \circ T\right)$$
$$= \sum_v \sum_\kappa p(v,\kappa)\log \frac{p(v,\kappa)}{p_t(v)p_s(\kappa)} \qquad (26)$$

where $H(I_t)$ and $H\left(I_s \circ T\right)$ are the marginal entropies of the target and transformed source images, and $p_t(v)$ and

$p_s(\kappa)$ are the marginal PDFs of the target and transformed source images, respectively:

$$H(I_t) = -\sum_v p_t(v) \log p_t(v) \quad \text{and}$$
$$H(I_s \circ T) = -\sum_\kappa p_s(\kappa) \log p_s(\kappa) \tag{27}$$

As a similarity measure, MI is maximized in image registration. A normalized version of MI may be used to obtain slightly more robust performance under variations in image overlap regions:

$$S_{NMI} = \frac{H(I_t) + H(I_s \circ T)}{H(I_t, I_s \circ T)} \tag{28}$$

This definition of normalized mutual information (NMI) is related to another measure called entropy correlation coefficient by the following formula:

$$S_{ECC} = \sqrt{2\left(1 - \frac{1}{S_{NMI}}\right)} \tag{29}$$

The calculation of similarity measures is extremely important for the success of a registration algorithm. This involves the choice of the histogram bin size and the computation of PDFs. For robust and smooth registration, the PDFs can be modeled by a mixture of Gaussian distributions and calculated through nonparametric Parzen window estimation.[7] This leads to analytical expressions for the derivatives of the similarity measures with respect to transformation parameters that are crucial for efficient optimization.

Correct and efficient implementation of image similarity metrics for optimization involves important considerations, such as local computations, and is key to the success of a registration algorithm, for example, Ref. [8].

Optimization Technique

The core of a conventional image registration algorithm, as defined by Eq. 1, is an optimization technique that involves a search over the transformation parameters to maximize a similarity measure or minimize a cost or distance function between the source and target images. Although it might be possible to find a closed-form solution for such an optimization problem, oftentimes iterative approaches such as Powell's method, downhill simplex, gradient descent, conjugate gradient, quasi-Newton, Levenberg–Marquardt, genetic or evolutionary algorithm, or simulated annealing are used. Deformable registration that involves the solution of PDEs is solved through finite differences or successive over-relaxation. Finite element methods are used to solve registration based on biomechanical or biophysical models.

In many applications, the correspondence bases, including image features and similarity measures, only provide a surrogate of a true measure of alignment between images; therefore, the solution of the registration problem may not always be considered a global optimum. Moreover, the parameter space of the transformation models for biomedical image registration is often high dimensional (typically ≥6 DOF). For this reason, the use of brute-force search or global optimization techniques such as evolutionary algorithm and simulated annealing is usually inefficient and often slow and costly. On the other hand, multi-resolution and multi-scale optimization frameworks may dramatically accelerate the convergence and improve the robustness of a registration algorithm. The use of these techniques as well as prior knowledge in the form of initial transformation should be seriously considered in any image registration task. Methods involving moments, based on the first- and second-order moments of the source and target images, may be used to match the center of gravity and the principal orientation of the images. This approach may be considered as a powerful initialization of rigid registration, but should be used very carefully if the fields of view of images only partially overlap. A high-dimensional nonrigid registration task may be divided into stages, in which the transformation complexity increases in each stage. Such a task may start with rigid registration and then evolve into affine and high-dimensional deformable registration.

Most of the widely used optimization techniques, including gradient descent, quasi-Newton, and Levenberg–Marquardt, require derivative calculations. While it is possible to numerically calculate the derivatives of a similarity measure or cost function with respect to transformation parameters, analytical expressions for the gradients of similarity measures or cost functions may dramatically accelerate and improve optimization. Depending on the size of the parameter space of the transformation as well as the size of the images, different computational methods and programming approaches may be used to balance between memory and CPU usage and speed of convergence. For example, efficient modeling and estimation of image PDFs through Gaussian mixture models and analytical calculation of their derivatives with respect to transformation parameters is essential to the development of an image registration algorithm that maximizes a statistical or an information-theoretic similarity measure.

Interpolation Method

Image registration involves geometric transformation of images, which requires image resampling. Image resampling includes interpolation. Interpolation is performed within each iteration of image registration to resample the transformed source image to the space of the target image to compute the correspondence basis. By definition, interpolation reconstructs a continuous image from its discrete samples by using an interpolation kernel function.[9]

When Nyquist criterion is fulfilled as per the sampling theory, the continuous image can be perfectly reconstructed by using the ideal interpolation kernel, which is the Sinc function. The Sinc function in image domain corresponds to a rectangular filter in the Fourier domain; however, perfect interpolation is not practical as the Sinc kernel is infinite and signal is not band limited. In practice, windowed Sinc function or other kernels with appropriate properties may be used.

Common interpolation kernels are separable and symmetric; therefore, a 1D kernel can be applied sequentially in all n dimensions of the image. The easiest and fastest interpolation is the nearest neighbor interpolation which corresponds to a step function that assigns the value of the closest point to each interpolated image point; however, linear, cubic, or B-spline kernels are often used as they provide more accurate estimations of the ideal Sinc interpolation kernel. The choice of an interpolation kernel between linear and higher order kernels involves a trade-off between the speed of interpolation and its accuracy. This is particularly important in registering two large images, where the source image is transformed and resampled in every iteration for the optimization of a similarity measure.

APPLICATIONS OF IMAGE REGISTRATION

In this section, some biomedical applications of image registration are discussed based on their implications in two main categories: intra-subject registration and intersubject registration.

Intra-subject Registration

Intra-subject registration, or the registration of images of the same subject, may be used for three main purposes: (1) to correct for motion that occurs during an imaging session, (2) to correct for variations in imaging at different scan sessions, and (3) to correct for variations in anatomy.

Motion Correction

In vivo medical imaging of live subjects may take time, so subjects may inadvertently move during a long imaging session. Moreover, imaging may be performed on moving organs such as heart or lung. Image registration may be used for motion correction. Registration-based motion correction can be performed retrospectively or prospectively. If sufficient computational power is available, many retrospective registration techniques may be efficiently implemented and used for prospective motion correction. Data from motion tracking devices may be used to guide or initialize image registration; otherwise, the transformation model parameters may be initialized by those of an identity transform.

Motion can be rigid or nonrigid. For example, head motion is often considered rigid, whereas cardiac and respiratory motions are nonrigid. Motion of the head and brain in many MRI neuroimaging studies is corrected using rigid registration. In many of these studies, such as functional MRI (fMRI), precise spatial correspondences between points in a time series of volumetric (3D) images is a crucial requirement for post-acquisition statistical analysis. An fMRI imaging session usually involves hundreds of fast 3D acquisitions as a time series (3D+time), so it is typically lengthy and may last between 5 and 15 min. Any subvoxel head motion can interfere with the detection of the small blood oxygenation level-dependent (BOLD) signal, which amounts to a small fraction of signal intensity changes in fMRI. Consequently, the estimation and correction of even small subvoxel motion is crucial for accurate fMRI analysis.

Large motion within an fMRI session destroys the signal, and so in addition to prevention of motion, often a prospective motion correction technique based on real-time rigid registration is used during acquisition to detect, discard, and repeat motion-corrupted scans and update the field of view (origin and rotation) for the rest of the scan. This is performed during fMRI acquisition and is followed by retrospective rigid registration for subvoxel motion correction prior to statistical analysis. Figure 2 shows the estimated parameters of a 6-DOF rigid transformation for 300 volumes acquired in an fMRI time series acquisition. Rigid registration with NCC similarity measure was used to compute these parameters using the realignment tool in the Statistical Parametric Mapping (SPM) software package.[a] Figure 2b shows the joint histogram of the first and last volumes in this fMRI time series prior and after retrospective registration.

Similar to fMRI, motion correction is useful in many other MRI studies. Multiple MR images may be acquired from the beginning to the end of an MRI scan session, and the patient may inadvertently move during the session or even within image acquisitions. Image registration may be used to correct motion and align images acquired during an MRI scan session. The correction of motion is particularly crucial in post-acquisition analysis of long scans such as diffusion-tensor imaging (DTI), in which spatial point-by-point correspondence between a reference 3D image and a large number of diffusion-sensitized 3D images (with different gradient directions) is used to detect and quantify the directional diffusivity of water molecules at each voxel. Figure 3 shows color fractional anisotropy (FA) maps of a DTI image corrupted by head motion, and then corrected through retrospective rigid registration using maximization of MI similarity measure between the diffusion-sensitized images and the reference image.

---

[a]http://www.fil.ion.ucl.ac.uk/spm/.

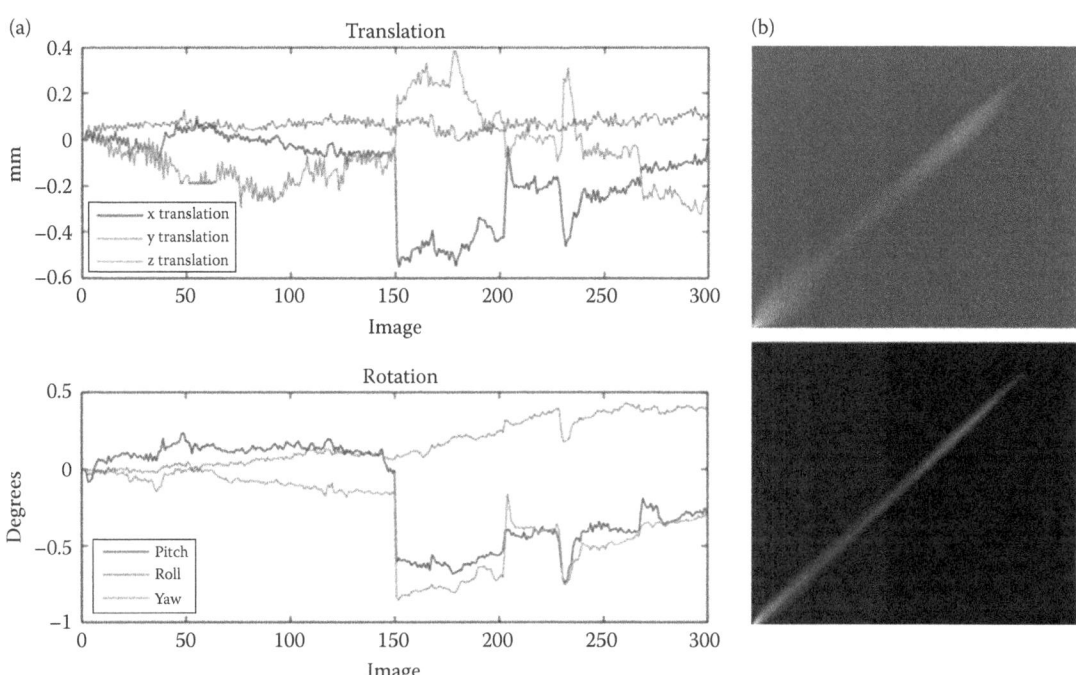

Fig. 2 Estimation and correction of motion in an fMRI time series using retrospective rigid registration based on NCC similarity measure using the realignment tool in the SPM software package: (a) transformation parameters (three translation and three rotation parameters) for the 300 acquired scans; (b) joint histogram of the first and last images in the time series prior and after registration in the top and bottom graphs, respectively

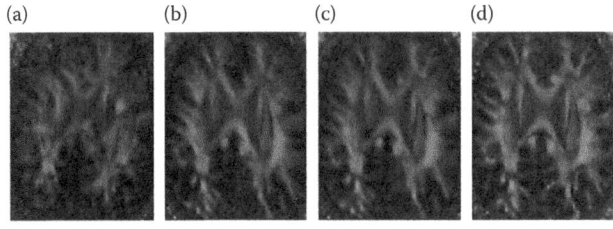

Fig. 3 The effect of registration-based motion correction on color FA maps obtained from diffusion tensor analysis of a diffusion-weighted (DW) MRI scan acquired while the subject moved between scans: (a) color FA of motion-corrupted scan; (b) color FA after motion correction but without correction of gradient directions; (c) color FA after motion correction and correction of gradient directions; (d) color FA of a DW-MRI scan of the same subject without motion. Part (d) is regarded as the gold standard. The similarity between (c) and (d) shows the success of registration-based motion correction in DW-MRI analysis

Motion correction is not limited to neuroimaging. In fact, the estimation and correction of motion in body, chest, and heart imaging is more challenging as the body and chest often undergo nonrigid motion. Figure 4 shows two chest MRI scans—one acquired at expiration and the other at inspiration—and the result of deformable registration to map the expiration image into the inspiration image. The large diffeomorphic deformation based on viscous fluid model along with NCC similarity metric was used to register these images. Image registration may be used along with cardiac and respiratory gating to correct motion between scans and reconstruct high-resolution 3D and 4D images of moving body organs. The application of such advanced, spatiotemporal image registration and reconstruction may provide a movie of cardiac motion in 3D, which is very useful in the evaluation of the function and structure of the heart.

Fast and efficient image registration may be integrated into an imaging protocol in the form of image-based navigation for real-time motion correction. Moreover, image registration may play a critical role in imaging uncooperative patients or patients who have difficulty staying still during scans, thus leading to the notion of motion-robust imaging. When a subject moves continuously, the most effective approach is to make imaging as fast as possible, similar to snapshot imaging with high shutter speed in digital photography. If image acquisition is sufficiently fast, a high-quality snapshot image of the subject is obtained. Nevertheless, most biomedical imaging techniques are slow and have relatively low SNR at high resolutions. Possible solutions include accelerated imaging through compressed sensing, minimizing the imaging time by restricting the image acquisition field of view and combining images through image registration afterwards.

In an application such as fetal MRI, in which the fetus moves intermittently and continuously, the problem is tackled through fast 2D slice acquisitions. Given a high chance that the fetus does not move fast at the exact time of a slice acquisition (which takes about 1 s), a high-quality high-resolution slice can be obtained. Hundreds to thousands of such slices are acquired in different angles and views in a fetal MRI scan session. These 2D slices provide a comprehensive survey of the fetal brain and body anatomy for clinical evaluation. The slices can then be combined together to generate a volumetric image of the fetus, but this requires image registration as the fetus moves between slice acquisitions. A 3D-to-3D, slice-to-volume registration[10] can be used to correct inter-slice motion and reconstruct a volumetric image from a collection of slice acquisitions in fetal MRI.[11] Figure 5 shows a sample outcome of this technique. A multistage registration framework is important here, since a high-quality reference volume for registration does not exist and is only estimated in each iteration of the algorithm. The registration is thus performed for stacks of slices at the beginning, and the transformation is gradually refined for each slice.

Variations in Imaging

The second scenario in intra-subject image registration deals with variations in imaging, which involves geometric and/or photometric variations. Regardless of the modality of the images and the type of imaging device or scanner used, images that are acquired for a subject in different scan sessions may have different geometry and may not be readily comparable. This can be due to differences in the coordinate systems, patient position, and different fields of view that result in differences in image origin, orientation, and spacing. To compare these images side by side and to combine information from different scans, the images should be brought into alignment in a common coordinate system. This can be done through image registration. Although this type of registration is often performed retrospectively, its integration into the imaging devices is useful.

The choice of the transformation model depends on the characteristics of the imaged organ or tissue. A rigid transformation is a valid assumption for the registration of head and brain images even if the source and target images are from different modalities. On the other hand, if the tissue or organ undergoes nonrigid motion or deformation, or one or both imaging modalities are affected by geometric distortions, other suitable transformation models may be used. For example, a combination of rigid and nonrigid transformations may be used to account for sectioning, morphological deformations, and stain variations in registering serial microscopic images in histology.[12,13]

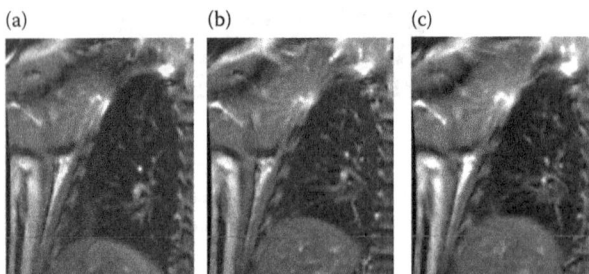

Fig. 4 Deformable registration of lung MRI scans of an adult subject acquired during inspiration and expiration: (a) image acquired at inspiration; (b) image acquired at expiration; (c) is (a) registered to (b). The line shows the amount of diaphragm motion that has been compensated by deformable registration. Deformable registration was performed using the viscous fluid model by maximizing the NCC similarity measure[8]

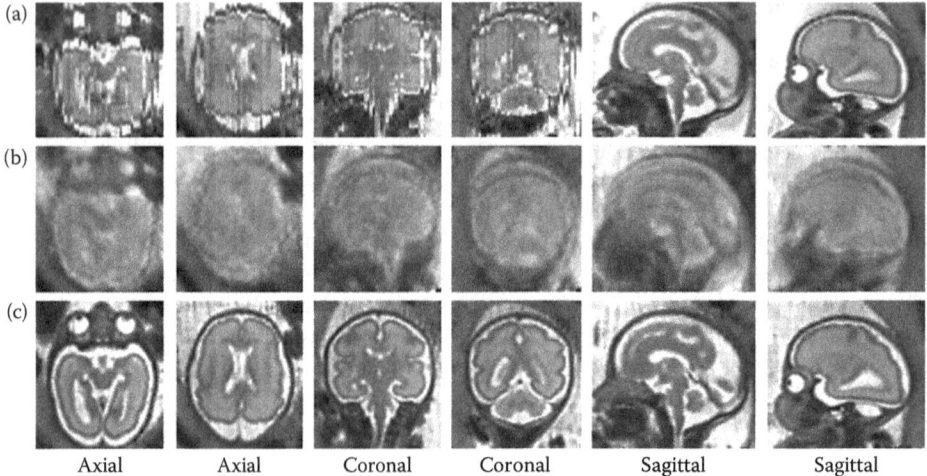

Fig. 5 Reconstruction of volumetric fetal brain MRI from slice acquisitions; each row shows two axial, two coronal, and two sagittal views of an image: (a) original sagittal single-shot fast spin echo (SSFSE) MRI scan of the fetal head; (b) Average of eight SSFSE scans acquired in the axial, coronal, and sagittal planes of the fetal head; (c) reconstructed volumetric image of the fetal head after iterations of 3D-to-3D, slice-to-volume motion correction, and super-resolution volume reconstruction.[11] Inter-slice motion artifacts are observed in the out-of-plane (i.e., axial and coronal) views of the image in (a). Inter-slice motion results in a blurred average image in (b). Registration-based slice motion correction results in a motion-free high-resolution volumetric image in (c) which can be used in further computer-aided analysis

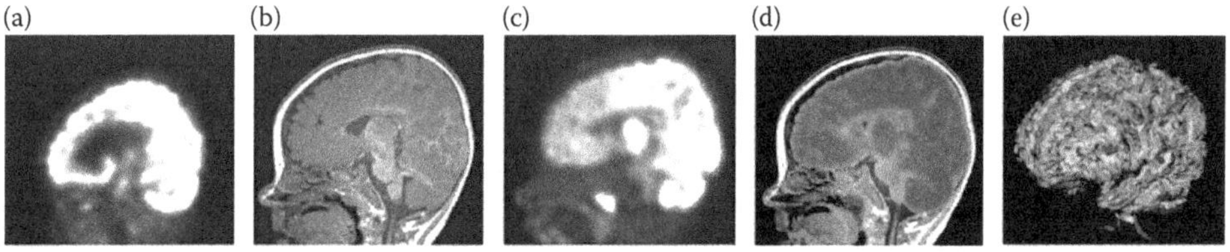

Fig. 6 Intermodality rigid registration of the brain images of a pediatric patient acquired with PET and MRI scanners in different sessions: (a) a PET scan; (b) an MRI scan; (c) the PET scan registered to the MRI scan; (d) color-coded registered PET scan overlaid on the MRI scan; (e) 3D-rendered models of the brain vasculature obtained from an MRA scan registered to a structural MRI scan from which a 3D surface mesh of the cortex and white matter is extracted and shown in this image. Image registration allows fusion of data from multiple scans and multiple modalities

The choice of the correspondence basis depends on the type of images. Statistical measures based on intensity differences, such as SSD, may be used for images of the same modality. Information-theoretic measures, such as MI, CC, or CR, will be appropriate choices for registering images of different modalities. Figure 6 shows brain images of a pediatric patient acquired on a PET scanner and an MRI scanner in different sessions. These images have been registered using a 6-DOF rigid transformation and CR similarity measure using the FLIRT registration tool in the FSL software package.[b] The figure shows that after registration, the PET image could be accurately overlaid on the MR image. Also, different MR images of this subject could be fused together via registration to provide a 3D model of the main vessels obtained from an MR angiography (MRA) scan and a 3D model of the cortex obtained from automatic segmentation of a T1-weighted structural MRI scan.

Variations in Anatomy

The anatomy of a subject, organ, or tissue may change over time between scans. Intra-subject anatomic variations may have different sources, for example, normal growth, aging, disease, atrophy, tumor growth, and surgery (e.g., resection). Appropriate transformation models and similarity measures are required to be used to account for such variations in image registration. For example, deformation fields obtained from deformable registration may be used to quantify the growth or aging process, or to quantify the progress or improvement of disease. Variations in anatomy may be quantified by local volume changes which are obtained from the Jacobian determinant of the deformation field at each point, see Eq. 16.

Deformable registration may also be used in image-navigated surgery to register intraoperative images to preoperative images for real-time detection of landmarks or anatomic regions. It may also be used to register postoperative and preoperative images for the evaluation of surgical outcomes. For example, fast deformable registration may be used to account for brain shift in registering preoperative and intraoperative brain images in craniotomy.

Intersubject Registration

Intersubject registration, or the registration of images of different subjects, may be used (1) to quantify the differences in anatomy between subjects, (2) to bring subjects into a standard spatial space in which group analysis can be performed (this process is called spatial normalization), (3) to construct atlases or standard templates that can be used as representatives of a population anatomy, and (4) for tasks such as atlas-based image segmentation.

Differences in Anatomy

Low-dimensional rigid or affine transformations may be used to register images of different subjects for side-by-side comparison. Variations in anatomy may be described by shape differences at voxel level through a process called voxel-based morphometry.[14] High-dimensional deformable registration may also be used to register images of different subjects. The deformation fields obtained from high-dimensional deformable registration may be used to quantify anatomic differences between subjects. These differences can be described by properties of the deformation field, such as the determinant of Jacobian defined in Eq. 16, in analyses such as deformation-based morphometry or tensor-based morphometry.

Spatial Normalization

Oftentimes pathologic or normal differences between subjects in a population can only be identified through statistical group analysis. The anatomic variability between subjects is a major problem in analyzing images of a group of subjects. The most intuitive way to address this problem is to spatially normalize the images, that is, to bring the images of all subjects into alignment in a common space. Because the anatomic variability between subjects is highly nonlinear, accurate spatial normalization typically requires nonrigid registration. Affine and low-dimensional nonrigid transformations are easier and faster to calculate,

---

[b]http://fsl.fmrib.ox.ac.uk/fsl/fslwiki/FSL.

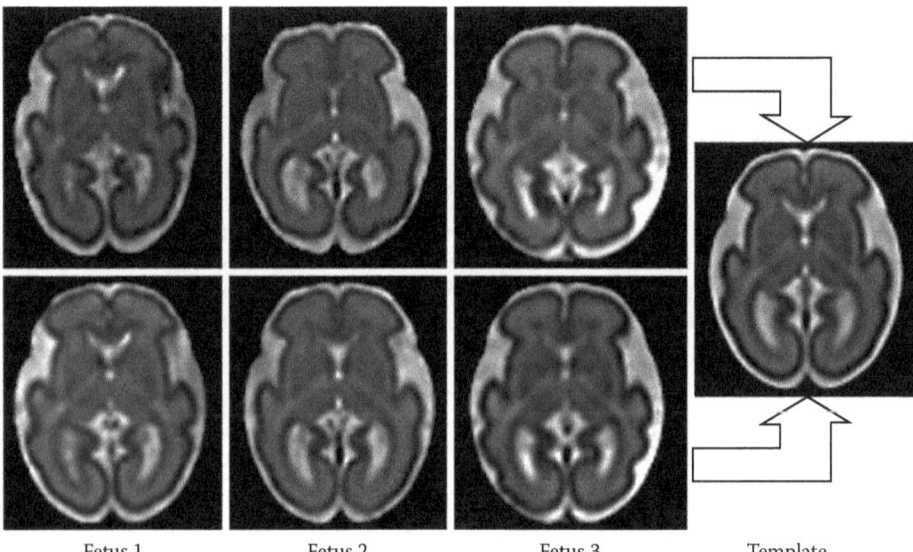

Fig. 7 Spatial normalization: brain MR images of three fetuses at 28 weeks' gestational age are spatially normalized to a 28-week template image (on the right). The original images are shown in the top row, and the spatially normalized images are shown in the bottom row. The majority of anatomic variability between the three brains is compensated after spatial normalization, so the images can be compared. The higher level of gyrification and cortical maturation in fetus 3, as compared to the other fetuses, is still evident after spatial normalization. The DOF of the deformation depends on the application

but high-dimensional deformations are usually preferred especially in applications such as functional and structural neuroimage analysis.[15,16] A considerable amount of work has been done and a vast number of software tools are available for deformable registration for neuroimage analysis and other applications.[c] For a literature review of deformable registration techniques, see Ref. [17].

An example of spatial normalization is shown in Fig. 7. This figure shows fetal brain MR images of three fetuses scanned at the approximate gestational age of 28 weeks. These images have been registered to a 28-week fetal brain MRI template (on the right) through symmetric deformable registration using viscous fluid model and the NCC similarity measure (using the ANTs registration tools[d]). It can be seen that the anatomic structures of the three fetuses are registered to the template; however, the higher level of gyrification and cortical maturation in the third fetus is still observable after spatial normalization by the topology-preserving large diffeomorphic deformable registration. The degree of deformable registration may be defined based on the nature and purpose of intersubject registration for spatial normalization through balancing between the external and internal forces that derive the deformations.

Atlas Construction

Another important application of intersubject registration is the construction of atlases that are average unbiased representatives of a population. The construction of good representatives of a population involves specialized strategies for population-based averaging of anatomy that generates local encoding of anatomic variability. Obviously, atlas construction requires intersubject image registration. The transformation models used can range from rigid, affine, or low-dimensional nonrigid transformations to high-dimensional large deformations to construct deformable atlases of the population anatomy.

Groupwise registration is the technique of choice to build an unbiased atlas from a sample set of images of subjects drawn from a population. This is achieved through iterative estimation of an unbiased template and symmetric diffeomorphic deformable registration of each subject to the estimated template. At first iteration, the template can be simply calculated as the average of all subject anatomies. The precision of the atlas in capturing anatomic details depends on the DOF of the deformation model. The choice of the similarity metric is also important. CC or CR is usually an appropriate choice. Figure 8 shows a 4D (spatiotemporal) atlas of the fetal brain constructed from volumetric brain MRI of 40 normal fetuses in the gestational age range of 26–36 weeks. The figure shows an average of all subjects on the left and the constructed atlas at 1-week intervals between 27 and 32 weeks' gestation.

---

[c]See the Insight Segmentation and Registration ToolKit (ITK): http://www.itk.org/, as a programming framework for the implementation of many image registration components and techniques, and the Neuroimaging Tools and Resources Clearinghouse (NITRC) for a list of tools: http://www.nitrc.org/. Elastix (http://elastix.isi.uu.nl/) and MIRTK (https://mirtk.github.io/) are good examples of general-purpose image registration toolkits.
[d]Advanced Normalization Tools (ANTs): http://picsl.upenn.edu/software/ants/.

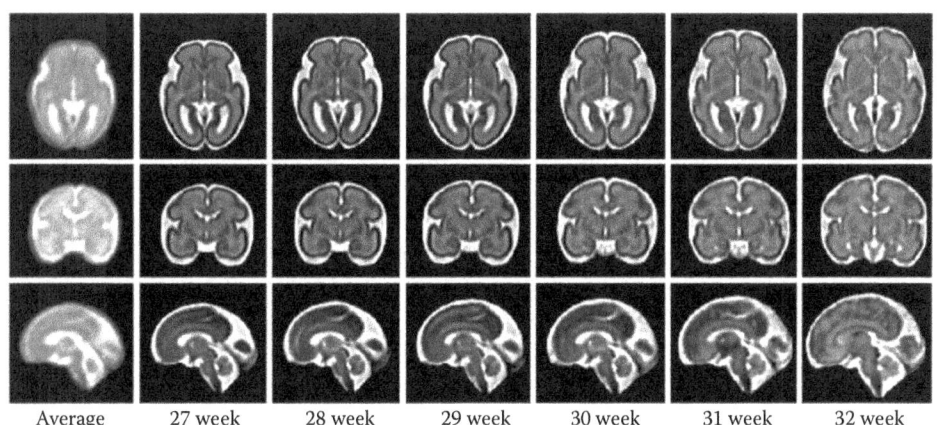

Fig. 8 An unbiased spatiotemporal atlas of the fetal brain constructed from reconstructed fetal brain MRI scans of 40 fetuses scanned between 26 and 36 weeks' gestational age; the image on the left shows a simple average of all 40 MRI scans after rigid registration but without any deformable registration. The atlas has been shown at 1-week intervals between 27 and 32 weeks' gestation. This atlas was constructed using an iterative procedure that involved kernel regression in time and symmetric diffeomorphic deformable registration of individual anatomies in space

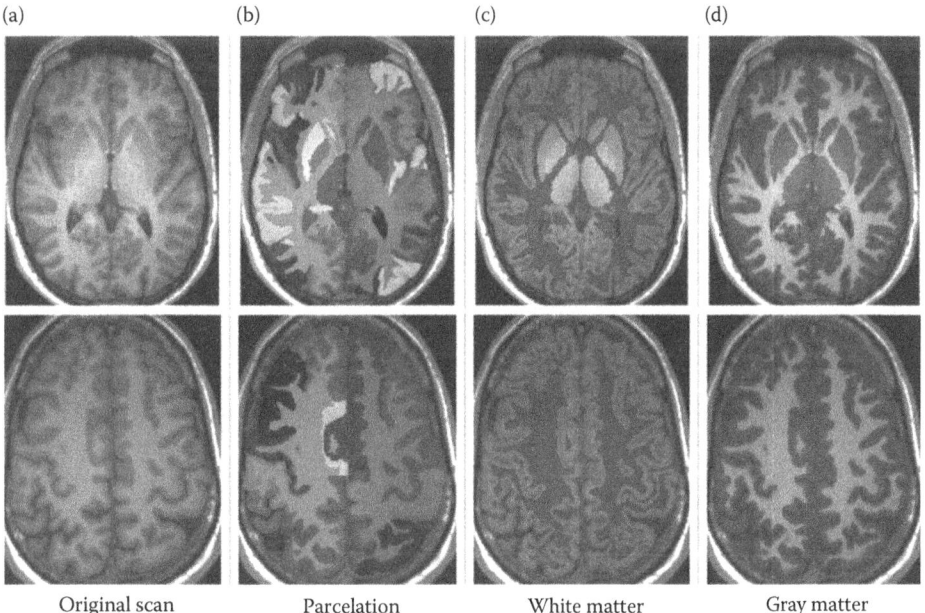

Fig. 9 Multi-atlas segmentation of an adult brain MRI scan using deformable registration and probabilistic simultaneous truth and performance level estimation for label fusion:[18] (a) original scan; (b) brain parcelation; (c) white matter segmentation; (d) gray matter segmentation, all overlaid on the original image

Atlas-Based Segmentation

Atlas-based segmentation relies on the registration of an atlas (source) image to a query (target) image to achieve segmentation of the query image through label propagation. A deformable transformation model is computed to map the anatomy of the atlas to the subject through registration, and the computed transformation is applied to the atlas labels to map them into the subject anatomy (label propagation). This is an intuitive approach to incorporate prior knowledge in the form of atlases into automatic segmentation. Registration errors translate to segmentation errors, but their effect can be reduced by using multiple atlases and label fusion, which can be done through robust label fusion methods. Figure 9 shows multi-atlas parcelation and segmentation of a brain MR image to anatomic regions, white matter, and gray matter through deformable registration, label propagation, and robust probabilistic label fusion.[18]

VALIDATION OF IMAGE REGISTRATION

Validation is necessary to ensure the fidelity and usefulness of image registration techniques, and includes considerations such as accuracy, robustness, reliability, resource requirements, computational complexity, and impact. The validation of biomedical image registration techniques, in terms of accuracy, robustness, and reliability, suffers from the lack of ground truth. It shares the same problems associated with finding appropriate correspondences for registration. The most promising similarity measures that are often used in image registration operate in the intensity domain and not in the spatial domain, thus do not provide useful information regarding the fidelity of registration, or the magnitude or distribution of registration errors. In the absence of ground truth, image registration may be validated using techniques such as (1) visual inspection and assessment, (2) gold standards obtained from controlled experiments, (3) surrogate measurements based on the quality of registration products or outcomes, and (4) test–retest reliability for the assessment of consistency and reliability.

Visual assessment of registration accuracy can be done through the assessment of segmented edges or contours, anatomic landmarks, subtracted images, image overlay, or checkerboard display of images. Extrinsic markers implanted on a number of subjects may be used to set up gold standard transformations for the validation of techniques. Alternatively controlled experiments with phantoms or with known motion or deformations may be used for validation. The quality of the outcomes or products of registration may be used as surrogate measures of registration accuracy. Examples of such surrogate measures are the quality of atlas-based segmentations obtained from label propagation using deformable registration,[16] quality of atlases in atlas construction, and the locality and focus of statistical parametric maps in brain functional localization.[15] Finally, test–retest reliability may be used to evaluate the consistency and reliability of registration and sensitivity analysis for the evaluation of robustness to scene and parameter variations. Although high-performance computing is now available at relatively low costs, the computational cost and resource requirements of a registration technique may also need to be critically assessed and validated especially in dealing with big data, such as images obtained from microscopic imaging or in large population studies. Efficient implementation of transformations, interpolation, and optimization is often crucial. Computational techniques that accelerate registration and reduce computational costs are very useful.[19]

CONCLUSION

This chapter has covered the structure of a biomedical image registration algorithm that conventionally consists of four main components: a transformation model, a correspondence basis (typically in the form of a distance, or similarity metric, or a cost function), an optimization technique, and an interpolation method. Various combinations of these components may be used for any specific application. The choices of the transformation model and the correspondence basis depend on geometric and photometric differences between the images, respectively. The geometric differences may vary between global rigid transformation to large local deformations. Photometric differences between multimodality images may require the use of robust similarity measures based on correlation or information-theoretic measures. The choice of an interpolation technique is established through a trade-off between speed and accuracy of registration. An optimization technique is highly dependent on the other components. Second-order gradient optimization techniques are typically used when computational or analytical derivatives of the cost function can be obtained for parametric transformations. Finite-element methods are used to solve PDEs associated with biomechanical and biophysical deformation models. Image registration is a complex task and can be computationally intensive and slow. In recent years, it has been shown that complex image registration tasks may be learned through unsupervised or supervised methods using sufficiently flexible models such as deep convolutional neural networks.[20,21] Such learning-based methods lead to much faster image registration as they rely on training phases and learned models to eliminate optimization at test time.

A review of applications shows how image registration constitutes a fundamental tool in many biomedical imaging applications. There is extensive literature on biomedical image registration, and the review of this literature and the related theories and techniques, which span across multiple disciplines including signal and image processing, computer graphics, computer vision, mathematics, statistics, information theory, continuum mechanics, and thermodynamics, is beyond the capacity of any single article. Among many valuable publications, only some references that cover different aspects of the field are cited here, and more can be found through related searches. These references include a book,[22] and review articles on medical image registration,[23,24] a review of image registration for image-guided intervention,[25] a survey of slice-to-volume medical image registration,[10] a survey of image registration techniques for neuroimaging and brain functional localization,[15] a review of medical image registration on multicore processing units and graphical processing units (GPUs),[19] an evaluation of nonrigid registration techniques in neuroimaging,[16] and a survey of deformable medical image registration techniques.[17]

REFERENCES

1. Ashburner, J.; Friston, K.J. Nonlinear spatial normalization using basis functions. Hum. Brain Mapping **1999**, 7 (4), 254–266.
2. Rueckert, D.; Sonoda, L.I.; Hayes, C.; Hill, D.L.; Leach, M.O.; Hawkes, D.J. Nonrigid registration using free-form

deformations: Application to breast MR images. IEEE Trans. Med. Imaging **1999**, *18* (8), 712–721.
3. Thirion, J.P. Image matching as a diffusion process: An analogy with Maxwell's demons. Med. Image Anal. **1998**, *2* (3), 243–260.
4. Vercauteren, T.; Pennec, X.; Perchant, A.; Ayache, N. Diffeomorphic demons: Efficient non-parametric image registration. NeuroImage **2009**, *45* (1), S61–S72.
5. Arun, K.S.; Huang, T.S.; Blostein, S.D. Least-squares fitting of two 3D point sets. IEEE Trans. Pattern Anal. Mach. Intell. **1987**, *9* (5), 698–700.
6. Besl, P.; McKay, N. A method for registration of 3-D shapes. IEEE Trans. Pattern Anal. Mach. Intell. **1992**, *14* (2), 239–256.
7. Wells, W.M.; Viola, P.; Atsumi, H.; Nakajima, S.; Kikinis, R. Multi-modal volume registration by maximization of mutual information. Med. Image Anal. **1996**, *1* (1), 35–51.
8. Avants, B.B.; Tustison, N.J.; Song, G.; Cook, P.A.; Klein, A.; Gee, J.C. A reproducible evaluation of ANTs similarity metric performance in brain image registration. Neuroimage **2011**, *54* (3), 2033–2044.
9. Lehmann, T.M.; Gönner, C.; Spitzer, K. Survey: Interpolation methods in medical image processing. IEEE Trans. Med. Imaging **1999**, *18* (11), 1049–1075.
10. Ferrante, E.; Paragios, N. Slice-to-volume medical image registration: A survey. Med. Image Anal. **2017**, *39*, 101–123.
11. Gholipour, A.; Estroff, J.A.; Warfield, S.K. Robust super-resolution volume reconstruction from slice acquisitions: Application to fetal brain MRI. IEEE Trans. Med. Imaging **2010**, *29* (10), 1739–1758.
12. Wang, C.W.; Gosno, E.B.; Li, Y.S. Fully automatic and robust 3D registration of serial-section microscopic images. Sci. Rep. **2015**, *5*, 15051.
13. Qu, L.; Long, F.; Peng, H. 3-d registration of biological images and models: Registration of microscopic images and its uses in segmentation and annotation. IEEE Sig. Process. Mag. **2015**, *32* (1), 70–77.
14. Ashburner, J.; Friston, K.J. Voxel-based morphometry—The methods. NeuroImage **2000**, *11* (6), 805–821.
15. Gholipour, A.; Kehtarnavaz, N.; Briggs, R.; Devous, M.; Gopinath, K. Brain functional localization: A survey of image registration techniques. IEEE Trans. Med. Imaging **2007**, *26* (4), 427–451.
16. Klein, A.; Andersson, J.; Ardekani, B.A.; Ashburner, J.; Avants, B.; Chiang, M.C.; Christensen, G.E.; Collins, D.L.; Gee, J.; Hellier, P.; Song, J.H.; Jenkinson, M.; Lepage, C.; Rueckert, D.; Thompson, P.; Vercauteren, T.; Woods, R.P.; Mann, J.J.; Parsey, R.V. Evaluation of 14 nonlinear deformation algorithms applied to human brain MRI registration. Neuroimage **2009**, *46* (3), 786–802.
17. Sotiras, A.; Davatzikos, C.; Paragios, N. Deformable medical image registration: A survey. IEEE Trans. Med. Imaging **2013**, *32* (7), 1153–1190.
18. Akhondi-Asl, A.; Warfield, S.K. Simultaneous truth and performance level estimation through fusion of probabilistic segmentations. IEEE Trans. Med. Imaging **2013**, *32* (10), 1840–1852.
19. Shams, R.; Sadeghi, P.; Kennedy, R.A.; Hartley, R.I. A survey of medical image registration on multicore and the GPU. IEEE Sig. Process. Mag. **2010**, *27* (2), 50–60.
20. Yang, X.; Kwitt, R.; Styner, M.; Niethammer, M. Quicksilver: Fast predictive image registration–a deep learning approach. NeuroImage **2017**, *158*, 378–396.
21. de Vos, B.D.; Berendsen, F.F.; Viergever, M.A.; Staring, M.; Išgum, I. End-to-end unsupervised deformable image registration with a convolutional neural network. In *Deep Learning in Medical Image Analysis and Multimodal Learning for Clinical Decision Support*; Cardoso, J.; Arbel, T.; Carneiro, G.; Syeda-Mahmood, T.; Tavares, J.M.R.S.; Moradi, M.; Bradley, A.; Greenspan, H.; Papa, J.P.; Madabushi, A.; Nascimento, J.C.; Cardoso, J.S.; Belagiannis, V.; Lu, Z.; Eds.; Springer, 2017, 204–212.
22. Hajnal, J.V.; Hill, D.L.G.; Hawkes, D.J. *Medical Image Registration*, Biomedical Engineering Series; Neuman, P.R.; Ed.; CRC Press: Boca Raton, FL, 2001.
23. Maes, F.; Vandermeulen, D.; Suetens, P. Medical image registration using mutual information. Proc. IEEE **2003**, *91* (10), 1699–1722.
24. Pluim, J.P.W.; Maintz, J.B.A.A.; Viergever, M.A. Mutual-information-based registration of medical images: A survey. IEEE Trans. Med. Imaging **2003**, *22* (8), 986–1004.
25. Markelj, P.; Tomaževič, D.; Likar, B.; Pernuš, F. A review of 3D/2D registration methods for image-guided interventions. Med. Image Anal. **2012**, *16* (3), 642–661.

Boolean Algebras

A. R. Bednarek
University of Florida, Gainesville, Florida, U.S.A.

Abstract
Boolean algebra, named after the 19th century mathematician and logician, George Boole, has contributed to many aspects of computer science and information science. In information science, Boolean logic forms the basis of most end-user search systems, from searches in online databases and catalogs, to uses of search engines in information seeking on the World Wide Web.
—**ELIS Classic**, from 1970

Keywords: Boolean algebra Boole; George.

INTRODUCTION

In this entry attention is focused on mathematical models of proven utility in the area of information handling, namely, Boolean algebras. Following some general comments concerning mathematical models, particular examples of Boolean algebras, serving as motivation for the subsequent axiomatization, are presented. Some elementary theorems are cited, particularly the very important representation theorem that justifies, in some sense, the focusing of attention on a particular Boolean algebra, namely, the algebra of classes, and applications more directly related to the information sciences are given.

Running the risk of redundancy, attention will be called to an often-repeated observation, but one of extreme importance in applications of mathematics to physical problems. Referring to Fig. 1, it is important to realize that when one constructs a mathematical model as a representation of a physical phenomenon, one is abstracting and, as a consequence, the model formulated is doomed to imperfection. That is, one can never formally mirror the physical phenomenon, and must always be satisfied with an imperfect copy. However, following the initial commitment to a model, the logic that one appeals to dictates the resultant theorems derived within the framework of the model. Of course, the depth of the theorems realized is limited by the sophistication of the model as well as the ingenuity of those who attempt to formulate the propositions within it. After theorems are derived within the framework of the model, they are interpreted relative to the physical situation that motivated the model.

It is not necessary to go very deeply into mathematics before facing the necessity of examining, in some detail, this cycle and developing a feeling for its power as well as its limitations. By way of example, almost any student of calculus encounters, in one form or another, the following problem:

> The deceleration of a ship in still water is proportional to its velocity. If the velocity is v_0 feet per second at the time the power is shut off, show that the distance S the ship travels in the next t seconds is $S = (v_0/k)[1 - e^{-kt}]$ where k is the constant of proportionality.

HISTORY

The desired equation relating the distance traveled to the time is easily arrived at by means of the calculus. However, a close look at the solution reveals a few puzzling aspects. When does the ship stop? The conclusion is that it never stops. How far does it go? The conclusion is that it goes no further than v_0/k, that is, the distance it travels is bounded. Sympathy is due the beginning student of the calculus who is puzzled by these observations, but, too often, we neglect to focus our attention on the source of the puzzlement. It really has nothing to do with the limit process that plays such an integral role in analysis, nor must we drag poor Zeno into the picture. This disturbing conclusion is not the consequence of any faulty mathematics, but is more directly related to the naïveté of the original model. If we say that the deceleration of a ship in still water (an idealization in itself) is proportional only to its velocity, then the conclusion that asserts itself is that the ship never stops but only goes a finite distance.

The usual remedy applied in such cases as the ship problem is to construct a more sophisticated model, that is, a model that takes into account more of the phenomena observed. For example, in the ship problem, the assertion that the deceleration is proportional only to the velocity

Boolean Algebras

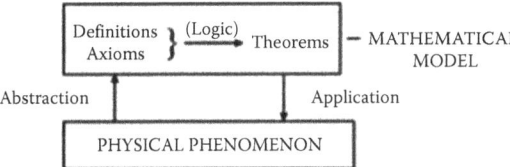

Fig. 1

might be amended to include friction in some way, resulting in an equation of greater complexity, the formulation and solution of which require a more general mathematical model. We might extend the preceding model to look like Fig. 2.

The great power of mathematics lies in its ability to reflect several different phenomena at one time, and the theorems derived within the framework of a single axiomatization of these varied phenomena will, in turn, be applicable to each of them. However, the trade-off that exists between generalization and depth must constantly be kept in mind. That is, it should be remembered that it is difficult to prove deep theorems in very general models. But when axioms are added to the model, the phenomena which the model reflects begin to be delimited, and certainly one does not wish to undermine the real power of mathematics, that is, its ability to treat a variety of situations at the same time.

Examples

We now turn our attention to an examination of some of the particular examples of the model that is the principal concern of this entry, Boolean algebras. One must keep in mind that the common characteristics of these models are precisely those that will constitute the elements of our later axiomatization. To avoid infinite regress a certain level of sophistication on the part of the reader, if not actual mathematical experience, is assumed.

Example 1 (A finite algebra)

The system considered in this example consists of the two digits, 0 and 1, and two binary operations of multiplication, "•," and addition, "+." The operations are defined by the multiplication and addition tables shown.

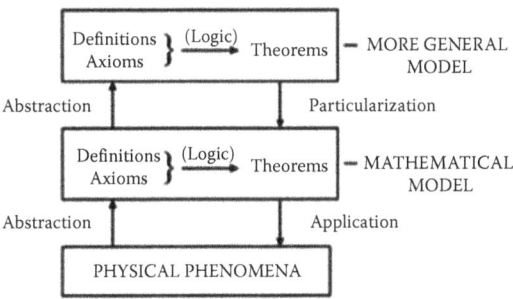

Fig. 2

If x, y, and z are any *variables* that are allowed to assume one of the two values 0 or 1, then the structure defined above has an algebra possessing (among others) the following properties:

$$x \cdot x = x \qquad x + x = x \tag{1a}$$

$$x \cdot y = y \cdot x \qquad x + y = y + x \tag{1b}$$

$$x \cdot (y \cdot z) = (x \cdot y) \cdot z \qquad x + (y + z) = (x + y) + z \tag{1c}$$

$$(x \cdot y) + x = x \qquad (x + y) \cdot x = x \tag{1d}$$

$$x \cdot (y + z) = (x \cdot y) + (x \cdot z) \qquad x + (y \cdot z) = (x + y) \cdot (y + z) \tag{1e}$$

Each of the above can be verified by a consideration of all the possible values of the variables.

If B is the collection consisting of the elements 0 and 1, and for each x in B x' is defined by $x' = 1$ if $x = 0$, and $x' = 0$ if $x = 1$, then:

$$x \cdot x' = 0 \qquad \text{and} \qquad x + x' = 1$$
$$0 \cdot x = 0 \qquad 1 \cdot x = x \qquad \text{and} \qquad 0 + x = x \qquad 1 + x = x \tag{1f}$$

We let $[B; \bullet, +, ']$ denote the system described in Example 1.

Example 2 (Algebra of propositions)

The elements in this example are *propositions*, that is, statements to which it is possible to assign one of the truth values "true" or "false." Two propositions p and q are defined to be equal if and only if they have the *same* truth value. We consider the two logical binary operations of *conjunction* and *disjunction* as well as the unary (operating on a single proposition as contrasted with a binary operation, which operates on pairs of propositions) operation of *negation*. The conjunction of the propositions p and q is denoted by pq and is the proposition corresponding to that obtained by applying the logical connective "and." The conjunction is defined to be true only if both p and q are true. Otherwise it is false. The disjunction of p and q, denoted by $p + q$, is the proposition corresponding to that obtained by applying the logical connective "or." The proposition $p + q$ is false if and only if both p and q are false. The negation of p, denoted by $\bar{p}$, is the proposition having truth values opposite those of p. It corresponds to the logical statement, "It is false that p."

All of the above can be summarized nicely by employing "truth tables" that give the truth values of compound statements, realized by applying the operations discussed to the truth values of the component propositions.

p	q	pq
T	T	T
T	F	F
F	T	F
F	F	F

Conjunction

p	q	$p+q$
T	T	T
T	F	T
F	T	T
F	F	F

Disjunction

p	$\bar{q}$
T	F
F	T

Negation

One can verify that, in view of the definitions above, if p, q, and r are any propositions, the following statements hold.

$$pp = p \qquad p + p = p \tag{2a}$$

$$pq = qp \qquad p + q = q + p \tag{2b}$$

$$p(qr) = (pq)r \qquad p + (q+r) = (p+q) + r \tag{2c}$$

$$(pq) + p = p \qquad (p+q)p = p \tag{2d}$$

$$p(q+r) = pq + pr \qquad p+(qr) = (p+q)(q+r) \tag{2e}$$

$$p\bar{p} \text{ is always false and } p + \bar{p} \text{ is always true} \tag{2f}$$

Denoting $p\bar{p}$ by 0 and $p + \bar{p}$ by 1 we have

$$0q = 0 \quad 1q = q \quad 0 + q = q \quad 1 + q = q$$

(*Note*: It is easy to see that $p\bar{p} = q\bar{q}$ and $p+\bar{p} = q+\bar{q}$ for any propositions p and q.)

We illustrate the employment of the truth table technique in the verification of part of the assertion in Eqs. 2e, namely, that $p + (qr) = (p+q)(p+r)$.

p	q	r	qr	$p + qr$	$p + q$	$p + r$	$(p+q)(p+r)$
T	T	T	T	T	T	T	T
T	T	F	F	T	T	T	T
T	F	T	F	T	T	T	T
T	F	F	F	T	T	T	T
F	T	T	T	T	T	T	T
F	T	F	F	F	T	F	F
F	F	T	F	F	F	T	F
F	F	F	F	F	F	F	F

Since the columns headed "$p + qr$" and "$(p + q)(p + r)$" have identical entries the propositions $p+qr$ and $(p + q)(p + r)$ have the same truth value and are therefore equal.

The structure described in Example 2 is denoted in the sequel by $[;,+,-]$.

Example 3 (Algebra of sets)

The term *set* is taken as undefined and used synonymously with class, aggregate, and collection. The objects that constitute a set E are called the *elements* of E. To denote the logical relation of "being an element of E" we use the notation $x \in E$. This is read: "x is an element of E." The denial of this relation is symbolized by $x \notin E$.

The notation of $E = \{xP(x)\}$ denotes the set E consisting of all x for which the proposition $P(x)$ is true. When the set under consideration is finite, it is often denoted by a simple listing of its elements; thus, $E = \{a,b\}$ is the set consisting of the elements a and b.

If $A = \{xP(x)\}$ and there are no elements which satisfy the proposition $P(x)$, A is said to be the *empty set*. The empty (or *null*) set is denoted by ϕ.

If the sets E and F have the property that every element of E is an element of F, E is called a *subset* of F; this is denoted by $E \subset F$. If the set E is a subset of F, but F is not a subset of E, then E is said to be a *proper subset* of F, or F *properly contains* E. The empty set ϕ is a subset of every set.

Two sets E and F are *equal*, written $E = F$, if and only if $E \subset F$ and $F \subset E$.

Given two sets E and F, we define the *union*, denoted by $E \cap F$, by the set equation:

$$E \cup F = \{x \mid x \in E \text{ or } x \in F\}$$

Similarly, the *intersection* of E and F, denoted by $E \cap F$, is defined by:

$$E \cap F = \{x \mid x \in E \text{ and } x \in F\}$$

In general, consideration centers on subsets of a fixed set often referred to as the *universal set*. In particular, if X is the universal set, we let (X) denote the set of all subsets of X. The set (X) is often called the *power set* of X. If $E \in (X)$, then the *complement* of E, denoted by E', is defined as $E' = \{xx \in X \text{ and } x \notin E\}$. If E and F are elements of (X), that is, subsets of X, then the *difference* of the sets E and F, denoted by $E-F$, is the set defined by:

$$E - F = \{x \mid x \in E \text{ and } x \notin F\}$$

It should be noted that $E - F = E \cap F'$.

It is often helpful to employ the schematics shown in Fig. 3 in visualizing the set-theoretic relations and operations defined above. The rectangular area represents the universal set X; subsets of X are denoted by areas within the rectangle.

Boolean Algebras

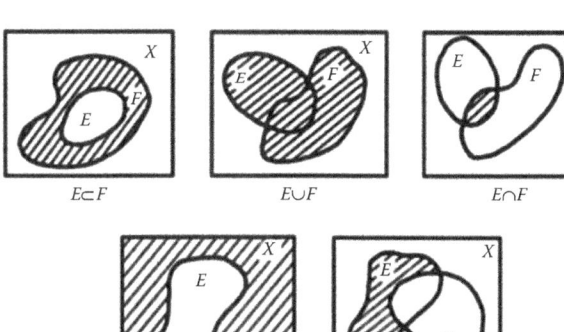

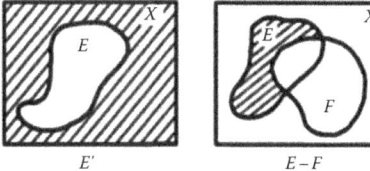

Fig. 3

Focusing our attention on a particular universal set X and its power set (X), it is easy to verify the following properties (in no sense exhaustive) of the algebra of sets, where E, F, and G are arbitrary subsets of X.

$$E \cap E = E \quad E \cup E = E \tag{3a}$$

$$E \cap F = F \cap E \quad E \cup F = F \cup E \tag{3b}$$

$$E \cap (F \cap G) = (E \cap F) \cap G \quad E \cup (F \cup G) = (E \cup F) \cup G \tag{3c}$$

$$(E \cap F) \cup E = E \quad (E \cup F) \cap E = E \tag{3d}$$

$$E \cap (F \cup G) = (E \cap F) \cup (E \cap G)$$
$$E \cup (F \cap G) = (E \cup F) \cap (E \cup G) \tag{3e}$$

$$E \cap E' = \phi \quad\quad E \cup E' = X$$
$$\phi \cap E = \phi \quad X \cap E = E \quad \phi \cup E = E \quad X \cup E = E \tag{3f}$$

In the sequel we denote the above algebra of sets by $[(X); \cap, \cup, ']$.

Axiomatization

In the preceding section we examined three structures possessing some common properties, namely, (1i), (2i), and (3i), where $i = 1, 2, 3, 4, 5, 6$. We abstract to construct the important mathematical (see Fig. 1) model called a *Boolean algebra*, named in honor of G. Boole who first studied it in 1847.[1,2]

A *Boolean algebra* is a set B with two binary operations $\wedge$ (cap) and $\vee$ (cup) and a unary operation (complementation) satisfying the following axioms:

$$x \wedge x = x \quad \text{and} \quad x \vee x = x \tag{Ia}$$

$$x \wedge y = y \wedge x \quad \text{and} \quad x \vee y = y \vee x \tag{Ib}$$

$$x \wedge (y \wedge z) = (x \wedge y) \wedge z \text{ and } x \vee (y \vee z)$$
$$= (x \vee y) \vee z \tag{Ic}$$

$$(x \wedge y) \vee x = x \text{ and } (x \vee y) \wedge x = x \tag{Id}$$

$$x \wedge (y \vee z) = (x \wedge y) \vee (x \wedge z) \text{ and}$$
$$x \vee (y \wedge z) = (x \vee y) \wedge (x \vee z) \tag{Ie}$$

B contains *distinct* elements 0 and 1 such that:

$$x \wedge x' = 0 \quad\quad \text{and} \quad x \vee x' = 1$$
$$0 \wedge x = 0 \quad 1 \wedge x = x \quad \text{and} \quad 0 \vee x = x \quad 1 \vee x = 1 \tag{If}$$

This is by no means the only axiomatization possible,[3] but it is probably the one most commonly used.

To emphasize the relationship between the above axiomatization and the preceding particularizations:

Boolean algebra	Example 1	Example 2	Example 3
B	set $\{0,1\}$	set $0°$ of all propositions	set $0°(X)$ of all subsets of a fixed set X
$\wedge$	•	conjunction	intersection
$\vee$	+	disjunction	union
$'$	$'$	negation	complementation
0	0	$p\bar{p}$	empty set ϕ
1	1	$p + \bar{p}$	universal set X

(Examples 1, 2, and 3) we identify in tabular form the corresponding structural elements.

We now prove a particular theorem to illustrate the generation of results within the framework of the model and their subsequent application.

Theorem. If $(B; \wedge, \vee, ')$ is a Boolean algebra, then for any x and y in B we have:

i. $x'' = x$
ii. $(x \wedge y)' = x' \vee y'$

Proof. First of all we prove that every element has only one complement. Suppose $\bar{x}$ is an element such that $x \wedge \bar{x} = 0$ and $x \vee \bar{x} = 0$. Then

$$\bar{x} = \bar{x} \wedge 1 = \bar{x} \wedge (x \vee x') = (\bar{x} \wedge x) \vee (\bar{x} \wedge x')$$
$$= 0 \vee (\bar{x} \wedge x') = \bar{x} \wedge x'$$

but

$$x' = x' \wedge 1 = x' \wedge (x \vee \bar{x}) = (x' \wedge x) \vee (x' \wedge \bar{x})$$
$$= 0 \vee (x' \wedge \bar{x}) = (x' \wedge \bar{x})$$

and since $\bar{x} \wedge x' = x' \wedge \bar{x}$, we have $\bar{x} \wedge x'$. We then apply the preceding by demonstrating that $(x \wedge y) \wedge (x' \vee y') = 0$ and $(x \wedge y) \vee (x' \vee y') = 1$, so that $(x \wedge y)' = x' \vee y'$.

$$(x \wedge y) \wedge (x' \vee y') = [(x \wedge y) \wedge x'] \vee [(x \wedge y) \wedge y']$$
$$= [(x \wedge y) \wedge x'] \vee [(x \wedge y) \wedge y']$$
$$= [y \wedge (x \wedge x')] \vee [x \wedge (y \wedge y')]$$
$$= [y \wedge 0] \vee [x \wedge 0] = 0 \vee 0 = 0$$
$$(x \wedge y) \wedge (x' \vee y') = [(x \wedge y) \vee x'] \vee y'$$
$$= [(x \vee x') \wedge (y \vee x')] \vee y' = [1 \wedge (y \wedge x')] \vee y'$$
$$= (x \vee x') \vee y' = (x' \vee y) \vee y'$$
$$= x' \vee (y \wedge y') = x' \vee 1 = 1$$

An interpretation (application) of this theorem in Example 2 yields the fact that the negation of the conjunction of two propositions is the disjunction of the negations of each of them. For example, "it is false that x is a positive integer and x is greater than or equal to 5" is logically equivalent to the proposition "x is not a positive integer or x is less than 5."

An interpretation of the above in Example 2 yields the set-theoretic equation

$$(E \cap F)' = E' \cup F'$$

that is, the complement of the intersection of two sets is equal to the union of their complements, as shown in Fig. 4.

We describe very briefly some of the more significant results and developments in the theory of Boolean algebras. For a comprehensive treatment of the subject, see Birkhoff[3] and Halmos[4] and their bibliographies.

Every Boolean algebra can be made into a ring with identity in which every element is multiplicatively idempotent; that is, $x^2 = x$ for every x. This is accomplished by defining addition and multiplication as follows:

$$x + y = (x \wedge y') \vee (x' \wedge y) \text{ and } xy = x \wedge y$$

Because rings are more familiar and more carefully studied many of the useful concepts can be translated into the context of Boolean algebras.

Conversely, if one starts with a ring with identity in which every element is idempotent (usually called a Boolean ring), defining $\wedge$ and $\vee$ by

$$x \wedge y = xy \quad x \vee y = x + y + xy$$

the Boolean ring is converted into a Boolean algebra.

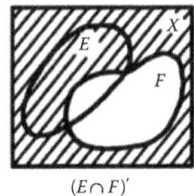 

$(E \cap F)'$ $E' \cup F'$

Fig. 4

Two Boolean algebras, B_1 and B_2, are said to be *isomorphic* if there exists a function $h: B_1 \to B_2$ that maps B_1 onto B_2 in such a way that distinct elements of B_1 are mapped onto distinct elements of B_2, and h preserves the operations; that is, $h(x \wedge y) = h(x) \wedge h(y)$; $h(x \vee y) = h(x) \vee h(y)$; and $h(x') = h(x)'$.

If X is a compact Hausdorff space, then the class of sets that are both open and closed forms a Boolean algebra. A topological space is totally disconnected if the only components (maximal connected sets) are points. There is a very important *representation theorem* in the theory of Boolean algebras, the Stone Representation Theorem (M.H. Stone[5]). If B is a Boolean algebra, then a compact totally disconnected Hausdorff space S exists such that B is isomorphic to the Boolean algebra of all open-closed subsets of S.

An Application

We consider here one modest application of the preceding to switching theory. Switching theory is concerned with circuits composed of elements that can assume a finite number of discrete states, most commonly two states. These circuits are modeled as described earlier, and the models are analyzed. This is an idealization; the models neglect such characteristics as stability, temperature effects, and transition times. The theory of Boolean algebras has played an important role in the analysis of these models for circuits made of binary (two-state) devices.

A *switching function* is a rule by which the output of a composite circuit can be ascertained from the states of its components. If the variables x, y, and z denote switches and each switch can assume one of the states, open or closed (0 or 1), then the function $w = x \wedge y$ describes the output of a series circuit containing the switches x and y. Similarly, $t = x \vee y$ is a function describing a parallel circuit containing the switches x and y. These components, along with the negation function (x' is a switch that is open whenever x is closed and closed whenever x is open), allow the construction and analysis of complex circuits. This analysis can be carried out by the use of truth tables, and the circuits can be indicated by a diagram, as shown in Fig. 5.

With this interpretation it is readily seen that the above is a Boolean algebra. For example, the verification of axiom (Id) involves the observation that the circuits in Fig. 6 are equivalent.

After observing that it is indeed a Boolean algebra, the machinery of that algebra may be used to synthesize

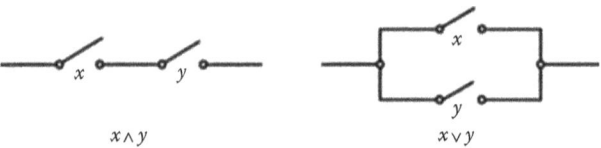

Fig. 5

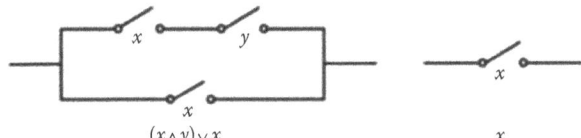

Fig. 6

circuits, consider questions of realizability, minimize circuitry, and so forth. We can only hint at the possible applications.[6]

REFERENCES

Boole, G. *The Mathematical Analysis of Logic*: Cambridge, 1847.
Boole, G. *An Investigation of the Laws of Thought*: London, 1854.
Birkhoff, G. *Lattice Theory*: Providence, RI, 1961, 155.
Halmos, P.R. *Lectures on Boolean Algebras*; Van Nostrand: Princeton, NJ, 1963.
Stone, M.H. The theory of representations for boolean algebras. Trans. Am. Math. Soc. **1936**, *40*, 37–111.
Flegg, H.G. *Boolean Algebra and Its Application*; Wiley: New York, 1964.

Cantor, Fuzzy, Near, and Rough Sets

James F. Peters
Computational Intelligence Laboratory, Electrical & Computer Engineering, University of Manitoba, Winnipeg, Manitoba, Canada

Sankar K. Pal
Machine Intelligence Unit, Indian Statistical Institute, Kolkata, India

Abstract
This entry provides an introduction to fuzzy sets, near sets, and rough sets. It continues with an overview of the relationship between sets and their linkage to Cantor sets. Applications to image processing and computational intelligence are also discussed.

Keywords: Cantor sets; Fuzzy sets; Near sets; Rough sets.

INTRODUCTION

This entry considers how one might utilize fuzzy sets, near sets, and rough sets, taken separately or taken together in hybridizations, in solving a variety of problems in image analysis. A brief consideration of Cantor sets[6,7] provides a backdrop for an understanding of several recent types of sets useful in image analysis. Fuzzy, near and rough sets provide a wide spectrum of practical solutions to solving image analysis problems such as image understanding, image pattern recognition, image retrieval and image correspondence, mathematical morphology, perceptual tolerance relations in image analysis and segmentation evaluation. Fuzzy sets result from the introduction of a membership function that generalizes the traditional characteristic function. The notion of a fuzzy set was introduced by L. Zadeh in 1965.[64] Sixteen years later, rough sets were introduced by Z. Pawlak in 1981.[39] A set is considered rough whenever the boundary between its lower and upper approximation is non-empty. Of the three forms of sets, near sets are newest, introduced in 2007 by J.F. Peters in a perception-based approach to the study of the nearness of observable objects in a physical continuum.[16,44,46,49,51,54,55]

This entry highlights a context for three forms of sets that are now part of the computational intelligence spectrum of tools useful in image analysis and pattern recognition. The principal contribution of this entry is an overview of the high utility of fuzzy sets, near sets and rough sets with the emphasis on how these sets can be used in image analysis, especially in classifying parts of digital images presented in this book.

CANTOR SET

To establish a context for the various sets utilized in this book, this section briefly presents the notion of a Cantor set. From the definition of a Cantor set, it is pointed out that fuzzy sets, near sets and rough sets are special forms of Cantor sets. In addition, this entry points to links between the three types of sets that are part of the computational intelligence spectrum. Probe functions in near set theory provide a link between fuzzy sets and near sets, since every fuzzy membership function is a particular form of probe function. Probe functions are real-valued functions introduced by M. Pavel in 1993 as part of a study of image registration and a topology of images.[38] Z. Pawlak originally thought of a rough set as a new form of fuzzy set.[39] It has been shown that every rough set is a near set (this is Theorem 4.8 in Ref. [45]) but not every near set is a rough set. For this reason, near sets are considered a generalization of rough sets. The contribution of this entry is an overview of the links between fuzzy sets, near sets and rough sets as well as the relation between these sets and the original notion of a set introduced by Cantor in 1883.[6]

> By a 'manifold' or 'set' I understand any multiplicity, which can be thought of as one, *i.e.*, any aggregate [*inbegriff*] of determinate elements which, can be united into a whole by some law.
> —*Foundations of a General Theory of Manifolds,*
> —*G. Cantor, 1883.*

> ... A set is formed by the grouping together of single objects into a whole.
> —*Set Theory*
> —*F. Hausdorff, 1914.*

In this mature interpretation of the notion of a set, G. Cantor points to a property or law that determines elementhood in a set and "unites [the elements] into a whole",[6] elaborated in Ref. [7], and commented on in Ref. [19]. In 1851, Bolzano[2] writes that "an aggregate so conceived that is indifferent to the arrangement of its members I call a *set*". At that time, the idea that a set could contain just one element or no elements (null set) was not contemplated. This is important in the current conception of a near set, since such a set must contain pairs of perceptual objects with similar descriptions and such a set is never null. That is, a set is a perceptual near set if, and only if it is never empty and it contains pairs of perceived objects that have descriptions that are within some tolerance of each other.

NEAR SETS

How Near
How near to the bark of a tree are drifting snowflakes, swirling gently round, down from winter skies?
How near to the ground are icicles,
slowly forming on window ledges?

–Fragment of a Philosophical Poem.
–Z. Pawlak & J.F. Peters, 2002.

The basic idea in the near set approach to object recognition is to compare object descriptions. Sets of objects X, Y are considered near each other if the sets contain objects with at least partial matching descriptions.

–Near sets. General theory about nearness of objects,
–J.F. Peters, 2007.

Set Theory Law 1 Near Sets

Near sets contain elements with similar descriptions.

Near sets are disjoint sets that resemble each other.[17] Resemblance between disjoint sets occurs whenever there are observable similarities between the objects in the sets. Similarity is determined by comparing lists of object feature values. Each list of feature values defines an object's description. Comparison of object descriptions provides a basis for determining the extent that disjoint sets resemble each other. Objects that are perceived as similar based on their descriptions are grouped together. These groups of similar objects can provide information and reveal patterns about objects of interest in the disjoint sets. For example, collections of digital images viewed as disjoint sets of points provide a rich hunting ground for near sets. For example, near sets can be found in the favite pentagona coral fragment in Fig. 1a from coral reef near Japan. If we consider the greyscale level, the sets X, Y in Fig. 1b are near sets, since there are many pixels in X with grey levels that are very similar to pixels in Y.

Near Sets and Rough Sets

Near sets are a generalization of rough sets. It has been shown that every rough set is, in fact, a near set but not every

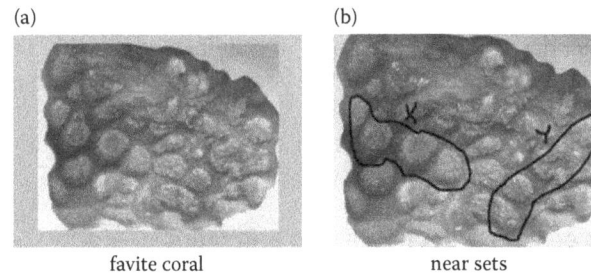

(a) favite coral (b) near sets

Fig. 1 Sample near sets

near set is a rough set Peters.[45] Near set theory originated from an interest in comparing similarities between digital images. Unlike rough sets, the near set approach does not require set approximation.[55] Simple examples of near sets can sometimes be found in tolerance classes in pairs of image coverings, if, for instance, a subimage of a class in one image has a description that is similar to the description of a subimage in a class in the second image. In general, near sets are discovered by discerning objects–either within a single set or across sets–with descriptions that are similar.

From the beginning, the near set approach to perception has had direct links to rough sets in its approach to the perception of objects[27,39] and the classification of objects.[39,41–43] This is evident in the early work on nearness of objects and the extension of the approximation space model (see, *e.g.*, Refs. [51,54]). Unlike the focus on the approximation boundary of a set, the study of near sets focuses on the discovery of affinities between perceptual granules such as digital images viewed as sets of points. In the context of near sets, the term *affinity* means *close relationship between perceptual granules (particularly images) based on common description*. Affinities are discovered by comparing the descriptions of perceptual granules, *e.g.*, descriptions of objects contained in classes found in coverings defined by the tolerance relation $\cong_{\mathbb{F},\varepsilon}$.

Basic Near Set Approach

Near set theory provides methods that can be used to extract resemblance information from objects contained in disjoint sets, i.e., it provides a formal basis for the observation, comparison, and classification of objects. The discovery of near sets begins with choosing the appropriate method to describe observed objects. This is accomplished by the selection of probe functions representing observable object features. A basic model for a probe function was introduced by M. Pavel[38] in the text of image registration and image classification. In near set theory, a probe function is a mapping from an object to a real number representing an observable feature value.[44] For example, when comparing fruit such as apples, the redness of an apple (observed object) can be described by a probe function representing colour, and the output of the probe function is a number representing the degree of redness. Probe functions provide a basis for describing and discerning affinities between objects as well as between groups of similar objects.[53] Objects that have, in some degree, affinities are considered near each other. Similarly, groups of objects (i.e. sets) that have, in some degree, affinities are also considered near each other.

Near Sets, Psychophysics and Merleau-Ponty

Near sets offer an ideal framework for solving problems based on human perception that arise in areas such as image processing, computer vision as well as engineering and science problems. In near set theory, perception is a combination of the view of perception in psychophysics[5,18] with a view of perception found in Merleau-Ponty's work.[23] In the context of psychophysics, perception of an object (i.e., in effect, our knowledge about an object) depends on sense inputs that are the source of signal values (stimularions) in the cortex of the brain. In this view of perception, the transmissions of sensory inputs to cortex cells senses are likened to probe functions defined in terms of mappings of sets of sensed objects to sets of real-values representing signal values (the magnitude of each cortex signal value represents a sensation) that are a source of object feature values assimilated by the mind.

Perception in animals is modelled as a mapping from sensory cells to brain cells. For example, visual perception is modelled as a mapping from stimulated retina sensory cells to visual cortex cells (see Fig. 2). Such mappings are called probe functions. A probe measures observable physical characteristics of objects in our environment. In other words, a probe function provides a basis for what is commonly called *feature extraction*.[11] The sensed physical characteristics of an object are identified with object features. The term *feature* is used in S. Watanabe's sense of the word,[63] *i.e.*, a feature corresponds to an observable property of physical objects. Each feature has a 1-to-many relationship to real-valued functions called probe functions representing the feature. For each feature (such as colour) one or more probe functions can be introduced to represent the feature (such as grayscale, or RGB values). Objects and sets of probe functions form the basis of near set theory and are sometimes referred to as perceptual objects due to the focus on assigning values to perceived object features.

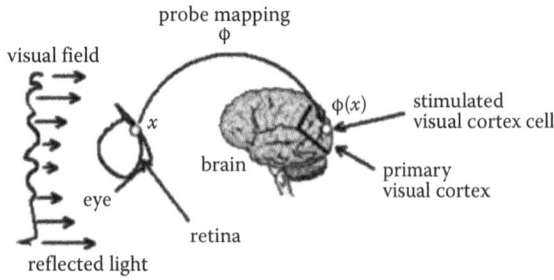

Fig. 2 Sample visual perception

Axiom 1 An object is perceivable if, and only if the object is describable.

In Merleau-Ponty's view,[23] an object is perceived to the extent that it can be described. In other words, object description goes hand-in-hand with object perception. It is our mind that identifies relationships between object descriptions to form perceptions of sensed objects. It is also the case that near set theory has been proven to be quite successful in finding solutions to perceptual problems such as measuring image correspondence and segmentation evaluation. The notion of a sensation in Poincaré[56] and a physical model for a probe function from near set theory[49,55] is implicitly explained by Zeeman[65] in terms of visual perception. That is, 'seeing' consists of mappings from sense inputs from sensory units in the retina of the eye to cortex cells of the brain stimulated by sense inputs. A sense input can be represented by a number representing the intensity of the light from the visual field (*i.e.*, everything in the physical world that causes light to fall on the retina.) impacting on the retina. The intensity of light from the visual field will determine the level of stimulation of a cortex cell from retina sensory input. Over time, varying cortex cell stimulation has the appearance of an electrical signal. The magnitude of cortex cell stimulation is a real-value. The combination of an activated sensory cell in the retina and resulting retina-originated impulses sent to cortex cells (visual stimulation) is likened to what Poincaré calls a sensation in his essay on separate sets of similar sensations leading to a perception of a physical continuum.[56] This model for a sensation underlies what is known as a probe function in near set theory.[45,55]

Definition 1 Visual Probe Function

Let O = {perceptual objects}. A perceptual object is something in the visual field that is a source of reflected light. Let R denote the set of reals. Then a *probe* ϕ is a mapping $\phi : X \rightarrow \Re$. For $x \in X$, $\phi(x)$ denotes an amplitude in a visual perception (see *e.g.*, Fig. 2).

In effect, a probe function value $\phi(x)$ measures the strength of a feature value extracted from each sensation. In Poincaré, sets of sensations are grouped together because they are, in some sense, similar within a specified distance, *i.e.*, tolerance. Implicit in this idea in Poincaré is the perceived feature value of a particular sensation that makes it possible for us to measure the closeness of an individual senation to other sensations.

A human sensation modelled as a probe measures observable physical characteristics of objects in our environment. The sensed physical characteristics of an object are identified with object features. In Merleau-Ponty's view, an object is perceived to the extent that it can be described.[23] In other words, *object description goes hand-in-hand with object perception*. It is our mind that identifies relationships between object descriptions to form perceptions of sensed objects. It is also the case that near set theory has been proven to be quite successful in finding solutions to perceptual problems such as measuring image correspondence and segmentation evaluation.

Axiom 2 Formulate object description to achieve object perception.

In a more recent interpretation of the notion of a near set, the nearness of sets is considered in the context of perceptual systems.[55] Poincaré's idea of perception of objects such as digital images in a

Cantor, Fuzzy, Near, and Rough Sets

physical continuum can be represented by means of *perceptual systems*, which is akin to but not the same as what has been called a *perceptual information system*.[49,55] A *perceptual system* is a pair $\langle O, \mathbb{F} \rangle$ where O is a non-empty set of *perceptual objects* and $\mathbb{F}$ is a non-empty, countable set of probe functions (see Def. 1).

Definition 2 Perceptual System[49]

A perceptual system $\langle O, \mathbb{F} \rangle$ consists of a sample space O containing a finite, non-empty set of sensed sample objects and a non-empty, countable set $\mathbb{F}$ containing probe functions representing object features.

The perception of physical objects and their description within a perceptual system facilitates pattern recognition and the discovery of sets of similar objects. In the near set approach to image analysis, one starts by identifying a perceptual system and the defining a cover on the sample space with an appropriate perceptual tolerance relation.

Method 1 Perceptual Tolerance

1. identify a sample space O and a set $\mathbb{F}$ to formulate a perceptual system $\langle O, \mathbb{F} \rangle$, and then
2. introduce a tolerance relation τ_ε that defines a cover on O.

Visual Acuity Tolerance

Zeeman[65] introduces a tolerance space (X, τ_ε), where X is the visual field of the right eye and ε is the least angular distance so that all points indistinguishable from $x \in X$ are within ε of x. In this case, there is an implicit perceptual system $\langle O, \mathbb{F} \rangle$, where $O := X$ consists of points that are sources of reflected light in the visual field and $\mathbb{F}$ contains probes used to extract feature values from each $x \in O$.

Sets of Similar Images

Consider $\langle O, \mathbb{F} \rangle$, where O consists of points representing image pixels and $\mathbb{F}$ contains probes used to extract feature values from each $x \in O$. Let $\mathcal{B} \subseteq \mathbb{F}$. Then introduce tolerance relation $\cong_{\mathcal{B},\varepsilon}$ to define a covers on $X, Y \subset O$. Then, in the case where X, Y resemble each other, i.e., $X \bowtie_{\mathcal{B},\varepsilon} Y$, then measure the degree of similarity (nearness) of X, Y (a publicly available toolset that makes it possible to complete this example for any set of digital images is available at Refs. [15,17]. See Table 1 (also, Refs. [48,49,55]) for details about the bowtie notation $\bowtie_{\mathcal{B},\varepsilon}$ used to denote resemblance between X and Y, i.e., $X \bowtie_{\mathcal{B},\varepsilon} Y$.

Tolerance Near Sets

In near set theory, the trivial case is excluded. That is, an element $x \in X$ is not considered near itself. In addition, the empty set is excluded from near sets, since the empty set is never something that we perceive, i.e., a set of perceived objects is never empty. In the case where one set X is near another set Y, this leads to the realization that there is a third set containing pairs of elements $x, y \in X \times Y$ with similar descriptions. The key to an understanding of near sets is the notion of a description. The description of each perceived object is specified a vector of feature values and each feature is represented by what is known as a probe function that maps an object to a real value. Since our main interest is in detecting similarities between seemingly quite disjoint sets such as subimages in an image or pairs of classes in coverings on a pair of images, a near set is defined in context of a tolerance space.

Definition 3 Tolerance Near Sets[49]

Let $\langle O, \mathbb{F} \rangle$ be a perceptual system. Put $\varepsilon \in \mathfrak{R}$, $\mathcal{B} \subset \mathbb{F}$. Let $X, Y \subset O$ denote disjoint sets with coverings determined by a tolerance relation $\cong_{\mathcal{B},\varepsilon}$. Sets X, Y are tolerance near sets if, and only if there are preclasses $A \subset X, B \subset Y$ such that $A \bowtie_{\mathcal{B},\varepsilon} B$.

Near Sets in Image Analysis

The subimages in Figs. 3b and 4b delineate tolerance classes (each with its own grey level) subregions of the

Table 1 Nomenclature

Symbol	Interpretation
O, X, Y	Set of perceptual objects, $X, Y \subseteq O, A \subset X, x \in X, y \in Y$,
$\mathbb{F}, \mathcal{B}$	Sets of probe functions, $\mathcal{B} \subseteq \mathbb{F}, \phi_i \in \mathcal{B}$,
$\phi_i(x)$	$\phi_i : X \to \mathfrak{R}$, i^{th} probe function representing feature of x,
$\phi_\mathcal{B}(x)$	$(\phi_1(x), \phi_2(x), ..., \phi_i(x), ..., \phi_k(x))$, description of x of length k, $\varepsilon \in \mathfrak{R}$ (reals) such that $\varepsilon \geq 0$,
$\|\cdot\|_2$	$= (\sum_{i=1}^{k} (\cdot_i)^2)^{\frac{1}{2}}, L_2$ (Euclidean) norm,
$\cong_{\mathcal{B},\varepsilon}$	$\{(x,y) \in O \times O : \|\phi(x) - \phi(y)\|_2 \leq \varepsilon\}$, tolerance relation,
$\cong_\mathcal{B}$	shorthand for $\cong_{\mathcal{B},\varepsilon}$,
$A \subset \cong_{\mathcal{B},\varepsilon}$	$\forall x, y \in A, x \cong_{\mathcal{B},\varepsilon} y$ (i.e., $A \cong_{\mathcal{B},\varepsilon}$ is a preclass in $\cong_{\mathcal{B},\varepsilon}$),
$\mathbb{C} \cong_{\mathcal{B},\varepsilon}$	tolerance class, maximal preclass of $\cong_{\mathcal{B},\varepsilon}$,
$X \bowtie_{\mathcal{B},\varepsilon} Y$	X resembles (is near) $Y \Leftrightarrow X \cong_{\mathcal{B},\varepsilon} Y$.

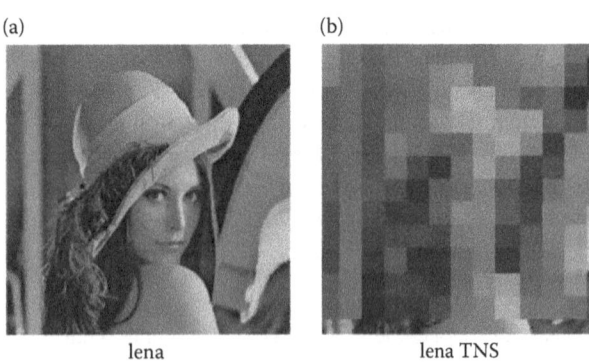

Fig. 3 Lena tolerance near sets (TNS)

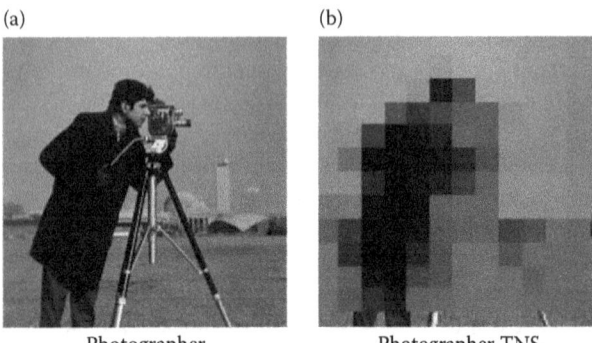

Fig. 4 Photographer tolerance near sets

original images in Figs. 3a and 4a. The tolerance classes in these images are dominated by ▪ (light grey), ▪ (medium grey) and ▪ (dark grey) subimages along with a few ▪ (very dark) subimages in Fig. 3b and many very dark ▪ subimages in Fig. 4b. From Def. 2, it can be observed that the images in Figs. 3a and 4a are examples of tolerance near sets, *i.e.*, $Image_{Fig. 4a} \bowtie_{\mathcal{B},\varepsilon} Image_{Fig. 3a}$). Examples of the near set approach to image analysis can be found in, *e.g.*, Refs. [13, 475–482; 14, 1–6; 9,10,12,15, 22,47,48,49,52,55].

From set composition Law 1, near sets are Cantor sets containing one or more pairs of objects (*e.g.*, image patches, one from each digital image) that resemble each other as enunciated in Def. 2, *i.e.*, $X, T \subset O$ are near sets if, and only if $X \bowtie_{\mathcal{B},\varepsilon} Y$).

FUZZY SETS

A fuzzy set is a class of objects with a continuum of grades of membership.
　　　　　　　　　　　　　–Fuzzy sets, Information and Control 8
　　　　　　　　　　　　　　　　　　　　　–L.A. Zadeh, 1965.

... A fuzzy set is characterized by a membership function which assigns to each object its grade of membership (a number lying between 0 and 1) in the fuzzy set.

　　　　　　　　　　　　　　–A new view of system theory
　　　　　　　　　　　　　–L.A. Zadeh, 20–21 April 1965.

Set Theory Law 2 Fuzzy Sets

Every element in a fuzzy set has a graded membership.

Notion of a Fuzzy Set

The notion of a fuzzy set was introduced by L.A. Zadeh in 1965.[64] In effect, a Cantor set is a fuzzy set if, and only if every element of the set has a grade of membership assigned to it by a specified membership function. Notice that a membership function $\phi : X \to [0,1]$ is a special case of what is known as a probe function in near set theory.

Near Fuzzy Sets

A fuzzy set X is a near set relative to a set Y if the grade of membership of the objects in sets X, Y is assigned to each object by the same membership function ϕ and there is a least one pair of objects $x, y \in X \times Y$ such that $\|\phi(x) - \phi(y)\|_2 \leq \varepsilon$, *i.e.*, the description of x is similar to the description y within some ε.

Fuzzy Sets in Image Analysis

Fuzzy sets have widely used in image analysis (see, *e.g.*, Refs. [8,12,21,26,28–34,36,60,62]). In the notion of fuzzy sets,[31,32] defined an image of $M \times N$ dimension and L levels as an array of fuzzy singletons, each with a value of membership function denoting the degree of having brightness or some property relative to some brightness level l, where $l = 0, 1, 2,..., L - 1$. The literature on fuzzy image analysis is based on the realization that the basic concepts of edge, boundary, region, relation in an image do not lend themselves to precise definition.

From set composition Law 2, it can be observed that fuzzy sets are Cantor sets.

ROUGH SETS

A new approach to classification, based on information systems theory,
given in this paper.... This approach leads to a new formulation
of the notion of fuzzy sets (called here the rough sets).
The axioms for such sets are given, which are the same
as the axioms of topological closure and interior.

　　　　　　　　–Classification of objects by means of attributes.
　　　　　　　　　　　　　　　　　　　　　–Z. Pawlak, 1981.

Set Theory Law 3 Rough Sets

Any non-empty set X is a rough set if, and only if the approximation boundary of X is not empty.

Rough sets were introduced by Z. Pawlak[39] and elaborated in Refs. [40–43]. In a rough set approach to classifying sets of objects X, one considers the size of the boundary region in the approximation of X. By contrast, in

Table 2 Pawlak indiscernibility relation and partition symbols

Symbol	Interpretation
$\sim_{\mathscr{B}}$	$= \{(x,y) \in X \times X \mid f(x) = f(y)\ \forall f \in \mathscr{B}\}$, indiscernibility, cf. [39],
$x/_{\sim\mathscr{B}}$	$x/_{\sim\mathscr{B}} = \{y \in X \mid y \sim_{\mathscr{B}} x\}$ elementary set (class),
$U/_{\sim\mathscr{B}}$	$U/_{\sim\mathscr{B}} = \{x/_{\sim\mathscr{B}} \mid x \in U\}$ quotient set.
$\mathscr{B}_*(X)$	$\mathscr{B}_*(X) = \bigcup_{x/_{\sim\mathscr{B}} \subseteq X} x/_{\sim\mathscr{B}}$ (lower approximation of X),
$\mathscr{B}^*(X)$	$\mathscr{B}^*(X) = \bigcup_{x/_{\sim\mathscr{B}} \cap X \neq \emptyset} x/_{\sim\mathscr{B}}$ (upper approximation of X).

a near set approach to classification, one does not consider the boundary region of a set. In particular, assume that X is a non-empty set belonging to a universe U and that $\mathbb{F}$ is a set of features defined either by total or partial functions. The lower approximation of X relative to $\mathscr{B} \subseteq \mathbb{F}$ is denoted by $\mathscr{B}_*(X)$ and the upper approximation of X is denoted by $\mathscr{B}^*(X)$, where

$$\mathscr{B}_*(X) = \bigcup_{x/_{\sim\mathscr{B}} \subseteq X} x/_{\sim\mathscr{B}},$$
$$\mathscr{B}^*(X) = \bigcup_{x/_{\sim\mathscr{B}} \cap X \neq \emptyset} x/_{\sim\mathscr{B}}.$$

The $\mathscr{B}$-boundary region of an approximation of a set X is denoted by $Bnd_{\mathscr{B}}(X)$, where

$$Bnd_{\mathscr{B}}(X) = \mathscr{B}^*(X) \setminus \mathscr{B}_*(X)$$
$$= \{x \mid x \in \mathscr{B}^*(X) \text{ and } x \notin \mathscr{B}_*(X)\}.$$

Definition 3 Rough Set[39]

A non-empty, finite set X is a rough set if, and only if $|\mathscr{B}^*(X) - \mathscr{B}_*(X)| = 0$.

A set X is roughly classified whenever $Bnd_{\mathscr{B}}(X)$ is not empty. In other words, X is a rough set whenever the boundary region $Bnd_{\mathscr{B}}(X) \neq \emptyset$. In sum, a rough set is a Cantor set if, and only if its approximation boundary is non empty. It should also be noted that rough sets differ from near sets, since near sets are defined without reference to an approximation boundary region. This means, for example, with near sets the image correspondence problem can be solved without resorting to set approximation.

Method 2 Rough Set Approach
1. Let $(U, \mathscr{B})$ denote a sample space (universe) U and set of object features $\mathscr{B}$.
2. Using relation $\sim\mathscr{B}$, partition the universe U.
3. Determine the size of the boundary of a set X.

Sample Non-Rough Set

Let $x \in U$. $x/_{\sim\mathscr{B}}$ (any elementary set) is a non-rough set.

Sample Rough Set

Any set $X \subset U$ where

$$\bigcup_{x/_{\sim\mathscr{B}} \subseteq U/_{\sim\mathscr{B}}} x/_{\sim\mathscr{B}} \neq X.$$

In other words, if a set X does not equal its lower approximation, then the set X is rough, *i.e.*, roughly approximated by the equivalence classes in the quotient set $U/_{\sim\mathscr{B}}$.

Rough Sets in Image Analysis

> The essence of our approach consists in viewing a digitized image as a universe of a certain information system and synthesizing an indiscernibility relation to identify objects and measure some of their parameters.
>
> – *Adam Mrozek and Leszek Plonka, 1993.*

In terms of rough sets and image analysis, it can be observed that A. Mrózek and L. Plonka were pioneers.[24] For example, he was one of the first to introduce a rough set approach to image analysis and to view a digital image as a universe viewed as a set of points. The features of pixels (points) in a digital image are a source of knowledge discovery. Using Z. Pawlak's indiscernibility relation, it is then a straightforward task to partition an image and to consider set approximation relative to interesting objects contained in subsets of an image. This work on digital images by A. Mrózek and L. Plonka appeared six or more years before the publication of papers on approximate mathematical morphology by Lech Polkowski[58] (see, also, Refs. [57,59]) and connections between mathematical morphology and rough sets pointed to by Isabelle Bloch.[1] The early work on the use of rough sets in image analysis has been followed by a number of articles by S.K. Pal and others.[3,4,20,25,35,37,50,61]

From set composition Law 3, it can be observed that rough sets are Cantor sets.

CONCLUSION

In sum, fuzzy sets, near sets and rough sets are particular forms of Cantor sets. In addition, each of these sets in the computational intelligence spectrum offer very useful approaches in image analysis, especially in classifying objects.

ACKNOWLEDGEMENTS

This research by James Peters has been supported by the Natural Sciences and Engineering Research Council of Canada (NSERC) grant 185986, Manitoba Centre of Excellence Fund (MCEF) grant, Canadian Centre of

Excellence (NCE) and Canadian Arthritis Network grant SRI-BIO-05, and Manitoba Hydro grant T277 and that of Sankar Pal has been supported by the J.C. Bose Fellowship of the Govt. of India.

BIBLIOGRAPHY

1. Bloch, L. On links between mathematical morphology and rough sets. Pattern Recognit. **2000**, *33* (9), 1487–1496.
2. Bolzano, B. *Paradoxien des unendlichen* (paradoxes of the infinite), trans. Steele, D.A.; Routledge and Kegan Paul: London, 1959.
3. Borkowski, M. *2d to 3d conversion with direct geometrical search and approximation spaces*; PhD thesis; Dept. Elec. Comp. Eng, 2007. Available at http://wren.ee.umanitoba.ca/.
4. Borkowski, M.; Peters, J.F. Matching 2d image segments with genetic algorithms and approximation spaces. In *Transactions on Rough Sets*, V(LNAI 4100), 2006, 63–101.
5. Bourbakis, N.G. Emulating human visual perception for measuring difference in images using an spn graph approach. IEEE Trans. Syst. Man Cybern. Part B **2002**, *32* (2), 191–201.
6. Cantor, G. Über unendliche, lineare punktmannigfaltigkeiten. Mathematische Annalen **1883**, *201*, 72–81.
7. Cantor, G. *Gesammelte abhandlungen mathematischen und philosophischen inhalts*; Zermelo, E.; Ed.; Springer: Berlin, 1932.
8. Deng, T.Q.; Heijmans, H.J.A.M. Grey-scale morphology based on fuzzy logic. J. Math. Imag. Vis. **2002**, *16*, 155–171.
9. Fashandi, H.; Peters, J.F.; Ramanna, S. L2 norm length-based image similarity measures: Concrescence of image feature histogram distances. In *Signal and Image Processing, International Association of Science & Technology for Development*, Honolulu, HI, 2009, 178–185.
10. Gupta, S.; Patnaik, K.S. Enhancing performance of face recognition systems by using near set approach for selecting facial features. J. Theor. Appl. Inf. Technol. **2008**, *4* (5), 433–441.
11. Guyon, I.; Gunn, S.; Nikravesh, M.; Zadeh, L.A. *Feature Extraction Foundations and Applications*; Springer: Berlin, 2006.
12. Hassanien, A.E.; Abraham, A.; Peters, J.F.; Schaefer, G.; Henry, C. Rough sets and near sets in medical imaging: A review. IEEE Trans. Info. Tech. Biomed. **2009**, *13* (6), 955–968. doi:10.1109/TITB.2009.2017017.
13. Henry, C.; Peters, J.F. Image pattern recognition using approximation spaces and near sets. In *Proceedings of 11th International Conference on Rough Sets, Fuzzy Sets, Data Mining and Granular Computing (rsfdgrc 2007), Joint Rough Set Symposium (jrs 2007)*. Lecture Notes in Artificial Intelligence 4482, Heidelberg, 2007, 475–482.
14. Henry, C.; Peters, J.F. Near set image segmentation quality index. In *Geobia 2008 Pixels, Objects, Intelligence Geographic Object Based Image Analysis for the 21st century*; University of Calgary: Alberta, 2008, 1–6.
15. Henry, C.; Peters, J.F. *Near Set Evaluation and Recognition (Near) System*; Technical Report, Computationa Intelligence Laboratory, University of Manitoba. UM CI Laboratory Technical Report No. TR–2009–015, 2009a.
16. Henry, C.; Peters, J.F. Perception-based image analysis. Int. J. Bio-Inspired Comput. **2009b**, *2* (2), in press.
17. Henry, C.; Peters, J.F. Near sets. Wikipedia, 2010. Available at http://en.wikipedia.org/wiki/Near_sets.
18. Hoogs, A.; Collins, R.; Kaucic, R.; Mundy, J. A common set of perceptual observables for grouping, figure-ground discrimination, and texture classification. IEEE Trans. Pattern Anal. Mach. Intell. **2003**, *25* (4), 458–474.
19. Lavine, S. *Understanding the Infinite*; Harward University Press: Cambridge, MA, 1994.
20. Maji, P.; Pal, S.K. Maximum class separability for rough-fuzzy c-means based brain mr image segmentation. Trans. Rough Sets **2008**, *IX* (LNCS-5390), 114–134.
21. Martino, F.D.; Sessa, S.; Nobuhara, H. Eigen fuzzy sets and image information retrieval. In *Handbook of Granular Computing*; Pedrycz, W.; Skowron, A.; Kreinovich, V.; Eds.; John Wiley & Sons, Ltd.: West Sussex, 2008, 863–872.
22. Meghdadi, A.H.; Peters, J.F.; Ramanna, S. *Tolerance Classes in Measuring Image Resemblance*. Intelligent Analysis of Images & Videos, KES 2009, Part II, Knowledge-Based and Intelligent Information and Engineering Systems, LNAI 5712, 2009, 127–134. ISBN:978-3-64-04591-2. doi:10.1007/978-3-642-04592-9_16.
23. Merleau-Ponty, M. *Phenomenology of Perception*; Smith, Gallimard, Paris and Routledge & Kegan Paul: Paris and New York. Trans. Colin Smith, 1945/1965.
24. Mrózek, A.; Plonka, L. Rough sets in image analysis. Found. Comput. Decis. Sci. **1993**, *18* (3–4), 268–273.
25. Mushrif, M.; Ray, A.K. Color image segmentation: Rough-set theoretic approach. Pattern Recognit. Lett. **2008**, *29* (4), 483–493.
26. Nachtegael, M.; Kerre, E.E. Connections between binary, grayscale and fuzzy mathematical morphologies. Fuzzy Sets Syst. **2001**, *124*, 73–85.
27. Orłowska, E. Semantics of vague concepts. Applications of rough sets. *Polish Academy of Sciences* 469. In *Foundations of Logic and Linguistics. Problems and Solutions*; Dorn, G.; Weingartner, P.; Eds.; Plenum Press: London, 1982, 1985, 465–482.
28. Pal, S.K. A note on the quantitative measure of image enhancement through fuzziness. IEEE Trans. Pattern Anal. Machine Intell. **1982**, *4* (2), 204–208.
29. Pal, S.K. A measure of edge ambiguity using fuzzy sets. Pattern Recognit. Lett. **1986**, *4* (1), 51–56.
30. Pal, S.K. Fuzziness, image information and scene analysis. In *An Introduction to Fuzzy Logic Applications in Intelligent Systems*; Yager, R.R.; Zadeh, L.A.; Eds.; Kluwer Academic Publishers: Dordrecht, 1992, 147–183.
31. Pal, S.K.; King, R.A. Image enhancement with fuzzy set. Electron. Lett. **1980**, *16* (10), 376–378.
32. Pal, S.K.; King, R.A. Image enhancement using smoothing with fuzzy set. IEEE Trans. Syst. Man Cybern. **1981**, *11* (7), 495–501.
33. Pal, S.K.; King, R.A.; Hashim, A.A. Image description and primitive extraction using fuzzy sets. IEEE Trans. Syst. Man Cybern. **1983**, *13* (1), 94–100.
34. Pal, S.K.; Leigh, A.B. Motion frame analysis and scene abstraction: Discrimination ability of fuzziness measures. J. Intell. Fuzzy Syst. **1995**, *3*, 247–256.

35. Pal, S.K.; Mitra, P. Multispectral image segmentation using rough set initialized em algorithm. IEEE Trans. Geosci. Remote Sens. **2002**, *11*, 2495–2501.
36. Pal, S.K.; Mitra, S. Noisy fingerprint classification using multi layered perceptron with fuzzy geometrical and textual features. Fuzzy Sets Syst. **1996**, *80* (2), 121–132.
37. Pal, S.K.; UmaShankar, B.; Mitra, P. Granular computing, rough entropy and object extraction. Pattern Recognit. Lett. **2005**, *26* (16), 401–416.
38. Pavel, M. *Fundamentals of Pattern Recognition*, 2nd Ed.; Marcel Dekker, Inc.: New York, 1993.
39. Pawlak, Z. Classification of objects by means of attributes. Pol. Acad. Sci. **1981a**, 429.
40. Pawlak, Z. Rough sets. Int. J. Comp. Inform. Sci. **1981b**, *11*, 341–356.
41. Pawlak, Z.; Skowron, A. Rough sets and boolean reasoning. Inf. Sci. **2007a**, *177*, 41–73.
42. Pawlak, Z.; Skowron, A. Rough sets: Some extensions. Inf. Sci. **2007b**, *177*, 28–40.
43. Pawlak, Z.; Skowron, A. Rudiments of rough sets. Inf. Sci. **2007c**, *177*, 3–27.
44. Peters, J.F. Near sets. General theory about nearness of objects. Appl. Math. Sci. **2007a**, *1* (53), 2609–2029.
45. Peters, J.F. Near sets. General theory about nearness of objects. Appl. Math. Sci. **2007b**, *1* (53), 2609–2029.
46. Peters, J.F. Near sets. Special theory about nearness of objects. Fundamenta Informaticae **2007c**, *75* (1–4), 407–433.
47. Peters, J.F. Discovering affinities between perceptual granules: L2 norm-based tolerance near preclass approach. In *Man-Machine Interactions, Advances in Intelligent & Soft Computing 59*; The Beskids: Kocierz Pass, 2009a, 43–55.
48. Peters, J.F. Tolerance near sets and image correspondence. Int. J. Bio-Inspired Comput. **2009b**, *1* (4), 239–445.
49. Peters, J.F. Corrigenda and addenda: Tolerance near sets and image correspondence. Int. J. Bio-Inspired Comput. **2010**, *2* (5). in press.
50. Peters, J.F.; Borkowski, M. K-means indiscernibility relation over pixels. In *Lecture Notes in Computer Science 3066*; Tsumoto, S., Slowinski, R., Komorowski, K.; Gryzmala-Busse, J.W., Eds.; Springer: Berlin, 2004, 580–585. doi:10.1007/b97961.
51. Peters, J.F.; Henry, C. Reinforcement learning with approximation spaces. Fundamenta Informaticae **2006**, *71*, 323–349.
52. Peters, J.F.; Puzio, L. Image analysis with anisotropic wavelet-based nearness measures. Int. J. Comput. Intell. Syst. **2009**, *3* (2), 1–17.
53. Peters, J.F.; Ramanna, S. Affinities between perceptual granules: Foundations and perspectives. In *Human-Centric Information Processing through Granular Modelling Sci 182*; Bargiela, A.; Pedrycz, W.; Eds.; Springer-Verlag: Berlin, 2009, 49–66.
54. Peters, J.F., Skowron, A.; Stepaniuk, J. Nearness of objects: Extension of approximation space model. Fundamenta Informaticae **2007**, *79* (3–4), 497–512.
55. Peters, J.F.; Wasilewski, P. Foundations of near sets. Inf. Sci. Int. J. **2009**, *179*, 3091–3109. doi:10.1016/j.ins.2009.04.018.
56. Poincaré, H. *La science et l'hypothèse*, Paris. Ernerst Flammarion. Later Ed., Champs sciences, Flammarion, 1968 & Science and Hypothesis, 1902, trans. J. Larmor, Walter Scott Publishing, London, 1905.
57. Polkowski, L. Mathematical morphology of rough sets. In Bulletin of the Polish Academy of Sciences Mathematics; Polish Academy of Sciences: Warsaw, 1993.
58. Polkowski, L. *Approximate Mathematical Morphology. Rough Set Approach. Rough and Fuzzy Sets in Soft Computing*; Springer—Verlag: Berlin, 1999.
59. Polkowski, L.; Skowron, A. Analytical morphology: Mathematical morphology of decision tables. Fundamenta Informaticae **1994**, *27*, 255–271.
60. Rosenfeld, A. Fuzzy digital topology. Inform. Contrl. **1979**, *40* (1), 76–87.
61. Sen, D.; Pal, S.K. Histogram thrsholding using fuzzy and rough means of association error. IEEE Trans. Image Process. **2009**, *18* (4), 879–888.
62. Sussner, P.; Valle, M.E. Fuzzy associative memories and their relationship to mathematical morphology. In *Handbook of Granular Computing*; Pedrycz, W.; Skowron, A.; Kreinovich, V.; Eds.; John Wiley & Sons, Ltd.: West Sussex, 2008, 733–753.
63. Watanabe, S. *Pattern Recognition: Human and Mechanical*; John Wiley & Sons: Chichester, 1985.
64. Zadeh, L.A. Fuzzy sets. Inf. Control **1965**, *201*, 72–81.
65. Zeeman, E.C. 1962. *The topology of the brain and the visual perception*. In *Topology of 3-Manifolds and Selected Topics*; Fort, K.M.; Ed.; Prentice Hall: New Jersey, 240–256.

Color

Tania Pouli
Technicolor, Rennes, France

Erik Reinhard
Technicolor Research and Innovation, Rennes, France

Douglas W. Cunningham
Brandenburg University of Technology, Cottbus, Germany

Abstract
This entry examines color information in images and discusses applications that rely on this information. The human visual system has special adaptations allowing it to take advantage of the color distribution in nature, and these adaptations are also discussed. Other topics include: The nature of color and its processing by the human visual system, color as a 3D space, color transfer, color space statistics, and consistency.

Keywords: Color space; Color; Trichromatic theory; Visual perception.

We have so far considered statistical regularities either without specifically including the notion of color, or deliberately ignoring color and assuming a single luminance channel. Although this is sufficient for studying many of the regularities within natural scenes, color deserves special treatment as it is one of the most informative aspects of the visual world. The use of color in images and art can be used to create a specific mood and to induce particular emotions, it can serve as a signal (e.g., to indicate fruit ripeness in nature or whether we can cross the road in manmade environments) and even give indication about the type of environment or time of the year.[1]

In general, images are formed as light is transduced to electric signals. This happens both in the retina as well as in electronic image sensors. Before that, light is emitted and usually reflected several times, encountering several different surfaces before it reaches an electronic or physiological sensor. This process is modeled by the rendering equation, given by:

$$L_o(\lambda) = L_e(\lambda) + \int_\Omega L_i(\lambda) f_r(\lambda) \cos(\Theta) d\omega \quad (1)$$

This equation models the behavior of light at a surface: light L_i impinging on a surface point from all directions Ω is reflected into an outgoing direction of interest, weighted by the bi-directional reflectance distribution function (BRDF) f_r. The amount of light L_o going into that direction is then the sum of the reflected light and the light emitted by the surface point (which is zero unless the surface is a light source). Here, we have specifically made all relevant terms dependent on wavelength λ to indicate that this behavior happens at all wavelengths, including all visible wavelengths. This means that L_o can be seen as a spectral distribution.

Light reaching our eyes is therefore carrying information about the intensity and direction of illumination but also about the materials and colors of objects that it has interacted with in its path. For a sensor—be it biological or electronic—to be able to perceive color, special adaptations are necessary. Ideally, to accurately represent color, the full spectrum of light should be captured. This, however, would require a prohibitive amount of processing.

To reduce the amount of data, the light spectrum can instead be sampled through cells with different sensitivities. The human retina, for instance, is populated by three types of cone photoreceptors, while the eyes of mantis shrimp contain at least eight visual pigments with narrowly tuned sensitivities[2] and many species of birds are tetrachromatic.[3] In all of these cases, although the precise form of sampling of the spectrum differs, the purpose is the same: to reduce the amount of data to be transmitted and processed while minimizing loss of information.

This remains an important goal in further stages of the visual system, with many processes relying on particular regularities in the color information of the environment, allowing us to quickly and effortlessly understand our environment. In this entry, we will review the main color processes of the human visual system and discuss how they relate to certain statistical regularities and patterns in natural environments. We will also look at the implications and applications of these regularities in imaging disciplines.

TRICHROMACY AND METAMERISM

The human visual system is well equipped for viewing color, and several different processes take place in our visual pathways that mediate its perception. Several models were proposed in the last centuries to account for the different aspects of color vision. One of the earliest theories suggests that new colors can be created by mixing three other colors.[4,5] This was determined by a color matching experiment where participants were asked to match given colors by mixing a set of light sources with adjustable intensities. Helmholtz found that three light sources were necessary and sufficient for matching any given color. This led to what is now known as the *Young-Helmholtz trichromatic theory*.

Within the human visual system, photopic vision is indeed mediated by three types of cone photoreceptors, each with a different peak sensitivity. Each cone type integrates the incident light according to a different weight function:

$$L = \int_\lambda L_o(\lambda)\bar{l}(\lambda)d\lambda \qquad (2a)$$

$$M = \int_\lambda L_o(\lambda)\bar{m}(\lambda)d\lambda \qquad (2b)$$

$$S = \int_\lambda L_o(\lambda)\bar{s}(\lambda)d\lambda \qquad (2c)$$

where $\bar{l}(\lambda)$, $\bar{m}(\lambda)$, and $\bar{s}(\lambda)$ are the weight functions (responsivities), which peak at wavelengths that are perceived roughly as red, green, and blue (Fig. 1), specifically at 565 nm, 545 nm, and 440 nm.[6,7] The letters L, M, and S stand for "long," "medium," and "short" wavelengths. A spectral distribution $L_o(\lambda)$ therefore gives rise to a triple of numbers (L, M, S), which represent the signal that is passed on from the photoreceptors to the remainder of the human visual system. Such triples are called *tristimulus values*.

An important property of this behavior is that different spectral distributions can integrate to the same tristimulus values since each receptor type responds to a wide range of wavelengths. Such spectral distributions are then necessarily perceived to be identical, which is known as *metamerism*.[8] Although metameric surfaces do not often occur in nature,[9] a crucial implication of metamerism is that with three carefully controlled light sources, the full spectrum (or nearly) can be simulated. Because of that, we are able to build display devices that emit light using three primaries rather than emulating the full spectral distribution, which has shaped the way digital images are stored and managed. Digital imaging relies on this precise process too, which is why digital images are also typically encoded using a triplet of values per pixel. Similar to our visual system, this allows images to be stored using a relatively small amount of data compared to what would be required for storing their full spectral distribution.

COLOR AS A 3D SPACE

The primaries used in display devices or digital imaging need not be the same as the responsivities of the cones: tristimulus values for one choice of primaries can be converted to tristimulus values for a different set of primaries, and although they do not represent the same spectral distributions, through metamerism they lead to the same percept. Consequently, the same color can be described in infinitely many ways by changing the set of primaries that it is defined upon. In effect, each tristimulus value corresponds to a point in a three-dimensional space, known as a *color space*, and each axis defines a channel. By shifting, scaling, and rotating the axes defining a color space, a different space can be constructed to achieve different goals.

This treatment of color allows for two different ways of studying and using chromatic information in images. A set of pixels in an image can be seen as a set of coordinates within the color space of choice, which can be manipulated as a three-dimensional manifold. Although this would ensure that relations between pixels remain unchanged, it bears no resemblance to the workings of the visual system and would require more complex imaging algorithms.

On the other hand, each channel can be treated as a separate entity, resulting in three one-dimensional distributions. This, of course, is a much simpler problem compared to its 3D counterpart, but it comes with its own set of limitations. If, for instance, we look at the response curves of the L, M, and S cones, it is easy to see that they have a large overlap. This, albeit necessary for the visual system, leads to a highly correlated color space. Effectively, a given value for one channel becomes a good predictor for the values of other channels. This is illustrated in Fig. 2 (top row) where the pixel values for pairs of channels are plotted for the LMS cone space. Most values sit near the diagonal, indicating that changes in one channel will affect the behavior of the other two as well.

OPPONENT PROCESSING

Fortunately, the issue of correlation between color channels can be easily resolved. If an appropriate transformation is

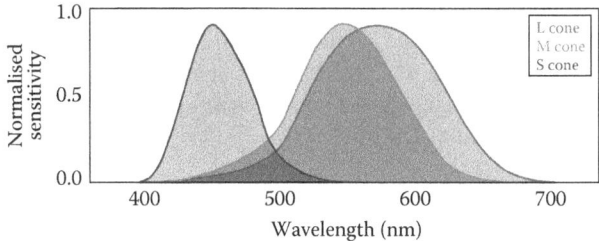

Fig. 1 The three cone photoreceptor sensitivity curves

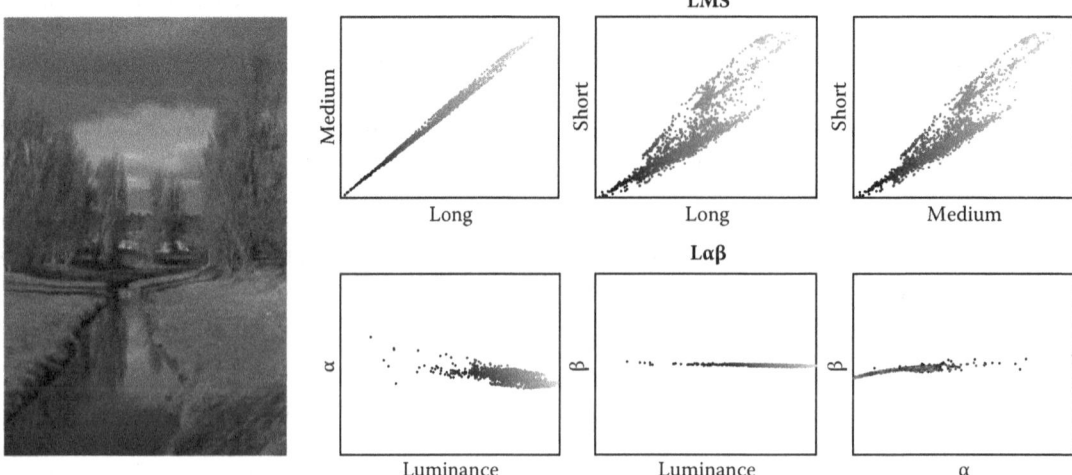

Fig. 2 For the image shown, corresponding pixel values are plotted for all pairs of channels in the LMS cone space (top) and the $l\alpha\beta$ opponent space (bottom). Values in the LMS space lie along the diagonal, indicating high correlation, while this is much less so the case with $l\alpha\beta$. (Seamills, UK, 2009)

chosen, the axes of the color space may be rotated such that data in one channel no longer predict the values in other channels, thus reducing correlation.

Further stages of the visual system employ one such solution known as the *opponent processing theory*.[10,11] It states that as proposed in the trichromatic theory, there are indeed three variables mediating color vision, but in this case, instead of assigning unique color sensations to each of them, pairs of opponent sensory qualities are assigned. These sensory qualities form pairs of red-green and blue-yellow, which are responsible for color information, while a black-white pair allows for luminance perception, illustrated in Fig. 3. This opponent nature of the visual system leads to sensations that are paired and mutually exclusive, explaining both why color blindness removes pairs of colors as well as why subjective experiences of reddish-green and yellowish-blue are not possible.

An additional implication of opponent processing in the visual system is that it reduces correlation for the input that it typically encounters. The LMS cone responses and in particular the L and M cones (approximately red and green) have a large overlap, meaning that a particular signal is likely to excite both cone types. It has been shown that an eigenvector decomposition of the LMS cone responses leads to a more efficient representation, with less redundancy. In addition, this representation matches the opponency in the visual system, suggesting that dimensionality reduction is its main purpose.[12]

To further link the opponent processing stages of the HSV to regularities in natural environments, Ruderman perfomed an interesting statistical experiment.[13] A set of spectral images of natural scenes was captured and converted to log LMS space, effectively simulating the photoreceptor responses to the same scenes. The transformed images were then subjected to Principal Components Analysis (PCA), which was applied on the color data. Each image pixel in this analysis was treated as a point in a 3D color space (log LMS) while spatial information was ignored. The results were striking: finding that the eigenvectors resulting from PCA closely resemble the opponency in the visual system.

The process followed for this experiment can be summarized as follows:

1. The input spectral images are converted to the LMS cone space.
2. A square patch of 128×128 pixels is selected from the center of each of the images.
3. Images are logarithmically compressed to ensure that values are more symmetrically spread within their coordinate system.
4. Values in each of the patches are centered around their mean. This operation is applied on each channel separately and it simulates a simplified von Kries adaptation step.[14]

$$\mathbf{L} = \log L - <\log L> \quad (3a)$$

$$\mathbf{M} = \log M - <\log M> \quad (3b)$$

$$\mathbf{S} = \log S - <\log S> \quad (3c)$$

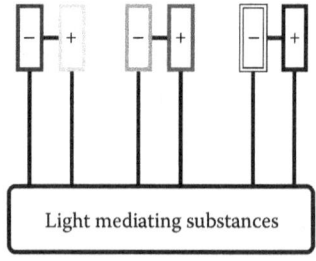

Fig. 3 The opponent processing scheme proposed by Hering[10] and later verified by Hurvich and Jameson[11]

Color

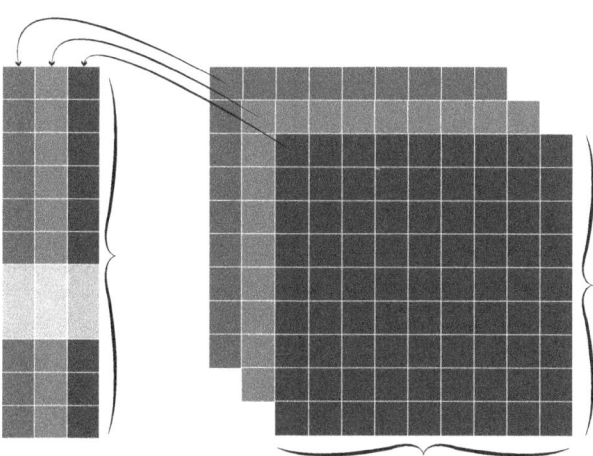

Fig. 4 Before processing image data with principal component analysis, images are reshaped such that each channel is now a single vector

5. Each of the three channels of each image patch is reshaped to a single column vector. These vectors are concatenated into a three-column matrix, containing 16,384 LMS triples for each image. This process is repeated for all images in an ensemble, and all subsequent matrices are appended to the end of the initial matrix. See Fig. 4 for an illustration.
6. Finally, principal component analysis (PCA) is performed on the resulting matrix, rotating the axes so that they are maximally decorrelated, producing a set of three components (eigenvectors) sorted according to the decreasing order of their respective eigenvalues. The mathematical details of that process are given in Section 7.1.

By starting in LMS cone space, the rotation yields a new color space, which surprisingly corresponds closely to the color opponency found in further stages of the visual system. Color opponent spaces are characterized by a single achromatic channel, typically encoding luminance or lightness, and two chromatic axes, which can be thought of as spanning red-green and yellow-blue (although the axes do not precisely correspond to these perceptual hues). The chromatic axes can have positive and negative values; a positive value can, for instance, denote the degree of redness, whereas a negative value in the same channel would denote the degree of greenness. A consequence of color opponency is that we are not able to simultaneously perceive an object to have green and red hues. Although we may describe objects as reddish-blue or yellowish-green, we never describe them as reddish-green. The same holds for yellow and blue.

The color opponency of the $l\alpha\beta$ space is demonstrated in Fig. 5, where the image is decomposed into its separate channels. The image representing the α channel has the β channel reset to zero and vice versa. We have retained the luminance variation here for the purpose of visualization.

Fig. 5 The top-left image is decomposed into the l channel of the $l\alpha\beta$ color space, shown in the top-right image, as well as $l+\alpha$ and $l+\beta$ channels in the bottom-left and bottom-right images. (Seamills, UK, 2009)

The image showing the luminance channel only was created by setting both the α and β channels to zero.

The conversion from LMS to the aforementioned decorrelated opponent color space $l\alpha\beta$ is given by:

$$\begin{bmatrix} L \\ \alpha \\ \beta \end{bmatrix} = \begin{bmatrix} \frac{1}{\sqrt{3}} & 0 & 0 \\ 0 & \frac{1}{\sqrt{6}} & 0 \\ 0 & 0 & \frac{1}{\sqrt{2}} \end{bmatrix} \begin{bmatrix} 1 & 1 & 1 \\ 1 & 1 & -2 \\ 1 & -1 & 0 \end{bmatrix} \begin{bmatrix} \log L \\ \log M \\ \log S \end{bmatrix} \quad (4)$$

while the inverse transform is given by:

$$\begin{bmatrix} \log L \\ \log M \\ \log S \end{bmatrix} = \begin{bmatrix} 1 & 1 & 1 \\ 1 & 1 & -1 \\ 1 & -2 & 0 \end{bmatrix} \begin{bmatrix} \frac{\sqrt{3}}{3} & 0 & 0 \\ 0 & \frac{\sqrt{6}}{6} & 0 \\ 0 & 0 & \frac{\sqrt{2}}{2} \end{bmatrix} \begin{bmatrix} L \\ \alpha \\ \beta \end{bmatrix} \quad (5)$$

Returning to Fig. 2, it is readily visible that such a transform is quite successful at reducing correlation in the color data. The bottom row shows plots of corresponding pixel values for all pairs of channels in the $l\alpha\beta$ color space, resulting in point clouds that are much less diagonal compared with their LMS counterparts, shown in the top row. Values in one channel no longer predict the behavior of the other two. In addition, the transform to $l\alpha\beta$ not only decorrelates data, but it effectively makes it independent.[13] These properties have an important implication in imaging: each channel of an image can now be processed separately without affecting the others. One particular application of this observation is discussed in the following section.

It is worth noting here that $l\alpha\beta$ is not the only color space that exhibits such statistical properties. Although the $l\alpha\beta$ space was explicitly constructed through statistical analysis and has strong links with the opponent processing steps in the visual system, other color spaces that are often used in imaging applications have similar properties. For instance, both CIELab and IPT color spaces have a nonlinear compression (albeit not logarithmic), and both encode color data in luminance channel and two opponent channels.[15,16] The opponency and decorrelation properties of these and several other color spaces will be discussed in "Color Space Statistics" section.

COLOR TRANSFER

Images can convey information not only through the depicted objects but also through the particular mood, color scheme, and composition of the scene. Artists can manipulate the color palette manually to change the appearance of an image and achieve specific effects but that can be a time-consuming process, requiring advanced image manipulation skills. An alternative solution for editing a given image is to find another image with the desired look and transfer properties from that image to our original one. If the property being transferred is color, this process is known as *color transfer*.

This of course is a non-obvious problem: the color in an image could be taken to mean anything from the overall appearance or color palette (warm or cool colors, dark or light mood, and so on) to the full color distribution in a chosen color space. Matters are further complicated since for any chosen definition for describing color within an image, an almost infinite number of ways to transfer it to another exist. Let us consider the images in Fig. 6, where the color palette of the green landscape is transferred to an autumn foliage scene. The approach taken here makes the whole image green overall but maintains a hint of red on the leaves. Another possible interpretation would be to make the red leaves green but preserve the yellow background and yet another would be to manipulate the source such that it has exactly the same color distribution as the target.

Fig. 6 A color transfer example. Here the green colors of the target image (bottom left) are transferred to the source autumn scene (top left) to obtain the result on the right. (Westonbirt Arboretum, Bristol, 2009)

The lack of a strict definition of what constitutes a successful transfer has led to an expansive collection of techniques, each approaching the problem in a slightly different way and focusing on slightly different application areas. Many of them though share a particular characteristic, which makes them interesting for us in the context of color statistics. Specifically, many of the existing color transfer techniques rely on the decorrelation afforded by opponent spaces such as the $l\alpha\beta$ space, discussed earlier, or CIELab to transfer properties between single channels at a time, rather than having to consider the full 3D color distribution. We will look at two examples of color transfer methods to demonstrate the flexibility that this fortunate regularity of the color content of images allows.

Color Transfer through Simple Moments

One of the earliest color transfer techniques takes advantage of the decorrelation property of $l\alpha\beta$ and shifts and scales the pixel values of the source image to match the mean and standard deviation from the target.[17] Each channel of the image is processed separately in $l\alpha\beta$, as it allows the transfer to take place independently in each channel, turning a potentially complex 3D problem into three much simpler 1D problems.

Input images are assumed to be given in the sRGB space and are first converted to $l\alpha\beta$. The colors of the source image can now be altered to approximate those of the target. This is achieved by shifting and scaling the source distribution such that it is aligned with that of the target. Figure 7 shows an example pair of source and

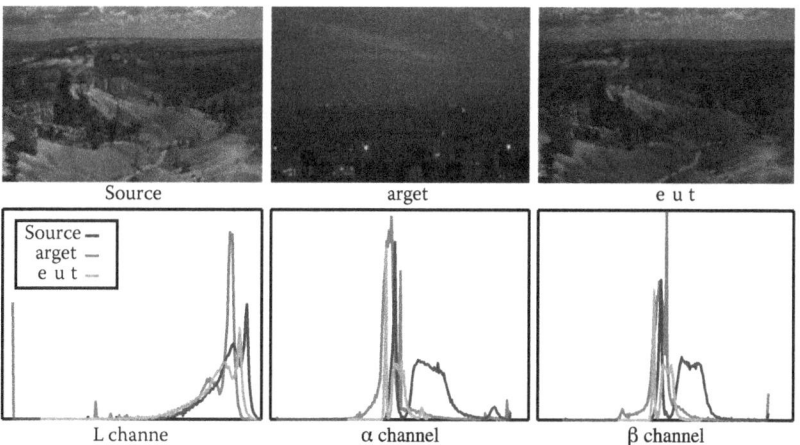

Fig. 7 An example pair of source and target images as well as their resulting output after using the color transfer technique by Reinhard et al.[17] are shown on the top row. The corresponding histograms are shown on the bottom row for the three channels of the $l\alpha\beta$ color space. (Left: Bryce Canyon, Utah, 2009; right: Paris, 2008)

target images as well as their corresponding histograms before and after transferring the colors between them.

To achieve this, the values of each channel of the source image are shifted to a zero mean by subtracting the mean of the source from each pixel. They are then scaled by the standard deviations of both images such that they acquire that of the target image, and finally, they are shifted to the mean of the target by adding its mean instead of the source mean that was originally subtracted:

$$l' = l_s - \mu_s \tag{6a}$$

$$l'' = \frac{\sigma_{t,l}}{\sigma_{s,l}} l' \tag{6b}$$

$$l_o = l'' + \mu_t \tag{6c}$$

Here, the subscripts s, t, and o correspond to the source, target, and output images and μ and σ are their respective means and standard deviations. Equations (6a)–(6c) describe the distribution transfer between source and target for the luminance channel only. This process effectively ensures that the distributions of the luminance values for the two images have the same first and second moment. The same process is repeated for the two chromatic channels of the $l\alpha\beta$ space to complete the color transfer. The images can then be converted back to their original display space.

Although this technique can be successful for a wide range of images, the quality of the results largely depends on the composition of the source and target images. Swatches may be used to supplement this technique and allow more control over color correspondences if the two images are too different. The user can in this case select pairs of swatches from the source and target images to indicate corresponding regions between them. The pixels in these swatches form clusters in the $l\alpha\beta$ space, allowing statistics to be computed separately for each of the swatches. Source pixels are transformed according to the statistics of each pair of swatches, leading to a number of possible renditions of the source image. To construct the final image, these different renditions are blended by considering the distance of each pixel's color to the center of each cluster. Distances are divided by the standard deviation of each cluster to account for different cluster sizes, and pixels are blended using inversely proportional weights to the normalized distances.

Despite the increased control that these swatches allow, in both the local and global versions of this technique the transfer of colors between the two images relies on first-order statistics, namely the mean and standard deviation of each channel. Such statistics can provide useful information about overall tendencies in a distribution but, more information and ideally higher-order analysis are necessary to capture more subtle variations in images.

An obvious next step is to use higher statistical moments in addition to the mean and standard deviation to achieve a better result. For instance, the skew and kurtosis of the distribution can be transferred to reshape the histogram of the source image to be closer to that of the target. Although such a process is conceptually elegant, unfortunately it is not as effective in practice. The mean and standard deviation of a distribution can easily be transferred analytically (Eqs. 6a–6c), while higher moments such as the skew and kurtosis require more complex solutions. Although a distribution can be uniquely described by its moments, arbitrary moments cannot be changed on demand, thus requiring optimization-based solutions. Even so, the improvements afforded by transferring further moments in this way are very minor.

Color Transfer through Higher-Order Manipulation

In many cases a small set of statistical moments is not sufficient to adequately describe subtleties of the color

Fig. 8 Histogram matching between two color images. (Westonbirt Arboretum, Bristol, 2009)

distribution of images. A much more faithful representation can be achieved by considering the full distribution of values in each channel. This process can be easily extended to a color image by repeating the histogram matching process for each of the three channels to achieve a result as shown in Fig. 8.

This process ensures that the transformed image will have exactly the same distribution as the selected target and therefore the same colors. Occasionally, however, the resulting image may be too harsh as the transfer can amplify artifacts that were previously invisible, indicating that higher-order properties of the image may need to be matched or preserved to achieve a successful result. Based on this observation, a wealth of color transfer methods have been developed, each relying on a different set of statistical properties of the images to transfer the color palette between them without otherwise affecting the appearance of the image.

Histogram Features at Different Scales

Since a full histogram match is likely to push the image appearance too far, one possibility is to only partially match the two histograms by taking advantage of their local structure.[18] One recent solution achieves that by considering features of the histograms in different scales. An example result from this method is shown in Fig. 9.

Consider, for instance, the image in Fig. 10; a pyramid of increasingly coarse scales can be constructed by filtering the original histogram with a Gaussian kernel of increasing size. Similar to image scale-space filtering, coarser levels preserve only large scale features of the histogram, which represent larger segments of the image. On the other hand, finer scales contain more details but each of the histogram features at those scales will correspond to smaller regions in the image.

The color palette between the two images is transferred by reshaping the source histogram so that it approaches that of the target. Rather than aim for an accurate match, though, where every subtlety of the target is matched, the source can be progressively transformed to the target by matching more and more scales, starting from the coarsest. When all scales including the finest are matched, then the result will be identical to the simpler histogram matching discussed earlier.

This leads to an interesting observation: such an approach is possible because nearby pixels in images tend to be similar. This is a recurring property of natural scenes, which we have discussed in the context of simple luminance statistics, edges, and higher-order properties. In this case, it means that by treating each feature of the

Fig. 9 An example result obtained with the histogram scale-space approach discussed here.[18] The input image (high dynamic range original—shown both tonemapped and linearly normalized) is transformed to match the color palette of the given reference. (Utah, USA, 2009)

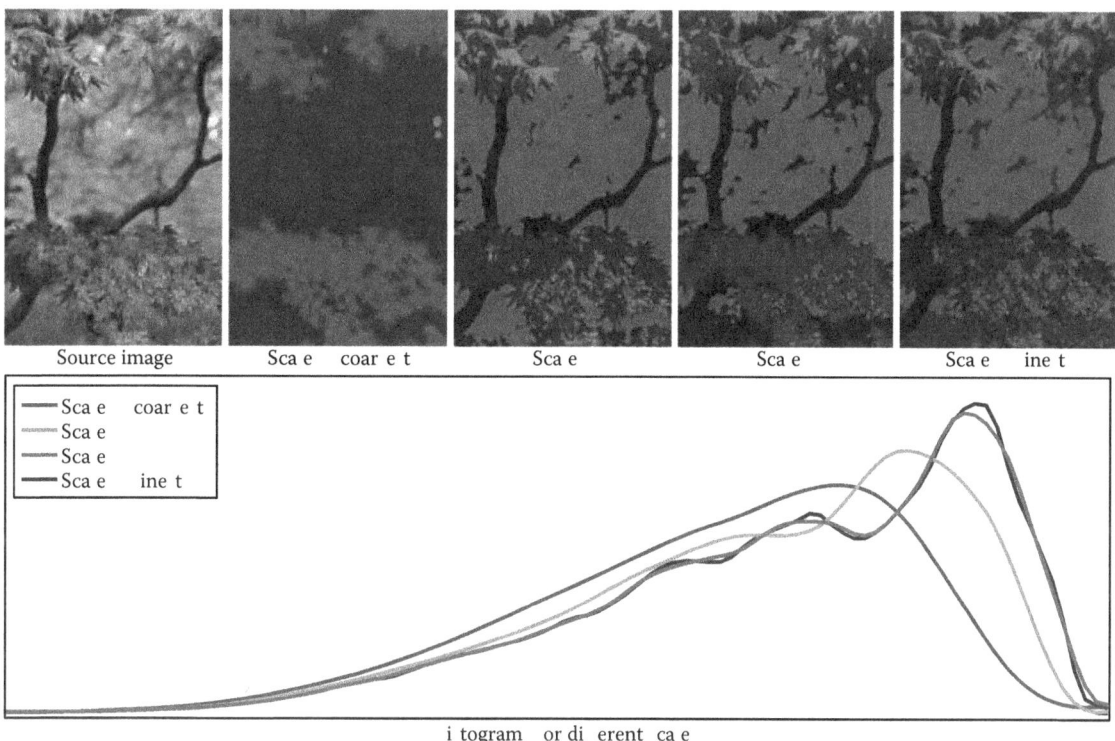

Fig. 10 The histogram of the top-left image, shown at the bottom, is progressively filtered to remove detail. The coarser features of the histogram correspond to large coherent regions of the image with similar colors. When examining the finer scales of the histogram, though, peaks in the histogram only contain a small number of pixels and therefore correspond to small image segments. (Westonbirt Arboretum, Bristol, 2009)

filtered histograms separately, we are inherently treating the image as a collection of segments, without ever having to explicitly segment it.

Although this approach is successful for many images, some transfers require relatively flat areas to acquire much more extreme color transitions, which may amplify noise or compression artifacts or simply change the appearance of the scene in terms of contrast. Edges are one of the main streams of information within visual data, and consequently, in an application such as color transfer it would be desired to preserve them. This, however, is not entirely straightforward as edge information is intricately linked to the pixel values in the image and therefore to color as well. Whether edges are recovered by adding local contrast back to the image after the transfer has taken place[18,19] or through an optimization scheme that aims to preserve the gradient distribution of the source image,[20] such methods present a tradeoff between accurate matching of the color distribution and preservation of edge content.

Color Transfer as a 3D Problem

Referring back to "Color as a 3D Space" section, we have seen how color spaces describe a three-dimensional space. Pixels in an image are in that case given as a triplet, denoting a point within that space. In all our discussion on color transfer so far, we have focused only on techniques that manipulate each of the image channels independently. Although this description of color provides the obvious advantage of simplifying a potentially complex 3D problem to a set of three 1D problems, it cannot capture all subtleties in the color distribution of images.

To maintain local color information and interrelations, the 3D color distribution of the two images needs to be treated as a whole; this is done so that the source 3D distribution will be reshaped to match or approximate that of the target. Unfortunately, translating processes, such as histogram matching or histogram reshaping, to more than one dimension is not straightforward and either requires an optimization-based solution or a way to simplify the problem to fewer dimensions.

In the latter case, a general approach that iteratively matches the two distributions through 1D projections may be employed, proposed by Pitié et al.[21,22] Specifically, at each iteration step, the 3D distributions of the source and target images are rotated using a random 3D rotation matrix and projected to the axes of their new coordinate system. Each 1D projection of the source is then matched to that of the target and the data is transformed back to its original coordinate system. This process is repeated with different rotations until convergence. An example result and the effect of an increasing number of iterations can be seen in Fig. 11.

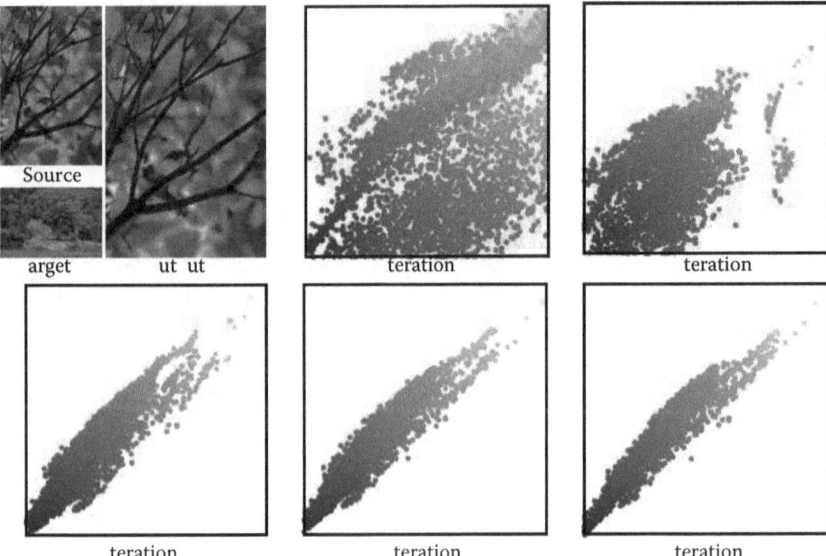

Fig. 11 The colors of the source image are progressively mapped to those of the target by iteratively rotating the 3D color distributions of both images to a randomly chosen coordinate system and matching the 1D marginal distributions for each of the axes

Whether the color distribution is matched by manipulating each of the color channels separately or by treating the image as a 3D color distribution, a few important observations arise. Although all the techniques discussed have at a high level the same overall goal, they employ widely different approaches to get there, offering a unique view into color processing.

Another issue of note is the lack of consensus on the choice of color space to be used for such processing. The simple moment transfer technique discussed earlier[17] as well as many other color transfer techniques[20,23–31] operate in Ruderman's $l\alpha\beta$ space,[13] while the histogram reshaping process uses *CIE Lab*.[19]

On the other hand, the iterative projection of 1D marginals begins in RGB and selects a random coordinate system in each iteration,[22] effectively transforming the image to a randomly selected color space at each step, while a similar dimensionality reduction process has been demonstrated in the *LC*h space.[32]

Other color spaces implicated in color transfer are YC_bC_rbr[33] and *Yu*v,[34,35] as well as color appearance models.[36] Finally, several authors suggest to compute a dedicated color space for each transfer, based on PCA.[37–41] The following section will attempt to shed some light on the issue of color space choice.

COLOR SPACE STATISTICS

Any RGB-like color space tends to be significantly correlated. This means that a high value in, say, the red channel suggests a good probability of finding high values in the green and blue channels for those pixels. In essence, this is a different way of saying that colors in natural scenes tend to be desaturated, or that images on average tend towards gray. This feature of natural images is explicitly exploited in photography, where the gray-world assumption is made to infer the color of the illumination in a scene, which is discussed in more detail in the following section.

The high correlation of RGB-like color spaces also means that for imaging applications such as color transfer, as we saw in the previous section, a different color space may be necessary to ensure that changes in one channel do not inadvertently affect the other channels and therefore lead to unpredictable results. A decorrelated space, such as the $l\alpha\beta$ space proposed by Ruderman et al.[13] allows images to be transformed to a coordinate space where values in one of their channels predict those in other channels as little as possible. Decorrelated color spaces are frequently used in image editing applications for this precise reason.

This is of course a desirable property as it reduces the dimensionality of the problem at hand as we have seen, but such spaces have been found to not fully decorrelate all images. The $l\alpha\beta$ space, for instance, is designed around the principal axes arising from the decorrelation of natural scenes. Consequently, converting an image to that space is equivalent to rotating its data so that it is aligned to the axes defining the $l\alpha\beta$ space. If the content of this image is significantly different to the types of images used for the initial derivation of that space, these axes may or may not correctly describe most of the variation in the new data. It is not clear, however, whether different scene categories or images possessing a higher dynamic range would lead to a different color space or how such differences might impact applications such as color transfer.

To answer this question, a study was done comparing various color spaces in terms of their decorrelation

Fig. 12 Sample images from the four image categories used for the color space analysis and the computation of PCA-based spaces

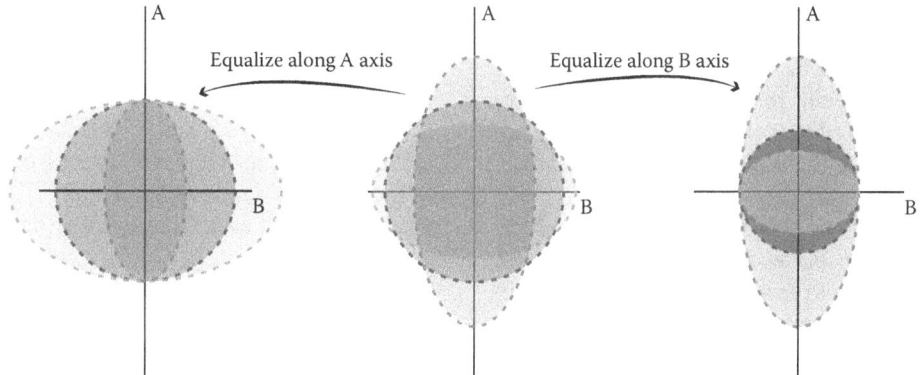

Fig. 13 Since different color spaces are defined along different ranges, normalizing along a single axis is unlikely to correctly allow for comparisons between the different spaces

properties.[42] Images from different scene categories were analyzed to assess whether color spaces such as $l\alpha\beta$ are in fact better at decorrelating particular scene types. Each of the image categories was also used to construct a custom PCA-based space following a similar process to that the $l\alpha\beta$ space discussed in "Trichromacy and Metamerism" section, based on the hypothesis that a color space constructed through decorrelation of a particular type of image would be better at reducing correlation when processing similar images. The study focused on both manmade and natural imagery, specifically categorizing images as daytime manmade scenes denoted by MD, indoors manmade scenes (MI), nighttime manmade scenes (MN), and daytime natural scenes (ND). Figure 12 shows a few sample images from each of the categories used.

To compare both existing and the per-scene type color spaces, correlation measures were employed as well as a short experiment where pairs of images were processed using a linear color transfer method[17] and the color transfer result was ranked as successful or not. In addition to $l\alpha\beta$ and the PCA-based spaces discussed, the following color spaces were also considered: CIELab (using both illuminants D65 and E), Yxy, Yuv, Yu'v', XYZ, RGB, and HSV.[16]

Color Space Normalization

Different color spaces are defined for different ranges. For instance, an RGB image may be defined between 0 and 1, while CIELab is not explicitly constrained. To ensure that results from different color spaces are comparable in such an analysis, data needs to be normalized. This, however, is not straightforward since the meaning of different channels, as well as the relations between them, vary with color space. Consequently, simply normalizing along each of the channels will lead to incorrect results, as illustrated in Fig. 13.

To resolve this issue, data can be normalized according to the volume of each color space, which is computed according to the range that each channel or axis is expected to acquire. To approximate the volume of each of the color spaces, the maximum and minimum achievable values for each channel are determined by using input RGB values constrained between 0 and 1. The computed volumes used for the normalization step are shown in Table 1.

Correlation Analysis

In Chapter 4, we saw how statistical moments can be computed for distributions. The second moment corresponds to the variance of a dataset I of size N and is computed as follows:

$$\sigma^2(I) = \sum_{p=1}^{N} \frac{(I(p)-c)^2}{N-1} \qquad (7)$$

Table 1 The volumes of different color spaces, approximated by a cuboid

Color Space	Volume
$l\alpha\beta$	6.4480
CIELab (**E**)	6.0447e + 03
CIELab (**D65**)	8.8841e + 03
Yxy	0.2646
Yuv	0.0879
Yu'v'	0.1318
HSV	0.8333
XYZ	1.0351
RGB	1.0000

where c is a constant around which the data is considered. Typically this can be taken as 0, if statistics of the data are computed with respect to the origin, or more frequently, c is set to the mean of the data μ.

The variance of a dataset provides information about the spread of the values within that set. To measure how more than one set varies with respect to each other, a more general formulation of this measure is necessary, known as *covariance*.

$$\text{Cov}(I,J) = \sum_{p=1}^{N} \frac{(I(p)-\mu_I)(J(p)-\mu_J)}{N-1} \quad (8)$$

For each color space, covariance matrices were computed corresponding to the four datasets shown in Fig. 12. Values along the diagonals of the covariance matrices contain the variance within each channel, while elements off the diagonal capture covariance values for pairs of channels:

$$\text{Cov}_s = \begin{bmatrix} \text{Cov}(s_1,s_1) & \text{Cov}(s_1,s_2) & \text{Cov}(s_1,s_3) \\ \text{Cov}(s_2,s_1) & \text{Cov}(s_2,s_2) & \text{Cov}(s_2,s_3) \\ \text{Cov}(s_3,s_1) & \text{Cov}(s_3,s_2) & \text{Cov}(s_3,s_3) \end{bmatrix} \quad (9)$$

where s = { $l\alpha\beta$, CIELab (E), CIELab (D65), Yuv, Yu'v', HSV, XYZ, RGB} and $s$ii refers to the i'hth channel of that space, with i = {1, 2, 3}. The covariance values for each channel pair were then averaged to a single number per image set, representing the ability of each color space to decorrelate different classes. These results are summarized in Fig. 14.

As would be expected, opponent spaces lead to the lowest covariance values, while the RGB and XYZ spaces are much more correlated. Interestingly, the choice of white point used for the conversion to the *CIELab* color space was found to drastically matter when assessing color spaces with the covariance measure. When the achromatic illuminant E was used, *CIELa*b resulted to much lower covariance values compared to using the bluish D65 illuminant.

In addition to standard color spaces, ensemble-specific spaces similar to $l\alpha\beta$ were constructed following the procedure detailed in "Opponent Processing" section for each of the image ensembles shown in Fig. 12. Since these color spaces are computed from specific datasets representing different images categories, one would expect that they would lead minimal covariance for the categories they were derived from.

Figure 15 summarizes the results for the covariance analysis of these ensemble-specific spaces. Although all four spaces lead to lower covariance overall compared to standard color spaces, only for the "natural day" (ND) and "manmade day" (MD) sets does the corresponding color space lead to better performance. On the other hand, the PCA-based space constructed from the "manmade night" (MN) set led to the worst performance overall. One hypothesis explaining these results is that more varied image sets offer a better sampling of the color gamut, therefore leading to a more descriptive decomposition.

To confirm whether the findings from the statistical analysis correspond with viewers' preferences, a short study was also conducted. Pairs of images from the four ensembles were transformed to one of the color spaces

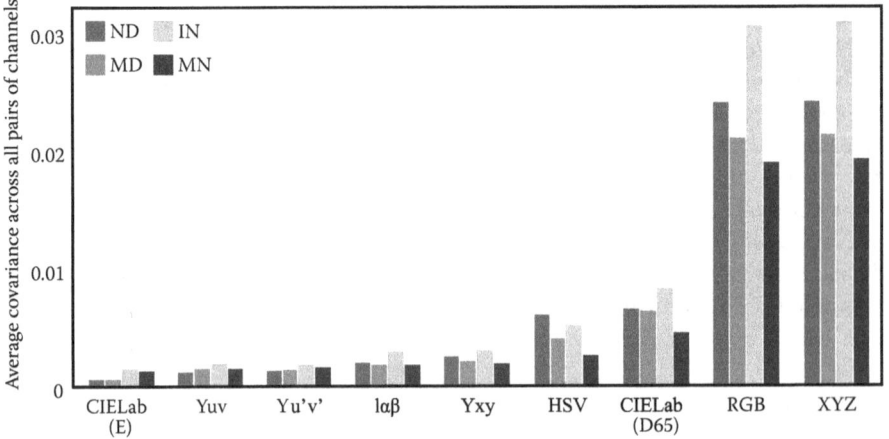

Fig. 14 Average covariance values for the color spaces tested for each image ensemble. Higher values suggest higher correlation and therefore indicate spaces less suitable for edits where each channel is processed separately

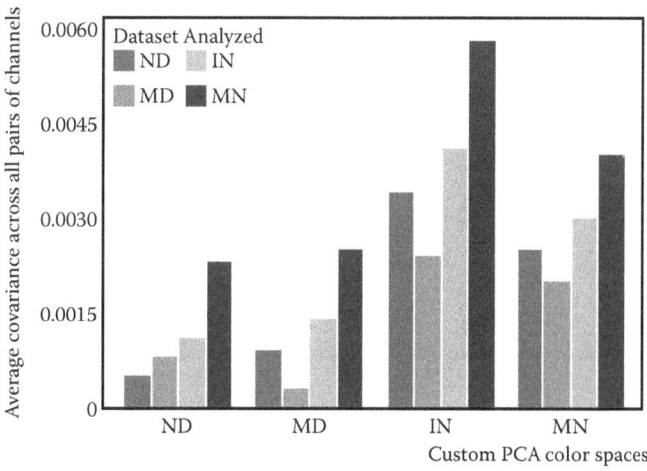

Fig. 15 Average covariance for the four PCA-based color spaces computed from different image ensembles. Images from each of the ensembles were transformed with all spaces to assess whether better decorrelation can be achieved when the color space is computed from similar images

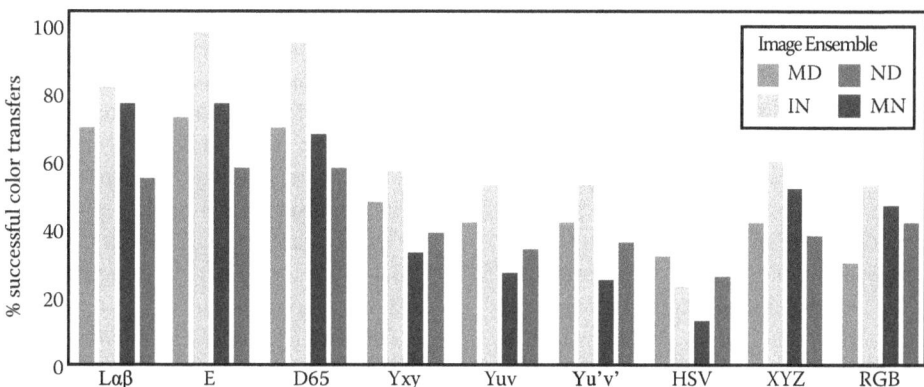

Fig. 16 Percentage of successful color transfers for each color space for each of the ensembles. E and D65 refer to CIELab with illuminants E and D65, respectively

discussed and then used as input for the color transfer algorithm presented in "Color Transfer through Simple Moments" section. In addition to the standard and PCA-based color spaces discussed in the previous section, PCA spaces were this time also computed from each of the individual input images. Participants were then shown the pair of input images as well as the color transferred result and asked to classify it as successful or not.

Interestingly, the color space rankings from this study, shown in Figs. 16 and 17 for the standard and PCA spaces, respectively, were very similar to the results of the previous analysis. The CIELab space led to the most successful color transfer, followed by the PCA-based space computed from the reference image for each transfer. Unlike the covariance analysis results, though, the choice of white point for the color space conversion seemed to matter less.

In summary, although one may argue that decorrelating individual images would be the most fine-grained approach to color transfer, it was found that similarly excellent results can be obtained by simply choosing the CIELab space. Thus, it does not appear to be necessary to expend computational resources to deriving either category-based color spaces or image-specific color spaces for the purpose of color transfer.

COLOR CONSTANCY AND WHITE BALANCING

The color of the light reaching our retinas is determined by the spectral distribution of the illumination in the scene and the reflectance properties of the surfaces it encounters in its path to our eyes. An intriguing property of our visual system is its ability to correctly assess the color of objects under varying illumination. A banana will be perceived as yellow whether it is seen under the orange cast of tungsten light or outside in sunlight, even though the precise properties of the light reflected from its surface will be very different in these two situations. In other words, the visual system is able to separate the effect of surface reflectance and the prevalent illumination in a scene, effectively

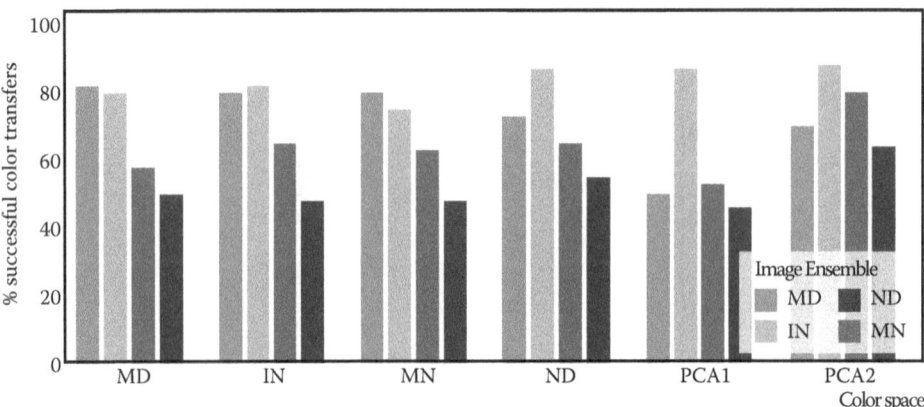

Fig. 17 Percentage of successful color transfers for each ensemble-specific color space for each of the ensembles. Results are also shown for per-image PCA, carried out in the color space derived from the source image (PCA1) or the reference image (PCA2)

discounting the color of the light source when assessing the color of an object.

This property is known as *color constancy* and was first experimentally demonstrated by Edwin H. Land in his "Color Mondrian" experiment.[43,44] A Mondrian-like display was constructed out of colored pieces of paper, which were illuminated by three narrow-spectrum red, green, and blue projectors that could be individually controlled, as illustrated in Fig. 18. In a typical experiment, the tristimulus values of light reflected from a patch of a particular color, such as the green patch (a) in the figure, were first measured and the projected light was then adjusted so that a different patch, such as the orange patch (b), reflected the same tristimulus values. Although in both cases the tristimulus values of the light reaching the retina were identical, viewers would consistently see the green patch as green and the orange one as orange when viewed as part of the complete display.

Unlike the visual system, though, cameras simply capture light exactly as it reaches the sensor. If a scene has a color cast due to the illumination, this cast will remain visible and noticeable on the image since the conditions where the image was captured are unlikely to match the display and viewing environment where it is then shown. Consequently, the ability to discount or remove the prevalent illumination from images would be a very desirable one for cameras. This is typically implemented in a process known as *white-balancing*, which aims to simulate the color constancy properties of the visual system.[44–46]

In color spaces such as the aforementioned LMS space, equal values of the three components denote achromatic colors. One way to achieve such neutral colors is to start with an equal energy spectrum, i.e., a spectral distribution which has the same value L_o for each wavelength λ. This could happen if a scene was illuminated by an equal-energy light source, a source that emitted the same energy at all wavelengths. The colors that a camera will capture, then, from such a scene can be attributed to the reflectance properties of the objects and surfaces in it.

In practice, a scene is illuminated by only one or at most a few light sources, with an emission spectrum that is off-white. In natural scenes, for instance, the main source of illumination is likely to be the sun, with evidence suggesting that this may even be encoded as a prior in human color constancy.[47] The reflectance properties in a scene, on the other hand, are much more variable, suggesting that the local variations in color in an image are likely due to different materials and reflectance properties. Despite this variation, a surprising finding is that when a collection of reflectances are averaged, the result ends up being a distribution function that is close to gray. This is known as the *gray-world assumption*.[48]

The implication of this finding is that if we were to average the colors of all pixels in an image, and the gray-world assumption holds, the average tristimulus value

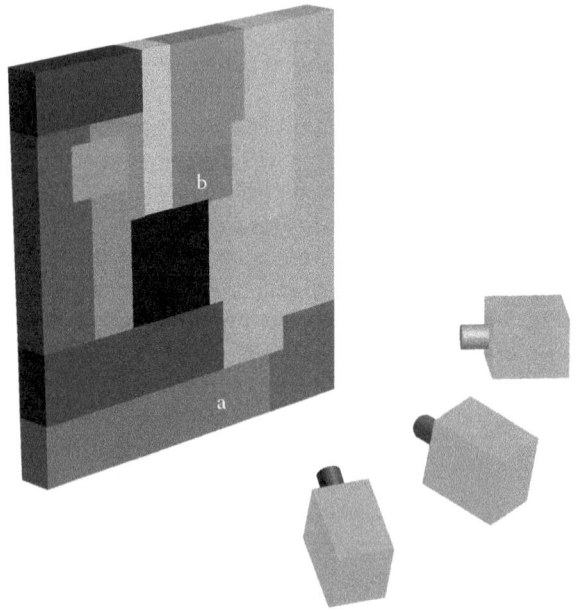

Fig. 18 An illustration of the experimental setup used to demonstrate color constancy

$(\bar{L}, \bar{M}, \bar{S})$ would be a good indicator of the color of the light source. Unfortunately, the gray-world assumption does not always hold. Figure 19a–c shows the result of averaging pixel values over single natural image, while 19d shows the average pixel value over a hundred natural images. As can be seen, in single images this assumption may or may not hold. In particular, if the surface colors in the scene are biased towards some specific saturated color, the average reflectance will not tend toward gray, and therefore extracted information about the illumination will likely also be biased.

The gray-world assumption is often used to aid white-balancing in imaging applications. After all, if we know the color of the illuminant, then we can correct all pixels by simply dividing all pixels by the average image value. Moreover, if we know that the display has a different white point, say $(L_{d,w}, M_{d,w}, S_{d,w})$, white-balancing can be implemented as follows:

$$L_{wb} = L \frac{L_{d,w}}{\bar{L}} \quad (10a)$$

$$M_{wb} = M \frac{M_{d,w}}{\bar{M}} \quad (10b)$$

$$S_{wb} = S \frac{S_{d,w}}{\bar{S}} \quad (10c)$$

The gray-world assumption is a statistical argument that is necessary to perform white-balancing on images in the absence of further information about the illuminant, given that white-balancing is by itself an underconstrained problem.[49] Note that this procedure is best applied in a perceptual color space, such as LMS, thereby mimicking chromatic adaptation processes that occur in the human visual system.[16] If an image is given in a different color space, most likely the RGB space, the image should first be converted to LMS.

The approximation of the illuminant can be improved by excluding the most saturated pixels from the estimation.[50] Alternatively, the image can be subjected to further statistical analysis to determine if the color distribution is due to colored surfaces or a colored light source.[51] Here the image is first converted to the CIELab color opponent space. Ignoring the lightest and darkest pixels, since they do not contribute to a reliable estimate, the remaining pixels are used to computed a two-dimensional histogram $F(a, b)$ on the two chromatic channels a and b. In each channel, the chromatic distributions are modeled with:

$$\mu_k = \int_k k F(a,b) dk \quad (11)$$

$$\sigma_k^2 = \int_k (\mu_k - k) F(a,b) dk \quad (12)$$

where $k=a, b$. These are the mean and variances of the histogram projections onto the a and b axes. In CIELab neutral colors lie around the $(a, b)=(0, 0)$ point. To assess how far the histogram lies from this point, the distance D can be computed:

$$D = \mu - \sigma \quad (13)$$

where $\mu = \sqrt{\mu_a^2 + \mu_b^2}$ and $\sigma = \sqrt{\sigma_a^2 + \sigma_b^2}$.

The measure $D_\sigma = D/\sigma$ can be used to assess the strength of the cast. If the spread of the histogram is small and lies far away from the origin, the image is likely to be dominated by strong reflectances rather than illumination.

Computational Color Constancy as the Minkowski Norm

It was found that while gray-world algorithms work well on some images, alternate solutions, such as the *white-patch algorithm*,[43] perform better on texturerich images. The white-patch algorithm assumes that the lightest pixels in an image depict a surface with neutral reflectance, so that its color represents the illuminant.

Both the gray-world and white-patch algorithms are special instances of the Minkowski norm:[52]

$$L_p = \left(\frac{\int f^p(x) dx}{\int dx} \right)^{1/p} = ke[0,1] \quad (14)$$

where $f(x)$ denotes the image at pixel x. The average of the image is computed for $p=1$, thereby implementing the gray-world assumption. The maximum value of the image is computed by substituting $p=\infty$, which represents the white patch algorithm, also known as Max-RGB.

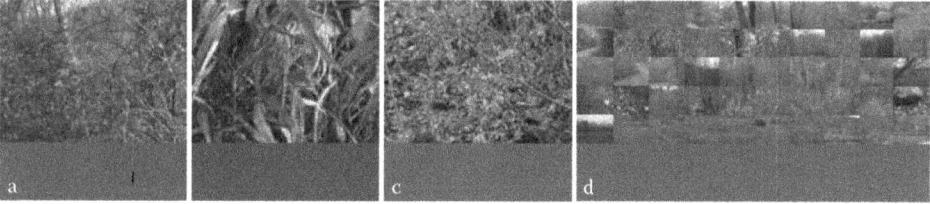

Fig. 19 The white points shown in the bottom row were estimated for three individual images, as well as for a collection of 128 images using the gray-world assumption. Although when a large set of images is averaged, the result will be approximately gray, this is not the case with individual images as the dominant colors of the scene are likely to bias the estimation (images taken from the Zurich Natural Images database; see Chapter 3 for more details)

Table 2 Different formulations based on the Minkowski norm can be used to obtain different color constancy algorithms (adapted from Ref. [54])

Algorithm	Symbol	Description
Gray-World	$e^{0,1,0}$	The average reflectance in a scene is assumed to be achromatic; therefore, any shifts from an achromatic average are due to illumination.[48]
White-Patch (Max-RGB)	$e^{0,\infty,0}$	The brightest patch in the scene is assumed to belong to a surface with neutral reflectance and therefore the color of that patch represents the color of the illumination.
Shades of Gray	$e^{0,p,0}$	A more general formulation of the gray-world and max-RGB algorithms that assumes that the pth norm of the scene is achromatic. Shown to perform best when p = 6.[52]
General Gray-World	$e^{0,p,\sigma}$	A more general version of the above, where a local region of scale σ is considered, which can be achieved by first filtering the image.
Gray-Edge	$e^{1,p,\sigma}$	The average of the derivative of the image is assumed to be achromatic.[53]
Max-Edge	$e^{1,\infty,\sigma}$	The maximum difference of reflectances in the scene is assumed to be achromatic.
Second-Order Gray-Edge	$e^{2,p,\sigma}$	The average of higher-order derivatives (second in this case) is assumed to be achromatic.[54]

A further generalized assumption can be made about images, which is that the average difference between two pixels evaluates to gray. This is known as the gray-edge assumption[53,54] and can be formulated as follows[55]:

$$\left(\int \left|\frac{\partial^n f^\sigma(x)}{\partial x^n}\right|^p dx\right)^{1/p} = k e^{n,p,\sigma} \quad (15)$$

Here, n is the order of the derivative, p is the Minkowski-norm, and $f^\sigma(x) = f \otimes G^\sigma$ is the Gaussian-blurred image where the size of the filter kernel is given by σ. With this formulation several color constancy algorithms can be constructed, shown in Table 2. Examples of some of these methods can be seen in Fig. 20.

Although different algorithms rely on different and increasingly complex statistics to correctly white-balance images, no single method has yet emerged as a universally effective solution.[46] This suggests that a single aspect of images may not be sufficient to determine the color of the illumination, especially in ambiguous scenarios. An overall white scene illuminated with a blue light may look substantially similar to a blue scene illuminated with a white light. Yet most color constancy algorithms will find it hard to distinguish between them. On the other hand, our visual system is capable of such a distinction in most cases, perhaps by relying on a variety of higher-order statistical information in scenes.[56–58]

White-Balance Algorithm Selection

Having noted that many color constancy and white-balancing algorithms exist, with none of them universally applicable, Gijsenij and Gevers use natural image statistics to classify an image, and then they use this classification to select the most appropriate white-balancing algorithm for the image at hand.[55,59] In particular, they use the finding that the distribution of edge responses in an image can be modeled by means of a Weibull distribution:[60]

$$f(x) = \frac{\gamma}{\beta}\left(\frac{x}{\beta}\right)^{\gamma-1} \exp\left(\frac{x}{\beta}\right)^\gamma \quad (16)$$

The parameters β and γ have meaning in this context. The contrast of an image (or image ensemble) is given by β, whereas γ is an indicator of grain size (i.e., the peakedness of the distribution). This means that higher values for β represent images with more contrast. Higher values for γ indicate finer textures.

Fitting a Weibull distribution involves computing a Gaussian derivative filter in both x and y directions. This results in the (β_x, β_y, γ_x, γ_y) set of parameters for each color channel. The Gaussian derivative filter can be first-, second-, or third-order. However, it was found that the order of the chosen filter is relatively unimportant: high correlations exist between the fitted parameters.[60]

It is now possible to fit Weibull parameters to the derivatives of a large number of images and correlate their values with the white-balancing algorithm that performs best for each image. The parameter space tends to form clusters where a specific algorithm tends to produce the most accurate result. This means that the Weibull distribution can be used to select the most appropriate white-balancing algorithm for each image individually.[55,59]

SUMMARY

In this Entry we looked at some of the regularities of color information in images and discussed a number of imaging applications that rely on them. The human visual system has special adaptations allowing it to take advantage of the

Fig. 20 Different color constancy algorithms based on the Minkowski norm were used to white-balance the image shown at the top left (a). The algorithms used are given in Table 2. The small rectangle below each image shows the white point estimated by each algorithm and the manually white-balanced image is given in (b). (Santorini, Greece, 2009)

color distribution in nature. One important such example is color opponency ("Opponent Processing" section). The visual system re-encodes information from the photoreceptors into three opponent channels, which effectively reduce redundancy. The same principle has been applied to images, leading to a number of color opponent spaces that rotate image data such that correlation between the three channels is minimized.

A consequence of the decorrelation abilities of such color spaces is that each channel can be processed separately without affecting the other two. One imaging application that relies on this particular property of color opponent spaces is color transfer, which was discussed in "Color Transfer" section. Given a pair of source and reference images, color transfer methods aim to re-color the source image using the color palette or distribution of the reference.

In "Color Space Statistics" section, we described a set of experiments for guiding the selection of a particular color space for color transfer as well as other image editing applications. Several standard color spaces were compared by means of a correlation analysis as well as a simple study. In addition, custom spaces were created through principal component analysis of large sets of images of the same category or even single images, finding that in most cases, the CIELab color space offers good decorrelation of color information in images.

Another important mechanism of the HVS that relies on color regularities of natural images is color constancy, which allows us to assess the colors of objects despite changing illumination. In the case of images, algorithms that are functionally similar to the color constancy of the visual system can be used to remove color shifts due to illumination. This process is then known as white-balancing. In "Color Constancy and White Balancing" section we discussed a number of such methods as well as a general formulation for many of them, which is based on the Minkowski norm. Finally, we showed how image statistics can guide the selection of white-balancing algorithm, increasing the likelihood of a correct white-point estimation.

REFERENCES

1. Webster, M.A.; Mizokami, Y.; Webster, S.M. Seasonal variations in the color statistics of natural images. Netw. Comput. Neural Syst. **2007**, *18* (3), 213–233.
2. Marshall, N.J.; Land, M.F.; King, C.A.; Cronin, T.W. The compound eyes of mantis shrimps (crustacea, hoplocarida, stomatopoda). II. Colour pigments in the eyes of stomatopod crustaceans: Polychromatic vision by serial and lateral filtering. Philos. Trans. R. Soc. Lond. B **1991**, *334* (1269), 57–84.
3. Bennett, A.; Cuthill, I. Avian color vision and coloration: Multidisciplinary evolutionary biology. Am. Nat. **2007**, *169* (S1), 1–6.
4. Young, T. The Bakerian lecture: On the theory of light and colors. Philos. Trans. R. Soc. Lond. **1802**, *92*, 12–48.
5. von Helmholtz, H.L.F. *Handbuch der Physiologischen Optik*. Leopold Voss, Leipzig, Germany, 1867.
6. Bowmaker, J.K.; Dartnall, H.J. Visual pigments of rods and cones in a human retina. J. Phys. **1980**, *298* (1), 501–511.
7. Stockman, A.; MacLeod, D.; Johnson, N. Spectral sensitivities of the human cones. J. Opt. Soc. Am. A **1993**, *10* (12), 2491–2521.
8. Fairchild, M.D. *Color Appearance Models*, 3rd Ed.; John Wiley & Sons: Chichester, 2013.
9. Foster, D.H.; Amano, K.; Nascimento, S.; Foster, M.J. Frequency of metamerism in natural scenes. J. Opt. Soc. Am. A **2006**, *23* (10), 2359–2372.
10. Hering, E. *Outlines of a Theory of the Light Sense (Translation from German: Zur Lehre vom Lichtsinne, 1878)*; Harvard University Press: Cambridge, MA, 1920.
11. Hurvich, L.M.; Jameson, D. The opponent process theory of color vision. Psychol. Rev. **1957**, *64* (6), 384–404.
12. Buchsbaum, G.; Gottschalk, A. Trichromacy, opponent colours coding and optimum colour information transmission in the retina. Proc. R. Soc. Lond. B **1983**, *220* (1218), 89–113.
13. Ruderman, D.L.; Cronin, T.W.; Chiao. C.C. Statistics of cone responses to natural images: Implications for visual coding. J. Opt. Soc. Am. A **1998**, *15* (8), 2036–2045.
14. Wyszecki, G.; Stiles, W.S. *Color Science: Concepts and Methods, Quantitative Data and Formulae*, 2nd Ed.; John Wiley & Sons: New York, 2000.
15. Ebner, F.; Fairchild, M.D. Development and testing of a color space (IPT) with improved hue uniformity. In *Proceedings of the 6th IS&T/SID Color Imaging Conference: Color Science, Systems and Applications*, 1998, 8–13.
16. Reinhard, E.; Khan, E.A.; Akyüz, A.O.; Johnson, G.M. *Color Imaging: Fundamentals and Applications*; A K Peters: Wellesley, 2008.
17. Reinhard, E.; Ashikhmin, M.; Gooch, B.; Shirley, P. Color transfer between images. IEEE Comput. Graph. Appl. **2001**, *21* (5), 34–41.
18. Pouli, T.; Reinhard, E. Progressive histogram reshaping for creative color transfer and tone reproduction. In *NPAR '10: Proceedings of the 8th International Symposium on Non-Photorealistic Animation and Rendering*, 2010, 81–90.
19. Pouli, T.; Reinhard, E. Progressive color transfer for images of arbitrary dynamic range. Comput. Graph. **2011**, *35* (1), 67–80.
20. Xiao, X.; Ma, L. Gradient-preserving color transfer. Comput. Graph. Forum **2009**, *28* (7), 1879–1886.
21. Pitié, F.; Kokaram, A.; Dahyot, R. N-dimensional probability density function transfer and its application to colour transfer. In *Proceedings of the IEEE International Conference on Computer Vision*, Vol. 2; Washington, DC, 2005, 1434–1439.
22. Pitié, F.; Kokaram, A.; Dahyot, R. Automated colour grading using colour distribution transfer. Comput. Vis. Image Underst. *107* (2), 1434–1439.
23. Greenfield, G.; House, D. Image recoloring induced by palette color associations. J. WSCG **2003**, *11* (1), 189–196.
24. Toet, A. Natural colour mapping for multiband nightvision imagery. Inform. Fusion **2003**, *4* (3), 155–166.
25. Yin, J.; Cooperstock, J.R. Color correction methods with applications to digital projection environments. J. WSCG **2004**, *12* (1–3), 241.
26. Li, Z.; Jing, Z.; Yang, X.; Sun, S. Color transfer based remote sensing image fusion using non-separable wavelet frame transform. Pattern Recognit. Lett. **2005**, *26* (13), 2006–2014.
27. Xu, S.; Zhang, Y.; Zhang, S.; Ye, X. Uniform color transfer. In *Proceedings of the IEEE International Conference on Image Processing*, 2005, 940–943.
28. Wang, C.; Huang, Y.; Huang, M. An effective algorithm for image sequence color transfer. Math. Comput. Model. **2006**, *44* (7–8), 608–627.
29. Luan, Q.; Wen, F.; Xu, Y.-Q. Color transfer brush. In *Proceedings of Pacific Graphics*, 2007, 465–468.
30. Zhao, Y.; Berns, R.S. Image-based spectral reflectance reconstruction using the matrix R method. Color Res. Appl. **2007**, *32* (5), 343–351.
31. Xiang, Y.; Zou, B.; Li, H. Selective color transfer with multi-source images. Pattern Recognit. Lett. **2009**, *30* (7), 682–689.
32. Neumann, L.; Neumann, A. Color style transfer techniques using hue, lightness and saturation histogram matching. In *Proceedings of the 1st Eurographics Conference on Computational Aesthetics in Graphics, Visualization and Imaging*, 2005, 111–122.
33. Kumar, R.; Mitra, S.K. Motion estimation based color transfer and its application to color video compression. Pattern Anal. Appl. **2007**, *11* (2), 131–139.
34. Levin, A.; Lischinski, D.; Weiss, Y. Colorization using optimization. ACM Trans. Graph. **2004**, *24* (3), 689–694.
35. Wang, L.; Zhao, Y.; Jin, W.; Shi, S.; Wang, S. Real-time color transfer system for low-light level visible and infrared

35. images in YUV color space. In *Proceedings of the SPIE*, Vol. 6567; 2007, 65671G.
36. Morovic, J.; Sun, P.-L. Accurate 3D image colour histogram transformation. Pattern Recognit. Lett. **2003**, *24* (11), 1725–1735.
37. Kotera, H.; Morimoto, T.; Saito, R. Object-oriented color matching by image clustering. In *Proceedings of the 6th IS&T/SID Color Imaging Conference*, 1998, 154–158.
38. Abadpour, A.; Kasaei, S. A fast and efficient fuzzy color transfer method. In *Proceedings of the 4th IEEE International Symposium on Signal Processing and Information Technology*, 2004, 491–494.
39. Kotera, H. A scene-referred color transfer for pleasant imaging on display. In *Proceedings of the IEEE International Conference on Image Processing*, 2005, 2, 5–8.
40. Xiao, X.; Ma, L. Color transfer in correlated color space. In *Proceedings of the 2006 ACM International Conference on Virtual Reality Continuum and Its Applications*, New York, 2006, 305–309, ACM.
41. Abadpour, A.; Kasaei, S. An efficient PCA-based color transfer method. J. Vis. Commun. Image Represent. **2007**, *18* (1), 15–34.
42. Reinhard, E.; Pouli, T. Colour spaces for colour transfer. In *IAPR Computational Color Imaging Workshop, volume 6626 of Lecture Notes in Computer Science*, (invited paper); Schettini, R.; Tominaga, S.; Trémeau, A.; Eds; Springer: Berlin, 2011, 1–15.
43. Land, E. The retinex theory of color vision. Sci. Am. **1977**, *237* (6), 108–128.
44. Foster, D.H. Color constancy. Vis. Res. **2011**, *51* (7), 674–700.
45. Ebner, M. *Color Constancy*; John Wiley & Sons: West Sussex, 2007.
46. Gijsenij, A.; Gevers, T.; van De Weijer, J. Computational color constancy: Survey and experiments. IEEE Trans. Image Process. **2011**, *20* (9), 2475–2489.
47. Delahunt, P.B.; Brainard, D.H. Does human color constancy incorporate the statistical regularity of natural daylight? J. Vis. **2004**, *4* (2).
48. Buchsbaum, G. A spatial processor model for object colour perception. J. Franklin Inst. **1980**, *310* (1), 1–26.
49. Funt, B.; Barnard, K.; Martin, L. Is machine colour constancy good enough? In *Proceedings of the 5th European Conference on Computer Vision*, Freiburg, 1998, 445–459.
50. Adams, J.; Parulski, K.; Spaulding, K. Color processing in digital cameras. IEEE Micro **1998**, *18* (6), 20–30.
51. Gasparini, F.; Schettini, R. Color balancing of digital photos using simple image statistics. Pattern Recognit. **2004**, *37* (6), 1201–1217.
52. Finlayson, G.; Trezzi, E. Shades of grey and colour constancy. In *Proceedings of the 12th IS&T/SID Color Imaging Conference*, 2004, 37–41.
53. van de Weijer, J.; Gevers, T. Color constancy based on the grey-edge hypothesis. In *Proceedings of the IEEE International Conference on Image Processing*, 2005, 722–725.
54. van de Weijer, J.; Gevers, T.; Gijsenij, A. Edge-based color constancy. IEEE Trans. Image Process. **2007**, *16* (9), 220–2214.
55. Gijsenij, A.; Gevers, T. Color constancy using natural image statistics. In *Proceedings of the IEEE International Conference on Computer Vision and Pattern Recognition*, 2007, 1–8.
56. Golz, J.; MacLeod, D.I. Influence of scene statistics on colour constancy. Nature **2002**, *415* (6872), 637–640.
57. Golz, J. The role of chromatic scene statistics in color constancy: Spatial integration. J. Vis. **2008**, *8* (13), 253.
58. Granzier, J.J.; Brenner, E.; Cornelissen, F.W.; Smeets, J.B. Luminance—Color correlation is not used to estimate the color of the illumination. J. Vis. **2005**, *5* (1).
59. Gijsenij, A.; Gevers, T. Color constancy using natural image statistics and scene semantics. IEEE Trans. Pattern Anal. Mach. Intell. **2011**, *33* (4), 687–698.
60. Geusebroek, J.-M.; Smeulders, A.W.M. A six-stimulus theory for stochastic texture. Int. J. Comput. Vis. **2005**, *62* (1–2), 7–16.

Color Image Denoising

Palungbam RojiChanu
Department of Electronics & Communication Engineering, National Institute of Technology
Nagaland, Dimapur, India

Kh. Manglem Singh
Department of Computer Science & Engineering, National Institute of Technology
Manipur, Imphal, India

Abstract

Vector median filters and its variants are traditional vector filters used for removing impulse noise from color images. Color images are vector signals and have strong correlation among their components. Use of quaternion in image processing applications is increasing its popularity. Quaternion treats a color image as a single unit by combining its components into a single hypercomplex number. This preserves the vector nature of the color signal. Based on the quaternion theory, a switching vector median filter is proposed. The distance between two color pixels is expressed in terms of their intensity and chromaticity differences, and the concept of rank condition filter is introduced for detecting the impulse. The color pixels in the filtering window are ranked based on the aggregated sum of color differences from other pixels in the window. The pixel with the least rank is known as vector median. The central pixel is considered as corrupted by impulse noise if its rank is larger than predefined healthy rank and the minimum color differences between the central pixel and its neighboring pixels in the four edge direction of the window is larger than a predefined threshold. The noisy pixel is replaced by vector median computed using the quaternion distance. The experimental results show that the proposed filter outperforms other well-known nonlinear filters in removing impulse noise. It combines the quaternion benefits of color representation and rank conditioned filters.

Keywords: Chromaticity; Correlation; Hypercomplex number; Rank condition filter; Vector nature.

INTRODUCTION

Image denoising is an important task of image processing which aims at suppressing noise while preserving the image details. Impulse noise is one kind of noise which appears as large energy spikes occurring for a short duration. Images are often corrupted by impulse noise due to certain factors such as faulty sensors, electronic interference, or errors in the communication channels. In case of color images, the presence of noise leads to color fluctuations which deviate pixel values from their ideal values thereby introducing errors.[1,2] Hence, many filtering techniques have been developed which can be broadly classified into marginal and vector filtering. Marginal filtering techniques process each channel independently that produce color artifacts while the vector filters treat every pixel as set of vectors. This introduces no new color and is able to preserve the strong spectral correlation existing among the channels.[3,4] Color images are nonstationary due to the presence of edges and fine details. Hence, nonlinear filters are preferred than linear filters.

Vector median filter (VMF)[5] is the basic nonlinear order statistics filter for removing impulse noise. It has the properties of median filter such as zero impulse response and good smoothening capabilities. Minkowski metrics such as Euclidean distance and city-block distance are used for calculation of magnitude differences of color samples. For every pixel, the distance between it and other pixels inside the filtering window is computed and the aggregated sum of distances is ordered in an ascending order. The pixels with larger aggregated distance are those which vary greatly from the pixel population. Hence, the pixel with minimum aggregated distance is given the lowest rank and represents the vector median. In vector directional filter (VDF),[6] the angle between two pixels is considered as the distance measure. The pixels with atypical direction is considered as an impulse and replaced by the VMF output. It is able to preserve the chromaticity better than the VMF since vector's direction corresponds to its chromaticity. Directional distance filter (DDF)[6,7] integrates the magnitude and directional processing of both the VMF and the VDF. For improving the filtering performance, weighted filters such as weighted VMF (WVMF),[8] weighted VDF (WVDF),[9,10] and weighted DDF (WDDF)[9,10] are designed. The weight assigned to pixels controls the filtering action and helps in designing an optimal filter by offering greater flexibility as compared with the median-based filters. If the central pixel is given more weight, then the resulting weighted filters are center-weighted VMF (CWVMF), center-weighted VDF (CWVDF), and center-weighted DDF (CWDDF).[11,12]

The traditional vector filters mentioned above have the major drawback that they filter the uncorrupted vector pixels without checking if they are actually an impulse or not. This leads to excessive blurring of the image. To overcome the problem, adaptive filters are designed which give an estimate of the filter kernel coefficients based on the local statistics obtained from a relatively smaller window. If the current pixel is detected as an impulse, it is filtered by the traditional basic vector filters else no action is performed. The resulting filters are adaptive VMF (AVMF),[13] adaptive VDF (AVDF), adaptive DDF (ADDF), adaptive CWVMF (ACWVMF), adaptive CWVDF (ACWVDF), and adaptive CWDDF (ACWDDF).[14,15] A switching action is performed and filtering is done only on pixels considered as corrupted. Some other common switching filters include the peer group filters (PGFs),[16] vector sigma filters,[17–19] entropy filters,[20] fuzzy-based filters,[21,22] and switching vector filters based on noncausal (NC) linear prediction techniques.[23]

In this article, the application of quaternion for removing impulse noise from color images is reviewed, and a new switching vector median filter is proposed. *RGB* color pixel is expressed in pure quaternion form which helps in improving the measurement of the color dissimilarity. The concept of quaternion and its properties are discussed in Section "Quaternions." Application of quaternion in image denoising is given in Section "Application of Quaternion in Color Image Denoising." The proposed work is presented in Section "Proposed Switching Vector Median Filter" with experimental results and discussions given in Section "Experimental Results and Discussions" followed by conclusion in the last section.

QUATERNIONS

Quaternions were discovered by Sir William Hamilton, a British mathematician in 1843. A quaternion q is a hyper complex number having a real part a and three imaginary parts b, c, and d. It is defined in a four-dimensional space ($\mathbb{R}^4$) in Cartesian form as follows:

$$q = a + bi + cj + dk \tag{1}$$

where i, j, and k are complex operators which follow Hamilton's rules.

$$i^2 = j^2 = k^2 = ijk = -1,$$
$$ij = k,\ jk = i,\ ki = j, \tag{2}$$
$$ji = -k, kj = -i, ik = -j.$$

The three complex operators are orthogonal with each other and signify the three-dimensional vectors.

$$i = (1, 0, 0),$$
$$j = (0, 1, 0),$$
$$k = (0, 0, 1).$$

A quaternion is considered as a hypercomplex number. It is split into "scalar" (S) and "vector" (V) parts:

$$q = Sq + Vq, Sq = a, Vq = bi + cj + dk. \tag{3}$$

Common Properties of Quaternions

Addition

Addition of quaternions follow component-wise rule. If $q_1 = a_1 + b_1 i + c_1 j + d_1 k$ and $q_2 = a_2 + b_2 i + c_2 j + d_2 k$ are two quaternions, then the addition is given by Jia[24] as follows:

$$\begin{aligned} q_1 + q_2 &= (a_1 + b_1 i + c_1 j + d_1 k) + (a_2 + b_2 i + c_2 j + d_2 k) \\ &= (a_1 + a_2) + (b_1 + b_2)i + (c_1 + c_2)j + (d_1 + d_2)k \\ &= (Sq_1 + Sq_2) + (Vq_1 + Vq_2) \end{aligned} \tag{4}$$

Quaternion addition is both associative and commutative.

Subtraction

Subtraction of two quaternions is given by:

$$\begin{aligned} q_1 - q_2 &= (a_1 + b_1 i + c_1 j + d_1 k) - (a_2 + b_2 i + c_2 j + d_2 k) \\ &= (a_1 - a_2) + (b_1 - b_2)i + (c_1 - c_2)j + (d_1 - d_2)k \\ &= (Sq_1 - Sq_2) + (Vq_1 - Vq_2) \end{aligned} \tag{5}$$

Multiplication

Multiplication of quaternions is similar with polynomials and follows the multiplicative properties of the Hamilton's rule. It is denoted by:

$$\begin{aligned} q_1 q_2 &= (a_1 + b_1 i + c_1 j + d_1 k)(a_2 + b_2 i + c_2 j + d_2 k) \\ &= (a_1 a_2 - b_1 b_2 - c_1 c_2 - d_1 d_2) + (a_1 b_2 - b_1 a_2 - c_1 d_2 - d_1 c_2)i \\ &\quad + (a_1 c_2 - b_1 d_2 - c_1 a_2 - d_1 b_2)j + (a_1 d_2 - b_1 c_2 - c_1 b_2 - d_1 a_2)k \end{aligned} \tag{6}$$

The above quaternion product has a more concise form:

$$q_1 q_2 = (Sq_1 Sq_2) - (Vq_1 \cdot Vq_2) + (Sq_1 Vq_2) + (Sq_2 Vq_1) + (Vq_1 \times Vq_2) \quad (7)$$

where "·" and "×" are "dot" and "cross" products of vector algebra. Here, multiplication is associative but not commutative. The multiplication causes rotation. Consider $ij = k$, it is the multiplication of i on the right by j causing a 90° rotation in the $\mathbb{R}^4$ by rotating the i axis into the k axis and k axis into the $-i$ axis.

Conjugate and Norm

Conjugate of a quaternion is given by:

$$q^* = (a + bi + cj + dk)^* = (a - bi - cj - dk). \quad (8)$$

Also, the magnitude or norm is expressed as:

$$|q| = \sqrt{a^2 + b^2 + c^2 + d^2}. \quad (9)$$

It is observed that $qq^* = q^*q = |q|^2$.

Inverse

The inverse of a quaternion q^{-1} is denoted by:

$$q^{-1} = \frac{q^*}{|q|^2}. \quad (10)$$

Pure Quaternion

A quaternion having a zero real part is called a pure quaternion. It is equivalent to an ordinary vector. This representation resembles the Argand diagram of the complex plane.

$$q = 0 + bi + cj + dk = bi + cj + dk. \quad (11)$$

Unit Quaternion

A quaternion with $|q| = 1$ is called a unit quaternion q_u. For a unit quaternion, $q_u^{-1} = q_u^*$. It can also be expressed as follows[25]:

$$q_u = R_u \cos\phi + P_u \sin\phi = \cos\phi + P_u \sin\phi \quad (12)$$

where $R_u = (1, 0, 0, 0)$ is the real part of unit quaternion, $P_u = (0, ip_2, jp_3, kp_4)$ represents the vector unit quaternion which is parallel to the vector component of q_u, and ϕ is an arbitrary real number. Its geometrical interpretation is used in specifying the quaternion angle of rotation. This is a very useful property and it is proved below:

$$\begin{aligned} |q_u|^2 &= q_u q_u^* = R_u \cos\phi + P_u \sin\phi (R_u \cos\phi + P_u \sin\phi)^* \\ &= R_u R_u^* \cos^2\phi + R_u P_u^* \sin\phi\cos\phi + P_u R_u^* \sin\phi\cos\phi \\ &\quad + P_u P_u^* \sin^2\phi \\ &= \cos^2\phi + \sin^2\phi = 1. \end{aligned} \quad (13)$$

Polar Form of Quaternion

Any quaternion can be expressed in polar form as follows:

$$q = |q| e^{\mu\theta} = |q|(\cos\theta + \mu\sin\theta) \quad (14)$$

where μ is a unit pure quaternion (a pure quaternion with unit magnitude). It is computed by

$$\mu = \frac{V(q)}{|V(q)|} = \frac{(bi + cj + dk)}{|bi + cj + dk|} = \frac{(i + j + k)}{\sqrt{1+1+1}} = \frac{(i + j + k)}{\sqrt{3}} \quad (15)$$

with the only exception of μ being undefined when $V(q) = 0$. It is also referred to as *eigenaxis* which identifies the direction in the three-dimensional ($\mathbb{R}^3$) space of the vector part.

Here, θ is the angle between the real part and three-dimensional imaginary parts. It is also known as the *eigenangle* or phase. It is computed as:

$$\theta = \begin{cases} \tan^{-1} \frac{\sqrt{b^2 + c^2 + d^2}}{a}, & a \neq 0 \\ \frac{\pi}{2}, & a = 0 \end{cases} \quad (16)$$

It is always positive and lies in the range $0 \leq \theta \leq \pi$.

Quaternion Rotation in Three-Dimensional Space

In order to define how quaternions are used to rotate vectors in three-dimensional space, it is important to discuss the product of a unit quaternion and a perpendicular vector quaternion. Consider a vector quaternion s_v perpendicular to a unit quaternion q_u ($s_v \perp q_u$). Their product is written as:

$$W = q_u s_v = (\cos\phi + P_u \sin\phi) s_v = (\cos\phi + \sin\phi P_u) s_v = \cos\phi s_v + \sin\phi P_u s_v. \quad (17)$$

Since P_u is the unit vector along vector parts of q_u, $P_u \| q_u$. The first part represents a vector $W_{v(1)} \| s_v$ so that $W_{v(1)} \perp q_u$. The second term is also a vector $W_{v(2)}$ since $s_v \perp q_u$. Moreover $W_{v(2)} \perp s_v$ and $W_{v(2)} \perp (q_u \| P_u)$. Here, the product W is the sum of two vectors, it is also a vector denoted as W_v. Both $W_{v(1)}$ and $W_{v(2)}$ lie in a plane perpendicular to q_u. Hence, the product W_v can be geometrically interpreted as rotation of vector s_v in the plane perpendicular to q_v by an angle ϕ. The axis of rotation is parallel to q_u.

Again, consider the product of W_v with q_u^{-1} denoted as follows:

$$T_v = W_v q_u^{-1} = W_v q_u^* = \cos\phi W_v + \sin\phi W_v P_u^* \\ = \cos\phi W_v - \sin\phi W_v P_u. \tag{18}$$

Using the concept of vector identity $W_v P_u = -P_u W_v$, the above equation is rewritten as:

$$T_v = \cos\phi W_v + \sin\phi P_u W_v. \tag{19}$$

This represents another rotation of the vector by an angle ϕ about q_u. Therefore, it is concluded that the operation

$$T_v = q_u s_v q_u^{-1} \tag{20}$$

signifies the rotation of the vector s_v by an angle 2ϕ about q_u.

Equation 20 can be used in describing rotation of any arbitrary vector v_v about q_u. v_v is decomposed into $v_v = p_v + s_v$ in which p_v is parallel $(p_v \| q_u)$, and s_v is vertical $(s_v \perp q_u)$ components along q_u. Then,

$$q_u v_v q_u^{-1} = q_u (p_v + s_v) q_u^{-1} = q_u p_v q_u^{-1} + q_u s_v q_u^{-1} = q_u p_v q_u^{-1} + T_v \tag{21}$$

where T_v is s_v rotated by an angle 2ϕ about q_u in an axis parallel to it. Since $p_v \| q_u$, the component p_v can be written as $p_v = zP_u$, where z is a real number and $P_u \| q_u$. Thus, the first term in Eq. 21 is denoted as:

$$q_u p_v q_u^{-1} = q_u zP_u q_u^{-1} = zq_u P_u q_u^{-1} = zP_u q_u q_u^{-1} = zP_u = p_v \tag{22}$$

Finally,

$$q_u v_v q_u^{-1} = p_v + T_v. \tag{23}$$

This operation can be interpreted as a rotation of vector v_v about q_u by an angle 2ϕ.[25]

Quaternion Representation of Color Pixel

A color pixel is defined in a color space based on the red (R), green (G), and blue (B) primaries and a white reference point in *RGB* model. To use a quaternion which is defined in $\mathbb{R}^4$ space to operate on a color vector in $\mathbb{R}^3$ space, a pure quaternion is used. An *RGB* color vector is represented by a pure quaternion whose imaginary parts represent the red, green, and blue component, respectively. This concept is supported by the coincidence between the three-space imaginary part of the quaternion and three-dimensional nature of the *RGB* color triplets. It is denoted by

$$q = ri + gj + bk, \tag{24}$$

where *r*, *g*, and *b* are the red, green, and blue amplitudes of the signal.

The gray line of the unit *RGB* color space is denoted by the unit pure quaternion $\mu = (i + j + k)/\sqrt{3}$. A pixel in the gray line is achromatic, where all the three components are equal $(r = g = b)$.[26]

Quaternion Unit Transform and Quaternion-Based Representation of Color Differences

A unit quaternion vector is defined as follows:

$$U = |q|e^{\mu\theta} = \cos\theta + \mu\sin\theta \\ = \cos\theta + \frac{1}{\sqrt{3}}(\sin\theta i + \sin\theta j + \sin\theta k). \tag{25}$$

Let $c = ri + gj + bk$ be a color pixel in *RGB* color space expressed in pure quaternion form. Now an operator called quaternion unit transform[27] is derived. It is the convolution among a quaternion function and unit vectors. It is expressed by:

$$UcU^* = \left[\cos\theta + \frac{1}{\sqrt{3}}\sin\theta(i+j+k)\right](ri+gj+bk) \\ \left[\cos\theta - \frac{1}{\sqrt{3}}\sin\theta(i+j+k)\right]. \tag{26}$$

UcU^* is the rotation of vector c by 2θ around μ axis which represents the gray line in the *RGB* space. For a unit quaternion, θ is $\pi/2$. Hence, UcU^* is a rotation of π radians. It plays an important role in impulse noise filtering. The color pixels c and its rotated form UcU^* are positioned at equal distances from the gray line μ in the opposite direction. Thus $c + UcU^*$ should lie on the gray line.

The simplification of quaternion rotational operator UcU^* is given in the work of Chi[25] as follows:

$$Y = UcU^* = \left((S(U))^2 - \|V(U)\|^2\right)c + 2(V(U).c)V(U) \\ + 2S(U)(V(U) \times c) \\ = (\cos^2\theta - \sin^2\theta)(ri + gj + bk) \\ + 2\left[\frac{1}{\sqrt{3}}\sin\theta(i+j+k)\cdot(ri+gj+bk)\right] \\ \left(\frac{1}{\sqrt{3}}\sin\theta(i+j+k)\right) + 2(\cos\theta \times (ri+gj+bk)) \\ = \cos2\theta(ri+gj+bk) + \frac{2\sin^2\theta(r+g+b)}{3}(i+j+k) \\ + \frac{\sin2\theta}{\sqrt{3}}\left[(b-g)i + (r-b)j + (g-r)k\right]. \\ \triangleq Y_{RGB} + Y_I + Y_\Delta, \tag{27}$$

where

$$Y_{RGB} = \cos2\theta(ri+gj+bk),$$

$$Y_I = \frac{2\sin^2\theta(r+g+b)}{3}(i+j+k),$$

$$Y_\Delta = \frac{\sin 2\theta}{\sqrt{3}}\left[(b-g)i+(r-b)j+(g-r)k\right],$$

denote the *RGB* space component, the intensity, and the color difference. Y_Δ is the projection of the tristimuli values in the Maxwell triangle which is rotated by 90°. It represents the chromaticity component. Also it is important to mention that any vector c with coordinates (r', g', b') in the unit cube of *RBG* color space can be decomposed into two orthogonal components c' and i'. i' is the main diagonal of the cube which represents the gray line. The later represents the chromaticity plane perpendicular to the gray line. Therefore, a color vector is decomposed into intensity component along gray axis (parallel) and chromaticity component (perpendicular).

When $\theta = \pi/4$, Y_{RGB} reduces to zero and the unit transform transfers the *RGB* space into hue–saturation–intensity (*HSI*) like space. Denoting $T = U|_{\theta \triangleq \frac{\pi}{4}}$, then:

$$G = TcT^* = UcU^*|_{\theta=\frac{\pi}{4}} = \frac{\mu}{3}(r+g+b) + \frac{1}{\sqrt{3}}$$
$$\left[(b-g)i+(r-b)j+(g-r)k\right] = Y_I + Y_\Delta \quad (28)$$

Similarly,

$$G' = T^*cT = \frac{\mu}{3}(r+g+b)$$
$$-\frac{1}{\sqrt{3}}\left[(b-g)i+(r-b)j+(g-r)k\right] = Y_I - Y_\Delta \quad (29)$$

For calculating the intensity and chromaticity component, the following operations are performed.

Adding Eqs. 28 and 29,

$$Y_I = \frac{1}{2}\left[TcT^* + T^*cT\right].$$

Subtracting Eq. 29 from Eq. 28,

$$Y_\Delta = \frac{1}{2}\left[TcT^* - T^*cT\right].$$

For two pixels q_1 and q_2, the intensity and chromaticity parts are computed as follows:

$$Y_{I(q_1)} = \frac{1}{2}\left[Tq_1T^* + T^*q_1T\right],$$

$$Y_{I(q_2)} = \frac{1}{2}\left[Tq_2T^* + T^*q_2T\right],$$

$$Y_{\Delta(q_1)} = \frac{1}{2}\left[Tq_1T^* - T^*q_1T\right], \text{ and}$$

$$Y_{\Delta(q_2)} = \frac{1}{2}\left[Tq_2T^* - T^*q_2T\right].$$

Therefore, the intensity difference between q_1 and q_2 is derived as:

$$d_1(q_1,q_2) = Y_{I(q_1)} - Y_{I(q_2)} = \frac{1}{2}\left[Tq_1T^* + T^*q_1T\right]$$
$$-\frac{1}{2}\left[Tq_2T^* + T^*q_2T\right]$$
$$= \frac{1}{\sqrt{3}}\left[r_1 + g_1 + b_1 - r_2 - g_2 - b_2\right]. \quad (30)$$

Similarly, the chromaticity difference between q_1 and q_2 is derived as:

$$d_2(q_1,q_2) = Y_{\Delta(q_1)} - Y_{\Delta(q_2)}$$
$$= \frac{1}{2}\left[Tq_1T^* - T^*q_1T\right] - \frac{1}{2}\left[Tq_2T^* - T^*q_2T\right]$$
$$= \frac{1}{\sqrt{3}}\Big[\left((b_1-g_1)-(b_2-g_2)\right)i \quad (31)$$
$$+ \left((r_1-b_1)-(r_2-b_2)\right)j$$
$$+ \left((g_1-r_1)-(g_2-r_2)\right)k\Big].$$

APPLICATION OF QUATERNION IN COLOR IMAGE DENOISING

A color image has three channels per pixel. It is a vector signal which depends on all its components for its existence. The common method of component-wise image processing ignores the three-dimensional vectorial nature of color image. The quaternion representation of color combines the three channels into a single hypercomplex number. It treats a color as a single entity and processes as a unit rather than three independent separate components. Hence, quaternion representation analyzes a color image as a vector field. Also quaternion processes color and texture information in combination. It plays an important role in segmentation.[28] The spatial interchannel and intrachannel relationships can also be maintained. It is found that while multiplying two quaternions, each component interacts every other component whose effect spreads over all other components. Moreover, the quaternion rotation solves the problem of gimbal lock found in Euler's rotation.

Early works of quaternion in color image edge detectors and filters are found in the studies of Sangwine[29] and Sangwine and Ell.[30] They are based on quaternion convolution which utilizes the quaternion rotational operator $U[\ ]U^*$, where [ ] denotes a space for a quaternion to be operated. Zhang and Karim[31] develop a gray-scale impulse detector based on the differences between the central pixel and its neighbors aligned with the four edge main directions inside the filtering window. The minimum absolute value of four convolutions is obtained by using one-dimensional Laplacian operators. If this value is bigger than a predefined threshold, filtering is performed. This idea is extended in color noise filtering using the quaternion rotation.[32] For pixels $q_1 = r_1i + g_1j + b_1k$ and $q_2 = r_2i + g_2j + b_2k$, if they have similar or approximate color values, $q_1 + Uq_2U^*$ must lie on or near the gray line. On the other hand, if they have large color difference, $q_1 + Uq_2U^*$ lie apart from the gray line. Denoting $q_3 = q_1 + Uq_2U^* = r_3i + g_3j + b_3k$, an estimate of the color difference is computed as follows:

$$d(q_1 + Uq_2U^*) = d(q_3)$$
$$= \left(r_3 - \frac{r_3 + g_3 + b_3}{3}\right)i + \left(g_3 - \frac{r_3 + g_3 + b_3}{3}\right)j \quad (32)$$
$$+ \left(b_3 - \frac{r_3 + g_3 + b_3}{3}\right)k,$$

where

$$r_3 = r_1 - r_2 + \frac{2}{3}(r_2 + g_2 + b_2),$$
$$g_3 = g_1 - g_2 + \frac{2}{3}(r_2 + g_2 + b_2),$$
$$b_3 = b_1 - b_2 + \frac{2}{3}(r_2 + g_2 + b_2).$$

The color difference between q_1 and q_2 is denoted by $|d(q_1 + Uq_2U^*)|$. For example, if $q_1 = 20i + 80j + 190k$ and $q_2 = 170i + 180j + 90k$, then $Q(q_3) = 187.08$.

This color difference measure is used in calculating the difference between the central pixel and neighbors in the four directions at 0°, 45°, 90°, and 135° inside the filtering window. The average color difference Val_h between the central pixel and other pixel values in the direction $h = (1, 2, 3, 4)$ for respective degrees is calculated as follows:

$$Val_h = \frac{1}{4}\sum_{j=1}^{4}\left|d(x_{(N+1)/2}, x_j)\right| \quad (33)$$

with $x_{(N+1)/2}$ being the central pixel. The minimum value Val is given by

$$Val = \min(Val_1, Val_2, Val_3, Val_4) \quad (34)$$

which can be used for impulse detection. If the central pixel is an isolated impulse, then all the Val_h values will be quite high giving a higher Val. If the central pixel is an impulse, its Val value will be large. On the other hand, if the central pixel is a healthy one, its Val will be quite small. Therefore the Val value is compared with a predefined threshold Tol for impulse detection. If $Val \geq Tol$, the central pixel is replaced by quaternion vector median filter $Q^{(VMF)}$ else identity operation is performed. Hence, the output of the filter can be represented by the following equation:

$$x^{QVMF} = \begin{cases} Q^{(VMF)}, & \text{if } Val \geq Tol \\ x_{(N+1)/2}, & \text{otherwise.} \end{cases} \quad (35)$$

A similar quaternion filter is proposed in the work of Jin et al.[33] If the central pixel is detected as corrupted, it is replaced by the output of *WVMF*. The algorithm is defined as follows:

$$x^{QVMF} = \begin{cases} Q^{(WVMF)}, & \text{if } Val \geq Tol \\ x_{(N+1)/2}, & \text{otherwise.} \end{cases} \quad (36)$$

In the work of Jin et al.,[34] a quaternion-based switching filter is designed. It utilizes both the quaternion color difference estimate of Eq. (32) and four edge direction method. The number of similar pixels π_h in each direction Γ_h^* is computed. A pixel in a direction Γ_h^* is considered as a similar pixel if the color difference between it and the vector median of the color samples in each direction Γ_h^* is smaller than the predefined threshold. Then the edge direction k with maximum number of similar pixels π_k is computed as follows:

$$k = \arg\max\{Card(\pi_h), h = 1, 2, 3, 4\}$$

with $Card(\pi_h)$ representing the cardinality of the set π_h.

The color samples in Γ_k^* are used for checking if the central pixel is corrupted or not. The switching method is defined as follows:

$$q_{output} = \begin{cases} x_{\frac{N+1}{2}}, & \text{if } Card(\pi_h) \geq m \text{ and } \left|d\left(x_{\frac{N+1}{2}}, x_k^{(VM)}\right)\right| < Tol \\ q^{VMF}, & \text{otherwise} \end{cases}$$
$$(37)$$

where m is an integer, q^{VMF} is the classical VMF output on central pixel $x_{(N+1)/2}$ and $d(x_{(N+1)/2}, x_k^{(VM)})$ is the quaternion color difference between $x_{(N+1)/2}$ and $x_k^{(VM)}$, ($x_k^{(VM)}$ being the vector median of the pixels in the direction Γ_k^* with maximum number of similar pixels).

The chromaticity distance measure is unable to differentiate pixels in the gray line in which all the three components are equal. If $q_1 = 0i + 0j + 0k$ and $q_2 = 255i + 255j + 255k$ are black and white pixels lying in the gray axis, respectively, then $|d(q_1 + Uq_2U^*)| = 0$. It considers both the pixels to be equivalent which is not true. To overcome this problem, the intensity difference is incorporated in the distance measure.[35] The color pixel difference measure $d(q_1, q_2)$ is the summation of the intensity $d_1(q_1, q_2)$ and chromaticity $d_2(q_1, q_2)$ differences. Using this distance measure, the difference between the central and neighboring pixels in the four edge direction are calculated. If an impulse is detected, it is filtered by the output of the VMF.

A two-stage quaternion-based switching filter is proposed by Wang et al.[36] for color images corrupted by mixture noise such as impulse and Gaussian noise. It combines quaternion switching filter (QSF) with a nonlocal mean filter for efficient filtering. Another two-stage switching filter based on quaternion is also designed by Jin et al.[37] First an effective color distance measure based on quaternion is developed. Then using the four edge directional method, the filter classifies pixel into most probably noisy and noise free ones. The probably noisy pixel is again checked for an impulse by utilizing the concept of peer group. If an impulse is detected, it is replaced by the output of weighted vector median filter.

For removal of random-valued impulse noise from video sequences, the concept of quaternion is also extended in the work of Jin et al.[38] In order to measure the color difference between color samples, the luminance and chromaticity differences are expressed in quaternion form. The presence of impulse is detected based on this color difference by utilizing the color vectors along the horizontal, vertical, and diagonal directions in current frame and adjacent frames on motion trajectory. The corrupted pixels are replaced by three-dimensional weighted vector median output while the healthy pixels are left intact. A similar filter for removal of impulse noise from video sequences is also presented in the work of Elaiyaraja et al.[39]

In the work of Yin et al.[40], quaternion wavelet transform is developed which is based on the quaternion algebra, quaternion Fourier transform, and Hilbert transform by using four real discrete wavelet transform. Its application in image denoising gives a higher peak signal-to-noise ratio (PSNR) and better visual effect compared to current denoising methods.

PROPOSED SWITCHING VECTOR MEDIAN FILTER

In this section, the proposed switching vector median filter based on the quaternion theory is described. For calculating the intensity and chromaticity differences between two color pixels, the expressions given in Eqs. 30 and 31 are used. Let us consider pixels in a 3×3 filtering window of Fig. 1. First, the color difference between two pixels is calculated. Then the sum of aggregated color differences denoted by δ_i assigned to a sample q_i, where i = 1, 2, 3…N with the remaining pixels inside the filtering window having N pixels is computed as follows:

$$\delta_i = \sum_j^N d(q_i, q_j), (1 \leq j \leq N) \quad (38)$$

q_1	q_2	q_3
q_4	q_5	q_6
q_7	q_8	q_9

Fig. 1 A 3×3 filtering window

If the aggregated distance measures δ_1, δ_2, δ_3…δ_N are sorted in ascending order by assigning rank, that is,

$$\delta_{(1)} \leq \delta_{(2)} \leq \cdots \leq \delta_{(N)}$$

then it also implies similar ordering to the input set q_1, q_2, q_3,…q_N, resulting in an ordered input sequence:

$$q_{(1)} \leq q_{(2)} \leq \cdots \leq q_{(N)}$$

The lowest ranked sample $q_{(1)}$ is the output of classical VMF and their closest neighborhoods have similar values of the quaternion distance measure calculated to other remaining input pixels. If $q_{(1)}$, $q_{(2)}$, $q_{(3)}$…$q_{(k)}$ are a set of k input ranked samples having the highest similarity to input set, it acts as a good measurement for detection of noise corruption of the central pixel.[41] If the rank of the central pixel lies at the extreme end, it is regarded as an impulse. On the other hand, if it ranges in the middle then the central pixel is possibly not an impulse. To improve the performance of VMF, rank-conditioned vector median filter (RCVMF) is developed by Singh and Prabin[42] which gives the output if the rank of the central pixel (r_c) is larger than a predefined rank of a healthy vector pixel (r_k) inside the filtering window. This is expressed as follows:

$$x_{RCVMF} = \begin{cases} q_{VMF}, & \text{if } r_c > r_k \\ q_{\frac{(N+1)}{2}}, & \text{otherwise} \end{cases} \quad (39)$$

The main advantage is that pixels with extreme rank which are likely to possess abnormal value are processed and pixel replacement is conditioned to their ranks. The proposed method incorporates the advantage of rank conditioned filter. In the first stage of noise detection, if the central pixel has a rank bigger than the predefined rank r_k, it is regarded as probably corrupted and consider for second stage of noise detection. In the second stage, a threshold mechanism is added for further detection of impulse. The central pixel whose rank is bigger than a predefined rank may not always be the corrupted pixel since it can have smaller distance between the pixel q_k corresponding to the predefined rank r_k. As inspired from Dong and Xu,[43] the four edge directional method is applied for detection of noise. The color differences between the central pixel and pixels in the four directions, that is, 0°, 45°, 90°, and 135° inside the filtering window are computed. This color difference is the summation of both the intensity and chromaticity differences. If q_1 and q_2 are two pixels, then the distance measure is given as follows:

$$\begin{aligned} d(q_1, q_2) &= d_1(q_1, q_2) + d_2(q_1, q_2) \\ &= \frac{1}{2}\left[Tq_1T^* + T^*q_1T\right] - \frac{1}{2}\left[Tq_2T^* + T^*q_2T\right] \\ &\quad + \frac{1}{2}\left[Tq_1T^* - T^*q_1T\right] - \frac{1}{2}\left[Tq_2T^* - T^*q_2T\right]. \end{aligned} \quad (40)$$

If $m = \{1, 2, 3,$ and $4\}$ denotes the four directions in 0°, 45°, 90°, and 135° respectively, then the corresponding color pixel differences are defined as follows:

$$D_1 = \frac{1}{2}\big(d(q_5,q_4) + d(q_5,q_6)\big)$$

$$D_2 = \frac{1}{2}\big(d(q_5,q_3) + d(q_5,q_7)\big)$$

$$D_3 = \frac{1}{2}\big(d(q_5,q_2) + d(q_5,q_8)\big)$$

$$D_4 = \frac{1}{2}\big(d(q_5,q_1) + d(q_5,q_9)\big)$$

Then the minimum color difference of the four direction $D = \min(D_1, D_2, D_3, D_4)$ is compared with a predefined threshold Tol. The central pixel is concluded as corrupted if D is bigger than the predefined threshold Tol and is replaced by the output of the VMF computed using the quaternion representation defined by:

$$q_{VMF} = \min_{q_h \in \{q_1, q_2, \ldots q_N\}} \left\{ \sum_{l=1}^{N} |q_l - q_h| \right\} \quad (41)$$

When the central pixel is on the edge, it is expected that at least one value among them is extremely small; if the central pixel is on the smooth area, the values are approximately equal to zero. On the other hand, if the current pixel is an impulse, all the four values will become very large. Hence, this second stage of noise detection further enhances the noise detection capability. Based on the above formulation, the proposed filter (PF) has the following form:

$$q_{PF} = \begin{cases} q_{VMF}, & \text{if } r_c > r_k \text{ and } D > Tol \\ q_5, & \text{otherwise} \end{cases} \quad (42)$$

Here, q_1, q_2, q_3, and q_4 take the previously filtered results, while filtering the central pixel q_5 to reduce the adverse effects of the noisy neighbors and q_5, q_6, q_7, and q_8 are the original pixels.

The PF combines the advantages of rank conditioned filters[41,42] such as edge preservation and minimization of outliers, and threshold mechanism in addition to the quaternion benefits of color representation. Rank conditioned filters always give the output as one of the pixels inside the filtering window. Thus, no new intermediate points are introduced which results in edge preservation. The nature of rank ordered data helps in limiting the effect of outliers since outliers tend to occupy the extreme rank in the ordered data. The other common method of distance measure such as Euclidean distance considers only the intensity component of color image. However, it is possible that there exists large color discontinuity among the pixels with only small intensity difference.[44] It becomes difficult to differentiate impulse from normal pixels in such situations. Hence, it is necessary to include the chromatic information in addition to the intensity measure. By expressing a color pixel in quaternion form, it is possible to retain orientation information relative to the mid-gray line axis. This corresponds to the chromatic or hue-related information. Quaternion treats the color pixel as a single entity rather than as separated chrominance and luminance components, or separated color space channels such as red, green, or blue. Thus, accuracy in high color information is achieved. This also preserve the nature of vector pixel.

EXPERIMENTAL RESULTS AND DISCUSSIONS

To show the filtering performance, the PF is compared with eight different filters using ten different color images of size 512 × 512. The filters are VMF,[5] AVMF,[13] entropy VMF (EVMF),[20] PGF,[16] ACWVMF,[14] and

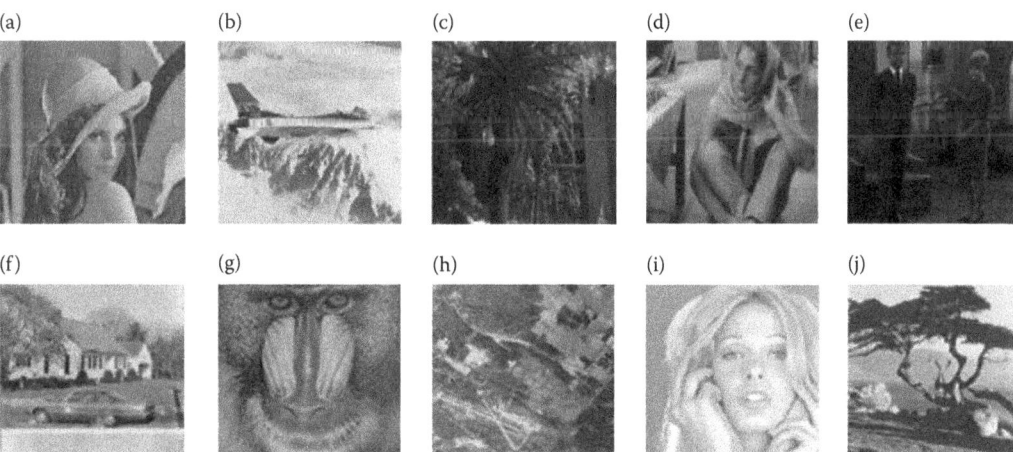

Fig. 2 Standard test images: (a) Lena; (b) Airplane; (c) Aptus; (d) Barbara; (e) Couple; (f) House; (g) Mandrill; (h) Miramar; (i) Tiffany; (j) Tree

fuzzy VMF (FVMF).[22] adaptive rank weighted switching filter (ARWSF),[45] and QSF.[34] The ten images are given in Fig. 2.

The impulse noise model proposed by Viero et. al.[1] is used for simulation. It has the following form:

$$q^k = \begin{cases} \{q^1 \quad q^2 \quad q^3\} \text{ with probability } 1-p \\ \{n^1, \quad q^2, \quad q^3\} \text{ with probability } p_1 p \\ \{q^1, \quad n^2, \quad q^3\} \text{ with probability } p_2 p \\ \{n^1, \quad n^2, \quad n^3\} \text{ with probability } p_a p \\ \{n^1, \quad n^2, \quad n^3\} \text{ with probability } p_a p \end{cases} \quad (43)$$

where n^k(k = 1, 2, 3) is either 0 or 255 in case of fixed-valued impulse noise having equal probability, whereas for random-valued impulse noise, it can take any value in the range [0, 255]: p represents the corruption probability of the pixel; p_1, p_2, and p_3 are the corruption probabilities of the channel; and $p_a = 1 - p_1 - p_2 - p_3$.

The objective evaluation criteria used for comparing other filters are mean absolute error (MAE), PSNR in dB, normalized color distance (NCD), and structural similarity index (SSIM). They are described in the sections that follow.

Mean Absolute Error

MAE is given by

$$MAE = \frac{1}{3 \times M_1 \times N_1} \sum_{i=1}^{M_1} \sum_{j=1}^{N_1} |q(i,j) - f(i,j)| \quad (44)$$

where $M_1 \times N_1$ is the size of the image, $q(i,j)$ and $f(i,j)$ are the original and filtered pixel values at (i,j) location.

Peak Signal-to-Noise Ratio

PSNR is given by

$$PSNR = 10 \log_{10}\left(\frac{I_{max}^2}{MSE}\right) \quad (45)$$

where I_{max} is the maximum pixel value of the original image, and MSE stands for mean squared error given as

$$MSE = \frac{1}{3 \times M_1 \times N_1} \sum_{i=1}^{M_1} \sum_{j=1}^{N_1} (q(i,j) - f(i,j))^2 \quad (46)$$

Normalized Color Distance

NCD is given by

$$\frac{\sum_{i=1}^{M_1} \sum_{j=1}^{N_1} \sqrt{(L^o(i,j) - L^x(i,j))^2 + (u^o(i,j) - u^x(i,j))^2 + (v^o(i,j) - v^x(i,j))^2}}{\sum_{i=1}^{M_1} \sum_{j=1}^{N_1} \sqrt{(L^o(i,j))^2 + (u^o(i,j))^2 + (v^o(i,j))^2}} \quad (47)$$

which is defined in the Lu*v* color space where $L^o(i,j)$, $u^o(i,j)$, $v^o(i,j)$ and $L^x(i,j)$, $u^x(i,j)$, $v^x(i,j)$ are values of the lightness and two chrominance components of the original image sample $q(i,j)$ and filtered image sample $f(i,j)$, respectively.

Structural Similarity Index

SSIM is given by

$$SSIM = \frac{(2\mu_x\mu_y + C_1)(2\mu_{xy} + C_2)}{(\mu_x^2 + \mu_y^2 + C_1)(\sigma_x^2 + \sigma_y^2 + C_2)} \quad (48)$$

which measures the similarity between two images. where, μ_x and μ_y are the means of the original and filtered images; σ_x^2 and σ_y^2 represent the corresponding covariance and variance of the original and filtered images, respectively; and C_1 and C_2 are the constants.[46]

The parameters which give the best compromised results used for all the vector filters is given in Table 1.

The filters are used for comparison due to the following reasons: VMF is the most common traditional vector filter which is used as a reference in many filtering methods. ACWVMF is an efficient filter with better filtering performance as compared with other filters. QSF is a good filter which is based on quaternion utilizing the four edge directional operators. AVMF and PGF are fast computational filters with good performance. EVMF are adaptive filters based on the local contrast entropy of the

Table 1 Parameters used for various filters

Filters	Parameters
AVMF	$T = 35$
FVMF	$\delta = 0.5$
PGF	$T = 60$
ACWVMF	$w = 3, T = 80$
ARWSF	$T = 65$
QSF	$T = 35$
Proposed	$r_h = 5, T = 35$

Table 2 Comparison of different filters in removal of fixed-valued impulse noise from Lena

Filter	Impulse noise ratio																				
	10%			20%			30%			40%			50%			60%			70%		
	NCD	PSNR	MAE	NCD	PSNR	MAE	NCD	PSNR	MAE	NCD	PSNR	MAE	NCD	PSNR	MAE	NCD	PSNR	MAE	NCD	PSNR	MAE
VMF	0.0216	31.76	3.72	0.0134	31.34	3.94	0.0189	30.82	4.18	0.0181	30.10	4.48	0.0189	29.32	4.84	0.0205	28.40	5.26	0.0223	27.34	5.81
AVMF	0.0055	34.95	1.00	0.0096	33.85	1.47	0.0088	32.67	1.98	0.0100	31.47	2.55	0.0113	30.33	3.16	0.0130	28.98	3.87	0.0215	27.69	4.69
EVMF	0.0078	33.41	2.57	0.0078	33.16	2.54	0.0084	32.41	2.69	0.0178	31.22	3.00	0.0187	29.87	3.45	0.0200	28.26	4.10	0.0218	26.64	4.97
FVMF	0.0188	32.13	3.69	0.0193	31.67	3.88	0.0198	31.05	4.13	0.0193	30.30	4.43	0.0202	29.48	4.81	0.0213	28.41	5.33	0.0220	27.29	6.03
PGF	0.0051	35.50	0.87	0.0062	34.01	1.39	0.0171	32.56	1.95	0.0179	31.10	2.57	0.0190	30.01	3.23	0.0202	28.58	3.98	0.0219	27.10	4.86
ACWVMF	0.0052	36.19	0.76	0.0042	34.85	1.22	0.0166	33.57	1.72	0.0173	32.15	2.29	0.0182	30.92	2.89	0.0193	29.43	3.61	0.0209	28.00	4.44
RWASF	0.0042	38.61	0.52	0.0156	35.16	1.06	0.0166	32.36	1.70	0.0179	29.71	2.52	0.0200	27.10	3.65	0.0233	24.55	5.29	0.0286	22.11	7.72
QSF	0.0047	35.73	0.83	0.0061	33.55	1.47	0.0081	31.56	2.23	0.0192	29.83	3.11	0.0210	28.26	4.15	0.0237	26.51	5.50	0.0270	25.03	7.08
Proposed	0.0040	37.92	0.59	0.0156	35.64	1.07	0.0069	34.02	1.63	0.0171	32.83	2.19	0.0180	31.22	2.80	0.0190	29.88	3.49	0.0205	28.16	4.32

Table 3 Comparison of different filters in removal of fixed-valued impulse noise from Airplane

Filter	Impulse noise ratio																				
	10%			20%			30%			40%			50%			60%			70%		
	NCD	PSNR	MAE	NCD	PSNR	MAE	NCD	PSNR	MAE	NCD	PSNR	MAE	NCD	PSNR	MAE	NCD	PSNR	MAE	NCD	PSNR	MAE
VMF	0.0086	28.15	3.28	0.0088	27.85	3.52	0.0096	27.49	3.57	0.0147	27.02	4.13	0.0163	26.42	4.56	0.0373	25.62	5.10	0.0383	24.68	5.79
AVMF	0.0191	28.77	1.38	0.0195	28.40	1.79	0.0200	27.99	2.21	0.0205	27.44	2.72	0.0363	26.78	3.30	0.0370	25.89	4.03	0.0380	24.86	4.91
EVMF	0.0170	28.72	2.36	0.0351	28.49	2.42	0.0352	28.08	2.64	0.0353	27.47	3.00	0.0355	26.66	3.53	0.0359	25.62	4.30	0.0365	24.45	5.28
FVMF	0.0089	28.20	3.36	0.0095	27.89	3.60	0.0105	27.52	3.89	0.0191	27.02	4.28	0.0213	26.36	4.77	0.0277	25.50	5.43	0.0361	24.56	6.31
PGF	0.0193	28.83	1.31	0.0197	28.37	1.77	0.0354	27.86	2.27	0.0358	27.16	2.84	0.0362	26.47	3.50	0.0370	25.54	4.29	0.0381	24.38	5.24
ACWVMF	0.0025	29.13	1.10	0.0136	28.77	1.49	0.0152	28.35	1.91	0.0357	27.82	2.40	0.0361	27.17	3.02	0.0368	26.38	3.71	0.0376	25.44	4.54
RWASF	0.0180	31.30	0.78	0.0344	29.98	1.36	0.0343	28.59	2.09	0.0343	26.92	3.08	0.0343	25.07	4.48	0.0346	23.14	6.46	0.0356	21.01	9.39
QSF	0.0349	30.48	1.12	0.0352	29.52	1.72	0.0357	28.42	2.44	0.0364	27.22	3.35	0.0370	25.92	4.48	0.0388	24.55	5.91	0.0409	23.19	7.77
Proposed	0.0346	31.36	0.79	0.0347	30.52	1.22	0.0348	29.56	1.71	0.0351	28.10	2.34	0.0352	28.07	2.85	0.0355	26.86	3.60	0.0359	25.80	4.48

Table 4 Comparison of different filters in removal of for fixed-valued impulse noise from Aptus

Filters	Impulse noise ratio																					
	10%			20%			30%			40%			50%			60%			70%		
	NCD	PSNR	MAE	NCD	PSNR	MAE	NCD	PSNR	MAE	NCD	PSNR	MAE	NCD	PSNR	MAE	NCD	PSNR	MAE	NCD	PSNR	MAE
VMF	0.0338	25.32	6.34	0.0557	25.01	6.68	0.0380	24.63	7.09	0.0408	24.16	7.59	0.0443	23.61	8.19	0.0486	22.89	8.97	0.0536	22.16	9.86
AVMF	0.0193	26.09	4.04	0.0232	25.62	4.64	0.0274	25.12	5.28	0.0322	24.52	6.03	0.0375	23.89	6.86	0.0438	23.09	7.87	0.0511	22.28	9.04
EVMF	0.0243	26.82	4.54	0.0250	26.58	4.65	0.0270	26.11	4.94	0.0304	25.48	5.45	0.0352	24.72	6.14	0.0416	23.77	7.07	0.0498	22.76	8.28
FVMF	0.0349	25.34	6.50	0.0373	24.94	6.93	0.0401	24.48	7.41	0.0434	23.96	7.99	0.0473	23.41	8.66	0.0521	22.73	9.50	0.0579	22.02	10.51
PGF	0.0172	26.30	3.68	0.0221	25.70	4.41	0.0272	25.13	5.17	0.0328	24.42	6.03	0.0391	23.67	6.98	0.0462	22.77	8.09	0.0538	22.03	9.33
ACWVMF	0.0147	26.88	3.17	0.0187	26.34	3.79	0.0232	25.75	4.47	0.0282	25.06	5.26	0.0338	24.35	6.13	0.0403	23.51	7.17	0.0483	22.59	8.58
RWASF	0.0108	28.21	2.05	0.0173	26.92	2.90	0.0254	25.55	4.12	0.0355	24.14	5.47	0.0481	22.75	7.11	0.0649	21.18	9.30	0.0877	19.63	12.15
QSF	0.0148	26.63	3.21	0.0211	25.73	4.20	0.0285	24.84	5.33	0.0369	23.94	6.61	0.0467	23.01	8.10	0.0580	22.06	9.85	0.0714	21.02	11.97
Proposed	0.0106	28.36	2.18	0.0156	27.42	2.91	0.0213	26.42	3.71	0.0284	25.21	4.66	0.0348	24.66	5.79	0.0425	23.73	6.93	0.0519	22.66	8.25

Table 5 Comparison of different filters in removal of fixed-valued impulse noise from Barbara

Filters	Impulse noise ratio																					
	10%			20%			30%			40%			50%			60%			70%		
	NCD	PSNR	MAE	NCD	PSNR	MAE	NCD	PSNR	MAE	NCD	PSNR	MAE	NCD	PSNR	MAE	NCD	PSNR	MAE	NCD	PSNR	MAE
VMF	0.0324	23.01	10.23	0.0329	22.92	10.39	0.0336	22.79	10.62	0.0360	22.67	10.84	0.0354	22.50	11.14	0.0471	22.23	11.53	0.0482	21.85	12.06
AVMF	0.0245	23.70	7.61	0.0296	23.51	8.07	0.0331	23.31	8.54	0.0343	23.07	9.08	0.0336	22.82	9.61	0.0469	22.46	10.33	0.0483	22.01	11.15
EVMF	0.0284	23.86	8.45	0.0285	23.89	8.14	0.0317	23.82	8.07	0.0324	23.61	8.25	0.0336	23.30	8.64	0.0457	22.81	9.30	0.0479	22.19	10.22
FVMF	0.0300	23.21	10.31	0.0308	23.11	10.54	0.0322	22.99	10.81	0.0353	22.82	11.14	0.0383	22.64	11.54	0.0396	22.34	12.09	0.0478	21.94	12.85
PGF	0.0265	24.04	6.63	0.0280	23.74	7.33	0.0434	23.44	8.03	0.0445	23.11	8.75	0.0456	22.78	9.46	0.0469	22.41	10.28	0.0487	21.90	11.19
ACWVMF	0.0188	24.18	6.64	0.0263	23.93	7.19	0.0301	23.68	7.7406	0.0314	23.40	8.34	0.0329	23.10	8.97	0.0330	22.74	9.87	0.0345	22.37	10.66
RWASF	0.0197	25.76	3.75	0.0389	24.95	4.91	0.0411	24.13	6.13	0.0439	23.27	7.47	0.0476	22.26	9.05	0.0528	21.11	10.97	0.0605	19.69	13.58
QSF	0.0206	24.22	6.07	0.0269	23.71	7.13	0.0315	23.26	8.19	0.0461	22.77	9.32	0.0485	22.23	10.59	0.0515	21.60	12.60	0.0553	20.88	13.85
Proposed	0.0169	25.35	4.67	0.0396	24.82	5.51	0.0409	24.25	6.35	0.0426	23.54	7.25	0.0303	23.27	8.44	0.0457	22.83	9.33	0.0477	22.49	10.24

Table 6 Comparison of different filters in removal of fixed-valued impulse noise from Couple

Filters	10%			20%			30%			Impulse noise ratio 40%			50%			60%			70%		
	NCD	PSNR	MAE	NCD	PSNR	MAE	NCD	PSNR	MAE	NCD	PSNR	MAE	NCD	PSNR	MAE	NCD	PSNR	MAE	NCD	PSNR	MAE
VMF	0.0226	39.99	1.08	0.0237	38.97	1.25	0.0253	37.83	1.46	0.0276	36.16	1.74	0.0302	34.71	2.25	0.0361	33.37	2.61	0.0392	31.98	3.03
AVMF	0.0048	42.28	0.29	0.0103	39.42	0.574	0.0150	37.59	0.88	0.0206	35.69	1.24	0.0279	34.54	1.65	0.0329	33.09	2.13	0.0408	31.61	2.72
EVMF	0.0156	42.34	0.64	0.0139	40.35	0.77	0.0168	38.14	0.98	0.0208	35.80	1.283	0.0229	33.62	1.68	0.0286	35.19	1.97	0.0298	33.62	2.36
FVMF	0.0213	40.66	1.25	0.024E	39.40	1.44	0.0266	38.06	1.66	0.0281	36.37	1.95	0.0309	34.67	2.30	0.0350	32.58	2.79	0.0457	30.77	3.43
PGF	0.0130	41.61	0.31	0.0115	38.55	0.62	0.0290	36.43	0.96	0.0345	34.60	1.35	0.0406	32.81	1.81	0.0479	30.58	2.39	0.0568	28.64	3.10
ACWVMF	0.0047	43.74	0.25	0.0103	40.561	0.52	0.0175	38.46	0.81	0.0227	36.36	1.16	0.0279	35.30	1.53	0.0354	32.37	2.07	0.0513	30.36	2.71
RWASF	0.0139	40.47	0.33	0.0206	36.53	0.72	0.0288	33.89	1.20	0.0455	31.36	1.84	0.0588	28.99	2.69	0.0765	26.60	3.87	0.1001	24.31	5.52
QSF	0.0131	42.05	0.31	0.0125	38.34	0.68	0.0193	35.78	1.12	0.0277	33.49	1.69	0.0471	31.49	2.39	0.0576	29.60	3.29	0.0710	27.79	4.46
Proposed	0.0128	43.03	0.27	0.0113	39.78	0.55	0.0169	38.56	0.88	0.0237	36.62	1.26	0.0416	35.41	1.70	0.0493	37.69	2.23	0.0589	30.52	2.91

Table 7 Comparison of different filters in removal of fixed-valued impulse noise from House

Filters	Impulse noise ratio																				
	10%			20%			30%			40%			50%			60%			70%		
	NCD	PSNR	MAE	NCD	PSNR	MAE	NCD	PSNR	MAE	NCD	PSNR	MAE	NCD	PSNR	MAE	NCD	PSNR	MAE	NCD	PSNR	MAE
VMF	0.0167	26.20	5.02	0.0166	25.86	5.37	0.0174	25.45	5.79	0.0198	25.00	6.28	0.0196	24.40	6.919	0.0207	23.72	7.64	0.0323	22.92	8.60
AVMF	0.0116	27.03	2.52	0.01328	26.59	3.09	0.0151	26.08	3.75	0.01648	25.53	4.46	0.0180	24.85	5.32	0.0198	24.08	6.29	0.0315	23.18	7.52
EVMF	0.0121	26.98	3.57	0.0258	26.75	3.66	0.0261	26.36	3.94	0.0267	25.82	4.38	0.0276	25.10	5.06	0.0287	24.21	5.96	0.0304	23.15	7.18
FVMF	0.0164	26.33	5.16	0.0174	25.97	5.55	0.0184	25.55	6.03	0.0259	25.06	6.58	0.0271	24.43	7.31	0.0281	23.71	8.20	0.0318	22.84	9.38
PGF	0.0114	27.18	2.30	0.0129	26.64	2.97	0.0265	26.06	3.70	0.0275	25.42	4.50	0.0286	24.67	5.42	0.0301	23.80	6.46	0.0320	22.93	7.75
ACWVMF	0.0099	27.59	1.85	0.0120	27.10	2.44	0.0259	26.59	3.09	0.0268	25.98	3.82	0.0278	25.28	4.66	0.0290	24.47	5.65	0.0308	23.66	6.95
RWASF	0.0093	28.25	1.23	0.0243	27.40	1.98	0.0252	26.45	2.8	0.0265	25.18	4.00	0.0283	23.69	5.53	0.0310	22.01	7.6	0.0354	20.09	10.74
QSF	0.0253	27.09	2.22	0.0264	26.24	3.17	0.0279	25.28	4.30	0.0297	24.34	5.59	0.0320	23.26	7.20	0.0350	22.15	9.16	0.0388	20.94	11.63
Proposed	0.0241	27.93	1.48	0.0247	27.28	2.11	0.0247	27.28	2.11	0.0263	26.06	3.66	0.0273	25.36	4.54	0.0286	24.52	5.51	0.0303	23.50	6.73

Color Image Denoising

Table 8 Comparison of different filters in removal of fixed-valued impulse noise from Mandrill

	Impulse noise ratio																				
	10%				20%				30%				40%				50%			
Filters	NCD	PSNR	MAE	NCD	PSNR	MAE	NCD	PSNR	MAE	NCD	PSNR	MAE	NCD	PSNR	MAE	NCD	PSNR	MAE			
VMF	0.0574	22.33	11.56	0.0551	22.14	11.97	0.0566	21.88	12.48	0.0540	21.62	13.03	0.0412	21.29	13.72						
AVMF	0.0473	22.75	8.89	0.0481	22.51	9.54	0.0437	22.22	10.28	0.0480	21.92	11.08	0.0358	21.56	12.00						
EVMF	0.0369	23.06	9.27	0.0337	22.98	9.24	0.0299	22.78	9.47	0.0451	22.49	9.93	0.0464	22.08	10.65						
FVMF	0.0572	22.29	11.86	0.0534	22.04	12.37	0.0595	21.76	12.95	0.0575	21.46	13.61	0.0469	21.10	14.38						
PGF	0.0413	22.96	8.01	0.0428	22.63	8.94	0.0462	22.28	9.90	0.0476	21.92	10.88	0.0493	21.51	11.96						
ACWVMF	0.0425	23.24	7.33	0.0422	22.95	8.11	0.0371	22.61	8.97	0.0305	22.28	9.87	0.0328	21.86	10.94						
RWASF	0.0154	24.18	4.83	0.0197	23.53	6.04	0.0422	22.76	7.43	0.0451	21.94	8.98	0.0489	20.97	10.85						
QSF	0.0336	23.47	6.50	0.0348	22.80	8.01	0.0384	22.18	9.54	0.0418	21.57	11.20	0.0519	20.93	13.05						
Proposed	0.0180	24.31	5.15	0.0235	23.75	6.24	0.0416	23.13	7.33	0.0437	22.39	8.56	0.0462	22.05	10.21						

	60%			70%		
Filters	NCD	PSNR	MAE	NCD	PSNR	MAE
VMF	0.0418	20.89	14.55	0.0543	20.37	15.65
AVMF	0.0385	21.11	13.08	0.0532	20.57	14.43
EVMF	0.0481	21.54	11.66	0.0506	20.87	13.02
FVMF	0.0489	20.70	15.29	0.0555	20.21	16.45
PGF	0.0512	21.00	13.16	0.0538	20.36	14.61
ACWVMF	0.0359	21.39	12.12	0.0484	20.79	13.86
RWASF	0.0538	19.88	13.13	0.0613	18.52	16.24
QSF	0.0555	20.27	15.08	0.0597	19.49	17.52
Proposed	0.0485	21.47	11.57	0.0513	20.72	13.09

Table 9 Comparison of different filters in removal of fixed-valued impulse noise from Miramar

										Impulse noise ratio											
	10%			20%			30%			40%			50%			60%			70%		
Filters	NCD	PSNR	MAE	NCD	PSNR	MAE	NCD	PSNR	MAE	NCD	PSNR	MAE	NCD	PSNR	MAE	NCD	PSNR	MAE	NCD	PSNR	MAE
VMF	0.0276	25.73	8.40	0.0287	25.41	8.78	0.0299	25.06	9.20	0.0313	24.65	9.70	0.0355	24.17	10.28	0.0482	23.62	10.96	0.0501	22.95	11.85
AVMF	0.0261	27.00	4.80	0.0273	26.52	5.52	0.0314	25.99	6.30	0.0312	25.41	7.15	0.0333	24.77	8.11	0.0473	24.10	9.15	0.0492	23.32	10.40
EVMF	0.0280	26.85	6.32	0.0297	26.76	6.22	0.0302	26.49	6.38	0.0430	26.03	6.75	0.0442	25.36	7.38	0.0457	24.50	8.29	0.0480	23.49	9.49
FVMF	0.0280	25.76	8.55	0.0290	25.42	8.95	0.0306	25.02	9.42	0.0315	24.55	9.98	0.0350	24.01	10.68	0.0424	23.4	11.50	0.0510	22.71	12.57
PGF	0.0268	27.73	3.79	0.0284	26.98	4.77	0.0434	26.24	5.79	0.0446	25.49	6.85	0.0460	24.7	7.96	0.0476	23.88	9.18	0.0496	23.13	10.52
ACWVMF	0.0108	27.99	3.54	0.0133	27.36	4.33	0.0211	26.74	5.19	0.0283	26.04	6.12	0.0304	25.32	7.14	0.0462	24.53	8.29	0.0483	23.77	9.76
RWASF	0.0228	30.36	1.93	0.0397	28.83	2.95	0.0412	27.30	4.10	0.0433	25.71	5.44	0.0461	24.05	7.10	0.0505	22.19	9.31	0.0574	20.22	12.34
QSF	0.0409	27.90	3.51	0.0424	26.72	4.79	0.0441	25.61	6.20	0.0463	24.57	7.76	0.0489	23.48	9.58	0.0523	22.39	11.70	0.0566	21.26	14.25
Proposed	0.0396	29.17	2.65	0.0405	28.15	3.58	0.0415	27.13	4.52	0.0428	26.40	5.73	0.0439	25.61	6.76	0.0455	24.64	7.93	0.0474	23.78	9.22

Table 10 Comparison of different filters in removal of fixed-valued impulse noise from Tiffany

Filters	Impulse noise ratio																			
	10%				20%				30%				40%				50%			
	NCD	PSNR	MAE		NCD	PSNR	MAE		NCD	PSNR	MAE		NCD	PSNR	MAE		NCD	PSNR	MAE	
VMF	0.0057	27.64	3.67		0.0059	27.56	3.79		0.0062	27.37	3.96		0.0065	27.14	4.15		0.0069	26.87	4.37	
AVMF	0.0026	28.47	1.35		0.0031	28.25	1.70		0.0036	27.99	2.07		0.0042	27.66	2.47		0.0049	27.28	2.91	
EVMF	0.0045	28.14	8.79		0.0044	28.09	2.73		0.0046	27.89	2.81		0.0049	27.53	3.00		0.0054	27.05	3.31	
FVMF	0.0057	27.74	3.65		0.0059	27.64	3.77		0.0062	27.51	3.92		0.0065	27.33	4.11		0.0069	27.08	4.35	
PGF	0.0024	28.55	1.25		0.0029	28.30	1.63		0.0036	27.91	2.04		0.0043	27.44	2.50		0.0051	27.02	2.98	
ACWVMF	0.0023	28.71	1.16		0.0028	28.48	1.51		0.0034	28.20	1.89		0.0040	27.81	2.28		0.0047	27.47	2.73	
RWASF	0.0019	29.10	0.90		0.0026	28.56	1.35		0.0034	27.78	1.90		0.0044	26.74	2.60		0.0058	25.44	3.52	
QSF	0.0023	28.65	1.20		0.0031	28.23	1.68		0.0039	27.70	2.24		0.0048	27.19	2.83		0.0059	26.60	3.53	
Proposed	0.0020	29.01	0.97		0.0026	28.66	1.37		0.0032	28.18	1.80		0.0039	27.92	2.24		0.0046	27.53	2.71	

Filters	60%			70%		
	NCD	PSNR	MAE	NCD	PSNR	MAE
VMF	0.0074	26.51	4.64	0.0080	26.24	5.07
AVMF	0.0057	26.81	3.39	0.0065	26.52	3.93
EVMF	0.0061	26.35	3.75	0.0071	25.39	4.36
FVMF	0.0075	26.72	4.66	0.0083	26.10	5.14
PGF	0.0060	26.40	3.53	0.0072	25.37	4.21
ACWVMF	0.0054	27.18	3.23	0.0063	26.67	3.77
RWASF	0.0078	23.88	4.82	0.0108	22.05	6.74
QSF	0.0073	25.82	4.39	0.0091	24.88	5.49
Proposed	0.0054	26.98	3.22	0.0063	26.52	3.78

Table 11 Comparison of different filters in removal of fixed-valued impulse noise from Tree

Filters	10%			20%			30%			40%			50%			60%			70%		
	NCD	PSNR	MAE	NCD	PSNR	MAE	NCD	PSNR	MAE	NCD	PSNR	MAE	NCD	PSNR	MAE	NCD	PSNR	MAE	NCD	PSNR	MAE
VMF	0.0079	34.38	2.27	0.0106	32.13	3.17	0.0100	32.14	3.05	0.0170	30.75	3.56	0.0186	29.10	4.22	0.0291	27.52	5.02	0.0314	25.85	6.04
AVMF	0.0118	37.70	0.638	0.0131	35.51	1.08	0.0141	33.65	1.61	0.0250	31.72	2.26	0.0263	29.81	3.04	0.0279	27.95	3.99	0.0302	26.17	5.19
EVMF	0.0235	36.6	1.37	0.0237	35.51	1.56	0.0242	33.93	1.89	0.0249	32.02	2.36	0.0261	29.97	3.02	0.0277	28.01	3.89	0.0302	26.02	5.07
FVMF	0.0085	34.83	2.52	0.0095	33.61	2.88	0.0108	32.23	3.33	0.0151	30.63	3.90	0.0172	28.83	4.64	0.0215	27.11	5.60	0.0317	25.40	6.85
PGF	0.0118	37.50	0.65	0.0132	35.01	1.15	0.0242	32.97	1.73	0.0253	30.96	2.43	0.0266	29.20	3.28	0.0284	27.37	4.29	0.0309	25.61	5.55
ACWVMF	0.0115	40.09	0.43	0.0126	37.60	0.81	0.0237	35.51	1.25	0.0244	33.36	1.81	0.0255	31.28	2.52	0.0270	28.99	3.37	0.0291	27.44	4.50
RWASF	0.0228	38.43	0.47	0.0233	34.80	0.98	0.0244	31.72	1.68	0.0260	28.99	2.62	0.0286	26.38	3.98	0.0330	23.77	6.00	0.0406	21.26	9.03
QSF	0.0230	37.73	0.60	0.0239	34.33	1.23	0.0253	31.30	2.13	0.0273	28.77	3.29	0.0303	26.47	4.82	0.0340	24.52	6.71	0.0399	22.50	9.41
Proposed	0.0227	40.04	0.42	0.0231	37.17	0.81	0.0236	34.90	1.30	0.0243	32.84	1.87	0.0253	30.90	2.57	0.0266	28.91	3.44	0.0287	26.69	4.59

Table 12 Comparison of filters based on average NCD, PSNR, and MAE for ten images corrupted with fixed-valued impulse noise

Filter	\multicolumn{3}{c}{10%}			\multicolumn{3}{c}{20%}			\multicolumn{3}{c}{30%}			\multicolumn{3}{c}{40%}			\multicolumn{3}{c}{50%}			\multicolumn{3}{c}{60%}			\multicolumn{3}{c}{70%}		
	NCD	PSNR	MAE	NCD	PSNR	MAE	NCD	PSNR	MAE	NCD	PSNR	MAE	NCD	PSNR	MAE	NCD	PSNR	MAE	NCD	PSNR	MAE
VMF	0.0234	28.45	5.55	0.0237	27.91	5.88	0.0245	27.54	6.14	0.0266	26.92	6.55	0.0267	26.23	7.04	0.0337	25.47	7.6280	0.0378	24.63	8.37
AVMF	0.0173	29.87	3.25	0.0197	29.01	3.74	0.0212	28.26	4.29	0.0242	27.44	4.90	0.0265	26.66	5.59	0.0313	25.75	6.4130	0.0370	24.82	7.39
EVMF	0.0198	29.67	4.78	0.0233	29.25	4.19	0.0234	28.60	4.38	0.0281	27.74	4.74	0.0296	26.77	5.28	0.0328	26.02	6.0290	0.0352	24.85	7.02
FVMF	0.0230	28.64	5.68	0.0241	28.16	5.99	0.0255	27.61	6.35	0.0282	26.95	6.79	0.0291	26.20	7.34	0.0324	25.32	8.0390	0.0388	24.38	8.95
PGF	0.0175	30.02	2.88	0.0188	28.99	3.49	0.0296	28.08	4.15	0.0315	27.16	4.87	0.0336	26.28	5.65	0.0361	25.23	6.5650	0.0393	24.14	7.66
ACWVMF	0.0123	30.77	2.62	0.0156	29.79	3.14	0.0214	28.94	3.70	0.0249	28.02	4.33	0.0270	27.20	5.05	0.0308	26.09	5.9090	0.0356	25.10	6.97
ARWSF	0.0139	31.46	1.67	0.0236	29.66	2.42	0.0283	28.05	3.31	0.0322	26.39	4.40	0.0367	24.67	5.79	0.0428	22.83	7.6750	0.0518	20.88	10.34
QSF	0.0213	30.39	2.55	0.0232	28.91	3.40	0.0263	27.59	4.37	0.0316	26.36	5.47	0.0369	25.16	6.78	0.0414	23.94	8.4190	0.0470	22.69	10.34
Proposed	0.0185	31.64	1.91	0.0228	30.31	2.53	0.0255	29.34	3.13	0.0288	28.19	3.97	0.0307	27.40	4.83	0.0347	26.75	5.7250	0.0379	25.08	6.76

Impulse noise ratio

Table 13 Ranking of filters based on average of PSNR, MAE, and NCD

Rank	Fixed-valued impulse noise					
	Filter	PSNR	Filter	MAE	Filter	NCD
1	PF	28.38	PF	4.12	ACWVMF	0.0239
2	ACWVMF	27.98	ACWVMF	4.53	AVMF	0.0253
3	EVMF	27.55	PGF	5.03	EVMF	0.0275
4	AVMF	27.40	AVMF	5.08	VMF	0.0281
5	PGF	27.12	ARWSF	5.09	PF	0.0284
6	FVMF	26.75	EVMF	5.20	FVMF	0.0287
7	VMF	26.73	QSF	6.73	PGF	0.0295
8	QSF	26.43	FVMF	7.01	QSF	0.0325
9	ARWSF	26.27	VMF	8.00	ARWSF	0.0328

Table 14 SSIM values on Lena

Noise	Fixed-valued impulse noise						
	VMF	AVMF	PGF	ACWVMF	ARWSF	QSF	Proposed
10%	0.9938	0.9970	0.9974	0.9978	0.9987	0.9975	0.9985
20%	0.9931	0.9962	0.9964	0.9969	0.9972	0.9959	0.9975
30%	0.9922	0.9950	0.9948	0.9959	0.9946	0.9935	0.9963
40%	0.9908	0.9934	0.9928	0.9943	0.9902	0.9903	0.9948
50%	0.9890	0.9914	0.9901	0.9925	0.9825	0.9860	0.9930
60%	0.9864	0.9883	0.9856	0.9894	0.9692	0.9790	0.9896
70%	0.9827	0.9842	0.9790	0.9853	0.9480	0.9704	0.9859
Average	0.9897	0.9922	0.9909	0.9932	0.9829	0.9875	0.9937

pixels, and FVMF are vector filters which deal with the imprecision in the image regions. Both are used for comparison with many existing filters. ARWSF is an efficient impulse noise reduction filter based on the rank weighted, aggregated pixel dissimilarity measures. All the filters except QSF and the PF are implemented using the Euclidean distance.

Tables 2–11 show the comparison of the PF with other filters in terms of PSNR, NCD, and MAE. The images are corrupted with fixed-valued impulse noise with noise density ranging from 10% to 70%. The average values of the parameters for the 10 images are compared in Table 12. It is observed that the average values of NCD, PSNR, and MAE of the PF at 20% impulse noise are 0.0228, 30.31, and 2.53 dB, whereas for VMF, the values are 0.0237, 27.91, and 5.88 dB. This shows the superiority of the proposed method. In Table 13, the ranking of the filters based on average values of NCD, PSNR, and MAE is given. The PF occupies the topmost rank in terms of PSNR and MAE, whereas ACWVMF has the lowest NCD. For SSIM, the values are depicted for Lena image in Table 14. It is seen that the PF has the highest SSIM with 0.9975 followed by ARWSF with a value of 0.9972, and ACWVMF occupies the third rank with a value of 0.9969 at 20% impulse noise ratio.

CONCLUSIONS

This article presents a brief review of the quaternion theory and its properties. The recent development in the application of quaternion in the field of image filtering corrupted by impulse noise is discussed in details. Based on the quaternion rotation theory, a new vector median filter is proposed. The color difference between two samples is expressed in terms of their intensity and chromaticity differences. Based on this concept, the impulse detection is developed. The current pixel is concluded as corrupted if its rank is larger than a predefined rank, and the minimum color difference between it and its neighboring pixels along the four edge direction is greater than predefined threshold. The PF shows better performance in terms of NCD, PSNR, MAE, and SSIM compared with many existing filters.

REFERENCES

1. Plataniotis, K.N.; Venetsanopoulos, A.N. *Color Image Processing and Applications*; Springer, 2000.
2. Pitas, I.; Venetsanopoulos, A.N. *Nonlinear Digital Filters Principles and Applications*; Kluwer Academic Publishers: Boston, MA, 1990.

3. Lukac, R.; Plataniotis, K.N. *Color Image Processing: Methods and Applications*; CRC Press: Boca Raton, FL, 2006.
4. Rastislav, R.; Plataniotis, K.N. A taxonomy of color image filtering and enhancement solutions. Adv. Imaging Electron Phys. **2006**, *140*, 187–264.
5. Astola, J.; Haavisto, P.; Neuvo, Y. Vector median filters. Proc. IEEE **1990**, *78* (4), 678–689.
6. Trahanias, P.E.; Karakos, D.; Venetsanopoulos, A.N. Directional processing of color images: Theory and experimental results. IEEE Trans. Image Process. **1996**, *5* (6), 868–880.
7. Karakos, D.G.; Trahanias, P.E. Combining vector median and vector directional filters: The directional-distance filters. In *International Conference on Image Processing*, Vol. 1; IEEE: Washington, DC, 1995.
8. Wichman R.; Oistamo, K.;, Liu, Q.; Grundstrom, M.; Neuvo, Y.A. Weighted vector median operation for filtering multispectral data. In *Proceedings of SPIE 1818, Visual Communications and Image Processing*, Boston, MA, 1992.
9. Lukac, R.; Plataniotis, K.N.; Smolka, B.; Venetsanopoulos, A.N. Selection weighted vector directional filters. Computer Vision Image Underst. **2004**, *94* (1), 140–167.
10. Lukac, R.; Plataniotis, K.N.; Smolka, B.; Venetsanopoulos, A.N. Generalized selection weighted vector filters. EURASIP J. Adv. Signal Process. **2004**, *12*, 1–16.
11. Celebi, M.E.; Kingravi, H.A.; Aslandogan, Y. A nonlinear vector filtering for impulse noise removal from color images. J. Electron. Imaging **2007**, *16* (3), 033008-21.
12. Ko, S.J.; Lee, Y.H. Center weighted median filters and their applications to image enhancement. IEEE Trans. Circuits Syst. **1991**, *38*, 984–993.
13. Lukac, R. Adaptive vector median filtering. Pattern Recogn. Lett. **2003**, *24* (12), 1889–1899.
14. Lukac, R.; Smolka, B. Application of the adaptive center-weighted vector median framework for the enhancement of cDNA microarray images. Int. J. Appl. Math. Computer Sci. **2003**, *13* (3), 369–384.
15. Lukac, R. Adaptive color image filtering based on center-weighted vector directional filters. Multidimens. Syst. Signal Process. **2004**, *15* (2), 169–196.
16. Smolka, B. Peer group switching filter for impulse noise reduction in color images. Pattern Recogn. Lett. **2010**, *31*, 484–495.
17. Lukac, R.; Smolka, B.; Plataniotis, K.N.; Venetsanopoulos, A.N. Vector sigma filters for noise detection and removal in color images. J. Visual Commun. Image Represent. **2006**, *17*, 1–26.
18. Lukac, R.; Smolka, B.; Plataniotis, K.N.; Venetsanopoulos, A.N. A variety of multichannel sigma filters. In *Proceedings of the SPIE*, Vol. 3146, 2003, 244–253.
19. Lukac, R.; Smolka, B.; Plataniotis, K.N.; Venetsanopoulos, A.N. Generalized adaptive vector sigma filters. In *Proceedings of 2003 International Conference on Multimedia and Expo.*; Vol. 1; Baltimore, MD, 2003, 537–540.
20. Lukac, R.; Smolka, B.; Plataniotis, K. N.; Venetsanopoulos, A.N. Entropy vector median filter. In *Iberian Conference on Pattern Recognition and Image Analysis*; Springer: Berlin and Heidelberg, 2003.
21. Lukac, R.; Plataniotis, K.N. Fuzzy vector filters for cDNA microarray image processing. Studies Fuzziness Soft Comput. **2008**, *242*, 67–82.
22. Plataniotis, K.N.; Androutsos, D.; Venetsanopoulos, A.N. Adaptive fuzzy systems for multichannel signal processing. Proc. IEEE **1999**, *87* (9), 1601–1622.
23. Singh, K. M.; Bora, P.K. Switching vector median filters based on non-causal linear prediction for detection of impulse noise. Imaging Sci. J. **2014**, *62* (6), 313–326.
24. Jia, Y.B. *Quaternions and Rotations*. 2008 Com S 477.577; University of Southern California: Los Angeles, 2013.
25. Chi, V. *Quaternion and Rotations in 3-space: How it Works. Tutorial on Quaternions and Rotations in 3-Space*; Microelectronic Systems Laboratory, Department of Computer Science, University of North Carolina: Chapel Hill, 1998.
26. Angulo, J. Quaternion colour representations and derived total orderings for morphological operators. In *Conference on Colour in Graphics, Imaging, and Vision*, Vol. 2008; Terrassa, 2008, 417–422.
27. Cai, C.; Mitra, S.K. A normalized color difference edge detector based on quaternion representation. *In Proceedings of 2000 International Conference on Image Processing*, Vol. 2; Vancover, BC, 10–13th September, 2000, 816–819.
28. Shi, L.; Funt, B. Quaternion color texture segmentation. Computer Vision Image Underst. **2007**, *107* (1), 88–86.
29. Sangwine, S.J. Colour image edge detector based on quaternion convolution. Electron. Lett. **1998**, *34* (10), 969–971.
30. Sangwine, S. J.; Ell, T.A. Colour image filters based on hypercomplex convolution. IEE Proc. Vision Image Signal Process. **2000**, *147* (2), 89–93.
31. Zhang S.; Karim M.A. A new impulse detector for switching median filters. IEEE Signal Process. Lett. **2002**, *9* (11), 360–363.
32. Jin, L.; Li, D. An efficient color-impulse detector and its application to color images. IEEE Signal Process. Lett. **2007**, *14* (6), 397–400.
33. Jin, L.; Liu, H.; Xu, X.; Song, E. Qauternion based color image filtering for impulse noise suppression. J. Electron. Imaging **2010**, *19* (4), 043003-12.
34. Jin, L.; Liu, H.; Xu, X.; Song, E. Color impulsive noise removal based on quaternion representation and directional vector order-statistics. Signal Process. **2011**, *91* (5), 1249–1261.
35. Geng, X.; Hu, X.; Xiao, J. Quaternion switching filter for impulse noise reduction in color image. Signal Process. **2012**, *92* (1), 150–162.
36. Wang, G.; Liu, Y.; Zhao, T. A quaternion-based switching filter for colour image denoising. Signal Process. **2014**, *102*, 216–225.
37. Jin, L.; Zhu, Z.; Xu, X.; Li, X. Two-stage quaternion switching vector filter for color impulse noise removal. Signal Process. **2016**, *128*, 171–185.
38. Jin, L.; Liu, H.; Xu, X.; Song, E. Quaternion-based impulse noise removal from color video sequences. IEEE Trans. Circuits Systems Video Technol. **2013**, *23* (5), 741–755.
39. Elaiyaraja, G.; Kumaratharan, N.; Prapau, C. R. Modified quaternion based impulse noise removal with adaptive threshold from color video sequences and medical images. Middle-East J. Sci. Res. **2015**, *23* (7), 1382–1389.
40. Yin, M.; Liu, W.; Shui, J.; Wu, J. Quaternion wavelet analysis and application in image denoising. Math. Probl. Eng. **2012**, *2012*, 1–21.

41. Hardie, R.C.; Barner, K.E. Rank conditioned rank selection filters for signal restoration. IEEE Trans. Image Process. **1994**, *3* (2), 192–206.
42. Singh, K.M.; Prabin, B. Adaptive vector median filter for removal of impulse noise from color images. IU-J. Electr. Electron. Eng. **2004**, *4* (1), 1063–1072.
43. Dong, Y.; Xu, S. A new directional weighted median filter for removal of random-valued impulse noise. IEEE Signal Process. Lett. **2007**, *14* (3), 193–196.
44. Cheng, C.M.; Pei, S.C. Subpixel color edge detection by using binary quaternion-moment preserving thresholding technique. In *Proceedings of 1998 International Symposium on Underwater Technology*, Tokyo, 17th April, 1998, 295–298.
45. Smolka, B.; Malik, K.; Malik, D. Adaptive rank weighted switching filter for impulsive noise removal in color images. J. Real-Time Image Process. **2015**, *10* (2), 289–311.
46. Wang, Z.; Bovik, A. C.; Sheikh, H. R.; Simoncell, E.P. Image quality assessment from error visibility to structural similarity. IEEE Trans. Image Process. **2004**, *13* (4), 600–612.

Color Image Segmentation Metrics

Majid Harouni and Hadi Yazdani Baghmaleki
Department of Computer Science, Dolatabad Branch, Islamic Azad University, Isfahan, Iran

Abstract
An automatic image segmentation procedure is an inevitable part of many image analyses and computer vision which deeply affect the rest of the system; therefore, a set of interactive segmentation evaluation methods can substantially simplify the system development process. This entry presents the state of the art of quantitative evaluation metrics for color image segmentation methods by performing an analytical and comparative review of the measures. The decision-making process in selecting a suitable evaluation metric is still very serious because each metric tends to favor a different segmentation method for each benchmark dataset. Furthermore, a conceptual comparison of these metrics is provided at a high level of abstraction and is discussed for understanding the quantitative changes in different image segmentation results.

Keywords: Analytical and empirical study; Color image segmentation; Error measurement; Segmentation evaluation; Similarity and distance metrics; Unsupervised evaluation.

INTRODUCTION

One of the most important current issues in image analysis and computer vision is evaluation of the segmentation methods of the systems. In fact, the task of image segmentation is to produce a separated image into homogenous and distinct regions, which has some information explicit in analyzing the image processing, and contributes to computerized data extraction. Some well-known examples of image segmentation use for image analysis and vision are scene analysis,[1,2] image retrieval and recognition,[3–6] natural or medical image understanding,[7,8] saliency detection,[9] object tracking systems,[10,11] and many more fields of study. The image segmentation methods can be categorized into seven major classes: region-based,[12,13] boundary-based,[14–16] edge detection-based,[17,18] template matching,[19,20] cluster-based,[21–23] threshold-based,[24–27] graph-based,[28,29] and the combination of the mentioned classes.[30–33] In all the working steps of the aforementioned systems, the image segmentation step has a substantial impact on the overall algorithmic performance; therefore, the process of selecting segmentation evaluation metrics should be considered carefully. The key problems with this selection are that the images are captured with poor contrast, low resolution, illumination variability, scale changes, color balances, complex subsurface structure and/or different texture details, various outdoor conditions, and variation of camera calibration and sensitivity. Generally, the human visual system is well formed to segment a complex and nonuniform image background into the desired objects and/or regions. Nevertheless, it is very difficult to segment such images into perceptually distinct regions. Hence, the evaluation of image segmentation results is necessary to effectively carry out the objectives of the research; this can help researchers to identify the strengths and weaknesses of the image segmentation methods, which are of paramount importance. Nevertheless, the measurement of segmentation performance methods and a standard comparison with other methods have always been the main principles for researchers.

Image segmentation evaluation problem has been reviewed substantially, which is still a continuing challenge in determining the best evaluation method with respect to optimality and computation time for the researchers. Besides, several different decision-making processes have been proposed to tackle this challenge. These processes has the following issues: (1) to efficiently exploit the ability of a proposed segmentation method in comparison with other image segmentation methods; (2) to define the proposed method is paramount in different image categories, for example, natural or medical images; and (3) to validate the selected parameterization biases of the proposed method for various image conditions are more effective in this research plan. This entry provides an extensive description of each image segmentation metrics, where a much more methodical study considers that quantitative evaluation methods would usefully complement and extend the qualitative image analysis. In general, image segmentation evaluation methods can be divided into two main categories:[34–37] empirical evaluation and analytical evaluation methods. The empirical methods analyze a segmentation method with respect to its findings and outputs, some of these methods are presented in the works of Bouthemy and François,[1] Said,[38] and Zhang;[39] instead, the analytical methods assess the method based on its complexity, functionality, utilities, etc.[7,21,23,29]

The rest of the entry is organized as follows: Section "Brief Background" presents the related work. Section "Types of Quantitative Metrics" gives an overview of the structure of image segmentation methods and presents the extended hierarchy of different image segmentation evaluation methods. Section "Discussion and Suggestion" provides the comparative performance analysis of various metrics. Finally, Section "Conclusions" concludes this entry.

BRIEF BACKGROUND

Image segmentation methods have been utilized in computer vision and image understanding to partition an image into its distinct regions.[10,25,26,38–40] A great variety of color image segmentation methods is presented; some are used for general images and some are planned for a specific purpose. Furthermore, the outputs of these methods can perceptually be classified as correct-segmentation, over-segmentation, and under-segmentation.[29,41–44] A color image dataset is rich in both homogeneous color regions and texture; in this case, the texture-based segmentation method would be proposed and the most challenging problems are as follows: (1) to segment the elements of texture regions in various shapes and sizes without losing desired details; (2) to reduce the processing time of the method; and (3) to evaluate the computational efficiency of the different algorithms. In the following section, a brief literature survey of related works on the third challenging problem is summarized. A color–texture image segmentation method was proposed based on a three-dimensional deformable surface model and energy function;[45] a set of quantitative evaluation metrics is applied to measure the quality of the proposed method in natural image scenes, that is, Probabilistic Rand Index (PRI), Normalized Probabilistic Rand (NPR) index, global consistency error (GCE), variation of information (VOI), and boundary displacement error (BDE). In the work of Mridula,[46] a hybrid color texture segmentation model was developed for pixel labeling of images by combining -level co-occurrence matrix feature extraction and Markov random field algorithm; the hybrid model was evaluated using misclassification error (MCE) metrics. In the proposed image segmentation method in Ref. [21] the different pixels or regions were extracted from labeling the pixel-level image features using support vector machine classification and the evaluation of the proposed method was used to measure the segmentation errors by three metrics: segmentation error rate (ER), local consistency error (LCE), and bidirectional consistency error (BCE). The performance of a lossy data compression was introduced as an image segmentation method in color–texture natural images and to assess the performance measurements of the method, four metrics were used, that is, PRI, GCE, VOI, and BDE.[7] The graph-based image segmentation processes have a big problem: finding the reasonable weights for the graph cuts implementation, like the previous one, the PRI, GCE, VOI, and BDE metrics were used to compare an unsupervised multilevel color image segmentation method with others.[47] A hybrid roof segmentation approach was presented for the color aerial images in the work of El Merabet et al.[48] Its qualitative image segmentation was evaluated by the mean value of VINET from the work of Vinet.[49] Following that different segmentation evaluation methods are employed when dealing with different studies, that is, object accuracy measurement: Jaccard index (JI),[36,50–52] similarity matching methods: Dice coefficient (Dice),[36,37,51] image contour matching methods: Hausdorff distance (HAUSD),[22,50] and/or contour-based evaluation: normalized sum of distances (NSD);[53,54] pixel difference between a segmented image and the ground-truth (GT) image: Hamming distance (HD),[36,55,56] Fowlkes–Mallows index (FMI)[57–59] and boundaries discrepancy between them: boundary Hamming distance (BHD),[60] polyline distance metric (PDM);[61–63] surface distance measurement methods: mean absolute surface distance (MASD);[64–66] statistic based methods: mutual information (MI)[34,67] and/or normalized mutual information (NMI);[68,69] texture-based methods: earth movers distance (EMD);[70–72] relevance-based metrics: recall and precision.[73,74] The conceptual framework of image understanding and computer vision systems will be discussed in the following subsections.

Conceptual Image and Vision Processing System Design

The overall framework of image analysis systems is shown in Fig. 1. Normally, the image acquisition stage is to enter the number of sequence images from a benchmark standard dataset or a real-time image capturing device. The image preprocessing stage can also be a preliminary stage of the systems, in which a raw image is normalized into desired image.[75] In the next stage, the desired image segmentation stage is to separate uniform regions in a desired image, and postprocessing may be used to enhance the regions. A set of local and global features from these regions of the desired image is extracted to describe the entrance image, which are used in the classification stage.

Hierarchical Structure in Color Image Segmentation Evaluation

The extended hierarchy of the color image segmentation evaluation process in the work of Zhang et al.[42] is shown in Fig. 2; this process can be divided into three main categories: subjective, objective, and hybrid subjective–objective evaluations. A point of note is that the performance measurement of

Color Image Segmentation Metrics

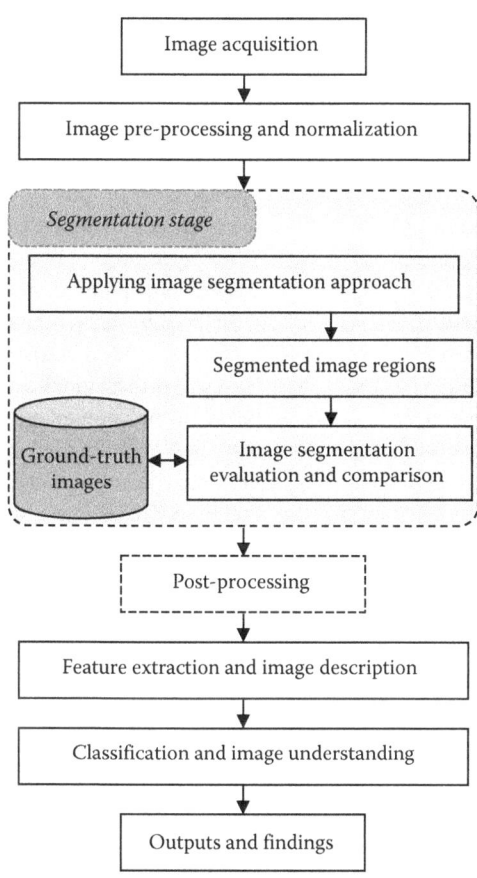

Fig. 1 An overal framework of computer vision and image undestanding system

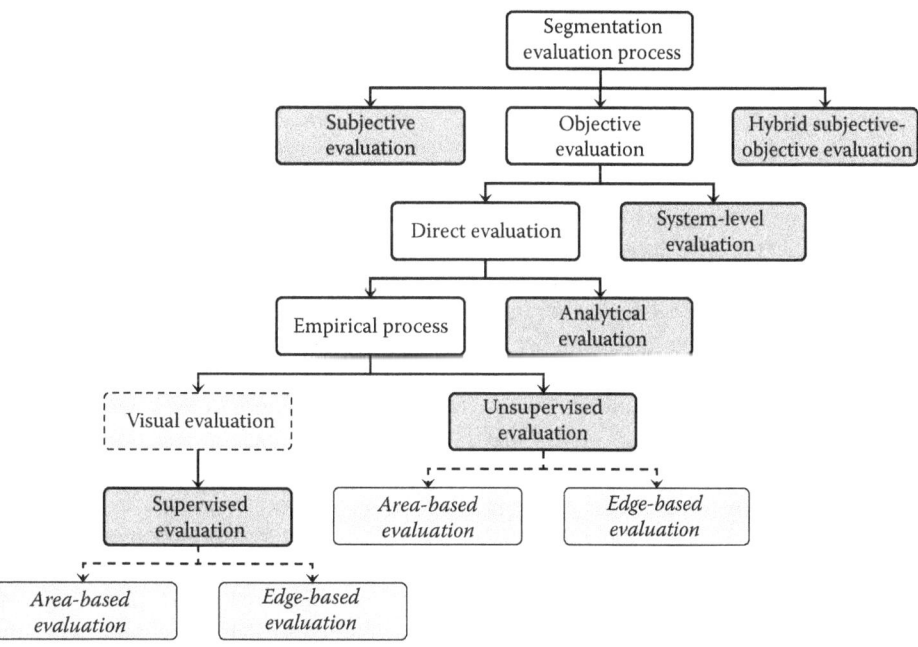

Fig. 2 A hierarchical structure for color image segmentation evaluation

image segmentation methods must meet strict criteria and are compared with other studies. The details of main categories and/or subcategories are presented as follows:

- **Subjective evaluation**: This evaluation of image segmentation algorithms is a frequently used process for the measurement of image (video) quality. In general, there are two main determinant factors that contribute to a more efficient evaluation of the segmentation algorithms: different experimental conditions, and different human observation experiences and skills. However, subjective evaluation is a time-consuming, energy-intensive, and error-prone process, which can cause an exceedingly tedious task for a viewer. Furthermore and also noteworthy, a quantitative–subjective evaluation method cannot reproduce the statistical properties of the segmented images.[34,37,50]
- **Objective evaluation**: This evaluation of image segmentation is based on quantitative criteria and analyzed by internal observation. In fact, the objective methods calculate appropriate metrics directly from the images; these metrics try to measure certain similarities and significant differences between a segmented image and its GT image. To prove the quality validation of a segmentation method, it is necessary to avoid the influence of human activities and factors, to provide consistency, and also to improve reliability in segmenting different types of images.[39] However, this kind of segmentation evaluation has a number of limitations when utilizing the hand-segmenting benchmark datasets as follows: a variety of file formats for hand-segmenting images, different GT segmentation images for a given image, and the datasets are too large, thus providing hardware with necessary processing capabilities.
  - *System-level evaluation:* Different evaluation metrics attempt to assess the impact of a designed segmentation method on the complete system. A well-designed segmentation method provides useful information on tracking performance analysis of the system; hence, this task is needed to investigate the significant accuracy improvement on an applicative system. However, it is very difficult in many cases to use a large number of manually segmented images to accomplish fairly accurate pixel-level evaluation criteria. Therefore, obtaining high system accuracy shows that the proposed segmentation method also has a high-performance design. Three criteria in the work of Tomalik[76] are presented for the impact of a segmentation method on the performance analysis of the system classification: (1) minimizing and avoiding false positives, (2) fitting as much as possible the segmented regions into their expected hand-segmented regions, and (3) the segmentation findings would be the most useful for expansion and further analysis.
  - *Analytical evaluation:* These methods are considered as result-independent methods to be analyzed based on the principles, properties, for example, processing scheme and algorithmic structure, and the processing complexity of image segmentation methods.[34] In theory, the image segmentation methods can be assessed analytically.[77]
  - *Visual evaluation:* In these methods, a benchmark dataset is separated into different image groups with the same subject matter,[78,79] in which an evaluation method is statistically established by analyzing the individual and group-level differences between the segmented images and their original images. The main goal of these methods is to show the quantification of the quality of an image segmentation results that can display more detailed visual information. Another goal is to analyze possible errors in the segmentation stage. On the other hand, the difficult problem is that the diversity in the image region types of different image databases can lead to lower segmentation performance. However, the crucial challenge is that the visual evaluation is still under development.
  - *Supervised evaluation:* In order to evaluate the accuracy of supervised segmentation methods, the segmentation results are used to compare between GT images and segmented images. In fact, the methods are directly based on region overlapping information; in which reviewers tend to consider the possibility of the standard quality measurement methods.[51] The supervised segmentation evaluation can be also divided into two subgroups: area-based and edge-based evaluations.[80] Briefly, area-based evaluation methods are derived from the intersection of segmented image pixels and hand-segmenting image pixels. On the other hand, edge-based evaluation methods often provide greater insight into analyzing the image segmentation error between the image edges. A more significant discussion can be found in Refs. [51,54,80,81].
  - *Unsupervised evaluation:* The unsupervised evaluation of color image segmentation methods, objectively in a quantitative manner, can be designed for measuring the region-by-region correlation and differences of segmented images in observed original images; which means that the error between these images can be obtained without the reference-segmented images;[82] hence, it becomes harder due to little information concerning the best consensus GT for original image dataset. The main issues addressed in an efficient segmentation for different images are as follows: (1) textures and/or regions are homogeneous, in which neighboring regions must be readily distinguishable from each other; (2) to have simple region interiors; and (c) region boundaries must not lose real boundaries.[42,83,84]

- *Hybrid objective–subjective evaluation*: A hybrid objective–subjective evaluation process is employed to measure the performance of an image segmentation approach with perceptual analysis. The main goal of objective evaluation metrics is to assess the performance of the quality of images. Subjective evaluation is needed to measure the image quality and provide the human visual comparison of an image details. It can be useful in understanding how humans perceive visual quality of images which can influence the intricacy of visual information.

TYPES OF QUANTITATIVE METRICS

Quantitative metrics are proposed to evaluate the performance of image segmentation methods using the intuitive knowledge. Typically, comparable segmentation results and/or errors will be necessary to help advance the state of the art. Therefore, the categorization of quantitative metrics into different performance groups is critical to support further development. The following subsections describe this categorization in more detail.

Region/Volume Based

The region/volume based measures are based on the segmented object regions that can provide an evaluation of outline matching in pixel-to-pixel autosegmentation and its reference images. Rather than considering performance metrics, these measures could be divided into two subgroups: statistical/probabilistic analysis and distance analysis.

Statistical/Probabilistic Analysis

In fact, the target of the statistical analysis is to provide a set of result values for comparison of a proposed segmentation method with other methods, where this work can calculate the overall probability of segmentation results for all image pixels.

a. Relevance measurement
   The notion of relevance measurement is a computation sequence for analyzing, in which segmented image pixels have been detected correctly in their respective regions of origin, that is, tagged GT data, or have not been detected correctly. For image segmentation, a confusion matrix for this experiment is shown in Fig. 3. As shown in Fig. 4, the basic concepts of the confusion matrix are defined as follows:
   – True positives denote the number of pixels, regions, and/or objects of the foreground of a segmented image, that is, region of interest, that are truly identified with its foreground GT image.

		Automatic segmentation result	
		Positive	Negative
Ground truth	Positive	$TP = \sum_{i=1}^{k} n_{(i,i)}$	$FN = \sum_{i=1}^{k}\sum_{j \neq i}^{k} n_{(i,j)}$
Ground truth	Negative	$FP = \sum_{j=1, i \neq j}^{k}\sum n_{(i,j)}$	$TN = \sum_{j=1}^{k}\sum_{i \neq j}^{k}\sum_{j \neq i}^{k} n_{(i,j)}$

Fig. 3 A confusion matrix of segmented images and their GT images

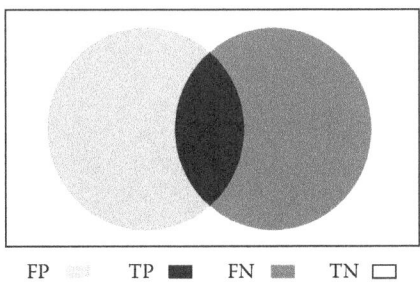

Fig. 4 The conception of confusion: segmented image pixels (left area), their respective GT image pixels (right area), and image background (white area)

- False positives denote the number of pixels, regions, and/or objects of the foreground of a segmented image that are falsely identified with its foreground GT image.
- True negatives denote the number of pixels, regions, and/or objects of the background of a segmented image that are truly identified with its background GT image.
- False negatives denote the number of pixels, regions, and/or objects of the background of a segmented image that are falsely identified with its background GT image.

Some well-known relevance metrics can be defined as presented in Table 1; these metrics are overlap based, so they indicate the ability of the system's performance to evaluate mis-segmented pixels and correct segmentation results, that is, target pixels.

The VD metric measured the volume of the foreground pixel regions on both segmented images and their reference images; to proof this point, after some algebraic manipulations, the Eq. VD is modified as follows:

$$
\begin{aligned}
VD &= \frac{|FN-FP|}{2TP+FN+FP} \\
&= \frac{\left|\left(\left|S_{GT}(f)\right|-TP\right)-\left(\left|S_{auto}(f)\right|-TP\right)\right|}{(FN+TP)+(FP+TP)} \\
&= \frac{\left\|S_{GT}(f)\right|-\cancel{TP}-\left|S_{auto}(f)\right|+\cancel{TP}\right|}{\underbrace{(FN+TP)}_{|S_{GT}(f)|}+\underbrace{(FP+TP)}_{|S_{auto}(f)|}} \\
&= \frac{\left\|S_{GT}(f)\right|-\left|S_{auto}(f)\right|\right|}{\left|S_{GT}(f)\right|+\left|S_{auto}(f)\right|} \quad \blacksquare
\end{aligned}
\quad (1)
$$

b. Similarity measurement

JJI: In the case of segmentation evaluation, most researchers identify the significance of a comparative study of similarity measures. One of these metrics that computes the similarity between the GT images and

Table 1 Quantitative relevance metrics derived from confusion matrix

Evaluation metrics	Symbols/names	Equations
True Negative Rate	TNR, specificity	$TNR = \dfrac{TN}{TN+FP}$ (1)
True Positive Rate	TPR, recall, sensitivity	$TPR = \dfrac{TP}{TP+FN}$ (2)
Positive Likelihood Ratio	PLR	$PLR = \dfrac{Sensitivity}{1-Specificity}$ (3)
Negative Likelihood Ratio	NLR	$NLR = \dfrac{1-Sensitivity}{Specificity}$ (4)
False Positive Rate	FPR	$FPR = \dfrac{FP}{FP+TN}$ (5)
False Negative Rate	FNR	$FNR = \dfrac{FN}{FN+TP}$ (6)
Precision	P, Reliability	$P = \dfrac{TP}{TP+FP}$ (7)
F-measure	F	$F = \dfrac{Precision \times Recall}{Precision + Recall}$ (8)
XOR	XOR	$XOR = \dfrac{FP+FN}{TP+FN}$ (9)
Accuracy	AC, Validity	$AC = \dfrac{TP+TN}{TP+TN+FP+FN}$ (10)
Error Probability	EP	$EP = 1 - AC$ (11)
Volumetric Distance	VD	$VD = \dfrac{|FN-FP|}{2TP+FN+FP}$ (12)
Volumetric Similarity	VS	$VS = 1 - VD$ (13)
Area Under Curve	AUC	$AUC = 1 - \dfrac{FPR+FNR}{2}$ (14)

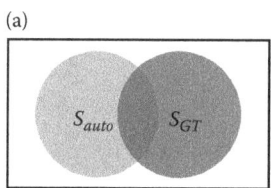 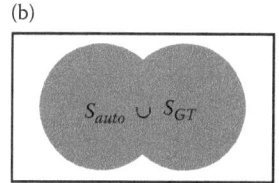 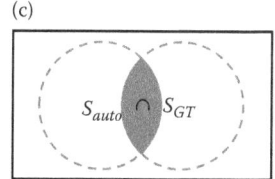

Fig. 5 Symmetric differences: (a) overlapped areas; (b) union operation; (c) intersect operation

its automatic segmented result is the JI. This metric is defined as follows:

$$JI = \frac{|S_{auto} \cap S_{GT}|}{|S_{auto} \cup S_{GT}|} \quad (2)$$

where S_{auto} is the automatic segmented images set, and S_{GT} is its GT images set. The vertical line brackets denote the absolute value of set's elements; "$\cap$" and "$\cup$" symbols are the intersection and union operations, respectively (Fig. 5). With increase in the similarity between two sets, the number of elements in the intersection set will increase and the cardinality of union set will decrease. However, in considering this equation, when similarity of two sets reach maximum, the JI value becomes 1. And in similarity absence, JI will be 0. Often, this metric is applied when evaluating an automatic segmentation method on image dataset that consists of just two main regions: foreground and background.

Dice coefficient: The dice coefficient is a statistical-based metric that evaluates the performance of a color image segmentation method by comparing its result set with the hand-segmented image set. The mathematical calculation of Dice coefficient is presented as follows:

$$DICE = \frac{2 * |S_{auto} \cap S_{GT}|}{|S_{auto}| + |S_{GT}|} \quad (3)$$

The cardinality value of the sum of S_{auto} and S_{GT} sets is static, which means by increasing or decreasing the similarity of two sets, the denominator remains fixed; therefore, the value of the fraction increase with increase in the size of intersection set. However, the range of Dice coefficient value is between 0 and 1, where the values denote the relative effect of dissimilarity and similarity pixels between two sets, respectively. Following this, the XOR metric is employed in calculating the mis-segmented pixels probability,[85,86] which can be differently defined as follows:

$$XOR = \frac{|S_{auto} \cup S_{GT}| - |S_{auto} \cap S_{GT}|}{|S_{GT}|} s \quad (4)$$

FMI: It has been presented essentially for computing the similarity between two clustering results; but in image segmentation evaluation, it intends to compare two segmented areas by Eq. 5. The value range of this metric is in interval [0 1], where 0 and 1 denote the mismatching and the perfect agreement between two segmented images, respectively.

$$FMI = \frac{|S_{auto} \cap S_{GT}|}{\sqrt{|S_{auto}| \cdot |S_{GT}|}} \quad (5)$$

Rand index (RI): It was originally presented for clustering quality evaluation. This measurement works based on the point pairs in cluster placement. In the segmentation evaluation process, both the point pairs and the clusters are replaced with the pixel pairs and the segments, respectively. If any pixel pair in automatic segmentation results is placed in the same segment, the corresponding pixels in its human-segmented image must be in the same region as well; also, if a specific pixel pair in one belongs to separated segments, the corresponding pixels must be in separated segments; in this term, the similarity will be increased. However, the RI is defined in Eq. 6. The RI value range is between 0 and 1.

$$RI(S_{auto}, S_{GT}) = \frac{1}{\binom{N}{2}} \sum_{\substack{i,j=1 \\ i \neq j}}^{N} \mathbb{F}\left[\left(P_i^{auto}, P_j^{auto}\right), \left(P_i^{GT}, P_j^{GT}\right)\right] \quad (6)$$

where P_i^{auto}, P_j^{auto} and P_i^{GT}, P_j^{GT} are pixel pairs in automatic segmented images and the corresponding pair in the GT images; also, $\mathbb{F}$ is identity function that is defined as follows:

$$\mathbb{F}\left[\left(P_i^{auto}, P_j^{auto}\right), \left(P_i^{GT}, P_j^{GT}\right)\right]$$
$$= \begin{cases} 1 & \text{if} \begin{cases} \left(\left(P_i^{auto} = P_j^{auto}\right) AND \left(P_i^{GT} = P_j^{GT}\right)\right) \\ OR \\ \left(\left(P_i^{auto} \neq P_j^{auto}\right) AND \left(P_i^{GT} \neq P_j^{GT}\right)\right) \end{cases} \\ 0 & \text{Otherwise} \end{cases}$$
$$(7)$$

PRI: Some image segmentation datasets are supervised, which have at least one GT segmented image for any existing image; in some cases, there are more than one GT for a given image. The RI is not appropriate for comparison of segmentation result with multiple human segmented images. Considering this condition, a modified RI called PRI is defined as follows:

$$PRI = \frac{1}{K}\sum_{j=1}^{K} RI\left(S_{auto}, S_{GT}^{j}\right) \quad (8)$$

where K is the number of existing GT segmented images of a given image, in which S_{GT}^{j} is the jth GT of this image. The range values of the segmentation results and their meaning are similar to RI metric.

NPR: Considering that the different aspects of image complexity is significant, it addresses its difficulty in evaluating segmentation result and describing more detailed information. In these cases, it is required to perform a more accurate evaluation of segmentation results by comparing them with different evaluation methods in a dataset. One of the ways to achieve this goal is to normalize the evaluation results by computing a baseline model; this model expresses the expected performance evaluation of an image segmentation method for a given dataset. A normalized evaluation that is presented in the work of Unnikrishnan et al.[87] is shown as follows:

$$NPR = \frac{PRI - EV}{MaxPR - EV} \quad (9)$$

where *PRI* is the current automatic segmented image, *MaxPR* is the maximum achieved *PRI* by applying the segmentation method through current image dataset, and the EV is computed as follows:

$$EV = \frac{1}{m}\sum_{j=1}^{m} PRI\left(S_{j}\right) \quad (10)$$

where m is the number of images in a benchmark dataset, and S denotes the automatic segmentation result of an image in the dataset.

Misclassification errors: This metric is introduced by Yasnoff et al.[88] to compute the percentage of displaced pixels in a given automatic segmented image by Eq. 11, regarding its GT image.

$$MCE = 1 - \frac{\left|S_{auto}(b) \cap S_{GT}(b)\right| + \left|S_{GT}(f) \cap S_{auto}(f)\right|}{\left|S_{auto}(b)\right| + \left|S_{auto}(f)\right|} \quad (11)$$

where $S_{auto}(b)\left(S_{auto}(f)\right)$ and $S_{GT}(b)\left(S_{GT}(f)\right)$ denote the background (foreground) pixels in the segmented and the GT images. The MCE value is in interval [0, 1], and it is similar to the other error metrics, lower value will be desirable.

Error rate: The error simply expresses the percent of pixels placed in a wrong region as shown in the following equation:

$$ER = \frac{N_{false} + N_{miss}}{N_{total}} \times 100 \quad (12)$$

where N_{false} and N_{miss} are the numbers of segmented pixels that have been located in the erroneous region, and N_{total} denotes the number of pixels in the hand-segmented image. In these terms, the value range of ER is in the interval between 0 and 100, where 0 value shows there is no similarity between segmentation results and corresponding GT image; also the value of 100 denotes the excellent segmentation results.

MI: It is a metric that analyzes the relationships between two sets of variable regions by measuring the amount of feature information of a set associated with the other set as expressed in Eq. 13 and presented in Refs.[89,90]

$$MI(S_{auto}, S_{GT}) = Ent(S_{auto}) + Ent(S_{GT}) - Ent(S_{auto}, S_{GT}) \quad (13)$$

where $Ent\left(S_{auto}\right)$ and $Ent\left(S_{GT}\right)$ denote the marginal entropy, and $Ent\left(S_{auto}, S_{GT}\right)$ is the joint entropy.

VOI: This metric has been basically proposed for clustering comparison. It can denote the measurement of missed and extra information rates in resemblance of the two image sets, which is based on their mutual information and marginal entropy as follows:

$$VOI(S_{auto}, S_{GT}) = Ent(S_{auto}) + Ent(S_{GT}) - 2MI(S_{auto}, S_{GT}) \quad (14)$$

NMI: As described earlier, the MI measures the amount of shared information between two image sets; but because of the upper bound problem, it seems to be necessary to normalize this metric to achieve a fixed maximum value as an upper bound limit. Eq. 15 shows how Strehl and Ghosh[91] introduced the NMI for solving this problem:

$$NMI = \frac{MI\left(S_{auto}, S_{GT}\right)}{\sqrt{Ent\left(S_{auto}\right).Ent\left(S_{GT}\right)}} \quad (15)$$

c. Consistency error measurements

Some quantitative evaluation methods are used for measuring the consistency between automatic segmented image and its GT image. It is suggested that there are three measures, that is, GCE, LCE, and BCE; to calculate them, it is necessary to compute local refinement error (LRE) as nonsymmetric formula in the following equation:

$$LRE(S_{auto}, S_{GT}, P) = \frac{\left|R(S_{auto}, P) \setminus R(S_{GT}, P)\right|}{\left|R(S_{auto}, P)\right|} \quad (16)$$

where $R(S_{auto}, P)$ denotes a region in segmented image that contain pixel P, and "\" symbolizes set difference. The meaning of other variables and operators are the same as previously mentioned. Now, LCE, GCE, and BCE can be computed. The value range of these metric is between 0 and 1, in which 0 value represents no mis-segmented region and 1 value means that there is no homogeny region between two image sets.

LCE: Some metrics provide the label refinement homogeneously through the image in order to rectify slight difference in graininess when comparing segmentations. The LCE is a measurement (Eq. 17) that shows the degree of refinement between automatic segmentation result and its GT.

$$LCE = \frac{1}{N} \sum_{i=1}^{N} MIN\{LRE(S_{auto}, S_{GT}, P_i), LRE(S_{GT}, S_{auto}, P_i)\} \quad (17)$$

where N is the total number of image pixels.

GCE: In this measurement, it can be assumed that any segmentation like S_i possesses a refinement of segmentation S_j (Eq. 18). As shown in Eq. 14, this metric reports maximum similarity, when all local refinements are placed in the same direction. It is shown that in the absence of accommodation, no similarity will be achieved.

$$GCE = \frac{1}{N} MIN \left\{ \sum_{i=1}^{N} LRE(S_{auto}, S_{GT}, P_i), \sum_{i=1}^{N} LRE(S_{GT}, S_{auto}, P_i) \right\} \quad (18)$$

BCE: When excellent segmentation refinement is achieved, it is necessary to measure the degree of refinement in order to provide more accurate quantitative analysis; in this situation, one may simply set the minimal LRE to a very low value, so that by enhancing segmentation method, the LCE will reduce slowly. The BCE is applied when LCE is very low, it is defined as follows:

$$BCE = \frac{1}{N} \sum_{i=1}^{N} MAX\{LRE(S_{auto}, S_{GT}, P_i), LRE(S_{GT}, S_{auto}, P_i)\} \quad (19)$$

Distance-Based Metrics

Accurate region contour is a key point in image segmentation evaluation; hence, some image segmentation measurements focus on distance between segmented object's contour and reference object's boundary. Some metrics work on physical distance[92,93] and few are based on logical distance.[94,95] In physical distance, there are several geometric distance measurement methods, for example, Euclidean distance, Manhattan distance, etc. And in logical distance, the value of distance can only be 0 or 1, where 0 means there is no distance between two objects and 1 shows that there is a distance. However, the most used physical distance measurement method is two-dimensional Euclidean distance, which is defined as follows:

$$d_{EQ}(P_1, P_2) = \sqrt{(x_{p1} - x_{p2})^2 + (y_{p1} - y_{p2})^2} \quad (20)$$

where $P_1(x_{p1}, y_{p1})$ and $P_2(x_{p2}, y_{p2})$ are the first and second points, respectively; also, a derived concept from Euclidean distance is a distance distribution signature (the so-called minimal Euclidean distance as shown in Eq. 21) that shows minimum Euclidean distance between a point and a set of points.

$$d_{\min}(p_s, B_t) = MIN\{d_{EQ}(p_s, p_t) | p_t \in B_t\} \quad (21)$$

where p_s and p_t symbolize the pixels that are located on the boundary of automatic segmented object and its GT boundary, respectively. Another derived concept is supremum–infimum distance that shows minimum Euclidean distance between two furthest points from two distinct images sets:

$$d_{\sup-\inf}(B_{auto}, B_{GT}) = MAX\{d_{\min}(p_i, B_{GT}) | p_i \in B_{auto}\} \quad (22)$$

where B_{auto}, and B_{GT} denote the automatic segmented object boundary and its human segmented GT boundary sets, respectively. It is citable that minimal Euclidean and supremum–infimum distance are not symmetric.

HAUSD: This metric measures the longest Euclidean distance between two sets of boundary pixels from automatic segmented and human segmented objects; its definition is given by the following equation:

$$HAUSD = MAX\{d_{\sup-\inf}(B_{auto}, B_{GT}), d_{\sup-\inf}(B_{GT}, B_{auto})\} \quad (23)$$

It is obvious that 0 value for the HAUSD shows that there is exhaustive matching between result and its reference. With increasing mismatching between two sets, the distance will increase too.

MASD: As explained earlier, the minimal Euclidean distance is not symmetric, because the minimal Euclidean distance from B_{GT} to B_{auto} is not equal to the distance from B_{auto} to B_{GT}; hence, to show the average of distances between two boundary pixel sets, the mean absolute surface distance was introduced in the following equation:

$$MASD = \frac{1}{2}\left(\frac{1}{N} \sum_{p_i \in B_{auto}}^{N} d_{\min}(p_i, B_{GT}) + \frac{1}{M} \sum_{p_j \in B_{GT}}^{M} d_{\min}(p_j, B_{auto}) \right) \quad (24)$$

In considering this equation, the minimum value for the metric is 0 which means superior matching is achieved but because of losing upper bound for mean absolute surface distance, there is no maximum value to denote the explicit dissimilarity.

NSD: The metric provided the basis for computing the distance between any misclassified pixel in automatic segmentation result and the contour of its referenced object's GT. To normalize the measurement, the sum of these distances is divided by the sum of the distances between any pixel in the result and the boundary of its reference. The NSD equation is defined as follows:

$$NSD = \frac{\sum_{p_i \in \{S_{auto} \triangle S_{GT}\}} d_{min}(p_i, B_{GT})}{\sum_{p_j \in \{S_{auto} \cup S_{GT}\}} d_{min}(p_j, B_{GT})} \quad (25)$$

Like other distance-based metrics, a lower value for distance denotes the perfect segmentation and a higher value refers to the undesired segmentation results. Hence, after the metric is normalized, the value range will be between 0 and 1.

Average Symmetric Surface Distance (ASD): Another distance-based metric is based on the ASD relating to two image segmentation results, which reports the average of the minimum Euclidean distance between the contours of the corresponding objects in both segmentations as follows:

$$ASD = \frac{\sum_{p_j \in B_{GT}} d_{min}(p_j, B_{auto}) + \sum_{p_i \in B_{auto}} d_{min}(p_i, B_{GT})}{|B_{auto}| + |B_{GT}|} \quad (26)$$

Obviously, the lesser value of ASD denotes more desirable segmentation results.

BDE: One of the most distinctive quantitative segmentation evaluation methods is the BDE that is focused on the distance-based features, where this metric is sensitive to the distance between the contours of an object in the automatic segmented image and the contours of the corresponding object in its GT image. In other words, it computes the average of the distance between any pixel in the boundary of an automatic segmented object and the nearest pixel in the contour of the reference object as follows:

$$BDE = \frac{1}{N} \sum_{p_i \in B_{auto}} d_{min}(p_i, B_{GT}) \quad (27)$$

where *N* is the number of the boundary pixels in an automatic segmented image. The lower bound of BDE is 0 which denotes perfect matching between the contours of both automatic segmented and GT; although, there is no upper bound, any high value means undesirable segmentation result.

HD: It is a metric based on logical distance originally presented to calculate the disagreements between two strings. In segmentation evaluation, it is used to enumerate the number of displaced pixels as shown in Eq. 28. To normalize the distance, it is divided by the total number of pixels; in this condition, the value range is given as [0, 1], where 0 and 1 denote absolute dissimilarity and excellent matching between a segmentation method result on an image and its GT.

$$HD = \frac{|S_{auto}(b) \cap S_{GT}(f)| + |S_{GT}(b) \cap S_{auto}(f)|}{|S_{auto}|} \quad (28)$$

BHD: This metric is a variation of HD that processes to the determination of the contour of automatic segmented object. When an accurate object boundary is involved, the common HD is not applicable anymore. The BHD calculates the mis-segmented boundary pixel on original boundary pixels in Eq. 29. The value range and its description are similar to the HD.

$$BHD = \frac{|B_{auto}(b) \cap B_{GT}(f)| + |B_{GT}(b) \cap B_{auto}(f)|}{|B_{GT}(f) \cup B_{auto}(f)|} \quad (29)$$

DISCUSSION AND SUGGESTION

The system-level evaluation of a proposed image segmentation method can show the impact of the designed method on a complete image analysis system; also, the outcome of the segmentation evaluation results will set a high standard for automatic qualitative evaluation, and it may provide desirable features of a precedent for future enhancements. In many situations, it is difficult to choose a set of segmentation metrics for a fair comparison with other existing methods. In order to show a state-of-the-art segmentation accuracy of the standard benchmark test datasets, it is necessary to consider and investigate the morphological features of the regions of the test images, where these regions are formed in different textural structures, natural conditions, and colors. The most commonly used metrics are summarized in Table 2.

There has been much effort to provide a common potential explanation and to find the best self-balancing stability among the selected metrics for the accuracy of a proposed image segmentation method, for example, natural image segmentation methods with respect to. PRI and VOI metrics;[96] unsupervised image segmentation methods with respect to PRI, VOI, GCE, and BDE;[47,97] parallelized segmentation methods with respect to HM and NSD; automatic brain tumor segmentation methods with respect to Dice and EP; and nuclei segmentation methods with respect to RI, JI, HAUSD, and NSD.[54] As shown in Table 3, the desired values for all the mentioned metrics are summarized, which means the direction of mean changes for each metric: higher means better ("↑") and smaller means better ("↓").

Table 2 Image segmentation evaluation methods

Metric name	Symbol	Region based			Boundary based			Binary segmentation	General segmentation	Brief description
		Distance	Relevance	Similarity	Distance	Relevance	Similarity			
True Negative Rate (Specificity)	TNR		✓					✓		A set of pixels/voxels as a nonobject
True Positive Rate (Recall, Sensitivity)	TPR		✓					✓		A number of correct overlapping pixels/voxels
Positive Likelihood Ratio	PLR		✓					✓		A subsegment location probability rate
Negative Likelihood Ratio	NLR		✓					✓		A non-subsegment location probability rate
False Positive Rate	FPR		✓					✓		A set of autosegment pixels/voxels labeled as nonobject
False Negative Rate	FNR		✓					✓		A set of hand-segment pixels/voxels labeled as nonobject
Precision	P		✓					✓		A measure of segmentation accuracy
Accuracy	AC		✓					✓		A ratio of correct overlapping over the number of all autosegment pixels/voxels
XOR	XOR		✓					✓		A ratio of the pixels/voxels mis-segmented over the number of all hand-segment pixels/voxels
F-measure	F		✓					✓		A harmonic mean of precision and recall rates
Error Probability	EP		✓					✓		The probability of mis-segmenting an object
Volumetric Distance	VD		✓					✓		The mean absolute volume difference between hand-autosegment and the sum of their volumes
Volumetric Similarity	VS							✓		A ratio of volumes of the hand-autosegments to show the similarity of them
Area Under Curve	AUC		✓					✓		A ratio of the segmentation performance that how well the correct and mis-segment are
Jaccard Index	JI			✓				✓		A ratio of how closely the autosegment overlaps the hand-segment pixels/voxels
Dice Coefficient	Dice			✓				✓		A ratio of overlapping over the average size of both hand and autosegment pixels/voxels
Fowlkes-Mallows Index	FMI			✓				✓		A geometric mean of precision and recall rates

(Continued)

Table 2 Image segmentation evaluation methods (*Continued*)

Metric name	Symbol	Region based Distance	Region based Relevance	Region based Similarity	Boundary based Distance	Boundary based Relevance	Boundary based Similarity	Binary segmentation	General segmentation	Brief description
Rand Index	RI			✓				✓	✓	A measure of the similarity between hand and autosegment pixels/voxels
Probabilistic Rand Index	PRI			✓				✓	✓	A measure of the similarity between auto-segment and a set of hand-segment pixels/voxels
Normalized Probabilistic Rand Index	NPR			✓				✓	✓	Measuring of similarity between autosegment methods and a set of their hand segments
Misclassification Errors	MCE		✓	✓				✓		Known as the similar mis-segmentation errors
Error Rate	ER		✓	✓				✓	✓	The ratio of the counts of the incorrect auto-segment pixels/voxels to the reference segments
Local Consistency Error	LCE			✓				✓	✓	A ratio of identical segmentations
Global Consistency Error	GCE			✓				✓	✓	A measure for ignoring over-segmentations
Bidirectional Consistency Error	BCE			✓				✓	✓	An approximation ratio for a minimization of a fit error over a set of hand segments
Mutual Information	MI			✓				✓	✓	A measure based on the concept of conditional entropy
Variation of Information	VOI			✓				✓	✓	A measure based on the concept of both conditional entropy and mutual information
Normalized Mutual Information	NMI			✓				✓	✓	Known as the dissimilarity or distance measure
Hausdorff Distance	HAUSD				✓			✓	✓	A measure of proximity between the autosegment pixels/voxels and its hand segment
Mean Absolute Surface Distance	MASD				✓			✓	✓	A measure for how much on average the two autosegment and its hand segment differ
Average Symmetric Surface Distance	ASD				✓			✓	✓	A measure for how close two segmented surfaces/contours are
Normalized Sum of Distances	NSD	✓			✓			✓	✓	A ratio of measuring the final distance between both hand and autosegment pixels/voxels
Boundary Displacement Error	BDE				✓			✓	✓	A displacement error measurement between nearest hand and autosegment boundaries
Hamming Distance	HD	✓		✓				✓		The number of pixels/voxels in disagreement of two hand and autosegment sets
Boundary Hamming Distance	BHD						✓	✓		A measure based on the concept of both Hamming distance and the object boundaries

Color Image Segmentation Metrics

Table 3 Balancing relationship values in the segmentation metrics

TNR and TPR										
↑↓	FPR, FNR									
↑↑	↓↑	P, AC								
↑↓	↓↓	↑↓	XOR							
↑↑	↓↑	↑↑	↓↑	F						
↑↓	↓↓	↑↓	↓↓	↑↓	EP, VD					
↑↑	↓↑	↑↑	↓↑	↑↑	↓↑	AUC, VS, JI, Dice, FMI, RI, PRI, NPR				
↑↓	↓↓	↑↓	↓↓	↑↓	↓↓	↑↓	MCE, ER, LCE, GCE, BCE			
↑↑	↓↑	↑↑	↓↑	↑↑	↓↑	↑↑	↓↑	MI		
↑↓	↓↓	↑↓	↓↓	↑↓	↓↓	↑↓	↓↓	↑↓	VOI	
↑↑	↓↑	↑↑	↓↑	↑↑	↓↑	↑↑	↓↑	↑↑	↓↑	NMI
↑↓	↓↓	↑↓	↓↓	↑↓	↓↓	↑↓	↓↓	↑↓	↓↓	↑↓ HAUSD, MASD, ASD, NSD, BDE, HD, BHD

Figure 6 illustrates a set of sample images, in which the objects were located and identified; there are possible changes in the autosegmented object location and structure that are intentionally placed within the location of the GT object. It should be noted that segmentation evaluation results are often based on the size and location of overlapping object areas, specifically in similarity-based computational methods; thus, different object areas and shapes with fixed-pixel sizes, that is, the autosegmented object (1,225 pixels), the hand-segmented object (4,900 pixels), and the entire background image (10,000 pixels), are drawn; and their average pixel intensity values are measured and compared in six different test sample images as shown in Table 4. Computation of the relevance metrics of such test images are presented in Fig. 7, where according to Table 3, the relationship values among the segmentation results are a quantitative representation of the expected trend. Due to the common boundaries in some sample images, that is, S3 and C3, the BHD is different from others (the gray cells in Table 4); the best evaluation results would be achieved when the number of common boundary pixels in both a machine-segmented object and its reference region are similar.

In order to investigate how the overlapped pixels, that is, intersection of the reference object and machine-segmented object, can influence the segmentation evaluation results, where the number of object pixels are approximately identical, a set of sample images associated with different rotation sequences are assumed and presented in Fig. 8. As shown in Fig. 9, the performance relevance measurements indicate that an object similarity interpretation of the segmentation method is based on the shapes and locations of both objects. Furthermore, Table 5 presents comparison of the segmentation evaluation results which shows the best scores depend on the equivalent fraction of the geometrical properties in the objects.

For another evaluation of color image segmentation methods, the original images with their four different hand-segmented images are shown in Fig. 10,[98] where there are value judgments in comparing these different images and the original image. As shown in Fig. 11, the proposed measures represent how much the human perceptions are fuzzy; to prove this, assume that three of the hand-segmented images are known as machine-segmented images (MSIs) with different number of segments. Both RI and PRI measures are approximately similar for both

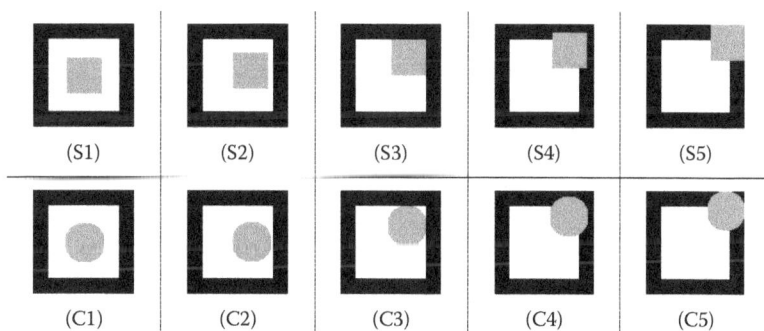

Fig. 6 Test sample images set: an automatic segmented area (gray), its hand-segmented area (white), and its image background (black)

Table 4 Performance evaluation results for the test sample images set

Image name	JI	Dice	FMI	RI	VOI	GCE	BHD	TNR	TPR	FNR	P	F	XOR	AC
S1, C1, S2, C2	0.25	0.40	0.50	0.54	1.26	0.18	1.00	1.00	0.25	0.75	1.00	0.20	0.75	0.63
S3	0.25	0.40	0.50	0.54	1.26	0.18	0.74	1.00	0.25	0.75	1.00	0.20	0.75	0.63
C3	0.25	0.40	0.50	0.54	1.26	0.18	0.85	1.00	0.25	0.75	1.00	0.20	0.75	0.63

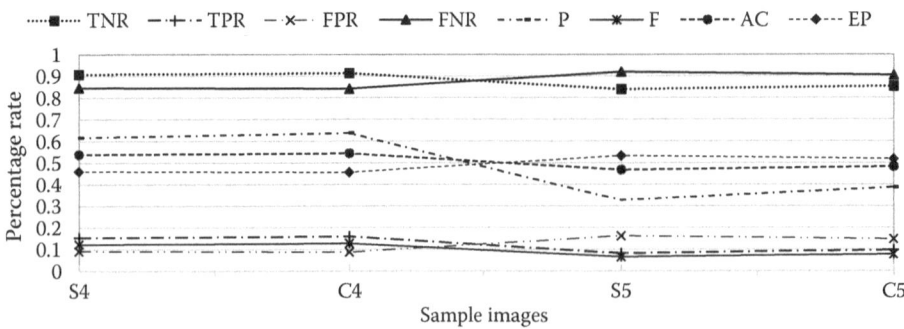

Fig. 7 The relevance measurements of segmentation results

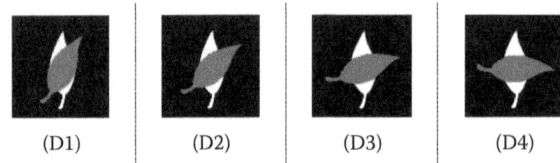

Fig. 8 Test sample images: an automatic segmented area (gray), its hand-segmented area (white), and its image background (black) with the same volumetric areas

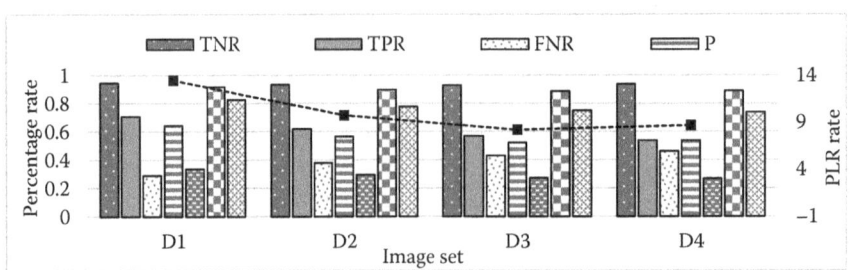

Fig. 9 The segmentation results of various relevance measurements

Table 5 Comparison of the evaluation results for the test sample images

Image name	JI	Dice	FMI	RI	VOI	GCE	HAUSD	BHD
D1	0.51	0.67	0.68	0.85	0.70	0.13	19.10	0.34
D2	0.42	0.59	0.59	0.82	0.80	0.15	23.71	0.42
D3	0.37	0.55	0.55	0.80	0.85	0.16	26.08	0.45
D4	0.37	0.54	0.54	0.81	0.83	0.16	26.00	0.46

images, in which the highest value of these metrics is obtained when an MSI contains almost all the boundaries in the hand-segmented image and/or both the test and reference images have the same large regions. The smaller value of the VOI presents a segmentation closer to its reference image (Table 6). Moreover, The GCE metric is used for evaluating a measure of overlap between image pixels of the two segmentations, and it can be defined as the minimization of the image error between the two segmentations.

CONCLUSION

In this entry, a survey of performance evaluation metrics for color image segmentation methods is given, which can be used as a starting point for the selection of the most appropriate metrics for assessing automatic image segmentation procedure. In addition, a set of comparative results for different metric groups is presented. The key finding of this comparative study is that the balancing relationships among the metrics are mutually interacted. Besides, the

Color Image Segmentation Metrics

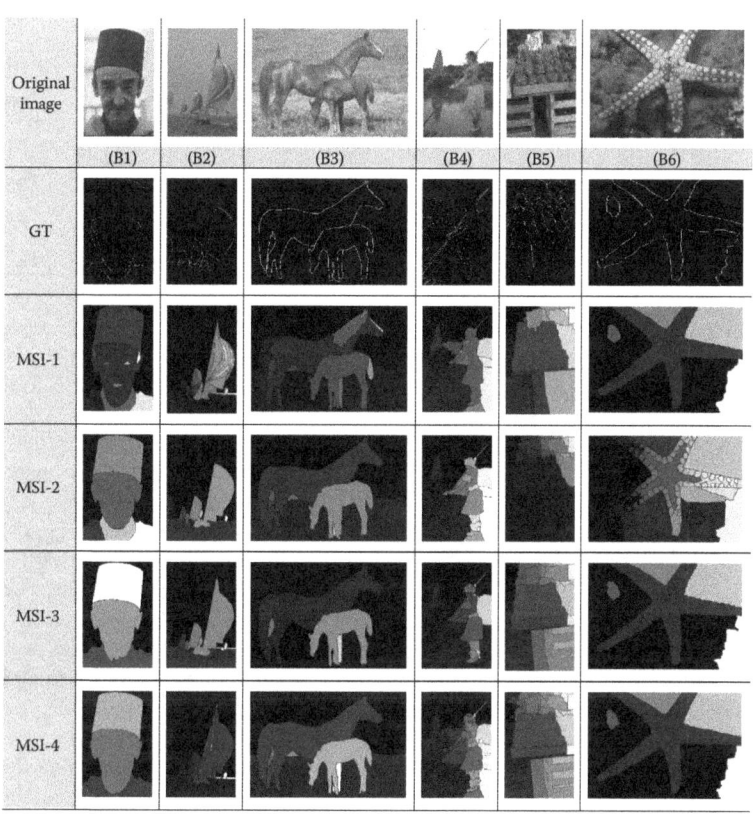

Fig. 10 Examples of the GT and original images; (B1–B6) assumed machine-segmented methods (MSI)

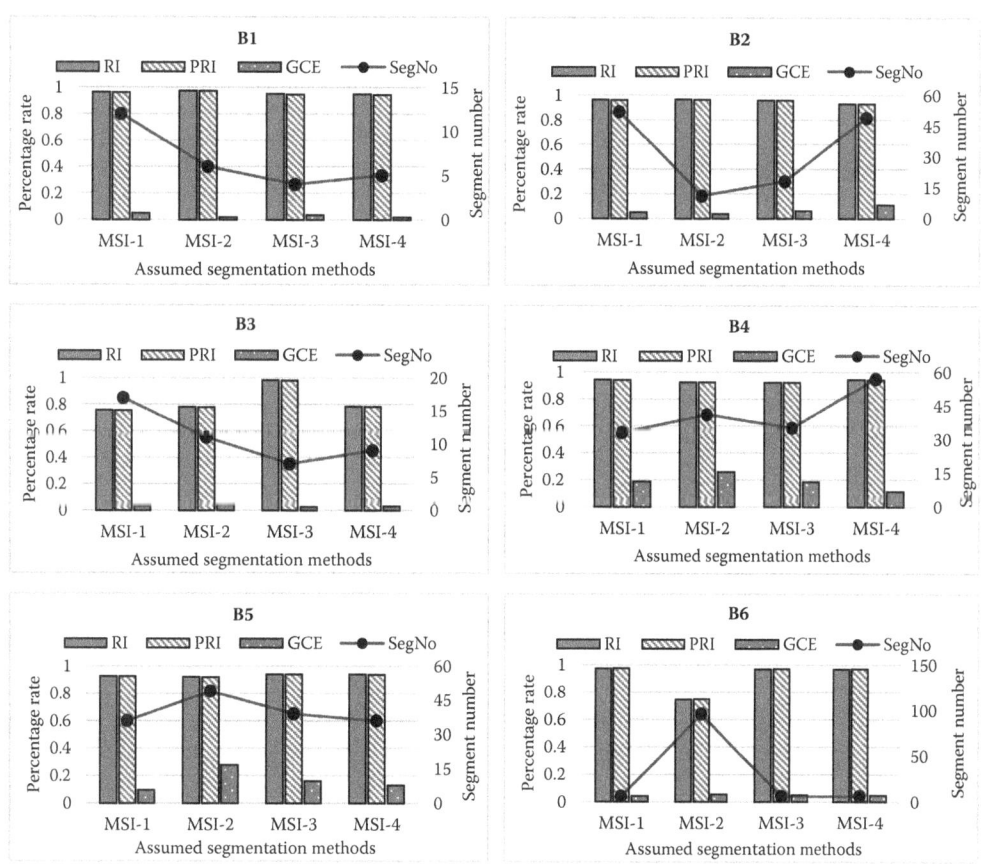

Fig. 11 Examples of the original images; (B1–B6) and assumed machine-segmented methods (MSI)

Table 6 The evaluation results for the test sample images

Methods	B1			B2			B3			B4			B5			B6		
	MI	NMI	VOI	MI	NMI	VOI	MI	NMI	VOI	MI	NMI	VOI	MI	NMI	VOI	MI	NMI	VOI
MSI-1	1.69	0.91	0.35	3.39	0.77	2.05	2.58	0.81	1.25	1.25	0.66	1.28	1.72	0.81	0.83	2.12	0.89	0.55
MSI-2	1.69	0.55	2.75	3.61	0.74	2.54	2.49	0.74	1.73	1.22	0.72	0.94	1.57	0.86	0.50	2.12	0.91	0.44
MSI-3	1.67	0.90	0.38	3.74	0.80	1.90	2.39	0.75	1.57	1.25	0.91	0.24	1.54	0.83	0.64	1.83	0.83	0.75
MSI-4	1.68	0.90	0.36	3.63	0.79	1.93	2.83	0.78	1.56	1.25	0.73	0.91	1.50	0.77	0.89	1.85	0.84	0.70

metrics have been categorized from two viewpoints: the first is focused on the importance of volume and boundary sizes; the second is focused on the features of the segmentation results, for example, relevance and/or overlapping, and distance and similarity between autosegmented image results and its reference image. Moreover, a potential benefit of analyzing the segmentation evaluation results is the ability to characterize the system development process.

REFERENCES

1. Bouthemy, P.; François, E. Motion segmentation and qualitative dynamic scene analysis from an image sequence. Int. J. Comput. Vision **1993**, *10* (2), 157–182.
2. Shirkhorshidi, A.S.; Aghabozorgi, S.; Wah, T.Y. A comparison study on similarity and dissimilarity measures in clustering continuous data. PLoS One **2015**, *10* (12), e0144059.
3. Jakóbczak, D. Modeling of high-dimensional data for applications of image segmentation in image retrieval and recognition tasks. In *Hybrid Soft Computing for Image Segmentation*; Springer International Publishing: Basel, 2016, 291–317.
4. Ozden, M.; Polat, E. A color image segmentation approach for content-based image retrieval. Pattern Recognit. **2007**, *40* (4), 1318–1325.
5. Harouni, M.; Mohamad, D.; Rasouli, A. Deductive method for recognition of on-line handwritten Persian/Arabic characters. In *The 2nd International Conference on Computer and Automation Engineering (ICCAE)*, Vol. 5; IEEE: Singapore, 2010, 791–795.
6. Lin, C.; Hsiao, M.; Lin, W. Object-based image segmentation and retrieval for texture images. Imaging Sci. J. **2015**, *63* (4), 220–234.
7. Yang, A.Y.; Wright, J.; Ma, Y.; Sastry, S.S. Unsupervised segmentation of natural images via lossy data compression. Comput. Vision Image Underst. **2008**, *110* (2), 212–225.
8. Tadeusiewicz, R.; Ogiela, M.R. *Medical Image Understanding Technology*; Springer: Heidelberg, 2004.
9. Bai, X.; Wang, W. Saliency-SVM: An automatic approach for image segmentation. Neurocomputing **2014**, *136*, 243–255.
10. Zhang, P.; Zhuo, T.; Xie, L.; Zhang, Y. Deformable object tracking with spatiotemporal segmentation in big vision surveillance. Neurocomputing **2016**, *204*, 87–96.
11. Mahalingam, T.; Mahalakshmi, M.; Eds.; Vision based moving object tracking through enhanced color image segmentation using haar classifiers. In *Trendz in Information Sciences & Computing (TISC)*; IEEE: Chennai, 2010.
12. Ko, B.; Byun, H. FRIP: A region-based image retrieval tool using automatic image segmentation and stepwise Boolean and matching. IEEE Trans. Multimedia **2005**, *7* (1), 105–113.
13. Hettiarachchi, R.; Peters, J.F. Voronoï region-based adaptive unsupervised color image segmentation. Pattern Recognit. **2017**, *65*, 119–135.
14. Zhu, G.; Zhang, S.; Zeng, Q.; Wang, C. Boundary-based image segmentation using binary level set method. Opt. Eng. **2007**, *46* (5), 050501.
15. Miranda, P.A.; Falcao, A.X.; Spina, T.V. Riverbed: A novel user-steered image segmentation method based on optimum boundary tracking. IEEE Trans. Image Process. **2012**, *21* (6), 3042–3052.
16. Harouni, M. *A smoothing embedded segmentation algorithm in recognition of online handwritten persian character*; Doctoral dissertation; Universiti Teknologi Malaysia, 2013.
17. Bellon, O.R.; Silva, L. New improvements to range image segmentation by edge detection. IEEE Signal Process Lett. **2002**, *9* (2), 43–45.
18. Brejl, M.; Sonka, M. Object localization and border detection criteria design in edge-based image segmentation: Automated learning from examples. IEEE Trans. Med. Imaging **2000**, *19* (10), 973–985.
19. Bhanu, B.; Fonder, S. Functional template-based SAR image segmentation. Pattern Recognit. **2004**, *37* (1), 61–77.
20. Zhang, H.; Wu, Q.J.; Nguyen, T.M. A robust fuzzy algorithm based on student's t-distribution and mean template for image segmentation application. IEEE Signal Process Lett. **2013**, *20* (2), 117–120.
21. Wang, X.-Y.; Wang, T.; Bu, J. Color image segmentation using pixel wise support vector machine classification. Pattern Recognit. **2011**, *44* (4), 777–787.
22. Kumar, V.; Chhabra, J.K.; Kumar, D. Automatic cluster evolution using gravitational search algorithm and its application on image segmentation. Eng. Appl. Artif. Intell. **2014**, *29*, 93–103.
23. Rasouli, P.; Meybodi, M.R. Cluster-based image segmentation using fuzzy markov random field. J. Comput. Rob. **2016**, *9* (2), 1–9.
24. Maitra, M.; Chatterjee, A. A hybrid cooperative–comprehensive learning based PSO algorithm for image segmentation using multilevel thresholding. Expert Syst. Appl. **2008**, *34* (2), 1341–1350.
25. Zhang, C.; Xie, Y.; Liu, D.; Wang, L. Fast threshold image segmentation based on 2D fuzzy fisher and random local optimized QPSO. IEEE Trans. Image Process. **2017**, *26* (3), 1355–1362.
26. Kaur, T.; Saini, B.S.; Gupta, S. Optimized multi threshold brain tumor image segmentation using two dimensional

minimum cross entropy based on co-occurrence matrix. In *Medical Imaging in Clinical Applications: Algorithmic and Computer-Based Approaches*; Dey, N.; Bhateja, V.; Ella Hassanien, A.; Eds.; Springer: Cham, 2016, 461–486.
27. Cuevas, E.; Sossa, H. A comparison of nature inspired algorithms for multi-threshold image segmentation. Expert Syst. Appl. **2013**, *40* (4), 1213–1219.
28. Peng, B.; Zhang, L.; Zhang, D. A survey of graph theoretical approaches to image segmentation. Pattern Recognit. **2013**, *46* (3), 1020–1038.
29. Heimowitz, A.; Keller, Y. Image segmentation via probabilistic graph matching. IEEE Trans. Image Process. **2016**, *25* (10), 4743–4752.
30. Tabb, M.; Ahuja, N. Multiscale image segmentation by integrated edge and region detection. IEEE Trans. Image Process. **1997**, *6* (5), 642–655.
31. Daoud, M.I.; Atallah, A.A.; Awwad, F.; Al-Najar, M.; Eds.; Accurate and fully automatic segmentation of breast ultrasound images by combining image boundary and region information. In *IEEE 13th International Symposium on, Biomedical Imaging (ISBI)*; IEEE: Prague, 2016.
32. Jones, J.L.; Xie, X.; Essa, E. Combining region-based and imprecise boundary-based cues for interactive medical image segmentation. Int. J. Numer. Methods Biomed. Eng. **2014**, *30* (12), 1649–1666.
33. Altarawneh, N.M.; Luo, S.; Regan, B.; Sun, C.; Jia, F. Global threshold and region-based active contour model for accurate image segmentation. Signal Image Process. **2014**, *5* (3), 1.
34. Cardoso, J.S.; Corte-Real, L. Toward a generic evaluation of image segmentation. IEEE Trans. Image Process. **2005**, *14* (11), 1773–1782.
35. Ortiz, A.; Oliver, G. On the use of the overlapping area matrix for image segmentation evaluation: A survey and new performance measures. Pattern Recognit. Lett. **2006**, *27* (16), 1916–1926.
36. Polak, M.; Zhang, H.; Pi, M. An evaluation metric for image segmentation of multiple objects. Image Vision Comput. **2009**, *27* (8), 1223–1227.
37. Yeghiazaryan, V.; Voiculescu, I. *An Overview of Current Evaluation Methods used in Medical Image Segmentation*; Technical Report CS-RR-15-08, Department of Computer Science; University of Oxford: Oxford, 2015.
38. Said, E.H. *Image segmentation for biometric identification systems*. Doctoral dissertation; West Virginia University Libraries: Morgantown, WV, 2007.
39. Zhang, Y.J. A survey on evaluation methods for image segmentation. Pattern Recognit. **1996**, *29* (8), 1335–1346.
40. Amer, A.; Ed.; Memory-based spatio-temporal real-time object segmentation for video surveillance. In *Electronic Imaging 2003; International Society for Optics and Photonics*; Santa Clara, CA, 2003.
41. Sharma, M. *Performance evaluation of image segmentation and texture extraction methods in scene analysis*; Master of Philosophy in Computer Science to the University of Exeter: Exeter, 2001.
42. Zhang, H.; Fritts, J.E.; Goldman, S.A. Image segmentation evaluation: A survey of unsupervised methods. Comput. Vision Image Underst. **2008**, *110* (2), 260–280.
43. Jaber, M. *Probabilistic framework for image understanding applications using bayesian networks*; Thesis; Rochester Institute of Technology, 2011.
44. Sousa, C.H.R.de. *Evaluating segmentation and classification approaches using rapideye data for vegetation mapping in Minas Gerais*, Brazil, 2013.
45. Krinidis, M.; Pitas, I. Color texture segmentation based on the modal energy of deformable surfaces. IEEE Trans. Image Process. **2009**, *18* (7), 1613–1622.
46. Mridula, J. *Feature based segmentation of colour textured images using Markov random field model*; Thesis; National Institute of Technology Rourkela: Odisha, 2011.
47. Yin, S.; Qian, Y.; Gong, M. Unsupervised hierarchical image segmentation through fuzzy entropy maximization. Pattern Recognit. **2017**, *68*, 245–259.
48. El Merabet, Y.; Meurie, C.; Ruichek, Y.; Sbihi, A.; Touahni, R. Building roof segmentation from aerial images using a lineand region-based watershed segmentation technique. Sensors **2015**, *15* (2), 3172–3203.
49. Chabrier, S.; Emile, B.; Rosenberger, C.; Laurent, H. Unsupervised performance evaluation of image segmentation. EURASIP J. Appl. Signal Process. **2006**, *1*, 217–217.
50. Stegmaier, J.; Otte, J.C.; Kobitski, A.; Bartschat, A.; Garcia, A.; Nienhaus, G.U.; Strähle, U.; Mikut, R. Fast segmentation of stained nuclei in terabyte-scale, time resolved 3D microscopy image stacks. PLoS One **2014**, *9* (2), e90036.
51. Randrianasoa, J.; Kurtz, C.; Gançarski, P.; Desjardin, E.; Passat, N. Supervised evaluation of the quality of binary partition trees based on uncertain semantic ground-truth for image segmentation. In *International Conference on Image Processing (ICIP)*, IEEE: Beijing, 2017, 3874–3878.
52. Beneš, M.; Zitova, B. Performance evaluation of image segmentation algorithms on microscopic image data. J. Microsc. **2015**, *257* (1), 65–85.
53. Bergeest, J.-P.; Rohr, K.; Eds.; Segmentation of cell nuclei in 3D microscopy images based on level set deformable models and convex minimization. In *IEEE 11th International Symposium on Biomedical Imaging (ISBI)*; IEEE: Beijing, 2014.
54. Chen, C.; Wang, W.; Ozolek, J.A.; Rohde, G.K. A flexible and robust approach for segmenting cell nuclei from 2D microscopy images using supervised learning and template matching. Cytometry Part A **2013**, *83* (5), 495–507.
55. Mohammed, M.M.; Badr, A.; Abdelhalim, M. Image classification and retrieval using optimized pulse-coupled neural network. Expert Syst. Appl. **2015**, *42* (11), 4927–4936.
56. Harouni, M.; Mohamad, D.; Rahim, M.S.M.; Halawani, S.M.; Afzali, M. Handwritten Arabic character recognition based on minimal geometric features. Int. J. Mach. Learn. Comput. **2012**, *2* (5), 578–582.
57. Mehri, M.; Gomez-Krämer, P.; Héroux, P.; Boucher, A.; Mullot, R. A texture-based pixel labeling approach for historical books. Pattern Anal. Appl. **2015**, 1–40.
58. McGuinness, K. *Image segmentation, evaluation, and applications*; Doctoral dissertation; Dublin City University: Dublin, 2010.
59. Vidyarthi, A.; Mittal, N.; Eds.; Brain tumor segmentation approaches: Review, analysis and anticipated solutions in machine learning. In *Systems Conference (NSC), 2015 39th National*; IEEE: Noida, 2015.
60. Kohli, P.; Torr, P.H. Robust higher order potentials for enforcing label consistency. Int. J. Comput. Vision **2009**, *82* (3), 302–324.

61. Molinari, F.; Zeng, G.; Suri, J.S. Inter-greedy technique for fusion of different segmentation strategies leading to high-performance carotid IMT measurement in ultrasound images. J. Med. Syst. **2011**, *35* (5), 905–919.
62. Molinari, F.; Krishnamurthi, G.; Acharya, U.R.; Sree, S.V.; Zeng, G.; Saba, L.. Nicolaides, A.; Suri, J.S. Hypothesis validation of far-wall brightness in carotid-artery ultrasound for feature-based IMT measurement using a combination of level-set segmentation and registration. IEEE Trans. Instrum. Meas. **2012**, *61* (4), 1054–1063.
63. Suri, J.; Poolla, A.; Ye, Z.; Bilhanan, A. SU-FF-I-12: Boundary Based Vs. Region-based segmentation techniques for breast lesion phantoms produced by Fischer's Full Field Digital Mammography Ultrasound System (FFDMUS): A novel tool for performance evaluation of LCD's and CRT's. Med. Phys. **2005**, *32* (6), 1906.
64. Gerig, G.; Jomier, M.; Chakos, M.; Eds.; Valmet: A new validation tool for assessing and improving 3D object segmentation. In *Medical Image Computing and Computer-Assisted Intervention–MICCAI 2001*; Utrecht, Springer: Berlin and Heidelberg, 2001.
65. Sluimer, I.; Prokop, M.; Van Ginneken, B. Toward automated segmentation of the pathological lung in CT. IEEE Trans. Med. Imaging **2005**, *24* (8), 1025–1038.
66. Zreik, M.; Leiner, T.; de Vos, B.D.; van Hamersvelt, R.W.; Viergever, M.A.; Išgum, I.; Eds.; Automatic segmentation of the left ventricle in cardiac CT angiography using convolutional neural networks. In *IEEE 13th International Symposium on, Biomedical Imaging (ISBI)*; IEEE: Prague, 2016.
67. Viergever, M.A.; Maintz, J.A.; Klein, S.; Murphy, K.; Staring, M.; Pluim, J.P. A survey of medical image registration–under review. Med. Image Anal. **2016**, *33*, 140–144.
68. Karadağ, Ö.Ö.; Senaras, C.; Vural, F.T.Y. Segmentation fusion for building detection using domain-specific information. IEEE J. Sel. Top. Appl. Earth Obs. Remote Sens. **2015**, *8* (7), 3305–3315.
69. Jiang, Y.; Zhou, Z.-H. SOM ensemble-based image segmentation. Neural Process. Lett. **2004**, *20* (3), 171–178.
70. Ni, K.; Bresson, X.; Chan, T.; Esedoglu, S. Local histogram based segmentation using the Wasserstein distance. Int. J. Comput. Vision **2009**, *84* (1), 97–111.
71. Rubner, Y.; Tomasi, C.; Eds.; Texture-based image retrieval without segmentation. In *The Proceedings of the Seventh IEEE International Conference on Computer Vision*; IEEE: Kerkyra, 1999.
72. Pratihar, S.; Begum, N.; Eds.; Understanding shape context by analysis of Farey ranks. In *5th International Conference on, Informatics, Electronics and Vision (ICIEV)*; IEEE: Dhaka, 2016.
73. Arbelaez, P.; Maire, M.; Fowlkes, C.; Malik, J. Contour detection and hierarchical image segmentation. IEEE Trans. Pattern Anal. Mach. Intell. **2011**, *33* (5), 898–916.
74. Shahedi, M.; Cool, D.W.; Bauman, G.S.; Bastian-Jordan, M.; Fenster, A.; Ward, A.D. Accuracy validation of an automated method for prostate segmentation in magnetic resonance imaging. J. Digital Imaging **2017**, 1–14.
75. Harouni, M.; Rahim, M.; Al-Rodhaan, M.; Saba, T.; Rehman, A.; Al-Dhelaan, A. Online Persian/Arabic script classification without contextual information. The Imaging Sci. J. **2014**, *62* (8), 437–448.
76. Tomalik, E. *Image-based microscale particle velocimetry in live cell microscopy*; Master of Science Thesis; Blekinge Institute of Technology: Sweden, 2013.
77. Olsén, C. *Towards automatic image analysis for computerised mammography*; Datavetenskap, Doctoral dissertation; Umea University: Sweden, 2008.
78. Huang, S.-C.; Ye, J.-H.; Chen, B.-H. An advanced single-image visibility restoration algorithm for real-world hazy scenes. IEEE Trans. Ind. Electron. **2015**, *62* (5), 2962–2972.
79. Zhou, J.; Gao, S.; Jin, Z. A new connected coherence tree algorithm for image segmentation. TIIS **2012**, *6* (4), 1188–1202.
80. Berber, T.; Alpkocak, A.; Balci, P.; Dicle, O. Breast mass contour segmentation algorithm in digital mammograms. Comput. Meth. Program. Biomed. **2013**, *110* (2), 150–159.
81. Rosenberger, C.; Chabrier, S.; Laurent, H.; Emile, B. Unsupervised and supervised image segmentation evaluation. Adv. Image Video Segmentation **2006**, 365–393.
82. Levine, M.D.; Nazif, A.M. Dynamic measurement of computer generated image segmentations. IEEE Trans. Pattern Anal. Mach. Intell. **1985**, *7* (2), 155–164.
83. Johnson, B.; Xie, Z. Unsupervised image segmentation evaluation and refinement using a multi-scale approach. ISPRS J. Photogramm. Remote Sens. **2011**, *66* (4), 473–483.
84. Borsotti, M.; Campadelli, P.; Schettini, R. Quantitative evaluation of color image segmentation results. Pattern Recognit. Lett. **1998**, *19* (8), 741–747.
85. Celebi, M.E.; Schaefer, G.; Iyatomi, H.; Stoecker, W.V.; Malters, J.M.; Grichnik, J.M. An improved objective evaluation measure for border detection in dermoscopy images. Skin Res. Technol. **2009**, *15* (4), 444–450.
86. Xie, F.; Bovik, A.C. Automatic segmentation of dermoscopy images using self-generating neural networks seeded by genetic algorithm. Pattern Recognit. **2013**, *46* (3), 1012–1019.
87. Unnikrishnan, R.; Pantofaru, C.; Hebert, M.; Eds.; A measure for objective evaluation of image segmentation algorithms. In *CVPR Workshops IEEE Computer Society Conference on, Computer Vision and Pattern Recognition-Workshops*, IEEE: San Diego, CA, 2005.
88. Yasnoff, W.A.; Mui, J.K.; Bacus, J.W. Error measures for scene segmentation. Pattern Recognit. **1977**, *9* (4), 217–231.
89. Wells, W.M.; Viola, P.; Atsumi, H.; Nakajima, S.; Kikinis, R. Multi-modal volume registration by maximization of mutual information. Med. Image Anal. **1996**, *1* (1), 35–51.
90. Viola, P.; Wells III, W.M. Alignment by maximization of mutual information. Int. J. Comput. Vision **1997**, *24* (2), 137–154.
91. Strehl, A.; Ghosh, J. Cluster ensembles—A knowledge reuse framework for combining multiple partitions. J. Mach. Learn. Res. **2002**, *3* (December), 583–617.
92. Zhang, H.; Zuo, W.; Wang, K.; Zhang, D. A snake-based approach to automated segmentation of tongue image using polar edge detector. Int. J. Imaging Syst. Technol. **2006**, *16* (4), 103–112.
93. Hausman, K.; Balint-Benczedi, F.; Pangercic, D.; Marton, Z.C.; Ueda, R.; Okada, K.; Beetz, M.; Eds.; Tracking-based interactive segmentation of textureless objects. In *IEEE International Conference on, Robotics and Automation (ICRA)*; IEEE: Karlsruhe, 2013.

94. Harouni, M.; Mohamad, D.; Rahim, M.S.M.; Halawani, S.M. Finding critical points of handwritten Persian/Arabic character. Int. J. Mach. Learn. Comput. **2012**, *2* (5), 573–577.
95. Mazaheri, M.; Mozaffari, S.; Eds.; Real time adaptive background estimation and road segmentation for vehicle classification. In *19th Iranian Conference on Electrical Engineering (ICEE)*; Tehran: IEEE, 2011.
96. Mobahi, H.; Rao, S.R.; Yang, A.Y.; Sastry, S.S.; Ma, Y. Segmentation of natural images by texture and boundary compression. Int. J. Comput. Vision **2011**, *95* (1), 86–98.
97. Wang, B.; Tu, Z.; Eds.; Affinity learning via self-diffusion for image segmentation and clustering. In *IEEE Conference on, Computer Vision and Pattern Recognition (CVPR)*; IEEE: Providence, 2012.
98. Martin, D.; Fowlkes, C.; Tal, D.; Malik, J.; Eds.; A database of human segmented natural images and its application to evaluating segmentation algorithms and measuring ecological statistics. In *ICCV 2001 Proceedings Eighth IEEE International Conference on, Computer Vision*; IEEE: Vancouver, BC, 2001.

Depth Statistics

Tania Pouli
Technicolor, Rennes, France

Erik Reinhard
Technicolor Research and Innovation, Rennes, France

Douglas W. Cunningham
Brandenburg University of Technology, Cottbus, Germany

Abstract

The statistics of 2D images can lead to powerful tools for image analysis. But in 3D environments, statistical analysis will give rise to even stronger regularities describing the structure within scenes. 2D images can be extracted from 3D ones through extraction. 3D images can be constructed from series of 2D images through other means. In this entry, these concepts are discussed and illustrated.

Keywords: 2D information; 3D information; Depth statistics.

The natural world has three spatial dimensions, and we need to percieve all three to function within it. Although the images formed on our retinas are two-dimensional spatial representations of the world, we employ a variety of cues in order to determine the geometry and depth of the environment around us.[1] Some of these cues, such as binocular disparity or motion parallax, require us to be able to move and perceive the 3D environment directly. Yet our ability to understand depth relationships in scenes is still present even when viewing a photograph, indicating that at least a subset of the necessary information is still present in the 2D representation.

Statistics of 2D images can lead to powerful tools for image analysis and understanding, as we have seen throughout this book. Strong links have been found between many statistical regularities in scenes and human perception. To form a 2D image, 3D information is projected onto a plane—be it our retinas or a camera sensor, effectively flattening depth information and removing empty space. In 3D environments, in contrast, most space is empty, making the information within them very sparse. Consequently, it is reasonable to expect that statistical analysis of the 3D information will give rise to even stronger regularities describing the structure within scenes.[2–4]

To study 3D scenes, the first step is to capture a representation of depth information within them. Typically, range scanners are used for this task, as they measure depth for each point in the scene. Figure 1 shows an example range image for a natural scene taken from the Brown Range Image Database,[5,6] visualized with a heat map (see Section 3.1.5 for a more detailed discussion on range and depth capture). Once range data is captured, it can be analyzed much in the same way as intensity images. By analyzing depth rather than intensity, the statistical properties of scene structures may be determined.

One of the earlier such studies analyzed images from the Brown Range Image Database using a series of simple statistical tools, including gradient and wavelet analysis.[5] Although many of the findings were similar to previously reported statistics from analyses of 2D image collections,[7] some interesting features were observed.

Gradient analysis of range images suggested that the structure of scenes as captured by such data is even sparser than would be obtained with 2D photographs. Figure 2 shows the logarithmic gradient distribution of the images from the Brown database for both the range and the intensity data. The log gradient distributions shown have a kurtosis of 5.1 for the vertical and 5.4 for the horizontal gradients, while the intensity distributions lead to kurtosis values of 3.2 and 3.3, respectively.

More striking results were found in the analysis of wavelet coefficients and bivariate statistics computed from the range data, further supporting the hypothesis that natural scenes are sparse and scale-invariant. Bivariate statistics can be used to study the correlations between pairs of pixels at a given distance:[5]

$$K(a,b \mid x) = Pr\{I(x_1) = a, I(x_2) = b \mid \|x_1 - x_2\| = r\}, \quad (1)$$

where x_1 and x_2 represent two different positions within the image I, which are at a distance r apart. Since images consist of mostly flat regions with discontinuities between them, it can be expected that nearby pixels are more likely to be correlated. In addition to the qualitative differences between the two representations, the bivariate distributions of the range data proved to be a better fit for a model

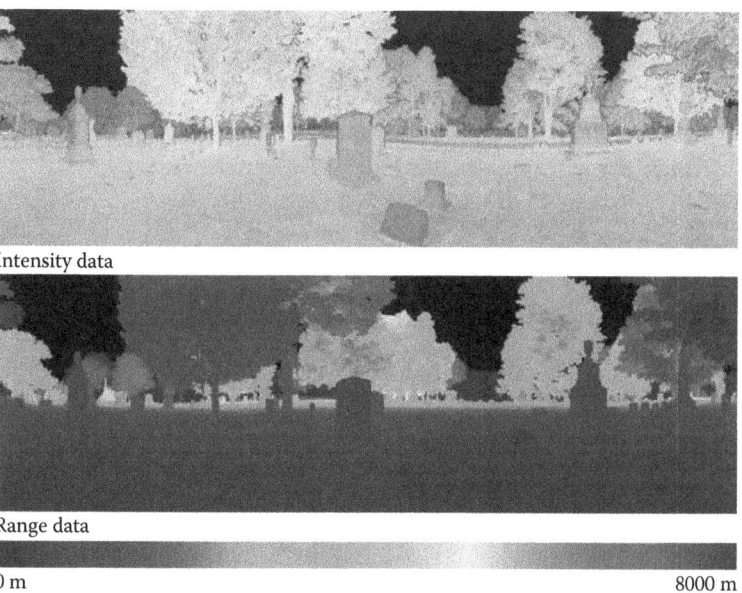

Fig. 1 An example image from the Brown Range Image Database.[5,6] The top image shows intensity data for the scene while the bottom image shows depth, here visualized with a heat map. Note that for both intensity and depth images, black pixels indicate areas where there is no information, such as sky areas or highly reflective surfaces

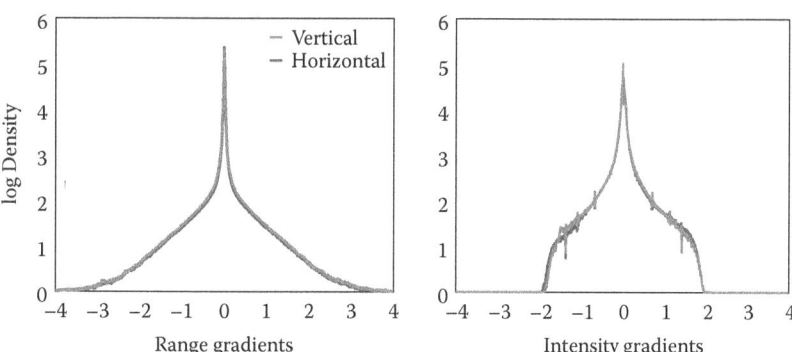

Fig. 2 Log gradient distributions for the range (left) and corresponding intensity data (right) for images in the Brown Range Image Database. The increased kurtosis in the range distributions suggests that depth information is sparser than 2D intensities of the same scenes

formalizing the sparseness and scale invariance properties of natural scenes. This model will be discussed in the following section.

THE "DEAD LEAVES" MODEL

One of the main goals of natural image statistics is to derive models that can robustly describe and predict images with natural characteristics. An important property of natural images is that they are highly non-Gaussian. Many of the statistical regularities of natural scenes are characterized by this recurring theme. For instance, gradient distributions both of individual scenes and image collections are highly kurtotic, with a sharp peak at zero. Similar distributions arise if we analyze wavelet coefficients as well as virtually any other transform we may apply to image data.

Given the non-Gaussian nature of images, what model can adequately describe natural scenes and their structure? A prevalent example is known as the "dead leaves" model, which has been repreatedly explored in the context of natural image formation.[8,9,6] Based on this model, images can be seen as consisting of sets of elementary objects ("leaves"), whose properties such as position and scale are drawn from a Poisson distribution, but which are independent of each other. Objects appear in layers and partially occlude each other, like fallen leaves would.

Figure 3 shows two example images generated with the dead leaves model using circular and square primitives and an image generated using a Gaussian model. All images are approximately scale-invariant, but the Gaussian model

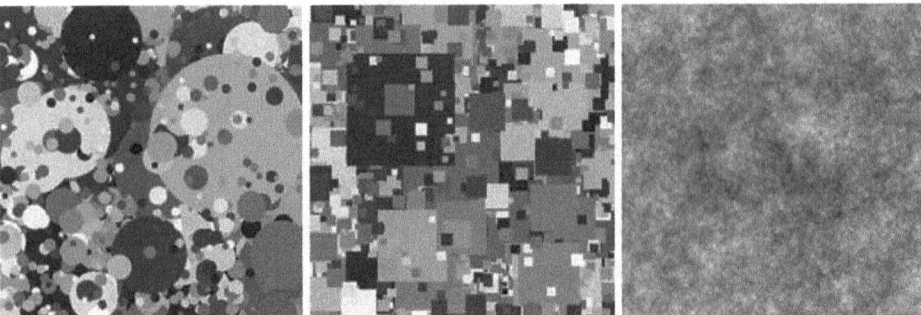

Fig. 3 The first two images were generated with the dead leaves model[6] using circular and square primitives, while the third image was created using a Gaussian model. All images were constructed such that they are approximately scale invariant

cannot generate the distinct structures that characterize natural scenes. Lee et al. showed that by using different primitives as in the two examples shown, or by varying additional parameters (e.g., using ellipsoid primitives with varying widths) this model can generate images that replicate the statistics of specific image classes, such as natural or manmade image categories.[6]

In the case of bivariate statistics (see Eq. 1), the dead leaves model can be expressed as:

$$K(a,b \mid x) = [1 - \lambda(x)]q(a)q(b) + 2\lambda(x)h_x(a+b)g_x(b-a) \quad (2)$$

where q represents the marginal distribution for a pixel, and h_x, g_x are distributions with predefined shapes. Specifically, h_x is similar to q, while g_x is concentrated at 0. The model effectively uses the parameter λ to control the probability of pixels a and b being part of the same object or belonging to different objects. If the bivariate distributions considered are looking at nearby pixels, a high λ value will be necessary, while larger separations are modeled with a lower value for this parameter. Despite its simplicity, the model can predict and replicate the statistical properties of natural scenes very effectively. More strikingly, when range data is analyzed instead of intensities, an even better fit can be achieved.

PERCEPTION OF SCENE GEOMETRY

In addition to providing useful insight into scene formation, statistics of range data can be linked to the way we perceive scene geometry. Figure 4 shows some visual illusions where properties such as length and orientation are often misperceived. Scene statistics have been related to increasingly complex perceptual oddities and visual illusions such as these.[10–14]

Although we live and function in a 3D world, we perceive the environment around us through 2D projections formed on our retinas. The loss of a dimension inevitably poses many ambiguitites that our visual system needs to resolve. However, in several cases, some aspects of the scene (such as distances or orientations) may be overestimated or underestimated. An interesting theory, which has gained growing support in recent years, explains these mismatches between real scene properties and their perceived counterparts by hypothesizing that we rely on statistical priors that relate the appearance of scenes in 2D projections with real 3D measurements.[15–17] A given 3D scene can lead to an infinite number of different 2D projections, all of which are mathematically valid. However, some are statistically more likely to occur in natural scenes than others, leading to a probability distribution for different aspects of scenes, which in turn may induce perceptual biases.

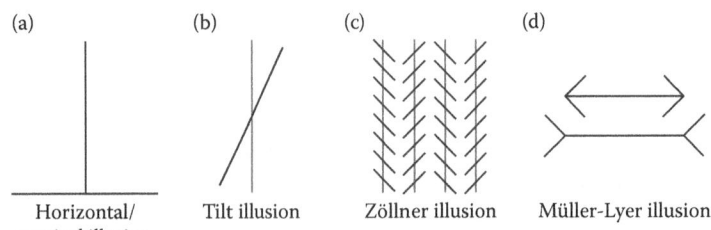

Fig. 4 Several oddities can be observed concerning the perception of lines of different orientations and configurations, many of which can be related to the statistics of 3D scenes. Some example illusions illustrating these mismatches between perception and actual scene configuration are shown here. (a) Horizontal/vertical illusion: both lines are the same length but typically the vertical line is perceived as longer. (b) Tilt illusion: the vertical line may appear to be slightly tilted counterclockwise. (c) Zöllner illusion: the vertical lines are parallel but appear tilted. (d) Müller-Lyer illusion: both horizontal lines are the same length but appear longer or shorted depending on the orientation of the arrowheads

Length Perception

One example is the perception of length and its relation to orientation: if the orientation of a stimulus in a scene is changed, so is its perceived length (see Fig. 4a). Thus, vertical distances or lengths appear longer than horizontal, even if they are identical, with the maximum perceived length occuring at 20 to 30 degrees from the vertical axis.[18–20] This relation is shown in blue in Fig. 5.

Howe et al. showed that the relation between lengths in 2D images and corresponding distances in 3D data correlates very strongly with perceived length.[10] To study this phenomenon, a set of range images was captured and 2D projections of parts of the 3D scenes were formed. Pairs of points were then randomly selected from within the 2D images and their corresponding distance in the range scenes was measured. These measurements were used to compute a ratio λ between the projected length l and the real 3D distance. Figure 5 shows in green the relation between λ and the orientation of the segments, which very closely approximates the perceptual relation (shown in blue).

Other relations between real scene properties and 2D projections have also been studied, further supporting the hypothesis that the visual system may internally employ probability distributions of scene properties to make judgments about them.[11–14,21] Many visual illusions indicate that intersecting lines at different orientations pose a challenge to our visual system.[22]

Orientation and Angle Perception

It has long been known that human observers tend to misjudge angles in images.[23,24] Specifically, acute angles are overestimated while obtuse angles are underestimated.[25] The magnitude of misperception for different angles is shown in Fig. 6.

The tilt illusion shown in Fig. 4b is one of the simplest line configurations that can cause orientation misperception consistent with these angle mismatches. In such a configuration, one of the lines (in this case, the vertical gray line) is considered in the context of another (the slanted black line). Because the acute angle subtended by the two lines is perceived as larger than it really is, the vertical gray line in this case will likely appear to be slightly tilted away from the obliquely oriented black line.

Similar to perceived length mismatches, evidence from studies of range images suggests that this perceptual oddity can also be explained by the statistics of natural scenes.[12] To study the relation between angles in 2D projections and their corresponding physical sources in the range data, intersecting line segments were detected in images using a template-based approach. Given that data, the probability of different configurations of intersecting lines were computed, revealing that 90-degree intersections were less likely than smaller or larger angles subtended by the two lines. Figure 6 (right) shows the cumulative probability of different angles as detected in this study.

Interestingly, the probability of different physical line configurations occurring in the range data was shown to be a strong predictor of the perceptual misjudgments of angles in images.[12] At a high level, this can be understood as having a prior expectation of what line configurations we are likely to encounter in our visual environment in order to resolve the ambiguity of the 2D projections on our retina. If this is true, then projected angles would be misperceived such that they fit with the probability of a given physical angle causing them. As can be seen in Fig. 6 (left), statistical findings support this hypothesis.

Specific mismatches like the ones discussed so far are not the only perceptual aspects that can be explained by the statistical structure in natural scenes. Statistical analyses of scenes have repeatedly provided evidence that a probabilistic model may underlie many perceptual oddities, especially in the context of scene geometry.[26,27,13,14,28] We refer the reader to[11] for a more detailed discussion of these phenomena.

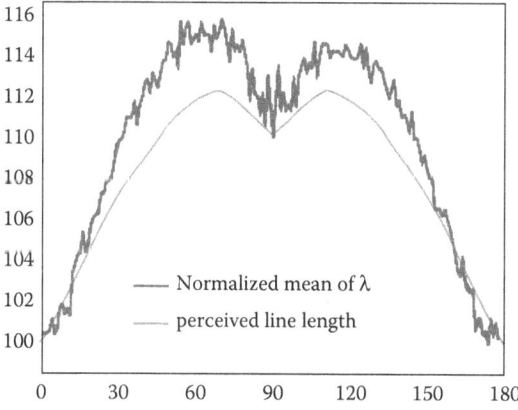

Fig. 5 The blue line shows the relation between the perceived length of lines of the same actual length in images and their orientation. Observers generally perceive vertical lines as longer, with the maximum perceived length at 20 to 30 degrees from the vertical axis. The red line shows the statistical relationship uncovered by Howe et al.[10] between orientation of edges and the ratio of their length in 2D images versus their actual 3D distance

CORRELATIONS BETWEEN 2D AND RANGE STATISTICS

The previous section looked at the links between statistical properties of range data and the way we perceive scenes. The goal of these studies is ultimately to understand human vision by determining what priors, if any, the visual system may use to resolve ambiguities between 3D physical scenes and their 2D projections. A similar problem is faced by computer vision and image processing: algorithms need to make decisions about the content of scenes based on 2D

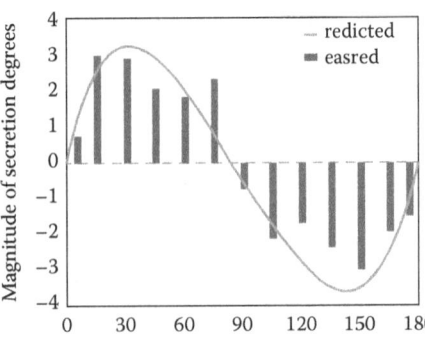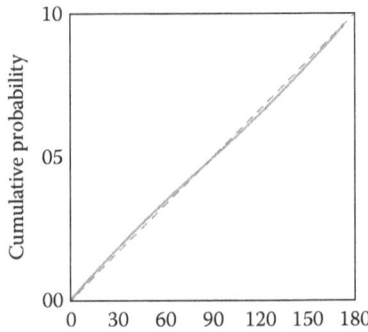

Fig. 6 Left: human observers tend to misperceive angles. Acute angles are overestimated while obtuse ones are underestimated. The gray bars show psychophysical measurements[25], while the green line shows the predicted misperception, based on the statistical analysis.[12] Right: the green line shows the cumulative probability of different angles occuring in physical, 3D scenes. If plotted in degrees rather than probability, the green line in the left plot is obtained[12]

representations of them. Since a given image may be generated by an infinite number of different configurations, some priors or assumptions are necessary to guide algorithms through this inherent ambiguity of 2D projections.

We have discussed several cases where 2D image statistics over large image sets have been used as priors for particular algorithms (e.g., see Sections 5.6 or 5.8). However, in these examples, priors serve as a representation of the structure that we would expect to find in a natural 2D image. On the other hand, the studies discussed in the earlier parts of this entry have focused on statistics of range data describing the 3D structure of scenes.

In contrast to either of these approaches, a small set of studies has focused on correlations between 2D statistics and their 3D counterparts. Kalkan et al. analyzed patches in 2D and 3D scenes in order to determine the physical causes of different types of structure in images.[29,30] Structures in 2D images can be categorized according to the following classes:[29]

- Homogeneous patches.
- Edge-like structures.
- Corners.
- Texture.

Figure 7 shows examples of these structures in an image.

Although it is possible to find patches that exemplify the characteristics of these different categories, most image patches are likely to display mixed characteristics

Fig. 7 Examples of the four types of image structure identified by Kalkan et al.[29]

and are therefore better described in a continuum, such as using the *intrinsic dimensionality* scheme.[31,32,33] This scheme was first introduced in image processing applications by Zetsche & Barth[34] and it classifies image patches according to the shape of their Fourier spectrum. A patch is categorized as:

- **i0D** if the spectrum is concentrated at the origin, as would be the case for homogeneous patches
- **i1D** if the spectrum forms a line, which is likely to occur if the patch contains an edge
- **i2D** if the spectrum is neither focused at the origin nor forming a line, which would be the case with corner or junction structures

As most patches will not perfectly fit within a single category, the intrinsic dimensionality of a patch can be defined using barycentric coordinates as a point in a triangle, as illustrated in Fig. 8. Please refer to Felsberg and Krüger[31] for details regarding the computation of this space.

Similar to 2D images, patches of 3D geometry also exhibit different types of structure. In some cases, such as for continuous surfaces, 3D structures can be directly correlated to corresponding 2D patches: a homogeneous patch in an image is likely to be caused by a smooth, continuous surface. When discontinuities occur, however, correlations between different patches are less straightforward. For instance, an edge in a 2D patch may be caused by a depth discontinuity where one object partially occludes another. It may also be due to texture on an otherwise continuous surface, or it may be caused by orientation discontinuities where two surfaces meet.

To determine whether different types of 2D structure can predict the underlying physical geometry, Kalkan et al. analyzed scenes where both range and chromatic data was captured.[29] Four types of geometry discontinuities were identified:

Surface continuity, where the underlying surface is homogeneous and does not change.
Regular gap discontinuity, occurring where a small set of surfaces meet or overlap.
Orientation discontinuity, occurring at corners where two surfaces with different orientations meet.
Irregular gap discontinuity, where the 3D structure cannot be described with a small set of surfaces and more complex interactions take place, e.g., tree branches or leaves.

These types of surface discontinuities are illustrated in Fig. 9 using simple surfaces.

Based on this classification and the intrinsic dimensionality scheme described earlier, corresponding patches in the range and 2D images were selected and categorized, allowing for correlations between 2D and 3D structures to be computed. As would be expected, continuous surface patches in 3D correlated strongly with homogeneous regions, and orientation discontinuities appeared as edges in images. But some less obvious correlations were found with other structure classes as well. Regular gap discontinuities were correlated with edges, and more so, with corner structures. Additionally, irregular gap discontinuities appeared mostly correlated with texture structures in their 2D counterparts.[29,30]

DEPTH RECONSTRUCTION

In Section 4.5, the "Dark-Is-Deep" paradigm was discussed. Extracting depth from two-dimensional projections of a scene is and has been an actively researched problem.[35–37] Since this is an underconstrained problem, statistical priors can be employed.

Human vision often relies on contextual information to resolve underconstrained problems such as recognizing objects or estimating the shape of an item. By considering not only the object in question but also the environment in which it is placed, contextual information can help constrain the space of possible solutions.[38,39] Recently, statistical priors have served to provide context in the problem of both object recognition[40–42] and geometric reconstruction of scenes from 2D images.[43–45]

In an attempt to estimate surface orientation in scenes, Hoiem et al. rely on statistical learning to divide the scene into oriented surfaces. Their method relies on two observations. First, most surfaces in outdoors images can be categorized as either sky, ground, or vertical surfaces perpendicular to the ground. Second, the orientation of surfaces in 3D can be determined from the appearance of corresponding image regions in 2D.[44,45] Although this approach does not recover accurate depth for each pixel in

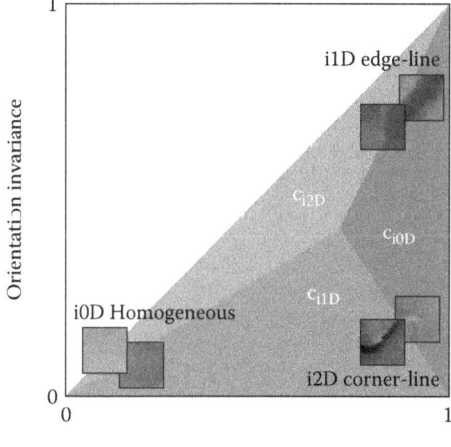

Fig. 8 A visualization of the intrinsic dimensionality space for image patches. The intrinsic dimensionality of a patch is given by barycentric coordinates ($c_{i0D} - c_{i1D}$). The axes can be understood as a representation of contrast of a patch (X-axis) and orientation invariance on the y-axis.[31,29]

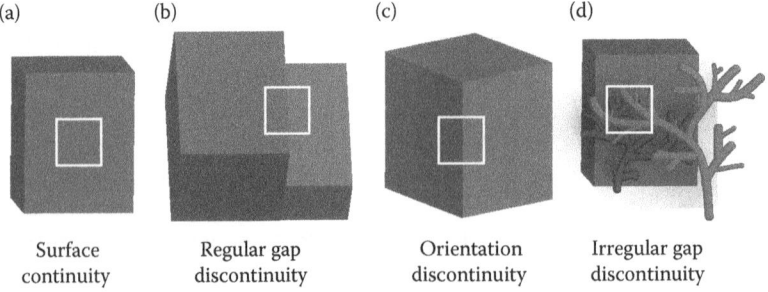

Fig. 9 Different types of patch structures in 3D geometry as identified by Kalkan et al.[29]

the scene, it successfully creates a geometric context from a single image as shown in Figs. 10 and 11.

Alternatively, instead of attempting to understand the scene as a whole, depth can be recovered from images by learning the relations between image structures and their corresponding depth. As we saw in Section 11.3, even when considered in isolation, image patches have been found to correlate to particular types of depth discontinuities.[30,29] Although these correlations are not strong enough to be used directly for reconstructing depth from an image, probabilistic learning models, such as the Markov random fields (MRF) have been used to achieve reasonably accurate reconstructions of 3D scenes.[43,46,47,48]

Saxena and colleagues combined both contextual information and local priors in a supervised learning approach for depth recovery from a single, monocular image.[49,50,51,46] Based on the observation that local features cannot be accurately interpreted without context, they used a set of multiscale features to train an MRF. Images were divided into patches, for which a depth value was computed from ground-truth 3D scanned data. Patches were then analyzed using two types of features that looked at absolute depth in each patch and relative depth relations between two patches.[49]

These examples demonstrate that even a single image contains sufficient information to recover a representation of the underlying scene geometry. Although the methods discussed do not aim to compute accurate depth maps, they show that well chosen statistical priors carry significant descriptive power.

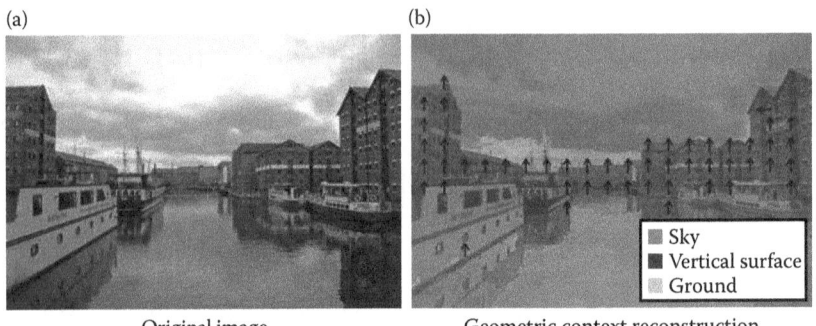

Fig. 10 Using a statistical learning approach, Hoiem et al. recover the geometric context of the scene, classifying surfaces in the image as sky, ground, and oriented surfaces[45] (Gloucester, UK, 2011)

Fig. 11 Using global and local features, Saxena et al. trained an MRF to reconstruct 3D geometry from a single image.[46] The image (a) can be mapped on the reconstructed geometry to form a virtual environment (b) that can be navigated interactively (Schwäbisch Hall, Germany, 2012)

REFERENCES

1. Howard, I.P.; Rogers, B.J. *Perceiving in Depth*; Oxford University Press: New York, 2012.
2. Potetz, B.; Lee, T.S. Statistical correlations between two-dimensional images and three-dimensional structures in natural scenes. J. Opt. Soc. Am. A **2003**, *20* (7), 1292–1303.
3. Su, C.-C.; Bovik, A.C.; Cormack, L.K. Natural scene statistics of color and range. In *Proceedings of the IEEE International Conference on Image Processing*, 2011, 257–260.
4. Wan, W.; Yang, Z. Statistics of three-dimensional natural scene structures. J. Vis. **2012**, *12* (9), 1203–1203.
5. Huang, J.; Lee, A.; Mumford, D. Statistics of range images. In *Proceedings of the IEEE International Conference on Computer Vision and Pattern Recognition*, Washington, DC, 2000, 324–331.
6. Lee, A.B.; Mumford, D.; Huang, J. Occlusion models for natural images: A statistical study of a scale-invariant dead leaves model. Int. J. Comput. Vis. **2001**, *41* (1), 35–59.
7. Huang, J.; Mumford, D. Statistics of natural images and models. In *Proceedings of the IEEE International Conference on Computer Vision and Pattern Recognition*, Vol. 1; Washington, DC, 1999; Vol. 1
8. Ruderman, D.L. Origins of scaling in natural images. Vis. Res. **1997**, *37* (23), 3385–3398.
9. Alvarez, L.; Gousseau, Y.; Morel, J.-M. The size of objects in natural and artifi-cial images. Adv. Imaging Electron Phys. **1999**, *111*, 167–242.
10. Howe, C.Q.; Purves, D. Range image statistics can explain the anomalous perception of length. Proc. Nat. Acad. Sci. **2002**, *99* (20), 13184–13188.
11. Howe, C.Q.; Purves, D. *Perceiving Geometry: Geometrical Illusions Explained by Natural Scene Statistics*; Springer: New York, 2005.
12. Howe, C.Q.; Purves, D. Natural-scene geometry predicts the perception of angles and line orientation. Proc. Nat. Acad. Sci. U. S. A. **2005**, *102* (4), 1228–1233.
13. Howe, C.Q.; Purves, D. The Müller-Lyer illusion explained by the statistics of image–source relationships. Proc. Nat. Acad. Sci. U. S. A. **2005**, *102* (4), 1234–1239.
14. Howe, C.Q.; Yang, Z.; Purves, D. The Poggendorff illusion explained by natural scene geometry. Proc. Nat. Acad. Sci. U. S. A. **2005**, *102* (21), 7707–7712.
15. Knill, D.C.; Richards, W. *Perception as Bayesian Inference*; Cambridge University Press: Cambridge, 1996.
16. Rao, R.P.; Olshausen, B.A.; Lewicki, M.S. *Probabilistic Models of the Brain: Perception and Neural Function*; MIT Press: Cambridge, MA, 2002.
17. Purves, D.; Lotto, R.B. *Why We See What We Do Redux: A Wholly Empirical Theory of Vision*; Sinauer Associates, Inc.: Sunderland, MA, 2011.
18. Pollock, W.T.; Chapanis, A. The apparent length of a line as a function of its inclination. Q. J. Exp. Psych. **1952**, *4* (4), 170–178.
19. Cormack, E.O.; Cormack, R.H. Stimulus configuration and line orientation in the horizontal-vertical illusion. Atten. Percept. Psycho. **1974**, *16* (2), 208–212.
20. Schiffman, H.R.; Thompson, J.G. The role of figure orientation and apparent depth in the perception of the horizontal-vertical illusion. Perception **1975**, *4* (1), 79.
21. Burge, J.; Fowlkes, C.C.; Banks, M.S. Natural-scene statistics predict how the figure–ground cue of convexity affects human depth perception. J. Neurosci. **2010**, *30* (21), 7269–7280.
22. Robinson, J.O. *The Psychology of Visual Illusion*; Dover Publications: Mineola, NY, 1998.
23. von Helmholtz, H.L.F. *Handbuch der Physiologischen Optik*; Leopold Voss: Leipzig, 1867.
24. Hering, E. *Outlines of a Theory of the Light Sense* (Translation from German: Zur Lehre vom Lichtsinne, 1878); Harvard University Press: Cambridge, MA, 1920.
25. Nundy, S.; Lotto, B.; Coppola, D.; Shimpi, A.; Purves, D. Why are angles misperceived? Proc. Nat. Acad. Sci. **2000**, *97* (10), 5592–5597.
26. Yang, Z.; Purves, D. A statistical explanation of visual space. Nat. Neurosci. **2003**, *6* (6), 632–640.
27. Yang, Z.; Purves, D. Image/source statistics of surfaces in natural scenes. Netw. Comput. Neural Syst. **2003**, *14* (3), 371–390.
28. Long, F.; Yang, Z.; Purves, D. Spectral statistics in natural scenes predict hue, saturation, and brightness. Proc. Nat. Acad. Sci. **2006**, *103* (15), 6013–6018.
29. Kalkan, S.; Wörgötter, F.; Krüger, N. Statistical analysis of local 3D structure in 2D images. In *Proceedings of the IEEE International Conference on Computer Vision and Pattern Recognition*, Vol. 1, 2006, 1114–1121.
30. Kalkan, S.; Wörgötter, F.; Krüger, N. First-order and second-order statistical analysis of 3D and 2D image structure. Netw. Comput. Neural Syst. **2007**, *18* (2), 129–160.
31. Felsberg, M.; Krüger, N. A probabilistic definition of intrinsic dimensionality for images. In *Pattern Recognition*; Springer, 2003, 140–147.
32. Krüger, N.; Felsberg, M. A continuous formulation of intrinsic dimension. In *Proceedings of the British Machine Vision Conference*, Vol. 2; 2003.
33. Felsberg, M.; Kalkan, S.; Krüger, N. Continuous dimensionality characterization of image structures. Image Vision Comput. **2009**, *27* (6), 628–636.
34. Zetzsche, C.; Barth, E. Fundamental limits of linear filters in the visual processing of two-dimensional signals. Vision Res. **1990**, *30* (7), 1111–1117.
35. Koenderink, J.J.; van Doorn, A.J. Shape from shading. In *The Visual Neurosciences*; Chalupa, L.M.; Werner, J.S.; Eds.; MIT Press: Cambridge, MA, 2003, 1090–1105.
36. Prados, E.; Faugeras, O. Shape from shading. In *Handbook of Mathematical Models in Computer Vision*; Paragios, N.; Chen, Y.; Faugeras, O.; Eds.; Springer: New York, 2006, 375–388.
37. Zhang, R.; Tsai, P.-S.; Cryer, J.E.; Shah, M. Shape from shading: A survey. IEEE Trans. Pattern Anal. Mach. Intell. **1999**, *21* (8), 690–706.
38. Palmer, S.E. The effects of contextual scenes on the identification of objects. Memory Cognition **1975**, *3* (5), 519–526.
39. De Graef, P.; Christiaens, D.; d'Ydewalle, G. Perceptual effects of scene context on object identification. Psychol. Res. **1990**, *52* (4), 317–329.
40. Torralba, A.; Murphy, K.P.; Freeman, W.T. Contextual models for object detection using boosted random fields. In *NIPS-17: Proceedings of the 2004 Conference on Advances in Neural Information Processing Systems*, 2004, 1401–1408.

41. Torralba, A. Contextual priming for object detection. Int. J. Comput. Vision **2003**, *53* (2), 169–191.
42. Torralba, A.; Sinha, P. Statistical context priming for object detection. In *Proceedings of the 8th IEEE International Conference on Computer Vision*, Vol. 1; 2001, 763–770.
43. Torralba, A.; Oliva, A. Depth estimation from image structure. IEEE Trans. Pattern Anal. Mach. Intell. **2002**, *24* (9), 1226–1238.
44. Hoiem, D.; Efros, A.A.; Hebert, M. Geometric context from a single image. In *Proceedings of the 10th IEEE International Conference on Computer Vision*, Vol. 1; 2005, 654–661.
45. Hoiem, D.; Efros, A.A.; Hebert, M. Recovering surface layout from an image. Int. J. Comput. Vision **2007**, *75* (1), 151–172, 267, 268.
46. Saxena, A.; Sun, M.; Ng, A.Y. Make3D: Learning 3D scene structure from a single still image. IEEE Trans. Pattern Anal. Mach. Intell. **2009**, *31* (5), 824–840.
47. Rajagopalan, A.; Chaudhuri, S. An MRF model-based approach to simultaneous recovery of depth and restoration from defocused images. IEEE Trans. Pattern Anal. Mach. Intell. **1999**, *21* (7), 577–589.
48. Rajagopalan, A.; Chaudhuri, S. Optimal recovery of depth from defocused images using an MRF model. In *Proceedings of the 6th IEEE International Conference on Computer Vision*, 1998, 1047–1052.
49. Saxena, A.; Chung, S.H.; Ng, A. Learning depth from single monocular images. In *NIPS-20: Proceedings of the 2006 Conference on Advances in Neural Information Processing Systems*; Weiss, Y.; Schölkopf, B.; Platt, J.; Eds.; MIT Press: Cambridge, MA, 2006, 1161–1168.
50. Saxena, A.; Sun, M.; Ng, A.Y. Learning 3-D scene structure from a single still image. In *Proceedings of the 11th IEEE International Conference on Computer Vision*, 2007, 1–8.
51. Saxena, A.; Chung, S.H.; Ng, A.Y. 3-D depth reconstruction from a single still image. Int. J. Comput. Vision **2008**, *76* (1), 53–69.

Digital Halftoning: Computational Intelligence-Based Approach

Arpitam Chatterjee and Kanai Chandra Paul
Department of Printing Engineering, Jadavpur University, Kolkata, India

Bipan Tudu
Department of Electronics & Instrumentation Engineering, Jadavpur University, Kolkata, India

Abstract

Digital halftoning is a process of representing continuous tone (contone) images comprising many tonal variations with devices that can support limited number of tones at output. A very common example may be printing a grayscale image consisting of different variations of gray shades with a black-and-white printer that can print only one color, that is, black on white paper. Historically, halftoning was invented for printing; however, with the advent of different digital displays, this is an indispensable process for digital displays with a limited number of tonal reproducibility. Considering the grayscale image reproduction, this can be framed as an optimization problem where the solutions are the optimized halftone patterns of the subjected continuous tone images. This entry presents the application of computational intelligence (CI) toward generation of optimized halftone patterns through minimization of visual cost function. The fundamental framework of applying CI is elaborated with three popular CI algorithms, namely, genetic algorithm, particle swarm optimization, and artificial bee colony optimization. However, the applications of binary versions of them are being presented due to the nature of the problem. The results are portrayed in detail with different qualitative mathematical evaluations. The visual and objective evaluations of the results in comparison with the output of standard digital halftoning techniques show the possible potential of applying CI techniques for obtaining improved halftones.

Keywords: Binary artificial bee colony optimization; Binary optimization; Computational intelligence; Digital halftoning; Genetic algorithm; Particle swarm optimization.

INTRODUCTION

Digital halftoning[1] is an essential process for devices such as printers, digital displays, and monitors, which are limited with their tonal reproducibility. This process fundamentally prepares a close representation of original continuous tone (contone) images with the limited number of tones that are available with the output devices. For example, a black-and-white printer works in only two states, whether *to print* or *not to print*. The halftoning process enables the devices to generate the entire image with lots of gray shades by different arrangement of black dots/pixels. Years back, this was achieved using physical screens and with the advancements of digital devices, this is achieved digitally. It is worth having a short review on some of the established techniques.

Broadly, halftoning techniques can be divided into three classes based on the conversion principle: point processing, neighborhood processing, and iterative processing. In case of point processing algorithms, whether the output pixel value will be black or white is governed by the intensity value of original input pixel. The major drawbacks of such algorithms include false contouring and sort of artificial image appearance. In case of neighborhood processing algorithms apart from the pixel under processing the neighboring pixels' original intensity values also govern the output halftone decision. The major drawback of error diffusion algorithms is objectionable low-frequency noises that created worm-like patterns. Finally, the iterative techniques, which are comparatively new and producing significant improvements, generate the halftone images based on some search-based techniques. The iterative approach of halftoning considers halftoning as a search problem, where search for the optimum halftone patterns is governed by the optimization algorithms. This paradigm is comparatively new but very promising specially in case of model-based halftoning. A consolidated list of some major works in the abovementioned classes of halftoning algorithms is shown in Table 1 for further reading.

Although halftoning was a very important process, the mathematical evaluation of halftones was not studied for long. Ulichney first significantly addressed the mathematical analysis of halftone patterns and also established three important statistical parameters, namely, pair correlation

Encyclopedia of Image Processing, First Edition
DOI: 10.1201/9781351110273-140000166
Copyright © 2018 by Taylor & Francis. All rights reserved.

Table 1 List of some major halftoning algorithms

Class of halftoning algorithm	Name of the algorithm	Proposed by
Point processing	Dispersed-dot ordered dithering	Bayer,[2] Judic et al.,[3] Ostromoukhov et al.,[4,5] Ulichney,[1] Knuth[6]
	CDOD	Holladay,[7] Adobe Inc.,[8] Kang,[9,10] Roetling[11]
Neighborhood processing	Error diffusion	Floyd and Steinberg,[12] Jarvice et al.,[13] Stucki,[14] Fan,[15] Eschbach,[16] Deitz[17]
Iterative processing	Search-based halftoning	Analoui and Allebach,[18] Velho and Gomes,[19] Scheermesser and Bryngdahl,[20] Allebach and Wong,[21] Eschbach,[22] Pappas,[23] Ulichney,[24] Lau et al.,[25] Sullivan et al.[26]

function, radially averaged power spectrum density (RAPSD), and anisotropy, toward statistical evaluation of halftones.[1] In general, the quality of the halftones depends on the distribution of minority pixels, that is, pixels with value 0 in bright regions and 1 in dark regions considering 0 representing black and 1 representing white pixels. The details of such parameters can be found in the work of Ulichney[1] and for halftones of fixed gray levels, two major desired noise properties can be observed, namely, blue noise[1] and green noise.[25] The former one is desired for halftones that are subjected to reproduction by devices such as monitors or LCD displays that can recognize individual pixels at output, whereas the latter one is desired for devices such as printers that cannot reproduce individual pixels at output rather print in terms of dots. In case of green noise, the clusters of minority pixels are obvious, but quality halftones will maintain the distribution of cluster centers with blues noise properties. In a very simple note, blue and green noise characteristics reflect the concentration of noise in high- and mid-frequency regions of human visual sensitivity. Among the different statistical parameters, RAPSD is popularly used as it shows significant regions: in case of blue noise halftone patterns with distinguishing (a) region with little or no low-frequency components, (b) region with flat high-frequency components, and (c) region with a sharp spectral peak at cutoff frequency f_b; in case of green noise halftone patterns with distinguishing (a) region with little or no low-frequency components, (b) region with high-frequency spectral components that diminish with increasing cluster size, and (c) region with a sharp spectral peak at f_g. The blue and green noise principal frequencies, that is, f_b and f_g, respectively, depend on the subjected gray level. The desired RAPSD plots are shown in Fig. 1. It is also noteworthy here that both Ulichney and Lau et al. proposed blue noise mask (BNM) and green noise mask (GNM) toward achieving high-quality halftones satisfying blue noise and green noise characteristics, respectively.

As mentioned previously, black-and-white output devices reproduce the original grayscale images by means of only two states, whether there will be a black or a white dot/pixel. This also provides the scope of forming the halftoning as a binary pattern generation process, where 1 represents white and 0 represents black. Such binary patterns can be generated adaptively as opposed to a fixed screen/weight kernels as used in case of many established techniques. Computational intelligence (CI)[27] comprises many algorithms which are principally motivated by biological behaviors present in nature. It has proven success in solving many complicated optimization problems.[28–31] This can be applied to generating halftone patterns as well by minimizing a problem-specific cost function. Among different CI algorithms, this contribution presents three popular techniques: genetic algorithm (GA)[32] motivated by biological crossover and mutation behavior in biological reproduction system, particle swarm optimization (PSO),[33] and artificial bee colony (ABC)[34] optimization, which are motivated by food-searching behavior of bird or fish swarm and honeybee, respectively. It is important to note that since the problem is shaped as finding optimized binary pattern as representation of halftones, the binary versions of those algorithms, that is, binary genetic algorithm (BGA),[35] binary particle swarm optimization (BPSO),[36] and binary artificial bee colony (bABC), are applied.

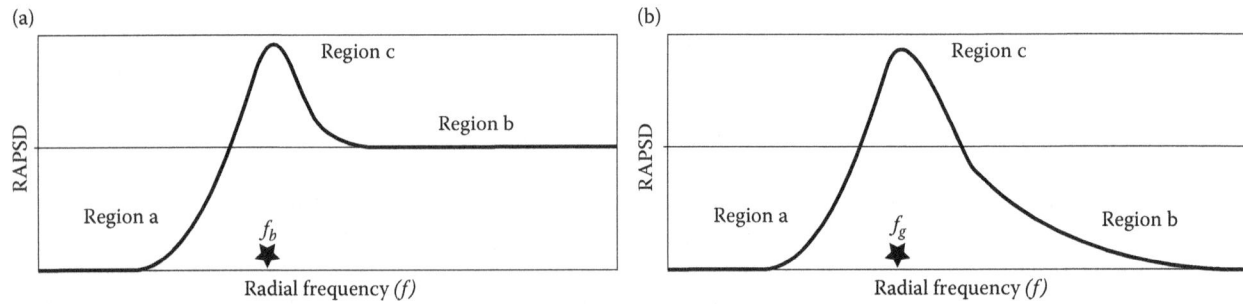

Fig. 1 Desired RAPSD for (a) blue noise and (b) green noise halftone patterns[1,25]

Table 2 Some popularly used HVS models

HVS model	CSF expression
Campbell HVS model[35]	$\text{CSF} = k\left(e^{-2\pi\alpha f} - e^{-2\pi\beta f}\right); \alpha = 0.012, \beta = 0.046$
Daly HVS model[36]	$\text{CSF} = \begin{cases} a(b+cf)\exp\left(-(cf)^d\right), \text{if } f > f_{\max} \\ 1, \text{otherwise} \end{cases}$ $a = 2.2, b = 0.192, c = 0.114, d = 1.1, f_{\max} = 6.6$
Mannos and Sakrison HVS model[37]	$\text{CSF} = a(b+cf)\exp\left(-(cf)^d\right);$ $a = 2.6, b = 0.0192, c = 0.114, d = 1.1$
Näsänen HVS model[38]	$\text{CSF} = \exp\left(-f/(c\log(L)+d)\right); c = 0.525, d = 3.91, L = 11$

The results are presented with three test images that cover images of lighter gray shades, balanced proportion of highlight, midtone, and shadows in the image and images consisting of darker shades of gray. The results are also evaluated against the objective image quality evaluation parameters and statistical analysis. One of the major drawbacks of applying CI techniques may be the high processing time. A pattern lookup table (p-LUT) approach[37] is indicated to address that.

OBJECTIVE FUNCTION FORMULATION

Inclusion of Human Vision System in Objective Function

Halftone is a visual representation of contone images that motivate the use of visual difference calculation between original and halftone. Further, this difference may be subjected to minimization. Thus, the different parameters in the cost function are calculated between human vision system (HVS) filtered original and halftone images or subimages denoted as g and h, respectively. The HVS filters are well studied, and different models of HVS filters have been proposed, while most popularly used HVS filters with their contrast sensitivity function (CSF)[38] are presented in Table 2. The general response characteristics of HVS filter can be presented in Fig. 2.

Parameters of Cost Function

As mentioned previously, search-based optimization techniques need an objective function or cost function to minimize. In this entry, the objective function φ is formulated using three parameters as described below. Note that the parameters are calculated between HVS filtered original and halftone image or subimage.

φ_1—the similarity between original and halftone images in terms of local tonal characteristics. This may

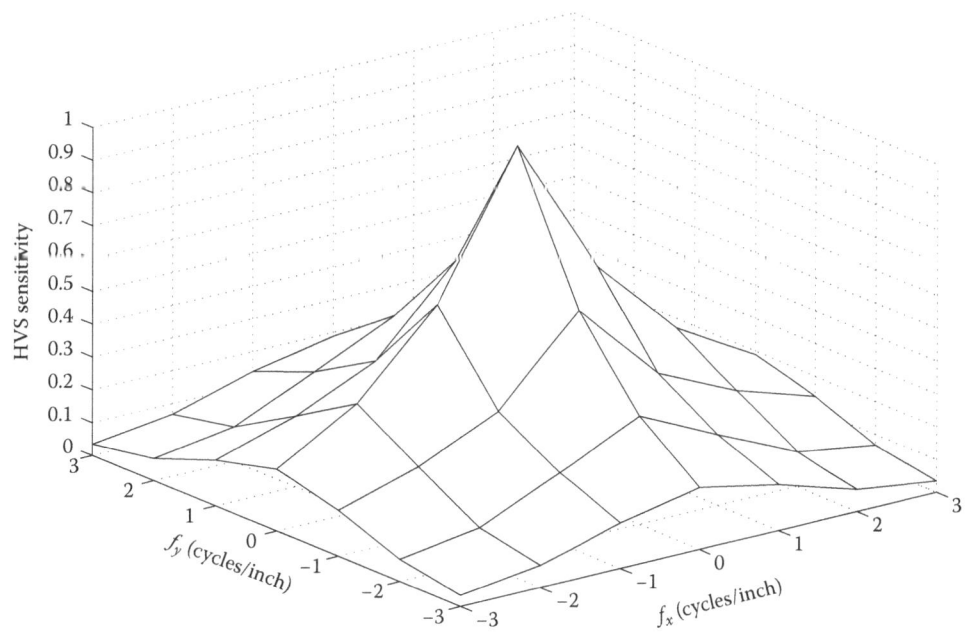

Fig. 2 The general response characteristics of HVS filters

be measured using popular image quality metric mean squared error (MSE)[39] between original grayscale and halftone image as described in the following equation:

$$\varphi_1 = \text{MSE}_{gh} = \frac{1}{N}\sum_{i=1}^{N}(g_i - h_i)^2 \tag{1}$$

where N is the total number of pixels in the subjected subimage, and $i = 1, 2, \ldots, N$ is the pixel index.

φ_2—the similarity between original and halftone images in terms of local structural characteristics. This can be measured using structural similarity index measure (SSIM).[40] But the conventional SSIM found to be insensitive to sub-pixel misregistration since it does not respect convex combination. Slightly modified form of SSIM, called sub-pixel SSIM (SPSSIM), as introduced and mathematically proven by Barkol et al.,[41] can address this limitation of SSIM and can be used to formulate φ_2 as follows:

$$\varphi_2 = 1 - \text{SPSSIM}_{gh} \tag{2}$$

φ_3—the third parameter can be related to homogeneity in the distribution of black-and-white pixels.[42] This can be obtained through a distance-based measure. The decision on minority pixel value depends on the image grayness; that is, the white pixels (pixels with value "1") in dark regions and the black pixels (pixels with value "0") in white regions as shown in the following equation:

$$p_{mn} = \begin{cases} 1, & \text{if } 0 \leq \mu_g < 0.5 \\ 0, & \text{otherwise} \end{cases} \tag{3}$$

where p_{mn} denotes the minority pixel value in the halftone representation, and μ_g is the local average intensity of the HVS filtered original image. The principal distance d_p between the minority pixels can be calculated from Eq. 4 based on approximation using square packing.

$$d_p = \begin{cases} \sqrt{\frac{1}{\mu_g}}, & \text{if } 0 \leq \mu_g < 0.5 \\ \sqrt{\frac{1}{(1-\mu_g)}}, & \text{otherwise} \end{cases} \tag{4}$$

Considering d_{mn} as the distance between minority pixel at m, n and its nearest minority pixel, the distortion measure to address blue noise characteristics can be addressed by parameter φ_3 as shown in the following equation:

$$\varphi_3 = \begin{cases} 0, & \text{if } d_{mn} \geq d_p \text{ and } h_{mn} = p_{mn} \\ 0, & \text{if } d_{mn} < d_p \text{ and } h_{mn} \neq p_{mn} \\ \left(\frac{d_p - d_{mn}}{d_p}\right)^2, & \text{otherwise} \end{cases} \tag{5}$$

where h_{mn} indicates the binary value of halftone image at pixel location m, n.

Finally, combining the above three parameters the objective function can be formulated from Eq. 6, where the weight associated to each parameter may be adjusted based on prior problem knowledge; however, here all the weights remained equal.

$$\varphi = w_1\varphi_1 + w_2\varphi_2 + w_3\varphi_3 \tag{6}$$

CI ALGORITHMS: BGA, BPSO, AND bABC

Binary Genetic Algorithm

GA is one of the most popular early developments of evolutionary search heuristics that mimics the genetic behaviors exist in nature. It has been successfully deployed to solve problems in diverse areas, for example, optimization, automatic programming, machine learning, economics, immune systems, ecology, population genetics, evolution, and learning. The advantages of GA include wide solution space, scope for high degree of parallelism, and can optimize problems with multi-objective functions, while it has limitations such as occurrence of premature convergence, cannot use gradient, and trouble in finding exact global optimum.

The fundamental steps of GA can be listed in Table 3. The binary version of BGA can be elaborated in Fig. 3.[43] As it is clear that in case of BGA, the probable solutions are coded as binary strings, and there are two identifiers one for solution and one for bit.

Binary Particle Swarm Optimization

PSO shares few commonalities with GA, for example, initiating search with randomly generated solution pool called initial population, searching in an iterative manner, and fitness evaluation at the end of iterations for each solution. But it is found to be more potential due to some remarkable differences with GA, and some of them are listed below.

- New solutions are generated by updating existing solutions instead of performing crossover and mutation operations.
- In case of GA, all the solutions share information among them and cause the whole population to move toward the optima. In contrast, for PSO, the information sharing is one way, that is, only information pertaining to the best solutions is shared for update. This contributes toward faster convergence.
- PSO does not involve any survival of the fittest dynamics like GA; hence, no solutions are discarded that provide better population stability.
- As PSO does not require such storage of solutions, it requires lesser memory compared to GA.

Table 3 Basic steps of GA search dynamics

Step 1: Defining GA parameters, cost function, fitness function, and variables.

Step 2: Generating initial population *P* randomly.

Step 3: Finding fitness of each fit$_i$ | $i = 1, 2, \ldots, P$ solution.

Step 4: Finding mates and construction of elite class. The solutions with better fitness are retained discarding the solutions that have lower fitness. The selection process is performed using mechanisms such as Roulette wheel selection, tournament selection, steady-state reproduction, and ranking and scaling.

Step 5: Mating between the solutions in elite group. The mating is performed in terms of crossover operation controlled by crossover probability (C_p).

Step 6: Performing mutation operation. Mutation alters one or more bits with a probability equal to mutation rate (μ).

Step 7: Checking convergence and declaring the optimized solution. In case the solution satisfies the stopping criteria, rest of the iteration cycles are terminated and optimized solution is declared. Otherwise, steps 3–7 are continued till the stopping criteria is met or the specified number of iterations is exhausted.

The basic steps of PSO can be presented in Table 4, while in BPSO the major steps of PSO remains unchanged except the velocity update mechanism. In this case, a probabilistic approach is adopted for velocity update.

The velocity of existing solution strings is modified in a bit-wise manner following Eq. 7, where the notations remain same as mentioned in step 5 in Table 4 but includes one more index *j* to denote the bit of individual solutions.

$$V_{i,j}^{t+1} = \varpi V_{i,j}^t + c_1 r_1 \left(l_{\text{best},i,j}^t - p_{i,j}^t \right) + c_2 r_2 \left(g_{\text{best},j}^t - p_{i,j}^t \right) \quad (7)$$

In binary space, the differences in Eq. 7 reduce to the Hamming distance and leave only three possibilities as shown in the following equation:

$$l_{\text{best},i,j}^t - p_{i,j}^t \text{ or } g_{\text{best},j}^t - p_{i,j}^t$$
$$= \begin{cases} 1, & \text{if } l_{\text{best},i,j}^t \text{ or } g_{\text{best},j}^t = 1 \text{ and } p_{i,j}^t = 0 \\ 0, & \text{if } l_{\text{best},i,j}^t \text{ or } g_{\text{best},j}^t = p_{i,j}^t = 0 \text{ or } 1 \\ -1, & \text{if } l_{\text{best},i,j}^t \text{ or } g_{\text{best},j}^t = 0 \text{ and } p_{i,j}^t = 1 \end{cases} \quad (8)$$

The sigmoid transform as shown in Eq. 9 is performed using equation, and the transformed value is utilized as the probability of changing the current bit values. This sigmoid transform also eliminates the boundary handling technique required for continuous PSO, where a velocity range [$V_{\max}, V_{\min}$] is necessary to ensure the particles are not flying out of the swarm.

$$s_{ij}^t = \frac{1}{1 + \exp(-V_{i,j}^t)} \quad (9)$$

The sigmoid transformed values obtained from Eq. 9 are compared against a uniformly generated random number $q_{i,j}^t \in [0,1]$, and the modified binary values of the bit are decided as shown in the following equation:

$$p_{i,j}^{t+1} = \begin{cases} 1, & \text{if } s_{i,j}^t \geq q_{i,j}^t \\ 0, & \text{if } s_{i,j}^t < q_{i,j}^t \end{cases} \quad (10)$$

In case of equality between $s_{i,j}^t$ and $q_{i,j}^t$, there is equal probability of changing the bit values to 1 and 0. Such cases can be addressed by random update of the bit values to 1 or 0.

Binary Artificial Bee Colony

ABC is a comparatively new development in the field of swarm intelligence that mimics the foraging behavior exists in honeybee swarms. It shares common features with GA and PSO, for example, initiating search with randomly generated population of solutions, but this algorithm does not include survival-of-the-fittest strategy like GA and provide better population stability, fitness evaluation at each cycle of search, memory-based search dynamics like PSO, having few control parameters, etc. One of the main differences of ABC algorithm from GA and PSO is its ability to perform local as well as global search consecutively and separately by involving different classes of bees. This feature provides advantages such as faster convergence performance, strong robustness, and high flexibility.[44]

The major steps of ABC can be described in Table 5, whereas in bABC, the binary version of ABC, the update dynamics follow the same bit-wise probabilistic fashion as described in Section "Binary Particle Swarm Optimization."

p-LUT Implementation

Application of above algorithms in halftoning can be accomplished in a pixel-wise fashion, where each image can be divided into subimages and each subimages to be subjected to optimization techniques for obtaining the optimized halftone pattern for the subjected subimage. This approach can be expensive in terms of processing time. The p-LUT approach may be a possible solution to that. In this approach, a lookup table comprising the possible halftone patterns for different gray levels is generated using the optimization techniques. The subimages of the original is then evaluated against the best matching pattern

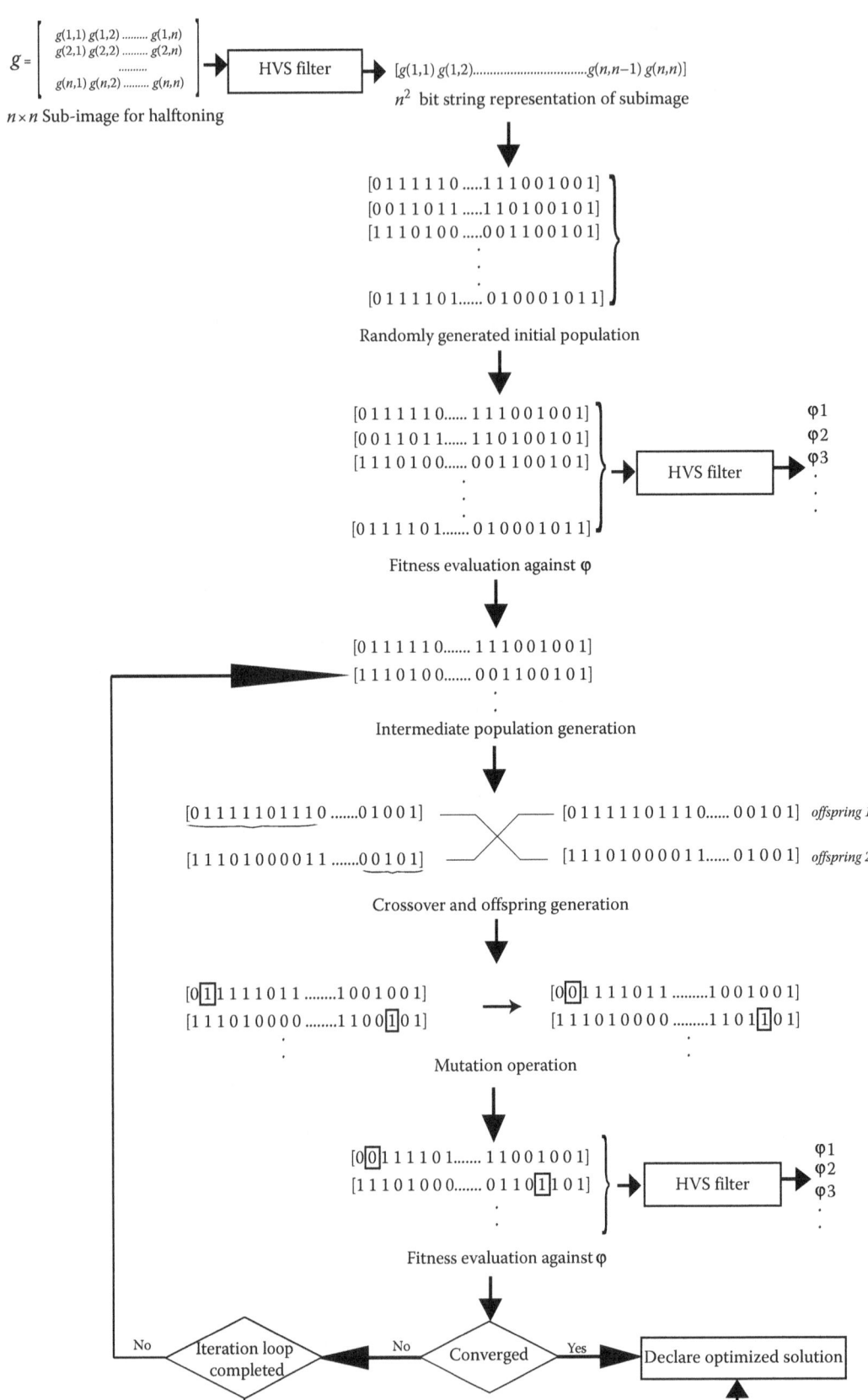

Fig. 3 BGA optimization to obtain $n \times n$ optimized halftone patterns

Table 4 Steps of PSO

Step 1: Defining PSO parameters, cost function, fitness function, and variables.

Step 2: Generating initial population P randomly.

Step 3: Evaluating fitness of solutions.

Step 4: Finding local and global best solutions l_{best} and g_{best}, respectively.

Step 5: Updating velocity and position of particles using following equations where V_i^t and X_i^t are the velocity and position for the ith particle in the swarm at the tth iteration, respectively. Similarly, V_i^{t+1} and X_i^{t+1} are the updated velocity and position at the $(t+1)^{th}$ iteration, respectively; $l_{best,i}^t$ represents the local best solution found by the ith particle at the tth iteration; g_{best}^t is the global best solution at the tth iteration; ϖ is a time-varying inertia that contributes toward the convergence speed of the optimization process; and c_1, c_2 in below equation are positive weight constants that control the relative impact of l_{best} and g_{best} positions on the velocity update.

$$V_i^{t+1} = \varpi V_i^t + c_1 r_1 \left(l_{best,i}^t - p_i^t \right) + c_2 r_2 \left(g_{best}^t - p_i^t \right)$$

$$X_i^{t+1} = X_i^t + V_i^{t+1}$$

Step 6: Evaluating the modified solution. The modified solution is retained in case that causes higher fitness than the existing solution; otherwise the existing solution is retained.

Step 7: Checking convergence and declaring the optimized solution; steps 4–7 are repeated if stopping criteria is not met or the given number of iterations is exhausted.

Table 5 Major steps of ABC algorithm

Step 1: Defining PSO parameters, cost function, fitness function, and variables.

Step 2: Initializing employed bees where half of the total N solutions in the initial population are considered as employed bees, and fitness of solution $fit_i \mid i = 1, 2, ..., SN$ is evaluated against the defined fitness function.

Step 3: Generating new solutions by employed bees where each employed bee performs search in its vicinity and generates new solution $v_i \mid i = 1, 2, \cdots, SN$ using the following equation

$$v_{ij} = x_{ij} + \eta_{ij} \left(x_{ij} - x_{kj} \right)$$

where j is the index for dimension of the optimization problem, η is a random number generated in $[-1, 1]$, and k is a randomly selected number in $[1, 2, ..., SN]$. The fitness of the new solution v_i is evaluated. The current solution x_i is replaced by the new solution v_i, in case it shows higher fitness.

Step 4: Assigning onlookers. The probability of the onlooker bees getting associated to the employed bees p_i is calculated as follows:

$$p_i = fit_i \bigg/ \sum_{i=1}^{SN} fit$$

Step 5: Generating new solution by the onlookers. New solutions by the onlookers are generated using equation mentioned under step 3, and fitness of the new solutions is evaluated.

Step 6: Generating new solutions by scout bees by the following equation, where r is a random number in $[0, 1]$ and x_j^{min}, x_j^{max} are the upper and lower bounds of the jth dimension in the solution space.

$$x_{ij} = x_j^{min} + r \left(x_j^{max} - x_j^{min} \right)$$

Step 7: Selecting an optimized solution once stopping criterion is reached or the defined number of iterations is exhausted.

from the p-LUT and used as the halftone representation of the subjected subimage. The pattern selection can be made based on different parameters of the cost function presented in Section "Parameters of Cost Function." The size of the patterns in p-LUT can be user specified, but since the grayness matching is performed based on the local mean intensity, a higher pattern size can cause loss of information fidelity. Also, the p-LUT construction itself requires considerable processing time, but once constructed the halftoning of individual images is considerably faster. The p-LUT approach is close to the clustered-dot ordered dithering (CDOD) method in terms of implementation, and hence, the target noise profile is green noise.

RESULTS

The results can be first studied with the halftoned ramp generated with different CI algorithms as shown in Fig. 4. The ramp presents halftone representation for nine distinguishable gray levels from 0.1 presenting dark and 0.9 presenting bright regions. This can be considered as comprehensive representation of the ability of the CI algorithms to produce different gray levels. The results in Fig. 4 were accomplished using pixel-wise optimization, and the results are reproduced at 300 dpi. The RAPSD plots for halftones of fixed gray level 0.5 are shown in Fig. 5, which can be used to assess the degree of

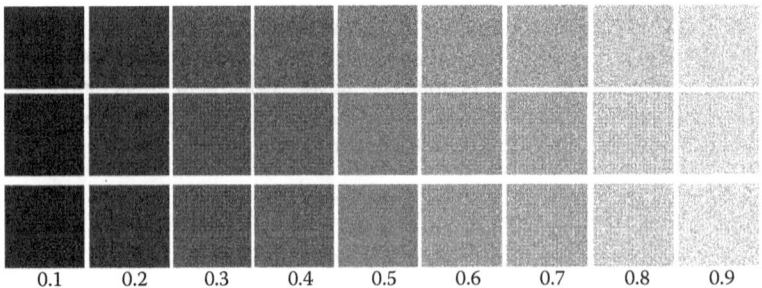

Fig. 4 Halftoned ramp generated using different CI algorithms: (top) BGA; (middle) BPSO; (bottom) bABC

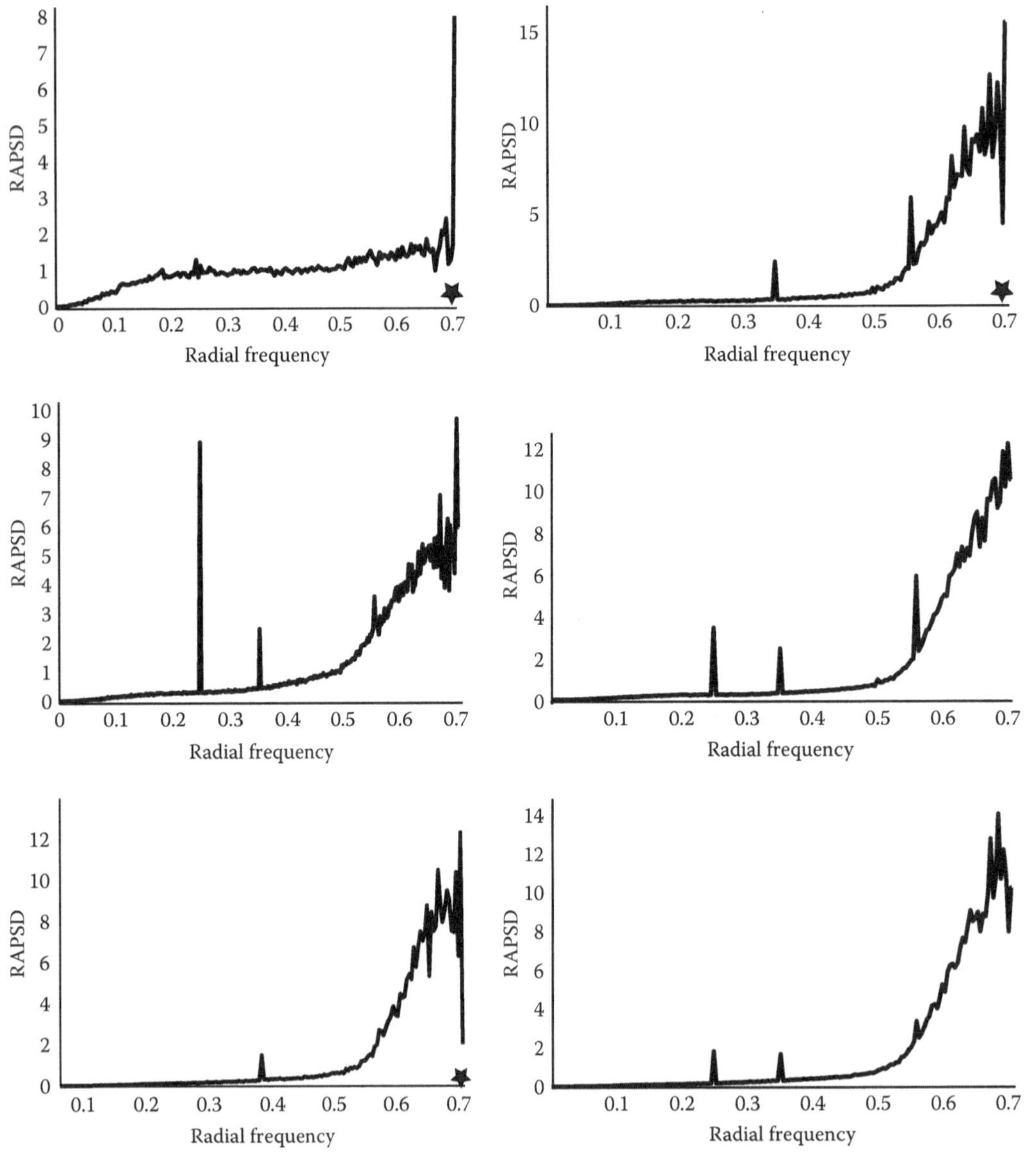

Fig. 5 RAPSD plots for fixed gray level 0.5 with CI algorithms using (top) pixel-wise and (bottom) p-LUT approach. The CI algorithms are (left) BGA, (middle) BPSO, and (right) bABC

conformity to the desired blue or green noise characteristics. In Fig. 5, the halftones of fixed gray level generated using both pixel-wise and p-LUT fashions are considered for RAPSD analysis. The results presented in this entry are generated as 1-bit tiff images so that the binary patterns can be viewed.

Results of CI-based halftoning using the p-LUT approach for the test images obtained from standard image databases[45,46] are shown in Figs. 6–8. Three images with different lightness are presented to assess the performance of images with different gray level distribution. The processing time, the main reason for introducing the p-LUT approach, was studied, and the same for different image sizes is shown in Table 6. It includes some of the established technique processing times as well for comparative analysis.

Finally, the output of halftones for real-life images obtained from two standard databases was evaluated against the standard image quality evaluation metrics. This may address the subjective nature of visual evaluation. The outcome of this objective evaluation is presented in Table 7, where some other standard techniques are also included to draw comparison. Among many such qualitative metrics, four popular metrics, namely, MSE,[39] visual signal-to-noise ratio (VSNR),[48] mean SSIM (MSSIM),[40] and feature similarity index measure (FSIM),[49] are shown here.

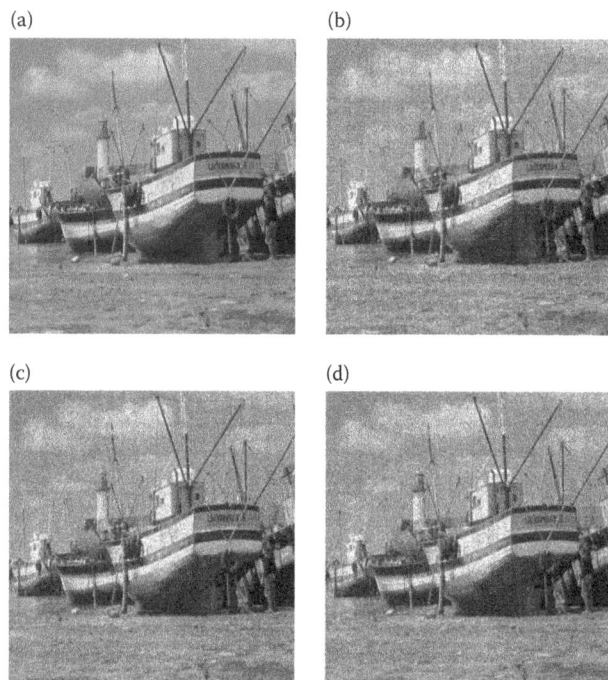

Fig. 7 Halftone of sample (a) image with balanced gray level distribution using (b) BGA, (c) BPSO, and (d) bABC generated p-LUT approach (reproduced at 600 dpi)

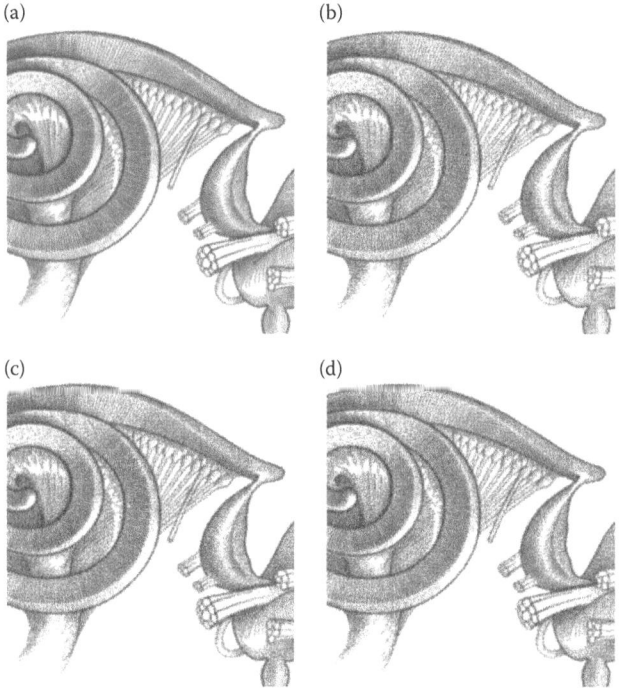

Fig. 6 Halftone of sample (a) bright image using (b) BGA, (c) BPSO, and (d) bABC generated p-LUT approach (reproduced at 600 dpi)

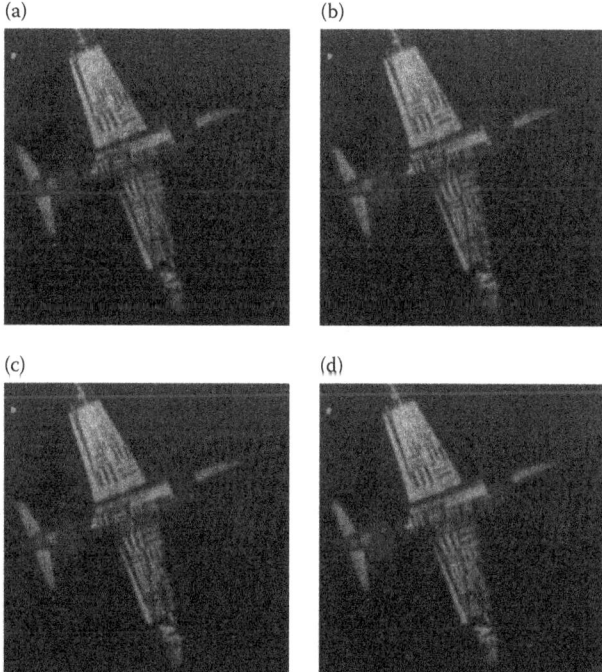

Fig. 8 Halftone of sample (a) dark image using (b) BGA, (c) BPSO, and (d) bABC generated p-LUT approach (reproduced at 600 dpi)

Table 6 Comparison of processing time in seconds using Windows PC with Intel® Core™ 2 Duo CPU, E7400, 2.8GHz, 4 GB RAM

Techniques	Image size (in pixels)				
	256 × 256	512 × 512	1024 × 1024	2048 × 2048	2560 × 2560
Error diffusion (ED)[12]	1.519	2.870	12.046	31.580	46.662
BNM[1]	0.727	1.301	4.165	5.701	9.019
GNM[25]	5.937	9.518	18.144	53.033	81.437
Block error diffusion (BED)[47]	3.324	5.348	12.940	44.441	69.315
p-LUT	0.595	1.895	5.201	21.367	37.293

Table 7 Objective evaluation against standard image quality metrics

	ED	BNM	GNM	BED	BGA	BPSO	bABC
For image database[45]							
MSE[39]	0.212	0.208	0.210	0.199	0.212	0.207	0.208
VSNR[48]	10.05	8.98	8.55	4.71	8.950	9.265	10.257
MSSIM[40]	0.9998	0.9997	0.9997	0.9994	0.9997	0.9997	0.9997
FSIM[49]	0.9218	0.9318	0.9370	0.9105	0.964	0.951	0.9627
For image database[46]							
MSE	0.168	0.153	0.154	0.243	0.154	0.152	0.156
VSNR	9.685	7.669	11.388	8.444	10.267	11.849	11.198
MSSIM	0.9998	0.9997	0.9997	0.9978	0.9998	0.9998	0.9998
FSIM	0.9378	0.9328	0.9836	0.9459	0.9808	0.9901	0.9872

DISCUSSIONS

Figure 4 clearly shows the potential of all three CI algorithms to generate distinguishable gray levels. Although all of them produce almost same quality halftones visually, the RAPSD plots shown in Fig. 5 reveal that they are qualitatively different as the noise distributions are different between the plots. Figure 5 also includes the RAPSD plots for the halftones obtained with the p-LUT approach. It is important to note that in case of p-LUT generated halftones, the noise concentration is comparatively more in mid-frequency regions. This reveals the fact that the pixel-wise implementation results are in higher degree of conformity with the blue noise properties, whereas the p-LUT approach results are with the green noise characteristics. The principal blue noise frequency (f_b) is denoted by a star mark in the frequency axis. This is the point where the peak should take place in the ideal blue noise scenario along with no or less noise concentration prior to f_b, whereas in case of green noise, the peak is expected to appear prior to f_b. In all the CI results, the peaks are appearing near or at the desired point (star mark), but the distribution to the lower frequencies is different which results in the varying amount of graininess in the results.

The halftone images in Figs. 6–8 clearly show the ability of the presented CI algorithms to reproduce faithfully the original images comprising different lightness distribution using binary pixels. The p-LUT approach also reduces the processing time considerably as shown in Table 6. The processing time may be considered acceptable in comparison with the standard techniques.

The results presented in Table 7 also reveal the fact that the CI-based techniques with p-LUT implementation perform competitively with the standard techniques. The four metrics shown here also indicates improvement in different aspects. The lower values in terms of MSE show the gray level-wise difference, whereas the higher values of VSNR indicate the better visual signal fidelity. In case of MSSIM higher values represent the better similarities between the original and halftone images in terms of structural information in the original image, whereas the higher values of FSIM reflect the ability of the CI algorithms toward better retention of original image features with the halftone images. The comparative results also show that among the different CI algorithms considered in this entry, BPSO and bABC outperform BGA. It is also important to note that the CI-based implementations also draw a trade-off between blue and green noise techniques; hence, it may be suitable for both type of devices that can and cannot reproduce the individual pixels at output.

CONCLUSION

The entry presented CI algorithm-based generation of halftone images for grayscale reproduction with devices that can reproduce limited number of tones at output. The visual cost function assesses the visual difference between the original and corresponding halftone representation. The CI algorithms are used to minimize the calculated difference, and the use of binary version of the CI algorithms enables generating the binary solutions. The steps and application of binary version for three popular CI algorithms, that is, BGA, BPSO, and bABC, have been presented in this entry. The results for fixed gray levels have been presented for visual assessments and also been analyzed statistically to portray the noise profile and degree of conformations to the desired noise profiles. The p-LUT approach for minimizing the high processing time with pixel-wise optimization framework is presented, and results of its application to the real-life images from standard databases have been shown. The halftones of real-life images have been evaluated objectively using popular image quality metrics and compared with the results of established techniques. The results obtained using CI algorithms are inspiring as they show significant improvement over the standard techniques under consideration. The work can be further extended to color images, more robust objective function for improvement, applications of multi-objective optimization concepts, etc. Finally, this entry may be a motivating step toward application of CI algorithms in digital halftoning.

ACKNOWLEDGMENTS

We acknowledge sincere thanks to publishers of our previously published papers for granting the permissions: Chatterjee, A.; Tudu, B.; Paul, K.C. Towards optimized binary pattern generation for grayscale digital halftoning: A binary particle swarm optimization (BPSO) approach. J. Vis. Commun. Image R **2012**, *23*, 1245–1259, Chatterjee, A.; Tudu, B.; Paul, K.C. Binary genetic algorithm-based pattern LUT for grayscale digital half-toning. Signal Image Video Process. **2013**, *7*, 377, Chatterjee, A.; Tudu, B.; Paul, K.C. Binary grayscale halftone pattern generation using binary artificial bee colony (bABC). Signal Image Video Process. **2013**, *7*, 1195.

REFERENCES

1. Ulichney, R.A. *Digital Halftoning*; MIT Press: Cambridge, MA, 1987.
2. Bayer, B.E. An optimum method for two level renditions of continuous-tone pictures. In *Proceedings of International Conference on Communications*; Conference Record 11–15, Seattle, Washington, DC, June 11–13, 1973.
3. Judice, C.N.; Jarvis, J.F.; Ninke, W.H. Using ordered dither to display continuous tone pictures on AC plasma panel. Proc. SID **1974**, *15*, 161–169.
4. Ostromoukhov, V.; Hersch, R.D. Halftoning by rotating non-Bayer dispersed dither array. Proc. SPIE **1995**, *2441*, 180–197.
5. Osromoukhov, V.; Hersch, R.D.; Amidror, I. Rotated dispersed dither: A new technique for digital halftoning. Comput. Graph. Proc. Ann. Conf. Series **1994**, 123–130.
6. Knuth, D.E. Digital halftones by dot-diffusion. ACM TOG **1987**, *4*, 245–273.
7. Holladay, T.M. An optimum algorithm for halftone generation for displays and hard copies. Proc. SID **1980**, *21*, 185–192.
8. Adobe Systems Inc. *PostScript™ Language Reference Manual Supplement*, 1995, 167–170.
9. Kang, H.R. Frequency analyses of microcluster halftoning. In *IS&T's NIP 13*, Seattle, Washington, DC, 1997, 492–501.
10. Kang, H.R. Microcluster line screens and frequency analyses. In *EI'98*, San Jose, CA, 1998, 24–30.
11. Roetling, P.G. Electronic Halftone Imaging System. US patent 4051536, 1977.
12. Floyd, R.W.; Steinberg, L. An adaptive algorithm for spatial gray-scale. Proc. SID **1976**, *17* (2), 75–78.
13. Jarvis, J.F.; Judice, C.N.; Ninke, W.H. A survey of techniques for the display of continuous-tone pictures on bilevel displays. Comput. Vision Graph. **1976**, *5*, 13–40.
14. Stucki, P. *Mecca: A Multiple-Error Correcting Computation Algorithm for Bilevel Image Hardcopy Reproduction*; Technical Report RZ1060; IBM Research Laboratory: Zurich, 1981.
15. Fan, Z. A simple modification of error diffusion weights. In *Proceedings of IS&T 46th Annual Conference*; Cambridge, MA, 1993, 113–115.
16. Eschbach, R. Reduction of artifacts in error diffusion by means of input-dependent weights. JEI **1993**, *2* (4), 352–358.
17. Dietz, H. *Randomized Error Distribution*; Industrial Associates at Brooklyn Polytechnic Institute: New York, 1985.
18. Analoui, M.; Allebach, J.P. Model based halftoning using direct binary search. In *Proceedings of the SPIE, Human Vision, Visual Processing, and Digital Display III*, August 27, 1992. doi:10.1117/12.135959.
19. Velho, L.; deMiranda Gomes, J. Digital haltoning with space filling curves. Comput. Graph. **1991**, *25*, 81–90.
20. Scheermesser, T.; Bryngdahl, O. Spatially dependent texture analysis and control in digital halftoning. J. Opt. Soc. Am. A **1996**, *13*, 2348–2354.
21. Allebach, J.P.; Wong, P.W. *Introduction to Digital Halftoning, Tutorial Notes, IS&T NIP 13*, Seattle, Washington, DC, November 2, 1997, 4.2–4.26.
22. Eschbach, R. Pulse-density modulation on rastered media: Combining pulse-density modulation and error diffusion. J. Opt. Soc. Am. A **1990**, *26*, 708–716.
23. Pappas, T.N. Digital halftoning techniques for printing. In *ICPS'94: The Physics and Chemistry in Imaging Systems—IS&T 47th Annual Conference*, Vol. 2; Rochester, NY, 1994, 468–471.
24. Ulichney, R. The void-and-cluster method for dither array generation. Proc. SPIE **1993**, *1913*, 332–343.
25. Lau, D.L.; Arce, G.R. *Modern Digital Halftoning*; Marcel Dekker: New York, 2001.
26. Sullivan, R.J.; Ray, L.A.; Miller, R. Design of minimum visual modulation halftone patterns. IEEE Trans. Syst. Man Cybern. **1991**, *21*, 33–38.

27. Eberhart, R.C.Y.; Shi, Y. *Computational Intelligence: Concepts to Implementations*; Morgan Kaufmann: Burlington, MA, 2007.
28. Yao, X. *Evolutionary Computation: Theory and Applications*; World Scientific Publishers: Singapore, 1999.
29. Smolinski, T.G. *Application of Computational Intelligence in Biology: Current trends and Open Problems*; Smolinski, T.G.; Milanova, M.G.; Hassanien, A.E.; Eds.; Springer: Berlin, 2008.
30. Palit, A.K.; Popovic, D. *Computational Intelligence in Time Series Forecasting: Theory and Engineering Applications*; Springer: London, 2005.
31. Wang, L.; Fu, X. *Data Mining with Computational Intelligence*; Wu, X.; Jain, L.; Eds.; Springer-Verlag: Berlin, 2005.
32. Holland, J.H. *Adaptation in Natural and Artificial Systems*; University of Michigan Press: Ann Arbor, MI, 1975.
33. Kennedy, J.; Eberhart, R. Particle swarm optimization. Proc. IEEE Int. Conf. Neural Networks **1995**, *4*, 1942–1948.
34. Karaboga, D. *An Idea Based on Honey Bee Swarm for Numerical Optimization*; Technical Report. TR06, Computer Engineering Department, Erciyes University: Kayseri, 2005.
35. Haupt, R.L.; Haupt, S.E. *Practical Genetic Algorithms*, 2nd Ed.; John Wiley & Sons: New York, 2004.
36. Kennedy, J.; Eberhart, R. A discrete binary version of the particle swarm algorithm. Proc. IEEE Int. Conf. Syst. Man Cybern. **1997**, *5*, 4104–4108.
37. Chatterjee, A.; Tudu, B.; Paul, K.C. Towards optimized binary pattern generation for grayscale digital halftoning: A binary particle swarm optimization (BPSO) approach. J. Vis. Commun. Image R. **2012**, *23*, 1245–1259.
38. Kim, S.H.; Allebach, J.P. Impact of HVS models on model-based halftoning. IEEE Trans. Image Process. **2002**, *11*, 258–269.
39. Wang, Z.; Bovik, A.C. Mean squared error: Love it or leave it? IEEE Signal Process. Mag. **2009**, 98–117.
40. Wang, Z.; Bovik, A.C.; Sheikh, H.R.; Simoncelli, E.P. Image quality assessment: From error visibility to structural similarity. IEEE Trans. Image Process. **2004**, *13*, 600–612.
41. Barkol, O.; Kogan, H.; Shaked, D.M.; Fischer, M. A robust similarity measure for automatic inspection. In *17th IEEE International Conference on Image Processing*, Hong Kong, September 2010, 2489–2492.
42. Wong, P.W. Entropy-constrained halftoning using multipath tree coding. IEEE Trans. Image Process. **1997**, *6*, 1567–1579.
43. Chatterjee, A.; Tudu, B.; Paul, K.C. Binary genetic algorithm-based pattern LUT for grayscale digital halftoning. Signal Image Video Process. **2013**, *7*, 377–388.
44. Yan, G.; Li, C. An effective refinement artificial bee colony optimization algorithm based on chaotic search and application for PID control tuning. J. Comput. Inform. Syst. **2011**, *7*, 3309–3316.
45. USC-SIPI Image Database. Available at http://sipi.usc.edu/database/database.php?volume=misc, (accessed in July 15, 2009).
46. Li, H.; Mould, D. Contrast-aware halftoning. Comput. Graph. Forum **2010**, *29*, 273–280. Available at http://onlinelibrary.wiley.com/doi/10.1111/j.1467-8659.2009.01596.x/suppinfo, (accessed in July 1, 2011).
47. Damera-Venkata, N.; Evans, B.L. FM halftoning via block error diffusion. IEEE Proc. Int. Conf. Image Process. **2001**, *2*, 1081–1084.
48. Chandler, D.M.; Hemami, S.S. VSNR: A wavelet-based visual signal-to-noise ratio for natural images. IEEE Trans. Image Process. **2007**, *16*, 2284–2298.
49. Zhang, L.; Zhang, L.; Mou, X.; Zhang, D. FSIM: A feature similarity index for image quality assessment. IEEE Trans. Image Process. **2011**, *20*, 2378–2386.

Dimensionality Reduction

Tania Pouli
Technicolor, Rennes, France

Erik Reinhard
Technicolor Research and Innovation, Rennes, France

Douglas W. Cunningham
Brandenburg University of Technology, Cottbus, Germany

Abstract
Images of scenes can often contain huge numbers of pixels. Yet much of this information, in terms of scene analysis, can be redundant. Therefore it is desirable to extract meaning form the image from as few pixel samples as possible. This situation is considered in the entry. In particular, principal component analysis is used. This entry introduces the relevant concepts for this type of analysis and provides useful examples.

Keywords: Dimensionality reduction; Principal component analysis; Scene analysis.

Photographs of scenes typically contain a very large number of pixels. It is not uncommon for modern cameras to have in excess of 15 megapixels, and high-end cameras can have 80 megapixels or more. The pixels in an image can be seen as observations of a set of processes that gave rise to this image. Light sources emit light, all objects in the scene reflect, absorb or transmit light, and the medium through which light travels may participate by scattering light. Some light will eventually pass through the optical system of the camera and will be recorded as pixels.

The number of observations made in a single image or a set of images is usually much larger than the number of processes (think of objects in the scene). It would be possible to see the pixels in an image, therefore, as random variables that sample the same random process. The question then is whether it would be possible to learn something about the underlying process that gave rise to the samples/pixels.

If the dimensionality of the underlying process is lower than the number of observations, then the data should form clusters in N-dimensional space. An example is shown in Fig. 1, where an image of a snowy scene is analyzed by plotting 500 randomly chosen pixels in linear sRGB space. As the process that gave rise to the scene is predominantly snowy weather, it is no surprise that the pixels lie more or less on a straight line in 3D sRGB color space. An example where the pixels stem from multiple processes is shown in Fig. 2. Here, the butterfly and its background lead to different clusterings of pixels in 3D space.

If clusters lie in lower-dimensional planes as in this example, then there exist various types of analysis that will be able to detect what the dimensionality of the underlying process was.[1] Principal Component Analysis (PCA) is perhaps the most well-known algorithm that accomplishes this.[2–4] There are many more techniques that can be used for dimensionality reduction, including Independent Component Analysis (ICA),[5] canonical correlation analysis,[5,6] linear discriminant analysis,[7] Fisher's linear discriminant,[8] topic models, and latent dirichlet allocation. Both PCA and ICA are special cases of blind source separation, which is a general collection of techniques that try to separate a set of signals from a mixed signal without prior knowledge of the source signals or the mixing process.[9]

PRINCIPAL COMPONENT ANALYSIS

Principal component analysis (PCA—also known as the Karhunen-Loéve transform or the Hotelling transform) is a common data mining and dimensionality reduction technique that takes as input a series of n d-dimensional observations. Its output consists of a new set of eigenvectors (or basis vectors), as well as their eigenvalues (or weighting values). PCA is also often used in many different application areas beyond data mining and dimensionality reduction, including data visualization, variance calculations, factor analysis, perceptual experiment data analysis, and of course, image statistics. The eigenvectors point to directions of maximum variance in the data. We begin by showing how eigenvectors and eigenvalues are computed.

The first step in applying PCA consists of choosing an appropriate set of n d-dimensional input vectors. This can, for example, be a set of images for which the most important "eigenimages" are of interest. Alternatively, it would

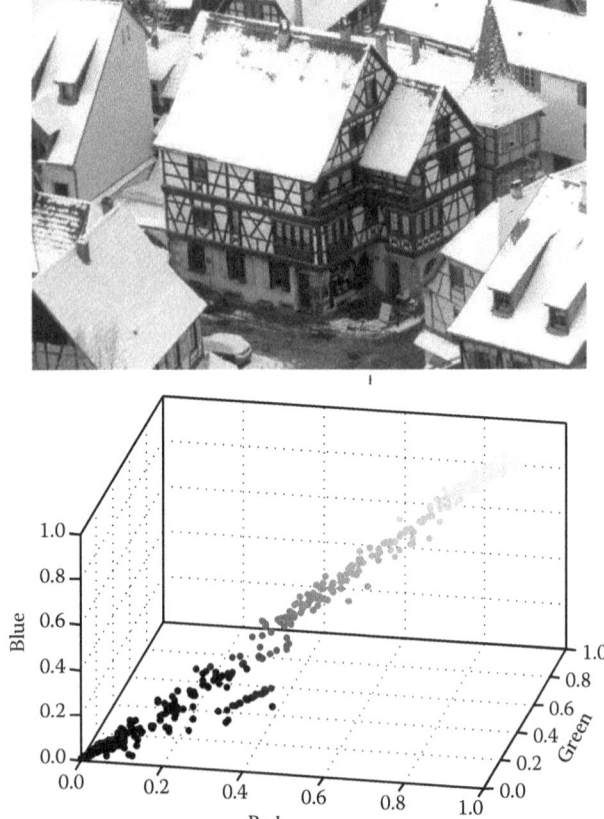

Fig. 1 We have randomly picked 500 pixels of the image shown at the top and plotted them in linear sRGB space. As a result of the snowy weather, most pixels ended up lying on a straight line in 3D space
Source: Keysersberg, Alsace, France, 2013.

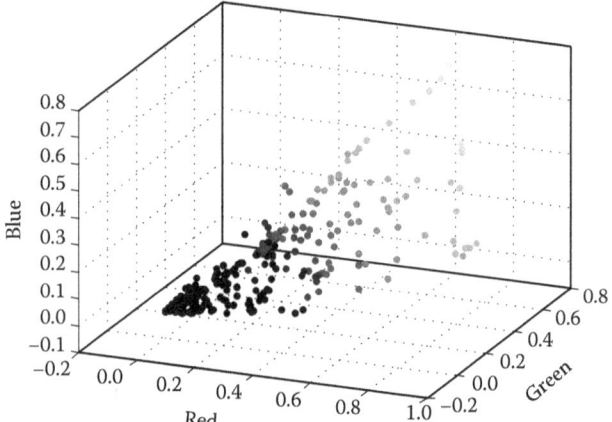

Fig. 2 We have randomly picked 500 pixels of the image shown at the top and plotted them in linear sRGB space. The butterfly and the background are distinctly different processes, leading to less straightforward clustering than the example shown in Fig. 1
Source: Changi Airport butterfly garden, Singapore, 2012.

be possible to use small patches within a set of images as the input vectors. Finally, one could use individual pixels in an image, or set of images, as n three-dimensional observations.

In the standard definition of PCA, one then creates a $d \times n$ matrix $\mathbf{M}$ by concatenating all the input vectors. The covariance between two dimensions across all input vectors $\mathbf{x}_1$ and $\mathbf{x}_2$ is given by:

$$\mathrm{cov}(\mathbf{x}_1, \mathbf{x}_2) = \frac{\sum_{i=1}^{n}(\mathbf{x}_{1,i} - \overline{\mathbf{x}}_1)(\mathbf{x}_{2,i} - \overline{\mathbf{x}}_2)}{n-1} \quad (1)$$

To facilitate the computation of a covariance matrix, the data is first centered so that the mean value in each of the d dimensions is zero, i.e., $\overline{\mathbf{x}}_1 = \overline{\mathbf{x}}_2 = \ldots \overline{\mathbf{x}}_d = 0$. The covariance computation then simplifies to:

$$\mathrm{cov}(\mathbf{x}_1, \mathbf{x}_2) = \frac{\sum_{i=1}^{n} \mathbf{x}_{1,i} \mathbf{x}_{2,i}}{n-1} \quad (2)$$

Note that the variance within a single dimension can be computed as $\mathrm{var}(\mathbf{x}_i) = \mathrm{cov}(\mathbf{x}_i, \mathbf{x}_i)$. The covariance matrix is then the covariance between each pair of dimensions in our matrix of input vectors M:

$$\mathbf{C} = \begin{bmatrix} \mathrm{cov}(\mathbf{x}_1, \mathbf{x}_1) & \mathrm{cov}(\mathbf{x}_1, \mathbf{x}_2) & \cdots & \mathrm{cov}(\mathbf{x}_1, \mathbf{x}_d) \\ \mathrm{cov}(\mathbf{x}_2, \mathbf{x}_1) & \mathrm{cov}(\mathbf{x}_2, \mathbf{x}_2) & \cdots & \mathrm{cov}(\mathbf{x}_2, \mathbf{x}_d) \\ \vdots & \vdots & \ddots & \vdots \\ \mathrm{cov}(\mathbf{x}_d, \mathbf{x}_1) & \mathrm{cov}(\mathbf{x}_d, \mathbf{x}_2) & \cdots & \mathrm{cov}(\mathbf{x}_d, \mathbf{x}_d) \end{bmatrix} \quad (3)$$

As the data has zero mean, this can be efficiently computed as a matrix multiplication:

$$\mathbf{C} = \frac{1}{n-1} \mathbf{M}\mathbf{M}^T \quad (4)$$

We then wish to compute the eigenvectors and eigenvalues of this square matrix. Each of the d eigenvectors has an eigenvalue that gives the length of its corresponding eigenvector. It is then possible to sort the eigenvectors according to their length. In that case, the longest eigenvector will point in the direction of largest variability in the data.

It is normally described as the first principal component. The second-longest eigenvector points in the direction that exhibits the second-most variability. In this manner, each subsequent eigenvector will explain less variability in the data. To achieve dimensionality reduction, it would be possible to ignore all dimensions of the data for which the corresponding eigenvalues are below some threshold.

The decomposition into eigenvalues and eigenvectors proceeds as follows. An eigenvector of a square matrix $\mathbf{C}$ is a nonzero vector $\mathbf{v}$ such that:

$$\mathbf{C}\mathbf{v} = \lambda \mathbf{v} \tag{5}$$

where λ is an eigenvalue. This can be rewritten as:

$$\mathbf{C}\mathbf{v} - \lambda \mathbf{v} = 0 \tag{6}$$

$$(\mathbf{C} - \lambda \mathbf{I})\mathbf{v} = 0 \tag{7}$$

where $\mathbf{I}$ is the identity matrix. There can only be a nonzero solution for $\mathbf{v}$ in the above equation if:

$$\det(\mathbf{C} - \lambda \mathbf{I}) = 0 \tag{8}$$

In other words, the determinant of $\mathbf{C} - \lambda \mathbf{I}$ should be zero. This equation, known as the *characteristic equation* (or secular equation), is essentially a polynomial equation of degree d in the variable λ. As we are applying eigenvalue decomposition on pixels, patches of pixels, or images, the matrix $\mathbf{C} - \lambda \mathbf{I}$ will be real-valued, meaning that there are at most n real-valued roots to the above polynomial equation. These could be computed numerically, giving a set of eigenvalues. By substituting these into Eq. 5 the eigenvectors could be computed.

In practice, eigendecomposition is achieved via a different route. There exist numerical algorithms to directly compute eigenvectors and eigenvalues,[10,11] including the QR algorithm[12–14] and Arnoldi iteration.[15]

Performing PCA directly on the covariance matrix of this dimensionality is often computationally infeasible. If small, say, 50×50 pixel, grayscale images are used as input to the PCA, each image is a 2500-dimensional vector, and accordingly the covariance matrix $\mathbf{C}$ has 6.25×10^6 elements.

There is, however, a trick based on linear algebra that can be used to simplify this: the rank (that is, the maximum number of linearly independent rows or columns) of the covariance matrix $\mathbf{C}$ is limited by the number of training examples. If there are n training examples, there will be at most $n-1$ eigenvectors with nonzero eigenvalues. Therefore, if the number of input feature vectors n is smaller than the dimensionality d of each vector, the principal component analysis can be performed on a relatively small $n \times n$ sized matrix, instead of the larger $d \times d$ sized one.

To achieve this, we calculate the covariance matrix $\mathbf{C}' = \mathbf{M}^T \mathbf{M}$. The eigenvalue decomposition then creates n eigenvectors $\mathbf{v}'_i$ and corresponding eigenvalues λ'_i. It can be proven that if $\mathbf{v}'$ is an eigenvector of $\mathbf{C}' = \mathbf{M}^T \mathbf{M}$, then $\mathbf{v} = \mathbf{M} \mathbf{v}'$ is an eigenvector of $\mathbf{C} = \mathbf{M} \mathbf{M}^T$, having the same eigenvalue λ. This can be seen as follows. Let us assume that $\mathbf{v}'$ is an eigenvector of $\mathbf{M}^T \mathbf{M}$ with eigenvalue λ. We then have:

$$(\mathbf{M}^T \mathbf{M})\mathbf{v}' = \lambda \mathbf{v}' \tag{9}$$

$$\mathbf{M}(\mathbf{M}^T \mathbf{M})\mathbf{v}' = \mathbf{M} \lambda \mathbf{v}' \tag{10}$$

$$(\mathbf{M}\mathbf{M}^T)(\mathbf{M}\mathbf{v}') = \lambda(\mathbf{M}\mathbf{v}')(\mathbf{C})(\mathbf{M}\mathbf{v}') = \lambda(\mathbf{M}\mathbf{v}') \tag{11}$$

This proves that $\mathbf{v} = \mathbf{M} \mathbf{v}'$ is an eigenvector of covariance matrix $\mathbf{C}$.

As stated before, components with a larger eigenvalue correspond to more important features in that they explain more of the variance in the data. Conversely, lower eigenvalues signify less important components. In order to identify how many dimensions might be needed for dimensionality reduction, it is possible to sort the eigenvalues. The amount of explained variance up to a dimension $d_k < d$, then, is the sum of all eigenvalues up to that point:

$$\sum_{i=1}^{d_k} \lambda_i \tag{12}$$

Depending on the desired fidelity of the smaller, reconstructed space, it is then possible choose an appropriate cut-off point α that selects the k most important dimensions:

$$k \mid \sum_{i=1}^{d_k} \lambda_i < \alpha \sum_{i=1}^{d} \lambda_i \tag{13}$$

The original stimuli or feature vectors (as well as any new stimulus) can then be reconstructed using the reduced set of corresponding eigenvectors. This representation offers the advantage that a relatively small number of components is sufficient to encode most of the information in the images.

Nevertheless, although the computed eigenvectors are orthogonal, statistical independence cannot be guaranteed. Natural images tend to be non-Gaussian and thus the decomposition offered by PCA can only decorrelate the data.[16] The main consequence of this is that meaningful information is only captured by the first few components, while further ones mostly correspond to less important features. That this still affords useful information is demonstrated in the following paragraph.

Whitening

One of the strengths of PCA is that it can be used to decorrelate data. The eigenvalues are a measure of the variance along each dimension. If we were to scale decorrelated data to have unit variance in each dimension, we would obtain whitened data, i.e., data that is uncorrelated and of

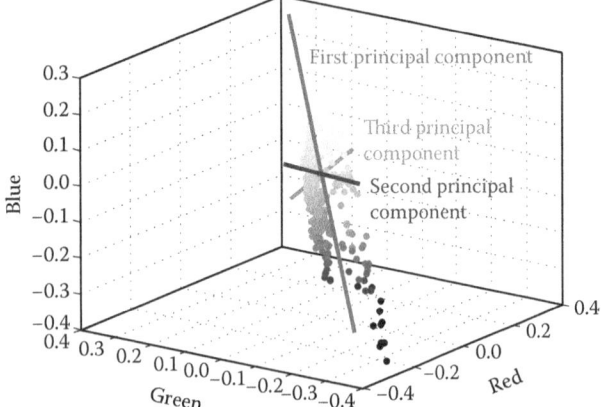

Fig. 3 PCA analysis on three-dimensional pixel data produces three directions, which are ordered according to how much variation each direction explains. (Dunedin, New Zealand, 2012)

Fig. 4 PCA run on 5000 randomly chosen patches taken from the high dynamic range image at the top. The first 25 principal components are shown in reading order for patches of size 25×25 (bottom left), 50×50 (bottom middle), and 100×100 (bottom right). (Westonbirt Arboretum, UK, 2009)

unit variance. This means that we would have removed all second-order effects from the data, "second-order" meaning variances and correlations. As such, we would be able to use the resulting images to study higher-order statistics. In "Independent Components Analysis" section, we demonstrate that removal of second-order statistics from the data is a necessary prerequisite for computing independent components.

PCA on Pixels

An example of PCA analysis of the pixels in an image is given in Fig. 3. The three-dimensional pixel data was analyzed, leading to three eigenvectors that are plotted in the figure. Note that the point cloud is stretched most in the direction of the first principal component. In this figure, we adjusted the length of the axes for the purpose of visualization; the actual length of the second and third principal components is significantly shorter than seen in this figure.

Applying PCA to single pixels of either individual images or ensembles of images is useful in color applications. In essence, it is possible to decorrelate the three color channels by running PCA on images in this manner. Decorrelation of color information appears to happen in human vision, and it has direct applications in image processing.

PCA on Patches

We can also apply PCA to small patches, drawn randomly from one or more images. The output is then an ordered set of patches whereby the first principal components represent the most common modes of variation in local image regions. An example is shown in Fig. 4, where the high dynamic range input image was transformed to a luminance-only image. We then subtracted the mean luminance and randomly selected 5000 patches of given size (see Fig. 4). The patches are then linearly transformed into one-dimensional vectors by concatenating the rows of the patch. As before, the resulting 5000 vectors are placed in a matrix that can then be subjected to PCA. The resulting basis vectors $\mathbf{v}_i$ are then reshaped into 2D patches $\mathbf{V}_i(x, y)$ for visualization. Figure 4 shows the first 25 principal components for different patch sizes ranging from 25×25 pixels to 100×100 pixels.

Irrespective of how large the patches are, the first principal component gives an average response, being mostly uniform in luminance distribution. The second and third principal components tend to encode horizontal and vertical gradients/edges. The fourth and fifth principal components appear to encode horizontal and diagonal lines. The remaining principal components show increasingly complex patterns, which are less meaningful.

The eigenvectors are only good features if the multidimensional data contains at least a few dimensions which carry a strong signal—if the data contains significant

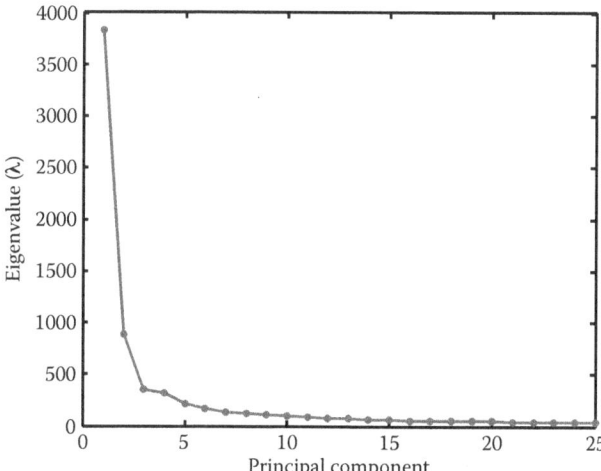

Fig. 5 The first 25 eigenvalues corresponding to the eigenvectors of size 100×100, which are shown in the bottom right of Fig. 4

levels of noise, then the interpretation of the eigenvectors becomes difficult. This can usually be checked by investigating how quickly the eigenvalues fall off as the dimensionality becomes larger: if the falloff is steep in the beginning and then tapers off, then the first few eigenvectors are potential candidates for "good" features. The relative importance of the eigenvectors of Fig. 4 are plotted in Fig. 5. We note that the falloff is very significant for the first few components and then tapers off, which is consistent with our ability to assign meaning to the first few components.

Results such as these are also found in studies whereby ensembles of natural images are taken as input,[17] although in this work the patches were windowed with a Gaussian window first. They observe that the first few principal components begin to resemble receptive fields of simple cells, for example in cats.[18]

A consequence of applying PCA to image patches is that in principle an image patch can be represented to a reasonable degree by the first few components. Effectively, a patch $I(x, y)$ could be represented by a linear superposition of basis functions. The principal components $V_i(x, y)$ in such a case act as the basis functions:[19]

$$I(x,y) = \sum_i w_i V_i(x,y) \qquad (14)$$

For human vision, the implication is that only a few different types of receptive fields can account for a relatively large proportion of our visual experience. While it is unlikely that the human visual system applies PCA to its input, certain receptive fields do bear some resemblance to the patches obtained by this method.

PCA, however, assumes that either the data has a Gaussian distribution (which is already violated, for instance, in the image of Fig. 2, where the point cloud is not ellipsoidal), or that linear pairwise correlations are the most important form of statistical dependence.[19] As many, if not all natural images have higher-order dependencies, PCA is limited in representing such images as basis functions, and the technique also does not fully explain the receptive field structure in the human visual system.

Thus, one of the main applications of PCA is in the realm of feature extraction in that it finds a good, new set of basis vectors (or features) for the data. For this interpretation to hold, however, the data must be distributed according to a single, Gaussian distribution. For non-Gaussian data, or for multimodal Gaussian data, PCA only decorrelates the axes. For highly clustered data, PCA may therefore not be a good method.

On the other hand, the results obtained with patch-based PCA do point into an interesting direction, which is that images may be analyzed with a small set of basis functions that capture the dominant trends in the image. This leads to an important concept in natural image statistics, which is that of sparse coding. It is now believed that the human visual system aims to preserve information, but to process and represent it sparsely, i.e., with as few neurons active as possible.[19,20] If the variability in the input signal can be represented by as few neurons as possible, then this has several advantages for organism: such neuronal activity is metabolically efficient,[21] it minimizes wiring length[22] and it increases capacity in associative memory.[23]

The concept of sparse coding returns in the discussion of Independent Component Analysis (ICA) in a later section.

PCA on Images

So far, we have discussed the application of PCA on pixels as well as on patches of images. It is also possible to apply PCA on entire images. Each image is then considered a separate feature vector, and are known as eigenimages. Such an approach can be useful if the set of images contain carefully constrained exemplars. Variations in illumination, reflectance properties, location, and orientation or the objects in the image can complicate processing severely.[24,25] Robustness is often a problem, in that the method does not handle outliers or noisy data very well.[26,27]

Nonetheless, various approaches have been proposed to overcome these limitations,[28] which have led to various applications, including illumination planning,[29] tracking of robot manipulators,[30] visual inspection,[31] and human face recognition,[32,33] the latter of which is discussed in the next section.

Eigenfaces

Perhaps the most famous example of PCA in the context of image analysis is the extraction of so-called *eigenfaces* within the computer vision domain,[33] the principles of which are illustrated in Fig. 6. The input consists of n images of $p \times p = d$ pixels. The eigenfaces are the

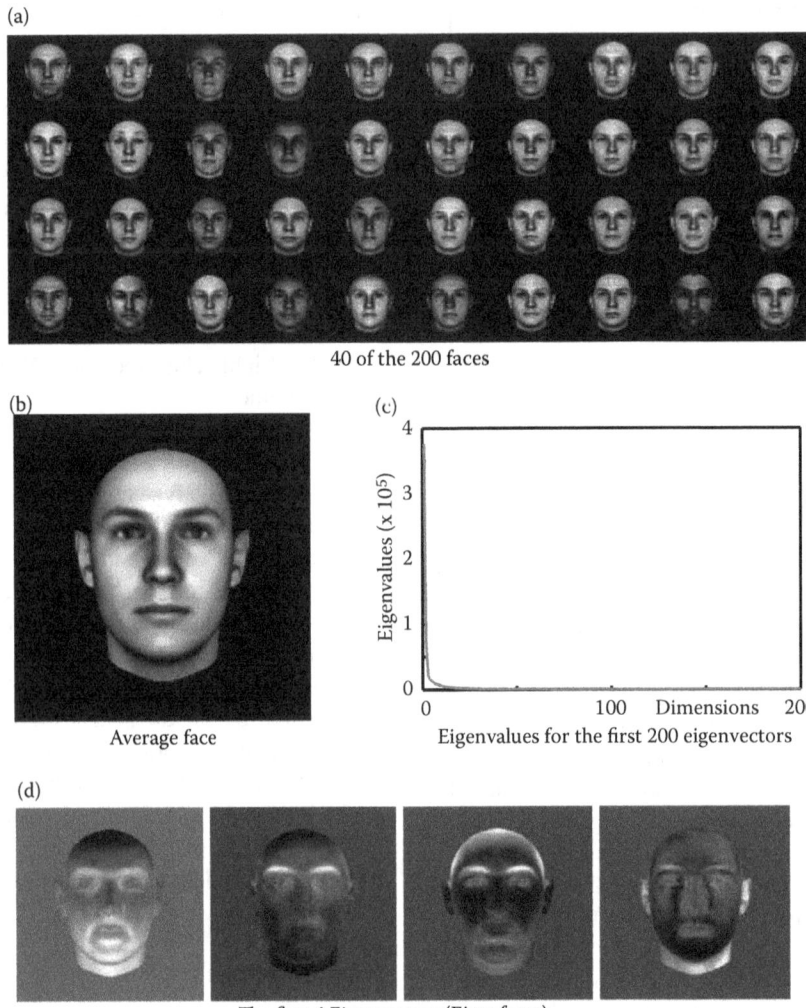

Fig. 6 PCA can be used to decompose a collection of face images into a set of components, which can be used for constructing novel faces or for face recognition. (a) Forty of the 200 faces from the MPI face database[69]—note that only the texture of the face varies but not its shape. (b) Average face of all 200 faces. (c) Plot of the eigenvalues of the PCA showing how quickly the values fall off as the number of dimensions increases. (d) First four eigenvectors (Principal components, or *eigenfaces*) showing where changes in the face texture happen
Source: Figure used with permission from the copyright holder Christian Wallraven.

eigenvectors resulting from the PCA decomposition of the input face images. Since PCA is a global method, each of the d pixels is treated as single dimension. This means that any difference in background, lighting, pose, and size of the faces will become encoded in the PCA eigenvectors, which usually is not desirable.

If one is interested only in the variation in appearance of the faces, then one possibility is to bring all faces into correspondence by warping them onto a common face coordinate system. Each face will then have the exact same shape with each pixel in the image exactly specifying the same facial features in all faces. The only variation left will then be the one according to the facial appearance (or texture) of the face or variations in illumination, which PCA can extract.

Figure 6 shows an example in which a database of 200 three-dimensional faces (100 male, 100 female) was brought into correspondence first.[34] The faces were then rendered onto the shape of the average face (Fig. 6a). Figure 6b shows how fast the eigenvalues fall off, indicating that the first few eigenvectors explain a large amount of variance. Furthermore, Fig. 6c–f show the first four eigenvectors, or eigenfaces: the first eigenface highlights the change in appearance from top to bottom, the second focuses on the eyebrows, the third on the cheeks, and the fourth on the chin and cheeks (the beard region). Although eigenfaces do not always correspond to such well-defined features, they are able to capture the main modes of variation in the images.

It has been shown that such a separation between form on the one hand and texture on the other hand corresponds well to human recognition performance characteristics.[35] Furthermore, the idea that faces are represented in a vector space (known as the *face space*[36]) might be compatible with a representation of the average face in a coordinate system defined by PCA dimensions.[35,37]

PCA has also been shown to be effective as a statistical approach in the reconstruction of 3D representations of human faces on the basis of a single 2D images.[38] In this work, a low-dimensional parameterization of head shape was learned from a large set of scans of 3D fases. A single 2D input image can then be mapped to 3D using this parameterization.

Finally, PCA has not only been applied to analysis of static faces, but also to the modeling of facial expressions. Indeed, applying PCA to a set of faces varying in identity and expressions yields an interesting separation of identity and expression axes, which is also compatible with behavioral and neurophysiological data, suggesting a large dissociation between the two types of information.[39]

INDEPENDENT COMPONENTS ANALYSIS

The orthogonality of the axes of PCA can be seen as a limitation. A more advanced, albeit computationally much more involved technique is Independent Component Analysis (ICA).[16,40–43] Rather than producing decorrelated axes that are orthogonal as achieved with PCA, independent components analysis finds axes that are more or less independent, but which are not necessarily orthogonal, as shown in Fig. 7.

Several ICA algorithms are known, including infomax,[44] extended infomax,[45] fastICA,[46] Jade,[46] kernel ICA,[47] and RADICAL.[48] Although the implementations vary, their goals are the same: represent multivariate data as a sum of components in such a way that the components reveal the underlying structure that gave rise to the observed signal.

The number of ICA algorithms available and their computational and algorithmic complexity make an in-depth discussion of this class of algorithms beyond the scope of this book. However, we will discuss the general principle as well as some of their implications for natural image statistics as well as for human vision.

While PCA measures covariances of pixels separated by varying distances, the components that are found constitute a linear transform. This means that although the data may be decorrelated, there is no guarantee that the resulting components are in any way independent. In particular, only if the input data happens to have a Gaussian distribution, then the resulting principal components are both decorrelated and independent.[49]

As we have seen, many of the statistics of natural images that are currently known point in the direction of high kurtosis, i.e., they are highly non-Gaussian. This has given rise to the use of ICA in examing natural image statistics, where image patches[43] or the color components of a set of pixels (as shown in Fig. 7) are calculated. ICA could also conceivably be used on sets of entire images

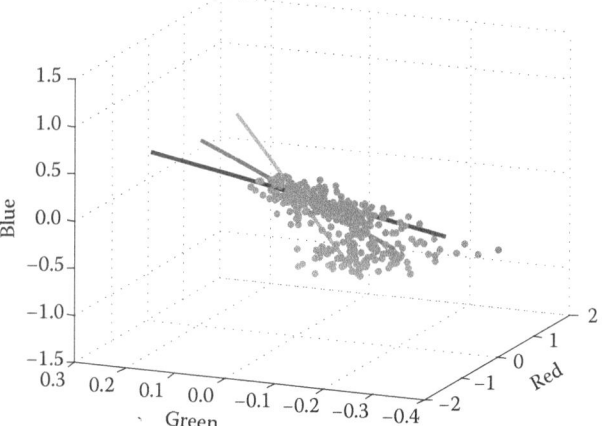

Fig. 7 In this plot, we have chosen 50,000 points from the image at the top, and applied ICA to the color channels of these points, revealing three new axes that are not orthogonal—the main difference with PCA. Note that we only plot 500 randomly chosen points here. Further, the axes were scaled for visualization; the pink axis does not correspond to a meaningful basis, it being drawn more than 3000 times longer than the other two axes. (Monument Valley, Arizona, 2012)

(although this would probably be prohibitively expensive computationally).

ICA algorithms tend to require a set of preprocessing steps to make the problem more tractable. In particular, the data need to be centred and whitened. In addition, data reduction is often applied. Data can be whitened by running the PCA algorithm first, as outlined in "Whitening" section. By keeping only the first n components, data reduction is achieved. This ensures that only those components are computed that will be meaningful to the problem being solved. Moreover, it speeds up the computations required to determine independent components.

Note that if the multivariate data being analyzed has a Gaussian distribution along each of its dimensions, there would be no higher-order statistical correlations available in the data. In that case, applying PCA would not only decorrelate the data but would also lead to the computation of independent axes. Thus, for ICA to be meaningful, there is the requirement that the distribution of data points is non-Gaussian.

Assuming for now that we are performing ICA on small image patches randomly drawn from a set of images, the pixels $\mathbf{I}(x, y)$ in the patch can be represented as the linear superposition of a weighted set of basis functions $\mathbf{V}_i$:

$$\mathbf{I}(x,y) = \sum_i w_i \mathbf{V}_i(x,y) \quad (15)$$

where w_i are the weights. This is the same equation as the one used for PCA (Eq. 14), albeit that the assumptions used for the construction of the basis functions are different. In particular, the assumptions made for ICA are:[16]

- The weights w_i can be seen as random variables that have a non-Gaussian distribution. They are also assumed to be statistically independent.
- The components $\mathbf{V}_i(x, y)$ are invertible.

Under these assumptions, and given a large enough set of image patches, it is possible to derive a set of components $\mathbf{V}_i$ without advance knowledge of any of the weights w_i. A property of linear systems is that the generative model described by Eq. 15 is equivalent to:

$$w_i = \sum_x \sum_y \mathbf{V}_i^{-1}(x,y) \mathbf{I}(x,y) \quad (16)$$

$$= \sum_x \sum_y \mathbf{Q}_i(x,y) \mathbf{I}(x,y) \quad (17)$$

This means that either set of equations can be solved, and the components computed for either system can be converted into the system by matrix inversion. Similar to the argumentation in "PCA on Patches" section, we can concatenate the values in the patches $\mathbf{Q}_i$ to form vectors $\mathbf{q}_i$, and similarly rearrange the image patch $\mathbf{I}(x, y)$ to vector $\mathbf{k} = k_1, \ldots, k_m$ so that we can write:

$$w_i = \sum_{j=1}^m q_{i,j} k_j = \mathbf{q}^T \mathbf{k} \quad (18)$$

We now have an expression that relates the independent components $\mathbf{q}_i$ to the weights w_i. Recall that these as yet unknown components could be combined by applying the weights to each corresponding component, and then we would obtain the original image patch, now represented by $\mathbf{k}$. The weights w_i can therefore be interpreted as random variables drawn from a certain probability density function (pdf). The probability density for each component $\mathbf{q}_i$ is then given by $p_i(w_i)$.

By the definition that the components we seek are independent, the multidimensional probability density of all the n weights combined is given by:

$$p(w_1,\ldots,w_n) = \prod_{i=1}^n p_i(w_i) \quad (19)$$

Unfortunately, we do not know the pdf of the weights w_i. Instead, we would like to find the pdf of the observed variable $\mathbf{k}$ (recall that this is a vector of pixel values that serves as input). Due to the linear transform applied in Eq. 18, this pdf is given by:

$$p(\mathbf{k}) = |\det(\mathbf{R})| \prod_{i=1}^n p_i(\mathbf{q}_i \mathbf{k}) \quad (20)$$

Here, the matrix $\mathbf{R}$ is the matrix that defines the linear transform in Eq. 18, i.e., $\mathbf{R} = \{q_{i,j}\}$.

Equation 20 constitutes the statistical model for independent component analysis. To estimate parameters given a statistical model, a technique known as *maximum likelihood estimation* is normally used. The unknown parameters are the independent components $\mathbf{q}_i$. If we are using a large number A of image patches, then we can construct an equally large number of vectors ($\mathbf{k}_1, \ldots, \mathbf{k}_A$). The likelihood is then the probability of observing these input vectors given the model parameters. The likelihood $L(\mathbf{q}_1, \ldots, \mathbf{q}_n)$ is given in log space by:

$$\log L(\mathbf{q}_1,\ldots,\mathbf{q}_n) = A \log|\det(\mathbf{R})| + \sum_{i=1}^n \sum_{a=1}^A \log p_i(\mathbf{q}_i^T \mathbf{k}) \quad (21)$$

It can be shown that if the features are uncorrelated, as they would be after whitening using PCA, then the determinant of $\mathbf{R}$ will be ± 1. As a result, the log-likelihood can be estimated as:

$$\log L(\mathbf{q}_1,\ldots,\mathbf{q}_n) = \sum_{i=1}^n \sum_{a=1}^A \log p_i(\mathbf{q}_i^T \mathbf{k}) \quad (22)$$

Of course, we would like to choose model parameters such that this log-likelihood is maximized. Maximum likelihood estimation can be performed by standard numerical optimization techniques,[50–53] although specific optimization techniques have been developed in the context of ICA.[16,46]

ICA ON NATURAL IMAGES

One could ask what the independent components are in the context of natural images. It could be argued that these are the objects that comprise a scene.[54] It is exactly these that provide the structure that gives rise to the pixel array that is analyzed by computer algorithms as well as by the human visual system. Both PCA and ICA, however, are generative models that are based on linear superposition. Thus, they cannot take into account translation and rotation of objects.[54]

However, it has been found that small image patches of natural image ensembles can be analyzed using ICA. It should be noted here that we are still free to choose our assumed probability density functions $p_i(w_i)$. So far, ICA is a general technique for blind source separation. To make ICA relate to natural images, the pdfs should be chosen such as to mimic a presumed feature of human vision. A reasonable assumption would be to argue that receptive fields in the human visual pathway are in some sense sparse, i.e., cells preferentially respond little if at all to most stimuli, and they respond vigorously to some stimuli to which they are tuned. Such sparse coding is likely to occur and helps, for instance, with metabolism: a cell that is dormant most of the time does not use precious energy.

The sparseness of a probability distribution function could, for instance, be measured by computing the kurtosis of the observed variables. Kurtosis is, in fact, a direct measure of how far a distribution deviates from being Gaussian. Larger values of kurtosis indicate a sparser distribution. It would therefore be possible to choose pdfs that maximize kurtosis. Several alternative functions have been proposed, including:[16]

$$\log p_i(w_i) = -2\log\cosh\left(\pi s / 2\sqrt{3}\right) - 4\sqrt{3}/\pi \qquad (23)$$

As an example, we have selected 15 natural images, shown in Fig. 8, and created a set of 50,000 randomly chosen patches of size 25×25 or 50×50 pixels. We then applied a fast ICA algorithm to this data.[46] A resulting set of 64 feature patches, corresponding to the basis functions $\mathbf{V}_i$ is shown in Fig. 9.

The feature detector weights, i.e., those corresponding to $\mathbf{V}_i^{-1} = \mathbf{Q}_i$, are shown in Fig. 10. As these are the patches that would be applied to the image to generate the encoding of an image, they could be compared to receptive fields in the human visual system. In that context, as argued by others,[55] it is interesting to note that such basis functions reveal structures that resemble those found in the human visual system.[55]

In particular, this technique yields Gabor-like patches—elongated structures that are localized in space, orientation,

Fig. 8 A set of 15 images used in the independent component analysis in "ICA on Natural Images" section. (Westonbirt Arboretum, UK, 2009 and 2011)

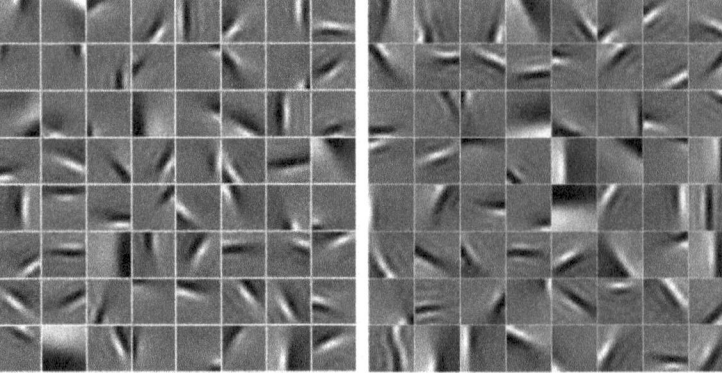

Fig. 9 Basis functions V_i computed using 50,000 patches of size 25×25 (bottom left) and 50×50 (bottom right), drawn from the 15 images presented in Fig. 8

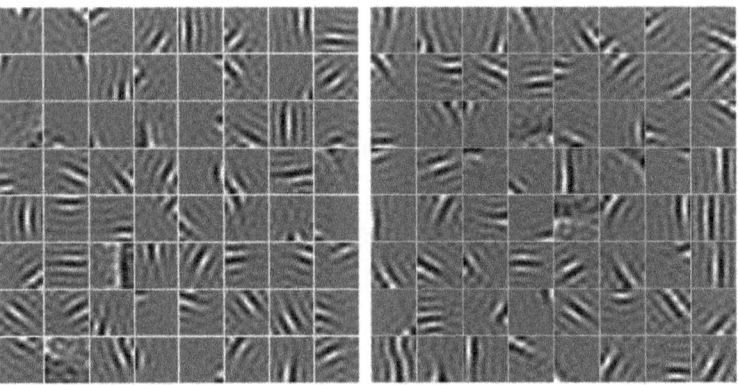

Fig. 10 Features $V_i^{-1} = Q_i$ computed using the 15 images presented in Fig. 8

and frequency. Their size, aspect ratio, spatial frequency bandwidth, receptive field length, and orientation tuning bandwidth are similar to those measured in cortical cells.[56] These results lend credence to the argument that the human visual system appears to represent natural images with independent variables, each having a highly kurtotic distribution that leads to a metabolically efficient sparse coding. A number of researchers have shown that the statistics of art can also be captured with very similar sparse coding, especially if a compressive nonlinearity is applied as a tone mapper.[57–60]

As an example, ICA decompositions have been used in object classification, specifically on images of human faces as well as flowers.[61]

GAUSSIAN MIXTURE MODELS

The analysis of image patches can be performed in many different ways, each revealing subtly different information about how the human visual system may function, and each offering different opportunities in imaging applications. We have seen in Section 5.2.8, for example, that image patches tend to recur nearby, both in scale and in space. In this entry, we have shown that the human visual system appears to analyze images by removing correlations, as evidenced by the analyzers that are revealed by applying principal component analysis to individual patches. This was taken one step further in the section on independent components analysis, showing that independence is often achieved, rather than mere decorrelation.

An alternative way to analyze image patches is by applying Gaussian Mixture Models.[62,63] A Gaussian mixture model is simply a sum of k (multivariate) Gaussian distributions, each with a weight factor w_j:

$$p(\mathbf{x}) = \sum_{i=1}^{k} w_j N(x \mid \mathbf{mu}_j, \Sigma_j) \qquad (24)$$

Here, the weights are prior probabilities and are therefore limited in range to $0 \leq w_j \leq 1$. Moreover, these weights sum to 1. See Appendix A.2 for the definition of multivariate Gaussian distributions. Given a set of observations, which can be individual pixels, patches, or entire images, a predetermined number of Gaussians can be fitted. This is

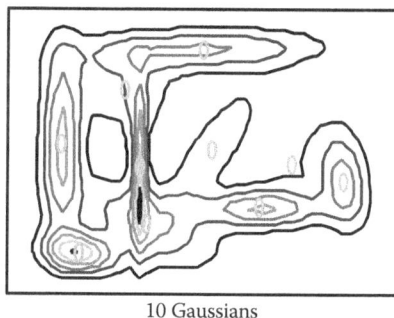
10 Gaussians

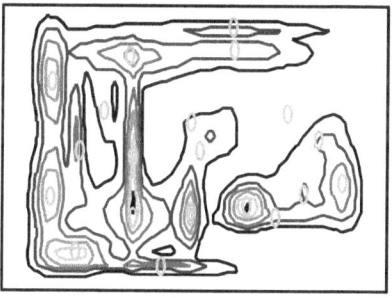
20 Gaussians

Fig. 11 The image at the top was fitted with 10 and 20 Gaussians. (Olympia, Greece, 2010)

accomplished with numerical optimization in the form of expectation maximization (EM).[64]

As an example, we have estimated an image with 10 and 20 Gaussians and plotted the resulting contours in Fig. 11. Here, a single color channel is treated as a 2D dataset of observations. The Gaussians therefore, to some extent, represent the image. The computational complexity of the algorithm is such that it would not be tractable to represent an image with sufficient Gaussians. This problem could be alleviated by employing Gaussian mixtures in a scale space approach, however.[65,66]

It is entirely possible to compute a Gaussian mixture for image patches as well as for individual pixels. The latter case has proven a useful representation in color transfer applications.[67] This application is discussed further in Section 10.4.

Application of Gaussian mixture models to image patches drawn from a natural image ensemble has produced interesting insights as well as applications in image segmentation and image querying,[68] as well as image restoration[62] and image denoising.[63,65]

One approach to image denoising would be to define the expected log-likelihood of all patches occurring in a given image $\mathbf{I}$.[62] This is computed by summing the log-likelihood of each (possibly overlapping) patch $\mathbf{I}_i$, after applying a given prior p.

$$E_p(\mathbf{I}) = \sum_i \log(p(\mathbf{I}_i)) \quad (25)$$

Note that this does not give the log probability of a full image, as patches $\mathbf{I}_i$ may overlap. Given a noisy image $\mathbf{J}$, the corruption can generally be modeled with:

$$\| \mathbf{AJ} - \mathbf{I} \|^2 \quad (26)$$

If we would wish to find an uncorrupted image $\mathbf{I}$ given a corrupted image $\mathbf{J}$, then this can be expressed as a minimization problem consisting of a likelihood term (Eq. 26) and a prior term (Eq. 25):

$$\arg\min \| \mathbf{AJ} - \mathbf{I} \|^2 - \lambda\, \mathrm{E}_p(\mathbf{I}) \quad (27)$$

where λ controls the relative weight of the prior term. The log-likelihood of a patch is given by:

$$\log p(\mathbf{I}_i) = \log\left(\sum_{k=1}^{K} w_k\, N(\mathbf{I}_j \mid \mathbf{mu}_j, \Sigma_j) \right) \quad (28)$$

where the number of Gaussian mixture components is K. This approach has been shown to produce favorable results for both patch restoration and image restoration tasks, as compared to various other techniques that include PCA and ICA.[62]

REFERENCES

1. Burges, C.J.C. Dimension reduction: A guided tour. Found. Trends Mach. Learn. **2009**, *2* (4), 275–365.
2. Pearson, K. On lines and planes of closest fit to systems of points in space. Philos. Mag. **1901**, *2* (11), 559–572.
3. Wold, S.; Esbensen, K.; Geladi, P. Principal component analysis. Chemom. Intell. Lab. Syst. **1987**, *2* (1), 37–52.
4. Jolliffe, I. Principal component analysis. In *Encyclopedia of Statistics in Behavioral Science*; Everitt, B.; Howell, D.; Eds.; John Wiley & Sons: New York, 2005.
5. Hardoon, D.R.; Szedmak, S.; Shawe-Taylor, J. Canonical correlation analysis: An overview with application to machine learning methods. Neural Comput. **2004**, *16* (12), 2639–2664.
6. Thompson, B. Canonical correlation analysis. In *Encyclopedia of Statistics in Behavioral Science*; Everitt, B.; Howell, D.; Eds.; John Wiley & Sons, 2005.
7. McLachlan, G.J. *Discriminant Analysis and Statistical Pattern Recognition*; John Wiley & Sons, 2004.
8. Fisher, R.A. The use of multiple measurements in taxonomic problems. Ann. Eugen. **1936**, *7* (2), 179–199.
9. Acharyya, R. *A New Approach for Blind Source Separation of Convolutive Sources: Wavelet Based Separation Using Shrinkage Function*; VDM Verlag Dr. Müller e.K: Saarbrücken, 2008.
10. Lanczos, C. An iteration method for the solution of the eigenvalue problem of linear differential and integral operators. J. Res. Natl. Bur. Stand. **1950**, *45* (4), 255–282.
11. Saad, Y. *Numerical Methods for Large Eigenvalue Problems*; Manchester University Press: Manchester, 1992.
12. Francis, J.G.F. The QR transformation, I. Comput. J. **1961**, *4* (3), 265–271.
13. Francis, J.G.F. The QR transformation, II. Comput. J. **1962**, *4* (4), 332–345.
14. Kublanovskaya, V.N. On some algorithms for the solution of the complete eigenvalue problem. USSR Comput. Math. Math. Phys. **1961**, *1* (3), 637–657.
15. Arnoldi, W.E. The principle of minimized iterations in the solution of the matrix eigenvalue problem. Q. Appl. Math. **1951**, *9* (1), 17–29.
16. Hyvärinen, A.; Hurri, J.; Hoyer, P.O. *Natural Image Statistics: A Probabilistic Approach to Early Computational Vision*; Springer: London, 2009.
17. Hancock, P.J.B.; Baddeley, R.; Smith, L. The principal components of natural images. Network: Comput. Neural Syst. **1992**, *3* (1), 61–70.
18. Orban, G.A. *Neuronal Operations in the Visual Cortex*; Springer-Verlag: Berlin, 1984.
19. Olshausen, B.A.; Field, D.J. Emergence of simple-cell receptive field properties by learning a sparse code for natural images. Nature **1996**, *381* (6583), 607–609.
20. Field, D.J. What is the goal of sensory coding? Neural Comput. **1994**, *6* (4), 559–601.
21. Baddeley, R.J. An efficient code in V1? Nature **1996**, *381* (6583), 560–561.
22. Földiák, P.; Young, M.P. Sparse coding in the primate cortex. In *The Handbook of Brain Theory and Neural Networks*; Arbib, M.A.; Ed.; MIT Press: Cambridge, MA, 1995, 895–989.
23. Baum, E.B.; Moody, J.; Wilczek, F. Internal representations for associative memory. Biol. Cybern. **1988**, *59* (4–5), 217–228.
24. Mel, B.W. SEEMORE: Combining color, shape, and texture histogramming in a neurally inspired approach to visual object recognition. Neural Comput. **1997**, *9* (4), 777–804.
25. Murase, H.; Nayar, S.K. Visual learning and recognition of 3-D objects from appearance. Int. J. Comput. Vis. **1995**, *14* (1), 5–24.
26. Huber, P.J. *Robust Statistics*; John Wiley & Sons: New York, 1981.
27. Rousseeuw, P.J.; Leroy, A.M. *Robust Regression and Outlier Detection*; John Wiley & Sons: Hoboken, NJ, 1987.
28. Leonardis, A.; Bischof, H. Robust recognition using eigenimages. Comput.Vis. Image Underst. **2000**, *78* (1), 99–118.
29. Murase, H.; Nayar, S.K. Illumination planning for object recognition using parametric eigenspaces. IEEE Trans. Pattern Anal. Mach. Intell. **1994**, *16* (12), 1219–1227.
30. Nayar, S.K.; Murase, H.; Nene, S.A. Learning, positioning, and tracking visual appearance. In *Proceedings of the IEEE International Conference on Robotics and Automation*, 1994, 3237–3244.
31. Yoshimura, S.; Kanade, T. Fast template matching based on the normalized correlation by using multiresolution eigenimages. In *Proceedings of the IEEE/RSJ/GI International Conference on Intelligent Robots and Systems (Advanced Robotic Systems and the Real World, IROS'94)*, Vol. 3; 1994, 2086–2093.
32. Beymer, D.; Poggio, T. Face recognition from one example view. In *Proceedings of the 5th IEEE International Conference on Computer Vision*, 1995, 500–507.
33. Turk, M.; Pentland, A. Eigenfaces for recognition. J. Cogn. Neurosci. **1991**, *3* (1), 71–86.
34. Blanz, V.; Vetter, T. A morphable model for the synthesis of 3D faces. In *SIGGRAPH '99: Proceedings of the 26th Annual Conference on Computer Graphics and Interactive Techniques*, 1999, 187–194.
35. Hancock, P.J.B.; Burton, A.M.; Bruce, V. Face processing: Human perception and principal components analysis. Mem. Cognit. **1996**, *24* (1), 26–40.
36. Valentine, T. *Cognitive and Computational Aspects of Face Recognition: Explorations in Face Space*; Routledge: London, 1995.
37. Leopold, D.A.; O'Toole, A.; Vetter, T.; Blanz, V. Prototype-referenced shape encoding revealed by high-level aftereffects. Nat. Neurosci. **2001**, *4* (1), 89–94.
38. Atick, J.J.; Griffin, P.A.; Redlich, A.N. Statistical approach to shape from shading: Reconstruction of three-dimensional face surfaces from single two-dimensional images. Neural Comput. **1996**, *8* (6), 1321–1340.
39. Calder, A.J.; Young, A.W. Understanding the recognition of facial identity and facial expression. Nat. Rev. Neurosci. **2005**, *6* (8), 641–651.

40. Hyvärinen, A. Survey on independent component analysis. Neural Comput. Surv. **1999**, *2* (4), 94–128.
41. Comon, P. Independent component analysis: A new concept? Signal Process. **1994**, *36* (3), 287–314.
42. Comon, P.; Jutten, C. *Handbook of Blind Source Separation: Independent Component Analysis and Applications*; Academic Press/Elsevier: Oxford, 2010.
43. Hyvärinen, A.; Karhunen, J.; Oja, E. *Independent Component Analysis*; John Wiley & Sons: New York, 2001.
44. Bell, A.J.; Sejnowski, T.J. An information-maximization approach to blind separation and blind deconvolution. Neural Comput. **1995**, *7* (6), 217–234.
45. Lee, T.-W.; Girolami, M.; Sejnowski, T.J. Independent component analysis using an extended infomax algorithm for mixed sub-Gaussian and super-Gaussian sources. Neural Comput. **1999**, *11* (2), 417–441.
46. Hyvärinen, A. Fast and robust fixed-point algorithms for independent component analysis. IEEE Trans. Neural Networks **1999**, *10* (3), 626–634.
47. Bach, F.R.; Jordan, M.I. Kernel independent component analysis. J. Mach. Learn. Res. **2002**, *3* (July), 1–48.
48. Learned-Miller, E.G.; Fisher III. J.W. ICA using spacings estimates of entropy. J. Mach. Learn. Res. **2003**, *4* 1271–1295.
49. Ruderman, D.L.; Cronin, T.W.; Chiao, C.C. Statistics of cone responses to natural images: Implications for visual coding. J. Opt. Soc. Am. A **1998**, *15* (8), 2036–2045.
50. Edgeworth, F.Y. On the probably errors of frequency-constants. J. R. Stat. Soc. **1908**, *71*, 381–397, 499–512, 651–678, 72, 81–90.
51. Fisher, R.A. On an absolute criterion for fitting frequency curves. Stat. Sci. **1997**, *12* (1), 39–41.
52. Fisher, R.A. On the mathematical foundations of theoretical statistics. Philos. Trans. R. Soc. A **1922**, *222*, 309–368.
53. Millar, R.B. *Maximum Likelihood Estimation and Inference: With Examples in R, SAS and ADMB*; John Wiley & Sons: Chichester, 2011.
54. Olshausen, B.A.; Field, D.J. Sparse coding with an overcomplete basis set: A strategy employed by V1? Vis. Res. **1997**, *37* (23), 3311–3325.
55. Bell, A.J.; Sejnowski, T.J. The 'independent components' of natural scenes are edge filters. Vis. Res. **1997**, *37* (23), 3327–3338.
56. van Hateren, J.H.; van der Schaaf, A. Independent component filters of natural images compared with simple cells in primary visual cortex. Proc. R. Soc. B: Biol. Sci. **1998**, *265* (1394), 359.
57. Hughes, J.M.; Graham, D.J.; Rockmore, D.N. Quantification of artistic style through sparse coding analysis in the drawings of Pieter Bruegel the elder. Proc. Natl. Acad. Sci. **2010**, *107* (4), 1279–1283.
58. Hughes, J.M.; Graham, D.J.; Jacobsen, C.R.; Rockmore, D.N. Comparing higher-order spatial statistics and perceptual judgements in the stylometric analysis of art. In *Proceedings of the European Signal Processing Conference (EUSIPCO)*, 2011, 1244–1248.
59. Koch, M.; Denzler, J.; Redies, C. 1/ f2 characteristics and isotropy in the Fourier power spectra of visual art, cartoons, comics, mangas, and different categories of photographs. PLoS One **2010**, *5* (8), e12268.
60. Olshausen, B.A.; DeWeese, M.R. The statistics of style. Nature **2010**, *463* (7284), 1027–1028.
61. Kanan, C.; Cottrell, G. Robust classification of objects, faces, and flowers using natural image statistics. In *Proceedings of the IEEE International Conference on Computer Vision and Pattern Recognition*, 2010, 2472–2479.
62. Zoran, D.; Weiss, Y. From learning models of natural image patches to whole image restoration. In *Proceedings of the IEEE International Conference on Computer Vision*, 2011, 479–486.
63. Zoran, D.; Weiss, Y. Natural images, Gaussian mixtures and dead leaves. In *NIPS-25: Proceedings of the 2012 Conference on Advances in Neural Information Processing Systems*, 2012, 1745–1753.
64. Dempster, A.P.; Laird, N.M.; Rubin, D.B. Maximum likelihood from incomplete data via the EM algorithm. J. R. Stat. Soc. B **1977**, *39* (1), 1–38.
65. Portilla, J.; Strela, V.; Wainwright, M.J.; Simoncelli, E.P. Image denoising using scale mixtures of Gaussians in the wavelet domain. IEEE Trans. Image Process. **2003**, *12* (11), 1338–1351.
66. Theis, L.; Hosseini, R.; Bethge, M. Mixtures of conditional Gaussian scale mixtures applied to multiscale image representations. PLoS One **2012**, *7* (7), e39857.
67. Tai, Y.; Jia, J.; Tang, C. Local color transfer via probabilistic segmentation by expectation-maximization. In *Proceedings of the IEEE International Conference on Computer Vision and Pattern Recognition*, Vol. 1; Washington, DC, 2005, 747–754.
68. Carson, C.; Belongie, S.; Greenspan, H.; Malik, J.; Blobworld: Image segmentation using expectation-maximization and its application to image querying. IEEE Trans. Pattern Anal. Mach. Intell. **2002**, *24* (8), 1026–1038.
69. Troje, N.F.; Bülthoff, H.H. Face recognition under varying poses: The role of texture and shape. Vis. Res. **1996**, *36* (12), 1761–1771.

Discrete Fourier Transforms

Dave K. Kythe
Security Leader & Advisor, Los Angeles, California, U.S.A.

Prem K. Kythe
University of New Orleans, New Orleans, Louisiana, U.S.A.

Abstract
The discrete Fourier transform (DFT) is one of the most important algorithms for processing digital signals, including images. It has many other applications as well in virtually all important engineering domains. In this article the DFT is introduced and many of its important properties are examined.

Keywords: Algorithm; DFT; Discrete fourier transform.

DISCRETE FOURIER TRANSFORM

The discrete Fourier transform (DFT) that transforms one function into another is called the *frequency domain representation*; it requires a discrete input function whose nonzero real or complex values are finite. Unlike the discrete-time Fourier transform (DIFT), the DFT only evaluates enough frequency components to reconstruct the finite segment that was analyzed. The DFT is used in processing information (data) stored in computers, and in signal processing and related fields where the frequencies in a signal are analyzed. An efficient computation of the DFT is provided by a fast Fourier transform (FFT) algorithm. Although DFT refers to a mathematical transformation while FFT refers to a specific family of algorithms for computing DFTs, the FFT algorithms commonly used to compute DFTs often mean DFT in common terminology, which has now become confusing by taking FFT as synonymous with DFT. For more details, see Kythe and Schäferkotter[1] [2005: 299 ff]. The DFT is defined as follows: Let a sequence of n nonzero complex numbers $\{x_0,\ldots,x_{n-1}\}$ be transformed into a sequence of m complex numbers $\{X_0,\ldots,X_{m-1}\}$ by the formula

$$X_j = \sum_{k=0}^{n-1} x_k e^{-2\pi jk\, i/m} \quad \text{for} \quad 0 \le j \le m-1;\ i=\sqrt{-1}, \tag{1}$$

where

$$x_k = \frac{1}{m}\sum_{j=0}^{m-1} X_j e^{2\pi kj\, i/m} \quad \text{for} \quad 0 \le k \le n-1. \tag{2}$$

Formula (1) is known as the DFT analysis equation and its inverse (2) the DFT synthesis equation, or the *inverse discrete Fourier transform* (IDFT). The DFT pair can be represented as

$$x_k \xrightarrow{DFT} X_j \quad \text{and} \quad X_j \xrightarrow{IDFT} x_k. \tag{3}$$

Example 1

Let $n = 4$, and define a nonperiodic sequence x_k by

$$x_k = \begin{cases} 2 & k=0, \\ 3 & k=1, \\ -1 & k=2, \\ 1 & k=3. \end{cases}$$

By (1), the 4-point DFT of x_k is

$$X_j = \sum_{k=0}^{3} x_k e^{-2\pi kj i/4} = 2 + 3e^{-\pi ji/2} - e^{-\pi ji} + e^{-3\pi ji/2}, \quad 0 \le j \le 3.$$

Thus, $X_0 = 2 + 3 - 1 + 1 = 5$; $X_1 = 2 - 3i + 1 + i = 3 - 2i$; $X_2 = 2 - 3 - 1 - 1 = -3$; $X_3 = 2 - 3i + 1 - i = 3 + 2i$. The vertical bar plots of the function x_k and its magnitude and phase are presented in Fig. 1.

Example 2

The n-point DFT of the nonperiodic sequence x_k of length n is defined by $x_k = \begin{cases} 1, & 0 \le k \le (n_1-1), \\ 0, & n_1 \le k \le n. \end{cases}$ Then

Discrete Fourier Transforms

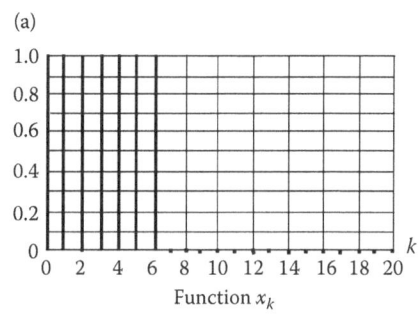

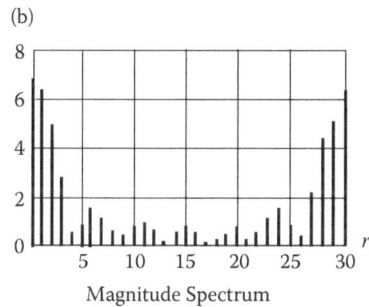
Magnitude Spectrum

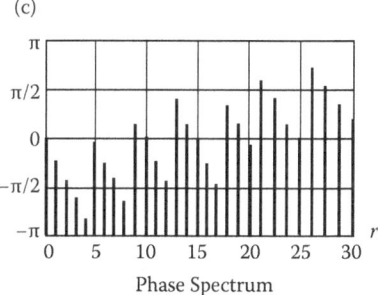
Phase Spectrum

Fig. 1 Magnification and phase spectra

$$X_j = \sum_{k=0}^{n_1-1} e^{2jk\pi\,i/n} = \begin{cases} n_1, & j=0, \\ \dfrac{1-e^{2n_1 j\pi\,i/n}}{1-e^{2j\pi\,i/n}}, & j \neq 0, \end{cases}$$

$$= \begin{cases} n_1, & j=0, \\ e^{-(n_1-1)\pi j/n}\dfrac{\sin(n_1 j\pi/n)}{\sin(j\pi/n)}, & j \neq 0. \end{cases}$$

The magnitude $|X_j| = \begin{cases} n_1 & \text{for } j=0, \\ \dfrac{\sin(n_1 j\pi/n)}{\sin(j\pi/n)} & \text{for } j \neq 0. \end{cases}$

The phase is given by $\phi(X_j) = \begin{cases} 0 & \text{for } j=0, \\ \sin(n_1 j\pi/n) & \text{for } j \neq 0. \end{cases}$

DFT IN MATRIX FORM

Expand (1) in terms of the time and frequency indices (k, j), and we get for $n = m$:

$$\begin{aligned}
X_0 &= x_0 + x_1 + x_2 \cdots + x_{n-1}, \\
X_1 &= x_0 + x_1 e^{-2\pi\,i/n} + x_2 e^{-4\pi\,i/n} + \cdots + x_{n-1} e^{-2(n-1)\pi\,i/n}, \\
X_2 &= x_0 + x_1 e^{-4\pi\,i/n} + x_2 e^{-8\pi\,i/n} + \cdots + x_{n-1} e^{-4(n-1)\pi\,i/n}, \\
&\vdots \\
X_{n-1} &= x_0 + x_1 e^{-2(n-1)\pi\,i/n} + x_2 e^{-4(n-1)\pi\,i/n} + \cdots + x_{n-1} e^{-2(n-1)^2 \pi\,i/n},
\end{aligned} \qquad (4)$$

which can be written in matrix form as

$$\begin{Bmatrix} X_0 \\ X_1 \\ X_2 \\ \vdots \\ X_{n-1} \end{Bmatrix} = \begin{bmatrix} 1 & 1 & 1 & \cdots & 1 \\ 1 & e^{-2\pi\,i/n} & e^{-4\pi\,i/n} & \cdots & e^{-2(n-1)\pi\,i/n} \\ 1 & e^{-4\pi\,i/n} & e^{-8\pi\,i/n} & \cdots & e^{-4(n-1)\pi\,i/n} \\ \vdots & \vdots & \vdots & & \vdots \\ 1 & e^{-2(n-1)\pi\,i/n} & e^{-4(n-1)\pi\,i/n} & \cdots & e^{-2(n-1)^2 \pi\,i/n} \end{bmatrix} \begin{Bmatrix} x_0 \\ x_1 \\ x_2 \\ \vdots \\ x_{n-1} \end{Bmatrix}. \qquad (5)$$

Similarly, for IDFT in matrix form we have

$$\begin{Bmatrix} x_0 \\ x_1 \\ x_2 \\ \vdots \\ x_{n-1} \end{Bmatrix} = \frac{1}{n} \begin{bmatrix} 1 & 1 & 1 & \cdots & 1 \\ 1 & e^{2\pi\,i/n} & e^{4\pi\,i/n} & \cdots & e^{2(n-1)\pi\,i/n} \\ 1 & e^{4\pi\,i/n} & e^{8\pi\,i/n} & \cdots & e^{2(n-1)\pi\,i/n} \\ \vdots & \vdots & \vdots & & \vdots \\ 1 & e^{2(n-1)\pi\,i/n} & e^{4\pi(n-1)\,i/n} & \cdots & e^{2(n-1)^2 \pi\,i/n} \end{bmatrix} \begin{Bmatrix} X_0 \\ X_1 \\ X_2 \\ \vdots \\ X_{n-1} \end{Bmatrix}. \qquad (6)$$

Example 3

In Example 2, we have $X = [5\ 3-2i\ -3\ 3+2i]$. Then

$$\begin{Bmatrix} x_0 \\ x_1 \\ x_2 \\ \vdots \\ x_{n-1} \end{Bmatrix} = \frac{1}{4} \begin{bmatrix} 1 & 1 & 1 & \cdots & 1 \\ 1 & e^{\pi i/2} & e^{\pi i} & e^{3\pi i/2} & \\ 1 & e^{\pi i} & e^{3\pi i/2} & e^{3\pi i} & \\ 1 & e^{3\pi i/2} & e^{3\pi i} & \cdots & e^{9\pi i/2} \end{bmatrix} \begin{Bmatrix} 5 \\ 3-2i \\ -3 \\ \vdots \\ 3+2i \end{Bmatrix}$$

$$= \begin{Bmatrix} 2 \\ 3 \\ -1 \\ 1 \end{Bmatrix}.$$

DFT BASIS FUNCTIONS

If we express (6) in the form

$$\begin{Bmatrix} x_0 \\ x_1 \\ x_2 \\ \vdots \\ x_{n-1} \end{Bmatrix} = \frac{1}{n} X_0 \begin{Bmatrix} 1 \\ 1 \\ 1 \\ \vdots \\ 1 \end{Bmatrix} + \frac{1}{n} X_1 \begin{Bmatrix} 1 \\ e^{2\pi i/n} \\ e^{4\pi i/n} \\ \vdots \\ e^{2(n-1)\pi i/n} \end{Bmatrix}$$

$$+ \frac{1}{n} X_2 \begin{Bmatrix} 1 \\ e^{4\pi i/n} \\ e^{8\pi i/n} \\ \vdots \\ e^{4(n-1)\pi i/n} \end{Bmatrix} + \cdots \quad (7)$$

$$+ \frac{1}{n} X_{n-1} \begin{Bmatrix} 1 \\ e^{2(n-1)\pi i/n} \\ e^{4(n-1)\pi i/n} \\ \vdots \\ e^{2(n-1)^2 \pi i/n} \end{Bmatrix},$$

then it is obvious that the DFT basis functions f_j are the columns of the right-hand square matrix B, i.e.,

$$f_j = \frac{1}{n} \begin{bmatrix} 1 & e^{2\pi j\,i/n} & e^{4\pi j\,i/n} & \cdots & e^{2(n-1)j\pi\,i/n} \end{bmatrix}^T,$$
for $0 \leq j \leq n-1$.

Eq. 7 represents a DT sequence as a linear combination of complex exponentials, which are weighted by the corresponding DFT coefficients. Such a representation can be used to analyze linear time-invariant systems.

PROPERTIES OF DFT

The properties of the m-point DFT are as follows:

(i) PERIODICITY. $X_j = X_{j+pm}$ for $0 \leq j \leq m - 1$, where $p \in \mathbb{R}^+$. In other words, the m-point DFT of an aperiodic sequence of length n, $n \leq m$, is periodic with period m.

(ii) LINEARITY. If x_k and y_k are two DT sequence with the m-point DFT pairs: $x_k \xrightarrow{DFT} X_j$ and $y_k \xrightarrow{DFT} Y_j$, then for any arbitrary constants a and b (which may be complex)

$$ax_k + by_k \xrightarrow{DFT} a_1 X_j + b Y_j. \quad (8)$$

(iii) ORTHOGONALITY. The column vectors f_j of the DFT matrix, defined in Eq. 7, form the basis vectors of the DFT and are orthogonal with respect to each other such that

$$f_j^h \cdot f_l = \begin{cases} m & \text{for } j = l, \\ 0 & \text{for } j \neq l, \end{cases}$$

where $\cdot$ denotes the dot product and h the Hermitian operation.

(iv) HERMITIAN SYMMETRY. This Hermitian symmetry implies that for the m-point DFT X_j of a real-valued aperiodic sequence x_k

$$X_j = \bar{X}_{m-j}, \quad (9)$$

where $\bar{X}_j$ is the complex conjugate of X_j. In other words, X_j is conjugate-symmetric about $j = m/2$. The magnitude $|X_{m-i}| + X_j|$, and the phase $\phi(X_{m-j}) = -\phi(X_j)$, i.e., the phase of the spectrum is odd.

(v) TIME SHIFTING. If $x_k \xrightarrow{DFT} X_j$, then for an m-point DFT and an arbitrary integer k_0

$$x_{k-k_0} \xrightarrow{DFT} e^{2k_0 j\pi i/m} X_j. \quad (10)$$

(vi) CIRCULAR CONVOLUTION. For two DT sequences x_k and y_k with the m-point DFT pairs: $x_k \xrightarrow{DFT} X_j$ and $y_k \xrightarrow{DFT} Y_j$, the circular convolution is defined by

$$x_k \otimes y_k \xrightarrow{DFT} X_j Y_j, \quad (11)$$

and

$$x_k y_k \xrightarrow{DFT} \frac{1}{m}\left[X_j \otimes Y_j\right], \quad (12)$$

where $\otimes$ denotes the circular convolution operation. In this operation the two sequences must be of equal length.

PARSEVAL'S THEOREM. If $x_k \xrightarrow{DFT} X_j$, then the energy E_x of an aperiodic sequence x_k of length n can be written in terms of its m-point DFT as

$$E_x = \sum_{k=0}^{n-1} |x_k|^2 = \frac{1}{m} \sum_{k=0}^{m-1} |X_j|^2. \qquad (13)$$

In other words, the DFT energy preserves the energy of the signal within a scale factor of m.

Example 4

Notation DFT index = j; DTFT frequency: $\omega_j = 2\pi j/n$; DFT coefficients X_j; DTFT coefficients X_ω. Using the DFT, calculate the DTFT of the DT decaying exponential sequence $x_k = 0.6^k u + k$. For time limitation, apply a rectangular window of length $n = 10$; then the truncated sequence is given by

$$x_k^w = \begin{cases} 0.6^k & \text{for } 0 \leq k \leq 9, \\ 0 & \text{elsewhere.} \end{cases}$$

The DFT coefficients are compared with the IDFT coefficients in Table 1.

$$0.6^k u_k \xrightarrow{CTFT} \frac{1}{1 - 0.6^{-\omega i}},$$

where CTFT refers to continuous-time Fourier transform.

Example 5

The DTFT of the aperiodic sequence $x_k = [2, 1, 0, 1]$ for $0 \leq k \leq 3$ is given by $X_j = [4, 2, 0, 2]$ for $0 \leq j \leq 3$. By mapping in the DTFT domain, the corresponding DTFT coefficients are given by $X_{\omega_j} = [4, 2, 0, 2]$ for $\omega_j = [0, 0.5\pi, \pi, 1.5\pi]$ rads/sec. On the other hand, if the DTFT is computed in the range $-\pi \leq \omega \leq \pi$, then the DTFT coefficients are given by $X_{\omega_j} = [4, 2, 0, 2]$ for $\omega_j = [\pi, -0.5\pi, 0, 0.5\pi]$ rads/sec.

Table 1 Comparison of DFT coefficients with IDFT coefficients

j	ω_j	X_j	X_ω
−5	−π	0.6212	0.6230
−4	−0.8π	0.6334 + 0.1504 i	0.6337 + 0.3297 i
−3	−0.6π	0.6807 + 0.3277 i	0.6337 + 0.1513 i
−2	−0.4π	0.8185 + 0.5734 i	0.8235 + 0.5769 i
−1	−0.2π	1.3142 + 0.9007 i	1.322 + 0.9062 i
0	0	2.4848	2.5000
1	0.2π	1.3142 − 0.9007 i	1.322 − +0.9062 i
2	0.4π	0.8185 − 0.5734 i	0.8235 − 0.5769 i
3	0.6π	0.6807 − 0.3277 i	0.6849 − 0.3297 i
4	0.8π	0.6334 − 0.1504 i	0.6373 − 0.1513 i

Note that the above DFT coefficients can be computed from the correspondence.

ZERO PADDING

To improve the resolution of the frequency axis ω in the DFT domain, a common practice is to append additional zero-valued samples to the DT sequences. This process, known as *zero padding*, is defined for an aperiodic sequence x_k of length n by

$$x_k^{zp} = \begin{cases} x_k & \text{for } 0 \leq k \leq n-1, \\ 0 & \text{for } n \leq k \leq m-1. \end{cases} \qquad (14)$$

Thus, the zero-padded sequence x_k^{zp} has an increased length of m. This improves the frequency resolution $\Delta\omega$ of the zero-padded sequence from $2\pi/n$ to $2\pi/m$.

PROPERTIES OF m-POINT DFT

Let the length of the DT sequence be $n \leq m$, and let the DT sequence be zero-padded with $m - n$ zero-valued samples. Then the properties of periodicity, linearity, orthogonality, Hermitian symmetry, and Parseval's theorem are the same as those given in § 4.

Example 6

(Circular convolution) Consider two aperiodic sequences $x_k = [0, 1, 2, 3]$ and $y_k = [5, 5, 0, 0]$ defined over $0 \leq k \leq 3$. We will use the property (11) to compute circular convolution as follows: Since $X_j = [6, -2 + 2i, -2, -2 - 2i]$ and $Y_j = [10, 5 - i, 0, 5 + 5i]$ for $0 \leq j \leq 3$, we have

$$x_k \otimes y_k \xrightarrow{DFT} [60, 20i, 0, -20i],$$

which after taking the inverse DFT yields

$$x_k \otimes y_k = [15, 5, 15, 25].$$

LINEAR CONVOLUTION VERSUS CIRCULAR CONVOLUTION

The linear convolution $x_k \star y_k$ between two time-limited DT sequences x_k and y_k of lengths n_1 and n_2, respectively, can be expressed in terms of the circular convolution $x_k \otimes y_k$ by zero-padding both x_k and y_k such that each sequence has length $N \geq (n_1 + n_2 - 1)$. It is known that the circular convolution of the zero-padded sequences is the same as that of the linear convolution. The algorithm for implementing the linear convolution of two sequences x_k and y_k is as follows:

STEP 1. Compute the N-point DFTs X_j and Y_j of the two time-limited sequences x_k and y_k, where the value of $N \geq n_1 + n_2 - 1$.

STEP 2. Compute the product $Z_j = X_j Y_j$ for $0 \leq j \leq N - 1$.

STEP 3. Compute the sequence z_k as the IDFT of Z_j. The resulting sequence z_k is the result of the linear convolution between x_k and y_k.

Example 7

Consider the DT sequences

$$x_k = \begin{cases} 2 & \text{for } k = 0, \\ -1 & \text{for } |k| = 1, \\ 0 & \text{otherwise,} \end{cases} \quad \text{and} \quad \begin{cases} 2 & \text{for } k = 0, \\ 3 & \text{for } |k| = 1, \\ -1 & \text{for } |k| = 2, \\ 0 & \text{otherwise.} \end{cases}$$

STEP 1. Since the sequences x_k and y_k have lengths $n_1 = 3$ and $n_2 = 5$, the value of $N \geq 5 + 3 - 1 = 7$; so we set $N = 7$. Then, zero-padding in x_k is $N - n_1 = 4$ additional zeros, which gives $x'_k = [-1, 2, -1, 0, 0, 0, 0]$; similarly, zero-padding in y_k is $N - n_2 = 2$ additional zeros, which gives $y'_k = [-1, 3, 2, 3, -1, 0, 0]$. The values of the DFT of x'_k and of y'_k are given in Table 2.

STEP 2. The value of $Z_j = X'_j Y'_j$ is shown in the fourth column of Table 2.

STEP 3. Taking the IDFT of Z_j gives

$$z_k = [0.998, -5, 5.001, -1.999, 5, -5.002, 1.001].$$

RADIX-2 ALGORITHM FOR FFT

Theorem 1.

For even values of N, the N-point DFT of a real-valued sequence x_k of length $m \leq N$ can be computed from the DFT coefficients of two subsequences: (i) x_{2k}, which contains the even-valued samples of x_k, and (ii) x_{2k+1}, which contains the odd-valued samples of x_k.

This theorem leads to the following algorithm to determine the N-point DFT:

STEP 1. Determine the $(N/2)$-point DFT G_j for $0 \leq j \leq N/2 - 1$ of the even-numbered samples of x_k.

STEP 2. Determine the $(N/2)$-point DFT H_j for $0 \leq j \leq N/2 - 1$ of the odd-numbered samples of x_k.

STEP 3. The N-point DFT coefficients X_j for $0 \leq j \leq k-1$ of x_k are obtained by combining the $(N/2)$ DFT coefficients of G_j and H_j using the formula $X_j = G_j + W_k^j H_j$, where $W_k^j = e^{-2\pi i/N}$ is known as the *twiddle factor*. Note that although the index $j = 0,\ldots, N - 1$, we only compute G_j and H_j over $0 \leq j \leq (N/2 - 1)$, and any outside value can be determined using the periodicity properties of G_j and H_j, which are defined by $G_j = G_{j+N/2}$ and $H_j = H_{j+N/2}$.

The flow graph for the above method for $N = 8$-point DFT is shown in Fig. 2.

In general, this figure computes two $(N/2)$-point DFTs along with N complex additions and N complex multiplications. Thus, $(n/2)^2 + N$ complex additions and $(N/2)^2 + N$ complex multiplications are needed. Since $(N/2)^2 +$

Table 2 Values of X'_j, Y'_j, Z_j for $0 \leq j \leq 6$

j	X'_j	Y'_j	Z_j
0	0	6	0
1	$0.470 - 0.589\,i$	$-1.377 - 6.031\,i$	$-4.199 - 2.024\,i$
2	$-5.440 - 2.384\,i$	$-2.223 + 1.070\,i$	$3.760 + 4.178\,i$
3	$-3.425 - 1.650\,i$	$-2.901 - 3.638\,i$	$3.933 + 17.247\,i$
4	$-3.425 + 1.650\,i$	$-2.901 + 3.638\,i$	$3.933 - 17.247\,i$
5	$-0.544 + 2.384\,i$	$-2.223 - 31.070\,i$	$3.760 - 4.178\,i$
6	$0.470 + 0.589\,i$	$-1.377 + 6.031\,i$	$-4.199 + 2.024\,i$

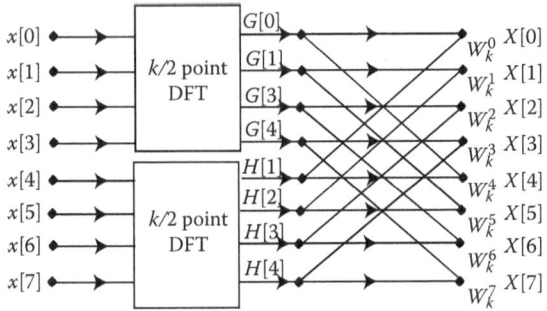

Fig. 2 Flow graph for 8-point DFT

$N < N^2$ for $N > 2$, the above algorithm is considerably cost effective.

In the case when N is a power of 2, the $(N/2)$-point DFTs G_j and H_j can be computed by the formula

$$G_j = \sum_{nu=0,1,2,\ldots}^{N/4-1} g_{2v} e^{-2vj\pi i/(N/4)} + W_{N/2}^j \sum_{v=0,1,2,\ldots}^{N/4-1} g_{2v+1} e^{-2vj\pi i/(N/4)}, \quad (15)$$

where the first summation will be denoted by G'_j and the second by G''_j. Similarly,

$$H_j = \sum_{nu=0,1,2,\ldots}^{N/4-1} h_{2v} e^{-2vj\pi i/(N/4)} + W_{N/2}^j \sum_{v=0,1,2,\ldots}^{N/4-1} h_{2v+1} e^{-2vj\pi i/(N/4)}, \quad (16)$$

where the first summation will be denoted by H'_j and the second by H''_j. Formula (15) represents the $(N/2)$-point DFT G_j in terms of two $(N/4)$-point DFTs of the even- and odd-numbered samples of g_k; similarly, formula (16) represents the $(N/2)$-point DFT H_j in terms of two $(N/4)$-point DFTs of the even- and odd-numbered samples of h_k. The flow graphs of these cases are presented in Fig. 3.

Thus, for example, the 2-point DFTs G'_0 and G'_1 can be expressed as

$$G'_0 = x_0 e^{-2vj\pi i/2}\Big|_{v=0,j=0} + x_4 e^{-vj\pi i/2}\Big|_{v=1,j=1} = x_0 + x_4,$$

and

$$G'_1 = x_0 e^{-2vj\pi i/2}\Big|_{v=0,j=1} + x_4 e^{-vj\pi i/2}\Big|_{v=1,j=1} = x_0 - x_4.$$

Discrete Fourier Transforms

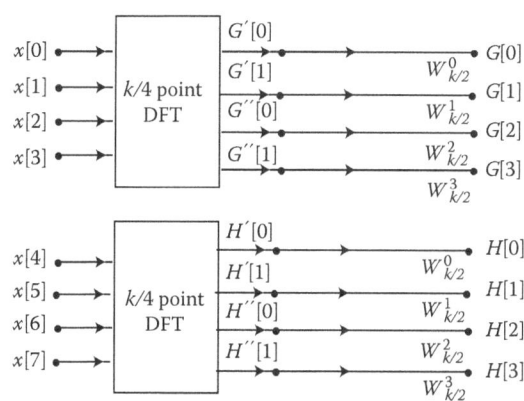

Fig. 3 Flow graph for $(N/2)$-point DFTs using $(N/4)$-point DFTs

Also note that in the flow graphs of Fig. 3, the twiddle factor for an 8-point DFT is

$$W_{N/2}^j = e^{-vj\pi i/(N/2)} = e^{-4vj\pi i/N} = W_N^{2j}.$$

REFERENCE

1. Kythe, P.K.; Scaferkotter, M.R. *Handbook of Computational Methods for Integration*; Chapman & Hall/CRC: Boca Raton, FL, 2005.

Facial Animation

Frederic I. Parke
Department of Visualization, College of Architecture, Texas A&M University, College Station, Texas, U.S.A.

Keith Waters
Kwaters Consulting, Boston, Massachusetts, U.S.A.

Abstract

In recent years, there has been dramatically increased interest in computer-based three-dimensional (3-D) facial character animation. Facial animation is not a new endeavor; initial efforts to represent and animate faces using computers go back more than 35 years. However, the recent explosion of activity in character animation has promoted a concurrent interest in facial animation. Our intent is to present the principles of facial animation to enable animation implementors to develop their own systems and environments.

Keywords: Animation; Herman Chernoff; Games industry; Medicine; Video teleconferencing; Avatars; Social robots; Believability flip; Uncanny valley.

The human face is interesting and challenging because of its familiarity. The face is the primary part of the body that we use to recognize individuals; we can recognize specific faces from a vast universe of similar faces and are able to detect very subtle changes in facial expression. These skills are learned early in life, and they rapidly develop into a major channel of communication. Small wonder, then, that character animators pay a great deal of attention to the face.

Human facial expression has been the subject of much investigation by the scientific community. In particular, the issues of universality of facial expression across cultures and the derivation of a small number of principal facial expressions have consumed considerable attention. *The Expression of the Emotions in Man and Animals*, published by Charles Darwin in 1872,[1] dealt precisely with these issues and sowed the seeds for a subsequent century to research, clarify, and validate his original theories. The value of this body of work, and of others in this field, requires no explanation in the context of facial animation.

The ability to model the human face and then animate the subtle nuances of facial expression remains a significant challenge in computer graphics. Despite a heavy reliance on traditional computer graphics algorithms such as modeling and rendering, facial modeling and animation are still being defined, without broadly accepted solutions. Facial animations often are developed with ad hoc techniques that are not easily extendible and that rapidly become brittle. Therefore, this entry presents a structured approach, by describing the anatomy of the face, working though the fundamentals of facial modeling and animation, and describing some state-of-the-art techniques.

ABOUT THIS ENTRY

Two-dimensional (2-D) facial character animation has been well defined over the years by traditional animation studios such as Disney Studios, Hanna-Barbera, and Warner Brothers. However, three-dimensional (3-D) computer-generated facial character animation is not as well defined. Therefore, this entry is focused principally on realistic 3-D faces.

The purpose of this entry is to provide a source for readers interested in the many aspects of computer-based facial animation. In this entry, we have tried to capture the basic requirements for anyone wanting to animate the human face, from key framing to physically based modeling. The nature of the subject requires some knowledge of computer graphics, although a novice to the subject can also find the entry an interesting resource about the face.

Clearly, the field of computer-generated facial animation is rapidly changing; every year, new advances are reported, making it difficult to capture the state of the art. However, it is clear that facial animation is a field whose time has come. The growth of increasingly complex computer-generated characters demands expressive, articulate faces. Most of the techniques employed today involve principles developed in the research community some years ago—in some instances, more than a couple of decades ago.

So why this surge of interest in computer-generated facial animation? There is no single reason, although we can point to several key influences. Perhaps the strongest interest comes from commercial animation studios, whose insatiable appetite for the latest and greatest visual effect is both enormous and endless. These studios are trendsetters who popularize new animation techniques. DreamWorks and Pixar are examples of such production studios where, for example, the movies *Shrek* and *The Incredibles* were produced. In addition, the advance in realism of video games has demanded expressive facial animation with high levels of realism.

Another key reason is the development of powerful interactive modeling and animation systems, such as Maya and 3D Studio. These systems dramatically ease the development of 3-D facial models. Improvements in surface scanner technology, such as optical laser scanners,[2] and motion capture systems have enabled many facial modeling and animation approaches based on surface and motion data from real faces. Also, overall advances in affordable computing power have made more sophisticated and computationally intensive modeling, animation, and rendering techniques broadly available.

Another intriguing influence is the advent of believable social agents. The construction of believable agents breaks the traditional mold of facial animation; agents have to operate in real time, bringing along a new set of constraints. While the basic algorithms used to animate real-time characters are concurrent with production animation, new tools have been developed to deal with issues such as lip synchronization and behavior interaction.

A BRIEF HISTORICAL SKETCH OF FACIAL ANIMATION

This section is a brief synopsis of key events that have helped shape the field, rather than a chronological account of facial animation. Most events in facial animation have been published in one form or another. The most popular forums have been the proceedings and course notes of the ACM SIGGRAPH conferences and other computer graphics journals and conference proceedings.

Historically, the first computer-generated images of 3-D faces were generated by Parke as part of Ivan Sutherland's computer graphics course at the University of Utah in early 1971. Parke began with very crude polygonal representations of the head, which resulted in a flip-pack animation of the face opening and closing its eyes and mouth. Several of these images are shown in Fig. 1.

While at the University of Utah, Henri Gouraud was also completing his dissertation work on his then-new smooth polygon shading algorithm. To demonstrate the effectiveness of the technique, he applied it to a digitized model of his wife's face. Parke used this innovative shading technique to produce several segments of fairly realistic facial

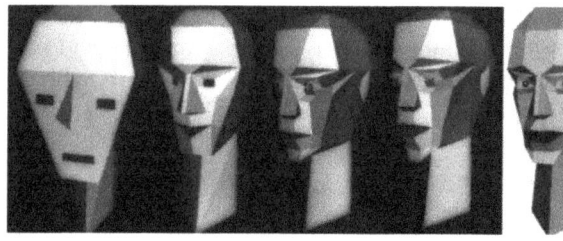

Fig. 1 Several of the earliest 3-D face models developed by Parke at the University of Utah in 1971

animation.[3] He did this by collecting facial expression polygon data from real faces using photogrammetric techniques and simply interpolating between expression poses to create animation. By 1974, motivated by the desire to quickly produce facial animation, Parke completed the first parameterized 3-D face model.[4]

In 1971, Chernoff first published his work using computer-generated 2-D face drawings to represent a k-dimensional space.[5] By using a simple graphical representation of the face, an elaborate encoding scheme was derived. Also in 1973, Gillenson at Ohio State University reported his work on an interactive system to assemble and edit 2-D line-drawn facial images, with the goal of creating a computerized photo identi-kit system.[6]

From 1974 through 1978, 3-D facial animation development was essentially dormant. However, during this period, the development of 2-D computer-assisted animation systems continued at the New York Institute of Technology, Cornell University, and later at Hanna-Barbera. These systems supported 2-D cartoon animation, including facial animation.

In 1980, Platt at the University of Pennsylvania published his master's thesis on a physically based muscle-controlled facial expression model.[7] In 1982, Brennan at MIT reported work on techniques for computer-produced 2-D facial caricatures.[8] Also at MIT in 1982, Weil reported on work using a video-disk-based system to interactively select and composite facial features.[9] Later at MIT, based on this work, Burson developed computer-based techniques for aging facial images, especially images of children.

In the mid-1980s, developments in facial animation took off once more. An animated short film, *Tony de Peltrie*, produced by Bergeron and Lachapelle in 1985, was a landmark for facial animation.[10] This was the first computer-animated short where 3-D facial expression and speech were a fundamental part of telling the story.

In 1987, Waters reported a new muscle model approach to facial expression animation.[11] This approach allowed a variety of facial expressions to be created by controlling the underlying musculature of the face. In 1988, Magnenat-Thalmann and colleagues also described an abstract muscle action model.[12] In 1987, Lewis and Parke,[13] and in 1988, Hill et al.[14] reported techniques for automatically synchronizing speech and facial animation.

Another groundbreaking animation short was *Tin Toy*, which received an Academy Award. Produced by Pixar, *Tin Toy* was an example of the capabilities of computer facial animation. In particular, a muscle model was used to articulate the facial geometry of the baby into a variety of expressions.[15]

The development of optical range scanners, such as the Cyberware™ optical laser scanner, provides a new wealth of data for facial animation.[2] In 1990, Williams reported the use of registered facial image texture maps as a means for 3-D facial expression animation.[16] By the late 1990s, large data sets of high-quality laser-scanned data were being used to create detailed morphable facial models by Blanz and Vetter.[17]

The new wave of enhanced image processing and scanning technology promised to usher in a new style of facial animation. In 1993, Lee et al.[18] described techniques to map individuals into a canonical representation of the face that has known physically based motion attributes.

Another growth area was in medicine, with a focus on surgical planning procedures and accurate simulation of face tissue dynamics. In 1988, Deng[19] and later Pieper[20] in 1991, used a finite-element model of skin tissue to simulate skin incisions and wound closure. More recently the finite-element approach has been applied to highly detailed biomechanical models of muscle and skin tissue derived from the Visible Human Project by Sifakis et al.[21]

Through the late 1990s, there was a surge of interest in facial analysis from video cameras. This interest was twofold: first, to provide the ability to track the human face to create lifelike characters, and second, to develop the ability to detect facial expression and thereby derive emotional states. There has been some success in both areas. Two popular techniques are model-based[22,23] and optical flow-based[24,25] techniques.

The late 1990s and early 2000s became a threshold for high-fidelity face capture and rendering for the film industry. Landmark films such as *The Lord of the Rings* (New Line Cinema, 2002), *The Matrix Reloaded* (Warner Bros., 2003), *The Polar Express* (Warner Bros., 2004), and *Monster House* (Sony Pictures, 2006) required face motion capture sessions of actors using markers and head gear. The capture sessions resulted in very large data sets, which had to be processed and rendered. Such techniques are referred to as *data-driven* facial animation and demand blending between more established modeling, rendering, and animation techniques, and alternative approaches.[26]

In the more recent past, the ability to create visual surrogates that are authentic enough to deceive observers into thinking they are real people is close at hand. Such techniques will most likely blend animation, modeling, and control with live captured data. How such surrogates will be used is speculative at this time; however, the film, games, medicine, and virtual online media are likely to be the first beneficiaries. The future is indeed bright for computer facial animation.

APPLICATION AREAS

By far the largest motivator, developer, and consumer of 3-D facial character animation is the animation industry itself. While the animation studios continue to shape how computers are used in animation, other emerging areas that influence animation are briefly mentioned below.

Games Industry

The games industry has experienced rapid recent development, due in part to increasing processor performance, coupled to more and more powerful graphics coprocessors. Such hardware advances have opened the door to more sophisticated real-time 3-D animation software that is different from the techniques employed in film production animation. High-quality real-time rendering is now commonplace. For example, texture mapping, as well as many special purpose effects such as hair rendering, skin reflectance mapping, environment mapping, as well as motion dynamics and multibody interaction, can be rendered on the fly to enhance the realism of game play.

While a maturing games industry has benefited from real-time performance, it also plays an important role in film production, where animated scenes can be blocked in and combined with live action well before final rendering is required. Within the context of facial animation, real-time sequencing and synchronization within the scene can save enormous amounts of time in film production.

It is clear that a number of opportunities for the next-generation facial animation techniques will be at the intersection of real-time performance and off-line, non-real-time production rendering. For the games industry, the challenges will be to combine lifelike performances with which players can interact; playing prerendered clips of a character will no longer be sufficient to provide a belief that the character is real and engaging. Engaging the player requires some sensing of user actions. For the face in particular, it will be increasingly important for the characters to engage in non-verbal communication, such as eye contact and the eye-gaze behaviors we experience in the real world.

The film industry can afford data-intensive capture and processing sessions to create *one-off* productions. In contrast, the games industry has to create algorithms that are used on the fly to fit the technical constraints of processors and rendering hardware. Nevertheless, compressing the large quantities of capture data down into manageable chunks, using principal component analysis (PCA), allows performance data to be used. Such techniques allow performance segments to be seamlessly stitched together into a real-time playback sequence.

Medicine

Computing in medicine is a large and diverse field. In the context of facial animation, two particular aspects are of interest: surgical planning and facial tissue surgical simulation. In both cases, the objective is to execute preoperative surgical simulation before the patient goes under the knife.

Craniofacial surgical planning involves the rearrangement of the facial bones due to trauma or growth defects.[27] Because this involves the rigid structures of bone tissue, the procedure essentially becomes a complex 3-D *cut-and-paste* operation. Computer models are typically generated from computer tomography scans of the head and the bone surfaces, generated from iso-surface algorithms such as the *marching cubes*.[28] More recently, the use of detailed data from the Visible Human Project has allowed models of the facial tissues—in particular, muscle and skin—to be identified and modeled.[29]

For facial tissue simulation, the objective is somewhat different. Here, the objective is to emulate the response of skin and muscle after they have been cut and tissue has been removed or rearranged.[30]

Video Teleconferencing

The ability to transmit and receive facial images is at the heart of video teleconferencing. Despite the rapid growth of available communication bandwidth, there remains a need for compression algorithms. One active research area is in model-based coding schemes and, in particular, algorithms applied to facial images.[31]

The components of a very low-bandwidth face video conferencing system are illustrated in Fig. 2. Each captured frame from video is analyzed by the encoder, with the assumption that the principal object in the scene is a human face. Computer vision algorithms are then used to extract and parameterize properties such as the shape, orientation, and motion of the head and face features.

These few parameters are compressed and transmitted to the decoder, where a 3-D model of the human head is synthesized to create a visual surrogate of the individual. As the head moves from frame to frame, new parameters are transmitted to the receiver and subsequently synthesized. This procedure is in contrast to existing video teleconferencing compression techniques that deal exclusively with compression and transmission of pixel-based image data.

While Fig. 2 represents teleconferencing in action, there remain a number of key initialization procedures for the encoder and decoder. Figure 3 illustrates an example of a 3-D canonical model mapped to an individual. As part of the initialization procedure for the encoder, features of the face must be accurately aligned, such that the mouth and eyes open in the correct location with respect to the image texture map.

One of the by-products of mapping images of individuals to canonical representations of the face is that any image can be used to create a novel character. This has resulted in some interesting opportunities to create avatars from animal—or non-animal—images.

Social Agents and Avatars

A developing area for facial animation is in user interfaces that have characters or agents. The principle of social agents lies in the ability of an agent to interact directly with the user. This ability can be as simple as a reactive behavior to some simple action such as searching for a file, or as complex as an embodiment or characterization of a personal assistant capable of navigating the Internet under voice commands and responding audibly and visually with a resulting find. Some themes include characters that display their activity state through facial expressions.

Ultimately, these agents will understand spoken requests, speak to the user, behave in real time, and respond with uncanny realism. These interfaces often are referred to as *social user interfaces* and are designed to supplement graphical user interfaces. For example, a character will appear to assist when you start a new application. If you hesitate or ask for help, the agent will reappear to provide you with further guidance. In many instances, these characters will be seen as active collaborators, with personalities of their own.

At first sight, building this type of interface appears to be straightforward: construct a character, build a set of behavior rules, and switch the character on. Unfortunately, it is not that simple. It is difficult enough to understand and model human-to-human behavior, let alone human-to-computer behavior. So by endowing a computer

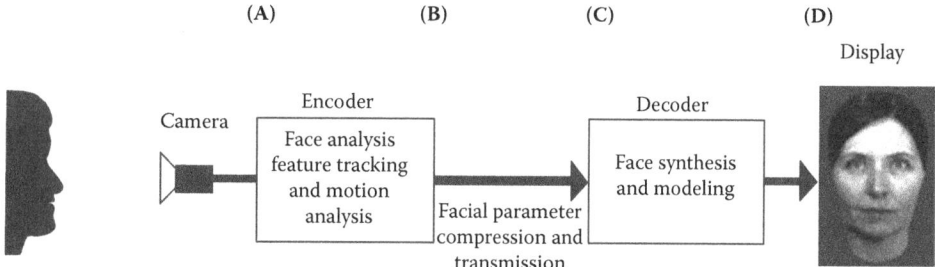

Fig. 2 A video teleconferencing protocol. A camera captures moving images of a face: (A) Face parameters are extracted. (B) The parameters are compressed and transmitted to a decoder. (C) They are reconstructed. (D) A visual surrogate is displayed on the receiving end. A two-way system replicates this sequence in the reverse direction

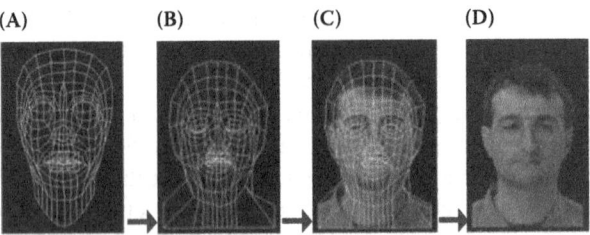

Fig. 3 A 3-D model created from a single image of an individual, precisely mapped to a canonical model: (A) the baseline canonical face, (B) and (C) the topology deformed to match an individual, and (D) the rendered 3-D visual surrogate

interface with some human characteristics, we turn on all our human responses. Most significantly, we expect the interface to behave like a human, rather than a computer. Bearing this in mind, a *useful* social interface, such as a computer-generated humanoid with a face, has yet to be seen. However, many academic and industrial labs are actively developing prototypes.

Social Robots

Robots present a new frontier for experiments to understand what makes us human. Not only is it possible to mimic human responses and behaviors, but new types of robots can serve as human surrogates. Unlike a computer-generated character that is constrained to a 2-D display, a physical embodiment of a robot has to move in the real world. This presents engineers with significant additional challenges. Nevertheless, the development of robot agents shares many of the underlying concepts developed for computer-generated 3-D characters.

Kismet is an early example of a social robot, developed by Cynthia Breazeal at MIT.[32] Kismet is capable of generating a range of facial expressions and emotions, as illustrated in Fig. 4. While Kismet is an exaggerated non-human character, with large eyes and ears, the design was carefully considered to ensure that it could participate in social interactions matching the robot's level of competence.[33]

Generating facial expressions for Kismet uses an interpolation technique over a 3-D *affect space*. The dimensions of this space correspond to arousal, valence, and stance (a, v, s). An emotional affect space, as defined by psychologists such as Russell,[34] maps well into Kismet's interpolation scheme, allowing the mixing of individual features of expressions. The specific (a, v, s) values are used to create a net emotive expression P_{net}, as follows:

$$P_{net} = C_{arousal} + C_{valence} + C_{stance} \qquad (1)$$

where $C_{arousal}$, $C_{valence}$, and C_{stance} vary within a specified range using a weighted interpolation scheme. Figure 5 illustrates where the expression of disgust can be located with respect to the 3-D affect space.

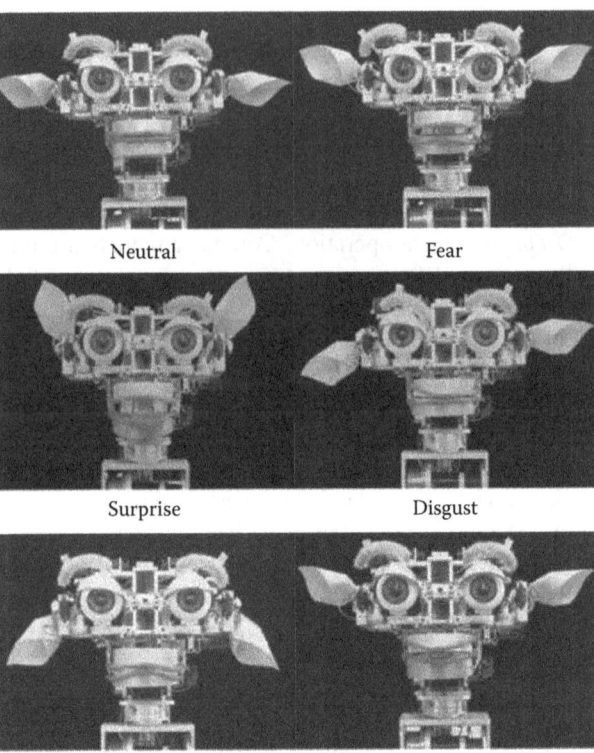

Fig. 4 Kismet generating a range of primary facial expressions from Mori.[35] Intermediate expressions are generated by blending the basis facial postures

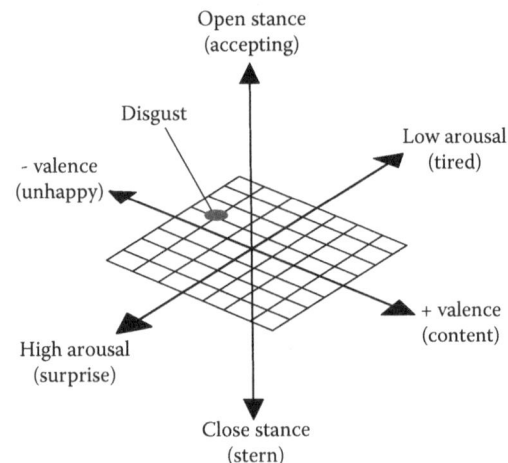

Fig. 5 The 3-D affect space used by Kismet

It remains a significant challenge to build an autonomous humanoid robot that can deceive a human into thinking it is real. The design of Kismet as a young, fanciful anthropometric creature with facial expressions that are easily recognizable to humans was carefully considered, ensuring that the expectations for Kismet's behavior were calibrated to its abilities, and therefore, not to fall into a believability trap.

Facial Animation

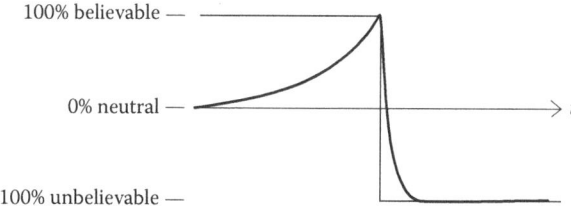

Fig. 6 The believability flip. At a certain point in time, when the character is perceived as no longer realistic, a flip occurs, and the character becomes completely unbelievable

THE BELIEVABILITY FLIP AND THE UNCANNY VALLEY

As the realism of animated characters improve—in terms of their visual and motion fidelity—there comes a point when our perception of a character identifies that something is not quite right. The character appears *too plastic*, the motion *lacks fluidity*, or the lip synchronization *looks strange*. At that point, we suspend our belief that the character is real. This is the *believability flip*, as illustrated in Fig. 6. This turning point has become a critical boundary which many of today's lifelike animated characters are attempting to overcome. This effect is even more pronounced when dealing with virtual surrogates of well-known personalities or people we know well. Once the flip occurs, there is no going back. It appears that we recalibrate with lower expectations and no longer respond to the character in the same way.

It is important to recognize that we have developed a deep and profound understanding of the visual patterns our faces create. This is because we as humans are highly social animals; we benefit from recognizing others in our groups, as well as deciphering emotional states from facial expressions. This has been confirmed by the discovery of cells in our brains that have been identified as exclusively targeting faces. Therefore, overcoming the *believability flip* for virtual surrogates will remain a profound challenge for some time.

The flip—when we realize the character is not real—varies based on the character; those closest to us are ultimately most familiar and are consequently very hard to synthesize, whereas an unfamiliar person is somewhat easier. In between are personalities that we might observe on television or in the newspapers; they may have some subtle traits that we recognize. Therefore, synthesis complexity is not evenly balanced.

In 1970, the Japanese roboticist Masahiro Mori coined the term *uncanny valley* as a concept of robotics.[38] It concerns the emotional response of humans to robots, as well as to other non-human entities. The hypothesis states that as a robot is made increasingly human-like in appearance and motion, our human response is increasingly empathetic until a point when there is a reversal and our response becomes strongly repulsive. Figure 7 illustrates

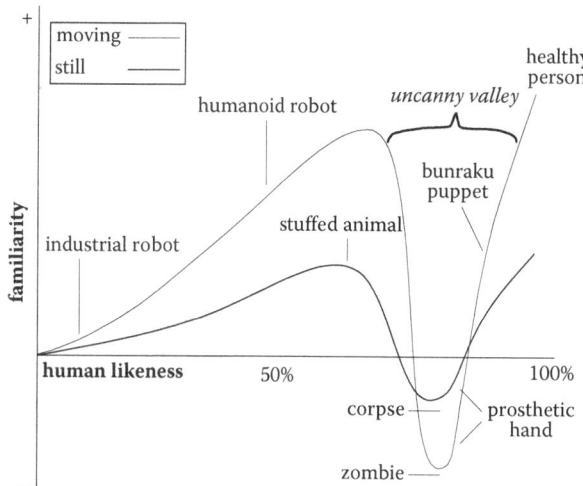

Fig. 7 The uncanny valley
Source: Adapted from Ezzat & Poggio.[38]

a familiarity response with respect to human likeness, and the *uncanny valley* is identified by the familiarity reversal.

While the discussion in this section appears to have a scientific basis, it should be recognized that there is little hard evidence either way. The rigorous exploration of human familiarity perception is relatively new, and there is much to learn. It is expected that scientific investigations will be able to shed light on the key components. For example, the formulations of questions on how to measure a synthetic character's performance could lead to some important discoveries about ourselves; this is especially true as computer-generated characters are now capable of exquisite visual fidelity.

A TURING TEST FOR FACES?

In 1950, Alan Turing, a renowned British mathematician who played a decisive role in the development of the computer during World War II, published a paper on computing machinery and intelligence[36] to consider the question "Can machines think?" He devised an imitation game to test if a machine can converse believably with a human. The test was along the following lines: a human judge engages in a natural language conversation with two other parties, one a human and the other a machine. If the judge cannot reliably tell which is which, then the machine is said to pass the test. Even today, the Turing test remains elusive and begs the question as to if and when a machine will be able to pass the test; Alan Turing predicted that by the year 2000, computers would have enough memory and processing power to pass the test. Today, computing resources are sufficiently plentiful to expose the problem as one of developing better software algorithms.

While the Turing test is philosophical in nature, there is an emerging need to practically test facial animation systems that attempt to mimic reality. For example, it is intriguing to suggest that an animated visual surrogate might be capable of deceiving us that it is real and thereby could be used as a replacement newscaster presenting the evening news. If so, what would be the criteria for passing such a test, especially in the light of our strong negative human response when we no longer believe the character is real?

A Visual Turing Test

A step toward testing the perception of face synthesis was carried out with Mary101,[37] where a surrogate was created from video resynthesis techniques.[38] The goal was to test human perception of a talking head to identify a) if people could distinguish between the surrogate and the real video images; and b) the intelligibility of lip reading.

Figure 8 illustrates frames from the tests. At the top are real images while the bottom illustrates frames of the virtual surrogate. The center images show the resynthesized face components. Their results indicate that a resynthesized talking head can approximate the video fidelity of a real person. The generation of a virtual newscaster capable of reading the evening news and being perceived as real is within reach. However, many challenges remain before such a surrogate could believably interact with humans.

Speech perception is easier to measure than the realism aspects of a surrogate, in part because there is a direct goal to understanding what was spoken. However, it is a common misconception that *lip reading* is all or nothing—is it possible to lip read or not? In fact, lip reading, more precisely speech reading,[39] varies enormously between people. Individuals who are hard of hearing rely to some degree on visual cues, typically preferring face-to-face conversations to assist a degraded auditory channel. On the other hand, the profoundly deaf depend exclusively on visual facial cues and therefore demand face-to-face communication, ideally in good lighting and close proximity. The rest of us, with adequate eyesight and hearing, also use facial cues to understand what is being said. This is why we have better comprehension when talking to one another face to face than when talking on the telephone.

As indicated by Mary101, the goal of creating a visually believable surrogate, capable of passing a series of visual perception tests that mimic a real person, can be attained. A visual Turing test for face behavior requires further definition. Perhaps this can be achieved through a set of step-wise face motion tests exercising clearly defined body, head, and face motions and actions.

RELATIONSHIP TO CONVENTIONAL ANIMATION

Computer animation is a direct outgrowth of conventional animation, and Disney Studios has had a significant influence on the animation industry over the years. Most of the hard lessons they learned through trial and error are directly applicable to computer animation, especially character animation. It could be argued that there are few differences between traditional animation techniques and those applied in computer animation, suggesting that computers are merely more powerful tools at the disposal of animators. This being essentially true, we have a great deal to learn from traditional animation.

Disney's Principles of Animation

Frank Thomas and Ollie Johnston outlined 12 principles of animation, which applied to the way Disney Studios

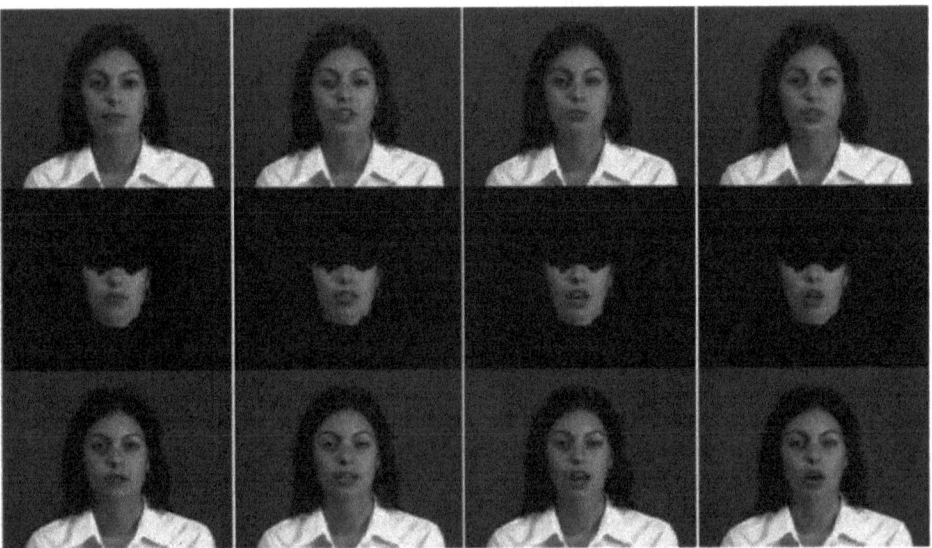

Fig. 8 Frames from Mary101.[37] At the top are real image sequences, in the center are regenerated face components, and at the bottom is the final composite

produces animation.[40] These "rules" are widely accepted as the cornerstone of any animation production and can be applied directly to the way computer character animation is produced.[41] What follows are brief descriptions of those principles, which can also be applied to facial animation.

Squash and Stretch

Squash and stretch is perhaps the most important aspect of how a character moves. Objects, such as a bouncing ball, will compress when they hit an immovable object, such as the floor, but they soon come back to their original shape, as illustrated in Fig. 9. A rule of thumb is that no matter how "squashy" or "stretchy" something becomes, its volume remains relatively the same.

If a character or object is in motion, it will undergo certain changes within its overall shape. For example, a cat character falling through space stretches in the direction of the fall and squashes, or "splats," when it reaches the ground. The scaling may seem extreme when viewed in a single frame, but in motion it is remarkable how much the squashing and stretching can be exaggerated while still retaining a natural look. This elasticity can be used to imply weight, mass, or other physical qualities. For example, the shape of an iron ball would not be affected by a drop to the ground, whereas a balloon full of water undergoes dramatic shape changes both as it is dropped and when it impacts the ground.

Complex models present complex squash and stretch issues. In a hierarchically defined model, squash and stretch are usually applied differently and at different times to the various model parts to achieve the illusion of mass and weight. Ideally, a flexible model is used, in which the shape of various parts can be appropriately changed by accelerations and impacts.

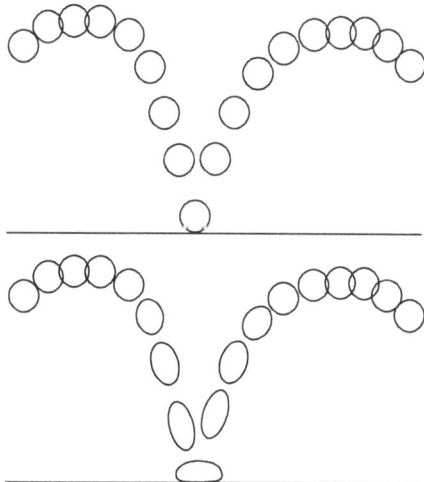

Fig. 9 In the motion of a ball bouncing, the ball can appear to have more weight if the drawings are closer together at the top of the arc. In the bottom illustration, a flattened ball on impact and elongation in acceleration and deceleration are the beginnings of squash and stretch

Anticipation

Anticipation is the act of hinting to the audience what is about to happen. This hint can be a broad physical gesture, or it can be as simple as a subtle change in facial expression. The key idea is not to allow any motion to come unexpectedly, unless that is the desired effect. For example, before a character zooms off, it gathers itself up, draws back in the opposite direction, and then moves rapidly off in the other direction.

These anticipatory moves do not necessarily imply why something is being done, but rather they clarify what is being done. Once a movement has been implied through anticipation, animating a vastly different move can be used to introduce an element of surprise. For example, a car coiling up, ready to shoot forward, but then zooming backward, could be considered a sight gag.

Staging

Staging is the actual location of the camera and characters within the scene. Staging is very important and should be done carefully. Principles of cinema theory come into play in the way that shots are staged. In general, there should be a distinct reason for the way that each shot in the film is staged. The staging should match the information that is required for that particular shot. The staging should be clear, and it should enhance the action. A common mistake in the design of computer-generated films is to make the staging too dynamic, simply because the computer has the capability to do so. As a consequence, the scenes become confusing, or else they distract from the action that is taking place.

One could easily write an entire paper on the meaning and importance of camera angles, lighting, and other film effects. Researching conventional film literature will enhance an animator's understanding of these theoretical film principles and is highly recommended.[42,43] However, the most basic advice for good staging is that the most important information required from a scene should be clear and uncluttered by unusual or poor staging.

Ease-In and Ease-Out

Newton's laws of motion state that no object with mass can start in motion abruptly, without acceleration. Even a bullet shot from a gun has a short period of acceleration. Only under the most unusual of circumstances does the motion of an object have an instantaneous start or stop. *Ease-in* and *ease-out* are the acceleration and deceleration, respectively, of an object in motion. Eases may be applied to any motion or attribute change, including translation, rotation, scaling, or change of color. How an object's motion eases helps define the weight and structure of the object.

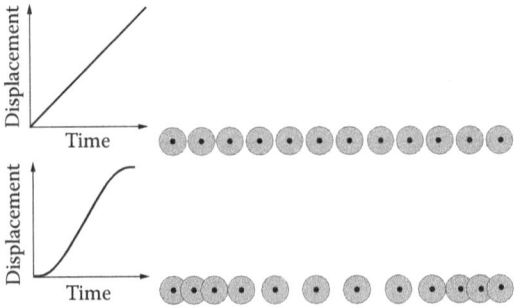

Fig. 10 The top profile illustrates a linear time displacement, while the bottom profile shows how the ease-in and ease-out can give the impression of acceleration and deceleration

An ease is used at the beginning or end of a move to soften the transition from an active state to a static state. Many animation systems offer a choice of eases, a common one being a cosine ease, as illustrated in Fig. 10 (bottom). The linear motion, as in Fig. 10 (top), is evenly spaced in time; all motion proceeds in a steady, predictable manner. However, linear motion does not lend itself to interesting animation, and thus it is the least desirable. Nonlinear eases are more widely used; their motion is fluid and more enjoyable. Being able to arbitrarily define eases for every action is best. Often, an interactive curve editor is used to graphically edit ease functions using combinations of spline curves, to allow for an infinite number of possible eases. Actually seeing the motion curve dip down to its rest position is sometimes as useful as seeing the animation in preview. The ability to interactively adjust the curves that determine the rate of animation or transition between key poses is crucial.

Follow-Through and Overlapping Action

If all the parts of a character stop or change motion at the same time, the effect is one of extreme rigidity. To impart a sense of fluidity, animators delay the movement of appendages. For example, consider a piece of animation in which a character falls on the ground. Letting the arms lag one or two frames behind the body impact imparts continuity and fluidity to the entire motion. This effect is called *follow-through*.

Overlapping action is also important when moving the camera through an environment or when moving a character through space. Early computer animation was typically composed of a move, a pause, a rotation, a pause, another move, another pause, and so on. This process quickly becomes tedious. A solution is to start the rotation before the move finishes, overlapping the action, instead of pausing. Follow-through is a common form of overlapping action. Rather than abruptly stopping an action after it has been completed, additional motion eases out along the same path of action. For example, a tennis swing is much more effective if the swing continues after the ball has been hit.

Arcs

Most motion is nonlinear; that is, an object usually follows some curved path. Rather than linearly interpolating from one key frame to the next, passing a curve through the keys gives a more dynamic look to the animation. If animation has been completely interpolated using splines, however, the motion may be too uniform—it will have no punch.

Any "oomph" lost by splining can be regained by editing the motion curves. Again, a function editor that gives an interactive graphic representation is ideal for defining motion curves. Most systems have a number of interpolation functions available to the animator. One issue with cubic interpolating splines is that although they keep slope continuity from key frame to key frame, they also tend to overshoot when confronted with sudden changes in velocity. Since animators usually intend key frames to represent extremes in motion, these overshoots can have undesired results. Feet go through the floor; fingers go through hands. Appropriate interactive control of motion curves is necessary in a production animation environment to allow specification of desired motion without tedious, iterative curve adjustments.

Secondary Motion

Secondary motion is the motion of objects or body parts that depend on primary motion. An example of secondary motion would be the motion of hair or the motion of clothing over the surface of a moving figure. In general, secondary motion is caused by the motion of a primary object. For example, the motions of floppy dog ears would be secondary motions caused by the motion of the dog's head and body.

Exaggeration

Exaggeration involves making the motion more dramatic than one would observe in the real world. If a scene is animated with little or no exaggeration, the motion will be dull and listless. Animators use exaggeration to *sell* the action or the movement of a character.

Exaggeration of motion is not always the way to go, but often exaggeration of motion characteristics is needed to create interesting animation. Exaggeration does not have to impart a cartoon feel to be effective. After the motion has been blocked out, it is up to the animator to decide which movements must be exaggerated to enhance the animation. Live action footage can be used for reference. The live action may be used to rough out the major movements, which are then subtly exaggerated to showcase aspects of the motion.

The exact amount of exaggeration that is required is difficult to judge; however, significant amounts of exaggeration can often be effective. One approach is to push the exaggeration until it is clearly too much and then back off a little.

Appeal

The characters should *appeal* to the audience in some way. This is not to say that the characters need to be cute, but rather that there should be some elements about the characters that make them interesting to watch. The audience should emotionally connect with the characters. They should love the heroine and hate the villain.

REFERENCES

1. Darwin, C. *Expression of the Emotions in Man and Animals*; J. Murray, London, 1872.
2. *4020/RGB 3D Scanner with Color Digitizer*, 1990.
3. Parke, F.I. *Computer Generated Animation of Faces*; Master's thesis; University of Utah: Salt Lake City, UT, 1972, UTEC-CSc-72-120.
4. Parke, F.I. *A Parametric Model for Human Faces*; PhD thesis; University of Utah: Salt Lake City, UT, 1974; UTEC-CSc-75-047.
5. Chernoff, H. *The Use of Faces to Represent Points in N-Dimensional Space Graphically*; Technical Report Project NR-042-993, Office of Naval Research, 1971.
6. Gillenson, M.L. *The Interactive Generation of Facial Images on a CRT Using a Heuristic Strategy*; PhD. thesis; Ohio State University, Computer Graphics Research Group: Columbus, OH, 1974.
7. Platt, S.M. *A System for Computer Simulation of the Human Face*; Master's thesis; The Moore School, University of Pennsylvania, 1980.
8. Brennan, S.E. *Caricature Generator*; Master's thesis, Massachusetts Institute of Technology: Cambridge, MA, 1982.
9. Weil, P. *About Face*; Master's thesis, Massachusetts Institute of Technology, Architecture Group, 1982.
10. Bergeron, P.; Lachapelle, P. Controlling facial expressions and body movements. In *Advanced Computer Animation, SIGGRAPH '85 Tutorials*; ACM: New York, 1985, 61–79.
11. Waters, K. A muscle model for animating three-dimensional facial expressions. SIGGRAPH '87, Comput. Graph. **1987**, *21* (4), 17–24.
12. Magnenat-Thalmann, N.; Primeau, N.E.; Thalmann, D. Abstract muscle actions procedures for human face animation. Vis. Comput. **1988**, *3* (5), 290–297.
13. Lewis, J.P.; Parke, F.I. Automatic lip-synch and speech synthesis for character animation. Proceedings Human Factors in Computing Systems and Graphics Interface '87, Canadian Human-Computer Communications Society: Toronto, ON, 1987, 143–147.
14. Hill, D.R.; Pearce, A.; Wyvill, B. Animating speech: An automated approach using speech synthesis by rules. Vis. Comput. **1988**, *3*, 277–289.
15. Parke, F.I.; Ed.; *State of the Art in Facial Animation, SIGGRAPH Course Notes 26*; ACM: New York, 1990.
16. Williams, L. Performance driven facial animation. SIGGRAPH '90, Comput. *Graph.* **1990**, *24* (4), 235–242.
17. Blanz, V.; Vetter, T.A morphable model for the synthesis of 3D faces. In *Proceedings of SIGGRAPH '99, Computer Graphics Proceedings, Annual Conference Series*; Rockwood, A.; Ed.; Addison Wesley Longman: Reading, MA, 1999, 187–194.
18. Lee, Y.; Terzopoulos, D.; Waters, K. Constructing physics-based facial models of individuals. In *Graphics Interface '93*; Canadian Human-Computer Communications Society: Toronto, ON, 1993, 1–8.
19. Deng, X.Q. *A Finite Element Analysis of Surgery of the Human Facial Tissue*; PhD. thesis; Columbia University: New York, 1988.
20. Pieper, S.D. *CAPS: Computer-Aided Plastic Surgery*; PhD. Thesis; Massachusetts Institute of Technology, Media Arts and Sciences, 1991.
21. Sifakis, E.; Neverov, I.; Fedkiw, R. Automatic determination of facial muscle activations from sparse motion capture marker data. Trans. Graph. **2005**, *24* (3), 417–425.
22. Yuille, A.L.; Cohen, D.S.; Hallinan, P.W. Feature extraction from faces using deformable templates. *IEEE Computer Society Conference on Computer Vision and Pattern Recognition (CVPR'89)*; IEEE Computer Society Press: San Diego, CA, 1989, 104–109.
23. Blake, A.; Isard, M. 3D position, attitude and shape input using video tracking of hands and lips. In *Proceedings of SIGGRAPH 94, Computer Graphics Proceedings, Annual Conference Series*; Glassner, A.; Ed.; ACM Press: New York, 1994, 185–192.
24. Black, M.J.; Yacoob, Y. Tracking and recognizing rigid and non-rigid facial motions using local parametric models of image motion. In *IEEE International Conference on Computer Vision*; IEEE Computer: Los Alamitos, CA, 1995, 374–381.
25. Essa, I.; Pentland, A. *A Vision System for Observing and Extracting Facial Action Parameters*; Technical Report 247; MIT Perceptual Computing Section: Cambridge, MA, 1994.
26. Deng, Z.; Neumann, U. *Data-Driven 3D Facial Animation*; Springer Verlag: New York, 2008.
27. Vannier, M.W.; Marsch, J.F.; Warren, J.O. Three-dimensional computer graphics for craniofacial surgical planning and evaluation. Proc. SIGGRAPH '83, Comput. Graph. **1983**, *17* (3), 263–273.
28. Lorensen, W.E.; Cline, H.E. Marching cubes: High resolution 3D surface construction algorithm. Proc. SIGGRAPH '87, Comput. *Graph.* **1987**, *21* (4), 163–169.
29. The Visible Human Project. 1994.
30. Larrabee, W. A finite element model of skin deformation. I. Biomechanics of skin and soft tissue: A review. Laryngoscope **1986**, *96*, 399–405.
31. Choi, C.S.; Harashima, H.; Takebe, T. *Highly Accurate Estimation of Head Motion and Facial Action Information on Knowledge-Based Image Coding*; Technical Report PRU90-68; Institute of Electronics, Information and Communication Engineers of Japan: Tokyo, 1990.
32. Breazeal, C.L. *Designing Sociable Robots*; The MIT Press: Cambridge, MA, 2002.
33. Breazeal, C.L.; Foerst, A. Schmoozing with robots: Exploring the original wireless network. In *Proceedings of Cognitive Technology*; IEEE Press: Los Alamitos, 1999, 375–390.
34. Russell, J.; Fernandez-Dols, J.; Eds.; *The Psychology of Facial Expression*; Cambridge University Press: Cambridge, 1979.

35. Mori, M. The uncanny valley. *Engery* **1970**, *7* (4), 33–35.
36. Turing, A.M. Computing machinery and intelligence. Mind **1950**, *59* (236), 433–460.
37. Geiger, G.; Ezzat, T.; Poggio, T. *AI Memo 2003-003: Perceptual Evaluation of Video-Realistic Speech*; Technical Report; MIT, 2003.
38. Ezzat, T.; Poggio, T. Visual speech synthesis by morphing visemes. Int. J. Comput. *Vis.* **2000**, *38* (1), 45–57.
39. Jeffers, J.; Barley, M. *Speechreading (Lipreading)*; Charles C. Thomas: Springfield, IL, 1971.
40. Thomas, F.; Johnson, O. *Disney Animation: The Illusion of Life*; Abbeville Press: New York, 1981.
41. Lassiter, J. Principles of traditional animation applied to 3D computer animation. In *SIGGRAPH '87 Tutorials, Course 21*; ACM: New York, 1987, 35–44.
42. Katz, S. *Shot by Shot*; Michael Wiese Productions, 1991.
43. Arijon, D. *Grammar of the Film Language*; Silman-James Press: Los Angeles, CA, 1976.

First Order Statistics

Tania Pouli
Technicolor, Rennes, France

Erik Reinhard
Technicolor Research and Innovation, Rennes, France

Douglas W. Cunningham
Brandenburg University of Technology, Cottbus, Germany

Abstract
Statistical analysis and manipulation of histograms can be used to derive powerful descriptors for images. For example, the first-order moments correspond to the centroids of the image, while second-order central moments can be used to determine the orientation of an image in a scene. This article introduces the concept of first-order statistics and shows how first-order they can been linked to such information as depth in scenes and material properties.

Keywords: First-order statistics; Image properties; Scene analysis.

Digital images consist of an array of pixels, each representing the average incoming illumination from a small part of the scene. Effectively, each pixel of a digital camera sensor integrates light over a small area, storing it as a single number. If we were to use a very low resolution sensor consisting of a single pixel, our digital image would simply be a single number representing the average light in the scene—this is, after all, what a photometer does. This is illustrated in Fig. 1.

Our visual system performs a similar task as photoreceptors in the retina, spatially sample light some creatures such as mollusks, however, sense light in a way more akin to our single-pixel camera example, suggesting that useful information may be found in images even at the single pixel level, without any spatial information.[1]

HISTOGRAMS AND MOMENTS

The distribution of pixel values in images can be represented with *histograms*. For a given intensity vector I, we define its histogram H with B bins of width V as follows:

$$H = \{(h(1), v(1)), \ldots, (h(B), v(B))\} \tag{1}$$

$$B = \left\lceil \frac{\max(I) - \min(I)}{V} \right\rceil \tag{2}$$

$$h(i) = \sum_{p=1}^{N} P(I(p), i), \quad i \in [1, B] \tag{3}$$

$$v(i) = \min(I) + (i-1)V \tag{4}$$

$$P(I(p), i) = \begin{cases} i = \left\lfloor \frac{I(p) - \min(I)}{V} + 1 \right\rfloor \\ 0 \quad \text{otherwise} \end{cases} \tag{5}$$

where H is the set of all pairs $(h(i), v(i))$ for all $i \in [1, B]$ corresponding to the number of elements and value of the ith bin of the histogram. $I(p)$ is the value of the pth pixel of vector I, which contains a total of N pixels, and $P(I(p), i)$ represents the probability of a pixel $I(p)$ belonging to a bin i.

Figure 2 shows the intensity histograms of some example images. By visual inspection, we can already surmize that the shape of a histogram depends on the content of the image: scenes with uniform illumination (top left) will typically have an approximately Gaussian shape (top right). If, on the other hand, there are shadows and highlights present (bottom left), the histogram will likely exhibit a bimodal distribution (two peaks corresponding to the different parts of the bottom-right image in Fig. 2).

Although such high-level characterization and understanding of histograms might be useful in photography, to derive an analytical description of the shape of an image histogram, properties of the intensity distribution may be computed, such as its mean and variance. *Statistical moments* are commonly employed to quantitatively describe the shape of a distribution. The kth moment of a distribution can be computed as follows:

$$m_k = \sum_{p=1}^{N} \frac{(I(p) - c)^k}{N} \tag{6}$$

Encyclopedia of Image Processing, First Edition
DOI: 10.1201/9781351110273-140000478
Copyright © 2018 by Taylor & Francis. All rights reserved.

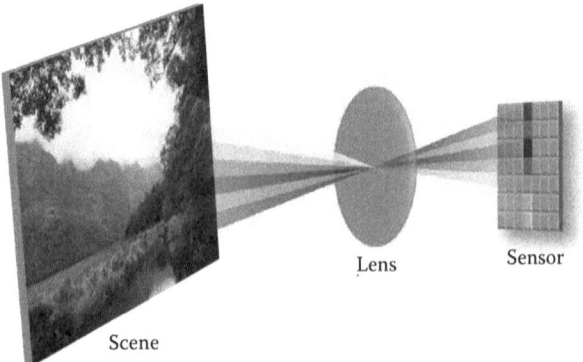

Fig. 1 Pixels on the camera sensor integrate light from the scene, effectively capturing the average incoming illumination over a small area of the scene

where c can be any constant. Generally, if $c = 0$, then the above equation computes the raw moments of the distribution, while setting $c = \mu$ gives us the central moments (i.e., centered at the mean). The first moment corresponds to the mean μ of the distribution and the second is the variance σ^2 (which is by itself the square of the standard deviation σ).

The meaning of further moments is less straightforward but the *skewness* S and *kurtosis* κ of a distribution relate to the third and fourth moments, respectively. More specifically, the skewness and the kurtosis are defined as:

$$S = \frac{m_3}{\sigma^3} \qquad (7)$$

$$k = \frac{m_4}{\sigma^4} \qquad (8)$$

respectively, where m_3 and m_4 are the third and fourth central moments, respectively. Skewness encodes asymmetry in the distribution while kurtosis closely relates to sparseness. Figure 3 shows some example distributions with specific values for mean, standard deviation, skewness, and kurtosis.

Image Moments and Moment Invariants

In our analysis so far, we have treated images as one-dimensional vectors of intensity values, whereby the intensity of a pixel p is given by $I(p)$. We can expand this definition to consider the horizontal and vertical position of a pixel and define its intensity value at position x, y as $I(x, y)$, where $x \in M$ and $y \in N$ and $M, N \in \mathbb{R}$ are the dimensions of the image. The moments of a 2D image $I(x, y)$ can now be defined as:

$$m_{j,k} = \sum_{x=1}^{M} \sum_{y=1}^{N} \left(x^j - c_x\right)\left(y^k - c_y\right) I(x, y) \qquad (9)$$

where j, k define the order of the moment in the horizontal and vertical direction, respectively. Setting c_x, c_y to 0 leads

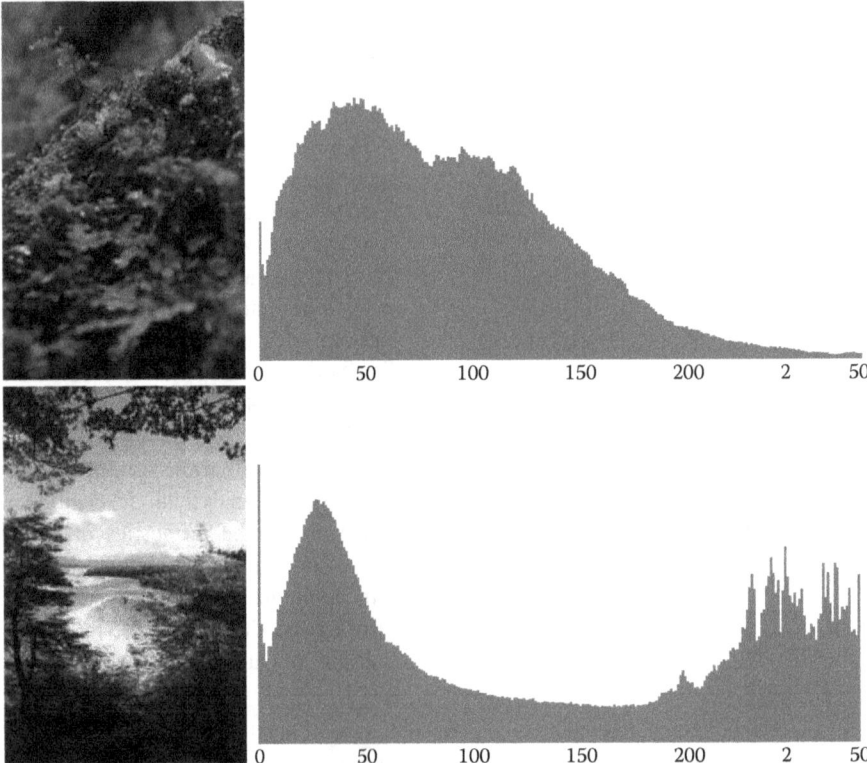

Fig. 2 The top image is uniformly illuminated, leading to an approximately Gaussian distribution. In the bottom image, parts of the scene are in the shade while other parts are brightly illuminated, leading to a bimodal distribution

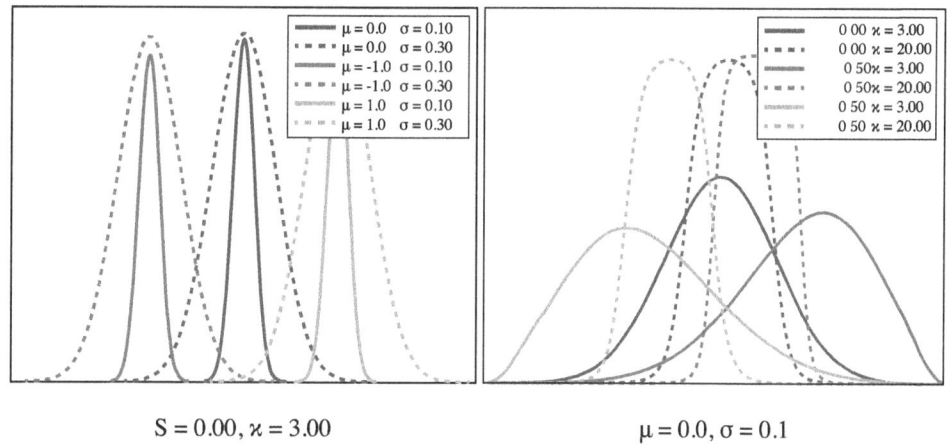

Fig. 3 Probability distributions with specific mean, standard deviation, skewness, and kurtosis values from sets of random numbers drawn from the Pearson system[2]

to the raw moments of the image, while $c_x = m_{1,0}/m_{0,0}$ and $c_x = m_{0,1}/m_{0,0}$ lead to central moments $m'_{i,j}$.

Similar to (6), the above definition can be normalized by the number of pixels or *area* of the image. Using the formulation in (9), the area $m_{0,0}$ can computed as:

$$m_{0,0} = \sum_{x=1}^{M} \sum_{y=1}^{N} I(x,y) \tag{10}$$

The first-order moments $m_{1,0}$ and $m_{0,1}$ correspond to the *centroids* of the image, while second-order central moments can be used to determine the orientation of an image.

Image moments offer a very powerful basis for representing and describing images. An image can be uniquely described and reconstructed by its (raw) moments—a property known as the *uniqueness theorem*[3] (although this is not a case for a general distribution). By manipulating or combining moments, image descriptors can be constructed that can characterize images uniquely up to certain transformations. That is they are *invariant* to specific transformations.

In the simplest case, we can consider the central image moments. As they effectively align the image centroid with the origin of the coordinate space, they are invariant to translations—they remain constant, irrespective of the position of the image (or segment of interest) within the coordinate space. In addition to translational invariance, moments can be combined and manipulated to define bases that are scale, rotation, and other property invariant.[3] Such bases are known as *moment invariants* and were first introduced in the pioneering work of Hu et al.,[4] where a set of moments $\phi_1,\ldots,\phi_7$ was introduced to obtain rotational invariance. These are the following:

$$\phi_1 = m'_{2,0} + m'_{0,2} \tag{11}$$

$$\phi_2 = \left(m'_{2,0} - m'_{0,2}\right)^2 + 4m'^2_{1,1} \tag{12}$$

$$\phi_3 = \left(m'_{3,0} - 3m'_{1,2}\right)^2 + \left(3m'_{2,1} - m'_{0,3}\right)^2 \tag{13}$$

$$\phi_4 = \left(m'_{3,0} + m'_{1,2}\right)^2 + \left(m'_{2,1} + m'_{0,3}\right)^2 \tag{14}$$

$$\begin{aligned}\phi_5 = &\left(m'_{3,0} - 3m'_{1,2}\right)\left(m'_{3,0} + m'_{1,2}\right) \\ &\left(\left(m'_{3,0} + m'_{1,2}\right)^2 - 3\left(m'_{2,1} + m'_{0,3}\right)^2\right) \\ &+ \left(3m'_{2,1} - m'_{0,3}\right)\left(m'_{2,1} + m'_{0,3}\right) \\ &\times \left(3\left(m'_{3,0} + m'_{1,2}\right)^2\right) - \left(m'_{2,1} + m'_{0,3}\right)^2\end{aligned} \tag{15}$$

$$\begin{aligned}\phi_6 = &\left(m'_{2,0} - m'_{0,2}\right)\left(\left(m'_{3,0} + m'_{1,2}\right)^2 - \left(m'_{2,1} + m'_{0,3}\right)^2\right) \\ &+ 4m'_{1,1}\left(m'_{3,0} + m'_{1,2}\right)\left(m'_{2,1} + m'_{0,3}\right)\end{aligned} \tag{16}$$

$$\begin{aligned}\phi_7 = &\left(3m'_{2,1} - m'_{0,3}\right)\left(m'_{3,0} + m'_{1,2}\right) \\ &\left(\left(m'_{3,0} + m'_{1,2}\right)^2 - 3\left(m'_{2,1} + m'_{0,3}\right)^2\right) \\ &- \left(m'_{3,0} - 3m'_{1,2}\right)\left(m'_{2,1} + m'_{0,3}\right) \\ &\times \left(3\left(m'_{3,0} + m'_{1,2}\right)^2 - \left(m'_{2,1} + m'_{0,3}\right)^2\right)\end{aligned} \tag{17}$$

Figure 4 shows the Hu moments for an image with different transformations. Scale and rotation does not affect the resulting moments, while blur does.

Since then, several different moment invariants have been proposed, initially for improved rotational invariance and later for achieving invariance to affine transforms,[5–7] convolution,[8,9] and even elastic transformations.[10] Because of the descriptive power of such bases, in conjunction with their ability to ignore image transforms or degradations,[11] moment invariants have found applications in fields as varied as character recognition,[6,12] forgery detection,[13] and mesh simplification.[14]

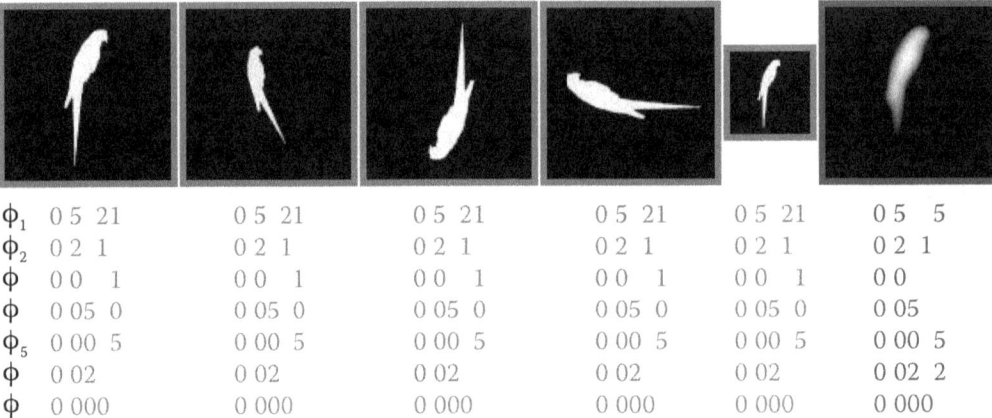

Fig. 4 Rotation and scaling transformations lead to the same Hu moments[4] while blurring the image does not. More recent moment invariants have been developed to handle blurring as well as other transforms[3]

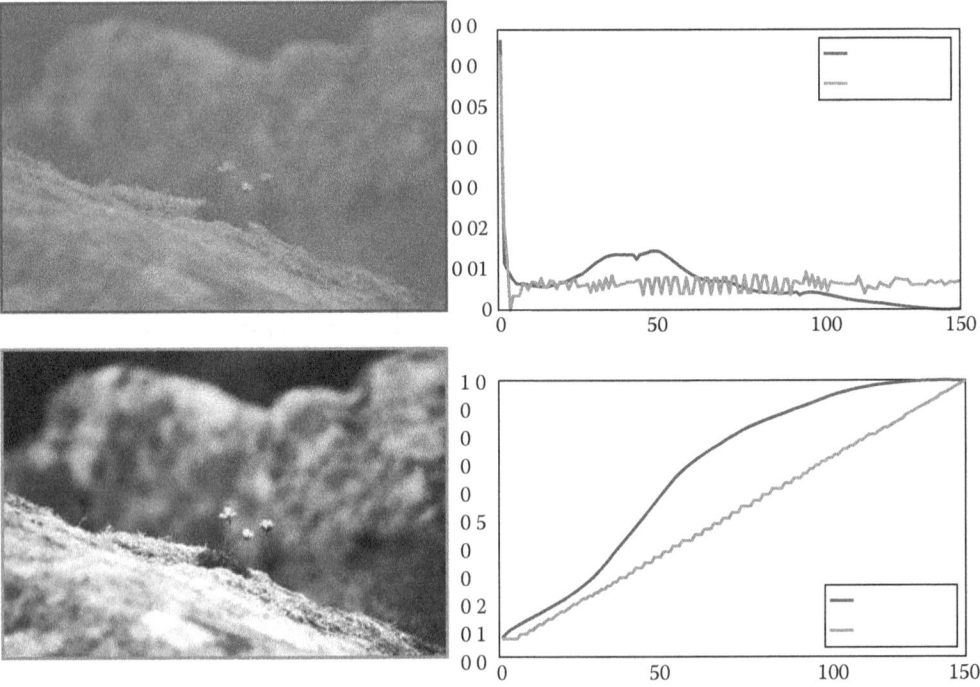

Fig. 5 Histogram equalization can increase the contrast of a low contrast image (top left) by reshaping the intensity distribution more equally. The equalized resulting image is shown at the bottom left

Histogram Adjustments

Despite their simplicity, first-order statistics have now found several applications in image processing. Studies have shown correlations between first-order statistical regularities in images and properties of the illuminant,[15,16] which has proven useful in areas such as white balancing. Moreover, transferring statistical moments between images in appropriate color spaces has been demonstrated in what is now known as color transfer.[17]

First-order statistics can also be computed on a single image basis. By manipulating the distribution of values within a single image, a variety of effects can be achieved. For instance, the contrast of an image that only covers a small portion of the available range of intensities can be increased by adjusting the pixel values such that the full range of intensities is more equally represented. This process is known as *histogram equalization* and an example can be seen in Fig. 5.

A more general version of histogram equalization is *histogram matching*, where the histogram of a source image (I_s) is matched to that of a given target (I_t). First, the cumulative histograms of the source and target are computed:

$$C_s(i) = \sum_{i=1}^{B} h_s(i) \qquad (18)$$

First Order Statistics

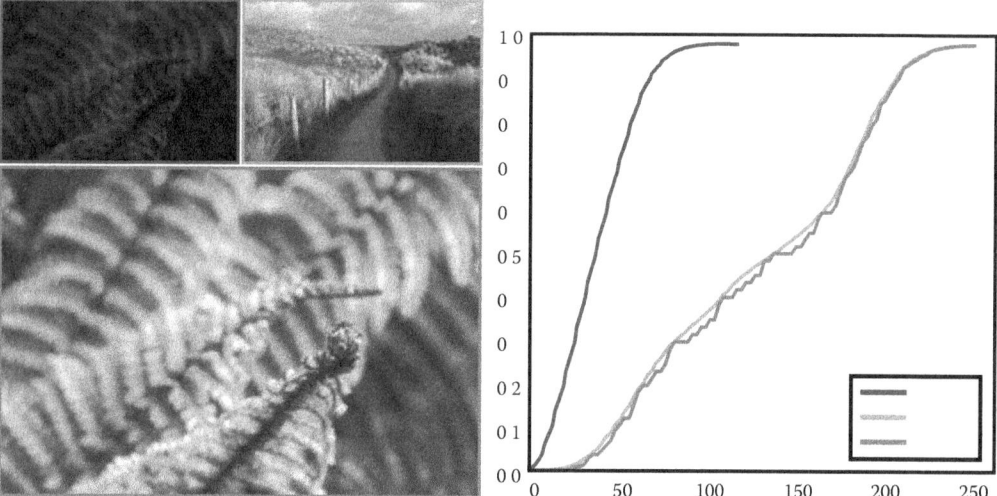

Fig. 6 The source image (top left, red) is matched to the target (top right, green) using histogram matching to produce the image at the bottom (blue). The corresponding cumulative histograms are shown

$$C_t(i) = \sum_{i=1}^{B} h_t(i) \qquad (19)$$

after which an image is matched to another according to these two cumulative histograms:

$$I_0(p) = v_t \left(C_t^{-1} \left(C_s \left(\frac{I(p) - \min(I) + 1}{V} \right) \right) \right) \qquad (20)$$

Here, a cumulative histogram C is defined as a function mapping a bin index to a cumulative count. The inverse function C^{-1} acts as a reverse lookup on the histogram, returning the bin index (and therefore the intensity value) corresponding to a given count. An example of this technique applied on a source-target pair of intensity images is shown in Fig. 6.

MOMENT STATISTICS AND AVERAGE DISTRIBUTIONS

Histogram moments have been used in the context of natural image statistics as a means to characterize images of specific classes. Depending on the content of images, different values will arise for the various moments as we will see in this section.

In a large-scale image statistics study, Huang and Mumford analyzed more than 4000 grayscale images of natural scenes (taken from the database created by J. H. van Hateren;[18] see Section 3.3.1) by computing central moments of logarithmic histograms.[19] The values found for the mean, standard deviation, skewness, and kurtosis were, respectively, $\mu = 0$, $\sigma = 0.79$, $S = 0.22$, and $\kappa = 4.56$. The value of the skewness shows that the distribution is not symmetric, which can be attributed, at least partly, to the presence of sky in many of the images, resulting in a bias towards higher intensities. In addition to that, the values of both the skewness and the kurtosis show that the distribution is non-Gaussian.

A less skewed distribution was found by Brady and Field[20] in their analysis of 46 logarithmically transformed natural images, while the linear images resulted in a distribution skewed towards darker values. Although no exact values were provided for the moments of the distributions, Fig. 7 shows the resulting histograms for both cases. As can be seen from these examples, although in both cases natural images were used and analyzed, the results can vary. Generally, the distribution of log intensities for natural images does not deviate far from Gaussian.

Results, however, do depend on the choice of images. In all the examples given, the images used to compute the distributions were captured using traditional imaging techniques and therefore were restricted in terms of the dynamic range they could represent (e.g., 8 bits in Ref. [21] and 12 bits in Refs. [19,20]). As discussed in Section 3.1.2, high dynamic range (HDR) imaging allows for scenes with more extreme illumination to be captured. HDR images have been analyzed and compared to traditional imagery in a small set of studies, indicating marked differences

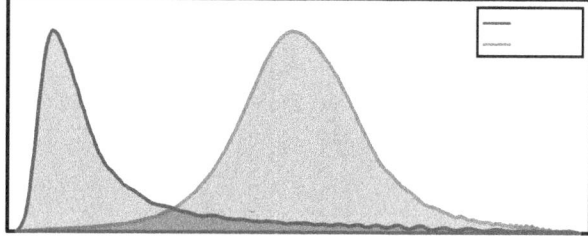

Fig. 7 The linear and log intensity histograms found by Brady and Field from their analysis of 46 natural images
Source: Adapted from the work of Brady and Field.[20]

between high dynamic range (HDR) and low dynamic range (LDR) ensembles.[21–23]

Dror et al. analyzed HDR environment maps to determine statistical regularities in real-world illumination.[22,23] Environment maps capture the full field of view from a particular viewpoint and represent it as a hemisphere (or sphere—see Section 3.1.3). The illumination arriving at the centre of the hemisphere is thus fully represented in such an image. Figure 8 shows the histograms for the two different datasets used in this study.

In a more recent study, Pouli et al. collected and analyzed images of four different scene categories for the purpose of determining the statistical differences between HDR and LDR imagery.[21] For each scene, a series of nine differently exposed images was captured to form the HDR image. The best-exposed image of the series was then selected to form the corresponding LDR ensemble, effectively constructing two sets depicting the same scenes but using different capture techniques (example images from these datasets can be seen in Fig. 6.14).

Figure 9 shows the average logarithmic histograms for the HDR and LDR sets discussed as well as the images from the HDR Photographic Survey.[24] Visual analysis of the histograms quickly shows considerable differences both between LDR and HDR as well as between scene classes. Unsurprisingly, given the 32-bit encoding of the HDR images, the quantization seen toward the left of the LDR ensembles is absent in the HDR histograms. On the other hand, long tails can be observed in the HDR distributions towards higher luminance values, which can be attributed to the presence of light sources or highlights that cannot be adequately represented with LDR imagery.

Looking at specific differences across the different image classes, a few more observations arise. In the LDR case, the average histograms of all datasets look very similar. When captured in HDR though, the same scenes lead to significant differences between categories.

Although the two capture methods lead to measurable differences, it is important to note that only qualitative conclusions can be drawn from this data. As the HDR images are linear but not radiometrically calibrated, their pixel values are correct up to a scaling factor, which may be different for each image. Consequently, to draw quantitative conclusions regarding distributions of radiances in real scenes, a fully calibrated HDR set would be necessary. To some extent, this is fulfilled by the HDR Photographic Survey.[24] However, the scenes in this collection were selected with the aim of depicting both deep shadows and bright highlights within the same image and thus their distributions may be biased.

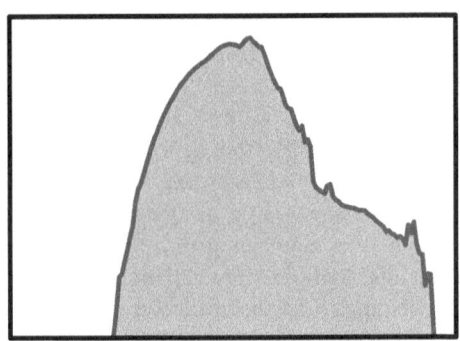 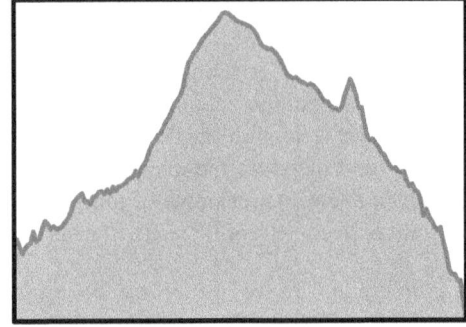

Fig. 8 The log luminance histograms for the environment map collections analyzed in the work of Dror et al.[22]

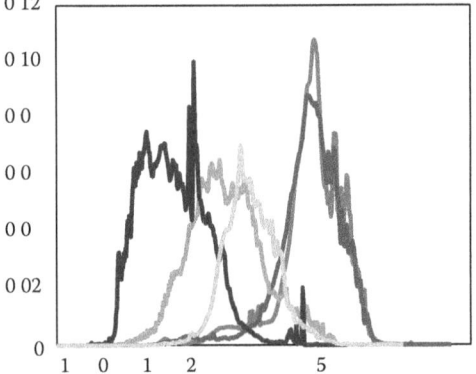 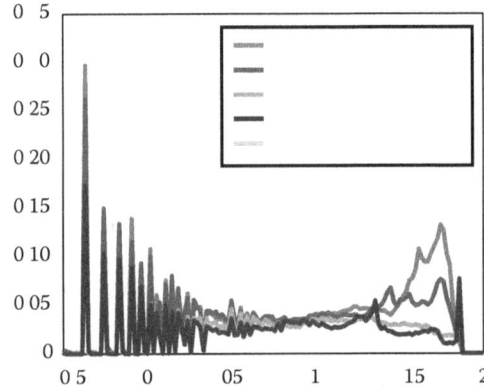

Fig. 9 Log histograms for the HDR and LDR datasets from the work of Pouli et al.[21] Different scene types lead to much larger differences when captured in HDR

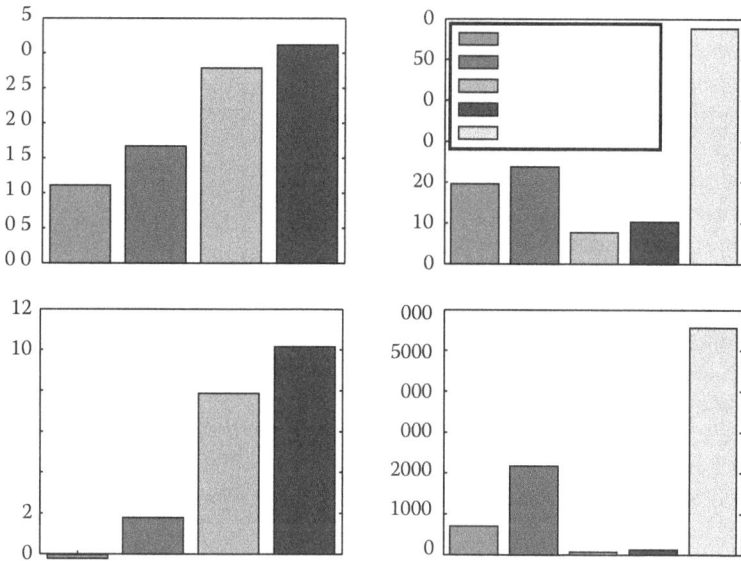

Fig. 10 Skewness and kurtosis values for the histograms in Fig. 9

MATERIAL PROPERTIES

Histograms of image categories, as we have seen, already hint at differences between scene types. In the general scenarios we have discussed so far, however, they do not offer sufficient discriminative power. For instance, although the average distributions for natural and manmade scenes may have measurable differences (e.g., Fig. 9), it would not be possible to accurately categorize the content of any given image as one or the other based solely on its histogram. If we shift our focus to more specific image classes or properties, however, histograms and their moments have been found to be more descriptive.

One notable example is the analysis of materials.[25–28] Human observers are very capable of making accurate judgements about material properties of objects and surfaces around them. Although the appearance of objects depends on complex interactions between light, material properties, and geometry, psychophysical evidence suggests that simple cues may be sufficient for assessing whether a surface is matte or glossy.[29] It is reasonable to expect that such cues may be directly or indirectly linked to statistical properties of images. For instance, Fig. 11 shows renderings of the same object but with different material properties and their corresponding histograms. The shape of the histogram changes significantly

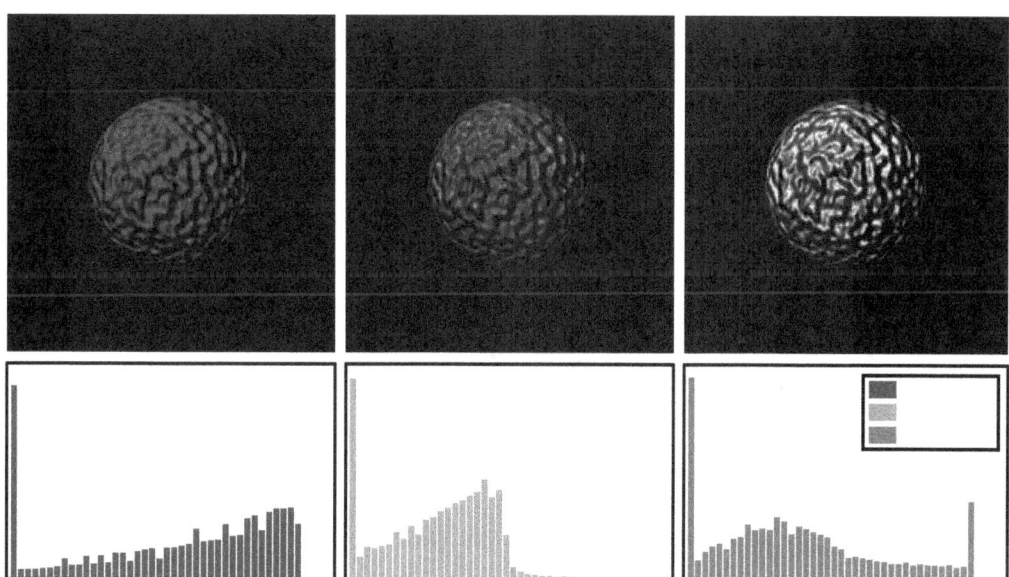

Fig. 11 The three images were generated in a 3D modeling program using different properties of the Phong material so that they look progressively glossier. Their log intensity histograms are shown in the bottom row. The histograms were computed only on the pixels within the sphere

with material changes even though both the lighting and geometry remain constant in the three images.

To determine whether a link between the perception of materials and statistical proprties exists, Motoyoshi et al.[30] studied the perceived surface qualities of various materials and analyzed their relation to the associated histograms and specifically their skewness. For patches of simple materials with mesostructure, such as stucco or crumpled paper, they found a correlation between histogram skewness and specular intensity, while an inverse relation was observed between skewness and the diffuse reflectance of the material. In addition to the physical properties of the materials, perceptual attributes of lightness and glossiness were also measured. Figure 12 shows the perceptual attributes and physical material properties in relation to the image skewness for a stucco material, while Fig. 13 gives the skewness for the images in Fig. 11.

Although skewness measurements correlate with material glossiness both physically and perceptually, it is important to note that such correlation only holds when certain assumptions are present.[31] Crucially, images are expected to depict surfaces of nearly constant *albedo* (the proportion of incident light reflected by a surface) and illumination. If these assumptions are violated, skewness measurements will likely not be indicative of specific material properties.

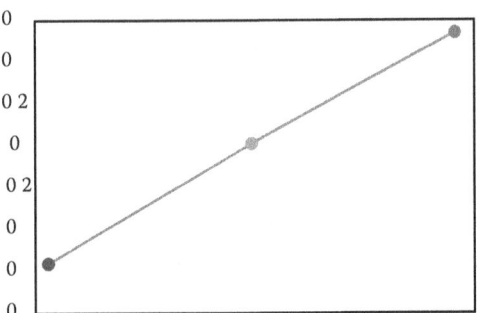

Fig. 13 Skewness measurements from the images shown in Fig. 11. Similar to the findings of Motoyoshi et al.,[30] skewness increases proportionally to the glossiness of the material

This is demonstrated in Fig. 14, where the image of the diffuse object (Fig. 11a) is inverted, therefore changing the skewness of its histogram to a value close to that of Fig. 11b ($s = 0.3273$). Despite the increased skewness, the object does not appear glossier, indicating that other aspects need to be considered, such as additional histogram moments or percentile statistics.[32]

NONLINEAR COMPRESSION IN ART

First-order statistics have also seen very widespread use in the study of paintings. Paintings form a very interesting category of images as they are generally abstracted or manipulated representations of real scenes, purposefully created for human viewers. Different creative styles can represent the same scene in totally different ways—from almost realistic to completely abstract—eliciting different reactions and emotions. Yet, despite art often representing a reduced, abstracted, or modified version of reality, we are not only capable of visually processing it but also find it aesthetically pleasing and engaging.

In an effort to understand the regularities in art and how the compare with real scenes, first-order (as well as higher-order) statistics have been extensively used to examine different properties of art.[33–38] This included various epochs of art, including properties of the illuminants,

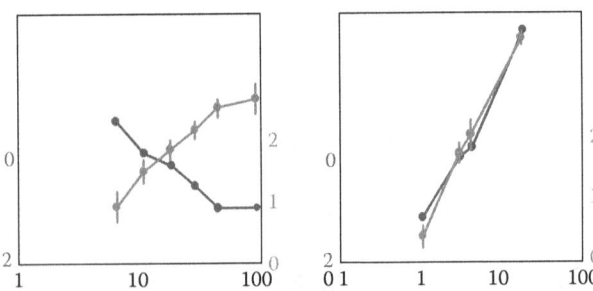

Fig. 12 As the diffuse reflectance of the surface increases, the lightness rating as perceived by the observers also increases while the corresponding histogram skewness decreases (left). An increase in specular intensity also results in an increased rating of glossiness as well as higher skewness value (right)
Source: Adapted from the work of Motoyoshi et al.[30]

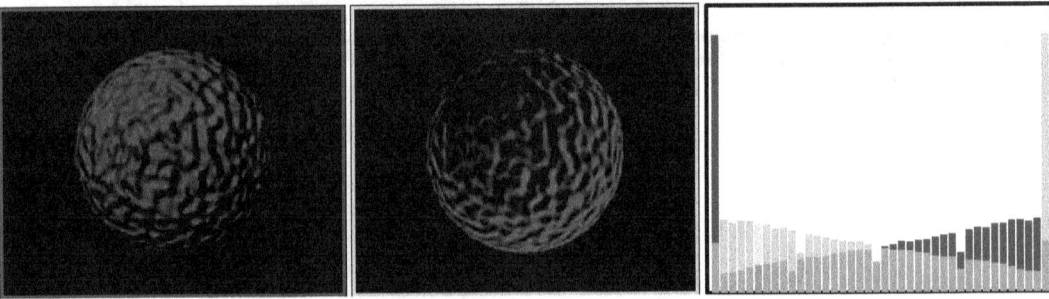

Fig. 14 The image of the diffuse object (a) is inverted to produce (b). Although this process changes the skewness of the distribution as the corresponding histogram is mirrored (c), it does not change the perceived glossiness of the object. Instead, the object appears as if it is illuminated from the opposite direction, indicating that skewness alone may not be a sufficient indicator for material glossiness

differences between epochs or cultures, and the relationship between image statistics and either the human visual system, personal preference, or real-world scenes.

One particularly interesting aspect of art studied in this work is the nonlinearity of intensities found in paintings compared to real-world illumination and the nonlinearities in the visual system.[33,35,37] As we have seen in Section 3.1.2, the range of illumination in natural scenes far exceeds the capabilities of modern cameras and displays, requiring nonlinear compression schemes to preserve detail in the scene while reducing its dynamic range to fit within more restricted devices. The human visual system faces a similar problem, as discussed in Section 2.3.2, employing a sigmoidal compression in the photoreceptors.[39]

Unsurprisingly, artists are also confronted with this issue, especially given the limited range that can be represented with paints on a canvas. Graham and colleagues compared a set of paintings to calibrated photographs of natural scenes to determine how the nonlinearity employed in paintings relates to natural scene illumination and the responses of photoreceptors.[33] Images in both sets were processed with a difference-of-Gaussians (DoG) filter, simulating ganglion responses, and with Gabor filters, simulating cortical responses. Further, images were analyzed both in linear and logarithmic (simulating the photoreceptor responses) domain. Skewness and kurtosis statistics were collected for each transform and image class (art or natural).

Interestingly, paintings tend to have a significantly lower skewness and sparseness than natural scenes, both when pixels are analyzed directly and after filtering with a DoG in the linear case, but this relation reverses in the log domain, as shown in Fig. 15. As the paintings already represent a compressed version of real-world intensities, they are less affected by the effect of the logarithmic compression.[33] Although nonlinearities in the way light is represented in art are common, no single function was found capable of modeling this compression across all art,[34] suggesting that much like tonemapping photographic images of high dynamic range, no single solution will work for all scenes.[40]

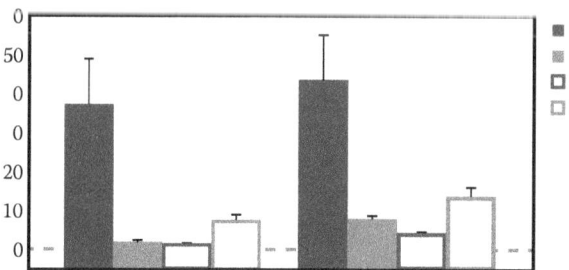

Fig. 15 Natural scenes exhibit much higher sparseness (kurtosis) in the linear case. However if intensities are logarithmically compressed, artistic images have sparser distributions than natural scenes
Source: Adapted from the work of Graham and Field.[33]

In addition to learning about artists' nonlinear compression, art has also been analyzed using statistics to estimate elements involved in the process of painting, including the position of light sources in Caravaggio-style paintings, whether optical projections were used, the position and properties of a virtual camera that would have taken the picture if it were a photograph, and a 3D version of the scene.[41-44] Finally, a number of researchers in addition to Graham have examined distinguishing art from different periods from either each other or from real scenes[45-48] or to digitally restore art[49] using first-order statistics combined with higher-order statistics. For a more detailed review of this area, we refer the reader to Ref. [3].

DARK-IS-DEEP PARADIGM

Photographs can capture in a two-dimensional array of pixels the appearance of a complex scene. Looking at an image, we can easily form an understanding about the shape of the depicted objects, the illumination, or the materials within the image. In the previous section we have seen how simple histogram statistics may be linked with material appearance in images. Substantial effort has been devoted to algorithms that can similarly recover information about the geometry and the estimation of the light sources present in the scene. Depth and shape, however, have been linked both perceptually and in terms of image reconstruction to the intensities of image pixels.[50-52]

The Shape from Shading (SFS) problem was first formulated by Horn in the context of computer vision and can be understood as the extraction or recovery of 3D geometry from a 2D image using shading information[53]—essentially relating luminance with depth. The intuitive simplicity of this idea has led to a remarkable amount of follow-up research, spanning several decades.[54] Although many of the SFS methods perform well in restricted situations, where the materials and illumination are well controlled, estimations tend to be less accurate in more general scenarios.

Human vision also relies on certain assumptions about the scene to interpret the shape of the objects. Mathematically, without prior information about the illumination in a scene, it is impossible to determine the geometry of an object. Consider, for instance, the spheres in Fig. 16; without knowledge of the location of the illumination we cannot accurately determine whether the shapes are convex or concave. Despite this ambiguity, the third shape is perceived as concave while the others appear convex. On the other hand, if all shapes are rotated by 90° as shown in Fig. 17 such a distinction cannot be made.

One of the main priors that the human visual system uses to determine shape is the "light from above" assumption[51] (although evidence suggests that a left bias also exists[55,56]). This is not surprising, as our visual system has evolved both genetically and personally in environments

Fig. 16 Although it is impossible to accurately determine whether these objects are concave or convex from this viewpoint, the third sphere is perceived as concave while the remaining objects are seen as convex. Our visual system employs priors to make such assessment relating both to the direction of the illumination and global convexity

Fig. 17 Unlike the objects in Fig. 16, it is harder to assess the convexity of these objects as illumination appears to come from either side

Fig. 18 The right image was created by randomly permuting the pixels of the image on the left, resulting in identical first-order statistics. (Bryce Canyon, USA, 2009)

where illumination typically comes from above. Further, Langer et al. have found evidence that a global convexity assumption exists when assessing shape.[57]

Although in such simple examples image intensity appears to be related to depth and geometry—at least to the extent that it conforms to the assumptions discussed—natural scenes consist of much more complex configurations (varying material properties, structure at different scales, translucency, and so on). To derive links between 2D image content and 3D depth, more complex statistics need to be considered.

SUMMARY

Statistical analysis and manipulation of image histograms can be used to derive powerful descriptors. Using well-chosen assumptions, first-order statistics have been linked to depth in scenes as well as material properties. However, these statistics do not capture spatial relations between pixels as the position of pixels is not considered. Figure 18 shows an example of two very different images that would result in identical first-order statistics when spatial information is ignored. The right image is constructed by randomly permuting the pixels of the left image, yet it appears unnatural—it carries no recognizable information.

As they only assess single-pixel properties, the histogram-based statistics discussed in this entry are the simplest to compute and interpret. To further explore the spatial relations between pixels and their associations with aspects of human vision, more complex transforms are necessary.

REFERENCES

1. Panda, S.; Provencio, I.; Tu, D.C.; Pires, S.S.; Rollag, M.D.; Castrucci, A.M.; Pletcher, M.T.; Sato, T.K.; Wiltshire, T.; Andahazy, M. Melanopsin is required for nonimage-forming photic responses in blind mice. Sci. Signal. **2003**, *301* (5632), 525.
2. Johnson, N.L.; Kotz, S.; Balakrishnan, N. *Continuous Univariate Distributions*, Vol. 1; John Wiley & Sons: New York, 1994.
3. Flusser, J.; Zitova, B.; Suk, T. *Moments and Moment Invariants in Pattern Recognition*; John Wiley & Sons: Chichester, 2009.
4. Hu, M.-K. Visual pattern recognition by moment invariants. IRE Trans. Inf. Theory **1962**, *8* (2), 179–187.
5. Flusser, J.; Suk, T. Pattern recognition by affine moment invariants. Pattern Recogn. **1993**, *26* (1), 167–174.
6. Flusser, J.; Suk, T. Affine moment invariants: A new tool for character recognition. Pattern Recogn. Lett. **1994**, *15* (4), 433–436.
7. Suk, T.; Flusser, J. Affine moment invariants of color images. In *Computer Analysis of Images and Patterns*; Springer: Berlin, 2009, 334–341.
8. Flusser, J.; Suk, T.; Saic, S. Recognition of blurred images by the method of moments. IEEE Trans. Image Process. **1996**, *5* (3), 533–538.

9. Suk, T.; Flusser, J. Combined blur and affine moment invariants and their use in pattern recognition. Pattern Recogn. **2003**, *36* (12), 2895–2907.
10. Flusser, J.; Kautsky, J.; Šroubek, F. Implicit moment invariants. Int. J. Comput. Vision **2010**, *86* (1), 72–86.
11. Flusser, J.; Suk, T. Degraded image analysis: An invariant approach. IEEE Trans. Pattern Anal. Mach. Intell. **1998**, *20* (6), 590–603.
12. Wong, W.-H.; Siu, W.-C.; Lam, K.-M. Generation of moment invariants and their uses for character recognition. Pattern Recogn. Lett. **1995**, *16* (2), 115–123.
13. Mahdian, B.; Saic, S. Detection of copy–move forgery using a method based on blur moment invariants. Forensic Sci. Int. **2007**, *171* (2), 180–189.
14. Tang, H.; Shu, H.; Dillenseger, J.-L.; Bao, X.D.; Luo, L.M. Moment-based metrics for mesh simplification. Comput. Graph. **2007**, *31* (5), 710–718.
15. Reinhard, E.; Khan, E.A.; Akyüz, A.O.; Johnson, G.M. *Color Imaging: Fundamentals and Applications. A K Peters*; A K Peters: Wellesley, MA, 2008.
16. Adams, J.; Parulski, K.; Spaulding, K. Color processing in digital cameras. IEEE Micro **1998**, *18* (6), 20–30.
17. Reinhard, E.; Ashikhmin, M.; Gooch, B.; Shirley, P. Color transfer between images. IEEE Comput. Graphics Appl. **2001**, *21* (5), 34–41.
18. van Hateren, J.H.; van der Schaaf, A. Independent component filters of natural images compared with simple cells in primary visual cortex. Proc. R. Soc. B Biol. Sci. **1998**, *265* (1394), 359.
19. Huang, J.; Mumford, D. Statistics of natural images and models. In *Proceedings of the IEEE International Conference on Computer Vision and Pattern Recognition*, Vol. 1; Washington, DC, 1999.
20. Brady, M.; Field, D.J. Local contrast in natural images: Normalisation and coding efficiency. Perception **2000**, *29* (9), 1041–1055.
21. Pouli, T.; Cunningham, D.; Reinhard, E. Statistical regularities in low and high dynamic range images. In *APGV '10: Proceedings of the 7th Symposium on Applied Perception in Graphics and Visualization*, New York, 2010, 9–16.
22. Dror, R.O.; Leung, T.K.; Adelson, E.H.; Willsky, A.S. Statistics of real-world illumination. In *Proceedings of the IEEE International Conference on Computer Vision and Pattern Recognition*, Vol. 2; Kauai, HI, 2001, 164–171.
23. Dror, R.O.; Willsky, A.S.; Adelson, E.H. Statistical characterization of realworld illumination. J. Vision **2004**, *4* (9), 821–837.
24. Fairchild, M.D. The HDR photographic survey. In *Proceedings of the 15th IS&T/SID Color Imaging Conference*, Vol. 15; The Society for Imaging Science and Technology, 2007, 233–238.
25. Fleming, R.W.; Dror, R.O.; Adelson, E.H. Real-world illumination and the perception of surface reflectance properties. J. Vision **2003**, *3* (5).
26. Fleming, R.W.; Jensen, H.W.; Bülthoff, H.H. Perceiving translucent materials. In *Proceedings of the First ACM Symposium on Applied Perception in Graphics and Visualization*, New York, 2004, 127–134.
27. Fleming, R.W.; Torralba, A.; Adelson, E.H. Specular reflections and the perception of shape. J. Vision **2004**, *4* (9), 798–820.
28. Motoyoshi, I.; Adelson, E.H. Luminance re-mapping for the control of apparent material. In *Proceedings of the 2nd Symposium on Applied Perception in Graphics and Visualization*, 2005, 165.
29. Beck, J.; Prazdny, S. Highlights and the perception of glossiness. Atten. Percept. Psychophys. **1981**, *30* (4), 407–410.
30. Motoyoshi, I.; Nishida, S.; Sharan, L.; Adelson, E.H. Image statistics and the perception of surface qualities. Nature **2007**, *447* (7141), 206–209.
31. Anderson, B.L.; Kim, J. Image statistics do not explain the perception of gloss and lightness. J. Vision **2009**, *9* (11), 10, 1–17.
32. Sharan, L.; Li, Y.; Motoyoshi, I.; Nishida, S.; Adelson, E.H. Image statistics for surface reflectance perception. J. Opt. Soc. Am. A **2008**, *25* (4), 846–865.
33. Graham, D.J.; Field, D.J. Statistical regularities of art images and natural scenes: Spectra, sparseness and nonlinearities. Spatial Vision **2007**, *21* (1–2), 1–2.
34. Graham, D.J.; Field, D.J. Global nonlinear compression of natural luminances in painted art. In *Proceedings of IS&T/SPIE Electronic Imaging*, Bellingham, WA, 2008, 68100K.
35. Graham, D.J.; Field, D.J. Variations in intensity statistics for representational and abstract art, and for art from the eastern and western hemispheres. Perception **2008**, *37* (9), 1341–1352.
36. Graham, D.J.; Friedenberg, J.D.; McCandless, C.H.; Rockmore, D.N. Preference for art: Similarity, statistics, and selling price. In *Proceedings of IS&T/SPIE Electronic Imaging. International Society for Optics and Photonics*, 2010, 75271A.
37. Graham, D.J.; Friedenberg, J.D.; Rockmore, D.N. Efficient visual system processing of spatial and luminance statistics in representational and nonrepresentational art. In *Proceedings of IS&T/SPIE Electronic Imaging*, Vol. 72401N; San Jose, CA, 2009.
38. Graham, D.J.; Friedenberg, J.D.; Rockmore, D.N.; Field, D.J. Mapping the similarity space of paintings: Image statistics and visual perception. Vis. Cognit. **2010**, *18* (4), 559–573.
39. Naka, K.I.; Rushton, W.A.H. S-potentials from luminosity units in the retina of fish (cyprinidae). J. Physiol. **1966**, *185* (3), 587–599.
40. Cadík, M.; Wimmer, M.; Neumann, L.; Artusi, A. Evaluation of HDR tone mapping methods using essential perceptual attributes. Comput. Graph. **2008**, *32* (3), 330–349.
41. Criminisi, A.; Stork, D.G. Did the great masters use optical projections while painting? Perspective comparison of paintings and photographs of renaissance chandeliers. In *Proceedings of the IEEE International Conference on Pattern Recognition*, Vol. 4; 2004, 645–648.
42. Kale, D.C.; Stork, D.G. Estimating the position of illuminants in paintings under weak model assumptions: An application to the works of two Baroque masters. In *Proceedings of IS&T/SPIE Electronic Imaging*, 2009, 72401.
43. Stork, D.G.; Johnson, M.K. Lighting analysis of diffusely illuminated tableaus in realist paintings: An application to detecting "compositing" in the portraits of Garth Herrick. In *Media Forensics and Security*, 2009, 72540.
44. Ziegaus, C.; Lang, E.W. Statistical invariances in artificial, natural and urban images. Zeitschrift für Naturforschung A **1998**, *53*, 1009–1021.

45. Wallraven, C.; Fleming, R.; Cunningham, D.; Rigau, J.; Feixas, M.; Sbert, M. Categorizing art: Comparing humans and computers. Comput. Graphics **2009**, *33* (4), 484–495.
46. Cutzu, F.; Hammoud, R.I.; Leykin, A. Distinguishing paintings from photographs. Comput. Vision Image Underst. **2005**, *100* (3), 249–273.
47. Leslie, L.; Chua, T.-S.; Jain, R. Annotation of paintings with high-level semantic concepts using transductive inference and ontology-based concept disambiguation. In *Proceedings of the 15th ACM Conference on Multimedia*, 2007, 443–452.
48. Marchenko, Y.; Chua, T.-S.; Aristarkhova, I. Analysis and retrieval of paintings using artistic color concepts. In *Proceedings of the IEEE International Conference on Multimedia and Expo*, 2005, 1246–1249s.
49. Pappas, M.; Pitas, I. Digital color restoration of old paintings. IEEE Trans. Image Process. **2000**, *9* (2), 291–294.
50. Todd, J.T.; Mingolla, E. Perception of surface curvature and direction of illuminant from patterns of shading. J. Exp. Psychol. Human. Perception Perform. **1983**, *9* (4), 583–595.
51. Ramachandran, V.S. Perception of shape from shading. Nature **1988**, *331* (6152), 163–166.
52. Langer, M.S.; Bülthoff, H.H. Depth discrimination from shading under diffuse lighting. Perception **2000**, *29* (6), 649–660.
53. Horn, B.K.P. *Shape from Shading: A Method for Obtaining the Shape of a Smooth Opaque Object from One View*; PhD Thesis; Massachusetts Institute of Technology, 1970.
54. Zhang, R.; Tsai, P.-S.; Cryer, J.E.; Shah, M. Shape from shading: A survey. IEEE Trans. Pattern Anal. Mach. Intell. **1999**, *21* (8), 690–706.
55. Mamassian, P.; Goutcher, R. Prior knowledge on the illumination position. Cognition **2001**, *81* (1), B1–B9.
56. McManus, I.C.; Buckman, J.; Woolley, E. Is light in pictures presumed to come from the left side? Perception **2004**, *33* (12), 1421–1436.
57. Langer, M.S.; Bülthoff, H.H. A prior for global convexity in local shape-fromshading. Perception **2001**, *30* (4), 403–410.

Floodplain Mapping

N. N. Ramaprasad and Priya Narayanan
Department of Geography, School of Earth Science, Central University of Karnataka, Kalaburagi, India

Abstract

Plain areas close to rivers that are vulnerable to frequent floods are floodplains. The floodplains depend on many factors such as catchment area, catchment response, rainfall pattern, evaporation, and recharge and flow rates. Floodplains are formed by continuous deposition of silts brought down from hills. Floodplain contains abundant minerals and fertile soil and easy to evolve. The urban floods are recent phenomenon where the covering of surfaces by impervious materials reduces the infiltration and increases runoff resulting in the flash flood in low-lying areas. Because of the fast catchment response, urban floods are highly unpredictable. The occupation of floodplains in developing countries is of great concern. Floodplain mapping involves careful study of the hydro-geomorphology of the terrain and is essential for planners and developers. Mapping floodplains with accuracy was a gigantic task with the conventional method and challenging until the availability of remotely sensed data. Presently, digital elevation models such as Shuttle Radar Topography Mission (SRTM), Advanced Spaceborne Thermal Emission and Reflection Radiometer (ASTER), and CartoDEM are widely used in interpreting the surface models. The aerial photographs can cover large areas. Creation of DEM using semiautomatic techniques and manual break lines is economical and accurate. Light detection and ranging uses a laser scanner mounted on an airborne platform and produces an accurate three-dimensional (3D) mapping. Laser pulses measure distances between the emitter and reflected objects. The by-products of the DEM are slope map, aspect, flow direction, and flow accumulation. The low-gradient areas close to rivers determine the floodplains. The logic of obtaining clustered cells of low-gradient areas along the rivers can help us locate floodplains.

Keywords: Aspect; Digital elevation models; Floodplain; Remote sensing; Slope.

INTRODUCTION

The floodplains are major components of a river system. The physical, chemical and biological processes of floodplains combine over a range of temporal and spatial scale contributing to environmental change. The impact of floods has significant socio-economic importance such that the relatively flat terrain with fertile land next to rivers and creeks attracts high percentage of the world's human population to dwell on floodplains at the peril of placing themselves in the hazardous situation.[1]

An extensive study on mapping of floodplains is critical as the flood management is impossible without proper scientific study of floodplains. The stormwater can be used to recharge ground water and can be converted from threat to resource. Proper preparedness in floodplains with the development of structural and nonstructural remedies at place, depending on the case history of floods can ease the life of human beings without any risks. Surface runoff occurs when the rainfall intensity exceeds the evaporation rate and infiltration capacity of the soil. The surface runoff flows due to gravity in the river channels. The subsurface flows, which were infiltrated also dissipate into river channels and water bodies. When the capacity of channel exceeds the combined effect of surface runoff and subsurface flow, the plain areas surrounding the river channels are inundated causing floods.

Floodplains

Floodplains are plain areas having a gentler slope less than 1% and land adjacent to a stream or river that extends from the river banks up to the base of the enclosing valley walls. They usually experience flooding during periods of high discharge. The soil usually consists of downhill eroded levees, silts, and sands deposited during floods. Levees are the primarily deposited heaviest materials, followed by finer materials of silt and sand. They are gradually building up to create the floor of the floodplain and when the drainage system has diverted in due course, the floodplain may become a level area of high fertility. They appear in similar to the floor of an old lake. The floodplain is not altogether flat. The floodplains have a gentle slope toward downstream, and often cover a distance from the side toward the center.[1] The width of floodplains depends on many factors such as the flow accumulation flow rate and the slope of the area. The floodplain is the most developed natural place of a river to dissipate its energy, where the fast-moving narrow channel stream dissipates to slow-moving wider channel submerging the floodplains. India alone has witnessed significant flood incidences in the major cities such as Mumbai, Chennai, and Bengaluru, each of them having a different topography. More often the affected homes in these cities are built up in the regions of floodplains.[2]

CLASSIFICATION OF FLOODPLAINS

Riverine Floodplains

The riverine flooding usually occurs at a distance from the catchment. The advanced, accurate warning system can minimize casualties. Dam failures or excess discharge from dams along with heavy storms, make the situation uncontrollable causing massive damages to the economic system. Riverine floods occur for a longer duration.

Urban Floodplains

Urban floods, also called as flash floods that last for a short period, are highly unpredictable because of small catchment areas and mainly caused due to land-use changes.[3] The binding of surface areas by impervious surfaces reduces the infiltration. The straightening of channels reduces the time of concentration that results in flooding the low-lying urban areas during heavy storms.

Back Water (Fluctuating Lake Levels) Floodplains

Water levels in the lakes can vary for short term (e.g., seasonal) or long term (e.g., on a yearly basis). Periodical rains cause high water levels for short periods of time. Long-term lake level fluctuations are a less-recognized phenomenon that can cause high water and subsequent flooding problems lasting for years or even decades. Although lakes exhibit fluctuating water levels, they do not change dramatically where outlet to downstreams provides a relatively natural balance between inflow and outflow. However, completely landlocked lakes without outlets are inadequate for maintaining a balance between inflow and outflow. These lakes, termed as "closed basin lakes," are particularly susceptible to dramatic fluctuations in water levels. They recede the runoff of storm water, flooding the upper catchment of lakes.

Coastal Floodplains

Coastal flooding and erosion are result of combined effect of storm surges and wave actions. Storm surge increases water surface height above normal tide levels, primarily because of low pressure over the ocean surface. The low pressure acts like a suction in a straw, creating a dome of water in the middle of the cyclone. In the central oceans, this dome of water sinks harmlessly and flows away. Nevertheless, as a storm hit nears coastal land, high winds in the storm push this dome of water toward the shore, the lifting sea floor blocks the water's escape, and it comes ashore as deadly storm surge.[4] An intense hurricane can spread a dome of water many miles wide and more than 25-feet-high barreling toward the shore as the storm hits land. Depending upon local topography it may inundate coastal areas for miles inland from the shoreline. Difference between the astrological tide level and observed storm tide determines the intensity of storm surge. Storm surge heights vary from a few feet to more than 25 feet. The height of the storm surge and corresponding coastal flood depends on many factors, including the strength, intensity, and speed of the storm. The storm surge is dependent on wind direction and velocity along the shoreline; the slope of the seafloor along the coastline; the configuration of the coast; and the astronomical tide. The storm surge is extremely detrimental when it passes on a shallowly sloped shore, as a result of high tide, over a densely populated developed country with little or no natural barrier islands, coral reefs, or coastal vegetation.

CHALLENGES IN MAPPING FLOODPLAINS

Managing the development of floodplains is a critical responsibility for regional and urban planners. The benefits of floodplains, including fertile agricultural land and accessible housing locations, mainly located in places of comfortable developments, must balance with the personal and economic security posed by floods. The floodplain although a vulnerable area for floods, it depends on many factors such as hydrological factor, land-use pattern, and intensity of rainfall. The frequency and return period is dependent on many factors. The historical evidence can only ease the designation of floodplains. However, for planning and management, mapping and analyzing floodplains were very necessarily required. Mapping floodplains with accuracy was a gigantic task with the conventional method and challenging until the availability of remotely sensed data.

DIGITAL ELEVATION MODEL

A digital elevation model (DEM) is a three-dimensional digital representation of a terrain surface created from terrain elevation data. The digital surface model (DSM) represents the face of the earth, including all objects on it (vegetation and buildings). The DEM and the digital terrain model (DTM) represent the bare soil surface without any objects such as plants and buildings. DEM is the representation of terrain as a raster in the array of rows and columns (a grid of squares, also known as an elevation map), where each pixel accounts for a portion of the earth's surface area, with an average height of the place recorded as the value of the pixel. Higher the resolution smaller the area it represents. The higher resolution of vertical and horizontal DEM will result in better topography that will be much more useful in determining catchment characteristics.

REMOTELY SENSED DIGITAL ELEVATION MODELS

The remotely sensed data can capture large areas with accuracy and can give bird's eye view of the area. The different optical sensor images from two different positions can resolve the geometry of the stereo image, and the position (height information) is extractable. The choice of the data sources depends on the scale of mapping and finer data for model parameterization. The primary method of generating DTMs for floodplain and channel demarcation, including airborne scanning laser altimetry light detection and ranging (LiDAR) and airborne interferometric synthetic aperture radar. The cost and time need to be balanced while selecting the data for hydrological modeling. A variety of sensors mounted on spacecraft remotely captured flooded areas and images of the earth including the height information. The advantages of active microwave systems are that they can be used to acquire the images during the flood events even in cloud coverages.

OPTICAL AND MICROWAVE REMOTE SENSORS

Microwave remote sensing of flood extent active microwave sensors, such as synthetic aperture radar, provide their source of energy and record the reflected energy returned from the imaged object. Microwave can penetrate clouds, emergent aquatic plants, and forests to detect standing water.[5–9] Cloud penetration is particularly important in capturing images for monitoring flood events, as they commonly occur during rains.[7,9–11] Inundated areas can be determined optimally by selecting the most suitable polarizations of the radar waves.

DIGITAL AERIAL PHOTOGRAMMETRY

DTMs can be produced using stereophotogrammetry applied to overlapping pairs of aerial photographs.[12] Aerial photos are captured in series along the fly line in different strips with adjoining photos having 60%–65% overlap in the flight direction and 20%–30% lateral overlap between the two strips. The photographs are scanned, digitized using a photogrammetric scanner. Precisely, the coordinates of fiducial marks in the camera are measured. The fiducial marks on the image are assigned the same coordinate system with camera calibration. In digital photographs, the photo coordinates are at once received. A relationship between the image coordinates and the ground space coordinates is defined by the onboard global positioning system (GPS) and the inertial navigation system (INS) to determine orientations. In the digital stereophotogrammetry instrument, the onboard coordinates can be directly fed to relate with the ground coordinate system. The ground elevations are observed in photogrammetry device from an overlapping stereo-pairs. Image matching technique is performed in semiautomatic and automatic modes. The automatic collection of the feature based on spatial relation is less time consuming than the manual collection. The three-dimensional (3D) ground space coordinates of points can then be determined and interpolated onto a regular grid. The model formed is essentially a DSM. The semiautomatic techniques are used to remove the noises. Additional break lines can be extracted directly in 3D using the stereo plotter. Depending upon the need, large-scale photography can be acquired. The vertical accuracy depends on the ground sample distance and the accuracy of ground control point. The skill of the operator enhances the accuracy and productivity.

LIGHT DETECTION AND RANGING

LiDAR is a laser technique that maps highly accurate and dense elevation data suitable for flood modeling.[13,14] A LiDAR airborne system uses a laser scanner mounted on an aircraft or helicopter platform. Pulses from the laser beams strike the earth's surface, and the reflected laser beam feature travels back toward the platform. The trip distance between the laser transmitter and the receiver via the surface component is computed using the trip time of the pulsation and the speed of light. The instantaneous position and orientation of the laser are known to use the GPS and INS systems on board the program. The 3D location of the ground feature stored in the WGS84 coordinate system and then transformed to the local map projection using the additional information on the scan angle and GPS base station. A high vertical accuracy of ±5 to 25 cm is possible to achieve with LiDAR technology. At average flight speeds and flying heights, the LiDAR can collect locations at a density of at least one point every 0.25–5 m depending upon laser characteristics and terrain elevations. The laser pulse may reflect from more than one part (multipath) of a ground feature in vegetated areas and buildings; it may also reflect from the top of the foliage and also from the ground below. Many LiDAR systems are equipped to collect both the first return and the last return, and in some systems, it is possible to obtain the complete reflected waveform. The intensity of the reflected pulse can also be used to analyze information about the surface feature imaged. For LiDAR data, filtering algorithms are applied to administer floodplain friction measurement and to resolve the problems encountered while integrating LiDAR data into a flood model.

COMPUTATION OF SLOPE AND ASPECT

The ratio of rise to run is defined as slope represented in either percentage or angles. The gradient can be calculated depending on the data available (coordinates and length, height difference). The magnitude of the land surface slope is given by the following equation as shown in Fig. 1.

$$\text{Slope} = \tan\theta = \frac{\text{Rise}}{\text{Run}}$$

In the 3D coordinate system, slope(degress) = atan $\sqrt{\left(\frac{dz}{dx}\right)^2 + \left(\frac{dz}{dy}\right)^2} \times 57.29578$

The units of z need to be in the same as that of x and y. The answer is a dimensionless quantity representing the gradient with the positive value representing uphill. When the slope is multiplied by 100, the vertical rise will be in percentage of horizontal distance. A measurement corresponding to that for the zenith angle described earlier gives the slope angle in degrees in the slope function.

a	b	c
d	e	f
g	h	i

Aspect identifies the downward slope direction. The partial derivatives evaluated in this equation are as positive toward the downhill slope. The average of central finite differences over each of the three rows of cells in the middle row counts twice as it appears in averages on each side.

$$\frac{dz}{dx} = \frac{(c + 2f + i) - (a + 2d + g)}{8 * x\_cell\_size}$$

$$\frac{dz}{dy} = \frac{(g + 2h + i) - (a + 2b + c)}{8 * y\_cell\_size}$$

$$\text{Slope} = \sqrt{\left(\frac{dz}{dx}\right)^2 + \left(\frac{dz}{dy}\right)^2}$$

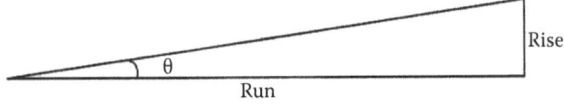

Fig. 1 Diagram showing the calculation of slope

In physical geography, aspect is a touchstone of the direction of the side of the ground surface, which is similar to the azimuth used in land surveying. By analogy with the azimuth, the function evaluates aspect as

$$\text{Aspect} = \text{atan2}\left(\frac{dz}{dx}, \frac{\Delta z}{\Delta y}\right)$$

To define slope in hydrology, we are concerned in the pathway of water running downhill. The gradient is taken as positive quantities if pointing downhill. The D8 (eight directions forming the central cell) flow direction is a continuously gridded elevation surface. Only eight discrete flow paths are possible. D8 slope is computed as

$$\text{D8 slope} = \frac{\text{Drop}}{\text{Run}}$$

where drop refers to the difference in height between the two cells, with downward considered positive, and the run is equal to the distance between the center of cells, the cell resolution(x) for flow along the coordinate directions and equal to √2(x) cell size for flow along the diagonals. From an elevation raster, for an arbitrary cell, we have its elevation value, plus elevation of its eight neighboring cells. The simplest approach to slope angle computation is to make use of x- and y-gradient filters. To get the real slope angle along path p, observe that both the x and y-gradient filters contribute to it. The normalized gradients are used to compute the slope aspect.

FLOODPLAIN MAPPING

The slope determined can be classified according to morphology. The DEM values can be used to find the average slope and to locate the floodplains. The slope obtained from LiDAR technology can help to find the accurate floodplain, but the covered drainages in urban areas make even the LiDER technology also complicated.Slope maps are classified based on the morphology defined in Table 1;[15] floodplains are the level grounds adjacent to the rivers. The rivers are demarcated from either available vector data of the area or using remotely sensed images. Figure 2 shows the LISS III image; Fig. 3 shows Cartosat DEM of the sample area obtained from Cartosat from longitude77°8'to 77°10' East and from latitude15°56' to 15°58' North. Based on the slope map, single cells may be ignored, and clustered cells of the low gradient may be selected adjoining the rivers. The careful observation of the DEM and historical evidence and the catchment area is necessary for determining the width of floodplains. However, depending upon the stage of the river, the buffer zone for selection of floodplains can be determined. From Fig. 4, we can quickly estimate the floodplains surrounding the river. Applying the hydro parameter with finer resolution DEM, the accurate mapping of floodplains is possible.

Table 1 Distribution of slopes based on Integrated Mission for Sustainable Development (IMSD) and National Remote Sensing Agency (NRSA) 1995 guidelines

Types of terrain	Slope (%)
Nearly level	0–1
Very gentle slope	1–3
Gentle slope	3–5
Moderate slope	5–10
Strong slope	10–15
Moderately steep to steep sloping	15–30
Very steep sloping	>30

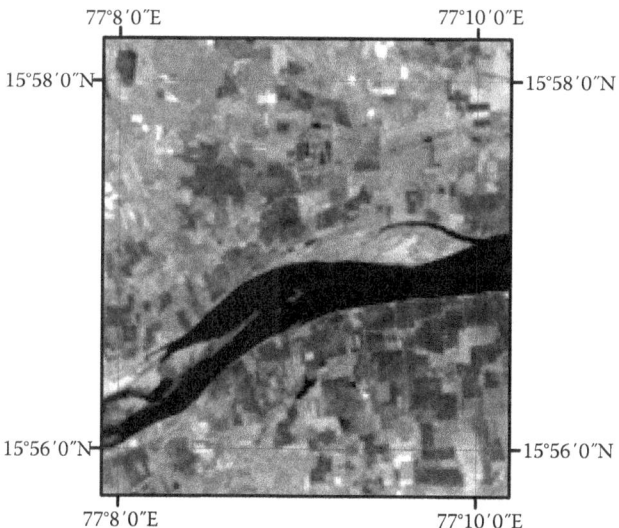

Fig. 2 LISS III image

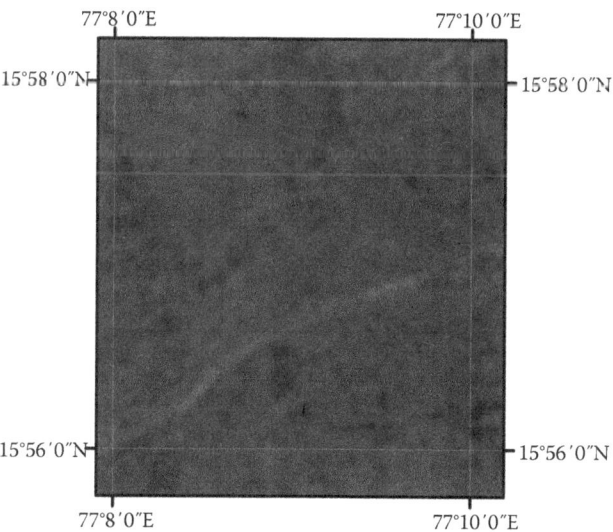

Fig. 3 Cartosat 1 DEM

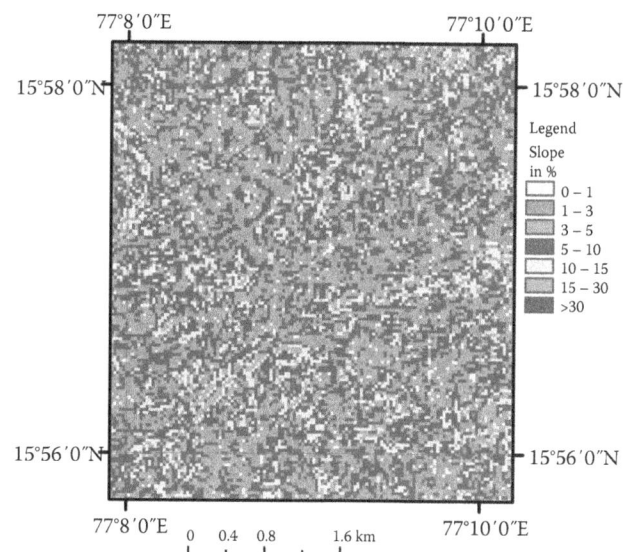

Fig. 4 Slope map from Cartosat 1 DEM

CONCLUSION

The remotely sensed data have many advantages than the conventional methods. By remote sensing, even hostile environments are accessible. They cover large areas, spaceborne satellites have high temporal resolution, and time series analysis is possible. The dynamic nature of floodplains requires time series data for its mapping and analysis. The photogrammetric techniques and filter algorithms enhance the image analysis. Future technologies are expected to improve accuracy by eliminating the delta factors as minimum as possible to obtain high accuracy. The accurate floodplain determination will assist planners and developers to bring structural and nonstructural remedies at these situations. The mitigation techniques adopted in floodplains can help in reinstating the natural resources and even downplay the catastrophe.

REFERENCES

1. Marriott, S.B.; Alexander, J. *Floodplains: Interdisciplinary Approaches*; The Geological Society: London, 1999.
2. Gupta, A.K.; Nair, S.S. Urban floods in Bangalore and Chennai: Risk management challenges and lessons for sustainable urban ecology. Curr. Sci. **2011**, *100* (11), 1638–1645.
3. Ramaprasad, N.N.; Narayanan, P. Vulnerability assessment of flood affected locations of Bangalore by using multi-criteria evaluation. Ann. GIS **2016**, *22* (2), 151–162.
4. Klemas, V. The role of remote sensing in predicting and determining coastal storm impacts. J. Coastal Res. **2009**, *25* (6), 1264–1275.
5. Horritt, M.S.; Mason, D.C.; Luckman, A.J. Flood boundary delineation from synthetic aperture radar imagery using a statistical active contour model. Int. J. Remote Sens. **2001**, *22* (13), 2489–2507.

6. Brivio, P.A.; Colombo, R.; Maggi, M.; Tomasoni, R. Integration of remote sensing data and GIS for accurate mapping of flooded areas. Int. J. Remote Sens. **2002**, *23* (3), 429–441.
7. Kiage, L.M.; Walker, N.D.; Balasubramanian, S.; Babin, A.; Barras, J. Application of Radarsat-1 synthetic aperture radar imagery to assess hurricane-related flooding of coastal Louisiana. Int. J. Remote Sens. **2005**, *26* (24), 5359–5380.
8. Townsend, P.A. Relationship between forest structure and the detection of flood inundation in forested wetlands using C-band SAR. Int. J. Remote Sens. **2002**, *23* (3), 443–460.
9. Klemas, V. Remote sensing of floods and flood-prone areas: An overview. J. Coastal Res. **2015**, *31* (4), 1005–1013.
10. Stevens, T.B. *Synthetic aperture radar for coastal flood mapping*; NASA Global Change Master Directory, in LSU Earth Scan Laboratory: Baton Rouge, LA, 2013.
11. Townsend, P.A.; Walsh, S.J. Modeling floodplain inundation using an integrated GIS with radar and optical remote sensing. Geomorphology **1998**, *21* (3–4), 295–312.
12. Wolf, P.R.; Dewitt, B.A. *Elements of Photogrammetry: With Applications in GIS*, 3rd Ed.; McGraw-Hill: Boston, MA, 2000.
13. Wehr, A.; Lohr, U. Airborne laser scanning: An introduction and overview. ISPRS J. Photogramm. Remote Sens. **1999**, *54* (1), 68–82.
14. Flood, M. Laser altimetry: From science to commercial mapping. Photogramm. Eng. Remote Sens. **2001**, *67* (11), 1209–1217.
15. Kumar, B.; Kumar, U. Integrated approach using RS and GIS techniques for mapping of ground water prospects in Lower Sanjai Watershed, Jharkhand. Int. J. Geomatics Geosci. **2010**, *1* (3), 587–598.

Forgery Detection in Digital Images

Zhili Zhou, Yi Cao, and Xingming Sun
Department of Computer Science and Information Engineering Nanjing University of Information Science and Technology, Nanjing, China

Ching-Nung Yang
Jiangsu Engineering Center of Network Monitoring & School of Computer and Software, National Dong Hwa University, Hualien, Taiwan

Abstract

With the increasing popularity of digital image and the wide use of image processing tools, a large number of digital images are replicated, transmitted, and redistributed via networks. The problem of copy forgeries has become more and more serious in various fields like entertainment, digital forensics, and journalism. Also, people become increasingly aware of their rights under copyright laws. Thus, the copyright protection of digital image is necessarily required. To prevent the illegal use of copyrighted images (original images), image copy forgery detection is developed to detect image copies, which are derived from an original image via various copy attacks such as geometric transformation and noise-like contamination. In the past two decades, a large number of image copy detection methods had been proposed. Those detection methods were categorized into four types: global feature-based detection, local feature-based detection, feature combination-based detection and learning feature-based detection. In this work, we review the key technologies of those detection methods, and meanwhile discuss their advantages and disadvantages, respectively. Furthermore, we also give related challenges and possible solutions.

Keywords: Copy forgery; Image copy detection; Image search; Near-duplicate retrieval; Partial-duplicate retrieval.

INTRODUCTION

Nowadays, due to the rapid development of Internet technology and the widespread use of personal digital image acquisition devices, thousands of digital contents (including digital image, audio, and video) are produced and uploaded to network. Because digital image is one of most popular and important digital media, it has a large percentage among the huge number of digital content. By using various image processing approaches, digital image is easily duplicated, modified, and redistributed over networks. Therefore, there are a lot of copy versions of copyrighted images. To prevent the illegal use of digital content, the copyright protection of digital image is necessarily required. The first step for achieving copyright protection is the copy forgery detection,[1] and thus, the technology of copy forgery detection has attracted a lot of attention.

The so-called image copies are defined as the images derived from the original image (a copyrighted image) by various copy attacks such as rescaling, rotation, shifting, cropping, change of intensity and contrast, and noise adding. From the literature,[2] there is a common consensus that the original image and its image copies come from the same digital resource. Actually, the definition of image copies is similar to that of near-duplicate images. The difference is that the near-duplicate images not only refer to the image copies of an original image, but also the similar images captured from the same scene or object by different conditions, such as different viewpoints, positions, and acquisition time. Thus, image copies can be regarded as a subset of near duplicates. Figure 1 demonstrates several examples of image copies and similar images for a given original image, respectively. Figure 1a is the original image. Figure 1b–d shows image copies, while Fig. 1e–h shows similar images. Although similar images and original are not from the same digital source, similar images look more visually similar to the original image than image copies. To achieve copyright protection, our aim is to detect image copies and exclude other irrelevant images.

This entry is organized as follows: In Section "Copy Forgery Detection Techniques," we first introduce two main copy forgery detection techniques, that is, watermarking and content-based copy detection. Sections "Feature Extraction" and "Image Matching" briefly review all previous technologies of two key steps of content-based copy detection: the feature extraction and the image matching. A detailed comparison between various content-based copy detection methods is demonstrated in

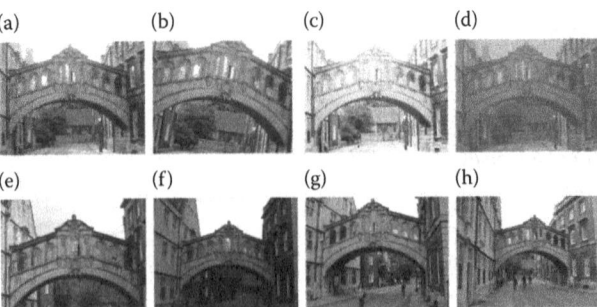

Fig. 1 Examples of image copies and similar images: (a) original images; (b–d) copy versions; and (e)–(f) similar images

Section "Comparison." Related challenges and possible solutions are given in Section "Concluding Remarks: Related Challenges and Possible Solutions."

COPY FORGERY DETECTION TECHNIQUES

There are two main copy forgery detection techniques: digital watermarking and content-based copy detection.[1] By the watermarking technology, we may embed the mark into a copyrighted image in an invisible manner. Afterward, the mark may be extracted to prove the ownership of an image. However, using watermarking technique to avoid copy forgery has two drawbacks. First, it is hard to implement the embedding process in real time for copyrighted images because the watermark has to be embedded into each copyrighted image before its distribution. This approach using watermark is impractical for some real-time scenarios, and thus limit its practical applications. Second, the embedded mark is vulnerable when the image is under some possible copy attacks during its transmission. Once the embedded mark is distorted seriously, it cannot be used to verify whether the image is a copy or not.

Instead of embedding a mark into the image, content-based copy detection relies on content-based image features. Because it is a passive technology, content-based copy detection attracts more and more attention for copy forgery detection. This detection technique is similar to image retrieval technique, and it includes two main steps: content-based feature extraction and image matching

based on extracted features. A content-based copy detection system is briefly illustrated by the following example. First, the system collects a large number of images, which are illegally used and suspected by the copyright owner. Then, the content-based features of each image are extracted. We store the extracted features and the URLs of the images (note: for the sake of reducing storage) to establish a large-scale image database. For an original image (a copyrighted image), the system also extracts its features in the same manner. Afterward, by image matching strategies with extracted features indexing, we search for the copy versions of the original image in this database. If there are image copies without usage permission, they may be regarded as illegal copies. Finally, the owner of the original image may make a further confirmation and consider whether to file lawsuit against illegal users. Figure 2 shows the flow chart of content-based copy detection.

Compared with watermarking technique, content-based copy detection does not require any additional information but image itself. The copy detection can be accomplished even though the image is already distributed. Therefore, content-based copy detection is a more promising technique for copy forgery detection. In this entry, we introduce two key parts of content-based copy detection techniques: the feature extraction and the image matching.

FEATURE EXTRACTION

Feature extraction is a crucial and basic step to detect image copies. Ideal image features should have high robustness against a variety of copy attacks, so that copies may be accurately detected under these attacks. Meanwhile, they should be discriminative enough to identify image copies from other irrelevant images. In addition, they are also expected to have good efficiency to extract and compare the features to ensure the copy detection. Therefore, how to achieve robustness, discriminability, and efficiency simultaneously in feature extraction is a challenging issue.

To date, many kinds of features have been proposed and explored for image copy detection. According to their characteristics, these features are divided into four categories on which various copy detection methods can be

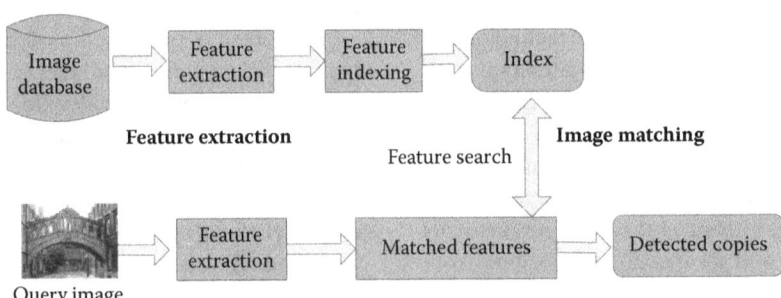

Fig. 2 The flow chart of content-based copy detection

implemented: global features, local features, combination of features, and learning features.

Global Features

In earlier years, researchers mainly focus on extracting global features for image copy detection. Global features are referred to as the features extracted from the whole or nearly whole image region. To the best of our knowledge, Chang et al.'s approach[3] was the first work for image copy detection. In Chang et al.'s approach, the wavelets and image color space are used to generate a global feature to detect image copies. This approach may detect the images after slight copy attacks, such as sharpening, softening, and re-quantization. However, it is not clear how to apply this approach to address big distortions such as histogram equalization, high contrast enhancement, and significant geometric transformations. To improve the robustness, a few global features based on different image division strategies were proposed for copy forgery detection. Kim[1] subdivides images into a number of rectangular blocks and then computes the ordinal measure of discrete cosine transform (DCT) coefficients of those blocks as global features. Kim's method can detect the copies generated by many image manipulations such as histogram equalization, contrast adjustment, color adjustment, and noise addition. Kim's method is also robust to 180° rotation transformation. However, it fails to resist rotation transformations with minor degree. To improve the robustness to rotation transformation, Wu et al.[4] and Zou et al.[5] extract global features by dividing image into elliptical and circular tracks, respectively. More specifically, images are first divided into N elliptical or circular tracks, and then the global features are generated by computing the average intensities of these tracks. As shown in Fig. 3, three image division strategies for feature extraction are illustrated.

To enhance the robustness, other global features, for example, Gabor texture-based descriptor,[6] edge-based signature,[7] and rotation invariant partition-based feature[8] are accordingly proposed. Although, these global features may detect the images after various content-preserved attacks including rescaling, rotation, nosing, and so on, they cannot easily detect the copies generated by content-discard attacks such as cropping and rotation with cropping. The main reason is that content-discard attacks lose some content of an image, and thus, global features extracted from an image, before and after these attacks, are quite different.

Local Features

Recently, more and more local features have been proposed for image copy detection. Instead of extracting one or several global features from each image, a local feature can be extracted from each local image region, and thus, hundreds to thousands of local features may be obtained from an image. These local features are extracted from different regions in an image. Even though the image is distorted by content-discard attacks and loses a part of image content, a number of regions still reserve image content, which can be successfully matched by their corresponding local features. Finally, local features are robust against content-discard attacks.

As one of most famous local features, scaling invariant feature transform (SIFT)[9] with its extension versions, such as principal component analysis on SIFT (PCA-SIFT),[10] speeded up robust feature (SURF),[11] and Affine-SIFT (ASIFT),[12] had been widely applied in various computer vision tasks, such as face recognition,[13,14] image retrieval,[15] and image registration.[16] Compared with global features, local features have relatively high robustness to many common kinds of image transformations and distortions. Generally, all of those features are composed of two components: keypoint and descriptor. To extract those features, hundreds of keypoints are first detected from each image by using various detectors such as Harris–Laplace detector,[17] difference of Gaussian (DOG) detector,[9] and Hessian-affine detector.[18] Then, the characteristic of surrounding region of each keypoint is captured to form a local image descriptor. The local feature-based copy detections generally work as follows. The local features are extracted and matched between images, and then the number of local matches is used to measure the image similarity to determine whether a test image is a copy version of a queried (an original) image. When they are directly used for copy detection such as in the work of Yan and Sukthankar,[10] we cannot achieve desirable performance for the following reasons. First, those features are extracted from relatively small local image regions, which do not contain enough spatial information, resulting in low discriminability. Then, it is difficult to effectively distinguish image copies from irrelevant images, especially similar images. Second, a trade-off is necessary between robustness and efficiency. Their robustness comes from the decrease of efficiency. The number of local features extracted from each image ranges from several hundreds to thousands and they are high-dimensional, and thus directly matching those local features between images is a quite time-consuming process. Figure 4 shows that the SIFT features are matched between the original image and its copy for image copy detection. Each line

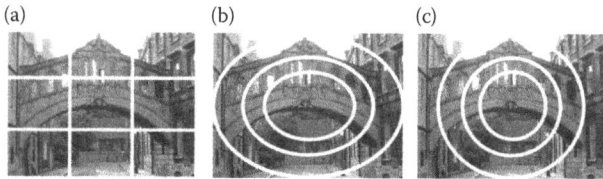

Fig. 3 Various image division strategies for feature extraction: (a) rectangular block division; (b) elliptical track division; and (c) circular track division

represents a local feature match between the two images. In Fig. 4, there are 220 pairs of SIFT feature matches, and then the number of matches, that is, 220, will be used as image similarity to implement copy detection. By setting a threshold of match number, we can determine whether a test image is a copy of the original image.

To improve the discriminability, Xu et al.[19] propose a novel local feature for copy detection. They first use DOG detector to detect image keypoints, and then, instead of using SIFT descriptor, use multi-resolution histogram descriptor[20] to capture the characteristics of relatively large size patches. Ling et al.[21] propose an affine-invariant descriptor based on polar-mapping and discrete Fourier transform (DFT) to describe local image regions. The affine-invariant local regions are detected by Hessian-affine detector,[18] and then polar-mapping is used to transform the regions. Afterward, 2-D DFT is performed on the transformed regions and the low-middle DFT magnitudes are used to form the final local descriptor. Ling et al.[22] propose another local descriptor, that is, multiscale-SIFT descriptor, to improve the discriminability of SIFT feature. The computing local descriptor only uses one support region surrounding each keypoint. The multiscale-SIFT descriptor selects four local patches centered at the same keypoint with four different sizes, and then computes the image gradients histogram for each patch. Afterward, all of these histograms are concatenated to form the final local descriptor.

Compared with the original SIFT feature, the above local features show better discriminability for copy detection. If matching the above features between images directly, it suffers from the computational complexity issue. To reduce the computational complexity, many quantization strategies have been proposed to generate quantized local features. Bag-of-visual words (BOW) quantization is one of the most popular approaches for feature quantization. By clustering a sample set of local features into a number of groups by using K-means algorithm, each group center is regarded as a visual word. For a given local feature, we assign it into its nearest group and use the corresponding visual word ID of the group to represent it. As a result, each high-dimensional local feature is compressed into a visual word, which is one-dimensional feature and thus is benefited to the efficiency of feature matching.

Robust image hashing is another technique, which can be used to quantize the local features. Without clustering process, it directly quantizes the local features into hash sequences. Ling et al.[23] accumulate every four successive elements of a 128-dimension SIFT feature to form 32-D vector V^s. Then, we use the sign of the difference between the value of each element in V^s and the mean value of all elements in the vector to form the final hash sequence. Using those quantized features may enhance computation efficiency. However, quantization errors reduce the discriminability of local features to some extent, and decrease the accuracy for copy forgery detection.

Fig. 4 SIFT features for copy forgery detection by using DOG detector: (a) white color points are the keypoints of an original image; (b) a copy version; (c) the lines represent the matched SIFT features between two images

Combination of Features

Apparently, it is not easy to achieve robustness, discriminability, and efficiency simultaneously by only using one single image feature. Therefore, some researches are dedicated to combining various features for copy detection. Zhou et al.[24] proposed a novel signature by combining both local keypoint and global distribution feature. Keypoints are first detected from a given image by using the Hessian-Affine detector, and then global distribution characteristics of these keypoints are adopted to form the finial signature. This signature is proven to be more discriminative and efficient compared with the local features, and robust against many kinds of noise adding attacks than the global features. Unfortunately, it has low robustness to some significant geometric attacks, such as cropping with a large percentage of image content, scaling with large factor, and shifting with many pixels.

Wang et al.[25] presented a copy detection method by combining both local and global features. Wang et al.'s approach consists of three steps: (i) extracting and matching local features; (ii) using the coordinates of the matched features, and estimating and using affine transformation to filter false matches; and (iii) generating and comparing local binary pattern (LBP) and color histogram of the areas by coordinates of the matched features between two images to further confirm the matching results. Zhou et al.[26] introduced another framework to combine local and global features for copy detection. The SIFT features are matched between images, followed by a global verification step in which the convex region-based context features of matched features are extracted and then compared to remove the false matches. Finally, the number of remaining matches is used for copy forgery detection.

As mentioned previously, the quantized local features allow efficient image matching, but the low discriminability of local features and the quantization errors lead to many false feature matches. Coarse-to-fine matching is a popular technology to reduce these false feature matches. In this technology, the quantized SIFT features are efficiently matched between images to obtain initial feature matching results. Generally, the geometric consistent information among the matched features or the configuration information around the matched features are used to filter false matches.

Jegou et al.[27] proposed the weak geometric consistency (WGC) scheme to verify the consistency of orientation and scale parameters of matched features to filter false matched features. With the additional assumption that correct matches will also be consistent in terms of translation transformation, Zhao et al.[28] proposed an improved WGC scheme by adding the translation consistency constraint. In order to fully capture geometric consistency information, the geometric consistency constraint strategy, that is, random sample consensus (RANSAC),[29] is widely used for filtering the false matches. It randomly samples a subset of matched features to estimate the affine transformations between images, and then it checks the geometric consistency of local matches to filter false matches. Due to the high computational complexity of iteration step for estimating potential affine transformations, RANSAC can only be used to remove false matches for some top-ranked candidate images. Without iteratively estimating the affine transformation, Zhou et al.[30] demonstrated a geometric coding scheme. It encodes the geometric relationships among features into three geo-maps, and then it compares these maps between images to filter geometrically inconsistent matches.

All the above methods are designed for near-duplicate image detection but not for copy forgery detection. Zhou et al.[2] described that many false matches between similar images cannot be effectively filtered by the geometric consistency verification, because there are many false matches that satisfy geometric consistency between similar images. The global context information of the falsely matched features between similar images is different to some extent, on which Zhou et al.[2] designed the global context verification scheme. After efficiently matching SIFT features between images, a global context descriptor, called overlapping region-based global context descriptor, sufficiently captures the global context information of matched features to filter false matches. Compared with geometric consistency verification, the global context verification scheme is more effective for filtering false matches in copy forgery detection.

In summary, compared with using only global features or local features for copy forgery detection, combining both features may achieve better performance. This hybrid a better choice for copy forgery detection.

Learning-based Features

As mentioned by Kim,[1] Zhou et al.,[2] and Hsiao et al.,[31] there are many kinds of copy attacks, such as rescaling, rotation, shifting, cropping, change of intensity and contrast, and noise adding. Although the existing features can accurately detect the images after one or several kinds of copy attacks, it is not easy to deal with all the attacks. To address this issue, Hsiao et al.[31] proposed an extended feature set for detecting image copies. In this framework, a number of virtual prior attacks are applied on copyrighted images to obtain many transformed images, and then features are generated from these transformed images and used as training feature dataset. For classifying copies and non-copies, three classifiers are then learned from the training feature dataset for copy forgery detection. Zhang et al.[32] proposed a novel copy forgery detection system based on convolutional neural networks (CNNs) to learn the features for copy detection task. More specifically, instead of manually extracting image features, the features are learned from the image pixels to determine whether two inputted images have a copy relationship or not. Through the learning process, the method using learning-based features achieves promising performance than the traditional methods in the aspect of robustness.

IMAGE MATCHING

Feature Indexing

If we match local features between each pair of images in an exhaustive manner, it will be very time consuming because of the high dimensionality and huge number of local features. Therefore, many feature indexing methods were proposed to expedite the feature matching process. The typical indexing method, for example, the k-d tree, has been widely used to index high-dimensional features. However, it cannot provide any speedup over the feature matching process if the dimensionality of features is more than 10.[33]

Lowe[9] used a best-bin-first (BBF) tree algorithm[34] to index features to obtain an approximate matching. In the BBF algorithm, it adopts a modified search ordering of the k-d tree algorithm, and the bins in feature space are searched in the order of their closest distance from the queried location. This algorithm may efficiently find the closest neighbor of the query location with high probability.

Locality-sensitive hashing (LSH)[35] is one of the most popular methods for indexing high-dimensional features. Each feature vector is mapped to a set of hash sequences using different hash functions. Then, a few corresponding hash tables are built to index these feature vectors. The LSH generally requires 100–500 bytes to index a descriptor with a large storage space. And, it is difficult to deal with large-scale image datasets, which usually contain billions of local descriptors. Lv et al.[36] proposed an improved indexing scheme, referred to as multi-probe LSH. Lv et al.'s multi-probe LSH outperforms the original LSH algorithm in the aspects of space and time efficiency, but it still suffers from the problem of memory usage.

In order to deal with large image datasets with reasonable memory usage, Sivic and Zisserman[37] proposed the BOW model to index local features. As mentioned in Section "Local Features," after quantizing local features into nearest visual words by the BOW quantization step, each local feature is replaced by its visual word. Thus, an image can be represented by the frequency histogram of visual words. To efficiently match the visual words between images, an inverted index file is established to index those visual words. Each visual word in the index file is followed by the IDs of the images that have the visual word and the corresponding frequencies of the visual word in the images.

Image Similarity Measurement

To measure the similarity between images, many similarity measurement strategies or distance functions were proposed. The Minkowski-like distance functions are used in Refs. [1,4,38] They are usually used for measuring the distance between two image feature vectors denoted as $f = [f(1), f(2), \ldots, f(i), \ldots, f(n)]$ and $f' = [f(1)', f(2)', \ldots, f(i)', \ldots, f(n)']$. The distance is defined as

$$D(f, f') = \left(\sum_{i=1}^{n} (f(i) - f(i)')^r \right)^{1/r} \qquad (1)$$

where r is a tunable parameter. For example, if $r = 1$, the distance is a Hamming (or L1 norm) distance; if $r = 2$, it can be regarded as Euclidean (or L2 norm) distance. The other typical distance functions including histogram intersection,[39] Chi-square statistic,[39] Canberra distance,[40] and Bhattacharyya distance[41] are defined, respectively, as follows:

Histogram intersection:

$$D(f, f') = 1 - \sum_{i=1}^{n} \min(f(i), f(i)'). \qquad (2)$$

Chi-square statistic:

$$D(f, f') = \sum_{i=1}^{n} \frac{(f(i) - f(i)')^2}{f(i) + f(i)'}. \qquad (3)$$

Canberra distance:

$$D(f, f') = \sum_{i=1}^{n} \frac{|f(i) - f(i)'|}{f(i) + f(i)'}. \qquad (4)$$

Bhattacharyya distance:

$$D(f, f') = -\log \sum_{i=1}^{n} \sqrt{f(i) \times f(i)'}. \qquad (5)$$

Beitao et al.[42] demonstrated that the above functions may be not effective enough for copy detection. They proposed the dynamic partial function (DPF) to measure image similarity. First, many features, such as color features, texture features, and shape features, are extracted from each image. Each image is then represented by a p-dimensional feature vector denoted as $(f_1, f_2, \ldots, f_p)$, where each element is an image feature. Instead of measuring image similarity by comparing the same kind of features between images, DPF dynamically selects the minimum values among the distances between many pairs of different features. The ith feature distance between images is defined as $\Delta d_i = |f_i - f_i'|$, where $1 \leq i \leq p$. If we define the set as the following:

$$\Delta_m = \text{the smallest } m \ \Delta d_i\text{'s of } (\Delta d_1, \Delta d_2, \ldots, \Delta d_p) \qquad (6)$$

where Δ_m denotes the set of the first m minimum distance values in the distances set $(\Delta d_1, \Delta d_2, ..., \Delta d_p)$. The PDF function between two images X and Y is given as follows:

$$D(X,Y) = \left(\sum_{\Delta d_i \in \Delta_m} \Delta d_i^r \right)^{1/r} \quad (7)$$

To overcome the "one-size-fits-all" problem in DPF, Qamra et al. proposed an enhanced DPF,[43] which contains thresholding, sampling, and weighting techniques. Experimental results demonstrate that the DPF function can achieve higher accuracy than the other traditional image similarity measurement strategies and distance functions.

When using local features for feature matching, the cardinality of the initial feature matches or the final feature matches after false match filtering is simply used to measure image similarity by $Sim(X,Y) = |S|$, where S is the set of matched features and $|S|$ is the cardinality. However, the cardinality of feature matches is often dependent on image scene complexity, and thus, the number of feature matches may range from a few to several thousands. Consequently, the similarity measurement leads to unstable performance.

To avoid this issue, Zhou et al. proposed an effective and fast similarity measurement strategy.[2] It randomly samples a set of feature matches with a certain number, denoted as *NR*, and then verifies the correctness of these matched features by comparing global context features. The number of the matches judged to be correct is denoted as *nr*. Finally, the ratio of the number of *NR* and *nr* is used to measure the similarity between the two images X and Y by $Sim(X,Y) = nr/NR$. Although only *NR* randomly sampled matches are verified, the similarity measurement strategy can maintain high accuracy level and improve the efficiency. In addition, this strategy can avoid the sensitive problem when using all the matches for similarity measurement, which has been proven by Zhou et al.[2]

Instead of using the number of feature matches, Zhao et al. presented a pattern learning (PL)-based similarity measurement strategy, referred to as pattern entropy (PE).[44] It computes image similarity by measuring the pattern regularity of patterns of feature matches based on the assumption that the patterns of near duplicate are coherently more regular and smooth than those of irrelevant images. Zhao and Ngo[45] proposed an improved PL-based similarity measurement strategy, that is, scale and rotation invariant pattern entropy, to achieve scale and rotation invariance for copy forgery detection.

COMPARISON

The following comparison demonstrates the advantages and disadvantages of each method and may help in understanding how to choose the detection technology for copy forgery detection in digital images. Here, we compare six typical copy detection methods in accuracy and efficiency. We adopt a well-known dataset, that is, Copydays dataset.[46] This dataset is composed of 3,212 digital images, which are generated from 157 original images. Each original image is suffered from three kinds of copy attacks: "JPEG compression," "cropping," and "strong." For each original image, it has nine copies generated by JPEG compression with nine various quality factors, nine cropped versions generated by cutting the image content from 10% to 80%, and two to six copies generated by "strong" attacks. The strong attacks include a variety of significant distortions, such as printing and scanning, blurring, painting, rotating, and their combinations. In addition, we also download 100K images from Internet as distracting images and add these images into the dataset. Finally, we choose 157 original images as query images and search for their copies in the dataset. For a given copy detection method, if its threshold changes, it will obtain many pairs of false positive and false negative rates. By using different threshold, we can plot points representing precision rate and recall rate on a two-dimensional graph to form a precision-recall curve. Then, we can use mean average precision (MAP), which represents the average precision across all different recall levels, to evaluate performance. Different from false positives/false negatives, the MAP may represent the overall performance. Also, the MAP is very popular for most of the recent copy detection methods, such as those described by Zhou et al.[2,26,30] and Zhao and Ngo.[45] Therefore, instead of simply using false positives/false negatives, we adopt the MAP to measure the detection performance of various copy detection methods. All the experiments are conducted on a standard PC (Intel Xeon 3.50G CPU and 32G RAM) with Windows 7 × 64 system.

Six typical copy detection methods compared in this entry include two global feature-based methods, one local feature-based method, two feature combination-based methods, and one learned feature (LF)-based method, which are listed and briefly described next.

Ordinal Measure Method

This method was proposed by Kim et al.[1] As described in Section "Global Features," images are subdivided into 8 × 8 rectangular blocks, and the ordinal measure of DCT coefficients of these blocks is computed as global features. Then, the image similarity is measured by using L1 norm distance between the global features. To speed up the image matching process, we adopt the K-means algorithm to cluster the global features and build inverted index file to index database images at the offline stage.

Elliptical Track Feature (ETF)-Based Method

Wu et al.[4] subdivide images into 16 elliptical tracks and compute the ratio of the mean intensity of each track to the mean intensity of whole image to generate the ETF. Then,

L1 norm distance between the ETFs is used to measure the image similarity. Also, we cluster the features by K-means algorithm and build inverted index file to expedite the image matching process.

BOW-Based Method

This method is based on the BOW model. The SIFT matches are obtained between images by using the BOW model, and then the number of these matches is used as the image similarity for copy detection.

RANSAC-Based Method

After obtaining the SIFT matches based on the BOW model, this method adopts RANSAC algorithm to verify the geometric consistency of SIFT matches to reduce false matches. The number of remaining matches is used as the image similarity for copy detection.

Geometric Coding (GC)-Based Method

Instead of using RANSAC to reduce false matches, this method adopts the geometric coding algorithm[30] for verifying SIFT matches to filter the false matches that are geometrically inconsistent. Then, the number of remaining matches is used as the image similarity.

LF-Based Method

Hsiao et al.[31] propose a new approach called the extended feature set for copy detection. In this method, the global features such as ordinal measure features[1] are extracted from the original images and their copies that are generated by the virtual prior attacks including various geometric transformations and image manipulations. Then, three pattern classification methods are used to learn and train the extracted features to obtain three classifiers for copy detection: the multivariate Gaussian, the Gaussian mixture model, and the support vector machine (SVM).

Table 1 lists the MAP values of all of these six copy detection methods. It is observed that the four methods (BOW-based method, RANSAC-based method, GC-based method, and LF-based method) achieve much higher MAP values than the former two. The reason is that the four use local features while the former two adopt global features, and the local features are more robust than the global features against many geometric transformations, such as cropping, rotation, and rotation with cropping, as mentioned in Section "Feature Extraction."

GC-based method and RANSAC-based method have the higher MAP values than BOW-based method. Because BOW-based method directly uses the number of the initial SIFT matches obtained by BOW model as the image similarity, BOW quantization decreases the detection performance and results in many false matches. Compared with BOW-based method, GC-based method and RANSAC-based method combine the local feature matching with additional verification steps, which can remove most of false matches that are geometrically inconsistent. Thus, GC-based and RANSAC-based methods can achieve a better accuracy performance than BOW-based method. Also, it is clear that LF-based method has the best detection accuracy among all of the six methods. Since the learning process is implemented on the large number of extended features by applying the priori simulated attacks, compared with the other conventional methods, the LF-based method achieves more promising performance for copy detection.

To compare the performance of these six copy detection methods against various copy attacks, we test four image categories from the Copydays dataset. Each category is generated by various copy attacks "JPEG compression," "cropping," and "strong," respectively. All the MAP values are listed in Table 2. Among the four categories, all methods achieve desirable results on the "original image" category, whereas all of them have relatively inferior results on the "strong" category. Intuitively, the "strong" category contains many kinds of significant copy attacks, which makes the copy detection much more challenging. Moreover, for the "JPEG compression" category, it is observed that the MAP values of global feature-based methods (ordinal measure method and ETF-based method) are slightly higher than those of the local feature-based method (BOW-based method) and the two feature combination-based methods (RANSAC-based and GC-based methods). That is mainly because the global features are more robust than the local features to the JPEG compression. On the contrary, for the "cropping" and "strong" categories, the local feature-based and feature combination-based methods perform much better than the global feature-based methods. The reason is that the "cropping" and "strong"

Table 1 MAP values of various copy detection methods

Copy detection methods	Ordinal measure	ETF based	BOW based	RANSAC based	GC based	LF based
MAP values	0.415	0.432	0.882	0.891	0.898	0.930

Table 2 MAP values of various copy detection methods using four image categories

Copy detection methods	Ordinal measure	ETF based	BOW based	RANSAC based	GC based	LF based
Original	0.932	0.927	0.959	0.959	0.959	0.967
JPEG	0.926	0.934	0.890	0.895	0.910	0.951
Cropping	0.212	0.208	0.918	0.924	0.931	0.943
Strong	0.122	0.143	0.786	0.786	0.793	0.812

content-discard attacks significantly change the global features. It can also be observed that LF-based method shows best robustness to these attacks among all of these methods. That is because this method extracts a large number of extended features by the priori simulated attacks and uses the learning process to obtain a copy detector, which makes the copy detection more robust to the possible copy attacks.

To test the efficiency of those methods, we evaluate their time cost of feature extraction and feature indexing (or feature training) at offline stage and their average detection time, which implies the time cost of searching for a given query image in the dataset at online stage. Table 3 lists the time cost in both the offline and online stages. It is observed that the global feature-based methods (ordinal measure method and ETF-based method) require the lesser time cost at both the online and offline stages than the local feature-based method (BOW-based method) and the feature combination-based methods (RANSAC-based and GC-based methods). This is because hundreds of high-dimensional local features could be extracted and matched between images, which require much more time. On the contrary, only single global feature with low dimensionality needs to be computed from each image, and thus, they are efficient to be extracted, indexed, and matched. Thus, the global feature-based methods are generally more efficient for copy detection. Moreover, the feature combination-based methods (RANSAC-based and GC-based methods) require more detection time than the local feature-based method (BOW-based method). The reason is that they adopt an additional verification step, in which the geometric consistency information is used to remove false matches that are geometrically inconsistent. In addition, the LF-based method requires most time at the offline stage, since a large number of extended features are required to be extracted and feature training is also a time-consuming process. Also, this method has relatively high detection time cost at the online stage.

From the above comparison, in summary, the global feature-based methods have the advantages of the better robustness to content-preserved attacks and the higher detection efficiency, while the local feature-based methods have the advantage of the better robustness to most content-discard attacks but with lower detection efficiency. The feature combination-based methods achieve the higher detection accuracy than the global feature-based and local feature-based methods at the expense of efficiency. The LF-based method is most robust to various copy attacks but with relatively low detection efficiency.

CONCLUDING REMARKS: RELATED CHALLENGES AND POSSIBLE SOLUTIONS

As illustrated in Section "Introduction," for a given image, the similar images may be more similar to the original one than the copy versions. As a result, it is difficult to effectively identify the copies from the image database, where there are a lot of similar images stored together with image copies. Although some copy detection methods have been proposed to address this issue, it is still not easy to identify image copies from the similar images. The reason is that the similar image and the copy version are quite similar to each other. Therefore, the big challenge is how to efficiently and correctly distinguish between similar images and image copies. For example, it is an intractable task to detect the similar images generated from the same scene or object by little different capturing conditions, such as viewpoint change with less than 10°.

There are some improvements on copy forgery detection, for example, capturing more detailed information, such as the properties of local features including orientation, scaling, and positions, to describe sufficiently the difference between similar images and image copies. Also, extracting global or local features from many samples of similar images and copies, and training the classifiers to discriminate similar images from image copies by using various machine learning techniques, such as SVM, extreme learning machine, and CNN, may be used to enhance the detection accuracy.

Another concern about copy detection is the so-called partial-duplicate image. As we know, partial-duplicate images are the images, and part of which are derived from the same original image by various geometric transformations including modifications of color, intensity and contrast, noising adding, and so on.[30] Generally, the original image shares a part of same image content with the partial-duplicate image. Compared with the image copies, the duplicate regions are much smaller, and thus,

Table 3 Time cost of various copy detection methods

		Ordinal measure	ETF based	BOW based	RANSAC based	GC based	LF based
Offline stage	Feature extraction	2.89×10^2	2.24×10^2	3.53×10^2	3.53×10^2	3.53×10^2	4.05×10^3
	Feature indexing (feature training)	3.92	2.35	2.96×10^2	2.96×10^2	2.96×10^2	1.93×10^3
Online stage	Average detection time (seconds)	0.10	0.08	0.14	6.70	0.28	1.35

there are only a small number of feature matches mixed with a lot of false matches between the images. Thus, it is difficult to effectively detect partial duplicates. Some partial-duplicate image detection methods have been proposed. Most of them use geometric consistency constraint or context verification to remove false matches to enhance detection accuracy. However, to ensure the robustness of geometric information constraints, the geometric information among features usually is extracted at online stage, and context information is not stable enough. To achieve high detection accuracy and meanwhile to have good robustness, we may extract more stable context information, such as the visually context information of local features, and then embed it into the index for filtering false matches.

In addition, since the recent copy detection methods are designed based on local features, the performance of those methods relies on the robustness of local features. However, the extraction of local features and feature indexing are relatively time-consuming and tedious, which degrade the detection efficiency at offline stage. Therefore, another challenge of copy forgery detection is to study more efficient feature extraction and feature indexing algorithms. Actually, the approximate computation of feature extraction and the multi-hierarchical feature indexing may be the possible solutions to improve the efficiency at offline stage.

ACKNOWLEDGMENTS

This work is supported by the National Key R&D Program of China under Grant 2018YFB1003205; in part by the National Natural Science Foundation of China under Grant 61602253, Grant U1536206, Grant U1405254, Grant 61772283, Grant 61672294, Grant61572258, Grant 61502242; in part by the Jiangsu Basic Research Programs-Natural Science Foundation under Grant BK20150925 and BK20151530; and in part by the Postgraduate Research & Practice Innovation Program of Jiangsu Province under Grant KYCX18_1016.

REFERENCES

1. Kim, C. Content-based image copy detection. Signal Process. Image Commun. **2003**, *18* (3), 169–184.
2. Zhou, Z.L.; Wang, Y.L.; Wu, Q.M.J.; Yang, C.N.; Sun, X.M. Effective and efficient global context verification for image copy detection. IEEE Trans. Inf. Forensics Secur. **2017**, *12* (1), 48–63.
3. Chang, E.Y.; Wang, J.Z.; Li, C.; Wiederhold, G. RIME: A replicated image detector for the world-wide web. Proc. Int. Soc. Opt. Eng. **1998**, *3527*, 58–67.
4. Wu, M.-N.; Lin, C.-C.; Chang, C.-C. Novel image copy detection with rotating tolerance. J. Syst. Softw. **2007**, *80* (7), 1057–1069.
5. Zou, F.; Li, X.; Xu, Z.; Ling, H.; Li, P. Image copy detection with rotation and scaling tolerance. Comput. Res. Dev. **2009**, *46* (8), 1349–1356.
6. Li, Z.; Liu, G.; Jiang, H.; Qian, X. Image copy detection using a robust Gabor texture descriptor. In *Proceedings of the First ACM Workshop on Large-Scale Multimedia Retrieval and Mining*, Beijing, 2009, 65–71.
7. Lin, C.-C.; Wang, S.-S. An edge-based image copy detection scheme. Fundamenta Informaticae **2008**, *83* (3), 299–318.
8. Zhou, Z.L.; Yang, C.N.; Chen, B.J.; Sun, X.M.; Liu, Q.; Wu, Q.M.J. Effective and efficient image copy detection with resistance to arbitrary rotation. IEICE Trans. Inf. Syst. **2016**, *E99D* (6), 1531–1540.
9. Lowe, D.G. Distinctive image features from scale-invariant keypoints. Proc. Int. J. Comput. Vision **2004**, *60* (2), 91–110.
10. Yan, K.; Sukthankar, R. PCA-SIFT: A more distinctive representation for local image descriptors. In *Proceedings of IEEE Conference Computer Vision Pattern Recognition*, Washington, DC, 2004, 506–513.
11. Bay, H.; Ess, A.; Tuytelaars, T.; Van Gool, L. Speeded-up robust features (SURF). Comput. Vision Image Underst. **2008**, *110* (3), 346–59.
12. Morel, J.M.; Yu, G.S. ASIFT: A new framework for fully affine invariant image comparison. SIAM J. Imag. Sci. **2009**, *2* (2), 438–469.
13. Geng, C.; Jiang, X.D. Face recognition using sift features. In *Proceedings of IEEE Conference Image Processing*, Cairo, 2009, 3313–3316.
14. Berretti, S.; Ben Amor, B.; Daoudi, M.; Del Bimbo, A. 3D facial expression recognition using SIFT descriptors of automatically detected keypoints. Visual Comput. **2011**, *27* (11), 1021–1036.
15. Zhang, S.; Tian, Q.; Lu, K.; Huang, Q.; Gao, W. Edge-SIFT: Discriminative binary descriptor for scalable partial-duplicate mobile search. IEEE Trans. Image Process. **2013**, *22* (7), 2889–2902.
16. Goncalves, H.; Corte-Real, L.; Goncalves, J.A. Automatic image registration through image segmentation and SIFT. IEEE Trans. Geosci. Remote Sens. **2011**, *49* (7), 2589–2600.
17. Mikolajczyk, K.; Schmid, C. Indexing based on scale invariant interest points. In *Proceedings of the 8th International Conference on Computer Vision*, Vancouver, BC, 2001, 525–531.
18. Mikolajczyk, K.; Schmid, C. Scale affine invariant interest point detectors. Int. J. Comput. Vision **2004**, *60* (1), 63–86.
19. Xu, Z.; Ling, H.F.; Zou, F.H.; Lu, Z.D.; Li, P. A novel image copy detection scheme based on the local multi-resolution histogram descriptor. Multimedia Tools Appl. **2011**, *52* (2–3), 445–463.
20. Hadjidemetriou, E.; Grossberg, M.D.; Nayar, S.K. Multi-resolution histograms and their use for recognition. IEEE Trans. Pattern Anal. Mach. Intell. **2004**, *26* (7), 831–847.
21. Ling, H.F.; Wang, L.; Zou, F.H.; Yan, W. Fine-search for image copy detection based on local affine-invariant descriptor and spatial dependent matching. Multimedia Tools Appl. **2013**, *52* (2–3), 551–568.
22. Ling, H.F.; Cheng, H.; Ma, Q.; Zou, F.H.; Yan, W. Efficient image copy detection using multiscale fingerprints. IEEE Multimedia **2012**, *19* (1), 60–69.

23. Ling, H.F.; Yan, L.Y.; Zou, F.H.; Liu, C.; Feng, H. Fast image copy detection approach based on local fingerprint defined visual words. Sig. Proc. **2013**, *93* (8), 2328–2338.
24. Zhou, Z.L.; Sun, X.M.; Chen, X.Y.; Chang, C.; Fu, Z.J. A novel signature based on the combination of global and local signatures for image copy detection. Secur. Commun. Netw. **2014**, *7* (11), 1702–1711.
25. Wang, Y.; Hou, Z.; Leman, K.; Pham, N.T.; Chua, T.; Chang, R. Combination of local and global features for near-duplicate detection. In *Proceedings of International Conference on Multimedia Modelling*, Taipei, 2011, 328–338.
26. Zhou, Z.L.; Sun, X.M.; Wang, Y.L.; Fu, Z.J.; Shi, Y.Q. Combination of SIFT feature and convex region-based global context feature for image copy detection. In *Proceedings of International Workshop Digital-Forensics Watermarking*, Taipei, 2014, 60–71.
27. Jegou, H.; Douze, M.; Schmid, C. Hamming embedding and weak geometric consistency for large scale image search. In *Proceedings of European Conference on Computer Vision*, Marseille, 2008, 304–317.
28. Zhao, W.L.; Wu, X.; Ngo, C.W. On the annotation of web videos by efficient near-duplicate search. IEEE Trans. Multimedia **2010**, *12* (5), 448–461.
29. Fischler, M.A.; Bolles, R.C. Random sample consensus: A paradigm for model fitting with applications to image analysis and automated cartography. Commun. ACM **1981**, *24* (6), 381–95.
30. Zhou, W.G.; Li, H.; Lu, Y.; Tian, Q. SIFT match verification by geometric coding for large-scale partial-duplicate web image search. ACM Trans. Multimedia Comput. Commun. Appl. **2013**, *9* (1), 1–18.
31. Hsiao, J.H.; Chen, C.S.; Chien, L.F.; Chen, M.S. A new approach to image copy detection based on extended feature sets. IEEE Trans. Image Process. **2007**, *16* (8), 2069–2079.
32. Zhang, J.; Zhu, W.; Li, B.; Hu, W.; Yang, J. Image copy detection based on convolutional neural networks. In *Proceedings of Chinese Conference on Computer Vision*, Chengdu, 2016, 111–121.
33. Friedman, J.H.; Bentley, J.L.; Finkel, R.A. An algorithm for finding best matches in logarithmic expected time. ACM Trans. Math. Softw. **1977**, *3* (3), 209–226.
34. Beis, J.S.; Lowe, D.G. Shape indexing using approximate nearest-neighbour search in high-dimensional spaces. In *Proceedings of IEEE Conference on Computer Vision and Pattern Recognition*, San Juan, PR, 1997, 1000–1006.
35. Indyk, P.; Motwani, R. Approximate nearest neighbors: Towards removing the curse of dimensionality. In *Proceedings of ACM Symposium on Theory of Computing*, Dallas, TX, 1998, 604–613.
36. Lv, Q.; Josephson, W.; Wang, Z.; Charikar, M.; Li, K. Multi-probe LSH: Efficient indexing for high-dimensional similarity search. In *Proceedings of International Conference on Very Large Data Bases*, Vienna, 2007, 950–961.
37. Sivic, J.; Zisserman, A. Video Google: A text retrieval approach to object matching in videos. In *Proceedings of IEEE International Conference on Computer Vision*, Nice, 2003, 1470–1477.
38. Wu, M.-N.; Lin, C.-C.; Chang, C.-C. A robust content-based copy detection scheme. Fundamenta Informaticae **2006**, *71* (2–3), 351–366.
39. Ahonen, T.; Hadid, A.; Pietikainen, M. Face recognition with local binary patterns. In *Proceedings of European Conference on Computer Vision*, Prague, 2004, 469–481.
40. Androutsos, D.; Plataniotiss, K.N.; Venetsanopoulos, A.N. Distance measures for color image retrieval. In *Proceedings of International Conference on Image Processing*, Chicago, IL, 1998, 770–774.
41. Sung-Hyuk, C.; Srihari, S.N. On measuring the distance between histograms. Pattern Recogn. **2002**, *35* (6), 1355–1370.
42. Beitao, L.; Chang, E.; Yi, W. Discovery of a perceptual distance function for measuring image similarity. Multimedia Syst. **2003**, *8* (6), 512–522.
43. Qamra, A.; Meng, Y.; Chang, E.Y. Enhanced perceptual distance functions and indexing for image replica recognition. IEEE Trans. Pattern Anal. Mach. Intell. **2005**, *27* (3), 379–391.
44. Ngo, C.W.; Zhao, W.L.; Jiang, Y.G. Fast tracking of near-duplicate keyframes in broadcast domain with transitivity propagation. In *Proceedings of ACM International Conference Multimedia*, Santa Barbara, CA, 2006, 845–854.
45. Zhao, W.L.; Ngo, C.W. Scale-rotation invariant pattern entropy for keypoint-based near-duplicate detection. IEEE Trans. Image Process. **2009**, *18* (2), 412–423.
46. Copydays dataset. Available at http://lear.inrialpes.tr/~jegou/data.php.

Fourier Analysis

Tania Pouli
Technicolor, Rennes, France

Erik Reinhard
Technicolor Research and Innovation, Rennes, France

Douglas W. Cunningham
Brandenburg University of Technology, Cottbus, Germany

Abstract
In a given image, the probability that neighboring pixels have similar values tends to be high. This phenomenon is described by the autocorrelation function, which gives the correlation between pixels at different points in image space. But the autocorrelation function tends to be computationally inefficient in the image domain. Instead, it is customary to compute the power spectrum in the frequency domain, and the use the Fourier transform to bring the results back into the image domain. This approach tends to be more computationally efficient, and takes advantage of the mathematical properties in both domains.

This entry introduces the relevant concepts for this type of analysis and provides useful examples.

Keywords: Autocorrelation function; Fourier analysis; Fourier transform; Frequency domain; Power specturm.

The interaction of light rays with an object systematically alters light according to the surface properties of the object. This means that the pattern of change across a set of light rays provides direct, reliable information about the object. This can also be seen by permuting the pixels of an image (Fig. 1). Although the exact same pixels are available in both images, the permutation has resulted in the loss of meaningful information. Thus, a statistical assessment of an image based on individual pixels alone will give only very partial insight into the structure of an image.

Of course, the structure in images is of central interest to the human visual system. It should not be surprising, then, that considerable attention has been directed toward investigating second- and even higher-order statistics. It has been shown, for example, that people can discriminate almost without effort between images that differ in their second-order statistics.[1,2] Second-order statistics are those that typically involve covariances or correlations between pairs of pixels.

In Chapter 5, we showed that looking at the difference between pairs of neighboring pixels provided a large amount of information about structure. Gradient analysis is, however, very limited in that it requires the two pixels to be neighbors. In this entry we address ways to analyze relationships between pixels that are not necessarily adjacent in images. In a given image, the probability that neighboring pixels have similar values tends to be high, whereas pixels some distance away have less ability to predict a pixel's color. This phenomenon is described by the autocorrelation function, which gives the correlation between pixels at different points in image space. It is explained in more detail in the next section.

Although the autocorrelation function could provide useful insights into image structure, it tends to be computationally inefficient to compute in image space. An equivalent analysis can be performed in frequency space by examining the power spectrum. This requires the use of Fourier transforms, which are introduced in a later section. The equivalence of the autocorrelation function and the power spectrum is known as the Wiener-Khintchine theorem, which states that these form a Fourier transform pair. This is explained in a later section.

The power spectrum encodes how much of each spatial frequency is available in the image. Spatial frequencies are measures of the number of oscillations (cycles) from dark to light per unit of length of the image (Fig. 2). In vision research it is common to express this as the number of cycles per degree (cpd)—effectively a measure of oscillations per unit of visual angle (α in Fig. 3). The two ways to express spatial frequencies convey the same information, although the latter assumes that the distance between the observer and the display is known. The importance of power spectral analysis is discussed in "Power Spectra" section.

Finally, Fourier decompositions return complex valued elements. The magnitude of the complex numbers gives

Fourier Analysis

Fig. 1 The pixels of the image on the left were permuted to create the image on the right, in the process destroying higher-order structure that prevents the image to be meaningfully interpreted. (Westonbirt Arboretum, UK, 2011)

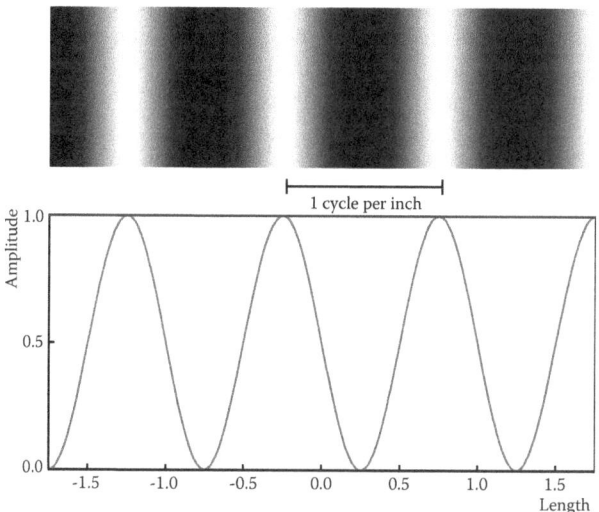

Fig. 2 A sine function encodes a single frequency, in this case at one cycle per inch

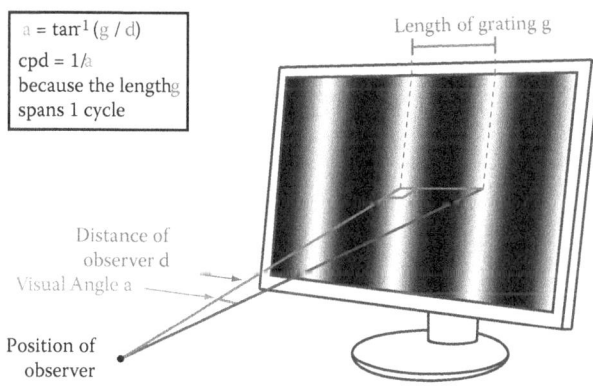

Fig. 3 The angular frequency (in cycles per degree, cpd) depends on the size g at which a sine wave is displayed as well as the distance d between the observer and the display

information of the frequency content, whereas the angle provides information regarding the position within the image, also known as *phase*. There have been attempts to discover statistical regularities in the phase content of images, but although most information that characterizes an image is known to be encoded into the phase spectrum, analyses of the phase structure of images have not been as overwhelmingly successful as frequency-domain analysis. Fourier-based phase structure is discussed further in "Phase Spectra" section.

AUTOCORRELATION

The autocorrelation function is a special case of cross-correlation. Cross-correlation is a measure of how similar an image is to a mask. Assume we have an image I and a mask M, with the mask having a size of $m_x \times m_y$ pixels, which is smaller than the size of the image ($n_x \times n_y$). Image and mask will have means μ_I and μ_M and standard deviations σ_I and σ_M, respectively.

Each pixel in the image now forms the center of a region the size of the mask. The similarity of the region with the mask can then be computed using the normalized cross-correlation $r(x, y)$:

$$r(x,y) = \sum_{x_s=-m_x/2}^{m_x/2-1} \sum_{y_s=-m_y/2}^{m_y/2-1} \frac{I(x+x_s, y+y_s) M(x_s + m_x/2, y_s + m_y/2)}{\sigma_I \sigma_M} \quad (1)$$

The normalized cross-correlation $r(x, y)$ produces values between −1 and 1. It will be close to 1 if the region around (x, y) closely resembles the pattern encoded by the mask. Such a measure is thus useful to detect specific patterns in images. An example is shown in Fig. 4. Here, the spikes show where the hexagonal template was matched best against the input image.

When such a cross-correlation is computed between the image and a displaced version of the same image, we measure autocorrelation as a function of displacement. If for a given displacement a high autocorrelation is found, then this could be interpreted as an indication that the image exhibits periodicity with a frequency commensurate with this displacement. The autocorrelation function r_a as a function of displacement (u, v) is given by:

Fig. 4 Cross-correlation between an image and a template yields values between −1 and 1 for each pixel. (Eden Project, Cornwall, UK, 2011)

$$r_a(u,v) = \sum_{x_s=0}^{n_x-1} \sum_{y_s=0}^{n_y-1} \frac{I(u+x_s, v+y_s) I(x_s, y_s)}{\sigma_I^2} \quad (2)$$

This function is essentially the cross-correlation of an image with itself, and is therefore a direct consequence of the formulation of the cross-correlation in Eq. 1. An example is shown in Fig. 5, where the autocorrelation was computed for each of the red, green, and blue channels separately.

The autocorrelation for the green channel is shown as a surface in Fig. 6. The zero displacement is shown in the middle of the surface as a large spike. For natural images, larger displacements tend to lead to smaller autocorrelation values. Note also that this image has produced some direction sensitivity, with the autocorrelation remaining at higher values in the vertical direction compared with similar horizontal displacements.

The computation of the autocorrelation function as in Eq. 2 is expensive to compute. In effect it has a time complexity of $O(n_x^2 \, n_y^2)$, which is not practical for decent-sized images. However, it can be shown that the autocorrelation function is equivalent to the power spectrum, which is computed in Fourier space. In fact, when applied to natural image ensembles, the power spectrum shows

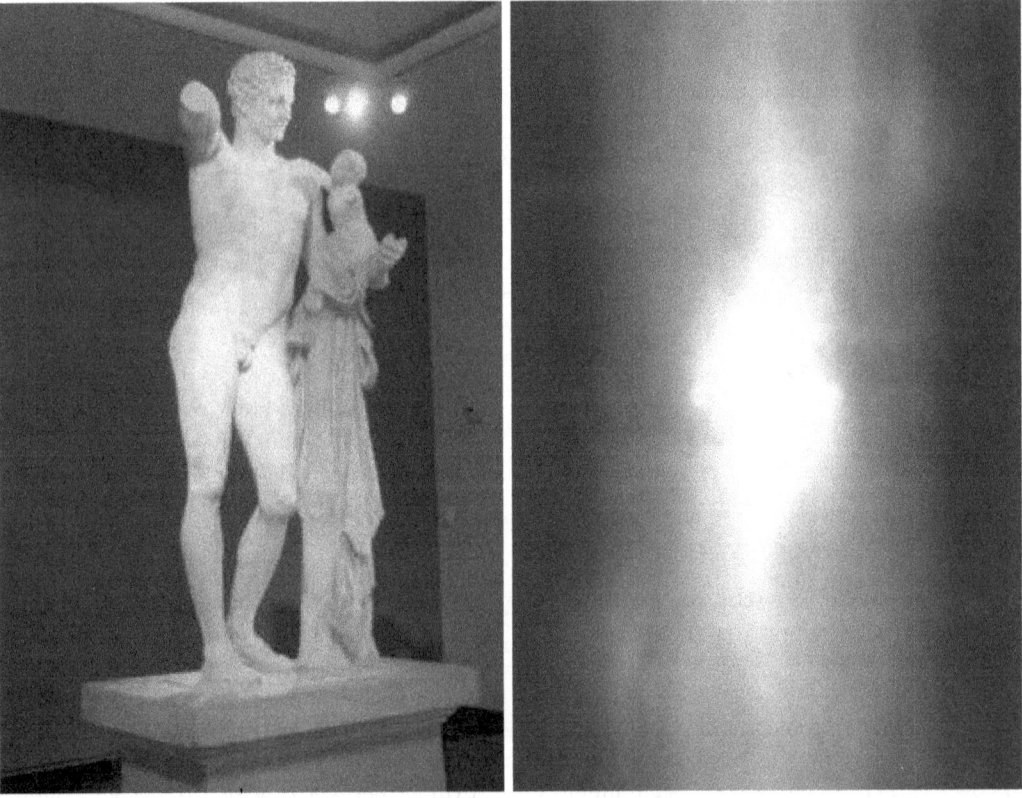

Fig. 5 The autocorrelation for the left image is shown on the right. The autocorrelation was computed for each of the red, green, and blue channels separately, leading to slight color variations in this image. (Olympia, Greece, 2010)

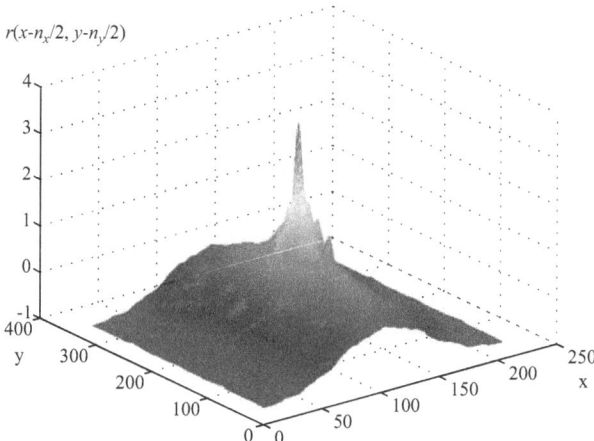

Fig. 6 The autocorrelation function for the green channel of the left image of Fig. 5 is shown here. The spike in the middle of the plot coincides with a zero displacement

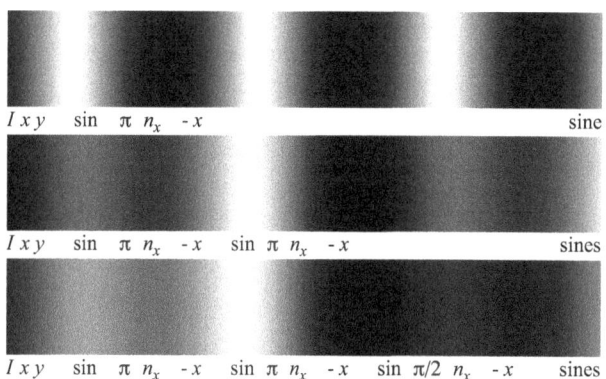

Fig. 7 Sines with commensurable periods are summed, leading to a new signal which is also periodic

striking statistical regularities. To discuss these findings, we first briefly explain Fourier transforms ("The Fourier Transform" section) and show that the autocorrelation and power spectrum constitute a Fourier transform pair ("The Wiener-Khintchine Theorem" section).

THE FOURIER TRANSFORM

The Fourier transform is exceptionally important in digital image processing, computer vision, and other fields, but it is also important in the statistical analysis of natural images. We therefore briefly review its construction and its properties. We begin by noting that sinusoids—sines and cosines—are periodic functions for all $k \in \mathbb{Z}$:

$$\sin(x + 2\pi k) = \sin(x) \tag{3}$$

$$\cos(x + 2\pi k) = \cos(x) \tag{4}$$

A sinusoidal function $g(x) = A \sin(2\pi f x + \theta)$ is periodic with frequency f, amplitude A, and phase θ. The period T_0 of this function is $1/f$. We can add sinusoids with different periods and if these periods are integer multiples of a base period (i.e., they are commensurable), then the resulting sum is also periodic. Figure 7 shows some examples.

We can add a large number of sinusoids together using this same progression, i.e., $g(x) = \sum_n A \sin(2\pi f x/2^n + \theta)$, as shown in Fig. 8. As each frequency is a harmonic, i.e., an integer multiple of a base frequency, a step edge is beginning to form, although there is still some undershoot and overshoot of the signal near the edge (known as the Gibbs phenomenon). In general, arbitrary periodic waveforms can be built by summing harmonically related sines and cosines.

Conversely, given an arbitrary periodic waveform, it is possible to find its trigonometric series representation, known as the trigonometric Fourier series of a signal. A trigonometric Fourier series is generally written as:

$$g(x) = a_0 + \sum_{n=1}^{\infty} a_n \cos(2\pi f n x) + \sum_{n=1}^{\infty} b_n \sin(2\pi f n x) \tag{5}$$

with $-\infty < x < \infty$. Thus, a specific signal $g(x)$ can be represented by the coefficients a_0, a_n, and b_n, which can be computed as follows:[3]

$$a_0 = f \int_{1/f} g(x) dx \tag{6}$$

$$a_0 = 2f \int_{1/f} g(x) \cos(2\pi f n x) dx \quad n \neq 0 \tag{7}$$

$$b_n = 2f \int_{1/f} g(x) \sin(2\pi f n x) dx \tag{8}$$

A Fourier series will converge to $g(x)$ if this function has continuous first and second derivatives over a period $T_0 = 1/f$, except possibly at a finite number of points where it may have finite jump discontinuities. A Fourier series is defined over a period $1/f$. Outside this interval, the Fourier series converges to a periodic extension of $g(x)$.

Fig. 8 Summing a large number of sinusoids (500 in this case) approximates a step edge

With the exception of certain special cases, the trigonometric Fourier series may alternatively be expressed in terms of cosines alone by noting that:

$$a_n \cos(2\pi f n x) + b_n \sin(2\pi f n x) = A_n \cos(2\pi f n x + \theta_n) \quad (9)$$

As a result, the trigonometric Fourier series is then:

$$g(x) = A_0 + \sum_{n=1}^{\infty} A_n \cos(2\pi f n x + \theta_n) \quad (10)$$

The Fourier coefficients are now given by the amplitudes A_n and phases θ_n, which go with frequencies fn. The coefficients A_n plotted against frequencies fn (where $n = 0, 1, 2, \ldots, \infty$) form the amplitude spectrum of the signal, whereas the coefficients θ_n plotted against frequencies fn (where $n = 1, 2, 3, \ldots, \infty$) give the phase spectrum.

A third form of the Fourier series is exponential in nature. We note that using Euler's formula, the cosine in (10) can be written as:

$$\cos(2\pi f n x + \theta_n) = \frac{1}{2} e^{i(2\pi f n x + \theta_n)} + \frac{1}{2} e^{-i(2\pi f n x + \theta_n)} \quad (11)$$

Thus, it is possible to represent a signal $g(x)$ as a sum of complex exponential terms:

$$g(x) = \sum_{n=-\infty}^{\infty} X_n e^{2\pi i f n x} \quad (12)$$

The complex coefficients X_n can be computed as follows:

$$X_n = \frac{1}{T_0} \int_{T_0} g(x) e^{-2\pi i f n x} dx \quad (13)$$

which defines the exponential Fourier series. Note that these coefficients directly relate to those of the trigonometric Fourier series, as follows:

$$A_n = 2|X_n| \quad (14)$$

$$\theta_n = \tan^{-1}\left(\frac{\operatorname{Im} X_n}{\operatorname{Re} X_n}\right) \quad (15)$$

This makes it convenient to calculate the exponential Fourier series and then derive amplitude and phase spectra according to these equations.

So far we have considered only one-dimensional signals $g(x)$. For images, the extention to two dimensions is as follows. The Fourier transform $F(u, v)$ as function of an image $I(x, y)$ is defined as:

$$F(u,v) = \sum_x \sum_y I(x,y) \exp(-2\pi i (xu + yv)) \quad (16)$$

Thus, an image can be transformed into an equivalent representation which is termed Fourier space or frequency space. The reasons why such a conversion represents frequencies will become apparent later in this section.

The inverse Fourier transform takes the frequency representation back to image space:

$$I(x,y) = \sum_u \sum_v F(u,v) \exp(2\pi i (xu + yv)) \quad (17)$$

With this mechanism, it is possible to convert an image to Fourier space, carry out calculations, and then transform the result back to image space. The reason that one might want to go to Fourier space is that some calculations become much more computationally efficient. Noting that for an image with n pixels the time complexity of the forward and backward transforms is $O(n \log n)$ for efficient Fast Fourier Transforms (FFTs), any calculation that costs $O(n \log n)$ in Fourier space, but more in image space, becomes amenable to treatment in this space.

An example is convolution, where in image space a sliding window is moved across the image. For every pixel, a set of neigboring pixels is weighted and summed according to a specified filter kernel δ:

$$I^{\text{conv}}(x,y) = \sum_i \sum_j I(i,j) \delta(x-i, y-j) = I \star \delta \quad (18)$$

where the "$\star$" is the convolution operator. Convolutions are very useful, for instance to blur an image. In this case, the filter kernel is often chosen to be Gaussian. Although the filter kernel could be capped, summing over only a small pixel neighborhood around each pixel, this would make the convolution approximate. In the formulation above, each pixel contributes to every other pixel, making the time complexity of this operation $O(n^2)$.

The convolution theorem states that a convolution could be equivalently expressed in the Fourier domain as a multiplication of the Fourier transform of the image and the Fourier transform of the filter kernel. Denoting the Fourier transform as $F[]$, the convolution theorem is given by:

$$F[I \star \delta] = c\, F[I] F[\delta] \quad (19)$$

where c is a normalizing constant. We can therefore transform both image and kernel, incurring a cost of $O(n \log n)$, followed by a multiplication that costs $O(n)$. Including the inverse Fourier transform results in a total cost of $O(n \log n)$, which is especially for large images significantly cheaper than $O(n^2)$.

At this point, it is interesting to note the mathematical similarity between convolution and cross-correlation. In 1D, the correlation between two functions f and g is given by:

$$r(x) = \int_{-\infty}^{\infty} f(\theta) g(x+\theta) d\theta \quad (20)$$

This is the one-dimensional case of the cross-correlation function given earlier in (1). Convolution in 1D is defined as:

$$(f \star g)(x) = \int_{-\infty}^{\infty} f(\theta) g(x-\theta) d\theta \quad (21)$$

Despite mathematical similarities, their interpretation and use is different. In signal processing for instance, convolutions are used to compute the output of a linear system given a specific input. Their Fourier space equivalent is a multiplication.

Correlation is used to compare the similarity of two signals, reaching its maximum for the displacement at which the two signals best match. As mentioned in the previous section, the autocorrelation function is a special case of cross-correlation, whereby a signal is compared to itself for different displacements. This function has the same time complexity as convolution and is therefore also expensive to compute in image space. However, it turns out that the autocorrelation function can also be expressed in Fourier space, an important finding discussed next.

THE WIENER-KHINTCHINE THEOREM

The reason to discuss the Fourier transform in some detail is that there is a direct link between computations in Fourier space and those in image space. While this is generally the case, here we refer specifically to the relation between the autocorrelation function ("Autocorrelation" section) and power spectra.[4] In other words, Fourier space gives us computationally convenient access to the Fourier equivalent of the autocorrelation function. Recall that the autocorrelation function $r_a(u, v)$ of an image $I(x, y)$ is the cross-correlation between the image and a displaced version of the same image as function of the displacement (u, v).

The Wiener-Khintchine theorem[5,6] states that the autocorrelation function r_a can be expressed as:

$$r_a = F^{-1}\left[|F|^2\right] \tag{22}$$

where F is the Fourier transform of the image, and F^{-1} denotes the inverse Fourier transform, as defined in Eq. 17. Given Eq. 14, we see that the amplitudes squared are the Fourier equivalent of the autocorrelation function. It is therefore possible to gain information about the autocorrelation function by analyzing the square of the amplitude spectrum, which is also known as the power spectrum. As discussed in the following section, natural images exhibit striking statistical regularities in their power spectra.

POWER SPECTRA

Images contain information at different spatial scales, from the large (e.g., the mountains in the distance of Fig. 9) to the very small (e.g., the grass on the left as well as the fine texture on the mountains due to vegetation). As is well known from Fourier analysis, sharp edges, such as at the silhouette of the mountains against the sky, can be described by a weighted sum of sinusoids, with higher frequencies being weighted less (see "The Fourier Transform" section and in particular Fig. 8). An examination of the relative power of the different spatial frequencies reveals several interesting trends, which are so prominent that many works in image statistics provide an analysis of the power spectrum.

Slope Computation

To compute the spectral slope of an ensemble of images, the following procedure can be used.[7] We begin by assuming that images are square. Then, the weighted mean intensity μ is subtracted to avoid leakage from the DC-component of the image, with μ defined as:

$$\mu = \frac{\sum_{(x,y)} L(x,y)w(x,y)}{\sum_{(x,y)} w(x,y)} \tag{23}$$

Here, $w(x, y)$ is a weight factor that is defined below.

Next, the images should be prefiltered to avoid boundary effects that are related to the Fourier transform. As the Fourier transform requires the signal to be periodic to give a correct answer, it should be used with care when applied to images. To see this, consider a cosine grating as shown in the top left of Fig. 10. This grating is aligned with the image and contains an exact number of periods between the left and right edges. The amplitude spectrum in Fourier space will show exactly two spikes, left and right of the center of the image (we have shifted this amplitude spectrum such that the zero frequency is in the middle of the image). These spikes represent the frequency at which this cosine occurs. Edge artifacts are carefully avoided in this image by making the signal truly periodic.

However, this can not be maintained in general, and even rotating the cosine by 45 degrees breaks the assumption of the image being periodic. As the left and right edges of the image no longer line up (the same is true for top and bottom images), the amplitude spectrum is no longer a pair of spikes. In fact, it is a pair of spikes augmented with various other structures that exist because opposite edges of the image are not lining up.

Fig. 9 A typical natural scene, containing details at various scales. (Milford Sound, New Zealand, 2012)

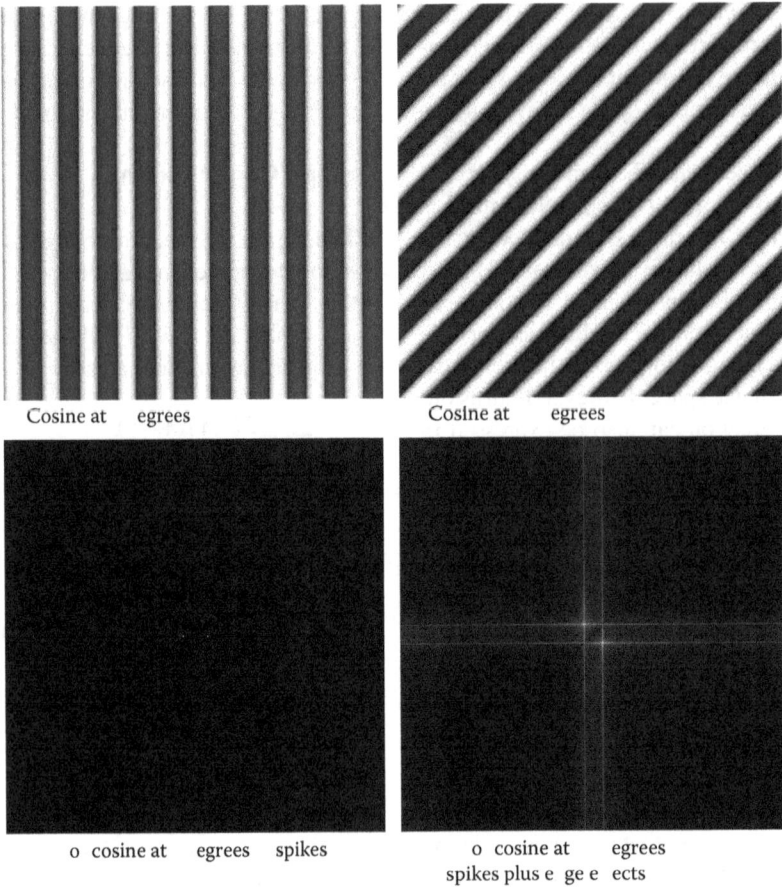

Fig. 10 The top row shows two oriented cosine gratings. Underneath are shown their amplitude spectra

Fig. 11 After windowing with a Kaiser-Bessel window (left), the amplitude spectrum shows much fewer edge artifacts (right)

In general, images are not periodic and this means that measures ought to be taken to minimize edge artifacts when analysing ensembles in Fourier space. This is normally accomplished by applying a window to the image first, which tapers pixel values to a constant value near the edges. This makes edges on opposite sides of the image have the same value and therefore simulates a periodic signal. The center of the image is largely unaffected to ensure that meaningful statistics can be computed. For the cosine example, the result of windowing is shown in Fig. 11.

There are many choices of window,[8] each with some advantages and disadvantages. A good trade-off between computability, side-lobe level, and main-lobe width is afforded by the circular Kaiser-Bessel window with parameter $\alpha = 2$.[8] It is computed as follows:

$$w(x, y) = \frac{I_0\left(\pi\alpha\sqrt{1.0 - \left(\frac{x^2+y^2}{(N/2)^2}\right)}\right)}{I_0(\pi\alpha)}: \quad 0 \leq \sqrt{x^2 + y^2} \leq \frac{N}{2} \quad (24)$$

Here, N is the window size. This weight function is then normalized by letting:

$$\sum_{(x,y)} w(x,y)^2 = 1 \quad (25)$$

Further, I_0 is the modified zero-order Bessel function of the first kind, computed as:

$$I_0(x) = \sum_{m=0}^{\infty} \frac{1}{m!\Gamma(m+1)}\left(\frac{x}{2}\right)^{2m} \quad (26)$$

and Γ is the gamma function, which is an extension of the factorial function:

$$\Gamma(z) = \int_0^{\infty} e^{-t} t^{z-1} dt \quad (27)$$

The windowed images are then Fourier transformed, and the power spectrum $P(u, v)$ can be computed as the square of the magnitudes:

$$P(u,v) = \frac{|F(u,v)|^2}{N^2} \quad (28)$$

where $F = F[I]$ is the Fourier transform of the image, and (u, v) are pixel indices in Fourier space. Two-dimensional frequencies can also be represented with polar coordinates (f, θ). where f is the spatial frequency and θ is the spatial orientation. They are computed as:

$$f = \sqrt{u^2 + v^2} \quad (29)$$

$$\theta = \tan^{-1}(v/u) \quad (30)$$

Conversion back to Cartesian coordinates is given by:

$$u = f\cos(\theta) \quad (31)$$

$$v = f\sin(\theta) \quad (32)$$

Although frequencies of up to $N/2$ cycles per image are computed, it would be better to use only half as many of the lowest frequencies. Higher frequencies may suffer from aliasing, noise, and low modulation transfer.[7]

The estimation of the spectral slope is then performed by fitting a straight line through the logarithm of these data points as a function of the logarithm of $1/f$.

Spectral Slope Analysis

The spectra of individual natural images are likely to vary, as shown in Fig. 12. This may play a role in the rapid detection of certain scenes or objects[9]). When averaging over a sufficiently large number of images (and across orientation θ), however, a clear pattern arises: the lower frequencies contain the most power, with power decreasing

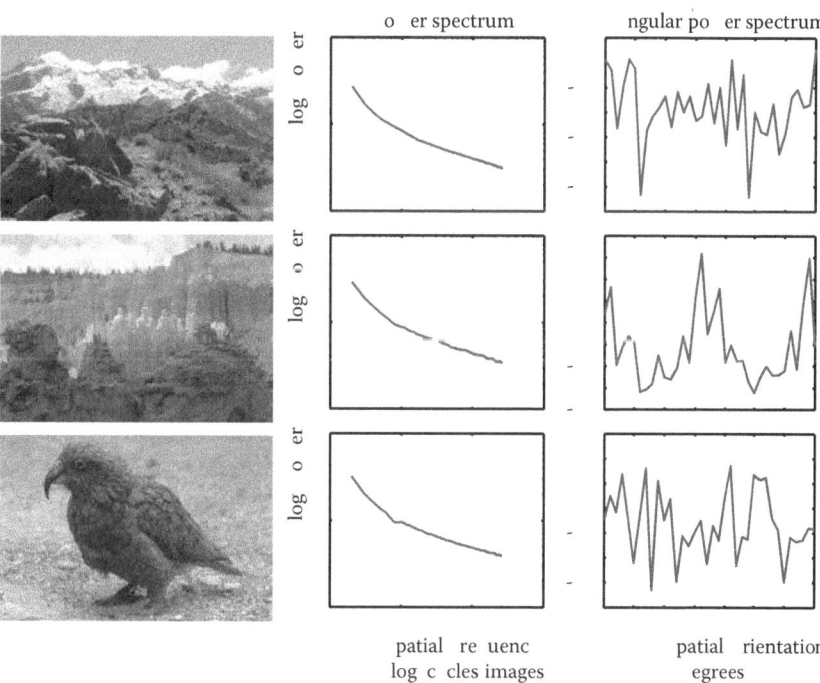

Fig. 12 The individual power spectra are shown here for three different images (middle column). The angular power spectra are shown on the right. (Top to bottom: Mt. Cook, New Zealand, 2012; Bryce Canyon, USA, 2009; Kea at Homer Tunnel, New Zealand, 2012)

as a function of frequency. In fact, on a log-log scale, amplitude as function of frequency lies approximately on a straight line. That is, the averaged spectrum tends to follow a power law, which can be well modeled with:

$$P(f) = \frac{1}{f^\beta} \qquad (33)$$

where P is the power as a function of frequency f, and β is the spectral slope.

Figure 13 shows the results of analysing three image ensembles, namely, a set of 133 natural images, drawn from the Van Hateren database[10] as well as two sets of synthetic images, whereby the first set of 30 images is of subjectively higher quality than the second set, which contains 18 images.[11] The natural image ensemble yields a slope of 1.88 with a standard deviation of 0.42, which, as discussed next, turns out to be typical for natural scenes. The high- and low-quality image ensembles produced somewhat steeper slopes, with especially the low-quality ensemble showing significantly less energy at high frequencies. This is thought to be because the subjectively low quality synthetic images have relatively little fine detail.

Several studies have reported values for the spectral slope for different image ensembles. While the particulars of the image ensembles vary considerably, they tend to focus on natural scenes, which usually means simply eliminating carpentered environments. The average power spectral slope varies from $\beta = 1.8$ to $\beta = 2.4$, with most values clustering around 2.0.[11–28] A summary of findings is listed in Table 1. Note that if the power spectrum has a slope of around $\beta \approx 2.0$, then the amplitude spectrum would have a slope of around $\alpha \approx 1.0$:

$$A(f) = \frac{1}{f^\alpha} \qquad (34)$$

Table 1 Spectral slopes for natural image ensembles (Starred studies were carried out on high dynamic range data)

Study	Ensemble size	$\beta \pm \sigma$
[12]	19	2.10±0.24
[13]	320	2.30
[14]*	95	2.29
[15]	6	2.00
[16]	85	2.20
[17]	20	2.20±0.28
[18]	117	2.13±0.36
[29]	216	1.96
[20]	29	2.22±0.26
[30]	95	2.22±0.24
[30]*	95	2.24±0.14
[21]	133	1.88±0.42
[22,23]	45	1.81
[24]	276	1.88±0.42
[25]	82	2.38
[26]	135	2.40±0.26
[27]	12,000	2.08
[28]	48	2.26

As α hovers around 1.0 in many studies, the amplitude spectrum is therefore often well characterized by $A(f) \approx 1/f$, which is why this particular statistical regularity, found in so many studies, is variously described as the $1/f$ property or $1/f$ statistic.

If the power as function of frequency is given by $P(f) \approx 1/f^2$, this leads to an interesting feature. By *Parceval's theorem* we have that the volume under the power spectrum is proportional to the variance of an image. This by itself relates to the available amount of contrast. We can therefore study contrast in terms of the power spectrum.

If we are moving through a natural environment, we do not expect the amount of contrast of the images formed on our retinas to vary considerably. Likewise, if we zoom in to a portion of a digital image, the contrast as measured by the variance in the image is not likely to change much. We may therefore expect the energy in different frequency bands to be roughly equal,[16] which points at a source of scale invariance as explained next.

A 2D image will have an energy E summed over all orientations as a function of frequency f given by:

$$E(f) = 2\pi f\, P(f) \qquad (35)$$

If we consider a range of frequencies between f_0 and nf_0, zooming in to a portion of the image will have the effect that this band of frequencies will shift to be between $a f_0$ and $an f_0$, where a is a measure of how much zoom has been applied. The energy between f_0 and $n f_0$ will also shift accordingly. To have an equal amount of energy at

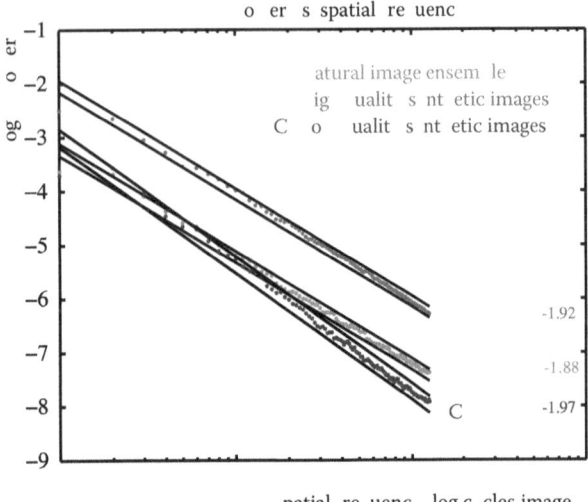

Fig. 13 The average power spectra for 133 natural images, 30 high-quality rendered images, and 18 lower-quality rendered images

all zoom levels, and thereby an equal amount of contrast irrespective of how much an image is zoomed in to, we require that:[16]

$$\int_{f_0}^{nf_0} 2\pi f\, P(f) df = K \qquad (36)$$

where K is a constant. This is indeed the case if $P(f)$ is chosen to be proportional to $1/f^2$, as in that case we have:

$$\int_{af_0}^{anf_0} 2\pi c \frac{1}{f} df = K \qquad (37)$$

where the constant c is introduced to reflect the fact that the power spectrum is proportional to $1/f^2$. Equation 37 evaluates to:

$$2\pi c \left|\log(f)\right|_{af_0}^{anf_0} = K \qquad (38)$$

$$2\pi c \left(\log(anf_0) - \log(af_0)\right) = K \qquad (39)$$

$$2\pi c \log\left(\frac{anf_0}{af_0}\right) = K \qquad (40)$$

$$2\pi c \log(n) = K \qquad (41)$$

The zoom level, given by parameter a, has been factored out of this equation, which means that the amount of energy remains constant irrespective of zoom level. This is of course because the power spectrum behaves according to $1/f^2$. This implies that an image with such a power spectral composition has constant variance at all scales. In other words, it is *scale invariant*.[15]

Thus, images tend to have features that appear at multiple scales, which is also a hallmark of fractal images (note that a simple linear transform of the spectral slope yields the image's fractal dimension[31]). In practice, this means that we should be able to zoom in and out of an image, or travel through a natural scene, and expect that the statistics will remain roughly constant. Thus, a scene imaged on one's retina or on the sensor of a camera $I(x, y)$ could be viewed from nearer or farther away, producing approximately $I(kx, ky)$, where k is a constant determining the amount of scaling.[32] Although this image is smaller or larger than $I(x, y)$, dependent on the value of k, it is a depiction of the same scene, and we would therefore expect it to exhibit the same statistics. This is borne out in the power spectrum of natural images, and it can also be seen in the analysis of wavelet coefficients, as discussed in Section 8.8.

As argued in "The Wiener-Khintchine Theorem" section, the power spectrum and the autocorrelation function form a Fourier transform pair. This means that the spectral slope of image ensembles can be interpreted as describing relations between pairs of pixels. Intuitively, this means that since the surface of a given object tends to be rather homogeneous, it is expected that neighboring pixels will be similar and that the farther apart two pixels are, the less similar they will be.

The $1/f$ property of images seems to arise from several sources. Edges, for example, show $1/f$ spectra.[33] Likewise, foliage and landscapes tend to exhibit fractal properties.[34] Further, the clustering of independent objects is such that the distribution of sizes in many scenes also tends to follow a power law.[17,35]

There is also some emerging evidence that the slope of the power spectrum is distinct for different scene types or objects.[27,28,30,36] For example, Huang and Mumford[29] examined 216 images, which had been painstakingly segmented into pixels representing 11 different categories, and found that although the image ensemble had an average slope of 1.96, there were systematic differences in the slopes across categories. Specifically, the slopes were 2.3, 1.8, 1.4, and 1.0 for manmade, vegetation, road, and sky pixels, respectively. Likewise, Webster and Miyahara[28] analyzed the power spectra and RMS-contrast of 48 natural scenes. They found significant differences in both spectral slope and contrast across the three scene types (2.15, 2.23, and 2.4 for the forest, close-up, and distant meadow scenes, respectively).

Finally, Pouli et al. compared three different manmade categories (indoors, night, and day scenes) against natural daytime scenes.[30] Example images from their ensembles are shown in Fig. 14. Prior to analysis, a window of 1024×1024 pixels was cropped from the middle of each image. This helps avoid photographer bias, as shown in the examples of Fig. 14. The number of images in each ensemble as well as slopes β and their variances are listed in Table 2. In this study, it was found that carpentered environments have on average steeper slopes than natural scenes: a two-tailed, independent measures t-test revealed that the difference between the manmade categories and the natural one is statistically significant ($t(668) = 2.64$, $p < 0.008$), with an average slope of $\beta = 2.32$ for the manmade image classes and $\beta = 2.22$ for the natural set.

Dynamic Range

In addition to examining different scene types, as noted above, Pouli et al. also investigated whether the $1/f$ statistic is dependent on the way the data is captured.[30] In particular, one might expect that high dynamic range capture techniques would allow the statistical assessment of scenes that cannot be captured with conventional imaging techniques. Many scenes have a range between light and dark that exceed the range captured by conventional cameras. This means that scenes would have to be well exposed before they can be captured, which may lead to bias in the image ensembles and therefore affect the statistical regularities computed on those images. With high dynamic range imaging techniques such limitations are removed (although in many cases there are other limitations, for instance, regarding how much movement can be present in a scene at capture time).

Fig. 14 Examples from four different classes of images, used to determine to what extent natural image statistics depend on scene type[30]

Pouli et al. assessed the $1/f$ behaviour of the four aforementioned scene types, namely natural images, manmade daytime images, nighttime images, and indoors images.[30] The comparison of the spectral slopes found for each category, as captured with both conventional and high dynamic range imaging techniques, is shown in Table 3.

Of note in this table is that there is a significant difference between the low dynamic range and high dynamic range capture techniques for the indoors and night scenes only. This means that both the carpentered and natural daytime scenes are equally well represented by both high dynamic range and low dynamic range images, suggesting that previous studies on natural image statistics remain valid, even if the ensembles were captured with conventional image capture techniques.

On the other hand, the same is not true for nighttime and indoors images. Here, the high dynamic range images especially produce much steeper slopes than for the natural image ensemble. This suggests that high dynamic range imaging is a particularly important tool for such image categories.

As spectral slopes obtained for natural images do not translate to other image categories, it appears that any

Table 2 For each of the LDR image ensembles from the study by Pouli et al.[30], the number of images N is shown, as well as spectral slope β and its variance σ^2

Data set	N	β	σ^2
Natural day	95	2.22	0.242
Manmade day	240	2.29	0.153
Indoors	125	2.44	0.121
Night	52	2.47	0.249

Table 3 Comparison of HDR and LDR image ensembles from the study by Pouli et al.[30] The number of images N is listed, as well as spectral slope β and its variance σ^2. The t-tests show that the difference between LDR and HDR ensembles is significant for the indoors and night ensembles

Data set	N	log HDR β (σ^2)	LDR β (σ^2)	t-test	p
Natural day	95	2.24 (0.144)	2.22 (0.242)	t(188)=0.390	>0.69
Manmade day	240	2.34 (0.099)	2.29 (0.153)	t(478)=1.683	>0.09
Indoors	125	2.61 (0.095)	2.44 (0.121)	t(248)=3.977	<0.00
Night	52	2.68 (0.152)	2.47 (0.249)	t(102)=2.362	<0.02

computer graphics applications that would make use of $1/f$ statistics would benefit from considering image type.

Dependence on Image Representation

As many studies have found similar power spectra, it is likely that this statistic is relatively robust against image distortions, for instance due to the choice of camera or lens system. In particular, several post-processing distortions were evaluated, leading to the following observations:[11]

File Formats. The choice of file format does not have an appreciable effect on the $1/f$ statistic. In particular, conversion from a lossless format to a lossy format such as GIF or JPEG does not significantly alter the spectral slope, with the exception of the smoothing parameter available in the JPEG format, which can destroy high frequency content.

Gamma Correction. To correct for display nonlinearities in cathode ray tube monitors, and for reason of backward compatibility also in liquid crystal displays, images should be gamma corrected. This involves applying a power function $I_d = I^{1/\gamma}$, where I_d denotes the pixel values that are sent to the display. The gamma value of typical displays hovers around 2.2–2.4. A similar nonlinearity is encoded in many color spaces, including the often-used sRGB space. This manipulation does not appear to affect the spectral slope significantly. Figure 15 shows the relationship between gamma correction value and the resulting spectral slope for the rendering shown in the same figure.

Aliasing. In image synthesis, geometry is normally sampled with point samples, which are then taken to represent small areas. If a single sample is taken per pixel, then this can lead to visible artifacts such as jagged edges and Moiré patterns, collectively known as *aliasing*. Multiple samples per pixel will ameliorate the effect, as more samples will provide a better estimate of the area represented by each pixel. Aliasing does have an effect on the spectral slope, and it has been found that at least 16 samples per pixel would be required to ensure that aliasing is not a factor affecting the $1/f$ statistic.

Rendering Parameters. When synthesizing an image, the choice of rendering parameters may affect the appearance of an image. For instance, rendering soft shadows instead of hard shadows (i.e., using area lights rather than point lights) could be thought to affect the power spectrum. However, it was found that this does not have a significant effect. Likewise, computationally expensive rendering techniques such as adding diffuse interreflection do not affect the power spectrum. In all, this suggests that rendering has less of an effect on the power spectrum than modeling, corroborating the idea that the $1/f$ behavior relates to the geometric composition of a scene more than the way it is lit.

Angular Dependence

When power spectra are not averaged over all orientations, it is clear that there is some variation as a function of angle: natural images tend to concentrate their power in horizontal and vertical angles.[24,33,35,37,38] Figure 16 shows the angular spectra for the aforementioned natural and synthetic image ensembles. Manmade structures especially tend to show strong vertical and horizontal lines, leading to similar angular dependence, even in single images (see Fig. 17).

Further, in examining over 6000 manmade and 6000 natural scenes, Torralba found that the slope varied as a function of both scene type and orientation (with slopes of 1.98, 2.02, and 2.22 for horizontal, oblique, and vertical angles in natural scenes and 1.83, 2.37, and 2.07 for manmade scenes).[27] Thus, the spectral slope may be useful in object or scene discrimination.[9] It is critical to mention that if two images have similar spectral slopes, then

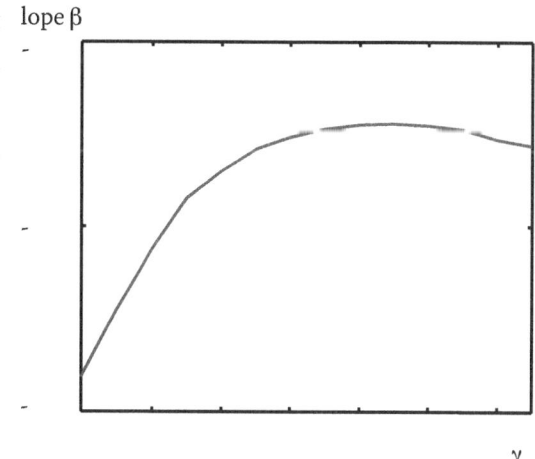

Fig. 15 The rendering on the left was subjected to gamma correction for a range of gamma values. The resulting spectral slope is plotted on the right

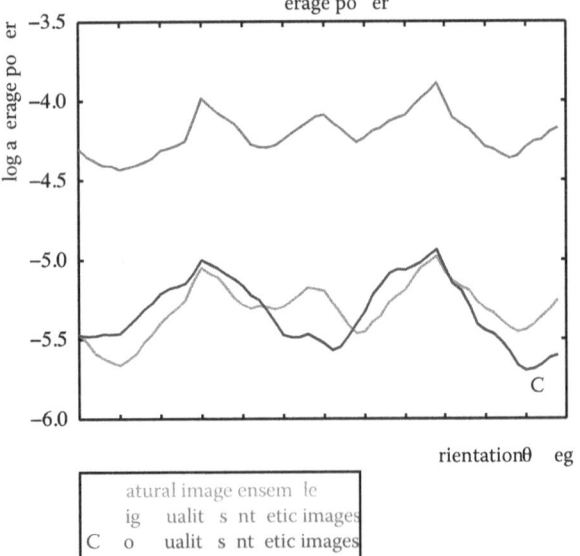

Fig. 16 Power as a function of spatial orientation for a natural image ensemble, and high- and low-quality rendered image ensembles

swapping their power spectra will not affect recognition as long as phase information is preserved.[26,39]

Temporal Dependence

It has also been shown that the pattern of change over time follows a power law. That is, if the contrast modulation over time for a given pixel is examined, the power spectra can also be modeled with $P = 1/f^{\alpha}$, where P is the power as function of frequency f, and α is the *temporal spectral slope*.[13,40,41]

Temporal spectral slopes between 1.2 and 2.0 have been reported for natural image sequences. The temporal spectral slope relates, perceptually, to the apparent speed and jitter of the scene and to some degree with the apparent "purposefulness" of the motion (the higher the slope is, the more persistent the motion will be).

1/f Failures

Although the $1/f$ statistical regularity found in so many studies is striking, this does not mean that it holds for all imagery. It is important to note that reliable image statistics emerge

Fig. 17 The distribution of power over spatial orientation tends to produce peaks at the horizontal and vertical direction for image ensembles. Here, a similar effect is shown for a single image of a manmade structure. (Parthenon, Athens, Greece, 2010)

when a large number of images are analyzed together. Specifically, individual images may deviate from any statistics observed in ensembles. This implies that these results should be used with care if they are somehow to be applied to individual images. In particular, the $1/f$ statistics are prone to failure for individual images, and then specifically at low frequencies, which correspond to large image scales.[42]

Further, some deviations from the predicted scale invariance may be observed at high frequencies. This can be attributed to noise in images, which manifests itself in high frequencies.[43]

PHASE SPECTRA

It can be argued that although statistical regularities are present in power spectra of natural images, much of the perceptually relevant information is encoded in phase spectra.[44,45] As an example, we have swapped the phase spectrum of two images, shown in Fig. 18. As can be seen, much of the image structure has swapped. Thus, the image structure is largely encoded into the phase spectrum. Second-order statistics such as the autocorrelation function (and therefore power spectra in Fourier space) as well as variance are insensitive to signal phase. They therefore do not adequately measure image structure.

To gain access to phase information without polluting the results with first-and second-order information (such as means, variances, and covariances), we can whiten the images first.[44,46] This amounts to adjusting the spectral slope to become flat. The autocorrelation function will therefore be zero everywhere, except at the origin. Alternatively, Principal Component Analysis (PCA) can be applied to whiten a signal; see Section 7.1. By removing the second moment from consideration, it is now possible to compute skewness and kurtosis on the whitened signal. The whitened skew S_w and whitened kurtosis κ_w are thus a measure of variations in the phase spectra.

The result of applying this procedure to a set of natural images leads to the conclusion that the whitened images are almost always positively skewed and are always positively kurtosed. In contrast, if the phase spectrum is randomized on the same images, the whitened skewness and kurtosis are close to zero.

While positive skewness and kurtosis of phase spectra points in the direction of the presence of statistical regularities in the phase spectrum, these results are relatively weak and do not easily explain aspects of human vision. Furthermore, they do not appear to have found employ in any graphics-related applications that we are aware of (although it may lead to higher-order constraints in texture synthesis algorithms). In the following section, however, we will discuss how extending the present analysis to be localized in space leads to further and stronger insights. Moreover, in Section 8.11 we will discuss a wavelet-based technique to help understand phase structure.

HUMAN PERCEPTION

There is an extremely large body of work examining the relationship between natural image statistics and human

Fig. 18 In this demonstration, the phases of the top two images are swapped to produce the bottom to images. The amplitude spectra are retained. Note that much of the image structure is located in the phase spectrum and has therefore been swapped. (Left: Methoni Castle, Greece, 2010; right: Horse Shoe Bend, Arizona, 2012)

perception. At the simplest level, our ability to discriminate two random phase textures based solely on changes in the spectral slope has been examined.[9,47,48,49] Humans are most sensitive to slopes around 2.8 to 3.2, which would represent an image with much less high spatial frequency content than natural images. There is some evidence (albeit controversial) for a second minimum near 1.2. Rainville and Kingdom examined the ability to detect symmetry for white noise images with different spectral slopes and found that one participant was best for slopes near 2.8, consistent with the image discrimination data.[50] The other participant was best for slopes between 1 and 2, consistent with the potential second minima.

Regardless, it is clear that humans are not maximally sensitive to changes in spectral slopes representing natural images. The reasons for this are still unclear, although several hypotheses have been forwarded including that the shift from 2 to 2.8 reflects blur perception.[51]

Instead of attempting to determine the tuning of the visual system by measuring discrimination, one can approach the problem more indirectly: one can estimate the sensitivity of the spatial perception mechanisms from an autocorrelation analysis of contrast sensitivity function (which has been referred to as a modulation transfer function of the human visual system).

It is generally accepted that human spatial perception is mediated by several, partially overlapping spatial frequency channels (at least seven at each orientation[51]). Since similar frequencies are likely to be processed by the same channel, the sensitivity to similar frequencies should be similar. The less similar the frequencies are, the less correlated their sensitivity thresholds should be.[52–55]

Billock[51] examined the correlation between contrast sensitivity thresholds as a function of the spatial frequency separation, and found that (for up to five octaves) the correlation functions were power laws with slopes ranging from 2.1 to 2.4. This held not only for static stimuli but also for slowly flickering stimuli (up to 1 Hz). These slopes are much more in line with the slopes found in natural images, suggesting that human spatial frequency mechanisms may be optimized for natural images. Interestingly, more rapid flicker yielded higher slopes (around 2.6). As mentioned above, higher slopes reflects an attenuation of the higher spatial frequencies. Billock suggested that the higher slopes for rapidly flickering images may represent motion deblurring.

In contrast, the discrimination of temporal spectral slopes appears to be more straightforward. Humans are most sensitive to differences in temporal spectral slope for slopes between 1.8 and 2.0, which is very similar to the range of slopes in natural image sequences.

The existence of spatial frequency channels in the human visual system are also implicated in lightness perception. Dakin and Bex have shown that if the amplitude of the response of these channels to natural stimuli is weighted according to their scale, i.e., with weights w_s proportional to $1/f^{-s}$, their combined response correlates well with the perception of lightness.[56] In particular, it can explain the Craik-O'Brien-Cornsweet[57] and White's illusions.[58]

FRACTAL FORGERIES

Thanks in large part to the seminal work of Mandelbrot,[34] many areas of computer graphics use fractals to synthesize textures, surfaces, objects, or even whole scenes.[59–61] A subset of this work focuses on fractal Brownian motion in general and fractal Brownian textures in specific, which bear striking resemblance to real surfaces and textures. Since the fractal dimension of such a fractal texture is a linear transform of the spectral slope, these works are essentially relying on the regularities in power spectra. Many of these techniques either explicitly or implicitly choose parameters so that the spectral slope will be similar to natural images. Perhaps the most famous of these synthetic scenes are the eerily familiar landscapes produced in Voss's "fractal forgeries".[62]

IMAGE PROCESSING AND CATEGORIZATION

As mentioned in "1/f Failures" section, despite the fact that a power law description clearly captures the regularities in large collections of images, individual images tend not to be 1/f. It has been suggested that differences in the spectral slope between parts of an image allow people to rapidly make some simple discriminations (e.g., the "pop-out" effect[1,2,9]). Others have speculated on the evolutionary advantage of being able to detect spectral slope.[9,51,63,64,65] Just as knowing about the statistics of natural images in general can inform us about how the human visual system works and how we might build more efficient computer vision and computer graphics algorithms, so too will an understanding of the cause of *variations* in the statistics provide insights.

A number of potential sources for 1/f patterns and their variations have been identified.[17,27,28,33,34,35,36] For example, Huang and Mumford[36] and Webster and Miyahara[28] found different average spectral slopes for different scene categories (both within as well as between images); at the very least, there seem to be differences between manmade, general vegetation, forest, meadow, road, and sky elements. It has also been shown that underwater scenes have different spectral slopes.[66] To help further distinguish between object or scene categories, one can look at the interaction between power spectra and other characteristics.[24,33,35,37,38] For example, a "spatial envelope" of a scene can be constructed from the interaction between power spectra and orientation combined with some information from Principal Components Analysis (PCA).[38] This envelope yields perceptual dimensions such as naturalness, roughness, and expansion. Similar categories tend to cluster together in this scene space. This approach was later extended to estimate absolute depth using the relationship between power spectra and orientation and some information from wavelets.[27]

In a related line of work, Dror and colleagues used a variety of natural image statistics to estimate the reflectance of an object (e.g., metal versus plastic) under conditions where the illumination is unknown.[67] They employed a wide range of image statistics from simple intensity distributions through oriented power spectra to wavelet coefficient distributions.

As noted in "Human Perception" section, it has been suggested that the differences between the average spectral slope of 2.0 and the peak of human sensitivity to changes in slope (at 2.8) is due to deblurring. Furthermore, it has been suggested that the higher slopes for an autocorrelation analysis of human contrast sensitivity are due to motion deblurring. In an indirect examination of this claim, Dror and colleagues examined the properties of Reichardt correlator motion detections.[68,69] While there is considerable evidence that the human visual system uses such correlators for the low-level detection of motion, it has also been shown using typical moving gratings that they signal temporal frequency and not velocity. Dror demonstrated that when stimuli that have natural image statistics are used, the response properties of Reichardt detectors are better correlated with velocity and suggest that they make a much better basis for the synthetic processing of motion than previously assumed

TEXTURE DESCRIPTORS

Increasingly, the spectral slope is being used as a low-dimensional descriptor of texture.[31,47,65,70,71] Perceptually, the primary effect of increasing the spectral slope is to increase the coarseness of the texture[47] (as shown in Fig. 19). Indeed, Billock and colleagues have suggested that a decent model of dynamic textures can be given with the equation $A(f) = K f_s^{-\beta} f_t^{-\alpha}$, where K is a constant, and f_s and f_t are the spatial and temporal frequencies, respectively.

Likewise, several researchers have suggested that the spectral slope might provide a good estimate of the blur in

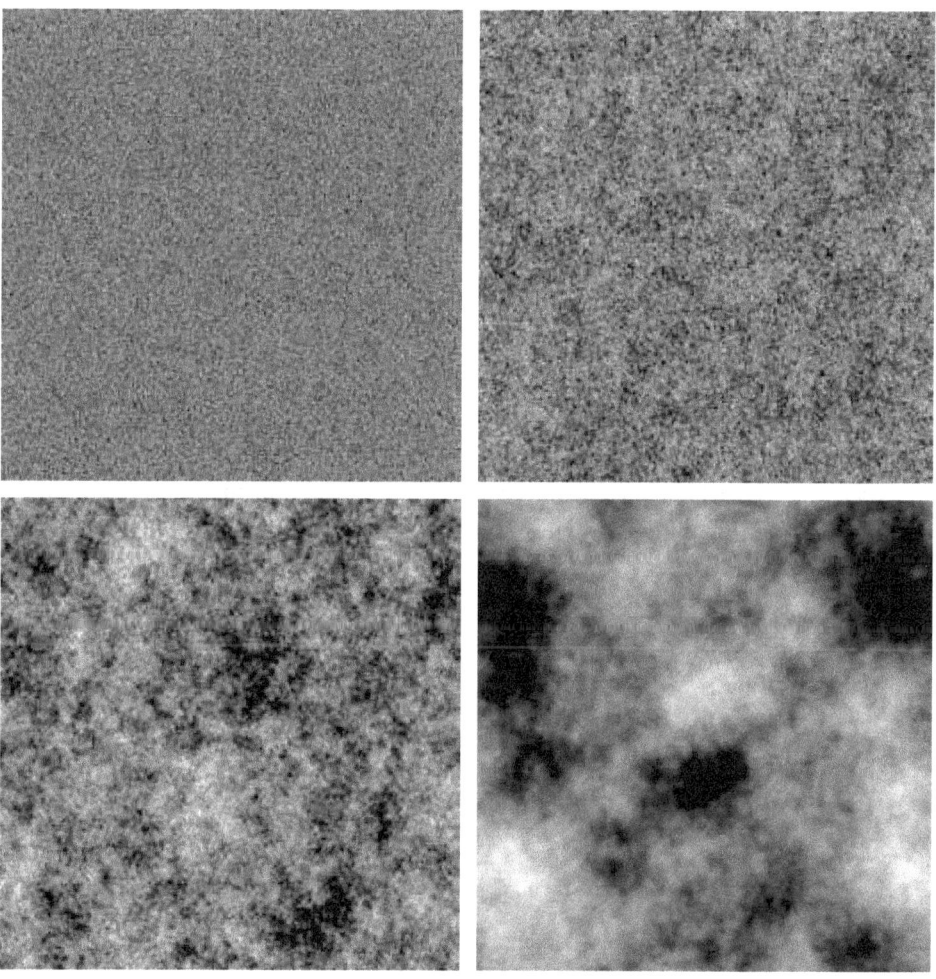

Fig. 19 Static, random phase patches produced by $1/f^{\beta}$ spatial-frequency filtering of random white noise. The values of the spectral slope are 0.8, 1.6, 2.4, and 3.2 for the top left, top right, bottom left, and bottom right, respectively

an image, either by using the slope directly[72,73] or by looking at the relative power at higher frequencies.[74,75] The spectral slope of an image is in fact often altered to synthetically blur an image.[76,77] Murray and colleagues, however, have shown that the perception of image blur is not well predicted by the relative energy at higher frequencies.[78] Changes in the relative phases must also be taken into account.

TERRAIN SYNTHESIS

The omnipresence of the $1/f$ statistic may be leveraged in applications that generate images involving parameter tuning. For example, in procedural plant modeling,[79] the well-known fact that plants tend to exhibit a strong degree of self-similarity is exploited. Many descriptive systems have been developed to take advantage of this characteristic, the most prominent of which are L-systems.[60] Other areas where the $1/f$ nature of natural images is employed is in displacement mapping[80,81] or solid texture generation.[82]

Here, we show the use of $1/f$ statistics in fractal terrain modeling. The simplest approach is to filter a white noise field with the desired spectral slope to produce the terrain height map.[61] In a more complex approach, called the *midpoint subdivision algorithm*,[83] a patch of terrain is iteratively subdivided while displacing the subdivided patches by a random amount. As the size of the subdivided patches decreases, the amplitude of the displacement is reduced. In particular, halving the size of the patch corresponds to a reduction in maximum displacement by a factor of k. Here, k is a user parameter that determines the roughness of the terrain.

Figure 20 shows a set of example terrains, created by varying k between 1.5 and 2.6. In each case the number of iterations was ten, resulting in terrains consisting of 524,288 triangles. The spectral slope of these images relates to parameter k and the number of iterations (up to ten) according to the plot in Fig. 21.

An interesting observation is afforded by showing these images to participants in an informal perceptual experiment. Simply asking observers to indicate the image which looks most natural resulted in the distribution shown in Figs. 22 and 23. As can be seen, there is good correspondence between what observers consider natural, and a $1/f$ slope of around $\beta = 1.86$. Note that this very closely corresponds to the average spectral slope that was found for natural images, as discussed in "Spectral Slope Analysis" section.

ART STATISTICS

A considerable amount of research has been conducted on the frequency statistics of art. Usually the goal is to examine the relationship between the real world and depictions of it. The knowledge gained is used to either support theories of how human vision works or for the (semi-) automatic analysis of art. For example, Graham and colleagues[84–88] examined the first- and second-order statistics of over 900 works from many different art epochs from both western and eastern cultures.

In one of their first works,[84] they compared the statistics of 124 paintings to a similar number of real-world images from Van Hateren's database[1]. The art images were uncompressed TIFF photographs (taken under controlled conditions by the museum photographer) of art works from the Herbert F. Johnson Museum of Art, Cornell University.[84] A patch of 818×818 pixels for each image was extracted randomly and examined for a variety of single pixel and frequency-based statistics. They found that the art works had a power spectral slope of 2.46 and the world images had a slope of 2.74. Although this value is above the average found by other labs, it is not that much higher and the difference can be explained by the exclusion of blurry images. The smaller slope value for art implies a relatively smaller amount of higher-frequency information (or the absense of blur in the real-world images).

A subsequent more detailed analysis revealed that beyond blur, the difference between art and photographs seems to be driven by two factors.[85] First, eastern artworks have shallower slopes than western artworks. Second, abstract art had shallower slopes (ca. 2.23) than either landscape or portrait paintings (both roughly 2.5). Furthermore, Graham and colleagues examine the luminance compression between real-world (which has a high dynamic range) and the very limited dynamic range (usually around 30:1) of paintings and found that painters use a compressive nonlinearity.[86,89] That is, they use a tone mapping algorithm. They also found that once a similar nonlinearity is applied to real-world scenes, both art and real-world scenes show similar sparsity.

In a similar line of research, Redies and colleagues examined monochrome versions of 200 western artworks from a variety of epochs excluding modern art. The exclusion of modern art allowed them to focus on representational art.[90–92] The statistics of the artworks were compared to those of pictures of household objects (179 images), plant parts and natural scenes (i.e., 408 natural images from the Van Hateren database), and scientific illustrations (209 images). They found a power spectral slope of 2.1 for the paintings, 2.1 for natural scenes, 2.9 for plant parts and objects (close-up views), and 1.6 for scientific illustrations. In short, the frequency statistics of representational western art seem similar to real-world scenes, but not to close-up views of real-world scenes.

There was an astonishingly consistent slope for the artworks: the spectral slopes did not vary significantly with changes in country of origin, century, painting technique,

---

[1]They excluded photos with an undefined amount of blur.

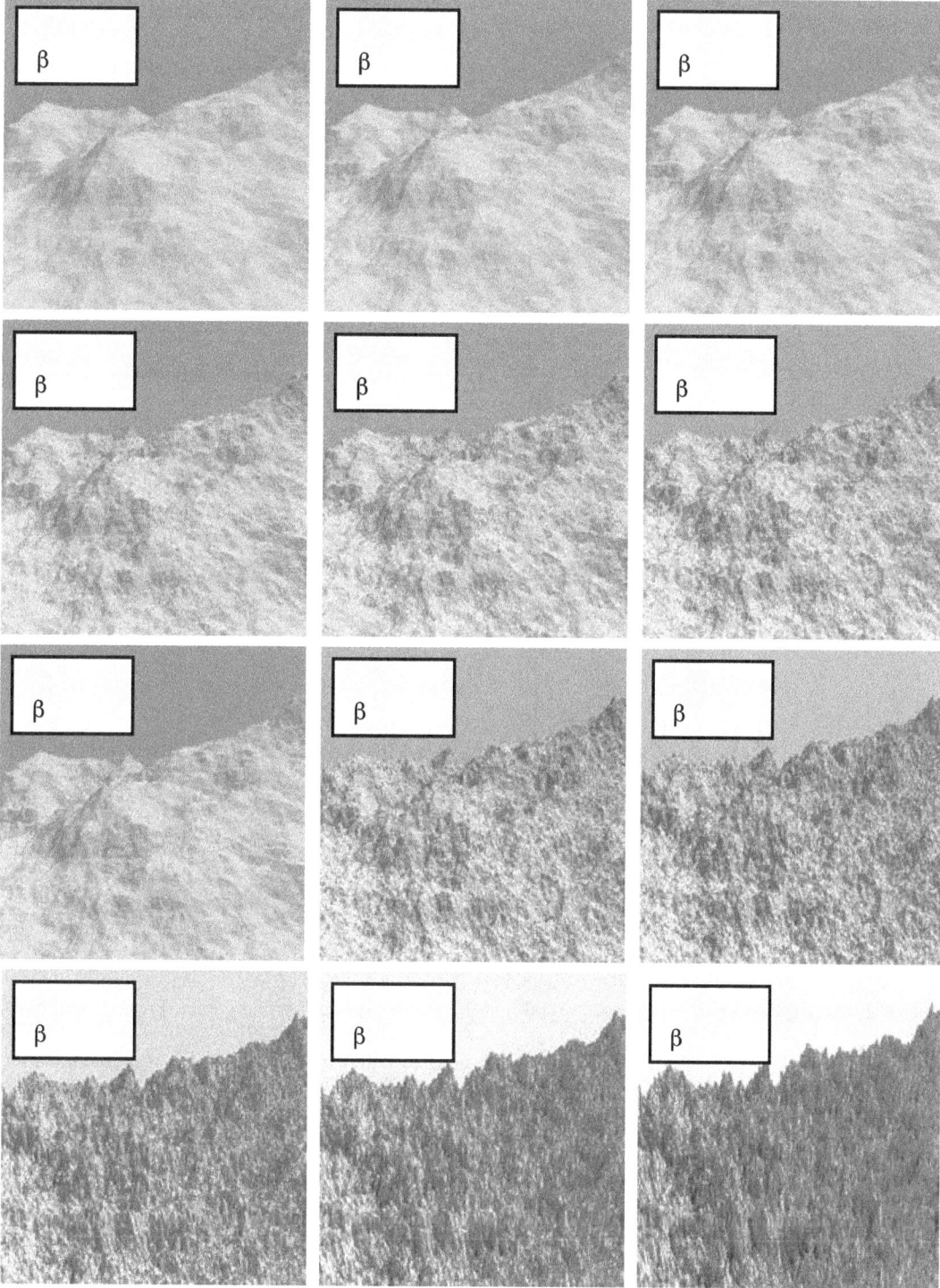

Fig. 20 A set of 12 fractal terrains, generated with parameter k, resulting in a spectral slope of β

or subject matter. Interestingly, they also found that representations of faces showed $1/f$ spectra even though photographs of faces are not $1/f$.[93] These and other results can be related to information from neuroscience for a theory on aesthetics.[91] It has been shown that political cartoons, comics, and Japenese mangas also have roughly similar power spectra (roughly 1.99, 2.04, and 2.08, respectively).[92] Perhaps most interesting is that the spectral slope of artworks does not vary much as a function of orientation, while it does for natural images (which can be the basis of a simple scene discriminator in photographs).

A large body of work has focused exclusively on Jackson Pollock's drip paintings.[94–101] While the bulk of the work has focused on descriptive statistics of

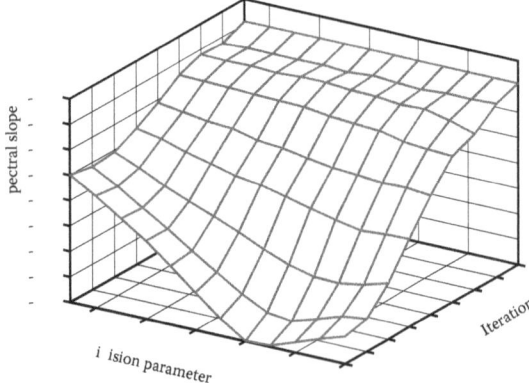

Fig. 21 The relation between terrain roughness parameter k, the number of iterations, and the spectral slope

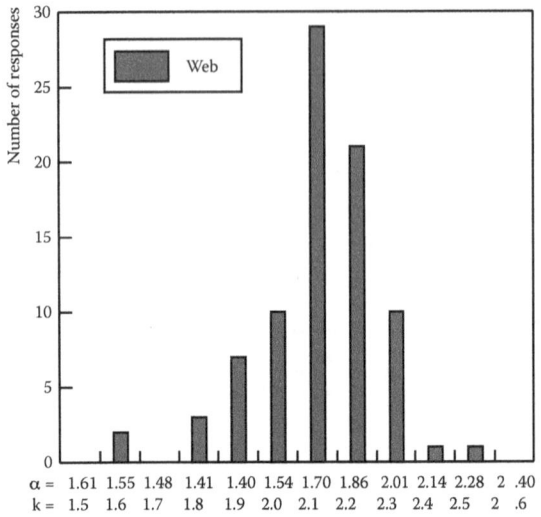

Fig. 22 Naturalness ratings for the images of Fig. 21. This experiment was carried out by means of a webpage, which showed images and asked viewers to send an e-mail to the author, stating the number of the image which appeared most natural

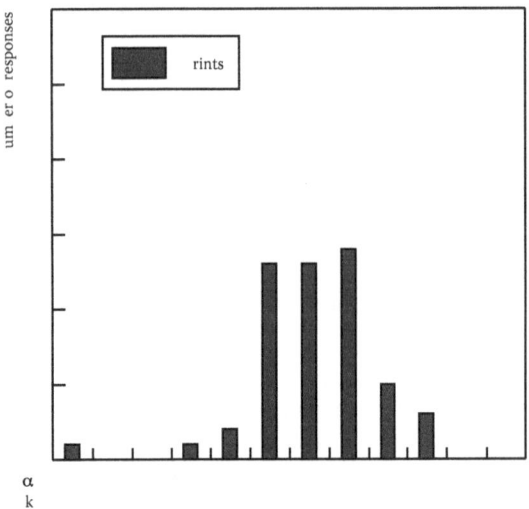

Fig. 23 Naturalness ratings for the images of Fig. 21. This experiment was carried out using high-quality prints, whereby participants were asked to select the print that appeared most natural

Pollock's work, some research addressed the issue of authenticating Pollock paintings.[96,98,102] Overall, it has been extensively shown that Pollock's artwork can be described well with a $1/f$ like power spectra. It has also been shown that the slope changed systematically over the course of Pollock's career.

REFERENCES

1. Julesz, B.; Caelli, T. On the limits of fourier decompositions in visual texture perception. Perception **1979**, *8* (1), 69–73.
2. Caelli, T. *Visual Perception: Theory and Practice*; Pergamon Press: Oxford, 1981.
3. Ziemer, R.E.; Tranter, W.H.; Fannin, D.R. *Signals and Systems: Continuous and Discrete*, 4th Ed.; Prentice Hall: Upper Saddle River, NJ, 1993.
4. Nikias, C.; Petropolu, A. *Higher-Order Spectra Analysis*; Prentice-Hall: Englewood Cliffs, NJ, 1993.
5. Wiener, N. Generalized harmonic analysis. Acta Math. **1930**, *55* (1), 117–258.
6. Khintchine, A. Korrelationstheorie der stationären stochastischen processe. Math. Ann. **1934**, *109* (1), 604–615.
7. van der Schaaf, A. *Natural Image Statistics and Visual Processing*; PhD thesis; Rijksuniversiteit Groningen: Groningen, 1998.
8. Harris, F.J. On the use of windows for harmonic analysis with the discrete fourier transform. Proc. IEEE **1978**, *66* (1), 51–84.
9. Billock, V.A.; Cunningham, D.W.; Tsou, B.H. What visual discrimination of fractal textures can tell us about discrimination of camouflaged targets. In *Human Factors Issues in Combat Identification*; Andrews, D.H.; Herz, R.P.; Wolf, M.B.; Eds.; Ashgate: Farnham, UK, 2008, 99–112.
10. van Hateren, J.H.; van der Schaaf, A. Independent component filters of natural images compared with simple cells in primary visual cortex. Proc. R. Soc. London B **1998**, *265* (1394), 359.
11. Reinhard, E.; Shirley, P.; Ashikhmin, M.; Troscianko, T. Second order image statistics in computer graphics. In *Proceedings of the 1st Symposium on Applied Perception in Graphics and Visualization*; ACM: New York, 2004, 99–106.
12. Burton, G.J.; Moorhead, I.R. Color and spatial structure in natural scenes. Appl. Opt. **1987**, *26* (1), 157–170.
13. Dong, D.W.; Atick, J.J. Statistics of natural time-varying images. Network: Comput. Neural Syst. **1995**, *6* (3), 345–358.
14. Dror, R.O.; Willsky, A.S.; Adelson, E.H. Statistical characterization of realworld illumination. J. Vision **2004**, *4* (9), 821–837.
15. Field, D.J. Relations between the statistics of natural images and the response properties of cortical cells. J. Opt. Soc. Am. A **1987**, *4* (12), 2379–2394.
16. Field, D.J. Scale-invariance and self-similar "wavelet" transforms: An analysis of natural scenes and mammalian visual systems. In *Wavelets, Fractals and Fourier Transforms*; Farge, M.; Hunt, J.C.R.; Vassilicos, J.C.; Eds.; Clarendon Press: Oxford, 1993, 151–193.

17. Field, D.J.; Brady, N. Visual sensitivity, blur and the sources of variability in the amplitude spectra of natural scenes. Vision Res. **1997**, *37* (23), 3367–3383.
18. van Hateren, J.H. A theory of maximizing sensory information. Biol. Cybern. **1992**, *68* (1), 23–29.
19. Huang, J.; Lee, A.; Mumford, D. Statistics of range images. In *Proceedings of the IEEE International Conference on Computer Vision and Pattern Recognition*, Washington, DC, 2000, 324–331.
20. Párraga, C.A.; Brelstaff, G.; Troscianko, T.; Moorehead, I.R. Color and luminance information in natural scenes. J. Opt. Soc. Am. A **1998**, *15* (3), 563–569.
21. Reinhard, E.; Ashikhmin, M.; Gooch, B.; Shirley, P. Color transfer between images. IEEE Comput. Graphics Appl. **2001**, *21* (5), 34–41.
22. Ruderman, D.L.; Bialek, W. Statistics of natural images: Scaling in the woods. Phys. Rev. Lett. **1994**, *73* (6), 814–817.
23. Ruderman, D.L. The statistics of natural images. Network: Comput. Neural Syst. **1994**, *5* (4), 517–548.
24. van der Schaaf, A.; van Hateren, J.H. Modelling the power spectra of natural images: Statistics and information. Vision Res. **1996**, *36* (17), 2759–2770.
25. Thomson, M.G.A.; Foster, D.H. Role of second- and third-order statistics in the discriminability of natural images. J. Opt. Soc. Am. A **1997**, *14* (9), 2081–2090.
26. Tolhurst, D.J.; Tadmor, Y.; Chao, T. Amplitude spectra of natural images. Ophthalmic Physiol. Opt. **1992**, *12* (2), 229–232.
27. Torralba, A.; Oliva, A. Statistics of natural image categories. Network: Comput. Neural Syst. **2003**, *14* (3), 391–412.
28. Webster, M.A.; Miyahara, E. Contrast adaptation and the spatial structure of natural images. J. Opt. Soc. Am. A **1997**, *14* (9): 2355–2366.
29. Huang, J.; Mumford, D. *Image Statistics for the British Aerospace Segmented Database*; Technical Report; Division of Applied Math, Brown University: Providence, RI, 1999.
30. Pouli, T.; Cunningham, D.; Reinhard, E. Statistical regularities in low and high dynamic range images. In *APGV '10: Proceedings of the 7th Symposium on Applied Perception in Graphics and Visualization*, New York, 2010, 9–16.
31. Cutting, J.; Garvin, J. Fractal curves and complexity. Percept. Psychophys. **1987**, *42*, 365–370.
32. Mumford, D.; Gidas, B. Stochastic models for generic images. Q. Appl. Math. **2001**, *59* (1), 85–112.
33. Switkes, E.; Mayer, M.J.; Sloan, J.A. Spatial frequency analysis of the visual environment: Anisotropy and the carpentered environment hypothesis. Vision Res. **1978**, *18* (10), 1393–1399.
34. Mandelbrot, B.B. *The Fractal Geometry of Nature*; W.H. Freeman and Co.: New York, 1983.
35. Ruderman, D.L. Origins of scaling in natural images. Vision Res. **1997**, *37* (23), 3385–3398.
36. Huang, J.; Mumford, D. Statistics of natural images and models. In *Proceedings of the IEEE International Conference on Computer Vision and Pattern Recognition*, Vol. 1; Washington, DC, 1999.
37. Baddeley, R.J.; Abbott, L.F.; Michael, C.A.; Booth, C.A.; Sengpiel, F.; Freeman, T.; Wakeman, E.A.; Rolls, E.T. Responses of neurons in primary and inferior temporal visual cortices to natural scenes. Proc. R. Soc. London B **1997**, *264* (1389), 1775–1783.
38. Oliva, A.; Torralba, A. Modeling the shape of the scene: A holistic representation of the spatial envelope. Int. J. Comput. Vision **2001**, *42* (3), 145–175.
39. Tadmor, Y.; Tolhurst, D.J. Both the phase and the amplitude spectrum may determine the appearance of natural images. Vision Res. **1993**, *33* (1), 141–145.
40. Billock, V.A.; Guzman, G.C.D.; Kelso, J.A.S. Fractal time and 1/f spectra in dynamic images and human vision. Physica D **2001**, *148* (1), 136–146.
41. Eckert, M.P.; Buchsbaum, G.; Watson, A.B. Separability of spatiotemporal spectra of image sequences. IEEE Trans. Pattern Anal. Mach. Intell. **1992**, *14* (12), 1210–1213.
42. Langer, M.S. Large-scale failures of f−α scaling in natural image spectra. J. Opt. Soc. Am. A **2000**, *17* (1), 28–33. 141.
43. Zoran, D.; Weiss, Y. Scale invariance and noise in natural images. In *Proceedings of the IEEE International Conference on Computer Vision*, 2009, 2209–2216.
44. Thomson, M.G.A. Higher-order structure in natural scenes. J. Opt. Soc. Am. A **1999**, *16* (7), 1549–1553.
45. McCotter, M.; Gosselin, F.; Sowden, P.; Schyns, P. The use of visual information in natural scenes. Visual Cogn. **2005**, *12* (6), 938–953. 142.
46. Thomson, M.G.A. Beats, kurtosis and visual coding. Network: Comput. Neural Syst. **2001**, *12* (3), 271–287.
47. Bllock, V.A.; Cunningham, D.W.; Havig, P.R.; Tsou, B.H. Perception of spatiotemporal random fractals: An extension of colorimetric methods to the study of dynamic texture. J. Opt. Soc. Am. A **2001**, *18* (10), 2404–2413.
48. Knill, D.C.; Field, D.J.; Kersten, D. Human discrimination of fractal images. J. Opt. Soc. Am. A **1990**, *7* (6), 1113–1123.
49. Tadmor, Y.; Tolhurst, D.J. Discrimination of changes in the second order statistics of natural and synthetic images. Vision Res. **1994**, *34* (4), 541–554.
50. Rainville, S.J.M.; Kingdom, F.A.A. Spatial-scale contribution to the detection of mirror symmetry in fractal noise. J. Opt. Soc. Am. A **1999**, *16* (9), 2112–2123.
51. Billock, V.A. Neural acclimation to 1/f spatial frequency spectra in natural images transduced by the human visual system. Physica D **2000**, *137* (3), 379–391.
52. Billock, V.A.; Harding, T.H. Evidence of a colour appearance model for colour management systems spatial and temporal channels in the correlational structure of human spatiotemporal contrast sensitivity. J. Physiol. **1996**, *490* (2), 509–517.
53. Owsley, C.; Sekular, R.; Siemsen, D. Contrast sensitivity throughout adulthood. Vision Res. **1983**, *23* (7), 689–699.
54. Peterzell, D.H.; Werner, J.S.; Kaplan, P.S. Individual differences in contrast sensitivity functions: Longitudinal study of 4-, 6- and 8-month-old human infants. Vision Res. **1995**, *35* (7), 9651–979.
55. Sekuler, R.; Wilson, H.R.; Owsley, C. Structural modeling of spatial vision. Vision Res. **1984**, *24* (7), 689–700.
56. Dakin, S.C.; Bex, P.J. Natural image statistics mediate brightness filling in. Proc. R. Soc. London B **2003**, *270* (1531), 2341–2348.
57. Cornsweet, T. *Visual Perception*; Academic Press: New York, 1970.

58. White, M. A new effect of patterns on perceived lightness. Perception **1979**, *8* (4), 413–416.
59. Deussen, O.; Colditz, C.; Coconu, L.; Hege, H. Efficient modeling and rendering of landscapes. In *Visualization in Landscape and Environmental Planning*, Bishop, I.; Lange, E.; Eds.; Taylor & Francis: Oxford, 2005.
60. Deussen, O.; Lintermann, B. *Digital Design of Nature: Computer Generated Plants and Organics*; Springer-Verlag: New York, 2005.
61. Peitgen, H.-O.; Saupe, D. *The Science of Fractal Images*; Springer-Verlag: New York, 1988.
62. Voss, R. Random fractal forgeries. In *Fundamental Algorithms for Computer Graphics*; Earnshaw, R.A.; Ed.; Springer: Berlin, 1985, 805–835.
63. Campbell, F.W.; Howell, E.R.; Johnson, J.R. A comparison of threshold and suprathreshold appearance of gratings with components in the low and high spatial frequency range. J. Physiol. **1978**, *284* (1), 193–201.
64. Hammett, S.T.; Bex, P.J. Motion sharpening: Evidence for the addition of high spatial frequencies to the effective neural image. Vision Res. **1996**, *36* (17), 2729–2733.
65. Rogowitz, B.; Voss, R. Shape perception and low-dimension fractal boundaries. Proc. SPIE **1990**, *1249*, 387–394.
66. Balboa, R.M.; Grzywacz, N.M. Power spectra and distribution of contrasts of natural images from different habitats. Vision Res. **2003**, *43* (24), 2527–2537.
67. Dror, R.O.; Adelson, E.H.; Willsky, A.S. Surface reflectance estimation and natural illumination statistics. In *Proceedings of the IEEE Workshop on Statistical and Computational Theories of Vision*; Vancouver, 2001.
68. Dror, R.O.; O'Carroll, D.C.; Laughlin, S.B. The role of natural image statistics in biological motion estimation. In *Proceedings of the IEEE Workshop on Biologically Motivated Computer Vision*, Seoul, 2000, 492–501.
69. Dror, R.O.; O'Carroll, D.C.; Laughlin, S.B. Accuracy of velocity estimation by Reichardt correlators. J. Opt. Soc. Am. A **2001**, *18* (2), 241–252.
70. Kube, P.; Pentland, A. On the imaging of fractal surfaces. IEEE Trans. Pattern Anal. Mach. Intell. **1988**, *10* (5), 704–707.
71. Taylor, R.P.; Spehar, B.; Wise, J.A.; Clifford, C.W.; Newell, B.R.; Hagerhall, C.; Purcell, T.; Martin, T.P. Perceptual and physiological responses to the visual complexity of fractal patterns. Nonlinear Dyn. Psychol. Life Sci. **2005**, *9*, 89–114.
72. Brady, N.; Bex, P.J.; Fredericksen, R.E. Independent coding across spatial scales in moving fractal images. Vision Res. **1997**, *37* (14), 1873–1883.
73. Tolhurst, D.J.; Tadmor, Y. Discrimination of changes in the slopes of the amplitude spectra of natural images: Band-limited contrast and psychometric functions. Perception **1997**, *26* (8), 1011–1025.
74. Marr, D.; Hildreth, E.C. Theory of edge detection. Proc. R. Soc. London B **1980**, *207* (1167), 187–217.
75. Mather, G. The use of image blur as a depth cue. Perception **1997**, *26* (9), 1147–1158.
76. Webster, M.A.; Georgeson, M.A.; Webster, S.M. Neural adjustments to image blur. Nat. Neurosci. **2002**, *5* (9), 839–840.
77. Vera-Diaz, F.A.; Woods, R.L.; Peli, E. Shape and individual variability of the blur adaptation curve. Vision Res. **2010**, *50* (15), 1452–1461.
78. Murray, S.; Bex, P.J. Perceived blur in naturally contoured images depends on phase. Front. Psychol. **2010**, *1* (185), 1–12.
79. Prusinkiewicz, P.; Lindenmayer, A. *The Algorithmic Beauty of Plants*; Springer-Verlag: New York, 1990.
80. Cook, R.L.; Carpenter, L.; Catmull, E. The Reyes image rendering architecture. In *SIGGRAPH '87: Proceedings of the 14th Annual Conference on Computer Graphics and Interactive Techniques*; 1987, 95–102.
81. Smits, B.; Shirley, P.; Stark, M.M. Direct ray tracing of displacement mapped triangles. In *Proceedings of the Eurographics Workshop on Rendering*; 2000, 307–318.
82. Perlin, K. An image synthesizer. In *SIGGRAPH '85: Proceedings of the 12th Annual Conference on Computer Graphics and Interactive Techniques*; 1985, 287–296.
83. Fournier, A.; Fussell, D.; Carpenter, L. Computer rendering of stochastic models. Commun. ACM **1982**, *25* (6), 371–384.
84. Graham, D.J.; Field, D.J. Statistical regularities of art images and natural scenes: Spectra, sparseness and nonlinearities. Spatial Vision **2007**, *21* (1–2), 1–2.
85. Graham, D.J.; Field, D.J. Variations in intensity statistics for representational and abstract art, and for art from the eastern and western hemispheres. Perception **2008**, *37* (9), 1341–1352.
86. Graham, D.J.; Friedenberg, J.D.; Rockmore, D.N. Efficient visual system processing of spatial and luminance statistics in representational and nonrepresentational art. In *Proceedings of IS&T/SPIE Electronic Imaging*, Vol. 72401N, 2009.
87. Graham, D.J.; Friedenberg, J.D.; McCandless, C.H.; Rockmore, D.N. Preference for art: Similarity, statistics, and selling price. In *Proceedings of IS&T/SPIE Electronic Imaging*; International Society for Optics and Photonics: Bellingham, WA, 2010, 75271A.
88. Graham, D.J.; Friedenberg, J.D.; Rockmore, D.N.; Field, D.J. Mapping the similarity space of paintings: Image statistics and visual perception. Visual Cogn. **2010**, *18* (4), 559–573.
89. Graham, D.J.; Field, D.J. Global nonlinear compression of natural luminances in painted art. In *Proceedings of IS&T/SPIE Electronic Imaging*, 2008, 68100K.
90. Redies, C.; Hasenstein, J.; Denzler, J. Fractal-like image statistics in visual art: Similarity to natural scenes. Spatial Vision **2007**, *21* (1–2), 97–117.
91. Redies, C. A universal model of esthetic perception based on the sensory coding of natural stimuli. Spatial Vision **2007**, *21* (1–2), 1–2.
92. Koch, M.; Denzler, J.; Redies, C. 1/ f 2 Characteristics and isotropy in the Fourier power spectra of visual art, cartoons, comics, mangas, and different categories of photographs. PLoS ONE **2010**, *5* (8), e12268.
93. Redies, C.; Hänisch, J.; Blickhan, M.; Denzler, J. Artists portray human faces with the Fourier statistics of complex natural scenes. Network: Comput. Neural Syst. **2007**, *18* (3), 235–248.
94. Taylor, R.P.; Micolich, A.P.; Jonas, D. Fractal analysis of Pollock's drip paintings. Nature **1999**, *399* (6735), 422.
95. Alvarez-Ramirez, J.; Ibarra-Valdez, C.; Rodriguez, E.; Dagdug, L. 1/ f-noise structures in Pollocks's drip paintings. Physica A **2008**, *387* (1), 281–295.

96. Coddington, J.; Elton, J.; Rockmore, D.; Wang, Y. Multifractal analysis and authentication of Jackson Pollock paintings. In *Proceedings of the SPIE: Computer Image Analysis in the Study of Art*, Vol. 6810; Stork, D.G.; Coddington, J.; Eds.; 2008, 68100F.
97. Irfan, M.; Stork, D.G. Multiple visual features for the computer authentication of Jackson Pollock's drip paintings: Beyond box counting and fractals. In *SPIE: Image Processing: Machine Vision Applications II*, Vol. 7251; 2009, 1–11.
98. Al-Ayyou, M.; Irfan, M.T.; Stork, D.G. Boosting multifeature visual texture classifiers for the authentication of Jackson Pollock's drip paintings. In *SPIE: Computer Vision and Image Analysis of Art II*, Vol. 7869; 2011, 1.
99. Mureika, J.R.; Dyer, C.C.; Cupchik, G.C. Multifractal structure in nonrepresentational art. Phys. Rev. E **2005**, *72* (4), 281–295.
100. Mureika, J.R.; Fairbanks, M.S.; Taylor, R.P. Multifractal comparison of the painting techniques of adults and children. In *SPIE: Computer Vision and Image Analysis of Art*, Vol. 7531; Stork, D.; Coddington, J.; Bentkowska-Kafel, A.; Eds.; 2010; 1–6.
101. Mureika, J.R.; Taylor, R.P. The abstract expressionists and les automatistes: A shared multi-fractal depth? Signal Process. **2013**, *93* (3), 573–578.
102. Taylor, R.P.; Guzman, R.; Martin, T.P.; Hall, G.D.R.; Micolich, A.P.; Jonas, D.; Scannell, B.C.; Fairbanks, M.S.; Marlow, C.A. Authenticating Pollock paintings using fractal geometry. Pattern Recognit. Lett. **2007**, *28* (6), 695–702.

Fuzzy Set Theory

Donald Kraft
Department of Computer Science, U.S. Air Force Academy, Colorado Springs, Colorado, U.S.A.

Gloria Bordogna
Italian National Research Council, Institute for the Dynamics of Environmental Processes, Dalmine, Italy

Gabriella Pasi
Department of Informatics, Systems and Communication, University of Studies of Milano Bicocca, Milan, Italy

Abstract

This entry presents a definition of fuzzy set theory and an overview of some applications to model flexible information retrieval systems. The entry focuses on a description of fuzzy indexing procedures defined to represent the varying significance of terms in synthesizing the documents' contents, the representation of structured documents so as to model a subjective view of document content, the definition of flexible query languages which allow the expression of soft selection conditions, and fuzzy associative retrieval mechanisms to model fuzzy pseudothesauri, fuzzy ontologies, and fuzzy categorizations of documents.

INTRODUCTION

The objective of this entry is to provide an overview of some applications of fuzzy set theory to design flexible information retrieval systems (IRSs). The term "flexible" implies that we consider IRSs that can represent and manage the uncertainty, vagueness, and subjectivity, which are characteristic of the process of information searching and retrieval.

Consider the notions that index terms offer only an approximate and incomplete view of a document's content, that query languages (such as those incorporating Boolean logic) do not usually allow users to express vague requirements for specifying selection conditions that are tolerant to imprecision, and that a document's relevance to the user's query is a subjective and an imprecise notion. We show how imprecision, vagueness, and subjectivity can be managed within the formal framework of fuzzy set theory. This means that retrieval mechanisms capable of both modeling human subjectivity and of estimating the partial relevance of documents to a user's needs can be designed.

The retrieval process is introduced as a fuzzy multicriteria decision-making (MCDM) activity in the presence of vagueness. Documents constitute the set of the alternatives described using weighted index terms. The query specifies a set of soft constraints on the document representations that are created via indexing. The retrieval mechanism performs a decision analysis in the presence of imprecision to rank the documents on the basis of their partial satisfaction of the soft constraints.

This entry is organized as follows: in the section on "Current Trends in IR," the current trends and key issues in IR are discussed. In the section on "Fuzzy Retrieval Models," an overview of the basic notions of fuzzy set theory to model flexible IRSs are presented. In the section on "Fuzzy Document Indexing," a description of the traditional fuzzy document representation is first illustrated. In addition, both a fuzzy representation of documents structured into logical sections that can be adapted to the subjective needs of a user and a fuzzy representation of HTML documents are presented. In the section on "Flexible Querying," a description of how the Boolean query language of IR can be extended so as to make it flexible and suitable to express soft constraints by capturing the vagueness of the user needs is presented. Both numeric and linguistic selection conditions are introduced to qualify term's importance, and it is shown how linguistic quantifiers are defined to specify soft aggregation operators of query terms. In the section on "Fuzzy Associative Mechanisms," a description of how fuzzy sets can serve to define associative mechanisms to expand the functionalities of IRSs are presented. The focus of current research trends in IR is on the semantic web, i.e., the capability to represent concepts and to model their semantic relationships: fuzzy sets provide notions that can be applied to this purpose allowing to model either fuzzy pseudothesauri and fuzzy ontologies and to build fuzzy categorizations of documents by fuzzy clustering techniques. In the section on "Fuzzy Performance Measures," fuzzy performance measures for IRSs are introduced and the conclusion summarizes the main contents of this entry.

CURRENT TRENDS IN IR

In this section the current trends and the key issues in IR are introduced.

Current Trends in IR

Some of the current trends in IR research run the gamut in terms of expanding the discipline both to incorporate the latest technologies and to cope with novel necessities. In terms of novel necessities, with the diffusion of the Internet and the heterogeneous characteristics of users of search engines, which can be regarded as the new frontier of IR, a new central issue has arisen, generally known as the semantic web. It mainly consists in expanding IRSs with the capability to represent and manage the semantics of both user requests and documents so as to be able to account for user and document contexts. This need becomes urgent with cross-language retrieval, which consists in expressing queries in one language, and retrieving documents written in another language, that is what commonly happens when submitting queries to search engines. Cross language retrieval not only implies new works on text processing, e.g., stemming conducted on a variety of languages, new models of IR such as the development of language models, but also the ability to match terms in distinct languages at a conceptual level, by modeling their meaning.

Another research trend of IR is motivated by the need to manage multimedia collections with nonprint audio elements such as sound, music, and voice, and video elements such as images, pictures, movies, and animation. Retrieval of such elements can include consideration of both metadata and content-based retrieval techniques. The definition of new IRSs capable to efficiently extract content indexes from multimedia documents, and to effectively retrieve documents by similarity or proximity to a query by example so as to fill the semantic gap existing between low-level syntactic index matching and the semantics of multimedia document and query are still to come.

In addition, modern computing technology, including storage media, distributed and parallel processing architectures, and improved algorithms for text processing and for retrieval, has an effect on IRSs. For example, improved string searching algorithms have improved the efficiency of search engines. Improved computer networks have made the Internet and the World Wide Web a possibility. Intelligent agents can improve retrieval in terms of attempting to customize and personalize it for individual users. Moreover, great improvements have been made in retrieval systems interfaces based on human–computer interface research.

These novel research trends in IR are faced by turning to technologies such as natural language processing, image processing, language models, artificial intelligence, and automatic learning.

Also fuzzy set theory can play a crucial role to define novel solutions to these research issues since it provides suitable means to cope with the needs of the semantic web,[1,2] e.g., to model the semantic of linguistic terms so as to reflect their vagueness and subjectivity and to compute degrees of similarity, generalization, and specialization between their meanings.

Key Issues in IR

Modeling the concept of relevance in IR is certainly a key issue, perhaps the most difficult one, and no doubt the most important one. What makes a document relevant to a given user is still not fully understood, specifically when one goes beyond topicality (i.e., the matching of the topics of the query with the topics of the document). Of course, this leads to the realization that relevance is gradual and subjective.

A second key issue is the representation of the documents in a collection, as well as the representation of users' information needs, especially for the purpose of matching documents to the queries at a "semantic" level. This implies introducing incompleteness, approximation, and managing vagueness and imprecision.

Finally, a key issue is how to evaluate properly an IRS's performance. Here, too, one sees imprecision.

IMPRECISION, VAGUENESS, UNCERTAINTY, AND INCONSISTENCY IN IR

Very often the terms imprecision, vagueness, uncertainty, and inconsistency are used as synonymous concepts. Nevertheless when they are referred to qualify a characteristic of the information they have a distinct meaning.[3] Since IR has to do with information, understanding the different meanings of imprecision, vagueness, uncertainty, and inconsistency allows to better understanding the perspectives of the distinct IR models defined in the literature.

Vagueness and imprecision are related to the representation of the information content of a proposition. For example, in the information request, "find *recent* scientific chapters dealing with the *early* stage of infectious diseases by HIV," the terms *recent* and *early* specify vague values of the publication date and of the temporal evolution of the disease, respectively. The publication date and the phase of an infectious disease are usually expressed as numeric values; their linguistic characterization has a coarser granularity with respect to their numeric characterization.

Linguistic values are defined by terms with semantics compatible with several numeric values on the scale upon which the numeric information is defined. Imprecision is just a case-limit of vagueness, since imprecise values have a full compatibility with a subset of values of the numeric reference scale.

There are several ways to represent imprecise and vague concepts. Indirectly, by defining similarity or proximity relationships between each pair of imprecise and vague concepts.

If we regard a document as an imprecise or vague concept, i.e., as bearing a vague content, a numeric value computed by a similarity measure can be used to express the closeness of any two pairs of documents. This is the way of dealing with the imprecise and vague document and query contents in the vector space model of IR. In this context the documents and the query are represented as points in a vector space of terms and the distances between the query and the documents points are used to quantify their similarity.

Another way to represent vague and imprecise concepts is by means of the notion of fuzzy set. The notion of a fuzzy set is an extension to normal set theory.[4] A set is simply a collection of objects. A fuzzy set (more properly called a fuzzy subset) is a subset of a given universe of objects, where the membership in the fuzzy set is not definite. For example, consider the idea of a person being middle-aged. If a person's age is 39, one can consider the imprecision of that person being in the set of middle-aged people. The membership function, μ, is a number in the interval [0, 1] that represents the degree to which that person belongs to that set. Thus, the terms *recent* and *early* can be defined as fuzzy subsets, with the membership functions interpreted as compatibility functions of the meaning of the terms with respect to the numeric values of the reference (base) variable. In Fig. 1, the compatibility function of the term *recent* is presented with the numeric values of the timescale measured in years. Note that here a chapter that has a publication date of the current year or 1 year previous is perfectly *recent*; however, the extent to which a chapter remains *recent* declines steadily over the next 2 years until chapters older than 3 years have no sense of being *recent*.

In the next sections we will see how the notion of fuzzy set has been used in the IR context to represent the vague concepts expressed in a flexible query for specifying soft selection conditions of the documents. Uncertainty is related to the truth of a proposition, intended as the conformity of the information carried by the proposition with the considered reality. Linguistic expressions such as "probably" and "it is possible that" can be used to declare a partial lack of knowledge about the truth of the stated information.

Further, there are cases in which information is affected by both uncertainty and imprecision or vagueness. For example, consider the proposition "probably document d

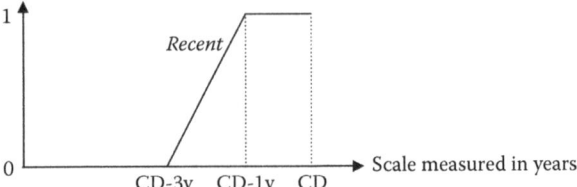

Fig. 1 Semantics of the term "recent" referring to the publication date of a scientific chapter. CD, current date; y, years

is relevant to query q." Possibility theory[5,6] together with the concept of a linguistic variable defined within fuzzy set theory[7] provide a unifying formal framework to formalize the management of imprecise, vague, and uncertain information.[8]

However, the same information content can be expressed by choosing a trade-off between the vagueness and the uncertainty embedded in a proposition. For example, one can express the content of the previous proposition by a new one "document d is more or less relevant to query q." In this latter proposition, the uncertain term *probably* has been eliminated, but the specificity of the vague term *relevant* has been reduced. In point of fact, the term *more or less relevant* is less specific than the term *relevant*. A dual representation can eliminate imprecision and augment the uncertainty, like in the expression "it is *not completely probable* that document d fully satisfies the query q."

One way to model IR is to regard it as an uncertain problem.[9]

On the basis of what has been said about the trade-off between uncertainty and vagueness to express the same information content, there are two alternative ways to model the IR activity. One possibility is to model the query evaluation mechanism as an uncertain decision process. Here the concept of relevance is considered binary (crisp) and the query evaluation mechanism computes the probability of relevance of a document d to a query q. Such an approach, which does model the uncertainty of the retrieval process, has been introduced and developed by probabilistic IR models.[10–12] Another possibility is to interpret the query as the specification of soft "elastic" constraints that the representation of a document can satisfy to an extent, and to consider the term *relevant* as a gradual (vague) concept. This is the approach adopted in fuzzy IR models.[9,13] In this latter case, the decision process performed by the query evaluation mechanism computes the degree of satisfaction of the query by the representation of each document. This satisfaction degree, called the retrieval status value (RSV), is considered as an estimate of the degree of relevance (or is at least proportional to the relevance) of a given document with respect to a given user query. An RSV of 1 implies maximum relevance; an RSV value of 0 implies absolutely no relevance. And, an RSV value in the interval [0, 1] implies an intermediate level or degree of relevance. For example, an RSV value of 0.5 could imply an average degree of relevance.

Inconsistency comes from the simultaneous presence of contradictory information about the same reality. An example of inconsistency can be observed when submitting the same query to several IRSs that adopt different representations of documents and produce different results. This is actually very common and often occurs when searching for information over the Internet using different search engines. To solve this kind of inconsistency, some fusion strategies can be applied to the ranked lists each search engine produces. In fact, this is what metasearch engines do.[14,15]

In this entry, we analyze the representation and management of vagueness as a means of improving the flexibility of IRSs. In particular, we will focus on the modeling of vagueness and in fuzzy IR models.

The document representation based on a selection of index terms is invariably incomplete. When synthesizing the content of a text manually by asking an expert to select a set of index terms, one introduces subjectivity in the representation. On the other hand, automatic full-text indexing introduces imprecision since the terms are not all fully significant in characterizing a document's content. However, these terms can have a partial significance that might also depend upon the context in which they appear, i.e., which document component.

In the query formulation, users often only have a vague idea of the information for which they are looking. Users therefore find it difficult to translate their needs into a precise request using a formal query language such as one employing Boolean logic.

A flexible IRS should be capable of providing more detailed and rich representations of documents and of interpreting vague queries in order to perform retrieval processes that tolerate, and account for, this vagueness.

FUZZY RETRIEVAL MODELS

Fuzzy retrieval models have been defined in order to reduce the imprecision that characterizes the Boolean indexing process, to represent the user's vagueness in queries, and to deal with discriminated answers estimating the partial relevance of the documents with respect to queries. Extended Boolean models based on fuzzy set theory have been defined to deal with one or more of these aspects.[16–24] Surveys of fuzzy extensions for IRSs and of fuzzy generalizations of the Boolean retrieval model can be found in Kraft[9] and Bordogna.[13]

Fuzzy "knowledge-based" models[25,26] and fuzzy associative mechanisms[27–30] have been defined to cope with the incompleteness that characterizes either the representation of documents or the users' queries. Miyamoto[31] illustrates a wide range of methods to generate fuzzy associative mechanisms.

It has been speculated that Boolean logic is passé, out of vogue. Yet, researchers have employed p-norms in the vector space model or Bayesian inference nets in the probabilistic model to incorporate Boolean logic into those models. In addition, the use of Boolean logic to separate a collection of records into two disjoint classes has been considered, e.g., using the one-clause-at-a time (OCAT) methodology.[32] Moreover, even now retrieval systems such as Dialog and Web search engines such as Google allow for Boolean connectives. It should come as no surprise, therefore, to see extensions of Boolean logic based upon fuzzy set theory for IR.

It is noteworthy that most of the research on fuzzy retrieval has been conducted by a relatively few scholars. Moreover, that research has focused upon theoretical models, focusing primarily upon text retrieval, so that precious little testing has to date been conducted.

Extensions of Fuzzy Boolean Retrieval Models

The fuzzy retrieval models have been defined as generalizations of the classical Boolean model. These allow one to extend existing Boolean IRSs without having to redesign them. This was first motivated by the need to be able to produce proper answers in response to the queries. In essence, the classical Boolean IRSs apply an exact match between a Boolean query and the representation of each document. This document representation is defined as a set of index terms. These systems partition the collection of documents into two sets, the retrieved documents and the rejected (non-retrieved) ones. As a consequence of this crisp behavior, these systems are liable to reject useful items as a result of too restrictive queries, as well as to retrieve useless material in reply to queries.[33]

The softening of the retrieval activity in order to rank the retrieved items in decreasing order of their relevance to a user query can greatly improve the effectiveness of such systems. This objective has been approached by extending the Boolean models at different levels. We shall consider those levels in order to model retrieval using a MCDM model that allows the expression of the users' queries as soft constraints and produce discriminated answers.

Fuzzy Techniques for Documents' Indexing

The aim here is to provide more specific and exhaustive representations of each document's information content. This means improving these representations beyond those generated by existing indexing mechanisms. We shall introduce the fuzzy interpretation of a weighted document representation and then later introduce a fuzzy representation of documents structured in logical sections that can be adapted to a user that has subjective criteria for interpreting the content of documents.[17] In this same vein, we shall describe an indexing procedure for HTML documents.[34]

Definition of Flexible Query Languages

The objective here is to define query languages that are more expressive and natural than classical Boolean logic. This is done in order to capture the vagueness of user needs as well as to simplify user–system interaction. This has been pursued with two different approaches. First, there has been work on the definition of soft selection criteria (soft constraints), which allow the specification of the different importance of the search terms. Query languages based on numeric query term weights with different semantics have been first proposed as an aid to define more expressive selection criteria.[18,21,22,24,35] An evolution of these approaches has been defined that introduces linguistic query weights, specified by fuzzy sets such as *important* or *very important*, in order to express the different vague importance of the query terms.[36] Second, there is the approach of introducing soft aggregation operators for the selection criteria, characterized by a parametric behavior which can be set between the two extremes of intersection (AND) and union (OR) as adopted in Boolean logic. Boolean query languages have been extended and generalized by defining aggregation operators as linguistic quantifiers such as *at least k* or *about k*.[16]

As we shall see, the incorporation of weighted document representations in a Boolean IRS is a sufficient condition to improve the system via a ranking capability. As a consequence of this extension, the exact matching that is employed by a classical Boolean IRS is softened using a partial matching mechanism that evaluates the degree of satisfaction of a user's query for each document. This degree of satisfaction is the RSV that is used for ranking.

Fuzzy Associative Mechanisms

These associative mechanisms allow to automatically generating fuzzy pseudothesauri, fuzzy ontologies, and fuzzy clustering techniques to serve three distinct but compatible purposes. First, fuzzy pseudothesauri and fuzzy ontologies can be used to contextualize the search by expanding the set of index terms of documents to include additional terms by taking into account their varying significance in representing the topics dealt with in the documents. The degree of significance of these associated terms depends on the strength of the associations with a document's original descriptors. Second, an alternative use of fuzzy pseudothesauri and fuzzy ontologies is to expand the query with related terms by taking into account their varying importance in representing the concepts of interest. The importance of an additional term is dependent upon its strength of association with the search terms in the original query. Third, fuzzy clustering techniques, where each document can be placed within several clusters with a given strength of belonging to each cluster, can be used to expand the set of the documents retrieved in response to a query. Documents associated with retrieved documents, i.e., in the same cluster, can be retrieved. The degree of association of a document with the retrieved documents does influence its RSV. Another application of fuzzy clustering in IR is that of providing an alternative way, with respect to the usual ranked list, of presenting the results of a search.

FUZZY DOCUMENT INDEXING

In order to increase the effectiveness of IRSs, the indexing process plays a crucial role. In fact, it is not sufficient to provide IRSs with powerful query languages or sophisticated retrieval mechanisms to achieve effective results if the representation of documents oversimplifies their information content.

Vector Space, Probabilistic, and Generalized Boolean Indexing

The vector space model and the probabilistic models generally adopt a weighted document representation, which has improved the Boolean document representation by allowing the association of a numeric weight with each index term.[10,33] The automatic computation of the index term weights is based on the occurrences count of a term in the document and in the whole archive.[37–39] In this case, the indexing mechanism computes for each document d and each term t a numeric value by means of a function F. An example of F which has the index term weight increasing with the frequency of term t in document d but decreasing with the frequency of the term in all the documents of the archive is given by

$$F(d,t) = tf_{dt} \times g(\text{IDF}_t) \qquad (1)$$

where
- tf_{dt} is a normalized term frequency, which can be defined as:

$$= tf_{dt} = \frac{\text{OCC}_{dt}}{\text{MAXOCC}_d}$$

- OCC_{dt} is the number of occurrences of t in d
- MAXOCC_d is the number of occurrences of the most frequent term in d
- IDF_t is an inverse document frequency which can be defined as: IDF_t

$$\text{IDF}_t = \log \frac{N}{\text{NDOC}_t}$$

- N is the total number of documents in the archive
- NDOC_t is the number of documents indexed by t
- g is a normalizing function

The computation of IDF_t is particularly costly in the case of large collections which are updated online.

The definition of such a function F is based on a quantitative analysis of the text, which makes it possible to model the qualitative concept of significance of a term in describing the information carried by the text. The adoption of weighted indexes allows for an estimate of the relevance, or of the probability of relevance, of documents to a query.[10,33]

Based on such an indexing function, and by incorporating Boolean logic into the query, the first fuzzy interpretation of an extended Boolean model has been to adopt a weighted document representation and to interpret it as a fuzzy set of terms.[40] From a mathematical point of view, this is a quite natural extension: the concept of the significance of index terms in describing the information content of a document can then be naturally described by adopting the function F, such as the one defined by Zadeh[5] as the membership function of the fuzzy set representing a document's being in the subset of concepts represented by the term in question. Formally, a document is represented as a fuzzy set of terms: $R_d = \sum_{t \in T} \mu_{Rd}/t$ in which the membership function is defined as $\mu_{Rd}: D \times T \to [0,1]$. In this case, $\mu_{Rd}(t) = F(d,t)$, i.e., the membership value, can be obtained by the indexing function F. We describe later that through this extension of the document representation, the evaluation of a Boolean query produces a numeric estimate of the relevance of each document to the query, expressed by a numeric score or RSV, which is interpreted as the degree of satisfaction of the constraints expressed in a query.

Fuzzy set theory has been applied to define new and more powerful indexing models than the one based on the function specified in Eq. 1. The definition of new indexing functions has been motivated by several considerations. First, these F functions do not take into account the idea that a term can play different roles within a text according to the distribution of its occurrences. Moreover, the text can be considered as a black box, closed to users' interpretation. Such users might naturally filter information by emphasizing certain subparts on the basis of their subjective interests. This outlines the fact that relevance judgments are driven by a subjective interpretation of the document's structure, and supports the idea of *dynamic* and *adaptive* indexing.[17,41] By adaptive indexing, we mean indexing procedures which take into account the users' desire to *interpret* the document contents and to "build" their synthesis on the basis of this interpretation.

Fuzzy Representation of Structured Documents

We also consider the synthesis of a fuzzy representation of structured documents that takes into account the user needs.[17] A document can be represented as an entity composed of sections (e.g., *title*, *authors*, *introduction*, and *references*). For example, a single occurrence of the term in the *title* indicates that the chapter is concerned with the concept expressed by the term, while a single occurrence in the *reference* suggests that the chapter refers to other publications dealing with that concept. The information role of each term occurrence depends then on the semantics of the subpart where it is located. This means that to the aim of defining an indexing function for structured documents the single occurrences of a term may contribute differently to the significance of the term in the whole document. Moreover, the document's subparts may have a different importance determined by the users' needs. For example, when looking for chapters written by a certain author, the most important subpart would be the *author name*; while when looking for chapters on a certain topic, the *title*, *abstract*, and *introduction* subparts would be preferred.

Of course, when generating an archive of a set of documents, it is necessary to define the sections which one wants to employ to structure each document. The decision of how to structure the documents, i.e., the type and number of sections, depends on the semantics of the documents and on the accuracy of the indexing module that one wants to achieve. A formal representation of a document will be constituted using a fuzzy binary relation: with each pair <section, term>, a significance degree in the interval [0, 1] is computed to express the significance of that term in that document section. To obtain the overall significance degree of a term in a document, i.e., the index term weight, these values are *dynamically* aggregated by taking into account the indications that a user explicits in the query formulation. Other non-fuzzy approaches have also introduced the concept of boosting factor to emphasize differently the contribution of the index terms occur-rences depending on the document sections to the overall index term weights. However these approaches compute *static* index term weights during the indexing process, without taking into account the user interpretation.

On the contrary, in the fuzzy approach the aggregation function is defined on two levels. First, the user expresses preferences for the document sections (the equivalent of the boosting factors), specifying those sections that the system should more heavily weight in order to take proper account of the evaluation of the relevance of a given document to that user's query. Second, the user should decide which aggregation function has to be applied for producing the overall significance degree. This is done by the specification of a linguistic quantifier such as *at least one, at least k*, or *all*.[42] By adopting this document representation, the same query can select documents in different relevance order depending on the user's indicated preferences.

An indexing model has been proposed by which the occurrences of a term in the different documents' sections are taken into account according to specific criteria, and the user's interpretation of the text is modeled.[17] During the retrieval phase, the user can specify the distinct importance (preference) of the sections and decide that a term must be present in *all* the sections of the document or in *at least a certain number* of them in order to consider the term fully significant. A section is a logical subpart identified by s_i, where $i \in 1,\ldots, n$ and n is the total number of the sections in the documents. We assume here that an archive contains documents sharing a common structure.

Formally, a document is represented as a fuzzy binary relation:

$$R_d = \sum_{(t,s) \in T \times S} \mu_d(t,s)/(t,s) \qquad (2)$$

The value $\mu_d(t, s) = F_s(d, t)$ expresses the significance of term t in section s of document d. A function $F_s: D \times T \to [0, 1]$ is then defined for each section s. The overall significance degree $F(d, t)$ is computed by combining the single significance degrees of the sections, the $F_s(d, t)$s, through an aggregation function specified by the user. This function is identified by a fuzzy linguistic quantifier such as *all*, *at least k*, or *at least 1*, which aggregates the significance degrees of the sections according to their importance values as specified by the user.

The criteria for the definition of F_s are based on the semantics of section s and are specified by an expert during the indexing of the documents. For example, for sections containing short texts or formatted texts, such as the *author* or *keywords*, a single occurrence of a term makes it fully significant in that section: in this case, it could be assumed that $F_s(d, t) = 1$ if t is present in s but 0 otherwise. On the other hand, for sections containing textual descriptions of variable length such as the *abstract* and *title* sections, $F_s(d, t)$ can be computed as a function of the normalized term frequency in the section as for example:

$$\mu_s(d,t) = tf_{dst} * IDF_t \qquad (3)$$

in which IDF_t is the inverse document frequency of term t [see definition (5)], tf_{dst} is the normalized term frequency defined as:

$$tf_{dst} = \frac{OCC_{dst}}{MAXOCC_{sd}}$$

in which OCC_{dst} is the number of occurrences of term t in section s of document d and $MAXOCC_{sd}$ is a normalization parameter depending on the section's length so as not to underestimate the significance of short sections with respect to long ones. For example, this normalization parameter could be computed as the frequency of the term with the highest number of occurrences in the section.

To simplify the computation of this value, it is possible to heuristically approximate it: during the archive generation phase, with an expert indicating the estimated percentage of the average length of each section with respect to the average length of documents ($PERL_s$). Given the number of occurrences of the most frequent term in each document d, $MAXOCC_d$, an approximation of the number of occurrences of the most frequent term in section s of document d is

$$MAXOCC_{sd} = PERL_s * MAXOCC_d$$

Term Significance

To obtain the overall degree of significance of a term in a document, an aggregation scheme of the $F_s(d, t)$s values has been suggested, based on a twofold specification of the user.[17] When starting a retrieval session, users can specify their preferences on the sections s by a numeric score $\alpha_s \in [0, 1]$ where the most important sections have an importance weight close to 1. Moreover, users can select a linguistic quantifier to specify the aggregation criterion; the quantifier can be chosen among *all* (the most restrictive one), *at least one* (the weakest one), or *at least k* which is associated with an intermediate aggregation criterion.

Within fuzzy set theory linguistic quantifiers used to specify aggregations are defined as ordered weighted averaging (OWA) operators.[43] When processing a query, the first step accomplished by the system for evaluating $F(d, t)$ is the selection of the OWA operator associated with the linguistic quantifier lq, OWA_{lq}. When the user does not specify any preferences on the documents' sections, the overall significance degree $F(d, t)$ is obtained by applying directly the OWA_{lq} operator to the values $\mu_1(d, t),\ldots,\mu_n(d, t)$:

$$F(d,t) = OWA_{lq}\left(\mu_1(d,t),\ldots,\mu_n(d,t)\right)$$

When distinct preference scores $\alpha_1,\ldots,\alpha_n$ are associated with the sections, it is first necessary to modify the values $\mu_1(d, t), \ldots, \mu_n(d, t)$ in order to increase the "contrast" between the contributions due to important sections with respect to those of less important ones. The evaluation of the overall significance degree $F(d, t)$ is obtained by applying the operator OWA_{lq} to the modified degrees $a_1,\ldots, a_n$: $F(d, t) = OWA_{lq}(a_1,\ldots, a_n)$.

We can now briefly sketch a comparison of the effectiveness of a system adopting a simple weighted representation versus a system with this structured weighted representation. In particular, the different rankings of two documents obtained by adopting the two different representations are outlined by an example. The two documents considered in the archive of CNR research projects contain the term "genoma." Table 1 shows the normalized frequency of "genoma" in the sections of the two documents; as it can be noticed, the term "genoma" has the same total number of occurrences in both documents. Since the normalization factors are the same, by applying F as defined in Eq. 1, the significance of "genoma" in both documents gets the same value $F(d_1, genoma) = F(d_2, genoma) = 0.8$. Table 2 shows the significance degrees for each section in which the term "genoma" occurs. These degrees are obtained using the fuzzy representation of structured documents; since the title and keywords sections are short texts, μ_{title} and $\mu_{keywords}$ are defined so as to take values in $\{0, 1\}$. After estimating that the objective section takes up averagely 30% of the documents' length, and the description section is around 40%, $\mu_{objective}$ and $\mu_{description}$ are defined.

When the user does not specify any criterion to aggregate the single degrees of the sections, a default aggregation operator is used.[16,17] Since no importance is specified to differentiate the contributions of the sections, all of them are assumed to have the same importance weight of 1. Notice that the document d_1, which contains "genoma" in the *keywords* and title sections is now considered more significant with respect to document d_2 that contains the term just in the *objectives* and *description* sections.

These results could be reversed if the user specifies that the presence of the term "genoma" in the *objectives* section is fundamental. Table 3 illustrates this situation, showing the modified degrees of significances of the sections when the user sets the aggregation criterion equal to at *least 1* and $\alpha_{objective} = 1$, $\alpha_{title} = \alpha_{keywords} = \alpha_{description} = 0.5$, and $\alpha_i = 0$ otherwise.

The fact that the user can explicate the preferences on the section and the aggregation criterion by a linguistic quantifier allows a subjective interpretation of document content

Table 1 Normalized frequency of "genoma" in the sections of the two documents

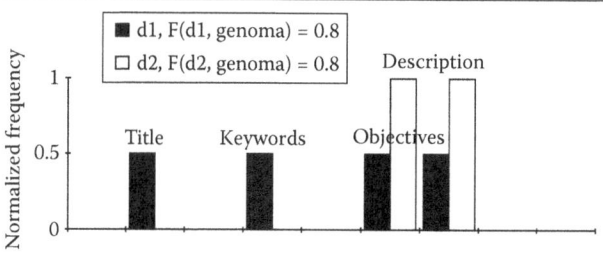

Table 2 Significance degrees of "genoma" in each section of the two documents

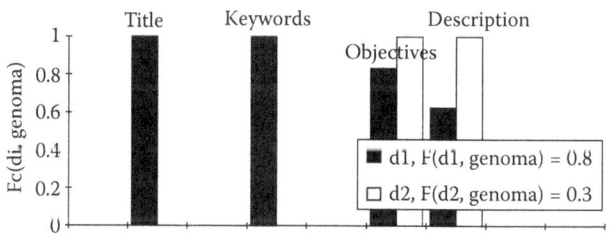

Table 3 Modified significance degrees of the term "genoma" in the documents sections

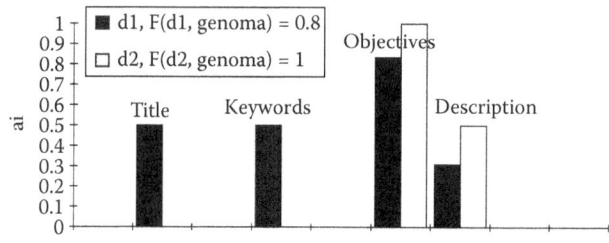

and gives the user the possibility of full control on the system behavior. This is not the case for other IR models, the probabilistic model (e.g., Bayesian updating of the probabilities as part of relevance feedback), and Rocchio's relevance feedback mechanism for the vector space model, or even the calculation of the rank for Web pages retrieved by Google using PageRank. In these models the retrieval criteria remain implicit and are not observable directly by the user.

Experimental Results

A comparison of the results produced by using the traditional fuzzy representation of documents and the fuzzy representation of structured documents can be found.[17] In this experiment, a collection of 2500 textual documents about descriptions of CNR research projects has been considered. The indexing module of the prototypal IRS named DOMINO, used for the experiment, has been extended in order to be able to recognize in the documents any structure simply by specifying it into a definition file. In this way it is not necessary to modify the system when dealing with a new collection of documents with a different structure. The definition of the documents sections has been made before starting the archive generation phase. During this phase it was also necessary to specify the criteria by which to compute the significance degrees of the terms in each section. Two kinds of sections have been identified: the "structured" sections, i.e., the research code, title, research leader, and the "narrative" sections, containing unstructured textual descriptions, i.e., the project description and the project objective. It has been observed that while the values of precision remain unchanged in the two versions of the system, the values of recall are higher by using the structured representation than those obtained by using the traditional fuzzy representation.

We illustrate another approach which produces a weighted representation of documents written in HTML.[34] An HTML document has a specific syntactic structure in which its subparts have a given format specified by the delimiting tags. In this context, tags are seen as syntactic elements carrying an indication of the importance of the associated text. When writing a document in HTML, an author associates varying importance to each of the different subparts of a given document by delimiting them by means of appropriate tags. Since a certain tag can be employed more than once, and in different positions inside the document, the concept of document subpart is not meant as a unique, adjacent piece of text. Such a structure is subjective and carries the interpretation of the document author. It can be applied in archives, which collect heterogeneous documents, i.e., documents with possibly different "logical" structures.

When generating a HTML document, an author exploits the importance weights associated with different subparts of the text. If characters of different fonts are used, it is assumed that the larger the font, the more important the information carried by the text. Moreover, to use boldface or italics for characters generally means the highlighting of a portion of the text. Tags constitute then indicators of the importance of documents' subparts.

An indexing function has been proposed which provides different weights for the occurrences of a given term in the document, depending on the tags by which they are delimited.[34] The overall significance degree $F(d, t)$ of a term t in a document d is computed by first evaluating the term significance in the different document tags, and then by aggregating these contributions. With each tag, a function $F_{tag}:D \times T \to [0, 1]$ is associated together an importance weight $\mu_{tag} \in [0, 1]$. Note that the greater is the emphasis of the text associated with a tag, the greater is its importance weight. A possible ranking of the considered tags has been suggested[34] in decreasing order of tag importance. The definition of such a list is quite subjective, although based on objective assumptions suggested by commonsense. These rankings include notion such as a larger font for the characters in a portion of text yields greater importance of that portion of text, or text in boldface or italics or appearing in a list can be assumed as having a higher importance. Of course, other orderings could be defined.

To simplify the hierarchy of the tags, we see that certain tags can be employed to accomplish similar aims, so one can group them into different classes. It is assumed that the members of a class have the same importance weight. Text not delimited by any tag is included into the lowest class. A simple procedure to compute numeric importance weights starting from the proposed ranking can be achieved. The definition of F_{tag} follows the same mechanism as the previous approach.[17] The following normalized frequency is now proposed:

$$F_{tag}(d,t) = \frac{NOCC_{tag\ dt}}{MAXOCC_{tag\ d}}$$

in which $NOCC_{tag\ dt}$ is the number of occurrences of term t inside tag in document d, and $MAXOCC_{tag\ d}$ is the number of occurrences of the most frequent term inside the tag.

Once the single significance degrees of a term into the tags have been computed, these have to be aggregated in order to produce an overall significance degree of the term into the document. In the aggregation all the significance degrees should be taken into account, so as to consider the contribution of each tag, modulated by their importance weights. To this aim a weighted mean can be adopted:

$$A\left(F_{tag_1}(d,t),\ldots,F_{tag_n}(d,t)\right) = \sum_{i=1,\ldots,n} F_{tag_i}(d,t) * w_i$$

in which $\sum_{i=1,\ldots,n} w_i = 1$. Starting from the list of tags in decreasing relative order of their importance, the numeric weights w_i are computed through a simple procedure. Assuming that tag_i is more important than tag_j iff $I<j$ (i and j being the positions of tag_i and tag_j respectively in the ordered list), the numeric importance weight w_i associated with tag_i can be computed as $w_i = (n-i+i)\big/\sum_{i=1,\ldots,n} i$.

In the computation of the overall significance degree $F(d, t)$, the inverse document frequency of term t could be taken into account:

$$F(d,t) = \left(\sum_{i=1,\ldots,n} F_{tag\ i}(d,t)^* w_i^* g(IDF_t) \right)$$

in which the definition of $g(IDF_t)$ is given in formula (5).

FLEXIBLE QUERYING

A flexible query language is a query language that incorporates some elements of natural language so users have a simple, powerful, and yet subjective mechanism by which to express their information needs. Flexibility can also be a characteristic of the query evaluation mechanism to allow a tuning of the query's concepts' semantics with respect to the document collection, the user's subjective statement of information need, and even the user's application domain. Linguistic variables provide a suitable framework to generalize, to represent, and to manage the linguistics of the query's concepts. Thus, this approach can be used to formalize the semantics of linguistic terms introduced in a Boolean query language.

Flexible query languages have been defined as generalizations of Boolean query languages that employ Boolean logic. Within the framework of fuzzy set theory, we have the idea of a softening of the rigid, crisp constraints of a Boolean condition being strictly true (a document has a keyword) or false (the document does not contain the keyword).

A flexible query consists of either or both of two soft components. First, there can be selection conditions interpreted as soft constraints on the significance of the index terms in each document representation. Second, there can be soft aggregation operators, which can be applied to the soft constraints in order to define compound selection conditions. The atomic selection conditions for individual terms are expressed by pairs <term, weight>, in which the weight can be a numeric value in the interval [0, 1] that is used to identify a soft constraint or the weight can be a linguistic value for the variable *importance*. The compound conditions for combining terms via Boolean logic are expressed by means of linguistic quantifiers used as aggregation operators.

Query Evaluation Mechanism

Query processing within retrieval can be interpreted as a decision-making activity. Its aim is to evaluate a set of alternatives or possible solutions, in this case a set of documents, based upon some criteria or selection conditions in order to select the optimal list (perhaps ranked) of documents in response to a user's query.

In the case of a Boolean query, the alternatives are the document representations as described based on the presence or absence of index terms or keywords. The selection conditions, as expressed by terms specified in a query, define a set of constraints requiring the presence or absence of these terms within a document's representation. These conditions are expressed connected by aggregation operators, i.e., the Boolean logic operators of AND, OR, and NOT. The decision process is performed through an exact matching function, which is strictly dependent on the system query language. This decision process evaluates the global satisfaction of the query constraints for each document representation. Relevance is modeled as a binary property of the documents with respect to the user's query.

Given a fuzzy approach to retrieval, query processing can be regarded as a decision activity affected by vagueness. In fact, the query can be seen as the specification of a set of soft constraints, i.e., vague selection conditions, that the documents can satisfy to a partial extent. The documents described through the significance degrees of the index terms constitute the alternatives. The query evaluation mechanism is regarded as fuzzy decision process that evaluates the degree of satisfaction of the query constraints by each document representation by applying a partial matching function. This degree is the RSV and can be interpreted as the degree of relevance of the document to the query and is used to rank the documents. Then, as a result of a query evaluation, a fuzzy set of documents is retrieved in which the RSV is the membership value. In this case the definition of the partial matching function is strictly dependent on the query language, specifically on the semantics of the soft constraints.

A wish list of requirements that a matching function of an IRS must satisfy has been proposed.[18,24] Included in this list is the separability property that the evaluation of an atomic selection condition for an individual term in a query should be independent of the evaluation of the other atomic components or their Boolean connectors. The matching function should be based solely upon a function evaluating atomic conditions. Following the calculation of these evaluations, one can then aggregate them based upon the Boolean operators in the query. It has been shown that this property guarantees a homomorphic mapping from the space of all single terms to the space of all possible Boolean queries using these terms.[44] This property has been considered widely within fuzzy retrieval models, especially in the definition of flexible query languages.

By designing the partial matching mechanism from the bottom-up the separability property is ensured. First, each atomic selection condition or soft constraint in the query is evaluated by a function E for a given document. Then the aggregation operators are applied to the results starting from the inmost operator in the query to the outermost operator by a function E^*. This E function evaluates the soft constraints associated with the query atoms on the fuzzy set R_d representing each document, where these soft constraints are defined as fuzzy subsets. The membership value $\mu_{atom}(i)$ is the degree of satisfaction of the soft constraint associated with the atomic query $atom$, i.e., $E(<atom>, d) = \mu_{atom}(F(d, t))$. In other words, E evaluates how well the term t, which has an indexing weight $F(d, t)$ for document d, satisfies the soft constraint specified by $atom$. The result of the evaluation is a fuzzy set, $\sum_{d \in D} \mu_{atom}(F(d,t))/d$ in which $\mu_{atom}(F(d, t))$ is interpreted as the RSV of document d with respect to the query atom.

The function $E^*: D \times Q \rightarrow [0, 1]$, where Q is the set of all the proper queries in the query language, evaluates the final RSV of a document, reflecting the satisfaction of the whole query. The definition of E^* depends strictly upon the structure of the query language, specifically upon the aggregation operators used to combine the atomic components. The AND connective is classically defined as the minimum (min) operator, the OR connective as the maximum (max) operator, and the NOT connective as the one-minus (1) or complement operator. These definitions preserve the idempotence property. A fuzzy generalization of the Boolean query structure has been defined in which the Boolean operators are replaced by linguistic quantifiers.[16] In this context, linguistic quantifiers are used as aggregation operators to determine the degree of satisfaction for the soft constraints. They allow to improve as well as to simplify the expressiveness of the Boolean query language.

Query Weights

To render a Boolean query language to be more user-friendly and more expressive, one can extend the atomic selection conditions by introducing query term weights.[22,23,45,46] An example of weighted query is the following: $<t_1, w_1>$ AND $(<t_2, w_2>$ OR $<t_3, w_3>)$ in which t_1, t_2, and t_3, are search terms with numeric weights w_1, w_2, and w_3 in the interval $[0, 1]$. These weights are implicitly given as being equal to 1 in the classical Boolean query language.

The concept of query weights raises the problem of their interpretation. Several authors have realized that the semantics of query weights should be related to the concept of the "importance" of the terms. Being well aware that the semantics of the query term weights influences the definition of the partial matching function, specifically of the E function, different semantics for the soft constraint imposed by a pair $<t, w>$ have been proposed in the literature trying to satisfy as much as possible properties of the wish list, in particular the separability property.

Early on, query weights were interpreted as a relative importance weight where the separability property does not hold. Two distinct definitions of E have been proposed for conjunctive and disjunctive queries, respectively.[22,47] Later, other models[23,24,46] used an interpretation of the query weights w as a threshold on the index term weight or as an ideal index term weight.[35,45]

Implicit Query Weights

The simplest extension of the Boolean model consists of the adoption of a weighted document representation but with a classical Boolean query language.[40] This retrieval mechanism ranks the retrieved documents in decreasing order of their significance with respect to the user query. In this case, an atomic query consisting of a single term t is interpreted as the specification of a pair $<t, 1>$ in which $w = 1$ is implicitly specified. The soft constraint associated with $<t, 1>$ is then interpreted as the requirement that the index term weight be "close to 1" and its evaluation is defined as $\mu_w(F(d, t)) = F(d, t)$. This means that the desired documents are those with maximum index term weight for the specified term t, i.e., index term weights closest to 1. This interpretation implies that the evaluation mechanism tolerates the under satisfaction of the soft constraint associated with $<t, 1>$ with a degree equal to $F(d, t)$.

Relative Importance Query Weights

Here, query weights are interpreted as measures of the "relative importance" of each term with respect to the other terms in the query.[22,47] This interpretation allows the IRS to rank documents so that documents are ranked higher if they have larger index term weights for those terms that have larger query weights. However, since it is not possible to have a single definition for the soft constraint μ_w that preserves the "relative importance" semantics independently of the Boolean connectors in the query, two distinct definitions of μ_w have been proposed, depending on the aggregation operators in the query. This approach, sadly, gives up the separability property. Two alternative definitions have been proposed for conjunctive and disjunctive queries.[22,47] The first proposal[22] yields

$\mu_w(F(d, t)) = w\, F(d, t)$ for disjunctive queries and
$\mu_w(F(d, t)) = \max(1, F(d, t)/w)$ for conjunctive queries;
while the second proposal[47] yields
$\mu_w(F(d, t)) = \min[w, F(d, t)]$ for disjunctive queries
$\mu_w(F(d, t)) = \max[(1 - w), F(d, t)]$ for conjunctive queries

Notice that any weighted Boolean query can be expressed in disjunctive normal form (DNF) so that any query can be evaluated by using one of these two definitions.

Threshold Query Weights

To preserve the separability property, an approach treating the query weights as thresholds has been suggested.[23,46] By specifying query weights as thresholds the user is asking to see all documents "sufficiently about" a topic. In this case, the soft constraint identified by the numeric query weight can be linguistically expressed as "more or less over w." Of course, the lower the threshold, the greater the number of documents retrieved. Thus, a threshold allows a user to define a point of discrimination between under- and over satisfaction.

The simplest formalization of threshold weights has been suggested as a crisp threshold:[23]

$$\mu_w(F(d,t)) = \begin{cases} 0 & \text{for } F(d,t) < w \\ F(d,t) & \text{for } F(d,t) \geq w \end{cases}$$

In this case, the threshold defines the minimally acceptable document. Due to its inherent discontinuity, this formalization might lead to an abrupt variation in the number of documents retrieved for small changes in the query weights. To remedy this, continuous threshold formalization has been suggested:[46]

$$\mu_w = (F(d,t)) = \begin{cases} P(w) * \frac{F(d,t)}{w} & \text{for } F(d,t) < w \\ P(w) + P(w) * \frac{(F(d,t)-w)}{(1-w)} & \text{for } F(d,t) \geq w \end{cases}$$

where $P(w)$ and $Q(w)$ might be defined as $P(w) = 1 + w/2$ and $Q(w) = 1 - w^2/4$.

For $F(d, t) < w$, the μ_w function measures the closeness of $F(d, t)$ to w; for $F(d, t) \geq w$, $\mu_w(F(d, t))$ expresses the degree of over satisfaction with respect to w, and under satisfaction with respect to 1.

Ideal Query Weights

Another interpretation for the query weights has been defined.[35,45] Here, the pair $<t, w>$ identifies a set of ideal or perfect documents so that the soft constraint μ_w measures how well $F(d, t)$ comes close to w, yielding:

$$\mu_w(F(d,t)) = e^{\ln(k)*(F(d,t)-w)^2}$$

The parameter k in the interval [0, 1] determines the steepness of the Gaussian function's slopes. As a consequence, k will affect the strength of the soft constraint close to w. So, the larger the value of k is, the weaker the constraint becomes. This parametric definition makes it possible to adapt the constraint interpretation to the user concept of close to w.[36] The retrieval operation associated with a pair $<t, w>$ corresponds in this model to the evaluation of a similarity measure between the importance value w and the significance value of t in R_d: $w \approx F(d, t)$.

Comparisons of These Query Weight Semantics

In order to analyze the results obtained by these different semantics associated with the query weight w, let us consider the archive represented by the fuzzy sets in Table 4. The rows are the documents, the columns are the terms, and the elements are the values of the index term

Table 4 Each row is a fuzzy set representing a document

	t_1	t_2	t_3	t_4
d_1	1	0.9	1	0.2
d_2	0.7	0.6	0.3	0.8

weights, i.e., an element of row d_i and column t_j is the value $F(d_i, t_j)$. Let us consider the query: $q = <t_1, 1>$ AND $<t_2, 0.6>$ AND $<t_4, 0.2>$, as represented in Table 5.

Table 6 yields the results of the evaluation of q for each of the query weight semantics, assuming that the AND connective is evaluated using the MIN operator.

Linguistic Query Weights

The main limitation of numeric query weights is their inadequacy in dealing with the imprecision which characterizes the concept of importance that they represent. In fact, the use of numeric query weights forces the user to quantify a qualitative and rather vague notion and to be aware of the weight semantics. Thus, a fuzzy retrieval model with linguistic query weights has been proposed[36] with a linguistic extension of the Boolean query language based upon the concept of a linguistic variable.[7] With this approach, the user can select the primary linguistic term "important" together with linguistic hedges (e.g., "very" or "almost") to qualify the desired importance of the search terms in the query. When defining such a query language the term set, i.e., the set of all the possible linguistic values of the linguistic variable *importance* must be defined. Such a definition depends on the desired granularity that one wants to achieve. The greater the number of the linguistic terms, the finer the granularity of the concepts that are dealt with. Next, the semantics for the primary terms must be defined. A pair $<t, important>$, expresses a soft constraint $\mu_{important}$ on the term significance values (the $F(d, t)$ values). The evaluation of the relevance of a given document d to a query consisting solely of the pair $<t, important>$ is based upon the evaluation of the degree of satisfaction of the associated soft constraint $\mu_{important}$.

Table 5 Query q (ANDed weighted pairs)

	t_1	t_2	t_4
q	1	0.6	0.2

Table 6 Results of query q in Table 5 referred to documents in Table 4

Query weight semantics	d_1	d_2
Ideal index term weight	0.3	0.6
Relative importance	0.8	0.6
Threshold on index term weight	0.2	0

The problem of giving a meaning to numeric weights reappears here in associating a semantic with the linguistic term *important*. The $\mu_{important}$ function is defined based on the ideal semantics of the numeric weight to yield:[36]

$$\mu_{important}(F(d,t)) = \begin{cases} e^{\ln(k)*(F(d,t)-i)^2} & \text{for } F(d,t) < i \\ 1 & \text{for } i \leq (d,t) \leq j \\ e^{\ln(k)*(F(d,t)-j)^2} & \text{for } F(d,t) j \end{cases}$$

We see that if $F(d, t)$ is less than the lower bound i or greater than the upper bound j, the constraint is under satisfied. The strength of the soft constraint $\mu_{important}$ depends upon both the width of the range $[i, j]$ and the value of the k parameter. The values i and j delimit the level of *importance* for the user. We note that as the value $|i - j|$ increases, the soft constraint becomes less precise. So, for the case of the ideal semantics of numeric query term weights, k determines the sharpness of the constraint in that as k increases, the constraint increases in fuzziness.

We can define the $\mu_{important}$ function based upon the threshold semantics to yield:[48]

$$\mu_{important} = (F(d,t)) = \begin{cases} \frac{1+i}{2} * e^{\ln(k)*(F(d,t)-i)^2} & \text{for } F(d,t) < i \\ \frac{1+F(d,t)}{2} & \text{for } i \leq F(d,t) \leq j \\ \frac{1+j}{2} * \left(1 + \frac{F(d,t)-j}{2}\right) & \text{for } F(d,t) > j \end{cases}$$

We note that this compatibility function is continuous and nondecreasing in $F(d, t)$ over the interval [0, 1]. For $F(d, t) < i$, $\mu_{important}$ increases as a Gaussian function. For $F(d, t)$ in the interval $[i, j]$, $\mu_{important}$ increases at a linear rate. For $F(d, t) > j$, $\mu_{important}$ still increases, but at a lesser rate. The compatibility functions of non-primary terms, such as *very important* or *fairly important*, are derived by modifying the compatibility functions of primary terms. This is achieved by defining each linguistic hedge as a modifier operator. For example, the linguistic hedges are defined as translation operators in[48] to yield

$$\mu_{very} \text{ important}(x) = \mu_{important}(x)$$

with $i_{very} = i + 0.2$ and $j_{very} = j + 0.2$ and $\forall x \in [0,1]$.

$$\mu_{averagely} \text{ important}(x) = \mu_{important}(x)$$

with $i_{averagely} = i - 0.3$ and $j_{averagely} = j - 0.3$ and $\forall x \in [0,1]$.

$$\mu_{minimally} \text{ important}(x) = \mu_{important}(x)$$

with $i_{minimally} = i - 0.5$ and $j_{minimally} = j - 0.5$ and $\forall x \in [0,1]$.

in which i and j are values in [0, 1] delimiting the range of complete satisfaction of the constraint $\mu_{important}$. With these definitions, any value $F(d, t)$ of the basic domain of the *importance* variable fully satisfies at least one of the constraints defined by the linguistic query terms.

In Herrara–Viedma[49] a query language with linguistic query weights having heterogeneous semantics have been proposed so as to benefit of the full potential offered of fuzzy set to model subjective needs.

Linguistic Quantifiers to Aggregate the Selection Conditions

In a classical Boolean query language, the AND and OR connectives allow only for crisp (non-fuzzy) aggregations which do not capture any of the inherent vagueness of user information needs. For example, the AND used for aggregating M selection conditions does not tolerate the no satisfaction of but a single condition which could cause the no retrieval of relevant documents. To deal with this problem, additional extensions of Boolean queries have been provided which involves the replacement of the AND and OR connectives with soft operators for aggregating the selection criteria.[33,50,51]

Within the framework of fuzzy set theory, a generalization of the Boolean query language has been defined based upon the concept of linguistic quantifiers that are employed to specify both crisp and vague aggregation criteria of the selection conditions.[16] New aggregation operators can be specified by linguistic expressions with self-expressive meaning, such as *at least k* and *most of*. They are defined to exist between the two extremes corresponding to the AND and OR connectives, which allow requests for *all* and *at least one of* the selection conditions, respectively. The linguistic quantifiers used as aggregation operators are defined by OWA operators.

Adopting linguistic quantifiers more easily and intuitively formulate the requirements of a complex Boolean query. For example, when desiring that *at least 2* out of the three terms "politics," "economy," and "inflation" be satisfied, one might formulate the Boolean query as:

(politics AND economy) OR (politics AND inflation) OR (economy AND inflation)

However, a simpler one can replace this,

at least 2(politics; economy; inflation)

This new query language via the nesting of linguistic quantifiers supports the expression of any Boolean query. For example the query

<image> AND (<processing> OR <analysis>) AND <digital>

can be translated into the new, more synthetic and clear formulation:

all (<image>, at least 1 of (<processing>, <analysis>), <digital>).

A quantified aggregation function can be applied not only to single selection conditions, but also to other quantified expressions. Then, the E^* function evaluating the entire query yields a value in [0, 1] for each document d in the archive D.

If S is the set of atomic selection conditions and Q is the set of legitimate Boolean queries over our vocabulary of terms, then the E^* function can be formalized by recursively applying the following rules

1. if $q \in S$ then $E^*(d, s) = d_w(F(d, t))$ in which $\mu_w(F(d, t))$ is the satisfaction degree of a pair $<t, w>$ by document d with w being either a numeric weight or a linguistic weight
2. if $q =$ quantifier $(q_1,\ldots,q_n)$ and $q_1,\ldots,q_n \in Q$ then

$$E^*(d,q) = \text{OWA}_{\text{quantifier}}\left(E^*(d,q_1),\ldots,E^*(d,q_n)\right)$$

3. $E^*(d, \text{NOT } q) = 1 - E^*(d, q)$

in which $\text{OWA}_{\text{quantifier}}$ is the OWA operator associated with *quantifier*.

The formal definition of the query language with linguistic quantifiers with the following quantifies has been generated[16]

- *all* replaces AND.
- *at least k* acts as the specification of a crisp threshold of value k on the number of selection conditions and is defined by a weighting vector $w_{\text{at least } k}$ in which $w_k = 1$, and $w_j = 0$, for $i \leq k$—noting that *at least 1* selects the maximum of the satisfaction degrees so that it has the same semantics of OR.
- *about k* is a soft interpretation of the quantifier *at least k* in which the k value is not interpreted as a crisp threshold, but as a fuzzy one so that the user is fully satisfied if k or more conditions are satisfied but gets a certain degree of satisfaction even if $k-1, k-2,\ldots,1$ conditions are satisfied—this quantifier is defined by a weighting vector $w_{\text{about } k}$ in which $w_i = \frac{i}{\sum_{j=1}^{k} j}$ for $i \leq k$, and $w_i = 0$ for $i > k$.
- *most of* is defined as a synonym of *at least* $\frac{2}{3} n$ in which n is the total number of selection conditions.

With respect to non-fuzzy approaches that tried to simplify the Boolean formulations, the fuzzy approach subsumes the Boolean language, allows reformulating Boolean queries in a more synthetic and comprehensible way, and improves the Boolean expressiveness by allowing flexible aggregations.

Other authors have followed these ideas by proposing alternative formalization of linguistic query weights and flexible operators based on ordinal labels and ordinal aggregations,[52] thus reducing the complexity of the evaluation mechanism.

FUZZY ASSOCIATIVE MECHANISMS

Associative retrieval mechanisms are defined to enhance the retrieval of IRSs. They work by retrieving additional documents that are not directly indexed by the terms in a given query but are indexed by other, related terms, sometimes called associated descriptors. The most common type of associative retrieval mechanism is based upon the use of a thesaurus to associate index or query terms with related terms. In traditional associative retrieval, these associations are crisp.

Fuzzy associative retrieval mechanisms obviously assume fuzzy associations. A fuzzy association between two sets $X = \{x_1,...,x_m\}$ and $Y = \{y_1, ..., y_n\}$ is formally defined as a fuzzy relation

$$f : X \times Y \to [0, 1]$$

where the value $f(x, y)$ represents the degree or strength of the association existing between the values $x \in X$ and $y \in Y$. In IR, different kinds of fuzzy associations can be derived depending on the semantics of the sets X and Y.

Fuzzy associative mechanisms employ fuzzy thesauri, fuzzy pseudothesauri, fuzzy ontologies, and fuzzy categorizations to serve three alternative, but compatible purposes: 1) to expand the set of index terms of documents with new terms; 2) to expand the search terms in the query with associated terms; and 3) to expand the set of the documents retrieved by a query with associated documents.

Fuzzy Thesauri

A thesaurus is an associative mechanism that can be used to improve both indexing and querying. It is well known that the development of thesauri is very costly, as it requires a large amount of human effort to construct and to maintain. In highly dynamic situations, i.e., volatile situations, terms are added and new meanings derived for old terms quite rapidly, so that the thesaurus needs frequent updates. For this reason, methods for automatic construction of thesauri have been proposed, named pseudothesauri, based on statistical criteria such as the terms' co-occurrences, i.e., the simultaneous appearance of pairs (or triplets, or larger subsets) of terms in the same documents.

In a thesaurus, the relations defined between terms are of different types. If the associated descriptor has a more general meaning than the entry term, the relation is classified as broader term (BT), while a narrower term (NT) is the inverse relation. Moreover, synonyms and near-synonyms are parts of another type of relationship associated by a related term (RT) connection.

The concept of a fuzzy thesaurus has been suggested,[27,31,53,54] where the links between terms are weighted to indicate the relative strengths of these associations. Moreover, fuzzy pseudothesauri are generated when the weights of the links are automatically computed by considering document relationships rather than concept relationships.[30,55]

The first work on fuzzy thesauri introduced the notion of fuzzy relations to represent associations between terms.[54,56] Let us look at a formal definition of a fuzzy thesaurus.[27,28] Consider T to be the set of index terms and C to be a set of concepts. Each term $t \in T$ corresponds to a fuzzy set of concepts $h(t)$:

$$h(t) = \{\langle c, t(c) \rangle \mid c \in C\}$$

in which $t(c)$ is the degree to which term t is related to concept c. A measure M is defined on all of the possible fuzzy sets of concepts, which satisfies:

$$M(\emptyset) = 0$$
$$M(C) < \infty$$
$$M(A) \leq M(B) \quad \text{if } A \subseteq B$$

A typical example of M is the cardinality of a fuzzy set. The fuzzy RT relation is represented in a fuzzy thesaurus by the similarity relation between two index terms, t_1 and $t_2 \in T$ and is defined as

$$s(t_1, t_2) = M[h(t_1) \cap h(t_2)] / M[h(t_1) \cup h(t_2)]$$

This definition satisfies the following:

- if terms t_1 and t_2 are synonymous, i.e., $h(t_1) = h(t_2)$, then $s(t_1, t_2) = 1$
- if t_1 and t_2 are not semantically related, i.e., $h(t_1) \cap h(t_2) = \emptyset$, then $s(t_1, t_2) = 0$
- $s(t_2, t_1) = s(t_1, t_2)$ for all $t_1, t_2 \in T$
- if t_1 is more +similar to term t_3 than to t_2, then $s(t_1, t_3) > s(t_1, t_2)$

The fuzzy NT relation, indicated as nt, which represents grades of inclusion of a narrower term t_1 in another (broader) term t_2, is defined as:

$$\text{nt}(t_1, t_2) = M\left[h(t_1) \cap h(t_2)\right] / M\left[h(t_1)\right]$$

This definition satisfies the following:

- if term t_1's concept(s) is completely included within term t_2's concept(s), i.e., $h(t_1) \subseteq h(t_2)$, then

$$\text{nt}(t_1, t_2) = 1$$

- if t_1 and t_2 are not semantically related, i.e., $h(t_1) \cap h(t_2) = \emptyset$, then nt$(t_1, t_2) = 0$
- if the inclusion of t_1's concept(s) in t_2's concept(s) is greater than the inclusion of t_1's concept(s) in t_3's concept(s); then nt$(t_1, t_2) >$ nt(t_1, t_3)

By assuming M as the cardinality of a set, s and nt are given as:

$$s(t_1, t_2) = \sum_{k=1}^{M} \min\left[t_1(c_k), t_2(c_k)\right] / \sum_{k=1}^{M} \max\left[t_1(c_k), t_2(c_k)\right]$$

$$\text{nt}(t_1, t_2) = \sum_{k=1}^{M} \min\left[t_1(c_k), t_2(c_k)\right] / \sum_{k=1}^{M} t_1(c_k)$$

A fuzzy pseudothesaurus can be defined by replacing the set C in the definition of $h(t)$ above with the set of documents D, with the assumption that $h(t)$ is the fuzzy set of documents indexed by term t. This yields

$$h(t) = \{(d, t(d)) \mid d \in D\}$$

in which $t(d) = F(d, t)$ is the index term weight defined above. F can be either a binary value defining a crisp representation, or it can be a value in [0, 1] to define a fuzzy representation of documents. The fuzzy RT and the fuzzy NT relations now are defined as:

$$s(t_1, t_2) = \sum_{k=1}^{M} \min[F(t_1, d_k), F(t_2, d_k)] \bigg/ \sum_{k=1}^{M} \max[F(t_1, d_k), F(t_2, d_k)]$$

$$\mathrm{nt}(t_1, t_2) = \sum_{k=1}^{M} \min[F(t_1, d_k), F(t_2, d_k)] \bigg/ \sum_{k=1}^{M} F(t_1, d_k)$$

Note that $s(t_1, t_2)$ and $\mathrm{nt}(t_1, t_2)$ are dependent on the co-occurrences of terms t_1 and t_2 in the set of documents, D. The set of index terms of document d, i.e., $\{t \mid F(d, t) \neq 0 \text{ and } t \in T\}$, can be augmented by those terms t_A which have $s(t, t_A) > \alpha$ and/or $\mathrm{nt}(t, t_A) > \beta$ for parameters α and $\beta \in [0, 1]$.

Suppose that in the definition of F we have the set T as a set of citations which are used to index documents, rather than a set of terms. In this case, a fuzzy association on citations can be defined through the fuzzy relations of s and/or nt. By using citations, a user may retrieve documents that cite a particular author or a particular reference. In addition, a keyword connection matrix has been proposed to represent similarities between keywords in order to reduce the difference between relationship values initially assigned using statistical information and a user's evaluation.[57] A new method is also proposed in which keywords that are attached to a document and broader concepts are hierarchically organized, calculating the keyword relationships through the broader concepts.

Moreover, a thesaurus can be generated based on the max-star transitive closure for linguistic completion of a thesaurus generated initially by an expert linking terms.[58] In addition, a probabilistic notion of term relationships can be employed by assuming that if one given term is a good discriminator between relevant and nonrelevant documents, then any term that is closely associated with that given term (i.e., statistically co-occurring) is likely to be a good discriminator, too.[10] Note that this implies that thesauri are collection-dependent.

One can also expand on Salton's[59] use of the $F(d, t)$ values. Salton[60] infers term relationships from document section similarities. On the other hand, one can manipulate the $F(d, t)$ values in order to generate co-occurrence statistics to represent term linkage weights.[61] Here, a synonym link is considered, defined as:

$$\mu_{\text{synonym}}(t_1, t_2) = \sum_{d \in D} \left[F(d, t_1) \leftrightarrow F(d, t_2) \right]$$

where $F(d, t_1) \leftrightarrow F(d, t_2) = \min[F(d, t_1) \to F(d, t_2), F(d, t_1) \leftarrow F(d, t_2)]$ and $F(d, t_1) \to F(d, t_2)$ can be defined in variety of ways. For instance, $F(d, t_1) \to F(d, t_2)$, the implication operator, can be defined as $[F(d, t_1)^c \vee F(d, t_2)]$, where $F(d, t_1)^c = 1 - F(d, t_1)$ is the complement of $F(d, t_1)$ and $\vee$ is the disjunctive (OR) operator defined as the max; or it can be defined as $\min(1, [1 - F(d, t_1) + F(d, t_2)])$. Here, a narrower term link (where term t_1 is narrower than term t_2, so term t_2 is broader than term t_1), is defined as:

$$\mu_{\text{narrower}}(t_1, t_2) = \sum_{d \in D} \left[F(d, t_1) \leftrightarrow F(d, t_2) \right]$$

Note that fuzzy narrower relationships defined between fuzzy sets can help the purpose of identifying generalization and specialization of topics, while the fuzzy similarity relationship between fuzzy sets can be of aid to identify similar topics. Thus they serve to build a labeled graph of relationships between concepts, regarded as fuzzy sets of terms, in the specific domain of the collection.

Fuzzy Clustering for Documents

Clustering in IR is a method for partitioning D, a given set of documents, into groups using a measure of similarity (or distance) which is defined on every pairs of documents. Grouping like documents together is not a new phenomenon, especially for librarians. The similarity between documents in the same group should be large, while the similarity between documents in different groups should be small.

A common clustering method is based on the simultaneous occurrences of citations in pairs of documents. Documents are clustered using a measure defined on the space of the citations. Generated clusters can then be used as an index for IR, i.e., documents which belong to the same clusters as the documents directly indexed by the terms in the query are retrieved.

Similarity measures have been suggested empirically or heuristically, sometimes analogously to the similarity measures for documents matched against queries.[33,38,62] When adopting a fuzzy set model, clustering can be formalized as a kind of fuzzy association. In this case, the fuzzy association is defined on the domain $D \times D$. By assuming $R(d)$ to be the fuzzy set of terms representing a document d with membership function values $d(t) = F(d, t)$ being the index term weights of term t in document d, the symmetric fuzzy relation s, as originally defined above, is taken to be the similarity measure for clustering documents:

$$s(d_1, d_2) =$$
$$\sum_{k=1}^{M} \min\left[d_1(t_k), d_2(t_k)\right] \Big/ \sum_{k=1}^{M} \max\left[d_1(t_k), d_2(t_k)\right]$$
$$= \sum_{k=1}^{M} \min\left[F(t_k, d_1), F(t_k, d_2)\right] \Big/ \sum_{k=1}^{M} \max\left[F(t_k, d_1), F(t_k, d_2)\right]$$

in which T is the set of index terms in the vocabulary and M is the number of index terms in T.

In fuzzy clustering, documents can belong to more than one cluster with varying degree of membership.[63] Each document is assigned a membership value to each cluster. In a pure fuzzy clustering, a complete overlap of clusters is allowed. Modified fuzzy clustering, also called soft clustering, uses thresholding mechanisms to limit the number of documents belonging to each cluster. The main advantage of using modified fuzzy clustering is the fact that the degree of fuzziness is controlled. The use of fuzzy clustering in IR have several applications, that span from unsupervised categorization of documents into homogenous overlapping topic categories, so as to offer users an overview of the contents of a collection or to organize the results of a search into labeled groups, thus allowing users to have an immediate view of what has been retrieved. With respect to crisp clustering, fuzzy clustering allows finding a document in several labeled groups, thus reflecting distinct interpretation of document's content.

FUZZY PERFORMANCE MEASURES

One problem with current criteria to measure the effectiveness of IRSs is the fact that Recall and Precision measures have been defined by assuming that relevance is a Boolean concept. In order to take into account the fact that IRSs rank the retrieved documents based on their RSVs that are interpreted either as a probabilities of relevance, similarity degrees of the documents to the query, or as degrees of relevance, Recall–Precision graphs are produced in which the values of precision are computed at standard levels of recall. Then the average of the precision values at different recall levels is computed to produce a single estimate.

Nevertheless, these measures do not evaluate the actual values of the RSVs associated with documents and do not take into account the fact that also users can consider relevance as a gradual concept. For this reason some authors have proposed some fuzzy measure of effectiveness. Buell and Kraft[46] proposed the evaluation of fuzzy recall and fuzzy precision, defined as follows:

$$\text{Fuzzy precision} = \frac{\sum_d \min(e_d, u_d)}{\sum_d u_d},$$

$$\text{Fuzzy recall} = \frac{\sum_d \min(e_d, u_d)}{\sum_d u_d}$$

where u_d is the user's evaluation of the relevance of document d (u_d can be binary or defined in the interval [0, 1]) and e_d is the RSV of document d computer by the IRSs. These measures take into account the actual values of e_d and u_d, rather than the rank ordering based in descending order on e_d.

These measures can be particularly useful to evaluate the results of fuzzy clustering algorithms.

CONCLUSIONS

This entry reviews the main objectives and characteristics of the fuzzy modeling of the IR activity with respect to alternative approaches such as probabilistic IR and vector space IR. The focus of the fuzzy approaches is on modeling imprecision and vagueness of the information with respect to uncertainty. The fuzzy generalizations of the Boolean Retrieval model have been discussed by describing the fuzzy indexing of structured documents, the definition of flexible query languages subsuming the Boolean language, and the definition of fuzzy associations to expand either the indexes or the queries, or to generate fuzzy clusters of documents. Fuzzy similarity and fuzzy inclusion relationships between fuzzy sets have been introduced that can help to define more evolved fuzzy IR models performing "semantic" matching of documents and queries, which is the current trend of research in IR.

REFERENCES

1. Tho, Q.T.; Hui, S.C.; Fong, A.C.M.; Cao, T.H. Automatic fuzzy ontology generation for semantic web. IEEE Trans. Knowl. Data Eng. **2006**, *18* (6), 842–856.
2. Sanchez, E. *Fuzzy Logic and the Semantic Web*; Elsevier: Amsterdam, **2006**.
3. Motro, A. Imprecision and uncertainty in database systems. In *Fuzziness in Database Management Systems*; Bosc, P., Kacprzyk, J., Eds.; Physica-Verlag: Heidelberg, 1995, 3–22.
4. Zadeh, L.A. Fuzzy sets. Inform. Control **1965**, 8, 338–353.
5. Zadeh, L.A. Fuzzy sets as a basis for a theory of possibility. Fuzzy Set Syst. **1978**, *1*, 3–28.
6. Dubois, D.; Prade, H. *Possibility Theory: An Approach to Computerized Processing of Uncertainty*; Plenum Press: New York, 1988.
7. Zadeh, L.A. The concept of a linguistic variable and its application to approximate reasoning, parts I, II. Inform. Sci. **1975**, 8, 199–249, 301–357.
8. Bosc, P. Fuzzy databases. In *Fuzzy Sets in Approximate Reasoning and Information Systems*; Bezdek, J., Dubois, D., Prade, H., Eds.; The Handbooks of Fuzzy Sets Series, Kluwer Academic Publishers: Boston, MA, 1999.
9. Kraft, D.; Bordogna, G.; Pasi, G. Fuzzy set techniques in information retrieval. In *Fuzzy Sets in Approximate Reasoning and Information Systems*; Bezdek, J.C., Dubois, D., Prade, H., Eds.; The Handbooks of Fuzzy Sets

Series; Kluwer Academic Publishers: Boston, MA, 1999; 469–510.
10. van Rijsbergen, C.J. *Information Retrieval*; Butterworths & Co. Ltd.: London, UK, 1979.
11. Fuhr, N. Models for retrieval with probabilistic indexing. Inform. Process. Manage. **1989**, *25* (1), 55–72.
12. Crestani, F.; Lalmas, M.; van Rijsbergen, C.J.; Campbell, I. Is this document relevant? Probably. ACM Comput. Surv. **1998**, *30* (4), 528–552.
13. Bordogna, G.; Pasi, G. The application of fuzzy set theory to model information retrieval. In *Soft Computing in Information Retrieval: Techniques and Applications*; Crestani, F., Pasi, G., Eds.; Physica-Verlag: Heidelberg, Germany, 2000.
14. Yager, R.R.; Rybalov, A. On the fusion of documents from multiple collections information retrieval systems. J. Am. Soc. Inform. Sci. **1999**, *49* (13), 1177–1184.
15. Bordogna, G.; Pasi, G.; Yager, R. Soft approaches to information retrieval on the WEB. Int. J. Approx. Reason. **2003**, *34*, 105–120.
16. Bordogna, G.; Pasi, G. Linguistic aggregation operators in fuzzy information retrieval. Int. J. Intell. Syst. **1995**, *10* (2), 233–248.
17. Bordogna, G.; Pasi, G. Controlling information retrieval through a user adaptive representation of documents. Int. J. Approx. Reason. **1995**, *12*, 317–339.
18. Cater, S.C.; Kraft, D.H. A generalizaton and clarification of the Waller-Kraft wish-list. Inform. Process. Manage. **1989**, *25*, 15–25.
19. Buell, D.A. A problem in information retrieval with fuzzy sets. J. Am. Soc. Inform. Sci. **1985**, *36* (6), 398–401.
20. Kraft, D.H. Advances in information retrieval: Where is that /#*%@^ record? In *Advances in Computers*; Yovits, M., Ed.; Academic Press: New York, 1985; 277–318.
21. Buell, D.A.; Kraft, D.H. A model for a weighted retrieval system. J. Am. Soc. Inform. Sci. **1981**, *32* (3), 211–216.
22. Bookstein, A. Fuzzy requests: An approach to weighted Boolean searches. J. Am. Soc. Inform. Sci. **1980**, *31* (4), 240–247.
23. Radecki, T. Fuzzy set theoretical approach to document retrieval. Inform. Process. Manage. **1979**, *15* (5), 247–260.
24. Waller, W.G.; Kraft, D.H. A mathematical model of a weighted Boolean retrieval system. Inform. Process. Manage. **1979**, *15*, 235–245.
25. Lucarella, D.; Zanzi, A. Information retrieval from hypertext: An approach using plausible inference. Inform. Process. Manage. **1993**, *29* (1), 299–312.
26. Lucarella, D.; Morara, R. FIRST: Fuzzy information retrieval system. J. Inform. Sci. **1991**, *17* (2), 81–91.
27. Miyamoto, S. Information retrieval based on fuzzy associations. Fuzzy Set Syst. **1990**, *38* (2), 191–205.
28. Miyamoto, S. Two approaches for information retrieval through fuzzy associations. IEEE Trans. Syst. Man Cybernet. **1989**, *19* (1), 123–130.
29. Murai, T.; Miyakoshi, M.; Shimbo, M. A fuzzy document retrieval method based on two-valued indexing. Fuzzy Set Syst. **1989**, *30* (2), 103–120.
30. Miyamoto, S.; Nakayama, K. Fuzzy information retrieval based on a fuzzy pseudothesaurus. IEEE Trans. Syst. Man Cybernet. **1986**, *SMC-16* (2), 278–282.
31. Miyamoto, S. *Fuzzy Sets in Information Retrieval and Cluster Analysis*; Kluwer Academic Publishers: Dordrecht, 1990.
32. Sanchez, S.N.; Triantaphyllou, E.; Kraft, D.H. A feature mining based approach for the classification of text documents into disjoint classes. Inform. Process. Manage. **2002**, *38* (4), 583–604.
33. Salton, G.; McGill, M.J. *Introduction to Modern Information Retrieval*; McGraw-Hill: New York, 1983.
34. Molinari, A.; Pasi, G. A fuzzy representation of HTML documents for information retrieval systems. In *Proceedings of the IEEE International Conference on Fuzzy Systems*, Vol. 1; New Orleans, September 8–12, 1996, 107–112.
35. Cater, S.C.; Kraft, D.H. TIRS: A topological information retrieval system satisfying the requirements of the Waller–Kraft wish list. In *Proceedings of the Tenth Annual ACM/SIGIR International Conference on Research and Development in Information Retrieval*, New Orleans, LA, June 1987, 171–180.
36. Bordogna, G.; Pasi, G. A fuzzy linguistic approach generalizing Boolean information retrieval: A model and its evaluation. J. Am. Soc. Inform. Sci. **1993**, *44* (2), 70–82.
37. Salton, G.; Buckley, C. Term weighting approaches in automatic text retrieval. Inform. Process. Manage. **1988**, *24* (5), 513–523.
38. Sparck Jones, K.A. *Automatic Keyword Classification for Information Retrieval*; Butterworths: London, 1971.
39. Sparck Jones, K.A. A statistical interpretation of term specificity and its application in retrieval. J. Doc. **1972**, *28* (1), 11–20.
40. Buell, D.A. An analysis of some fuzzy subset applications to information retrieval systems. Fuzzy Sets Syst. **1982**, *7* (1), 35–42.
41. Berrut, C.; Chiaramella, Y. Indexing medical reports in a multimedia environment: The RIME experimental approach. In *ACM-SIGIR 89*, Boston, MA, 1986; 187–197.
42. Zadeh, L.A. A computational approach to fuzzy quantifiers in natural languages. Comput. Math. Appl. **1983**, *9*, 149–184.
43. Yager, R.R. On ordered weighted averaging aggregation operators in multi criteria decision making. IEEE Trans. Syst. Man Cybernet. **1988**, *18* (1), 183–190.
44. Bartschi, M. Requirements for query evaluation in weighted information retrieval. Inform. Process. Manage. **1985**, *21* (4), 291–303.
45. Bordogna, G.; Carrara, P.; Pasi, G. Query term weights as constraints in fuzzy information retrieval. Inform. Process. Manage. **1991**, *27* (1), 15–26.
46. Buell, D.A.; Kraft, D.H. Performance measurement in a fuzzy retrieval environment. In *Proceedings of the Fourth International Conference on Information Storage and Retrieval*, Oakland, CA, May 31–June 2; ACM/SIGIR Forum; **1981**, *16* (1), 56–62.
47. Yager, R.R. A note on weighted queries in information retrieval systems. J. Am. Soc. Inform. Sci. **1987**, *38* (1), 23–24.
48. Kraft, D.H.; Bordogna, G.; Pasi, G. An extended fuzzy linguistic approach to generalize Boolean information retrieval. J. Inform. Sci. Appl. **1995**, *2* (3), 119–134.
49. Herrera-Viedma, E.; Lopez-Herrera, A.G. A model of an information retrieval system with unbalanced fuzzy

linguistic information. Int. J. Intell. Syst. **2007**, *22* (11), 1197–1214.
50. Paice, C.D. Soft evaluation of Boolean search queries in information retrieval systems. Inform. Technol. Res. Develop. Appl. **1984**, *3* (1), 33–41.
51. Sanchez, E. Importance in knowledge systems. Inform. Syst. **1989**, *14* (6), 455–464.
52. Herrera, F.; Herrera-Viedma, E. Aggregation operators for linguistic weighted information. IEEE Trans. Syst. Man Cybernet. Part A Syst. Hum. **1997**, *27* (5), 646–656.
53. Neuwirth, E.; Reisinger, L. Dissimilarity and distance coefficients in automation-supported thesauri. Inform. Syst. **1982**, *7* (1), 47–52.
54. Radecki, T. Mathematical model of information retrieval system based on the concept of fuzzy thesaurus. Inform. Process. Manage. **1976**, *12* (5), 313–318.
55. Nomoto, K.; Wakayama, S.; Kirimoto, T.; Kondo, M. A fuzzy retrieval system based on citation. Syst. Control **1987**, *31* (10), 748–755.
56. Reisinger, L. On fuzzy thesauri. In *COMPSTAT 1974*; Bruckman, G.; et al.; Eds.; Physica Verlag: Vienna, Austria, 1974; 119–127.
57. Ogawa, Y.; Morita, T.; Kobayashi, K. A fuzzy document retrieval system using the keyword connection matrix and a learning method. Fuzzy Set Syst. **1991**, *39* (2), 163–179.
58. Bezdek, J.C.; Biswas, G.; Huang, L.Y. Transitive closures of fuzzy thesauri for information-retrieval systems. Int. J. Man Mach. Stud. **1986**, *25* (3), 343–356.
59. Salton, G. *Automatic Text Processing: The Transformation, Analysis and Retrieval of Information by Computer*; Addison Wesley: Boston, MA, 1989.
60. Salton, G.; Allan, J.; Buckley, C.; Singhal, A. Automatic analysis, theme generation, and summarization of machine-readable texts. Science **1994**, *264*, 1421–1426.
61. Kohout, L.J.; Keravanou, E.; Bandler, W. *Information retrieval system using fuzzy relational products for thesaurus construction*. In *Proceedings IFAC Fuzzy Information*, Marseille, 1983, 7–13.
62. Salton, G.; Bergmark, D. A citation study of computer science literature. IEEE Trans. Prof. Commun. **1979**, *22* (3), 146–158.
63. Bezdek, J.C. *Pattern Recognition with Fuzzy Objective Function Algorithms*; Plenum Press: New York, 1981.

Gradients, Edges, and Contrast

Tania Pouli
Technicolor, Rennes, France

Erik Reinhard
Technicolor Research and Innovation, Rennes, France

Douglas W. Cunningham
Brandenburg University of Technology, Cottbus, Germany

Abstract
In this entry various topics related to scene analysis are discussed such as gradient determination, edge detection, linear space and scale considerations, image contrast, image deblurring, superresolution, inpainting. A very approachable mathematical treatment is given, thoroughly illustrated with examples. Historical background information is also given.

Keywords: Edge detection; Gradient determination; Image contrast; Image deblurring; Inpainting; Superresolution.

The information contained within a single pixel is essentially just a color, which is normally represented with three values. We can derive useful information from the analysis of individual pixel values. We can even perform some very elegant image manipulations with the information contained in single pixels. Nonetheless, the information obtained by analyzing single pixels is still very limited. In particular, a single pixel cannot, in principle, tell us anything about the structures depicted in the image. Does the image, say the one in Fig. 4.18, depict a bird? Is there a tree present, and if so, where does it start or stop? Is the tree closer to us than the bird or farther away? The information present in a single pixel is not enough to answer such questions.

As Figs. 1.1 and 4.18 demonstrated, simply rearranging the location of the pixels completely eliminates the visible structure. While this demonstration again emphasizes that single pixels by themselves do not tell us anything about structure, it also suggests a way to get at structural information. Pixels do in fact have one more critical piece of information: their relative location. The change in color or luminance between pixels placed at specific locations is the key to discovering more about the content of an image.

REAL-WORLD CONSIDERATIONS

Since we usually try to infer something about the real-world scenes depicted in images, it will be worthwhile to spend a few minutes thinking about the real world. From a physical point of view, every point of humanity's natural environment is filled with some form of matter. Some of this matter is in solid form (e.g., rocks, trees, glass), some in liquid form (e.g., ink, milk, water), and sometimes in gaseous form (e.g., fog, clouds, air). We tend to call spatially localized regions of similar material "objects," with gaseous media such as air being a special case of object. In other words, the real world is completely filled with objects. Thus, the surface of one object *always* abuts against the surface of another object, which means that a surface is really just a transition region between two kinds of matter.

Perceptual Consequences

Light is structured by its interaction with objects. In most cases, the surface of an object (and occasionally, such as in subsurface scattering, a very short distance under the surface) is solely responsible for the altering of the properties of light rays.[1] Since objects are spatially localized regions of similar material, neighboring points on an object's surface will tend to affect light in a similar way. Likewise, different objects are usually made of different materials and thus will tend to affect light in different ways.

Perceptually, then, it can be said that the visual world consists solely of surfaces, the properties of which we infer from how they structure light.[2] As a simple example, we can reverse the observations given earlier: if two neighboring points affect light in the same way, it is very likely that they come from the same object. If they affect light (significantly) differently, then they probably belong to different objects.

Image Space Consequences

Obviously, images are constructed from light rays that have previously interacted with real-world objects. It is not surprising, then, that some of the same light structures used by the visual system to infer properties of the real world are also present in images. For an image, therefore, we have the observation that two-dimensional images consist entirely of surfaces, and the place where two surfaces meet is called an edge. If two neighboring points in an image affect light in the same way, it is very likely that they come from the same surface. If they affect light (significantly) differently, then they probably belong to different surfaces. Thus, by looking for strong changes in luminance or color between neighboring points, we can find edges in an image.

The next part of this entry examines methods for looking at luminance changes. Following this, methods for looking specifically at edges will be examined.

GRADIENTS

There are two common methods for determining the magnitude of the change in the luminance function: using the first derivative and using the second derivative. The simplest approach to calculating the first derivative is to compare a given pixel with some subset of its neighbors, which provides a description of how the luminance in the image changes as a function of spatial direction. There are four primary methods for calculating a discrete approximation of the gradient: forward difference, backward difference, central difference, and the Söbel operator. The first three can be derived from a Taylor polynomial around the pixel of interest x.

The Forward Difference Method

To derive an estimate of the gradient at a given pixel location, we can start with a Taylor polynomial, where we try to calculate the value at the next point in a function based on the current value and the derivatives of the function at the current point. More specifically, we can employ a weighted sum of the derivatives, which in one spatial dimension leads to:

$$I(x+h) = \sum_{n=0}^{\infty} \frac{I^{(n)}(x)}{n!} h^n \tag{1a}$$

$$= I(x) + hI'(x) + \frac{h^2}{2!} I''(x) + \ldots \tag{1b}$$

where $I(x)$ is the value at pixel x, $I(x+h)$ is the value at $x+h$ with h being some constant, and $I^{(n)}(x)$ is the nth-order derivative of I evaluated at point x. Rewriting Eq. 1b to isolate the first derivative I' yields:

$$I'(x) = \frac{I(x+h) - I(x) - \sum_{n=2}^{\infty} \frac{h^n}{n!} I^{(n)}(x)}{h} \tag{2}$$

Given that gradients are effectively measures of the derivative of a function, we could ignore the higher-order terms, and rather than compute derivatives, we can compute gradients. If we additionally assume that the step between pixels h is 1, then we obtain the forward difference method for calculating the gradient $\mathbf{D} = (D_x, D_y)$ at a pixel (see Fig. 1). Since we are dealing with a two-dimensional image, horizontal and vertical differences are calculated separately:

$$D_x(i,j) = I(i+1,j) - I(i,j) \tag{3a}$$

$$D_y(i,j) = I(i,j+1) - I(i,j) \tag{3b}$$

where $I(i, j)$ is the luminance for pixel (i, j), D_x is the horizontal gradient and D_y is the vertical gradient. Commonly, gradients are computed in log space, which has the advantage that they represent contrasts, as pixel ratios become pixel differences:

$$D_x(i,j) = \log I(i+1,j) - I(i,j) \tag{4a}$$

$$D_y(i,j) = \log I(i,j+1) - I(i,j) \tag{4b}$$

We can construct a pair of convolution kernels to represent the forward difference operator:

$$\frac{\partial}{\partial x} = \begin{bmatrix} -1 & 1 \end{bmatrix} \tag{5a}$$

$$\frac{\partial}{\partial y} = \begin{bmatrix} -1 \\ 1 \end{bmatrix} \tag{5b}$$

A vector valued gradient image (dI_x, dI_y) can then be computed by applying these convolution kernels to the input image I in log space:

$$dI_x = \log I * \frac{\partial}{\partial x} \tag{6a}$$

$$dI_y = \log I * \frac{\partial}{\partial y} \tag{6b}$$

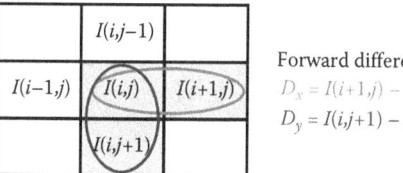

Fig. 1 Forward difference method for calculating gradients

It is important to note that the high-order terms are omitted. This means that this method yields only an approximation of the first derivative—albeit a rather accurate one for small values of h. The error can be estimated from the truncated terms and has the order $O(h)$.

The Backward Difference Method

The backward difference method is essentially the same, with the exception that the left pixel $I(i-1, j)$ is compared to the pixel of inters $I(i, j)$ rather than $I(i+1, j)$ (see Fig. 2). Since the step h between pixels is now negative, every odd-powered derivative in the series also becomes negative:

$$I(x-h) = \sum_{n=0}^{\infty} \frac{I^{(n)}(x)}{n!}(-h)^n \quad (7a)$$

$$= I(x) - hI'(x) + \frac{h^2}{2!}I''(x) - \frac{h^3}{2!}I''(x) + \ldots \quad (7b)$$

Rewriting Eq. 7b to isolate the first derivative I' yields:

$$I'(x) = \frac{I(x) - I(x-h) - \sum_{n=2}^{\infty} \frac{(-h)^n}{b!} I^{(n)}(x)}{h} \quad (8)$$

The gradients are therefore computed in horizontal and vertical directions using:

$$D_x = \log I(i, j) - \log I(i-1, j) \quad (9a)$$

$$D_y = \log I(i, j) - \log I(i, j-1) \quad (9b)$$

where, similar to the previous section, we have also added the log space computation.

The convolution kernel for the backward difference operator is the same as for the forward difference. The two operators differ merely in which of the cells is the pixel of interest. Just as with the forward difference, the high-order terms are truncated. The error, then, still has the order $O(h)$.

The Central Difference Method

Combining the forward and the backward difference methods yields a more accurate and robust technique called the central difference method, which is illustrated in Fig. 3:

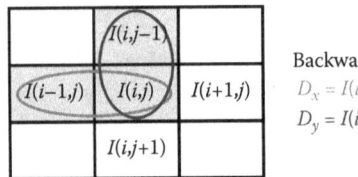

Fig. 2 Backward difference method for calculating gradients

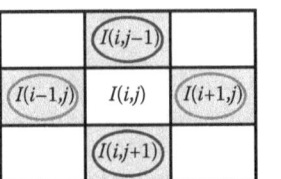

Fig. 3 Central difference method for calculating gradients

$$I(x+h) - I(x-h)$$
$$= \left(I(x) + hI'(x) + \frac{h^2}{2!}I''(x) + \frac{h^3}{3!}I'''(x) + \ldots \right)$$
$$- \left(I(x) - hI'(x) + \frac{h^2}{2!}I''(x) - \frac{h^3}{3!}I'''(x) + \ldots \right) \quad (10a)$$

$$= 2\left(hI'(x) + \frac{h^3}{3!}I'''(x) + \ldots \right) \quad (10b)$$

Rewriting Eq. 10b to isolate the first derivative I' yields:

$$\frac{I(x+h) - I(x-h)}{2} = hI'(x) + \frac{h^3}{3!}I'''(x) + \ldots \quad (11)$$

Ignoring the higher-order terms, assuming h is 1, moving into log space, and computing the horizontal and vertical gradients separately yields:

$$D_x(i, j) = \left(\log I(i+1, j) - \log I(i-1, j) \right)/2 \quad (12a)$$

$$D_y(i, j) = \left(\log I(i, j+1) - \log I(i, j-1) \right)/2 \quad (12b)$$

We likewise construct a pair of convolution kernels for the central difference operator:

$$\frac{\partial}{\partial x} = \begin{bmatrix} -1/2 & 0 & 1/2 \end{bmatrix} \quad (13a)$$

$$\frac{\partial}{\partial y} = \begin{bmatrix} -1/2 \\ 0 \\ 1/2 \end{bmatrix} \quad (13b)$$

These convolution kernels can then be applied according to Eqs. 6a and 6b to compute the gradient image.

Since the second derivative term in the forward difference method cancels out the second derivative term in the backward difference method, the error is of order $O(h^2)$. Although the central difference method is more accurate, it has the disadvantage that oscillating functions can yield a zero derivative.

The Söbel Operator

The forward and backward difference methods both incorporated a single neighboring pixel into the computation. The central difference examines two neighbors and is more

Once again, these kernels are convolved with the logged input image to produce the vector-valued gradient image as per Eqs. 6a and 6b.

Second Derivative Methods

Intuitively, the place of greatest change in luminance will be important. The places of greatest change in the luminance function correspond to the maxima in the first derivative. Since it is not always easy in practice to uniquely identify maxima, many applications take advantage of the fact that maxima in the first derivative will show up as zero crossings in the second derivative. Figure 5 visualizes the relationship between the luminance function and its derivatives. The top of the figure shows a simple horizontal sinusoid. The lower part of the graph shows the luminance function, as well as its first and second derivatives. The vertical bars represent the location of maximal luminance changes.

There are several common ways of estimating the second derivative. The most common is the Laplacian, which is the divergence of the gradient operator. The Laplacian ∇^2 can be calculated as $\nabla^2 = \frac{\partial^2}{\partial x^2} + \frac{\partial^2}{\partial y^2}$, or as a single convolution

$$\nabla^2 I = I * \begin{bmatrix} 0 & 1 & 0 \\ 1 & -4 & 1 \\ 0 & 1 & 0 \end{bmatrix} \quad (15)$$

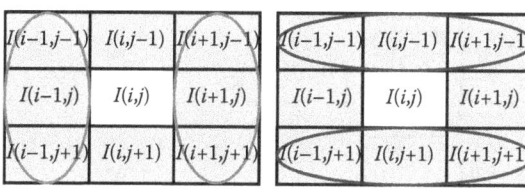

Fig. 4 Söbel operator for calculating gradients

accurate. The Söbel operator additionally includes diagonal neighbors (Fig. 4) and is the most accurate of the basic methods. To calculate a gradient using the Söbel operator, convolution kernels with the following weights are applied to the image:

$$\frac{\partial}{\partial x} = \begin{bmatrix} -1 & 0 & 1 \\ -2 & 0 & 2 \\ -1 & 0 & 1 \end{bmatrix} \quad (14a)$$

$$\frac{\partial}{\partial y} = \begin{bmatrix} 1 & 2 & 1 \\ 0 & 0 & 0 \\ -1 & -2 & -1 \end{bmatrix} \quad (14b)$$

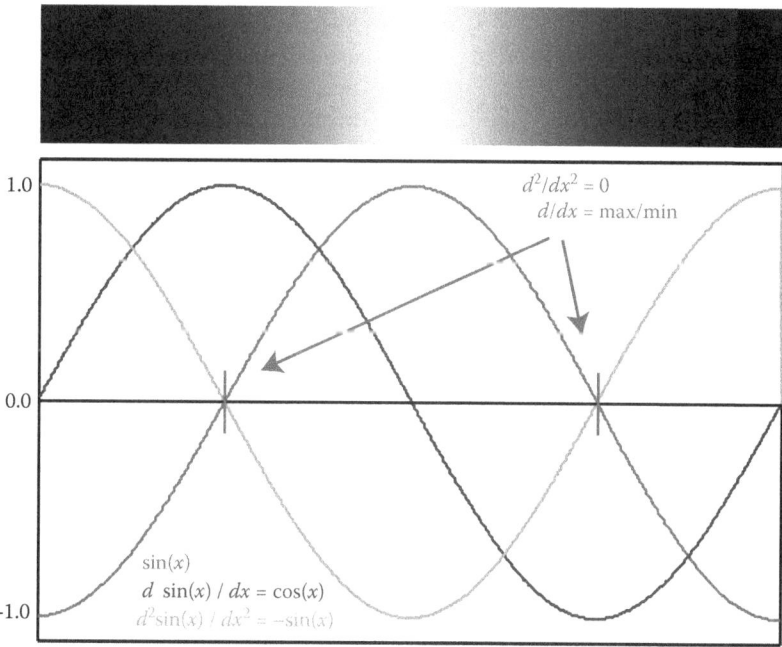

Fig. 5 A sinusoidal texture gradient. The luminance function and its first and second derivatives (both normalized to [−1, 1]) are plotted. Note that where the second derivative is 0 the luminance function has maximum variation, i.e., the first derivative is either at its maximum or minimum

This is a 2D convolution, which is separable, meaning that it can also be written as the sum of two 1D convolutions. In that case, a straightforward pair of convolution kernels would be:

$$\frac{\partial^2}{\partial x^2} = \begin{bmatrix} 1 & -2 & 1 \end{bmatrix} \quad (16a)$$

$$\frac{\partial^2}{\partial y^2} = \begin{bmatrix} 1 \\ -2 \\ 1 \end{bmatrix} \quad (16b)$$

The Laplacian of an image is then given by:

$$\nabla^2 I = I * \frac{\partial^2}{\partial x^2} + I * \frac{\partial^2}{\partial y^2} \quad (17)$$

Wherever the Laplacian of an image becomes small, the gradient of the image will be either at a maximum or a minimum.

Gradient Magnitude

In all cases, it is common to ignore the direction of the gradient, and instead calculate the mean gradient magnitude at a given location:

$$D(i,j) = \sqrt{D_x(i,j)^2 + D_y(i,j)^2} \quad (18)$$

In some cases, however, one may choose to keep the vertical and horizontal gradient magnitudes separate (as, for instance, in Levin's motion deblurring technique[3]):

$$D(i,j) = \begin{bmatrix} |D_x(i,j)| \\ |D_y(i,j)| \end{bmatrix} \quad (19)$$

Gradient Statistics

Although gradients can carry considerable information, especially about the texture of objects and surfaces, little is known about the statistics of gradients in natural images. It is known, however, that the gradient histogram has a very sharp peak at zero (see Fig. 6) and then falls off very rapidly (with long tails at the higher gradients).[5-10] The distribution can be modeled as:

$$\exp(-x^\alpha) \quad (20)$$

with $\alpha < 1$.[11-13] The reason for the specific shape of the gradient distribution seems to be precisely the observations outlined earlier: images contain many large surfaces that tend to be smooth and somewhat homogeneous. Such surfaces will have gradient magnitudes near zero. There

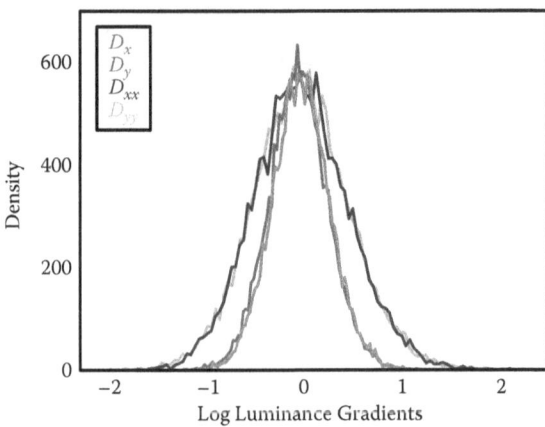

Fig. 6 Gradient distributions for a collection of natural images: D_x and D_y are the horizontal and vertical gradients (first derivative of the luminance function), respectively, and D_{xx} and D_{yy} are the horizontal and vertical second derivatives[4]

will, of course, be a few high-contrast edges, which will yield very high gradients.[14] Interestingly, similar objects tend to cluster together, which means that the transition from one object to another will be similar in any direction. This is reflected in the symmetry of the gradient distribution around the central peak. It has also been shown that the gradient histogram is roughly invariant to changes in scale.[5,8,15]

Single-Image Gradient Statistics

Instead of assessing average statistics over ensembles, it is possible to compute gradient statistics on a single image. Specifically, it is interesting to look at small patches and assess the probability of such patches recurring elsewhere in the image.[16] To characterize a 5 × 5 patch **p** centered at pixel location (x, y), first its mean gradient magnitude $\bar{D}_\mathbf{p}$ is computed:

$$\bar{D}_\mathbf{p}(x,y) = \sum_{i=-2}^{2} \sum_{j=-2}^{2} D(i+x, j+y) \quad (21)$$

Then, an empirical density d is computed for each patch within a neighborhood of radius r around the patch. This neighborhood has an area $A = \pi r^2$:

$$d(\mathbf{p}; r) = \sum_{\mathbf{p}_j \in A} \frac{G\left(\|\mathbf{p} - \mathbf{p}_j\|_2^2\right)}{A} \quad (22)$$

where G is a Gaussian kernel. We can then calculate an average density for each mean gradient magnitude:

$$\bar{d}(r, D_\mathbf{p}) = \frac{\sum_{D_{\mathbf{p}_j} = D_\mathbf{p}} d(\mathbf{p}_j; r)}{n} \quad (23)$$

where n is the number of elements over which we sum. The average number of matching nearest neighbors N with a distance r can then be computed as:

$$N(r, D_p) = A \bar{d}(r, D_p) \quad (24)$$

Calculations such as these, if carried out on a large set of images, reveal that within each image, smooth patches occur frequently, whereas structured patches, i.e., those with sharp gradients, occur less often. Moreover, patches tend to occur in clusters, meaning that a given patch is more likely to occur nearby than farther away. However, the distance over which a patch has a matching counterpart depends on its gradient content. In particular, the higher the mean gradient magnitude, the larger the search space r should be. This notion has been empirically quantified with the following expression, relating search radius to mean gradient magnitude:

$$r(D_p) = a + b \exp(D_p/10) \quad (25)$$

It was found that the variables a and b depend on the average number of matching neighbors N as follows:

$$a = 5 \cdot 10^{-3} N + 0.09 N^{0.5} - 0.044 \quad (26)$$

$$b = 7.3 \cdot 10^{-4} + 0.3234 N^{0.5} - 0.35 \quad (27)$$

The interpretation of these equations is that a patch has a certain amount of contrast, quantified by its mean gradient magnitude D_p. The search radius to find a given number N of matching patches for this patch is then given by Eq. 25.

Conversely, for a given patch and search radius, it is possible to calculate the number of matching neighbors N that are likely to be found:

$$N(r, D_p) = \left(\frac{-b + \sqrt{b^2 - 4ac}}{2a} \right)^2 \quad (28)$$

where

$$a = 0.001 \left(5 + 0.73 \exp(D_p/10) \right) \quad (29)$$

$$b = 0.1 \left(0.9 + 3.24 \exp(D_p/10) \right) \quad (30)$$

$$c = -0.1 \left(0.44 + 3.5 \exp(D_p/10) + r \right) \quad (31)$$

Applications which might benefit from knowing the search radius to find similar patches include, for example, image denoising.[16] The Non-Local Means denoising algorithm replaces the center pixel in a patch with the mean value of center pixels found in similar patches elsewhere in the image.[17] The standard algorithm limits the search space to a relatively small and fixed neighborhood. By making the search space adaptive as described earlier, however, it is possible to find better candidate patches and therefore obtain better results. More recently, it was shown that denoising results may benefit from a combination of single-image statistics and ensemble statistics.[18]

It was also discovered that matching patches can be found at different spatial scales. In other words, an image can be repeatedly downsampled by some factor, leading to a stack of images (an image pyramid). Searching matching patches between scales will find strong candidates in nearby scales at nearby locations. In essence, this means that patches are self-similar across scales. Local self-similarity, detected within a single image, has been used in an image and video upscaling application.[19]

Finally, we note that the likelihood of similar patches occurring drops off with distance. This is also seen with single pixel values, which can be computed with the autocorrelation function, or equivalently with the Fourier transform.

EDGES

As stated in a previous section, an edge is the meeting of two surfaces. It follows, then, that an edge contains information about both surfaces and as such provides a highly localized, very dense set of insights into the surfaces—or objects—on either side.

Definition of an Edge

The term *edge* has many definitions, which can differ considerably for different application domains. Given the observations already mentioned, though, it is clear that an edge is essentially a region of rapid or even maximal change in some underlying function. Quite often, edges are regions where there is a discontinuity. For the real world, the function can be seen as the type of matter across spatial location, and the edges of an object—its surface—are discontinuities in type of matter. For images, the underlying function is image luminance or color across spatial location, and edges are large, sudden changes in luminance. For light field photography, the underlying function is either radiance or intensity across light rays.[20] For 3D volume recordings, the function is often material density across spatial location. Note that it is even possible to extend this definition of edges to concepts, with edges in conceptual space. For our present purposes, however, we will restrict ourselves to color or luminance magnitude functions (Fig. 7).

Edge Detection Processes

Since both the first derivative and second derivative methods for calculating gradients examine neighboring pixels

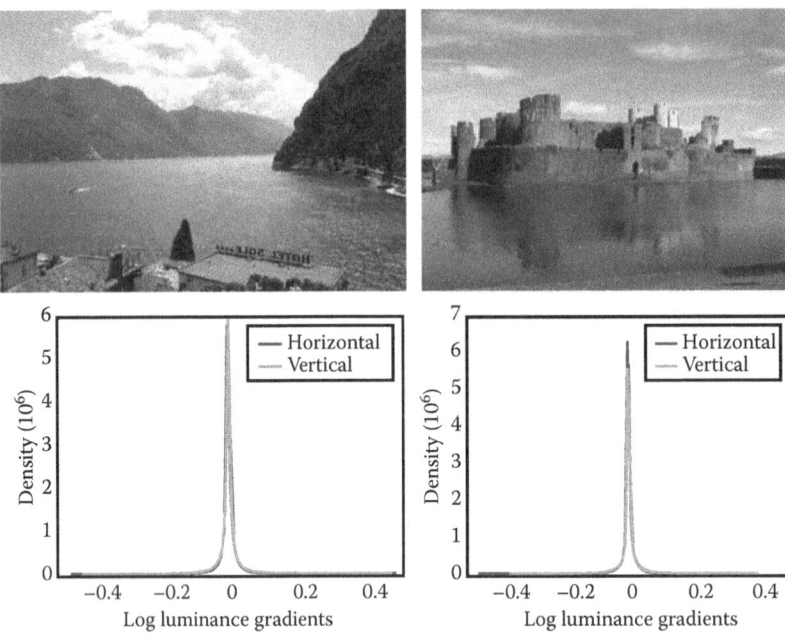

Fig. 7 Two color images and the gradient histograms of grayscale versions of the images, calculated using the central difference method. (Lago di Garda, Italy, 2012; Caerphilly Castle, UK, 2010)

to determine how much the luminance function changes (making them very local), they would seem to be natural choices for discovering edges. All we need to do is to decide precisely what we mean by a *large* change in luminance. At first blush, this would seem to be simple: we simply look for zero crossings if we are using a second derivative or really big gradients if we are using a first derivative. In practice, it is not so easy. For example, we could take the two images from Fig. 7. Both images have large, relatively homogenous regions such as lakes, the sky, and buildings. There are also a few high-contrast edges, such as the edges of the mountains or the silhouette of the buildings. The gradient histograms for these pictures are shown in the second row of Fig. 7 with a focus on the gradients with strengths between −0.5 and 0.5. As expected, there is a sharp peak at zero, confirming that there are many pixels in relatively homogeneous regions. The plots also exhibit heavy tails, showing that there are some strong edges.

Figure 8 shows the effects of two different threshold values (a low threshold in the top row and a higher threshold in the bottom row. Neither result is really satisfactory. The top row has too many edges, especially in the water,

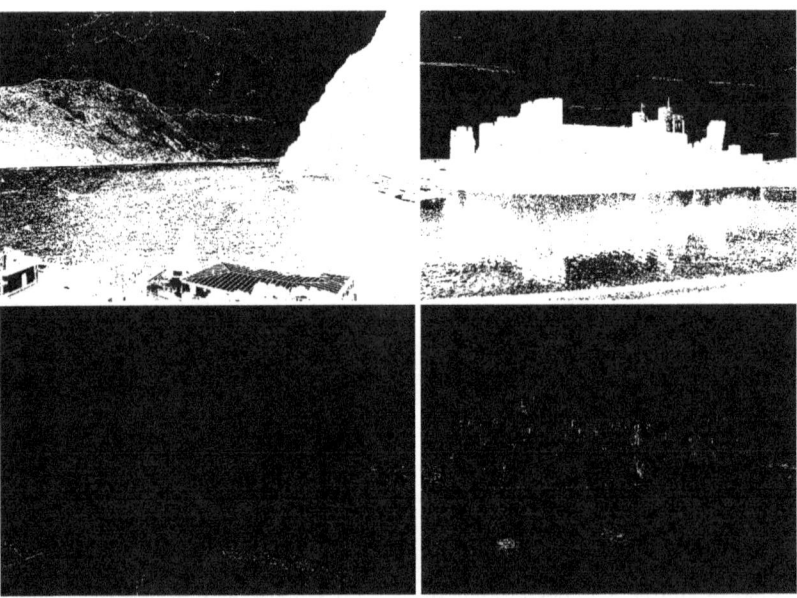

Fig. 8 The gradient magnitudes that are above a low (top) or high (bottom) threshold are shown in white

mountains, and buildings. The bottom row, on the other hand, is missing large, important edges (such as part of the castle or mountains) and yet still has some edges in the water. Although there are many reasons for these and similar difficulties, one central issue is that no surface is perfectly homogeneous. Thus, small local inhomogeneities will lead to small local fluctuations in luminance and therefore in the first derivative. This is particularly true for textured objects. Likewise, the presence of noise will lead to spurious, large changes in the gradient.

To help reduce the effect of noise and highly local edges, many edge detection procedures add a number of steps and constraints. The first step that is often taken is to reduce the local fluctuations by preprocessing the image with a local smoothing operator like a Gaussian filter. The top row in Fig. 9 shows the gradient histograms of the two images from Fig. 7 after they have been smoothed with a Gaussian filter. The bottom row shows the gradients in the image that are above the same low threshold used in the top row of Fig. 8. As expected, the histograms have become even more spiked around zero, leading to higher kurtosis.

Finally, if we remember that edges are merely the meeting places of two surfaces or objects and that an object is a spatially localized region of similar matter, it becomes clear that edges have another useful property: they are spatially extended. In other words, edges will cover many neighboring pixels, especially if they are important edges. Thus, one can use the presence or absence of an edge in a neighboring pixel to strengthen or weaken the response of an edge detector at a given pixel. This is part of the reason that local smoothing operators help to find "important" edges (anything that is extended over a larger space will be partly spared by a low-band pass filter). For a more detailed discussion of edge detection methods, we recommend several standard computer vision books, including Shapiro and Stockman's[21] and Forsyth and Ponce's.[22]

Edge Statistics

Edges play a central role in most theories of human visual perception. In fact, many theories focus solely on edges, ignoring any surface or texture qualities. This can be seen, for example, in the first computational theory of visual perception.[23] In Marr's theory, the human visual system first extracts important primitives in an image, which should reflect geometric structure as well as illumination effects, such as highlights. In practice, these features are extracted using the zero crossings of a Laplacian of a 2D Gaussian, which yields edges.

Although the features are sometimes modified to make bars or blobs, nearly all features in the implementation of Marr's theory can be considered to be "edges with some width or length." These (edge-like) features jointly form the *primal sketch*. The edges are then grouped into coherent units, and some depth information for the newly-formed units is extracted to make the 2.5D sketch. Naturally, further processing occurs at subsequent stages of the model. For the present purposes, though, we can see that essentially everything in the image is discarded with the exception of edges. Note that Marr's primal sketch can be formalized into a scale-invariant representation of edges,[24] leading to a sparse representation that models the probability distribution of edges in natural images.

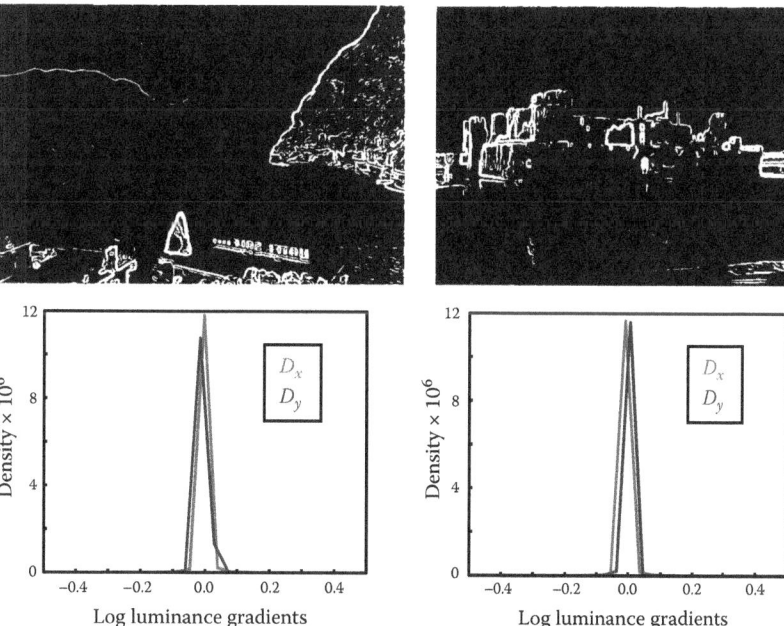

Fig. 9 The effect of filtering. The gradients (top) that are above the same low threshold used in Fig. 5.8 and the gradient histograms (bottom) are shown for the two images after they have been smoothed with a Gaussian filter

Arguably, nearly all modern computational theories of human visual perception follow Marr's basic model to a surprising degree, especially with regard to the focus on edges. This emphasis is to some degree justifiable, since most physiological studies on the neural basis of early visual processing show that one of the first steps in the visual cortex is to extract edges.[25,26] Of course, subsequent stages extract color, motion, depth, and many other useful features, but edges do seem to be a critical first stage.

Given this central role that edges play, it is very surprising that there are very few empirical studies on the statistics of edges in natural images. There are, of course, many theoretical discussions of what edges in an image *should* do, going back as far as the Gestalt psychologists, who had a lot to say about edges and their role in perceptual organization (for more on the Gestalt laws, please see Ref. [27]). Very little work, however, has been done what what edges *actually* do.

While it is true that a considerable amount of work has been done in the machine learning and computer vision fields on learning to detect edges and segment images using a wide variety of filters and that many of these methods rely on the (learned) statistics in image corpuses, most of the statistical descriptions of edge variations in an image corpus are hidden in various classifiers (see, for example, Konishi et al.[28]). Thus, these studies do not provide any *explicit* information about the statistics of gradients or edges.

A few studies have examined the statistics of edges as well as specific edge properties such as collinearity (e.g., Refs. [29,30]). Much of this work has focused on finding empirical evidence showing that the various Gestalt laws of perceptual organization can be derived from image statistics. The most intense focus has been on the law of similarity and the law of good continuation. The first law essentially states that similar elements will be grouped together into a coherent whole.

The second law has two major components. First, in a refinement of the law of similarity, it states that elements with similar orientations will be grouped together. Second, and more critically, it asserts that when several elements intersect, elements that have similar orientations will be joined into one continuous unit. In other words, the human visual system will try wherever possible to join neighboring edges into long, smooth contours.

Empirical evidence for the these laws comes from Geisler and colleagues,[29] who extracted edges (using a multistage procedure based on Gabor functions; from 20 images representing a wide range of natural (i.e., not manmade) scenes. After extracting the edges, they examined the geometrical relationship between all pairs of edges, focusing on the distance between edge centers, the orientation difference, and the direction of the second edge relative to the orientation of the first edge. Among other things, they found that regardless of the separation of two edges or their orientation, the two edges will most likely have the same orientation. This is even more the case when the two edge segments are near to each other and when they are collinear. They also found that for similarly oriented edges, regardless of how far apart they were, there is a high probability that the second edge is nearly co-circular (i.e., tangent to the same circle). This is reflected in the Gestalt law of good continuation. Using a very different methodology, Krüger and Wörtgötter[30] found very similar results.

Gradients and edges detection processes have also been applied to the study of art. Several researchers have examined the statistics of various categories of art and compared them to each other as well as to real scenes (including comparing a painting of a specific scene to a photograph of that exact scene), in part to learn about how humans represent or encode visual information[31–33] They have found that, in general, the statistics of most art are very similar to those of real-world scenes.

LINEAR SCALE SPACE

As was mentioned in the last section, one of the first steps in edge detection is often to smooth the image with a Gaussian filter in order to remove the very local, very high-frequency edges. How much the image is blurred will determine which edges get removed. Some edges like the leaves of a tree are very small, with lots of rapid changes in direction and thus only visible in the higher-frequency range. These disappear quickly with little blurring. Other edges, like the trunk of a tree, are larger, with fewer rapid changes and as such, they are visible in the high, middle, and possibly even lower frequencies. Thus, if we want to capture the leaves or fine texture details, we should blur very little. Of course, since both the coarse edges and the fine edges are visible with little blur, it will be hard to decide with little blur if an edge is a fine detail or coarse without additional processing. If, on the other hand, we want to focus on the larger, more spatially extended, coarser structure in an image, then blurring a lot will make these easier to see since they will remove the finer edges.

So, what do we do if we want to clearly see the coarse structure as well as the fine details at the same time? Or to remove the coarse and only focus on the fine detail? One way to do this is to look at the image at a number of different blur levels simultaneously. This is the core of *scale-space theory*, in which an image is represented as a family of one-parameter, smoothed images. The parameter is the size of the smoothing function used. Scale spaces have been applied to an incredible variety of fields in computer vision and computer graphics, including image segmentation, image denoising, medical imaging, optical flow,[a]

---

[a] In the human vision literature, this concept is often referred to as *optic flow*. Here we follow the computer vision literature and refer to this phenomenon as *optical flow*.

texture analysis, shape-from-texture, object recognition, image matching, even temporal changes (for a recent overview, see Ref. [34]).

In 1959, Taizo Iijima axiomatically derived scale space.[35] Since the original article was in Japanese, it was not widely read. A little over twenty years later, the idea was independently (re-)introduced by Witkin,[36] who focused on one-dimensional signals. The idea was immediately seized upon by the psychophysicist Jan Koenderink,[37] who applied it to image analysis and used it to explain many aspects of human visual perception.

Tony Lindeberg was central in further developing the mathematical and theoretical basis of scale space, as well as determining a number of its properties. A very thorough overview of the various methods for deriving linear scale space—including discrete versions of the operators—can be found in Lindeberg's book.[38] In brief, by assuming a few very general qualities, it can be shown that the Gaussian kernel is the only possible choice for the smoothing kernel. Perhaps the most critical property of the smoothing kernel is that the smoothed versions should be simplifications of the original signal. This criterion (often called *causality*) means that no new properties should be introduced by the smoothing (see Fig. 10). Other critical properties include that the operator not require any special knowledge of the structures in the image (e.g., it should be linear, scale invariant, isotropic, and shift invariant).

Fig. 10 shows on the left side a simple one-dimensional signal, at several different scales (the original is at the bottom). Notice that the curve becomes smoother the more it is blurred. That is, higher-frequency information is being increasingly suppressed. On the right, the zero crossings in the second derivative are shown as a function of increasing blur. Several critical things can be seen here. First, no new zero crossings are introduced as the width of the blur kernel is increased. Second, and perhaps more critically, the zero crossings form paths through scale space.[36] The nature of features across scales is referred as *deep structure* and can be used to infer rather interesting properties of edges in images.

One issue with scale-space approaches that can be seen from this figure is that dependent on the scale at which a signal is analyzed, zero crossings are detected at different spatial locations. In other words, edges drift across scale space. This is generally an undesirable property of scale-space approaches. This can, however, be overcome by processing the signal with edge-preserving or edge avoiding filters such as anisotropic diffusion[39] or the bilateral filter.[40,41] Edge-preserving scale-space decompositions can also be obtained by carefully designing filters that can be steered to smooth edges of specific strengths while avoiding others[42–44]. Such methods have found many applications in visual computing disciplines as they allow images to be processed at different scales, while minimizing the artefacts caused by traditional scale-space approaches.[45]

For a two-dimensional image $I(x, y)$, the family of one-parameter, smoothed images can be represented as (see the leftmost column of Fig. 11):

$$I(x,y;t) = g(x,y;t) * I(x,y;0) \tag{32}$$

for $t > 0$ where t is the scale parameter and g is the Gaussian kernel:

$$g(x, y; t) = \frac{1}{(2\pi t)} e^{\frac{x^2+y^2}{2t}} \tag{33}$$

The semicolon implies that the convolution is only applied over the variables before the semicolon. The value t defines the scale. Koenderink proved that convolution of the image with a Gaussian filter is the general solution to the heat diffusion equation (for a homogeneous medium with uniform conductivity[37]):

$$\partial_t I = \nabla^2 I \tag{34}$$

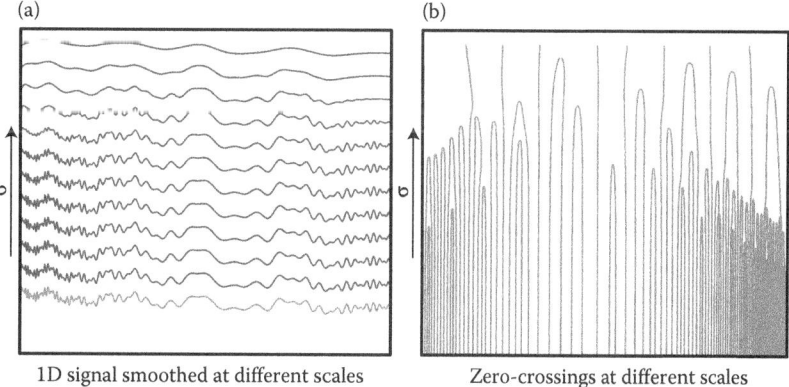

Fig. 10 Scale-space example for a one-dimensional signal. (a) A one-dimensional signal (bottom) is convolved with a Gaussian kernel of increasing width. (b) The zero crossings in the second-order derivative are shown as a function of increasing kernel size σ. Notice that no new zero crossings are introduced by the blur and that the zero crossings form paths across the blur levels. The change of features as a function of blur level is referred to as *deep structure*

In other words, the different scales can be thought of as how heat (or in this case, luminance) would spread over time with the scale parameter t representing time. As the second and third columns of Fig. 11 show, detecting zero crossings of the second derivative in image space leads at appropriately chosen scales to an edge detector. It has also been shown that, given specific assumptions, scale space can be seen as a special case of wavelets.

The N-Jet and Feature Detectors

Linear scale spaces become really useful for image statistics once one realizes that differentiation commutes with convolution:

$$\frac{\partial}{\partial x}(I * g) = I * \frac{\partial g}{\partial x} \quad (35)$$

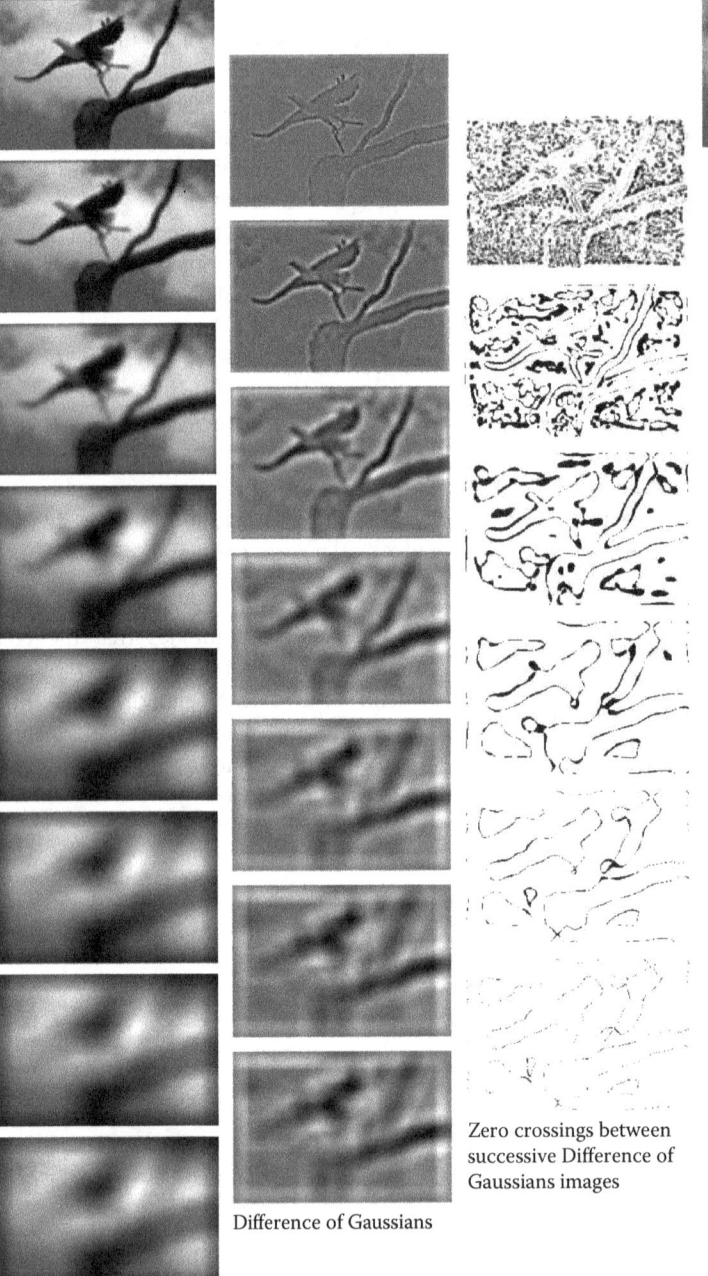

Input image

Zero crossings between successive Difference of Gaussians images

Difference of Gaussians

Gaussian filted images

Fig. 11 An input image was filtered several times with increasing blur kernels (first column). Pairs of successively filtered images where then subtracted to create a difference-of-Gaussians stack (second column). Then, pairs of difference of Gaussian images were tested to detect zero crossings for each pixel (third column), effectively forming the 2D equivalent of Fig. 5.10. (Falknerei Potzberg, Germany, 2013)

Thus, rather than convolving the image with a Gaussian and then calculating the derivative, we can convolve the original image with a Gaussian derivative operator. This naturally leads to a multiscale consideration of gradients as a form of feature detection.

As we saw in Eq. 1b, the Taylor expansion shows how the luminance for a given pixel is related to its neighbors based on a weighted combination of derivatives of increasing order. Similarly, scale space uses a set of derivatives up to a given order. More specifically, the set of derivatives up to the order N (at a given scale t) is referred to as the *local N-Jet*. The set of derivatives up to a given order N for all scales is the *multiscale local N-Jet*.

By combining Gaussian derivatives, a number of feature detectors can be readily constructed. In order to ensure that the detectors have certain properties, however, it is first necessary to construct a local coordinate system. For an edge detector, for example, it would be useful if the local coordinate system were aligned with the edge, thus ensuring rotational invariance of the detector. Since—as was shown earlier (see Fig. 5)—the first derivative is at a maximum for edges, it is possible to define the axes in terms of the first derivative and thereby ensuring that the edge detector is aligned with the edge. We can do this by defining one axis v of the local coordinate system as the direction of the gradient (and thus the perpendicular to the edge) and the other axis u as perpendicular to the direction of the gradient.

Now that we have an edge-aligned coordinate system, edges can be located by looking for maxima in the first-order derivative. This is often done using the second-order derivative, which crosses zero at maxima and minima in the first-order derivative. To determine if a given zero crossing is a maxima or a minima, we can look at the third derivative, which will be less than zero for maxima. Thus, the 3-Jet carries information about edges. Since the first derivative perpendicular to the gradient (that is, along u) is by definition zero, the equations simplify considerably within the local coordinate system. An edge can be defined as those points where the second-order derivative along the axis v is zero and the third is less than zero:

$$I''_v = 0 \tag{36a}$$

$$I'''_v < 0 \tag{36b}$$

As Lindeberg[38] pointed out, due to the discrete nature of an image, the actual zero crossing in the second-order derivative might be between pixels. Interpolating in the second-order derivative to find zero crossings using the constraint that the third-order derivative needs to be negative is an easy way to obtain subpixel edge detection.

Likewise, ridge detectors can be constructed by looking for zero crossings in I_{uv} that satisfy the condition that $I''^2_u - I''^2_v < 0$. Blob detectors are constructed from maxima or minima in the Laplacian. Junction detectors use a measure of curvature that can be constructed by combinations of Gaussian derivative operators. By selecting the scale(s) at which the feature detector's response is strongest, it is possible to automatically detect the proper scale(s) for examining a given feature.[38]

Implications for Human Perception

A considerable amount of work has been done connecting different properties of scale spaces to aspects of the human visual system. For example, Koenderink has suggested that the early stages of the human visual system use multiscale local N-Jets to represent the input.[46] It has also been shown that the structure of receptive fields in the retina can be modeled using a Gaussian scale space.[47] Likewise, the "center-surround" nature of receptive fields in early visual areas can also be modeled using the Laplacian of the Gaussian. More intriguingly, the layout of receptive field sizes on the retina (a linear increase in receptive field size as a function of distance from the retina) is precisely the layout that is predicted if the retina were trying to create a scale-space representation.[47] Finally, the fact that there are massive downward connections from higher visual areas to lower visual areas is consistent with scale-space theory,[34] and it suggests that the lower area receptive fields may be modified (something for which there is some physiological evidence; see, for example, Ref. [48]). Once nonlinear scale spaces are allowed, even more aspects of human vision (in particular, higher visual processing) can be explained. For more on the relationship between scale space and human vision, we recommend.[34]

Scale-Space Statistics

Several researchers have created created a wide variety of feature detectors using scale-space theory and tested them on natural images (e.g., Refs. [49,50,37,51,52]). Interestingly, a large amount of this work is directly interested in deriving 3D shape information from the 2D image information, usually by focusing on the distortions that shape information should undergo during perspective projection. Although these works present statistics on the recovered depth, edge, and orientation values, they focus on the accuracy of the detectors and less on the statistical analysis of natural images.

Salden and colleagues have provided extensive discussions of how one can use scale space to construct complete feature representation systems that are appropriate for natural images (see, for example, Ref. [53]). Although the statistical regularities of those features in natural images remains an interesting prospect for future research, there is some research on the statistical properties of the feature detector. Lindeberg,[38] for example, extensively examined how one might extract specific structures from images, including testing these detectors on sample images. In this work he also derived the statistical properties that one should expect from the detectors, then he examined synthetic images to see if the feature detectors functioned as expected. For example, he determined how the number of extrema should decrease as a function of scale (the density of extrema is essentially inversely proportional to the scale parameter). The expected pattern was more or less obtained. Since noise data was used, however, this study is more focused on the statistics of the detector than of images.

Edge statistics were assessed in scale space by Pedersen and colleagues.[54] They examined Van Hateren's database using a 3-Jet to determine edge statistics. In particular, they point out that when constructing an edge detector, at least three dimensions need to be described. Specifically, any given edge will have an orientation, a position within the receptive field, and a blur or scale level. They show that ideal step edges form a two-dimensional manifold in this space that depends solely on the orientation and the ratio of the position to the scale. They then construct a nine-dimensional 3-Jet space, consisting of the two first-order derivatives (I_x and I_y), the three second-order derivatives (I_{xx}, I_{yy}, and I_{yx}) and the four third-order derivatives and project the edge manifold into this space. In this theoretical part, they show that the ideal step edges still trace a 2D manifold.

The data is then whitened and contrast normalized. Whitening is essentially the removal of first- and second-order correlations. This can be achieved with the aid of principal components analysis. As a result, the covariance matrix of each nine-dimensional data point becomes the identity matrix. Contrast normalization subsequently takes these data points and divides each point by its norm.

After whitening and contrast normalization of the data (to eliminate the effects of lighting and thus be able to focus on geometry), they show that all 3-Jets of ideal step edges will lie on a eight-dimensional sphere in the nine-dimensional space, and thus the distance between and two 3-Jets can be measured by their angular separation. Finally, in the empirical part, they measure the angular distance between the ideal edge manifold and the measured 3-Jets, obtaining the full probability curve for all points in the image. They found that most of the points in an image are very close to the edge manifold, regardless of the scale! In other words, the 3-Jets of the points in the images are clustered near the ideal edge manifold. Furthermore, the edge statistics are roughly scale invariant. They conclude that natural images are extremely sparse, with most of the points in an image clustering around low-dimensional structures like edges, blobs, ridges, and junctions.

CONTRAST IN IMAGES

An alternative method of assessing relations between pixels is through the computation of contrast. Recall in particular Weber's contrast:

$$C_{\text{Weber}} = \frac{L - L_b}{I_b} \tag{37}$$

This measure is intended for uniform patches on uniform backgrounds. It can be extended to measure contrast in images in the form of center-surround processing and is typically computed as a difference of Gaussians. First, we apply to the image a Gaussian filter twice, but with different spatial extents, leading to L^{center} and L^{surround}. These are then subtracted and optionally normalized to yield a contrast measure for each pixel:[55]

$$C_{\text{DoG}} = \frac{L^{\text{center}} - L^{\text{surround}}}{L^{\text{surround}}} \tag{38}$$

This computation gives rise to the filter profile shown in Fig. 2.9. If we apply such processing to an image, then we obtain results such as shown in Fig. 12.

In essence, such center-surround processing can be seen as a method to extract edges from an image; it serves as a rudimentary edge detector. Such processing also occurs in the human visual system. It removes spatial correlations[56–58] and effectively returns a zero response in regions where there are no contrasts. This can also be seen as removing image components that are predictable and therefore uninformative. Conversely, the information that is left is unpredictable and therefore informative.[59] Alternatively, it can be understood by saying that surfaces are consistent;[60] edges form discontinuities that are transmitted to the brain. Anything in between edges is thought to be inferred by the human visual system from the edges alone, a process known as *filling-in*.[61–66]

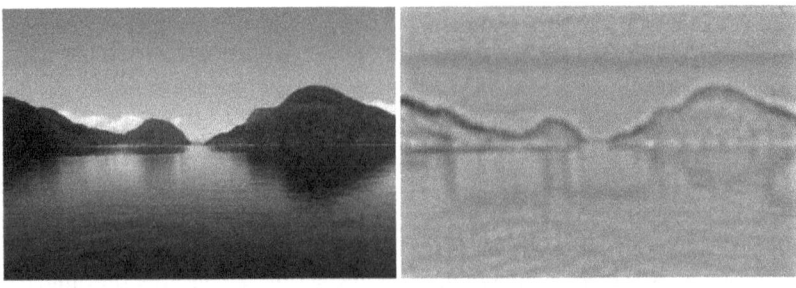

Fig. 12 Example of center-surround processing. Here, we applied separate processing to each of the three RGB color channels. The filter parameters were $\sigma^{\text{center}} = 6.25$ and $\sigma^{\text{surround}} = 12.5$ pixels. (Doubtful Sound, New Zealand, 2012)

As gradients compute differences between neighboring pixels and center-surround processing computes differences between pixel areas, one would expect that histograms of the magnitude of difference of Gaussian responses follow the same behavior as gradient histograms. That this is indeed the case is shown in Fig. 13, where a center-surround computation was carried out on two image ensembles. The 15 images of the first ensemble are shown in Fig. 14, whereas the second ensemble is shown in Fig. 7.8. The first ensemble consists of 15 landscape images, whereas the second ensemble consists of 15 high dynamic range forest scenes, which do not contain significant amounts of sky.

The forest image ensemble produces a longer tail for positive gradients, which is likely due to the fact that these images were captured in high dynamic range. The landscape image ensemble was linearized prior to the calculations. The shape of both plots suggests that a highly kurtotic distribution is present, consistent with the idea that center-surround processing removes unimportant information. High dynamic range imaging, which is a closer representation of light available in scenes than afforded by conventional images, appear to exhibit higher kurtosis, an issue also seen in moment statistics.

It has been suggested that negative center-surround responses to natural images are more numerous than positive responses.[67] The difference between the number of negative versus positive responses was measured to be as high as 50%. This could be an explanation for the fact that there are twice as many OFF pathways in the human visual system as there are ON pathways Unfortunately, we were not able to reproduce this result. For the HDR forest ensemble the ratio between positive and negative responses is 1:1.15. In other words, the negative responses are 15% more numerous. Given that landscape images tend to have bright sky near the top of the image, with darker ground and foreground areas, we carried out the same experiment on the 15 landscape images of Fig. 14. Here, the ratio is 1:1.025, albeit that in this case the positive responses were slightly more numerous (by 2.5%).

So far, we have relatively arbitrarily fixed the size of the center Gaussian to $\sigma^{center} = 6.25$ pixels and the surround to $\sigma^{surround} = 1.5$ pixels. There is of course no reason to believe that these values are in some sense optimal. Further, human vision is sensitive to contrasts at different scales as modeled by the contrast sensitivity function This has led to the development of contrast measurements that take different spatial scales into account, which can be obtained by first band-pass filtering the image.[68] Typically, center-surround processing at different spatial scales begins by creating a stack of Gaussians, To create differences of Gaussians at different scales, pairs of neighboring scales are simply subtracted.

Finally, note that contrast as well as luminance values vary greatly across an image. Moreover, contrast and luminance values tend to vary independently. Interestingly, human vision appears to process contrast and luminance separately, showing yet another form of adjustment to statistical regularities in natural scenes.[69]

This also affects adaptation in the human visual system, as saccadic eye movements will continuously cause the retina to focus on different patches with correspondingly different local contrasts and luminances.[70] This means that human vision does not tend to approach a state of full adaptation, as the retina is presented with different luminance and contrast statistics every few hundred milliseconds or so. This is important to maintain vision. Consider a situation whereby every photoreceptor is fully adapted to its input. In that case, the semi-saturation constant σ of Equation (2.1) would equal its input I, and as a result, the response of every photoreceptor would be the same, namely, half-maximal. If all photoreceptors emit the same signal, vision would essentially be blind! Non-stationary statistics combined with saccadic eye movements prevent this from happening.

Of course, motion of objects in scenes will also cause significant deviations from the zero response and are therefore particularly salient.

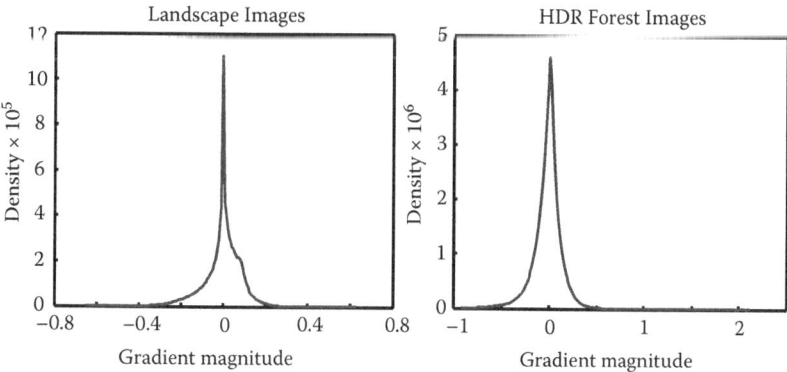

Fig. 13 Magnitude distribution of difference-of-Gaussian processing, performed on the linearized landscape images of Fig. 5.14 (left) and the high dynamic range forest images of Figure 7.8 (right)

Fig. 14 An ensemble of 15 images containing landscapes. (Various, South Island, New Zealand, 2012)

IMAGE DEBLURRING

When taking a photograph of a scene, it is not uncommon that either the camera or an object in the scene moves. The longer the aperture is open, the more likely this is to be the case. As a result, all or part of the image will be blurred. A number of approaches for sharpening an image have been proposed. One type of approach, *blind motion deconvolution*, essentially treats the photograph as the result of a convolution between an unknown sharp image and an equally unknown blurring kernel. The goal is to estimate the blur kernel so that it can be deconvolved with the blurred image to yield the sharp image. Naturally, this is an underspecified problem, so additional constraints are needed; recently, a number of researchers have employed natural image statistics to provide them.

For example, gradient distributions derived from natural images can be used to estimate the blur kernel.[71]

They can be effectively modeled with a hyper-Laplacian distribution:[72]

$$p(x) \propto \exp\left(-\alpha |x|^\beta\right) \tag{39}$$

The heavy-tailed distribution of gradient histograms was shown to be best modeled if parameter β was set to approximately 2/3.[72] Hyper-Laplacian distributions have also been successfully applied to applications such as transparency separation,[73] image segmentation[74] and superresolution[75]

Of course, other image statistics approaches are possible, for instance, by utilizing the typical $1/f^\beta$ distribution of power spectra which would hold for sharp images but not necessarily for blurred ones.[76] An interaction between power spectra and wavelets can also be used[77,78].

All of these approaches assume camera motion—that is, that there is a single blur kernel for the entire image. In an alternate approach, the gradient structure can be used to find those pixels that are blurred and segment them from the rest of the image, as demonstrated by Levin.[3] Specifically, a correspondence is found between the gradient distribution and the blur present in the image, allowing the discovery of the blur kernel for a given image.

One primary feature of motion blurring is the attenuation of higher frequencies. This shows in the slope of the power spectra (by increasing β) as well as in gradient histograms (in particular by removing the longer tails at the higher gradients). Levin attempts to recover the blur kernel by applying different blurs to the image to find the one that produces a gradient histogram that matches the blurred one. This requires a non-blurred image or image region. Using a completely different image tends not to produce reliable results (due to differences in the underlying gradient histogram). Since much terrestrial motion tends to be horizontal, the majority of the effects of motion blurring are also horizontal. Thus, the vertical gradients can, under some circumstances, be used to calculate the blur of the horizontal components. In a further refinement, Shan et al.[79] employ an iterative alternation between optimizing the blur kernel and the sharp image using priors from natural image statistics.

SUPERRESOLUTION

There is a growing field of work that aims to infer information from an image that is not actually present in the image. Image deblurring is one example, where sharpness is recreated. One core application in this area is superresolution, which essentially tries to increase the resolution of an image by intelligently guessing what the missing higher-frequency information is. For example, Fattal notes that there is a difference between shadow edges and texture edges.[80] Not only are humans sensitive to these differences, but information about what type of edge is present at a pixel can be found in the gradient structure. The unique intensity falloff found at the edge of a shadow—the penumbra—seems to be used by human vision to find shadows and separate illumination changes from reflectance changes.[81]

Following this insight, Fattal et al. examined 15 indoor images, looking at the statistics of pixels in a small area to see if the sharpness of an edge in the high-resolution image could be estimated from the low-resolution version of the image. To do this, they calculated a gradient field (using central differences) and then searched for the nearest edge (along the gradient) for each pixel. Finally, they determined how much the luminance changes along that edge, how far the edge is from the pixel, and how sharp the edge is. They found a clear dependency between the three measures and the continuity profiles exhibited by the image intensities, and they were able to use this information to improve the quality of a superresolution technique.

In a different approach, Tappen et al.[75] use the gradient distribution to fill in the holes produced when a high-resolution image is produced from a low-resolution image. They also apply the same procedure to yield a full-color image from an achromatic image.

INPAINTING

No imaging system is perfect. All imaging systems contain intrinsic noise and occasionally yield results with holes (sometimes due simply to self-occlusion). This is also true of the human visual system. The "blind spot," which is the section of the retina where the optic nerve leaves the eye, has no photoreceptors. Yet we do not see a blank spot there. Aside from that, there are many situations whereby it is desirable to remove an object from an image—for instance, a photograph shot through a wire fence may benefit from the removal of the fence. Distracting objects may also need to be removed, requiring algorithms to guess what may have been behind those objects.

There is a range of computer graphics algorithms for "hole filling" or image completion, which can be captured under the title *inpainting*.[82,83] Inpainting techniques can be divided into several classes based on the central algorithm. The three typical categories are texture synthesis, methods relying on partial differential equations, and exemplar-based techniques.

The first category, texture synthesis, is generally considered to be the oldest form of inpainting, likely originating with Fournier et al.[84] These methods have as their goal the synthesis of a new image patch that is perceptually similar to a user-defined patch. For example, the technique developed by Heeger and Bergen[85] generates stochastic textures using a straightforward model of the human visual system. Specifically, their method takes a white noise patch—of arbitrary size—and alters it using, in part, a pyramid-based histogram matching algorithm so that the first-order statistics of a series of oriented spatial-frequency filters are similar to those of the patch.

The second class of approaches is based on the theory of variational methods and partial differential equations (PDE). One of the first of these was proposed by Bertalmio et al.,[82] who used an iterative procedure, where edges (detected using gradients) at the border of the missing region are extended into the missing region. Chan and Shen[86] extended this model using the theory of total variation, as well as an anistropic diffusion based on the strength of the gradients. Bertalmio et al.[87] also proposed a hybrid approach, which uses PDE-like methods to fill in the structure of an image and then applies texture synthesis methods to finish by filling in the texture.

Finally, there are the examplar approaches, which use image statistics to find optimal replacements for the missing texture, usually by assuming that a patch similar to the missing bit can be found somewhere in the remaining image. For example, Hirani and Totsuka[88] use a combination of spectral and spatial information to find the replacement patch. The seminal work of Efros et al.[89,90] has done much to define the field of exemplar-based inpainting. Levin[91] suggested that the missing section can be filled in based on the gradients at the boundary region as well as some term that maximizes the match to the global gradient histogram.

Likewise, Criminisi et al.[83] included information from gradients in the algorithm that decides which regions should be completed first. They then select a small patch of pixels around the region to be completed and then find similar patches elsewhere in the image. Image editing can also proceed entirely in the gradient domain. In that case, an image can be reconstructed by solving the Poisson equation. This allows gradients to be directly interpolated into the missing region.[92,93] Alternatively, the patch-copying procedure can be treated as a global optimization of a Markov random field.(Fig. 15)[94]

Fig. 15 An example of inpainting using Crimini's[83] algorithm. After regions from an image (top left) are deleted (top right), the hole is filled (bottom) using this and similar images. (Castilla y Leon, Spain, 2010)

REFERENCES

1. Reinhard, E.; Khan, E.A.; Akyüz, A.O.; Johnson, G.M. *Color Imaging: Fundamentals and Applications*; Peters, A.K.; Ed.; Wellesley, 2008.
2. Gibson, J.J. *The Ecological Approach to Visual Perception*; Lawrence Erlbaum: Hillsdale, NJ, 1979.
3. Levin, A. Blind motion deblurring using image statistics. In *NIPS-20: Proceedings of the Conference on Advances in Neural Information Processing Systems*, Vol. 19; 2006, 841–848.
4. Pouli, T.; Cunningham, D.W.; Reinhard, E. Image statistics and their applications in computer graphics. In *Eurographics State-of-the-Art Reports*, 2010, 97.
5. Ruderman, D.L.; Bialek, W. Statistics of natural images: Scaling in the woods. Phys. Rev. Lett. **1994**, *73* (6), 814–817.
6. Ruderman, D.L. The statistics of natural images. Netw. Comput. Neural Syst. **1994**, *5* (4), 517–548.
7. Huang, J.; Mumford, D. Statistics of natural images and models. In *Proceedings of the IEEE International Conference on Computer Vision and Pattern Recognition*, Vol. 1; Washington, DC, 1999.
8. Huang, J.; Lee, A.; Mumford, D. Statistics of range images. In *Proceedings of the IEEE International Conference on Computer Vision and Pattern Recognition*, Washington, DC, 2000, 324–331.
9. Dror, R.O.; Leung, T.K.; Adelson, E.H.; Willsky, A.S. Statistics of real-world illumination. In *Proceedings of the IEEE International Conference on Computer Vision and Pattern Recognition*, Vol. 2; Kauai, HI, 2001, 164–171.
10. Dror, R.O.; Willsky, A.S.; Adelson, E.H. Statistical characterization of realworld illumination. J. Vision **2004**, *4* (9), 821–837.
11. Levin, A.; Zomet, A.; Weiss, Y. Learning to perceive transparency from the statistics of natural scenes. In *NIPS-15: Proceedings of the 2002 Conference on Advances in Neural Information Processing Systems*; MIT Press: Cambridge, MA, 2002.
12. Mallat, S.G. A theory for multiresolution signal decomposition: The wavelet representation. IEEE Trans. Pattern Anal. Mach. Intell. **1989**, *11* (7), 674–693.
13. Simoncelli, E.P. Statistical models for images: Compression, restoration and synthesis. In *Proceedings of the 31st Asilomar Conference on Signals, Systems and Computers*, 1997, 673–678.
14. Balboa, R.M.; Grzywacz, N.M. Occlusions and their relationship with the distribution of contrasts in natural images. Vision Res. **2000**, *40* (19), 2661–2669.
15. Field, D.J. Relations between the statistics of natural images and the response properties of cortical cells. J. Opt. Soc. Am. A **1987**, *4* (12), 2379–2394.
16. Zontak, M.; Irani, M. Internal statistics of a single natural image. In *Proceedings of the IEEE International Conference on Computer Vision and Pattern Recognition*, 2011, 977–984.
17. Buades, A.; Coll, B.; Morel, J.-M. A non-local algorithm for image denoising. In *Proceedings of the IEEE International Conference on Computer Vision and Pattern Recognition*, Vol. 2; 2005, 60–65.
18. Mosseri, I.; Zontak, M.; Irani, M. Combining the power of internal and external denoising. In *Proceedings of the IEEE International Conference on Computational Photography*; 2013.
19. Freedman, G.; Fattal, R. Image and video upscaling from local self-examples. ACM Trans. Graph. **2011**, *30* (2), 12.
20. Georgiev, T.; Intwala, C.; Babakan, S.; Lumsdaine, A. Unified frequency domain analysis of lightfield cameras. In *Proceedings of the 13th European Conference on Computer Vision*; 2008, 224–237, 100.
21. Shapiro, L.; Stockman, G.C. *Computer Vision*; Prentice Hall: Upper Saddle River, NJ, 2001, 102.
22. Forsyth, D.A.; Ponce, J. *Computer Vision: A Modern Approach*, 2nd Ed.; Prentice Hall: Upper Saddle River, NJ, 2011.
23. Marr, D. *Vision: A Computational Investigation into the Human Representation and Processing of Visual Information*; W H Freeman and Company: San Fransisco, CA, 1982.
24. Pedersen, K.S.; Lee, A.B. Toward a full probability model of edges in natural images. In *Proceedings of the 7th European Conference on Computer Vision*; Springer, 2002, 328–342.
25. Hubel, D.H.; Wiesel, T.N. Receptive fields and functional architecture of the monkey striate cortex. J. Physiol. **1968**, *195* (1), 215–243.
26. Thompson, R.F. *The Brain: A Neuroscience Primer*, 3rd Ed.; Worth Publishers: New York, 2000.
27. Köhler, W. *Gestalt Psychology*; Liveright: New York, 1929.
28. Konishi, S.; Yuille, A.; Coughlan, J.; Zhu, S.-C. Statistical edge detection: Learning and evaluating edge cues. IEEE Trans. Pattern Anal. Mach. Intell. **2003**, *25* (1), 57–74.
29. Geisler, W.S.; Perry, J.S.; Super, B.J.; Gallogly, D.P. Edge co-occurrence in natural images predicts contour grouping performance. Vision Res. **2001**, *41* (6), 711–724.
30. Krüger, N.; Wörgötter, F. Multi-modal estimation of collinearity and parallelism in natural image sequences. Netw. Comput. Neural Syst. **2002**, *13* (4), 553–576.
31. Graham, D.J.; Friedenberg, J.D.; Rockmore, D.N. Efficient visual system processing of spatial and luminance statistics in representational and nonrepresentational art. In *Proceedings of IS&T/SPIE Electronic Imaging*, Vol. 72401N; 2009.
32. Graham, D.J.; Friedenberg, J.D.; Rockmore, D.N.; Field, D.J. Mapping the similarity space of paintings: Image statistics and visual perception. Vis. Cogn. **2010**, *18* (4), 559–573.
33. Redies, C.; Amirshahi, S.A.; Koch, M.; Denzler, J. PHOG-derived aesthetic measures applied to color photographs of artworks, natural scenes and objects. In *European Conference on Computer Vision, Workshops and Demonstrations*, 2012, 522–531.
34. ter Haar Romeny, B.M. *Front-End Vision and Multi-Scale Image Analysis*; Kluwer Academic Publisher: Dordrecht, 2011.
35. Weickert, J.; Ishikawa, S.; Imiya, A. Linear scale-space has first been proposed in Japan. J. Math. Imaging Vis. **1999**, *10* (3), 237–252.
36. Witkin, A.P. Scale-space filtering. In *Proceedings of the 8th International Joint Conference on Artificial Intelligence*, 1983, 1019–1022.

37. Koenderink, J.J. The structure of images. Biol. Cybern. **1984**, *50* (5), 363–370.
38. Lindeberg, T. Linear scale space. In *Geometry-Driven Diffusion in Computer Vision*; ter Haar Romeny; Ed.; Kluwer Academic Publishers: Dordrecht, 1994, 1–77.
39. Perona, P.; Malik, J. Scale-space and edge detection using anisotropic diffusion. IEEE Trans. Pattern Anal. Mach. Intell. **1990**, *12* (7), 629–639.
40. Tomasi, C.; Manduchi, R. Bilateral filtering for gray and color images. In *Proceedings of the IEEE International Conference on Computer Vision*, Washington, DC, 1998, 839–846.
41. Paris, S.; Durand, F. A fast approximation of the bilateral filter using a signal processing approach. Int. J. Comput. Vis. **2009**, *81* (1), 24–52.
42. Farbman, Z.; Fattal, R.; Lischinski, D.; Szeliski, R. Edge-preserving decompositions for multi-scale tone and detail manipulation. ACM Trans. Graph. **2008**, *27* (3), 67.
43. Fattal, R. Edge-avoiding wavelets and their applications. ACM Trans. Graph. **2009**, *28* (3), 22.
44. Paris, S.; Hasinoff, S.W.; Kautz, J. Local Laplacian filters: Edge-aware image processing with a Laplacian pyramid. ACM Trans. Graph. **2011**, *30* (4), 68.
45. Trentacoste, M.; Mantiuk, R.; Heidrich, W.; Dufrot, F. Unsharp masking, countershading and halos: Enhancements or artifacts? Comput. Graph. Forum **2012**, *31* (2), 555–564.
46. Koenderink, J.J. The brain a geometry engine. Psychol. Res. **1990**, *52* (2–3), 122–127.
47. Lindeberg, T.; Florack, L. *Foveal scale-space and the linear increase of receptive field size as a function of eccentricity*. Technical report, ISRN KTH NA/P–94/27– SE, 1994.
48. Angelucci, A.; Bressloff, P.C. Contribution of feedforward, lateral and feedback connections to the classical receptive field center and extra-classical receptive field surround of primate v1 neurons. Prog. Brain Res. **2006**, *154*, 93–120.
49. Gårding, J.; Lindeberg, T. Direct estimation of local surface shape in a fixating binocular vision system. In *Proceedings of the 3rd European Conference on Computer Vision*; Eklund, J.-O.; Ed.; Volume 800 of Lecture Notes in Computer Science; Springer: Berlin, 1994, 365–376.
50. Jones, D.G.; Malik, J. A computational framework for determining stereo correspondence from a set of linear spatial filters. Image Vis. Comput. **1992**, *10* (10), 395–410.
51. Lindeberg, T.; Gårding, J. Shape from texture from a multi-scale perspective. In *Proceedings of the IEEE International Conference in Computer Vision*, 1993, 683–691.
52. Malik, J.; Rosenholtz, R. A differential method for computing local shape-fromtexture for planar and curved surfaces. In *Proceedings of the IEEE International Conference on Computer Vision and Pattern Recognition*, 1993, 267–273.
53. Salden, A.H.; ter Haar Romeny, B.M.; Viergever, M.A. Differential and integral geometry of linear scale-spaces. J. Math. Imaging Vis. **1998**, *9* (1), 5–27.
54. Pedersen, K.S.; Lee, A.B. Toward a full probability model of edges in natural images. In *Proceedings of the 8th European Conference on Computer Vision*; Springer: Heidelberg, 2002, 328–342.
55. Tadmor, Y.; Tolhurst, D.J. Calculating the contrasts that retinal ganglion cells and LGN neurones encounter in natural scenes. Vision Res. **2000**, *40* (22), 3145–3157.
56. Barlow, H.B. Possible principles underlying the transformations of sensory images. In *Sensory Communication*; Rosenblith, W.A.; Ed.; MIT Press: Cambridge, MA, 1961, 217–234.
57. Srinivasan, M.; Laughlin, S.; Dubs, A. Predictive coding: A fresh view of inhibition in the retina. Proc. R. Soc. Lond. B **1982**, *216* (1205), 427–459.
58. Atick, J.J.; Redlich, A.N. Towards a theory of early visual processing. Neural Comput. **1990**, *2* (3), 308–320.
59. Tkacik, G.; Prentice, J.S.; Victor, J.D.; Balasubramanian, V. Local statistics in ˇnatural scenes predict the saliency of synthetic textures. Proc. Natl. Acad. Sci. **2010**, *107* (42), 18149–18154.
60. Grimson, W. Surface consistency constraints in vision. Comput. Vision Graph. Image Process. **1983**, *24* (1), 28–51.
61. Walls, G. The filling-in process. Am. J. Optom. **1954**, *31* (7), 329–340.
62. Krauskopf, J. Effect of retinal stabilization on the appearance of heterochromatic targets. J. Opt. Soc. Am. **1963**, *53* (6), 741–744.
63. Grossberg, S.; Mingolla, E. Neural dynamics of form perception: Boundary adaptation, illusory figures, and neon color spreading. Psychol. Rev. **1985**, *92* (2), 173–211.
64. Elder, J.H. Are edges incomplete? Int. J. Comput. Vision **1999**, *34* (2/3), 97–122.
65. Neumann, H.; Pessoa, L.; Hansen, T. Visual filling-in for computing perceptual surface properties. Biol. Cybern. **2001**, *85* (5), 355–369.
66. Dakin, S.C.; Bex, P.J. Natural image statistics mediate brightness filling in. Proc. R. Soc. Lond. B **2003**, *270* (1531), 2341–2348.
67. Balasubramanian, V.; Sterling, P. Receptive fields and functional architecture in the retina. J. Physiol. **2009**, *587* (12), 2753–2767.
68. Peli, E. Contrast in complex images. J. Opt. Soc. Am. A **1990**, *7* (10), 2032–2040.
69. Mante, V.; Frazor, R.A.; Bonin, V.; Geisler, W.S.; Carandini, M. Independence of luminance and contrast in natural scenes and in the early visual system. Nat. Neurosci. **2005**, *8* (12), 1690–1697.
70. Frazor, R.A.; Geisler, W.S. Local luminance and contrast in natural images. Vision Res. **2006**, *46* (10), 1585–1598.
71. Fergus, R.; Singh, B.; Hertzmann, A.; Roweis, S.; Freeman, W. Removing camera shake from a single photograph. ACM Trans. Graph. **2006**, *25* (3), 787–794.
72. Krishnan, D.; Fergus, R. Fast image deconvolution using hyper-Laplacian priors. In *NIPS-23: Proceedings of the 2009 Conference on Advances in Neural Information Processing Systems*, Vol. 22; 2009, 1–9, 115.
73. Levin, A.; Weiss, Y. User assisted separation of reflections from a single image using a sparsity prior. IEEE Trans. Pattern Anal. Mach. Intell. **2007**, *29* (9), 1647–1654.
74. Heiler, M.; Schnorr, C. Natural image statistics for natural image segmentation. Int. J. Comput. Vis. **2005**, *63* (1), 5–19.
75. Tappen, M.F.; Russell, B.C.; Freeman, W.T. Exploiting the sparse derivative prior for super-resolution and image demosaicing. In *Proceedings of the 3rd International Workshop on Statistical and Computational Theories of Vision at ICCV*, 2003.

76. Caron, J.N.; Namazi, N.M.; Rollins, C.J. Noniterative blind data restoration by use of an extracted filter function. Appl. Opt. **2002**, *41* (32), 6884–6889.
77. Jalobeanu, A.; Blanc-Féraud, L.; Zerubia, J. Estimation of blur and noise parameters in remote sensing. In *Proceedings of the IEEE International Conference on Acoustics, Speech and Signal Processing*, 2002, 249–256.
78. Neelamani, R.; Choi, H.; Baraniuk, R. ForWaRD: Fourier-wavelet regularized deconvolution for ill-conditioned systems. IEEE Trans. Signal Process. **2004**, *52* (2), 418–433.
79. Shan, Q.; Jia, J.; Agarwala, A. High-quality motion deblurring from a single image. ACM Trans. Graph. **2008**, *27* (3), 1–10.
80. Fattal, R. Image upsampling via imposed edge statistics. ACM Trans. Graph. **2007**, *26* (3), 95.
81. Goldstein, E. *Sensation and Perception*, 9th Ed.; Wadsworth Cengage Learning: Belmont, CA, 2013.
82. Bertalmio, M.; Sapiro, G.; Caselles, V.; Ballester, C. Image inpainting. In *SIGGRAPH'00: Proceedings of the 27th Annual Conference on Computer Graphics and Interactive Techniques*, 2000, 417–424.
83. Criminisi, A.; Perez, P.; Toyama, K. Region filling and object removal by exemplar-based inpainting. IEEE Trans. Image Process. **2004**, *13* (9), 1200–1212.
84. Fournier, A.; Fussell, D.; Carpenter, L. Computer rendering of stochastic models. Commun. ACM **1982**, *25* (6), 371–384.
85. Heeger, D.J.; Bergen, J.R. Pyramid-based texture analysis/synthesis. In *SIGGRAPH'95: Proceedings of the 22nd Annual Conference on Computer Graphics and Interactive Techniques*; ACM: New York, 1995, 229–238.
86. Chan, T.; Shen, J. Local inpainting models and TV inpainting. SIAM J. Appl. Math. **2001**, *62* (3), 1019–1043.
87. Bertalmio, M.; Vese, L.; Sapiro, G.; Osher, S. Simultaneous structure and texture image inpainting. IEEE Trans. Image Process. **2003**, *12* (8), 882–889.
88. Hirani, A.N.; Totsuka, T. Combining frequency and spatial domain information for fast interactive image noise removal. In *SIGGRAPH '96: Proceedings of the 23rd Annual Conference on Computer Graphics and Interactive Techniques*, 1996, 269–276.
89. Efros, A.A.; Leung, T.K. Texture synthesis by non-parametric sampling. In *Proceedings of the 7th IEEE International Conference on Computer Vision*, Vol. 2; 1999, 1033–1038.
90. Efros, A.A.; Freeman, W.T. Image quilting for texture synthesis and transfer. In *SIGGRAPH '01: Proceedings of the 28th Annual Conference on Computer Graphics and Interactive Techniques*, 2001, 341–346.
91. Levin, A.; Zomet, A.; Weiss, Y. Learning how to inpaint from global image statistics. In *Proceedings of the 9th IEEE International Conference on Computer Vision*, Washington, DC, 2003, 305–312.
92. Pérez, P.; Gangnet, M.; Blake, A. Poisson image editing. ACM Trans. Graph. **2003**, *22* (3), 313–318.
93. Shen, J.; Jin, X.; Zhou, C.; Wang, C.C.L. Gradient based image completion by solving Poisson equation. Comput. Graph. **2007**, *31* (1), 119–126.
94. Komodakis, N.; Tziritas, G. Image completion using efficient belief propagation via priority scheduling and dynamic pruning. IEEE Trans. Image Process. *16* (11), 2649–2661.

Grayscale Morphological Image Enhancement

Jorge Domingo Mendiola Santibañez
Center of Research and Technological Development in Electrochemistry (CIDETEQ), Querétaro, México; Facultad de Ingeniería, Autonomous University of Querétaro (UAQ), Querétaro, México

Iván Ramón Terol Villalobos
Center of Research and Technological Development in Electrochemistry (CIDETEQ), Querétaro, México

Abstract

In the present work, several specialized transformations to enhance contrast are exposed. They are built with morphological operators to reinforce contours, increase the intensities of certain regions, improve images with bad illumination, and intensify contrast. Also, two methods to measure the quality of the output images are depicted: one quantifies contrast, and the other one determines the quality of the resultant contours.

Keywords: Bad illumination; Contrast enhancement; Contrast mappings; Contrast measure; Edges quality; Rational morphology; Top hats; Weber's law.

INTRODUCTION

In this study, several transformations to enhance contrast are presented. They are built with the following morphological operators: (a) morphological erosion and dilation,[1] (b) morphological opening and closing,[1] (c) morphological gradients,[2] (d) opening and closing by reconstruction,[3] and (e) top hats.[1]

Several of the aforementioned operators are briefly explained in Section "Basic Morphological Transformations" in order to understand the ideas implemented through several equations. The beginning of the study of morphological contrast is provided in Section "First Formal Work in Morphological Contrast," where the idea of Kramer and Bruckner to enhance contrast is presented.[4] This concept was used by Meyer and Serra to introduce the contrast mapping notion.[5] The *morphological slope filters* (MSFs)[6] are outlined in Section "Morphological Slope Filters." These operators are formalized using the contrast mapping notion where instead of utilizing a proximity criterion, a gradient criterion is used to select the primitives, the morphological erosion, dilation, or the original function. An important characteristic of MSFs is that the erosion and dilation are used in different ways with the purpose of reinforcing the contours of regions of low or high contrast. The sequential application of the MSFs is presented in Section "Sequential Morphological Slope Filters," and these transformations permit to obtain intermediate results compared with MSFs.[7] Another approach on MSFs is provided in Section "Weighted Morphological Slope Filters," where the proximity criterion is weighted by an increasing function with the purpose of considering the intensity levels of the gradient transformation. Initially, the weighted MSFs were studied in the work of Terol-Villalobos,[8] and an extension was developed in the work of Mendiola-Santibañez and Terol-Villalobos.[9] So far, the contrast is enhanced by reinforcing the image contours utilizing a gradient criterion; however, in Sections "Morphological Multiscale Contrast Approach Consistent with Human Vision Perception" and "Contrast Enhancement Using Connected Top Hats," the gradient criterion is replaced by a top-hat criterion bringing in consequence the improvement of the contrast considering the size of the regions into the image.[10] During this analysis, the authors found that Weber's law can be expressed in terms of the opening by reconstruction giving place to a new technique based on the rational morphology.[11,12] The new methodology enhances the flat zones, and in consequence, the contours of the image are sharpened. This proposal is depicted in Section "Rational Multiscale Contrast." The rational morphology introduced in the work of Kogan et al.[13] and the final expressions obtained in the works of Espino-Gudiño et al.[11] and Peregrina-Barreto et al.[12] are similar to the equations corresponding to the retinex method.[14,15] Other formulation to improve contrast is presented in Section "An Approach to Enhance Contrast on Images with Poor Lighting," where Weber's law is applied directly to improve the visual perception[16] of the processed image. On the other hand, it is very common to enhance contrast, but in few occasions, it is quantified. In Section "Morphological Contrast Measure," a solution to this problem is presented.[17–20] When contrast is enhanced, many times it is necessary to compare results, and one way to do this is to have an index capable of indicating the quality of the edges of the improved images; for this, in Section "Other Criterion to Measure the Quality of an Enhanced Image," a solution to this problem is presented.[9] The last section corresponds to the conclusions.

BASIC MORPHOLOGICAL TRANSFORMATIONS

In mathematical morphology,[1] increasing and idempotent transformations are frequently used. A transformation T is increasing if for two sets X, Y such that $X \subset Y \Rightarrow T(X) \subset T(Y)$. In the gray-level case, the inclusion is substituted by the usual order, that is, let f and g be two functions, then $f \leq g \Leftrightarrow f(x) \leq g(x)$ for all x. Then, a transformation T is increasing if for all pair of functions f and g, with $f \leq g \Rightarrow T(f) \leq T(g)$. In other words, increasing transformations preserve the order. A second important property is the idempotence notion. A transformation T is idempotent if and only if $T(T(f)) = T(f)$. The use of both properties plays a fundamental role in the theory of morphological filtering. Morphological transformations complying with these properties are known as morphological filters.[1] The basic morphological filters are the morphological opening $\gamma_{\mu B}(f)$ and closing $\varphi_{\mu B}(f)$ using a given structuring element μB. In this study, a square structuring element is employed, where B denotes the basic structuring element of size $\mu = 1$ with 3×3 pixels, which contains its origin in the center. While $\breve{B}$ is the transposed set with respect to its origin, $\breve{B} = \{-x : x \in B\}$, $\mu \in Z$ is a size parameter, and Z stands for the set of the integers. Let $f: Z^2 \to Z$ be the input image and x be a point on the definition domain D_f. The morphological opening $\gamma_{\mu B}$ is expressed as follows:

$$\gamma_{\mu B}(f)(x) = \delta_{\mu B}\left(\varepsilon_{\mu B}(f)\right)(x)$$

by duality, the morphological closing $\varphi_{\mu B}$ is defined as

$$\varphi_{\mu B}(f)(x) = [\gamma_{\mu B}(f^c)(x)]^c$$

with $f^c(x) = 255 - max(f(x))$, where f^c represents the complement function, $max(f(x))$ the maximum intensity level, $\varepsilon_{\mu B}(f)(x) = \wedge\{f(y) : y \in \mu \breve{B}_x\}$ and $\delta_{\mu B}(f)(x) = [\varepsilon_{\mu B}(f^c)(x)]^c$ are the morphological erosion and dilation, respectively, and $\wedge$ represents the *inf* operator. To simplify the notation, the next expression is considered, $\mu B = \mu$.

Structurally, the morphological opening $\gamma_{\mu B}(f)$ allows the elimination of those regions where the structuring element does not fit giving as result a smoothed image. The morphological closing $\varphi_{\mu B}(f)$ has the same behavior but in the complement of the image.

Morphological Gradients

The morphological gradient gm_μ, the external gradient $grade_\mu$, and the internal gradient $gradi_\mu$ are edge detectors.[2] These transformations are written as follows:

$$gm_\mu(f)(x) = \delta_\mu(f)(x) - \varepsilon_\mu(f)(x)$$
$$grade_\mu(f)(x) = \delta_\mu(f)(x) - f(x)$$
$$gradi_\mu(f)(x) = f(x) - \varepsilon_\mu(f)(x)$$

Morphological gradients detect important intensity changes within the processed image, which correspond to the image contours.

Morphological Top Hat

The top-hat transformation is the arithmetic difference between the original function and the opened function or the difference between the closed function and the original function:[1]

$$Thw_\mu(f) = f - \gamma_\mu(f)$$
and
$$Thb_\mu(f) = \varphi_\mu(f) - f$$

Because the opening is anti-extensive, $f \geq \gamma_\mu(f)$, and the closing is extensive, $f \leq \varphi_\mu(f)$, the functions $Thw_\mu(f)$ and $Thb_\mu(f)$ are nonnegatives. These transformations were proposed by Meyer.[1] Generally, these transformations are followed by a threshold operation for obtaining a binary image containing the structures with a given size and contrast. Indeed, the top hat leads to a size distribution involving contrast of the image. Both transformations, morphological gradients and top-hat transformations, play a main role into image segmentation.

Structurally, the white top hat $Thw_\mu(f)$ corresponds to those regions where the structuring element does not fit. The black top hat $Thb_\mu(f)$ has the same behavior of the white top hat but in the complement.

Transformations by Reconstruction

Transformations by reconstruction are built by means of geodesic transformations.[3] The geodesic dilation $\delta_f^1(g)$ is given by $\delta_f^1(g) = f \wedge \delta(g)$ with $g \leq f$. When the marker g is equal to the erosion of the original function $\varepsilon_\mu(f)$, and the geodesic dilation is iterated until the idempotence is reached, the opening by reconstruction $\tilde{\gamma}_\mu(f)$ is obtained. Formally, this is written as follows:

$$\tilde{\gamma}_\mu(f)(x) = \lim_{n \to \infty} \delta_f^n\left(\varepsilon_\mu(f)\right)(x)$$

By duality, the closing by reconstruction $\tilde{\varphi}_\mu(f)$ is expressed as follows:

$$\tilde{\varphi}_\mu(f)(x) = \left(\tilde{\gamma}_\mu(f^c)(x)\right)^c$$

Structurally, the opening by reconstruction $\tilde{\gamma}_\mu(f)$ is interpreted as follows: those regions supporting the morphological erosion will be reconstructed similar to the input image, while those components eliminated by the erosion will be merged with the background. The same behavior presents the closing by reconstruction $\tilde{\varphi}_\mu(f)$ but in the complement of the image.

MORPHOLOGICAL CONTRAST ENHANCEMENT TRANSFORMATIONS

First Formal Work on Morphological Contrast

The first formal work on morphological contrast was reported by Meyer and Serra,[5] where they proposed a framework theory for morphological contrast enhancement based on the activity lattice structure. In their work, the original idea of Kramer and Bruckner's transformation[4] was used. This transformation sharpens the transitions between the object and background and changes the gray value of the original image at each point by selecting the value of the closest function, the dilation or erosion. A similar transformation to the Kramer and Bruckner operator is presented as follows:

$$KB^{\delta,\varepsilon}(f)(x) = \begin{cases} \delta_\mu(f)(x) & \text{if } \delta_\mu(f)(x) - f(x) < f(x) - \varepsilon_\mu(f)(x) \\ \varepsilon_\mu(f)(x) & \text{otherwise} \end{cases} \quad (1)$$

Some problems in $KB^{\delta,\varepsilon}$ are the oscillations and jumps produced when it is iterated. An example of Eq. 1 is provided in Fig. 1b. The original image is displayed in Fig. 1a. This image was obtained from the Internet Brain Segmentation Repository (IBSR) developed by the Centre for Morphometric Analysis (CMA) at the Massachusetts General Hospital (www.nitrc.org). Due to the morphological erosion and dilation take the size $\mu = 5$, the edges of the output image are sharpened considerably.

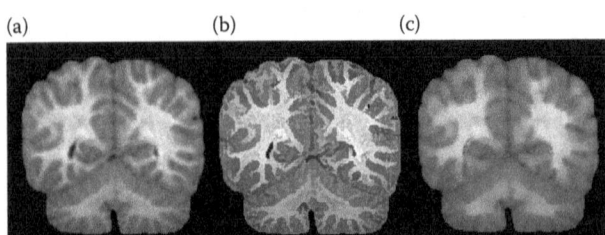

Fig. 1 Kramer and Bruckner's transformation. (a) Original image taken from the IBSR data set developed by the CMA at the Massachusetts General Hospital (www.nitrc.org). The database contains 18 volumes, and the image corresponds to the slice 39 of the volume 16; (b) Kramer and Bruckner's transformation, $KB\delta\varepsilon$ with $\mu = 5$; (c) contrast mapping of two states proposed by Meyer and Serra, $\kappa^{\varphi,\gamma}$ with $\mu = 2$

Meyer and Serra introduced the contrast mappings in the way suggested by Kramer and Bruckner, but the hypotheses are modified. They not only assume that the transformations are extensive, and anti-extensive, but also that the transformations must be idempotent. The use of this last hypothesis to build contrast operators avoids the risk of degrading the image. Formally, this is expressed as follows:[5]

$$\kappa^{\varphi,\gamma}(f)(x) = \begin{cases} \varphi_\mu(f)(x) & 0 \le \rho(x) \le \beta \\ \gamma_\mu(f)(x) & \beta < \rho(x) \le 1 \end{cases} \quad (2)$$

where $\rho(x)$ represents a proximity criterion given by:

$$\rho(x) = \frac{\varphi_\mu(f)(x) - f(x)}{\varphi_\mu(f)(x) - \gamma_\mu(f)(x)} \text{ and } \beta \in [0,1].$$

An example of Eq. 2 is presented in Fig. 1c. The morphological opening and closing take the size $\mu = 2$. Notice that the output image is smoothed because the opening eliminates narrow regions and the closing works similar but in the complement.

Morphological Slope Filters

Other way of attenuating the image degradation problem in the Kramer and Bruckner transformation was proposed by Terol-Villalobos.[6] In his work, the dilation and erosion transformations are also used in separate way to build a class of nonincreasing filters called *MSFs*. Applying the notion of toggle mappings, the next two operators were introduced in the work of Terol-Villalobos:[6]

$$\xi_\phi^{\varepsilon_1}(f)(x) = \begin{cases} \varepsilon(f)(x) & gradi(f(x)) \le \phi \\ f(x) & gradi(f(x)) > \phi \end{cases}$$

$$\xi_\phi^{\delta_1}(f)(x) = \begin{cases} \delta(f)(x) & grade(f(x)) \le \phi \\ f(x) & grade(f(x)) > \phi \end{cases}$$

where ϕ is a threshold parameter, and $gradi(f)$, $grade(f)$, $\varepsilon(f)$, and $\delta(f)$ are computed considering $\mu = 1$. When $\xi_\phi^{\varepsilon_1}(f)$ and $\xi_\phi^{\delta_1}(f)$ are iterated, the next result is obtained:

$$\xi_\phi^{\varepsilon_2}(f) = \xi_\phi^{\varepsilon_1}\left[\xi_\phi^{\varepsilon_1}(f)\right] \quad and \quad \xi_\phi^{\delta_2}(f) = \xi_\phi^{\delta_1}\left[\xi_\phi^{\delta_1}(f)\right]$$

At the nth step, when stability is reached ($n \to \infty$),

$$\xi_\phi^{\varepsilon_\infty}(f) = \xi_\phi^{\varepsilon_1}\left[\xi_\phi^{\varepsilon_n}(f)\right] = \xi_\phi^{\varepsilon_n}(f)$$

and

$$\xi_\phi^{\delta_\infty}(f) = \xi_\phi^{\delta_1}\left[\xi_\phi^{\delta_n}(f)\right] = \xi_\phi^{\delta_n}(f) \quad (3)$$

These families of contrast mappings are defined as *MSFs*. The case of filter $\xi_\phi^{\varepsilon_n}(f)(x)$ will be analyzed as follows; however, similar words can be mentioned by the operator $\xi_\phi^{\delta_n}(f)(x)$ but in the complement. In Fig. 2, an image illustrates the performance of filter $\xi_\phi^{\varepsilon_\infty}$ for a 1D signal. Original signal is given in Fig. 2a. The structuring element B is a line with three points and the origin is located in its center. D_R denotes the definition domain. The morphological erosion size $\mu = 1$ is presented with a dotted line. Also, there are the reference points x', x'', y', and p respectively. In Fig. 2b, the internal gradient, $gradi(f)$, is computed and those slopes less than $\phi = 1$ will be eroded; this is illustrated in Fig. 2c. The procedure is repeated in Fig. 2c–e. In Fig. 2e, the idempotence has been reached.

Filter $\xi_\phi^{\varepsilon_n}(f)$ produces a well-defined contrast because those zones into the original image f fulfilling $gradi_B(f) \leq \phi$ will be attenuated by means of the morphological erosion, that is, $\xi_\phi^{\varepsilon_n}(f) = \varepsilon_B(f)$. These regions are zones of weak contrast or weak gradient. While the zones of great contrast in f satisfy that $gradi_B(f) > \phi$, these zones are maintained unchanged, that is, $\xi_\phi^{\varepsilon_\infty}(f) = (f)$.

On the other hand, the output image $\xi_\phi^{\varepsilon_\infty}$ does not contain regions with slopes less than ϕ. Due to this characteristic, $\xi_{\phi_1}^{\varepsilon_\infty}$ and $\xi_{\phi_1}^{\delta_\infty}$ accomplish the next relation:

$$\text{for } \phi_1 \leq \phi_2, \begin{cases} \xi_{\phi_1}^{\varepsilon_\infty}(f)(x) \geq \xi_{\phi_2}^{\varepsilon_\infty}(f)(x) \\ \xi_{\phi_1}^{\delta_\infty}(f)(x) \leq \xi_{\phi_2}^{\delta_\infty}(f)(x) \end{cases}$$

In this way, an order is established on the MSFs according to the parameter ϕ. An example of Eq. 3 is presented in Fig. 3a considering $\xi_\phi^{\delta_\infty}$ with $\phi = 10$. The original image can be observed in Fig. 1a. The image in Fig. 3a presents several regions where the contrast has been over-enhanced.

Sequential Morphological Slope Filters

The MSFs applied sequentially rendering a selection of features at each level of the sequence of filters. The next expressions indicate how to apply these transformations:[7]

$$\xi_{\phi_1,\phi_2\cdots\phi_{m-1}}^{\varepsilon_\infty}(f) = \xi_{\phi_{m-1}}^{\varepsilon_\infty}\cdots\xi_{\phi_2}^{\varepsilon_\infty}\xi_{\phi_1}^{\varepsilon_\infty}(f)$$
$$\xi_{\phi_1,\phi_2\cdots\phi_{m-1}}^{\delta_\infty}(f) = \xi_{\phi_{m-1}}^{\delta_\infty}\cdots\xi_{\phi_2}^{\delta_\infty}\xi_{\phi_1}^{\delta_\infty}(f) \qquad (4)$$

Analyzing for $\xi_{\phi_1,\phi_2\cdots\phi_{m-1}}^{\varepsilon_\infty}$, if we apply a family of filters with parameters ϕ_i between ϕ_1 and ϕ_{m-1}, the output image obtained contains some great contrast regions of the levels $\phi_1, \phi_2, \ldots, \phi_m$ used for its computation, but with a greater contrast than ϕ_{m-1}. This can be observed from the following:

$$\xi_{\phi_m}^{\varepsilon_\infty} \leq \xi_{\phi_{m-1},\phi_m}^{\varepsilon_\infty} \leq \xi_{\phi_{m-1}}^{\varepsilon_\infty} \leq \cdots \leq \xi_{\phi_2}^{\varepsilon_\infty} \leq \xi_{\phi_1,\phi_2}^{\varepsilon_\infty} \leq \xi_{\phi_1}^{\varepsilon_\infty}$$

The last relation indicates that intermediate results between $\xi_{\phi_m}^{\varepsilon_\infty}(f)$ and $\xi_{\phi_{m-1}}^{\varepsilon_\infty}(f)$ are obtained if these filters are applied sequentially, that is, $\xi_{\phi_{m-1},\phi_m}^{\varepsilon_\infty}$. An example of Eq. 4 is presented in Fig. 3b, where the filter applied $\xi_{\phi_n\cdots\phi_1}^{\delta_n}$ produces an intermediate result with respect to the filter $\xi_\phi^{\delta_\infty}(f)$ with $\phi = 10$ presented in Fig. 3a, bringing in consequence the correction of those regions where a contrast saturation exists.

Weighted Morphological Slope Filters

The gradient criterion within the MSFs does not consider the intensity level of the original image when the contrast

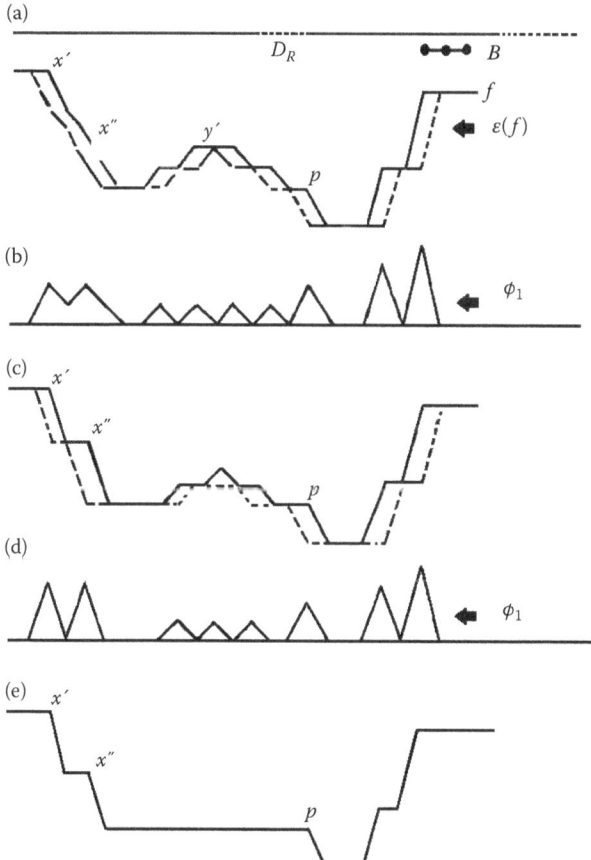

Fig. 2 MSFs. (a) Original function; (b) internal gradient of the original function; (c) $\xi_\phi^{\varepsilon_1}(f)$; (d) $gradi_B(\xi_\phi^{\varepsilon_1}(f))$; (e) $\xi_\phi^{\varepsilon_n}(f)$ until stability

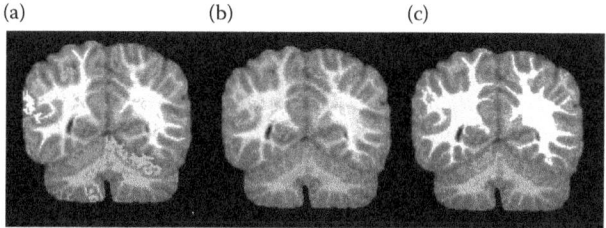

Fig. 3 MSF development. The original image is located in Fig. 1a. (a) $\xi_\phi^{\delta_n}(f)$ with $\phi = 10$; (b) $\xi_{\phi_n\cdots\phi_1}^{\delta_n}(f)$ with $\phi_1 = 1$, $\phi_2 = 2$, $\phi_3 = 3, \ldots, \phi_{10} = 10$; (c) $\xi_{\phi_1,\cdots,\phi_{10}d}^{\delta_\infty}(f)$ with $\phi_1 = 1, \ldots, \phi_{10} = 10$, $\chi_d(t) = A/{1 + Be^{k(t-d)}}$; $d = 60$; $t = f(x)$

of the processed image is modified. This problematic situation was considered in the work of Terol-Villalobos,[8] and the given solution consisted in weighting the gradient criterion by the intensity levels of the original image. In general way, the weighted MSFs are expressed as follows:[9]

$$\xi^{\varepsilon_1}_{\phi,\chi_d}(f)(x) = \begin{cases} \varepsilon(f)(x) & \chi_d(x)\,gradi(f(x)) \leq \phi \\ f(x) & \chi_d(x)\,gradi(f(x)) > \phi \end{cases}$$

$$\xi^{\delta_1}_{\phi,\chi_d}(f)(x) = \begin{cases} \delta(f)(x) & \chi_d(x)\,grade(f(x)) \leq \phi \\ f(x) & \chi_d(x)\,grade(f(x)) > \phi \end{cases}$$

At the nth step, when stability is reached ($n \to \infty$),

$$\xi^{\varepsilon_\infty}_{\phi,\chi_d}(f) = \xi^{\varepsilon_1}_{\phi,\chi_d}\left[\xi^{\varepsilon_n}_{\phi,\chi_d}(f)\right] = \xi^{\varepsilon_n}_{\phi,\chi_d}(f)$$

and

$$\xi^{\delta_\infty}_{\phi,\chi_d}(f) = \xi^{\delta_1}_{\phi,\chi_d}\left[\xi^{\delta_n}_{\phi,\chi_d}(f)\right] = \xi^{\delta_n}_{\phi,\chi_d}(f) \quad (5)$$

Definition 1.

Let $t = f(x)$ be the intensity level at the point x, and let $0 \leq \chi(t) \leq 1$ be an increasing function in this interval with: χ: [0,255] → [0,1]. The weighting function is defined as $\chi(t - d)$, where $d \in Z$ represents a displacement parameter, also if $t - d \leq 0$, $\chi(t - d) = 0$. When $(t - d) > 0$, $\chi(t - d)$ becomes a monotonic increasing function. In order to reduce the notation, the following expression is utilized, $\chi_d(t) = \chi(t - d)$.

If the displacement parameter d is increased, fewer regions of the internal gradient will be treated; a similar situation occurs if the external gradient is utilized. This situation is expressed as follows for the internal gradient case:

$$\chi_{d=0}(t) \geq \chi_{d=1}(t) \geq \cdots \geq \chi_{d=255}(t)$$

$$\chi_{d=0}(t)gradi(f) \geq \chi_{d=1}(t)gradi(f) \geq \cdots \geq \chi_{d=255}(t)gradi(f)$$

To enhance the regions in certain intervals of intensities d_1 to d_2 with $d_1 \leq d_2$, the weighting function must be applied in the following way:

$$\chi_{d=0}(t)[u(t - d_1) - u(t - d_2)]$$

where $u(t)$ represents the unitary step. An example of weighting function is, $\chi_d(t) = A/{1 + Be^{k(t-d)}}$ with A, B, k, and d constants, and $t = f(x)$. An example of Eq. 5 is presented in Fig. 3c where the displacement parameter $d = 60$ causes that those regions with an intensity level greater than 60 will be processed. The other constants took the values $A = 1$, $B = 1$, and $k = 0.02663$.

Contrast Enhancement Using Connected Top Hats

In this section, a study of contrast mappings is presented, but instead of using a proximity criterion for selecting the pattern, a contrast criterion given by the top-hat transformation will be used.[10] The interest in utilizing this type of criterion consists in knowing the contrast introduced in the output image. This idea comes from the notion of MSFs, where at each point of the gradient of the output image computed by this class of filters the gray level is greater than a given parameter ϕ or equal to zero. However, the top-hat criterion seems to be more interesting than the gradient one. We know that the contrast of an object depends on the luminance of its surroundings. Then the gradient transformation is a good criterion to decide how to increase the contrast in the image. However, contrast is a more complex notion and it depends on different parameters. Among the different parameters, the size of the region plays a main role. Thus, the top-hat transformation which selects a region according to its size and its gray-level difference with the neighboring regions must permit a better control of the contrast in the output image. Consider the next contrast mapping of two states:

$$W^2_{\mu,\beta}(f)(x) = \begin{cases} \varphi_\mu(f)(x) & \rho(x) < \beta \\ \gamma_\mu(f)(x) & \rho'(x) \leq 1 - \beta \end{cases}$$

where the proximity criterion is expressed in terms of the top-hat transformations:

$$\rho(x) = \frac{Thb_\mu(f)(x)}{Thb_\mu(f)(x) + Thw_\mu(f)(x)}$$

$$\rho'(x) = \frac{Thw_\mu(f)(x)}{Thb_\mu(f)(x) + Thw_\mu(f)(x)}$$

and

$$\rho(x) + \rho'(x) = \frac{Thb_\mu(f)(x)}{Thb_\mu(f)(x) + Thw_\mu(f)(x)} + \frac{Thw_\mu(f)(x)}{Thb_\mu(f)(x) + Thw_\mu(f)(x)} = 1$$

Observe that $\rho(x)$ represents the top hat on dark regions, whereas $\rho'(x)$ represents the top hat on bright ones; both top-hat values are normalized to the top-hat transformations on dark and bright regions. Also, the parameter β plays the role of a threshold for selecting the primitives. When $\rho(x)$ is between 0 and β, the closing affects the output image, while if $\rho(x)$ is between 0 and $1 - \beta$, the opening is used. Thus, to select the primitives, one uses the smallest values of the top-hat transformation and not the highest values as it is the strict definition of the top-hat transformation. For instance, when β is equal to 0.5, the rule to choose the primitive for sharpening the images is based on the smallest value between $\rho(x)$ and $\rho'(x)$. Indeed, the smallest top-hat value between the top hat working with dark structures and that working with bright structures is chosen to select the primitive. For β equal to 0, the contrast operator becomes $W^2_{\mu,\beta}(f) = \gamma_\mu$ (only the dark gray levels are enhanced), while for β equal to 1, one has the operator $W^2_{\mu,\beta}(f) = \varphi_\mu$ that works with bright structures. Contrary to the contrast operator that uses the morphological opening and closing as primitives, two operators using the opening and closing in separated ways are defined by the following relationships:

$$\kappa^\gamma_{\mu,\phi}(f)(x) = \begin{cases} \gamma_\mu(f)(x) & [f - \gamma_\mu(f)](x) \leq \phi \\ f(x) & [f - \gamma_\mu(f)](x) > \phi \end{cases}, \quad (6)$$

$$\kappa^\varphi_{\mu,\phi}(f)(x) = \begin{cases} \varphi_\mu(f)(x) & [\varphi(f) - f](x) \leq \phi \\ f(x) & [\varphi(f) - f](x) \leq \phi \end{cases}$$

The first operator works on bright structures, whereas the second one works on the dark regions. To enhance the image, the regions verifying the top-hat criterion are not modified, and the regions with low contrast (the contrast criterion is not verified) are attenuated to increase the contrast of those that remain intact. Consider the first operator using the original and the opened images as patterns. To enhance the image, the opened image serves as the background for detecting the regions of size μ with a contrast greater than the parameter ϕ.

The use of a contrast criterion to build these operators allows the classification of the points in the domain of definition of f in two sets:

- A set $S_{\mu,\phi}(f)$ composed of the regions of high contrast, where for all point $x \in S_{\mu,\phi}(f)$

$$[f - \gamma_\mu(f)](x) > \phi \quad \text{for } \kappa^\gamma_{\mu,\phi}(f) \quad \text{and}$$
$$[\varphi(f) - f](x) > \phi \quad \text{for } \kappa^\varphi_{\mu,\phi}(f)$$

- A set $S^c_{\mu,\phi}(f)$ composed of weak contrast zones (the complement of $S_{\mu,\phi}(f)$), where for all point $x \in S^c_{\mu,\phi}(f)$

$$[f - \gamma_\mu(f)](x) \leq \phi \quad \text{for } \kappa^\gamma_{\mu,\phi}(f) \quad \text{and}$$
$$[\varphi_\mu(f) - f](x) \leq \phi \quad \text{for } \kappa^\varphi_{\mu,\phi}(f)$$

The relevance of this classification consists in knowing the contrast introduced in the output image computed by the contrast operators. We will see in Property 1 that the contrast introduced in the output image by the operators $\kappa^\gamma_{\mu,\phi}$ or $\kappa^\varphi_{\mu,\phi}$ is strictly established. Then, one will say that the output image has a well-defined contrast according to the following definition.

Definition 2.

An image is said to have a well-defined contrast if one can classify its points in two sets according to a top-hat criterion; a set of points where the top-hat value is equal to zero, and a set of points where the top-hat value is greater than a parameter ϕ.

Now, observe in Eq. 6 that the symbols of the morphological opening γ_μ and the morphological closing φ_μ are used as patterns to define the contrast operators. Nevertheless, in their place, other types of openings and closings can be used, for example, the opening and closing by reconstruction.

In general, the two-state contrast operator $\kappa^\gamma_{\mu,\phi}$ using the opening and the original function as patterns will be analyzed in this work. However, similar properties and comments can be made when the closing is used as primitive $\kappa^\varphi_{\mu,\phi}$.

An example of Eq. 6 utilizing the opening by reconstruction as primitive instead of the morphological opening is presented in Fig. 4. Original image is located in Fig. 4a. The Fig. 4b represents the output image considering a scale 16, whereas scales 8 and 16 can be seen in Fig. 4c. Another example taking the scales 16 and 32 is given in Fig. 4d.

Now, let us show some interesting properties of these operators by studying how the contrast is modified in the input image in order to obtain the output one. The case of the contrast mapping using the opening and the original function as primitives is studied. By construction, this contrast mapping is an anti-extensive transformation. Thus, for all function f, the following ordering relation is verified:

$$\gamma_\mu(f) \leq \kappa^\gamma_{\mu,\phi} \leq f$$

Since the morphological opening is a strong filter, one obtains $\gamma_\mu(\kappa^\gamma_{\mu,\phi}) = \gamma_\mu(f)$. Remember that a morphological filter Ψ is said to be strong if it satisfies the following relationship:

$$\Psi(f) = \Psi(f \wedge \Psi(f)) = \Psi(f \vee \Psi(f))$$

This expression implies the following robustness condition: $\forall f, g$ such that $f \wedge \Psi(f) \leq g \leq f \vee \Psi(f) \Rightarrow \Psi(f) = \Psi(g)$. Then, by applying this last relation to the opening and using $g = \kappa^\gamma_{\mu,\phi}$, $\gamma_\mu(f) \leq \kappa^\gamma_{\mu,\phi} \leq f \Rightarrow \gamma_\mu(\kappa^\gamma_{\mu,\phi}(f)) = \gamma_\mu(f)$. This means that the top hat of the output image $\kappa^\gamma_{\mu,\phi}(f)$ is given by $\kappa^\gamma_{\mu,\phi}(f) - \gamma_\mu(\kappa^\gamma_{\mu,\phi}(f)) = \kappa^\gamma_{\mu,\phi}(f) - \gamma_\mu(f)$. Then, two-state contrast mappings based on a contrast criterion have the following property:

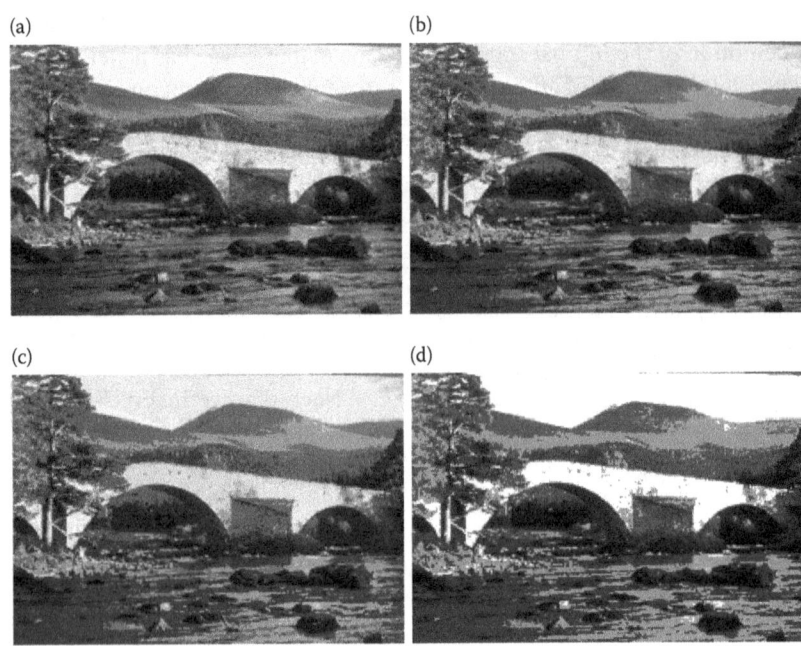

Fig. 4 Morphological contrast multiscale using top hat. (a) Original image; (b) output image using a scale 16; (c) output images using two scales, 8 and 16; (d) output image using two scales 16 and 32

Property 1.

The output image computed by $\kappa^{\gamma}_{\mu,\phi}(f)$ has a well-defined contrast. For all point x of its domain of definition, the top-hat transformation value of $\kappa^{\gamma}_{\mu,\phi}(f)(x)$ is

$$\left[\kappa^{\gamma}_{\mu,\phi}(f) - \gamma_{\mu}\left(\kappa^{\gamma}_{\mu,\phi}(f)\right)\right](x) > \phi \quad or$$
$$\left[\kappa^{\gamma}_{\mu,\phi}(f) - \gamma_{\mu}\left(\kappa^{\gamma}_{\mu,\phi}(f)\right)\right](x) = 0$$

Observe that by definition of the operator $\kappa^{\gamma}_{\mu,\phi}(f)$, the output image is composed by gray-level zones given by $\gamma_{\mu}(f)$ or by f. Thus, for all point $x \in S_{\mu,\phi}(f)$,

$$\left[\kappa^{\gamma}_{\mu,\phi}(f) - \gamma_{\mu}\left(\kappa^{\gamma}_{\mu,\phi}(f)\right)\right](x) = [f - \gamma_{\mu}(f)](x) > \phi$$

and, for all points $x \in S^c_{\mu,\phi}(f)$, $[\kappa^{\gamma}_{\mu,\phi}(f) - \gamma_{\mu}(\kappa^{\gamma}_{\mu,\phi}(f))](x) = [\gamma_{\mu}(f) - \gamma_{\mu}(f)](x) = 0$

Thus, the two-state contrast operators based on a top-hat criterion not only classify the high-contrast ($S_{\mu,\phi}(f)$) and weak-contrast ($S^c_{\mu,\phi}(f)$) regions of the input image but also impose a well-defined contrast to the output image. From this previous analysis, observe that $S_{\mu,\phi}([\kappa^{\gamma}_{\mu,\phi}(f)]) = S_{\mu,\phi}(f)$, which indicates that the transformation is idempotent. This situation is expressed formally in Property 2.

Property 2.

The two-state contrast operator $\kappa^{\gamma}_{\mu,\phi}(f)$ is an idempotent transformation. For all function f, one has that $\kappa^{\gamma}_{\mu,\phi}\kappa^{\gamma}_{\mu,\phi}(f) = \kappa^{\gamma}_{\mu,\phi}(f)$.

Morphological Multiscale Contrast Approach Consistent with Human Vision Perception

In this section, multiscale contrast algorithms for enhancing contrast are presented.[11] Here, the notion of Weber's law is utilized to derive a criterion for selecting the primitives. According to Weber's law, if the luminance of an object L is just noticeably different from its surrounding L_S or background, then the ratio $|L-L_S|/L_S$ is a constant between 0.02 and 0.03. If the luminance L is considered as the gray-level intensities of the original image f and its opening value as the background L_S, then $|L-L_S|/L_S$ can be rewritten as $[f - \tilde{\gamma}_{\mu}(f)](x)/\tilde{\gamma}_{\mu}(f)(x)$ or $f(x)/\tilde{\gamma}_{\mu}(f)(x) - 1$. This ratio can be used as a human visual criterion to build contrast operators. Thus, the next two toggle mappings are presented as follows:

$$\kappa^{\gamma}_{\mu,\phi}(f)(x) = \begin{cases} \tilde{\gamma}_{\mu}(f)(x) & \text{if } f(x)/\gamma_{\mu}(f)(x) \leq \phi \\ f(x) & \text{otherwise} \end{cases},$$

$$\kappa^{\varphi}_{\mu,\phi}(f)(x) = \begin{cases} \tilde{\varphi}_{\mu}(f)(x) & \text{if } \tilde{\varphi}_{\mu}(f)(x)/f(x) \leq \phi \\ f(x) & \text{otherwise} \end{cases} \quad (7)$$

Observe that, by construction, the operators $\kappa^{\gamma}_{\mu,\phi}(f)$ and $\kappa^{\varphi}_{\mu,\phi}(f)$ are anti-extensive and extensive, respectively. On the other hand, both operators are idempotent transformations. Thus, all point x of the output image is classified in either of two classes. Those verifying the contrast criterion:

$$\kappa_{\mu,\phi}^{\gamma}(f)(x) = f(x) \Rightarrow \frac{\kappa_{\mu,\phi}^{\gamma}(f)(x)}{\tilde{\gamma}_{\mu}\left(\kappa_{\mu,\phi}^{\gamma}(f)(x)\right)} = \frac{f(x)}{\tilde{\gamma}_{\mu}(f)(x)} > \phi$$

and those belonging to the background:

$$\kappa_{\mu,\phi}^{\gamma}(f)(x) = \tilde{\gamma}(f)(x) \Rightarrow \frac{\kappa_{\mu,\phi}^{\gamma}(f)(x)}{\tilde{\gamma}_{\mu}\left(\kappa_{\mu,\phi}^{\gamma}(f)(x)\right)} = \frac{\tilde{\gamma}_{\mu}(f)(x)}{\tilde{\gamma}_{\mu}(f)(x)} = 1 \leq \phi$$

This implies that the output image has a well-defined contrast according to the contrast criterion. Now, to generate a multiscale processing method, some properties are needed. Among them, *causality and edge preservation* are the most important. Causality implies that coarser scales can only be caused by what happened at finer scales. The derived images contain less and less details, some structures are preserved, and others are removed from one scale to the next. Particularly, the transformations should not create new structures at coarser scales. On the other hand, if the goal of image enhancement is to provide an image for image segmentation, edge preservation is required, which means that contours must remain sharp and not displaced. Thus, to generate a multiscale contrast algorithm, a family of contrast operators $\{\kappa_{\mu_k,\phi}^{\gamma}\}$ is applied by composition according to the scale (size) parameter. Let us analyze the composition of two contrast operators $\kappa_{\mu_2,\phi}^{\gamma}\kappa_{\mu_1,\phi}^{\gamma}(f)$ at scales μ_1 and μ_2 with $\mu_1 < \mu_2$. The first operator, as expressed above, defines two regions composed by the zones verifying $f(x)/\tilde{\gamma}_{\mu_1}(f)(x) > \phi$, as well as those serving as background for detecting the first ones $(f(x)/\tilde{\gamma}_{\mu_1}(f)(x) \leq \phi)$. For this last case, the output image will change the value of the criterion by 1, that is, for all regions verifying $f(x)/\tilde{\gamma}_{\mu_1}(f)(x) \leq \phi$, then $\frac{\kappa_{\mu_1,\phi}^{\gamma}(f)(x)}{\tilde{\gamma}_{\mu_1}\left(\kappa_{\mu_1,\phi}^{\gamma}(f)\right)(x)} = 1$. By applying the second operator $\kappa_{\mu_2,\phi}^{\gamma}(f)$ to the image $\kappa_{\mu_1,\phi}^{\gamma}(f)$, one has that: $\forall x$ such that $\frac{\kappa_{\mu_1,\phi}^{\gamma}(f)(x)}{\tilde{\gamma}_{\mu_1}\left(\kappa_{\mu_1,\phi}^{\gamma}(f)\right)(x)} > \phi$, then $\frac{\kappa_{\mu_1,\phi}^{\gamma}(f)(x)}{\tilde{\gamma}_{\mu_2}\left(\kappa_{\mu_1,\phi}^{\gamma}(f)\right)(x)} > \phi$, since $\tilde{\gamma}_{\mu_2}(f) < \tilde{\gamma}_{\mu_1}(f)$. This means that the regions verifying the contrast criterion for $\kappa_{\mu_1,\phi}^{\gamma}(f)$ are preserved when the second operator is applied. Then, $\kappa_{\mu_2,\phi}^{\gamma}(f)$ extracts high-contrast regions from $\tilde{\gamma}_{\mu_1}(f)$ verifying $\tilde{\gamma}_{\mu_1}(f)/\tilde{\gamma}_{\mu_2}(f) > \phi$. The output image $\kappa_{\mu_2,\phi}^{\gamma}\kappa_{\mu_1,\phi}^{\gamma}(f)$ will be composed by high-contrast regions of f and $\tilde{\gamma}_{\mu_1}(f)$, and some regions of the two backgrounds $\tilde{\gamma}_{\mu_1}(f)$ and $\tilde{\gamma}_{\mu_2}(f)$.

Property 3.

The composition of a family of contrast operators $\{\kappa_{\mu_k,\phi}^{\gamma}\}$, with $\mu_1 < \mu_2 < \cdots < \mu_n$, generates a multiscale contrast approach. This means that the structures at scale μ_i of the composition are preserved. For a given μ_i, such that $1 \leq i \leq n$, and for all point x verifying the contrast criterion,

$$\frac{\kappa_{\mu_i,\phi}^{\gamma}\cdots\kappa_{\mu_2,\phi}^{\gamma}\kappa_{\mu_1,\phi}^{\gamma}(f)(x)}{\tilde{\gamma}_{\mu_i}\left(\kappa_{\mu_i,\phi}^{\gamma}\cdots\kappa_{\mu_2,\phi}^{\gamma}\kappa_{\mu_1,\phi}^{\gamma}(f)\right)(x)} > \phi$$

the ratio for the nth scale $\frac{\kappa_{\mu_n,\phi}^{\gamma}\cdots\kappa_{\mu_2,\phi}^{\gamma}\kappa_{\mu_1,\phi}^{\gamma}(f)(x)}{\tilde{\gamma}_{\mu_i}\left(\kappa_{\mu_n,\phi}^{\gamma}\cdots\kappa_{\mu_2,\phi}^{\gamma}\kappa_{\mu_1,\phi}^{\gamma}(f)\right)(x)}$ is also greater than ϕ, then $\kappa_{\mu_n,\phi}^{\gamma}\cdots\kappa_{\mu_2,\phi}^{\gamma}\kappa_{\mu_1,\phi}^{\gamma}(f)(x) = \kappa_{\mu_i,\phi}^{\gamma}\cdots\kappa_{\mu_2,\phi}^{\gamma}\kappa_{\mu_1,\phi}^{\gamma}(f)(x)$.

On the other hand, no new structures are introduced at finer scales. For all point x such that

$$\frac{\kappa_{\mu_i,\phi}^{\gamma}\cdots\kappa_{\mu_2,\phi}^{\gamma}\kappa_{\mu_1,\phi}^{\gamma}(f)(x)}{\tilde{\gamma}_{\mu_i}\left(\kappa_{\mu_i,\phi}^{\gamma}\cdots\kappa_{\mu_2,\phi}^{\gamma}\kappa_{\mu_1,\phi}^{\gamma}(f)\right)(x)} = 1, \text{ then}$$

$$\frac{\kappa_{\mu_n,\phi}^{\gamma}\cdots\kappa_{\mu_2,\phi}^{\gamma}\kappa_{\mu_1,\phi}^{\gamma}(f)(x)}{\tilde{\gamma}_{\mu_i}\left(\kappa_{\mu_n,\phi}^{\gamma}\cdots\kappa_{\mu_2,\phi}^{\gamma}\kappa_{\mu_1,\phi}^{\gamma}(f)\right)(x)} = 1$$

In other words, for all point x of the output image, it has a well-defined contrast at each scale of the composition of the family $\{\kappa_{\mu_k,\phi}^{\gamma}\}$. Moreover, the composition of contrast operators not only preserves a well-defined contrast at each scale but also increases the contrast at finer scales as expressed by the following property.

Property 4.

In a composition of a family of contrast operators $\{\kappa_{\mu_k,\phi}^{\gamma}\}$ with $\mu_1 < \mu_2 < \cdots < \mu_n$, the following relation can be established. For a given μ_i, such that $1 \leq i \leq n$ and for all point x,

$$\frac{\kappa_{\mu_n,\phi}^{\gamma}\cdots\kappa_{\mu_2,\phi}^{\gamma}\kappa_{\mu_1,\phi}^{\gamma}(f)(x)}{\tilde{\gamma}_{\mu_i}\left(\kappa_{\mu_n,\phi}^{\gamma}\cdots\kappa_{\mu_2,\phi}^{\gamma}\kappa_{\mu_1,\phi}^{\gamma}(f)\right)(x)} \geq \frac{\kappa_{\mu_i,\phi}^{\gamma}\cdots\kappa_{\mu_2,\phi}^{\gamma}\kappa_{\mu_1,\phi}^{\gamma}(f)(x)}{\tilde{\gamma}_{\mu_i}\left(\kappa_{\mu_i,\phi}^{\gamma}\cdots\kappa_{\mu_2,\phi}^{\gamma}\kappa_{\mu_1,\phi}^{\gamma}(f)\right)(x)}$$

Similar properties can be expressed for a family of contrast operators $\{\kappa_{\mu_k,\phi}^{\varphi}\}$. An example of filter $\kappa_{\mu_n,\phi}^{\gamma}\cdots\kappa_{\mu_2,\phi}^{\gamma}\kappa_{\mu_1,\phi}^{\gamma}(f)(x)$ is presented in Fig. 5. Original image is located in Fig. 5a. Parameter ϕ is considered with a value of 1.067 with the following scales: Fig. 5b with a scale 16; Fig. 5c with scales 5 and 40; and Fig. 5d with scales 5, 40, and 80.

Rational Multiscale Contrast

When working with images with a wide range of scene brightness, as for example when strong highlights and deep shadows appear in the same image, the methods previously presented do not allow the enhancement of this type of images. Thus, the idea to avoid gray-level saturation consists in locally applying the logarithm transformation. Therefore, in this section, a method based on morphological contrast mappings that improves the contrast in a locally logarithmic way is explained.[12] The approach based on morphological contrast enhancement

Fig. 5 Morphological contrast multiscale based on Weber's law considering $\varnothing = 1.067$. (a) Original image; (b) output image using one scale 5; (c) output image using two scales, 5 and 40; (d) output image using three scales, 5, 40, and 80

presented in this section is derived from the well-known method called retinex.[14] In particular, multiscale retinex (MSR) has been shown to be an effective approach to enhance image contrast. The main idea in the retinex theory addresses the problem of separating illumination from reflectance dealing with the compensation of illumination effects on images. MSR is explained easily from the single-scale retinex (SSR) case. For SSR, one has at a point of coordinates (x, y):

$$R(x,y,c) = \log\{f(x,y)\} - \log\{F(x,y,c) * f(x,y)\}$$
$$= \log \frac{f(x,y)}{F(x,y,c) * f(x,y)}$$

where $R(x,y,c)$ is the output of image f, $*$ denotes convolution, and $F(x,y,c)$ is the Gaussian function $F(x,y,c) = Ke^{-(x^2+y^2)/c^2}$ with K selected so that $\iint F(x,y,c)dx\,dy = 1$; constant c is the scale. The form of the retinex expression is similar to the difference-of-Gaussian function widely used in natural vision science to model receptive fields of individual neurons and perceptual processes. The logarithmic function transforms the subtractive inhibition into an arithmetic division. This permits, under some assumptions of the spatial distribution of the source illumination, the computation of the reflectance ratio of the scene. In the work of Kogan et al.,[13] the single-scale retinex is extended to a multiscale version (MSR). The MSR is the weighted sum of several SSRs with different scales:

$$R_M(x,y,w,c) = \sum_{n=1}^{N} w_n R(x,y,c_n)$$

where R_M is the MSR result, $w = (w_1, w_2, \ldots, w_n)$, w_n is the weight of the nth SSR, and $c = (c_1, c_2, \ldots, c_n)$ where c_n is the scale of the nth SSR with $\sum_{n=1}^{N} w_n = 1$.

Now, instead of using the convolution of the Gaussian to enhance the regions in a given scale, the opening by reconstruction will be applied. On the other hand, Weber's law is used to build rational morphological operators for detecting the principal regions and to enhance them. Barnard and Funt[14] proposed the rational functions of the ratio of morphological transformations as follows:

$$R = \frac{\sum_{i=1}^{m} a^i \left(\frac{MT_1}{MT_2}\right)^i}{\sum_{j=0}^{n} b^j \left(\frac{MT_1}{MT_2}\right)^j}$$

where a and b are constants. In particular, they investigated the case where $m=1$ and $n=0$ and take the transformations MT1 and MT2 as the morphological dilation and erosion, respectively, which simplifies R to:

$$R = \frac{a_1}{b_0}\left(\frac{MT_1}{MT_2}\right)$$

Finding the ratio of this operation yields useful information about the contours. In the present work, one uses the ratio between the original image and the opened image. Then, Weber's law is taken into account and to enhance the regions the opened image serves as background to identify the principal regions of the original image. In a single scale, one has at point x:

$$R(x,\mu) = \frac{f(x)}{\tilde{\gamma}_\mu(f)(x)}$$

In a multiscale situation, the opened image $\tilde{\gamma}_{\mu_{i+1}}(f)$ at scale μ_{i+1} serves as the background for detecting and enhancing structures at scale μ_i of the image $\tilde{\gamma}_{\mu_i}(f)$. Considering that $\tilde{\gamma}_0(f) = f$, the multiscale connected algorithm is given by the updated equation:

$$R_{M_\gamma}(x,\mu) = \sum_{n=1}^{N} \frac{\gamma_{\mu_{n-1}}(f)(x)}{\gamma_{\mu_n}(f)(x)} \quad \text{with} \quad \tilde{\gamma}_0(f) = \tilde{\gamma}_{\mu_0}(f) = f \quad (8)$$

This rational multiscale contrast algorithm enables the enhancement of the structures at different scales. Observe that the ratio $\frac{\tilde{\gamma}_{\mu_{n-1}}(f)(x)}{\tilde{\gamma}_{\mu_n}(f)(x)}$ is directly linked with the top-hat transformation. In fact, the application of the log function to this ratio yields:

$$\log \frac{\tilde{\gamma}_{\mu_{n-1}}(f)(x)}{\tilde{\gamma}_{\mu_n}(f)(x)} = \log\{\tilde{\gamma}_{\mu_{n-1}}(f)(x)\} - \log\{\tilde{\gamma}_{\mu_n}(f)(x)\}$$

$$= \tilde{\gamma}_{\mu_{n-1}}(\log(f))(x) - \tilde{\gamma}_{\mu_n}(\log(f))(x)$$

This means that the ratio $\frac{\tilde{\gamma}_{\mu_{n-1}}(f)(x)}{\tilde{\gamma}_{\mu_n}(f)(x)}$ is equivalent to applying a logarithm to the original image, before the top hat is computed, then the anti-log is applied. To show the multiscale characteristics of this approach, observe that ratio $\frac{\tilde{\gamma}_{\mu_{n-1}}(f)(x)}{\tilde{\gamma}_{\mu_n}(f)(x)}$ of openings by reconstruction is equivalent to a difference of the two openings. Thus, from property 4, it does not introduce structures smaller than the scale $n-1$. Because of the use of reconstruction filters, no new contours are introduced in the processed image. An example of Eq. 8 is presented in Fig. 6. In this case, the image is processed at color level using the color space introduced in the work of Espino-Gudiño et al.[11] Original image is presented in Fig. 6a. This image has bright and dark regions, and in Fig. 6b,c, the information in dark zones is recovered avoiding an over illumination. The parameters used are 1 and 1.125 taking three scales. The MSR method is illustrated in Fig. 6d, where the color has been modified considerably.

When an image shows complex illumination conditions with dark and light regions, contrast improvement becomes much harder to achieve. The aim is to obtain equilibrium to produce general contrast enhancement in this type of image. A solution to this problem consists in obtaining a combination of rational multiscale processes with transformations by reconstruction, since R_{M_γ} treats light regions, then R_{M_φ} deals with dark regions:[12]

$$R_{M_\varphi}(x,\mu) = \sum_{n=1}^{N} \frac{\tilde{\varphi}_{\mu_n}(f)(x)}{\tilde{\varphi}_{\mu_{n-1}}(f)(x)} \quad \text{with} \quad \tilde{\varphi}_0(f) = \tilde{\varphi}_{\mu_0}(f) = f \quad (9)$$

Once R_{M_γ} and R_{M_φ} are available, they must be combined through a convex linear combination of the two images in the so-called morphological rational contrast operator (MRCO). This operator is mathematically expressed in Eq. 10, where a and b are constants that together add to 1.0, and MRCO is the resulting contrast-enhanced image.[12]

$$MRCO(x) = R_{M_\gamma}(x,\mu) \cdot a + R_{M_\varphi}(x,\mu) \cdot b \quad (10)$$

An Approach to Enhance Contrast on Images with Poor Lighting

In this section, some operators to enhance and normalize the contrast in gray-level images with poor lighting are introduced.[16] Contrast operators are based on the logarithm function in a similar way to Weber's law. The image f is divided into n blocks w^i of size $l_1 \times l_2$. Each block is a subimage of the original image. The minimum and maximum intensity values in each subimage are denoted as:

$$m_i = \wedge w^i(x) \quad \forall x \in D_{w^i} \subseteq D$$

$$M_i = \vee w^i(x) \quad \forall x \in D_{w^i} \subseteq D$$

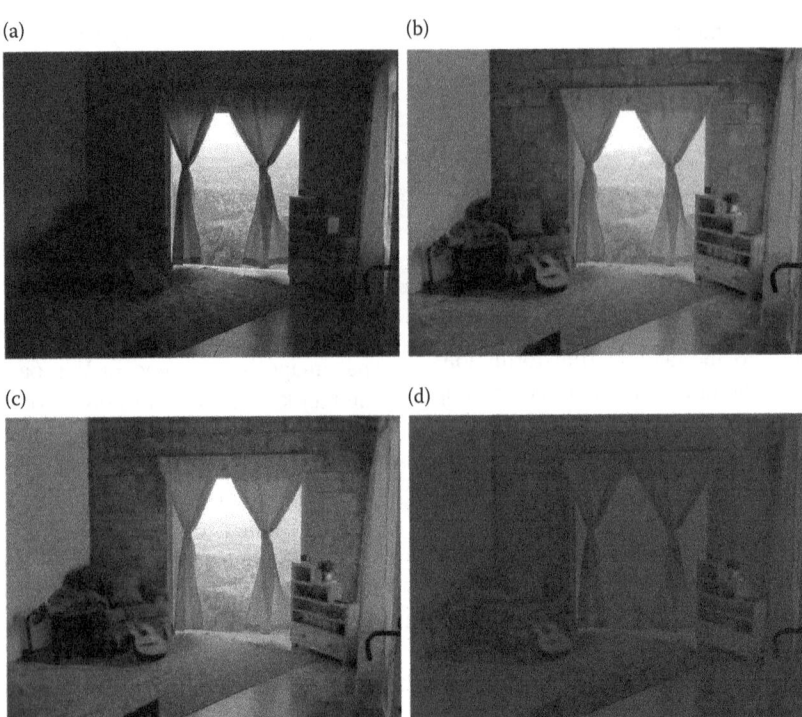

Fig. 6 (a) Original image; (b) three-scale rational multiscale contrast algorithm with criterion 1.0; (c) three-scale rational multiscale contrast algorithm with criterion 1.125; (d) MSR

where D and D_{w^i} represent the definition domain. For each analyzed block, maximum (M_i) and minimum (m_i) values are used to determine the background criteria τ_i in the following way:

$$\tau_i = \frac{m_i + M_i}{2} \quad \forall i = 1, 2, \ldots, n$$

In the one-dimensional case, the following expression is obtained:

$$v(x) = \begin{cases} \tau_1 & x \leq L_1 \\ \tau_2 & L_1 < x \leq L_2 \\ \tau_3 & L_2 < x \leq L_3 \\ \vdots \\ \tau_n & L_{n-1} < x \leq L_n \end{cases} \quad (11)$$

The value of τ_i represents a division line between clear ($f > \tau_i$) and dark ($f \leq \tau_i$) intensity levels. Once τ_i is calculated, this value is used to select the background parameter associated with the analyzed block. An expression to enhance the contrast is presented as follows:

$$\Gamma_{\tau_i}(f) = \begin{cases} k_i \log(f+1) + M_i & f \leq \tau_i \\ k_i \log(f+1) + m_i & \text{otherwise} \end{cases} \quad (12)$$

Note that the background parameter depends on the τ_i value. If $f \leq \tau_i$ (dark region), the background parameter takes the value of the maximum intensity (M_i) within the analyzed block, and the minimum intensity (m_i) value otherwise. Also, the unit was added to the logarithm function in equation to avoid indetermination. On the other hand, the constant k_i in the equation is obtained as follows:

$$k_i = \frac{255 - m_i^*}{\log(256)} \quad \forall i = 1, 2, \ldots, n$$

with

$$m_i^* = \begin{cases} m_i & f > \tau_i \\ M_i & f \leq \tau_i \end{cases}$$

The equation corresponding to $\Gamma_{\tau_i}(f)$ works similar to a contrast mapping; this modifies the intensity values depending on certain criterion. The criterion to modify the contrast is given by τ_i. On the other hand, M_i and m_i values are used as background parameters to improve the contrast depending on the τ_i value, because the background is different for clear and dark regions. In this way, the transformation $\Gamma_{\tau_i}(f)$ fulfills the next properties.

Property 5.

(a) It is a nonincreasing transformation, that is, for any two images f_1 and f_2 with $f_1 \leq f_2$, $\Gamma_{\tau_i}(f_1) \geq \Gamma_{\tau_i}(f_2)$. (b) It is not an idempotent transformation, that is, $\Gamma_{\tau_i}\Gamma_{\tau_i}(f) \neq \Gamma_{\tau_i}(f)$. (c) It is an extensive transformation, that is, $f \leq \Gamma_{\tau_i}(f)$. (d) It is possible to classify the definition domain of f in two sets: the set S_{τ_i} composed by high-contrast areas (for every point $x \in S_{\tau_i}, f > \tau_i$) and the set $S^c_{\tau_i}$ composed by low-contrast areas (for every point $x \in S^c_{\tau_i}, f \leq \tau_i$). (e) The composition of contrast mappings using Eq. 12 will result in lighter images for each iteration, reaching a limit imposed by the value of the highest level of intensity of the image, maxint = 255 in our particular case, that is, $\underbrace{\Gamma_{\tau_i}...\Gamma_{\tau_i}(\Gamma_{\tau_i}(f))}_{n \ times} \to$ maxint. (f) If image f is subdivided into smaller blocks each time, the background function $b(x)$ tends to be similar to the original function f.

An example of Eqs. 11 and 12 is provided in Fig. 7. The original image is displayed in Fig. 7a. Figure 7b illustrates the block division; in this case, four blocks were taken. The values assigned to τ are as follows: $\tau_1 = 10$, $\tau_2 = 1$, $\tau_3 = 2$, and $\tau_4 = 1$. The enhanced image by applying Eq. 12 is shown in Fig. 7c.

On the other hand, given that, maximum and minimum values are analyzed for each block, an extension using morphological operators is presented as follows. Let $I_{max}(x)$ and $I_{mix}(x)$ be the the maximum and minimum intensity values taken from one set of pixels contained in a window (B) of elemental size (3 × 3 elements), $x \in D$ (D represents the definition domain). Notice that the window corresponds to the structuring element B. For the sake of simplicity, let us consider $I_{max}(x) = max\{f(x+b) : b \subseteq B\}$ and $I_{min}(x) = min\{f(x+b) : b \subseteq B\}$, $x \in D$.

Then, a new expression is derived:

$$\tau(x) = \frac{I_{min}(x) + I_{max}(x)}{2}$$

where $I_{max}(x)$ and $I_{min}(x)$ values correspond to the morphological dilation and erosion defined by the order statistical filters. Thus, the last equation is expressed as follows:

$$\tau(x) = \frac{\varepsilon_\mu(f)(x) + \delta_\mu(f)(x)}{2}$$

Notice that τ_i was substituted by $\tau(x)$, since $\tau(x)$ has a local character given by the structuring element μB. In this way, the contrast operator in Eq. 12 is written as:

$$\Gamma_{\tau(x)}(f) = \begin{cases} k_{\tau(x)}\log(f+1) + \delta_\mu(f)(x) & f \leq \tau(x) \\ k_{\tau(x)}\log(f+1) + \varepsilon_\mu(f)(x) & \text{otherwise} \end{cases} \quad (13)$$

and $\quad k_{\tau(x)} = \dfrac{255 - \tau(x)}{\log(256)}$

On the other hand, it is desirable to obtain a function that resembles the image background without dividing the original image into blocks, and without using the morphological erosion and dilation, since these morphological transformations generate new contours when the structuring element is increased. In mathematical morphology, there is a class of transformations that allows the filtering of the

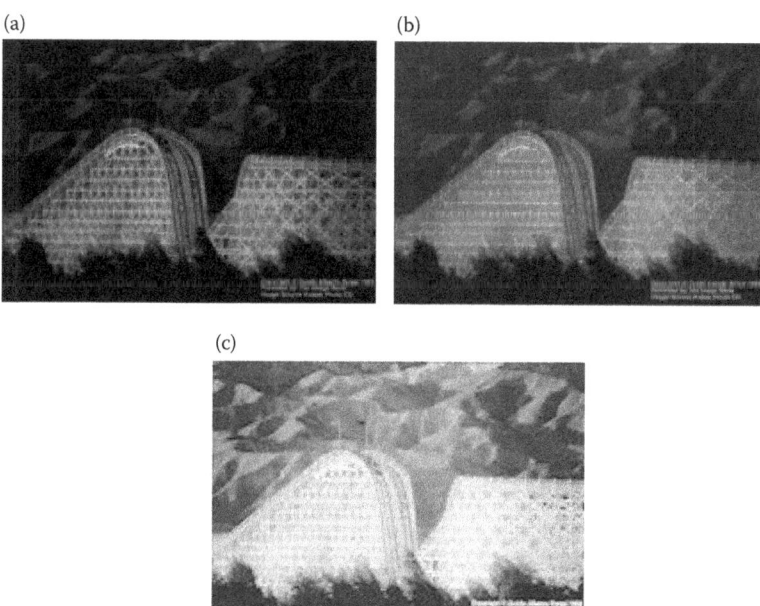

Fig. 7 Image background detection using block approach and contrast enhancement. (a) Original image; (b) image divided into four blocks. The values of τ for each block are as follows: $\tau_1 = 10$, $\tau_2 = 1$, $\tau_3 = 2$, and $\tau_4 = 1$; (c) enhanced image after applying Eq. 12

image without generating new components; these transformations are called transformations by reconstruction. In our case, the opening by reconstruction is our choice because it touches the regional minima and merges the regional maxima. This characteristic allows the modification of the altitude of regional maxima when the size of the structuring element increases. This effect can be used to detect the background criterion $\tau(x)$:

$$\tau(x) = \tilde{\gamma}_\mu(f)(x)$$

Eq. 12 is maintained, and only the way to detect the background is modified. When considering the opening by reconstruction to detect the background, one further operation is necessary to detect the local information given by the original function (image extremes are contained in the opening by reconstruction because of its behavior). The morphological transformation proposed for this task is the erosion size $\mu = 1$:

$$b(x) = \varepsilon\left[\tilde{\gamma}_\mu(f)\right](x)$$

Given that the morphological erosion tends to generate new information when the structuring element is enlarged, the image background was computed by using only the morphological erosion size $\mu = 1$.

Thus, the following expression derived from Eq. 13 is utilized to enhance the contrast in images with poor lighting:

$$\zeta_\mu(f) = k(x)\log(f+1) + \varepsilon\left[\tilde{\gamma}_\mu(f)\right]$$
$$\text{and}\quad k(x) = \frac{\text{maxint} - \varepsilon\left[\tilde{\gamma}_\mu(f)\right]}{\log(\text{maxint}+1)} \tag{14}$$

An interesting property is obtained from the behavior of the opening by reconstruction. This property is called multi-background and is presented as follows:

Property 6 (multi-background).

For all $\mu_1 > 0$, $\mu_2 > 0$, and $\mu_1 < \mu_2$, such that $\tilde{\gamma}_{\mu_1}(f) \geq \tilde{\gamma}_{\mu_2}(f)$, then $\varepsilon_{\mu_1}\left[\tilde{\gamma}_{\mu_1}(f)\right](x) \geq \varepsilon_{\mu_1}\left[\tilde{\gamma}_{\mu_2}(f)\right](x)$.

An example of Eq. 14 is presented in Fig. 8. The original image is located in Fig. 8a. In Fig. 8b, the background image is computed considering an opening by reconstruction size $\mu = 3$, and posteriorly Eq. 14 is applied, the output image is displayed in Fig. 8c.

Morphological Contrast Measure

It is common to enhance the contrast in the processed images; however, it is not easy identifying which image has the better contrast. A solution to this problem is given below.[17]

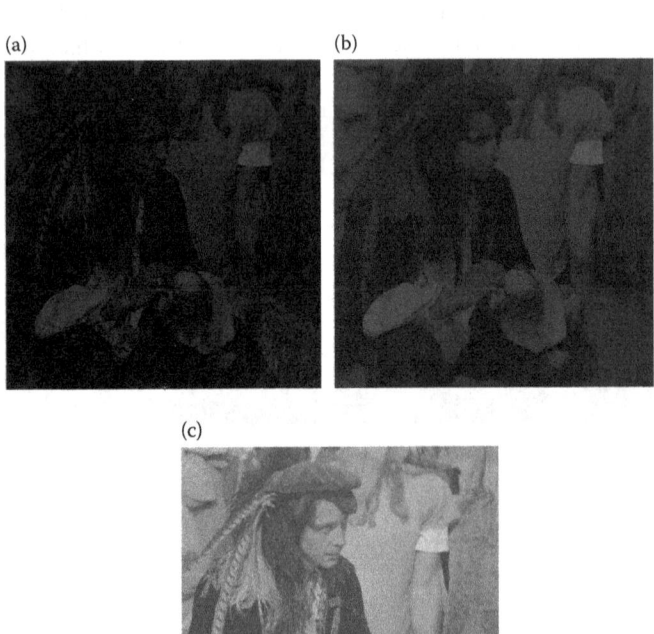

Fig. 8 Image with bad illumination. (a) Original image; (b) background image considering $\mu = 3$; (c) enhanced images obtained from Eq. 14

Weber's law can be expressed as follows:

$$C = k \log L + b, \quad L > 0$$

where C is the perceived contrast, L is the luminance, and k and b are constants, b being the uniform background. Background is a parameter from which a change or modification in the levels of luminance can be perceived by the human eye. Therefore, if the background image is modified, then the perceived contrast can be altered. The background will be computed as follows:

$$b(x) = \varepsilon\left[\tilde{\gamma}_\lambda(f)\right](x)$$

where λ is a size criterion, ε is an erosion size 1, and $\tilde{\gamma}_\lambda$ represents the opening by reconstruction. Consider an output image h which was obtained by the processing of a certain operator, that is, $h = \psi(f)$ where f is the original image. Now the purpose is to quantify the high contrast from image h. For this, consider certain square structuring element μB_x centered at its origin x. The structuring element μB_x moves over the output image h and for each set of pixels intersected by the structuring element, the maxima and minima intensity levels, that is, $I_{max}(\mu B_x)$ and $I_{min}(\mu B_x)$ will be detected. The luminance L could be expressed as follows:

$$L = \frac{I_{max}(\mu B_x)}{I_{min}(\mu B_x)} \text{ with } I_{min}(\mu B_x) \neq 0 \text{ and } I_{max}(\mu B_x) \neq 0$$

Notice that parameter L is detected locally. By substituting the luminance L in Weber's law, the following equation is obtained:

$$C_{\mu,\lambda}(h)(x) = k \log\left(\frac{I_{max}(\mu B_x)}{I_{min}(\mu B_x)}\right) + \varepsilon\left[\tilde{\gamma}_\lambda(h)\right](x), \text{ with}$$
$$I_{max}(\mu B_x) \neq 0 \text{ and } I_{min}(\mu B_x) \neq 0$$

However, $I_{max}(\mu B_x)$ and $I_{min}(\mu B_x)$ represent the dilation and erosion at the point x, because they are rank filters; therefore, $I_{max}(\mu B_x) = \delta_\mu(h)(x)$ and $I_{min}(\mu B_x) = \varepsilon_\mu(h)(x)$. Thus,

$$C_{\mu,\lambda}(h)(x) = k \log\left(\frac{\delta_\mu(h)(x)}{\varepsilon_\mu(h)(x)}\right) + \varepsilon\left[\tilde{\gamma}_\gamma(h)\right](x), \text{ with}$$
$$\delta_\mu(h)(x) \neq 0 \text{ and } \varepsilon_\mu(h)(x) \neq 0$$

Notice that $\log(\delta_\mu(h)) = \delta_\mu(\log(h))$ and $\log(\varepsilon_\mu(h)) = \varepsilon_\mu(\log(h))$, because the structuring element is flat. The previous equation is rewritten as follows:

$$C_{\mu,\lambda}(h)(x) = k\, gradm_\mu(\log(h))(x) + \varepsilon\left[\tilde{\gamma}_\lambda(h)\right](x), \; h(x) \neq 0$$

To have a measure of the contrast of the whole image, the sum of the values of $\Theta_{\mu,\lambda}$ is carried out, and it is denoted as $\Theta_{\mu,\lambda}$:

$$\Theta_{\mu,\lambda} = \sum_{\substack{0 \leq i \leq m \\ 0 \leq j \leq n}} C_{\mu,\lambda}(h)(x_{i,j})$$

with $h(x_{i,j}) \neq 0$, $h(x_{i,j})$ denotes the intensity level in the point $x_{i,j}$, whereas m and n correspond to the dimensions of the processed image. The k parameter is calculated by considering the maximum intensity level within the image; in this case, it is 255:

$$k = \frac{255 - \varepsilon\left[\tilde{\gamma}_\lambda(h)\right](x)}{\log 255}$$

The equation corresponding to $\Theta_{\mu,\lambda}$ will be divided by the volume of the original image f, where $Vol(f) = \sum_{\substack{0 \leq i \leq m \\ 0 \leq j \leq n}} f(x_{i,j})$, this is to work without large numbers. Consequently, the next index allows to have a global contrast measure:

$$\chi_{\mu,\lambda} = \frac{\Theta_{\mu,\lambda}}{Vol(f)} \tag{15}$$

with $Vol(f) \neq 0$

The images in Figs. 1a–c and 3a–c are considered to measure the contrast. The computed indexes $\chi_{\mu,\lambda}$ for these images are presented in Table 1. According to the values in the table, the image presenting the best contrast corresponds to the picture in Fig. 3a.

Other Criterion to Measure the Quality of an Enhanced Image

Several transformations reported in the literature permit to enhance the contrast by distributing the intensity of the pixels or stretching and shrinking the gray levels producing a large modification of the edges of the processed images. One way of quantifying edge preservation is presented as follows, where a morphological index to compare contours at pixel level between the original and output images is applied. The morphological contour preservation index (MCPI) is applied.[9] This index allows

Table 1 Morphological contrast measure

Image	$\chi_{\mu,\lambda}$
Figure 1b	0.97593
Figure 1c	1.03550
Figure 3a	1.04532
Figure 3b	1.04022
Figure 3c	1.02578

Table 2 Morphological contour preservation index

Image	MCPI
Figure 1a	1
Figure 1b	1.053
Figure 1c	0.963
Figure 3a	−0.004
Figure 3b	0.699
Figure 3c	0.603

to obtain a measure to evaluate the quality of the contours of the output image with respect to those of the original image. For obtaining MCPI, it is necessary to compute the edge preservation parameter (EPP). The EPP is obtained from the convolution (denoted as *) of two kernels of size 3×3 with the morphological internal gradient. These kernels allow to detect horizontal, G_h, and vertical, G_v, changes at each point in the internal gradient. EPP is expressed as follows:

$$EPP = \sum_{x \in D_{gradi(f)}} G_h(x) + \sum_{x \in D_{gradi(f)}} G_v(x)$$

$$\text{with } G_h = \begin{bmatrix} 1 & 1 & 1 \\ 0 & 0 & 0 \\ -1 & -1 & -1 \end{bmatrix} * gradi_\mu(f)$$

$$\text{and } G_v = \begin{bmatrix} -1 & 0 & 1 \\ -1 & 0 & 1 \\ -1 & 0 & 1 \end{bmatrix} * gradi_\mu(f)$$

where $D_{gradi(f)}$ represents the definition domain of the internal gradient. The MCPI is obtained as follows:

$$MCPI = \frac{EPP_{processed}}{EPP_{original}} \quad (16)$$

$EPP_{Processed}$ and $EPP_{original}$ correspond to the processed and original images. The best MCPI value must be nearest to 1.

The images in Figs. 1a–c and 3a–c will be used to compute the MCPI represented in Eq. 16. The set of indexes is presented in Table 2. According to the values in the table, the image with the contours of better quality corresponds to the picture in Fig. 1b.

CONCLUSION

Several transformations to enhance contrast have been depicted as well as interesting properties fulfilled by the contrast operators. Weber's law plays an important role because several transformations were developed considering it. Also, two approaches to enhance contrast in images with poor illumination were presented. The first one utilizes the rational morphology concepts, whereas the second one deals with Weber's law. On the other hand, a method to quantify the contrast in enhanced images was presented together with a criterion to measure the quality of the resultant contours. Finally, all the transformations depicted in this work are morphological transformations.

REFERENCES

1. Serra, J. *Mathematical Morphology*, Vol. I; Academic: London, 1982.
2. Rivest, J.; Soille, P.; Beucher, S. Morphological gradients. J. Electron. Imaging **1993**, *2* (4), 326–336.
3. Vincent, L. Morphological grayscale reconstruction in image analysis: Applications and efficient algorithms. IEEE Trans. Image Process. **1993**, *2* (2), 176–201.
4. Kramer, H.P.; Bruckner, J.B. Iteration of non-linear transformations for enhancement of digital image. Pattern Recognit. **1975**, *7* (4), 53–58.
5. Meyer, F.; Serra, J. Contrast and activity lattice. Signal Process. **1989**, *16* (4), 303–317.
6. Terol-Villalobos, I.R. Nonincreasing filters using morphological gradient criteria. Opt. Eng. **1996**, *35*, 3172–3182.
7. Terol-Villalobos, I.R.; Cruz-Mandujano, J.A. Contrast enhancement and image segmentation using a class of morphological nonincreasing filters. J. Electron. Imaging **1998**, *7*, 641–654.
8. Terol-Villalobos, I.R. Morphological image enhancement and segmentation. In *Advances in Imaging and Electron Physics*; Hawkes, P.W.; Ed.; Academic Press: Burlington, NJ, 2001, Vol. 118, 207–273.
9. Mendiola-Santibañez, J.D.; Terol-Villalobos, I.R. Image enhancement and segmentation using weighted morphological connected slope filters. J. Electron. Imaging **2013**, *22* (2), 023022–023022.
10. Terol-Villalobos, I.R. Morphological connected contrast mappings based on top-hat criteria: A multiscale contrast approach. Opt. Eng. **2004**, *43* (7), 1577–1595.
11. Espino-Gudiño, M.; Terol-Villalobos, I.R.; Santillan, I. Morphological multiscale contrast approach for gray and color images consistent with human visual perception. Opt. Eng. **2007**, *46* (6), 067003.
12. Peregrina-Barreto, H.; Herrera-Navarro, A.; Morales-Hernández, L.; Terol-Villalobos, I.R. Morphological rational operator for contrast enhancement. J. Opt. Soc. Am. A **2011**, *28*, 455–464.
13. Kogan, R.G.; Agaian, S.S.; Lentz, K.P. Visualization using rational morphology and magnitude reduction. In *Proceedings of SPIE Nonlinear Image Processing IX*, Vol. 3387; 1998, 153–163.
14. Barnard, K.; Funt, B. Investigations into multi-scale retinex. In *Color Imaging: Vision and Technology*; MacDonald, L.W.; Luo, M.R.; Eds.; John Wiley & Sons: New York, 1999, 9–17.
15. Jobson, D.J.; Rahman, Z.; Woodell, G.A. A multiscale retinex for bridging the gap between color images and the

human observation of scenes. IEEE Trans. Image Process. **1997**, *6* (7), 965–976.

16. Jiménez-Sánchez, A.R.; Mendiola-Santibañez, J.D.; Terol-Villalobos, I.R.; Herrera-Ruíz, G.; Vargas-Vázquez, D.; García-Escalante, J.J.; Lara-Guevara, A. Morphological background detection and enhancement of images with poor lighting. IEEE Trans. Image Process. **2009**, *18* (3), 613–623.

17. Jiménez-Sánchez, A.R.; Santillán, I.; Resendiz, J.R.; Gonzalez-Gutierrez, C.A.; Mendiola-Santibañez, J.D. Morphological contrast index based on the Weber's law. Int. J. Imaging Syst. Technol. **2012**, *22*, 137–144.

18. Mendiola-Santibañez, J.D.; Terol-Villalobos, I.R.; Herrera-Ruiz, G.; Fernández-Bouzas, A. Morphological contrast measure and contrast enhancement: One application to the segmentation of brain MRI. Signal Process. **2007**, *87* (9), 2125–2150.

19. Mendiola-Santibañez, J.D.; Terol-Villalobos, I.R.; Santillán-Méndez, I.M. Determination of adequate parameters for connected morphological contrast mappings through morphological contrast measures. In *Advances in Imaging and Electron Physics*, Vol. 161; Hawkes, P.W.; Ed.; Burlington, NJ: Academic Press, 2010, 55–88.

20. Jiménez Sánchez, A.R.; Mendiola-Santibañez, J.D.; Herrera-Ruíz, G.; Santillán, I. Índice de contraste morfológico basado en el análisis de los contornos y el fondo de la imagen. Computación y Sistemas **2012**, *16* (1), 99–110.

Image and Scene Analysis

James F. Peters
Computational Intelligence Laboratory, University of Manitoba, Winnipeg, Manitoba, Canada

Abstract
To facilitate image and scene analysis, a digital image can be viewed as a set of points (pixels) susceptible to the whole spectrum of mathematical structures. These structures include image neighborhoods, image clusters, image segments, image tessellations, collections of image segment centroids, sets of points nearest a particular image point such as a region centroid, image regions gathered together as near sets, adjacent image regions, and the geometry of polygonal image regions. The fact that structured images reveal hidden information in images is the main motivation for structuring an image. Structured images are then analyzed in terms of their component parts such as subsets of image regions, local neighborhoods, regional topologies, nearness and remoteness of sets, local convex sets, and mappings between image structures. The combination of traditional image analysis strategies and the more recent topology of digital images leads to a more complete view of digital images. In addition, the structured view of images facilitates the detection of hidden image patterns in image scenes as well as the solution of pattern recognition problems common in the study of digital images.

Keywords: Digital image; Geometry; Near sets; Set of points; Topology of digital images; Voronoi tessellation.

INTRODUCTION

Image analysis focuses on various digital image measurements (e.g., pixel size, pixel adjacency, pixel feature values, pixel neighborhoods, pixel gradient, closeness of image neighborhoods). Three standard region-based approaches in image analysis are isodata thresholding (binarizing images), watershed segmentation (computed using a distance map from foreground pixels to background regions), and non-maximum suppression (finding local maxima by suppressing all pixels that are less likely than their surrounding pixels).[60] In image analysis, object and background pixels are associated with different adjacencies (neighborhoods).[1] There are two basic types of neighborhoods: adjacency neighborhoods[46,22] and topological neighborhoods.[33,16] Using different geometries, an adjacency neighborhood is defined by pixels adjacent to a given pixel. Adjacency neighborhoods are commonly used in edge detection in digital images. A topological neighborhood is defined by finding all pixels with feature vectors that match or are similar to the feature vector of a given pixel (the neighborhood "center") within a prescribed radius. Unlike an adjacency neighborhood, a topological neighborhood can have holes in it, that is, pixels with feature vectors that do not match the neighborhood center and are not part of the neighborhood. A *feature vector* is a one-dimensional array of numbers that are feature values of a pixel. Examples of feature vectors are as follows:

Black-and-white (binary) image: (0) (black pixel), (1) (white pixel).
Greyscale image pixels: (50) (grey pixel), (200) (almost white pixel) on a scale of greyscale pixel values from 0 to 255.
Color image pixels: (50, 50, 50) (matching brightness of red, blue, and green pixel channel values), (50, 150, 200) (dark red, medium blue, and bright green pixel channel values) on a scale of color brightness pixel channel values from 0 to 255.
Greyscale edge pixel: (50, 30) (grey pixel with a gradient of 30°), (200, 45) (almost white pixel with a gradient of 45°).

With scene analysis, the focus shifts from image analysis measurements to a study of image objects and image regions and the relationships between image objects, object classes, and surrounding image regions.[8] Image regions and objects are found by considering the similarities between the feature vectors of pixels in subimages. Digital image segments and topological neighborhoods of pixels are examples of image regions of interest in scene analysis. In terms of geometry, scene analysis is a study of shapes in 2D regions (e.g., polygons and convex hulls) and 3D shapes (polyhedra) that can be extracted from plane polyhedral pictures. A polyhedron is a solid figure (e.g., cube, cylinder, pyramid, dodecahedron) with plane faces.[59]

What Is a Digital Image?

A digital image is obtained from a continuous (analog) view of a visual field by sampling and quantizing visible light. As a result, a digital camera superimposes a rectangular grid on an analog image and assigns an average brightness to each grid element.

There are two basic types of digital images: raster images and vector images. Raster images are arrays of tiny units called pixels, the default standard for digital camera, cell phone, digital microscope, notepad, laptop, and desktop computer displays. Raster image formats include .jpeg, .jpg (Joint Photographic Experts Group—supports RGB and CMYK color spaces, commonly used for photographic images by digital cameras and on the web, stores images as bitmaps at a resolution of 8 bits per color channel); .tiff, .tif (Tagged Image File Format, commonly used by scanners and other imaging devices, supports bit depths up to 16 and 32 bits per color channel); and .png (Portable Network Graphics, commonly used for graphics and photographic images on the web, stores color bitmaps at a resolution of 8 or 16 bits per color channel). Vector images consist of points and lines connecting the points, commonly used in gaming designs and satellite photography. This brief introduction to image and scene analysis considers only raster images. Vector image formats include .svg (Scalable Vector Graphics, designed for 2D vector graphics on the web), .eps (Encapsulated Postscript), and .pdf (Adobe PDF Format). Raster images can be converted to vector images.[51]

A *digital image* is a discrete representation of visual field objects that have spatial (layout) and intensity (color or grey tone) information. A *greyscale image* is an image containing pixels that are visible as black or white or grey tones (intermediate between black and white). From an appearance point of view, a *greyscale digital image* is represented by a 2D light intensity function $I(x, y)$, where x and y are spatial coordinates and the value of I at (x, y) is proportional to the intensity of light that impacted on an optical sensor and recorded in the corresponding picture element (pixel) at that point. The term *binary image* refers to a black-and-white image, *greyscale image* an image with varying shades of black and white (grey tones), and *color image* (or *multidimensional image*) for an image with one or more colors. If we have a multicolor image, then a pixel at (x, y) is 1×3 array, and each array element indicates a red, green, or blue brightness of the pixel in a color band (or color channel). A greyscale digital image I is represented by a single 2D array of numbers, and a color image is represented by a collection of 2D arrays, one for each color band or channel.

The smallest part in a digital image is called a *picture point* or *pixel* or *point sample*. A *pixel* is a physical point in a raster image.

MScript 1.

`PixelValue[`   `, {1, 1}, "Byte"]`
{61, 89, 31}

The terms *picture point*, *point*, and *pixel* are used interchangeably.

MScript 2.

`im =`
`GreyIm = ColorConvert[im, "Grayscale"]`

Each pixel has information that represents the response of an optical sensor to a particle of light (photon) reflected by a part of an object within a field of view (also field of vision).

A *visual field* or field of view is the extent of the observable world that is seen (part of scene in front of a camera) at any given moment. In terms of a digitized optical sensor value, a *point sample* is a single number in a greyscale image or a set of three numbers in a color image. For example, the byte values of the pixel in the (1,1) position in a tree nymph butterfly (genus *Vanessa*, species *atalanta*, family Nymphalidae) color image are given in MScript 1 (Mathematica® script that uses the built-in `PixelValue` function).

It is a straightforward task to convert a color image to a greyscale image using either Mathematica or MATLAB®. For example, in MScript 2, the butterfly image is converted to greyscale and that greyscale image is assigned to the variable `GreyIm`.

MScript 2.0

```
With[{img = ColorConvert[, "Grayscale"]},
Module[{w, h},
{w,h} = ImageDimensions[img];
Manipulate[ListLinePlot[PixelValue[img,
{All, row}], PlotRange → {0, 1}],
{row, 1, h,1}]]]
```

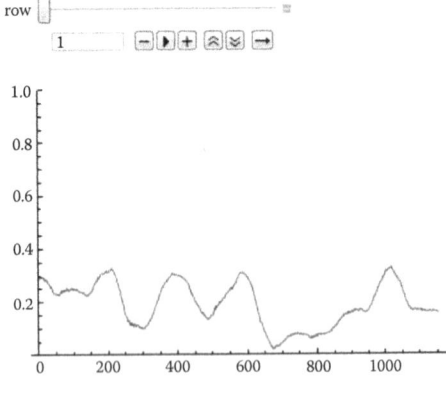

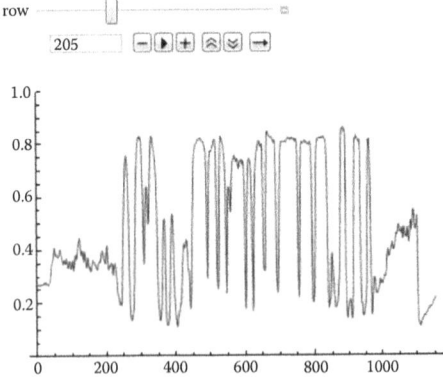

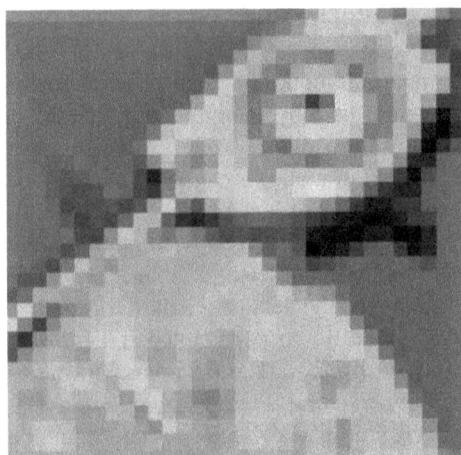

Gaining access and comparing the distributions of pixel values in a digital image are important in digital image analysis. For example, the greyscale image GreyIm is a 1152×656 image with 656 rows. The distribution of pixel values can be profitably examined during image analysis. This is done in a straightforward manner in MScript 2.0. The output for MScript 2.0 includes a row slider, which makes it possible to view the varying greylevel intensities in each row. Recall that a pixel value equal to or close to zero corresponds to black or dark grey pixels, and pixel values equal to or close to 1 (maximum intensity) correspond to white or nearly white pixels. For example, notice that the intensities in the first plot matching the corresponding dark pixels uniformly across the top of greyscale butterfly image. By contrast, in row 303 (roughly the middle of this image), most of the pixel intensities are at about 0.80 (in the white or nearly white range).

In raster images, it is common to use the little square model, which represents a pixel as a geometric square.

MScript 3.

Rasterize[ , RasterSize → 32

The tiny square model of raster images is illustrated in MScript 3 in terms of a picture of a rock containing fossils that average 1 mm in diameter. In this script, Rasterize specifies the number of pixels in an image square (in this case, each square is 32 pixels wide in a low-resolution image). As the raster size increases, the tiny squares tend to disappear. To see this, try RasterSize → 300 to obtain a high-resolution image. In photography, a *visual field* is the part of the physical world that is visible through a camera at a particular position and orientation in space. A visual field is identified with a view cone or angle of view.

Visualizing a Digital Image

The channels in a color image can be visualized as a histogram that displays a collection of bins, each with a particular color brightness level or with a greylevel intensity. It is possible to control the number of bins that are displayed. In the following examples, 32 bins are selected to make it easier to discern different color brightness levels. Color bins and greylevel intensity bins are combined in MScript 4.

MScript 4.

ImageHistogram[im, 32, Joined → False]

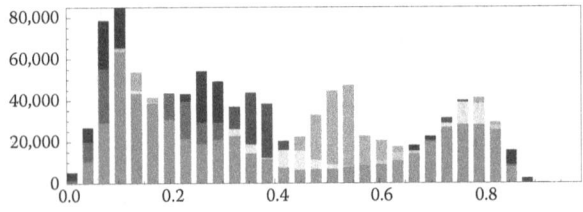

In MScript 4, im is the wood nymph butterfly image. The vertical scale on the output indicates the number of pixels that have a particular color brightness level or greylevel intensity. The horizontal scale gives the brightness (intensity) levels from 0 to 1. To display separate histograms for each color channel, use the Appearance → "Separated" option to obtain.

Image and Scene Analysis

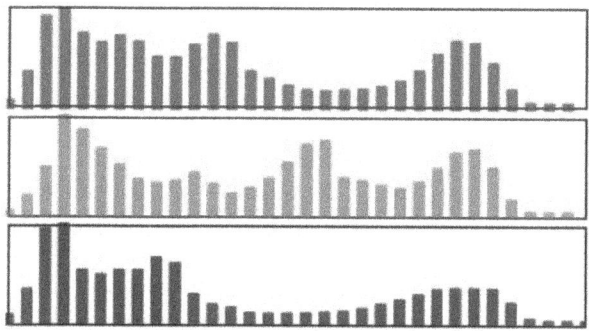

To display just the greylevel intensity bins in the greyscale butterfly image, try MScript 5.

MScript 5.

```
ImageHistogram[
ColorConvert[im,
"Grayscale"], 32, Joined → False]
```

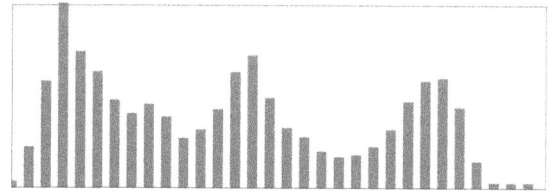

Distance between Feature Vectors

Every normed vector space results in a metric space. The norm of a vector is a measure of its length. For example, consider the feature vector $x = (x_1, ..., x_m)$ in $\mathbb{R}^m$, where each x_i is feature value for an image pixel. The norm $\|x\|$ of x is defined by:

$$\|x\| = \sqrt{x_1^2 + \cdots + x_m^2}.$$

Let $\mathbb{R}$ be the set of reals and let vectors x, y be in $\mathbb{R}^m$. Then the Euclidean distance $d_E : \mathbb{R}^m \times \mathbb{R}^m \to \mathbb{R}$ is defined by:

$$d_E(x, y) = \|x - y\| = \sqrt{\sum_{i=1}^{m}(x_i - y_i)^2}.$$

For more details about Euclidean distance, see the work of Deza and Deza.[9] A simpler metric to work with and implement is the Taxicab (or City Block) metric, defined by:

$$d(x, y) = \sum_{i=1}^{m}|x_i - y_i|.$$

Neighborhoods of Image Pixels

There are two types of neighborhoods (spherical and description based). In general, a *spherical neighborhood of a point x* in a nonempty set X is a set of points that are sufficiently near x. One point y belongs to a neighborhood of point x, provided y is sufficiently near x. In this case, *sufficient nearness of points x and y* is defined in terms of the fact that the distance between x and y is less than some number $\varepsilon \in (0, \infty]$. Distance is measured by some distance function. For a more general view of neighborhoods, see the work of Krantz.[21] The following image is used to illustrate the two types of neighborhoods in a digital image

Let $\varepsilon \in (0, \infty]$ be a bound on distances of points from a neighborhood center $x \in X$. A *neighborhood* of a point x (denoted $N_{x,\varepsilon}$ or simply by N_x, when ε is understood) is defined by:

$$N_x = \{y \in X : d(x, y) < \varepsilon\}.$$

In a digital image, pixels p, p' are adjacent, provided there are no pixels in between p and p'. For adjacent pixels p, p', the taxicab distance $d(p, p') = 1$. For example, let p, p' be at locations $(1, 1), (1, 2)$, respectively, and:

$$d(p, p') = |1 - 1| + |1 - 2| = 1.$$

For example, consider the following binary image containing a total of 25 pixels.

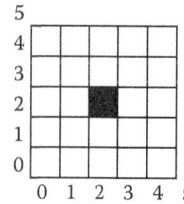

In this tiny 5×5 image, there is one black pixel at $(2, 2)$ surrounded by white pixels. Let the neighborhood center be at the $(2, 2)$. Then:

$$N_{p,1} = \{y \in X : d(p, y) < 1\} = \{+\}.$$

Notice that an ordinary neighborhood of a point is defined *spatially* in terms of the distance between points (pixels), not the distance between pixel feature vectors. In the following image showing the shadows of a plant, a sample image neighborhood with $\varepsilon = 150$ is represented by a green disk.

In this example, the distance between each neighborhood pixel and the neighborhood center is less than 150 pixels. This is an example of an open image neighborhood.

A *descriptive neighborhood* of a picture point is a set of picture points, each with a description that matches the description of the neighborhood center (also called neighborhood focal point). Let g be a set of picture points in a digital image and each pixel in g be represented by a feature vector $\phi(p)$ (feature vectors provide precise descriptions of picture points) and $\varepsilon > 0$ (neighborhood upper bound). Also, let $\phi(p)$, $\phi(p')$ denote the feature vectors of neighborhood center pixel p and another neighborhood pixel p', respectively. Assume that each pixel is described by a feature vector $(x_1,..., x_i,..., x_n)$, where x_i is the ith pixel feature value. Then, a descriptive neighborhood of x denoted by $N_{\Phi(x),\varepsilon}$ is defined by:

$$N_{\phi(p),\varepsilon} = \left\{ p' \in X : \|\phi(p) - \phi(p')\| \leq \varepsilon \right\}.$$

An example of a descriptive neighborhood with center x and radius $\varepsilon = 0.5$ is shown next.

In descriptive neighborhood of an image g, let pixel $p = g(x, y)$ at location (x, y) in g. Then, for example, each feature vector $\phi(p)$ for the neighborhood center p is defined by
 color = r, g, b channel values,
 G_x = gradient in the x-direction,
 G_y = gradient in the y-direction,
 $\theta = \arctan\left[\dfrac{G_y}{G_x}\right]$ (gradient),

$$h = \sum_{x,y} \frac{g(x,y)}{1+|x-y|} \text{ (homogeneity)},$$

$$e = \sum_{i,j} g(x,y)^2 \text{ (energy)}.$$

Let $g(x, y, 1)$, $g(x, y, 2)$, and $g(x, y, 3)$ denote the red, green, and blue color channel brightness values, respectively, for each pixel $g(x, y)$ in the image g. Then, let the feature vector that describes each pixel p in g be:

$$\phi(p) = \big(g(x,y,1), g(x,y,2), g(x,y,3), \theta, h(x,y), e(x,y)\big).$$

The angle θ (pixel gradient orientation) is an example of a shape descriptor. A *shape descriptor* is an expression that quantifies a shape characteristic of image regions and contours. Homogeneity and energy are the examples of image texture descriptors. A *texture descriptor* is an expression that describes elementary patterns that are repeated in an image. Let $g(i, j)$ denote the intensity of a pixel at location (i, j) in a greyscale image g. *Homogeneity* is a measure of the closeness of the distribution of the pixel intensities $g(i, j)$ in the greylevel co-occurrence matrix for an image g. *Energy* equals the sum of the squared pixel intensities in a local neighborhood. For more about shape and texture descriptors and near sets in image analysis. [15,29,30,33,37–39,43]

Unlike a spherical neighborhood, a descriptive neighborhood includes a nonuniform distribution of pixels that reflects the nonuniform distributions of pixels with feature vectors such that the distance between the feature vector for the center pixel p and any other neighborhood pixel p' is less than ε. This is an example of an open descriptive neighborhood, since the boundary pixels are not included in the neighborhood. For more about descriptive neighborhoods and image features, see the work of Peters.[33]

Image Clustering

Image clustering is a form of dimensionality reduction. An *image cluster* is a grouping of image pixels. The basic approach is to organize the pixels in an image into bins called clusters. Membership of a pixel in a cluster is determined by the proximity of the pixel intensity to the cluster centroid (or mean value of pixel intensities in a cluster). So, instead of an image with n different pixel intensities, we obtain an image with n/k intensities, so that each pixel is labeled with the intensity of one of the k clusters. For example, using MScript 6, a plant image is organized as a collection of eight clusters.

MScript 6.

ClusteringComponents[, 8]
//Colorize

Notice the singular color of the pixels in each cluster in the plant image. The color of a pixel is its cluster label. Also, notice that cluster pixels are often not adjacent to each other but instead scattered in different locations in the image. This is a common occurrence. The paradigm to remember in looking for clusters in an image is that every pixel in a cluster is descriptively near the centroid of the cluster. For a detailed introduction to clustering techniques, see the work of Sonka et al.[53]

The white edges on a black canvas can be inverted to obtain more distinct black edges on a white background.

Image Edge Detection

Edge detection is a fundamental part of image analysis. In a digital image, edges are groups of pixels where the intensity (or brightness) of each edge changes abruptly. Each pixel has a gradient orientation (angle) determined by comparing each pixel intensity with the intensities of neighboring pixels. The default edge detection method was introduced by John Canny in his MSc thesis at MIT.[3,42,53]

A good place to start in experimenting with finding edges in an image is to isolate the edges of the clusters found using MScript 6. For example, MScript 7 finds the edges in any image using the default Canny edge detector on plant clusters. For a topological view of clusters.[28]

Compare the results of edge detection on image clusters with straightforward edge detection on the original plant image. This is done in MScript 8. The result demonstrates the economy of edge detection in terms of all dominate edge pixels, that is, many lesser edge pixels with lower intensity changes are discarded using Canny's approach to edge detection.

MScript 8.

EdgeDetect[   ]

Scene Analysis

The foundations for scene analysis are built on the pioneering work by Rosenfeld on digital topology[18,44–48] (later called digital geometry[17]) and others.[11,19,22,24,25] The work on digital topology runs parallel with the introduction of computational geometry by Shamos[50] and Preparata,[40,41] building on the work on spatial tessellations by Voronoï,[56,57] and others.[4,13,14,23,27,55]

MScript 7.

EdgeDetect[   ]

To analyze image scenes, it is necessary to identify the objects in an image. Such objects can be viewed geometrically as collections of connected edges (e.g., skeletonizations or edges belonging to shapes or edges in polygons) or sets of pixels that are in some sense near each other or set of points near a fixed point (called a Voronoï region[10]).

Skeletonization

This section briefly introduces the skeletonization of scenes, reducing a complex scene to a skeleton view of the dominant arcs and line segments in an image. An *image skeleton* is a union of arcs and line segments that are centrally located in different parts of an image. The process of shrinking an image into a skeleton is called *thinning* or *skeletonization*.[17]

This can be accomplished using either the MATLAB `bwmorph` function or Mathematica `SkeletonTransform` function on a binary image. Consider, for example, the following tree nymph butterfly image.

For example, after converting the color image for the tree nymph to a binary image, the `bwmorph` function produces the following result.

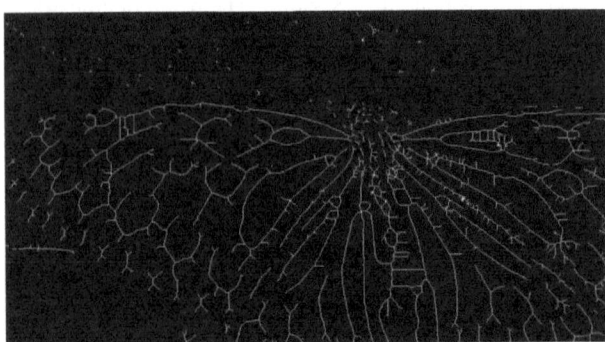

As a result, at least four main objects are visible in tree nymph skeleton, which are as follows:

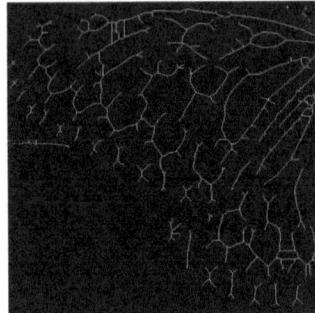

left wing:

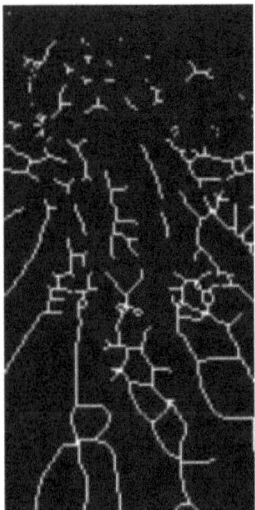

tree nymph body: :

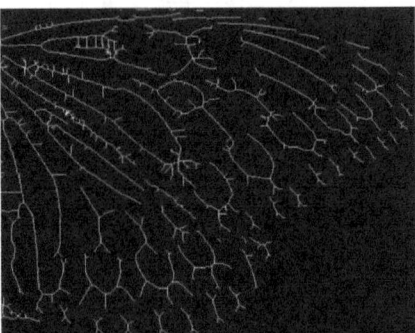

right wing: :

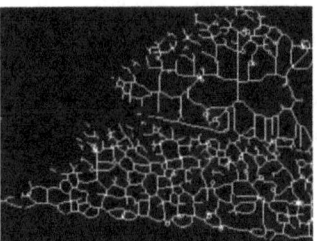

tree leaf: :

Notice that very fine-grained details can be observed in each part of the tree nymph skeletonization. Also, notice that the edges in the tree nymph wings reveal large number of either partial or complete convex sets. A set is *convex*, provided the line segment connecting each pair points in the set belongs to the set. In the upper left tree nymph wing, the bounded region ▱ is an example of a convex set. Since there is interest in classifying digital images, comparing bounded regions in the skeletonizations of pairs of images provides a basis for image classification.

The choice of the granularity of image skeletonization depends on the application. For example, in studies of microscope images of biological samples or medical images, there is advantage in seeing the finer skeleton edge details that can be derived by various means. There are many cases such as electrical circuit images or satellite images where coarse-grained skeletonization is sufficient.

Voronoï Diagrams

A *Voronoï diagram* is a subdivision of an image into regions that are closer to a generator element (e.g., some image centroid) than to any other such element.[7, §5.3, 537] Let X be a digital image and S be a subset of X. A pixel in S is called a *site* to distinguish the pixels in S from other pixels in an image. The important thing is to note that the sites in S are selected in some fashion from the pixels in an image. The pixels in S are used to construct Voronoï regions. For example, let z be a fixed pixel in S and y be any pixel in S. Then a Voronoï region V_z of site $z \in S$ is defined by:

$$V_z = \left\{ x \in X : \|x - z\|_{y_j \in S} \leq \|x - y\| \right\}.$$

The Voronoï region is also denoted by V_{zi}, where z_i is the ith element of the set of sites S. The size of the set S is denoted by $|s|$. The pixel y_j ranges over S so that $1 \leq j \leq |S|$. Hence, $1 \leq i \leq |S|$. The set $\left\{ V_{z_i} \right\}_{i=1}^{|S|}$ is a *Voronoï tessellation* or *Voronoï diagram* of an image X. A *tessellation* of a plane image has the appearance of a mosaic, a covering of a plane image with a repeated shape without gaps or overlaps. A *Delaunay triangulation* results from connecting the nearest neighbors of the region sites in a Voronoï diagram. For details, see, for example, the works of Du et al.,[10] Frank and Hart,[12] Agrell,[2] Wang et al.,[58] Chynoweth and Sewell,[6] and Peters.[31,32,34] For an example of Voronoï tessellation in plant sciences and in chromosome analysis, see the works of Korn[20] and Srisang et al.,[54] respectively. Chen[5] points out that Delaunay triangulation is the most popular of the different forms of triangulation.

A straightforward way to achieve a Voronoï diagram and Delaunay tessellation on a digital image reduces to a few basic steps.

1° Segment an image by some means. *Segmentation* of an image results from partitioning an image to obtain disjoint sets of homogeneous pixels called segments.
2° Find the centroid of each image segment, that is, find a segment pixel with an intensity that is nearest the average intensity of the segment pixels. The set of all centroids of image segments is an example of set of sites S in image.
3° Find the Voronoï region of each site in S. This leads to a Voronoï diagram on an image.
4° Construct a Delaunay tessellation on an image. To do this, draw line segments between the sites of adjacent Voronoï regions in an image.

Mathematical morphology (MM) offers an effective way to segment an image. It is a mainstay in the study of image segmentation. MM was introduced by Matheron (founder of geostatistics[26]) and Serra[49] during the 1960s (Matheron was Serra's PhD supervisor). Two basic operations (erosion and dilation) establish the foundations of MM. Digital images are viewed as sets of objects (dark areas) that are either thinned (eroded) or expanded (dilated). Combinations of erosion and dilation operations lead to closing and opening operations. Eventually, what is known as *watershed* segmentation was introduced as a means of delineating the boundaries of image objects as closed contours. A good introduction to watershed segmentation is given by Solomon and Breckon.[52]

For example, MScript 9 can be used to produce a watershed segmentation on a digital image (in this case, the image contains the archway shadows in the Rzvaniye Mosque in Şanlıurfa, southeastern Turkey; this image is used because there is a clear-cut separation between the archway and its surroundings). Watershed segmentation provides a basic threshold leading to the study of various structures in digital images (e.g., convex groupoids,[35] monoids,[36] Delaunay triangles, and Voronoï regions).

MScript 9.

```
Image[WatershedComponents[
```

```
Threshold[ColorNegate[         ],
0.58]], "Bit"]
```

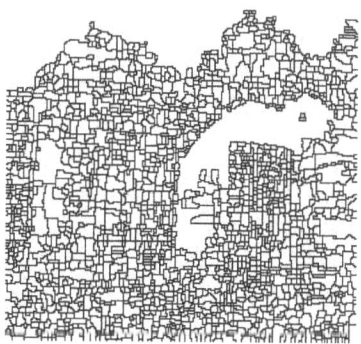

After finding the Voronoï region of the centroid of each image segment, then construct a Delaunay tessellation draw line segments between the sites of adjacent Voronoï regions in an image. The centroid of a bounded region may or may not coincide with image segment centroids. In this sample, Voronoï diagram on the archway shadows, each bounded region in the diagram consists of all points that are nearest the bounded region centroid. In scene analysis, a collection of tessellation polygons with a vertex in common are examples of near convex sets. In the analysis of a scene, maximal collections of near convex sets define regions of interest in a scene.

Except for the image borders, the centroid approach to constructing a Voronoï diagram results in a tessellation of the interior of an image. Based on the centroids in a watershed segmentation of the archway shadows example, the following Voronoï diagram is found. A sample Voronoï diagram on the arch image is shown in the following image.

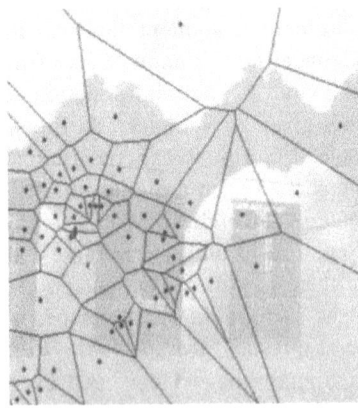

It is then possible to superimpose Voronoï diagram on the archway image. Using MScript 10, the original image and its Voronoï diagram are combined.

MScript 10.

ImageCompose[,

{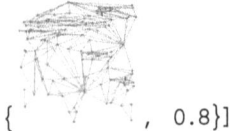, 0.8}]

After constructing a Voronoï diagram, it is a straightforward task to construct a Delaunay tessellation on an image. A sample Delaunay tessellation on the archway image is shown next.

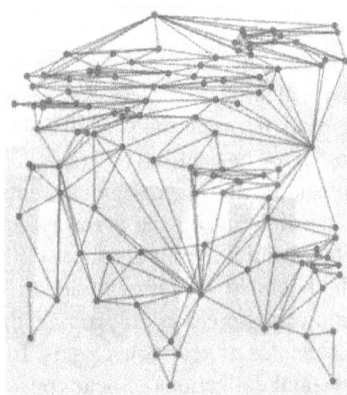

The Delaunay tessellation can then be superimposed on the archway image.

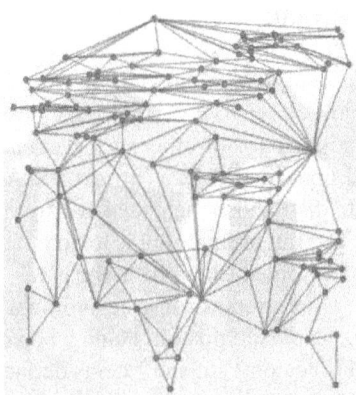

In terms of scene analysis, it is helpful to experiment with the percentage of the overlay (Voronoï diagram) that is blended with the original image. In this example, the overlay is set at 80%, and then at 1.0 (100%) to achieve an accurate rendition of the centroid-based Voronoï diagram on the archway image. After an image has been tessellated with a collection of Voronoï regions that are convex sets, the tessellated image can be compared with other tessellated images. This leads to the classification of tessellated image scenes based on factors such as the tessellation granularity, Voronoï region area, and Voronoï region shape (i.e., polygons such as the commonly occurring triangles and quadrilaterals in a Voronoï tessellation).

REFERENCES

1. Aberra, T.A. *Topology Preserving Skeletonization of 2d and 3d Binary Images*; Master's thesis; Technische Universität Kaiserslautern, Kaiserslautern, 2004. Supervisors: K. Schladitz, J. Franke.
2. Agrell, E. *Models and Algorithms for Image-Based Analysis of Microstructures*; PhD thesis; Chalmers University of Technology: Chalmers, 1997.
3. Canny, J.F. *Finding Edges and Lines in Images*; Master's thesis; MIT Artificial Intelligence Laboratory, 1983.
4. Chan, M.T. *Topical Curves and Metric Graphs*; PhD thesis; University of California: Berkeley, CA, 2012. Supervisor: B. Sturmfels.
5. Chen, L.M. *Digital and Discrete Geometry: Theory and Algorithms*; Springer: Berlin, 2014, xvii+322 p.
6. Chynoweth, S.; Sewell, M.J. Mesh duality and legendre duality. Proc. R. Soc. Lond. A **1990**, *428* (1875), 351–377.
7. Cornea, N.D.; Silver, D. Curve-skeleton properties, applications, and algorithms. IEEE Trans. Vis. Comput. Graph. **2007**, *13* (3), 530–548.
8. Dalal, N. *Finding People in Images and Videos*; PhD thesis; Grenoble, Institut National Polytechnique de Grenoble, 2006. Supervisors: B. Triggs, C. Schmid.
9. Deza, E.; Deza, M.-M. *Dictionary of Distances*; Elsevier: Amsterdam, 2006, 391, ISBN:0-444-52087-2.

10. Du, Q.; Faber, V.; Gunzburger, M. Centroidal Voronoi tessellations: Applications and algorithms. SIAM Rev. **1999**, *41* (4), 637–676.
11. Eckhardt, U.; Latecki, L.J. Topologies for the digital spaces Z^2 and Z^3. Comput. Vis. Image Underst. **2003**, *90* (3), 295–312.
12. Frank, N.P.; Hart, S.M. A dynamical system using the Voronoi tessellation. Amer. Math. Mon. **2010**, *117* (2), 92–112.
13. Gardner, M. On tessellating the plane with convex polygon tiles. Sci. Am. **1975**, *233*, 116–119.
14. Grünbaum, B.; Shepherd, G.C. Tilings with congruent tiles. Bull. Am. Math. Soc. **1980**, *3* (3), 951–973.
15. Henry, C.; Peters, J.F. Arthritic hand-finger movement similarity measurements: Tolerance near set approach. Comput. Math. Methods Med. **2011**, 14, Article ID 569898.
16. Henry, C.J. Neighbourhoods, classes and near sets. Appl. Math. Sci. **2011**, *5* (35), 1727–1732.
17. Klette, R.; Rosenfeld, A. *Digital Geometry: Geometric Methods for Digital Picture Analysis*; Morgan-Kaufmann Publishers: Amsterdam, 2004.
18. Kong, T.Y.; Roscoe, A.W.; Rosenfeld, A. Concepts of digital topology. Special issue on digital topology. Topol. Appl. **1992**, *46* (3), 219–262.
19. Kong, T.Y.; Rosenfeld, A. *Topological Algorithms for Digital Image Processing*; North-Holland: Amsterdam, 1996.
20. Korn, R.W. Window patterns in Lithops. Int. J. Plant Sci. **2011**, *172* (9), 1101–1109.
21. Krantz, S.G. *A Guide to Topology*; The Mathematical Association of America: Washington, DC, 2009, 106, ISBN:978-0-88385-346-7.
22. Kronheimer, E.H. The topology of digital images. Topol. Appl. **1992**, *42* (3), 279–303.
23. Lai, R. *Computational Differential Geometry and Intinsic Surface Processing*; PhD thesis; University of California Los Angeles: Los Angeles, CA, 2010. Supervisors: T.F. Chan, P. Thompson, M. Green, L. Vese.
24. Latecki, L. Topological connectedness and 8-connectedness in digital pictures. CVGIP: Image Underst. **1993**, *57*, 261–262.
25. Latecki, L.; Conrad, C.; Gross, A. Preserving topology by a digitization process, J. Math. Imaging Vis. **1998**, *8*, 131–159.
26. Matheron, G. *Estimating and Choosing*; Springer: Berlin, 1989.
27. Gavrilova, M.L.; Ed.; *Generalized Voronoi Diagrams: A Geometry-Based Approach to Computational Intelligence*; Springer: Berlin, 2008, xv+304 p.
28. Naimpally, S.A.; Peters, J.F. *Topology with Applications. Topological Spaces via Near and Far*; World Scientific: Singapore, 2013, xv+277 p.
29. Peters, J.F. Local near sets: Pattern discovery in proximity spaces. Math. Comput. Sci. **2013**, *7* (1), 3–9.
30. Peters, J.F. Nearness of sets in local admissible covers. Theory and application in micropalaeontology. Fund. Inform. **2013**, *126* (4), 433–444.
31. Peters, J.F. Proximal Delaunay triangulation regions, 2014 (6260), 1–4, arXive 1411.
32. Peters, J.F. Proximal Voronoï Regions, 2014, (3570), 1–4, arXive 1411.
33. Peters, J.F. *Topology of Digital Images. Visual Pattern Discovery in Proximity Spaces*, Vol. 63; Intelligent Systems Reference Library Springer: Berlin, 2014, xv+411 p.
34. Peters, J.F. Proximal Voronoï regions, convex polygons, & Leader uniform topology. Adv. Math. **2015**, *4*, 1–5.
35. Peters, J.F.; İnan, E.; Öztürk, M.A. Klee-Phelps Convex Groupoids, 2014 (0934), 1–4, arXiv 1411.
36. Peters, J.F.; İnan, E.; Öztürk, M.A. Monoids in proximal banach spaces. Int. J. Algebra **2014**, *8* (18), 869–872.
37. Peters, J.F.; Naimpally, S.A. Applications of near sets. Am. Math. Soc. **2012**, *59* (4), 536–542.
38. Peters, J.F.; Wasilewski, P. Foundations of near sets. Inform. Sci. **2009**, *179* (18), 3091–3109.
39. Poli, G.; Llapa, E.; Cecatto, J.R.; Saito, J.H.; Peters, J.F.; Ramanna, S.; Nicoletti, M.C. Solar flare detection system based on tolerance near sets in a GPU–CUDA framework. Knowl.-Based Syst. **2014**, *70*, 345–360.
40. Preparata, F.P. Convex hulls of finite sets of points in two and three dimensions. Commun. Assoc. Comp. Mach. **1977**, *2* (20), 87–93.
41. Preparata, F.P. *Steps into Computational Geometry*; Technical Report, Coordinated Science Laboratory; University of Illinois: Urbana-Champaign, 1977.
42. Prince, S.J.D. *Computer Vision: Models, Learning, and Inference*; Cambridge University Press: Cambridge, 2012, xvii+580 p.
43. Ramanna, S.; Meghdadi, A.H.; Peters, J.F. Nature-inspired framework for measuring visual image resemblance: A near rough set approach. Theor. Comput. Sci. **2011**, *412* (42), 5926–5938.
44. Rosenfeld, A. Distance functions on digital pictures. Pat. Rec. **1968**, *1* (1), 33–61.
45. Rosenfeld, A. *Digital Picture Analysis*; Springer-Verlag: Berlin, 1976, xi+351 p.
46. Rosenfeld, A. Digital topology. Am. Math. Mon. **1979**, *86* (8), 621–630.
47. Rosenfeld, A.; Kak, A.C. *Digital Picture Processing*, Vol. 1; Academic Press, Inc.: New York.
48. Rosenfeld, A.; Kak, A.C. *Digital Picture Processing*, Vol. 2; Academic Press, Inc.: New York, 1982, xii+349 p.
49. Serra, J. *Image Analysis and Mathematical Morphology*; Academic Press Inc.: London, 1982.
50. Shamos, M.I. *Computational Geometry*; PhD thesis; Yale University: New Haven, CT, 1978. Supervisors: D. Dobkin, S. Eisenstat, M. Schultz.
51. Sharma, O. *A Methodology for Raster to Vector Conversion of Colour Scanned Maps*; Master's thesis; Department of Geodesy and Geomatics Engineering, 2006.
52. Solomon, C.; Breckon, T. *Fundamentals of Digital Image Processing: A Practical Approach with Examples in Matlab*; Wiley-Blackwell: Oxford, 2011, x+328 p.
53. Sonka, M.; Hlavac, V.; Boyle, R. *Image Processing, Analysis and Machine Vision*, 3rd Ed.; Cengage Learning: Boston, MA, 2008, 829, ISBN-10: 0-495-24438-4.
54. Srisang, W.; Jaroensutasinee, K.; Jaroensutasinee, M. Segmentation of overlapping chromosome images using computational geometry. Walailak J. Sci. Technol. **2006**, *3* (2), 181–194.
55. Toussaint, G.T. Computational geometry and morphology. In *Proceedings of the First International Symposium for*

Science on Form; Kato, Y.; Takaki, R.; Toriwaki, J.; Eds.; KTK Scientific Publishers : Tokyo, 1986, 395–403.

56. Voronoi, G. Sur une fonction transcendante et ses applications à la sommation de quelque séries. Ann. Sci. Ecole Norm. Sup. **1904**, *21* (3), 207–267.

57. Voronoi, G. Nouvelles applications des parametres continus à la théorie des formes quadratiques. J. Reine Angew. Math. **1908**, *134*, 19–287.

58. Wang, J.; Ju, L.; Wang, X. An edge-weighted centroidal Voronoi tessellation model for image segmentation. IEEE Trans. Image Process. **2009**, *18* (8), 1844–1858.

59. Whiteley, W. *Rigidity of Molecular Structures: Generic and Geometric Analysis*; Kluwer/Plenum: New York, 1999.

60. Wirjadi, O. *Models and Algorithms for Image-Based Analysis of Microstructures*; PhD thesis, Technische Universität Kaiserslautern: Kaiserslautern, 2009. Supervisor: K. Berns.

Image Enhancement

S. Palanikumar
Information Technology, Noorul Islam Centre for Higher Education, Thuckalay, India

Abstract

Image enhancement is the process by which an input image is preprocessed to fit for the application. Enhancement includes improving contrast, brightness, and edge or reducing the noise of the image. The principal objective of enhancement is to process an image so that the result is more suitable than the original image for a specific application. Consequently, the enhancement methods are application specific and are often developed empirically. Image enhancement can be done in either spatial domain or frequency domain. Contrast stretching, image negative, gray-level slicing, bit-plane slicing, compression of dynamic range, histogram equalization, and histogram specification are some of the spatial domain techniques. Spatial domain techniques modify the gray-level value using the transformation function. Low-pass filter, high-pass filter, and band-pass filter are some of the frequency domain techniques. First, frequency domain methods transform the image from spatial domain to frequency domain using the transformation function, and then the modification is done in the frequency domain itself. Finally, the image is brought to spatial domain using inverse transformation function. Spatial domain processing methods include point, mask, and global operations, but frequency domain operations, by nature of the frequency transforms, are global operations. Of course, frequency domain operations can become mask operations, based only on a local neighborhood, by performing the transform on small blocks instead of the entire image. Image enhancement has played and will continue to play an important role in different fields such as medical, industrial, military, and scientific applications.

Keywords: Enhancement; Equalization; Filter; Frequency; Grayscale; Histogram; Image; Intensity; Spatial.

The primary objective of image enhancement is to process an image so that the result is more suitable than the original image for a specific application.

Pratt[1] defines image enhancement as a collection of techniques that seek to improve the visual appearance of an image or to convert the image to a form better suited for analysis by a human or a machine.

An image enhancement is basically anything that makes it easier or better to visually interpret an image. In some cases, like "low-pass filtering," the enhanced image can actually look worse than the original, but such an enhancement was likely performed to help the interpreter see low-spatial-frequency features among the usual high-frequency clutter found in an image. Also, an enhancement is performed for a specific application. This enhancement may be inappropriate for another purpose, which would demand a different type of enhancement.

Image enhancement procedures are used to process degrading images of unrecoverable objects. In image enhancement, the goal is to accentuate certain image features for subsequent analysis or for image display. Enhancement process does not increase the inherent information content in the data. It simply emphasizes certain specified image characteristics. Enhancement algorithms are generally interactive and application dependent. Examples include contrast and edge enhancement, pseudo coloring, noise filtering, sharpening, and magnifying. Image enhancement is useful in feature extraction, image analysis, and visual information display.

There are many techniques available for image enhancement. Most of the work related to palm print identification system uses histogram equalization (HE) as enhancement technique. But the traditional HE technique has various drawbacks. In order to overcome those drawbacks, a suitable technique based on HE is needed.

Figure 1 shows the standard lena image, and its histogram equalized image is shown in Fig. 2. In the histogram equalized image, the contrast is improved well. Figure 3 shows the histogram of lena image, and histogram of enhanced image using HE is shown in Fig. 4. In Fig. 3, the histogram is not distributed well in the bright region. In Fig. 4, the histogram is uniformly distributed in all gray-level ranges.

IMAGE ENHANCEMENT IN SPATIAL DOMAIN

Spatial domain refers to the image plane itself. In this method, enhancement is based on direct manipulation of

Encyclopedia of Image Processing, First Edition
DOI: 10.1201/9781351110273-140000133
Copyright © 2018 by Taylor & Francis. All rights reserved.

Fig. 1 Lena image

Fig. 2 Enhanced image using HE

pixels in an image. The enhancement process is denoted mathematically by a transformation function as follows:

$$s = T(r) \qquad (1)$$

where T is the transformation that maps a pixel value r into a pixel value s.

$g(x, y) = T(f(x, y))$ where $f(x, y)$ is the input image and $g(x, y)$ is the enhanced image. T is an operator on f, defined over some neighborhood of (x, y). The gray levels range from 0 to 255 for an 8-bit environment. In general, 0 to $L-1$, L is 2^n, where n is the number of bits used to represent the image pixel.

i. Image negative
Image negative means reversing the intensity levels of an image.[2] The transfer function is given in the following equation:

$$s = L - 1 - r \qquad (2)$$

For an 8-bit image, if input gray levels are 5, 100, and 200, then the values of output gray levels are as follows:

$$s = 256 - 1 - 5, 256 - 1 - 100, 256 - 1 - 200$$

where $s = 250, 155, 55$
Log transformations
The log transformation is given as follows:

$$s = c \ \log(1 + r) \qquad (3)$$

where c is the constant and $r \geq 0$.

ii. Power-law transformation
The power-law transformation is given as follows:

$$s = cr^\gamma \qquad (4)$$

where c and γ are positive constants.
This method is also known as gamma correction.

iii. Contrast stretching
Contrast stretching is used to increase the dynamic range of the gray levels of the image.

iv. Gray-level slicing
Gray level slicing is used to highlight a specific range of gray levels in an image.

v. Bit-plane slicing
Bit-plane slicing is used to highlight the contribution made by specific bits to the total appearance of an image.

vi. Histogram processing
The histogram of a digital image with gray levels in the range $[0, L-1]$ is a discrete function $h(r_k) = n_k$, where r_k is the kth gray level and n_k is the number of pixels in the image having gray level r_k. Histograms are the basis for numerous spatial domain processing techniques. Histogram manipulation can be used effectively for image enhancement.

Contrast is understood as a combination of the range of intensity values effectively used within a given image and the difference between the image's maximum and minimum pixel values. In some digital images, the features of interest occupy only a relatively narrow range of the gray scale. One might use a point operation to expand the contrast of the features of interest, so that they occupy a larger portion of the displayed gray-level range. This is known as contrast enhancement or contrast stretching.

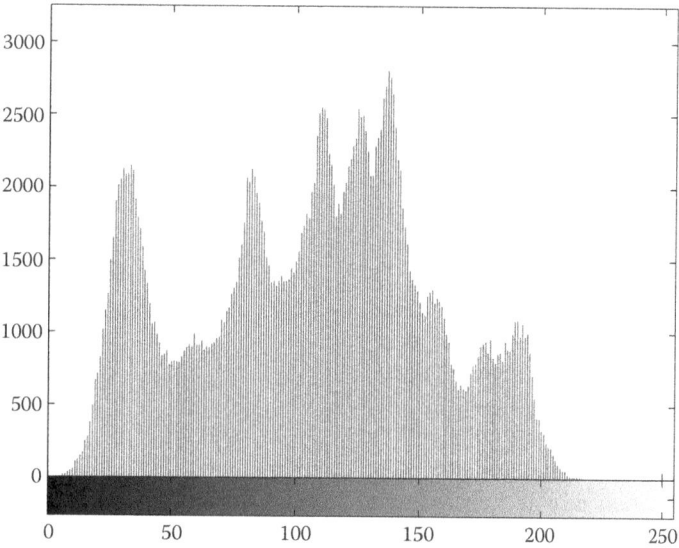

Fig. 3 Histogram of lena image

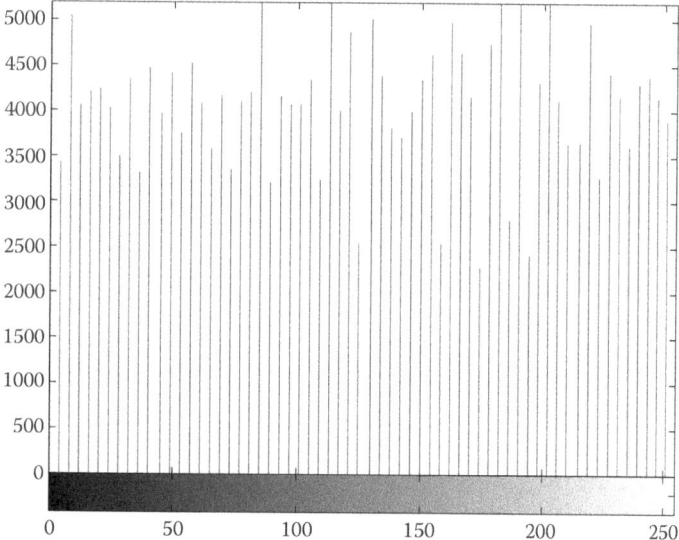

Fig. 4 Histogram of enhanced image

vii. Histogram modification

The histogram of pixel intensities in an image of a natural scene that has been linearly quantized is usually highly skewed toward darker levels. Detail is often imperceptible in the darker regions when such an image is viewed on a conventional display with only 8-bit pixels. Many image processing operations result in changes to the image's histogram. The class of histogram modifications which we consider here includes operations where the changes to pixel levels are computed so as to change the histogram in a particular way.

viii. Histogram stretching

The simplest form of histogram modification is histogram stretching. For example, if the image is underexposed, its values would only occupy the lower part of the dynamic range. You can perform the histogram stretching for an 8-bit image using HE-based enhancement.

ix. HE-based enhancement

The gray-level histogram is a function showing the relationship between gray level and the number of pixels in the image with that gray level. It is a plot of gray level and its probability. The technique used for obtaining a uniform histogram is known as histogram linearization[3] or HE.

HE is a widely used method of image contrast enhancement in a variety of applications due to its simple function and effectiveness.[4] It is a common technique for enhancing the appearance of images.

It involves finding a grayscale transformation function that creates an output image with a uniform histogram (or nearly so). Suppose there is an image which is predominantly dark, then its histogram would be skewed toward the lower end of the gray scale and all the image details are compressed into the dark end of the histogram. If the gray levels are stretched at the dark end, they produce a more uniformly distributed histogram. The image would then be much clearer. Grayscale transformation function can be determined assuming the gray levels to be continuous and normalized to lie between 0 and 1.

A transformation T is found that maps gray values r in the input image F to gray values $s=T(r)$ in the transformed image. It is assumed that T is single valued and monotonically increasing, and $0 \leq T(r) \leq 1$ for $0 \leq r \leq 1$ and the inverse transformation from s to r is given by $r=T^{-1}(s)$. If one takes the histogram for the input image and normalizes it, the area under the histogram is 1, having probability distribution for gray levels in the input image $P_r(r_k)$.

The probability distribution of gray levels in the input image is first calculated by Gonzalez and Woods[3] as follows:

$$P_r(r_k) = \frac{n_k}{N} \qquad (5)$$

where n_k is the number of pixels having gray level k, and N is the total number of pixels in the image. A plot of $P_r(r_k)$ versus r_k is called as a histogram. The transformation is represented as follows:

$$s_k = T(r_k) = \sum_{i=0}^{k} \frac{n_i}{N} = \sum_{i=0}^{k} P_r(r_k) \qquad (6)$$

where $0 \leq r_k \leq 1$, the index $k=0, 1, 2, \ldots, L-1$ and $0 \leq s_k \leq 1$. The values of s_k will have to be scaled up by $L-1$ and rounded to the nearest integer so that the output values of this transformation will range from 0 to $L-1$. Thus, the discretization and rounding of s_k to the nearest integer will mean that the transformed image will not have a perfectly uniform histogram. HE may not always produce desirable results, particularly if the given histogram is very narrow. It can produce false edges and regions. It can also increase image graininess and patchiness.

○ Conventional HE

The traditional HE technique is described as follows:

Consider the input image as **X**. Based on the histogram H(**X**), the probability density function (PDF) of the image is defined as follows:

$$p(k) = n_k/N = n_k/(n_0 + n_1 + \cdots + n_{L-1}) \text{ for } \\ k = 0, 1, \ldots L-1 \qquad (7)$$

where n_k is the number of pixels that have the gray level k that appears on the input image **X**, and N is the total number of pixels in the input image. Note that $p(k)$ is associated with the histogram of the input image which represents the number of pixels that have a specific intensity X_k. The graphical appearance of PDF is known as the histogram.

From the PDF in Eq. 7, the cumulative distribution function (CDF) is defined as follows:

$$c(k) = \sum_{j=0}^{k} p(j) \text{ for } k = 0, 1 \ldots L-1 \qquad (8)$$

Note that $c(L-1)=1$ from Eqs. 1 and 2.

Based on the CDF, HE now maps an input gray level X_k into an output gray level $f(k)$, where $f(k)$, which is commonly called as *level transformation function*, is defined as follows:

$$f(k) = X_0 + (X_{L-1} - X_0) \cdot c(k) \qquad (9)$$

where X_0 is the minimum gray level, and X_L is the maximum gray level. Thus, HE remaps the input image into the entire dynamic range $[X_0, X_{L-1}]$.

○ Bin underflow and bin overflow (BUBO)

A global HE-based enhancement method that uses the BUBO mechanism is given by Yang et al.[5] The PDF of the image is thresholded using a lower (underflow) threshold and an upper (overflow) threshold, where C_{BU} and C_{BO} are bin underflow and bin overflow thresholds. HE is performed using the thresholded histogram. BUBO is an effective technique for contrast enhancement, but in some cases, the method still cannot expand gray-level distribution to expand dynamic range of input image. However, we can add the function of adjusting parameter into BUBO to expand image for more dynamic range, and the method still needs to adjust different parameter for input images. To overcome the drawbacks of the traditional HE methods, the BUBO method puts constraint to avoid over enhancement as defined in the following equation:

$$P_{BUBO}(k) = \begin{cases} C_{BO}, & \text{if } P(k) > C_{BO} \\ P(k), & \text{if } C_{BU} \leq P(k) \leq C_{BO} \\ C_{BU}, & \text{if } P(k) < C_{BU} \end{cases} \qquad (10)$$

$$C_{BU} = (1-\alpha)/N$$
$$C_{BO} = (1-\alpha)/N$$

where α varies from 0 to infinity, and N is the total number of pixels in the given input image. It cannot expand gray-level distribution to expand dynamic range of input image. To expand image for more dynamic range, the method still needs and adjust different parameters for input images.

- Weighted threshold histogram equalization (WTHE)

The WTHE enhancement method described by Wang and Ward[6] performs HE based on a modified histogram. Each original probability density value $p(k)$ is replaced by a weighted and thresholded PDF value $p_{wt}(k)$ yielding

$$\Delta k = (L-1) \times p_{wt}(k) \qquad (11)$$

$p_{wt}(k)$ is obtained by applying a transformation function $TR(k)$ to $p(k)$ as follows:

$$p_{wt}(k) = TR(p(k)) \qquad (12)$$

The transfer function is given as follows:

$$TR(p(k)) = \begin{cases} p_u & \text{if } p(k) > p_u \\ \left[\dfrac{p(k)-p_l}{p_u-p_l}\right] \times p_u & \text{if } p_l \leq p(k) \leq p_u \\ 0 & \text{if } p(k) < p_l \end{cases} \qquad (13)$$

The original PDF is clamped at an upper threshold p_u and a lower threshold p_l, and it transforms all values between the upper and lower thresholds using a normalized power law function with index $r > 0$. The increment for each intensity level is decided by the transformed histogram. The increment can be controlled by adjusting the index r of the power law transformation function. To give an example, when $r < 1$, the power law function will give a higher weight to the low probabilities in the PDF than the high probabilities. Therefore, with $r<1$, the less-probable levels are "protected" and over-enhancement is less likely to occur. Besides the weighting mechanism described above, the PDF is also thresholded at an upper limit p_u. As a result, all levels whose PDF values are higher than p_u will have their increment clamped at a maximum value $\Delta_{max} = (L-1)p_u$. Such upper clamping further avoids the dominance of the levels with high probabilities when allocating the output dynamic range. In our algorithm, the value of p_u is decided by the following equation:

$$p_u = v \cdot P_{max}, 0 < v \leq 1 \qquad (14)$$

where P_{max} is the peak value (highest probability) of the original PDF, and the real number v defines the upper threshold normalized to P_{max}. It can be seen from Eq. 6 that when $r=1$, $p_u=1$, and $p_l=0$, the WTHE method reduces to the original HE. Some other global HE-base methods, such as the BUBO method, can also be considered special cases of the WTHE method. Our experiments show that for a large variety of images the value of v can be kept constant while achieving satisfactory effect of enhancement. After obtaining the weighted thresholded PDF $p_{wt}(k)$, the equalization process is similar to the traditional HE. The CDF is first obtained by the following equation:

$$C_{wt}(k) = \sum_{m=0}^{k} p_{wt}(m), \quad \text{for } k = 0,1,\ldots L-1 \qquad (15)$$

and HE procedure is then performed as follows:

$$G(i,j) = W_{out} \times C_{wt}(F(i,j)) + M_{adj} \qquad (16)$$

where W_{out} is the dynamic range of the output image. M_{adj} is the mean adjustment factor that compensates for the mean change after enhancement. For a simple case, W_{out} is equal to the full range $[0, L-1]$, and $M_{adj}=0$. $W_{out}=\min(L-1, G_{max}.W_{in})$ in which W_{in} is the dynamic range of the input image and G_{max} is a preset maximum gain of dynamic range. We usually set G_{max} in the range of 1.5 to 3.0.

- Adaptively increasing value of histogram equalization (AIVHE) contrast enhancement

The AIVHE method proposed by Ching-His and Hong-Yang[7] reshapes the original PDF to obtain new PDF to prevent a significant change in the gray levels. It also provides a mechanism of adjustment to contrast enhancement by means of adaptive constraint parameter $\alpha(k)$ for adjustment automatically, which is determined by the initial value γ and user defined parameter β. AIVHE divides the original PDF into upper and lower blocks on the basis of P_{bas}. A value of maximum threshold P_h is then be set to restrict the variation of the $P_{AIVHE}(k)$, and then limit the value of $P_{AIVHE}(k)$ be not greater than P_h. AIVHE reshapes original PDF, and the $P_{AIVHE}(k)$ is obtained as follows:

$$P_{AIVHE}(k) = \begin{cases} P_h, & \text{if } P(k) \geq P_h \\ P(k) - \alpha(k)(p(k)-P_{bas}) \times \beta & \text{if } P_{bas} < P(k) < P_h \\ P(k) - \alpha(k)(P_{bas}-p(k)) \times \beta & \text{if } P(k) \leq P_{bas} \end{cases} \qquad (17)$$

where P_{bas} is set to the average PDF, P_h is set as double of P_{bas}, β is the enhancement parameter adjusted by user, and $\alpha(k)$ is the adaptive constraint parameter for adjustment automatically. The initial value for β is a real number in the range of [0, 1]. The function of HE is produced when β is set to zero, and P_{bas} is the mean value of the maximum and minimum value of $P(k)$. Dark and bright regions stretching are controlled by γ at $\alpha(k)$. Whole contrast enhancement effect of the image is controlled by β. The effective constraint parameter can be calculated as follows:

$$\alpha(k) = \left(1 - (X_m - k)/X_m\right)^2 \times (1-\gamma) + \gamma \quad \text{if,} \quad k \leq X_m$$
$$\left(1 - (k - X_m)/((L-1) - X_m)\right)^2 \times (1-\gamma) + \gamma \quad \text{if,} \quad k > X_m$$
(18)

where X_m is the mean brightness, γ is the real number, and the initial value in the range of [0, 1]. It decides the distribution of pixels in dark and bright regions. By the $P_{AIVHE}(k)$, the cumulative distributive function $C_{AIVHE}(k)$ is found. $C_{AIVHE}(k)$ is normalized and then the output image is as follows:

$$f(k) = (L-1) \times (C_{AIVHE}(k))/(C_{AIVHE}(k)(L-1))$$
(19)

where $f(k)$ is the input/output transfer function, $L-1$ is the maximum gray level, k is the kth gray level, and $C_{AIVHE}(k)$ is the cumulative density function. Furthermore, AIVHE method enhances the contrast but the brightness of the palm print is not preserved.

○ Brightness preserving bi-histogram equalization (BBHE)
The BBHE method is developed by Kim.[8] First, the BBHE decomposes an input image into two sub-images based on the mean of the input image. One of the sub-images is the set of samples less than or equal to the mean, whereas the other one is the set of samples greater than the mean. Then, the BBHE equalizes the sub-images independently based on their respective histograms with the constraint that the samples in the formal set are mapped into the range from the minimum gray level to the input mean, and the samples in the latter set are mapped into the range from the mean to the maximum gray level. Denoted by X_m, the mean of the image X and assume that $X_m \in \{X_0, X_1, \ldots X_{L-1}\}$. Based on the mean, the input image is decomposed into two sub-images X_L and X_U as follows:

$$X = X_L \cup X_U \quad (20)$$

where

$$X_L = \{X(i,j) \mid X(i,j) \leq X_m, \forall X(i,j) \in X\}$$

and

$$X_U = \{X(i,j) \mid X(i,j) > X_m, \forall X(i,j) \in X\}$$

Note that the sub-image X_L is composed of $\{X_0, X_1, \ldots, X_m\}$ and the other sub-image X_U is composed of $\{X_{m+1}, X_{m+2}, \ldots, X_{L-1}\}$.

○ Dualistic sub-image histogram equalization (DSIHE)
The DSIHE method is introduced by Wan et al.[9] First, the image is decomposed into two equal area sub-images based on its original probability density function. Then the two sub-images are equalized, respectively. At last, we get the result after the processed sub-images are composed into one image. In other words, DSIHE separates the histogram based on gray level with cumulative probability density equal to 0.5 instead of the mean.

○ Minimum mean brightness error bi-histogram equalization (MMBEBHE)
Using input mean as the threshold level to separate the histogram does not guarantee maximum brightness preservation. The brightness preservation in MMBEBHE given by Chen and Ramli[10] is based on an objective measurement referred as AMBE. It is defined as the absolute difference between the input and the output means. Lower AMBE implies better brightness preservation. BBHE that set the threshold level, and X_T as input mean does not guarantee minimum AMBE. The threshold level should be chosen based on the resulting AMBE and not fixed to the input mean. MMBEBHE is formally defined by the following procedures:
  I. Calculate the AMBE for each of the threshold level.
  II. Find the threshold level, X_T that yield minimum MBE.
  III. Separate the input histogram into two based on the X_T found in step 2 and equalize them independently as in BBHE.

○ Recursive mean separate histogram equalization (RMSHE)
In typical HE, no mean separation is performed and thus no brightness preservation. In BBHE, the mean separation is done once and thus achieves certain extends of brightness preservation. HE is equivalent to RMSHE described by Chen and Ramli[11] with recursion level $r = 0$. BBHE is equivalent to RMSHE with recursion level $r = 1$.
When $r = 0$, no decomposition occurs. This case is simply equivalent to conventional HE.

When $r = 1$, the input histogram H(**X**) is decomposed into two sub-histograms $H_L(\mathbf{X})$ and $H_U(\mathbf{X})$ based on the input mean X_M. Obviously, this case is the same as BBHE. When $r = 2$, $H_L(\mathbf{X})$ is divided further into $H_{LL}(\mathbf{X})$ and $H_{LU}(\mathbf{X})$ based on a new mean X_{ML}, and $H_U(\mathbf{X})$ is also divided further into $H_{UL}(\mathbf{X})$ and $H_{UU}(\mathbf{X})$ based on another new mean X_{MU}. Here, X_{ML} is the mean of the sub-histogram $H_L(\mathbf{X})$, whereas X_{MU} is the mean of the sub-histogram $H_U(\mathbf{X})$. Similar procedures can be carried out when r is greater than 2. It is claimed that as the recursion level r increases, the mean of the output image comes near to the input mean X_M.

- Recursive sub-image histogram equalization (RSIHE)
  RSIHE is the generalization of DSIHE introduced by Sim et al.[12] so the median-based histogram segmentation occurs many times recursively in RSIHE. It also shares similar recursive framework with RMSHE except that for histogram segmentation, RSIHE uses the medians of sub-histograms instead of the means of sub-histograms.

x. Histogram specification
It is useful sometimes to be able to specify the shape of the histogram that we wish the processed image to have. The method used to generate a processed image that has a specified histogram is called histogram matching or histogram specification.

xi. Global/local processing
If the pixels are modified by a transformation function based on the gray-level content of the entire image, then it is called as global processing. Instead of the entire image, if the transformation function is based on square or rectangular neighborhood, then it is called as local processing.

xii. Enhancement using arithmetic/logic operations
Arithmetic/logic operations involving images are performed on a pixel by pixel basis between two or more images (this includes the logic operation NOT, which is performed on a single image). Logical AND, OR, and NOT operations can be done. Performing the NOT operation on a black, 8-bit pixel produces a white pixel. Intermediate values are processed the same way, changing all 1s to 0s and vice versa. The AND OR operations are used for masking. That is for selecting sub-images in an image.

xiii. Image subtraction
The difference between two images $f(x, y)$ and $h(x, y)$ can be expressed as follows:

$$g(x,y) = f(x,y) - h(x,y) \tag{21}$$

It is obtained by computing the differences between all pairs of corresponding pixels from f and h.

xiv. Image averaging
Image averaging is done by adding all the images and dividing the result by total number of images. Addition is performed by adding corresponding pixels from each image.

xv. Spatial filters
In spatial filtering, the filtering operation is done directly on the pixels of an image. Some neighborhood operations work with the values of the image pixels in the neighborhood and the corresponding values of a sub-image that has the same dimensions as the neighborhood. The sub-image is called a filter, mask, kernel, template, or window.

- Smoothing spatial filters
  Smoothing filters are used for blurring and for noise reduction. Blurring is used in preprocessing steps, such as removal of small details from an image prior to object extraction, and bridging of small gaps in lines or curves.

  i. Order statistics filters (mean, median)
     Order statistics filters are nonlinear spatial filters whose response is based on ordering or ranking the pixels contained in the image area encompassed by the filter, and then replacing the value of the center pixel with the value determined by the ranking result.

  ii. Median filter
      In median filter, center pixel value is replaced by the median of the gray levels in the neighborhood of the pixel.

  iii. Weighted median filter
       It is a type of median filter. In this filter, all the pixel values are sorted out, and then some weight is assigned to each pixel. These filters are also used to remove noise. This method is efficient when the window size is small.

  iv. Center weighted median filter
      It is a type of weighted median filter. Here, all the pixel values are sorted out and the center value is taken. Weighted median filter gives more weight only to the central value of each window. In high noise condition, it produces the blurring effect.

  v. Tristate median filter
     This filter is the combination of standard median filter and center weighted median filter. It helps to reduce impulse noise. Even at high noise condition it preserves image details.

  vi. Adaptive weighted median filter
      This filter is used to remove the speckle noise in images. It is based on the weighted median, which originates from the well-known median filter through the

introduction of weight coefficients. By adjusting the weight coefficients and consequently the smoothing characteristics of the filter according to the local statistics around each point of the image, it is possible to suppress noise while edges and other important features are preserved.

vii. Max filter
In max filter, center pixel value is replaced by maximum value of the gray levels in the neighborhood of the pixel.

viii. Min filter
In min filter, center pixel value is replaced by minimum value of the gray levels in the neighborhood of the pixel.

○ Sharpening spatial filters
The purpose of sharpening is to highlight fine detail in an image or to enhance details that has been blurred.

i. First-order filter
The basic definition of the first-order derivative of a one-dimensional function $f(x)$ is the difference given by Gonzalez and Woods:[13]

$$\frac{\partial f}{\partial x} = f(x+1) - f(x) \quad (22)$$

First-order derivatives generally produce thicker edges in an image. First-order derivatives generally have a stronger response to gray-level step.

ii. Second-order filter
A second-order derivative as the difference is given as follows:

$$\frac{\partial^2 f}{\partial x^2} = f(x+1) + f(x-1) - 2f(x) \quad (23)$$

Second-order derivatives have a stronger response to fine detail, such as thin lines and isolated points. They produce a double response at step changes in gray level.

iii. Laplacian filter
Laplacian for an image $f(x, y)$ of two variables is defined as follows:

$$\nabla^2 f = \frac{\partial^2 f}{\partial x^2} + \frac{\partial^2 f}{\partial y^2} \quad (24)$$

$$\frac{\partial^2 f}{\partial x^2} = f(x+1, y) + f(x-1, y) - 2f(x, y)$$

$$\frac{\partial^2 f}{\partial y^2} = f(x, y+1) + f(x, y-1) - 2f(x, y)$$

$$\nabla^2 f = [f(x+1, y) + f(x-1, y) + f(x, y+1) + f(x, y-1)] - 4f(x, y)$$

Laplacian mask is given as follows:

$$\begin{array}{ccc} 0 & 1 & 0 \\ 1 & -4 & 1 \\ 0 & 1 & 0 \end{array}$$

iv. Unsharp masking
The unsharp masking is given as follows:

$$f_s(x, y) = f(x, y) - \overline{f}(x, y) \quad (25)$$

where $f_s(x, y)$ denotes the sharpened image, and $\overline{f}(x, y)$ is a blurred version of $f(x, y)$.

v. High-boost filtering
A generalization of unsharp masking is called high-boost filtering. The transfer function is as follows:

$$f_{hb}(x, y) = Af(x, y) - \overline{f}(x, y) \quad (26)$$

where $A \geq 1$ and $\overline{f}(x, y)$ is a blurred version of $f(x, y)$.

vi. Robert operator

a. $\begin{array}{cc} -1 & 0 \\ 0 & -1 \end{array}$

b. $\begin{array}{cc} 0 & -1 \\ -1 & 0 \end{array}$

These two operators are called Robert operators.

vii. Sobel operator

a. $\begin{array}{ccc} -1 & -2 & -1 \\ 0 & 0 & 0 \\ 1 & 2 & 1 \end{array}$

b. $\begin{array}{ccc} -1 & 0 & 1 \\ -2 & 0 & 2 \\ -1 & 0 & 1 \end{array}$

These two operators are called Sobel operators.

IMAGE ENHANCEMENT IN FREQUENCY DOMAIN

Image enhancement in the frequency domain is obtained by multiplying the Fourier transform of the input image by the filter transfer function and taking the inverse transform of the multiplied result.

i. Low-pass filter
Low-pass filter is a filter that passes low-frequency signals but attenuates (reduces the amplitude of) signals with frequencies higher than the cutoff frequency.

ii. High-pass filter
High-pass filter is a filter that passes high frequencies well but attenuates (i.e., reduces the amplitude of) frequencies lower than the filter's cutoff frequency. The actual amount of attenuation for each frequency is a design parameter of the filter. It is sometimes called a low-cut filter or bass-cut filter.

iii. Band-pass filter
Band-pass filter is a device that passes frequencies within a certain range and rejects (attenuates) frequencies outside that range, that is, band-pass filter passes only the band frequency.

iv. Notch filter
The filter is called notch filter since it is a constant function with a hole (notch) at the origin.
The filter transfer function is given as follows:

$$H(u,v) = \begin{cases} 0 & \text{if } (u,v) = \left(\dfrac{M}{2}, \dfrac{N}{2}\right) \\ 1 & \text{otherwise} \end{cases} \quad (27)$$

v. Smoothing frequency domain filters
Edges and other sharp transitions (such as noise) in the gray levels of an image contribute significantly to the high-frequency content of its Fourier transform. Hence smoothing is achieved in the frequency domain by attenuating a specified range of high-frequency components in the transform of a given image.

○ Ideal low-pass filter
The simplest low-pass filter that cuts off all high-frequency components of the Fourier transform that are at a distance greater than a specified distance D_0 from the origin of the transform. This type of filter is called as ideal low-pass filter. It has the following transfer function:

$$H(u,v) = \begin{cases} 1 & \text{if } D(u,v) \leq D_0 \\ 0 & \text{if } D(u,v) > D_0 \end{cases} \quad (28)$$

where D_0 is the specified nonnegative quantity. $D(u, v)$ is the distance from point (u, v) to the origin of the frequency rectangle.

○ Butterworth low-pass filters
The transfer function of a Butterworth low-pass filter of order n, and with cutoff frequency at a distance D_0 from the origin, is defined as in the following equation:

$$H(u,v) = \dfrac{1}{1 + \left[D(u,v)/D_0\right]^{2n}} \quad (29)$$

where $D(u, v)$ is given as follows:

$$D(u,v) = \left[(u - M/2)^2 + (v - N/2)^2\right]^{1/2}$$

where M and N are the sizes of the image.

BLPF transfer function does not have a sharp discontinuity that establishes a clear cutoff between passed and filtered frequencies.

○ Gaussian low-Pass filters
The transfer function of a Gaussian low-pass filter with cutoff frequency at a distance D_0 from the origin is defined as follows:

$$H(u,v) = e^{-D^2(u,v)/2D_0^2} \quad (30)$$

where $D(u, v)$ is as follows:

$$D(u,v) = \left[(u - M/2)^2 + (v - N/2)^2\right]^{1/2}$$

where M and N are the sizes of the image.

vi. Sharpening frequency domain filters
The high-pass filtering process attenuates the low-frequency components without disturbing high-frequency information in the Fourier transform.
The transfer function of the high-pass filter is obtained from the transfer function of the low-pass filter as follows:

$$H_{hp}(u,v) = 1 - H_{lp}(u,v) \quad (31)$$

where

$H_{hp}(u,v)$ — transfer function of the high-pass filter

$H_{lp}(u,v)$ — transfer fuction of the low-pass filter

○ Ideal high-pass filters
A two-dimensional ideal high-pass filter transfer function is given as follows:

$$H(u,v) = \begin{cases} 0 & \text{if } D(u,v) \leq D_0 \\ 1 & \text{if } D(u,v) > D_0 \end{cases} \quad (32)$$

where D_0 is the cutoff frequency and $D(u, v)$ is given as follows:

$$D(u,v) = \left[(u - M/2)^2 + (v - N/2)^2\right]^{1/2}$$

○ Butterworth high-pass filters
The transfer function of the Butterworth high-pass filter of order n and with cutoff frequency locus at a distance D_0 from the origin is as follows:

$$H(u,v) = \dfrac{1}{1 + \left[D_0/D(u,v)\right]^{2n}} \quad (33)$$

○ Gaussian high-pass filters
The transfer function of the Gaussian high-pass filter (GHPF) with cutoff frequency locus at a distance D_0 from the origin is given as follows:

$$H(u,v) = 1 - e^{-D^2(u,v)/2D_0^2} \qquad (34)$$

- Unsharp masking
 Unsharp masking is done by obtaining the high-pass filtered image by subtracting from the low-pass filtered version. The transfer function is given as follows:

$$H_{hp}(u,v) = 1 - H_{lp}(u,v) \qquad (35)$$

- High-boost filtering
 Transfer function of high-boost filtering is given as follows:

$$H_{hb}(u,v) = (A-1) + H_{hp}(u,v) \qquad (36)$$

where $A \geq 1$.

- Homoorphic filtering
 In homomorphic filtering, first logarithm of the original image is taken. Then, DFT is applied. After applying DFT, the transfer function $H(u, v)$ is multiplied and inverse DFT of the result is taken. Finally, the enhanced image is obtained by taken exponential of the final result.

MULTISPECTRAL IMAGE ENHANCEMENT

In multispectral imaging, there is a sequence of I images $U_i(m, n)$, $i = 1, 2, 3 \ldots I$, where I is typically between 2 and 12. It is desired to combine these images to generate a single or few images that are representative of their features. There are three common methods of enhancing such images as follows:

i. Intensity ratio
 The intensity ratio is given as follows:

$$R_{i,j}(m,n) \triangleq \frac{u_i(m,n)}{u_j(m,n)}, i \neq j \qquad (37)$$

where $u_i(m, n)$ represents the intensity and is assumed to be positive. This method gives $I^2 - I$ combinations for the ratios, the few most suitable of which are chosen by visual inspection. Sometimes, the ratios are defined with respect to the average image $\left(\frac{1}{I}\right) \sum_{i=1}^{I} u_i(m,n)$ to reduce the number of combinations.

ii. Log ratio
 The log ratio is given in Eq. 38,

$$L_{i,j} \triangleq \log R_{i,j} = \log u_i(m,n) - \log u_j(m,n) \qquad (38)$$

The log ratio $L_{i,j}$ gives a better display when the dynamic range of $R_{i,j}$ is very large, which can occur if the spectral features at a spatial location are quite different.

iii. Principal components
 For each (m, n), define the $I \times 1$ vector
 The autocorrelation of ensemble of vectors is given as follows:

$$u(m,n) = \begin{bmatrix} u_1(m,n) \\ u_1(m,n) \\ \vdots \\ u_I(m,n) \end{bmatrix} \qquad (39)$$

The $I \times 1$ KL transform of $u(m, n)$, denoted by ϕ, is determined for the autocorrelation matrix of the ensemble of vectors $\{u_i(m, n), i = 1, 2 \ldots I\}$. The rows of ϕ, which are eigenvectors of the autocorrelation matrix, are arranged in decreasing order of their associated eigenvalues. Then for any $I_0 \leq I$, the images $v_i(m, n)$, $i = 1, 2 \ldots I_0$ are obtained from the KL transformed vector as follows:

$$v(m,n) = \phi_{u(m,n)} \qquad (40)$$

They are the first I_0 components of the multispectral images.

FALSE COLOR, TRUE COLOR, AND PSEUDO COLOR

i. False color
 False color implies mapping a color image into another color image to provide a more striking color contrast to attract the attention of the viewer.
ii. True color
 Joshi[14] discussed about true color mode. True color representation mode attempts to display a colored view of a scene, keeping as faithfully as possible to the colors in the original.
iii. Pseudo color
 Pseudo color refers to mapping a set of images $u_i(m, n)$, $i = 1, 1 \ldots I$ into a color image. Usually, the mapping is determined such that different features of the data set can be distinguished by different colors. Thus, a large data set can be presented comprehensively to the viewer.

COLOR IMAGE ENHANCEMENT

Color image enhancement may require improvement of color balance or color contrast in a color image. Enhancement of color images become a more difficult task not only because of the added dimension of the data but also due to the added complexity of color perception. The input color coordinates of each pixel are independently transformed into another set of color coordinates, where the image in each coordinate is enhanced by its own monochrome or grayscale image enhancement algorithm.

APPLICATION OF IMAGE ENHANCEMENT

i. Image enhancement on forensic application
The problem with collecting forensic evidence at a crime scene is that the evidence is often masked behind backgrounds. This makes it difficult for extracting key components from the evidence. Often times, the background color of the crime scene can overpower the faint detail of the evidence. Types of evidence that this can occur on is with finger prints and shoe prints at the crime scene. To correct this problem, image enhancement techniques can be used to obtain the relevant information that is needed for the investigators.

ii. Underwater image enhancement
Iqbal et al.[15] described an approach based on slide stretching algorithm to enhance the underwater images in which the clarity of the images is degraded by light absorption and scattering and as a result one color dominates the image. This involves contrast stretching of RGB algorithm which is applied to equalize the color contrast in images. Further, saturation and intensity stretching of HSI(Hue Saturation Intensity) is used to increase true color and solve the problem of lighting in images. Interactive software is developed for under water image enhancement and quality of images which is statistically illustrated through histograms.

iii. Image enhancement of low light scenes with infrared flash images
Matsui et al.[16] developed a technique for enhancing an image using near infrared flash images. In this technique, near infrared flash images are effectively used in removing annoying effects in images of dimly lit environments like image noise and motion blur where dual bilateral filters are used to decompose the color image into a large-scale image and a detail image.

iv. Medical image enhancement
Medical imaging technologies play more and more important roles not only in the diagnosis and treatment of diseases but also in disease prevention, health checkup, major disease screening, health management, early diagnosis, disease severity evaluation, choice of treatment methods, treatment effect evaluation, and rehabilitation. The status of medical imaging technologies has increased continuously in health care applications.

Due to its ability to make the diagnosis and treatment of diseases, image guided surgery, and other medical links more timely, accurate, and efficient, medical image enhancement has become a routine task. Through producing excellent tissue uniformity, optimized contrast, edge enhancement, artifact elimination, intelligent noise reduction, and so forth, cutting-edge image enhancement helps doctors to accurately interpret medical images, a crucial foundation for better diagnosis and treatment.

v. Image enhancement in remote sensing
Digital image processing plays a vital role in the analysis and interpretation of remotely sensed data. Especially data obtained from satellite remote sensing, which is in the digital form, can best be utilized with the help of digital image processing. Image enhancement and information extraction are two important components of digital image processing. Image enhancement techniques help in improving the visibility of any portion or feature of the image, suppressing the information in other portions or features. Information extraction techniques help in obtaining the statistical information about any particular feature or portion of the image.

vi. Biometric image enhancement
Biometrics is a rapidly evolving technology that is being widely used in forensics, such as criminal identification and prison security, and it has the potential to be used in a large range of civilian application areas. Biometrics can be used to prevent unauthorized access to automatic teller machines, cellular phones, smart cards, personal computers, workstations, and computer networks. In automobiles, biometrics can replace keys with keyless entry devices. The major areas of palm print application are described as follows: palm print can be used for identification cards issued by government like passports, national identification cards, voter cards, driving licenses, social services, and so on. It can be used for airport security, boarding passes, and commercial driving licenses.

Personal identification system is essential in day-to-day life. Biometrics plays a vital role in this field in which fingerprint, iris, face, voice, and palm are used to recognize individuals. The input image may be distorted due to reasons such as wet skin, poor illumination, and discontinued lines and ridges Also, the input image may be of low contrast. This may affect the process of personal identification. Hence, it is necessary to acquire noise-free fine details from the palm. This can be easily done once the image is enhanced. So image enhancement is a very important step in the process of personal identification. Palanikumar et al.[17] described a method for palm print enhancement.

vii. Image and video enhancement in display system
To improve the visual quality and remove artifact in the input signal, image enhancement algorithms are used. These improve the image and video quality and provide a pleasant view to the user.

CONCLUSION

Image enhancement plays a crucial role in preprocessing of images. Several techniques for preprocessing including spatial and frequency domain methods are discussed here. Several types of HE-based methods are discussed. HE-based methods are mostly used in medical image preprocessing. Few types of color image enhancement methods are also described. Finally, applications of image enhancement in various fields are given.

REFERENCES

1. Pratt, W.K. *Digital Image Processing*, 4th Ed.; PIKS Scientific Inside: New Jersey, 2012; Jain, A.K. *Fundamentals of Digital Image Processing*; Prentice Hall of India Learning Pvt. Ltd.: New Delhi, 2010.
2. www.eie.polyu.edu.hk/~enyhchan/imagee.pdf.
3. Gonzalez, R.C.; Woods, R.E. *Digital Image Processing*, First ISE reprint; Addison-Wesley: Reading, MA, 1998.
4. Menotti, D.; Najman, L.; Facon, J.; Araujo, A.A. Multi-histogram equalization methods for contrast enhancement and brightness preserving. IEEE Trans. Consum. Electron. **2007**, *53* (3), 1186–1194.
5. Yang, S.; Oh, J.; Park, Y. Contrast enhancement using histogram with bin underflow and bin overflow. In *Proceedings of International Conference on Image Processing*, Barcelona, Spain, 2003, 881–884.
6. Wang, Q.; Ward, R.K. Fast image/video contrast enhancement based on weighted thresholded histogram equalization. IEEE Trans. Consum. Electron. **2007**, *53* (2), 757–764.
7. Lu, C.-H.; Hsu, H.-Y.; Wang, L. A new contrast enhancement technique by adaptively increasing the value of histogram. In *IEEE International Workshop on Imaging Systems and Techniques*, 2009, 407–411.
8. Kim, Y. Contrast enhancement using brightness preserving bi-histogram equalization. IEEE Trans. Consum. Electron. **1997**, *43* (1), 1–8.
9. Wan, Y.; Chen, Q.; Zhang, B. Image enhancement based on equal area dualistic sub-image histogram equalization method. IEEE Trans. Consum. Electron. **1999**, *45* (1), 68–75.
10. Chen, S.; Ramli, A.R. Minimum mean brightness error bi-histogram equalization in contrast enhancement. IEEE Trans. Consum. Electron. **2003**, *49* (4), 1310–1319.
11. Chen, S.; Ramli, A.R.; Contrast enhancement using recursive mean-separate histogram equalization for scalable brightness preservation. IEEE Trans. Consum. Electron. **2003**, *49* (4), 1301–1309.
12. Sim, K.S.; Tso, C.P.; Tan, Y.Y. Recursive sub-image histogram equalization applied to gray-scale images. Pattern Recognit. Lett. **2007**, *28*, 1209–1221.
13. Gonzalez, R.C.; Woods, R.E. *Digital Image Processing*, 2nd Ed.; Fourth Indian Reprint; Pearson Education: New Delhi, 2003.
14. Joshi, M.A. *Digital Image Processing an Algorithmic Approach*; Prentice Hall of India learning Pvt. Ltd.: New Delhi, 2006.
15. Iqbal, K.; Salam, R.A.; Osman, A.; Tailb, A.Z. Underwater image enhancement using an integrated colour model. IAENG Int. J. Comput. Sci. **2007**, *34* (2), 239–244.
16. Matsui, S.; Okabe, T.; Shimano, M.; Sato, Y. Image enhancement of low light scenes with near-infrared flash images. IPJ Trans. Comput. Vision Appl. **2010**, *2*, 215–223.
17. Palanikumar, S.; Sasikumar, M.; Rajeesh, J. Palmprint enhancement using recursive histogram equalization. Imaging Sci. J. **2013**, *61*, 447–457.

Image Inpainting Evolution: A Survey

Bombaywala Md. Salman R.
Uka Tarsadia University, Bardoli, Surat, India

Chirag N. Paunwala
Sarvajanik College of Engineering and Technology, Surat, India

Abstract

Image inpainting is the process of reconstructing or altering the image in an undetectable manner. The goals and applications of inpainting are numerous—from restoration of damaged paintings and photographs to the removal of selected objects. Inspired by the professional art restorers, image inpainting techniques have evolved in the recent past. This entry presents the evolution of image inpainting techniques classified in different categories. For each category, appropriate classification have been made based on the fundamental technique used for inpainting. The pioneer techniques for each category are described in detail. Possible variations in each category are described for clear understanding of a reader. Appropriate comparison has been done in order to highlight the advantages of the techniques over other.

Keywords: Diffusion; Exemplar; Inpainting; Sparse; Texture synthesis; Total variation.

INTRODUCTION

Image inpainting is the process of restoring the images or the damaged area in the images in such a manner that the restoration or the change is undetectable. Roots of the technique lie way back in time where restorations of images were done by skilled artists. The purpose to retouch the image extended from paintings to photographs and video. In the digital domain, the problem of inpainting appeared as "error concealment" in telecommunications, where the need was to fillin image blocks that were lost during data transmission. While inpainting is related to noise removal on a broader sense, it is fundamentally a different problem. Apart from filling in the blank area in the image, inpainting finds its application in various fields like restoration of image from scratches and texts, object removal for editing, loss recovery of damaged images in transmission, etc.(see Fig. 1).

The concept of inpainting was first introduced by Bertalmio et al.[1] Since then, various algorithms and techniques have been developed. Apart from restoration, removal, and replacement, Chan and Shen[2] introduced its application in digital zoom-in which is useful in variety of applications such as image super resolution and data compression. As edges represent important information in images, inpainting has found its application in edge coding–decoding and compression based on edge information.[2] For compression, repeated patterns are drawn from an image, and Liu et al.[3] showed how inpainting can be used with image patches for compression. Inpainting is also used in extracting images from superimposed text. Modification in images has created severe problems with the security of digital images. The domain of digital forensic and investigation has emerged in recent years[4] to deal with this authenticity problems. Red-eye effect due to flash light passing through pupil can be detected and corrected using inpainting in digital photographs[5] (see Fig. 2). Image inpainting targets to fill a completely blank area in the image by using the information from the known part, so as to complete the image with visual plausibility. Image inpainting methods assume a similarity between pixels in known part and unknown part of the image. This similarity is utilized locally and globally to obtain the inpainted image.

The existing image inpainting methods can be divided into three categories. The first category of methods uses partial differential equations (PDEs) to diffuse the information from the exterior of the unknown region toward the interior, thus imitating the methods used by professional restorers. The information is propagated with smoothness constraints, analogous to the heat flow in structures. This flow of heat can be modeled by a differential equation using multivariate unknown functions and its partial derivative—PDE. The first category inpainting algorithm was pioneered by Bertalmio et al.[1] where they used the differential equations and inpainted structures using a repetitive algorithm. The major drawback of this approach was the number of diffusion steps required in between inpainting steps which was somewhat overcomed by fast digital inpainting algorithm derived by Oliveira et al.[6] at the cost of quality. Chan and Shen[2] later derived mathematical model based on the Ref. [7] total variation

Encyclopedia of Image Processing, First Edition
DOI: 10.1201/9781351045636-140000131
Copyright © 2018 by Taylor & Francis. All rights reserved.

Fig. 1 Scratch removal and text removal examples of image inpainting

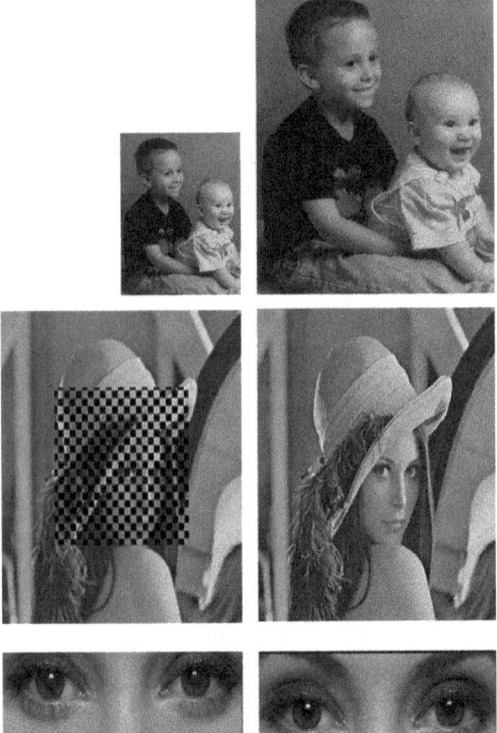

Fig. 2 Digital image zoom-in (top), image compression (middle), and red eye removal (bottom) examples of image inpainting

(TV) model, which was again improved by them with curvature-driven diffusion (CDD), solving the connectivity problem. These methods are well suited for completing straight structures, curved structures, and inpainting small regions. All these PDE-based methods have a common problem of inability of restoring texture information and tendency to blur large holes (see Fig. 3).

The second category of inpainting methods is concentrated on textures and minute details in the image. Synthesizing texture from the limited source aims at creating a texture that has similar appearance and structural properties to the original sample. Efros and Leung[8] developed a method which aimed to maintain local structure as much as possible. Texture synthesis methods can be divided into two categories. The first category is single pixel based, in which one pixel is synthesized at a time.[8] It is obvious that these methods are more time consuming. To overcome this drawback, patch-based methods, in which blocks of texture images also known as exemplars—available from the known part of image, are used to estimate blocks of pixels instead of single pixel at a time.[9–11] With the advent of sparse representations and optimization theories, sparsity-based techniques, the second category, have been developed for solving the inpainting problem.[12–14] It is assumed that the image can be represented sparsely in a given basis and that the known and unknown parts also have sparse representations. As shown in Fig. 4, exemplar based-methods outperform PDE-based methods when inpainting textures in an image. However, the computational cost of these methods is more as the search for the best match has to be made locally or globally.

The third category of image inpainting methods is the hybrid category which utilizes methods from both the above categories to incorporate their inherent advantages. Natural images contain structures (edges, corners, etc.) as well as textures (homogeneous patterns). Hybrid approaches separate the image into structure and texture components and use appropriate method for inpainting each component. Once inpainted; the components are combined to obtain the final result. Various methods based on bounded variational functions,[2] PDE,[1] discrete cosine transform (DCT), and many more were developed to separate the structure and texture components. The hybrid methods are well suited for natural images and provide the best results. The hybrid methods can inpaint structure as well as texture much better compared to other two categories.

General Inpainting Problem

Consider an image to be inpainted—I can be assumed to be made up of two major components. A—let us call the region which is to be inpainted as the target/unknown region Ω and the remaining part as the source/known region S. Hence, $I = \Omega \cup S$. For every pixel location (x,y) in image I, the pixel $p(x,y)$ will be value representing gray value in case of grayscale image and $p(x,y)$ will be a vector containing the values of R,G, and B components in case of color image.

The target of inpainting is to estimate the color/grayscale values of all the pixels belonging to Ω by using

Image Inpainting Evolution: A Survey

Fig. 3 Result of inpainting large object by using the diffusion-based technique. Notice the blur introduced by diffusion process

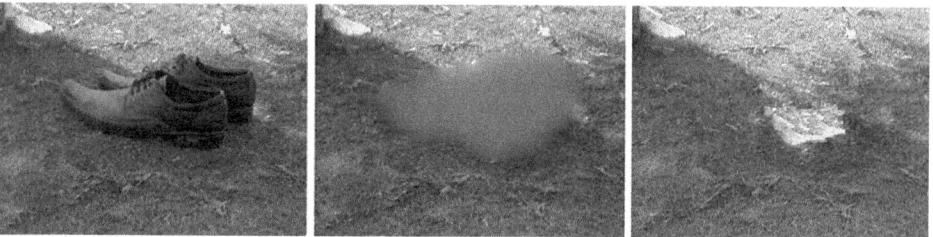

Fig. 4 Result of inpainting large object by using the diffusion-based technique (center) and exemplar-based method (right)

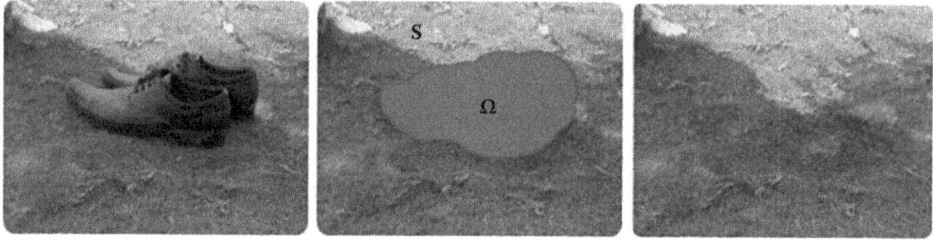

Fig. 5 The inpainting problem. Original image (left), target region (Ω), and source region (S), and inpainted image (right)

the values of pixels located in the region S (see Fig. 5) in such a manner that the final image looks natural to human eye.

FIRST CATEGORY: PDE-BASED DIFFUSION METHODS

The process of diffusion of information in the target region can be imitated by the flow of heat along the structure. The process of regularization diffuses the pixel values from the source to the target using PDEs or by minimizing a function representing a global variation in the image. The process of regularization propagates the information by smoothening and hence tends to remove the high frequency details in the image. This is undesirable for obtaining plausible visual output. Hence, a constraint must be imposed on the amount and direction of smoothening. If the pixel is located along the boundary of a structure, then the smoothening must be applied along the boundary and not across the boundary. If the pixel is located in a homogeneous region, smoothening can be done in omni direction. It therefore becomes necessary to extract the boundary and edges information of the image before applying PDE-based diffusion methods for the problem of image inpainting.

Anisotropic Diffusion

Since amount of smoothening needs to be controlled with respect to the image characteristics, isotropic diffusion (diffusion same in all the directions) is not a good approach for inpainting problem. Perona and Malik[14] presented the concept of anisotropic diffusion by introducing a diffusion coefficient in the heat equation. For an image I, the diffusion equation was defined as

$$\frac{\partial I}{\partial t} = div\big(c(x,y,t)\nabla I\big) = c(x,y,t)I + \nabla c \cdot \nabla I \qquad (1)$$

where *div*() stands for divergence which indicates how fast the image intensities changes in *x* and *y* directions; ∇ and Δ represent the gradient and Laplacian operators, respectively; and *c* is the diffusion coefficient which controls the rate of diffusion and is generally chosen as a function of the image gradient so that the image edges are preserved. The goal of *c* is to avoid smoothing across the region boundaries. Two functions were proposed for computing the diffusion coefficient:

$$c(\|\nabla I\|) = e^{-(\|\nabla I\|/K)^2} \quad (2)$$

and

$$c(\|\nabla I\|) = \frac{1}{1+(\|\nabla I\|/K)^2} \quad (3)$$

where *K* controls the edge sensitivity and is chosen experimentally. The function *c* will have a value of 1 in the smooth regions, while it will have a value close to zero around the edges. This way the diffusion process will be constrained and details of the images can be preserved during regularization (smoothing).

Diffusion-Based Image Inpainting

Bertalmio et al.[1] pioneered the use of anisotropic diffusion for the process of image inpainting. The authors propagated image Laplacians from the source region toward the interior of target region iteratively. The propagation direction was determined by the direction of isophotes estimated by finding the normal to image gradient at each point. The isophotes gives the direction of minimum change in pixel intensities. After every few steps of iteration, anisotropic diffusion was applied to periodically curve the lines so that they do not overlap.

As shown in Fig. 6, consider the original image is *I*, target region is denoted as Ω, the boundary of target region is denoted as $\Delta\Omega$, and the source region is denoted as *S*. Obviously, $S = I - \Omega$.

The algorithm works in an iterative way by taking the input image *I* and generating the set of images $I_1, I_2, I_3, \ldots, I_n$ in such a way that the new image is the improvement over the previous image. The process is described by the following equation:

$$I^{n+1}(i,j) = I^n(i,j) + \Delta t I_t^n(i,j), \forall (i,j) \in \Omega \quad (4)$$

where the super index *n* denotes the inpainting time. (*i,j*) are the pixel coordinates, Δt is the rate of improvement, and $I_t^n(i,j)$ is the update of image $I^n(i,j)$. The preceding equation runs only inside the region to be inpainted, Ω. The update for each iteration $I_t^n(i,j)$ can be calculated as follows:

$$I_t^n(i,j) = \left(\delta \vec{L}^n(i,j) \cdot \frac{\vec{N}(i,j,n)}{|\vec{N}(i,j,n)|}\right) |\nabla I^n(i,j)| \quad (5)$$

where $\delta \vec{L}^n(i,j)$ is an estimate of the smoothness of the image at pixel (*i,j*) and is calculated as the neighbors of the discrete implementation of the Laplacian of the pixel *i,j*.

$$\delta \vec{L}^n(i,j) := (L^n(i+1,j) - L^n(i-1,j),$$
$$L^n(i,j+1) - L^n(i,j-1)) \quad (6)$$

$$L^n(i,j) = I_{xx}^n(i,j) + I_{yy}^n(i,j) \quad (7)$$

The next term deals with the direction of the isophotes obtained by calculation perpendicular to the gradient at a point. We know that the gradient vector gives the direction of the largest rate of change while its 90° rotation will give the direction of smallest change. Hence, vector $\vec{N}(i,j,n) = \nabla^{\perp} I^n(i,j)$ gives the direction of isophotes. The gradient vector $\nabla I^n(i,j)$ can be calculated as $(I_x^n(i,j), I_y^n(i,j))$, and the orthogonal to this vector is $(-I_y^n(i,j), I_x^n(i,j))$. The normalized version of the equation can be calculated as follows:

$$\frac{\vec{N}(i,j,n)}{|\vec{N}(i,j,n)|} := \frac{(-I_y^n(i,j), I_x^n(i,j))}{\sqrt{(I_x^n(i,j))^2 + (I_y^n(i,j))^2}} \quad (8)$$

The last term of the equation is calculated as slope limited version of the norm of the gradient of the image. The equation is as follows:

$$|\nabla I^n(i,j)| = \begin{cases} \sqrt{(I_{xbm}^n)^2 + (I_{xfM}^n)^2 + (I_{ybm}^n)^2 + (I_{yfM}^n)^2}; when \beta^n > 0 \\ \sqrt{(I_{xbM}^n)^2 + (I_{xfm}^n)^2 + (I_{ybM}^n)^2 + (I_{yfm}^n)^2}; when \beta^n < 0 \end{cases} \quad (9)$$

where

$$\beta^n(i,j) = \left(\delta \vec{L}^n(i,j) \cdot \frac{\vec{N}(i,j,n)}{|\vec{N}(i,j,n)|}\right) \quad (10)$$

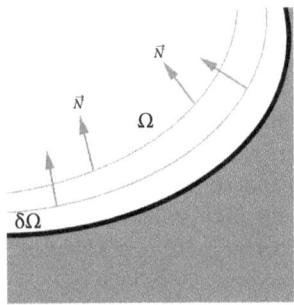

Fig. 6 Propagation direction ($\vec{N}$) is the normal to the signed distance to the boundary of the region to be inpainted

Image Inpainting Evolution: A Survey

Here, β is the projection of $\delta\vec{L}$ onto the vector $\vec{N}$. Sub indexes b and f denote the backward and forward differences, respectively, and m and M denote the minimum and maximum respectively, between derivative and zero. The equation is used to prevent too small or too big numbers. To ensure correct direction field, diffusion process is applied at regular intervals during the process of inpainting. That is after every few steps of inpainting, anisotropic diffusion[14] is done to periodically curve the lines so that they do not overlap. Anisotropic diffusion also removes noise from the image without removing the edge information. The equation for the continuous time diffusion can be discretized and used for the same:

$$\frac{\partial I}{\partial t}(x,y,t) = g_e(x,y)\kappa(x,y,t)|\nabla I(x,y,t)|, \forall (x,y) \in \Omega^e \quad (11)$$

where Ω^e is a dilation of Ω with a ball of radius e, κ is the Euclidean curvature of isophotes of I, and $g_e(x,y)$ is a smooth function in Ω^e such that $g_e(x,y) = 0$ in $\Delta\Omega^e$ and $g_e(x,y) = 1$ in Ω. The algorithm first runs one step of anisotropic diffusion on the whole image. This is to prevent noise from influencing the calculation of isophote lines. Proceeding this, A steps of inpainting followed by B steps of diffusion are applied. The process is continued till the process of inpainting converges, that is, image Laplacian remains constant in the direction of isophotes.

The above mentioned method requires iterative calculations which are time-consuming. Efforts to improve the Bertalmio et al.[1] method resulted into a quick and simple inpainting algorithm presented by Oliveira et al.[6] They used the idea of "diffusion barriers." Diffusion barriers are boundaries for the diffusion process inside Ω. Process starts with Ω with clearance of color information. Ω is convolved repeatedly with diffusion kernel. The boundary is one pixel wide. The kernel used was a weighted average kernel as shown in Fig. 7. The center of the kernels are zero so that only neighborhood pixels contribute toward the results. Fig. 8 shows the result of inpainting using the algorithm presented by Oliveira et al.[6]

Diffusion-based inpainting is suitable to inpaint small holes, smooth regions in an image, and inpainting strong structures. Although provision is made to preserve edges, after few diffusion iterations, the image tends to become blurred specially while filling large holes. These methods are not well suitable for inpainting textures and details in an image.

Fig. 7 Two diffusion kernels used with the algorithm: $* = 0.073235$, $\# = 0.176765$, $\blacklozenge = 0.125$

ORIGINAL IMAGE

IMAGE AFTER INPAINTING

Fig. 8 Result of inpainting using the Oliveira et al.[6] algorithm

Total Variational Image Inpainting

The problem of image inpainting can also be viewed as a variational problem by considering the image as a function of a bounded variation (BV). The Rudin et al.[7] TV model designed for denoising was extended to image inpainting application. The idea is to find a function of BV such that the TV inside the target region is minimized. Restoration model should follow locality, restoration of narrow broken edges, and robustness to noise.

Inpainting problem, in this case, as shown in Fig. 9, is to inpaint domain D with estimation of image information u on $(E \cup D)$, where E is an extended ring of inpainting domain. Now, an unconstrained variational inpainting problem is solved by Eq. 12. The first integral represents the regularization term, while the second term checks the similarity of reconstructed image with the input image based on the known region. The Euler–Lagrange equation for J_λ is given in Eq. 13, where λ_e is the extended Lagrange multiplier and is zero for pixels inside inpainting domain D.

$$J_\lambda[u] = \int_{E \cup D} |\nabla u| dx dy + \frac{\lambda}{2} \int_E |u - u^0|^2 dx dy \qquad (12)$$

$$\frac{\partial u}{\partial t} = 0 = \nabla \cdot \frac{\nabla u}{|\nabla u|} + \lambda_e(u^0 - u) \qquad (13)$$

This TV inpainting model can restore sharp edges but it fails to follow connectivity principle. The vision psychology shows that human observers prefer the connected result.[15] This problem is shown in Fig. 10. In attempt to improve this problem, same authors proposed modification[15] in the TV model and the resulted method is called "curvature-driven diffusion (CDD)." The modified TV inpainting equation is as follows:

$$-\nabla \left(\frac{g(|\kappa|)}{|\nabla u|} \nabla u \right) + \lambda_e(u^0 - u) = 0 \qquad (14)$$

where scalar curvature κ can be calculated as follows:

$$\kappa = \nabla \left[\frac{\nabla u}{|\nabla u|} \right] \qquad (15)$$

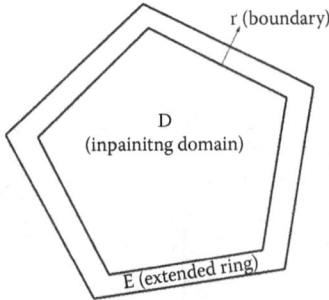

Fig. 9 Inpainting problem for the variational approach

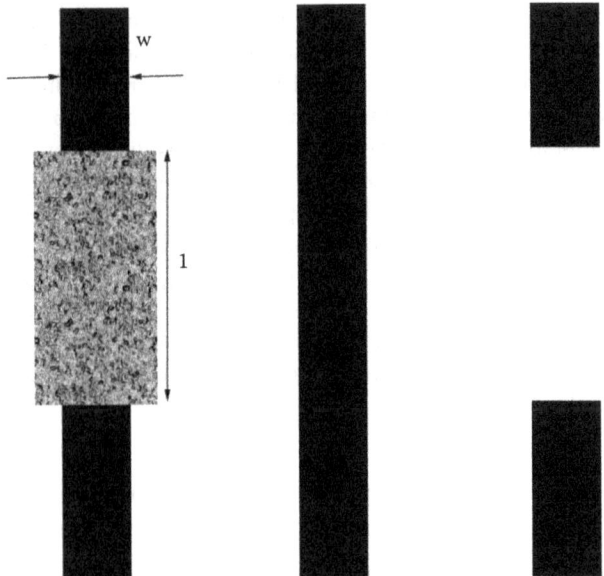

Fig. 10 What is behind the box (when $l > w$)? (left), answer by most humans (middle), answer by TV inpainting model (right)

and g is an annihilator of large curvatures and stabilizer for small curvatures. Results of TV model and CDD model are shown in Figs. 11 and 12 with comparison in terms of number of iterations (Q), PSNR (P), and SSIM (S). Figs. 11 and 12 show results of inpainting using TV and CDD algorithm, respectively. Number of iterations required to get visually good results are much higher in case of TV than CDD. Moreover, it can be seen that PSNR value may be obtained quite low while getting good result, and this is not the case for evaluation using SSIM. Though connectivity problem is solved, TV and CDD models, as in case of diffusion-based techniques, these methods cannot recover texture information.

The PDE-based techniques provide satisfactory results to remove noise and small holes in an image. However, these methods have been suffering from a common problem of texture restoration. For large holes in an image, the inpainting results tend to produce blurs. The methods are well suited for completing straight structures and curved structures.

SECOND CATEGORY: EXEMPLAR-BASED METHODS

The second category of image inpainting method developed from the very idea of synthesizing textures. The texture synthesis problem is a little different from the inpainting problem. As shown in Fig. 13, texture synthesis aims at creating a texture from a given sample such that the created texture has similar appearance and structural properties.

Texture synthesis method proposed by Efros and Leung[8] developed the pixel-based synthesis technique

Image Inpainting Evolution: A Survey

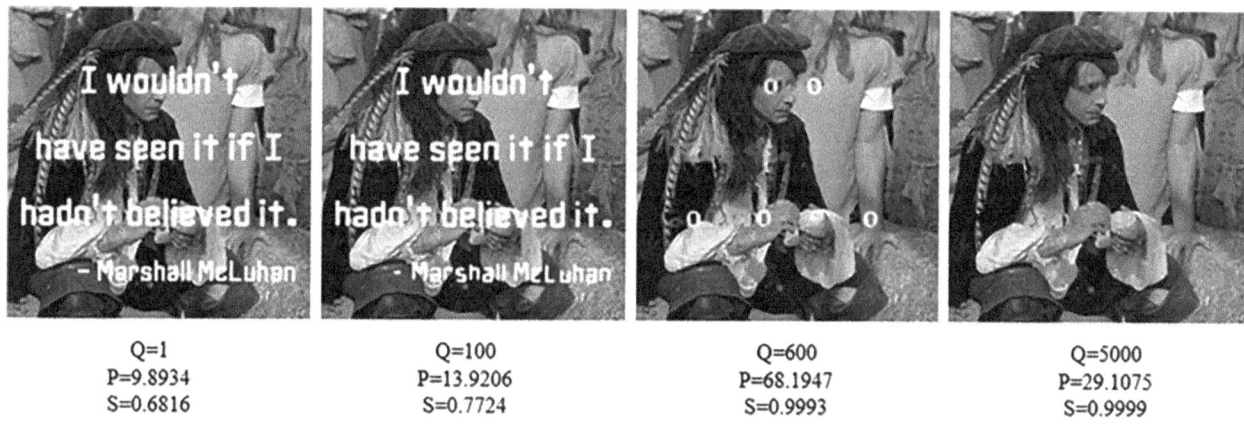

Fig. 11 Inpainting of image using the TV algorithm

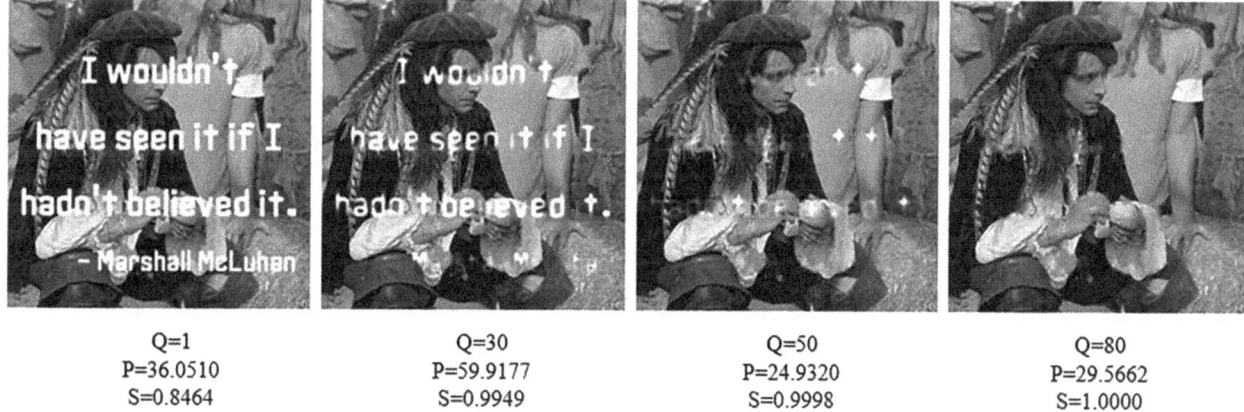

Fig. 12 Inpainting of pirate image using the CDD algorithm

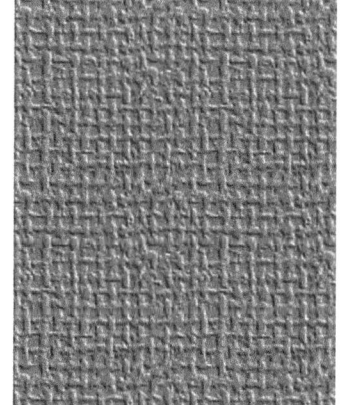

Fig. 13 The texture synthesis problem. Sample texture (left); synthesized texture (right)

which depend on Markov random field modeling. The method uses an already available texture as a source to generate artificial texture by expanding from borders. It runs pixel by pixel by copying the central pixel of the best-matched patch found from the source texture based on the known neighborhood of the pixel to be synthesized.

Texture synthesis method can be applied to image inpainting problem by considering the target region, the area to be synthesized, and the known part as the known texture.

Given an image I, the texture synthesis method proceeds as follows. As shown in Fig. 14, for a pixel $p(x,y)$ at location (x,y), let ψ_p be the patch around the pixel p. This patch contains a known part ψ_p^S and an unknown part ψ_p^U. The method searches for a patch ψ_m around a pixel m such that the known part of both the patches ψ_p and ψ_m matches the most. The pixel p is then replaced by m. Hence, the image is created pixel by pixel from known to unknown part.

Since the method uses the pixel-by-pixel approach, it is computationally costly and complex. The complexity can be reduced by reducing the search space for best match.[16] In order to overcome such a drawback, methods to synthesize entire patch rather than a single pixel from the source region[17] are used. In such a case, instead of copying a pixel m, the entire patch ψ_m is copied in place of ψ_p. The method introduces artifact at boundaries between the patches of textures. To overcome this, Efros and Freeman[9] proposed a method of image quilting which finds the optimal patch (seam) in the overlapped region of neighboring patches by minimizing the boundary cut error.

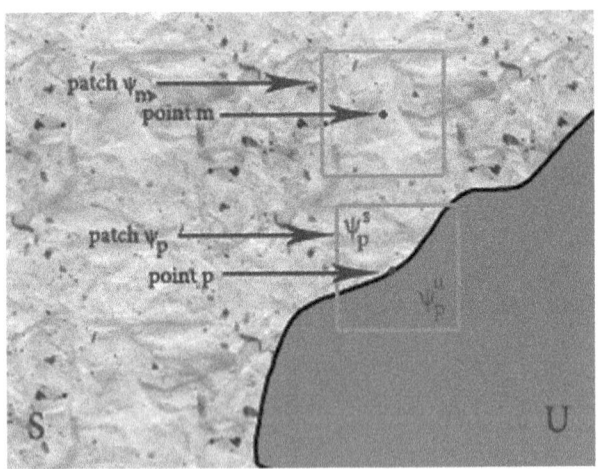

Fig. 14 Nomenclature used for exemplar-based image inpainting

The methods that use the patch-based approach for synthesizing are now referred to as exemplar-based inpainting. Since they replace the entire patch, the methods are faster than the pixel-by-pixel approach.

Exemplar-Based Image Inpainting

Exemplar-based image inpainting approach was pioneered by Criminisi et al.[11] The authors identified that patch-based texture synthesis techniques inherit the ability to inpaint structures as well as textures. However, there was no priority assigned to order the filling process. The authors presented an approach to prioritize the candidate patches such that edges, borders, and other structural information within the image to maintain during the process of inpainting. As shown in Fig. 15, the algorithm searches for a patch with the best match, and the entire patch is replaced to obtain the inpainted image.

The algorithm requires the target region to be marked in advance. Once the region is marked, the boundary $\Delta\Omega$ between the source region S and the target region Ω is identified (use edge detection/high-pass filters). For every point p on the detected boundary, a confidence term and a data term is calculated according to Eqs. 16 and 17, and final priority is calculated as per Eq. 18. The point having the maximum priority is chosen as the initial point for starting the inpainting procedure.

$$C(p) = \frac{\Sigma_{q \in \psi_p \cap (I-\Omega)} C(q)}{|\psi_p|} \quad (16)$$

$$D(p) = \frac{|\nabla I^\perp \bullet n_p|}{\alpha} \quad (17)$$

$$P(p) = C(p) \bullet D(p) \quad (18)$$

Equation 16 represents the confidence term which ensures that the selected patch has enough number of known pixels, so that a proper and justified match can be found from the source region. Equation 17 represents the data term which ensures that the point p lies on the most prominent structure. The function depends on the strength of isophote hitting the boundary which ensures that linear structures are inpainted first. Priority is calculated by multiplying confidence and data terms.

Once the priorities are calculated, the algorithm works iteratively, filling in the patches from boundary toward the center of the target region. For every patch defined on $\Delta\Omega$, a candidate patch is found from the source region by calculating the similarity of the known part ψ_p^s using the sum of squared differences (SSD). The patch with highest similarity is chosen to inpaint the target patch. The values of confidence and data terms are then updated at the fill front, and the algorithm repeats until all the pixels of target region are inpainted.

The result of inpainting used by Criminisi et al.[11] is shown in Fig. 16e. As can be seen in Fig. 16a–c, the

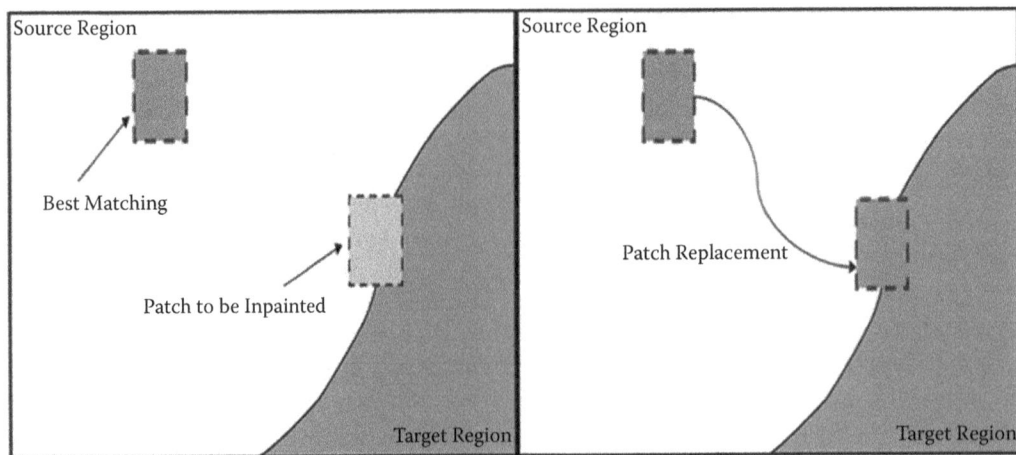

Fig. 15 Exemplar-based image inpainting problem

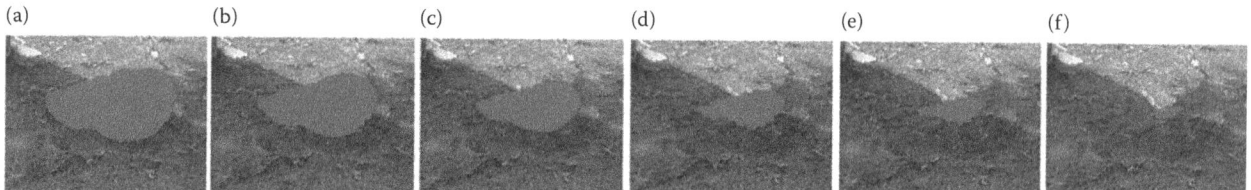

Fig. 16 Result of exemplar-based image inpainting. Notice how the inpainting starts from the boundary of the target region and proceeds toward the center (a)–(e). (f) Result of diffusion-based inpainting

inpainting approach starts from the boundary and propagates toward the center of the target region. Many variations are done to the above mentioned method to obtain better results in terms of quality and processing time.

Variation in Patch Filling Order

Cheng et al.[18] observed that the confidence term proposed by Criminisi et al.[11] tends to approach zero value quickly as the process of filling proceeds, which makes the computed priority almost undistinguishable. Hence, authors proposed addition of data term and confidence term. The authors also found that the data term was stable enough while the confidence term decays too fast, and hence, instead of using the original confidence term, they used a regularized version.

Wu and Ruan[19] presented a data term which used TV model to calculate the isophotes across $\Delta\Omega$ as opposed to the conventional technique which calculates the data term along the direction of isophotes. The data term presented is given as follows:

$$D(p) = \frac{1}{|\nabla u|}|u_{\xi\xi}| \quad (19)$$

where $u_{\xi\xi}$ represents the second-order directional derivative of image in the direction along the edge, and $|\nabla u|$ is the change of linear structure along the normal direction. This data term considers the extent of edge and hence is able to recover linear structure with more precision.

Isophotes are calculated on the basis of gradients of the image which may give false result in the presence of noise. To overcome this, Goswami and Paunwala[20] proposed an infinite symmetrical elliptical filter-based data term. This data term smoothes out the noise inherent and finds structure information better compared to the above mentioned methods.

LeMeur et al.[21] proposed a tensor-based data term based on Di Zenzo's structure tensor computed as

$$J = \sum_{i=1}^{n} \nabla I_i \nabla I_i^T \quad (20)$$

on each color channel of the image I. There are several advantages of structure tensor over the gradient field. Structure tensor improves the robustness to noise and local singularities. Based on the eigenvalues of the structure tensor variation, the degree of anisotropy of a local region is calculated. The tensor based data term is then given as

$$D(p_x) = \alpha + (1-\alpha)\exp\left(-\frac{\eta}{(\lambda_1 - \lambda_2)^2}\right) \quad (21)$$

where $\eta \geq 0$ and $\alpha \in [0,1]$. On the flat region ($\lambda_1 \approx \lambda_2$), no direction is favored for the propagation. When $\lambda_1 \gg \lambda_2$, it indicates the presence of a structure and the data terms tend to be 1.

An image consists of both structure and texture components, with structure components being sparsely distributed within the image compared to texture components. For example, two corners of a table will be very sparse as compared to the texture on the table. Taking this into consideration, Xu and Sun[12] introduced a new data term based on sparsity measured as

$$\rho(p) = \|w_{p_x}\|_2 \sqrt{\frac{N_s}{N}} \quad (22)$$

where N_s and N are the number of valid candidate patches in a search window defined around the point p. A large value of ρ represents a small similarity with neighbors, which indicates that the point belongs to a structure.

In order to overcome the problem of confidence term being decaying to zero, Hesabi and Mahdavi[13] improved the scheme proposed by Xu and Sun[12] by modifying the priority term as follows:

$$P(p) = \alpha T_{[\xi,1]}(\rho(p)) + \beta T'_{[\gamma,1]}(C(p)) \quad (23)$$

where all the terms are defined similar to as proposed by Xu and Sun.[12] The transformation T' is a linear transformation to control the decreasing rate of confidence term. The priority term is also made additive rather than multiplicative as multiplication is sensitive to extreme values.

Variation in Methods for Best Match Patch

The most widely used metric to search for a best match patch in the source region is SSD. However, SSD favors to copy the pixels of uniform regions. To overcome this, Bugeau et al.[22] introduced a weighted Bhattacharya distance.

$$d_{(SSD,BC)}(\psi_t,\psi_s) = d_{SSD}(\psi_t,\psi_s) \cdot d_{BC}(\psi_t,\psi_s) \quad (24)$$

The term d_{BC} is the modified Bhattacharya distance given by

$$d_{BC}(\psi_t,\psi_s) = \sqrt{1 - \Sigma_k \sqrt{h_1(k)h_2(k)}} \quad (25)$$

where h_1 and h_2 are the histograms of patches ψ_t and ψ_s, respectively.

Bhattacharya distance gave a zero value if the distributions of the patches are same irrespective of their spatial domain nature. In order to overcome this, LeMeur and Guillemot[23] proposed a variant of weighted Bhattacharya distance by multiplying the SSD with $(1 + d_{BC})$. Under this condition, even if two patches have same distribution, the distance between them will be equal to SSD.

Other distance metrics such as hamming distance and normalized cross correlation can also be used to find the best matching pair of patches. The detail comparison of above distance metrics can be found in the work of Ralekar et al.[24] Other metrics are also introduced which work well in the presence of geometric transformation such as flipping, scaling, and rotation.[10,25]

Exemplar-based methods carryout an exhaustive search among the known part of the image to find a best patch match. This brings in a high computational cost and slows down the process of inpainting. In order to overcome these limitations, different techniques were developed. Taking the coherent nature of images into consideration, it was observed that most of the time the patch similar to the target patch is found in the local vicinity. Hence, Anupam et al.[26] searched for the best patch in a local neighborhood and found the best match on the basis of variance calculation for the patches having the same mean squared error (see Fig. 17). The results presented shows significant improvement in the execution speed.

Xu and Sun[12] also considered a defined neighborhood for best match search and considered the parameters related to that neighborhood only for determining the best matching patch. Kwok et al.[27] proposed a fast exemplar-based method which decomposed the exemplar using DCT and then only selects the few coefficients which are most significant to evaluate the best match. Then the best match was found using the search array data structure and base-score array. The algorithm showed a significant improvement in the speed of execution. Performance of the abovementioned K-nearest neighbor (k-NN) techniques was further improved by introducing faster and approximate NN searches. A detail discussion regarding such methods can be found in the work of Kumar et al.[28]

A randomized search algorithm PatchMatch was introduced by Barnes et al.[29] which makes use of image coherency and uses the dependencies among the queries to find the best patch. PatchMatch algorithm is divided into two parts: a random/prior initialization, and search and propagation.

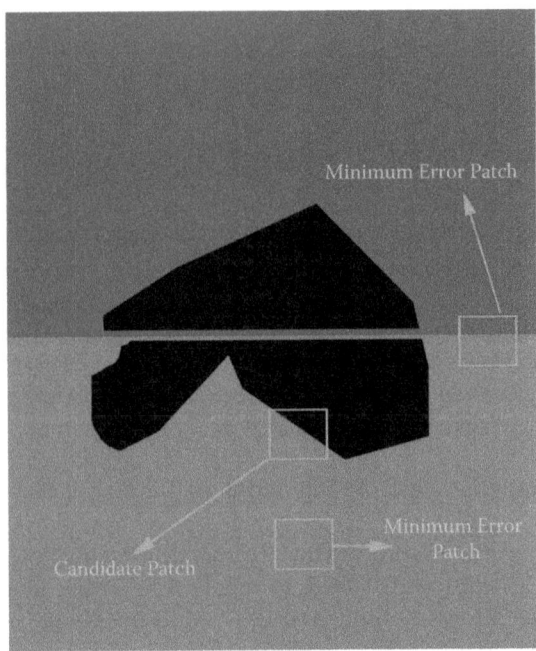

Fig. 17 Patches with same mean square error. Selecting the incorrect patch may not produce the most visually plausible result

As shown in Fig. 18, the algorithm iteratively executes the propagation step to update the offsets of patches obtained from a local random search. The algorithm was shown to be much faster than the tree-based approach.

The PatchMatch algorithm was further generalized by Barnes et al.[30] by finding k-NN as opposed to one; searching across the scales, rotations, and translations; and matching using arbitrary descriptors and distances instead of only SSD.

Sparsity-Based Image Inpainting

With the advent of sparse representation and optimization theories, sparse-based inpainting techniques were developed. A sparse representation of a signal makes use of dictionaries formed using atoms which are the elementary signals representing the template/signal. Combination of atoms can construct complex signals and images. In case of images, the image X is considered to be sparse in a given basis stored in a dictionary D in the form of atom a. Hence, the image can be represented as $X = Da$ (see Fig. 19). The dictionary can be initialized by directly learning from the source region[14,31] or can be learned by using different dictionary learning methods after a random initialization.

If a degraded image is given by G, the aim of sparse inpainting will be to represent G by using minimum number of atoms from dictionary D. Many solutions[32] exist for finding solution to such a problem. The patches are searched on the basis of known part as earlier, by searching for sparse vectors which best approximate the input patch as $\psi_p = Da_p$. The obtained linear combination of atoms thus obtained is then used to approximate the unknown part of the patch.

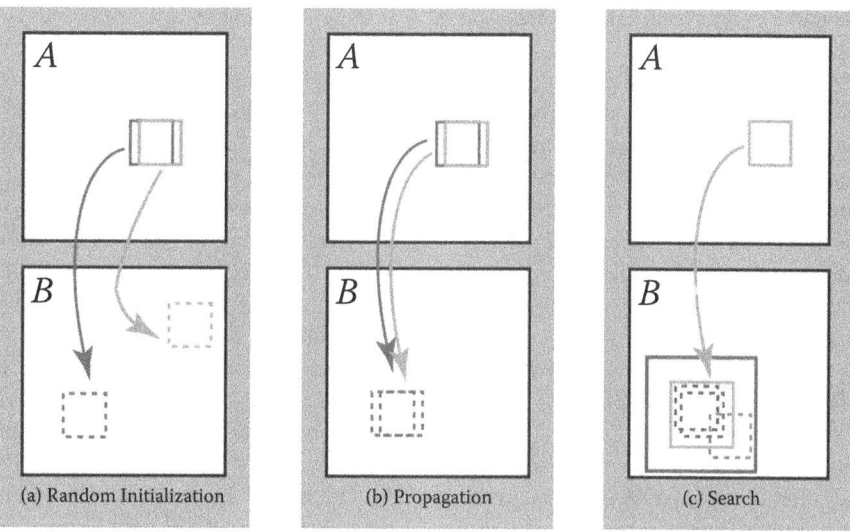

Fig. 18 Phases of the randomized nearest neighbor algorithm: (a) patches initially have random assignments; (b) patches check above and left neighbors to see if they will improve the patch mapping, propagating good matches; (c) patch searches randomly for improvements in concentric neighborhoods

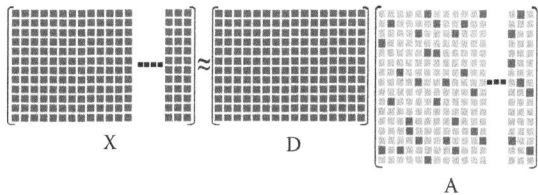

Fig. 19 Basic setup to understand dictionary learning

Shen, Hu, Zhang, et al.[14,31] presented a patch wise image inpainting algorithm using image signal sparse representation over a redundant dictionary constructed by directly sampling from the source region. For sparse representation, the Lasso regression method was used. Given the dictionary $x = [x_1, x_2, \ldots x_N]$ and the input signal $y = [y_1, y_2, \ldots y_p]$, the Lasso algorithm estimates the coefficients β of a signal over a given dictionary as follows:

$$\hat{\beta} = \arg\min\left(\|y - x\beta\|_2^2 + \lambda\|\beta\|_1\right) \quad (26)$$

The term $\|\beta\|_1$ encourages the sparsity, and the parameter λ controls the trade-off between the reconstruction error and the sparsity.

Xu and Sun[12] used sparse representation to obtain image inpainting by combining N best-matched patches linearly. Considering P and P' be the matrix of unknown and known pixels of the patch, respectively, the aim is to find Ψ_p', the estimate of Ψ_p, which can be represented as follows:

$$\psi_p' = \sum_{q=1}^{N} \alpha_q \psi_q \quad (27)$$

The combination coefficients α's are inferred by minimizing a constrained optimization problem in the framework of sparse representation. Two constraints were introduced on the appearance of the patch Ψ_p'. The first constraint stated that the approximated patch should be similar over already known pixels, and the second constraint stated that the newly filled pixels should be in local consistency with neighboring patched in appearance. The constraint can be mathematically stated as follows:

$$E = \left\|D\psi_p' - \psi_T\right\|^2 < \varepsilon, \text{where}, D = \begin{bmatrix} \overline{P} \\ \sqrt{\beta}P \end{bmatrix} \text{ and } \psi_T$$
$$= \begin{bmatrix} \overline{P}\psi_p \\ \sqrt{\beta}P\sum w_{p,pj}\psi_{pj} \end{bmatrix} \quad (28)$$

The second constraint is that the patch must be estimated by the sparsest linear combination of patches from the known region, that is, $\vec{\alpha}$ must have very small number of nonzero entries. The linear combination coefficients $\vec{\alpha}$ can be inferred by optimizing the following constrained optimization problem:

$$min\left\{\|\vec{\alpha}\|_0\right\} \text{ s.t. } \left\|D\psi_p' - \psi_T\right\|^2 < \varepsilon. \quad (29)$$

The optimization problem was solved by developing a greedy fashion similar to matching pursuit algorithm.[33]

As mentioned earlier, Xu and Sun[12] use the sparsity-based data term in order to calculate the patch filling order. The algorithm was applied to a variety of natural images for text removal, object removal, and block completion. The authors also made combination of different existing techniques[34] with their novel approach, and comparison was shown in the literature.

Exemplar-based methods are well suited to generate textures in an image and hence provides satisfactory results to inpaint textures and small holes. The methods also provide good results in case of larger holes. However, the quality structure reconstruction depends on the sequence of inpainting. Variations in exemplar-based methods tend to produce implausible results at the cost of increased complexities.

THIRD CATEGORY: HYBRID METHODS

Diffusion-based methods work well for small holes and tend to produce blur effects on inpainting large holes. They are well suited to reconstruct the missing part of a structure while the methods fail to recreate textures. On the other hand, exemplar-based methods perform well for texture recreation while they fail to preserve edges or structures. A natural image consists of textures as well as structures. Use of a single category method for the purpose of inpainting may lead to implausible results. In order to overcome such a problem, the image is first decomposed into texture and structure part, and then suitable inpainting methods are used to inpaint the corresponding sub image separately.

Bertalmio et al.[35] decomposed image using two functions. First function is of BV and represents the image structures, while the second function captures the texture part of the image. The structural part of the image contains the low-frequency components and hence results in sketchy output, whereas the textural part of the image contains high-frequency components, and hence results in an image containing all the edges and fine details. The structural image is inpainted using the diffusion method,[1] whereas the textural image is inpainted using the texture synthesis method.[8] Finally, both the images are added to generate the final inpainted output.

Wavelets have inherent property of decomposing image into different frequency components. Also the multi resolutional property can be used to improve the inpainting results. Zhang and Dai[36] used the wavelets to decompose the image into structure and texture sub images. The level scaling (LL) band of wavelet decomposition is considered as structural part, whereas the remaining three bands are combined to obtain the final textural sub image. The structural subimage is inpainted using the CDD method,[15] whereas the textural sub image is inpainted using the texture synthesis method.[8]

Cho and Bui[37] used the multi-resolution and the decoupling properties of wavelets to the advantage of the inpainting algorithm. Data are separated into low-frequency scaling and high-frequency wavelet coefficients, which make it possible to analyze structure as well as texture. In the proposed algorithm, first wavelet and scaling coefficients are estimated after decomposition of image. The coefficient of wavelet sub-band is dependent on the neighboring coefficient of the same sub-band as well as on the corresponding coefficient of the other sub-band. Hence, inter- and intrascale dependencies are considered. The algorithm works on the coarsest LL sub-band by using TV-based inpainting method.[3]

After inpainting the scaling sub-band, three wavelet sub-bands in the coarsest level are filled simultaneously by using exemplar-based approach.[11] Once all the coefficients are estimated, inverse transform for one level is performed to obtain the approximation of the next finer scale. The process is continued till all the scales have been processed. The results show a significant improvement over other inpainting approaches.

The hybrid approach of inpainting utilizes the advantages of both the PDE-based and exemplar-based methods to produce better results compared to former techniques. The cost of quality is the increase in complexity of the algorithm.

APPLICATIONS OF IMAGE INPAINTING

Various image processing problems can be solved by applying appropriate image inpainting algorithms. Few problems that can be addressed belong to image restoration, object removal, red eye removal, texture synthesis, image resizing, image compression, etc.

Image restoration deals with the problem of recovering the original image from different degradations (see Fig. 20). The degradations can be scratches in old document and photographs due to aging or folding. It can also be a reflection of flash light of camera or unwanted glare of light from the object. Since the degraded area is generally not large, diffusion-based methods gives satisfactory results. However, other methods can be used in order to restore image depending on the nature of degradation. For instance, if the degraded area is large, one must use exemplar-based method as diffusion-based methods will introduce implausible blur.

Object removal deals with the problem of removing an object from an image thereby reconstructing the underlying information in a plausible manner (see Fig. 21). The object to be removed is generally of significant size, and hence, diffusion-based techniques are not preferred for such a problem. Exemplar-based techniques, sparsity-based approaches, or hybrid approaches provide satisfactory results for object removal.

Image inpainting has also been applied to the problem of image compression and transmission loss concealment. In image compression problem, the repeated patterns in an image and small information chunks are deliberately omitted prior to transmission or storage. Since inpainting methods can accurately recreate the missing information, the bandwidth for transmission or storage space required can

 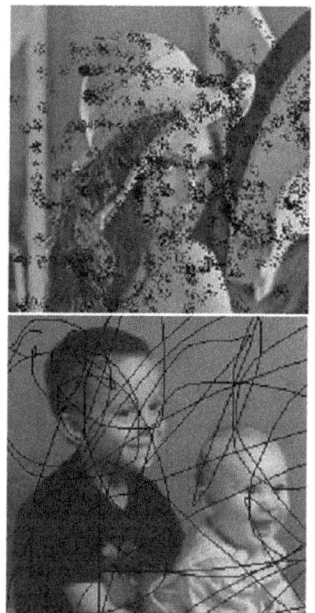 

Fig. 20 Image inpainting for image restoration

Fig. 21 Image inpainting for object removal

be significantly reduced. Errors in transmission create a similar situation at the receiver, where some random parts of an image go missing. Image inpainting can be applied to recover the original image in such cases.

CONCLUSION AND FUTURE SCOPE

Image inpainting has been widely studied in the field of image processing since past decades. The method is applicable to various applications extending from object removal to image restoration. Different categories described in this entry can be identified and used for an appropriate problem on hand. Some methods offer time advantages at the cost of result quality and vice versa.

Image inpainting algorithms can be directly applied to individual frames of a video to obtain the video inpainting. However, the temporal dimension, change of scene, motion of objects, motion of camera, etc. limit the direct application of the former.

REFERENCES

1. Bertalmio, M.; Sapiro, G.; Caselles, V.; Ballester, C. Image in painting. In *Proceedings of the 27th Annual Conference on Computer Graphics and Interactive Techniques*, SIGGRAPH '00, July 23–28; ACM Press/Addison-Wesley Publishing Co.: New York, 2000, 417–424.
2. Chan, T.F.; Shen., J.H. Mathematical models for local non-texture inpainting. SIAM J. Appl. Math. **2002**, *62* (3), 1019–1043.
3. Liu, D.; Sun, X.; Wu, F. Inpainting with image patches for compression. J. Vis. Commun. Image Represent. **2012**, *23* (1), 100–113.
4. Huynh, T.K.; Huynh, K.V.; Le-Tien, T.; Nguyen, S.C. A survey on image forgery detection techniques. In *The 2015 IEEE RIVF International Conference on Computing & Communication Technologies—Research, Innovation, and Vision for Future (RIVF)*, Can Tho, Vietnam, January 25–28; IEEE, 2015, 71–76.
5. Yoo, S.; Park, R.H. Red-eye detection and correction using inpainting in digital photographs. IEEE Trans. Consum. Electron. **2009**, *55* (3), 1006–1014.

6. Oliveira, M.M.; Bowen, B.; McKenna, R.; Chang, Y.S. Fast digital image inpainting. In *Proceedings of the International Conference on Visualization, Imaging and Image Processing VIIP 2001*; Marbella, Spain, September 3–5; Hamza, M.H.; Ed.; Acta Press: Marbella, Spain, 2001, 261–266.
7. Rudin, L.I.; Osher, S.; Fatemi, E. Nonlinear total variation based noise removal algorithms. Physica D **1992**, *60* (1), 259–268.
8. Efros, A.A.; Leung, T.K. Texture synthesis by non-parametric sampling. In *The Proceedings of the Seventh IEEE International Conference on Computer Vision*; Kerkyra, Greece, September 20–27; IEEE Computer Society: Washington, DC, 1999, Vol. 2, No. 2, 1033.
9. Efros, A.A.; Freeman, W.T. Image quilting for texture synthesis and transfer. In *Proceedings of the 28th Annual Conference on Computer Graphics and Interactive Techniques*; SIGGRAPH '01, Los Angeles, CA, August 12–16; ACM Press: New York, 2001, 341–346.
10. Drori, I.; Cohen-Or, D.; Yeshurun, H. Fragment-based image completion. ACM Trans. Graph. **2003**, *22* (3), 303–312.
11. Criminisi, A.; Perez, P.; Toyama, K. Region filling and object removal by exemplar based image inpainting. IEEE Trans. Image Process. **2004**, *13* (9), 1200–1212.
12. Xu, Z.; Sun, J. Image inpainting by patch propagation using patch sparsity. IEEE Trans. Image Process. **2010**, *19* (5), 1153–1165.
13. Hesabi, S.; Mahdavi-Amiri, N. A modified patch propagation-based image inpainting using patch sparsity. In *16th CSI International Symposium on Artificial Intelligence and Signal Processing AISP 2012*; Shiraz, Iran, May 2–3; IEEE: Washington, DC, 2012, 43–48.
14. Perona, P.; Malik, J. Scale-space and edge detection using anisotropic diffusion. IEEE Trans. Pattern Anal. Mach. Intell. **1990**, *12* (7), 629–639.
15. Chan, T.F.; Shen, J.H. Non-texture inpainting by curvature-driven diffusions (CDD). J. Vis. Commun. Image Represent. **2001**, *12* (4), 436–449.
16. Ashikhmin, M. Synthesizing natural textures. In *Proceedings of the Symposium on Interactive 3D Graphics 2001*, North Carolina, USA, March 19–21; ACM: New York, 2001, 217–226.
17. Liang, L.; Liu, C.; Xu, Y.Q.; Guo, B.; Shum, H.Y. Real-time texture synthesis by patch-based sampling. ACM Trans. Graph. **2001**, *20* (3), 127–150.
18. Cheng, W.H.; Hsieh, C.W.; Lin, S.K.; Wang, C.W.; Wu, J.L. In *Proceedings of International Conference on Computer Graphics, Imaging and Vision 2005*, Beijing, China, July 26–29, 2005, 64–69.
19. Wu, J.; Ruan, Q. Object removal by cross isophotes exemplar-based inpainting. In *18th International Conference on Pattern Recognition ICPR'06*, Hong Kong, August 20–24; IEEE Computer Society: Washington, DC, 2006, 810–813.
20. Goswami, P.; Paunwala, C. Exemplar-based image inpainting using ISEF for priority computation. In *International Conference on Circuits, Systems, Communication and Information Technology Applications (CSCITA'14)*; Mumbai, India, April 4–5; IEEE, 2014, 75–80.
21. Le Meur, O.; Gautier, J.; Guillemot, C. Exemplar-based inpainting based on local geometry. In *International Conference on Image Processing ICIP'11*; Brussels, Belgium, September 11–14; IEEE, 2011, 3401–3404.
22. Bugeau, A.; Bertalmío, M.; Caselles, V.; Sapiro, G. A comprehensive framework for image inpainting. IEEE Trans. Image Process. **2010**, *19* (10), 2634–1645.
23. Le Meur, O.; Guillemot, C. Super-resolution-based inpainting. In *Proceedings of the 12th European conference on Computer Vision ECCV'12*; Florence, Italy, October 7–13; Springer Verlag: Berlin and Heidelberg, Germany, 2012, 554–567.
24. Ralekar, C.; Dhondse, S.; Mushrif, M.M. Image reconstruction by modified exemplar based inpainting. In *International Conference on Communications and Signal Processing ICCSP'15*, Chengdu, China, October 10–11; IEEE, 2015, 1005–1010.
25. Zhang, Y.; Xiao, J.; Shah, M. Region completion in a single image. In *Proceedings of the 2004 Eurographics*; Nice, France, July 8–10; ACM: New York, 2004.
26. Anupam; Goyal, P.; Diwakar, S. Fast and enhanced algorithm for exemplar based image inpainting. In *Fourth Pacific-Rim Symposium on Image and Video Technology 2010*, Singapore, November 14–17; IEEE, 2010, 325–330.
27. Kwok, T.H.; Sheung, H.; Wang, C.C.L. Fast query for exemplar-based image completion. IEEE Trans. Image Process. **2010**, *19* (12), 3106–3115.
28. Kumar, N.; Zhang, L.; Nayar, S. What is a good nearest neighbors algorithm for finding similar patches in images? In *Proceedings of the 10th European Conference on Computer Vision Part II ECCV '08*; Marseille, France, October 12–18; Forsyth, D.; Torr, P.; Zisserman, A.; Eds.; Springer Verlag: Berlin and Heidelberg, Germany, 2008, 364–378.
29. Barnes, C.; Shechtman, E.; Finkelstein, A.; Goldman, D.B. PatchMatch: A randomized correspondence algorithm for structural image editing. ACM Trans. Graph. **2009**, *28* (3), 1–11.
30. Barnes, C.; Shechtman, E.; Goldman, D.B.; Finkelstein, A. The generalized patchmatch correspondence algorithm. In *Proceedings of the 11th European Conference on Computer Vision Conference on Computer Vision Part III ECCV'10*; Heraklion, Crete, Greece, September 5–11; Daniilidis, K.; Maragos, P.; Paragios, N.; Eds.; Springer Verlag: Berlin and Heidelberg, Germany, 2010, 29–43.
31. Shen, B.; Hu, W.; Zhang, Y.; Zhang, Y.J. Image inpainting via sparse representation. In *IEEE International Conference on Acoustics, Speech and Signal Processing,* Taipei, 19–24 April; IEEE: New York, 2009, 697–700.
32. Elad, M. *Sparse and Redundant Representations: From Theory to Applications in Signal and Image Processing*, 1st Ed.; Springer Publishing Company Incorporated: New York, 2010.
33. Mallat, S.G.; Zhang, Z. Matching pursuits with time-frequency dictionaries. IEEE Trans. Signal Process. **1993**, *41* (12), 3397–3415.
34. Wong, A.; Orchard, J. A nonlocal-means approach to exemplar-based inpainting. In *15th IEEE International Conference on Image Processing*; San Diego, CA, October 12–15; IEEE, 2008, 2600–2603.
35. Bertalmio, M.; Vese, L.; Sapiro, G.; Osher, S. Simultaneous Structure and Texture Image Inpainting. IEEE Trans. Image Process. **2003**, *12* (8), 882–889.

36. Zhang, H.; Dai, S. Image inpainting based on wavelet decomposition. Procedia Eng. **2012**, *29*, 3674–3678.
37. Cho, D.; Bui, T.D. Image inpainting using wavelet-based inter- and intra-scale dependency. In *19th International Conference on Pattern Recognition 2008, Tampa, Florida, December 8–11*; IEEE Computer Society: Washington, DC, 2008, 1–4.

BIBLIOGRAPHY

1. Qin, C.; Cao, F.; Zhang, X.P. Efficient image inpainting using adaptive edge-preserving propagation. Imaging Sci. J. **2011**, *59* (4), 211–218.
2. Chen, H.; Hagiwara, I. Image reconstruction based on combination of wavelet decomposition, inpainting and texture synthesis. Int. J. Image Graph. **2009**, *9* (1), 51–65.
3. Ou, J.; Chen, W.; Pan, B.; Li, Y. A new image inpainting algorithm based on DCT similar patches features. In *12th International Conference on Computational Intelligence and Security, CIS 2016*, Wuxi, China, December 16–19; IEEE, 2016, 152–155.
4. Shi, J.; Qi, C. Sparse modeling based image inpainting with local similarity constraint. In *IEEE International Conference on Image Processing, ICIP 2013*; Melbourne, Australia, September 15–18, 2013; IEEE, 2014, 1371–1375.
5. Kojenkine, N.; Hagiwara, I.; Savchenko, V. Software tools using CSRBFs for processing scattered data. Comput. Graph. **2003**, *27* (2), 311–319.
6. Jidesj, P.; George, S. Gauss curvature-driven image inpainting for image reconstruction. J. Chin. Inst. Eng. **2014**, *37* (1), 122–133.
7. Li, S.J.; Yao, Z.A. Image inpainting algorithm based on partial differential equation technique. Imaging Sci. J. **2013**, *61* (3), 292–300.
8. Tae-O-Sot, S.; Nishihara, A. DCT inpainting with patch shifting scheme. In *Proceedings of Annual Summit and Conference, APSIPA 2911*, Xi'an, China, October 18–21, 2011.
9. Liu, Y.; Caselles, V. Exemplar-based image inpainting using multiscale graph cuts. IEEE Trans. Image Process. **2013**, *22* (5), 1699–1711.
10. Guleryuz, O.G. Nonlinear approximation based image recovery using adaptive sparse reconstructions and iterated denoising-part I: Theory. IEEE Trans. Image Process. **2006**, *15* (3), 539–554.

Image Interpolation

Xin Li
Department of Computer Science and Electrical Engineering, West Virginia University, Morgantown, West Virginia, United States

Abstract
Image interpolation refers to computational approaches of obtaining high-resolution images from their low-resolution representation. It is closely related to the field of single-image super-resolution and image upsampling. This chapter reviews and compares two classes of image interpolation methods: model based versus learning based. It highlights the most recent advances in both categories and contains some experimental comparison on the same benchmark data set.

Keywords: Image interpolation; Image upsampling; Learning-based methods; Model-based methods; Single-image super-resolution.

INTRODUCTION

Image interpolation, also known as image scaling or image upsampling, refers to the technique of obtaining a higher resolution image from its low-resolution (LR) version. Applications related to image interpolation are diverse—from consumer electronics (e.g., the so-called "digital zooming" feature offered by most cameras or camcorders) and entertainment (e.g., whenever you click a "full screen" button while playing a PowerPoint slide or watching YouTube video) to medical imaging[1] and geostatistics.[2] The need for image interpolation is often driven by the necessity of developing a cost-effective computational alternative to the optical method of obtaining high-resolution (HR) images. Therefore, the fundamental trade-off between the cost and the performance is at the root of all image interpolation techniques.

In the past decades, the art of image interpolation has advanced rapidly. Early developments[3] directly leveraged mathematical tools such as cubic splines into signal processing applications related to image interpolation and digital filtering. These are the foundation of the class of nonadaptive interpolation techniques including bilinear and bicubic, which are still widely used due to their computational efficiency. The class of adaptive interpolation techniques (e.g., edge-directed interpolation[4–7]) became feasible in the late 1990s, thanks to the fast advance of computing technologies. Spatial adaptation has been shown to be a sensible idea of delivering interpolated images with higher subjective qualities at the price of increased computational complexity.

More recently, there are two flurries of ideas that have received increasingly more attention. On the one hand, recognizing the importance of exploiting self-similarity has spurred a lot of interest in developing nonlocal image processing algorithms including interpolation. Most recent works by Dong et al.[8,9] have convincingly shown that exploiting nonlocal self-similarity is particularly useful for the class of images containing abundant regular edges and textures. On the other hand, there has been a new trend of shifting from model-based (deriving analytical models underlying image data first) to data-driven or learning-based approach toward image interpolation/upsampling or super-resolution (SR).[10–12] The availability of training data raises new opportunities as well as challenges.

Where will the field of image interpolation go next? From our own perspective, there are still plenty of open problems in the field and those with practical significance could be tackled in the short term. First, as the technology of high-definition television evolves fast, there is a mismatch between the old LR video source and new display device. Therefore, it is a timely topic to develop computationally efficient interpolation/SR algorithms for image sequences. Explicit search-based motion estimation has always been the computational bottleneck in any attempt to exploit motion-related temporal dependency. Second, learning-based approaches could benefit from stronger a priori knowledge dictated by specific applications—for instance, in biomedical imaging, the content of imaging will be known to be a certain part of human body; in biometrics, the training data can be readily obtained from existing database. How to develop interpolation/SR algorithms tailored for a certain class of imagery deserves further study.

The rest of this chapter is organized as follows: We first provide a historical review of image interpolation in Section "Historical Review of Image Interpolation," which has to be selective and biased. Then, we detail on model-based interpolation including nonadaptive and adaptive methods

Encyclopedia of Image Processing, First Edition
DOI: 10.1201/9781351110273-140000467
Copyright © 2018 by Taylor & Francis. All rights reserved.

in Section "Model-Based Interpolation: Where Adaptation Matters." In Section "Learning-Based SR/Upsampling: When Seeing Is Believing," we lay out insights and ideas behind learning-based approaches and highlight the main technical challenges as well. We discuss the connection and difference between image interpolation and SR in Section "Image Interpolation versus SR: What Is the Difference? We make some concluding remarks and discuss future research directions in Section "Conclusions."

HISTORICAL REVIEW OF IMAGE INTERPOLATION

According to Meijering,[2] the problem of interpolation (for 1D data) could trace back to the works of Newton and Lagrange. Under the context of digital communication and signal processing, interpolation was studied by C. Shannon, H. Nyquist, and J. M. Whittaker in their famous sampling theorem for band-limited signals. While in the field of numerical analysis, interpolation was studied by Schoenberg under the disguise of approximation problem,[13] and the concept of spline was developed for interpolation purpose.[14] The connections between these two lines of research can be found in a tutorial review[15] where higher order kernels are found to possess considerably better low-pass properties than linear interpolators.

In the 1970s, the field of digital image processing started to develop, so did image interpolation. One of the first applications related to image interpolation was the geometrical rectification of digital images obtained from the first Earth Resources Technology Satellite launched in 1972. Cubic convolution-based interpolation was based on a sinc-like kernel composed of piecewise cubic polynomials. Its two-dimensional extension—bicubic interpolation—is still of wide use nowadays. A more sophisticated interpolation technique based on cubic B-spline was developed in 1978[3] and has remained influential. A comparative study among various competing interpolation techniques was conducted by Parker et al.[16] in 1983.

In the late 1980s and the early 1990s, information technology experienced an era of booming, which brought about the revolution of communication and computing. As digital images become ubiquitous in various multimedia applications, there has been a growing interest in obtaining more fundamental understanding of image signals and developing more powerful image processing algorithms. The importance of spatial adaptation has been long recognized for the problem of image interpolation, but there still lacked a principled approach toward spatially adaptive interpolation. Intuitive ideas such as those based on explicit edge detection[4,5] are often limited by the ad hoc definitions of edges in the literature.

A breakthrough in adaptive interpolation was made in the so-called new edge-directed interpolation (NEDI),[6] where edges were implicitly defined by the local covariance characteristics of image data. The key idea behind NEDI is a geometric duality between LR and HR covariances—if they are associated with the same edge, they should be correlated. Such covariance-based adaptation dated back to the classical Wiener filtering[17]; so NEDI can be interpreted as a multi-resolution extension of Wiener filtering to support the interpolation task. Since the publication of NEDI, numerous variations and improvements have been developed, for example, real-time artifact-free implementation[18] and iterative optimized implementation.[7,19]

In the last decade, new ideas borrowed from other fields have stimulated further advances in image interpolation and its closely related field of SR.[20] Due to space limitations, we will only review two examples here. The first is to leverage the idea of exploiting nonlocal self-similarity from image denoising[21] to image interpolation. Several nonlocal interpolation algorithms[8,9] have been developed along this line of research. Second, learning-based approaches toward image upsampling/SR have benefited from the use of sparse coding[11] and exploitation of self-similarity.[12] Subjective quality improvements have become more convincing in the situation of large magnification ratios (e.g., above 4).

MODEL-BASED INTERPOLATION: WHERE ADAPTATION MATTERS

In this section, we review four representative model-based interpolation methods: two nonadaptive (bilinear and bicubic) and two adaptive (NEDI and soft-decision estimation for adaptive image interpolation [SAI]). Generally speaking, adaptive methods can deliver higher quality interpolated images but at the price of increased computational complexity.

Bilinear Interpolation

One of the conceptually simplest but still widely used interpolation methods is called bilinear interpolation. It is a 2D extension of 1D linear interpolation, that is, from fitting lines to fitting planes on uniformly spaced data points. Mathematically, the formula of linear interpolation is straightforward: for any pair of data points $x(n)$ and $x(n + 1)$, one can obtain the interpolated value at the location $x(n + a)$; $0 < a < 1$ by $x(n + a) = a \times x(n + 1) + (1 - a) \times x(n)$. This method is called linear because it assumes a constant rate of change between $x(n)$ and $x(n + 1)$. Figure 1 illustrates two equivalent ways of generalizing the above interpolation formula from 1D to 2D. If one interpolates along the row directions first, the two intermediate interpolated values will be

$$x(m, n + a) = a \times x(m, n + 1) + (1 - a) \times x(m, n) x(m + 1, n + a)$$
$$= a \times x(m + 1, n + 1) + (1 - a) \times x(m + 1, n)$$

(1)

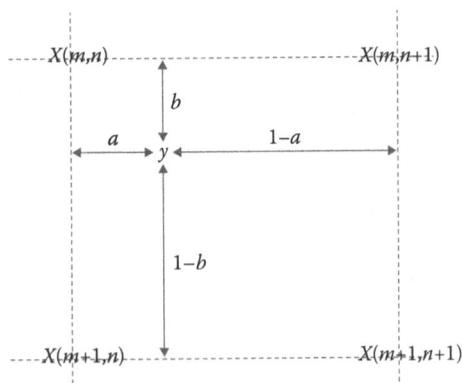

Fig. 1 Illustration of bilinear (left) and bicubic (right) interpolation

and then one can interpolate along the column direction using $x(m, n + a)$ and $x(m + 1, n + a)$ as pivots:

$$x(m+a, n+a) = b \times x(m, n+a) + (1-b) \times x(m+1, n+a) \qquad (2)$$

Substituting Eq. 1 in Eq. 2, we obtain the final formula of bilinear interpolation:

$$\begin{aligned} x(m+a, n+a) = &(1-a)(1-b) \times x(m, n) \\ &+ (1-a)b \times x(m+1, n) \\ &+ a(1-b) \times x(m, n+1) \\ &+ ab \times x(m+1, n+1) \end{aligned} \qquad (3)$$

It is straightforward to verify that if one interpolates along the column direction first (i.e., using $x(m + a, n)$ and $x(m + a, n + 1)$ as pivots), exactly the same result as Eq. 3 will be derived.

The main advantage of bilinear interpolation is its computational efficiency. Note that Eq. 3 admits parallel implementation, that is, instead of processing one pixel at a time, one can process the whole image in one pass. Under MATLAB®, the bilinear formula admits vectorized implementation—the last three terms on the right side of Eq. 3 are nothing but shifted versions of the original image (e.g., as implemented by *circshift*). Extra caution will be needed for handling pixels around image borders (e.g., one might use function *padarray* to avoid the introduction of boundary artifacts).

Bicubic Interpolation

Similar to the introduction of bilinear method, it is more convenient to define cubic interpolation for 1D case first and then extend it to 2D. The basic idea behind cubic interpolation is to fit a third-order polynomial (instead of straight lines) to uniformly spaced data points. Let $f(x) = ax^3 + bx^2 + cx + d$ denote the model, and its first-order derivative is given by $f'(x) = 3ax^2 + 2bx + c$. To determine four unknown parameters a, b, c, and d, we need to have a set of four linear equations:

$$f(0) = d, f(1) = a+b+c+d, f'(0) = c, f'(1) = 3a+2b+c \qquad (4)$$

from which we can obtain:

$$\begin{aligned} a &= 2f(0) - 2f(1) + f'(0) + f'(1), \\ b &= -3f(0) + 3f(1) - 2f'(0) - f'(1), \\ c &= f'(0), d = f(0) \end{aligned} \qquad (5)$$

Note that for consecutive values of p_0, p_1, p_2, and p_3, respectively, at $x = -1$, $x = 0$, $x = 1$, and $x = 2$, we can estimate:

$$f(0) = p_1, f(1) = p_2, f'(0) = \frac{p_2 - p_0}{2}, f'(1) = \frac{p_3 - p_1}{2} \qquad (6)$$

Substituting Eq. 6 into Eq. 5, we reach the formula of cubic interpolation:

$$\begin{aligned} a &= -\frac{1}{2}p_0 + \frac{3}{2}p_1 - \frac{3}{2}p_2 + \frac{1}{2}p_3, \\ b &= p_0 - \frac{5}{2}p_1 + 2p_2 - \frac{1}{2}p_3, \\ c &= -\frac{1}{2}p_0 + \frac{1}{2}p_2, d = p_1 \end{aligned} \qquad (7)$$

When compared against linear interpolation, note that cubic involves a total of four data points instead of two.

Based on the same principle of separability, 2D bicubic interpolation can be decomposed into two sequential 1D building blocks:

$$f(x,y) = f\big(f(p_{00},p_{01},p_{02},p_{03},y), f(p_{10},p_{11},p_{12},p_{13},y), \\ f(p_{20},p_{21},p_{22},p_{23},y), f(p_{30},p_{31},p_{32},p_{33},y), x\big). \qquad (8)$$

or equivalently:

$$f(x,y) = f\big(f(p_{00},p_{10},p_{20},p_{30},x), f(p_{01},p_{11},p_{21},p_{31},x), \\ f(p_{02},p_{12},p_{22},p_{32},x), f(p_{03},p_{13},p_{23},p_{33},x), y\big). \qquad (9)$$

Unlike bilinear interpolation involving four nearest neighbors, bicubic interpolation requires 16 surrounding points and is therefore computationally more demanding (Fig. 1). However, interpolated images by bicubic usually contain fewer artifacts than those by bilinear due to the adoption of higher order polynomials.

NEDI and ICBI

Both bilinear and bicubic interpolations are nonadaptive methods because the same model is assumed to fit everywhere in the image. Such lack of adaptation is desirable from the computational point of view but often produces unpleasant artifacts in the interpolated images. Spatial adaptation is the key to achieve better subjective quality especially around the areas of important image structures such as edges. An intuitively appealing idea is to interpolate only along the edge but never across it (because interpolating across edges tends to blur them and inconsistency might create artifacts). Despite the conceptual simplicity of this idea, computational implementation is not straightforward. For example, if one attempts explicit edge detection like that in the works of Jensen and Anastassiou[4] and Allabech and Wong,[5] the obtained interpolation results are not always consistent because edge detection decision is binary and seldom error free.

Inspired by the influential work of Wiener filtering, NEDI[6] proposed to achieve spatially adaptive interpolation via estimating local covariances. It is easy to observe that directly estimating HR covariance characteristics is not feasible due to the missing data at interpolated locations. However, based on the scale invariant of edge orientation, if one assumes that edge orientation is implicitly embedded into local covariances, it is possible to replace the unknown HR covariances (marked by dashed arrows in Fig. 2) by their LR counterparts (marked by solid arrows in Fig. 2). Such geometric duality can be verified for two sublattices with an interpolation factor of two—the first is samples along diagonal directions (odd; odd), and the second is the quincunx sublattice located at (even; odd); (odd; even) positions as shown in Fig. 2. The adaptation of local interpolation coefficients can then be solved by a standard least square optimization procedure.

The main strength of NEDI is its capability of delivering high-quality interpolated images especially for the class of images with abundant edges (e.g., digital reproduction of artistic works). However, since the calculation of local covariances has to be done on a block-by-block basis, it defies parallel implementation and requires higher computational cost. More recently, a computationally efficient implementation of similar idea to NEDI was developed in iterative curvature-based interpolation (ICBI).[18] Unlike the original setting of Wiener filtering in NEDI, ICBI uses local curvature as the guidance for adapting the interpolation coefficients, which enjoys lower computational complexity. It has been reported in the work of Giachetti and Asuni[18] that running time reduction can be achieved in terms of orders of magnitude.

SAI and NARM

More recently, as computing technologies quickly advance, it becomes feasible to further enhance/optimize the quality of interpolated images without worrying about the computational burden. Along this line of research, iterative optimization of model parameters such as those involved in NEDI has been studied in SAI[7] and nonlocal autoregressive modeling (NARM).[9] In SAI,[7] the missing data and unknown local covariances are treated as peer hidden variables and alternatively updated in an expectation maximization-like fashion. Such iterative optimization does not guarantee the convergence to a global optimum, but it is a sensible idea to enforce the consistency of local covariance estimates.

In NARM,[9] the idea of local adaptation is further connected with exploiting nonlocal self-similarity in images. In other words, to estimate some second-order statistics such as covariances, one can not only count on the data in the spatial neighborhood (i.e., through sliding windows) but also resort to image patches that are spatially distant but perceptually similar (i.e., those patches close to the exemplar one lying along the same manifold). Such nonlocal extension of the underlying model has shown to be particularly effective for the class of image structures including regular edges and textures. A bilateral variance estimation-based interpretation can be found in a related work.[8] However, the computational complexity of NARM is significantly higher than that of previous model-based approaches.

At the end of this section, we use some experimental results to demonstrate the performance of competing interpolation methods. Figures 3 and 4 include the comparison of interpolated images for four competing methods: bilinear, bicubic, NEDI, and SAI. It takes bilinear/bicubic

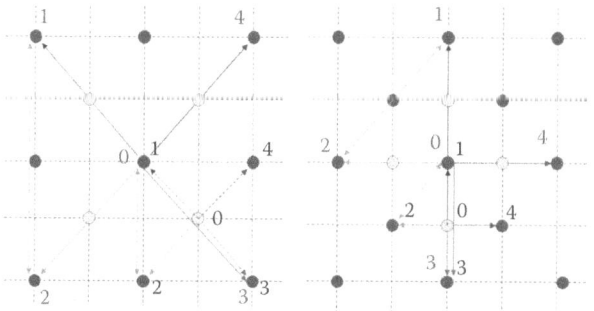

Fig. 2 Geometric duality between LR and HR covariances in the first step (left: interpolating diagonal values at (odd, odd) locations) and the second step (right: interpolating the other half missing values at (odd, even) and (even, odd) locations). The original LR image is assumed to be overlapped with the samples at (even, even) locations

Fig. 3 Subjective quality comparison among different interpolation methods (magnification ratio=4): (a) bilinear; (b) bicubic; (c) NEDI; (d) SAI

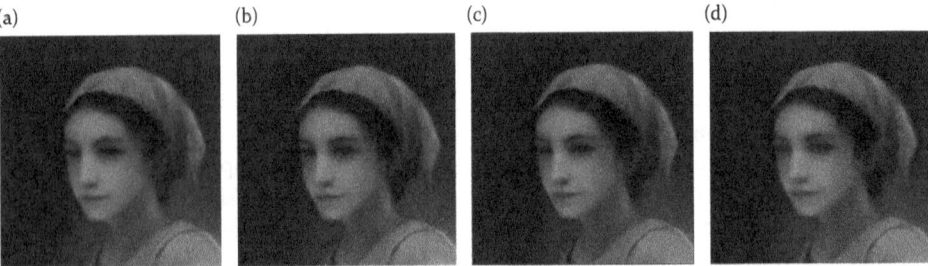

Fig. 4 Subjective quality comparison among different interpolation methods (magnification ratio=4): (a) bilinear; (b) bicubic; (c) NEDI; (d) SAI

interpolation a fraction of second to magnify an image sized by 320 × 240 by a factor of four. By contrast, the running time of NEDI and SAI is in the order of seconds, but one can observe that dramatic visual quality improvements can be achieved at the price of higher computational complexity.

LEARNING-BASED SR/UPSAMPLING: WHEN SEEING IS BELIEVING

In this section, we review four representative learning-based approaches. Unlike model based, learning-based approaches assume some training data (e.g., pair of LR and HR images) available, which can be used to learn how to interpolate/upsample test data (a separate collection of LR images) in a more principled way.

Example-Based SR

In the example-based SR,[12] a Markov network model is adopted to characterize the intrinsic relationship between LR and HR image patches. It is possible to derive optimal HR patches using belief propagation algorithms though their computational complexity is high. A more computationally tractable one-pass solution was discussed in the work of Freedman and Fattal[12] as well. More specifically, to estimate a missing HR patch, one can use its associated LR patch as an anchor and find the best matcher (i.e., find the nearest neighbor in the patch space) from the given training set. Then the corresponding HR patch to the best matcher will be used as the approximate solution. It has been experimentally found that such one-pass approximation can produce comparable results to the full-scale belief propagation algorithm.

SR via Neighborhood Embedding

An improved idea based on example-based SR is to replace the nearest neighbor finding by k-nearest neighbor (kNN) finding and exploit the local subspace constraint in the patch space. In the so-called SR via neighborhood embedding,[22] the local subspace constraint on the patch manifold was learned by a locally linear embedding (LLE) method.[23] Let x^q denote a target LR patch and its neighboring patches be $x^p \in N_q$. The strategy of LLE is to solicit an optimal linear reconstruction of the target patch from its kNNs, namely, to minimize

$$E = \left\| x^q - \sum_{x^p \in N_q} w_{qp} x^p \right\|^2 \qquad (10)$$

subject to $\sum_{x^p \in N_q} w_{qp} = 1$. The above constrained linear least square optimization problem admits the closed-form solution. Then the derived weighting coefficients w_{qp}'s can be applied to the corresponding HR patches to obtain the pursued HR patch. Despite the visual quality improvement, it has been found that neighborhood embedding might not achieve the lowest mean square error (MSE), which indicates that MSE metric is not a reliable indicator of subjective quality.

SR via Sparse Coding

A more sophisticated approach toward learning-based SR is to make a connection with the strategy of sparse coding. Spare coding refers to the idea of representing a vector by a small number of so-called atoms or elements in a dictionary. The sparsity constraint could be enforced by the total number of nonzero coefficients in w_{qp}'s; however, such l_0-based optimization is non-convex and computationally infeasible. There has been a recurring theme of replacing l_0-norm by its convex l_1-counterpart, which admits computationally efficient solutions. The so-called theory of compressed sensing[24] and numerous fast algorithms have been developed for the l_1-optimization problem.

In the work of Yang et al.,[11] two dictionaries for the LR and HR image patches are jointly trained, and the similarity of sparse representations between the LR and HR image patch pair with respect to their own dictionaries is enforced. Similar to neighborhood embedding, the linear representation of an LR image patch is applied with the HR image patch dictionary to generate an HR image patch; however, unlike neighborhood embedding, the sparsity constraint enforced with respect to the linear representation of an LR image patch makes[11] not only subjectively more pleasant but also objectively reaches lower MSE than other competing approaches. In particular, it has been shown in the work of Yang et al.[11] that such sparse coding-based approach is particularly suitable for the SR of certain constrained class of images such as face hallucination.

Local Self-Example-Based Upsampling

One of the fundamental weaknesses of learning-based approaches is the dependency of the interpolated/SR image quality upon training sets. It has been experimentally observed that artifacts could arise from the mismatch between training and test sets, but there is no systematic and quantitative study about the impact of such mismatch to the best of our knowledge. For the same reason, it is often difficult to compare different learning-based approaches toward interpolation/SR because varying training sets are often adopted by different researchers.

In a recent work,[12] an appealing idea of using an LR image itself as training example was proposed. The key insight behind the so-called local self-example (LSE) training is the self-similarity of local patches in photographic images—for any patch in smooth, edge, or texture area, it can find a similar patch in a scaled version of the original image as long as the scaling factor a is kept close to unity. Such small scaling factor is critical because it has been shown in the work of Freedman and Fattal[12] that patch similarity assumption is only valid for that range (typically $1 < a < 1.25$). Therefore, if one wants to achieve a larger scaling factor such as two, the trick is to decompose it into a product of smaller scaling factors, for example, $2 = 8/7 \times 7/6 \times 6/5 \times 5/4$. For each individual upsampling step, it follows a similar implementation to that of the work of Freeman et al.,[10] except that the training set is nothing but the upsampled version of the LR image itself.

We have used the same set of test images to compare the performance of learning-based SR/upsampling methods. Figures 5 and 6 include the comparison of interpolated images for four competing methods: example-based SR,[10] neighborhood embedding,[22] sparse coding,[11] and LSE.[12] All the source codes of those methods can be downloaded at the author's reproducible research website http://www.csee.wvu.edu/~xinl/source.html. It can be observed that they all achieve comparable visual quality in the reconstructed images (arguably LSE produces the sharpest after further zoom-in) (Fig. 6).

IMAGE INTERPOLATION VERSUS SR: WHAT IS THE DIFFERENCE?

In the literature, there has been a confusion between the two terms: image interpolation ands single-image SR. Both techniques take an LR image as the input and produce an HR image as the output; so, apparently they are related, but is there any fundamental difference? The answer is positive. In this section, we attempt to articulate this difference and highlight the importance of following a fair protocol in experimental comparisons.

Image interpolation assumes that LR image is directly a downsampled version of HR image. Consequently, it is

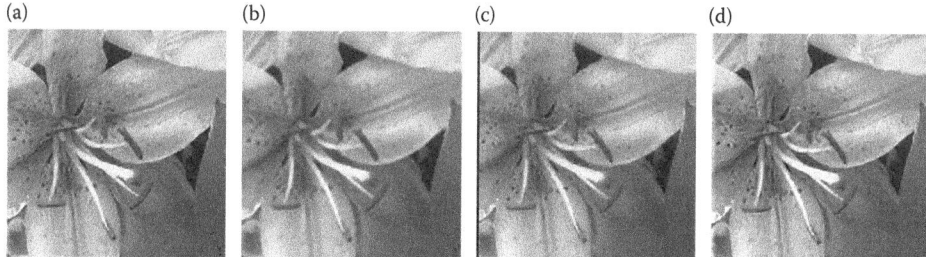

Fig. 5 Subjective quality comparison among different SR/upsampling methods (magnification ratio=4): (a) example-based SR; (b) neighborhood embedding; (c) sparse coding; (d) LSE

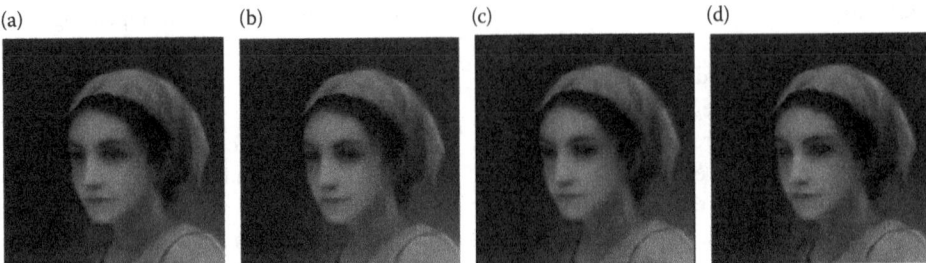

Fig. 6 Subjective quality comparison among different interpolation methods (magnification ratio=4): (a) example-based SR; (b) neighborhood embedding; (c) sparse coding; (d) LSE

Fig. 7 Conceptual difference between two similar problem formulations: image interpolation (left) versus image SR (right). In other words, SR can be viewed as a combination of image interpolation and blind image deconvolution in the absence of the knowledge about PSF

known as a priori knowledge that a subset of HR pixels is already given by the LR image as ground truth. By contrast, single-image SR assumes that there is an anti-aliasing filtering before downsampling. This anti-aliasing filter is often called point spread function (PSF) in the literature of optics, but its impact on LR image is profound—the interpolated image will not be the final result any more but its blurred version. In other words, the existence of anti-aliasing filter dictates that SR is decomposed into two subproblems: image interpolation and image deconvolution. These two subproblems have to be solved simultaneously in single-image SR (Fig. 7).

For the above reasons, it is only fair to compare interpolation/SR algorithms that have used the same protocol to produce LR images. Especially when objective quality metrics such as MSE/ peak signal-to-noise-ratio (PSNR) are used, LR images have to be produced by the same procedure—if anti-aliasing filter is used, the same PSF should be adopted by all SR reconstruction methods; if the knowledge about anti-aliasing filter is prohibited during HR image reconstruction (i.e., image deconvolution becomes blind), the same constraint should be applied to all competing approaches. In a more practical setting, only LR images are given and no ground-truth HR images are available, such situation calls for the study of blind image quality assessment, which is still an open problem as of today.

CONCLUSIONS

In this chapter, we have reviewed recent advances in the field of image interpolation and its related area called single-image SR/upsampling. Model- and learning-based approaches are discussed and compared with the same set of test images. We also highlight the subtle difference between these two similar problems and suggest that whether an anti-aliasing filter used could have a significant impact on the experimental results especially objective quality metrics such as PSNR/ structural similarity index (SSIM).

Looking down the path ahead, we believe that constant increase of computational resources will shift the focus of interpolation/SR from still images to moving pictures. How to achieve an improved trade-off between the cost and the performance for video interpolation/SR will deserve more systematic studies in the future. In particular, since video compression is inevitable to support efficient storage and transmission, joint investigation of video compression and SR seems a fruitful area.

REFERENCES

1. Lehmann, T.; Gonner, C.; Spitzer, K. Survey: Interpolation methods in medical image processing. IEEE Trans. Med. Imag. **1999**, *18* (11), 1049–1075.
2. Meijering, E. Chronology of interpolation: From ancient astronomy to modern signal and image processing. Proc. IEEE **2002**, *90* (3), 319–342.
3. Hou, H.S.; Andrews, H.C. Cubic splines for image interpolation and digital filtering. IEEE Trans. Acoust. Speech Signal Process. **1978**, *26*, 508–517.
4. Jensen, K.; Anastassiou, D. Subpixel edge localization and the interpolation of still images. IEEE Trans. Image Process. **1995**, *4* (3), 285–295.

5. Allabech, J.; Wong, P. Edge directed interpolation. In *International Conference on Image Processing*, Vol. 3; IEEE: Lausanne, 1996, 707–710.
6. Li, X.; Orchard, M. New edge directed interpolation. IEEE Trans. Image Process. **2001**, *10*, 1521–1527.
7. Zhang, X.; Wu, X. Image interpolation by adaptive 2-D autoregressive modeling and soft-decision estimation. IEEE Trans. Image Process. **2008**, *17* (6), 887–896.
8. Dong, W.; Shi, G.; Li, X. Nonlocal image restoration with bilateral variance estimation: A low-rank approach. IEEE Trans. Image Process. **2013**, *22* (2), 700–711
9. Dong, W.; Zhang, L.; Lukac, R.; Shi, G. Sparse representation based image interpolation with nonlocal autoregressive modeling. IEEE Trans. Image Process. **2013**, *22* (4), 1382–1394.
10. Freeman, W.T.; Jones, T.R.; Pasztor, E.C. Example-based super-resolution. IEEE Comput. Graph. Appl. **2002**, *22*, 56–65.
11. Yang, J.; Wright, J.; Huang, T.; Ma, Y. Image super-resolution via sparse representation. IEEE Trans. Image Process. **2010**, *19* (11), 2861–2873.
12. Freedman, G.; Fattal, R. Image and video upscaling from local self-examples. ACM Trans. Graph. (TOG) **2011**, *30* (2), 12.
13. Schoenberg, I.J. Contributions to the problem of approximation of equidistant data by analytic functions. Quart. Appl. Math. **1946**, *4* (2), 45–99.
14. Schoenberg, I.J. *Cardinal Spline Interpolation*; SIAM: Philadelphia, PA, Vol. 12, 1973.
15. Crochiere, R.E.; Rabiner, L. Interpolation and decimation of digital signals—A tutorial review. Proc. IEEE **1981**, *69* (3), 300–331.
16. Parker, J.A.; Kenyon, R.V.; Troxel, D. Comparison of interpolating methods for image resampling. IEEE Trans. Med. Imag. **1983**, *2* (1), 31–39.
17. Wiener, N. Extrapolation, interpolation, and smoothing of stationary time series with engineering applications. J. Amer. Stat. Assoc. **1949**, *47* (258), 319–321.
18. Giachetti, A.; Asuni, N. Real-time artifact-free image upscaling. IEEE Trans Image Process. **2011**, *20* (10), 2760–2768.
19. Liu, X.; Zhao, D.; Xiong, R.; Ma, S.; Gao, W.; Sun, H. Image interpolation via regularized local linear regression. IEEE Trans. Image Process. **2011**, *20* (12), 3455–3469.
20. Park, S.; Park, M.; Kang, M. Super-resolution image reconstruction: A technical overview. IEEE Signal Process. Mag. **2003**, *20*, 21–36.
21. Dabov, K.; Foi, A.; Katkovnik, V.; Egiazarian, K. Image denoising by sparse 3-D transform-domain collaborative filtering. IEEE Trans. Image Process. **2007**, *16*, 2080–2095.
22. Chang, H.; Yeung, D.-Y.; Xiong, Y. Super-resolution through neighbor embedding. In *Proceedings of the 2004 IEEE Computer Society Conference on Computer Vision and Pattern Recognition*, Vol. 1; IEEE: Washington, DC, 2004, I–I.
23. Roweis, S.T.; Saul, L.K. Nonlinear dimensionality reduction by locally linear embedding. Science **2000**, *290* (5500), 2323–2326.
24. Candès, E.J.; Romberg, J.K.; Tao, T. Robust uncertainty principles: Exact signal reconstruction from highly incomplete frequency information. IEEE Trans. Inform. Theory **2006**, *52* (2), 489–509.

Image Processing and Measurement

John C. Russ
Professor Emeritus, Department of Materials Science and Engineering, College of Engineering, North Carolina State University, Raleigh, North Carolina, U.S.A.

Abstract

Images provide important information in scientific, technical, and forensic situations, in addition to their role in everyday life. Extracting information from images acquired by digital cameras involves image processing to correct colors, reduce noise, and correct for non-uniform illumination or non-planar views. Enhancement of image details is generally accomplished by reducing the contrast of other information in the image, so that (for example) lines and edges that make measurements of structure are more accessible. The processing steps use a variety of computer algorithms and may be performed on the pixel array, or in a different space, e.g., by using a Fourier transform. Some applications, especially forensic ones, require simple comparisons, but for object identification, classification, or correlations, quantitative measurements of color or density, position, size, and shape are needed. Several possible measurement quantities are available for each category, particularly shape, for which a variety of dimensionless ratios, Fourier or wavelet coefficients, and invariant moments may be used. Interpretation of the measurements depends on the nature of the image and of the specimen or scene, for instance, whether it consists of discrete objects on a surface, a section through a complex structure, or a projection through a three-dimensional space.

Keywords: Image processing; Image measurement; Object measurement; Color; Size; Shape; Position; Stereology.

INTRODUCTION

Human beings depend to a high degree on images to gather information about their world, and to organize and understand that information. This dependence extends to scientific, technical, and forensic analysis as well, and to scales that include the microscopic and astronomical, aided by a broad variety of instruments designed to use infrared light, X-rays, radar, sound waves, and so on.

Human vision is not a quantitative tool, and is easily fooled by illusions and distracted by extraneous or random background features. Measurement requires a direct comparison to appropriate standards (rulers, protractors, color scales, etc.). Consequently, the design of instruments and computer algorithms that collect, process, and analyze images is a key part of acquiring quantitative data for many scientific, technical and forensic activities.

Image processing is done for two principal reasons: to improve visual appearance for a human observer, including printing and transmission, and to prepare images for measurement and for analysis of the features and structures that they reveal. Image processing methods can be considered under two principal categories: the correction of defects or limitations in acquisition, and the enhancement of important details. Image processing may alter the values or locations of pixels (picture elements) to produce another image. Image analysis, on the other hand, extracts numerical measurement information from the picture.

It is important to understand that the scale of an image (micrometers, feet, miles, or light years) matters little, as does the type of signal used to form the image. Most processing and measurement tools are equally applicable to a broad variety of images, and may be used in a very wide range of applications.

CORRECTION OF DEFECTS 1: COLOR ADJUSTMENT

Digital cameras and earth-observing satellites capture color images. Color correction should be the first operation performed if it is required. Compensation for variations in illumination can be made in several ways. The best results require capturing an image of known color standards under the same lighting, or having sufficient independent knowledge of the characteristics of the light source and the physics of the instrumentation.

With standards, a tristimulus matrix can be calculated that corrects for the overlap in the wavelength ranges of the filters used to form red, green, and blue (RGB) signals that are typically stored. In some cases, a simpler

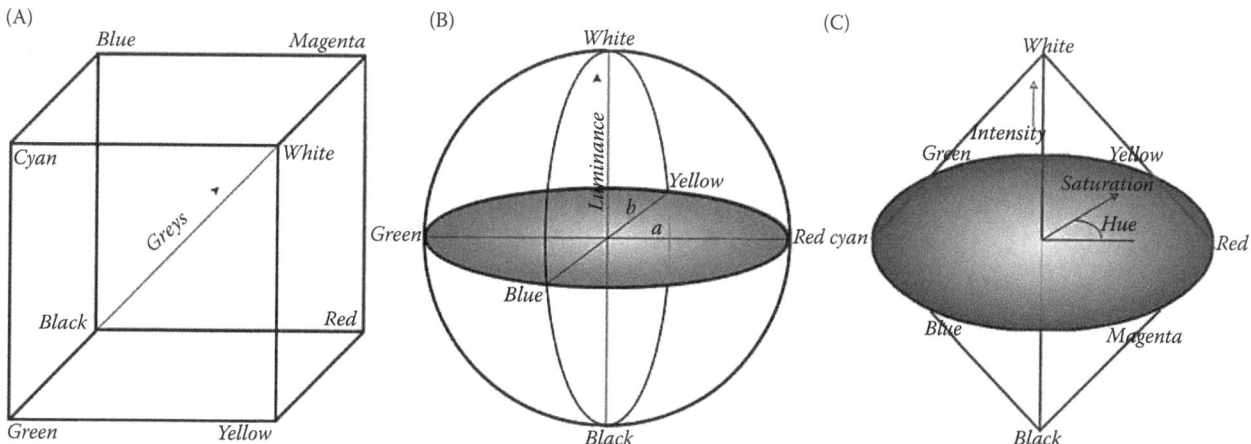

Fig. 1 Color spaces: (A) cubic RGB; (B) spherical L•a•b; (C) biconic HSI

and more approximate approach is used in which neutral gray objects are located and the RGB values are adjusted to be equal. This constructs adjustment curves for each color channel which are then applied throughout the image.

Most cameras and computers store and display color images as RGB values for each pixel, but for most processing and measurement purposes, other color spaces are more useful. L•a•b and HSI (Hue, Saturation, Intensity) color coordinates are often used as shown in Fig. 1. In the L•a•b space, which may be represented as a sphere with orthogonal axes, L is the luminance, or brightness, while the "a" and "b" axes are red–green and blue–yellow.

HSI space is more complicated, with H or Hue represented as an angle on the color wheel from red to yellow, green, cyan, blue, magenta, and back to red, while S is saturation, or the amount of color (e.g., the difference between gray, pink, and red), and I (also called V for value or B for brightness) is the intensity. This space may be represented as a cylinder, cone or bicone. In the bicone shown in the figure, saturation is reduced to zero at the ends. The color saturation at maximum intensity can be increased only by reducing some color contribution, and likewise at the dark end, saturation can be increased only by increasing intensity. Conversion from one color space to another is performed in software as necessary.

CORRECTION OF DEFECTS 2: NOISE REDUCTION

Noise is generally any part of the image that does not represent the actual scene, but arises from other sources. These may include the statistics of charge production in the detector, thermal or electronic noise in the amplifier and digitization process, electrical interference in transmission, vibration of the camera or flickering of the light source, and so on. The two principal kinds of noise are random and periodic; they are treated in different ways, under the assumption that they can be distinguished from the important details. Random or speckle noise usually appears as fluctuations in the brightness of neighboring pixels, and is treated in the spatial domain of the pixels, while periodic noise involves larger-scale variations and is best dealt with using the Fourier transform of the image.

Figure 2 shows an image with significant random noise, visible as variations in pixels in the uniform background above the cat's head. It arises primarily from the amplification required, because the photo was taken in dim light. The most common, but generally poor, approach used for random noise reduction is a Gaussian blur, which replaces each pixel value with the weighted average of the pixels in a small neighborhood. This reduces the noise as shown, but also blurs detail and shifts edges. It is identical to a low-pass filter in Fourier space that keeps low frequencies and reduces the high frequencies (variations over a short distance) that constitute the pixel-to-pixel noise variations, but which are also needed to define edges, lines, and boundaries. Extensions of the Gaussian model may adjust the weights applied to the neighboring pixels based on their difference in value or the direction of the local brightness gradient.

Median filters replace each pixel with the mean value found by ranking the pixel values in the neighborhood according to the brightness (all of the examples in the figure use a neighborhood with a radius of 3 pixels). The median filter is a non-linear operation that has no equivalent in Fourier space. This filter and variations that combine partial results from multiple neighborhoods, or use vectors for color images, are widely used and do a better job of preserving details such as lines and edges while reducing random noise. More computationally complex filters such as the non-local means filter[1] produce even better results. This works by replacing each pixel with a weighted average of all pixels in the image, based on the similarity of their neighborhoods.

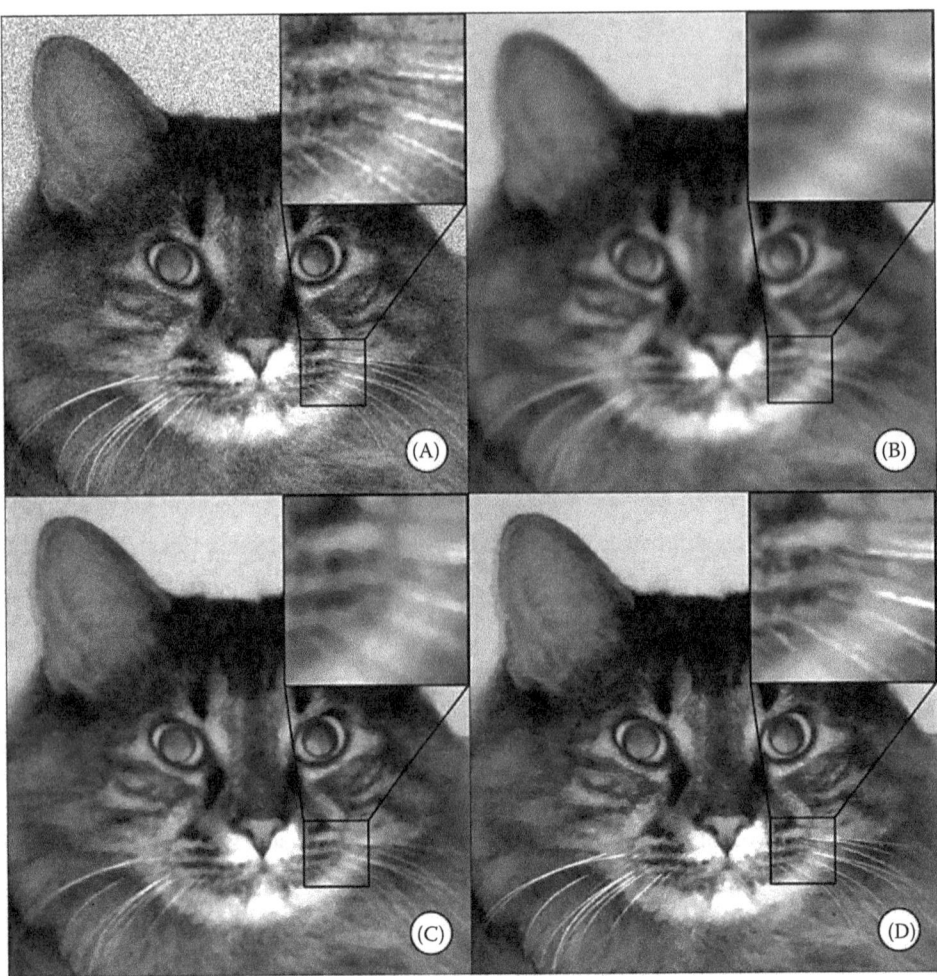

Fig. 2 Random noise reduction: (A) original; (B) Gaussian smooth; (C) median filter; (D) non-local means filter

Figure 3 shows an example of periodic noise. In the Fourier transform, this appears as "spikes" at radii corresponding to the frequency (inverse of the spacing of the lines) and at angles that correspond to their orientation. Removal of the spikes and calculating the inverse Fourier transform restores the image with the noise removed but all other information, which is composed of different frequencies and orientations, intact.

CORRECTION OF DEFECTS 3: NON-UNIFORM ILLUMINATION

A key assumption behind most methods for selecting features for measurement is that an object should have the same color and brightness wherever it happens to lie in the field of view. In some controlled situations, such as microscopy and laboratory setups, uniform illumination can be achieved. In real-world imagery, including crime scene photographs and satellite imaging of a curved planet, it may be difficult or impossible to do so. There are several ways to adjust the resulting image to correct the non-uniformity.

Figure 4 shows the preferred approach—recording an image of the background or substrate with the objects of interest removed. This background can then be subtracted or divided into the original to remove the variation. The choice of subtraction or division depends on how the camera recorded the brightness, as described below. When recording a background image is not practical, it may be possible to model the background by fitting a smooth function, typically a polynomial, to multiple points in the image that are known or assumed to be the same, or in some cases to calculate a background based on independent knowledge of the circumstances (such as the lighting of a spherical planet by the sun).

In other cases, it may be possible to "remove" the objects of interest by a morphological procedure called an opening. As shown in Fig. 5, replacing each pixel by its brightest neighbor, and repeating the operation until the dark letters are removed, and then reversing the operation and replacing each pixel by its darkest neighbor to restore the position of the edges and creases, produces a background image that can be subtracted.

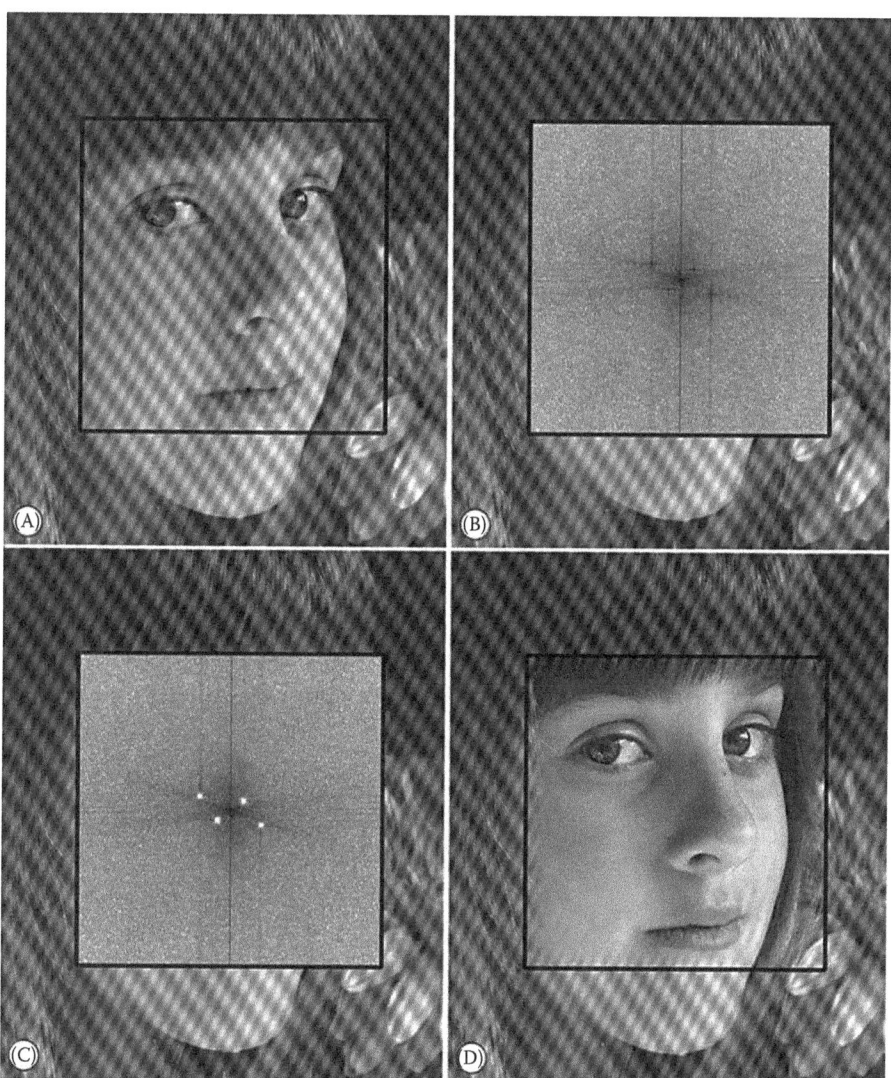

Fig. 3 Periodic noise removal: (A) original; (B) Fourier power spectrum of outlined region; (C) removal of spikes; (D) retransformed result

CORRECTION OF DEFECTS 4: GEOMETRIC DISTORTION

Measurements are most straightforwardly performed when the image shows the subjects of interest in a normal view of a flat surface. Transforming an image taken at an angle, or of a curved surface, requires knowing the geometry and performing a correction as shown in Fig. 6. Including rulers in images, and locating fiducial marks, is a critical step to enable this procedure and is standard practice in forensic imaging. Pixel values are interpolated from those in the original to generate the corrected image.

ENHANCEMENT 1: HISTOGRAM ADJUSTMENTS

After the corrective steps shown above, it is often useful to make adjustments to contrast and brightness. This is done by referring to the image histogram, a plot showing the number of pixels as a function of brightness. For color images, there may be a histogram for each channel, but adjustments, like all of the enhancement operations, should be performed on the brightness, luminance, or intensity values leaving the color information unchanged. Attempting to make adjustments to the RGB channels, for example, would alter the relative amounts producing new and strange colors in the resulting image.

When the brightness range captured in the image does not cover the full available dynamic range, a linear stretch of the values can be applied (Fig. 7B). It is important not to push pixel values beyond the black and white limits, causing them to be clipped to those values and data to be irretrievably lost. After the contrast expansion, there are just as many possible brightness values that have no pixels, as shown by the gaps in the histogram, but they are uniformly distributed across the brightness range rather than being collected at one or both ends of the histogram. Linear stretching is not the only possibility. Figure 7 shows

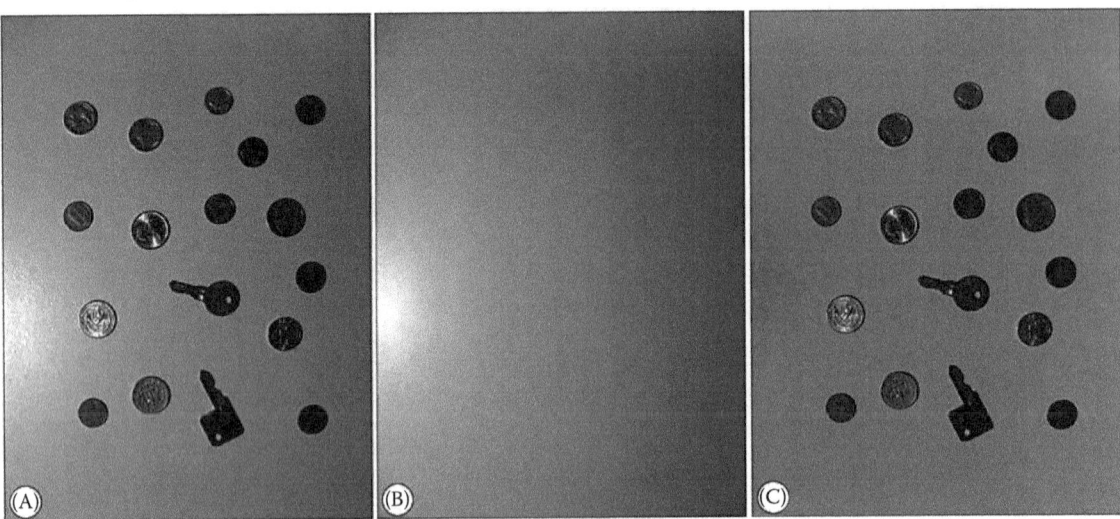

Fig. 4 Correction of non-uniform illumination: (A) original image; (B) image of the background with the objects removed; (C) after subtracting the background

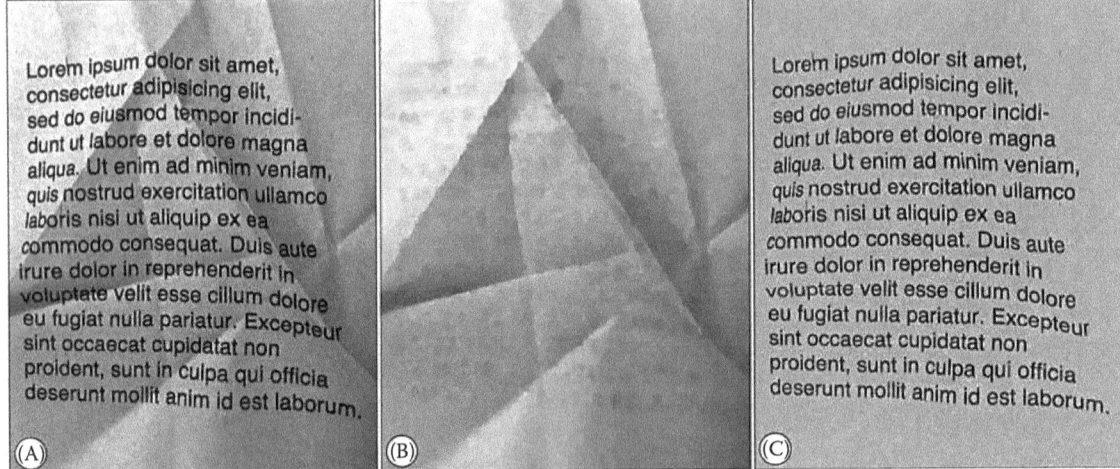

Fig. 5 Generating a background by morphological opening: (A) original; (B) removal of the letters; (C) subtracted result

Fig. 6 Geometric correction of the original image (A) produces a normal view (B) in which measurement of wear marks can be made to identify the individual tire. Similar procedures are used for footprints

Fig. 7 Histogram modification: (A) original image with limited brightness range; (B) linear stretch; (C) inverting the range to produce a negative image; (D) adjusting gamma to stretch the dark range and compress the bright values; (E) histogram equalization; (F) homomorphic compression

several other possibilities, with the resulting histogram shown for each case.

Adjusting the "gamma" value (Fig. 7D) changes the midgray point in the histogram and can expand the contrast for either the bright or dark portion of the image by compressing the values at the opposite end of the range. Rather than this manual adjustment, applying histogram equalization (Fig. 7E) adjusts values so that the histogram is as nearly uniform as possible, and all levels of brightness are represented by equal areas of the image. This is shown in the cumulative histogram, shown in Fig. 7E, which becomes a straight line. Equalization is often useful for comparing images taken under different lighting conditions. A more computationally intensive approach is the homomorphic transformation, which is applied in Fourier space by adjusting the amplitudes of dominant frequencies. In Fig. 7F, the details in both the bright and dark regions are clearly evident.

ENHANCEMENT 2: SHARPENING DETAIL

Human vision locates lines and edges in images as places where the brightness changes abruptly, and from these forms a mental sketch of the scene. Increasing the local contrast at steps, or narrowing the distance over which the change occurs, makes the image appear sharper. The simplest approach to this is the Laplacian filter, which calculates the difference between each pixel and the average value of its neighbors. A more flexible routine, called the unsharp mask and implemented in many programs, subtracts a Gaussian smoothed copy of the image from the original. This is a high-pass filter (it removes low frequencies or gradual variations in brightness, and "passes" or keeps the high frequencies) and may equivalently be performed using the Fourier transform. The results of all these "detail extracting" routines are typically added back to the original for viewing.

The difference of Gaussians or DoG filter is the most flexible of such an approach, which calculates the difference between two copies of the image which have been smoothed with Gaussians having different radii.[2] This is a band-pass filter that selects a range of frequencies, and can enhance detail while suppressing high-frequency noise as well as low-frequency variations. It is shown in Fig. 8B. Using similar logic but calculating the difference between median values in different-size neighborhoods (Fig. 8C) requires more computation but is superior in its ability to avoid haloes around edges. Local equalization (Fig. 8D) performs histogram equalization within a local neighborhood and keeps the new value only for each central pixel; this emphasizes the fine detail by increasing the difference, either positive or negative, between each pixel and its local neighbors.

DETAIL ENHANCEMENT 3: DEFINING EDGES

In addition to the visual enhancement of images, edges and boundaries are important for the measurement of features. Defining their position may be performed using several different approaches. The most common, the Sobel filter,[3] replaces the value of each pixel with the magnitude of the local gradient of pixel brightness, as shown in Fig. 9B. A different approach, the variance filter (Fig. 9C), calculates the statistical variance of pixel values in a neighborhood, responding strongly to local changes. Both of these produce broad lines because of the size of the neighborhood used for the calculation. The Canny filter (Fig. 9D) begins with the gradient but keeps only those pixels with the maximum value in the gradient direction, producing single-pixel-wide lines that mark the most probable location of the boundary.[4]

In addition to marking the location of boundaries, the brightness gradient vector has a direction that can be used to measure the orientation of edges. Figure 10 shows the use of the Sobel gradient vector to mark cellulose fibers used in papermaking with gray values proportional to

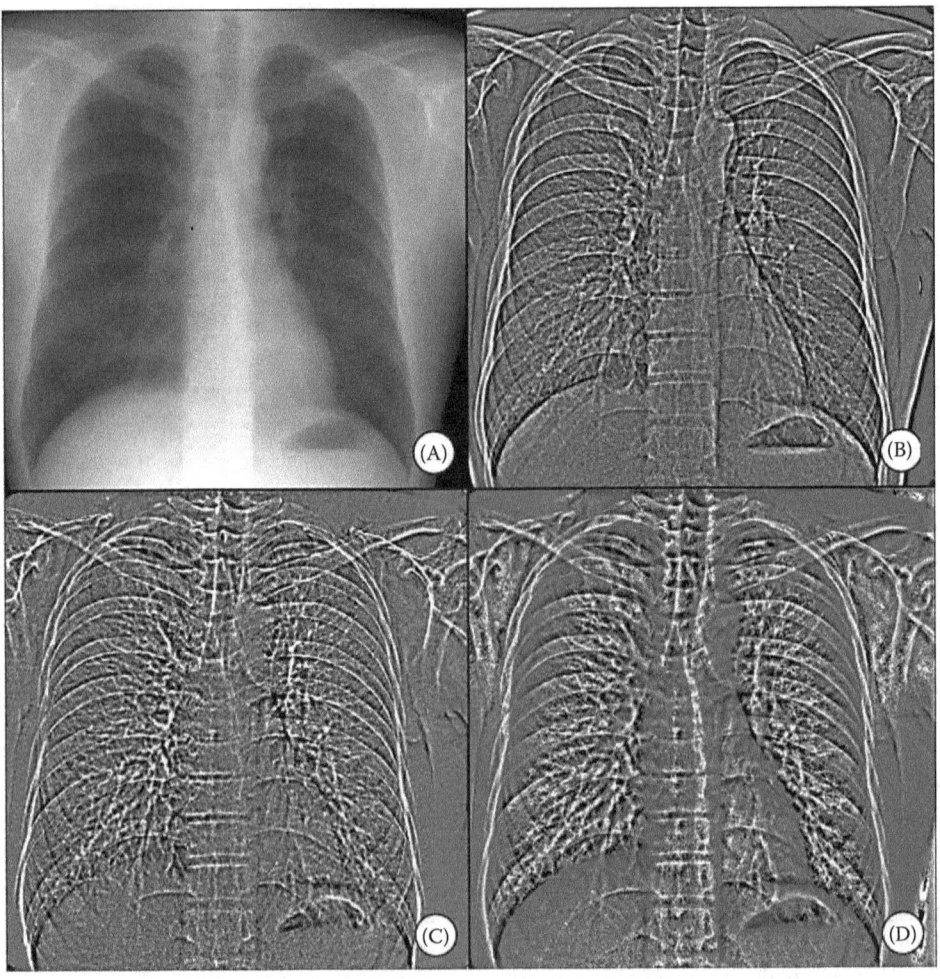

Fig. 8 Local detail enhancement: (A) original; (B) difference of Gaussians; (C) difference of medians; (D) local equalization

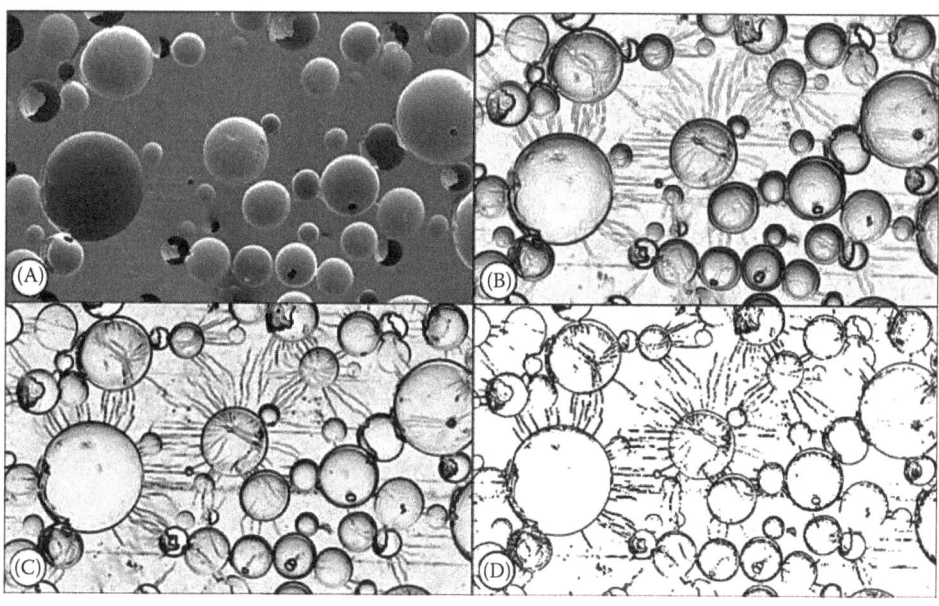

Fig. 9 Section through bubbles in a polymer: (A) original; (B) Sobel filter; (C) variance filter; (D) Canny filter

the local angle. A histogram of values, shown as a compass plot, indicates the non-isotropic distribution of fiber orientations.

DETAIL ENHANCEMENT 4: REVEALING TEXTURE

Features in images are not always distinguished by differences in brightness or color, or by outlined boundaries. Another criterion can be texture, which can be understood as a local variation in brightness or color. Figure 11 shows an example: the fat in the cheese does not have a distinct brightness, but has a "smooth" appearance while the surrounding matrix is highly textured.

Processing the image to replace each pixel value with the result from calculating various statistical properties of the local neighborhood can convert the image to one in which the regions have a unique brightness and can be isolated for measurement. The most commonly used properties are the range (difference between the brightest and darkest value) or the variance of the pixel values. In the figure, the fractal dimension has been calculated; this is a more complex calculation that fits the slope (on log–log axes) of the variance as a function of the size of the neighborhood. The resulting difference in brightness allows outlining the boundaries of the fat regions, so that their volume fraction and surface area can be determined using stereological relationships as explained below.

DETAIL ENHANCEMENT 5: PRINCIPAL COMPONENTS

RGB color images, and satellite images covering multiple wavelengths, may be processed using principal components analysis (also known as the Hotelling or Karhunen–Loève transform) to obtain one or more new color channels as a combination of the existing ones that can provide optimum contrast for the details in a particular image.

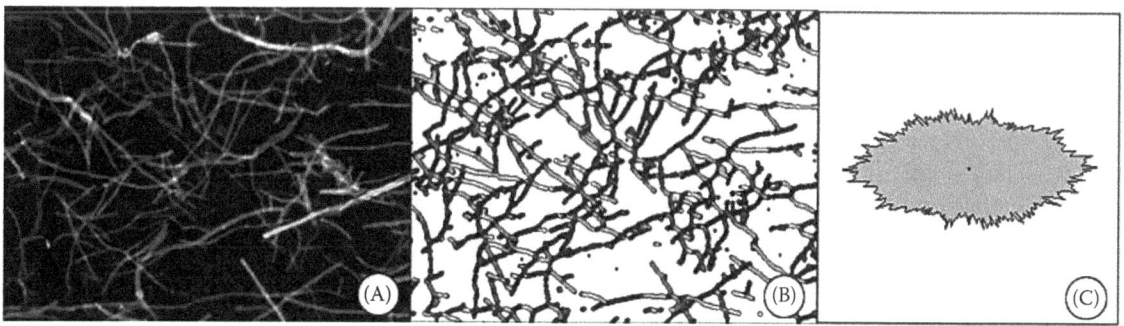

Fig. 10 Measuring fiber orientations: (A) original; (B) gray values along each fiber indicate the local compass angle; (C) rose plot of values shows a nearly 3:1 preferred orientation in the horizontal direction

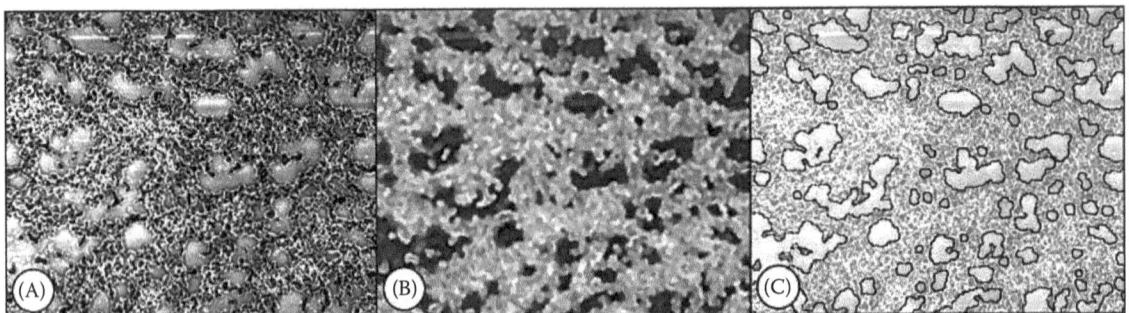

Fig. 11 Microscope image of a section of cheese, showing the smooth areas: (A) original; (B) fractal dimension as described in the text; (C) resulting outlines superimposed on the fat regions for measurement of volume and surface area

This can be visualized as a rotation of the color coordinate axes as shown in Fig. 12. The original image is a fingerprint on a check which has an imprinted texture pattern. In the original RGB channels the minutiae in the print are difficult to discern. Plotting each pixel's RGB values in a three-dimensional (3D) graph shows correlation, and fitting a plane to the data produces the maximum dispersion of the values and hence the greatest contrast.

Using the position of each pixel's point along the new principal component axes results in the images shown that separate the fingerprint from the printed background pattern. The third axis, which is perpendicular to the plane, generates an image with little contrast, containing primarily the random noise in the original image.

DETAIL ENHANCEMENT 7: IMAGE COMBINATIONS

The example of subtracting a recorded background image was shown above. There are other situations in which two or more images of the same scene may be acquired, for instance, using different wavelength bands, or different lighting, or different camera focus. Processing an image may also produce an additional representation (for example, the Gaussian blurred copy that is subtracted to produce the unsharp mask result).

Arithmetic operations between images are performed pixel by pixel, with scaling and offset applied to keep the resulting values within the permitted range (for single-byte images this is 0...255, but some programs accommodate many different bit depths and normalize all of them to 0...1 using real numbers rather than integers).

Either subtraction or division is used for removing background, depending on whether the acquisition device responds logarithmically (like film and vidicon cameras) or linearly (solid-state detectors, but the electronics may convert the result to logarithmic in order to mimic film). Division is used to ratio one wavelength band to another, compensating for variations in illumination and (for example) the curvature of the earth. Addition may be used to superimpose difference-of-Gaussian or edge-delineation results on the original image for visual enhancement. Multiplication is less often used, but is applied in graphics applications, for example, to superimpose texture on smooth regions. In addition, mathematical operations include keeping whichever pixel value is greater or smaller, and for black-and-white or "binary" thresholded images, the various Boolean operations (AND, OR, Exclusive-OR, and their combinations) are useful for combining various selections and information.

When a series of images acquired with different focal planes are captured, they can be combined to keep whichever pixel value at each location gives the sharpest focus, resulting in an extended focal depth. The pixel value selected may be the one with the highest local contrast or variance in its neighborhood. Figure 13 shows an example, with a map indicating the original image from which each pixel in the composite was selected.

DETAIL ENHANCEMENT 8: DECONVOLUTION

When the Hubble telescope was first launched, a fabrication error in the curvature of the primary mirror caused the images to be out-of-focus. Several years later, a replacement optical package was installed that compensated for the incorrect primary curvature, restored the focal sharpness, and increased the amount of light directed to the instrument package. But in the interim, sharp images were obtained by deconvolution using computer software. If the point spread function (PSF) of the optics, which is simply the recorded image produced by point of light like a star, can be either calculated or measured, it can be used to remove much of the blur introduced in image capture, either due to the optics or due to motion.

Figure 14 shows an example. The process is usually performed in Fourier space, with the most basic algorithm (Wiener deconvolution) dividing the transform of the blurred image by that of the PSF, plus a small scalar constant that depends on the amount of noise present.[5] Other methods include iterative techniques that may try

Image Processing and Measurement

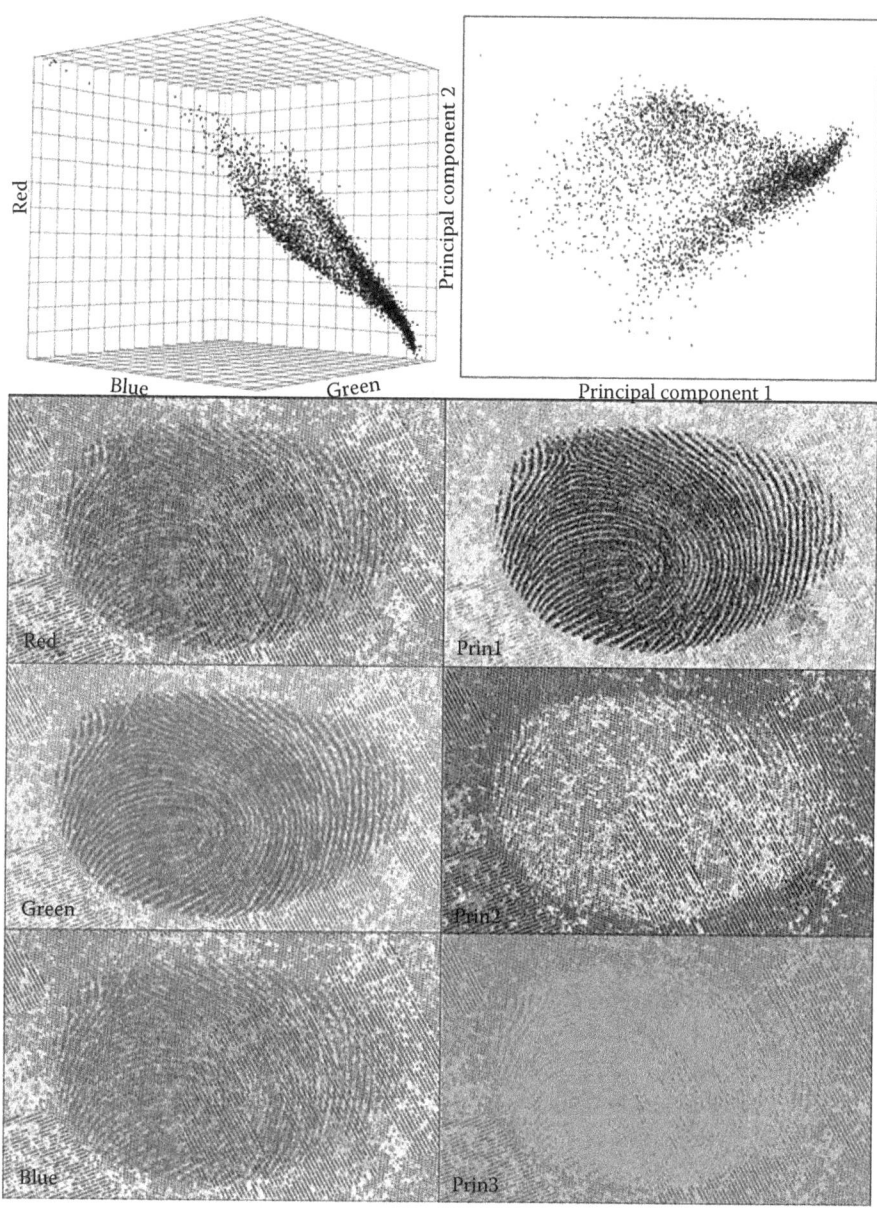

Fig. 12 Principal components: the red, green, and blue channels of the original image do not distinguish the fingerprint; the images formed using the principal component values separate the fingerprint from the printed background

to determine the PSF from the image itself (e.g., Lucy–Richardson deconvolution[6,7]). The results are never as good as a perfectly focused original image, because the noise is increased and not all of the blur can be removed. However, improvement over the original blurred image can be great, and for images such as forensic evidence may be critical.

DETAIL ENHANCEMENT 8: CROSS-CORRELATION

Cross-correlation is used to align images, and also to locate a target in a scene. It is often used for aerial surveillance, machine and robotic vision, and for finding faces in images. It is frequently carried out using Fourier transforms, but for small targets may be applied in the spatial or pixel domain. It is easy to visualize the process as having the target image on a transparent film and sliding it across all locations in the scene image to find a match. The result is another image in which each pixel records a measure of the similarity of that location to the target. Figure 15 shows an example. Searching for the target particle shape finds all of the occurrences with high matching scores, in spite of the different contrast for single particles versus those in groups, while ignoring the background texture of the filter and objects present with other sizes or shapes.

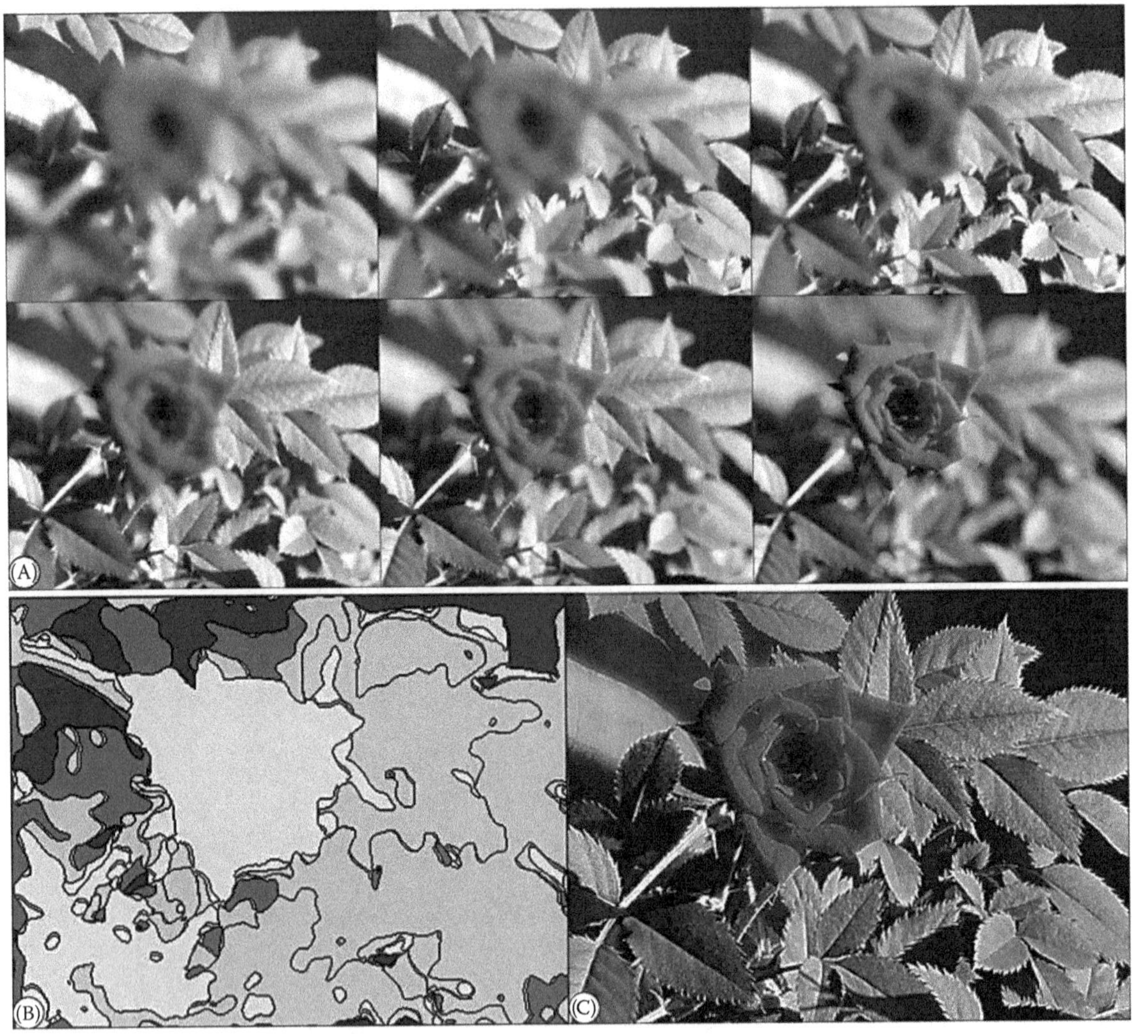

Fig. 13 A series of images taken with different focal settings (A), the extended focus composite produced by selecting the pixel with the greatest local variance (C), and a map showing the source of each pixel (B)

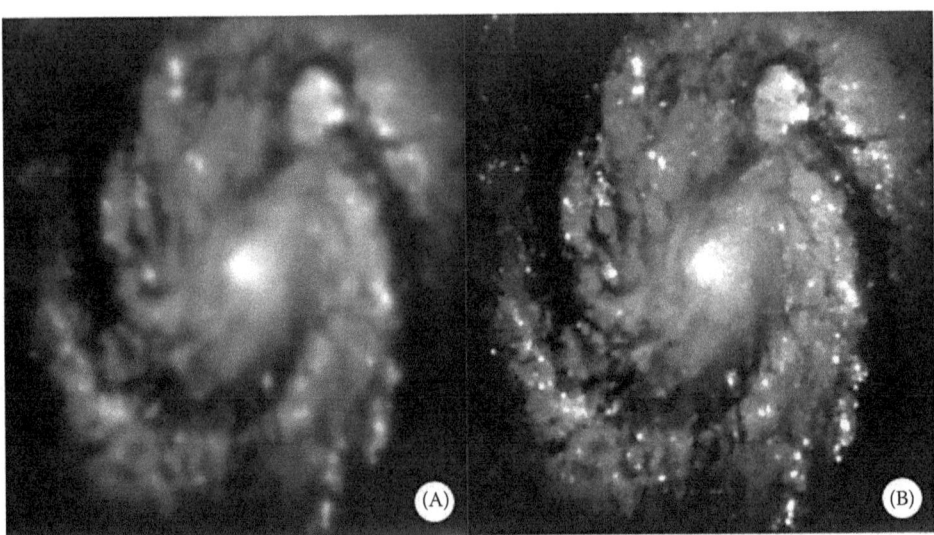

Fig. 14 Deconvolution: (A) original blurred image and (B) deconvolved result, which reveals fine details

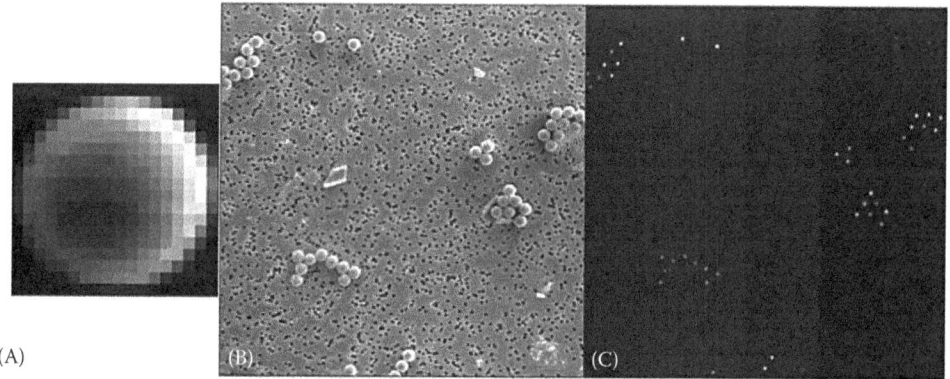

Fig. 15 Cross-correlation: (A) enlarged target image showing individual pixels; (B) image of filter with particles on a complex background; (C) cross-correlation result marking the particle locations

BINARY IMAGES 1: THRESHOLDING (AUTOMATIC)

Except for manual measurements on images, in which a human marks points using a mouse and the computer reports distances, most image measurements are performed after thresholding or segmentation to delineate the objects, structures, or other features of interest. Manual measurements are generally suspect, because of non-reproducibility and the possible influence of expectation or desire. For the same reason, manual thresholding, although often used, is not a preferred approach.

Thresholding selects pixels based on some defining characteristics as belonging to the features of interest. The process may identify all pixels at once as part of one or another of several classes of structure, or simply erase as background those which are not part of the structure of current interest. The simplest of all types of thresholding is based on the brightness histogram of the image, as shown in Fig. 16A.

A peak in the histogram indicates that many pixels have similar brightnesses, which may indicate that they represent the same type of structure. Placing thresholds "between peaks" may distinguish the features of current interest. In the example in the figure, the bright peak corresponds to the paper but there is no dark peak representing the ink. Instead, a statistical test is used to select the threshold value (marked with an arrow) that is used to (hopefully) isolate the printed characters for measurement and ultimately identification.

The test illustrated is one of the most widely used, producing often satisfactory and at least reproducible results. It uses the Student's t-test to compare the values of pixels above and below each possible threshold setting and selects the one that produces the greatest value of t. This indicates that the two groups are most different and distinguishable.[8,9] However, the statistical test makes the tacit assumption that the two populations have Gaussian or normal distributions, which is rarely the case. There are a variety of other statistical tests, which use entropy, fuzzy weighting of values, and other means, and which produce somewhat different threshold settings.

A different approach to automatic threshold setting uses not only the value of the pixels but also those of their immediate neighbors. The logic behind the test is that pixels within features, or within background, should be similar to their neighbors, while those along borders should not. A co-occurrence matrix that counts the number of

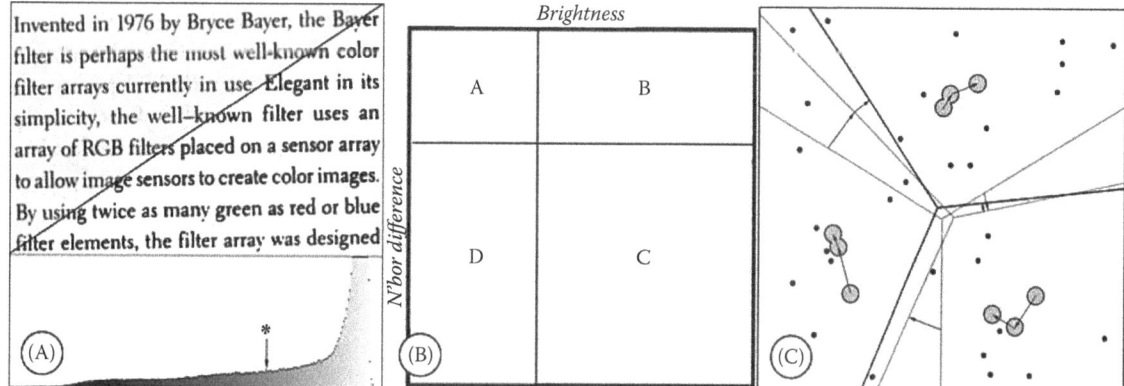

Fig. 16 Methods for automatic thresholding: (A) based on histogram statistics; (B) using a co-occurrence matrix; (C) iterative k-means cluster analysis

pixels with each value along one direction and the number with each average neighbor value along the other is used, as indicated schematically in Fig. 16B. In the figure, the areas marked A and C are the more uniform regions, corresponding to features and the background in which pixels are similar to their neighbors. Those marked B and D represent the borders where they are different. Iteratively adjusting thresholds to maximize the total counts or the entropy in A and C and minimize those in B and D is a more computer-intensive approach, but is superior in performance.[10]

Another powerful method is indicated in Fig. 16C. The k-means procedure[11] is particularly appropriate for color or multichannel images (the figure shows just two dimensions, but the method generalizes directly to any number). The values of all pixels are plotted and the method searches for clusters. An initial set of k locations are selected arbitrarily, and all pixel points that are closest to each location are temporarily given that class identity. The mean of each class is then used as the next proposed cluster center, and the procedure is repeated. This causes some points to change identity, and the cluster boundaries and cluster means to change. The procedure continues until no further changes take place.

BINARY IMAGES 2: THRESHOLDING (INTERACTIVE)

Other approaches to thresholding or segmentation are sometimes useful, and may involve some degree of human interaction. Figure 17A shows the use of an edge-marking procedure as described above to outline each object. Since there are also some lines and edges drawn within the objects (e.g., the pumpkin ridges and stems), it becomes necessary for the operator to select those lines that are the object boundaries, but which have been located automatically.

Human selection also operates in the seed-fill or region-growing approach, shown in Fig. 17B. Marking an initial point (indicated by an asterisk in the figure) begins the process. Then every neighboring point is examined and ones that are similar are added to the growing region. This continues until no further neighboring points are added. The resulting selection is outlined in the figure. The test for similarity can be a fixed range of color or brightness, or it may be based on the statistics of the growing region, or may be weighted toward the values of the pixels near the local expanding boundary. The most common problem with region growing is that it may "escape" from the feature and become free to spread across background or other objects.

Figure 17C illustrates the active contour approach. It begins with a manually drawn outline, which then contracts until it is stopped by the borders of the object (active contours that expand from an inner outline can also be used). The stopping may be based on color or brightness, gradient, or other criterion. Active contours can bridge over gaps where the border is indistinct because the shrinking criterion seeks to minimize the energy in the boundary, based on its length and curvature.[12] Active contours may be called "snakes," and when applied in 3D are referred to as "balloons."

These are not the only approaches used for thresholding and segmentation. Top-down split and merge segmentation examines the histogram for the image and, if it is not uniform by some statistical test, divides the area into parts. Each of these is examined similarly and divided, and the process continues. At each iteration, adjacent regions with different previous parents are compared and joined if they are similar. The final result reaches the level of individual pixels and produces a set of regions. Other computer-intensive methods include fuzzy approaches to cluster analysis that weight pixels by how different they are from the cluster mean, and neural net approaches which begin with the entire array of pixel values as input.

BINARY IMAGES 3: MORPHOLOGICAL PROCESSING

Thresholded images are an often imperfect delineation of the features or structures of interest. Random variations in pixel values may cause some individual errors, boundaries may be poorly defined if the finite size of pixels straddle them and have intermediate values, and some pixels may

Fig. 17 Additional segmentation methods: (A) edge delineation; (B) region growing; (C) active contours

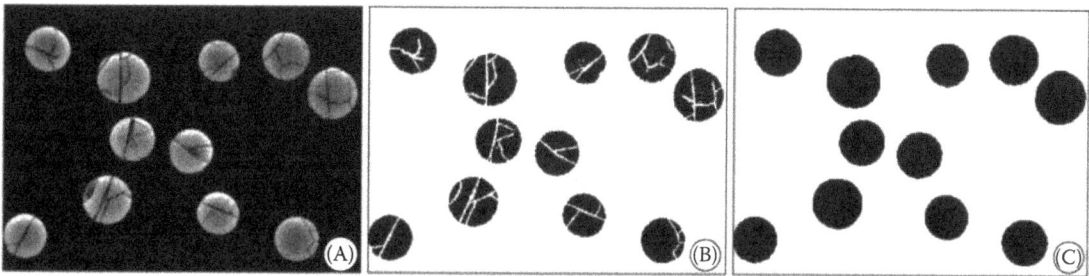

Fig. 18 Applying a closing: (A) original image showing cross-sections of glass fibers; (B) thresholded image showing cracks; (C) filling the cracks with a closing

have values that are the same as those within the structures of interest. These flaws are usually small in dimension (often single pixels) and are dealt with by morphological operations of erosion and dilation, which remove or add pixels according to the identity of their neighbors.

Dilation in its simplest form adds background pixels that are adjacent to a feature boundary, and erosion removes feature pixels that are adjacent to background. Since each of these changes the size of the object, they are usually used in combination. Figure 18 shows an example, in which a closing (the sequence of dilation followed by erosion) is able to fill internal gaps without changing the external dimensions of the fibers. The opposite sequence, erosion followed by dilation, is called an opening and is used to remove background noise or speckle.

Continued erosion with a rule that a pixel may not be removed if it causes an object to divide into two parts generates the feature skeleton. An alternative method assigns to each pixel within a feature a value that measures its straight line distance to the nearest background point. The ridges in this Euclidean distance map (EDM) define the skeleton and their values form the medial axis transform, which is often useful for measurement purposes.

In the example in Fig. 19, the number of end points in the skeleton (pixels with only one neighbor) identifies the number of teeth in the gear. In other cases, the number of node points (pixels with more than two neighbors) measure network connectivity. Euler's rule for the topology of skeletons in two-dimensional images is (number of loops − number of segments + number of ends + number of nodes = 1).

The EDM is also used to separate touching features, as shown in Fig. 20. The watershed segmentation method considers "rain" falling on the EDM and proceeds downhill from the peaks to locate points that would receive runoff from more than one initial peak. These locations mark watershed boundaries and are removed, leaving separated features for measurement. The method works for mostly convex features that have only a single peak in their EDM, with overlaps less than their radii.

MEASUREMENTS 1: PHOTOGRAMMETRY

Dimensions and spatial arrangements of objects in 3D scenes can be determined from measurements on images. In some cases, such as accident reconstruction, image measurements are used to construct detailed 3D models. Sometimes measurement is based on multiple images taken from different positions, for example, stereo pair images, employing trigonometry. But even single images often can be accurately interpreted to determine 3D information.

Fig. 19 A gear, with its skeleton superimposed, and the Euclidean distance map

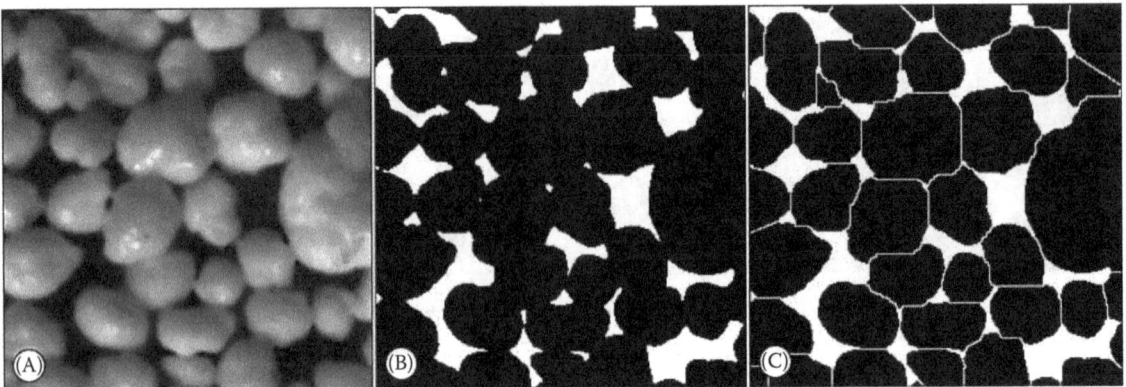

Fig. 20 Watershed segmentation: (A) original image of sand grains touching each other; (B) thresholded image; (C) after watershed segmentation

For example, knowing the location and lens specification of a surveillance camera makes it possible to determine the height of a person from the image. This can be done trigonometrically, but a scaled drawing of the geometry also provides a solution and is easier to explain, for instance, to a non-technical jury. An even simpler method, called "reverse projection," requires taking a suitable measuring ruler to the scene and recording its image using the same camera and geometry, and then superimposing the two images as shown in Fig. 21 so that the height or other dimension can be read directly.

Another forensic example is the measurement of a blood spatter pattern. The elongation of each droplet gives the angle and direction from which it arrived at the surface (a wall, floor, table, etc.). The intersection point of lines projected back in the indicated directions locates the point in space where the droplets originated, which is the exit wound from a gunshot and hence determines the location of the victim when shot.

MEASUREMENTS 2: STEREOLOGY

Sections through 3D samples are typically imaged in various kinds of light and electron microscopes, and are also produced by tomographic imaging using light, X-rays, sound, neutrons, and many other signals. The features revealed in these section images do not directly show the size or even the number of objects present in the space, because the sampling plane may pass through any portion of the object, not necessarily showing its full extent. However, it is possible using rules derived from geometric

Fig. 21 Reverse projection measurement used to measure the height of a bank robber
Source: Image courtesy of George Pearl, Atlanta Legal Photo Services, Atlanta, GA.

probability to infer many important structural parameters, including the volume fraction, surface area, length, curvature, number, and connectivity of the objects.

This field is known as stereology (from the Greek for study of three-dimensional space). Many of the rules and procedures are simple to apply and involve counting of "events"—the intersection of the structure(s) of interest with properly designed grids of lines or points—rather than the measurement of dimensions. The key to using stereological relationships is understanding that a section plane intersects a volume to produce an area, intersects a surface to generate a line, and intersects a linear feature producing points. In all cases, the dimension of the structure of interest is one greater than the evidence found in the image.

For example, the volume fraction of a structure is measured by the fraction of points in a regular or random grid that fall on the structure. The surface area per unit volume is equal to two times the number of intersections that a line grid makes with the surface, divided by the total length of the line, or to $(4\pi/3)$ times the length of the boundary line divided by the image area. The length of a linear structure per unit volume is two times the number of intersection points divided by the image area. In all cases, care is needed in the design of grids and the sectioning techniques used in order to produce unbiased results. This somewhat specialized topic is well covered in texts such as that by Baddeley and Vedel Jensen.[13]

MEASUREMENTS 3: FEATURE BRIGHTNESS, SIZE, AND LOCATION

The measurements of individual features in images fall generally into four groups: brightness or color, location, size, and shape. It is also important in many cases to count the number of features present. Figure 22 shows an image of rice grains captured using a desktop flatbed scanner. Some of the rice grains intersect the edges of the image, indicating that this is a sample of a larger field of objects. One unbiased procedure for counting the number per unit area is to count as one-half those grains that intersect the edges, since the other "half" count would be obtained if the adjacent field of view was measured.

For measurement purposes, the edge-intersecting grains cannot be used, as their dimension is unknown. Since large objects are more likely to intersect an edge, the bias in a measured size distribution such as the one shown in the figure can be compensated by counting each measurable grain with a weighting function equal to

$$(Wx \bullet Wy) / ((Wx - Fx) \bullet (Wy - Fy)), \qquad (1)$$

where Wx and Wy are the dimensions of the image in the x and y directions, and Fx and Fy are the projected or box dimensions of each object in those directions. For very small features, this weight is nearly 1, but for large features, it is greater than one to compensate for other similar-size objects that would have intersected the borders of the image and have been excluded from the measurements.

The distribution of the length of the rice grains is used, for example, to determine that the sampled rice has a small percentage of short grains and can be sold as "long-grain" rice. There are many other useful measures of size, such as area (which may or may not include internal holes and peripheral indentations), the radii of the maximum inscribed and minimum circumscribed circles, and the perimeter.

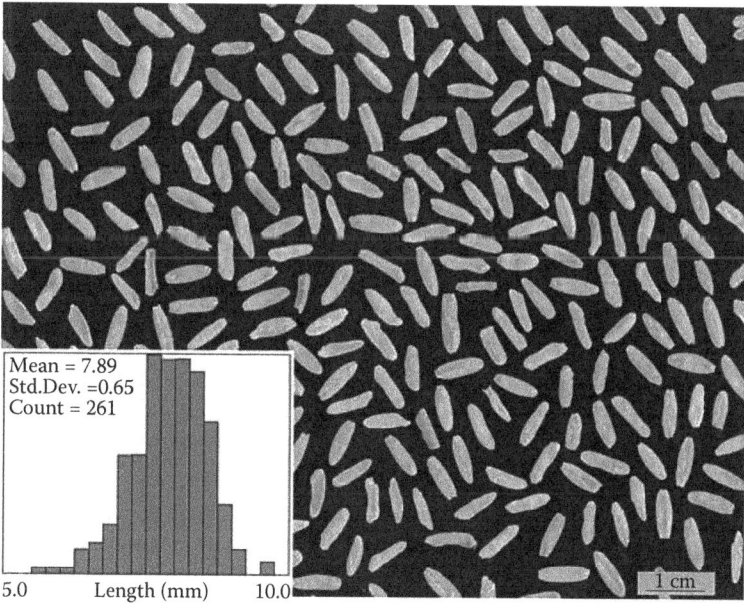

Fig. 22 Measurement of the distribution of lengths of rice grains

Perimeter is the most difficult measurement to determine properly. It may be calculated using the center-to-center path through the boundary pixels, or along their outer edges, or by fitting smooth curves, and these all give slightly different results. More importantly, the perimeter depends on the pixel size and resolution of the original image, and in many cases, as the magnification increases, the resolution reveals more and more irregularities, so that the perimeter is not a well-defined concept. Indeed, the rate at which perimeter varies with resolution is one of the ways to determine the fractal dimension of a shape.[14]

Pixel brightness values can be calibrated to measure density and other object parameters, but the values recorded in the RGB channels cannot be used to measure color in the sense of a spectrophotometer. This is because the filters used in cameras cover ranges of wavelengths so that different combinations of intensity and wavelength can produce identical results. This is also true for satellite images, which record many bands with each one covering a range of visible or infrared wavelengths.

The location of objects can be determined as their centroids, which may be weighted by density determined from the pixel values. Location may also be based on the center of the circumscribed or inscribed circles in some cases; the latter location is the only one guaranteed to lie within the boundary of the object. One use of location data for a collection of objects is determining whether the objects are clustered, randomly arranged, or self-avoiding. Cacti in the desert are naturally self-avoiding, as they compete for water and nutrients. People cluster in cities (and stars cluster in galaxies). Raindrops fall as separate events and their impacts are random. A comparison of the mean nearest neighbor distance between features with the square root of (image area/number of features) reveals these trends.[15] The measured value is less than the calculated test value for clustering, and greater for self-avoidance.

MEASUREMENTS 4: FEATURE SHAPE

Shape is a difficult concept to describe, and humans generally resort to nouns rather than adjectives ("… shaped like a …"). "Round" may mean "like a circle" (or a sphere or cylinder) but might also mean without indentations and sharp corners. "Skinny" and "bent" generally have meaning only by comparison to other forms. Putting numbers to shape description is complicated as well. The simplest and most widely used approach to measuring shape uses dimensionless ratios of size measurements. Table 1 lists a few as examples, but it should be understood that various names are assigned to these relationships with no consistency, and that it is possible to have shapes that are visually entirely different that share values for one or several of these ratios.

To illustrate the use of dimensionless ratios, a collection of leaves from various trees was used.[16] Figure 23 shows representative examples (not at the same scale), with a plot of the values for three of the shape factors that are able to identify the various species based on shape alone. The regions occupied by the points in each class are irregular, and improved results can be obtained by using linear discriminant analysis to calculate canonical variables, which are linear combinations of the measured parameters. This produces the plot shown in the figure, in which each class is represented by a spherical region centered on the mean value with a radius of two standard deviations.

Other methods for shape description can also distinguish all of these classes. The principal ones in use are harmonic coefficients and moments. The former is based on the periphery of the feature, for example, expressing the point coordinates along the boundary in complex form $(x + iy)$. A Fourier transform of the boundary then represents the shape as a series of terms, and the amplitudes can be used as numeric shape descriptors.[16,17] Instead of a Fourier transform, a wavelet transform may also be used.

Moments, on the other hand, use all of the interior pixel coordinates as well, which can be an advantage if the boundary is poorly defined, or when the shape consists of multiple parts (e.g., an animal paw print). There are invariant moments[18,19] that may be used to describe shape. Both the harmonic coefficients and the moment values can be used in subsequent statistical analysis for comparison and correlation.

MEASUREMENTS 5: DATA ANALYSIS

Measurements on objects and structures obtained from images are typically used for descriptive statistics and classification, and for correlation with object history or function. The common statistical parameters (mean, standard deviation, etc.) are convenient but make the tacit assumption that the values are normally distributed, which is not always the case (especially rarely so for shape parameters).

Non-parametric comparison between data sets using Mann–Whitney or Kolmogorov–Smirnov statistics are

Table 1 A few dimensionless ratios that may be used to describe shape

$$\text{Radius ratio} = \frac{\text{inscribed diameter}}{\text{circumscribed diameter}}$$

$$\text{Roundness} = \frac{4 \cdot \text{area}}{\pi \cdot \text{max diameter}^2}$$

$$\text{Form factor} = \frac{4\pi \cdot \text{area}}{\text{perimeter}^2}$$

$$\text{Aspect ratio} = \frac{\text{max caliper dimension}}{\text{min caliper dimension}}$$

$$\text{Solidity} = \frac{\text{area}}{\text{convex area}}$$

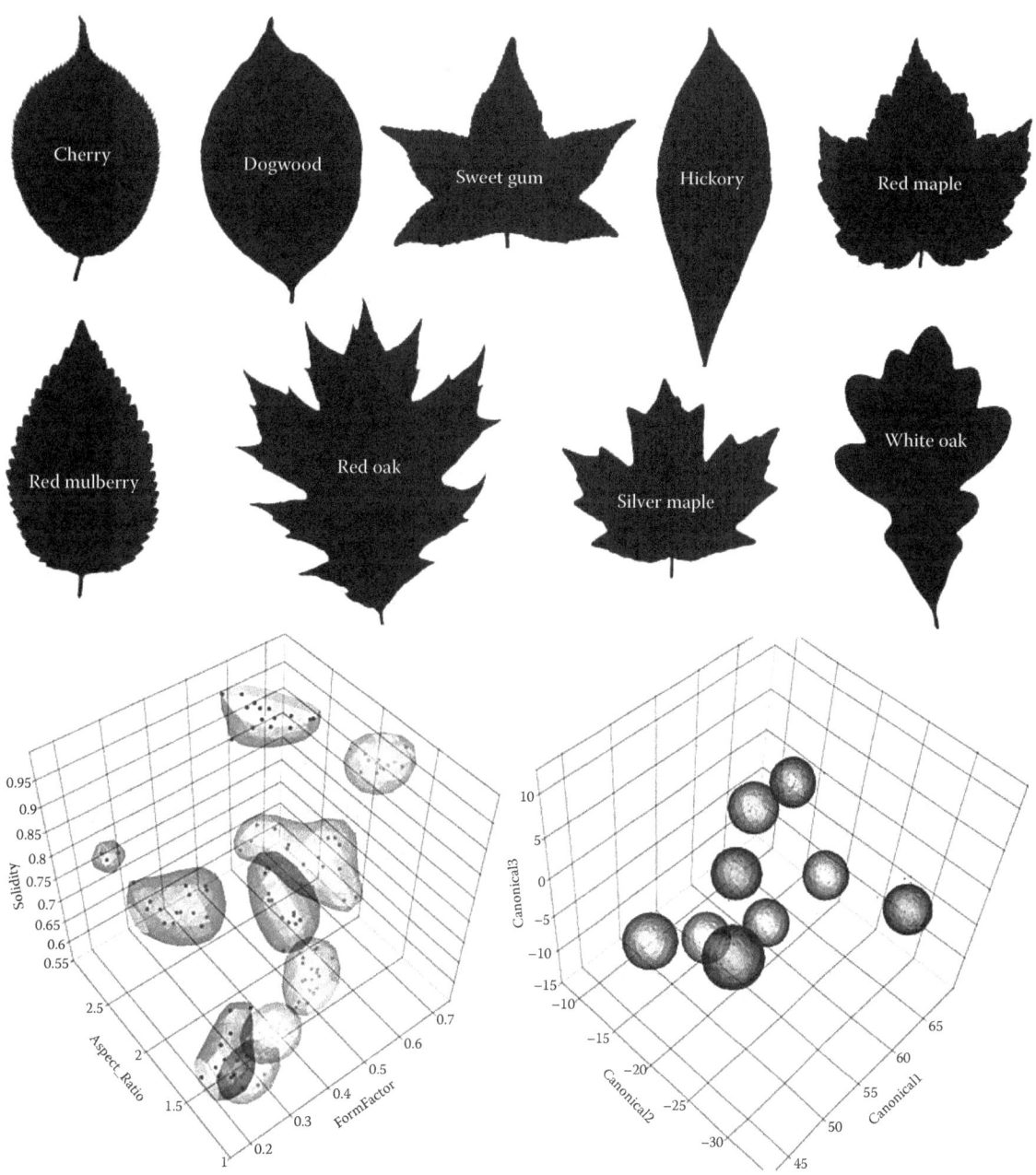

Fig. 23 Measurement of shape using leaves: several dimensionless shape factors are shown that can distinguish the classes, as well as the canonical parameters calculated by linear discriminant analysis

preferred, as they yield meaningful probabilities whether the data are normal or not. Likewise, correlation based on rank order (Spearman's correlation) is preferred over the usual Pearson's correlation if relationships may be nonlinear. The interpretation of the r-squared value is the same in both cases.

Classification based on measurements such as those shown in Fig. 23 may use linear discriminant analysis, neural nets, fuzzy cluster analysis, or k-nearest neighbor tests. These are standard tools for treating data, not limited to measurements from images, and are well covered in most statistics texts.

A particular interest for image analysis is database searching. Landmark methods, such as the Automated Fingerprint Identification Service (AFIS), work by using the relative location of multiple points. For fingerprints, these are minutiae such as the gaps, ends, and bifurcations of ridgelines in the print. A list of 12–16 such landmarks can call up the 10 or so most similar fingerprints on file, for a human to compare. Similar use of human judgment of a small number of "most like" selections found by automatic search algorithms is used in medical diagnosis, such as in the analysis of Pap smears and mammograms.

An elusive goal for image analysis is "query by example" in which the presentation of an image is used to locate other images of similar objects. The problem is that with a few exceptions such as finding paintings with the same predominant color(s), it is not easy for computer algorithms to decide what it is that the presenter believes to be the important characteristics of the example image. Online Internet searches for images work using the words in accompanying text, not the contents of the images themselves.

ACKNOWLEDGMENTS

The explanations and topics covered, and the examples shown, are from *The Image Processing Handbook* (John C. Russ, CRC Press, 2011). More detailed information, additional examples and comparisons of algorithms, and extended references are also available there.

REFERENCES

1. Buades, A.; Coll, B.; Morel, J.M. A non-local algorithm for image denoising. Comput. Vis. Pat. Recog. **2005**, *2*, 60–65.
2. Marr, D.; Hildreth, E. Theory of edge detection. Proc. R. Soc. Lond. B Biol. Sci. **1980**, 207 (1167), 187–217.
3. Sobel, I.E. *Camera Models and Machine Perception*; PhD thesis; Publ. AIM-121,; Stanford University Electrical Engineering Department, 1970.
4. Canny, J. A computational approach to edge detection. IEEE Trans. Pat. Anal. Mach. Intell. **1986**, PAMI-6, 679–698.
5. Pratt, W.K. Generalized Wiener filter computation techniques. IEEE Trans. Comput. **1972**, *C-21*, 636–641.
6. Richardson, W.H. Bayesian-based iterative method of image restoration. J. Opt. Soc. Am. **1972**, *62*, 55–59.
7. Lucy, L.B. An iterative technique for the rectification of observed distributions. Astronom. J. **1974**, *79*, 745.
8. Otsu, N. A threshold selection method from gray-level histograms. IEEE Trans. Syst. Man Cybernet. **1979**, *9* (1), 62–66.
9. Trussell, J. Comments on "Picture thresholding using an iterative selection method". IEEE Trans. Syst. Man Cybernet. **1979**, *9* (5), 311.
10. Pal, N.R.; Pal, S.K. Entropic thresholding. Signal Process. **1989**, *16*, 97–108.
11. Hartigan, J.A. *Clustering Algorithms*; John Wiley & Sons: New York, 1975.
12. Mumford, D.; Shah, J. Optimal approximation by piecewise smooth functions and associated variational problems. Commun. Pure Appl. Math. **1989**, *42*, 577–685.
13. Baddeley, A.; Vedel Jensen, E.B. *Stereology for Statisticians*; Chapman and Hall/CRC: Boca Raton, FL, 2005.
14. Mandelbrot, B.B. *The Fractal Geometry of Nature*; W. H. Freeman: San Francisco, CA, 1982.
15. Schwarz, H.; Exner, H.E. The characterization of the arrangement of feature centroids in planes and volumes. J. Microsc. **1983**, *129*, 155.
16. Neal, F.B; Russ, J.C. *Measuring Shape*; CRC Press: Boca Raton, FL, 2012.
17. Lestrel, P.E.; Ed.; *Fourier Descriptors and Their Applications in Biology*; Cambridge University Press: Cambridge, 1997.
18. Hu, M.K. Visual pattern recognition by moment invariants. IEEE Trans. Inform. Theory **1962** *IT-8*, 179–187.
19. Flusser, J.; Suk, T. Pattern recognition by affine moment invariants. Pat. Recogn. **1993**, *26* (1), 167–174.

Image Retrieval

Vittorio Castelli
T.J. Watson Research Center, IBM, Yorktown Heights, New York, U.S.A.

Abstract
We describe approaches and techniques for indexing and retrieving still images from multimedia databases. We specifically emphasize content-based image retrieval (CBIR), a class of techniques where the user composes queries that specify the content of the desired images. After a brief overview of digital image formats, we analyze different approaches to content specification: in terms of low-level visual features, of objects, and of metadata. We then describe a general progressive framework that combines these approaches. We finally conclude the entry with an overview of common applications of image repositories and digital libraries, such as medical imaging, remote-sensing imaging, and data for the oil industry.

INTRODUCTION

During the first decade of the twenty-first century digital images have rapidly supplanted traditional film-based images in many application areas. The most visible example is consumer photography, where, due to technological advancements, the digital medium has overtaken the traditional film. Digital cameras with high-quality optics and high-resolution sensors are much smaller than old 35 mm cameras. Cellular phones often have built-in digital cameras with resolutions that match that of a typical mid-range computer monitor. External USB hard disks with 1 Tb of capacity or more are available at a fraction of the cost of a personal computer, allowing consumers to store large numbers of digital pictures. Distributing digital images is also very easy: telecommunication companies offer services to share images acquired via cellular phone cameras, while social networking sites provide intuitive interfaces for posting personal photographs on the Internet. Accessing images over the Web has become almost instantaneous, thanks to the widespread availability of broadband connectivity for home use, over the phone lines (using DSL, i.e., Digital Subscriber Line, technology), or over the cable-television networks.

Digital imagery has quickly supplanted traditional imagery in scientific fields, particularly in radiology. Several types of radiological images have always been acquired through electronic sensors—magnetic resonance imaging (MRI), TAC, and Positron emission tomography (PET), for example. The availability of high-resolution sensors and especially of high-resolution, high-contrast computer displays has made it possible in many cases to replace traditional, film-based x-ray imagery with digital images. Technological advances are making digital radiography appealing even in applications where high-resolution images are required, such as for mammography.

Advances in imaging techniques have also enabled the reproduction of precious documents, historical artifacts, and figurative art masterpieces with sufficient wealth of details to be usable not only by the general public, but even by scholars. This trend was pioneered in the mid-1990s, for example, by the Vatican Digital Library initiative;[1] more recent development include the digital rendering of Michelangelo's David statue.[2]

Quite interestingly, techniques for *retrieving* digital images from large collections or from the Internet have only partially kept up with the pace of the digital imagery explosion. The most widely used search engines can retrieve images through keyword-based searches and build image indexes based on the HTML tags of the images and on keywords that appear in the text in the proximity of the images. Specific scientific fields have developed metadata standards to describe information on how images are acquired and on their content. However, the ultimate goal, the ability of retrieving images by specifying the desired content, is still a research topic and, despite the numerous advances in the field, has proven, so far, to be an elusive achievement.

In this entry, we provide an overview of still-image retrieval by content. We start section "Introduction" by discussing how images are represented in digital format. In section "Image Formats and Data Compression" we introduce the concept of content-based retrieval of still-image data (CBIR, for content-based image retrieval). Section "Query Specification" introduces the definition of *objects* as the building blocks for content representation within an image. In section "Content Representation and Objects" we identify different abstraction levels at which objects can be specified. We discuss how simple objects can be defined as connected regions that are homogeneous with respect to pixel-level, feature-level, semantic-level, and metadata-level characteristics. We describe how information can be efficiently represented at these different levels, how a user can specify content, and what mechanisms can be used to perform the search. Simple objects can also be defined simultaneously at multiple abstraction levels, and aggregated to form composite objects. The semantics of both types of objects, and the techniques required to search for them are the subject of section "Defining Content at Multiple Abstraction Levels."

In section "Progressive Search at Multiple Abstraction Levels" we briefly introduce a different perspective on image retrieval, affective image retrieval, where the image is treated as a signifier and the user specifies the desired signified. For example, a user could ask the system to retrieve images that convey "happiness," that describe "democracy," or that illustrate "Memorial Day."

In section "Effective Image Retrieval" we then discuss specific applications of the techniques described in the entry to Digital Libraries. We analyze specific examples of digital libraries of scientific data: medical image databases, repositories of remotely sensed images, and databases used by the oil industry.

IMAGE FORMATS AND DATA COMPRESSION

Data compression techniques are commonly used in image databases to reduce the required storage space. As we shall see in later sections it is sometimes possible to use properties of the compression algorithms for indexing and retrieval purposes. Hence, we briefly review some fundamental concepts of image compression.

The goal of source coding (data compression) is to produce a representation of the original signal which requires fewer bits. Compression is accomplished by reducing the redundancy present in the original data, and possibly by selectively discarding some information. Methods that discard information are called *lossy*, and the remaining ones are called *lossless*. In (gray scale) images, there are two main sources of redundancy: the similarity of spatially close pixels, and the nonuniformity in the overall intensity distribution.

Neighboring pixels commonly have similar brightness (intensity), and it is often possible to rather accurately estimate a pixel value from those of the surrounding ones. Hence, only the difference between the predicted and original values needs to be encoded, and this difference in general can be represented by fewer bits than the original grayscale value. Two main classes of approaches exist to reduce spatial redundancy. The first operates in the *spatial domain*, by exploiting local redundancy. A typical example is the lossless mode of JPEG standard: here, the image is scanned line by line from left to right, the value of each pixel is predicted using the values of the closest previously scanned pixels and the difference between predicted and actual values is computed (this approach is also known as predictive coding). The second class of approaches operates in a *transform domain*. Natural images have higher energy in the lower frequencies of the spatial spectrum, that describe slower intensity variations, than in the higher frequencies. The *two-dimensional Fourier transform*,[3] [Chapter 8] hence concentrates most of the image energy in a few low-frequency coefficients. Each Fourier coefficient is generated from the entire image, and is not easily predicted from neighboring coefficients, hence the transform effectively reduces spatial redundancy. The *two-dimensional block discrete cosine transform* (DCT), used in the JPEG standard[4] is closely related to the Fourier transform, but is more local in nature: the image is first blocked into 8 × 8 squares, and each block is transformed separately. Again, most of the energy is concentrated in a small subset of coefficients, and coefficients cannot be effectively predicted from the values of their neighbors because their statistical dependence is weak. The *wavelet transform*,[5] used in the JPEG2000 standard,[6] relies on a high-pass filter (H) and a low-pass filter (L). The image rows are filtered separately, and the result is filtered column by column. This operation produces a subband, which is commonly identified by the used filters: for instance, if rows are filtered with the high-pass filter and the columns with the low-pass filter, the resulting subband is denoted by HL. Hence, there are four possible subbands, LL, HL, LH, and HH, depicted in Fig. 1. Subbands are downsampled by retaining each other row and each other column. A ℓ level wavelet transform repeats the described operation on the LL subband $\ell - 1$ times. The transform yields both spatial and spectral information. The lower-frequency subband (LL at level ℓ) is well localized in frequency and each coefficient depends on a large number of image pixels. The higher-frequency subband (HH at level 1) is well localized in space, but contains information on roughly the top half of the frequency spectrum. Most transform coefficients at the higher levels have values close to zero, while most of the energy is in the lower-frequency subbands. Figure 1 illustrates this concept. The LL subband is a smaller version of the original image, the LH subband captures edges aligned with the horizontal axis, the LH subband captures edges aligned with the vertical axis, and the HH subband captures diagonal edges.

From both wavelet transform and block DCT it is easy to obtain a multiresolution representation of the image, that is, a sequence of increasingly smaller and coarser approximations to the original. As we shall see in subsequent entries, this property can be advantageously used during search.

The second cause of redundancy is present in most images: some of the intensity values are more common than others, and data compression can be achieved by encoding them with shorter codewords. Numerous techniques exist

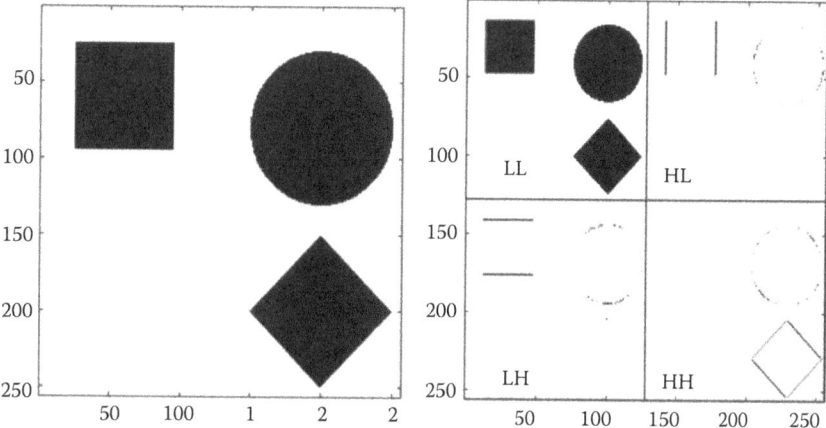

Fig. 1 A simple image (left) and its two-dimensional wavelet transform (right)

to accomplish this task, such as Shannon coding, Huffman coding, Arithmetic coding, and Lempel-Ziv-like codes, etc. An introduction can be found in Cover.[7] In general, higher intensity redundancy results in better compression. Spatial redundancy reduction almost invariably yields a significant increase in the intensity redundancy, and for this reason favors compressibility.

With lossless techniques, images typically compress by a factor of 2–3. If higher compression ratios are desired, information must be selectively discarded to improve compressibility. Two mechanisms are available, thresholding and quantizing,[8] which are usually applied after the spatial redundancy reduction and always before the intensity redundancy reduction. Thresholding changes to zero all values that are close to it. For example, in predictive coding, all small prediction errors could be set to zero. Quantizing (via *scalar quantization*, SQ) means reducing the number of allowed intensity values, by partitioning the intensity scale into regions (bins), assigning a representative intensity value to each bin, and changing each pixel to the representative value of the bin into which it falls. Both thresholding and quantization can be used simultaneously.

A substantial amount of research has been devoted to a general class of quantization schemes, called *vector quantization* (VQ).[8,9] VQ has its root in the well-known result in Information Theory that states that quantizing *independent* Gaussian random variables together yields lower distortion at the same rate than quantizing them separately (see, e.g., Problem 13.1 in Cover and Thomas[10]). Further improvements can be achieved if the random variables are not independent, as it is typically the case for neighboring image pixels. VQ compression of an image consists of: tessellating the image into groups of neighboring pixels having the same size, say d (e.g., dividing the image into 2×2 pixel regions); partitioning the d-dimensional space of pixel groups into bins; selecting a representative group of d pixels for each bin; and mapping each group of pixels from the image to the representative group of the bin into which it falls.

If thresholding, VQ, or SQ are used, either on the image or on its transform,[11,12] it is generally not possible to recover the original data from its compressed version. To select the thresholding and quantization parameters, often a measure of similarity between the original and reconstructed image is used, such as the mean squared error. Compression ratios of 10 to 1 or better are possible with lossy schemes, without appreciable deterioration of the image quality. However, since these techniques nevertheless introduce visual artifacts, their application to scientific data is rare, and essentially limited to remote browsing of the images over slow networks. By combining a lossy scheme with lossless encoding of the difference between the original and reconstructed image, it is possible to obtain a lossless scheme which can also be used for image retrieval.

SEARCHING IMAGE REPOSITORIES BY CONTENT

In traditional relational databases, queries are expressed using a highly structured language, such as SQL. This is possible because the information managed by the database is itself highly structured and can be stored, for instance, in one or more tables. Images, like many other types of multimedia data, are by nature unstructured. Therefore, they are difficult to organize using relational or object-relational databases and to retrieve using a structured query language. No universal solution exists to the problem of organizing and searching data in image databases: information retrieval from image repositories is a field of research that is currently in the process of reaching maturity.

Ideally, multimedia repositories should be searched by allowing the user to specify the desired content of the data to be retrieved. However, the old proverb "a picture is worth a thousand words" describes well the difficulties that a user encounters when attempting to specify exactly what the retrieved images should contain. The almost universal approach to overcome this difficulty is to rely on similarity-based search. The large class of heterogeneous methodologies developed for searching multimedia databases is commonly known in the field as *content-based retrieval* (CBR).[13] The main challenges posed by content-based search are automatic extraction and indexing of the content (to which this entry is devoted), query specification, and display of the results. In practice, however, the problem is very hard. In particular, a chasm exists between what search features can be extracted by existing systems from the user's query, and what content the user is actually interested in. This is called *semantic gap*,[14] and was well characterized by Smeulders et al.,[15] as the "lack of coincidence between the information that one can extract from the visual data and the interpretation that the same data have for a user in a given situation." Most of the research in CBR has been devoted to map information that can be extracted automatically from images, from associated metadata, and from context information (such as text near images embedded into documents) to user's semantic.

QUERY SPECIFICATION

There are numerous query specifications styles that are appropriate for image databases. The early CBR systems, such as QBIC,[13] VisualSeek,[16] and Virage,[17] used an example-based interface where the user would construct queries by supplying one or more examples. Graphical Query Languages offer an appealing alternative; however they have a steep learning curve and limited support for multimedia datatypes and relationships. A drag-and-drop interface supporting a quasinatural query language that combines English-like sentences with multimedia data, such as images, has been proposed in Bergman et al.[18] and used in conjunction with the SPIRE system[19] in a variety of application scenarios. This interface supports the composition of the type of queries described in this entry. Intelligent user interfaces for CBIR are described in Vermilyer.[20]

Many CBR systems rely on a paradigm for interaction with the user called *Relevance Feedback*.[21] In relevance feedback, the user issues a query and, upon receiving a set of results, inspects them, selects those that are most relevant and submits them to the system. The system appropriately modifies the query based on the user's feedback. The turn-taking dialog between user and system continues until the user is satisfied with the results.

CONTENT REPRESENTATION AND OBJECTS

In this entry, we define the content of images in terms of *objects*. We distinguish between simple objects, which can be thought as the building blocks of an image, and composite objects, collections of simple objects grouped by spatial (or temporal) relations. Therefore, we will not distinguish between a "scene" and the image of a complex object with multiple parts: our concept of composite object encompasses both.

Simple Objects and Attributes

A *simple object* is a connected region of an image which is homogeneous with respect to some specific characteristics. An entire image and any connected portion of an image are therefore simple objects, defined by constraints on pixel locations.

Objects have *attributes*. We distinguish *defining* attributes, that are used in the definition of a particular type of object, from *non-defining* attributes, that do not characterize the type of object. For example, a forested area can be defined as a region of a remotely sensed image having specific spectral reflectance and textural characteristics: in this case texture and spectral reflectance are defining attributes, while the location of the forest and its size in pixels are not.

Attributes can be *numerical* or *categorical*. Surface area and location are examples of numerical attributes. The type of a lesion observed in a MRI scan of the brain is a categorical attribute. Categorical attributes can be further characterized as *sharp* or *fuzzy*. An attribute of a geological stratum in a formation microscanner imager (FMI) is its thickness. Thickness can be defined as a sharp attribute, and measured for instance in feet, or as a fuzzy attribute, taking fuzzy values such as "thin," "medium," and "thick." A stratum 4 ft thick could also be characterized by membership values of .6 to the "thin" category, .4 to the "medium" category and 0 to the "thick" category. Attributes can also be either *deterministic* or *probabilistic*. Consider measuring the area of a lake in square meters (thus, we define it as a numerical, sharp attribute) using a remotely sensed image: while some pixels clearly fall entirely within the lake, some contain both dry land and water. Hence, the lake surface cannot be measured exactly, and only an estimate can be produced. This estimate can be treated as a random variable, and confidence intervals can be constructed from the image, for example by counting the number of pixels containing both water and dry land. Similarly, consider classifying automatically a forested area using a remotely sensed image, and assume that the statistical classifier decides that with probability .9 the forest is an evergreen forest, and with probability .1 is deciduous. Here the class label (a categorical attribute) is probabilistic rather than deterministic.

It is important to note the distinction between fuzzy attributes and probabilistic attributes. In the forest example, the classifier is telling us that of all forested areas looking like the one being analyzed, 90% are evergreen forests, and 10% are deciduous forests. This is different from saying that 90% of the trees in the patch of vegetation are evergreen and the remaining 10% are deciduous. The former characterization is probabilistic, the latter is fuzzy.

Finally, we note that attributes can be either *scalar* or *vector-valued*. The size (in number of pixels) of an object is an example of scalar attribute, its color histogram (section "Color Descriptors and Color Matching") is a vector-valued attribute.

Composite Objects

Simple objects are not sufficient to characterize the content of images. A *composite object* or *compound object* is a collection of simple objects that satisfy a set of spatial or temporal relations. A river delta as seen in a FMI image can be defined as a group of well-defined strata, arranged in a specific order.

Like attributes, spatial and temporal relations can themselves be sharp (for instance, "within n pixels") or fuzzy (for instance, "near"). Composite objects provide a powerful paradigm for specifying the content of images, especially those used in scientific applications.

Examples of object-based image representations are the paradigm adopted in SPIRE,[19,22] described in this entry, the *blobworld* representation,[23,24] and the framework used in VisualSEEk.[16] The blobworld approach, based on segmenting images with an Expectation-Maximization (EM) algorithm[25] applied to combined color and texture features, is well-tailored toward identifying objects in photographic images, provided that they stand out from the background. Each object is efficiently represented by replacing it with a "blob"—an ellipse identified by its centroid and its scatter matrix. In VisualSEEk, the query is specified in terms of color regions and their spatial organization.

DEFINING CONTENT AT MULTIPLE ABSTRACTION LEVELS

Attributes of image objects, or, more generally, of multimedia data, can be represented, described and searched at different levels of abstraction. We can readily identify four such levels, namely raw-data, feature, semantic, and metadata level. Simple objects can be defined at each level, or simultaneously at multiple levels.

Searching Images at the Raw Data Level

Digital images are two-dimensional arrays of pixels. Each pixel is represented as one or more numeric values denoting the intensities in different bands of the electromagnetic spectrum at its specific location. A pixel-level object can be defined as a connected portion of an image. Two pixel-level objects are equal if they have the same size and shape, and if the pixels in the corresponding positions have identical values.

Similarity between two pixel-levels objects having the same size and shape is defined in terms of similarity between the values of the pixels in corresponding positions. A pixel-level object having n pixels can be represented as a n-dimensional point, by making each pixel correspond to a different coordinate. The difference between two pixel-level objects can then be defined in terms of the Euclidean distance between the corresponding points. Alternatively, a pixel-level object can be represented as an n-dimensional vector, starting at the origin and ending at the above defined n-dimensional point. In this case, the similarity between two objects can be defined using their inner product.

The SPIRE system[19] supports image search at the numerous levels, including the raw-data level.

Progressive pixel-level retrieval

Computing the Euclidean distance (or the inner product) between a query object and all the subsets having identical shape and size, of all the images in the database, is a very expensive operation. Matching can be based on cross-correlation, or the correlation coefficient.[3] [Chapter 20] Here, image representation can significantly reduce the computational complexity. Cross-correlation is still an expensive operation in the pixel domain. However, multiplying the Fourier transforms of two images (a much faster operation) is equivalent to computing the cross-correlation in the pixel domain. In Vaidyanathan,[26] the author notices that cross-correlating the corresponding subbands of two images and adding the results produces a subsampled version of the cross-correlation. By noticing that the wavelet transform concentrates most of the energy in the lower-frequency subbands, and that a similar property holds for the low-frequency components of block-DCT, Li, Turek and Feig[27] and Castelli et al.[28] concluded that the results of Vaidyanathan,[26] can be well approximated by just considering the cross-convolution of the lower-frequency subbands (or DCT coefficients). Local maxima in this approximation suggest possible location of matches. Once candidate locations are identified, the search can be refined by computing and adding contributions from higher-frequency subbands. Large computational savings are achieved by refining the search only around candidate matches rather than on the entire image.

Rather than being a general mechanism, raw-level search has specific applications. It is extremely useful to identify distinctive objects across a time series of images or of images acquired from different instruments, hence it is useful for coregistration purposes. For example, remotely sensed images of the same region acquired by the same instrument vary over time because of slight differences in the orbit, altitude, and orientation of the platform. To coregister the images, ground control points are identified, using pixel-level techniques, and their positions used to compute the warping mapping between the images.

Analogously, in medical imaging, distinctive anatomical features must be precisely identified to compute the exact position of the tissues or organs of interest.

Features

The term *feature* denotes an aspect of an image that can be captured by a numeric or, more rarely, categorical quantity (called a descriptor) computed from the raw data. By definition, features do not have an immediate semantic meaning. In the literature, the term feature is often used to denote the descriptor.

Typical image features describe color, texture, and spatial properties, such as shape. We distinguish between *global features*, that represent an image in its entirety, and *local features* that capture localized properties. Global feature descriptors are often used for photographic images, but have in general poor retrieval qualities. It is likely that the use of global features originated with one of the early CBR systems, QBIC,[13] which represented each image with a color feature vector, a texture feature vector, and a shape feature vector. In reality, when a new image is ingested in QBIC, the operator is supposed to manually outline a region of interest, from which the feature descriptors are extracted, and that becomes the subject of the image. Local feature descriptors are much more useful for indexing purposes. They are either extracted by first segmenting the image into homogeneous regions and computing features for each region (as in the blobworld,) or by dividing the image into a large number of small, overlapping windows (e.g., overlapping), and computing features for each window.[28] The advantages of the first approach are that it produces a smaller data volume, and that an appropriate segmentation scheme often produces results that are acceptable from the viewpoint of the user. The advantage of the second approach is the increased flexibility, since objects are not predefined and can be extracted to better match the query.

We now discuss two of the most commonly encountered classes of features in image databases: color and texture. In each subsection, we briefly describe how the category is used, what descriptors are commonly used and how similarity is captured. We conclude each subsection with remarks on specific applications, and, when appropriate, on how to speed up the search by relying on properties of image compression schemes.

Color Features

The use of color for retrieval

Color is commonly used to index images,[29] and is one of the main features used in early CBIR systems, such as QBIC[13] and Virage.[17] It is a very important feature for photographic images and works of art, and histological images. It has limited applicability to radiological imaging, which is almost universally displayed in gray scale (and only rarely using false-colors, which are not useful in indexing), in oil-exploration imagery (where, again, false coloring is used), and in satellite imagery, where the multispectral information is used to extract semantics and not used directly as a low-level property.

Color descriptors and color matching

Color features for similarity retrieval are usually variations of the color histogram. The first step in the feature computation is a selection of the color space. For typical digital photographs, where each pixel is represented as the superposition of three basic colors, red, green, and blue, the RGB space is the most natural choice. Here, each pixel value corresponds to a point in a three-dimensional space, whose coordinates are the red, green, and blue intensity. Pixels of images compressed with the JPEG image compression standard can be immediately represented in the YCrCb space. The natural color space for television data (such as newscasts, sport events, movies, etc.) depends on the standard: NTSC uses the YIQ space, PAL and SECAM use the YUV standard. None of the mentioned color spaces is perceptually uniform, in the sense that Euclidean distances between points in the color space do not capture well the subjective similarity between colors. Hence, numerous other color spaces have been used with moderate success by different authors, such as the hue-saturation-value (HSV) space,[29] the Munsell color space,[30] the opponent-axis space,[31] or the CIE-LUV space.[32]

The second step in the color histogram computation is quantization: the color space is subdivided into cells, a representative color is selected for each cell, and all the pixel values falling in a specific cell are replaced by the representative value. The quantized color space usually contains anywhere from 64 to 4096 different colors.

The color histogram of an image is finally computed by counting how many pixels fall in each cell of the quantized color space. The counts can be normalized (divided by the size of the image), in which case they sum to 1. A non-normalized color histogram having b bins can be represented as a point in a b-dimensional space, the ith coordinate of which is equal to the number of counts in the ith bin. If the color histogram is normalized, it lies on a b–dimensional simplex, which is a $(b - 1)$-dimensional surface.

Different functions on the color histogram spaces have been proposed to capture similarity between images. These functions are usually distances between the points representing the histograms. While the choice of the color space and of the number of quantization bins appear to affect the quality of the retrieval, the choice of the similarity function is less important, and most commonly used ones are essentially interchangeable.

Image representation can be combined with color feature extraction. For instance, research has been done on how to approximate the histograms of images compressed with the JPEG standard with the color histogram of the DC components.[33]

Further Uses of Color

Color is a powerful means for conveying meaning and emotions. Itten[34] codified a psychological theory on the use of the color in art to convey semantics. It is unsurprising, then, that color has been used to retrieve images of paintings from collections of art works. Colombo, Del Bimbo, and Pala[35] analyze the color content of photographs of paintings, and derive measures of harmonic accordance, contrast of luminance, contrast of saturation, and contrast of warmth. As a response to a query, the system returns the most similar paintings with respect to one or more of the listed color quantities. Del Bimbo et al.[36] extract color and shape using a multiresolution representation of the image. By using a low-resolution approximation (which can be easily obtained from the wavelet transform or the block-DCT), the system can make judgments about the overall color composition of the painting and use them during retrieval.

Texture Features

Definition and use

Texture describes the local intensity variations in an image. In natural images, texture is a property of individual surfaces, and is an important visual cue especially for categorizing materials and substances: water, wood, grass, snow, cloth, fur, brick walls, sand, etc. are just few examples of entities that humans can identify even in black and white pictures based on texture.

Texture is extremely important in scientific imagery, where commonly expert use textural characteristics to interpret the data. For example, in remotely sensed images, erosion patterns, types of tree associations, and types of crops can be analyzed in terms of texture. In well-bore and core images, the stratigraphical analysis is exquisitely textural in nature. In some types of medical images, such as mammograms, texture is a powerful indicator of pathology.

Texture descriptors

Texture descriptors are generally extracted from grayscale images. For color images, descriptors are computed from the overall intensity map (computed by adding the square of the pixel values in each color band), from each individual color band, or from selected color bands. Color textures have been the subject of rather few studies.[37–39]

There are three main classes of texture descriptors.

Spatial-domain descriptors are computed directly from the image (or the intensity map). Several feature sets belong to this class, and here we describe the most common ones.

The *gray-level differences histogram*[40] counts how many times the differences between pixel values in a window and the value at the center occur within an image or a region. Statistical properties of the histogram, such as its mean, variance, central moments, entropy, etc., are used to represent texture properties, such as overall brilliance, intensity variations, etc.

The *co-occurrence matrix* counts how many times a pair of intensity values occurs in pixels having a fixed distance in a predefined direction. Corresponding texture descriptors[41–43] are derived properties of the matrix, such as the entropy or the moments, and have been related to visual properties, such as regularity, homogeneity, directionality, and periodicity of the texture.

Tamura's features[44] were selected to capture specific characteristics of the human visual system, and describe texture in terms of coarseness, directionality, roughness, regularity, contrast, and line-likeness. Studies show that, in practice, coarseness, contrast, and directionality alone yield most of the discriminatory ability of the feature set, and that adding the remaining three features improves retrieval results only marginally. Tamura's features capture very well high-level characteristics of texture, but fail to provide fine-level discrimination.

Transform-domain descriptors are computed from a transform of the image. When an image is compressed with a transform-based method, the corresponding texture descriptors can be obtained at a very moderate computational cost.

Wavelet-based descriptors[45] capture the local spatial frequency content of the image. $2^k \times 2^k$ portion of an image is described to a good degree of approximation by $3d + 1$ subblocks of a d-levels wavelet transform: one in each of the three high-frequency subbands (HL, LH, HH) at each level, and one in the low frequency subbands at level d. A block at level ℓ has size $2^{k-\ell} \times 2^{k-\ell}$. Smith and Chang[45] suggest computing the wavelet transform of the image, conceptually dividing the image into nonoverlapping square blocks, and considering for each block the corresponding $3d + 1$ portions of the transform. The mean value and the variance of the coefficients within each transform portion are then computed, and concatenated to produce a texture feature vector. The authors also suggest an algorithm for merging adjacent image blocks to segment the image into regions of homogeneous texture.

Gabor-based features[46] are among the most discriminating texture descriptors available. The (even, symmetric) Gabor filters are uncorrelated Gaussians with variances σ_x and σ_y in the x- and y-directions, modulated by a sinusoid of frequency ϕ varying along the x-axis, and rotated by an angle θ, hence they are defined by four parameters. In the spatial frequency domain, they correspond to two Gaussians having variances $1/(2\pi\sigma_x)$ and $1/(2\pi\sigma_y)$, centered at ϕ and $-\phi$ respectively, in a reference frame rotated by θ. Their main property is that they trade off optimally localization in the spatial domain and in the frequency/orientation domain. Hence, the magnitudes of the Gabor coefficients give a good indication of directionality and periodicity of the texture. A texture feature vector is constructed by selecting a group of filters at various orientations, scale (variances), and frequencies, computing the corresponding Gabor coefficients, and concatenating them. Experiments show that Gabor features are competitive with other texture descriptor sets in a wide variety of application domains,[45,47,48] and have been successfully used for texture discrimination and classification[49] and image segmentation.[50]

Random-field models describe texture as a spatial stochastic process. Technically, in this context a spatial stochastic process is a probability measure on the collection of possible images.[51] Practically, a random field is a probability distribution over sets of neighboring pixels. Markov random fields (MRF) are a special class of processes specified through the conditional distribution of the value of a pixel given the values of specific neighbors. Due to their effectiveness, they have been used for a long time for description and representation,[52–54] classification,[55–57] synthesis,[58,59] compression,[59] and segmentation of texture.[60–64] Gaussian MRF are particularly appealing, due to the simplicity of the model: here pixel values can be written as the sum of a linear combinations of the values of their neighbors and of a correlated Gaussian noise. Simultaneous autoregressive models (SAM), where the Gaussian noise is white (i.e., uncorrelated), are even simpler, and have been widely used for image retrieval field. Typical texture features obtained from random field models are the parameters of the model itself: for example, a SAM texture descriptor would consist of the vector of coefficients used in the linear combination and of the variance of the white Gaussian noise. It is also possible to simultaneously fit different models to the data (characterized, for instance, by different neighborhoods), and use the index selected model a further descriptor.

Texture Similarity and Texture-Based Retrieval

Similarity (or better, dissimilarity) between texture feature vectors is usually measured by means of a distance function. Let **x** and **y** denote two such vectors. The most commonly used metric is the Euclidean distance, defined as

$$D^{(2)}(\mathbf{x},\mathbf{y}) = \sqrt{\sum_{i=1}^{d}(\mathbf{x}[i]-\mathbf{y}[i])^2}$$

It is rotationally invariant and weights all the features equally. This last property is undesirable when the ranges of the different feature descriptors vary significantly. For example, the variance of the gray scale difference histogram is often much larger than the mean, and would have a significantly bigger effect in determining similarity. Furthermore, additional flexibility is required when the system is allowed to learn the relative importance of different features from the user input and feedback. Metrics that can be used to satisfy both requirements are the weighted Minkowsky distances and the generalized Euclidean distance. The former is defined as

$$D^p(\mathbf{x},\mathbf{y}) = \left[\sum_{i=1}^{d} w_i |\mathbf{x}[i]-\mathbf{y}[i]|^p\right]^{1/p}$$

where

p is a positive number
w_i are weights
d is the length of the feature vector
The latter is defined as

$$D(\mathbf{x},\mathbf{y}) = [(\mathbf{x}-\mathbf{y})^T K(\mathbf{x}-\mathbf{y})]^{1/2}$$

where **K** is a positive definite matrix having determinant equal to 1. Further details on similarity measures for retrieval can be found in Santini and Jain.[65]

Progressive texture retrieval

As previously described, several texture features can be extracted from image transforms, and therefore are easily computed when the images are compressed using a transform-based scheme. Texture features extracted from different levels of the multiresolution pyramid can also be used to speed up search.[40] Here, an optimal starting level in the pyramid is identified using the examples provided with the query. Pruning of the search space is then performed by texture matching at this starting level: note that higher starting levels are characterized by fewer homogeneous regions within images, and hence yield faster matching. However, a higher starting level introduces a larger approximation in the matching, as details are lost, and therefore the selectivity threshold for the pruning must be relaxed accordingly. Since the two effects are contrasting, there is always an optimum starting level from the speed viewpoint. Only the regions that pass the pruning stage are retained and further analyzed at the immediately finer level of the multiresolution pyramid. The process is repeated until the full-resolution level is reached. Tenfold increase in retrieval speed have been observed with this technique.

Shape Features

At a first glance, searching images repositories by specifying the shape of desired contained objects appears to be a natural and intuitive approach to CBR. In reality, however, it is a problematic endeavor at best. First, most animate objects are not rigid and their shapes can vary substantially as they move. Even for rigid objects, perspective and angle of view can radically change the shapes recorded in an image. Additionally, shape is not invariant with respect to rotations of the image. In spite of these difficulties, shape-based image retrieval has been an active and often successful field of investigation. Another difficulty with shape features is that they rely heavily on accurate images segmentation, which is largely unsolved. If objects cannot be precisely segmented, shape features can be misleading or even meaningless. In contrast, color or texture features are not so sensitive to inaccurate segmentation. For example, if an object is segmented into two pieces, the colors from the two pieces still reflect the object, but the partial shapes can be arbitrary and dramatically different from the true shape.

Indexing images by shape requires solving three separate classes of problems. The first is *shape extraction*: images must be analyzed to extract shapes of objects. The literature in this field is very large, and image retrieval has borrowed techniques from other disciplines, such as robotic vision. The wide spectrum of approaches to shape extraction range from accurate segmentation, to approximate representation of shape regions, to extraction of low-level features that capture aspects of shape, such as edges.

The second problem is *shape representation*: the shape must be described by numeric quantities that can be used for retrieval. Numerous descriptors have been proposed in the literature, which can be classified along several dimensions. A first distinction is between *global* and *local* descriptors. Global descriptors capture properties of the shape as a whole, for example its elongation or the overall smoothness of the contour. Local descriptor capture properties of selected regions of the shape, for example, the presence of sharp angles in the contour. A second distinction is between *complete* and *partial* descriptors. Complete descriptors attempt to retain all properties of the shape, while partial descriptor capture only characteristics that are deemed salient for retrieval. A third distinction is between *interior* and *boundary* representations. Interior methods describe a shape, for example, by means of approximations or using a sequence of approximations. Boundary methods describe properties of the contour line, for example, by means of piecewise polynomial lines. Other dimensions distinguish between *rotationally invariant* and *rotationally variant* features, *single-scale* and *multi-scale* features, *composition-of-part* and *deformation* methods.

The third problem is *shape matching*. This is typically accomplished by means of an appropriate distance function. Unfortunately, there is no universal shape similarity metric: different feature are matched with different, specific similarity functions. A detailed discussion of these metrics, a more in-depth view of the categorization reported in this section can be found in Kimia,[66] which is an excellent introduction to shape-based retrieval.

Shape-based techniques have had only limited application in photographic image retrieval, where often segmented shapes are associated with texture- and color-based representations of the interior.[67] In contrast, shape has been successfully used in several specialized application areas, such as medical imaging;[68] biometric identification;[69] and retrieval of logos, trademarks, and watermarks.[70] A survey on recent advances in CBR of three-dimensional shapes can be found in Icke.[71]

Searching Images at the Semantic Level

Semantic content characterization

Searching images at the semantic level is one of the most difficult tasks faced by CBIR systems. The main challenges are representing semantics, extracting it automatically from the images, and indexing it. These problems are particularly severe in databases of photographic image, where the subjectivity of the user perception plays a major role. For example, a specific image could be described as "an outdoors scene," "a picture from an outdoors sport event," "a picture from a golf tournament," or "a picture of the 18th hole at the St. Andrews golf club in the year 2000, taken while Tiger Woods is putting to win the tournament and the golf grand slam." Scientific data, however, is often more amenable to automatic extraction of semantic content. We first discuss photographic images, and later scientific imagery.

Jaimes and Chang[72] proposed a scheme to represent multimedia content which uses 10 abstraction levels, divided into two groups. The first group contains levels related to percept, corresponds to the pixel and feature levels, and divides into different levels global and local content. The second group, containing six levels (generic object, generic scene, specific object, specific scene, abstract object, abstract scene,) provides a very good framework for describing semantic content in photographic images and video, and can be successfully used as guideline for constructing indexes. Referring to the example of the previous paragraph, the first two descriptions would be indexed at the generic scene, while the last belongs to the specific scene level.

Semantic content extraction

While researchers have proposed solutions on how to organize semantic content for search purposes, the automatic extraction of semantic content from photographic images and video is still an unsolved problem and will remain such at least for the next few years. Some of the difficulties encountered are very similar to those raised by the automatic object recognition problem, to which the computer vision discipline has devoted decades of efforts. Specific algorithms exist to identify particular classes of objects within images, such as deciduous trees,[73] naked people and horses;[74] specific types of scenes, such as sunsets, outdoors, and indoors images;[75] and some very high level semantics such as warmth, and harmony of a painting.[35] Automatic image annotation is a form of semantic context extraction: Li and Wang[76,77] propose methods for automatically associating a collection of labels to images.[78]

Still-image retrieval at the semantic level can pose challenges beyond those related to object recognition: users, for example, might look for images with a specific purpose in mind, for example, to evoke desired emotions (e.g., images that evoke sadness, images with dramatic effect, etc.), which is discussed in the section "Affective Image Retrieval."

Following Liu et al.,[79] we classify the technical approaches to semantic content extraction into five categories:

1. *Defining high-level concepts through object ontology.* Methods of this class rely on small-size vocabularies of simple descriptors and a mechanism for mapping low-level features extracted from the image into the vocabulary. An example of such descriptor could be "bright-green region." High-level semantic concepts are expressed in terms of the descriptors: a "lawn" could be described as a "bright-green region located in the bottom half of a photograph." The main limitation of the approach lies in how to appropriately design ontologies for specific tasks.
2. *Using supervised or unsupervised learning.* In the case of supervised learning, a training set consisting of examples labeled with high-level semantic labels is used to train a classifier. The classifier learns to associate low-level features with semantic concepts. In unsupervised learning, the low-level features representations of a large collection of images are partitioned into similar groups, and groups are associated to semantic concepts. The similarity measure is defined on the feature space, or on the space of probability distributions over the feature space, such as in Li and Wang.[77]

    The main limitation of supervised learning methods is their reliance on large manually labeled training sets. The main limitation of unsupervised learning methods is that the correspondence between groups and semantic concepts is incidental.
3. *Interacting with the user via relevance feedback.* Semantic concepts are learned on the fly by a classifier through an iterative refinement process. During a turn of the iterative refinement process, the user is shown candidate examples picked by the classifier, marks some as positive (in some cases, as negative); the classifier then updates its internal model using the user feedback. The main limitation of the approach lies in the limited number of examples that a user can reasonably label.
4. *Matching content through semantic templates.* A semantic template is the "signature" of a semantic concept extracted from a collection of representative images. Unlike when using supervised learning, the designer of semantic concepts must often have a good understanding of the underlying features.
5. *Combining image features and features from surrounding text.* This approach is well suited for retrieval of images from the World Wide Web, which are embedded within HTML pages. Currently, the analysis of the surrounding text is often limited to keyword selection, but advances in natural language processing make the approach increasingly appealing. An important challenge arises from the difficulty of determining whether the surrounding text is actually descriptive of the image content.

Scientific data offers numerous opportunities for automatic semantic content extraction. Experts often interpret the imagery by relying on low-level visual cues, which might be captured by a computer. Additionally, a great deal of the semantic content can be described in terms of objects and their spatial relations, there are no scenes, and the abstract levels of Jaimes and Chang are irrelevant. We briefly discuss three cases: remotely sensed images, oil well-bore images, and medical images.

One of the main uses of multispectral remotely sensed images is to identify and distinguish different types of land cover. The applications are numerous: from management of forestry to the identification of diseases in crops, to crop yield prediction, to environmental monitoring. Different types of land cover have different spectral reflectance: for instance, vegetation reflects in the green part of the spectrum, absorbs in the red (hence, leaves and grass look green), and is highly reflective in the near infrared; barren terrain, on the other hand, has moderate to high reflectance in all the visible and near-infrared spectrum. Automatic classifiers can be constructed to label the individual pixels, identify connected regions and produce semantic objects such as "forests," "urban areas," "bodies of water," etc. These classifiers are usually specific to an instrument, a particular geographical region and a given time of the year.

Well bore data for the oil industry contain both image-like information (acquired, for instance, by the FMI instrument) and one-dimensional data, acquired by appropriate log instruments. Bulk lithology (the type of rock) can be inferred from the log data, and classifiers can automate the task. The well can then be partitioned into labeled strata and rock formations, defined as associations of strata satisfying relative position constraints, can easily be searched.

Radiological imagery has recently received substantial attention as a prime candidate for automatic extraction of semantic content. In addition to the data acquired by the medical instrument, each image is analyzed by a trained radiologist, who produces a reading containing semantic information. This reading is then stored in the medical information system and linked to the image, together with additional metadata, as mandated by the standards with which the repository complies. An ontology based approach for retrieving semantic content from medical image repositories is reported in by Wei and Barnaghi,[80] while Lehmann et al.[81] describes a strategy for automatic image categorization according to a large number of semantic classes. An overview of recent results related to CBIR from medical images can be found in Müller et al.[82]

Progressive semantic retrieval

Progressive techniques that rely on properties of compression schemes exist to speed up the daunting task of labeling the tens of gigabytes produced daily by an instrument. Progressive classification[83] analyzes the multiresolution pyramid. It uses a different classifier for each level of the pyramid. Starting from an appropriate level, the appropriate classifier decides whether each pixel corresponds to a semantically homogeneous region at full resolution, in which case it labels the entire region, or not, in which case it marks the pixel. The marked pixels are then analyzed at the immediately finer resolution level, using an analogous classifier. The process terminates when the full-resolution level is reached and all marked pixels are labeled. This approach is not only several times faster than pixel-wise classification, but, under general condition, it is also more accurate.

Metadata

Metadata is the highest content abstraction level. It corresponds to information that cannot be inferred or extracted from the image itself, or that is associated to the image in a manual fashion. The date and time of a photographic image, the names of the people appearing in it, the author of a painting and his biography, the location of a well, the name and medical history of a patient, the satellite used to acquire an image are examples of metadata.

Metadata is either structured, and is characterized by the presence of predefined fields whose values have well-specified types, or unstructured, for example textual captions of images. Standards exist or are emerging to regulate structured metadata. They are always specific to a particular application domain, such as medicine, geographical information, remotely sensed data, etc.

Structured metadata is amenable to management using a traditional database. Unstructured metadata can be indexed using information retrieval methodologies. Both cases are beyond the scope of this entry.

PROGRESSIVE SEARCH AT MULTIPLE ABSTRACTION LEVELS

Only the simplest multimedia queries are expressed in terms of a single attribute. These queries are called *atomic* by Fagin.[84] They retrieve the simplest simple objects. For example, one could ask a photographic database for pictures of the current president of the United States (semantic query), a museum digital catalog for paintings by Renoir (metadata query), a fashion archive for fabric having a certain mix of colors (feature-level query). The repository would probably answer the query by returning a large number of results. More specific queries that return a smaller, better defined set of images, are expressed in terms of multiple attributes. For instance, the user might want to retrieve images of paintings by Renoir having as subject scenes from "la Grenouillère," or images of a tumor (metadata-level) having specific size and contour characteristics (feature level).

To support multiple-attribute queries, an image repository has to solve several problems, related to both the semantics of the query and its execution. In this section we discuss these classes of problems and describe solutions proposed in the literature.

The Semantics of Combining Multiple Attributes

Image and multimedia databases must support approximate queries: the user can only provide an approximate description of the desired content, and ask the repository to return the images that best match the specification. It is acceptable to ask for images of gray cars, and have the system return a ranked list of three images, containing respectively a gray sedan, a "metallic silver" convertible and a "silver frost" SUV. The similarity search paradigm yields a significant amount of flexibility, but at the same time complicates the interpretation of queries. Consider asking a photographic image database for pictures of "red cars." In a traditional database, this query would be expressed in SQL as select image_id where subject='car' and color='gray', and the result would be a list of image identifiers containing gray cars. In the example, only the gray sedan would be returned.

In a multimedia database, the equalities in the constraints are substituted by similarity functions, and the "and" connective becomes a function that combines the similarity values of the two constraints. Early CBIR systems, such as QBIC, allowed the user to combine similarities with respect to color, texture, and shape. The user selects the importance of the three features using a graphical user interface, the similarity of images to the query is then computed separately for each feature, and the three resulting scores are combined by means of a weighted average with coefficients proportional to the importance of the features. This approach is simple, but lacks flexibility.

The query framework implemented in the Garlic system[85,86] solves the problem by treating similarity scores as fuzzy membership functions.[84,89] Scores obtained from matching individual constraints are normalized between 0 (no match) and 1 (perfect match), and combined using fuzzy Boolean connectives. The simplest forms of fuzzy AND and OR are respectively, the minimum and the maximum of the connected scores, while the negation (NOT) corresponds to subtracting the score from 1. Hence, if the color "silver frost" matches gray with a score of .8 and an SUV matches a car with a score of .9, the silver frost SUV matches the query for a gray car with an overall score of .8, the minimum of the two.

This framework can conceptually be extended to composite objects,[90] by noting that relations between simple objects produce sharp or fuzzy scores that can be treated in the same way as the object attribute scores.

Searching at Multiple Levels of Abstractions

Fagin[84,89] proposes an algorithm that executes queries containing m constraints on attributes. The algorithm assumes that the search engine can return the top k results in response to atomic queries, and that can compute the score of a database item with respect to an atomic query. The algorithm first evaluates in parallel the m atomic queries, returning for each the smallest set of top results such that there are k distinct database items that appear in each result set. Clearly, when the atomic queries are combined with the minimum function, or with a function which is monotonic in the scores, these k database items constitute the result set L. The algorithm then combines, for each item of L, the scores of the individual atomic query, to produce its score. The scored results are finally sorted in decreasing order and returned to the user.

A sequential processing algorithm for retrieving composite objects is described by Li et al.,[90] and consists of three procedures. The first procedure consists of linearizing the description of a composite object into a set of subgoals. The ordering is a function of the dependence between simple objects and of the availability of precomputed indexes for executing atomic queries. The result of the step is a chain of subtasks. The second procedure manages the computation of sets of L results from each subtask. The first time the procedure is invoked on a subtask, it produces the best L matches, the second time it produces the next best L matches, etc. The third procedure controls the execution flow in a dynamic programming fashion: it starts from the first task in the chain, and retrieves the best k matches (out of the block of L items retrieved by the second procedure), then it executes the second task, and retrieves the best k matches for the subproblem consisting of the first and second task. The computation continues by subsequently adding tasks, and keeping track of the scores of the individual subtasks and of the current set of results. It is possible that the score of the kth partial result be smaller than the scores of the worst result currently produced by a particular subtask: if this is the case, the algorithm might be ignoring relevant database items, and a backtracking is invoked, that retrieves further objects from the offending subtask.

Further enhancements of the algorithm are presented in by Li et al.[91] where several fuzzy relations are discussed, and details are given on the execution flow control.

When ordering the subtasks, it pays to consider the abstraction level at which they operate. Metadata atomic queries are usually faster to execute and more restrictive than queries at any other level. They are in general followed by semantic queries, feature-level queries, and raw-data queries in the order. Further optimization can be performed in a query-dependent fashion: the system can collect statistics on how effective different types of queries are at pruning the search space, and use the information while staging the query execution.[92]

AFFECTIVE IMAGE RETRIEVAL

Up to this point, we have concerned ourselves with the problem of describing the visual content of images for the purpose of search and retrieval. However, since the early days of mankind, imagery has been used as a powerful way of describing concepts that transcend the pictorial representation. The image acts as a signifier and refers to a signified which could be an emotion (e.g., sadness, tranquility, etc.), an abstract concept (e.g., portrait paintings often convey information on the personality of the sitter), an event or a recurrence (e.g., the Vietnam War, the Declaration of Independence, etc.). Paintings have also been used to convey philosophical or theological teaching (e.g., the background of the Mona Lisa captures Leonardo's view of a dynamic, ever changing nature, and contains clear references to the author's fascination with hydrodynamics and hydraulics). Affective image indexing and retrieval is a discipline that lies at the intersection of affective computing[93] and CBIR; its goal is to support queries containing the specification of the emotions evoked by the desired images.

Most approaches to affective image indexing and retrieval rely on the correlation between low-level perceptual cues and the emotions they cause.[35] For example Li et al.,[94] map texture features to a thesaurus of affective concepts; Wu et al.[95] analyze the usefulness of color, text, shape, and their combination for affective image retrieval, while Bianchi-Berthouze and Kato[96] describe a system that interactively creates a Kansei user model based on low-level features extracted from Web images.

IMAGE REPOSITORIES AND DIGITAL LIBRARIES

Early research in still-image retrieval was often pursued in connection with the field of *multimedia digital libraries*. Digital libraries are organized collections of multimedia data, providing storage, retrieval, and transmission functionalities. They are used to manage text,[97] music and audio,[98] images,[30] video,[99] and other forms of electronic content. During their development, early digital libraries posed numerous challenges in all fields of computer science. At the hardware and system software levels, new computer architectures have been invented to efficiently store and transmit large amount of data; requirements have been imposed on operating systems to provide the desired quality of service and data integrity; and new assumptions on how data is accessed and modified have guided the design of file systems that efficiently manage multimedia files. At the application level, new systems for acquiring digital content have been developed; novel content representation models have been devised; algorithms for efficiently searching large collections of data have been explored; and simple, yet powerful user interfaces have been investigated to specify the desired content and to represent the returned results.

In recent years, efforts have been pursued in different directions that often depend on the application field. For example, in remote sensing the focus has shifted from creating large repositories containing data from a large variety of sources to supporting federations of data as service providers. For example, the Federation of Earth Science Information Partners, or ESIP—http://www.esipfed.org/—, originated from a NASA grant, is composed of distributors of satellite and ground-based data sets, providers of data products and technologies, organizations that develop tools for Earth Science, and strategic funding partners. As a consequence, problems of metadata standardization and of interoperability of data sets and data products derived from the data sets have taken the precedence over the original still-image retrieval problems. Other fields of investigation include collaborative learning,[100] data fusion, digital library federation, leveraging new computation paradigms such as grid computing,[101] supporting e-Science,[102] all of which are beyond the scope of this entry. The present section contains an overview of the early work in the area of digital libraries for image data, where a substantial emphasis was devoted to image retrieval.

The ground work in the general area of digital libraries was sponsored by the Digital Libraries Initiative (DLI). The first studies were conducted at Carnegie Mellon,[99] U.C. Berkeley,[103] U.C. Santa Barbara,[104] University of Illinois at Urbana-Champaign,[97] University of Michigan,[105] and Stanford University.[106] The field of research has since seen a proliferation of projects. Early image digital libraries organized photographic images, catalogs of museums and art galleries,[107] fingerprints,[69] medical data,[108,109] geographically referenced data,[104] satellite images,[28] etc. We briefly review some of these applications.

Medical Image Databases

Medical imaging[110] is one of the most powerful diagnostic tools available. X-ray radiographs are essentially the only radiological images acquired using film, in analog format: most other modalities acquire data in digital format, hence can be managed by a digital library. Digital radiography and digital mammography are becoming increasingly popular, and we will see in the near future filmless radiology departments. Enabling technologies in digital radiology include: advances in sensor quality (resolution, signal-to-noise ratio), which make the quality of digital images comparable to that of film-based images; improvements in high-resolution, high-contrast flat-panel displays, which increase the productivity of radiologist; and high-speed connections to the Internet, that enable teleradiology, whereby a specialist can diagnose images acquired in multiple hospitals or imaging clinics.

Medical imaging is characterized by a variety of modalities, which are suited to investigating different types of properties, and yield data in different formats. Anatomy, physiology, biochemistry, and spatial properties of the body and its organs can be studied with appropriate radiological instruments.

X-ray radiography is the most common form of medical imaging. It directly measures the opacity of the body or of contrast media to electromagnetic radiation having wavelengths in the 100 to 0.01 Å range. The part of the body to be imaged is placed between a source that produces a large, non-diverging beam of x-rays and a sensor array, which records the intensity of the transmitted radiation. A digital radiograph is large: for example, a digital mammogram is a gray-scale image whose typical size is 4000×4000 pixels, and where each pixel is represented as a 12 or 14 bit number: hence, a single mammogram corresponds to 32 megabytes of data.

Computed tomography (CT) produces indirectly images of slices of the body. A series of digital radiographs of the same thin area of the body are acquired from different angles. The resulting set of projection is then analyzed by a computer which reconstructs a slice, whose typical size is 512×512 pixels, each represented by 2 bytes. During an examination, several slices are acquired, and about 20 megabytes of data are generated.

MRI measures the amount of water present in tissues, by aligning the spins of the hydrogen atoms in a thin slice of the body to a strong magnetic field, tilting them, and measuring the variations in magnetic field while the spins realign. A computer analyzes the signal and produces a 256×256 pixels image, where each pixel requires one or two bytes. Numerous slices are typically acquired during a single examination, and 10–20 megabytes of data are generated. Functional MRI is a novel technique that measures activity of organs, typically the brain. Angiographic MRI is used to image blood vessels. Diffusion MRI images the diffusion of liquids in tissues, and is used in the diagnosis of ischemic strokes.

PET measures positrons emitted by radioactive dies which are injected in the body and distribute themselves within the target organ in proportion to the physiological activity of its various parts. PET therefore measures the distribution of the source of radiation within the body.

Ultrasounds are extremely useful for imaging soft tissues, but cannot penetrate bones. Their use is mostly limited to the abdomen and the heart. During an ultrasound examination, the apparatus measures the reflectance (rather than the transparency) of the tissues to the ultrasounds emitted by a source. Doppler ultrasound is a special technique that allows to measure the blood flow within desired organs, such as the liver, and is the only type of medical image to be displayed in color for diagnostic purposes.

Several other diagnostic imaging techniques exist, including single photon emission computed tomography (SPECT), magnetic source imaging (MSI), digital subtraction angiography (DSA), electrical impedance tomography (EIT), electrical source imaging (ESI), etc.

A typical radiology department can easily generate several gigabytes of data a day. The data is heterogeneous in nature: even within the same modality, scanners produced by different manufacturers generate data with different characteristics and formats. The two main problems in this field are dealing with the large number of different equipment types (from scanners, to display workstations, to communication networks, large storage subsystems, database management systems, etc.) that form a picture archiving and communication system (PACS), and managing the sheer volume and the different formats of the data. Recently the ARC-NEMA Digital Imaging and Communications in Medicine (DICOM) standard has specified a non-proprietary digital image format, a specific data file structure and protocols for the interchange of biomedical images and associated medical information.

Medical image databases have an incredible potential in diagnostic medicine, in medical research, and in education; however, their use in such applications is still limited. For example, we can envision that future medical image databases will be powerful differential diagnosis tools: the radiologist, facing an unclear case, will be able to retrieve and consult all the images containing a similar lesion stored in the data repository, but there still are no commercial systems supporting this capability. In medical research, we auspicate that image features will be widely included in clinical studies. Education is probably the application where image databases are starting to realize their potential: for example, the Uniformed Service University of Health Sciences, the U.S. federal government health sciences university, operates an online medical image database called MedPix™, which incorporates peer-reviewed radiological teaching files. MedPix contains images with associated textual information and provides image and textual search tools. MedPix is a prominent example of the radiological resources available on the World Wide Web. A survey of these Web-based resources can be found in Schiller and Fink.[111]

Query-by-content from medical image databases has mostly relied on metadata, texture, and shape features, since most medical images are typically in gray scale. In a recent study,[81] the authors investigate the use of the combination of texture features with scaled representation of the images to categorize medical images for CBR purpose. They report that, even with global features and with a large number of categories,[81] they were able to achieve 85% classification accuracy and the correct class was within the top ten guesses in 98% of the cases. Specialized applications of texture in the medical domain include the analysis and enhancement of mammograms. A mammogram is essentially a texture image that rarely contains objects with a well-defined contour. Digital enhancement of mammographical images should increase the visual difference between normal tissue and abnormalities,[112] and filters that match the specific texture of abnormalities should be used.[113] Similarly, texture features are extremely valuable inputs to automatic classifiers of medical images.[114] We conclude this section by directing the reader interested in CBIR in the medical domain to Müller et al.,[82] which contains a broad overview of the topic as well as an extensive bibliography.

Remotely Sensed Image Databases

Remotely sensed images provides us with a wealth of information that find applications in meteorology, earth sciences, environmental studies, urban planning, forestry management, agriculture, education, and, of course, law enforcement and defense.[115]

Instruments are carried on satellites (platforms) that orbit the earth. Often several instruments are carried on a single platform. Geostationary satellites are on an equatorial orbit, while lower-altitude platforms are usually on a quasipolar orbit. An instrument consists of one or a few parallel rows of sensors. Each row contains from a few dozens to a few tens of thousands sensors, each of which acquires a pixel of data, requiring one or two bytes. For polar-orbiting satellites, the rows are parallel to the surface of the earth and perpendicular to the direction of motion of the platform. At predefined intervals, each row of sensors acquires data from a long and narrow strip of the surface of the earth, and produces a line in the image. Between intervals, the platform moves along its orbit, and at different acquisition times the field of view of the instrument covers a different strip. The imaging process is therefore somewhat analogous to that of a desktop scanner.

Satellite images represent different quantities. Most instruments acquire data reflected data in one or more spectral bands. Instrument having long sensor rows usually acquire data in few spectral bands. The LANDSAT Thematic Mapper (TM), for example, acquires data in 6 spectral bands from blue to mid-infrared and in one thermal band, at 30 m resolution, and has more than 6000 pixels per line. Spectrometers, on the other hand, acquire data simultaneously in hundreds of narrow spectral bands, but typically have shorter sensor rows. For example, the CASI instrument has 288 spectral channels and 512 pixels per line.

Few instruments image emissions in the far infrared portion of the electromagnetic spectrum, and essentially measure the temperature of the surface of the earth, of the oceans, or of strata in the atmosphere. Synthetic aperture radars (SARs) measure the reflectance of the surface to microwave or short wave emissions generated on the satellite, as well as the distance between the surface and the satellite, and they yield images that can be used to produce elevation maps or to study certain types of land covers, such as ice.

Satellite instruments image the surface at a wide variety of resolutions. In some meteorological satellites, a pixel corresponds to several square miles on the surface. The NOAA AVHRR instrument acquires data having resolution of about 1 km. The LANDSAT Multi-Spectral Scanner is a medium resolution instrument with resolution of 79 m on the ground. The French SPOT 1 and 2 have resolutions of 10 m. Until recently, satellites capable of acquiring high-resolution images have had only military applications. More recent commercial satellites can acquire images with impressive details: for example the QuickBird satellite owned by DigitalGlobe is capable of 60 cm (roughly 2 ft) resolution in panchromatic mode and of 240 cm in multispectral mode. Images from QuickBird form part of the high-resolution images available via the popular Web-based application Google Earth mapping service. Some images from the Google Earth mapping service have an even higher resolution; at the time of this writing these images are actually aerial photos, rather than satellite images.

Numerous properties of satellite imagery pose challenges to the image database technology. Orbiting instruments generate data at an impressive rate: for example, the Earth Observing System satellites produce about 300 gigabytes a day. The data is collected at few sites, and its distribution in electronic format is difficult due to its sheer volume. The image content is extremely dense, and different information is useful in different application fields: hence satellite data management requires powerful indexing methodologies.

Image Databases for the Oil Industry

The oil industry is a major producer and user of image and volumetric data. Using seismic techniques, large three-dimensional models of geological formations are constructed. From the types of strata and their formations, it is possible to infer the presence of oil, the extension of the reservoir, and to determine the best strategies for drilling. The resolution of such three-dimensional data is in the order of meters or of tens of meters, and each model corresponds to a surface area of tens or hundreds of square km, and to depths of several km.

The relatively poor resolution of such data makes it inadequate for fine-tuning the drilling process. In some occasions, the stratum containing oil is only a few feet deep, and the drill has to be steered right into the stratum. Different types of images are used for this purpose. During the drilling process, it is possible to extract portions of the core, which are then cut, polished, and photographed. Similarly, packs of instruments are lowered to the bottom of the well and slowly retrieved. Some of the instruments measure global properties of the surrounding strata every few feet, such as the gamma ray emission, while others have arrays of sensors that are pressed against the walls of the well, and produce high-resolution measurements of properties such as electric conductivity. The formation FMI has 196 such sensors that measure the electrical resistivity along the circumference of the bore at depth intervals of 0.1 in. FMI data is usually represented as a false-color image having 196 columns and tens of thousands (or more) rows.

In some cases, microphotographs of core samples are used to assess the yield potentials of an oil field.

Besides the large amount of data, image databases for the oil industry face the challenge of data fusion: three-dimensional data, well-bore images, and microphotographs are often used in conjunction to make operative decisions. Powerful indexing techniques are also needed: for example, when combining three-dimensional seismic data (which provides a global, low-resolution view of the oil field) with images from the different well bores (which provide a sparse, highly localized, and high-resolution characterization), the analyst is often interested in determining the exact depths of specific geological formations at all well locations, to improve the accuracy of the model. This is currently a slow, labor-intensive process that requires manually matching hard copies of FMI images, tens of meters long. Automatic extraction and indexing of the data can substantially simplify the task.[48,116]

CONCLUSIONS

We have defined a framework for defining, representing, and searching the content of image repositories at different abstraction levels. We have discussed how raw-data, feature, and semantic level descriptions can be extracted from the images and automatically indexed. We have noted how the combination of image representation (compression) and processing yield significant speedups in content extraction at each abstraction level.

Simple objects, defined as connected image regions which are homogeneous with respect to specific characteristics, form the atomic unit of content retrieval. They can be defined and searched at one or more abstraction levels. Composite objects are sets of simple objects satisfying spatial or temporal constraints. We have discussed the semantics of simple and composite objects and described appropriate retrieval methodologies.

There are numerous open problems in the field. The automatic extraction of semantic content from photographic images is probably the most complex. The investigation of new high-dimensional indexing structures supporting flexible metrics is also an open area of research. Improvements in the interaction between the user and the system where the system learns how individual users tend to formulate queries are needed to make large digital libraries easier to use. Standards need to be defined, to allow a search engine to simultaneously query multiple repositories, and combine the contained information. Finally, a better infrastructure, with faster communication lines, is needed to remotely query large image repository.

ACKNOWLEDGMENTS

The author would like to thank Dr. Chung-Sheng Li, Dr. Lawrence D. Bergman, Dr. Yuan-Chi Chang, Dr. John R. Smith, and Dr. John J. Turek for the years of cooperation that led to the development of the SPIRE system, Dr. Ian Bryant, Dr. Peter Tilke, Dr. Barbara Thompson, Dr. Loey Knapp, and Dr. Nand Lal, for their comments and suggestions, and for defining applications scenarios for our technology.

REFERENCES

1. Mintzer, F.; et al.; Toward on-line, worldwide access to Vatican library materials. IBM J. Res. Dev. **1996**, *40* (2), 139–162.
2. Koller, D.; et al.; Protected interactive 3D graphics via remote rendering. In *Proceedings of the 31st International Conference on Computer Graphics and Interactive Techniques, SIGGRAPH2004*; New York; ACM Press: New York, 2004, 695–703.
3. Pratt, W.K. *Digital Image Processing,* 2nd Ed.; John Wiley & Sons: New York, 1991.
4. Pennebaker, W.; Mitchell, J.L. *JPEG Still Image Data Compression Standard*; Van Nostrand Reinhold: New York, 1993.
5. Shensa, M.J. The discrete wavelet transform: Wedding the Á trous and Mallat algorithms. IEEE Trans. Sig. Proc. **1992**, *40* (10), 2110–2130.
6. Lee, D.T. JPEG 2000: Retrospective and new developments. Proc. IEEE **2005**, *93* (1), 32–41.
7. Cover, T.J.; Thomas, J.A. *Elements of Information Theory,* 2nd Ed.; John Wiley & Sons: New York, 2006.
8. Gray, R.M.; Neuhoff, D.L. Quantization. *IEEE Trans. Info. Theory Commem. Issue.* **1998**, *44* (5), 2325–2383.
9. Gersho, A.; Gray, R.M. *Vector Quantization and Signal Compression*; Kluwer: Boston, MA, 1992.
10. Cover, T.J.; Thomas, J.A. *Elements of Information Theory*; John Wiley & Sons: New York, 1991.
11. Cosman, P.; Gray, R.M.; Vetterli, M. Vector quantization of image subbands: A survey. IEEE Trans. Image Process. **1996**, *5* (2), 202–225.
12. Shapiro, J.M. Embedded image coding using zerotrees of wavelet coefficients. IEEE Trans. Signal Process. **1993**, *41* (12), 3445–3462.
13. Niblack, W.; et al.; The QBIC project: Querying images by content using color texture, and shape. In *Proceedings of the SPIE—The International Society for Optical Engineering, Storage and Retrieval for Image and Video Databases*, Vol. 1908; 1993, 173–187.
14. Hare, J.S.; et al.; Bridging the semantic gap in multimedia information retrieval: Top-down and bottom-up app roaches. In *Mastering the Gap: From Information Extraction to Semantic Representation,* 3rd European Semantic Web Conference, Bouquet, P.; Brunelli, R.; Chanod, J.-P.; Niederée, C.; Stoermer, H.; Eds.; Budva, Montenegro, 2006.
15. Smeulders, A.; et al.; Content-based image retrieval at the end of the early years. IEEE Trans. Pattern Anal. Mach. Intell. **2000**, *22* (12), 1449–1380.
16. Smith, J.R.; Chang, S.-F. VisualSeek: A fully automated content-based image query system. In *Proceedings of the ACM Multimedia '96*, Boston, MA, November 18–22, 1996, 87–98.
17. Bach, J.R.; et al.; The Virage image search engine: An open framework for image management. In *Storage and Retrieval for Still Image Video Databases, Proceedings of the SPIE—The International Society for Optical Engineering*, Vol. 2670; 1996, 76–87.
18. Bergman, L.D. et al.; Drag-and-drop multimedia: An interface framework for digital libraries. IJODL **1999**, *2* (2/3), 170–177, (Special Issue on User Interfaces for Digital Libraries).
19. Bergman, L.D.; et al.; SPIRE, a digital library for scientific information. IJODL **2000**, *3* (1), 85–99, (Special Issue on Tradition of Alexandrian Scholars).
20. Vermilyer, R. Intelligent user interface agents in content-based image retrieval. In *Proceedings of the 2006 IEEE Southeast Con*, Memphis, TN, 2006, 136–142.
21. Zhou, X.S.; Huang, T.S. Relevance feedback in image retrieval: A comprehensive review. Multimedia Syst. **2003**, *8* (3), 536–544 (Special Issue on Content-Based Image Retrieval).
22. Castelli, V.; et al.; Search and progressive information retrieval from distributed image/video databases: The SPIRE project. In *Proceedings of the ECDL '98*, Crete, Greece, September, 1998.
23. Belongie, S.; et al.; Color- and texture-based image segmentation using EM and its application to content-based image retrieval. In *Proceedings of the Sixth International Conference on Computer Vision*, January 1998.
24. Carson, C.; et al.; Region-based image query. In *Proceedings of the IEEE CVPR '97 Workshop on Content-Based Access of Image and Video Libraries*, Santa Barbara, CA, 1997.
25. Dempster, A.P.; Laird, N.M.; Rubin, D.B. Maximum likelihood from incomplete data via the EM algorithm. J. Roy. Stat. Soc. B **1977**, *39* (1), 1–38.
26. Vaidyanathan, P.P. Orthonormal and biorthonormal filter banks as convolvers, and convolutional coding gain. *IEEE Trans. Signal Process.* **1993**, *41* (6), 2110–2130.
27. Li, C.-S.; Turek, J.J.; Feig, E. Progressive template matching for content-based retrieval in earth observing satellite image databases. Proc. SPIE Photonic East **1995**, *2606*, 134–144.
28. Castelli, V.; et al.; Progressive search and retrieval in large image archives. IBM J. Res. Dev. 1998, *42* (2), 253–268.
29. Smith, J.R.; Chang, S.-F. Tools and techniques for color image retrieval. In *Proceedings of the SPIE Storage and Retrieval for Still Image Video Databases*, San Jose, CA, February, 1996, Vol. 2670, 426–637.
30. Niblack, W.; et al.; The QBIC project: Querying images by content using color texture, and shape. IBM Res. J. **1993**, *9203* (81511).
31. Swain, M.J.; Ballard, D.H. Color indexing. Int. J. Comput. Vis. **1991**, *7* (1).
32. Gray, R.S. *Content-Based Image Retrieval: Color and Edges*; Technical Report 95–252, Department of Computer Science, Dartmouth University: Dartmouth, MA, 1995.
33. Feig, E.; Li, C.-S. Computing image histograms from compressed data. In *Proceedings of the SPIE Electronic*

Imaging and Multimedia Systems, Beijing, China, 1996, Vol. 2898, 118–124.
34. Itten, J. *Kunst der Farbe*; Otto Maier Verlag: Ravensburg, Germany, 1961 (in German).
35. Colombo, C.; Del Bimbo, A.; Pala, P. Semantics in visual information retrieval. IEEE Multimedia **1999**, *6* (3), 38–53.
36. Del Bimbo, A.; et al.; Visual querying by color perceptive regions. Pattern Recogn. **1998**, *31* (9), 1241–1253.
37. Gagalowicz, A.; Ma, S.D.; Tournier-Lasserve, C. Efficient models for color textures. In *Proceedings of the IEEE International Conference on Pattern Recognition, ICPR '86*, 1986, 412–414.
38. Hernandez, O.J.; Khotanzad, A. An image retrieval system using multispectral random field models, color, and geometric features. In *Proceedings of the 33rd Applied Imagery Pattern Recognition Workshop*, October, 2004, 251–256.
39. Yu, H.; Li, M.; Zhang, H.; Feng, J. Color texture moment for content-based image retrieval. In *Proceedings of the IEEE International Conference Image Processing, ICIP '02*, Rochester, NY, June, 2002, 929–932.
40. Li, C.-S.; Chen, M.-S. Progressive texture matching for earth observing satellite image databases. In *Proceedings of the SPIE on Multimedia Storage and Archiving Systems*, Boston, MA, November 18–19, 1996, Vol. 2916, 150–161.
41. Davis, L.S.; Johns, S.; Aggarwal, J.K. Texture analysis using generalized co-occurrence matrices. IEEE Trans. Pattern Anal. Mach. Intell. **1979**, *1* (3), 251–259.
42. Haralick, R.M.; Shanmugam, K.; Dinstein, I. Texture features for image classification.; IEEE Trans. Syst. Man. Cybernet. **1973**, *3*, 610–621.
43. Parkkinen, J.; Selkainaho, K.; Oja, E. Detecting texture periodicity from the co-occurrence matrix. Pattern Recogn. Lett. **1990**, *11*, 43–50.
44. Tamura, H.; Mori, S.; Yamawaki, T. Texture features corresponding to visual perception. IEEE Trans. Syst. Man. Cybern. **1978**, *8* (6), 460–473.
45. Smith, J.R.; Chang, S.-F. Quad-tree segmentation for texture-based image query. In *Proceedings of the ACM Multimedia '94*, San Francisco, CA, October 15–20, 1994, 279–286.
46. Jain, A.K.; Farrokhnia, F. Unsupervised texture segmentation using Gabor filters. Pattern Recogn. **1991**, *24* (12), 1167–1186.
47. Li, C.-S., Castelli, V. Deriving texture feature set for content-based retrieval of satellite image database. In *Proceedings of the IEEE International Conference on Image Processing, ICIP '97*, Santa Barbara, CA, October 26–29, 1997, 567–579.
48. Li, C.-S.; et al.; Comparing texture feature sets for retrieving core images in petroleum applications. In *Proceedings of the SPIE Storage and Retrieval for Image and Video Databases VII*, Vol. 3656, San Jose, CA, January 1999, 2–11.
49. Ma, W.Y.; Manjunath, B.S. Texture features and learning similarity. In *Proceedings of the IEEE Computer Vision and Pattern Recognition, CVPR '96*, San Francisco, CA, June 18–20, 1996, 425–430.
50. Ma, W.Y.; Manjunath, B.S. Edge flow: A framework of boundary detection and image segmentation. In *Proceedings of the IEEE Computer Vision and Pattern Recognition, CVPR '97*, 1997, 744–749.
51. Guyon, X. *Random Fields on a Network: Modeling, Statistics, and Applications*; Springer Verlag: New York, 1995.
52. Chen, C.C.; Dubes, R.C. Experiments in fitting discrete Markov random fields to textures. In *Proceedings of the IEEE Computer Vision and Pattern Recognition, CVPR '89*, 1989, 298–303.
53. Hassner, M.; Sklansky, J. Markov random field models of digitized image texture. In *Proceedings of the IEEE International Conference on Pattern Recognition, ICPR '78*, 1978, 538–540.
54. Hassner, M.; Sklansky, J. The use of Markov random fields as models of texture. Comput. Graphics Image Process. **1980**, *12*, 357–370.
55. Chellappa, R.; Chatterjee, S. Classification of textures using Gaussian Markov random fields. IEEE Trans. Acoust. Speech Signal Process **1985**, *33*, 959–963.
56. Cohen, F.S.; Fan, Z.; Attali, S. Automated inspection of textile fabrics using textural models. IEEE Trans. Pattern Anal. Mach. Intell. **1991**, *13* (8), 803–808.
57. Solberg, A.H.S.; Taxt, T.; Jain, A.K. A Markov random field model for classification of multisource satellite imagery. IEEE Trans. Geosci. Remote Sens. **1996**, *34* (1), 100–113.
58. Chellappa, R.; Chatterjee, S.; Bagdazian, R. Texture synthesis and compression using Gaussian-Markov random field models. IEEE Trans. Syst. Man Cybernet. **1985**, *15* (2), 298–303.
59. Chellappa, R.; Kashyap, R.L. Texture synthesis using 2-D noncausal autoregressive models. IEEE Trans. Acoust. Speech Signal Process. **1985**, *33*, 194–203.
60. Bouman, C.A.; Shapiro, M. A multiscale random field model for Bayesian image segmentation. IEEE Trans. Image Process. **1994**, *3* (2), 162–177.
61. Chen, J.L.; Kunda, A. Automatic unsupervised texture segmentation using Hidden Markov Model. In *Proceedings of the IEEE ICASSP '93*, 1993, 21–24.
62. Goktepe, M.; Yalabik, N.; Atalay, V. Unsupervised segmentation of gray level Markov model textures with hierarchical self organizing maps. In *Proceedings of the IEEE International Conference on Pattern Recognition, ICPR '96*, Vienna, Austria, 1996, D7M.3.
63. Hansen, F.R.; Elliott, H. Image segmentation using simple Markov field models. Comput. Graphics Image Process. **1982**, *20*, 101–132.
64. Noda, H.; Shirazi, M.N.; Kawaguchi, E. A MRF model-based method for unsupervised textured image segmentation. In *Proceedings of the IEEE International Conference on Pattern Recognition, ICPR '96*, 1996, B94.2
65. Santini, S.; Jain, R. Similarity measures. IEEE Trans. Pattern Anal. Mach. Intell. **1999**, *21* (9), 871–883.
66. Kimia, B.B. Shape representation for image retrieval. In *Image Databases, Search and Retrieval of Digital Imagery*; Castelli, V.; Bergman, L.D.; John Wiley & Sons: New York, 2002, Chap. 13, 345–372.
67. Carson, C.; et al.; *Blobworld: A System for Region-Based Image Indexing and Retrieval (Long Version)*; Technical Report UCB/CSD-99-1041, EECS Department, University of California: Berkeley, CA, 1999.
68. Antani, S.; Lee, D.J.; Longa, L.R.; Thoma, G.R. Evaluation of shape similarity measurement methods for spine

x-ray images. J. Vis. Commun. Image Repres. **2004**, *15* (3), 285–302.
69. Jain, A.; Lin, H.; Bolle, R. On-line fingerprint verification. IEEE Trans. Pattern Anal. Mach. Intell. **1997**, *19* (4), 302–314.
70. Eakins, J.P.; Boardman, J.M.; Shields, K. Retrieval of trade mark images by shape feature-the ARTISAN project. In *Proceedings of the IEE Colloquium on Intelligent Image Databases*, May, 1996, 1–6.
71. Icke, I. *Content Based 3d Shape Retrieval a Survey of State of the Art*; Pattern Recognition Laboratory, The Graduate Center, City University of New York, 2004. Available at http://www.cs.gc.cuny.edu/icke/academic/2ndexam.pdf.
72. Jaimes, A.; Chang, S.-F. A conceptual framework for indexing visual information at multiple levels. In *Proceedings of the SPIE—The International Society for Optical Engineering, Internet Imaging*, San Jose, CA, January, 2000, Vol. 3964, 2–15.
73. Haering, N.; da Vitoria Lobo, N. Features and classification methods to locate deciduous trees in images. Comput. Vis. Image Und. July/August **1999**, *75* (1/2), 133–149.
74. Forsyth, D.A.; Fleck, M.M. Body plans. In *Proceedings of the IEEE Computer Vision and Pattern Recognition, CVPR '97*, 1997.
75. Vailaya, A.; et al.; Content-based hierarchical classification of vacation images. In *Proceedings of the IEEE International Conference on Multimedia Computing and Systems*, Florence, Italy, June, 7–11, 1999, 518–523.
76. Li, J.; Wang, J.Z. Automatic linguistic indexing of pictures by a statistical modeling approach. IEEE Trans. Pattern Anal. Mach. Intell. **2003**, *25* (9), 1075–1088.
77. Li, J.; Wang, J.Z.; Real-time computerized annotation of pictures. IEEE Trans. Pattern Anal. Mach. Intell. **2008**, *30* (6), 985–1002.
78. Li, J.; Wang, J.Z. Alpir™ automatic photo tagging and visual image search. Online Demo.
79. Liu, Y.; et al.; A survey of content-based image retrieval with high-level semantics. Pattern Recogn. **2007**, *40* (1), 262–282.
80. Wei, W.; Barnaghi, P.M. *Semantic support for medical image search and retrievalBIEN'07: Proceeding of the Fifth IASTED International Conference*, Anaheim, CA, 2007; ACTA Press: Anaheim, CA, 2007, 315–319.
81. Lehmann, T.; et al.; Automatic categorization of medical images for content-based retrieval and data mining. Comput. Med. Imag. Grap. **2005**, *29*, 2–3, 143–155.
82. Müller, H.; et al.; A review of content-based image retrieval systems in medical applications—Clinical benefits and future directions. Int. J. Med. Inform. **2003**, *73* (1), 1–23.
83. Castelli, V.; et al.; Progressive classification in the compressed domain for large EOS satellite databases. In *Proceedings of the IEEE ICASSP '96*, May 1996, Vol. 4, 2201–2204.
84. Fagin, R. Fuzzy queries in multimedia database systems. In *Proceedings of the 17th ACM Symposium on Principles of Database Systems, PODS '98*, Seattle, WA, June 1–3, 1998; ACM Press: New York, 1998, 1–10.
85. Carey, M.J.; et al.; Towards heterogeneous multimedia information systems: The Garlic approach. In *Proceedings of the Fifth International Workshop on Research Issues in Data Engineering: Distributed Object Management*, 1995, 124–131.
86. Cody, W.; et al.; Querying multimedia data from multiple repositories by content, the Garlic project. In *Proceedings of the Third Working Conference on Visual Database Systems*; 1995.
87. Kilr, G.J.; Yuan, B. *Fuzzy Sets and Fuzzy Logic, Theory and Applications*; Prentice Hall: Upper Saddle River, NJ, 1995.
88. Zadeh, L.A. Fuzzy sets. Inform. Control **1965**, *8* (3), 338–353.
89. Fagin, R. Combining fuzzy information from multiple systems. In *Proceedings of the 15th ACM Symposium on Principles of Database Systems, PODS '96*, Montreal, Quebec, Canada, June 3–5, 1996; ACM Press: New York, 1996, 216–226.
90. Li, C.-S.; et al.; Sequential processing for content-based retrieval of composite objects. In *Proceedings of the SPIE Storage and Retrieval for Image and Video Databases VI*; San Jose, CA, January, 24–30, 1998, Vol. 3312, 2–13.
91. Li, C.-S.; et al.; Framework for efficient processing of content-based fuzzy Cartesian queries. In *Proceedings of the SPIE Storage and Retrieval for Media Databases*, 2000, Vol. 3972, 64–75. Also available as IBM Research Report RC21640, January 11, 2000.
92. Li, C.-S.; et al.; Progressive content-based retrieval of image and video with adaptive and iterative refinement. U.S. Patent 05,734,893, March 31, 1998.
93. Tao, J.; Tan, T. Affective computing: A review. In *Affective Computing and Intelligent Interaction*, Vol. 3784; Lecture Notes in Computer Science; Springer: Berlin/Heidelberg, 2005, 239–247.
94. Li, H.; Li, J.; Song, J.; Chen, J. Fuzzy mapping from image texture to affective thesaurus. In *Bio-Inspired Computational Intelligence and Applications*, Vol. 4688; Lecture Notes in Computer Science; Springer: Berlin/Heidelberg, 2007, 357–367.
95. Wu, Q.; Zhou, C.; Wang, C. Content-based affective image classification and retrieval using support vector machines. In *Affective Computing and Intelligent Interaction*, Vol. 3784; Lecture Notes in Computer Science; Springer: Berlin/Heidelberg, 2005, 239–247.
96. Bianchi-Berthouze, N.; Kato, T. K-dime: An adaptive system to retrieve images from the web using subjective criteria DNIS '00 In *Proceedings of the International Workshop on Databases in Networked Information Systems*; Springer-Verlag: London, 2000, 157–172.
97. Schatz, B.; et al.; Federating diverse collections of scientific literature. IEEE Comput. Mag. **1996**, *29* (5), 28–36.
98. Li, V.O.K.; Wanjiun, L. Distributed multimedia systems. Proc. IEEE **1997**, *85* (7), 1063–1108.
99. Wactlar, H.D.; Kanade, T.; Smith, M.A.; Stevens, S.M. Intelligent access to digital video: Informedia project. IEEE, Comput. Mag. **1996**, *29* (5), 46–52.
100. Collins, L.M.; et al.; Collaborative eScience libraries. Int. J. Digit. Libr. **2007**, *7* (1), 31–33.
101. Candela, L.; et al.; Diligent: Integrating digital library and grid technologies for a new Earth observation research infrastructure. Int. J. Digit. Libr. **2007**, *7* (1), 59–80.
102. Digital Library Goes e-Science (DLSci06). Workshop Held in Conjunction with ECDL 2006, Alicante, Spain, September 17–22, 2006.

103. Wilensky, R. Towards work-centered digital information services. IEEE Comput. Mag. **1996**, *29* (5), 37–43.
104. Smith, T.R. A digital library for geographically referenced materials. IEEE Comput. Mag. **1996**, *29* (5), 54–60.
105. Atkins, D.E.; et al.; Toward inquiry-based education through interacting software agents. IEEE Comput. Mag. May **1996**, *29* (5), 69–76.
106. Paepcke, A.; et al.; Using distributed objects for digital library interoperability. IEEE Comput. Mag. **1996**, *29* (5), 61–68.
107. Mintzer, F. Developing digital libraries of cultural content for internet access. IEEE Commun. Mag. **1999**, *37* (1), 72–78.
108. D'Alessandro, M.P.; et al.; The Iowa Health Book: Creating, organizing and distributing a digital medical library of multimedia consumer health information on the internet to improve rural health care by increasing rural patient access to information. In *Proceedings of the Third Forum on Research and Technology Advances in Digital Libraries ADL '96*, 1996, 28–34.
109. Lowe, H.J.; et al.; The image engine HPCC project, a medical digital library system using agent-based technology to create an integrated view of the electronic medical record. In *Proceedings of the 3rd Forum on Research and Technology Advances in Digital Libraries, ADL '96*, Washington, DC, May 1996, 45–56.
110. Macowsky, A. *Medical Imaging Systems*; Prentice Hall: Englewood Cliffs, NJ, 1983.
111. Schiller, A. Fink, G. Radiology resources on the net: So many images, so little time. Emerg. Med. News **2004**, *26* (3), 38–40.
112. Petrick, N.; et al.; An adaptive density-weighted contrast enhancement filter for mammographic breast mass detection. IEEE Trans. Med. Imaging **1996**, *15* (1), 59–67.
113. Strickland, R.N.; Han, H. Wavelet transform matched filters for the detection and classification of microcalcifications in mammography. In *Proceedings of the IEEE International Conference on Image Processing, ICIP '95*, Washington, DC, October, 23–26, 1995, Vol. 1, 422–425.
114. Christoyianni, I.; Dermatas, E.; Kokkinakis, G. Neural classification of abnormal tissue in digital mammography using statistical features of the texture. In *Proceedings of the Sixth IEEE International Conference on Electronics, Circuits and Systems, ICECS '99*, September 1999, 117–120.
115. Richards, J.A. *Remote Sensing Digital Image Analysis, an Introduction*, 2nd Ed.; Springer-Verlag: New York, 1993.
116. Bergman, L.D.; et al.; PetroSPIRE: *A multi-modal content-based retrieval system for petroleum applications*; In *Multimedia Storage and Archiving System* Boston, MA, 1999, Vol. 3846.
117. Castelli, V.; Bergman, L.D.; Eds.; *Image Databases, Search and Retrieval of Digital Imagery*; John Wiley & Sons: New York, 2002.
118. Chen, Y.; Li, J.; Wang, J.Z. *Machine Learning and Statistical Modeling Approaches to Image Retrieval*; Springer, 2004.
119. Feng, D.D.; Siu, W.C.; Zhang, H.; Eds.; *Multimedia Information Retrieval and Management*; Springer: Berlin, 2003.
120. Marques, O.; Furht, B. *Content-Based Image and Video Retrieval (Multimedia Systems and Applications)*; Springer: Berlin, 2002.
121. Wilhelmus, A.; Smeulders, M.; Jain, R.; Eds.; *Image Databases and Multi-media Search*; World Scientific: Singapore, 1997.
122. Sagarmay, D.; Ed.; *Multimedia Systems and Content-Based Image Retrieval*; Idea Group, Inc., 2004.
123. Eakins, J.; Graham, M. Content-based image retrieval. Available at http://www.jisc.ac.uk/publications/publications/contentimagefinalreport.aspx 39, JISC, (accessed in October 1999).
124. Datta, R.; Joshi, D.; Li, J.; Wang, J.Z. Image retrieval: Ideas, influences, and trends of the new age. ACM Comput. Surv. April **2008**, *40* (2), 1–60.
125. Venters, C.C.; Cooper, M. A review of content-based image retrieval systems, 2000. Available at http://www.jisc.ac.uk/publications/publications/contentreviewfinalreport.aspx.
126. Kherfi, M.L.; Ziou, D.; Bernardi, A. Image retrieval from the World Wide Web: Issues, techniques, and systems. ACM Comput. Surv. **2004**, *36* (1), 35–67.

Bibliography

We conclude this entry by listing additional resources for the interested readers. In recent years, several books have been published on the topic of still-image repositories, search, and retrieval. A general introduction can be found in Chen,[117] while Feng[118] describes a Machine Learning approach to CBIR. Numerous books on multimedia repositories contain sections on image collections, including Marques[119] and Wilhelmus.[120] In addition to these publications, which can be used as introductions to the topic or as textbooks, several collections of selected papers have appeared in the literature.[121,122] Numerous articles review current results in CBIR,[123,124] CBIR systems,[125] shape-based retrieval,[71] semantic-level retrieval,[79] as well as for applications such as CBIR for medical imaging[82] and retrieval from the World Wide Web.[126]

Image Secret Sharing

WeiQi Yan and Jonathan Weir
Queen's University, Belfast, U.K.

Mohan S. Kankanhalli
National University of Singapore, Singapore

Abstract
Visual cryptography allows visual information to be encrypted in such a way that decryption can be done via sight reading. In this entry, a variation of the technique for color images introduced. The security of secret sharing is built upon the computation of polynomials based on moving lines. This version of the technique can be used for sharing compressed-domain images.

Keywords: Color image; Compression; Cryptography; Visual cryptography; Secret image; Security.

INTRODUCTION

Cryptography plays a very vital enabling role in our modern computing infrastructure. Almost all real-world applications require keys (such as passwords) for the purposes of confidentiality, authentication, and nonrepudiation. The strength of such cryptographic applications is based on the secrecy of a key. Therefore, the loss of a key can lead to disastrous consequences. Thus, many cryptographers have tackled the following problem:

Suppose a secret s (a key) is divided into $n > 1$ parts (called secret shares) and it satisfies these properties:

1. The secret key s can be easily restored from k ($k \leq n$) shares.
2. The secret key s cannot be restored from $k-1$ (or less) shares.
3. The size of each share is not more than the size of the secret key s.

Such a scheme is referred to as a (k, n) threshold cryptography scheme or a secret sharing system.[2,17] It provides a backup mechanism to the secret key and it provides protection against the loss of a key. Secret sharing is also regarded as a mechanism to transfer secret information by public channels in cryptography.[4] Blakley based his secret sharing scheme on hyperplanes[17] and Shamir provided a solution based on the Lagrange interpolation.[2] Asmuth and Bloom scheme is based on the Chinese Remainder Theorem.[5] The details of these methods are available in Refs. [1,12]. These traditional secret sharing schemes primarily concentrate on bit strings and do not take the specific content of these bits into account. However, with the increasing emphasis on security and digital rights management of multimedia data, the connection between multimedia and cryptography is becoming stronger. In this context, we present our novel ideas on color image sharing in which we utilize the concept of secret sharing from cryptography and employ it to protect a secret color image. As we shall see, the ideas cannot be directly applied so we need to take into account that the data under consideration describes color images and is not any generic bit stream. In our scheme, a secret color image is divided into n shares. Each share is an innocuous image totally unrelated to the secret image. We utilize k (or more) shares in order to perfectly reconstruct the secret image. However, having access to $k - 1$ (or less) shares will not reveal the secret color image.

We envisage several useful applications for a color image sharing scheme. Suppose we have a secret color image that we desire to protect. If we employ traditional cryptographic techniques, then we need to encrypt the image and store the image on a secure server. We then need to pay attention on the security of the key used for encryption. This server would then become a single focus of attack from a potential adversary. However, with an image sharing scheme, we can divide the information in the image into several shares and keep them on separate servers. This would allow for a lot more redundancy in the protection since breaking one server will not reveal the secret image. Another application would be that of data hiding. Suppose we would like to transmit a secret image over a noisy and insecure channel. We could divide the image information into several shares that are basically innocuous images. These images could be transmitted and at the other end, the secret image could be reconstructed from the threshold number of shares. Another useful application would be that of a military command and control

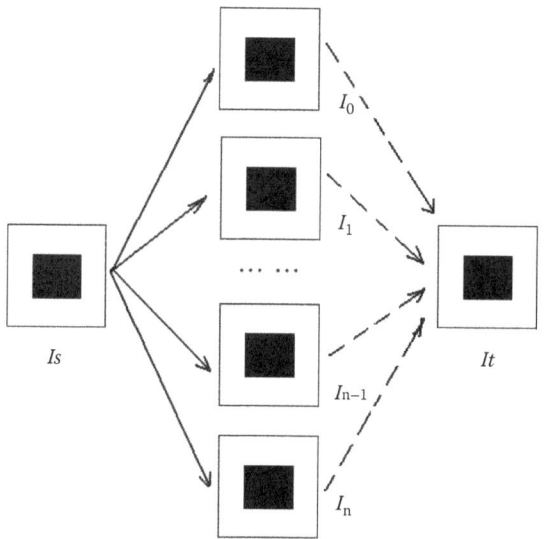

Fig. 1 Principle of image sharing

system based on the Clark–Wilson security model.[11] Suppose we want the battlefield plans to be made only if k out of n commanders agree. In which case, we could divide the battle terrain map into n shares and distribute it to the commanders. Only if k of them get together can they restore the terrain map and agree to a battle plan. In general, our scheme would be advantageous in any situation where a group of mutually suspicious individuals or processes need to cooperate and every threshold subset of the group needs to be given the veto power.

The problem of color image sharing is formally defined as follows. In order to transfer a color image $\mathcal{I}$ through a public channel securely, the information of the color image is divided into n pieces and embedded into images I_i, $(i = 1, 2, …, n)$, and we call the images I_i, $(i = 1, 2, …, n)$ as shares. With the knowledge of any $k (k<n)$ shares I_i, $(i = 1, 2, …, k)$, the restoration of the original color I image is easy; with the knowledge of any $k − 1 (k < n)$ shares, the restoration of the original image $\mathcal{I}$ is impossible (i.e., any image is equally likely to be reconstructed).

In Fig. 1, the left-most image is the original color image that we desire to keep as a secret and it is divided into several blocks (subimages). We process each of these blocks into n shares in order to create the subimages shares. If required, we can compose these shares I_i, $(i, = 1, 2, …, n)$ together and compute the blocks in the right-most image. Block-based processing is done for breaking correlation. Note that our scheme will require a minimum of k shares in order for I_r to be exactly equal to I_s. If less than k shares are used, then I_s cannot be restored. The threshold is indispensable in color image sharing.

STATE OF THE ART

Applying visual cryptography techniques to color images is a very important area of research because it allows the use of natural color images to secure some types of information. Due to the nature of a color image, this again helps to reduce the risk of alerting someone to the fact that information is hidden within it. It should also allow high quality sharing of these color images. Color images are also highly popular and have a wider range of uses when compared to other image types. Many of the techniques presented within this section use halftone technologies on the color images in order to make them work with visual cryptography. That is why color visual cryptography is presented within this section.

In 1996, Naor and Shamir published a second article on visual cryptography "Visual Cryptography II: Improving the Contrast via the Cover Base".[23] The new model contains several important changes from their previous work; they use two opaque colors and a completely transparent one.

The first difference is the order in which the transparencies are stacked. There must be an order to correctly recover the secret. Therefore, each of the shares needs to be predetermined and recorded so recovery is possible. The second change is that each participant has c sheets, rather than a single transparency. Each sheet contains red, yellow, and transparent pixels. The reconstruction is done by merging the sheets of participant I and participant II, i.e., put the i-th sheet of II on top of the i-th sheet of I and the $(i + 1)$-th of I on top of the i-th of II.

The two construction methods are monochromatic construction and bichromatic construction. In the monochromatic construction, each pixel in the original image is mapped into c subpixels and each participant holds c sheets. In each of participant I sheets, one of the subpixels is red and the remaining $c−1$ subpixels are transparent. In each of participant II sheets, one of the subpixels is yellow, the other $c−1$ subpixels are transparent. The way the sheets of participant I and II are merged is by starting from the sheet number 1 of participant I, then putting sheet number 2 of participant II on top of it, then sheet number 2 of participant I on top of that, and so on.

The order in which subpixels of participant I are colored red constitutes a permutation π on $\{1, …, c\}$ and the order which the subpixels of participant II are colored yellow constitutes a permutation σ. π and σ are generated as follows: π is chosen uniformly at random from the set of all permutations on c's elements. If the original pixel is yellow, then $\pi = \sigma$, therefore each red subpixel of the i-th sheet of participant I will be covered by a yellow subpixel of the same position of the i-th sheet of participant II. If the original pixel is red, then $\sigma(i) = \pi(i + 1)$ for $1 \leq i \leq c − 1$ and $\sigma(c) = \pi(1)$, therefore each yellow subpixel of the i-th sheet of participant II will be covered by a red subpixel of the same position of the $(i + 1)$-th sheet of participant I except the c-th sheet. In practice, the first sheet of participant I is not necessarily stored since it is always covered by other sheets.

A very primitive example of color image sharing appeared in Ref. [24]. In this example, each pixel of the color secret image is expanded to a block of 2 × 2 subpixels. Each one of these blocks is filled with red, green, blue, and white (transparent) colors, respectively. Taking symmetries into account, 24 different possibilities for the combination of two pixels can be obtained. It is claimed that if the subpixels are small enough, the human visual system will average out the different possible combinations to 24 different colors. To encrypt a pixel of the colored image, round the color value of that pixel to the nearest representable color. Select a random order for the subpixels on the first share and select the ordering on the second share such that the combination produces the required color.

The advantage of this scheme is that it can represent 24 colors with a resolution reduction of 4, instead of 242 = 576. The disadvantage is that the 24 colors are fixed once the basic set of subpixel colors is fixed.

Another primitive scheme was also presented[29] and extended more recently.[34] Verheul and Van Tilborg's scheme provides a c-color (k, n)-threshold scheme. This scheme uses the black pixel to superimpose on the result of two color pixels, superimposition, if they give a resultant color that is not in the original color palette. This can be achieved by making sure the superimposed color pixels result in a noncolor palette color, one of which is changed to a black pixel or by ensuring that one of the color pixels is changed to black before the superimposing operation.[10] Yang and Laih improve on the pixel expansion aspect of the Verheul and Van Tilborg scheme and their (n, n)-threshold scheme is optimal since they match the following lower bound placed on pixel expansion, formulated in Ref. [10]:

$$m \geq \begin{cases} c \cdot 2^{n-1} - 1 & \text{if } n \text{ is even} \\ c \cdot 2^{n-1} - c + 1, & \text{if } n \text{ is odd} \end{cases} \quad (1)$$

Hou et al.[18] proposed a novel approach to share color images based on halftoning. With this halftone technology, different gray levels can be simulated simply by altering the density of the printed dots. Within bright parts of the image the density is sparse, while in the darker parts of the image, it is dense. This is very helpful in the visual cryptography sense because it is able to transform a grayscale image into a black and white image. This allows for traditional visual cryptography techniques to be applied. Similarly, the color decomposition method is used for color images, which also allows the proposed scheme to retain all the advantages of traditional visual cryptography, such as no computer participation required for the decryption/recovery of the secret.

Hou himself also provided one of the first color decomposition techniques to generate visual cryptograms for color images.[19] Using this color decomposition, every color within the image can be decomposed into one of three primary colors: cyan, magenta, or yellow. This proposal is similar to traditional visual cryptography with respect to the pixel expansion that occurs. One pixel is expanded into a 2 × 2 block where two color pixels are stored along with two transparent (white) pixels.

However, Leung et al.[21] examined the security of Hou's[19] scheme, and while the scheme is secure for a few specific two-color secret images, the security cannot be guaranteed for many other cases.

Improving this pixel expansion and also working out the optimal contrast of color visual cryptography schemes have been investigated.[10] In the paper, they prove that contrast-optimal schemes are available for color visual cryptography (VC) and then further go on to prove the optimality with regard to pixel expansion.

A lossless recovery scheme outlined by[20] considers halftoning techniques for the recovery of color images within visual cryptography. The scheme generates high quality halftone shares that provide lossless recovery of the secrets and reduces the overall noise in the shares without any computational complexity. Their proposed method starts by splitting the color channels into its constituent parts, cyan (C), magenta (M), and yellow (Y). Each channel has grayscale halftoning applied to it. Error diffusion techniques discussed in Ref. [35] are then applied to each halftone channel. A circularly symmetric filter is used along with a Gaussian filter. This provides an adequate structure for the dot placement when constructing the shares.

Efficiency within color visual cryptography[25] is also considered which improves on the work done by[3,34] The proposed scheme follows Yang and Laih's color model. The model considers the human visual system's effect on color combinations out of a set of color subpixels. This means that the set of stacked color subpixels would look like a specific color in original secret image. As with many other visual cryptography schemes, pixel expansion is an issue. However Shyu's scheme has a pixel expansion of $\lceil log_2 c \rceil$, which is superior to many other color visual cryptography schemes especially when c, the number of colors in the secret image becomes large. An area for improvement however would be in the examination of the difference between the reconstructed color pixels and the original secret pixels. Having high quality color VC shares would further improve on the current schemes examined within this survey, this includes adding a lot of potential for visual authentication and identification.

Chang et al.[6] present a scheme based on smaller shadow images, which allows color image reconstruction when any authorized k shadow images are stacked together using their proposed revealing process. This improves on the following work,[32] which presents a scheme that reduces the shadow size by half. Chang et al.'s technique improves on the size of the share in that, as more shares are generated for sharing purposes, the overall size of those shares decreases.

In contrast to color decomposition, Yang and Chen[33] propose an additive color mixing scheme based on probabilities. This allows for a fixed pixel expansion and improves on previous color secret sharing schemes. One problem with this scheme is that the overall contrast is reduced when the secrets are revealed.

In most color visual cryptography schemes, when the shares are superimposed and the secret is recovered, the color image gets darker. This is due to the fact that when two pixels of the same color are superimposed, the resultant pixel gets darker. Cimato et al.[9] examine this color darkening by proposing a scheme that has to guarantee that the reconstructed secret pixel has the exact same color as the original. Optimal contrast is also achieved as part of their scheme. This scheme differs from other color schemes in that it considers only 3 colors when superimposing, black, white, or one pixel of a given color. This allows for perfect reconstruction of a color pixel, because no darkening occurs, either by adding a black pixel or by superimposing two colors that are identical, that ultimately results in a final darker color.

A technique that enables visual cryptography to be used on color and grayscale images is developed in progressive color visual cryptography.[13] Many current state-of-the-art visual cryptography techniques lead to the degradation in the quality of the decoded images, which makes it unsuitable for digital media (image, video) sharing and protection. In Ref. [13], a series of visual cryptography schemes have been proposed that not only support gray-scale and color images, but also allow high quality images including that of perfect (original) quality to be reconstructed.

The annoying presence of the loss of contrast makes traditional visual cryptography schemes practical only when quality is not an issue which is relatively rare. Therefore, the basic scheme is extended to allow visual cryptography to be directly applied on grayscale and color images. Image halftoning is employed in order to transform the original image from the grayscale or color space into the monochrome space which has proved to be quite effective. To further improve the quality, artifacts introduced in the process of halftoning have been reduced by inverse halftoning.

With the use of halftoning and a novel microblock encoding scheme, the technique has a unique flexibility that enables a single encryption of a color image but enables three types of decryptions on the same ciphertext. The three different types of decryptions enable the recovery of the image of varying qualities. The physical transparency stacking type of decryption enables the recovery of the traditional VC quality image. An enhanced stacking technique enables the decryption into a halftone quality image. A progressive mechanism is established to share color images at multiple resolutions. Shares are extracted from each resolution layer to construct a hierarchical structure; the images of different resolutions can then be restored by stacking the different shared images together.

The advantage is that this scheme allows for a single encryption, multiple decryptions paradigm. In the scheme, secret images are encrypted/shared once, and later, based on the shares, they can be decrypted/reconstructed in a plurality of ways. Images of different qualities can be extracted, depending on the need for quality as well as the computational resources available. For instance, images with loss of contrast are reconstructed by merely stacking the shares; a simple yet effective bit-wise operation can be applied to restore the halftone image; or images of perfect quality can be restored with the aid of the auxiliary look-up table. Visual cryptography has been extended to allow for multiple resolutions in terms of image quality. Different versions of the original image of different qualities can be reconstructed by selectively merging the shares. Not only this, a spatial multiresolution scheme has been developed in which images of increasing spatial resolutions can be obtained as more and more shares are employed.

This idea of progressive visual cryptography has recently been extended[14] by generating friendly shares that carry meaningful information and that also allows decryption without any computation at all. Purely stacking the shares reveals the secret. Unlike Refs. [7,13] which require computation to fully reconstruct the secret, the scheme proposed in Ref. [15] has two types of secrets, stacking the transparencies reveals the first, but computation is again required to recover the second-level secret. Fang's scheme is also better than the polynomial sharing method proposed in Ref. [26] by Thien and Lin. The method proposed in Ref. [26] is only suitable for digital systems and the computational complexity for encryption and decryption is also a lot higher.

Currently, one of the most robust ways to hide a secret within an image is by typically employing visual cryptography. The perfect scheme is extremely practical and can reveal secrets without computer participation. Recent state-of-the-art watermarking[8] can hide a watermark in documents that require no specific key in order to retrieve it. We take the idea of unseen visible watermarks and apply a secure mask to them and incorporate it for use within the VC domain, thus improving the overall security that is currently one of its weaknesses.

Weir et al.[31] also provide a mechanism for secret sharing using color images as a base. The technique relies on visual cryptography as a mechanism for sharing the secret. Many smaller secrets can be embedded within a color image and a final share can be created in order to reveal all of the secrets. The color image is visually similar to the original image before embedding due to the high Peak-Signal-to-Noise Ratio (PSNR) achieved after embedding.

Another recent novel application for color image sharing and using color images for secret sharing was presented by Weir and Yan.[30] Using Google Maps, along with its Street View implementation, personally identifiable information can be obscured using visual cryptography techniques and can be accurately recovered by authorized individuals who may need the information. Specifically for law enforcement agencies who many find this type of information helpful for a particular case. This type of practical application is very important for the progression of visual cryptography, which presents a unique way of using these techniques in a real-world situation.

APPROACHES FOR IMAGE SHARING

Shamir's Secret Sharing Scheme

Shamir's secret sharing scheme is based on the Lagrange interpolation.[2] Given a set of points (x_i, y_i) $(i=0, 1, 2, ..., k)$, the Lagrange interpolation polynomial $L^k(x)$ can be constructed using:

$$f^k(x) = \sum_{i=0}^{k} y_i \prod_{j=0; i \neq j}^{k} \frac{x - x_i}{x_j - x_i} \qquad (2)$$

Given a secret, it can be easily shared using this interpolation scheme. If $GF(q)$ denotes a Galois field ($q > n$), the following polynomial is constructed by choosing proper coefficients $a_0, a_1, ..., a_k$ from $GF(q)$, which satisfy:

$$f^k(x) = s^* + \sum_{i=0}^{k} a_i x^i \qquad (3)$$

where s^* is the secret key. The coefficients are randomly chosen over the integers $(0, q)$ and the details are provided in Ref. [2]. Suppose $s_i = f(a_i), (i=0, 1, \cdots, k)$, each S_i is known as a *share* and they all can be delivered to different persons.

Now we would like to reconstruct the original secret. Suppose k people have provided their shares s_i, $(i = 0, 1, \cdots, k)$. The following Lagrange interpolation polynomial is utilized to reconstruct the original secret:

$$P^k(x) = \sum_{i=0}^{k} s_i \prod_{j=0; i \neq j}^{k} \frac{a - a_i}{a_j - a_i} \qquad (4)$$

where addition, subtraction, multiplication, and division are defined over $GF(q)$.

$$P^k(a_i) = s_i; i=0, 1, 2, \cdots, k; s^* = P^k(0); \qquad (5)$$

Thus, we can obtain the original secret s^*.[1,2]

Color Image Sharing Based on the Lagrange Interpolation

When an image is treated as a secret, we can share the secret based on the Lagrange interpolation. We consider the image to be a matrix; and Lagrange interpolation is generalized for matrices.

We assume a grayscale image corresponds to a matrix A. Given matrix $A_i = (d_{w,h}^i)_{W \times H}$, where w and h are integer ($w = 1, 2, ..., W; h = 1, 2, ..., H; i = 0, 1, ..., k$), k is the number of shares. We define the matrix operations of Lagrange polynomials with degree k in the Galois field:

$$L^k(x) = \sum_{i=0}^{k} A_i \prod_{j=0; i \neq j}^{k} \frac{x - x_i}{x_j - x_i} \qquad (6)$$

where (x_i, A_i) is the feature point, $A_i ((i = 0,1,..., k))$ are matrices of size $W \times H$, the elements being nonnegative, x_i is a real number and $x_i \leq x_j$ ($i \leq j$).

Our novel idea for color image sharing using Lagrange interpolation is the following. We utilize the secret image and a few ($<n$) other chosen innocuous images to build the Lagrange interpolation polynomial. We then construct new images based on this interpolation to obtain a total of n images. We now can use all the innocuous images and the new reconstructed images as the shares of the secret image. And the secret image is not distributed but it can be always reconstructed using the requisite number of shares. So our novelty is in the construction of new shares based on the secret image and some chosen innocuous images. We now provide the details of the scheme. In Fig. 2, assume that the secret image that we desire to share is at the position $x = 0$.

In Fig. 2a, we position an innocuous image at x_2. We can now compute many new shares with the parameters in the interval $(x_2 - d, x_2 + d)$, $d > 0$, (we have used $d_{max} = \frac{x_2 - x_0}{4}$). Without loss of generality, we select the parameter at the position x_1 as our newly created share (based on 0 and x_2). We now consider the images at x_2 and x_1 as the shares of the secret image at 0. For the reconstruction of the secret, we can collect the two shares, we calculate the coefficients of Lagrange interpolation polynomial first and then compute the image at position 0. In Fig. 2b, we put the shares in x_1, x_2 and obtain the secret color image at 0; and

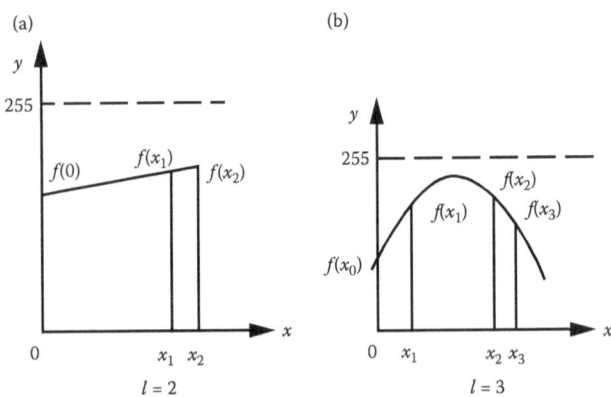

Fig. 2 Image sharing based on the Lagrange interpolation in (a) and (b)

Image Secret Sharing

(a) Original image and shares 1 and 2

(b) Original image and shares 1, 2, and 3

Fig. 3 Experimental results of image sharing based on the Lagrange interpolation in (a) and (b)

in Fig. 2b, we put the shares at x_1, x_2, x_3 and restore the secret image at position 0. For a color image, this operation can be performed for each color channel. Figure 3 shows our experimental results for the Lagrange interpolation scheme of a color image.

In Fig. 3a, the original color image is the secret image to be shared, at least two shares are needed to reconstruct the original image, i.e., it is a (2, n) scheme, thus $k = 2$. Thus, share 2 has been generated using the original color image and share 1 (which is an innocuous image). Now, share 1 and share 2 can be distributed independently and the original color image can be reconstructed anytime if we obtain both the shares. In Fig. 3b, the original color image is the secret color image and at least three shares are required to restore the secret color image. This is because $k = 3$, therefore it is a (3, n) secret image sharing scheme. Notice that the secret image is somewhat visible in the generated shares. We will fix this problem later with the block-based approach.

However, the Lagrange polynomial based image sharing has a potential practical problem. It cannot yield too many shares. This is because the Lagrange polynomial curve with a high degree has severe oscillations once the degree of the polynomial is greater than nine.[16] The consequence of this oscillations phenomenon is that it is impossible to constrain the pixel values to lie between 0 and 255. As a result, the resultant shares are not proper images anymore as shown in Fig. 4.

In Fig. 4, the resultant image Fig. 4a is the original secret image, Fig. 4b is the innocuous image and Fig. 4c is the generated share while the rest of the shares are all-white images. We use a polynomial of degree 8 for generating Fig. 4c. Figure 4c has obvious color overflow problems because some pixel values are more than 255 and some are less than zero, hence the image quality is severely degraded. Note that this is not a problem in Shamir's original secret sharing scheme because they consider the secret to be shared as a binary integer, and thus a share can take on any value. In our case, this binary integer has some constraints because they denote image pixel values. In order to overcome this serious limitation, it is obvious that some form of a piece-wise polynomial interpolation is required in order to bound the degree of the polynomial and thus constrain the oscillations. We have developed a new color image sharing scheme based on moving lines, which is a rational implicit curve.

(a) Original image   (b) Shares 1   (c) Shares 2

Fig. 4 The image sharing by using a high degree polynomial interpolation in (a)-(c)

Color Image Sharing Based on Moving Lines

It is well known that the equation of a line in the homogenous form in projective geometry[22] is:

$$aX + bY + cW = 0 \qquad (7)$$

where a, b, and c are not all zero, (X, Y, W) are the homogenous coordinates of points whose Cartesian coordinates are:

$$(x, y) = (X/W, Y/W) \qquad (8)$$

It is obvious that X, Y, W cannot all be zero. Let P denote the triple (X, Y, W) and L denote the triple (a, b, c). We refer to the line L that is:

$$\{(X, Y, W) \mid L \cdot P = (a,b,c) \cdot (X, Y, W) \\ = aX + bY + cW = 0\} \qquad (9)$$

Thus, we can see that a point P lies on a line $L = (a, b, c)$ only and if only $P \cdot L = 0$, where $P \cdot L$ is the dot product.

Now we consider the line L containing two points $P_1 = (X_1, Y_1, W_1)$ and $P_2 = (X_2, Y_2, W_2)$, and also the point P at which two lines $L_1 = (a_1, b_1, c_1)$ and $L_2 = (a_2, b_2, c_2)$ intersect. Because of the duality principle (of points and lines), we have these cross products:

$$L = P_1 \times P_2, \ P = L_1 \times L_2 \qquad (10)$$

A homogeneous point whose coordinates are functions of a variable t (i.e., it is parameterized by a variable t) is denoted as:

$$P[t] = (X[t], Y[t], W[t]) \qquad (11)$$

which actually is the rational curve:

$$x = \frac{X[t]}{W[t]}; \ y = \frac{Y[t]}{W[t]}; \qquad (12)$$

If the functions are of the following form:

$$X[t] = X_i \phi_i[t]; Y[t] = Y_i \phi_i[t]; W[t] = W_i \phi_i[t] \qquad (13)$$

With $\{\phi_i[t]\}$ being a given set of blending function, then Eq. 11 defines a curve:

$$P[t] = \sum P_i \phi_i[t] \qquad (14)$$

where the homogeneous points are $P_i = (X_i, Y_i, W_i)$. Likewise,

$$L[t] = (a[t], b[t], c[t]) \qquad (15)$$

denotes the family of lines $a[t]x + b[t]y + c[t] = 0$.

In order to obtain the intersection of two pencils, we notice the following four lines $L_{0,0}$, $L_{0,1}$, $L_{1,0}$, and $L_{1,1}$ from which two pencils are defined:

$$L_0[t] = L_{0,0}(1-t) + L_{0,1}t, L_1[t] = L_{1,0}(1-t) + L_{1,1}t \qquad (16)$$

The points, at which they intersect for parameter values $t = 0$, Δt, $2\Delta t$, ..., 1 are on a curve. The parameter $t = 0$, Δt, $2\Delta t$, ..., 1 is adaptively given and it may take any real number value in the interval [0,1] for a piece of curve and it also can be extended to the infinite interval [−∞, +∞] for the whole curve. The curve turns out to be a conic section, which can be expressed as a rational Bernstein–Bezier curve $P[t]$ as follows:

$$P[t] = L_0[t] \times L_1[t] \qquad (17)$$

It is clear that $P[t]$ is a quadratic rational Bezier curve[27,28] whose control points are:

$$P_0 = L_{0,0} \times L_{1,0}, P_1 = \frac{L_{0,0} \times L_{1,1} + L_{0,1} \times L_{1,0}}{2}, P_2 = L_{0,1} \times L_{1,1} \qquad (18)$$

The graph of a quadratic curve generated by moving lines is shown in Fig. 5a.

In general, the curve comprising of $L_{0,0}$, $L_{0,1}$, $L_{0,2}$, and $L_{1,0}$, $L_{1,1}$ is:

$$L_0[t] = L_{0,0}(1-t)^2 + 2L_{0,1}(1-t)t + L_{0,2}t^2, L_1[t] \\ = L_{1,0}(1-t) + L_{1,1}t \qquad (19)$$

then $P[t] = L_0[t] \times L_1[t]$, $t \in [0, 1]$ is a cubic rational Bezier curve.[16]

Without loss of generality, the moving lines consisting of $L_{0,0}, L_{0,1}, \ldots, L_{0,p}$, and $L_{1,0}, L_{1,1}, \ldots, L_{1,q}$ are:

$$L_0[t] = \sum_{i=0}^{p} L_{i,0} B_i^p(t), \quad L_1[t] = \sum_{i=0}^{q} L_{i,1} B_i^q(t) \qquad (20)$$

then $P[t] = L_0[t] \times L_1[t]$, $t \in [0, 1]$ represents a rational Bezier curve of degree $p + q$ as shown in Fig. 5b.[16]

As in the case of Lagrange interpolation, we now construct the scheme of image sharing based on moving lines as shown in Fig. 6. In Fig. 6, the x-coordinate depicts the given position of the original secret color image, the y-coordinate indicates the values of the gray-scale pixel value, t is the parameter for moving lines generation, and x^* is the position of the new computed share. For image sharing, we compute the shares using the above implicit curve. We now present the detailed steps for our moving lines based on the color image sharing scheme.

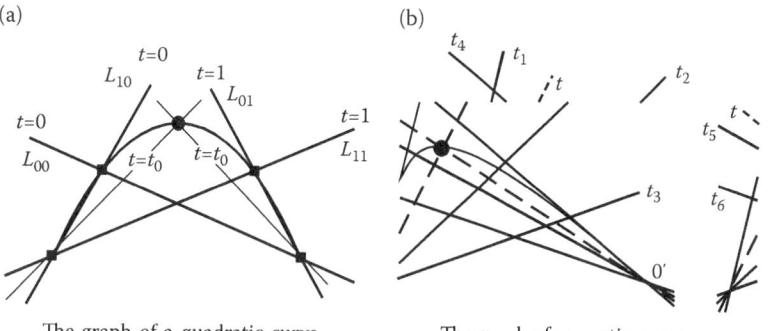

Fig. 5 Intersection of two pencils of lines in (a) and (b)

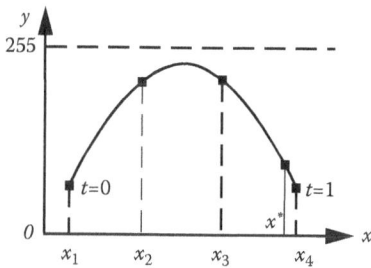

Fig. 6 Image sharing scheme based on moving lines

Steps of sharing a color image:

1. Suppose we are given one secret image I_0 to be shared, $k + 1$ images $I_0, I_1, \ldots, I_k$ as the innocuous images and a group of corresponding parameters $x_0, x_1, \ldots, x_k$ ($x_0 \leq x_1 \leq \ldots \leq x_k$) as input.
2. Calculate an arbitrary new share corresponding to a parameter within the interval of the given parameters (x_1, x_k).
3. The images (greater than k) $I_0, I_1, \ldots, I_{i-1}, I_p, I_i, I_{i+1}, \ldots, I_k$; $0 \leq i - 1 \leq p \leq i \leq k$ and their corresponding parameters $x_0, x_1, \ldots, x_{i-1}, x_p, x_i, x_{i+1}, \ldots, x_k$ ($x_0 \leq x_1 \leq \ldots \leq x_{i-1} \leq x_p \leq x_{i+1} \leq \ldots \leq x_k$) are the output computed image shares.

Steps of reconstructing the color image:

1. Select k image shares $I_0, I_1, \ldots, I_k$ and their corresponding position parameters $x_0, x_1, \ldots, x_k$ ($x_0 \leq x_1 \leq \ldots \leq x_k$) and the position parameter of the color image x_0 as the input.
2. Calculate the implicit curve corresponding to pixels in each share according to the moving lines scheme $P[t] = L_0[t] \times L_1[t]$, $t \in [0, 1]$.
3. Reconstruct the secret color image by the scheme of moving lines at the position of x_0 as output.

Note that each of the steps has to be applied separately for each color channel. The use of a rational curve has several advantages, such as it does not have the oscillations phenomenon of polynomials with a high degree; a rational curve is easy to be controlled and is able to express conic curves. Also a polynomial curve with a high degree cannot guarantee that the interpolation values can be constrained to a given range.[16] Thus, when a rational curve is employed to share a color image, the scheme based on moving lines yields more shares than that of the one based on the Lagrange interpolating polynomial. With more shares, the secret color image can be shared in a more secure manner with greater flexibility. However, this greater flexibility comes at an increased cost of computation compared to that of the Lagrange interpolations scheme.

IMPROVED ALGORITHM

If we carefully examine Fig. 3, we can see that the profile of the secret image is visible in the constructed image shares. The reason is that the correlation of the secret image is not broken during the image sharing process. So far, we selected only one X-position parameter for the secret image, thus only one share (i.e., the newly created one) is closely related to the secret image and the other shares are totally independent. Thus, the innocuous images do not contribute towards image hiding. We now modify the earlier approach slightly to make all the shares involved in data hiding by utilizing a block-based approach with multiple parameters:

Steps of block-based image sharing:

1. Divide the secret image into blocks.
2. Designate different parameters (i.e., different innocuous image position) for each block.
3. Share the secret image using the earlier approach.
4. Write down the parameters and positions embedded in the corresponding shares of each block for secret restoration.

The restoration handling is the inverse procedure of the encoding procedure.

In order to clearly explain the scheme, we illustrate the improved approach in Fig. 7. In Fig. 7, the secret image is divided into four blocks, each block is embedded into the

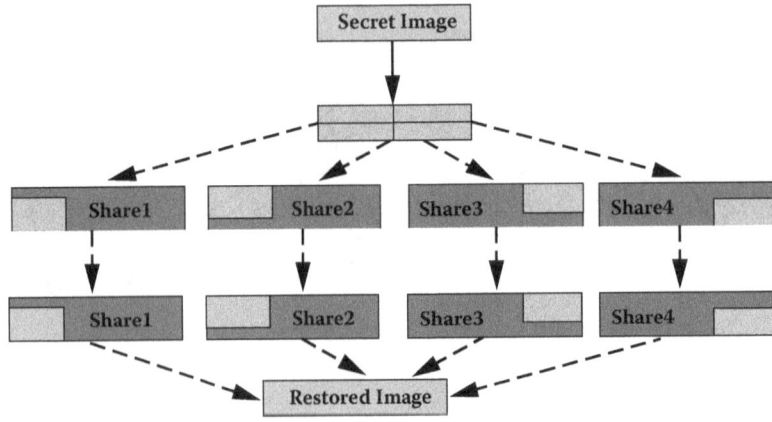

Fig. 7 Improved algorithm of image sharing

Fig. 8 The experimental results of image sharing by moving lines

Fig. 9 The experimental results of image sharing by moving lines

given innocuous image share at a different spatial position. The secret image can be reconstructed by applying the scheme block-wise again. This method effectively breaks the correlation of the original secret image with the secret image being divided into many blocks and these blocks being hidden in different shares at different locations. Thus, the secret image is no longer visible in the shares.

EXPERIMENT AND EVALUATION

Both of the algorithms proposed in this entry can theoretically be employed for image sharing, but actually they are different in terms of practical implementation. Color image sharing based on Lagrange interpolation allows for only a limited number of shares due to the oscillations phenomenon. Color image sharing based on moving lines can theoretically generate an unlimited number of shares. In practice, generating more shares will require more computation. Thus, as a trade-off, we usually share a color image into at most five shares.

We now present some experimental results on images of size 128×128. The shares and original secret image based on quadric curves of moving lines are shown in Fig. 8, which is a $(2, n)$ sharing example. In Fig. 9, we illustrate a $(3, n)$ scheme based on cubic curves of moving lines.

Figure 9 is the result of $(3, n)$ scheme based upon moving lines to share a given original image, with share 1 and share 2 being innocuous. Share 3 is the new computed share. By using shares 1, 2, 3, and their parameters, the original color image can be restored.

Figure 10 is the result of a $(4, n)$ scheme based on a moving line to share a color image by given original secret image, share 1, share 2, and share 3. Share 4 is the new computed share. When shares 1, 2, 3, 4, and their parameters

Fig. 10 The experimental results of image sharing by moving lines

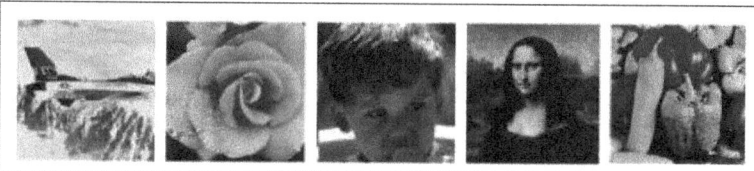

Fig. 11 Breaking the correlation of neighboring blocks in an image

are used for the moving lines based scheme, the original color image can be perfectly restored.

For a sensitive application, it is better to compute the shares by performing block-by-block (e.g., 8 × 8 pixel blocks) processing. The advantage of such an approach is that distributing the reconstructed blocks into the various innocuous images can break the correlation of the secret image in the image share. Thus, instead of taking $n - 1$ innocuous images and creating the n-th share (which can possibly reveal some correlations), we can take n innocuous images and distribute the secret image blocks within them. In fact, the secret color images need not be of the same size as that of the shares. Actually, if the shares have a larger size, the hiding of the color image will be more secure.

Figure 11 shows an example of breaking the correlation between neighboring blocks of the secret image. In this $(4, n)$ image sharing scheme, we have divided the original image into four equal-sized rectangular blocks and have shared these blocks using a quadric curve generated by the moving lines technique. What is different here from the earlier examples (where we created a whole new image as a share) is that we embed the information of each rectangular block into one of the image shares. The embedded information of share 1 is in the lower-left corner, that of share 2 is in the upper-left position while that of share 3 is embedded is in the upper-right region, and that of share 4 is in the lower-right corner. Thus, the information of the original secret image is shared among all the image shares. While this example is a simple illustration of how the correlation can be broken, it is clear that more sophisticated secret image subdivision and sharing schemes can be devised using the same principle.

CONCLUSION

In this entry, we introduce the novel concept of color image sharing based on Lagrange interpolation and moving lines.

The security of secret sharing is built upon the computation of polynomials. Given enough shares, the coefficients of the interpolating polynomial can be exactly determined. However, if an attacker does not possess the threshold number of shares and the image shares, he will not be able to reconstruct the polynomial. Thus, he will not able to restore the secret image properly.

Secret sharing based on Lagrange interpolation is often utilized to share binary strings, but it is difficult to use for color image sharing since it yields only a limited number of shares. We therefore have developed a new color image sharing scheme based on moving lines that does not have that limitation. An implicit curve generated by moving lines is a rational curve; we believe it can therefore be directly applied for sharing compressed-domain images.

Our future work will focus on further investigation of color image sharing in the compressed domain. The possible research directions in the future are:

1. Color image sharing based on compressed domain processing. Ideas related to DCT, DFT, and DWT can be employed in color image sharing.
2. Color image sharing based on visual cryptography. If visual cryptography could be implemented in color images, we could share color images with visual cryptography. Since visual cryptography is perfectly secure, color image sharing based on visual cryptography will be extremely robust.

BIBLIOGRAPHY

1. Salomaa, A. *Public-Key Cryptography*; Springer: Berlin Heidelberg, **1990**.
2. Shamir, A. How to share a secret. Commun. ACM **1979**, *22* (11), 612–613.
3. Blundo, C.; De Bonis, A.; De Santis, A. Improved schemes for visual cryptography. Des. Codes Cryptogr. **2001**, *24* (3), 255–278.

4. Liu, C. *Introduction to Combinatorial Mathematics*; McGraw-Hill: New York, 1968.
5. Asmuth, C.A.; Bloom, J. A modular approach to key safeguarding. IEEE Trans. Inf. Theory **1983**, *29*, 208–210.
6. Chang, C.-C.; Lin, C.-C.; Lin, C.-H.; Chen, Y.-H. A novel secret image sharing scheme in color images using small shadow images. Inf. Sci. **2008**, *178* (11), 2433–2447.
7. Chen, S.-K.; Lin, J.-C. Fault-tolerant and progressive transmission of images. Pattern Recognit. **2005**, *38* (12), 2466–2471.
8. Chuang, S.-C.; Huang, C.-H.; Wu, J.-L. Unseen visible watermarking. In *ICIP (3)*, 261–264, 2007.
9. Cimato, S.; De Prisco, R.; De Santis, A. Colored visual cryptography without color darkening. Theor. Comput. Sci. **2007**, *374* (1–3), 261–276.
10. Cimato, S.; De Prisco, R.; De Santis, A. Optimal colored threshold visual cryptography schemes. Des. Codes and Cryptogr. **2005**, *35* (3), 311–335.
11. Gollmann, D. *Computer Security*; John Wiley & Sons: Chichester, 1999.
12. Denning, D.E. *Cryptography and Data Security*; Addison-Wesley: London, 1982.
13. Duo, J.; Yan, W.-Q.; Kankanhalli, M.S. Progressive color visual cryptography. SPIE J. Electron. Imaging **2005**, *14* (3).
14. Fang, W.-P. Friendly progressive visual secret sharing. Pattern Recognit. **2008**, *41* (4), 1410–1414.
15. Fang, W.-P.; Lin, J.-C. Visual cryptography with extra ability of hiding confidential data. J. Electron. Imaging **2006**, *15* (2), 023020.
16. Farin, G. *Curves and Surfaces for Computer Aided Geometric Design: A Practical Guide*, 3rd Ed.; Academic Press: Boston, 1992.
17. Blakley, G.R. Safeguarding cryptographic keys. In *Proceedings of AFIPS 1979 National Computer Conference*, Aelington, vol. 48, 313–317, 1979.
18. Hou, Y.C.; Chang, C.Y.; Tu, S.F. Visual cryptography for color images based on halftone technology. In *Proceedings of the 5th World Multiconference on Systemics, Cybernetics, and Informatics (SCI 2001)*, Orlando, FL, vol. XIII, 441–445.
19. Hou, Y.-C. Visual cryptography for color images. Pattern Recognit. **2003**, *36*, 1619–1629.
20. Prakash, N.K.; Govindaraju, S. Visual secret sharing schemes for color images using halftoning. In *Proceedings of Computational Intelligence and Multimedia Applications*, 3, 174–178, December 2007.
21. Leung, B.W.; Ng, F.Y.; Wong, D.S. On the security of a visual cryptography scheme for color images. Pattern Recognit. **2009**, *42* (5), 929–940.
22. Penna, M.A.; Patterson, R.R. *Projective Geometry and Its Applications to Computer Graphics*; Prentice Hall: New Jersey, 1986.
23. Naor, M.; Shamir, A. Visual cryptography II: Improving the contrast via the cover base. In *Proceedings of the International Workshop on Security Protocols*, 1997, 197–202; Springer-Verlag: London.
24. Rijmen, V.; Preneel, B. Efficient color visual encryption for shared colors of benetton. In *EUCRYPTO '96*, 1996.
25. Shyu, S.J. Efficient visual secret sharing scheme for color images. Pattern Recognit. **2006**, *39* (5), 866–880.
26. Thien, C.-C.; Lin, J.-C. An image-sharing method with user-friendly shadow images. IEEE Trans. Circuits Syst. Video Technol. **2003**, *13* (12), 1161–1169.
27. Sederberg, T.W.; Chen, F.L. Implicitization using moving curves and surfaces. In *Proceedings of SIGGRAPH'95*; ACM Press: Los Angeles, 1995, 301–308.
28. Sederberg, T.W.; Saito, T.; Qi, D.X.; Klimaszewski, K.S. Curve implicitization using moving lines. Comput. Aided Geom. Des. **1994**, (11), 687–706.
29. Verheul, E.R.; Van Tilborg, H.C.A. Constructions and properties of k out of n visual secret sharing schemes. Des. Codes Cryptogr. **1997**, *11* (2), 179–196.
30. Weir, J.; Yan, W. Resolution variant visual cryptography for street view of Google maps. In *IEEE International Symposium on Circuits and Systems, 2010*. ISCAS 2010, May 2010.
31. Weir, J.; Yan, W.Q.; Crookes, D. Secure mask for color image hiding. In *Third International Conference on Communications and Networking in China, 2008*. ChinaCom 2008, 1304–1307, August 2008.
32. Yang, C.-N.; Chen, T.-S. New size-reduced visual secret sharing schemes with half reduction of shadow size. In *ICCSA (1)*, 19–28, 2005.
33. Yang, C.-N.; Chen, T.-S. Colored visual cryptography scheme based on additive color mixing. Pattern Recognit. **2008**, *41* (10), 3114–3129.
34. Yang, C.-N.; Laih, C.-S. New colored visual secret sharing schemes. Des. Codes Cryptogr. **2000**, *20* (3), 325–336.
35. Zhou, Z.; Arce, G.R.; Di Crescenzo, G. Halftone visual cryptography. IEEE Trans. Image Process. **2006**, *15* (8), 2441–2453.

Image Super Resolution

Mita C. Paunwala and Amisha J. Shah
Electronics and Communication Department, C. K. Pithawala College of Engineering and Technology, Surat, India

Abstract
The resolution of an image is one of the main parameters of image quality. Higher resolution image inherits detailed information about the image. Therefore, higher resolution is desired and often required in both civilian and military applications. High resolution (HR) is usually referred as a large number of pixels within a given size of the image. However, the image will be severely degraded due to shot noise as the pixel density increases. Hence, the limit exists on the pixel density within an image. An alternate hardware-based solution is to increase the chip size. This leads to reduced charge transfer rate and increased image capturing time. Hence, this method is not advisable for commercial applications. Thus, finding an effective way to increase image resolution is a matter of importance. When, resolution of an image cannot be improved by replacing hardware devices, either due to constraints of increased cost and size or both, one can resort to image super resolution (SR) algorithms. The main aim of image SR is to enhance the size of an image without upsetting the inherent information. Thus, the term image SR refers to a process of reconstructing an HR image from single or multiple degraded low-resolution (LR) images. The existing SR techniques have successfully demonstrated ways to enhance image quality. This chapter essentially explores survey on image SR for both single and multiple LR images.

Keywords: Image resolution; Interpolation; Iterative back projection; Learning; Projection onto convex set.

INTRODUCTION

In the present era, digital imaging has become a part of our life. There is growing demand to obtain better image quality by increasing image resolution and by adding more functionality features. High resolution (HR) means high-pixel density within an image, which is very important for various military and civilian applications.

The spatial resolution of an image can be increased by reducing the pixel size (i.e., accommodating more number of pixels per unit area). A pixel corresponds to a detector sensor in an imaging device. Most digital imaging devices use a charge-coupled device or complementary metal oxide semiconductor sensor. The photosensitive detectors integrate the incident light and accordingly generate various intensity levels. The reduce pixel size reduces the amount of light and therefore degrades the image quality. The remedy of this problem is to increase the number of pixels. Accommodation of more pixels increases chip size and chip capacitance which, in turn, reduces the charge transfer rate, and increases the cost of imaging device. Thus, this approach is not an effective solution. An effective and economical solution is to use signal processing algorithms. The process of obtaining an HR image from the observed LR images using a signal processing algorithm is called super resolution (SR).

An HR image can provide minute details of the captured scene that can be critical for human interpretation. Problems motivating SR arise in a number of imaging areas, such as military and civilian applications, where the camera used for imaging may introduce physical constraints on the resolution of the image data. Target reorganization, identification, and detection systems are some of the military applications that require the best quality of an image. Similarly, vehicle license plate readers, surveillance monitors, and medical imaging applications are few examples of civilian applications where image resolution plays an important role. Usually, satellite images are of low resolution (LR) due to wide area coverage.

The methods developed to produce an HR image may differ by observation model used to formulate the image SR problem. The most commonly used observation model is represented in Section "Imaging Model."

Imaging Model

An imaging model provides the relationship between an original HR image and a set of LR images of the same

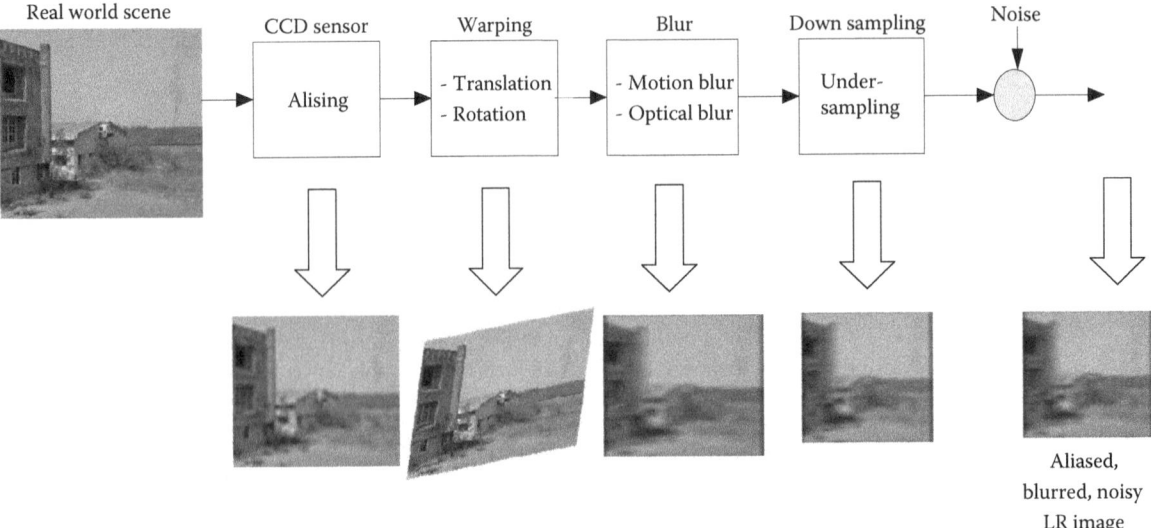

Fig. 1 LR image generation from a real-world image referring Eq. 1

scene, which plays an important role in SR reconstruction techniques. The imaging model which affects the real world scene is shown in Fig. 1. The real-world scene is affected by aliasing and motion blur effects. Aliasing effect may occur when the resolving power in acquisition system is pushed too near the pixel resolution. Such aliasing effect may appear as jagged edges in an image. Motion blur takes place because of limited shutter speed and/or relative motion between the scene and the imaging system. The imaging model, with consideration of these effects, can be mathematically represented as follows:[1]

$$y_k = DB_k M_k x + n_k, \quad 1 \leq k \leq p \tag{1}$$

where y is the observed LR image and x is the desired HR image. Aliasing effect is represented by subsampling matrix D. The motion is considered using warp matrix M, and the blur is considered as a blur matrix B. n and p represent the noise of the sensor and the number of LR images, respectively.

SR techniques can be broadly classified on the basis of the domain selected and number of input LR frames. According to the domain of operation, the SR techniques can be classified as frequency domain methods or spatial domain methods. Further, the SR techniques can also be configured as a multi-frame or a single-frame SR approach, according to the number of LR frames available. The multi-frame SR algorithms can be further distinguished by their operating principles as shown in Fig. 2.

Acquisition of LR Images

As defined earlier, image SR is a technique to reconstruct an HR image from a series of LR images. Multiple images of the same scene can be acquired by using multiple sensors or a single sensor as shown in Fig. 3.

The multiple LR images must be aliased and have a sub-pixel shift as shown in Fig. 4b. These images contain some new information that can be used in SR reconstruction. If the LR images have a shift of integer units, then there is no new information for SR reconstruction as shown in Fig. 4c.

FREQUENCY DOMAIN SR APPROACH

Frequency domain approaches transform the input LR image(s) to the frequency domain and then estimate the HR image in the same domain. The estimated HR image is then transformed back to the spatial domain. The SR process generates the missing high frequency components such that it minimize aliasing, blur, and noise. Aliasing effect is well analyzed in the frequency domain. Fourier domain easily describes planer shifts and possibly planar rotation and scale.

Fourier transform-based iterative method is introduced by Santis and Gori.[2] Fourier transform extends the spectrum of a given signal and therefore increase its resolution. Same algorithms were later reintroduced by Walsh and Nielsen-Delaney[3] in a non-iterative form, based on singular value decomposition (SVD). Tsai and Huang[4] first introduced the multi-image SR algorithm in the frequency domain. This algorithm was developed by considering only relative motion between LR images acquired by the Landsat 4 satellite and avoiding noise and blur available in the imaging system. Later on, Kim et al.[5] extended this approach by using weighted least squares formula for a blurred and noisy image with the pre assumption that all LR images have the same blur and the same noise characteristics. This method was further modified by Kim and

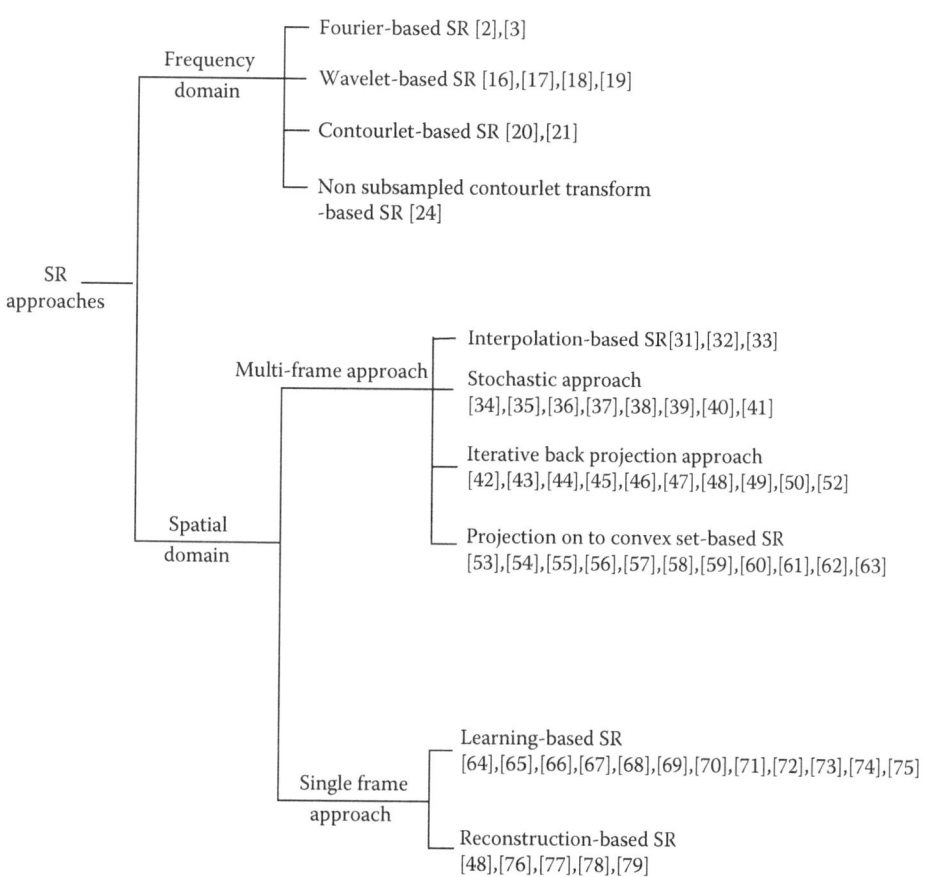

Fig. 2 Classification of SR approach

Su[6] with consideration that each LR image has different blurs. They used the Tikhonov regularization method to overcome the ill-posed problem which results from blur operator.

Multi-image SR algorithm is possible if at least one of the parameters involved in the imaging model, such as zoom, motion, multiple images from different sensors,[7,8] blur, different channels of a color image,[7] and multiple aperture,[9,10] changes from one LR image to another. Therefore, an accurate registration step is required to compensate such changes in the multi-image SR algorithm before the actual reconstruction. However, in certain environments, we cannot guarantee the performance of the registration algorithms. In such case, the error caused by an inaccurate registration should be considered in the reconstruction process. Most of the SR algorithms model the registration error as an additive Gaussian noise. Bose et al.[11,12] prepared system matrix W to consider the error generated by inaccurate registration and proposed the total least squares method to minimize the error. Ng and Bose[13] analyzed displacement errors on the convergence rate of the iteration. Here, LR images are acquired from multiple cameras which are shifted from each other by a known sub-pixel displacement. Such displacement causes registration error along the boundary and after that reduced by linear convergence of the conjugate gradient method. Vandewalle et al.[14] proposed an algorithm in the frequency domain, where the low frequency alias free part is used to precisely register a set of aliased images. Rhee and Kang[15] introduced a method to reduce memory requirement and computational cost by using discrete cosine transform (DCT) instead of DFT. They also apply multichannel adaptive regularization parameters to overcome ill-posed problems such as under determined cases or insufficient motion information cases.

The wavelet transform (WT) is used to decompose the input image into structurally correlated sub images. Demirel et al.[16] used discrete wavelet transform (DWT) along with stationary wavelet transform (SWT) to reconstruct an HR image, in which interpolation is first done in a high-frequency sub-band obtained through DWT and after that improved by using interpolated high-frequency sub-band of SWT. Kim et al.[17] proposed a learning-based method with wavelet synthesis. The approach relies on the fact that the relationship between an image and the LL band of DWT is closely correlated to the relationship between an HR image and its LR image. In this approach, efficient training sets are prepared by using LH band and a transposed HL band of LR image. These training sets are utilized to estimate the wavelet coefficients of LH band and the HL band for better estimation of HR image. Image SR using wavelet-based fusion

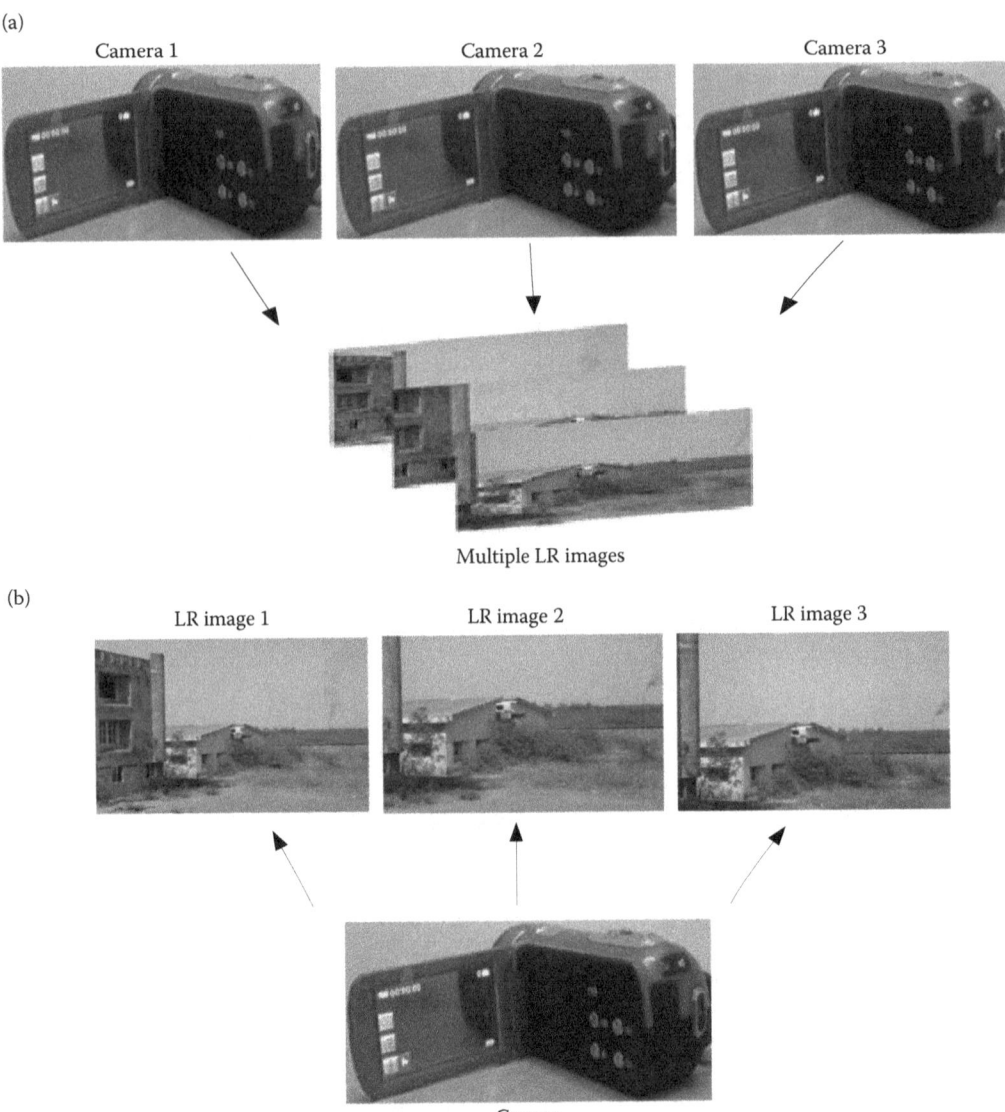

Fig. 3 (a) Several cameras acquire still images of the same scene which are combined to produce an HR image; (b) a video camera records a dynamic scene

is proposed by Liyakathunisa.[18] Here the LR, sub-pixel-shifted LR images are registered using Affine transformation. A wavelet-based fusion is performed and the noise is removed using a soft threshold. This algorithm is computationally fast. Zheng et al. proposed an algorithm for video sequence SR using wavelet-based nonlocal means (NLM).[19] Here, the HR sub-bands HL and LH are estimated from HL and LH sub-bands of LR by (NLM-SR), while the LL sub-band is directly obtained from the LR image. To speed up the computation, the sub-band HH is directly filled out with zeros.

DWT helps to preserve information along discontinuities when they are at orientations of 0°, 45°, and/or 90° only. The drawback of WT is that it cannot offer the better directional information, especially when an image contains contours/curved edges. Contourlet transform (CT) offers high directional information and can be used in image SR. CT-based single-image SR is proposed by Jiji et al.[20] Here, an HR image is reconstructed through a learning process in CT domain using a basic principle that, in a CT pyramid, every coefficient at the finer level can be related to the coefficient at the next coarser level for the similar orientation. Self-similarity-based, single-image SR is proposed by Park.[21] In his algorithm, an input LR image is partitioned into low-frequency (LF) and high-frequency components, and the SR image is reconstructed by searching similar LR and HR patches in LF and high-frequency domains, respectively.

CT causes the Gibbs effect along the edges as it is shift variant.[22] Unlike CT, nonsubsampled contourlet transform (NSCT) is more suitable for image SR as it is shift-invariant transform.[23] An adaptive NSCT decomposition-based learning approach is proposed by Shah and Gupta[24] Here, the directional decomposition is made adaptive using a

Image Super Resolution

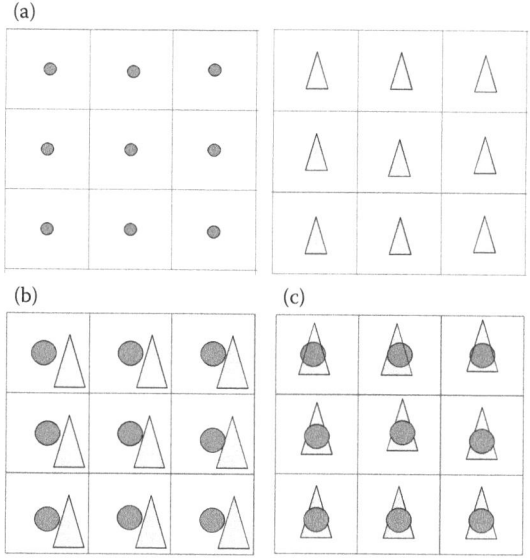

Fig. 4 (a) Two LR images with (b) sub-pixel shift between them and (c) integer point shift

classifier. With the aid of a classifier, an optimum benefit of directionality is achieved, which in turn reduces the computation time and complexity.

Extreme sensitivity to model errors, requirement of pure translational motion, and linear space invariant blur during image acquisition are challenges of frequency domain multi-frame SR algorithms. Moreover, addition of prior information in order to regularize the ill-posed SR problem is considerably difficult using these algorithms.

The frequency domain methods are capable of reducing hardware complexity and also convenient for parallel implementation. However, the observation model is restricted to only global translational motion and blur. It is also difficult to apply the spatial domain a priori knowledge for regularization because of the lack of data correlation in the frequency domain.

SPATIAL DOMAIN SR APPROACH

The spatial domain techniques consider nonideal sampling, motion, optical/motion blur, and artifacts due to compression. In addition, it allows the use of a priori constraints which results in extrapolation during the reconstruction process. Thus, spatial domain approaches overcome the difficulties of the frequency domain methods. SR techniques can be classified as multi-image SR techniques and single-image SR techniques based on the available input LR images as discussed earlier.

Spatial Domain Multi-Frame SR Approach

The basic premise for increasing the spatial resolution in SR techniques is the availability of multiple LR images captured from the same scene.[25] Multi-image SR techniques fuse the information from multiple LR images to produce an HR image. Interpolation, iterative back projection (IBP), projection onto convex set (POCS), and stochastic-based methods are most widely used in multi-frame SR techniques.

Interpolation-Based Approach

This is the most intuitive approach to generate HR image. It consists of three basic stages: (i) registration, (ii) interpolation, and (iii) restoration as shown in Fig. 5. First of all, a relative motion between sub-pixel shifted LR images is estimated using a image registration. This results in a motion-compensated image having nonuniformly distributed samples. There are various techniques of registration in literature.[14,26]

Interpolation is the process of determining the values of a function between its samples. A uniformly spaced up sampled image is obtained through interpolation. Neighbor interpolation, bilinear interpolation, bicubic interpolation, and cubic spline interpolation[27] are basic interpolation techniques. The disadvantage of these basic techniques is that they filter out the pixels across edges which blurs the resultant image. Numerous edge-directed interpolation methods are proposed by Wong and Siu.[28–30] At last, the effect of sensor point spread function blur and noise is removed by using image restoration.

A geostatistical interpolation technique called Kriging interpolation is proposed by Panagiotopoulou et al.[31] Here, a gradient-based algorithm is used to estimate relative motion for each frame followed by Kriging interpolation. The interpolation considers both the distance and the degree of variation between known data points for estimation of unknown areas to construct a uniformly spaced HR grid. Finally, Wiener filtering is used to reconstruct an HR image. The proposed Kriging interpolation procedure requires at least 18 frames in order to generate a satisfactory HR image. Furthermore, the computational complexity is also higher due to nonlinear interpolation.

A fast interpolation method based on optimized resampling filter coefficients is proposed by Gillman.[32]

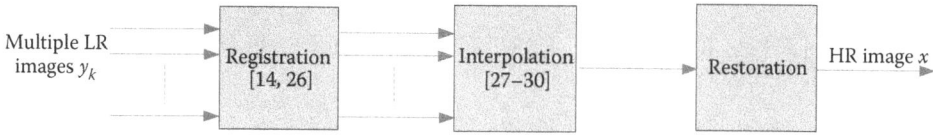

Fig. 5 The steps followed by the interpolation-based approach for SR

He used tenth-order filter to generate compatible results. The quality of the final HR image depends on both the LR image content and the sub-pixel shift of each input LR image.

Zhou[33] proposed an interpolation-based SR approach using multi-surface fitting to consider the spatial structure information optimally. In this approach, each set of LR pixels is fitted with one surface, and the resultant SR image is generated by fusing the multisampling values on these surfaces.

The interpolation-based SR methods are simple in computation but propagates an error to the later processing steps due to inaccurate registration. In addition, the researchers had still not considered noise and blurring effect in this approach. These loopholes limit the approach to generate an optimal estimated HR image. Furthermore, the lack of prior constraint forces this approach to use a special treatment for reducing aliasing. It should also be noted that this method is not able to handle the single-frame SR problem as it cannot reconstruct the high-frequency components, that is, edge information.

Stochastic Approach

Unlike the interpolation-based SR approach, the stochastic approach allows the use of prior constraint necessary for optimal solution of the ill-posed SR inverse problem. In this technique, an unknown HR image is estimated based on the LR measurements as well as assumption of noise or motion models. A popular family of estimators is the maximum likelihood (ML) type estimators from many of the available estimators.[34,35] To find the ML estimation of SR image, Eq. 1 can be rewritten as follows:

$$\hat{\underline{X}} = \underset{\underline{X}}{ArgMin} \left[\sum_{k=1}^{N} \rho\left(y_k, DB_k M_k \underline{X}\right) \right] \sqrt{b^2 - 4ac} \qquad (2)$$

where ρ measures the "distance" between the model (i.e., $DB_k M_k X$) and the measurements (i.e., y_k).[34] This estimator is extremely sensitive to noise present in the input LR images. Furthermore, its performance depends on the available LR images. The ML estimator requires minimum 100 LR images to optimally restore HR image.[35]

The estimated SR image can be regularized by a priori probability distribution on an SR image. The maximum a posteriori (MAP) estimator of X can maximize the prior probability distribution $P(X|y_k)$ with respect to X. The MAP estimator also allows the use of a priori image model $P(X)$. Thus, the MAP estimator can be represented as follows:

$$X = ArgMax\left[\ln P\left(y_1, y_2, \ldots, y_R | X\right)\right] + \ln P(X) \qquad (3)$$

MAP-based SR algorithm is proposed by Liu et al.[36] They considered an independent motion between LR images as well as identical zero mean Gaussian noise in each LR image. Hyperspectral images (HSIs) are widely used in agriculture, geology, meteorology, etc. The spatial resolution of such HSI is very poor. A MAP-based SR algorithm for such HSI is proposed by Zhang.[37] Here, the principal component analysis (PCA) is used for motion estimation as well for image reconstruction. This approach also reduces the computational load and noise by using different SR techniques for different groups of components. Using the Markov random field prior, $P(x)$ is described by a Gibbs prior whose probability density is defined as follows:

$$\begin{aligned} P(x) &= \frac{1}{Z} \exp[-U(x)] \\ &= \frac{1}{Z} \exp\left[-\sum_{c \in s} \phi_c(x)\right] \end{aligned} \qquad (4)$$

where Z is a normalizing constant, $U(x)$ is the energy function, and $\phi_c(x)$ is the derivative of the image. The non-convex (gradient discontinuities) property of the Gibbs prior blurs the edges in the resultant SR image. This limits the use of Gibbs prior. A discontinuity preserving MAP-based SR method was proposed by Schultz and Stevenson.[38] They used a Huber-Markov Gibbs prior to reconstruct an SR image from an LR video sequence. This prior offers good smoothness, but still it is lenient to the step edges. L2 norm-based Tikhonov prior is an edge-preserving regularization term.[39] It can be represented as follows:[40]

$$\gamma_T(x) = \left\| \Gamma \underline{X} \right\|_2^2 \qquad (5)$$

where Γ is a high-pass operator such as derivative or Laplacian. The Tikhonov prior causes over smoothness as it is a quadratic regularization term with a linear solution, which strongly penalize large edges in images.[39] A total variation (TV) prior overcomes the limitation of the Tikhonov prior as it is based on a differential L1 norm which considers the magnitude of the gradient as follows[40]:

$$\gamma_{TV}(X) = \left\| \nabla X \right\|_1 \qquad (6)$$

Aly and Dubois[41] proposed an SR approach in which they used TV regularizer along with a data fidelity term to yield a cost function. This cost function is minimized using a level set-based method which results in a unique SR image.

A regularization term compensates the missing measurement information with some general prior information about the desirable HR solution.[34] In such case, the optimization problem can be defined as follows:

$$\hat{\underline{X}} = \underset{\underline{X}}{ArgMin}\left[\sum_{k=1}^{N} \rho\left(y_k, DB_k M_k \underline{X}\right) + \lambda \gamma(\underline{x})\right] \qquad (7)$$

Table 1 Various regularization priors

Sr. No.	Type	Features
1	Gibbs prior	• It performs over smoothing. • The processed image results blurred.
2	Huber–Markov prior	• It offers local smoothing. • It is still lenient to the step edges.
3	Tihkhonov	• It is a high-pass operator (Laplacian). • Resultant image will be blurred.
4	Total variation	• It uses L1 norm of the magnitude of gradient. • Edges are preserved in the resultant image.
5	Bilateral total variation	• It is computationally fast. • Edges are preserved in the resultant image.

where $\gamma(x)$ is a regularization parameter. Aforementioned various regularization parameters are summarized in Table 1.

The main advantages of this technique are follows: (i) it allows the use of a priori knowledge about the solution, and (ii) it offers flexible and robust modeling of noise characteristics.

Iterative Back Projection-Based Approach

The resolution enhancement of an image through the IBP approach is initially proposed by Irani-Peleg[42] and Komatsu et al.[43] In this approach, each HR image pixel is a sum of different projections of the same LR image area, determined by the image blurring and displacement. The HR image obtained through IBP method can be expressed by the following equation:

$$f_{i+1} = f_i + c \sum_{k=1}^{n} \left(D_{BP}^{(k)} \right)^2 \left(g^{(k)} - g_i^{(k)} \right) \quad (8)$$

where $g^{(k)}$ is the kth LR image, f_i is the SR image obtained after the ith iteration, and $g_i^{(K)}$ is the decimated image obtained through f_i. The term $(D_{BP}^{(k)})2$ is the back projection matrix that creates a projection from the difference $(g_i^{(k)} - g_i^{(k)})$ in an HR image, and c is a normalizing constant.

The theory of signal extrapolation, where the entire signal can be find out from the available subset of the signal, has been applied to perform image super resolution. Papoulis[44] and Gerchberg[45] (PG) independently demonstrated the method of iterative signal (1-D) extrapolation. Vandewalle et al.[46] used the PG algorithm in a modified form to get SR image from multiple LR registered images. Chatterjee et al.[47] restricted the use of PG algorithm for a single LR image. The classical PG method was not able to deliver a good result in the presence of blur and noise of LR image. Chatterjee et al. calculate an error between the LR image and the simulated LR image which is added to the corresponding block of pixels in the obtained SR image. Thus mathematically,

$$z_i = z_i + P(\varepsilon_i) \quad (9)$$

where z_i is the estimated SR image vector obtained after the ith iteration, P is the back projection factor, and ε_i is an error calculated as follows:

$$\varepsilon_i = y - DHz_i \quad (10)$$

where y is the input LR image, D is the decimation matrix, and H is the blurring matrix.

The original IBP method is suffering from chessboard effect/ringing effect, especially at edges. Dai et al.,[48] Dong et al.,[49] and Lin et al.[50] proposed different approaches to overcome these artifacts. A bilateral back projection algorithm is proposed by Dai,[48] in which the bilateral filtering is integrated with a back projection method to reduce the jaggy artifacts. On the other side, Dong et al.[49] employed the nonlocal image redundancy to improve the quality of images. However, the computation complexity of updating the reconstruction error in each step is high. Lin et al.[50] proposed the edge-directed interpolation method with IBP to overcome the effect of chessboard.

In recent years, the rate of video acquisition by LR imaging devices is increasing. Hence, SR of such video is also required.[51,52] Hsieh et al.[51] introduced a video SR technique based on motion-compensated IBP considering motion between the two consecutive frames of the video. They used block matching algorithm to determine motion between two frames of a video which cannot be accurate as the results are highly sensitive to block size. Zhang et al.[52] has proposed the critical point filter-based image matching to overcome limitations of the block-based image matching algorithm. In spite of its simplicity, the IBP method is suffering from the problem of slow convergence.

POCS Approach

Similar to IBP, POCS is also an iterative method that allows the use of prior knowledge in the reconstruction of HR image. This approach performs restoration and interpolation simultaneously during the estimation of HR image.

The use of set theory in the reconstruction of SR image was first proposed by Stark et al.[53] They used the convex projection method to remove blur spot present in an image, produced due to image-plane detector arrays of imaging device. They consider each detector of size 4×2 (in pixels). This enforced to have a minimum 8-pixel angular displacement between two LR images. In addition, this approach was able to generate a satisfactory SR image for noiseless input LR image but not for noisy LR image. Tekalp et al.[54] extended the algorithm for compressed LR images.

Assuming that the motion information is provided, the imaging model is given as follows:[55]

$$g_d(m_1,m_2,k) = \sum_{n_1,n_2} f(n_1,n_2)h(n_1,-m_1,n_2-m_2,k) + v(m_1,m_2) \quad (11)$$

where $h(\cdot)$ is a linear blur mapping of the HR source image $f(n_1, n_2, k)$ to the kth measured LR image $g_d(m_1, m_2, k)$, and $v(m_1, m_2)$ is an additive white Gaussian noise. A closed, convex constraint set for pixels in the estimated HR image $\tilde{f}(n_1,n_2,k)$ is as follows:

$$c_r = \tilde{f}(n_1,n_2,k) : |r^{(\tilde{f})}(m_1,m_2,k)| \leq \delta_0(m_1,m_2,k) \quad (12)$$

where the value of $g_d(m_1, m_2, k)$ at each pixel (m_1, m_2) constrained such that its associated residual r is bounded by δ_0. The bound δ_0 is determined from the noise statistics. The projection of an estimate $\tilde{f}(n_1,n_2,k)$ onto c_r is defined as follows:

$$P_R(m_1,m_2,k)[\tilde{f}(n_1,n_2,k)]$$
$$= \tilde{f}(n_1,n_2,k) \begin{cases} \dfrac{(r^{(\tilde{f})}(.) - \delta_0(.))h(n_1,n_2,..)}{\sum_{o_1}\sum_{o_2}h^2(o_1,o_2,..)}, & r^{(\tilde{f})}(.) > \delta_0(.) \\ 0, & r^{(\tilde{f})}(.) \leq \delta_0(.) \\ \dfrac{(r^{(\tilde{f})}(.) + \delta_0(.))h(n_1,n_2,..)}{\sum_{o_1}\sum_{o_2}h^2(o_1,o_2,..)}, & r^{(\tilde{f})}(.) < -\delta_0(.) \end{cases}$$
(13)

where the function argument "." is meant to be m_1, m_2, k. Note that the blurring function h is assumed to be slightly different in each frame depending on the small motion between neighboring frames which usually is a small fraction of a pixel.

The amplitude constraint should always be satisfied which is termed as $0 \leq \tilde{f}(n_1,n_2,k) \leq 255$. The estimation of HR image $f(n_1, n_2, k)$ is obtained from images $g_d(m_1, m_2; k)$ ($q = k - 1, k, k + 1$) through the projections in the backward and forward directions.[55,56]

A space varying blur and sensor noise is considered by Patti et al.[57] The same approach is then extended by Tekalp et al.[54] for multiple moving objects in the scene. As an iterative algorithm, the POCS-based SR approach may generate a ringing effect across the edges. Patti and Altunbasak[58] used higher order interpolation method to reduce the ringing effect.

Enhanced resolution of MRI heart images is very helpful for proper diagnosis. An application of Wavelet-based POCS approach for cardiovascular MRI image enhancement is proposed by Hsu.[59] Wheeler et al.[60] proposed a POCS-based SR algorithm in the frequency domain in order to correct the effect of blurring and associated aliasing. They used resampling of SR image in the spatial domain as it is difficult to formulate rotation of LR images in the frequency domain.

Human face image super-resolution techniques have a variety of applications such as surveillance, recognition, transmission. Yu et al.[61] proposed a novel approach for face image SR in two stages. In the first stage, the HR image is estimated from their smoothed and down sampled LR images through POCS and in the second stage, the residual between an original HR image and a reconstructed image is recompensated.[61] They used a data set of 300 HR face images, in which 280 images are used for training and 20 for testing. The LR images are generated by smoothing and down sampling the corresponding HR image. In this process, the images are aligned with the two eyes manually.

Su[62] proposed a hybrid SR algorithm. In this novel approach, an HR intermediate image sequence is obtained from the sparse representation of the LR image sequence. The final HR image is obtained by fusing these intermediate HR image sequence using POCS method at the cost of computational complexity.

As discussed earlier, an SR image reconstructed using POCS method is suffering from edge artifacts. Shen et al.[63] proposed a computationally complex SR approach in the space–frequency domain to reduce the edge artifacts.

In summary, simplicity is an advantage of POCS. Nevertheless, the disadvantages of nonunique solution, slow convergence and a high computational cost hide its simplicity.

Learning-Based Singe-Frame SR Approach

Learning-based technique estimates high-frequency details from a training set of HR images that encode the relationship between HR and LR images. These approaches synthesize missing details based on patch similarities between the input LR image and the examples in the training set. The basic concept of the patchwise learning-based approach is depicted in Fig. 6.

In the patchwise learning technique, all given images are represented as a set of patches and the dictionary (LR/HR dictionary) is therefore formed by pairs of LR and HR patches. A best match is chosen from the LR dictionary for super resolving a test LR patch. The HR patches corresponding to the chosen LR patches are considered to reconstruct a final HR image. Thus, the learning-based approach contains a training step in which the relationship between some HR examples and their LR counterparts are learned. This learned knowledge is then incorporated into a priori term of the reconstruction. Example-based or learning-based SR algorithm has been proposed recently as a very attractive solution for single-image SR.

The dictionary (training data set) which is used to reconstruct an HR image may not include LR patches similar to an input LR patch. This limits recovery of high frequency details. Kim et al.[64] used a residual training data set to

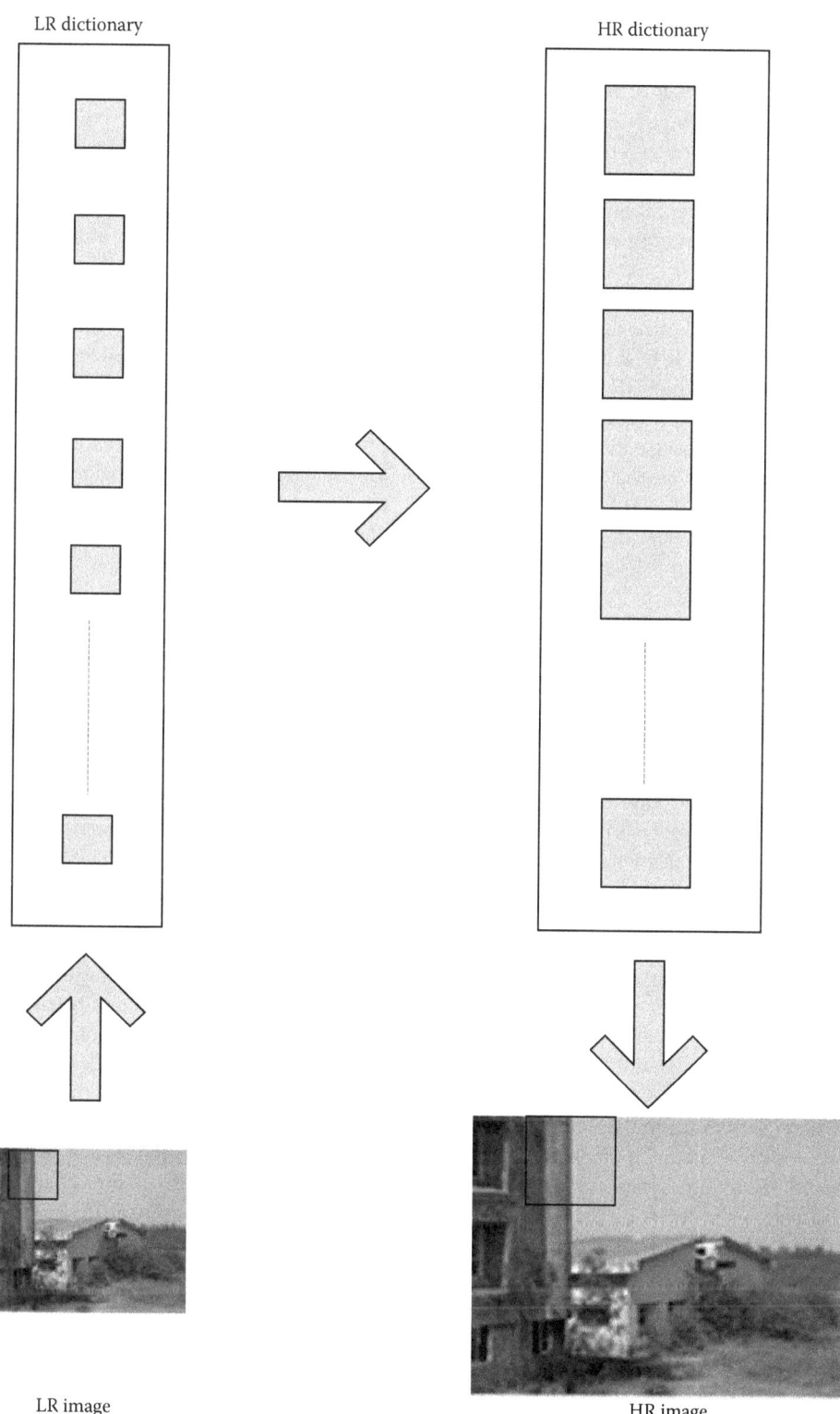

Fig. 6 Reconstruction of HR image through a patchwise learning process that uses a set of low-resolution image patches (LR dictionary) and their corresponding HR image patches (HR dictionary)

overcome this limitation. Further, in this work, LR patches are selected adaptively to efficiently use the training data.

In the learning-based SR approach, it is important to properly estimate high frequency components else presence of outlier patches cause artifacts in the resultant SR image. Kim et al.[65] proposed an efficient algorithm that uses a robust error norm with global and local patch constraints to efficiently reject the outliers.

The SR approach based on dictionary leaning has high computation time. Qinlan et al.[66] proposed an improved example-based single-image SR method that reduces computational complexity by classifying the high-frequency patches of the LR image into different classes. They used an example-based image SR algorithm for only edge area of LR image. Another fast converging SR algorithm was proposed by Yang et al.[67] They used both external examples and self-examples.

Neighbor embedding (NE) is also a learning method that has stimulated a lot of interest in machine learning. NE-based approach[68–70] selects "K best candidates" from the LR dictionary. The corresponding HR patches from the HR dictionary are considered to construct an unknown HR image using a basic assumption that a patch in the LR input image and the corresponding HR unknown patch share similar neighborhood structures.[70] NE-based SR is inadequate for complex/high-dimensional data. Tang et al.[71] overcome this limitation by using a local learning-based SR approach.

The conventional learning approach assumes that the training data set represents a similar class of objects. Hence, the performance of these conventional approaches degrades when the input LR image contains different objects or textures. This discrepancy can be overcome by using a hyper sphere of support vector data description (SVDD) as a classifier. Ogawa[72] proposed a single-image SR approach where HR image is reconstructed using hyper spheres of the SVDD obtained from training examples. However, this method suffers from some artifacts around edges due to use of block-based process. Zhang et al.[73] used an ordinary dictionary as well as residual high frequency dictionary to reconstruct high frequency details in the estimated HR image.

A self- learning approach for SR is proposed by Yang,[74] which does not require the collection of training low and HR image data in advance. Yang has used support vector regression (SVR) with image sparse representation to accurately model relationship between input LR image and generated HR image. The problem of learning the mapping functions between LR and HR images based on LR and HR dictionary is addressed by Jiang et al.[75] In this work, the input LR image patch is individually super resolves by learning the prior information using anchored neighborhood regression followed by the nonlocal median filter to improve the performance of the proposed approach.

In summary, learning-based image SR is one of the popular methods that allow the use of priors to infer missing details in LR image. However, the performance of this method highly depends on the size of the dictionary. Huge dictionary improves quality of the resultant SR image at the cost of computation. Also, the quality of final SR image degrades with higher zooming factor. Furthermore, it is difficult to use learning-based SR approach for a video where real time processing is required.

Reconstruction Approach

The reconstruction-based SR approach had rapidly gained prominence. It provides a powerful theoretical framework for the inclusion of a priori constraints, necessary to improve the reconstruction quality of the ill-posed SR problem. This approach can be used for single-frame as well as for multi-frame LR images. This method can be broadly classified as deterministic and stochastic regularization approach. The stochastic approach for multiple LR images is discussed earlier.

The deterministic regularized SR approach solves the inverse SR problem by using the prior information about the solution which can be used to make the problem well posed. Constrained least square (CLS) is the most common deterministic regularization approach. The cost function, used to obtain an SR image using this approach, is defined as follows:

$$\hat{x} = \underset{x}{ArgMin} \left[\sum_{k=1}^{N} \left\| y_k - M_k x \right\|_2^2 + \lambda \left\| Lx \right\|_2^2 \right] \quad (14)$$

where N is the number of LR images, the operator L is usually a high-pass filter, and M_k is the degradation matrix for the LR image y_k. The Lagrange multiplier λ is the regularization parameter, which controls the balance between the data fidelity term and the smoothness constraint.

Reconstruction-based approaches estimate an HR image by enforcing priors in the up sampling process. Such priors are commonly incorporated into a back projection framework to reduce artifacts around edges, but it suffers from the chessboard and ringing effect. A novel prior called steering kernel regression total variation (SKRTV) is proposed by Li et al.[76] This prior helps to explore the local structural regularity properties in natural images and thus preserves the structural smoothness and edge sharpness. On the other side, Dai et al.[48] has integrated the bilateral filtering into the back projection method to smoothen the pixels which are nearby in both space and feature domains and thus reduces the edge artifacts.

An edge-directed SR approach is proposed by Wang et al.[77] Here a sharp HR gradient field is estimated by an adaptive self-interpolation algorithm. Prior information about the image sparsity can also be used to reconstruct an HR image. The key feature of an algorithm proposed by Wang et al.[77] is a displacement field calculated from the gradient magnitude. Afterward, this is used to interpolate the Bicubic up-sampled image. In this work, the algorithm is able to preserve image details along the silent edges at the cost of computation time. Mallat et al.[78] has proposed an SR algorithm based on sparse mixing estimators. Here mixing coefficients are calculated using an appropriate regularity and a sparse signal representation. Furthermore, the computational complexity is reduced by orthogonal matching pursuit algorithm. One more novel prior is proposed by Zhang et al.[79] in order to preserve sharp edges. They used a nonlocal steering kernel regression (NLSKR) model to solve the ill-posed single-image SR problem. Resolution enhancing technologies are necessary for medical imaging also, where diagnosis or analysis from a low-resolution image is extremely difficult. Wang et al.[80] and Isaac et al.[81] proposed different algorithms for magnetic resonance imaging (MRI).

The biggest advantage of the reconstruction-based SR approach is that it provides unique solution, but its performance may degrade with large magnification factor.

CONCLUSION AND DISCUSSIONS

In most of the applications higher resolution is desired to get detailed information of the captured image. An economical and effective solution to obtain an HR image is the use of image SR algorithms. This chapter explores the classification of various SR algorithms with pros and cons. In reality, it is not fair to compare different SR algorithms as they are developed for different applications with different model and assumptions.

The aforementioned literature survey concludes that the incorporation of noise model, a priori information and/or regularization term with SR algorithms in frequency domain is more difficult than the spatial domain. Furthermore, in the spatial domain, an HR image can be reconstructed more easily from the multiple LR images as compared to the single LR image.

Lack of prior constrain and error due to inaccurate registration limits the use of interpolation-based SR approach. Also generation of SR image using the interpolation approach is easier for multiple LR images as compared to the single LR image. The IBP- and POCS-based SR algorithms perform equally for both single and multiple LR images, but slow convergence and nonunique solution hide their simplicity. On the other side, stochastic and reconstruction-based SR approaches give a unique solution by allowing the use of appropriate a priori term.

REFERENCES

1. Elad, M.; Feuer, A. Restoration of a single super-resolution image from several blurred, noisy, and under-sampled measured images. IEEE Trans. Image Process. **1997**, *6* (12), 1646–1658.
2. Santis, P.D.; Gori, F. On an iterative method for super-resolution. J. Mod. Opt. **1975**, *22* (8), 691–695.
3. Walsh, D.O.; Nielsen-Delaney, P.A. A direct method for super-resolution. J. Opt. Soc. Am. A **1994**, *11*, 572–579.
4. Tsai, R.; Huang, T. Multiframe image restoration and registration. In *Advances in Computer Vision and Image Processing*, Vol. 1; Tsai, R.Y.; Huang, T.S.; Eds.; JAI Press Inc.: Stamford, CT, 1984, 317–339.
5. Kim, S.P.; Bose, N.K.; Valenzuela, H.M. Recursive reconstruction of high resolution image from noisy under-sampled multi-frames. IEEE Trans. Acoust. Speech Signal Process. **1990**, *38*, 1013–1027.
6. Kim, S.P.; Su, W.Y. Recursive high-resolution reconstruction of blurred multi-frame images. IEEE Trans. Image Process. **1993**, *2*, 534–539.
7. Zomet, A.; Peleg, S. Multi-sensor super-resolution. In *Proceedings of 6th IEEE Workshop on Applications of Computer Vision*; IEEE Computer Society: Washington, DC, 2002, 27–31.
8. Shechtman, E.; Caspi, Y.; Irani, M. Space-time super-resolution. IEEE Trans. Pattern Anal. Mach. Intell. **2005**, *27* (4), 531–545.
9. Komatsu, T.; Aizawa, K.; Saito, T. Resolution enhancement using multiple apertures. In *Super-Resolution Imaging*; Chaudhuri, S.; Eds.; Kluwer Academic: Norwell, MA, 2002, 171–193.
10. Faramarzi, E.; Bhakta, V.R.; Rajan, D.; Christensen, M.P. Super resolution results in panoptes, an adaptive multi-aperture folded architecture. In *Proceedings of IEEE International Conference on Image Processing*, Hong Kong, 2010, 2833–2836.
11. Bose, N.K.; Kim, H.C.; Zhou, B. Performance analysis of the TLS algorithm for image reconstruction from a sequence of under sampled noisy and blurred frames. In *Proceedings ICIP-94, IEEE International Conference Image Process*; IEEE: Austin, TX, 1994, Vol. 3, 571–575.
12. Ng, M.; Koo, J.; Bose, N. Constrained total least squares computations for high resolution image reconstruction with multi sensors. Int. J. Imaging Syst. Technol. **2002**, *12*, 35–42.
13. Ng, M.K.; Bose, N.K. Analysis of displacement errors in high-resolution image reconstruction with multi sensors. IEEE Trans. Circuits Syst. I **2002**, *49*, 806–813.
14. Vandewalle, P.; Susstrunk, S.; Vetterli, M. A frequency domain approach to registration of aliased images with application to super-resolution. EURASIP J. Appl. Sig. Process. **2006**, Article ID 71459. doi:10.1155/ASP/2006/71459, 14.
15. Rhee, S.H.; Kang, M.G. Discrete cosine transform based regularized high resolution image reconstruction algorithm. Opt. Eng. **1999**, *38* (8), 1348–1356.
16. Demirel, H.; Jafari, G.A. Image resolution enhancement by using discrete and stationary wavelet decomposition. IEEE Trans. Image Process. **2011**, *20* (5), 1458–1460.
17. Kim, C.; Choi, K.; Hwang, K.; Ra, J. Learning-based super-resolution using a multi-resolution wavelet approach. In *Proceedings of International Workshop on Advance Image Technology IWAIT*, Seoul, Korea, 2009.
18. Syed, L.; Ananthashayana, V.K. Super resolution blind reconstruction of low resolution images using wavelets based fusion. World Acad. Sci. Eng. Technol. **2008**, *40*, 177–181.
19. Zheng, H.; Bouzerdoum, A.; Phung, S.L. Wavelet based nonlocal-means super-resolution for video sequences. In *Proceedings of 2010 IEEE 17th International Conference on Image Processing*, Hong Kong, 2010, 2817–2820.
20. Jiji, C.V.; Chaudhuri, S. Single-frame image super-resolution through contourlet learning. EURASIP J. Appl. Signal Process. **2006**, 1–11. Article ID 73767.

21. Park, S.; Lee, O.; Kim, J. Self-similarity based image super-resolution on frequency domain. In *Signal and Information Processing Association Annual Summit and Conference (APSIPA)*, Kaohsiung, Taiwan, 2013, 1–4.
22. Wang, P.; Tian, H.; Zheng, W. A Novel Image Fusion Method Based on FRFT-NSCT; Hindawi Publishing Corporation Mathematical Problems in Engineering, 2013. Article ID 408232. doi:10.1155/2013/408232.
23. Lu, Y.; Do, M. A new contourlet transform with sharp frequency localization. In *IEEE Proceedings of International Conference on Image Processing*, Atlanta, GA, 2006, 1629–1632.
24. Shah, A.; Gupta, S.B. Adaptive directional decomposition in non sub sample contourlet transform domain for single image super resolution. Int. J. Multimedia Tools Appl. **2015**, *74* (14), 8443–8467.
25. Park, S.C.; Park, M.K.; Kang, M.G. Super-Resolution Image Reconstruction: A Technical Overview. IEEE Signal Processing Magazine, 2003, 21–36.
26. Cortelazzo, L.; Lucchese, G.M. A noise-robust frequency domain technique for estimating planarro to translations. IEEE Trans. Signal Process **2000**, *48*, 1769–1786.
27. Hou, H.S.; Andrews, H. Cubic splines for image interpolation and digital filtering. IEEE Trans. Acoustic Speech Signal Process. **1978**, *26* (6), 508–517.
28. Wong, C.S.; Siu, W.C. Further improved edge-directed interpolation and fast EDI for SDTV to HDTV conversion. In *Proceedings of European Signal Processing Conference*, Aalborg, North Denmark, 2010, 23–27.
29. Yun, Y.; Bae, J.; Kim, J. Adaptive multidirectional edge directed interpolation for selected edge regions. In *Proceedings of Region 10 Conference (TENCON 2011)*, Bali, November 2011, 385–388.
30. Zhou, D.; Shen, X.; Dong, W. Image zooming using directional cubic convolution interpolation. IET Image Process. **2012**, *6* (6), 627–634.
31. Panagiotopoulou, A.; Anastassopoulos, V. Super-resolution image reconstruction employing Kriging interpolation technique. In *EURASIP Conference Focused on Speech and Image Processing, Multimedia Communications and Services*, Maribor, Slovenia, 2007, 144–147.
32. Gilman, A.; Bailey, D.G.; Marsland, S. Least-squares optimal interpolation for fast image super-resolution. In *IEEE International Symposium on Electronic Design, Test and Application*, Vietnam, 2010, 29–34.
33. Zhou, F.; Yang, W.; Liao, Q. Interpolation-based image super resolution using multisurface fitting. IEEE Trans. Image Process. **2012**, *21* (7), 3312–3318.
34. Farsiu, S.; Robinson, M.D.; Elad, M.; Milanfar, P. Fast and robust multiframe super resolution. IEEE Trans. Image Process. **2004**, *13* (10), 1327–1344.
35. Capel, D.; Zisserman, A. Super-resolution enhancement of text image sequences. In *Proceedings 15th International Conference on Pattern Recognition ICPR*, Barcelona, Spain, 2000, Vol. 1, No. 1, 600–605.
36. Liu, X.; Song, D.; Dong, C.; Li, H. MAP-based image super-resolution reconstruction. World Acad. Sci. Eng. Technol. **2008**, *37*, 208–211.
37. Zhang, H.; Zhang, L.; Shen, H. A super-resolution reconstruction algorithm for hyper- spectral images. Elsevier J. Sig. Process. **2012**, *92*, 2082–2096.
38. Schulz, R.R.; Stevenson, R.L. Extraction of high-resolution frames from video sequences. IEEE Trans. Image Process. **1996**, *5*, 996–1011.
39. Greeshma, T.R.; Ameeramol, P.M. Bayesian MAP Model for Edge Preserving Image Restoration: A Survey. In *Conference on Advances in Computational Techniques (CACT)*; Springer: Greater Noida, India, 2011, 14–18.
40. Farsiu, S.; Robinson, M.D.; Elad, M.; Milanfar, P. Fast and robust multiframe super resolution. IEEE Trans. Image Process. **2004**, *13* (10), 1327–1344.
41. Aly, H.A.; Dubois, E. Image up-sampling using total-variation regularization with a new observation model. IEEE Trans. Image Process. **2005**, *14* (10), 1647–1659.
42. Irani, M.; Peleg, S. Super resolution from image sequence. In *Proceedings of 10th International Conference on Pattern Recognition (ICPR)*; IEEE Computer Society Press: NJ, 1990, 115–120.
43. Komatsu, K.A.T.; Igarashi, T.; Saito, T. Very high resolution imaging scheme with multiple different-aperture cameras. Sig. Process. Image Commun., Special Issue on Digital High Definition Television **1993**, *5* (6), 511–526.
44. Papoulis, A. A new algorithm in spectral analysis and band-limited extrapolation. IEEE Trans. Circuits Syst. CAS-22 **1975**, 735–742.
45. Gerchberg, R.W. Super-resolution through error energy reduction. J. Modern Optics **1974**, *21* (9), 709–720.
46. Vandewalle, P.; Susstrunk, S.; Vetterli, M. Superresolution images reconstructed from aliased images. In *SPIE Visual Communication and Image Processing Conference*; Ebrahimi, T.; Sikora, T.; Eds.; Lugano, Switzerland, 2003, 1398–1405.
47. Chatterjee, P.; Mukherjee, S.; Chaudhuri, S.; Seetharaman, G. Application of Papoulis- Gerchberg method in image super-resolution and inpainting. Comput. J. **2009**, *52* (1), 80–89.
48. Dai, S.; Han, M.; Wu, Y.; Gong, Y. Bilateral back-projection for single image super Resolution. IEEE Proc. ICME **2007**, 1039–1042.
49. Dong, W.; Zhang, L.; Shi, G.; Wu, X. Nonlocal back-projection for adaptive image enlargement. In *IEEE International Conference on Image Processing*, Genova, Italy, 2005, 349–352.
50. Lin, C.K.; Wu, Y.H.; Yang, J.F.; Liu, B.D. An iterative enhanced super-resolution system with edge-dominated interpolation and adaptive enhancements. EURASIP J. Adv. Sig. Process **2015**, *18* (79).
51. Hsieh, C.; Huang, Y.P.; Chen, Y.; Fuh, C.S. Video super-resolution by motion compensated iterative back-projection approach. J. Inf. Sci. Eng. **2011**. *27*, 1107–1122.
52. Yixiong, Z.; Mingliang, T.; Kewei, Y.; Zhenmiao, D. Video super-resolution reconstruction using iterative back projection with critical-point filters based image matching. Adv. Multimedia **2015**, Article ID *285969*, 10 pages.
53. Stark, H.; Oskoui, P. High resolution image recovery from image plane arrays using convex projections. J. Opt. Soc. Am. A **1989**, *6*, 1715–1726.
54. Tekalp, A.M.; Ozkan, M.K.; Sezan, M.I. High-resolution image reconstruction from lower-resolution image sequences and space varying image restoration. In *Proceedings of IEEE International Conference on Acoustics,*

Speech and Signal Processing (ICASSP), San Francisco, CA, March 3, 1992, 169–172.
55. Pratt, W.K. *Digital Image Processing*, 2nd Ed.; John & Wiley Sons: New York, 1991.
56. Youla, D.C. Generalized image restoration by the method of alternating orthogonal projections. IEEE Trans. Circuits Syst. **1978**, *CAS-25*, 694–702.
57. Patti, A.J.; Sezan, M.I.; Tekalp, A.M. Super resolution video reconstruction with arbitrary sampling lattices and nonzero aperture time. IEEE Trans. Image Process. **1997**, *6* (8), 1064–1076.
58. Patti, A.J.; Altunbasak, Y. Artifact reduction for set theoretic super resolution image reconstruction with edge adaptive constraints and higher-order interpolants. IEEE Trans. Image Process. **2001**, *10* (1), 179–186.
59. Hsu, J.T.; Yen, C.C.; Li, C.C.; Sun, M.; Tian, B.; Kaygusuz, M. Application of Wavelet-based POCS Super-resolution for Cardiovascular MRI Image Enhancement. In *IEEE Proceedings of the Third International Conference on Image and Graphics*, Hong Kong, China, 2004, 572–575.
60. Wheeler, F.W.; Hoctor, R.T.; Barrett, E.B. Super-Resolution Image Synthesis using Projections onto Convex Sets in the Frequency Domain; SPIE, 2005.
61. Yu, H.; Xiang, M.; Hua, H.; Chun, Q. *Face Image Super-resolution through POCS and Residue Compensation*; VIE, 2008, 494–497.
62. Xiaoqing, S.; Shutao, L. Multi-frame image super-resolution reconstruction based on sparse representation and POCS. Int. J. Digital Content Technol. Appl. **2011**, *5* (8), 127–135.
63. Shen, W.; Fang, L.; Chen, X.; Xu, H. Projection onto convex sets method in space-frequency domain for super resolution. J. Comput. **2014**, *9* (8), 1959–1966.
64. Kim, C.; Choi, K.; Beom, R.J. Improvement on learning based super resolution by adopting residual information and patch reliability. In *IEEE International Conference on Image Processing*, Genova, 2009, 1197–1200.
65. Kim, C.; Cho, K.; Lee, H.; Hwang, K.; Beom, R.J. Robust learning based super resolution. In *IEEE 17th International Conference on Image Processing*; Hong Kong, 2010, 2017–2020.
66. Qinlan, X.; Hong, C.; Huimin, C. Improved example-based single image super resolution. In *IEEE International Conference on Image Processing*; Genova, 2010, 1204–1207.
67. Yang, J.; Lin, Z.; Cohen, S. Fast image super-resolution based on in-place example regression. In *IEEE Conference on Computer Vision and Pattern Recognition (CVPR)*, Portland, OR, 2013, 1059–1066.
68. Fan, W.; Dit-Yan, Y. Image hallucination using neighbor embedding over visual primitive manifolds. In *IEEE Conference on Computer Vision and Pattern Recognition (CVPR)*, Minneapolis, MN, 2007, 1–7.
69. Chan, T.M.; Zhang, J.; Pu, J.; Huang, H. Neighbor embedding based super-resolution algorithm through edge detection and feature selection. Pattern Recognit. Lett. **2009**, *30* (5), 494–502.
70. Bevilacqua, M.; Roumy, A.; Guillemot, C.; Marie, L.; Alberi, M. Low-complexity single image super-resolution based on nonnegative neighbor embedding. In *British Machine Vision Conference*, Surrey, 2012, 1–10.
71. Tang, Y.; Yan, P.; Yuan, Y.; Li, X. Single-image super-resolution via local learning. Int. J. Mach. Learn. Cyb. **2011**, 15–23. doi:10.1007/s13042-011-0011-6.
72. Ogawa, T.; Haseyama, M. Adaptive single image super resolution approach using support vector data description. EURASIP J. Adv. Sig. Process. **2011**, Article ID 852934. doi:10.1155/2011/852934.
73. Zhanga, J.; Zhaob, C.; Xiongb, R.; Mab, S.; Zhaoa, D. Image super-resolution via dual dictionary learning and sparse representation. In *IEEE International Symposium on Circuits and Systems (ISCAS)*, Seoul, 2012, 1688–1691.
74. Yang, M.C.; Chiang, Y.; Wang, F. A self-learning approach to single image super resolution. IEEE Trans. Multimed. **2013**, *15* (3), 498–508.
75. Jiang, J.; Ma, X.; Chen, C.; Lu, T.; Wang, Z.; Ma, J. Single image super-resolution via locally regularized anchored neighborhood regression and nonlocal means. IEEE Trans. Multimed. **2016**. doi:10.1109/TMM.2016.2599145.
76. Li, L.; Xie, Y.; Hu, W.; Zhang, W. Single image super-resolution using combined total variation regularization by split Bregman iteration. Neurocomputing **2014**, *142*, 551–560.
77. Wang, L.; Xiang, S.; Meng, G.; Wu, H.; Pan, C. Edge-directed single-image super resolution via adaptive gradient magnitude self-interpolation. IEEE Trans. Circuits Syst. Video Technol. **2013**, *23* (8), 1289–1299.
78. Mallat, S.; Yu, G. Super-resolution with sparse mixing estimators. IEEE Trans. Image Process. **2010**, *19* (11), 2889–2900.
79. Kaibing, Z.; Xinbo, G.; Jie, L.; Hongxing, X. Single image super-resolution using regularization of non-local steering kernel regression. Signal Process. **2016**, *123*, 53–63.
80. Wang, Y.; Li, J.; Fu, P. Medical image super-resolution analysis with sparse representation. In *International Conference on Intelligent Information Hiding and Multimedia Signal Processing (IIH-MSP)*, Greece, July 2012.
81. Isaac, J.; Kulkarni, R. Super resolution techniques for medical image processing. In *International Conference on Technologies for Sustainable Development (ICTSD)*, India, February 2015.

Information Sets for Edge Detection

Shaveta Arora
Department of EECE, The NorthCap University, Gurgaon, India

Vignesh Kothapalli
Department of EECE, Electronics and Electrical Engineering, Indian Institute of Technology Guwahati, India

Madasu Hanmandlu
CSE Department, MVSR Engineering College, Nadergul, Hyderabad, India

Gaurav Gupta
School of Natural, Applied and Health Sciences, Wenzhou-Kean University, China

Abstract

In most of the computer vision applications, edge detection is the initial step for performing high-level tasks such as object recognition and scene analysis. The traditional algorithms cannot meet the desired accuracy and robustness of these applications. Information set theory is utilized for defining edge strength measures that help in finding the robust edges in the presence of noise. Information sets are derived from fuzzy sets by the application of Hanman–Anirban entropy. Fuzzy sets represent the vagueness present in an image using membership functions (MFs), whereas information sets represent the overall uncertainty in an image by linking the information source (any property/attribute) values with the corresponding MF values. Information set theory is already applied successfully to image processing applications such as enhancement of underexposed and overexposed images, face recognition, and gait recognition. The effectiveness of this theory is demonstrated here through the nonderivative smallest univalue segment assimilating nucleus (SUSAN) edge detector and Sobel fractional gradient by extracting localized features such as the fine edges in the image.

Keywords: Edge detection; Fractional derivative; Fuzzy sets; Information sets; Membership function; SUSAN edge detector.

INTRODUCTION

Any image can be characterized by certain cues such as edges, color and texture, which are essential for the analysis of its constituent parts. Intensity changes in images are generally observed mainly because of the lack of coherence in the elements of structure (objects, humans, surroundings, documents, etc.) or surface. Other reasons can be attributed to the illumination bound and reflectance property of the material. But, spatial localization is an important property of the intensity changes that leads to the formulation of rules for combining zero crossing segments from different channels of an image into a particular description as explained in the work of Marr and Hildreth.[1] The boundaries where these intensity changes occur are termed as edges of an image, which help us in providing skeletal view of the subject. The main aim of edge detection is to determine those discontinuities in pixel intensities, which are generally the high-frequency components of an image. But, noise also being a high-frequency component poses a problem while determining the edge pixels. The output of edge detection results in significant reduction in the amount of data to be processed for higher level image processing tasks.

The usage of fuzzy logic has witnessed an exponential growth in almost all tasks of image processing as fuzzy logic facilitates the representation of inherent uncertainties in the image information. These uncertainties can be either local in the case of edge detection, segmentation, and recognition or global in the case of image enhancement tasks.[2]

Motivated by the need to represent the uncertainty in the fuzzy sets using the entropy functions, a novel approach toward information processing called "Information set theory" was introduced in 2011.[3] The successful application of this theory includes the fusion of images of different resolutions using the singular value decomposition method, image enhancement, face recognition, iris recognition, and edge detection.[4–7]

This article illustrates the effectiveness of the information sets on two edge detection techniques, namely, smallest univalue segment assimilating nucleus (SUSAN)[8] and the Sobel-based fractional gradient filter.[9] The SUSAN edge detector is a nonderivative method for fast and accurate detection of edges and corners in images. The Sobel based fractional gradient edge detection algorithm makes use of Sobel derivative function and modifies it into a fractional-order derivative comprising an infinite series. The origin of fractional derivatives dates back to Cauchy, Riemann, Liouville, and Letnikov in the nineteenth century. A detailed explanation of fractional calculus and its effectiveness in the fields of physics, mechanics, electronics, biology, time and frequency domain system identification, etc., can be found in the works of Loverro,[10] Oldham and Spanier.[11]

The edge detectors Sobel and SUSAN are good at preserving the connectivity of the edge pixels but have a limitation of ignoring the weak edges. To overcome this issue, we apply information set theory to these detectors by utilizing different membership functions (MFs) for the intensification of weak edges that contain some hidden information.

Traditional approaches determine the edge pixels either as the optimal points of the first-order derivative or as the zero crossings of the second-order derivative.[12] A few of these are as follows: Sobel, Prewitt, Canny, Robert, and Laplacian of Gaussian (LOG) edge detectors. Traditional edge detectors have been combined with fuzzy logic to expedite the logical reasoning based on approximate values.[16] For instance, Sobel edge detector is combined with interval type 2 fuzzy logic system as an alternative to applying filtering on the images.[13] Whereas, anisotropic diffusion-driven edge detection proposed by Maiseli and Gao makes use of an iterative approach to detect edges in severely degraded images.[14] Edge detection has also found its way in the extraction of edges of deep geological bodies through interpretation of the "potential field"[15] data.[16]

SUSAN APPROACH

USAN Area Computation and Its Fuzzification

Using SUSAN principle, univalue segment assimilating nucleus (USAN) area, which accounts for the number of pixels having similar brightness as that of center pixel, is computed first. For this, a circular mask (having a center pixel known as the "nucleus") of 37 pixels as shown in Fig. 1 is placed on each image pixel. The intensity of each pixel within this mask is compared with the intensity of nucleus using Eq. 1, and USAN area s is calculated by summing these 37 compared outputs in equation 2.

$$C(r, r_0) = e^{-\left(\frac{I(r)-I(r_0)}{t}\right)^6} \quad (1)$$

$$s(r_0) = \Sigma_r C(r, r_0) \quad (2)$$

Here, r and r_0 are the positions of the pixel within the mask and of nucleus, respectively. $I(r)$ and $I(r_0)$ are the intensities of the pixel within the mask and nucleus, respectively. t is the threshold for the edge strength. Lower value of $t < 20$ is preferred for synthetic images, and $20 < t < 50$ is found suitable for real-life images. It determines the edges of minimum strength and provides an easy way of controlling the edge map.

According to SUSAN principle, s attains the maximum value of 37 when the nucleus lies in the flat region of image, becomes half of the maximum very near a straight edge, and falls even further inside a corner.[8]

To expedite the process of simulation of edge detection, we make use of a histogram-based approach. To plot histogram, $h(s)$, that is, the total number of pixels having USAN area s is calculated. The frequency of occurrence of the USAN area s denoted by (s) in the whole image is given as

$$p(s) = h(s) / (\Sigma h(s); \quad x = 1, 2, \ldots, 37 \quad (3)$$

satisfying the condition $\Sigma p(s) = 1$. The histogram is a plot of USAN area (s) versus $p(s)$.

Formation of a Fuzzy Set

Fuzzification deals with the uncertainty associated with the detection of an edge pixel. Fuzzy MF indicates how strongly a pixel belongs to an edge pixel in the range 0–1. Therefore, a fuzzy set is formed from the USAN area (s) and its MF, $\mu(s)$, as follows:

$$S = \bigcup \{s, \mu(s)\}; \quad s = 1, 2, \ldots, 37 \quad (4)$$

FRACTIONAL DERIVATIVE APPROACH

Fraction derivative deals with derivatives of arbitrary orders (including complex orders). For instance, let us consider a variable A raised to a positive integral exponent b. We can perceive this as multiplying A with itself $|b|$ number of times. Instead, if we think of raising A to a fractional exponent, the multiplication is quite difficult to contemplate. For instance, it is quite challenging to imagine the physical meaning of $X^{2.73}$. We can emphatically say that this has a finite value that can be determined by the infinite series expansion.[10,11]

THE GAMMA FUNCTION

The generalization of the factorial function for real numbers leads to the gamma function, expressed as follows:

$$\Gamma(z) = \int_0^\infty e^{-u} u^{z-1} du, \quad \text{for all } z \in R \quad (5)$$

Fig. 1 Circular mask

The gamma function of any natural number z is expressed as the product of that number minus one and its function value is given by

$$\Gamma(z+1) = z\Gamma(z), \text{ When } z \in N. \qquad (6)$$

For the negative integer values of z, we observe that the gamma function value tends to infinity but can be defined for the non-integer values as shown in Fig. 2.[10]

Grunwald–Letnikov Fraction Derivative

To start with, let us first consider the general expression for a first-order derivative:

$$f'(x) = \lim_{h \to 0} \frac{f(x+h) - f(x)}{h} \qquad (7)$$

Repeating this process again on $f'(x)$ gives us the second-order derivative as

$$f''(x) = \lim_{h \to 0} \frac{f'(x+h) - f'(x)}{h}$$
$$= \lim_{\substack{h1 \to 0}} \frac{\lim_{h2 \to 0} \frac{f(x+h1+h2) - f(x+h1)}{h2} - \lim_{h2 \to 0} \frac{f(x+h2) - f(x)}{h2}}{h1} \qquad (8)$$

On selecting the same value for all of $h1$, $h2$, that is, $h1 = h2 = h$. The second-order expression simplifies to

$$f''(x) = \lim_{h \to 0} \frac{f(x+2h) - 2f(x+h) + f(x)}{h^2} \qquad (9)$$

For the nth derivative, the expression can be written as a summation:

$$d^n f(x) = \lim_{h \to 0} \frac{1}{h^n} \sum_{m=0}^{n} (-1)^m \binom{n}{m} f(x - mh) \qquad (10)$$

When the binomial term is represented using the gamma function instead of the factorial, the above expression can be generalized for nonintegral values of $n = \alpha$ where $\alpha \in R$.

Thus, the generalized expression for the Grunwald–Letnikov fractional derivative of order α is given by

$$d^\alpha f(x) = \lim_{h \to 0} \frac{1}{h^n} \sum_{m=0}^{\frac{u-l}{h}} (-1)^m \frac{\Gamma(\alpha+1)}{m!\Gamma(\alpha-m+1)} f(x - mh) \qquad (11)$$

where u and l represent the upper and lower limits of differentiation i.e., $l < \alpha \leq u$.[10]

Sobel Fractional Derivative Masks

An edge point in an image results from a discontinuity/irregularity in the intensities of adjacent pixels. It is marked with a high-gradient magnitude value. The gradient of an image function (x, y) at location (x, y) is given by

$$\nabla f(x, y) = [G_x, G_y]^T \qquad (12)$$

where $G_x = f(x+1, y) - f(x, y)$ and $G_y = f(x, y+1) - f(x, y)$ are the gradients along X and Y directions, respectively. Sobel edge detector uses two masks of size 3×3 centered at the pixel location (x, y) to compute the gradients along X and Y axes, which are as follows:

$$G_x = f(x+1, y-1) - f(x-1, y-1) + 2f(x+1, y)$$
$$- 2f(x-1, y) + f(x+1, y+1) - f(x-1, y-1) \qquad (13)$$

$$G_y = f(x-1, y+1) - f(x-1, y-1) + 2f(x, y+1)$$
$$- 2f(x, y-1) + f(x+1, y+1) - f(x-1, y-1) \qquad (14)$$

Assuming the difference between any two pixel locations along X and Y axes be Δx and Δy, respectively,[11] we set $\Delta x = \Delta y = 2$. Then the gradient components are mathematically expressed as follows:

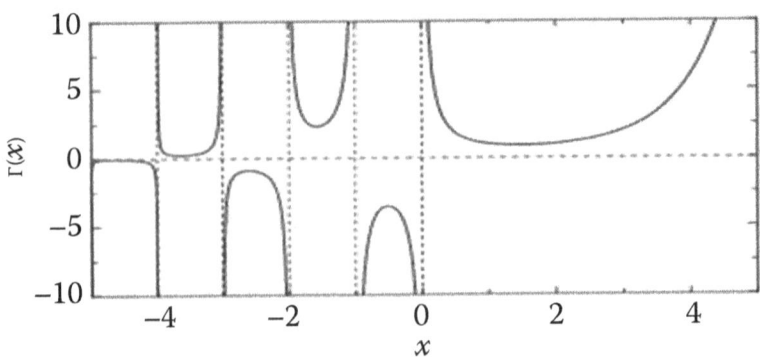

Fig. 2 Gamma function

$$G_x = \frac{1}{2}\left(\frac{\partial f(x+1,y-1)}{\partial x} + 2\frac{\partial f(x+1,y)}{\partial x} + \frac{\partial f(x+1,y+1)}{\partial x}\right) \quad (15)$$

$$G_y = \frac{1}{2}\left(\frac{\partial f(x-1,y+1)}{\partial y} + 2\frac{\partial f(x,y+1)}{\partial y} + \frac{\partial f(x+1,y+1)}{\partial y}\right) \quad (16)$$

We now generalize the order of derivative from an integer to a fraction to get the fractional-order gradient equations:

$$G_x^\alpha = \frac{1}{2}\left(\frac{\partial^\alpha f(x+1,y-1)}{\partial x^\alpha} + 2\frac{\partial^\alpha f(x+1,y)}{\partial x^\alpha} + \frac{\partial^\alpha f(x+1,y+1)}{\partial x^\alpha}\right) \quad (17)$$

$$G_y^\alpha = \frac{1}{2}\left(\frac{\partial^\alpha f(x-1,y+1)}{\partial y^\alpha} + 2\frac{\partial^\alpha f(x,y+1)}{\partial y^\alpha} + \frac{\partial^\alpha f(x+1,y+1)}{\partial y^\alpha}\right) \quad (18)$$

We next apply the Grunwald–Letnikov fractional derivative on the image functions in Eqs. 17 and 18 leading to

$$(\nabla^\alpha f)_{i,j} = \left((\Delta_1^\alpha f)_{i,j}, (\Delta_2^\alpha f)_{i,j}\right) \quad (19)$$

$$\text{where } (\Delta_1^\alpha f)_{i,j} = \sum_{k=0}^{T-1}(-1)^k \binom{\alpha}{k} f_{i-k,j},$$

$$(\Delta_2^\alpha f)_{i,j} = \sum_{k=0}^{T-1}(-1)^k \binom{\alpha}{k} f_{i,j-k}$$

$$\binom{\alpha}{k} = \frac{\Gamma(\alpha+1)}{k!\,\Gamma(\alpha-k+1)}$$

and $1 \leq i \leq M$, $1 \leq j \leq N$—M and N being the rows and columns, respectively. The fractional derivatives in X and Y directions are as follows:

$$G_x^\alpha = \frac{1}{2}\begin{bmatrix} f(x+1,y-1) - \alpha f(x,y-1) + \\ \frac{\alpha^2-\alpha}{2}f(x-1,y-1) + \\ \ldots + (-1)^{T-1}C_{T-1}^\alpha f(x-T+2,y-1) + \\ + 2f(x+1,y) - 2\alpha f(x,y) + \\ (\alpha^2-\alpha)f(x-1,y) + \\ \ldots + 2*(-1)^{T-1}C_{T-1}^\alpha f(x-T+2,y) + \\ + f(x+1,y+1) - \alpha f(x,y+1) \\ + \frac{\alpha^2-\alpha}{2}f(x-1,y+1) + \\ \ldots + (-1)^{T-1}C_{T-1}^\alpha f(x-T+2,y+1) \end{bmatrix} \quad (20)$$

$$G_y^\alpha = \frac{1}{2}\begin{bmatrix} f(x-1,y+1) - \alpha f(x-1,y) + \\ \frac{\alpha^2-\alpha}{2}f(x-1,y-1) + \\ \ldots + (-1)^{T-1}C_{T-1}^\alpha f(x-1,y-T+2) + \\ + 2f(x,y+1) - 2\alpha f(x,y) + \\ (\alpha^2-\alpha)f(x,y-1) + \\ \ldots + 2*(-1)^{T-1}C_{T-1}^\alpha f(x,y-T+2) + \\ + f(x+1,y+1) - \alpha f(x+1,y) + \\ \frac{\alpha^2-\alpha}{2}f(x+1,y-1) + \\ \ldots + (-1)^{T-1}C_{T-1}^\alpha f(x+1,y-T+2) \end{bmatrix} \quad (21)$$

Converting the above relations into the fractional Sobel masks with $T = 3$ here, T represents the order/size of the masks. Leads to two 3×3 operators, G_x and G_y.

G_y Operator

$(\alpha^2-\alpha)/4$	$-\alpha/2$	$1/2$
$(\alpha^2-\alpha)/2$	$-\alpha$	1
$(\alpha^2-\alpha)/4$	$-\alpha/2$	$1/2$

G_x Operator

$(\alpha^2-\alpha)/4$	$(\alpha^2-\alpha)/2$	$(\alpha^2-\alpha)/4$
$-\alpha/2$	$-\alpha$	$-\alpha/2$
$1/2$	1	$1/2$

The magnitude of the fractional-order gradient[9] is expressed as follows:

$$\nabla^\alpha F = \text{mag}(\nabla^\alpha f) = [(G_x^\alpha)^2 + (G_y^\alpha)^2]^{1/2} \quad (22)$$

Fuzzification of Fractional Gradient

Fuzzification converts a crisp input into the fuzzy variable lying in the interval (0, 1) and is represented by an MF. Here, the fractional-order gradient magnitude calculated using Eq. 22 is fuzzified by plotting its histogram. Fuzzification of these gradient magnitude values helps detect edge pixels in the proximity of the pixels with the maximum magnitude. The probability p of fractional gradient magnitude of $(\nabla^\alpha F)$ denoted by v is computed from

$$p(v) = \frac{\text{hist}(v)}{\sum_{i=0}^{v\max}\text{hist}(i)} \quad (23)$$

where hist(v) denotes the number of pixels having the value of v. Before proceeding to the concept of information set, we recall the definition of fuzzy set. A pair of the fractional gradient magnitude and its MF value is an element in our fuzzy set:

$$V = \bigcup\{v, \mu(v)\} \qquad (24)$$

where $\mu(v)$ is its membership function.

USE OF INFORMATION SET CONCEPT

According to the information theory, entropy is represented as

$$H = \Sigma p(-\log p) \qquad (25)$$

where (p) is the probability and $(-\log p)$ is the logarithmic gain.[17] It may be noted that Pal and Pal entropy function replaces this logarithmic gain by the exponential gain[18] resulting in

$$H = \Sigma p e^{(1-p)} \qquad (26)$$

The representation of uncertainty in the gradient values of pixel intensities is necessary for edge detection. The effectiveness of an edge detector lies in detecting an optimum number of edges and their strengths. For detecting strong edges, both possibilistic and probabilistic uncertainties are essential for which the Hanman–Anirban entropy function is immensely suitable as it can represent both types of uncertainties. This entropy function in the probabilistic domain is defined as

$$H = \Sigma p e^{-(ap^3 + bp^2 + cp + d)} \qquad (27)$$

where a, b, c, and d are real-valued parameters, and p represents probability. To represent the possibilistic uncertainty, we replace the probability values (p) with the possibility values (z), that is, the gradient values. By referring to both p and z as the information source values, we can use Eq. 27 to compute the probabilistic uncertainty with p as the information source value and possibilistic uncertainty with z as the information source value. We now convert the exponential gain function in Eq. 27 into the Gaussian membership function (GMF) by assuming the distribution of gradient values as Gaussian. With the choice of parameters in Eq. 27 as $p = z$, $a = 0$, $b = \dfrac{1}{2\sigma^2}$, $c = -\dfrac{2m}{2\sigma^2}$, $d = \dfrac{m^2}{2\sigma^2}$ the exponential gain function becomes the GMF:

$$\mu(z) = e^{-(z-m)^2/2\sigma^2} \qquad (28)$$

where m is the mean and σ^2 is the variance in z. In view of this substitution, the Hanman–Anirban entropy function takes the form as:

$$H = \Sigma z \mu(z) \qquad (29)$$

The entropy function now represents the possibilistic uncertainty. A set of values $\{z\mu(z)\}$ constitutes the information set and each $z\mu(z)$ is the information value. The mean information of information values is the expectation of the information values:

$$H_m = \sum_{Z=1}^{L} z\mu(z) p(z) \qquad (30)$$

where (z) is the probability of z such that $\sum_{Z=1}^{L} p(z) = 1$. Thus, Eq. 30 contains the possibilistic uncertainty as given by $\mu(z)$ and probabilistic uncertainty as given by $p(z)$.

APPLICATION OF SUSAN PRINCIPLE

The symmetric Gaussian membership function (SGMF) that depends on two parameters, mean m_s and standard deviation σ_s, is given by

$$\mu_1(s) = e^{-\left(\dfrac{(s-m_s)^2}{2\sigma_s^2}\right)} \qquad (31)$$

As per SUSAN principle, the mean of this function should take the half of maximum USAN area, s_{max}, that is, $m_s = s_{max}/2$ and σ_s is calculated as

$$\sigma_s = \sqrt{\dfrac{1}{2} \dfrac{\sum_{s=1}^{L} \left((s_{max}/2) - s\right)^4 p(s)}{\sum_{s=1}^{L} \left((s_{max}/2) - s\right)^2 p(s)}} \qquad (32)$$

In contrast to SGMF, $m_s = s_{max}$ in GMF in the works of Hanmandlu et al.[19,20] Both the functions are plotted in Fig. 3b for the original image in Fig. 3a. The values of the standard deviation and mean defined in SGMF give rise to more number of edges than GMF.[19,20] We will now modify SGMF to obtain sharp edges by using a sigmoid function.

Sigmoid membership function (SMF) is a two-parameter function and is applied to SGMF, $\mu_1(s)$, as

$$\mu_2(s) = \dfrac{1}{1 + e^{-\tau(\mu_1(s) - \mu_c)}} \qquad (33)$$

where τ is the intensification and μ_c is the crossover. For better demarcation of the edges, the value of τ is chosen such that $\tau > 15$. Figure 3c illustrates SGMF and SMF at different values of crossover μ_c. The graph clearly shows the intensification of SGMF except for $\mu_c = 0.9$. To enhance the weak edges, an optimal value of μ_c is required. This will be discussed in the "Effect of Parameters on Edge Strength Factor—SUSAN" section.

APPLICATION OF FRACTIONAL DERIVATIVES

We now make a small modification in GMF from the standpoint of edges because the presence of an edge pixel

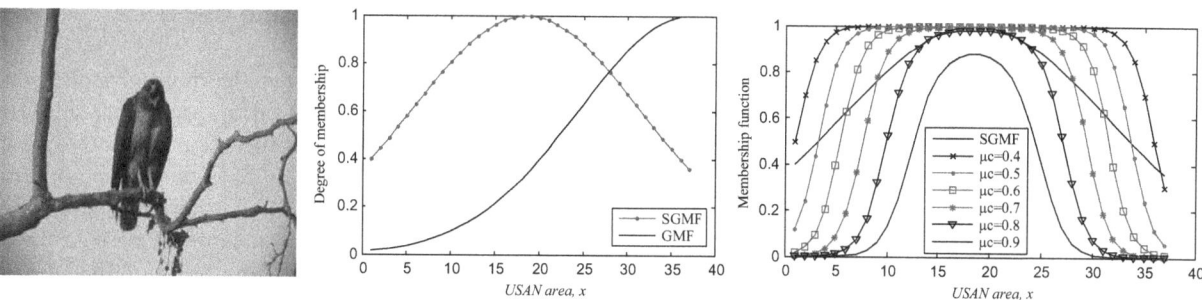

Fig. 3 (a) Original image: "42049"; (b) SGMF and GMF; (c) SGMF and SMF at $\tau = 20$ with different values of μ_c [7]

arises when its fractional gradient magnitude attains the maximum, that is, v_{max}. In view of this, GMF is written as

$$\mu_1(v) = e^{-\frac{(v-(v_{max}))^2}{2\sigma^2}} \quad (34)$$

The variance σ_v is taken as the fuzzifier calculated from:

$$\sigma_v = \sqrt{\frac{1}{2} \frac{\sum_{v=1}^{L} \left((v_{max}/2) - v\right)^4 p(v)}{\sum_{v=1}^{L} \left((v_{max}/2) - v\right)^2 p(v)}} \quad (35)$$

The use of fuzzifier in GMF results in more edges than obtained with the actual standard deviation. A sigmoid function is devised to modify the GMF to get sharper edges. Let us denote the MF $\mu(v)$ before and after the intensification by two symbols, $\mu_1(v)$ and $\mu_2(v)$, respectively. The S-shape of SMF allows it to act as an intensifier that can strengthen the weak edges. It is defined as

$$\mu_2(v) = \frac{1}{1 + e^{-r(\mu_1(v) - \mu_c)}} \quad (36)$$

where r is the intensification and μ_c is the crossover. The value of r is considered to be greater than 16 or in general to be higher so that we get better demarcation of edges. Moreover, with the increasing values of r, the image tends to be binary and the thresholding requirement can be eliminated to some extent. To intensify the weak edges, we use Eq. 33 by selecting an optimal value of crossover μ_c and intensification r. From the simulations, the value of $r = 25$ is found to give the best results.[9]

QUANTITATIVE MEASURES OF ANALYSIS

For ascertaining the performance of an edge detector, some quantitative measures are necessitated. As mentioned before, we bank upon the edge strength factor to find the optimal value of the crossover parameter in SMF.

Humans cannot evaluate the quality and accuracy of an edge map quantitatively. In order to quantify the performance of the proposed edge detector, there is a need to define some quantitative measures. Apart from edge strength factor introduced in this work, two other measures, namely, Pratt's figure of merit (*FoM*) and the structural similarity (SSIM) index, are used.

Pratt's Figure of Merit

To validate the performance of the proposed edge detector and to compare the performance of the edge preservation ability of different methods of edge detection, *FoM*, is used. This measure is based on the difference between the actual and the ideal edge pixels[21] and is defined as

$$\text{FoM} = \frac{1}{\max(I_I, I_A)} \sum_{i=1}^{I_A} \frac{1}{(1 + \alpha' d_i^2)} \quad (37)$$

where I_I and I_A are the number of ideal and actual detected edge pixels, respectively; d_i is the distance between a pixel declared as the edge point and the nearest ideal edge pixel; and α' is a scaling constant set to 1/9. A value of *FoM* close to 1 implies that there is less difference between an ideal edge map and the actual edge map.

The Structural Similarity Index

SSIM[22] is a full-reference metric which is used to measure the similarity between the two images. The SSIM index can be viewed as a quantitative measure between two images such that one of the images is of perfect quality. It is based on the computation of three terms: luminance comparison (x, y), contrast comparison $c(x, y)$, and structure comparison $s(x, y)$. The overall index is multiplicative of these three terms:

$$\text{SSIM}(x, y) = [l(x, y)]^\alpha \cdot [c(x, y)]^\beta \cdot [s(x, y)]^\gamma \quad (38)$$

where $\alpha = \beta = \gamma = 1$ for equal importance of the three terms. From the definition of SSIM, if the two images X and Y are equal then SSIM = 1.

Edge Strength Factor

The uncertainty in the edge segments is caused due to the variation in the edge strengths. The edge magnitude together with the MF value facilitates the computation of edge strength expressed as $\{(s\mu(s))^2\}$. This edge strength with respect to the crossover changes to $\{(\mu(s) - \mu_c)^2 s\}$. With the use of MF μ_1, the edge strength becomes:

$$\mathrm{ES}_1 = \frac{1}{L} \sum_{s=1}^{L} \left(\mu_1(s) - \mu_c\right)^2 s \tag{39}$$

The mean edge information is computed from:

$$\mathrm{ES}_{m1} = \frac{1}{L} \sum_{s=1}^{L} \left(\mu_1(s) - \mu_c\right) s \tag{40}$$

The edge strength with the use of SMF now becomes:

$$\mathrm{ES}_2 = \frac{1}{L} \sum_{s=1}^{L} \left(\mu_2(s) - \mu_c\right)^2 s \tag{41}$$

The mean edge information is:

$$\mathrm{ES}_{m2} = \frac{1}{L} \sum_{s=1}^{L} \left(\mu_2(s) - \mu_c\right) s \tag{42}$$

The ratio of mean edge information to the edge strength defined as edge quality factor is:

$$\mathrm{EQ}_1 = \left| \frac{\mathrm{ES}_{m1}}{\mathrm{ES}_1} \right| \tag{43}$$

The edge quality factor after the intensification appears as:

$$\mathrm{EQ}_2 = \left| \frac{\mathrm{ES}_{m2}}{\mathrm{ES}_2} \right| \tag{44}$$

The Edge Strength Factor is the ratio of quality factors before and after intensification defined as:

$$\mathrm{ES}_f = \frac{\mathrm{EQ}_1}{\mathrm{EQ}_2} \tag{45}$$

This factor helps us in obtaining a suitable edge map containing the maximum edge information. The computation is repeated for (v) as well from Eqs. 39–45.

Optimal Crossover

The edge strength factor facilitates the computation of the optimal value of crossover for SMF, $\mu_2(s)$. The crossover μ_c is learned at the maximum edge strength factor ES_f. Figure 4 shows the graph of ES_f vs μ_c. The maximum value of ES_f is marked here, and the corresponding μ_c is called the optimal crossover.

Figure 5 gives the flowchart of the proposed algorithm named Algorithm 1, which is outlined here.

Proposed Algorithm Using USAN

The proposed edge detection algorithm is applied on the three components of RGB color space separately. Histogram of the USAN area is fuzzified using SGMF, which in turn is fuzzified by SMF. Selection of an appropriate MF and its parameter is essential for the success of edge detection. An edge strength measure is derived from the information sets, which will provide the optimal crossover for the optimal edge map. Finally, the obtained edge maps of the individual components are concatenated to get the resultant edge image.

Algorithm 1

1. Separate the three components R, G, and B of colored RGB image.
2. Place the circular mask on each pixel of the individual component and compute the USAN area at this pixel using Eqs. 1 and 2.
3. Calculate the histogram of the USAN area (s) using Eq. 3.
4. Calculate the standard deviation using Eq. 32 and fuzzify s as SGMF using Eq. 31.
5. Fuzzify SGMF to SMF using Eq. 33. For this, the optimal crossover in the range [0.5–0.9] is required to find by maximizing the value of edge strength factor ES_f using Eq. 45 as shown in Fig. 4.
6. Defuzzify the output to convert the histogram-based fuzzy output into the spatial domain edge map.
7. Concatenate the three components of the edge map and obtain the binary edge map after applying Otsu's thresholding.[23]

THE PROPOSED EDGE DETECTOR USING FRACTIONAL DERIVATIVE MASK

The flowchart of the fractional derivative-based algorithm is explained using Fig. 6.

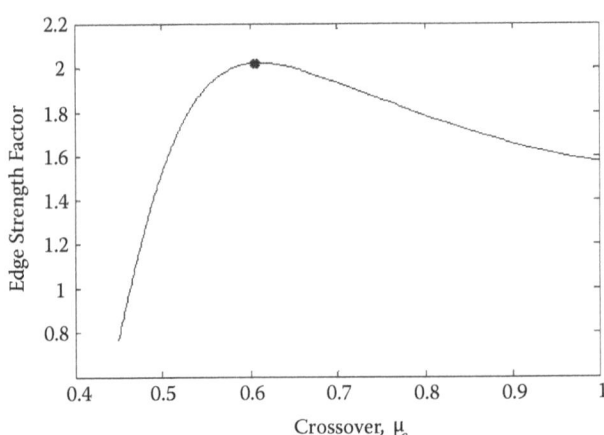

Fig. 4 A plot of ES_f versus μ_c[7]

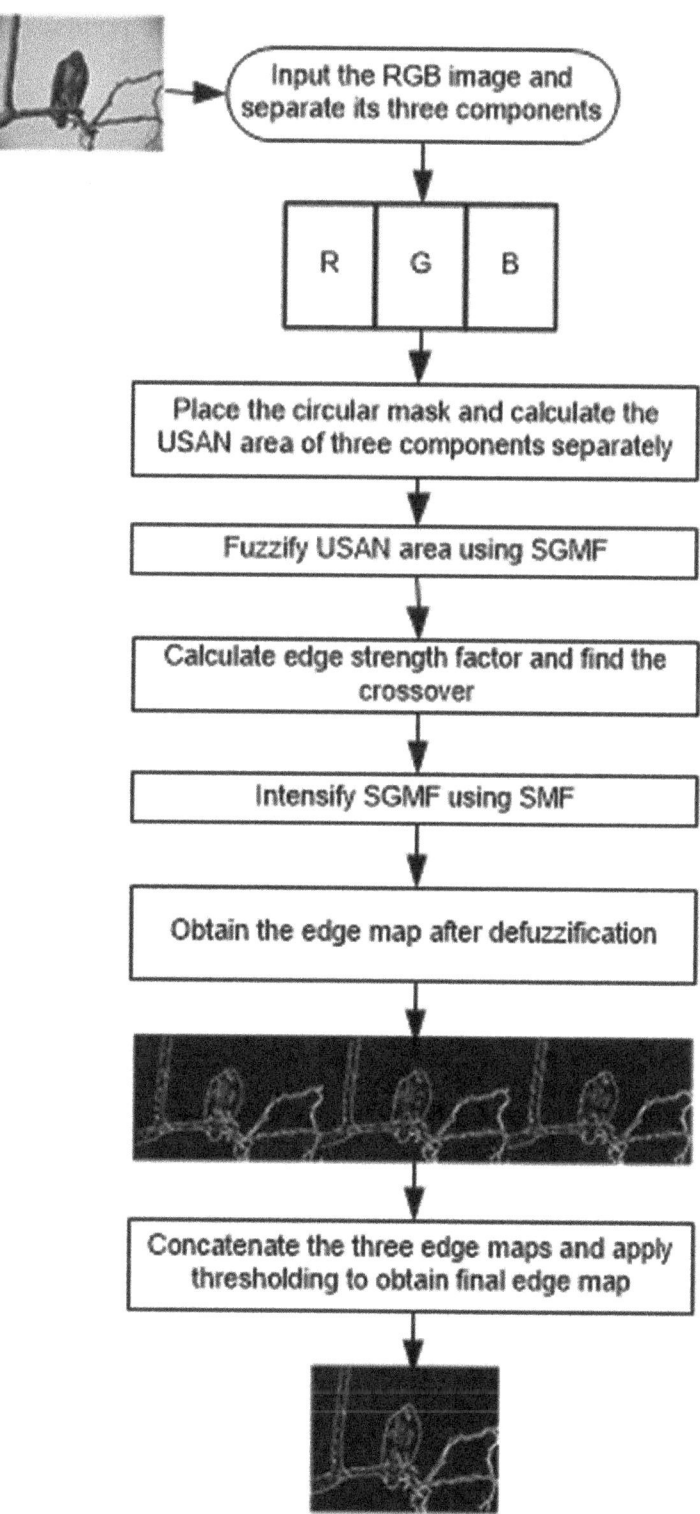

Fig. 5 Flowchart of the SUSAN-based algorithm[7]

EXPERIMENTAL RESULTS AND DISCUSSIONS

We have simulated the proposed information set-based edge detector on different types of natural and standard images. To evaluate its performance, we have used images from Berkeley segmentation database[24] comprising varying levels of complexities and is widely used by researchers. This database includes the ground truth images as a result of the combined judgment of five to ten human observers. Both quantitative and qualitative comparisons are performed on this database in terms of accuracy and robustness to noise.

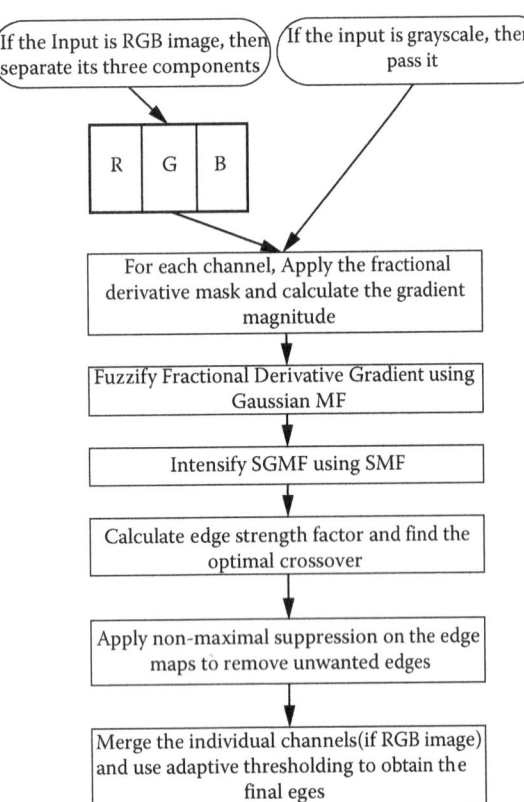

Fig. 6 Flowchart of the fractional derivative-based algorithm

Qualitative Comparison Using SUSAN

A comparison of six images is shown row-wise in Fig. 7—first row: the original image; second row: the ground truth; third row: the result of Canny edge detector; fourth row: the result of Sobel's; fifth row: the result of Prewitt's; sixth row: the result of Robert's; seventh row: the results of SUSAN; eighth row: the result of fuzzy (GMF); ninth row: the result of fuzzy (JADE); tenth row: the proposed results with the optimal value of crossover, μ_c; and the eleventh row: the proposed results with fixed $\mu_c = 0.8$. The parameters t and τ are kept constant as $t = 30$ and $\tau = 20$.

In the edge maps of Canny and Sobel edge detectors in Fig. 7e and f, there is a loss of connectivity. Prewitt and Robert edge detectors provide the similar edge maps in Fig. 7a. SUSAN edge detector produces thicker edges along with noise than those of other edge detectors. The two fuzzy-based edge detectors provide weak edges and less connectivity in some images as in Fig. 7d where we can see the missed background edges. It is observed that the most connected edges are obtained by the proposed (optimal) method. More parameters are required to be adjusted in the fuzzy-based algorithms where GMF and bell-shaped MF are considered, whereas the proposed edge detector needs only the crossover adjustment for the SMF. It produces continuous and fine edges even in the areas of the image with the least contrast.

Qualitative Comparison Using Fractional Derivatives

Figure 8 shows a comparison of the proposed and other edge detectors on six images. In this figure, the first set of images is the original color images. The second set corresponds to the grayscale images. The third set depicts the ground truth images. The fourth set shows the results of Sobel edge detection, whereas the fifth set shows the results of Canny edge detector. The sixth set depicts the mapping of Prewitt edge detector. The seventh set shows the result of Robert operator. Finally, the eighth set presents the result of the proposed optimal fractional derivative-based edge detection. Here, we consider the optimal value of crossover for better edge detection. In Canny and LOG edge maps, we can see the presence of random connectivity and detection of more random thick edges. In Prewitt, Roberts, and Sobel edge maps, there is a loss of connectivity as well as edge information. The proposed edge detector maintains a balance between the detection of edges and the connectivity as observed from the results.

Simulations on Noisy Images—SUSAN

The edge maps of all the edge detectors are also obtained for the images corrupted by 10% and 20% impulse noise density as shown in Figs. 9a–j and 10a–j, respectively. The proposed edge detector gives the best performance in terms of edge localization in the presence of noise and provides the maximum value of FoM among all the edge detectors compared.

Simulations on Noisy Images—Fractional Derivative

To test the performance on the noisy images, we have added the standard Gaussian noise to the images as shown in Figs. 11 and 12. As can be observed that the edges are least affected by the noise while the edge mapping and the highest *FoM* values are achieved by our edge detector among all the edge detectors compared.

Quantitative Comparison Using SUSAN

Figure 13 shows the histogram of *FoM* for 100 test images which lies in the range of 0–1 for all nine types of edge detectors presented. This histogram depicts the number of images (y-axis) that falls in the range given on the x-axis. The more the number of images close to 1, the better is the similarity of the edge detector to the ground truth. It can be noted from this graph that the maximum number of images yields the value of *FoM* close to 1 for the proposed method compared to others as the histogram is aligned toward the right having a peak at 0.7–0.8, whereas SUSAN edge detector shows the least similarity having a peak in the range of 0.2–0.3.

Information Sets for Edge Detection

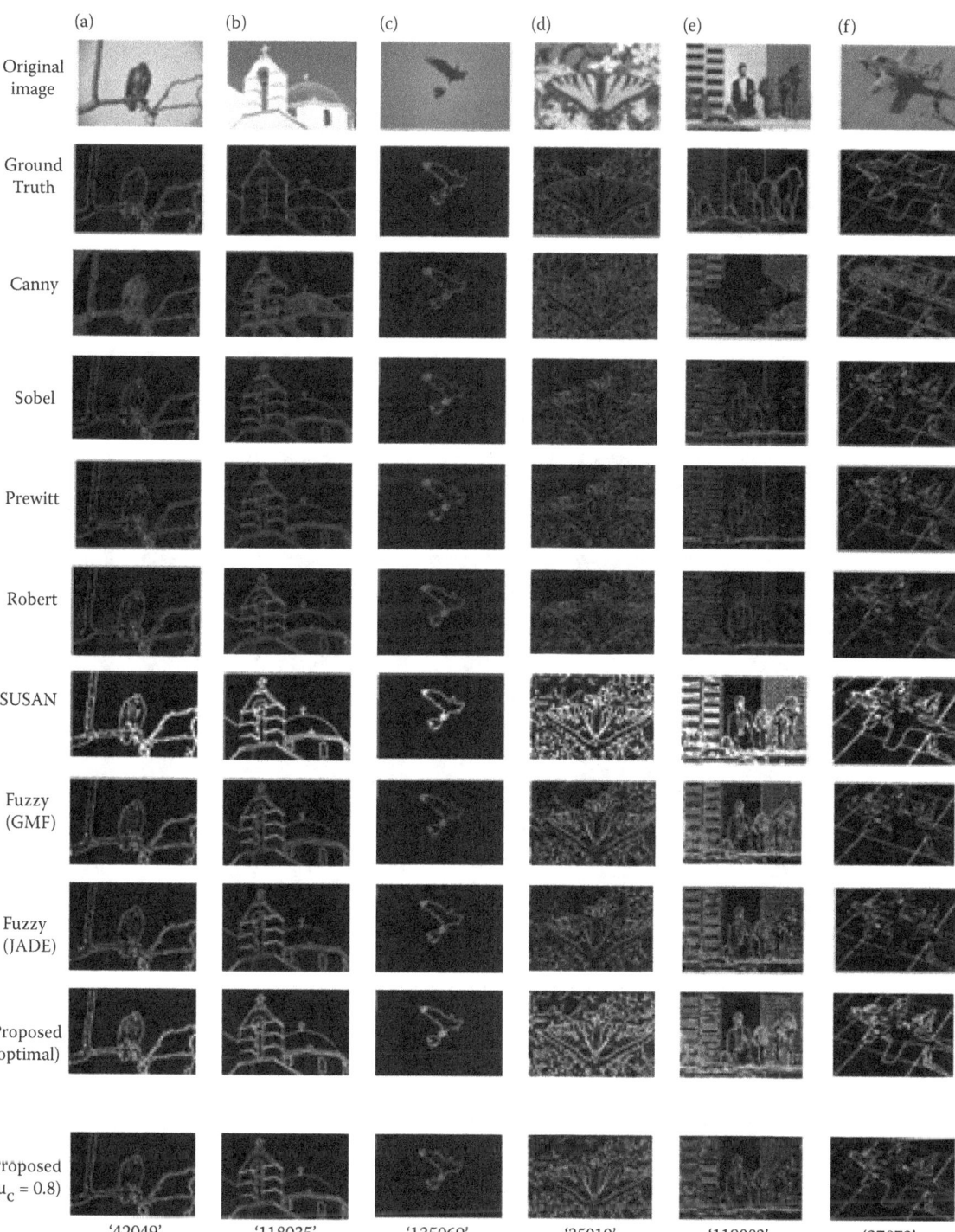

Fig. 7 Row-wise comparison of different approaches on images (a–f)[7]

Figure 14a and b show the graphs of the average value of *FoM* and SSIM on 100 images by different edge detectors. It is observed from Fig. 14a that the proposed edge detector provides sufficiently high value of *FoM* among all. Figure 14b also shows the highest SSIM index among all the edge detectors. These values indicate that the proposed edge detector is a good match with the ground truth images.

FoM Measure Comparison Using Fractional Derivative

Figure 15 depicts the average values of *FoM* that range from 0 to 1 as a bar chart on six edge detectors. The value of 1 indicates 100% similarity. The proposed edge detector has the highest *FoM* values and is a better match with the ground truth images than others.

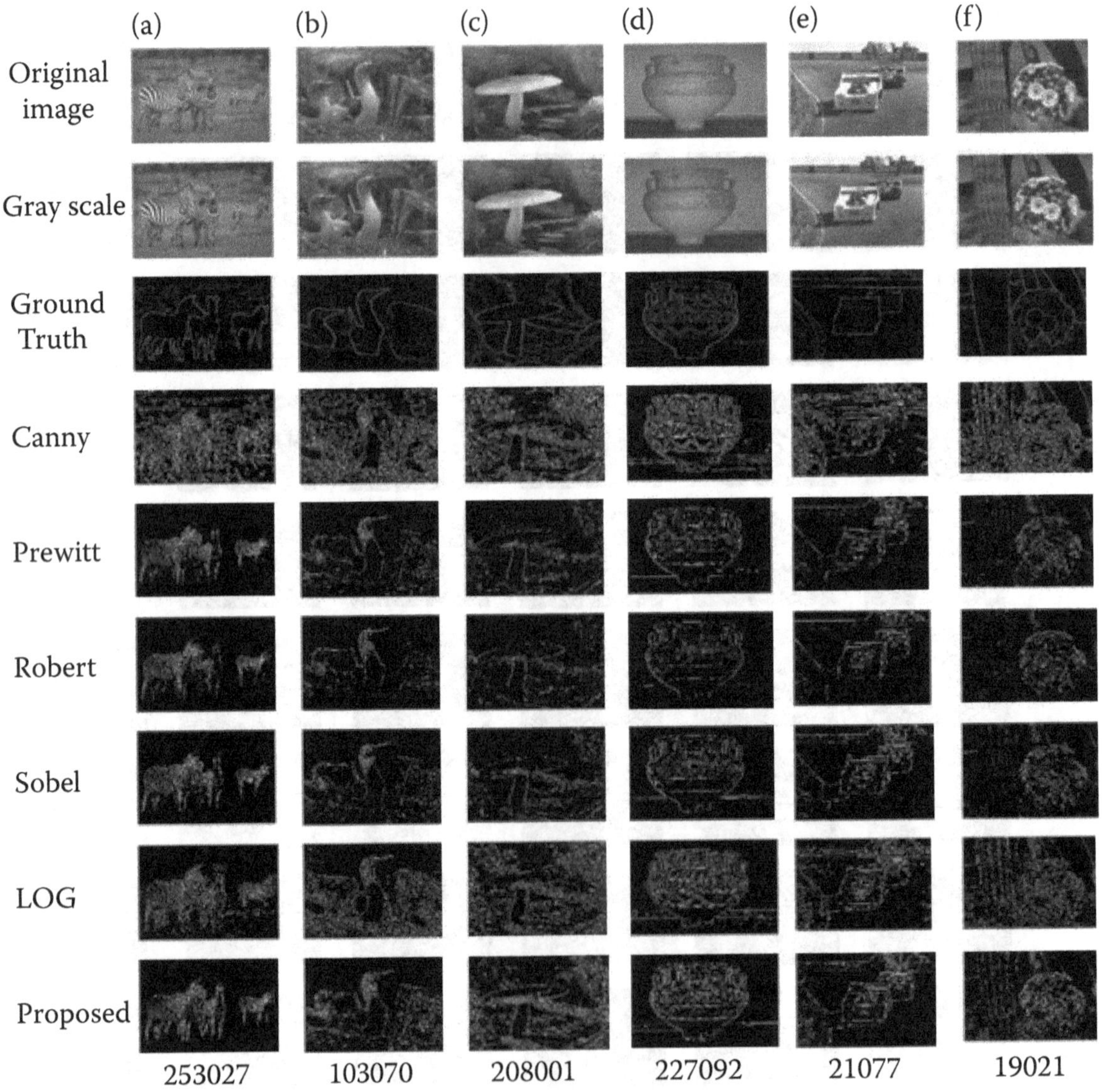

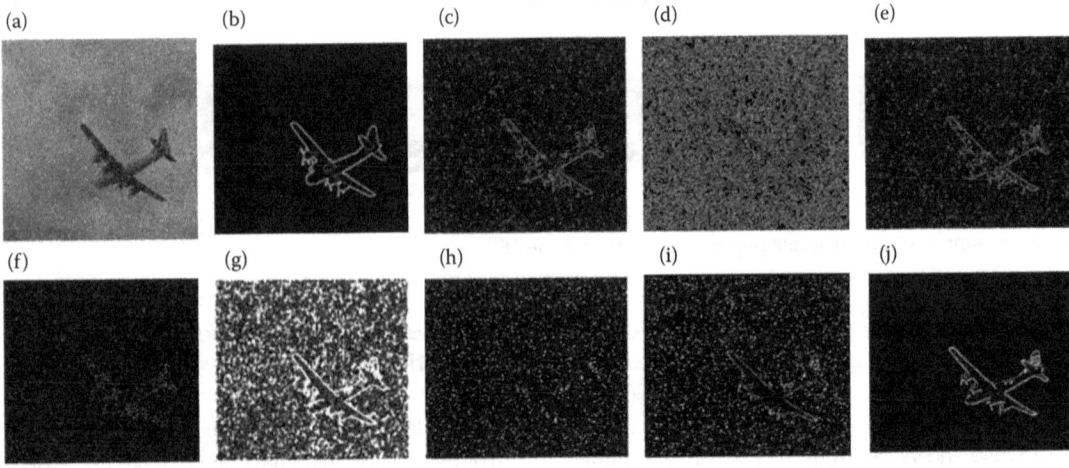

Fig. 8 Comparison of edge detectors on images (a) 253027, (b) 103070, (c) 208001, (d) 227092, (e) 21077, and (f) 19021[9]

Fig. 9 (a) Original image "3096" with noise density of 10%; (b) ground truth; (c) Sobel, *FoM* = 0.47; (d) Canny, *FoM* = 0.07; (e) Prewitt, *FoM* = 0.47; (f) Robert, *FoM* = 0.67; (g) SUSAN, *FoM* = 0.05; (h) fuzzy (GMF), *FoM* = 0.21; (i) fuzzy (JADE), *FoM* = 0.16; (j) proposed, *FoM* = 0.86[7]

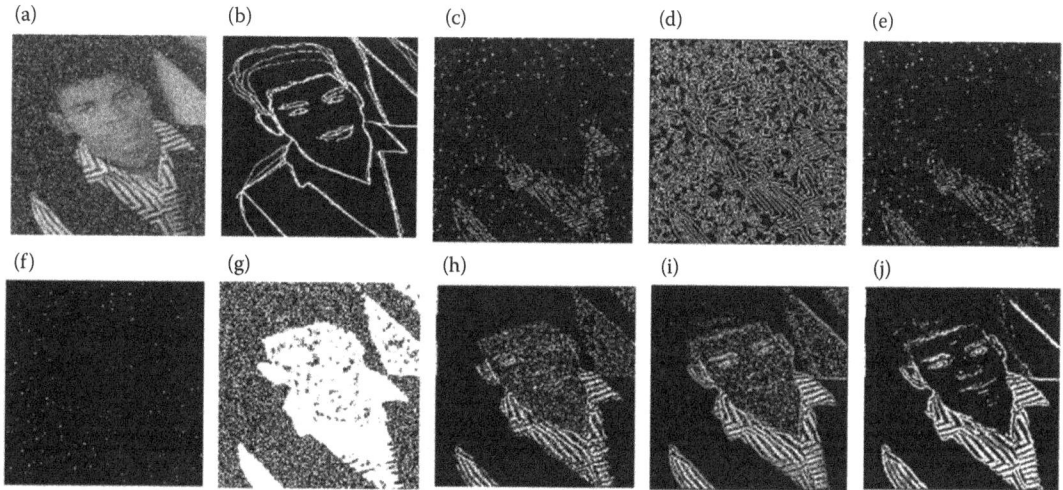

Fig. 10 (a) Original image "302008" with noise density of 20%; (b) ground truth; (c) Sobel, $FoM = 0.54$; (d) Canny, $FoM = 0.40$; (e) Prewitt, $FoM = 0.53$; (f) Robert, $FoM = 0.23$; (g) SUSAN, $FoM = 0.16$; (h) fuzzy (GMF), $FoM = 0.56$; (i) fuzzy (JADE), $FoM = 0.37$; (j) proposed, $FoM = 0.57$[7]

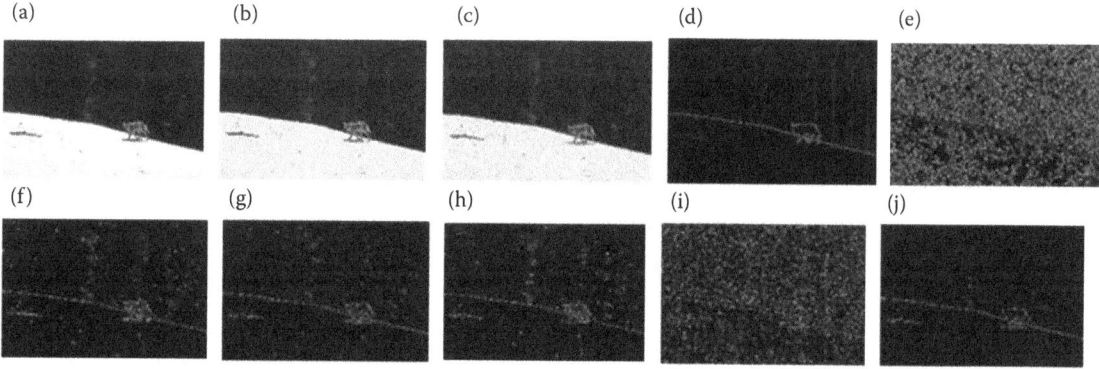

Fig. 11 Edge maps corrupted with standard Gaussian noise: (a) original image "167062"; (b) noisy color image; (c) noisy grayscale image; (d) ground truth; (e) Canny, $FoM = 0.03$; (f) Prewitt, $FoM = 0.49$; (g) Roberts, $FoM = 0.71$; (h) Sobel, $FoM = 0.48$; (i) LoG ($FoM = 0.05$) (j) proposed (optimal), $FoM = 0.88$

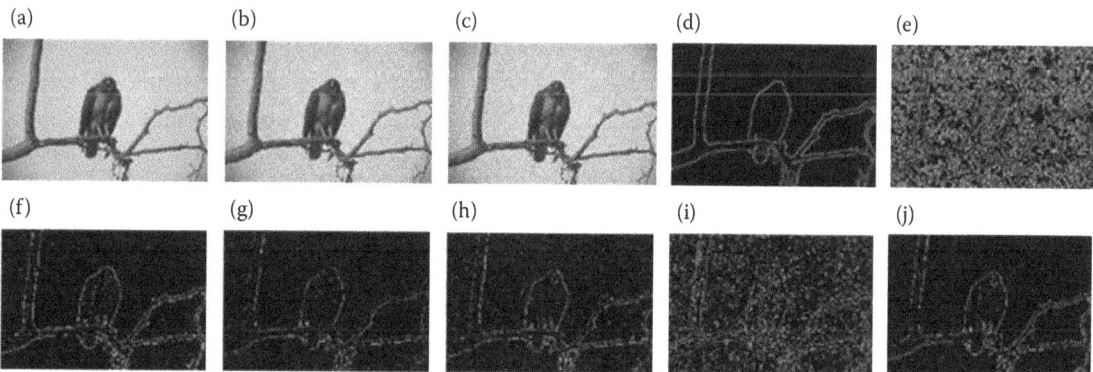

Fig. 12 Edge maps corrupted with standard Gaussian noise: (a) original image "42049"; (b) noisy color image; (c) noisy grayscale image; (d) ground truth; (e) Canny, $FoM = 0.15$; (f) Prewitt, $FoM = 0.82$; (g) Roberts, $FoM = 0.78$; (h) Sobel, $FoM = 0.80$; (i) LoG ($FoM = 0.2$) (j) Proposed (optimal), $FoM = 0.86$

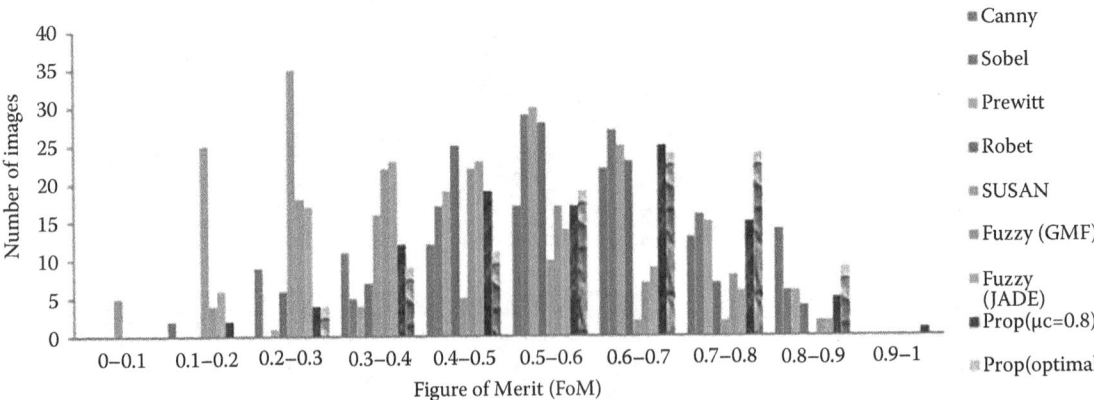

Fig. 13 Histograms of *FoM* of the proposed (optimal) method with Canny, Sobel, Prewitt, Robert, SUSAN, Fuzzy (GMF), Fuzzy (JADE), and proposed ($\mu_c = 0.8$)[7]

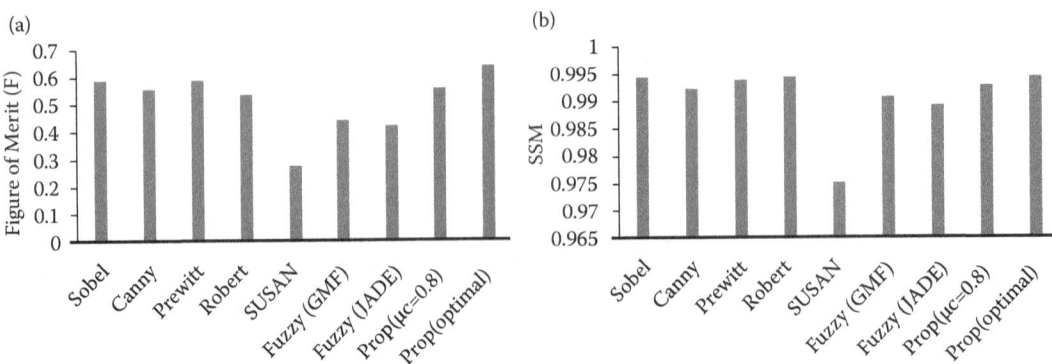

Fig. 14 (a) Average values of *FoM* and (b) SSIM of 100 images for different methods[7]

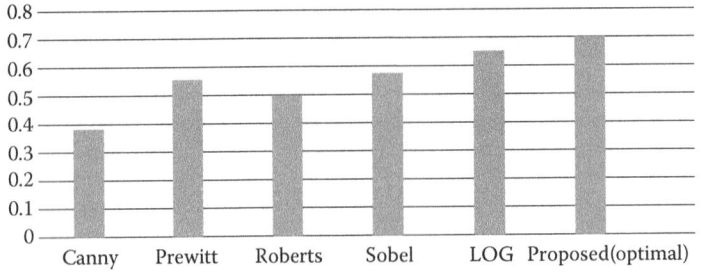

Fig. 15 Average values of *FoM*[9]

Effect of Parameters on Edge Strength Factor—SUSAN

Effect of Crossover on Edge Strength Factor (ES$_f$)

The quality of an edge map is greatly influenced by proper selection of crossover (μ_c) and threshold (*t*). To obtain the optimal edge map, edge strength factor and FoM are calculated for different images by varying μ_c and *t*. The effect of changing the crossover and threshold on the information set-based measure and edge strength factor is discussed in this section.

To examine the effect of change in the crossover (μ_c) on the edge strength factor, first the threshold (*t*) is kept constant say at 35. Figure 16 illustrates the ideal edge maps of the original image "42049" using different values of crossover, μ_c. Table 1 presents the values of ES$_f$ and *FoM* for different values of μ_c. Both ES$_f$ and *FoM* follow the same behavior, that is, first increasing then decreasing. It is apparent that the values of $\mu_c < 0.5$ yield low values of ES$_f$ as well as *FoM* and result in hazy, low contrast, and thick edge map. The value of ES$_f$ is found to be the maximum at one point after $\mu_c > 0.5$. The maximum value of ES$_f$ provides an edge map with the maximum number of edges. Increasing

Information Sets for Edge Detection

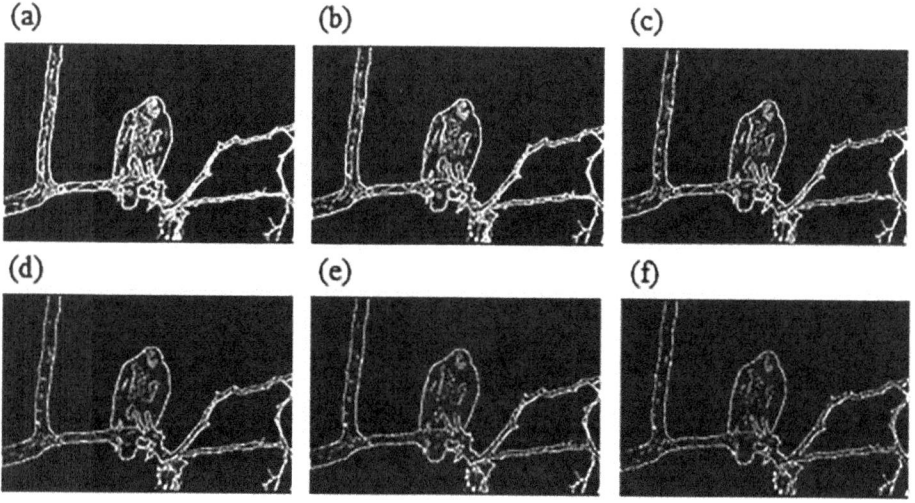

Fig. 16 Ideal edge maps of SMF *with* (a) $\mu_c = 0.4$, (b) $\mu_c = 0.5$, (c) $\mu_c = 0.6$, (d) $\mu_c = 0.7$, (e) $\mu_c = 0.8$, and (f) $\mu_c = 0.9$, at $t = 35$[7]

Table 1 Crossover μc versus ES_f and *FoM*

μ_c	t	ES_f	*FoM*
0.4	35	0.29	0.5
0.5	35	2.07	0.61
0.6	35	2.23	0.72
0.7	35	2	0.87
0.8	35	1.8	0.9
0.9	35	1.67	0.88

μ_c, 0.5 onwards, the edges become thinner and beyond 0.8, it leads to the loss of information and broken edges.

Effect of Threshold on Edge Strength Factor (ES_f)

To examine the effect of change in the threshold (t) on the edge strength factor, the crossover (μ_c) is kept constant say at 0.8. Figure 17 shows the edge maps of the given image for different values of t. It indicates that the lower the value of threshold, the more the number of edges in the edge map. For $t = 20$–45, the optimal crossover, μ_c, is found at the maximum value of ES_f as recorded in Table 2, which also shows that at a fixed value of μ_c, both ES_f and FoM present the same behavior. This is found true for every image we simulated.

Table 2 Threshold t versus ES_f and FoM

	Optimal μ_c			Fixed μ_c		
t	μ_c	ES_f	*FoM*	μ_c	ES_{00}	*FoM*
20	0.58	2.13	0.54	0.8	1.79	0.75
25	0.55	2.32	0.55	0.8	1.8	0.86
30	0.54	2.38	0.6	0.8	1.82	0.91
35	0.53	2.43	0.6	0.8	1.81	0.9
40	0.53	2.47	0.72	0.8	1.8	0.89
45	0.53	2.51	0.79	0.8	1.8	0.88

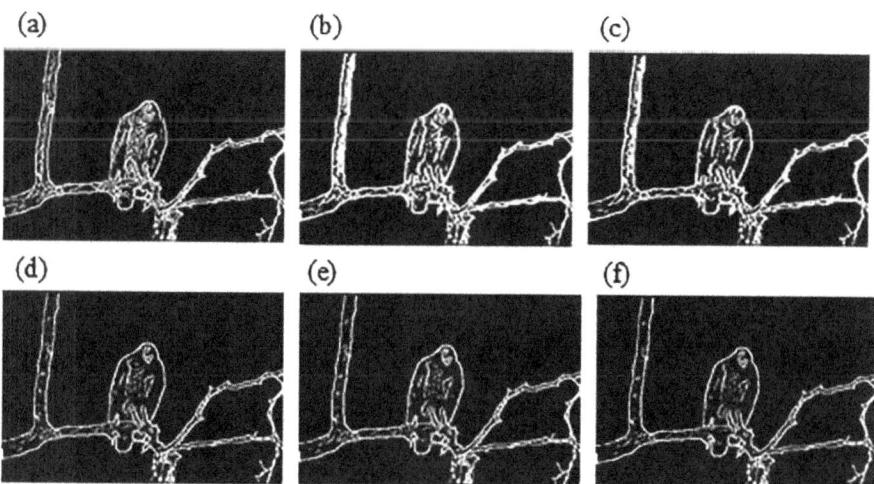

Fig. 17 Ideal edge maps of SMF with (a) $t = 20$, (b) $t = 25$, (c) $t = 30$, (d) $t = 35$, (e) $t = 40$, and (f) $t = 45$[7]

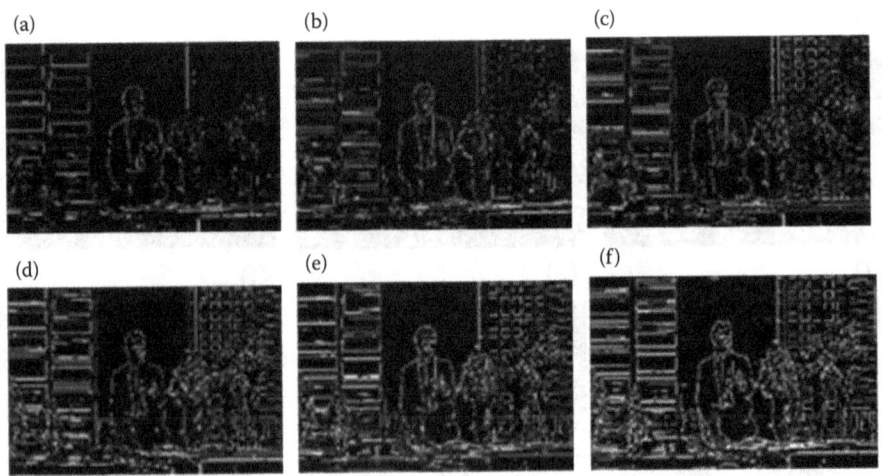

Fig. 18 Edge maps of SMF with (a) $\mu_c = 0.5$, (b) $\mu_c = 0.7$, (c) $\mu_c = 0.9$, (d) $\mu_c = 1$, (e) $\mu_c = 1.25$, and (f) $\mu_c = 1.5$

Table 3 Effect of crossover on *FoM*

Alpha (α)	Order (mask)	μ_c	FoM
1	3	0.5	0.5
1	3	0.7	0.66
1	3	0.9	0.71
1	3	1	0.75
1	3	1.25	0.67
1	3	1.5	0.65

Effect of parameters on *FoM*—Fractional Derivative

Effect of Change in Crossover Parameter on FoM

To examine the effects of change in crossover, we fix the value of α and the orders of masks as 1 and 3. We will see subsequently how they affect *FoM*. Figure 18 shows the edge maps for test image "119082" for the varying values of crossover.

Table 3 shows the average values of *FoM* at different values of crossover for all the test images. From the bar graph analysis, we find that the maximum average value of *FoM* is around 0.7. So, we look for the crossover point for which *FoM* is around 0.7.

Therefore, the crossover lies in the range of approximately 0.8–1.1 for maximum *FoM*.

Effect of Change in Alpha on FoM

Alpha (α) is the order of the fractional derivative. Generally, we take α as 1. For the sake of analysis, we have considered the test image "296059." Changing α affects the edges in the following ways as shown in Fig. 19. The quantitative results can be observed in Table 4.

On increasing the value of α, we increase the order of fractional derivative. Thus, there is a chance that it is affected by noise. Thus, we get more edge information at the cost of noise.

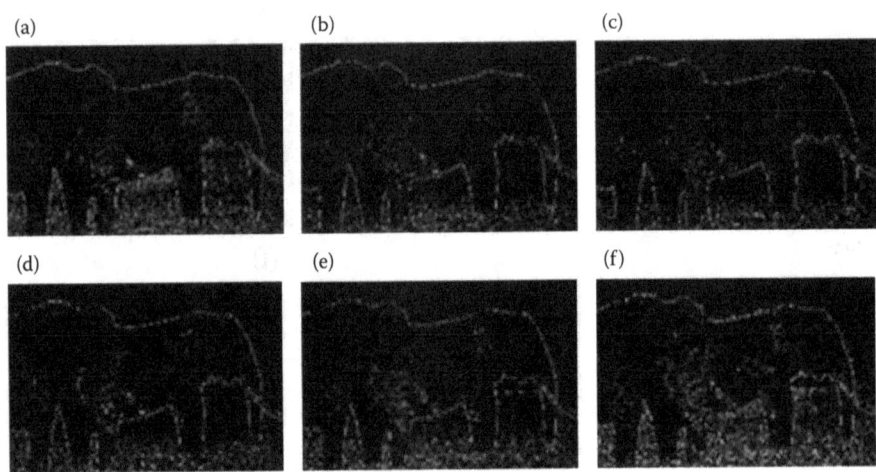

Fig. 19 Edge maps of SMF with (a) $\alpha = 0.8$, (b) $\alpha = 0.9$, (c) $\alpha = 1$, (d) $\alpha = 1.1$, (e) $\alpha = 1.2$, and (f) $\alpha = 1.5$

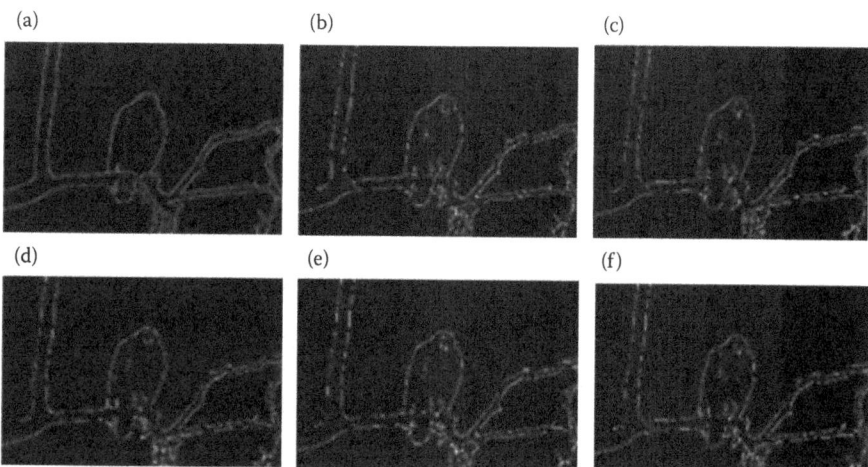

Fig. 20 Edge maps of SMF with (a) ground truth and (b) order(mask) = 3 (c) 4 (d) 5 (e) 6 and (f) 7

Table 4 Effect of alpha on *FoM*

Alpha (α)	Order (mask)	μ_c	*FoM*
0.8	3	1	0.54
0.9	3	1	0.66
1	3	1	0.75
1.1	3	1	0.73
1.2	3	1	0.64
1.5	3	1	0.45

Table 5 Effect of the size of mask on *FoM*

Alpha (α)	Order (mask)	μ_c	*FoM*
1	3	1	0.83
1	4	1	0.8
1	5	1	0.79
1	6	1	0.76
1	7	1	0.75

Effect of Change in the Size of Mask on FoM

To analyze the variations in *FoM* due to change in the size of mask as in Fig. 20, we choose a random test image "42029." The other two parameters are kept constant for maximum information. The quantitative variations can be observed in Table 5.

We observe that there is a change in *FoM* arising out of loss of information and retention of noise due to large masks.

CONCLUSIONS

The proposed edge detectors are developed using the information set-based theory, and they make use of the edge strength measure to extract the edge information. The edge strength factor is devised to determine the optimal value of the crossover for the detector. The modified SUSAN edge detector gives better results with this single unknown parameter, whereas other methods are dependent on several parameters. Experimental results show that the modified SUSAN edge detector is more efficient in locating the sharp edges than the other traditional and fuzzy-based detectors. It provides robust edges for the complex and noisy images. These advantages make this edge detector more suitable to biometrics, biomedical applications, features identification, contour detection, surveillance, etc. However, in the proposed fractional derivative edge detector, we use the fractional derivative-based Sobel mask where each fractional derivative uses infinite series for gradient calculation. Nevertheless, we restrict the length of series according to the size of the mask used. We adjust the parameters of the mask to get the maximum edge information by means of the information set theory. The experimental results show that the proposed edge detector outperforms the traditional edge detectors. The drawback of this edge detector is its moderate performance on highly noisy images. Our future work will be to make the detector robust to the noise.

REFERENCES

1. Marr, D.; Hildreth, E. Theory of edge detection. Proc. R Soc. B Biol. Sci. **1980**, *207* (1167), 187–217. Available at http://rspb.royalsocietypublishing.org/content/207/1167/187.
2. Chaira, T.; Ray, A.K. *Fuzzy Image Processing and Applications with Matlab*; CRC Press: Boca Raton, FL and New York, 2009, 237.
3. Hanmandlu, M. Information sets and information processing. Def. Sci. J. **2011**, *61* (5), 405–407.
4. Mamta; Hanmandlu, M. Robust authentication using the unconstrained infrared face images. Expert Syst. Appl. **2014**, *41* (14), 6494–6511. doi:10.1016/j.eswa.2014.03.040.
5. Arora, P.; Hanmandlu, M.; Srivastava, S. Gait based authentication using gait information image features.

Pattern Recognit. Lett. **2015**, *68*, 336–342. doi:10.1016/j.patrec.2015.05.016.

6. Hanmandlu, M.; Arora, S.; Gupta, G.; Singh, L. Underexposed and overexposed colour image enhancement using information set theory. Imaging Sci. J. **2016**, *64* (6), 321–333. doi:10.1080/13682199.2016.1215063.

7. Arora, S.; Hanmandlu, M.; Gupta, G. Edge detection of digital color images using information sets. J. Electron Imaging **2016**, *25* (6), 1–10. doi:10.1117/1.JEI.25.6.061607.

8. Smith, S.M.; Brady, J.M. SUSAN—A new approach to low level image processing. Int. J. Comput. Vis. **1997**, *23* (1), 45–78. doi:10.1023/A:1007963824710.

9. Kothapalli, V.; Arora, S.; Hanmandlu, M. Edge detection using fractional derivatives and information sets. J. Electron. Imaging **2018**, *27* (5), 051226.

10. Loverro, A. *Fractional calculus: History, definitions and applications for the engineer.* Report, 2004. http://nd.edu/~msen/Teaching/UnderRes/FracCalc.pdf

11. Oldham, K.B.; Spanier, J. *Theory and Applications of Differentiation and Integration of Arbitrary Order. The Fractional Calculus*; Dover Publications Inc.: New York, 2006.

12. Gonzalez, R.C.; Woods, R.E. *Digital Image Processing*; 3rd Ed.; Pearson Education India: New Delhi 2009, 954.

13. Melin, P.; Mendoza, O.; Castillo, O. An improved method for edge detection based on interval type-2 fuzzy logic. Expert Syst. Appl. **2010**, *37* (12), 8527–8535. doi:10.1016/j.eswa.2010.05.023.

14. Maiseli, B.J.; Gao, H. Robust edge detector based on anisotropic diffusion-driven process. Inf. Process Lett. **2016**, *116* (5), 373–378. doi:10.1016/j.ipl.2015.12.003.

15. Yao, Y.; Huang, D.; Yu, X.; Chai, B. Edge interpretation of potential field data with the normalized enhanced analytic signal. Acta Geod. Geophys. **2016**, *51* (1), 125–136. doi:10.1007/s40328-015-0120-x.

16. Gao, W.; Zhang, X.; Yang, L.; Liu, H. An improved Sobel edge detection. In *2010 3rd International Conference on Computer Science and Information Technology*; IEEE, Beijing, China, 2010, 67–71

17. Shannon, C.E. Mathematical theory of communication. Bell Syst. Tech. J. **1948**, *27* (3), 379–423.

18. Pal, N.R.; Pal, S.K. Entropy: A new definition and its applications. IEEE Trans. Syst. Man Cybern. **1991**, *21* (5), 1260–1270. Available at http://ieeexplore.ieee.org/lpdocs/epic03/wrapper.htm?arnumber=120079.

19. Hanmandlu, M.; Verma, O.P.; Gangwar, P.; Vasikarla, S. Fuzzy edge and corner detector for color images. In *Sixth International Conference on Information Technology: New Generations*; IEEE: Las Vegas, NV, 2009, 1301–1306. Available at http://ieeexplore.ieee.org/lpdocs/epic03/wrapper.htm?arnumber=5070806.

20. Hanmandlu, M.; Kalra, R.R.; Madasu, V.K.; Vasikarla, S. Area based novel approach for fuzzy edge detection. In *TENCON IEEE Region 10 Conference*; IEEE: Singapore, 2006, 1–4. Available at http://ieeexplore.ieee.org/lpdocs/epic03/wrapper.htm?arnumber=4142435.

21. Abdou, I.E.; Pratt, W.K. Quantitative design and evaluation of enhancement/thresholding edge detectors. Proc. IEEE **1979**, *67* (5), 753–763. Available at http://ieeexplore.ieee.org/lpdocs/epic03/wrapper.htm?arnumber=1455594.

22. Wang, Z.; Bovik, A.C.; Sheikh, H.R.; Simoncelli, E.P. Image quality assessment: From error visibility to structural similarity. IEEE Trans. Image Process. **2004**, *13* (4), 600–612. Available at http://ieeexplore.ieee.org/lpdocs/epic03/wrapper.htm?arnumber=1284395.

23. Otsu, N. A threshold selection method from gray-level histograms. IEEE Trans. Syst. Man Cybern. **1979**, *9* (1), 62–66. Available at http://ieeexplore.ieee.org/lpdocs/epic03/wrapper.htm?arnumber=4310076.

24. The Berkeley Segmentation Dataset and Benchmark. Available at www.eecs.berkeley.edu/Research/Projects/CS/vision/bsds/

JPEG: Symmetric Exponential Quantization

Yuri A. Gadzhiev
Dagestan State Technical University, Makhachkala, Russia

Abstract
A systematic way to construct JPEG quantization tables using the assumption of exponential dependence of discrete cosine transform coefficients on their polynomial positions is considered. The experiment shows that this approach has some advantages over quantization used in the standard JPEG package by both image visual quality and achieved compression ratio.

Keywords: DCT; JPEG; Quantization.

INTRODUCTION

It is evident that the quality of JPEG-like (lossy) image compression depends very highly on used quantization. This was apparently one of the main reasons to define separately the quantization table of the processing algorithm as it is in JPEG. Anyway, the JPEG implements three predetermined quantization tables by an one for each of three YCbCr color components. In due time, these tables had been announced as a result of numerous experiments by a visual comparison of obtained with them images and achieved at that compression and now are used in JPEG by default. Despite this, everyone can be sure that the results leave much to be desired. First, the tables are nonsymmetric that causes various images to depend on their spatial orientation. We think it is obvious that an image should not change under its quarter-turn. So, any used quantization table is necessarily to be a symmetrical one. In addition, a systematic approach should be used as the basis for the construction of the quantization table, not merely manual fitting the quantizers' numeric values with a visual estimation of the results. One such approach, which we use further, can be connected with a look at the quantization process simply as at a way to set apart of the others the accuracy of each coefficient at the terms of a given source data functional decomposition.

QUANTIZED TRANSFORM CONCEPT

Suppose we have a source data vector $x = (x_1, x_2, \ldots, x_N)$. We always may want to consider its components $x_1, x_2, \ldots, x_N$ as the values of some function $x(t)$ at some points $t_1, t_2, \ldots, t_N$ of some argument t, so that $x_1 = x(t_1), x_2 = x(t_2), \ldots, x_N = x(t_N)$. If so, then we can always approximate this function $x(t)$ by means of N some others, known to us linearly independent functions $f_1(t), f_2(t), \ldots, f_N(t)$, so that the approximation equation

$$x(t) = c_1 f_1(t) + c_2 f_2(t) + \cdots + c_N f_N(t)$$

becomes an equality at the points $t_1, t_2, \ldots, t_N$. Now a new N-component vector $c = (c_1, c_2, \ldots, c_N)$, obtained by means of such an approximation, completely replaces $x = (x_1, x_2, \ldots, x_N)$, because we always can go from the former to the latter by a calculation of the form

$$x_i = c_1 f_1(t_i) + c_2 f_2(t_i) + \cdots + c_N f_N(t_i)$$

for $i = 1, \ldots, N$. Such a conversion from x to c, determined by the functions $f_1(t), f_2(t), \ldots, f_N(t)$, is usually called discrete f-transform of x, where f is a family name of the approximate functions $f_1(t), f_2(t), \ldots, f_N(t)$.

Each ith component c_i of c shows "the degree of presence" of ith function $f_i(t)$ in the source function $x(t)$, so that the greater the absolute value of c_i, the greater a contribution of the function $f_i(t)$ into x. Without going into intricacies of detailed analysis, we may note that when we fix the precision of each c_i on one fixed level, then we fix on the same level (more precisely, on the same level multiplied by the value $d_i = \sqrt{\Sigma_j f_i(t_j)^2}$ which is equal to unity for any orthonormal transform) the precision of each x_i obtained as the result of calculation by known c. However, the system of functions $\{f_i(t)\}$ is chosen usually so that as many of the coefficients c_i as possible were minimized by their absolute values. If so, then the precision of smaller coefficients can be chosen less than the precision of the greater ones. Slightly simplifying, assume, for instance, that some two of c, to say for definiteness c_i and c_j, are such that $c_i \approx (1/10) c_j$. Then, if the precision of c_j is $10^{-p} c_j$ and the precision of c_i is

$10^{-(p-1)}c_i$, then the precision of $(c_i + c_j)$ is still approximately

$$\frac{1}{2}\left(10^{-p}c_j + 10^{-(p-1)}c_i\right) \approx \frac{1}{2}\left(10^{-p}c_j + 10^{-(p-1)}\frac{1}{10}c_j\right) = 10^{-p}c_j.$$

In terms of the quantization, it means that we can use for the coefficient c_i a quantizer (divisor) q_i ten times as large as the quantizer q_j without considerable reduction of summary precision. It shows that main idea of the quantization, that is to use larger quantizers for smaller coefficients and vice versa. The question is how larger must be these quantizers.

EXPONENTIAL QUANTIZATION

First, we should note that many interesting works on the optimization of the quantization process are already done: using a general approach, like represented in;[1–9] for medical imaging, such as,[10–12] including standardization of quantization for ultrasonic images;[11] using some image specifics, like,[13–15] or specific algorithms like.[16] However, at the base of our method lies only an assumption, justified by research works of Krupinski and Purczynski[17] and Lam and Goodman,[18] on the exponential character of the distribution of discrete cosine transform coefficients.

Suppose the coefficients mentioned in Section "Quantized Transform Concept" are ordered by their expected absolute values so that $i < j$, where i, j are the indices of some two coefficients, implies $|\tilde{c}_i| \leq |\tilde{c}_j|$, where tilde marks an expectation in the sense of expected value of a coefficient considered as a random variable. We shall call quantization *exponential* one when the quantizers used under this quantization are defined by the equality

$$q_i = p^{i-1}, \quad (1)$$

where p is a constant numerical parameter greater than 1.

Now we will try to construct an exponential quantization for JPEG. Used in JPEG functions $f_n(t), n = 1,\ldots,N$ are the functions of the form $f_n(t) = k_n \cdot \cos[(n-1)t], n = 1,\ldots,N$, where factors k_n are chosen such that if it is possible that all the system of functions $f_n(t), n = 1,\ldots,N$ becomes orthonormal at used points $t_1,\ldots,t_N$. It is out of scope to discuss why exactly cosine functions were chosen for JPEG. However, we have to remark that the frequency functions generally are the best for the data we can perceive, including visual images and sound. Possibly, it is due to the fact that such data have a wave nature, possibly not. Cosine functions have some advantages over other trigonometric ones, such as the absence of unwanted Gibbs effect and closeness of the corresponding transform matrix to the optimal one by a feature of the source data decorrelation. So, with the above specified functions, we have a representation of the form

$$\begin{aligned}x(t) = &c_1 k_1 + c_2 k_2 \cos(t) + c_3 k_3 \cos(2t) \\ &+ \cdots + c_N k_N \cos[(N-1)t].\end{aligned} \quad (2)$$

JPEG utilizes the above specified functions $f_n(t), n = 1,\ldots,N$ at the points $t_m = \frac{\pi}{N}\left(m - \frac{1}{2}\right), m = 1,\ldots,N$ of interval $[0, \partial]$ where transform matrix T is defined by the equality

$$T = (CD)^{-1},$$

where C is cosine matrix with components

$$C_{m,n} = \cos[(n-1)t_m], (m = 1,\ldots,N; n = 1,\ldots,N),$$

and D is a diagonal matrix such that $D_{1,1} = 1/\sqrt{N}$ and $D_{n,n} = \sqrt{2/N}$ for all other $n > 1$.

Therefore, if we want to use some quantization q for coefficients $c_n k_n, n = 1,\ldots,N$, from (2), then for coefficients $c_n, n = 1,\ldots,N$, from (2), we should use an equivalent quantization of the form

$$q' = D^{-1}q. \quad (3)$$

For $N = 8$ with $p = 1.27$, using (3) and (1), we obtain

$$q'^T = [2.828 \; 2.54 \; 3.226 \; 4.097 \; 5.203 \; 6.608 \; 8.392 \; 10.658]$$

and quantization table Q' of rounded off components of quantization table $Q := q' \cdot q'^T$ obtains the form

$$Q' = \begin{bmatrix} 8 & 7 & 9 & 12 & 15 & 19 & 24 & 30 \\ 7 & 6 & 8 & 10 & 13 & 17 & 21 & 27 \\ 9 & 8 & 10 & 13 & 17 & 21 & 27 & 34 \\ 12 & 10 & 13 & 17 & 21 & 27 & 34 & 44 \\ 15 & 13 & 17 & 21 & 27 & 34 & 44 & 55 \\ 19 & 17 & 21 & 27 & 34 & 44 & 55 & 70 \\ 24 & 21 & 27 & 34 & 44 & 55 & 70 & 89 \\ 30 & 27 & 34 & 44 & 55 & 70 & 89 & 114 \end{bmatrix} \quad (4)$$

Let us note in conclusion that the value of the parameter p in (1) determines both the compressed image quality and achieved compression level. The value 1.27, which was used above, has been taken rather arbitrarily, only to obtain the results as possibly close to usually obtained by JPEG with its standard quantization table* (for details, see Section "Comparison of Exponential and Standard* JPEG Quantizations").

COMPARISON OF EXPONENTIAL AND STANDARD* JPEG QUANTIZATIONS

Here, we are using (4) as a luminance quantization table with the program cjpeg.exe, the standard JPEG compressor of JPEG designers, which allows defining the quantization table in a separate text file. The results of processing test images both with (4) and usual JPEG quantization table*

JPEG: Symmetric Exponential Quantization

Table 1 Results for a comparison between the standard JPEG* and exponential (4) quantizations

File name	File size (byte) with normal JPEG quantization	File size (byte) with exponential quantization (4)	PSNR** (dB) of Y-component with normal JPEG quantization	PSNR (dB) of Y-component with exponential quantization (4)
4.1.01.jpg	9 947	9 448	37.81	38.00
4.1.02.jpg	9 930	9 488	38.38	38.49
4.1.03.jpg	6 002	5 724	40.38	40.58
4.1.04.jpg	10 342	9 806	39.19	39.11
4.1.05.jpg	10 198	9 888	38.17	38.36
4.1.06.jpg	16 038	16 005	33.52	34.13
4.1.07.jpg	6 333	6 145	42.31	42.41
4.1.08.jpg	8 507	8 259	40.69	40.89
4.2.01.jpg	31 972	29 270	40.06	39.74
4.2.02.jpg	37 346	35 538	37.36	37.53
4.2.03.jpg	77 244	77 171	31.34	32.01
4.2.04.jpg (Lena)	37 788	35 742	37.83	37.91
4.2.05.jpg	38 679	36 708	38.48	38.63
4.2.06.jpg	52 482	51 599	34.28	34.58
4.2.07.jpg	41 308	39 059	36.29	36.30
5.1.09.jpg	11 651	11 654	34.03	34.29
5.1.10.jpg	18 297	18 457	32.61	33.41
5.1.11.jpg	6 709	6 100	40.50	40.46
5.1.12.jpg	8 672	8 107	37.60	37.98
5.1.13.jpg	10 880	10 675	37.50	38.05
5.1.14.jpg	14 693	14 659	34.44	35.08
5.2.08.jpg	42 027	40 053	36.86	36.97
5.2.09.jpg	58 828	58 285	34.12	34.76
5.2.10.jpg	64 494	64 132	32.19	32.82
5.3.01.jpg	162 440	155 795	36.10	36.22
5.3.02.jpg	201 569	202 936	33.14	33.62
7.1.01.jpg	41 220	40 288	36.23	36.30
7.1.02.jpg	22 108	19 850	38.99	39.15
7.1.03.jpg	41 708	41 245	35.30	35.43
7.1.04.jpg	39 737	38 283	36.91	36.78
7.1.05.jpg	52 567	53 086	33.37	33.79
7.1.06.jpg	52 458	53 114	33.49	33.91
7.1.07.jpg	50 398	50 705	33.83	34.16
7.1.08.jpg	34 311	32 985	36.31	36.31
7.1.09.jpg	47 590	47 767	34.58	34.90
7.1.10.jpg	40 528	39 287	36.75	36.64
7.2.01.jpg	123 375	117 919	35.76	35.80
boat.512.jpg	43 417	41 501	35.66	35.78
elaine.512.jpg	38 215	37 632	34.18	34.25
Gray21.512.jpg	8 112	7 650	52.89	52.24
house.jpg	48 009	46 448	36.80	37.17

(Continued)

Table 1 Results for a comparison between the standard JPEG* and exponential (4) quantizations (*Continued*)

File name	File size (byte) with normal JPEG quantization	File size (byte) with exponential quantization (4)	PSNR** (dB) of Y-component with normal JPEG quantization	PSNR (dB) of Y-component with exponential quantization (4)
numbers.512.jpg	67 965	66 296	32.70	32.96
ruler.512.jpg	68 652	66 361	37.38	37.73
testpat.1k.jpg	84 328	82 418	42.51	<u>42.06</u>
Overall	1 899 074	1 853 538		

*Under the standard quantization table, there is meant a table of the form

8	6	5	8	12	20	26	31
6	6	7	10	13	29	30	28
7	7	8	12	20	29	35	28
7	9	11	15	26	44	40	31
9	11	19	28	34	55	52	39
12	18	28	32	41	52	57	46
25	32	39	44	52	61	60	51
36	46	48	49	56	50	52	50

which is used in cjpeg.exe for the luminance component on default.
**PSNR is the "peak signal-to-noise ratio" calculated as usual.

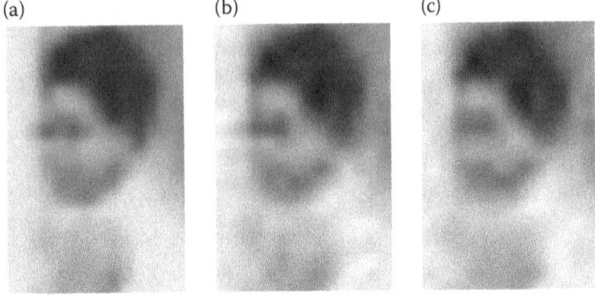

Fig. 1 Three fivefold magnified small fragments of the image 5.1.12: (a) a fragment of the original image; (b) JPEG with symmetric exponential quantization (4); (c) JPEG with the standard JPEG quantization table

Fig. 2 Three threefold magnified fragments of the image 4.1.03: (a) a fragment of the original image; (b) JPEG with symmetric exponential quantization (4); (c) JPEG with the standard JPEG quantization table

are shown in Table 1. The test images used were from the Miscellaneous Test Images Database of the Signal and Image Processing Institute at the University of Southern California, available at the URL http://sipi.usc.edu/database/misc.zip.

We can see from Table 1 that the exponential quantization used in the experiments gains overall 45,536 byte in the compression, and, at the same time, in the 37 cases out of all 44 ones by the obtained image's quality. The results when used exponential quantization yet concedes to the standard JPEG quantization are labeled in Table 1 by underlying.

However, there are cases which undoubtedly show the advantages of used exponential quantization over the standard nonsymmetric one by the image visual quality. So, it is seen in Fig. 1 that the latter leads to essential visual image distortions (Fig. 1c), which do not take place in the image (Fig. 1b) obtained from the JPEG image quantized with the table (4). Figure 2 also shows that the exponential quantization (4) gives rise to some less noticeable artifacts (Fig. 2b) than the standard JPEG quantization (Fig. 2c).

CONCLUSION

We have considered an approach for constructing a JPEG quantization table. The method is universal and allows making a table with a level of quantization given by the value of p in (1). The advantages of the method, in comparison with others, are as follows: its relative simplicity, possibility to achieve higher compression with lower visual image distortion and, finally, its universality in relation to a kind of used transform and a size of the transform block. When using along with JPEG this method does not require to modify the JPEG software since JPEG packages allow defining the used quantization tables distinctly. Performed experiments show that the method allows obtaining better results than the generally used standard JPEG quantization by both achieved image compression and quality.

REFERENCES

1. Battiato, S.; Mancuso, M.; Bosco, A.; Guarnera, M. Psychovisual and statistical optimization of quantization tables for DCT compression engines. In *Proceedings of the 11th International Conference on Image Analysis and Processing*; IEEE, 2001.
2. Fong, W.C.; Chan, S.C.; Ho, K.L. Designing JPEG quantization matrix using rate-distortion approach and human visual system model. In *Proceedings of the 1997 IEEE International Conference on Communications, ICC'97*, Vol. 3; Montreal: Towards the Knowledge Millennium; IEEE, 1997.
3. Jiang, Y.; Pattichis, M.S. JPEG image compression using quantization table optimization based on perceptual image quality assessment. In *Conference Record of the Forty Fifth Asilomar Conference on Signals, Systems and Computers (ASILOMAR)*; IEEE, 2011.
4. Minguillon, J.; Pujol, J. Uniform quantization error for Laplacian sources with applications to JPEG standard. In *SPIE's International Symposium on Optical Science, Engineering, and Instrumentation*; International Society for Optics and Photonics, 1998.
5. Sherlock, B.G.; Nagpal, A.; Monro, D.M. A model for JPEG quantization. In *Proceedings of the 1994 International Symposium on Speech, Image Processing and Neural Networks. ISSIPNN'94*; IEEE, 1994.
6. Shohdohji, T.; Hoshino, Y.; Kutsuwada, N. Optimization of quantization table based on visual characteristics in DCT image coding. Comput. Math. Appl. **1999**, *37* (11), 225–232.
7. Wang, C.Y.; Lee, S.M.; Chang, L.W. Designing JPEG quantization tables based on human visual system. Signal Process. Image Commun. **2001**, *16* (5), 501–506.
8. Watson, A.B. DCT quantization matrices visually optimized for individual images. In *IS&T/SPIE's Symposium on Electronic Imaging: Science and Technology*; International Society for Optics and Photonics, 1993.
9. Watson, A.B. Perceptual optimization of DCT color quantization matrices. In *Proceedings of the 1994 IEEE International Conference on Image Processing. ICIP-94*, Vol. 1; IEEE, 1994.
10. Batista, L.V.; Melcher, E.U.K.; Carvalho, L.C. Compression of ECG signals by optimized quantization of discrete cosine transform coefficients. Med. Eng. Phys. **2001**, *23*, 127–134.
11. Hamamoto, K. Standardization of JPEG quantization table for medical ultrasonic echo images. In *Proceedings of the 6th IEEE International Conference on Electronics, Circuits and Systems, ICECS'99*, Vol. 2; IEEE, 1999.
12. Karson, T.H.; Chandra, S.; Morehead, A.J.; Stewart, W.J.; Nissen, S.E.; Thomas, J.D. JPEG compression of digital echocardiographic images: Impact on image quality. J. Am. Soc. Echocardiogr. **1995**, *8* (3), 306–318.
13. Jeong, G.M.; Kang, J.H.; Mun, Y.S.; Jung, D.H. JPEG quantization table design for photos with face in wireless handset. In *Advances in Multimedia Information Processing-PCM 2004*; Springer, 2005, 681–688.
14. Watson, A.B. DCTune: A technique for visual optimization of DCT quantization matrices for individual images. In *Sid International Symposium Digest of Technical Papers*, Vol. 24; Society For Information Display, 1993.
15. Wu, S.W.; Gersho, A. Rate-constrained picture-adaptive quantization for JPEG baseline coders. In *Proceedings of the 1993 IEEE International Conference on Acoustics, Speech, and Signal Processing. ICASSP-93*, Vol. 5; IEEE, 1993.
16. Costa, L.F.; Veiga, A.C.P. Identification of the best quantization table using genetic algorithms. In *Proceedings of the IEEE Pacific Rim Conference on Communications, Computers and Signal Processing, PACRIM 2005*; IEEE, 2005.
17. Krupinski, R.; Purczynski, J. Modeling the distribution of DCT coefficients for JPEG reconstruction. Signal Process. Image Commun. **2007**, *22* (5), 439–447.
18. Lam, E.Y.; Goodman, J.W. A mathematical analysis of the DCT coefficient distributions for images. IEEE Trans. Image Process. **2000**, *9* (10), 1661–1666.

JPEG: YCgCb Color Space vs. YCbCr

Yuri A. Gadzhiev
Dagestan State Technical University (DSTU), Makhachkala, Russia

Abstract
YCbCr color space is widely used in image processing. A color space called YCgCb is proposed for using JPEG instead of YCbCr. Performed experiment shows that with JPEG it often provides some better compression (approximately 1%–5%) with as a rule some better visual quality, measured both objectively, by PSNR between compressed and original images, and subjectively, by a visual estimation of compressed image.

Keywords: Color; JPEG; Transform; YCbCr; YCgCr.

INTRODUCTION

The R, G, B color components of usual colored image are often highly correlated with each other. That is why before compressing an image, it is usually first subjected to some linear transform of its R, G, B components which removes this correlation. By means of such a transform, given by its square matrix $\begin{pmatrix} A_R & A_G & A_B \\ B_R & B_G & B_B \\ C_R & C_G & C_B \end{pmatrix}$, each ith pixel of image, originally given by pixel's RGB components $\begin{pmatrix} R_i \\ G_i \\ B_i \end{pmatrix}$, is represented by a new pixel's ABC components $\begin{pmatrix} A_i \\ B_i \\ C_i \end{pmatrix}$ so that

$$\begin{pmatrix} A_i \\ B_i \\ C_i \end{pmatrix} = \begin{pmatrix} A_R & A_G & A_B \\ B_R & B_G & B_B \\ C_R & C_G & C_B \end{pmatrix} \cdot \begin{pmatrix} R_i \\ G_i \\ B_i \end{pmatrix}$$

and

$$\begin{pmatrix} R_i \\ G_i \\ B_i \end{pmatrix} = \begin{pmatrix} R_A & R_B & R_C \\ G_A & G_B & G_C \\ B_A & B_B & B_C \end{pmatrix} \cdot \begin{pmatrix} A_i \\ B_i \\ C_i \end{pmatrix},$$

where

$$\begin{pmatrix} R_A & R_B & R_C \\ G_A & G_B & G_C \\ B_A & B_B & B_C \end{pmatrix} = \begin{pmatrix} A_R & A_G & A_B \\ B_R & B_G & B_B \\ C_R & C_G & C_B \end{pmatrix}^{-1}$$

is a matrix of inverse transform. In order to uncorrelate new components, it is required that a pixel with equal RGB components $\begin{pmatrix} R_i \\ G_i \\ B_i \end{pmatrix} = \begin{pmatrix} C \\ C \\ C \end{pmatrix}$ was always transferred into a vector with even two zero components. If such are the last two components, then it will have a form $\begin{pmatrix} D \\ 0 \\ 0 \end{pmatrix}$.

That is, for arbitrary constant C, a well-decorrelating transform always does:

$$\begin{pmatrix} A_R & A_G & A_B \\ B_R & B_G & B_B \\ C_R & C_G & C_B \end{pmatrix} \cdot \begin{pmatrix} C \\ C \\ C \end{pmatrix} = \begin{pmatrix} D \\ 0 \\ 0 \end{pmatrix}$$

and its inverse, respectively, does:

$$\begin{pmatrix} R_A & R_B & R_C \\ G_A & G_B & G_C \\ B_A & B_B & B_C \end{pmatrix} \cdot \begin{pmatrix} D \\ 0 \\ 0 \end{pmatrix} = \begin{pmatrix} C \\ C \\ C \end{pmatrix}.$$

From the first equality, we have $A_R + A_G + A_B = \dfrac{D}{C}$, and therefore, when a transform is normalized so that $A_R + A_G + A_B = 1$, then $D = C$, and from the second equality, it now follows that $R_A = G_A = B_A = 1$, that is, in this, normalized, case

$$\begin{pmatrix} A_R & A_G & A_B \\ B_R & B_G & B_B \\ C_R & C_G & C_B \end{pmatrix} \cdot \begin{pmatrix} C \\ C \\ C \end{pmatrix} = \begin{pmatrix} C \\ 0 \\ 0 \end{pmatrix}$$

and consequently,

$$\begin{pmatrix} A_R & A_G & A_B \\ B_R & B_G & B_B \\ C_R & C_G & C_B \end{pmatrix}^{-1} = \begin{pmatrix} 1 & R_B & R_C \\ 1 & G_B & G_C \\ 1 & B_C & B_C \end{pmatrix}.$$

All the color transforms used in practice are normalized so as to satisfy to the last two equalities. Color space obtained as a result of such transform is always different from the original RGB color space.

YCgCb COLOR SPACE

The color spaces of various kinds are now intensively studied and designed in very many parts of science and technologies. Some of them are the so-called ICA color space[1] supposed to be used for the purpose of pattern recognition; a new perception-based one[2] proposed for the illumination-invariant image processing; a widely used CIELAB color space;[3] a set of YCbCr-like color spaces proposed for the purposes of face recognition, like that described in the work of De Dios and Garcia;[4] a few specific color spaces used for various purposes, like that described in the works of Kasson et al.[5] and Paschos;[6] and many others, provided by the video standards such as those described in the works of Sullivan et al.[7] and Manjunath et al.,[8] the main objective of which is to decorrelate original RGB colors to attain better video compression.

Note that in principle, we can use any discrete functional transform with well-expressed decorrelation property as a decorrelate color transform. Such might be discrete Fourier transform, cosine, Haar's, Walsh, and so on. So, for example, discrete cosine transform of first type leads to the well-known YCoCg color space. Indeed, provided that $f_n(x) = \cos(nx)$, $x_k = \dfrac{k\pi}{2}$, and $k, n \in \{0,\ldots,2\}$, the transform matrix defined as an inverse of the matrix $F = \|f_n(x_k)\|$ is:

$$F^{-1} = \begin{pmatrix} 0.25 & 0.5 & 0.25 \\ 0.5 & 0 & -0.5 \\ 0.25 & -0.5 & 0.25 \end{pmatrix}$$

while

$$F = \begin{pmatrix} 1 & 1 & 1 \\ 1 & 0 & -1 \\ 1 & -1 & 1 \end{pmatrix}.$$

This, except for the sign of Cg, is the well-known YCoCg color transform.[9]

Nevertheless, still one of the most well known and intensively used, which now is even an essential part of the HEVC standard (see the work of Sullivan et al.,[7] part IV, par. A), is YPbPr or YCbCr color space, which can be defined, as it is done in Ref.[10] by the following equations:

$$Y(R,G,B) = K_R \cdot R + (1 - K_R - K_B) \cdot G + K_B \cdot B,$$

$$C_B(R,G,B) = \frac{1}{2} \cdot \frac{B - Y(R,G,B)}{1 - K_B},$$

$$C_R(R,G,B) = \frac{1}{2} \cdot \frac{R - Y(R,G,B)}{1 - K_R},$$

where $K_R, K_B > 0$, and $K_R + K_B < 1$. From this follows that if each of the R, G, B components is within the range $[0\ldots1]$, the value of $Y(R, G, B)$ is also within the same range $[0\ldots1]$. Used in the second and third equalities factors of the form $\dfrac{1}{2} \cdot \dfrac{1}{1 - K_x}$ are required "to restore the signal excursion of the color-difference signals to unity (i.e., +0.5 to −0.5)." [11, par. 2.5.2] That is, such a factor does a renormalization of the function $C_X(R, G, B)$, so that if all its R, G, B arguments are within the range $[0\ldots1]$, the value of C_X will be within the range $[-0.5 \cdots +0.5]$.

When the coefficients are set to the values $K_R = 0.299$, $K_B = 0.114$, the corresponding transform matrix has a form

$$\begin{bmatrix} Y(1,0,0) & Y(0,1,0) & Y(0,0,1) \\ C_B(1,0,0) & C_B(0,1,0) & C_B(0,0,1) \\ C_R(1,0,0) & C_R(0,1,0) & C_R(0,0,1) \end{bmatrix} \qquad (1)$$

$$= \begin{bmatrix} 0.299 & 0.587 & 0.114 \\ -0.169 & -0.331 & 0.5 \\ 0.5 & -0.419 & -0.081 \end{bmatrix}$$

with its inverse

$$\begin{bmatrix} Y(1,0,0) & Y(0,1,0) & Y(0,0,1) \\ C_B(1,0,0) & C_B(0,1,0) & C_B(0,0,1) \\ C_R(1,0,0) & C_R(0,1,0) & C_R(0,0,1) \end{bmatrix}^{-1} \qquad (2)$$

$$= \begin{bmatrix} 1 & 0 & 1.402 \\ 1 & -0.344 & -0.714 \\ 1 & 1.772 & 0 \end{bmatrix},$$

which is set by the recommendation of ITU[10] as a standard of YCbCr color transform.

Now we define our YCgCb color space similarly by the following equations:

$$Y(R,G,B) = (1 - K_G - K_B) \cdot R + K_G \cdot G + K_B \cdot B,$$

$$C_G(R,G,B) = \frac{1}{2} \cdot \frac{G - Y(R,G,B)}{1 - K_G},$$

$$C_B(R,G,B) = \frac{1}{2} \cdot \frac{B - Y(R,G,B)}{1 - K_B}.$$

Provided for $K_G = K_B = \frac{1}{3}$, it leads to the following transform matrix:

$$\begin{bmatrix} Y(1,0,0) & Y(0,1,0) & Y(0,0,1) \\ C_G(1,0,0) & C_G(0,1,0) & C_G(0,0,1) \\ -C_B(1,0,0) & -C_B(0,1,0) & -C_B(0,0,1) \end{bmatrix} \rightarrow \begin{bmatrix} \frac{1}{3} & \frac{1}{3} & \frac{1}{3} \\ \frac{-1}{4} & \frac{1}{2} & \frac{-1}{4} \\ \frac{1}{4} & \frac{1}{4} & \frac{-1}{2} \end{bmatrix}, \quad (3)$$

with an inverse:

$$\begin{bmatrix} Y(1,0,0) & Y(0,1,0) & Y(0,0,1) \\ C_G(1,0,0) & C_G(0,1,0) & C_G(0,0,1) \\ -C_B(1,0,0) & -C_B(0,1,0) & -C_B(0,0,1) \end{bmatrix}^{-1} \rightarrow \begin{bmatrix} 1 & \frac{-4}{3} & \frac{4}{3} \\ 1 & \frac{4}{3} & 0 \\ 1 & 0 & \frac{-4}{3} \end{bmatrix} \quad (4)$$

EXPERIMENTAL RESULTS

Here we perform, applied to JPEG, an experimental comparison of YCgCb with YCbCr. To do it, we replaced two program modules in the JPEG package responsible for the color transforms—Jccolor.c and Jdcolor.c—so that they could perform YCgCb color transform instead of their original YCbCr one. The modification concerned only final calculations—all numbers from (1) and (2) were replaced in programs with the numbers from (3) and (4). Algorithms and computing methods rest the same. The result of an experiment on Miscellaneous Test Images Database of the Signal and Image Processing Institute, University of Southern California (available at http://sipi.usc.edu/database/misc.zip) is shown in Table 1.

Table 1 The results of using YCbCr and YCgCb color transforms with JPEG

RGB colored test image	Image size (pixels)	JPEG file size (byte) with YCbCr	JPEG file size (byte) with YCgCb	PSNR (dB) of the JPEG image with YCbCr	PSNR (dB) of the JPEG image with YCgCb
4.1.01	256 × 256	9,947	9,901	32.71	32.65
4.1.02	—	9,930	9,694	34.06	34.08
4.1.03	—	6,002	5,937	37.88	37.83
4.1.04	—	10,342	10,142	32.04	32.35
4.1.05	—	10,198	9,797	31.44	31.63
4.1.06	—	16,038	15,476	29.30	29.48
4.1.07	—	6,333	6,190	34.48	34.53
4.1.08	—	8,507	8,344	33.10	33.21
4.2.01	512 × 512	31,972	28,749	33.23	33.27
4.2.02	—	37,346	33,000	31.03	31.55
4.2.03	—	77,244	74,800	26.21	26.47
4.2.04 (Lena)	—	37,788	36,090 (see Fig. 1)	33.21	33.36 (see Fig. 1)
4.2.05	—	38,679	37,156	32.61	32.98
4.2.06	—	52,482	50,145	28.66	28.83
4.2.07	—	41,308	39,573	30.29	30.56
House	—	48,009	45,828	30.39	30.31
Overall		442,125	420,822 (−4.82%)		

Fig. 1 Lena image decompressed from JPEG created under proposed YCgCb color transform (compressed file size 36,090 bytes, PSNR 33.36)

We can see in this table that proposed YCgCb color transform in all cases shows better results by the bitrate and in 13 cases from 16 it shows better results by the visual quality calculated as PSNR between compressed and original images. Well-known image Lena (4.2.04) compressed into JPEG with proposed YCgCb color transform (compressed file size 36,090 bytes, PSNR 33.36) is shown in Fig. 1. Its visual learning under a magnification which was performed by us showed that it has overall some less expressed DCT-transform's artifacts than JPEG image obtained under standard JPEG YCbCr color transform. Those three cases where YCgCb color transform was worse by PSNR than YCbCr are marked in Table 1 by means of underlining.

All shown experiments were performed with utilities CJPEG and DJPEG, release 6a of February 7, 1996, compiled under Windows 7 with bcc32, and then being launched with their default modes. Used in experiments JPEG package (of the same release) modified by us for using YCgCb color transform is available at the URL http://spritesoft.narod.ru/utilities/JpegYCgCb/JpegYCgCb.zip. It can be freely downloaded and used in any experiments not contradicted to the license of the JPEG package.

CONCLUSION

Earlier we performed a comparison between using with JPEG common YCbCr color space and introduced YCgCb one. As a result, we see that YCgCb color transform slightly improves visual quality of the JPEG compressed images with a gain from 1% to 5% of their bitrate.

REFERENCES

1. Liu, C.; Yang, J. ICA color space for pattern recognition. IEEE Trans. Neural Netw. **2009**, *20* (2), 248–257.
2. Chong, H.; Gortler, S.; Zickler, T. A perception–based color space for illumination-invariant image processing. ACM Trans. Graph. **2008**, *27* (3), 8, Article 61.
3. Connolly, C.; Fliess, T. A study of efficiency and accuracy in the transformation from RGB to CIELAB color space. IEEE Trans. Image Process. **1997**, *6* (7), 1046–1048.
4. De Dios, J.J.; Garcia, N. Face detection based on a new color space YCgCr. In *Proceedings of International Conference on Image Processing*, ICIP 2003, Barcelona, Spain, vol. 3; IEEE, 2003, 909–912.
5. Kasson, J.M.; Nin, S.I.; Plouffe, W.; Hafner, J.L. Performing color space conversions with three-dimensional linear interpolation. J. Electron. Imaging **1995**, *4* (3), 226–250.
6. Paschos, G. Perceptually uniform color spaces for color texture analysis: An empirical evaluation. IEEE Trans. Image Process. **2001**, *10* (6), 932–937.
7. Sullivan, G.J.; Ohm, J.-R.; Han, W.-J.; Wiegand, T. Overview of the high efficiency video coding (HEVC) standard. IEEE Trans. Circuits Syst. Video Technol. **2012**, *22* (12), 1649–1668.
8. Manjunath, B.S.; Ohm, J.-R.; Vasudevan, V.V.; Yamada, A. Color and texture descriptors. IEEE Trans. Circuits Syst. Video Technol. **2001**, *11* (6), 703–715.
9. Malvar, H.; Sullivan, G. YCoCg-R: A color space with RGB reversibility and low dynamic range. *JVT PExt Ad Hoc Group Meeting*, Trondheim, 22–24 July 2003.
10. YCbCr. Wikipedia article. Available at https://en.wikipedia.org/wiki/YCbCr.
11. Studio Encoding Parameters of Digital Television for Standard 4:3 and Wide-Screen 16:9 Aspect Ratios. Recommendation ITU-R BT.601-7 (03/2011). Available at https://www.itu.int/dms_pubrec/itu-r/rec/bt/R-REC-BT.601-7-201103-I!!PDF-E.pdf.

Large-Scale Remote Sensing Image Processing

Fang Huang
School of Resources and Environment, University of Electronic Science and Technology of China, Chengdu, China; Institute of Remote Sensing Big Data, Big Data Research Center, University of Electronic Science and Technology of China, Chengdu, China

Jian Taoc
Texas Engineering Experiment Station and High Performance Research Computing, Texas A&M University, College Station, Texas, U.S.A.

Abstract

With the rapid development of information technology and remote sensing (RS) sensor technology, modern RS data sources have been enriched. Due to improvements in spatial resolution, the file size of a single RS image has increased from megabytes to gigabytes. At the same time, the demand for more accurate RS data and shorter processing times presents a challenge for researchers in the field. How to obtain Earth observation data rapidly through RS technology and effectively obtain useful information from these large quantities of RS images have become very hot research subjects. In particular, it is now necessary to address the challenge of high performance computing (HPC) for large-scale RS data processing. Parallel computing based on computer clusters has been the mainstream approach and the most efficient way to enhance overall algorithm performance. With the evolution of computing technologies, some researchers have also explored new HPC technologies based on multiple cores, graphics processing units, Intel Many Integrated Cores, cloud computing, and big data computing platforms to process large-scale RS images.

Keywords: Cloud computing; Heterogeneous computing; High performance computing; Large-scale data processing; Parallel algorithm; Parallel computing; Processing; Real-time image processing; RS big data; Task scheduling.

REMOTE SENSING AND REMOTE SENSING IMAGE

Remote sensing (RS) is regarded as the science and art of identifying, observing, and measuring an object without coming into direct contact with it.[1] RS systems integrate various sensing devices, such as cameras, scanners, radiometers, and radars, to collect, process, and distribute large amounts of data.[2] As RS technologies have many advantages such as high object recognition ability, wide range of coverage, short data acquisition cycles, and diverse acquisition channels, they have been widely applied in agriculture, forestry, geology, weather, hydrology, military, environmental protection, and many other fields.

RS data contain reflected or emitted radiation from surfaces in different portions of the electromagnetic spectrum including visible, ultraviolet, reflected infrared, thermal infrared, microwave, and so on. Thus, most RS data fall into one of three main categories: visible spectral RS, thermal infrared RS, and microwave RS. Additionally, based on the altitude of the sensors, RS can be divided into space RS, satellite RS, and ground RS. RS can also be divided into resource RS and environmental RS from the difference of the research object, and so on.

Multiband or multispectral data consist of sets of radiation data that individually cover intervals of continuous wavelengths within some finite portion(s) of the electromagnetic spectrum. Each interval makes up a band or channel. The data are used to produce images of the Earth's surface and/or atmosphere to serve as input to sophisticated RS applications and analyses. RS images are produced by radiation from ground areas that are used as a sample for a larger region. The measured radiation varies depending on the reflectance, absorption, and emittance properties of various ground objects. The sampling area varies from a square meter to a square kilometer depending on a given sensor's position and accuracy.

Nowadays, hyperspectral sensors and unmanned aerial vehicles are making accurate and precise measurements of individual materials using spectrometers operating from space. The resulting data set produces a detailed spectral signature.

RS IMAGE PROCESSING

The primary objective of an RS application is to extract physical quantities of a surface of interest from RS data and obtain thematic information and domain knowledge related to geosciences. In other words, the very first step in the workflow of an RS application is RS image processing.

During image acquisition, many factors can cause geometric deformation. Such factors include changes in the height and attitude of the sensors, atmospheric refraction, curvature of the Earth, topography, rotation of the Earth, and the structure and performance of the sensors. Geometric deformation causes physical objects in an RS image to differ from real-world objects in certain projections, thus the geometric deformation of an RS image produces geometric or position distortions that necessitate the radiometric and geometric calibration of RS data. As RS data are basically images, the algorithms and operations used in image processing are applicable and commonly used in RS applications. A detailed description of RS image processing procedures and methodologies can be found in the works of Jong and Meer,[3] Schowengerdt,[4] and Chen.[5] The basic RS image processing involved in satellite imagery include spectral transforms (such as various vegetation indices, principal components and contrast enhancement, independent component analysis, vertex component analysis, convolution, and Fourier filtering), multi-resolution image pyramids and scale-space techniques (such as wavelets and image spatial decomposition), thematic classification (using traditional statistical approaches, neural networks, or fuzzy classification methods), image modeling, and image fusion.[2]

LATEST DEVELOPMENTS IN RS AND RS IMAGE PROCESSING

The latest developments in RS technologies show the following characteristics: three alls (all-weather, all-day, and all-global); three highs (high resolution, high spectral resolution, and high temporal resolution); and three multiples (multi-platform, multi-sensor, and multi-angle). With the rapid development of information and RS technologies, modern RS data sources have been enriched. Due to improvements in spatial resolution, the file size of a single RS image has increased from megabytes to gigabytes.

How to rapidly obtain Earth observation (EO) data through RS technology and effectively obtain useful information from the images has become a very hot research subject. At the same time, there has been an increase in demand for more accurate RS data and shorter processing times. The usage of high performance computing (HPC) for large-scale RS data processing has become a necessity to address these challenges.

LARGE-SCALE RS IMAGE PROCESSING USING HPC TECHNOLOGIES

There are some data- or computation-intensive operations in RS image processing. As a matter of fact, large-scale RS image processing is one of the most important tasks in large-scale RS analysis, applications, and simulations. The rapid development of HPC technologies makes large-scale RS image processing a reality, and many RS algorithms and applications in HPC systems have been studied and developed.

When using HPC for large-scale RS data processing, one needs to focus on the following aspects:

1. *Adoption of HPC technologies*. Currently, there are many HPC technologies that can be adopted for RS data processing. These technologies include computer clusters, multi-core processing, many-core processing, heterogeneous computing, cloud computing/big data methods, and so on. Each technology has its own advantages and disadvantages. Any decision to be made should be based on the specific requirements and characteristics of a given application, as well as the size of the calculation.
2. Regardless of the considered HPC technology, one needs to overcome common technical problems, such as mass data storage, fast and efficient I/O, data partitioning, data indexing, and visualization.
3. Based on the characteristics of the application itself, it is usually necessary to either design or/and implement a corresponding parallel algorithm and system framework with the particular HPC platform in mind. When designing or/and implementing parallel algorithms/systems, the parallel programming patterns, data or task parallel methods should be considered as well.

ROADMAP OF LARGE-SCALE RS IMAGE PROCESSING USING HPC TECHNOLOGIES

In the geosciences field, many researchers use various HPC techniques to improve the performance of their algorithms, applications, and/or simulations. Parallel computing based on computer clusters has been the mainstream approach and the most efficient way to enhance overall algorithm performance. With the evolution of computing technologies, some researchers have also explored new HPC technologies based on multiple cores, cloud computing, and graphics processing units (GPUs), Intel Many Integrated Cores (MICs), and big data computing platforms.

However, for a large-scale RS data processing application, it is important to choose the most suitable HPC technology and the right platform to meet one's specific needs. Furthermore, some issues need to be taken into consideration first when designing and implementing the desired approach using HPC computing technology: (1) It is helpful to decompose complex RS data processing into isolated tasks which can be executed independently and simultaneously on multiple processors,[6] and (2) it is important to enable dynamic task scheduling and load balancing in the distribution algorithm.[7]

Large-Scale RS Image Processing with Parallel Cluster Computing

Parallel computing is a major form of HPC, which refers to any computational technique that solves a large problem faster than is possible using stand-alone, off-the-shelf systems. Parallel computing has had a tremendous impact on a variety of areas, ranging from computational simulations for scientific and engineering applications to commercial applications in data mining and transaction processing. In essence, parallel processing entails dividing a problem into multiple tasks that can be executed within several processes upon a number of processors. How this is accomplished depends on the hardware and software environment.[8]

Until now, parallel computing based on clusters is the mainstream approach and the most efficient way to process large-scale data. Message passing interface (MPI) is a standard for implementing parallel programming and is a library rather than a programming language. Implementation of the MPI standard is done via MPICH or OpenMPI, both of which are dedicated to providing an MPI implementation for common parallel platforms, including symmetric multiprocessors, massively parallel processors, Beowulf clusters, and so on. It also provides a platform for MPI implementation research and for developing new and better parallel programming environments.

Processing large-scale RS data can be classified as a task-level parallelism problem. For applications like this, the common way to use parallel computing is to divide the tasks and distribute them across all computing units. In the present case, RS data is divided and distributed to computing nodes and processed simultaneously. A flowchart of the parallel processing steps is demonstrated in Fig. 1 (taken from the work of Huang et al.[9,10]).

Different nodes in the cluster will start sequentially. The computing nodes are divided into master and workers. All the instructions and the work of task decomposition are controlled by the master, and the master will collect the processing results for each process. When the workers receive the instructions and subtasks, they will process the assigned subtasks simultaneously. If the workers finish a subtask, they will wait for instructions from the master either to shut down or to continue to process the next computing subtask. The operation of the workers will depend on their assigned subtasks, or in other words, the number that needs to be processed in one subtask distribution will affect the efficiency of a parallel module. In actual operation, the assigned number of subtasks is determined by

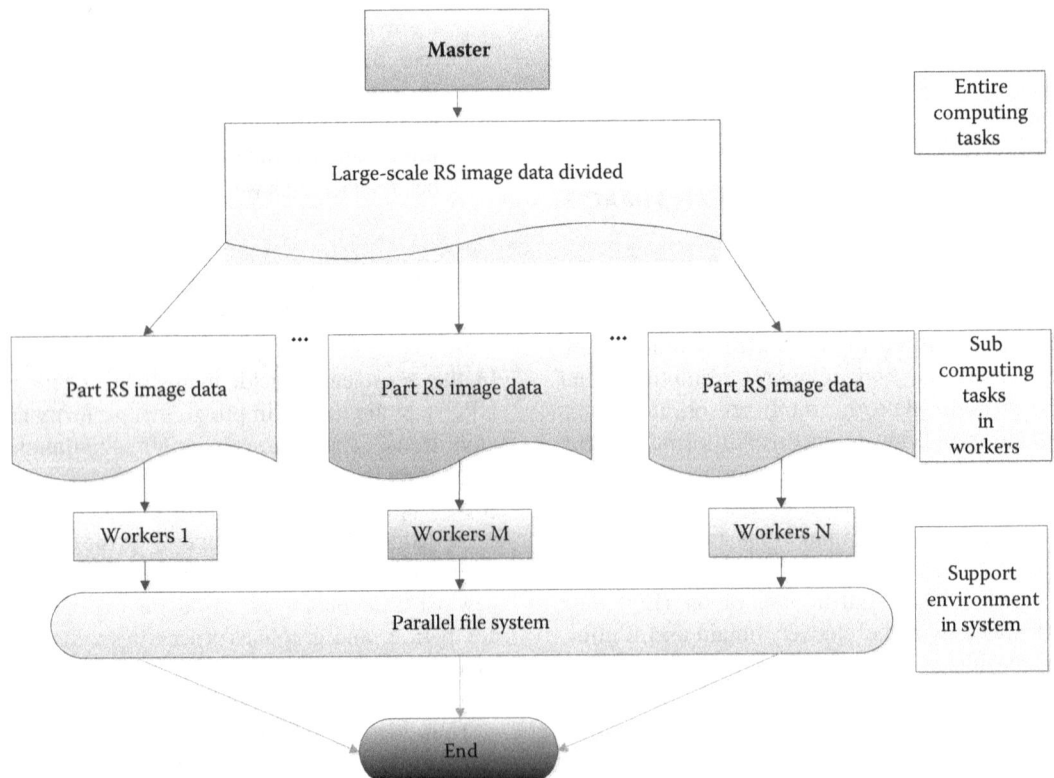

Fig. 1 Large-scale RS image data processing using cluster-based parallel computing

the number of RS image data to be processed, the average elapse time of each RS image, and the number of computing nodes involved.

Working principle of the master's side. The communication portion on the master's side is implemented using a *while* loop (illustrated in Fig. 2) with the exit condition that no subtasks remain to be sent to the workers. The master will receive the sent state information from any workers; if it is an idle tag from one worker and there are still some subtasks to be processed, the master will send the "keep working" tag to the worker, followed by the subtask information. Otherwise, it will send the "shut down" instruction to tell the worker to exit its loop and wait to terminate. If the sent tag information includes "result sending" information, the master will receive the results from that worker. The information received by the master is the status of the subtask processing, which is 0 if the subtask has been processed successfully or a nonzero number otherwise. After all subtasks have been completed, the master will exit the loop.

Working principle of the workers' side. To stay coordinated with the master, the communication portion on the worker side is also a *while* loop (Fig. 3). A worker first sends its current state information, that is, idle, working, or sending result, to the master. Next, it will wait for instructions from the master. Following the command from the master, the worker will either shut down or stay alive. If it keeps on working, it will receive process instructions and start. If a worker finishes processing, either by obtaining the correct results or by terminating due to some error, the results will be sent to the master. The master will resolve the results and send the corresponding prompts to the user.

Large-Scale RS Image Processing with Grid/Cloud Computing

In some RS applications, such as global-scale land surface temperature retrieval[9] and essential land variables, and so on, tremendous computing resources are required to process the vast amount of RS data. How to shorten the processing time and substantially increase the efficiency becomes vital for a successful large-scale RS image processing application.

Grid computing technologies that have evolved over the last decade promise to make the creation of an environment feasible for RS applications that can handle hundreds of distributed databases, heterogeneous computing resources, and simultaneous users.[11,12] In grid computing, computers throughout the world are loosely coupled via the Internet to form one or more virtual supercomputers. Each computer is one node on the interconnected grid. There are two unique features of such a virtual supercomputer: (1) It has high-performance and high-throughput computing capabilities, and (2) it takes advantage of the computing resources that are not fully used.[13,14] Grid computing enables the sharing of the computing, the storage, information, and knowledge resources. Grid computing technologies have been widely used in medical image processing, numerical simulations, RS image processing, and many other fields.[11,12]

Leveraging the grid computing technologies in RS image processing has been an effective approach to speed up the processing of massive RS data with multiple loosely coupled computing systems. Meanwhile, with the grid computing model, RS data can be systematically shared in a wide area network which helps to reduce the storage requirement due to data redundancy.

To address the performance issues of large-scale RS image processing, it is important to investigate new numerical algorithms and methodologies to create suitable grid middle-wares based on grid computing technologies. Considerable research in this direction has been carried out, and many software packages have been developed, including Globus and Condor.

In recent years, with the advent of "cloud computing" and related information technologies, RS image processing has found a new solution. Cloud computing has evolved from grid computing, and it is a type of internet-based computing that provides shared computer processing resources and data to computers and other devices on demand. By using cloud computing, scientists can easily access a great deal of computing resources to process RS data. Additionally, the cost of cloud computing is relatively inexpensive. Most cloud computing service providers adopt the pay-as-you-go system, which saves research expenditure. Moreover, cloud computing presents a universal and

```
While loop, until there is no running job
{
    Receive workers' state;
    If the worker is in Idle state and there is still has subtask
        then send the subtask to that worker;
    Else
        then send Shutdown instruction to that worker;
    If the worker is in result state
        then receive the results;
    Output the accomplishment message;
}
```

Fig. 2 Master's communication procedure in the loop

```
While loop, until the subtask has finished
{
    Send current state to Master;
    Receive Master's instruction;
    If receive the Shutdown instruction
        then the worker will exit the loop;
    else
        Receive the subtask;
        Process the current subtask in serial operations;
        Send results to Master after sending the result state tag;
}
```

Fig. 3 Workers' communication procedure in the loop

ubiquitous mechanism to use HPC facilities for the non-specialists. One of the core technologies is the universal parallel computing paradigm, that is, MapReduce. When using the MapReduce programming mode, researchers who are not familiar with traditional HPC can also design and implement their own parallel programs to solve their scientific or engineering problems.

When performing large-scale RS image processing utilizing the MapReduce mode, there is a common flow (Fig. 4). In addition to the open source cloud platforms such as Hadoop and Spark, there are many commercial cloud platforms such as Window Azure, Amazon Web services, and Alibaba Cloud, and many researchers have successfully carried out RS image processing on these platforms.[15,16]

Large-Scale RS Image Processing with Heterogeneous Computing

Recently, due to a rapid increase in the computing capacity of accelerators, such as GPUs and Intel Xeon Phi (Intel MIC architecture), the use of accelerators for big data processing has become a hot research topic in various fields. Many studies of GPUs have been conducted in the geosciences field.[17–19] One thing to note is that different computing platforms, that is, GPU and MIC, require different programming models and tools.

A heterogeneous computing system integrates GPUs and Intel Xeon Phi etc. acceleration components into a conventional computing system to implement computing tasks together with the CPU. Heterogeneous computing integrates each heterogeneous platform in an asynchronous manner by utilizing separate resources for computing or task scheduling, thereby maximizing the overall efficiency of a computing system by assigning tasks based on considerations of the capacities of each computing device. Indeed, heterogeneous computing is playing an increasingly important role in big data processing, and we envision the rapid adoption of heterogeneous computing for large-scale spatial data interpolation processing using improved algorithms. Until now, few studies have addressed the application of heterogeneous computing in the geosciences field.

Generally speaking, heterogeneous computing is suitable for designing and implementing a parallel algorithm.[20] To process large-scale RS data, it is better to adopt a load balancing strategy to fully use the multiple different computing devices. For a given large-scale RS image data processing application, the task partition granularity and the load balancing problem must be addressed when there are multiple computing devices. The main thread needs to distribute the overall task to multiple devices for processing, before collecting and combining the computing results to obtain the final product. In addition, it needs to use the lock mechanism to implement the dynamic load balancing strategy to compute the task distribution schedule. The detailed implementation of the load balancing scheduling strategy is illustrated in Fig. 5.

Large-Scale RS Image Processing in Big Data Areas

In the EO field, the amount of information that can be obtained from observed RS data is increasing, where the data structure is becoming more and more complicated, and the timeliness requirements are getting stronger and stronger. Large-scale RS image processing has obvious similarities with big data. That is, EO-based RS typically shows the same four-dimensional space–time features that big data also possesses. In the context of big data, the development of RS technology also poses great challenges to traditional data processing methods. "RS big data" is a necessity that conforms to the development of RS in the age of big data. RS big data is a utilization of RS technology under the guidance and technical support of big data, which is the value of big data in RS and its related fields. RS big data contains a full range of knowledge of the Earth's surface, where the fundamental purpose of interpreting RS big data is to master the knowledge and find the hidden laws after the data acquisition.

There are several aspects and features of RS big data that need to be discussed: the huge volume and velocity of the RS data, the diversity of the RS data, the complexity of the RS data especially with regard to higher dimensionality,[21] and also the conversion and transmission

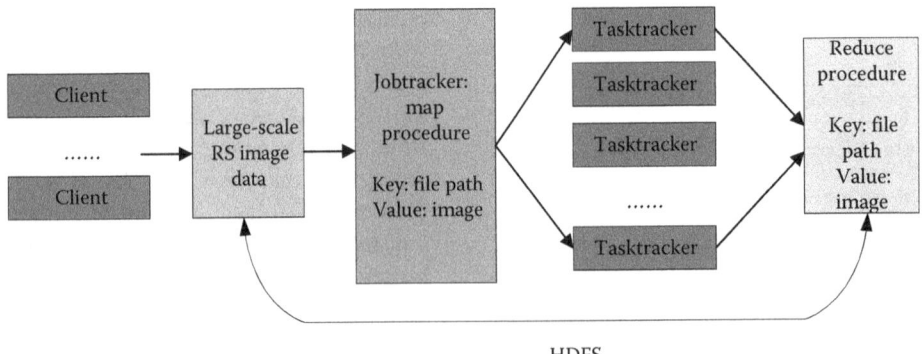

Fig. 4 Large-scale RS image data processing using MapReduce

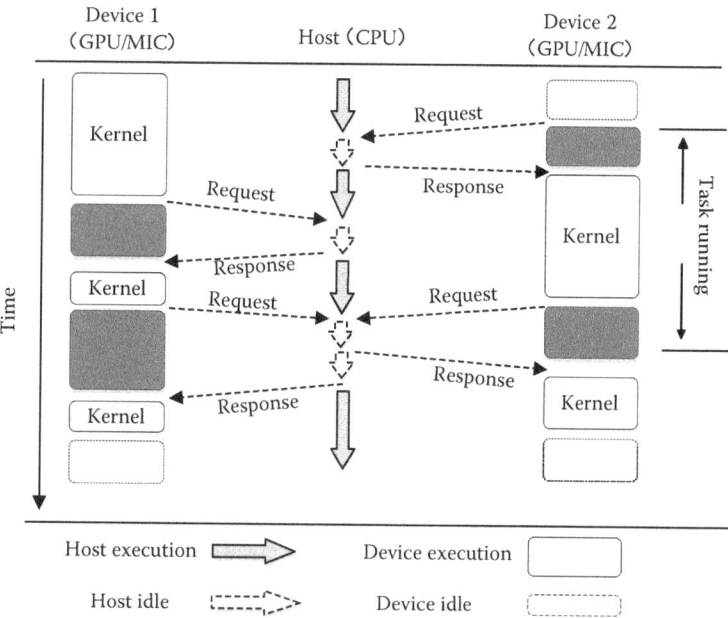

Fig. 5 Strategy for load balancing among multiple devices

of the RS data. The evolving applications based on RS big data toward economic analyses include the following: (1) analysis of land use changes based on high-resolution RS data, including buildings, vegetation, water and other temporal and spatial changes; (2) based on light RS data to analyze the gross domestic product of an area/country, its population density, water and electricity consumption, and to create global poverty zone thematic maps; and (3) based on hyperspectral RS data, monitor and assess crop growth and yield, etc.

At present, the main focus of RS big data is on the expression, understanding, mining of the RS big data, and so on. The use of artificial intelligence and deep learning methods will be very important for the development of RS big data.[22,23]

ACKNOWLEDGMENTS

This study was supported mainly by Hubei Provincial Key Laboratory of Intelligent Geo-information Processing (China University of Geosciences) (Grant No. KLI-GIP2016A03), Key Laboratory of Spatial Data Mining and Information Sharing of the Ministry of Education, Fuzhou University (Grant No. 2017LSDMIS03), the Fundamental Research Funds for the Central Universities (Grant No. ZYGX2015J111), and also the National Science Foundation of the United States (Award No. 1251095, 1723292). Thanks to the Taylor & Francis Group for granting us the permission to use the following material: Fang Huang, Zhou, J.; Tao, J.; Tan, X.; Liang, S.; Cheng, J. PMODTRAN: A parallel implementation based on MODTRAN for massive remote sensing data processing. Int. J. Digit. Earth. **2016**, *9* (9), 819–834. We also thank Multidisciplinary Digital Publishing Institute (MDPI) AG for use of Huang, F.; Bu, S.; Tao, J.; Tan, X. OpenCL implementation of a parallel universal kriging algorithm for massive spatial data interpolation on heterogeneous systems. Int. J. Geo-Inf. **2016**, *5* (6), 96.

REFERENCES

1. Lillesand, T.M. *Remote Sensing and Image Interpretation*; John Wiley & Sons: New Delhi, India, 2006.
2. Petcu, D.; Zaharie, D.; Neagul, M.; Panica, S.; Frincu, M.; Gorgan, D.; Stefanut, T.; Bacu, V. Remote sensed image processing on grids for training in earth observation, Chapter 08. In *Image Processing*; Chen, Y.-S.; Ed.; InTech: Rijeka, 2009.
3. Jong, S.M.D.; Meer, F.D.V.D. *Remote Sensing Image Analysis. Including the Spatial Domain*; Springer: Netherlands, 2006.
4. Schowengerdt, R.A. *Remote Sensing, Models and Methods for Image Processing*, 3rd Ed.; Academic Press, Elsevier Inc.: Burlington, MA, 2007.
5. Chen, C.H. *Image Processing for Remote Sensing*; CRC Press, Taylor & Francis Group: Boca Raton, FL, 2007.
6. Akhter, S.; Honda, K.; Chemin, Y.; Amin, M.A. Experiments on distributed remote sensing data (modis and aster) processing using optima cluster. In *NECSEC 2005*, 2005.
7. Plaza, A.J.; Chang, C.-I. *High Performance Computing in Remote Sensing*; Chapman & Hall CRC: Boca Raton, FL, 2007, 496.
8. Mineter, M.J.; Dowers, S.; Gittings, B.M. Towards a HPC framework for integrated processing of geographical data: Encapsulating the complexity of parallel algorithms. Trans. GIS **2000**, *4* (3), 245–261.

9. Huang, F.; Zhou, J.; Tao, J.; Tan, X.; Liang, S.; Cheng, J. PMODTRAN: A parallel implementation based on MODTRAN for massive remote sensing data processing. Int. J. Digit. Earth **2016**, *9* (9), 1–16.
10. Huang, F.; Liu, D.; Tan, X.; Wang, J.; Chen, Y.; He, B. Explorations of the implementation of a parallel IDW interpolation algorithm in a Linux cluster-based parallel GIS. Comput. Geosci. **2011**, *37* (4), 426–434.
11. Ian, F.; Carl, K. *The Grid: Blueprint for a New Computing Infrastructure*; Morgan Kaufmann Publishers Inc.: San Francisco, CA, 1999, 677.
12. Foster, I.; Kesselman, C.; Tuecke, S. The anatomy of the grid: Enabling scalable virtual organizations. Int. J. High Perform. Comput. Appl. **2001**, *15* (1), 200–222.
13. Buyya, R.; Abramson, D.; Giddy, J. Nimrod/G: An architecture for a resource management and scheduling system in a global computational grid. In *Proceedings Fourth International Conference/Exhibition on High Performance Computing in the Asia-Pacific Region*, 14–17 May 2000, 2000, Beijing, China, Vol. 1, 283–289.
14. Goux, J.P.; Kulkarni, S.; Yoder, M.; Linderoth, J. Master–worker: An enabling framework for applications on the computational grid. Cluster Comput. **2001**, *4*, 63–70.
15. Deqiang, G.; Keping, D.; Yonghua, Q.; Yuzhen, Z.; Linli, L. Remote sensing algorithm platform in windows azure. In *20th International Conference on Geoinformatics*, 15–17 June 2012, Hong Kong, China, 2012, 1–6.
16. Tan, X.; Di, L.; Deng, M.; Huang, F.; Ye, X.; Sha, Z.; Sun, Z.; Gong, W.; Shao, Y.; Huang, C. Agent-as-a-service-based geospatial service aggregation in the cloud: A case study of flood response. Environ. Modell. Softw. **2016**, *84*, 210–225.
17. Liu, P.; Yuan, T.; Ma, Y.; Wang, L.; Liu, D.; Yue, S.; Kołodziej, J. Parallel processing of massive remote sensing images in a GPU architecture. Comput. Inform. **2014**, *33* (1), 197–217.
18. Cheng, T. Accelerating universal Kriging interpolation algorithm using CUDA-enabled GPU. Comput. Geosci. **2013**, *54*, 178–183.
19. Chen, D.; Li, D.; Xiong, M.; Bao, H.; Li, X. GPGPU-aided ensemble empirical-mode decomposition for EEG analysis during anesthesia. IEEE Trans. Inf. Technol. Biomed. **2010**, *14* (6), 1417–1427.
20. Huang, F.; Bu, S.; Tao, J.; Tan, X. OpenCL implementation of a parallel universal kriging algorithm for massive spatial data interpolation on heterogeneous systems. Int. J. Geo-Inf. **2016**, *5*, 96.
21. Ma, Y.; Wu, H.; Wang, L.; Huang, B.; Ranjan, R.; Zomaya, A.; Jie, W. Remote sensing big data computing: Challenges and opportunities. Future Gener. Comput. Syst. **2015**, *51*, 47–60.
22. Wang, L.; Zhang, J.; Liu, P.; Choo, K.-K.R.; Huang, F. Spectral–spatial multi-feature-based deep learning for hyperspectral remote sensing image classification. Soft Comput. **2017**, *21* (1), 213–221.
23. Liu, P.; Choo, K.-K.R.; Wang, L.; Huang, F. SVM or deep learning? A comparative study on remote sensing image classification. Soft Comput. **2017**, *21* (23), 7053–7065.

Markov Random Fields

Tania Pouli
Technicolor, Rennes, France

Erik Reinhard
Technicolor Research and Innovation, Rennes, France

Douglas W. Cunningham
Brandenburg University of Technology, Cottbus, Germany

Abstract
Images often need to be analyzed with respect to context, either in space or time, or depending on many factors. One way to do this is through probabilistic models, such as Markov random fields (MRFs). These rely on estimated parameters of some experimenter-specified model for the scene. This entry introduces the basic concepts necessary to understand MRFs and provide a few insights into several statistical regularities that can be detected with them.

Keywords: Markov random field; Parameter estimation; Scene analysis.

There are many thousands of visual illusions, most of which serve to show that the human visual system does not, in fact, perform some form of inverse physics but instead uses a wide variety of assumptions, simplifications, and heuristics (see Section 2.5.5). Perhaps the most common feature of nearly all visual processing—from simple edge detection through object recognition and up to aesthetic judgment—is that these processes are *context dependent*. That is, the same visual signal can be interpreted in many different ways depending on the information right next to it (either spatially or temporally), on our previous experience, and on a host of other factors.

For example, our recognition of an edge is based not just on the local luminance change from one pixel to the next but on the other luminance changes in the image (i.e., only maxima in the gradient field are edges; see Chapter 5). Context dependency is encoded at a very early stage in visual processing, as the center-surround nature of receptive fields shows (see Chapter 2). In addition to this local context dependency, there is also evidence that global context dependency is encoded early: it has been repeatedly demonstrated that information that is clearly outside of a cell's receptive field strongly influences the response of that cell, even though by definition a cell should not be able to react to information outside its receptive fields (see, for example, Ref. [1]).

For edges and many other low-level processes, the "outside influences" come largely from cells that are immediate neighbors (the so-called *near-field*), again a decent reflection of real-world statistics. This influence is often implemented with simple lateral inhibitions. Some influence, however, comes from cells that are very far away, and this is believed to occur either through the massive feedback projections from higher visual areas down to lower visual areas or through chains of lateral connections (i.e., propagated influence). In short, to fully describe the statistics of real-world scenes, as well as how humans process the real world, one cannot look solely at individual cells, regions, or pixels. It is critical that at least the local neighborhood be examined. In the ideal case, global influences must also be taken into account.

One excellent method for embedding this context dependency in visual computing applications is through *Markov random fields* (MRFs). MRFs are based on Ising's work in the physics of condensed matter in the 1920s.[2,3] After Ising, MRFs continued to be developed and expanded, at first mostly by physicists and mathematicians. The seminal paper from Geman and Geman in 1984[4] brought MRFs to the field of image processing (for a historical review of MRFs, please see Ref. [5]).

Since the 1980s, MRFs have been successfully applied to an incredible range of problems in computer graphics and computer vision, including denoising and general image restoration, image segmentation, edge detection, optical flow, shape detection, depth segmentation, and texture analysis (for more on these and other examples, see, for example, Ref. [6]). Although most MRFs focus on highly local context dependencies similar to the effect of retinal cells on their immediate neighbors and other low-level visual tasks, modeling global dependencies is also possible. Further, it is also possible to use MRFs to learn the statistical dependencies in image sets (including natural image sets). This often takes the form of estimating the parameters of some experimenter-specified model

Encyclopedia of Image Processing, First Edition
DOI: 10.1201/9781351110273-140000483
Copyright © 2018 by Taylor & Francis. All rights reserved.

(generally using Maximum Pseudo Likelihood Estimation or contrastive divergence).

In this chapter, we focus on the basic concepts necessary to understand MRFs and provide a few insights into several statistical regularities that can be detected with them. For further reading, we direct the reader to any of the many excellent books an Markov random fields and their myriad applications, such as Refs. [6 and 7].

IMAGE INTERPRETATION

In some areas of psychology, perception is thought of as a form of hypothesis testing (see, for example, Ref. [8]). That is, given (ambiguous) information at a specific spot on the retina (sometimes in vision research, this is called the *proximal stimulus*), the visual systems poses hypotheses or guesses as to what the real-world object or event (called the *distal stimulus*) might be. If necessary, the visual system then searches for information. If there is not enough information in the current proximal stimulus to confirm or disprove the hypothesis, then new information is acquired (e.g., the person moves to get a better view). In other words, vision is sometimes thought of as the process of inference where a set of proximal stimuli (the retinal excitations) are assigned a value from a (limited) set of interpretations. For example, the visual system might try to determine if the information at a given cell came from an edge or not. Likewise, cells higher in the visual system might try to determine if the stimulation at a given cell came from a tree trunk, from leaves, from the ground, or from the sky.

Within the field of image analysis using Markov random fields, this process of inference is often called *labeling*. More formally, let S be a set of m observations or *sites*:

$$S = \{1,...,m\} \tag{1}$$

where the numbers are indexes to individual observations. These might be cells on the retina, pixels in an image, segmented regions in an image, and so on. In a two-dimensional $n_1 \times n_2$ image, the set can also be written as:

$$S = \{(i,j) | 1 \leq i \leq n_1, 1 \leq j \leq n_2\} \tag{2}$$

Note that connectivity between sites is important in MRFs, but this is usually represented separately. The global order of the sites, on the other hand, is not important. Thus, 2D images can be represented with a single parameter h, where $h = \{1, 2,..., m\}$ with $m = n_1 \times n_2$.

Labeling also requires the definition of a set of possible interpretations that a site may have. This list of categories or *labels* can be nominal (and therefore contain only information about identity), such as edge versus non-edge; ordinal (containing information about the identify and ranking of categories), such as relative depth ordering (closest surface, next closest surface,..., most distant surface); or even interval (containing information about the identity, ranking, and relative distance between categories), such as the intensity at a pixel (0,..., 255).

The class of operations that are possible on the labels is strongly dependent on the nature of the labels. For nominal label sets, it is only possible to determine if two sites have the same label or not. For ordinal and interval label sets, it is possible to define operations like similarity and distance. If the labels are discrete, the set of M labels can be written as:

$$\mathcal{L} = \{1,...,M\} \tag{3}$$

with the indices pointing to individual categories or labels. If the labels are continuous (such as a floating point representation of luminance), then the set $\mathcal{L}$ can be thought of as a line in $\mathbb{R}$ or some portion of it:

$$\mathcal{L} = [X_1, X_2] \subset \mathbb{R} \tag{4}$$

with X_1 and X_2 being the lower and upper bounds of the set, respectively.

We can then interpret the image. More specifically, we assign each site $i \in S$ a value f from the set of labels $\mathcal{L}$. Commonly the label at a given site i is denoted with f_i. The set $f = \{f_1, f_2,..., f_m\}$ of labels for all m sites can be called a labeling of the sites in S in terms of the labels in $\mathcal{L}$. If all sites can be assigned a single label then we have a *mapping* f, which can be written as:

$$f : S \to \mathcal{L} \tag{5}$$

Depending on the nature of the set of labels, we can create a wide variety of maps. If the labels are {edge, non-edge} then the mapping f_i is an *edge map*. If $\mathcal{L}$ consists of distances from the viewer, then f is a *depth map*. In the area of Markov random fields, a map is often called a *configuration*.

If all sites have the same set of labels $\mathcal{L}$, then the set of all possible configurations $\mathcal{F}$—which is also called the *configuration space*—is the Cartesian product of the set of labels $\mathcal{L}$ by itself m times, with m being the number of sites. If $\mathcal{L}$ is discrete, then the configuration space is combinatorial: with m sites and M labels, there are M^m possible mappings. For example, if there is only one label, then there is only one configuration (or interpretation of the data) regardless of the number of sites: $1^m = 1$. If there are two labels, then with two sites there are four logical possible configurations: both sites have label 1 ($f = \{1,1\}$), both sites have label 2 ($f = \{2,2\}$), site 1 has label 1 and site 2 has label 2 ($f = \{1,2\}$), or site 1 has label 2 and site 2 has label 1 ($f = \{2,1\}$).

GRAPHS

As was mentioned earlier, most visual processes are context-dependent and can be modeled by looking at the influence of sites that are immediately adjacent to the site of interest. This, of course, requires some formal notion of adjacency. In essence, i and j can be thought of as being connected to one another. One way of explicitly representing this is with a graph $G = (S, E)$, where E is the set of edges (i.e., the a set of adjacent sites). If two sites i, $j \in V$ are neighbors, then they will have an edge $(i, j) \in E$ between them. One important property of adjacency in the context of Markov random fields is that, like neighbors in real life, the relationship is mutual. If the site i is adjacent to the site j, then j must be adjacent to i. This is represented with an *undirected* graph, which is equivalent to saying if $(i, j) \in E$, then $(j, i) \in E$. Note that, as in real life, a site cannot be its own neighbor $(i,i) \notin E$.

Neighborhood Systems

Now that we have a method for formalizing context, we need to define what we mean by immediate or local context. The neighborhood $\mathcal{N}_i$ of a site i is defined as the set of all sites that share an edge with i:

$$\mathcal{N}_i = \{ j \mid \forall j \in S, (i,j) \in E \} \qquad (6)$$

The complete set of all neighborhoods, then, is a *neighborhood system*:

$$\mathcal{N} = \{ \mathcal{N}_i \mid \forall i \in S \} \qquad (7)$$

In practice, this is made easier by the fact that most images are spatially regular (in fact, they form a lattice). Thus, the set of neighbors for a given site is visually obvious. There are two common neighborhood systems in lattices, depending on whether diagonals are allowed. The simplest case is a first-order neighborhood (also called a 4-neighborhood), where we do not look at diagonals (see Fig. 1). For example, the 4-neighborhood of the site $S_{u,v}$ consists of:

$$\mathcal{N}_{u,v}^1 = \{ S_{u-1,v}, S_{u+1,v}, S_{u,v-1}, S_{u,v+1} \} \qquad (8)$$

The superscript denotes that we are using a first-order neighborhood. In a second-order neighborhood (also called an 8-neighborhood), diagonal neighbors are also considered (see Fig. 1). Thus, the 8-neighborhood of site $S_{u,v}$ consists of:

$$\mathcal{N}_{u,v}^2 = \{ S_{u-1,v}, S_{u+1,v}, S_{u,v-1}, S_{u,v+1}, S_{u-1,v-1}, S_{u-1,v+1}, S_{u+1,v-1}, S_{u+1,v+1} \} \qquad (9)$$

It should be mentioned that pixels at the edges of an image have fewer neighbors. Often, this is avoided in practice by "wrapping" an image (so that in an $n_1 \times n_2$ image, pixel $I(0,y)$ is connected to $I(n_1,y)$).

If the graphs are irregular and there is no explicit concept of connection (for example, as might occur in an edge map), we need to define some distance function $dist(i, j)$ between sites i and j. All sites that fall within a given threshold distance are included in the neighborhood. For example, if the location of the sites can be interpreted spatially, then an acceptable distance function would be to take a circle centered on the site of interest and all cells that are within a threshold radius are considered to be within the local neighborhood. Of course, care needs to be taken when defining both the distance function and the threshold to ensure that the neighborhood range completely includes the initial site itself (for example, since most edges are spatially extended, it is important to ensure that a given edge in an edge map is within its own neighborhood calculations). If the graph cannot be spatially interpreted, then some other form of distance function needs to be defined.

Cliques

The final term that is needed is a *clique*, which is a subgraph of G in which all the elements are neighbors of each other. Cliques may be unary (containing just the site itself), binary (pairs of neighbors), triplets of neighbors, and so on. The set of all unary cliques will be referred to as C_1, the set of all binary cliques as C_2, and so on. The set of all cliques C is given by:

$$C = C_1 \cup C_2 \cup \ldots \qquad (10)$$

Fig. 1 Neighborhoods: first-order or 4-neighborhood (left) and second-order or 8-neighborhood (right)

Fig. 2 The possible cliques in a 4-neighborhood: single or unary cliques (left), pairwise or binary cliques (middle), and three-way cliques (right)

Critically, cliques are ordered, so that (i, j) is not the same as (j, i). Additionally, a site can be in more than one clique. Thus, a single site is in its own unary clique as well as possibly in one or more binary cliques. Possible cliques within a 4-neighborhood in a lattice are shown in Fig. 2. These include the sites themselves (unary cliques), horizontal and vertical pairs (binary cliques), and the three-way cliques. A *maximal clique* is a clique where it is not possible to add any more sites while still retaining complete connectedness.

PROBABILITIES AND MARKOV RANDOM FIELDS

Some of all the possible configurations f in the configuration space $\mathscr{F}$ are more likely than others. Markov random fields capture this intuition more concretely and allow us to model contextual dependencies. For an image I whose pixels make up the set S of m sites, we can define a set $F = \{F_1, F_2, \ldots, F_m\}$ of random variables that are the actual mappings from S to $\mathscr{L}$. Each site's random variable F_i takes one of the possible mappings f_i. The probability that the actual mapping at a specific site F_i is a particular possible mapping for that site f_i is given by $p(F_i = f_i)$, or $p(f_i)$ for short. Following standard notation, the $(F = f)$ term will be shortened everywhere to f. The set or family F is called a random field.

If there were no context dependencies, then the probability that a given site has a given interpretation $p(f_i)$ would depend only on the site itself. That is, this probability would be conditionally independent of all other sites j:

$$p(f_i \mid f_j) = p(f_i) \tag{11}$$

where f_j is the mapping at all other sites $j \neq i$. As we have seen, however, this is generally not the case and therefore context dependencies need to be modeled. The random field F is a Markov random field if the following two conditions are met:

Positivity. The probability of a mapping occurring is greater than zero for all possible mappings in the configuration space:

$$p(f) > 0 \qquad \forall f \in \mathscr{F} \tag{12}$$

Markovianity. The probability that a site i takes a specific value f_i is dependent only on the local neighborhood of that site $\mathscr{N}_i$:

$$p(f_i \mid f_{\{S-i\}}) = p(f_i \mid \mathscr{N}_i) \tag{13}$$

where $\{S - i\}$ is all of S except i.

Gibbs Distributions

The Hammersley-Clifford theorem[9] states that if the above two assumptions are met then we can write the joint probability $p(f)$ as a Gibbs distribution:

$$p(f) = \frac{1}{Z} e^{-\frac{1}{T} U(f)} \tag{14}$$

where Z is the normalizing constraint called the partition function, T is a global parameter called the *temperature* (often assumed to be 1), and $U(f)$ is the prior energy. The partition function is given by:

$$Z = \sum_{f \in \mathscr{F}} e^{-\frac{1}{T} U(f)} \tag{15}$$

and ensures that the sum of all probabilities equals unity:

$$\sum_{f \in \mathscr{F}} p(f) = 1 \tag{16}$$

The prior energy $U(f)$ is given by:

$$U(f) = \sum_{c \in C} V_c(f) \tag{17}$$

where $V_c(f)$ denotes the prior potentials (the values of subgraphs of G), which depend only on f_i, with $i \in c$ (the labels for sites within a clique). Lower energy states $U(f)$ are more likely. Note that the temperature T controls the sharpness of the distribution. The higher T is, the more evenly distributed all configurations are. As T approaches zero, the distribution concentrates around the global energy minima.

In some cases, it can be convenient to rewrite the energy function as a sum of terms, with each term relating to a clique size (single, pair, triple, etc.):

$$U(f) = \sum_{i \in C_1} V_1(f_i) + \sum_{i,j \in C_2} V_2(f_i, f_j) + \sum_{i,j,k \in C_3} V_3(f_i, f_j, f_k) + \cdots \tag{18}$$

where C_1 contains the unary cliques, C_2 contains the binary cliques, C_3 contains the triple cliques, and so on. Likewise, V_1 denotes the potentials of the unary cliques, and so on.

Auto-Models

The simplest MRF model would be to restrict our considerations solely to cliques up to size two. If the set of sites S is a two-dimensional image and we use only a first-order neighborhood, then the auto-model is the same as the Ising model. In these cases, Eq. 18 can be rewritten as:

$$U(f) = \sum_{i \in S} V_1(f_i) + \sum_{i \in S} \sum_{j \in \mathcal{N}_i} V_2(f_i, f_j) \quad (19)$$

Note that here we sum over *all* unary and binary cliques, and that each site is in a unary clique and four binary cliques. Combining Eq. 19 with Eqs. 14 and 15, and assuming T is 1 gives the conditional probability of a site based on its neighbors:

$$P(f_i | f_{\mathcal{N}_i}) = \frac{\exp\left(-V_1(f_i) - \sum_{j \in \mathcal{N}_i} V_2(f_i, f_j)\right)}{\sum_{f_i \in \mathcal{L}} \exp(-V_1(f_i) - V_2(f_i, f_j))} \quad (20)$$

In the original Ising model, there were two states $\mathcal{L} = \{-1, +1\}$. The Potts model provides an important extension to the Ising model by allowing the set of labels to have more than two states. Ising also ignored unary cliques (for physically based reasons that were relevant to the specifics of the phenomenon he was studying), leaving only binary term:

$$U(f) = \sum_{c \in C} V_c(f) = -\beta \sum_{(i,j) \in C_2} f_i f_j \quad (21)$$

where β captures the interaction between neighbors. If $\beta > 0$, then neighbors will tend to have the same value. One can generate images using an Ising model, with different values for β giving different images. It is also possible to use different β for horizontal and vertical cliques, yielding anisotropic images. Interestingly, as was shown in Chapter 5, the difference between pairs of neighbors is an approximation of the gradient, so the binary terms are essentially modeling the first derivative properties of the image.

Since the joint probability $p(f)$ measures the probability that a given configuration occurs, we can bias the outcome towards specific configurations by setting the clique potentials carefully. Thus, much of the interest in MRFs is based on the choice of potential functions (using prior knowledge about interactions between neighboring sites and labels) so that the desired behavior arises.

MAP-MRF

MRFs and their connections with Gibbs distributions provide us with the ability to calculate the probability of different configurations. As mentioned, this is interesting for texture generation. A simple Ising or Potts model can produce convincing textures for different settings of the available parameters.

Within image analysis, however, MRFs present some limitations. As stated earlier, problems in low-level vision or image analysis can be thought of as inference or hypothesis testing problems: we have a given image and we want to interpret it. The probability of a given configuration $p(f)$ simply tells us how likely any given configuration is in principle (with lower energy configurations being more likely). Often, however, we would like to know which of the many possible configurations applies to our current image. MRFs by themselves have no means for incorporating real observations and thus are insufficient to answer this question. For this reason MRFs are often embedded within a Bayesian framework.

Although a full treatment of Bayesian statistics is beyond the scope of this book,[a] a few comments are required. An example of how Bayesian inference can be used in the context of MRFs is as follows. Assume that there is a real-world from which we take some picture or observation. Due to the process of image capture, however, the information from the real-world gets slightly corrupted. If we know what the real-world scene might have contained and have some idea of how the information was corrupted, we could make some plausible guesses about what the scene really contained based on the (corrupted) image. In other words, there is a known dataset (e.g., the observed pixel values) d and the possible interpretations of that dataset (the labels $\mathcal{L}$). The knowledge about how likely different sets of interpretations or configurations are captured by $p(f)$, as we saw earlier. In Bayesian statistics, this is called the *a priori* or prior. Knowledge about how the information was corrupted during the observation process is captured by the *likelihood* $p(d|f)$. Note that this is very much application-dependent. Given this information we can calculate how likely any given interpretation or configuration f is given the observed data:

$$p(f|d) = \frac{p(d|f)p(f)}{p(d)} \quad (22)$$

This is called the *a posterior* or posterior. Note that the denominator $p(d)$ is a constant since the image is known, but its calculation is generally intractable. Fortunately, this is usually not a problem as we shall see. Since we want to know which interpretation is most likely, it is common to calculate the maximum *a priori* or MAP estimate:

$$f^* = argmax_{f \in \mathcal{F}} p(f|d) \quad (23)$$

---

[a]The interested reader is directed to[6] for more on Bayesian statistics in image processing. See also the appendix for a brief discussion on Bayes' rule.

Since $p(d)$ is constant, it is not needed for the MAP, which therefore becomes:

$$f^* = argmax_{f \in \mathcal{F}} p(d|f) p(f) \qquad (24)$$

In sum, when using MRFs in a Bayesian framework, we need to define a neighborhood system, then define the allowable cliques, and then define the clique potentials. This gives us the prior $p(f)$. We then derive the likelihood energy based on assumptions on the underlying distributions and processes involved in generating the dataset. Finally, we derive the posterior energy and find the maximum.

APPLICATIONS

We present two examples where MRFs have been successfully used to detect statistical regularities in the image. For more on what regularities MRFs have been used to model, please see Ref. [6].

Image Restoration

One of the classic examples of how an MRF can be used is simple image restoration. There are many, many different versions of this procedure. The one we present here is based on Ref. [10]. We assume that we have an image, represented as a two-dimensional spatial lattice, and that we only look at the first-order neighborhood. The set of sites $S = 1,..., m$ contains the pixel locations, while the set of labels $\mathcal{L} = 0,..., 255$ contains possible 8-bit pixel luminances. The possible cliques are the sites themselves and pairs of vertically or pairs of horizontally adjacent pixels. The binary clique potential is $V_2(f_i, f_j) = \beta_g (f_i - f_j)$ where β is a scalar and $g(.)$ is a function that penalizes differences between f_i and f_j (i.e., it penalizes smoothness violations). Further, the assumption is made that the "correct" image was corrupted by adding white noise[b] $d_i = F_i + \mu_i$ where d_i is the observed pixel value, F_i is the correct pixel value, and μ_i is a zero-mean Gaussian distribution with variance σ^2. Thus, the likelihood is:

$$p(d|f) = \left(\frac{1}{2\pi\sigma^2}\right)^{|S|/2} \exp\left(-\frac{1}{2\sigma^2} \sum_{i \in S} (d_i - f_i)^2\right) \qquad (25)$$

where $|S|$ is the number of data points. Using an Ising model and ignoring the unary terms gives:

---

[b]Note that the noise in real imaging systems is often not white. In real restoration processes usually a Poisson distribution is assumed.[4] Interestingly, there is some evidence that electronic noise—and thus the corrupting factor between real scene and observation—is not white, but the power spectrum actually follows a power law[26,27]. Using a power law rather than white noise in the likelihood might improve the performance of MRFs in image restoration.

$$p(f) = \frac{1}{Z} \exp\left(-\beta \sum_{i,j} f_i f_j\right) \qquad (26)$$

which gives the *a posteriori* distribution of:

$$p(f|d) = \frac{1}{Z} e^{-\left(\beta \sum_{i,j} f_i f_j\right) - \left(\frac{1}{2\sigma^2}\right) \sum_{i \in S}(d_i - f_i)^2} \qquad (27)$$

To find the solution the energy function in the exponent is minimized:

$$U(f) = -\left(\beta \sum_{i,j} f_i f_j\right) - \left(\frac{1}{2\sigma^2}\right) \sum_{i \in S}(d_i - f_i)^2 \qquad (28)$$

Object Segmentation

Another classic application of MAP-MRF is to segment an object from the rest of an image (part of the classic graph cut problem). Here, we present Boykov and Jolly's algorithm,[11] which uses an auto-model. Specifically, the user selects a few pixels that are a part of the object of interest and a few pixels that are definitely not on the object of interest. The unary term focuses on the how well the remaining pixels match a description of the object or of the background. Although a number of potential functions are named, the one used in this implementation assigns a value to each pixel based on how well it matches the log of two histograms (one each for object and background pixels, respectively). The binary term governs smoothness, again with a list of potentially useful functions. In the specific implementation, a simple penalty was used:

$$V_2(f_i) = \exp\left(-\frac{(I_i - I_j)^2}{2\sigma^2} \cdot \frac{1}{\text{dist}(i,j)}\right) \qquad (29)$$

for $i \neq j$. This gives us the following energy function:

$$U(f) = \lambda \cdot \sum_{i \in S} V_1(f_i) + \sum_{i \in S} \sum_{j \in \mathcal{N}_i} V_2(f_i, f_j) \qquad (30)$$

where $\lambda \geq 0$ is a constant that determines the relative weight of the unary term. Minimizing the resulting energy function yields a rather successful segmentation of the object.

COMPLEX MODELS AND PATCH-BASED REGULARITIES

Most of the MRF models contain relatively simple underlying functions. There are, however, several methods for combining multiple simple models to make a more powerful one. By combining different classes of models and learning their parameters, it should be possible to ensure that we can adequately represent real-world statistics.

Perhaps the most promising approach to combining simple models is the *Products of Experts* (PoE) technique. Here, we present the approach, followed by how it can be used with MRFs (a procedure called *Field of Experts*) to learn the statistical regularities of image patches.

Products of Experts

High-dimensional datasets can be modeled with PoE.[12] The basic idea is to create several straightforward models called experts, each of which is trained to model a part of the dataset (such as one of the many dimensions in a high-dimensional dataset) and then multiply the experts together (and then normalizing) to model the full dimensionality of the dataset. The multiplication overcomes some limitations of mixture models (see Chapter 7). The experts essentially model how likely a given image is given a specific set of models $p(d|\theta_n)$ where d is the observed data (i.e., the image) and θ_n represents the parameters of model n. The joint probability is then given by:

$$p(d|\theta_1,\ldots,\theta_n) = \frac{\Pi_n p(d|f_n,\theta_n)}{\Sigma_c \Pi_n p(c|f_n,\theta_n)} \quad (31)$$

where c indexes all possible vectors in the dataspace (e.g., all possible images). Using machine-learning techniques (including sampling and contrastive divergence to make the problem computationally tractable), it is possible to learn the parameters of the individual experts on a training dataset. The trained experts can be used (with some limitations) to process new data. For example, a large number of simple experts, each of which is a mixture of a uniform distribution and a Gaussian aligned to a single axis in the dataset, can give rise to edge detectors when trained on a training set of images that consist of a single edge each.[12]

Since natural image statistics tend to be highly kurtotic, using Student-*t* distributions as the basis of the experts is a viable alternative.[13] This procedure is called a *Product of T-Distribution* (PoT) model. Learning the parameters of a set of Student-*t*-based experts on 5 × 5 pixel image patches from a database of natural images has produced filters that look much like those obtained in sparse encoding approaches.

Fields of Experts

Most MRFs focus on first-order neighborhoods and use simple models, meaning they can only explicitly capture simple, very local image structures.[14,15] It is clear, however, that both longer-range and more complex interactions are very important in images and need to be modeled. One approach to include larger ranges of context using more complex models is to use Product of Experts in a Markov random field, which is called a *Field of Experts* (FoE) for short. In essence, the experts of a PoE are learned, as mentioned earlier, and then used as priors in MRFs with very large, overlapping cliques. Student-*t* distributions and a variant on the L1 norm have both been shown to be effective as the basis of such experts.[14,15]

The Fields of Experts were tested on the Berkeley Segmentation Benchmark.[16] For computational reasons, the size of the image was limited to being between three and five times as large as the patches. To test the approach, 20,000 image patches were randomly extracted (usually each 5 × 5 pixels large). These were subsequently converted to grayscale. They then created subsets of 200 patches each (again, for computational reasons) and used a different subset at each training step. There were usually 5,000 training steps in the experiments. Visual inspection of the recovered filters does not immediately lead to any recognizable structure. They do not look like edge detectors or any of the usual filters found in sparse encoding approaches (or even in PoE approaches). Roth et al. suggest that this is due to the overlap of the cliques.

A direct comparison of the models obtained under different circumstances (e.g., clique sizes, training steps, etc.) is not possible since the partition function cannot be evaluated. Instead, the effectiveness of different FoEs on different tasks, such as image denoising and inpainting can be assessed. The FoE model performs very well, showing that these models can capture some regularities in an image.

STATISTICAL ANALYSIS OF MRFS

Using MRFs within a Bayesian framework is an elegant way to perform many computer vision tasks or even to model the human perceptual system. It is amazingly flexible and can, in principle, measure a truly astonishing variety of statistical regularities. A part of this flexibility is due to the fact that the researcher is free to specify the underlying model. Once a model is specified, it is then possible to use a set of images as a training dataset to estimate the parameters of the model underlying the MRF.

In most cases, however, it is unclear how well the model has represented the images or what features it has modeled.[17] This is largely because the partition function is computationally intractable. Thus, in most cases, the success of learning image parameters is tested by using the MRF to perform some task such as image denoising or image segmentation.

Zhu and Mumford,[18] however, suggested that an appropriate way to see what the models are modeling is to take advantage of the generative nature of MRFs. That is, as mentioned earlier, it is possible to use a (trained) MRF to synthesize textures. To examine the statistics of their MRF, Zhu and Mumford trained it on 44 real-world images and then generated several images from their MRF—a process they called sampling the MRF. Finally, they examined the statistical moments of the marginal distributions of both real and synthesized images. As usual, the intensity histograms of real-world images had high kurtosis and heavy tails and the statistics were (mostly) scale invariant. The synthesized images neither visually resembled real-world images nor did their statistics. Although the intensity histograms of the synthetic images had flat tails, they had only small spikes. Zhu and Mumford suggested that this was due to the size of the filters used. Adjusting the filter size improved the similarity to real-world images somewhat.

Later, several researchers simultaneously revived the idea of using a trained MRF to synthesize images and then examine their statistical moments.[19–24] As an example, Lyu and Simoncelli[21] trained an MRF based on Fields of Gaussian Scale Mixtures using five images (each 512×512 pixels) using 5×5 pixel neighborhoods. The marginal statistics of the samples turn out to be clearly non-Gaussian but not as heavy-tailed as real-world images. The joint statistics were also close but note quite the same as real-world images.

Ranzato and colleagues[22] augmented the MRF to have two latent variables rather than one. The first set of latent variables models pixel intensities. The second set models image-specific pixel covariances. They call this a gated MRF. They use a Product of Student-t model (PoT)[25] as the base MRF and they also add the ability to model mean pixel intensities and not just pixel pair differences. After training, they sampled the model (with patches of 160×160 pixels) and assessed how similar the intensity histograms of synthesized images and random images were to either random images or real-world images using the KullbackLeibler divergence. This is a measure of how different two probability distribution functions are, as explained in the appendix. They found that adding mean pixel intensities greatly improves the quality of the generated images. Interestingly, not accounting for the mean pixel intensity yields intensity histograms that are closer to random images than real ones.

Schmidt and colleagues suggest that a more flexible underlying model, a variant of Gaussian scale mixtures, might yield better results than traditional ones.[24] They examined intensity histograms and multiscale derivatives using the Kullback-Leibler divergence. They found that for both pairwise MRFs and Fields of Experts, this model yields a higher kurtosis than the older models. We used this procedure on a set of 15 images (shown in Fig. 7.8) using pairwise MRF and obtained the results shown in Fig. 3. We also used a 3×3 clique Fields of Experts on the same set of images and obtained the results shown in Fig. 4. Note that the experts shown in this figure have a broad base and a narrow peak, showing heavy-tailed behavior. On of the applications of this approach is image inpainting, a result of which is shown in the bottom-right of Fig. 5. Note that the Fields of Experts improve the quality of the result somewhat relative to pairwise MRFs.

Schmidt and colleagues also find that the Bayesian minimum mean squared error estimate (MMSE) was

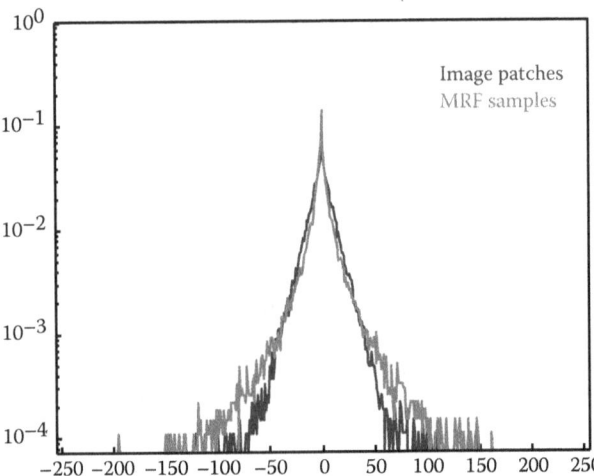

Fig. 3 Filter marginals computed on an MRF model using a variant of Gaussian scale mixtures as the underlying model.[24] They are compared with the derivative statistics of the image patches themselves. The images used to generate these results are shown in Fig. 7.8

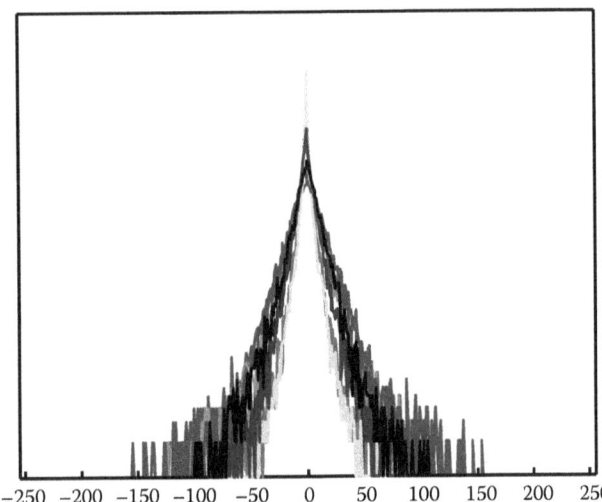

Fig. 4 Eight filter marginals computed on a 3×3 clique FoE model[24]

better than MAP. As their approach lends itself not only to analysis but also applies in a generative setting, Schmidt et al. showed its applicability in several scenarios, including image inpainting (a result obtained with this method is given in the bottom-left of Fig. 5). Subsequently, they further extended the model and showed that the synthesized images more accurately matched the multiscale derivative statistics, random filter statistics, and joint feature statistics of of real-world images.[17]

Several other researchers have shown that MRFs trained on patches do reasonably well but are not perfect. Karklin and Lewicke[19] examined the properties of MRFs based on multivariate Gaussians trained on 20×20 pixel patches from real-world images and compared the responses to

Fig. 5 Inpainting result using pairwise MRFs (bottom left) and FoE (bottom right), generated with the method of Schmidt et al.[24] (Dunedin, New Zealand, 2012)

those of simulated neurons. They found that the properties are similar to those in visual areas V1 and V2 in the human visual system. Likewise, Osindero and Hinton[23] examined the performance of a model that has several hidden layers, each with its own MRF (conditioned on the variables in the layer above). They used 150,000 patches each of 20 × 20 pixels to train their model. They generated 10,000 sample images and found that the intensity histograms for the real-world patches have a high kurtosis (8.3). The version of their model that includes lateral connections yields images with a kurtosis of 7.3, while the version without lateral connections has a kurtosis of 3.4.

REFERENCES

1. Schwabe, L.; Obermayer, K.; Angelucci, A.; Bressloff, P.C. The role of feedback in shaping the extra-classical receptive field of cortical neurons: A recurrent network model. J. Neurosci. **2006**, *26* (36), 9117–9129, 207.
2. Ising, E. *Beitrag zur Theorie des Ferro- und Paramagnetismus*; PhD thesis; University of Hamburg, 1924, 208.
3. Niss, M. History of the Lenz-Ising model 1920–1950: From ferromagnetic to cooperative phenomena. Arch. Hist. Exact Sci. **2005**, *59* (3), 267–318, 208.
4. Geman, S.; Geman, D. Stochastic relaxation, Gibbs distributions, and the bayesian restoration of images. IEEE Trans. Pattern Anal. Mach. Intell. **1984**, *6* (6), 721–741, 203, 208, 216.
5. Kindermann, R.; Snell, J.L. *Markov Random Fields and their Applications*, Vol. 1; American Mathematical Society: Providence, RI, 1980, 208.
6. Li, S.Z. *Markov Random Field Modeling in Image Analysis*; Springer, 2009, 208, 215, 216.
7. Blake, A.; Kohli, P.; Rother, C. *Markov Random Fields for Vision and Image Processing*; MIT Press: Cambridge, MA, 2011, 208.
8. Gregory, R.L. Perceptual illusions and brain models. Proc. Royal Soc. Lond. B **1968**, *171* (1024), 279–296, 208.
9. Hammersley, J.M.; Clifford, P.E. Markov random fields on finite graphs and lattices. Unpublished manuscript, 1971, 213.
10. Geman, S.; Graffigne, C. Markov random field image models and their applications to computer vision. In *Proceedings of the International Congress of Mathematicians*, Vol. 1; AMS: Providence, RI, 1986, 2, 216.
11. Boykov, Y.Y.; Jolly, M.-P. Interactive graph cuts for optimal boundary & region segmentation of objects in N-D images. In *Proceedings of the 8th IEEE International Conference on Computer Vision*, Vol. 1; 2001, 105–112, 217.
12. Hinton, G.E. Training products of experts by minimizing contrastive divergence. Neural Comput. **2002**, *14* (8), 1771–1800, 218.
13. Welling, M.; Hinton, G.E.; Osindero, S. Learning sparse topographic representations with products of student-t distributions. In *NIPS-15: Proceedings of the 2002 Conference on Advances in Neural Information Processing Systems*, 2002, 1359–1366, 218.
14. Roth, S.; Black, M.J. Fields of experts: A framework for learning image priors. In *Proceedings of the IEEE International Conference on Computer Vision and Pattern Recognition*, Washington, DC, 2005, 860–867, 7, 219.
15. Roth, S.; Black, M.J. Fields of experts. Int. J. Comput. Vision **2009**, *82* (2), 205–229, 219.
16. Martin, D.; Fowlkes, C.; Tal, D.; Malik, J. A database of human segmented natural images and its application to

evaluating segmentation algorithms and measuring ecological statistics. In *Proceedings of the IEEE International Conference in Computer Vision*, Vol. 2; 2001, 416–423, 219.

17. Gao, Q.; Roth, S. How well do filter-based MRFs model natural images? In *Pattern Recognition*; Pinz, A.; Pock, T.; Bischof, H.; Leberl, F.; Eds. Volume 7476 of Lecture Notes in Computer Science; Springer: Berlin, 2012, 62–72.

18. Zhu, S.; Mumford, D. Prior learning and Gibbs reaction-diffusion. IEEE Trans. Pattern Anal. Mach. Intell. **1997**, *19* (1), 1236–1250.

19. Karklin, Y.; Lewicki, M.S. Emergence of complex cell properties by learning to generalize in natural scenes. Nature **2008**, *457* (7225), 83–86.

20. Levi, E. *Using Natural Image Priors—Maximizing or Sampling?* Master's thesis; The Hebrew University, 2009.

21. Lyu, S.; Simoncelli, E.P. Modeling multiscale subbands of photographic images with fields of Gaussian scale mixtures. IEEE Trans. Pattern Anal. Mach. Intell. **2009**, *31* (4), 693–706.

22. Ranzato, M.; Mnih, V.; Hinton, G.E. Generating more realistic images using gated MRF's. In *NIPS-23: Proceedings of the 2010 Conference on Advances in Neural Information Processing Systems*, 2010, 2002–2010.

23. Osindero, S.; Hinton, G.E. Modeling image patches with a directed hierarchy of Markov random fields. In *NIPS-22: Proceedings of the 2008 Conference on Advances in Neural Information Processing Systems*, vol. 20, 2008, 1121–1128.

24. Schmidt, U.; Gao, Q.; Roth, S. A generative perspective on MRFs in low-level vision. In *Proceedings of the IEEE International Conference on Computer Vision and Pattern Recognition*, 2010, 1751–1758.

25. The, Y.W.; Welling, M.; Osindero, S.; Hinton, G.E. Energy-based models for sparse overcomplete representations. J. Mach. Learn. Res. **2003**, *4*, 1235–1260.

26. Keshner, M. 1/f noise. Proc. IEEE **1982**, *70* (3), 212–218.

27. Savilli, M.; Lecoy, G.; Nougier, J. *Noise in Physical Systems and 1/f Noise*; Elsevier: New York, 1983.

Mathematical Topics from 3-D Graphics

Fletcher Dunn
Valve Software, Bellevue, Washington, U.S.A.; Computer Science and Engineering, University of North Texas, Denton, Texas, U.S.A.

Abstract
This entry introduces a number of mathematical issues that arise when creating 3-D graphics on a computer. It presents an extremely brief and high-level overview of the subject matter, focusing on topics for which mathematics plays a critical role. The authors try to pay special attention to those topics that, from their experience, are glossed over in other sources or are a source of confusion to beginners.

Keywords: Rendering; Graphics; Bidirectional reflectance distribution function (BRDF); Colorimetry; Radiometry.

This entry alone is not enough to teach you how to get some pretty pictures on the screen. However, it should be used parallel with (or preceding!) some other course, book, or self-study on graphics, and we hope that it will help you breeze past a few traditional sticky points. You will not find much to help you figure out which DirectX or OpenGL function calls to make to achieve some desired effect. These issues are certainly of supreme practical importance, but alas, they are also in the category of knowledge that Robert Maynard Hutchins dubbed "rapidly aging facts," and we have tried to avoid writing an entry that requires an update every other year when ATI releases a new card or Microsoft a new version of DirectX. Luckily, up-to-date API references and examples abound on the Internet, which is a much more appropriate place to get that sort of thing. (API stands for application programming interface. In this entry, API will mean the software that we use to communicate with the rendering subsystem.)

One final caveat is that since this is an entry on math for video games, we will have a real-time bias. This is not to say that the entry cannot be used if you are interested in learning how to write a raytracer; only that our expertise and focus is in real-time graphics.

HOW GRAPHICS WORKS

We begin our discussion of graphics by telling you how things *really* work, or perhaps more accurately, how they really *should* work, if we had enough knowledge and processing power to make things work the right way. The beginner student is to be warned that much introductory material (especially tutorials on the Internet) and API documentation suffers from a great lack of perspective. You might get the impression from reading these sources that diffuse maps, Blinn–Phong shading, and ambient occlusion are "the way images in the real world work," when in fact you are probably reading a description of how one particular lighting model was implemented in one particular language on one particular piece of hardware through one particular API. Ultimately, any down-to-the-details tutorial must choose a lighting model, language, platform, color representation, performance goals, etc.—as other entries will have to do. (This lack of perspective is usually purposeful and warranted.) However, we think it is important to know which are the fundamental and timeless principles, and which are arbitrary choices based on approximations and trade-offs, guided by technological limitations that might be applicable only to real-time rendering, or are likely to change in the near future. So, before we get too far into the details of the particular type of rendering most useful for introductory real-time graphics, we want to take our stab at describing how rendering *really* works.

We also hasten to add that this discussion assumes that the goal is photorealism, simulating how things work in nature. In fact, this is often not the goal, and it certainly is never the only goal. Understanding how nature works is a very important starting place, but artistic and practical factors often dictate a different strategy than just simulating nature.

The Two Major Approaches to Rendering

We begin with the end in mind. The end goal of rendering is a bitmap, or perhaps a sequence of bitmaps if we are producing an animation. You almost certainly already know that a bitmap is a rectangular array of colors, and each grid entry is known as *pixel*, which is short for "picture element." At the time we are producing the image, this bitmap is also known as the *frame buffer*, and often there is additional postprocessing or conversion that happens when we copy the frame buffer to the final bitmap output.

How do we determine the color of each pixel? That is the fundamental question of rendering. Like so many challenges in computer science, a great place to start is by investigating how nature works.

We see light. The image that we perceive is the result of light that bounces around the environment and finally enters the eye. This process is complicated, to say the least. Not only is the physics of the light bouncing around very complicated (actually, almost everybody approximates the true physics of light by using simpler geometric optics), but so are the physiology of the sensing equipment in our eyes and the interpreting mechanisms in our minds. Speaking of equipment, there are also many phenomena that occur in a camera but not the eye, or as a result of the storage of an image on film. These effects, too, are often simulated to make it look as if the animation was filmed. Thus, ignoring a great number of details and variations (as any introductory entry must do), the basic question that any rendering system must answer for each pixel is "What color of light is approaching the camera from the direction corresponding to this pixel?"

There are basically two cases to consider: Either we are looking directly at a light source and light traveled directly from the light source to our eye, or (more commonly) light departed from a light source in some other direction, bounced one or more times, and then entered our eye. We can decompose the key question asked previously into two tasks. This entry calls these two tasks *the rendering algorithm*, although these two highly abstracted procedures obviously conceal a great deal of complexity about the actual algorithms used in practice to implement it.

The rendering algorithm

- *Visible surface determination.* Find the surface that is closest to the eye, in the direction corresponding to the current pixel.
- *Lighting.* Determine what light is emitted and/or reflected off this surface in the direction of the eye.

At this point, it appears that we have made some gross simplifications, and many of you no doubt are raising your metaphorical hands to ask "What about translucency?" "What about reflections?" "What about refraction?" "What about atmospheric effects?" Please hold all questions until the end of the presentation.

The first step in the rendering algorithm is known as *visible surface determination*. There are two common solutions to this problem. The first is known as *raytracing*. Rather than following light rays in the direction that they travel from the emissive surfaces, we trace the rays backward, so that we can deal only with the light rays that matter: the ones that enter our eye from the given direction. We send a ray out from the eye in the direction through the center of each pixel to see the first object in the scene this ray strikes. Then we compute the color that is being emitted or reflected from that surface back in the direction of the ray. A highly simplified summary of this algorithm is illustrated by Table 1.

Actually, it is probably not a good idea to think of pixels as having a "center," as they are not really rectangular blobs of color, but rather are best interpreted as infinitely small point samples in a continuous signal. The question of which mental model is best is incredibly important,[1,2] and is intimately related to the process by which the pixels are combined to reconstruct an image. On CRTs, pixels were definitely not little rectangles, but on modern display devices such as LCD monitors, "rectangular blob of color" is not too bad of a description of the reconstruction process. Nonetheless, whether pixels are rectangles or point samples, we still might not send a single ray through the center of each pixel, but rather we might send several rays ("samples") in a smart pattern, and average them together them in a smart way. The other major strategy for visible surface determination, the one used for real-time rendering at the time of this writing, is known as *depth buffering*. The basic plan is that at each pixel we store not only a color value, but also a depth value. This depth buffer value records the distance from the eye to the surface that is reflecting or emitting the light used to determine the color for that pixel. As illustrated in Table 1, the "outer loop" of a raytracer is the screen-space pixel, but in real-time graphics, the "outer loop" is the geometric element that makes up the surface of the scene.

The different methods for describing surfaces are not important here. What *is* important is that we can project the surface onto screen-space and map them to screen-space pixels through a process known as *rasterization*. For each pixel of the surface, known as the *source fragment*, we compute the depth of the surface at that pixel and compare it to the existing value in the depth buffer, sometimes known as the *destination fragment*. If the source fragment we are currently rendering is farther away from the camera than the existing value in the buffer, then whatever we rendered before this is obscuring the surface we are now rendering (at least at this one pixel), and we move on to the next pixel. However, if our depth value is closer than the existing value in the depth buffer, then we know this is the closest surface to the eye (at least of those rendered so far) and so we update the depth buffer with this new, closer depth value. At this point, we might also proceed to step 2 of the rendering algorithm (at least for this pixel)

Table 1 Pseudocode for the raytracing algorithm

```
for (each x, y screen pixel) {
    // Select a ray for this pixel
    Ray ray = getRayForPixel (x, y) ;

    // Intersect the ray against the geometry. This will
    // not just return the point of intersection, but also
    // a surface normal and some other information needed
    // to shade the point, such as an object reference,
    // material information, local S, T coordinates, etc.
    // Don ' t take this pseudocode too literally.
    Vector3 pos, normal ;
    Object *obj ; Material *mtl ;
    if (rayIntersectScene (ray, pos, normal, obj, mtl)) {

        // Shade the intersection point. (What light is
        // emitted / reflected from this point towards the camera ?)
        Color c = shadePoint (ray, pos, normal, obj, mtl) ;

        // Put i t into the frame buffer
        writeFrameBuffer (x, y, c) ;

    } else {

        // Ray missed the entire scene. Just use a generic
        // background color at this pixel
        writeFrameBuffer (x, y, backgroundColor) ;

    }

}
```

Table 2 Pseudocode for forward rendering using the depth buffer

```
// Clear the frame and depth buffers
fillFrameBuffer (backgroundColor) ;
fillDepthBuffer (infinity);

// Outer loop iterates over all the primitives (usually triangles)
for (each geometric primitive) {

    // Rasterize the primitive
    for (each pixel x, y in the projection of the primitive) {

        // Test the depth buffer, to see if a closer pixel has
        // already been written.
        float primDepth = getDepthOfPrimitiveAtPixel (x, y) ;
        if (primDepth>readDepthBuffer (x, y)) {

            // Pixel of this primitive is obscured, discard it
            continue ;
        }

        // Determine primitive color at this pixel.
        Color c = getColorOfPrimitiveAtPixel (x, y) ;

        // Update the color and depth buffers
        writeFrameBuffer (x, y, c) ;
        writeDepthBuffer (x, y, primDepth) ;
    }
}
```

and update the frame buffer with the color of the light being emitted or reflected from the surface that point. This is known as *forward rendering*, and the basic idea is illustrated by Table 2.

Opposed to forward rendering is *deferred rendering*, an old technique that is becoming popular again due to the current location of bottlenecks in the types of images we are producing and the hardware we are using

Table 3 Pseudocode for deferred rendering using the depth buffer

```
// Clear the geometry and depth buffers
clearGeometryBuffer () ;
fillDepthBuffer (infinity);

// Rasterize all primitives into the G-buffer
for (each geometric primitive) {
    for (each pixel x, y in the projection of the primitive) {

        // Test the depth buffer, to see if a closer pixel has
        // already been written.
        float primDepth = getDepthOfPrimitiveAtPixel (x, y) ;
        if (primDepth>readDepthBuffer (x, y)) {

            // Pixel of this primitive is obscured, discard it
            continue ;
        }

        // Fetch information needed for shading in the next pass.
        MaterialInfo mtlInfo ;
        Vector3 pos, normal ;
        getPrimitiveShadingInfo (mtlInfo, pos, normal) ;

        // Save it off into the G-buffer and depth buffer
        writeGeometryBuffer (x, y, mtlInfo, pos, normal) ;
        writeDepthBuffer (x, y, primDepth) ;
    }
}

// Now perform shading in a 2nd pass, in screen space
for (each x, y screen pixel) {
    if (readDepthBuffer (x, y) == infinity) {

        // No geometry here. Just write a background color
        writeFrameBuffer (x, y, backgroundColor) ;

    } else {

        // Fetch shading info back from the geometry buffer
        MaterialInfo mtlInfo ;
        Vector3 pos, normal ;
        readGeometryBuffer (x, y, mtlInfo, pos, normal) ;

        // Shade the point
        Color c = shadePoint (pos, normal, mtlInfo) ;

        // Put it into the frame buffer
        writeFrameBuffer (x, y, c) ;
    }
}
```

to produce them. A deferred renderer uses, in addition to the frame buffer and the depth buffer, additional buffers, collectively known as the *G-buffer* (short for "geometry" buffer), which holds extra information about the surface closest to the eye at that location, such as the 3-D location of the surface, the surface normal, and material properties needed for lighting calculations, such as the "color" of the object and how "shiny" it is at that particular location. (Later, we see how those intuitive terms in quotes are a bit too vague for rendering purposes.) Compared to a forward renderer, a deferred renderer follows our two-step rendering algorithm a bit more literally. First, we "render" the scene into the G-buffer, essentially performing only

visibility determination—fetching the material properties of the point that is "seen" by each pixel but not yet performing lighting calculations. The second pass actually performs the lighting calculations. Table 3 explains deferred rendering in pseudocode.

Before moving on, we must mention one important point about why deferred rendering is popular. When multiple light sources illuminate the same surface point, hardware limitations or performance factors may prevent us from computing the final color of a pixel in a single calculation, as was shown in the pseudocode listings for both forward and deferred rendering. Instead, we must use multiple passes, one pass for each light, and *accumulate* the reflected light from each light source into the frame buffer. In forward rendering, these extra passes involve rerendering the primitives. Under deferred rendering, however, extra passes are in image space, and thus depend on the 2-D size of the light in screen-space, not on the complexity of the scene! It is in this situation that deferred rendering really begins to have large performance advantages over forward rendering.

Describing Surface Properties: The BRDF

Now let us talk about the second step in the rendering algorithm: lighting. Once we have located the surface closest to the eye, we must determine the amount of light emitted directly from that surface, or emitted from some other source and reflected off the surface in the direction of the eye. The light directly transmitted from a surface to the eye—for example, when looking directly at a light bulb or the sun—is the simplest case. These *emissive* surfaces are a small minority in most scenes; most surfaces do not emit their own light, but rather they only reflect light that was emitted from somewhere else. We will focus the bulk of our attention on the nonemissive surfaces.

Although we often speak informally about the "color" of an object, we know that the perceived color of an object is actually the light that is entering our eye, and thus can depend on many different factors. Important questions to ask are: What colors of light are incident on the surface, and from what directions? From which direction are we viewing the surface? How "shiny" is the object? Further relevant questions that should influence what color we write into the frame buffer could be asked concerning the general viewing conditions, but these issues have no bearing on the light coming into our eye; rather, they affect our perception of that light. So a description of a surface suitable for use in rendering does not answer the question "What color is this surface?" This question is sometimes meaningless—what color is a mirror, for example? Instead, the salient question is a bit more complicated, and it goes something like, "When light of a given color strikes the surface from a given incident direction, how much of that light is reflected in some other particular direction?" The answer to this question is given by the *bidirectional reflectance distribution function*, or BRDF for short. So rather than "What color is the object?" we ask, "What is the distribution of reflected light?"

Symbolically, we write the BRDF as the function f(**x**, $\hat{\omega}_{in}$, $\hat{\omega}_{out}$, λ). (Remember that ω and λ are the lowercase Greek letters omega and lambda, respectively.) The value of this function is a scalar that describes the relatively likelihood that light incident at the point **x** from direction $\hat{\omega}_{in}$ will be reflected in the outgoing direction $\hat{\omega}_{out}$ rather than some other outgoing direction. As indicated by the boldface type and hat, $\hat{\omega}$ might be a unit vector, but more generally it can be any way of specifying a direction; polar angles are another obvious choice and are commonly used. Different colors of light are usually reflected differently; hence the dependence on λ, which is the color (actually, the wavelength) of the light.

Although we are particularly interested in the incident directions that come from emissive surfaces and the outgoing directions that point toward our eye, in general, the entire distribution is relevant. First of all, lights, eyes, and surfaces can move around, so in the context of creating a surface description (for example, "red leather"), we do not know which directions will be important. But even in a particular scene with all the surfaces, lights, and eyes fixed, light can bounce around multiple times, so we need to measure light reflections for arbitrary pairs of directions.

Before moving on, it is highly instructive to see how the two intuitive material properties that were earlier disparaged, color and shininess, can be expressed precisely in the framework of a BRDF. Consider a green ball. A green object is green and not blue because it reflects incident light that is green more strongly than incident light of any other color. (Here and elsewhere, we use the word "color" in a way that is technically a bit dodgy, but is OK under the assumptions about light and color made in most graphics systems.) For example, perhaps green light is almost all reflected, with only a small fraction absorbed, while 95% of the blue and red light is absorbed and only 5% of light at those wavelengths is reflected in various directions. White light actually consists of all the different colors of light, so a green object essentially filters out colors other than green. If a different object responded to green and red light in the same manner as our green ball, but absorbed 50% of the blue light and reflected the other 50%, we might perceive the object as teal. Or if most of the light at all wavelengths was absorbed, except for a small amount of green light, then we would perceive it as a dark shade of green. To summarize, a BRDF accounts for the difference in color between two objects through the dependence on λ: any given wavelength of light has its own reflectance distribution.

Next, consider the difference between shiny red plastic and diffuse red construction paper. A shiny surface reflects incident light much more strongly in one particular direction compared to others, whereas a diffuse surface scatters light more evenly across all outgoing directions. A perfect reflector, such as a mirror, would reflect all the light from one incoming direction in a single outgoing direction, whereas a perfectly diffuse surface would reflect light equally in all outgoing directions, regardless of the direction of incidence. In summary, a BRDF accounts for the difference in "shininess" of two objects through its dependence on $\hat{\omega}_{in}$ and $\hat{\omega}_{out}$.

More complicated phenomena can be expressed by generalizing the BRDF. Translucence and light refraction can be easily incorporated by allowing the direction vectors to point back into the surface. We might call this mathematical generalization a *bidirectional surface scattering distribution function* (BSSDF). Sometimes, light strikes an object, bounces around inside of it, and then exits at a different point. This phenomenon is known as *subsurface scattering* and is an important aspect of the appearances of many common substances, such as skin and milk. This requires splitting the single reflection point **x** into $\mathbf{x}_{in}$ and $\mathbf{x}_{out}$, which is used by the BSSDF. Even volumetric effects, such as fog and subsurface scattering, can be expressed, by dropping the words "surface" and defining a *bidirectional scattering distribution function* (BSDF) at any point in space, not just on the "surfaces." Taken at face value, these might seem like impractical abstractions, but they can be useful in understanding how to design practical tools.

By the way, there are certain criteria that a BRDF must satisfy in order to be physically plausible. First, it does not make sense for a negative amount of light to be reflected in any direction. Second, it is not possible for the total reflected light to be more than the light that was incident, although the surface may absorb some energy so the reflected light can be less than the incident light. This rule is usually called the *normalization constraint*. A final, less obvious principle obeyed by physical surfaces is *Helmholtz reciprocity*: if we pick two arbitrary directions, the same fraction of light should be reflected, no matter which is the incident direction and which is the outgoing direction. In other words,

Helmholtz reciprocity $\quad f(\mathbf{x}, \hat{\omega}_1, \hat{\omega}_2, \lambda) = f(\mathbf{x}, \hat{\omega}_2, \hat{\omega}_1, \lambda)$

Due to Helmholtz reciprocity, some authors do not label the two directions in the BRDF as "in" and "out" because to be physically plausible, the computation must be symmetric.

The BRDF contains the complete description of an object's appearance at a given point, since it describes how the surface will reflect light at that point. Clearly, a great deal of thought must be put into the design of this function. Numerous lighting models have been proposed over the last several decades, and what is surprising is that one of the earliest models, Blinn–Phong, is still in widespread use in real-time graphics today. Although it is not physically accurate (nor plausible: it violates the normalization constraint), we study it because it is a good educational stepping stone and an important bit of graphics history. Actually, describing Blinn–Phong as "history" is wishful thinking—perhaps the most important reason to study this model is that it still is in such widespread use! In fact, it is the best example of the phenomenon we mentioned at the start of this entry: particular methods being presented as if they are "the way graphics work."

Different lighting models have different goals. Some are better at simulating rough surfaces, others at surfaces with multiple strata. Some focus on providing intuitive "dials" for artists to control, without concern for whether those dials have any physical significance at all. Others are based on taking real-world surfaces and measuring them with special cameras called goniophotometers, essentially sampling the BRDF and then using interpolation to reconstruct the function from the tabulated data. The notable Blinn–Phong model discussed in the entry "*3-D Graphics: Standard Local Lighting Model*" (see pp. 66–74) is useful because it is simple, inexpensive, and well understood by artists. Consult the sources in the suggested reading for a survey of lighting models.

A Very Brief Introduction to Colorimetry and Radiometry

Graphics is all about measuring light, and you should be aware of some important subtleties, even though we will not have time to go into complete detail here. The first is how to measure the color of light, and the second is how to measure its brightness.

In your middle school science classes, you might have learned that every color of light is some mixture of red, green, and blue (RGB) light. This is the popular conception of light, but it is not quite correct. Light can take on any single frequency in the visible band, or it might be a combination of any number of frequencies. Color is a phenomenon of *human perception* and is not quite the same thing as frequency. Indeed different combinations of frequencies of light can be perceived as the same color—these are known as *metamers*. The infinite combinations of frequencies of light are sort of like all the different chords that can be played on a piano (and also tones between the keys). In this metaphor, our color perception is unable to pick out all the different individual notes, but instead, any given chord sounds to us like some combination of middle C, F, and G. Three color channels is not a magic number as far as physics is concerned, it is peculiar to human vision. Most other mammals have only two different types of receptors (we would call them "color blind"), and fish, reptiles, and birds have *four* types of color receptors (they would call *us* color blind).

However, even very advanced rendering systems project the continuous spectrum of visible light onto some discrete basis, most commonly, the RGB basis. This is a ubiquitous simplification, but we still wanted to let you know that it is a simplification, as it does not account for certain phenomena. The RGB basis is not the only color space, nor is it necessarily the best one for many purposes, but it is a very convenient basis because it is the one used by most display devices. In turn, the reason that this basis is used by so many display devices is due to the similarity to our own visual system. Hall[3] does a good job of describing the shortcomings of the RGB system.

Since the visible portion of the electromagnetic spectrum is continuous, an expression such as $f(\mathbf{x}, \hat{\omega}_{in}, \hat{\omega}_{out}, \lambda)$ is continuous in terms of λ. At least it should be in theory. In practice, because we are producing images for human consumption, we reduce the infinite number of different λs down to three particular wavelengths. Usually, we choose the three wavelengths to be those perceived as the colors red, green, and blue. In practice, you can think of the presence of λ in an equation as an integer that selects which of the three discrete "color channels" is being operated on.

KEY POINTS ABOUT COLOR

- To describe the spectral distribution of light requires a continuous function, not just three numbers. However, to describe the human perception of that light, three numbers are essentially sufficient.
- The RGB system is a convenient color space, but it is not the only one, and not even the best one for many practical purposes. In practice, we usually treat light as being a combination of red, green, and blue because we are making images for human consumption.

You should also be aware of the different ways that we can measure the intensity of light. If we take a viewpoint from physics, we consider light as energy in the form of electromagnetic radiation, and we use units of measurement from the field of *radiometry*. The most basic quantity is *radiant energy*, which in the SI system is measured in the standard unit of energy, the *joule* (J). Just like any other type of energy, we are often interested in the rate of energy flow per unit time, which is known as *power*. In the SI system power is measured using the *watt* (W), which is one joule per second (1 W = 1 J/sec). Power in the form of electromagnetic radiation is called *radiant power* or *radiant flux*. The term "flux," which comes from the Latin *fluxus* for "flow," refers to some quantity flowing across some cross-sectional area. Thus, radiant flux measures the total amount of energy that is arriving, leaving, or flowing across some area per unit time.

Imagine that a certain amount of radiant flux is emitted from a 1 m² surface, while that same amount of power is emitted from a different surface that is 100 m². Clearly, the smaller surface is "brighter" than the larger surface; more precisely, it has a greater flux per unit area, also known as *flux density*. The radiometric term for flux density, the radiant flux per unit area, is called *radiosity*, and in the SI system, it is measured in watts per meter. The relationship between flux and radiosity is analogous to the relationship between force and pressure; confusing the two will lead to similar sorts of conceptual errors.

Several equivalent terms exist for radiosity. First, note that we can measure the flux density (or total flux, for that matter) across any cross-sectional area. We might be measuring the radiant power emitted from some surface with a finite area, or the surface through which the light flows might be an imaginary boundary that exists only mathematically (for example, the surface of some imaginary sphere that surrounds a light source). Although in all cases we are measuring flux density, and thus the term "radiosity" is perfectly valid, we might also use more specific terms, depending on whether the light being measured is coming or going. If the area is a surface and the light is arriving on the surface, then the term *irradiance* is used. If light is being emitted from a surface, the term *radiant exitance* or *radiant emittance* is used. In digital image synthesis, the word "radiosity" is most often used to refer to light that is leaving a surface, having been either reflected or emitted.

When we are talking about the brightness at a particular point, we cannot use plain old radiant power because the area of that point is infinitesimal (essentially zero). We can speak of the flux *density* at a single point, but to measure flux, we need a finite area over which to measure. For a surface of finite area, if we have a single number that characterizes the total for the entire surface area, it will be measured in flux, but to capture the fact that different locations within that area might be brighter than others, we use a function that varies over the surface that will measure the flux density.

Now we are ready to consider what is perhaps the most central quantity we need to measure in graphics: the intensity of a "ray" of light. We can see why the radiosity is not the unit for the job by an extension of the ideas from the previous paragraph. Imagine a surface point surrounded by an emissive dome and receiving a certain amount of irradiance coming from all directions in the hemisphere centered on the local surface normal. Now imagine a second surface point experiencing the same amount of irradiance, only all of the illumination is coming from a single direction, in a very thin beam. Intuitively, we can see that a ray along this beam is somehow "brighter" than any one ray that is illuminating the first surface point. The irradiance is somehow "denser." It is denser *per unit solid area*.

The idea of a solid area is probably new to some readers, but we can easily understand the idea by comparing it to angles in the plane. A "regular" angle is measured (in radians) based on the length of its projection onto the unit circle. In the same way, a solid angle measures the *area* as projected onto the unit sphere surrounding the point. The SI unit for solid angle is the *steradian*, abbreviated "sr." The complete sphere has 4π sr; a hemisphere encompasses 2π sr.

By measuring the radiance per unit solid angle, we can express the intensity of light at a certain point as a function that varies based upon the direction of incidence. We are very close to having the unit of measurement that describes the intensity of a ray. There is just one slight catch, illustrated by Fig. 1, which is a close-up of a very thin pencil of light rays striking a surface. On the top, the rays strike the surface perpendicularly, and on the bottom, light rays of the same strength strike a different surface at an angle. The key point is that the area of the top surface is smaller than the area of the bottom surface; therefore, the irradiance on the top surface is larger than the irradiance on the bottom surface, despite the fact that the two surfaces are being illuminated by the "same number" of identical light rays. This basic phenomenon, that the angle of the surface

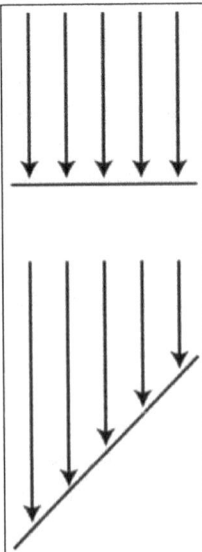

Fig. 1 The two surfaces are receiving identical bundles of light, but the surface on the bottom has a larger area, and thus has a lower irradiance

Table 4 Common radiometric terms

Quantity	Units	SI unit	Rough translation
Radiant energy	Energy	J	Total illumination during an interval of time
Radiant flux	Power	W	Brightness of a finite area from all directions
Radiant flux density	Power per unit area	W/m²	Brightness of a single point from all directions
Irradiance	Power per unit area	W/m²	Radiant flux density of incident light
Radiant exitance	Power per unit area	W/m²	Radiant flux density of emitted light
Radiosity	Power per unit area	W/m²	Radiant flux density of emitted or reflected light
Radiance	Power per unit projected area, per unit solid angle	W/(m² sr)	Brightness of a ray

Table 5 Units of measurement from radiometry and photometry

Radiometric term	Photometric term	SI photometric unit
Radiant energy	Luminous energy	talbot, or lumen second (lm s)
Radiant flux	Luminous flux, luminous power	lumen (lm)
Irradiance	Illuminance	lux (lx = lm/m²)
Radiant exitance	Luminous emittance	lux (lx = lm/m²)
Radiance	Luminance	lm/ (m² sr)

causes incident light rays to be spread out and thus contribute less irradiance, is known as *Lambert's law*. We have more to say about Lambert's law in the entry "*3-D Graphics: Standard Local Lighting Model*" (see pp. 66–74) but for now, the key idea is that the contribution of a bundle of light to the irradiance at a surface depends on the angle of that surface.

Due to Lambert's law, the unit we use in graphics to measure the strength of a ray, *radiance*, is defined as the radiant flux per unit *projected area*, per unit solid angle. To measure a projected area, we take the actual surface area and project it onto the plane perpendicular to the ray. (In Fig. 1, imagine taking the bottom surface and projecting it upward onto the top surface). Essentially this counteracts Lambert's law.

Table 4 summarizes the most important radiometric terms.

Whereas radiometry takes the perspective of physics by measuring the raw energy of the light, the field of *photometry* weighs that same light using the human eye. For each of the corresponding radiometric terms, there is a similar term from photometry (Table 5). The only real difference is a nonlinear conversion from raw energy to perceived brightness.

Throughout the remainder of this entry, we try to use the proper radiometric units when possible. However, the practical realities of graphics make using proper units confusing, for two particular reasons. It is common in graphics to need to take some integral over a "signal"—for example, the color of some surface. In practice, we cannot do the integral analytically, and so we must integrate numerically, which boils down to taking a weighted average of many samples. Although mathematically we are taking a weighted average (which ordinarily would not cause the units to change), in fact what we are doing is *integrating*, and that means each sample is really being multiplied by some differential quantity, such as a differential area or differential solid angle, which causes the physical units to change. A second cause of confusion is that, although many signals have a finite nonzero domain in the real world, they are represented in a computer by signals that are nonzero at a single point. (Mathematically, we say that the signal is a multiple of a Dirac delta.) For example, a real-world light source has a finite area, and we would be interested in the radiance of the light at a given point on the emissive surface, in a given direction. In practice, we imagine shrinking the area of this light down to zero while holding the radiant flux constant. The flux density becomes infinite in theory. Thus, for a real area light, we would need a signal to describe the flux density, whereas for a point light, the flux density becomes infinite and we instead describe

Mathematical Topics from 3-D Graphics

the brightness of the light by its total flux. We will repeat this information when we talk about point lights.

KEY POINTS ABOUT RADIOMETRY

- Vague words such as "intensity" and "brightness" are best avoided when the more specific radiometric terms can be used. The scale of our numbers is not that important and we do not need to use real-world SI units, but it is helpful to understand what the different radiometric quantities measure to avoid mixing quantities together inappropriately.
- Use radiant flux to measure the total brightness of a finite area, in all directions.
- Use radiant flux density to measure the brightness at a single point, in all directions. Irradiance and radiant exitance refer to radiant flux density of light that is incident and emitted, respectively. Radiosity is the radiant flux density of light that is leaving a surface, whether the light was reflected or emitted.
- Due to Lambert's law, a given ray contributes more differential irradiance when it strikes a surface at a perpendicular angle compared to a glancing angle.
- Use radiance to measure the brightness of a ray. More specifically, radiance is the flux per unit projected angle, per solid angle. We use projected area so that the value for a given ray is a property of a ray alone and does not depend on the orientation of the surface used to measure the flux density.
- Practical realities thwart our best intentions of doing things "the right way" when it comes to using proper units. Numerical integration is a lot like taking a weighted average, which hides the change of units that really occurs. Point lights and other Dirac deltas add further confusion.

The Rendering Equation

Now let us fit the BRDF into the rendering algorithm. In step 2 of our rendering algorithm, we are trying to determine the radiance leaving a particular surface in the direction of our eye. The only way this can happen is for light to arrive from some direction onto the surface and get reflected in our direction. With the BRDF, we now have a way to measure this. Consider all the potential directions that light might be incident upon the surface, which form a hemisphere centered on **x**, oriented according to the local surface normal $\hat{n}$. For each potential direction $\hat{\omega}_{in}$, we measure the color of light incident from that direction. The BRDF tells us how much of the radiance from $\hat{\omega}_{in}$ is reflected in the direction $\hat{\omega}_{out}$ toward our eye (as opposed to scattered in some other direction or absorbed). By summing up the radiance reflected toward $\hat{\omega}_{out}$ over all possible incident directions, we obtain the total radiance reflected along $\hat{\omega}_{out}$ into our eye. We add the reflected light to any light that is being *emitted* from the surface in our direction (which is zero for most surfaces), and voila, we have the total radiance. Writing this in math notation, we have the *rendering equation.*

The Rendering Equation

$$L_{out}(x, \hat{\omega}_{out}, \lambda) = L_{emis}(x, \hat{\omega}_{out}, \lambda) \\ + \int_{\Omega} L_{in}(x, \hat{\omega}_{in}, \lambda) f(x, \hat{\omega}_{in}, \hat{\omega}_{out}, \lambda)(-\hat{\omega}_{in} \cdot \hat{n}) d\hat{\omega}_{in} \quad (1)$$

As fundamental as Eq. 1 may be, its development is relatively recent, having been published in SIGGRAPH in 1986 by Kajiya.[4] Furthermore, it was the *result* of, rather than the cause of, numerous strategies for producing realistic images. Graphics researchers pursued the creation of images through different techniques that seemed to make sense to them before having a framework to describe the problem they were trying to solve. And for many years after that, most of us in the video game industry were unaware that the problem we were trying to solve had finally been clearly defined. (Many still are.)

Now let us convert this equation into English and see what the heck it means. First of all, notice that **x** and λ appear in each function. The whole equation governs a balance of radiance at a single surface point **x** for a single wavelength ("color channel") λ. So this balance equation applies to each color channel individually, at all surface points simultaneously.

The term $L_{out}(\mathbf{x}, \hat{\omega}_{out}, \lambda)$ on the left side of the equals sign is simply "The radiance leaving the point in the direction $\hat{\omega}_{out}$." Of course, if **x** is the visible surface at a given pixel, and $\hat{\omega}_{out}$ is the direction from **x** to the eye, then this quantity is exactly what we need to determine the pixel color. But note that the equation is more general, allowing us to compute the outgoing radiance in any arbitrary direction $\hat{\omega}_{out}$ and for any given point **x**, whether or not $\hat{\omega}_{out}$ points toward our eye.

On the right-hand side, we have a sum. The first term in the sum $L_{emis}(\mathbf{x}, \hat{\omega}_{out}, \lambda)$, is "the radiance emitted from **x** in the direction $\hat{\omega}_{out}$" and will be nonzero only for special emissive surfaces. The second term, the integral, is "the light reflected from **x** in the direction of $\hat{\omega}_{out}$." Thus, from a high level, the rendering equation would seem to state the rather obvious relation:

$$\begin{pmatrix} \text{Total radiance} \\ \text{towards } \hat{\omega}_{out} \end{pmatrix} = \begin{pmatrix} \text{Radiance emitted} \\ \text{towards } \hat{\omega}_{out} \end{pmatrix} \\ + \begin{pmatrix} \text{Radiance reflected} \\ \text{towards } \hat{\omega}_{out} \end{pmatrix}$$

Now let us dig into that intimidating integral. (By the way, if you have not had calculus, just replace the word "integral" with "sum," and you will not miss any of the main points of this section.) We have actually already discussed how it works when we talked about the BRDF, but let us repeat it with different words. We might rewrite the integral as

$$\begin{pmatrix} \text{Radiance reflected} \\ \text{towards } \hat{\omega}_{out} \end{pmatrix} = \int_{\Omega} \begin{pmatrix} \text{Radiance incident from } \hat{\omega}_{in} \\ \text{and reflected towards } \hat{\omega}_{out} \end{pmatrix} d\hat{\omega}_{in}$$

Note that symbol Ω (uppercase Greek omega) appears where we normally would write the limits of integration. This is intended to mean "sum over the hemisphere of possible incoming directions." For each incoming direction $\hat{\omega}_{in}$, we determine how much radiance was incident in this incoming direction and got scattered in the outgoing direction $\hat{\omega}_{out}$. The sum of all these contributions from all the different incident directions gives the total radiance reflected in the direction $\hat{\omega}_{out}$. Of course, there are an infinite number of incident directions, which is why this is an integral. In practice, we cannot evaluate the integral analytically, and we must sample a discrete number of directions, turning the "$\int$" into a "Σ."

Now all that is left is to dissect the integrand. It is a product of three factors:

$$\begin{pmatrix} \text{Radiance incident from } \hat{\omega}_{in} \\ \text{and reflected towards } \hat{\omega}_{out} \end{pmatrix} = L_{in}(x, \hat{\omega}_{in}, \lambda) f(x, \hat{\omega}_{in}, \hat{\omega}_{out}, \lambda)(-\hat{\omega}_{in} \cdot \hat{n})$$

The first factor denotes the radiance incident from the direction of $\hat{\omega}_{in}$. The next factor is simply the BRDF, which tells us how much of the radiance incident from this particular direction will be reflected in the outgoing direction we care about. Finally, we have the *Lambert factor*. As discussed previously, this accounts for the fact that more incident light is available to be reflected, per unit surface area, when $\hat{\omega}_{in}$ is perpendicular to the surface than when at a glancing angle to the surface. The vector $\hat{n}$ is the outward-facing surface normal; the dot product $-\hat{\omega}_{in} \cdot \hat{n}$ peaks at 1 in the perpendicular direction and trails off to zero as the angle of incidence becomes more glancing. We discuss the Lambert factor once more in the entry "*3-D Graphics: Standard Local Lighting Model*" (see pp. 66–74).

In purely mathematical terms, the rendering equation is an *integral equation*: it states a relationship between some unknown function $L_{out}(x, \hat{\omega}_{out}, \lambda)$, the distribution of light on the surfaces in the scene, in terms of its own integral. It might not be apparent that the rendering equation is recursive, but L_{out} actually appears on both sides of the equals sign. It appears in the evaluation of $L_{in}(x, \hat{\omega}_{in}, \lambda)$, which is precisely the expression we set out to solve for each pixel: what is the radiance incident on a point from a given direction? Thus, to find the radiance exiting a point x, we need to know all the radiance incident at x from all directions. But the radiance incident on x is the same as the radiance leaving from *all other surfaces visible to* x, in the direction pointing from the other surface toward x.

To render a scene realistically, we must solve the rendering equation, which requires us to know (in theory) not only the radiance arriving at the camera, but also the entire distribution of radiance in the scene in every direction at every point. Clearly, this is too much to ask for a finite, digital computer, since both the set of surface locations and the set of potential incident/exiting directions are infinite. The real art in creating software for digital image synthesis is to allocate the limited processor time and memory most efficiently, to make the best possible approximation.

The simple rendering pipeline we present in the entry "*3-D Graphics: Real-Time Graphics Pipeline*" (see pp. 51–60) accounts only for direct light. It does not account for indirect light that bounced off of one surface and arrived at another. In other words, it only does "one recursion level" in the rendering equation. A huge component of realistic images is accounting for the indirect light—solving the rendering equation more completely. The various methods for accomplishing this are known as *global illumination* techniques.

This concludes our high-level presentation of how graphics works. Although we admit we have not yet presented a single practical idea, we believe it is very important to understand what you are trying to approximate before you start to approximate it. Even though the compromises we are forced to make for the sake of real time are quite severe, the available computing power is growing. A video game programmer whose only exposure to graphics has been OpenGL tutorials or demos made by video card manufacturers or books that focused exclusively on real-time rendering will have a much more difficult time understanding even the global illumination techniques of today, much less those of tomorrow.

REFERENCES

1. Heckbert, P.S.What are the coordinates of a pixel? In *Graphics Gems*; Glassner, A.S.; Ed.; Academic Press Professional: San Diego, CA, 1990, 246–248. available at http://www.graphicsgems.org/.
2. Smith, A.R. *A Pixel Is Not a Little Square, a Pixel Is Not a Little Square, a Pixel Is Not a Little Square! (and a Voxel Is Not a Little Cube)*; Technical Report, Technical Memo 6, Microsoft Research, 1995. Available at http://alvyray.com/memos/6pixel.pdf.
3. Hall, R. *Illumination and Color in Computer Generated Imagery*; Springer-Verlag: New York, 1989.
4. Kajiya, J.T. The rendering equation. In *SIGGRAPH '86: Proceedings of the 13th Annual Conference on Computer Graphics and Interactive Techniques*; ACM: New York, 1986, 143–150.

Missing Depth Data In Painting

Ju Shen
Department of Computer Science, University of Dayton, Dayton, Ohio, U.S.A.

Sen-ching Samson Cheung
Department of Electrical and Computer Engineering, University of Kentucky, Lexington, Kentucky, U.S.A.

Chen Chen
Center for Research in Computer Vision, University of Central Florida, Orlando, Florida, U.S.A.

Ruixu Liu
Department of Electrical and Computer Engineering, University of Dayton, Dayton, Ohio, U.S.A.

Abstract

With the remarkable development of active depth-sensing technology, red, green, blue-depth (RGB-D)-based systems have been widely adopted in different fields, such as virtual reality and telecommunication on mobile devices. For a given depth image, a depth value is provided per pixel, which can significantly facilitate the three-dimensional (3D) estimation procedure and enable many applications from 3D scanning to motion tracking to activity recognition. Despite its potential uses and popularity, the quality of the depth measurements of modern depth sensors is far from perfect. There are often many pixels with erroneous values or missing values in the acquired depth map. The uncertainty in depth measurements among these sensors can significantly degrade the performance of any subsequent vision processing. One possible solution is to infer or correct the measured pixel values by using imaging in-painting technique. However, due to the different nature of RGB and depth acquisition, traditional in-painting methods often yield unsatisfactory results on the depth map. In this entry, we will take the Kinect (v1) sensor as an example and introduce a probabilistic model to capture various types of uncertainties in the depth measurement process among structured light systems. The key idea is to utilize the correlation between color and depth channels to classify scene objects from the depth image to different layers according to their distance to the camera. Then these layers are used to guide the inference procedure of those missing or erroneous pixels.

Keywords: Color-depth sensing; Depth imaging; Image completion; Image in-painting.

PROBLEM DEFINITION

Microsoft Kinect represents a typical stereo red, green, blue-depth (RGB-D) system that successfully achieves real-time depth measurement with the early version based on structured light principle and later using time-of-flight mechanism (Kinect v2). In this entry, we will focus on the structured light depth sensors that acquire the scene depth by periodically emitting infrared (IR) patterns that can be captured by an IR complementary metal-oxide semiconductor (CMOS) sensor. According to the shape changes of the received IR pattern, depth distribution of the captured scene can be estimated. With the known extrinsic parameters from calibration, each depth pixel can be aligned with its corresponding color pixel that is obtained from the RGB camera. Thus, the capture scene's three-dimensional (3D) geometry and texture can be reconstructed using the RGB-D data. Figure 1 demonstrates the 3D reconstruction result by using a Kinect, which enables an arbitrary view generation using the RGB-D input.

Despite the promising applications of RGB-D systems, depth image from modern structured light depth sensors is still far from perfect. Noises and missing depth pixels often present in the acquired depth map that can significantly affect subsequent performance. According to the process of missing depth generation, we classify them into two categories. The first category is referred as randomly distributed "small holes" on the depth map. Such type of missing depth is usually produced by random noises from sensors, absorption, poor reflection, or even shadow reflection of the light patterns, for example, objects with darker colors, specular surfaces, or fine-grained surfaces such as human hair often cause poor depth measurements as shown in Fig. 2.

For the second category of missing depth, we define it as large contiguous regions of missing and erroneous

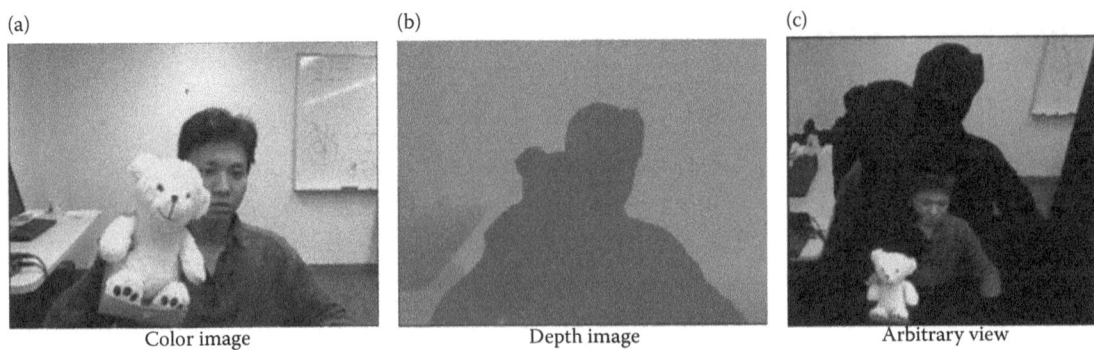

Fig. 1 Reconstructed arbitrary view by using the color and depth image input from Kinect

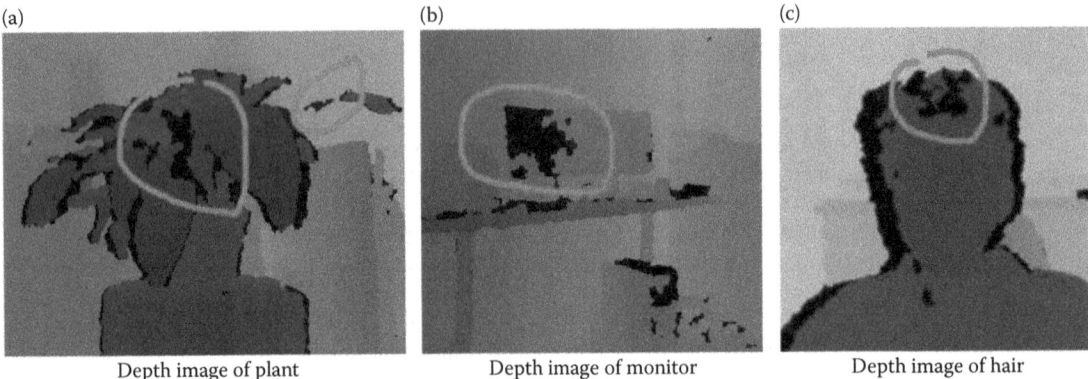

Fig. 2 Missing depth values from a particular object surface: plant leaves, monitor, and hair

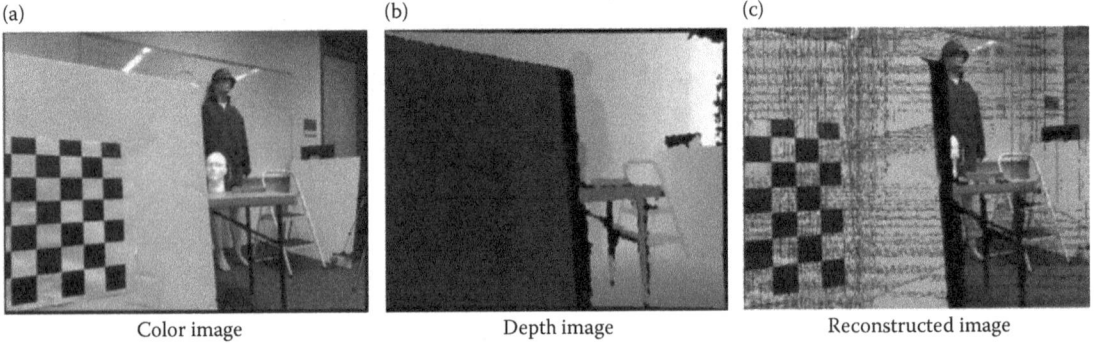

Fig. 3 Large contiguous region of missing depth along object boundary

depth. Such type of missing depth is usually caused by the imaging principle of structured light sensors as demonstrated in Fig. 3. The checkerboard produces a contiguous region of missing depth along its boundary. This missing region makes the head model, and the person's arms have no depth values although their corresponding textures are available in the RGB image. The unsatisfactory result is yielded in the reconstructed image as shown in Fig. 3c.

Such missing regions often present along the object boundaries, where background regions were visible to the IR camera but are occluded by foreground objects and receive no structured light patterns. This represents a major source of depth error caused by the disparity between the projector and the sensor. Sometimes, they can be orders of magnitude different from their true values. Inadequate calibration between the RGB and depth camera or roundoff error during normalization can further affect the estimated depth values. Surface orientation also plays a role in depth measurements—as the surface normal deviates from the principal axis of the IR camera, the accuracy of the depth measurement declines and becomes unreliable near depth discontinuities, for example, the book surface in Fig. 4b.

For the first class of missing depth values, they can be inferred by surrounding available depth pixels in the neighboring regions. In addition, the corresponding RGB information can be used to steer the depth completion by using

Fig. 4 Depth map in-painting by joint bilateral filter[1]

techniques such as joint bilateral filter.[1] Figure 4 shows the result of applying joint bilateral filter to infer the missing depth that produces a plausible depth map compared to the original input. Such type of missing depth can usually be reasonably recovered by using interpolation-based solutions. However, for the second type of large contiguous missing depth, generic image denoising and complete algorithms fail to take into account the unique problems of structured light RGB-D systems. Noticeable artifacts are often yielded by directly applying image in-painting-based approaches.

To tackle with those large contiguous region of missing depth values, in this entry, we introduce a layer-based approach for depth denoising and completion.[2] The basic idea is to label each pixel by a layer index, which steers the completion process, which infers the missing depth using available pixels while preserve well-defined depth edges. A stochastic framework is developed to separate the depth image into multiple layers and combines multiple RGB-D system noise models to robustly determine the depth layer label. The goal is to identify and remove erroneous pixels and complete the remaining missing values on the depth image to facilitate any subsequent RGB-D applications.

EXISTING METHODS

In the literature, a number of algorithms have been developed to denoise depth image using super-resolution and image in-painting. A common theme among these works is to rely on information obtained from the companion color images to predict missing depth values. The use of color information for depth enhancement is based on the assumption that certain correlation exists between depth continuality and color image consistency.[1] While providing useful cue for interpolation, this assumption does not always hold as color edges and depth edges do not necessarily coincide with each other. Garro et al.[3] presented an interpolation scheme for depth super-resolution. A high-resolution RGB camera was used to guide the up-sampling process on the depth image. To interpolate the missing depth pixel, the scheme used neighboring depth pixels mapped onto the same color segment as the target pixel. This method relied strongly on the extrinsic alignment between the color and depth image. However, noisy depth values along object boundaries may map into a wrong color segment and propagate its effect to other pixels in the segment. A similar segmentation-based method can also be found in the work of Park et al.,[4] where a nonlocal means filtering-based approach was used to regularize depth maps and maintain fine detail and structure. Wang et al.[5] proposed a stereoscopic in-painting algorithm to jointly complete missing texture and depth by using two pairs of RGB and depth cameras. Regions occluded by foreground were completed by minimizing an energy function. The system required an additional pair of color and depth cameras to achieve the goal (Fig. 5).

Various probabilistic frameworks are often used in modeling depth measurements, fusing depth, and color

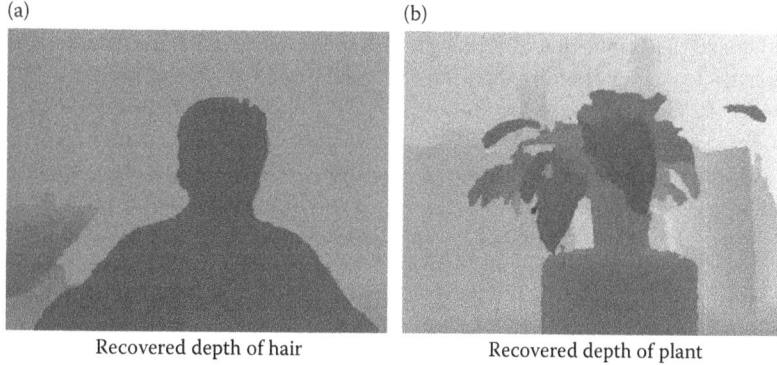

Fig. 5 Depth map completion results with noticeable artifact by image in-painting techniques[12]

information, and predicting missing values. Diebel et al.[6] demonstrated the use of Markov random field (MRF) in the super-resolution of depth data using high-resolution color data. However, their work provided little insight in modeling the sources of error in the depth sensor. Similarly, a low-resolution depth image was iteratively refined through the use of a high-resolution color image.[7] Bilateral filter was applied to a cost function based on depth probabilities. A final high-resolution image was produced by a winner-takes-all approach on the cost function. These approaches work well for the super-resolution problem where missing depth pixels are uniformly distributed. Depth images obtained by structured light sensors often have large contiguous regions of missing depth measurements, which cannot be handled by such approaches. A recent depth fusion paper proposed by Zhang et al. aimed at capturing full-frame depth by adaptively adjusting the contribution from photometric stereo and completed the depth in an edge-preserving manner.[8] Three additional LEDs are used as assistance to fulfill the task. The system was cumbersome as an additional pair of color and depth cameras was needed. In Section "Method," we introduce the layer base solution that does not require additional sensors or environmental setup to achieve automatic depth denoising and completion.

METHOD

We assume that the dynamic 3D environment consists of two parts: *a static background* and *a number of dynamic foreground objects*. This is a configuration typically seen in a home or office environment with a small number of individuals moving in front of the device. We also assume that the RGB and depth cameras are extrinsically aligned and temporally synchronized. After an initial step of offline training on background-only frames, our online algorithm consists of two main phases: layer labeling followed by depth denoising and completion. *In the first phase*, each pixel of the incoming frame is labeled by different layers via a probabilistic framework that incorporates a data measurement model and a smoothing neighborhood model based on available observations. Maximum a posteriori (MAP) estimation is used in the labeling to prevent blurring along depth discontinuities. *In the second phase*, the labels estimated in the first phase are used to steer the removal of outlier and the completion of missing depth values from the background model or from neighboring depth values with the same labels. The robust labeling allows us to preserve the shape of object boundary and prevent noise propagation across objects with significant depth differences.

Layer Assignment

The multilayered RGB-D measurement process is described by probabilistic graphical model as shown in Fig. 6. Let G be the input of the two-dimensional (2D) color and depth images. At each pixel location $s \in G$, X_s denotes the latent random integer variable indicating which layer the pixel belongs to. X_s can assume any value in the set $\{-n', \ldots -2, -1\} \cup \{1, 2, \ldots, n\}$. Negative numbers represent layers from the static background, whereas positive numbers represent layers in the foreground. The larger the layer number is, the closer it is to the camera. The number of background layers n' and the number of foreground layers n are determined based on the observed data and are the same for all pixels. The approaches to estimate n and n' will be described in Section 3.1.1.

X_t represents one of the closest neighbor to the target pixel X_s in the 2D space. All the labels over the entire image thus form an MRF, and the spatial relationship between adjacent labels is governed by an edge factor $\psi(X_s, X_t, f_{st})$, where f_{st} is the measured similarity between the two pixels. Each layer label X also has its evidence potential function $\varphi(X_s)$ based on the measurement of Bayesian network (BN) shown in the lower half of Fig. 6. As all the measurements are made at the same pixel, the subscript s is omitted, and $\varphi(X)$ is defined as follows:

$$\varphi(X) \triangleq P(I_d \mid M, \theta, X) \cdot P(I_c \mid X) \cdot P(M \mid X) \cdot P(X)$$

where

$$P(I_d \mid M, \theta, X) = \int_D P(D \mid X) \int_{Z_d} P(I_d \mid Z_d, M) \cdot P(Z_d \mid D, \theta)$$

I_c represents the observed color values. D represents the true but unobserved depth values. D is corrupted by an additive Gaussian noise that produces a noisy measurement Z_d. The noise variance is determined by the closeness measurement θ to the nearest edge. Due to the missing depth problem, Z_d may not be directly observable. We thus introduce an observable depth indicator random variable M which is 1 if the depth value is observed and 0 otherwise. Combining these two random variables results in the observable depth value $I_d = M \cdot Z_d$. Using this probabilistic model, layer labeling can be formulated as a MAP or MAP problem:

$$X^{\text{MAP}} \triangleq \operatorname{argmax}_{XG} \sum_s \log \varphi(X_s) + \sum_{(s,t) \in G} \log \psi(X_s, X_t, f_{st})$$

The choice of parameterization allows $\psi(X_s, X_t, f_{st})$ and $\varphi(X_s)$ to be computed. While the complexity of the exact solution to the MAP problem is exponential in the image size, which is known to be NP hard. To improve the computation speed, various approximation algorithms have been proposed for a global optimization, such as graph cuts or loopy belief propagation.[9] In this entry, we used max-product-based loopy belief propagation to approximate the inference. One major reason of choosing it than graph cuts is that belief propagation is more efficient in processing video sequence:

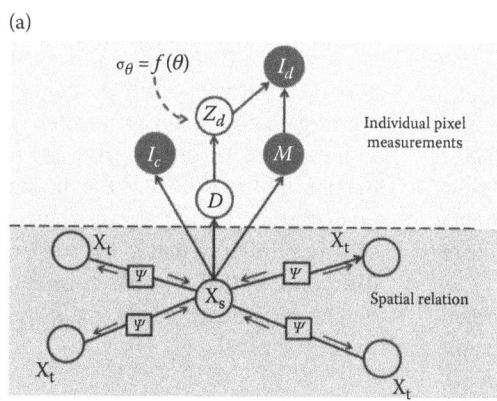

Fig. 6 RGB-D model: the spatial relation at the bottom of the figure (blue background) specifies the constraints between neighboring labels while the individual pixel measurement on the top describes the measurement process. Darken nodes are actual measurements, white nodes are hidden variables, and square nodes are factors

the output of previous frame can be used as the initial value of messages for current frame, which makes the convergence speed improved.

Based on the designed graphical model, we first perform the individual pixel measurement by estimating the four layer distributions $P(X)$, $P(M|X)$, $P(D|X)$, and $P(I_c|X)$ according to both offline training data and online input. The depth distribution $P(D, X) = P(X) \cdot P(D|X)$ is modeled as a mixture of Gaussian (MOG) model, whereas the color distribution $P(I_c, X) = P(X) \cdot P(I_c|X)$ is modeled as multiple color histograms on the quantized HSV (hue, saturation, and value) space, one for each layer. The observable depth indicator distribution $P(M, X) = P(X) \cdot P(M|X)$ is modeled by Bernoulli distribution for each layer. In fact, the parameter estimation for the depth indicator is a simpler version of the color distribution. This leads to our focus on computing the depth and color distributions.

The static background model can be learned through offline training on a set of pre-scanned RGB-D images to obtain the parameters for the background depth layers, which are marked as negative values. There are two phases of the estimation: global estimation and local adaptation. During the global estimation, all the pixels with both color and depth measurements will be aggregated to estimate a single pair of $P(D, X)$ and $P(I_c, X)$ using the expectation–maximization (EM) approach. $P(D, X)$ is initialized by K-means, and $P(I_c, X)$ is initialized as an uniform distribution. It is important to note that the concept of layers is based only on depth but not on color. As such, the EM process is primarily driven by the depth data in the sense that the E-step only estimates the layer posterior $P(X|D)$ for the depth but not the color. During the M-step, we use the depth data to update the estimates for the layer prior to $P(X)$ and the depth layer conditional $P(D|X)$, and only use the color data to update the color layer conditional $P(I_c|X)$ using the posterior probability $P(X|D)$ of the colocated depth pixel. As a result, two pixels with the same color value can have different contributions to different layers. Different number of layers are tested, and the optimal number n' is determined by using the Bayesian information criterion on the depth data.[10] The example in Fig. 1 shows the results after two separate layers are identified and obtain the global color and depth models. In the second phase, the global distributions are adapted to each individual pixel by using only the temporal data at that pixel location. Sequential exponential weighing scheme is used for the adaption. For example, the local mean for layer $X_s = i$ at location s is updated by a new depth value d_{new} as follows:

$$\mu^{(t+1)} := \left(1 - \lambda P^{(t)}\left(X_s = i \mid D_s = d_{new}\right)\right) \cdot \mu^{(t)} \\ + \lambda P^{(t)}\left(X_s = i \mid D_s = d_{new}\right) \cdot d_{new}$$

t represents the iteration step, and λ controls the rate of adaptation which is empirically set to 0.3. All the other parameters are updated in a similar fashion. Similar to the global distributions, the two layer conditional probabilities are updated based on the corresponding color or depth data. The layer prior is updated using the depth data, if available, or the color data, if the depth data is missing. After the adaptation, the parameters can better describe the characteristics of the local pixel where the local depth distribution $P_s(D)$ and the color posterior $P_s(X|I_c)$ clearly indicate that this pixel is more likely to belong to layer-1.

The foreground layer distributions are estimated online. To deliver a real-time response and cope with fast-moving objects, the foreground distributions are estimated for every frame. As no temporal data is maintained, we only estimate global distributions using an approach identical to that of the background global distribution. The training data are obtained based on only those pixels with valid depth measurements and very low background posterior probability, that is, $\max_{i=1,\ldots,n}$, $P_s(X_s = -i|D_s) < \varepsilon$ for a small fixed ε. To obtain the full range of $P(X)$, $P(D|X)$, and $P(I_c|X)$, we also need prior probabilities for foreground and background. We simply set them to be equally likely for our experiments though better performance may be possible with more foreground training data.

Next, the noisy depth measurement Z_d is modeled based on an additive Gaussian model with a noise standard deviation σ_θ that reflects the uncertainty in the depth measurement. As argued in Section "Existing Methods," erroneous depth measurements occur predominantly near object boundaries. To model this effect, we apply an edge detector on the depth map and use the spatial distance θ to the closest depth edge as a reliability measure. The noise variance σ_θ is modeled as a deterministic logistic function. This simple model is easy to compute, though a more sophisticated one incorporating surface normal, texture, and color can be used in a similar fashion. For the spatial MRF, the edge potential $\psi(X_s, X_t)$ is defined based on the similarity in color and depth between the neighboring pixel. The similarity strength is a feature based on how close the color and depth of the neighboring pixels are from the target pixel. The color similarity ratio C_{st} and the depth similarity ratio D_{st} are defined in a similar metric.[11] We use a parameter $\alpha \in [0,1]$ to indicate a weight trade-off between depth and color information. If the depth measurements are reliable, most of the weight should be assigned to depth values as they are more reliable for foreground/background labeling; if the depth measurements are unreliable, they should not be used at all in computing the edge potential. Similar to noise term computation, α is defined as the logistic function of the two pixels involved distance.

Depth Image Completion

With the guidance of layer classification, pixels with erroneous depth measurements can be identified efficiently by comparing with pixels from the same layers. These pixels usually have significant depth difference from other pixels in the neighborhood. As most measurement errors occur around object boundaries, it is imperative not to mistake true depth discontinuities as wrong depth values. To determine if a depth pixel is an outlier, we robustly estimate the depth distribution in the neighborhood around the pixel via a Random Sample Consensus (RANSAC)-like procedure. First, we only consider depth values in the neighborhood that share the same label as the target pixel. Then, multiple small sets of random sample pixels are drawn, and a Gaussian distribution is estimated for each set. If only a small fraction of the neighborhood can be fit within two standard deviations from the mean of a sample distribution, this distribution is likely to contain outlier samples and is thus discarded. Among those that survive the robustness test, the one with the smallest variance is used, and the target depth pixel is declared an outlier if it is beyond two standard deviations from the mean. The outlier depth pixel will join the rest of the missing depth pixels and will be completing using a joint color–depth bilateral filtering scheme similar to that in the work of Kopf et al.[1] The only difference is that we only consider the contributions from neighboring depth pixels that have the same layer label as the center pixel.

DEPTH COMPLETION RESULTS

In this section, we demonstrate the depth denoising and completion results by applying the described method to different scenarios. Figure 7 shows a simple example with one foreground and one background layer. From the input depth image, one can see that the background wall is clearly visible in the color image. However, from the depth image, there are a number of missing depth regions especially along the boundary of the hair area due to the rapid spatial changes in depth layers. By applying the layer labeling algorithm, accurate segmentation can be achieved by identifying the optimal number of layers and correctly assigning each pixel to the associate layer, as the layer mask image shown in Fig. 7c. Guided by this segmentation mask, it corrects and completes the depth values as shown in Fig. 7d.

Figure 8 presents a more complex example with two foreground and two background layers. The two foreground layers are the head model and the cup (layer 2) as well as the books and the doll (layer 1). The two background layers are the white board and the table (layer −1) as well as the rest of the region (layer −2) as shown in Fig. 8a. Notice that each layer can have objects of varying depths as shown in Fig. 8b. The curves and straight lines from different objects are effectively preserved in the output result as shown in Fig. 8c.

More challenging foreground object is shown in Fig. 9, where the plant leaves lead to significant measurement error of the depth map. The wrongly assigned depth values move some of the background pixels to the foreground or vice versa. This error is clearly noticeable that a significant number of foreground pixels are assigned with background pixels. For example, in the depth image in Fig. 9b, a certain portion of the tree leave depth is wiped out and assigned with the depth values of the background wall. Furthermore, due to the considerable difference between the foreground and background pixels, large contiguous missing depth pixels are produced along foreground object boundaries. Steered by correct layer labeling, the depth completion can effectively preserve the shape of object boundaries and prevent noise propagation across objects with significant depth differences.

Missing Depth Data In Painting

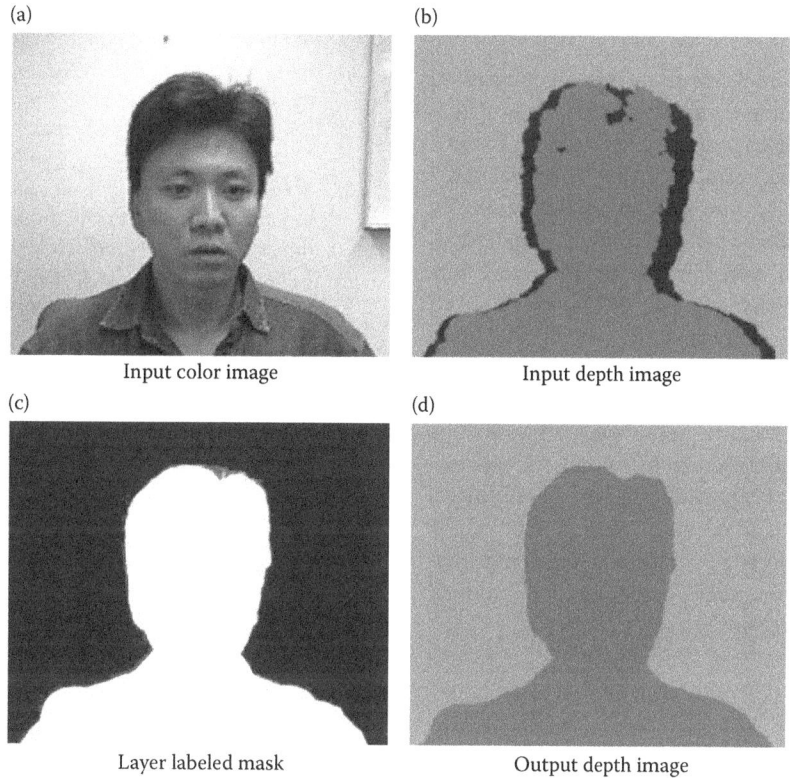

Fig. 7 The input and final output depth image after denoising and completion

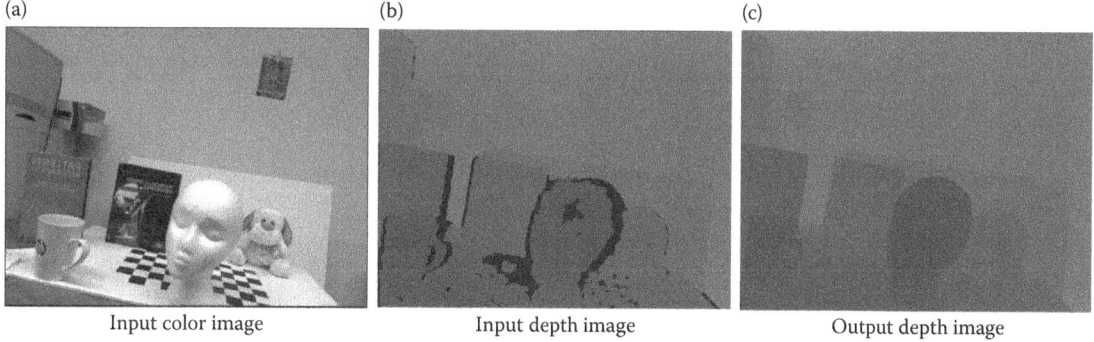

Fig. 8 Scene with multiple foreground and background layers

Fig. 9 Depth completion on more complex foreground object (plant)

SUMMARY

In this entry, we introduce a depth denoising and completion scheme by combining color–depth correlation, background modeling, spatial smoothness, and measurement error models pertinent to structured light systems. A probabilistic graphical model is used to fuse all these different factors together with the key latent variable being the depth layer at each pixel. The depth layer labeling is formulated as a MAP problem, and an MRF attuned to the uncertainty in depth measurements is used to spatially smooth the labeling process. Driven by the obtained depth layer labels, depth noise is removed and depth interpolation is performed using a bilateral filter. There could be some weaknesses to this design. For example, additional depth features such as surface normal and texture may provide a more accurate uncertainty model in depth measurements. More computationally efficient approaches for depth layer labeling could be further investigated, for example, parallel computing can be applied to the layer labeling procedure.

REFERENCES

1. Kopf, J.; Cohen, M.F.; Lischinski, D.; Uyttendaele, M. Joint bilateral upsampling. ACM Trans. Graph. **2007**, *26* (3), 96
2. Shen, J.; Cheung, S.-C.S. Layer depth denoising and completion for structured-light RGB-D cameras. In *IEEE Conference on Computer Vision and Pattern Recognition (CVPR)*, Portland, OR, 2013.
3. Garro, V.; dal Mutto, C.; Zanuttigh, P.; Cortelazzo, G. A novel interpolation scheme for range data with side information. In *Conference for Visual Media Production (CVMP)*, London, 2009.
4. Park, J.; Kim, H.; Tai, Y.; Brown, M.; Kweon, I. High quality depth map upsampling for 3d-tof cameras. In *International Conference on Computer Vision (ICCV)*, Barcelona, 2011.
5. Wang, L.; Jin, H.; Yang, R.; Gong, M. Stereoscopic in-painting: Joint color and depth completion from stereo images. In *IEEE Conference on Computer Vision and Pattern Recognition (CVPR)*, Anchorage, AK, 2008.
6. Diebel, J.; Thrun, S. An application of Markov random fields to range sensing. Adv. Neural Inf. Process. Syst. **2006**, *18*, 291–295.
7. Yang, Q.; Yang, R.; Davis, J.; Nister, D. Spatial-depth super resolution for range images. In *IEEE Conference on Computer Vision and Pattern Recognition (CVPR)*, Minneapolis, MN, 2007.
8. Zhang, H.Q.; Ye, M.; Yang, R.; Matsushita, Y.; Wilburn, B.; Yu, H. Edge-preserving photometric stereo via depth fusion. In *IEEE Conference on Computer Vision and Pattern Recognition (CVPR)*, Providence, RI, 2012.
9. Freeman, W.; Pasztor, E.; Carmichael, O. Learning low-level vision. Int. J. Comput. Vision (I-JCV) **2000**, *40*, 25–47.
10. Bhat, H.S.; Kumar, N. *On the Derivation of the Bayesian Information Criterion*; Technical Report, University of California: Merced, 2010.
11. Rother, C.; Blake, A.; Kolmogorov, V. Grabcut—Interactive foreground extraction using iterated graph cuts. In *Proceedings of ACM SIGGRAPH*, Los Angeles CA, 2004.
12. Camplani, M.; Salgado, L. Efficient spatio-temporal hole filling strategy for kinect depth maps. In *Three-Dimensional Image Processing (3DIP) and Applications*, Burlingame, CA, 2012.

Modern Cryptography

Debdeep Mukhopadhyay and Rajat Subhra Chakraborty
Indian Institute of Technology Kharagpur, West Bengal, India

Abstract
Cryptography is used to protect information from illegal access and it involves developing schemes (ciphers) which allow secret data to be shared over insecure channels. This entry provides a brief introduction to cryptographic concepts, history, standards and basic algorithms.

Keywords: Cryptography; Cypher; Security.

INTRODUCTION

The art of keeping messages secret is cryptography, while cryptanalysis is the study attempted to defeat cryptographic techniques. *Cryptography* is used to protect information from illegal access. It largely encompasses the art of building schemes (*ciphers*) which allow secret data exchange over insecure channels.[1] The need of secured information exchange is as old as civilization itself. It is believed that the oldest use of cryptography was found in non-standard hieroglyphics carved into monuments from Egypt's Old Kingdom. In 5 B.C. the Spartans developed a cryptographic device, called *scytale* to send and receive secret messages. The code was the basis of transposition ciphers, in which the letters remained the same but the order is changed. This is still the basis for many modern day ciphers. The other major ingredient of many modern-day ciphers is substitution ciphers, which was used by Julius Caesar and is popularly known as Caesar's shift cipher. In this cipher, each plaintext character was replaced by the character 3 places to the right in the alphabet set modulo 26. However, in the last three decades cryptography has grown beyond designing ciphers to encompass also other activities like design of signature schemes for signing digital contracts. Also the design of cryptographic protocols for securely proving one's identity has been an important aspect of cryptography of the modern age. Yet the construction of encryption schemes remains, and is likely to remain, a central enterprise of cryptography.[2] The primitive operation of cryptography is hence *encryption*. The inverse operation of obtaining the original message from the encrypted data is known as *decryption*. Encryption transforms messages into representation that is meaningless for all parties other than the intended receiver. Almost all cryptosystems rely upon the difficulty of reversing the encryption transformation in order to provide security to communication.[3] *Cryptanalysis* is the art and science of breaking the encrypted message. The branch of science encompassing both cryptography and cryptanalysis is cryptology and its practitioners are cryptologists. One of the greatest triumph of cryptanalysis over cryptography was the breaking of a ciphering machine named Enigma and used during Worldwar II. In short cryptology evolves from the long-lasting tussle between the cryptographer and cryptanalyst.

For many years, many fundamental developments in cryptology outpoured from military organizations around the world. One of the most influential cryptanalytic papers of the twentieth century was William F. Friedman's monograph[4] entitled *The Index of Coincidence and its Applications in Cryptography*. For the next fifty years, research in cryptography was predominantly done in a secret fashion, with few exceptions like the revolutionary contribution of Claude Shannon's paper *"The Communication Theory of Secrecy Systems"*, which appeared in the *Bell System Technical Journal* in 1949.[5]

However, after the world wars cryptography became a science of interest to the research community. The *Code Breakers* by David Kahn produced the remarkable history of cryptography.[6] The significance of this classic text was that it raised the public awareness of cryptography. The subsequent development of communication and hence the need of privacy in message exchange also increased the impetus on research in this field. A large number of cryptographers from various fields of study began to contribute leading to the rebirth of this field. Horst Fiestel[7] began the development of the US Data Encryption Standard (DES) and laid the foundation of a class of ciphers called as private or symmetric key algorithms. The structure of these ciphers became popular as the Fiestel Networks in general. Symmetric key algorithms use a single key to both encrypt and decrypt. In order to establish the key between the sender and the receiver they required to meet once to decide the key. This problem commonly known as the key exchange problem was solved by Martin Hellman and Whitfield Diffie[8] in 1976 in their ground-breaking

Encyclopedia of Image Processing, First Edition
DOI: 10.1201/9781351110273-140000474
Copyright © 2018 by Taylor & Francis. All rights reserved.

paper *New Directions in Cryptography*. The developed protocol allows two users to exchange a secret key over an insecure medium without any prior secrets. The work not only solved the problem of key exchange but also provided the foundation of a new class of cryptography, known as the public key cryptography. As a result of this work the RSA algorithm, named after the inventors Ron Rivest, Adi Shamir, and Leonard Adleman, was developed.[9] The security of the protocol was based on the computational task in factoring the product of large prime numbers.

Cryptology has evolved further with the growing importance of communications and the development in both processor speeds and hardware. Modern-day cryptographers have thus more work than merely jumbling up messages. They have to look into the application areas in which the cryptographic algorithms have to work. The transistor has become more powerful. The development of the VLSI technology (now in submicrons) have made the once cumbersome computers faster and smaller. The more powerful computers and devices will allow the complicated encryption algorithm run faster. The same computing power is also available to the cryptanalysts who will now try to break the ciphers with both straight forward brute force analysis, as well as by leveraging the growth in cryptanalysis. The world has thus changed since the DES was adopted as the standard cryptographic algorithm and DES was feeling its age. Large public literature on ciphers and the development of tools for cryptanalysis urged the importance of a new standard. The National Institute for Standards and Technology (NIST) organized a contest for the new Advanced Encryption Standard (AES) in 1997. The block cipher Rijndael emerged as the winner in October 2000 because of its features of security, elegance in implementations and principled design approach. Simultaneously Rijndael was evaluated by cryptanalysts and a lot of interesting works were reported. Cryptosystems are inherently computationally complex and in order to satisfy the high throughput requirements of many applications, they are often implemented by means of either VLSI devices or highly optimized software routines. In recent years such cryptographic implementations have been attacked using a class of attacks which exploits leaking of information through side-channels like power, timing, intrusion of faults etc. In short as technology progresses new efficient encryption algorithms and their implementations will be invented, which in turn shall be cryptanalyzed in unconventional ways. Without doubt cryptology promises to remain an interesting field of research both from theoretical and application point of view.

CRYPTOGRAPHY: SOME TECHNICAL DETAILS

The aim of the cryptographer is to find methods to secure and authenticate messages. The original message is called the plaintext and the encrypted output is called the ciphertext. A secret key is employed to generate the ciphertext from the plaintext. The process of converting the plaintext to the cipher text is called encryption and the vice versa is called decryption. The cryptographer tries to keep the messages secret from the attacker or intruder. A cryptosystem is a communication system encompassing a message source, an encryptor, an insecure channel, a decryptor, a message destination and a secure key transfer mechanism. The scenario of a cryptographic communication is illustrated in Fig. 1. The encryptor uses a key K_a and the decryptor used a key K_b, where depending on the equality of K_a and K_b there are two important classes of cryptographic algorithms.

The sender and the receiver are often given the names of **Alice** and **Bob**, while the untrusted channel is being observed by an adversary whom we name as **Mallory**. She has access to the ciphertexts, and is aware of the encryption and decryption algorithm. The goal of the attacker Mallory is to ascertain the value of the decryption key K_b, thus obtaining the information which he is not supposed to know. The attacker or cryptanalyst is a powerful entity who studies the cipher and uses algebraic and statistical techniques to attack a cryptographic scheme.

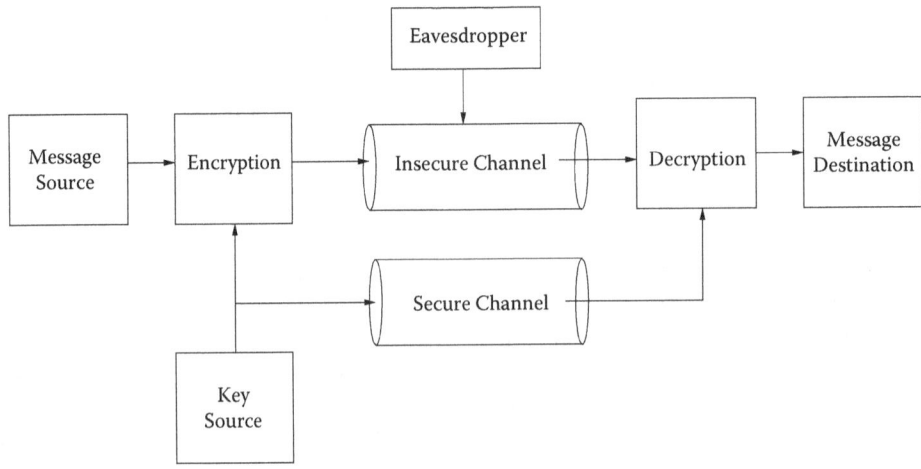

Fig. 1 Secret key cryptosystem model

A cryptanalytic attack is a procedure through which the cryptanalyst gains information about the secret decryption key. Attacks are classified according to the level of a-priori knowledge available to the cryptanalyst.

A **Ciphertext-only attack** is an attack where the cryptanalyst has access to ciphertexts generated using a given key but has no access to the corresponding plaintexts or the key. A **Known-plaintext attack** is an attack where the cryptanalyst has access to both ciphertexts and the corresponding plaintexts, but not the key.

A **Chosen-plaintext attack** (CPA) is an attack where the cryptanalyst can choose plaintexts to be encrypted and has access to the resulting ciphertexts, again their purpose being to determine the key.

A **Chosen-ciphertext attack** (CCA) is an attack in which the cryptanalyst can choose ciphertexts, apart from the challenge ciphertext and can obtain the corresponding plaintext. The attacker has access to the decryption device.

In case of CPA and CCA, adversaries can make a bounded number of queries to its encryption or decryption device. The encryption device is often called as oracle: meaning it is like a black-box without details like in an algorithm of how an input is transformed or used to obtain the output. Although this may seem a bit hypothetical, but there are enough real life instances where such encryption and decryption oracles can be obtained. Thus security analysis with the existence of such oracles is imperative.

The attacks are measured against a worst case referred to as the brute-force method. The method is a trial-and-error approach, whereby every possible key is tried until the correct one is found. Any attack that permits the discovery of the correct key faster than the brute-force method, on average, is considered successful. An important principle known as the *Kerckhoff's principle* states that the secrecy of a cipher must reside entirely in the key. Thus an enemy will have a complete knowledge of the cipher but shall not know the key. A secured cryptographic scheme should withstand the attack of such a well-informed adversary.

Formal definition of a cryptosystem is stated below for the sake of completeness:

Definition 1

A cryptosystem is a five-tuple (P, C, K, E, D), where the following are satisfied:

1. *P is a finite set of possible plaintexts*
2. *C is a finite set of possible ciphertexts*
3. *K, the key space, is a finite set of possible keys*
4. *$\forall K_a, K_b \in K$, there is an encryption rule $e_{K_a} \in E$ and a corresponding decryption rule $d_{K_b} \in D$. Each chosen pair $e_{K_a}: P \to C$ and $d_{K_b}: C \to P$ are invertible functions, ie. $\forall x \in P, d_{K_b}(e_{K_a}(x)) = x$.*

Example 1

Let $P = C = Z_{26}$. Let $K_a = K_b = k \in Z_{26}$. We can define a cryptosystem as follows $\forall x, y \in Z_{26}$:

$$e_k(x) = (x + k) \bmod 26 \quad d_k(x) = (y - k) \bmod 26$$

Example 2

Let $P = C = \{0, 1\}^{128}$. Let $K_a = K_b = k \in \{0, 1\}^{128}$. We can define a cryptosystem as follows $\forall x, y \in \{0, 1\}^{128}$:

$$e_K(x) = (x \oplus k) \quad d_K(x) = (y \oplus k)$$

Here the operator $\oplus$ is a bitwise operation and is a self-invertible operation. Not all ciphers have K_a and K_b same. In fact, depending on their equality we have two important dichotomy of ciphers which are explained next:

- Private-key (or symmetric) ciphers: These ciphers have the same key shared between the sender and the receiver. Thus, referring to Fig. 1 $K_a = K_b$.
- Public-key (or asymmetric) ciphers: In these ciphers we have $K_a \neq K_b$. The encryption key and the decryption keys are different.

These types differ mainly in the manner in which keys are shared. In symmetric-key or private-key cryptography both the encryptor and decryptor use the same key. Thus, the key must somehow be securely exchanged before secret key communication can begin (through a secured channel, Fig. 1). In public key cryptography the encryption and decryption keys are different.

In such algorithms we have a key-pair, consisting of:

- Public key, which can be freely distributed and is used to encrypt messages. In the Fig. 1, this is denoted by the key K_a.
- Private key, which must be kept secret and is used to decrypt messages. The decryption key is denoted by K_b in the Fig. 1.

In the public key or asymmetric ciphers, the two parties – namely Alice and Bob–are communicating with each other and have their own key pair. They distribute their public keys freely. Mallory has the knowledge of not only the encryption function, the decryption function, and the ciphertext, but also has the capability to encrypt the messages using Bob's public key. However, he is unaware of the secret decryption key, which is the private key of the algorithm. The security of these classes of algorithms rely on the assumption that it is mathematically hard or complex to obtain the private key from the public informations. Doing so would imply that the adversary solves a mathematical problem which is widely believed to be difficult. It may be noted that we do not have any proofs for their hardness, however we are unaware of any efficient techniques to solve them. The elegance of constructing these ciphers lies in the fact that the public keys and private keys still have to be related in the sense, that they perform the invertible operations to obtain the message back. This is achieved through a class of magical functions, which are called *one-way* functions. These functions are easy to compute in one direction, while computing the inverse from the output is believed to be a difficult problem. We shall discuss this in more details in a following section. However, first let us see an example for this class of ciphers.

Example 3

This cipher is called the famous RSA algorithm (Rivest Shamir Adleman). Let $n = pq$, where p and q are properly chosen and large prime numbers. Here the proper choice of p and q are to ensure that factorization of n is mathematically complex. The plaintexts and ciphertexts are $P = C = Z_n$, the keys are $K_a = \{n, a\}$ and $K_b = \{b, p, q\}$, st $ab \equiv 1 \mod \phi(n)$. The encryption and decryption functions are defined as, $\forall x \in P$, $e_{K_a}(x) = y = x^a \mod n$ and $d_{K_b}(y) = y^b \mod n$.

The proof of correctness of the above algorithm follows from the combination of the Fermat's little theorem and the Chinese Remainder Theorem (CRT). The algorithm is correct if $\forall x \in P$, we have:

$$x^{ab} \equiv x \mod n$$

It suffices to show that:

$$x^{ab} \equiv x \mod p \quad x^{ab} \equiv x \mod q \quad (1)$$

It may be observed that since $gcd(p, q) = 1$ we have from the Extended Euclidean Algorithm (EEA) $1 = (q^{-1} \mod p)q + (p^{-1} \mod q)p$. Thus, from Eq. 1 applying CRT we have $x^{ab} \equiv x((q^{-1} \mod p)q + (p^{-1} \mod q)p) \mod n = x$.

If $x \equiv 0 \mod p$, then it is trivial that $x^{ab} \equiv x \mod p$.

Otherwise if $x \not\equiv 0 \mod p$, $x^{p-1} \equiv 1 \mod p$. Also, since $ab \equiv 1 \mod \phi(n)$ and $\phi(n) = (p-1)(q-1)$ we have $ab = 1 + k(p-1)(q-1)$ for some integer k.

Thus we have, $x^{ab} = xx^{k(p-1)(q-1)} \equiv x \mod p$. Likewise, we have $x^{ab} \equiv x \mod q$. Combining the two facts, by CRT we have that $x^{ab} \equiv x \mod n$. This shows the correctness of the RSA cipher.

It may be observed that the knowledge of the factors of p and q help to ascertain the value of the decryption key K_b from the encryption key K_a. Likewise, if the decryption key K_b is leaked, then the value of n can be factored using a probabilistic algorithm with probability of success at least 0.5.

Another kind of public key ciphers is the ElGamal cryptosystem, which is based on another hard problem which is called the *Discrete Log Problem* (DLP).

Consider a finite mathematical group $(G, .)$. For an element $\alpha \in G$ of order n, let:

$$< \alpha > = \{\alpha^i : 0 \leq i \leq n - 1\}$$

The DLP problem is to find the unique integer i st. $\alpha^i = \beta$, $0 \leq i \leq n - 1$. We denote this number as $i = \log_\alpha \beta$ and is referred as the *Discrete Log*.

Computing Discrete Log, is thus the inverse computation of a modular exponentiation operation. We have efficient algorithms for computing the modular exponentiation, by the square and multiply algorithm, however it is generally difficult to compute the DLP for properly chosen groups. Thus, the modular exponentiation is a potential one-way function having applications in public key cryptography.

We define one such cryptosystem, known as the ElGamal cipher.

Example 4

Let p be a prime, st. computing DLP in $(Z_p^*, .)$ is hard. Let $\alpha \in Z_p^*$ be a primitive element, and define the plaintext set as $P = Z_p^*$ and the ciphertext set as $C = Z_p^* \times Z_p^*$. The key set is defined as $K = (p, \alpha, a, \beta): \alpha^a \equiv \beta \mod p$.

For a given $k \in K$, $x \in P$, $c \in C$ and for a secret number $r \in Z_{p-1}$, define $c = e_k(x, r) = (y_1, y_2)$, where $y_1 = \alpha^r \mod p$, and $y_2 = x\beta^r \mod p$. This cryptosystem is called as the ElGamal cryptosystem. The decryption is straightforward: for a given ciphertext, $c = (y_1, y_2)$, where $y_1, y_2 \in Z_p^*$, we have $x = (y_1^a)^{-1}(y_2)$.

The plaintext x is thus masked by multiplying it by β^r in the second part of the ciphertext, y_2. The hint to decrypt is transmitted in the first part of the ciphertext in the form of α^r. It is assumed that only the receiver who has the secret key a can compute β^r by raising α^r to the power of a, as $\beta \equiv \alpha^a \mod p$. Then decrypting and obtaining back x as one just needs to multiply the multiplicative inverse of β^r with y_2.

Thus one can observe that the ElGamal cipher is randomized, and one can for the same plaintext x obtain $p - 1$ ciphertexts, depending on the choice of r.

An interesting point to note about the hardness of the DLP is that the difficulty arises from the modular operation. As otherwise, α^i would have been monotonically increasing, and one can apply a binary search technique to obtain the value of i from a given value of α and $\beta = \alpha^i$. However, as the operations are performed modular p, there is no ordering among the powers, a higher value of i can give a lower value of the α^i. Thus in the worst case, one has to do brute force search among all the possible $p - 1$ values of i to obtain the exact value (note that there is a unique value of i). Hence the time complexity is $O(p)$. One can try to use some storage and perform a time-memory trade-off.

An attacker can pre-compute and store all possible values of (i, α^i), and then sort the table based on the second field using an efficient sorting method. Thus the total storage required is $O(p)$ and the time to sort is $O(p \log p)$. Given a value of β, now the time to search is $O(\log p)$. Sometimes for complexity analysis of DLP, we neglect the value of $\log$, and then thus in this case the time complexity is reduced to $O(1)$ while the memory complexity is increased to $O(p)$. However there are developments in cryptanalysis which allows us to solve the DLP in time-memory product of $O(\sqrt{p})$, but the study of these algorithms is beyond the scope of this text.

Public (or asymmetric) and Private (or symmetric) key algorithms have complementary advantages and disadvantages. They have their specific application areas. Symmetric key ciphers have higher data throughput but the key must remain secret at both the ends. Thus in a large network there are many key pairs that should be managed. Sound cryptographic practice dictates that the key should be changed frequently for each communication session. The throughputs of the most popular public-key encryption methods are several orders of magnitude slower than the best known symmetric key schemes. In a large network the number of keys required are considerably smaller and needs to be changed less frequently. In practice thus public-key cryptography is used for efficient key management while symmetric key algorithms are used for bulk data encryption.

In the next subsection, we highlight an application of public key systems to achieve key-exchanges between two parties. The famous protocol known as the Diffie-Hellman key-exchange is based on another hard problem related closely with the DLP. This is called as the Diffie-Hellman Problem (DHP) and the key-exchange is called as the Diffie Hellman (DH) key-exchange.

In this exchange, Alice and Bob (see Fig. 2) agree upon two public elements, p and g. Alice has a secret element a, and Bob has a secret element b, where $a, b \in Z_{p-1}$. Alice computes $x_1 \equiv g^a \bmod p$, while Bob computes $x_2 \equiv g^b \bmod p$ and then exchanges these informations over the network. Then Alice computes $x_2^a \bmod p$, while Bob computes $x_1^b \bmod p$, both of which are the same. Apart from the agreement (which is quite evident), the most important question is of the secrecy of the agreed key, i.e. the untrusted third party should not be able to compute the agreed key, which is numerically $x_1^b \equiv x_2^a \equiv g^{ab} \bmod p$.

Thus the eavesdropper has to compute this value from the public information of g and p and the exchanged information of $x_1 \equiv g^a \bmod p$ and $x_2 \equiv g^b \bmod p$. This problem is known as the *Computational Diffie Hellman Problem*

(CDH). As can be observed this problem is related to the DLP: if one can solve the DLP he can obtain the values of a or b and can solve the CDH problem as well. The other direction is however not so straightforward and is beyond the current discussion.

The classical DH key-exchange can be subjected to simple man-in-the-middle (MiM) attacks. As an interceptor *Eve* can modify the value x_1 from Alice to Bob and hand over Bob a modified value of $x_1' \equiv g^t \bmod p$, for some arbitrarily chosen $t \in Z_{p-1}$. Similarly, she also modifies x_2 received from Bob into $x_2' = x_1' \equiv g^t \bmod p$. However, Alice and Bob are unaware of this attack scenario and goes ahead with the DH key-exchange and computes the keys as $g^{ta} \bmod p$ and $g^{tb} \bmod p$, respectively. They use these keys to communicate with each other. However, the future messages that are encrypted with these keys can all be deciphered by Eve as she also can compute these keys using the exchanged values of x_1 and x_2 and the public values g and p. This simple attack obviates the use of other reinforcements to the classical DH Key-exchange, like encrypting the exchanged messages by symmetric or asymmetric ciphers. Thus for an end-to-end security interplay of symmetric and asymmetric ciphers is very important. However, the objective of this text is to understand the design challenges of these primitives on hardware.

One of the important class of symmetric algorithms is block ciphers, which are used for bulk data encryption. In the next section, we present an overview of block cipher structures. As an important example, we present the Advanced Encryption Standard (AES), which is the current standard block cipher. The AES algorithm uses finite field arithmetic and the underlying field is of the form $GF(2^8)$. Subsequently in a later section, we describe a present-day public key cipher, namely elliptic curve cryptosystem. These ciphers rely on the arithmetic on elliptic curves which can be defined over finite fields over characteristic 2 and primes.

BLOCK CIPHERS

Block ciphers are encryption algorithms that encrypt n bits of plaintext using an m bits of the key (m and n may be different) and produces an n bits of the ciphertext. Figure 3 shows a top-level diagram of a block cipher. As can be observed that the plaintext is divided into the **Block Length**, which is a block of size n bits. Each block is transformed by the encryption algorithm to result in an n bits of the ciphertext. The plaintext block P_i is thus processed by the key K, resulting in the ciphertext block $C_i = E_K(P_i)$.

The encryption algorithm is used in several **modes** to obtain the ciphertext blocks. The most naïve way of doing the operation is called **Electronic Code Book** (ECB). In this mode, as shown in Fig. 4, each block P_i gets encrypted independent of another block P_j, where $i \neq j$.

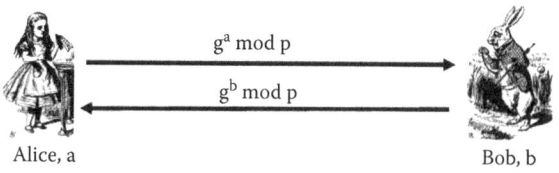

Fig. 2 The Diffie Hellman key exchange

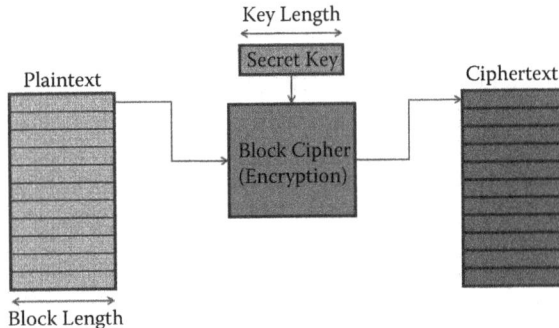

Fig. 3 Block Cipher: Encryption

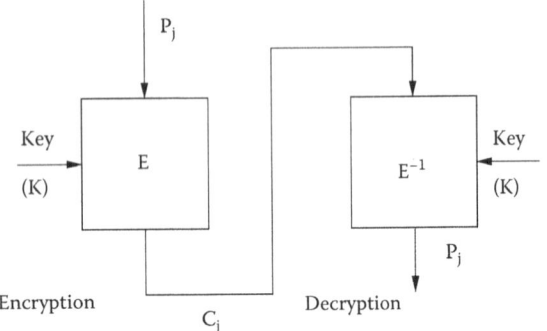

Fig. 4 Electronic Code Book

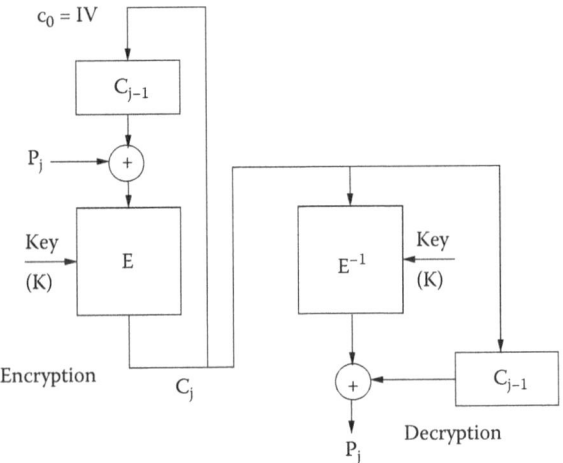

Fig. 5 Cipher Block Chaining

However, this is not a secured form of encryption and is not used for most applications. A popular and secured mode of encryption is called as the **Cipher Block Chaining** (CBC).

In this mode, as shown in Fig. 5 the cipher of a block, C_{j-1} is XORed with the next plaintext block, P_j. Thus the ciphertext for the next block is $C_j = E_K(P_j \oplus C_{j-1})$. This indicates that the output of the jth instance depends on the output of the previous step. Thus although, as we shall see in the following sections, that the block ciphers have an iterated structure, there is no benefit from pipelining. More precisely, the reason is that the next block encryption cannot start unless the encryption of the previous block is completed. However, there are other modes of ciphers, like **counter mode** and *Output Feedback* (OFB) where pipelining provides advantage.

Inner Structures of a Block Cipher

In order to understand the design aspects of a block cipher, it is important to know what they are comprised of.

The block ciphers of the present day have typically blocks of size 128 bits, while the keys are of size 128, 192, or 256 bits. For lightweight applications, there are some ciphers which have keys of length 80 bits. The choice of the key size is very important for security against brute force attacks, and is referred to as the security margin of the cipher. However, the longer the key implies that the cipher design has a larger overhead, in terms of hardware area, time, power. Further it may be noted that a cipher with a large key size is not necessarily more secured. For example, it is widely believed that AES-128 is the most secured among its other variants with key sizes 192 and 256 bits.

The block cipher is typically made of further subdivisions or transformations. The transformations are often called *rounds* of the cipher (refer Fig. 6). A block cipher has, say, N_r number of rounds. The input key, which is the *secret key* is transformed by the *key-scheduling algorithm*, to generate the N_r round keys. The input key is often used as the *whitening key* and is mixed with the plaintext block, P_j. Typically the key mixing is performed through bit-wise XOR between the plaintext block and the input key. Subsequently, each round operates on the message, and the message *state* gets updated due to each transformation. The transformations of a round is achieved by further sub-operations, which make up the rounds. The round keys, computed by the key-scheduling algorithm also are mixed with the present state of the message, typically through bit-wise *XOR*. After the N_r rounds, the final state is returned as the ciphertext.

The round of a cipher is made of further components which provide the cipher much needed **confusion** and **diffusion**. Classically, diffusion hides the relation between the ciphertext and the plaintext. On the other hand, confusion obscures the relation between the ciphertext and the key. The objective of both the steps are to make the task of the cryptanalyst harder.

The round comprises of three distinct operations:

1. **Addition with Round Key**: The message state is typically XORed with the round key.
2. **D-box**: It is a key-less transformation called as the diffusion box, or D-box. It provides diffusion to the cipher. This step is typically a linear transformation wrt. the XOR operation. Hence it can be expressed in terms of the input using only *XOR* gates. Thus often they are easy to implement and hence can

Modern Cryptography

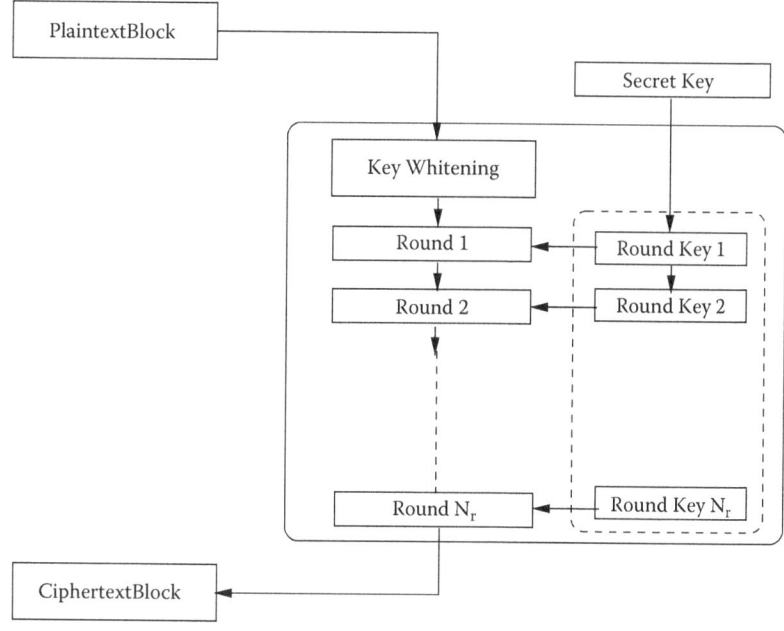

Fig. 6 Structure of a Block Cipher

be applied on larger block lengths as the resource requirement typically is less.

3. **S-Box**: It is generally a key-less transformation, commonly referred to as the substitution box, or S-Box. It provides the much needed confusion to the cipher as it makes the algebraic relations of the ciphertext bits in terms of the message state bits and the key bits more complex. The S-Boxes are typically non-linear wrt. the XOR operations. These transformations require both **XOR** and also **AND** gates. These transformations are mathematically complex and pose a large over-head. Hence they are often performed in smaller chunks. The hardware required also grows fast with the input size and thus requires special techniques for implementation.

The rounds combine the diffusion and the substitution layers suitably for achieving security. In the following section we present the design of the AES algorithm, to illustrate the construction of a block cipher.

The Advanced Encryption Standard

In 1997 the National Institute of Standards and Technology (NIST) initiated the selection for the next generation standard block cipher after DES. One of the primary reasons being the shorter key size of DES (56 bits for encryption) was becoming more and more inadequate for providing security. Efforts to extend DES by cascading instances of DES also was not successful owing to the existence of *meet-in-the-mddle* attacks. Thus 3-DES was evaluated to provide security corresponding to 2 rounds of DES (112 bits), as opposed to the expected security provided by a 168-bit key. Moreover, DES was not very efficient for implementations because of its rather unexplained S-Box design. On November 26, 2011 a cipher designed by Belgian inventors, Rijmen and Daemen, was selected as the Advanced Encryption Standard (AES).

AES is thus a block cipher which works on $GF(2^8)$. Although Rijndael was originally designed to support plaintext blocks and the key blocks of 128, 192, or 256 bits, the adopted AES cipher has a restricted plaintext block of side 128 bits.

The AES algorithm receives a plaintext block as input and produces the ciphertext block after several rounds of the cipher. The cipher algorithm explains the input plaintext, the intermediate blocks and the final ciphertext blocks as states, denoted by matrices with elements in $GF(2^8)$. We next explain some of the notations that have been used in the subsequent exposition:

The state matrix of the Rijndael cipher has Nb 32-bit words, $4 \leq Nb \leq 8$, thus the block length is $32Nb$. For AES, as stated before the block length is 128, thus $Nb = 4$. The key block is parameterized by Nk, which denotes the number of columns of size 32 bits. The range of Nk is $4 \leq Nk \leq 8$. For AES the key length can be either 128, 192, or 256, thus $Nk = 4$, 6, or 8. The number of rounds of the cipher is denoted by Nr, which varies with the size of the key.

The state matrix for AES is as follows:

$$\mathbb{S} = \begin{pmatrix} b_{0,0} & b_{0,1} & b_{1,0} & b_{1,1} \\ b_{1,0} & b_{1,1} & b_{1,2} & b_{1,3} \\ b_{2,0} & b_{2,1} & b_{2,2} & b_{2,3} \\ b_{3,0} & b_{3,1} & b_{3,2} & b_{3,3} \end{pmatrix}$$

The state Σ comprises of 16 bytes, indicated by $b_{i,j}$, where $0 \le i, j \le 15$. Each of the bytes are elements of $GF(2^8)$.

Algorithm 1

The AES function Cipher

Input: byte in[4,Nb], word w[Nb(Nr+1)]

Output: byte out[4,Nb]

```
1   byte state[4,Nb]
2   state = in
3   AddRoundKey(state, w[0:Nb-1])
4   for round = 1 to Nr-1 do
5       SubBytes(state)
6       ShiftRows(state)
7       MixColumns(state)
8       AddRoundKey(state, w[round*Nb:(round+1)*Nb-1])
9   end
10  SubBytes(state)
11  ShiftRows(state)
12  AddRoundKey(state, w[Nr*Nb:(Nr+1)*Nb-1])
13  out=state
```

The state matrices of AES undergo transformations through the rounds of the cipher. The plaintext is of 128 bits and are arranged in the state matrix, so that each of the 16 bytes are elements of the state matrix. The AES key can also be arranged in a similar fashion, comprising of *Nk* words of length 4 bytes each. The input key is expanded by a **Key-Scheduling** algorithm to an expanded key *w*. The plaintext state matrix (denoted by **in**), is transformed by the round keys which are extracted from the expanded key *w*. The final cipher (denoted by **out**) is the result of applying the encryption algorithm, **Cipher** on the plaintext, **in**. In the next two sections, we present the round functions and the key scheduling algorithm respectively.

The AES Round Transformations

The AES **Cipher** receives as an input the plaintext, denoted by the byte **in[4,Nb]**, while the output is denoted by **out[4,Nb]**. The plaintext is stored in the state matrix, denoted by the byte array **state**. The key is stored in a key matrix, **w** which is mixed with the plaintext by XORing. This step is often referred to as the key whitening. The plaintext is subsequently transformed by *Nr* rounds. Each of the first *Nr* − 1 rounds have round transformations, namely **SubBytes**, **ShiftRows**, **MixColumns**, and **AddRoundKey**. In the last round only the transformations SubBytes, ShiftRows, and AddRoundKey are present. Each of the *Nr* + 1 rounds thus require a share of the key which is stored in the key **w[Nb(Nr+1)]**, generated via the key-scheduling algorithm.

The bytes of the state matrix are elements of $GF(2^8)$ and are often written in hexadecimal notation. For example an element, $a(x) = x^7 + x + 1$, can be encoded in binary as 10000011, where the ones denote the corresponding coefficient in $GF(2)$. The element in hexadecimal is denoted as {13}. Likewise, an element in $GF(2^8)$ encoded as 10110011 is expressed as {F3}. As described before, the field is generated by using the following irreducible poynomial as the reduction polynomial:

$$m(X) = X^8 + X^4 + X^3 + X + 1$$

Thus the extension field $GF(2^8)$ is created and the elements of the field are expressible as polynomials $\in GF(2)[X]/\langle m(X) \rangle$. Each non-zero element has a multiplicative inverse, which can be computed by the Euclidean inverse algorithm. This forms the basis of what is known as the SubBytes step of the algorithm.

SubBytes

The **SubBytes** step is a non-linear byte-wise function. It acts on the bytes of the state and subsequently applies an affine transformation on the cipher (Fig. 7). The step is based on the computation of finite field inverse, which is as follows:

$$x' = \begin{cases} x^{-1} & \text{if } x \ne 0 \\ 0 & \text{otherwise} \end{cases}$$

The final output is computed as $y = A(x') + B$, where A and B are fixed matrices defined as follows:

$$A = \begin{pmatrix} 1 & 0 & 0 & 0 & 1 & 1 & 1 & 1 \\ 1 & 1 & 0 & 0 & 0 & 1 & 1 & 1 \\ 1 & 1 & 1 & 0 & 0 & 0 & 1 & 1 \\ 1 & 1 & 1 & 1 & 0 & 0 & 0 & 1 \\ 1 & 1 & 1 & 1 & 1 & 0 & 0 & 0 \\ 0 & 1 & 1 & 1 & 1 & 1 & 0 & 0 \\ 0 & 0 & 1 & 1 & 1 & 1 & 1 & 0 \\ 0 & 0 & 0 & 1 & 1 & 1 & 1 & 1 \end{pmatrix} \quad (2)$$

and the value of B vector is:

$$(\mathbf{B})^t = \begin{pmatrix} 0 & 1 & 1 & 0 & 0 & 0 & 1 & 1 \end{pmatrix} \quad (3)$$

Here, $(B)^t$ represents the transpose of B, and the left most bit is the LSB.

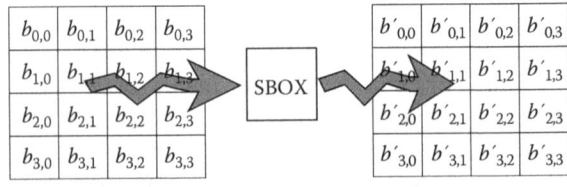

Fig. 7 SubBytes transformation

The **InvSubBytes** step operates upon the bytes in the reverse order. It is defined as:

$$X = Y^{-1}A^{-1} + D \quad (4)$$

where D is an 8 × 1 vector-matrix. The matrix A^{-1} is defined as:[10]

$$A^{-1} = \begin{pmatrix} 0 & 0 & 1 & 0 & 0 & 1 & 0 & 1 \\ 1 & 0 & 0 & 1 & 0 & 0 & 1 & 0 \\ 0 & 1 & 0 & 0 & 1 & 0 & 0 & 1 \\ 1 & 0 & 1 & 0 & 0 & 1 & 0 & 0 \\ 0 & 1 & 0 & 1 & 0 & 0 & 1 & 0 \\ 0 & 0 & 1 & 0 & 1 & 0 & 0 & 1 \\ 1 & 0 & 0 & 1 & 0 & 1 & 0 & 0 \\ 0 & 1 & 0 & 0 & 1 & 0 & 1 & 0 \end{pmatrix} \quad (5)$$

and the value of D vector is:

$$(\mathbf{D})^t = \begin{pmatrix} 0 & 0 & 0 & 0 & 0 & 1 & 0 & 1 \end{pmatrix} \quad (6)$$

Here, $(D)^t$ represents the transpose of D, and the left most bit is the LSB.

The SubBytes is the only non-linear layer of the cipher. The other two operations ShiftRows and MixColumns are linear and provide fast diffusion of disturbances in the cipher.[10]

ShiftRows

In the operation **ShiftRows**, the rows of the **State** are cyclically left shifted over different offsets. We denote the number of shifts of the 4 rows by c_0, c_1, c_2 and c_3. The shift offsets c_0, c_1, c_2 and c_3 depend on N_b. The different values of the shift offsets are specified in Table 1.[10]

Table 1 Shift offsets for different block lengths

N_b	c_0	c_1	c_2	c_3
4	0	1	2	3
5	0	1	2	3
6	0	1	2	3
7	0	1	2	4
8	0	1	3	4

The **InvShiftRows** operation performs circular shift in the opposite direction. The offset values for InvShiftRows are the same as ShiftRows (Table 1). ShiftRows implementations do not require any resource as they can be implemented by rewiring.

MixColumns

The **MixColumns** transformation operates on each column of **State** (X) individually. Each column of the state matrix can be imagined as the extension field $GF(2^8)^4$. For $0 \leq j \leq Nb$ a column of the state matrix S is denoted by the polynomial:

$$s_j(X) = s_{3,j}X^3 + s_{2,j}X^2 + s_{1,j}X + s_{0,j} \in GF(2^8)[X]$$

The transformation for MixColumns is denoted by the polynomial:

$$m(X) = \{03\}X^3 + \{01\}X^2 + \{01\}X + \{02\} \in GF(2^8)[X]$$

The output of the MixColumns operation is obtained by taking the product of the above two polynomials, $s_j(X)$ and $m(X)$ over the field $GF(2^8)^4$, with the reduction polynomial being $X^4 + 1$.

Thus the output can be expressed as a modified column, computed as follows:

$$s'_j(X) = (s_j(X) * m(X)) \bmod (X^4 + 1), 0 \leq j < Nb$$

The transformation can also be viewed as a linear transformation in $GF(2^8)^4$ as follows:

$$\begin{bmatrix} s'_{0,j} \\ s'_{1,j} \\ s'_{2,j} \\ s'_{3,j} \end{bmatrix} = \begin{bmatrix} \{02\} & \{03\} & \{01\} & \{01\} \\ \{01\} & \{02\} & \{03\} & \{01\} \\ \{01\} & \{01\} & \{02\} & \{03\} \\ \{03\} & \{01\} & \{01\} & \{02\} \end{bmatrix} \begin{bmatrix} s_{0,j} \\ s_{1,j} \\ s_{2,j} \\ s_{3,j} \end{bmatrix} \quad (7)$$

In case of **InvMixColumns**, the inverse of the same polynomial is used. If $m^{-1}(X)$ is defined as a function of the transformation of InvMixColumns that operates on **State** X, then.

In matrix form the InvMixColumns transformation can be expressed as:

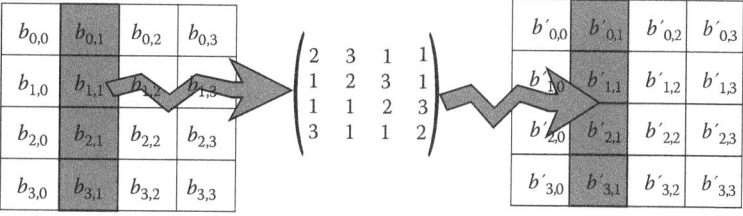

Fig. 8 MixColumn transformation

$$\begin{bmatrix} s''_{0,j} \\ s''_{1,j} \\ s''_{2,j} \\ s''_{3,j} \end{bmatrix} = \begin{bmatrix} \{0E\} & \{0B\} & \{0D\} & \{09\} \\ \{09\} & \{0E\} & \{0B\} & \{0D\} \\ \{0D\} & \{09\} & \{0E\} & \{0B\} \\ \{0B\} & \{0D\} & \{09\} & \{0E\} \end{bmatrix} \begin{bmatrix} s_{0,j} \\ s_{1,j} \\ s_{2,j} \\ s_{3,j} \end{bmatrix} \quad (8)$$

AddRoundKey

Let the input state of a particular round of the cipher **round** be denoted by s. The columns of the state are denoted by $s_0, s_1, \ldots, s_{Nb-1}$. The function **AddRoundKey(state, w[round*Nb,(round+1)*Nb-1])** is denoted as:

$$s_j = s_j \oplus w[round * Nb + j], 0 \leq j < Nb$$

Here $\oplus$ is bit-wise XOR operation. Thus the words of the round key are combined with the state through a mod 2 addition (bitwise XOR). The objective of the key mixing step is to make every round states after the key mixing independent of the previous rounds, assuming that the round keys are generated by an efficient key-scheduling algorithm, which is detailed next.

Key-Scheduling in AES

The algorithm **Key-Scheduling** or **Key-Expansion** takes a Rijndael key **key** and generates the round keys of the ciphers. The input key is a byte-array of length $4Nk$, while the expanded key is a word array of length $Nb(Nr + 1)$. The round keys are mixed in the cipher via application of XOR.

The pseudocode presented in Algorithm 6 explains the generation of the round keys from the input key in AES. The pseudocode uses the functions, **word**, **SubWord** and **RotWord**.

The function Word just concatenates its arguments. The input to SubWord is a word, which is transformed by the SubBytes transformations. Consider the input to SubWord is the word (b_0, b_1, b_2, b_3), where each of the b_is are bytes. Each byte, b_i is transformed by the SubBytes transformation, thus resulting in $d_i = SubBytes(b_i)$. Thus the output of the SubWord is (d_0, d_1, d_2, d_3), after the application of the SubBytes transformation on each of the bytes.

Algorithm 2

The AES KeyExpansion function, KeyExpansion

Input: Nk, byte key[4,Nk]

Output: word word w[Nb(Nr+1)]

```
1    word temp
2    i=0
3    while (i<Nk) do
4      w[i]=word(key[4*i],key[4*i+1],key[4*i+2],key[4*i+3])
5      i=i+1
6    end
7    i=Nk
8    while (i < Nb*(Nr+1)) do
9      temp=w[i-1]
10     if (i mod Nk=0) then
11       temp=SubWord(RotWord(temp)) XOR Rcon[i/Nk]
12     end
13     else if (Nk>6 and i mod Nk = 4) then
14       temp = SubWord(temp)
15     end
16     w[i]=w[i-Nk] XOR temp
17     i=i+1
18   end
```

The input to the RotWord is also a word (b_0, b_1, b_2, b_3). The output is (b_1, b_2, b_3, b_0), which is nothing but the bytewise left cyclic rotation applied on the input word.

Finally, the round constant, abbreviated as **Rcon[n]** = $(\{02\}^n, \{00\}, \{00\}, \{00\})$. The round constants are added to the round keys to provided asymmetry to the key expansion algorithm and protect against certain class of attacks.

RIJNDAEL IN COMPOSITE FIELD

Rijndael involves arithmetic in $GF(2^8)$ elements. As discussed in *Section 1.12* the operations can be expressed in composite fields, exploiting isomorphism properties.

A wide variety of techniques have evolved for implementing the AES algorithm with various objectives. Isomorphism properties and use of subfield arithmetic helps to obtain compact circuits for the AES operations. The techniques proposed[11] presents a method of efficiently expressing the inverse in $GF(2^8)$ using inverse computations in the subfield $GF(2^4)$.

Expressing an Element of $GF(2^8)$ in Subfield

The AES algorithm uses the particular Galois field of 8-bit bytes, where the bits are coefficients of a polynomial. The multiplications are performed modulo an irreducible polynomial $q(X) = X^8 + X^4 + X^3 + X + 1$, while the additions of the coefficients are performed using modulo 2 arithmetic. This representation, as described earlier is called the polynomial representation. If A is a root of the polynomial then the standard polynomial basis of the field is denoted by $1, A, A^2, \ldots, A^7$. Following the notations as introduced[12], we use uppercase Roman letters for specific elements of $GF(2^8)$ or its isomorphic field $GF(2^4)^2$. Lowercase Greek letters are used for the subfield $GF(2^4)$.

An element in $GF(2^8)$ is mapped to an element in the composite field $GF(2^4)^2$. The converted element is expressed as a linear polynomial in y over $GF(2^4)$. Thus, $G \in GF(2^4)^2$ is expressed as $G = \gamma_1 Y + \gamma_0$. The multiplications are defined modulo an irreducible polynomial $r(Y) = Y^2 + \tau Y + \mu$. All the coefficients are in the field $GF(2^4)$ and the pair (γ_1, γ_0) represents G in terms of a polynomial basis $(y, 1)$ where y is one root of $r(Y)$.

Alternately, one can use the normal basis also for $GF(2^4)^2$, which is (y, y^{16}) using the roots of $r(Y)$. Note that $r(Y) = Y^2 + \tau Y + \mu = (Y + y)(Y + y^{16})$. Here, $\tau = y + y^{16}$ is the trace, while $\mu = (y)(y^{16})$ is the norm of Y.[a]

Inversion of an Element in Composite Field

The most complex operation is the finite field inverse, which forms the basis of the AES S-Box. Direct computation of the inverse of an eight degree polynomial modulo the irreducible eight degree polynomial is difficult. However, an efficient technique was proposed[11]. As outlined previously, the inverse of a $GF(2^8)$ element is computed by converting it into an isomorphic composite field $GF(2^4)^2$. The element in the composite field is expressed as a polynomial of the first degree with coefficients from $GF(2^4)$.[13]

Let us assume that the element in $GF(2^4)^2$ whose multiplicative inverse is to be computed is denoted as $\gamma_1 Y + \gamma_0$. The operations are performed modulo the polynomial $r(Y) = Y^2 + \tau Y + \mu$.

The multiplication modulo $Y^2 + \tau Y + \mu$ is:

$$(\gamma_1 Y + \gamma_0)(\delta_1 Y + \delta_0) = \gamma_1 \delta_1 Y^2 + (\gamma_1 \delta_0 + \gamma_0 \delta_1)Y + \gamma_0 \delta_0$$
$$= \gamma_1 \delta_1 (\tau Y + \mu) + (\gamma_1 \delta_0 + \gamma_0 \delta_1)Y + \gamma_0 \delta_0$$
$$= (\gamma_1 \delta_0 + \gamma_0 \delta_1 + \gamma_1 \delta_1 \tau)Y + (\gamma_0 \delta_0 + \gamma_1 \delta_1 \mu)$$

Let, $(\gamma_1 Y + \gamma_0)^{-1} = (\delta_1 Y + \delta_0) \bmod (Y^2 + \tau Y + \mu)$. Rearranging we have $(\gamma_1 Y + \gamma_0)(\delta_1 Y + \delta_0) \equiv 1 \bmod (Y^2 + \tau Y + \mu)$. Thus, using the product and equating to 1 by matching the coefficients we can write the following simultaneous equation:

$$\gamma_1 \delta_0 + \gamma_0 \delta_1 + \gamma_1 \delta_1 \tau = 0$$
$$\gamma_0 \delta_0 + \gamma_1 \delta_1 \mu = 1$$

We solve the above equations to compute the values of δ_0 and δ_1:

$$\delta_0 = (\gamma_0 + \gamma_1 \tau)(\gamma_0^2 + \gamma_0 \gamma_1 \tau + \gamma_1^2 \mu)^{-1}$$
$$\delta_1 = \gamma_1 (\gamma_0^2 + \gamma_0 \gamma_1 \tau + \gamma_1^2 \mu)^{-1}$$

The computations can also be similarly reworked if the basis is normal. Considering the normal basis of (Y, Y^{16}). Since both the elements of the basis are roots of the polynomial $Y^2 + \tau Y + \mu = 0$, we have the following identities which we use in the equations of the multiplication and inverse of the elements in the composite field:

$$Y^2 = \tau Y + \mu$$
$$1 = \tau^{-1}(Y^{16} + Y)$$
$$\mu = (Y^{16})Y$$

Thus the multiplication modulo $Y^2 + \tau Y + \mu$ in the normal basis is:

$$(\gamma_1 Y^{16} + \gamma_0 Y)(\delta_1 Y^{16} + \delta_0 Y)$$
$$= \gamma_1 \delta_1 Y^{32} + (\gamma_1 \delta_0 + \gamma_0 \delta_1)(Y^{16}Y) + \gamma_0 \delta_0 Y^2$$
$$= \gamma_1 \delta_1 (\gamma^2 + Y^2) + \mu(\gamma_1 \delta_0 + \gamma_0 \delta_1) + \gamma_0 \delta_0 Y^2$$
$$= Y^2(\gamma_1 \delta_1 + \gamma_0 \delta_0) + [\gamma_1 \delta_1 \tau^2 + \mu(\gamma_1 \delta_0 + \gamma_0 \delta_1)]$$
$$= (\tau Y + \mu \tau^{-1}(Y^{16} + Y))(\gamma_1 \delta_1 + \gamma_0 \delta_0)$$
$$\quad + [\gamma_1 \delta_1 \tau^2 + \mu(\gamma_1 \delta_0 + \gamma_0 \delta_1)](\tau^{-1}(Y^{16} + Y))$$
$$= [\gamma_1 \delta_1 \tau + \theta]Y^{16} + [\gamma_0 \delta_0 \tau + \theta]Y,$$

where $\theta = (\gamma_1 + \gamma_0)(\delta_1 + \delta_0)\mu\tau^{-1}$

Thus, if $(\gamma_1 Y^{16} + \gamma_0 Y)$ and $(\delta_1 Y^{16} + \delta_0 Y)$ are inverses of each other, then we can equate the above product to $1 = \tau^{-1}(Y^{16} + Y)$. Equating the coefficients we have:

$$\delta_0 = \left[\gamma_1 \gamma_0 \tau^2 + \left(\gamma_1^2 + \gamma_0^2\right)\mu\right]^{-1} \gamma_1$$
$$\delta_1 = \left[\gamma_1 \gamma_0 \tau^2 + \left(\gamma_1^2 + \gamma_0^2\right)\mu\right]^{-1} \gamma_0$$

These equations show that the inverse in the field $GF(2^8)$ can be reduced to the inverse in the smaller field $GF(2^4)$ along with several additional operations, like addition, multiplication, and squaring in the subfield. The inverse in the subfield can be stored in a smaller table (as compared to a table to store the inverses of $GF(2^8)$). The operations in $GF(2^4)$ can be in turn expressed in the sub-subfield $GF(2^2)$. The inverses in the sub-subfield $GF(2^2)$ is the same as squaring.

Depending on the choices of the irreducible polynomials, the level of decompositions and the choices of the basis of the fields, the complexity of the computations differ and is a subject of significant research.

The Round of AES in Composite Fields

Like the SubByte, the entire round of AES (and the entire AES algorithm) can be expressed in the composite fields. In Ref. [14], the authors develop the entire round of AES in composite fields. It must be kept in mind that though there is a gain in terms of compact representations and efficient computations in the subfield and sub-subfields, for further decompositions, there is an accompanied cost involved. The cost comes from the transformation of the

---

[a] In order to map an element in $GF(2^8)$ to an element in the composite field $GF((2^4)^2)$, as discussed in *section 1.12.1* the element is multiplied with a transformation matrix, T.

elements between the various field representations. Hence, it is the designer's job to study these transformations and decide a final architecture which optimizes this trade-off efficiently.

The above point also implies that performing the inverse in composite fields for computing the S-Box operation, imply a continuous overhead of the transformation of the elements from $GF(2^8)$ to $GF(2^4)^2$, and vice versa. Hence, it is worth-while to explore techniques to represent the entire AES in the composite field representation. This minimizes the overhead in the transformations among the different field representations, being performed once at the beginning and finally at the end.

The Rijndael round transformations in subfield are defined as follows Consider the transformation **T** maps an element from $GF(2^8)$ to $GF(2^4)^2$. The **T**, as discussed before represents a transformation matrix an 8×8 binary matrix, which operates on each byte of the 4×4 state matrix of AES. Denote the AES state by S, where each element is denoted by b_{ij}, where $0 \leq i, j \leq 3$. Thus, an element in $x \in GF(2^8)$ is mapped to $T(x) \in GF(2^4)^2$. Now let us consider each of the round transformations one by one:

1. **SubBytes Transformation**: This operation has two steps:
   (a). **Inverse**: $b'_{i,j} = (b_{i,j})^{-1}$. In the composite field, we have $T(b'_{i,j}) = (T(b_{ij}))^{-1}$. Note that the inverse on the RHS of the above equation is in $GF(2^4)^2$. The computation of the inverse is as explained above in section 2.4.1.
   (b). **Affinne:** $b''_{i,j} = A(b'_{i,j}) + B$. Here A and B are fixed matrices as discussed in section 2.3.3.1. In the composite field, $T(b''_{i,j}) = T(A(b'_{i,j})) + T(B) = T A T^{-1}(T((b'_{i,j}))) + T(B)$. Thus the matrices of the SubBytes operations needs to be changed by applying the transformation matrix **T**.
2. **ShiftRows**: This step remains the same as this is a mere transposition of bytes and the field transformation to the composite field is localized inside a byte.
3. **MixColumns**: This step essentially involves multiplication of a column of the state matrix with a row of the Mix Column matrix. As can be observed from Eq. 6, all the rows of the Mix Column matrix are permutations of ({01}, {01}, {02}, {03}). If we denote the ith row of this matrix as (m_0, m_1, m_2, m_3), and the jth column of the state matrix as $(s_{0,j}, s_{1,j}, s_{2,j}, s_{3,j})$, then the $(i, j)^{th}$ element of the state matrix corresponding to the output of the Mix Column is:

$$s'_{i,j} = m_0 s_{0,j} + m_1 s_{1,j} + m_2 s_{2,j} + m_3 s_{3,j}$$

Thus in the composite field the above transformation is,

$$T(s'_{i,j}) = T(m_0)T(s_{0,j}) + T(m_1)T(s_{1,j}) + T(m_2)T(s_{2,j}) + T(m_3)T(s_{3,j})$$

Here the additions in either the original field or the composite field are all in characteristic-2 field, they are bitwise XORs in both the representations.

4. **Add Round Key**: The operation is $s'_{i,j} = s_{i,j} + k_{i,j}$, where $k_{i,j}$ is a particular byte of the round key. In the composite field, thus this transformation is $T(s'_{i,j}) = T(s_{i,j}) + T(k_{i,j})$. Again the addition is a bitwise XOR.

This implies that the round keys also need to be computed in the composite field. Hence, similar transformations also needs to be performed on the key-scheduling algorithm.

In the next section, we present an overview on a popular public key encryption algorithm, known as **Elliptic Curve Cryprography** (ECC), which leads to much efficient implementations compared to the older generation algorithms like RSA, ElGamal etc.

ELLIPTIC CURVES

Let us start with a puzzle: What is the number of balls that may be piled as a square pyramid and also rearranged into a square array? The number is more than one.

The answer to this simple question can be solved by assuming that the height of the pyramid is denoted by the integer x and the dimension of the sides of the rearranged square is denoted by the integer y.

Since the number of balls in both the arrangements are the same, we have the following equation:

$$y^2 = 1^2 + 2^2 + \cdots + x^2 = \frac{x(x+1)(2x+1)}{6}$$

Some discrete values of y are plotted wrt. x and is depicted in Fig. 9. Curves of this nature are commonly called **Elliptic Curves**: these are curves which are quadratic wrt. y and cubic wrt. x. It may be observed that the curve has two distinct regions or lobes, as it is often referred to as curves of genus 2. Also since, the curve is quadratic wrt. y, the curve is symmetric over the x-axis.

We next present a method by Diophantus of Alexandria, who lived around 200 A.D. to determine non-trivial points on the curve. This method uses a set of known points to find an unknown point on the curve. Let us start with two trivial points: (0,0) and (1,1). Clearly, both these two points do not indicate a solution to the puzzle.

Now the equation of a straight line between these two points is: $y = x$. Since the equation of the curve is cubic wrt. x, the straight line must intersect the curve on a third point

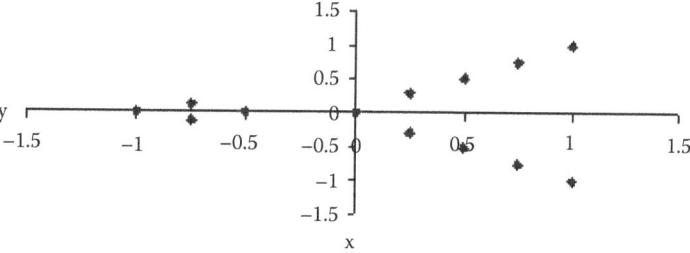

Fig. 9 Plot of y vs x

(the points may not be distinct though!). In order to obtain the third point, we substitute $y = x$ in the equation of the curve, $y^2 = x(x + 1)(2x + 1)/6$, and we obtain:

$$x^3 - \frac{3}{2}x^2 + \frac{1}{2}x = 0$$

We know that $x = 0$ and 1 are two roots of the equation. From the theory of equations, thus if the third root of the equation is $x = \alpha$, we have $0 + 1 + \alpha = \frac{3}{2} \Rightarrow \alpha = \frac{1}{2}$. Since the point on the curve is $y = x$, we have $y = \frac{1}{2}$.

Thus $\left(\frac{1}{2}, \frac{1}{2}\right)$ is a point on the curve. Since the curve is symmetric over the x-axis, $\left(\frac{1}{2}, -\frac{1}{2}\right)$ is also another point on the curve. However these points also do not provide a solution as they are not integral.

Now, consider a straight line through $\left(\frac{1}{2}, -\frac{1}{2}\right)$ and $(1, 1)$. The equation of this line is $y = 3x - 2$, and intersecting with the curve we have:

$$x^3 - \frac{51}{2}x^2 + \cdots + = 0$$

Thus, again the third root $x = \beta$ can be obtained from $1 + \frac{1}{2} + \beta = \frac{51}{2} \Rightarrow \beta = 24$. The corresponding y value is 70, and so we have a non-trivial solution of the puzzle as 4900.

Through this seemingly simple puzzle, we have observed an interesting *geometric* method to solve an *algebraic* problem. This technique forms the base of the geometric techniques (the **chord-and-tangent** rule) in Elliptic Curves.

An Elliptic curve over a field K is a cubic curve in two variables, denoted as $f(x, y) = 0$, along with a rational point, which is referred to as the point at infinity. The field K is usually taken to be the complex numbers, reals, rationals, algebraic extensions of rationals, p-adic numbers, or a finite field. Elliptic curves groups for cryptography are examined with the underlying fields of F_p (where $p > 3$ is a prime) and F_{2^m} (a binary representation with 2^m elements).

A general form of the curve is introduced next. However the curve equation used for implementation is often transformed forms of this curve using the properties of the underlying field K.

Definition 2

An elliptic curve E over the field K is given by the Weierstraß equation mentioned in Eq. 9. *The generalized Weierstraß equation is:*

$$E : y^2 + a_1 xy + a_3 y = x^3 + a_2 x^2 + a_4 x + a_6 \quad (9)$$

with the constant coefficients $a_1, a_2, a_3, a_4, a_6 \in K$ and $\Delta \neq 0$, where Δ is the discriminant of E and is defined as:

$$\Delta = -d_2^2 d_8 - 8d_4^3 - 27d_6^2 + 9d_2 d_4 d_6$$
$$d_2 = a_1^2 + 4a_2$$
$$d_4 = 2a_4 + a_1 a_3$$
$$d_6 = a_3^2 + 4a_6$$
$$d_8 = a_1^2 a_6 + 4a_2 a_6 - a_1 a_3 a_4 + a_2 a_3^2 - a_4^2$$

This equation, known as the generalized *Weierstraß* equation defines the Elliptic Curve E over the field K. It may be noted that if E is defined over K, it is also defined over any extension of the field K. If L is any extension of K, then the set of L-rational points on E is defined as:

$$E(L) = \{(x, y) \in L \times L : y^2 + a_1 xy + a_3 y$$
$$= x^3 + a_2 x^2 + a_4 x + a_6 = 0\} \cup \{\infty\}$$

where ∞ is the point of infinity.

Point at ∞ is a point at the top of the y-axis, and is also at the bottom of the y-axis. We conceptualize this by thinking that the ends of the y-axis are wrapped around and meet in the back. However when working with finite fields, there is no meaningful ordering of the points. Thus the point at ∞ is also conceptualized as the intersecting point of two vertical lines. By symmetry, if they meet at the top they also meet in the bottom. Also from another sense, two parallel lines intersect at only one point, thus implying that the top point at ∞ is the same as that in the bottom.

Simplification of the Weierstraß Equation

Two elliptic curves E_1 and E_2 defined over K are said to be *isomorphic* over K if change of variables transform one form to the other. However the change of variables should be *admissible* depending on the underlying field.

More precisely, consider two elliptic curve equations:

$$E_1 : y^2 + a_1xy + a_3y = x^3 + a_2x^2 + a_4x + a_6$$
$$E_2 : y^2 + a_1'xy + a_3'y = x^3 + a_2'x^2 + a_4'x + a_6'$$

If there exists $u, r, s, t \in K$, $u \neq 0$, such that the change of variables:

$$(x, y) \to (u^2x + r, u^3y + u^2sx + t)$$

transform equation E_1 into equation E_2. We next present those simplifications for different characteristics for K.

Characteristic of K is neither 2 nor 3: The admissible change of variables are:

$$(x, y) \to \left(\frac{x - 3a_1^2 - 12a_2}{36}, \frac{y - 3a_1x}{216} - \frac{a_1^3 + 4a_1a_2 - 12a_3}{24}\right)$$

transforms E to the curve:

$$y^2 = x^3 + ax + b$$

where $a, b \in K$. The discriminant of the curve is $\Delta = -16(4a^3 + 27b^2)$.

Characteristic of K is 2: If $a_1 \neq 0$, then the admissible change of variables are:

$$(x, y) \to \left(a_1^2 x + \frac{a_3}{a_1}, a_1^3 y + \frac{a_1^2 a_4 + a_3^2}{a_1^3}\right)$$

transforms the curve E to the form:

$$y^2 + xy = x^3 + ax^2 + b$$

where a and $b \in K$. The discriminant of the curve is $\Delta = b$.

If $a_1 = 0$, then the admissible change of variables are as follows:

$$(x, y) \to (x + a_2, y)$$

This transforms the curve E to the form:

$$y^2 + cy = x^3 + ax + b$$

where $a, b, c \in K$. The discriminant of the curve is $\Delta = c^4$.

Characteristic of K is 3: Similar simplification of the curve can be done for curves of characteristic 3 also using the admissible change of variables:

$$(x, y) \to \left(x + \frac{d_4}{d_2}, y + a_1x + a_1x + a_1\frac{d_4}{d_2} + a_3\right)$$

where $d_2 = a_1^2 + a_2$ and $d_4 = a_4 - a_1a_3$, transforms E to the curve:

$$y^2 = x^3 + ax^2 + b$$

where $a, b, c \in K$. The discriminant of the curve is $\Delta = -a^3b$. If $a_1^2 = -a_2$, then the admissible change of variables:

$$(x, y) \to (x, y + a_1x + a_3)$$

transforms E to the curve:

$$y^2 = x^3 + ax + b$$

where $a, b, c \in K$. The discriminant of the curve is $\Delta = -a^3$.

Singularity of Curves

For an elliptic curve defined as $y^2 = f(x)$ defined over some K, singularity is the point (x_0, y_0) where there are multiple roots. This can be alternately stated by defining $F(x, y) = y^2 - f(x)$ and evaluating when the partial derivatives vanish wrt. both x and y.

$$\frac{\delta F}{\delta x}(x_0, y_0) = \frac{\delta F}{\delta y}(x_0, y_0) = 0$$
$$or, -f'(x_0) = 2y_0 = 0$$
$$or, f(x_0) = f'(x_0) = 0$$

Thus f has a double root at the point (x_0, y_0). Usually we assume that Elliptic Curves do not have singular points. Let us find the condition for the curve defined as $y^2 = x^3 + Ax + B$, defined over a field K with appropriate characteristics.

Thus, we have:

$$3x^2 + A = 0 \Rightarrow x^2 = -A/3$$

Also we have,

$$x^3 + Ax + B = 0 \Rightarrow x^4 + Ax^2 + Bx = 0$$
$$\Rightarrow (-A/3)^2 + A(-A/3) + Bx = 0 \Rightarrow x = -\frac{2A^2}{9B}$$
$$\Rightarrow 3\left(\frac{2A^2}{9B}\right)^2 + A = 0 \Rightarrow 4A^3 + 27B^2 = 0$$

Thus, the criteria for non-singularity of the curve is $\Delta = 4A^3 + 27B^2 \neq 0$. For elliptic curve cryptography usually the curves do not have singularity.

The Abelian Group and the Group Laws

In this section we show that addition laws can be defined on the points of the elliptic curve so that they satisfy the conditions required for a mathematical group. The essential requirements for a group operation is that the operations have to be associative, there should be a neutral or identity element, and every element should have an inverse on the elliptic curve. The rules are commonly called as chord-and-tangent rules (useful to conceptualize when the elliptic curve is defined over real numbers), also known as the **double** and **add** rules. Further the group is abelian, implying that the operations are commutative. The operation of addition is realized by two distinct operations (unlike over a finite field where a single operation is used): addition when the two points are distinct, and doubling when the points are same.

We summarize the properties of the addition operations (doubling is a special case of the addition operation when the two points are same). The addition is denoted by the symbol + below for an elliptic curve $E(K)$, where K is some underlying field.

Given two points $P, Q \in E(K)$, there is a third point, denoted by $P + Q \in E(K)$, and the following relations hold for all $P, Q, R \in E(K)$

- $P + Q = Q + P$ (commutative)
- $(P + Q) + R = P + (Q + R)$ (associativity)
- $\exists O$, such that $P + O = O + P = P$ (existence of an identity element, O)
- $\exists (-P)$ such that $-P + P = P + (-P) = O$ (existence of inverses)

For cryptography, the points on the elliptic curve are chosen from a large finite field. The set of points on the elliptic curve form a *group* under the addition rule. The point at infinity, denoted by O, is the identity element of the group. The operations on the elliptic curve, i.e., the group operations are *point addition, point doubling* and *point inverse*. Given a point $P = (x, y)$ on the elliptic curve, and a positive integer n, *scalar multiplication* is defined as

$$nP = P + P + P + \cdots P (n \text{ times}) \quad (10)$$

The *order* of the point P is the smallest positive integer n such that $nP = O$. The points $\{O, P, 2P, 3P, \cdots (n-1)P\}$ form a group generated by P. The group is denoted as $<P>$.

The security of ECC is provided by the Elliptic Curve Discrete Logarithm problem (ECDLP), which is defined as follows : *Given a point P on the elliptic curve and another point $Q \in <P>$, determine an integer k ($0 \leq k \leq n$) such that $Q = kP$.* The difficulty of ECDLP is to calculate the value of the scalar k given the points P and Q. k is called the discrete logarithm of Q to the base P. P is the generator of the elliptic curve and is called the basepoint.

The ECDLP forms the base on which asymmetric key algorithms are built. These algorithms include the elliptic curve Diffie-Hellman key exchange, elliptic curve ElGamal public key encryption and the elliptic curve digital signature algorithm.

Next we define the above operations and the underlying computations for elliptic curves of characteristic 2, which is the object of focus of this textbook.

Elliptic Curves with Characteristic 2

For elliptic curves defined on characteristic 2 fields we have the alternate definition as follows:

Definition 3

An elliptic curve E over the field $GF(2^m)$ is given by the simplified form of the Weierstraß equation mentioned in Eq. 9. The simplified Weierstraß equation is:

$$y^2 + xy = x^3 + ax^2 + b \quad (11)$$

with the coefficients a and b in GF (2^m) and b≠0.

Equation 11 can be rewritten as

$$F(x, y) : y^2 + x^3 + xy + ax^2 + b = 0 \quad (12)$$

The partial derivatives of this equation are:

$$\frac{\delta F}{dy} = x$$
$$\frac{\delta F}{dx} = x^2 + y \quad (13)$$

If we consider the curve given in Eq. 11, with $b = 0$, then the point (0, 0) lies on the curve. At this point $\delta F/dy = \delta F/dx = 0$. This forms a *singular point* and cannot be included in the elliptic curve group, therefore an additional condition of $b \neq 0$ is required on the elliptic curve of Eq. 11. This condition ensures that the curve is *non singular*. Hence for the rest of the text we will assume $b \neq 0$, ie. the curve in Eq. 11 is a *non-singular curve*.

The set of points on the elliptic curve along with a special point O, called the *point at infinity*, form a group under addition. The identity element of the group is the point at infinity (O). The arithmetic operations permitted on the group are point inversion, point addition and point doubling which are described as follows.

Point Inversion: Let P be a point on the curve with coordinates (x_1, y_1), then the inverse of P is the point $-P$ with coordinates ($x_1, x_1 + y_1$). The point $-P$ is obtained by drawing a vertical line through P. The point at which the line intersects the curve is the inverse of P.

Let $P = (x_1, y_1)$ be a point on the elliptic curve of Eq. 11. To find the inverse of point P, a vertical line is drawn passing through P. The equation of this line is $x = x_1$. The point at which this line intersects the curve is the inverse $-P$. The coordinates of $-P$ is (x_1, y_1'). To find y_1', the point of intersection between the line and the curve must be found. Equation 12 is represented in terms of its roots p and q as shown below.

$$(y - A)(y - B) = y^2 - (p + q)y + pq \quad (14)$$

The coefficients of y is the sum of the roots. Equating the coefficients of y in Eqs. 12 and 14.

$$p + q = x_1$$

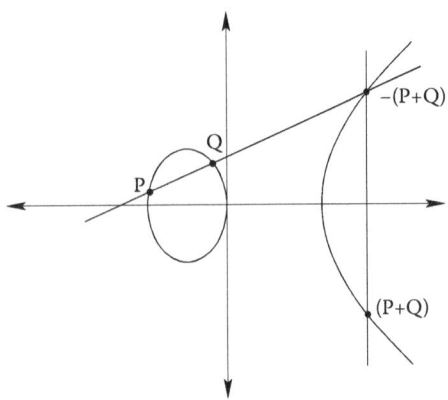

Fig. 10 Point addition

One of the roots is $q = y_1$, therefore the other root p is given by:

$$p = x_1 + y_1$$

This is the y coordinate of the inverse. The inverse of the point P is therefore given by $(x_1, x_1 + y_1)$.

Point Addition: Let P and Q be two points on the curve with coordinates (x_1, y_1) and (x_2, y_2). Also, let $P \neq \pm Q$, then adding the two points results in a third point $R = (P + Q)$. The addition is performed by drawing a line through P and Q as shown in Fig. 10. The point at which the line intersects the curve is $-(P + Q)$. The inverse of this is $R = (P + Q)$. Let the coordinates of R be (x_3, y_3), then the equations for x_3 and y_3 is

$$\begin{aligned} x_3 &= \lambda^2 + \lambda + x_1 + x_2 + a \\ y_3 &= \lambda(x_1 + x_3) + x_3 + y_1 \end{aligned} \quad (15)$$

where $\lambda = (y_1 + y_2)/(x_1 + x_2)$. If $P = -Q$, then $P + (-P)$ is O.

The derivation of the co-ordinates can be done from simple principles of co-ordinate geometry. Let $P = (x_1, y_1)$ and $Q = (x_2, y_2)$ be two points on the elliptic curve. To add the two points, a line (l) is drawn through P and Q. If $P \neq \pm Q$, the line intersects the curve of Eq. 11 at the point $-R = (x_3, y_3')$. The inverse of the point $-R$ is $R = (P + Q)$ having coordinates (x_3, y_3).

The slope of the line l passing through P and Q is given by:

$$\lambda = \frac{y_2 - y_1}{x_2 - x_1}$$

Equation of the line l is:

$$\begin{aligned} y - y_1 &= \lambda(x - x_1) \\ y &= \lambda(x - x_1) + y_1 \end{aligned} \quad (16)$$

Substituting y from 16 in the elliptic curve Eq. 11 we get,

$$(\lambda(x - x_1) + y_1)^2 + x(\lambda(x - x_1) + y_1) = x^3 + ax^2 + b$$

This can be rewritten as

$$x^3 + (\lambda^2 + \lambda + a)x^2 + \cdots = 0 \quad (17)$$

Equation 17 is a cubic equation having three roots. Let the roots be p, q, and r. These roots represent the x coordinates of the points on the line that intersect the curve (the point P, Q and $-R$). Equation 17 can be also represented in terms of its roots as:

$$\begin{aligned} (x - p)(x - q)(x - r) &= 0 \\ x^3 - (p + q + r)x^2 \cdots &= 0 \end{aligned} \quad (18)$$

Equating the x^2 coefficients of Eqs. 18 and 17 we get:

$$p + q + r = \lambda^2 + \lambda + a \quad (19)$$

Since $P = (x_1, y_1)$ and $Q = (x_2, y_2)$ lie on the line l, therefore two roots of Eq. 17 are x_1 and x_2. Substituting $p = x_1$ and $q = x_2$ in Eq. 19 we get the third root, this is the x coordinate of the third point on the line which intersects the curve (i.e., $-R$). This point is denoted by x_3, and it also represents the x coordinate of R.

$$x_3 = \lambda^2 + \lambda + x_1 + x_2 + a \quad (20)$$

The y coordinate of $-R$ can be obtained by substituting $x = x_3$ in Eq. 16. This point is denoted as y_3'.

$$y_3' = \lambda(x_3 + x_1) + y_1 \quad (21)$$

Reflecting this point about the x axis is done by substituting $y_3' = x_3 + y_3$. This gives the y coordinate of R, denoted by y_3.

$$y_3 = \lambda(x_3 + x_1) + y_1 + x_3 \quad (22)$$

Since we are working with binary finite fields, subtraction is the same as addition. Therefore:

$$\begin{aligned} x_3 &= \lambda^2 + \lambda + x_1 + x_2 + a \\ y_3 &= \lambda(x_3 + x_1) + y_1 + x_3 \\ \lambda &= \frac{y_2 + y_1}{x_2 + x_1} \end{aligned} \quad (23)$$

Point Doubling: Let P be a point on the curve with coordinates (x_1, y_1) and $P \neq -P$. The double of P is the point $2P = (x_3, y_3)$ obtained by drawing a tangent to the curve through P. The inverse of the point at which the tangent intersects the curve is the double of P (Fig. 11). The equation for computing $2P$ is given as

$$\begin{aligned} x_3 &= \lambda^2 + \lambda + a = x_1^2 + \frac{b}{x_1^2} \\ y_3 &= x_1^2 + \lambda x_3 + x_3 \end{aligned} \quad (24)$$

where $\lambda = x_1 + (y_1/x_1)$.

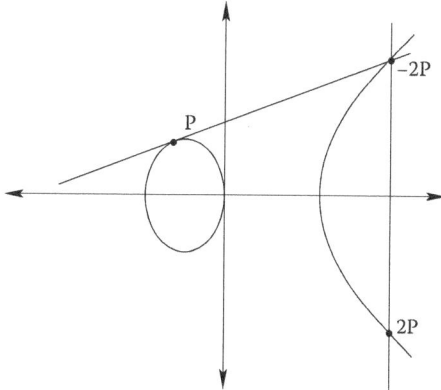

Fig. 11 Point doubling

Let $P = (x_1, y_1)$ be a point on the elliptic curve. The double of P, ie. $2P$, is found by drawing a tangent t through P. This tangent intersects the curve at the point $-2P = (x_3, y_3')$. Taking the reflection of the point $-2P$ about the X axis gives $2P = (x_3, y_3)$.

First, let us look at the tangent t through P. The slope of the tangent t is obtained by differentiation of Eq. 11.

$$2y \frac{dy}{dx} + x \frac{dy}{dx} + y = 3x^2 + 2ax$$

Since we are using modular 2 arithmetic:

$$x \frac{dy}{dx} + y = x^2$$

The slope dy/dx of the line t passing through the point P is given by:

$$\lambda = \frac{x_1^2 + y_1}{x_1} \quad (25)$$

The equation of the line t can be represented by the following:

$$y + y_1 = \lambda(x + x_1) \quad (26)$$

This gives:

$$y = \lambda(x + x_1) + y_1$$
$$y = \lambda x + c \text{ for some constant } c$$

To find x_3 (the x coordinate of $-2P$), substitute for y in Eq. 11.

$$(\lambda x + c)^2 + x(\lambda x + c) = x^3 + ax + b$$

This equation can be rewritten as:

$$0 = x^3 + (\lambda^2 + \lambda + a)x + \cdots \quad (27)$$

This equation is cubic and has three roots. Of these three roots, two roots must be equal since the line intersects the curve at exactly two points. The two equal roots are represented by p. The sum of the three roots is $(\lambda^2 + \lambda + a)$, similar to Eq. 18. Therefore:

$$p + p + r = \lambda^2 + \lambda + a$$
$$r = \lambda^2 + \lambda + a$$

The dissimilar root is r. This root corresponds to the x coordinate of $-2P$ ie. x_3. Therefore:

$$x_3 = \lambda^2 + \lambda + a$$

To find the y coordinate of $-2P$, ie. y_3', substitute x_3 in Eq. 26. This gives:

$$y_3' = \lambda x_3 + \lambda x_1 + y_1$$
$$y_3' = \lambda x_3 + x_1^2$$

To find y_3, the y coordinate of $2P$, the point y_3' is reflected on the x axis. From the point inverse equation:

$$y_3 = \lambda x_3 + x_1^2 + x_3$$

To summarize, the coordinates of the double are given by Eq. 28:

$$\begin{aligned} x_3 &= \lambda^2 + \lambda + a \\ y_3 &= x_1^2 + \lambda x_3 + x_3 \\ \lambda &= x_1 + \frac{y_1}{x_1} \end{aligned} \quad (28)$$

The fundamental algorithm for ECC is the *scalar multiplication*, which can be obtained using the basic double and add computations as shown in Algorithm 3. The input to the algorithm is a *basepoint* P and a m bit scalar k. The result is the scalar product kP, which is equivalent to adding the point P k times.

As an example of how Algorithm 3 works, consider $k = 22$. The binary equivalent of this is $(10110)_2$. Table 2 below shows how $22P$ is computed.

Each iteration of i does a doubling on Q if k_i is 0 or a doubling followed by an addition if k_i is 1. The underlying operations in the addition and doubling equations use the finite field arithmetic discussed in the previous section. Both point doubling and point addition have 1 inversion (I) and 2 multiplications (M) each (from Eqs. 15 and 24), neglecting squaring operations which are free in characteristic 2. From this, the entire scalar multiplier for the m bit

Table 2 Scalar multiplication using double and add to find $22P$

i	k_i	Operation	Q
3	0	Double only	$2P$
2	1	Double and add	$5P$
1	1	Double and add	$11P$
0	0	Double only	$22P$

scalar k will have $m(1I+2M)$ doublings and $\frac{m}{2}(1I+2M)$ additions (assuming k has approximately $m/2$ ones on an average). The overall expected running time of the scalar multiplier is therefore obtained as:

$$t_a \approx (3M+\frac{3}{2}I)m \qquad (29)$$

For this expected running time, finite field addition and squaring operations have been neglected as they are simple operations and can be considered to have no overhead to the run time.

Algorithm 3

Double and Add algorithm for scalar multiplication

Input: Basepoint $P=(x,y)$ and Scalar $k=(k_{m-1}, k_{m-2} \cdots k_0)_2$, where $k_{m-1}=1$
Output: Point on the curve $Q=kP$

```
1    Q = P
2    for i = m − 2 to 0 do
3        Q = 2 · Q
4        if k_i = 1 then
5            Q = Q + P
6        end
7    end
8    return Q
```

Projective Coordinate Representation

The complexity of a finite field inversion is typically eight times that of a finite field multiplier in the same field.[15] Therefore, there is a huge motivation for an alternate point representation which would require lesser inversions. The two point coordinate system (x,y) used in Eqs. 11, 15, and 24 discussed in the previous section is called *affine representation*. It has been shown that each affine point on the elliptic curve has a one to one correspondence with a unique equivalence class in which each point is represented by three coordinates (X, Y, Z). The three point coordinate system is called the *projective representation*.[16] In the projective representation, inversions are replaced by multiplications. The projective form of the Weierstraß equation can be obtained by replacing x with X/Z^c and y by Y/Z^d. There are several projective coordinates systems proposed. The most commonly used projective coordinate system are the *standard* where $c=1$ and $d=1$, the *Jacobian* with $c=2$ and $d=3$ and the *López-Dahab (LD) coordinates*[16] which has $c=1$ and $d=2$. The LD coordinate system[17] allows point addition using *mixed coordinates*, i.e., one point in affine while the other in projective.

Replacing x by X/Z and y by Y/Z^2 in Eq. 11 results in the LD projective form of the Weierstraß equation.

$$Y^2 + XYZ = X^3 + aX^2Z^2 + bZ^4 \qquad (30)$$

Let $P=(X_1, Y_1, Z_1)$ be an LD projective point on the elliptic curve, then the inverse of point P is given by $-P = (X_1, X_1Z_1+Y_1, Z_1)$. Also, $P+(-P) = O$, where O is the point at infinity. In LD projective coordinates O is represented as $(1, 0, 0)$.

The equation for doubling the point P in LD projective coordinates[17] results in the point $2P=(X_3, Y_3, Z_3)$. This is given by the following equation.

$$\begin{aligned} Z_3 &= X_1^2 \cdot Z_1^2 \\ X_3 &= X_1^4 + b \cdot Z_1^4 \\ Y_3 &= b \cdot Z_1^4 \cdot Z_3 + X_3 \cdot \left(a \cdot Z_3 + Y_1^2 + b \cdot Z_1^4\right) \end{aligned} \qquad (31)$$

The equations for doubling require 5 finite field multiplications and zero inversions.

The equation in LD coordinates for adding the affine point $Q=(x_2, y_2)$ to P, where $Q \neq \pm P$, is shown in Equation 2.32. The resulting point is $P+Q=(X_3, Y_3, Z_3)$.

$$\begin{aligned} A &= y_2 \cdot Z_1^2 + Y_1 \\ B &= x_2 \cdot Z_1 + X_1 \\ C &= Z_1 \cdot B \\ D &= B^2 \cdot (C + a \cdot Z_1^2) \\ Z_3 &= C^2 \\ E &= A \cdot C \\ X_3 &= A^2 + D + E \\ F &= X_3 + x_2 \cdot Z_3 \\ G &= (x_2 + y_2) \cdot Z_3^2 \\ Y_3 &= (E + Z_3) \cdot F + G \end{aligned} \qquad (32)$$

Point addition in LD coordinates thus requires 9 finite field multiplications and zero inversions. For an m bit scalar with approximately half the bits one, the running time expected is given by Equation 2.33. One inversion and 2 multiplications are required at the end to convert the result from projective coordinates back into affine.

$$\begin{aligned} t_{ld} &\approx m(5M + \frac{9M}{2}) + 2M + 1I \\ &= (9.5m + 2)M + 1I \end{aligned} \qquad (33)$$

The LD coordinates require several multiplications to be done but have the advantage of requiring just one inversion. To be beneficial, the extra multiplications should have a lower complexity than the inversions removed.

SCALAR MULTIPLICATIONS: LSB FIRST AND MSB FIRST APPROACHES

The scalar multiplication algorithm is at the heart of the ECC systems. Several optimization techniques have been evolved to implement this operation efficiently. In this section, we compare a simple but interesting and effective variation of the scalar multiplication algorithms. Consider the algorithm stated in Algorithm 3 for performing scalar multiplication using double and add. The algorithm parses the scalar bits from the left, and is often referred to as the **MSB first** algorithm. For an m-bit length scalar, this algorithm requires m-doubling operations and on an average $(m-1)/2$ additions. However in the following we consider a variation of the algorithm where the scalar is read from the LSB.

Algorithm 4

Double and Add algorithm for scalar multiplication (LSB First)

Input: Basepoint $P = (x, y)$ and Scalar $k = (k_{m-1}, k_{m-2} \cdots k_0)_2$, where $k_{m-1} = 1$

Output: Point on the curve $Q = kP$

1 $Q = 0, R = P$
2 for $i = 0$ to $m - 1$ do
3   if $k_i = 1$ then
4      $Q = Q + R$
5   end
6    $R = 2R$
7 end
8 return Q

The working of the algorithm is self evident. However we can observe that compared to the Algorithm 4, the LSB first algorithm has the opportunity of parallelism. However it requires two variables R and Q. In the following section, we present another trick called as the Montgomery's ladder for efficient implementation of the scalar multiplications. The algorithm also has consequences in the **side channel analysis** of the hardware implementations derived from these algorithms.

MONTGOMERY'S ALGORITHM FOR SCALAR MULTIPLICATION

Let $P = (x_1, y_1)$ be a point on the curve: $y^2 + xy = x^3 + ax^2 + b$, where $(x, y) \in GF(2^m) \times GF(2^m)$. It is evident that $-P = (x_1, x_1 + y_1)$. We restate the equations for $R = P + Q = (x_3, y_3)$ as follows:

$$x_3 = \begin{cases} (\frac{y_1+y_2}{x_1+x_2})^2 + (\frac{y_1+y_2}{x_1+x_2}) + x_1 + x_2 + a & \text{if } P \neq Q; \\ x_1^2 + \frac{b}{x_1^2} & \text{if } P = Q. \end{cases}$$

$$y_3 = \begin{cases} (\frac{y_1+y_2}{x_1+x_2}) + (x_1+x_3) + x_3 + y_1 & \text{if } P \neq Q; \\ x_1^2 + (x_1 + \frac{y_1}{x_1})x_3 + x_3 & \text{if } P = Q. \end{cases}$$

Neglecting squaring and addition operations, as they are cheap, point addition and doubling each has one inversion and two multiplication operations. It is interesting to note that the x-coordinate of the doubling operation is devoid of any y-coordinate, it works only using x-coordinates. However, the x-coordinate of the addition operation naïvely needs the y-coordinate. If both the operations, namely addition and doubling can be performed with only one coordinate, say the x-coordinate, then the entire scalar multiplication can be performed without storing one of the coordinates. This can lead to a compact hardware implementation, and each of these coordinates is quite a large value and typically stored in a register.

Before explaining how we can perform the addition without the y-coordinate we present a technique for performing the scalar multiplication, which is referred to as the **Montgomery's Ladder**.

Montgomery's Ladder

Algorithm 5 presents the Montgomery's ladder for performing point multiplication. In this algorithm, like the LSB first algorithm, there are two variables. The variables are initialized with the values $P_1 = P$ and $P_2 = 2P$. The algorithm parses the key bits from the MSB; depending on the present key bit being one **point addition** is performed on P_1, and **point doubling** is performed on P_2. On the contrary, if the key bit is zero, **point addition** is performed on P_2, while **point doubling** is performed on P_1. Thus at every iteration, both addition and doubling are performed, making the operations uniform. This helps to prevent simple side channel analysis, like simple power attacks (SPA).

Algorithm 5

Montgomery's Ladder for scalar multiplication

Input: Basepoint $P = (x, y)$ and Scalar $k = (k_{m-1}, k_{m-2} \cdots k_0)_2$, where $k_{m-1} = 1$

Output: Point on the curve $Q = kP$

1 $P_1 = P, P_2 = 2P$
2 for $i = m - 2$ to 0 do
3   if $k_i = 1$ then
4      $P_1 = P_1 + P_2, P_2 = 2P_2$
5   end
6   else
7      $P_2 = P_1 + P_2, P_1 = 2P_1$
8   end
9 end
10 return P_1

Apart from this, we also an interesting property: the difference between $P_2 - P_1 = P$ throughout the scalar multiplication. The **invariance property** was found to very useful in designing fast scalar multiplication circuits without any pre-computations.

Faster Multiplication on EC without Pre-computations

We have previously seen that the *x*-coordinate of the double of a point P can be performed without the *y*-coordinate. The *x*-coordinate of the addition of two points P_1 and P_2 can also be similarly performed using only the *x*-coordinate, using the **invariance property**: $P = P_2 - P_1$ throughout the scalar multiplication in Algorithm 5. The following results help us to understand the technique to do so.

Theorem 1

Let $P_1 = (x_1, y_1)$ and $P_2 = (x_2, y_2)$ be points on the ECC curve, $y^2 + xy = x^3 + ax^2 + b$, where $(x, y) \in GF(2^m) \times GF(2^m)$. Then the *x*-coordinate of $P_1 + P_2$, x_3 can be computed as:

$$x_3 = \frac{x_1 y_2 + x_2 y_1 + x_1 x_2^2 + x_2 x_1^2}{(x_1 + x_2)^2}$$

The result is based on the fact that the characteristic of the underlying field is 2, and that the points P_1 and P_2 are on the curve.

The next theorem expresses the *x*-coordinates of the $P_1 + P_2$ in terms of only the *x*-coordinates of the P_1, P_2 and that of $P = P_2 - P_1$.

Theorem 2

Let $P = (x, y)$, $P_1 = (x_1, y_1)$ and $P_2 = (x_2, y_2)$ be elliptic points. Let $P = P_2 - P_1$ be an invariant. Then the *x*-coordinate of $P_1 + P_2$, x_3 can be computed in terms of the *x*-coordinates as:

$$x_3 = x + \left(\frac{x_1}{x_1 + x_2}\right)^2 + \left(\frac{x_1}{x_1 + x_2}\right)$$

Thus the *x*-coordinates of both the sum and doubling can be computed storing only the *x*-coordinates of the respective points. The next theorem shows how to observe the *y*-coordinates after the computation of the scalar product.

Theorem 3

Let $P = (x, y)$, $P_1 = (x_1, y_1)$ and $P_2 = (x_2, y_2)$ be elliptic points. Assume that $P_2 - P_1 = P$ and x is not 0. Then the *y*-coordinates of P_1 can be expressed in terms of P, and the *x*-coordinates of P_1 and P_2 as follows:

$$y_1 = (x_1 + x) \frac{(x_1 + x)(x_2 + x) + x^2 + y}{x} + y$$

Using these theorems, one can develop Algorithm 6 for performing scalar multiplications. Note that the algorithm uses only the x-coordinates of the points P_1 and P_2, and the coordinate of the point P, which is an invariant.

Algorithm 6

Detailed working of Montgomery's Ladder for scalar multiplication

Input: Basepoint $P = (x, y)$ and Scalar $k = (k_{m-1}, k_{m-2} \cdots k_0)_2$, where $k_{m-1} = 1$
Output: Point on the curve $Q = kP$

1   **if** $k = 0$ *or* $x = 0$ **then**
2   **return** $(0, 0)$
3   **end**
4    $x_1 = x, x_2 = x^2 + \frac{b}{x^2}$
5   **for** $i = m - 2$ **to** 0 **do**
6    $t = \frac{x_1}{(x_1 + x_2)}$
7   **if** $k_i = 1$ **then**
8    $x_1 = x + t^2 + t, x_2 = x_2^2 + \frac{b}{x_2^2}$
9   **end**
10   **else**
11   $x_2 = x + t^2 + t, x_1 = x_1^2 + \frac{b}{x_1^2}$
12   **end**
13   **end**
14   $r_1 = x_1 + x; r_2 = x_2 + x$
15   $y_1 = \frac{r_1(r_2 + x^2 + y)}{x} + y$
16   **return** (x_1, y_1)

The number of operations required in the scalar multiplication can be observed by counting the number of multiplications, inversions, squaring, and additions required. As can be observed that each of the loops require two multiplications and two inversions, thus accounting for $2(m - 2)$ multiplications and inversions. Further, each of the loops also requires 4 additions, and 2 squaring operations. This accounts for $4(m - 2)$ additions and $2(m - 2)$ squarings to be performed. Outside the loop also there are some computations to be performed. There is one inverse, four multiplications, six additions, and two squarings to be performed. It may be pointed out that since the inverse is a costly operation in finite fields, we minimize them at the cost of other operations. Like in the Montgomery's ladder, we compute the inverse of x, and then evaluate the inverse of x^2 by squaring x^{-1} rather than paying the cost of another field inversion. This simple trick helps to obtain efficient architectures!

Using Projective co-ordinates to Reduce the Number of Inversions

The number of inversions can be reduced by using Projective Co-ordinates as discussed in section 2.5.5. Using the projective co-ordinates the transformed equations are:

Table 3 Computations for performing ECC scalar multiplication (projective vs affine coordinates)

Computations	Affine coordinates	Projective coordinates
Addition	$4k+6$	$3k+7$
Squaring	$2k+2$	$5k+3$
Multiplication	$2k+4$	$6k+10$
Inversion	$2k+1$	1

$$X_3 = \begin{cases} (xZ_2 + (X_1Z_2))(X_2Z_1) & \text{if } P \neq Q; \\ X_1^4 + bZ_1^4 & \text{if } P = Q. \end{cases}$$

$$Z_3 = \begin{cases} (X_1Z_2 + X_2Z_1)^2 & \text{if } P \neq Q; \\ (Z_1^2 X_1^2) & \text{if } P = Q. \end{cases}$$

Each of the above steps thus require, no inversions, four multiplications, three additions, and five squaring operations.

The final conversion from projective to affine coordinates however requires inversions and can be performed using the following equations:

$$x_3 = X_1/Z_1$$
$$y_3 = (x + X_1/Z_1)[(X_1 + xZ_2) + (x^2 + y)(Z_1Z_2)](xZ_1Z_2)^{-1} + y$$

This step reduces the number of inversions, by computing $(xZ_1Z_2)^{-1}$, and then obtaining Z_1^{-1} by multiplying with xZ_2. The required number of multiplication is thus ten, however the inversions required is only one.

In Table 3, we present a summary of the two design techniques that we have studied: namely affine vs projective coordinates for implementation of the scalar multiplication using Montgomery's ladder. Before going into the design, we shall develop some other ideas on hardware designs on FPGAs.

CONCLUSIONS

The entry presents an overview on modern cryptography. It starts with the classification of ciphers, presenting the concepts of symmetric and asymmetric key cryptosystems. The entry details the inner compositions of block ciphers, with a special attention to the AES algorithm. The composite field representation of the Advanced Encryption Standard (AES) algorithm is used in several efficient implementations of the block cipher, hence the present entry develops the background theory. The entry subsequently develops the underlying mathematics of the growingly popular asymmetric key algorithm, the Elliptic Curve Cryptosystems (ECC). The entry discusses several concepts for efficient implementations of the ECC scalar multiplication, like LSB first and MSB first algorithms, the Montgomery ladder. These techniques are used for developing efficient hardware designs for both AES and ECC.

REFERENCES

1. Huang, J.; Mumford, D. *Image statistics for the British Aerospace segmented database*; Technical Report, Division of Applied Math, Brown Univeristy, 1999.
2. Criminisi, A.; Perez, P.; Toyama, K. Region filling and object removal by exemplar-based inpainting. IEEE Trans. Image Process. **2004**, *13* (9), 1200–1212.
3. Huang, S.-J.; Hsieh, C.-T. Coiflet wavelet transform applied to inspect power system disturbance-generated signals. IEEE Trans. Aerosp. Electron. Syst. **2002**, *38* (1), 204–210.
4. Clark, J.; Zhang, C.; Wallace, A. Image aquisition using fixed and variable triangulation. In *Proceedings of the 5th IET International Conference on Image Processing and its Applications*, 1995, 539–543.
5. Hughes, J.M.; Graham, D.J.; Jacobsen, C.R.; Rockmore, D.N. Comparing higher-order spatial statistics and perceptual judgements in the stylometric analysis of art. In *Proceedings of the European Signal Processing Conference (EUSIPCO)*, 2011, 1244–1248.
6. DeVore, R.A.; Lucier, B.J. Fast wavelet techniques for near-optimal image processing. In *Proceedings of the IEEE Military Communications Conference*, New York, 1992, 48.3.1–48.3.7.
7. Chevreul, M.E. *The Principles of Harmony and Contrast of Colors and Their Applications to the Arts*; Schiffer: West Chester, PA, 1987.
8. Canters, F.; Decleir, H. *The World in Perspective: A Directory of World Map Projections*; John Wiley & Sons: New York, 1989.
9. Hart, W.M., Jr. The temporal responsiveness of vision. In *Adler's Physiology of the Eye, Clinical Application*; Moses, R.A.; Hart, W.M.; Eds.; The C. V. Mosby Company: St. Louis, MO, 1987.
10. Calow, D.; Lappe, M. Local statistics of retinal optic flow for self-motion through natural sceneries. Netw. Comput. Neural Syst. **2007**, *18* (4), 343–374.
11. von Helmholtz, H.L.F. *Handbuch der Physiologischen Optik*; Leopold Voss: Leipzig, 1867.
12. Boynton, R.M. *Human Color Vision*; Holt, Rinehart, and Winston: New York, 1979.
13. Helga Kolb, R.N.; Fernandez, E. *Webvision: The Organization of the Retina and Visual System*, 2010. Available at http://webvision.med.utah.edu/.
14. Adams, J.; Parulski, K.; Spaulding, K. Color processing in digital cameras. IEEE Micro **1998**, *18* (6), 20–30.
15. Hinton, G.E. Training products of experts by minimizing contrastive divergence. Neural Comput. **2002**, *14* (8), 1771–1800.
16. Gao, Q.; Roth, S. How well do filter-based MRFs model natural images? In *Pattern Recognition*; Pinz, A.; Pock, T.; Bischof, H.; Leberl, F.; Eds.; Volume 7476 of Lecture Notes in Computer Science; Springer: Berlin, 2012.
17. Fleming, R.W.; Torralba, A.; Adelson, E.H. Specular reflections and the perception of shape. J. Vision **2004**, *4* (9), 798–820.

Modern Hardware Design Practices

Debdeep Mukhopadhyay and Rajat Subhra Chakraborty
Indian Institute of Technology Kharagpur, West Bengal, India

Abstract
Cryptography is used in many applications such as mobile communications, automatic teller machines (ATMs), digital signatures, and online banking. But the real-time processing of these applications requires optimized and high performance implementations of the ciphers. In these cases dedicated hardware provides significant opportunities for processing speed up due to the inherent parallelism. This article introduces one technology for implementing real-time cryptographic process, namely, the Field Programmable Gate Array (FPGAs). FPGAs are reconfigurable platforms to build hardware that are widely used in industry.

Keywords: Cyptography; FPGA; Real-time processing; Reconfigurable hardware.

INTRODUCTION

With the growth of electronics, there has been tremendous growth in the applications requiring security. Mobile communications, automatic teller machines (ATMs), digital signatures, and online banking are some probable applications requiring cryptographic principles. However, the real-time processing required in these applications obviates the necessity of optimized and high performance implementations of the ciphers.

There has been lot of research in the design of cryptographic algorithms, both on software platforms as well as dedicated hardware environments. While conventional software platforms are limited by their parallelisms, dedicated hardware provides significant opportunities for speed up due to their parallelism. However they are costly, and often require off-shore fabrication facilities. Further, the design cycle for such Application Specific Integrated Circuits (ASICs) are lengthy and complex. On the contrary, the Field Programmable Gate Arrays (FPGAs) are reconfigurable platforms to build hardware. They combine the advantages of both hardware: (in extracting parallelism, and achieving better performance); and software: (in terms of programmability). Thus these resources are excellent low cost, high performance devices for performing design exploration and even in the final prototyping for many applications. However, designing in FPGAs is tricky, and as what works for ASIC libraries does not necessarily work for FPGAs. FPGAs have a different architecture, with fixed units in the name of Look-up-Tables (LUTs) to realize the basic operations, along with larger inter-connect delays. Thus, the designs need to be carefully analyzed to ensure that the utilizations of the FPGAs are enhanced, and the timing constraints are met. In the next section, we provide an outline of the FPGA architecture.

FPGA Architecture

FPGAs are reconfigurable devices offering parallelism and flexibility, on one hand, while being low cost and easy to use on the other. Moreover, they have much shorter design cycle times compared to ASICs. FPGAs were initially used as prototyping device and in high performance scientific applications, but the short time-to-market and on site reconfigurability features have expanded their application space. These devices can now be found in various consumer electronic devices, high performance networking applications, medical electronics and space applications. The reconfigurability aspect of FPGAs also makes them suited for cryptographic applications. Reconfigurability results in flexible implementations allowing operating modes, encryption algorithms and curve constants etc. to be configured from software. FPGAs do not require sophisticated equipment for production, they can be programmed in house. This is a distinct advantage for cryptography as no third party is involved, thus increasing trust in the hardware circuit (reducing chances of IP theft, IC cloning, counterfeiting, insertion of Trojans etc.).

There are two main parts of the FPGA chip [1] : the input/output (I/O) blocks and the core. The I/O blocks are located around the periphery of the chip, and are used to provide programmable connectivity to the chip. The core of the chip consists of programmable logic blocks and programmable routing architectures. A popular architecture for the core called island style architecture is shown in Fig. 1. *Logic blocks*, also called *configurable logic blocks* (CLB), consists of logic circuitry for implementing logic. Each CLB is surrounded by routing channels connected through switch blocks and connection blocks. A *switch block* connects wires in adjacent channels through programmable switches. A *connection block* connects the

Modern Hardware Design Practices

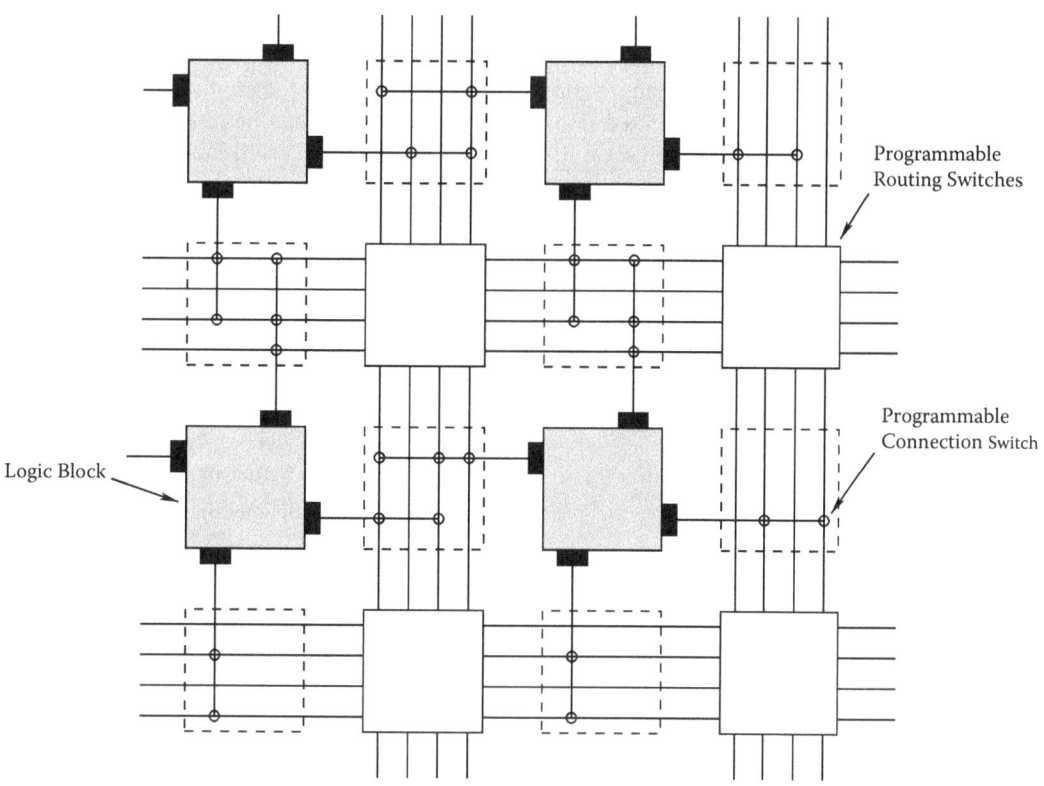

Fig. 1 FPGA island-style architecture

wire segments around a logic block to its inputs and outputs, also through programmable switches. Each logic block further contains a group of *basic logic elements* (BLE). Each BLE has a *look up table* (LUT), a storage element and combinational logic as shown in Fig. 2. The storage element can be configured as a edge triggered D-flipflop or as level-sensitive latches. The combinational logic generally contains logic for carry and control signal generation.

The LUTs can be configured to be used in logic circuitry. If there are m inputs to the LUT then any m variable Boolean function can be implemented. The LUT mainly

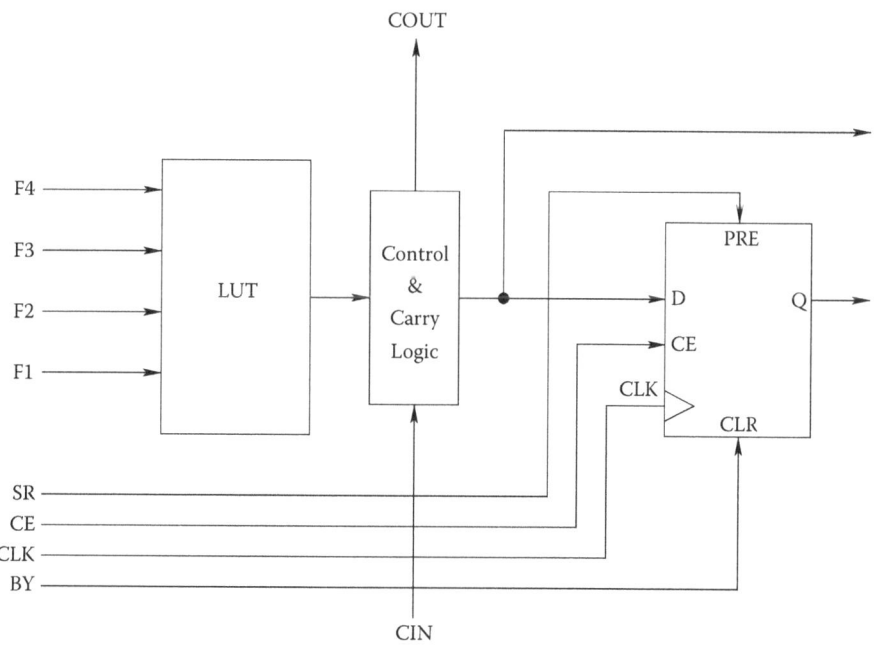

Fig. 2 FPGA logic block

contains memory to store truth tables of Boolean functions, and multiplexers to select the values of memories. There have been several studies on the best configuration for the LUT. A larger LUT would result in more logic fitted into a single LUT, and hence lesser critical delay. However, a larger LUT would also indicate larger memory and bigger multiplexers, hence larger area. Most studies show that a 4-input LUT provides the best area-delay product, though there have been few applications where a 3-input LUT [2] and 6-input LUT [3] is more beneficial. Most FPGA manufacturers, including Xilinx[a] and Altera[b] have an increasing number of inputs, namely 4 and 6. However, for a given device the number of inputs are fixed.

The FPGA Design Flow

The Field Programmable Gate Arrays (FPGAs) are programmable devices which can be configured through a design flow as entailed next. The design flow is a semi-automated or automated sequence of steps that translates the conception of the design to an actual hardware realization on a target platform, in our case an FPGA device. The sequence of steps starts with the hardware description of the design using a *Hardware Description Language* (HDL). We describe the inputs to the design flow subsequently:

1. **HDL description:** The design is described through an HDL specification. The most commonly known languages are Verilog and VHDL. Today, even higher-level languages are used like SystemC, MatLab, or Simulink to make the design faster and easier. The HDL describes the design at what is known at the *Register Transfer Level* (RTL), which describes the flow of signals or data between hardware registers and the operations that are performed on them. The hardware registers are conceptually the lowest level of abstraction to the designer, in this case. The HDL design can be designed in two alternative styles, *behavorial* and *structural*. The former style is using more high-level constructions, and based on the manner or behavior in which the described hardware functions. The latter style is based on the actual instantiations of hardware elements or sub-blocks. Thus a structural style requires an architectural knowledge of the designer, and is often preferred for architectural exploration for design trade-offs, performance etc. In real life, the HDL codes are developed in a mixed fashion. For example, in verilog the designs are modeled as encapsulations called as *modules*. The entire design is in reality a hierarchy of these modules and can be depicted often in the form of a tree-like structure. The root node of the tree, is the top-level of the design, which can be thought of to be made of sub-modules, which can be in turn broken into submodules. Finally, the design hierarchy terminates at the leaf levels, which are often simple descriptions, like shifters, flip-flops etc. A popular way of developing the design, is to make the leaf-level behavorial, while as we go up the tree the HDLs become more structural, the top-level being entirely structural. Such an approach improves the programmability, readibilty, and testability of the design, and often helps in an efficient design exploration for better performance.

2. **Constraints:** The HDLs described above are mapped into actual hardware through a sequence of automated steps, namely synthesis, mapping, translation, and routing. However, along with the HDLs the designer also provides some *constraints* to the CAD tools which convert the HDL to an actual hardware. The constraints are typically the desired clock frequency (f_{clk}), and various delays which characterize a given design. The delays could be the following:
   — **Input delay:** This delay is from the input pad[c] to the register, which holds the input signal.
   — **Register to Register Delay:** The synthesis tool assumes that all combinational paths in the design are to be performed in a single clock period. Thus this component of the delay, which describes the delay of a combinational path between two registers, help to compute the critical path of the design. The critical path of the design gives an upper bound to the value of f_{clk}, hence the delay specifications should be carefully provided. Following are some important delay constraints which needs special mention:

   (a) **Set-up time:** It is the minimum time that the synchronous data must arrive before the active clock edge in a sequential circuit.

   (b) **Hold Time:** It is the minimum time that the synchronous data should be stable after the active clock edge.

   (c) **False Path:** The analyzer considers all combinational paths that are to be performed in a single clock cycle. However in the circuits there be paths that are never activated. Consider Fig. 3 consisting of two sub-circuits separated by the dashed line. First consider the portion on the right-side of the dashed line and the signal transitions showing how the path (the critical path), $g1 \rightarrow g2 \rightarrow g3 \rightarrow g4 \rightarrow g5$ can get sensitized. Next consider the other portion of the circuit and note that

---

[a]http://www.xilinx.com.
[b]http://www.altera.com.

[c]Pads are dedicated terminals through which the design communiciates with the external world.

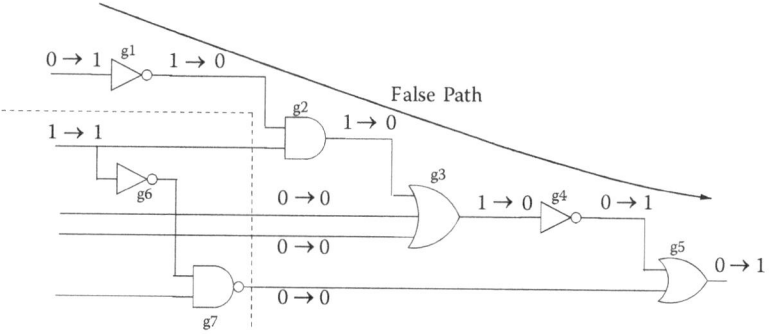

Fig. 3 False path in a circuit

due to the presence of the gates $g6$ and $g7$, this path becomes a false path as no input condition can trigger this path. In the example shown, the inverter ($g6$) and the NAND gate ($g7$) ensure that an input of logic one to the inverter, results in the NAND gate producing an output logic one, thus making the output of the circuit logic one much earlier. When the input to the inverter is logic zero, the mentioned path is again false. Thus to obtain a proper estimate of f_{clk}, the designer or the CAD tool should properly identify these false paths.

(d) **Multi-cycle Path:** There are some paths in a design which are intentionally designed to require more than one clock signals to become stable. Thus the set-up and hold-time violation analysis for the overall circuit should be done by taking care of such paths, else the timing reports will be wrongly generated. Consider the Fig. 4, showing an encryption hardware circuit. The selection of the two multiplexers, MUX-A and MUX-B are the output of a 3-stage circular shifter made of 3-DFFs (D-Flip Flop) as shown in the diagram. The shifter is initially loaded with the value (1, 0, 0), which in 3 clock cycles makes the following transitions: (1, 0, 0) → (0, 1, 0) → (0, 0, 1) → (1, 0, 0). Thus at the start the input multiplexer (MUX-A) selects the plaintext input and passes the result to the DFF. The DFF subsequently latches the data, while the encryption hardware performs the transformation on the data. The output multiplexer (MUX-B) passes the output of the encryption hardware to the DFF as the ciphertext, when the select becomes one in the third clock cycle, i.e., two clock cycles after the encryption starts. Meanwhile it latches the previous ciphertext, which gets updated every two clock cycles. Thus the encryption circuit has two clock cycles to finish its encryption operation. This is an example of a multi-cycle path, as the combinational delay of the circuit is supposed to be performed in more than one clock cycles. This constraint also should be detected and properly kept in mind for a proper estimation of the clock frequency.

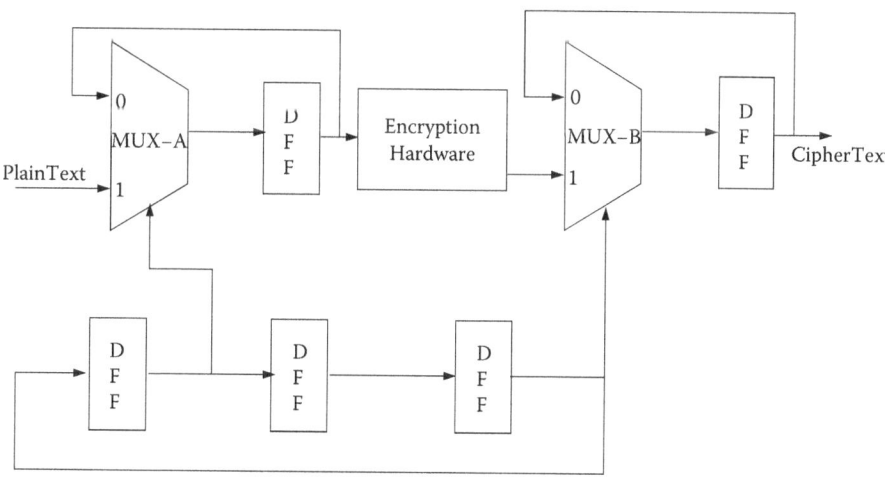

Fig. 4 Multi-cycle path in a circuit

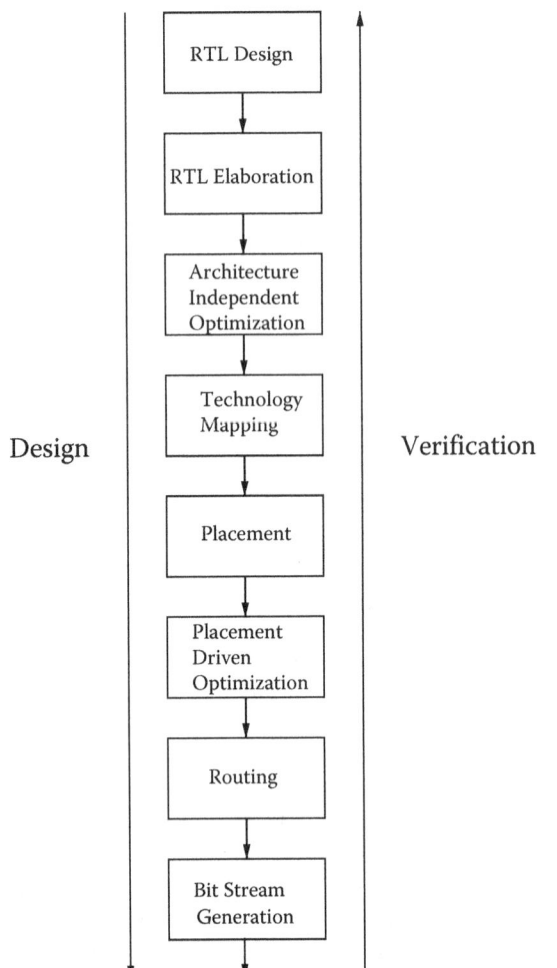

Fig. 5 The FPGA design flow

The other important design choices are the type of FPGA device, with different cost, performance and power consumptions. Generally, the designer starts with the lower end FPGAs and iteratively depending on the complexity of the design chooses higher end platforms.

Figure 5 depicts the typical design flow for FPGA design. As may be observed that the design flow is top-down starting from the RTL design, RTL elaboration, Architecture Independent Optimizations, Technology Mapping (Architecture Dependent Optimizations), Placement, Placement Driven Optimizations, Routing and Bit Stream Generation. In the design world along with this flow, the verification flow also goes on hand in hand. However it may be noted that the verification flow proceeds in the opposite direction: answering queries like is the RTL elaboration equivalent to the RTL design? In the following description we describe the individual steps in the flow in more details.

1. **RTL Design:** This step involves the description of the design in a HDL language, like verilog. The step involves the architecture planning for the design into sub-modules, understanding the data-path and control-path of the design and in developing the RTL codes for the sub-modules. This step also involves the integration of the sub-modules to realize the complete design. Testing individual sub-modules and the complete design via test-benches, also written in a high-level language, often verilog, system verilog etc. is also an integral part of this step.

2. **RTL Elaboration:** This step involves the inferring of data-path to be realized by special components internal to the FPGA, like adders with dedicated fast-carry chains, specially designed multipliers, etc. The control-path elements on the other hand get elaborated into state machines, or Boolean equations.

3. **Architecture Independent Optimization:** This step involves various optimization methods which are not related to the underlying architecture of the FPGA platform. For data-path optimization constant propagation, strength reduction, operation sharing, and expression optimization are some popular techniques. On the other hand for control-path optimizations, finite state machine encoding and state minimization are some well known methods. The combinational circuits are optimized exploiting don't-care logic present in the circuit.

4. **Technology Mapping:** In this step, the various elements of the design are optimally assigned to the resources of the FPGAs. Hence this step is specific to the FPGA device, and depends on the underlying architecture. Depending on the platform the data-path elements get inferred to adders, multipliers, memory elements embedded in the device. The control-path elements and the elements in the control-path, which are not inferred to special embedded elements are realized in the FPGA logic block. The performance of the implemented design, both area and delay, depends on the architecture of the LUTs of the FPGA logic block. We shall discuss later, that the number of inputs to the LUTs can be suitably used to advantage to have high-performance implementations. Thus these optimizations are specific to the underlying architecture and depends on the type of the FPGAs being used.

5. **Placement:** Placement in FPGA decides the physical locations and inter connections of each logic block in the circuit design, which becomes the bottleneck of the circuit performance. A bad placement can increase the interconnects which leads to significant reduction in performance.

6. **Placement-Driven Optimization:** In order to reduce the interconnect delay, and to improve the performance of the design, the initial placement is incrementally updated through logic restructuring, rewiring, duplication etc.

7. **Routing:** Global and detailed routing are performed to connect the signal nets using restricted routing resources which are predesigned. The routing

resources used are programmable switches, wire segments which are available for routing, and multiplexers.
8. **Bit-stream Generation:** This is the final step of the design flow. It takes the routed design as input, and produces the bit-stream to program the logic and interconnects to implement the design on the FPGA device.

MAPPING AN ALGORITHM TO HARDWARE: COMPONENTS OF A HARDWARE ARCHITECTURE

The conversion of an algorithm to an efficient hardware is a challenging task. While functional correctness of the hardware is important, the main reason for designing a hardware is performance. Thus, one needs to consider all opportunities for a high-performance design: namely by reducing the critical path of the circuit, by making it more compact, thus ensuring that the resources of the FPGA platform are used efficiently. Hence, in order to develop an efficient implementation, one needs to look into the components of a hardware and understand the *architecture* of the design. Figure 6 describes the important components of a digital hardware design. As can be observed, the three most important parts of the architecture are the data-path elements, the control-path block, and the memory unit.

- The *data-path* elements are the computational units of the design. The data-paths are central to the performance of a given circuit, and have a dominating effect on the overall performance. Thus the data-path elements need to be properly optimized and carefully designed. However it is not trivial, as there are numerous equivalent circuit topologies and various designs have different effect on the delay, area and power consumption of the device. Also one has to decide whether the data-path elements will be combinational or sequential units, depending on the underlying application and its constraints. Examples of common data-path elements are registers, adders, shifters etc. These data-path elements often form the components of the Arithmetic Logic Unit (ALU) of a given design.
- The *control-path* elements, on the other hand, sequences the data flow through the data-path elements. Hence the input data is processed or transformed by the data-path elements, which are typically combinational. On the other hand, the data is switched and cycled through the data-path elements by the control unit, which is typically a sequential design. The control signals generated by the sequential controller are often dependent on the states, or sometimes on both the state or partial outputs from the datapath. The former form of controller is known as the **Moore** machine, while the latter as the **Mealy** machine.

Algorithm 1 Binary gcd Algorithm

Input: Integers u and v
  Output: Greatest Common Divisor of u and v: $z = gcd(u, v)$
1 **while** $(u! = v)$ **do**
2   **if** *u and v are even* **then**
3         $z = 2gcd(u/2, v/2)$
4   **end**
5   **else if** *(u is odd and v is even)* **then**
6         $z = gcd(u, v/2)$
7   **end**
8   **else if** *(u is even and v is odd)* **then**
9         $z = gcd(u/2, v)$
10   **end**
11   **else**
12   **if** $(u \geq v)$ **then**
13             $z = gcd((u - v)/2, v)$
14   **end**
15   **else**
16             $z = gcd(u, (v - u)/2)$
17   **end**
18   **end**
19 **end**

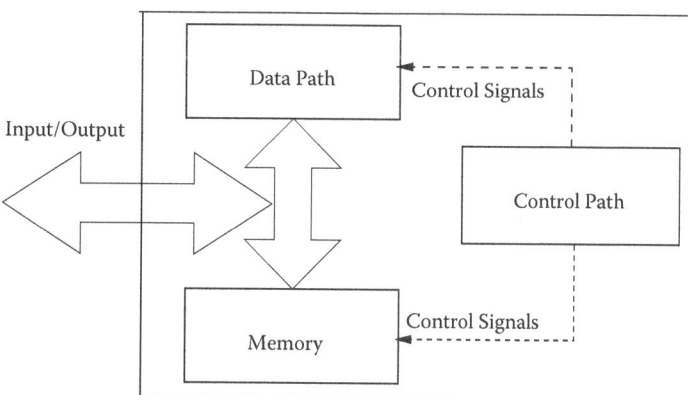

Fig. 6 Important components of an architecture

Whatever be the design type of the controller, a key to a good design is to comprehend the effective split between the data-path and control-path elements. We illustrate this concept with the help of a case study in the next section.

CASE STUDY: BINARY GCD PROCESSOR

Consider the following algorithm (algorithm 11) for computing the greatest common divisor (gcd) of two given integers. The algorithm is commonly known as the binary Euclidean gcd algorithm, and is an improvement on the classical school book Euclidean algorithm for computing gcd.

In order to illustrate the development of a separate data and control paths, we shall take the example of the above gcd algorithm. The objective of the present exercise is to develop a special-purpose hardware that computes the greatest common divisor of two positive integers, X and Y. We assume that $gcd(0, 0) = 0$. We shall present a step by step approach to realize the hardware.

1. **Identification of the states of the algorithm**: The pseudo-code of the algorithm is expressed in an HDL-like language in (algorithm 12). The pseudo-code identifies the essential states or simple stages of the algorithm. It may be noted that in each state, the intended hardware is expected to perform certain computations, which are realized by the computation or data path elements. The pseudo-code shows that there are six states of the design, denoted by S_0 to S_5.

Algorithm 2 HDL like description of the Binary gcd Algorithm

Input: Integers u and v
   Output: Greatest Common Divisor of u and v: $z = gcd(u, v)$
  1 register X_R, Y_R;
  2 $X_R = u$; $Y_R = v$; $count = 0$;   /* State 0 */
  3 **while** $(X_R ! = Y_R)$ **do**
  4   **if** $(X_R[0] = 0$ and $Y_R[0] = 0)$ **then** /* State 1 */
  5   X_R = right shift (X_R)
  6   Y_R = right shift (Y_R)
  7         $count = count + 1$
  8   **end**
  9   **else if** $(X_R[0] = 1$ and $Y_R[0] = 1)$ **then** /* State 2 */
10   Y_R = right shift (Y_R)
11   **end**
12   **else if** $(X_R[0] = 0$ and $Y_R[0] = 1)$ **then** /* State 3 */
13   X_R = right shift (X_R)
14   **end**
15   **else**   /* State 4 */
16   **if** $(X_R \geq Y_R)$ **then**
17   X_R = right shift$(X_R - Y_R)$
18   **end**
19   **else**
20   Y_R = right shift$(Y_R - X_R)$
21   **end**
22 **end**
23 **while** $(count > 0)$ **do**   /* State 5 */
24   X_R = left shift (X_R)
25     $count = count - 1$
26 **end**

---

2. **Identification of the data path elements**: As evident from the pseudo-code, the data path elements required for the gcd computation are *subtracter*, *complementer*, *right shifter*, *left shifter* and *counter*. The other very common data path element is the *multiplexer* which are required in large numbers for the switching necessary for the computations done in the datapath. The *selection* lines of the multiplexers are configured by the control circuitry, which is essentially a state machine.

3. **Identification of the state machine of the control path**: The control path is a sequential design, which comprises of the state machine. In this example, there is a six state machine, which receives inputs from the computations performed in the data path elements, and accordingly performs the state transitions. It also produces output signals which configures or switches the data path elements.

4. **Design of the data path architecture**: The data path of the gcd processor is depicted in Fig. 7. The diagram shows the two distinct parts: data path and control path for the design. The data-path stores the values of X_R and Y_R (as mentioned in the HDL-like code) in two registers. The registers are loadable, which means they are updated by an input when they are enabled by an appropriate control signal (e.g., $load\_X_R$ for the register X_R). The values of the inputs u and v are initially loaded into the registers X_R and Y_R through the input *multiplexer*, using the control signals $load\_uv$. The least bits of X_R and Y_R are passed to the controller to indicate whether the present values of X_R and Y_R are even or not. The next iteration values of X_R and Y_R are updated, by feeding back the register values (after necessary computations) through the input *multiplexer* and this is controlled by the signals $update\_X_R$ and $update\_Y_R$. The computations on the registers X_R and Y_R are division by 2, which is performed easily by the two *right-shifters*, and subtraction and comparison for equality, both of which are performed by a *subtracter*. The values stored in X_R and Y_R are compared using a subtracter, which indicates to the controller the events $(X_R! = Y_R)$ and $(X_R \geq Y_R)$ by raising appropriate flag signals. In the case, when $X_R < Y_R$ and the subtraction $Y_R - X_R$ is to be performed, the result is complemented. The next iteration values of X_R and Y_R are loaded either after the subtraction or directly, which is controlled by the signals $load\_X_R\_after\_sub$ and $load\_Y_R\_after\_sub$.

Modern Hardware Design Practices

The circuit also includes an *up down counter*, which is incremented whenever both X_R and Y_R are even values. Finally, when $X_R = Y_R$ the result is obtained by computing $2^{count}(X_R)$, which is obtained by using a *left-shifter* and shifting the value of X_R, until the value of *count* becomes zero.

5. **Design of the state machine for the controller**: The state machine of the controller is depicted in Table 3.3. As discussed there are six states of the controller, and the controller receives four inputs from the data path computations, namely $(X_R! = Y_R), X_R[0], Y_R[0]$, and $X_R \geq Y_R$ respectively. The state transitions are self-explanatory and can be easily followed by relating the table and the data path diagram of Fig. 7. The state machine is an example of a Mealy machine.

ENHANCING THE PERFORMANCE OF A HARDWARE DESIGN

One of the primary goals of developing a hardware architecture is *performance*. But the term performance has several implications, depending on the application at hand. For certain applications, speed may be of utmost importance, while for others it may be the area budget of the design is of primary concern.

In general, if any standard book of computer architecture is referred to we obtain several definitions of performance. We revise certain definitions here and consider some more variants of these. To start with, the performance of a hardware design is often stated through its critical path, as that limits the clock frequency. In a combinational circuit, the critical path is of primary concern and a circuit which has a better optimized critical path, ie. a smaller critical delay is faster. On the other hand for a sequential circuit, it is also important to know the number of clock cycles necessary to complete a computation. Like in the previous example of the gcd processor, the number of clock cycles needed is proportional to the number of bits in the larger argument. However, the number of clock cycles required is not a constant, and varies with the inputs. Thus one may consider the average number of clock cycles needed to perform the computation. Let the fastest clock frequency be denoted by f_{max} and the average number of clock cycles is say denoted by cc_{avg}, then the total computation time for the gcd processor is obtained by $t_c = \frac{cc_{avg}}{f_{max}}$. Another important metric is the throughput of the hardware, denoted by $\tau = N_b/t_c = \frac{N_b f_{max}}{cc_{avg}}$, where N_b is the number of bytes of data being simultaneously processed.

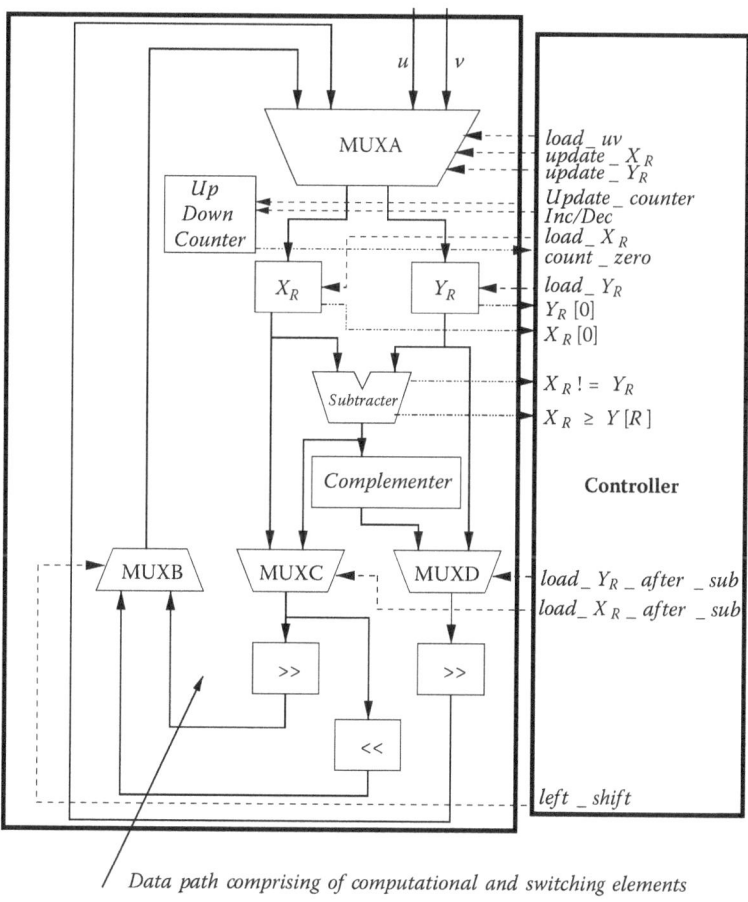

Data path comprising of computational and switching elements

Fig. 7 Data path of a gcd processor

Table 1 State transition matrix of the controller

Present State	Next state					Output signals									
	0__	100_	110_	101_	111_	load uv	update X_R	update Y_R	load X_R	load Y_R	load_X_R after_sub	load_Y_R after_sub	Update counter	Inc/Dec	left shift	count zero
S_0	S_5	S_1	S_2	S_3	S_4	1	0	0	1	1	0	0	0	–	–	–
S_1	S_5	S_1	S_2	S_3	S_4	0	1	1	1	1	0	0	1	1	–	–
S_2	S_5	S_1	S_2	S_3	S_4	0	0	1	0	1	0	0	0	–	–	–
S_3	S_5	S_1	S_2	S_3	S_4	0	1	0	1	0	0	0	0	–	–	–
$S_4 (X_R \geq Y_R)$	S_5	S_1	S_2	S_3	S_4	0	1	0	1	0	1	0	0	–	–	–
$S_4 (X_R < Y_R)$	S_5	S_1	S_2	S_3	S_4	0	0	1	0	1	0	1	0	–	–	–
S_5	S_5	S_5	S_5	S_5	S_5	0	0	0	0	0	0	0	1	0	1	0

The other important aspect of hardware designs is the resource consumed. In context to FPGAs the resources largely comprise of slices, which are made of LUTs and flipflops. As discussed, the LUTs have typically fixed number of inputs. In order to improve the performance of a hardware design, it requires to customize the design for the target architecture to ensure that the resource used is minimized. The smallest programmable entity on an FPGA is the lookup table As an example, Virtex-4 FPGAs have LUTs with four inputs and can be configured for any logic function having a maximum of four inputs. The LUT can also be used to implement logic functions having less than four inputs, two for example. In this case, only half the LUT is utilized the remaining part is not utilized. Such a LUT having less than four inputs is an *under-utilized LUT*. For example, the logic function $y=x_1+x_2$ under utilizes the LUT as it has only two inputs. Most compact implementations are obtained when the utilization of each LUT is maximized. From the above fact it may be derived that the minimum number of LUTs required for a q bit combinational circuit is given by Eq. 1.

$$\#LUT(q) = \begin{cases} 0 & \text{if } q=1 \\ 1 & \text{if } 1<q\leq 4 \\ \lceil q/3 \rceil & \text{if } q>4 \text{ and } q \bmod 3 = 2 \\ \lfloor q/3 \rfloor & \text{if } q>4 \text{ and } q \bmod 3 \neq 2 \end{cases} \quad (1)$$

The delay of the q bit combinational circuit in terms of LUTs is given by Eq. 2, where D_{LUT} is the delay of one LUT.

$$DELAY(q) = \lceil \log_4(q) \rceil * D_{LUT} \quad (2)$$

The percentage of under-utilized LUTs in a design is determined using Eq. 3. Here, LUT_k signifies that k inputs out of 4 are used by the design block realized by the LUT. So, LUT_2 and LUT_3 are under utilized LUTs, while LUT_4 is fully utilized.

$$\%UnderUtilizedLUTs = \frac{LUT_2 + LUT_3}{LUT_2 + LUT_3 + LUT_4} * 100 \quad (3)$$

It may be stressed that the above formulation provides minimum number of LUTs required and not an exact count. As an example, consider $y = x_5x_6x_1 + x_5x_6x_2 + x_1x_2x_3 + x_2x_3x_4 + x_1x_3x_5$. Observe that the number of LUTs required is 3, and the formula says that the minimum is 2-LUTs. Our analysis and experiments, show that the above formulation although provides a lower bound, matches quite closely with the actual results. Most importantly the formulation helps us to perform design exploration much faster, which is the prime objective of such a formulation.

The number of LUTs required to implement a Boolean function is the measure of the area of a function. The above formulation can also be generalized for any k-input LUT. A k input LUT (k-LUT) can be considered a black box that can perform any functionality of a maximum of k variables. If there is a single variable, then no LUT is required. If there are more than k variables, then more than one k–LUTs are required to implement the functionality. The lower bound of the total number of k input LUTs for a function with x variables can thus be similarly expressed as;

$$lut(x) = \begin{cases} 0 & \text{if } x \leq 1 \\ 1 & \text{if } 1 < x \leq k \\ \lfloor \frac{x-k}{k-1} \rfloor + 2 & \text{if } x > k \text{ and } (k-1) \nmid (x-k) \\ \frac{x-k}{k-1} + 1 & \text{if } x > k \text{ and } (k-1) \mid (x-k) \end{cases} \quad (4)$$

Delay in FPGAs comprises of LUT delays and routing delays. Analyzing the delay of a circuit on FPGA platform is much more complex than the area analysis. By experimentation we have found that for designs having combinational components, the delay of the design varies linearly with the number of LUTs present in the critical path. Figure 8 shows this linear relationship between the number of LUTs in the critical path and the delay in multipliers of different sizes. Due to such linear relationship, we can consider that the number of LUTs in the critical path is a measure of actual delay. From now onwards, we use the term LUT delay to mean the number of k–LUTs present in the critical path.

For an x variable Boolean function, number of k–LUTs in the critical path is denoted by the function $maxlutpath(x)$ and is thus expressed as:

$$maxlutpath(x) = \lceil \log_k(x) \rceil. \quad (5)$$

We will use Eqs. 4 and 5 for estimating area and delay of the architecture proposed in Fig. 7. In the following section, we present gradually the estimation of the hardware blocks as required in the data-path.

MODELLING OF THE COMPUTATIONAL ELEMENTS OF THE GCD PROCESSOR

The important datapath elements as shown in the Fig. 7 are *multiplexers* and *subtracter*.

Modeling of an Adder

In the architecture as depicted in Fig. 7, the most important components to influence the delay are the integer adders/subtracters. As on hardware platforms, both adder and subtracters have same area and delay, we consider a general

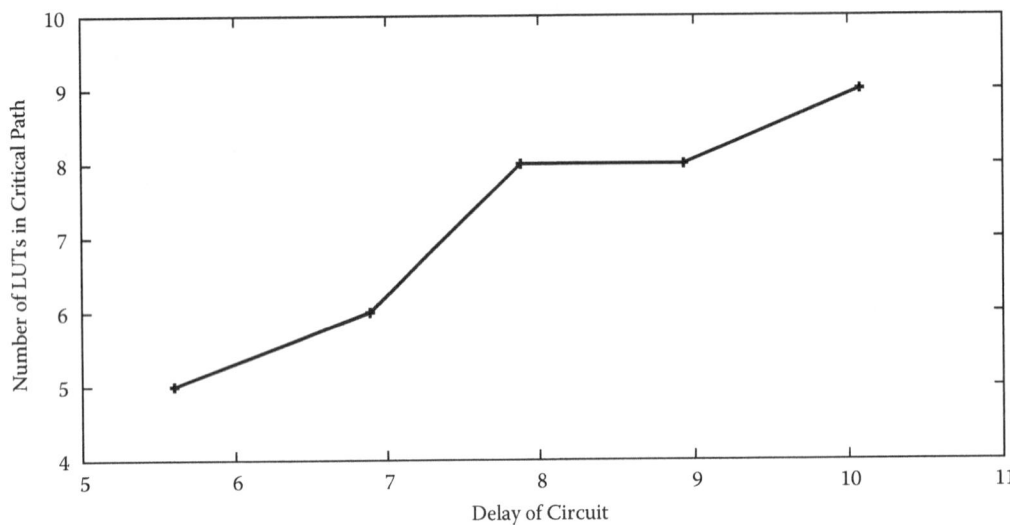

Fig. 8 LUTs in critical path vs. delay for a combinational multiplier

structure. The carry propagation is a major challenge in designing efficient integer adders on hardware platforms. There are several methods for realizing fast adder topologies. On FPGA platforms, carry chain based adders are very popular due to their low-cost and optimized carry propagation techniques. Here in this discussion, we consider such a carry chain–based integer adder available in most common FPGAs.

Internal diagram of such an adder is shown in Fig. 9. For fast carry propagation, a dedicated chain of MUXCY is provided in FPGAs. For an m bit adder, the carry propagates through m number of cascaded MUXCY. Dedicated

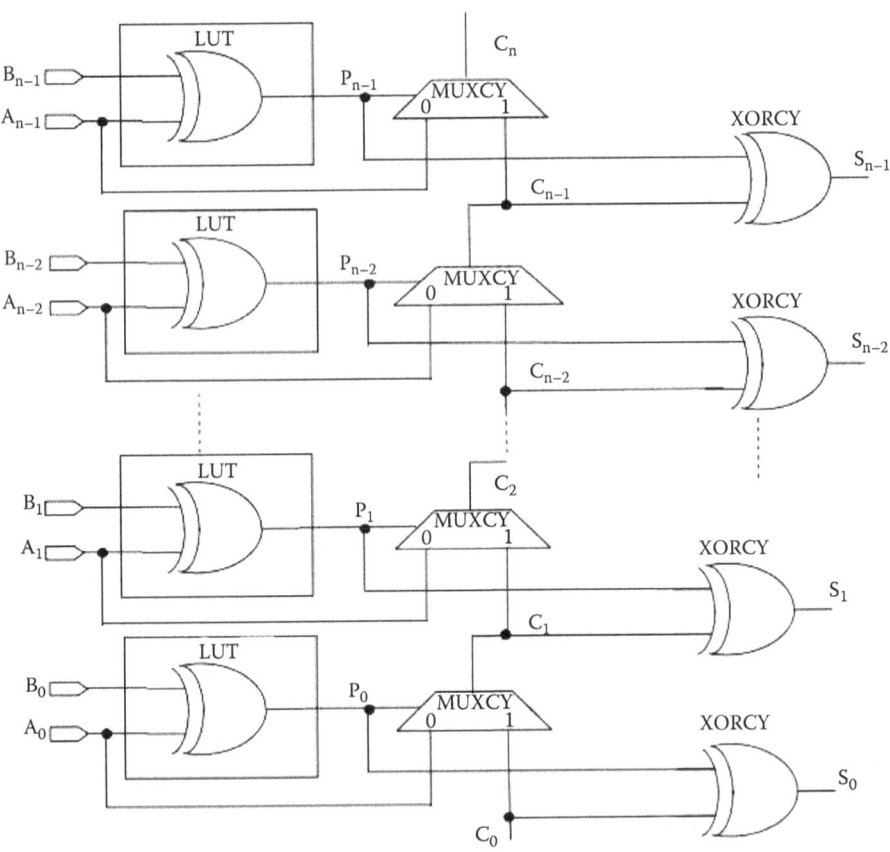

Fig. 9 Adder with carry chain

carry chains are much faster than generic LUT based fabric in FPGA, hence carry propagation delay is small. Since these MUXCY are used only for fast carry propagation, and other blocks present are constructed of LUTs, we need to scale the delay of MUXCY circuits for comparing delay of an adder with any other primitive. Let us consider that delay of a MUXCY is s times lesser than that of a LUT. This scaling factor s depends on device technology. For Xilinx Virtex IV FPGAs, $s \approx 17$. So, for an m bit adder, we can say that the LUT delay of the carry chain is $\lceil m/s \rceil$. Since the delay of the adder is determined by the delay of the carry chain, we can consider that the delay of the adder

$$D_{add} = \lceil m/s \rceil. \quad (6)$$

Likewise the delay of a subtracter can also be approximated similarly as:

$$D_{sub} = \lceil m/s \rceil. \quad (7)$$

The architecture of the gcd circuit also requires a complementer, which is obtained in the usual 2's complement sense. This requires also a subtracter, and hence has similar area and delay requirements as above.

Modeling of a Multiplexer

For a $2^t:1$ MUX, there are t selection lines, thus the output of a 2^t input MUX is a function of $2^t + t$ variables. So the total LUT requirements to implement the output functionality is $lut(2^t + t)$. For $GF(2^m)$, each input line to the MUX has m bits and the output has m bits. Thus the total LUT requirement for a 2^s input MUX, is given by

$$\# LUT_{MUX} = m \times lut(2^t + t). \quad (8)$$

Delay of the MUX in terms of LUTs is equal to the *maxlutpath* of $2^t + t$ variables and is given by,

$$D_{MUX} = maxlutpath(2^t + t). \quad (9)$$

If $2^{t-1} < number\ of\ inputs < 2^t$, then estimations in Eqs. 8 and 9 for 2^t inputs give an upper bound. Practically, the values in this case are slightly lesser than the values for 2^t inputs in Eqs. 8 and 9, and the difference can be neglected.

Total LUT Estimate of the gcd Processor

From Fig. 7, we can observe that the total number of LUTs in the gcd processor is the sum of the LUTs in the multiplexers, namely MUX_A, MUX_B, MUX_C and MUX_D, and the subtracter along with the complementer, which is also another subtracter. The state machine (control block in Fig. 7) consumes very few LUTs and is not considered in the overall LUT count. Thus the total number of k–LUTs in the entire circuit is,

$$\# LUT_{gcd} = 2LUT_{Subtractor} + LUT_{MUX_A} + LUT_{MUX_B} + LUT_{MUX_C} + LUT_{MUX_D}$$

Delay Estimate of the gcd Processor

In Fig. 7 we can observe that the critical path of the design goes through the path: subtracter $\rightarrow$ complementer $\rightarrow$ $MUX_D \rightarrow MUX_B \rightarrow MUX_A$. Hence the total delay using Eqs. 6, 7, and 9 can be approximated as:

$$D_{PATH} = 2D_{sub} + D_{MUX_D} + D_{MUX_B} + D_{MUX_A}$$
$$\approx 2\lceil m/s \rceil + 1 + 1 + 1$$
$$nonumber \approx 3 + 2\lceil m/s \rceil$$

Note that the last part of the equation, namely the delay of MUX_A comes from the fact that the multiplexer is made of two smaller 2-input multiplexers in parallel: one input writing into the register X_R and the other into the register Y_R.

EXPERIMENTAL RESULTS

The above design was synthesized using Xilinx ISE tools and targeted on a Virtex-4 FPGA. The objective of the experiments was to study the above dependence on the LUT utilization and to estimate the critical path delay for the circuit. It may be noted that while the estimation of the LUT utilization matches quite closely, the objective of the delay is to observe the trend, as an exact delay optimization is not aimed at. We vary the bit length of the gcd processor and repeat the experiments to study the scalability of the design. The estimations help in design exploration as we are able to estimate the dependence and thus tweak the architecture in a more planned manner and most importantly before actually implementing the hardware.

LUT Utilization of the gcd Processor

Figure 10 shows the plots for the LUT utilization of the gcd processor designed as described in the previous section. The results are furnished with two settings of the FPGA tool: one with the hierarchy on (Fig. 10 (a)) and the other flattened (Fig. 3.10 (b)). It may be observed that the theoretical estimates and the actual resource utilization match quite closely.

Delay Estimates for the gcd Processor

Figure 11 shows the plots for the critical path delays for the gcd processor with varying bit sizes. The plots are shown for both for hierarchy on and hierarchy off, as shown in Fig. 11(a) and Fig. 11(b), respectively.

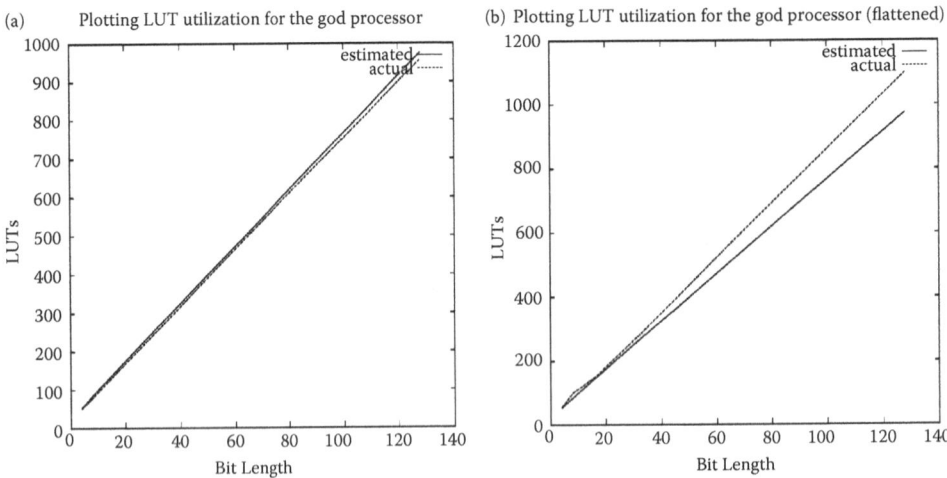

Fig. 10 LUT utilization of the gcd processor (both theoretical and actual) (a) LUT utilization of the gcd processor (hierarchy on) (b) LUT utilization of the gcd processor (hierarchy flattened)

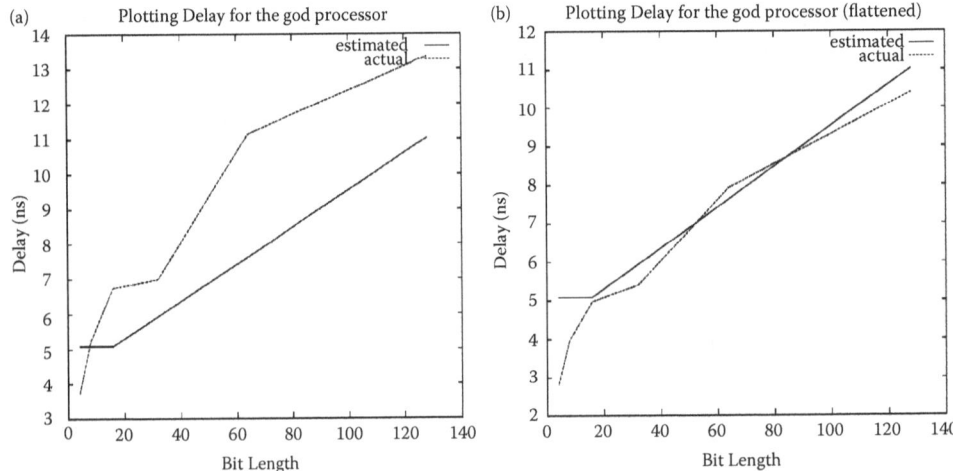

Fig. 11 Delay Modeling of the gcd processor (both theoretical and actual) (a) Critical Path Delay of the gcd processor (hierarchy on) (b) Critical Path Delay of the gcd processor (hierarchy flattened)

CONCLUSIONS

Both the estimates for the LUTs and the critical path show that the designer can make estimates of the performance ahead of the design. Thus, the theoretical model may be used as a guideline for design exploration. Again it may be noted that the estimates are not exact, and provides approximations to the exact requirements but nevertheless, can be used for design explorations in an analytic way. We show subsequently in the design of finite field circuits and an Elliptic Curve Crypto-processor how to leverage these models and framework for designing efficient architectures.

REFERENCES

1. Chen, D.; Cong, J.; Pan, P. FPGA design automation: A survey. Found. Trends Electron. Des. Autom. **2006**, *1* (3), 139–169.
2. Horiyama, T.; Nakanishi, M.; Kajihara, H.; Kimura, S. Folding of logic functions and its application to look up table compaction. ICCAD **2002**, 694–697.
3. Hutton, M.; Schleicher, J.; Lewis, D.; Pedersen, B.; Yuan, R.; Kaptanoglu, S.; Baeckler, G.; Ratchev, B.; Padalia, K.; Bourgeault, M.; Lee, A.; Kim, H.; Saini, R. Improving FPGA performance and area using an adaptive logic module. LNCS **2004**, *3203*, 135–144.

Multibiometric Fusion and Template Protection Strategies

Aarohi Vora and Chirag N. Paunwala
Electronics and Communication Department, Sarvajanik College of Engineering and Technology, Surat, India

Mita Paunwala
Electronics and Communication Department, C. K. Pithawala College of Engineering and Technology, Surat, India

Abstract

In recent scenario, the ability to reliably identify individuals in real time is a fundamental requirement in majority of the applications such as forensics, government, and commercial domain. Biometrics refers to the automated measurement of ubiquitous and unique measureable characteristics of an individual, which distinguishes an individual from the mass. Biometric technologies often are just a part of the security system, but it has a crucial role in the layering of increased security in the system. The design of a multibiometrics system in real-time applications is an exciting research due to the limitations of unimodal biometric systems. Though biometric system provides security, it is unsafe due to various attacks on it. The two most security threats to a biometric authentication system are spoofing and template breaching in the database. Thus, it is required to address the issues that make the system unsafe and provide the solution. The critical issues such as template protection and fusion strategies in the design of such systems are discussed in this entry. After reviewing multibiometric approaches, challenges, and integration strategies, some open research problems are suggested.

Keywords: Biometric systems; Classifier-based fusion; Cryptosystem; Feature transformation template protection; Fusion strategies; Rule-based fusion; Template protection scheme.

INTRODUCTION

In today's scenario, the advancement of technological transformations that have taken place in the field of digital media, e-commerce, banking system, and physical access systems has aroused the necessity for developing a sustainable secured system to a great extent. Security systems used for verification are categorized into three classes: (1) knowledge-based systems (password, PIN number, etc.), (2) token-based systems (voter card, passport, radio frequency identification tags, etc.), and (3) biometric-based systems (face, fingerprint, iris, etc.). Knowledge- and token-based systems can be spoofed easily by social engineering, dictionary attacks, forging the identity of the genuine person, etc. Thus, the increased need of privacy and security has given a call to develop a system based on anatomical and behavioral characteristics called biometric of the user. Biometric technologies have become the foundation of an extensive array of highly secure identification and personal verification solutions.

A biometric system is defined as automated methods of uniquely identifying an individual based on a physiological or behavioral characteristic. Biometric information should be universal, divergent, eternal/stable, quantifiable, consistent, and acceptable by most of the civilians and possess anti-spoof characteristic. Most widely used leading biometric technologies are face, fingerprint, hand geometry, iris, retina, signature, voice, palm print, gait, keystroke dynamics, ear, veins, knuckles, etc. The choice of biometric depends on the design considerations of the application because not a single biometric satisfies all the seven characteristics to the same markup level. Biometric systems operate in two distinct phases: (1) enrollment and (2) authentication as shown in Fig. 1.

In the enrollment phase, the biometric system operates by acquiring biometric data of an individual from sensor, extracting a feature set from the acquired data generating template from extracting feature and storing the template in the database against the identity (I) associated with the biometric.

In an authentication phase, the biometric is sensed and template is generated from it, in a manner similar to the enrollment phase. The representation, thus extracted is matched against the biometric representation(s) previously stored in the system to identify/verify the identity of the person.

In the verification mode, the system validates a person's identity by comparing the captured biometric data with

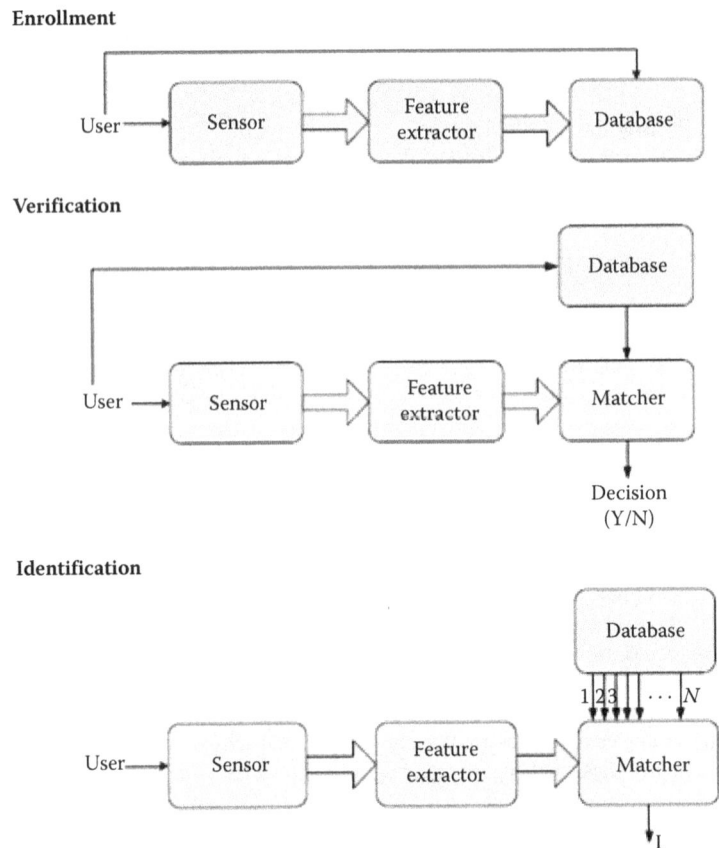

Fig. 1 Phases of biometric system operation

his/her own biometric template stored in the system database. In the identification mode, the system recognizes an individual by searching the templates of all the users in the database for a match.

All the users of a unimodal system must possess the biometric trait. It is possible that a subset of users does not possess a particular biometric. Thus, there is a large failure to enroll (FTE) rate associated with that unimodal biometric system. Furthermore, unimodal biometric systems suffer from several limitations which are as follows: (1) noisy data is captured through the sensor, (2) biometric templates of the same person vary during enrollment and authentication, (3) biometric templates of different people have similarities to a great extent, (4) all users to be enrolled in the database do not possess the trait during enrollment, and (5) intruder spoofs the template.

Multibiometric systems provide a variety of solutions to mitigate the problems associated with unibiometric systems. Multibiometric systems improve many of the limitations of unimodal biometric systems because the different biometric sources usually compensate for the inherent limitations of the other sources. Multibiometric system integrates evidences from multiple biometric sources and provides significant improvement in the recognition accuracy. Multibiometric system design depends on various factors such as sources of information, acquisition and processing architecture, level of information fusion, and fusion methodology.

Issues and Challenges in the Design of Biometric System

Metrics that determine the reliability of a biometric system are false acceptance rate (FAR), false rejection rate (FRR), equal error rate (EER), genuine acceptance rate (GAR), FTE, receiver operating characteristic (ROC) curve, etc. FAR measures the number of impostors that are falsely accepted by the system. FRR measures the number of authenticate people that are falsely rejected by the system. For a reliable multibiometric system, FAR and FRR of the system must be as low as possible. But it is not possible to reduce both the errors simultaneously, and hence, an optimum point has to be decided where both of them are equal which is nothing but the EER of the system. GAR is another metric to measure the performance of the system, which calculates the amount of genuinely accepted people by the system. ROC curve has two interpretations: (1) plot of threshold versus FAR and FRR and (2) plot of FAR versus GAR. If the plot of threshold versus FAR and FRR is examined, then the graph of FAR will increase and FRR will decrease, and hence, EER has to be obtained for optimum solution. When the plot of FAR versus GAR is observed, the higher value of GAR with lower value of FAR is accepted as a better performance region for a biometric system.

Open Research Problems

The design issues in the real-time biometric system and some open research problems are as follows.

- The first and foremost issue related to a practical biometric system is that the two feature sets originating from the same biometric trait of a user are not identical during each acquisition. This results in error that limits the system accuracy. The major factors affecting the accuracy of a biometric system are inter-user similarity (overlap of the biometric samples from two different individuals in the feature space), non-universality (people with hand-related disabilities, manual workers with many cuts and bruises on their fingertips, people with eye-related disabilities), noisy sensor data (e.g., a noisy fingerprint image due to smearing and residual deposits, a blurred iris image due to loss of focus, a face image captured with poor illumination, a noisy palm print due to overpressure on sensor), and lack of invariant representation. Due to these factors, the error rates associated with biometric systems designed are higher than the tolerance value decided for a particular application.[1] This indicates that biometric systems have nonzero error rates, and there is scope for improving the accuracy of biometric systems.
- Another challenge in the design of a biometric system is the dependence of system performance on the size of the database. In the case of a biometric verification system, the size of the database is not an issue, but in case of large-scale identification systems, the query feature set is matched with N identities enrolled in the system. Sequentially comparing the query with all the N templates is not a computationally effective solution but very much time consuming. This process affects the FAR of the system adversely and significantly reduces the throughput (number of queries that can be processed per unit time) of the system for large value of N. Hence, efficient scaling of system database is essential. Scaling can be done by a filtering or indexing process, where the database is narrowed down based on extrinsic factors (gender, ethnicity, age) or intrinsic factors (fingerprint pattern class). The search is confined to a smaller fraction of the database that is likely to contain the true identity of the user. There is not much research done in this domain, and hence, this is still an active area of research in the biometrics community.
- Multibiometric system design is a challenging problem because it is very difficult to predict the optimal sources of biometric information and the optimal fusion strategy for a particular application. Although utilizing more sources of evidence increases the accuracy, it may reduce the throughput of the system, and it is hard to find the optimal trade-off between the two. Hence, the accuracy of a multibiometric system degrades if an appropriate technique is not followed for consolidating the evidence provided by the different sources. Due to these reasons, information fusion in biometrics is still an active area of research.
- One more challenge in designing a biometric system is the level of security of the templates stored in the database against attacks devised specifically to hinder or degrade their operation. An explicit problem of system robustness against attacks is guaranteeing the security of biometric templates. Attacks on the template can lead to the following four vulnerabilities: (1) The templates can be used for crossmatching across different databases to covertly track a person without the person's consent; (2) a template can be replaced by an impostor's template; (3) a physical spoof can be created from the template; and (4) the stolen template can be replayed to the matcher to gain unauthorized access.[1] Due to these reasons, biometric templates should not be stored in plaintext form. Foolproof techniques are required to securely store or transmit the templates for remote applications such as ATM transactions and visa verification.
- Researchers not only have to address the issues related to reducing error rates but also have to look at the ways to enhance the usability of biometric systems. Solutions to advance the state of the art in biometrics include the design of new sensors that can acquire the biometric traits of an individual in a more reliable, convenient, and secure manner; the development of invariant representation schemes; robust and efficient matching algorithms; fusion schemes; template security; and privacy enhancement of biometric systems.

This entry addresses two critical issues in the design of a multibiometric system: template security and fusion methodology.

ISSUES, CHALLENGES, AND SOLUTION OF FUSION SCHEMES

Fusion in biometrics helps to expand the feature space used to represent individuals.[1] This increases the number of users that can be effectively enrolled in a personal identification system. Fusion of different modalities in the multibiometric system is a very critical stage. The levels of consolidation of the data from different modalities must be chosen very critically in order to enhance the recognition and performance of the system.

In general, there are mainly four levels of consolidation of data in a multibiometric system: sensor-level fusion, feature-level fusion, score-level fusion, and decision-level fusion as shown in Fig. 2, which are widely utilized in most of the research work.[1–4]

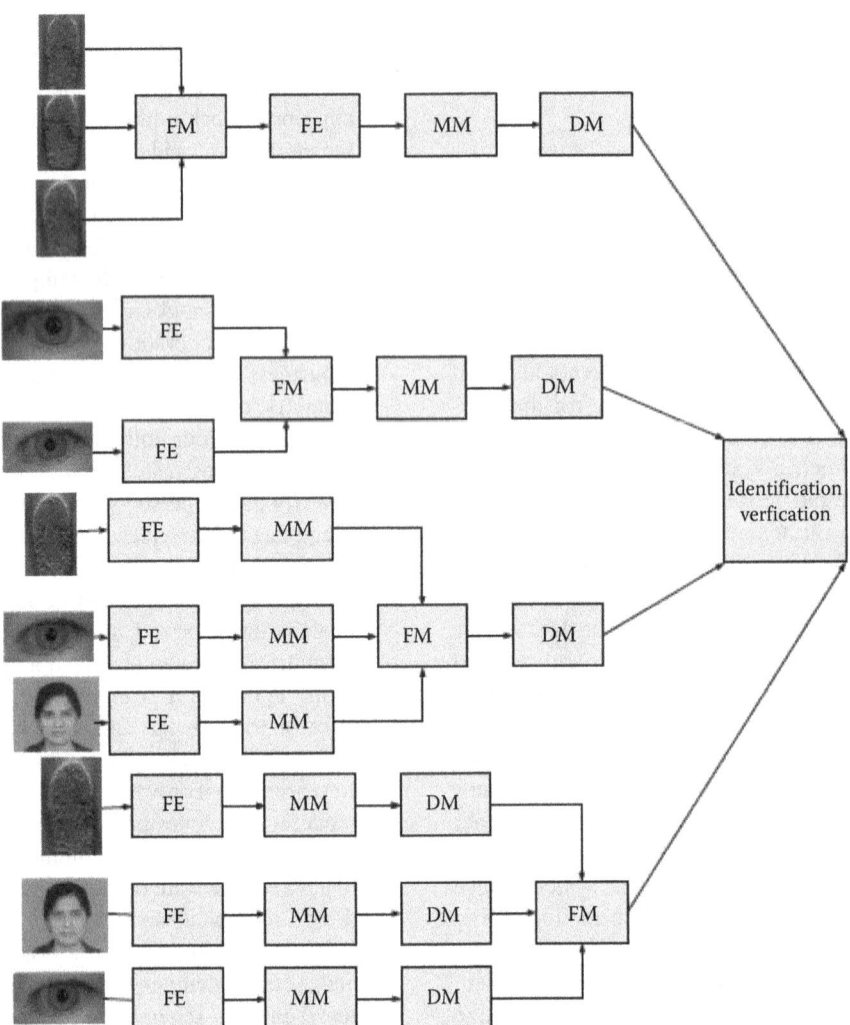

Fig. 2 Levels of fusion. Here FM, fusion module; FE, feature extraction module; MM, matching module; and DM, decision module

Sensor-Level Fusion

The biometric traits obtained from the sensors are directly fused with each other as shown in Fig. 3. Such type of fusion is possible only when different sensors capture the same biometric trait and then fuse it.

Al-Osaimi et al.[5] use a spatially optimized sensor-level fusion of 3D shape and texture for face recognition. The proposed approach by author optimizes fusion by modeling the variations caused by facial expressions and illumination changes and separating them from interpersonal discriminative information. A spatially optimized data-level fusion can outperform fusion at the score level under a similar experimental setup.

Kisku et al.[6] used an image fusion for face and palm print images using wavelet decomposition at low level. Fusion is accomplished with two high-resolution biometric images and scale-invariant feature transform operator. Finally, identity is verified by probabilistic relational graph with posteriori attributes matching between a pair of fused images. The performance of fusion at the low level is found to be superior with "maximum" fusion rule when it is compared with the other three fusion methods based on the three different wavelet-based fusion rules: "up-down" (UD) wavelet fusion rule, "down-up (DU) wavelet fusion rule, and "mean" wavelet fusion rule. Multisensor biometric fusion based on the maximum fusion rule produces 98.81% accuracy, whereas biometric fusion based on mean fusion rule, DU fusion rule, and UD fusion rule produces 97.43%, 96.27%, and 89.93% accuracy, respectively.

The researchers, Noushath et al.,[7] fused face and palm print modalities at sensor level using the wavelet-based image fusion scheme. According to their observation, sensor-level fusion is not performing satisfactorily, and in fact its performance is worse than unimodal counterparts. This may be due to the fact that the sensor-level fusion produces a kind of image that lacks adequate discriminatory information.

Tharwat et al.[8] performed sensor-level fusion by fusing images of ear and finger knuckle. Features are extracted using discrete cosine transform (DCT), discrete wavelet transform (DWT), and linear discriminant analysis (LDA)

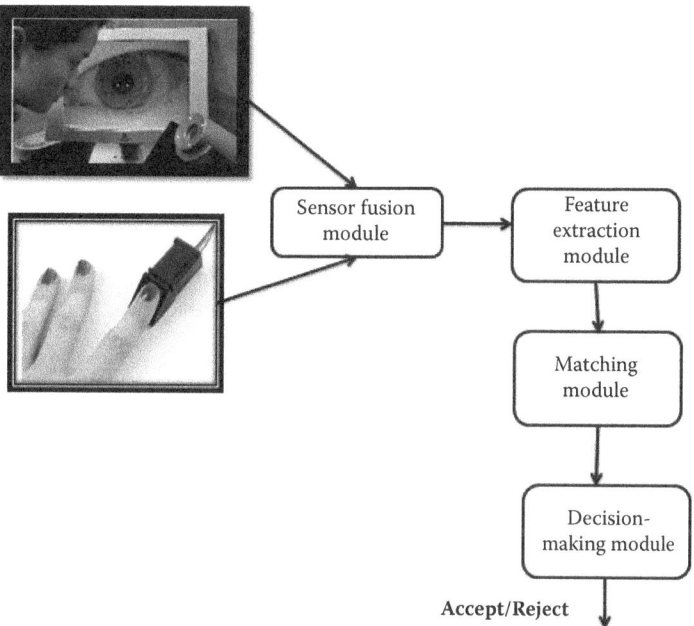

Fig. 3 Sensor level fusion

methods. They achieve the recognition rate of almost 98.82% using LDA method compared to DCT and DWT.

The raw data has the highest amount of information, but if contaminated by noise, it will degrade the performance of the system.[1,3] Hence, sensor-level fusion is not widely used in the multibiometric system.

Feature-Level Fusion

The feature templates/vectors generated from each of the modalities are integrated in a fusion block. When the feature sets are homogeneous, that is, if the multiple traits to be fused are from the same modality, fused vector can be calculated as a weighted average of individual feature templates of each trait. For n modalities, the fused template F using the weighted average technique is defined as[1,2]

$$F = \frac{1}{n}\sum_{k=0}^{n} w_k F_k \qquad (1)$$

where w_k and F_k are the weight parameters and feature templates, respectively. When the feature sets to be fused are not homogeneous, that is, if the multiple traits to be fused are from different modalities, the fused vector can be calculated by concatenating the feature templates of each trait.

The fused template F using the concatenation technique[9–20] is defined as

$$F = \left[F_1 \| F_2 \| F_3 \ldots \| F_n \right] \qquad (2)$$

Besides the concatenation technique, fused template can be performed using Fourier transform.[21]

The fused template F using Fourier transform technique is defined as:[21]

$$F = FT(F_1\ F_2\ F_3 \ldots F_n) \qquad (3)$$

The issue that arises with this level of fusion is that when the feature templates of different modalities to be fused are not compatible with each other, the recognition performance of the system will degrade as the data will be falsely integrated. For example, if the fusion of minutiae vectors and Eigen coefficients of faces is to be performed, then both the feature sets are not compatible with each other, and hence, fused vector formed will be faulty, resulting in failure of integration. Fusion at this level also suffers from the curse of dimensionality.

Score-Level Fusion

The most widely used level of fusion is the match score-level fusion. The matching module compares the template generated with the template already stored in the database and generates genuine and impostor score. This score indicates the proximity of input template with the stored template.

There are two types of score: similarity score and distance score.[2] Fusion requires homogeneity in score type. If an iris matcher produces distance score and a fingerprint matcher produces similarity score, then any one of the score type has to be transformed into other type to have homogeneity of score type. Furthermore, normalization in the score is required if the range of score generated by each matcher is wide.

The richest and optimal information carrying data of a biometric system is score vector. Match scores can easily be accessed and integrated, as they are always compatible with each other unlike feature templates. The techniques used for match score-level fusion are classified into two categories: (1) rule-based method[2–4,22–33] and (2) classification-based method.[2–4,34–43]

Rule-based method comprises linear sum fusion, weighted sum rule, product rule, min rule, max rule, and median rule. In sum rule, the match scores generated from different modalities are fused linearly. Let S_k, k = 0,1,2,... n, be the score vector of different n modalities. Then the fused score vector S using linear sum rule is given as:[2–4,22,23]

$$S = \sum_{k=0}^{n} S_k \quad (4)$$

All the biometric modalities do not possess equal qualities, that is, some may be stronger than the other in terms of its recognition performance. In such case, it is wiser to use weighted some rule where the effect of weaker biometric on the performance of the system is minimized, and the effect of strong biometric is maximized comparatively by assigning higher value of weight. Let w_k, k = 0,1,2, ..., n, be the weight parameters of n different model. The fused score vector S using linear weighted sum rule (LWSR) is given as:[24–29]

$$S = \sum_{k=0}^{n} S_k w_k \quad (5)$$

Another variant of rule-based method is a product rule fusion, which is expressed as a product of their score vectors. The fused score vector S using product rule is given as:[2,4,30,31]

$$S = \prod_{k=0}^{n} S_k \quad (6)$$

Similarly, the fused score vector S using min,[2,4,30,31] max,[2,4,30,31] and median[2,4,30,31] rules is given, respectively, as follows:

$$S = \min(S_1, S_2, \ldots S_k) \quad (7)$$

$$S = \max(S_1, S_2, \ldots S_k) \quad (8)$$

$$S = \mathrm{median}(S_1, S_2, \ldots S_k) \quad (9)$$

LWSR does not require any training session for its implementation. Hence, it is less time consuming and computationally efficient. Fusion using LWSR results in higher performance compared to other methods. Thus, LWSR fusion is a widely used rule-based fusion technique as it is computationally efficient, simple, and robust.

Classification-based methods treat the score vectors from different modalities as feature vectors. This category of methods includes several state-of-the-art algorithms such as k-nearest neighbors (KNNs),[32–34] support vector machines (SVM),[32,33,35–41] Bayesian classifier,[2,4] decision trees,[3,4,32,33] LDA,[3,4,32,33,42,43] etc., for fusion of scores. A classifier is used to construct a separation boundary between the genuine and impostor user in a verification system.

In KNN classifier technique, the distances of the test sample from k-nearest reference points are calculated and then the sample is assigned to a class, which has the majority of nearest neighbors. This procedure of calculating distances is tedious, and hence, it is time consuming.[3,4,32,33] Decision tree method categorizes the biometric samples according to a series of tests on a specific attribute of the data. These hierarchical tests lead to a particular class.[3,4,32,33] Each of the tested attributes is found based on maximizing the information gain at the particular node. The advantage of this method is that it provides direct insight into the predictive structure.

KNN and decision tree operating thresholds are not adjustable because their outputs are not a score but a class label, which is threshold independent. Although, in SVM and LDA, operating thresholds are also nonadjustable, these algorithms can be modified to generate a score value but not a class label. So a threshold can be used to classify these biometrics samples associated with confidence value.

Linear SVM constructs a separation boundary. The distance from boundary to the nearest data points (support vectors) on each side is maximized from a given set of training samples. A nonlinear SVM can be built by applying this algorithm in a transformed feature space. This feature space is created through a kernel function to project the samples to higher dimensional space. Radial basis function kernels are widely used in the literature as they improve recognition performance of the multibiometric system to a great extent.[32,33,35–41] Instead of using the output class label by SVM, the signed distance from the tested sample to the SM separating surface is used as output score. There are three widely used solvers for SVM: quadrature programming (QP), least squares (LS), and sequential minimal optimization (SMO). Performance accuracy of LS and SMO solvers has been proved to be better than QP. After training is carried out, the decision function is modified to get the fused score **S** as given as[32,33,35–41]

$$S = \sum_{i=1}^{N} \sum \lambda_i y_i K(X_i, X_T) + b^* \quad (10)$$

where N is the number of test samples, K() is the kernel function, λ is the Lagrange multiplier, y is the test sample, and b is the bias value. The decision threshold parameter is changed to reach different working points on ROC plot. This modification allows the comparison of different fusion methods in terms of their ROC plots along with their performance curves. The SVM has been widely researched and reported to have the best fusion performance compared to the methods including the decision-level fusion approach, sum rule, KNNs, and decision tree.

In recent years, theories of sparse representation and compressive sensing have emerged as powerful tools for efficiently processing data.[44,45] This has led to resurgence in interest in the principles of sparse representation and compressive sensing for biometrics recognition. The sparse representation-based multimodal system first sparsely encodes the individual modalities query samples independently to derive sparsity-based match scores and then combines them with the sum rule fusion. Suppose the multimodal system to be designed consists of face and iris modalities and their dictionaries are A^f and A^i, respectively. Given the face and iris query samples are y^f and y^i, the sparse coding problem is formulated as follows:

$$\hat{\alpha}^f = \operatorname{argmin} \|\alpha^f\|_1 \quad \text{s.t.} \|y^f - A^f \alpha^f\|_2 \prec \varepsilon \quad (11)$$

$$\hat{\alpha}^i = \operatorname{argmin} \|\alpha^i\|_1 \quad \text{s.t.} \|y^i - A^i \alpha^i\|_2 \prec \varepsilon \quad (12)$$

After the sparse match scores are obtained, they are fused using sum rule given in Eq. 5. SR-based multimodal systems have been proved to give better recognition performance compared to various state-of-the-art algorithms.

Decision-Level Fusion

In the decision-level fusion, the decision provided by each matcher of different modalities whether accepted/rejected is integrated.[2,4] The algorithms such as AND/OR rules,[1,2,4,46–50] majority voting,[1,2,4,51,52] Bayesian decision function,[1,2,4,53–56] and Dempster–Shafer theory of evidence,[1,2,4,57,58] are widely used to consolidate the decisions rendered by individual systems. Most of the commercial biometric systems provide only the final decision as the output, and hence, this level of fusion is often only the feasible solution for designing multibiometric system.[2] The simplest fusion technique in a multibiometric system involves combining decisions output by the different matchers using AND/OR rules.[1,2,4,46–50] The output of the AND rule is "match" only when all the biometric matchers agree that the input sample matches with the template, whereas the output of the OR rule is "match" as long as at least one matcher decides that the input sample matches with the template. The issue with the simple AND/OR rule is that when the AND rule is used, FAR of multibiometric system will tend to reduce and FRR will increase and the reverse is true for OR rule. Hence, the recognition performance of the multibiometric system will degrade.

In the majority voting rule,[1,2,4,51,52] decision is considered final if two or more matchers have same decision. One of the limitations of majority voting rule is that the recognition performance of the multibiometric system degrades if the matchers of different individual modalities do not have same recognition accuracy.

Bayesian decision fusion[1,2,4,53–56] transforms the output decision labels of each individual matcher into continuous probability values. This technique deals with finding out posterior probabilities of the modalities. This fusion results in misclassification of data labels if a priori probabilities are not properly predicted.

The Dempster–Shafer theory of evidence[1,2,4,57,58] applies degree of belief to the output of individual matchers. Let an iris matcher has a reliability of 0.90, that is, the output of the matcher is reliable 90% of the time and unreliable 10% of the time, and the matcher outputs "accept" decision. In such case, the algorithm assigns 0.90° of belief to the "accept" decision and 0.10° of belief to the "reject" decision. The degree of belief indicates that there is no reason to believe that the input does not match successfully against the template. Thus, the Dempster–Shafer technique improves the recognition accuracy of the multibiometric system in case of the decision-level fusion.

The research done by different authors for fusion of biometric evidences at various levels is summarized in Table 1.

ISSUES, CHALLENGES, AND SOLUTION OF TEMPLATE PROTECTION SCHEMES

Template leakage from the database is a serious threat to the security of the biometric system.[1] Template leakage leads to serious issues such as manipulated template data and can be replayed by impostor to get unauthorized access to the system. Template protection is a serious issue in the design of biometric system. Multibiometric systems have multiple templates to be stored in database, and hence, this issue is more critical in the multibiometric system.

Biometric authentication depends on unequal match between enrolled and query templates, and hence, handling large intra-user variability becomes the most challenging issue of template protection scheme. The template protection scheme adopted must have three basic properties: cryptographic security, performance, and revocability. It means that given a secured template, it must be computationally hard to find biometric features that match the secured template and also the scheme designed should not degrade the recognition performance of the multibiometric system.[1] Furthermore, it must be difficult for the intruder to identify from where the template is derived. There are three main approaches to address this issue: data

Table 1 Reference works at different levels of fusion

S. No.	Level of fusion	Reference papers
1	Sensor-level fusion	[5–8]
2	Feature-level fusion	[9–21]
3	Score-level fusion	[22–45]
4	Decision-level fusion	[46–58]

hiding scheme,[1,59–68] feature transformation scheme,[1,69–77] and cryptography scheme.[1,78–91]

Data Hiding Scheme

Data hiding schemes involve hiding the critical information inside a carrier. The watermarking technique in the multibiometric scenario hides one biometric data in another biometric data or some confidential information (personal ID information) in the compressed domain in order to increase the security of the biometric system.[1,59–68]

The main issue with the use of watermarking technique in the multibiometric system is the robustness and embedding capacity to be optimized simultaneously. Fingerprint template, embedded into a face template, must not alter the features of face; otherwise, the matching performance of the system will degrade to a greater extent, which limits the embedding capacity. This challenge offers one more research area for researcher.

Standard Encryption Algorithms

Another efficient method of template protection is to encrypt the template using standard cryptographic algorithms as shown in Fig. 4. In this method, enrolled template is stored in the database using encryption algorithms. During authentication, the stored template is decrypted first and then matched with the query template.[1,69]

Hence, the major gain of this method is that no matcher algorithms need to be redesigned, but the key management and vulnerability to template theft are the major cons of this technique. Hence, in order to overcome this issue, feature transformation scheme and biometric cryptosystem are developed.

Feature Transformation Scheme

In this technique, transformation function which is derived from a random key is applied to a biometric template, and then it is stored in the database[1,69–77] as shown in Fig. 5. During authentication, the same transformation function is applied to query template, and matching is performed in the transformed domain. There are basically two transformation schemes depending upon the characteristics of transformation function: invertible transforms and non-invertible transforms.

Invertible Transform Scheme

The transformation function designed is invertible in nature, and hence, there is a possibility that the intruder gets access to the key and stored template to recover the original biometric template.[1,69–77] Thus, the key management remains a major issue with this scheme. The matching algorithms must be redesigned to match templates in transformed domain. Random multi-space quantization technique and homomorphic encryption are the widely used invertible transform schemes for biometric template security.

Non-invertible Transform Scheme

In this scheme, the transformation function is pre-image resistant and non-invertible, and hence, it is computationally hard to invert a transformed template even if the key is known. This technique provides higher security than the invertible transform.[1,69–77] One type of non-invertible transformation is a function that leaves the biometric template in the original feature space even after the transformation. Such techniques are known as cancellable biometrics.[1,69–77]

The major issue of this approach is the trade-off between recognition performance and security of the transformation function. The choice of transformation function must be done critically based on the characteristics of the biometric features used in a specific application.

Biometric Cryptosystems

Biometric cryptosystems store helper data or secure sketch, that is, some public information about biometric features

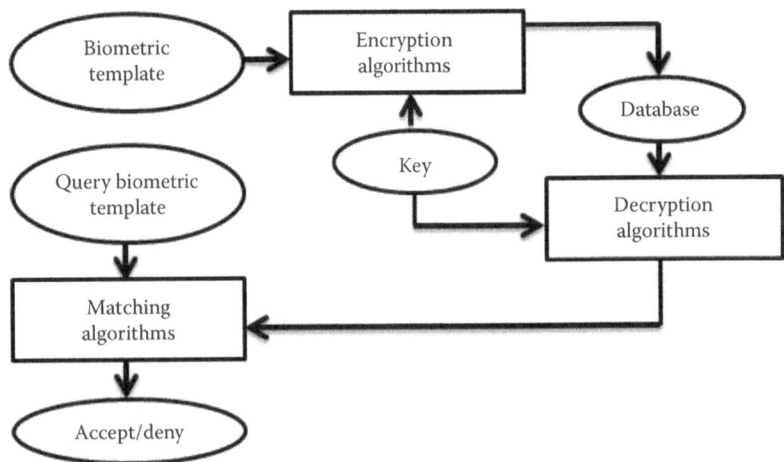

Fig. 4 Standard encryption scheme for template protection

Multibiometric Fusion and Template Protection Strategies

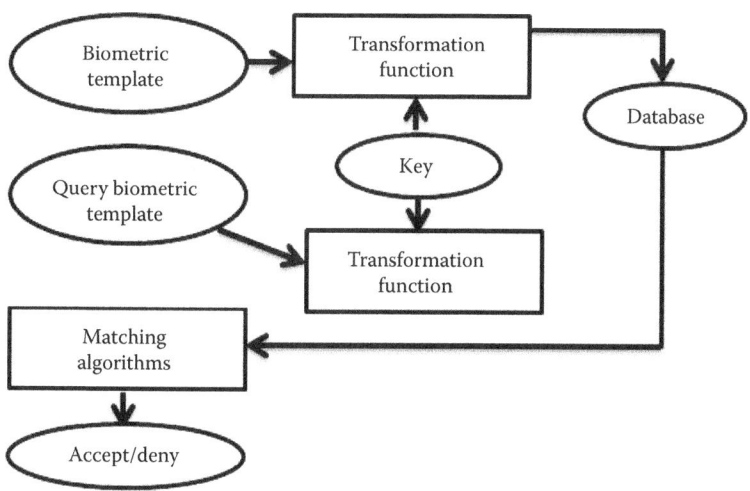

Fig. 5 Feature transformation scheme for template protection

in the database.[1,78–91] This helper data is used to extract the cryptographic key from query template during authentication. Matching is performed by verifying the validity of the extracted key.

The secure sketch is not secret, and hence, it should not reveal important information about original biometric template or the cryptographic key. Hence, biometric cryptosystems address the challenging issue of key management and template protection simultaneously.

There are two types of biometric cryptosystems depending on how the helper data is computed: key binding and key generation systems.

Key Binding Technique

In this cryptosystem, the secure sketch is obtained by binding a cryptographic key with the biometric template[1,78–91] as shown in Fig. 6.

In this technique, a single block will embed the key and the template, and this is stored in the database as a secure sketch. Matching is performed by recovering key from the helper data and query biometric template and then validating the key.[1,78–91] Indirect matching based on the error correction scheme needs some specific matchers, which can possibly lead to a reduction in the matching accuracy.

Key Generation Technique

In this technique, the helper data is first computed from the biometric template as shown in Fig. 7. Thereafter, the cryptographic key is generated from the helper data and the query biometric features. The cryptographic key generated should have uniform random distribution.[1,78–91]

The fuzzy extractor scheme generates a uniformly random cryptographic key from the biometric features, and hence, it addresses the issue of key stability. The

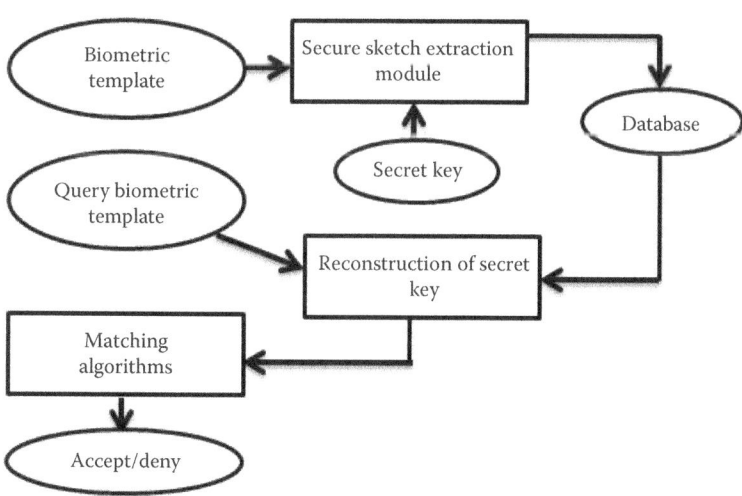

Fig. 6 Key binding biometric cryptosystem

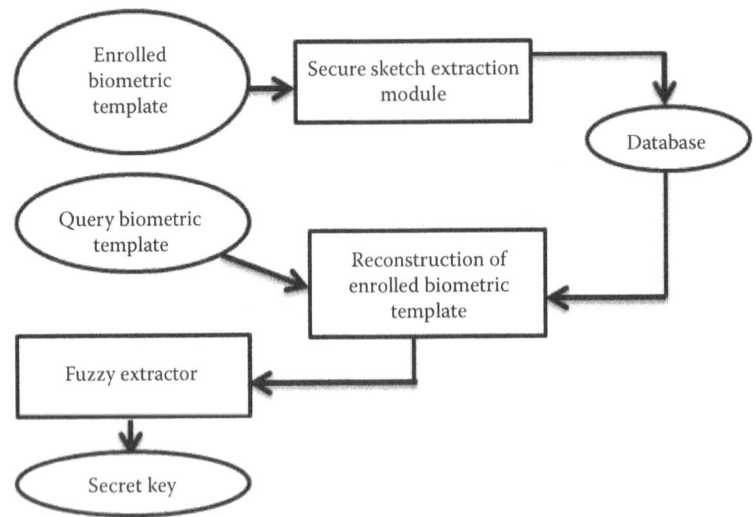

Fig. 7 Key generation biometric cryptosystem

Table 2 Reference work for different template protection schemes

S. No.	Template protection schemes	Reference papers
1	Data hiding scheme	[1,59–68]
2	Standard encryption	[1,69]
3	Feature transformation scheme	[1,69–77]
4	Cryptosystem	[1,78–91]

nonuniformity issue is handled by applying cryptographic hash functions to the biometric template. The research work done by various authors in the domain of biometric template protection is shown in Table 2.

CONCLUSION

Biometric system is the major need of today's society as every individual needs his/her confidential information to be secured. There has been a drastic increase in the security breaching, and hence, most of the security systems are now including biometrics as their major part. However, there are some issues related to the biometric systems also. The main issues with the biometric authentication system are template breaching, spoofing, recognition performance, and population coverage. These issues are addressed by implementing multibiometric systems and template protection schemes.

The main issue with the multibiometric system is to choose an optimal fusion technique to get higher performance accuracy. The amount of information reduces as one moves from a sensor- to decision-level fusion. The widely utilized fusion is at the match score level. Depending upon the nature of application and the amount of security required, one can choose the fusion scheme.

Various template protection schemes are watermarking, encryption, cryptosystem, and feature transformation method. Out of these schemes, cryptosystems and feature transformation are widely used as they address the issue of key management and key stability as well as maintain the secrecy of the template to a high extent.

In recent scenario, sparse content of the biometric features is widely explored to optimally reduce the template size. The intra-class variation and interclass similarity issues are widely addressed recently. The template protection scheme assumes that the probability distribution of biometric features should be uniform, but that is not the case in practical scenario, and hence, this is also an active area of research. Addressing liveliness of a biometric trait is also a challenging area to focus on.

REFERENCES

1. Jain, A.; Ross, A.; Nandakumar, K. *Introduction to Biometrics*, 1st Ed.; Springer Science and Business Media, LLC: New York, 2011.
2. Jain, A.; Ross, A.; Nandakumar, K. *Handbook of Multibiometrics*, 1st Ed.; Springer: New York and London, 2006.
3. Ross, A.; Jain, A. Information fusion in biometrics. Pattern Recogn. Lett. **2003**, *24* (13), 2115–2125.
4. Atrey, P.; Hossain, M.; Saddik, A.; Kankanhalli, M. Multimodal fusion for multimedia analysis: A survey. Appeared Multimedia Syst. **2010**, *16* (6), 345–379.
5. Al-Osaimi, F.; Bennamoun, M.; Mian, A. Spatially optimized data-level fusion of texture and shape for face recognition. IEEE Trans. Image Process. **2012**, *21* (2), 859–872.
6. Kisku, D.; Rattani, A.; Gupta, P.; Sing, J. Biometric sensor image fusion for identity verification: A case study with wavelet-based fusion rules graph matching. In *Conference*

on *Technologies for Homeland Security*; IEEE: Boston, MA, 2009, 433–439.

7. Noushath, S.; Imran, M.; Jetly, K.; Rao, A.; Kumar, G. Multimodal biometric fusion of face and palmprint at various levels. In *IEEE International Conference on Advances in Computing, Communications and Informatics (ICACCI)*; IEEE: Mysore, 2013, 1793–1798.

8. Tharwat, A.; Ibrahim, A.; Ali, H. Multimodal biometric authentication algorithm using ear and finger knuckle images. In *Seventh International Conference on Computer Engineering & Systems (ICCES)*; IEEE: Cairo, 2012, 176–179.

9. Ramalho, M.; Singh, S.; Correia, P.; Soares, L. Secure multi-spectral hand recognition system. In *19th European Signal Processing Conference*; IEEE: Barcelona, 2011, 2269–2273.

10. Ahmad, M.; Ilyas, M.; Isa, M.; Ngadiran, R.; Darsono, A. Information fusion of face and palmprint multimodal biometrics. In *IEEE Region 10 Symposium*; IEEE: Kuala Lumpur, 2014, 635–639.

11. Almayyan, W.; Own, H.; Zedan, H. A comparative evaluation of feature level based fusion schemes for multimodal biometric authentication. In *11th International Conference on Hybrid Intelligent Systems (HIS)*; IEEE: Melacca, 2011, 22–27.

12. Azom, V.; Adewumi, A.; Tapamo, J.R. Face and Iris biometrics person identification using hybrid fusion at feature and score-level. In *Pattern Recognition Association of South Africa and Robotics and Mechatronics International Conference (PRASA-RobMech)*; IEEE: Port Elizabeth, 2015, 207–212.

13. Brown, D.; Bradshaw, K. A multi-biometric feature-fusion framework for improved uni-modal and multi-modal human identification. In *Symposium on Technologies for Homeland Security (HST)*; IEEE: Waltham, MA, 2016, 1–6.

14. Dandawate, Y.H.; Inamdar, S.R. Fusion based multimodal biometric cryptosystem. In *International Conference on Industrial Instrumentation and Control (ICIC)*; IEEE: Pune, 2015, 1484–1489.

15. Daniel, D.M.; Monica, B. A data fusion technique designed for multimodal biometric systems. In *10th International Symposium on Electronics and Telecommunications*; IEEE: Timisoara, 2012, 155–158.

16. Joshi, S.C.; Kumar, A. Design of multimodal biometrics system based on feature level fusion. In *10th International Conference on Intelligent Systems and Control (ISCO)*; IEEE: Coimbatore, 2016, 1–6.

17. Parmar, P.A.; Degadwala, S.D. A feature level fusion fingerprint indexing approach based on MV and MCC using SVM classifier. In *International Conference on Communication and Signal Processing (ICCSP)*; IEEE: Melmaruvathur, 2016, 1024–1028.

18. Raghavendra, R. PSO based framework for weighted feature level fusion of face and palmprint. In *Eighth International Conference on Intelligent Information Hiding and Multimedia Signal Processing*; IEEE: Piraeus, 2012, 506–509.

19. Rattani, A.; Kisku, D.R.; Bicego, M.; Tistarelli, M. Robust feature-level multibiometric classification. In *Biometrics Symposium: Special Session on Research at the Biometric Consortium Conference*; IEEE: Baltimore, MD, 2006, 1–6.

20. Rattani, A.; Kisku, D.R.; Bicego, M.; Tistarelli, M. Feature level fusion of face and fingerprint biometrics. In *First International Conference on Biometrics: Theory, Applications, and Systems*; IEEE: Crystal City, VA, 2007, 1–6.

21. Patel, H.; Panuwala, C.; Vora, A. Hybrid feature level approach for Multi-biometric cryptosystem. In *International Conference on Wireless Communications, Signal Processing and Networking (WiSPNET)*; IEEE: Chennai, 2016, 1087–1092.

22. Gao, G.; Zhang, L.; Yang, J.; Zhang, L.; Zhang, D. Reconstruction based finger-knuckle-print verification with score level adaptive binary fusion. IEEE Trans. Image Process. **2013**, *22* (12), 5050–5062.

23. Kabir, W.; Ahmad, M.O.; Swamy, M.N.S. A new anchored normalization technique for score-level fusion in multimodal biometric systems. In *International Symposium on Circuits and Systems (ISCAS)*; IEEE: Montreal, QC, 2016, 93–96.

24. Xu, Y.; Fei, L.; Zhang, D. Combining left and right palmprint images for more accurate personal identification. IEEE Trans. Image Process. **2015**, *24* (2), 549–559.

25. Kelkboom, E.J.C.; Zhou, X.; Breebaart, J.; Veldhuis, R.N.J.; Busch, C. Multi-algorithm fusion with template protection. In *3rd International Conference on Biometrics: Theory, Applications, and Systems*; IEEE: Washington, DC, 2009, 1–8.

26. Awalkar, K.V.; Kanade, S.G.; Jadhav, D.V.; Ajmera, P.K. A multi-modal and multi-algorithmic biometric system combining iris and face. In *International Conference on Information Processing (ICIP)*; IEEE: Pune, 2015, 496–501.

27. Chaudhary, S.; Nath, R. A multimodal biometric recognition system based on fusion of palmprint, fingerprint and face. In *International Conference on Advances in Recent Technologies in Communication and Computing*; IEEE: Kottayam, Kerala, 2009, 596–600.

28. Vora, A.; Paunwala, C.; Paunwala, M. Improved weight assignment approach for multimodal fusion. In *International Conference on Circuits, Systems, Communication and Information Technology Applications (CSCITA)*; IEEE: Mumbai, 2014, 70–74.

29. Telgad, L.R.; Deshmukh, D.P.; Siddiqui, A.M.N. Combination approach to score level fusion for multimodal biometric system by using face and fingerprint. In *International Conference on Recent Advances and Innovations in Engineering (ICRAIE-2014)*; IEEE: Jaipur, 2014, 1–8.

30. Zhang, Y.; Sun, D.; Qiu, Z. A novel method for fusion operators evaluating at score-level fusion in biometric authentication. In *10th International Conference on Signal Processing Proceedings*; IEEE: Beijing, 2010, 1339–1342.

31. Proença, H. Ocular biometrics by score-level fusion of disparate experts. IEEE Trans. Image Process. **2014**, *23* (12), 5082–5093.

32. Theodoridis, S.; Koutroumbas, K. *Pattern Recognition*, 2nd Ed.; Elsevier, Academic Press: San Diego, CA, 2003.

33. Bishop, C. *Pattern Recognition and Machine Learning*, 1st Ed.; Springer-Verlag: New York, 2006.

34. Imran, M.; Rao, A.; Kumar, G.H. A new hybrid approach for information fusion in multibiometric systems. In *Third*

National Conference on Computer Vision, Pattern Recognition, Image Processing and Graphics; IEEE: Hubli, Karnataka, 2011, 235–238.

35. Vora, A.; Paunwala, C.; Paunwala, M. Statistical analysis of various kernel parameters on SVM based multimodal fusion. In *Annual IEEE India Conference (INDICON)*; IEEE: Pune, 2014, 1–5.
36. Vora, A.; Paunwala, C.; Paunwala, M. Nonlinear SVM fusion of multimodal biometric system. In *International Conference on Communication and Computing track (ICCC)*; Elsevier: Bangalore, 2014, 204–210.
37. Mezai, L.; Hachouf, F. Score level fusion algorithm using differential evolution and proportional conflict redistribution rule. In *International Conference on Systems, Man, and Cybernetics (SMC)*; IEEE: Budapest, 2016, 1051–1056.
38. Horng, S.J.; Chen, Y.H.; Run, R.S.; Chen, R.J.; Lai, J.L.; Sentosal, K.O. An improved score level fusion in multimodal biometric systems. In *International Conference on Parallel and Distributed Computing, Applications and Technologies*; IEEE: Higashi Hiroshima, 2009, 239–246.
39. Kang, B.J.; Park, K.R. Multimodal biometric method based on vein and geometry of a single finger. IET Comput. Vis. **2010**, *4* (3), 209–217.
40. Ejarque, P.; Hernand, J. Score bi-Gaussian equalisation for multimodal person verification. IET Sig. Process. **2009**, *3* (4), 322–332.
41. Elmir, Y.; Elberrichi, Z.; Adjoudj, R. Score level fusion based multimodal biometric identification (Fingerprint & voice). In *6th International Conference on Sciences of Electronics, Technologies of Information and Telecommunications (SETIT)*; IEEE: Sousse, 2012, 146–150.
42. Marfella, L.; Marasco, E.; Sansone, C. Liveness-based fusion approaches in multibiometrics. In *Workshop on Biometric Measurements and Systems for Security and Medical Applications (BIOMS) Proceedings*; IEEE: Salerno, 2012, 1–7.
43. Ren, C.; Yin, Y.; Ma, J.; Yang, G. A novel method of score level fusion using multiple impressions for fingerprint verification. In *International Conference on Systems, Man and Cybernetics*; IEEE: San Antonio, TX, 2009, 5051–5056.
44. Shekhar, S.; Patel, V.M.; Nasrabadi, N.M.; Chellappa, R. Joint sparse representation for robust multimodal biometrics recognition. IEEE Trans. Pattern Anal. Mach. Intell. **2015**, *36* (1), 113–126.
45. Nair, S.A.H.; Aruna, P.; Sakthivel, K. Analysis of sparse representation and PCA fusion methods for Iris and Palmprint biometric features. In *International Conference on Communication and Signal Processing*; IEEE: Melmaruvathur, 2014, 562–566.
46. Toh, A.K.; Yau, W. Combination of hyperbolic functions for multimodal biometrics data fusion. IEEE Trans. Syst. Man Cybern. Part B **2004**, *34* (2), 1196–1209.
47. Tao, Q.; Veldhuis, R. Hybrid fusion for biometrics: Combining score-level and decision-level fusion. In *Computer Society Conference on Computer Vision and Pattern Recognition Workshops*; IEEE: Anchorage, AK, 2008, 1–6.
48. Yang, B.; Busch, C.; Groot, K.; Xu, H.; Veldhuis, R. Decision level fusion of fingerprint minutiae based pseudonymous identifiers. In *International Conference on Hand-Based Biometrics*; IEEE: Hong Kong, 2011, 1–6.
49. Sudhamani, M.J.; Venkatesha, M.K.; Radhika, K.R. Fusion at decision level in multimodal biometric authentication system using Iris and Finger Vein with novel feature extraction. In *Annual IEEE India Conference (INDICON)*; IEEE: Pune, 2014, 1–6.
50. Kumar, A.; Hanmandlu, M.; Sanghvi, H.; Gupta, H. Decision level biometric fusion using ant colony optimization. In *International Conference on Image Processing*; IEEE: Hong Kong, 2010, 3105–3108.
51. Sanches, T.; Antunes, J.; Correia, P.L. A single sensor hand biometric multimodal system. In *15th European Signal Processing Conference*; IEEE: Poznan, 2007, 30–34.
52. Barbu, T.; Ciobanu, A.; Luca, M. Multimodal biometric authentication based on voice, face and iris. In *E-Health and Bioengineering Conference (EHB)*; IEEE: Iasi, 2015, 1–4.
53. Osadciw, L.; Varshney, P.; Veeramachaneni, K. Improving personal identification accuracy using multisensor fusion for building access control applications. In *Fifth International Conference on Information Fusion*; IEEE: Annapolis, MD, 2002, 1176–1183.
54. Kanhangad, V.; Kumar, A.; Zhang, D. Comments on an adaptive multimodal biometric management algorithm. IEEE Trans. Syst. Man Cybern. Part C **2008**, *38* (6), 841–843.
55. Tao, Q.; Veldhuis, R. Robust biometric score fusion by naive likelihood ratio via receiver operating characteristics. IEEE Trans. Inform. Foren. Secur. **2013**, *8* (2), 305–313.
56. Acharya, S.; Fridman, A.; Brennan, P.; Juola, P.; Greenstadt, R.; Kam, M. User authentication through biometric sensors and decision fusion. In *47th Annual Conference on Information Sciences and Systems (CISS)*; IEEE: Baltimore, MD, 2013, 1–6.
57. Mezai, L.; Hachouf, F. Score-level fusion of face and voice using particle swarm optimization and belief functions. IEEE Trans. Human-Mach. Syst. **2015**, *45* (6), 761–772.
58. Nguyen, K.; Denman, S.; Sridharan, S.; Fookes, C. Score-level multibiometric fusion based on Dempster–Shafer theory incorporating uncertainty factors. IEEE Trans. Human-Mach. Syst. **2015**, *45* (1), 132–140.
59. Paunwala, M.; Patnaik, S. Biometric template protection with DCT-based watermarking. Mach. Vis. Appl. **2014**, *25* (1), 263–275.
60. Paunwala, M.; Patnaik, S. Augmenting security of biometric with watermarking. Imaging Sci. J. **2014**, *62* (4), 189–196.
61. Vatsa, M.; Singh, R.; Mitra, P.; Noore, A. Digital watermarking based secure multimodal biometric system. In *International Conference on Systems, Man and Cybernetics*; IEEE: The Hague, Netherlands, 2004, Vol. 3, 2983–2987.
62. Ghany, K.K.A.; Hassan, G.; Hassanien, A.E.; Hefny, H.A.; Schaefer, G.; Ahad, M.A.R. A hybrid biometric approach embedding DNA data in fingerprint images. In *International Conference on Informatics, Electronics & Vision (ICIEV)*; IEEE: Dhaka, 2014, 1–5.
63. Abdullah, M.A.M.; Dlay, S.S.; Woo, W.L.; Chambers, J.A. A framework for Iris biometrics protection: A marriage between watermarking and visual cryptography. IEEE Access **2016**, *4* (1), 10180–10193.
64. Islam, M.R.; Sayeed, M.S.; Samraj, A. Multimodality to improve security and privacy in fingerprint authentication

system. In *International Conference on Intelligent and Advanced Systems*; IEEE: Kuala Lumpur, 2007, 753–757.
65. Islam, M.R.; Sayeed, M.S.; Samraj, A. Biometric template protection using watermarking with hidden password encryption. In *International Symposium on Information Technology*; IEEE: Kuala Lumpur, 2008, 1–8.
66. Lalithamani, N.; Sabrigiriraj, M. Embedding of iris data to hand vein images using watermarking technology to improve template protection in biometric recognition. In *International Conference on Electrical, Computer and Communication Technologies (ICECCT)*; IEEE: Coimbatore, 2015, 1–7.
67. Li, C.; Wang, Y.; Ma, B. Protecting biometric templates using LBP-based authentication watermarking. In *Conference on Pattern Recognition*; IEEE: Nanjing, 2009, 1–5.
68. Majumder, S.; Devi, K.J.; Sarkar, S.K. Singular value decomposition and wavelet-based iris biometric watermarking. IET Biometrics **2013**, *2* (1), 21–27.
69. Rathgeb, C.; Uhl, A. A survey on biometric cryptosystems and cancelable biometrics. EURASIP J. Inform. Secur. **2011**, *1* (3), 1–25.
70. Butt, M.; Damer, N. Helper data scheme for 2D cancelable face recognition using bloom filters. In *IWSSIP Proceedings*; IEEE: Dubrovnik, 2014, 271–274.
71. Yang, H.; Jiang, X.; Kot, A.C. Generating secure cancelable fingerprint templates using local and global features. In *2nd International Conference on Computer Science and Information Technology*; IEEE: Beijing, 2009, 645–649.
72. Chandra, E.; Kanagalakshmi, K. Cancelable biometric template generation and protection schemes: A review. In *3rd International Conference on Electronics Computer Technology*; IEEE: Kanyakumari, 2011, 15–20.
73. Patel, V.; Ratha, N.; Chellappa, R. Cancelable biometrics: A review. IEEE Sig. Process. Mag. **2015**, *32* (5), 54–65.
74. Kanade, S.; Petrovska-Delacretaz, D.; Dorizzi, B. Cancelable iris biometrics and using Error Correcting Codes to reduce variability in biometric data. In *Conference on Computer Vision and Pattern Recognition*; IEEE: Miami, FL, 2009, 120–127.
75. Paul, P.P.; Gavrilova, M. Multimodal biometrics using cancelable feature fusion. In *International Conference on Cyberworlds*; IEEE: Santander, 2014, 279–284.
76. Paul, P.P.; Gavrilova, M. Rank level fusion of multimodal cancelable biometrics. In *13th International Conference on Cognitive Informatics and Cognitive Computing*; IEEE: London, 2014, 80–87.
77. Kaur, H.; Khanna, P. Biometric template protection using cancelable biometrics and visual cryptography techniques. Multimedia Tools Appl. **2016**, *75* (23), 16333–16361.
78. Nandakumar, K.; Jain, A.; Pankanti, S. Fingerprint-based fuzzy vault: Implementation and performance. IEEE Trans. Inform. Foren. Secur. **2007**, *2* (4), 744–757.

79. Jain, A.; Nandakumar, K.; Nagar, A. Biometric template security. EURASIP J. Adv. Sig. Process., Special Issue on Advanced Signal Processing and Pattern Recognition Methods for Biometrics, **2008**, 1–17.
80. Li, C.; Hu, J.; Pieprzyk, J.; Susilo, W. A new biocryptosystem-oriented security analysis framework and implementation of multibiometric cryptosystems based on decision level fusion. IEEE Trans. Inform. Foren. Secur. **2015**, *10* (6), 1193–1206.
81. Lee, Y.J.; Park, K.R.; Lee, S.J.; Bae, K.; Kim, J. A new method for generating an invariant iris private key based on the fuzzy vault system. IEEE Trans. Syst. Man Cybern. Part B **2008**, *38* (5), 1302–1313.
82. Al-Saggaf, A.A.; Ghouti, L.; Acharya, H.S. Biometric cryptosystem with renewable templates. In *National Workshop on Information Assurance Research*; IEEE: Riyadh, Kingdom of Saudi Arabia, 2012, 1–5.
83. Bansal, D.; Sofat, S.; Kaur, M. Fingerprint fuzzy vault using hadamard transformation. In *International Conference on Advances in Computing, Communications and Informatics (ICACCI)*; IEEE: Kochi, 2015, 1830–1834.
84. Feng, Y.C.; Yuen, P.C.; Jain, A. A hybrid approach for generating secure and discriminating face template. IEEE Trans. Inform. Foren. Secur. **2010**, *5* (1), 103–117.
85. Ghany, K.K.A.; Hefny, H.A.; Hassanien, A.E.; Ghali, N.I. A hybrid approach for biometric template security. In *International Conference on Advances in Social Networks Analysis and Mining*; IEEE: Istanbul, 2012, 941–942.
86. Mai, G.; Lim, M.H.; Yuen, P.C. Fusing binary templates for multi-biometric cryptosystems. In *7th International Conference on Biometrics Theory, Applications and Systems (BTAS)*; IEEE: Arlington, VA, 2015, 1–8.
87. Li, C.; Hu, J. A security-enhanced alignment-free fuzzy vault-based fingerprint cryptosystem using pair-polar minutiae structures. IEEE Trans. Inform. Foren. Secur. **2016**, *11* (3), 543–555.
88. Nandakumar, K.; Jain, A.K.; Pankanti, S. Fingerprint-based fuzzy vault: Implementation and performance. IEEE Trans. Inform. Foren. Secur. **2007**, *2* (4), 744–757.
89. Nagar, A.; Nandakumar, K.; Jain, A.K. Multibiometric cryptosystems based on feature-level fusion. IEEE Trans. Inform. Foren. Secur. **2012**, *7* (1), 255–268.
90. Yuan, L. Multimodal cryptosystem based on fuzzy commitment. In *17th International Conference on Computational Science and Engineering*; IEEE: Chengdu, 2014, 1545–1549.
91. Nguyen, T.H.; Wang, Y.; Nguyen, T.N.; Li, R. A fingerprint fuzzy vault scheme using a fast chaff point generation algorithm. In *International Conference on Signal Processing, Communication and Computing (ICSPCC 2013)*; IEEE: Kunming, 2013, 1–6.

Optical Character Recognition (OCR)

Matthew C. Mariner
George A. Smathers Libraries, University of Florida, Gainesville, Florida, U.S.A.

Abstract
This entry is a basic overview of the technology and uses of optical character recognition (OCR). OCR is a technology used to capture text from printed page and convert it to a machine-readable format. The history of OCR's major milestones and implementations is covered as well as the most widespread current applications. The realm of OCR is split into two basic categories: OCR hardware and OCR software. Their respective uses and advantages are covered so as to outline the form from one system to another. The hurdles of OCR technology, such as manuscript capture and archaic printing practices, are discussed and possible solutions based on current research presented.

Keywords: Algorithm; American standard code for information interchange (ASCII); Capture; Detection; Intelligent character recognition (ICR); Mimeograph; Multiline optical character reader (MLOCR); Recognition.

INTRODUCTION

Optical character recognition (OCR) is a blanket term used to identify a number of processes by which written or typed text is electronically translated into machine-readable text. The technology is "optical" in that it, similar to human readers, scans the document for characters that are relevant to some output goal. Some applications are inherently mechanical and literally look at the document with a high-speed lens; others assess the image of a document after it has been scanned by another device. OCR has been present in the business and governmental world for more than 60 yr, but is a recent development in the realm of academic and public libraries. Now used by thousands of universities and institutions for capturing searchable and editable text, OCR devices and software are essential in the rapidly developing world of digital libraries and textual preservation. Through the advent of OCR, academic institutions can scan the pages of a book, recognize its text, load the images online, and allow users to instantly search for keywords or proper names without having to manually "flip" through each page of the work. The research value of such a technology is likely to be profound: permitting text comparisons and analyses never before possible.

Understanding the history, key concepts, and variety of uses of OCR are not necessary for OCR use. However OCR is a very complex and mathematically intensive science, and its history, concepts, and usage provide context for its current and evolving applications. This entry will provide a beginner's understanding of a highly important and advanced field.

BRIEF HISTORY OF OCR

Optical character recognition as a readily available and practical technology has only existed for a handful of decades. In its earliest form, OCR was implemented by means of photodetection, mechanical devices, and bulky templates. Gustav Tauschek, an Austrian born pioneer in information technology famous for developing many improvements to early calculators and punch-card devices, filed the first known patent for an OCR machine in 1929. Tauschek's device used a photodetector to recognize text placed behind a template; when both the text and the template matched exactly, light projected through the bottom of the device was obstructed and could not reach the photodetector.[1]

Tauschek's device paved the way for more complex and practical solutions. After a period of 20 years, when rudimentary OCR machines relied heavily on light sensitivity and restrictive templates, the 1950s introduced the so-called "Fabulous" era of not only OCR advancement, but of computer technology as a whole. In 1951, David Shepherd, a 27-year-old research scientist employed by the U.S. Department of Defense, developed out of his home a simple but heraldic OCR solution. Known as "Gismo," Shepherd's machine was capable of recognizing 23 letters of the Latin alphabet, so long as they were produced cleanly by a standard typewriter.[2] "Gismo" created a media stir among the ranks of technology journalists, allowing others with greater vision and resources to capitalize.

Jacob Rabinow, founder of Intelligent Machines Research Corporation (IMR), was one such visionary. His corporation is widely known for its marketing of the very first commercial OCR machines. Based on Vannevar Bush's "rapid selector," an early microfilm reader, Rabinow's OCR device was complex enough to process thousands of texts and support the needs of a major corporation. In fact, Rabinow's device was sold to *Reader's Digest* magazine for the processing of sales reports into punched cards for the magazine's sales department—an implementation acknowledged as the first commercial use of OCR.[2] The second execution of OCR use by a large organization was made by the United States Postal Service (USPS).[2]

As the uses of OCR permeated the business and governmental world, standards of practice became necessary to streamline a very complex mathematical process. By 1970, approximately 50 manufacturers had developed and were actively marketing over 100 OCR machines and systems. The main source of confusion for developers and their machines was the variety of fonts and character sets used by their customers. In 1960, the American Standards Institute (ANSI) X3A1 Committee on Automatic Data Processing was established.[2] The goal of the committee was to design and implement the use of a standard font for document creation. After 6 yr of study and research, the committee adopted a standard character set known as the United States of America Standards Institute (USASI)-A font, more commonly known as "OCR-A" (Fig. 1).[2] This font standard was crucial in the standardization of business document generation practices, as well as the creation of OCR devices designed specifically for the reading of documents using the OCR-A font.

With the new USASI-A standard in place, OCR devices remained relatively unchanged for several decades. They advanced in terms of accuracy, speed, and volume handling, but were still gigantic machines far too expensive to be used by small companies and the average academic institution. What truly catapulted OCR into mainstream usage was the rapid development of computer technologies in the late 1970s and 1980s. As computers became smaller, less expensive, and exponentially more powerful, OCR technologies proved equally adaptable. OCR devices shrunk from giant, hand-fed monstrosities to hand-held and tabletop scanners capable of reading thousands of different fonts and many non-Latin character sets like Cyrillic and Greek.

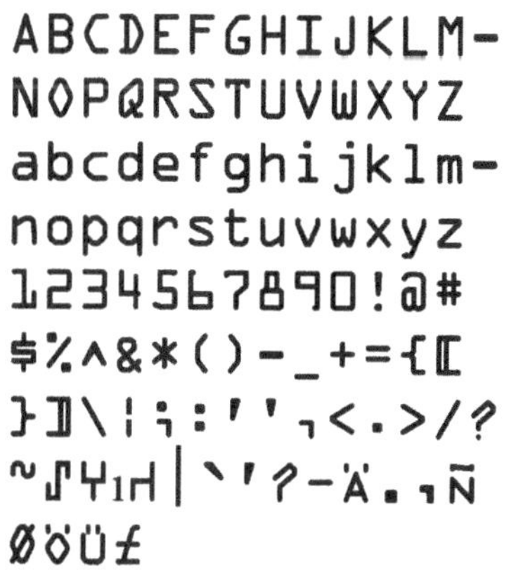

These developments cut costs in business and government, and created new niches in the scholarly community. OCR as a tool for scholarly research has been in use since the early 1990s and is currently one of the most important facets of digital library functionality. Rather than having to rely on complex OCR machines, universities and other research organizations can afford software-based OCR systems. OCR software, which handles the recognition and output processes, is dependent on high-quality scans of documents. Most OCR software also relies on lexicons from various languages and a sophisticated zoning feature that encapsulates a suspected character and determines what it is. Depending on the original document's complexity, the results can be nearly flawless text files. These are frequently embedded in parent images and used to search documents for keywords instantly.[3]

HOW OCR WORKS

At its core, OCR is a mathematical enterprise. OCR technology employs a series of algorithms[1] to determine the existence of characters on a document and to match them to a set of predetermined characters, or a lexicon. For example, were this entry to be scanned and the images run through OCR, each letter, punctuation mark, and number would be suspected and recognized against a lexicon of characters. The output would then be encoded in machine-readable text, i.e., Unicode, and consequently seen by a text editor as human-readable. OCR's sophisticated process has three main components: detection, recognition, and output.

Detection

Depending on whether an OCR machine or OCR software is being used, the act of detection occurs at different points. OCR software works from the digital images of scanned documents. In most cases, flatbed scanners are used to capture the page images, but other solutions such as high-speed scanners and digital cameras are often used depending on the size and format of the document.

In order to increase the readability of the document by the OCR engine, all pages of a document must be scanned at a reasonably high resolution. Monochromatic and grayscale images are typically scanned at 300 dots per inch (dpi), while documents containing highly detailed color plates are scanned at 600 dpi or higher. Generally, the higher the dpi and clearer the image, the less time the OCR engine will have to spend determining the actual size and shape of a particular character. This attention to

Fig. 1 All characters represented by OCR-A

scanning standards is especially important when dealing with mimeographed documents (see section on "Problems and Issues") and text with non-Western characters like Chinese and Arabic. OCR software detects characters by recognizing white space around a suspect, thus determining where it ends and begins. Poorly scanned documents can corrupt this process and cause the OCR engine to incorrectly recognize a character or miss it altogether. Documents scanned at a lower dpi can still be text-recognized, albeit with a significant risk of degradation in the output quality.

In order to more accurately match the original, scanned images are typically formatted as tagged image files (TIFs). This is a lossless file format that best preserves the detail and quality of the document. TIF images are also ideal for preservation in that their expected lifespan of usability as well as their amenability to applications is long and wide. They make ideal files for OCR and are, in a sense, the root of digitization.

Recognition

Once a set of images is prepared, the recognition part of the OCR process can be performed. Typically, each image—representing a single page of a book or document—is recognized separately one item at a time. Accuracy is determined by the overall software design and its specific purpose, e.g., Chinese OCR, and by the original document quality. In older OCR systems, characters were captured and matched against the program's cache of recognizable characters. This caused an unreliable result pattern that was responsible for OCR's reputation as inaccurate and impractical.

Today's OCR engines utilize various algorithms to analyze features of characters, rather than whole character instantiation. These algorithms are designed to evaluate stroke edge, lines of discontinuity between characters, and white space. A suspect is found by the analysis of white space around dark space—or, rather, paper space around a typed character. Once this suspect is found and its features are evaluated, the collected data is then matched to a database of known characters and a guess is offered. This guess, because of the advanced algorithms borrowed from neural network technology, is almost always correct. Many programs go a step beyond this and utilize a revolutionary polling process whereby results are attained through votes. Such OCR programs are equipped with several algorithms that each offer a guess as to what character a particular suspect is; a vote is then cast, the average guess is calculated, and a final decision is reached.

There are a variety of factors that can confound these outcomes (see section on "Problems and Issues") and impact the OCR engine's ability to generate correct text. Accuracy levels, measured in percentages, typically surpass 99% when dealing with well-scanned, plain-print documents. If a document is scanned poorly at low dpi or the original document is rife with ink runs or sloppy typesetting, the accuracy level will drop significantly. Levels of 100% are possible[1] but require that documents be nearly perfect in terms of legibility and format, and that the OCR engine be trained over a period of time with similar documents. For academic uses, and specifically those of libraries, this margin of error is acceptable as text output is mostly used as a basis for text-searchability in digitized artifacts.

Accuracy is a complex idea and not all OCR applications are capable of or even require the same level. Consumer-grade OCR software typically offer high levels of accuracy, but nothing compared to the innovation inherent in government-contracted OCR solutions. This is not a problem; as the average consumer who may simply be employing OCR to capture a term paper or handwritten notes does not require the highest level of accuracy. Government agencies and commercial enterprises where sensitive, personal data is being captured using OCR demands the absolute zenith of accuracy with the smallest margin of error possible.

Output

Once text has been successfully detected and recognized, output files with machine and human-readable characters are generated. Depending on the program being used, output varies. Traditionally, plain American Standard Code for Information Interchange (ASCII) encoded texts are generated as plain text (txt) files capable of being read by anyone via a simple text editor or word processor. These files do not possess much formatting but preserve the text in an easily accessible and preservable form.

Much of the OCR software currently available has the ability to encode files in a variety of encodings including Unicode and its many mapping methods (e.g., UTF-7, 8, 16); ISO/IEC 10646; and simple ASCII. The benefit of encoding in one character set vs. another can be determined by the user's needs. Actual file formats able to be generated by OCR software range from simple text files to Microsoft Word documents, Adobe PDF documents, and many other text and image types.

Once a file type and encoding have been determined, and the ensuing output generated, the text and its images are ready to be preserved and presented. The text files associated with these page images are most often used in libraries to make items text-searchable. In other words, a user of a particular library's online digital library may access a book, and within the constraints of the search field, find any keyword, phrase, or proper name within the pages of that item. Rather than rely on manually generated transcripts of texts, researchers and library patrons can simply link straight to a page containing the keyword or phrase for which they are searching. The actual text file can also be made available for download and edited for research purposes.

OCR TECHNOLOGIES

OCR technologies can be broken down into two basic categories: hardware- and software-based. The latter of the two is the more prominent in the academic world, while the former is found on-site in many large corporations and government agencies. The following sections will overview both categories focusing on the specific uses of these solutions and the typical requirements of their users.

Hardware-Based OCR

Originally, hardware-based OCR was the only solution for those in need of rapid, efficient document processing. Machines were large, expensive, and required a great deal of maintenance. Today hardware-based systems are still in use, but function far more efficiently and cost the user less space and money. Hardware-based systems are implemented when repeated data must be extracted from a large amount of like forms. As previously mentioned in this entry, *Reader's Digest* magazine implemented the first OCR system for such a use. Even today with the advent of electronic forms, hardcopies of older forms or official documents still circulate the corporate and governmental world. These forms contain data that must be harvested quickly and without error.

One of the most well-known implementations of hardware-based OCR is the device used by the USPS to sort and distribute letters and parcels. The USPS utilizes a device called a multiline optical character reader (MLOCR). MLOCRs are similar to the large OCR devices used in the 1960s and 1970s, but are vastly more efficient and designed specifically for large amounts of input that must be sorted. An MLOCR works like any other system in that it detects, recognizes, and produces output. In terms of the USPS, however, an MLOCR scans across many lines on a piece of flat mail for an easily recognizable ZIP code. If the ZIP code is recognized, a barcode representing the postal code is printed on the letter, which is then sorted to its appropriate bin. The shortcoming of the USPS MLOCR, like many OCR devices, is its inability to read handwritten addresses and ZIP codes. If the MLOCR encounters such a letter, a special barcode is applied and the mail is redirected to another area for further processing by more advanced human-operated machines.[4]

Not all hardware OCR systems are comprised of hulking machinery like MLOCRs. Many commercially available and intuitive systems exist that can fit on a user's desk or in the palm of his or her hand. One of the more popular and easy-to-use devices is a pen reader. This OCR device is roughly the size of a fountain pen and is generally capable of reading any typewritten text the scanning element is "drawn" across. The input is typically transferred instantly to the user's personal computer where it can be edited and archived. Pen readers are as efficient as most other hardware solutions but benefit from their portability and facility for capturing and recognizing small sections of a document as opposed to the entire thing.

Between pen readers and large-scale devices like MLOCRs are desktop document readers, medium-sized devices used primarily for quick, medium-volume document capture. Larger businesses might employ several of these devices as a way to process official documents without compromising their validity. Government agencies use document readers to scan and capture images and text from passports and photographic identification cards. These devices are very effective at capturing certain templates of information like passports or official forms that occur in thousands of instances and do not differ greatly in format and quality.

Software-Based OCR

In the academic world, OCR software is the most widely used OCR solution. The software itself can be as expensive as machines, but in many cases can be modified and tailored to meet the specific needs of the user. For academic institutions and specialized libraries, this malleability affords them the ability to have absolute control over the imaging and text-capture processes.

OCR software exists in several echelons of complexity and ability, as well as expense and manipulability. Many commercially available software packages specialize in only one language, while others do a generally acceptable job at recognizing a few hundred. Similarly, some software costs a little over $100 and can be perfectly acceptable for use by a small business or for personal scholarship, but be lacking in flexibility. Many academic libraries opt for open-source OCR because the software's programming can be modified to either improve features or add new ones.

Program modifications are especially important when an academic library, with on-staff programmers, wants to implement a system that can be changed to suit shifting research foci without having to purchase additional specialized systems. Because dedicated software systems focus entirely on recognition and output of text, the sophisticated multi-algorithm voting programs function in a fully dedicated capacity.

Most software OCR systems allow users to specify not only the output type for recognized text, but the encoding as well. Depending on the program design, output can be a variety of Microsoft Word document types, LaTeX (a typesetting format easily integrated with XML), XML, HTML, rich text, plain text, and many more. This allows users to recognize any text and seamlessly export the program's results into a variety of formats for any purposes without having to use dedicated conversion programs.

Within the realm of OCR software a few specialized applications exist. These programs recognize characters like any other OCR system, but are highly specialized in fields that have many nonliteral symbols and marks.

Music OCR

One of the more revolutionary—and troublesome—technologies is optical music recognition or Music OCR. Music OCR systems are used to capture and recognize the musical notations on sheet music and printed scores. The interpretation of sheet music by Music OCR allows the recognized symbols to be placed into an editable format like MusicXML or even a MIDI file, and altered or preserved.

Most Music OCR systems are able to achieve accuracy rates comparable to basic text OCR software, but face some major issues. If a piece of sheet music is simple, with successive notes and no nonstandard symbols or accompaniments, then Music OCR can achieve 99% accuracy. However, if a piece of sheet music contains complex notation or staccato marks (simple dots), artifacts on the page can be mistaken for those symbols creating skewed, inaccurate output. This problem is not unique to Music OCR, but is especially destructive to sheet music recognition since the misinterpretation of a symbol early on in a score can render the entire piece unusable. Postrecognition editing and spot-checking is crucial in maximizing the accuracy of Music OCR output.[5]

Mathematical formula recognition

Like Music OCR, mathematical formula OCR is specialized for users with a specific symbology. Symbols used in expressing equations and mathematical formulae are not included in most encoders and character sets utilized by basic OCR systems. A special field exists to cater to the needs of engineers and mathematicians. Because simple text files are unable to express certain characters, mathematical formula OCR software typically allows the user to save recognized text in more versatile formats like XML, HTML, and LaTeX. The latter being a complex multi-layered format used to design scientific documents with nonstandard expressions. Essentially, mathematical OCR allows users the ability to recognize documents with both text and complex, multiline formulae and output the results to usable, editable files amenable to these expressions.[6]

PROBLEMS AND ISSUES

With all its benefits, there are a few things that OCR technologies are still unable to do, or, rather, do well. Generally when dealing with Western texts, the problems most often encountered are columns and breaks. Columns occur when two separate islands of text are next to one another, but do not influence the other. The most basic of OCR solutions are able to correct for this by recognizing the white space between columns and not mistaking adjacent islands as connected text. Breaks occur when non-text items such as graphs, tables, and figures interrupt the flow of a document, often creating even more confusion by themselves being partially textual. These problems are largely taken care of by more advanced programming and more highly trained OCR handlers. This section will overview the three most pressing concerns in the OCR development community and how these issues are key factors in academic institutions and research libraries.

Handwriting and Intelligent Character Recognition

The recognition of handwritten documents is, perhaps, the largest hurdle for developers of OCR systems. Currently, the technology to recognize and decipher an essentially infinite variety of letter shapes does not exist. Furthermore, cursive handwriting poses an even greater problem when the interconnection of letters is taken into account. OCR typically relies on the identification of white space around characters to separate them for determination of size, shape, and curvature. Cursive characters, however, are usually connected to the previous and following characters, thus making an entire word appear as a single character.[7]

While the majority of documents recognized by corporate entities contain simple, typed characters, academic libraries often find themselves at a loss when digitizing manuscripts. Thousands of historically important handwritten documents are digitized each year but not text-recognized, limiting their usability by scholars and patrons alike. Even though the documents are still of use in their digitized, preserved state, their contents are unsearchable and uneditable, making it difficult for researchers not familiar with the language of said documents to analyze them. Handwritten documents, in many cases, must still undergo a rigorous and time-consuming transcription process. The expense of doing so is compounded by the language and legibility of the document.

To combat the complexities of plain, handwritten text, a cousin of OCR has been under development since the early 1990s. Intelligent character recognition (ICR) is a system based on neural network technology that recognizes handwritten and cursive characters.[8] Handwritten characters are so troublesome because, unlike machine-printed text, they vary by writer and can often appear in several forms within the same word. For example, an individual may write the letter "E" in a certain way five times in a row, but will deviate due to any number of environmental or emotional factors, thus creating a new variant within the same body of work. Handwritten characters are also spaced unevenly, sometimes so close they appear connected; sometimes so distant a computer would not recognize their relation. These factors and more contribute to the many complications faced by ICR systems.

ICR, like OCR, recognizes text character-by-character and starts by segmenting words into their component characters, then matching them against a lexicon. Where ICR differs from OCR, is in its ability to both learn patterns and account for context. Like human readers, ICR has the ability to make a judgment call on an obscure character based on its surroundings. For example, a human reader, if unable to clearly discern whether a character is a "U" or a "V," will make a decision based on the context. ICR manages to perform similar judgments when faced with unclear suspects. Incorporated in ICR software is a lexicon of possible values for a specific field that allows the program to correct itself. As with OCR, however, the process of lexicon-based corrections occurs after a judgment has been made during the recognition phase.

With the ability to learn patterns, ICR can become more efficient over various periods of exposure to similar samples. Where this skill is most useful, at least in the academic world, is in the processing of a certain individual's personal correspondence or notes. When an ICR system is exposed to one writer's style of expressing characters over a period of time, it becomes familiar with that style and can make better judgments when faced with obscure or uncertain suspects. In the business world, this skill is used most often to identify characters a form or application, where characters are corralled by the constraints of the form into relatively equidistant stables.

Cursive text, however, is still the bane of OCR. Many developers have tried to combat the complexities of cursive text by taking a holistic approach. Holistic recognition, rather than character-based analytical recognition, takes into account the entire word. The system finds whole words as suspects and attempts to match them to a lexicon. This approach has proved problematic due to the sheer size of a lexicon needed to attain any level of acceptable accuracy, and the exorbitant space needed to accommodate new words and proper names.[7]

ICR is capable of recognizing cursive text, but cannot do so with the same accuracy OCR delivers when recognizing machine-printed text. The algorithms used to recognize cursive text are designed specifically to deal with the connections between cursive characters. The world "clear" (Fig. 2), e.g., might, because of the proximity between "c" and "l" appear as the word "dear" (Fig. 3). To determine which word it has encountered, ICR might access its lexicon to search for any words that begin with a combination of "c" and "l"; otherwise it will default to "dear." ICR will then create a log of such suspects, alerting the user to possibly spurious results. As such, ICR is still dependent on postrecognition editing by humans.

The OCR of manuscripts and cursive texts is far more fathomable than it was 20 years ago, but research continues to discover more obstacles with every one it surmounts. There is no clear-cut solution, but as technology advances in other spheres, and microcomputing power and efficiency increases, OCR will benefit.

Document Quality

The most widespread issue facing OCR is the quality of the document being recognized. Readability of a document

Fig. 2 Ambiguous cursive word segmented and recognized as "clear"

Fig. 3 Ambiguous cursive word segmented and recognized as "dear"

can be affected by many factors including age of the original compounded by physical degradation; the original having been produced using a manual typewriter causing characters to show variation in pressure and position; original having been carbon copied; and most often, the original having been a low-quality photocopy. These factors and more contribute to the majority of OCR errors. In the academic library world, these problems arise more often because of the antiquity and environmental degradation of the items scanned.

On a per-character level, the most common errors are broken characters and collided characters. Broken characters occur for many reason, but most often because of repeated photocopying of an original, where the photocapture of a character becomes less and less clear, or because of the inherently low level of quality created by mimeograph machines. This is rarely a problem today, as modern printers—either using laser or ink-jet printing—can maintain the darkness and integrity of characters without

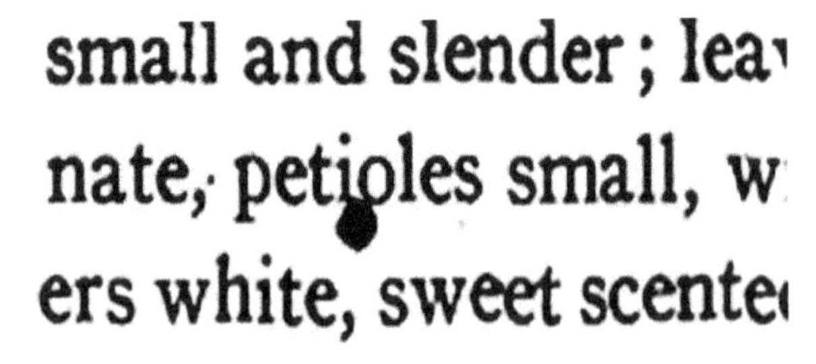

Fig. 4 Ink blot artifact present in mimeographed text
Source: From University of Florida agricultural document *The Mandarin Orange Group* by H. Harold Hume, 1903, http://www.uflib.ufl.edu/UFDC/UFDC.aspx?c=ufir&b=UF00027526&v=00001. H. Harold Hume. "The Mandarin Orange Group." Page image. *University of Florida Digital Collections.* 17 August 2007. http://www.uflib.ufl.edu/UFDC/UFDC.aspx?c=ufir&b=UF00027526&v=00001.

error for long periods of time. When mimeographed, characters, either because of imperfections in the device or poor distribution of ink across rollers, often lose sections or are left out entirely (Fig. 4). This problem is compounded by the quality of the original—often produced using a manual typewriter—as adjacent characters are sometimes too close and appear as one.[9]

OCR accuracy can potentially be affected by interpreting a disfigured character as a completely different one, or simply missing it altogether. In order to correct the errors in OCR output caused by document quality, users have the option to either correct post-OCR, or compensate pre-OCR. Postediting techniques vary by institution, but generally involve a labor-intensive comparison between the original document and the output generated. An individual might have to hand-correct thousands of misread characters, much in the same way a copyeditor would edit a columnist's article. While such a practice certainly strives for quality, in most cases it simply is not practical nor economically feasible.

Certain algorithms can be employed as part of an OCR system's programming to identify and correct errors likely caused by document degradation, but their use is limited and the software expensive. In most cases it makes more sense, fiscally, to allow these errors to exist as part of an allowed margin of error. Many of today's advanced OCR software packages are able to produce such accurate output despite these errors and hand correction is simply not feasible. For certain items that are degraded in particular areas, and if the priority for such action exists, pre-editing is advised. Using photoediting software, users can "touch-up" characters that exhibit the aforementioned errors. If the letter "M" is missing its middle portion, the user can simply draw it in. Also, artifacts caused by photocopying and mimeography can be erased, removing any possibility of blotches and stains being mistaken for characters.[10]

As far as text-searchability is concerned, a certain amount of character misreads is acceptable and, in most cases, does not affect the overall usability of a document.

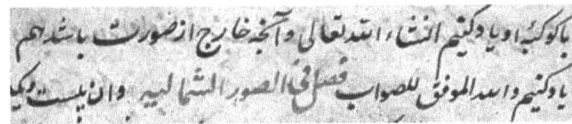

Fig. 5 Two lines of unsegmented, handwritten Arabic calligraphy
Source: Cropped image from *Ajā'ib al-makhlūqāt wa-gharā'ib al-mawjūdāt (Marvels of Things Created and Miraculous Aspects of Things Existing)* by al-Qazwīnī d. 1283/682; retrieved from the Islamic Medical Manuscripts at the National Library of Medicine, http://www.nlm.nih.gov/hmd/arabic/natural_hist5.html. al-Qazwi_ni_d. "Aja_'ib al-makhlu_qa_t wa-ghara_'ib al-mawju_da_t (Marvels of Things Created and Miraculous Aspects of Things Existing)". Page image. *Islamic Medical Manuscripts at the National Library of Medicine.* 20 October 2007. http://www.nlm.nih.gov/hmd/arabic/natural_hist5.html.

Generally, it is up to the institution to decide what level of pre- or postediting a document needs in order to generate the most suitable OCR output.

Foreign Language OCR

Recognition of uniform Western (Latin, Greek, Cyrillic, etc.) characters is widely considered a solved problem. Non-Western characters, especially Chinese, Japanese, and Arabic, are still an oft-researched and highly challenging field. Western characters tend to exhibit a uniform squareness that is most amenable to OCR systems, while many non-Western sets are more ornate, with greater variance in stroke and angle.

Arabic, especially, because of its connected, calligraphic nature, poses the greatest challenge (Fig. 5). Because of the complexities of developing Arabic OCR, commercially available systems are expensive and limited. Nevertheless, estimates indicate that there are upwards of 1 billion readers and users of Arabic script. Despite the commercial viability of such a demand, the technological problems of recognizing Arabic characters abound. Similar to the recognition problems plaguing cursive English text, Arabic is rife with segmentation problems. While it is clear to human readers where one character ends and another begins, a machine has much more difficulty in performing such intelligent separations.[11]

Beyond the lack of separation between characters, an Arabic word can contain one or more subwords; characters themselves can be represented in as many as four different forms, their meaning often dependent on position relative to other characters. These hurdles are not insurmountable by any means, but pose a great challenge to mathematicians and programmers concerned with OCR development. Ultimately it will be intelligent OCR, much like the systems used to recognize Western cursive texts and handwriting, which conquers Arabic script. However, because of the multilayered nature of Arabic script, preprocessing would become essential to creating any semblance of usable output. A system dedicated to identifying the exact direction of strokes from one endpoint to another on an entire line is seen as a possible solution to these issues. These vectors, combined with a database of all possible character classes is key in bringing Arabic OCR to the same level of accuracy and usability as Western-based systems.

CONCLUSION

At its beginning, OCR was a curiosity with little impetus for application. As with most things during the rapid advancement of computer technology, OCR quickly found its place in the growing digital world. Now it is an absolutely essential step in the process of digitization and has found integration in numerous peripheral programs. Aside from simply saving time and money for its users and benefiters, OCR enhances the research value of a document. The text is opened up to searching, data mining, and rapid translation, allowing it to be shared more widely and with greater benefit to the user. While OCR does suffer from a handful of technological limitations created by the current level of neural network sophistication and microcomputing as a whole, several subordinate systems have evolved which help to lessen the effect of these borders. In time these limitations will be lifted, and OCR will likely outgrow its current formats, extending into untapped fields and perhaps others, not yet defined.

REFERENCES

1. Mori, S.; Suen, C.Y.; Yamamoto, K. Historical review of OCR research and development. Proc. IEEE **1992**, *80* (7), 1029–1058.
2. Schantz, H.F. *The History of OCR: Optical Character Recognition*; Recognition Technologies Users Association: Manchester, VT, 1982.
3. Arora, J. Building digital libraries: An overview. DESIDOC Bull. Inform. Technol. **2001**, *21* (6), 3–24.
4. http://www.usps.com/history/history/his3_5.htm, (accessed in October 2007).
5. Bellini, P.; Bruno, I.; Nesi, P. Assessing optical music recognition tools. Comput. Music J. **2007**, *31* (1), 68–93.
6. Uchida, S.; Nomura, A.; Suzuki, M. Quantitative analysis of mathematical documents. Int. J. Doc. Anal. Recog. **2005**, *7* (4), 211–218.
7. Arica, N.; Yaruman-Vural, F.T. Optical character recognition for cursive handwriting. IEEE Trans. Pattern Anal. Mach. Intell. **2002**, *24* (6), 801–813.
8. Fischetti, M. The write type. Sci. Am. **2007**, *296* (6), 96.
9. Hartley, R.T.; Crumpton, K. Quality of OCR for degraded text images. In *Proceedings of the Fourth ACM Conference on Digital Libraries,* Berkeley, CA, 1999; Association for Computing Machinery: New York, 1999, 228–229.
10. Bieniecki, W.; Grabowski, S.; Rozenberg, W. Image preprocessing for improving OCR accuracy. In *MEMSTECH' 2007*, Lviv-Polyana, Ukraine, May 23–26, 2007.
11. Al-Ali, M.; Ahmad, J. Optical character recognition system for Arabic text using cursive multi-directional approach. J. Comput. Sci. **2007**, *3* (7), 549–555.

Optical Coding Schemes: Applications

Wing C. Kwong
Hofstra University, Hempstead, New York, U.S.A.

Guu-Chang Yang
National Chung Hsing University, Taichung City, Taiwan

Abstract
In this entry various potential applications of optical coding along the seven basic optical coding types are discussed.

Keywords: Coding applications; Optical coding.

As hardware technologies mature, optical coding has been gathering attention and proposed for various applications.[1,2,3,4,5,6,7–19] Experimental OCDMA testbeds at 10 Gbit/s have been successfully demonstrated. For example, Brés et al.[3] demonstrated an incoherent, tunable, wavelength-time-coding testbed supporting 8 simultaneous users by means of four-code keying with the carrier-hopping prime codes Hernandez et al.[20] demonstrated a coherent, spectral-phase-coding testbed, supporting 32 simultaneous users by transmitting 8 Walsh codewords in two time slots and two polarizations with forward error correction. Both coherent and incoherent (or synchronous versus asynchronous) coding have their pros and cons. Which one to use depends on the application and operation environment. In fact, they can be used together to complement their deficiencies. For example, coding-based passive optical networks have been recently proposed with the downlink traffic transported by coherent (or synchronous) codes and the uplink traffic carried by incoherent (or asynchronous) codes.[6,7]

Desirable features of optical coding, in particular 2-D spectral-temporal amplitude coding, include dynamic bandwidth assignment, efficient in bursty traffic, asynchronous and uncoordinated statistical multiple-user access, high scalability for supporting more users by simply adding codewords, flexible code-cardinality enlargement by means of increasing the number of wavelengths and/or time slots independently, performance degradation gracefully under heavy traffic, potential data obscurity, and the support of multiple bit rates, variable QoS, multimedia services. It is predictable that optical coding, in particular O-CDMA, is slowly replacing OTDMA and WDMA as means of contention resolution in optical systems, and gradual upgrade strategies from O-TDMA or WDMA to O-CDMA are being sought. For example, O-CDMA has been proposed to incorporate into various network applications, such as local and metropolitan area networks, burst-mode switching, ring networks, passive optical networks, optical interconnects, optical wireless, optical interconnects, optical microarea networks, and O-CDMA-to-WDM gateways for long-haul WDM backbone.[1,2,3,5,7,8,10–12,14,16] In addition, optical coding finds other applications, such as IP-routing, in-service monitoring, and fiber fault surveillance in optical networks and sensor identification in fiber-sensor systems, which use optical codes for the purpose of address or user identification.[9,13,15,19]

Last but not least, optical coding theory has unexpectedly found application in the area of preventing four-wave-mixing crosstalk in high-capacity, long-haul, repeaterless, WDM lightwave systems due to fiber nonlinearities. While Forghieri et al.[21,22] proposed the use of unequal channel spacings in order to prevent four-wave-mixing crosstalk from falling in wavelength channels and solved for the channel spacings by means of integer linear programming with computer exhaustive search, Kwong et al.[23,24] formulated optimal solutions algebraically by recognizing that the problem was identical to the construction 1-D temporal-amplitude codes with ideal autocorrelation sidelobes.

SUMMARY

In this entry, various coding techniques and enabling hardware technologies of the seven major categories of optical coding schemes were investigated. Supporting the transmission of multirate, multimedia services by means of multiple-length codes, and increasing bit rate by means of multicode and shifted-code keying were also reviewed. AWG- and FBG-based programmable encoder/decoder designs were investigated. Finally, the potential applications of optical coding were discussed.

Among these major optical coding schemes, 2-D spectral-temporal amplitude (or wavelength-time) coding is found to be the most advantageous because of 1) simplicity: asynchronous access (no need of global clock), little scheduling, supporting bursty traffic, gradual performance degradation under heavy load, and less sensitive to fiber nonlinearities; 2) ease of implementation: tunable and supporting dynamic services, such as variable QoS and multiple data rates; 3) larger code size and flexible coding in wavelength and time independently; 4) more functionalities: supporting multiple bit rates and QoS by varying code length and weight, better obscurity by means of hopping in wavelength and time, and supporting multicode and shifted-code keying for better spectral efficiency and code obscurity; and 5) better scalability and compatibility: trading bandwidth for scalability, compatible with WDM technology, and supporting overlay and gradual upgrade strategies.

REFERENCES

1. Yang, G.-C.; Kwong, W.C. *Prime Codes with Applications to CDMA Optical and Wireless Networks*; Artech House: Norwood, MA, 2002.
2. Prucnal, P.R. (ed.) *Optical Code Division Multiple Access: Fundamentals and Applications*; Taylor & Francis Group: Boca Raton, FL, 2006.
3. Brés, C.-S.; Glesk, I.; Prucnal, P.R. Demonstration of an eight-user 115-Gchip/s incoherent OCDMA system using supercontinuum generation and optical time gating. IEEE Photonic. Technol. Lett. **2006**, *18* (7), 889–891.
4. Brés, C.-S.; Prucnal, P.R. Code-empowered lightwave networks. J. Lightwave Technol. **2007**, *25* (10), 2911–2921.
5. Stok, A.; Sargent, E.H. System performance comparison of optical CDMA and WDMA in a broadcast local area network. IEEE Commun. Lett. **2002**, *6* (9), 409–411.
6. Hu, H.-W.; Chen, H.-T.; Yang, G.-C.; Kwong, W.C. Synchronous Walsh-based bipolar-bipolar code for CDMA passive optical networks. J. Lightwave Technol. **2007**, *25* (8), 1910–1917.
7. Lundqvist, H.; Karlsson, G. On error-correction coding for CDMA PON. J. Lightwave Technol. **2005**, *23* (8), 2342–2351.
8. Menendez, R.C.; Toliver, P.; Galli, S.; Agarwal, A.; Banwell, T.; Jackel, J.; Young, J.; Etemad, S. Network applications of cascaded passive code translation for WDM-compatible spectrally phase-encoded optical CDMA. J. Lightwave Technol. **2005**, *23* (10), 3219–3231.
9. Yeh, C.; Chi, S. Optical fiber-fault surveillance for passive optical networks in S-band operation window. Opt. Express **2005**, *13* (14), 5494–5498.
10. Meenakshi, M.; Andonovic, I. Code-based all optical routing using two-level coding. J. Lightwave Technol. **2006**, *24* (4), 1627–1637.
11. Khattab, T.; Alnuweiri, H. Optical CDMA for all-optical sub-wavelength switching in core GMPLS networks. J. Sel. Areas Commun. **2007**, *25* (5), 905–921.
12. Farnoud, F.; Ibrahimi, M.; Salehi, J.A. A packet-based photonic label switching router for a multirate all-optical CDMA-based GMPLS switch. IEEE J. Sel. Top. Quantum Electron. **2007**, *13* (5), 1522–1530.
13. Fathallah, H.A.; Rusch, L.A. Code division multiplexing for in-service out-of-band monitoring of live FTTH-PONs. J. Opt. Networking **2007**, *6* (7), 819–829.
14. Ghaffari, B.M.; Matinfar, M.D.; Salehi, J.A. Wireless optical CDMA LAN: Digital design concepts. IEEE Trans. Commun. **2008**, *56* (12), 2145–2155.
15. Rad, M.M.; Fathallah, H.A.; Rusch, L.A. Fiber fault monitoring for passive optical networks using hybrid 1-D/2-D coding. IEEE Photonic. Technol. Lett. **2008**, *20* (24), 2054–2056.
16. Sowailem, M.Y.S.; Morsy, M.H.S.; Shalaby, H.M.H. Employing code domain for contention resolution in optical burst switched networks with detailed performance analysis. J. Lightwave Technol. **2009**, *27* (23), 5284–5294.
17. Beyranvand, H.; Salehi, J.A. All-optical multiservice path switching in optical code switched GMPLS core network. J. Lightwave Technol. **2009**, *27* (12), 2001–2012.
18. Deng, Y.; Wang, Z.; Kravtsov, K.; Chang, J.; Hartzell, C.; Fok, M.P.; Prucnal, P.R. Demonstration and analysis of asynchronous and survivable optical CDMA ring networks. J. Opt. Commun. Networking **2010**, *2* (4), 159–165.
19. Rad, M.M.; Fathallah, H.A.; Rusch, L.A. Fiber fault PON monitoring using optical coding: Effects of customer geographic distribution. IEEE Trans. Commun. **2010**, *58* (4), 1172–1181.
20. Hernandez, V.J.; Cong, W.; Hu, J.; Yang, C.; Fontaine, N.K.; Scott, R.P.; Ding, Z.; Kolner, B.H.; Heritage, J.P.; Yoo, S.J.B. A 320-Gb/s capacity (32-user × 10 Gb/s) SPECTS O-CDMA network testbed with enhanced spectral efficiency through forward error correction. J. Lightwave Technol. **2007**, *25* (1), 79–86.
21. Forghieri, F.; Tkach, R.W.; Chraplyvy, A.R.; Marcuse, D. Reduction of four-wave mixing crosstalk in WDM systems using unequally spaced channels. IEEE Photonic. Technol. Lett. **1994**, *6* (6), 754–756.
22. Forghieri, F.; Gnauck, A.H.; Tkach, R.W.; Chraplyvy, A.R.; Derosier, R.M. Repeaterless transmission of eight channels at 10 Gb/s over 137 km (11 Tb/s-km) of dispersion-shifted fiber using unequal channel spacing. IEEE Photonic. Technol. Lett. **1994**, *6* (11), 1374–1376.
23. Kwong, W.C.; Yang, G.-C. An algebraic approach to the unequal-spaced channel-allocation problem in WDM lightwave systems. IEEE Trans. Commun. **1997**, *45* (3), 352–359.
24. Chang, K.-D.; Yang, G.-C.; Kwong, W.C. Determination of FWM products in unequal-spaced-channel WDM lightwave systems. J. Lightwave Technol. **2000**, *18* (12), 2113–2122.

Optical Coding Schemes: Basics

Wing C. Kwong
Hofstra University, Hempstead, New York, U.S.A.

Guu-Chang Yang
National Chung Hsing University, Taichung City, Taiwan

Abstract

Since the mid-1980s, there have been many advances in coding-based optical systems and networks. There are two main categories of optical coding schemes— synchronous and asynchronous, with synchronization occurring in either time or wavelength. Both approaches have certain advantages and disadvantages. But, asynchronous coding, in general, allows simultaneous users to access the same optical transmission medium independently with no wait time, scheduling, or coordination. Based on the choices of signaling format, detection method, synchronization requirement, coding domain, and code format, optical coding schemes can be generally classified into seven main categories:

1. Temporal amplitude coding (1-D)
2. Temporal phase coding (1-D)
3. Spectral amplitude coding (1-D)
4. Spectral phase coding (1-D)
5. Spatial-temporal amplitude coding (2-D)
6. Spectral-temporal amplitude coding (2-D)
7. Spatial/polarization-spectral-temporal coding (3-D)

In this entry, 1-D temporal amplitude coding is discussed.

Keywords: 1-D coding; Amplitude coding; Optical coding; Temporal coding.

Figure 1 outlines the basic configuration of a typical coding-based optical system and network. It consists of multiple stations (or users) linked to a shared optical medium via optical fibers or free space.[1,2] The shared medium usually consists of an optical multiplexing/demultiplexing device, such as a star coupler. The device is used to combine optical codewords from and then distribute to all stations. Each station consists of a pair of optical transmitter and receiver, whose structures depend on the coding scheme in use.

In incoherent on-off keying (OOK) modulation, a user sends out an optical codeword corresponding to the address (signature) codeword of its intended receiver for a data bit 1, but nothing for a data bit 0. Figure 2a shows the timing diagrams of a continuous stream of optical clock pulses of width T_c and repetition rate $1/T$. Figure 2b shows an example of low-bandwidth nonreturn-to-zero electrical data bits of period T at one station in Fig. 1. Electro-optic OOK conversion is performed using the voltages of data bits to control the opening and closing of the intensity modulator in the transmitter.[1,2] Every data bit 1, which carries a high voltage, will close the switch and let pass one clock pulse, resulting in a high-bandwidth data-modulated optical signal, as shown in Fig. 2c. Assume that 1-D time-spreading binary (0, 1) codewords are used as the address codewords of the stations. To accommodate the 1s and 0s of the codewords, each bit period T is subdivided into a number of time slots (or so-called chips) of width T_c, giving code length $N = T/T_c$. Following the data-bit pattern in this illustration, Fig. 2d shows the corresponding time-spreading binary (0, 1) codewords generated by an optical encoder. Finally, these codewords are multiplexed with the codewords from all stations at the star coupler, as exemplified in Fig. 2e.

At each receiver, an optical decoder, which operates as an inverse filter of its corresponding optical encoder, correlates its own address (signature) codeword with any received codeword. Assuming chip synchronism, the decoder output is written as a discrete periodic correlation function, according to Section 1.5. If a codeword arrives at the correct destination, an autocorrelation function with a high peak is generated. Otherwise, the codeword is treated as interference and a cross-correlation function results. So, it is necessary to maximize the autocorrelation peaks but to minimize the cross-correlation functions in order

Optical Coding Schemes: Basics

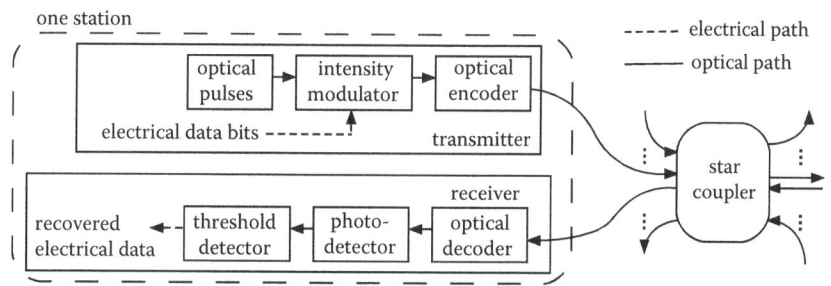

Fig. 1 A typical coding-based optical system model
Source: Reproduced with permission from Guu-Chang Yang and Wing C. Kwong, *Prime Codes with Applications to CDMA Optical and Wireless Networks*, Norwood, MA: Artech House, Inc., 2002. © 2002 by Artech House, Inc.

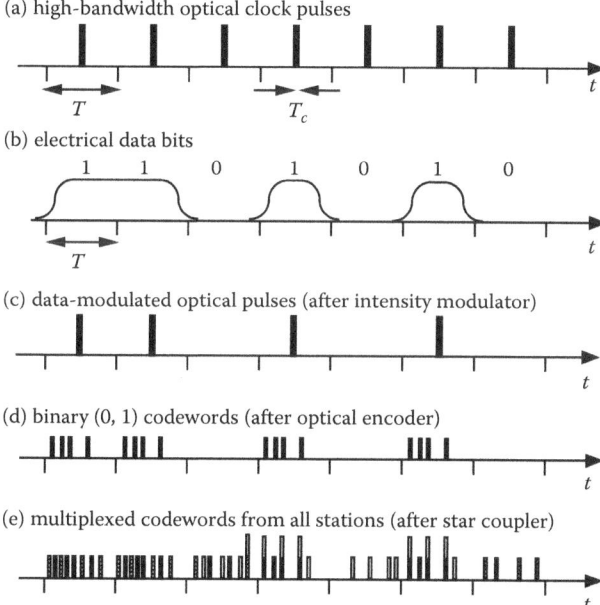

Fig. 2 Signal formats at various stages of an optical transmitter
Source: Reproduced with permission from Guu-Chang Yang and Wing C. Kwong, *Prime Codes with Applications to CDMA Optical and Wireless Networks*, Norwood, MA: Artech House, Inc., 2002. © 2002 by Artech House, Inc.

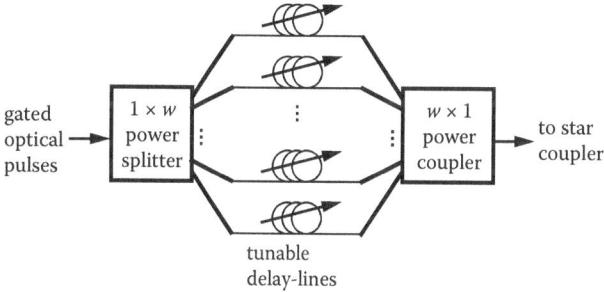

Fig. 3 Tunable optical encoder in a parallel coding configuration[2,3]

to optimize the discrimination between the correct codewords and interfering codewords. An electrical pulse is generated whenever an autocorrelation peak is detected at the threshold detector. Such a pulse indicates the reception of a data bit 1; otherwise, a data bit 0 is recovered.

Prucnal et al.[3] proposed the first-of-its-kind tunable incoherent optical encoder, in which a parallel configuration of fixed fiber-optic delay-lines and intensity modulators was used, for temporal amplitude coding. Figure 3 shows the block diagram of a revised version of such a parallel design. The encoder consists of a $1 \times w$ optical power splitter, a set of w electronically tunable (fiber-optic or waveguide) delay-lines, and a $w \times 1$ optical power coupler, where w is the code weight (or number of pulses) of the 1-D time-spreading binary (0, 1) codes in use. At the encoder input, the power splitter divides an incoming gated optical pulse, which represents the transmission of a data bit 1, into w pulses. These w pulses are delayed individually by their own tunable delay-lines, according to the locations of the mark chips (or binary 1s) in the address codeword of the intended receiver. Finally, these pulses are combined by the coupler to form the desired codeword for transmission.

As fixed delay-lines were assumed in Prucnal's original design,[3] tunability was achieved by first generating pulses in all possible time-delays and then selecting only w of them in according to the address codeword of the intended receiver. So, the original setup required the use of a $1 \times N$ power splitter, a set of N fixed (fiber-optic or waveguide) delay-lines, N intensity modulators (one per line), and an $N \times 1$ power coupler, where N is the code length. Each of these N delay-lines takes a distinct value from the set of $\{0, T_c, 2T_c, \ldots, (N-1)T_c\}$, where T_c is the chip-width. The w properly delayed pulses are selected by closing w intensity modulators.

The setup of the decoder is identical to that of the encoder, except that only w fixed-delay paths are needed if a fixed-address-receiver configuration is assumed. The decoder correlates incoming codewords by accumulating the reversely delayed pulses to give a high autocorrelation peak if there exists a codeword matching the decoder's address codeword. Otherwise, a low cross-correlation function results and is considered as interference.

The advantage of this parallel coding configuration is the capability of performing incoherent optical processing without the need for fast-response photodetectors or high-speed electronics. Also, it can universally generate any 1-D time-spreading binary (0, 1) code, However, this kind of parallel structure creates a very stringent power requirement because each optical pulse is split into many pulses in the encoders and decoders. In addition, the number of delay-lines adds up to a huge number, and this kind of parallel coding devices is usually bulky and lossy.

Figure 4 shows the block diagram of a tunable incoherent optical encoder in a serial coding configuration.[2,4,5] This setup improves the power, size, and delay-length requirements of the parallel configuration. The encoder consists of a series of 2×2 optical switches, which are connected with two separate (fiber-optic or waveguide) delay-lines between two adjacent switches. Each pair of delay-lines generates a differential time-delay of one chip-width T_c. The DC-bias voltages of the 2×2 optical switches are individually controlled so that each switch is configured to operate in either 3-dB or bar state independently. At an encoder input, the power splitter divides an incoming gated optical pulse (of width T_c), which represents the transmission of a data bit 1, into two pulses. The amount of differential time-delays is accumulated as these two pulses pass through a series of bar-state switches. The bar state allows optical pulses at the two inputs of a switch to directly exit the corresponding outputs without any change. After the proper amount of differential time-delays has been created, both pulses are combined at a 3-dB-state switch and then split into four pulses—two pulses at each output. Functioning like a 2×2 passive coupler, a 3-dB state switch duplicates optical pulses arriving at its two inputs and makes them available at the two outputs. The process repeats until the desired number of pulses, which are properly arranged according to the address codeword of the intended receiver, is generated. If there are n switches in the 3-dB state, a time-spreading binary (0, 1) codeword with 2^n pulses can be generated. For example, if the second, sixth, and thirteenth switches are set to the 3-dB state, 2, 6, and 13 chips of differential time-delays are accumulated to generate the codeword 1010001010000101000101.

The setup of the optical decoder is similar to that of the encoder, except that the 3-dB-state 2×2 optical switches are now replaced by 2×2 passive couplers if a fixed-address-receiver configuration is assumed. Only $n+1$ 2×2 passive couplers are needed, and the delay-lines between two couplers are used to generate the differential time-delays of two groups of pulses directly. For example, to have an address (signature) codeword of 1010001010000101000101 in the decoder, four 2×2 passive couplers are needed, and the delay-lines between two couplers are arranged to generate the differential time-delays of 2, 6, and 13 chips, correspondingly.

While this serial configuration improves the power, size, and delay-length requirements of the parallel configuration, it can only generate certain families of 1-D time-spreading binary (0, 1) codes that have replicative pulse separations, such as 1-D even-spaced codes, 2^n codes, and 2^n prime codes in Section 3.5.[2,4,5]

Making use of the block structure of some optical codes, such as the 1-D prime codes the power splitting/combining loss and number of optical switches in the serial coding configuration can be reduced substantially. For instance, each of the prime codewords (in Section 3.1) of length p^2 and weight p can be divided into p blocks, and each block has a single one (or a pulse) and $p-1$ 0s, where p is a prime number. If all p pulses in a prime code can be generated by a laser with a repetition rate p/T, the power loss created by pulse splitting and combining at the optical splitter and switches can be avoided, where T is the bit period, as explained in the following.

Shown in Fig. 5 is an improved serial tunable encoder. It consists of a series of $L+1$ 2×2 optical switches, which are connected with two separate (fiber-optic or waveguide) delay-lines between two adjacent switches, where $L = \lceil \log_2 p \rceil$ and $\lceil \cdot \rceil$ is the ceiling function.[2,4,5] The DC-bias voltages of the 2×2 optical switches are individually controlled so that each switch is configured to operate in either cross or bar state independently. While the bar state allows an optical pulse at an input of a switch to directly exit the corresponding output, the cross state allows the pulse to cross over to the opposite output. The differential time-delays are all distinct and assigned with values equal to the product of the chip-width T_c and consecutive powers of 2. This design is based on the tunable O-TDMA coder proposed by Prucnal et al.,[6,7] in which any discrete time delay of $\{0, T_c, 2T_c,..., (p-1)T_c\}$ chips can be generated by setting the 2×2 optical switches in either cross or bar states, accordingly. The last 2×2 optical switch is used to route the properly delayed pulse to the output of the encoder. With this last switch in place, the use of intensity modulators for performing electro-optic conversion of data bits (see Figs. 1 and 2) can be eliminated. This is because all p pulses within a bit

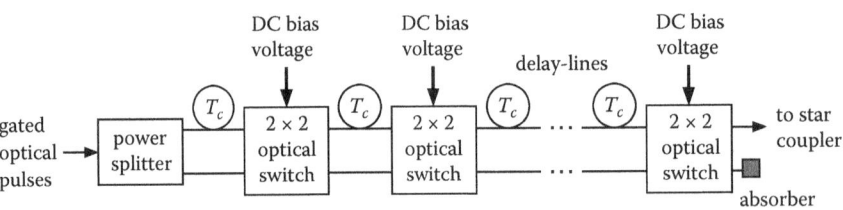

Fig. 4 Tunable optical encoder in a serial coding configuration[2,4,5]

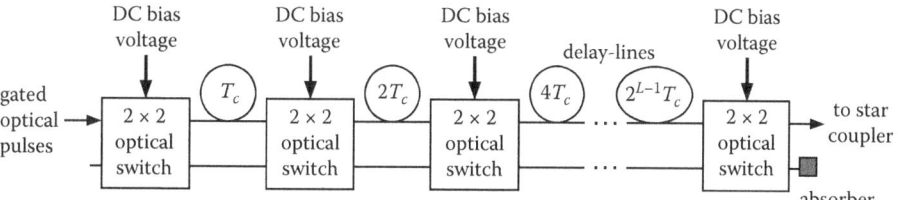

Fig. 5 Tunable incoherent optical encoder in an improved serial coding configuration[2,4,5]

period can be simply routed to the unused output of the last switch for every data bit 0. Similarly, as in the 2^n prime codewords in Section 3.5, a block may have no pulse but all p 0s, any optical pulse within such a block can also be routed to this unused output.

Because optical pulses are now entering this improved serial encoder at a rate of p/T, p-fold increases in the speed of electronics and the repetition rate of lasers are required, as compared to the serial design in Fig. 4. Nevertheless, there is no power loss due to splitting and combining of optical pulses, and only $\lceil \log_2 p \rceil + 1$ optical switches are required in the improved serial design, resulting in substantial cost savings and size reduction, and making the design more suitable for waveguide implementation.

1-D TEMPORAL PHASE CODING

Temporal amplitude coding can only accommodate 1-D binary (0, 1) codes and is restricted to the use of incoherent processing and detection because optical intensity is used for transmission. This kind of incoherent system is usually asynchronous in nature because these optical codes, such as the prime codes in Chapter 3, are designed to operate without system-wise time synchronization. Due to non-scheduled transmission and nonzero mutual interference, temporal amplitude coding supports only a limited number of subscribers and even fewer simultaneous users before a rapid deterioration of code performance occurs. By introducing 0 or π phase shifts to optical pulses, temporal phase coding supports orthogonal bipolar (−1, +1) codes, such as maximal-length sequences and Walsh codes,[8,9] with zero (in-phase) cross-correlation functions. Rather than using OOK, temporal phase coding transmits a conjugated form of the bipolar codewords in use for data bit 0s, resulting in better code performance than temporal amplitude coding.[10–15] However, this kind of coherent system is usually synchronous in nature because bipolar codes require system-wise synchronization in order to maintain code orthogonality. Also, the need for phase preservation in optical fibers often hinders the development of temporal phase coding.

Figure 6 shows a typical tunable coherent encoder and decoder for temporal phase coding.[10–15] Assume that bipolar codes of length and weight of both equal to N are used. A narrow optical pulse is first split into N pulses.

These pulses are then delayed by fixed (fiber-optic or waveguide) delay-lines, and 0 or π phase shifts are introduced by tunable phase modulators. These phase-coded pulses are finally recombined to form a temporal-phase codeword. In this scheme, a station (or user) transmits the address (bipolar) codeword of its intended receiver for a data bit 1, but a data bit 0 is conveyed by the same codeword with conjugated phase shifts. In a receiver, phase tracking and correlation of arriving bipolar codewords (from all stations) with the receiver's address codeword are usually done by optical heterodyne detection in a phase-locked loop, as shown in Fig. 6b.[10,12,13] The recovery of data bits is performed by a balanced photodetector and an electronic, integrating, sampling, and thresholding circuit.

Another type of temporal-phase-coding device consists of superstructured fiber Bragg gratings.[16,17] Phase modulation is performed by segments of (0 or π) phase-shifted fiber Bragg gratings placed inside a piece of optical fiber. The locations of the segments determine the spacings of the pulses and, in general, are not easily tunable.

1-D SPECTRAL PHASE CODING

Because spectral phase coding transmits phase-modulated codes in the wavelength domain, the code length is independent of data rate, and codes are inherently synchronous as long as the coded spectra all align to a common wavelength reference plane. In this kind of scheme, each station (or user) is assigned an orthogonal bipolar codeword, such as maximal-length sequences and Walsh codewords, as its address signature. A user transmits the (bipolar) address codeword of its intended receiver for a data bit 1, but a data bit 0 is conveyed by the same codeword with conjugated phase shifts. As shown in Fig. 7, a narrow, broadband optical pulse is first dispersed in wavelengths by gratings.[18–21] Phase coding is then performed by passing spectral components through a phase mask, in which the pixels can be made electronically programmable with the use of liquid-crystal-based phase modulators. The phase-coded spectral components are finally recombined by (inverse) gratings to form a spectral-phase codeword. The length of the bipolar codes in use is determined by the wavelength resolutions of the gratings and phase mask. The decoder has the same setup as the encoder but with a conjugated phase mask.

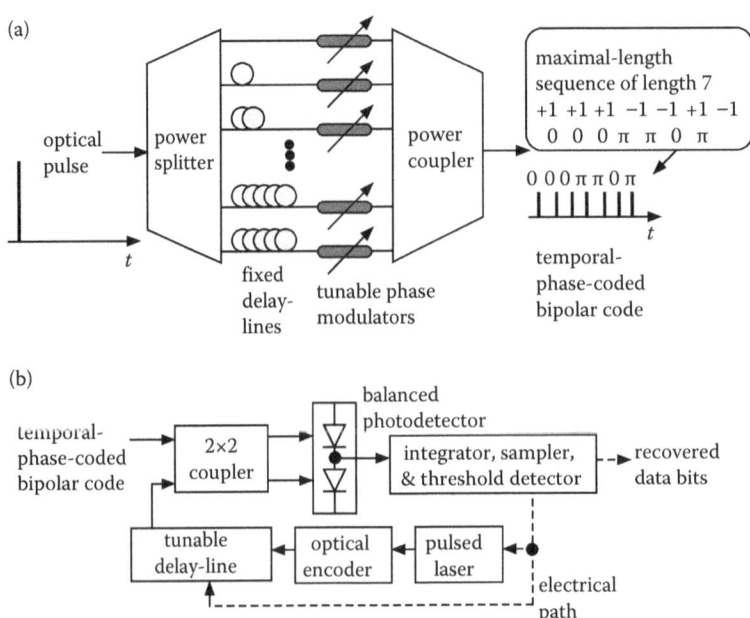

Fig. 6 Temporal phase coding: (a) tunable encoder; (b) tunable decoder[10–15]

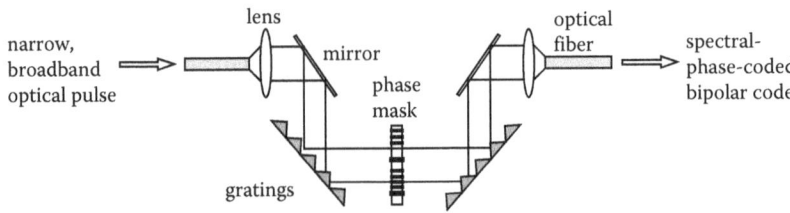

Fig. 7 Spectral phase coding in free space[18–21]

In addition to free-space, spectral phase coding can also be performed in waveguide and fiber. In Fig. 8a, spectral spreading is done by a wavelength-division-multiplexing (WDM) demultiplexer (DEMUX), such as thin-film filters, arrayed-waveguide-grating devices, holographic Bragg reflectors, and micro-disk resonators.[22–28] Phase coding is then performed by passing the spectral components through tunable phase modulators. The phase-coded spectral components are finally recombined by a wavelength multiplexer (MUX) to form a spectral-phase codeword.

Similarly, based on the time-domain spectral-phase encoder by Gao et al. in Ref. [20] spectral spreading can be done by placing segments of fiber Bragg gratings (FBGs) with different center wavelengths inside a piece of optical fiber, as shown in Fig. 8b. Those spectral components matching the center wavelengths get reflected back to the fiber input and routed through an optical circulator to a rapidly tunable phase modulator. Because different FBGs are placed at different locations of the fiber, time delays (or time spreading) are introduced to the reflected wavelengths. The phase modulator then performs 0 or π phase shifts to the time-spreading wavelengths one-by-one, according to the (bipolar) address codeword of the intended receiver. A second piece of optical fiber with reversed placement of FBGs as those of the first piece of optical fiber is used to (time) realign the phase-coded spectral components, finally forming the desired spectral-phase codeword.

1-D SPECTRAL AMPLITUDE CODING

In spectral amplitude coding, OOK is assumed as each data bit 1 is conveyed by a bipolar (−1, +1) codeword, but data 0s are not transmitted.[14,15,29–33] In addition, amplitude coding means that optical intensity is used for transmission. So, only the "+1" elements of bipolar codewords are transmitted in wavelengths but the "−1" code elements are not. The three kinds of spectral-phase encoders in "1-D Spectral Phase Coding" section can be modified for spectral amplitude coding. For the free-space encoder, amplitude masks with transparent and opaque pixels are used in lieu of phase masks. For the waveguide-type encoder, intensity modulators replace phase modulators, as shown in Fig. 9a. For the fiber-based encoder,[29,32,33] only fiber Bragg gratings (FBGs) with center wavelengths matching those of the "+1" elements of the bipolar codeword in use exist inside the two pieces of optical fiber, as shown in Fig. 9b. The wavelengths matching those of the "+1" code elements

Optical Coding Schemes: Basics

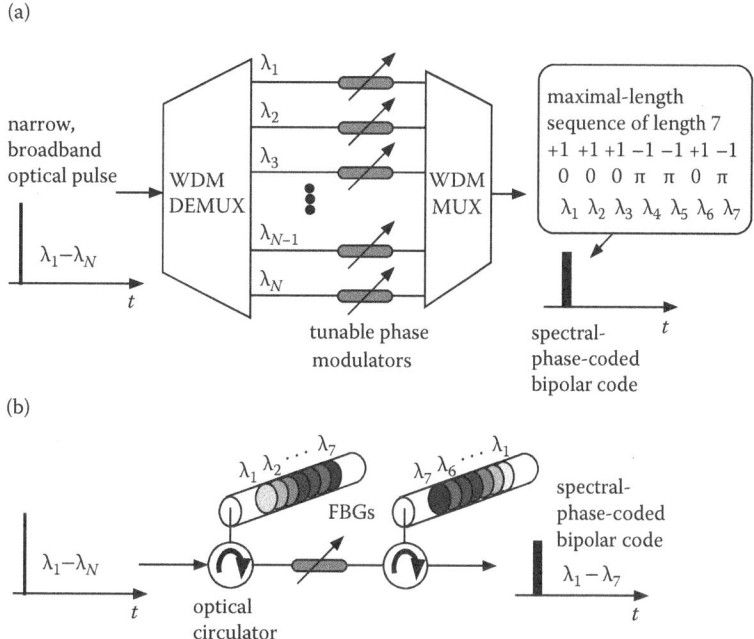

Fig. 8 Spectral phase coding in (a) waveguide and (b) fiber[20–25]

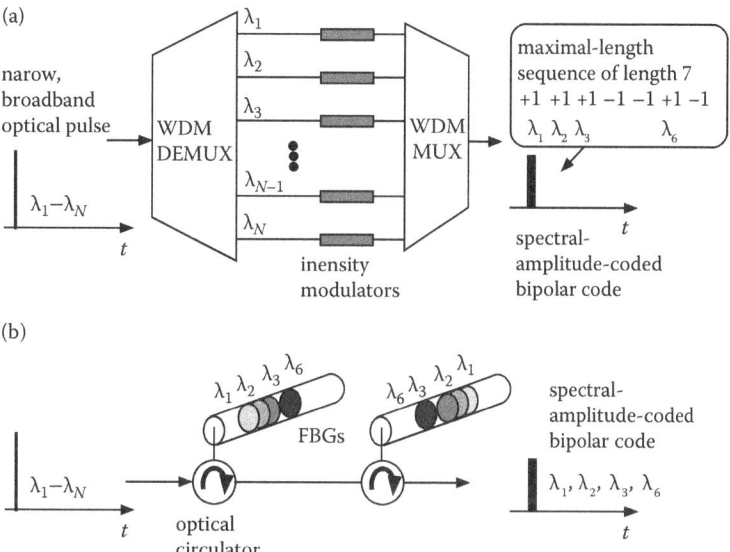

Fig. 9 Spectral amplitude coding in (a) waveguide and (b) fiber

are reflected back to the input of the first piece of optical fiber and then routed to the second piece of optical fiber for (time) realignment.

Because spectral amplitude coding only transmits the "+1" elements of the bipolar codewords, an optical differential receiver, which consists of a "true" decoder, a "conjugated" decoder, and a pair of balanced photodetectors, is designed to emulate the "−1" code elements.[15,27,30,32,33] The decoders, which function as inverse filters, are made of the same structure as the encoders. The true decoder is used to collect incoming wavelengths that match those wavelengths of the "+1" elements of its (bipolar) address codeword, and the conjugated decoder collects the rest of the incoming wavelengths. The "−1" code-elements are emulated at the balanced photodetectors, which allow the electrical current obtained from the true decoder being subtracted by the electrical current obtained from the conjugated decoder.

For example, the spectral-amplitude differential receiver for the maximal-length sequence (+1, +1, +1, −1, −1, +1, −1) of length 7 is shown in Fig. 10. The true decoder detects wavelengths λ_1, λ_2, λ_3, and λ_6, where λ_i is the i th wavelength used to carry the i th sequence element that is equal to "+1." The conjugated decoder then detects wavelengths

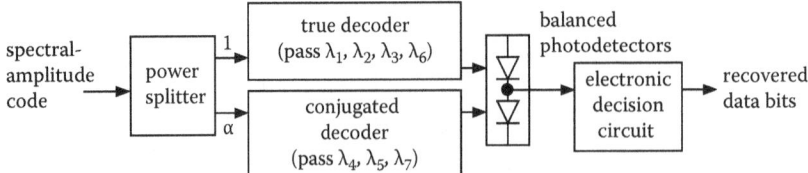

Fig. 10 Spectral-amplitude differential receiver for the maximal-length sequence (+1, +1, +1, −1, −1, +1, −1), where α is the output ratio of the power splitter, which is equal to 1 when orthogonal bipolar codes are in use[15,29,30,32,33]

station	address code	bit	transmitted signal
1	$\lambda_1\ \lambda_2\ \lambda_3\ 0\ 0\ \lambda_6\ 0$	1	$\lambda_1\ \lambda_2\ \lambda_3\ 0\ 0\ \lambda_6\ 0$
2	$0\ \lambda_2\ \lambda_3\ \lambda_4\ 0\ 0\ \lambda_7$	0	$0\ 0\ 0\ 0\ 0\ 0\ 0$
3	$\lambda_1\ 0\ \lambda_3\ \lambda_4\ \lambda_5\ 0\ 0$	1	$\lambda_1\ 0\ \lambda_3\ \lambda_4\ \lambda_5\ 0\ 0$
4	$0\ \lambda_2\ 0\ \lambda_4\ \lambda_5\ \lambda_6\ 0$	1	$0\ \lambda_2\ 0\ \lambda_4\ \lambda_5\ \lambda_6\ 0$
5	$0\ 0\ \lambda_3\ 0\ \lambda_5\ \lambda_6\ \lambda_7$	1	$0\ 0\ \lambda_3\ 0\ \lambda_5\ \lambda_6\ \lambda_7$
6	$\lambda_1\ 0\ 0\ \lambda_4\ 0\ \lambda_6\ \lambda_7$	0	$0\ 0\ 0\ 0\ 0\ 0\ 0$
7	$\lambda_1\ \lambda_2\ 0\ 0\ \lambda_5\ 0\ \lambda_7$	1	$\lambda_1\ \lambda_2\ 0\ 0\ \lambda_5\ 0\ \lambda_7$

multiplexed signal = $3\lambda_1\ 3\lambda_2\ 3\lambda_3\ 2\lambda_4\ 4\lambda_5\ 3\lambda_6\ 2\lambda_7$

	receiver 1		receiver 2	
	autocorrelation	intensity	cross-correlation	intensity
true decoder output	$3\lambda_1\ 3\lambda_2\ 3\lambda_3\ 0$ $0\ 3\lambda_6\ 0$	12	$0\ 3\lambda_2\ 3\lambda_3\ 2\lambda_4$ $0\ 0\ 2\lambda_7$	10
cojugated decoder output	$0\ 0\ 0\ 2\lambda_4$ $4\lambda_5\ 0\ 2\lambda_7$	8	$3\lambda_1\ 0\ 0\ 0$ $4\lambda_5\ 3\lambda_6\ 0$	10
balanced photo-detector output		4		0

Fig. 11 Example of the spectral-amplitude decoding process with zero mutual interference, where the maximal-length sequences of length 7 are used

λ_4, λ_5, and λ_7. Assume that the cyclic-shifted versions of this maximal-length sequence are used as the other bipolar codewords; they support at most 7 possible subscribers (or stations). Figure 11 illustrates an example of the autocorrelation and cross-correlation processes in the receivers of stations 1 and 2, respectively. In this example, representing the transmission of data bit 1s, the spectral-amplitude (address) codewords of stations 1, 3, 4, 5, and 7 are being transmitted simultaneously, multiplexed, and distributed to the receivers of all stations. While the receivers of stations 1, 3, 4, 5, and 7 expect to see autocorrelation functions, the receivers of stations 2 and 6 will see cross-correlation functions because no matching (address) codeword is transmitted to these two receivers for data bit 0s in OOK. So, the wavelengths arriving at receivers 1 and 2 at one time instant are $3\lambda_1$, $3\lambda_2$, $3\lambda_3$, $2\lambda_4$, $4\lambda_5$, $3\lambda_6$, and $2\lambda_7$. For receiver 1, the total number of wavelengths detected at the upper photodetector is 12 (from $3\lambda_1 + 3\lambda_2 + 3\lambda_3 + 3\lambda_6$) and that at the lower photodetector is 8 (from $2\lambda_4 + 4\lambda_5 + 2\lambda_7$). The balanced photodetectors give an autocorrelation peak of $12 − 8 = 4$ units of electrical current. For receiver 2, the upper photodetector sees ten wavelengths (from $3\lambda_2 + 3\lambda_3 + 2\lambda_4 + 2\lambda_7$) and the lower photodetector also sees 10 wavelengths (from $3\lambda_1 + 4\lambda_5 + 3\lambda_6$), resulting in $10 − 10 = 0$ unit of electrical current—zero cross-correlation value—at the output of the balanced photodetectors.

In addition to bipolar $(−1, +1)$ codes, spectral amplitude coding also supports the use of binary $(0, 1)$ codes, but these codes need to have a low, fixed in-phase cross-correlation function λ_c.[32,33] Because spectral coding does not involve time spreading, the correlation process is not a function of time anymore. So, the cross-correlation function is called in-phase and has one value only. If the amount of mutual interference caused by each interfering codeword at the output of the true decoder is λ_c, the conjugated decoder will output $w − \lambda_c$.[32] By adjusting the splitting ratio of the power splitter in Fig. 10 to $\alpha = \lambda_c/(w − \lambda_c)$, the net interference at the output of the conjugated decoder will then become $\alpha(w − \lambda_c) = \lambda_c$. As a result, the mutual interference seen at both decoders can completely be cancelled out at the balanced photodetectors, even through pseudo-orthogonal binary $(0, 1)$ codes, such as the synchronous prime codes in Chapter 4, are used, as long as their in-phase cross-correlation functions are equal to a constant number.

Instead of OOK, a variation of spectral amplitude coding conveys data bit 0 by transmitting the conjugate wavelengths of bipolar codewords.[34] In this scheme, the "−1" elements of bipolar codewords are transmitted in wavelengths but

Optical Coding Schemes: Basics

the "+1" elements are not whenever data bit 0s are conveyed. By transmitting both data bit 1s and 0s, this scheme results in code-performance improvement, similar to that of spectral phase coding. However, only restricted families of bipolar codes, such as Walsh codes and "balanced" maximal-length sequences,[34–38] which have code lengths equal to some even numbers and code weights equal to half of the code lengths, are used in order to maintain zero mutual interference.

2-D SPATIAL-TEMPORAL AMPLITUDE CODING

To lessen the long-code-length and large-bandwidth-expansion problems in 1-D temporal amplitude coding, spatial-temporal amplitude coding uses 2-D binary (0, 1) codewords to carry data bit 1s in space and time simultaneously.[39–45] The spatial domain is provided by multiple transmission channels, such as free space, multiple fibers, or a multiple-core fiber.

Figure 12 shows an example of spatial-temporal-amplitude-coding systems with multiple optical fibers and star couplers.[39,40] Assume that there are M stations and each of them is assigned a 2-D binary (0, 1) codeword of size $L \times N$ as its address signature, where L is the number of spatial channels (provided by optical fibers and star couplers, in this example) and N is the number of time slots.[2,45] A narrow optical pulse, which represents the transmission of a data bit 1, is first split into L pulses; each pulse is time-delayed to one of the N time slots, according to the address codeword of the intended receiver. These L pulses are then conveyed separately via their own optical fibers and multiplexed with the corresponding pulses from other stations at the star couplers. The decoders in the receivers reverse the process to give correlation functions.

Figure 13 illustrates an example of encoding and decoding processes of 2-D binary pixels in spatial-temporal amplitude coding.[41,43] With the support of 2-D binary (0, 1) codes,[2,45] the technique is suitable for parallel transmission and simultaneous access of multiple digitized 2-D images. There is no need to perform the bottleneck-prone parallel-to-serial or serial-to-parallel conversion because the pixels of every digitized image are transmitted in parallel. The concept also applies to free-space optics, in lieu of the multicore fiber or fiber bundle.[42] In this example, 2-D images of 2×2 binary pixels are transmitted, and each black pixel is encoded at an optical encoder with a 4×4 2-D unipolar codeword, forming an array of 8×8 coded pixels. These encoded pixels are then multiplexed with the corresponding encoded pixels from other stations at an optical

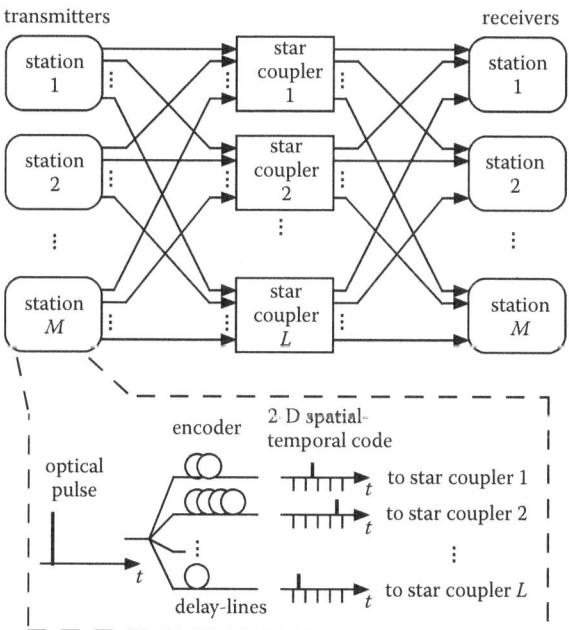

Fig. 12 Spatial-temporal amplitude coding in multiple fibers and star couplers
Source: Reproduced with permission from Guu-Chang Yang and Wing C. Kwong, *Prime Codes with Applications to CDMA Optical and Wireless Networks*, Norwood, MA: Artech House, Inc., 2002. © 2002 by Artech House, Inc.

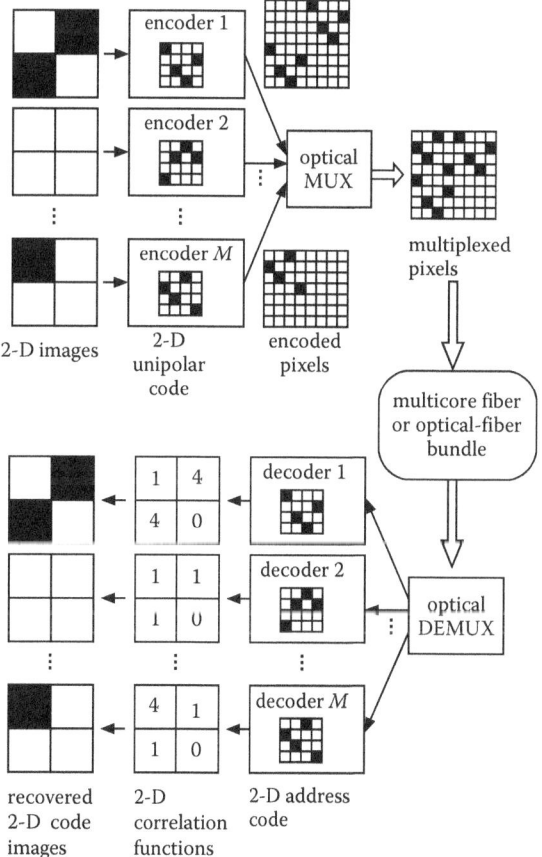

Fig. 13 Example of the encoding and decoding processes of 2-D binary pixels in spatial-temporal amplitude coding, where black squares represent dark pixels and the number in each square of the 2-D correlation functions represents the darkness level after correlation

MUX. Afterward, the multiplexed arrays of 8×8 encoded pixels are transmitted in parallel via multicore fiber, optical-fiber bundle, or free space, and later distributed to all stations by an optical DEMUX. If an array of coded pixels arrives at the correct decoder, an autocorrelation function, which is represented by a high (darkness-level) number, results. Otherwise, a cross-correlation function, which is represented by a low number, results and is treated as interference. By threshold-detecting the darkness levels, the binary pixels of the transmitted 2-D image are recovered.

2-D SPECTRAL-TEMPORAL AMPLITUDE CODING

If deployment of multicore fiber or optical-fiber bundle is not possible or too complicated, one alternative is to use a hybrid WDM-coding scheme,[2,46,47] in which 1-D time-spreading codewords are used in conjunction with multiple wavelengths. Every codeword is reusable and can be sent out simultaneously at different wavelengths, if needed. The selection of codewords of a specific wavelength at a receiver is done by an optical WDM filter placed in front of the optical decoder. The code length and, in turn, the speed of hardware are reduced due to the lessening in the number of simultaneous users in each wavelength. However, unless there is a central controller coordinating the wavelength usage among simultaneous users, wavelengths cannot usually be utilized evenly, and the scheme fails to achieve the optimal performance, as discussed in Section 5.1.[2,47]

Another approach, spectral-temporal amplitude (or wavelength-time) coding, imbeds multiple wavelengths within 1-D time-spreading codewords as the second coding dimension.[2,4,22–24,26,35–38,47,49–52] The scheme involves fast wavelength hopping, and the wavelength-hop takes place at every pulse of a time-spreading codeword, instead of having the same wavelength for all pulses within each codeword as in the aforementioned hybrid WDM-coding scheme. Using the same number of wavelengths and code length, 2-D wavelength-time codes, such as the prime codes in Chapter 5, have a larger cardinality than and can perform as well as the 1-D time-spreading codes used in the hybrid WDM-coding scheme.

Figure 14a shows a waveguide-based, spectral-temporal-amplitude encoder, which consists of WDM MUX/DEMUX devices, such as thin-film filters, arrayed-waveguide-grating (AWG) devices, holographic Bragg reflectors, and micro-disk resonators,[22–28] combined with tunable time-delay elements. A narrow, broadband laser pulse, which represents the transmission of a data bit 1, is split into L pulses of distinct wavelengths, where L is the number of wavelengths used in the wavelength-time codes. These multiwavelength pulses are time-delayed by tunable delay-lines, according to the address codeword of the intended receiver. If a wavelength is not used in the codeword, it will be blocked by an intensity modulator along that wavelength path. This encoder is integrable by fabricating the components all in the waveguide. The all-fiber approach in Fig. 14b places segments of fiber Bragg gratings (FBGs) with different center wavelengths inside a piece of optical fiber. The locations of these wavelength segments, which are usually not tunable, determine the spacings of the corresponding wavelength pulses.[16,17,37,53,54] More examples on AWG- and FBG-based encoder/decoder designs that are tunable are given in Section 2.10.

THREE-DIMENSIONAL CODING

To further improve the number of simultaneous users and subscribers, higher coding dimension can be achieved by

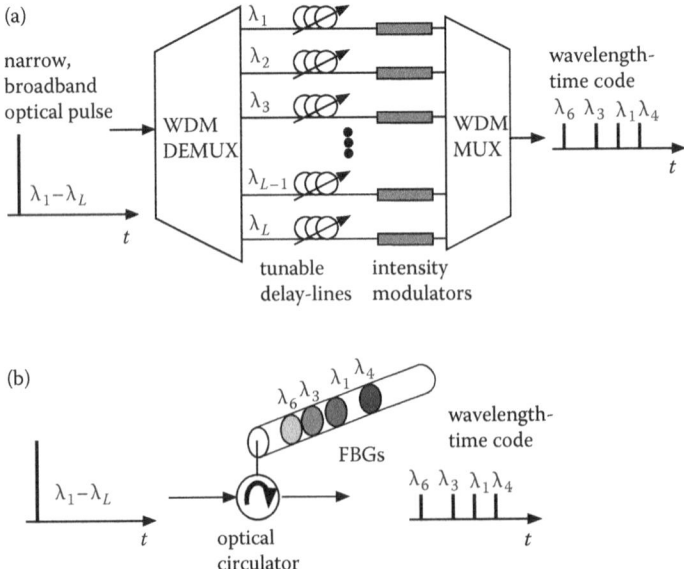

Fig. 14 Spectral-temporal-amplitude encoders in (a) waveguide and (b) fiber[23,24,34,53,54]

Optical Coding Schemes: Basics

combining temporal, spatial, and spectral coding.[55,56] Furthermore, the use of the two polarizations of optical field to replace the use of multiple fibers in the spatial domain has been proposed.[57] 3-D unipolar codes are needed to support this kind of coding schemes. The prime codes that support 3-D coding are studied in Chapter 8.[2]

MULTIRATE AND MULTIPLE-QOS CODING

So far in this entry, only one type of service with identical bit rate is assumed to exist in coding-based optical systems and networks. However, future systems are expected to support a variety of services, such as data, voice, and video, with different bit rates, QoSs, and even priorities. It is known that coherent (phase) coding schemes have better performance than incoherent (amplitude) coding schemes because the former allow the use of orthogonal bipolar codes.[23] Nevertheless, coherent coding requires strict phase control with the use of special optical fibers in order to maintain code orthogonality. The code choices in coherent coding are mostly limited to bipolar codes, such as maximal-length sequences and Walsh codes. This kind of code has restrictive code cardinality, which is about the same as the code length, giving a hard limit on the number of possible subscribers. Also, bipolar codes are very sensitive to any change in the code structure, and code orthogonality can be easily destroyed due to induced phase fluctuations during transmission in optical fiber. On the other hand, the unipolar codes used in incoherent coding schemes have totally opposite characteristics. First of all, they are less sensitive to phase changes, and regular optical fiber can be used because optical intensity is transmitted. Second, with pseudo-orthogonality (or nonzero cross-correlations), the unipolar code structure is more flexible and less restrictive in the relationships among code cardinality, weight, and length. Unipolar codes allow trading among performance, number of simultaneous users, and number of subscribers. Third, there exist unipolar codes for 1-D, 2-D, and even 3-D coding, while bipolar codes are all 1-D. Fourth, special unipolar codes can be designed with multiple lengths and variable weights without sacrificing the cross-correlation function. These features support multirate, multimedia services in incoherent coding systems and networks with different bit rates, QoSs, and priorities.[2,23,58–62] The prime codes that support this kind of coding schemes are studied in Chapter 7.

To illustrate an application of multilength coding, three types of multirate, multimedia services (i.e., digitized voice, data, and video) are assumed in Fig. 15.[2,23] Real-time video transmission, which has a continuous traffic pattern, is usually assigned the highest priority and requires the highest bit rate, whereas the bursty voice service requires the lowest priority and bit rate. Assume that video-service bit rate $1/T_v$ is an integer multiple of the data-service bit rate $1/T_d$ (i.e., $1/T_v = r_1/T_d$), which, in turn, is a multiple of the voice-service bit rate $1/T_s$ (i.e., $1/T_d = r_2/T_s$), where r_1 and r_2 are the expansion factors. For the service (i.e., video) with the highest bit rate, code length N is assumed, where $T_v = NT_c$ and T_c is the pulse-width (or chip-width). Because the same chip-width T_c is used for all services, the medium-rate service (i.e., data) requires optical codewords of r_1 times longer than that of the highest-rate service in order to support the rate $1/T_d = 1/(r_1 T_v) = 1/(r_1 NT_c)$. Similarly, the lowest-rate service (i.e., voice) requires codewords of $r_1 r_2$ times longer than that of the video service in order to support the rate

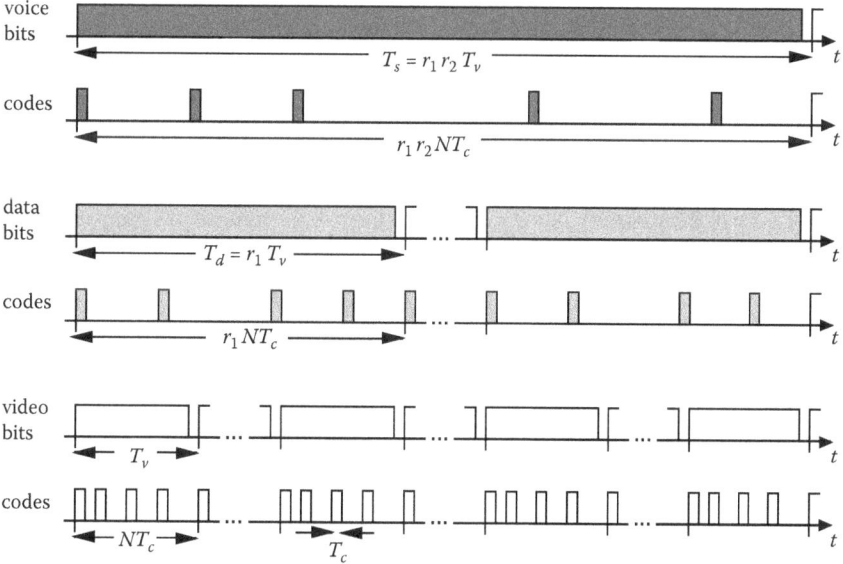

Fig. 15 Timing diagrams of three types of multirate, multimedia services supported by multilength coding
Source: Reproduced with permission from Guu-Chang Yang and Wing C. Kwong, *Prime Codes with Applications to CDMA Optical and Wireless Networks*, Norwood, MA: Artech House, Inc., 2002. © 2002 by Artech House, Inc.

$1/T_s = 1/(r_1 r_2 T_v) = 1/(r_1 r_2 N T_c)$. To support these three types of services, 1-D or 2-D unipolar codewords of lengths N, $r_1 N$, and $r_1 r_2 N$ are constructed with the same maximum cross-correlation function that is independent of code length.[2,23,58–62] The shortest codewords are then assigned to the real-time services (i.e., video) with the highest bit-rate and priority, whereas the longest codewords are for the voice services. Because the analyses in Chapter 7 show that the shortest codewords have the best performance, the QoS of critical real-time video transmission is guaranteed. This unique priority feature, however, cannot be found in conventional single-length coding schemes. In addition, one system clock and lasers with the same pulse-width can be used for all types of services in this multilength approach, simplifying system hardware and timing requirements.

Furthermore, two multirate asynchronous O-CDMA schemes were proposed by Maric and Lau.[63] For example, in the parallel-mapping multiple-code scheme,[63,64] each user is assigned multiple 1-D codewords. If a user needs to transmit at a rate of M times the basic bit rate, every M serial bits are first converted into M parallel bits. Then, each parallel bit 1 is conveyed by one of the assigned M codewords, but nothing is transmitted for a bit 0. As a result, the number of codewords that are transmitted at the same time ranges from 0 to M, depending on the user's bit rate and the number of parallel bit 1s after the serial-to-parallel conversion. Because of the need of transmitting many codewords simultaneously, optical codes with huge cardinality are required in this scheme. Nevertheless, the scheme is still asynchronous in nature because user-to-user synchronization is not needed, even though multiple codewords are simultaneously transmitted by every user.

MULTICODE KEYING AND SHIFTED-CODE KEYING

To support higher bit-rate transmission without increasing the speed of optics and electronics, three methods of multiple-bit-per-symbol transmission have been proposed.[65–70] In pulse-position modulation (PPM) coding, each bit period is divided into 2^m nonoverlapping PPM frames.[65,68] Each user is assigned one distinct (address) codeword and all m serial data bits are converted into one of 2^m possible symbols. A symbol is transmitted by placing the codeword entirely inside one of the 2^m PPM frames designated for that symbol. As illustrated in Fig. 16, every two serial data bits are grouped to form one of the four possible symbols and, in turn, the symbol is conveyed by transmitting the codeword entirely within one of the four PPM frames. As a result, the total number of time slots is increased by a factor of 2^m in this nonoverlapping PPM scheme and so is the transmission bandwidth.

Another method of transmitting symbols is by means of multicode keying,[64,70,71] in which each user is assigned 2^m distinct codewords to represent m serial data bits per symbol. One of these codewords is conveyed each time in order to represent the transmission of one of the 2^m symbols. Figure 17 shows an example of four-code keying, in which every two serial data bits are grouped to form one of the four possible symbols and, in turn, the symbol is conveyed by one of the four distinct codewords. This multicode-keying approach does not need system-wise synchronization but only needs the communicating transmitter-receiver pair be synchronized, the same requirement as any asynchronous OOK coding scheme anyway. However, multicode keying requires an 2^m-fold increase in the number of codewords, all with the same low cross-correlation function. Details about the cross-correlation requirements and the prime codes that are suitable for multicode keying can be found in Section 5.6.

Without the need for huge code cardinality, shifted-code keying assigns each user with one codeword and its $2^m - 1$ (time or wavelength) shifted copies to represent the 2^m symbols of m serial data bits per symbol.[64,69] Figure 18 shows an example of shifted-code keying with 4 symbols, in which 4 time positions (within a bit period) are used as the transmission start-time of a codeword. This scheme is different from the nonoverlapping PPM scheme[65,68] in such a way that no increase in the number of time slots or the bandwidth expansion is needed. The overlapping PPM scheme [66,67] belongs to a case of the shifted-code-keying scheme. Shifted-code keying does not require system-wise synchronization or a

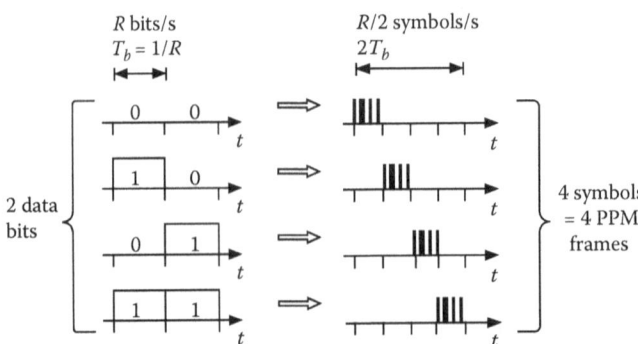

Fig. 16 Example of PPM coding with 4 symbols, represented by 4 PPM frames, where T_b is the bit period and R is the bit rate[65,66,68]

Optical Coding Schemes: Basics

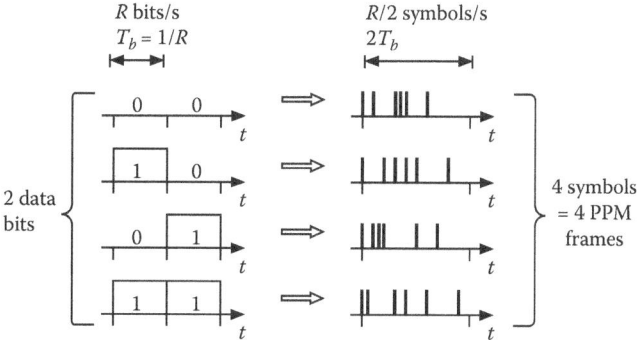

Fig. 17 Example of multicode keying with 4 symbols, represented by 4 distinct codewords, where T_b is the bit period and R is the bit rate[64,70,71]

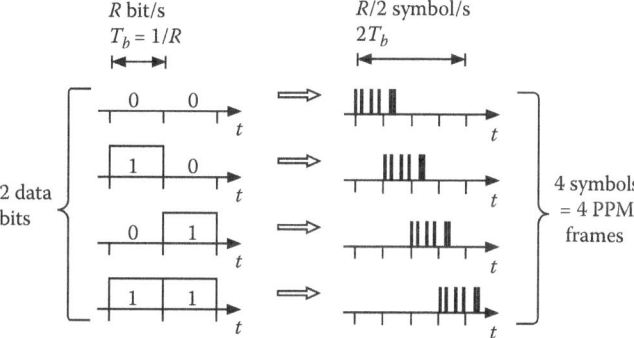

Fig. 18 Example of shifted-code keying with 4 symbols, represented by shifting a codeword to one of the four time positions, where T_b is the bit period and R is the bit rate[64,69]

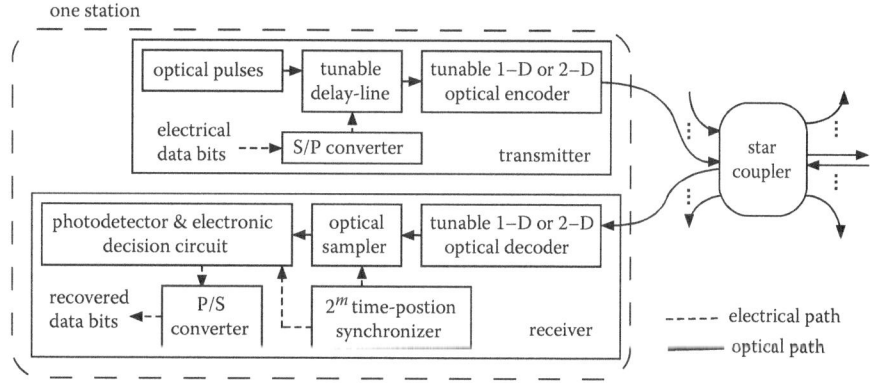

Fig. 19 Tunable transmitter and receiver for shifted-code keying with time-shifted codewords

2^m-fold increase in code cardinality. Depending on time or wavelength shifts, two tunable transmitter-receiver designs are given in Figs. 19 and 20.[69] The prime codes that are suitable for shifted-code keying can be found in Section 5.3.

Shown in Fig. 19 is a tunable transmitter and receiver for shifted-code keying with 1-D or 2-D time-spreading codewords and their time-shifted copies.[69] The transmitter consists of a serial-to-parallel (S/P) data-to-symbol converter, a tunable delay-line, and a tunable 1-D or 2-D optical encoder. The S/P data-to-symbol converter groups every m serial data bits to form one of the 2^m symbols. A narrow laser pulse is then delayed to one of the 2^m time positions by the tunable delay-line; the amount of time-delay depends on which symbol is transmitted. The delay-line only needs to be tuned as fast as the symbol rate and consists of the improved serial encoder in Section 2.1.[6,7] This optical pulse is then passed through the tunable optical encoder to form the address codeword of its intended receiver with the proper amount of time shift. The receiver consists of a tunable 1-D or 2-D optical decoder, an optical sampler, a 2^m time-position synchronizer, a photodetector, an electronic decision circuit, and a P/S symbol-to-data converter. Because only one time-shifted codeword is transmitted during one symbol period, the intended receiver will see at most one autocorrelation peak per symbol period, but the peak's time position depends on which symbol is received. Time-gated by the 2^m time-position synchronizer, the optical sampler inspects the existence of such a peak at these 2^m positions. The optical sampler consists of optical interferometric devices, such as terahertz optical asymmetric demultiplexers and nonlinear optical loop mirrors.[72,73] These devices can periodically generate an optical sampling window of size as narrow as several picoseconds by means of ultrafast optical nonlinear effects. So, they can be used to gate very narrow optical features, such as autocorrelation peaks, which appear at most once in each symbol period. Optical signals, such as noise and cross-correlation functions, falling outside the sampling windows are dropped. The received symbol is then determined by the electronic decision circuitry after photodetection. Finally, the data bits are recovered at the P/S symbol-to-data converter.

Shown in Fig. 20 is a tunable transmitter and receiver for shifted-code keying with 2-D wavelength-time codewords and their wavelength-shifted copies.[69] These 2-D wavelength-time codewords are assumed to have L distinct wavelengths, and each wavelength is used at most once per codeword, such as the wavelength-shifted carrier-hopping prime codes in Section 5.3. Because L wavelengths can generate at most L wavelength-shifted copies of a codeword, 2^m-ary shifted-code keying requires $2^m \leq L$. The transmitter consists of an S/P data-to-symbol converter, an $L \times 1$ optical router, and a tunable 2-D optical encoder. The encoder, which generates wavelength-shifted codewords, consists of a $1 \times L$ power splitter, a set of L tunable delay-lines, and an $L \times L$ arrayed-waveguide-gratings (AWG) device with periodic-wavelength assignment.[27,35–38] Wavelength periodicity means that exit wavelengths at the AWG output ports are rotatable, depending on which input port is injected with a laser pulse. Assume that the wavelengths of the pulses at output ports 1, 2, 3, and 4 are $\lambda_1, \lambda_2, \lambda_3$, and λ_4, respectively, when input port 1 of a 4×4 AWG device is injected with a broadband optical pulse. The wavelengths of the pulses are then rotated up once and become $\lambda_2, \lambda_3, \lambda_4$, and λ_1, at output ports 1, 2, 3, and 4, respectively, when input port 2 is injected with the pulse. In this encoder, a narrow, broadband laser pulse is first split into L pulses at the power splitter, and these pulses are delayed by the tunable delay-lines, according to the address codeword of its intended receiver. All possible wavelength shifts of the codeword are performed at the $L \times L$ AWG device. Depending on which symbol is transmitted, only one of the AWG output ports, which has the proper wavelength-shifted codeword, is picked by the $L \times 1$ optical router.

The receiver consists of one tunable 2-D optical decoder, a bank of L photodetectors, an electronic decision circuit, and a P/S symbol-to-data converter. The decoder, which has an identical setup as the encoder, is used to reverse the amounts of time delay and wavelength shift introduced by its corresponding encoder. Because only one wavelength-shifted codeword is transmitted during one symbol period, the intended receiver will see at most one autocorrelation peak per symbol period at one of the L output ports of the optical decoder. By identifying which output port has the autocorrelation peak, the received symbol is finally determined by the electronic decision circuitry after photodetection. Finally, the data bits are recovered at the P/S symbol-to-data converter.

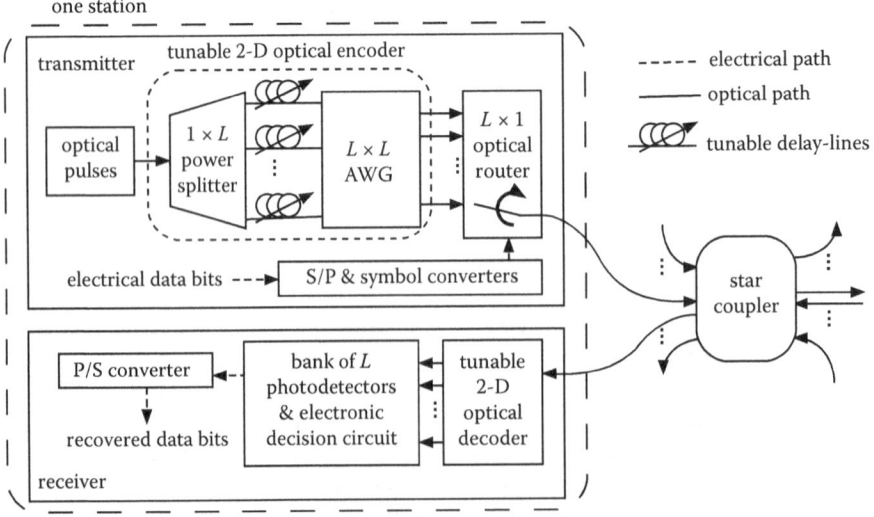

Fig. 20 Tunable transmitter and receiver for shifted-code keying with wavelength-shifted codewords

REFERENCES

1. Prucnal, P.R.; Santoro, M.A.; Fan, T.R. Spread spectrum fiber-optic local area network using optical processing. J. Lightwave Technol. **1986**, *4* (5), 547–554.
2. Yang, G.-C.; Kwong, W.C. *Prime Codes with Applications to CDMA Optical and Wireless Networks*; Artech House: Norwood, MA, 2002.
3. Prucnal, P.R.; Santoro, M.A.; Sehgal, S.K. Ultra-fast all-optical synchronous multiple access fiber networks. IEEE J. Selected Areas Commun. **1986**, *4* (9), 1484–1493.
4. Kwong, W.C.; Yang, G.-C.; Zhang, J.-G. 2n prime-sequence codes and coding architecture for optical code-division multiple-access. IEEE Trans. Commun. **1996**, *44* (9), 1152–1162.
5. Yang, G.-C.; Kwong, W.C. On the construction of 2n codes for optical code-division multiples-access. IEEE Trans. Commun. **1995**, *43* (2–4), 495–502.
6. Prucnal, P.R.; Krol, M.; Stacy, J. Demonstration of a rapidly tunable optical time-division multiple-access coder. IEEE Photon. Technol. Lett. **1991**, *3* (2), 170–172.
7. Deng, K.-L.; Runser, R.J.; Toliver, P.; Glesk, I.; Prucnal, P.R. A highly-scalable, rapidly-reconfigurable, multicasting-capable, 100 Gb/s photonic switched interconnect based upon OTDM technology. J. Lightwave Technol. **2000**, *18* (12), 1892–1904.
8. Lam, A.W.; Tantaratana, S. *Theory and Application of Spread-Spectrum Systems—A Self Study Course*; IEEE Press: Piscataway, NJ, 1994.
9. Dinan, E.H.; Jabbari, B. Spreading codes for direct sequence CDMA and wideband CDMA cellular networks. IEEE Commun. Mag. **1998**, *36* (9), 48–54.
10. Huang, W.; Nizam, M.H.M.; Andonovic, I.; Tur, M. Coherent optical CDMA (OCDMA) systems used for high-capacity optical fiber networks—System description, OTDMA comparison, and OCDMA/WDMA networking. J. Lightwave Technol. **2000**, *18* (6), 765–778.
11. Marhic, M.E. Coherent optical CDMA networks. J. Lightwave Technol. **1993**, *11* (5/6), 895–864.
12. Huang, W.; Andonovic, I.; Tur, M. Decision-directed PLL for coherent optical pulse CDMA systems in the presence of multiuser interference, laser phase noise, and shot noise. J. Lightwave Technol. **1998**, *16* (10), 1786–1794.
13. Andonovic, I.; Tur, M.; Huang, W. Coherent optical pulse CDMA systems based on coherent correlation detection. IEEE Trans. Commun. **1999**, *47* (2), 261–271.
14. Nguyen, L.; Dennis, T.; Aazhang, B.; Young, J.F. Experimental demonstration of bipolar codes for optical spectral amplitude CDMA communication. J. Lightwave Technol. **1997**, *15* (9), 1647–1653.
15. Lam, C.F.; Tong, D.T.K.; Wu, M.C.; Yablonovitch, E. Experimental demonstration of bipolar optical CDMA system using a balanced transmitter and complementary spectral encoding. IEEE Photon. Technol. Lett. **1998**, *10* (10), 1504–1506.
16. Teh, P.C.; Petropoulos, P.; Ibsen, M.; Richardson, D.J. A comparative study of the performance of seven- and 63-chip optical code-division multiple-access encoders and decoders based on superstructured fiber Bragg gratings. J. Lightwave Technol. **2000**, *19* (9), 1352–1365.
17. Teh, P.C.; Petropoulos, P.; Ibsen, M.; Richardson, D.J. Phase encoding and decoding of short pulses at 10 Gb/s using superstructured fiber Bragg gratings. IEEE Photon. Technol. Lett. **2001**, *13* (2), 154–156.
18. Sardesai, H.P.; Chang, C.-C.; Weiner, A.M. A femtosecond code-division multiple-access communication system test bed. J. Lightwave Technol. **1998**, *16* (11), 1953–1964.
19. Cong, W.; Yang, C.; Scott, R.P.; Hernandez, V.J.; Fontaine, N.K.; Kolner, B.H.; Heritage, J.P.; Yoo, S.J.B. Demonstration of 160- and 320-Gb/s SPECTS O-CDMA network testbeds. IEEE Photon. Technol. Lett. **2006**, *18* (15), 1567–1569.
20. Gao, Z.; Wang, X.; Kataoka, N.; Wada, N. Rapid reconfigurable OCDMA system using single-phase modulator for time-domain spectral phase encoding/decoding and DPSK data modulation. J. Lightwave Technol. **2011**, *29* (3), 348–354.
21. Hernandez, V.J.; Cong, W.; Hu, J.; Yang, C.; Fontaine, N.K.; Scott, R.P.; Ding, Z.; Kolner, B.H.; Heritage, J.P.; Yoo, S.J.B. A 320-Gb/s capacity (32-user×10 Gb/s) SPECTS O-CDMA network testbed with enhanced spectral efficiency through forward error correction. J. Lightwave Technol. **2007**, *25* (1), 79–86.
22. Yu, K.; Shin, J.; Park, N. Wavelength-time spreading optical CDMA system using wavelength multiplexers and mirrored fiber delay lines. IEEE Photon. Technol. Lett. **2000**, *12* (9), 1278–1280.
23. Prucnal, P.R. (Ed.) *Optical Code Division Multiple Access: Fundamentals and Applications*; Taylor & Francis Group: Boca Raton, FL, 2006.
24. Brés, C.-S.; Glesk, I.; Prucnal, P.R. Demonstration of an eight-user 115-Gchip/s incoherent OCDMA system using supercontinuum generation and optical time gating. IEEE Photon. Technol. Lett. **2006**, *18* (7), 889–891.
25. Huang, Y.-K.; Baby, V.; Prucnal, P.R.; Greiner, C.M.; Iazikov, D.; Mossberg, T.W. Integrated holographic encoder for wavelength-hopping/time-spreading optical CDMA. IEEE Photon. Technol. Lett. **2005**, *17* (4), 825–827.
26. Brés, C.-S.; Prucnal, P.R. Code-empowered lightwave networks. J. Lightwave Technol. **2007**, *25* (10), 2911–2921.
27. Takahashi, H.; Oda, K.; Toba, H.; Inoue, Y. Transmission characteristics of arrayed waveguide N×N wavelength multiplexer. J. Lightwave Technol. **1995**, *13* (3), 447–455.
28. Djordjev, K.; Choi, S.-J.; Dapkus, R.D. Microdisk tunable resonant filters and switches. IEEE Photon. Technol. Lett. **2002**, *14* (6), 828–830.
29. Huang, J.-F.; Hsu, D.-Z. Fiber-grating-based optical CDMA spectral coding with nearly orthogonal M-sequence codes. IEEE Photon. Technol. Lett. **2000**, *12* (9), 1252–1254.
30. Lin, C.-H.; Wu, J.; Tsao, H.-W.; Yang, C.-L. Spectral amplitude-coding optical CDMA system using Mach-Zehnder interferometers. J. Lightwave Technol. **2005**, *23* (4), 1543–1555.
31. Zaccarin, D.; Kavehrad, M. Performance evaluation of optical CDMA systems using non-coherent detection bipolar codes. J. Lightwave Technol. **1994**, *12* (1), 96–105.
32. Wei, Z.; Shalaby, H.M.H.; Ghafouri-Shiraz, H. Modified quadratic congruence codes for fiber Bragg-grating-based spectral-amplitude-coding optical CDMA systems. J. Lightwave Technol. **2001**, *19* (9), 1274–1281.

33. Wei, Z.; Ghafouri-Shiraz, H. Unipolar codes with ideal in-phase cross-correlation for spectral amplitude-coding optical CDMA systems. IEEE Trans. Commun. **2002**, *50* (8), 1209–1212.
34. Zeng, F.; Wang, Q.; Yao, J. Sequence-inversion-keyed optical CDMA coding/decoding scheme using an electrooptic phase modulator and fiber Bragg grating arrays. IEEE J. Sel. Top. Quantum Electron. **2007**, *13* (5), 1508–1515.
35. Kwong, W.C.; Yang, G.-C.; Liu, Y.-C. A new family of wavelength-time optical CDMA codes utilizing programmable arrayed waveguide gratings. IEEE J. Selected Areas Commun. **2005**, *23* (8), 1564–1571.
36. Kwong, W.C.; Yang, G.-C.; Chang, C.-Y. Wavelength-hopping time-spreading optical CDMA with bipolar codes. J. Lightwave Technol. **2005**, *23* (1), 260–267.
37. Hsieh, C.-P.; Chang, C.-Y.; Yang, G.-C.; Kwong, W.C. A bipolar-bipolar code for asynchronous wavelength-time optical CDMA. IEEE Trans. Commun. **2006**, *54* (7), 1190–1194.
38. Hu, H.-W.; Chen, H.-T.; Yang, G.-C.; Kwong, W.C. Synchronous Walsh-based bipolar-bipolar code for CDMA passive optical networks. J. Lightwave Technol. **2007**, *25* (8), 1910–1917.
39. Hui, J.Y.N. Pattern code modulation and optical decoding: A novel code division multiplexing technique for multifiber network. IEEE J. Selected Areas Commun. **1985**, *3* (3), 916–927.
40. Park, E.; Mendez, A.J.; Garmire, E.M. Temporal/spatial optical CDMA networks—Design, demonstration, and comparison with temporal networks. IEEE Photon. Technol. Lett. **1992**, *4* (10), 1160–1162.
41. Kitayama, K. Novel spatial spread spectrum based fiber optic CDMA networks for image transmission. IEEE J. Selected Areas Commun. **1994**, *12* (5), 762–772.
42. Hassan, A.A.; Hershey, J.E.; Riza, N.A. Spatial optical CDMA. IEEE J. Selected Areas Commun. **1995**, *13* (3), 609–613.
43. Kwong, W.C.; Yang, G.-C. Image transmission in multicore-fiber code-division multiple-access networks. IEEE Commun. Lett. **1998**, *2* (10), 285–287.
44. Yeh, B.-C.; Lin, C.-H.; Yang, C.-L.; Wu, J. Noncoherent spectral/spatial optical CDMA system using 2-D diluted perfect difference codes. J. Lightwave Technol. **2009**, *27* (13), 2420–2432.
45. Yang, G.-C.; Kwong, W.C. Two-dimensional spatial signature patterns. IEEE Trans. Commun. **1996**, *44* (2), 184–191.
46. Perrier, P.A.; Prucnal, P.R. Wavelength-division integration of services in fiber-optic networks. Inter. J. Digital Analog Cabled Systems **1988**, *1* (3), 149–157.
47. Yang, G.-C.; Kwong, W.C. Performance comparison of multiwavelength CDMA and WDMA+CDMA for fiber-optic networks. IEEE Trans. Commun. **1997**, *45* (11), 1426–1434.
48. Mendez, A.J.; Gagliardi, R.M.; Hernandez, V.J.; Bennett, C.V.; Lennon, W.J. High-performance optical CDMA system based on 2-D optical orthogonal codes. J. Lightwave Technol. **2004**, *22* (11), 2409–2419.
49. Tančevski, L.; Andonovic, I. Hybrid wavelength hopping/time spreading schemes for use in massive optical networks with increased security. J. Lightwave Technol. **1996**, *14* (12), 2636–2647.
50. Fathallah, H.; Rusch, L.A.; LaRochelle, S. Passive optical fast frequency-hop CDMA communications system. J. Lightwave Technol. **1999**, *17* (3), 397–405.
51. Kwong, W.C.; Yang, G.-C.; Baby, V.; Brès, C.-S.; Prucnal, P.R. Multiple-wavelength optical orthogonal codes under prime-sequence permutations for optical CDMA. IEEE Trans. Commun. **2005**, *53* (1), 117–123.
52. Baby, V.; Glesk, I.; Runser, R.J.; Fischer, R.; Huang, Y.-K.; Brés, C.-S.; Kwong, W.C.; Curtis, T.H.; Prucnal, P.R. Experimental demonstration and scalability analysis of a four-node 102-Gchip/s fast frequency-hopping time-spreading optical CDMA network. IEEE Photon. Technol. Lett. **2005**, *17* (1), 253–255.
53. Chen, L.R.; Benjamin, S.D.; Smith, P.W.E.; Sipe, J.E. Applications of ultrashort pulse propagation in Bragg gratings for wavelength-division multiplexing and code-division multiple access. IEEE J. Quantum Electron. **1998**, *34* (11), 2117–2129.
54. Chen, L.R. Flexible fiber Bragg grating encoder/decoder for hybrid wavelength-time optical CDMA. IEEE Photon. Technol. Lett. **2001**, *13* (11), 1233–1235.
55. Yeh, B.-C.; Lin, C.-H.; Wu, J. Noncoherent spectral/time/spatial optical CDMA system using 3-D perfect difference codes. J. Lightwave Technol. **2009**, *27* (6), 744–759.
56. Kim, S.; Yu, K.; Park, N. A new family of space/wavelength/time spread three-dimensional optical code for OCDMA networks. J. Lightwave Technol. **2000**, *18* (4), 502–511.
57. McGeehan, J.E.; Nezam, S.M.R.M.; Saghari, P.; Willner, A.E.; Omrani, R.; Kumar, P.V. Experimental demonstration of OCDMA transmission using a three-dimensional (time-wavelength-polarization) codeset. J. Lightwave Technol. **2005**, *23* (10), 3282–3289.
58. Kwong, W.C.; Yang, G.-C. Double-weight signature pattern codes for multicore-fiber code-division multiple-access networks. IEEE Commun. Lett. **2001**, *5* (5), 203–205.
59. Kwong, W.C.; Yang, G.-C. Design of multilength optical orthogonal codes for optical CDMA multimedia networks. IEEE Trans. Commun. **2002**, *50* (8), 1258–1265.
60. Kwong, W.C.; Yang, G.-C. Multiple-length, multiple-wavelength optical orthogonal codes for optical CDMA systems supporting multirate, multimedia services. IEEE J. Selected Areas Commun. **2004**, *22* (9), 1640–1647.
61. Kwong, W.C.; Yang, G.-C. Multiple-length extended carrier-hopping prime codes for optical CDMA systems supporting multirate, multimedia services. J. Lightwave Technol. **2005**, *23* (11), 3653–3662.
62. Baby, V.; Kwong, W.C.; Chang, C.-Y.; Yang, G.-C.; Prucnal, P.R. Performance analysis of variable-weight, multilength optical codes for wavelength-time O-CDMA multimedia systems. IEEE Trans. Commun. **2007**, *55* (7), 1325–1333.
63. Maric, S.V.; Lau, V.K.N. Multirate fiber-optic CDMA: System design and performance analysis. J. Lightwave Technol. **1998**, *16* (1), 9–17.
64. Chen, H.-W.; Yang, G.-C.; Chang, C.-Y.; Lin, T.-C.; Kwong, W.C. Spectral efficiency study of two multirate schemes for optical CDMA with/without symbol synchronization. J. Lightwave Technol. **2009**, *27* (14), 2771–2778.
65. Shalaby, H.M.H. Performance analysis of optical synchronous CDMA communication systems with PPM signaling. IEEE Trans. Commun. **1995**, *43* (2), 624–634.

66. Shalaby, H.M.H. A performance analysis of optical overlapping PPM-CDMA communication systems. J. Lightwave Technol. **1999**, *17* (3), 426–433.
67. Kim, J.Y.; Poor, H.V. Turbo-coded optical direct-detection CDMA system with PPM modulation. J. Lightwave Technol. **2001**, *19* (3), 312–323.
68. Kamakura, K.; Yashiro, K. An embedded transmission scheme using PPM signaling with symmetric error-correcting codes for optical CDMA. J. Lightwave Technol. **2003**, *21* (7), 1601–1611.
69. Narimanov, E.; Kwong, W.C.; Yang, G.-C.; Prucnal, P.R. Shifted carrier-hopping prime codes for multicode keying in wavelength-time O-CDMA. IEEE Trans. Commun. **2005**, *53* (12), 2150–2156.
70. Chang, C.-Y.; Yang, G.-C.; Kwong, W.C. Wavelength-time codes with maximum cross-correlation functions of two for multicode keying optical CDMA. J. Lightwave Technol. **2006**, *24* (3), 1093–1100.
71. Chang, C.-Y.; Chen, H.-T.; Yang, G.-C.; Kwong, W.C. Spectral efficiency study of quadratic-congruence carrier-hopping prime codes in multirate optical CDMA system. IEEE J. Selected Areas Commun. **2007**, *25* (9), 118–128.
72. Lee, J.-H.; Teh, P.C.; Petropoulos, P.; Ibsen, M.; Richardson, D.J. A grating-based OCDMA coding-decoding system incorporating a nonlinear optical loop mirror for improved code recognition and noise reduction. J. Lightwave Technol. **2002**, *20* (1), 36–46.
73. Sokoloff, J.P.; Prucnal, P.R.; Glesk, I.; Kane, M. A terahertz optical asymmetric demultiplexer (TOAD). IEEE Photon. Technol. Lett. **1993**, *5* (7), 787–790.

Optical Coding Schemes: Hardware

Wing C. Kwong
Hofstra University, Hempstead, New York, U.S.A.

Guu-Chang Yang
National Chung Hsing University, Taichung City, Taiwan

Abstract

With the advances in optical hardware technology, the large bandwidth expansion required by optical codes can now be accommodated. In this article, particular hardware designs of coding devices, based on arrayed waveguide gratings and fiber Bragg gratings are investigated. With the advances in optical hardware technology, the large bandwidth expansion required by optical codes can now be accommodated. In this entry, particular hardware designs of coding devices, based on arrayed waveguide gratings and fiber Bragg gratings are investigated.

Keywords: Arrayed waveguide grating; Fiber bragg grating; Optical coding; Optical hardware; Wavelength-aware hard-limiting detector.

WAVELENGTH-AWARE HARD-LIMITING DETECTOR

A hard-limiter can be placed at the front end of an optical decoder to prevent interference from becoming heavily localized in small sections of a cross-correlation function.[1,2,3] In concept, the hard-limiter can also equalize the interference strength at nonempty chip positions and eliminate the near-far problem due to unequalized powers of received codewords caused by different transmission distances. However, a fast-response, all-optical hard-limiter that can work with ultrashort optical pulses is still under research. To alleviate this deficiency, Baby et al.[4] proposed a wavelength-aware hard-limiting detector for 2-D wavelength-time codes, as shown in Fig. 1. This detector is placed at the output of an optical sampler, which is, in turn, connected to the output of a 2-D optical decoder in a receiver. The role of the optical sampler is to periodically generate an optical sampling window of size as narrow as several picoseconds in order to gate autocorrelation peaks, but to remove noise and cross-correlation functions that fall outside the sampling window.[5,6] The opening of the sampling window is synchronized to the expected time locations of the autocorrelation peaks, which appear at most once in the same time location per bit period, after the correlation process at the optical decoder. With the use of an optical sampler, slower-response photodetectors and electronics can now be used in the wavelength-aware hard-limiting detector. This is because the detector only needs to discriminate the energy strength of the gated signal exiting the optical sampler, but a precise opto-electronic conversion—the actual shape—of the narrow gated signal is not necessary. If strong energy is detected, an autocorrelation peak is decided and a data bit 1 is recovered.

To provide 2-D hard-limiting functionality, the wavelength-aware hard-limiting detector consists of a wavelength DEMUX, a bank of photodetectors and electronic hard-limiters, and an electronic decision circuit. In this setup, the L wavelengths of the gated signal are individually photodetected and hard-limited. Because each wavelength is only counted once after hard-limiting, the probability of getting a decision error is reduced if the strong energy seen at the gated signal is caused by one or more wavelengths being repeated many times in the cross-correlation function. The setup is particularly useful for 2-D wavelength-time codes that use every wavelength exactly once in each codeword, such as the carrier-hopping prime codes. This is because the gated signal always contains all L wavelengths after hard-limiting if it contains an autocorrelation peak. Finally, by recombining and threshold-detecting the hard-limited signals at the electronic decision circuit, a data bit 1 will be recovered if there still exists an autocorrelation peak.

FIBER BRAGG GRATINGS

Fiber Bragg gratings (FBGs), which periodically vary the refractive index in an optical fiber core, can be used to build in-fiber distributed Bragg (wavelength) reflectors.[7,8–11] This type of reflector can be used as a WDM device, such as in-fiber wavelength filters and sensors. Shown in Fig. 2 is a FBG-based, 2-D tunable wavelength-time encoder introduced by Chen.[11] Tunability is achieved using an optical

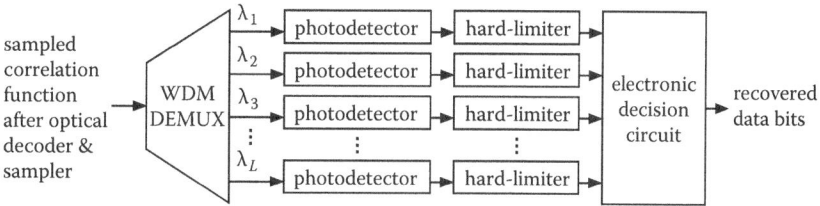

Fig. 1 Wavelength-aware hard-limiting detector for 2-D wavelength-time codes[4]

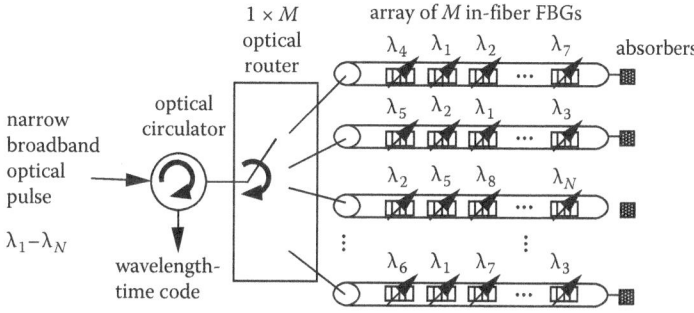

Fig. 2 2-D tunable wavelength-time encoder with in-fiber FBGs[11]

router with an array of multiple FBG fibers. While M different rows are designed with different FBG arrangements for generating M different wavelength-time codewords, a $1 \times M$ optical router is used to route an incoming narrow, broadband optical pulse to one of these rows. The optical circulator routes the generated codeword to the encoder output.

When a narrow, broadband optical pulse is injected into a section of FBGs, the input wavelength, which matches the Bragg wavelength $\lambda_B = 2n_{\mathrm{eff}}\Lambda$ of the FBGs, will be reflected, but the rest of the wavelengths will pass through the FBGs, where Λ is the resonance grating period and n_{eff} is the effective index of the unmodified mode.[7,8,11] Because there are multiple sections of distinct-Bragg-wavelength FBGs within a piece of optical fiber, the location of each section determines the time position of a certain wavelength pulse in a wavelength-time codeword. In addition to the setup in Fig. 2, a limited degree of tunability can also be achieved by introducing equal strain to the FBGs in a piece of optical fiber such that these FBGs see the same amount of Bragg-wavelength shift. The amount of wavelength shift in a FBG is determined by $\Delta\lambda_{\mathrm{shift}} = 0.8\lambda_B(\Delta L/L)$, where L is the FBG length and ΔL is the amount of fiber stretch. For example, a typical section of FBGs is of length $L = 10\,\mathrm{mm}$, and the maximum amount of strain that can be created is about $\Delta L = 0.1\,\mathrm{mm}$, giving $\Delta L/L = 0.01$. At $\lambda_B \approx 1550$ nm, the corresponding wavelength shift is about $\Delta\lambda_{\mathrm{shift}} \approx 12.4$ nm. Assume that the spacing between adjacent wavelengths is 0.8 nm (or 100 GHz in frequency), the number of available wavelengths obtained by straining the FBGs can be as high as $12.4/0.8 = 15$. By mounting in-fiber FBGs onto a piezoelectric transducer for strain induction, different wavelength-shifted codewords, such as those in the shifted carrier-hopping prime codes, are generated.

ARRAYED WAVEGUIDE GRATINGS

Shown in Fig. 3 is a tunable AWG-based coding device that can be used to generate 1-D spectral-phase codes, 1-D spectral-amplitude codes, and 2-D spectral-temporal-amplitude codes. This device features four kinds of tunability. The first kind of tunability is on the spreading and rotation of the spectral components of broadband optical pulses. It is achieved by the use of an optical router in conjunction with the wavelength periodicity of the AWG.[12–15,16] For an $M \times M$ AWG device with periodic-wavelength assignment, exit wavelengths at the output ports are rotatable, depending on which input port is injected with a broadband optical pulse. The $1 \times M$ optical router is used to route an incoming pulse to one of the M input ports of the AWG device. The second kind of tunability is supported by tunable delay-lines along each wavelength path for temporal coding. The third kind of tunability is provided by phase modulators for phase coding. The fourth kind of tunability is due to the use of intensity modulators for preserving or dropping optical pulses in some wavelength paths. The properly wavelength-shifted, time-delayed, phase-modulated, and/or intensity-modulated pulses are finally combined at the WDM MUX to generate the desired 1-D spectral-phase codeword, 1-D spectral-amplitude codeword, or 2-D spectral-temporal-amplitude codeword.

For an $M \times M$ AWG device with wavelength periodicity, the exit wavelengths, denoted $\lambda_{(i+j)\bmod M}$, at output port j

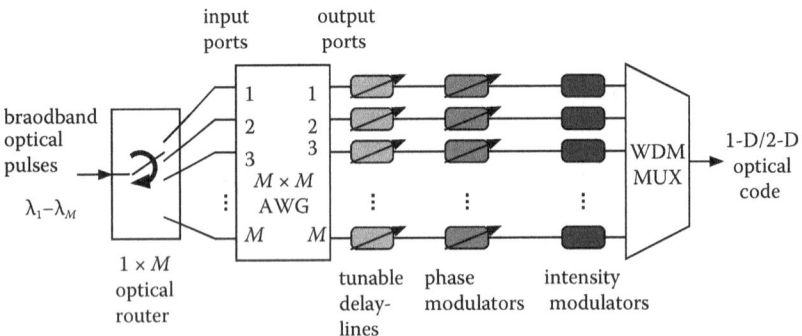

Fig. 3 Tunable AWG-based encoder for 1-D spectral (phase and amplitude) coding and 2-D spectral-temporal coding

can be represented as a modulo-M addition determined by which input port i is injected with a broadband optical pulse, where i and $j \in [1, M]$. For example, consider an 8×8 AWG device. Assume that the exit wave-lengths at the output ports $j = \{1, 2, 3, 4, 5, 6, 7, 8\}$ are $\{\lambda_1, \lambda_2, \lambda_3, \lambda_4, \lambda_5, \lambda_6, \lambda_7, \lambda_8\}$, respectively, when input port $i = 1$ is injected with an optical pulse. Then, the exit wavelengths at the output ports $j = \{1, 2, 3, 4, 5, 6, 7, 8\}$ are rotated once and become $\{\lambda_2, \lambda_3, \lambda_4, \lambda_5, \lambda_6, \lambda_7, \lambda_8, \lambda_1\}$, respectively, when input port $i = 2$ receives the optical pulse. Table 1 tabulates the wavelength at each output port of this 8×8 AWG device as a function of which input port (or row) is applied with a broadband optical pulse. By turning the optical router to the desired input port and tapping into the appropriate output ports, the spectral components of an optical pulse can be separated and rotated rapidly.

To generate 1-D spectral-amplitude codes based on the maximal-length sequence $(+1, +1, +1, -1, -1, +1, -1)$ of length 7, the encoder is configured with $M = 7$, all delay-lines being set to give zero time-delay, and the phase modulators giving zero phase shift. In this configuration, every "+1" element in the maximal-length sequence is transmitted with a wavelength that is determined by the "+1" element's position. The "−1" elements are not transmitted, which is done by turning off the intensity modulators along the corresponding wavelength path. So, the resulting spectral-amplitude codeword C_1 contains $\lambda_1\, \lambda_2\, \lambda_3\, \lambda_6$. There exist totally 7 orthogonal spectral-amplitude codewords $\{C_1, C_2, C_3, ..., C_7\}$ by cyclic-shifting this maximal-length sequence, as given in Table 2.[17] Codeword C_i is generated when the 1×7 optical router is turned to input port $i \in$[18,19] of the 7×7 AWG device.

To generate spectral-amplitude codes based on the "+1" elements of other bipolar codes, such as Walsh codes and "modified" maximal-length sequences of length M,[20,14,15] wavelength periodicity of the AWG is not useful. (The modified maximal-length sequences of length N are constructed by padding a "−1" to the ends of every maximal-length sequences of length $N - 1$ in order to have the same number of "+1" and "−1" in each sequence.) By setting the $1 \times M$ optical router to one input port and all delay-lines to zero time shift, code generation is tunable through the use of the intensity modulators to select which wavelengths to preserve or drop.[21] Also given in Table 2 are Walsh code of length 8, whereas 8 wavelengths are

Table 1 Periodic-wavelength assignment of an 8×8 AWG output port

	1	2	3	4	5	6	7	8
input port 1	λ_1	λ_2	λ_3	λ_4	λ_5	λ_6	λ_7	λ_8
input port 2	λ_2	λ_3	λ_4	λ_5	λ_6	λ_7	λ_8	λ_1
input port 3	λ_3	λ_4	λ_5	λ_6	λ_7	λ_8	λ_1	λ_2
input port 4	λ_4	λ_5	λ_6	λ_7	λ_8	λ_1	λ_2	λ_3
input port 5	λ_5	λ_6	λ_7	λ_8	λ_1	λ_2	λ_3	λ_4
input port 6	λ_6	λ_7	λ_8	λ_1	λ_2	λ_3	λ_4	λ_5
input port 7	λ_7	λ_8	λ_1	λ_2	λ_3	λ_4	λ_5	λ_6
input port 8	λ_8	λ_1	λ_2	λ_3	λ_4	λ_5	λ_6	λ_7

Table 2 Transmission wavelengths of spectral-amplitude code based on the maximal-length sequences of length 7 and walsh code of length 8

	Maximal-length sequences	Wavelength usage
C_1	$(+1, +1, +1, -1, -1, +1, -1)$	$\lambda_1\, \lambda_2\, \lambda_3\, \lambda_6$
C_2	$(-1, +1, +1, +1, -1, -1, +1)$	$\lambda_2\, \lambda_3\, \lambda_4\, \lambda_7$
C_3	$(+1, -1, +1, +1, +1, -1, -1)$	$\lambda_1\, \lambda_3\, \lambda_4\, \lambda_5$
C_4	$(-1 +1, -1, +1, +1, +1, -1)$	$\lambda_2\, \lambda_4\, \lambda_5\, \lambda_6$
C_5	$(-1, -1, +1, -1, +1, +1, +1)$	$\lambda_3\, \lambda_5\, \lambda_6\, \lambda_7$
C_6	$(+1, -1, -1, +1, -1, +1, +1)$	$\lambda_1\, \lambda_4\, \lambda_6\, \lambda_7$
C_7	$(+1, +1, -1, -1, +1, -1, +1)$	$\lambda_1\, \lambda_2\, \lambda_5\, \lambda_7$
	Walsh codes	**Wavelengths usage**
C_1	$(+1, -1, +1, -1, +1, -1, +1, -1)$	$\lambda_1\, \lambda_3\, \lambda_5\, \lambda_7$
C_2	$(+1, +1, -1, -1, +1, +1, -1, -1)$	$\lambda_1\, \lambda_2\, \lambda_5\, \lambda_6$
C_3	$(+1, -1, -1, +1, +1, -1, -1, +1)$	$\lambda_1\, \lambda_4\, \lambda_5\, \lambda_8$
C_4	$(+1, +1, +1, +1, -1, -1, -1, -1)$	$\lambda_1\, \lambda_2\, \lambda_3\, \lambda_4$
C_5	$(+1, -1, +1, -1, -1, +1, -1, +1)$	$\lambda_1\, \lambda_3\, \lambda_6\, \lambda_8$
C_6	$(+1, +1, -1, -1, -1, -1, +1, +1)$	$\lambda_1\, \lambda_2\, \lambda_7\, \lambda_8$
C_7	$(+1, -1, -1, +1, -1, +1, +1, -1)$	$\lambda_1\, \lambda_4\, \lambda_6\, \lambda_7$

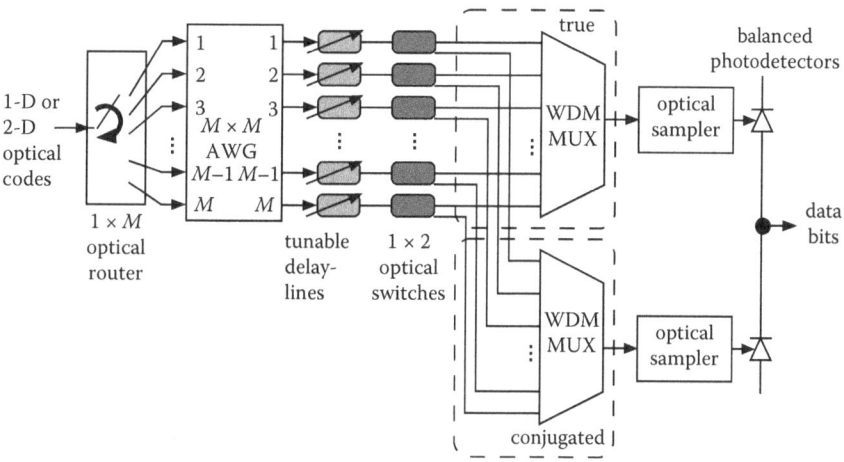

Fig. 4 Tunable AWG-based optical balanced receivers for 1-D spectral (phase and amplitude) coding and 2-D spectral-temporal coding

used to represent the "+1" elements in the codewords in accordance to the "+1" element's positions.

Similarly, for 1-D spectral phase coding of bipolar codes, the phase modulators come into play in order to add π phase shifts to the "−1" elements of the codes. For 2-D spectral-temporal amplitude coding, time spreading is achieved by tunable delay-lines in addition to the wavelength selection by the intensity modulators. Variations of this tunable spectral encoder have been proposed for various optical coding schemes in.[12–15]

Figure 4 shows a tunable optical balanced receiver that can be used to correlate 1-D spectral-phase codes, 1-D spectral-amplitude codes, and 2-D spectral-temporal-amplitude codes. The receiver consists of a $1 \times M$ optical router, an $M \times M$ AWG device, a set of M tunable delay-lines and 1×2 optical switches, two WDM MUXs, two optical samplers, and a pair of balanced photodetectors. Because each receiver is assigned with one 1-D or 2-D spectral codeword as its address signature, the top WDM MUX is used to collect incoming wavelengths that match those wavelengths in its assigned (address) spectral codeword, and the bottom WDM MUX collects the rest of the wavelengths arriving at the receiver. For the case of 1-D spectral amplitude coding, the "+1" elements of a bipolar codeword, such as the maximal-length sequences, are conveyed by wavelengths, the "−1" elements are emulated through subtraction at the balanced photodetectors. This is done by routing the wavelengths corresponding to the "−1" elements of the address codeword to the bottom WDM MUX by properly setting the 1×2 optical switches. The tunable delay-lines are all set to give zero time-delay. Wavelength tunability is achieved by the $1 \times M$ optical router, which routes incoming 1-D spectral-amplitude codewords to one of the input ports of the $M \times M$ AWG device for the proper amount of wavelength rotations. Optical samplers are used to improve the signal-to-interference ratio and reduce the bandwidth requirement of the balanced photodetectors. If one of the arriving codewords matches the address codeword of the receiver, an autocorrelation peak results and is then passed to an electronic hard-limiter/regenerator for data-bit-1 recovery.

For other bipolar codes, such as Walsh codes and modified maximal-length sequences, tunability is achieved via the use of 1×2 optical switches to route the wavelengths corresponding to the "+1" and "−1" elements of the address codeword to the true and conjugated decoders, respectively. Also, the tunable delay-lines are all set to give zero time-delay. For 2-D spectral-temporal amplitude coding, time despreading is performed by the tunable delay-lines. The unused wavelengths are routed to the bottom WDM MUX and then ignored by disabling the lower optical sampler.

REFERENCES

1. Baby, V.; Brés, C.-S.; Glesk, I.; Xu, L.; Prucnal, P.R. Wavelength aware receiver for enhanced 2D OCDMA system performance. Electron. Lett. **2004**, *40* (6), 385–387.
2. Kwong, W.C.; Yang, G.-C.; Baby, V.; Brés, C.-S.; Prucnal, P.R. Multiple-wavelength optical orthogonal codes under prime-sequence permutations for optical CDMA. IEEE Trans. Commun. **2005**, *53* (1), 117–123.
3. Kwong, W.C.; Yang, G.-C. Multiple-length extended carrier-hopping prime codes for optical CDMA systems supporting multirate, multimedia services. J. Lightwave Technol. **2005**, *23* (11), 3653–3662.
4. Hsu, C.-C.; Yang, G.-C.; Kwong, W.C. Hard-limiting performance analysis of 2-D optical codes under the chip-asynchronous assumption. IEEE Trans. Commun. **2008**, *56* (5), 762–768.
5. Lee, J.-H.; Teh, P.C.; Petropoulos, P.; Ibsen, M.; Richardson, D.J. A grating-based OCDMA coding-decoding system incorporating a nonlinear optical loop mirror for improved code recognition and noise reduction. J. Lightwave Technol. **2002**, *20* (1), 36–46.
6. Sokoloff, J.P.; Prucnal, P.R.; Glesk, I.; Kane, M. A terahertz optical asymmetric demultiplexer (TOAD). IEEE Photonic. Technol. Lett. **1993**, *5* (7), 787–790.

7. Prucnal, P.R; (Ed.); *Optical Code Division Multiple Access: Fundamentals and Applications*; Taylor & Francis Group: Boca Raton, FL, 2006.
8. Chen, L.R.; Benjamin, S.D.; Smith, P.W.E.; Sipe, J.E. Applications of ultrashort pulse propagation in Bragg gratings for wavelength-division multiplexing and code-division multiple access. IEEE J. Quantum Electron. **1998**, *34* (11), 2117–2129.
9. Teh, P.C.; Petropoulos, P.; Ibsen, M.; Richardson, D.J. A comparative study of the performance of seven- and 63-chip optical code-division multiple-access encoders and decoders based on superstructured fiber Bragg gratings. J. Lightwave Technol. **2000**, *19* (9), 1352–1365.
10. Teh, P.C.; Petropoulos, P.; Ibsen, M.; Richardson, D.J. Phase encoding and decoding of short pulses at 10 Gb/s using superstructured fiber Bragg gratings. IEEE Photonic. Technol. Lett. **2001**, *13* (2), 154–156.
11. Chen, L.R. Flexible fiber Bragg grating encoder/decoder for hybrid wavelength-time optical CDMA. IEEE Photonic. Technol. Lett. **2001**, *13* (11), 1233–1235.
12. Kwong, W.C.; Yang, G.-C.; Liu, Y.-C. A new family of wavelength-time optical CDMA codes utilizing programmable arrayed waveguide gratings. IEEE J. Sel. Areas Commun. **2005**, *23* (8), 1564–1571.
13. Kwong, W.C.; Yang, G.-C.; Chang, C.-Y. Wavelength-hopping time-spreading optical CDMA with bipolar codes. J. Lightwave Technol. **2005**, *23* (1), 260–267.
14. Hsieh, C.-P.; Chang, C.-Y.; Yang, G.-C.; Kwong, W.C. A bipolar-bipolar code for asynchronous wavelength-time optical CDMA. IEEE Trans. Commun. **2006**, *54* (7), 1190–1194.
15. Hu, H.-W.; Chen, H.-T.; Yang, G.-C.; Kwong, W.C. Synchronous Walsh-based bipolar-bipolar code for CDMA passive optical networks. J. Lightwave Technol. **2007**, *25* (8), 1910–1917.
16. Takahashi, H.; Oda, K.; Toba, H.; Inoue, Y. Transmission characteristics of arrayed waveguide $N \times N$ wavelength multiplexer. J. Lightwave Technol. **1995**, *13* (3), 447–455.
17. Huang, W.; Nizam, M.H.M.; Andonovic, I.; Tur, M. Coherent optical CDMA (OCDMA) systems used for high-capacity optical fiber networks—System description, OTDMA comparison, and OCDMA/WDMA networking. J. Lightwave Technol. **2000**, *18* (6), 765–778.
18. Prucnal, P.R.; Santoro, M.A.; Fan, T.R. Spread spectrum fiber-optic local area network using optical processing. J. Lightwave Technol. **1986**, *4* (5), 547–554.
19. Brackett, C.A.; Heritage, J.P.; Salehi, J.A.; Weiner, A.M. Optical telecommunications system using code division multiple access. US Patent 4,866,699. Issue date: September 12, 1989.
20. Dinan, E.H.; Jabbari, B. Spreading codes for direct sequence CDMA and wideband CDMA cellular networks. IEEE Commun. Mag. **1998**, *36* (9), 48–54.
21. Lam, C.F.; Tong, D.T.K.; Wu, M.C.; Yablonovitch, E. Experimental demonstration of bipolar optical CDMA system using a balanced transmitter and complementary spectral encoding. IEEE Photonic. Technol. Lett. **1998**, *10* (10), 1504–1506.

Perceptual Image and Video Quality

Deepti Ghadiyaram, Todd Goodall, Lark Kwon Choi, and Alan C. Bovik
Laboratory of Image and Video Engineering, The University of Texas at Austin, Austin, Texas, U.S.A.

Abstract
Owing to the explosive growth in the volume of photographs and videos in the recent years, finding effective and efficient ways to control and monitor their perceptual quality has become a topic of significant interest. Great strides have been made in understanding human vision and in developing advanced perceptual image and video quality measures. In this entry, we discuss the fundamental principles that have thus far motivated the design of a broad array of sophisticated, biologically inspired image and video quality assessment models. We also present the tremendous potential of these powerful models in delivering high-quality visual content to the end user.

Keywords: Digital images and videos; Human visual system; Natural scene statistics; Objective quality; Perceptual quality; Visual perception.

INTRODUCTION

We live in a world obsessed with taking images, recording, and streaming videos. Visual media is increasingly pervasive everywhere: entertainment and social networks to news reports; the film industry along with social media websites has an almost bottomless supply of visual content. The Internet offers many venues such as Flickr, YouTube, Vimeo, Instagram, and Vine for publishing images and videos that reflect individualized creativity of the increasingly knowledgeable consumer base. Such ubiquitousness of images and videos has led to rapid, synergistic advances in technology by camera and mobile device manufacturers, allowing consumers to efficiently capture and store high-resolution images and videos. This increasing demand for visual data also necessitates the development of accurate metrics that understand and even estimate its quality in light of the evolving standards and devices.

A vast majority of the captured digital images (and videos) are by amateur photographers whose unsure hands and eyes could potentially introduce annoying artifacts during the capture process. Furthermore, every digital image passes through various stages during its acquisition, storage, and transmission, and thus could suffer from a wide variety of spatial and temporal distortions. Blocking, overexposure, underexposure, ringing, noise, compression, and blurring are some of the examples of commonly occurring spatial distortions whereas ghosting, smearing, and flickering are some examples of temporal distortions that afflict videos.

Blocking effects are discontinuities between the boundaries of adjacent logical blocks algorithmically formed as part of certain block-based compression operations such as JPEG, MPEG, and H.264. Ringing artifacts usually appear as a ripple around areas with sharp transitions in a signal such as high-contrast edges, commonly introduced due to signal compression. Blurring is the loss of sharpness and spatial detail in images and could be caused by several factors during or after image acquisition. Sensor noise and low illumination could sometimes cause additive noise in images, which presents itself as a grainy texture.

Pertaining to temporal distortions, ghosting appears as a motion-blurred remnant around moving objects. One of the many causes for a temporal blurring artifact is a non-instantaneous exposure while moving objects are being captured. Flickering is the noticeable discontinuity between consecutive frames as a result of abrupt changes in the luminance caused by low frame rate, severe coding artifacts, or object motion.

Irrespective of their source, any kind of distortion that afflicts images (or videos) could potentially degrade an end user's quality of experience (QoE). The dramatic shift toward capturing images and videos and the increase in over-the-top video streaming services in the last few years despite the limited capacity of wireless networks has propelled the immense need to understand and measure QoE. Studying the impact of various spatial and temporal distortions on human perception and identifying their presence in images and videos are therefore critical for understanding human perceptual visual quality, thereby aiding in addressing the end goal of delivering better quality visual content to end users.

Perceptual Quality and Objective Quality Assessment

The word "quality" is touted as a measure of excellence, but there are several distinct connotations to it in the image and video processing literature. For instance, "perceptual quality" refers to the subjective quality of a visual stimuli as perceived by the observer. By contrast, "aesthetic quality" refers to the appreciation of beauty and possibly the artistic quality of the visual stimuli. In this entry, however, we exclusively deal with perceptual quality.

Given that the ultimate receiver of any visual media is the human eye, there is considerable interest in the research community to deeply understand the functionalities of our visual system that assist us to evaluate the perceptual fidelity of visual media. These neurobiological findings have the potential to motivate the development of appropriate and successful biologically inspired computational models for quality assessment.

Subjective assessment of perceptual quality of a given set of images (and videos) involves polling a large number of human populace for their opinion scores. Though having humans in the loop aids in gathering authentic data, given the tremendous surge in the volume of visual media content across the Internet, gathering opinion scores becomes laborious and cumbersome. Thus, *objective algorithms* that automatically and accurately determine image quality, without involving humans, are more viable solutions. The predictive capability of objective quality assessment (QA) algorithms can be used to improve the perceptual quality of visual signals by integrating them into processes that perceptually optimize the signal capture process or that modify transmission rates to ensure good quality across wired or wireless networks. Such "quality-aware" strategies might in turn ensure that end users receive a satisfactory QoE.

However, the subtle and highly subjective nature of perceptual visual quality makes objective quality assessment extremely challenging. Designing accurate objective quality assessment models requires a deep understanding of a broad swathe of topics such as the perceptual processing in the human visual system (HVS), the regularities in natural image and video statistics, the influence of different distortions on these statistics, and the architecture of powerful machine learning-driven models. The aim of this entry is to briefly introduce each of these topics. We start by discussing the fundamentals of HVS and proceed to introducing the remarkable statistical regularities observed in natural images and videos in Section "Human Visual Processing and Natural Scene Statistics." We also discuss the design of a few state-of-the-art visual quality assessment models and introduce a few benchmark image and video quality databases in Sections "Perceptual-Based Objective QA Models" and "Subjective Image and Video Quality Databases," respectively, and conclude with an array of unaddressed research problems pertaining to visual quality in Section "Future Directions of Quality Assessment."

HUMAN VISUAL PROCESSING AND NATURAL SCENE STATISTICS

Human Visual System

Substantial strides have been made toward understanding and modeling low-level visual processing in the human system.[1] While high-level cognitive factors (such as semantic content and attention) can certainly affect a human's perception of quality, it is widely accepted that distortion sensing (of still images at least) is largely pre-attentive[2] and dominated by low-level visual signal processing.

As light from the outside world falls onto the photoreceptors in the human retina, each visual signal is locally encoded by multiple sensory neurons. First, the bipolar cells near the surface of the retina relay the gradient potentials from the photoreceptors to the ganglion cells. There is considerable evidence that local center-surround excitatory–inhibitory processes occur at the receptive fields of the ganglion cells,[3] thus providing a band-pass response to the input luminance (Fig. 1). These ganglion cells are interconnected, and together, they compute a retinal contrast signal that can simply be approximated as:

$$A(x,t) = \frac{f(x,t) - \bar{f}(x,t)}{\bar{f}(x,t)} \qquad (1)$$

This spatial contrast signal is directed to the lateral geniculate nucleus (LGN) (depicted in Fig. 2). LGN decomposes these two center-surround contrast response signals from each retina, the temporal responses of which can be modeled using a set of difference of temporal gamma filters. This operation yields a set of lagged and unlagged temporal responses. The retina decorrelates the spatial signal, whereas the LGN decorrelates the retinal signal temporally. These four response signals

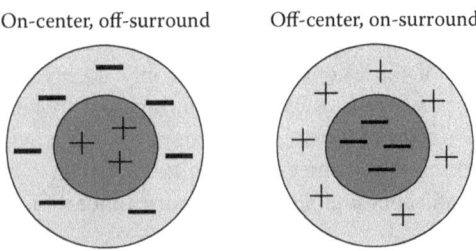

Fig. 1 Depiction of on-center and off-center excitatory–inhibitory responses of ganglion cells in the retina

provide a simple and complete "dictionary" necessary to describe the visual input signal[4] for later parts of the visual pathway. Even though the model assumes separability and linearity, it closely predicts the behavior observed in HVS, thus reinforcing the "efficient coding" hypothesis, which we describe later.

A bridge connecting LGN to the primary visual cortex, also known as visual area 1 (V1) transmits these four response signals to the simple cells. These *simple cells* in area V1 can be modeled as collectively providing a large bank of quadrature pairs of log-gabor type responses. *Complex cells* can be simply modeled as adding the local half-square rectified responses of simple cells and further normalizing them. The collective output of complex cells mimics a filter bank of log-gabor filters tuned for various spatiotemporal orientations of visual stimuli across scales.

After area V1, the flow of information along the visual pathway is often broadly modeled as a split into the ventral and dorsal streams. The ventral stream ("What Pathway") mostly follows the pathway from V1 to the temporal lobe via visual area 2 (V2) and visual area 4 (V4), and corresponds to object recognition and shape representation. The dorsal stream ("Where Pathway") follows the pathway from V1 to middle temporal visual area (MT) via V2 and corresponds to motion computation of object locations and trajectories including the control of eyes and arms.

Natural Scene Statistics

The development of HVS is strongly dependent on early visual stimulation and the statistics of the surrounding visual environment. Although statistically analyzing the image environment could lead to a deeper understanding of human visual processing, there is no way to collect enough data to fully characterize the same.

Nevertheless, the statistics of real-world natural images, generally referred to as *Natural Scene Statistics* (NSS), have been deeply studied for the past several years.[5,6] NSS models are based on the principled observation that good-quality real-world photographic images exhibit certain perceptually relevant statistical regularities. Despite the tremendous diversity of natural images in terms of content and capture processes, these statistical regularities are remarkably consistent. For instance, the amplitude spectra of the spatial Fourier transforms of natural images obey an approximate reciprocal law. This is a statistically self-similar phenomenon and is invariant to scale.

Another powerful statistical regularity is founded on the thesis that the firings of sensory neurons along the visual pathways carry efficient representations of the visual information. Olshausen and Field[7] thus conjectured that natural images also have such an efficient and *sparse* representation that can be exploited by the visual system. The sparse codes (illustrated in Fig. 3) derived from natural images provide minimal reconstruction error while preserving information. They strongly resemble Gabor filters, or the receptive field profiles of 2D simple cells in primary visual cortex. Another analysis showed that the principal (and independent) spatial (and spatiotemporal) components of natural time-varying images strongly resemble the simple cell responses in the visual cortex.

Another useful statistical regularity of natural images surfaces when subjected to a spatial linear band-pass filters such as a difference of Gaussian (DoG) or a predictive coding filter (Eq. 5).[8] The empirical probability distributions of these filtered responses can be reliably modeled using a Gaussian probability distribution.

For example, given an image's intensity map I of size $M \times N$, a divisive normalization operation[5] yields a normalized luminance coefficients (NLCs) map:

$$\text{NLC}(i,j) = \frac{I(i,j) - \mu(i,j)}{\sigma(i,j) + 1}, \qquad (2)$$

where

$$\mu(i,j) = \sum_{k=-3}^{3} \sum_{l=-3}^{3} w_{k,l} I_{k,l}(i,j), \qquad (3)$$

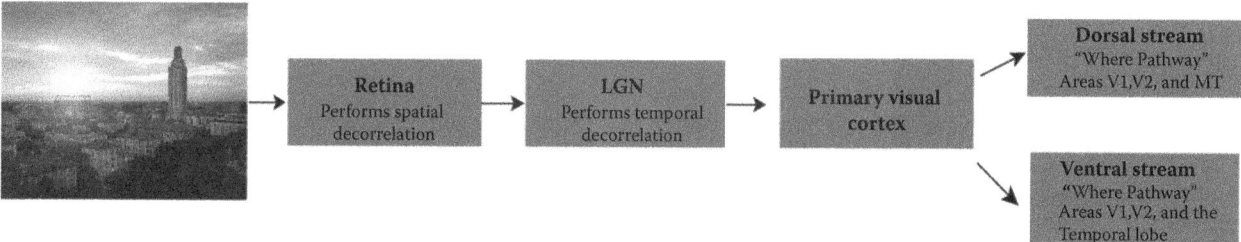

Fig. 2 The big picture of the visual pathway. The ganglion cells in the retina spatially decorrelate the incoming visual input, and the cells in the LGN temporally decorrelate the resulting spatial signal. This spatiotemporally decorrelated signal is transmitted to area V1 for further processing. After V1, the two streams split to perform two main categories of processing (popularly known as the What and Where pathways) in the HVS

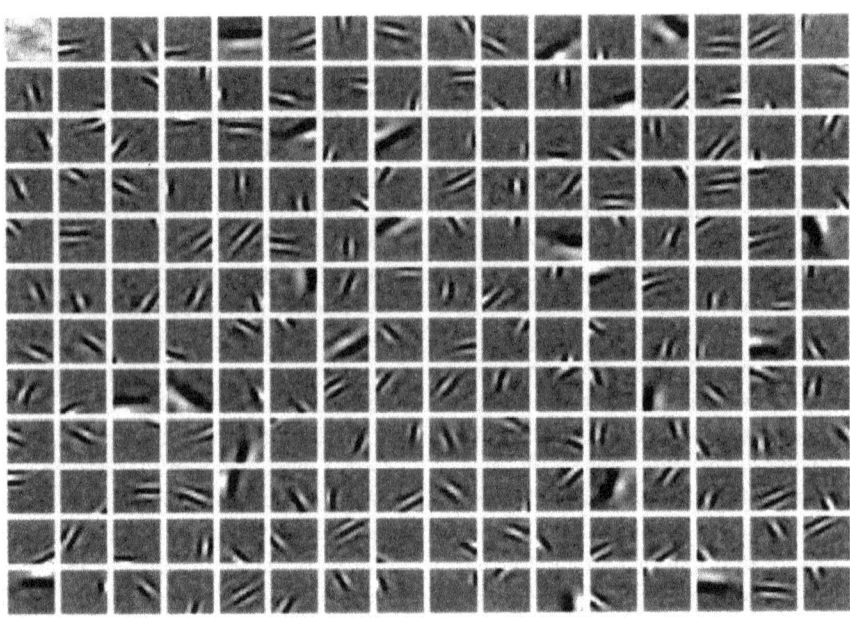

Fig. 3 Illustration of a few sparse spatial codes derived on natural image data. These codes strongly resemble the receptive field profiles (2D impulses responses) of 2D simple cells in primary visual cortex

and

$$\sigma(i,j) = \sqrt{\sum_{k=-3}^{3}\sum_{l=-3}^{3} w_{k,l}\left[I_{k,l}(i,j) - \mu(i,j)\right]^2}, \quad (4)$$

where $i \in 1,2\ldots M$, $j \in 1,2\ldots N$ are spatial indices and w is a 2D circularly symmetric Gaussian weighting function. This *debiasing* and *divisive normalization* process mimics the normalization operation performed by complex cells.

Similarly, a good filter to model the center-surround excitatory–inhibitory processes that occur at various stages of visual processing is the 2D difference of isotropic Gaussian filters (DoG):

$$\text{DoG} = \frac{1}{\sqrt{2\pi}}\left(\frac{1}{\sigma_1}e^{\frac{-(x^2+y^2)}{2\sigma_1^2}} - \frac{1}{\sigma_2}e^{\frac{-(x^2+y^2)}{2\sigma_2^2}}\right), \quad (5)$$

where $\sigma_2 = 1.5\sigma_1$.

Figure 4 depicts a pristine natural image, its NLC map, and their corresponding histograms of intensity values along with the scatter plots of horizontally adjacent pixels. The white noise-like scatter plot of the NLC map of I is indicative of the decorrelation of the pixels, which contrasts from the near-linear correlation between the plot of I. As illustrated in Fig. 5, the empirical probability density function of $I''(x,y)$ (obtained from applying DoG on I) of natural images also closely follows a Gaussian-like distribution.

PERCEPTUAL-BASED OBJECTIVE QA MODELS

Perceptual-Based versus Metric-Based Models

Objective quality assessment algorithms can be divided into two classes based on the kind of information extracted from the visual signal—metric based and perceptual based. The former category refers to the class of algorithms that are mostly designed with a focus on making direct measurements of the contained image (video) artifacts and/or analyzing certain distortion-specific changes to the spatial structures in the image (video).

However, the most efficient QA algorithms to date are perceptual based and are founded on the basis of NSS. It has been observed by vision scientists and image engineers that certain perceptually relevant statistical laws obeyed by natural scenes (introduced in Section "Natural Scene Statistics") are violated by the presence of common distortions. Pertaining to images, if they are singly distorted, that is, if images contain only one of the few synthetically introduced distortions, then the natural statistics of such distorted images make it possible to determine the presence and identify the type of distortion as well.

For example, the systematic Gaussian-like behavior of the histogram of the normalized coefficients in Eq. 2 of natural, undistorted images is modified by the presence of common

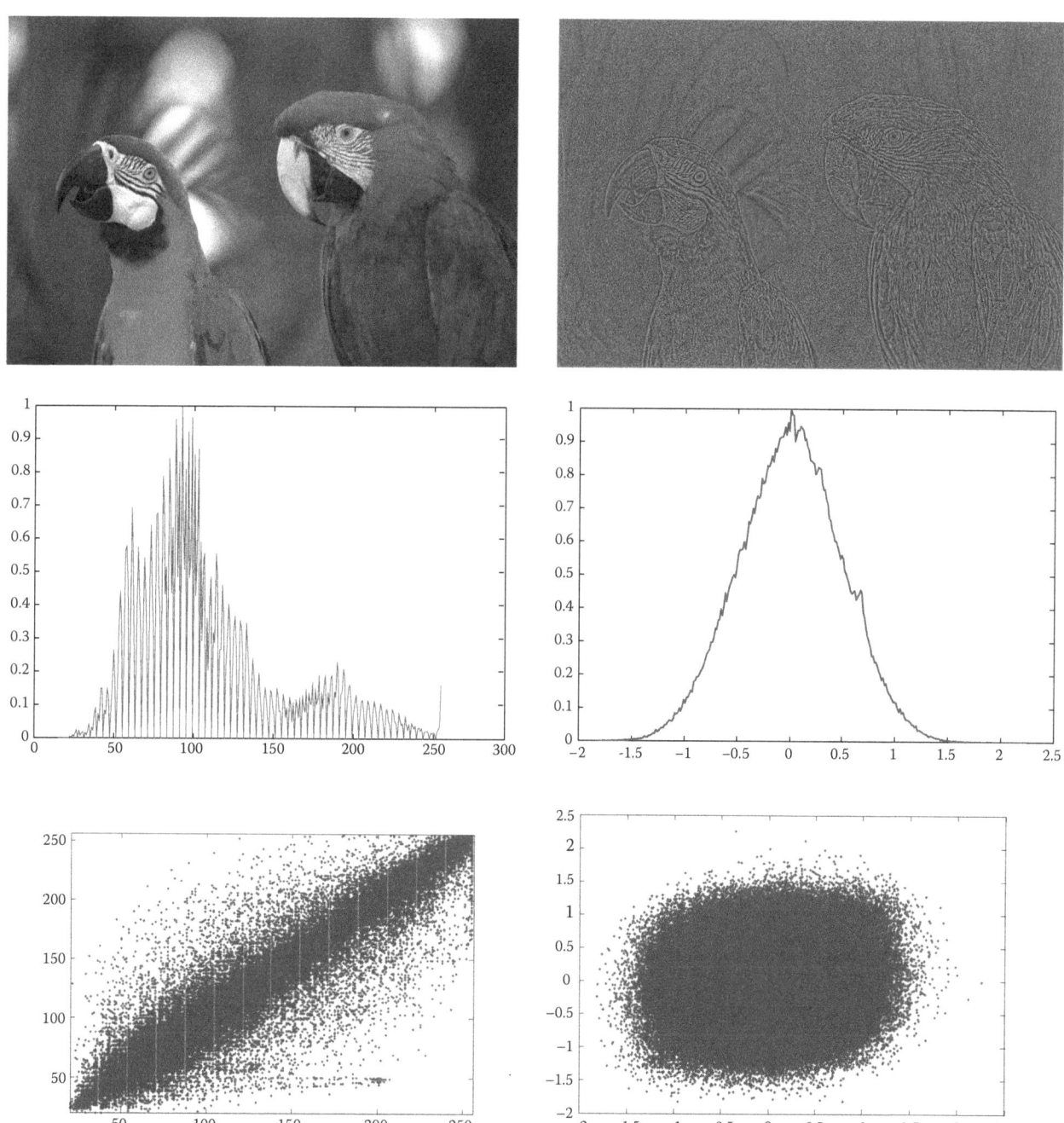

Fig. 4 A natural undistorted image (shown in the upper left) when processed by applying the debiasing and normalization produces a decorrelated NLC map (shown in the upper right). The histogram of the intensity values of the NLC map (middle right) follows a Gaussian distribution. The scatter plots (lower row) contrast the highly correlated natural image and the nearly decorrelated NLC map

image distortions, as illustrated in Fig. 6. Furthermore, each distortion characteristically affects the image distribution (Fig. 6a). Effectively quantifying these deviations is crucial for being able to predict the perceptual quality of that image. A generalized Gaussian distribution (GGD) with zero mean is given by

$$f(x;\alpha,\sigma^2) = \frac{\alpha}{2\beta\Gamma(1/\alpha)} e^{-\left(\frac{|x|}{\beta}\right)^\alpha}, \qquad (6)$$

where

$$\beta = \sigma\sqrt{\frac{\Gamma(1/\alpha)}{\Gamma(3/\alpha)}}, \qquad (7)$$

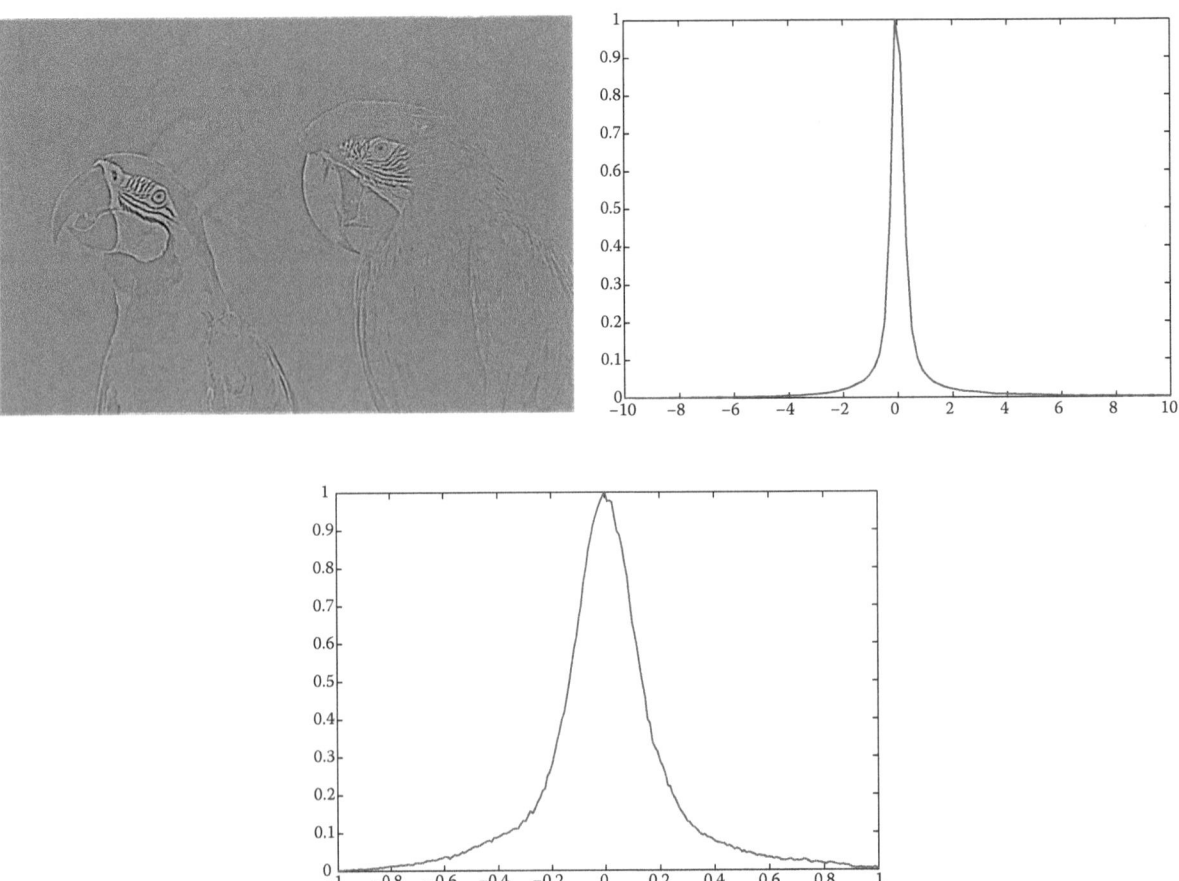

Fig. 5 (Left) The residual when the DoG filter is applied on a natural image shown in Fig. 4a. (Middle) The histogram of the *DoG* residual. (Right) The debiased and normalized residual also closely follows a Gaussian distribution

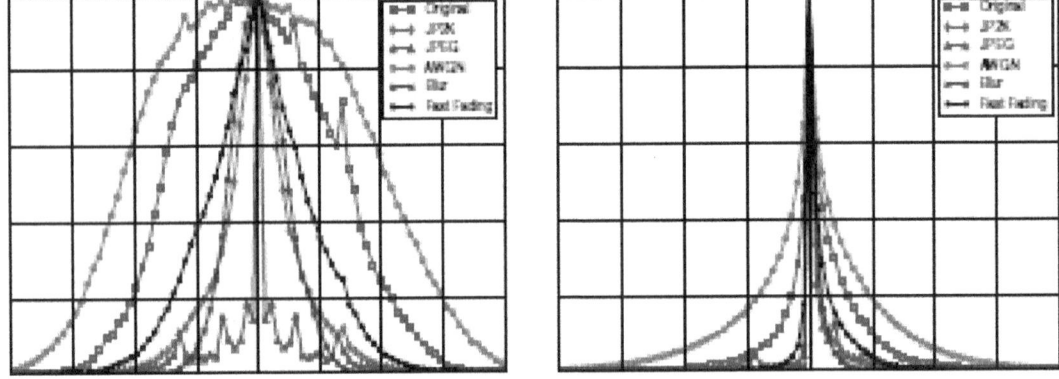

Fig. 6 Histogram of (left) the NLC maps of a naturally undistorted image and its various distorted versions. (Right) The paired products of NLC maps of a naturally undistorted image and its various distorted versions

and $\Gamma(\cdot)$ is the gamma function:

$$\Gamma(a) = \int_0^\infty t^{a-1} e^{-t} dt \quad a > 0 \qquad (8)$$

can be effectively used to model a broad spectrum of singly distorted image statistics. The parameter controls the "shape" of the distribution while controls its variance.

State-of-the-art NSS-based image quality assessment (IQA) models exploit this statistical unnaturalness of images in the presence of distortions using the estimates of the shape and scale parameters to identify and/or assess distortions (Section "No-Reference Visual Quality Assessment").

Furthermore, these spatial distortions also affect the statistical relationships between neighboring pixels thereby introducing unnatural dependencies. The pairwise products of the neighboring normalized coefficients along four orientations, horizontal, vertical, main diagonal, and secondary diagonal, are defined as

$$H(i,j) = \text{NLC}(i,j)\text{NLC}(i,j+1) \tag{9}$$

$$V(i,j) = \text{NLC}(i,j)\text{NLC}(i+1,j) \tag{10}$$

$$D1(i,j) = \text{NLC}(i,j)\text{NLC}(i+1,j+1) \tag{11}$$

$$D2(i,j) = \text{NLC}(i,j)\text{NLC}(i+1,j-1), \tag{12}$$

where $i \in 1,2\ldots M$, $j \in 1,2\ldots N$ are spatial indices.

An asymmetric GGD model given by

$$f(x;v,\sigma_l^2,\sigma_r^2) = \begin{cases} \dfrac{v}{(\beta_l+\beta_r)\Gamma(1/v)}\exp\left(-\left(\dfrac{-x}{\beta_l}\right)^v\right) & x<0 \\[2ex] \dfrac{v}{(\beta_l+\beta_r)\Gamma(1/v)}\exp\left(-\left(\dfrac{x}{\beta_r}\right)^v\right) & x>0 \end{cases} \tag{13}$$

where

$$\beta_l = \sigma_l\sqrt{\dfrac{\Gamma(1/\alpha)}{\Gamma(3/\alpha)}} \tag{14}$$

$$\beta_r = \sigma_r\sqrt{\dfrac{\Gamma(1/\alpha)}{\Gamma(3/\alpha)}}, \tag{15}$$

where η is given by

$$\eta = (\beta_r - \beta_l)\dfrac{\Gamma(2/v)}{\Gamma(1/v)}, \tag{16}$$

captures the shapes and distortion-driven asymmetries in the distributions of these pairwise products. These parameters ($v, \sigma_l^2, \sigma_r^2$) are highly reliable indicators of distortions contained in the images as shown in Fig. 6b and play a critical role in IQA models (Section "No-Reference Visual Quality Assessment"). These kinds of features are also important in video quality assessment (VQA) models, when applied on frame differences.[9]

Regardless of the types and severities of the contained distortions, image and VQA models can be broadly divided into three categories, depending on the amount of information about the reference signal that is available to the algorithm: full-reference (FR), reduced-reference (RR), and no-reference (NR) or "blind."

Full-Reference Visual Quality Assessment

An FR quality assessment algorithm assumes that a pristine signal is available to it, thus allowing a full comparison between pristine and distorted signals. Requiring the reference signal is favorable for explicitly interpreting the fidelity of the distorted signal by measuring the mutual information of the two signals. One of the most attractive and computationally inexpensive signal fidelity measures is the mean squared error (MSE) between the pristine and distorted signals. MSE is usually converted into a peak signal-to-noise ratio (PSNR) measure and widely used in the image and video processing literature. However, MSE (or PSNR) does not incorporate any information about the transmitter, channel, or most importantly, the final receiver. Thus, developing sophisticated models to evaluate the perceptual signal fidelity by exploiting our current understanding of biological visual information processing mechanisms is of great interest.

One of the FR IQA algorithms that is proven to be highly effective is the (Structural Similarity) SSIM index.[10] The fundamental idea underlying the original SSIM approach is that natural images are highly structured with strong neighbor dependencies, and these dependencies carry critical information about the structures of the objects in the visual scene. It is further based on the premise that the HVS is highly adapted to extract such structural information from visual scenes. Thus, a measure of signal fidelity should incorporate the perceived differences in the structural information between pristine and distorted signals, in addition to the nonstructural illumination and contrast changes.

Specifically, suppose that x and y are two spatially aligned patches extracted from the pristine image and its distorted variant, respectively. The SSIM framework separates the task of measuring the similarity between the pristine and distorted images into three comparisons: luminance $l(x, y)$, contrast $c(x, y)$, and structure $s(x, y)$.

$$S(x,y) = l(x,y)c(x,y)s(x,y) \tag{17}$$

Despite its simplicity, the SSIM index performs remarkably well across a wide variety of images and distortion types and has been extensively used for benchmarking numerous applications such as image fusion, compression, and visual surveillance. A brief overview of several other top-performing FR IQA models such as visual information fidelity (VIF), visual signal-to-noise ratio (VSNR), most apparent distortion (MAD), and feature similarity (FSIM) can be found in the work of Bovik.[11]

While extensions of SSIM to video have been shown to perform well, the leading FR VQA tool to date is the Motion-based Video Integrity Evaluation (MOVIE) index.[12] MOVIE uses a model of visual area V5, also known as visual area MT (middle temporal) of the extra-striate cortex[13] and works by estimating a temporal distortion map and a spatial distortion map. Local estimates of motion produce 3D vectors oriented in space and time. The frequency-transformed input contains planes corresponding and orthogonal to these 3D vectors. The misalignments of these aforementioned planes between reference and distorted video provide a *perceptual temporal distortion map*. A bank of spatial Gabor filters across scale and orientation are used to compare the reference and distorted video and generate a *perceptual spatial distortion map*. These two distortion maps are combined multiplicatively and over a video, the coefficient of variation is computed to provide a single quality score.

RR Visual Quality Assessment

The availability of a pristine signal along with its distorted version is impractical in many real-world scenarios, thus posing a serious limitation in terms of the applicability of FR quality assessment models to several "quality-centric" applications and analysis. NR QA techniques lie on the other end of this spectrum of information availability as they are not based on any additional information except for the distorted signal whose quality needs to be ascertained (Section "No-Reference Visual Quality Assessment"). RR quality assessment algorithms fall somewhere in the middle of this spectrum, where, in addition to the distorted signal, a small fraction of information pertaining to the reference signal is also made available to the algorithm. Reduced reference entropic differencing (RRED)[9] is one of the most successful frameworks of RR IQA algorithms. RRED adopts an information theoretic approach where natural image approximations of the distorted image are considered. Specifically, wavelet coefficients are first obtained by a steerable pyramid decomposition of reference and distorted images. Next, a distorted image is *approximated* to follow a Gaussian scale mixture (GSM) distribution by projecting it on to the space of natural images. The problem of quality assessment is approached from the perspective of measuring the average difference between the scaled entropies of wavelet coefficients of the reference and projected distorted images. A family of algorithms are proposed in the work of Soundararajan and Bovik[9] depending on the sub-band in which the quality computation is carried out, and the amount of information is transmitted from the reference image. Thus, the quality prediction power is evaluated in relation to the amount of side information extracted and transmitted from the reference image, and it could be generally concluded that the prediction power is improved with the increase in the availability of side information.

The central idea of RRED has been extended to the RR VQA problem wherein a steerable pyramid decomposition of video frame differences produces sub-bands covering different scales and orientations. Local blocks are selected in each sub-band, and a GSM model is used to estimate conditional entropy, spatially and temporally. These entropies are then combined to obtain a single quality index. The resulting spatiotemporal RRED index performs quite competitively with top-performing FR VQA models.

NR Visual Quality Assessment

Objective blind or NR quality assessment refers to the scenario where the only information that is available to the algorithm is the distorted signal whose quality is to be predicted. Blind quality assessment is certainly the most challenging as well as the most interesting problem with a potential for being integrated into various real-time applications. In reality, the possibility of a reference image (or video) being available is rare, and thus it becomes extremely important to rapidly develop robust NR QA algorithms with high predictive capabilities.

Past and ongoing research on developing new blind IQA models has been largely devoted to extracting handcrafted, low-level image descriptors that are independent of the image content. There are several models proposed in the past that assume a particular kind of distortion such as blur, ringing, and blocking, and extract distortion-specific features. A comprehensive summary of all such models can be found in the work of Bovik.[11]

The development of NR IQA models based on NSS that do not make assumptions a priori on the contained distortion is also experiencing a surge of research. Tang et al.[14] proposed an approach which combines features derived using NSS with certain texture, blur, and noise statistics. Moorthy and Bovik's DIIVINE (Distortion Identification-based Image Verity and Integrity Evaluation) index deploys summary statistics derived from an NSS wavelet coefficient model, using a two-stage framework for QA: distortion identification followed by distortion-specific QA. Another complementary single-stage approach, BLIINDS-II, operates in the discrete cosine transform (DCT) domain where a small number of features are computed and yet achieves highly competitive performance. Codebook Representation for No-Reference Image Assessment (CORNIA) is an unsupervised technique and is not an NSS-based model as it builds several distortion-specific code words to compute image quality.

The NR IQA Blind/Referenceless Image Spatial QUality Evaluator (BRISQUE)[15] index utilizes the GGD model (Eq. 6) fit to its normalized coefficients with two distortion scene statistical features and pair product NSS feature maps modeled by Eq. 13 with four features ($\mu, \sigma_l, \sigma_r, \lambda$) measured in four directions, thus resulting in 18 features. These features, when measured at two scales, yield a total of 36 features extracted from each image. These features are used to train a support vector regressor (SVR) and predict quality scores on test data.

Naturalness image quality evaluator (NIQE)[16] is a recently proposed unsupervised technique driven by NSS-based features that neither requires any exposure to distorted images a priori nor involves training on subjective human opinion scores. Given an image, NIQE first computes the empirical distribution of its normalized coefficients in the spatial domain and compares it against that of a large corpus of representative pristine images. This resulting measure of closeness of a given image to *naturalness* represents the NIQE. It thus affords more generality and competes well with top-performing IQA algorithms.

NR VQA algorithms are only very recently being developed. The current top-performing NR VQA algorithm is called Video BLind Image Integrity Notator using DCT-Statistics (Video BLIINDS).[17] Video BLIINDS works by modeling the behavior of DCTs of the frame differences, behavior of motion vectors, coherency measurements, a flicker feature, and a feature extracted using the NIQE. To provide an index, difference mean opinion scores (DMOS) and features are used to train a SVR and given an input test video, the learner automatically predicts a quality score, without requiring a reference.

Most of these are supervised techniques, and they generally follow a machine learning paradigm, where image/video features are first extracted and a kernel function is learned that maps these features to ground-truth subjective quality scores. Consequently, several benchmark databases such as the LIVE Image Quality database,[18] the TID2008 database,[19] and the LIVE VQA database[20] have been designed to contain several human opinion scores on images (and videos) corrupted by only one of a few synthetically introduced distortions and are extensively utilized by the quality assessment research community. Despite their encouraging performance on the standard benchmark databases, the caveat of most of the earlier models is that they require being sufficiently trained on the distortion types that we wish to test them on. We refer the reader to the work of Bovik[11] for a detailed summary of these models.

Temporal Visual Masking toward VQA

Understanding and modeling perceptual motion statistics in videos is of high interest since they account for the visibility of temporal distortions, and thus perceptual quality. Very recently, Suchow and Alvarez demonstrated a striking "motion silencing" illusion,[21] in which salient temporal changes of an object's luminance, color, size, and shape appear to cease in the presence of large, coherent object motion. This dramatic phenomenon has potential implications for understanding temporal distortion perception. Although the mechanisms underlying this phenomenon remain unknown, a quantitative, spatiotemporal flicker detector model[22] was developed. The model predicts when human subjects would judge that motion silencing has occurred in the presented stimulus, as a function of the velocity of the stimulus, flicker frequency, and spacing. The high correlation between the subjective human opinions and the predictions of the model suggests that the model may be a starting point for better understanding the neural processing involved in motion silencing in the visual pathway.

A consistent physiological and computational model that detects motion silencing might be useful to probe the related motion perception effects, such as distortion visibility in compressed videos. A recent human study investigated the effects of speed of coherent object motions in naturalistic videos on flicker visibility. The results suggest that the visibility of flicker distortions is strongly reduced when the speed of the coherent motion is large, and the influence is more pronounced when the quality of the video is poor. Further understanding the effects of motion silencing on perceptual temporal distortions is an open research problem.

Stereoscopic (3D Image and Video) Quality Assessment

Assessing the perceptual quality of stereoscopic images and videos is a topic in which research is currently in an early stage. Conventional 3D displays cause issues such as visual discomfort and fatigue, mismatches between vergence and accommodation, depth compression, and perceived acuity degradation, all affecting an end user's viewing experience.

As 3D visual input propagates through the HVS, area V1 produces the cyclopean image, and the dorsal pathway comprehends depth (Fig. 2). The cyclopean image is formed from fusing the right and left views by nontrivially accounting for binocular rivalry and differences in visual perception between the two views. This task, however, is extremely challenging given our limited understanding of the formation of the cyclopean image in HVS by accounting for the display geometry, the presumed fixation, vergence, and accommodation.

Recent understandings of the visual pathways have led to better quality assessment models that account for some of these nontrivialities to a certain extent. The development of such QA models is very important because they could help us understand the perceptual quality of the 3D visual content and improve its quality thereby reduce stress and fatigue of the end user.

3D stereoscopic IQA models aim to estimate the quality of the true cyclopean image formed within an observer's mind when a stereo image pair is stereoscopically presented. Aside from the aforementioned challenges with regard to estimating a cyclopean image, these models absolutely must estimate disparity using left and right views. The quality of the estimated cyclopean image directly depends on the estimated disparity whose accuracy, despite the existence of several disparity estimation algorithms, is limited by external factors such as display device position and small depth of field, thereby making the problem more elusive.

Cyclopean multiscale (MS)-SSIM[23] is the state of the art in FR 3D IQA. This model first estimates reference and distorted cyclopean images and then computes an SSIM index. Chen et al. recently proposed an NR 3D IQA that competes favorably with several FR models. In 3D VQA, the addition of the temporal dimension introduces motion and parallax motion. This compounds the issues encountered in static 3D IQA models. Even without motion, the time-varying nature of a video introduces complex dependencies between estimated cyclopean frames and disparity maps. Though a few FR 3D VQA models have been proposed to tackle this problem, much further research needs to be done.

SUBJECTIVE IMAGE AND VIDEO QUALITY DATABASES

Importance of Representative Databases

As mentioned earlier in Section "Perceptual Quality and Objective Quality Assessment," given that the ultimate receivers of visual media are humans, the only reliable way to understand and predict the effect of distortions on a typical person's viewing experience is to capture opinions from a large sample of human subjects, which is termed *subjective image/video quality assessment*. These subjective scores are vital for understanding human perception of quality. They are also crucial for designing and evaluating reliable quality assessment models that are consistent with subjective human evaluations, regardless of the type and severity of the distortions.

Existing legacy image and video quality databases have played an important role in advancing the field of quality prediction. These benchmark databases have been designed to contain a small set of high-quality real-world photographs (or videos), each corrupted by only one of a few synthetically introduced distortions, for example, images corrupted by simulated camera sensor noise, or MPEG-2 compressed videos. (Section "Image Content in the Benchmark IQA Databases"). Current top-performing IQA/VQA models are designed, trained, and evaluated based only on the statistical perturbations observed on such "singly" distorted datasets. This might result in quality prediction models that inadvertently assume that every image/video has a single distortion that most objective viewers could agree upon.

However, the unsure eyes and hands of most amateur photographers frequently lead to occurrences of annoying visual artifacts, which are usually mixtures of several possible distortions. The unrepresentativeness of the legacy benchmark databases challenges the robustness, scalability, and applicability of the current QA models in several user-centric visual media applications. It is thus desirable to design challenging databases containing a large number of authentically distorted images and videos of different quality "types," mixtures, and distortion severities, and a wide variety of visual content. Such challenging datasets could ensure that state-of-the-art quality assessment algorithms are validated on more representative data, thus furthering efforts toward building better QA algorithms.

Image Content in the Benchmark IQA Databases

1. *2D IQA databases:* Most of the top-performing IQA models have been extensively evaluated on two benchmark databases: the LIVE IQA database[18] designed in 2005 and the TID2008 database[19] designed and released in 2008. The LIVE IQA database, one of the first comprehensive databases, consists of 779 images, much larger than the small databases that existed at the time of its introduction. This legacy database contains 29 pristine reference images and models five distortion types—jp2k, jpeg, Gaussian blur, white noise, and fast-fading noise, of varying severities.[18]

   The TID2008 database is larger, consisting of 25 reference and 1,700 distorted images over 17 distortion categories, each again of varying severities. We refer the reader to the work of Ponomarenko et al.[19] for more details on the categories, severities, and distortion simulation methods followed by the creators of this database. TID2013 is a very recently introduced image quality database with an end goal to include the peculiarities of color distortions in addition to the 17 spatial distortions included in TID2008. It consists of 3,000 images and includes seven new types of distortions, thus modeling a total of 24 distortions.

   To overcome the limitations of the restrictive, singly distorted 2D IQA datasets and to create a holistic resource for designing the next generation of robust, perceptually aware image assessment models, a unique and challenging database called the LIVE Blind Authentic Image Quality Challenge database was recently introduced,[24] containing images afflicted by large number of diverse, authentic distortion mixtures. All of the images in this database were captured using a wide variety of mobile device cameras. Since these images are naturally distorted as opposed to being artificially distorted post-acquisition from pristine reference images, they often contain mixtures of multiple distortions, creating an even broader spectrum of image impairments. Figure 7 illustrates a few images from this database.

2. *3D IQA databases:* LIVE 3D IQA database is one of the first publicly available databases containing high-quality stereo and depth images. These stereo images are distorted both symmetrically and asymmetrically and annotated with subjective human ratings.

Video Content in the Benchmark VQA databases

1. *2D VQA databases:* The LIVE Video Quality database[20] is one of the first modern public-domain comprehensive VQA databases. This dataset contains a total of 160 videos in a lossless YUV 4:2:0 planar format. Of the 160 videos, 10 are reference with each reference distorted in 15 ways to test MPEG-2 and H.264 compression along with streaming H.264

Fig. 7 Sample images that illustrate few spatial distortions such as (left) overexposure, (middle left) noise, (middle right) underexposure, and (right) motion blur, introduced during the capture process
Source: LIVE Blind Image Quality Challenge database.[24]

through error-prone Internet Protocol (IP) and wireless networks.

The LIVE Mobile VQA database expands on the LIVE VQA database to include combinations of temporal dynamics, rate adaptations, and frame freezes. This dataset contains a total of 210 videos in a lossless YUV 4:2:0 planar format. Of the 210 videos, 10 are reference with each distorted in 20 ways.

2. *3D VQA databases:* The IRCCyN/IVC NAMA3DS1-COSPAD1 3D Video Quality database is one of the few freely available 3D VQA databases. Ten video sequences are recorded using a twin-lens camera without the need for stereo adjustment as necessary for most other databases. Some video sequences are shot without compression where possible, but others use H.264 with high-profile mode with an average bit rate of 21 Mbps. Degradations to these videos include H.264 compression using quality settings, JPEG2k, downsampling, image sharpening, and combinations of downsampling and sharpening.

Test Methodologies

1. *Laboratory settings:* All the images and videos in the databases listed in Sections "Image Content in the Benchmark IQA Databases" and "Video Content in the Benchmark VQA Databases" contain quality ratings obtained by conducting subjective studies, mostly in controlled laboratory settings. Two popular ITU recommended test methodologies are adopted by the creators of most of the quality assessment databases: *single-stimulus* and *double-stimulus* methodologies. Under the single-stimulus methodology, stimulus is singly presented at any time and an observer's perceived quality of that stimulus is recorded. This method does not rely on the availability of reference stimuli and thus is of significant practical value. By contrast, under the double-stimulus methodology, both the reference stimulus and one of its distorted versions are presented to an observer, and a comparative assessment of the perceived quality of the stimuli is recorded. Single-stimulus methodology aids in gathering absolute quality ratings on a calibrated scale of choice, whereas pairwise comparisons require adopting a sophisticated approach to accurately generate the preferential subjective quality rankings.

   The opinion scores contained in the TID2008 database[19] were obtained from 838 observers by conducting batches of large-scale subjective studies, whereby a total of 256,000 comparisons of the visual quality of distorted images were performed. Although this is a large database, some of the test methodologies that were adopted do not abide by the ITU recommendations. Conversely, the LIVE IQA database[18] was created following the ITU recommended single-stimulus methodology and computes quality difference scores to address user biases.

   On the other hand, most of the VQA databases were designed by the following single-stimulus methodology, which includes presenting the hidden reference. For example, each video in the LIVE VQA database is assessed by 38 subjects, whereas 18 subjects rated each video in LIVE mobile VQA database. Additionally, for the LIVE mobile VQA database, a continuous temporal quality score was captured for the entire duration of all the videos. The IRCCyN 3D video quality database also followed the single-stimulus methodology, and the videos were rated by a total of 29 participants.

2. *Online studies (web and mobile applications)*: The experimental environment in the studies conducted in laboratory settings is stringently controlled, involving small, nonrepresentative subject samples (typically graduate and undergraduate students). However, in reality, the highly variable ambient conditions and the wide array of display devices on which a user might potentially view images and videos will have a considerable influence on her perception of quality.

This greatly motivates the recent interest in conducting IQA studies on the Internet, which can enable access to a much larger and more diverse subject pool while allowing for more flexible study conditions. The creators of the LIVE Blind Authentic Image Challenge database[24] have designed a sophisticated crowdsourcing system and conducted an online human study to obtain a very large number of human opinion scores. This human-powered framework (illustrated in Fig. 8) proved to be an effective way to gather over 280,000 human opinion scores on 1,163 naturally distorted images from over 7,000 distinct subjects, and the authors expect to collect at least 350,000 subjective judgments overall. The opinion scores gathered from this massive human study on a specific set of images chosen randomly have very high correlation values with the scores obtained on the same image set gathered via a study conducted in a laboratory. This high correlation advocates the veracity of the online system proposed in the work of Ghadiyaram and Bovik[24] in gathering reliable human opinion scores.

FUTURE DIRECTIONS OF QUALITY ASSESSMENT

In this section, we briefly touch upon some current quality assessment issues where research is still in its early stages.

Bitstream-Based VQA

Since streaming video services are increasingly demanding of bandwidth, bitstream-based VQA algorithms have been the focus of current research. Although VQA models described in Section "Perceptual-Based Objective QA Models" capture the essence of perceptual video quality, the encoded bitstream needs to be decoded before applying these models, which is a computationally intensive process. Hence, it is highly desirable to design low-complexity, bitstream-based quality assessment models which do not require complete decoding of videos (Fig. 9).

Color Quality Assessment

Most of the current top-performing IQA models process just the luminance component of images. A small body of recent work extends models designed for grayscale images to chromatic channels as an attempt to address color quality. This is however insufficient as it is extremely important to deeply understand the implications of errors in color reproductions and distortions in the chromatic channels on perceptual quality.[25,26] Given the pervasiveness of color images and the existence of several image enhancement-based mobile applications, this lack of robust quality assessment models for color distortion perception is a pressing concern and an interesting topic for future research.

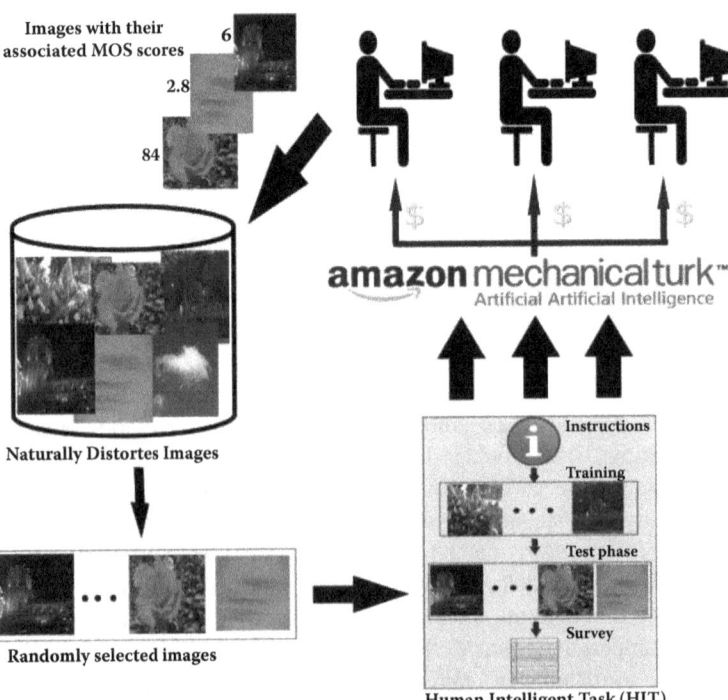

Fig. 8 Illustrating how the crowdsourcing system proposed in the work of Ghadiyaram and Bovik[24] packages the task of rating images as a human intelligence task and disperses it on Amazon's Mechanical Turk

Fig. 9 Sample video frames that illustrate some of the distortions modeled in LIVE VQA database.[20] (a) MPEG-2 compressed frame; (b) H.264 compressed frame; (c) IP loss simulated frame; (d) wireless loss simulated frame

Perceptually Optimized Image Acquisition

Despite the significant improvement of cameras in smartphones and tablets, automated tools to fix the acquisition parameters still remain imperfect. Smartphone cameras, owing to their limited form factor, lead to a higher incidence of obvious artifacts compared to single-lens reflex (SLR) cameras. Current smartphone cameras also lack a way to perceptually control and optimize acquisition parameters such as camera International Standards Organization (ISO), shutter speed, focus settings, and the like.

Thus, highly reliable perceptual mechanisms founded on the recently developed, efficient quantitative perceptual IQA/VQA models offer a great potential for improved mobile camera performance during image and video acquisition. However, such perceptual capabilities have not been integrated with many of the existing mobile cameras to date and is an exciting direction of research.

Time-Varying Video Quality for the Wireless Infrastructure

Perceptual IQA/VQA for the wireless video infrastructure remains an unsolved and largely unaddressed problem. The explosion of wireless video and the paucity of available network bandwidth are leading to a "wireless spectrum crunch" as data usage continues to double each year. Chen et al.[27] recently proposed a promising blind measure to predict the quality of videos containing distortions produced by carefully simulating the effect of varying bitrates over time and over wireless networks. Progress remains to be made on this problem in order to design robust models, which could be deployed throughout the wireless network infrastructure to address the network constraints perceptually.

Quality Assessment in Different Image Modalities

In addition to the aforementioned topics, applying quality assessment to other image modalities could open up interesting applications and new challenges. For instance, infrared (IR) images suffer from distortions such as nonuniformity, halo effects, and hotspots. These co-occurring distortions are different from those that occur in natural images. Detecting the types and amounts of distortions (or mixtures of distortions) is practically an unsolved problem for this modality. Recent work[28] suggests that the scene statistics of IR images, derived in the same spirit as on natural images, strongly correlate with human perceptual quality.

CONCLUSION

In any case, there is an exceedingly growing awareness of the importance of understanding, predicting, and monitoring the perceptual quality of images and videos that are delivered to human viewers. The leaps of progress made thus far in vision science and image engineering continue to push the boundaries of achievable perceptual quality prediction. This would eventually lead to the design of biologically inspired automatic quality assessment algorithms that are robust to the varied image and video content, distortions, and display devices. These automated tools in turn have tremendous practical and industrial significance and could help in delivering the best possible visual content to the end users.

REFERENCES

1. Carandini, M.; Demb, J.B.; Mante, V.; Tolhurst, D.J.; Dan, Y.; Olshausen, B.A.; Gallant, J.L.; Rust, N.C. Do we know what the early visual system does? J. Neurosci. **2005**, *25* (46), 10577–10597.
2. Moorthy, A.K.; Bovik, A.C. Visual importance pooling for image quality assessment. IEEE J. Sel. Top. Signal Process. **2009**, *3* (2), 193–201.
3. Kuffler, S.W. Discharge patterns and functional organization of mammalian retina. J. Neurophysiol. **1953**, *16* (1), 37–68.
4. Dong, D.W.; Atick, J.J. Temporal decorrelation: A theory of lagged and nonlagged responses in the lateral geniculate nucleus. Netw. Comput. Neural Syst. **1995**, *6* (2), 159–178.
5. Ruderman, D.L. The statistics of natural images. Netw. Comput. Neural Syst. **1994**, *5* (4), 517–548.
6. Mandelbrot, B. *The Fractal Geometry of Nature*; Freeman: New York, 1982.
7. Olshausen, B.A.; Field, D.J. Sparse coding with an overcomplete basis set: A strategy employed by V1? Vis. Res. **1997**, *37* (23), 3311–3325.
8. Wilson, H.R.; Bergen, J.R. A four mechanism model for threshold spatial vision. Vis. Res. **1979**, *19* (1), 19–32.
9. Soundararajan, R.; Bovik, A.C. RRED indices: Reduced reference entropic differencing for image quality assessment. IEEE Trans. Image Process. **2012**, *21* (2), 517–526.
10. Wang, Z.; Bovik, A.C.; Sheikh, H.R.; Simoncelli, E.P. Image quality assessment: From error visibility to structural similarity. IEEE Trans. Image Process. **2004**, *13* (4), 600–612.
11. Bovik, A.C. Automatic prediction of perceptual image and video quality. IEEE Proc. **2013**, *101* (9), 2008–2024.
12. Seshadrinathan, K.; Bovik, A.C. Motion-tuned spatiotemporal quality assessment of natural videos. IEEE Signal Process. **2010**, *19* (2), 335–350.
13. Movshon, J.A.; Newsome, W.T. Visual response properties of striate cortical neurons projecting to area MT in Macaque monkeys. J. Neurosci. **1996**, *16* (23), 7733–7741.
14. Tang, H; Joshi, N.; Kapoor, A. Blind image quality assessment using semi-supervised rectifier networks. In *Proceedings of IEEE International Conference on Computer Vision Pattern Recognition*; IEEE: Washington, DC, 2014.
15. Mittal, A.; Moorthy, A.K.; Bovik, A.C. No-reference image quality assessment in the spatial domain. IEEE Trans. Image Process. **2012**, *21* (12), 4695–4708.
16. Mittal, A.; Soundararajan, R.; Bovik, A.C. Making a completely blind image quality analyzer. IEEE Sig. Proc. Lett. **2013**, *20* (3), 209–212.
17. Saad, M.; Bovik, A.C.; Charrier, C. Blind prediction of natural video quality. IEEE Trans. Image Process. **2014**, *23* (3), 1352–1365.
18. Sheikh, H.R.; Sabir, M.F.; Bovik, A.C. A statistical evaluation of recent full reference image quality assessment algorithms. IEEE Trans. Image Process. **2006**, *15* (11), 3440–3451.
19. Ponomarenko, N.; Lukin, V.; Zelensky, A.; Egiazarian, K.; Carli, M.; Battisti, F. TID2008-A database for evaluation of full-reference visual quality assessment metrics. Adv. Modern Radio Electron. **2009**, *10* (4), 30–45.
20. Seshadrinathan, K.; Soundararajan, R.; Bovik, A.C.; Cormack, L.K. Study of subjective and objective quality assessment of video. IEEE Trans. Image Process. **2010**, *19* (6), 1427–1441.
21. Suchow, J.W.; Alvarez, G.A. Motion silences awareness of visual change. Curr. Biol. **2011**, *21* (2), 140–143.
22. Choi, L.K.; Bovik, A.C.; Cormack, L.K. A flicker detector model of the motion silencing illusion. J. Vis. **2012**, *12* (9), 777.
23. Chen, M.-J.; Su, C.-C.; Kwon, D.-K.; Cormack, L.K.; Bovik, A.C. Full-reference quality assessment of stereopairs accounting for rivalry. Image Commun. **2013**, *28* (9), 1143–1155.
24. Ghadiyaram, D.; Bovik, A.C. Crowdsourced study of subjective image quality. In *Proceedings of Annual Asilomar Conference on Signals, Systems and Computers,* Pacific Grove, CA, November 2–5, 2014.
25. Zhang, X.; Wandell, B. Color image fidelity metrics evaluated using image distortion maps. Signal Process. **1998**, *70*, 201–214.
26. Rajashekar, U.; Wang, Z.; Simoncelli, E.P. Perceptual quality assessment of color images using adaptive signal representation. In *SPIE Conference on Human Vision and Electronic Imaging XV*, Vol. 75271, San Jose, CA, 2010, 75271L.
27. Chen, C.; Choi, L.K.; de Veciana, G.; Caramanis, C.; Heath, R.W.; Bovik, A.C. Modeling the time-varying subjective quality of HTTP video streams with rate adaptations. IEEE Trans. Image Process. **2014**, *23* (5), 2206–2221.
28. Goodall, T.; Bovik, A.C. No-reference task performance prediction on distorted LWIR images. In *IEEE Southwest Symposium on Image Analysis Interpretation*, San Diego, CA, April 6–8, 2014.

Perceptual Image Enhancement

Deepti Ghadiyaram, Todd Goodall, Lark Kwon Choi, and Alan C. Bovik
Laboratory of Image and Video Engineering, The University of Texas at Austin, Austin, Texas, U.S.A.

Abstract

Enhancing image quality has become an important issue as the volume of digital images increases exponentially and the expectation of high-quality images grows insatiably. Digitized images commonly suffer from poor visual quality due to distortions, low contrast, deficiency of lighting, defocusing, and atmospheric influences such as fog, severe compression, and transmission errors. Hence, image enhancement is indispensable for better perception, interpretation, and subsequent analysis. Since humans are generally regarded as the final arbiter of the visual quality of the enhanced images, perceptual image enhancement has been of great interest. In this article, we discuss actively evolving perceptual image enhancement research. It discusses the principles of the human visual system that have been used in perceptual image enhancement algorithms and then presents modern image enhancement models and applications on the perceptual aspects of human vision.

Keywords: Adaptive image compression; Contrast enhancement; Denoising; Face recognition; Human visual system; Image defogging; Image enhancement; Image quality; Mammogram; Natural scene statistics.

INTRODUCTION

The advent of digital imaging and the high-speed Internet have opened up a new era in photography and image processing. Amateur users easily capture, store, edit, and share images, whereas professional users including researchers effectively examine details, identify objects, interpret findings, and present results. As the global volume of digital images increases exponentially from mobile devices to large screens, the expectation of high-quality images grows insatiably. Digitized images commonly suffer from poor image quality, particularly due to distortions (e.g., blocking, ringing, blur, Gaussian noise), lack of contrast (e.g., variant illumination), deficiency in focusing (e.g., motion effects), lighting (e.g., under or overexposure), adverse atmospheric influences (e.g., fog), and transmission errors (e.g., packet loss). Hence, image enhancement is frequently applied to improve the overall visual quality of an image for better perception, interpretation, and subsequent analysis.

Since humans are generally regarded as the final arbiter of the visual quality of enhanced images, understanding how the human visual system (HVS) tunes to natural environments and how to achieve visual perception has been studied for decades among vision scientists and engineers. The HVS promptly and reliably recognizes objects in a blurred or noisy image and penetrates the variability in shape, color, and shading of objects under the drastically changing illuminant conditions. From retina to the primary visual cortex, the HVS efficiently codes the visual signal using multiscale/multiorientation subband decomposition and a divisive normalization process. In addition, the HVS imposes spatiotemporal visual masking effects. When visual signals are superimposed on similar orientations, spatiotemporal frequencies, motion, or color, the local high-frequency energy in an image can strikingly reduce the visibility of other signals such as noise.

Growing interest in the HVS has inspired developing perceptual image enhancement models whose successful performance has also shown that accounting for the HVS is an essential part of perceptual image enhancement. Perceptually optimized denoising, deblurring, and distortion-blind image repair models enhance the visual quality of distorted images based on the HVS and of natural scene statistics (NSS). The multiscale Retinex (MSR) with color restoration model, called MSRCR, combines models of the color constancy property of human vision to provide both enhanced contrast and sharpness. More examples of image enhancement models that are based on perceptual image processing can be found, such as contrast enhancement using Gaussian mixture models (GMMs), fusion-based approaches, and tone mapping operations (TMOs) to achieve high dynamic range (HDR) using structural fidelity and statistical naturalness.

Perceptual image enhancement techniques vary depending on enhancement purposes and applications. Denoising and deblurring models alleviate visual distortions, while contrast and visibility enhancement algorithms allocate new luminance or chrominance values to emphasize edges based on global or local neighborhood. TMOs adjust the color appearance of HDR natural scenes for display mediums that have a limited dynamic range. These perceptual image enhancement techniques are used in a variety of applications from improvement of daily snapshots on mobile devices to robust face recognition in security, visibility enhancement in bad weather scenes, diagnostic analysis of medical imaging such as mammogram, and adaptive image/video compression.

To introduce actively evolving perceptual image enhancement research, we discuss image enhancement models and applications based on the perceptual aspects of human vision. Section "Principles of HVS" summarizes the principles of the HVS that have been used in perceptual image enhancement. Section "Perceptual Image Enhancement Models" presents modern perceptual image enhancement models. Section "Perceptual Image Enhancement Applications" describes practical applications of perceptual image enhancement algorithms. Section "Conclusion" concludes this entry with future directions.

PRINCIPLES OF HVS

Spatial and Temporal Contrast Sensitivity Function

The HVS, having many interconnecting neurons with hidden functions, is often treated like a black box. Measuring spatial and temporal acuity is important for determining the amount of stimulus contrast necessary to elicit a criterion level of response from the neuron. The perceived contrast is a key for visual perception since the visual information in the HVS is represented in terms of contrast rather than the absolute level of light. The spatial contrast threshold indicates the amount of contrast for the retina to detect a spatial pattern, whereas the temporal contrast threshold is the required temporal contrast to sense a time-varying signal. Contrast sensitivity is the inverse of the contrast threshold.

The spatial contrast sensitivity function (CSF) depicted in Fig. 1a describes how the retina responds to contrast as a function of frequency.[1] Retinal cone cells are most sensitive to stimuli at 4–8 cycles per degree (cpd), whereas nothing is visible beyond approximately 60 cpd. Perceptual image enhancement methods mostly concentrate on signals in the visible range of spatial frequencies, discarding imperceptible signals based on the CSF. The retina is composed of a layer of ganglion cells, which can be modeled as having circularly symmetric center-surround excitatory and inhibitory, on and off receptive fields. Differences in light between the center and periphery of these receptive fields enable neurons to fire at different rates. In addition, contrast sensitivity is a function of viewing distance. When close to a viewing screen, individual pixels become visible to a viewer, whereas at normal viewing distance, individual pixels become indistinguishable. Quantifying

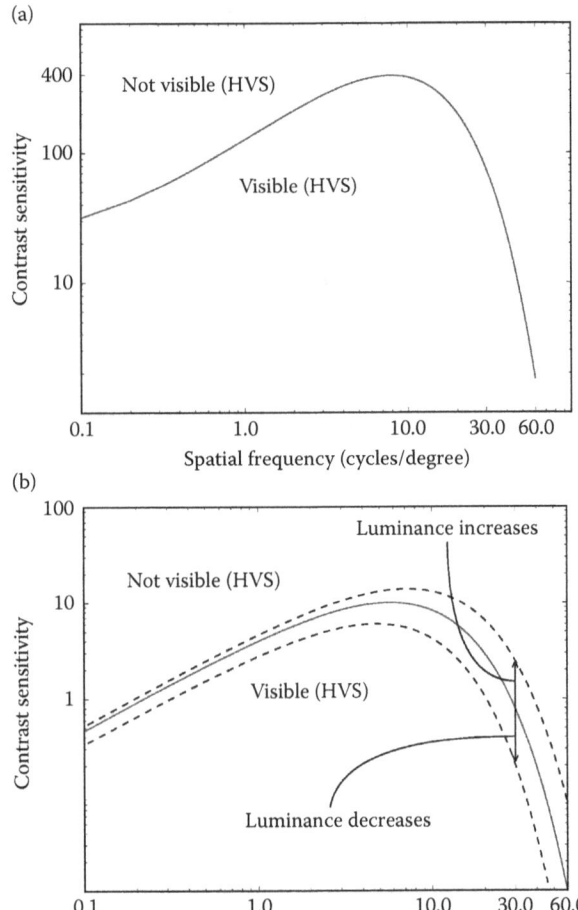

Fig. 1 (a) Spatial contrast sensitivity function; (b) temporal contrast sensitivity functions

the perceived contrast of an image is important to design perceptually enhanced images and displays.

The temporal CSF provides a summary of the temporal response characteristics in the HVS.[1] The highest contrast sensitivity is in the range of 8–15 Hz under sufficient lighting conditions, whereas visual stimuli over 50 Hz become undetectable. This temporal sensitivity varies with lighting conditions, so the temporal CSF curve shifts up or down depending on the integrated amount of light at the retina, as shown in Fig. 1b. For example, the HVS reduces the contrast sensitivity for higher frequency changes in low light. The lateral geniculate nucleus (LGN) decomposes the contrast signal from the retina into lagged and unlagged temporal responses, which can be modeled as temporally differenced outputs of gamma filters. The characteristics of the temporal CSF are also useful to develop perceptual video quality enhancement algorithms.

Multiscale and Multiorientation Decomposition

The natural environment is inherently multiscale and multiorientation. As a result, the HVS necessarily adjusts objects at various scales, orientations, and distance by efficiently coding the visual signals. The visual signals sensed by the eye are decomposed into band-pass channels as they pass through the visual cortex of the brain. The receptive fields of simple cells in the primary visual cortex (area V1) consist of adjacent excitatory and inhibitory regions, where they are elongated and oriented, and hence respond to stimuli in some orientations better than others. This orientation selectivity enables simple cells to detect the orientation of a stimulus. By using a wide range of orientations, lobe separations, and sizes, simple cells can be modeled as collectively providing a large number of a quadrature pair log-Gabor filter type responses; however, each simple cell is also sensitive to precise position of the visual stimuli within the receptive field. Complex cells are less dependent on stimulus position. They can be modeled as rectifying the responses of simple cells and further as normalizing them. The collective output of complex cells is often modeled using a filter bank of log-Gabor filters tuned to various spatiotemporal orientations of visual stimuli across scales. By using multiscale and multiorientation cell responses, the later stages of the HVS are able to efficiently encode both shape representations and object trajectories.

To mimic these multichannel characteristics of the HVS, several multiresolution image representations have been used for image analysis: the Gaussian and Laplacian pyramid, the wavelet decomposition, and Gabor filter banks. The Gaussian pyramid provides a representation of the same image at multiple scales using low-pass filtering and decimation, whereas the Laplacian pyramid presents a set of detailed band-pass filtered images at different scales. The construction of a Gaussian pyramid involves continuous two-dimensional Gaussian low-pass filtering and subsampling ($\downarrow$) by a factor of 2. The Laplacian pyramid is then computed as the difference between the original image and the low-pass filtered image after upsampling ($\uparrow$) by a factor of 2. Figure 2a explains these processes. The wavelet decomposition represents an image in subbands having different scales and orientations. As shown in Fig. 2b, the LL (low, low) subband is a low-resolution version of the original image, and the HL (high, low), LH (low, high), and HH (high, high) subbands, respectively, contain details with vertical, horizontal, and diagonal orientations. The Gabor filter banks more closely approximates the response of the HVS, where the basis functions are Gaussian functions modulated by sinusoidal waves. As can be seen in Fig. 2c, Gabor filters have both frequency-selective and orientation-selective properties. Due to its optimally localized space–time–frequency characteristics, Gabor filters are widely used in texture analysis, motion analysis, object recognition (e.g., fingerprint enhancement and iris recognition), etc.[2] These natural and efficient multiscale/multiorientation decompositions are also extensively used for perceptual image enhancement models.

Visual Masking

Visual masking refers to the inability of the HVS to detect a stimulus, the target, when another stimulus, the mask, is superimposed on the target in a similar orientation, spatiotemporal frequency, motion, color, or other attributes.

Contrast (or Texture) Masking

Contrast masking indicates the variation of the detection thresholds of the target as a function of the contrast of the mask. The presence of local high-frequency energy in an image strikingly reduces the perceptual significance of the distortions.[3] For example, JPEG blocking artifacts are evident on the lady's face and neck, but far less visible or eliminated in the flora regions as shown in Fig. 3a. Figure 3b is another example, where the image is uniformly distorted by the same level of additive Gaussian noise. Distortions are highly visible on the smooth regions (e.g., face), but nearly imperceptible on the more textured areas (e.g., hair and scarf).

Heeger's model[4] explains how contrast masking occurs in the HVS. As processed visual stimuli pass through the eyes to visual cortex, V1 neurons execute multiscale and multiorientation decompositions, computing local energy responses to the visual stimulus. These responses, in the form of contrast energy, are then normalized by local summations of neighboring responses. In this manner, neurons with separate spatiotemporal tunings respond relative to each other. This nonlinearity is commonly called the divisive normalization. The computational divisive normalization transform model is depicted in Fig. 4.

Luminance Masking

The HVS may be modeled as performing a transform following processing by the retinal ganglion cells where the processed input is subtracted and then divided by the local mean response. This operation accomplishes an adaptive gain control and entropy reduction for signal coding. This transform causes masking of luminance changes in proportion to the background luminance. The Weber–Fechner law states that the ratio of changes in luminance to the average luminance must be larger than a specific constant to be visible.

Temporal Masking and Silencing

Temporal masking in videos is analogous to spatial contrast masking in images, but is made more perceptually complex by the addition of motion. Models of temporal masking have been developed using heuristics for video compression. The commonly modeled case of sparse, global scene changes is one special case of temporal masking. A striking motion silencing illusion was recently introduced in which local salient temporal changes of objects in luminance, hue, size, and shape are masked in the presence of large, coherent object motions.[5] Although

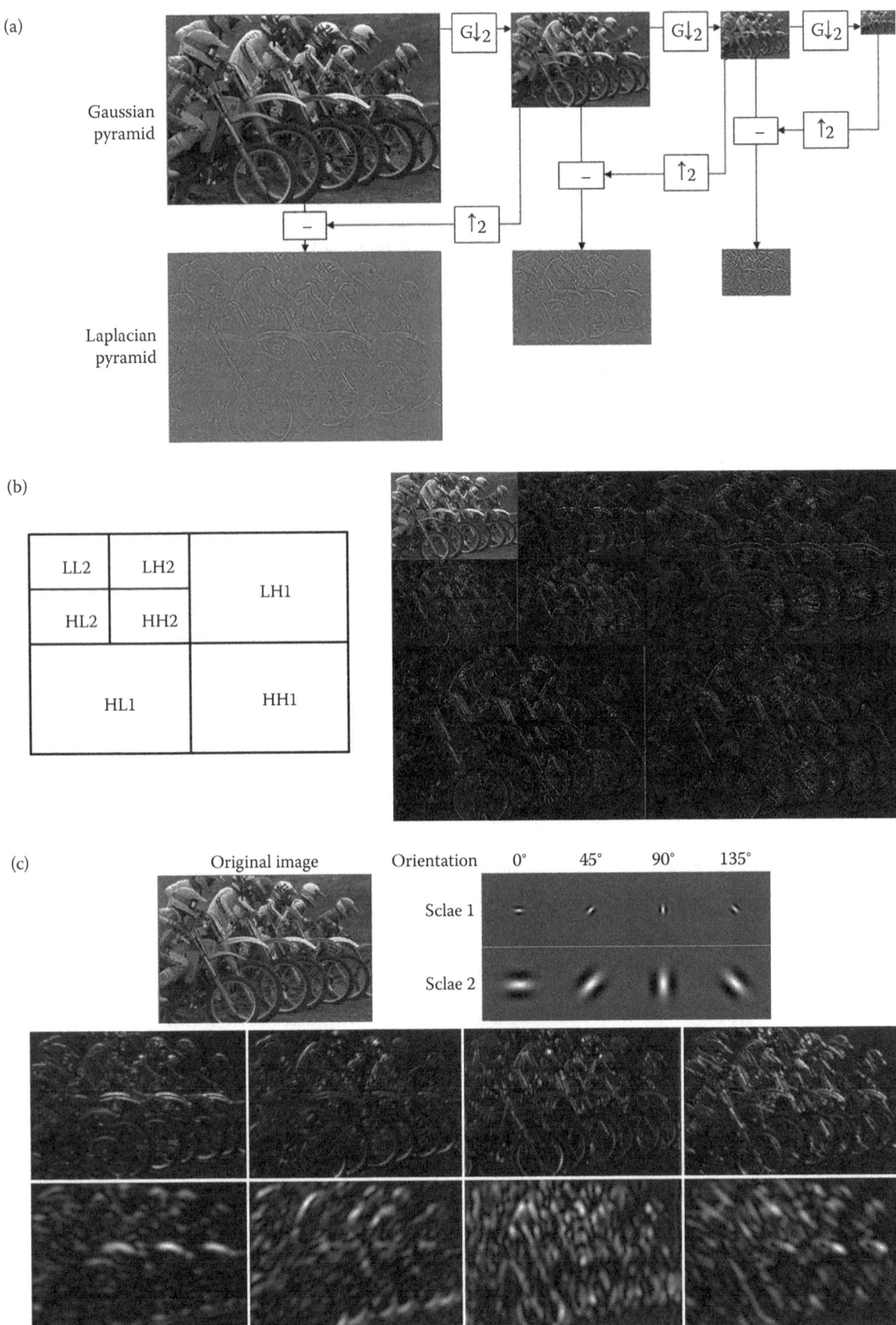

Fig. 2 Multiresolution image representations: (a) Gaussian and Laplacian pyramid; (b) wavelet decomposition; (c) Gabor filters and filtered images

Perceptual Image Enhancement

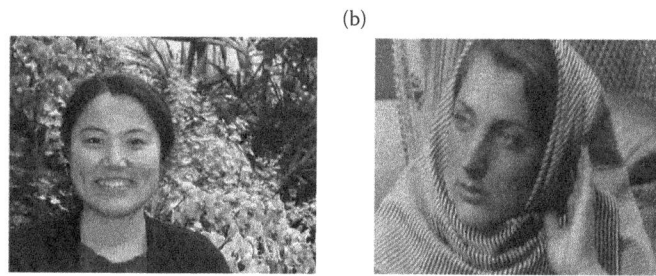

Fig. 3 Contrast masking of (a) JPEG artifacts and (b) additive white Gaussian noise

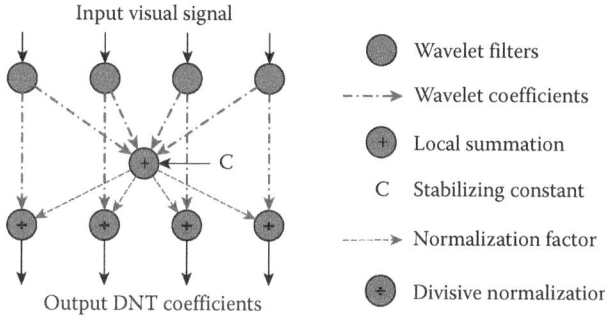

Fig. 4 Illustration of the computational divisive normalization transform (DNT) model

the underlying mechanisms of silencing remain unknown, Choi et al.[6] have developed a filter-based quantitative spatiotemporal flicker detector model that accurately predicts human judgments of silencing as a function of object velocity and object change rate. They suggest that the band-pass responses of cortical neurons may be implicated in whether changes in moving objects are perceived.

Modeling Statistical Regularities of Natural Images

The development of the HVS is strongly dependent on early visual stimulation and the statistics of the surrounding visual environment. The early stages of vision, from the retina to the primary cortex, are constrained to process local points of the visual field using groups of neurons representing localized responses. The output of these neurons is determined by natural statistical regularities. Hence, modeling the statistical regularities of natural images is a modern approach to model the HVS.

The primary goal of the HVS is to efficiently reduce redundancy while maximizing information of incoming visual stimuli using efficient statistical measures. Similar to the remarkably efficient coding of visual signals into Gabor-like complex cell responses by cortical neurons in area V1, the first few local principal (and independent) components of natural images are also observed to maximally decorrelate the image content. The principal component analysis filters yielding the highest response variance are found to be oriented edges or their derivatives, thus resembling the oriented receptive fields found in the cortex. This suggests that a decorrelation operation is accomplished by HVS to reduce the redundancy.

The highly structured statistical regularities exhibited by natural images are well modeled by spatial linear band-pass filters such as difference of Gaussian (DoG), Laplacian of Gaussian (LoG), Gabor, or a predictive coding filter. DoG models the center-surround excitatory–inhibitory processes that occur at various stages of visual processing, while LoG measures the second derivative of an image after Gaussian smoothing. LoG filters are useful for finding edges and blobs. It has been observed that the empirical probability distributions of the responses of these filters can be reliably modeled using a Gaussian scale mixture (GSM) probability distribution. The GSM model can be used to capture both the marginal and the joint distributions of the wavelet coefficients across subbands and to extract enhanced image by removing noisy signals. It has also been observed that the non-Gaussian long-tailed histograms of natural scenes imply the feasibility of 'sparse coding' by visual neurons. A large class of image enhancement algorithms is based on modeling of statistical regularities of natural images.

PERCEPTUAL IMAGE ENHANCEMENT MODELS

Denoising, Deblurring, and Blind Repair of Natural Images

Perceptual Denoising

Visual signals are often degraded by noise during image acquisition and transmission, so the development of efficient denoising techniques is highly desirable to enhance perceived visual quality. The statistical regularities of natural images have been exploited by applying wavelet decomposition to the signal with a set of multiscale band-pass oriented filters. The observed marginal distributions of wavelet coefficients are highly kurtotic and long-tailed. In addition to regularities in marginal distributions, it has been observed that higher order dependencies exist between neighboring subbands across scales and orientations. These observations were successfully incorporated by Portilla et al.[7] in a model to denoise images corrupted by independent additive Gaussian noise of known variance.

When an image is decomposed into oriented subbands at multiple scales, the neighborhood includes coefficients from other subbands (i.e., at nearby scales and orientations) as well as from the same subband. The coefficients within each local neighborhood around a reference coefficient of a pyramid subband are modeled using a GSM[7]: A random vector x is a GSM if it can be expressed as the product of two independent components

$$x = \sqrt{z}u, \qquad (1)$$

where u is a zero-mean Gaussian vector with covariance matrix C_u, and z is called a mixing multiplier. Assuming that the original image is corrupted by independent additive white Gaussian noise, and then a noisy neighborhood coefficient vector y can be characterized as follows:

$$y = x + w = \sqrt{z}u + w, \qquad (2)$$

where w is a noise coefficient vector with covariance matrix C_w.

The image denoising method decomposes the image into pyramid subbands at different scales and orientations; denoises each subband, except the low-pass residual band; and inverts the pyramid transform, obtaining the denoised image. The group of neighboring coefficients constitutes a sliding window that moves across the wavelet subband. At each step, the reference center coefficient, x_c, of the window is estimated (i.e., denoised). Therefore, a denoising method is to estimate x_c of a spatial location x from a noisy neighborhood y. The Bayes least square estimator is used to compute the conditional mean as follows[7]:

$$E\{x_c \mid y\} = \int_0^\infty E\{x_c \mid y, z\} p\{z \mid y\} dz, \qquad (3)$$

where $E\{x \mid y, z\} = zC_u(zC_u + C_w)^{-1}y$, from a local linear wiener estimate, and C_u can be estimated from the observed noisy covariance matrix by $C_u = C_y - C_w$. The posterior density $p(z \mid y)$ can be predicted by Bayes' rule: $p(z \mid y) \propto p(y \mid z)p_z(z)$, where $p_z(z) \propto 1/z$. $p(y \mid z)$ can be calculated as a Gaussian function of zero-mean and covariance $C_y = zC_u + C_w$. These estimates in different subbands are used to reconstruct the denoised image. On a set of images corrupted by simulated additive white Gaussian noise of known variance, this model is shown to introduce fewer artifacts and to better preserve edges.

Another model in the wavelet domain but designed for video denoising was proposed.[8] Since there exist strong correlations between wavelet coefficients across adjacent video frames, grouping wavelet coefficients along temporal directions based on a GSM framework reinforces the statistical regularities of the coefficients and enhances the performance of Bayesian signal denoising algorithm.

A GSM was used to capture the local correlations between the spatial and temporal wavelet coefficients, and such correlations were strengthened with a noise-robust motion compensation process, resulting in competitive performance of denoising.[8] The keys of the spatiotemporal GSM (ST-GSM) denoising method were to include temporal neighborhoods in the wavelet coefficient vector and to correctly estimate global motion in the presence of noise. Figure 5 shows the visual effects of the denoising algorithm for the *Suzie* sequence. ST-GSM shows quite effective performance at suppressing background noise while maintaining the edge and texture details, where the structural similarity (SSIM) maps of the corresponding frames were used to verify the denoising enhancement.

Perceptual Deblurring

One of the most common artifacts in digital photo and videography is blurring, which could be caused by a number of factors—camera shake, long exposure, poor focus, etc. This artifact can be modeled as a blur kernel convolved with image intensities, and thus recovering an unblurred image signal from a single blurred signal translates to signal deconvolution. The heavy-tailed property of the empirical distributions of natural images has been successfully used for a number of image enhancement models including deconvolution. The NSS of derivative distributions of unblurred images is exploited by Fergus et al.[9] The authors adopted a Bayesian approach using the aforementioned scene statistics as a prior to find a blur kernel, then reconstructed the deblurred image using the estimated kernel via a standard deconvolution method.

The algorithm proposed by Fergus et al.[9] takes a blurred input image **B** and few more manual inputs from the user: a rectangular patch, rich in edge structure, within the blurred image; the maximum size of the blur kernel; and an initial 3×3 blur kernel. The blurry image **B** was assumed to have been generated by convolution of a blur kernel **K** with a latent sharp image **L** plus sensor noise **N** at each pixel:

$$B = K \otimes L + N, \qquad (4)$$

where $\otimes$ denotes discrete image convolution (with nonperiodic boundary conditions). In the first step, the blur kernel **K** is inferred using multiscale approach. Specifically, let **P** denote the blurred patch selected by the user. The blur kernel **K** and the latent patch image L_p are estimated by varying image resolution in a coarse-to-fine manner in order to avoid local minima. This estimation problem is posed as a search for the max-marginal blurring kernel, by finding the highest probability, guided by a prior on the statistics of **L**. In the second step, using this estimated kernel, a standard deconvolution (e.g., Richardson–Lucy) algorithm is applied to estimate the unblurred image **L**.[9]

Figure 6 shows that the algorithm significantly sharpens a blurred image. The result suggests that applying natural

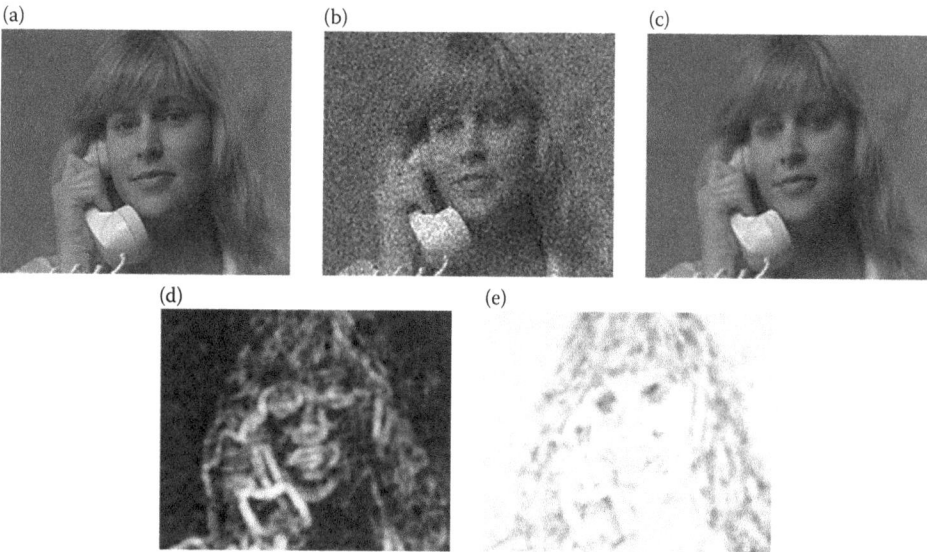

Fig. 5 Denoising results of Frame 100 in *Suzie* sequence corrupted with noise standard deviation $\sigma = 20$. (a) Original image; (b) noisy image; (c) ST-GSM denoised sequence; (d) SSIM quality map corresponding to (b); (e) SSIM quality map corresponding to (c). A brighter region indicates a larger SSIM value, high image quality in SSIM quality map. Images are generated using an algorithm [8]

image priors and advanced statistical techniques greatly assists in recovering an unblurred to blurred signal from a blurry image. Although ringing artifacts occur near saturated regions and regions of significant object motion, the algorithm achieves superior performance when applied on a large number of real images with varying degrees of blur.

Perceptually Optimized Blind Repair of Natural Images

A distortion-blind, perceptually optimized GEneral-purpose No-reference Image Improver (GENII) enhances images corrupted by any of JPEG compression, JPEG2000 compression, additive white noise, or Gaussian blur distortions.[10] The GENII framework operates based on NSS features extracted from a blind image quality assessment (IQA) algorithm in the wavelet or spatial domains. Since natural images possess highly reliable statistical regularities that are measurably modified by the presence of distortions, NSS features can be used to identify unknown distortions by type and severity. Once a distortion type has been identified, the extracted features are utilized to predict the perceptual quality of the image using a quality assessment process. If the predicted quality falls below a certain threshold, a trained regression module such as support vector regression is queried to estimate perceptually optimized parameters of that distortion. GENII then proceeds to invoke an appropriate image repair algorithm such as deringing, deblocking, denoising, or deblurring models based on the predicted distortion type. The repaired intermediate image is then passed back into the loop in order to evaluate visual quality and identify its distortion type. This loop continues until the obtained intermediate image has the highest possible quality or until a specific finite number of repair iterations have been performed.[10] An illustration of the GENII framework including a distorted image and the corresponding repaired image is shown in Fig. 7.

As a working exemplar model of the GENII framework, GENII-1 is aligned with distortion-specific image repair algorithms. For example, the block matching three-dimensional (3D) (BM3D) algorithm and the trilateral filter are used for denoising and deringing, respectively, but other repair algorithms can be adopted. GENII-1 was tested to evaluate the performance on 4,000 distorted images. GENII-1 performs quite well predicting and ameliorating the distortions present in the image, even when the image has multiple distortions.

Perceptual Image Contrast Enhancement

MSR with Color Restoration

Retinex is a model of the lightness and color perception of human vision. As the word "Retinex" implies, this model relates to the neurophysiological functions of individual neurons in the primate retina, LGN, and cerebral cortex of the brain as a form of a center-surround spatial structure.[11] The Retinex algorithm uses multiscale analysis to decompose the image into different frequency bands, and then enhances its desired global and local frequencies. Retinex image processing achieves sharpening, local contrast enhancement, color constancy, and natural rendition of HDR lightness and color. The Retinex concept evolved from single-scale Retinex into a full-scale image enhancement algorithm, called the MSRCR.[11] MSRCR improves the visibility of color and details in shadows by using color constancy. Here, the constancy refers to the resilience of perceived color and lightness to spatial and spectral illumination

Fig. 6 *Top*: A scene with a blur. The patch selected by the user is indicated by the gray rectangle. *Bottom*: The deblurred image. Close-up images of the selected patch are shown on the right. Images are generated using an algorithm [9]

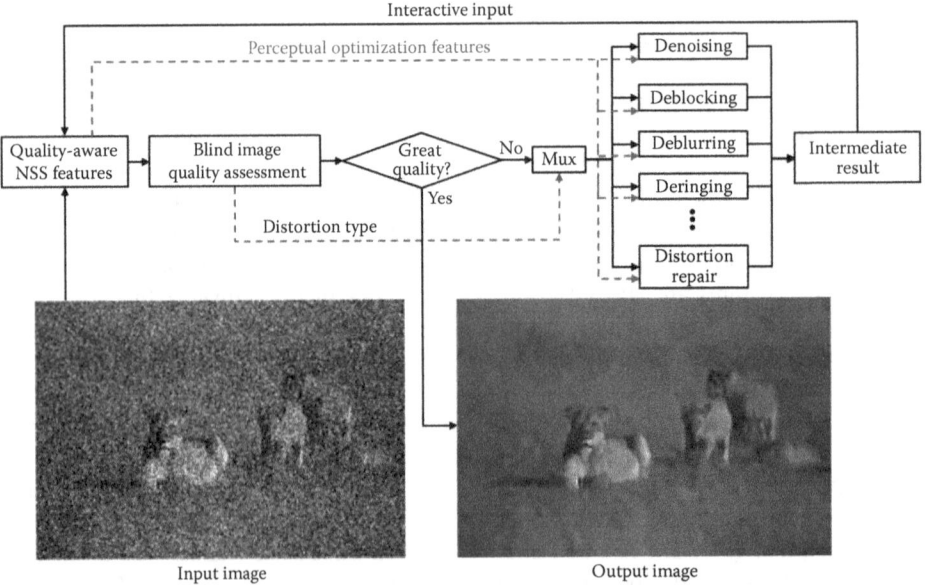

Fig. 7 An illustration of the GENII framework. NSS features are used to predict the distortion type, the severity, the visual quality, and the distortion parameters that serve as inputs to a possible non-blind repair algorithm. The intermediate result is fed back to the system until the best possible quality is achieved. Images are provided by Moorthy et al.[10]

variations. MSRCR is applied not only to everyday images, but also to lunar, medical X-ray, mammographic, CT, and magnetic resonance images to enhance contrast and sharpness and to reduce the bandwidth requirement, as desired in telemedicine applications.

MSRCR algorithm consists of MSR and color restoration as follows[11]:

$$R_{MSRCRi}(x,y) = \alpha_i(x,y) \sum_{i=1}^{K} W_k \{\log I_i(x,y) - \log[F_k(x,y) \otimes I_i(x,y)]\}, \quad (5)$$

$$\alpha_i(x,y) = \log\left[NI_i(x,y) / \sum_{n=1}^{N} I_n(x,y)\right], \quad (6)$$

$$F_k(x,y) = \kappa \exp[-(x^2 + y^2)/\sigma_k^2], \quad (7)$$

$$\kappa = 1 / \sum_x \sum_y F(x,y), \quad (8)$$

where index i is the ith spectral band in an image I, (x_1, x_2) is the pixel location, and $\otimes$ is the convolution operator. $\alpha_i(x_1, x_2)$ is the color restoration coefficient in the ith spectral band, N is the number of spectral band (typically, $N = 1$ for grayscale images and $N = 3$ for RGB color images), F_k is the kth Gaussian surround function, while K is the number of scales. σ_k is the standard deviation of the Gaussian kernels. MSRCR performs a log operation on the ratio of each pixel in each spectral band to both spatial and spectral averages. This suppression of spatial and spectral lighting variations mimics human perception of color and lightness.

MSRCR has been tested on a large number of images and has consistently proven to be better than conventional contrast enhancement methods. Results show that MSRCR automatically improves images with diverse degrees of visual deficit. For example, to produce more realistic enhanced images, MSRCR slightly sharpens the dimed original image or removes the effects of the shadows for the moderately underexposed images. Further, MSRCR reduces the underexposure and completely eliminate the effect of dark shadows by enhancing the details in the dark zones as shown in Fig. 8. MSRCR image processing is patented (e.g., USPTO #5,991,456, #6,843,125, #6,842,543, and CAN #PCT/US1997/007996) and available for licensing to commercial purposes.

Histogram Equalization Approach Using GMM

Adaptive image equalization using the GMM is another contrast enhancement method based on the HVS.[12] For decades, histogram modification techniques are generally regarded as the most straightforward and intuitive contrast enhancement approaches due to their simplicity. Global histogram equalization, local histogram equalization, brightness preserving histogram equalization with maximum entropy, the flattest histogram specification with accurate brightness preservation, histogram modification framework, and contrast enhancement based on genetic algorithm are representative examples of contrast enhancement methods using histogram modification. However, the aforementioned techniques cause problems when the histogram has spikes, often resulting in a hashed, noisy appearance of the output image or undesirable checkerboard effects. Further, they often suffer from empirical

Fig. 8 *Top*: Original images. *Bottom*: Output images using the MSRCR algorithm. Images are generated using a demo version of TruView® PhotoFlair

parameter settings. To solve these problems and to improve the visual quality of input images, the automatic image equalization and contrast enhancement algorithm using the GMM is proposed.[12]

Celik and Tjahjadi[12] explained that human eyes are more sensitive to widely scattered fluctuations than small variations around dense data. Hence, in order to increase the contrast while retaining image details, dense data with low standard deviation should be spread, whereas scattered data with high standard deviation should be compacted. To achieve this, the GMM-based algorithm[12] partitions the histogram distribution of an input image into a mixture of different Gaussian components as shown in Fig. 9b. The significant intersection points are selected from all the possible intersections between the Gaussian components. Then, the contrast enhanced image is generated by transforming the pixel's gray levels in each input interval to the appropriate output gray-level interval according to the dominant Gaussian component and the cumulative distribution function of the input interval. Since the HVS is more sensitive to sudden changes in widely scattered data and less sensitive to smooth changes in densely scattered data, the Gaussian components with small variances are weighted with smaller values than the Gaussian components with larger variances and vice versa.[12] Further, the gray-level distributions are used to weight the components in the mapping of the input interval to the output interval as shown in Fig. 9c. For color images, the luminance component is processed for contrast enhancement, while the chrominance components are preserved in $L*a*b*$ color space.

Performance evaluation using absolute minimum mean brightness, discrete entropy, and edge-based contrast on 300 gray and color images of the Berkeley image database shows that the proposed algorithm outperforms other histogram modification algorithms without any parameter tuning, even under diverse illumination conditions.[12] The proposed algorithm attains drastic dynamic range compression while preserving fine details without yielding artifacts such as halos, gradient reversals, or loss of local contrast.

Image Fusion

Global contrast enhancement is limited in its ability to improve local details, while local enhancement approaches can suffer from block discontinuities (e.g., checkerboard effects), noise amplification, and unnatural image modifications. A fusion-based contrast enhancement solves these problems and balances the effects of local and global contrast.[13]

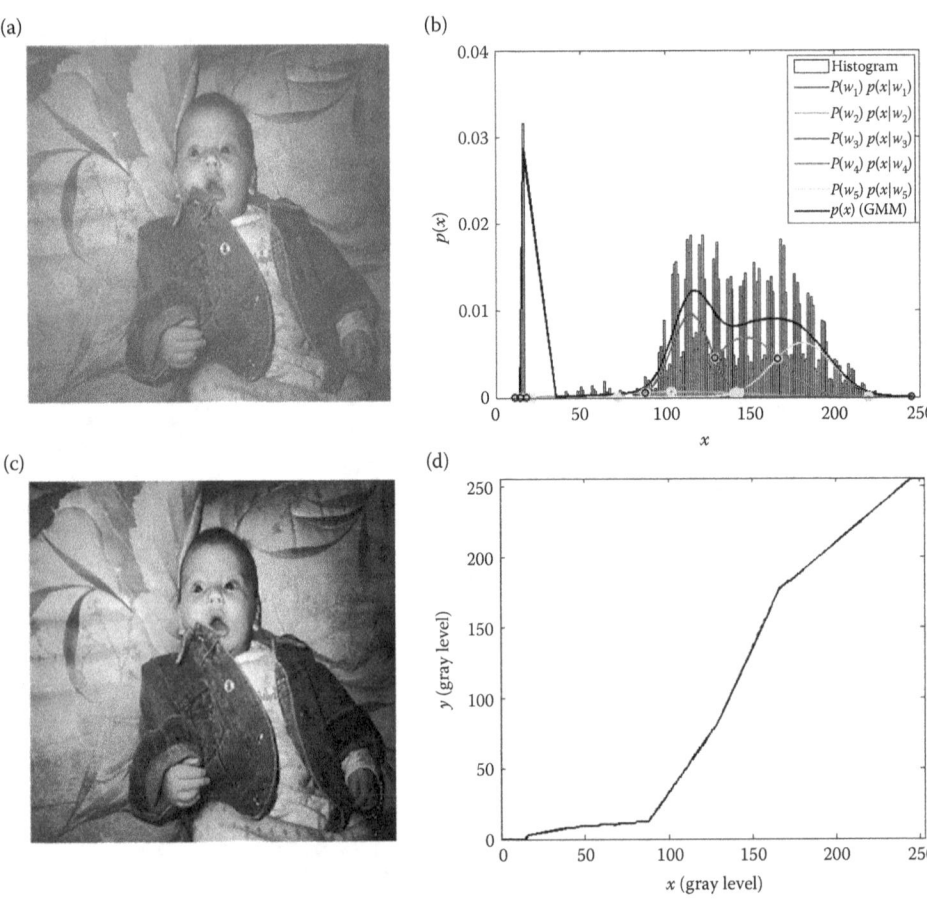

Fig. 9 (a) Gray-level input image; (b) histogram and GMM fit of (a); (c) data mapping between the input and the output; (d) output image. Images are provided by Celik and Tjahjadi[12]

A fusion-based contrast enhancement involves extracting the most informative regions from the input images and filtering the local areas of input images with weighted sum to achieve the contrast enhanced output. Since the HVS is more sensitive to local contrast changes such as edges, and since multiscale decomposition provides convenient localization of contrast changes, multiresolution-based image fusion approaches are widely used in practice. Specifically, Laplacian multiscale refinement achieves a seamless, halo-free fused image. Several input images and corresponding normalized weight maps are decomposed via the Gaussian and the Laplacian pyramid, respectively, and then they are blended into a fused pyramid as follows[13]:

$$F_l = \sum_{k=1}^{N} G_l\{\overline{W}_k\} L_l\{I_k\}, \tag{9}$$

$$\overline{W}_k = W_k / \sum_{k=1}^{N} W_k, \tag{10}$$

where k is the index of input images I, N is the number of input images, W_k is a weight map of the input image at index k, $\overline{W}_k$ is a normalized weight map to ensure that the sum of W_k to unity, and l is the number of pyramid levels. l can be various but is recommended to be over six in order to eliminate halo effects. $G_l\{\cdot\}$ and $L_l\{\cdot\}$ represent the Gaussian and the Laplacian decomposition at pyramid levels l, respectively. Operations are performed successively for each level, in a bottom-up manner. Finally, a contrast enhanced image J is achieved by the Laplacian pyramid reconstruction as follows[13]:

$$J = \sum_l F_l \uparrow^n, \tag{11}$$

where $\uparrow^n$ is the upsampling operator with factor $n = 2^{l-1}$.

The challenging part of the fusion-based approaches is the design of the weight maps. The weight maps vary depending on applications, but are largely based on visual quality features or salience dictated by the particular vision task. For example, the scalar weight map

$$W_k = (W_k^{chr})^\alpha (W_k^{sat})^\beta (W_k^{con})^\gamma (W_k^{sal})^\delta, \tag{12}$$

where W^{chr}, W^{sat}, W^{con}, and W^{sal} denote chrominance, saturation, contrast, and saliency respectively, whereas the exponents α, β, γ, and δ represent impact portions between weighting features. Results over a wide range of test images show that the fusion-based approach balances local and global contrasts without block discontinuities and artifacts amplifications, while preserving the original image appearance. Fusion-based contrast enhancement techniques are widely used to improve image focus, image exposure, HDR photography, and image segmentation. Figure 10 shows some example results of the fusion-based contrast enhancement.

Fig. 10 (a, c) Original images; (b, d) output images of the fusion-based contrast enhancement. Images are generated using an algorithm [13]

Tone Mapping and HDR Image Enhancement

The dynamic range of the natural environment is extremely large. The HVS efficiently adapts colors, luminance, and tones in the real world into the best appearance, while standard acquisition and display devices suffer from visualization of the real world due to their limited low dynamic ranges (LDRs). To effectively visualize the natural environment, various TMOs and HDR image enhancement techniques have been suggested. In spite of significant progress on developing TMOs, further improvement of HDR image enhancement are challenging due to the difficulty of perceptual IQA for HDR images.[14] Recently, a perceptually relevant tone-mapped image quality index (TMQI)[14] and a new optimized TMO[15] based on TMQI were introduced, which have shown to correlate well with human judgments on a wide range of tone-mapped images.

Yeganeh and Wang[14] explain that a good tone-mapped image should achieve a high performance both in multiscale structural fidelity and in statistical naturalness. Since the HVS extracts structural information from the visual scene, multiscale signal fidelity is a good predictor of perceptual quality. In addition, the deviation from the NSS of an input image can be used as an indicator of image naturalness in terms of perceptual quality.[14] TMQI not only provides an overall quality score of the tone-mapped image but also creates multiscale quality maps that reflect the structural fidelity variations across scale and space. Let X and Y be the HDR and the tone-mapped LDR image, respectively. The TMQI computation is expressed as follows[14]:

$$\text{TMQI}(X,Y) = a[S(X,Y)]^\alpha + (1-a)[N(Y)]^\beta, \quad (13)$$

$$S(X,Y) = \frac{1}{M}\sum_{i=1}^{M} S_{local}(x_i, y_i), \quad (14)$$

$$S_{local}(x,y) = \frac{2\sigma'_x \sigma'_y + C_1}{\sigma'^2_x + \sigma'^2_y + C_1} \times \frac{2\sigma_{xy} + C_2}{\sigma_x \sigma_y + C_2}, \quad (15)$$

$$\sigma' = \frac{1}{\sqrt{2\pi}\theta_\sigma} \int_{-\infty}^{\sigma} \left[-\frac{(t-\tau_\sigma)^2}{2\theta_\sigma^2}\right] dt, \quad (16)$$

$$N(Y) = \frac{1}{K} P_m P_d, \quad (17)$$

where S and N denote the structural fidelity and statistical naturalness, respectively. The parameters α, β determine the sensitivities of the two factors, and $0 \le a \le 1$ adjusts the relative importance between them. x and y denote two local image patches extracted from X and Y. S_{local} means the SSIM-motivated local structural fidelity, whereas σ_x, σ_y, and σ_{xy} indicate the local standard deviations and cross correlation between the two corresponding patches. C_1 and C_2 are stability constants. τ_σ is a contrast threshold, and $\theta_\sigma = \tau_\sigma/3$. The statistical naturalness N is modeled as the product of two density functions of image brightness and contrast. K is a normalization factor given by $K = \max\{P_m, P_d\}$, where P_m and P_d are a Gaussian density function and a Beta-density function, respectively.

A new TMO approach[15] based on TMQI is an iterative algorithm that searches for the best solution in the image space. Specifically, the algorithm adopts a gradient ascent method to improve the structural fidelity S, and then solves a constrained optimization problem to enhance the statistical naturalness N. The two-step iteration alternately continues until convergence to an optimized TMQI. Performance evaluation is executed on a database of 14 HDR images including various contents such as humans, landscapes, architectures, as well as indoor and night scene. Results show that both the structural fidelity and naturalness enhancement steps significantly improve the overall quality of the tone-mapped image as can be seen in Fig. 11.

Fig. 11 *Top*: Initial images created by gamma corrections. *Bottom*: Output images of the tone mapped and HDR image enhancement. Images are provided by Ma et al.[15]

PERCEPTUAL IMAGE ENHANCEMENT APPLICATIONS

Robust Face Recognition under Variable Illumination Conditions

Face recognition plays a central role in a variety of applications such as security and surveillance systems. Various face recognition methods including eigenface, fisherface, independent component analysis, and Bayesian face recognition have been suggested in the past decades. However, due to the difficulties in handling the lighting conditions in practice, recognizing human faces in variable illuminations, especially in the side lighting effects, is challenging. To achieve robust face recognition under varying illuminations, an adaptive region-based image enhancement method is proposed.[16]

The first step of the algorithm of Du and Ward[16] is segmenting an image into differently lit regions according to its local homogeneous illumination conditions. The region segmentation is based on an edge map of the face image obtained using two-level wavelet decomposition. Adaptive region-based histogram equalization (ARHE) is then regionally applied to enhance the contrast and to minimize variations under different lighting conditions based on the spatial frequency responses of the HVS. The variations of illumination generally lie in the low-frequency band. Next, since the high-frequency components become more substantial in recognition of details with poor illuminations, the algorithm enlarges the high-frequency coefficients at different scales so as to enhance the edges, which is called edge enhancement (EdgeE). At the end, an enhanced image is reconstructed using the inverse discrete wavelet transform with the modified coefficients. Finally, the reconstructed enhanced image is used for face recognition. Figure 12 shows the examples of face images under various illumination conditions and corresponding enhanced images using an algorithm.[16]

The experimental results using the ARHE and EdgeE on a variety of face recognition databases (e.g., Yale B, Extended Yale B, and Carnegie Mellon University Pose, Illumination, and Expression) show that the method significantly improves the face recognition rate when the face images include illumination variations. The simplicity and generality of the algorithm enable the approach to be applied directly to any single image without any lighting assumption, without any prior information about the 3D face geometry, and without any data fitting steps.[16]

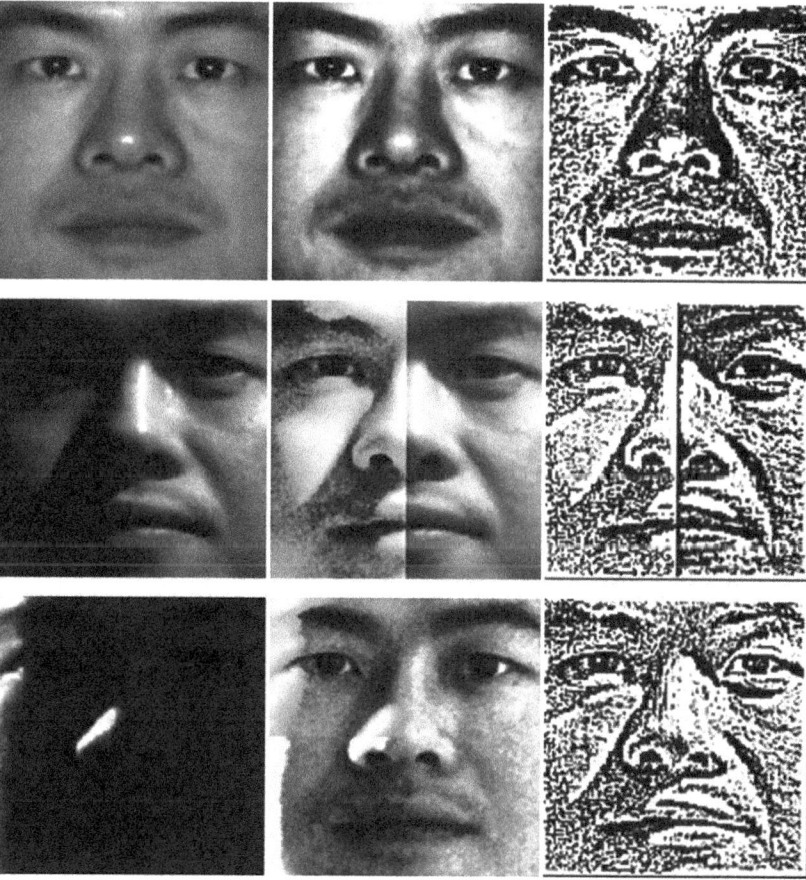

Fig. 12 *First column*: Original evenly, unevenly, and badly illuminated face images. *Second column*: Enhanced images by ARHE. *Third column*: Enhanced images by ARHE+EdgeE. Images are provided by Du and Ward[16]

Perceptual Image Defogging and Scene Visibility Enhancement

The perception of outdoor natural scenes is important for successfully conducting visual activities such as object detection, recognition, and navigation. In bad weather, the absorption or scattering of light by atmospheric particles such as fog, haze, and mist can seriously degrade scene visibility. As a result, objects in images captured under such conditions suffer from low contrast, faint color, and shifted luminance. Since degraded visibility can cause serious operator misjudgments in vehicles guided by camera images and can induce erroneous sensing in surveillance systems, image defogging designed to enhance scene visibility of foggy images is important.[17]

Recently a referenceless, perceptual image defogging and visibility enhancement algorithm was introduced[18] based on a space domain NSS model and multiscale "fog aware" statistical features.[17] Here, "referenceless" means that the perceptual defogging algorithm does not require multiple foggy images, different degrees of polarization, pre-detected salient objects in a foggy scene, side geographical information obtained from an onboard camera, a depth-dependent transmission map, training on human-rated judgments, or content-dependent assumptions such as smoothness of airlight layers, smoothness of a depth map, or substantial variations of color and luminance in a foggy scene.[18] Fog aware statistical features include the variance of local mean subtracted contrast normalized (MSCN) coefficients, the variance of the vertical product of MSCN coefficients, the sharpness, the coefficient of variance of sharpness, the Michelson contrast (or contrast energy), the image entropy, a dark channel prior, the color saturation in HSV color space, and the colorfulness.

The defogging and visibility enhancer utilizes statistical regularities observed on foggy and fog-free images to predict perceptual fog density of the input foggy image, then extract the most visible information from three processed images: one white balanced and two contrast enhanced images. Chrominance, saturation, saliency, perceptual fog density, fog aware luminance, and contrast weight maps are applied to effectively filter the most visible areas of the three preprocessed images and to smoothly blend them via Laplacian multiscale refinement.[18]

Evaluation on a variety of foggy images in subjective and objective comparisons shows that the perceptual image defogging algorithm produces more naturalistic, clear contrast of edges, and vivid color restoration after defogging. Figure 13 shows the examples of foggy images and their corresponding defogged, visibility enhanced images.

Nonlinear Unsharp Masking for Mammogram Enhancement

Digital mammography is the most common technique to detect breast cancer since it yields better image quality, uses a lower X-ray dose, and offers faster diagnosis with confident interpretation.[19] Mammogram enhancement algorithms generally focus on the contrast improvement of specific regions and details in mammograms, and then apply a threshold in order to separately visualize them from their surroundings. Nonlinear unsharp masking (NLUM) method combining the nonlinear filtering and unsharp masking (UM) technique is one of the powerful mammogram enhancement algorithms.[19]

A traditional UM technique is to subtract a low-pass filtered image from the original image. The UM improves visual quality by adding a scaled high-frequency part of the signal to its original; however, it can simultaneously amplify noise and over-enhances steep edges. NLUM enhances the

Fig. 13 *Top*: Original foggy images. *Bottom*: Defogged, visibility enhanced images using the perceptual image defogging algorithm. Images are provided by Choi et al.[18]

fine details of mammographic images by integrating the nonlinear filtering with UM. The original mammogram $I(m,n)$ is filtered by a nonlinear filter, then normalized and combined with the original mammogram using the arithmetic addition and multiplication in order to obtain an enhanced mammogram. Mathematically, the enhanced mammogram is expressed by Panetta et al.[19]:

$$E(m,n) = A_1 I(m,n) + A_2 \frac{F(m,n)}{|F|_{max}} I(m,n), \qquad (18)$$

where A_1 and A_2 are scaling factors, and $|F|_{max}$ is the maximum absolute value of the nonlinear filtered mammogram, $F(m,n) = w_0 I_0 + w_1 I_1 + w_2 I_2$, where w_0, w_1, and w_2 are weight coefficients working as high-pass filters, and I_0, I_1, I_2 are exponentially center-weighted mean filter of the original mammogram I, respectively. The second derivative-like measure of enhancement (SDME) is applied to optimize the NLUM parameters, obtaining the best enhancement result. Specifically, the NLUM parameters are decided where the SDME curve reaches the local maxima by changing the scaling factor and exponential gain coefficients.[19]

Results tested on a variety of 322 mammograms from the mini-MIAS database show that NLUM achieves excellent performance for improving the local contrast of specific regions and fine details such as microcalcifications. The abnormal regions are more recognizable in the enhanced mammograms. It was also observed that NLUM helps doctors to diagnose breast cancers successfully. For detailed examples, we refer to the work of Panetta et al.[19]

Adaptive Perceptual Image and Video Compression

Most existing approaches in image and video compression have focused on developing reliable methods to minimize mathematically tractable and easy to measure distortion metrics. Some of these non-perceptual techniques sometimes fail to correlate well with the perceived quality of the reconstructed images, especially at lower bit rates.[20] As a new effort to overcome these limitations, certain properties of the HVS have been used in the context of signal compression in order to minimize perceptually redundant information while preserving visual quality.

Höntsch and Karam[20] proposed a discrete cosine transform-based, locally adaptive, and perceptual image coder (PIC) based on the contrast sensitivity and visual masking effects of the HVS. This coder locally and adaptively discriminates between signal components based on their perceptual relevance. Specifically, the coder attempts to conceal coding distortions using the computed local spatial detection thresholds, thereby discarding the signal components which are imperceptible to the human receiver.

The coder proposed in the work of Hontsch and Karam[20] decomposes a given image into several subbands that vary in frequency and orientation. The subbands at varying local frequencies and orientations are used to adaptively compute local distortion sensitivity profiles in the form of distortion thresholds (i.e., perceptual weights). These thresholds are then used to adaptively control the quantization and dequantization stages of the coding system by adapting the quantization reconstruction levels to the varying characteristics of the visual data. The nature of this approach is that the process depends on image content and the thresholds are derived locally. These thresholds are needed both at the encoder and at the decoder in order to reconstruct the coded visual data. To avoid the need to transmit certain amount of side information, locally available masking characteristics of the visual input were estimated only at the decoder.[20]

The authors compared the bit rates (in bits per pixel) obtained using their locally adaptive perceptual image compression (APIC) with a nonlocally adaptive Safranek–Johnston method (called a PIC) for perceptually lossless compression. We reproduced the results presented by the authors in their work in Table 1 and Fig. 14. As shown in Table 1 and Fig. 14, APIC successfully removes a large number of perceptually redundant information in images and results in superior performance when compared to existing nonlocally adaptive schemes.

Table 1 Bit rates for the PIC and the APIC for (a) perceptually-lossless compression and (b) near transparent compression with a step size multiplier of 2.0

(a)			
	Bit rate (bits per pixel)		Compression gain (%)
Image	PIC	APIC	
Baboon	1.57	1.24	21
Actor	0.99	0.55	45
Lighthouse	0.92	0.61	34
Lena	0.79	0.54	31
Boat	0.90	0.66	27

(b)			
	Bit rate (bits per pixel)		Compression gain (%)
Image	PIC	APIC	
Baboon	1.05	0.85	18
Actor	0.71	0.37	47
Lighthouse	0.67	0.41	39
Lena	0.54	0.35	34
Boat	0.64	0.44	32

Fig. 14 Results of coding *Actor* and *Lenna* with a step size multiplier of 2.0. (a, d) Original images at 8 bits per pixel (bpp); (b, e) PIC at 0.706 and 0.635 bpp; (c, f) APIC at 0.372 and 0.351 bpp, respectively. The images are generated using an algorithm [20]

CONCLUSION

We have discussed actively evolving perceptual image enhancement models and applications based on the principles of the HVS. Since the HVS has remarkable abilities to analyze visual stimuli, understanding how the HVS tunes to natural environments and achieves stable, clear visual perception leads to the development of successful perceptual image enhancement methods. Among numerous properties of the HVS, CSF, multiscale and multiorientation decomposition, visual masking, and modeling of the NSS have played crucial roles in the development of algorithms that alleviate visual distortions, that increase contrast and visibility, and that adjust HDR natural scenes to display mediums. Perceptually inspired image enhancement models have been extensively applied to practical applications for better perception, interpretation, and subsequent analysis such as robust face recognition, image defogging, mammogram diagnosis, and image/video compression. In summary, understanding and modeling the HVS is fundamental to developing modern image enhancement models, and perceptual image enhancement is a promising direction for future research.

Although we introduced representative perceptual image enhancement models and applications, many emerging image enhancement topics remain in an early stage. Enhancement of color quality is an important area, yet there are not well-accepted perceptual quality predictions of color images. Progress remains to be made on color perception including color constancy and oppenency. Enhancement of stereoscopic 3D images is also a motivating topic. The role of depth and visual comfort are worth exploring for 3D image enhancement. As mobile cameras become very widely used, enhancement of pictures acquired using mobile devices such as image stabilization, color balance, and noise reduction under a low-light condition are pressing areas for future work on perceptual image enhancement.

REFERENCES

1. Robson, J.G. Spatial and temporal contrast-sensitivity functions of the visual system. J. Opt. Soc. Am. **1966**, *56* (8), 1141–1142.
2. Bovik, A.C.; Clark, M.; Geisler, W.S. Multichannel texture analysis using localized spatial filters. IEEE Trans. Pattern Anal. Mach. Intell. **1990**, *12* (1), 55–73.
3. Bovik, A.C. Automatic prediction of perceptual image and video quality. Proc. IEEE **2013**, *101* (9), 2008–2024.
4. Heeger, D.J. Normalization of cell responses in cat striate cortex. Visual Neurosci. **1992**, *9* (02), 181–197.
5. Suchow, J.W.; Alvarez, G.A. Motion silences awareness of visual change. Curr. Biol. **2011**, *21* (2), 140–143.
6. Choi, L.K.; Bovik, A.C.; Cormack, L.K. A flicker detector model of the motion silencing illusion. J. Vision **2012**, *12* (9), 777.
7. Portilla, J.; Strela, V.; Wainwright, M.J.; Simoncelli, E.P. Image denoising using scale mixtures of Gaussians in the wavelet domain. IEEE Trans. Image Process. **2003**, *12* (11), 1338–1351.
8. Varghese, G.; Wang, Z. Video denoising based on a spatio-temporal Gaussian scale mixture model. IEEE Trans. Circuits Syst. Video Technol. **2010**, *20* (7), 1032–1040. Available at https://ece.uwaterloo.ca/~z70wang/research/stgsm/.

9. Fergus, R.; Singh, B.; Hertzmann, A.; Roweis, S.T.; Freeman, W.T. Removing camera shake from a single photograph. ACM Trans. Graph. **2006**, *25* (3), 787–794. Available at http://cs.nyu.edu/~fergus/research/deblur.html.
10. Moorthy, A.K.; Mittal, A.; Bovik, A.C. Perceptually optimized blind repair of natural images. Sig. Process. Image Commun. **2013**, *28* (10), 1478–1493.
11. Rahman, Z.; Jobson, D.J.; Woodell, G.A. Retinex processing for automatic image enhancement. J. Electron. Imaging **2004**, *13* (1), 100–110.
12. Celik, T.; Tjahjadi, T. Automatic image equalization and contrast enhancement using Gaussian mixture modeling. IEEE Trans. Image Process. **2012**, *21* (1), 145–156.
13. Mertens, T.; Kautz, J.; Reeth, F.V. Exposure fusion: A simple and practical alternative to high dynamic range photography. Comput. Graph. Forum **2009**, *28* (1), 161–171.
14. Yeganeh, H.; Wang, Z. Objective quality assessment of tone-mapped images. IEEE Trans. Image Process. **2013**, *22* (2), 657–667. Available at https://ece.uwaterloo.ca/~z70wang/research/tmqi/.
15. Ma, K.; Yeganeh, H.; Zeng, K.; Wang, Z. High dynamic range image tone mapping by optimizing tone mapped image quality index. In *IEEE International Conference on Multimedia and Expo*, Chengdu, 2014, 1–6.
16. Du, S.; Ward, R.K. Adaptive region-based image enhancement method for robust face recognition under variable illumination conditions. IEEE Trans. Circuits Syst. Video Technol. **2010**, *20* (9), 1165–1175.
17. Choi, L.K.; You, J.; Bovik, A.C. Referenceless perceptual fog density prediction model. In *IS&T/SPIE Human Vision Electronic Imaging XIX*, San Francisco, CA, 2014, 90140H.
18. Choi, L.K.; You, J.; Bovik, A.C. Referenceless prediction of perceptual fog density and perceptual image defogging. IEEE Trans. Image Process. **2015**, *24* (11), 3888–3901. Available at http://live.ece.utexas.edu/research/fog/fade_defade.html.
19. Panetta, K.; Zhou, Y.; Agaian, S.; Jia, H. Nonlinear unsharp masking for mammogram enhancement. IEEE Trans. Inf. Technol. Biomed. **2011**, *15* (6), 918–928.
20. Hontsch, I.; Karam, L.J. Locally adaptive perceptual image coding. IEEE Trans. Image Process. **2000**, *9* (9), 1472–1483.

Photographic Image Characteristics

Miroslav Mikota and Mile Matijević
Faculty of Graphic Arts, University of Zagreb, Zagreb, Croatia

Abstract

Photography as a medium is characterized with the high level of iconicity and is trusted more than other image information types. The development and dominance of digital photography system have in the technical sense substantially influenced the approach to the creation of a photographic image. Furthermore, the manipulation of digital data enabled even more precise image data transfer and the influence on perception while maintaining iconic character of the image. However, the transformations of digital media in several procedures such as RAW into JPG, JPG into WebP, WebP into JPG, resolution change, document compression, and color mode also influence the image data potentially leading to degradation of image data. It is necessary to take those changes into account for the photographic image to maintain the author's intent and characteristics. Understanding of those changes also enables their targeted application.

Keywords: Color space; Digital photographic system; File format; Image characteristics; Photographic image.

INTRODUCTION

The appearance of digital photography led to the significant changes in the creation of photographic images in relation to classical photography systems. The basic change is in the photosensitive media used for imaging and in the fact that the light-sensitive sensor of the digital camera is not the media in which the image is stored.

The image in the digital photographic system is created on the photosensitive sensor of specific characteristics. The digital data is then stored on media (most commonly memory card of specific type—Compact Flash, Secure Digital, etc.). This digital record is characterized by sensor properties, chosen camera options, and file format specifics. Those options change technical and syntactic–semantic characteristics of the photographic image. The photographer must be aware of digital record manipulations in order for the photographic image to deliver the appropriate message to the viewer.

The digital photograph is stored in appropriate file format, depending on the digital camera options and the photographer.

The most common file format is JPEG (Joint Photographic Expert Group), rarely TIFF (Tagged Image File Format) and RAW file formats.[1] It uses loose compression. The intention of this format is to keep as much visually important details, and reject irrelevant ones, considering that the impression of the photograph is more influenced by small changes in lightness than the chromaticity. The efficiency of compression is dependent on the motif. It is less efficient in the high detail motifs.

In general, compared to TIFF format, the JPEG file is 10–20 times smaller depending on the selected compression level. JPEG compression is performed in three stages: (i) discrete cosine transform creates JPEG-specific 8-point "blocks," (ii) quantization shuffles data with the definition of the file size and image quality loss, and (iii) compression and encoding of the results.[2]

RAW file formats are not universal. They depend on the manufacturer of the camera. Most common are crw (Canon), nef (Nikon), mrw (Minolta), ptx (Pentax), raf (Fuji), orf (Olympus), and dng (Adobe, Leica, Hasselblad). Preediting of the RAW file format can redefine certain selected parameters of the imaging process in the digital camera.

Preprocessing of the RAW file defines the parameters usually attributed by the central processing unit of the digital camera (according to the settings set by the photographer). The first stage of preprocessing is the correction of exposition. Exposition is changed for all or no hue values. The second stage of preprocessing is the adjustment of white balance. Available white balance settings usually correspond to those available on the camera, with the possibility of manual white point selection, as software white balance definition.

The preprocessing of RAW photo enables the color general adjustment according to the preset settings on the camera prior to shooting, standard settings, or the change in contrast, saturation, and hue. The preprocessing can influence sharpness, noise, and graininess of the image, color space (usually sRGB, Adobe RGB), color depth or change it into gray scale. Preprocessed photo is saved into

JPEG or rarely TIFF file formats. The RAW file format can be used again for new preprocessing of the same photo.[3,4]

Depending on the used software, preprocessing enables multiple image manipulation options.

Irrelevant to its creation, the 8-bit (by R, G, B channel) JPEG passes through the processing which changes its properties. The changes should not create communication noise, as changing the message of the author-photographer to the viewer. Even though the color space of human vision is larger than the camera, screen, or print color space, the human brain can accept it as real. This acceptance of narrow color space enables the acceptance of photography as a media which presents the "realistic" reality.

The most commonly used color space today is sRGB. It was defined by Microsoft and HP in 1996 in an effort to standardize monitor colors. This color space is quite narrow in the cyan–green area. As the digital camera sensors can "read" those colors, camera manufacturers also enable wider Adobe RGB color space. As most of the exit units such as monitors, printers, and projectors display sRGB color space, the usual color space in imaging software is redefined in sRGB.

Color space redefining is called rendering. There are several color rendering intents: absolute, relative, saturation, and perceptual rendering. Absolute rendering renders the colors between color spaces while maintaining their colorimetric values. Color outside the spectrum (such as rendering from Adobe RGB to sRGB) is rendered into maximal saturation of narrower color space. The same principle is used in relative color rendering while taking into account the shift of the white color in comparison to the white in starting color space.

Colorimetric color rendering (from Adobe RGB to sRGB) losses the differences of colors outside the narrower color space and creates problems in fine tone gradients. This rendering is generally not used for Adobe RGB to sRGB rendering unless the colors in the photos are damped so that the photo does not contain relevant information in saturated colors.

Saturation rendering transforms maximally saturated colors of one color space into maximally saturated colors of the second color space while disregarding differences in hue, lightness, and saturation. It is not appropriate for rendering from Adobe RGB into sRGB color space in the area of digital photography. The exception can be the photos based on specific colors on the border of unnatural colors. Rendering of photos from Adobe RGB into sRGB is usually done with perception intent also called photographic rendering.

The basic idea of perceptional rendering is keeping the perceptional impression of the image as close to the original even though colorimetric changes do occur.[5,6] Less saturated colors are altered less than high saturated colors, which are compressed into smaller color space while maintaining the color difference in the narrower color space. This is the only reversible rendering.

In order to enable rendering, color engine must be set—image 145. Although the rendering should be standard, different color engines are available (such as Apple, Microsoft, Agfa, Linotype, Kodak, and Adobe). Since the standard of digital photography is Adobe Photoshop, Adobe ACE is taken as the standard color engine. The alternative is Microsoft ICM, with minimal differences in rendering result.

Color depth defines the number of colors that can be presented.

The depth of 8-bit per channel (total of 24 bit) divides each channel into 256 (2^8) colors, or 16.8 million colors ($256 \times 256 \times 256$). 16.8 million colors are considered the basic number for good color reproduction quality.

The depth of 16-bit per channel (total of 48 bit) enhances the number of colors the image program can process. It also enlarges file size significantly and limits the processing of such files in the computer software. Because of this, the 48-bit color depth is mostly used for scanning of classical photos which are processed with "levels" and "curves" as the basic color management and rendered into 24-bit depth for further processing.

Digital record of the photographic image is a sequence of pixels creating an image (bitmap). A number of pixels per unit of length is called a resolution, and it is usually defined as the number of pixels per inch or dots per inch (dpi) and should be related to the image size. Image size for photograph is defined as 1:1 (record:print) in resolution of 180–300 dpi.

This image definition makes sense only if the original contains an appropriate number of pixels. For the maximal standard format for classic photography of 50×60 cm, the number of pixels can be doubled at most.

Larger pictures are usually viewed from a larger distance and allow for the greater enlargement of total pixel number without reduction of image perception quality from a larger distance. This is appropriate for format definition, resolution, and additional cropping. Change in size is defined as size and resolution in menu image size. Downsampling of the image is performed best with Bicubic Sharper option and upscaling with Bicubic Smoother.

Digital photos can be defined as a single channel or multichannel. Single channel modes are meant for black-and-white photographs and multichannel for color photographs. The main channel for black-and-white photos is gray scale where every pixel is defined with values from 0 to 255, where 0 is black and 255 is white. Tone difference is separated into three areas for easier processing: shadows, mid-tones, and highlights. Color processing is performed in RGB mode. It is default color mode for digital photography and screen preview. In RGB, the color is defined with red, green, and blue primary colors with values from 0 to 255 for each channel.

The grayscale photo can be recorded in RGB color mode. RGB grayscale or color image can be rendered in grayscale mode. The primary way is rendering from RGB to grayscale mode. The alternative is rendering into Lab mode and selection of L channel, or separation of R, G, or B channel in RGB mode. Separation of R, G, or B channel can simulate color filters for black-and-white photography. Green channel is most often used. The red channel can be used for portrait photography. For borderline, black-and-white and color effects or application of color on black-and-white photo document stays in RGB mode and "desaturated." Other conversion methods can be used, such as Image–Adjustments–Black and White. Simulation of color filters can be performed, and the image is converted to black-and-white mode.

Digital photograph also led to the images with a high dynamic range (HDR) of photography. Development of digital photographic systems in the 1980s and the 1990s led to development of HDR photographs. First, HDR format was introduced in the year 1985 by Gregory Ward (Radiance RGBE). In 1995, Steve Mann and Rosalind Picard presented a mathematical formula for HDR generation (global HDR) which is the basis for a modern digital HDR photograph.

In 1997, Paul Debovec introduced HDR image generation based on the merging of differently exposed pictures of the same scene. Tone mapping then creates the HDR image displayable on exit units of lower dynamic range. This type of image generation introduced the term HDR photography.

This technique became the basis of multiple HDR-generating programs. The most known one is Photomatix. From the year 2005, the possibility of generation HDR images from differently exposed images is built in into Adobe Photoshop CS2.[7]

There is no unique definition of HDR or HDR photography. The theory of HDR photography defines it as the one containing all the tones present in the recorded scene. Disregarding the technical problems of generation of such photograph in comparison to the potential dynamic range of the scene (17 apertures) and human perception of the scene (14 apertures), it is clear that although technically superior, visually appears surreal. This is one of the basic characteristics of HDR photography. As an artistic expression, HDR photograph maintains this surreal appearance. If such definition is applied, then the HDR file must have the possibility of infinite tone and color record. This is possible in the usage of 32-bit color depth with the so-called floating reference point.[7,8]

Standard photography considers HDR as any photography with higher dynamic range than a standard photograph. There are no strict boundaries since there is no standard dynamic range of taken photographs. It is dependent on the camera, mostly on the imaging sensor and chosen sensitivity.

HDR is also defined as a photograph created by merging two or more differently exposed photographs of the same scene. If we accept that the basic element of HDR photography is the expansion of the dynamic range of the photograph and not the whole dynamic range of the scene, then this definition is only partially correct. The expansion of the dynamic range of the scene can be achieved from one photograph with appropriate preprocessing. Definition of HDR mentions fake or pseudo-HDR photograph, most commonly when one JPEG photograph is used for generation of HDR photograph.

However, without regard for HDR definition, the problem is the display of the photo with higher dynamic range. Taking the photo with the photographic camera creates the image of certain dynamic range. This range is quite smaller in comparison to the dynamic range of the scene. Assuming the image is created with correct exposition elements reproduces optimal tones in mid areas, but loses the details in highlights and shadows. If the scene is taken preexposed, recording takes shadow areas of the image with loss of data in other areas. Underexposure records highlight with a tonal loss in other areas. When digital records of these photographs "meld" into one, a document is created with a good resolution of tones in highlights, midtones, and shadows. Two or more photos are combined into one, the more commonly even number of photos in 0.5 or 1 aperture with AEB (Automatic Exposure Bracketing) recording option.

The alternative is a recording of the one photo processed to create an entire exposition range. A single photograph is processed into one correctly exposed, one underexposed, and one preexposed. The corrected photos are then merged into a single HDR photo. The processing of files needs to be done in a RAW file format that uses the full dynamic range of the photographic camera. This range is still narrower than the range of the scene. This process is sometimes referred as pseudo-HDR photography, and in certain cases, especially in creative purposes, can be achieved with JPEG photography.

The process is based on loading recorded photographs in specialized software, such as Photomatix—image 165 or the corresponding Photoshop option—"Merge to HDR." This results with the 32-bit record with floating white reference point containing the information from all the merged photographs into a new HDR photograph.

In order for the HDR photo to be stored into 16-bit TIFF or 8-bit JPEG, mapping of tones must be performed. In essence, the linear tonal record of HDR document is compressed into a narrower tonal range that can be reproduced. Technically, the relatively narrow dynamic color range is used to create an appearance of a wider dynamic range. This phase gives the opportunity of generating multiple finished photographs from a single HDR record.[9]

Contemporary visual communications are increasingly relying on web technologies while moving away from conventional methods and techniques for distributing various content, for example, text or images. Taking such a situation into consideration, the proliferation of photography as a media cannot be ignored. It is undeniable that photography has left an indelible imprint on means of communication even until today. However, until the 2010s, creating and publishing images (especially photographs) with a goal of reaching a large number of consumers was not easily available to everyone. Mentioned trends are accompanied by a significant decrease of photographic equipment prices but also with an appearance of mobile phones equipped with digital cameras that give their users a possibility of creating high-quality photographic images and instantly publishing them to reach a large number of people. All these mentioned trends make a solid foundation for developing new, more suitable file formats for digital photography. Although existing and most widespread file formats (JPEG in particular) are optimized for a wide spectrum of media platforms, there is still a need for development of such a format which would enable a more rational usage of data storage and transfer capacities while preserving an optimal quality and technical characteristics of digital photographic images.

One of the formats showing the ambition to endanger the position of existing conventional formats lead by JPEG is WebP (Web Photo) developed by One2 Technologies and taken over by the Internet giant Google. Its development is based on the idea that a new format should be created to be used in web technologies, enabling a major reduction in file sizes, thus making it possible to save data storage and reduce data traffic.

Due to the fact that the existing file formats used for digital photographic images are based on a different paradigm and intended to be used in a different kind of media, they are proving to be insufficient when it comes to contemporary web technologies. New solutions are expected to offer a lesser file size while preserving the image quality, especially when it comes to reproduction of color and image resolution. The mentioned solution in the form of WebP file format has come from the development of a file format for video files called WebM (Web Movie) which is based on algorithms used for coding VP8 video format which uses Resource Interchange File Format for creating and storing blocks of data. Taking into consideration that video is basically a series of consecutive image frames, it seems quite reasonable that a development of a new video format is linked to a file format for digital photographic images. Restructuring the blocks of data done in the conversion to WebP image format from conventional ones enables up to a 40% reduction of file sizes which ensures a more efficient data handling, and it was also announced that such a format would keep colors and image resolution, as well as other technical characteristics, virtually intact. Other announcements considering the WebP format include an introduction of ICC profiles, further development of transparency feature, and compatibility with a majority of web browsers and software make it a serious option for different media and cross-media platforms.

Problems associated with the application of WebP format are mainly connected to the fact that the photographer while using the Internet and presenting his photographs to public often is not even aware of the conversion from JPEG to WebP. It should be mentioned that a number of published papers indicate strongly that there are changes in the image resolution and sharpness as well as there are changes in color reproduction, especially near the achromatic point. Having all that said, there are concerns considering the reliability of transfer information and messages. Due to modern trends in the proliferation of digital photography, this effect could prove to be most problematic in the field of portrait photography which is a form of "selfie," the most present form of photography on the Internet today. After all, the last phase in any photographic system is the realization of photography through output units (either electronic or printed form), which can influence the most on the quality of photography and can be the source of the biggest information noise.[10]

If this problem is observed from the point of the dynamic range of digital photographic image and its realization technique, contemporary displays have a dynamic range up to 9.5 EV (exposure value), which is significantly less than the dynamic range of standard photography. A possible solution is found in the technology of the so-called HDR displays based on LED technology, so they could achieve a dynamic range bigger than 14 EV which correlates with a dynamic range of human vision while watching an open scenery. However, when discussing a dynamic range of displays and projectors, one should have in mind that it is based on contrast and not the dynamic range directly and the contrast (e.g., 750:1) correlates to watching in ideal conditions and is changed depending on the conditions of watching; therefore, there could be a link to the noise inherent in the digital camera sensor.

While discussing printing techniques for digital photography, the expanding of dynamic range is attempted in various ways, depending on the technique. Although techniques that use a true multitone subtractive (CMY) synthesis of color could result in photographic prints with wide dynamic range, the biggest advances in this field are being made with high fidelity drop-on-demand inkjet technology (whether it is piezoelectric or thermal bubble jet). The development of new printing machines for digital photography is followed by a big emphasis put on new ink formulations, for example, HDR inks which use HDR pigments. This way, a variety of colors that can be printed are increased leading to the expansion of gamut (new colors from additive color synthesis are being used) and an advancement in the reproduction of light colors (using photo cyan and photo magenta). All of these features lead to bigger bit depth and fine gradient in gray area. Those new approaches to digital photography printing require new substrates with specific reception layers, thus achieving bigger color density (Dmax of black goes up to 2.6) which in theory gives contrast round 400:1 and dynamic range roughly around 8.6 EV.

Having all of these in mind, it is obvious that a digital photographic image goes through a series of changes to its technical characteristics between its acquisition and realization which could possibly cause a noise in communication and changes in semantic and syntactic characteristics of images.[11,12]

DYNAMIC RANGE

Dynamic range stands for a range between the lightest and the darkest tone. Although this definition of dynamic range seems rather simple, the very meaning of dynamic range varies depending on the field it is being used in.[3] When it comes to the dynamic range of a certain scenery, it basically comes to a ratio of maximal and minimal luminance. If an open scene is being watched during a sunny day, the dynamic range is around 100,000:1 and human vision has a dynamic range going around 10,000:1. The dynamic range of a certain scenery can be even larger because it can be determined by different lighting conditions while human vision adapts to those conditions. The dynamic range of a digital camera is determined by its sensor, and it represents the ratio between the strongest and the weakest signals it can generate. The dynamic range of a digital camera is expressed in EV; if a certain camera has the dynamic range of 10 EV, it means its dynamic range is in fact 1024 (2^{10}).

The dynamic range of a camera is directly connected to the bit depth which is expressed in bits for every channel, namely, R (red), G (green) and B (blue) channel. The 8-bit sensor can distinguish 256 t per channel (2^8), 10-bit can distinguish 1,024 t (2^{10}), 12-bit does the same for 4,096 t (2^{12}) and a 14-bit sensor distinguishes 16,384 t (2^{14}). In theory, it is possible to achieve a dynamic range of 8, 10, 12, or 14 EV; however, the signal noise inherent in the imaging system decreases that value. It is obvious that a true dynamic range depends on each imaging device or camera individually.

When it comes to output units, there are differences between ways of defining the dynamic range. In displays, it is defined as a ratio of luminance between the lightest and the darkest pixel it shows while the dynamic range of printed photographic images is defined by maximum color density (e.g., if the color density value is 2, then the dynamic range of the print is 100:1, 100 being 10^2). It should be noticed that the dynamic range defined in such a manner shows information about contrast but gives none of it about the ability of the system to differentiate tones and colors. Although ways of defining dynamic range differ one from another, it can be used in a specific photographic system to define the problem of tone and color reproduction from image acquisition to its realization.

CIE LUV SYSTEM, CIE LAB SYSTEM, AND COLOR DISTANCE ΔE

CIE (Commission Internationale de l'Eclairage) Luv and CIE Lab systems were represented in 1976. CIE Luv is a dominant color space for color reproduction in computer monitors, televisions, etc. Lab system is used for printing photographs and for printing. Lab system represents a unique color space with equilized spatial and visual differences of color. CIE Lab color space enables the color viewing in CIE Lab chromaticity diagram and is determined by rectangular coordinates of L, a, and b defined as follows[13,14]:

$$L^* = 116 \left(\frac{Y}{Y_n}\right)^{0.333} - 16, \quad (1)$$

$$a^* = 500 \left[\left(\frac{X}{X_n}\right)^{0.333} - \left(\frac{Y}{Y_n}\right)^{0.333}\right], \quad (2)$$

$$b^* = 200 \left[\left(\left(\frac{Y}{Y_n}\right)^{0.333} - \left(\frac{Z}{Z_n}\right)^{0.333}\right)\right], \quad (3)$$

where tristimulus values X_n, Y_n, and Z_n relate to nominal stimuli of white object with standard lightning type.

The difference of color (ΔE) in Lab system is originally specified with the following formula:[13,14]

$$\Delta E^*_{ab} = \left[(\Delta L^*)^2 + (\Delta a^*)^2 + (\Delta b^*)^2\right]^{0.5} \quad (4)$$

This formula was developed, and in the year 1994, it was defined as a mathematical formula for measuring the color difference (ΔE_{94}). Further formula development for color difference was made in 2000 when (ΔE_{00}) was released.[13,14]

$$\Delta E^*_{94} = \sqrt{\left(\frac{\Delta L^*}{k_L S_L}\right)^2 + \left(\frac{\Delta C^*_{ab}}{k_C S_C}\right)^2 + \left(\frac{\Delta H^*_{ab}}{k_H S_H}\right)^2} \quad (5)$$

$$S_L = 1 \quad (6)$$

$$S_C = 1 + 0.045 \cdot C^*_{ab} \quad (7)$$

$$S_H = 1 + 0.015 \cdot C^*_{ab} \quad (8)$$

$$\Delta E_{00} = \left[\left(\frac{\Delta L'}{k_L S_L}\right)^2 + \left(\frac{\Delta C'_{ab}}{k_C S_C}\right)^2 + \left(\frac{\Delta H'_{ab}}{k_H S_H}\right)^2 + R_T \left(\frac{\Delta C'_{ab}}{k_C S_C}\right)\left(\frac{\Delta H'_{ab}}{k_H S_H}\right)\right]^{0.5} \quad (9)$$

where

$$L' = L^* \quad (10)$$

$$a' = (1+G)a^* \quad (11)$$

$$b' = b^* \quad (12)$$

$$C'_{ab} = \left[(a')^2 + (b')^2\right]^{0.5} \quad (13)$$

$$h' = \tan^{-1}\left(\frac{b'}{a'}\right) \quad (14)$$

$$G = 0.5\left\{1 - \left[\frac{\left(\overline{C^*_{ab}}\right)^7}{\left(\overline{C^*_{ab}}\right)^7 + 25^7}\right]^{0.5}\right\} \quad (15)$$

where $\overline{C^*_{ab}}$ is arithmetic middle C^*_{ab} of values for paired samples.

$$\Delta L' = L'_b - L'_s \quad (16)$$

$$\Delta C'_{ab} = C'_{ab,b} - C'_{ab,s} \quad (17)$$

$$\Delta H'ab = \left[2\left(C'_{ab,b} C'_{ab,s}\right)^{0.5} \sin\left(\frac{\Delta h'_{ab}}{2}\right)\right] \quad (18)$$

where variables b and s relate to comparison of samples and reference samples, which are given by

$$S_L = 1 + \frac{0.015\left(\overline{L'} - 50\right)^2}{\left[20 + \left(\overline{L'} - 50\right)^2\right]^{0.5}} \quad (19)$$

$$S_C = 1 + 0.045\overline{C'_{ab}} \quad (20)$$

$$S_H = 1 + 0.015\overline{C'_{ab}}T \quad (21)$$

where

$$T = 1 - 0.17\cos\left(\overline{h'_{ab}} - 30°\right) + 0.24\cos\left(2\overline{h'_{ab}}\right) \\ + 0.32\cos\left(3\overline{h'_{ab}} + 6°\right) - 0.20\cos\left(4\overline{h'_{ab}} - 63°\right) \quad (22)$$

$$R_T = -\sin(2\Delta\Theta)R_C \quad (23)$$

$$\Delta\Theta = 30\exp\left\{-\left[\frac{\overline{h'_{ab}} - 275°}{25}\right]^2\right\} \quad (24)$$

$$R_C = 2\left(\frac{\overline{C'_{ab}}^7}{\overline{C'_{ab}}^7 + 25^7}\right)^{0.5} \quad (25)$$

$\overline{L'}, \overline{C'_{ab}}, \overline{h'_{ab}}$ are arithmetic middles of paired samples of single values.

RESOLUTION AND SHARPNESS ASSESSMENT

The final quality of a digital photographic image depends on the ability of the used system to differentiate individual parts of an object. This ability depends on the technical characteristics of a photographic film or any photographic media used in classical system. Quality of digital photograph is dependent on sensor size and resolution, chosen file format, lens sharpness, and all the features of output units and materials used for printing.

The overall assessment of image resolution and sharpness is done by taking photographs of appropriate test charts with ISO 12233 being the most commonly used method. This chart, besides containing fields for assessing the resolving power of a camera (the so-called Modulation Transfer Function chart), has some other elements—to ease focusing and to detect errors inherent in camera's optics.

To test the resolving power while taking photos (one should test both the lens and the used photographic media), this chart is being photographed in a 1:1 scale, and then the next step is to determine the last field in which one can clearly separate two lines. The number read right under it is multiplied by 100 and gives a value to resolving power through the ratio of lines per width and picture height. This procedure is being done in both directions—vertical and horizontal.

To assess the resolving power in image printing, the same ISO 12233 chart is being saved to a file format with no compression (e.g., TIFF) or JPEG with minimal compression and printed afterward. The following procedure is the same as explained previously.

VISUAL TECHNICAL ASSESSMENT OF AN IMAGE QUALITY

The visual technical image quality assessment is used to compare original and reproduced digital photographic images. Different methods can be used to compare images made with the same technique of realization (e.g., comparing different prints) or even images across different media (e.g., comparing prints to images on displays). Several viewing techniques can be used to compare images on various types of displays to printed images under different illuminants and luminance levels. There are five techniques defined by Braun et al. in 1996[15]—memory, successive-binocular, simultaneous-binocular, simultaneous-haploscopic, and successive-Ganzfeld-haploscopic viewing. Choosing which technique to use is dependent on the goal of research and conditions in which the viewing should take place.

The memory viewing technique implies that a viewer is given two images separately without the possibility to return to the previous one (thus the name—memory). This technique has proved to be the best to simulate the consummation of photographic images.

Successive binocular viewing technique is also based on the comparison of a sample and an original, but it allows the viewer to switch from one image to another at any time and as much as he or she wants; however, it is not possible to view them at the same time.

Simultaneous-binocular viewing technique is based on the simultaneous comparison of the original and the reproduction of a photographic image. Compared to color measurement methods, this technique gives the most similar results.

Simultaneous-haploscopic viewing technique gives the opportunity for the research participants to view both images simultaneously with a barrier between their eyes, keeping one eye on the original while the other eye is viewing the reproduction.

Successive-Ganzfeld-haploscopic viewing technique defines the method as showing the images to the participant exactly the same as the previous technique, but not simultaneously.

DIGITAL PHOTOGRAPHIC IMAGE SIMILARITY INDEX

While processing a digital photographic image and especially when changing the file format or color space, there is a deformation in the structure of digital image, thus causing a degradation of information transferred using the image and especially causing noise in the communication channel. All those changes can be tracked and examined by using various methods—similarity index included.

Generally there are three basic approaches to this subject: FR (full reference), RR (reduced reference), and NR (no reference) methods. FR methods compare and determine differences between two images, one being the original and the other being a reproduction. RR methods use a smaller part of compared images to determine the difference, whereas NR methods use no original to assess the quality of the reproduction. Due to the fact that the goal is to get objective results of the comparison, the FR methods are predominantly used; thus, they can be used along with other viewing techniques. Among the FR methods, there are several ways to determine similarity between images, for example, peak-signal-to-noise ratio and mean square error, but these methods show insufficient correlation to visual technical assessment methods. Other contemporary methods include feature similarity, visual-signal-to-noise ratio, and visual information fidelity. An alternative is using SSIM (structural similarity index measurement) which uses lightness, contrast, and structural information between two images (Fig. 1).[16]

FORENSICS AND INTELLECTUAL PROPERTY RIGHTS IN DIGITAL PHOTOGRAPHY

Digital image analysis enables digital photographic forensics and determining changes in the original JPEG which could be considered a forgery when it comes to defining the authors' genuinity of the photographic image.

In such a manner, determining the level of lumination areas on the same optical distance with difference luminance can be found. The method is based on a conversion of digital signal to binary based on predefined thresholds. Determining the origin of each pixel is based on light intensity threshold; if that value would be 1, the entire image surface would be seen as black, while if that value would be 0, it would be white. Defining the right threshold depends on the type of photographic image being analyzed. The starting value is usually 0.5, and it decreases or increases depending on the results. This kind of analysis shows suspicious areas which show unusual light intensities. Deviations in color values found in HSV color space (hue, saturation, value) can also detect elements that do not originate from the analyzed image, in other words elements that were added to the photographic image. The conversion to binary format or to HSV color space is usually done in an image analysis software.

The confirmation of results can be made with an analysis of JPEG blocks and their characteristics, while further methods include the analysis of the digital image file structure through the EXIF file.

The combination of given methods gives the opportunity to determine if there were some later interventions to the digital photographic image and if the image can be considered genuine or a forgery.[17,18]

HISTOGRAM IN A DIGITAL PHOTOGRAPHIC IMAGE

A histogram represents a graphic representation of tone distribution in a digital photographic image or, in other words, the distribution of pixel for each and every tone

Fig. 1 Comparison of the original photography in JPG (left) and converted WebP (right), where SSIM is 0.2

Fig. 2 Histogram of the original photography in JPG format

Fig. 3 Histogram of the photography in a certain RAW format

found in the image. In the case of 8-bit RGB image, each channel can have 256 different tones (from 0 representing black to 255 representing full intensity) thus resulting in an image made of 16,777,216 t (colors). The left side of the histogram represents dark areas, whereas the right side represents light areas. If the image is overexposed during photographing, most of the pixels are shown on the right side of the histogram, whereas in the case of underexposure, they are mostly located on the left. If the image has balanced tones and colors, with proper lighting, its histogram is balanced throughout the entire area with just some peaks. This is why histogram can be considered as a useful tool for objective analysis of tones and colors, and this analysis can be done for the entire image or for each of the channels separately (Figs. 2 and 3).

This entry describes the basic elements that influence the changes in digital photographic image through various phases of digital photographic system. It has also shown methods that make assessments of these changes in digital photography in a way to detect noise in communication channel or even false information transfer and interpretation between a photographer and a final photography consumer. This entry has mostly portrayed photography in its traditional form of two-dimensional and static media, but nowadays photography is becoming much more and boundaries between static and moving image or two-dimensional and three-dimensional media are being erased.

REFERENCES

1. Allen, E.; Triantaphillidou, S. (Eds.) *The Manual of Photography*, 10th Ed.; Elsevier/Focal Press: Oxford, Burlington, MA, 2011, 566.
2. Solomon, C.; Breckon, T. *Fundamentals of Digital Image Processing: A Practical Approach with Examples in Matlab*; John Wiley & Sons, Ltd.: Chichester, 2010.
3. Eismann, K.; Duggan, S.; Grey, T.; McClelland, D. *Real World Digital Photography*, 3rd Ed.; Peachpit Press: Berkeley, CA, 2011, 575.
4. Mikota, M.; Kulčar, R.; Jecić, Z. Digital shooting in area of applied and art photography. In *9th International Conference on Printing, Design and Graphic Communications Blaž Baromić 2005*; Bolanča, Z.; Mikota, M.; Ed.; Zagreb; Senj, Ljubljana: Grafički fakultet Sveučilišta; Ogranak Matice hrvatske; Inštitut za celulozo in papir, 2005, 79–86.

5. Mikota, M.; Pavlović, I.; Matijević, M. Effect of the changes in the RGB digital image channel on the perception of fashion photography while retaining its iconicity. J. Text Cloth Technol. **2015**, *64*, 1–2, 19–24.
6. Mikota, M.; Matijevic, M.; Pavlovic, I. Colour reproduction analysis of portrait photography in cross-media system: Image on the computer monitor–electrophotographic printing. Imaging Sci J. **2016**, *64* (6), 299–304.
7. Langford, M. *Langford's Advanced Photography*; Elsevier/Focal Press: Amsterdam and Boston, MA, 2008.
8. Mikota, M.; Matijević, M.; Pavlović, I. Realizacija HDR fotografije Hi Fi DOD bubble jet ispisom na optimalnoj podlozi za ispis. Teh Glas. **2013**, *7* (3), 252–257.
9. Mikota, M. HDR photography–New challanges in realization and photographic image reproduction. In *Proceedings 15th International Conference on Printing, Design and Graphic Communications Blaž Baromić 2011*; Mikota, M.; Ed.; HDG: Zagreb, 2011, 88–98.
10. Matijević, M. Impact of JPEG-WebP conversion on the characteristics of the photographic image. Teh Vjesn—Tech Gaz. **2016**, *23* (2). Available at http://hrcak.srce.hr/index.php?show=clanak&id_clanak_jezik=231164&lang=en, (accessed in March 31, 2017).
11. Periša, M.; Mrvac, N.; Mikota, M. The visual grammar of photographic images produced by media convergence. In *Communication Management Forum 2015 Proceedings Zagreb*; Ciboci, L.; Ed.; Edward Bernays: Zagreb, 2015, 647–667.
12. Peres, M.R.; Ed.; *The Focal Encyclopedia of Photography: Digital Imaging, Theory and Applications History and Science*, 4th Ed.; Elsevier: Amsterdam, 2007, 846.
13. Kuehni, R.G. *Color Space and its Divisions: Color Order from Antiquity to the Present*; Wiley-Interscience: Hoboken, NJ, 2003, 408.
14. Sharma, G. *Digital Color Imaging Handbook*; CRC Press: Boca Raton, FL, 2003.
15. Braun, K.M.; Fairchild, M.D.; Alessi, P.J. Viewing techniques for cross-media image comparisons. Color Res. Appl. **1996**, *21* (1), 6–17.
16. Žeželj, T.; Mikota, M.; Tomiša, M.; Maričević, M. The degradation of photographic image by changing the compression and file format. In *International Conference Materials, Tribology, Recycling 2016*; Žmak, I.; Fabijanić Aleksandrov, T.; Ćorić, D.; Eds.; Hrvatsko društvo za materijale i tribologiju: Zagreb, 2016, 458–463.
17. Johnson, M.K.; Farid, H. Exposing digital forgeries in complex lighting environments. IEEE Trans. Inf. Forensics Secur. **2007**, *2* (3), 450–461.
18. Johnson, M.K.; Farid, H. Exposing digital forgeries by detecting inconsistencies in lighting. In *Proceedings of the 7th Workshop on Multimedia and Security*; ACM: New York, 2005, 1–10.

Raindrop Imaging

J. R. Saylor
Department of Mechanical Engineering, Clemson University, Clemson, South Carolina, U.S.A.

Abstract
Obtaining the diameter of a raindrop from a digital image of that drop is critical to several aspects of precipitation science. Initially this image processing task seems straightforward: under proper illumination conditions, the quality of the drop image can be quite good; deviations from sphericity are typically small; the image outline is smooth and close to circular, facilitating the rejection of noise and extraneous objects, which tend to be rough. However, typical image processing algorithms result in several problems and inaccuracies in the process of obtaining an absolute measurement of the drop diameter from the image. Most of these problems pertain to the depth of field of the entire imaging system, which is determined by both the physical optics and the digital imaging algorithms used. A large depth of field is desired in order to obtain a large number of raindrop images, which is key to obtaining converged statistics on drop size and shape. However, a large depth of field can result in significant variations in the magnification ratio for the optical system along the optical axis, which can result in errors in droplet diameter measurement, an important characteristic of raindrops. The image processing algorithms used to convert grayscale images of raindrops into measurements of drop diameter and shape can help or hinder the task of obtaining a large number of raindrop images and measuring their diameters accurately. This entry describes these raindrop imaging problems and discusses the strengths and weaknesses of specific image processing approaches within the context of those problems.

Keywords: Drop imaging; Edge detection; Histogram methods; Rain; Raindrop; Thresholding.

INTRODUCTION

Rain is a key aspect of the global water cycle and is therefore a critical part of our understanding of the global climate. Problems associated with the prediction of rain are well known to laymen and scientists alike. Perhaps less well known are the profound difficulties in simply knowing how much rain has fallen on a particular region. Point measurements can be made using standard volumetric rain gauges, tipping bucket rain gauges, as well as a variety of optical methods. However, rain rates can vary significantly with space and time, preventing accurate measurement of global rainfall using these single-point measurements. Practically speaking, radar measurement of rain is the only feasible means by which the amount of rain falling on the planet can be estimated. Precipitation radars that are ground based, such as the U.S. NEXRAD system (WSR-88D)[1] or satellite-based radar systems such as the NASA Tropical Rainfall Measurement Mission system,[2] have the potential for providing a complete picture of rainfall on the planet.

Using ground-based radar as an example, the radar waves are scattered back from the raindrops, and time-gated measurements of these signals can be used to get a radar reflectivity factor, Z, at every radial and azimuthal point within the radar's circular measurement area. Following the treatment from Doviak and Zrnic,[3] Z is related to rain as follows:

$$Z = \int_0^\infty N(D) D^6 \, dD \quad (1)$$

where D is the drop diameter and $N(D)$ is the drop size distribution, a measure of the number of drops per unit volume per unit drop size. The rain rate R can be obtained from Eq. 1 by writing R in terms of $N(D)$ as:

$$R = (\pi/6) \int_0^\infty D^3 N(D) w_t(D) \, dD \quad (2)$$

where $w_t(D)$ is the terminal velocity for a drop of diameter D. Equation 2 reveals that if $N(D)$ can be obtained from measurements of Z, then R can be obtained. To do this, the drop size distribution is typically modeled as a decaying exponential, a formulation first suggested by Marshall and Palmer:[4]

$$N(D) = N_0 e^{-\Lambda D} \quad (3)$$

Substituting Eq. 3 into Eq. 1 gives an equation for Z in terms of Λ and N_0. To close this system, N_0 is typically assumed to be a constant, allowing each measurement of Z to give a value for Λ. This value of Λ along with the constant for N_0 is then used to give $N(D)$, which is then substituted into Eq. 2 allowing computation of R at each measurement location and time. The value for N_0 is obtained in a variety of ways. For example, Marshall and Palmer[4] simply recommended a value of $N_0 = 8 \times 10^3$ m^{-3} mm^{-1} or by relating N_0 to some characteristic of the rain event.

The strength of the previous approach is that R can be obtained over very large regions, potentially over the entire planet, something that cannot be done with point measurements. The drawback of this approach is the assumptions used in modeling $N(D)$, including the assumption of an exponential function and errors incurred in developing a value for N_0. Consequently, the accuracy of measurements of R for the planet is determined in large part by the accuracy of $N(D)$. As a result, obtaining measurements of $N(D)$ to then reveal the accuracy of the function presented in Eq. 3 and/or to determine better ways for obtaining N_0 are critical.

The previous approach is taken when using single polarization radars, which represent the majority of precipitation radar installations. Dual polarization radars yield a vertical and horizontally polarized measurement of Z at each measurement cell in the range of the radar, that is, (Z_v, Z_h). Because there are two measurements obtained at each cell, such measurements enable extraction of two unknowns, that is, both Λ and N_0 from the radar data thereby eliminating the need for assumptions regarding the means for estimating a value for N_0. However, these radar installations are rare, and even with such radars, measurements of $N(D)$ are still needed to determine the validity of the exponential form of $N(D)$ and to improve upon this functional form.

All of the above motivate the need for measurements of $N(D)$ which, historically, have been obtained using the Joss–Waldvogel disdrometer (JWD),[5,6] a momentum-based method wherein drops impacting a Styrofoam cone are translated to an electric signal that is processed to yield drop diameter. Additionally, methods using line scan cameras, such as the 2DVD system,[7,8] have been used to obtain image slices of a falling drop that are then reconstructed to generate a complete two-dimensional image that is then used to obtain raindrop diameter.

With the advent of two-dimensional detectors, the ability to obtain $N(D)$ using digital imaging has enabled methods that benefit from the ever-improving resolution of commercially available digital cameras. This method was pioneered by Bliven at NASA Wallops (personal communication) using a setup similar to that illustrated in Fig. 1, where a digital camera faces a uniform light source, and the camera is focused on a location somewhere between the two. In this setup, raindrops falling between the camera and the light appear as silhouettes with the drop appearing dark on a bright background. The drops also have a bright spot in their center, which is an image of the light source as seen through the drop, i.e., with the drop

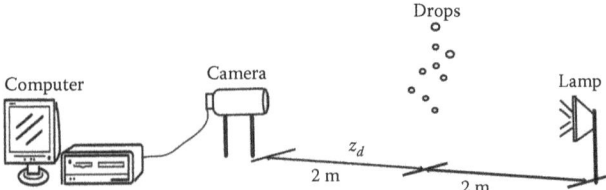

Fig. 1 Schematic illustration of raindrop imaging method for collecting images of the drop silhouette
Source: From Saylor and Sivasubramanian.[15]

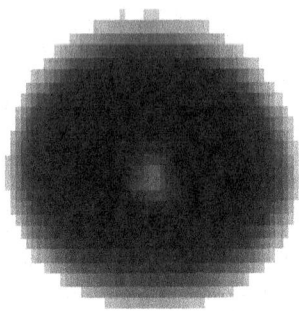

Fig. 2 Sample raindrop image is taken from the apparatus similar to that shown in Fig. 1. Note that the white spot in the middle of the raindrop is an image of the light source as seen through the drop

acting as a lens. A sample of such an image is presented in Fig. 2.

To obtain measurements of $N(D)$ using a digital camera setup like that shown earlier requires that a large number of drops should be measured so that $N(D)$, a statistical measure, is well converged. Moreover, the measurements should be obtained over a relatively short period of time, since $N(D)$ is itself a function of time. Satisfying the need for measuring a large number of raindrops over a small period of time requires a large depth of field. Large depths of field enable the measurement of large numbers of raindrops per second. However, a consequence of this concerns the magnification ratio:

$$M = \frac{d}{D} \qquad (4)$$

where D is the raindrop diameter and d is the diameter of the drop image on the digital camera sensor, both measured in the same units. Since drops closer to the camera appear larger than those farther away, M varies along the optical axis z. This error can be reduced by making the depth of field small, but as noted earlier, a large depth of field is inherently needed in this application to enable measurement of a large number of drops. Hence, variations of M with z will typically be large and can cause large errors in measurements of D.

A means for correcting the variation in M and thereby enabling a large depth of field along with accurate measurements of D is to develop a method for determining z_d, the location of the raindrop along the optical axis z and then using a value of M specific to that location. This would require that some characteristic of the raindrop image provides information related to z_d. Once this is done, it must be confirmed that this method is independent of the drop diameter, that is, that the functional relationship between z_d and the characteristic does not vary with D. Next, a method is described for doing precisely this. However, it is also shown that the performance of this approach varies with the digital image processing methods used.

DROP HOLES

As noted earlier and shown in Fig. 2, the image of a drop in silhouette exhibits a bright spot or "hole" in the center, which is an image of the light source as seen through the drop. This image of the light source is clearest when the drop is in the focal plane and becomes progressively blurred as the drop moves away from the focal plane.

This is shown in Fig. 3, which shows a sequence of images obtained using a glass sphere located at a range of positions along the optical axis, including the focal plane at 2 m from the camera and locations closer to and farther from the camera than the focal plane. In this sequence of images, the glass sphere takes the place of a drop. This is a deliberate choice used to evaluate this method since it allows precise location of the "drop" as well as precise knowledge of the drop diameter, which is quite difficult to do with actual drops. Additionally, the glass was MgF_2, which has an index of refraction very close to that of water. With all image processing algorithms that can be used to convert these grayscale images into binary images, the size of the white spot in the center of each sphere becomes smaller with the distance from the focal plane. The ratio of the spot size to the outer diameter of the sphere is

$$\alpha = \frac{d_s}{d} \qquad (5)$$

where d_s is the diameter of the spot, which can be used as a means to determine z_d, the location of the sphere along the optical axis. This is shown in Fig. 4 which is a plot of α versus z_d for glass spheres obtained from the work of Jones et al.[9]

This plot clearly shows how a measurement of α can be used to determine the z location at which the image was obtained, which in turn can be used to obtain the magnification ratio M (with suitable calibration) allowing a more accurate measurement of the actual drop/sphere diameter. In Fig. 4, each image was made binary using a threshold. This threshold was determined using a histogram clustering algorithm, where the image histogram was divided into two classes, and the centroid of each class in the histogram was averaged to define the threshold. However, many different image processing methods can be used, and this can have a significant impact on this overall approach, as is now explored.

Figure 5 shows a mosaic of glass sphere images. The left most column is the grayscale image of a MgF_2 sphere obtained at a range of z_d, and the remaining columns show the binary image obtained using a range of thresholding algorithms which were: double thresholding,[11] entropy,[12] a fixed threshold, the iterative thresholding method,[11,13] and a moment preservation technique.[14] As the figure shows, the range of z_d for which a hole in the image is observed differs with each algorithm. This range is the effective depth of field of the entire imaging system, since within the range where holes are observed, M can be determined from α. This is true regardless of how blurry the original grayscale image might be. Hence, an opportunity exists for increasing depth of field simply through the choice of the digital image processing algorithm. This will be further explored below.

Although the earlier approach works well, it also raises a secondary problem, which is that of a variable depth of field. As shown in Fig. 6, the different thresholding algorithms cited earlier also give a depth of field, which, while controllable by the choice of algorithm, exhibits a depth of field that varies with the drop diameter itself.

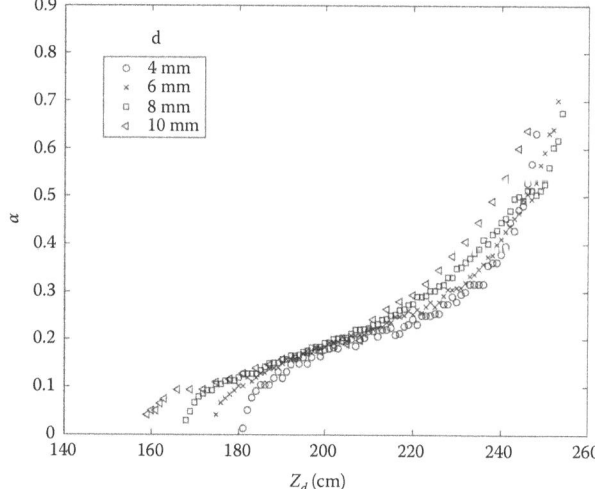

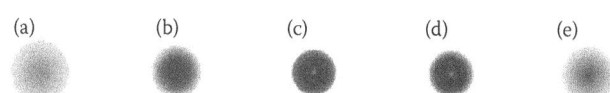

Fig. 3 Images of a 3-mm MgF_2 sphere obtained using the setup in Fig. 1. For these images, the camera and light source were separated by 4 m with the camera focused at 2 m. The distances of the spheres from the camera are (a) 186 mm, (b) 196 mm, (c) 200 mm (in focus), (d) 203 mm, and (e) 213 mm

Fig. 4 Plot of α versus z_d. The ratio of the diameter of the spot in the drop image to the diameter of the drop can be used to locate the position of the drop in the depth of field, regardless of the diameter. Note here that the "drops" are actually glass spheres
Source: From Jones et al.[9]

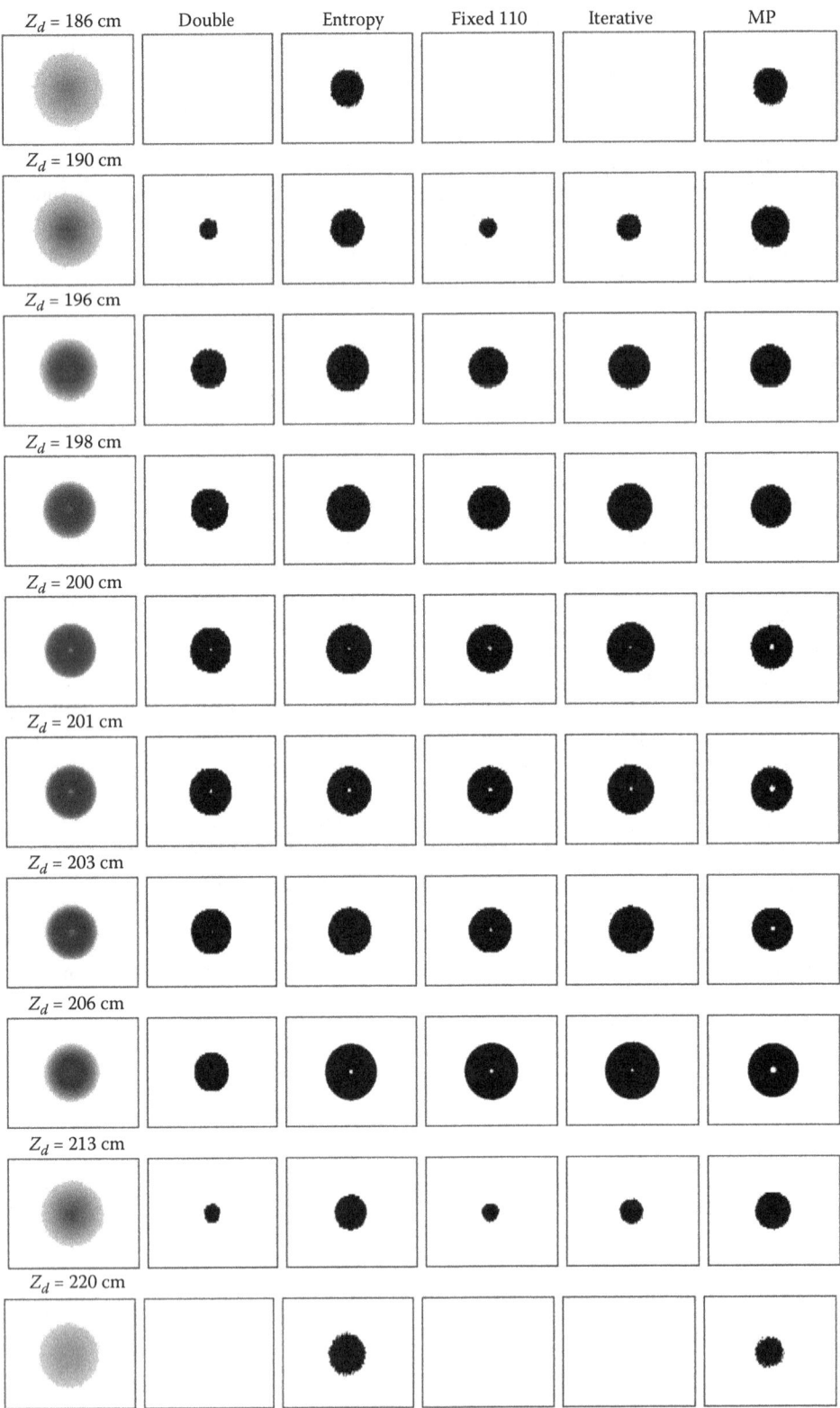

Fig. 5 Mosaic of images of glass spheres. The first column is the actual grayscale image of the glass sphere. The distance of the drop from the camera, z_d, is noted above each of these grayscale images, with the camera focused at $z_d = 200$ cm. The remaining five columns are the resulting binary images obtained using the indicated computer algorithm. The glass spheres were made of MgF$_2$
Source: From Saxena and Saylor.[10]

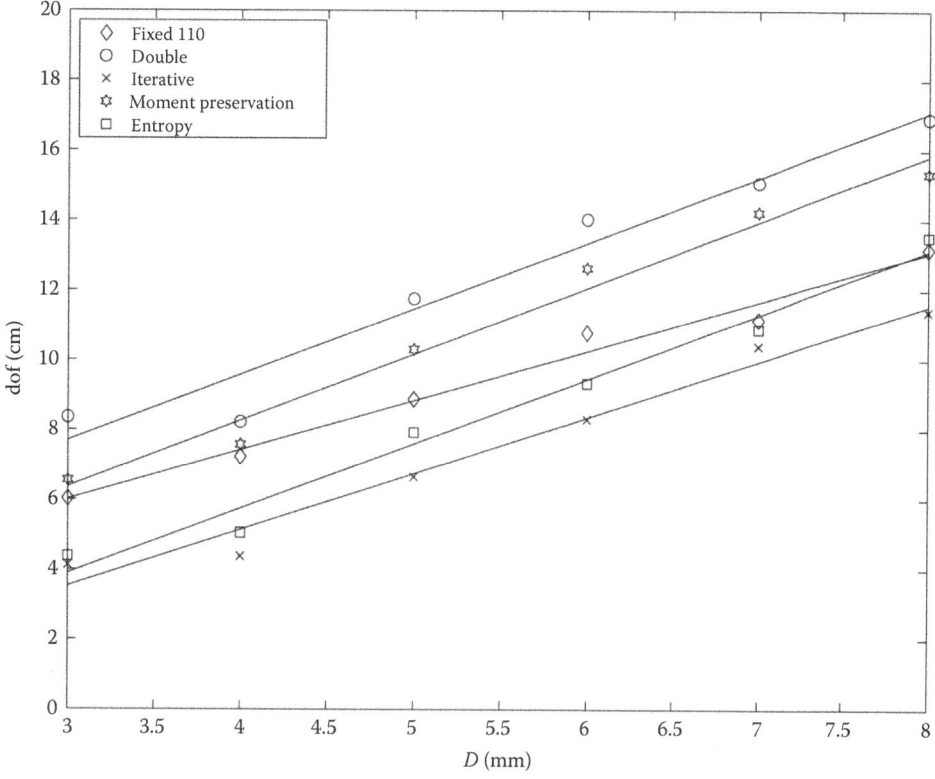

Fig. 6 Plot of effective depth of field (of the optical system plus the image processing algorithms) versus actual diameter
Source: From Saxena and Saylor.[10]

In theory, this problem can be worked around. If one obtains a relationship between α and z_d, the value of z_d can then be used to give the raindrop diameter D. Once this is known, a relationship between D and the depth of field enables accurate computation of $N(D)$. However, this process involves three transformations, each with its own uncertainty. A better approach would be one where there is less dependency on the relevant variables.

Saylor and Sivasubramanian[15] showed that using edge detection methods prior to thresholding gave better results. Specifically, the variation of M with z_d was quite small. This is illustrated in Fig. 7 which shows plots of measured drop diameter versus z_d, where the measured drop diameter is obtained using a value of M for the focal plane. As the figure shows, there is very little variation in the measured diameter over the entire depth of field. This is true for a range of drop diameters. Hence, even if one did not know the location of the drop in the depth of field, little error would be incurred if a single value of M was used.

Fig. 8 shows a plot of the diameter obtained by using several edge detection algorithms with a fixed magnification ratio versus the actual diameter. As the figure shows, all of the methods show excellent linearity in the relation between measured diameter and actual diameter. The Hueckel method[16,17] shows especially good behavior in that the plot of measured versus actual diameter is essentially identical to that of a unity slope line with zero intercept, showing that this approach can be used with excellent accuracy to get drop diameter.

However, using this approach, it is still the case that the depth of field varies with the drop diameter, as can be seen by examining Fig. 7 and as shown using a single plot in Fig. 9, below. However, the variation in depth of field with drop diameter is monotonic if one uses the Hueckel edge detection algorithm, and therefore, once drop diameter is obtained, the depth of field for that diameter bin is easily obtained from a functional fit to data such as that plotted in Fig. 9.

Field Comparisons

The studies cited previously consist of laboratory investigations using glass spheres. This approach is important and necessary since field studies of actual raindrops do not enable a comparison between a measured drop diameter and a known drop diameter. However, once an image processing method is developed in the laboratory, comparison with field data is needed. This was done in fact by Saxena and Saylor[10] for the thresholding methods. Specifically, rain data obtained using a raindrop imaging approach similar to that shown in Fig. 1 was obtained on several days during June and July of 2004. Measurements of rain rate were obtained using a JWD in parallel. Each of the thresholding methods explored was used to obtain $N(D)$, and

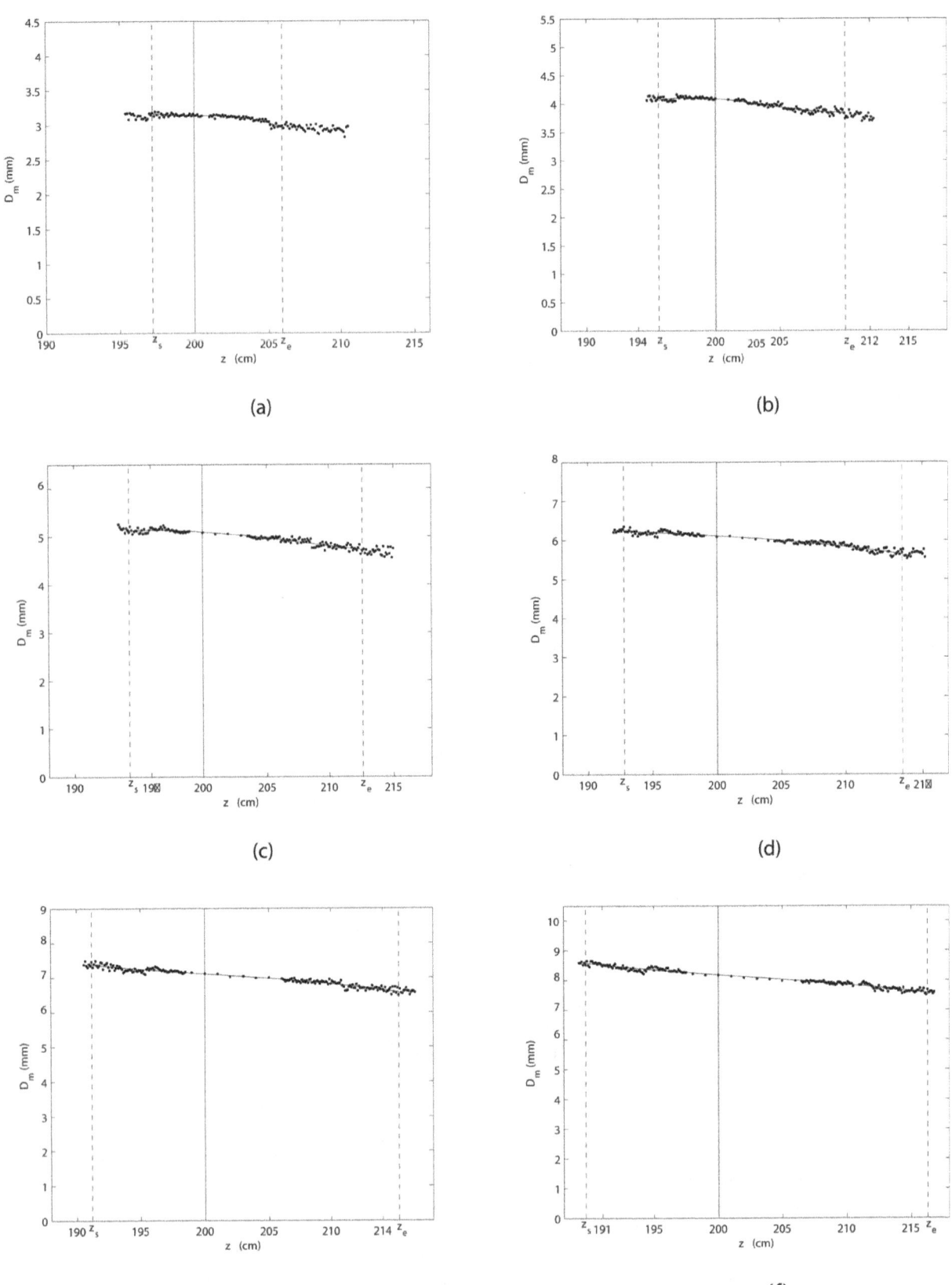

Fig. 7 Plots of measured diameter versus distance from the camera for six sphere diameters: (a) 3 mm; (b) 4 mm; (c) 5 mm; (d) 6 mm; (e) 7 mm; (f) 8 mm. In all plots, the measured diameter was obtained using the Hueckel[16,17] algorithm
Source: From Saylor and Sivasubramanian.[15]

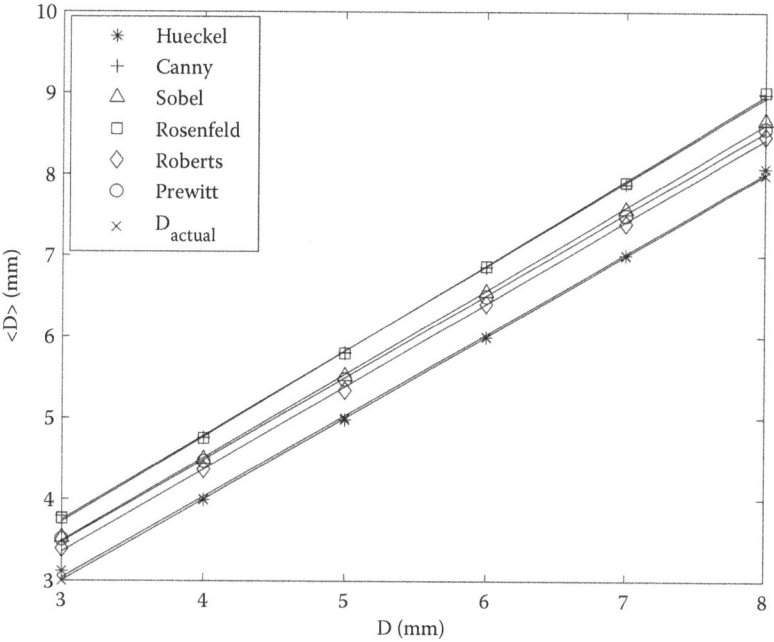

Fig. 8 Plot of measured diameter versus actual diameter for the same spheres used in Fig. 7, but for several image processing algorithms, including the Hueckel algorithm, which displays the best performance
Source: From Saylor and Sivasubramanian.[15]

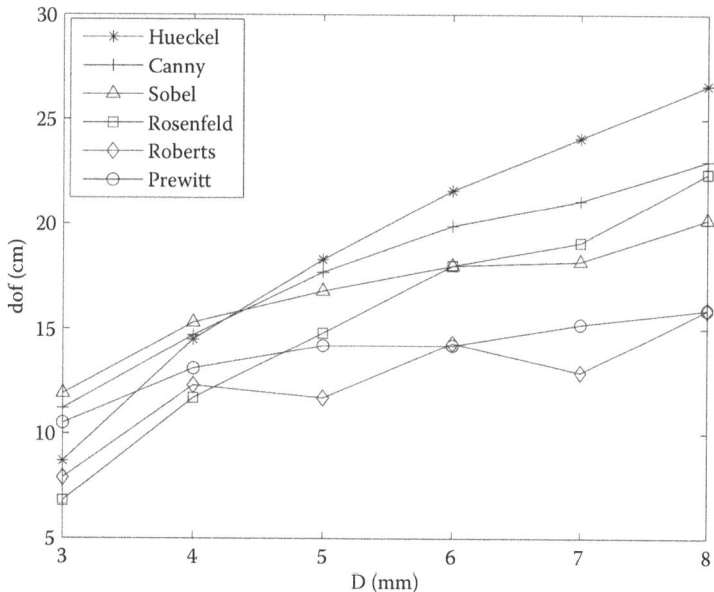

Fig. 9 Plot of effective depth of field versus actual sphere diameter for each of the image processing algorithms referred to in Fig. 8
Source: From Saylor and Sivasubramanian.[15]

that function was then used to obtain R using Eq. 2. All of the algorithms explored showed good agreement with the rain rate obtained using the JWD. In particular, the double thresholding algorithm performed especially well, and a plot of the rain rate obtained using this algorithm R_A is plotted against that obtained using the JWD, R_J, in Fig. 10.

It should be noted that the JWD has several measurement uncertainties. Chief among them is the fact that the impact velocity on the disdrometer head may deviate from terminal velocity due to vertical winds. Such deviations cause an error in drop diameter measurement using this device, since it relies on the combination of the measured signal (momentum) and the terminal velocity function to obtain D and hence $N(D)$. Of course, all the current methods use the terminal velocity function through Eq. 2; however, obtaining R using the JWD requires use of this

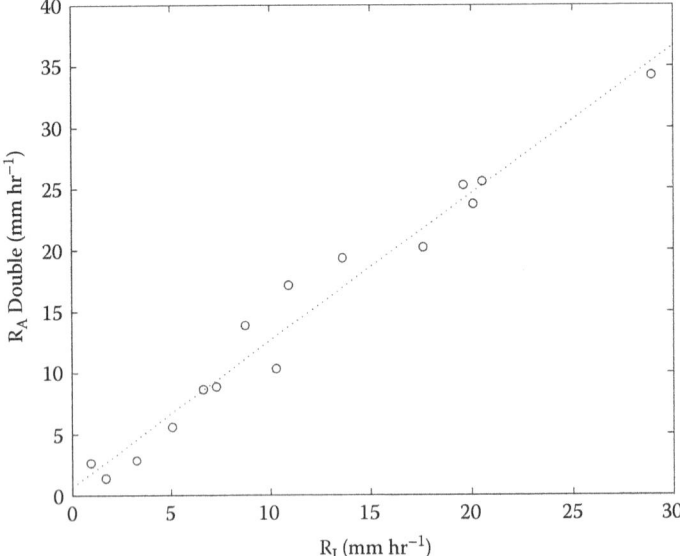

Fig. 10 Plot of rain rate obtained using raindrop imaging along with a thresholding algorithm (double thresholding) versus the rain rate obtained using the JWD
Source: From Saxena and Saylor.[15]

function twice and therefore introduces a greater level of uncertainty. This being the case, though the comparisons shown in Fig. 10 suggest an excellent performance of the methods described herein, future work may be profitably focused on comparing R obtained with the current approach to that obtained using a simple volumetric gage, or using a tipping bucket rain gauge, though even these devices have uncertainties associated with splash drops.

Other Issues

The laboratory work described earlier used perfect spheres. However, actual raindrops greater than approximately a millimeter in diameter exhibit measurable oblateness. To determine if this is an issue, Jones et al.[9] investigated ellipsoidal glass spheres and found that the method of using α to determine z_d and therefore M worked for these ellipsoidal-shaped drops, just as it did for spheres. Some of these ellipsoids deviated from sphericity in a way that was significantly larger than that expected from raindrops, illustrating the robustness of this method. This study also revealed that the index of refraction of the glass used affects the α versus z_d relationship. Magnesium fluoride is a glass that is transparent in the visible region and has an index of refraction very close to that of liquid water, $n = 1.38$,[18] compared with 1.33 for water. MgF_2 was used in the majority of the works cited earlier and is the material that should be used in future work focused on the methods described herein.

The dynamic range and spatial resolution of commercially available digital cameras is growing at a vigorous rate and at modest price ranges, suggesting that imaging of raindrops will enjoy a natural growth in resolution and accuracy as time passes. Hence, using digital imaging of raindrops to obtain an improved understanding of $N(D)$ is likely the best approach in future work. What are especially needed are further field studies, in particular those that use the edge detection approach, specifically the Hueckel algorithm and where the resulting rain rates are compared to those obtained by tipping bucket rain gauges and volumetric measurements.

CONCLUSION

Optical imaging of raindrops can provide important statistical characterizations of drop size and shape, which are important in and of themselves, and which can improve the ability of radars to extract rain rate from radar data. Key to raindrop imaging are the digital image processing algorithms, which are needed to automate the sizing of raindrops in raindrop imagery. It is important to properly select and use these algorithms to avoid several errors that are inherent in this approach, but which can be accounted for.

REFERENCES

1. Bonewitz, J.D. The NEXRAD program—An overview. Bull. Am. Meteorol. Soc. **1981**, *62* (6), 944–944.
2. Simpson, J.; Adler, R.F.; North, G.R. A proposed tropical rainfall measuring mission (TRMM) satellite. Bull. Am. Meteorol. Soc. **1988**, *69* (3), 278–295.
3. Doviak, R.J.; Zrnić, D.S. *Doppler Radar and Weather Observations*, 2nd Ed.; Academic Press: San Diego, CA, 1993.
4. Marshall, J.S.; Palmer, W.M. The distributions of raindrops with size. J. Meteorol. **1948**, *9*, 327–332.

5. Joss, J.; Waldvogel, A. A raindrop spectrograph with automatic analysis. Pure Appl. Geophys. **1967**, *68*, 240–246.
6. Joss, J.; Waldvogel, A. Comments on "Some observations on the Joss-Waldvogel rainfall disdrometer". J. Appl. Meteor. **1977**, *16*, 112–113.
7. Schönhuber, M.; Urban, H.E.; Poiares-Baptista, J.P.V.; Randeu, W.L.; Riedler, W. Weather radar versus 2D-video-distrometer data. In *Proceedings of the Third International Symposium on Hydrological Applications of Weather Radars*, Sao Paulo, 1995.
8. Schöhuber, M.; Urban, H.E.; Poaires Baptista, J.P.V.; Randeu, W.L.; Riedler, W. Weather radar versus 2D-video-distrometer data. In *Weather Radar Technology for Water Resources Management*; Braga Jr., B.; Massambani, O. Eds.; UNESCO: Paris, 1997, 159–171.
9. Jones, B.K.; Saylor, J.R.; Bliven, L.F. Single-camera method to determine the optical axis position of ellipsoidal drops. Appl. Opt. **2003**, *42* (6), 972–978.
10. Saxena, D.D.; Saylor, J.R. Use of thresholding algorithms in the processing of raindrop imagery. Appl. Opt. **2006**, *45* (12), 2672–2688.
11. Jain, R.; Kasturi, R.; Schunck, B.G. *Machine Vision*; McGraw Hill: Boston, MA, 1995.
12. Kapur, N.J.; Sahoo, P.K.; Wong, A.K. A new method for gray-level picture thresholding using the entropy of the histogram. Comput. Vis. Graph. Image Process. **1985**, *29*, 273–285.
13. Leung, C.K.; Lam, F.K. Performance analysis for a class of iterative image thresholding algorithms. Pattern Recogn. **1996**, *29*, 1523–1530.
14. Tsai, W. Moment-preserving thresholding: A new approach. Comput. Vis. Graph. Image Process. **1985**, *29*, 377–393.
15. Saylor, J.R.; Sivasubramanian, N.A. Edge detection methods applied to the analysis of spherical raindrop images. Appl. Opt. **2007**, *46* (22), 5352–5367.
16. Hueckel, M.H. An operator which locates edges in digitized pictures. J. Assoc. Comput. Mach. **1971**, *18*, 113–125.
17. Hueckel, M.H. A local visual operator which recognizes edges and lines. J. Assoc. Comput. Mach. **1973**, *20*, 634–647.
18. Lucas, J.; Smektala, F.; Adam, J.L. Fluorine in optics. J. Fluor. Chem. **2002**, *114*, 113–118.

Real Time Object Detection and Tracking

Bhaumik Vaidya and Chirag N. Paunwala
Electronics and Communication Department, Sarvajanik College of Engineering and Technology, Surat, India

Abstract

Object detection and tracking is an ongoing research topic in computer vision that makes efforts to detect, recognize, and track objects through a series of frames. It has been found that object detection and tracking in the video sequence is a challenging task and time consuming process. In this entry, various feature-based and silhouette-based object detection techniques are discussed along with their pros and cons. Object tracking techniques such as absolute difference, background subtraction, optical flow, and mean shift are explained in detail. The challenges in implementing these algorithms on hardware and few implementations of object detection algorithms are discussed in the last part of the article.

Keywords: Background subtraction; FAST; LBP; Object detection; Optical flow; SIFT; Silhouette-based detection; SURF; Tracking.

INTRODUCTION

During the past few years, there has been a rapid advancement in computer vision technologies. One area in computer vision that has attained great focus from researchers around the world is object detection, recognition, and tracking.

In many computer vision systems, object detection is the first step in building a larger system. A lot of information can be derived from the detected object: (i) object can be classified into a particular class, (ii) it can be tracked in image sequence, and (iii) more information about the scene or other object inferences can be derived from the detected object.

Object tracking can be defined as the task of detecting objects in every frame of the video and establishing correspondence between these detected objects from one frame to another in all frames of the video in the video file. The next part of the entry consists of a detailed description of object detection and object tracking methods with their pros and cons and applications.

APPLICATIONS OF OBJECT DETECTION AND TRACKING

Many factors such as the application of machine learning algorithms to solve computer vision problem, the development of new mathematical models for representation, and an increase in the computational power of processors have helped in rapid advancement of object detection applications in recent years.[1]

Object detection and tracking has many applications such as video surveillance system to track suspicious activities, events, persons, or group of persons in real time;[2,3] and intelligent traffic system to track vehicle,[4–7] to detect traffic signs,[8] and visual traffic surveillance (Fig. 1).

Object detection is essential in autonomous robots to give them information about their surroundings and planning for their navigation.[9] It can be used for real-time human–computer interaction based on face and hand gesture recognition;[10] industrial assembly and quality control in production lines;[11] and biomedical image analysis, for example, to detect cervical vertebra in X-ray images, breast cancer, or brain tumor.[12] It can be used in facial expression analysis which includes marketing, perceptual user interfaces, human–robot interaction, drowsy driver detection, telenursing, pain assessment, analysis of mother–infant interaction, social robotics, facial animation, and expression mapping for video gaming, among others.[13] It is also useful for pedestrian detection or head pose estimation in automatic driver assistance system,[14] image retrieval in search engine, and photo management.[15] Nowadays, object detection is widely used in consumer electronics specifically in smartphones.[16]

The object detection system can be categorized by the number of classes of objects they recognize accurately in challenging conditions posed by camera view. The categories are single class from a single view, for example, face detection from frontal views as shown in Fig. 2; single class from multiple views, for example, human tracking system in visual surveillance[17] The example for this is shown by taking video from PETS database[18] as shown in Fig. 3; multiple class from single view, for example, detection of

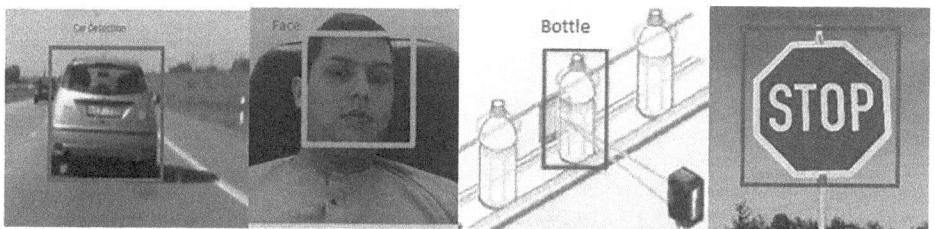

Fig. 1 Object detection application

Fig. 2 Single-class single-view object detection

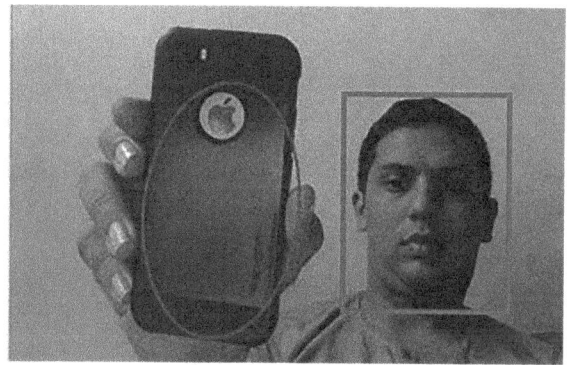

Fig. 4 Multiple-class single-view object detection

Fig. 3 Single-class multiple-view object detection

Fig. 5 Multiple-class multiple-view object detection

humans and mobile in a single scene as shown in Fig. 4; and multiple class from multiple views, for example, intelligent traffic management system to detect vehicles, pedestrians,[19,20] traffic sign, etc. from a video taken from PETS database[18] as shown in Fig. 5.

CHALLENGES IN OBJECT DETECTION

Object detection is challenging task because images in real life is affected by many factors such as noisy image, illumination changes (Fig. 6), dynamic backgrounds (Fig. 7), shadowing effect (Fig. 8), and camera jitter (Fig. 9).

Object detection is difficult when object is rotated, scaled, under occlusion condition and cluttering. Many applications require detecting more than one object class. If a large number of classes are being detected, the processing speed becomes an important issue, as well as the kind of classes that the system can handle without accuracy loss.[21]

Most methods cannot handle multiple views or large pose variations apart from deformable part-based methods. Efficiency and computational power are still the issues for implementation of some techniques in real-time applications. By using a specialized hardware (e.g., field

Fig. 6 Illumination problem

(a) Waving Trees   (b) Waving Calendar   (c) Flowing Water

Fig. 7 Dynamic background

Fig. 8 Shadowing effect

programmable gate array [FPGA], graphics processing unit [GPU], application specific integrated circuit [ASIC]), some methods can run in real time.[21]

Fig. 9 Camera jitter

OBJECT DETECTION STRATEGIES

This section mainly deals with different types of object detection techniques.

Feature-Based Object Detection

The block diagram of feature-based object detection system is shown in Fig. 10. As shown in block diagram, feature points or key points are derived first from the input image that can describe unique features of the image and that are invariant to all transformation. Subsequently, invariant feature vector representation, also called region descriptors, are derived for each feature point, each representing the image information available in a local neighborhood around one interest point. Object recognition can then be performed by matching the region descriptors to the model database.

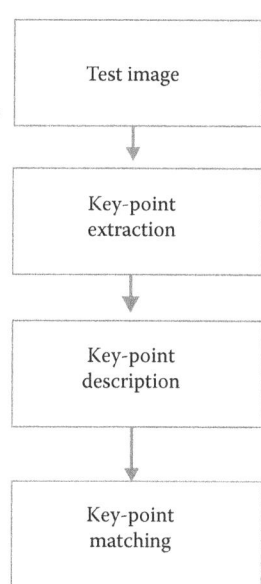

Fig. 10 Block diagram of feature-based methods for object detection

Features detected can be global features or local features. Global features such as color, texture, and histogram aim to describe an image as whole, whereas local features detect key points in an image and describe them. Global features are much faster, are easy to compute, and requires small amount of memory, but they can only be used when some amount of information related to object of interest is available. Also global features are sensitive to clutter and occlusion.[22] Local features require significant amount of memory because of large amount of image features, but they result in better performance.[22] To reduce the memory problem, dimensionality of vectors can be reduced and compact, localized key points can be found. For a good feature detection algorithm, the features found should be robust, accurate, and repeatable.

This section describes various feature-based algorithms for object detection with their pros and cons.

Harris Corner Detection

Corners are regions in the image with large variation in intensity in all the directions. In Harris corner detection, the difference in intensity for a displacement of (u, v) is found in all directions. Window function can be either a rectangular window or Gaussian window. It is used to give weights to pixels underneath.[23] It can be expressed in mathematical form as follows:

$$E(u, v) = \sum_{x,y} w(x,y) * \left[I(x+u, y+v) - I(x,y) \right]^2 \quad (1)$$

This function $E(u, v)$ should be maximize for corner detection. That means, the second term should be maximum. After applying Taylor expansion to Eq. 1 and using some mathematical steps, a final equation is derived as follows:

$$E(u, v) \approx [u \ v] * M \begin{bmatrix} u \\ v \end{bmatrix} \quad (2)$$

where

$$M = \sum w(x,y) \begin{bmatrix} I_x I_x & I_x I_y \\ I_x I_y & I_y I_y \end{bmatrix} \quad (3)$$

Here, I_x and I_y are the local image derivatives in x and y directions, respectively. After this, cornerness measure C is found for each pixel using an equation, which will determine if a window contains a corner or not.

$$R = \det(M) - k\left[\mathrm{trace}(M)\right]^2 \quad (4)$$

where

$$\det(M) = \lambda_1 \lambda_2 \quad (5)$$

$$\mathrm{trace}(M) = \lambda_1 + \lambda_2 \quad (6)$$

λ_1 and λ_2 are the eigenvalues of M, and k is an adjusting parameter. Eigenvalues are computationally expensive, so Harris suggested using a measure that combines the two eigenvalues in a single measure. So the value of cornerness measure decide whether a region is corner, edge, or flat. For smaller $|R|$, the region is flat. For $R < 0$, the region is edge. When R is large, the region is a corner. Figure 11 shows the output obtained using Harris detector.

Harris corner detectors exhibit high-speed performance and requires less memory, so it is suitable for real-time video processing applications specifically in the embedded environment. However, it is not scale invariant and not robust to noise.

Features from Accelerated Segment Test Algorithm

Features from accelerated segment test (FAST) is a corner detector in which corner points are detected by applying a segment test to every pixel by considering a circle of 16 pixels around the candidate pixel.[24] A pixel p in the image is selected which is to be identified as an interest point or not. Let its intensity be I_p. Select appropriate threshold value t. Now the pixel p is classified as a corner if there exists a set of n contiguous pixels in the circle of 16 pixels which are all brighter than $I_p + t$, or all darker than $I_p - t$.

A high-speed test was proposed to exclude a large number of non-corners. This test examines only the four pixels at positions 1, 9, 5, and 13 (First, 1 and 9 are tested if they are too brighter or darker. If so, then 5 and 13 are checked.)[24] If p is a corner, then at least three of these must be brighter than $I_p + t$ or darker than $I_p - t$. If this is not the case, then p cannot be a corner. The full segment

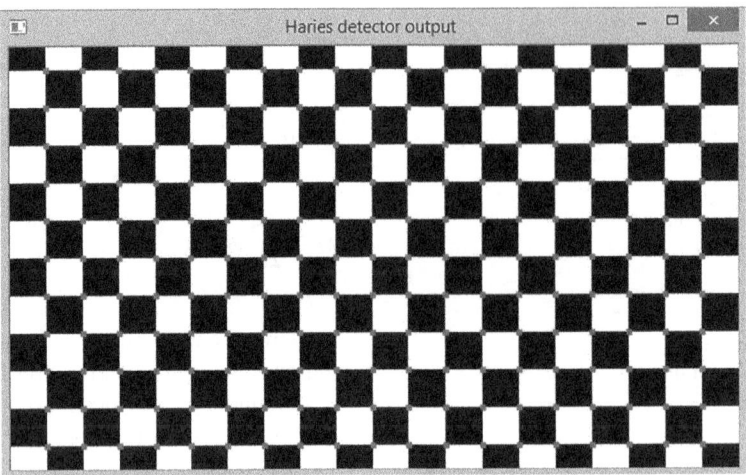

Fig. 11 Output of Harris corner detector

test criterion can then be applied to the passed candidates by examining all the pixels in the circle.

This detector itself exhibits high performance, but multiple features are detected adjacent to one another. It does not reject as many candidates for $n < 12$. The choice of pixels is not optimal because its efficiency depends on the ordering of the questions and the distribution of corner appearances. Multiple features are detected adjacent to one another. An improvement for addressing these limitations is achieved with a machine learning approach and non-maximal suppression.[25] The results of FAST with and without non-maximal suppression are shown in Fig. 12. It can be observed that non-maximal suppression considerably reduces the features adjacent to each other.

Scale-Invariant Feature Transform

A corner in a small image within a small window is flat when it is zoomed in the same window but still will be a corner when the image will be rotated. So corner detectors described earlier are not only rotation invariant but also scale invariant. To overcome this, a new feature detection and description algorithm is developed which is called scale-invariant feature transform (SIFT).[26] The flowchart for SIFT is shown in Fig. 13 below.

As shown in the figure, there are mainly five steps involved in SIFT feature detection, description, and matching.

1. Scale-space extrema detection

   To detect key points with different scale in SIFT, scale-space filtering is used. Laplacian of Gaussian (LoG) is found for the image with various σ values. LoG is basically a blob detector which detects blobs in various sizes due to change in σ. In short, σ acts as a scaling parameter. Gaussian kernel with low σ gives high value for small corners, whereas Gaussian kernel with high σ fits well for larger corners.

Fig. 12 Output using FAST detector with and without maximal suppression

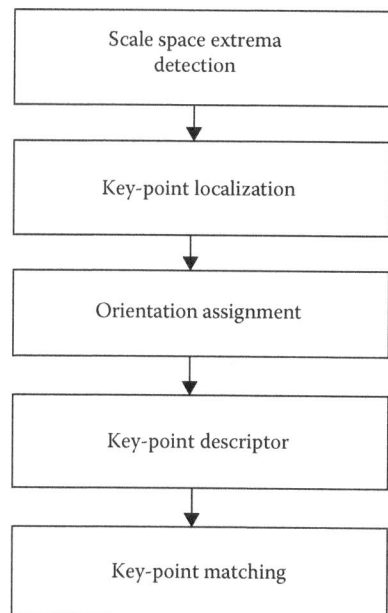

Fig. 13 Algorithmic flowchart for SIFT

To optimize LoG operation, SIFT uses difference of Gaussians (DoG). Mathematically, it can be shown that LoG can be approximately equal to DoG. DoG is obtained as the DoG blurring of an image with two different σ's: σ and $k\sigma$.

$$\frac{\partial G}{\partial \sigma} = \sigma \Delta^2 G = \frac{G(x,y,k\sigma) - G(x,y,\sigma)}{k\sigma - \sigma} \quad (7)$$

where

$$G(x, y, k\sigma) = \frac{1}{2\pi(k\sigma)^2} e^{-\left(\frac{x^2+y^2}{2k^2\sigma^2}\right)} \quad (8)$$

This process is done for different octaves of the image in Gaussian Pyramid.

Once this DoG are found, images are searched for local extrema over scale and space. For example, one pixel in an image is compared with its eight neighbors as well as nine pixels in next scale and nine pixels in previous scales. If it is a local extrema, it is a potential key point. It basically means that key point is best represented in that scale. Regarding different parameters, there are some empirical data values which can be summarized as follows: number of octaves = 4, number of scale levels = 5, initial σ = 1.6, and $k = \sqrt{2}$ as optimal values.[26]

2. Key point localization

To refine the key points found in the previous stage and get more accurate results, further computations are made. In this step, key points with low contrast and edge key points that were detected in the previous stage because of DoG sensitiveness to it are removed. Taylor series expansion of scale space is used to remove low contrast point. If the intensity of Taylor series expansion at this extrema is less than a contrast threshold value, then it is rejected.

To remove edges 2 × 2, Hessian matrix (H) is used to compute the principal curvature. From Harris corner detector, it is known that for edges, one eigenvalue is larger than the other. This concept is used here. If this ratio is greater than the edge threshold, the key point is discarded.

3. Orientation assignment

To make SIFT rotation invariant, a neighborhood is taken around the key point location depending on the scale, and the gradient magnitude and direction are calculated in that region. An orientation histogram with 36 bins covering 360° is created. It is weighted by gradient magnitude and Gaussian-weighted circular window. The highest peak in the histogram is taken, and any peak above 80% of it is also considered to calculate the orientation. It creates key points with same location and scale, but different directions. It contributes to stability of matching.

4. Key point descriptor

The description stage of SIFT starts by taking a 16×16 neighborhood around the key point using its scale to select the level of Gaussian blur for the image. It is divided into 16 sub-blocks of 4×4 size. For each sub-block, eight-bin orientation histogram is created. So a total of 128 bin values are available. It is represented as a vector to form key point descriptor. Finally, it is normalized to gain invariance to changes in illumination.

5. Key point matching

Key points between two images are matched by brute force matching or by identifying their nearest neighbors. But in some cases, the second closest match may be very near to the first. It may be due to noise or some other reasons. In that case, the ratio of the closest distance to the second closest distance is taken. If it is >0.8, they are rejected.

The result for SIFT algorithm for object detection is shown in Fig 14.

SIFT gives good result in case of transformation and changes in illumination but it is slow. SIFT is more suitable in case of images that are effected by translation, rotation, and scaling.[26] To overcome that, speeded up robust features (SURF) are developed.

Speeded-Up Robust Features

The SURF scheme is designed as an efficient and fast alternative to SIFT. The flowchart of SURF is shown in Fig. 15.

SURF approximates LoG with computation based on a simple 2-D box filter. The main advantage of this approximation is that convolution with box filter can be easily

Fig. 14 SIFT object detection results

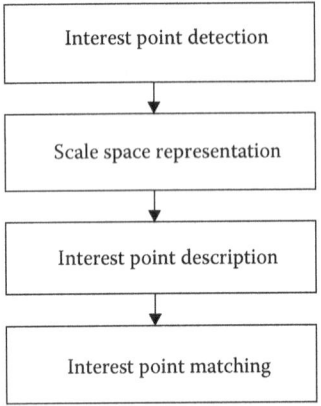

Fig. 15 Flowchart for SURF

calculated with the help of integral images. And it can be done in parallel for different scales. Also the SURF rely on determinant of Hessian matrix for both scale and location.[27] The approximated determinant of Hessian can be expressed as follows:

$$\mathrm{Det}\left(H_{\mathrm{approx}}\right) = D_{xx}D_{yy} - \left(wD_{xy}\right)^2 \tag{9}$$

where w is a relative weight for the filter response and used to balance the expression for the determinant. For orientation assignment, SURF uses wavelet responses in horizontal and vertical directions using integral image approach for a neighborhood of size 6s, where s is the scale at which interest point is detected. Adequate Gaussian weights are also applied to it. The dominant orientation is estimated by calculating the sum of all responses within a sliding orientation window of angle 60°.

For feature description, SURF uses Haar wavelet responses in horizontal and vertical directions in a neighborhood of size 20s×20s, where s is the scale at which interest point is taken. Interest region is divided into 4 × 4 subregions. For each subregion, horizontal and vertical wavelet responses are taken which is denoted by d_x and d_y. These responses are again weighted with Gaussian window centered at interest point. Then, they are summed up for each subregion and a feature vector is formed as follows:

$$v = \left(\sum d_x, \sum d_y, \sum |d_x|, \sum |d_y|\right) \tag{10}$$

This is computed for all 4×4 subregions resulting in SURF feature descriptor with total 64 dimensions. The lower the dimension, the higher the speed of computation and matching. For more distinctiveness, SURF feature descriptor has an extended 128 dimension version. The sums of d_x and $|d_x|$ are computed separately for $d_y < 0$ and $d_y \geq 0$. Similarly, the sums of d_y and $|d_y|$ are split up according to the sign of d_x, thereby doubling the number of features.

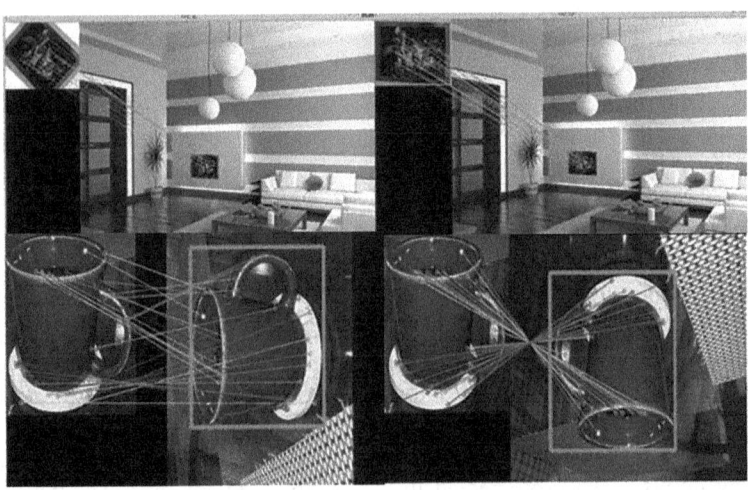

Fig. 16 Result of SURF for object detection

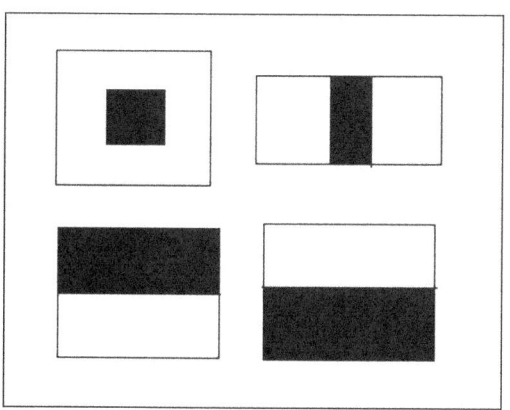

Fig. 17 Rectangular Haar-like features

The sign of the Laplacian can be used to distinguish bright blobs on dark backgrounds from the reverse situation without adding any computational cost. The results of SURF algorithm for object detection is shown in Fig. 16. It can be seen that SURF is rotation invariant and scale invariant.

SURF has a higher processing speed compared to SIFT because it uses a 64-dimensional feature vector compared to SIFT which uses 128-dimensional feature vector. SURF is good at handling images with blurring and rotation, but not good at handling viewpoint change and illumination change.[27]

Haar-Like Features

The principle of Haar-like features is based on detection of features encoding of some information about the class to be detected. They are adjacent rectangles in a particular position of an image. Figure 17 shows the types of Haar-like features depending on the number of adjacent rectangles. There are three types of Haar-like features: edge feature (Fig. 17 two bottom squares), line feature (Fig. 17 upper left square), and center-surround feature (Fig. 17 upper right square).[28]

The idea behind the Haar-like feature selection algorithm is simple. It lies on the principle of computing the difference between the sum of white pixels and the sum of black pixels. The main advantage of this method is the fast sum computation using the integral image. It is called Haar-like because it is based on the same principle of Haar wavelets.

The result of face detection using the Haar-like feature is shown in Fig. 18.

Local Binary Pattern

Local binary pattern (LBP) compute a local representation of textures. This local representation is constructed by comparing each pixel with its surrounding neighborhood of pixels. The first step in constructing the LBP texture descriptor is to convert the image to gray scale. The LBP algorithm slides its processing window over the grayscale image for evaluating the successive stages of the cascade algorithm by scoring their features. Each feature is described by 3×3 neighboring rectangular areas.

The value of each feature is computed by comparing the central area with the neighboring area around it (eight neighbors). The result is in the form of 8-bit binary value called LBP as can be seen in Fig. 19. A number of features represent a stage of the cascade algorithm. Every feature has positive and negative weights associated with it. For the case where the feature is in consistence with the object to be detected, the positive weight is added to the sum. For the case where the feature is inconsistent with the object, the negative value is added to the sum. The sum is then compared to the threshold of the stage. If the sum is below the threshold, the stage fails and the cascade terminates early, and thus, the processing window moves to the next window. If the sum is above the threshold, the next stage of the cascade is attempted. In general, if no stage rejects a candidate window, it is assumed that the object has been detected.

In order to avoid the redundancy of computing the integral of rectangles, the integral images are calculated to speed up the calculation of the feature.

Binary Robust Independent Elementary Features

SURF and SIFT feature descriptors take lots of memory because they have high-dimension feature vectors. But all these dimensions may not be needed for actual matching. It can be compressed using several methods such as PCA and LDA. Even other methods such as locality sensitive hashing is used to convert these SIFT descriptors in floating point numbers to binary strings. These binary strings are used to match features using Hamming distance. This provides better speedup because finding hamming distance is just applying XOR and bit count, which are very fast in modern CPUs with SSE instructions. But here also the descriptors are found first then hashing is applied, which does not solve the problem on memory.[29]

Binary robust independent elementary features (BRIEF) provide a shortcut to find the binary strings directly without finding descriptors. It takes smoothened image patch

Fig. 18 Face detection using Haar-like features

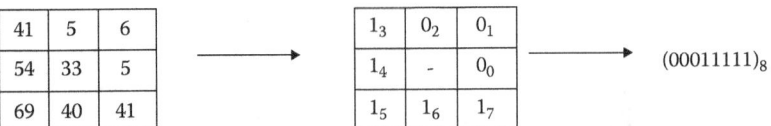

Fig. 19 Method to calculate LBP

and selects a set of $n_d(x, y)$ location pairs in a unique way. Then, some pixel intensity comparisons are done on these location pairs. For example, let the first location pairs be p and q. If $I(p) < I(q)$, then its result is 1, else it is 0. This is applied for all the n_d location pairs to get an n_d-dimensional bit string.[29]

One important point is that BRIEF is a feature descriptor, it does not provide any method to find the features. So other feature detectors such as SIFT and SURF have to be used. In short, BRIEF is a faster method for feature descriptor calculation and matching. It also provides high recognition rate unless there is large in-plane rotation.[29]

Feature Matching

For feature matching, many methods such as brute force matching, Fast Library for Approximate Nearest Neighbors, and fast directional chamfer matching can be used. Brute force matcher is simple. It takes the descriptor of one feature in first set and is matched with all other features in second set using some distance calculation such as L1 norm, L2 norm, or hamming distance. And the closest one is returned. It contains a collection of algorithms optimized for fast nearest neighbor search in large datasets and for high-dimensional features. It works faster than brute force matcher for large datasets.

Contour- or Silhouette-Based Object Detection

Different appearances and articulations of the object are difficult to model without a very large training set for feature-based method. Humans use many visual cues to recognize an object like shape information retrieved from image contours. Contour-based segmentation lead to a robust, flexible, and efficient system, invariant to object appearance such as color, texture, and illumination, without the need for learning from a large set of images.[30]

In this method, hand-drawn sketches can be used as the model(s), which can preserve the significant amount of discriminative object structural information required for liable detection. The sketch model is then segmented into multiple parts, which are expected to capture certain genuine object parts. It is important to identify those parts, which are likely to describe certain actual parts of an object and therefore is expected to be aligned to the choices of a human observer. For part decomposition, different types of algorithms can be used based on convexity measures of an object. Each part is segmented into contour fragments and then matched using different matching algorithms such as fast directional chamfer matching (FDCM). Figure 20 indicates the flowchart and the result of one example of silhouette-based object detection and tracking system.

From the given input image or sketch, shape segmentation and part decomposition is done automatically by using concept of convexity defect[31] and information regarding parts is stored in the database on the server side. On the client side from a test video, the first frame is extracted and parts are matched using FDCM. Then using SURF, key points are extracted and matched. As can be seen from the results in the flowchart, deformation in the parts can be easily handled by this method.

Comparison between Feature-Based and Contour-Based Approaches

Contour based methods are robust, flexible, and invariant to object deformation and they require no training. While they have the limitation that it is not so easy to distinguish the object contours from other edges due to the presence of object and/or background texture. Also the gradient along the object boundary may become very weak in certain portions due to the presence of a matching background.[32,33] Feature-based methods are more popular because of its accuracy and speed but they require large training set to deal with different appearances and articulation of the object.[30]

OBJECT TRACKING

Object detection involves locating objects in frames of a video sequence, whereas tracking is the process of locating moving objects over a period of time. There are two key steps in object tracking: detection of moving objects and tracking of such objects from frame to frame.

Tracking objects can be complex due to loss of information caused by projection of the 3D world on a 2D image, noise in images, complex object motion, nonrigid or articulated nature of objects, partial and full object occlusions, complex object shapes, scene illumination changes, and real-time processing requirements.[34] Many object tracking strategies have been designed that can overcome some of the above issues and can work in real life. All techniques are described one by one in Section "Absolute Difference Method."

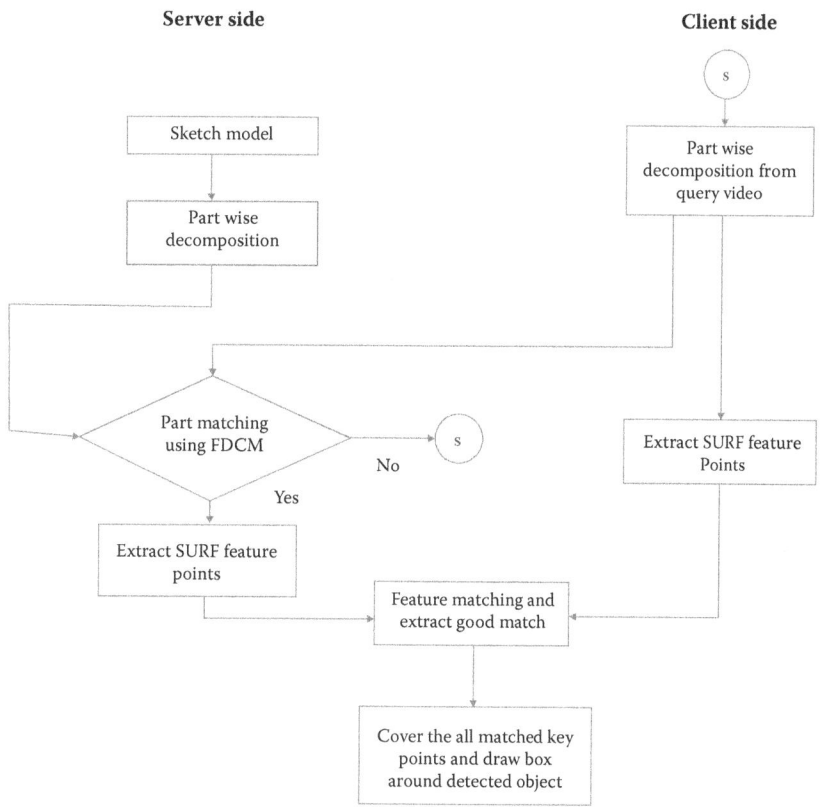

Fig. 20 Flowchart of silhouette-based object detection and tracking system

Absolute Difference Method

In this method, a pixel-by-pixel difference in consecutive frames or in sequence of images is calculated. It is basically image subtraction. It can be mathematically represented as follows:

$$D(t) = I(t_i) - I(t_j) \qquad (11)$$

where $I(t_i)$ and $I(t_j)$ are the pixel intensity values in two different frames. Difference of intensities are stored in D. The value of D gives an image frame with changed and unchanged regions. It gives better result for only dynamic scene changes; in other words, this method fail to detect a nonmoving object in the scene. This method also fails when the object is moving slowly because there will not be much difference in the pixel map. This method is easy to understand but involves lots of computations per frame. So it will be hard to use this method for real-time applications.

Background Subtractions

Absolute difference method completely fails to detect objects that are moving slowly or not moving at all. For that conditions, background subtraction method is used. As the name suggests, background subtraction is the process of separating out foreground objects from the background in a sequence of video frames. The major steps in a background subtraction method are preprocessing, background modeling, foreground detection, and data validation.

i. Preprocessing is done to remove any kind of noise in the image.
ii. Background modeling is very important for this method. In basic model, background is modeled using an average, a median, or a histogram analysis over time.[35] In this case, once the model is computed, the pixels of the current image are categorized as foreground by thresholding the difference between the background image and the current frame. The statistical models are more robust models to illumination changes and for dynamic backgrounds.[35] They can be categorized as Gaussian models,[36] support vector models,[37,38] and estimation models.[39,40] Gaussian is the easiest way to represent a background.[36] It expects the history over the time of pixel's intensity values. But a single model cannot deal with the dynamic background, for example, waving trees and rippling water. To overcome this problem, the mixture of Gaussians (MoG) or Gaussian mixture model is used.[36]

The second category utilizes more sophisticated statistical models, for example, support vector machine (SVM)[37] and support vector regression (SVR).[38] In estimation models, the filter is used to estimate the background. The filter may be a

Wiener filter,[39] Kalman filter,[40] or Chebyshev filter.[41] The Wiener filter works well for periodically changing pixels and it produces a larger value of the threshold for random changes that are utilized in the foreground detection. The main advantage of the Wiener filter is that it lessens the uncertainty of a pixel value by representing how it varies with time. A drawback occurs when a moving object corrupts the history values. The Chebyshev filter slowly updates the background for changes in lighting and scenery while making the use of a small memory footprint with low-computational complexity. In Cluster model, K-means models, codebook models, neural network models can be used.[42] Table 1 shows the comparison between different background modeling techniques in the challenging environment.

iii. After modeling background, foreground is detected by taking the absolute difference between the current frame and the modeled background.
iv. Threshold is used to validate that object is present or not. The background initialization and updating of background after specific time is necessary and important in this method.

The result for background subtraction using two different methods is shown in Fig. 21.

Optical Flow Technique

Optical flow is the pattern of apparent motion of image objects between two consecutive frames caused by the movement of object or camera. It is a 2D vector field where each vector is a displacement vector showing the movement of points from first frame to second. Optical flow has many applications in areas such as structure from motion, video compression, and video stabilization. Optical flow works on the assumptions that the pixel intensities of an object do not change between consecutive frames and neighboring pixels have similar motion.[43]

Consider a pixel $I(x, y, t)$ in the first frame at time t. It moves by distance (d_x, d_y) in the next frame taken after time d_t. So, from the assumption in calculating the optical flow, it can be concluded that

Table 1 Comparison of different algorithms for background modeling

No	Challenges	Solution
1	Noisy image	Clusters models: K-means, codebook[42]
		Features: blocks, clusters[58]
2	Camera jitter	Statistical models: MoG[36]
		Advanced statistical models: DMM
		Fuzzy models
3	Camera automatic adjustments	Background maintenance: MoG[36]
		Features: edges
4	Illumination changes	Filter models: Wiener filter[39]
		Sparse models
		Features: textures
5	Bootstrapping	Background initialization: consecutive frames
		Cluster models: K-means, codebook[42]
		Features: blocks
6	Moved background objects	Background maintenance: MoG[36]
7	Inserted background objects	Background maintenance: MoG[36]
8	Dynamic backgrounds	Fuzzy models
		Features: texture
		Features: histogram
9	Beginning moving objects	Background maintenance: MoG[36]
10	Sleeping foreground objects	Background maintenance: MoG[36]

$$I(x, y, t) = I(x+dx, y+dy, t+dt) \qquad (12)$$

If Taylor series approximation is taken of the right-hand side, common terms are removed and equation is divided by d_t to get the following equation:

$$f_x u + f_y v + f_t = 0 \qquad (13)$$

where

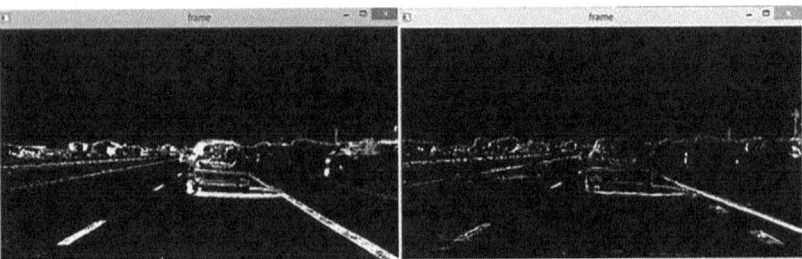

Fig. 21 Background subtraction using the KNN subtraction technique and the adaptive Gaussian mixture model

Real Time Object Detection and Tracking

$$f_x = \frac{\partial f}{\partial x}; f_y = \frac{\partial f}{\partial y} \tag{14}$$

$$u = \frac{dx}{dt}; v = \frac{dy}{dt} \tag{15}$$

The preceding equation is called optical flow equation. In this equation, f_x and f_y are the image gradients. Similarly, f_t is the gradient along time. But (u, v) are unknown. Two unknown variables cannot be solved with one equation. So, several methods are provided to solve this problem and one of them is Lucas–Kanade. It takes a 3×3 patch around the point. All the nine points have the same motion. An image gradient (f_x, f_y, f_t) is found for these nine points. Now from the assumptions that pixel intensities do not change and neighboring pixels have the same motion, problem becomes solving nine equations with two unknown variables which is overdetermined. A better solution is obtained with the least squares fit method. The final solution which is two-equation, two-unknown problem and solved to get the solution is given as follows:

$$\begin{bmatrix} u \\ v \end{bmatrix} = \begin{bmatrix} \sum_i f_{x_i}^2 & \sum_i f_{x_i} f_{y_i} \\ \sum_i f_{x_i} f_{y_i} & \sum_i f_{y_i}^2 \end{bmatrix}^{-1} \begin{bmatrix} -\sum_i f_{x_i} f_{t_i} \\ -\sum_i f_{y_i} f_{t_i} \end{bmatrix} \tag{16}$$

The idea of an optical flow is simple: given some points to track, the optical flow vectors of those points are received and solved for u and v, but it fails when there is large motion. So, the concept of image pyramids is used. When we go up in the pyramid, small motions are removed and large motions become small motions. After applying Lucas–Kanade at every scale, optical flow along with the scale information is received. The output of optical flow algorithm for two different videos is shown in Figs. 22 and 23. Figure 22 indicates the optical flow output of two different frames of traffic on road. Green arrows on frame indicates magnitude and direction of vectors. Results are color coded for better visualization. Direction corresponds to the hue value of the image. Magnitude corresponds to the value plane.

Figure 23 indicates optical flow for face tracking from webcam.

Dense Optical Flow Technique (Gunner Farneback's Optical Flow)

Lucas–Kanade method computes optical flow for a sparse feature set. Sparse techniques only need to process some pixels from the whole image, while dense techniques process all the pixels. Dense techniques are slower but can be more accurate. An example of a dense optical flow algorithm is Gunner Farneback's optical flow. It computes the optical flow for all the points in the frame.[44] The results for algorithms are shown in Fig. 24. Results are color coded for better visualization. Direction corresponds to the hue value of the image. Magnitude corresponds to the value plane.

Meanshift and Camshift Method

The meanshift algorithm is an efficient approach to track objects whose appearance is defined by histograms. For a given set of points (it can be a pixel distribution like histogram) and a small window, you have to move that window to the area of maximum pixel density:[45]

Algorithm 1: Meanshift method

Do the iteration till
   {
   C1 = Starting window
   C1_o = center of window
   C1_r = centroid of window
   If (C1_o == C1_r)
      Break;
   else
      Move to Next window such that C2_o = C1_r where C2_o is center of new window
   }

So in this method, the histogram back-projected image and initial target location is passed. When the object moves, obviously the movement is reflected in histogram back-projected image. As a result, the meanshift algorithm moves window to the new location with maximum density.[45]

One problem with the meanshift algorithm is window always has the same size when object is farther away or very close to camera. This problem can be solved by using camshift (continuously adaptive meanshift).[46]

In this method, the meanshift is applied first. Once the meanshift converges, it updates the size of the window using the following equation:

$$s = 2 * \sqrt{\frac{M_{00}}{256}}$$

It also calculates the orientation of best-fitting ellipse to it. Again, it applies the meanshift with new scaled search window and previous window location. The process is continued until required accuracy is met. The result for mobile tracking using camshift is shown in Fig. 25. To improve the accuracy of tracking further, camshift is used with different features such as SIFT features, color feature, and texture information, or Camshift algorithm is combined with the Kalman filter.

Fig. 22 Optical flow output for two different frames of road

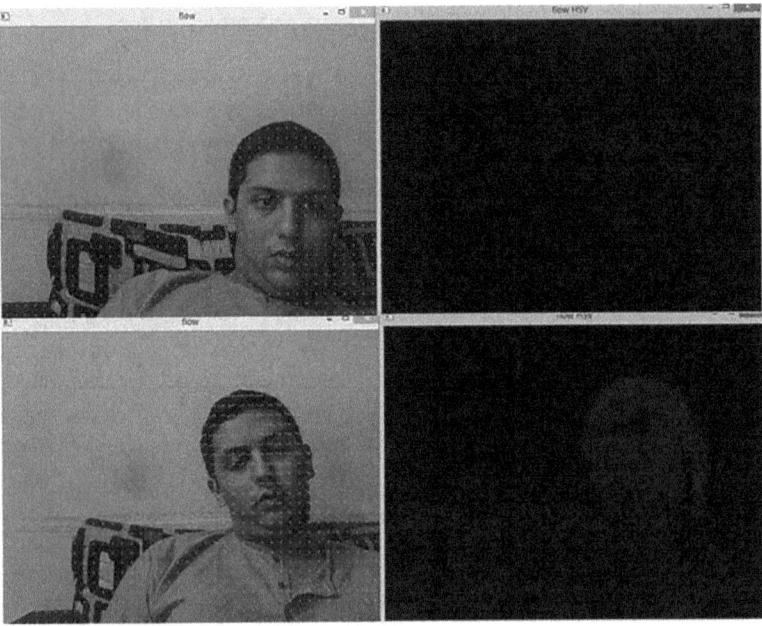

Fig. 23 Face tracking using optical flow

HARDWARE IMPLEMENTATION CHALLENGES OF OBJECT DETECTION AND TRACKING METHODS

Normally, general-purpose processors and specialized architectures such as digital signal processors do not provide the flexibility required to achieve the real-time performance of object detection and tracking algorithms, so there is always need of dedicated hardware architectures for these algorithms. Most of the applications of object detection and tracking are associated with real-time performance constraints. The problem is further aggravated in an embedded systems environment, where most of these applications are deployed. The high computational complexity and power consumption makes implementing an embedded object detection system with real-time performance a challenging task.[47] Consequently, there is a strong need for dedicated hardware architectures capable of delivering high detection accuracy within an acceptable processing time given the available hardware resources. Almost all algorithms discussed above for object detection and tracking have inherent parallelism associated with it. If hardware with more parallel resources is used to implement this algorithms, there can be considerable speed up. Same way it can also be implemented on graphical processing units with multiple cores with high computational tasks divided between multiple cores.

Fig. 24 Result of dense optical flow for two frames in a video

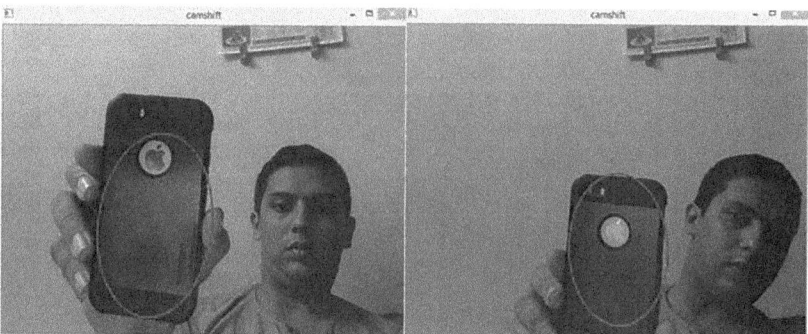

Fig. 25 Mobile tracking using camshift

The platforms with more number of DSP blocks (embedded multipliers and accumulators) and more amount of memory are ideally suitable for these algorithms. While implementing these algorithms, there are strict timing constraints, memory constraints, and resource sharing constraints. The data rate requirements of real-time applications impose a strict timing constraint. At video rates, all required processing for each pixel must be performed at the pixel clock rate (or faster).[48] This generally requires low-level pipelining to meet this constraint. Another form of timing constraint occurs when asynchronous processes need to synchronize to exchange data or synchronization within multiple clock domains. Some operations require images to be partially or wholly buffered. While current high-capacity devices have sufficient on-chip memory to buffer a single image, in most applications it is poor use of this valuable resource to simply use it as an image buffer. For this reason, image frames and other large datasets are more usually stored in off-chip memory.[48] The simplest way to increase memory bandwidth is to have multiple parallel memory systems. If each memory system has separate address and data connections to the FPGA, then each can be accessed independently and in parallel. A further approach to increasing bandwidth is to increase the word width of the memory. Each memory access will then read or write several pixels. This is effectively connecting multiple banks in parallel but using a single set of address lines common to all banks. Every hardware platform will have a limited amount of resources, so there will always be a need to share resources among different processes.

The majority of the published implementations of object detection are on CPU and GPU platforms. The implementation by Benenson et al.[49] achieves higher throughput on a GPU at 100 fps using histogram gradient approach presented by Dollar et al.[50] but with a resolution of 640×480 pixels. For higher throughput and resolution,[51] FPGA-based implementations have been reported recently. Different parts of the detector are implemented on different platforms: HOG feature extraction is divided between an FPGA and a CPU, and SVM classification is done on a GPU.[52] It can process 800×600 pixels at 10 fps for single-scale detection. The entire HOG-based detector is implemented on

FPGA by Mizuno et al.[53] and can process 1080 HD video (1920×1080 pixels) at 30 fps. However, the implementation only supports a single image scale. An ASIC version of this design is presented by Takagi et al.,[54] with dual cores to enable voltage scaling for power consumption of 40.3 mW for 1080 HD video at 30 fps, but still only supports a single image scale. It should be noted that these hardware implementations have relatively large on-chip memory sizes (e.g., Takagi et al.[54] uses 1.22 Mbit on ASIC, Hahnle et al.[55] uses 7 Mbit on FPGA), which contributes to increased hardware cost. Thus, from the discussion, it can be concluded that it is hard to satisfy all the desired requirements for accurate and robust object detection in embedded systems, which include real-time, high resolution (1,080 HD), high frame rate (>30 fps), multiple image scale, low power and low hardware costs.[56,57] There has to be trade-off according to the application. Performance requirements can vary quite a lot depending on application. Consider the scenarios of analyzing the footage of a security camera at the restricted area in a busy airport and demographic analysis of visitors in a retail store.

In the second scenario, algorithm can run at just five frames per second (FPS), but in the first scenario, a higher FPS is critical. The first scenario also places a much greater demand for accuracy. So, after knowing the performance requirement, suitable device and implementation technique can be chosen.

CONCLUSION

Due to high-speed computers and low-cost cameras, the demand for object detection and tracking algorithms that works in real time is increasing day by day. Although great progress has been observed in the last years, we are still far from achieving human-level performance. Object detection and tracking is an important task in computer vision field. Object detection is the first stage in many computer vision applications. This entry focused on the methods of real-time object detection. A comparison of different object detection methods in challenging environment is discussed. For object detection, SURF works well in embedded applications with its high accuracy of results with low feature size and high speed. Object tracking involves detecting object in every frame, tracing its path from frame to frame, and analyzing its behavior. For object tracking, combination of optical flow and meanshift techniques work well in almost all conditions. Hardware implementation constraints along with trade-offs in designing algorithms for real-time applications are also discussed.

REFERENCESS

1. Rodrigo, V.; Javier, R. Object detection: Current and future directions. Front. Rob. AI **2015**, *2*, 29.
2. Kalirajan, K.; Sudha, M. Moving object detection for video surveillance. Sci. World J. **2015**, *2015*, 10. doi:10.1155/2015/907469.
3. Haritaoglu, I.; Harwood, D.; Davis, L. W4: Real time surveillance of people and their activities. IEEE Trans. Pattern Anal. Mach. Intell. **2000**, *22* (8), 809–830.
4. Beymer, D.; McLauchlan, P.; Coifman, B.; Malik, J. A real-time computer vision system for measuring traffic parameters. In *IEEE Computer Society Conference on Computer Vision and Pattern Recognition*, San Juan, PR, 1997.
5. Bramberger, M.; Pflugfelder, R.P.; Rinner, B.; Schwabach, H.; Strobl, B. Intelligent traffic video sensor: Architecture and applications. In *Telecommunications and Mobile Computing Workshop on Wearable Computing*, Graz, Austria, March 2003.
6. Koller, D.; Weber, J.; Malik, J. Towards real time visual based tracking in cluttered traffic scenes. In *IEEE Intelligent Vehicles Symposium*, Paris, 1994, 201–206.
7. Pflugfelder, R. *Visual Traffic Surveillance Using Real-Time Tracking*; Diploma Thesis; Technical Report PRIP-TR-071, PRIP, TU Wien, 2002.
8. Yang, Y.; Luo, H.; Xu, H.; Wu, F. Towards real-time traffic sign detection and classification. In *IEEE 17th International Conference on Intelligent Transportation Systems (ITSC)*, Qingdao, October 8–11, 2014.
9. Pauly, M.; Surmann, H.; Finke, M.; Liang, N. Real-Time object detection for autonomous robots. In *Autonome Mobile Systeme*; Wörn, H.; Dillmann, R.; Henrich, D.; Eds.; Informatik aktuell; Springer: Berlin and Heidelberg, 1999.
10. Azad, R.; Azad, B.; Nabil, K.; Shahram, J. Real-time human-computer interaction based on face and hand gesture recognition. Int. J. Found. Comput. Sci. Technol. **2014**, *4* (4), 37–48.
11. Herakovic, N. Robot vision in industrial assembly and quality control processes. In *Robot Vision*; Ude, A.; Ed.; 2010, In Tech. Available at http://www.intechopen.com/books/robotvision/robot-vision-in-industrial-assembly-and-quality-control-processes.
12. Lecron, F.; Benjelloun, M.; Mahmoudi, S. Descriptive image feature for object detection in medical images. In *Image Analysis and Recognition*, ICIAR 2012. Lecture Notes in Computer Science; Campilho, A.; Kamel, M.; Eds.; Springer: Berlin and Heidelberg, Vol. 7325, 2012.
13. De la Torre, F.; Cohn, J.F. *Facial Expression Analysis. Visual Analysis of Humans*; Springer: London, 2011, 377–409.
14. Geronimo, D.; Lopez, A.M.; Sappa, A.D.; Graf, T. Survey of pedestrian detection for advanced driver assistance systems. IEEE Trans. Pattern Anal. Mach. Intell. **2010**, *32* (7), 1239–1258.
15. Kumar, N.; Belhumeur, P.; Nayar, S. Facetracer: A search engine for large collections of images with faces. In *Computer Vision—ECCV 2008. ECCV 2008*, Vol. 5305, Lecture Notes in Computer Science; Forsyth, D.; Torr, P.; Zisserman, A.; Eds.; Springer: Berlin and Heidelberg, 2008, 340–353.
16. Jeong, K.; Moon, H. Object detection using FAST corner detector based on smartphone platforms. In *First ACIS/JNU International Conference on Computers, Networks,*

Systems and Industrial Engineering (CNSI), Jeju Island, 2011, 111–115

17. Blauensteiner, P.; Kampel, M. Visual surveillance of an airport's Apron—An overview of the AVITRACK project. 2004.
18. PETS. *Benchmark Data*, 2009. Available at http://www.cvg.reading.ac.uk/PETS2009/a.html.
19. Daniele, T. FPGA-based pedestrian detection under strong distortions. In *Proceedings of the IEEE Conference on Computer Vision and Pattern Recognition Workshops*, Boston, MA, 2015, 66–71.
20. Roig, G.; Boix, X.; Shitrit, H.B.; Fua, P. Conditional random fields for multi-camera object detection. In *IEEE International Conference on Computer Vision (ICCV)*, Barcelona, 2011, 563–570.
21. Bouwmans, T. Traditional and recent approaches in background modeling for foreground detection: An overview. Sci. Direct J. Comput. Sci. Rev. **2014**, *11–12*, 31–66.
22. Hassaballah, M.; Abdelmgeid, A.; Hammam, A. Image features detection, description and matching. In *Image Feature Detectors and Descriptors: Foundations and Applications*, Part of the Studies in Computational Intelligence; Awad, A.I.; Hassaballah, M.; Eds.; Springer: Cham, 2016, 630. doi:10.1007/978-3-319-28854-3_2.
23. Harris, C.; Stephens, M. A combined corner and edge detector. In *Proceedings of the 4th Alvey Vision Conference*; University of Manchester: Manchester, 1988, 147–151.
24. Rosten, E.; Drummond, T. Machine learning for high-speed corner detection. In *Computer Vision—ECCV 2006*, Vol. 3951, ECCV 2006. Lecture Notes in Computer Science; Leonardis, A.; Bischof, H.; Pinz, A.; Eds.; Springer: Berlin and Heidelberg, Germany, 2006, 430–443.
25. Rosten, E.; Porter, R.; Drummond, T. Faster and better: A machine learning approach to corner detection. IEEE Trans. Pattern Anal. Mach. Intell. **2010**, *32*, 105–119.
26. Lowe, D. Distinctive image features from scale-invariant keypoints. Int. J. Comput. Vision **2004**, *60*, 1–28.
27. Bay, H.; Ess, A.; Tuytelaars, T.; Van Gool, L. Speeded-up robust features (SURF). Comput. Vis. Image Underst. **2008**, *110* (3), 346–359.
28. Guennouni, S.; Ahaitouf, A.; Mansouri, A. A comparative study of multiple object detection using haar-like feature selection and local binary patterns in several platforms Modell. Simul. Eng. **2015**, *2015*, 8. doi:10.1155/2015/948960.
29. Michael, C.; Lepetit, V.; Strecha C.; Fua, P. BRIEF: Binary robust independent elementary features. In *11th European Conference on Computer Vision (ECCV)*, LNCS; Springer: Heraklion, 2010.
30. Bhattacharje, S.; Mittal, A. Part-base deformable object detection with a single sketch. Comput. Vision Image Underst. **2015**, *139*, 73–87.
31. Gopalan, R.; Turaga, P.; Chellappa, R. Articulation invariant representation of non-planer shapes. In *Proceedings of the European Conference of Computer Vision*; Verlag: Berlin and Heidelberg, 2010.
32. Dickinson, S. Object representation and recognition. In *What is Cognitive Science?*; Lepore, E.; Pylyshyn, Z.; Eds.; Basil Black Well Publishers: New Brunswick, NJ, 1999, 172–207.
33. Treiber, M. *An Introduction to Object Recognition*; Springer: London.
34. Hemalatha, C.; Muruganand, S.; Maheswaran, R. A survey on real time object detection, tracking and recognition in image processing. Int. J. Comput. Appl. (0975–8887) **2014**, *91* (16), 38–42.
35. Bouwmans, T.; El-Baf, F.; Vachon, B. Statistical background modeling for foreground detection: A survey. Handb. Pattern Recognit. Comput. Vision **2010**, *4* (2), 181–199.
36. Grimson, W.; Stauffer, C. Adaptive background mixture models for real-time tracking. In *IEEE Conference on Computer Vision and Pattern Recognition*, Fort Collins, CO, 1999, 246–252.
37. Lin H.; Liu T.; Chuang, J. A probabilistic SVM approach for background scene initialization. In *International Conference on Image Processing*, Rochester, NY, September 2002, 893–896.
38. Wang, J.; Bebis, G.; Miller, R. Robust video-based surveillance by integrating target detection with tracking. In *IEEE Workshop on Object Tracking and Classification beyond the Visible Spectrum in Conjunction*, New York, June 2006.
39. Toyama, K.; Krumm, J.; Brumitt, B.; Meyers, B. Wallflower: Principles and practice of background maintenance. In *International Conference on Computer Vision*, Kerkyra, September 1999, 255–261.
40. Ridder, C.; Munkelt, O.; Kirchner, H. *Adaptive Background Estimation and Foreground Detection using Kalman-filtering*; Bavarian Research Center for Knowledge-Based Systems: Erlangen, 1995, 193–199.
41. Chang, R.; Gandhi, T.; Trivedi, M. Vision modules for a multi sensory bridge monitoring approach. In *IEEE Intelligent Transportation Systems Conference*, Washington, WA, October 2004, 971–976.
42. Kim, K.; Chalidabhongse, T.; Hanuood, D.; Davis, L. Background modeling and subtraction by codebook construction. In *IEEE International Conference on Image Processing*, Singapore, 2004.
43. Barron, J.; Fleet, D.; Beauchemin, S. Performance of optical flow techniques. Int. J. Comput. Vision **1994**, *12*, 43–77.
44. Farnebäck, G. Two-frame motion estimation based on polynomial expansion. Image Anal. **2003**, *2749*, 363–370.
45. Gorry, B.; Chen, Z.; Hammond, K.; Wallace, A.; Michaelson, G. Using Mean-Shift Tracking algorithms for real-time tracking of moving images on an autonomous vehicle tested platform. In *Conference of the World Academy of Science Engineering and Technology*, Venice, November 2007, 23–25.
46. Bradski, G. Computer vision face tracking for use in a perceptual user interface. Intel Technol. J. Q2, **1998**, *1*, 1–15.
47. Jinwook, O.; Gyeonghoon, K.; Injoon, H.; Junyoung, P.; Seungjin, L.; Joo-Young, K.; Jeong-Ho, W.; Hoi-Jun, Y. Low-power, real-time object-recognition processors for mobile vision systems. IEEE Micro **2012**, *32* (6), 38–50.
48. Bailey, D.G. *Design for Embedded Image Processing on FPGAs*, 1st Ed.; John Wiley & Sons (Asia) Pte Ltd.: Singapore, 2011.
49. Benenson, R.; Mathias, M.; Timofte, R.; Van Gool, L. Pedestrian detection at 100 frames per second. In *Proceedings IEEE Conference on Computer Vision and Pattern Recognition*, Providence, RI, 2012, 2903–2910.

50. Dollar, P.; Belongie, S.; Perona, P. The fastest pedestrian detector in the west. In *Proceedings of the British Machine Vision Conference*, *Aberystwyth, GBR*, 2010, 68.1–68.11.
51. Wei, Z.; Zelinsky, G.; Samaras, D. Real-time accurate object detection using multiple resolutions. In *Proceedings IEEE International Conference on Computer Vision*, *Rio de Janeiro*, Brazil, 2007, 1–8.
52. Bauer, S.; Kohler, S.; Doll, K.; Brunsmann, U. FPGA-GPU architecture for kernel SVM pedestrian detection. In *Proceedings IEEE Computer Society Conference on Computer Vision and Pattern Recognition Workshops*, San Francisco, CA, 2010, 61–68.
53. Mizuno, K.; Terachi, Y.; Takagi, K.; Izumi, S.; Kawaguchi; H.; Yoshimoto, M. Architectural study of HOG feature extraction processor for real-time object detection. In *Proceedings IEEE Workshop on Signal Processing Systems*, Quebec City, QC, 2012, 197–202.
54. Takagi, K.; Mizuno, K.; Izumi, S.; Kawaguchi, H.; Yoshimoto, M. A sub-100-milliwatt dual-core HOG accelerator VLSI for real-time multiple object detection. In *Proceedings IEEE International Conference on Acoustics, Speech and Signal Processing*, Vancouver, BC, 2013, 2533–2537.
55. Hahnle, M.; Saxen, F.; Hisung, M.; Brunsmann, U.; Doll, K. FPGA-based real-time pedestrian detection on high-resolution images. In *Proceedings IEEE Conference on Computer Vision and Pattern Recognition Workshops*, Portland, OR, 2013, 629–635.
56. Dollár, P.; Belongie, S.J.; Perona, P. The fastest pedestrian detector in the west. Br. Mach. Vision Conf. **2010**, *2* (3), 7.
57. Qasaimeh, M. *Fpga-based parallel hardware architecture for real-time object classification*; A Thesis Presented to the Faculty of the American University of Sharjah, June 2014.
58. Bove, V.M. Jr.; Sridharan, S.; Butler, D.E. Real time adap-adaptive foreground/background segmentation. EURASIP **2005**, *14*, 2292–2304.

Reflectance Modeling

Ariyo Kanno
Graduate School of Science and Engineering, Yamaguchi University, Ube, Japan

Abstract
In clear shallow water (e.g., coral reefs), a fraction of the visible light from the sun penetrates into water, gets scattered in water or reflected at the bottom, and reaches the optical sensor on satellites and aircraft. Reflectance models, by approximating the in-water radiative transfer processes, formulate the fraction (reflectance) as a function of variables including water depth, bottom reflectance, and water optical properties. This entry introduces two types of reflectance models commonly used for shallow water remote sensing of these variables.

Keywords: Bottom type; Passive remote sensing; Radiative transfer; Underwater; Water depth.

INTRODUCTION

Visible light from the sun is not only abundant in the earth's surface but also penetrable in water for some distance. In clear shallow water (e.g., coral reefs), some of the downwelling sunlight penetrates into water, gets scattered in water or reflected at the bottom, and leaves the water upward to the air, before getting detected by the optical sensor on satellites and aircraft (Fig. 1). The ratio of the energy of the water-leaving light to the downwelling sunlight, the reflectance, depends on water depth, bottom reflectance, and water optical properties. The modeling of the dependency enables the passive remote sensing of these variables, which are fundamental information to coastal zone managements and shallow water environmental managements.

The next section describes the basics of the propagation of visible light in shallow water. The following sections introduce two common models (in the form of equations) that have been applied for the remote sensing of shallow water. An emphasis is put on the older and simpler model (what we call Lyzenga's model).

PROPAGATION OF VISIBLE LIGHT IN SHALLOW WATER

The propagation of visible light in deep water is characterized by four processes: reflection and refraction at the surface and absorption and scattering in water. Consider a collimated beam of sunlight hitting the ocean surface from the air. A small fraction of the light energy is reflected at the surface back into the air. The rest enters into water and starts traveling downward in a direction determined by refraction. The light energy traveling in the direction largely exponentially decreases with the travel distance due to the interactions with water: absorption and scattering. Here, absorption is the transformation of the light energy into the internal energy of water. Scattering is the change of the traveling direction. The rates of these interactions depend on the optical properties of water (e.g., absorption coefficient, scattering coefficient, scattering phase function) as well as the wavelength of the light.

For shallow water, another process, reflection at the bottom, needs to be considered as well. A certain fraction (bottom reflectance) of the light energy hitting the water bottom is reflected upward. The direction of the reflected light depends on that of the incident light to some extent. When this dependency can be neglected and the directional distribution of the reflected light is isotropic, the bottom is called as Lambertian reflector.

In the viewpoint of the optical sensor flying over shallow water, the energy of visible light coming from a direction (or observed for a pixel) has four main components as shown in Fig. 1: the energy of the sunlight reflected once at the bottom, that scattered once in the water, that reflected once at the surface, and that scattered once in the atmosphere. The first two components are utilized in shallow water remote sensing to estimate water depth and the optical properties of the bottom and water.

LYZENGA'S MODEL

Overview

What we call Lyzenga's model in this entry is a simple model that relates the reflectance of optically shallow water to the water depth, bottom reflectance, and effective attenuation coefficient. The model can be transformed into

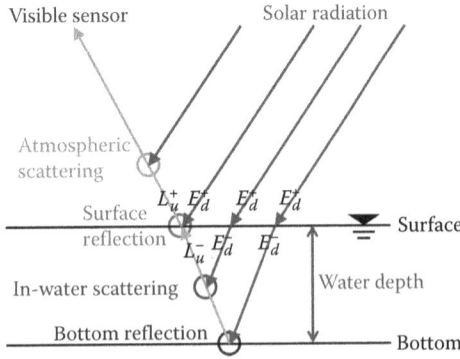

Fig. 1 Schematic view of radiance components observed by visible sensor over optically shallow water[1]

a linear form, which decomposes a remotely sensed quantity into the contributions of water depth and that of bottom reflectance. This enables the remote sensing of water depth and bottom type by linear transformations of multiple wavelength bands. Thus, Lyzenga's model is the most widely used reflectance model for the multispectral remote sensing of water depth and bottom type. Because the effective attenuation coefficient is not an IOP (inherent optical property) of water, Lyzenga's model is not suitable for remote sensing of water optical properties.

The derivation of the model by Lyzenga[2] on the basis of the radiative transfer theory presumes homogeneous optical properties of the water and a level Lambertian bottom with homogeneous reflectance. It also neglects the internal reflection effects and simplifies the multiple scatterings in water. The inaccuracy caused by the ignorance of the internal reflection effects is significant only when water depth is small and bottom reflectance is large as quantified by radiative transfer simulations.[3]

Model

In Lyzenga' model, as it appears in Ref. [4], the subsurface reflectance R^- of optically shallow water is modeled as follows:

$$R^- \equiv \frac{\pi L_u^-}{E_d^-} = r_v + \left(r_b - r_v\right) \cdot e^{-k \cdot h}. \quad (1)$$

where L_u^- and E_d^- are the upwelling radiance and downwelling irradiance just below the water surface, respectively; r_v is the subsurface reflectance of infinitely deep water due to in-water volume scattering; r_b is the bottom reflectance; k is the effective attenuation coefficient (positive; not IOP); and h is the water depth. According to Eq. 1, as h increases, the contribution of bottom reflection ($r_b \cdot e^{-k \cdot h}$) to R^- exponentially decreases, whereas that of in-water volume scattering ($r_v \cdot (1 - e^{-k \cdot h})$) increases. The simplicity of the model is achieved by approximating the depth dependencies of the two components using the same exponential function ($e^{-k \cdot h}$), which is first proposed by Lyzenga.[2] As a result, we can bundle them as $(r_b - r_v) \cdot e^{-k \cdot h}$, which shows that R^- decreases with increasing h in usual case, where r_b is larger than r_v.

By ignoring the effect of internal reflection at the surface, the reflectance just above the surface (R^+) can be modeled in the same form with R^- as follows:

$$R^+ \equiv \frac{\pi L_u^+}{E_d^+} = R_\infty^+ + \left(R_b^+ - R_\infty^+\right) \cdot e^{-k \cdot h}. \quad (2)$$

where L_u^+ and E_d^+ are the upwelling radiance and downwelling irradiance just above the water surface, respectively; R_b^+ is the contribution of bottom-reflected light and depends on the bottom reflectance and surface transmittance; and R_∞^+ is defined as

$$R_\infty^+ \equiv \lim_{h \to \infty} R^+ \quad (3)$$

and includes the contribution of both the in-water volume scattering and the surface reflection of incident sunlight.

In some literature, the reflectance at the sensor (on a satellite or an aircraft) is modeled implicitly based on Eq. 2.[2] The model form at the sensor level is still the same with Eq. 2 because the atmospheric effects are approximately absorbed by the additive term (R_∞^+) and the multiplicative factor ($R_b^+ - R_\infty^+$) of Eq. 2.[4] Therefore, in this entry, we refer to any of the models in the form of Eqs. 1 and 2 as Lyzenga's model.

Applications

Lyzenga's model has been used in many researches for remote sensing of water depth and bottom type (e.g., Refs. [2,4–9]). The beauty of this model is that it can be transformed to the linear form

$$X \equiv \log\left[R^+ - R_\infty^+\right] = \log\left[R_b^+ - R_\infty^+\right] - k \cdot h \quad (4)$$

where X can be derived from the remotely sensed imagery (by using the average R^+ of the deep water pixels as a substitute for R_∞^+), and $\log\left[R_b^+ - R_\infty^+\right]$ depends on R_b^+ but not on h. This form has been used in remote sensing of water depth to estimate h from X (e.g., Refs. [4,6,7]) and in remote sensing of bottom type to remove the effect of h from X (e.g., Refs. [8,9]).

In these applications, multiple wavelength bands of X are usually combined, if available, to estimate the water depth or bottom type without the difficulty of estimating k or $\log\left[R_b^+ - R_\infty^+\right]$. The simplest way is the linear transformations of multiple bands of X.[1,4,7,8]

Example

As an example application, Fig. 2 shows the water depth distribution estimated from a multispectral satellite image of coral reef by using Lyzenga's model. For each shallow

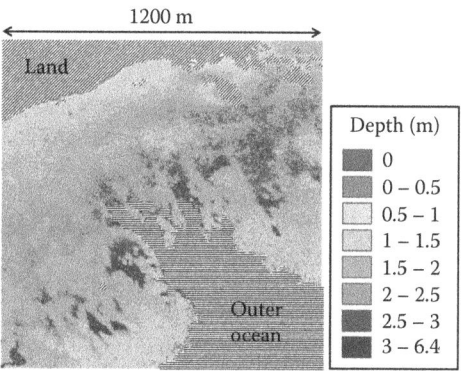

Fig. 2 Water depth distribution estimated from a multispectral satellite image (QuickBird, with three visible bands of 2.4 m resolution) of coral reef by using a method based on Lyzenga's model
Source: The result was reproduced from the work of Kanno et al.[10] with permission. This figure includes copyrighted material of DigitalGlobe, Inc., All Rights Reserved.

water pixel, the water depth was estimated by the following linear transformation of the multiple (three in this case) visible bands of X:

$$h = \beta_0 + \beta_1 X_1 + \beta_2 X_2 + \beta_3 X_3. \quad (5)$$

where X_i ($i = 1, 2, 3$) is the X of band i. The coefficients $\beta_0, \ldots, \beta_3$ were calibrated by least squares using 100 pixels with known depth (in situ depth soundings).

The water depth map shown in Fig. 2 has a high resolution of 2.4 m inherited from the satellite image. It represents the complex bottom topography of the reef. This resolution of bathymetry is not feasible by conventional ship soundings because of the shallow and complex bottom topography. Still, such maps are needed, for example, as an input to the fine-scale numerical simulations of currents, waves, water qualities, and ecosystems of coral reefs. Although the topography in deep area cannot be mapped by the approach, it is relatively smooth and easily measured by ship soundings.

Accuracy can be a problem. According to a cross validation using 501 pixels with known depth (min: 0.572 m, max: 3.32 m, mean: 1.96 m, standard deviation: 0.572 m) in the area, the root mean square error of the depth estimation was 0.316 m when 100 pixels were used for calibration. A more complex method using a nonlinear function of the three bands of X, described in Section 3.2 of the work of Kanno et al.[6] performed significantly better in this case, with a root-mean-square error of 0.253 m.

LEE'S MODEL

Overview

What we call Lee's model here is a more complex but a more accurate reflectance model than Lyzenga's model. It is essentially obtained by relating the effective attenuation coefficient in Lyzenga's model to IOPs of water.

Lee's model cannot be linearized like Lyzenga's model and is usually solved by numerical optimization. The model is often combined with spectral models of the bottom reflectance and water optical properties to simultaneously estimate the water depth, bottom reflectance, and water optical properties from hyperspectral imagery.

The derivation of the model by Lee et al.[11] starts from Lyzenga's model and hence their assumptions are similar. One difference is that the derivation of Lee's model does not neglect the internal reflection effects. Another important difference is that Lee's model has some empirical coefficients, which were fitted to the radiative transfer simulations of limited cases.

Model

Lee's model was first presented by Lee et al.[11] but soon modified by Lee et al.[12] to correct for errors and achieve more generality. In the modified version, the subsurface reflectance is modeled as

$$R^- = R_\infty^- \left(1 - \exp\left\{ -[a + b_b] \cdot \left[\frac{1}{\cos\theta_w} + \frac{D_u^C}{\cos\theta} \right] \cdot h \right\} \right)$$
$$+ r_b \exp\left\{ -[a + b_b] \cdot \left[\frac{1}{\cos\theta_w} + \frac{D_u^B}{\cos\theta} \right] \cdot h \right\}. \quad (6)$$

where θ_w is the subsurface solar zenith angle, θ is the subsurface viewing angle, a is the absorption coefficient of water, and b_b is the backscattering coefficient. Note that the coefficients a and b_b are IOPs. These coefficients are further decomposed but the decompositions are skipped here. D_u^C and D_u^B are the optical path elongation factors for scattered light and bottom-reflected light, respectively, and further modeled as

$$D_u^C = 1.03 \left(1 + 2.4 \frac{b_b}{a + b_b} \right)^{0.5} \quad (7)$$

and:

$$D_u^B = 1.04 \left(1 + 5.4 \frac{b_b}{a + b_b} \right)^{0.5}. \quad (8)$$

The variable R_∞^- in Eq. 6 is further modeled as:

$$R_\infty^- = \left(0.084 + 0.170 \frac{b_b}{a + b_b} \right) \cdot \frac{b_b}{a + b_b}. \quad (9)$$

The reflectance just above the surface (R^+) in the case of nadir view is provided as

$$R^+ = \frac{0.5 R^-}{1 - 1.5 R^-}, \quad (10)$$

taking into account the effect of the internal reflection.

By comparing Eq. 6 with Eq. 1, it can be said for subsurface reflectance (R^-) that Lee's model is obtained by modeling the effective attenuation coefficient (k) of Lyzenga's model using IOPs. The coefficient is approximated as a product of the optical path length and $a + b_b$. This approximation itself is traditional as was used in developing Lyzenga's model.[2] However, Lee's model discriminated the optical path lengths for the light scattered in-water and that for the light reflected at the bottom, and achieved better accuracy at the cost of simplicity.

Applications

Unlike Lyzenga's model, Lee's model cannot be linearized. From Eq. 6, it is obvious that the model can be explicitly inverted for bottom reflectance (r_b) but not for water depth (h). Thus, it is usually used with numerical optimization techniques.

Lee's model is often combined with the spectral models of the bottom reflectance and water optical properties (often parameterized for the target site) to simultaneously estimate the water depth, bottom reflectance, and water optical properties (e.g., Refs. [12–15]). The estimation is done by numerically optimizing the model unknown parameters to minimize the difference between the reflectance spectra calculated by the model and that observed. Some applications further introduced additional techniques such as linear unmixing (e.g., Refs. [15–18]) and spatial smoothing[19] to enhance the accuracy. Because the number of unknowns in the simultaneous estimation is large, hyperspectral imagery is more suitable than multispectral imagery to avoid the statistical problem of overfitting.

CONCLUSION

This entry introduced two famous models (in the forms of equation) of the shallow water reflectance of visible light. They relate the reflectance just above/below surface to water depth, bottom reflectance, and optical properties of water. Lyzenga's model is simple and Lee's model is more detailed and concrete. Both can be applied for any of the single-band, multispectral, or hyperspectral remote sensing. However, probably due to historical reasons, Lyzenga's model is widely used in multispectral remote sensing and Lee's model is popular in hyperspectral remote sensing.

REFERENCES

1. Kanno, A.; Tanaka, Y. Modified Lyzenga's method for estimating generalized coefficients of satellite-based predictor of shallow water depth. IEEE Geosci. Remote Sens. Lett. **2012**, *9* (4), 715–719.
2. Lyzenga, D.R. Passive remote-sensing techniques for mapping water depth and bottom features. Appl. Opt. **1978**, *17* (3), 379–383.
3. Kanno, A.; Tanaka, Y.; Sekine, M. Validation of shallow-water reflectance model for remote sensing of water depth and bottom type by radiative transfer simulation. J. Appl. Remote Sens. **2013**, *7* (1), 073516–073516.
4. Lyzenga, D.R.; Malinas, N.R.; Tanis, F.J. Multispectral bathymetry using a simple physically based algorithm. IEEE Trans. Geosci. Remote Sens. **2006**, *44* (8), 2251–2259.
5. Philpot, W.D. Bathymetric mapping with passive multispectral imagery. Appl. Opt. **1989**, *28* (8), 1569–1578.
6. Kanno, A.; Koibuchi, Y.; Isobe, M. Shallow water bathymetry from multispectral satellite images: Extensions of lyzenga's method for improving accuracy. Coastal Eng. J. **2011**, *53* (04), 431–450.
7. Lyzenga, D.R. Shallow-water bathymetry using combined lidar and passive multispectral scanner data. Int. J. Remote Sens. **1985**, *6* (1), 115–125.
8. Lyzenga, D. Remote sensing of bottom reflectance and water attenuation parameters in shallow water using aircraft and landsat data. Int. J. Remote Sens. **1981**, *2* (1), 71–82.
9. Armstrong, R.A. Remote-sensing of submerged vegetation canopies for biomass estimation. Int. J. Remote Sens. **1993**, *14* (3), 621–627.
10. Kanno, A.; Koibuchi, Y.K.; Takeuchi, W.; Isobe, M. A generalized satellite-based method of water depth mapping with a semiparametric optical model. J. Remote Sens. Soc. Jpn. **2009**, *29* (3), 459–470 (in Japanese with English Abstract).
11. Lee, Z.P.; Carder, K.L.; Mobley, C.D.; Steward, R.G.; Patch, J.S. Hyperspectral remote sensing for shallow waters. I. A semianalytical model. Appl. Opt. **1998**, *37* (27), 6329–6338.
12. Lee, Z.P.; Carder, K.L.; Mobley, C.D.; Steward, R.G.; Patch, J.S. Hyperspectral remote sensing for shallow waters: 2. Deriving bottom depths and water properties by optimization. Appl. Opt. **1999**, *38* (18), 3831–3843.
13. Lee, Z.; Carder, K.L.; Chen, R.F.; Peacock, T.G. Properties of the water column and bottom derived from airborne visible infrared imaging spectrometer (aviris) data. J. Geophys. Res.-Oceans **2001**, *106* (C6), 11639–11651.
14. McIntyre, M.L., Naar, D.F.; Carder, K.L.; Donahue, B.T.; Mallinson, D.J. Coastal bathymetry from hyperspectral remote sensing data: Comparisons with high resolution multibeam bathymetry. Mar. Geophys. Res. **2006**, *27* (2), 128–136.
15. Brando, V.E.; Anstee, J.M.; Wettle, M.; Dekker, A.G.; Phinn, S.R.; Roelfsema, C. A physics based retrieval and quality assessment of bathymetry from suboptimal hyperspectral data. Remote Sens. Environ. **2009**, *113* (4), 755–770.
16. Giardino, C.; Bartoli, M.; Candianai, G.; Bresciani, M.; Pellegrini, L. Recent changes in macrophyte colonisation patterns: An imaging spectrometry-based evaluation of southern lake garda (northern Italy). J. Appl. Remote Sens. **2007**, *1* (1), 011509.

17. Goodman, J.A.; Ustin, S.L. Classification of benthic composition in a coral reef environment using spectral unmixing. J. Appl. Remote Sens. **2007**, *1* (1), 011501.
18. Klonowski, W.M.; Fearns, P.; Lynch, M.J. Retrieving key benthic cover types and bathymetry from hyperspectral imagery. J. Appl. Remote Sens. **2007**, *1* (1), 011505.
19. Filippi, A.M.; Kubota, T. Introduction of spatial smoothness constraints via linear diffusion for optimization-based hyperspectral coastal ocean remote-sensing inversion. J. Geophys. Res. Oceans **2008**, *113*, C03013.

Remote Sensing Image Fusion

Christine Pohl
Institute of Computer Science, University of Osnabrueck, Osnabrueck, Germany

J. L. van Genderen
Department of Earth Observation Science, Faculty of Geoinformation Science and Earth Observation (ITC), University of Twente, Enschede, The Netherlands

Abstract

This entry describes the combination of two or more remote sensing images using image fusion techniques. In the field of data fusion, image fusion is performed at pixel level in contrast to feature or decision fusion. Prior to the fusion process, the input imagery is harmonized in order to ensure a data match on a pixel-by-pixel basis. Further optional image improvements are possible prior to the fusion process, depending on what type of features is supposed to be highlighted in the final product. For the fusion at a pixel level, a variety of fusion algorithms and methods are available. They are described in this entry and grouped into categories of image fusion techniques to provide a better overview. It also illustrates the benefit of image fusion using various examples from land cover/land use mapping, flood and oil palm plantation monitoring, and others. The need of a standardized quality assessment in terms of qualitative and quantitative evaluation is discussed, which is followed by conclusions.

Keywords: Data integration; Image fusion; Multisensor; Multitemporal; Pansharpening; Quality assessment; Remote sensing.

INTRODUCTION

Remote sensing image fusion is the combination of two or more remote sensing images of different characteristics producing a new image using a certain algorithm. The aim is to provide imagery with improved interpretation potential for visual and computer-assisted image exploitation. The idea is to enhance objects of interest that are not identifiable or extractable from a single image alone. Apart from being able to extract more information from fused images, the quality of the contained information is supposed to be improved. Quality refers to the improvement of

- the number of objects identifiable
- higher accuracy in the interpretation
- reduced ambiguity
- improved reliability
- more robustness in the operation

and originates from the combination of different views on the object. Electromagnetic waves of different wavelengths interact differently with the object of interest. Therefore, the combination of diverse sensors operating with different wavelengths allows various perspectives of the object leading to a better description.

Remote sensing image fusion is an established scientific field serving the improvement of remote sensing images for improved interpretation and information extraction. Since the availability of operational satellite systems in the early 1970s, providing multiple sensor type data on a regular basis, it has been evolving from a research domain to an operational application. One of the first applications using the combination of different observations taken from different domains of the electromagnetic spectrum is geology. The characterization of geological units, structures, and rock types along with their contained minerals is most successful if different bands or images are combined. In particular, the combination of active (radar) with passive (optical) remote sensing data is very powerful in revealing geological information. The utilization of special band ratios allows the mapping of minerals and their distribution, even if the Earth's surface is covered by dense vegetation. Also, other relevant remote sensing applications strongly benefit from image fusion. A very popular example is the so-called *pansharpening*. Here, lower spatial resolution multispectral imagery is fused with a high spatial resolution layer, mostly a panchromatic channel or high spatial resolution synthetic aperture radar (SAR) data. High spatial resolution multispectral images are of great benefit to many remote sensing applications, such as the monitoring of urbanization or deforestation processes, coastal zones,

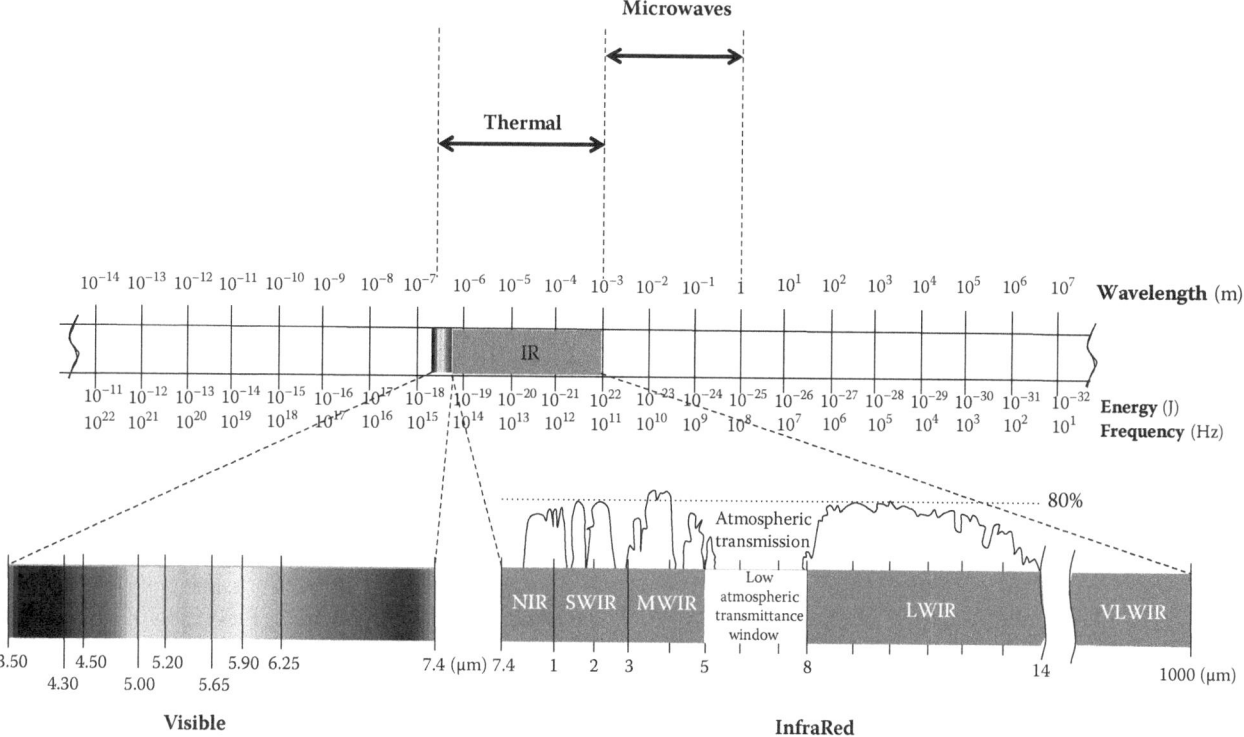

Fig. 1 Electromagnetic spectrum and sensor types categorized by their respective wavelengths

agricultural crops, and natural disasters. Since the acquisition of such imagery is physically not possible, image fusion is a valuable solution to this problem.

Remote Sensing Data Used in Image Fusion

There are various types of sensors providing remote sensing images on a regular basis. Especially since the availability of long-term remote sensing programs, such as Landsat (U.S. longest continuous space-based Earth observation) jointly operated by the North American Space Association (NASA) and United States Geological Survey (USGS) and Copernicus operated by European Space Agency (ESA) with its Sentinel satellites and contributing missions, the joint use of multimodal images has become popular. In addition to the availability of multisensor data, nowadays a lot of imagery can be obtained free of charge (open data policy of Landsat and Sentinel).

Among the sensors operationally acquiring images on a regular basis, there are passive and active systems. We distinguish visible (VIS) and infrared and thermal infrared (TIR) in the passive domain as well as SAR and light detection and ranging (LiDAR) in the active domain. The sensors are categorized according to the wavelengths used for Earth observation. Figure 1 illustrates the different sensor systems in the electromagnetic spectrum. Passive remote sensing operates with the radiation from the sun reflected at the Earth's surface or emitted from it. Active remote sensing emits its own radiation and records backscattered energy arriving at the sensor.

Optical Remote Sensing

Multispectral sensors acquire images with sensors sensitive in the VIS (red, green, blue), infrared, and TIR range of the spectrum. Typical image bands are red, green, blue, near-infrared (NIR), short-wave infrared (SWIR), mid-infrared (MIR), and TIR (compare with Table 1). Often optical sensors also provide a high spatial resolution panchromatic channel with a broader bandwidth.

A combination of bands as color composite already provides interesting features for visual interpretation as shown in Fig. 2. This Sentinel-2 scene subset, acquired on February 16, 2017, in the area of Berlin, Germany, displays settlements, forest, agricultural areas with and without vegetation, as well as water features according to their spectral differences in the individual bands. The composite shown here is a false color composite, built by its bands 8 (NIR),

Table 1 Typical spectral bands of optical sensors

Spectral range (μm)	Type of band
0.4–0.7	VIS[a]
0.7–1.5	NIR[a]
1.5–2.5	SWIR[a]
3.0–5.0	MIR[b]
8.0–14.0	TIR[c]

[a]Dominated by reflected solar energy.
[b]Reflected solar energy and surface emission.
[c]Dominated by surface emission.

Fig. 2 Sentinel-2 false color composite example

4 (red), and 3 (green), which means vegetation is displayed in red color and water in blue tones.

Multispectral remote sensors collect the sum of reflected radiation, emissions from the ground, and path radiance. Hyperspectral sensors operate similarly, only that their number of bands is significantly higher (in the range of several tens or hundreds of channels) with a small bandwidth for each band. This allows the study of biophysical variables of plants or the material of man-made objects rather than just the differentiation of objects based on their multispectral signature. The hyperspectral sensor acquires a continuous signature rather than discrete measurements using only a few selected spectral bands in the entire spectrum.

All objects emit thermal radiation if their temperature is higher than the absolute zero at −273°C. Thermal remote sensing measures the emitted radiation of the Earth's surface as radiant temperature. The region between 3 and 35 μm is the coverage of TIR. However, in remote sensing, the atmospheric window in the region of 8–14 μm is most common. A key parameter to be considered in thermal remote sensing is the emissivity of objects. This is compared to a blackbody, an artificial construct representing the ideal object that absorbs and emits all incident energy. Depending on the wavelength and temperature, the maximum spectral radiance varies. Thermal remote sensing uses the fact that the emitted radiation of objects depends on the composition and geometry of the observed surface.

Active Remote Sensing

A constraint of optical remote sensing is cloud cover, which hinders the acquisition of data. Therefore, the use of radar remote sensing gained importance. With the technical improvement of radar sensors in terms of orbit positioning accuracy and spatial resolution, SAR contributes important information in an operational context. Radar data is complementary to optical imagery since it contains information on surface roughness, moisture content, shape and geometry of the objects, and texture. Due to its own energy source, it can acquire images independent of daylight. Its rather long wavelength leads to the fact that it can penetrate clouds.

Table 2 Typical radar bands used in remote sensing

Radar band	Wavelength (cm)
X	2.5–3.75
C	3.75–7.5
L	15–30
P	100

Radar sensors operate with a radar antenna. The spatial resolution of the system depends on the physical antenna length, which, of course, is limited. Multiple acquisitions at different positions along the orbit create a synthetically enlarged radar antenna (a principle called SAR) and therefore lead to higher spatial resolution. Nowadays, operational sensors, such as the Sentinel-1 C-band SAR (wavelength 5.4 cm), provide up to 5 m ground sampling distance. A radar image contains the intensity at which the sent radar pulse is received back at the antenna along with its phase. Both data components are used for interpretation. The phase of the received signal is important in radar interferometry, where two acquisitions from different locations in the orbit observing the same object on the ground are used to calculate the phase difference, which then leads to the object height. Since the signal is controllable, other properties of the electromagnetic wave can be influenced and serve object interpretation. Here polarization is a main feature. It is important to know that longer wavelengths in radar remote sensing allow deeper penetration into the object (e.g., tree canopies or Earth's surface). Typical radar bands are listed in Table 2.

LiDAR operates with transmitted coherent light pulses that are reflected by the observed object. The collected backscattered energy is detected by a receiver and processed into intensity images as well as height information to create 3D models of the surface observed. LiDAR remote sensing entered operational status. Most height models are nowadays created from 3D point clouds collected by LiDAR sensors, which are currently mostly operated on airborne platforms. It is therefore also called airborne laser scanning in the literature. A big advantage is the acquisition of multiple pulse returns, which lead to detailed information on object structures. In forest monitoring, for example, it is possible to obtain information on canopy, branches, trunk, and ground.

IMAGE FUSION TECHNIQUES

It is obvious that the combination of diverse sensor types provides a more complete view of the object of interest.

Therefore, image fusion is an important option in remote sensing image processing. For pixel level, there are six groups of fusion techniques, which have been developed over the last decades, namely, component substitution (CS), numerical methods, statistical image fusion, multi-resolution approaches (MRAs), hybrid techniques, and others.[1]

CS techniques usually transfer the data into a different coordinate system, replace one component in the new coordinate system by an additional image channel, and return the data into the original data space. The resulting image contains the fused image. Examples of CS techniques are intensity–hue–saturation (IHS) fusion or principal component substitution (PCS) fusion.

Numerical methods combine image channels using mathematical operations, such as addition, subtraction, multiplication, and division. A famous technique is the Brovey transform (BT), which uses sum, differencing, and ratio in a certain combination.

The group of statistical fusion techniques comprises the principal component analysis (PCA), regression variable substitution (RVS), or a least-square fit. The fusion takes place using a statistical relationship between the channels to be combined. These techniques often lead to satisfactory results because they consider statistical relationships and characteristics of the particular data. They are rather independent from user interaction, which makes them valuable in an automation process.

The large group of MRAs decomposes the images into multiple channels resulting from different frequency contents. They are based on a multi-scale pyramid concept. The coefficients for the transformation of the images describe the local gray-level variation in a neighborhood window.

In the recent years, hybrid fusion techniques became more and more popular. They combine the advantages of different fusion techniques while overcoming constraints when using only one technique alone. They allow the enhancement of particular features in the imagery in order to produce an optimized fused image in the end. Increase in computer power, available memory, and hard disk space have fostered the popularity of these methods.

The following sections explain the techniques within these six groups in more detail.

Component Substitution Fusion

The techniques contained in this group are rather popular due to their simplicity. CS techniques transform an image with a certain number of bands into another data space, for example, color space: red–green–blue (RGB) to IHS. One component is replaced by another image (channel) containing the desired information to be integrated into the fused image. A typical example is the implementation of IHS transform for pansharpening. After CS, the bands are reverse transformed into the original data space.

Intensity Hue Saturation Fusion

This technique can be said to be the most popular fusion technique. The remote sensing image is transformed from the RGB color space into IHS in order to separate the individual color characteristics as perceived by the human eye into its components. There are various IHS models based on different geometric shapes, such as the cylindrical, cone, or spherical models. Therefore, the mathematical equations to transform RGB into IHS vary. This can lead to different results when using different software. The cylindrical color model is rather common. The related mathematical expressions are represented in the following equations:

$$\begin{pmatrix} I \\ v_1 \\ v_2 \end{pmatrix} = \begin{pmatrix} \frac{1}{\sqrt{3}} & \frac{1}{\sqrt{3}} & \frac{1}{\sqrt{3}} \\ \frac{1}{\sqrt{6}} & \frac{1}{\sqrt{6}} & -\frac{2}{\sqrt{6}} \\ \frac{1}{\sqrt{2}} & -\frac{1}{\sqrt{2}} & 0 \end{pmatrix} * \begin{pmatrix} R \\ G \\ B \end{pmatrix} \quad (1)$$

$$I = \frac{(R+G+B)}{3} \quad (2)$$

$$H = tan^{-1}\left(\frac{v_2}{v_1}\right) \quad (3)$$

$$S = \sqrt{v_1^2 + v_2^2} \quad (4)$$

In remote sensing image fusion, IHS fusion is mostly applied by extracting the intensity component and replacing it with a different image of either different texture (e.g., SAR) or higher spatial resolution (e.g., panchromatic band). The reverse IHS transform produces the actual fused image. An improvement of fused image quality can be achieved by matching the data range of the image to replace the intensity component to the intensity itself. The mathematical expression to reverse the three components into the fused image is provided in the following equation:

$$\begin{pmatrix} R \\ G \\ B \end{pmatrix} = \begin{pmatrix} \frac{1}{\sqrt{3}} & \frac{1}{\sqrt{6}} & \frac{1}{\sqrt{2}} \\ \frac{1}{\sqrt{3}} & \frac{1}{\sqrt{6}} & -\frac{1}{\sqrt{2}} \\ \frac{1}{\sqrt{3}} & -\frac{2}{\sqrt{6}} & 0 \end{pmatrix} * \begin{pmatrix} I \\ v_1 \\ v_2 \end{pmatrix} \quad (5)$$

Further variations of the IHS fusion can be implemented either by substituting other color components, that is, hue or saturation, or by modifying the components before reverse transforming the data. This is done, for example, by contrast stretching as implemented in the color contrast stretch.

The limitation of three channels for this technique has been resolved with modifications of this technique developed by different scientists. An example is the generalized IHS introduced by Tu et al.[2] Their approach uses the difference image between the panchromatic band to be fused and the calculated intensity component. Since there is always a trade-off between the preservation of spectral quality and the increase in spatial resolution, modifications of the IHS transformation including the trade-off parameter have been introduced about 10 years ago.[3,4] Other improvements to the traditional IHS transform include the possibility of edge-enhancing features[5,6] and a spectrally adjusted IHS.[7] The improvement is achieved by two approaches: (1) the newly introduced high spatial resolution panchromatic band is histogram-matched to the intensity channel, and (2) the introduction of weights, adjusting the contribution of multispectral and panchromatic information to the fused image.

Principal Component Substitution

Originally developed to decorrelate redundant measurements, the PCA or Karhunen–Loève transform produces independent orthogonal components. Most of the data variance is comprised in the first three components, while the information content reduces drastically from one to the next component. In remote sensing image fusion, this transformation serves, similarly to IHS, as a component substitution technique (PCS), where the first principal component (PC1) is replaced with the new image to be introduced. This could be, for example, a higher spatial resolution panchromatic image. The number of principal components equates to the number of channels of the original multiband image. In hyperspectral remote sensing, the PCA is applied to reduce data redundancy and data volume. In image fusion, the common practice is the replacement of PC1 with the histogram-matched panchromatic image for pansharpening. An inverse transformation returns the fused image in its original data space.

Each pixel contains two coordinates in the feature space per channel containing the gray value. If plotted in feature space the normal distributed measurements form clusters. These can be described by an ellipse in the 2D case. For an n-dimensional case it becomes a hyper-ellipsoid. The new coordinate system is designed in a way that the origin is placed in the center of the cluster, while PC1 is directed along the main axis of the ellipse (long axis), the second orthogonal to the first (short axis of ellipse), and so on. Mathematically, the eigenvector forms the direction, and the eigenvalue builds the length. In the new coordinate system, the data is represented uncorrelated. In the literature, a standardized PCA and a non-standardized PCA are differentiated: The standardized case uses the correlation values, and the non-standardized PCA is based on the covariance. The processing steps are as follows:

1. Calculation of the covariance/correlation matrix
2. Determination of the eigenvalues, vectors
3. Formation of the principal components

Equations 6 and 7 show the calculation of the cluster means μ_A and the variance of image band A for k number of pixels. j stands for a particular pixel; A_j is the gray value of that pixel.

$$\mu_A = \frac{\sum_{j=1}^{k} A_j}{k} \tag{6}$$

$$\sigma_A^2 = \frac{\sum_{j=1}^{k} (A_j - \mu_A)^2}{k-1} \tag{7}$$

The covariance C_{AB} between two image bands A and B (μ_B is the cluster mean for band B) is calculated using the following equation:

$$C_{AB} = \frac{\sum_{j=1}^{k} (A_j - \mu_A)(B_j - \mu_B)}{k} \tag{8}$$

The actual PCA transform is performed with the eigenvector matrix E and its transposed form (E^T) resulting in the eigenvalue matrix V, containing only diagonal elements using the following equation:

$$V = E\ C\ E^T \tag{9}$$

The eigenvalues are organized according to the variance from largest to smallest. Then the principal component values P_e for each PC with the number e result from the following equation:

$$P_e = \sum_{i=1}^{n} p_i E_{ie} \tag{10}$$

with i being the band and n being the number of input bands containing pixel value p.

Gram-Schmidt Fusion

This is a patented image fusion technique called Gram–Schmidt (GS) spectral sharpening.[8] It is a generalization of the PCS. It contains an additional step where a panchromatic band is simulated (PAN') following Eq. 11, where each multispectral band MS_i ($i = \{1, 2, ..., n\}$ with n being the number of multispectral MS bands and w_i being weighting factors. The Gram–Schmidt transform (GST) produces an orthogonal set of normalized vectors. For image fusion, PAN' is replaced by the gain and bias-adjusted high spatial resolution panchromatic band. Again the original panchromatic image is histogram-matched to the first GS component, and a reverse GST produces the fused image.

$$PAN' = \sum_{i=1}^{n} w_i\ MS_i \tag{11}$$

Variations of GS fusion use minimum mean square error (MMSE) minimizing weights or spectral sensitivity curves.[9] For MMSE, a modulation transfer function (MTF) serves to filter high spatial resolution data to improve spatial detail. Based on the injection model, spatial detail is inserted into the lower resolution multispectral image.[9]

Numerical Fusion Techniques

This group of fusion techniques operates with mathematical combinations of images. They are rather straightforward and simple to implement. Commonly known combinations are difference images or ratios, which are highly used in change detection and enhancement of particular objects, such as vegetation or soil. A very popular and successful pansharpening strategy is the high-pass filter fusion process. Here the ratio of the spatial resolution between the two image data defines the size of the filter kernel. This filter kernel is used to high-pass filter the panchromatic image. Then it is added to the multispectral bands including some weighting factor. The weighting factor depends on the correlation between the multispectral and panchromatic bands The kernel size is calculated from the image ratio as follows:

$$R = \frac{PR_{MS}}{PR_{PAN}} \quad (12)$$

where PR_{MS} and PR_{PAN} are the pixel resolutions of the multispectral and panchromatic images, respectively. The optimal kernel size is then 2R.

High-pass filtered (HPF) resolution merge came along with the IHS transform in the early 1980s.[10] HPF is a numerical method that uses high-pass filtering to enhance spatial detail. The fusion process introduces this spatial detail to the fused image MS_{fused}. The following equation shows the mathematical approach:

$$MS_{fused} = MS + PAN\_HPFw \quad (13)$$

with MS being the original multispectral image and PAN_HPF being the enhanced panchromatic image multiplied by the calculated weighting factor w.

Similarly to the adding of image bands or images, the data can also be multiplied. In this case, the data needs to be rescaled to fit the original data range.

Image differencing and rationing lead to image differences for change detection. Using ratios, slight variations are even further enhanced in the fused image. An additional effect is the elimination of illumination differences based on topography variations. Pansharpening through rationing can be achieved by

$$DN_f = DN_P - \frac{DN_{XSi}}{DN_{synP}} \quad (14)$$

with DN_f being the fused image. DN_P is the panchromatic, high spatial resolution image, and DN_{XSi} represents the multispectral image (ith band). The equation contains a synthesized panchromatic band calculated as average from the multispectral bands that overlap the data range of the panchromatic channel.

The most famous and common numeric method is the so-called BT. It normalizes the multispectral bands using a ratio and uses multiplication for image fusion. A further development of BT is the colour-normalized (CN) spectral sharpening allowing for more than three image bands to be fused. The original purpose of BT was a contrast improvement for better visual interpretation of remote sensing images. The following equation describes the algorithm for $i=\{1, 2, 3\}$ number of bands to obtain BT, $i=\{1, 2, 3,…, n\}$ for CN:

$$Fused_i = \frac{MS_i}{\sum_i^n MS_i} PAN \quad (15)$$

Statistical Fusion Methods

Statistical parameters describe the relationship between different images or channels. These parameters can be used to produce a fused image. Depending on the local correlation between the input images, information is adaptively inserted into the resulting fused image.[11] This approach is commonly known as *spatially adaptive image fusion*. Also RVS establishes a linear relationship between the image channel that is supposed to replace another to form the fused image. Strictly speaking, the PCA should be contained in this group of methods. PCS was already discussed in a in the section on Principal Component Analysis. Here, it should be mentioned that the PCA can be used to fuse images by inserting all image bands to be considered for the fused image as layer stack to produce fused PCs. Sometimes individual bands are excluded, such as the thermal channel of Landsat or Sentinel-2 because the information contained in TIR is rather different to the spectral reflectance in VIS or NIR bands. Especially in geology, PCS and PCA are successfully implemented.[12]

A linear relationship using correlation is the regression fusion approach. It is adaptive using local variations. The following equation illustrates the mathematical expression behind this fusion technique:

$$XS_f = a_i + b_i * PAN \quad (16)$$

with a_i being the bias and b_i being the scaling calculated by a least-square approach from the resampled multispectral and the high-resolution panchromatic band.

The first commercial fusion tool based on statistical measures is FuzeGo. Originally developed at the University of New Brunswick as UNB Pansharp, it is now available to perform pansharpening of optical images (MS and PAN) or the integration of optical and radar data.[13] The foundation of this approach is a statistical relationship calculated from a least-square fit to produce the optimal gray value for the fused PAN/MS image. It is anticipated to correlate the frequency range of the high spatial resolution PAN with the MS image bands. Weight factors allow adaptive processing to adjust the resulting fused image optimally for individual spatial detail in different regions. The big advantage of this software is its independence from operator interaction while producing robust high-quality results.

Multi-resolution Approaches

A computationally intensive but effective group of fusion techniques are the MRAs. They comprise the various forms of wavelets used for image transformation to extract individual frequencies to inject spatial detail into the multispectral image. There are many varieties of wavelet transforms (WTs), among which spatial resolution by structure injection is performed. Examples are as follows:

- Multi-scale transform. [14]
- Multi-scale decompositions.
- Multi-resolution wavelet decomposition. [15,16]
- Bilateral decomposition. [17]
- Pyramid transform and high-frequency injection. [18,19]

WT, discrete wavelet transform, additive wavelet transform, contourlet transform, and curvelet transforms belong to this fusion category too.

MRAs decompose multispectral images with respect to local frequency content. At different levels of the scale pyramid, fine to coarse resolutions of the original image are modeled. Each input image delivers a coarse spatial resolution image (LL) and three detailed high spatial resolution images (horizontal: LH, vertical: HL, and diagonal: HH). The coefficients of the WT are used to fuse the data. The fusion is done by either addition or substitution of the MS planes and the panchromatic image. The processing flow is illustrated in a recent book by Pohl and van Genderen.[20]

Hybrid Image Fusion

Recently due to increased computer power and experiences, different fusion algorithms are combined to overcome negative effects of individual techniques. Limitations could be a restricted number of bands (e.g., three) or distortions of the spectral content of the fused image. The resulting hybrid techniques produce better quality in the fused image. Spectral content is preserved while introducing features of the addition image, such as high spatial resolution of panchromatic data or texture from radar imagery. There are popular combinations, that is, IHS-BT, WT-IHS, WT-PCA, or BT-WT. Also, Ehlers fusion combines two elements of image fusion: fast Fourier transform and IHS. The following equation shows the mathematical equation for IHS-BT. Workflows of the abovementioned hybrid fusion techniques can be found in the book by Pohl and van Genderen:[20]

$$\begin{pmatrix} R'_{IHS-BT} \\ G'_{IHS-BT} \\ B'_{IHS-BT} \end{pmatrix} = \frac{PAN}{I+k(Pan-I)} \begin{pmatrix} R+k(PAN-I) \\ G+k(PAN-I) \\ B+k(PAN-I) \end{pmatrix} \quad (17)$$

with R'_{IHS-BT}, G'_{IHS-BT}, and B'_{IHS-BT} representing three fused channels for red, green, and blue, respectively. k is a weighting factor with a value between 0 and 1 regulating the degree of saturation stretch.[21]

WT-IHS helps to reduce spectral distortion in IHS transformation fusion. It uses WT to decompose the produced intensity component of IHS, resulting in a high spatial resolution multispectral image. Using WT-PCA, spatial detail is injected into PC_1.[22] First, the PCA produces the uncorrelated principal components. A histogram match fits the panchromatic channel to PC_1. Addition is used to inject the spatial detail extracted by the WT to PC_1. An inverse PCA produces the actual fused image.

Spatiotemporal Fusion Algorithms

With the availability of long-term observations from space, time series analysis gained importance. Sometimes significant acquisitions at certain points in time are impossible due to cloud cover or other circumstances, such as revisit times of the satellite. To solve this problem, the missing data is simulated using spatiotemporal image fusion. Popular combinations of data are, for example, MODIS and Landsat. After the correction of atmospheric effects and calibration, surface reflectance values are obtained. These can be inserted into the spatial and temporal adaptive reflectance fusion model (STARFM) or its derivatives.[23,24] This is performed on the basis that changes in surface reflectance over time are consistent at different spatial resolutions. STARFM uses neighboring pixels to predict missing data for the center pixel with a moving window. The approach also uses weights to account for the distance, as well as spectral and temporal differences in relation to the center pixel to be predicted. A further development based on the STARFM idea is the spatial temporal adaptive algorithm for mapping reflectance change.[24] It detects changes in reflectance integrating the tasseled cap transform. The tasseled cap transform produces a set of vegetation variables using a new coordinate system, that is, brightness, greenness, and wetness, important for crop monitoring. The enhanced version of STARFM can accommodate other optical remote sensing data. The process is conducted in three steps based on two pairs of high and low spatial resolution images acquired at two different dates. Prior to assigning weights w_i, the algorithm uses a search window at the size of the low spatial resolution data for pixels resembling the center pixel.

A linear regression based on the relationship between these pixels in both images produces the conversion coefficients to predict the high spatial resolution image for the missing date using the following equation:

$$F(x,y,t_p) = H(x,y,t_j) + \sum_{i=1}^{N} w_i \times v_i \times (L(x_i,y_i,t_p) - L(x_i,y_i,t_j)) \quad (18)$$

where H represents the reflectance values of the high spatial resolution image; L indicates the values in the low spatial resolution image at location (x, y); i stands for the ith similar pixel of N total number of similar pixels in a chosen window; and $j = \{1, 2\}$ specifies the point in time t (selected date).[25]

REMOTE SENSING APPLICATIONS OF IMAGE FUSION

Remote sensing applications are manifold. In the context of image fusion, authors are concerned with urbanization, change detection, geology, forestry, vegetation, agriculture, hazards, and land cover mapping in general.

Urban Applications

Due to the complexity of urban structures, image fusion plays an important role in urban remote sensing. Urban areas contain a mix of land cover types and are difficult to be interpreted from one satellite image alone. Another aspect is the fact that urban areas are increasing worldwide due to the population growth and the trend to move to cities to earn a living. Therefore, urban areas are quite dynamic. Image fusion is applied at global and local scale. Globally of interest are the urban growth and the development of urban sprawl. Locally and regionally fused images are used for planning, maintenance, and resource management. Another component that gains importance is the third dimension. The height nowadays provided by LiDAR is integrated in many remote sensing studies. Therefore, the fusion of remote sensing images with height derived from LiDAR for urban mapping is a popular application. Also, pansharpening is important due to the need for high spatial detail.

Agriculture

The second most important application of fused imagery is agriculture. The need for yield prediction and a proper monitoring of the growth cycles of agricultural crops call for multisensor data. The increased spatial and spectral resolution of fused images with the possibility to acquire the data at certain points in time if using multisensor observations is of great importance. The latter is achieved with the spatiotemporal fusion approach. Arithmetic combinations of different channels to produce physically relevant variables are established fields of image fusion.

Land Use/Land Cover Mapping

This group of applications is the largest in the field. Together with geology, it belongs to the first application that used remote sensing image fusion. In the beginning, especially pansharpening led to improved mapping results. But later on the combination of optical and radar data gained more and more importance for two reasons: (1) The two types of data are complementary and provide different information, and (2) radar sensors can acquire images under cloudy weather conditions, which make them a valuable data source in time series. State of the art is the use of hybrid fusion approaches for accurate mapping at various scales.

Change Detection

Strongly linked to the previous application is the monitoring of changes in our environment. Image fusion is very important because of the time component as mentioned earlier. In addition, a multisensor look at the object provides much better descriptions of the types of changes that occurred. In change detection image differencing or rationing are most popular. Hybrid fusion is valuable too since it allows more flexibility in the focus on different objects to be observed.

Geology

This application was the first to value the integration of very diverse data sets, such as passive and active remote sensing data. The spectral signature of optical images together with the texture and topography features of radar images are valuable sources in the interpretation of rock types, geological structures, and minerals. Even though the physical properties observed with optical and radar sensors are very different, the combination of both leads to better results for the interpretation, which is mostly done visually.

Natural Hazards and Disasters

Hazard and disaster monitoring is of utmost importance to protect life. With the climate change and the pressure of overpopulation along with its anthropogenic changes of our natural environment, more and more often people are threatened by disasters. Multimodal images are valuable sources of information to assess the extent of a disaster, provide information to mitigate the disaster, and later on study the possibility to prevent further disasters in the future. Image fusion is common to increase spatial, spectral, and temporal resolution of the observations. In addition, the use of optical and SAR remote sensing is popular due to their different aspects and availability. Examples are flood monitoring or the detection of landslides and volcano eruptions.

SELECTION OF SUITABLE FUSION TECHNIQUES

Many different aspects influence the resulting quality of a fusion approach. The object of interest within a certain application defines quality. This could be the necessity of preserving the spectral content of remote sensing images for further interpretation, that is, classification. Alternatively, the user is looking for spatial detail along with spectral information in order to visually exploit the image. The choice of an appropriate fusion technique depends on many factors. It largely alters with the type and quality of accessible data, the image processing software and user capabilities, the storage capacity, the application and desired information, etc. It requires knowledge about the sensors, the algorithm, and the application for which the information is provided. Combining the available number of satellites and sensors with the options to fuse the images, a large amount of possibilities is obtained.[26]

Issues in Remote Sensing Image Fusion

Issues to be considered are shown in Fig. 3. The key to a successful fusion is the desired application. Agriculture looks for different objects than geology. The availability of certain data types limits the selection too. Furthermore, it is necessary to decide on the processing level. This entry refers to pixel-level fusion. However, there are other more advanced processing levels, that is, feature or decision fusion. With increasing processing level, more and more information contained in the original data is lost, but value is added to the data. All issues are exclusively interdependent, meaning that the selection of one issue has an influence on all other issues. As a result, it is necessary to define the type of objects and features that are supposed to be extracted or visualized in the data. This will lead to certain criteria the fusion process has to fulfill. In the end, it is an iterative approach where different techniques will be tested to obtain the best result.

Image Fusion Quality Assessment

Fused image quality assessment is an active research field. Commonly quantitative and qualitative evaluation is applied. For the quantitative analysis, quality parameters, indices, and protocols are available. Their application is rather diverse and not unified. The qualitative approach using visual interpretation leads to even more diversity since the human interpretation is rather subjective and biased. The motivation to compare the performance of different solutions is manifold. First, users require to identify an optimized fused image to extract as much as accurate information as possible for their application. Second, fusion algorithm developers intend to improve the existing techniques

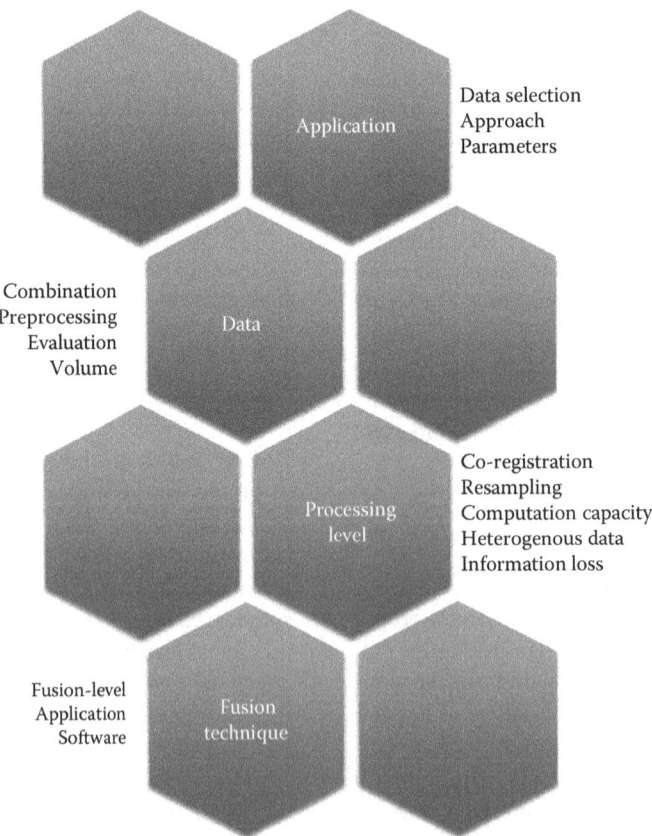

Fig. 3 Issues in data fusion

and eliminate drawbacks that require quality assessment. Third, each fused result needs to be accompanied by quality criteria to provide the necessary metadata for further processing. Quality assurance is an important aspect in the operational use of remote sensing data, especially if interpretation results are used in a legal context.[27] The common quantitative quality assessment measures and approaches to qualitative assessment are listed in the sections that follow.

Quantitative Quality Assessment

Many users of remote sensing image fusion use statistical measures to assess the quality of their fused images. Examples are presented in Table 3. A complete list can be found in the works of Pohl and van Genderen[20] and Li et al.[28] Since an individual parameter does not allow for a description of the global, regional, and local achievement of a fusion process, more complex quality descriptions were developed. A formal framework for the quantitative assessment of fused images was first established in 1997.[29] First, the fused image should be as similar as possible to the original multispectral image, which means that the fused image needs to be compared to the original image at its lower spatial resolution. Second, the ideal image is an image that resembles a theoretically sensed image with an ideal sensor (high spatial and high spectral resolution). Since such a sensor is physically impossible,

Table 3 Selected quantitative quality measures

Quality index	Meaning of abbreviation	Main quality assessed
CC	Correlation coefficient	Spatial or spectral
σ	Standard deviation	Contrast
$\Delta\mu$	Mean bias	Spectral
RMSE	Root mean square error	General
RASE	Relative average spectral error	Global spectral quality
ERGAS	Erreur Relative Global Adimensionnelle de Synthèse	Global spectral quality
UIQI[1]	Universal image quality index	Global spectral and spatial
(M)SSIM	(Mean) Structure similarity index measure	Luminance, contrast, and structure
SSQM	Structural similarity quality metric	Spatial
Q_{PS}	Quality pansharpening	Global spectral and spatial
QNR	Quality with no reference	Global spectral and spatial

the protocol "Quality with No Reference" (QNR) has been developed. Later the research team produced Khan's protocol[30] comprising two indices (spectral and spatial).

Zhou's protocol is a fusion quality assessment protocol where spectral and spatial qualities are evaluated separately.[31] The spectral assessment is done band by band and results in an average value of image differences between the original and fused images. The spatial assessment is done on HPF images, namely, the panchromatic image and the fused image. A Laplace filter is used to extract the high-frequency details of the imagery. The actual quality evaluation is done by calculating the correlation coefficient (CC) between the filtered images. However, other researchers have found limitations in this protocol, which is why other protocols were designed.

In most cases, users do not have access to an appropriate reference image to evaluate the quality of the fused image. This is why the quality assessment without reference image was developed.[32] This protocol is particularly interesting for pansharpening. It contains two separate indices describing the spatial and spectral distortions in the fused image. The calculation is based on the universal image quality index (UIQI).[33] Its advantage is the good performance on low and high spatial resolution imagery.[34] For the practical implementation, the panchromatic image is resampled to the resolution of the multispectral image. Then spatial and spectral differences are calculated. The final QNR index is derived from a multiplication of the spatial and spectral distortion indexes produced. The QNR index allows a modification to concentrate either on spatial or on spectral improvement, depending on the user's preference. Often users investigate spectral and spatial parameters independently.[35]

Khan's protocol takes the development further.[30] It integrates Wald's, Zhou's, and the QNR protocols. Comprising the consistency criteria of Wald, the introduction of high-frequency components from the panchromatic image to the fused image using an HPF as in Zhou's protocol and the definition of the spectral distortion from the QNR protocol, Khan's protocol assesses the overall quality of fused images, considering the MTF. The MTF is a measure of quality, representing the sharpness and the quality of edges and lines and is determined in the frequency domain.[36] Khan's protocol integrates the Q4 index[37] to obtain a comparable quantitative value. The results are a spatial index ($Spatial\_DI$) and a spectral distortion index ($Spectral\_DI$).

According to the literature, quantitative fused image quality assessment should be done using the QNR and Khan's protocol.[20,35] However, it has not been investigated so far, how reliable these protocols perform. For this purpose, a study was performed at the University of Osnabrueck. The two resulting values from Khan are transformed into one value ($Khan_{total}$) following the calculation of the QNR protocol (compare Eq. 19):

$$Khan_{total} = (1 - Spectral\_DI) * (1 - Spatial\_DI) \quad (19)$$

The comparison of Khan's with the QNR protocol led to the fact that their performance needs to be interpreted based on their implementation. The results vary with land cover types and selected evaluation site size. The intercomparison of results from different experiments carried out on different data and locations is rather impossible. It can be stated that both protocols support the user in identifying the best algorithm for different land cover types. Another finding is the fact that fusion results deteriorate if the wavelength range of the panchromatic image is located outside the range of the multispectral image bands. The protocols are not suitable for the detection of artifacts. This has to be done qualitatively, that is, visually (see Section "Visual Quality Assessment"). However, it also has to be stated that the protocols lead to contradictory results, especially in the agricultural areas.

Visual Quality Assessment

According to the literature, visual inspection is a must.[38–40] Due to its subjectivity, it is a problematic quality evaluation approach. Published remote sensing image fusion research contains the qualitative evaluation but mostly very superficially. No effort has been spent on establishing criteria and workflows for visual quality assessment of fused images. For the purpose of standardizing qualitative fused image quality assessment, a study was designed. With the support of the international scientific and professional community criteria, relevant information was collected using a questionnaire. From the collected information, a protocol for evaluating fused image quality visually was created. In order to provide a quantitative value of the visual quality, weighting factors and values were introduced.

Using the protocol, the user is guided through the process of fused image assessment from global via regional to local features. Criteria such as sharpness, color preservation, and object recognition are used to judge the quality of the images. This leads to a repeatable process with unified values leading to an objectified visual quality assessment result that can be used for comparison.

The first step in the procedure development required a compilation of important criteria to be used in visual quality assessment. Important quality criteria are considered during the assessment as listed in Table 4.

The criteria are prioritized in the order of spatial improvement (sharpness, spatial quality), color preservation

Table 4 Visual quality criteria in fused images

Spatial	Spectral
No pixelized lines	Color preservation
Spatial resolution	Color transition
Recognizable spatial objects	Contrast
Sharpness	Natural appearance
Artifacts	Brightness
Texture	Saturation

Table 5 Visual quality aspects in the protocol

Image area	Aspect category	Aspect
Global		General impression
	Spatial	Sharpness (resolution)
	Spectral	Natural appearance
		Brightness
Regional	Spatial	Sharpness (resolution)
		Artifacts
	Spectral	Color preservation
		Contrast of large areas
		Color distortions
	Objects	Groups of objects
		Large objects
Local	Spatial	Details
		Sharpness (edges)
		Artifacts
	Spectral	Color preservation
		Color distortions
		Hue and brightness
		Contrast of small areas
		Color transitions
	Objects	Individual objects
		Object details
		Homogenous areas
		Heterogeneous areas

(spectral quality), and object recognition. Based on the importance of the individual aspects, weighting factors are introduced in the protocol. As a result, the weighted criteria in the order of their importance received a value based on the answers to the questions posed. In total, 23 aspects listed in the protocol are subdivided into spatial-, spectral-, and object-level criteria on a global, a regional, and a local scale, respectively. These aspects are shown in Table 5.

The final protocol contains six parts, namely an introduction explaining the context, a description of possible applications of the protocol, a glossary with explanations of the used terminology, an overview on the structure, the questions to be followed (evaluation protocol), and an external, digital answer sheet with an automated calculation of the values and final score. In the external answer sheet, the process to define quality contains drop-down menus to simplify the usage of the sheet. This standardizes and facilitates the evaluation. In addition, it speeds up the process.[27]

CONCLUSIONS

With the information contained in this chapter, a broad overview has been given in the background, mathematical theory, and applications of remote sensing image fusion. It has been shown that it is an active research field with operational components gaining importance with the availability of long-term remote sensing archives and the multitude of operational sensors in space today. For further literature study, it is recommended to consult two books[20,35] and an international journal focusing on the subject: *International Journal of Image and Data Fusion* published by Taylor & Francis.

ACKNOWLEDGMENT

The original idea to write a book and this entry on remote sensing image fusion came from my mentor and friend Prof. Dr. John L. van Genderen, emeritus professor at the University of Twente in the Netherlands. The study on quality assessment was carried out with my master's students, Johanna Moellmann (quantitative protocols comparison) and Kevin Fries (visual assessment protocol). All contributions are gratefully acknowledged.

REFERENCES

1. Pohl, C.; van Genderen, J. Structuring contemporary remote sensing image fusion. Int. J. Image Data Fusion **2015**, *6* (1), 3–21.
2. Tu, T.-M., Su, S.-C.; Shyu, H.-C.; Huang, P.S. A new look at IHS-like image fusion methods. Information Fusion **2001**, *2* (3), 177–186.
3. Te-Ming, T., Wen-Chun, C.; Chien-Ping, C.; Huang, P.S.; Jyh-Chian, C. Best tradeoff for high-resolution image fusion to preserve spatial details and minimize color distortion. IEEE Geosci. Remote Sens. Lett. **2007**, *4* (2), 302–306.
4. Choi, M. A new intensity-hue-saturation fusion approach to image fusion with a Tradeoff Parameter. IEEE Trans. Geosci. Remote Sens. **2006**, *44* (6), 1672–1682.
5. Leung, Y.; Liu, J.; Zhang, J. An improved adaptive intensity-hue-saturation method for the fusion of remote sensing images. IEEE Geosci. Remote Sens. Lett. **2014**, *11* (5), 985–989.
6. Rahmani, S.; Strait, M.; Merkurjev, D.; Moeller, M.; Wittman, T., *An Adaptive IHS Pan-Sharpening Method*. IEEE Geosci. Remote Sens. Lett. **2010**, *7* (4), 746–750.
7. Te-Ming, T.; Huang, P.S.; Chung-Ling, H.; Chien-Ping, C. A fast intensity-hue-saturation fusion technique with spectral adjustment for IKONOS imagery. IEEE Geosci. Remote Sens. Lett. **2004**, *1* (4), 309–312.
8. Laben, C.A.; Bernard, V.; Brower, W. Process for enhancing the spatial resolution of multispectral imagery using pan-sharpening. U.S. Patent, Editor, Laben et al.: US & International, 2000, 9. Available at http://www.google.com/patents/US6011875
9. Garzelli, A.; Nencini, F.; Capobianco, L. Optimal MMSE pan sharpening of very high resolution multispectral images. IEEE Trans. Geosci. Remote Sens. **2008**, *46* (1), 228–236.
10. Schowengerdt, R.A. Reconstruction of multispatial, multispectral image data using spatial frequency content. Photogramm. Eng. Remote Sens. **1980**, *46*, 1325–1334.
11. Price, J.C. Combining multispectral data of differing spatial resolution. IEEE Trans. Geosci. Remote Sens. **1999**, *37* (3), 1199–1203.
12. Yésou, H.; Besnus, Y.; Rolet, J. Extraction of spectral information from Landsat TM data and merger with SPOT panchromatic imagery—A contribution to the study of geological structures. ISPRS J. Photogramm. Remote Sens. **1993**, *48* (5), 23–36.
13. Zhang, Y.; Mishra, R.K. From UNB PanSharp to Fuze Go—The success behind the pan-sharpening algorithm. Int. J. Image Data Fusion **2014**, *5* (1), 39–53.
14. Tai, G.M.; Nipanikar, S.I. Implementation of image fusion techniques for remote sensing application. Int. J. Emerging Technol. Adv. Eng. **2015**, *5* (6), 109–113.
15. Chibani, Y.; Houacine, A. The joint use of IHS transform and redundant wavelet decomposition for fusing multispectral and panchromatic images. Int. J. Remote Sens. **2002**, *23* (18), 3821–3833.
16. Fanelli, A.; Leo, A.; Ferri, M. Remote sensing images data fusion: A wavelet transform approach for urban analysis. *Remote Sensing and Data Fusion over Urban Areas, IEEE/ISPRS Joint Workshop* 2001, Rome, 2001.
17. Jianwen, H.; Shutao, L. Fusion of panchromatic and multispectral images using multiscale dual bilateral filter. In *18th International Conference on Image Processing (ICIP)*; IEEE: Brussels, 2011.
18. Rong, K.; Wang, S.; Yang, S.; Jiao, L. Pansharpening by exploiting sharpness of the spatial structure. Int. J. Remote Sens. **2014**, *35* (18), 6662–6673.
19. Schowengerdt, R.A. Reconstruction of multi-spatial, multi-spectral image data using spatial frequency content. Photogramm. Eng. Remote Sens. **1980**, *45* (10), 1325–1334.
20. Pohl, C.; van Genderen, J.L. *Remote Sensing Image Fusion: A Practical Guide*. CRC Press: Baton Rouge, 2016.
21. Su, Y.; Lee, C.H.; Tu, T.M. A multi-optional adjustable IHS-BT approach for high resolution optical and SAR image fusion. Chung Cheng Ling Hsueh Pao/Journal of Chung Cheng Institute of Technology **2013**, *42* (1), 119–128.
22. Gonzalez-Audicana, M.; Saleta, J.L.; Catalan, R.G.; Garcia, R. Fusion of multispectral and panchromatic images using improved IHS and PCA mergers based on wavelet decomposition. IEEE Trans. Geosci. Remote Sens. **2004**, *42* (6), 1291–1299.
23. Gao, F.; Hilker, T.; Zhu, X.; Anderson, M.; Masek, J.; Wang, P.; Yang, Y. Fusing Landsat and MODIS data for vegetation monitoring. IEEE Geosci. Remote Sens. Magazine **2015**, *3* (3), 47–60.
24. Hilker, T.; Wulder, M.A.; Coops, N.C.; Linke, J.; McDermid, G.; Masek, J.G.; Gao, F.; White, J.C. A new data fusion model for high spatial- and temporal-resolution mapping of forest disturbance based on Landsat and MODIS. Remote Sens. Environ. **2009**, *113* (8), 1613–1627.
25. Tewes, A.; Thonfeld, F.; Schmidt, M.; Oomen, R.J.; Zhu, X.; Dubovyk, O.; Menz, G.; Schellberg, J. Using RapidEye

26. Pohl, C.; Zeng, Y. Development of a fusion approach selection tool. In *2015 International Workshop on Image and Data Fusion. The International Archives of the Photogrammetry, Remote Sensing and Spatial Information Sciences*; ISPRS: Kona, HI, 2015.
and MODIS Data Fusion to Monitor Vegetation Dynamics in Semi-Arid Rangelands in South Africa. Remote Sens. **2015**, *7* (6), 6510–6534.
27. Pohl, C.; Moellmann, J.; Fries, K. Standardizing quality assessment of fused remotely sensed images. In *International Workshop in Image and Data Fusion*; ISPRS: Wuhan, 2017, 7.
28. Li, S.; Li, Z.; Gong, J. Multivariate statistical analysis of measures for assessing the quality of image fusion. Int. J. Image Data Fusion **2010**, *1* (1), 47–66.
29. Wald, L.; Ranchin, T.; Mangolini, M. Fusion of satellite images of different spatial resolutions—Assessing the quality of resulting images. Photogramm. Eng. Remote Sens. **1997**, *63* (6), 691–699.
30. Khan, M.M.; Alparone, L.; Chanussot, J. Pansharpening quality assessment using the modulation transfer functions of instruments. IEEE Trans. Geosci. Remote Sens. **2009**, *47* (11), 3880–3891.
31. Zhou, J., Civco, D.L.; Silander, J.A. A wavelet transform method to merge Landsat TM and SPOT panchromatic data. Int. J. Remote Sens. **1998**, *19*(4), 743–757.
32. Alparone, L.; Aiazzi, B.; Baronti, S.; Garzelli, A.; Nencini, F.; Selva, M. Multispectral and panchromatic data fusion assessment without reference. Photogramm. Eng. Remote Sens. **2008**, *74* (2), 193–200.
33. Wang, Z.; Bovik, A.C. A universal image quality index. IEEE Signal Process. Lett. **2002**, *9* (3), 81–84.
34. Khan, M.M.; Chanussot, J.; Siouar, B.; Osman, J. Using QNR index as decision criteria for improving fusion quality. In *Advances in Space Technologies, 2008. 2nd International Conference on ICAST 2008*. IEEE, 2008.
35. Alparone, L.; Aiazzi, B.; Baronti, S.; Garzelli, A. *Remote Sensing Image Fusion*. Signal and image Processing Earth Observations Series; Chen, C.H., Ed.; CRC Press: New York, 2015, 342.
36. Thomas, C.; Wald, L. A MTF-based distance for the assessment of geometrical quality of fused products. In *9th International Conference on Information Fusion*, 2006.
37. Alparone, L., Baronti, S.; Garzelli, A.; Nencini, F. A global quality measurement of pan-sharpened multispectral imagery. IEEE Trans. Geosci. Remote Sens. **2004**, *1* (4), 313–317.
38. Toet, A.; Franken, E.M. Perceptual evaluation of different image fusion schemes. Displays **2003**, *24* (1), 25–37.
39. Pohl, C.; van Genderen, J.L. Review article multisensor image fusion in remote sensing: Concepts, methods and applications. Int. J. Remote Sens. **1998**, *19* (5), 823–854.
40. Chavez, P.S.; Bowell, J.A. Comparison of the spectral information content of Landsat Thematic Mapper and SPOT for three different sites in the Phoenix, Arizona Region. ISPRS J. Photogramm. Remote Sens. **1988**, *54* (12), 1699–1708.

Remotely Sensed Images: Lossy Compression

Lluís Pesquer
Grumets Research Group, CREAF, Universitat Autònoma de Barcelona, Catalonia, Spain

Alaitz Zabala
Grumets Research Group, Dep Geografia, Universitat Autònoma de Barcelona, Catalonia, Spain

Abstract
The huge increase in remotely sensed data produced over the previous decades provides an excellent test bed for compression methodologies. While this being the case, image compression also needs to be considered carefully to avoid lowering the quality of data fed to applications relying on image analysis. The present work reviews main topics of image compression, focusing on quality indicators for the preservation of main remote sensing (RS) image characteristics: spatial, temporal, and spectral dimensions. These quality analyses are covered in selected research application case studies. Additionally, standardization and interoperability issues are discussed in depth. This entry offers RS a first guide in image compression.

Keywords: Lossy compression; Quality indicators; Remote sensing; Spatial pattern preservation; Temporal and spectral image series.

INTRODUCTION

We have entered into the era of big data,[1] and we lack a better method to manage and analyze the large volumes of data that are becoming available. Obviously, big data affects many fields and disciplines: economics, physics, medicine, ecology, etc., and the huge amounts of geospatial data generated in previous decades also belong to this flood of digital information. Images are one of the most important sources of geospatial information, and counting among their multiple uses, images are exploited by remote sensing (RS) and geographic information systems (GIS) to predict climate change, forecast weather, conduct land cover mapping and analysis, for disaster management and prevention, in a vast range of applications in agriculture and forestry, in urban management, and to control and monitor the environment. The large increase in the amount of digital information clearly provides enormous application potential, but there is also an important handling problem and a growing need for compression formats. The storage, access, and transmission of these large volumes of information could be problematic, and compression techniques could be a part of the solution.

Compression techniques are methods which reduce the number of symbols used to represent source information.[2] Thus, they imply reducing the amount of space needed to store information or the amount of time necessary to transmit it. There are different types of data compression methods; the main methods are detailed in Section "Image Compression Methods" and some of them are used for remotely sensed images. Most of these methods can be classified as lossless (completely reversible) or lossy (with some loss of information) compressions (see Section "Image Compression Methods" for definitions).

Lossless compression is usually incapable of achieving the high compression requirements of many storage and distribution applications,[3] which can only be achieved by lossy compression algorithms.

As lossy compression introduces irreversible alterations in the original remotely sensed information, rigorous studies are needed to understand the effects and consequences of this manipulation. To this end, in the present work, we review papers from different fields analyzing the effect of compression when using RS images for multiple objectives: spectral analysis, digital classification, texture analysis, stereoscopy, geostatistics, and multivariate regression, among others.

A brief review of image compression methods is presented in Section "Image Compression Methods" (mainly focused on lossy techniques). Section "Remotely Sensed Images" explains the main characteristics of RS images that are relevant for compression purposes in order to understand some interesting issues that will be elaborated in the following sections. Next, the entry focuses on topics relevant to remotely sensed images and their compression techniques, emphasizing the perspectives of standardization and interoperability. Following these review sections, we consider selected research issues and current applications in case studies in order to illustrate the previous theoretical sections. The key issue, directly or indirectly treated during these case studies, is the usefulness and quality of compressed RS images: missing values, metadata and standardization, and different dimensionalities and corresponding redundancies. RS users need a wide range of quality indicators to be convinced that compressed remotely sensed images (using standardized methods) preserve the necessary information for their studies and applications. These issues should be prominent within the main research scopes of future contributions to lossy compression methodologies in RS.

IMAGE COMPRESSION METHODS

User-Side Compression

Image compression methods may be broadly divided into[4] the following:

- Lossless compression: bit-preserving or reversible compression.
- Lossy compression: irreversible compression.

Main lossless data compression techniques are as follows:

- Run-length encoding (RLE) is probably the simplest compression algorithm that replaces sequences of the same data values by a count number and a single value.
- Burrows–Wheeler transform is normally used for textual data, it is based on block sorting, generating a new order of characters (similar characters tend to come together) for a more efficient RLE coding.
- Lempel Ziv (LZ) and several variations are dictionary compression methods, often used for textual data, but some of them can be applied to image formats: deflate for Portable Network Graphics (PNG), Lempel Ziv Welch (LZW) for Tagged Image File Format (TIFF), and others with more universal purposes: Lempel Ziv Renau for ZIP and LZX for Cabinet and Microsoft Compiled HTML Help (CHM) files.
- Huffman coding is a statistical base method that produces the shortest possible average code length depending on the source symbols and their frequencies. The generated corresponding codes are mapped into a binary tree. New adaptive (e.g., Faller-Gallager-Knuth [FGK] algorithm) versions improve the original one.
- Arithmetic coding as JBIG, the standard method from the Joint Bi-level Image Experts Group.

This is only a brief list of lossless compression techniques, as this work is focused on lossy compression methods. Of course, we can additionally include particular parameterizations of some lossy compression methods such as JPEG and JPEG 2000 in this list of lossless compression methods.

The most used lossy compression methods for RS images are as follows:

- Karhunen–Loeve transform (KLT): It is an optimal statistical block-based transform based on a principal component analysis (PCA).
- Discrete cosine transform: It transforms the image into elementary frequency components. JPEG is based on this transformation.
- Discrete Fourier transform: It converts the discrete signal in the space/time domain to a signal in its discrete frequency domain representation.
- Discrete wavelet transform (DWT): splits frequency band of image into different subbands. JPEG 2000 and Consultative Committee for Space Data Systems (CCSDS) Image data compression (CCSDS 122.0-B-1) are based on this transformation. Also, some specific versions have been developed, such as directional lifting wavelet transform (DLWT), or combined to a Tucker decomposition (DWT-TD).
- Set partitioning in hierarchical trees (SPIHT) and some variations as set partitioned embedded block (SPECK), both are scalable coding techniques that allow progressive transmission.

Usually, in case of remotely sensed images, multidimensional modes of these compression techniques are more useful than two dimensional versions, since remotely sensed images have significant redundancies in extra spatial dimensions (see dimensional properties in Image Dimensions section).

For a detailed review on compression methods applied to RS images, refer to the works of Lillesand et al.[4] and Kou.[2]

OnBoard Compression

The amount of RS data we are acquiring is growing, not only because the currently existing satellites keep sending images but also because new instruments are collecting more information due to the increasing spatial, spectral, and temporal resolutions. This amount of data needs to be handled not only by the final user, but also on space-borne remote platforms with limited availability of bandwidth to transmit data to the ground in real time. Compression algorithms need to be adapted to the restricted onboard memory and computation resources, and usually there is sought a balance between the complexity of the algorithm and its performance to be easily implemented by hardware or by software. Main alternatives of onboard compression are cited; the interested reader is referred to the bibliography for more detailed information.

Regarding standard compression methods, the CCSDS has been developing space data handling standards for onboard compression since 1982. Three standards have been approved by the Committee:

- CCSDS 122.0-B-1: image data compression (IDC).
- CCSDS 121.0-B-2: lossless data compression (LDC).
- CCSDS 123.0-B-1: lossless multispectral and hyperspectral image compression (MHIC).

Additionally, several state-of-the-art approaches have been presented in the literature for onboard compression methods, mainly prediction based or transform based. Transform coding computes a linear transform of the data to achieve energy compaction and hence transmit few carefully chosen transform coefficients. Some examples of transform-based algorithms are JPEG 2000, CCSDS-IDC, KLT, and some low-complexity approximations such as pairwise orthogonal transform or fast approximate KLT, JPEG-XR or one based on PCA. Predictive coding predicts the pixel values using a mathematical model and then only encodes the prediction error. Adaptive linear prediction is often used, but other methods have been devised as well, for example, based on edge detection or vector quantization. Examples of predictive methods are differential pulse code modulation and JPEG-LS. Performance evaluation of some of these algorithms can be found in the work of Motta et al.[5] or in the bibliography. Some other approaches, less used, are those based on lookup tables, clustering, or projections.

Standards in Compression Methods

One of the main decisions in compression is related to the compression algorithm and/or compression format to be used. Of course, the features of each particular algorithm need to be taken into account to choose the best algorithm for each application. That being said, sometimes compression formats or algorithms are available with similar characteristics. The range of possibilities usually includes standard algorithms with public specifications developed by international standardization organizations, to proprietary formats without public documentation and usually only some compression and decompression software (or dynamic link libraries, DLLs) available. Some examples of the first group (public) are well-known standards such as JPEG, JPEG 2000, standards of the CCSDS group, or even; Geography Markup Language (GML). Examples of the second (proprietary) are Enhanced Compressed Wavelet (ECW; ERDAS) or Multi-resolution Seamless Image Database (MrSID; Lizardtech). Somewhere in-between there are proprietary formats that have open specifications such as TIFF (Adobe Systems) and IMG (ERDAS Imagine), which have become de facto standards.

Traditionally, in the RS community, there has been a great reluctance to employ lossy image compression due to the importance of retaining all the information for ulterior digital analysis of the images. On the other hand, the capabilities of modern sensors, which increase the size of images, as well as the growing number of Earth observation missions, translate to an enormous demand for storage capacity. The need for an efficient image compression technique becomes acute for the user who has to manage this enormous image volume.

The new spatial data infrastructures (SDIs) paradigm, developed over recent years, promotes the establishment of web data services, usually in terms of the Open Geospatial Consortium's standards such as Web Map Service, Web Coverage Service, Web Map Tile Service, or Web Processing Service. To use these Web Services, especially WCS and WPS, it is necessary to standardize data compression and transmission formats in SDI environments in order to achieve the desired interoperability. Thus, the use of standard compression formats is unavoidable even if it implies slightly worse compression than that obtained with novel but nonstandard algorithms recently published.

REMOTELY SENSED IMAGES

RS can be defined as science and technology that aims to obtain remote data through sensors, as well as the processing and analysis applied to observations (from the space, from the air, or from the ground) of the Earth, the Universe, seabed, etc.[6]

The production of remotely sensed information from air transport platforms began many decades ago with aerial photography. The first RS information coming from satellite platforms was in the 1960s with the first meteorological satellite TIROS and in the 1970s with the LANDSAT land use satellite. With these early uses, the production of remotely sensed information was not very large, and there were not particular demands on real time (or near real time) distribution. For example, Landsat-1 (in 1972) had 5 bands with $3,240 \times 2,340$ samples, and Landsat-8 (in 2013) had 12 bands with $7,611 \times 7,381$ samples in the multispectral bands, doubling the number of samples in the panchromatic band.

Nowadays, remote sensors produce enormous volumes of data of different types including hyperspectral, multispectral satellite images, aerial images, radar, etc. and also some real-time applications which require very quick transmission capabilities. Thus, compression techniques become very useful. Remotely sensed images differ from other kind of images (medical, real or art photography, etc.) in their own specific characteristics, mainly spatial autocorrelation, as well as other correlations in time and spectral dimensions.

Spatial autocorrelation (correlation of a variable with itself over space) is a consequence of the first law of geography: *Everything is related to everything else, but near things are more related than distant things.*[7] For compression purposes, spatial autocorrelation introduces an interesting redundancy that compression methods can exploit. However, some indicators of spatial autocorrelation, such as the Moran index or variogram analysis,[8] are relevant for characterizing spatial patterns; thus, some regionalized variables need specific methodologies in order to preserve spatial properties of remotely sensed images. This point will be deeply analyzed in Section "Spatial Pattern Alterations."

In order to improve the compression performance procedures, it is important to know the technical specifications from each remotely sensed image (some examples are summarized in Tables 1 and 2) to be able to better treat autocorrelation, spectral redundancy, temporal redundancy, range and precision of radiometric values, etc.

Image Dimensions

Spatial Dimension

Due to spatial autocorrelation, the first image dimension that is usually considered is spatial dimension, and it is related to spatial resolution and geographic scene extension. Spatial resolution is defined as the size of the smallest object that can be separately identified in an image, and it is a measure of the fineness of detail of the image. Mainly, but not in all cases, it can be considered as the ground projection of the pixel size. As pixels are typically square, spatial resolution is generally expressed as the side length of a pixel. It is determined by the sensors' instantaneous field of view and may range from very different values, from kilometers (Meteosat second generation [MSG] images) to centimeters (GeoEye, DigitalGlobe's WorldView, or Pléiades). The spatial resolution jointly with the landscape fragmentation determines the redundancy in the spatial domain of the images. Figure 1 depicts different representations of the same area with images having different spatial resolution.

Spectral Dimension

Regarding spectral dimension, remotely sensed images are classified into panchromatic, multispectral, and hyperspectral. Panchromatic is a single image band formed by a wide range of wavelengths in the visible spectrum. Multispectral images are composed of a moderate number of bands, usually (depending on authors) less than 64 (or less than 10). Hyperspectral are composed of a large number of bands (more than 64 or more than 10, depending of the criterion for multispectral). Some authors consider two additional properties: the width (or narrowness) and continuity of the bands to better define hyperspectral with respect to multispectral images (wider and more discrete bands). Figure 2 demonstrates the difference between multispectral and hyperspectral images.

The characteristics of selected examples of multispectral and hyperspectral images are detailed in Tables 1 and 2, respectively.

This selection aims to illustrate the wide range of values within the main specifications that are involved in possible redundancy in different dimensions: spatial, temporal,

Table 1 Selection of different multispectral remotely sensed images in order to show several contrasting configurations

Platform/sensor	Spatial resolution (m)	Number of bands	Spectral width average (nm)	Radiometric resolution (bits)	Revisit time (days)
Landsat-5 TM	30, 120	7	136, 2,100	8	16
MODIS	250, 500, 1,000	36	22, 188, 367	12	1, 2
SPOT-5 HRG	5, 10, 20	5	110, 230	8	2–3, 25
ASTER	15, 30, 60, 90	14	68, 490	8, 12	4–16
Meteosat-9 SEVIRI	1,000, 3,000	12	400, 1,055	12	0.01 (15 min)
RapidEye-REIS	6.5	5	65	12	5–6
Sentinel-2 MSI	10, 20, 60	13	50	12	5

Table 2 Selection of different hyperspectral remotely sensed images in order to show several contrasting configurations and illustrate the main differences with respect to multispectral images

Platform/sensor	Spatial resolution (m)	Number of bands	Spectral width average (nm)	Radiometric resolution (bits)	Revisit time (days)
AIRS	2,300, 13,500	2,382	4		0.5
AVIRIS	2,000	224	10	12, 16	On demand
CASI-550	3	288	7.6	14	On demand
DAIS-7915	15	79	10, 16		On demand
EO-1 Hyperion	30	220	10	12	16

Fig. 1 Comparison of images (SPOT on the left, Landsat on the right) with different spatial resolution (10 and 30 m) over the same area (Barcelona harbor, 41° 22′ 17″ N–2° 10′ 54″ E)

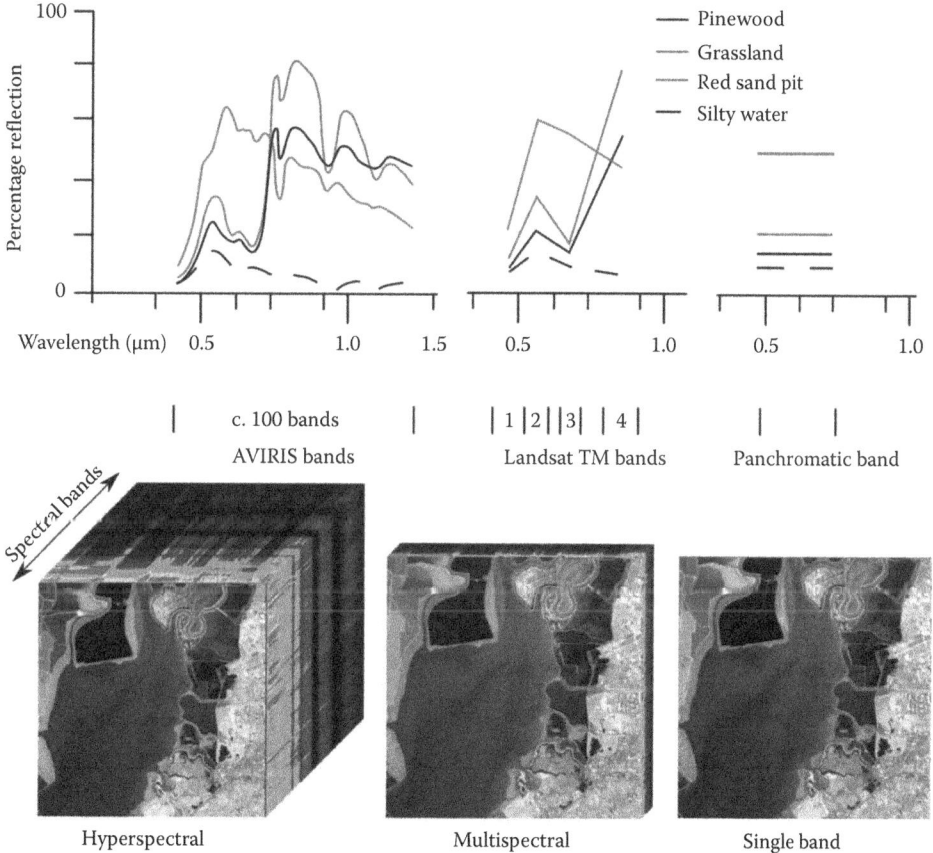

Fig. 2 Comparison of hyperspectral and multispectral images. The graphics show how land covers are represented in an AVIRIS hyperspectral image (100 bands), in a Landsat multispectral image (first four bands), and in a panchromatic (B&W) image (single band)
Source: DART Project Flickr photo stream, under Creative Commons Attribution-Non Commercial 2.0 license, modified by the authors. Website of the project: www.dartproject.info/.

spectral, radiometric, etc. The performance and parameterization of the chosen compression method should depend on the image characteristics.

Time Dimension

Deforestation, drought monitoring, land use change, and classification methods for land cover map generation are some of the remotely sensed image applications that are based on two-time properties: revisit time (also called temporal resolution) and temporal dynamic of the studied area. Regarding its implication for compression, time redundancy (or change along time) is analyzed and modeled by compression methods. Usually, high temporal resolution image series present more redundancy than low temporal resolution image series, of course depending on the particular velocity of land changes in a specific study region.

Another important consideration is that, in many cases, high temporal resolution images present low/medium spatial resolutions, for example, MSG (temporal resolution of 15 min, see Table 1) has between 1 and 3 km of spatial resolution; on the other hand, Moderate Resolution Imaging Spectroradiometer (MODIS) (daily product) has an spatial resolution from 250 m to 1 km, depending on the radiometric (spectral) band.

Finally, in order to maximize redundancy and thus take advantage of compression methods, it is important to use images encoding physical variables, such as reflectance or surface temperature, instead of digital numbers (DNs) captured by a sensor. These magnitudes are obtained through radiometric (atmospheric+topographic) correction methods[9] or through some type of normalization of the temporal image series. These methodologies reduce the different illumination conditions caused by the solar position according to the relief or different atmospheric conditions (water vapor, aerosols, etc.), and obviously, this harmonization improves compression methods.

Metadata for Remotely Sensed Images

Metadata describe the content, quality, condition, and other characteristics of data, and they are essential for exchanging, cataloging, searching, and consistently using geospatial information. Metadata for RS images have special needs to properly document topics that are not needed in other types of data (such as platform and mission information, multiband images, processing algorithms, and special quality indicators) or that need special consideration when referring to RS data (such as reference system, spatial extension, bounding box, pixel size, and resolution).

Several international standards have been developed to define the metadata elements that should be included when documenting geographical information in general and RS in particular. The most widely used are the standards from the International Organization for Standardization Technical Committee 211 (ISO TC-211), such as ISO 19115-1:2014 and its RS extension ISO 19115-2:2009 (under review). Special attention needs to be given to standards describing quality (ISO 19138 and the revisions being considered in ISO 19157, 19158, and 19159) as this information (e.g., image processing and algorithms, quality indicators regarding positional accuracy, compression method and its parameters) is of paramount importance to properly use RS images and their derived products.[10] Metadata may be encoded in several ways, XML being the most widely used (following ISO/TS 19139:2007 and ISO/TS 19139-2:2012, which will be reviewed by the new ISO 19115-3).

Moreover, metadata is very important in RS images to allow automatic (or semiautomatic) processing of images. This is a very important issue especially in a distributed and automatic processing chain such as those following OGC WPS standards. Examples of processes taking advantage of metadata are as follows: automatic geometric correction, which can be performed if some parameters are available in metadata files; acquisition date and time information, which is used in radiometric correction with consideration of illumination conditions; etc. To effectively use metadata within a WPS, it is necessary that the metadata are embedded on the image file itself or, at least, directly available in a standard format (XML following ISO guidelines). OpenGeospatial Standards "GML Application Schema—Coverages"[11] and "GML in JPEG 2000 (GMLJP2) Encoding Standard Part 1: Core"[12] describe how to encode a metadata XML file within the JPEG 2000 file as well as how to distribute JPEG 2000 in a WCS, for example, for use in a WPS.

Pesquer et al.[13] describe an application with a WPS chain of two processes for Landsat images, where JPEG 2000 is one of the possible output formats for the first step and later serves as input for the second step. In this implementation, metadata are embedded in an XMLBox in one reserved JPEG 2000 section, following GMLJP2·standard (see Fig. 3). These RS processes are geometric correction and radiometric (atmospheric+topographic) correction, and the role of specific metadata is essential in order to apply the best parameter configuration for both processes and achieve optimized results.

Missing Values

In a raster data model, missing values can represent different concepts and situations: no information, out of study region, error value, etc. Missing values (also referred as NODATA) are not always represented by the same value, for example, it may be −999 in an elevation raster (SRTM DEM) and 0 for a radiometric value in DN units. The main rule is that this value should be out of the range of valid values (−999 is not a realistic elevation, but it could be a valid value in bathymetry maps).

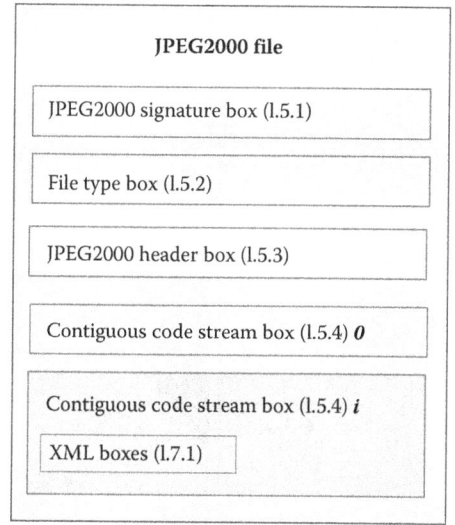

Fig. 3 Location of XML boxes in the file structure file of the JPEG 2000 format

Dealing with NODATA in Compression Systems

Most compression systems are not devised to consider these regions separately from the rest of the image, sometimes causing a loss in the coding efficiency and in the post-processing applications. The consideration of this value as a "normal" value has two main shortcomings for the coding process: first, the final coding performance could be penalized by the overhead produced by these NODATA regions, and second, it could generate errors in the post-processing applications, for instance, in a digital image classification process or when computing a slope in digital elevation models.

There are two main reasons for the extra coding cost: the cost of coding the sharp edges and the cost of coding the coefficients belonging to the NODATA regions. Sharp edges are produced by the difference between the values of the images and the value used to represent the missing values. If the missing value is near the range of valid image values, the coding efficiency will improve. However, the drawback is that the recovered image will have similar values on the border between the valid data and NODATA areas, which may imply turning valid values into invalid ones (e.g., vegetation indices with values out of the theoretic range −1, 1).

To identify original NODATA areas on the recovered image, it is necessary to include the NODATA mask on the compressed file, but this is not enough as this does not deal with the extra coding cost of images with NODATA values.

Several approaches within the JPEG 2000 framework can be developed to reduce the penalization in the encoding process produced by the presence of NODATA regions. The first technique (average data region [ADR]) is carried out as simple preprocessing and the second technique (shape-adaptive JPEG 2000, SA-JPEG 2000) modifies the coding system to avoid the regions without information.

This second approach SA-JPEG 2000 is similar to binary set splitting with k-d trees (BISK), entirely devised to support shape-adaptive processing, and thus, it is ideal for the coding of NODATA.

Effects on Coding Performance

Experimental results, performed on data from real applications and different scenarios, suggest that the proposed approaches can achieve, for example, for SA-JPEG 2000, a signal-to-noise ratio (SNR) improvement of about 8 dB. For images with a high amount of data percentage, the difference between the SA-JPEG 2000 and ADR is meaningless, for example, for the SPOT dataset (Fig. 4a). However, when the data region decreases, the difference grows, and SA-JPEG 2000 clearly outperforms ADR for all bit rates.

In addition to the data percentage within the image, how these data are distributed is also important for the final coding performance. In images with similar percentage of data region, if most data regions are narrow and not very connected, the non-shape-adaptive approaches like ADR are penalized.

Moreover, as explained before, when data percentage in the image is higher, differences between ADR and SA-JPEG 2000 become less important, and therefore, the penalization of the ADR approach due to the fragmentation of the NODATA areas is also less pronounced. For example, for a Landsat TM dataset (Fig. 4b), results for ADR are comparable to those obtained by SA-JPEG 2000 (Table 3) due to the higher proportion of data in the image (67%), in spite of the disconnected NODATA regions.

Effects on Image Classification

In addition to coding performance, it is necessary to evaluate the suitability of the images for post-processing

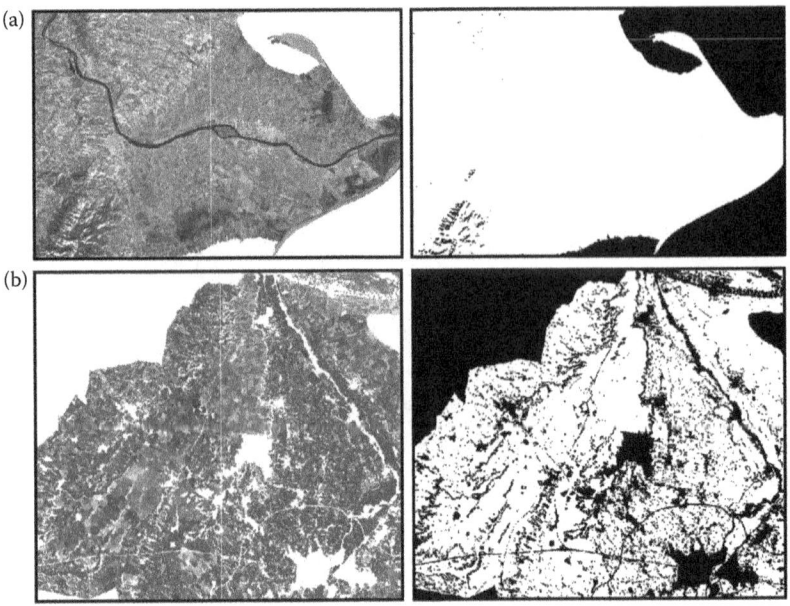

Fig. 4 (a) SPOT dataset over Ebro river Delta (NE Spain, 41° 43′ 3″ N–0° 40′ 58″ E): 03-10-2008, dimensions 1593 × 1033 × 4, 79% data region (b) Landsat-5 Thematic Mapper (TM) dataset over Lleida (NE Spain, 41° 42′ 50″ N–0° 31′ 45″ E): 17-06-2004, dimensions 1517 × 1311 × 6, 76% data region. For both datasets, an RGB composition (NIR, red, and green bands) and a data (white) and NODATA (black) region are presented

applications in RS and GIS scenarios. Higher compression performance does not necessarily yield a higher quality RS image for a given application, so it is necessary to evaluate encoding approaches not only according to their coding performance but also according to the effects produced on the product obtained after image coding.

One of the most common uses for remotely sensed images is to obtain thematic classifications. Several approaches can be developed to obtain them. Three classification methodologies (discriminant analysis, multitemporal maximum likelihood, and a hybrid classifier) were applied to classify three available datasets (SPOT, Landsat TM, and AVIRIS).

Moreover, in a post-processing application such as a digital classification, the best classification results are obtained when the SA-JPEG 2000 and ADR approaches are applied. It is noticeable that when a maximum likelihood classification is applied over a Landsat-5 TM multitemporal set of five images, ADR obtains, in general, similar results for overall accuracy but better results for classified areas as compared to SA-JPEG 2000 for the same bit rates, even with a lower SNR energy. The achievement of similar results is due to the higher proportion of data on the image (67%), in spite of the disconnected NODATA regions.

Final Considerations

Two important points are worth noting from these results. The first is that SA-JPEG 2000 yields the best coding performance, even outperforming BISK. Nevertheless, it does not generate a compliant code stream. The second point is that the best compliant approach for all the experiments is the ADR, whose coding performance can be equivalent to shape-adaptive approaches depending on the image. Regarding the results of post-processing classification performed over encoded images, ADR and SA-JPEG 2000 obtain similar results, especially if data region proportion is high or if NODATA regions are few and highly connected.[14]

Table 3 Coding performance for five Landsat-5 TM images used to obtain a 2004 crop map over Lleida (NE Spain)

	Landsat-5 TM (five dates)—(1517 × 1311 × 6)—data region: 67%			
Bit rate	**JPEG 2000**	**ADR**	**SA-JPEG 2000**	**BISK**
0.0625	6.56 (−11.70)	**18.25**	17.49 (−0.77)	17.16 (−1.09)
0.125	7.83 (−11.99)	**19.82**	19.59 (−0.24)	18.93 (−0.89)
0.25	9.65 (−12.17)	21.68 (−0.14)	**21.82**	21.00 (−0.82)
0.5	12.62 (−11.95)	24.02 (−0.54)	**24.56**	23.83 (−0.73)
1	18.11 (−10.43)	27.59 (−0.96)	**28.54**	27.95 (−0.59)
1.5	22.73 (−9.40)	30.68 (−1.45)	**32.14**	31.93 (−0.21)

Note: Results report the SNR Energy (in dB) for different bit rates (in bits per sample per band). The best coding performance is printed in bold, and the difference between this best coding approach and others is depicted in parentheses.

Table 4 Summary of desktop software that implement one or several lossy compression methods

Name	Website	Main format
BOI	www.gici.uab.cat/BOI/	JPEG 2000
ERDAS ER Mapper	www.erdas.com.ar/productos_imagine.htm	ECW
ESA compression tool and WhiteDwarf	www.esa.int/Our_Activities/Space_Engineering/ Onboard_Data_Processing/Data_compression_tools	CCSDS 121 (LDC), 122 (IDC) and 123 MHIC
GeoExpress	www.lizardtech.com/products/geoexpress/	MrSID
Jasper	www.ece.uvic.ca/~frodo/jasper/	JPEG 2000
Kakadu	www.kakadusoftware.com	JPEG 2000
LEADTOOLS	www.leadtools.com/sdk/compression/	JPEG 2000, LZW, ZIPLIB
MapImagery	www.mapimagery.com/ecw/	ECW
SPIHT image compression	www.openjpeg.org/	JPEG 2000
Szip	www.hdfgroup.org/doc_resource/SZIP/	CCSDS 121 (LDC)
TER/Emporda/Delta	www.gici.uab.cat/TER/ gici.uab.es/GiciWebPage/emporda.php gici.uab.es/GiciWebPage/delta.php	CCSDS 122 (IDC) / 123 (MHIC) / 122 (IDC)
UNL	http://hyperspectral.unl.edu/	CCSDS 122 (IDC)

Note: The aim of this collection is to indicate the links and main format specifications of these specialized compression software and main implementations in GIS and RS software. With the provided links, users can find many utilities for processing the remotely sensed images, before and after specific compression procedures.

SOFTWARE

Most RS software and GIS, for example, ArcGIS (www.arcgis.com), ENVI (www.exelisvis.com/ProductsServices/ENVIProducts/ENVI.aspx), GRASS (http://grass.osgeo.org/), and MiraMon (www.creaf.uab.cat/MiraMon) support some standard (JPEG 2000) or/and proprietary (ECW, MrSID) lossy compression formats. However, we will need to use special software in order to exploit advanced options and all capabilities, and also for specific cases of multi-dimensional compression. Table 4 shows a brief summary description of a selection of specialized software.

In addition to those in Table 4, there are some particular lossy compression utilities or library functions developed in high-level languages such as MATLAB®, for example, www.ux.uis.no/~karlsk/ICTools/ictools.html, or www.cipr.rpi.edu/research/SPIHT/spiht3.html, and there are also some lossy compression online tools as web services (e.g.,: https://kraken.io/web-interface).

QUALITY EVALUATION

Different compression goals can have different requirements: storage (compression ratio, CR), computational efforts (time processing, memory demands, etc.), and quality evaluation. Most applications for remotely sensed images (except onboard compression, which was discussed in a previous section) have quality requirements, to be analyzed in this section. In addition, higher compression performance does not necessarily mean higher quality for a given RS application[15]; thus, the possible malicious effects on the post-processing stages, such as classification methods, have to be carefully studied too.

Compression Metrics

The main compression measures used are as follows:

- CR: It is defined as the relation between the size of the original noncompressed image and the size of the compressed image. For example, a CR of 10:1 means that 10 bytes of the original image is reduced to 1 byte in the compressed image.
- Compression factor (CF): It is defined as dividing the size of the original noncompressed image by the compressed image. For example, 10 is the corresponding CF value for 10:1 in CR.
- Compression gain (CG): It is defined as the percent log ratio between the reference size image and the compressed size image.
- Bits per pixel (Bpp): It is defined as the ratio of the bit depth of the original image and the CF. For example, a 1:10 CR compression of a bitmap image of 24 bits (8 bits RGB) has a Bpp of 2.4.

Distortion Measurements

There are different global evaluators that compare the quality of a compressed image with respect to the original (noncompressed image), otherwise understood as a measure of the level of distortion caused or loss of information.

In the following mathematical expressions, x_i denotes a pixel from the original image (before compression) and $\hat{x}_i$ a pixel from the decoded image (after coding and decoding compression procedures). n is the number of samples analyzed, for example, for two-dimensional images, it represents the number of rows multiplied by number of columns, and in case of three-dimensional images, this is multiplied by the number of bands. The index i goes over the number of samples n; thus, these formulas can be applied to any dimensionality.

- Mean absolute error (MAE): $MAE = \frac{1}{n}\sum_{i=1}^{n}|x_i - \hat{x}_i|$.
- Mean squared error (MSE): $MSE = \frac{1}{n}\sum_{i=1}^{n}(x_i - \hat{x}_i)^2$.
- Root mean squared error (RMSE): $RMSE = \sqrt{MSE}$.
- $SNR = 10 \cdot \log_{10}\frac{\sigma^2}{MSE}$

where σ^2 is the variance of the original image.

- Peak SNR (PSNR): $PSNR = 10 \cdot \log_{10}\frac{\max|x_i|}{MSE}$. Probably, this is the most used quality indicator for lossy compression techniques for many types of images, including remotely sensed ones.
- Correlation coefficient.
- Multiscale and multilevel distortion: They are formed by different procedures detailed in the work of Jiang et al.[16]
- Mean phase error.
- Spectral angle mapper (SAM): $SAM = arcos\frac{\langle x,\hat{x}\rangle}{|x|^2 \cdot |\hat{x}|^2}$

where $\langle x,\hat{x}\rangle$ is the scalar product between both images.

- Spectral information divergence (SID): $SID = D(x\,|\,\hat{x}) + D(\hat{x}\,|\,x)$.

where $D(x\,|\,\hat{x})$ is the Kullback–Leibler information, measure of the relative entropy between both images, which is as follows:

$$D(x\,|\,\hat{x}) = \sum_{i=1}^{n} p_k \log\left(\frac{p_k}{q_k}\right),$$

where p and q are the vectors on the spectral dimension. It measures the probabilistic discrepancy between spectral signatures.

- The area under the ROC curve: for binary responses, this represents the relation between true positive rates (positives in both images) versus false positive rates (negatives in the original image but positives in the compressed image).

Refer to the work of Serra-Sagristà and Aulí-Llinàs[3] and general-scope books in the bibliography section to consult the specific details of most of these indicators.

Specific Preservations

Most of these distortion evaluators are used not only for remotely sensed images but for any kind of image. Additionally, they are global, meaning that they represent, for example, average values for the complete image, without explaining the corresponding spatial, time, or radiometric distribution. However, other indicators are particular to some image applications, for example, geospatial information or radiometric properties.

Depending on the RS application, corresponding images need to minimize the distortion in some particular aspects. In the next sections, we exhibit two examples for different preservation purposes: spatial pattern alterations and quality assessment of classification.

Spatial Pattern Alterations

One of the main properties of different remotely sensed images is the spatial resolution as seen in Tables 1 and 2. Depending on the spatial pattern of a particular landscape, spatial resolution could imply spatial redundancy, for example, as in Fig. 5, in large and homogeneous monocropping extensions, or, in other cases, a high spatial resolution could be essential for describing very fragmented landscapes.

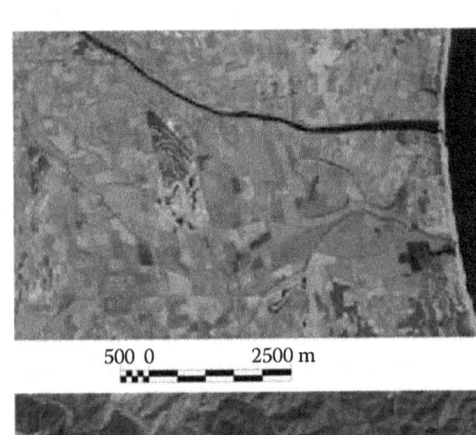

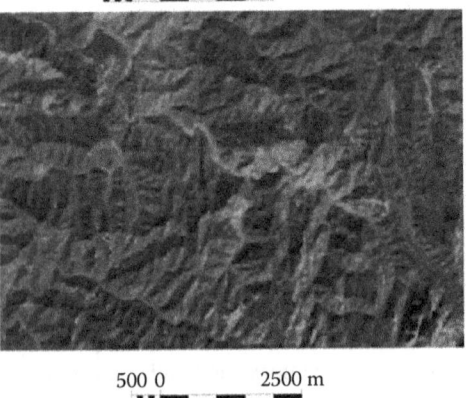

Fig. 5 (Top) Landsat image (4-5-3 RGB band composition) corresponding to a fragmented landscape (different kind of crops) implying a high variability image. (Bottom) A homogeneous forestry landscape

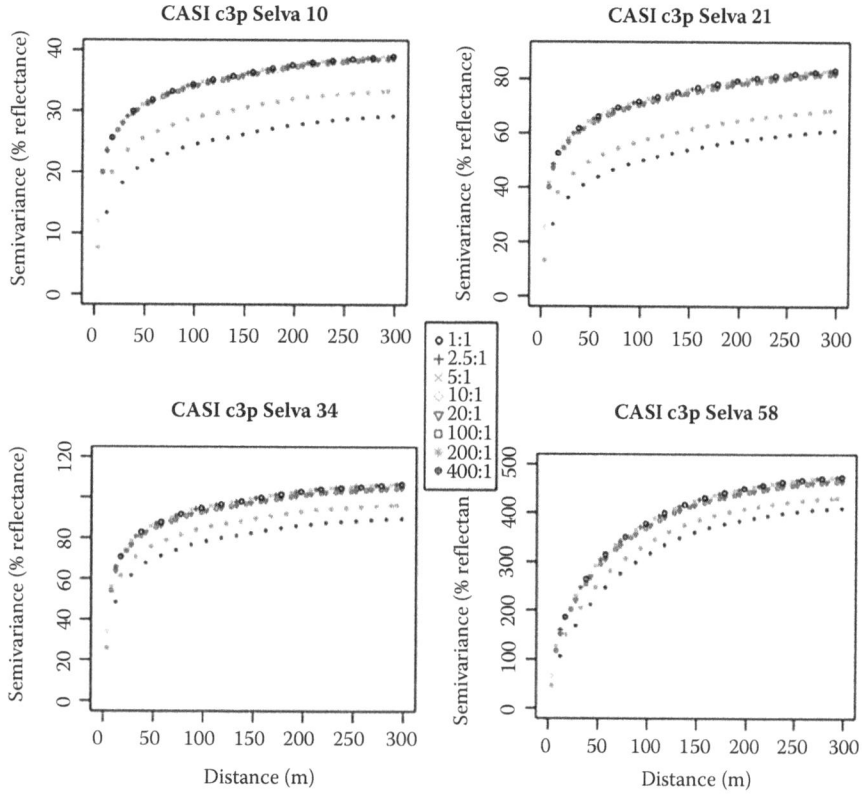

Fig. 6 Comparison of variogram results for a wide range of CRs on the JPEG 2000 3D-DWT compression method in the four selected CASI bands on 18-05-2007 (c3p scene) for the Selva study region, a mountainous landscape in the northeast Iberian Peninsula

If we explore the most usual distortion indicators, detailed in Section "Distortion Measurements," from a spatial point of view, we can assert that they do not specifically analyze spatial interactions between (near and mid-distance) neighboring pixels.

Exploring, describing, and analyzing the spatial variation in images is one of the main application of geostatistics in RS. Geostatistics is a subset of statistics specialized in the analysis, interpretation, and inference of geographically referenced data[17] that can provide indicators for describing spatial patterns, optimal sampling designs for ground surveys, techniques for downscaling images, measurements for spatial autocorrelation, tools for estimating continuous variables, and data for radiometric coregionalization analysis. The variogram plots this dependence on the spatial variance and it allows the spatial patterns to be analyzed. It is based on the geostatistical assumption that considers as constant (non-regionalized) the mean as well as the variance of the differences between regions separated by a given distance and direction.

As an example, Fig. 6 shows a variogram comparison study for a three-dimensional DWT compression method (JPEG 2000 format). In the figure, black dots correspond to the variogram (spatial pattern) of the original image (1:1 CR), and different symbols and colors representing several CRs are depicted. These plots (each for a different band of a hyperspectral image, four bands are studied) show that compressions up to a CR of 100:1 preserve quite well the variogram structure; however, a CR of 200:1 and a CR of 400:1 introduce significant alterations. Pesquer et al.[8] describe in detail more cases of this kind of geostatistical analysis.

Therefore, variogram analysis is a very suitable technique in order to analyze spatial alterations in remotely sensed images caused by lossy compression methods, and it offers a more specialized quality evaluation than distortion measurements, usually applied to any kind of image.

Quality Assessment on Classification

One of the main applications of RS images is the generation of Land Use/Land Cover (LULC) maps through digital image classification. In spite of the spectacular CRs reached by lossy compression algorithms, there has been little quantitative analysis on the implications of these compressions for classification, especially in real agroforestry management applications (i.e., multitemporal analysis to take advantage of the vegetation dynamics) with a regional approach (using images that cover a wide area). Indeed, previous research has generally focused on compression and has explored modifications in compression techniques to improve them; however, only some of these studies address the effect of this compression on classification.

Moreover, none of these articles used multitemporal analyses, and they all used small images (the widest area was 512 pixels × 512 pixels).

Pixel-by-Pixel Classification over Forest Areas

Several Landsat images and other topo-climatic variables (NDVI, average precipitation, solar radiation annual average, maximum temperature in July, minimum temperature in April, and minimum temperature in April) are combined to classify a legend of 15 forest communities (up to species level) using a hybrid classification method. Images have been compressed using JPEG and JPEG 2000 compression standards.

The results show that the classifications of images compressed using JPEG 2000 has less salt and pepper effect than classifications obtained with the original images. In addition, higher CRs (without exceeding the maximum recommended) can get even better results than those obtained with the original images. The maximum level of compression recommended depends on the fragmentation of the image (higher fragmentation accepts less compression) and compression method (JPEG 2000 allows greater compression than JPEG with the same results): for non-fragmented areas a CR of 10:1 or 20:1 can be applied, but for fragmented areas only JPEG 2000 at a maximum CR of 3:33 to 5:1 is advised.[18]

Pixel-by-Pixel Classification over Crop Areas

The study of an agricultural landscape uses a multitemporal approach with Landsat images to classify six types of crops using a hybrid classifier, as well as the minimum distance and maximum likelihood classifiers. Images have been compressed using JPEG, JPEG 2000, and 3D JPEG 2000 compression standards.

As in forest areas, results show that JPEG 2000 eliminates salt and pepper effect on the obtained classification, producing a better visual outcome. If the classification is performed using the multitemporal approach in a low fragmentation area, it is possible to use a CR of 20:1 (using JPEG) or 100:1 (using JPEG 2000), and if the area is more fragmented, the recommendation is not to compress more than a CR of 10:1. If the classification is performed using a single date, it is only recommended to use JPEG 2000 3D, from 10:1 to 100:1 (depending on the fragmentation of the area) (see Fig. 7).[19]

Object-Based Classification over Urban Areas

Image classification has traditionally employed pixel-based methods based on statistical measurements of individual pixel values to classify each pixel with supervised, unsupervised or hybrid classifiers. Nevertheless, with the improvement in spatial resolution of remote images (such as the obtained from digital airborne cameras or from high-resolution satellite sensors), the number of articles on geographic object-based image analysis (GEOBIA) techniques has increased.

To evaluate the effects of lossy compression in the field of object-based classification images over similar urban landscapes (characterized by low-density urban areas), images have been compressed using the JPEG 2000 compression standard and then submitted to segmentation and classification. The images that have been used are four aerial color (RGB) orthophotos (1 m resolution) and one Quickbird four-band image (including NIR and RGB bands).

The main conclusion of this study is that at high compression levels (40:1 or higher), poor accuracies are obtained (e.g., 55.40%) when image segmentation is applied to obtain categorical maps. Although for all uncompressed images similar overall accuracies were obtained (e.g., 69.40%), in more fragmented areas the decrease in accuracy at the first compression level was more significant than in the less fragmented areas. Similar overall accuracies were obtained for classifications of less fragmented areas in the uncompressed image and in the first three compression levels (up to 20:1), and accuracy decreases more significantly if more compression (40:1 or higher) is applied.[20]

CONCLUSION

Nowadays, lossy compression applied to remotely sensed images is not widely used; however the huge amount of data produced by a large number of sensors is exponentially increasing. Users of remotely sensed images are aware that compression is needed. Nevertheless, they are more worried about the effect of the compression on the final image applications (classification, retrieval of quantitative variables such as temperature and leaf area index) than about obtaining a very high compression performance (e.g., using nonstandard, recently published algorithms). Moreover, the specific characteristics of RS images (and its different spatial, temporal and spectral redundancies) lead to the need of compression algorithms that correctly preserve the relevant information for the user.

Thus, rigorous studies that analyze the quality of images compressed and studies that verify that their applications do not reduce their capabilities, coherence, quality of results, etc., are needed. After the introductory sections, this work has collected different studies that present results pertaining to quality, ranges of parameterization for several compression techniques, recommendations, etc. in different types of RS applications for environmental studies. These studies emphasize the specific characteristics in the spatial, time, and radiometric domains of remotely sensed images and their corresponding requirements for correct application of optimal compression techniques.

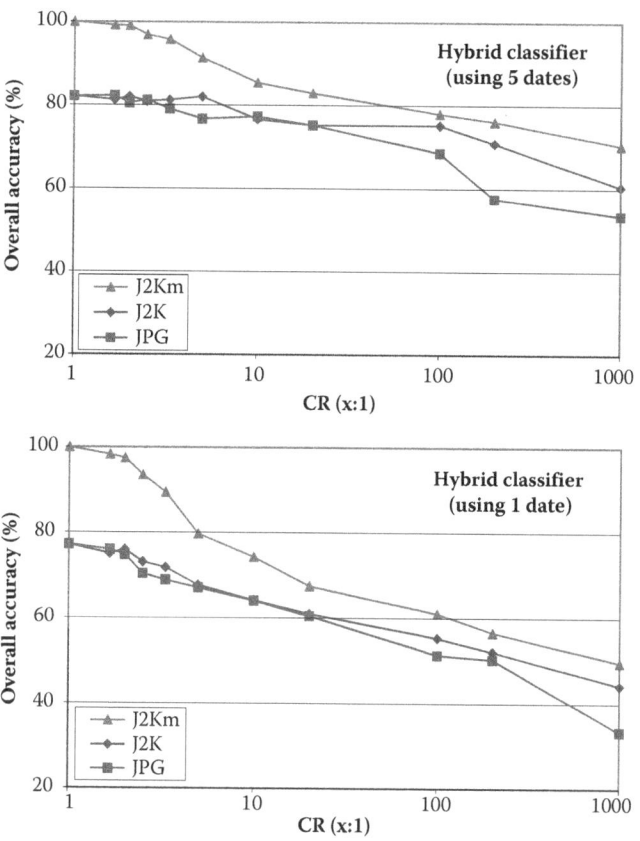

Fig. 7 Overall accuracy in relation to the original classification of the classifications obtained using the full set of images (left) and only one image (right), using the hybrid classifier

ACKNOWLEDGMENTS

This work was partially supported by the Catalan Government under Grant SGR2009-1511, the Spanish Ministry of Economy and Competitiveness, and the European regional development fund under Grant CGL2012-33927.

REFERENCES

1. Agrawal, D.; Bernstein, P.; Bertino, E.; Davidson, S.; Dayal, U.; Franklin, M.; et al. Challenges and Opportunities with Big Data—A Community White Paper Developed by Leading Researchers across the United States. Computing Research Association, 2012. Executive Summary.
2. Kou, W. *Digital Image Compression: Algorithms and Standards*; Kluwer Academic Publishers: Boston, MA, 1995.
3. Serra-Sagristà, J.; Aulí-Llinàs, F. Remote sensing data compression. In *Computational Intelligence for Remote Sensing*; Graña, M.; Duro, R.J.; Eds.; Springer Verlag: Berlin, 2008, 27–61.
4. Rabbani, M.; Jones, P.W. In *Digital Image Compression Techniques*; O'Shea, D.C.; Ed.; SPIE Optical Engineering Press: Bellingham, WA, 1991.
5. Faria, L.N.; Fonseca, L.M.G.; Costa, M.H.M. Performance evaluation of data compression systems applied to satellite imagery. J. Electr. Comput. Eng. **2012**, *2012*, 18, Article number 471857.
6. Pons, X.; Arcalís, A. *Diccionari terminològic de Teledetecció*; Enciclopèdia Catalana i Institut Cartogràfic de Catalunya: Barcelona, 2012.
7. Tobler, W.R. A computer model simulation of urban growth in the Detroit region. Econ. Geogr. **1970**, *46* (2), 234–240.
8. Pesquer, L.; Pons, X.; Cortés, A.; Serral, I. Spatial pattern alterations of JPEG2000 lossy compression in remote sensing images: Massive variogram analysis in high performance computing. J. Appl. Remote Sens. **2013**, Article number 73595.
9. Richards, J.; Jia, X. *Remote Sensing Digital Image Analysis: An Introduction*, 4th Ed.; Springer Verlag: Berlin, 2005.
10. Yang, K.; Blower, J.; Bastin, L.; Lush, V.; Zabala, A.; Masó, J.; Cornford, D.; Díaz, P.; Lumsden, J. An integrated view of data quality in earth observation. Philos. Trans. R. Soc. A **2013**, *371* (1983), 20120072.
11. Bauman, P. *OGC GML Application Schema—Coverages (1.0) 2010-10-27* Document: OGC 09–146r1, Open Geospatial Consortium, 2010.
12. Colaiacomo, L.; Masó, J.; Devys, E. *GML in JPEG 2000 (GMLJP2) Encoding Standard Part 1: Core (OGC 08–085r4)*. Open Geospatial Consortium, 2014.
13. Pesquer, L.; Masó, J.; Moré, G.; Pons, X., Peces, J.; Doménech, E. Interoperable service (WPS) of Landsat image processing. Revista de Teledetección **2012**, *37*, 51–56.

14. Zabala, A.; González-Conejero, J.; Serra-Sagrista, J.; Pons, X. JPEG2000 encoding of images with NODATA regions for remote sensing applications. J. Appl. Remote Sens. **2010**, *4* (1), 041793.
15. Penna, B.; Tillo, T.; Magli, E.; Olmo, G. Transform coding techniques for lossy hyperspectral data compression. IEEE Trans. Geosci. Remote Sens. **2007**, *45* (5), 1408–1421.
16. Jiang, H.; Yang, K.; Liu, T.; Zhang, Y. Remote sensing image compression assessment based on multilevel distortions. J. Appl. Remote Sens. **2014**, *8* (1), 083680.
17. Goovaerts, P. *Geostatistics for Natural Resources Evaluation*; Oxford University Press: New York, 1997.
18. Zabala, A.; Pons, X. Effects of lossy compression on remote sensing image classification of forest areas. Int. J. Appl. Earth Obs. Geoinf. **2011**, *13* (1), 43–51.
19. Zabala, A.; Pons, X. Impact of lossy compression on mapping crop areas from remote sensing. Int. J. Remote Sens. **2013**, *34* (8), 2796–2813.
20. Zabala, A.; Cea, C.; Pons, X. Segmentation and thematic classification of color orthophotos over non-compressed and JPEG 2000 compressed images. Int. J. Appl. Earth Obs. Geoinf. **2012**, *15*, 92–104.

BIBLIOGRAPHY

1. Barni, M. *Document and Image Compression*; Taylor & Francis: New York, 2006.
2. Campbell, J.B. *Introduction to Remote Sensing*, 2nd Ed.; Taylor & Francis: London, 1996.
3. Huang, B. *Satellite Data Compression*; Springer Science+Business Media: New York, 2011.
4. Lillesand, T.; Kiefer, R.W.; Chipman, J. *Remote Sensing and Image Interpretation*, 4th Ed.; John Wiley & Sons: New York, 2000.
5. Motta, G.; Rizzo, F.; Storer, J.A. *Hyperspectral Data Compression*; Springer Science+Business Media: New York, 2006.
6. O'Sullivan, D.; Unwin, D. *Geographic Information Analysis*; John Wiley & Sons: Hoboken, NJ, 2002.
7. Peisheng, Z.; Liping, D. *Geospatial Web Services: Advances in Information Interoperability*; Information Science Reference: New York, 2010.
8. Sinha, A.K.; Arctur, D.; Jackson, I.; Gundersen, L.C. *Societal Challenges and Geoinformatics*; Geological Society of America: Denver, CO, 2011.
9. Taubman, D.; Marcellin, M. *JPEG2000: Image Compression Fundamentals, Standards and Practice*; Kluwer Academic Publishers: Boston, MA, 2001.

Scattering Features of Complex Mediums

P. K. Choudhury
Institute of Microengineering and Nanoelectronics, Universiti Kebangsaan Malaysia, Bangi, Malaysia

N. Iqbal
Institute of Microengineering and Nanoelectronics, Universiti Kebangsaan Malaysia, Bangi, Malaysia; Department of Physics, University of Gujrat, Hafiz Hayat Campus, Pakistan

Abstract

Image processing techniques involve the scattering characteristics of objects in order to extract the information. This entry focuses on the scattering response of complex mediums composed of metal and chiral medium-coated objects, and chiro-ferrites. Physical optics and image theory have been exploited. The tuning feature of the scatterer would be useful in stealth technology and other applications.

Keywords: Complex mediums; Electromagnetic waves; Scattering.

INTRODUCTION

Image processing techniques involve the scattering characteristics of objects in order to extract the information. In this stream, the shooting and bouncing ray (SBR) tracing is a well-known asymptotic method used at high-frequency values. It is basically exploited for scattering-related problems for the determination of radar cross section (RCS) and formation of image of temporarily shaped electrically large objects.[1] This technique was first reported by Ling et al. to estimate the RCS of arbitrarily shaped cavities.[2] Further, the XPATCH software was developed for the calculation of RCS, which is based on the SBR.[3]

The procedure adopted for SBR method has three steps: ray tracing, field tracking, and physical optics (PO). In ray tracing algorithm, Persistence of Vision Ray (POV-Ray) tracer (a computer graphic software) efficiently amalgamates the ray tracing module,[4] wherein the information related to the geometry of target is added into the POV-Ray software. A virtual aperture composed of small grids in the direction perpendicular to the incidence ray is then developed. The incidence plane wave can be modeled by sending geometrical optics (GO) rays toward the target. In this way, the incidence ray from the aperture to target is defined, thereby determining its point of incidence on the target; the reflected ray can be obtained by using the GO technique. This procedure (the reflected ray being considered as the incidence ray) is repeated until it leaves the target. Finally, the scattered field can be obtained by using the field tracking and PO techniques. Once the scattered field is obtained, the RCS can then be determined easily.[5]

Some of the well-known existing phenomena in nature, such as the light from the moon, the color of the sky, and the rainbow are the results of the reflection, refraction, and scattering. The interaction of light with particles would determine their characteristics through the study of scattering response (of particles). In other words, the size, shape, and some other properties of particles can be evaluated through the intensity and polarization of the scattered light. In particular, to measure the size and refractive index (RI) of particles, many techniques have been reported. For example, the diameter of wire can be determined upon exploiting the phenomenon of diffraction. The other reported techniques include scattering model, Fraunhofer diffraction, and Lorenz–Mie theory (LMT).[6,7] Among several approaches to study the interaction of light with particles, the scattering of plane waves by regular-shaped particles (such as spheroid, ellipsoid, circular, and elliptical cylinders of infinite length) uses Maxwell's equations and the separation-of-variable method.[8–10] The approach of GO remains suitable for evaluating the scattering response of spherical particles or cylinders of infinite length. On the other hand, for irregular-shaped particles, where the theory of GO is rarely used, suitable numerical approaches have been developed[11–13] to tackle the problems.

The demand for new materials and structures of optical guides has been exponentially increasing as these play important roles in R&D perspective through their physical and chemical characteristics.[14–17] Within the context, bianisotropic mediums, particularly chiral mediums, are the key milestones in the advancement of electromagnetic (EM) features of mediums. Louis Pasteur discovered that some chemical substances make mirror images despite having the identical molecular structure.[18] These *chiral* objects cause alterations of the polarization state of light due to handedness (or chirality) of molecules.[16] Further, the right-handed or left-handed substances represent imbalance in the symmetry of their molecules. Later, Drude[19] presented the constitutive relations for such mediums in which the chirality parameter measures the handedness (or the optical activity). Chirality also provides information on circular dichroism—the feature due to which chiral medium splits a linearly polarized light into left- and right-circularly polarized waves. Also, the chirality parameter (of medium) determines the absorption strengths for these waves while passing through the medium. Interestingly, the fabrication of chiral mediums can be achieved through helices,[18] metal cranks,[20] twisted Swiss rolls,[21] and chiral honeycomb distribution.[22]

The uniaxial anisotropic material is a special case of bianisotropic material in which the medium has only one particular direction (instead of three). These are easier to fabricate and can be designed by putting small helices directed along the z-direction in a host medium.[15] Also, the electric or magnetic z-directed polarization can be produced in these by suitably applied external fields.[23]

The idea of nihility medium, which possesses null-valued constitutive parameters, was put forward in 2001.[24] These have the constitutive relations as $D(r, w)=0$ and $B(r, w)=0$, which yield $\nabla \times E(r, w)=0$. Later on, Tretyakov presented the generalized form of nihility medium as assumed to be the mixture of nihility and chiral inclusions of the same handedness.[25] In these, the real parts of both permittivity and permeability become zero at certain frequency, and the chirality remains nonzero. The energy can propagate through nihility mediums provided $\kappa^2 \leq \varepsilon\mu$—the condition that is required for lossless mediums.[26] Further, these mediums exhibit interesting feature of having negative refraction. Similar to chiral mediums, two waves exist in chiral nihility mediums as well, the one of which is the backward wave, as observed in the case of Veselago medium.[27] The use of chiral nilility mediums in applications related to the group velocity dispersion communication technology was also reported in the literature.[28]

The property of nihility can also be achieved through nonreciprocal and gyrotropic chiral materials as well, where the condition of negative refraction can be met at certain frequencies.[29] Qiu et al.[30] analyzed the nihility effect on chiral medium, where the values of both the permittivity and permeability are assumed to be approximately zero, instead of *exactly* zero. Gyrotropic features can be found in chiro-ferrite mediums and can be exploited in many optical applications, for example, dispersion compensation in optical communication systems.[31]

SCATTERING DUE TO COMPLEX-STRUCTURED OBJECTS

The scattering of light remains greatly important in many research arena, namely signal processing, telecommunications, and microfluidics. Regarding the use of classical theory to model plane wave scattering due to the infinitely large cylinder, LMT remains greatly important. This is, however, applicable only in the cases of objects having regular geometrical shapes, such as the infinitely long circular or elliptical cylinders. According to this theory, the incidence, scattered, and transmitted fields can be written in terms of special eigenfunctions, and the expansion coefficients of the scattered and transmitted fields are calculated by using the boundary conditions. After the evaluation of these constants, all the other physical entities (related to scattering) can be determined. In this connection, one may look at the research reports pivoted to scattering by spherical particles[32] and infinitely long cylinders.[33,34]

The study of scattering characteristics of objects remains useful in radar- and antenna-related applications.[35] Within the context, Wait[36] reported the scattering of plane waves by dielectric cylinder under the assumption of oblique incidence. The study revealed that both the transverse electric (TE) and transverse magnetic (TM) polarized waves remain present in the scattered field under the incidence waves of either TE or TM kind—the phenomenon which is contrary to a normal incidence of light.

Considering the EM behavior of some specialized forms of complex mediums, the use of conducting sheath helix structures was greatly dealt with in the context of guidance of waves.[37–41] The exploitation of such helical structures in achieving slow- and fast-wave propagation was reported in the literature.[42,43] Also, sheath[44] and tape[45] helical structures were tried with liquid crystal-based guides to control the evanescent fields so that the sensitivity of structures can be modified. Apart from these, the implementation of double-helix geometries in liquid crystal optical guides was also investigated in the context of controlling the dispersion and power sustainment features.[46] Within the context, the sheath helix kind of geometry has been found to be prudent to engineer the pattern of scattering characteristics, which are suitable for various optical applications. According to Chen,[47] conducting sheath helix structure behaves as perfectly conducting cylinder in the situation when its orientation remains perpendicular to the axial direction. The study determined bi-scattering echo width, backscattering echo width, current density, and surface charge density due to the sheath helix when illuminated by light waves.

Keeping in mind the role of scattering of waves in image processing techniques, in this section, the scattering response of different kinds of objects, which are assumed to be infinitely long and composed of various materials with different forms of cross-sectional shapes (namely the circular and elliptical), is briefly touched upon.

Case of Chiral Cylinder

The incidence of parallel-polarized plane waves may be considered normally on the surface of circular chiral cylinder (of radius a) in free space.[8] While using

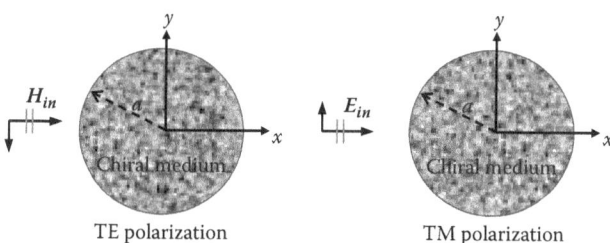

Fig. 1 Plane wave incidence on circular chiral cylinder

the symmetric property, fields corresponding to the perpendicular polarization can also be evaluated. The structures corresponding to both the cases, that is, the incidence of TE (perpendicularly polarized) and TM (parallel polarized) waves are shown in Fig. 1. The EM treatment of such situations essentially needs the use of cylindrical polar coordinate system (r, ϕ, z) due to the geometry of cylinder. Two regions are assumed in the whole system, namely, the incidence and scattered fields in the free space (region I) and the transmitted field in the chiral circular cylinder (region II). The field components in these regions can be written as

$$E_{in}^1 = \tfrac{1}{k_0^2} \sum_{n=-\infty}^{n=\infty} j^n \nabla \times \nabla \times \left[\hat{z} J_n(k_0 r) e^{jn\phi} \right] \quad (1)$$

$$H_{in}^1 = \tfrac{1}{j\omega\mu_0} \sum_{n=-\infty}^{n=\infty} j^n \nabla \times \left[\hat{z} J_n(k_0 r) e^{jn\phi} \right] \quad (2)$$

$$E_s^1 = -\tfrac{1}{k_0} \sum_{n=-\infty}^{n=\infty} j^n$$
$$\left[jC_n \nabla \times \left\{ \hat{z} H_n(k_0 r) e^{jn\phi} \right\} + \tfrac{1}{k_0} B_n \nabla \times \nabla \times \left\{ \hat{z} H_n^1(k_0 r) e^{jn\phi} \right\} \right] \quad (3)$$

$$H_s^1 = -\tfrac{1}{j\omega\mu_0} \sum_{n=-\infty}^{n=\infty} j^n$$
$$\left[\tfrac{1}{k_0} jC_n \nabla \times \nabla \times \left\{ \hat{z} H_n^1(k_0 r) e^{jn\phi} \right\} + B_n \nabla \times \left\{ \hat{z} H_n^1(k_0 r) e^{jn\phi} \right\} \right] \quad (4)$$

$$E_t^2 = \tfrac{1}{k_0} \sum_{n=-\infty}^{n=\infty} j^n D_n$$
$$\left[\nabla \times \left\{ \hat{z} J_n(\gamma_{1c} r) e^{jn\phi} \right\} + \tfrac{1}{k_0} \nabla \times \nabla \times \left\{ \hat{z} J_n(\gamma_{1c} r) e^{jn\phi} \right\} \right] \quad (5)$$

$$H_t^2 = \tfrac{1}{k_0} \sum_{n=-\infty}^{n=\infty} j^n F_n$$
$$\left[\nabla \times \left\{ \hat{z} J_n(\gamma_{2c} r) e^{jn\phi} \right\} - \tfrac{1}{k_0} \nabla \times \nabla \times \left\{ \hat{z} J_n(\gamma_{2c} r) e^{jn\phi} \right\} \right] \quad (6)$$

In these equations, the subscripts i, s, and t represent the incidence, scattered, and transmitted waves, respectively, and the superscripts 1 and 2 stand for the regions $r > a$ (i.e., the free space) and $r < a$ (i.e., inside the chiral cylinder), respectively. Also, $J_n(\cdot)$ and $H_n^1(\cdot)$ are, respectively, Bessel and Henkel functions in cylindrical coordinates system, and B_n, C_n, D_n, and F_n are the unknown constants which can be determined upon using the boundary conditions (i.e., the tangential components of both electric and magnetic fields are continuous). After the evaluation of unknown constants, the scattering and extinction efficiencies can be stated as[6]

$$Q_{sca} = \tfrac{2}{ak_0} \Re \sum_{n=-\infty}^{n=\infty} \left(|B_n|^2 + |C_n|^2 \right) \quad (7)$$

$$Q_{ex} = \tfrac{2}{ak_0} \Re \sum_{n=-\infty}^{n=\infty} \left(B_n + C_n \right) \quad (8)$$

where $\Re$ represents the real part and k_0 is the free space wave number. Within the context, one may consider scattering due to multilayer chiral cylinder of circular cross section in free space; the central region (of cylinder) being composed of either perfectly electrical conducting (PEC) or dielectric medium. The incidence of (parallel or perpendicularly polarized) light happens normally on the surface of cylinder. The scattered field can be determined by using the eigenfunction expansion method under the usage of cylindrical coordinate system. The right- and left-circularly polarized fields exist in each chiral layer, which can be written in terms of unknown constants that can be determined by using the boundary conditions. Finally, the echo widths corresponding to the co- and cross-field components can be evaluated. The numerical results reveal that the plane of polarization of light gets significantly altered due to the chirality of medium.[48] It has been reported before that the chirality parameter plays a determining role in controlling the RCS.[49]

Chiral Nihility Medium-Based Cylinder

One may consider the normal incidence of uniform plane waves over an infinitely long nihility cylinder having circular cross section. Assuming the cylindrical coordinate system (r, ϕ, z), the length of cylinder can be taken parallel to the z-axis. A parallel-polarized plane wave falls normally on the surface of cylinder. For such a boundary value problem, the incidence and scattered field components can be determined, and the scattering and extinction efficiencies may be expressed as follows:[50]

$$Q_{scat} = \tfrac{2}{k_0 a} \sum_{n=-\infty}^{\infty} |A_n|^2 \quad (9)$$

$$Q_{ext} = \tfrac{2}{k_0 a} \Re \left(\sum_{n=-\infty}^{\infty} A_n \right) \quad (10)$$

Here, R represents the real part, and A_n and k_0 are the unknown constant and free space wave number,

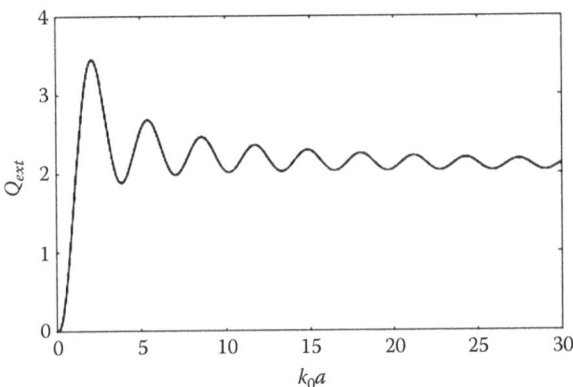

Fig. 2 Plot of extinction efficiency as a function of the dimension of nihility cylinder[50]

respectively. The values of Q_{scat} and Q_{ext} can be evaluated numerically by using the expansion coefficients.

Now, according to the definition of nihility medium, the RI (of medium) becomes vanishing as its material parameters (ε, μ) are null valued.[24] In such a case, the expansion coefficients may be expressed as

$$A_0 = -\frac{J_1(k_0 a)}{H_1(k_0 a)} \qquad (11)$$

$$A_n = -\frac{J_n(k_0 a)}{H_n(k_0 a)} \qquad n \neq 0 \qquad (12)$$

The scattering and extinction efficiencies become equal (i.e., $Q_{scat} = Q_{ext}$) because

$$\left|\frac{J_1(\xi_n)}{H_1(\xi_n)}\right|^2 = \Re\left|\frac{J_1(\xi_n)}{H_1(\xi_n)}\right| \qquad (13)$$

Figure 2 illustrates the plot of extinction efficiency as a function of the size of object.[50] Nihility mediums, under certain circumstances, behave similarly to nonabsorptive materials, and the impedance becomes matched with that of the free space. Similar results can be obtained under the situation of perpendicular polarization of the incidence field.

Applications in Invisibility

Let us consider the case of anisotropic cylinder of circular cross section, which is composed of core and shell, and is illuminated by parallel-polarized plane wave. The field components in each layer of it can be determined in terms of unknown expansion coefficients, the values of which can be obtained by matching the boundary conditions at the layer interfaces. The scattering and extinction efficiencies can be determined using Eqs. 9 and 10.

For this purpose, the effective limit theory under the conditions of long-wavelength limit (i.e., $k_0 b \ll 1$ and $k_s b \ll 1$, where b is the radius of shell, and k0 and ks are wave numbers of free space and shell, respectively) may be implemented. Also, the dominant terms in the effective scattering width are n=0 and n=1. Upon using some approximations, as in Refs. [51–53] along with the long-wavelength limit, one can determine the scattering efficiency as[54]

$$Q_s = \pi^5 \left(\frac{b}{\lambda}\right)^3 \left(\left|\frac{\mu_0 - \mu_{eff}}{\mu_0}\right|^2 = 2\left|\frac{\varepsilon_0 - \varepsilon_{eff}}{\varepsilon_0 - \varepsilon_{eff}}\right|^2\right) \qquad (14)$$

with μ_{eff} and ε_{eff} being the effective permeability and permittivity, respectively.

Here, it is worth noting that, in the long-wavelength limit, the fields become decoupled and the cylinder can be viewed as isotropic, as in the case of scattering efficiency observed in Eq. 14. Furthermore, in achieving invisibility, some constraints in the core–shell ratio must be fulfilled.[54] The results depicted in Fig. 3 reveal that the effective permittivity and scattering width can be controlled through the use of suitable values of core/shell anisotropy. However, it is apparent from the plots that the echo width becomes minimum for a certain ratio of core–shell dimension, especially in the case of variation in anisotropy of the shell (instead of the core), so that good transparency can be achieved. As such, the anisotropy plays a significant role in achieving invisibility.

Considering the importance of the phenomenon of scattering in varieties of scientific and technological applications, such as image processing, antennas, radars,

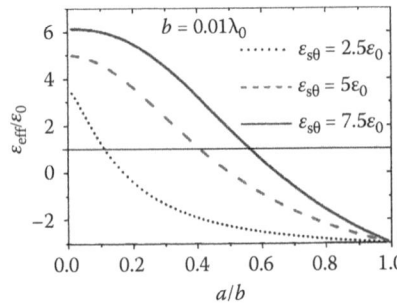
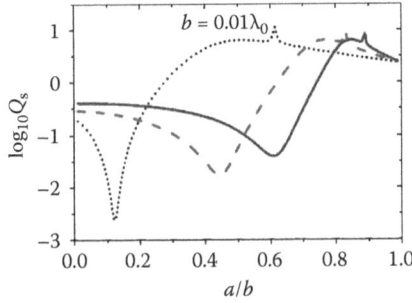

Fig. 3 Plots of (a) effective permittivity and (b) scattering efficiency as a function of core–shell ratio for different values of anisotropic parameter of the shell[54]

and communications, a few kinds of objects are touched upon in this entry with respect to the investigation of their scattering features. In this stream, the dependence of scattering characteristics on the TE and TM polarizations of the incidence waves, the oblique angle of incidence, the orientation of conducting sheath helix structure, and the kind of scattering medium is investigated. For this purpose, the emphasis is given on the mathematical modeling of scattering system/configuration and the numerically obtained results with respect to bi-, back-, and forward-scattering features with respect to the co- and cross-polarized incidence fields. The scattering objects considered in this entry generally include complex-structured mediums illuminated with polarized plane waves under the oblique angle of incidence. Different kinds of scattering configurations are now discussed in the sections that follow.

CONDUCTING SHEATH HELIX AND L-NIHILITY-BASED SILVER METAL CYLINDER

As stated earlier, the artificially designed metamaterials having null-valued permittivity and permeability developed new interests among the researchers. In the constitutive parameters, either or both (permittivity and permeability) can assume vanishing values. The mediums with near-zero RI contain small wave number—the feature that results in small variation in phase over longer dimension of medium.[55] Such properties find applications in highly directive antennas. Studies of these materials include EM energy tunneling in channel at ultrathin sub-wavelength dimension. Further, these are greatly useful in impedance-matched EM components.

Considering the case of silver metal, the real part of its permittivity becomes zero at certain frequency, and the medium then becomes the one with L-nihility.[56,57] With this background, let us first consider the case of silver metal cylinder having circular cross section, which is coated with dielectric material and embedded with conducting sheath helix structure of varieties of orientation.[58] Exploiting the concept of L-nihility, co- and cross-polarized components of scattered fields are analyzed under two states of polarization.[59] The numerically obtained results are compared with those deduced from the case of scattering due to PEC cylinder only. The results reveal that the complex-structured medium with sheath helix orientation parallel to the axial direction becomes identical to the kind of scatterer having the PEC cylinder.

Polarization States

Let us consider a circular cylinder having the cross-sectional radius a, and made of silver metal, as shown in Fig. 4. A dielectric coating is assumed to be over the cylinder, which makes the radius b of the outer surface of the structure. Now, the whole medium is finally wrapped

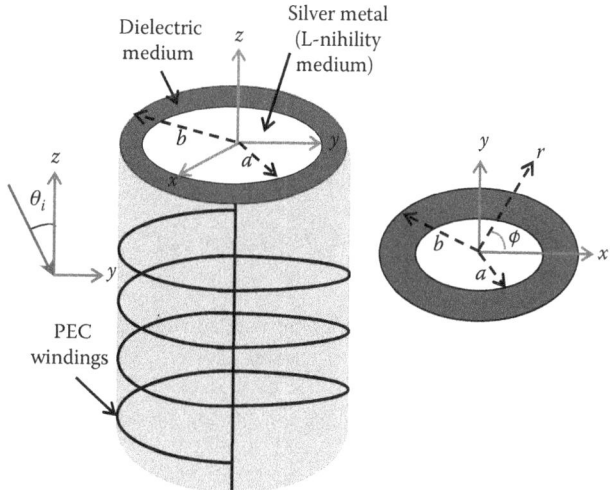

Fig. 4 Schematic of the scatterer

with PEC sheath helix structure, the orientation of which would vary between the two extremes determined by the pitch angle values 0° and 90°. The structure essentially needs the cylindrical polar coordinate system (r, ϕ, z) for analysis; the length of cylinder is taken to be extending along the z-direction, as illustrated in Fig. 4. As the figure depicts, the scatterer is assumed to be composed of three layers, namely, the silver metal $(r<a)$, the dielectric $(a<r<b)$, and the free space $(r>b)$.

The total field inside the free space becomes the superposition of the incidence (E_{0i}, H_{0i}) and scattered (E_{0s}, H_{0s}) fields, whereas within the dielectric and silver metal regions, these are denoted by (E_1, H_1) and (E_2, H_2), respectively. A plane wave is assumed to fall obliquely on the surface of scatterer structure at a certain value of incidence angle θ_i making with the z-axis. The cases of parallel and perpendicular polarizations of incidence waves are separately touched upon.

Parallel Polarization (TM)

We consider the incidence of plane waves with parallel polarization on the surface of cylinder (Fig. 4); the E- and H-fields in this case may be expanded in the form of cylindrical waves as follows:

$$E^i_{0z} = E_0 \sin\theta_i e^{jk_0 z \cos\theta_i} \sum_{n=-\infty}^{n=\infty} j^{-n} J_n(k_0 r \sin\theta_i) e^{jn\phi} \quad (15)$$

$$H^i_{0\phi} = \tfrac{E_0}{j\eta_0} e^{jk_0 z \cos\theta_i} \sum_{n=-\infty}^{n=\infty} j^{-n} J'_n(k_0 r \sin\theta_i) e^{jn\phi} \quad (16)$$

The scattered field components in the free space region consist of both the TM and TE polarizations and can be written as:

$$E^s_{0z} = E_0 \sin\theta_s e^{-jk_0 z \cos\theta_s} \sum_{n=-\infty}^{n=\infty} j^{-n} \chi_{1n} H^2_n(k_0 r \sin\theta_s) e^{jn\phi} \quad (17)$$

$$H_{0\phi}^s = \frac{E_0}{j\eta_0} e^{-jk_0 z \cos\theta_s} \sum_{n=-\infty}^{n=\infty} j^{-n} \mathcal{X}_{1n} H_n^{2'}(k_0 r \sin\theta_s) e^{jn\phi} \quad (18)$$

$$H_{0z}^s = \frac{E_0}{j\eta_0} e^{-jk_0 z \cos\theta_s} \sum_{n=-\infty}^{n=\infty} j^{-n} \mathcal{X}_{2n} H_n^2(k_0 r \sin\theta_s) e^{jn\phi} \quad (19)$$

$$E_{0\phi}^s = \frac{E_0}{\sin\theta_s} e^{-jk_0 z \cos\theta_s} \sum_{n=-\infty}^{n=\infty} j^{-n} \mathcal{X}_{2n} H_n^{2'}(k_0 r \sin\theta_s) e^{jn\phi} \quad (20)$$

Now, the E- and H-field components inside the dielectric region ($a < r < b$) can be expressed as

$$E_{1z} = E_0 \sin\theta_r e^{jk_1 z \cos\theta_r}$$
$$\sum_{n=-\infty}^{n=\infty} j^{-n} \left[\mathcal{X}_{3n} H_n^2(k_1 r \sin\theta_r) + \mathcal{X}_{4n} H_n^1(k_1 r \sin\theta_r) \right] e^{jn\phi} \quad (21)$$

$$H_{1\phi} = \frac{E_0}{j\eta_1} e^{jk_1 z \cos\theta_r}$$
$$\sum_{n=-\infty}^{n=\infty} j^{-n} \left[\mathcal{X}_{3n} H_n^{2'}(k_1 r \sin\theta_r) + \mathcal{X}_{4n} H_n^{1'}(k_1 r \sin\theta_r) \right] e^{jn\phi} \quad (22)$$

$$H_{1z} = -\frac{E_0}{j\eta_1} e^{jk_1 z \cos\theta_r}$$
$$\sum_{n=-\infty}^{n=\infty} j^{-n} \left[\mathcal{X}_{5n} H_n^2(k_1 r \sin\theta_r) + \mathcal{X}_{6n} H_n^1(k_1 r \sin\theta_r) \right] e^{jn\phi} \quad (23)$$

$$E_{1\phi} = -\frac{E_0}{\sin\theta_r} e^{jk_1 z \cos\theta_r}$$
$$\sum_{n=-\infty}^{n=\infty} j^{-n} \left[\mathcal{X}_{5n} H_n^{2'}(k_1 r \sin\theta_r) + \mathcal{X}_{6n} H_n^{1'}(k_1 r \sin\theta_r) \right] e^{jn\phi} \quad (24)$$

Similarly, the E- and H-field components inside the metallic region ($r < a$) can be written as:[60]

$$E_{2z} = E_0 \sin\theta_r e^{-(jk_{2s} z \cos\psi_{2s} + 2p_s)}$$
$$\sum_{n=-\infty}^{n=\infty} j^{-n} \mathcal{X}_{7n} J_n(k_{2s} r \sin\psi_{2s}) e^{jn\phi} \quad (25)$$

$$H_{2\phi} = \frac{E_0 \omega \varepsilon_{2s} k_{2s} \sin\psi_{2s} \sin\theta_r}{j(\omega^2 \mu_0 \varepsilon_{2s} - k_{2s}^2 \cos^2\psi_{2s})} e^{-(jk_{2s} z \cos\psi_{2s} + 2p_s)}$$
$$\sum_{n=-\infty}^{n=\infty} j^{-n} \mathcal{X}_{7n} J_n'(k_{2s} r \sin\psi_{2s}) e^{jn\phi} \quad (26)$$

$$H_{2z} = -\frac{E_0}{j\eta_{2s}} e^{-(jk_{2s} z \cos\psi_{2s} + 2p_s)}$$
$$\sum_{n=-\infty}^{n=\infty} j^{-n} \mathcal{X}_{8n} J_n(k_{2s} r \sin\psi_{2s}) e^{jn\phi} \quad (27)$$

$$E_{2\phi} = -\frac{E_0 k_{2s} \sin\psi_{2s}}{j\eta_{2s}(j\omega\varepsilon_{2s} - k_{2s}^2 \cos^2\psi_{2s})} e^{-(jk_{2s} z \cos\psi_{2s} + 2p_s)}$$
$$\sum_{n=-\infty}^{n=\infty} j^{-n} \mathcal{X}_{8n} J_n'(k_{2s} r \sin\psi_{2s}) e^{jn\phi} \quad (28)$$

In the preceding equations, E_0, η_0, and k_0 are the amplitude of incidence electric field, wave impedance, and wave number in free space, respectively. $J_n(\cdot)$, $H_n^1(\cdot)$, and $H_n^2(\cdot)$ indicate Bessel function of the first kind, and Henkel function of the first and second kinds, respectively; prime being the representation of differentiation with respect to the argument. The quantities $\mathcal{X}_{1n}, \mathcal{X}_{2n}, \mathcal{X}_{3n}, \mathcal{X}_{4n}, \mathcal{X}_{5n}, \mathcal{X}_{6n}, \mathcal{X}_{7n}$ and $\mathcal{X}_{8n}$ are the unknown constants, and θ_s and θ_r represent angles of scattering and refraction in the free space and dielectric regions, respectively. Further, $k_1 (= \omega\sqrt{\mu_0 \varepsilon_1})$, $\eta_1 (= \sqrt{\mu_0/\varepsilon_1})$, and ε_1 are, respectively, the wave number, wave impedance, and relative permittivity inside the dielectric medium, and $\eta_{2s} (= \sqrt{\mu_0/\varepsilon_{2s}})$, ε_{2s}, and σ_s are the respective values of wave impedance, relative permittivity, and conductivity of silver metal. The other used terms have meanings as follows:

$$\sin\theta_t = \left(\frac{jk_1}{\alpha_2 + j\beta_2}\right) \sin\theta_r \quad (29)$$

$$\cos\theta_t = (1 - \sin^2\theta_t)^{1/2} = s_1 e^{j\xi_s} \quad (30)$$

$$p_s = s_1(\alpha_2 \cos\xi_s - \beta_2 \sin\xi_s) \quad (31)$$

$$q_s = s_1(\alpha_2 \sin\xi_s - \beta_2 \cos\xi_s) \quad (32)$$

$$k_{2s} = \{(k_{1s} \sin\theta_r)^2 + q_s^2\}^{1/2} \quad (33)$$

$$\psi_{2s} = \tan^{-1}\left(\frac{k_{1s} \sin\theta_r}{q_s}\right) \quad (34)$$

$$\alpha_{2s} = \omega\sqrt{\mu_0 \varepsilon_{2s}} \left[\frac{1}{2}\left\{\sqrt{1 + \left(\frac{\sigma_s}{\omega\varepsilon_{2s}}\right)^2} - 1\right\}\right]^{1/2} \quad (35)$$

$$\beta_{2s} = \omega\sqrt{\mu_0 \varepsilon_{2s}} \left[\frac{1}{2}\left\{\sqrt{1 + \left(\frac{\sigma_s}{\omega\varepsilon_{2s}}\right)^2} - 1\right\}\right]^{1/2} \quad (36)$$

$$\eta_{2s} = \sqrt{\frac{j\omega\mu_0}{\sigma_s + j\omega\varepsilon_{2s}}} \quad (37)$$

Now, one may implement the boundary conditions at the interfaces $r = a$ and $r = b$; the corresponding equations may be expressed as:

$$\left. \begin{array}{ll} E_{1z} - E_{2z} = 0, & E_{1\phi} - E_{2\phi} = 0 \\ H_{1z} - H_{2z} = 0, & H_{1\phi} - H_{2\phi} = 0 \end{array} \right|_{r=a} \quad (38)$$

$$\left. \begin{array}{c} (E_{0z}^i + E_{0z}^s)\sin\psi + E_{0\phi}^s \cos\psi = 0, \quad E_{1z}\sin\psi - E_{1\phi}\cos\psi = 0 \\ (E_{0z}^i + E_{0z}^s - E_{1z})\cos\psi - (E_{0\phi}^s + E_{1\phi})\sin\psi = 0 \\ (H_{0z}^s - H_{1z})\sin\psi + (H_{0\phi}^i + H_{0\phi}^s - H_{1\phi})\cos\psi = 0 \end{array} \right|_{r=b}$$
$$(39)$$

By using the values of field components in the previously mentioned equations in the boundary conditions, one gets eight simultaneous equations, which can be stated in the matrix form given below; the solution to those equations results in obtaining the values of unknown constants used earlier.

$$
\begin{bmatrix} 0 \\ 0 \\ 0 \\ 0 \\ a_5 \\ 0 \\ a_7 \\ a_8 \end{bmatrix} = \begin{bmatrix} 0 & 0 & a_{13} & a_{14} & 0 & 0 & a_{17} & 0 \\ 0 & 0 & a_{23} & a_{24} & 0 & 0 & 0 & a_{28} \\ 0 & 0 & 0 & 0 & a_{35} & a_{36} & 0 & a_{38} \\ 0 & 0 & a_{43} & a_{44} & 0 & 0 & a_{47} & 0 \\ a_{51} & a_{52} & 0 & 0 & 0 & 0 & 0 & 0 \\ 0 & 0 & a_{63} & a_{64} & a_{65} & a_{66} & 0 & 0 \\ a_{71} & a_{72} & a_{73} & a_{74} & a_{75} & a_{76} & 0 & 0 \\ a_{81} & a_{82} & a_{83} & a_{84} & a_{85} & a_{86} & 0 & 0 \end{bmatrix} \begin{bmatrix} \mathcal{X}_{1n} \\ \mathcal{X}_{2n} \\ \mathcal{X}_{3n} \\ \mathcal{X}_{4n} \\ \mathcal{X}_{5n} \\ \mathcal{X}_{6n} \\ \mathcal{X}_{7n} \\ \mathcal{X}_{8n} \end{bmatrix}
$$

(40)

The symbols used in Eq. 40 have specific meanings as given in the Appendix-A.

Perpendicular Polarization (TE)

The TE polarization of incidence wave is now assumed. The field components in each layer of the scattering object (as in Fig. 4) can be determined by using the duality transformation.[10] The boundary conditions in the case of TE polarization may be modified as follows:

$$
\left. \begin{array}{ll} E_{1z} - E_{2z} = 0, & E_{1\phi} - E_{2\phi} = 0 \\ H_{1z} - H_{2z} = 0, & H_{1\phi} - H_{2\phi} = 0 \end{array} \right|_{r=a}
$$

(41)

$$
\left. \begin{array}{c} E_{0z}^s \sin\psi + \left(E_{0\phi}^i + E_{0\phi}^s\right)\cos\psi = 0, \quad E_{1z}\sin\psi - E_{1\phi}\cos\psi = 0 \\ \left(E_{0z}^s - E_{1z}\right)\cos\psi - \left(E_{0\phi}^i + E_{0\phi}^s - E_{1\phi}\right)\sin\psi = 0 \\ \left(H_{0z}^i + H_{0z}^s - H_{1z}\right)\sin\psi + \left(H_{0\phi}^s - H_{1\phi}\right)\cos\psi = 0 \end{array} \right|_{r=b}
$$

(42)

Upon substituting the values of field components in the preceding equations related to the boundary conditions, one gets eight equations, which can be written in the matrix form as

$$
\begin{bmatrix} 0 \\ 0 \\ 0 \\ 0 \\ b_5 \\ 0 \\ b_7 \\ b_8 \end{bmatrix} = \begin{bmatrix} 0 & 0 & 0 & 0 & b_{15} & b_{16} & 0 & b_{18} \\ 0 & 0 & b_{23} & b_{24} & 0 & 0 & b_{27} & 0 \\ 0 & 0 & 0 & 0 & b_{35} & b_{36} & b_{37} & 0 \\ 0 & 0 & 0 & 0 & b_{45} & b_{46} & 0 & b_{48} \\ b_{51} & b_{52} & 0 & 0 & 0 & 0 & 0 & 0 \\ 0 & 0 & b_{63} & b_{64} & b_{65} & b_{66} & 0 & 0 \\ b_{71} & b_{72} & b_{73} & b_{74} & a_{75} & a_{76} & 0 & 0 \\ b_{81} & b_{82} & b_{83} & b_{84} & a_{85} & a_{86} & 0 & 0 \end{bmatrix} \begin{bmatrix} \mathcal{X}'_1 \\ \mathcal{X}'_2 \\ \mathcal{X}'_3 \\ \mathcal{X}'_4 \\ \mathcal{X}'_5 \\ \mathcal{X}'_6 \\ \mathcal{X}'_7 \\ \mathcal{X}'_8 \end{bmatrix}
$$

(43)

In the preceding equation, $\mathcal{X}'_1, \mathcal{X}'_2, \mathcal{X}'_3, \mathcal{X}'_4, \mathcal{X}'_5, \mathcal{X}'_6, \mathcal{X}'_7$, and $\mathcal{X}'_8$ are the unknown constants with the prime indicating distinction from the constants used in the case of TM-polarized incidence. The other used symbols have meanings as stated in the Appendix-B.

The echo width is defined as the ratio of scattered power to the incidence power and can be explicitly stated for the co- and cross-polarized field components under the far-zone condition as follows:

$$
\frac{\sigma_{co}}{\lambda} = \frac{2}{\pi} \left| \sum_{n=-\infty}^{n=\infty} \left(\cos\Gamma_s \mathcal{X}_{1n} + \sin\Gamma_s \mathcal{X}'_{1n} \right) e^{jn\phi} \right|^2
$$

(44)

$$
\frac{\sigma_{cross}}{\lambda} = \frac{4}{\pi} \left| \sum_{n=-\infty}^{n=\infty} \left(\cos\Gamma_s \mathcal{X}_{2n} + \sin\Gamma_s \mathcal{X}'_{2n} \right) e^{jn\phi} \right|^2
$$

(45)

In the preceding equations, the quantity Γ_s can take values as $0°$ and $90°$ for the cases of TM- and TE-polarized incidence, respectively. The constants $\mathcal{X}_{1n}, \mathcal{X}'_{1n}, \mathcal{X}_{2n}$, and $\mathcal{X}'_{2n}$ can be determined from the solutions of simultaneous equations given in the matrix. Finally, the results corresponding to the co- and cross-polarized fields in both the TM and TE cases are obtained from the previous equations, which are then compared with the respective situation when the scatterer is in the form of PEC cylinder having a radius b.[10]

Scattering Analysis of the Structure

For the analytical purpose of the system configuration shown in Fig. 4, the parametric values of dimension are taken to be $a=0.6\lambda$ and $b=0.8\lambda$, which are the inner and outer radii of cylinder, respectively. The relative permittivity of silver metal has the form

$$
\varepsilon_2 = 5.7 + 0.4j - \left(\frac{9q}{2\pi c \hbar}\right)^2 \lambda^2
$$

(46)

where q is electronic charge (i.e., 1.6022×10^{-19} Coulomb), c is the speed of light in free space, and $\hbar$ is Planck's constant (i.e., 1.0546×10^{-34} J·s). For the L-nihility cylinder,[57] the free space wavelength is assumed to be $\lambda = 329$ nm so that $Re(\varepsilon) = 0$. This is because, according to the definition of L-nihility cylinder, the value of wavelength λ is chosen in a way that the real part of ε_2 becomes zero (i.e., $\Re(\varepsilon_2) = 0$ at $\lambda = 329$ nm, with R being the representation of the real part). The RI n_1 of dielectric medium is assumed to be 1.54.

Now, the RCS (or the echo width σ/λ) is analyzed, assuming the co- and cross-polarized field components under the consideration of TE- and TM-polarized incidence waves. For this purpose, the variations in helix pitch angle and oblique angle of incidence are taken into account. We assume that the width of sheath helix medium used in the structure approaches zero (i.e., the wire is too thin to have appreciable amount of surface current) and also the two successive helical windings are isolated from each other. The presence of conducting sheath helix in the structure yields some useful results. As the helix pitch angle ψ approaches 0° (i.e., the helical wraps become perpendicular to the direction of wave propagation), the echo width becomes weak. On the other hand, when ψ becomes close to 90°, the scatterer acts as a perfectly conducting cylinder.[47] The feature of anisotropic objects remains that the mutual coupling of different polarization occurs at its surface, which is contrary to the far-zone region.

In this work, the analysis is carried out in two ways—first, the echo width is studied against the azimuthal angle ϕ under the assumption of three different values of oblique angle of incidence while keeping the other parameters fixed, and second, the echo width is plotted against the incidence angle θ_i while considering the same values for the other parameters. Also, the values of azimuthal and incidence angles are taken in the ranges as $0° \leq \phi \leq 360°$ and $0° \leq \theta_i \leq 75°$, respectively.

Bistatic Echo Width as a Function of Azimuthal Angle

The normalized echo widths σ/λ for different values of helix pitch angle ψ and the angle of incidence θ_i are evaluated, and compared with those obtained for the system containing scatterer composed of PEC cylinder only. In order to plot the normalized echo width, we consider three different values of pitch angle of the sheath helix structure, namely, 0°, 45°, and 90°. The results with respect to the normalized echo widths are illustrated corresponding to the two situations—first, when the scattering of waves takes place under the configuration of the object depicted in Fig. 4, and second, when the scattering happens due to the PEC nature of cylinder.

Figure 5 exhibits the bi-scattering echo width obtained for the TE and TM polarizations under the assumption of three different values of oblique angle of incidence θ_i, namely, 15°, 30°, and 45°; the respective results are illustrated in Fig. 5a–c. It is noticed that the echo width gets altered with the variation in incidence angle; it approaches the maximum corresponding to the value of θ_i as 30°, and in the case of TM polarization (Fig. 5b). The echo width becomes the smallest in the case of $\theta_i = 15°$ for the same state of polarization (i.e., the TM). As to the case of TE polarization, the maximum normalized echo width is noticed corresponding to $\theta_i = 15°$, and it decreases with the increase in the angle of incidence. It is also found that the echo width due to the PEC cylinder remains close to that obtained from the structure (of Fig. 4) having the helix

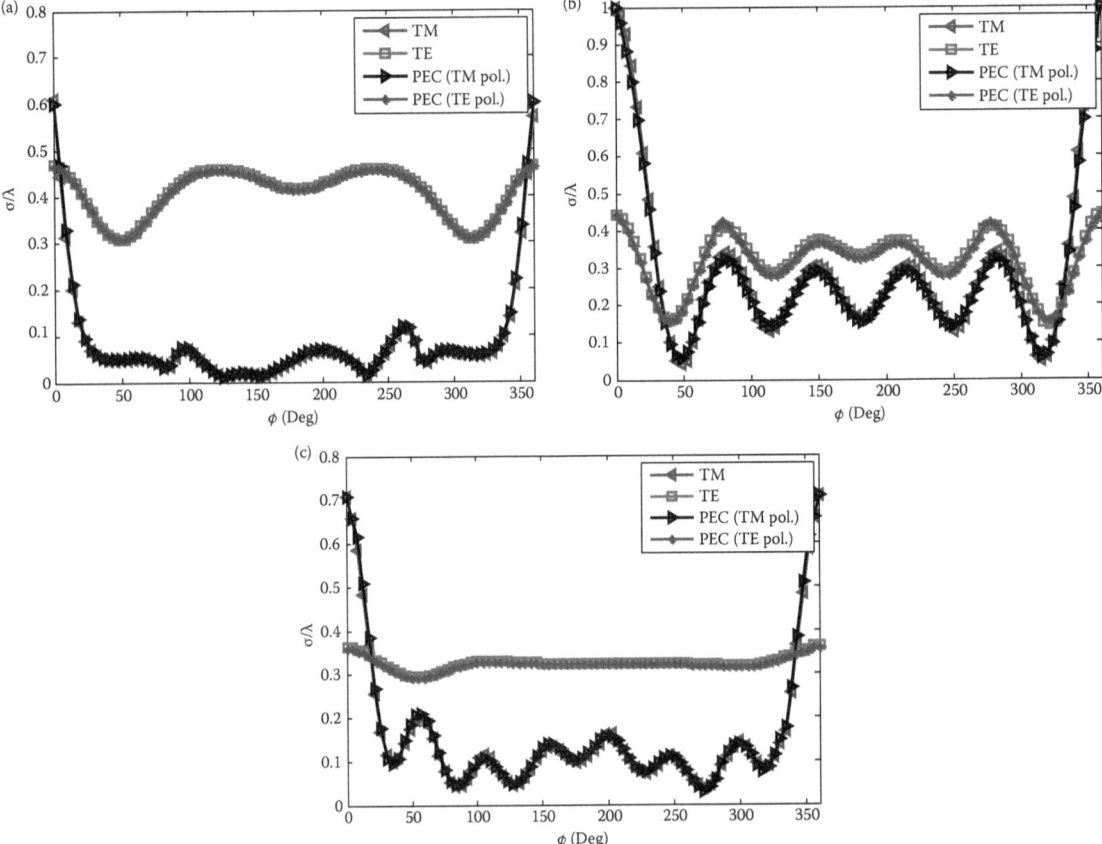

Fig. 5 Plots of the normalized echo width σ/λ against ϕ corresponding to the angles of incidence θ_i as (a) 15°, (b) 30°, and (c) 45° under the assumption of ψ as 90° and only the PEC kind of cylinder

pitch angle value as 90°. As such, the structure of L-nihility cylinder behaves similarly to the PEC cylinder when the conducting sheath helix (present on its surface) remains oriented parallel to the axial direction (i.e., $\psi=90°$).

Now, in the context of impact on the normalized echo width due to the other values of helix pitch angle, the co- and cross-polarized waves under the consideration of TM and TE polarizations are taken into account. In this stream, the results in Fig. 6a and b correspond to the case of TM polarization, whereas those in Fig. 6c and d represent the TE-polarized case. Further, two values of the angle of incidence θ_i, namely, 15° and 45°, are assumed along with two values of the helix pitch angle ψ as 0° and 45°. It is clearly noticed that the scatterer structure under consideration with the helix pitch angle value other than 90° no more remains identical to a PEC cylinder—the feature drawn from the characteristics of the normalized echo width obtained in this case. Here, we consider the results obtained from the structure in Fig. 4 only, whereas those for the PEC cylinder are not incorporated.

From the plots of TM polarization case, it is observed that the normalized echo width remains maximum corresponding to the co-polarized component under the incidence angle $\theta_i=15°$ and the helix pitch angle $\psi=45°$; the case of cross-polarization component under the same conditions of incidence and pitch angles exhibits small echo width. In this situation, the increase of θ_i to 45° (keeping pitch angle ψ fixed to 45°) results in increase in echo width for the co-polarized field components. In the case of TE polarization, the results corresponding to the plots of σ/λ are illustrated in Fig. 6c and d. It is observed from these graphs that the echo width for the co-polarized case becomes high in the situation of helix pitch angle as 0° and the incidence angle as 15°. Thus, the variations in incidence angle along with the sheath helix orientation significantly affect the scattering cross section. Upon comparing the results obtained due to both the TE and TM polarizations, it is noticed that the effect due to the TM-polarized field on the normalized echo width remains more dominant under various situations of incidence and pitch angles than what has been achieved in the case of TE-polarized incidence field.

Bistatic Echo Width as a Function of Incidence Angle

Now, the normalized bi-scattering echo width is plotted against the incidence angle θ_i upon keeping the azimuthal angle ϕ fixed. Figure 7a and b, respectively, represent the

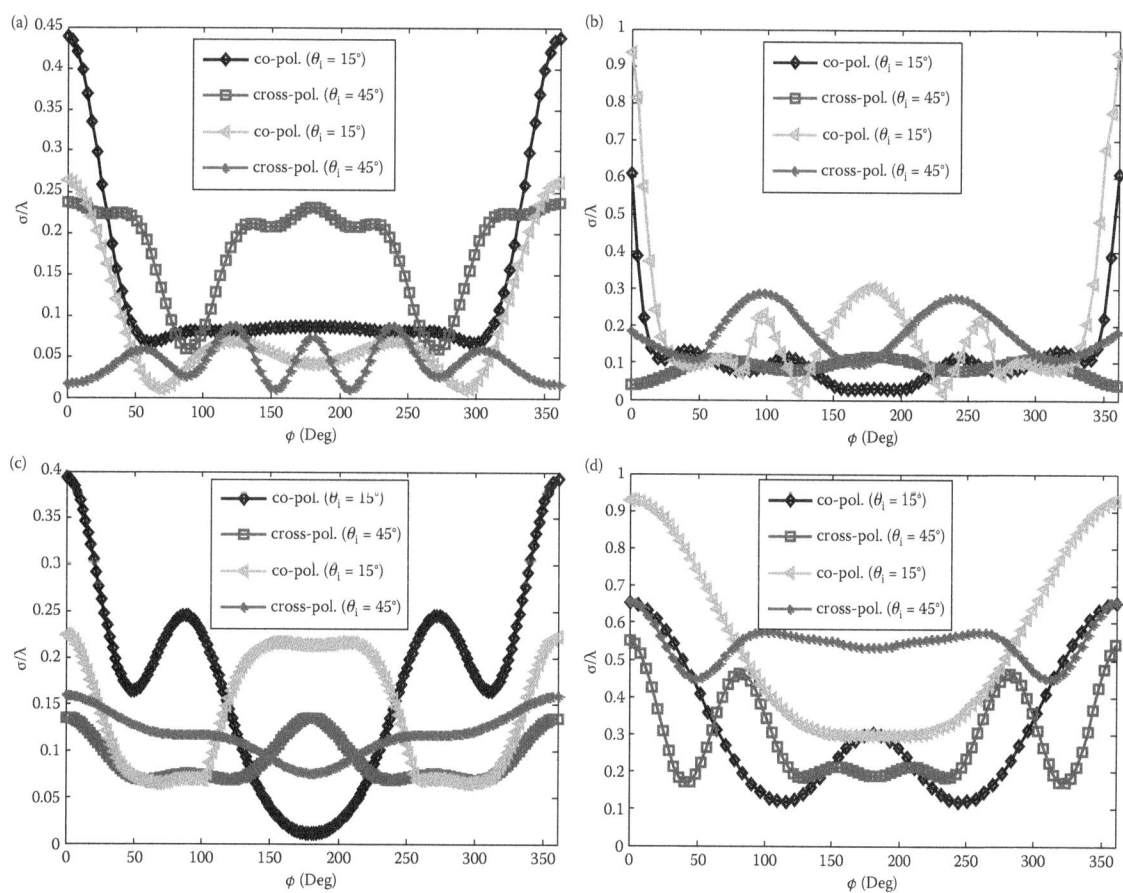

Fig. 6 Plots of normalized echo width σ/λ against ϕ corresponding to (a) $\psi=0°$, (b) $\psi=45°$ under the TM-polarized incidence field, and (c) $\psi=0°$, (d) $\psi=45°$ under the TE-polarized incidence

plots of the normalized bi-scattered echo widths for the values of ϕ as 0° and 90°; in each case, computations are made corresponding to both the TM- and TE-polarized incidence fields. The normalized echo widths are determined for two situations: (i) when the conducting sheath helix structure assumes orientation parallel to the axial direction and (ii) the case of scatterer structure containing the PEC cylinder only.

The results demonstrate that the echo width remains high corresponding to $\phi=0°$ under the situation of TM polarization compared to the case of TE-polarized incidence. This is in contrast to the situation corresponding to $\phi=90°$ when large echo width is observed in the case of TE-polarized incidence. As such, it can be inferred that the scatterer structure of Fig. 4, under the helix pitch angle value 90°, behaves as PEC cylinder irrespective of the values of incidence and azimuthal angles.

Apart from the case of helical wraps being parallel to the direction of wave propagation, the other values (namely, $\psi=0°$ and 45°) of sheath helix orientation are also taken into account. Considering the co- and cross-polarized field components along with both the states of polarization (i.e., the TE and the TM), the normalized echo width is evaluated; the corresponding results are illustrated in Fig. 8. It is observed that, in the situation of TM polarization, the echo width for the co-polarized components remains dominant at $\phi=0°$ and $\psi=0°$, and also the cross-polarized component is noticed to be the maximum under similar conditions (Fig. 8a and b). In the case of TE-polarized incidence, the situation generally seems to be reversed (to the case of TM polarization), as becomes clear from the plots in Fig. 8c and d.

Current Density Distributions

The current density distribution along the sheath helix structure mounted over the L-nihility silver metal cylinder coated with dielectric medium can be evaluated.[61] For this purpose, a uniform plane wave of parallel polarization is assumed to fall obliquely on the surface of scatterer. The following condition in which the magnetic field becomes discontinuous across the sheath helix is applied; the condition is

$$\boldsymbol{K}(b,\phi) = \hat{r} \times \left[\left(\boldsymbol{H}_0^i + \boldsymbol{H}_0^s \right) - \left(\boldsymbol{H}_1 \right) \right]_{r=b} \tag{47}$$

Upon substituting the values of magnetic field components, which were determined earlier under the situation of parallel-polarized incidence wave, one can determine the amplitude and phase of current distribution numerically along the sheath helix structure (Fig. 4).

For numerical computations of the normalized current density distribution (NCDD), two different values of the angle of incidence θ_i (namely 30° and 45°) and three different values of the sheath helix orientation ψ (namely 0°, 45°, and 90°) are taken into account. The parametric values of scatterer structure with respect to the dimension (i.e., the inner and outer radii) are kept the same as considered before (i.e., $a=0.6\lambda$ and $b=0.8\lambda$). The real part of relative permittivity becomes zero corresponding to the wavelength value $\lambda=329$ nm (which is essentially by the definition of L-nihility medium, as described before in Section "Chiral Nihility Medium-Based Cylinder"), and the dielectric RI n_1 is assumed to be 1.54. The current density distribution is evaluated as a function of ϕ in the range $0° \leq \phi \leq 180°$; the case of $\phi=0°$ corresponds to the forward scattering and $\phi=180°$ shows the backscattering. The direction of current can be expressed along the unit tangent to the sheath helix as

$$\hat{T} = \frac{\hat{\phi} + \varphi \hat{z}}{\left(1+\varphi^2\right)^{1/2}} \tag{48}$$

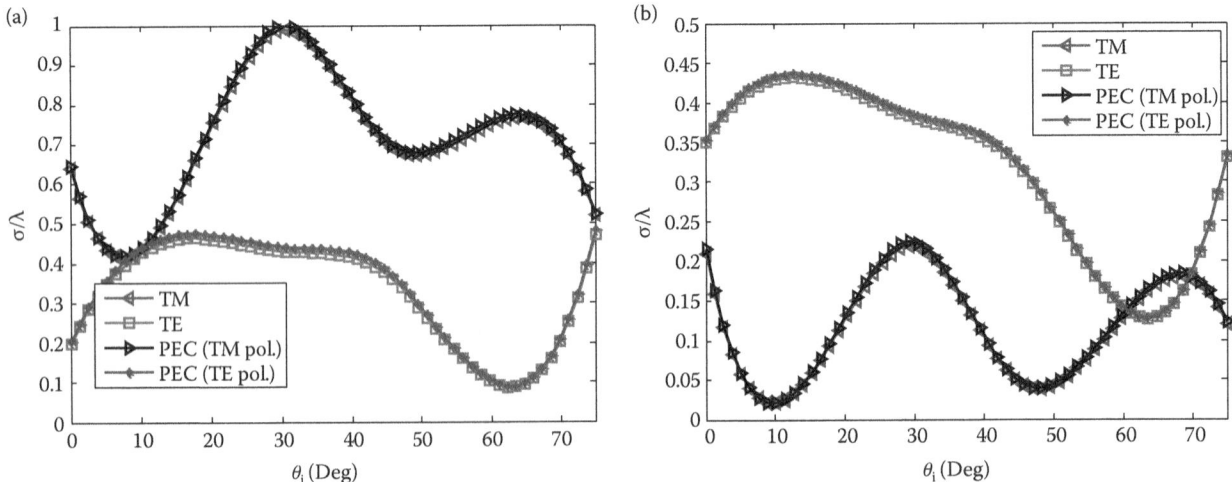

Fig. 7 Plots of normalized echo width σ/λ against θ_i corresponding to the azimuthal angle values (a) $\phi=0°$ and (b) $\phi=90°$ when the scatterer structure has $\psi=90°$, and only the PEC kind of cylinder is considered

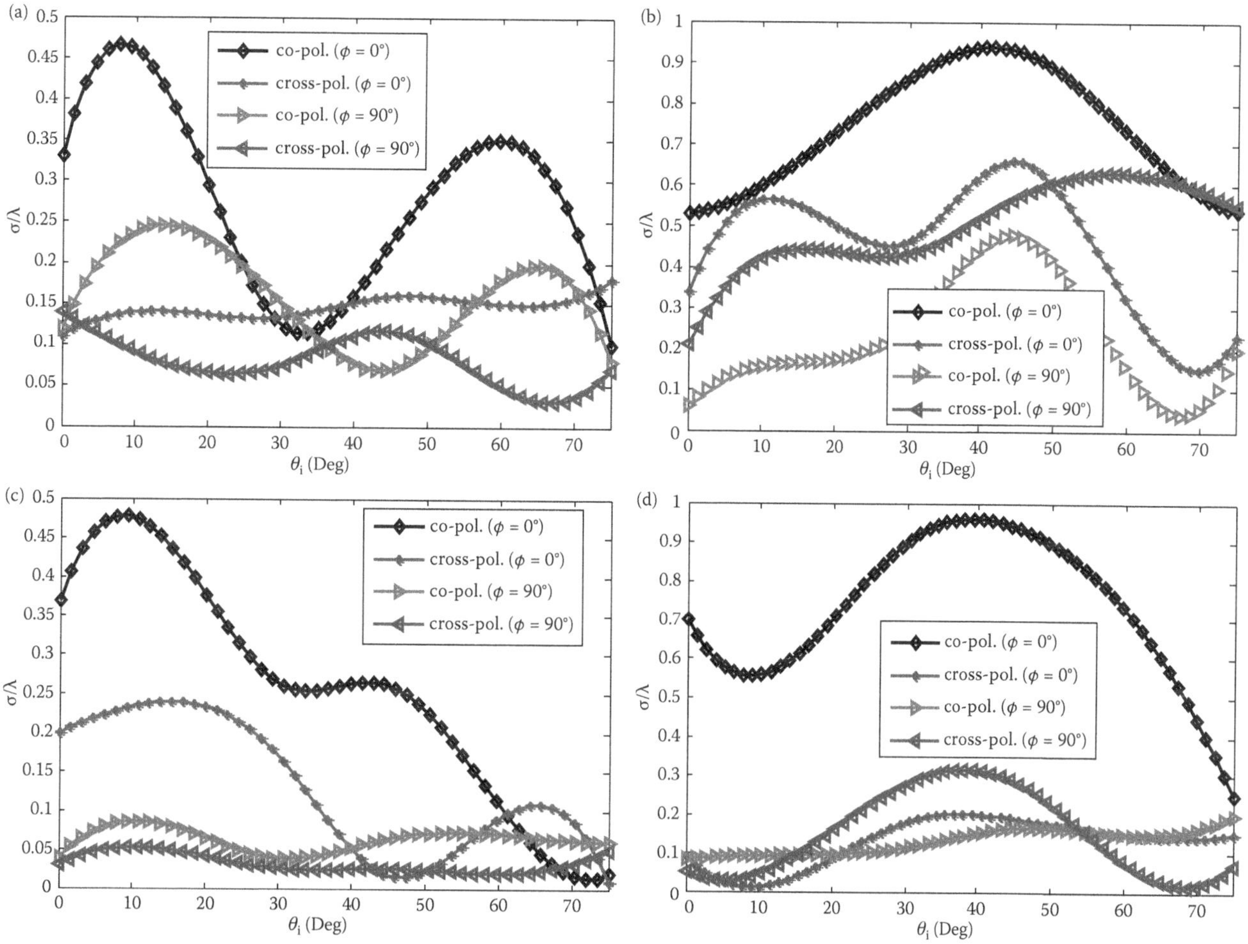

Scattering Features of Complex Mediums

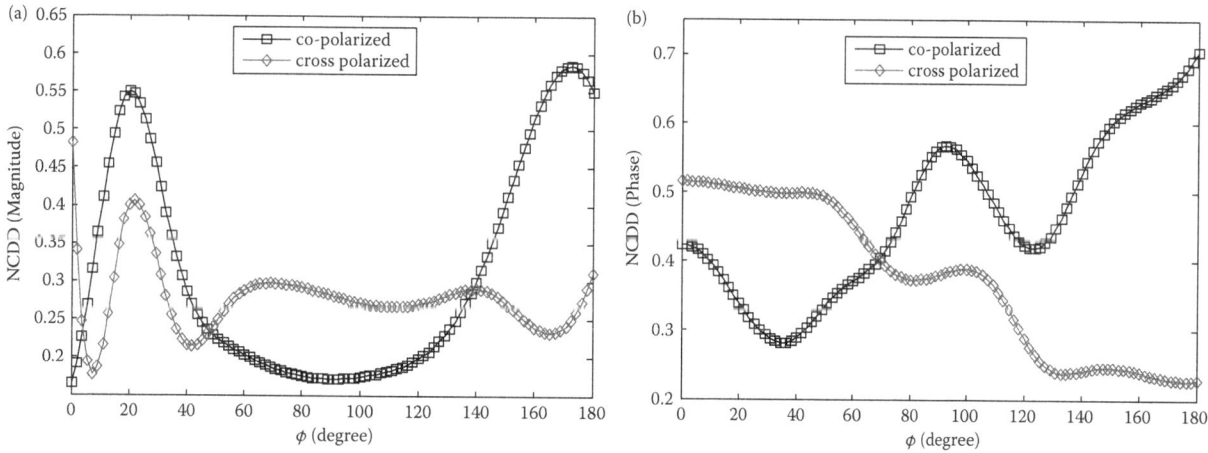

Fig. 8 Plots of normalized echo width σ/λ against θ_i corresponding to the helix pitch angle (a) $\psi=0°$, (b) $\psi=45°$ under the TM polarization and (c) $\psi=0°$, (d) $\psi=45°$ under the TE polarization

Fig. 9 Plots of NCDD (a) magnitude and (b) phase versus ϕ corresponding to $\theta_i=30°$ and $\psi=0°$

with $\varphi=p/2\pi b=\tan\psi$.[47] The amplitude and phase of the normalized current density are discussed individually.

Figure 9a and b, respectively, illustrate the plots of magnitude and phase of NCDD corresponding to the co- and cross-polarized waves against the angle ϕ along the sheath helix structure when the incidence angle is $\theta_i=30°$ and the helix orientation angle is $\psi=0°$. The case of parallel polarization of the incidence wave is considered in this case. Figure 9a shows two peaks in the amplitude pattern of NCDD with the maxima at ϕ values as 20° and 173° in

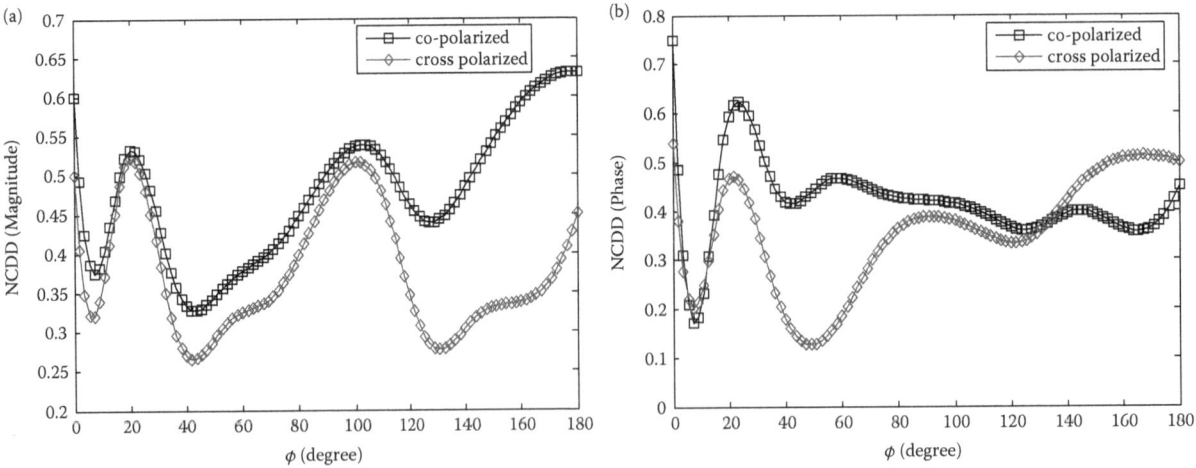

Fig. 10 Plots of NCDD (a) magnitude and (b) phase versus ϕ corresponding to $\theta_i = 30°$ and $\psi = 45°$

the entire range of the azimuthal angle; this is obtained corresponding to the co-polarized component. The lobes in distribution patterns occur due to interference between the forward and backward traveling waves. The NCDD remains minimum corresponding to $\phi = 90°$. The cross-polarized wave exhibits decrease in NCDD with the increase in ϕ.

The magnitude and phase properties of NCDD are depicted in Fig. 10a and b, respectively, corresponding to the case of $\psi = 45°$. It is noticed that the increase of pitch angle ψ causes the magnitude of current density to increase by a very small value, but it becomes more irregular compared to the situation with $\psi = 0°$ (Fig. 9a). The co-polarized incidence corresponds to larger magnitude of NCDD, whereas the phase of NCDD remains more uniform in this case (Fig. 10b) compared to the case of $\psi = 0°$ (Fig. 9b). As such, the increase of helix pitch causes stronger interference between the forward and backward waves. In this case too, the co-polarized incidence exhibits stronger backscattering of waves compared to the cross-polarized one.

Considering the case of PEC sheath helix orientation with $\psi = 90°$, the ϕ-dependence of NCDD amplitude and phase is illustrated in Fig. 11a and b, respectively. Figure 11a shows that, corresponding to the co-polarized wave, the amplitude of NCDD exhibits nearly sinusoidal variation with peaks at $\phi \approx 20°$, $\phi \approx 100°$, and $\phi \approx 175°$. In this case, the current density magnitudes are larger than those observed before in Fig. 9a (for $\psi = 0°$) and Fig. 9b (for $\psi = 45°$). However, the amplitude pattern for the cross-polarized wave remains very similar to that observed in the case of $\psi = 0°$ (Fig. 9a); the only existing difference is that the amplitude maxima becomes a little lesser now. The existing lobes exhibit the backscattering to be extremely strong in this case. Thus, a helix orientation parallel to the axis of cylinder causes large scattering—the feature that would be of great use in optical applications.

The phase plots of NCDD are shown in Fig. 11b. It is noticed that the helix pitch angle has great roles to alter the phase corresponding to both the kinds of incidence polarization. Upon comparing Figs. 9b, 10b, and 11b, it is found that

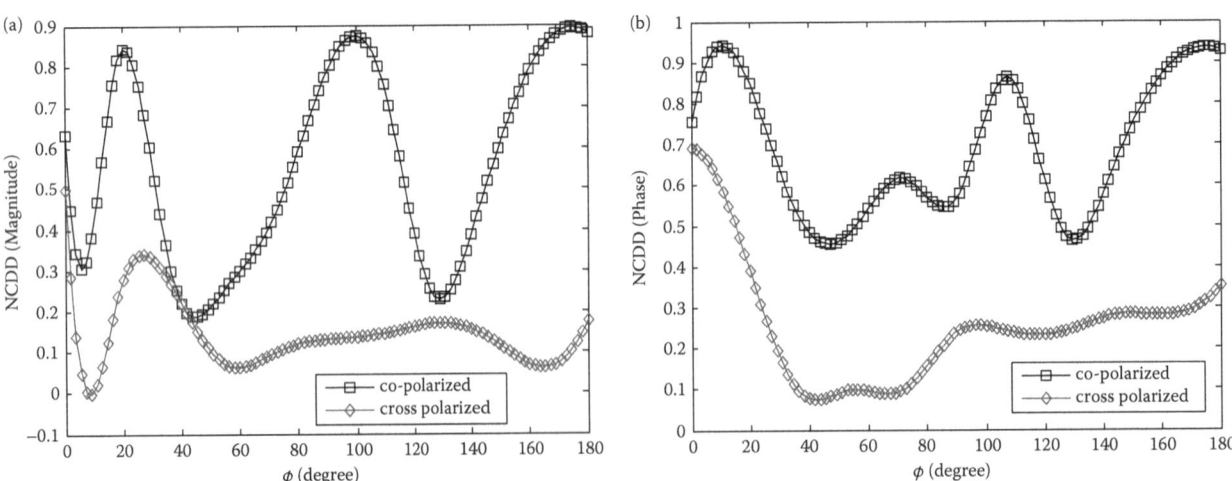

Fig. 11 Plots of NCDD (a) magnitude and (b) phase versus ϕ corresponding to $\theta_i = 30°$ and $\psi = 90°$

the phase of NCDD generally increases with the increase in helix pitch. The same happens with the NCDD magnitude as well. This is because $\psi=90°$ yields larger amplitude than what was noticed for lower values of pitch angle.

Increasing the incidence angle θ_i to 45°, the NCDD magnitude and phase plots corresponding to 0°, 45°, and 90° pitch angles are depicted in Figs. 12–14, respectively. It is noticed from Fig. 12a that, corresponding to $\psi=0°$, the amplitude pattern of NCDD for the co-polarized wave is almost inversed compared to that observed in Fig. 9a. A broadband maxima, centered at $\phi\approx90°$, appears in this figure corresponding to the co-polarized incidence. The cross-polarized wave results in larger amplitude maxima near ($\phi\approx85°$) compared to that in Fig. 9a. Thus, corresponding to 0° helix pitch angle, the increase in θ_i generally results in enhancement of the NCDD amplitude. The NCDD phase also shows significant alterations due to the increase in θ_i for the co- as well as cross-polarized incidence waves (Fig. 12b).

The NCDD magnitude and phase plots corresponding to $\theta_i=45°$ and $\psi=45°$ are shown in Fig. 13a and b, respectively. It is generally observed that the amplitude and phase increase with the increase in pitch angle. This is further justified from Fig. 14, which corresponds to the situation of $\theta_i=45°$ and $\psi=90°$, where more increase in amplitude and phase of the NCDD is noticed. The phase of NCDD is also greatly affected due to the increase in ψ.

Thus, it is generally noticed that the features of NCDD patterns remain greatly dependent on the helix pitch angle of the scatterer structure. The NCDD amplitude becomes larger corresponding to larger helix pitch angle. Also, the increase in incidence angle generally causes increase in the current density, and hence, the scattered radiation. Therefore, the increase in incidence angle causes to alter the scattering behavior of the object. Large scattering of waves remains vital in many antenna-related applications and medical diagnostics.

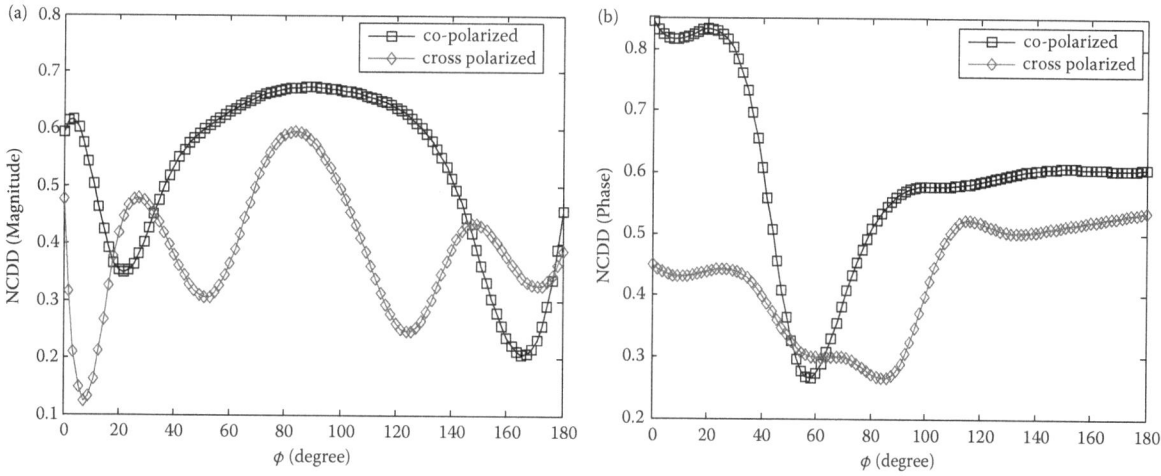

Fig. 12 Plots of NCDD (a) magnitude and (b) phase versus ϕ corresponding to $\theta_i=45°$ and $\psi=0°$

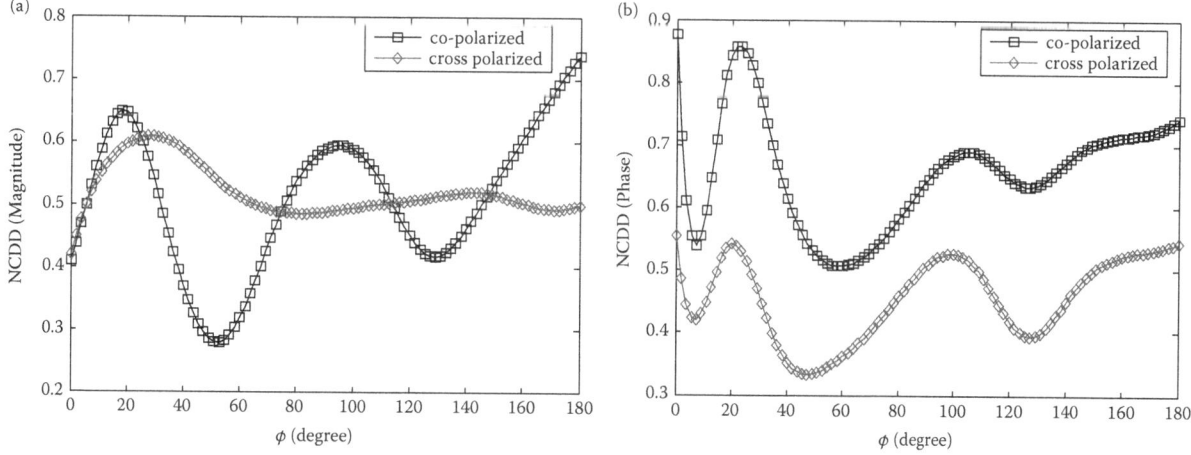

Fig. 13 Plots of NCDD (a) magnitude and (b) phase versus ϕ corresponding to $\theta_i=45°$ and $\psi=45°$

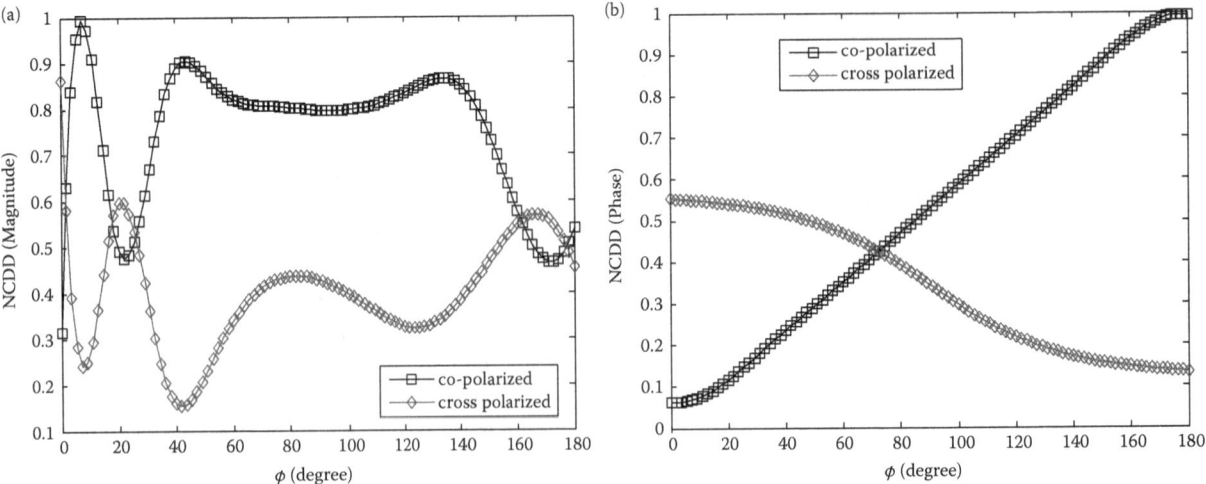

Fig. 14 Plots of NCDD (a) magnitude and (b) phase versus ϕ corresponding to $\theta_i=45°$ and $\psi=90°$

CHIRAL-COATED TWISTED CYLINDRICAL MEDIUMS

As described before, the negative index metamaterials (NIMs) are of new class. These exhibit unusual thought-provoking features and phenomena, and find prudent and novel applications, such as sub-wavelength imaging,[62–64] slow-light devices,[65] and cloaking.[66] NIMs require a condition through which both the permittivity and permeability simultaneously become negative—a stringent condition that is harder to achieve through the use of conventional mediums. Within the context, the NIM condition can be met through exploiting chiral metamaterial, which is an alternative way, as proposed in the work of Tretyakov et al.[67] Chiral NIMs exhibit fascinating EM properties, for example, two circularly polarized waves propagating though such mediums possess different RIs and cause negative reflection as well.[68]

Chiral medium-based guides of various forms were reported in the literature emphasizing the dispersion and power confinement properties.[69–74] However, in the context of scattering by such mediums, this section deals with the evaluation of echo width. In particular, a coating of chiral material over dielectric cylinders of elliptical/circular cross section, mounted with conducting sheath helix structure, is taken into account as scattering object.[75] The normalized bi-scattering echo width is computed numerically under parallel polarization of the oblique incidence wave. The results are obtained corresponding to variations in helix pitch angle, chiral admittance, and the condition of negative refraction. As such, this section basically describes comparative studies of scattering characteristics of the aforementioned kind(s) of object having circular and elliptical cross-sectional geometries; these are now discussed individually in the subsections that follow.

The Case of Elliptical Dielectric Cylinder

Figure 15 illustrates the schematic of twisted dielectric elliptical cylinder composed of silica glass, which is coated with linear, homogeneous, isotropic, and nonmagnetic chiral medium, and wrapped with a PEC sheath helix structure. It is assumed that the lengths of semimajor and semiminor axes of the inner layer are a and b, respectively, whereas those of the outer layer are a_1 and b_1. The elliptical coordinate system (ξ, η, z) is used for the analyses, where η and ξ lie in the respective ranges as $0° \leq \eta \leq 360°$ and $0 \leq \xi \leq \infty$.

The elliptical coordinate system is related to the rectangular coordinate system (x, y, z) through the coordinate transformation equations

$$x = q\xi\eta \tag{49}$$

$$y = q\sqrt{\xi^2-1}\sqrt{\eta^2-1} \tag{50}$$

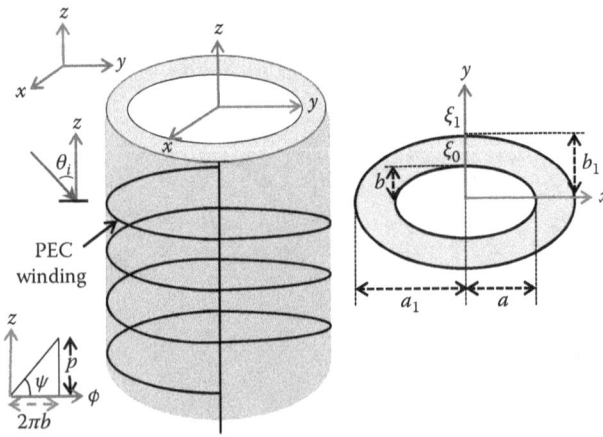

Fig. 15 Schematic of the scatterer structure having elliptical cross section

Scattering Features of Complex Mediums

In the preceding equations, q is the semi-focal length of elliptical cross section. The parametric coordinate of the inner dielectric medium is assumed to be as $\xi = \xi_0^e$, whereas that of the PEC sheath helix is as $\xi = \xi_1^e$. It is noteworthy that the PEC sheath helix encapsulates both (i.e., the inner dielectric and outer chiral) layers. For the simplicity of analyses, both the mediums are assumed to be nonmagnetic in nature, and therefore, the permeability values of these remain equivalent to those of the free space.

We consider the parallel-polarized (i.e., TM) $\left(H_{0z}^i = 0\right)$ incidence plane waves along the direction parallel to the xy-plane, with θ_i being the angle of incidence, as shown in Fig. 15. The tangential components of EM fields in the region outside the cylindrical structure are composed of incidence and scattered fields. Due to the anisotropic nature of PEC sheath helix structure, the scattered fields, and those inside the helix structure, consist of both the TM and TE polarizations. Now, the axial and transverse components of the E- and H-fields in each region can be expressed as

$$E_{0z}^{ie} = \sum_{n=0}^{n=\infty} Ae_n Re_n^{(1)}(\xi, k_0 q) \sin(n\eta) \tag{51}$$

$$H_{0\eta}^{ie} = -\frac{j\omega\varepsilon_0}{k_0^2 \mathbb{C}_s} \sum_{n=0}^{n=\infty} Ae_n Re_n^{(1)'}(\xi, k_0 q) \sin(n\eta) \tag{52}$$

$$E_{0z}^{se} = \sum_{n=0}^{n=\infty} Be_n Re_n^{(4)}(\xi, k_0 q) \sin(n\eta) \tag{53}$$

$$H_{0\eta}^{se} = -\frac{j\omega\varepsilon_0}{k_0^2 \mathbb{C}_s} \sum_{n=0}^{n=\infty} Be_n Re_n^{(1)'}(\xi, k_0 q) \sin(n\eta) \tag{54}$$

$$E_{0z}^{se} = \frac{j}{\eta_0} \sum_{n=0}^{n=\infty} Ce_n Re_n^{(4)}(\xi, k_0 q) \sin(n\eta) \tag{55}$$

$$E_{0\eta}^{se} = -\frac{\omega\mu_0}{k_0^2 \mathbb{C}_s \eta_0} \sum_{n=0}^{n=\infty} Ce_n Re_n^{(4)'}(\xi, k_0 q) \sin(n\eta) \tag{56}$$

$$E_{1z}^{e} = \sum_{n=0}^{n=\infty} \left[De_n Re_n^{(3)}(\xi, k_{1R}^2 q) + Ee_n Re_n^{(4)}(\xi, k_{1L}^2 q) \right] \sin(n\eta) \tag{57}$$

$$H_{1\eta}^{e} = -\frac{j\mathbb{C}_6}{\mathbb{C}_s} \sum_{n=0}^{n=\infty} \left[De_n Re_n^{(3)'}(\xi, k_{1R}^2 q) - Ee_n Re_n^{(4)'}(\xi, k_{1L}^2 q) \right] \sin(n\eta) \tag{58}$$

$$H_{1z}^{e} = -\frac{1}{j\eta_1} \sum_{n=0}^{n=\infty} \left[Fe_n Re_n^{(3)}(\xi, k_{1R}^2 q) + Ge_n Re_n^{(4)}(\xi, k_{1L}^2 q) \right] \sin(n\eta) \tag{59}$$

$$E_{1\eta}^{e} = -\frac{\mathbb{C}_4}{\mathbb{C}_s \eta_1} \sum_{n=0}^{n=\infty} \left[Fe_n Re_n^{(3)'}(\xi, k_{1R}^2 q) - Ge_n Re_n^{(4)'}(\xi, k_{1L}^2 q) \right] \sin(n\eta) \tag{60}$$

$$E_{2z}^{e} = \sum_{n=0}^{n=\infty} Le_n Re_n^{(1)}(\xi, k_2 q) \sin(n\eta) \tag{61}$$

$$H_{2\eta}^{e} = -\frac{j\omega\varepsilon_2}{k_2^2 \mathbb{C}_s} \sum_{n=0}^{n=\infty} Le_n Re_n^{(1)'}(\xi, k_2 q) \sin(n\eta) \tag{62}$$

$$H_{2z}^{e} = -\frac{j\omega\varepsilon_2}{k_2^2 \mathbb{C}_s} \sum_{n=0}^{n=\infty} Le_n Re_n^{(1)'}(\xi, k_2 q) \sin(n\eta) \tag{63}$$

$$E_{2\eta}^{e} = -\frac{\omega\mu_0}{k_2^2 \mathbb{C}_s \eta_2} \sum_{n=0}^{n=\infty} Pe_n Re_n^{(1)'}(\xi, k_2 q) \sin(n\eta) \tag{64}$$

In these equations, n is the azimuthal mode index, the superscripts i and s, respectively, represent the incidence and scattered fields, and the superscript e stands for the elliptical cross section. Also, the subscripts 0, 1, and 2, respectively, indicate the situations in free space, chiral, and dielectric mediums (of the structure in Fig. 15). Further, the terms $Re_n^{(1)}, Re_n^{(2)}, Re_n^{(3)}$, and $Re_n^{(4)}$ represent the radial Mathieu functions of the first, second, third, and fourth kinds, respectively,[76] with the prime indicating the differentiation with respect to the argument. The symbols $Be_n, Ce_n, De_n, Ee_n, Fe_n, Ge_n, Pe_n$, and Le_n are the unknown constants in the three different regions of the scatterer structure under consideration. The meanings of other symbols are made explicit in Appendix-C.

We now apply the boundary conditions[77] at the two interfaces—the dielectric/chiral interface (at the boundary $\xi = \xi_0^e$) and the chiral/free space interface (at the boundary $\xi = \xi_1^e$). At the dielectric–chiral interface, the tangential components of electric/magnetic fields remain continuous, and at the chiral-free space boundary, the tangential components along the direction of helix are zero, whereas those along the direction perpendicular to helix are continuous. The use of these boundary conditions yields a set of eight equations, which may be written in a matrix form as follows:

$$\begin{pmatrix} 0 \\ 0 \\ 0 \\ 0 \\ c_5 \\ 0 \\ c_7 \\ c_8 \end{pmatrix} = \begin{pmatrix} 0 & 0 & c_{13} & c_{14} & 0 & 0 & c_{17} & 0 \\ 0 & 0 & 0 & 0 & c_{25} & c_{26} & 0 & c_{28} \\ 0 & 0 & 0 & 0 & c_{35} & c_{36} & 0 & c_{38} \\ 0 & 0 & c_{43} & c_{44} & 0 & 0 & c_{47} & 0 \\ c_{51} & c_{52} & 0 & 0 & 0 & 0 & 0 & 0 \\ 0 & 0 & c_{63} & c_{64} & c_{65} & c_{66} & 0 & 0 \\ c_{71} & c_{72} & c_{73} & c_{74} & c_{75} & c_{76} & 0 & 0 \\ c_{81} & c_{82} & c_{83} & c_{84} & c_{85} & c_{86} & 0 & 0 \end{pmatrix} \begin{pmatrix} Be_n \\ Ce_n \\ De_n \\ Ee_n \\ Fe_n \\ Ge_n \\ Pe_n \\ Le_n \end{pmatrix} \tag{65}$$

In the preceding equation, the used symbols have meanings as given in Appendix-C. Upon solving this equation, the unknown constants can be determined.

The Case of Circular Dielectric Cylinder

We now consider the situation in which the cylindrical structure in Fig. 15 has circular cross section. In this case, the cylindrical polar coordinate system (r, ϕ, z) is to be used for the analyses. The parametric values of the inner and outer radii of circular cross section are taken to be r_0 and r_1, respectively. By adopting the similar procedure, as described earlier for the case of elliptical cylinder, the longitudinal and transverse components of the E- and H-fields under the consideration of parallel polarization in each layer of the structure can be determined as follows:

$$E_{0z}^{ic} = E_0 \sin\theta_i e^{jk_0 z \cos\theta_i} \sum_{n=-\infty}^{n=\infty} j^{-n} J_n(k_0 r \sin\theta_i) e^{jn\phi} \qquad (66)$$

$$H_{0\phi}^{ic} = \frac{E_0}{j\eta_0} e^{jk_0 z \cos\theta_i} \sum_{n=-\infty}^{n=\infty} j^{-n} J_n'(k_0 r \sin\theta_i) e^{jn\phi} \qquad (67)$$

$$E_{0z}^{sc} = E_0 \sin\theta_s e^{-jk_0 z \cos\theta_s} \sum_{n=-\infty}^{n=\infty} j^{-n} \mathbb{C}_{1n} H_n^2(k_0 r \sin\theta_s) e^{jn\phi} \qquad (68)$$

$$H_{0\phi}^{sc} = \frac{E_0}{j\eta_0} e^{-jk_0 z \cos\theta_s} \sum_{n=-\infty}^{n=\infty} j^{-n} \mathbb{C}_{1n} H_n^{2\prime}(k_0 r \sin\theta_s) e^{jn\phi} \qquad (69)$$

$$H_{0z}^{sc} = \frac{E_0}{j\eta_0} e^{-jk_0 z \cos\theta_s} \sum_{n=-\infty}^{n=\infty} j^{-n} \mathbb{C}_{2n} H_n^2(k_0 r \sin\theta_s) e^{jn\phi} \qquad (70)$$

$$E_{0\phi}^{sc} = \frac{E_0}{\sin\theta_s} e^{-jk_0 z \cos\theta_s} \sum_{n=-\infty}^{n=\infty} j^{-n} \mathbb{C}_{2n} H_n^{2\prime}(k_0 r \sin\theta_s) e^{jn\phi} \qquad (71)$$

$$E_{1z}^c = E_0 \sum_{n=-\infty}^{n=\infty} j^{-n} \Big[\sin\theta_{r1} e^{jk_1 R z \cos\theta_{r1}} \{\mathbb{C}_{3n} H_n^2(k_{1R} r \sin\theta_{r1}) \\
 + \mathbb{C}_{4n} H_n^1(k_{1R} \rho \sin\theta_{r1})\}\} + \sin\theta_{r2} e^{jk_1 L z \cos\theta_{r2}} \\
 \{\mathbb{C}_{5n} H_n^2(k_{1L} r \sin\theta_{r2}) + \mathbb{C}_{6n} H_n^1(k_{1L} r \sin\theta_{r2})\} \Big] e^{jn\phi} \qquad (72)$$

$$E_{1\phi}^c = E_0 \sum_{n=0}^{\infty} j^{-n} \Big[k_{1R} \sin^2\theta_{r1} e^{jk_1 R z \cos\theta_{r1}} (\mathbb{C}_4 \mathbb{C}_+ - \mathbb{C}_2) \{\mathbb{C}_{3n} H_n^{2\prime}(k_{1R} r \sin\theta_{r1}) + \mathbb{C}_{4n} H_n^{1\prime}(k_{1R} r \sin\theta_{r1})\} \\
+ k_{1L} \sin^2\theta_{r2} e^{jk_1 L z \cos\theta_{r2}} (\mathbb{C}_4 \mathbb{C}_- - \mathbb{C}_2) \{\mathbb{C}_{5n} H_n^{2\prime}(k_{1L} r \sin\theta_{r2}) + \mathbb{C}_{6n} H_n^{1\prime}(k_{1L} r \sin\theta_{r2})\} \\
+ \frac{n}{r} \sin\theta_{r1} e^{jk_1 R z \cos\theta_{r1}} (-\mathbb{C}_1 + j\mathbb{C}_3 \mathbb{C}_+) \{\mathbb{C}_{3n} H_n^2(k_{1R} r \sin\theta_{r1}) + \mathbb{C}_{4n} H_n^1(k_{1R} r \sin\theta_{r1})\} \\
+ \frac{n}{r} \sin\theta_{r2} e^{jk_1 L z \cos\theta_{r2}} (-\mathbb{C}_1 + j\mathbb{C}_3 \mathbb{C}_-) \{\mathbb{C}_{5n} H_n^2(k_{1L} r \sin\theta_{r2}) + \mathbb{C}_{6n} H_n^1(k_{1L} r \sin\theta_{r2})\} \Big] e^{jn\phi} \qquad (73)$$

$$H_{1z}^c = jE_0 \sum_{n=-\infty}^{n=\infty} j^{-n} \Big[\mathbb{C}_+ \sin\theta_{r1} e^{jk_1 R z \cos\theta_{r1}} \{\mathbb{C}_{3n} H_n^2(k_{1R} r \sin\theta_{r1}) + \mathbb{C}_{4n} H_n^1(k_{1R} r \sin\theta_{r1})\} \\
+ \mathbb{C}_- \sin\theta_{r2} e^{jk_1 L z \cos\theta_{r2}} \{\mathbb{C}_{5n} H_n^2(k_{1L} r \sin\theta_{r2}) + \mathbb{C}_{6n} H_n^1(k_{1L} r \sin\theta_{r2})\} \Big] e^{jn\phi} \qquad (74)$$

$$H_{1\phi}^c = E_0 \sum_{n=0}^{\infty} j^{-n} \Big[-jk_{1R} \sin^2\theta_{r1} e^{jk_1 R z \cos\theta_{r1}} (\mathbb{C}_2 \mathbb{C}_+ + \mathbb{C}_6) \{\mathbb{C}_{3n} H_n^{2\prime}(k_{1R} r \sin\theta_{r1}) + \mathbb{C}_{4n} H_n^{1\prime}(k_{1R} r \sin\theta_{r1})\} \\
- jk_{1L} \sin^2\theta_{r2} e^{jk_1 L z \cos\theta_{r2}} (\mathbb{C}_2 \mathbb{C}_- + \mathbb{C}_6) \{\mathbb{C}_{5n} H_n^{2\prime}(k_{1L} r \sin\theta_{r2}) + \mathbb{C}_{6n} H_n^{1\prime}(k_{1L} r \sin\theta_{r2})\} \\
+ \frac{n}{r} \sin\theta_{r1} e^{jk_1 R z \cos\theta_{r1}} (j\mathbb{C}_5 - \mathbb{C}_1 \mathbb{C}_+) \{\mathbb{C}_{3n} H_n^2(k_{1R} r \sin\theta_{r1}) + \mathbb{C}_{4n} H_n^1(k_{1R} r \sin\theta_{r1})\} \\
+ \frac{n}{r} \sin\theta_{r2} e^{jk_1 L z \cos\theta_{r2}} (j\mathbb{C}_5 - \mathbb{C}_1 \mathbb{C}_-) \{\mathbb{C}_{5n} H_n^2(k_{1L} r \sin\theta_{r2}) + \mathbb{C}_{6n} H_n^1(K_{1L} r \sin\theta_{r2})\} \Big] e^{jn\phi}. \qquad (75)$$

$$E_{2z}^c = E_0 \sin\theta_t e^{jk_2 z \cos\theta_t} \sum_{m=0}^{m=\infty} j^{-n} \mathbb{C}_{7n} J_n(k_2 r \sin\theta_t) e^{jn\phi} \qquad (76)$$

$$H_{2\phi}^c = \frac{E_0}{j\eta_2} e^{jk_2 z \cos\theta_t} \sum_{n=-\infty}^{n=\infty} j^{-n} \mathbb{C}_{7n} J_n'(k_2 r \sin\theta_t) e^{jn\phi} \qquad (77)$$

$$H_{2z}^c = \frac{-E_0}{j\eta_2} \sin\theta_t e^{jk_2 z \cos\theta_t} \sum_{m=0}^{m=\infty} j^{-n} \mathbb{C}_{8n} J_n(k_2 r \sin\theta_t) e^{jn\phi} \qquad (78)$$

$$E_{2\phi}^c = \frac{E_0}{\sin\theta_t} e^{jk_2 z \cos\theta_t} \sum_{n=-\infty}^{n=\infty} j^{-n} \mathbb{C}_{8n} J_n'(k_2 r \sin\theta_t) e^{jn\phi} \qquad (79)$$

In the preceding equations, the superscript c represents the circular cross section of the scaterrer and $\mathbb{C}_{1n}$, $\mathbb{C}_{2n}$, $\mathbb{C}_{3n}$, $\mathbb{C}_{4n}$, $\mathbb{C}_{5n}$, $\mathbb{C}_{6n}$, $\mathbb{C}_{7n}$, and $\mathbb{C}_{8n}$ are the unknown constants. Further, the quantities θ_{r1} and θ_{r2} are the angles of refraction of the left- and right-handed waves, respectively, in the chiral medium. The meanings of other used symbols are explicitly mentioned in Appendix-D.

Now, after using the field components, as shown in the previous equations, and applying the boundary conditions at the two aforementioned interfaces in the circular cross section of the scattering object, eight equations are obtained; the solution to those equations can be used to determine the values of unknown constants. The set of equations can be written in matrix form as follows:

$$\begin{pmatrix} 0 \\ 0 \\ 0 \\ 0 \\ d_5 \\ 0 \\ d_7 \\ d_8 \end{pmatrix} = \begin{pmatrix} 0 & 0 & d_{13} & d_{14} & 0 & 0 & d_{17} & 0 \\ 0 & 0 & 0 & 0 & d_{25} & d_{26} & 0 & d_{28} \\ 0 & 0 & 0 & 0 & d_{35} & d_{36} & 0 & d_{38} \\ 0 & 0 & d_{43} & d_{44} & 0 & 0 & d_{47} & 0 \\ d_{51} & d_{52} & 0 & 0 & 0 & 0 & 0 & 0 \\ 0 & 0 & d_{63} & d_{64} & d_{65} & d_{66} & 0 & 0 \\ d_{71} & d_{72} & d_{73} & d_{74} & d_{75} & d_{76} & 0 & 0 \\ d_{81} & d_{82} & d_{83} & d_{84} & d_{85} & d_{86} & 0 & 0 \end{pmatrix} \begin{pmatrix} \mathbb{C}_{1n} \\ \mathbb{C}_{2n} \\ \mathbb{C}_{3n} \\ \mathbb{C}_{4n} \\ \mathbb{C}_{5n} \\ \mathbb{C}_{6n} \\ \mathbb{C}_{7n} \\ \mathbb{C}_{8n} \end{pmatrix}$$

(80)

wherein the used symbols have meanings as given in Appendix-D.

Finally, the obtained values of expansion coefficients in both the circular and elliptical cases (of the scattering object) are used for the evaluation of bistatic echo widths, as illustrated earlier.

Scattering Features of Elliptical/Circular Cylinders

Let us consider that parallel (TM-) polarized plane waves fall upon the surface of elliptical/circular dielectric cylindrical scatterers, coated with chiral medium, at an incidence angle $\theta_i = 45°$. The RI values n_1 and n_2 (of the chiral and dielectric mediums, respectively) are taken as 2.0 and 1.445, respectively, and the operating wavelength is kept as $\lambda = 1.55\,\mu m$. For computations, two different values of chiral admittance ξ_c are assumed to be as 0.001 S and 0.01 S. The dimensional feature (in terms of semimajor and semiminor axes) of the elliptical cylinder is defined by the parametric values $a = 0.8\lambda$ and $b = 0.75\lambda$ corresponding to the inner layer, and $a_1 = 0.82\lambda$ and $b_1 = 0.77\lambda$ for the outer layer. In the case of cylinder having circular cross section, the radii of the inner and outer layers are denoted by $a = 0.8\lambda$ and $a_1 = 0.82\lambda$, respectively. Further, in both the cases of cylindrical geometries, the sheath helix orientation is assumed to have four different values of pitch angle, namely, 0°, 30°, 60°, and 90°.

Figure 16a and b illustrate the plots of bi-scattering echo width as a function of azimuthal angle ϕ for two different values of chirality admittance as $\xi_c = 0.001$ S and $\xi_c = 0.01$ S, while considering the elliptical cross section of the scatterer. It is observed that the highest value of echo width is attained in the range of ϕ as 100°–120° corresponding to the case of $\xi_c = 0.001$ S, and with the helix pitch angle value as $\psi = 60°$. Upon increasing the chirality admittance of coating (over the dielectric cylinder), the scattering patterns remain almost similar, but the echo width increases at the pitch angle 90°.

Keeping the operating conditions fixed, as in the case of scatterer composed of elliptical cylinder, the circularly cylindrical scatterer is now taken into account. Considering the chirality admittance values as $\xi_c = 0.001$ S and $\xi_c = 0.01$ S, the results with respect to ϕ-dependence of echo width are depicted in Fig. 17a and b, respectively. It is observed that the maximum values of echo width (or the scattering cross section) are observed with the pitch angle values as 60° and 90° corresponding to $\xi_c = 0.001$ S. The increase in chirality admittance results in increase of echo width.

Upon comparing these results with the ones obtained for the elliptical cylinder, it is noticed that the scatterer having circular cross section exhibits higher echo width, which would be attributed to the uniform distribution of medium in space. In other words, the anisotropy in shape

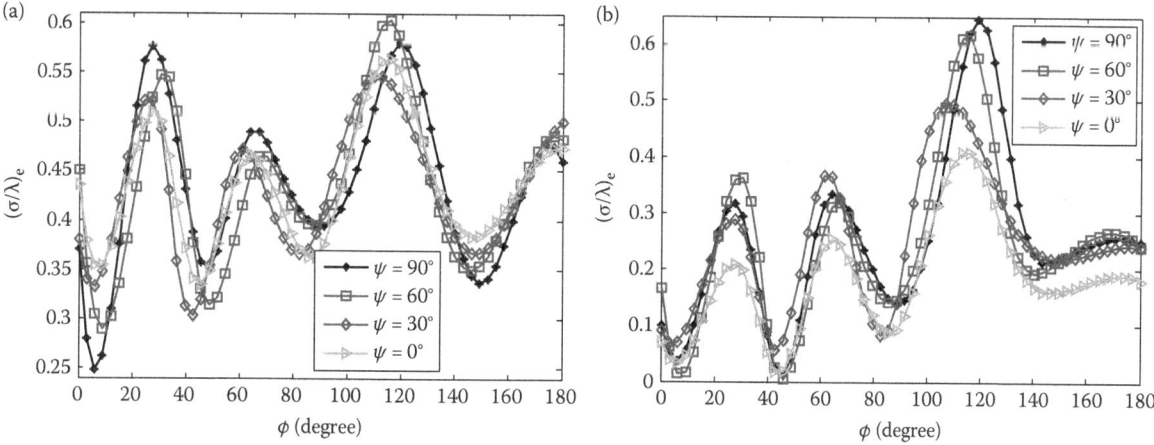

Fig. 16 The normalized bi-scattering echo width versus ϕ corresponding to (a) $\xi_c = 0.001$ S and (b) $\xi_c = 0.001$ S in the case of elliptical scatterer

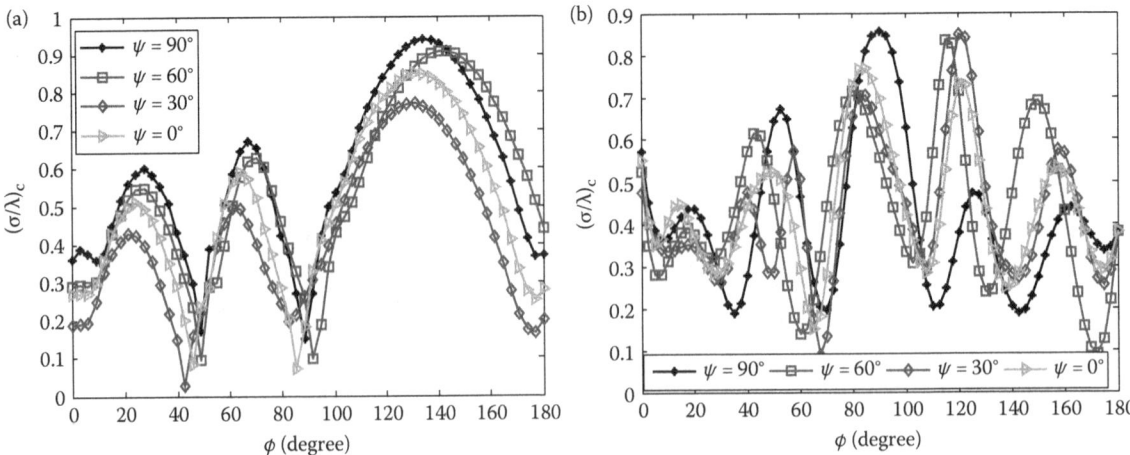

Fig. 17 The normalized bi-scattering echo width versus ϕ corresponding to (a) ξ_c=0.001 S and (b) ξ_c=0.001 S in the case of circular scatterer

in the case of elliptical scatterer causes reduced scattering cross section. It is found that the two maxima, corresponding to two different pitch angle values, are obtained in two different situations of chirality admittance in the case of elliptical scatterer, which get enhanced further in this case (i.e., the scatterer having a circular cross section).

We now focus on the effect of introducing negative refraction (through chiral medium) on the normalized echo width. It is known that, in chiral mediums, two circularly polarized plane waves propagate with different phase velocities, and their RI values are represented as $n_{\pm} = \sqrt{\varepsilon\mu} \pm \kappa$. Furthermore, for the transportation of positive energy in naturally occurring chiral mediums, the weak condition over chirality $\left(\sqrt{\varepsilon\mu} > |\kappa|\right)$ should hold. It has been reported before that the strong chirality condition can be achieved at particular frequencies.[78] With the use of such negative refraction condition, the obtained results corresponding to both the cases of elliptical and circular cross-sectional scatterers, operated under the same value of incidence angle (i.e., θ_i=45°) and helix pitch angles ψ, are illustrated in Fig. 18a and b, respectively. The results clearly indicate that the echo width maxima is achieved in the case of elliptical cylinder corresponding to the situation of sheath helix orientation parallel to the axial direction (i.e., ψ=90°); the patterns of echo width, however, remain the same for both the kinds of cross-sectional geometries of the scatterer. As such, the use of negative refraction enhances the bi-scattering echo width, and it becomes higher particularly with the helix pitch angle 90°.

CHIRAL/CHIRO-FERRITE-COATED TWISTED PEC CYLINDER

Chiro-ferrite material is the kind of Faraday chiral medium, which can be modeled by combining chiral objects with magnetically biased ferrites.[31,79] The purpose of the use of such material is to control the effects due to chirality. In this section, we consider the case of PEC cylinder with a coating of chiral/chiro-ferrite medium, and loaded with conducting sheath helix structure.[79] The results corresponding to the bi-and mono-scattered echo widths are determined under the consideration of parallel-polarized plane waves of oblique incidence, and taking

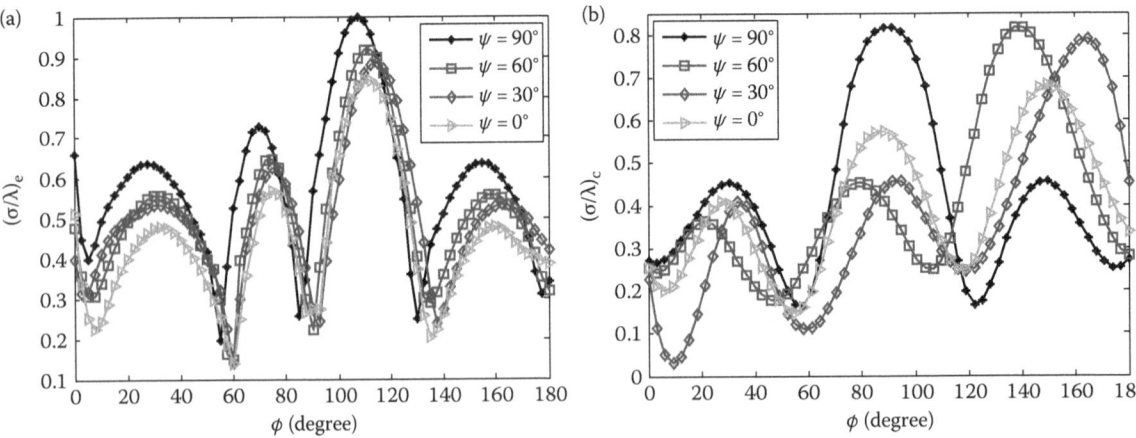

Fig. 18 The normalized bi-scattered echo width against ϕ for (a) elliptical and (b) circular scatterers in the case of negative RI condition

Scattering Features of Complex Mediums

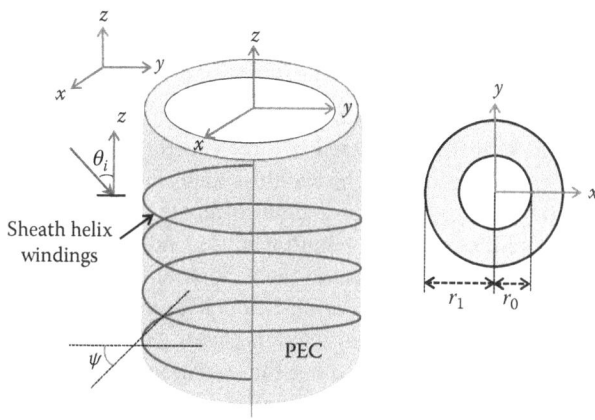

Fig. 19 Schematic of chiro-ferrite medium-coated PEC cylinder loaded with conducting sheath helix

into account the coating of both the kinds of materials, as stated earlier.

The structural geometry of the scatterer is shown in Fig. 19, where a chiro-ferrite coating exists over a PEC cylinder having radius r_0. Such a coated cylinder has the radius r_1, which is wrapped with a conducting sheath helix structure. As such, the helical windings assume the radius as r_1, and these are mounted with different angles of orientation. For the PEC cylinder, the tangential components of electric field at its surface become vanishing, and also no fields exist inside (the cylinder).

The constitutive relations for chiro-ferrite mediums are prescribed as[80]

$$\mathbf{D} = \varepsilon \mathbf{E} + j\xi_c \mathbf{B} \tag{81}$$

$$\mathbf{H} = j\xi_c \mathbf{E} + \overline{\overline{\mu}}^{-1} \cdot \mathbf{B} \tag{82}$$

where ξ_c and $\overline{\overline{\mu}}$ are the chirality admittance and permeability tensor (in gyrotropic form), respectively; $\overline{\overline{\mu}}$ is given as

$$\overline{\overline{\mu}} = \begin{bmatrix} \mu & -j\kappa_1 & 0 \\ j\kappa_1 & \mu & 0 \\ 0 & 0 & \mu_z \end{bmatrix} \tag{83}$$

where the term κ_1 is gyrotropic in nature.

We consider the oblique incidence of uniform parallel-polarized $\left(H_{0z}^i = 0\right)$ plane waves on the surface of cylinder. The longitudinal and transverse field components in each layer of the scatterer can be written as

$$E_{0z}^i = E_0 \sin\theta_i e^{jk_0 z \cos\theta_i} \sum_{n=-\infty}^{n=\infty} j^{-n} J_n\left(k_0 r \sin\theta_i\right) e^{jn\phi} \tag{84}$$

$$E_{0z}^s = E_0 \sin\theta_s e^{-jk_0 z \cos\theta_s} \sum_{n=-\infty}^{n=\infty} j^{-n} A_n H_n^2\left(k_0 r \sin\theta_s\right) e^{jn\phi} \tag{85}$$

$$H_{0z}^s = \tfrac{E_0}{j\eta_0} e^{-jk_0 z \cos\theta_s} \sum_{n=-\infty}^{n=\infty} j^{-n} B_n H_n^2\left(k_0 r \sin\theta_s\right) e^{jn\phi} \tag{86}$$

$$E_{1z}^t = E_0 \sum_{n=-\infty}^{n=\infty} j^{-n} \begin{bmatrix} \sin\theta_{r1} e^{jk_{2+}z\cos\theta_{r1}}\left\{C_n H_n^2\left(k_{2+}r\sin\theta_{r1}\right)+D_n H_n^1\left(k_{2+}r\sin\theta_{r1}\right)\right\} + \\ \sin\theta_{r2} e^{jk_{2-}z\cos\theta_{r2}}\left\{E_n H_n^2\left(k_{2-}r\sin\theta_{r2}\right)+F_n H_n^1\left(k_{2-}r\sin\theta_{r2}\right)\right\} \end{bmatrix} e^{jn\phi} \tag{87}$$

$$H_{1z}^t = jE_0 \sum_{n=-\infty}^{n=\infty} j^{-n} \begin{bmatrix} q_+ \sin\theta_{r1} e^{jk_{2+}z\cos\theta_{r1}}\left\{C_n H_n^2\left(k_{2+}r\sin\theta_{r1}\right)+D_n H_n^1\left(k_{2+}r\sin\theta_{r1}\right)\right\} + \\ q_- \sin\theta_{r2} e^{jk_{2-}z\cos\theta_{r2}}\left\{E_n H_n^2\left(k_{2-}r\sin\theta_{r2}\right)+F_n H_n^1\left(k_{2-}r\sin\theta_{r2}\right)\right\} \end{bmatrix} e^{jn\phi} \tag{88}$$

In these equations, the incidence, scattered, and transmitted fields (as identified by the respective use of superscripts i, s, and t) are denoted by the use of symbols E_{0z}^i, $\left(E_{0z}^s, H_{0z}^s\right)$, and $\left(E_{1z}^t, H_{1z}^t\right)$, respectively. Also, the subscripts 0 and 1 on the left-hand side represent the situations in the free space and coating regions (i.e., the chiro-ferrite medium of width as $r_1 - r_0$) of the scatterer, respectively. The quantity E_0 represents the amplitude of the incidence E-field, and the terms A_n, B_n, C_n, D_n, E_n, and F_n denote the unknown constants, which can be determined by using the suitable boundary conditions. Furthermore, θ_i and θ_s are, respectively, the incidence and scattering angles in the free space region, whereas θ_{r1} and θ_{r2} are those of the refracted right- and left-circularly polarized waves, respectively, in the middle layer region. The meanings of symbols $k_{2\pm}$ and $q_\pm$ are given in Appendix-E.

Now, applying the boundary conditions at the PEC interface—the tangential components of electric fields become vanishing; however, the boundary conditions for conducting sheath helix (at the interface $r = r_1$) remain the same as given in Eq. 39. Upon implementing the boundary conditions, six equations are obtained, which can be represented in a matrix form as follows:

$$\begin{bmatrix} 0 \\ 0 \\ e_3 \\ 0 \\ e_5 \\ e_6 \end{bmatrix} = \begin{bmatrix} 0 & 0 & e_{13} & e_{14} & e_{15} & e_{16} \\ 0 & 0 & e_{23} & e_{24} & e_{25} & e_{26} \\ e_{31} & e_{32} & 0 & 0 & 0 & 0 \\ 0 & 0 & e_{43} & e_{44} & e_{45} & e_{46} \\ e_{51} & e_{52} & e_{53} & e_{54} & e_{55} & e_{56} \\ e_{61} & e_{62} & e_{63} & e_{64} & e_{65} & e_{66} \end{bmatrix} \begin{bmatrix} A_n \\ B_n \\ C_n \\ D_n \\ E_n \\ F_n \end{bmatrix} \tag{89}$$

The meanings of the used symbols in the preceding equation are illustrated in Appendix-E. Upon solving the equations (in the preceding matrix), the values of the aforementioned unknown constants can be determined.

Scattering Features

The features of bi-scattered and backscattered echo widths σ/λ are evaluated by using the definition illustrated elsewhere.[81] It is noteworthy that the bi-scattered echo width is described as the ratio of the total scattered power to the input power. In the case of backscattered echo width, the positions of both the source and observer remain the same. Also, the observations are developed under the far-zone conditions.

For the analysis of the scattering object in Fig. 19, the dimensions of the scatterer are chosen by the use of parametric coordinates as $r_0 = 0.8\lambda$ and $r_1 = 0.9\lambda$, which are, respectively, the inner (i.e., the PEC cylinder) and outer (i.e., the conducting sheath helix structure) radii of the cylindrical scatterer confined inside the sheath helix; $\lambda = 1.55\,\mu m$ is the free space wavelength. The relative permittivity value of chirro-ferrite material is taken to be as $\varepsilon_2 = 13.8$, and the values of the elements of permeability tensor are taken as $\kappa_1 = 0.4\mu_0$, $\mu = 0.65\mu_0$, and $\mu_z = \mu_0$.

As to the sheath helix structure, four different values of helix pitch angle are assumed, namely $\psi = 0°$, 30°, 60°, and 90°. The oblique angle of incidence is taken as $\theta_i = 45°$ under the assumption of parallel (TM) polarization of plane wave. The bi- and backscattered echo widths are plotted against the azimuthal angle ϕ, which is specified for the interval of one complete cycle (i.e., $0° \leq \phi \leq 360°$), and can be measured from the x-axis (of the coordinate). The chiro-ferrite and chiral coating mediums assume the same chiral admittance ξ_c, and the results are obtained corresponding to three different values of it, namely, $\xi_c = 0.001\,S$, $\xi_c = 0.002\,S$, and $\xi_c = 0.003\,S$. Furthermore, the chiro-ferrite medium can be modified to a chiral one by choosing the values of elements of its permeability tensor as $\kappa_1 = 0$ and $\mu_z = \mu = \mu_0$.

The obtained results with respect to the ϕ-dependence of echo width are discussed in three categories as follows: first, the normalized echo width is analyzed corresponding to the cases of chiral and chiro-ferrite coatings under the alterations of chiral admittance; second, the impact of gyrotropy of the medium on the echo width under the situation of chiro-ferrite coating is touched upon; and third, the mono-scattering echo width is investigated in the case of chiro-ferrite coating under the variations in gyrotropy and chirality admittance parameters.

Effect of Chiral Admittance

The variation in sheath helix pitch angle ψ, and its effects on the normalized echo width, as a function of azimuthal angle ϕ is analyzed. For instance, four different values of ψ are taken into account, namely, 0°, 30°, 60°, and 90°. Figure 20a and b, respectively, illustrates the obtained results with a coating of chiro-ferrite and chiral mediums over the PEC cylinder; in both the kinds of coating (i.e., the chiro-ferrite and chiral mediums), the chirality admittance remains the same as $\xi_c = 0.001\,S$. It is noticed that the amplitude of echo width follows oscillatory trend as a function of azimuthal angle, and it is greatly affected upon altering the orientation of conducting sheath helix structure. The maximum value of echo width is observed corresponding to the pitch angle values 60° and 90° in the situation of chiral coating; the use of chiro-ferrite coating, however, increases the echo width maxima to a little extent.

Upon increasing the value of chirality admittance to 0.002 S, the results with respect to echo widths due to both the aforementioned kinds of coating are illustrated in Fig. 21a and b. It is noticed that the echo width maxima for both the cases increases with an increase in the

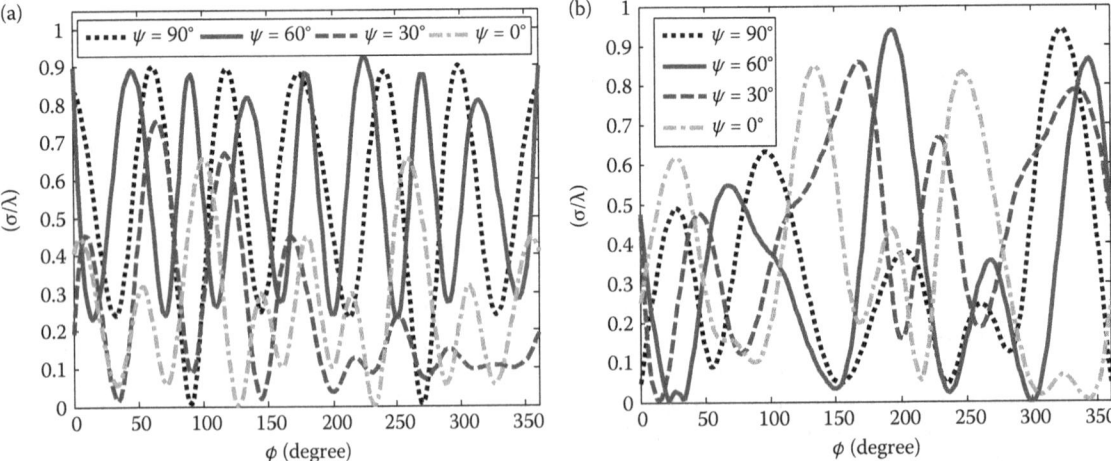

Fig. 20 The normalized echo width versus ϕ with the coating of (a) chiral and (b) chiro-ferrite mediums having the chirality admittance $\xi_c = 0.001\,S$

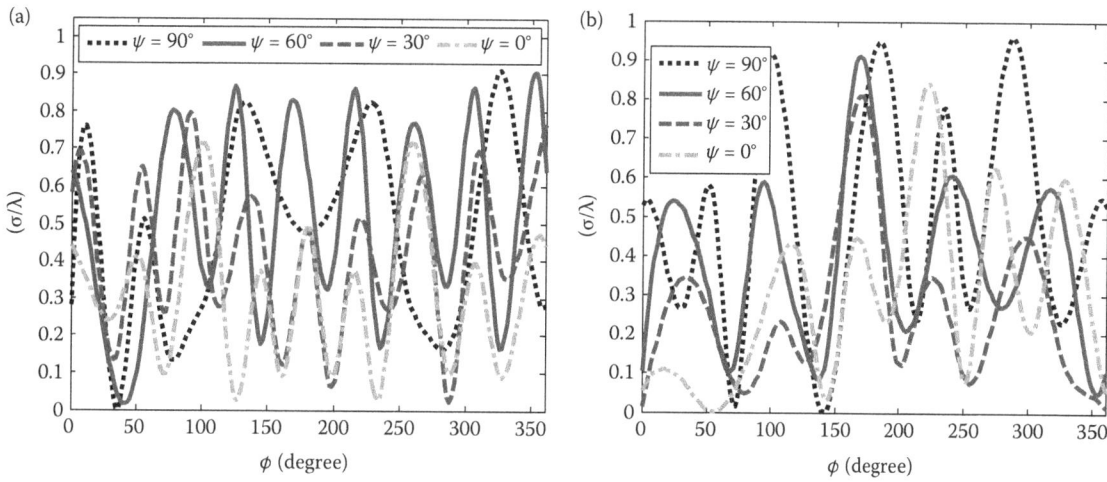

Fig. 21 The normalized echo width versus ϕ with the coating of (a) chiral and (b) chiro-ferrite mediums having the chirality admittance $\xi_c = 0.002$ S

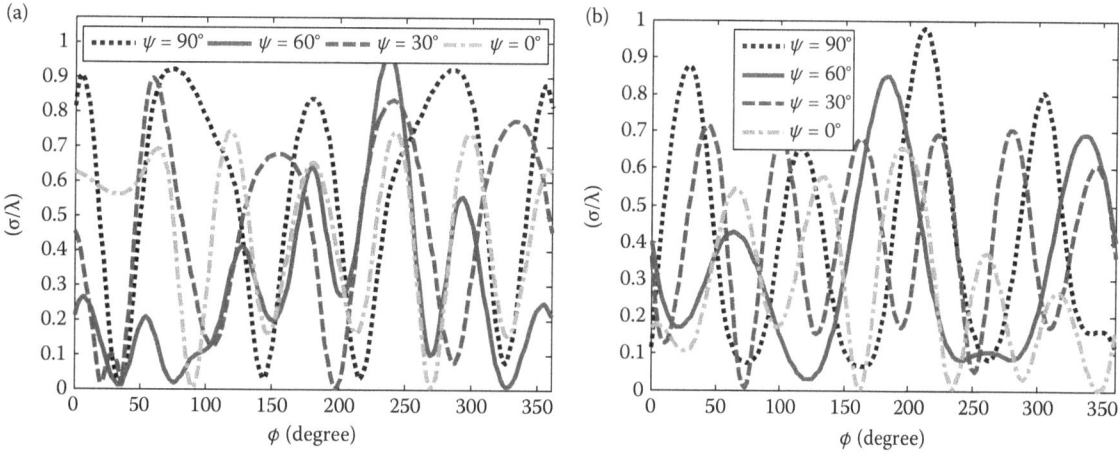

Fig. 22 The normalized echo width versus ϕ with the coating of (a) chiral and (b) chiro-ferrite mediums having the chirality admittance $\xi_c = 0.003$ S

value of helix pitch angle. A further increase in chirality admittance value to 0.003 S causes more increase in the respective echo width maxima (Fig. 22a and b).

It is finally observed that, in all the three situations of chirality admittance, the echo width values obtained in the case of chiro-ferrite coating remain higher than that achieved when the chiral coating remains in use, with a value of pitch angle close to 90°. Eventually, the presence of conducting sheath helix structure at a certain orientation modifies the scattering behavior of the coated PEC cylinder toward the simple PEC one.

Effect of Gyrotropy

The effect of varying the value of gyrotropy (of the chiro-ferrite medium) on the echo width is investigated assuming the chirality admittance ξ_c as 0.001 S; the other operating conditions are kept the same. Figure 23a and b illustrates the ϕ-dependence of echo width corresponding to two different values of gyrotropy, namely, $\kappa_1 = 0.45\mu_0$ and $0.5\mu_0$. Also, for each situation, four different kinds of sheath helix orientations are assumed, as used before. It is apparent from the results that the increase in gyrotropy value yields increased echo width maxima; the amplitude of maxima increases when the orientation of sheath helix remains parallel to the axial direction of the scattering object.

Features of Backscattering

The discussion so far covered the effects of chirality admittance (of chiral/chiro-ferrite mediums) and gyrotropy (of the chiro-ferrite medium) on the bi-scattering echo width. The features of backscattering due to the PEC cylinder, coated with either chiral or chiro-ferrite mediums, have also been touched upon. The corresponding results with

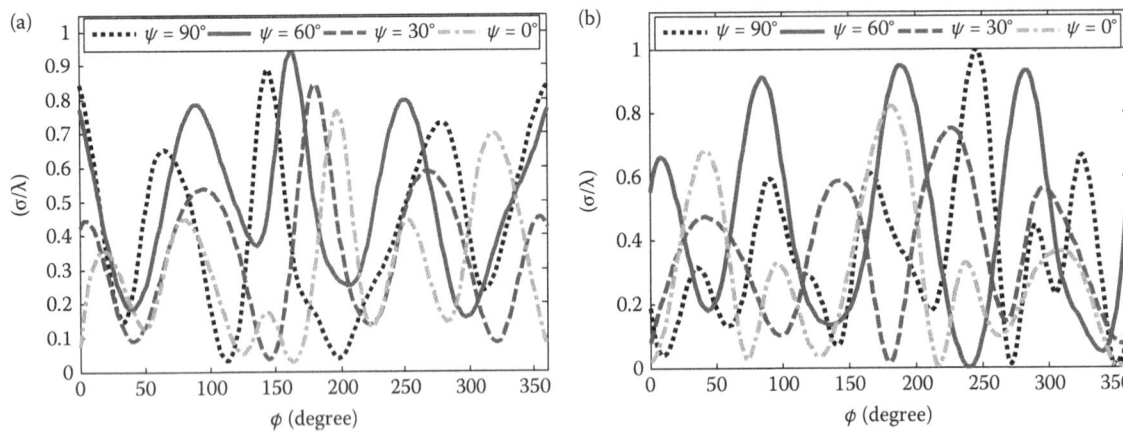

Fig. 23 The echo width versus ϕ plots when the PEC cylinder is coated with chiro-ferrite medium having the parametric values as $\xi_c=0.001$ S, $\varepsilon_1=13.8$, $\mu=0.65\mu_0$, $\mu_z=\mu_0$, and (a) $\kappa_1=0.45\mu_0$ and (b) $\kappa_1=0.5\mu_0$

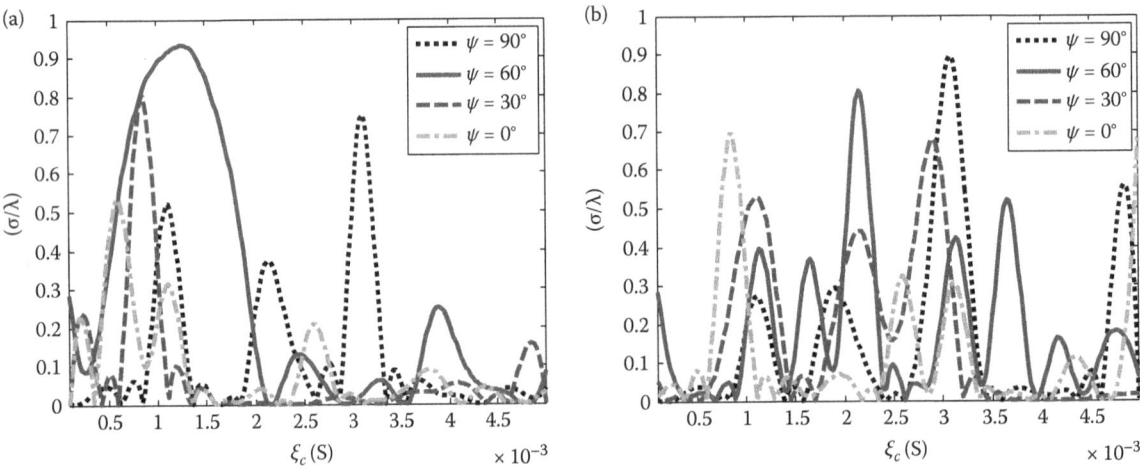

Fig. 24 The normalized backscattering echo width versus chirality admittance ξ_c with the parametric values as $\varepsilon_1=13.8$, $\mu_z=\mu_0$, and (a) chiral medium with $\kappa_1=0$, $\mu=\mu_0$ and (b) chiroferrite medium with $\kappa_1=0.4\mu_0$, $\mu=0.65\mu_0$

respect to the dependence of echo width on the chirality admittance (of the medium) are illustrated in Fig. 24a and b. Upon comparing the results of both types of coating mediums in the specified range of chirality admittance, it is found that the use of chiro-ferrite medium as a coating over the PEC cylinder yields larger backscattering. Also, reduced backscattering is observed with the increase in chirality admittance in the case of scatterer being coated with simple chiral medium.

Figure 25 illustrates backscattering as a function of gyrotropy in the case of PEC cylinder coated with chiro-ferrite medium. It is observed that the perpendicular (to the axis of the PEC cylinder) orientation of conducting sheath helix structure exhibits the maximum amount of backscattering. As such, the variations in gyrotropy as well as the orientation of sheath helix structure would cause alterations in backscattering-related features of the medium—the theme of the scatterer that would be greatly prudent in antenna-related applications.

UNIAXIAL ANISOTROPIC CHIRAL MEDIUM-COATED TWISTED PEC CYLINDER

Uniaxial anisotropic chiral mediums (UACMs) are the subclass of bianisotropic materials, which can be realized through the use of helixes and omega-shaped objects. The arrangement of helixes or omega-shaped particles in periodic order introduces uniaxial symmetry in the material. Furthermore, the chirality remains only in one direction in the case of UACMs. Propagation of EM waves through complex-structured guides under different forms of boundary conditions was investigated before.[82–84] However, in this section, we emphasize on the investigation of scattering response by the UACM-coated PEC cylinder having circular cross section. Similar to the structural configurations of scatterers discussed before, in this section, the scatterer under consideration has the mounted sheath helix structure at the outer surface of UACM; the impact of altering the sheath helix orientation on the scattering

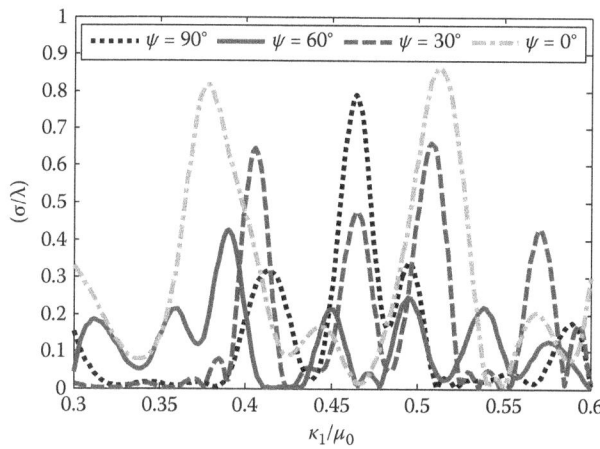

Fig. 25 The normalized backscattering echo width versus κ_1/μ_0 with the parametric values as $\varepsilon_1 = 13.8$, $\mu_z = \mu_0$, $\xi_c = 0.001$ S, and $\mu = 0.65\mu_0$ in the case of chiro-ferrite medium

$$\boldsymbol{D} = \left[\varepsilon_t \bar{I}_t + \varepsilon_z \hat{z}\hat{z}\right] \times \boldsymbol{E} + j\kappa\hat{z}\hat{z} \times \boldsymbol{H} \quad (90)$$

$$\boldsymbol{B} = \left[\mu_t \bar{I}_t + \mu_z \hat{z}\hat{z}\right] \times \boldsymbol{H} - j\kappa\hat{z}\hat{z} \times \boldsymbol{E} \quad (91)$$

where the permittivities ε_t and ε_z assume negative and positive values, the permeabilities are as $\mu_t > 0$, $\mu_z > 0$, and $\bar{I}_t = \hat{x}\hat{x} + \hat{y}\hat{y}$ is the unit dyadic in the direction transverse to the direction of z-axis; the subscript t is the representation of the transverse component of a vector or dyadic. Also, the chirality is such that $\kappa > 0$.

Now, exploiting the constitutive relations for UACM,[85] the longitudinal and axial field components can be determined upon using Maxwell's equations. Finally, the E- and H-field components in all the three different layers of the configuration (in Fig. 26) may be stated as

$$E_{0z}^i = E_0 \sin\theta_i e^{jk_0 z \cos\theta_i} \sum_{n=-\infty}^{n=\infty} j^{-n} J_n(k_0 r \sin\theta_i) e^{jn\phi} \quad (92)$$

$$H_{0\phi}^i = \frac{E_0}{j\eta_0} e^{jk_0 z \cos\theta_i} \sum_{n=-\infty}^{n=\infty} j^{-n} J_n'(k_0 r \sin\theta_i) e^{jn\phi} \quad (93)$$

$$E_{0z}^s = E_0 \sin\theta_s e^{-jk_0 z \cos\theta_s} \sum_{n=-\infty}^{n=\infty} j^{-n} \mathfrak{C}_{1n} H_n^2(k_0 r \sin\theta_s) e^{jn\phi} \quad (94)$$

$$H_{0\phi}^s = \frac{E_0}{j\eta_0} e^{-jk_0 z \cos\theta_s} \sum_{n=-\infty}^{n=\infty} j^{-n} \mathfrak{C}_{1n} H_n^{2'}(k_0 r \sin\theta_s) e^{jn\phi} \quad (95)$$

$$H_{0z}^s = \frac{E_0}{j\eta_0} e^{-jk_0 z \cos\theta_s} \sum_{n=-\infty}^{n=\infty} j^{-n} \mathfrak{C}_{2n} H_n^2(k_0 r \sin\theta_s) e^{jn\phi} \quad (96)$$

$$E_{0\phi}^s = \frac{E_0}{\sin\theta_s} e^{-jk_0 z \cos\theta_s} \sum_{n=-\infty}^{n=\infty} j^{-n} \mathfrak{C}_{2n} H_n^{2'}(k_0 r \sin\theta_s) e^{jn\phi} \quad (97)$$

cross section will remain one of the issues of investigation. In short, the bistatic echo width, along with other scattering-related features, such as the extinction and scattering efficiencies, are analyzed under the assumption of parallel-polarized oblique incidence of wave on the surface of scatterer.

For the analytical purpose, let us consider a PEC cylinder having radius r_0, which is coated with UACM on its outer surface. This makes the overall cylindrical structure to assume a radius r_1 (of the outer surface), and therefore, the thickness of the used coating becomes $r_1 - r_0$. The schematic of the structure is illustrated in Fig. 26. In this figure, a conducting sheath helix structure (with adjustable pitch angle) exists at the UACM–free space interface.

Figure 27 illustrates the three different ways of realizing uniaxial medium in which the chirality lies only in one direction. The constitutive relations for such mediums are given as follows[85]:

$$E_{1z}^t = E_0 \sum_{n=-\infty}^{n=\infty} j^{-n} \begin{bmatrix} \sin\theta_{t1} e^{j\mathfrak{C}_{s+} z \cos\theta_{t1}} \{\mathfrak{C}_{3n} H_n^2(\mathfrak{C}_{s+} r \sin\theta_{t1}) + \mathfrak{C}_{4n} H_n^1(\mathfrak{C}_{s+} r \sin\theta_{t1})\} + \\ \sin\theta_{t2} e^{j\mathfrak{C}_{s-} z \cos\theta_{t2}} \{\mathfrak{C}_{5n} H_n^2(\mathfrak{C}_{s-} r \sin\theta_{t2}) + \mathfrak{C}_{6n} H_n^1(\mathfrak{C}_{s-} r \sin\theta_{t2})\} \end{bmatrix} e^{jn\phi} \quad (98)$$

$$H_{1z}^t = \frac{jE_0}{\eta_1} \sum_{n=-\infty}^{n=\infty} j^{-n} \begin{bmatrix} \mathfrak{C}_+ \sin\theta_{t1} e^{j\mathfrak{C}_{s+} z \cos\theta_{t1}} \{\mathfrak{C}_{3n} H_n^2(\mathfrak{C}_{s+} r \sin\theta_{t1}) + \mathfrak{C}_{4n} H_n^1(\mathfrak{C}_{s+} r \sin\theta_{t1})\} + \\ \mathfrak{C}_- \sin\theta_{t2} e^{j\mathfrak{C}_{s-} z \cos\theta_{t2}} \{\mathfrak{C}_{5n} H_n^2(\mathfrak{C}_{s-} r \sin\theta_{t2}) + \mathfrak{C}_{6n} H_n^1(\mathfrak{C}_{s-} r \sin\theta_{t2})\} \end{bmatrix} e^{jn\phi} \quad (99)$$

$$E_{1\phi}^t = E_0 \sum_{n=-\infty}^{n=\infty} j^{-n} \begin{bmatrix} \frac{\kappa_0 n}{p\mathfrak{C}_{1c}^2} \sin\theta_{t1} e^{j\mathfrak{C}_{s+} z \cos\theta_{t1}} \{\mathfrak{C}_{3n} H_n^2(\mathfrak{C}_{s+} r \sin\theta_{t1}) + \mathfrak{C}_{4n} H_n^1(\mathfrak{C}_{s+} r \sin\theta_{t1})\} + \\ \frac{\kappa_0 n}{r\mathfrak{C}_{1c}^2} \sin\theta_{t2} e^{j\mathfrak{C}_{s-} z \cos\theta_{t2}} \{\mathfrak{C}_{5n} H_n^2(\mathfrak{C}_{s-} r \sin\theta_{t2}) + \mathfrak{C}_{6n} H_n^1(\mathfrak{C}_{s-} \rho \sin\theta_{t2})\} - \\ \frac{\omega\mu_1}{\eta_1 \mathfrak{C}_{1c}^2} \mathfrak{C}_+ \mathfrak{C}_{s+} \sin^2\theta_{t1} e^{j\mathfrak{C}_{s+} z \cos\theta_{t1}} \{\mathfrak{C}_{3n} H_n^{2'}(\mathfrak{C}_{s+} r \sin\theta_{t1}) + \mathfrak{C}_{4n} H_n^{1'}(\mathfrak{C}_{s+} r \sin\theta_{t1})\} - \\ \frac{\omega\mu_1}{\eta_1 \mathfrak{C}_{1c}^2} \mathfrak{C}_-^2 \mathfrak{C}_{s-} \sin^2\theta_{t2} e^{j\mathfrak{C}_{s-} z \cos\theta_{t2}} \{\mathfrak{C}_{5n} H_n^{2'}(\mathfrak{C}_{s-} r \sin\theta_{t2}) + \mathfrak{C}_{6n} H_n^{1'}(\mathfrak{C}_{s-} r \sin\theta_{t2})\} \end{bmatrix} e^{jn\phi} \quad (100)$$

$$H^t_{1\phi} = E_0 \sum_{n=-\infty}^{n=\infty} j^{-n} \begin{bmatrix} \frac{jk_0 n}{\eta_1 r \mathfrak{C}_{1c}^2} \mathfrak{C}_+ \sin\theta_{t1} e^{j\mathfrak{C}_{s+}z\cos\theta_{t1}} \{\mathfrak{C}_{3n} H_n^2(\mathfrak{C}_{s+}r\sin\theta_{t1}) + \mathfrak{C}_{4n} H_n^1(\mathfrak{C}_{s+}r\sin\theta_{t1})\} \\ + \frac{jk_0 n}{\eta_1 r \mathfrak{C}_{1c}^2} \mathfrak{C}_- \sin\theta_{t2} e^{jS_-z\cos\theta_{t2}} \{\mathfrak{C}_{5n} H_n^2(\mathfrak{C}_{s-}r\sin\theta_{t2}) + \mathfrak{C}_{6n} H_n^1(\mathfrak{C}_{s-}r\sin\theta_{t2})\} \\ + \frac{j\omega\varepsilon_t}{\mathfrak{C}_{1c}^2} \mathfrak{C}_{s+} \sin^2\theta_{t1} e^{j\mathfrak{C}_{s+}z\cos\theta_{t1}} \{\mathfrak{C}_{3n} H_n^{2\prime}(\mathfrak{C}_{s+}r\sin\theta_{t1}) + \mathfrak{C}_{4n} H_n^{1\prime}(\mathfrak{C}_{s+}r\sin\theta_{t1})\} + \\ \frac{j\omega\varepsilon_t}{\mathfrak{C}_{1c}^2} \mathfrak{C}_{s-} \sin^2\theta_{t2} e^{j\mathfrak{C}_{s-}z\cos\theta_{t2}} \{\mathfrak{C}_{5n} H_n^{2\prime}(\mathfrak{C}_{s-}r\sin\theta_{t2}) + \mathfrak{C}_{6n} H_n^{1\prime}(\mathfrak{C}_{s-}r\sin\theta_{t2})\} \end{bmatrix} e^{jn\phi} \quad (101)$$

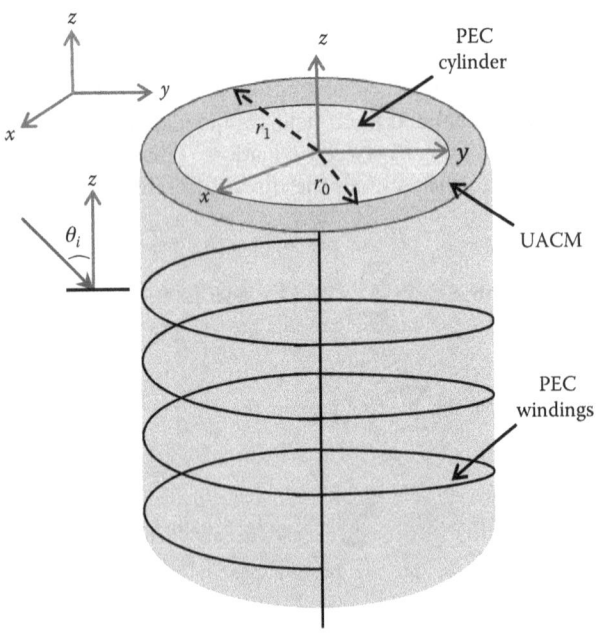

Fig. 26 Schematic of the UACM-coated scatterer

In the preceding equations, the subscripts 0 and 1 denote the free space and UACM situations, and the superscripts i, s, and t stand for the incidence, scattered, and transmitted fields, respectively. Also, E_0 is the incidence wave amplitude, and C_{1n}, C_{2n}, C_{3n}, C_{4n}, $C5_n$, and C_{6n} are the unknown constants, the values of which can be determined by using the suitable boundary conditions. Furthermore, the quantities θ_i, θ_s, θ_{t1}, and θ_{t2} are the respective angles of incidence, scattered, and left-, and right-circularly polarized transmitted waves. The other used symbols are defined as follows:

$$\mathfrak{C}^2_{s\pm} = \frac{\mathfrak{C}_{1c}^2}{2}\left[\frac{\varepsilon_z}{\varepsilon_t} + \frac{\mu_z}{\mu_1} \pm \sqrt{\left(\frac{\varepsilon_z}{\varepsilon_t} + \frac{\mu_z}{\mu_1}\right)^2 + 4\kappa_u^2 \frac{\varepsilon_z \mu_z}{\varepsilon_t \mu_1}}\right] \quad (102)$$

$$\mathfrak{C}_\pm = \frac{1}{\kappa_u}\left(\frac{\mathfrak{C}_{s\pm}^2}{\mathfrak{C}_{1c}^2} - \frac{\varepsilon_z}{\varepsilon_t}\right)\sqrt{\frac{\varepsilon_z \mu_z}{\varepsilon_t \mu_1}} \quad (103)$$

$$\mathfrak{C}_{1c}^2 = \omega^2 \varepsilon_t \mu_1 - k_0^2 \quad (104)$$

The boundary conditions at the two interfaces (PEC and sheath helix) are now implemented for the evaluation of the values of unknown constants. This gives six equations—two from the PEC interface (where the tangential components become vanishing) and the rest four come from the conducting sheath helix interface (where the tangential components of E-field along the circumferential direction are zero, whereas those of the E- and H-fields along the axial direction are continuous).

The expressions for bi-scattering echo width and the scattering and extinction efficiencies are represented as σ, Q_{scat}, and Q_{ext}, respectively, and can be written as follows[50]:

$$\sigma = \frac{2\lambda}{\pi}\left|\sum_{n=-\infty}^{n=\infty} \mathfrak{C}_{1n} e^{jn\phi}\right|^2 + \frac{4\lambda}{\pi}\left|\sum_{n=-\infty}^{n=\infty} \mathfrak{C}_{2n} e^{jn\phi}\right|^2 \quad (105)$$

$$Q_{scat} = \frac{2}{k_0 a}\sum_{n=-\infty}^{\infty}\left[\left|\mathfrak{C}_{1n}\right|^2 + \left|\mathfrak{C}_{2n}\right|^2\right] \quad (106)$$

$$Q_{ext} = -\frac{2}{k_0 a}\Re\left(\sum_{n=-\infty}^{\infty}\left[\mathfrak{C}_{1n} + \mathfrak{C}_{2n}\right]\right) \quad (107)$$

Here, R indicates the real part of quantity. These expressions can be used to evaluate the quantities of interest,

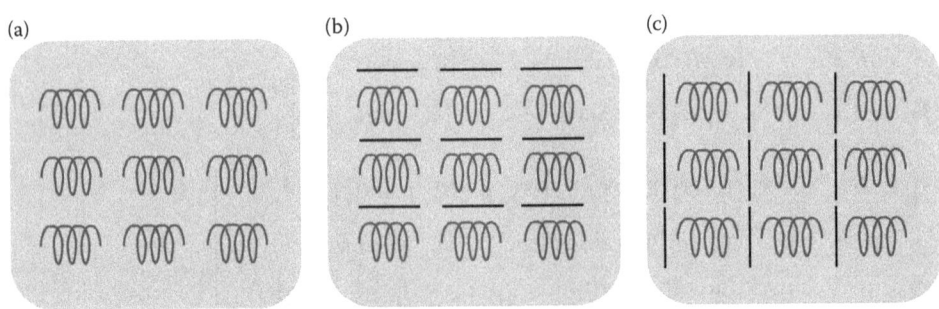

Fig. 27 UACM formation due to suitably oriented copper springs along the z-axis in a host medium in three different situations: (a) $\varepsilon_t > 0$ and $\varepsilon_z > 0$; (b) $\varepsilon_t > 0$ and $\varepsilon_z < 0$; (c) $\varepsilon_t < 0$ and $\varepsilon_z > 0$

as stated earlier. In this stream, the effects of uniaxial chiral medium parameters, and that of the orientation of conducting sheath helix structure, on the scattering echo width and extinction efficiency are touched upon for investigations.

Scattering Features of the Medium

For the computational purpose, the dimensional parameters of the structure are chosen as $r_0=0.8\lambda$ and $r_1=0.9\lambda$; λ being the free space wavelength, which is taken to be 1.55 μm. The chirality parameter is chosen as $\kappa=2.0$, the incidence angle is kept as $\theta_i=45°$, and the sheath helix orientations are assumed to be of the forms $\psi=0°$, 30°, 60°, and 90°. Apart from these, three different sets of material parameters are defined as (i) Type I—the ordinary dielectric medium (with $\varepsilon_t=\varepsilon_z=2\varepsilon_0$), (ii) Type II—the negative uniaxial medium (with $\varepsilon_t=2\varepsilon_0$, $\varepsilon_z=-2\varepsilon_0$), and (c) Type III—the positive uniaxial medium (with $\varepsilon_t=-2\varepsilon_0$, $\varepsilon_z=2\varepsilon_0$).

Now, assuming Type I kind of material coated over the PEC cylinder, the echo width is plotted as a function of azimuthal angle ϕ, as illustrated in Fig. 28a. The echo width maxima are obtained corresponding to the helix pitch angles $\psi=0°$ and 30°. As such, the scattering of waves may be increased by tuning the structure through proper adjustments of the sheath helix orientation. Furthermore, the kind of material in use yields higher value of forward scattering (than the backscattering) corresponding to all forms of helix orientation, except the case of $\psi=0°$, when it (the forward scattering) is reduced. It is worth noting at this point that the forward scattering and backscattering correspond to the situations of azimuthal angle values $\phi=0°$ and $\phi=180°$, respectively.

Figure 28b depicts the dependency of extinction efficiency on the parameter ε_z that is assumed in the range $\varepsilon_0 \leq \varepsilon_z \leq 2\varepsilon_0$. It is observed that, corresponding to lower values of permittivity ε_z, the perpendicular orientation of helix structure (i.e., $\psi=0°$) yields relatively large scattering, and the efficiency goes on decreasing with the increase in ε_z. Furthermore, for the chosen scatterer dimension, it is generally noticed that the extinction efficiency remains low corresponding to other larger values of helix pitch, and it becomes almost uniform within the chosen range of medium permittivity value corresponding to the helix orientation with $\psi=30°$.

Considering the coating of Type II kind of (negative) uniaxial material, a little larger value of echo width is obtained for $\phi \approx 5°-15°$ corresponding to the parallel orientation of sheath helix structure (i.e., $\psi=90°$; Fig. 29a) compared to what was noticed while Type I material coating was in use (Fig. 28a). However, the value of echo width is generally reduced in the case of Type II medium. Further, the forward scattering echo width remains dominant compared to the backscattering one.

Figure 29b describes the permittivity dependency of scattering efficiency in the case of Type II kind of uniaxial material. It is observed that, with the use of such materials, the value of scattering efficiency becomes very high (nearly 100%) corresponding to the permittivity value as $\varepsilon_z \approx 1.3$ and with the perpendicular orientation of helix structure. In the case of parallel orientation, the efficiency becomes lower (down to ~65%) corresponding to $\varepsilon_z \approx 1.45$, which is similar to the results obtained in the case of Type I medium. Further, the helix pitch angle value $\psi=30°$ also results in large scattering efficiency. As such, the negative UACM yields large scattering efficiency compared to the ordinary kind of uniaxial material, when coated over the PEC cylinder under the nonabsorption condition (of the coating medium).

In the case of Type III coating medium (composed of positive uniaxial material), the obtained results corresponding to echo widths are illustrated in Fig. 30a. In this situation, the echo width is enhanced compared to what was observed in Fig. 29a; it becomes high with the perpendicular orientation of sheath helix structure compared to the parallel one. However, the scattering efficiency remains less (in this case) than when Type II kind of coating material is in use—the feature which is observed corresponding to both the extreme kinds of sheath helix orientation (Fig. 30b).

In the case of UACMs, weak backscattering is generally observed, which can be reduced to zero so that it becomes invisible for monostatic radar. Such feature was found in

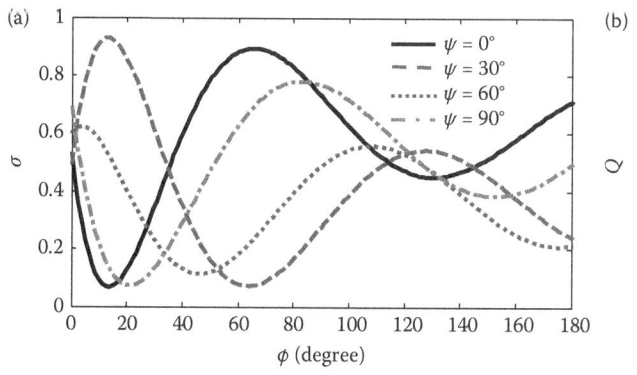
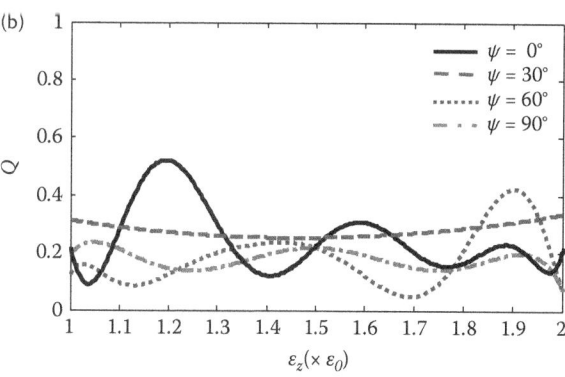
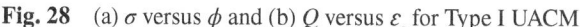

Fig. 28 (a) σ versus ϕ and (b) Q versus ε_z for Type I UACM

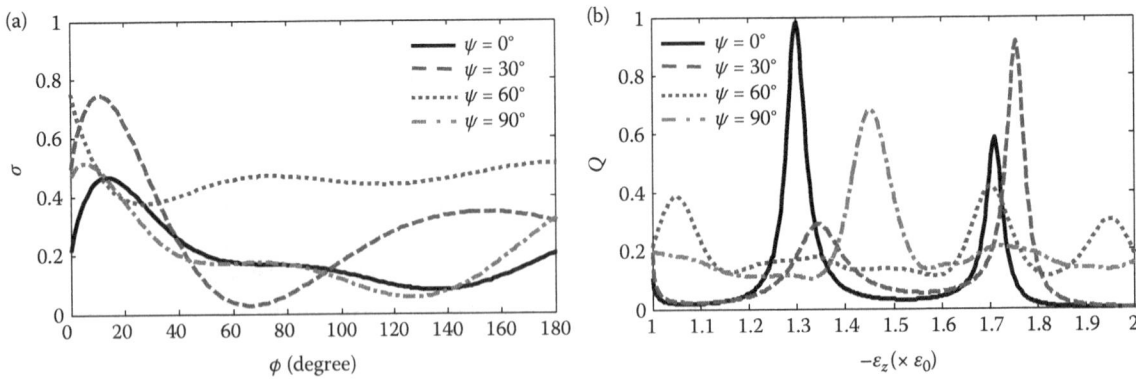

Fig. 29 (a) σ versus ϕ and (b) Q versus ε_z for Type II UACM

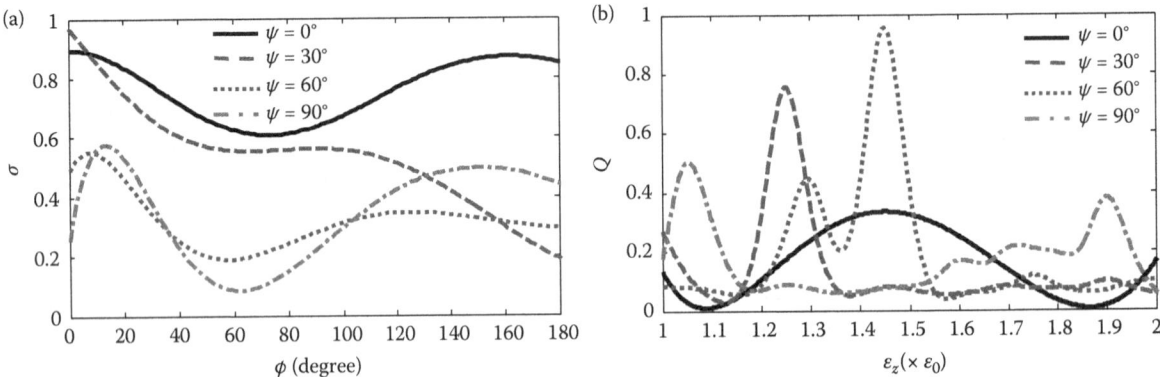

Fig. 30 (a) σ versus ϕ and (b) Q versus ε_z for Type III UACM

the case of DB sphere as well. But it is difficult to meet the boundary conditions in the DB case.[86] Further, the scattering efficiency remains equivalent to the extinction efficiency—the feature that shows the nondispersive property of scatterer. This kind of study was also found in the work of Lakhtakia and Geddes III [50]; however, the medium involved did not support the propagation of EM waves.

Many researchers reported achieving the features such as transparency, cloaking, or invisibility.[87–89] The relevant studies still remain challenging. In this entry, however, emphasis is given on the effect of introducing conducting sheath helix structure, along with the usage of varieties of complex mediums. It is expected that such investigations would be useful in attaining the required scattering characteristics due to the tuning flexibility in the scatterer.

CONCLUSION

This entry is devoted to the study of light scattering features of cylindrical objects having varieties of configurations as well as material properties, so far as the EM characteristics are concerned. In this connection, investigations of silver metal cylinder due to L-nihility, and coated with dielectric medium and encapsulated by conducting sheath helix structure, were initially touched upon. It was found that such a kind of scatterer, under the perpendicular orientation of sheath helix structure, becomes identical to a simple PEC cylinder. Further, under the parallel-polarized and perpendicularly polarized excitations of field, it was noticed that the scattering response (as a function of azimuthal angle) generally remains high in the former case, compared to the latter one, under different situations of the angle of incidence and the orientation of sheath helix. It is investigated that the current density remains greatly dependent on the structural parameters of the scatterer, and its magnitude becomes proportional to the conducting helix pitch and incidence angles.

The structures having circular as well as elliptical cross-sectional dielectric cylinders, coated with nonmagnetic isotropic chiral mediums and surrounded by sheath helix, are also investigated with respect to scattering characteristics. The results reveal that the scattering echo width remains strongly dependent on the chirality admittance of the coating medium as well as the geometrical cross section and helix pitch angle. For example, the backscattering is significantly reduced in the elliptical case under certain conditions of helix orientation. As such, the tuning of backscattered light can be achieved through suitable structural configurations of the scatterer. Furthermore, the negative refraction properties in the elliptical case enhance the echo width, which is useful in remote sensing applications.

The coating of chiro-ferrite medium over circular PEC cylinder, encapsulated by the sheath helix structure, was also taken into account. In this situation, larger bi-scattered echo width was noticed compared to the case of coating of simple chiral medium (over the PEC cylinder). Moreover, the perpendicular orientation of sheath helix structure results in higher echo width, as it behaves simply like a PEC cylinder in this case.

In the case of coating of UACM over the PEC cylinder, wrapped with conducting sheath helix structure, it was observed that the efficiency of scattering gets enhanced corresponding to the negative kind of UACM. The absorptive characteristic of scatterer was also briefly touched upon. The tuning feature of scatterer would be useful in stealth technology and other relevant applications.

Appendix A

In Eq. 40, the used symbols have meanings as

$$a_{13} = \sin\theta_r \, e^{jk_1 z \cos\theta_r} H_n^2(k_1 a \sin\theta_r);$$

$$a_{14} = \sin\theta_r \, e^{jk_1 z \cos\theta_r} H_n^1(k_1 a \sin\theta_r);$$

$$a_{17} = -\sin\theta_r \, e^{-(jk_{2s} z \cos\psi_{2s} + 2p_s)} J_n(k_2 a \sin\psi_2);$$

$$a_{23} = \tfrac{1}{\sin\theta_r} e^{jk_1 z \cos\theta_r} H_n^{2'}(k_1 a \sin\theta_r);$$

$$a_{24} = \tfrac{1}{\sin\theta_r} e^{jk_1 z \cos\theta_r} H_n^{1'}(k_1 a \sin\theta_r);$$

$$a_{28} = \frac{k_{2s} \sin\psi_{2s}}{j\eta_{2s}(j\omega\varepsilon_{2s} - k_{2s}^2 \cos^2\psi_{2s})} e^{-(jk_{2s} z \cos\psi_{2s} + 2p_s)} J'_n(k_{2s} a \sin\psi_{2s});$$

$$a_{35} = -\tfrac{1}{j\eta_1} e^{jk_1 z \cos\theta_r} H_n^2(k_1 a \sin\theta_r);$$

$$a_{36} = -\tfrac{1}{j\eta_1} e^{jk_1 z \cos\theta_r} H_n^1(k_1 a \sin\theta_r);$$

$$a_{38} = \tfrac{1}{j\eta_{2s}} e^{-(jk_{2s} z \cos\psi_{2s} + 2p_s)} J_n(k_{2s} a \sin\psi_{2s});$$

$$a_{43} = \tfrac{1}{j\eta_1} e^{jk_1 z \cos\theta_r} H_n^{2'}(k_1 a \sin\theta_r);$$

$$a_{44} = \tfrac{1}{j\eta_1} e^{jk_1 z \cos\theta_r} H_n^{1'}(k_1 a \sin\theta_r);$$

$$a_{47} = -\frac{\omega\varepsilon_{2s} k_{2s} \sin\psi_{2s} \sin\theta_r}{j(\omega^2 \mu_0 \varepsilon_{2s} - k_{2s}^2 \cos^2\psi_{2s})} e^{-(jk_{2s} z \cos\psi_{2s} + 2p_s)} J'_n(k_2 a \sin\psi_2);$$

$$a_{51} = \sin\theta_s \, e^{-jk_0 z \cos\theta_s} H_n^2(k_0 b \sin\theta_s) \sin\psi;$$

$$a_{52} = \tfrac{2}{\sin\theta_s} e^{-jk_0 z \cos\theta_s} H_n^2(k_0 b \sin\theta_s) \cos\psi;$$

$$a_5 = \sin\theta_i \, e^{jk_0 z \cos\theta_i} J_n(k_0 b \sin\theta_i) \sin\psi;$$

$$a_{63} = \sin\theta_r \, e^{jk_1 z \cos\theta_r} H_n^2(k_1 b \sin\theta_r) \sin\psi;$$

$$a_{64} = \sin\theta_r \, e^{jk_1 z \cos\theta_r} H_n^1(k_1 b \sin\theta_r) \sin\psi;$$

$$a_{65} = \tfrac{1}{\sin\theta_r} e^{jk_1 z \cos\theta_r} H_n^{2'}(k_1 b \sin\theta_r) \cos\psi;$$

$$a_{66} = \tfrac{1}{\sin\theta_r} e^{jk_1 z \cos\theta_r} H_n^{1'}(k_1 b \sin\theta_r) \cos\psi;$$

$$a_{71} = \sin\theta_s \, e^{-jk_0 z \cos\theta_s} H_n^2(k_0 b \sin\theta_s) \cos\psi;$$

$$a_{72} = -\tfrac{1}{\sin\theta_s} e^{-jk_0 z \cos\theta_s} H_n^{2'}(k_0 b \sin\theta_s) \sin\psi;$$

$$a_{73} = -\sin\theta_r \, e^{jk_1 z \cos\theta_r} H_n^2(k_1 b \sin\theta_r) \cos\psi;$$

$$a_{74} = -\sin\theta_r \, e^{jk_1 z \cos\theta_r} H_n^1(k_1 b \sin\theta_r) \cos\psi;$$

$$a_{75} = \tfrac{1}{\sin\theta_r} e^{jk_1 z \cos\theta_r} H_n^{2'}(k_1 b \sin\theta_r) \sin\psi;$$

$$a_{76} = \tfrac{1}{\sin\theta_r} e^{jk_1 z \cos\theta_r} H_n^{1'}(k_1 b \sin\theta_r) \sin\psi;$$

$$a_7 = \sin\theta_i \, e^{jk_0 z \cos\theta_i} J_n(k_0 b \sin\theta_i) \cos\psi;$$

$$a_{81} = \tfrac{1}{j\eta_0} e^{-jk_0 z \cos\theta_s} H_n^{2'}(k_0 b \sin\theta_s) \cos\psi;$$

$$a_{82} = \tfrac{1}{j\eta_0} e^{-jk_0 z \cos\theta_s} H_n^2(k_0 b \sin\theta_s) \sin\psi;$$

$$a_{83} = -\tfrac{1}{j\eta_1} e^{jk_1 z \cos\theta_r} H_n^{2'}(k_1 b \sin\theta_r) \cos\psi;$$

$$a_{84} = -\tfrac{1}{j\eta_1} e^{jk_1 z \cos\theta_r} H_n^{1'}(k_1 b \sin\theta_r) \cos\psi;$$

$$a_{85} = \tfrac{1}{j\eta_1} e^{jk_1 z \cos\theta_r} H_n^2(k_1 b \sin\theta_r) \sin\psi;$$

$$a_{86} = \tfrac{1}{j\eta_1} e^{jk_1 z \cos\theta_r} H_n^1(k_1 b \sin\theta_r) \sin\psi;$$

$$a_8 = \tfrac{1}{j\eta_0} e^{jk_0 z \cos\theta_i} J'_n(k_0 b \sin\theta_i).$$

Appendix B

In Eq. 43, the used symbols have meanings as

$$b_{15} = -e^{jk_1 z \cos\theta_r} H_n^2(k_1 a \sin\theta_r);$$

$$b_{16} = -e^{jk_1 z \cos\theta_r} H_n^1(k_1 a \sin\theta_r);$$

$$b_{18} = -\sin\theta_r \, e^{-(jk_2 z \cos\psi_2 + 2p)} J_n(k_2 a \sin\psi_2);$$

$$b_{23} = e^{jk_1 z \cos\theta_r} H_n^{2'}(k_1 a \sin\theta_r);$$

$$b_{24} = e^{jk_1 z \cos\theta_r} H_n^{1'}(k_1 a \sin\theta_r);$$

$$b_{27} = -\frac{\omega\mu_0 k_2 \sin\psi_2 \sin\theta_r}{\eta_2(\omega^2\mu_0\varepsilon_2 - k_2^2 \cos^2\psi_2)} e^{-(jk_2 z \cos\psi_2 + 2p)} J'_n(k_2 a \sin\psi_2);$$

$$b_{35} = \tfrac{1}{j\eta_1 \sin\theta_r} e^{jk_1 z \cos\theta_r} H_n^{2'}(k_1 a \sin\theta_r);$$

$$b_{36} = \tfrac{1}{j\eta_1 \sin\theta_r} e^{jk_1 z \cos\theta_r} H_n^{1'}(k_1 a \sin\theta_r);$$

$$b_{37} = \tfrac{\sin\theta_r}{j\eta_2} e^{-(jk_2 z \cos\psi_2 + 2p)} J_n(k_2 a \sin\psi_2);$$

$$b_{45} = \tfrac{1}{j\eta_1 \sin\theta_r} e^{jk_1 z \cos\theta_r} H_n^{2'}(k_1 a \sin\theta_r);$$

$$b_{46} = \tfrac{1}{j\eta_1 \sin\theta_r} e^{jk_1 z \cos\theta_r} H_n^{1'}(k_1 a \sin\theta_r);$$

$$b_{48} = \frac{k_2 \sin\psi_2}{(j\omega\mu_0 - k_2^2 \cos^2\psi_2)} e^{-(jk_2 z \cos\psi_2 + 2p)} J'_n(k_2 a \sin\psi_2);$$

$$b_{51} = e^{-jk_0 z \cos\theta_s} H_n^{2'}(k_0 b \sin\theta_s) \cos\psi;$$

$$b_{52} = e^{-jk_0 z \cos\theta_s} H_n^2(k_0 b \sin\theta_s) \sin\psi;$$

$$b_5 = e^{jk_0 z \cos\theta_i} J'_n(k_0 b \sin\theta_i) \cos\psi;$$

$$b_{63} = e^{jk_1 z \cos\theta_r} H_n^{1'}(k_1 b \sin\theta_r) \cos\psi;$$

$$b_{64} = e^{jk_1 z \cos\theta_r} H_n^{1'}(k_1 b \sin\theta_r) \cos\psi;$$

$$b_{65} = e^{jk_1 z \cos\theta_r} H_n^2(k_1 b \sin\theta_r) \sin\psi;$$

$$b_{66} = e^{jk_1 z \cos\theta_r} H_n^1(k_1 b \sin\theta_r) \sin\psi;$$

$$b_{71} = -e^{-jk_0 z \cos\theta_s} H_n^{2'}(k_0 b \sin\theta_s) \sin\psi;$$

$$b_{72} = e^{-jk_0 z \cos\theta_s} H_n^2(k_0 b \sin\theta_s) \cos\psi;$$

$$b_{73} = e^{jk_1 z \cos\theta_r} H_n^{2\prime}(k_1 b \sin\theta_r) \sin\psi;$$

$$b_{74} = -e^{jk_1 z \cos\theta_r} H_n^{1\prime}(k_1 b \sin\theta_r) \sin\psi;$$

$$b_{75} = e^{jk_1 z \cos\theta_r} H_n^2(k_1 b \sin\theta_r) \sin\psi;$$

$$b_{76} = -e^{jk_1 z \cos\theta_i} J_n'(k_1 b \sin\theta_r) \sin\psi;$$

$$b_7 = \sin\theta_i \, e^{jk_0 z \cos\theta_i} J_n(k_0 b \sin\theta_i) \cos\psi;$$

$$b_{81} = \tfrac{1}{j\eta_0} \sin\theta_s \, e^{-jk_0 z \cos\theta_s} H_n^2(k_0 b \sin\theta_s) \sin\psi;$$

$$b_{82} = \tfrac{1}{j\eta_1 \sin\theta_s} e^{-jk_0 z \cos\theta_s} H_n^{2\prime}(k_0 b \sin\theta_s) \cos\psi;$$

$$b_{83} = -\tfrac{1}{j\eta_1} \sin\theta_r \, e^{jk_1 z \cos\theta_r} H_n^2(k_0 b \sin\theta_s) \sin\psi;$$

$$b_{84} = -\tfrac{1}{j\eta_1} \sin\theta_r \, e^{jk_1 z \cos\theta_r} H_n^1(k_0 b \sin\theta_s) \sin\psi;$$

$$b_{85} = \tfrac{1}{j\eta_1 \sin\theta_r} e^{jk_1 z \cos\theta_r} H_n^{2\prime}(k_1 b \sin\theta_r) \cos\psi;$$

$$b_{86} = \tfrac{1}{j\eta_1 \sin\theta_r} e^{jk_1 z \cos\theta_r} H_n^{1\prime}(k_1 b \sin\theta_r) \cos\psi;$$

$$b_8 = \tfrac{1}{j\eta_0} \sin\theta_i \, e^{jk_0 z \cos\theta_i} J_n(k_0 b \sin\theta_i).$$

Appendix C

The meanings of the used symbols in Eq. 65 are as follows:

$$c_{13} = \mathrm{Re}_m^{(3)}(\xi_0^e, k_{1R} q);$$

$$c_{14} = \mathrm{Re}_m^{(4)}(\xi_0^e, k_{1R} q);$$

$$c_{17} = -\mathrm{Re}_m^{(1)}(\xi_0^e, k_2 q);$$

$$c_{25} = -\frac{\mathbb{C}_4}{\mathbb{C}_s \eta_1} \mathrm{Re}_m^{(3)\prime}\!\left(\xi_0^e, k_{1R} q\right);$$

$$c_{26} = \frac{\mathbb{C}_4}{\mathbb{C}_s \eta_1} \mathrm{Re}_m^{(4)\prime}\!\left(\xi_0^e, k_{1L} q\right);$$

$$c_{28} = \frac{\omega\mu_0}{k_2^2 \mathbb{C}_s \eta_2} \mathrm{Re}_m^{(1)\prime}\!\left(\xi_0^e, k_2 q\right);$$

$$c_{35} = -\tfrac{1}{j\eta_1} \mathrm{Re}_m^{(3)}(\xi_0^e, k_{1R} q);$$

$$c_{36} = -\tfrac{1}{j\eta_1} \mathrm{Re}_m^{(4)}(\xi_0^e, k_{1L} q);$$

$$c_{38} = -\tfrac{1}{j\eta_2} \mathrm{Re}_m^{(1)}(\xi_0^e, k_2 q);$$

$$c_{43} = -\frac{j\mathbb{C}_6}{\mathbb{C}_s} \mathrm{Re}_m^{(3)\prime}\!\left(\xi_0^e, k_{1R} q\right);$$

$$c_{44} = \frac{j\mathbb{C}_6}{\mathbb{C}_s} \mathrm{Re}_m^{(4)\prime}\!\left(\xi_0^e, k_{1L} q\right);$$

$$c_{47} = \frac{j\omega\varepsilon_2}{\mathbb{C}_s k_2^2} \mathrm{Re}_m^{(1)\prime}\!\left(\xi_0^e, k_2 q\right);$$

$$c_{51} = \mathrm{Re}_m^{(4)}(\xi_1^e, k_0 q) \sin\psi;$$

$$c_{52} = -\frac{\omega\mu_0}{k_0^2 \mathbb{C}_s \eta_0} \mathrm{Re}_3^{(4)\prime}\!\left(\xi_1^e, k_0 q\right) \cos\psi;$$

$$c_5 = A e_m \mathrm{Re}_m^{(1)}(\xi_1^e, k_0 q) \sin\psi;$$

$$c_{63} = \mathrm{Re}_m^{(3)}(\xi_1^e, k_{1R} q) \sin\psi;$$

$$c_{64} = \mathrm{Re}_m^{(4)}(\xi_1^e, k_{1L} q) \sin(m\eta) \sin\psi;$$

$$c_{65} = \frac{\mathbb{C}_4}{\mathbb{C}_s \eta_1} \mathrm{Re}_m^{(3)\prime}\!\left(\xi_1^e, k_{1R} q\right) \cos\psi;$$

$$c_{66} = -\frac{\mathbb{C}_4}{\mathbb{C}_s \eta_1} \mathrm{Re}_m^{(4)\prime}(\xi_1^e, k_{1L} q) \cos\psi;$$

$$c_{71} = \mathrm{Re}_m^{(4)}(\xi_1^e, k_0 q) \cos\psi;$$

$$c_{72} = \frac{\omega\mu_0}{k_0^2 \mathbb{C}_s \eta_0} \mathrm{Re}_m^{(4)\prime}\!\left(\xi_1^e, k_0 q\right) \sin\psi;$$

$$c_{73} = -\mathrm{Re}_m^{(3)}(\xi_1^e, k_{1R} q) \cos\psi;$$

$$c_{74} = -\mathrm{Re}_m^{(4)}(\xi_1^e, k_{1L} q) \cos\psi;$$

$$c_{75} = -\frac{\mathbb{C}_4}{\mathbb{C}_s \eta_1} \mathrm{Re}_m^{(3)\prime}\!\left(\xi_1^e, k_{1R} q\right) \sin\psi;$$

$$c_{76} = \frac{\mathbb{C}_4}{\mathbb{C}_s \eta_1} \mathrm{Re}_m^{(4)\prime}\!\left(\xi_1^e, k_{1L} q\right) \sin\psi;$$

$$c_7 = -A e_m \mathrm{Re}_m^{(1)}(\xi_1^e, k_0 q) \cos\psi;$$

$$c_{81} = -\frac{j\omega\varepsilon_0}{k_0^2 \mathbb{C}_s} \mathrm{Re}_m^{(4)\prime}\!\left(\xi_1^e, k_0 q\right) \cos\psi;$$

$$c_{82} = \tfrac{j}{\eta_0} \mathrm{Re}_m^{(4)}(\xi_1^e, k_0 q) \sin\psi;$$

$$c_{83} = \frac{j\mathbb{C}_6}{\mathbb{C}_s} \mathrm{Re}_m^{(3)\prime}\!\left(\xi_1^e, k_{1R} q\right) \cos\psi;$$

$$c_{84} = -\frac{j\mathbb{C}_6}{\mathbb{C}_s} \mathrm{Re}_m^{(4)\prime}(\xi_1^e, k_{1L} q) \cos\psi;$$

$$c_{85} = \tfrac{1}{j\eta_1} \mathrm{Re}_m^{(3)}(\xi_1^e, k_{1R} q) \cos\psi;$$

$$c_{86} = \tfrac{1}{j\eta_1} \mathrm{Re}_m^{(4)}(\xi_1^e, k_{1L} q) \sin\psi;$$

$$c_8 = -\tfrac{j\omega\varepsilon_0}{k_0^2 s} A e_m \mathrm{Re}_m^{(1)\prime}(\xi_1^e, k_0 q) \cos\psi;$$

$$k_{1R}^2 = p_1 + p_4 + \tfrac{1}{2}\{(p_1 - p_4)^2 - 4 p_2 p_3\}^{1/2};$$

$$k_{1L}^2 = p_1 + p_4 - \tfrac{1}{2}\{(p_1 - p_4)^2 - 4 p_2 p_3\}^{1/2};$$

$$p_1 = \frac{\omega}{\Delta_2}\{\mathbb{C}_4(\varepsilon_1 + \mu_0 \xi_1^2) + \mu_0 \xi_1 \mathbb{C}_2\};$$

$$p_2 = \frac{\omega\mu_0}{\Delta_s}(\mathbb{C}_2 - \mathbb{C}_4 \xi_1);$$

$$p_3 = \frac{\omega}{\Delta_2}\{\mathbb{C}_2(\varepsilon_1 + \mu_0 \xi_1^2) + \mu_0 \xi_1 \mathbb{C}_6\};$$

$$p_4 = \frac{\omega\mu_0}{\Delta_2}(\mathbb{C}_2 \xi_1 + \mathbb{C}_6);$$

$$\mathbb{C}_1 = \frac{1}{\Delta_1}\{-k_0^3 + 2\omega^2 \xi_1^2 k_0 \mu_0^2 + \omega^2 \mu_0 \varepsilon_1 k_0\};$$

$$\mathbb{C}_2 = \frac{1}{\Delta_1}\left(\omega\mu_0\xi_1 k_0^2 + \omega^3\varepsilon_1\xi_1\mu_0^2\right);$$

$$\mathbb{C}_3 = \frac{1}{\Delta_1} 2\omega^2\xi_1 k_0\mu_0^2;$$

$$\mathbb{C}_4 = \frac{1}{\Delta_1}\left(-\omega\mu_0 k_0^2 + \omega^3\mu_0^2\varepsilon_1\right);$$

$$\mathbb{C}_5 = -\frac{\omega}{\Delta_1}\left(2\omega\varepsilon_1\xi_1 k_0\mu_0 + 2\omega\xi_1^3 k_0\mu_0^2\right);$$

$$\mathbb{C}_6 = \frac{1}{\Delta_1}\left(\varepsilon_1 + \mu_0\xi_1^2\right)\left(\omega k_0^2 - \omega^3\mu_0\varepsilon_1\right);$$

$$\Delta_1 = k_0^4 - 4\omega^2\xi_1^2 k_0^2\mu_0^2 - 2\omega^2 k_0^2\mu_0\varepsilon_1 + \omega^4\mu_0^2\varepsilon_1^2;$$

$$\Delta_2 = \mathbb{C}_2^2 + \mathbb{C}_4\mathbb{C}_6;$$

$$Ae_n = \sqrt{8\pi}\, j^n E_0 Se_n(\cos\theta_i, k_0 q) / Ne_n(k_0 q);$$

$$Ne_n(k_0 q) = \int_0^{2\pi} \left[Se_n(\eta, k_0 q)\right]^2 dv \quad \text{with} = \cos v;$$

$$\mathbb{C}_s = q\left(\cos h^2\xi - \cos^2\eta\right)^{1/2};$$

$$¥_+ = \frac{p_1 - k_{1R}^2}{p_2};$$

$$¥_- = \frac{p_1 - k_{1L}^2}{p_2}.$$

Here, $Se_m(\cdot)$ is the angular Mathieu function[76] and v is the azimuthal angle that takes values between 0 and 2π.

Appendix D

The meanings of the used symbols in Eq. 80 are as follows:

$$d_{13} = \sin\theta_{r1} e^{jk_{1R}z\cos\theta_{r1}} H_n^2(k_{1R}r_0\sin\theta_{r1});$$

$$d_{14} = \sin\theta_{r1} e^{jk_{1R}z\cos\theta_{r1}} H_n^1(k_{1R}r_0\sin\theta_{r1});$$

$$d_{15} = \sin\theta_{r2} e^{jk_{1L}z\cos\theta_{r2}} H_n^2(k_{1L}r_0\sin\theta_{r2});$$

$$d_{16} = \sin\theta_{r2} e^{jk_{1L}z\cos\theta_{r2}} H_n^1(k_{1L}r_0\sin\theta_{r2});$$

$$d_{17} = -\sin\theta_t e^{jk_2 z\cos\theta_t} J_n(k_2 r_0\sin\theta_t);$$

$$d_{23} = k_{1R}\sin^2\theta_{r1} e^{jk_{1R}z\cos\theta_{r1}}\left(\mathbb{C}_4\mathbb{C}_+ - \mathbb{C}_2\right)H_n^{2'}\left(k_{1R}r_0\sin\theta_{r1}\right)$$
$$+\frac{n}{r_0}\sin\theta_{r1} e^{jk_{1R}z\cos\theta_{r1}}\left(-\mathbb{C}_1 + j\mathbb{C}_3\mathbb{C}_+\right)H_n^2\left(k_{1R}r_0\sin\theta_{r1}\right);$$

$$d_{24} = k_{1R}\sin^2\theta_{r1} e^{jk_{1R}z\cos\theta_{r1}}\left(\mathbb{C}_4\mathbb{C}_- - \mathbb{C}_2\right)H_n^{1'}\left(k_{1R}r_0\sin\theta_{r1}\right)$$
$$+\frac{n}{r_0}\sin\theta_{r1} e^{jk_{1R}z\cos\theta_{r1}}\left(-\mathbb{C}_1 + j\mathbb{C}_3\mathbb{C}_-\right)H_n^1\left(k_{1R}r_0\sin\theta_{r1}\right);$$

$$d_{25} = k_{1L}\sin^2\theta_{r2} e^{jk_{1R}z\cos\theta_{r2}}\left(\mathbb{C}_4\mathbb{C}_- - \mathbb{C}_2\right)H_n^{2'}\left(k_{1L}r_0\sin\theta_{r2}\right)$$
$$+\frac{n}{r_0}\sin\theta_{r2} e^{jk_{1L}z\cos\theta_{r2}}\left(-\mathbb{C}_1 + j\mathbb{C}_3\mathbb{C}_+\right)H_n^2\left(k_{1L}r_0\sin\theta_{r2}\right);$$

$$d_{26} = k_{1L}\sin^2\theta_{r2} e^{jk_{1L}z\cos\theta_{r2}}\left(\mathbb{C}_4\mathbb{C}_- - \mathbb{C}_2\right)H_n^{1'}\left(k_{1L}r_0\sin\theta_{r2}\right)$$
$$+\frac{n}{r_0}\sin\theta_{r2} e^{jk_{1L}z\cos\theta_{r2}}\left(-\mathbb{C}_1 + j\mathbb{C}_3\mathbb{C}_-\right)H_n^1\left(k_{1L}r_0\sin\theta_{r2}\right);$$

$$d_{28} = -\frac{E_0}{\sin\theta_t} e^{jk_2 z\cos\theta_t} J_n'(k_2 r_0\sin\theta_t);$$

$$d_{33} = \mathbb{C}_+ \sin\theta_{r1} e^{jk_{1R}z\cos\theta_{r1}} H_n^2\left(k_{1R}r_0\sin\theta_{r1}\right);$$

$$d_{34} = \mathbb{C}_+ \sin\theta_{r1} e^{jk_{1R}z\cos\theta_{r1}} H_n^1\left(k_{1R}r_0\sin\theta_{r1}\right);$$

$$d_{35} = \mathbb{C}_- \sin\theta_{r2} e^{jk_{1L}z\cos\theta_{r2}} H_n^2\left(k_{1L}r_0\sin\theta_{r2}\right);$$

$$d_{36} = \mathbb{C}_- \sin\theta_{r2} e^{jk_{1L}z\cos\theta_{r2}} H_n^1\left(k_{1L}r_0\sin\theta_{r2}\right);$$

$$d_{38} = \frac{\sin\theta_t e^{jk_2 z\cos\theta_t}}{j\eta_2} J_n(k_2 r_0\sin\theta_t);$$

$$d_{43} = -jk_{1R}\sin^2\theta_{r1} e^{jk_{1R}z\cos\theta_{r1}}\left(\mathbb{C}_4\mathbb{C}_+ + \mathbb{C}_6\right)H_n^{2'}\left(k_{1R}r_0\sin\theta_{r1}\right)$$
$$+\frac{n}{r_0}\sin\theta_{r1} e^{jk_{1R}z\cos\theta_{r1}}\left(j\mathbb{C}_5 - \mathbb{C}_1\mathbb{C}_+\right)H_n^2\left(k_{1R}r_0\sin\theta_{r1}\right);$$

$$d_{44} = -jk_{1R}\sin^2\theta_{r1} e^{jk_{1R}z\cos\theta_{r1}}\left(\mathbb{C}_4\mathbb{C}_+ + \mathbb{C}_6\right)H_n^{1'}\left(k_{1R}r_0\sin\theta_{r1}\right)$$
$$+\frac{n}{r_0}\sin\theta_{r1} e^{jk_{1R}z\cos\theta_{r1}}\left(j\mathbb{C}_5 - \mathbb{C}_1\mathbb{C}_+\right)H_n^1\left(k_{1R}r_0\sin\theta_{r1}\right);$$

$$d_{45} = -k_{1L}\sin^2\theta_{r2} e^{jk_{1L}z\cos\theta_{r2}}\left(\mathbb{C}_4\mathbb{C}_- + \mathbb{C}_6\right)H_n^{2'}\left(k_{1L}r_0\sin\theta_{r2}\right)$$
$$+\frac{n}{r_0}\sin\theta_{r2} e^{jk_{1L}z\cos\theta_{r2}}\left(j\mathbb{C}_5 - \mathbb{C}_1\mathbb{C}_-\right)H_n^2\left(k_{1L}r_0\sin\theta_{r2}\right);$$

$$d_{46} = -k_{1L}\sin^2\theta_{r2} e^{jk_{1L}z\cos\theta_{r2}}\left(\mathbb{C}_4\mathbb{C}_- + \mathbb{C}_6\right)H_n^{1'}\left(k_{1L}r_0\sin\theta_{r2}\right)$$
$$+\frac{n}{r_0}\sin\theta_{r2} e^{jk_{1L}z\cos\theta_{r2}}\left(j\mathbb{C}_5 - \mathbb{C}_1\mathbb{C}_-\right)H_n^1\left(k_{1L}r_0\sin\theta_{r2}\right);$$

$$d_{47} = -\frac{E_0}{j\eta_2} e^{jk_2 z\cos\theta_t} J_n'(k_2 r_0\sin\theta_t);$$

$$d_{51} = \sin\theta_s\, e^{-jk_0 z\cos\theta_s} H_n^2(k_0 r_1\sin\theta_s)\sin\psi;$$

$$d_{52} = \frac{E_0}{\sin\theta_s} e^{-jk_0 z\cos\theta_s} H_n^{2'}(k_0 r_1\sin\theta_s)\cos\psi;$$

$$d_5 = \sin\theta_i\, e^{jk_0 z\cos\theta_i} J_n(k_0 r_1\sin\theta_i)\sin\psi;$$

$$d_{63} = \sin\theta_{r1} e^{jk_{1R}z\cos\theta_{r1}} H_n^2\left(k_{1R}r_1\sin\theta_{r1}\right)\sin\psi$$
$$+\left[k_{1R}\sin^2\theta_{r1} e^{jk_{1R}z\cos\theta_{r1}}\left(\mathbb{C}_4\mathbb{C}_+ - \mathbb{C}_2\right)H_n^{2'}\left(k_{1R}r\sin\theta_{r1}\right)\right.$$
$$\left.+\frac{n}{r}\sin\theta_{r1} e^{jk_{1R}z\cos\theta_{r1}}\left(-\mathbb{C}_1 + j\mathbb{C}_3\mathbb{C}_+\right)H_n^{2'}\left(k_{1R}r_1\sin\theta_{r1}\right)\right]\cos\psi;$$

$$d_{64} = \sin\theta_{r1} e^{jk_{1R}z\cos\theta_{r1}} H_n^1\left(k_{1R}r_1\sin\theta_{r1}\right)\sin\psi$$
$$+\left[k_{1R}\sin^2\theta_{r1} e^{jk_{1R}z\cos\theta_{r1}}\left(\mathbb{C}_4\mathbb{C}_+ - \mathbb{C}_2\right)H_n^{1'}\left(k_{1R}r_1\sin\theta_{r1}\right)\right.$$
$$\left.+\frac{n}{r}\sin\theta_{r1} e^{jk_{1R}z\cos\theta_{r1}}\left(-\mathbb{C}_1 + j\mathbb{C}_3\mathbb{C}_+\right)H_n^{2'}\left(k_{1R}r_1\sin\theta_{r1}\right)\right]\cos\psi;$$

$$d_{65} = \sin\theta_{r2} e^{jk_{1L}z\cos\theta_{r2}} H_n^2(k_{1L}r_1\sin\theta_{r2})\sin\psi$$
$$+\left[k_{1L}\sin^2\theta_{r2} e^{jk_{1L}z\cos\theta_{r2}}\left(\mathbb{C}_4\mathbb{C}_- - \mathbb{C}_2\right)H_n^{2'}\left(k_{1L}r_1\sin\theta_{r2}\right)\right.$$
$$\left.+\frac{n}{r}\sin\theta_{r2} e^{jk_{1L}z\cos\theta_{r2}}\left(-\mathbb{C}_1 + j\mathbb{C}_3\mathbb{C}_-\right)H_n^{2'}\left(k_{1L}r_1\sin\theta_{r2}\right)\right]\cos\psi;$$

$$d_{66} = \sin\theta_{r2} e^{jk_{1L}z\cos\theta_{r2}} H_n^1(k_{1L}r_1'\sin\theta_{r2})$$
$$+\left[k_{1L}\sin^2\theta_{r2} e^{jk_{1L}z\cos\theta_{r2}}(\mathbb{C}_4\mathbb{C}_- - \mathbb{C}_2)H_n^{1'}(k_{1L}r_1'\sin\theta_{r2})\right.$$
$$\left.+\frac{n}{r}\sin\theta_{r2} e^{jk_{1L}z\cos\theta_{r2}}(-\mathbb{C}_1 + j\mathbb{C}_3\mathbb{C}_-)H_n^1(k_{1L}r_1'\sin\theta_{r2})\right]\cos\psi;$$

$$d_{71} = \sin\theta_s e^{-jk_0z\cos\theta_s} H_n^2(k_0r_1'\sin\theta_s)\cos\psi;$$

$$d_{72} = -\frac{E_0}{\sin\theta_s} e^{-jk_0z\cos\theta_s} H_n^{2'}(k_0r_1'\sin\theta_s)\sin\psi;$$

$$d_{73} = -\sin\theta_{r1} e^{jk_{1R}z\cos\theta_{r1}} H_n^2(k_{1R}r_1'\sin\theta_{r1})\cos\psi$$
$$+\left[k_{1R}\sin^2\theta_{r1} e^{jk_{1R}z\cos\theta_{r1}}(\mathbb{C}_4\mathbb{C}_+ - \mathbb{C}_2)H_n^{2'}(k_{1R}r_1'\sin\theta_{r1})\right.$$
$$\left.+\frac{n}{r}\sin\theta_{r1} e^{jk_{1R}z\cos\theta_{r1}}(-\mathbb{C}_1 + j\mathbb{C}_3\mathbb{C}_+)H_n^2(k_{1R}r_1'\sin\theta_{r1})\right]\sin\psi;$$

$$d_{74} = -\sin\theta_{r1} e^{jk_{1R}z\cos\theta_{r1}} H_n^1(k_{1R}r_1'\sin\theta_{r1})\cos\psi +$$
$$\left[k_{1R}\sin^2\theta_{r1} e^{jk_{1R}z\cos\theta_{r1}}(\mathbb{C}_4\mathbb{C}_+ - \mathbb{C}_2)H_n^{1'}(k_{1R}r_1'\sin\theta_{r1})\right.$$
$$\left.+\frac{n}{r}\sin\theta_{r1} e^{jk_{1R}z\cos\theta_{r1}}(-\mathbb{C}_1 + j\mathbb{C}_3\mathbb{C}_+)H_n^1(k_{1R}r_1'\sin\theta_{r1})\right]\sin\psi;$$

$$d_{75} = -\sin\theta_{r2} e^{jk_{1L}z\cos\theta_{r2}} H_n^2(k_{1L}r_1'\sin\theta_{r2})\cos\psi$$
$$+\left[k_{1L}\sin^2\theta_{r2} e^{jk_{1L}z\cos\theta_{r2}}(\mathbb{C}_4\mathbb{C}_- - \mathbb{C}_2)H_n^{2'}(k_{1L}r_1'\sin\theta_{r2})\right.$$
$$\left.+\frac{n}{r}\sin\theta_{r2} e^{jk_{1L}z\cos\theta_{r2}}(-\mathbb{C}_1 + j\mathbb{C}_3\mathbb{C}_-)H_n^2(k_{1L}r_1'\sin\theta_{r2})\right]\sin\psi;$$

$$d_{76} = -\sin\theta_{r2} e^{jk_{1L}z\cos\theta_{r2}} H_n^1(k_{1L}r_1'\sin\theta_{r2})\cos\psi$$
$$+\left[k_{1L}\sin^2\theta_{r2} e^{jk_{1L}z\cos\theta_{r2}}(\mathbb{C}_4\mathbb{C}_- - \mathbb{C}_2)H_n^{1'}(k_{1L}r_1'\sin\theta_{r2})\right.$$
$$\left.+\frac{n}{r}\sin\theta_{r2} e^{jk_{1L}z\cos\theta_{r2}}(-\mathbb{C}_1 + j\mathbb{C}_3\mathbb{C}_-)H_n^1(k_{1L}r_1'\sin\theta_{r2})\right]\sin\psi;$$

$$d_7 = \sin\theta_i e^{jk_0z\cos\theta_i} J_n(k_0r_1'\sin\theta_i)\cos\psi;$$

$$d_{81} = \frac{E_0}{j\eta_0} e^{jk_0z\cos\theta_s} H_n^{2'}(k_0r\sin\theta_s)\cos\psi;$$

$$d_{82} = \frac{E_0}{j\eta_0} e^{jk_0z\cos\theta_s} H_n^2(k_0r\sin\theta_s)\sin\psi;$$

$$d_{83} = -j\mathbb{C}_+ \sin\theta_{r1} e^{jk_{1R}z\cos\theta_{r1}} H_n^2(k_{1R}r_1'\sin\theta_{r1})\sin\psi$$
$$+\left[jk_{1R}\sin^2\theta_{r1} e^{jk_{1R}z\cos\theta_{r1}}(\mathbb{C}_2\mathbb{C}_+ + \mathbb{C}_6)H_n^{2'}(k_{1R}r_1'\sin\theta_{r1})\right.$$
$$\left.+\frac{n}{r_1}\sin\theta_{r1} e^{jk_{1R}z\cos\theta_{r1}}(j\mathbb{C}_5 - \mathbb{C}_1\mathbb{C}_+)H_n^2(k_{1R}r_1'\sin\theta_{r1})\right]\cos\psi;$$

$$d_{84} = -j\mathbb{C}_+ \sin\theta_{r1} e^{jk_{1R}z\cos\theta_{r1}} H_n^1(k_{1R}r_1'\sin\theta_{r1})\sin\psi$$
$$+\left[jk_{1R}\sin^2\theta_{r1} e^{jk_{1R}z\cos\theta_{r1}}(\mathbb{C}_2\mathbb{C}_+ + \mathbb{C}_6)H_n^{1'}(k_{1R}r_1'\sin\theta_{r1})\right.$$
$$\left.+\frac{n}{r_1}\sin\theta_{r1} e^{jk_{1R}z\cos\theta_{r1}}(j\mathbb{C}_5 - \mathbb{C}_1\mathbb{C}_+)H_n^1(k_{1R}r_1'\sin\theta_{r1})\right]\cos\psi;$$

$$d_{85} = -j\mathbb{C}_- \sin\theta_{r2} e^{jk_{1L}z\cos\theta_{r2}} H_n^2(k_{1L}r_1'\sin\theta_{r2})\sin\psi$$
$$+\left[jk_{1L}\sin^2\theta_{r2} e^{jk_{1L}z\cos\theta_{r2}}(\mathbb{C}_2\mathbb{C}_+ + \mathbb{C}_6)H_n^{2'}(k_{1L}r_1'\sin\theta_{r2})\right.$$
$$\left.+\frac{n}{r_1}\sin\theta_{r2} e^{jk_{1L}z\cos\theta_{r2}}(j\mathbb{C}_5 - \mathbb{C}_1\mathbb{C}_-)H_n^2(k_{1L}r_1'\sin\theta_{r2})\right]\cos\psi;$$

$$d_{86} = -j\mathbb{C}_- \sin\theta_{r2} e^{jk_{1L}z\cos\theta_{r2}} H_n^2(k_{1L}r_1'\sin\theta_{r2})\sin\psi$$
$$+\left[jk_{1L}\sin^2\theta_{r2} e^{jk_{1L}z\cos\theta_{r2}}(\mathbb{C}_2\mathbb{C}_+ + \mathbb{C}_6)H_n^{1'}(k_{1L}r_1'\sin\theta_{r2})\right.$$
$$\left.+\frac{n}{r_1}\sin\theta_{r2} e^{jk_{1L}z\cos\theta_{r2}}(j\mathbb{C}_5 - \mathbb{C}_1\mathbb{C}_-)H_n^1(k_{1L}r_1'\sin\theta_{r2})\right]\cos\psi;$$

$$d_8 = \frac{E_0}{j\eta_0} e^{jk_0z\cos\theta_i} J_n'(k_0r_1'\sin\theta_i)\cos\psi.$$

In these equations, $M_\pm = \yen_\pm$

Appendix E

The meanings of the used symbols in Eq. 89 are as follows:

$$e_{13} = \sin\theta_{r1} e^{jk_{2+}z\cos\theta_{r1}} H_n^2(k_{2+}r_0\sin\theta_{r1});$$
$$e_{14} = \sin\theta_{r1} e^{jk_{2+}z\cos\theta_{r1}} H_n^1(k_{2+}r_0\sin\theta_{r1});$$
$$e_{15} = \sin\theta_{r2} e^{jk_{2-}z\cos\theta_{r2}} H_n^2(k_{2-}r_0\sin\theta_{r2});$$
$$e_{16} = \sin\theta_{r2} e^{jk_{2-}z\cos\theta_{r2}} H_n^1(k_{2-}r_0\sin\theta_{r2});$$

$$e_{23} = \sin\theta_{r1} e^{jk_{2+}z\cos\theta_{r1}}$$
$$\left\{k_{2+}\sin\theta_{r1}(\mathbb{Q}_d q_+ - \mathbb{Q}_b)H_n^{2'}(k_{2+}r_0\sin\theta_{r1})\right.$$
$$\left.+\frac{n}{r_0}(-\mathbb{Q}_a + j\mathbb{Q}_c q_+)H_n^2(k_{2+}r_0\sin\theta_{r1})\right\};$$

$$e_{24} = \sin\theta_{r1} e^{jk_{2+}z\cos\theta_{r1}}$$
$$\left\{k_{2+}\sin\theta_{r1}(\mathbb{Q}_d q_+ - \mathbb{Q}_b)H_n^{1'}(k_{2+}r_0\sin\theta_{r1})\right.$$
$$\left.+\frac{n}{r_0}(-\mathbb{Q}_a + j\mathbb{Q}_c q_+)H_n^1(k_{2+}r_0\sin\theta_{r1})\right\};$$

$$e_{25} = \sin\theta_{r2} e^{jk_{2-}z\cos\theta_{r2}}$$
$$\left\{k_{2-}\sin\theta_{r2}(\mathbb{Q}_d q_- - \mathbb{Q}_b)H_n^{2'}(k_{2-}r_0\sin\theta_{r2})\right.$$
$$\left.+\frac{n}{r_0}(-\mathbb{Q}_a + j\mathbb{Q}_c q_-)H_n^2(k_{2-}r_0\sin\theta_{r2})\right\};$$

$$e_{26} = \sin\theta_{r2} e^{jk_{2-}z\cos\theta_{r2}}$$
$$\left\{k_{2-}\sin\theta_{r2}(\mathbb{Q}_d q_- - \mathbb{Q}_b)H_n^{1'}(k_{2-}r_0\sin\theta_{r2})\right.$$
$$\left.+\frac{n}{r_0}(-\mathbb{Q}_a + j\mathbb{Q}_c q_-)H_n^1(k_{2-}r_0\sin\theta_{r2})\right\};$$

$$e_{31} = \sin\theta_s e^{-jk_0z\cos\theta_s} H_n^2(k_0r_1'\sin\theta_s)\sin\psi;$$
$$e_{32} = \frac{1}{\sin\theta_s} e^{-jk_0z\cos\theta_s} H_n^{2'}(k_0r_1'\sin\theta_s)\cos\psi;$$
$$e_3 = \sin\theta_i e^{jk_0z\cos\theta_i} J_n(k_0r_1'\sin\theta_i)\sin\psi;$$

$$e_{43} = \sin\theta_{r1} e^{jk_{2+}z\cos\theta_{r1}}$$
$$\left[H_n^2(k_{2+}r_1'\sin\theta_{r1})\sin\psi + \left\{k_{2+}\sin\theta_{r1}(\mathbb{Q}_d q_+ - \mathbb{Q}_b)H_n^{2'}\right.\right.$$
$$\left.\left.(k_{2+}r_1'\sin\theta_{r1}) + \frac{n}{r_1}(-\mathbb{Q}_a + j\mathbb{Q}_c q_+)H_n^2(k_{2+}r_1'\sin\theta_{r1})\right\}\cos\psi\right];$$

$$e_{44} = \sin\theta_{r1} e^{jk_{2+}z\cos\theta_{r1}}$$
$$\left[H_n^1\left(k_{2+}r_1\sin\theta_{r1}\right)\sin\psi + \left\{k_{2+}\sin\theta_{r1}\left(\mathbb{Q}_d q_+ - \mathbb{Q}_b\right)H_n^{1\prime}\left(k_{2+}r_1\sin\theta_{r1}\right)\right.\right.$$
$$\left.\left. + \frac{n}{r_1}\left(-\mathbb{Q}_a + j\mathbb{Q}_c q_+\right)H_n^1\left(k_{2+}r_1\sin\theta_{r1}\right)\right\}\cos\psi\right];$$

$$e_{45} = \sin\theta_{r2} e^{jk_{2-}z\cos\theta_{r2}}$$
$$\left[H_n^2\left(k_{2-}r_1\sin\theta_{r2}\right)\sin\psi + \left\{k_{2-}\sin\theta_{r2}\left(\mathbb{Q}_d q_- - \mathbb{Q}_b\right)\right.\right.$$
$$\left.\left. H_n^{2\prime}\left(k_{2-}r_1\sin\theta_{r2}\right) + \frac{n}{r_1}\left(-\mathbb{Q}_a + j\mathbb{Q}_c q_+\right)H_n^2\left(k_{2-}r_1\sin\theta_{r2}\right)\right\}\cos\psi\right];$$

$$e_{46} = \sin\theta_{r2} e^{jk_{2-}z\cos\theta_{r2}}$$
$$\left[H_n^1\left(k_{2-}r_1\sin\theta_{r2}\right)\sin\psi + \left\{k_{2-}\sin\theta_{r2}\left(\mathbb{Q}_d q_- -_b\right)H_n^{1\prime}\left(k_{2-}r_1\sin\theta_{r2}\right)\right.\right.$$
$$\left.\left. + \frac{n}{r_1}\left(-\mathbb{Q}_a + j\mathbb{Q}_c q_-\right)H_n^1\left(k_{2+}r_1\sin\theta_{r2}\right)\right\}\cos\psi\right];$$

$$e_{51} = \sin\theta_s e^{-jk_0 z\cos\theta_s}H_n^2(k_0 r_1 \sin\theta_s)\cos\psi;$$
$$e_{52} = -\frac{1}{\sin\theta_s}e^{-jk_0 z\cos\theta_s}H_n^{2\prime}(k_0 r_1 \sin\theta_s)\sin\psi;$$

$$e_{53} = \sin\theta_{r1} e^{jk_{2+}z\cos\theta_{r1}}$$
$$\left[-H_n^2\left(k_{2+}r_1\sin\theta_{r1}\right)\cos\psi + \left\{k_{2+}\sin\theta_{r1}\left(\mathbb{Q}_d q_+ - \mathbb{Q}_b\right)H_n^{2\prime}\right.\right.$$
$$\left.\left. \left(k_{2+}r_1\sin\theta_{r1}\right) + \frac{n}{r_1}\left(-\mathbb{Q}_a + j\mathbb{Q}_c q_+\right)H_n^2\left(k_{2+}r_1\sin\theta_{r1}\right)\right\}\sin\psi\right];$$

$$e_{54} = \sin\theta_{r1} e^{jk_{2+}z\cos\theta_{r1}}$$
$$\left[-H_n^1\left(k_{2+}r_1\sin\theta_{r1}\right)\cos\psi + \left\{k_{2+}\sin\theta_{r1}\left(_d q_+ -_b\right)H_n^{1\prime}\left(k_{2+}r_1\sin\theta_{r1}\right)\right.\right.$$
$$\left.\left. + \frac{n}{r_1}\left(-_a + j_c q_+\right)H_n^1\left(k_{2+}r_1\sin\theta_{r1}\right)\right\}\sin\psi\right];$$

$$e_{55} = \sin\theta_{r2} e^{jk_{2-}z\cos\theta_{r2}}$$
$$\left[-H_n^2\left(k_{2-}r_1\sin\theta_{r2}\right)\cos\psi + \left\{k_{2-}\sin\theta_{r2}\left(\mathbb{Q}_d q_- - \mathbb{Q}_b\right)\right.\right.$$
$$\left.\left. H_n^{2\prime}\left(k_{2-}r_1\sin\theta_{r2}\right) + \frac{n}{r_1}\left(-\mathbb{Q}_a + j\mathbb{Q}_c q_-\right)H_n^2\left(k_{2-}r_1\sin\theta_{r2}\right)\right\}\sin\psi\right];$$

$$e_{56} = \sin\theta_{r2} e^{jk_{2-}z\cos\theta_{r2}}$$
$$\left[-H_n^1\left(k_{2-}r_1\sin\theta_{r2}\right)\cos\psi + \left\{k_{2-}\sin\theta_{r2}\left(\mathbb{Q}_d q_- - \mathbb{Q}_b\right)H_n^{1\prime}\left(k_{2-}r_1\sin\theta_{r2}\right)\right.\right.$$
$$\left.\left. + \frac{n}{r_1}\left(-\mathbb{Q}_a + j\mathbb{Q}_c q_-\right)H_n^1\left(k_{2-}r_1\sin\theta_{r2}\right)\right\}\sin\psi\right];$$

$$e_5 = \sin\theta_i e^{jk_0 z\cos\theta_i} J_n(k_0 r_1 \sin\theta_i)\cos\psi;$$
$$e_{61} = \frac{1}{j\eta_0} e^{-jk_0 z\cos\theta_s} H_n^{2\prime}(k_0 r_1 \sin\theta_s)\cos\psi;$$
$$e_{62} = \frac{E_0}{j\eta_0} e^{-jk_0 z\cos\theta_s} H_n^2(k_0 r_1 \sin\theta_s)\sin\psi;$$

$$e_{63} = \sin\theta_{r1} e^{jk_{2+}z\cos\theta_{r1}}$$
$$\left[-jq_+ H_n^2\left(k_{2+}r_1\sin\theta_{r1}\right)\sin\psi + \left\{jk_{2+}\sin\theta_{r1}\left(\mathbb{Q}_b q_+ + \mathbb{Q}_f\right)\right.\right.$$
$$\left.\left. H_n^{2\prime}\left(k_{2+}r_1\sin\theta_{r1}\right) + \frac{n}{r_1}\left(j\mathbb{Q}_e + \mathbb{Q}_a q_+\right)H_n^2\left(k_{2+}r_1\sin\theta_{r1}\right)\right\}\cos\psi\right];$$

$$e_{64} = \sin\theta_{r1} e^{jk_{2+}z\cos\theta_{r1}}$$
$$\left[-jq_+ H_n^1\left(k_{2+}r_1\sin\theta_{r1}\right)\sin\psi + \left\{jk_{2+}\sin\theta_{r1}\left(\mathbb{Q}_b q_+ + \mathbb{Q}_f\right)\right.\right.$$
$$\left.\left. H_n^{1\prime}\left(k_{2+}r_1\sin\theta_{r1}\right) + \frac{n}{r_1}\left(j\mathbb{Q}_e + \mathbb{Q}_a q_+\right)H_n^2\left(k_{2+}r_1\sin\theta_{r1}\right)\right\}\cos\psi\right];$$

$$e_{65} = \sin\theta_{r2} e^{jk_{2-}z\cos\theta_{r2}}$$
$$\left[-jq_- H_n^2\left(k_{2-}r_1\sin\theta_{r2}\right)\sin\psi + \left\{jk_{2-}\sin\theta_{r2}\left(\mathbb{Q}_b q_- + \mathbb{Q}_f\right)\right.\right.$$
$$\left.\left. H_n^{2\prime}\left(k_{2-}r_1\sin\theta_{r2}\right) + \frac{n}{r_1}\left(j\mathbb{Q}_e + \mathbb{Q}_a q_-\right)H_n^2\left(k_{2-}r_1\sin\theta_{r2}\right)\right\}\cos\psi\right];$$

$$e_{66} = \sin\theta_{r2} e^{jk_{2-}z\cos\theta_{r2}}$$
$$\left[-jq_- H_n^1\left(k_{2-}r_1\sin\theta_{r2}\right)\sin\psi + \left\{jk_{2-}\sin\theta_{r2}\left(\mathbb{Q}_b q_- + \mathbb{Q}_f\right)\right.\right.$$
$$\left.\left. H_n^{1\prime}\left(k_{2-}r_1\sin\theta_{r2}\right) + \frac{n}{r_1}\left(j\mathbb{Q}_e + \mathbb{Q}_a q_-\right)H_n^1\left(k_{2-}r_1\sin\theta_{r2}\right)\right\}\cos\psi\right];$$

$$e_6 = \frac{1}{j\eta_0} e^{jk_0 z\cos\theta_i} J_n'(k_0 r_1 \sin\theta_i)\cos\psi;$$

$$k_{2\pm} = \left[\left\{\mathbb{Q}_1 + \mathbb{Q}_4 \pm \sqrt{\left(\mathbb{Q}_1 - \mathbb{Q}_4\right)^2 - 4\mathbb{Q}_2 \mathbb{Q}_3}\right\}/2\right]^{1/2};$$

$$q_\pm = \frac{1}{\mathbb{Q}_2}\left(\mathbb{Q}_1 - k_{2\pm}^2\right);$$

$$\mathbb{Q}_1 = \frac{\omega}{\wp_2}\left\{\mathbb{Q}_d\left(\varepsilon_1 + \mu_z \xi_c^2\right) - \mu_z \xi_c \mathbb{Q}_b\right\};$$

$$\mathbb{Q}_2 = \frac{\omega}{\wp_2}\mu_z\left(\mathbb{Q}_b - \mathbb{Q}_d \xi_c\right);$$

$$\mathbb{Q}_3 = \frac{\omega}{\wp_2}\left[\mathbb{Q}_d\left(\varepsilon_1 + \mu_z \xi_c^2\right) - \mathbb{Q}_f \mu_z \xi_c\right];$$

$$\mathbb{Q}_4 = \frac{\omega}{\wp_2}\mu_z\left(\mathbb{Q}_b \xi_c + \mathbb{Q}_f\right);$$

$$\mathbb{Q}_a = \frac{1}{\wp_1}\left[-k_0^3 + 3\omega\kappa\xi_c k_0^2 + 2\omega^2 \xi_c^2 k_0\left(\mu^2 - \kappa^2\right) + \omega^2 \mu \varepsilon_1 k_0\right];$$

$$\mathbb{Q}_b = \frac{1}{\wp_1}\left[\omega k_0\left(\mu \xi_c k_0 + \omega \varepsilon_1 \kappa\right) + \omega^3 \varepsilon_1 \xi_c \left(\mu^2 - \kappa^2\right)\right];$$

$$\mathbb{Q}_c = \frac{1}{\wp_1}\left[2\omega^2 \xi_c k_0\left(\mu^2 - \kappa^2\right) + \omega\kappa k_0^2\right];$$

$$\mathbb{Q}_d = \frac{1}{\wp_1}\left[-\omega\mu k_0^2 + \omega^3 \varepsilon_1\left(\mu^2 - \kappa^2\right)\right];$$

$$\mathbb{Q}_e = -\frac{\omega}{\wp_1}\left[\xi_c k_0\left(2\omega\varepsilon_1 \mu + \xi_c \kappa k_0\right)\right.$$
$$\left. + 2\omega\xi_c^3 k_0\left(\mu^2 - k^2\right) + \omega^2 \varepsilon_1^2 \kappa\right];$$

$$\mathbb{Q}_f = \frac{1}{\wp_1}\left[\omega k_0^2\left(\varepsilon_1 + \mu \xi_c^2\right) - 2\omega^2 \varepsilon_1 \xi_c \kappa k_0\right.$$
$$\left. - \omega^3 \xi_c^2 \varepsilon_1\left(\mu^2 - \kappa^2\right) - \omega^3 \varepsilon_1^2 \mu\right];$$

$$\wp_1 = k_0^4 - 4\omega^2 \xi_c^2 k_0^2 (\mu^2 - \kappa^2) - 2\omega^2 k_0^2 \mu \varepsilon_1$$
$$+ \omega^4 \varepsilon_1^2 (\mu^2 - \kappa^2) - 4\omega \xi_c k_0^3 \kappa.$$
$$\wp_2 = \mathbb{Q}_b^2 + \mathbb{Q}_d \mathbb{Q}_f;$$

REFERENCES

1. Ling, H.; Chou, R.C.; Lee, S.W. Shooting and bouncing rays: Calculating the RCS of an arbitrarily shaped cavity. IEEE Trans. Antennas Propag. **1989**, *37* (2), 194–205.
2. Baldauf, J.; Lee, S.W.; Lin, L.; Jeng, S.K.; Scarborough, S.M.; Yu, C.L. High frequency scattering from trihedral corner reflectors and other benchmark targets: SBR versus experiments. IEEE Trans. Antennas Propag. **1991**, *39* (9), 1345–1351.
3. Volakis, J.L. XPATCH—A high-frequency electromagnetic scattering prediction code and environment for complex three-dimensional objects. IEEE Antennas Propag. Mag. **1994**, *36* (1), 65–69.
4. Auer, S.; Hinz, S.; Bamler, R. Ray-tracing simulation techniques for understanding high-resolution SAR images. IEEE Trans. Geosci. Remote Sens. **2010**, *48* (3), 1445–1456.
5. Xie, Z.; Xiaojian, X. Radar cross-section calculations and radar image formations using shooting and bouncing rays. In *Proceedings of the IEEE Image and Signal Processing (CISP)*, Hangzhou, 2, 804–808, 2013.
6. van de Hulst, H.C. *Light Scattering by Small Particles*; Dover: New York, 1957.
7. Lebrun, D.; Belaid, S.; Ren, K.F. Enhancement of wire diameter measurements: Comparison between Fraunhofer diffraction and Lorenz-Mie theory. Opt. Eng. **1996**, *35* (4), 946–950.
8. Bohren, C.F. Scattering of electromagnetic waves by an optically active cylinder. J. Colloid Interface Sci. **1978**, *66* (1), 105–109.
9. Bohren, C.F.; Huffman, D.R. *Absorption and Scattering of Light by Small Particles*; John Wiley & Sons: New York, 2007.
10. Balanis, C.A. *Advanced Engineering Electromagnetics*; John Wiley & Sons: New York, 2012.
11. Yang, P.; Liou, K.N. Geometric-optics-integral-equation method for light scattering by nonspherical ice crystals. Appl. Opt. **1996**, *35* (33), 6568–6584.
12. Mishchenko, M.; Hovenier, J.; Travis, L. *Light Scattering by Nonspherical Particles*; Academic Press: San Diego, CA, 2000.
13. Tsalamengas, J. Exponentially converging Nystro/Spl Uml/M's methods for systems of singular integral equations with applications to open/closed strip- or slot-loaded 2-D structures. IEEE Trans. Antennas Propag. **2006**, *54* (5), 1549–1558.
14. Choudhury, P.K.; Singh, O.N. Some multilayered and other unconventional lightguides. In *Electromagnetic Fields in Unconventional Structures and Materials*; Singh, O.N.; Lakhtakia, A.; Eds.; John Wiley & Sons: New York, 2000, 289–357.
15. Choudhury, P.K.; Singh, O.N. Electromagnetic materials. In *Encyclopedia of RF and Microwave Engineering*; Chang, K.; Eds.; John Wiley & Sons: New York, 2005, 1216–1231.
16. Choudhury, P.K.; Dey, K.K.; Basu, S. Micro- and nanoscale structures/systems and their applications in certain directions: A brief review. In *Nanoscale Spectroscopy with Applications*; Musa, S.M.; Eds.; CRC Press: Boca Raton, FL, 2013, 41–91.
17. Choudhury, P.K. Liquid crystal optical fibers and their applications—A brief overview. In *New Developments in Liquid Crystals and Applications*; Choudhury, P.K.; Eds.; Nova: New York, 2013, 229–266.
18. Lindell, I.V.; Sihvola, A.H.; Tretyakov, S.A.; Viitanen, A.J. *Electromagnetic Waves in Chiral and Bi-Isotropic Media*; Artech House: London, 1994.
19. Cho, K. Dispersion relation in chiral media: Credibility of Drude-Born-Fedorov equations. arXiv preprint arXiv:1501.01078, 2015.
20. Molina-Cuberos, G.J.; Garcia-Collado, A.J.; Margineda, J.; Nunez, M.J.; Martin, E. Electromagnetic activity of chiral media based on crank inclusions. IEEE Microw. Wirel. Compon. Lett. **2009**, *19* (5), 278–280.
21. Pendry, J.B. A chiral route to negative refraction. Science **2004**, *306* (5700), 1353–1355.
22. Kopyt, P.; Damian, R.; Celuch, M.; Ciobanu, R. Dielectric properties of chiral honeycombs—Modelling and experiment. Compos. Sci. Technol. **2010**, *70* (7), 1080–1088.
23. Lindell, I.V.; Viitanen, A.J. Plane wave propagation in uniaxial bianisotropic medium. Electron. Lett. **1993**, *29* (2), 150–152.
24. Lakhtakia, A. An electromagnetic trinity from "negative permittivity" and "negative permeability". Int. J. Infrared Millimeter Waves **2001**, *22* (12), 1731–1734.
25. Tretyakov, S.; Nefedov, I.; Sihvola, A.; Maslovski, S.; Simovski, C. Waves and energy in chiral nihility. J. Electromagn. Waves Appl. **2003**, *17* (5), 695–706.
26. Baqir, M.A.; Choudhury, P.K. Slow- and fast-waves through chiral/chiral nihility slab waveguides. J. Electromagn. Waves Appl. **2014**, *28* (18), 2229–2242.
27. Veselago, V.G. The electrodynamics of substances with simultaneously negative values of ε and μn. Soviet Physics Uspekhi **1968**, *10* (4), 509–514.
28. Iqbal, N.; Choudhury, P.K. Tailoring of group velocity dispersion in dual core chiral nihility waveguide. J. Mod. Opt. **2015**, *62* (17), 1419–1426.
29. Qiu, C.-W.; Zouhdi, S.; Tretyakov, S.; Li, L.-W. Possibilities for chiral nihility to achieve negative-index materials. Proc. Metamater. **2007**, *2007*, 213–214.
30. Qiu, C.-W.; Burokur, N.; Zouhdi, S.; Li, L.-W. Chiral nihility effects on energy flow in chiral materials. J. Opt. Soc. Am. A **2008**, *25* (1), 55–63.
31. Iqbal, N.; Choudhury, P.K. On the chiroferrite medium-based waveguide dispersion compensator. IEEE Photon. Technol. Lett. **2017**, *29* (1), 715–718.
32. Hovenac, E.A.; Lock, J.A. Assessing the contributions of surface waves and complex rays to far-field Mie scattering by use of the Debye series. J. Opt. Soc. Am. A **1992**, *9* (5), 781–795.
33. Lock, J.A.; Adler, C.L. Debye-series analysis of the first-order rainbow produced in scattering of a diagonally incident plane wave by a circular cylinder. J. Opt. Soc. Am. A **1997**, *14* (6), 1316–1328.

34. Li, R.; Han, X.E.; Ren, K.F. Generalized Debye series expansion of electromagnetic plane wave scattering by an infinite multilayered cylinder at oblique incidence. Phys. Rev. E **2009**, *79* (3), 036602.
35. Hamid, A.-K.; Cooray, F.R. Scattering by a perfect electromagnetic conducting elliptic cylinder. Prog. In Electromagn. Res. Lett. **2009**, *10*, 59–67.
36. Wait, J.R. Scattering of a plane wave from a circular dielectric cylinder at oblique incidence. Can. J. Phys. **1955**, *33* (5), 189–195.
37. Lim, K.Y.; Choudhury, P.K.; Yusoff, Z. Chirofibers with helical windings—An analytical investigation. Optik **2010**, *121* (11), 980–987.
38. Ghasemi, M.; Choudhury, P.K. On the sustainment of optical power in twisted clad dielectric cylindrical fibers. J. Electromagn. Waves Appl. **2013**, *27* (11), 1382–1391.
39. Ghasemi, M.; Choudhury, P.K. A revisit to the propagation through conducting helix loaded dielectric elliptic optical waveguides. Int. J. Microw. Opt. Technol. **2014**, *9* (1), 119–123.
40. Baqir, M.A.; Choudhury, P.K. Twisted clad microstructured optical fibers: Revisited. Appl. Phys. B **2014**, *117* (1), 481–486.
41. Ghasemi, M.; Choudhury, P.K. On the control of optical confinement in sheath helix loaded dielectric elliptical fiber. J. Mod. Opt. **2014**, *61* (18), 1509–1518.
42. Choudhury, P.K.; Kumar, D. On the slow-wave helical clad elliptical fibers. J. Electromagn. Waves Appl. **2010**, *24* (14,15), 1931–1942.
43. Baqir, M.A.; Choudhury, P.K. On the fast-waves in dispersive core twisted clad waveguides. IEEE Antennas Wirel. Propag. Lett. **2016**, *15*, 1735–1738.
44. Ghasemi, M.; Choudhury, P.K. Propagation through complex structured liquid crystal optical fibers. J. Nanophoton. **2014**, *8* (1), 083997-1–083997-13.
45. Ghasemi, M.; Choudhury, P.K. Conducting tape helix loaded radially anisotropic liquid crystal clad optical fiber. J. Nanophoton. **2015**, *9* (1), 093592-1–0093592-15.
46. Ghasemi, M.; Choudhury, P.K. On the conducting sheath double-helix loaded liquid crystal optical fibers. J. Electromagn. Waves Appl. **2015**, *29* (12), 1580–1592.
47. Chen, C.-L. Scattering by a helical sheath. IEEE Trans. Antennas Propag. **1966**, *14* (3), 283–290.
48. Kluskens, M.S.; Newman, E.H. Scattering by a multilayer chiral cylinder. IEEE Trans. Antennas Propag. **1991**, *39* (1), 91–96.
49. Tanaka, M.; Kusunoki, A. Depolarization properties of a chiral coated dielectric cylinder with application to RCS reduction. In *Proceedings of 1993 Asia Pacific Microwave Conference (APMC'93)*; IEEE: Hsinchu, Taiwan, 1993, 93. doi:10.1109/APMC.1993.468664.
50. Lakhtakia, A.; Geddes III, J.B. Scattering by a nihility cylinder. Int. J. Electron. Commun. **2007**, *61* (1), 62–65.
51. Jackson, J. *Classical Electrodynamics*. John Wiley & Sons: New York, 1999.
52. Luk'yanchuk, B.; Ternovsky, V. Light scattering by a thin wire with a surface-plasmon resonance: Bifurcations of the Poynting vector field. Phys. Rev. B **2006**, *73* (23), 235432-1–235432-12.
53. Wu, Y.; Li, J.; Zhang, Z.-Q.; Chan, C.T. Effective medium theory for magnetodielectric composites: Beyond the long-wavelength limit. Phys. Rev. B **2006**, *74* (8), 085111-1–085111-9.
54. Ni, Y.; Gao, L.; Qiu, C.-W. Achieving invisibility of homogeneous cylindrically anisotropic cylinders. Plasmonics **2010**, *5* (3), 251–258.
55. Mahmoud, A.M.; Engheta, N. Wave-matter interactions in epsilon-and-mu-near-zero structures. Nature Commun. **2014**, *5*, 5638-1–5638-7.
56. Naqvi, Q.A.; Mackay, T.G.; Lakhtakia, A. Optical refraction in silver: Counterposition, negative phase velocity and orthogonal phase velocity. Eur. J. Phys. **2011**, *32* (4), 883–893.
57. Mahmood, A.; Illahi, A.; Syed, A.; Choudhury, P.K. The effects on waves in silver metal due to L-nihility. J. Mod. Opt. **2013**, *60* (16), 1332–1336.
58. Choudhury, P.K.; Ghasemi, M.; Baqir, M.A. Twisted clad optical guides—Concept, features and applications. In *Optical Fibers: Technology, Communications and Recent Advances*; Ferreira, M.F.S.; Eds.; Nova: New York, 2017, 183–210.
59. Iqbal, N.; Choudhury, P.K.; Menon, P.S. Scattering from silver metal cylinder due to L-nihility coated with conducting sheath helix embedded dielectric medium. J. Electromagn. Waves Appl. **2015**, *29* (10), 1354–1374.
60. Chen, H.C. A coordinate-free approach to wave reflection from a uniaxially anisotropic medium. IEEE Trans. Microw. Theory Tech. **1983**, *31* (4), 331–336.
61. Iqbal, N.; Choudhury, P.K. Current density distribution along conducting sheath helix encapsulated silver metal cylinder coated with dielectric material under L-nihility and upon oblique incidence. Waves Random Complex Med. **2016**, *26* (4), 581–591.
62. Pendry, J.B. Negative refraction makes a perfect lens. Phys. Rev. Lett. **2000**, *85* (18), 3966–3969.
63. Shelby, R.A.; Smith, D.R.; Schultz, S. Experimental verification of a negative index of refraction. Science **2001**, *292* (5514), 77–79.
64. Engheta, N. An idea for thin subwavelength cavity resonators using metamaterials with negative permittivity and permeability. IEEE Antennas Wirel. Propag. Lett. **2002**, *1*, 10–13.
65. Tsakmakidis, K.L.; Boardman, A.D.; Hess, O. 'Trapped Rainbow' storage of light in metamaterials. Nature **2007**, *450*, 397–401.
66. Cai, W.; Chettiar, U.K.; Kildishev, A.V.; Shalaev, V.M. Optical cloaking with metamaterials. Nature Phot. **2007**, *1*, 224–227.
67. Tretyakov, S.; Sihvola, A.; Jylhä, L. Backward-wave regime and negative refraction in chiral composites. Photonics Nanostruct. **2005**, *3* (2,3), 107–115.
68. Zhang, C.; Cui, T.J. Negative reflections of electromagnetic waves in a strong chiral medium. Appl. Phys. Lett. **2007**, *91* (19), 194101-1–194101-3.
69. Choudhury, P.K.; Yoshino, T. Dependence of optical power confinement on core/cladding chiralities in chirofibers. Microw. Opt. Technol. Lett. **2002**, *32* (5), 359–364.
70. Choudhury, P.K.; Yoshino, T. Characterization of the optical power confinement in a simple chirofiber. Optik **2002**, *113* (2), 89–96.
71. Choudhury, P.K. On the propagation of electromagnetic waves through parabolic cylindrical chiroguides with small flare angles. Microw. Opt. Technol. Lett. **2002**, *33* (6), 414–419.

72. Choudhury, P.K. Partial electromagnetic wave guidance by a parabolic cylindrical chiroboundary with small flare angle. Optik **2002**, *113* (4), 177–180.
73. Iqbal, N.; Baqir, M.A.; Choudhury, P.K. Waves in microstructured conducting sheath helix embedded optical guides with chiral nihility and chiral materials. J. Nanomat. **2014**, *2014*, Article ID 362739.
74. Iqbal, N.; Choudhury, P.K. On the power distributions in elliptical and circular helically designed chiral nihility core optical fibers. J. Nanophoton. **2016**, *10* (1), 016008-1–016008-12.
75. Iqbal, N.; Choudhury, P.K.; Menon, P.S. Scattering of waves by chiral medium coated dielectric elliptical/circular cylinders with PEC sheath helix loading. J. Electromagn. Waves Appl. **2016**, *30* (11), 1504–1518.
76. Sebak, A.-R. Scattering from dielectric-coated impedance elliptic cylinder. IEEE Trans. Antennas Propag. **2000**, *48* (10), 1574–1580.
77. Iqbal, N.; Baqir, M.A.; Choudhury, P.K. Dispersion features of conducting sheath helix embedded elliptical and circular fibers with chiral nihility core. J. Nanomat. **2015**, *2015*, Article ID 912569.
78. Chen, J.; Wang, Y.; Jia, B.; Geng, T.; Li, X.; Feng, L.; Qian, W.; Liang, B.; Zhang, X.; Gu, M. Observation of the inverse Doppler effect in negative-index materials at optical frequencies. Nature Phot. **2011**, *5*, 239–245.
79. Iqbal, N.; Choudhury, P.K. Scattering due to twisted PEC cylinder coated with chiral/chiroferrite mediums. Optik **2016**, *127* (17), 7030–7039.
80. Hasanovic, M.; Mei, C.; Lee, J.K.; Arvas, E. Frequency-domain solution to electromagnetic scattering from dispersive chiroferrite materials. ACES J. **2013**, *28* (7), 565–572.
81. Ghaffar, A.; Yaqoob, M.; Alkanhal, M.A.; Sharif, M.; Naqvi, Q.A. Electromagnetic scattering from anisotropic plasma-coated perfect electromagnetic conductor cylinders. Int. J. Electron. Commun. **2014**, *68* (8), 767–772.
82. Baqir, M.A.; Choudhury, P.K. Propagation through uniaxial anisotropic chiral waveguide under DB-boundary conditions. J. Electromagn. Waves Appl. **2013**, *27* (6), 783–793.
83. Baqir, M.A.; Choudhury, P.K. On the propagation and sensitivity of optical fiber structure with twisted clad DB interface under slow-wave approximation. Opt. Commun. **2015**, *338*, 511–516.
84. Baqir, M.A.; Choudhury, P.K. Investigation of uniaxial anisotropic chiral metamaterial waveguide with perfect electromagnetic conductor loading. Optik **2015**, *126* (11,12), 1228–1232.
85. Cheng, Q.; Cui, T.J. Negative refractions in uniaxially anisotropic chiral media. Phys. Rev. B **2006**, *73* (11), 113104-1–113104-4.
86. Sihvola, A.; Wallen, H.; Yla-Oijala, P.; Taskinen, M.; Kettunen, H.; Lindell, I.V. Scattering by DB Spheres. IEEE Antennas Wirel. Propag. Lett. **2009**, *8*, 542–545.
87. Leonhardt, U.; Tyc, T. Broadband invisibility by non-Euclidean cloaking. Science **2009**, *323* (5910), 110–112.
88. Lai, Y.; Chen, H.; Zhang, Z.-Q.; Chan, C. Complementary media invisibility cloak that cloaks objects at a distance outside the cloaking shell. Phys. Rev. Lett. **2009**, *102* (9), 093901-1–093901-4.
89. Chen, X.; Luo, Y.; Zhang, J.; Jiang, K.; Pendry, J.B.; Zhang, S. Macroscopic invisibility cloaking of visible light. Nature Commun. **2011**, *2*, 176-1–17-6.

Semantic Processing

Phivos Mylonas
Department of Informatics, Ionian University, Corfu, Greece

Abstract
Semantic processing of multimedia content focuses in principle on the analysis of digital audiovisual content according to its high-level characteristics or entities to be derived in an (semi-)automated manner by suitable computational equipment. The notion of semantics is crucial in the process, since a "correct" machine-processable interpretation will allow for content providers to efficiently manipulate content and provide meaningful services to people in accordance with their individual standards, tastes, and preferences. The overall goal remains, of course, to contribute to the bridging of the gap between the semantic nature of user/people's needs and raw multimedia content. The herein discussed approach analyzes visual content (such as still images or video sequences) and its associated textual annotation, in order to extract the underlying semantics and construct a meaningful semantic index, based on a unified knowledge model. Content of interest may then be retrieved from the semantics by carrying out semantic interpretation and expansion. All described processes are based on a semantic processing methodology, employing fuzzy algebra and principles of taxonomic knowledge representation, illustrating the semantic unification. As a result, the overall contribution of semantic processing to the improvement of multimedia content understanding and retrieval effectiveness is of great importance for the entire research community.

Keywords: Content analysis; Knowledge management; Multimedia semantics; Visual context.

INTRODUCTION

The task of multimedia content indexing and retrieval has been influenced during the last decade by the important progress in numerous fields, such as digital content production, archiving and standardization, multimedia database management, multimedia signal processing, analysis and coding, computer vision, artificial and computational intelligence, human–computer interaction, and information retrieval. One major obstacle, though, multimedia retrieval systems still need to overcome in order to gain widespread acceptance is the so-called *semantic gap*[1,2] (Fig. 1).

This refers to the extraction of the semantic content of multimedia entities, the interpretation of user information needs and requests, as well as to the matching between the two. This obstacle becomes even harder when attempting to access vast amounts of multimedia information stored in different audiovisual (a/v) archives and represented in different formats. Among them, digital photographs and video sequences are the most demanding and complex data structures, due to their large amounts of spatiotemporal interrelations; video understanding is a key step toward more efficient manipulation of visual media, presuming semantic information extraction. Current and evolving international standardization activities, such as MPEG-4,[3] MPEG-7,[4] and MPEG-21[5] for video, or JPEG-2000[6] for still images, deal with aspects related to a/v content and metadata coding and representation. Syntactic description seems to be well in hand in MPEG-7, but fleshing out the semantic description has not yet received the required attention. It becomes clear among the research community dealing with image processing and content-based retrieval, that the results to be obtained will be ineffective, unless major focus is given to the semantic information level, defining what most users desire to retrieve. Thus, in order to close the loop between the user and available content, the extraction of information at a semantic level is required.

In recent years, several research activities emerged in the direction of *knowledge acquisition and modeling*, capturing knowledge from raw information and multimedia content in distributed repositories to turn poorly structured information into machine-processable knowledge.[7–11] A second direction is *knowledge sharing and use*, combining semantically enriched information with context, so as to provide inferencing for decision support and collaborative use of trusted knowledge between organizations.[12,13] Finally, in the *intelligent content* vision, multimedia objects integrate content with metadata and intelligence and learn to interact with devices and networks.[14] It is becoming apparent in all the above research fields that integration of diverse, heterogeneous, and distributed—preexisting—multimedia content will only be feasible through the design of *mediator systems*. In Biskup et al.,[15] for

Encyclopedia of Image Processing, First Edition
DOI: 10.1201/9781351045636-140000375
Copyright © 2018 by Taylor & Francis. All rights reserved.

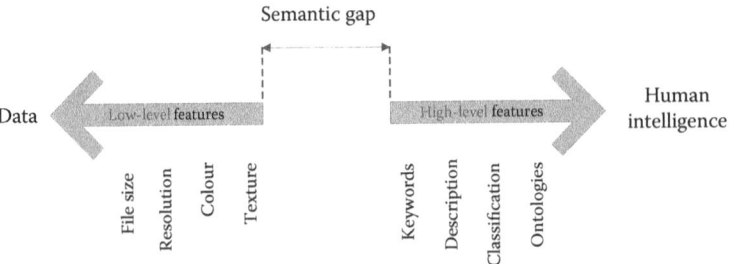

Fig. 1 The semantic gap

instance, a multimedia mediator is designed to provide a well-structured and controlled gateway to multimedia systems, focusing on schemas for semi-structured multimedia items and object-oriented concepts, while Altenschmidt et al.[16] focuses on security requirements of such mediated information systems. On the other hand, Brink et al.[17] deals with media abstraction and heterogeneous reasoning through the use of a unified query language for manually generated annotation, again without dealing with content or annotation semantics. A semantically rich retrieval model is suggested in the work of Glöckner and Knoll,[18] based on fuzzy set theory with domain-specific methods for content analysis and allowing natural language queries. Finally, Cruz and James[19] focus on the design of a single intuitive interface supporting visual query languages to access distributed multimedia databases.

In principle, the term *semantic processing* refers to a standalone, integrated approach, offering user-friendly and highly informative access to heterogeneous, distributed multimedia (audiovisual) pieces of information (archives). Focusing on a unified semantic analysis of digital multimedia information, as well as of their user-related needs, queries, and profiles, it contributes toward bridging of the gap between the semantic nature of initial user needs and raw multimedia content. As a result, it serves as a mediator between users and audiovisual archives, providing access to a/v content characterized by *semantic phrasing of the request*, *unified handling*, and *personalized response*. The core contribution of this task relies on the fact that it provides the missing link between low-level a/v features and high-level semantics that underlie in video and still images, on the one hand, and the purely semantic needs of users, on the other hand. To achieve this within the framework of semantic processing, we typically retrieve a/v content and associated textual annotation from participating a/v archives and perform visual and textual analysis to extract the underlying semantics and construct a semantic index, based on a unified knowledge model. We may then accept user queries, and, carrying out semantic interpretation and expansion, retrieve a/v content from the index, similarly to traditional text retrieval. Personalized ranking may be also supported, while user profiles are automatically generated and updated by monitoring and analyzing usage history. All above processes are typically based on a common semantic processing methodology, employing fuzzy algebra and principles of novel taxonomic knowledge representations.

OVERVIEW

Architecture

The general system architecture of a semantic processing framework is briefly depicted in Fig. 2. Connections between its main subsystems are to be identified, that is, a single *user interface* provides a user-friendly access to all participating archives, whereas the a/v *archive interfaces* are responsible for the communication between the main system and each a/v archive. A suitably constructed database is typically used to store the knowledge of the system, the semantic index, and the user profiles. The main framework consists typically of three subsystems: (a) *semantic unification*, (b) *searching*, and (c) *personalization*. The semantic unification subsystem constructs and updates the semantic index, whereas the personalization subsystem updates potential user profiles. The searching subsystem typically analyzes user queries, carries out matching with the semantic index, and returns retrieved content to the end users.

Data Models

Since the above framework is aimed to operate as a mediator between the end user and diverse a/v archives, the mapping of archive content to a uniform data model is of crucial importance. The specification of the model itself is a challenging issue, as it needs to be descriptive enough, to adequately and meaningfully serve user queries, and at the same time, abstract and general enough, to accommodate the mapping of the content of any a/v archive at a semantic level. In the following, we provide the overview of such a data model, consisting mainly out of two components: (a) the *knowledge model* and (b) the *semantic index* (Fig. 3).

Knowledge Model

The knowledge model contains all semantic information used in the system. It supports structured storage of

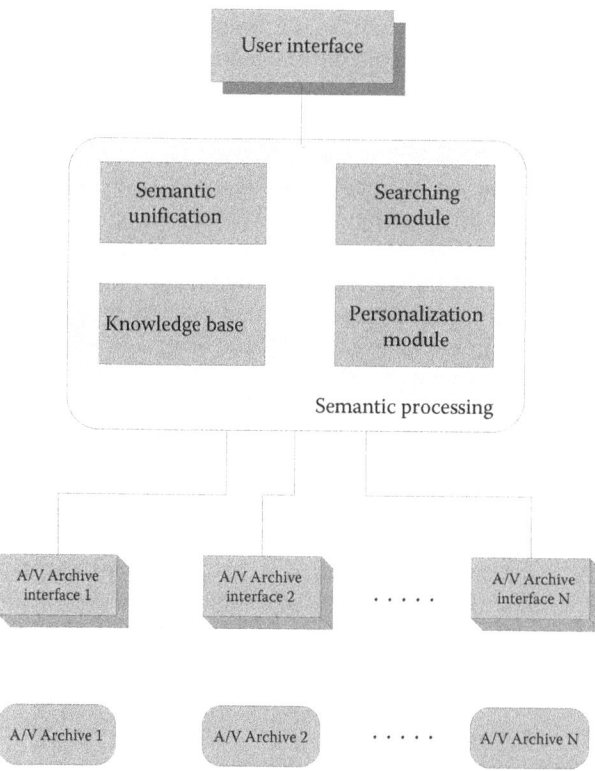

Fig. 2 General system architecture of a semantic processing framework

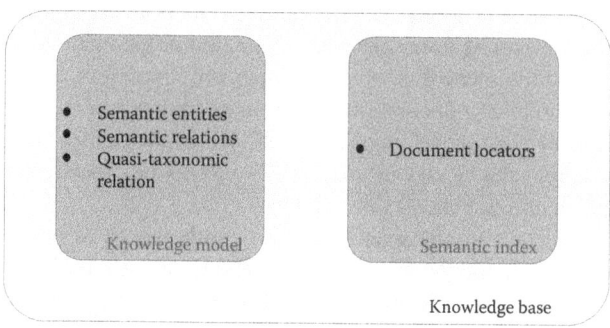

Fig. 3 Overview of the knowledge base data model

semantic entities and relations that experts have defined for indexing and retrieval purposes. Among other actions, it allows expanding a user query by looking for synonyms or related concepts. Three main types of information are introduced in the model:

1. *Semantic entities*: Entities such as thematic categories, objects, events, concepts, agents, and semantic places and times.
2. *Semantic relations*: Relations linking semantic entities, for example, "part of," "specialization of," and so on.
3. A *quasi-taxonomic relation*: A taxonomic knowledge representation to interpret the meaning of a multimedia item, composed of several elementary relations, also referred to as a taxonomy.

The knowledge model is manually constructed for a limited application set of specific multimedia content categories using the experts' assessment.

Semantic Index

The semantic index is used to collect the results of multimedia content analysis in order to support unified access to archives. The index contains sets of information locators for each semantic entities, identifying which semantic entities have been associated to each available multimedia item. It is used for fast and uniform retrieval of content related to the semantic entities specified in, or implied by, the query and the user profile. Content locators associated to index entities may link to complete a/v content, objects, still images, or other video decomposition units that may be contained in the a/v databases.

KNOWLEDGE REPRESENTATION

The theoretical basis on top of which the framework under discussion is constructed derives from careful selection and definition of an appropriate knowledge model. A typical model contains a set of semantic entities and semantic relations between them, which form the basic elements toward semantic interpretation. Through this knowledge representation, a detailed content description of all potential multimedia archives, such as still images, is established in a unified manner. Due to the fact that relations among real-life entities are always a matter of degree, and are, therefore, best modeled using fuzzy relations instead of crisp ones, the best approach followed is based on a formal methodology founded on fuzzy relational algebra and the exploitation of contextual information.[20]

The Notion of Context

It is common knowledge that the term *context* can take on many meanings and there is, in general, no solid definition that satisfies and covers the many aspects the term is used. The long history of the term appearance and usage varies from diverse areas of computer science to even philosophical and medical approaches.[21] In the field of computer science, the interest in contextual information is of great importance in artificial intelligence, information retrieval, and image and video analysis.[22] The nature of the applications in these fields is the one that signifies context, mostly dominated by rapid changes in the user's context. An indicative example is formed by handheld and ubiquitous computing.[24] Still, effective use of available contextual information within multimedia applications remains an open and challenging problem, although several researchers have tried to categorize context-aware applications in general according to subjective criteria, thus resulting in classes of applications.[25]

A fundamental problem tackled via access to and processing of contextual information is the bridging of two fundamental gaps in the literature: the *semantic* and the *sensory gaps*.[22] As already mentioned earlier, the *semantic gap* forms an issue inherent in most developments of multimedia systems and applications and may be described as the gap between the high-level semantic descriptions humans ascribe to images and the low-level features that machines can automatically parse. Given, for instance, the raw digital image of a wolf (Fig. 4), image analysis may extract feature or vectors (so-called descriptors) that focus on the segmented particles, salient regions, color histograms, etc.[23] Still, semantic processing of multimedia content includes and may advance research toward steps including (Fig. 5) a prototypical combination of image descriptors, extraction of semantic concepts, and assignment of symbolic names to them, as well as the utilization and exploitation of metadata information and/or identification of higher level entities inter- and intra-relations.[23]

As broader as the image domain gets, the wider the gap between the feature description and the semantic interpretation is. Narrowing down the domain to a specialized image one results into a smaller gap between features and their semantic, so domain-specific models may help in tackling the problem. However, the latter are not sufficient, but are considered as the first step toward an efficient approach to the problem. To illustrate all of the above, consider, for example, a picture of a man tossing a blue ball to a dog on the beach (Fig. 6a), which would be "seen" by a vision system as a series of moving color regions or the analysis of a medical image, which would be tackled only as raw numerical medical data (Fig. 6b). Contextual information, such as the relationships between the man, the dog, the location where the ball is being thrown and the significance of this event to the person taking the picture, in the first case, or qualitative information, such as the perspective of the image, or even the age of the associated patient, in the second case, are all gone.

The *sensory gap* is described as the gap between an object and the computer's ability to sense and describe that object. For example, for some computational systems, a *car* ceases to be a *car* if there is a tree in front of it, effectively dividing the car in two parts from the machine's perspective. Characteristics different in nature and texture may determine the demands of the search and retrieval methods, for example, possible presence or absence of occlusion, illumination, and clutter. In other words, the sensory gap can be thought as the gap between the object in the world and the information in a computational description derived from a recording of the particular scene. The latter clearly yields uncertainty in what is known about the state of the object and is particularly poignant when a precise knowledge of the capturing conditions is missing. For instance, considering the famous Mona Lisa painting, there might be a sensory gap in the sense of the inability to record the scene due to, for example, too few colors or pixels, too low light conditions, too small memory, or too few frames per second imposed by the hardware (sensor) that attempts the video capture (Fig. 7) prior to the depiction of the captured content to its end user.[26]

All in all, whenever there is a gap between an object and a computer's ability to sense and describe the object, the sensory gap is present. And this is also the case when an infinite number of different "signals" can be produced by the same object and different objects can produce similar signals, like it is the case of an image of the same object taken from different viewpoints (Fig. 8). Human perceptual machinery excels at recognizing when different "signal patterns" are the same object or when similar patterns are different ones, but the problem is difficult for computers and clearly lies in their ability to "understand" this process.[27]

Still, it is contextual knowledge that may enable computational systems to bridge these semantic and sensory gaps. With the advent of all kind of new multimedia-enabled devices and multimedia-based systems, new opportunities arise to infer the media semantics. Contextual metadata are capable of playing the important role of a "semantic mediator." Toward that scope, two aspects of context seem to have special salience in most multimedia applications: *where* and *when*, that is, *spatial* and *temporal* context. By taking into account the spatial and to lesser extend the

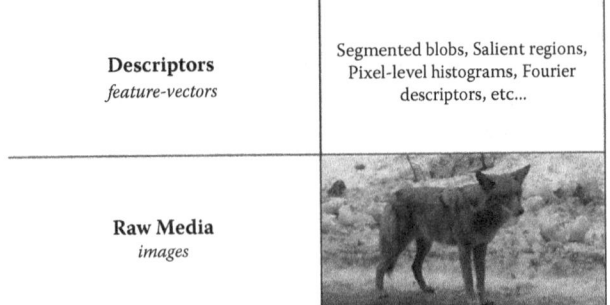

Fig. 4 Raw media and primitive image analysis example

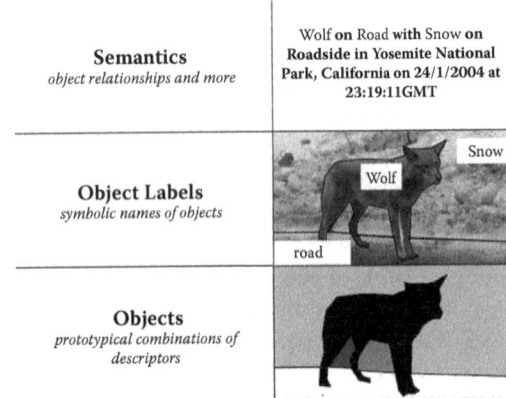

Fig. 5 Semantic processing of the previous example

Semantic Processing

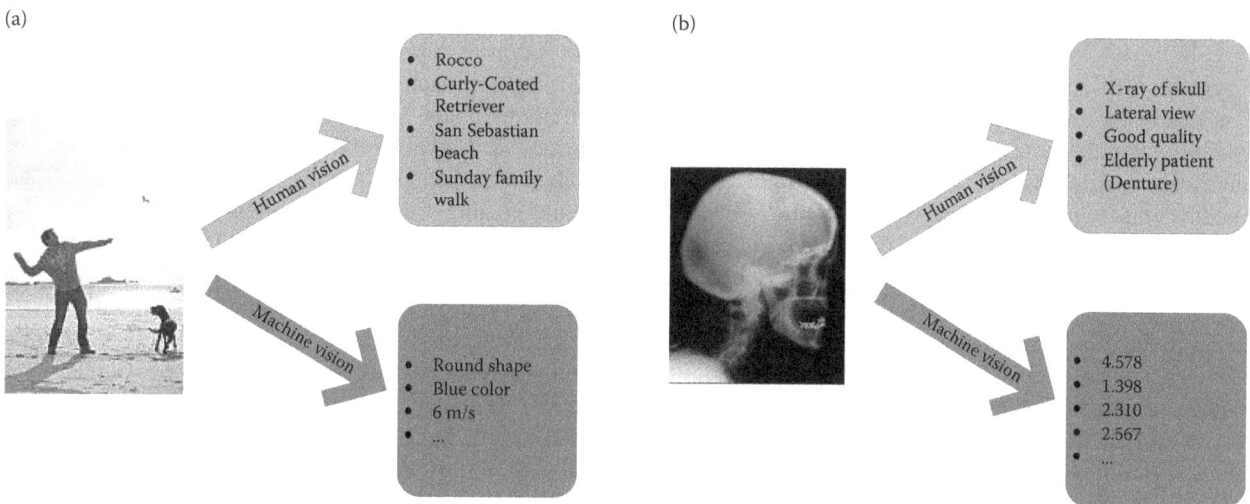

Fig. 6 Human versus machine vision interpretation: (a) a man tossing a ball to a dog; (b) a medical image of a human skull

Fig. 7 The sensory gap—limitations imposed by sensors

Fig. 8 The sensory gap—limitations imposed by different viewpoints

temporal context, we are able to gather and incorporate user interaction to adjust and add the extra information needed for interpretation. It is common knowledge though that context itself appears in various forms and modifications and researchers commonly emphasize distinctions between different types of context. Each context type illustrates different aspects of it and consequently has very little in common with the others.[28] Depending on the specific objectives of the task at hand, different formalizations of what is meant by context have been developed. For modern knowledge-based multimedia systems, context has a rather precise meaning and usage. In this framework, a constant enhancement of offered capabilities and functionalities is introduced, based on the always increasing contextual aspect of information provided by their end users. Context is exploited in pursue of a more efficient personalized approach, bringing the end user in the center of the application's interest.

Visual Context Identification

In general, the task of suitable visual context definition and identification is very important in the integration of a multimedia content-based semantic processing system. This is mainly due to the fact that all knowledge required for multimedia content analysis is thought to be context-sensitive, resulting in a specific need for formal definitions of context structures, prior to any static or dynamic context detection and analysis. The first objective formed within this task is the definition of the suitable aspect of context at hand, providing conceptual and audiovisual information. The latter will be used for context tailoring in several stages through the life cycle of the system's content, such as creation and analysis, consumption, and user interaction.

We may identify two related types of context with respect to its usage and applicability within multimedia systems, namely, the *context of content analysis* and the *context of use*. *Context of content analysis* refers to the context during the phase of analysis, including tasks such as knowledge-assisted analysis and reasoning. It is intended to be used to aid the extraction of semantic metadata both at the level of simple concepts and at the level of composite events and higher level concepts. In this manner, it forms the main employment of visual context in the semantic processing framework, for example, by performing scene classification to detect whether an image or video clip represents *city* or *landscape* content (Fig. 9), essentially aiding the analysis process to detect and recognize specific concepts or objects in the content. On the other hand, *context of use* is related to the use of content by a system's application modules, such as search/retrieval and personalization. In this case, given the multimedia content and metadata, contextual information from an external source are provided, consisting mainly of information about the particular user, network and client device, and so on. Ultimately, improved design and development of these applications is achieved, including retrieval, search, browsing, sharing, and management of content.

In multimedia computing applications, the aspects of context, that are thought to be the most suitable and appropriate for research and progress, are the ones of visual context described above. Therefore, from now on they may be presented under a common approach, summarized in the notion of *visual context*. Visual context forms a rather classical approach to context, tackling it from the scope of environmental or physical parameters in multimedia applications. Different architectures, conceptual approaches, and models support dynamic and adaptive modeling of visual context. One of the main objectives in the field is the combination of context parameters extracted from low-level visual features with higher level concepts, like fuzzy set theory, to support reasoning. Specifically, the context description supports fuzziness in order to face the uncertainty introduced by content analysis or the lack of knowledge. This context representation also supports audiovisual information (e.g., lighting conditions and information about the environment) and is separately handled by visual context models. The second objective is visual context analysis, that is, to take into account the extracted/recognized concepts during content analysis in order to find the specific context, express it in a structural description form, and use it for improving or continuing the content analysis, indexing, and searching procedures, as well as for content personalization purposes.

In terms of *knowledge-assisted content analysis and processing*, a set of core functionalities of the multimedia application is defined, regarding the way such a system is expected to execute knowledge-assisted image analysis functions automatically or in a supervised mode, either to detect or to recognize parts of content. Additionally, it is thought to generate or assist end users classify their contents and metadata, through suggestions or sorting being performed in a sophisticated way, making quite naturally implicit use of context analysis functionalities. For example, in a face recognition scenario, visual clues help the system detect the right person. Issues relating more to the automatic creation of metadata even after analysis, for example, through inference, make use of context as different sources of information (different analysis modules, textual inputs) are also integrated.

As far as *retrieval* is concerned, a set of core functionalities of a multimedia search and retrieval system may also be defined; there are many distinct aspects suggested and commented by users, regarding the way of performing searches, the type of searches they expect to have and

Fig. 9 City/landscape classification

the constraints they imagine. Organizing multimedia data into meaningful categories marked by end users as being important could, for example, exploit contextual information. Additionally, several use case scenarios illustrate a system's capability of learning a user's interest in order to adapt its behavior to the assumed interests, taking into account *personalization* aspects. With respect to the learning functionality, users are in general concerned about privacy, while the adaptive system behavior can also be commented in terms of controllability. Retrieval is strongly related to context, when tackling textual query analysis, search by semantic, visual, or metadata similarity, semantic grouping, browsing and rendering of retrieved content, personalization, and relevance feedback. However, user browsing capabilities, together with retrieval capabilities, suppose detection of common metadata, which is not related to visual context. Another form of context, dealing mostly with the semantic part of the analysis would be more useful in this case.[29] In any case, search by visual similarity may benefit by the use of visual context information, as in the cases of scene classification and object detection.

It should have been clear by now that visual context can be clearly exploited in *multimedia content processing*. During the phase of image metadata generation from one or more content analysis application modules, scene classification and/or object detection techniques can be of use, providing the necessary contextual information, for example, information about indoor/outdoor scenery at the metadata level. The same applies to knowledge-assisted image/video analysis and metadata generation; techniques and methodologies can be helpful in implementing scene classification and object detection. Moreover, when classifying and sorting content, degrees of confidence can be obtained by taking into consideration contextual information. The latter is useful for extraction of the semantics of content and detection of repeated content. It can help deal with identification of semantically similar content, as well as analyze relative metadata information stored or derived automatically from images. All aspects of the so far presented contextual information assist toward transparency and automation of knowledge extraction, providing the means to gather additional meaningful information in ontology mapping, image/video analysis, and metadata generation processes.

Visual Context in Image Analysis

By visual context in the sequel, we shall refer to all information related to the visual scene content of a still image (or video sequence) that may be useful for its analysis. Image analysis deals with a few well-known research problems shortly presented in the following, whereas visual context is mostly related to two of the problems in image analysis, namely, *scene classification* and *object detection* (Fig. 10). *Scene classification* forms a top-down approach,

where low-level visual features are employed to globally analyze the scene content and classify it in one of a number of predefined categories, for example, indoor/outdoor, city/landscape, and so on. On the other hand, *object detection/recognition* is a bottom-up approach that focuses on local analysis to detect and recognize specific objects in limited regions of an image, without explicit knowledge of the surrounding context, for example, recognize a building or a tree. These two major fields of image analysis actually comprise a chicken-and-egg problem as, for instance, detection of a building in the middle of an image might imply a picture of a city with a high probability, whereas pre-classification of the picture as "city" would favor the recognition of a building versus a tree. Solution to the above problem can be dealt through modeling of visual concept descriptors in one or more domain ontologies and ontology learning/visual concept detection techniques.

Another topic in the field of image processing is the automatic detection of important or interesting regions in an image; topic that has been tackled by a number of researchers over the past 20 years.[30,31] An example methodology is illustrated in the work of Milanese,[32] where a computational model of *visual attention* by combining knowledge about the human visual system with computer vision techniques is developed. The notion of region segmentation is aimed at generating regions of homogeneous properties such as color and texture, which are the basic units and also an efficient intermediate representation of the scene, for higher level reasoning and interpretation. This topic is considered in general as something that a human observer performs with relative ease, but at the same time it is difficult for an automated system to understand and make higher level analysis tractable. Moreover, *region-based segmentation* aids significantly in the process of reasoning on spatial relationships; a task that becomes meaningful and useful this way. Of course, the segmented regions need further processing in order to be able to provide semantically meaningful information. In this scope they need to be classified, if applicable, into semantic object classes that are encountered frequently in

Fig. 10 Scene classification versus object detection tasks: A single test image is being used to either identify an open-country scene and a polo event or to detect interesting objects like the human rider and the horses

images, such as sky, cloud, grass, and tree. The purpose is to provide information about the existence of a few typical semantic object classes, which tend to belong either to the foreground or background, as well as an estimate of the scene category. The latter proves to be very useful for triggering top-down reasoning.

In *content-based image search and retrieval*, more and more researchers are looking beyond low-level color, texture, and shape features in pursuit of more effective searching methods. Natural object detection in indoor or outdoor scenes, that is, identifying key object types such as sky, grass, foliage, water, and snow, can facilitate content-based applications, ranging from image enhancement to coding or other multimedia applications. However, a significant number of misclassifications usually occur because of the similarities in color and texture characteristics of various object types and the lack of context information, which is a major limitation of individual object detectors. Toward the solution to the latter problem, an interesting approach is the one presented in the work of Luo et al.[33] A spatial context-aware object-detection system is proposed, initially combining the output of individual object detectors in order to produce a composite belief vector for the objects potentially present in an image. Subsequently, spatial context constraints, in the form of probability density functions obtained by learning, are used to reduce misclassification by constraining the beliefs to conform to the spatial context models.

Other attempts in the area include the one proposed in the work of Naphade and Huang,[34] where a list of semantic objects, including sky, snow, rock, water, and forest, is used in a framework for semantic indexing and retrieval of video. As already expected, color has been one of the central features of existing work on natural object detection. For example, in the work of Saber et al.,[35] *color classification* is utilized in order to detect sky. In the context of content-based image retrieval, Smith and Li[36] assumed that a blue extended patch at the top of an image is likely to represent clear sky. An exemplar-based approach is presented more recently that uses a combination of color and texture features to classify subblocks in an outdoor scene as sky or vegetation, assuming correct image orientation.[38] The latter brings up the issue of utilizing context orientation information in object class detection algorithms, a task that is generally avoided due to the fact that such contextual information is not always available and the performance of the algorithms is more than adequate despite this shortcoming.

So far, none of the above methods and techniques utilize context in any form. This tends to be the main drawback of these individual object detectors, since they only examine isolated strips of pure object materials, without taking into consideration the context of the scene or individual objects themselves. This is very important and also extremely challenging even for human observers. The notion of visual context is able to aid in the direction of natural object detection methodologies, simulating the human approach to similar problems. Many object materials can have the same appearance in terms of color and texture, while the same object may have different appearances under different imaging conditions (e.g., lighting, magnification). However, one important trait of humans is that they examine all the objects in the scene before making a final decision on the identity of individual objects. The use of visual context forms the key for this unambiguous recognition process, as it refers to the relationships among the location of different objects in the scene. In this manner, it is useful in many cases to reduce the ambiguity among conflicting detectors and eliminate improbable spatial configurations in object detection. As already discussed, visual context may be either spatial or temporal; *spatial context* is associated with spatial relationships between objects or regions in a still image or video sequence, whereas *temporal context* is associated with temporal relationships between objects, regions, or scenes in the case of a video sequence. In the sequel, discussion will be restricted to spatial context. One can identify two types of spatial contextual relationships that exist in natural images: (a) relationships that exist between co-occurrence of objects in natural images and (b) relationships that exist between spatial locations of certain objects in an image (Fig. 11).[37]

The definition of spatial context is an important issue for the notion of visual context in general. In order to be able to use context in applications, a mechanism to sense the current context—when thought as location, identities of nearby people or objects and changes to those objects—and deliver it to the application is crucial and must be present. A significant distinction exists between methods trying to determine location in computing applications and research fields. Most of the existing approaches tend to restrict themselves, trying to infer the location where the image was taken (i.e., camera location); inferring the location of what the image was taken of (i.e., image content location) is a rather difficult and more complex task tackled by much less approaches.[39] In the work of Davis,[22] this challenge is addressed by leveraging regularities in a given user's and in a community of users' photo-taking behaviors. Suitable weights, based on past experience and intuition, are chosen in order to assist in the process of location-determining features and then adjusted through a process of trial and error. An example describing the notion behind the method considers the following: it seems rather intuitive that if two pictures are being taken in the same location within a certain time frame (e.g., a few minutes for pedestrian users), they are probably in or around the same location.

Another factor to be considered is the intersection of spatial (and temporal) metadata in determining the location of image content. For example, patterns of being in certain locations at certain times with certain people will help determine the probability of which building in an area a user might be in. Information on whether this particular

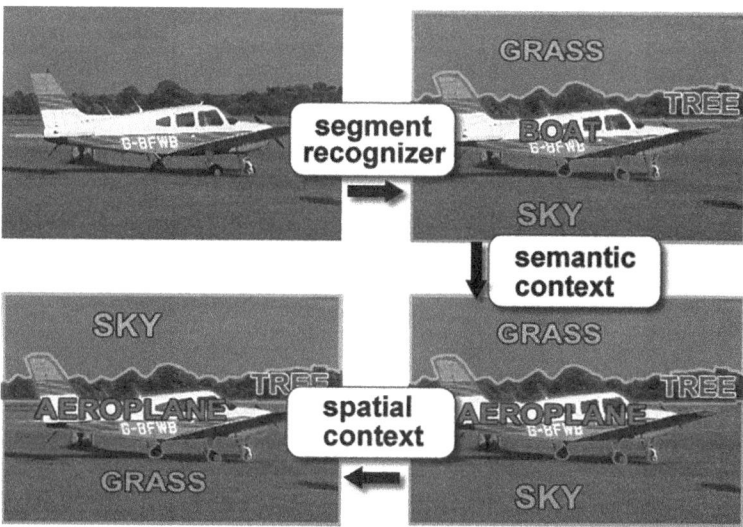

Fig. 11 Utilization of spatial context in semantic processing

building is the place he/she works in can also be derived in such a case. Rule-based constraint and inference engines can also be used to aid reasoning, as well as machine learning algorithms to learn from past performance to optimize and adjust the relative importance of the various location-determining features. Taking the process a step further into the field of context modeling transforms the problem into how to represent contextual information in a way that can help bridging the gap between applications using contextual information and the deployment of context-aware services. The development of such applications requires tools that are based on clearly defined context models. A simple approach is to use a plain model with context being maintained by a set of environment variables. Of course, visual context information can also come from an overall description of the whole scene. In that case, we are referring to the so-called *scene context*. In a number of studies, the context provided by a real-world scene has been claimed to have a mandatory, perceptual effect on the identification of individual objects in such a scene. This claim has provided a basis for challenging widely accepted data-driven models of visual perception. The so far discussed visual context, defined by the normal relationships among the locations of different materials in the scene without knowing exactly what the scene type is, is referred to as spatial context, and it is the one that is going to be used mostly in a multimedia system application. In the sequel, visual context analysis is discussed in relation to the problems of scene classification and object detection. With the increase in the number and size of digital archives and libraries, there is a need for automated, flexible, and reliable image search and retrieval algorithms, as well as for image and video database indexing. *Scene classification* provides solutions in the means of suitable applications for the problem. The ultimate goal is to classify scenes based on their content. However, scene classification remains a major open challenge. Most solutions proposed so far, such as those based on color histograms and local texture statistics,[40,41] lack the ability to capture a scene's global configuration, which is critical in perceptual judgments of scene similarity.

Initiatives have been taken in the field, whose main features can be summarized in the use of qualitative spatial and photometric relationships within and across regions. The emphasis on such qualitative measures leads to enhanced generalization abilities that are critical in achieving better coherence and efficiency for the final output/application. However, these similarity measures are rather inadequate for the problem at hand, if not combined together with additional information from other sources. And that is so, mainly because the previously defined similarity measures often produce results incongruent with human expectations, if the goal is to find images from a given object/scene class, such as snowy mountains or waterfalls. For example (Fig. 12), using color histograms to find the most similar images to a water scene at sunset could possibly return pictures of money, molten liquids, or even a watermelon! Obviously, all these images have the same overall gold color, although they differ in great degree in their semantic content.

Common standard approaches to object detection usually look at local pieces of the image in isolation when deciding if the object is present or not at a particular location. Of course, this is suboptimal and can be easily illustrated in the following example: consider the problem of finding a table in an office. A table is typically covered with other objects; indeed almost none of the table itself may be visible, and the parts that may be visible, such as its edge, are fairly generic features that may occur in many images. However, the table can be identified using contextual cues of various kinds. Of course, this problem is not restricted to tables or occluded objects: almost any object, when seen at a large enough distance, becomes impossible to recognize without using

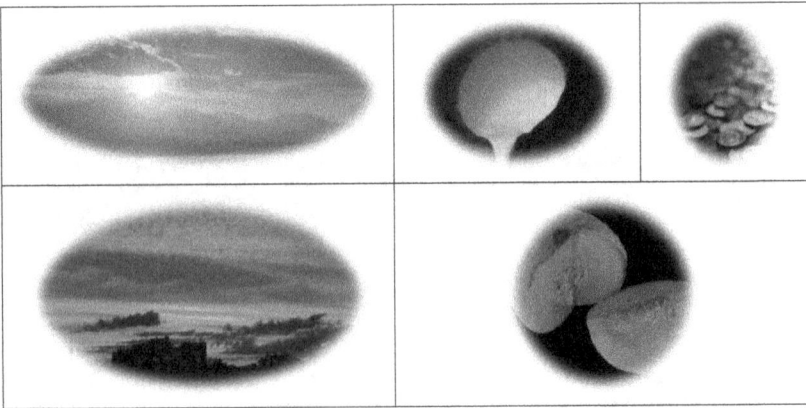

Fig. 12 All five images share a similar color histogram

context. Without doubt, there is a large amount of work in the field of *object detection and recognition* as well. However, techniques utilized in the field have usually positive results only in case of objects that have well-defined boundaries. Consequently, such strategies are not well suited for complex scenes, especially those that consist mostly of natural objects. The main difference between scene classification and object recognition techniques relies in the latter statement. Given these difficulties inherent in individual object recognition, scene classification approaches usually classify scenes without first attempting to recognize their components. This kind of strategy is also supported by psychophysical evidence showing that humans may holistically classify visual stimuli before recognizing the individual parts.[42] Also, efforts have been made in using scene classification to facilitate object detection and vice versa.[43] In general, scene classification methodology is characterized by the following three principles:

- The significance of the global scene configuration.
- The use of qualitative measurements.
- The sufficiency of low spatial frequency information.

Of course, careful work has to be done to define what images might be indoor, what images might be outdoor, and what kind of images might not be classified, notably close-up scenes, portraits, faces, animals, flowers, etc. The most addressed semantic image classification problem in the literature is the one of *indoor/outdoor classification*. One domain where automatic image classification is viable is personal photo collections, where specific categories of indoor or outdoor scenes can be identified (e.g., see Fig. 13).

The main features which are important in scene classification are *color* and *texture* and to a lesser extent, *shape*. Taking color into consideration, local color descriptors are in general detected in a straightforward manner. Clues such as blue sky, green plants, or red-tinted indoor scenes are the most commonly used.[44] Furthermore, faster and more efficient global color histograms are also utilized toward that goal.[45] In addition to color, texture and shape features are also of great importance. These features are capable of detecting regions with typical high frequencies in outdoor images, such as grass, leaves, sky, and water, and with typical vertical structures in indoor scenes, like wall corners or furniture, respectively. *City/landscape classification* (Fig. 9) is stated to be easier than indoor/outdoor classification.[44] And that is so because city images are mainly characterized by continuous vertical and horizontal edges, whereas outdoor images are typically dominated by short edges in all directions. Usually,

Fig. 13 Indoor/outdoor classification

either polygonized edges of minimal length or direction coherence measures are used in order to take into account the presence of line-shaped, straight, continuous edges in city images.

Now, as already discussed in this section, several approaches of analyzing the content of images exist in the literature and many aspects of context are identified aiding in the process of image analysis. The main task of related research work and one of the main goals in the field is the effective combination of local and global information, toward implementing robust methods to use in typical image processing problems and techniques. It should be clear by now that visual context can play a key role in the procedure of combining this information; context should actually stand in the middle, being able to handle both types of information and providing the means to achieve better coherence and reliable research results. In order to achieve the latter, appropriate visual context models should be selected and designed in a straightforward and productive manner, utilizing the variations of the particular aspects of visual context; in the following sections we shall focus on such a model.

Fuzzy Taxonomic Relations

It is common knowledge that retrieval systems based on terms suffer from the problematic mapping of terms to semantic entities.[46] Specifically, as more than one term may be associated to the same entity and more than one entity may be associated to the same term, the processing of query and index information is not trivial. In order to overcome such problems, one should work directly with *semantic entities*, rather than terms. In the sequel, we will denote by $S = \{s_1, s_2, \ldots, s_n\}$, the set of semantic entities that are known. A knowledge representation model may consist of the definitions of these semantic entities, together with their textual descriptions, that is, their corresponding terms, as well as a set of *relations* among the semantic entities. The objective is to construct a model in which the context determines the intended meaning of each word, and a word used in different context may have different meanings. An initial formal definition of such a model may be given as follows:

$$M = \{S, \{R_i\}\}, \quad i = 1\ldots n$$

$$R_i : S \times S \to \{0,1\}, \quad i = 1\ldots n$$

where M is the knowledge model and R_i is the ith relation among the semantic entities. Although any type of relation may be included, the two main categories are *taxonomic* (i.e., *ordering*) and *compatibility* (i.e., *symmetric*) relations. Compatibility relations fail to assist in the determination of the context of a query; the use of ordering relations

is necessary for such tasks.[1] Thus, a main challenge is the meaningful exploitation of information contained in taxonomic relations.

In addition, a knowledge model, in order to be highly descriptive, needs to contain a large number of distinct and diverse in relations among semantic entities. As a result, available information will be divided among them, making each one of them inadequate to fully describe a context. Thus, more than one such relation may need to be combined to provide a view of the knowledge that suffices for context definition and estimation. In order to overcome such problems, fuzzy semantic relations have been proposed for the modeling of real life information.[1] In particular, several commonly encountered relations that can be modeled as *fuzzy ordering relations* can be combined for the generation of a meaningful, fuzzy, quasi-taxonomic relation. More formally, a new knowledge model M_F is thus constructed, denoting the fuzziness in the approach in comparison to the knowledge model presented above and summarized in the following:

$$M_F = \{S, \{R_i\}\}, \quad i = 1\ldots n$$

$$r_i = F(R_i) : S \times S \to [0,1], \quad i = 1\ldots n$$

where F denotes the fuzzification of the relations R_i. The existence of many relations has led to the need for utilization of more relations for the generation of an adequate *taxonomic relation T*. Based on the relations r_i, we construct the following relation:

$$T = Tr^t(\bigcup_i r_i^{p_i}), \quad p_i \in \{-1,1\}, \quad i = 1\ldots n$$

where $Tr^t(A)$ is the sup-t transitive closure of some relation A, and the role of p_i is depicted by the specific definition of each relation used in the construction of T. Depending on the semantics of the relation definition (e.g., order of arguments a, b in Table 1), some relations may need to be inversed before being used in the construction of T. The transitivity of relation T, a required property in order for it to be taxonomic, was not implied by the above definition as the union of transitive relations is not necessarily transitive. For the purpose of analyzing multimedia content descriptions, relation T may be generated with the use of the following fuzzy taxonomic relations, whose semantics are defined in MPEG-7 and summarized in Table 1.

Based on the above fuzzy relations, T is a new semantic relation that is calculated as follows:[47]

$$T = Tr^t\left(Sp \cup P^{-1} \cup Ins \cup Pr^{-1} \cup Pat \cup L \cup Ex\right)$$

Based on the semantics of the participating relations, it is easy to see that T is ideal for the determination of the

Table 1 Fuzzy taxonomic relations used for generation of T

Name	Symbol	Meaning	Example	
			a	b
Part	$P(a,b)$	b is a part of a	Human body	Hand
Specialization	$Sp(a,b)$	a is a generalization of b	Vehicle	Car
Example	$Ex(a,b)$	b is an example of a	Player	Jordan
Instrument	$Ins(a,b)$	b is an instrument of a	Music	Drums
Location	$L(a,b)$	b is the location of a	Concert	Stage
Patient	$Pat(a,b)$	b is a patient of a	Course	Student
Property	$Pr(a,b)$	b is a property of a	Jordan	Star

topics that an entity may be related to, as well as for the estimation of the common meaning, that is, the context, of a set of entities. All relations used for the generation of T are partial ordering relations. Still, there is no evidence that their union is also antisymmetric, a property which is required for it to be taxonomic. Quite the contrary, T may vary from being a partial ordering to being an equivalence relation. This is an important observation, as true semantic relations also fit in this range (total symmetricity, as well as total antisymmetricity often have to be abandoned when modeling real life). Still, the semantics of the used relations indicate that T is very close to antisymmetric. Therefore, we categorize it as quasi-ordering or *quasi-taxonomic*.

Taxonomic Context Model

When using a taxonomic knowledge representation to interpret the meaning of a multimedia entity, it is the context of a term that provides its truly intended meaning. In other words, the true source of information is the co-occurrence of certain entities and not each one independently. Thus, the common meaning of terms should be used in order to best determine the entities to which they should be mapped. We will refer to this as their *taxonomic context*; in general, term *context* refers to whatever is common among a set of entities. Relation T will be used for the detection of the context of a set of entities, as explained in the remaining of this subsection.

The fact that relation T is (almost) an ordering relation allows us to use it in order to define, extract, and use the context of a set of entities in general. Relying on the semantics of relation T, we define the context $K(s)$ of a single entity $s \in S$ as the set of its antecedents in relation T, where S is the set of all entities. More formally, $K(s) = T(s)$, following the standard superset/subset notation from fuzzy relational algebra. Assuming that a set of entities $A \subseteq S$ is crisp, that is, that all considered entities belong to the set with degree one, the context of the set, which is again a set of entities, can be defined simply as the set of their common antecedents:

$$K(A) = \bigcap K(s_i), \quad s_i \in A$$

Obviously, as more entities are considered, the context becomes narrower, that is, it contains less entities and to smaller degrees, as illustrated in Fig. 14: $A \supset B \to K(A) \subseteq K(B)$.

When the definition of context is extended to the case of fuzzy sets of entities, this property must still hold. Moreover, we demand that the following are satisfied as well, basically because of the nature of fuzzy sets:

- $A(s) = 0 \Rightarrow K(A) = K(A - \{s\})$, that is, no narrowing of context.
- $A(s) = 1 \Rightarrow K(A) \subseteq K(s)$, that is, full narrowing of context.
- $K(A)$ decreases monotonically with respect to $A(s)$.

Taking this into consideration, we demand that, when A is a normal fuzzy set, the "considered" context $\mathcal{K}(s)$ of s, that is, the entity's context when taking its degree of participation to the set into account, is low when the degree of participation $A(s)$ is high or when the context of the crisp entity $K(s)$ is low.

Therefore, $cp(\mathcal{K}(s)) = cp(K(s)) \cap (S \cdot A(s))$, where cp is an involutive fuzzy complement and $S \cdot A(s)$ is a fuzzy set defined as $[S \cdot A(s)]_{(x)} = A(s) \forall x \in S$. By applying De Morgan's law, we obtain the following:

$$\kappa(s) = K(s) \cup cp(S \cdot A(s))$$

Then the set's context is easily calculated as follows:

$$K(A) = \bigcap_i \kappa(s_i), \quad s_i \in A$$

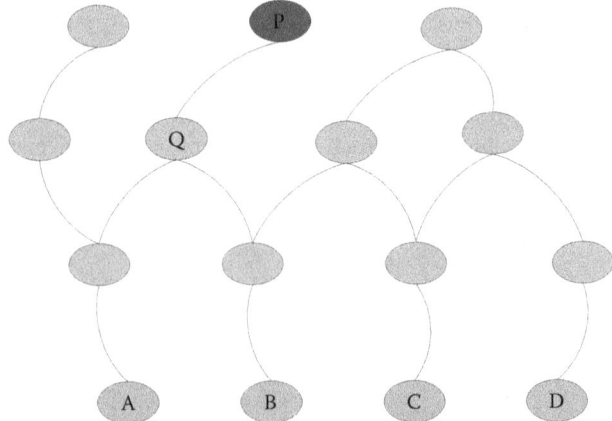

Fig. 14 As more entities are considered, the context it contains less entities and to smaller degrees: Considering only the first two leaves from the left (A, B), the context contains two entities (P, Q), whereas considering all the leaves (A, B, C) narrows the context to just one common ascendant (P)

Considering the semantics of the T relation and the process of context determination, it is easy to realize that when the entities in a set are highly related to a common meaning, the context will have high degrees of membership for the entities that represent this common meaning. Therefore, the height of the context $h(K(A))$ may be used as a measure of the semantic correlation of entities in set A. We refer to this measure as *intensity* of the context. The intensity of the context also represents the degree of relevance of the entities in the set.

SEMANTIC INDEXING

One of the main aspects of semantic processing tasks relies in the achievement of unified access to a/v content. The latter is accomplished through mapping of all a/v content and metadata to a semantically unified index, which is then used to serve user queries. In principle, the semantic index is constantly enriched and adapted to archive content changes. The main novelty introduced by this process is the fact that it allows an integrated detection of concepts both in multimedia content (i.e., images and videos) and text (i.e., in the form of metadata). Additionally, it sets the basis for efficient, intelligent clustering methods, for example, for the materialization of automatic thematic categorization of multimedia entities, using an appropriate knowledge model and the notions of semantic entities/relations and hierarchical contextual taxonomy. Thus, visual content analysis directly maps raw multimedia video to high-level semantic entities, integrating several algorithms and techniques toward an efficient image processing.

Detection of Semantic Entities

Very different strategies may be followed in order to detect the sought semantic entities. In general, techniques for semantic entities detection can be divided into two major fields: detection of semantic entities when dealing with multimedia content and textual semantic entities extraction when tackling text documents. In both cases, different content (e.g., digital images vs. text) is considered and as a result different analysis techniques are required, however, a uniform approach to their semantic handling is followed and same representations are obtained. The role of this semantic unification is to correlate the multimedia content descriptions provided by the a/v archives with the semantic entities stored in the knowledge base, so that user queries can be issued and handled at a semantic level. The result contains the correlations between multimedia content items and semantic entities. The built-in knowledge about entities permits robustness and uncertainty handling, whereas updating of knowledge ensures adaptation to environmental changes.

Visual Content Analysis

In a text document, the topic and semantics of the content are most often explicitly specified or at least contained within the document in a textual form. In a digital image, on the other hand, the entities to be indexed are not directly encountered in the image; recognizable features must be extracted and matched to the ones found in the knowledge base. Furthermore, abstract concepts, such as "sports" or "arts," are not directly encountered and must be inferred from concrete objects and events, as well as features that are not attributed to a particular object or event, such as light.[48] There are several options to implement this, even in the more complex case of video sequences, such as shot detection, key-frame extraction, and object localization and tracking techniques, so as to prepare a meaningful syntactic description of multimedia content.[49] In this case, following such video preprocessing techniques, we may achieve a hierarchical, spatiotemporal partition of the video sequence into meaningful time entities or shots. Thus, detected objects are then linked together to form adjacency graphs, that are then matched to the so-called *description graphs*, that model, in principal, a complex semantic entity in a combination of simpler ones[50] and are stored among objects, events, and other entities in the knowledge of the system. In this entry, we shall not examine more closely these problems encountered in visual content analysis, since it is not focusing on image processing and is more oriented toward the semantic interpretation of multimedia content.

Semantic Annotation Interpretation

The result of all (pre-)processing techniques mentioned in the previous section, including shot partitioning, key-frame extraction, and object detection and tracking, may be extended and integrated in a useful digital content annotation application; its annotation will have been semantically interpreted, that is, mapped to semantic entities and will be stored again within the semantic index. During this process, a query may be issued to each a/v archive for all content items that have not been indexed or whose description has been updated. The textual annotation contained in the MPEG-7 compliant description of each such item may be analyzed and semantic entities may be identified through matching with their definitions in the knowledge model. Links between detected semantic entities and the item in question may then be added to the index; weights may also be added depending on the location of each entity in the description and the degree of entity's matching.

CONCLUSIONS

The core contribution of *semantic processing* in image processing tasks is the provision of uniform access to heterogeneous a/v archives. This is accomplished by mapping all a/v content and metadata to a semantic index used to serve user queries, based on a common underlying knowledge model. A key aspect in these developments has been the exploitation of semantic metadata. In the following years, multimedia content management tasks are going to become even more complex. As it becomes obvious day by day, multimedia content itself will soon be a commodity, making the use of semantic metadata essential. Content providers will have to understand the benefits obtained from the systematic generation and exploitation of semantic information; service providers will have to accept them as the basis on which to build new services; and the producers of software tools for end users will redirect their imagination toward more appropriate integration of application software with content, taking advantage of semantic metadata. These developments clearly present some challenging prospects in technological, economic, standardization, and business terms and constitute semantic interpretation of multimedia content a key aspect of the conducted research activities.

REFERENCES

1. Dorai, C.; Venkatesh, S. Computational media aesthetics: Finding meaning beautiful. IEEE Multimedia **2001**, *8* (4), 10–12.
2. Smeulders, A.W.M.; Worring, M.; Santini, S.; Gupta, A.; Jain, R. Content-based image retrieval at the end of the early years. IEEE Trans. Pattern Anal. Mach. Intell. **2000**, *22*, 1349–1380.
3. Battista, S.; Casalino, F.; Lande, C. MPEG-4: A multimedia standard for the third millenium Part 2. IEEE Multimedia **2000**, *7* (1), 76–84.
4. Sikora, T. The MPEG-7 visual standard for content description—An overview. IEEE Trans. Circuits Syst. Video Technol. Special issue on MPEG-7, **2001**, *11* (6), 696–702.
5. MPEG-21 Overview v.3, ISO/IEC JTC1/SC29/WG11 N4511, December 2001.
6. ISO/IEC JTC1/SC29/WG1 N1646R: JPEG 2000 Part I Final Committee Draft Version 1.0, 2000.
7. Ramesh Naphade, M.; Kozintsev, I.V.; Huang, T.S. A factor graph framework for semantic video indexing. IEEE Trans. Circuits Syst. Video Technol. **2002**, *12* (1), 40–52.
8. Mich, O.; Brunelli, R.; Modena, C.M. A survey on video indexing. J. Visual Commun. Image Represent. **1999**, *10*, 78–112.
9. Yang, Y.; Zha, Z.-J.; Gao, Y.; Zhu, X.; Chua, T.-S. Exploiting web images for semantic video indexing via robust sample-specific loss. IEEE Trans. Multimedia **2014**, *16* (6), 1677–1689.
10. Naphide, H.R.; Huang, T.S. A probabilistic framework for semantic video indexing, filtering, and retrieval. IEEE Trans. Multimedia **2001**, *3* (1), 141–151.
11. Jun, W.; Worring, M. Efficient genre-specific semantic video indexing. IEEE Trans. Multimedia **2012**, *14* (2), 291–302.
12. Euzenat, J.; Fensel, D.; Lara, R.; Gómez-Pérez, A. Knowledge web: Realising the semantic web, all the way to knowledge enhanced multimedia documents. In *Proceedings of European Workshop on the Integration of Knowledge*; Semantics and Digital Media Technology (EWIMT): London, 2004, 25–26.
13. Nikolopoulos, S. *Semantic Multimedia Analysis Using Knowledge and Context*; PhD thesis; Queen Mary University of London, 2012. Available at http://qmro.qmul.ac.uk/jspui/handle/123456789/3148.
14. Kompatsiaris, I.; Avrithis, Y.; Hobson, P; Strinzis, M.G. Integrating knowledge, semantics and content for user-centred intelligent media services: The aceMedia project. In *Proceedings of Workshop on Image Analysis for Multimedia Interactive Services (WIAMIS)*, Lisboa, 2004, 21–23.
15. Biskup, J.; Freitag, Y.; Karabulut; Sprick, B. A mediator for multimedia systems. In *Proceedings 3rd International Workshop on Multimedia Information Systems*, Como, 1997.
16. Altenschmidt, C.; Biskup, J.; Flegel, U.; Karabulut, Y. Secure mediation: Requirements, design, and architecture. J. Comput. Secur. **2003**, *11* (3), 365–398.
17. Brink, A.; Marcus, S.; Subrahmanian, V. Heterogeneous multimedia reasoning. IEEE Comput. **1995**, *28* (9), 33–39.
18. Glöckner, I.; Knoll, A. Natural language navigation in multimedia archives: An integrated approach. In *Proceedings of the 7th ACM International Conference on Multimedia*, Orlando, FL, 1999, 313–322.
19. Cruz, I.; James, K. A user-centered interface for querying distributed multimedia databases. In *Proceedings of the 1999 ACM SIGMOD International Conference on Management of Data*, Philadelphia, PA, 1999, 590–592.
20. Klir, G.; Yuan, B. *Fuzzy Sets and Fuzzy Logic, Theory and Applications*; Prentice Hall: Upper Saddle River, NJ, 1995.
21. Winograd, T. Architectures for context. Hum.-Comput. Interact. **2001**, *16* (2), 401–419.
22. Davis, M.; Good, N.; Sarvas, R. From context to content: Leveraging context for mobile media metadata. In *Workshop on Context Awareness at the Second International Conference on Mobile Systems, Applications, and Services (MobiSys 2004)*, Boston, MA, 2004.
23. Raieli, R. *Multimedia Information Retrieval: Theory and Techniques*, Chandos Information Professional Series; Chandos Publishing: New Delhi, 2013.
24. Weiser, M. Some computer science issues in ubiquitous computing. Special Issue, Comput.-Augmented Environ. CACM **1993**, *36* (7), 74–83.
25. Schilit, B.; Adams, N.; Want, R. Context-aware computing applications. In *Proceedings of IEEE Workshop on Mobile Computing Systems and Applications*, Palo Alto Research Center: Santa Cruz, CA, 1994.
26. Lux, M. *Visual Information Retrieval*. Klagenfurt University: Graz, May 2009.
27. Glushko, R. *Information Organization & Retrieval*; School of Information, University of California: Berkeley, CA, October 6, 2014.
28. Edmonds, B. The pragmatic roots of context 119–132. In *Proceedings of the 2nd International and Interdisciplinary Conference on Modeling and Using Context (CONTEXT-99)*; Bouquet, P.; Serafini, L.; Brézillon, P.;

29. Wallace, M.; Akrivas, G.; Mylonas, Ph.; Avrithis, Y.; Kollias, S. Using context and fuzzy relations to interpret multimedia content. In *Proceedings of 3rd International Workshop on Content-Based Multimedia Indexing (CBMI)*; IRISA: Rennes, 2003.
30. Zhao, J.; Shimazu, Y.; Ohta, K.; Hayasaka, R.; Matsushita, Y. An outstandingness oriented image segmentation and its applications. In *Proceedings of 4th International Symposium on Signal Processing and Its Applications*, Gold Coast, QLD, 1996.
31. Osberger, W.; Maeder, A.J. Automatic identification of perceptually important regions in an image. In *Proceedings of IEEE International Conference on Pattern Recognition*, Brisbane, QLD, 1998.
32. Milanese, R. *Detecting Salient Regions in an Image: From Biology to Implementation*; PhD thesis; University of Geneva, Geneva, 1993.
33. Luo, J.; Singhal, A.; Zhu, W. Natural object detection in outdoor scenes based on probabilistic spatial context models. In *Proceedings of IEEE International Conference on Multimedia and Expo*, Baltimore, MD, 2002.
34. Naphade, M.; Huang, T.S. A factor graph framework for semantic indexing and retrieval in video. In *CVPR Workshop on Content-based Image and Video Retrieval*, 2000.
35. Saber, E.; Tekalp, A.M.; Eschbach, R.; Knox, K. Automatic image annotation using adaptive colour classification. CVGIP: Graph. Models Image Process. **1996**, *58*, 115–126.
36. Smith, J.R.; Li, C.-S. Decoding image semantics using composite region templates. In *Proceedings IEEE International Workshop on Content-based Access of Image and Video Database*, Santa Barbara, CA, 1998.
37. Galleguillos, C.; Rabinovich, A.; Belongie, S. Object categorization using co-occurrence, location and appearance. In *IEEE Conference on Computer Vision and Pattern Recognition (CVPR)*, Anchorage, AK, 23–28, 2008.
38. Vailaya, A.; Jain, A. Detecting sky and vegetation in outdoor images. In *Proceedings of SPIE*, Vol. 3972, January 2000.
39. Kalantidis, Y.; Tolias, G.; Avrithis, Y.; Phinikettos, M.; Spyrou, E.; Mylonas, Ph.; Kollias, S. VIRaL: Visual Image Retrieval and Localization. Multimedia Tools Appl. **2011**, *51* (2), 555–592.
40. Ashley, J.; Flickner, M.; Lee, D.; Niblack, W.; Petkovic, D. *Query by Image Content and Its Applications*. IBM Research Report, RJ 9947 (87906) Computer Science/Mathematics, March, 1995.
41. Smith, J.R.; Chang, S. Local color and texture extraction and spatial query. In *Proceedings of IEEE International Conference on Image Processing*, 1996.
42. Tanaka, J.W.; Farah, M. Parts and wholes in face recognition. Q. J. Exp. Psychol. **1993**, *46A* (2), 225–245.
43. Murphy, K.; Torralba, A.; Freeman, B. Using the forest to see the trees: A graphical model relating features, objects, and scenes. In *Advances in Neural Information Processing Systems 16 (NIPS 2003)*.
44. Vailaya, A.; Figueiredo, M.; Jain, A.; Zhang, H.-J. Content-based hierarchical classification of vacation images. In *Proceedings IEEE International Conference on Multimedia Computing and Systems*, Florence, 1999, 7–11.
45. Stauder, J.; Gouzien, G.; Chupeau, B.; Vigouroux, J.R.; Kijak, E. Semantic image browsing using hidden categories and confidence values. In *Storage and Retrieval for Media Databases 2003*, Santa Clara, CA, January 20–24, 2003.
46. Salembier, P.; Smith, J.R. MPEG-7 multimedia description schemes. IEEE Trans. Circuits Syst. Video Technol. **2001**, *11* (6), 748–759.
47. Wallace, M.; Akrivas, G.; Mylonas, Ph.; Avrithis, Y.; Kollias, S. Using context and fuzzy relations to interpret multimedia content. In *Proceedings of the Third International Workshop on Content-Based Multimedia Indexing (CBMI)*; IRISA: Rennes, 2003.
48. Zhao, R.; Grosky, W.I. Narrowing the semantic gap-improved text-based web document retrieval using visual features. IEEE Trans. Multimedia, Special Issue on Multimedia Database **2002**, *4* (2), 189–200.
49. Tsechpenakis, G.; Akrivas, G.; Andreou, G.; Stamou, G.; Kollias, S. Knowledge-assisted video analysis and object detection. In *Proceedings of European Symposium on Intelligent Technologies, Hybrid Systems and their Implementation on Smart Adaptive Systems*, Albufeira, Portugal, 2002.
50. Giro, X.; Marques, F. Semantic entity detection using description graphs. In *Workshop on Image Analysis for Multimedia Application Services (WIAMIS'03)*, London, 2003.

Sparse Modeling

Irina Rish
IBM, Yorktown Heights, New York, United States

Genady Ya. Grabarnik
St. John's University, Queens, New York, United States

Abstract
In this entry, we primarily focus on continuous sparse signals, following the developments in modern sparse statistical modeling and compressed sensing. Clearly, no single entry can possibly cover all aspects of these rapidly growing fields. Thus, our goal is to provide a reasonable introduction to the key concepts and survey major recent results in sparse modeling and signal recovery, such as common problem formulations arising in sparse regression, sparse Markov networks and sparse matrix factorization, several basic theoretical aspects of sparse modeling, state-of-the-art algorithmic approaches, as well as some practical applications. We start with an overview of several motivating practical problems that give rise to sparse signal recovery formulations.

A common question arising in a wide variety of practical applications is how to infer an unobserved high-dimensional "state of the world" from a limited number of observations. Examples include finding a subset of genes responsible for a disease, localizing brain areas associated with a mental state, diagnosing performance bottlenecks in a large-scale distributed computer system, reconstructing high-quality images from a compressed set of measure-ments, and, more generally, decoding any kind of signal from its noisy encoding, or estimating model parameters in a high-dimensional but small-sample statistical setting.

The underlying inference problem is illustrated in Fig. 1, where $x = (x_1, ..., x_n)$ and $y = (y_1, ..., y_m)$ represent an n-dimensional unobserved state of the world and its m observations, respectively. The output vector of observations, y, can be viewed as a noisy function (encoding) of the input vector x. A commonly used inference (decoding) approach is to find x that minimizes some *loss function* $L(x; y)$, given the observed y. For example, a popular probabilistic *maximum likelihood* approach aims at finding a parameter vector x that maximizes the likelihood $P(y|x)$ of the observations, i.e., minimizes the negative log-likelihood loss.

However, in many real-life problems, the number of unobserved variables greatly exceeds the number of measurements, since the latter may be expensive and also limited by the problem-specific constraints. For example, in computer network diagnosis, gene network analysis, and neuroimaging applications, the total number of unknowns, such as states of network elements, genes, or brain voxels, can be on the order of thousands, or even hundreds of thousands, while the number of observations, or samples, is typically on the order of hundreds. Therefore, the above maximum-likelihood formulation becomes underdetermined, and additional *regularization* constraints, reflecting specific domain properties or assumptions, must be introduced in order to restrict the space of possible solutions. From a Bayesian probabilistic perspective, regularization can be viewed as imposing a *prior* $P(x)$ on the unknown parameters x, and maximizing the posterior probability $P(x|y) = P(y|x)P(x)/P(y)$.

Perhaps one of the simplest and most popular assumptions made about the problem's structure is the solution *sparsity*. In other words, it is assumed that only a relatively small subset of variables is truly important in a specific context: e.g., usually only a small number of simultaneous faults occurs in a system; a small number of nonzero Fourier coefficients is sufficient for an accurate representation of various signal types; often, a small number of predictive variables (e.g., genes) is most relevant to the response variable (a disease, or a trait), and is sufficient for learning an accurate predictive model. In all these examples, the solution we seek can be viewed as a sparse high-dimensional vector with only a few nonzero coordinates. This assumption aligns with a philosophical principle of parsimony, commonly referred to as *Occam's razor*, or *Ockham's razor*, and attributed to William of Ockham, a famous medieval philosopher, though it can be traced back perhaps even further, to Aristotle and Ptolemy. Post-Ockham formulations of the principle of parsimony include, among many others, the famous one by Isaac Newton: "We are to admit no more causes of natural things than such as are both true and sufficient to explain their appearances."

Encyclopedia of Image Processing, First Edition
DOI: 10.1081/E-ECST2-120060006
Copyright © 2018 by Taylor & Francis. All rights reserved.

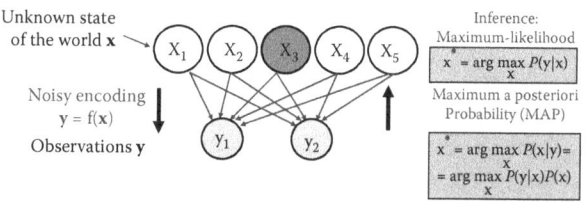

Fig. 1 Is it possible to recover an unobserved high-dimensional signal x from a low-dimensional, noisy observation y? Surprisingly, the answer is positive, provided that x has some specific structure, such as (sufficient) sparsity, and the mapping y = f(x) preserves enough information in order to reconstruct x

Statistical models that incorporate the parsimony assumption will be referred to as *sparse models*. These models are particularly useful in scientific applications, such as biomarker discovery in genetic or neuroimaging data, where the interpretability of a predictive model, e.g., identification of the most-relevant predictors, is essential. Another important area that can benefit from sparsity is signal processing, where the goal is to minimize signal acquisition costs while achieving high reconstruction accuracy; as we discuss later, exploiting sparsity can dramatically improve cost-efficiency of signal processing.

From a historical perspective, sparse signal recovery problem formulations can be traced back to 1943, or possibly even earlier, when the *combinatorial group testing* problem was first introduced by Dorfman.[1] The original motivation behind this problem was to design an efficient testing scheme using blood samples obtained from a large population (e.g., on the order of 100,000 people) in order to identify a relatively small number of infected people (e.g., on the order of 10). While testing each individual was considered prohibitively costly, one could combine the blood samples from groups of people; testing such combined samples would reveal if at least one person in the group had a disease. Following the inference scheme in Fig. 1, one can represent the health state of the i-th person as a Boolean variable x_i, where $x_i = 0$ if the person is healthy, and $x_i = 1$ otherwise; the test result, or measurement, y_j for a group of people G_j is the *logical-OR* function over the variables in the group, i.e., $y_j = 0$ if and only if all $x_i = 0$, $i \in G_j$, and 1 otherwise. Given an upper bound on the number of sick individuals in the population, i.e., the bound on sparsity of x, the objective of group testing is to identify all sick individuals (i.e., nonzero x_i), while minimizing the number of tests.

Similar problem formulations arise in many other diagnostic applications, for example, in computer network fault diagnosis, where the network nodes, such as routers or links, can be either functional or faulty, and where the group tests correspond to end-to-end transactions, called network probes, that go through particular subsets of elements as determined by a routing table.[2] (In the next section, we consider the network diagnosis problem in more detail, focusing, however, on its continuous rather than Boolean version, where the "hard faults" will be relaxed into performance bottlenecks, or time delays.) In general, group testing has a long history of successful applications to various practical problems, including DNA library screening, multiple access control protocols, and data streams, just to name a few. For more details on group testing, see the classical monograph by Du and Hwang,[3] as well as various recent publications.[4–6]

During the past several decades, half a century since the emergence of the combinatorial group testing field, sparse signal recovery is experiencing a new wave of intense interest, now with the primary focus on continuous signals and observations, and with particular ways of enforcing sparsity, such as using l_1-norm regularization. For example, in 1986, Santosa and Symes[7] proposed an l_1-norm-based optimization approach for the linear inversion (deconvolution) of band-limited reflection seismograms. In 1992, Rudin et al.[8] proposed *total variation* regularizer, which is closely related to l_1-norm, for noise removal in image processing. In 1996, the seminal paper by Tibshirani[9] on LASSO, or the l_1-norm regularized linear regression, appeared in the statistical literature, and initiated today's mainstream application of sparse regression to a wide range of practical problems. Around the same time, the *basis pursuit*[10] approach, essentially equivalent to LASSO, was introduced in the signal processing literature, and breakthrough theoretical results of Candès et al.[11] and Donoho[12] gave rise to the exciting new field of *compressed sensing* that revolutionized signal processing by *exponentially* reducing the number of measurements required for an accurate and computationally efficient recovery of sparse signals, as compared to the standard Shannon–Nyquist theory. In recent years, compressed sensing attracted an enormous amount of interest in signal processing and related communities, and generated a flurry of theoretical results, algorithmic approaches, and novel applications.

MOTIVATING EXAMPLES

Computer Network Diagnosis

One of the central issues in distributed computer systems and networks management is fast, real-time diagnosis of various faults and performance degradations. However, in large-scale systems, monitoring every single component, i.e., every network link, every application, every database transaction, and so on, becomes too costly, or even infeasible. An alternative approach is to collect a relatively small number of overall performance measures using end-to-end transactions, or *probes*, such as *ping* and *traceroute* commands, or end-to-end application-level tests, and then make inferences about the states of individual components. The area of research within the systems management field that focuses on diagnosis of network issues from indirect observations is called *network tomography*, similarly to medical tomography, where health issues are diagnosed based on inferences made from tomographic images of different organs.

In particular, let us consider the problem of identifying network performance bottlenecks, e.g., network links responsible for unusually high end-to-end delays, as discussed, for example, by Beygelzimer et al.[13] We assume that $y \in R^m$ is an observed vector of end-to-end transaction delays, $x \in R^n$ is an unobserved vector of link delays, and A is a *routing matrix*, where $a_{ij} = 1$ if the end-to-end test i goes through the link j, and 0 otherwise; the problem is illustrated in Fig. 2. It is often assumed that the end-to-end delays follow the noisy linear model, i.e.,

$$y = Ax + \epsilon \qquad (1)$$

where ϵ is the observation noise that may reflect some other potential causes of end-to-end delays, besides the link delays, as well as possible nonlinear effects. The problem of reconstructing x can be viewed as an *ordinary least squares* (OLS) regression problem, where A is the design matrix and x are the linear regression coefficients found by minimizing the least-squares error, which is also equivalent to maximizing the conditional log-likelihood log P(y|x) under the assumption of Gaussian noise ϵ:

$$\min_x \| y - Ax \|_2^2$$

Since the number of tests, m, is typically much smaller than the number of components, n, the problem of reconstructing x is underdetermined, i.e., there is no unique solution, and thus some regularization constraints need to be added. In case of network performance bottleneck diagnosis, it is reasonable to expect that, at any particular time, there are only a few malfunctioning links responsible for transaction delays, while the remaining links function properly. In other words, we can assume that x can be well approximated by a *sparse* vector, where only a few coordinates have relatively large magnitudes, as compared to the rest.

Neuroimaging Analysis

We now demonstrate a different kind of application example which arises in medical imaging domain. Specifically, we consider the problem of predicting mental states of a person based on brain imaging data, such as, for example, functional magnetic resonance imaging (fMRI). In the past decade, neuroimaging-based prediction of mental states became an area of active research on the intersection between statistics, machine learning, and neuroscience. A mental state can be cognitive, such as looking at a picture versus reading a sentence,[14] or emotional, such as feeling happy, anxious, or annoyed while playing a virtual-reality videogame.[15] Other examples include predicting pain levels experienced by a person,[16,17] or learning a classification model that recognizes certain mental disorders such as schizophrenia,[18] Alzheimer's disease,[19] or drug addiction.[20]

In a typical "mind reading" fMRI experiment, a subject performs a particular task or is exposed to a certain stimulus, while an MR scanner records the subject's blood-oxygenation-level-dependent (BOLD) signals indicative of changes in neural activity, over the entire brain. The resulting full-brain scans over the time period associated with the task or stimulus form a sequence of three-dimensional images, where each image typically has on the order of 10,000–100,000 subvolumes, or *voxels*, and the number of time points, or time repetitions (TRs), is typically on the order of hundreds.

As mentioned above, a typical experimental paradigm aims at understanding changes in a mental state associated with a particular task or a stimulus, and one of the central questions in the modern multivariate fMRI analysis is whether we can predict such mental states given the sequence of brain images. For example, in a recent pain perception study,[21] the subjects were rating their pain level on a continuous scale in response to a quickly changing thermal stimulus applied to their back via a contact probe. In another experiment, associated with the 2007 Pittsburgh Brain Activity Interpretation Competition,[22] the task was to predict mental states of a subject during a videogame

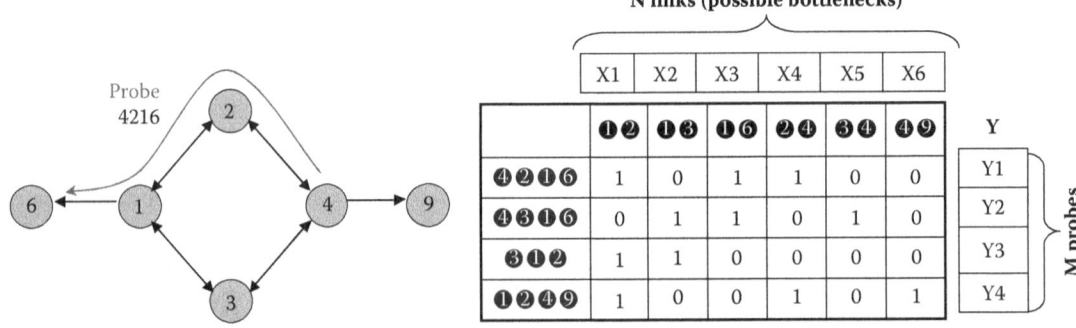

Fig. 2 Example of a sparse signal recovery problem: diagnosing performance bottleneck(s) in a computer network using end-to-end test measurements, or *probes*

session, including feeling annoyed or anxious, listening to instructions, looking at a person's face, or performing a certain task within the game.

Given an fMRI data set, i.e., the BOLD signal (voxel activity) time series for all voxels, and the corresponding time series representing the task or stimulus, we can formulate the prediction task as a linear regression problem, where the individual time points will be treated as independent and identically distributed (i.i.d.) samples—a simplifying assumption that is, of course, far from being realistic, and yet often works surprisingly well for predictive purposes. The voxel activity levels correspond to predictors, while the mental state, task, or stimulus is the predicted response variable. More specifically, let $A_1, \ldots, A_n$ denote the set of n predictors, let Y be the response variable, and let m be the number of samples. Then, $A = (a_1 | \ldots | a_n)$ corresponds to an $m \times n$ fMRI data matrix, where each a_i is an m-dimensional vector of the i-th predictor's values, for all m instances, while the m-dimensional vector y corresponds to the values of the response variable Y, as it is illustrated in Fig. 3.

As it was already mentioned, in biological applications, including neuroimaging, interpretability of a statistical model is often as important as the model's predictive performance. A common approach to improving a model's interpretability is *variable selection*, i.e., choosing a small subset of predictive variables that are *most relevant* to the response variable. In neuroimaging applications discussed above, one of the key objectives is to discover brain areas that are most relevant to a given task, stimulus, or mental state. Moreover, variable selection, as well as a more general *dimensionality reduction* approach, can significantly improve generalization accuracy of a model by preventing it from overfitting high-dimensional, small-sample data common in fMRI and other biological applications.

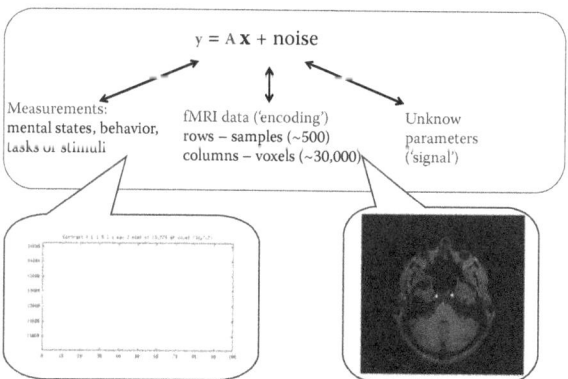

Fig. 3 Mental state prediction from functional MRI data, viewed as a linear regression with simultaneous variable selection. The goal is to find a subset of fMRI voxels, indicating brain areas that are most relevant (e.g., most predictive) about a particular mental state

A simple approach to variable selection, also known in the machine-learning community as a *filter-based* approach, is to evaluate each predictive variable independently, using some univariate relevance measure, such as, for example, correlation between the variable and the response, or the mutual information between the two. For example, a traditional fMRI analysis approach known as General Linear Models (GLMs)[23] can be viewed as filter-based variable selection, since it essentially computes individual correlations between each voxel and the task or stimulus, and then identifies brain areas where these correlations exceed a certain threshold. However, such mass-univariate approach, though very simple, has an obvious drawback, as it completely ignores multivariate interactions, and thus can miss potentially relevant groups of variables that individually do not appear among the top-ranked ones. Perhaps one of the most well-known illustrations of a multiway interaction among the variables that cannot be detected by looking at any subset of them, not only at the single variables, is the parity check (logical XOR) function over n variables; the parity check response variable is statistically independent of each of its individual inputs, or any subset of them, but is completely determined given the full set of n inputs. As it was demonstrated by Haxby et al.[24] and others (see, for example, recent work by Rish et al.[25]), highly predictive models of mental states can be built from voxels with submaximal activation that would not be discovered by the traditional GLM analysis. Thus, in recent years, multivariate predictive modeling became a popular alternative to univariate approaches in neuroimaging. Since a combinatorial search over all subsets of voxels in order to evaluate their relevance to the target variable is clearly intractable, a class of techniques, called *embedded methods*, appears to be the best practical alternative to both the univariate selection and the exhaustive search, since it incorporates variable selection into multivariate statistical model learning.

A common example of embedded variable selection is *sparse regression*, where a cardinality constraint restricting the number of nonzero coefficients is added to the original regression problem. Note that in case of linear, or OLS, regression, the resulting sparse regression problem is equivalent to the sparse recovery problem introduced in the network diagnosis example.

Compressed Sensing

One of the most prominent recent applications of sparsity-related ideas is *compressed sensing*, also known as *compressive sensing*, or *compressive sampling*,[11,12] an extremely popular and rapidly expanding area of modern signal processing. The key idea behind compressed sensing is that the majority of real-life signals, such as images, audio, or video, can be well approximated by sparse vectors, given some appropriate basis, and that exploiting the sparse signal structure can dramatically reduce the signal acquisition cost; moreover, accurate signal reconstruction can be achieved in a *computationally efficient* way, by using *sparse optimization* methods.

Traditional approach to signal acquisition is based on the classical Shannon–Nyquist result stating that in order to preserve information about a signal, one must sample the signal at a rate which is at least twice the signal's *bandwidth*, defined as the highest frequency in the signal's spectrum. Note, however, that such classical scenario gives a worst-case bound, since it does not take advantage of any specific structure that the signal may possess. In practice, sampling at the Nyquist rate usually produces a tremendous number of samples, e.g., in digital and video cameras, and must be followed by a compression step in order to store or transmit this information efficiently. The compression step uses some basis to represent a signal (e.g., Fourier and wavelets) and essentially throws away a large fraction of coefficients, leaving a relatively few important ones. Thus, a natural question is whether the compression step can be combined with the acquisition step, in order to avoid the collection of an unnecessarily large number of samples. Fourier and wavelet bases are two examples commonly used in image processing, though, in general, finding a good basis that allows for a sparse signal representation is a challenging problem, known as *dictionary learning*.

As it turns out, the above question can be answered positively. Let $s \in R^n$ be a signal that can be represented sparsely in some basis B, i.e., $s = Bx$, where B is an $n \times n$ matrix of basis vectors (columns), and where $x \in R^n$ is a sparse vector of the signal's coordinates with only $k \ll n$ nonzeros. Though the signal is not observed directly, we can obtain a set of linear measurements:

$$y = Ls = LBx = Ax \qquad (2)$$

where L is an $m \times n$ matrix, and $y \in R^m$ is a set of m measurements, or samples, where m can be much smaller than the original dimensionality of the signal, hence the name "compressed sampling." The matrix $A = LB$ is called the *design* or *measurement matrix*. The central problem of compressed sensing is reconstruction of a high-dimensional sparse signal representation x from a low-dimensional linear observation y, as it is illustrated in Fig. 4A. Note that the problem discussed above describes a *noiseless* signal recovery, while in practical applications there is always some noise in the measurements. Most frequently, Gaussian noise is assumed which leads to the classical linear, or OLS, regression problem, discussed before, though other types of noise are possible. The noisy signal recovery problem is depicted in Fig. 4B, and is equivalent to the diagnosis and sparse regression problems encountered in the sections "Computer Network Diagnosis" and "Neuroimaging Analysis," respectively.

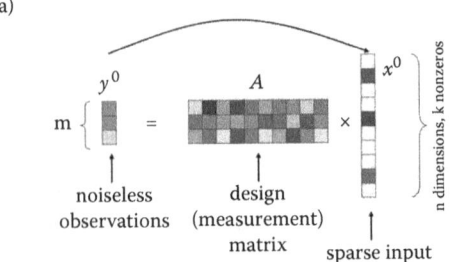

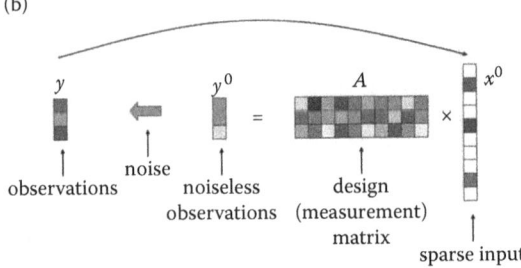

Fig. 4 Compressed sensing—collecting a relatively small number of linear measurements that allow for an accurate reconstruction of a high-dimensional sparse signal: (A) noiseless case, (B) noisy case

SPARSE RECOVERY IN A NUTSHELL

The following two questions are central to all applications that involve sparse signal recovery: *when* is it possible to recover a high-dimensional sparse signal from a low-dimensional observation vector? And, *how* can we do this in a computationally efficient way? The key results in sparse modeling and compressed sensing identify particular conditions on the design matrix and signal sparsity that allow for an accurate reconstruction of the signal, as well as optimization algorithms that achieve such reconstruction in a computationally efficient way.

Sparse signal recovery can be formulated as finding a minimum-cardinality solution to a constrained optimization problem. In the noiseless case, the constraint is simply $y = Ax$, while in the noisy case, assuming Gaussian noise, the solution must satisfy $\|y - y^*\|_2 \leq \varepsilon$, where $y^* = Ax$ is the (hypothetical) noiseless measurement, and the actual measurement is ε-close to it in l_2-norm (Euclidean norm). The objective function is the cardinality of x, i.e., the number of nonzeros, which is often denoted $\|x\|_0$ and called l_0-norm of x (though, strictly speaking, l_0 is not a proper norm). Thus, the optimization problems corresponding to noiseless and noisy sparse signal recovery can be written as follows:

$$\min_x \|x\|_0 \text{ subject to } y = Ax \qquad (3)$$

$$\min_x \|x\|_0 \text{ subject to } \|y - Ax\|_2 \leq \epsilon \qquad (4)$$

In general, finding a minimum-cardinality solution satisfying linear constraints is an NP-hard combinatorial problem.[26] Thus, an approximation is necessary to achieve computational efficiency, and it turns out that, under certain conditions, approximate approaches can recover the exact solution.

Perhaps the most widely known and striking result from the compressed sensing literature is that, for a random design matrix, such as, for example, a matrix with i.i.d. Gaussian entries, with high probability, a sparse n-dimensional signal with at most k nonzeros can be reconstructed exactly from only m = O(k log(n/k)) measurements.[11,12] Thus, the number of samples can be *exponentially smaller* than the signal dimensionality. Moreover, with this number of measurements, a computationally efficient recovery is possible by solving a convex optimization problem:

$$\min_x \|x\|_1 \text{ subject to } y = Ax \quad (5)$$

where $\|x\|_1 = \sum_{i=1}^{n} |x_i|$ is the l_1-norm of x. The above problem can be reformulated as a linear program and thus easily solved by standard optimization techniques.

More generally, in order to guarantee an accurate recovery, the design matrix does not necessarily have to be random, but needs to satisfy some "nice" properties. The commonly used sufficient condition on the design matrix is the so-called *restricted isometry property* (*RIP*),[11] which essentially states that a linear transformation defined by the matrix must be almost isometric (recall that an isometric mapping preserves vector length), when restricted to any subset of columns of certain size, proportional to the sparsity k.

Furthermore, even if measurements are contaminated by noise, sparse recovery is still *stable* in a sense that recovered signal is a close approximation to the original one, provided that the noise is sufficiently small, and that the design matrix satisfies certain properties such as RIP.[11] A sparse signal can be recovered by solving a "noisy" version of the above l_1-norm minimization problem:

$$\min_x \|x\|_1 \text{ subject to } \|y - Ax\|_2 \leq \epsilon \quad (6)$$

The above optimization problem can be also written in two equivalent forms (see, for example, Section 3.2 of Borwein et al.[27]): either as another constrained optimization problem, for some value of bound t, uniquely defined by ε:

$$\min_x \|y - Ax\|_2^2 \text{ subject to } \|x\|_1 \leq t \quad (7)$$

or as an unconstrained optimization, using the corresponding Lagrangian for some appropriate Lagrange multiplier λ uniquely defined by ε, or by t:

$$\min_x \frac{1}{2}\|y - Ax\|_2^2 + \lambda \|x\|_1 \quad (8)$$

In the statistical literature, the latter problem is widely known as LASSO regression,[9] while in signal processing, it is often referred to as *basis pursuit*.[10]

STATISTICAL LEARNING VERSUS COMPRESSED SENSING

Finally, it is important to point out similarities and differences between *statistical* and *engineering* applications of sparse modeling, such as learning sparse models from data versus sparse signal recovery in compressed sensing. Clearly, both statistical and engineering applications involving sparsity give rise to the same optimization problems that can be solved by the same algorithms, often developed in parallel in both statistical and signal processing communities.

However, statistical learning pursues somewhat different goals than compressed sensing, and often presents additional challenges:

- Unlike compressed sensing, where the measurement matrix can be constructed to have desired properties (e.g., random i.i.d. entries), in statistical learning, the design matrix consists of the observed data, and thus we have little control over its properties. Thus, matrix properties such as RIP are often not satisfied; also note that testing RIP property of a given matrix is NP-hard, and thus computationally infeasible in practice.
- Moreover, when learning sparse models from real-life data sets, it is difficult to evaluate the accuracy of sparse recovery, since the "ground-truth" model is usually not available, unlike in the compressed sensing setting, where the ground truth is the known original signal (e.g., an image taken by a camera). An easily estimated property of a statistical model is its predictive accuracy on a test data set; however, predictive accuracy is a very different criterion from the *support recovery*, which aims at correct identification of nonzero coordinates in a "ground-truth" sparse vector.
- While theoretical analysis in compressed sensing is often focused on sparse *finite-dimensional* signal recovery and the corresponding conditions on the measurement matrix, the analysis of sparse statistical models is rather focused on *asymptotic consistency* properties, i.e., decrease of some statistical errors of interest with the growing number of dimensions and samples. Three typical performance metrics include: 1) *prediction error*—predictions of the estimated model must converge to the predictions of the true model in some norm, such as l_2-norm; this property is known as model *efficiency*; 2) *parameter estimation error*—estimated parameters must converge to the true parameters, in some norm such as l_2-norm; this property is called *parameter estimation consistency*; and 3) *model-selection error*—the sparsity pattern, i.e., the location of nonzero coefficients, must converge to the one of the true model; this property is also known as *model selection consistency*, or *sparsistency* (also, convergence of the *sign pattern* is called *sign consistency*).

- Finally, recent advances in sparse statistical learning include a wider range of problems beyond the basic sparse linear regression, such as sparse generalized linear models, sparse probabilistic graphical models (e.g., Markov and Bayesian networks), as well as a variety of approaches enforcing more complicated structured sparsity.

SUMMARY AND BIBLIOGRAPHICAL NOTES

In this entry, we introduced the concepts of sparse modeling and sparse signal recovery, and provided several motivating application examples, ranging from network diagnosis to mental state prediction from fMRI and to compressed sampling of sparse signals. As it was mentioned before, sparse signal recovery dates back to at least 1943, when combinatorial group testing was introduced in the context of Boolean signals and logical-OR measurements.[1] Recent years have witnessed a rapid growth of the sparse modeling and signal recovery areas, with a particular focus on continuous sparse signals, their linear projections, and l_1-norm regularized reconstruction approaches, triggered by the breakthrough results of Candès et al.[11] and Donoho,[12] on high-dimensional signal recovery via l_1-based methods, where the number of measurements is logarithmic in the number of dimensions—an exponential reduction when compared to the standard Shannon–Nyquist theory. Efficient l_1-norm-based sparse regression, such as LASSO[9] in statistics and its signal processing equivalent, *basis pursuit*,[10] are now widely used in various high-dimensional applications.

In the past years, sparsity-related research has expanded significantly beyond the original signal recovery formulation, to include sparse nonlinear regression, such as GLMs, sparse probabilistic networks, such as Markov and Bayesian networks, sparse matrix factorization, such as dictionary learning, sparse principal component analysis (PCA) and sparse nonnegative matrix factorization (NMF), and other types of sparse settings.

Due to the enormous amount of recent developments in sparse modeling, a number of important topics remain beyond the scope of this entry. One example is the *low-rank matrix completion*—a problem appearing in a variety of applications, including collaborative filtering, metric learning, multitask learning, and many others. Since the rank minimization problem, similarly to l_0-norm minimization, is intractable, it is common to use its convex relaxation by the *trace norm*, also called the *nuclear norm*, which is the l_1-norm of the vector of singular values. For more details on low-rank matrix learning and trace norm minimization, see, for example, Fazel et al.,[28] Srebro et al.,[29] Bach,[30] Candès and Recht,[31] Toh and Yun,[32] Negahban and Wainwright,[33] Recht et al.,[34] Rohde and Tsybakov,[35] Mishra et al.,[36] and references therein. Another area we are not discussing here in detail is *sparse Bayesian learning*,[37–40] where alternative priors, beyond the Laplacian (equivalent to the l_1-norm regularizer), are introduced in order to enforce the solution sparsity. Also, besides several applications of sparse modeling that we will discuss herein, there are multiple others that we will not be able to include, in the fields of astronomy, physics, geophysics, speech processing, and robotics, just to name a few.

For further references on recent developments in the field, as well as for tutorials and application examples, we refer the reader to the online repository available at the Rice University website (http://dsp.rice.edu/cs), and to other online resources (e.g., http://nuit-blanche.blogspot.com). Several recent books focus on particular aspects of sparsity; for example, Elad[41] provides a good introduction to sparse representations and sparse signal recovery, with a particular focus on image processing applications. A classical textbook on statistical learning by Hastie et al.[42] includes, among many other topics, introduction to sparse regression and its applications. Also, a recent book by Bühlmann and van de Geer[43] focuses specifically on sparse approaches in high-dimensional statistics. Moreover, various topics related to compressed sensing are covered in several recently published monographs and edited.[44–46]

REFERENCES

1. Dorfman, R. The detection of defective members of large populations. Ann. Math. Stat. **1943**, *14* (4), 436–440.
2. Rish, I.; Brodie, M.; Ma, S.; Odintsova, N.; Beygelzimer, A.; Grabarnik, G.; Hernandez, K. Adaptive diagnosis in distributed systems. IEEE Trans. Neural. Netw. **2005**, *16* (5), 1088–1109.
3. Du, D.; Hwang, F. *Combinatorial Group Testing and Its Applications*. 2nd Ed.; World Scientific Publishing Co., Inc.: River Edge, NJ, 2000.
4. Gilbert, A.; Strauss, M. Group testing in statistical signal recovery. Technometrics **2007**, *49* (3), 346–356.
5. Atia, G.; Saligrama, V. Boolean compressed sensing and noisy group testing. IEEE Trans. Inf. Theory **2012**, *58* (3), 1880–1901.
6. Gilbert, A.; Hemenway, B.; Rudra, A.; Strauss, M.; Wootters, M. Recovering simple signals. In *Information Theory and Applications Workshop ITA*, San Diego, CA, 2012, 382–391.
7. Santosa, F.; Symes, W. Linear inversion of band-limited reflection seismograms. SIAM J. Sci. Stat. Comput. **1986**, *7* (4), 1307–1330.
8. Rudin, L.; Osher, S.; Fatemi, E. Nonlinear total variation based noise removal algorithms. Physica D **1992**, *60*, 259–268.
9. Tibshirani, R. Regression shrinkage and selection via the Lasso. J. R. Stat. Soc. B **1996**, *58* (1), 267–288.
10. Chen, S.; Donoho, D.; Saunders, M. Atomic decomposition by basis pursuit. SIAM J. Sci. Comput. **1998** *20* (1), 33–61.
11. Candès, E.; Romberg, J.; Tao, T. Robust uncertainty principles: Exact signal reconstruction from highly incomplete

frequency information. IEEE Trans. Inf. Theory **2006a**, *52* (2), 489–509.
12. Donoho, D. Compressed sensing. IEEE Trans. Inf. Theory **2006a**, *52* (4), 1289–1306.
13. Beygelzimer, A.; Kephart, J.; Rish, I. Evaluation of optimization methods for network bottleneck diagnosis. In *Proceedings of the Fourth International Conference on Autonomic Computing (ICAC)*, Washington, DC, 2007.
14. Mitchell, T.; Hutchinson, R.; Niculescu, R.; Pereira, F.; Wang, X.; Just, M.; Newman, S. Learning to decode cognitive states from brain images. Mach. Learn. **2004**, *57*, 145–175.
15. Carroll, M.; Cecchi, G.; Rish, I.; Garg, R.; Rao, A. Prediction and interpretation of distributed neural activity with sparse models. NeuroImage. **2009** *44* (1), 112–122.
16. Rish, I.; Cecchi, G.; Baliki, M.; Apkarian, A. Sparse regression models of pain perception. In: *Brain Informatics*, Yao, Y., Sun, R., Poggio, T., Liu, J., Zhong, N., Huang, J.; Eds.; Springer: Berlin, 2010, 212–223.
17. Cecchi, G.; Huang, L.; Hashmi, J.; Baliki, M.; Centeno, M.; Rish, I.; Apkarian, A. Predictive dynamics of human pain perception. PLoS Computat. Biol. **2012**, *8*, 10.
18. Rish, I.; Cecchi, G.; Heuton, K. Schizophrenia classification using fMRI-based functional network features. In *Proceedings of SPIEMedical Imaging*, February 2012a.
19. Huang, S.; Li, J.; Sun, L.; Liu, J.; Wu, T.; Chen, K.; Fleisher, A.; Reiman, E.; Ye, J. Learning brain connectivity of Alzheimer's disease from neuroimaging data. In *Proceedings of Neural Information Processing Systems (NIPS)*, Vol. 22; 2009, 808–816.
20. Honorio, J.; Ortiz, L.; Samaras, D.; Paragios, N.; Goldstein, R. Sparse and locally constant Gaussian graphical models. In *Proceedings of Neural Information Processing Systems (NIPS)*, 2009, 745–753.
21. Baliki, M.; Geha, P.; Apkarian, A. Parsing pain perception between nociceptive representation and magnitude estimation. J. Neurophysiology **2009**, *101*, 875–887.
22. Pittsburgh EBC Group PBAIC Homepage: http://www.ebc.pitt.edu/2007/competition.html 2007.
23. Friston, K.; Holmes, A.; Worsley, K.; Poline, J.-P.; Frith, C.; Frackowiak, R. Statistical parametric maps in functional imaging: A general linear approach. Hum. Brain Map. **1995**, *2* (4), 189–210.
24. Haxby, J.; Gobbini, M.; Furey, M.; Ishai, A.; Schouten, J.; Pietrini, P. Distributed and overlapping representations of faces and objects in ventral temporal cortex. Science **2001**, *293* (5539), 2425–2430.
25. Rish, I.; Cecchi, G.; Heuton, K.; Baliki, M.; Apkarian, A. V. Sparse regression analysis of task-relevant information distribution in the brain. In *Proceedings of SPIE Medical Imaging*, February 2012b.
26. Natarajan, K. Sparse approximate solutions to linear systems. SIAM J. Comput. **1995**, *24*, 227–234.
27. Borwein, J.; Lewis, A.; Borwein, J.; Lewis, A. *Convex Analysis and Nonlinear Optimization: Theory and Examples*. Springer: New York, 2006.
28. Fazel, M.; Hindi, H.; Boyd, S. A rank minimization heuristic with application to minimum order system approximation. In *Proceedings of the 2001 American Control Conference*; IEEE, Vol. 6; 2001, 4734–4739.
29. Srebro, N.; Rennie, J.; Jaakkola, T. Maximum-margin matrix factorization. In *Proceedings of Neural Information Processing Systems (NIPS)*, Vol. 17, Vancouver, BC, 2004, 1329–1336.
30. Bach, F. Consistency of trace norm minimization. J. Machine Learn. Res. **2008c**, *9*, 1019–1048.
31. Candès, E.; Recht, B. Exact matrix completion via convex optimization. Found. Comput. Math. **2009**, *9* (6), 717–772.
32. Toh, K.-C.; Yun, S. An accelerated proximal gradient algorithm for nuclear norm regularized least squares problems. Pacific J. Optim. **2010**, *6*, 615–640.
33. Negahban, S.; Wainwright, M. Estimation of (near) low-rank matrices with noise and high-dimensional scaling. Ann. Stat. **2011**, *39* (2), 1069–1097.
34. Recht, B.; Fazel, M.; Parrilo, P. Guaranteed minimum-ranksolutions of linear matrix equations via nuclear norm minimization. SIAM Rev. **2010**, *52* (3), 471–501.
35. Rohde, A.; Tsybakov, A. Estimation of high-dimensional low-rank matrices. Ann. Stat. **2011**, *39* (2), 887–930.
36. Mishra, B.; Meyer, G.; Bach, F.; Sepulchre, R. Low-rank optimization with trace norm penalty. SIAM J. Opt. **2013**, *23* (4), 2124–2149.
37. Tipping, M. Sparse Bayesian learning and the Relevance Vector Machine. J. Mach. Learn. Res. **2001**, *1*, 211–244.
38. Wipf, D.; Rao, B. Sparse Bayesian learning for basis selection. IEEE Trans. Signal Process. **2004**, *52* (8), 2153–2164.
39. Ishwaran, H.; Rao, J. Spike and slab variable selection: Frequentist and Bayesian strategies. Ann. Stat. **2005**, *33* (2), 730–773.
40. Ji, S.; Xue, Y.; Carin, L. Bayesian compressive sensing. IEEE Trans. Signal Process. **2008**, *56* (6), 2346–2356.
41. Elad, M. *Sparse and Redundant Representations: From Theory to Applications in Signal and Image Processing*; Springer: New York, 2010.
42. Hastie, T.; Tibshirani, R.; Friedman, J. *The Elements of Statistical Learning: Data Mining, Inference, and Prediction*, 2nd Ed. Springer-Verlag: New York, 2009.
43. Bühlmann, P.; van de Geer, S. *Statistics for High-Dimensional Data: Methods; Theory and Applications*. Springer: Berlin, 2011.
44. Eldar, Y.; Kutyniok, G.; Eds.; *Compressed Sensing: Theory and Applications*. Cambridge University Press: Cambridge, 2012.
45. Foucart, S.; Rauhut, H. *A Mathematical Introduction to Compressive Sensing*. Springer: New York, 2013.
46. Patel, V.; Chellappa, R. *Sparse Representations and Compressive Sensing for Imaging and Vision*: Springer Briefs in Electrical and Computer Engineering; Springer: New York, 2013.

The Human Visual System

Tania Pouli
Technicolor, Rennes, France

Erik Reinhard
Technicolor Research and Innovation, Rennes, France

Douglas W. Cunningham
Brandenburg University of Technology, Cottbus, Germany

Abstract
This entry introduces concepts related to the human visual system and human visual perception. First, relevant radiometric and photometric terms are introduced. Then, the processing of visual signals in the retina and visual cortex are discussed. Finally, implications for human visual processing are considered.

Keywords: Human visual system; Retina; Visual cortex; Visual perception.

In this entry we briefly discuss the human visual system and human visual perception. We begin by defining relevant radiometric and photometric terms. Then, the human visual system is described by following light as it enters the eye and is transduced into an electrical signal by the photoreceptors. We then follow the path this signal takes through several layers of cells in the retina and onwards to the visual cortex. Although much is known about cortical visual processing, there is even more that remains unknown. We therefore describe cortical processing only at a very abstract level and only sofar as is relevant in the context of natural image statistics.

In essence, the nature of human vision is such that natural images can be observed, interpreted, and understood at a level necessary to allow humans to function in a manner appropriate to their environment. Thus, this entry concludes with a discussion of the implications of visual processing.

RADIOMETRIC AND PHOTOMETRIC TERMS

To be able to describe light as it is observed by humans or captured and displayed by various technologies, several important terms need to be defined. Radiometric terminology describes light as energy. Radiometry is the science of measuring optical radiation, which occurs in a range of wavelengths between 10 nm and 10^5 nm and includes ultraviolet, visible, and infrared radiation. Radiometric quantities are listed in Table 1.

Photometry parallels radiometry, with the important exception that quantities are weighted according to the spectral sensitivity curve of the human visual system (given in Fig. 1) and therefore only extends between around 400 and 800 nm. Photometric quantities are listed in Table 2.

In human vision one further quantity is often used, which is the Troland (td), a measure of retinal illuminance. The reason for this is that the light that hits the retina has passed through the pupil, which has a diameter dependent on the amount of light incident. The Troland accounts for pupil size by multiplying luminance (in cd/m^2) by the size of the pupil (in mm^2).

HUMAN VISION

The human visual system forms a significant portion of the brain and is dedicated to processing visual information. Light enters the eye and is eventually transduced into an electrical signal by the photoreceptors. Through several layers of cells with associated processing, a heavily transformed signal is passed to the lateral geniculate nucleus (LGN), a part of the brain that can be thought of as acting as a relay station. After a small amount of further processing, the signal travels from the LGN to the visual cortex where it is processed by a multitude of modules, each responsible for increasingly abstract and high-level tasks.[1] The routes that the signal takes from the eyes to the brain and each of these modules are known as the visual pathways.

Human vision can be studied in many different ways. Visual psychophysics, for instance, presents meticulously designed light patterns (called stimuli) while asking observers to perform certain tasks.[2–6] Task performance is measured a large number of times for a sufficiently large

The Human Visual System

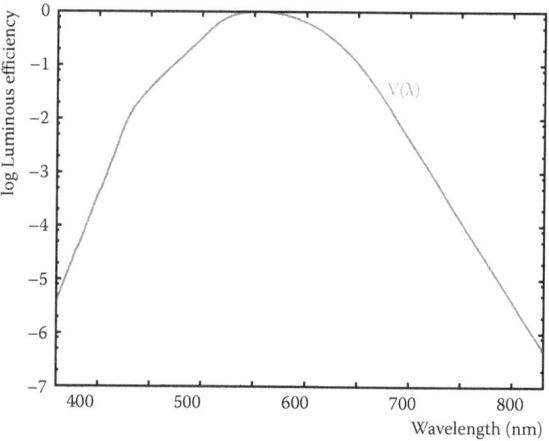

Fig. 1 The photopic luminous efficiency curve

Table 1 Radiometric quantities

Quantity	Unit	Description
Radiant energy (Q_e)	J (Joule)	Total energy of a beam of radiation
Radiant flux (P_e)	J/s = W (Watt)	Rate of change of radiant energy
Radiant exitance (M_e)	W/m^2	Radiant flux emitted from a delta surface area
Irradiance (E_e)	W/m^2	Radiant flux striking a delta surface area
Radiant intensity (I_e)	W/sr	Flux emitted into a given direction
Radiance (L_e)	W/m^2/sr	Flux per area emitted into a given direction

Table 2 Photometric quantities

Quantity	Unit	Radiometric equivalent
Luminous energy (Q_v)	lm s	Radiance energy
Luminous flux (P_v)	lm (lumen)	Radiant flux
Luminous exitance (M_v)	lm/m^2 = lx (lux)	Radiant exitance
Illuminance (E_v)	lm/m^2 = lx (lux)	Irradiance
Luminous intensity (I_v)	lm/sr = cd (candela)	Radiant intensity
Luminance (L_v)	lm/m^2/sr = cd/m^2	Radiance

number of observers using systematically altered versions of the stimuli, which then allows the experimenter to infer something about the inner workings of the human visual system.

In neuroscience, a different approach is taken, one that involves measuring cell responses using electrodes.[7] This cannot normally be done on humans, and is therefore typically performed on animals with visual systems that are thought to be comparable to those found in humans. Many aspects of low-level visual processing have been uncovered in this manner, especially relating to receptive fields of a variety of specific cells. The receptive field of a cell has several critical characteristics, such as the location on the retina or in the environment to which is responds and the specific size, shape, and pattern of light that needs to be shone onto that cell for it to respond maximally. It is reasonable to expect that cells in the retina have simpler receptive fields than those in the visual cortex.

Recent advances in measuring cell responses have made it possible to record and measure groups of cells.[8,9] Nonetheless, it remains difficult to obtain a view of how the visual system as a whole responds to stimuli. Visual psychophysics is only able to indirectly infer visual processing, while cell recordings often include only single cells. The field of natural image statistics has emerged as another tool for helping to understand human vision. The thought is that the human visual system over a long period has evolved to make sense of natural scenes, and as such is in some sense optimized for this task. It is therefore a good idea to understand the statistical regularities of natural scenes as a further tool to help understand human vision. Hence the emergence of natural image statistics as a field of study.

In our opinion, natural image statistics can also be useful in designing visual algorithms—for instance, in computer graphics, computer vision, and digital image processing. In this entry following the visual pathway, we give an overview of the human visual system and its structure, as well as some of its idiosyncrasies. We note that by and large the retina is easier to measure than any of the later stages. As a result, more is known about the retina. Consequently, this entry focuses mostly on the processing encountered in the human eye.

THE EYES

The eyes form the first part of a long chain of processing that allows humans to "see" the world.[10] The eyes sit in sockets that allow them to rotate, a feature that is useful for two reasons. First, it allows features of interest to be projected onto the fovea where vision has the highest acuity. Second, it allows humans to follow moving objects without head movement.

Figure 2 shows an annotated cross section of the human eye. Light enters the eye through the pupil, travels through the ocular media, and finally hits the retina. The retina is a layer of neural sensors lining the back of the eye. These sensors transduce light into a signal that is transmitted to the brain. Several layers of neurons are present in the retina (shown in Fig. 3). Only the photoreceptors (i.e., the rods and cones) are sensitive to light.

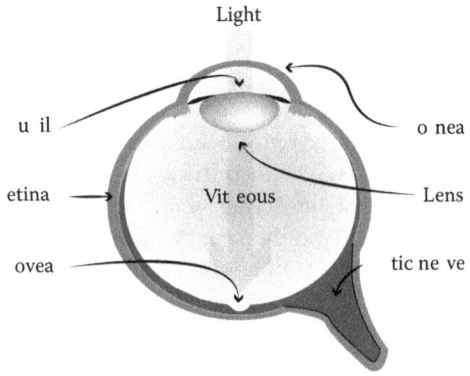

Fig. 2 A simplified cross section of the human eye
Source: Adapted from Ref. [10].

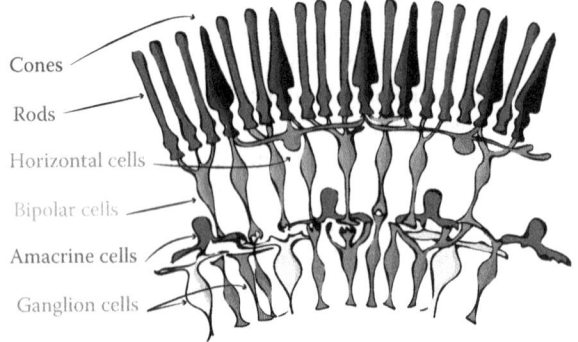

Fig. 3 The human retina consists of several layers of neurons that mediate vision
Source: Adapted from Ref. [19].

Optical Media and the Retina

Light enters the eye through the cornea and then passes through the aqueous humor, a cavity filled with clear liquid, before reaching the iris and the lens. The visual field over which light can enter the eye extends to about 60 degrees nasally and 95 degrees temporally. Vertically, the field of view extends 60 degrees above and 75 degrees below the horizontal meridian. Both eyes together therefore achieve a field of view of almost 180 degrees.

The iris contains muscles to control the diameter of the pupil and thereby helps regulate how much light passes through.[a] The lens is attached to muscles that control its shape and is used to focus on objects located at different distances to the eye. Note, however, that the cornea performs most of the focusing due to its curvature and refractive index ($n = 1.376$), while the lens contributes the remainder.[3]

After passing through the lens, light traverses the vitreous humor before reaching the retina, a neurosensory layer able to transduce light into signals and transmit those to the brain. The retina consists of several layers of cells as well as several regions with different properties. Light first passes through all layers of the retina before being transduced by the photoreceptors. The photoreceptors pass their electrical signal to a layer of bipolar cells, which in turn transmit the signal to the ganglion cells. Although the ganglion cells are located in the retina, they synapse in the LGN. Their axons are collectively called the *optic nerve bundle*. In between the photoreceptors and the ganglion cells, there are two layers of lateral connections, formed by horizontal cells and amacrine cells. As such, there is already a substantial amount of visual processing occurring within the retina. This processing is discussed further in the following sections.

As mentioned, the retina consists of a number of regions that perform somewhat different functions and are anatomically different. The central part of the retina contains the fovea, an area of 1.5 mm in diameter. In the center of this region is a smaller disk of 0.35 mm diameter, called the foveola, which is where the density of photoreceptors is highest. As a result, this is where visual acuity is highest as well. The region outside the fovea is the peripheral retina. This region also contains the optic disk, a small area known as the blind spot.

Photoreceptors

Two types of photoreceptors exist in our eyes, namely, cones and rods. These exist in different densities and fulfill different purposes.[11] Rods are sensitive to low light levels and allow us to detect contrast, brightness, and motion—but not color—in these conditions. The range of illumination that rod photoreceptors operate in is termed *scotopic*. Cones, on the other hand, operate at higher light levels, known as *photopic* conditions. In addition to contrast, motion, and brightness, cones are involved in seeing color (see Fig. 4). There exists a range of light levels under which both rods and cones are active. This range is called *mesopic*. The responsivities to different wavelengths for the rod and cone systems are known as the *scotopic* and *photopic luminous efficiency*, which are plotted in Fig. 5.

Photoreceptor Types

Three types of cones exist in the human eye and are classified according to the wavelengths that they are most sensitive to, namely, short, medium, and long (referred to as S, M, and L). These translate to peak sensitivities of approximately 440 nm, 545 nm, and 565 nm, respectively, corresponding roughly to blue, green, and red light, although as shown in Fig. 6 significant overlap exists between the L, M, and S cone sensitivities.[12,13]

As each photoreceptor integrates over a wide range of wavelengths, there are many combinations of colored light that would lead to the same photoreceptor output. For

---

[a]Recent research has shown that photo-responsive ganglion cells regulate this mechanism.[318]

The Human Visual System

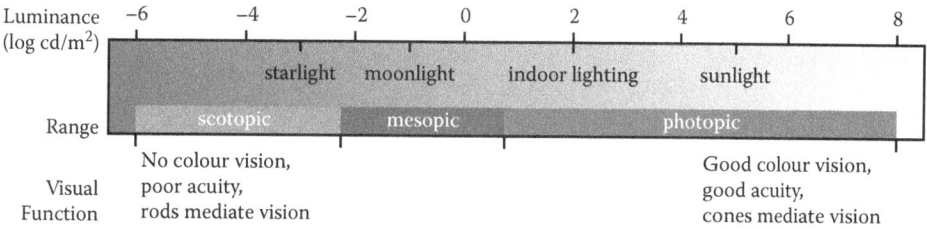

Fig. 4 The ranges of luminance where the different types of photoreceptors are operational. Rods operate in low light conditions, known as scotopic, while cones cover the brighter range of illumination known as photopic. Some overlap exists between the two photoreceptor types. This is known as the mesopic range
Source: Adapted from Ref. [104].

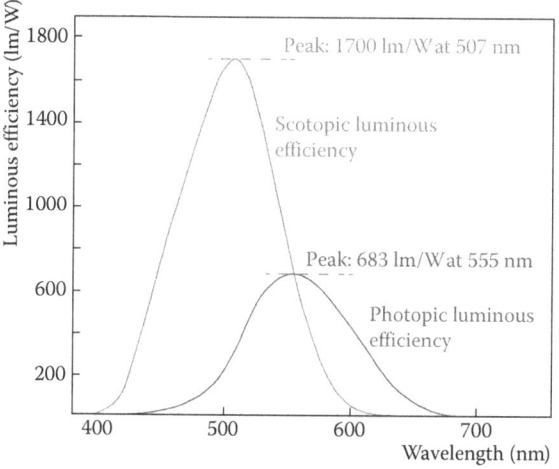

Fig. 5 Photopic and scotopic luminous efficiency curves

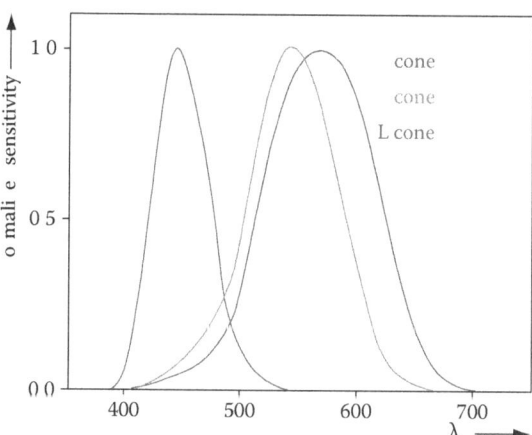

Fig. 6 The sensitivities of the three cone types for different wavelengths
Source: Adapted from Ref. [12].

instance, supposing we could change a monochromatic light from one wavelength to another and also change its intensity, it should be possible to produce a new stimulus that leaves the output of a given cone unchanged. If two different spectral distributions lead to the same response of all three photoreceptor types, they are visually equivalent to humans and are referred to as *metamers*. A second consequence of this integration is that a single photoreceptor type is not able to distinguish between different colors. For color vision to occur, the outputs of multiple cone types must be combined.[14] Processing in the retina does indeed achieve this, as the S, M, and L cones provide sufficient information to the remainder of the visual system to mediate color vision. Rods exist in only one type, and as a result color vision is impaired or even absent under scotopic lighting conditions.

Adaptation

A central property of photoreceptors, as well as many if not all cells involved in neural activity, is that of adaptation.[15] Photoreceptors can simultaneously transduce a range of illumination that spans about four orders of magnitude.[16] Our environment has illumination levels that span from starlight to direct sunlight, a range of about ten orders of magnitude (see Fig. 4). Adaptive processes allow photoreceptors to function over this much greater range. We experience its effect on a daily basis: it is sufficient to walk into a dark room from bright sunlight to realize that our eyes adjust to allow us to see under the new conditions. Effectively, our visual system adapts to the prevailing lighting conditions so that we can still distinguish contrasts in the scene.[17]

Adaptive processes give rise to both light and dark adaptation as well as chromatic adaptation. A demonstration of the latter is shown in Fig. 7, where an image is shown with a yellow and a cyan filter applied. To recover the correct colors of the image, it is possible to adapt the retina to the yellow and cyan color casts. This is achieved by staring at the fixation cross at the top of the image for at least 30 seconds. The image below also has a fixation cross. By focusing on this fixation cross after the adaptation period, the image should regain its normal coloring.

Light Sensitivity

The light sensitivity of rods and cones is mediated by photosensitive pigments that react with light. As light hits the photoreceptors, the photosensitive pigments break down (bleaching). In the case of rods, these photopigments

Fig. 7 Adaptation is local, as can be observed in this demonstration. To see the correct colors of the image below, first fixate on the upper cross for at least 30 seconds to allow the retina to locally adapt. By then focusing on the bottom fixation cross, the image should appear normally colored
Source: Keysersberg, Alsace, France, 2013.

are completely depleted in lighting conditions above the mesopic range, while cones are not fully depleted even in very bright conditions.[18] Photopigment bleaching causes the electrical current present in a photoreceptor to change, which results in a neural response transferred to the brain. Through this transduction process, the retina can translate light energy into a neural response, which effectively is what allows us to see.[19]

A final set of mechanisms occurring in the photoreceptors prevents them from saturating while adapting to the enormous range of illumination in nature.[20] Several studies have shown that the photoreceptors do not respond linearly to the full range of illumination.[18,21–23] This non-linearity was first studied by Naka and Rushton[24] and provides a very effective mechanism for the compression of dynamic range. Intensities in the middle of the range lead to approximately logarithmic responses while as the intensity is increased, the response tails off to a maximum. After that maximum is reached, further increases in intensity will not lead to a corresponding increase in

The Human Visual System

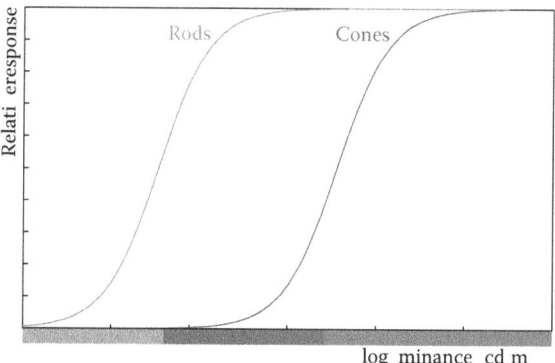

Fig. 8 Photoreceptors respond nonlinearly to increasing luminance levels. The Naka-Rushton equation models this nonlinearity, producing S-shaped curves known as sigmoids

response of the photoreceptors, ensuring that saturation will not occur.[18]

Figure 8 shows the nonlinear response of the rod and cone cells to increasing luminance values. The response for both types of photoreceptors can be modeled with what is known as the Naka-Rushton equation:

$$\frac{V}{V_{max}} = \frac{I^n}{I^n + \sigma^n} \quad (1)$$

where V is the response at an intensity I, V_{max} is the peak response at saturation, and σ is the intensity necessary for the half-maximum response (also known as the semi-saturation constant). The exponent n controls the slope of the function and is generally reported to be in the range of 0.7 to 1.0.[18] This functional form is known as a sigmoid because it forms an S-shaped curve when plotted on log-linear axes. In this model adaptation can be accounted for by changing the value of σ. Moreover, this value could be used to model temporal adaptation.

Note that the response function of photoreceptors can be linked to natural image statistics. It has been shown that the joint probability density function of various components that make up images (illumination, reflectance, and textured regions) leads to a lightness scale[b] as function of illuminance that is remarkably similar to the sigmoidal response function of photoreceptors that are shown in Fig. 8.[25] More recently, lightness perception has been shown to directly correlate with this sigmoidal response function.[26] Finally, temporal adaptation is implicated in the enhancement of efficient gain control in the context of natural images,[27] as well as redundancy reduction.[28–30]

---

[b]The official definition of lightness is as follows: lightness is the attribute of visual sensation according to which the area in which the visual stimulus is presented appears to emit more or less light in proportion to that emitted by a similarly illuminated area perceived as the "white" stimulus.

Photoreceptor Mosaic

Photoreceptors are located in the retina with a somewhat unexpected layout.[31] The S cones are sparse in the fovea and absent in the foveola. In the peripheral retina, S cones are packed at regular distances.[32] The L and M cones, on the other hand, are essentially randomly distributed over the retina, although the ratio between the two increases toward the periphery.[14]

The ratio between the L and M cones varies significantly between individuals,[33,34] although this has only a minor effect on the perception of color.[35,36] It has been postulated that this may be the result of plasticity in the human visual system, whereby later processing in essence compensates for the individual's precise photoreceptor mosaic.[37] Similar plasticity could help adjust an individual's visual system to long-term prevailing viewing conditions.

Given that at any given location in the retina there exists at most one cone, color vision is necessarily also a spatial process: different cone types need to be compared, and different cones are necessarily located in different positions. This places a limitation on the visual acuity of color vision.

Horizontal and Bipolar Cells

The retina consists of several layers of neural cells. The photoreceptors make up the first layer. As discussed above, their output is transmitted to a layer of bipolar cells, which subsequently convey the signal to the ganglion cells. These in turn carry the signal through the optic nerve to the LGN.

Bipolar cells take their input primarily from the photoreceptors and form the start of several pathways. Some bipolar cells respond to light against a darker background, forming the start of a so-called ON pathway. Others respond to dark spots against a lighter background, and are part of the OFF pathway. One could argue that the existence of separate ON and OFF pathways helps encode polarity. There are approximately two times more retinal OFF pathways, which encode negative contrasts, than ON pathways, which encode positive contrasts.[38] From the bipolar cells onwards, electrical signals are transported in the form of discrete spike trains. The firing rate of such spike trains encodes the quantity transmitted. One can think of spike trains as encoding only positive numbers. To also encode negative numbers (for instance, those corresponding to dark patches against a light background), a separate pathway needs to be constructed. Both L and M cone types are involved in both ON and OFF pathways, which can be either chromatic or achromatic.[39] The S cones are only involved in separate chromatic ON pathways.[40]

In between the photoreceptors and the bipolar cells sits a further layer of cells that provide lateral connectivity. These cells are known as horizontal cells. These cells tend to have wide receptive fields, meaning that they connect to a large number of photoreceptors and bipolar cells over a significant spatial extent.

Due to the connectivity of horizontal cells, they are able to change the input to bipolar cells dependent on signals carried from photoreceptors some distance away. Horizontal cells provide inhibitory (opponent) signals to bipolar cells. This means that if their input from photoreceptors is large, they reduce the input signal available to the bipolar cells, and vice versa. Horizontal cells, therefore, form the mechanism by which bipolar cells are able to respond to light patches on a dark background or dark patches on a light background.

Thus, the opponent processing of the horizontal cells causes the receptive fields of bipolar cells to be of the center-surround variety. A reasonable way to think of such receptive fields is as if both center and surround are obtained by maximally responding to a Gaussian-blurred version of the image, although the size of the Gaussian filter for the surround is wider than the center. As shown in Fig. 9, this leads to a receptive field that resembles the classical Mexican hat shape.

To visualize the responses carried by center-surround bipolar cells, we have blurred a grayscale image twice and subtracted the result, shown in Fig. 10. The size of the receptive field was arbitrarily chosen and does not reflect actual processing in the visual system. Note that circularly symmetric center-surround receptive fields respond most strongly to circular features matched in size to the filter width. Further, this type of center-surround processing also responds to edges.

Amacrine and Ganglion Cells

Between the bipolar and ganglion cells lies a further layer of cells providing lateral connectivity. This layer of amacrine cells also provides connections between ON and OFF pathways as well as between rod and cone pathways. There exists a large number of different amacrine cells, each with different functionality. They appear to be involved in processes that allow the visual system to adapt to its environment.

As a result of the multitude of amacrine cell types, the receptive fields of ganglion cells are significantly varied.[41] They exist with center-surround receptive fields and may also show color opponency.[42] The latter happens, for instance, when the ganglion cell is stimulated by a pathway originating with an L cone, and it is inhibited by a pathway that started with an M cone. Thus, the ganglion cells effectively encode a very different color space than the cones that approximately correspond to red, green, and blue light. In fact, many ganglion cells in the fovea are color opponent, encoding in their signal the proportion of red against green, or alternatively the amount of yellow against blue.

Color opponency and spatial center-surround processing can occur simultaneously in a process called *double opponency*. Here, if the center of the receptive field is red, then the surround will be green and vice versa. Likewise, a yellow center will be flanked by a blue surround and vice versa. The result of such opponency is simulated in Fig. 11.

Although the opponency is normally denoted as red-green and yellow-blue, note that these cardinal color directions do not completely correspond to perceptual experience. Later modules in the visual cortex are thought to perform further processing to arrive at the percepts of hues in terms of red, green, yellow, and blue.[43-45] Finally, there exist ganglion cells that are selective for direction and motion.

THE LATERAL GENICULATE NUCLEUS AND CORTICAL PROCESSING

The lateral geniculate nucleus (LGN) is the dominant recipient of signals transmitted from the retina, and it is thought of as a relay station, effectively broadcasting the signal to modules in the visual cortex. The receptive fields of neurons in this area resemble those seen for ganglion cells, showing spatial and chromatic opponency. The LGN is organized in layers,[46] termed *magnocellular*, *parvocellular*, and *koniocellular*.

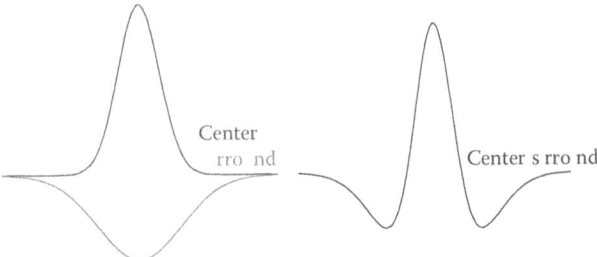

Fig. 9 Cross-section of a center-surround receptive field. This receptive field was created by subtracting two Gaussian profiles from each other

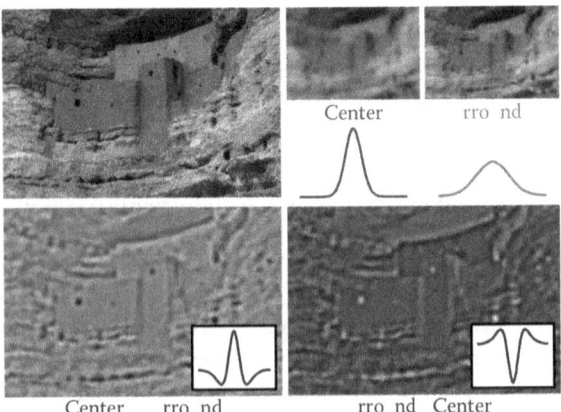

Fig. 10 The receptive field images at the bottom were processed by subtracting two differently blurred achromatic versions of the input image
Source: Montezuma's Castle, Arizona, 2012.

The Human Visual System

Fig. 11 Double opponency demonstrated on an image of a pair of lorikeets. Shown here are red-green, green-red, yellow-blue, and blue-yellow double opponent results. Note that these images are visualizations only; the results were created at an arbitrary spatial scale and they were computed in sRGB color space rather than LMS. The ratios between different cone types were approximately taken into account
Source: Jurong Bird Park, Singapore, 2012.

The magnocellular layers carry an achromatic center-surround signal. The parvocellular layers mediate red-green color opponency, especially in the fovea.[47,48] This layer does not appear to receive input from S cones. The koniocellular layers transmit several differentiated types of information, including blue-yellow color opponency.[46,49] The organization of the layers in the LGN is retinotopic, in that the topology of cells reflects that of the retina. Thus, nearby features in an image projected onto the retina will be processed by nearby cells in the LGN. There is evidence that the ON and OFF pathways are augmented with lagged and non-lagged responses, i.e., some signals are transmitted more rapidly than others.[50] This would allow the calculation of temporal changes and could form the basis for motion detection. There is also some evidence that the spatial properties of some LGN cells change as a function of time (usually, the spatial organization changes in a manner that looks like motion in a single direction[51]).

Most of the signal transmitted by the LGN arrives in the visual cortex in a module named V1, although some of the

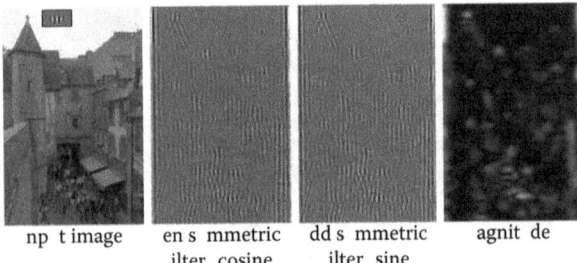

Fig. 12 A simulation of direction-selective receptive fields. An even symmetric Gabor patch (a windowed cosine, see inset, top left) was applied to the left image, yielding the second image. An odd symmetric Gabor filter was applied to the image as well. As an aside, the magnitude of the two receptive fields leads to a response shown on the right
Source: Mont St. Michel, France, 2007.

koniocellular layers transmit to the extrastriate cortex and bypass V1 altogether.[52–54] Nonetheless, V1 is the primary entry point for visual signals in the brain. From there, processing passes through more than 30 visually responsive cortical areas.[1]

Area V1 receives signals from the lateral geniculate cortex, and consists of six different layers. Some of these layers are subdivided into sublamina.[55–57] There are further structures called blobs or patches. It is thought that the structure of V1 relates to its function. Many cells in this area are orientationsensitive,[58] a simulation of which is shown in Fig. 12. This means that their receptive fields are formed by elongated structures rather than the circularly symmetric receptive fields in the retina and LGN.[59–61] Cells may also be responsive to motion as well as changes to scale[62] or to binocular signals.[63,64] The retinotopic layout of the LGN is preserved in area V1.[65] Area V1, like most, if not all, modules in the visual cortex has both feed-forward as well as feed-back connections with many other processing units.[66]

Other cortical areas, such as V2, V3, V4, and MT, are implicated in a variety of different functions. Area MT, for instance, is known to be involved in the perception of motion, while V4 may take part in color perception.

IMPLICATIONS OF HUMAN VISUAL PROCESSING

The substrate of visual processing is organized in a way that may not be considered obvious. A small selection of features known to exist in the human visual system were discussed in previous sections. For instance, we have presented ON and OFF pathways, color opponency, and center-surround processing, as well as combinations of these. As a result, the signal that the brain receives is not a straightforward image, but one that is already heavily processed. The organization and structure of the human visual system leads to several important characteristics,[67] some of which are discussed in the following sections.

Visual Acuity

Visual acuity is defined as the resolving power of the human visual system.[68] The ability to detect and perceive fine detail is affected by various aspects of vision, including diffraction and aberrations in the ocular media, as well as photoreceptor density.[69] Note in particular that photoreceptor density varies across the retina, cones being packed most densely in the fovea. This means that visual acuity is highest in the fovea. Here, a single cone subtends approximately 28 minutes of arc, which means that this is also theoretically the highest resolving power that cones could achieve. This would lead to a resolving power of about 60 cycles per degree.[70] The rod system has lower visual acuity, pooling rod responses to increase light sensitivity.

However, visual acuity may be limited by diffraction in the ocular media due to the edge of the pupil. A point light source would project according to a point spread function (PSF) on the retina. Diffraction would cause this point spread function to consist of concentric rings of light, i.e., an Airy disk pattern. To be able to separate a point source from a second point source some distance away, the point spread functions of both light sources should not overlap. Two point lights are considered sufficiently separated if the center of one PSF lies on the first trough of the second PSF. This is formalized by Raleigh's criterion, which links wavelength of light λ and the diameter of the pupil d:

$$a = 1.22 \frac{\lambda}{d} \quad (2)$$

Here, a is the angular radius of the first ring of the Airy disk pattern.

Further factors affecting visual acuity include refractive error, which occurs if light refracted through the ocular media is not focused sharply on the retina. Visual acuity is also dependent on the amount of illumination present in the scene[71] and the corresponding state of adaptation of the eye.

Under certain circumstances, visual acuity may be significantly higher than could be expected from the limitations imposed by diffraction and cone spacing.[72] This is generally known as *hyperacuity*. An example of when this occurs is during assessment of the alignment of two line segments—for instance, when reading a caliper. This can be done with an accuracy of about five to ten times higher than the normal resolving power of the human visual system. In essence, a set of photoreceptors are involved in localizing the line segments. Assuming the cones are of the L and M variety, their placement is randomized, which may help in such localization tasks.

In some sense, hyperacuity achieves for human vision what superresolution does in image processing. The randomization of cones helps with anti-aliasing. One implication for the study of natural image statistics may be that image resolution needs to be chosen high enough to anticipate the perception of edges and their relative alignment.

Temporal Resolution

To be able to detect motion, the eye must be able to detect rates of change in the image that falls on the retina. Some amount of time is needed to transduce light and propagate the resulting signal to the brain. This poses a limit on how fast or slow a signal can change and still be perceived as fluent motion.[73] For instance, if the rate of change is too slow, then the sensation of fluent motion is lost, and the stimulus is seen as flicker. At photopic light levels, the eye is most sensitive to flicker at frequencies of around 15 to 20 Hz.[74]

To detect two flashes of light separated in time, rods require an integration time of 100 ms, while cones require 10 to 15 ms. Thus, the rod system is slower to help improve light sensitivity.

Contrast

Contrast involves the relative values between points or regions in an image and can be measured in many different ways. For a region in an image, it would, for instance, be possible to compute root mean square contrast, which is the standard deviation of a set of pixels:

$$C_{RMS} = \sqrt{\frac{\sum_{n \in N}(L_n - \mu_L)^2}{N}} \qquad (3)$$

It would also be possible to normalize this measure of contrast by the mean μ_L:

$$C_{norm} = \frac{C_{RMS}}{\mu_L} \qquad (4)$$

However, it can be argued that for complex images a more involved measure would be appropriate—for instance, by computing the contrast on band-pass filtered images.[75]

If an image/stimulus is a small feature on a uniform background, then the background luminance L_b will be close to the average luminance. In that case, Weber contrast is appropriate. It is defined as:

$$C_{Weber} = \frac{L - L_b}{L_b} \qquad (5)$$

For sinusoidal gratings, a different contrast measure is commonly used, a measure known as Michelson contrast.

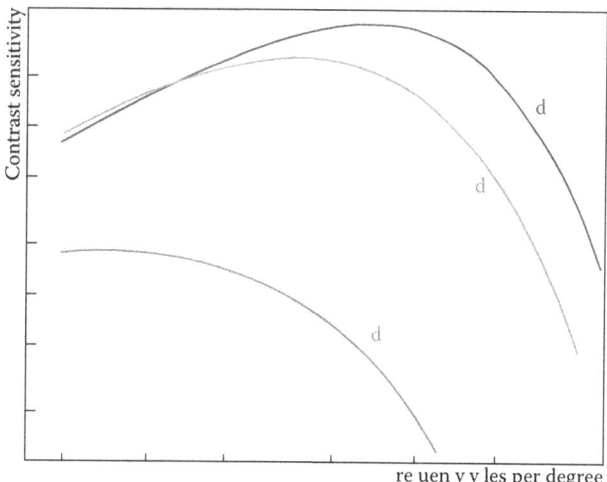

Fig. 13 Contrast sensitivity functions for different light levels.[73,105]

This measures the differences between the peaks and troughs of the grating:

$$C_{Michelson} = \frac{L_{max} - L_{min}}{L_{max} + L_{min}} \qquad (6)$$

Humans are sensitive to contrast in images (whether natural or psychophysical stimuli). This sensitivity depends on the frequency at which the contrast occurs, as well overall light levels. More recently, it has been shown that contrast sensitivity in natural scenes also depends on the distribution of local edges.[76] It would be possible to plot threshold contrast against frequency, which leads to the well-known contrast sensitivity function; this function is plotted for different light levels in Fig. 13.[77] Note that for low light levels, the response is low-pass, whereas at high light levels the response is band-pass.

Color Processing

Under photopic lighting conditions, the three types of cones are active, each integrating over the visible spectrum over a slightly different range and with different peak sensitivities. Thus, human vision is trichromatic under these conditions.[78,79] As a result, in each region of the fovea where there are multiple cone types active, color vision can emerge. However, as noted, different spectra may lead to the same three integrations, which means that the output of the cones will be identical and these different spectra cannot be differentiated. These spectra are said to be *metameric*.

Trichromacy has a second implication, which is that almost any color can be matched by mixing three other colors. This can be achieved, for instance, by matching colors on a bipartite field, which is in essence a disk partitioned into two halves. One of the halves displays the color to be matched, for instance, by means of projection. The other

Fig. 14 Two-tone images of human faces are easily recognized if lit from above (left). The inverted version of this image (right) is not easily recognized. Images such as these are therefore good candidates for accurate color matching experiments.[81]

half is illuminated by three projectors, each projecting a given color by an amount that is user-controllable.

This method can also be used to infer the luminance of a colored patch. Here, the colored patch is held fixed while the other half displays achromatic white in an amount controlled by the user. The amount of light that is necessary to make the border between the left and right half of the bipartite field minimally distinct then represents the luminance of the colored patch.[80]

When carrying out such experimentation on a display, significantly higher accuracy can be achieved by replacing the bipartite field with a two-tone image of a human face.[81] The accuracy gain stems from two characteristics of human vision. First, humans appear to make the assumption that objects, including faces, are predominantly illuminated from above (see Fig. 14). Second, humans are particularly adept at recognizing human faces, but only if illuminated from above.

Color opponent theory predicts that colors are perceived along three different dimensions, namely, light-dark, red-green, and yellow-blue.[82] This is consistent with perceptual experience[83] and after-images, as well as what is known about retinal processing. There is also a strong link between color opponent processing in the retina and statistical attributes of natural scenes.

Visual Illusions

Human vision solves a complicated problem, which is to makes sense of the world around us. One of complications stems from the fact that scene understanding and navigation requires a mental image of a three-dimensional world that must be derived from a pair of two-dimensional retinal images.[84] That this is a complicated problem is well known and is, for instance, exemplified by the difficulty of figuring out shape from shading in computer vision–related applications.[85,86]

It appears that human vision makes many assumptions that help in image understanding. Figure 14 already showed an example, namely, that face recognition is directly facilitated by ensuring that light comes from above. When this assumption is broken, face recognition becomes significantly more difficult. Similarly, a human face is more easily recognized when it is positioned right-side up.

The human visual system appears to assume that the scene being observed does not contain anything special, i.e., there are no accidental viewpoints, light comes from above and, possibly above all, the scene is in some sense natural. When these expectations are broken, then the scene may be viewed as having artefacts or, alternatively, a visual illusion may arise.[87,88]

In computer graphics, for example, achieving visual realism has been difficult for many years.[89] Only recently have modeling and rendering systems become available that add enough dirt, grime, and general detail that rendered images have become difficult to differentiate from photographs. Leaving out such detail, or omitting certain light paths because they are computationally costly to render, may lead to images that are generally scene as plastic or fake.

Visual illusions occur when expectations are broken in specific ways. It can be argued that human vision is betting on what is likely to be true about the world given the evidence available.[88] Thus, the study of visual illusions may lead to insights into human visual processing.[90] There are too many visual illusions to list in this book; moreover, many of them involve very simple line drawings. Illusions that can be reproduced in photographs or can be demonstrated as shaded surfaces are less numerous.

The most common illusion that can be seen in a photograph is perhaps the cafe wall illusion, discovered in Bristol by Richard Gregory[91] (see Fig. 15). Here, a set of rectangular tiles are positioned such that the horizontal grout lines appear to be oriented differently. Due to the camera angle, however, these lines should (and do!) all converge at the same point. In three dimensions, these lines are parallel.

Fig. 15 Cafe wall illusion
Source: Bristol, UK, 2006.

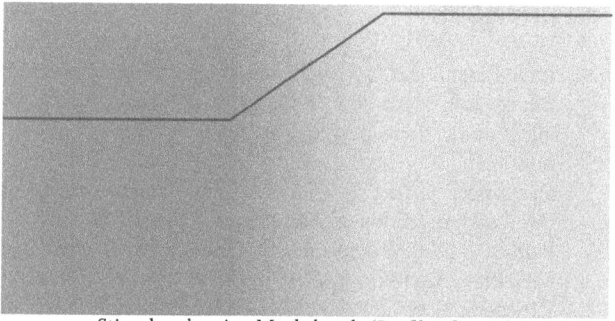

Stimulus showing Mach-bands (Profile of ramp)

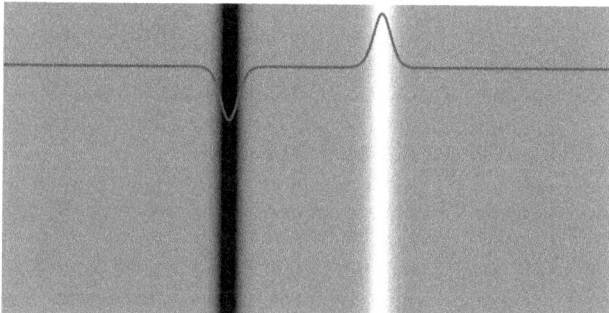

Exaggerated response to ramp (Profile of response)

Fig. 16 C^0 continuity may lead to Mach-bands, seen as under- or over-shoot where gradients are discontinuous

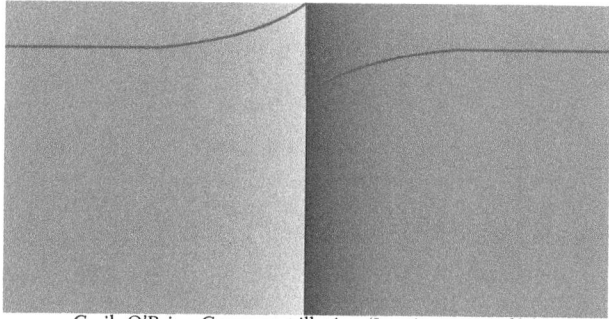

Craik-O'Brien-Cornsweet illusion (Luminance profile)

Masking the middle part reveals that the left and right parts have the same luminance

Fig. 17 The Craik-O'Brien-Cornsweet illusion. In the top panel, the left and right quarter of the image will appear to have a different gray level. Masking the center half of the image reveals that the actual luminance levels are identical

The center-surround organization found in the retina is thought to give rise to several interesting phenomena. One consequence of center-surround processing is the occurrence of Mach-bands.[92,93] In essence, wherever a uniform luminance gradient abruptly changes there may be the perception of either under or over-shoot. An example is shown in Fig. 16 (top) where a linear ramp abuts two uniform areas. At the transition point there is C^0 continuity, but a discontinuity in gradient. Processing such an image with a center-surround filter shows the cause for the perceived under- and over-shoot as seen in Fig. 16 (bottom).

A further consequence of center-surround processing is the Craik-O'Brien-Cornsweet illusion, shown in Fig. 17. This figure contains a step edge flanked by ramps that smoothly vary their gradients.[94–97] As a result, a step edge is seen, suggesting that the left part of the image is lighter than the right part. The shape of the ramps is chosen such that they generate little to no response from center-surround mechanisms in the retina (see Fig. 18). As the step edge in the center does evoke a response from these mechanisms, the resulting percept is similar as that which would be generated by an actual step edge flanked by uniform regions.

A consequence of this type of retinal processing is that the visual cortex receives a signal that highly emphasizes edges. Perceiving the structure of a scene in between edges may be based on a process called filling-in,[98–101] suggesting that the brain completes its perception of a scene predominantly on the basis of edges. One may therefore expect edges and their statistics to be of crucial importance in human vision.

Fig. 18 The Craik-O'Brien-Cornsweet illusion passed through a center-surround filter. Note that there is virtually no response for most of the image, except near the edge

An alternative suggestion is that filling-in is aided by other statistical regularities in natural images, specifically the nature of the power spectrum,[102] which is known to give a relation between power and frequency of $1/f^2$.

That the Craik-O'Brien-Cornsweet illusion may have yet a different and higher level explanation can be seen in Fig. 19.[103] Here, the edge of the image is changed to suggest curvature in depth, and this appears to enhance the effect. It may be that this simple change to the figure increases the probability that the luminance profile is due to illumination, and that the human visual system is able to detect this. This effect can be enhanced further by embedding it

Fig. 19 The Craik-O'Brien-Cornsweet illusion becomes stronger if it is placed in context. Here, for instance, the shape of the image suggests a 3D surface, which enhances the effect[103]

A 3D stimulus showing a Cornsweet-type profile

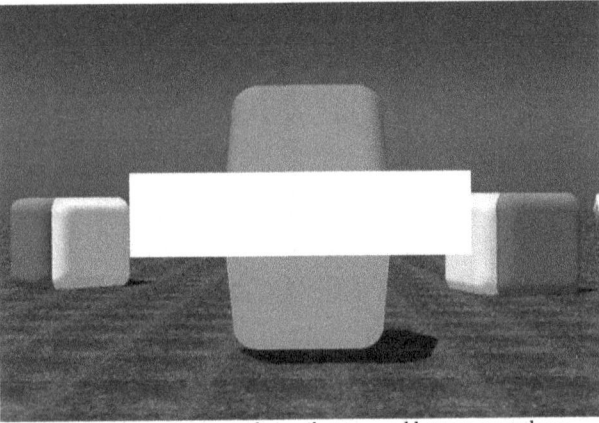

Masking the center part shows that top and bottom parts have identical luminance

Fig. 20 The Craik-O'Brien-Cornsweet illusion becomes stronger if it is placed in context. Here, for instance, the shape of the image suggests a 3D surface, which enhances the effect[103]
Source: Image used by permission from Dale Purves.

into a scene that affords further 3D cues, as demonstrated in Fig. 20. Incidentally, the strength of this illusion also relates to the assumption that light comes from above. Thus, while center-surround processing may be involved, it is by no means the only component of human visual processing that contributes to this variant of the illusion.

REFERENCES

1. van Essen, D.C.; Anderson, C.H.; Felleman, D.J. Information processing in the primate visual system: An integrated systems perspective. Science **1992**, *255* (5043), 419–423.
2. Gescheider, G.A. *Psychophysics: The Fundamentals*, 3rd Ed.; Lawrence Erlbaum Associates: Mahwah, NJ, 1997.
3. Palmer, S.E. *Vision Science: Photons to Phenomenology*; MIT Press: Cambridge, MA, 1999.
4. Kingdom, F.A.A.; Prins, N. *Psychophysics: A Practical Introduction*; Academic Press / Elsevier: London, 2009.
5. Cunningham, D.; Wallraven, C. *Experimental Design: From User Studies to Psychophysics*; AK Peters: Natick, MA, 2011.
6. Snowden, R.; Thompson, P.; Troscianko, T. *Basic Vision: An Introduction to Visual Perception*, 2nd Ed.; Oxford University Press: Oxford, 2012.
7. Hubel, D.H.; Wiesel, T.N. Ferrier lecture: Functional architecture of macaque monkey visual cortex. Proc. R. Soc. London B **1977**, *198* (1130), 1–59.
8. Ohki, K.; Chung, S.; Ch'ng, Y.H.; Kara, P.; Reid, R.C. Functional imaging with cellular resolution reveals precise micro-architecture in visual cortex. Nature **2005**, *433* (7026), 597–603.
9. Kerr, J.N.D.; Denk, W. Imaging in vivo: Watching the brain in action. Nat. Rev. Neurosci. **2008**, *9* (3), 195–205.
10. Reinhard, E.; Khan, E.A.; Akyüz, A.O.; Johnson, G.M. *Color Imaging: Fundamentals and Applications*; A K Peters: Wellesley, MA, 2008.
11. Forrester, J.; Dick, A.; McMenamin, P.; Lee, W. *The Eye: Basic Sciences in Practice*; W B Saunders: London, 2001.
12. Bowmaker, J.K.; Dartnall, H.J. Visual pigments of rods and cones in a human retina. J. Physiol. **1980**, *298* (1), 501–511.
13. Stockman, A.; MacLeod, D.; Johnson, N. Spectral sensitivities of the human cones. J. Opt. Soc. Am. A **1993**, *10* (12), 2491–2521.
14. Solomon, S.G.; Lennie, P. The machinery of colour vision. Nat. Rev. Neurosci. **2007**, *8* (4), 276–286.
15. Webster, M.A. Adaptation and visual coding. J. Vision **2011**, *11* (5), 1–23.
16. Kunkel, T.; Reinhard, E. A reassessment of the simultaneous dynamic range of the human visual system. In *Proceedings of the 7th Symposium on Applied Perception in Graphics and Visualization*; ACM: New York, 2010, 17–24.
17. Wandell, B.A. *Foundations of Vision*; Sinauer Associates, Inc.: Sunderland, MA, 1995.
18. Dowling, J.E. *The Retina: An Approachable Part of the Brain*; Belknap Press: Cambridge, MA, 1987.
19. Kolb, H.; Fernandez, E.; Nelson, R. *Webvision: The Organization of the Retina and Visual System*, 2010. Available at http://webvision.med.utah.edu/.
20. Dowling, J.E.; Ripps, H. Adaptation in skate photoreceptors. J. Gen. Physiol. **1972**, *60* (6), 698–719.
21. Hood, D.C.; Finkelstein, M.A. Comparison of changes in sensitivity and sensation: Implications for the response-intensity function of the human photopic system. J. Exp. Psychol. Hum. Percept. Perform. **1979**, *5* (3), 391–405.

22. Adelson, E.H. Saturation and adaptation in the rod system. Vision Res. **1982**, *22* (10), 1299–312.
23. Valeton, J.M.; van Norren, D. Light adaptation of primate cones: An analysis based on extracellular data. Vision Res. **1983**, *23* (12), 1539–1547.
24. Naka, K.I.; Rushton, W.A.H. S-potentials from luminosity units in the retina of fish (cyprinidae). J. Physiol. **1966**, *185* (3), 587–599.
25. Richards, W. A lightness scale from image intensity distributions. Appl. Opt. **1982**, *21* (14), 2569–2604.
26. Allred, S.R.; Radonjic, A.; Gilchrist, A.L.; Brainard, D.H. Lightness perception in high dynamic range images: Local and remote luminance effects. J. Vision **2012**, *12* (2).
27. Sinz, F.; Bethge, M. Temporal adaptation enhances efficient contrast gain control on natural images. PLoS Comput. Biol. **2013**, *9* (1).
28. Barlow, H.; Földiák, P. Chapter 4: Adaptation and decorrelation in the cortex. In *The Computing Neuron*; Mitchinson, G.; Ed.; Addison-Wesley Longman Publishing Co., Inc.: New York, 1989, 54–72.
29. Barlow, H.B. A theory about the functional role and synaptic mechanism of visual after-effects. In *Vision: Coding and Efficiency*; Blakemore, C.; Ed.; Cambridge University Press: Cambridge, 1990, 363–375.
30. Wainwright, M.J.; Schwartz, O.; Simoncelli, E.P. Natural image statistics and divisive normalization: Modeling non-linearities and adaptation in cortical neurons. In *Statistical Theories of the Brain*; Rao, R.; Olshausen, B.; Lewicki, M.; Ed.; MIT Press: Cambridge, MA, 2002, 203.
31. Ahnelt, P.K. The photoreceptor mosaic. Eye **1998**, *12* (3b), 531–540.
32. Williams, D.R.; Roorda, A. The trichromatic cone mosaic in the human eye. In *Color Vision: From Genes to Perception*; Gegenfurtner, K.R.; Sharpe, L.T.; Eds.; Cambridge University Press: Cambridge, 1999, 113–122.
33. Mollon, J.D.; Bowmaker, J.K. The spatial arrangement of cones in the primate retina. Nature **1992**, *360* (6405), 677–679.
34. Roorda, A.; Williams, D.R. The arrangement of the three cone classes in the living human eye. Nature **1999**, *397* (6719), 520–522.
35. Brainard, D.H.; Roorda, A.; Yamauchi, Y.; Calerone, J.B.; Metha, A.; Neitz, M.; Neitz, J.; Williams, D.R.; Jacobs, G.H. Functional consequences of the relative numbers of L and M cones. J. Opt. Soc. Am. A **2000**, *17* (3), 607–614.
36. Pokorny, J.; Smith, V.C.; Wesner, M. Variability in cone populations and implications. In *From Pigments to Perception*; Valberg, A.; Lee, B.B.; Eds.; Plenum: New York, 1991, 23–34.
37. Neitz, J.; Carroll, J.; Yamauchi, Y.; Neitz, M.; Williams, D.R. Color perception is mediated by a plastic neural mechanism that is adjustable in adults. Neuron **2002**, *35* (4), 783–792.
38. Ahmad, K.M.; Klug, K.; Herr, S.; Sterling, P.; Schein, S. Cell density ratios in a foveal patch in macaque retina. Visual Neurosci. **2003**, *20* (2), 189–209.
39. Wässle, H.; Boycott, B.B. Functional architecture of the mammalian retina. Physiol. Rev. **1991**, *71* (2), 447–480.
40. Kouyama, N.; Marshak, D.W. Bipolar cells specific for blue cones in the macaque retina. J. Neurosci. **1992**, *12* (4), 1233–1252.
41. Dacey, D.M.; Peterson, B.B.; Robinson, F.R.; Gamlin, P.D. Fireworks in the primate retina: In vitro photodynamics reveals diverse LGN-projecting ganglion cell types. Neuron **2003**, *37* (1), 15–27.
42. Conway, B.R.; Chatterjee, S.; Field, G.D.; Horwitz, G.D.; Johnson, E.N.; Koida, K.; Mancuso, K. Advances in color science: From retina to behavior. J. Neurosci. **2010**, *30* (45), 14955–14963.
43. De Valois, R.L.; De Valois, K.K. A multi-stage color model. Vision Res. **1993**, *33* (8), 1053–1065.
44. De Valois, R.L.; De Valois, K.K. On a three-stage color model. Vision Res. **1996**, *36* (6), 833–836.
45. De Valois, R.L.; De Valois, K.K.; Switkes, E.; Mahon, L. Hue scaling of isoluminant and cone-specific lights. Vision Res. **2000**, *37* (7), 885–897.
46. Dobkins, K.R. Moving colors in the lime light. Neuron **2000**, *25* (1), 15–18.
47. Dacey, D.M.; Lee, B.B. Cone inputs to the receptive field of midget ganglion cells in the periphery of the macaque retina. Invest. Ophthalmol. Vis. Sci. **1997**, *38*, S708.
48. Dacey, D.M.; Lee, B.B. Functional architecture of cone signal pathways in the primate retina. In *Color Vision: From Genes to Perception*; Gegenfurtner, K.R.; Sharpe, L.; Eds.; Cambridge University Press: Cambridge, 1999, 181–202.
49. Hendry, S.H.C.; Reid, R.C. The koniocellular pathway in primate vision. Annu. Rev. Neurosci. **2000**, *23* (1), 127–153.
50. Dong, D.W.; Atick, J.J. Temporal decorrelation: A theory of lagged and nonlagged responses in the lateral geniculate nucleus. Network: Comput. Neural Syst. **1995**, *6* (2), 159–178.
51. DeAngelis, G.C.; Ohzawa, I.; Freeman, R.D. Receptive-field dynamics in the central visual pathways. Trends. Neurosci. **1995**, *18* (10), 451–458.
52. Rodman, H.R.; Sorenson, K.M.; Shim, A.J.; Hexter, D.P. Calbindin immunoreactivity in the geniculo-extrastriate system of the macaque: Implications for the heterogeneity in the koniocellular pathway and recovery from cortical damage. J. Comp. Neurol. **2001**, *431* (2), 168–181.
53. Sincich, L.C.; Park, K.F.; Wohlgemuth, M.J.; Horton, J.C. Bypassing V1: A direct geniculate input to area MT. Nat. Neurosci. **2004**, *7* (10), 1123–1128.
54. Yukie, M.; Iwai, E. Direct projection from the dorsal lateral geniculate nucleus to the prestriate cortex in macaque monkeys. J. Comp. Neurol. **1981**, *201* (1), 81–97.
55. Brodmann, K. *Vergleichende Localisationslehre der Großhirnrhinde*; Barth Verlag: Leipzig, 1909. Translated by L. J. Gaery, Localisation in the Cerebral Cortex, London, 1994.
56. Lund, J.S. Organization of neurons in the visual cortex, area 17, of the monkey (Macaca mulatta). J. Comp. Neurol. **1973**, *147* (4), 455–496.
57. Bonhoeffer, T.; Grinvald, A. Iso-orientation domains in cat visual cortex are arranged in pinwheel-like patterns. Nature **1991**, *353* (6343), 429–431.
58. Hubel, D.H.; Wiesel, T.N. Receptive fields and functional architecture of the monkey striate cortex. J. Physiol. **1968**, *195* (1), 215–243.
59. Gur, M.; Kagan, I.; Snodderly, D.M. Orientation and direction selectivity of neurons in V1 of alert monkeys:

Functional relationships and laminar distributions. Cerebral Cortex **2005**, *15* (8), 1207–1221.

60. Hawken, M.J.; Parker, A.J.; Lund, J.S. Laminar organization and contrast sensitivity of direction-selective cells in the striate cortex of the old world monkey. J. Neurosci. **1988**, *8* (10), 3541–3548.
61. Ringach, D.L.; Shapley, R.M.; Hawken, M.J. Orientation selectivity in macaque V1: Diversity and laminar dependence. J. Neurosci. **2002**, *22* (13), 5639–5651.
62. Wang, C.; Yao, H. Sensitivity of V1 neurons to direction of spectral motion. Cerebral Cortex **2011**, *21* (4), 964–973.
63. Hubel, D.H.; Wiesel, T.N. Laminar and columnar distribution of geniculocortical fibers in the macaque monkey. J. Comp. Neurol. **1972**, *146* (4), 421–450.
64. Lund, J.S.; Boothe, R.G. Interlaminar connections and pyramidal neuron organization in the visual cortex, area 17, of the macaque monkey. J. Comp. Neurol. **1975**, *159* (3), 305–334.
65. Erwin, E.; Baker, F.H.; Busen, W.F.; Malpeli, J.G. Relationship between laminar topology and retinotopy in the rhesus lateral geniculate nucleus: Results from a functional atlas. J. Comput. Neurol. **1999**, *407* (1), 92–102.
66. Barone, P.; Batardiere, A.; Knoblauch, K.; Kennedy, H. Laminar distribution of neurons in extrastriate areas projecting to visual areas V1 and V4 correlates with the hierarchical rank and indicates the operation of a distance rule. J. Neurosci. **2000**, *20* (9), 3263–3281.
67. Thompson, W.B.; Fleming, R.W.; Creem-Regehr, S.H.; Stefanucci, J.K. *Visual Perception from a Computer Graphics Perspective*; CRC Press/AK Peters: Boca Raton, FL, 2011.
68. Kalloniatis, M.; Luu, C. Visual acuity. In *Webvision: The Organization of the Retina and Visual System*, Salt Lake City, UT, 2011. Available at http://webvision.med.utah.edu/book/part-viii-gabacreceptors/visual-acuity/.
69. Smith, G.; Atchison, D.A. *The Eye and the Visual Optical Instruments*; Cambridge University Press: New York, 1997.
70. Campbell, F.W.; Green, D.G. Optical and retinal factors affecting visual resolution. J. Physiol. **1965**, *181* (3), 576–593.
71. Riggs, L.A. Visual acuity. In *Vision and Visual Perception*; Graham, C.H.; Ed.; John Wiley & Sons, Inc.: New York, 1965.
72. Williams, D.R.; Coletta, N.J. Cone spacing and the visual resolution limit. J. Opt. Soc. Am. A **1987**, *4* (8), 1514–1523.
73. Kalloniatis, M.; Luu, C. Temporal resolution. In *Webvision: The Organization of the Retina and Visual System*, Salt Lake City, UT, 2007. Available at http://webvision.med.utah.edu/book/part-viii-gabacreceptors/temporal-resolution/.
74. Hart, Jr, W.M. The temporal responsiveness of vision. In *Adler's Physiology of the Eye, Clinical Application*; Moses, R.A.; Hart, W.M.; Eds.; The C. V. Mosby Company: St. Louis, MO, 1987.
75. Peli, E. Contrast in complex images. J. Opt. Soc. Am. A **1990**, *7* (10), 2032–2040.
76. Bex, P.J.; Solomon, S.G.; Dakin, S.C. Contrast sensitivity in natural scenes depends on edge as well as spatial frequency structure. J. Vision **2009**, *9* (10), 1–19.
77. Campbell, F.W.; Robson, J.G. Application of Fourier analysis to the visibility of gratings. J. Physiol. **1968**, *197* (3), 551–566.
78. Young, T. The Bakerian lecture: On the theory of light and colors. Philos. Trans. R. Soc. London **1802**, *92*, 12–48.
79. von Helmholtz, H.L.F. *Handbuch der Physiologischen Optik*; Leopold Voss: Leipzig, 1867.
80. Boynton, R.M. *Human Color Vision*; Holt, Rinehart, and Winston: New York, 1979.
81. Kindlmann, G., Reinhard, E.; Creem, S. Face-based luminance matching for perceptual colormap generation. In *Proceedings of IEEE Visualization*, 2002, 309–406.
82. Hering, E. *Outlines of a Theory of the Light Sense (Translation from German: Zur Lehre vom Lichtsinne, 1878)*; Harvard University Press: Cambridge, MA, 1920.
83. Hurvich, L.M.; Jameson, D. The opponent process theory of color vision. Psychol. Rev. **1957**, *64* (6), 384–404.
84. Koenderink, J.J.; van Doorn, A.J. Shape from shading. In *The Visual Neurosciences*; Chalupa, L.M.; Werner, J.S.; Eds.; MIT Press: Cambridge, MA, 2003, 1090–1105.
85. Zhang, R.; Tsai, P.-S.; Cryer, J.E.; Shah, M. Shape from shading: A survey. IEEE Trans. Pattern Anal. Mach. Intell. **1999**, *21* (8), 690–706.
86. Prados, E.; Faugeras, O. Shape from shading. In *Handbook of Mathematical Models in Computer Vision*; Paragios, N.; Chen, Y.; Faugeras, O.; Eds.; Springer: New York, 2006, 375–388.
87. Gregory, R.L. *Eye and the Brain: The Psychology of Seeing*, 3rd Ed.; McGraw-Hill: New York, 1978.
88. Gregory, R.L. *Seeing Through Illusions*; Oxford University Press: Oxford, 2009.
89. Reinhard, E.; Efros, A.; Kautz, J.; Seidel, H.-P. On visual realism of synthesized imagery. Proc. IEEE **2013**, *101* (9).
90. Gregory, R.L. Knowledge in perception and illusion. Philos. Trans. R. Soc. London B **1997**, *352* (1358), 1121–1128.
91. Gregory, R.L.; Heard, P. Border locking and the café wall illusion. Percept. **1979**, *8* (4), 365–380.
92. Chevreul, M.E. *The Principles of Harmony and Contrast of Colors and Their Applications to the Arts*; Schiffer: West Chester, PA, 1987.
93. Mach, E. *The Analysis of Sensations and the Relation of the Physical to the Psychical*; Dover: New York, 1959.
94. O'Brien, V. Contour perception, illusion and reality. J. Opt. Soc. Am. **1958**, *48* (2), 112–119.
95. Sherwood, S.L. (Ed.) *The Nature of Psychology: A Selection of Papers, Essays and Other Writings by Kenneth J W Craik*; Cambridge University Press: Cambridge, 1966.
96. Cornsweet, T. *Visual Perception*; Academic Press: New York, 1970.
97. Kingdom, F.A.A.; Moulden, B. Border effects on brightness: A review of findings, models and issues. Spatial Vision **1988**, *3* (4), 225–262.
98. Elder, J.H. Are edges incomplete? Int. J. Comput. Vision **1999**, *34* (2/3), 97–122.
99. Grossberg, S.; Mingolla, E. Neural dynamics of form perception: Boundary adaptation, illusory figures, and neon color spreading. Psychol. Rev. **1985**, *92* (2), 173–211.
100. Krauskopf, J. Effect of retinal stabilization on the appearance of heterochromatic targets. J. Opt. Soc. Am. **1963**, *53* (6), 741–744.
101. Walls, G. The filling-in process. Am. J. Optom. **1954**, *31* (7), 329–340.

102. Dakin, S.C.; Bex, P.J. Natural image statistics mediate brightness filling in. Proc. R. Soc. London B **2003**, *270* (1531), 2341–2348.
103. Purves, D.; Shimpi, A.; Lotto, R.B. An empirical explanation of the Cornsweet effect. J. Neurosci. **1999**, *19* (19), 8542–8551.
104. Ferwerda, J.A.; Pattanaik, S.; Shirley, P.; Greenberg, D.P. A model of visual adaptation for realistic image synthesis. In *SIGGRAPH '96: Proceedings of the 23th Annual Conference on Computer Graphics and Interactive Techniques*, 1996, 249–258.
105. Lamming, D. Contrast sensitivity. In *Vision and Visual Dysfunction*, Vol. 5; Cronly-Dillon, J.; Ed.; Macmillan Press: London, 1991.

Thermal Infrared Remote Sensing

Juan Carlos Jimenez, Jose Antonio Sobrino, Guillem Soria, Yves Julien, Drazen Skokovic, and Jose Gomis-Cebolla
Global Change Unit, Image Processing Laboratory, University of Valencia, Valencia, Spain

Abstract

Thermal infrared (TIR) remote sensing techniques for Earth observation (EO) purposes collect the TIR radiation emitted by our planet's surface in a number of natural or urban landscapes using sensor on board a variety of platforms (such as satellites or airplanes). TIR remote sensing data is essential for understanding land surface processes and land–atmosphere interactions, and it is used in a broad range of applications in a number of fields, such as oceanography, geology, climatology, and hydrology. The analysis and information extraction from thermal imagery acquired by EO sensors requires image processing techniques that are common to other branches of remote sensing but also specific techniques that are common to the TIR radiation nature. This entry provides a basic description of TIR remote sensing concepts, corrections, and algorithms for the retrieval of land surface temperature and emissivity.

Keywords: Emissivity; Infrared; Temperature; Thermal.

GENERAL OVERVIEW

The term "thermal infrared remote sensing", or simply "thermal remote sensing," refers to remote sensing techniques aimed at the measurement of thermal infrared (TIR) radiation from objects. It may be considered as a particular branch of general remote sensing techniques, so most of the general concepts related to remote sensing science can be also applied to thermal remote sensing. For the specific purpose of Earth observation (EO), in which this description is focused, it uses sensors on board a variety of platforms (satellites, airplanes, etc.) with spectral bands capturing the TIR radiation emitted by our planet's surface in a number of natural or urban landscapes.

A clear consensus on wavelength limits for the definition of the TIR spectral range and its different subregions does not exist in the literature, but it covers approximately a wavelength range between 3 and 100 µm. However, two main regions are considered to measure the emitted radiance by EO sensors: (i) 3–5 µm, denoted as midwave infrared (MIR), and (ii) 8–14 µm, denoted as longwave infrared (LWIR).

Traditionally less attention has been paid to TIR remote sensing compared to the exploitation of remote sensing in the visible and near-infrared (VNIR) range. However, TIR data is required to understand the land surface processes as well as those processes occurring at the land–atmosphere interface. The surface energy balance can be rigorously analyzed only if TIR data is available, since most of the energy fluxes explicitly depend on magnitudes directly related to TIR data.

Approximately, 80% of the thermal energy registered by a sensor in the region 10.5–12.5 µm is emitted by the land surface,[1] so the land surface temperature (LST) is the main variable to be extracted from the TIR signal. Because the thermal energy radiated by a body (e.g., a natural surface) is modulated by the spectral emissivity, this variable has also a key role in TIR remote sensing. Therefore, LST and land surface emissivity (LSE) are considered the primary variables to be retrieved from a set of spectral bands in the TIR region. It is worth to mention that LST does not depend on the spectral range but LSE is a spectral magnitude.

Applications of TIR remote sensing include a number disciplines, such as oceanography, geology, climatology, and hydrology, among others. LST plays an important role in the estimation of heat fluxes and evapotranspiration (e.g., water management purposes over agricultural areas), water stress detection over agricultural fields, orchards, or forests. Knowledge of emissivity spectra is required for accurate estimations of LST, but it also allows identification of minerals (mineral mapping),[2] and it is also a good indicator of land cover changes.[3] A review of applications is presented by Sobrino et al.[4]

BRIEF HISTORICAL BACKGROUND

The discovery of infrared radiation is attributed to Frederick William Herschel in 1800, who discovered a type of invisible radiation at longer wavelengths than red light. Herschel discovered infrared radiation when sunlight was passing through a prism and he observed an increase in temperature over a thermometer placed just beyond the red end of the visible spectrum. Since then the development in infrared technology has substantially increased, both in terms of detectors for measuring infrared radiation as well as powerful computers to process and extract information from thermal imagery.

In the framework of EO science, NASA launched in 1978 the Heat Capacity Mapping Mission, which is considered the first mission providing TIR remote sensing data at spatial resolutions around 600 m. The Landsat series in the 1980s (beginning in Landsat-4) provided also the first high-resolution thermal images (120 m) for natural resources studies. Currently, a number of platforms at geostationary or polar orbits include TIR sensors with daily and sub-daily temporal resolutions. However, most of these sensors provide TIR data at the kilometric scale, and thermal imagery at high spatial resolution is not well resolved with the exception of Landsat8/TIRS and Terra/ASTER.

BASIC PHYSICS OF THERMAL RADIATION

Radiometric Magnitudes

TIR radiation is a part of the electromagnetic radiation and of the wave propagation between the source and the detector. Therefore, the basic radiometric magnitudes used in remote sensing techniques also apply to TIR remote sensing, namely, radiant energy, radiant flux, radiant flux density, radiant intensity, and radiance. In particular, the spectral radiance (L_λ) is the main radiometric magnitude, since it is the magnitude being measured by TIR sensors. Spectral radiance is defined as the radiant flux in a given solid angle element through a perpendicular surface to the propagation direction and for a given wavelength. The commonly used unit for spectral radiance in the TIR region is $Wm^{-2}sr^{-1}\mu m^{-1}$.

Basic Laws of Radiation

The three basic laws of radiation given by the Planck's law, the Stefan–Boltzmann law, and the Wien's displacement law are applied to the TIR wavelengths to relate the measured spectral radiance with the temperature of the body. Hence, Wien's displacement law ($\lambda_{max} T = 2.8975 \cdot 10^{-3}$ m·K) applied to typical environmental temperatures (~300 K) shows that maximum emission is produced at wavelengths of around 10 μm, which justifies in part why TIR sensors are commonly working around this spectral range. The Stefan–Boltzmann law states that the total emissive power of a body (without any consideration of the spectral distribution) is a function of T^4 (where T is the temperature of the body). This result is especially important to convert hemispherical and broadband measurements of TIR radiation to temperature. However, Planck's law is the most important law when working with TIR remotely sensed data, since it allows the conversion of the measured spectral radiance in a given direction (B_λ) to temperature for a blackbody using the following equation:

$$B_\lambda(T) = \frac{c_1 \lambda^{-5}}{\exp\left(\dfrac{c_2}{\lambda T}\right) - 1} \quad (1)$$

with $c_1 = 1.19104 \cdot 10^8$ $W\mu m^4 m^{-2} sr^{-1}$ and $c_2 = 14387.7$ μm K. Because a blackbody is Lambertian, spectral radiance can be converted to spectral emittance (M_λ) using a π factor, $M_\lambda(T) = \pi B_\lambda(T)$.

Emissivity and Kirchhoff's Law

A blackbody is a theoretical material that absorbs all the radiant energy that strikes it, and it also radiates all of its energy in a spectral distribution pattern given by the Planck's law (Eq. 1). However, real objects do not behave as perfect blackbodies. The departure of a real object from the condition of blackbody can be measured through a spectral thermal property called emissivity (ε). The emissivity for a blackbody is 1, whereas real objects accomplish $\varepsilon < 1$. This consideration implies that relationships between radiance and temperature given by Stefan–Boltzmann and Planck's laws need to be corrected by the emissivity factor.

Another useful approach adopted in TIR remote sensing is the so-called Kirchoff's law. When a radiation source is surrounded by other radiation source, it emits radiation but also absorbs radiation. If we assume that a radiation equilibrium is accomplished (emitted energy equals absorbed energy at any wavelength), then the absorptivity of the surface equals the emissivity of the surface at the same temperature. For opaque surfaces, this results lead to a relationship between emissivity and reflectivity:

$$\varepsilon_\lambda = 1 - \rho_\lambda \quad (2)$$

The preceding equation is a fundamental approach in radiometry, since emissivity values can be obtained from reflectivity measurements.

THE RADIATIVE TRANSFER EQUATION

The radiative transfer equation (RTE) provides a relationship between the at-surface parameters and the measurements at the sensor level. The at-sensor registered radiance is the value directly extracted from a remote sensing image, so the RTE allows a physically based retrieval of the surface parameters LST and LSE from the image data. Because of the atmospheric layer between the sensor and the surface, the RTE also includes the atmospheric parameters influencing the emitted radiance which finally arrives to the sensor at a certain altitude. Radiative transfer theory applied in a number of situations has been addressed in the classical books of Chandrasekhar[5] and Lenoble.[6] In the case of TIR radiation, in which scattering processes are usually neglected, the RTE can be expressed as follows:

$$L^{sen} = L^{sur}\tau + L^{up} \quad (3)$$

where L^{sen} is the at-sensor radiance, L^{sur} is the radiance coming from the surface (at-surface radiance), τ is the atmospheric transmissivity, and L^{up} is the upwelling atmospheric radiance or path radiance. These magnitudes are spectral magnitudes, so they depend on the wavelength (or sensor band). Also, these magnitudes depend on the observation angle. For a given surface temperature (T_s), the at-surface radiance is dominated by the emission term, but it also includes the reflected atmospheric radiance which can be related to the surface emissivity via Kirchhoff's law:

$$L^{sur} = \varepsilon B(T_s) + (1-\varepsilon)L^{down} \quad (4)$$

where L^{down} is the down-welling atmospheric radiance (hemispherical flux divided by π). Equations 3 and 4 finally lead to the RTE used in thermal remote sensing:

$$L^{sen} = \left[\varepsilon B(T_s) + (1-\varepsilon)L^{down}\right]\tau + L^{up} \quad (5)$$

The preceding equation provides the main expression used in thermal remote sensing to develop LST and LSE retrieval algorithms with a physical basis. Ideally, the magnitudes involved in Eq. 5 refers to a given pixel (x,y) in the image. However, for practical purposes, the atmospheric terms are usually assumed to be constant for the whole image or a certain region or window of pixels. Figure 1 illustrates the different terms of the RTE contributing to the at-sensor registered radiance.

ATMOSPHERIC AND EMISSIVITY CORRECTIONS ON TIR IMAGERY

Radiative Transfer Codes and Spectral Libraries

The RTE given by Eq. 5 shows that retrieval of LST (Ts) requires the knowledge of the surface emissivity (ε) and the atmospheric parameters (τ, L^{up}, L^{down}). The estimation of the atmospheric parameters over the whole atmospheric layer requires computation of radiative transfer terms over thin layers in order to obtain the total atmospheric column magnitude. This complex task is addressed by the so-called radiative transfer codes, for example, the MODTRAN code,[7] considered one of the state-of-the-art radiative transfer codes in remote sensing techniques. MODTRAN usually uses input measurements of atmospheric soundings (e.g., pressure, air temperature and air humidity at certain altitudes) and provides the different spectral terms involved in the RTE. Radiative transfer codes are key software in order to simulate a number of atmospheric and surface conditions and to develop retrieval algorithms from the simulated datasets.

Although some radiative transfer codes also include a set of surface emissivity spectra, it is recommended to employ spectral libraries with a more complete dataset of emissivity spectral for both natural and manmade surfaces. These emissivity spectra are mostly derived from laboratory measurements of hemispheric reflectivity, which in turn is converted to emissivity from Kirchhoff's law. An example of these spectral libraries is the ASTER spectral library[8] (http://speclib.jpl.nasa.gov), which includes a number of emissivity spectra for soils, rocks, vegetation, water, snow, and several manmade materials. Another example of spectral library is the MODIS spectral library (http://www.icess.ucsb.edu/modis/EMIS/html.em.html). Figure 2 shows some emissivity spectra samples extracted

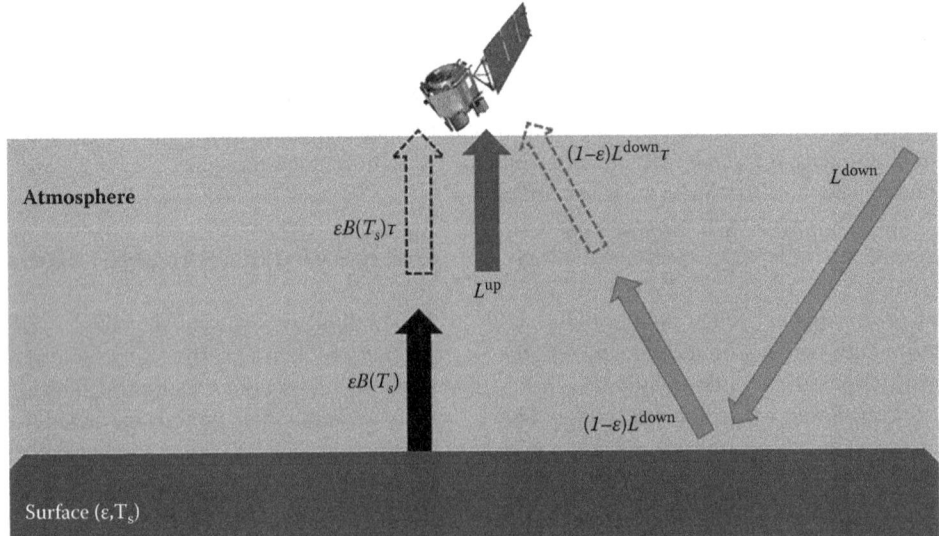

Fig. 1 The different terms contributing to the at-sensor registered radiance and related by the RTE adapted to the TIR (Eq. 5)

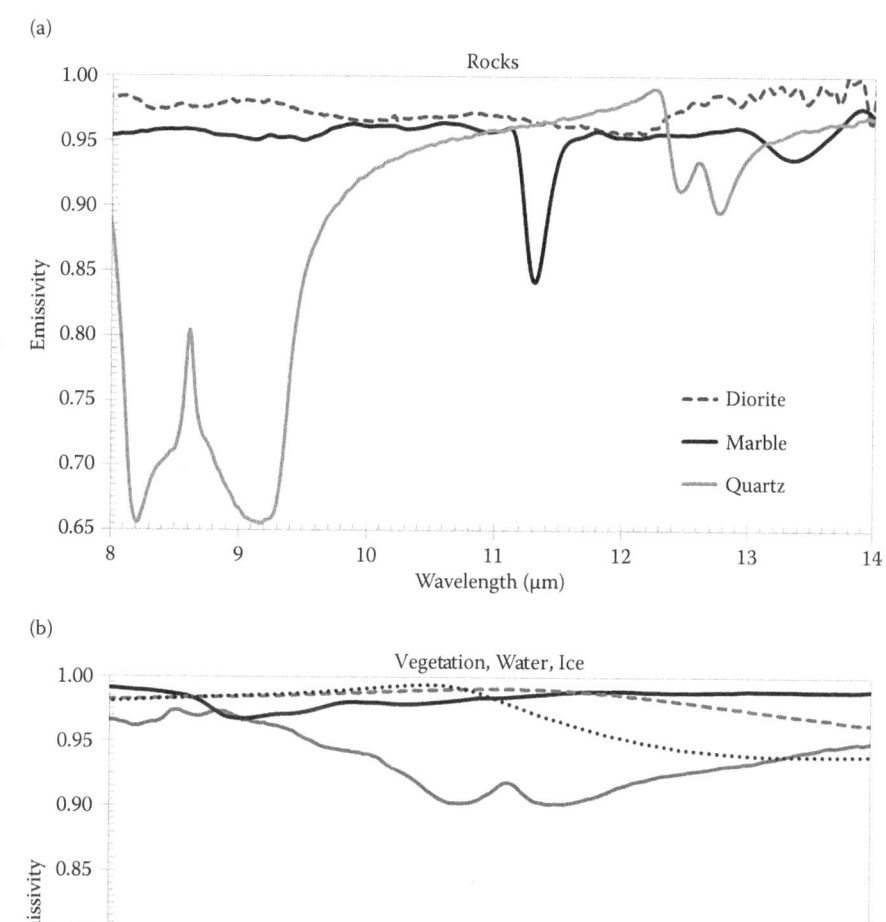

Fig. 2 Emissivity spectra for some natural samples ((a) rocks; (b) vegetation, water, and ice)

from the ASTER spectral library for the TIR spectral range between 8 and 14 µm. Different materials show different spectral features. Water and green vegetation show almost a flat spectrum with high emissivity values, so in some cases a near blackbody behavior can be considered for these surfaces.

MODTRAN Simulations

Next figures show an example of atmospheric and emissivity correction effects on the TIR signal measured by the sensor and the retrieved LST.

Figure 3 shows the surface emission term overplotted to the spectral atmospheric transmissivity for a standard atmospheric profile. The following conclusions can be drawn from Fig. 3:

i. The MIR region (3–5 µm) and the LWIR region (8–14 µm) typically show high atmospheric transmissivity values, the so-called atmospheric windows.
ii. Radiance emitted by the surface is considerably higher in the LWIR than in the MIR. This result justifies the selection of sensor bands in the LWIR for LST retrieval purposes, since the signal-to-noise ratio in the MIR is worse than in the LWIR. In the case of daytime images acquired in the MIR region, the contribution from the solar reflection should be also taken into account in the RTE. However, the MIR region is useful for the detection of hot temperature events because it avoids the sensor saturation problems. The MIR region is also useful for the characterization of some materials because some

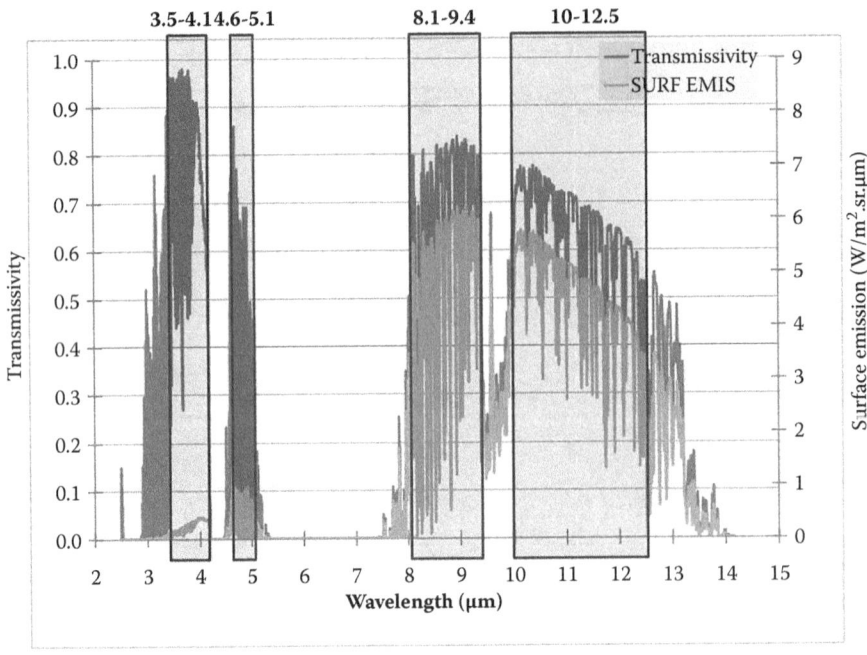

Fig. 3 Atmospheric transmissivity and surface emission simulated with MODTRAN radiative transfer code for a given standard atmospheric profile

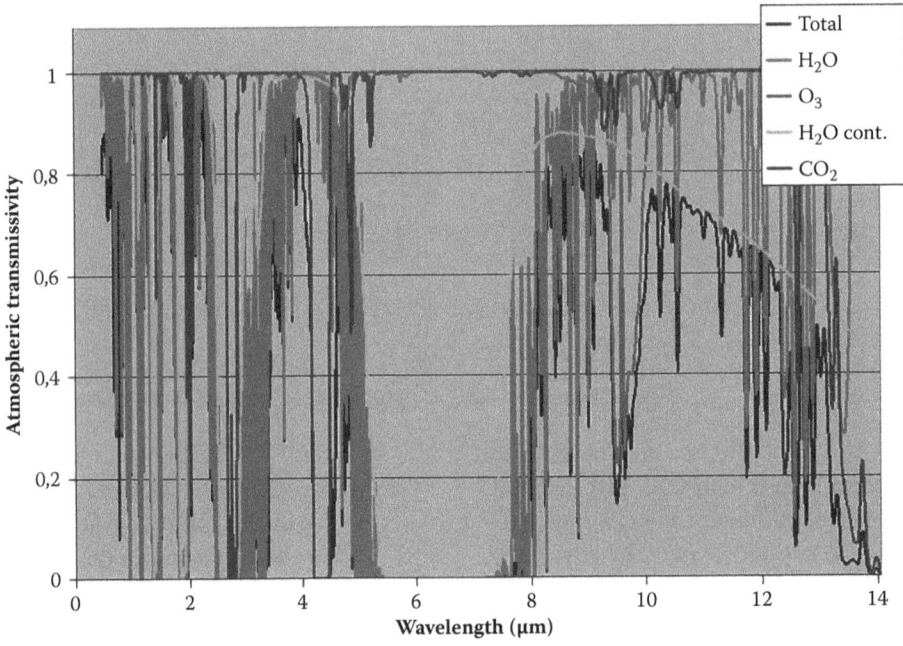

Fig. 4 Atmospheric transmissivity for the main atmospheric constituents

samples show also characteristic spectral features in the MIR.

iii. The TIR region shows two atmospheric windows, approximately between 8–9 μm and 10–12.5 μm. Surface emissivity is typically higher and with lower spectral variation in the second atmospheric windows, so most TIR sensors use the 10–12.5 μm to retrieve LST from the measured signal. This region is usually referred as the "split-window" region.

In order to better understand the location of the atmospheric windows, Fig. 4 shows the contribution to the atmospheric absorption of main atmospheric constituents.

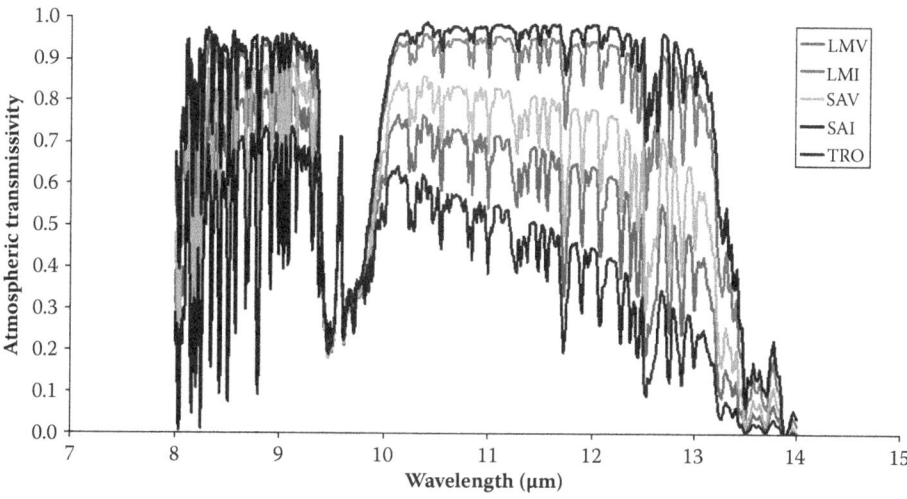

Fig. 5 Atmospheric transmissivity for standard atmospheres included in MODTRAN with different total atmospheric water vapor content. LMV, midlatitude summer; LMI, midlatitude winter; SAV, subarctic summer; SAI, subarctic winter; TRO, tropical

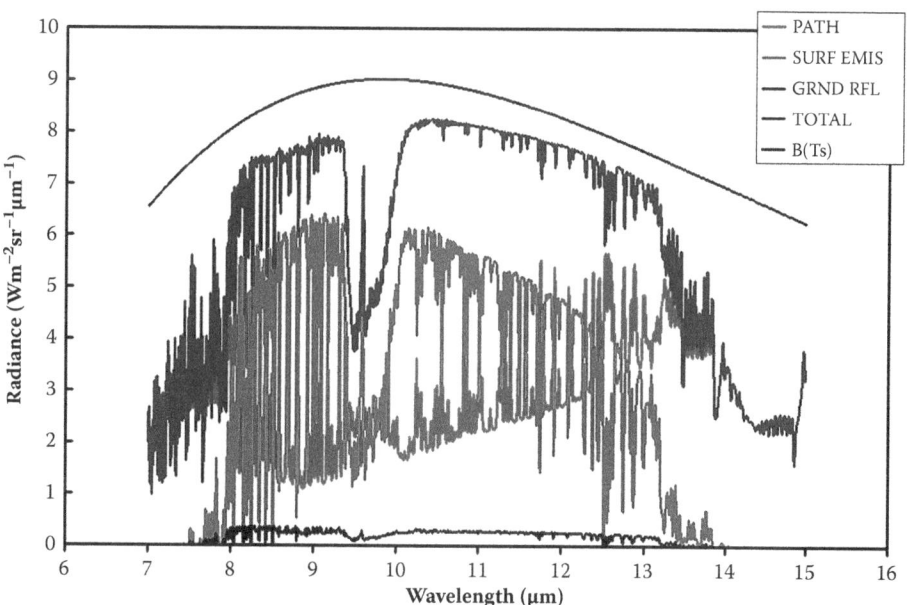

Fig. 6 Components of the RTE simulated with MODTRAN code. PATH, upwelling atmospheric radiance or path radiance; SURF EMIS, surface emission attenuated by the atmosphere; GRND RFL, down-welling atmospheric radiance reflected by the surface and attenuated by the atmosphere; TOTAL, total radiance registered by the sensor; B(Ts), radiance for a blackbody at temperature Ts

The atmospheric water vapor is the main contributor to the atmospheric effect in the TIR region, whereas the ozone shows a strong atmospheric absorption around 9.6 μm. Figure 5 shows the atmospheric transmissivity for standard atmospheres included in MODTRAN. The most humid atmospheres (tropical, midlatitude summer) provide the lowest values of atmospheric transmissivity because of its high atmospheric water vapor content, whereas dry atmospheres (subarctic winter) provide the highest atmospheric transmissivity.

The different terms involved in the RTE (Eqs. 3–5) are graphed in Fig. 6. It can be observed that the emission term has the major contribution to the total radiance measured by the sensor, followed by the atmospheric path radiance. The reflected term has a low contribution to the total radiance, and it is almost negligible for surfaces with emissivity values near 1.

METHODS FOR LST AND LSE RETRIEVAL FROM THERMAL IMAGERY

Several LST and LSE retrieval methods have been published in the literature in the last few decades (see the reviews by Li et al.[9,10]) Algorithms with a physical basis are developed from the RTE given in Eq. 5. However, the inversion of LST (Ts) and LSE (ε) from Eq. 5 is undetermined because there is always one more unknown than the number of measurements. For a given sensor with N spectral bands, and provided that atmospheric parameters are known, there are always N equations and $N+1$ unknowns (N spectral emissivities and one temperature). Therefore, at least one additional degree of freedom must be constrained to solve the set of equations. The different LST and LSE retrieval methods constrain this additional degree of freedom in different ways. Over surfaces with a well-known surface emissivity (e.g., sea or green vegetation), the LST can be directly inverted from the RTE.

Different classification schemes can be considered to describe the LST/LSE retrieval methods. In terms of LST retrieval, a convenient classification relies in the number of spectral bands required to resolve the undetermined problem, whereas in the case of LSE retrieval, one may consider methods for absolute emissivity values retrieval or methods for relative emissivity values retrieval.

Relative Emissivity

Relative methods for emissivity retrieval assume a constant value (or a similar approach) for a reference spectral band. The values retrieved with these methods refer to relative emissivity values and not the actual ones. However, the relative emissivity spectrum preserves the spectral shape of the absolute emissivity spectrum, so it is still valid in many applications for the identification of characteristic spectral features. The following published methods are included in this category: reference channel,[11] alpha coefficients,[12] and normalized emissivity method.[13]

Absolute Emissivity from Vegetation Indices and Classification-Based Approaches

Surface emissivity can be independently estimated by using data measured in other spectral ranges than the TIR. This is the case of emissivity retrieval methods based on relationships between surface emissivity and vegetation indices.[14] These approaches use VNIR bands to estimate the surface emissivity because vegetation indices are usually computed from bands located in the red and near infrared, for example, the normalized difference vegetation index. This is a useful approach over agricultural areas where pixels are a mixture of bare soil and vegetation.

Other approaches for TIR independent surface emissivity characterization are classification-based or spectral unmixing-based approaches, in which emissivity values are assigned to each class or end member. These approaches require a previous knowledge of the emissivity to be assigned to each land cover.

LST Estimation from Single-Channel Algorithms

LST can be retrieved from TIR data acquired with a single band whenever the emissivity is known or has been obtained from independent data. In this case, LST can be directly inverted from the RTE (Eq. 5) or single-channel algorithms also based on the RTE. This approach has been used for the Landsat series, with sensors with only one TIR band (except for the case of Landsat-8 with two TIR bands). A few examples of these algorithms can be found in Qin et al.,[15] Jiménez-Muñoz and Sobrino,[16] Jiménez-Muñoz et al.,[17] and Cristóbal et al.[18]

LST Estimation from Two-Channel or Dual-Angle Techniques

These techniques require TIR data acquired with two spectral bands (for the case of two-channel algorithms) or TIR data acquired with one single band under two different view angles (dual-angle techniques). In both cases, techniques are based on the "differential absorption" concept:[19] the difference between the TIR radiance (or brightness temperatures) measured at two different wavelengths is sensitive to the atmospheric conditions, so this difference can be used to compensate for the atmospheric effects and then to recover the surface temperature. When the two TIR bands are located in the 10.5–12 µm, the concept "split-window" algorithm is typically used. The same concept can be applied to the difference between the radiance or brightness temperature measured at one particular wavelength under two different view angles. Split-window algorithms have been successfully employed to retrieve LST from remotely sensed data with a number of sensors (e.g., AVHRR, ATSR, SEVIRI, MODIS). Some examples of these techniques can be found in the works of Sobrino et al.,[20] Kerr et al.,[21] and Sobrino et al.[22,23]

These techniques also require a previous knowledge (or an independent estimation) of the surface emissivity for the two TIR spectral bands. In the case of dual-angle algorithms, the angular dependence of the surface emissivity should be also known or parameterized.

LST and LSE Retrieval from Multispectral TIR Data

LST and LSE can be simultaneously retrieved from TIR data acquired with a sensor with multispectral capabilities (three or more TIR bands). This is the case of the temperature and emissivity separation (TES) algorithm originally developed to generate the LST and LSE products of the ASTER sensor.[24] TES algorithm is composed of three modules: NEM, RATIO, and MMD. The NEM module provides a first estimation of relative emissivities and a first guess of the surface temperature. The RATIO module normalizes the spectral emissivities estimated with the NEM module, and finally the MMD module recovers the absolute emissivities and temperature using an empirical relationship between the minimum emissivity and the spectral contrast. TES algorithm has been also applied to polar-orbiting and geostationary satellites.[25,26]

CONCLUSIONS

Thermal remote sensing is recognized as a powerful tool for many environmental applications spanning different topics such as water management, hot spot and geothermal anomalies detection (including fires and volcanoes), urban heat island monitoring, soil composition, and security and surveillance topics. At global scale, collection of TIR data allows for the retrieval of land and sea surface temperature for the monitoring of the Earth's climate. Retrieval algorithms and radiative transfer codes are currently mature enough to extract biophysical and geophysical variables from the TIR signal. However, a detailed atmospheric characterization at the image acquisition time and accurate surface emissivity estimations are still the main challenges to be faced, although significant improvements have been performed in the last years. Future TIR missions aimed at providing data at high spectral, spatial, and temporal resolutions would open new challenges for the development of robust retrieval algorithms and end-user applications.

REFERENCES

1. Czajkowski, K.P.; Goward, S.N.; Mulhern, T.; Goetz, S.J.; Walz, A.; Shirey, D.; Stadler, S.; Prince, S.D.; Dubayah, R.O. Estimating environmental variables using thermal remote sensing. In *Thermal Remote Sensing in Land Surface Processes*; Quattrochi, D.A.; Luvall, J.C.; Eds.; CRC Press: Boca Raton, FL, 2004, 11–32.
2. Vaughan, R.G.; Calvin, W.M.; Taranik, V. SEBASS hyperspectral thermal infrared data: Surface emissivity measurement and mineral mapping. Remote Sens. Environ. **2003**, *85*, 48–63.
3. French, A.N.; Schmugge, T.J.; Ritchie, J.C.; Hsu, A.; Jacob, F.; Ogawa, K. Detecting land cover change at the Jornada Experimental Range, New Mexico with ASTER emissivities. Remote Sens. Environ. **2008**, *112* (4), 1730–1748.
4. Sobrino, J.A.; Del Frate, F.; Drusch, M.; Jiménez-Muñoz, J.C.; Manunta, P.; Regan, A. Review of thermal infrared applications and requirements for future high-resolution sensors. IEEE Trans. Geosci. Remote Sens. **2016**, *54* (5), 2963–2972.
5. Chandrasekhar, S. *Radiative Transfer*; Dover Publications Inc.: New York, 1960.
6. Lenoble, J. *Atmospheric Radiative Transfer*; A. Deepak Publishing: Hampton, VA, 1993.
7. Berk, A.; Anderson, G.P.; Acharya, P.K.; Chetwynd, J.H.; Bernstein, L.S.; Shettle, E.P.; Matthew, M.W.; Adler-Golden, S.M. *MODTRAN4 User's Manual*; Air Force Research Laboratory: Hanscom AFB, MA, 1999.

8. Baldridge, A.M.; Hook, S.J.; Grove, C.I.; Rivera, G. The ASTER spectral library version 2.0. Remote Sens. Environ. **2009**, *113*, 711–715.
9. Li, Z.-L.; Tang, B.-H.; Wu, H.; Ren, H.; Yan, G.; Wan, Z.; Trigo, I.F.; Sobrino, J.A. Satellite-derived land surface temperature: Current status and perspectives. Remote Sens. Environ. **2013**, *131*, 14–37.
10. Li, Z.-L.; Wu, H.; Wang, N.; Qiu, S.; Sobrino, J.A.; Wan, Z.; Tang, B.-H.; Yan, G. Land surface emissivity retrieval from satellite data. Int. J. Remote Sens. **2013**, *34* (9/10), 3084–3127.
11. Kahle, A.B.; Madura, D.P.; Soha, J.M. Middle infrared multispectral aircraft scanner data analysis for geological applications. Appl. Opt. **1980**, *19*, 2279–2290.
12. Kealy, P.S.; Hook, S.J. Separating temperature and emissivity in the thermal infrared multispectral scanner data: Implications for recovering land surface temperatures. IEEE Trans. Geosci. Remote Sens. **1993**, *31* (6), 1155–1164.
13. Gillespie, A.R. Lithologic mapping of silicate rocks using TIMS. In *Proceedings of the TIMS Data User's Workshop*; JPL Publications, California Institute of Technology: Pasadena, CA, 1985, 86-38, 29–44.
14. Sobrino, J.A.; Jiménez-Muñoz, J.C.; Sòria, G.; Romaguera, M.; Guanter, L.; Moreno, J.; Plaza, A.; Martínez, P. Land surface emissivity retrieval from different VNIR and TIR sensors. IEEE Trans. Geosci. Remote Sens. **2008**, *46* (2), 316–327.
15. Qin, Z.; Karnieli, A.; Berliner, P. A mono-window algorithm for retrieving land surface temperatura from Landsat TM data and its application to the Israel-Egypt border region. Int. J. Remote Sens. **2001**, *22* (18), 3719–3746.
16. Jiménez-Muñoz, J.C.; Sobrino, J.A. A generalizad single-channel method for retrieving land surface temperature from remote sensing data. J. Geophys. Res. **2003**, *108* (D22), doi:10.1029/2003JD003480.
17. Jiménez-Muñoz, J.C.; Cristóbal, J.; Sobrino, J.A.; Sòria, G.; Ninyerola, M.; Pons, X. Revision of the single-channel algorithm for land surface temperature retrieval from Landsat thermal-infrared data. IEEE Trans. Geosci. Remote Sens. **2009**, *47* (1), 339–349.
18. Cristóbal, J.; Jiménez-Muñoz, J.C.; Sobrino, J.A.; Ninyerola, M.; Pons, X. Improvements in land surface temperature retrieval from the Landsat series thermal band using water vapor and air temperature. J. Geophys. Res. **2009**, *114* (D08103). doi:10.1029/2008JD010616.
19. McMillin, L.M. Estimation of sea surface temperature from two infrared window measurements with different absorption. J. Geophys. Res. **1975**, *80*, 5113–5117.
20. Sobrino, J.A.; Li, Z.-L., Stoll, M.P., Becker, F. Multichannel and multi-angle algorithms for estimating sea and land surface temperature with ATSR data. Int. J. Remote Sens. **1996**, *17*, 2089–2114.
21. Kerr, Y.H.; Lagouarde, J.P.; Nerry, F.; Ottle, C. Land surface temperature retrieval techniques and applications: Case of the AVHRR. In *Thermal Remote Sensing in Land Surface Processes*; Quattrochi, D.A.; Luvall, J.C.; Eds.; CRC Press: Boca Raton, FL, 2004, 33–109.
22. Sobrino, J.A.; El Kharraz, J.; Li, Z.-L. Surface temperature and water vapour retrieval from MODIS data. Int. J. Remote Sens. **2003**, *24* (24), 5161–5182.

23. Sobrino, J.A.; Sòria, G.; Prata, A.J. Surface temperature retrieval from along track scanning radiometer 2 data: Algorithms and validation. J. Geophys. Res. **2004**, *109* (D11101). doi:10.1029/2003JD004212.
24. Gillespie, A.; Rokugawa, S.; Matsunaga, T.; Cothern, J.S.; Hook, S.; Kahle, A.B. A temperature and emissivity separation algorithm for advanced spaceborne thermal emission and reflection radiometer (ASTER) images. IEEE Trans. Geosci. Remote Sens. **1998**, *36*, 1113–1126.
25. Hulley, G.C.; Hook, S.J. Generating consistent land surface temperature and emissivity products between ASTER and MODIS data for earth science research. IEEE Trans. Geosci. Remote Sens. **2011**, *49*, 1304–1315.
26. Jiménez-Muñoz, J.C.; Sobrino, J.A.; Mattar, C.; Hulley, G.; Göttsche, F.-M. Temperature and emissivity separation from MSG/SEVIRI data. IEEE Trans. Geosci. Remote Sens. **2014**, *52* (9), 5937–5951.

BIBLIOGRAPHY

Quattrochi, D.A.; Luvall, J.C. *Thermal Remote Sensing in Land Surface Processes*; CRC Press: Boca Raton, FL, 2004.

Sobrino, J.A. *Teledetección*; Servei de Publicacions de la Universitat de València: Valencia, 2000.

Three Dimensional Intensity-Curvature Measurement

Carlo Ciulla
University of Information Science and Technology, Ohrid, Macedonia

Abstract

The three-dimensional intensity-curvature measurement approaches (ICMAs) are signal post-processing techniques. The objective of the study conducted in this entry is twofold: (i) the presentation of the ICMAs and (ii) the use of the ICMAs to image the vasculature of the human brain. The concept at the root of the calculation of the ICMAs is the intensity-curvature term. The concept entails merging the value of the signal intensity with the sum of second-order partial derivatives with respect to the spatial coordinates of the model function fitted to the image data, and this sum is called classic-curvature. The three-dimensional ICMAs are (i) the intensity-curvature functional (ICF), (ii) the intensity-curvature term before interpolation (ICTBI), (iii) the intensity-curvature term after interpolation (ICTAI), and (iv) the resilient curvature (RC). The ICF is the ratio between ICTBI and ICTAI. The RC is the sum of second-order partial derivatives of the signal calculated equating the ICTBI to the ICTAI. The formulation of the ICMAs strictly depends on the model function fitted to the image data. The model function used is the cubic trivariate Lagrange polynomial, and the three-dimensional ICMAs were calculated fitting the polynomial to three-dimensional magnetic resonance imaging (MRI) volumes of the human brain. The MRI volumes were collected using T1-weighted, T2-weighted, and fluid-attenuated inversion recovery pulse sequences. The use of the ICMAs in this entry is to highlight human brain vessels.

Keywords: Human brain vessels; Intensity-curvature functional; Intensity-curvature term after interpolation; Intensity-curvature term before interpolation; Magnetic resonance imaging; Model polynomial function; Resilient curvature.

INTRODUCTION

This entry reports a novel study aimed to show that through magnetic resonance imaging (MRI), post-processing techniques is possible to image the vasculature of the human brain. The main theme of the entry is to introduce the three-dimensional intensity-curvature measurement approaches (ICMAs). The ICMAs are resampling techniques featured by the intensity-curvature term (which is a numerical value), and which replaces the image pixel intensity. The sum of second-order partial derivatives of the Hessian of the model function fitted to the pixel (classic-curvature [CC]) is used to begin the mathematical procedure which merges together (i) the value of the image pixel intensity (which can be the original image pixel intensity or the value resampled at the intra-pixel coordinate through the model polynomial function) with (ii) the CC. The objective of the study departs from a due literature review. The study continues with the mathematical presentation of the ICMAs and with the display of the ICMA images. The main goal is to highlight the vasculature of the human brain detected with MRI.

LITERATURE

Normally, the vasculature of the human brain is imaged with magnetic resonance angiography[1] and, in recent times, also with susceptibility-weighted imaging.[2] Imaging the vasculature of the human brain requires the task of vessel measurement, which is commonly referred to as vesselness measure.[3] The vesselness measure makes use of the Gaussian as the mean to obtain the second-order partial derivatives of an image. The Hessian is the matrix of second-order partial derivatives calculated using the Gaussian. The calculation of the eigenvalues and eigenvectors of the Hessian hence provides three scalars (the eigenvalues), which are used to calculate geometric ratios. And such ratios are then combined together into a formulation called vesselness function.[3] The use of the Gaussian as a preprocessing step in vesselness measures does not suppress the image noise and may determine the diffusion of two or more tubular objects into one another. Hence, in order to overcome the aforementioned limitation, there exists approaches which calculate the gradient vector field, which preserves the edges, and thus helps the detection and the measurement of the tubular objects.[4] What is relevant to the elucidation of the difference between the work of Frangi et al.[3] and the work presented here is the choice made in order to calculate the differential operators of an image. The Hessian of the Gaussian[3] versus the second-order partial derivatives of the model polynomial function is fitted to the MRI data.[5] An alternative approach to the calculation of the entries of the Hessian matrix of the Gaussian is the calculation of the matrix coefficients as the derivatives of Lindeberg's γ-parametrized normalized Gaussian kernels.[6] The aforementioned approach proved to be a successful complement to an image processing methodology developed with the purpose to segment the human brain vasculature from proton density-weighted MRI.[7] There is a straight difference between (i) the use of the model polynomial function to fit the MRI data and to calculate the second-order partial derivatives of the function and (ii) the use of the Gaussian to calculate the second-order derivatives of the image. In this entry, the CC, which is defined as the sum of second-order partial derivatives of the model polynomial function, is the approximation of the second-order derivative on an image. It is also relevant to report that in order to approximate image differentiation, other research works make use of derivative filters[8] (the calculation makes use of calculus), compact finite differences,[9] and the Sobel operator.[10] However, the approach based on model function fitting is different. As far as regards fitting the model polynomial function to the image pixels,[11] the study presented in this entry has the same line of thought of Haralick's work[12] (it uses calculus to obtain the second-order partial derivatives). The intersection between Haralick's[12] work and the work here presented is thus in the model polynomial function fitting approach. More specifically, the partial derivatives are used in edge-finding algorithms.[13] However, the mathematical procedure used for the calculation of the ICMAs is featured by the intensity-curvature term, which is obtained by merging (i) the sum of second-order partial derivatives of the model function fitted to the image data and (ii) the image pixel intensity. The intensity-curvature term is novel in the literature.[5] However, fitting a model function to the image pixels is certainly not new.[12,14]

ICMAS OF THE CUBIC TRIVARIATE LAGRANGE POLYNOMIAL

Let Eq. 1 define the cubic trivariate Lagrange polynomial[15] with ω_1 and ω_2 postulated in Eqs. 2 and 3, respectively. The voxel to resample is $f(0, 0, 0)$. The voxels neighboring $f(0, 0, 0)$ appear in the definition of ω_1 and ω_2.

$$g_4(x, y, z) = f(0, 0, 0) + \omega_1 \cdot a$$
$$= \left[(x+y+z)^3 + \left(\frac{1}{2}\right)(x+y+z)^2 \right.$$
$$\left. + (x+y+z) + 1\right] \quad (1)$$
$$+ \omega_2 \cdot a \left[(x+y+z)^2 + 2(x+y+z) + 1\right]$$

$$\omega_1 = \begin{bmatrix} f(-1,0,1)+f(-1,0,0)+f(1,0,0)+f(1,0,-1) \\ +f(-1,-1,1)+f(-1,-1,0)+f(1,-1,0)+f(1,-1,-1) \\ +f(-1,1,1)+f(-1,1,0)+f(1,1,0)+f(1,1,-1) \end{bmatrix} \quad (2)$$

$$\omega_2 = \begin{bmatrix} f(-1,0,1)+f(0,0,1)+f(1,0,1)+f(-1,0,0) \\ +f(1,0,0)+f(-1,0,-1)+f(0,0,-1) \\ +f(1,0,-1)+f(-1,-1,1)+f(0,-1,1) \\ +f(1,-1,1)+f(-1,-1,0)+f(0,-1,0) \\ +f(1,-1,0)+f(-1,-1,-1)+f(0,-1,-1) \\ +f(1,-1,-1)+f(-1,1,1)+f(0,1,1)+f(1,1,1) \\ +f(-1,1,0)+f(0,1,0)+f(1,1,0)+f(-1,1,-1) \\ +f(0,1,-1)+f(1,1,-1) \end{bmatrix} \quad (3)$$

The curvature that this study uses is the CC.[15] The CC is defined as the sum of the second-order partial derivatives with respect to the dimensional variables (x, y, and z) of the model function $(g_4(x, y, z))$ fitted to the MRI data and is termed as $\Omega(x, y, z)$. The CC $\Omega(x, y, z)$ of the model function $g_4(x, y, z)$ is defined as follows:

$$\Omega(x,y,z)$$
$$= \begin{bmatrix} \left(\frac{\partial^2(g_4(x,y,z))}{\partial x^2}\right) + \left(\frac{\partial^2(g_4(x,y,z))}{\partial y^2}\right) \\ + \left(\frac{\partial^2(g_4(x,y,z))}{\partial z^2}\right) + \left(\frac{\partial^2(g_4(x,y,z))}{\partial x \partial y}\right) \\ + \left(\frac{\partial^2(g_4(x,y,z))}{\partial y \partial x}\right) + \left(\frac{\partial^2(g_4(x,y,z))}{\partial x \partial z}\right) \\ + \left(\frac{\partial^2(g_4(x,y,z))}{\partial y \partial z}\right) + \left(\frac{\partial^2(g_4(x,y,z))}{\partial z \partial x}\right) \\ + \left(\frac{\partial^2(g_4(x,y,z))}{\partial z \partial y}\right) \end{bmatrix} \quad (4)$$
$$= 9 \cdot \{\omega_1 \cdot a[6(x+y+z)+1] + 2\omega_2 \cdot a\}$$

The intensity-curvature functional (ICF) is the ratio of the intensity-curvature term before interpolation (ICTBI) (at the numerator: termed as $E_0(x, y, z)$) to the intensity-curvature term after interpolation (ICTAI) (at the denominator: termed as $E_{IN}(x, y, z)$). The ICTBI is calculated as the antiderivative of the product between the signal intensity and the value of the CC at the origin of the pixel coordinate system. The ICTAI is calculated as the antiderivative of the product between the model function and the value of the CC at the intra-pixel location (x, y). The signal resilient to interpolation (SRI) is the signal intensity recalculated from the solution of the equation between the ICTBI and the ICTAI.

The ICTBI $E_0(x, y, z)$ of $g_4(x, y, z)$ is defined as follows:

$$E_0(x,y,z) = \iiint f(0,0,0) \cdot \Omega(x,y,z)_{(0,0,0)} \, dx \, dy \, dz$$
$$= f(0,0,0) \cdot 9 \cdot \{\omega_1 \cdot a + 2\omega_2 \cdot a\} \cdot xyz \tag{5}$$

The ICTAI $E_{IN}(x, y, z)$ of $g_4(x, y, z)$ is defined as follows:

$$E_{IN}(x,y,z) = \iiint g_4(x,y,z) \cdot \Omega(x,y,z)_{(x,y,z)} \, dx \, dy \, dz$$

$$= 9 \cdot \begin{cases} f(0,0,0) \cdot \{\omega_1 \cdot a \\ \left[6\left(\left(\frac{1}{2}\right)yzx^2 + \left(\frac{1}{2}\right)xzy^2 + \left(\frac{1}{2}\right)xyz^2\right) + xyz\right] \\ + 2\omega_2 \cdot a \, xyz\} + 6(\omega_1 a)^2 \Pi^{(4)} + \left[4(\omega_1 a)^2 \right. \\ \left. + 2\omega_2 \omega_1 \cdot a^2 + 6\omega_2 \omega_1 \cdot a^2 \right] \Pi^{(3)} + \\ \left[\left(\frac{6}{4}\right)(\omega_1 a)^2 + \frac{1}{2}(\omega_1 a)^2 + \omega_2 \omega_1 \cdot a^2 + 12\omega_2 \omega_1 \cdot a^2 \right. \\ \left. + \omega_1 \omega_2 a^2 + 2(\omega_2 a)^2 \right] \Pi^{(2)} + \\ \left[6(\omega_1 a)^2 + \left(\frac{1}{4}\right)(\omega_1 a)^2 + \left(\frac{1}{2}\right)\omega_2 \omega_1 \cdot a^2 \right. \\ \left. + 6\omega_2 \omega_1 \cdot a^2 + 2\omega_1 \omega_2 a^2 + 4(\omega_2 a)^2 \right] \Pi^{(1)} + \\ \left[(\omega_1 a)^2 + 2\omega_2 \omega_1 \cdot a^2 + \omega_1 \omega_2 a^2 + 2(\omega_2 a)^2 \right] xyz \end{cases} \tag{6}$$

Equations 7–10 are posited in order to simplify the preceding equation:

$$\Pi^{(1)} = \left(\frac{x^2 yz}{2} + \frac{xy^2 z}{2} + \frac{xyz^2}{2}\right) \tag{7}$$

$$\Pi^{(2)} = \left(\frac{x^3 yz}{3} + \frac{xy^3 z}{3} + \frac{xz^3 y}{3} + \frac{x^2 y^2 z}{2} + \frac{x^2 yz^2}{2} + \frac{xy^2 z^2}{2}\right) \tag{8}$$

$$\Pi^{(3)} = \begin{pmatrix} \frac{x^4 yz}{4} + \frac{x^2 y^3 z}{6} + \frac{x^2 yz^3}{6} + \frac{x^3 y^2 z}{3} + \left(\frac{1}{3}\right)x^3 yz^2 \\ + \frac{x^2 y^2 z^2}{4} + \frac{x^3 y^2 z}{6} + \frac{xy^4 z}{4} + \frac{xy^2 z^3}{6} \\ + \left(\frac{1}{3}\right)x^2 y^3 z + \frac{x^2 y^2 z^2}{4} + \frac{xy^3 z^2}{3} + \frac{x^3 yz^2}{6} \\ + \frac{xy^3 z^2}{6} + \frac{xyz^4}{4} + \frac{x^2 y^2 z^2}{4} + \left(\frac{1}{3}\right)x^2 yz^3 \\ + \left(\frac{1}{3}\right)xy^2 z^3 \end{pmatrix} \tag{9}$$

$$\Pi^{(4)} = \begin{pmatrix} \frac{x^5 zy}{5} + \frac{x^3 y^3 z}{9} + \frac{x^3 yz^3}{9} + \frac{x^4 y^2 z}{4} + \frac{x^4 z^2 y}{4} \\ + \frac{x^3 y^2 z^2}{6} + \frac{xy^5 z}{5} + \frac{x^3 y^3 z}{9} + \frac{xy^3 z^3}{9} + x^2 y^4 z \\ + \frac{x^2 z^2 y^3}{6} + \frac{xy^4 z^2}{4} + \frac{x^3 yz^3}{9} + \frac{xz^3 y^3}{9} + \frac{xz^5 y}{5} \\ + \left(\frac{1}{6}\right)x^2 y^2 z^3 + \left(\frac{1}{4}\right)x^2 yz^4 + \left(\frac{1}{4}\right)xy^2 z^4 \\ + \frac{x^4 y^2 z}{4} + \left(\frac{1}{4}\right)x^2 y^4 z + \left(\frac{1}{6}\right)x^2 y^2 z^3 + \left(\frac{4}{9}\right)x^3 y^3 z \\ + \frac{x^3 y^2 z^2}{3} + \left(\frac{1}{3}\right)x^2 y^3 z^2 + \frac{x^4 z^2 y}{4} + \frac{x^2 y^3 z^2}{6} \\ + x^2 yz^4 + \frac{x^3 y^2 z^2}{3} + \left(\frac{4}{9}\right)x^3 yz^3 + \left(\frac{4}{12}\right)x^2 y^2 z^3 \\ + \frac{x^3 y^2 z^2}{6} + \frac{xy^4 z^2}{4} + \left(\frac{1}{4}\right)xy^2 z^4 + \left(\frac{1}{3}\right)x^2 y^3 z^2 \\ + \left(\frac{4}{12}\right)x^2 y^2 z^3 + \left(\frac{4}{9}\right)xy^3 z^3 \end{pmatrix} \tag{10}$$

The ICF $\Delta E(x, y, z)$ of $g_4(x, y, z)$ is defined as follows:

$$\Delta E(x,y,z) = \frac{E_0(x,y,z)}{E_{IN}(x,y,z)} \tag{11}$$

The resilient curvature (RC) is the sum of second-order partial derivatives with respect to the spatial coordinates of the SRI, and the SRI is defined as per equation 126 in Chapter 11 of earlier works[15] $\left(\rho(x,y,z) = \left\{\left[\Phi_1 \Lambda_{(4)}\right] + \left[\Phi_2 \Lambda_{(3)}\right] + \left[\Phi_3 \Lambda_{(2)}\right] + \left[\Phi_4 \Lambda_{(1)}\right] + \left[\Phi_5\right]\right\} / \left\{\Lambda_6\right\}\right)$.

The $RC(x, y, z)$ of $g_4(x, y, z)$ is defined as follows:

$$RC(x,y,z) = \begin{bmatrix} \left(\frac{\partial^2 \rho(x,y,z)}{\partial x^2}\right) + \left(\frac{\partial^2 \rho(x,y,z)}{\partial y^2}\right) + \left(\frac{\partial^2 \rho(x,y,z)}{\partial z^2}\right) \\ + \frac{\partial\left(\frac{\partial \rho(x,y,z)}{\partial y}\right)}{\partial x} + \frac{\partial\left(\frac{\partial \rho(x,y,z)}{\partial x}\right)}{\partial y} \\ + \frac{\partial\left(\frac{\partial \rho(x,y,z)}{\partial y}\right)}{\partial z} + \frac{\partial\left(\frac{\partial \rho(x,y,z)}{\partial z}\right)}{\partial y} \\ + \frac{\partial\left(\frac{\partial \rho(x,y,z)}{\partial z}\right)}{\partial x} + \frac{\partial\left(\frac{\partial \rho(x,y,z)}{\partial x}\right)}{\partial z} \end{bmatrix} \tag{12}$$

In the preceding equation, the calculation of the second-order partial derivatives with respect to the dimensional variables (x, y, and z) demands the calculations illustrated in Eqs. 13, 40, and 42 through 48:

$$\left(\frac{\partial^2(\rho(x,y,z))}{\partial x^2}\right)$$

$$= \frac{\left\{\begin{bmatrix}\left(\frac{\partial^2(N_L)}{\partial x^2}\right)\cdot[\Lambda_6] + \left(\frac{\partial(N_L)}{\partial x}\right)\cdot\left(\frac{\partial(\Lambda_6)}{\partial x}\right)\\ -\left(\frac{\partial(N_L)}{\partial x}\right)\left(\frac{\partial(\Lambda_6)}{\partial x}\right)\end{bmatrix}\cdot[\Lambda_6]^2 - \left\{\begin{bmatrix}\left(\frac{\partial(N_L)}{\partial x}\right)\cdot\Lambda_6\\ -N_L\cdot\left(\frac{\partial(\Lambda_6)}{\partial x}\right)\end{bmatrix}\cdot\left\{2\cdot\Lambda_6\cdot\left(\frac{\partial(\Lambda_6)}{\partial x}\right)\right\}\right\}\right\}}{[\Lambda_6]^4} \quad (13)$$

Equations 14–19 simplify the formulation of Eq. 20.

$$\Phi_1 = 6(\omega_1 a)^2 \quad (14)$$

$$\Phi_2 = \left[4(\omega_1 a)^2 + 2\omega_2\omega_1\cdot a^2 + 6\omega_2\omega_1\cdot a^2\right] \quad (15)$$

$$\Phi_3 = \left[\left(\frac{6}{4}\right)(\omega_1 a)^2 + \left(\frac{1}{2}\right)(\omega_1 a)^2 + \omega_2\omega_1\cdot a^2 + 12\omega_2\omega_1\cdot a^2 + \omega_1\omega_2 a^2 + 2(\omega_2 a)^2\right] \quad (16)$$

$$\Phi_4 = \left[6(\omega_1 a)^2 + \left(\frac{1}{4}\right)(\omega_1 a)^2 + \left(\frac{1}{2}\right)\omega_2\omega_1\cdot a^2 + 6\omega_2\omega_1\cdot a^2 + 2\omega_1\omega_2 a^2 + 4(\omega_2 a)^2\right] \quad (17)$$

$$\Phi_5 = \left[(\omega_1 a)^2 + 2\omega_2\omega_1\cdot a^2 + \omega_1\omega_2 a^2 + 2(\omega_2 a)^2\right] \quad (18)$$

$$\Phi_6 = \left[\omega_1\cdot a + 2\omega_2\cdot a\right] \quad (19)$$

$$N_L = \left\{\Phi_1\cdot\Lambda_{(4)} + \Phi_2\cdot\Lambda_{(3)} + \Phi_3\cdot\Lambda_{(2)} + \Phi_4\cdot\Lambda_{(1)} + \Phi_5\right\} \quad (20)$$

The second-order partial derivatives of the preceding equation are calculated in Eqs. 22, 30, and 32–38. Equations 23–27 are posited to simplify the calculation of the first- and second-order partial derivatives of N_L.

$$\left(\frac{\partial(N_L)}{\partial x}\right) = \left\{\Phi_1\cdot\left(\frac{\partial(\Lambda_{(4)})}{\partial x}\right) + \Phi_2\cdot\left(\frac{\partial(\Lambda_{(3)})}{\partial x}\right) + \Phi_3\cdot\left(\frac{\partial(\Lambda_{(2)})}{\partial x}\right) + \Phi_4\cdot\left(\frac{\partial(\Lambda_{(1)})}{\partial x}\right)\right\} \quad (21)$$

$$\left(\frac{\partial^2(N_L)}{\partial x^2}\right) = \left\{\Phi_1\cdot\left(\frac{\partial^2(\Lambda_{(4)})}{\partial x^2}\right) + \Phi_2\cdot\left(\frac{\partial^2(\Lambda_{(3)})}{\partial x^2}\right) + \Phi_3\cdot\left(\frac{\partial^2(\Lambda_{(2)})}{\partial x^2}\right) + \Phi_4\cdot\left(\frac{\partial^2(\Lambda_{(1)})}{\partial x^2}\right)\right\} \quad (22)$$

$$\Lambda_{(1)} = \Pi^{(1)}/(x\,y\,z) \quad (23)$$

$$\Lambda_{(2)} = \frac{\Pi^{(2)}}{(x\,y\,z)} \quad (24)$$

$$\Lambda_{(3)} = \frac{\Pi^{(3)}}{(x\,y\,z)} \quad (25)$$

$$\Lambda_{(4)} = \frac{\Pi^{(4)}}{(x\,y\,z)} \quad (26)$$

$$\Lambda_6 = \left\{\Phi_6 - \left\{\omega_1\cdot a\left[6\left(\left(\frac{1}{2}\right)x + \left(\frac{1}{2}\right)y\right) + \left(\frac{1}{2}\right)z + 1\right] + 2\omega_2\cdot a\right\}\right\} \quad (27)$$

$$\left(\frac{\partial^2(\Lambda_6)}{\partial x^2}\right) = \left(\frac{\partial^2(\Lambda_6)}{\partial y^2}\right) = \left(\frac{\partial^2(\Lambda_6)}{\partial z^2}\right)$$
$$= \left(\frac{\partial^2(\Lambda_6)}{\partial x\partial y}\right) = \left(\frac{\partial^2(\Lambda_6)}{\partial y\partial x}\right) = \left(\frac{\partial^2(\Lambda_6)}{\partial x\partial z}\right) \quad (28)$$
$$= \left(\frac{\partial^2(\Lambda_6)}{\partial z\partial x}\right) = \left(\frac{\partial^2(\Lambda_6)}{\partial y\partial z}\right) = \left(\frac{\partial^2(\Lambda_6)}{\partial z\partial y}\right) = 0$$

Equation 28 is relevant to Eqs. 13, 40, 42–48. Let us posit

$$\left(\frac{\partial(N_L)}{\partial y}\right) = \left\{\Phi_1\cdot\left(\frac{\partial(\Lambda_{(4)})}{\partial y}\right) + \Phi_2\cdot\left(\frac{\partial(\Lambda_{(3)})}{\partial y}\right) + \Phi_3\cdot\left(\frac{\partial(\Lambda_{(2)})}{\partial y}\right) + \Phi_4\cdot\left(\frac{\partial(\Lambda_{(1)})}{\partial y}\right)\right\} \quad (29)$$

Three Dimensional Intensity-Curvature Measurement

$$\left(\frac{\partial^2(N_L)}{\partial y^2}\right) = \left\{\Phi_1 \cdot \left(\frac{\partial^2(\Lambda_{(4)})}{\partial y^2}\right) + \Phi_2 \cdot \left(\frac{\partial^2(\Lambda_{(3)})}{\partial y^2}\right) + \Phi_3 \cdot \left(\frac{\partial^2(\Lambda_{(2)})}{\partial y^2}\right) + \Phi_4 \cdot \left(\frac{\partial^2(\Lambda_{(1)})}{\partial y^2}\right)\right\} \tag{30}$$

$$\left(\frac{\partial(N_L)}{\partial z}\right) = \left\{\Phi_1 \cdot \left(\frac{\partial(\Lambda_{(4)})}{\partial z}\right) + \Phi_2 \cdot \left(\frac{\partial(\Lambda_{(3)})}{\partial z}\right) + \Phi_3 \cdot \left(\frac{\partial(\Lambda_{(2)})}{\partial z}\right) + \Phi_4 \cdot \left(\frac{\partial(\Lambda_{(1)})}{\partial z}\right)\right\} \tag{31}$$

$$\left(\frac{\partial^2(N_L)}{\partial z^2}\right) = \left\{\Phi_1 \cdot \left(\frac{\partial^2(\Lambda_{(4)})}{\partial z^2}\right) + \Phi_2 \cdot \left(\frac{\partial^2(\Lambda_{(3)})}{\partial z^2}\right) + \Phi_3 \cdot \left(\frac{\partial^2(\Lambda_{(2)})}{\partial z^2}\right) + \Phi_4 \cdot \left(\frac{\partial^2(\Lambda_{(1)})}{\partial z^2}\right)\right\} \tag{32}$$

$$\left(\frac{\partial^2(N_L)}{\partial x \partial y}\right) = \left\{\Phi_1 \cdot \left(\frac{\partial^2(\Lambda_{(4)})}{\partial x \partial y}\right) + \Phi_2 \cdot \left(\frac{\partial^2(\Lambda_{(3)})}{\partial x \partial y}\right) + \Phi_3 \cdot \left(\frac{\partial^2(\Lambda_{(2)})}{\partial x \partial y}\right) + \Phi_4 \cdot \left(\frac{\partial^2(\Lambda_{(1)})}{\partial x \partial y}\right)\right\} \tag{33}$$

$$\left(\frac{\partial^2(N_L)}{\partial y \partial x}\right) = \left\{\Phi_1 \cdot \left(\frac{\partial^2(\Lambda_{(4)})}{\partial y \partial x}\right) + \Phi_2 \cdot \left(\frac{\partial^2(\Lambda_{(3)})}{\partial y \partial x}\right) + \Phi_3 \cdot \left(\frac{\partial^2(\Lambda_{(2)})}{\partial y \partial x}\right) + \Phi_4 \cdot \left(\frac{\partial^2(\Lambda_{(1)})}{\partial y \partial x}\right)\right\} \tag{34}$$

$$\left(\frac{\partial^2(N_L)}{\partial x \partial z}\right) = \left\{\Phi_1 \cdot \left(\frac{\partial^2(\Lambda_{(4)})}{\partial x \partial z}\right) + \Phi_2 \cdot \left(\frac{\partial^2(\Lambda_{(3)})}{\partial x \partial z}\right) + \Phi_3 \cdot \left(\frac{\partial^2(\Lambda_{(2)})}{\partial x \partial z}\right) + \Phi_4 \cdot \left(\frac{\partial^2(\Lambda_{(1)})}{\partial x \partial z}\right)\right\} \tag{35}$$

$$\left(\frac{\partial^2(N_L)}{\partial z \partial x}\right) = \left\{\Phi_1 \cdot \left(\frac{\partial^2(\Lambda_{(4)})}{\partial z \partial x}\right) + \Phi_2 \cdot \left(\frac{\partial^2(\Lambda_{(3)})}{\partial z \partial x}\right) + \Phi_3 \cdot \left(\frac{\partial^2(\Lambda_{(2)})}{\partial z \partial x}\right) + \Phi_4 \cdot \left(\frac{\partial^2(\Lambda_{(1)})}{\partial z \partial x}\right)\right\} \tag{36}$$

$$\left(\frac{\partial^2(N_L)}{\partial y \partial z}\right) = \left\{\Phi_1 \cdot \left(\frac{\partial^2(\Lambda_{(4)})}{\partial y \partial z}\right) + \Phi_2 \cdot \left(\frac{\partial^2(\Lambda_{(3)})}{\partial y \partial z}\right) + \Phi_3 \cdot \left(\frac{\partial^2(\Lambda_{(2)})}{\partial y \partial z}\right) + \Phi_4 \cdot \left(\frac{\partial^2(\Lambda_{(1)})}{\partial y \partial z}\right)\right\} \tag{37}$$

$$\left(\frac{\partial^2(N_L)}{\partial z \partial y}\right) = \left\{\Phi_1 \cdot \left(\frac{\partial^2(\Lambda_{(4)})}{\partial z \partial y}\right) + \Phi_2 \cdot \left(\frac{\partial^2(\Lambda_{(3)})}{\partial z \partial y}\right) + \Phi_3 \cdot \left(\frac{\partial^2(\Lambda_{(2)})}{\partial z \partial y}\right) + \Phi_4 \cdot \left(\frac{\partial^2(\Lambda_{(1)})}{\partial z \partial y}\right)\right\} \tag{38}$$

Equations 22, 30, and 32–38 are posited to simplify the calculation of the first- and second-order partial derivatives as illustrated in Eqs. 13 and 39–48. It follows that:

$$\left(\frac{\partial(\rho(x,y,z))}{\partial y}\right) = \frac{\left\{\left(\frac{\partial(N_L)}{\partial y}\right) \cdot \Lambda_6 - N_L \cdot \left(\frac{\partial(\Lambda_6)}{\partial y}\right)\right\}}{[\Lambda_6]^2} \tag{39}$$

$$\left(\frac{\partial^2(\rho(x,y,z))}{\partial y^2}\right) = \frac{\left\{\left\{\left(\frac{\partial^2(N_L)}{\partial y^2}\right)\cdot[\Lambda_6] + \left(\frac{\partial(N_L)}{\partial y}\right)\cdot\left(\frac{\partial(\Lambda_6)}{\partial y}\right) - \left(\frac{\partial(N_L)}{\partial y}\right)\cdot\left(\frac{\partial(\Lambda_6)}{\partial y}\right)\right\}\cdot[\Lambda_6]^2 - \left\{\left(\frac{\partial(N_L)}{\partial y}\right)\cdot\Lambda_6 - N_L\cdot\left(\frac{\partial(\Lambda_6)}{\partial y}\right)\right\}\cdot\left\{2\cdot\Lambda_6\cdot\left(\frac{\partial(\Lambda_6)}{\partial y}\right)\right\}\right\}}{[\Lambda_6]^4} \qquad (40)$$

$$\left(\frac{\partial(\rho(x,y,z))}{\partial z}\right) = \frac{\left\{\left(\frac{\partial(N_L)}{\partial z}\right)\cdot\Lambda_6 - N_L\cdot\left(\frac{\partial(\Lambda_6)}{\partial z}\right)\right\}}{[\Lambda_6]^2} \qquad (41)$$

$$\left(\frac{\partial^2(\rho(x,y,z))}{\partial z^2}\right) = \frac{\left\{\left\{(\partial^2(N_L)/\partial z^2)\cdot[\Lambda_6] + (\partial(N_L)/\partial z)\cdot(\partial(\Lambda_6)/\partial z) - \left(\frac{\partial(N_L)}{\partial z}\right)\cdot\left(\frac{\partial(\Lambda_6)}{\partial z}\right)\right\}\cdot[\Lambda_6]^2 - \left\{\left(\frac{\partial(N_L)}{\partial z}\right)\cdot\Lambda_6 - N_L\cdot\left(\frac{\partial(\Lambda_6)}{\partial z}\right)\right\}\cdot\left\{2\cdot\Lambda_6\cdot\left(\frac{\partial(\Lambda_6)}{\partial z}\right)\right\}\right\}}{[\Lambda_6]^4} \qquad (42)$$

$$\left(\frac{\partial^2(\rho(x,y,z))}{\partial x\partial y}\right) = \frac{\left\{\left\{\left(\frac{\partial\left(\frac{\partial(N_L)}{\partial x}\right)}{\partial y}\right)\cdot[\Lambda_6] - \left(\frac{\partial(N_L)}{\partial y}\right)\cdot\left(\frac{\partial(\Lambda_6)}{\partial x}\right) + \left(\frac{\partial(N_L)}{\partial x}\right)\cdot\left(\frac{\partial(\Lambda_6)}{\partial y}\right)\right\}\cdot[\Lambda_6]^2 - \left\{\left(\frac{\partial(N_L)}{\partial x}\right)\cdot\Lambda_6 - N_L\cdot\left(\frac{\partial(\Lambda_6)}{\partial x}\right)\right\}\cdot\left\{2\cdot\Lambda_6\cdot\left(\frac{\partial(\Lambda_6)}{\partial y}\right)\right\}\right\}}{[\Lambda_6]^4} \qquad (43)$$

$$\left(\frac{\partial^2(\rho(x,y,z))}{\partial y\partial x}\right) = \frac{\left\{\left\{\left(\frac{\partial\left(\frac{\partial(N_L)}{\partial y}\right)}{\partial x}\right)\cdot[\Lambda_6] - \left(\frac{\partial(N_L)}{\partial x}\right)\cdot\left(\frac{\partial(\Lambda_6)}{\partial y}\right) + \left(\frac{\partial(N_L)}{\partial y}\right)\cdot\left(\frac{\partial(\Lambda_6)}{\partial x}\right)\right\}\cdot[\Lambda_6]^2 - \left\{\left(\frac{\partial(N_L)}{\partial y}\right)\cdot\Lambda_6 - N_L\cdot\left(\frac{\partial(\Lambda_6)}{\partial y}\right)\right\}\cdot\left\{2\cdot\Lambda_6\cdot\left(\frac{\partial(\Lambda_6)}{\partial x}\right)\right\}\right\}}{[\Lambda_6]^4} \qquad (44)$$

$$\left(\frac{\partial^2(\rho(x,y,z))}{\partial z \partial x}\right) = \frac{\left\{\left[\left(\frac{\partial\left(\frac{\partial(N_L)}{\partial z}\right)}{\partial x}\right) \cdot [\Lambda_6] - \left(\frac{\partial(N_L)}{\partial x}\right) \cdot \left(\frac{\partial(\Lambda_6)}{\partial z}\right) + \left(\frac{\partial(N_L)}{\partial z}\right) \cdot \left(\frac{\partial(\Lambda_6)}{\partial x}\right)\right] \cdot [\Lambda_6]^2 - \left\{\left(\frac{\partial(N_L)}{\partial z}\right) \cdot \Lambda_6 - N_L \cdot \left(\frac{\partial(\Lambda_6)}{\partial z}\right)\right\} \cdot \left\{2 \cdot \Lambda_6 \cdot \left(\frac{\partial(\Lambda_6)}{\partial x}\right)\right\}\right\}}{[\Lambda_6]^4} \qquad (45)$$

$$\left(\frac{\partial^2(\rho(x,y,z))}{\partial x \partial z}\right) = \frac{\left\{\left[\left(\frac{\partial\left(\frac{\partial(N_L)}{\partial x}\right)}{\partial z}\right) \cdot [\Lambda_6] - \left(\frac{\partial(N_L)}{\partial z}\right) \cdot \left(\frac{\partial(\Lambda_6)}{\partial x}\right) + \left(\frac{\partial(N_L)}{\partial x}\right) \cdot \left(\frac{\partial(\Lambda_6)}{\partial z}\right)\right] \cdot [\Lambda_6]^2 - \left\{\left(\frac{\partial(N_L)}{\partial x}\right) \cdot \Lambda_6 - N_L \cdot \left(\frac{\partial(\Lambda_6)}{\partial x}\right)\right\} \cdot \left\{2 \cdot \Lambda_6 \cdot \left(\frac{\partial(\Lambda_6)}{\partial z}\right)\right\}\right\}}{[\Lambda_6]^4} \qquad (46)$$

$$\left(\frac{\partial^2(\rho(x,y,z))}{\partial y \partial z}\right) = \frac{\left\{\left[\left(\frac{\partial\left(\frac{\partial(N_L)}{\partial y}\right)}{\partial z}\right) \cdot [\Lambda_6] - \left(\frac{\partial(N_L)}{\partial z}\right) \cdot \left(\frac{\partial(\Lambda_6)}{\partial y}\right) + \left(\frac{\partial(N_L)}{\partial y}\right) \cdot \left(\frac{\partial(\Lambda_6)}{\partial z}\right)\right] \cdot [\Lambda_6]^2 - \left\{\left(\frac{\partial(N_L)}{\partial y}\right) \cdot \Lambda_6 - N_L \cdot \left(\frac{\partial(\Lambda_6)}{\partial y}\right)\right\} \cdot \left\{2 \cdot \Lambda_6 \cdot \left(\frac{\partial(\Lambda_6)}{\partial z}\right)\right\}\right\}}{[\Lambda_6]^4} \qquad (47)$$

$$\left(\frac{\partial^2(\rho(x,y,z))}{\partial z \partial y}\right) = \frac{\left\{\left[\left(\frac{\partial\left(\frac{\partial(N_L)}{\partial z}\right)}{\partial y}\right) \cdot [\Lambda_6] - \left(\frac{\partial(N_L)}{\partial y}\right) \cdot \left(\frac{\partial(\Lambda_6)}{\partial z}\right) + \left(\frac{\partial(N_L)}{\partial z}\right) \cdot \left(\frac{\partial(\Lambda_6)}{\partial y}\right)\right] \cdot [\Lambda_6]^2 - \left\{\left(\frac{\partial(N_L)}{\partial z}\right) \cdot \Lambda_6 - N_L \cdot \left(\frac{\partial(\Lambda_6)}{\partial z}\right)\right\} \cdot \left\{2 \cdot \Lambda_6 \cdot \left(\frac{\partial(\Lambda_6)}{\partial y}\right)\right\}\right\}}{[\Lambda_6]^4} \qquad (48)$$

MATERIALS, METHODS, RATIONALE, AND OBJECTIVES

Ten subjects participated in the study. Compliance with the Declaration of Helsinki is assured because the MRI scans were collected after proper administration of the informed consent to the patient and in agreement with the ethical committees of Skopje City General Hospital. Table 1 reports the MRI acquisition parameters. The strength of the magnet was 1.5 T.

The model polynomial fitted to the MRI data was the cubic trivariate Lagrange function, and it was chosen from an array of seven available polynomials because the formula is parametric in the constant "a" (see Eq. 1), and thus allows tuning of the outward show of the ICMA images. To merge the value of the image pixel intensity with the sum of second-order partial derivatives of the model function is the concept at the root foundation of the development of the ICMAs. The ICMAs studied in this entry are (i) the ICF, (ii) the ICTBI, (iii) the ICTAI, and (iv) the RC. The rationale of the study is described here. T1-weighted MRI, T2-weighted MRI, and fluid-attenuated inversion recovery (FLAIR) MRI are MRI modalities. The T1-weighted MRI modality is commonly used to image the structure of the human brain. The T2-weighted MRI modality is used mainly to image fat and water in the human brain. The FLAIR MRI modality[16] is used mainly to suppress the MRI signal resulting from fluids of the human brain such as the cerebrospinal fluid. T1-weighted MR images were used to identify human brain vessels. Once the vessels were identified, the allied pulse sequences T2-weighted MRI and FLAIR MRI[16] were processed to obtain the ICMAs. The expectation was to highlight the human brain vessels in the ICMA images. Thus, the rationale of the study is the following: (i) to present the three-dimensional ICMAs and (ii) to elucidate if it is possible to highlight human brain vessels in the ICMA images of T2-weighted MRI and FLAIR MRI.

RESULTS

This section presents the ICMAs calculated from the T2-weighted MRI and the FLAIR MRI (the allied pulse sequences) when fitting the cubic trivariate Lagrange polynomial to the image data. The organization of the section mirrors the rationale of the study. Slices of interest were chosen from the MRI volumes acquired with the allied pulse sequences. The criterion adopted was that of identifying human brain vessels in the slices. Then the vessel structures in the human brain were studied with the ICMAs of T2-weighted MRI and FLAIR MRI. The ICMA images provide highlight of the vasculature and also additional and complementary structural information about the vessels.

Table 1 Recording parameters of the MR images

MRI acquisition	T1-weighted	T2-weighted	FLAIR
Time to echo (ms)	2.59	96	84
Repetition time (ms)	7	5,050 4,720	9,000
Field of view (mm)	280 × 280	204 × 250 193 × 250 171 × 250	203 × 250 165 × 250
Pixel matrix size	410 × 512	196 × 320 186 × 320 165 × 320	208 × 320 170 × 320
Number of slices	9, 16	27, 28	28

Preliminary Study

A T2-weighted MRI image is shown in Fig. 1a, the CC in Fig. 1b calculated when fitting the cubic bivariate Lagrange polynomial model function, and the Gaussian-filtered images in Fig. 1c–e. It is worth noting that the four images are smooth. However, the CC is obtained through the solid mathematical procedure, which descends from algebra and calculus. Due to mention that in Fig. 1, the significant and rapid loss of spatial resolution of the image in Fig. 1e (versus the images in Fig. 1c and d) is consequential to the use of the 2×2 Gaussian mask in Fig. 1c, the 3×3 Gaussian mask in Fig. 1d, and the 4×4 Gaussian mask in Fig. 1e. The CC image (see Fig. 1b) shows smoothing comparable to the image in Fig. 1c. Figure 1 also presents the convolution between the T2-weighted MRI image and the CC image. Such convolution (see Fig. 1f) can be compared with the ICF images (see Fig. 1g, h). The rationale of the comparison is to ascertain if convolving the image intensity with the second-order derivative of the image (of which the CC is the representation) is not the same as the calculation of the ICF. Figure 1f–h demonstrates the difference in image space.

ICMAs of the MR images

In the top and bottom rows, from left to right, Fig. 2 presents the T2-weighted MRI, the ICF of the T2-weighted MRI image, and the RC of the T2-weighted MRI image. The images are labeled with the slice numbers: 15 and 16. The arrows indicate the vasculature highlighted in both MR and ICMA images. See in the first row of images the vessel structure inside the ellipse, which is visible in the T2 image and also clearly demarcated in the ICF and RC images. The RC labeled "16" shows the vessel structure projection of three consecutive slices and this is consequential to the convolution of voxels calculated with Eqs. 2 and 3 and thus provides with a view similar to the maximum intensity projection (MIP), which is used in other works[17] to present the structures of multiple slices of the image volume in a single slice.

Three Dimensional Intensity-Curvature Measurement

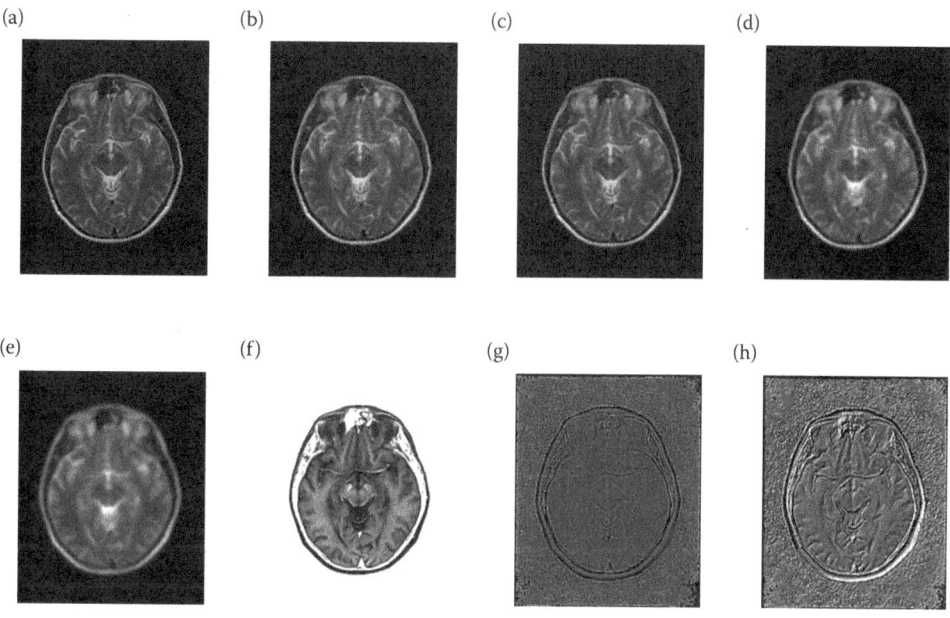

Fig. 1 (a) Two-dimensional T2-weighted MRI image (subject 1); (b) CC of the T2 image. Gaussian of the T2 image: 2×2 mask in (c); 3×3 mask in (d); 4×4 mask in (e). (f) Convolution between the T2-weighted MRI image and the CC image. (g and h) ICF of the T2-weighted MRI, which was obtained when fitting the cubic bivariate Lagrange model polynomial function (see Fig. 1g) and when fitting the bivariate linear model polynomial function (see Fig. 1h)

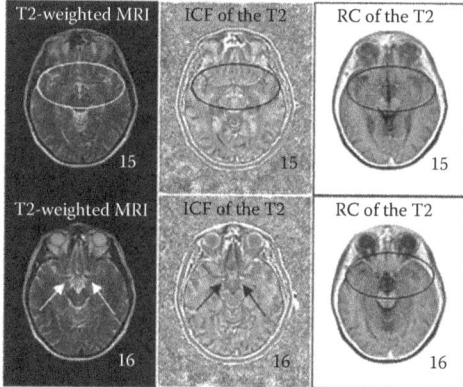

Fig. 2 Subject 1: two slices of interest (15 and 16) were chosen from the T2-weighted MRI volume, and they are presented in the leftmost images of the figure. The ICF images are displayed in the middle column. The RC images are located in the rightmost column. The vessel structures are identifiable in the T2-weighted MRI, and they are highlighted in the ICMA images (see the arrows and inside the ellipses)

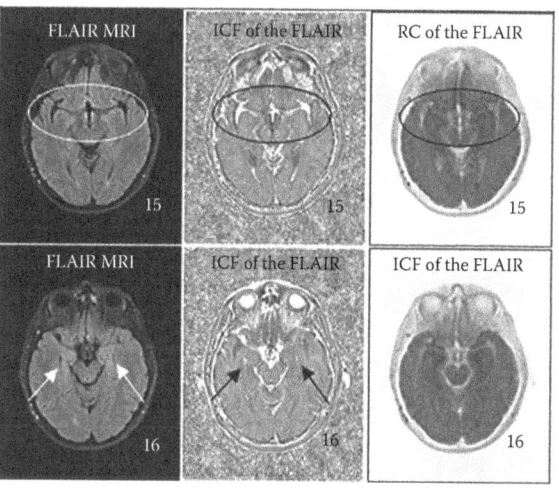

Fig. 3 Subject 1: the FLAIR imaging modality of the slices of interest (15 and 16) is presented in the leftmost column. The middle column displays the ICF images. The right column displays the RC images

In Fig. 3, the structure inside the ellipse in the FLAIR image labeled "15" is reproduced in the ICF and RC images (see inside the ellipse). In the ICF image (see the bottom row), the arrows indicate highlighting of presumed vessel-like structures.

Figure 4 sets as an example of the capability of the ICTBI and the ICTAI to highlight the vessel structures (see inside the ellipses). It is worth noting that the vessel structures are more demarcated in the ICTBI than they are in the ICTAI.

The images labeled "15" in Fig. 5 set as an example of how the ICTBI and the ICTAI can reproduce the brain structures of the FLAIR image (presented in Fig. 3). Whereas, the images labeled "16" show highlight of presumed vessel-like structures as indicated by the arrows. As regards the task of highlighting the vessel structures, the observation of Figs. 2–5 makes it possible to elucidate on the feature of the ICF of the T2-weighted MR images and of the ICTBI of the T2-weighted MR images. The feature is the highlight of human brain vessels and it is object of additional study in Figs. 10 and 11.

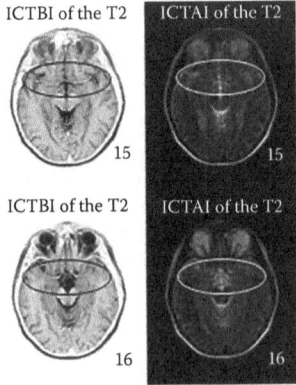

Fig. 4 Subject 1: the ICTBI of the T2 images displayed in Fig. 2 is presented in the left column. The images in the right column present the ICTAI of the T2 images located in Fig. 2

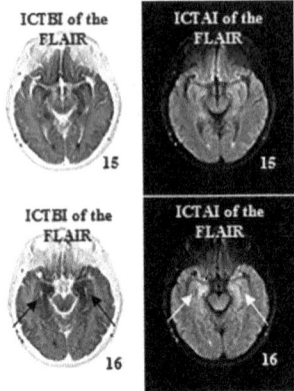

Fig. 5 Subject 1: the images presented in the left column are the ICTBI of the FLAIR images displayed in Fig. 3. The ICTAI of the same FLAIR images seen in Fig. 3 is placed in the right column

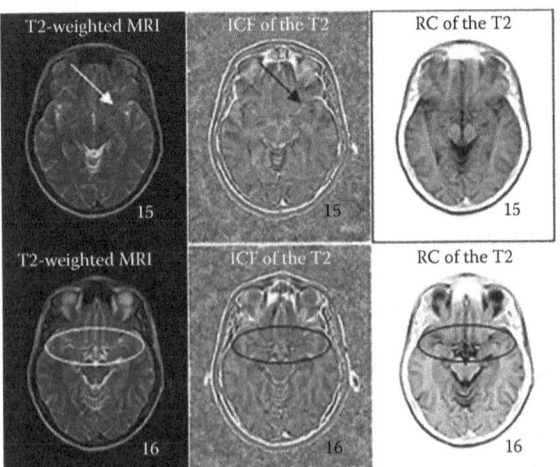

Fig. 6 Subject 2: two T2-weighted MRI slices of interest (15 and 16) are presented in the leftmost column. The ICF images of the T2-weighted MRI are displayed in the central column. The RC images of the T2-weighted MRI are located in the rightmost column

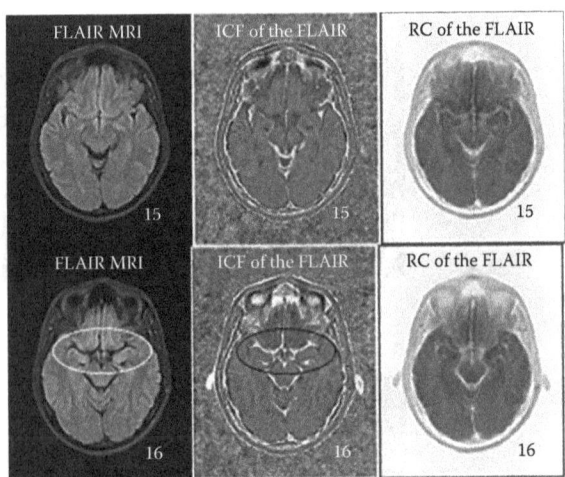

Fig. 7 Subject 2: two slices of interest of the FLAIR volume (15 and 16) are presented in the leftmost column. The ICF images are displayed in the central column. The RC images are located in the rightmost column

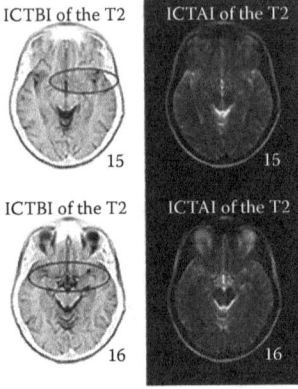

Fig. 8 Subject 2: the left column presents slices of the ICTBI calculated from the T2-weighted MRI displayed in Fig. 6. The vessels structures (see inside the ellipses) are more discernible and visible in the ICTBI images than they are in the T2-weighted MRI slices in Fig. 6. The right column displays the ICTAI of the T2 images. The ICTAI images offer less visibility of the vessel structures than the ICTBI images

In Fig. 6, slice 15, the arrows indicate the vessel structures in the T2-weighted MRI image. The vessels are well visible in the ICF image. The vessel structure inside the ellipse in the T2-weighted MRI (slice 16) is well reproduced in the ICF and RC images. It is worth noting that the outward show of the RC in slice 16 presents the structures of slice 15, thus displaying a view similar to the MIP.

In Fig. 7, the vessel structures are visible in slice 16 (see inside the ellipse), and they are well reproduced in the ICF image. Likewise noted in Fig. 6, the RC in slice 15 acts similarly to a MIP, because it contains brain structures located in slice 16.

Figure 8 illustrates the major finding reported in this entry: the capability of the ICTBI of the T2-weighted

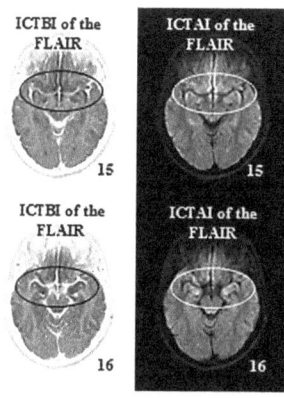

Fig. 9 Subject 2: the left column shows the ICTBI images (slices 15 and 16) of the FLAIR imaging modality presented in Fig. 7. The vessel structures are highlighted and well visible (see inside the ellipses of the ICTBI images). The right column displays the ICTAI of the FLAIR slices located in Fig. 7

MRI image to highlight the visibility of vessel structures more clearly than the T2-weighted MRI image. Compare slices 15 and 16 in Fig. 8 with the T2-weighted MR images located in Fig. 6 (slices 15 and 16). Also clarifies that the ICTAI is a smoother reproduction of the image from which is calculated; hence, the ICTAI of the T2-weighted MRI image does not allow significant highlights.

Figure 9 reports on another finding of this work, which is the aptitude of the ICTAI of the FLAIR imaging modality to highlight vessel structures. In fact, it can be observed that the ICTAI of the FLAIR offers more structural information as regards the visibility of the vessels. The aforementioned observation is elicited through the comparison of the ICTAI images presented in Fig. 8 with the ICTAI images presented in Fig. 9.

In Figs. 10 and 11, it can be summarized that the ICTBI of the T2-weighted MRI does highlight human brain vessels more than the T2-weighted imaging modality does. As regards the behavior of the ICF of the T2-weighted MRI, it is similar to the behavior revealed by the ICTBI of the T2-weighted MRI (see for instance in Fig. 10g the highlight of vessel structures in both of ICF and ICTBI). Finally, the ICTBI of the FLAIR imaging modality does not highlight the vessels like the ICTBI of the T2-weighted MRI does.

Figures 12 and 13 present the comparison between the ICMAs and other methodologies. The CC offers mild

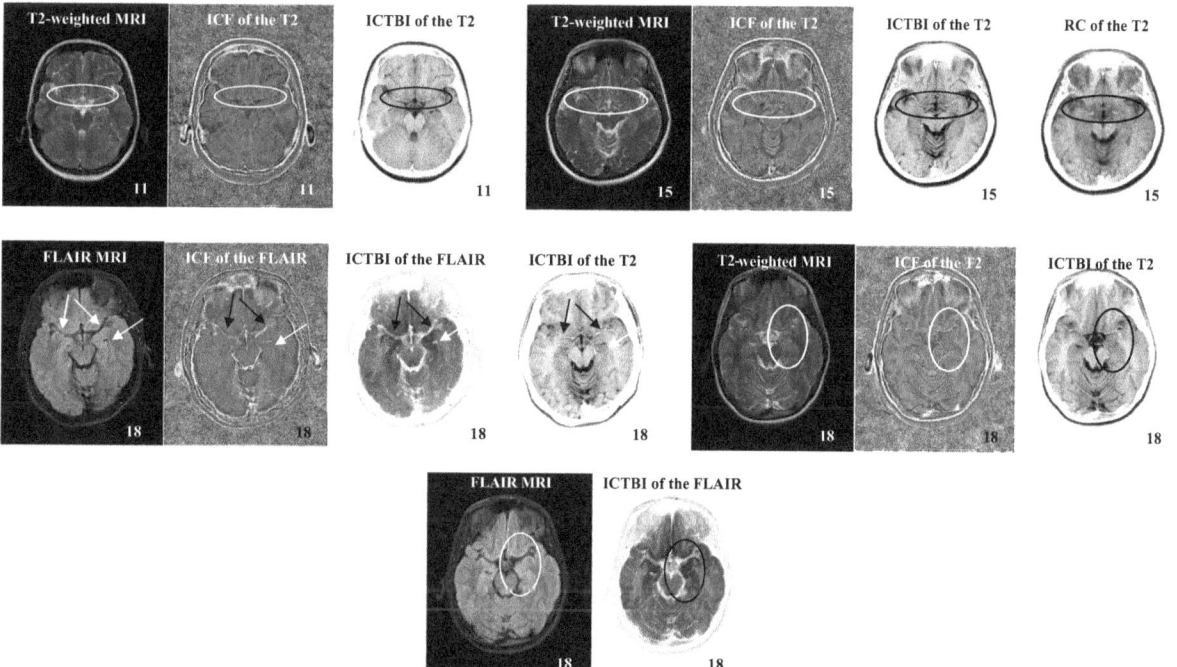

Fig. 10 Subject 3: the slice number is 11. T2-weighted MRI, and the ICF and ICTBI of the T2-weighted MRI are shown. The vessel in the T2-weighted MRI (see inside the ellipse) becomes more visible in the ICF and ICTBI images. Subject 4: the slice number is 15. T2-weighted MRI, and the ICF, ICTBI, and RC of the T2-weighted MRI are shown. The maximum visibility of the vessel object of study is obtained in the ICTBI image (see inside the ellipses). Subject 5: the slice number is 18. FLAIR image, the ICF and ICTBI of the FLAIR image, and the ICTBI of the T2-weighted MRI are shown. Both ICF and ICTBI show highlights on the presumed vessel structure (see arrows). However, as regards the highlight of the vessels, the ICTBI of the T2-weighted MRI is more effective than the FLAIR image (see the dark arrows in the ICTBI image). Subject 6: the slice number is 18. T2-weighted MRI, the ICF and ICTBI of the T2-weighted MRI, the FLAIR image, and the ICTBI of the FLAIR are shown. Inside the ellipses is visible a vessel structure which is highlighted in the ICF image of the T2-weighted MRI and more so in the ICTBI image of the T2-weighted MRI. The FLAIR and its ICTBI do not show the aforementioned vessel structure, thus the latter is a negative case

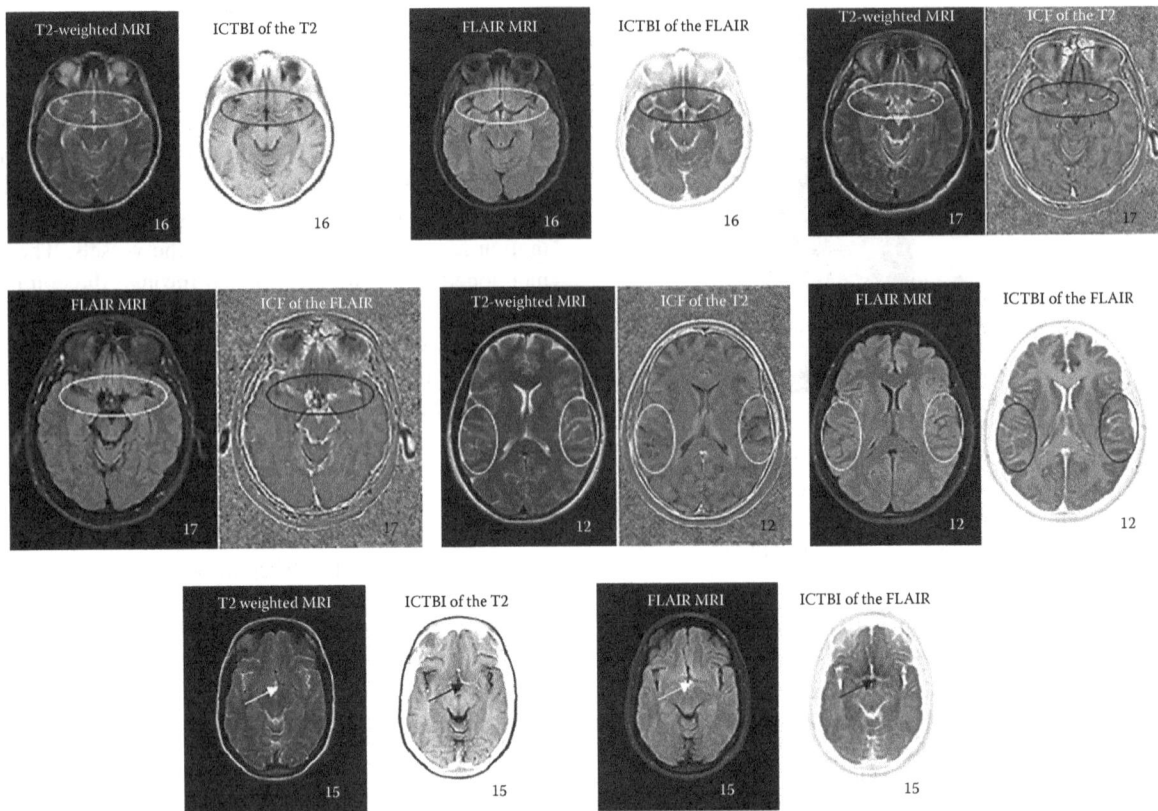

Fig. 11 Subject 7: the slice number is 16. T2-weighted MRI and the resultant ICTBI, and the FLAIR image and the resultant ICTBI are shown. The highlight of the vessel structure (see inside the ellipses) is more accentuated and detailed in the ICTBI of the T2-weighted image. Subject 8: the slice number is 17. T2-weighted MRI and the resultant ICF, and the FLAIR image and the resultant ICF are shown. The ICF images reproduce the vasculature (see inside the ellipses). The vessels in the ICF of the T2-weighted MRI are, however, more demarcated. Subject 9: the slice number is 12. T2-weighted MRI and the resultant ICF, and the FLAIR image and the resultant ICTBI are shown. The emphasis is on the presumed vessel-like structures (see inside the ellipses). The ICF shows a well-defined visibility of the vessel-like structures, whereas in the ICTBI of the FLAIR image, the structures are less visible. However, this case is positive both for the ICF of the T2-weighted image and the ICTBI of the FLAIR image. Subject 10: the slice number is 15. T2-weighted MRI and the resultant ICTBI, and the FLAIR image and the resultant ICTBI are shown. The emphasis is on the vessel-like structure located in the T2-weighted image and the demarcated highlight displayed by the resultant ICTBI. The same structure is less accentuated in the FLAIR image and also in the resultant ICTBI (hence this is a negative case), whereas the ICTBI of the T2-weighted image shows a positive case

smoothing and no significant loss of spatial resolution. The ICF offers better visibility of the vessels compared to the high-pass filtered MRI signal (see arrows in Figs. 12 and 13). The ICTBI of the T2-weighted MRI is featured by notable vessel visibility, whereas the ICTBI of the FLAIR and the inverted FLAIR is similar (see regions inside the ellipses in Figs. 12 and 13). The ICTAI and the RC are smoother than the T2-weighted MRI and the FLAIR MRI (see Fig. 13b). The RC images act as MIP of three consecutive slices.

DISCUSSION

Historical Perspective on Image Processing of MRI of the Human Brain

The notion of the common anatomical space of the human brain was introduced in 1988.[18,19] In 1991, the statistical parametric mapping (SPM) project was conceived and used to study positron emission tomography (PET) brain images in order to obtain estimates of regional cerebral blood flow (rCBF)[20] SPM is used by the scientific community in order to relate human brain function to human brain structure (among other uses). PET was the object of study when the SPM was invented. Nowadays, PET is a diagnostic modality functioning on physics principles different from those which MRI is based on. The endeavor of relating the structure to the function of the human brain had been addressed by the international consortium for brain mapping (ICBM), which was founded in 1992, through the development of the probabilistic atlas of the human brain.[21,22] The ICBM[23] was created in 1995 after the establishment of the Human Brain Project in 1992. The aim of the Human Brain Project was also the development of the probabilistic atlas.[24] In 1998, ascended the quest for the determination of the tissue type of every voxel of MRI data. Such quest can be satisfied with the

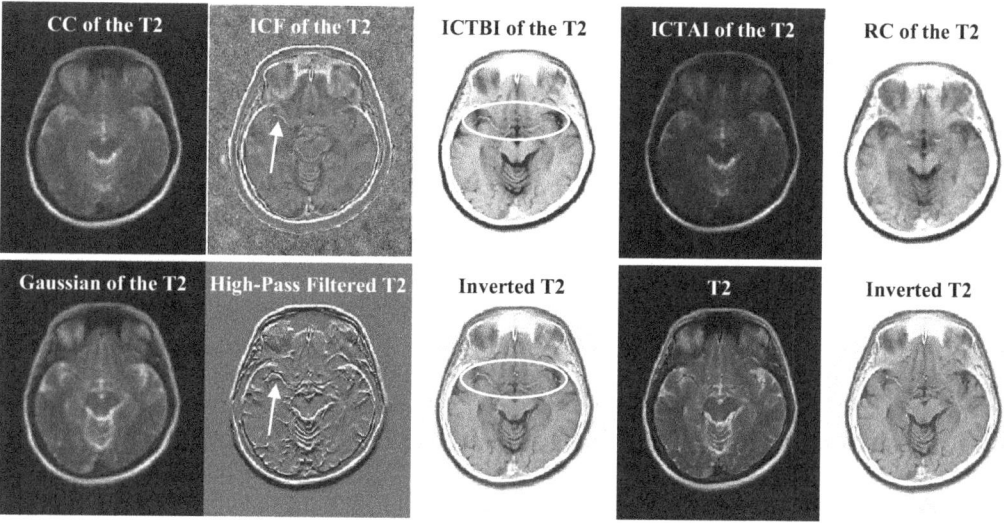

Fig. 12 T2-weighted MRI of subject 4: evidence of the difference between the ICMAs and other existing methods. The CC is compared to Gaussian filtering (3×3 mask). The ICF of the T2-weighted image is related to the high-pass filtered T2-weighted MRI image. The ICTBI and the RC are paralleled to the inverted T2-weighted MRI. And, the ICTAI is compared to the T2-weighted MRI

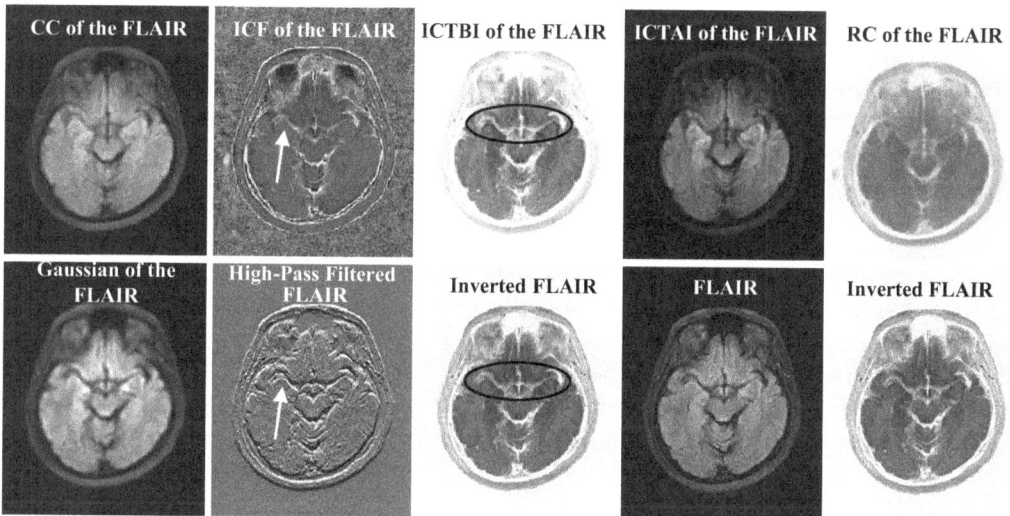

Fig. 13 FLAIR MRI of subject 4: evidence of the difference between the ICMAs and other existing methods. Likewise in Fig. 12, the comparisons are (i) the CC versus Gaussian filtering (3×3 mask), (ii) the ICF versus the high-pass filtered FLAIR MRI, (iii) the ICTBI of the FLAIR and the RC of the FLAIR versus the inverted FLAIR MRI, and (iv) the ICTAI of the FLAIR versus the FLAIR MRI

digital 3D brain phantom. Nonetheless, the digital 3D brain phantom can be useful to validate MRI image processing algorithms.[25] In 2001, the size of the probabilistic atlas was 7,000 subjects' brains, and the most relevant and important implications were to study the variance of the human brain as the function of age to validate MRI image processing algorithms and to study human brain diseases.[23] Indeed in the beginning of the new century, MRI[26,27] was recognized as the most powerful noninvasive in vivo imaging and diagnostic technology of the human brain.[28] Thus, the use of MRI brought to the attention of the scientific community a vast array of scientific challenges which were in need of solution. In addition to the challenges, MRI expanded and diversified into several allied imaging modalities. The scientific challenges were (i) image registration[29–31] and co-registration,[32] or more generally nonlinear spatial normalization.[33] Spatial normalization is the task of mapping an individual brain volume into a standardized space to allow congruency of analysis of images of the same modality collected at different times (registration) and of different imaging modalities (co-registration), and to allow quantitative analysis of human brain activations into a standardized domain.[34] (ii) Image segmentation which is the task of extraction of the human brain from the skull and the subsequent removal of non-brain structures.[35] (iii) Image smoothing, and more in general, any type of image processing step such as filtering.[36] As the list of scientific challenges is unveiled, it is due of very high consideration the massive effort of the scientific community in devising human brain image processing algorithms (for both qualitative and quantitative analyses of human brain images) and their implementation in software freely available to the scientific community. For an appraisal of human brain image processing algorithms, the reader is referred to the review provided in the work of Cui et al.[37] Bearing in mind that because of its variety, and depending on the scientific task at hand, it can be inferred that the list of image processing algorithm domains then continues with (iv) statistical image analysis,[38,39] among which some of the most popular techniques such as voxel-based morphometry (VBM)[40,41] and cortical surface-based analysis[42,43] have gained recognition and routine use in clinical applications.[44] VBM allows congruent statistical analysis of cerebral volume and tissue type concentration between different subjects' MRI scans. Cortical surface-based analysis renders the highly folded structure of the human cortex, made of gyri and sulci, into a parametric surface such as the sphere to allow congruent analysis among subjects' cortical surface, topography, and human brain activation patterns. Recall that voxel-by-voxel statistical image analysis was initiated in 1990,[45] 1991,[46] and 1998.[18] Statistical image analysis also encompasses principal component analysis (PCA), independent component analysis (ICA), and scale sub-profile model approach (SSM).[47] As regards the MRI allied techniques, a comprehensive list with references is given elsewhere[48] whereas in this entry, emphasis and attention is devoted to susceptibility-weighted imaging (SWI) which was initiated in 1997[49] and ascended in 2004, and is nowadays a promising and distinguished MRI imaging modality used to study the human brain vasculature.[2,50]

Study of Human Brain Vasculature

The discovery of deoxyhemoglobin as a natural tracer able to determine contrast in gradient echo MRI pulse sequences[49] and the recognition that phase images contain important information about the vasculature of the human brain[2] were the two milestones which determined a new course of research in MRI as regards the study of the human brain vasculature through the use of SWI.[2] SWI is based on gradient echo scans (T2* imaging) and uses two sources of data: (i) the SWI filtered phase image and (ii) the merged SWI magnitude image.[51] To obtain the SWI filtered phase image, the phase image, originally collected at the time of the MRI acquisition, is unwrapped and high-pass filtered.[2,51] The merged SWI magnitude image is obtained through masking the MRI magnitude image with the SWI filtered phase image.[2,51] It is noteworthy to report what is possible to achieve with the SWI filtered phase image: (i) to image the vasculature, (ii) to enhance white/gray matter contrast, (iii) to enhance water/fat contrast, and (iv) to identify iron[52] and other brain structures that are not background tissue.[2,53] In addition, the SWI filtered phase images allow the creation of susceptibility maps of the human brain which then yield venograms.[54] It is worth mentioning that using the merged SWI magnitude image, it is possible to create venograms determined through MIPs across slices.[2,51] Within the context of the study of the human brain vasculature through MRI image post-processing, the ICMAs are novel techniques. The ICMAs descend from the study of the intensity-curvature image content and were originally presented as viable two-dimensional image processing algorithms for the first time in 2014,[55] and they were dedicated in 2015 to the study of human brain tumors.[5] The study of the properties of the two-dimensional ICMAs releases the following characterization. While displaying the appearance of high-pass filtered signal, the CC and the ICF can be used as filter masks.[56,57] The SRI is a filter, and the RC is able (in one step) to invert, smooth, and magnify the grayscale of the image.[58] This entry presents the novelty of the study of the ICMAs obtained when processing the full MRI volume versus the previous studies[55–57] which had analyzed single slices of MRI volumes. And, this entry also reports the novelty that the ICMAs can be engaged in the study of human brain vessels.

CONCLUSION

The three-dimensional ICMAs are novel image processing techniques, the theoretical basis of which is the intensity-curvature concept. The concept is to merge the image intensity with the sum of second-order partial derivatives with respect to the dimensional variables of the polynomial model function fitted to the image data. The aforementioned sum is called CC.

The findings of this research are detailed here. The superiority of the ICTBI and the ICF of the T2-weighted MRI over the T2-weighted MRI in the task of highlighting human brain vessels have been discussed. The propensity of the ICTAI of the FLAIR imaging modality to highlight vessel structures has been explained. The aptitude of the ICTAI of the FLAIR to offer more structural information, as regards the visibility of the vessels, than the ICTAI of the T2-weighted MRI has been discussed. The behavior of the ICTAI of the T2-weighted MRI, which is inclined to offer a smoother version of the T2 image, has been described. The ability of the RC image to provide with a view similar to the MIP has been explained. In conclusion, merging the image intensity with the CC of the image allows to calculate images (the ICMAs) which can extract features from MRI. Future work should extend the investigation on the ICMA-based visualization of the human brain vessels to different MRI modalities.

ACKNOWLEDGMENTS

The author expresses sincere gratitude to Dr. Dimitar Veljanovski and Dr. Filip A. Risteski because of the availability of the MR images. Dr. Dimitar Veljanovski and Dr. Filip A. Risteski are affiliated with the Department of Radiology, General Hospital 8-mi Septemvri, Boulevard 8th September, Skopje, 1000, Republic of Macedonia. The author expresses sincere gratitude also to Professor Ustijana Rechkoska Shikoska for the help provided to coordinate the human resources. Professor Ustijana Rechkoska Shikoska is affiliated with the University of Information Science and Technology, Partizanska BB, Ohrid, 6000, Republic of Macedonia.

REFERENCES

1. Haacke, E.M.; Brown, R.F.; Thompson, M.; Venkatesan, R. *Magnetic Resonance Imaging: Physical Principles and Sequence Design*; John Wiley & Sons: New York, 1999.
2. Haacke, E.M.; Xu, Y.; Cheng, Y.-C.N.; Reichenbach, J.R. Susceptibility weighted imaging SWI. Magn. Reson. Med. **2004**, *52* (3), 612–618.
3. Frangi, A.F.; Niessen, W.J.; Vincken, K.L.; Viergever, M.A. Multiscale vessel enhancement filtering. In *International Conference on Medical Image Computing and Computer-Assisted Intervention*, Vol. 1496, Lecture Notes in Computer Science; Wells, W.M.; Colchester, A.; Delp, S.L.; Eds.; Springer Verlag: Berlin, 1998, 130–137.
4. Bauer, C.; Bischof, H. A novel approach for detection of tubular objects and its application to medical image analysis. In *Joint Pattern Recognition Symposium*; Springer-Verlag: Berlin, 2008, 163–172.
5. Ciulla, C.; Veljanovski, D.; Rechkoska, S.U.; Risteski F.A. Intensity-curvature measurement approaches for the diagnosis of magnetic resonance Imaging brain tumors. J. Adv. Res. **2015**, *6* (6), 1045–1069.
6. Lindeberg, T. Edge detection and ridge detection with automatic scale selection. Int. J. Comput. Vision **1998**, *30* (2), 77–116.
7. Descoteaux, M.; Collins, L.; Siddiqi, K. Geometric flows for segmenting vasculature in MRI: Theory and validation. In *International Conference on Medical Image Computing and Computer-Assisted Intervention*; Springer Verlag: Berlin and Heidelberg, 2004, 500–507.
8. Farid, H.; Simoncelli, E.P. Differentiation of discrete multidimensional signals. IEEE Trans. Image Process. **2004**, *13* (4), 496–508.
9. Lele, S.K. Compact difference schemes with spectral-like resolution. J. Comput. Phys. **1992**, *103*, 16–42.
10. Cha, Y.; Kim, S. The error-amended sharp edge (EASE) scheme for image zooming. IEEE Trans. Image Process. **2007**, *16* (6), 1496–1505.
11. Haralick, R. Edge and region analysis for digital image data. Comput. Graph. Image Process. **1980**, *12*, 60–73.
12. Haralick, R.M. Digital step edges from zero crossing of second directional derivatives. IEEE Trans. Pattern Anal. Mach. Intell. **1984**, *PAMI-6* (1), 58–68.
13. Haralick, R.; Watson, L. A facet model for image data. Comput. Graph. Image Process. **1981**, *15*, 113–129.
14. Prewitt, J. Object enhancement and extraction. In *Picture Processing and Psychopictorics*; Lipkin, B.; Rosenfeld, A.; Eds.; Academic: New York, 1970, 75–149.
15. Ciulla, C. *Signal Resilient to Interpolation: An Exploration on the Approximation Properties of the Mathematical Functions*; CreateSpace Publisher: North Charleston, SC, 2012. www.createspace.com.
16. Hajnal, J.V.; Bryant, D.J.; Kabuboski, L.; Pattany, P.M.; De Coene, B.; Lewis, P.D.; Pennock, J.M.; Oatridge, A.; Young, I.R.; Bydder G.M. Use of fluid attenuated inversion recovery FLAIR pulse sequences in MRI of the brain. J. Comput. Assisted Tomogr. **1992**, *16* (6), 841–844.
17. Haacke, E.M.; Tang, J.; Neelavalli, J.; Cheng, Y.C.N. Susceptibility mapping as a means to visualize veins and quantify oxygen saturation. J. Magn. Reson. Imaging **2010**, *32* (3), 663–676.
18. Fox, P.T.; Mintun, M.A.; Reiman, E.M.; Raichle, M.E. Enhanced detection of focal brain responses using intersubject averaging and change-distribution analysis of subtracted PET images. J. Cerebral Blood Flow Metab. **1998**, *80* (5), 642–653.
19. Talairach, J.; Tournoux, P. *Co-planar Stereotaxic Atlas of the Human Brain: 3-Dimensional Proportional System: An Approach to Cerebral Imaging*; Thieme Medical Publishers: New York, 1988.
20. Friston, K.J.; Frith, C.D.; Liddle, P.F.; Frackowiack, R.S.J. Comparing functional (PET) images: The assessment of significant changes. J. Cerebral Blood Flow Metab. **1991**, *110* (4), 690–699.
21. Mazziotta, J.; Toga, A.; Evans, A.; Fox, P.; Lancaster, J.; Zilles, K.; Woods, R.; Paus, T.; Simpson, G.; Pike, B.; Holmes, C.; Collins, L.; Thompson, P.; MacDonald, D.; Iacoboni, M.; Schormann, T.; Amunts, K.; Palomero-Gallagher, N.; Geyer, S.; Parson, L.; Narr, K.; Kabani, N.; Le Goualher, G.; Boomsma, D.; Cannon, T.; Kawashima, R.; Mazoyer, B. A probabilistic atlas and reference system for the human brain: International Consortium for Brain Mapping (ICBM). Philos. Trans. Royal Soc. B Biol. Sci. **2001**, *356* (1412), 1293–1322.
22. Thompson, P.M.; Toga, A.W. Detection, visualization and animation of abnormal anatomic structure with a deformable probabilistic brain atlas based on random vector field transformations. Med. Image Anal. **1997**, *1* (4), 271–294.
23. Mazziotta, J.C.; Toga, A.W.; Evans, A.; Fox, P.; Lancaster, J. A probabilistic atlas of the human brain: Theory and rationale for its development the international consortium for brain mapping (ICBM). Neuroimage **1995**, *2* (2PA), 89–101.
24. Huerta, M.; Koslow, S.; Leshner, A. The human brain project: An international resource. Trends Neurosci. **1993**, *16*, 436–438.

25. Collins, D.L; Zijdenbos, A.P.; Kollokian, V.; Sled, J.G.; Kabani, N.J.; Holmes, C.J.; Evans, A.C. Design and construction of a realistic digital brain phantom. IEEE Trans. Med. Imaging **1998**, *17* (3), 463–468.
26. Lauterbur, P.C. Image formation by induced local interactions: Examples of employing nuclear magnetic resonance. Nature **1973**, *242* (5394), 190–191.
27. Mansfield, P. *Proton Magnetic Resonance Relaxation in Solids by Transient Methods*; PhD thesis; Queen Mary College, University of London: London, 1962.
28. Duncan, J.S.; Ayache, N. Medical image analysis: Progress over two decades and the challenges ahead. IEEE Trans. Pattern Anal. Mach. Intell. **2000**, *22* (1), 85–106.
29. Collins, D.L.; Neelin, P.; Peters, T.M.; Evans, A.C. Automatic 3D intersubject registration of MR volumetric data in standardized Talairach space. J. Comput. Assisted Tomogr. **1994**, *18* (2), 192–205.
30. Woods, R.P.; Grafton, S.T.; Holmes, C.J.; Cherry, S.R.; Mazziotta, J.C. Automated image registration: I. General methods and intrasubject, intramodality validation. J. Comput. Assisted Tomogr. **1998**, *22* (1), 139–152.
31. Maintz, J.A.; Viergever, M.A. A survey of medical image registration. Med. Image Anal. **1998**, *2* (1), 1–36.
32. Woods, R.P.; Grafton, S.T.; Watson, J.D.; Sicotte, N.L.; Mazziotta, J.C. Automated image registration: II. Intersubject validation of linear and nonlinear models. J. Comput. Assisted Tomogr. **1998**, *22* (1), 153–165.
33. Ashburner, J.; Friston, K.J. Nonlinear spatial normalization using basis functions. Hum. Brain Mapp. **1999**, *7* (4), 254–266.
34. Fox, P.T. Spatial normalization origins: Objectives, applications, and alternatives. Hum. Brain Mapp. **1995**, *3*, 161–164.
35. Zhang, Y.; Brady, M.; Smith, S. Segmentation of brain MR images through a hidden Markov random field model and the expectation maximization algorithm. IEEE Trans. Med. Imaging **2001**, *20* (1), 45–57.
36. Gerig, G.; Kubler, O.; Kikinis, R.; Jolesz, F.A. Nonlinear anisotropic filtering of MRI data. IEEE Trans. Med. Imaging **1992**, *11* (2), 221–232.
37. Cui, Z.; Zhao, C.; Gong, G. Parallel workflow tools to facilitate human brain MRI post-processing. Front. Neurosci. **2015**, *9* (171), 1–7.
38. Ashburner, J. SPM: A history. Neuroimage **2012**, *62* (2), 791–800.
39. Friston, K.J.; Ashburner, J.; Frith, C.D.; Poline, J.B.; Heather, J.D.; Frackowiak, R.S.J. Spatial registration and normalization of images. Hum. Brain Mapp. **1995**, *3*, 165–189.
40. Martin, P.; Bender, B.; Focke, N.K. Post-processing of structural MRI for individualized diagnostics. Quant. Imaging Med. Surg. **2015**, *5*, 188–203.
41. Good, C.D.; Johnsrude, I.S.; Ashburner, J.; Henson, R.N.A.; Friston, K.J.; Frackowiak, R.S.J. A voxel-based morphometric study of ageing in 465 normal adult human brains. Neuroimage **2001**, *14*, 21–36.
42. Dale, A.M.; Fischl, B.; Sereno, M.I. Cortical surface-based analysis: I. Segmentation and surface reconstruction. Neuroimage **1999**, *9* (2), 179–194.
43. Fischl, B.; Sereno, M.I.; Dale, A.M. Cortical surface-based analysis: II. Inflation, flattening, and a surface-based coordinate system. Neuroimage **1999**, *9* (2), 195–207.
44. Mechelli, A.; Price, C.J.; Friston, K.J.; Ashburner, J. Voxel-based morphometry of the human brain: Methods and applications. Curr. Med. Imaging Rev. **2005**, *1* (2), 105–113.
45. Friston, K.J.; Frith, C.D.; Liddle, P.F.; Dolan, R.J.; Lammertsma, A.A.; Frackowiak, R.S. The relationship between global and local changes in PET scans. J. Cerebral Blood Flow Metabo. **1990**, *10*, 458–466.
46. Friston, K.J.; Frith, C.D.; Liddle, P.F.; Frackowiak, R.S.J. Comparing functional (PET) images: The assessment of significant change. J. Cerebral Blood Flow Metabo. **1991**, *11*, 690–699.
47. Petersson, K.M.; Nichols, T.E.; Poline, J.B.; Holmes, A.P. Statistical limitations in functional neuroimaging. I. Non-inferential methods and statistical models. Philos. Trans. Royal Soc. B Biol. Sci. **1999**, *354* (1387), 1239–1260.
48. Ciulla, C.; Shikoska, U.R.; Veljanovski, D.; Risteski, F.A. The intensity-curvature of human brain vessels detected with magnetic resonance imaging. Int. J. Appl. Sci. Biotechnol. **2017**, *5* (3), 326–335.
49. Reichenbach, J.R.; Venkatesan, R.; Schillinger, D.J.; Kido, D.K.; Haacke, E.M. Small vessels in the human brain: MR venography with deoxyhemoglobin as an intrinsic contrast agent. Radiology **1997**, *204* (1), 272–277.
50. Haacke, E.M.; Reichenbach, J.R. *Susceptibility Weighted Imaging in MRI: Basic Concepts and Clinical Applications*; John Wiley & Sons: New York, 2014.
51. Haacke, E.M.; Mittal, S.; Wu, Z.; Neelavalli, J.; Cheng, Y.C. Susceptibility-weighted imaging: Technical aspects and clinical applications, part 1. Am. J. Neuroradiol. **2009**, *30* (1), 19–30.
52. Haacke, E.M.; Ayaz, M.; Khan, A.; Manova, E.S.; Krishnamurthy, B.; Gollapalli, L.; Ciulla, C.; Kim, I.; Petersen, F.; Kirsch, W. Establishing a baseline phase behavior in magnetic resonance imaging to determine normal vs. abnormal iron content in the brain. J. Magn. Reson. Imaging **2007**, *26*, 256–264.
53. Liu, C.; Li, W.; Tong, K.A.; Yeom, K.W.; Kuzminski, S. Susceptibility-weighted imaging and quantitative susceptibility mapping in the brain. J. Magn. Reson. Imaging **2015**, *42* (1), 23–41.
54. Haacke, E.M.; Tang, J.; Neelavalli, J.; Cheng, Y.C.N. Susceptibility mapping as a means to visualize veins and quantify oxygen saturation. J. Magn. Reson. Imaging **2010**, *32* (3), 663–676.
55. Ciulla, C.; Rechkoska, S.U.; Bogatinoska, D.C.; Risteski F.A.; Veljanovski, D. Biomedical image processing of magnetic resonance imaging of the pathological human brain: An intensity-curvature based approach. In *AICT 2014 : The Tenth Advanced International Conference on Telecommunications, Ohrid, Macedonia, September 9–12, 2014*; Bogdanova, M.A.; Gjorgjevikj, D.; Eds.; ICT ACT: Ohrid, Republic of Macedonia, 2014, 56–65.
56. Ciulla, C.; Risteski, F.A.; Veljanovski, D.; Rechkoska, U.S.; Adomako, E.; Yahaya, F. A compilation on the contribution of the classic-curvature and the intensity-curvature

functional to the study of healthy and pathological MRI of the human brain. Int. J. Appl. Pattern Recognit. **2015**, *2* (3), 213–234.

57. Ciulla, C.; Veljanovski, D.; Risteski, F.A.; Rechkoska, S.U. On the filtering properties of classic-curvature and intensity-curvature functional images: Applications in magnetic resonance imaging. Int. J. Appl. Pattern Recognit. **2016**, *3* (1), 77–98.

58. Ciulla, C.; Veljanovski, D.; Rechkoska, S.U.; Risteski, F.A. On the properties of the intensity-curvature measurement approaches: The signal resilient to interpolation and the resilient curvature. Int. J. Innov. Comput. Appl. **2016**, *7* (2), 91–118.

Time and Motion

Tania Pouli
Technicolor, Rennes, France

Erik Reinhard
Technicolor Research and Innovation, Rennes, France

Douglas W. Cunningham
Brandenburg University of Technology, Cottbus, Germany

Abstract

This entry examines temporal properties of images with a focus on motion. For example, it covers, optical flow, gradient based motion detection, motion blur, and statistical based object detection. It also introduces scenarios where these techniques can be applied in visual computing.

Keywords: Motion; Optical flow; Time series.

The world is ever changing. Heraclitus of Ephesus stated that "Everything flows, nothing stands still."[a] He is also the source of the more well-known saying, "You could not step twice into the same river; for other waters are ever flowing on to you."[b] This constant flux and change of the world is an important source of information.

Despite the impressive range of visual abilities in different animals, there are no known motion-blind species.[1] It seems that, biologically, motion perception is one of the most basic visual abilities. This can be especially seen in those species with limited spatial vision: they use motion to compensate for the lack of retinal receptors. The *Copilia quadrata*, for instance, moves its spotlight retina in a particular pattern, giving it one-dimensional vision. Similarly, the carnivorous sea snail and the jumping spider move their eyes so that the one-dimensional strip of visual receptive cells can sense two spatial dimensions.

Since many of the changes that occur in the world are not random, temporal flux is a potential source of information not just for organic visual systems, but also both the synthesis and computational analysis of image sequences. This entry examines some temporal properties with a focus on motion. It also introduces some scenarios where these findings have been applied in visual computing.

THE STATISTICS OF TIME

All of the statistics in this book so far have been applied to one or more spatial locations at a given point in time. Other than technical reasons, there is nothing preventing us from performing the temporal equivalent: comparing different points in time for a given spatial location. Although such purely temporal statistics are certainly possible, it is much more common to examine both space and time simultaneously.

We could compare just two points in time. Rather than subtracting two spatially neighboring pixels to get a spatial gradient it is possible to subtract the intensity value at one pixel for two temporally neighboring images in an image sequence. This would give a measurement of the temporal contrast modulation. Likewise, by varying the separation (either in space or time) of the pairs of pixels, we could examine the statistical regularities at different frequencies (spatial or temporal, respectively).

In perhaps the first spatiotemporal analysis of image sequences, a series of works examined the spatiotemporal properties of natural image sequences that were not photographs. Specifically, they examined multispectral satellite image sequences (LANDSAT multispectral scanner sequences) and were able to detect such regularities as seasonal changes.[2–5]

For video sequences of terrestrial-based natural scenes, the first spatiotemporal analysis seems to be that of Watson and Ahumada,[6] who suggested that time be considered like space, and thus image sequences should be represented as a spatiotemporal volume $I(x, y, t)$, where $I(x, y)$ is a given image and t is the time at which that image was taken. They then proposed that the frequency transform of these spatiotemporal volumes be examined.

Starting with an analysis from first principles, they showed that different forms of motion within an image sequence should trace very specific paths within the spatiotemporal volume. For example, a vertical line moving

---
[a] As quoted by Plato [379], p. 344.
[b] As quoted by Plato [379], pp. 344–345.

horizontally (a common stimulus in psychophysics) should trace a diagonal path through the spatiotemporal volume. Interestingly, in the frequency domain, the path lies along a straight line in f_s, f_t (where f_s is the frequency domain transform of x and y and f_t is the frequency domain transform of t). The slope of this line is then $-1/r$ where r is the horizontal speed. Thus, higher spatial frequencies will have higher temporal frequencies, and higher velocities will have shallower slopes. Moreover, leftwards motion will show up in the odd quadrants while rightwards motion will show up in the even quadrants in Fourier space. Physical analysis of real images containing a vertical line moving horizontally confirmed the theoretical analysis. Moreover, the critical sampling frequency for which discretely presented motion should be seen as continuous was shown to be a linear function of velocity.

Finally, Watson and Ahumada also demonstrated that the spatiotemporal volume can be used to create motion detectors.[6] The idea is to use pairs of simple cells with one of the cell's responses being adjusted by a hyperbolic filter. The response is modified once in the spatial domain and once in the temporal domain. The signals of the two cells are then added. Note that the sensor only detects direction of motion, not speed. The speed is obtained by examining the temporal frequency response of the sensors. Eckert and colleagues examined the spatiotemporal power spectra of 14 image sequences (256 × 256 pixels × 64 frames at 30 frames per second, with no scene cuts), and found that all spectra were separable (with an index of separability of 0.98).[7] The power spectra fit the form:

$$P(k,f) = \frac{(ab)/(4\pi^3)}{\left((a/2\pi)^2 + k^2\right)^{3/2}\left((b/2\pi)^2 + f^2\right)} \quad (1)$$

with k representing the radial spatial frequency, f representing temporal frequency, and a and b representing model parameters that define the spatiotemporal bandwidth of the signal.

Similarly, Adelson and Bergen[8] suggested that the statistical regularities of spatiotemporal sequences can be examined with any traditional volume analysis. They also created a motion detector (the energy detector) that is formally identical to the elaborate Reichardt Detector. Many others have since used spatiotemporal volumes to analyze or manipulate image sequences, examining slices, tunnels, and other features in the volumes (see, e.g., Refs. [9–12]).

Dong and Atick examined the spatiotemporal properties of a set of 1,049 commercial movie clips (each clip was 64 × 64 pixels large and 64 frames long, with a temporal resolution of 24 frames per second) and 320 clips from homemade movies (each at 64 × 64 × 256 with a temporal resolution of 60 frames per second[c][13]).

---

[c] They also tried to avoid scene cuts in the clips, as these do not represent natural motion.

Their basic measure was the spatiotemporal correlation $R(x, y, t)$ between two pixels. In a first analysis, the correlation between two pixels at the same point in space but at two different points in time (a separation of 33 ms) was examined. Notice that this correlation is related to a pure temporal gradient. They found that the value of the pixel at the two times was highly correlated ($R = 0.9$). In other words, whatever the intensity was at that pixel at time 1, there was a high probability it was the same 33 ms later. The greater the difference in intensity between two frames was, the lower the probability for that change to happen. In fact, a visual inspection of the probability distribution suggests that it is strikingly similar to the spatial gradient distribution, once again revealing a high kurtosis.

Dong and Atick then examined the two-point correlation matrix for many spatial and temporal separations, as well as the results in the frequency domain.[13] They found that for any given temporal frequency, the spatial frequency power spectrum follows the usual power law:

$$1/f_s^a \quad (2)$$

where f_s is the spatial frequency. Interestingly, the value of the slope a varied as a function of temporal frequency. At low temporal frequencies they found that the exponent a tends to 2 while at higher temporal frequencies a approaches 1.

The temporal frequencies showed the same behavior: for a given spatial frequency, the temporal frequency power spectrum also follows a power law:

$$1/f_t^b \quad (3)$$

where f_t is the temporal frequency. Interestingly, it was found that the exponent b varies as function of spatial frequency. The slope b tends to 2 for low spatial-frequencies and b approaches 1 for higher spatial frequencies. The temporal slope can be thought of as being related to the persistance or purposefulness of motion.[14] A slope of 0.5 represents pure Brownian motion: the changes from frame 1 to frame 2 are completely uncorrelated with the changes from frame 2 to frame 3. Slopes greater than 0.5 represent persistent motion: any change that occurred from frame 1 to frame 2 is likely to occur again from frame 2 to frame 3. Exponents smaller than 0.5 are anti-persistent (the opposite motion is likely to occur).

Critically, if the correlation is multiplied by a power of f_s (such as f_s^{m+1}) and plotted as a function of the ratio of temporal to spatial frequency f_t/f_s, all the curves line up.[13] Otherwise stated:

$$R(f_s, f_t) \approx \frac{1}{f_s^{(m+1)}} F(f_t/f_s) \quad (4)$$

where m is a constant and $F(f_t/f_s)$ is a function of the ratio of spatial and temporal frequencies. In both the theoretical analysis and the experimental estimates (using real image sequences), the value of m is around 2.3, which is consistent with the slope of the power spectra for static images. Theoretically this function, including the value of m, can be derived by assuming a specific distribution of relative velocities from objects at many depths.[13] This result can be linked to human vision, in that the receptive fields of neurons are often coupled in space and time, thereby reflecting the real-world spatiotemporal statistics. As such, receptive fields should not be characterized by their spatial and temporal properties separately but by the ratio of temporal to spatial frequency f_t/f_s.[15]

The properties of natural image sequences have been compared to the visual systems of both humans as well as flies. In one study, the spatial properties of 117 photographs, using patches of 128×128 pixels, were analyzed, revealing the typical $1/f_s^b$ power spectrum, with $b = 2.13$.[16] Note that if an observer were to move in a straight line through a static environment that was uniformly cluttered with objects, the predicted power spectrum would be $1/f$ for a range of observer speeds,[16] which is at the lower end of the range measured by Dong and Atick.[13] The spatiotemporal statistics were used to derive the receptive field structure of cells early in the fly's visual system.

Alternatively, it is possible to capture digital recordings of what a single retinal cell would see by having people wear a specially designed photosensor-based recording system on their head as they walk around.[17,18] Signals captured in this manner were also found to conform to the $1/f$ property of the temporal power spectrum, confirming the predicted value for observers moving through a static environment. In addition, by processing the captured signal, the responses of various stages of the visual system were simulated. Figure 1 shows the differences between direct measurements of this time sequence and (simulated) photoreceptor responses for the same data in terms of intensity distribution and power spectrum.[18]

Independent component analysis can also be performed on image sequences.[19] A database of 216 clips taken from television signals, each 128×128 pixels large and 4,800 frames long (192 seconds at 25 frames per second) was used for this purpose. The first 288 independent components (ICs) for a set of patches, each $12 \times 12 \times 12$ large, were calculated. The ICs resemble edges (or bars) moving at a constant velocity, with the direction being perpendicular to the orientation of the bar. Interestingly, this is very similar to the spatiotemporal receptive fields of LGN neurons.[20] The related independent component filters (ICFs; which can be used to filter an image sequence) also resemble other aspects of the human visual cortex (especially in their spatial properties). Interestingly, they found a significant correlation between spatial frequency and velocity, with low spatial frequency ICFs being tuned to faster movement than higher spatial-frequency ICFs and fast-moving ICFs mainly being encoded at low spatial frequencies.

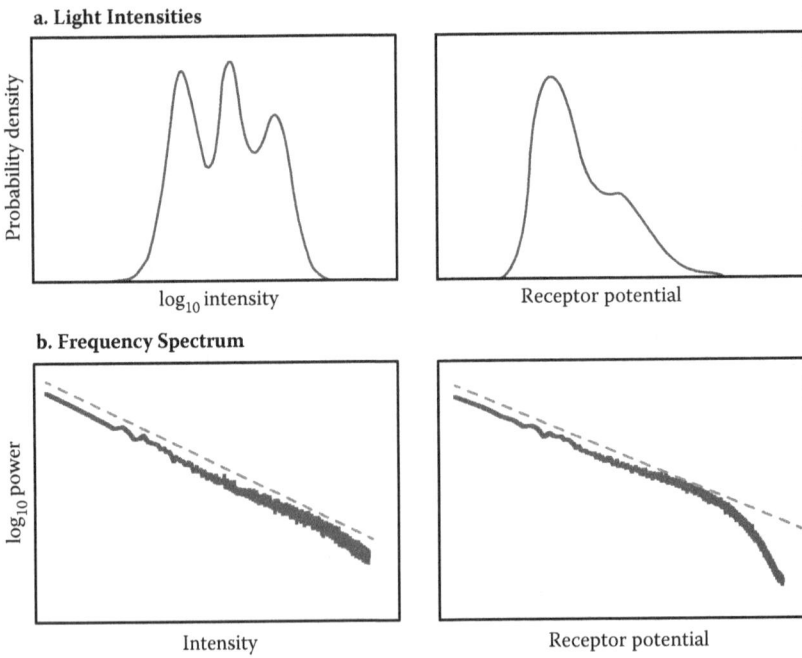

Fig. 1 (a) Probability distributions of the light intensity (left) and the corresponding photoreceptor responses for a time sequence simulating the input to a single photoreceptor. (b) Average power spectrum for consecutive sections from the same time sequence, following approximately a $1/f^t$ behavior, with $t = -1$. In the photoreceptor-processed spectrum, higher frequencies are filtered by the low-pass behavior of the photoreceptors and as such deviate from $1/f$.[18]

MOTION

One of the strongest statistical regularities that arise when we look at spatiotemporal changes is that of motion. Motion has been the subject of intense interdisciplinary study for at least the last 150 years. A large number of theories exist about what kinds of motion there are, what causes motion, what properties motion has, and how motion can be processed. For a recent overview on the perception of motion and some of the related theoretical and philosophical issues, the reader is directed to.[21–24]

Before we can examine the statistical regularities of motion, we first need to extract it. That is, we need some form of spatiotemporal transform that takes image sequences and extractes or emphasizes motion information. We have already mentioned the one from Watson and Ahumada[6] as well as the one from Adelson and Bergen.[8] The very first motion detector, though, was developed a century earlier by Sigmund Exner.[25] Although Exner's model is no longer used, it provides an excellent (and remarkably accurate) starting point for subsequent work on motion detection. Here, we present the basics of the two most common forms of motion detection. Nearly all analyses of the statistical regularities of motion start with one of these. We then examine a number of motion regularities.

Correlation-Based Motion Detection

In the 1950s, Werner Reichardt examined the behavior and underlying neurophysiology of the beetle and the fly, with a focus on motion perception.[26] He suggested that a good motion detector should see a given object as it moves through two different points in space at two different times. Thus, it needs two temporal gradient sensors separated by a specific distance in space and one of them should respond with a specific temporal delay. The two signals are multiplied together to obtained the motion signal. Since both receptors must see the *same* change in the environment, this form of motion detector is called a *correlator type*.

The basic version of the Reichardt motion detector allows detection of motion at a specific speed in a specific direction.[26] If an object is moving in the right direction at the right speed, then the change in illumination caused by the object passing over the first receptor cell will hit the multiplication point just after the object passes the second receptor, causing the motion detector to emit a signal. By adding a second set of pathways connecting the same two sensors to a second multiplier, and then subtracting the signals from the two multipliers, we can construct an elaborate Reichardt detector, which is capable of detecting motion at a given speed in the two opposed directions. Technically, the Reichardt detector measures temporal frequency,[22] but Dror has shown that due to the statistics of natural images this is a very reliable and consistent estimation of velocity.[27] It has been shown that humans use correlator-type detectors[28,29] (so do flies[30]), that the tuning is done by changing spatial preferences,[31,32] and that the detector can explain many low-level visual illusions.[33,34] There exist several computational implementations of the elaborate Reichardt detector (e.g., Ref. [35]).

Gradient-Based Motion Detection

In examining the encoding of television signals, Lamb and colleagues[36] noted that the correlation approach requires many multiplications, and that any increase in the number of velocities that should be detected required the addition of a new multiplication unit.[d] To create a more efficient encoding of television signals, Lamb and colleagues developed a model for measuring velocity directly from spatial and temporal gradients. First, the temporal derivative of local luminance ($\partial I(x,t)/\partial t$) and the spatial derivative of local luminance ($\partial I(x,t)/\partial x$) are calculated. Then the two signals are divided. Gradient-based detectors have the advantage that they are not dependent on pattern properties. The disadvantages include the fact that for low spatial derivatives, any noise in temporal derivatives is amplified.

APPLICATIONS USING STATISTICAL MOTION REGULARITIES

Once motion has been extracted, a number of statistical regularities are noticeable. The perception literature is full of many cases of motion-based perceptual processes and motion illusions, such as structure-from-motion, color-from-motion, depth-from-motion (e.g., motion parallax), heading and spatial orientation from optical flow (see "Statistics of Real-World Motion and Retinal Flow" section), and so on. For an overview of the perception of motion and some of the many motion-based statistical regularities that people can sense.[24]

There are also a number of applications that take advantage of spatiotemporal regularities. These applications use a wide variety of approaches from simple gradients up to scale space and Markov random fields (for example, noise reduction,[37,38] feature tracking,[39] camera motion estimation,[40] video retrieval,[41] motion interpretation,[42–44] and object segmentation[45]). Several others are discussed in this section.

Specular Highlights

It has been shown that the pattern of change over time can be used to extract specular highlights. Using knowledge of how caustics change in real-world scenes and some simple geometry, specular highlights can be removed from

---

[d]It has subsequently been shown that the multiplication phase can be implemented using a combination of linear and nonlinear filters and does not explicitly require multiplication.[125]

video sequences.[46] Similarly, a "temporal color space" can be defined, where each pixel in an image sequence is represented in a four-dimensional space, with three color dimensions (R, G, and B) and one temporal dimension.[47] Given knowledge of how specular reflections change as a function of time, and that the specular highlights generally have the same spectral distribution as the illuminant, specular highlights can then be removed and the color of the illuminant can be estimated in an image sequence.

Markov Random Fields

Many researchers use Markov random fields (MRFs) with temporal data. For example, it can be shown that with three assumptions, namely that all luminance changes are due to translational motion, that Gaussian noise corrupts the observation, and that slower motions are more likely to occur than faster ones, the trained MRF model interprets several motion illusions in a manner similar to humans.[48]

MRFs can be used to track people in videos using three underlying models (one each for edges, ridges, and optical flow).[49] After training the models on 150 images and short sequences, results were shown to be better when all three models were used than for any of the three models individually.

Black and colleagues used four generative models within an MRF context to detect motion in general.[50,51] The four models reflect changes in form (i.e., including translational and rotational motion), changes in illumination, changes in specular reflections, and a catch-all for all other changes (which they call iconic changes). Note that the iconic changes are domain-dependent and thus require a separate model for each set of domain-specific change (e.g., a model for detecting eye blinks needs to be represented separately from a model for mouth motion). They tested the models on complex, real-world scenes and showed that once again it works best when all four models are used simultaneously.

People Detection and Biological Motion

People are remarkably sensitive to the motion of other people. In fact, Johansson showed that by placing dots on the joints of people and showing just the motion of those dots (usually a set of about 15 points) is sufficient for people to recognize how many people are present in the video and what they are doing.[52,53] It has since been shown that these motion signals are also sufficient to tell many other things, including what gender a walking person is (for an overview, please see Ref. [24]). This field has come to be known as biological motion. A number of models of how people process biological motion signals have been proposed (see, e.g., Refs. [54–57]).

In an interesting variation, it was suggested that crowds of people can be tracked by following corners.[58] In essence, once corners are detected, it is assumed that only the corners on people move (background corners do not move) and that each person has on average the same number of corners. Those assumptions allow humans to detect how many people are present. A measurement of the average speed for walking in a sequence allows the detection of running later on.

Motion Blur

All of the photons that hit a given cell within a specific temporal window are integrated and are perceptually treated as if they all arrived at exactly the same instant in time.[23,34] The temporal integration window is believed to be about 80 ms in bright conditions and around 150 ms when we are in the dark and our eyes are fully dark-adapted. When an object is stationary, all photons from a specific location on the object hit the same receptor cell for the duration of the temporal integration window. The result is that the retinal cell receives enough photons to respond and the retinotopic location of the spot on the object is unambiguous.

If the object is moving fast, however, light from a given spot on its surface will sequentially hit a number of neighboring retinal cells during the temporal integration window. This means that this spot will be seen by the retina as existing at one instant in time at multiple locations. Thus, directly after processing by the retinal receptors, the object will seem to be longer in the direction of motion than it should be. Moreover, the visual system will not be able to resolve the higher spatial frequencies, since they are spread out. Cameras exhibit similar phenomena relating to blur since the shutter speed of modern cameras is not zero.

The human visual system, however, is built such that it can at least partially compensate for this motion-based spreading. Once the visual system detects that the object is moving, it automatically performs some motion deblurring.[23] Several computational approaches to removing motion blur from images have been developed (see Section 5.6). Motion blur is also routinely applied to computer-generated sequences, often with the aid of high dynamic range imaging.[59]

Shape from Dynamic Occlusion

As discussed by Reichardt, a moving object disappears from one location and appears at another location. Obviously, when the object disappears from a location that location is not empty. One of two things can happen: either another part of the surface of the moving object shows up (which consequently correlates with a disappearance elsewhere) or some background texture suddenly appears.

Likewise, at the leading edge of the figure, background texture will suddenly disappear. The specific pattern of background texture appearances and disappearances at the edges of a moving object is referred to as *dynamic occlusion*. A moment of thought will show that the appearance and disappearance of the background will not have a match elsewhere and thus will not be picked up as a motion signal.

These temporal contrast modulations that do not come from motion provide a wealth of information. Obviously, they indicate that there is or was an occluding object at that location.[60,61] It also provides reliable information that disappearing background still exists.[62] Finally, the collection of dynamic occlusion signals in an image sequence specify the shape, velocity, and degree of transparency of the moving object.[62–82] For a review, see Ref. [83]. The visual process responsible for seeing a form from dynamic occlusion changes is referred to as Spatiotemporal Boundary Formation (SBF). The class of changes that can result in the perception of a moving bounded form extends beyond the simple accretion and deletion of texture elements; the class includes changes in an element's color, orientation, or location.[80]

Shipley and Kellman[81] provided a mathematical proof showing that the orientation of a small portion of the moving object's contour (a *local edge segment* or LES) may be recovered from changes of three texture elements, as long as the elements are not spatially collinear and all three changes occur within the temporal integration window. If the velocity of the LES is known, only two element changes are needed. Cunningham and colleagues[70,84] modified and extended the proof to show how the global shape and the global velocity can also be recovered from six element transformations. They then tested the model on synthetic and real-world images, with decent results. Most of the failures of the algorithm on real-world images seems to lie in the implementation of the Reichard detectors, which are needed to separate the motion-based and non-motion-based temporal gradients.

OPTICAL FLOW

Perhaps the most well-known and widely used motion-based technique in visual computing is optical flow. Optical flow has its origins in James J. Gibson's work describing the information that a moving animal receives.[85] Optical flow is in principle described by the temporal gradient field for an image sequence. Essentially, the change in location of each point on each object in a scene is measured. In practice, this can be performed on an image directly, rather than on the real-world scene. If both the observer and the real-world scene are perfectly stationary, then there will be no difference between the locations of any objects (or their two-dimensional image projections) at any two points in time and the vector field will be zero everywhere.

If the observer is moving forward (while looking forward), every point in the image will move towards the borders of the image (see the left panel of Fig. 2). This yields a radially expanding vector field; the center of the field is the point towards which we are traveling.[86] If the observer is looking forward but traveling to the left, all points in the image will move to the right of the image, creating a vector field with a large global translation component. In principle, optical flow can be used to determine the direction and speed of motion of a moving observer, since the same pattern of optical flow is obtained as when the scene is moving and the observer is static. All that is necessary is relative motion between the observer and the scene. Although humans can use optical flow for determining direction of travel, they often do not seem to use it in real-world conditions.[87]

Optical flow is also useful in object perception. If an otherwise static scene has a single moving object in it, this will show up as a local patch of high vector values in an empty optical flow vector field. Likewise, if an object within a scene has a different motion than its surrounding scene, this will also show up as a characteristic pattern within the optical flow field[85] (see Ref. [88] for an example of how this can be used in a computational framework). Of particular interest is the fact that at the edge of a moving object there will be discontinuities in the flow field. Humans can use these discontinuities to detect the shape, location, relative depth, global motion, and transparency of

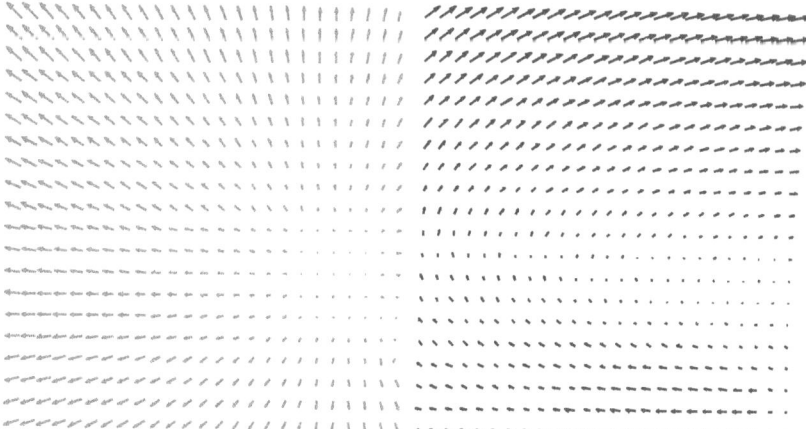

Fig. 2 Two examples of flow fields. The field on the left shows a radially expanding flow, which is likely to be due to forward motion within the scene. The movement within the right field is most likely due to camera rotation

the moving object (as discussed in "Shape from Dynamic Occlusion" section).

Many optical flow algorithms are gradient-based. Given that an image sequence is represented as a spatiotemporal volume $I(x, y, t)$, where (x, y) are spatial indices and t is the temporal index (see "The Statistics of Time" section), it is assumed that the total derivative of the image intensities in both space and time is zero at all times, i.e.:

$$I_x(x,y,t)v_x + I_y(x,y,t)v_y + I_t(x,y,t) = 0 \quad (5)$$

Here, the subscripts indicate partial derivatives and $\mathbf{v} = (v_x, v_y)$ is the optical flow, which is always in the image plane and normal to the spatial image orientation.[89] The implicit assumption made here is that any changes in intensity are due to translation and not to changes in illumination and reflectance. This equation would allow one to estimate an optical flow field $\mathbf{v}$, for instance by defining an error term:

$$E(\mathbf{v}) = \left(I_x(x,y,t)v_x + I_y(x,y,t)v_y + I_t(x,y,t)\right)^2 \quad (6)$$

It is then possible to compute a linear least-squares estimate of $\mathbf{v}$ by setting the gradient of this error term to zero:[89]

$$\nabla E(\mathbf{v}) = \mathbf{A}\mathbf{v} + \mathbf{b} = 0 \quad (7)$$

In this formulation, the matrix $\mathbf{A}$ is defined as:

$$\mathbf{A} = \begin{bmatrix} I_x^2 & I_xI_y \\ I_xI_y & I_y^2 \end{bmatrix} \quad (8)$$

and vector $\mathbf{b}$ is given by:

$$\mathbf{b} = \begin{bmatrix} I_xI_t \\ I_yI_t \end{bmatrix} \quad (9)$$

This would give the following solution:

$$\mathbf{v} = -\mathbf{A}^{-1}\mathbf{b} \quad (10)$$

However, computing the inverse of $\mathbf{A}$ is problematic, as this matrix is singular and therefore has a determinant that is always zero. To allow optimization using this function, additional constraints can be added to the error term. The problem can for instance be regularized by adding a global smoothness constraint, as suggested by Horn and Schunck.[90]

An alternative solution is to consider a neighborhood of pixels and select the velocity that is most consistent with the translations observed for all pixels within the patch. Given a patch with n pixels, the error term then becomes:[91]

$$E(\mathbf{v}) = \sum_{k=1}^{n} w_i \left(I_x(x_k,y_k,t)v_x + I_y(x_k,y_k,t)v_y + I_t(x_k,y_k,t)\right)^2 \quad (11)$$

Following the same procedure as above, the error term can be written as:

$$\nabla E(\mathbf{v}) = \sum_{i=1}^{n} w_i \left(\mathbf{A}_i\mathbf{v} + \mathbf{b}_i\right) \quad (12)$$

The solution to this problem is then:

$$\mathbf{v} = -\left(\sum_{i=1}^{n} w_i \mathbf{A}_i\right)^{-1} \sum_{i=1}^{n} w_i \mathbf{b}_i \quad (13)$$

Note that there is still a chance that the sum of matrices $\mathbf{A}_i$ is singular. In the following, we discuss a probabilistic method that would reduce the chances of creating a singular matrix.

Probabilistic Optical Flow

There are many sources of variation that cause spatiotemporal gradients in images. These include noise, quantization, lack of precision of the gradient operators (see Section 5.2 for a discussion), illumination variations, the possibility of multiple motions within a region, and areas of low contrast, as well as the aperture problem. As mentioned earlier, the latter is an issue that relates to both gradient-based algorithms as well as motion-sensitive neurons in the human visual system.[92] A single neuron receives input over a spatially localized receptive field (an aperture). Within this aperture, the direction of the motion is necessarily ambiguous (see "Shape from Dynamic Occlusion" section for more on this topic). There will be many patterns, directions, and speeds of movement that will elicit the same response within the neuron.

Despite all these sources of spatiotemporal variation, optical flow algorithms are intended to detect motion. Gradient-based algorithms will suffer from low accuracy whenever gradients are due to any of the other sources. To account for this, optical flow detection can be restated in terms of a probabilistic framework.[89] In this case, the probability of a vector field $\mathbf{v}$ given an observed image gradient $\nabla I = (I_x, I_y, I_t)$ can be constructed:

$$P(\mathbf{v} | \nabla I) \quad (14)$$

The different types of uncertainty can be accounted for by the introduction of Gaussian noise terms bfn_1 and n_2, accounting effectively for spatial and temporal sources of uncertainty. Including these noise terms in to Eq. 5, and denoting (I_x, I_y) as $\mathbf{I}_s$, we can write:

$$\mathbf{I}_s(\mathbf{v} - \mathbf{n}_1) + I_t = n_2 \quad (15)$$

This describes the conditional probability $P(I_t|\mathbf{v},\mathbf{I}_s)$. Using Bayes' rule (see the Appendix), we obtain:

$$P(\mathbf{v}|\mathbf{I}_s,I_t) = \frac{P(I_t|\mathbf{v},\mathbf{I}_s)P(\mathbf{v})}{P(I_t)} \quad (16)$$

Simoncelli and colleagues choose a zero-mean Gaussian as the prior distribution of $P(I_t)$[89] with covariance Λ_p. If the variance of $\mathbf{n}_1$ is set to (σ_1,σ_1) and the variance of n_2 is σ_2, then the covariance and mean of this distribution are given by:

$$\Lambda_v = \left(\frac{\mathbf{A}}{\sigma_1\|\mathbf{I}_s\|^2 + \sigma_2} + \Lambda_p^{-1}\right)^{-1} \quad (17)$$

$$\mu_v = -\frac{\Lambda_v \mathbf{b}}{\sigma_1\|\mathbf{I}_s\|^2 + \sigma_2} \quad (18)$$

Since this distribution is Gaussian, the mean μ_v is also the maximum *a posteriori* (MAP) estimate. This approach ensures that the matrix inversion is possible. This statistical approach can be extended to image patches, in which case the covariance and mean become:

$$\Lambda_v = \left(\sum_{i=1}^{n}\frac{w_i\mathbf{A}}{\sigma\|\mathbf{I}_s\|^2 + \sigma_2} \Lambda_p^{-1}\right)^{-1} \quad (19)$$

$$\mu_v = -\Lambda_v \sum_{i=1}^{n}\frac{w_i\mathbf{b}}{\sigma_1\|\mathbf{I}_s\|^2 + \sigma_2} \quad (20)$$

Finally, note that the division by $\sigma_1\|\mathbf{I}_s\|^2 + \sigma_2$ acts as a nonlinear gain control, which has been shown to improve performance.[89] A similar normalization was discussed in the context of the reduction of dependencies between wavelet coefficients in Section 8.9.

Statistics of Real-World Motion and Retinal Flow

Optical flow in image sequences, as well as optical flow of retinal projections, can arise due to self-motion of the observer/camera or due to objects moving within the environment. Optical flow due to self-motion generates much information about the direction of movement and the structure of the environment, as well as the presence of possible objects and obstacles.[85,93] The speed of the flow is also important. As an example, consider moving directly forward. In this case, the optical flow of the projected environment on the retina (or the camera sensor) will radiate outward from the point toward which the observer is traveling. If any object within this expanding field generates a stationary flow field, then that object is on a collision course with the observer/camera. A car coming out of a side street at a speed that would lead to a collision, for instance, would remain stationary on the retina.

It appears that such information is used by animals.[94] Further, there are specialized regions in the brain that process optical flow. These regions take their input from simple motion detectors earlier in the human visual system. It can be hypothesized that statistical regularities may play a role in the analysis taking place in these brain regions.[95,96] This leads to two ways in which the statistics of optical flow can be assessed. The first is to directly study motion signals,[95–98] whereas the second is to pass motion signals through basic motion detectors first and then study the emerging statistical regularities.[99–101]

To study motion signals directly, it is possible to derive flow fields from range databases,[95,96,98] such as the Brown Range Image Database (see Section 3.3.13).[102,103] In this manner average velocities can be analyzed. As it turns out, looking forward in natural scenes causes the horizon to roughly separate images into upper and lower halves. On average, the distance between the viewer and the nearest point will be farther in the upper part of the image than in the lower part. As a consequence, egocentric motion will cause objects located in the upper half of the scene to move more slowly across the retina than the lower half, an effect which can be detected in optical flows derived from range data.[95] The variance in retinal speed is lower in the lower half of the visual field, indicating more uniform retinal speed, whereas this variance is higher in the upper half of the visual field. However, it appears that the distribution of retinal speed can be modeled well with a log-Gaussian distribution.[96]

The distribution of retinal speed depends to a large extent on the distribution of depth values in a natural scene, showing a near-inverse relationship. However, this changes dramatically for scenes that are non-natural.[96] For natural scenes, the statistical relation between depth and retinal speed may help inform the human visual system about relative depth in the scene, especially in the lower half of the visual field.

The distribution of the directions of retinal flow is close to Gaussian for positions in the lower half of the visual field, whereas extreme non-Gaussian distributions occur near the horizontal median. In this region, the distribution becomes heavy-tailed, i.e., with a high kurtosis.[96] The direction of the optical flow as projected on the retina is strongly dependent on the direction of movement of the observer, especially in the lower half of the visual field.

Finally, it was found that the retinal speed and the direction of the optical flow are largely uncorrelated,[96] especially after gaze stabilization. This allows the human visual system to encode direction and speed independently in cortical area MT.[104]

Statistics of Optical Flow

Rather than estimate the optical flow as it would arise after projection onto the retina, it would be possible to directly assess the statistics of optical flow, which is for instance of interest in the application of video retrieval.[105] This is for instance possible by subjecting optical flow to principal components analysis[106] (discussed previously in Section 7.1). In such a model, motion is decomposed into a set of basis motions, which can then be ordered in order of prevalence. General motion can then be described with a small number of basis motions. In addition, the model coefficients can be derived directly from image derivatives, and there is therefore no requirement to compute dense image motion first. Such a model can be applied effectively to domain-specific problems. An example is the analysis and representation of the motion of mouths or human gait.[106] Moreover, it is possible to incorporate such models within a Bayesian inference framework to reliably estimate image motion in the vicinity of occlusion boundaries.[107]

The statistics of optical flow are interesting also for the purpose of designing optical flow algorithms. As argued earlier in "Optical Flow" section, gradient-based optical flow algorithms require a matrix inversion, which requires the matrix to be non-singular. This has given rise to the formulation of smoothness constraints (see Roth and Black for a discussion[108]), formulating optical flow as a Bayesian inference problem using an appropriate posterior distribution. Using Bayes' rule, this can be split into a data term and a smoothness term (see "Probabilistic Optical Flow" section).

Many smoothness constraints have been formulated, usually enforcing local smoothness by being based on nearest-neighbor differences or other local measures of flow gradient.[108] Roth and Black argue that such smoothness contraints can model piecewise smooth flow, but that they cannot not model more complex flows.[108] Global constraints, on the other hand, could account for more arbitrary flows. The overall insight appears to be that, dependent on the stimuli chosen, motion as captured in optical flow fields can be characterized as "slow and smooth"[109] or "mostly slow and smooth, but sometimes fast and discontinuous".[108]

Analyzing optical flow data derived from the Brown Range Image Database, Roth and Black[108] come to several conclusions regarding the velocity of optical flow:

- Vertical velocity follows a roughly Laplacian distribution.
- Horizontal velocity is more prevalent than vertical motion and therefore leads to a broader distribution.
- The heavy-tailed nature of motion is due to camera translations rather than rotations.

The distribution of flow orientations also tends to favor horizontal directions.

Spatial derivative histograms of optical flow show the characteristic central peak and long tails associate with sparseness and high kurtosis. This suggests that optical flow is often smooth, but on occasion it exhibits much stronger motions. Horizontal and vertical flow gradients are largely independent, as can be determined by analyzing joint statistics of horizontal and vertical flow gradients. This independence can be confirmed by analyzing the flow gradients with PCA.[106,108] Temporal derivative histograms show similar heavy-tailed behavior, which can be attributed to the translational component of motion.[108]

These statistical findings have been used to form priors for the calculation of optical flow with the aid of Markov random fields, and especially using the Fields of Experts model,[108] which was discussed in Section 9.6.2.

REFERENCES

1. Braddick, O. Motion perception. In *Encyclopedia of Perception*; Goldstein, E.B.; Ed.; SAGE Publications, Inc.: Thousand Oaks, CA, 2010, 572–578.
2. Ingebritsen, S.; Lyon, R. Principal components analysis of multitemporal image pairs. Int. J. Remote Sens. **1985**, *6* (5), 687–696.
3. Lodwick, G. Measuring ecological changes in multitemporal landsat data using principal components. In *Proceedings of the International Symposium on Remote Sensing of the Environment*, 1979, 1131–1141.
4. Lodwick, G. A computer system for monitoring environmental change in multitemporal landsat data. Can. J. Remote Sens. **1981**, *7*, 24.
5. Byrne, G.F.; Crapper, P.F.; Mayo, K.K. Monitoring landcover change by principal component analysis of multitemporal landsat data. Remote Sens. Environ. **1980**, *10* (3), 175–184.
6. Watson, A.B.; Ahumada, A. *A Look at Motion in the Frequency Domain*, Vol. 84352; National Aeronautics and Space Administration, Ames Research Center: Moffett Field, CA, 1983.
7. Eckert, M.P.; Buchsbaum, G.; Watson, A.B. Separability of spatiotemporal spectra of image sequences. IEEE Trans. Pattern Anal. Mach. Intell. **1992**, *14* (12), 1210–1213.
8. Adelson, E.H.; Bergen, J.R. Spatiotemporal energy models for the perception of motion. J. Opt. Soc. Am. A **1985**, *2* (2), 284–299.
9. Ngo, C.-W.; Pong, T.-C.; Chin, R.T. Detection of gradual transitions through temporal slice analysis. In *Proceedings of the IEEE International Conference on Computer Vision and Pattern Recognition*, Vol. 1, 1999.
10. Ngo, C.-W.; Pong, T.-C.; Zhang, H.-J. Motion analysis and segmentation through spatio-temporal slices processing. IEEE Trans. Image Process. **2003**, *12* (3), 341–355.
11. Ngo, C.-W.; Pong, T.-C.; Zhang, H.-J.; Chin, R.T. Motion characterization by temporal slices analysis. In *Proceedings of the IEEE International Conference on Computer Vision and Pattern Recognition*, Vol. 2, 2000, 768–773.
12. Ristivojevic, M.; Konrad, J. Joint space-time image sequence segmentation: Object tunnels and occlusion volumes. In *Proceedings of the IEEE International Conference on Acoustics, Speech, and Signal Processing*, Vol. 3, 2004, 9–12.

13. Dong, D.W.; Atick, J.J. Statistics of natural time-varying images. Network: Comput. Neural Sys. **1995**, *6* (3), 345–358.
14. Billock, V.A.; Guzman, G.C.D.; Kelso, J.A.S. Fractal time and 1/f spectra in dynamic images and human vision. Physica D **2001**, *148* (1), 136–146.
15. Dong, D.W.; Atick, J.J. Temporal decorrelation: A theory of lagged and nonlagged responses in the lateral geniculate nucleus. Network: Comput. Neural Sys. **1995**, *6* (2), 159–178.
16. van Hateren, J.H. Theoretical predictions of spatiotemporal receptive fields of fly LMCs, and experimental validation. J. Comp. Physiol. A **1992**, *171*, 151–170.
17. van Hateren, J.H. Processing of natural time series of intensities by the visual system of the blowfly. Vision Res. **1997**, *37* (23), 3407–3416.
18. van Hateren, J.H.; van der Schaaf, A. Temporal properties of natural scenes. In *Electronic Imaging: Science & Technology*, 139–143, 1996.
19. van Hateren, J.H.; Ruderman, D.L. Independent component analysis of natural image sequences yields spatio-temporal filters similar to simple cells in primary visual cortex. Proc. R. Soc. London B **1998**, *265* (1412), 2315–2320.
20. DeAngelis, G.C.; Ohzawa, I.; Freeman, R.D. Receptive-field dynamics in the central visual pathways. Trends Neurosci. **1995**, *18* (10), 451–458.
21. Burr, D.C.; Thompson, P. Motion psychophysics: 1985–2010. Vision Res. **2011**, *51* (13), 1431–1456.
22. Clifford, C.W.G.; Ibbotson, M.R. Fundamental mechanisms of visual motion detection: Models, cells and functions. Prog. Neurobiol. **2003**, *68* (6), 409–437.
23. Goldstein, E. *Sensation and Perception*, 9th Ed.; Wadsworth Cengage Learning: Belmont, CA, 2013.
24. Thompson, W.B.; Fleming, R.W.; Creem-Regehr, S.H.; Stefanucci, J.K. *Visual Perception from a Computer Graphics Perspective*; CRC Press/AK Peters: Boca Raton, FL, 2011.
25. Exner, S. Entwurf zu Einer Physiologischen Erklärung der Psychischen Erscheinungen. I. Teil [Draft of a Physiological Explanation of Mental Impressions. Part I.]. Deutike, Leipzig, Wein, 1894.
26. Hassenstein, B.; Reichardt, W. Systemtheoretische Analyse der Zeit, Reihenfolgen, und Vorzeichenauswertung bei der Bewegungsperzepion des Rüsselkäfers Chlorophanus. Z. Naturforsch. **1956**, *11*, 513–524.
27. Dror, R.O.; O'Carroll, D.C.; Laughlin, S.B. The role of natural image statistics in biological motion estimation. In *Proceedings of the IEEE Workshop on Biologically Motivated Computer Vision*, Seoul, 2000, 492–501.
28. Heeger, D.; Boynton, G.; Demb, J.; Seidemann, E.; Newsome, W. Motion opponency in visual cortex. J. Neurosci. **1999**, *19* (16), 7162–7174.
29. Qian, N.; Andersen, R. Transparent motion perception as detection of unbalanced motion signals. II. Physiology. J. Neurosci. **1994**, *14* (12), 7367–7380.
30. Haag, J.; Denk, W.; Borst, A. Fly motion vision is based on Reichardt detectors regardless of the signal-to-noise ratio. Proc. Nat. Acad. Sci. U.S.A. **2004**, *101* (46), 16333–16338.
31. Anderson, S.J.; Burr, D.C. Spatial and temporal selectivity of the human motion detection system. Vision Res. **1985**, *125* (8), 1147–1154.
32. Burr, D.C.; Ross, J. Contrast sensitivity at high velocities. Vision Res. **1982**, *22* (4), 479–484.
33. Burr, D.C. Motion perception, elementary mechanisms. In *The Handbook of Brain Theory and Neural Networks*, Arbib, M.A.; Ed.; MIT Press: Cambridge, MA, 2003.
34. Burr, D.C.; Ross, J. Visual processing of motion. Trends Neurosci. **1986**, *9*, 304–306.
35. Crowder, N.A.; Dawson, M.; Wylie, D. Temporal frequency and velocitylike tuning in the pigeon accessory optic system. J. Neurophysiol. **2003**, *90* (3), 1829–1841.
36. Limb, J.; Murphy, J. Estimating the velocity of moving images in television signals. Comput. Graphics Image Process. **1975**, *4* (4), 311–327.
37. Dubois, E.; Sabri, S. Noise reduction in image sequences using motioncompensated temporal filtering. IEEE Trans. Commun. **1984**, *32* (7), 826–831.
38. Zlokolica, V.; Pizurica, A.; Philips, W. Noise estimation for video processing based on spatio-temporal gradients. IEEE Signal Process Lett. **2006**, *13* (6), 337–340.
39. Fusiello, A.; Trucco, E.; Tommasini, T.; Roberto, V. Improving feature tracking with robust statistics. Pattern Anal. Appl. **1999**, *2* (4), 312–320.
40. Hirschmuller, H.; Innocent, P.R.; Garibaldi, J.M. Fast, unconstrained camera motion estimation from stereo without tracking and robust statistics. In *ICARCV '02: Proceedings of the 7th IEEE International Conference on Control, Automation, Robotics and Vision*, Vol. 2, 2002, 1099–1104.
41. Ho, Y.-H.; Lin, C.-W.; Chen, J.-F.; Liao, H.-Y. Fast coarse-to-fine video retrieval using shot-level spatio-temporal statistics. IEEE Trans. Circuits Syst. Video Technol. **2006**, *16* (5), 642–648.
42. Laptev, I. *Local Spatio-Temporal Image Features for Motion Interpretation*. PhD Thesis, KTH, 2004.
43. Laptev, I.; Lindeberg, T. Space-time interest points. In *Proceedings of the IEEE International Conference on Computer Vision*, 2003, 432–439.
44. Laptev, I.; Lindeberg, T. Velocity adaptation of space-time interest points. In *Proceedings of the 17th IEEE International Conference on Pattern Recognition*, 2004, 52–56.
45. Spagnolo, P.; Orazio, T.; Leo, M.; Distante, A. Moving object segmentation by background subtraction and temporal analysis. Image Vision Comput. **2006**, *24* (5), 411–423.
46. Swaminathan, R.; Kang, S.B.; Szeliski, R.; Criminisi, A.; Nayar, S.K. On the motion and appearance of specularities in image sequences. In *Proceedings of the 8th European Conference on Computer Vision, volume 2350 of Lecture Notes in Computer Science*; Heyden, A.; Sparr, G.; Nielsen, M.; Johansen, P.; Eds.; Springer: Berlin, 2002.
47. Sato, Y.; Ikeuchi, K. Temporal-color space analysis of reflection. J. Opt. Soc. Am. A **1994**, *11* (11), 2990–3002.
48. Weiss, Y.; Simoncelli, E.P.; Adelson, E.H. Motion illusions as optimal percepts. Nat. Neurosci. **2002**, *5* (6), 598–604.
49. Sidenbladh, H.; Black, M.J. Learning image statistics for Bayesian tracking. In *Proceedings of the 8th IEEE International Conference on Computer Vision*, Vol. 2, 2001, 709–716.

50. Black, M.J.; Fleet, D.J.; Yacoob, Y. A framework for modeling appearance change in image sequences. In *Proceedings of the 6th IEEE International Conference on Computer Vision*, 660–667, 1998.
51. Black, M.J.; Fleet, D.J.; Yacoob, Y. Robustly estimating changes in image appearance. Comput. Vision Image Understanding **2000**, *78* (1), 8–31.
52. Johansson, G. Visual perception of biological motion and a model for its analysis. Percept. Psychophys. **1973**, *14* (2), 201–211.
53. Johansson, G. Visual motion perception. Sci. Am. **1975**, *232* (6), 76–88.
54. Casile, A.; Giese, M.A. Critical features for the recognition of biological motion. J. Vision **2005**, *5* (4), 348–360.
55. Cutting, J.E.; Proffitt, D.R.; Kozlowski, L.T. A biomechanical invariant for gait perception. J. Exp. Psycholo. Hum. Percept. Perform. **1978**, *4* (3), 357–372.
56. Giese, M.A.; Poggio, T. Neural mechanisms for the recognition of biological movements. Nat. Rev. Neurosci. **2003**, *4* (3), 179–192.
57. Troje, N.F. Decomposing biological motion: A framework for analysis and synthesis of human gait patterns. J. Vision, **2002**, *2* (5).
58. Albiol, A.; Silla, M.J.; Albiol, A.; Mossi, J.M. Video analysis using corner motion statistics. In *Proceedings of the IEEE International Workshop on Performance Evaluation of Tracking and Surveillance*, 2009, 31–38.
59. Debevec, P.; Malik, J. Recovering high dynamic range radiance maps from photographs. In *SIGGRAPH '97: Proceedings of the 24th Annual Conference on Computer Graphics and Interactive Techniques*; ACM: New York, 1997, 369–378.
60. Michotte, A. A propos de la permanence phénoménale: Faits et théories (On phenomenal permanence: Facts and theories). Acta Psychol. **1950**, *7*, 298–322.
61. Michotte, A.; Thinès, G.; Crabbé, G. Les compléments amodaux des structures perceptives [amodal completion and perceptual organization]. In *Studia Psychologica*; Publications Universitaires de Louvain: Louvain, 1964.
62. Gibson, J.J.; Kaplan, G.A.; Reynolds, Jr., H.N.; Wheeler, K. The change from visible to invisible: A study of optical transitions. Percept. Psychophys. **1969**, *5* (2), 113–116.
63. Anderson, G.J.; Cortese, J.M. 2-D contour perception resulting from kinetic occlusion. Percept. Psychophys. **1989**, *46* (1), 49–55.
64. Bradley, D.R.; Lee, K. Animated subjective contours. Percept. Psychophys. **1982**, *32* (4), 393–395.
65. Bruno, N.; Bertamini, M. Identifying contours from occlusion events. Percept. Psychophys. **1990**, *48* (4), 331–342.
66. Bruno, N.; Bertamini, M.; Domini, F. Amodal completion of partly occluded surfaces: Is there a mosaic stage? J. Exp. Psychol. Hum. Percept. Perform. **1997**, *23* (5), 1412–1426.
67. Bruno, N.; Gerbino, W. Illusory figures based on local kinematics. Percept. **1991**, *20* (2), 259–274.
68. Cicerone, C.M.; Hoffman, D.D. Color from motion: Dichoptic activation and a possible role in breaking camouflage. Percept. **1997**, *26* (11), 1367–1380.
69. Cicerone, C.M.; Hoffman, D.D.; Gowdy, P.D.; Kim, J.S. The perception of color from motion. Percept. Psychophys. **1995**, *57* (6), 761–777.
70. Cooke, T.; Cunningham, D.W.; Wallraven, C. Local processing in spatiotemporal boundary formation. In *Proceedings of the 7th Tübingen Perception Conference*, 2004, 65.
71. Cunningham, D.W.; Shipley, T.F.; Kellman, P.J. The dynamic specification of surfaces and boundaries. Percept. **1998**, *27* (4), 403–416.
72. Cunningham, D.W.; Shipley, T.F.; Kellman, P.J. Interactions between spatial and spatiotemporal information in spatiotemporal boundary formation. Percept. Psychophys. **1998**, *60* (5), 839–851.
73. Hine, T. Subjective contours produced purely by dynamic occlusion of sparse-points array. Bull. Psychon. Soc. **1987**, *25* (3), 182–184.
74. Kellman, P.J.; Cohen, M.H. Kinetic subjective contours. Percept. Psychophys. **1984**, *35* (3), 237–244.
75. Kellman, P.J.; Palmer, E.M.; Shipley, T.F. Effects of velocity in dynamic object completion. Invest. Ophthalmol. Vis. Sci. **1998**, Supplement, 39, S855.
76. Palmer, E.; Kellman, P.J.; Shipley, T.F. Spatiotemporal relatability in dynamic object completion. Invest. Ophthalmol. Vis. Sci. **1997**, *38* (4), 256.
77. Prazdny, K. Illusory contours from inducers defined solely by spatiotemporal correlation. Percept. Psychophys. **1986**, *39* (3), 175–178.
78. Rock, I.; Halper, F. Form perception without a retinal image. Am. J. Psychol. **1969**, *82* (4), 425–440.
79. Shipley, T.F.; Kellman, P.J. Optical tearing in spatiotemporal boundary formation: When do local element motions produce boundaries, form, and global motion? Spat. Vis. **1993**, *7* (4), 323–339.
80. Shipley, T.F.; Kellman, P.J. Spatiotemporal boundary formation: Boundary, form, and motion perception from transformations of surface elements. J. Exp. Psychol. Gen. **1994**, *123* (1), 3–20.
81. Shipley, T.F.; Kellman, P.J. Spatiotemporal boundary formation: The role of local motion signals in boundary perception. Vision Res. **1997**, *37* (10), 1281–1293.
82. Stappers, P.J. Forms can be recognized from dynamic occlusion alone. Percept. Mot. Skills **1989**, *68* (1), 243–251.
83. Shipley, T.F.; Cunningham, D.W. Perception of occluding and occluded objects over time: Spatiotemporal segmentation and unit formation. In *From Fragments to Objects: Segmentation and Grouping in Vision*, Shipley, T.F.; Kellman, P.J.; Eds.; Elsevier Science: Oxford, 2001, 557–585.
84. Cunningham, D.W.; Graf, A.B.A.; Bülthoff, H.H. A relative encoding approach to modeling spatiotemporal boundary formation. J. Vision **2002**, *2* (7), 704.
85. Gibson, J.J. *The Perception of the Visual World*; Houghton Mifflin: Boston, MA, 1950.
86. Gibson, J.J. *The Ecological Approach to Visual Perception*; Lawrence Erlbaum: Hillsdale, NJ, 1979.
87. Harris, J.M.; Bonas, W. Optic flow and scene structure do not always contribute to the control of human walking. Vision Res. **2002**, *42* (13), 1619–1626.
88. Pless, R.; Brodsky, T.; Aloimonos, Y. Detecting independent motion: The statistics of temporal continuity. IEEE Trans. Pattern Anal. Mach. Intell. **2000**, *22* (8), 768–773.
89. Simoncelli, E.P.; Adelson, E.H.; Heeger, D.J. Probability distributions of optical flow. In *Proceedings of the IEEE International Conference on Computer Vision and Pattern Recognition*, 1991, 310–315.

90. Horn, B.K.P.; Schunck, B.G. Determining optical flow. Artif. Intell. **1981**, *17* (1), 185–203.
91. Lucas, B.D.; Kanade, T. An iterative image registration technique with an application in stereo vision. In *Proceedings of the 7th International Joint Conference on Artificial Intelligence*, 1981, 674–679.
92. Wallach, H. Über visuell warhgenomme Bewegungsrichtung. Psychologische Forshcung **1935**, *20* (1), 325–380.
93. Gibson, J.J. *The Senses Considered as Perceptual Systems*; Houghton Mifflin: Boston, MA, 1966.
94. Lappe, M. (Ed.) *Neuronal Processing of Optic Flow*, Vol. 44; Academic Press: San Diego, CA, 1999.
95. Calow, D.; Krüger, N.; Wörgötter, F.; Lappe, M. Statistics of optic flow for selfmotion through natural scenes. Dyn. Percept. **2004**, 133–138.
96. Calow, D.; Lappe, M. Local statistics of retinal optic flow for self-motion through natural sceneries. Network: Comput. Neural Sys. **2007**, *18* (4), 343–374.
97. Ivins, J.; Porrill, J.; Frisby, J.; Orban, G. The "ecological" probability density function for linear optic flow: Implications for neurophysiology. Percept. **1999**, *28* (1), 17–32.
98. Roth, S.; Black, M.J. On the spatial statistics of optical flow. In *Proceedings of the 10th IEEE International Conference on Computer Vision*, Vol. 1, 2005, pages 42–49.
99. Fermüller, C.; Shulman, D.; Aloimonos, Y. The statistics of optical flow. Comput. Vision Image Understanding **2001**, *82* (1), 1–32.
100. Kalkan, S.; Calow, D.; Wörgötter, F.; Lappe, M.; Krüger, N. Local image structures and optic flow estimation. Network: Comput. Neural Sys. **2005**, *16* (4), 341–356.
101. Zanker, J.; Zeil, J. Movement-induced motion signal distributions in outdoor scenes. Network: Comput. Neural Sys. **2005**, *16* (4), 357–376.
102. Huang, J.; Lee, A.; Mumford, D. Statistics of range images. In *Proceedings of the IEEE International Conference on Computer Vision and Pattern Recognition*, Washington, DC, 2000, 324–331.
103. Lee, A.B.; Mumford, D.; Huang, J. Occlusion models for natural images: A statistical study of a scale-invariant dead leaves model. Int. J. Comput. Vision **2001**, *41* (1), 35–59.
104. Rodman, H.R.; Albright, T.D. Coding of visual stimulus velocity in area MT of the macaque. Vision Res. **1987**, *27* (12), 2035–2048.
105. Fablet, R.; Bouthemy, P. Motion recognition using nonparametric image motion models estimated from temporal and multiscale co-occurrence statistics. IEEE Trans. Pattern Anal. Mach. Intell. **2003**, *25* (12), 1619–1624.
106. Fleet, D.J.; Black, M.J.; Yacoob, Y.; Jepson, A.D. Design and use of linear models for image motion analysis. Int. J. Comput. Vision **2000**, *36* (3), 171–193.
107. Fleet, D.J.; Black, M.J.; Nestares, O. Bayesian inference of visual motion boundaries. In *Exploring Artificial Intelligence in the New Millennium*, Lakemeyer, G.; Nebel, B.; Eds.; Morgan Kaufmann: San Francisco, CA, 2002, 139–174.
108. Roth, S.; Black, M.J. On the spatial statistics of optical flow. Int. J. Comput. Vision **2007**, *74* (1), 33–50.
109. Weiss, Y.; Adelson, E.H. *Slow and Smooth: A Bayesian Theory for the Combination of Local Motion Signals in Human Vision*. Technical Report, AI Memo 1624, MIT AI Lab, Cambridge, MA, 1998. 284.

Topology in Geographic Information System

Prasanna Kumar
Mathematics, Birla Institute of Technology and Science—Goa Campus, Zuarinagar, India

Abstract
Topology is an important branch of mathematics which conceived out of properties of real line, Euclidean spaces, and the continuous functions on these spaces. Because of this, it has strong and wide applications in geographic information systems (GISs). Most topological structures fit well into GIS, with the dynamics of real-life problems within and across spatial proximities. In the geographic framework, the spatial relationships are the central points for the objects generating them. These structures are identified through topo-geographic and metric-geographic functionalities and the corresponding topologies emerged out of it: mostly object or feature flavored or both. The modeling capabilities are explained and illustrated in detail in many of the papers in this direction. Most prominent theories developed for representing the geographical objects through their topological characteristics will facilitate an integrated visualization of the problem in hand. The abstractions of characterized objects are justified through greater works that can be seen widely in the literature.

Due to the diversity in the geography, location, distance, proximity are more or less contextual. This kind of uncertainty and imprecision are handled effectively by integrating topology and GIS as mentioned earlier. In this entry, an attempt is made to provide references starting from the dawn of using topology in GIS to some of the most recent ones. In fact, we present the literature centered around abstraction and conceptualization of topological features in the geographic objects and the subsequent applications in an expository way.

Keywords: Data model; GIS; Spatial objects; Spatial relationship; Topology.

INTRODUCTION

Topology is popularly known as an axiomatic branch of mathematics. Even then, it has wide applications across engineering and science branches. The idea of topology evolved from the mysterious behavior of the real line in many aspects, Euclidean space, and the associated continuous functions. Hence, one can find its applications in geographic information systems (GISs) in abundance. Topological functionalities get better understanding in GIS, which involve the dynamics of problems within and across locations and proximities. In the geographic setup, these concepts remain focal points forever for the objects they represent. The notions of bisection, continuity, homeomorphism, diffeomorphism, compactness, connectedness, separability, and denseness are the key tools for any GIS researcher today, like simple Euclidean geometry were to cartographers of last few centuries. The advantage of topological structures is that information storage could be reduced by storing boundaries between adjacent regions exactly once. Here, adjacency relations are crucial, and therefore, topological maps improve the data entry and map production.

As mentioned earlier, this survey article explains about the interplay between GIS and topology. Topological content will play multiple roles and cover a large range of expectations of qualitative features, which experience topology and translates it into GIS. There is always a requirement for computer-based representations of geospatial information that parallel the structure of related geographic space. The distance, proximity, context, perspective, dimension, and neighborhood are modeled accordingly in GIS. We are familiar with open intervals in the real line or open discs in the plane. The open intervals and the open circles are the open sets in standard topologies on the real line and the plane, respectively. Any sets can be an open set if we define the topology appropriately. Thus, a set may have several topologies defined on it. If we wish to measure the similarities and differences between information units, an appropriate metric can be found to do so. So, properties of metric spaces can be used accordingly. Since every metric space is a topological space, the knowledge from topological domain can be effectively used in the development of the theory. When a topology is induced by a metric, we have some additional structures that are useful. Such spaces are known as metrizable spaces. But there are topological spaces that are not induced by metric, and we need to cautiously deal with such spaces.

The metric space dynamics is most of the time insufficient as a formal base for distance measures in geographic spaces. Therefore, classical theory of topological representations of the spatial structures associated with the neighborhoods have seen refinements time and again. The locality and neighborhood in non-Euclidean metric spaces are also studied widely. On the other hand, geometry is a prominent aspect of geographic information that changes with the change in reference systems. The construction of topologies in geographic spaces from a nonmetric function is a new discovery in the field of GIS.

Let us present the results appeared in the literature on applications of topology in GIS in the sections that follow.

ENTRY OF CLASSICAL TOPOLOGY INTO GIS

Topology,[1] like many other branches of mathematics, is fascinating but an abstract subject. As mentioned earlier, it has plenty of applications across various engineering and science branches.[2] It is more or less accepted that topology has entered into the dictionary of GIS through a book by McDonnell and Kemp[3] formally. But informally, the existence and use of topology in GIS literature goes back to 1970s. First and foremost, GIS researchers who used topological tools are Chrisman,[4] Peucker and Chrisman,[5] Corbett,[6,7] Goodchild,[8] and Freeman.[9] They mainly used the properties of metric and topological spaces in many analytical applications. Blackmore[10] went little further and applied these concepts to do error analysis in spatial databases. Carter's[11] book mentions the use of topology in computer mappings. Another important material is a monograph by Worboys,[12] which unravels the initial developments in this direction.

Bonham and Carter,[13] Chou,[14] and Chrisman[15] have also given a nice presentation of the topic in their monographs which covers the proceedings in GIS topology of next two to three decades after Chrisman' work,[4] the outcomes in view of the topological data structures and their advantages in terms of adjacency, connectivity, and mappings. The results of analysis of this kind were presented in a paper by Goodchild.[16] Egenhofer[17] introduced the point set topology in a different way in GIS and developed a mechanism[18] that vastly reduces the number of individual topological relations necessary to describe a total configuration. This model is suitable for a multilevel treatment of spatial objects. Clementini[19] did a comparative analysis of different methods for representing topological relationships.

A map driven by a topology always contains space filling and nonoverlapping regions. Hence, cartographic nontopological data structures have been hardly used by mainstream GIS software. Nontopological data structures are conformal to planar graph theory. A data structure can represent proximal and directional spatial relations in metric forms,[20] in addition to its topological shaping. This extension permits a larger set of functional connections between geographical features that are to be explicitly represented. Detailed explanations are given in a paper by Theobald.[21]

SPATIAL RELATIONSHIPS AND TOPOLOGICAL MAPS

Even though it was formally added to the GIS dictionary in 1995,[3] much before Eigenhofer[22,23] had introduced the point-set topological relations between spatial objects which was extended further[24] for modeling conceptual neighborhoods of topological line–region relations. Several models of this kind[25,26] on topological relations have appeared in the literature. The common feature of these models is that they provide a computational basis for spatial reasoning being a trade-off between the formal ground needed by an information system and the corresponding geographic space.[27,28] Clementini[29] structured approximate topological relations in graphs having multifold interpretations.

The practical requirements in GIS have led to the investigation of formal methods of describing spatial relations.[30] After an introduction to the basic notions of topology, an extended theory of topological spatial relations between sets is developed by Eigenhofer,[31] in which the relations are defined in terms of the intersections of the boundaries and interiors of two sets. These relations correspond to some of the standard topological spatial relations between sets such as equality, disjointness, and containment in the interior.

Since points and polygons can be precisely, imprecisely, or randomly recognized within a spatial information system,[32] its treatment is not easy. The analysis in[33,34] clarifies some unresolved issues of the uncertain relations between imprecise regions[35,36]. Min Deng et al.[37] established some reasoning of topological relations between imprecise regions. A locational error model for spatial features in vector-based GIS has been proposed in the paper by Leung.[38] The need of topological tools in handling the previous explained problems is elaborately presented in an article by Reed.[39]

Quite a number of papers[40–42] have discussed the topological relations between multilevel spatial objects. There are many topological properties, and it is inadequate to only consider the empty and non-empty invariants[43] because other topological properties, such as connectivity, compactness, first fundamental group, and subspace topology, can help to identify the topological relations in the use of GIS. In a paper by Liu,[44] the topological relations between GIS objects have been extended by considering properties including connectivity and first fundamental group. Such models can be applied for designing and implementing GIS[45] mechanism in its own way.

MORE QUALITATIVE DESCRIPTION USING TOPOLOGY

Topological relativity is of interest in GIS theory. One way of assessing the spatial reasoning is to analyze commonplace terms from natural language with respect to conceptual neighborhood graphs, the sequence choice for topological relations.[46] Dube and Egenhofer[47] have analyzed sixteen English-language spatial prepositions for region–region relations for their corresponding topological relations, each of which was found to represent a convex subset within the conceptual neighborhood graph of the region–region relations. Nested neighborhood graphs are also used as reference systems[48] for topological relations with compound spatial objects. They support the identification of complete sets of feasible relations by incrementally building relations from the base set of topological relations between simple regions, lines, and points. From an application point of view, spatial usages require by far more complex geometric structures than simple points, lines, and regions that can be found in GIS. Some GIS and database researchers have recognized this shortcoming and begun to incorporate more complex spatial data types into their systems.[49,50] Complex points composed of a finite set of isolated points, complex lines representing spatially embedded graphs possibly consisting of several connected components, and complex regions consisting of several components where each component possibly contains holes were handled effectively by Limen et al.[51] and Schmeider.[52] Schneider and Thomas[53] have studied topological relationship between complex spatial objects based on the well-known nine-intersection model. Some of the shapefiles that shape GIS data models were designed by Strand.[54]

GIS deals geographical objects ranging from physical landscapes to towns and transport systems.[55] Such objects, exactly located in space, can easily be handled by modern GIS,[56–57] yet form only a small proportion of all the possible geographical objects.[58] The book authored by Burrough[59] exposes the essential complexity of the world and stresses the point that current GISs do not adequately address problems as diverse as the resolution of crime between national boundaries, or the interpretation of views of people from different cultures.[60–61] On the other hand, the paper by Li et al.[62] set up the basic topology with a different topological structure of spatial objects. The basic topological components of a spatial entity in topological paradox[63] associated with positional error[64] are the other important aspects of it.

In one of their papers, Gotts et al.[65] describe the topological aspect of a logic-based, artificial intelligence approach to understand the qualitative description of spatial properties and relations, with reasoning. This technique is popularly known as region connection calculus theory.[65]

METRIC-INDUCED MODELING

Naive Geography's popular phrasing "Topology matters, metric refines" perpetuates the properties that provide scope for improvement in purely topological structures. Metric-induced[66–67] computational models with metric information in terms of splitting ratios and closeness measures have appeared in GIS literature.

The paper by Egenhofer and Dube[68] defines a comprehensive set of 11 metric refinements that apply to the eight coarse topological relations between two regions that the 9-intersection and the region connection calculus develop the allowable values for each metric refinement. It is shown that any topological relation between two regions can be derived uniquely from the union of at most three such refinement specifications. Such interrelation hybridizations have been found useful for the size of regions[69–70] and direction relations.[71–72] Mereotopologies have traditionally been defined in terms of the intersection of point sets representing the regions. These schemes function well for purely metric-induced topological models. Two other significant papers[73–74] explore the idea of a distance-based interpretation for mereotopology.

Spatial relations are the basis for querying GIS, and such query languages use natural language-like terms. To bridge the gap between computational models for spatial relations and people's use of spatial terms in their natural languages, some models for the geometry of spatial relations[75–76] were calibrated for a set of 59 English-language spatial predicates. The models distinguish topological and metric properties. Brisson[77] investigated data structures for representing and manipulating d-dimensional geometric objects for arbitrary $d \geq 1$. A new representation is given for such cell objects, which provides direct access to topological structure, ordering information among cells, the topological dual, and boundaries. The paper[77] also focuses on the problem of 3D data modeling within GIS.

HANDLING COMPLEXITY OF 3D DATA VIA TOPOLOGY

The approach presented by Pigot[78] gives new techniques to manage the diversity and the complexity of 3D data in GIS. This approach uses two levels of modeling in order to describe the whole 3D scene. The first level uses a 3D topology composed of nodes, edges, and faces. Improved version of this model can be seen in the paper by Ramos.[79]

The need for GIS data models that allow for 3D spatial query, analysis, and visualization[80–81] explored a new way of representing the topological relationships among 3D geographical features such as buildings and their internal partitions or subunits.[82–83] The 3D topological data model is called the combinatorial data model. It is a logical data model that simplifies the complex topological relationships among 3D features through a hierarchical network structure called the node-relation structure.[84] It is modeled using graph theoretic techniques. These kinds of models were implemented with real data for evaluating its effectiveness for performing 3D spatial queries[85] and visualization. Using the geometric network obtained from the transformations, spatial queries based upon the complex connectivity relationships between 3D urban objects are implemented[86] using Dijkstra algorithm. The paper by Birkin et al.[87] presents the results of an experimental implementation of a 3D network data model (GNM) using GIS data[88–89]. Requirements for topological functionality in 3D were then grouped and categorized[90–91] and handled accordingly. Other significant papers[92–94] conclude by suggesting that these needs can be used as a basis for the implementation of topology in 3D. As we all know, ArcGIS and TIGER[95] also implemented topology in its toolkits.

RECENT ADVANCES

Let us review some recent developments in topology supported GIS. One thing we need to be clear; topology is not the solution to every GIS problem, it is just one of its ingredients. The more we structure these ingredients, the better is for GIS. In this direction, Prasanna Kumar[96] published a paper on perspective topology. This entry discusses a new concept of perspective neighborhood associated with an unusual nonmetric function and a topology generated by the basis involving these neighborhoods. A brief description of a framework model for application[96] is also presented at the end of this entry. The second paper by Prasanna Kumar[97] also introduced two new topologies: one is object oriented and the second one is feature oriented. The important structural properties and modeling capabilities are also illustrated with a simple model. The applications of this model are also briefly discussed.

Qualitative locations describe spatial objects by relating them to a frame of reference with qualitative relations. Most models only formalize spatial objects, frames of reference, and their relations at one scale, thus limiting their applicability in location changes of spatial objects across scales. A topology-based, multi-scale qualitative location model is proposed[98] to represent the multiple representations of the same objects in relation with the frames of reference at different levels. Multi-scale regional partitions are the frames of reference at multiple levels of scale and formalized to relate multiple representations of the same objects to the multiple frames of reference.[99]

Dube et al.[100] considered the nineteen planar discrete topological relations that apply to regions bounded by a digital Jordan curve. Here, metrics are developed to determine the topological relations. Two sets of five such metrics were found to be minimal and sufficient to uniquely identify each of the 19 topological relations. Dube et al.[101] developed another formalism to construct topologically distinct configurations based on simple regions. As an extension to the compound object model, this paper contributes a method for constructing a complex region and also provides a mechanism to determine the corresponding complement of such a region.

The topological relations can be modified and updated parametrically. Case studies evaluating the topological relations between 3D objects are performed in a paper by Yu et al.[102] The model can express and compute the topological relations between objects in a symbolic and geometry-oriented way. The method can also support topological relation series computation between objects with location or shape changes.

CityGML, as the standard for the representation and exchange of 3D city models, contains rich information in terms of geometry, semantics, topology, and appearance. CityGML adopts the XLink approach to represent topological relationships between different geometric aggregates or thematic features, but limited to shared objects. The paper by Li et al.[103] proposes a two-level model for representing 3D topological relationships in CityGML: high-level (semantic-level) topology between semantic features and low-level (geometric-level) topology between geometric primitives. Five topological relationships are adopted in this model: touch, in, equal, overlap, and disjoint. An application domain extension, called TopoADE, is proposed for the implementation of the topological model.

Rolf et al.[104] explored the practicability and analyzed the quality of searches for Wikipedia pages of topologically related administrative divisions in Switzerland and Scotland via Linked Data. The quality of searches in the English and German versions of DBpedia is compared, as is that of searches in GeoNames and DBpedia using DBpedia's links to GeoNames or a manually created list of links. It turns out that live searches are practicable with acceptable performance.

Changes in river plane shapes are called river planform changes (RPCs), and such changes can impact sustainable human development. RPCs can be identified through field surveys—a method that is highly precise but time consuming, or through remote sensing and GIS, which are less precise but more efficient. Leng et al.[105] developed a combinatorial reasoning mechanism based on topological and metric relations that can be used to classify RPCs. This approach may reduce the abrupt and wrong information caused by varying river water levels.

Describing the geometry of a spatial object using the OpenGIS Simple Features Specification requires only simple features: the interior, boundary, and exterior of a spatial object are defined. The paper by Jingwei et al.[106] has proposed a comprehensive model, the 27-intersection model (27IM), which considers both the dimensions and the number of intersections. Also some propositions are presented in the same paper to exclude relations that the 27IM cannot implement.

CONCLUSION

We have discussed some significant developments in GIS in view of topology. The notions from a wide range of papers have been presented comprehensively for the first time. The main aim of this work was to compile how topological properties can be used to explore the hidden features of the geographic space. The properties such as connectedness, compactness, and the subsidiary features of a general topology can be defined in GIS similar to the notions on general topological space, but it is more dependent on the structure of basis and open sets associated with it. The standard notion of topology in GIS centers around explicit representation of adjacent spatial relations and involves various geographic features.

The advantages attributed to topological data structures have become clear now, and therefore GIS users need to adequately understand the topological data structures and use them appropriately. Of course, further work is required to discover the deeper applications of the topology and their impact on GIS.

This collection of classic of somewhat exhaustive literature is presented after an intensive and comprehensive study of the topology integrated GIS by the author. At the same time, it is quite essential to understand the kind of exclusions that occurs and the case of noncoverage of any articles. We hope that a knowledge of these results provides a powerful tool for the discovery of deep and important results in this subject. Improved methods are the fundamentally important area for the future work. The topological formalization of realistic criteria will boost further research and the development of more applicable general-purpose GIS techniques.

REFERENCES

1. Munkers, J.R. *Topology*; Prentice Hall: Englewood Cliffs, NJ, 2000.
2. Alexandroff, P. *Elementary Concepts of Topology*; Dover Publishing Inc.: New York, 1961.
3. McDonnell, R.; Kemp, K. *International GIS Dictionary*; John Wiley & Sons: Cambridge, 1995.
4. Chrisman, N. Topological data structures for geographic representation. In *Proceedings of the International Symposium on Computer-Assisted Cartography: Auto-Carto II*; American Congress on Surveying and Mapping: Bethesda, MD, 1975, 346–351.
5. Peucker, T.K.; Chrisman, N. Cartographic data structures. Am. Cartographer **1975**, *2* (1), 55–69.
6. Corbett, J.F. Topological principles in cartography. In *Proceedings of the International Symposium on Computer-Assisted Cartography: Auto-Carto II*; American Congress on Surveying and Mapping: Bethesda, MD, 1975, 61–65.
7. Corbett, J.F. *Topological principles in cartography*. Technical paper 48, US Department of Commerce; Bureau of the Census; Washington, DC, 1979.
8. Goodchild, M.F. Statistical aspects of the polygon overlay problem. In *Proceedings of the First International Advanced Study Symposium on Topological Data Structures for Geographic Information Systems*; Laboratory for Computer Graphics and Spatial Analysis, Graduate School of Design; Harvard University: Harvard, MA, 1977.
9. Freeman, J. The modeling of spatial relations. Comput. Graphics Image Process. **1975**, *4* (2), 156–171.
10. Blakemore, M. Generalization and error in spatial databases. Cartographica **1984**, *21* (2 & 3), 131–139.
11. Carter, J.R. *Computer Mapping: Progress in the '80s*; Association of American Geographers: Washington, DC, 1984.
12. Worboys, M.F. *GIS: A Computing Perspective*; Taylor & Francis: London, 1995.
13. Bonham-Carter, G.F. *Geographic Information Systems for Geoscientists: Modeling with GIS*; Pergamon Press: Ottawa, ON, 1996.
14. Chou, Y.H. *Exploring Spatial Analysis in Geographic Information Systems*; Onword Press: Sante Fe, NM, 1997.
15. Chrisman, N. *Exploring Geographic Information Systems*; John Wiley & Sons: New York, 1990. Clarke, K.C. *Analytical and Computer Cartography*; Prentice Hall: Englewood, CO, 1997.
16. Goodchild, M.F. A spatial analytical perspective on geographical information systems. Int. J. Geogr. Inf. Sci. **1987**, *1* (4), 327–334.
17. Egenhofer, M.J.; Franzosa, R. Point-set topological spatial relations. Int. J. Geogr. Inf. Syst. **1991**, *5* (2), 161–174.
18. Egenhofer, M.J.; Clementini, E.; Di Felice, P. Topological relations between regions with holes. Int. J. Geogr. Inf. Syst. **1994**, *8* (2), 129–144.
19. Clementini, E.; Di Felice, P. A comparison of methods for representing topological relationships. Inf. Syst. **1995**, *20* (3), 149–178.
20. Worboys, M.F. Metrics and topologies for geographic space. In *Proceedings of 7th International Symposium on Spatial Data Handling*, Delft, August 12–16, 1996.
21. Theobald, D.M. Topology revisited: Representing spatial relations. Int. J. Geogr. Inf. Sci. **2001**, *15* (8), 689–705.
22. Egenhofer, M.; Herring, J. *Categorizing Binary Topological Relations between Regions, Lines and Points in Geographic Databases*; Technical Report 01/1991, University of Maine: Orono, ME, 1991, 28.
23. Egenhofer, M. A model for detailed binary topological relationships. Geomatica **1993**, *47* (3), 261–273.
24. Egenhofer, M.J.; Mark, D. Modeling conceptual neighborhoods of topological line- region relations. Int. J. Geogr. Inf. Syst. **1995**, *9* (5), 555–565.
25. Clementini, E.; Di Felice, P. A model for representing topological relationships between complex geometric features in spatial database. Inf. Sci. **1996**, *90* (1–4), 121–136.
26. Clementini, E.; Di Felice, P.; van Oosterom. A small set of format topological relationships for end-user interaction. In *Advances in Spatial Databases-Third International Symposium, SSD'93*, Lecture notes in Computer Science 692; Abel, D.; Oai, B.C.; Eds.; Springer-Verlag: Singapore, 1993, 277–295.
27. Cui, Z.; Cohn, A.G.; Randell, D.A. Qualitative and topological relationships in spatial databases. In *Advances in Spatial Databases-Third International Symposium, SSD*

'93, Lecture notes in Computer Science 692; Abel, D.; Oai, B.C.; Eds.; Springer-Verlag: Singapore, 1993, 296–315.
28. Clementini, E.; Felice, P.; Oosterom, P. A small set of formal topological relationships suitable for end-user interaction. In *Proceedings of the International Symposium Ssd'93*, Lecture Notes in Computer Science, Singapore, June 23–25, 1993, 277–295.
29. Clementini, E.; Di Felice, P. Approximate topological relations. Int. J. Approx. Reason. **1996**, *16* (2), 173–204.
30. Egenhofer, M.; Mark, D. Naive geography. In *Proceedings of the COSIT'95*, Semmering, September 21–23, 1995, 1–15.
31. Egenhofer, M.; Franzosa, R. On the equivalence of topological relations. Int. J. Geogr. Inf. Sci. **1995**, *9* (2), 133–152.
32. Chen, X. Spatial relationships between uncertain sets. In *International Archives of Photogrammetry and Remote Sensing*; XVIII Congress: Vienna, 1996, 105–110.
33. Leung, Y.; Yan, J.P. Point-in-polygon analysis under certainty and uncertainty. GeoInformatica **1997**, *1* (1), 93–114.
34. Winter, S. Uncertain topological relations between imprecision regions. Int. J. Geogr. Inf. Sci. **2000**, *14* (5), 411–430.
35. Billen, R.; Zlatanova, S.; Mathonet, P.; Boniver, F. The dimensional model: A framework to distinguish spatial relationships. In *Proceedings of the ISPRS Symposium on Geospatial Theory, Processing and Applications*, July 9–10, Ottawa, ON, 2002.
36. Chen, S.P.; Zhou, C.H. Conformity information source. Earth Inf. Sci. **2003**, *3* (1), 1–3.
37. Deng, M.; Chen, X.; Kusanagi, M.; Phien, H.N. Reasoning of topological relations between imprecise regions. Geogr. Inf. Sci. **2004**, *10* (1), 73–81.
38. Leung, Y.; Yan, J.P. A locational error model for spatial features. Int. J. Geogr. Inf. Sci. **1998**, *12* (6), 607–620.
39. Reed, C. GIS users shouldn't forget about topology. GeoWorld **1999**, *12* (4), 12.
40. Renz, J.; Rauh, R.; Knauff, M. Towards cognitive adequacy of topological spatial relations. In *Spatial Cognition II, Integrating Abstract Theories, Empirical Studies, Formal Methods, and Practical Applications*; Springer-Verlag: London, 2000, 184–197.
41. Chen, X.; Deng, M. Complete description of topological relations between spatial regions. In *2nd International Symposium on Spatial Data Quality*; Shi, W.Z.; Goodchild, M.; Fisher, 2003, 51–60.
42. Clough P.D.; Joho, H.; Purves, R. Judging the spatial relevance of documents for GIR. In *Advances in Information Retrieval*, Lecture Notes in Computer Science, Vol. *3936*; Lalmas, M.; Mac-Farlane, A.; R€uger, S.; Tombros, A.; Tsikrika, T.; Yavlinksy, A.; Eds.; Springer: Berlin, 2006, 548–552.
43. Deng, M.; Cheng, T.; Chen, X.; Li, Z. Multi-level topological relations between spatial regions based upon topological invariants. Geoinformatica **2007**, *11* (2), 239–267.
44. Liu, K.; Shi, W. Extended model of topological relations between spatial objects in geographic information systems. Int. J. Appl. Earth Obs. Geoinf. **2007**, *9* (3), 264–275.
45. Kurata, Y. The 9+ -intersection: A universal framework for modeling topological relations. In *Geographic Information Science*; Springer: Berlin/Heidelberg, 2008, 181–198.

46. Alboody, A.; Sedes, F.; Inglada, J. Modeling topological relations between uncertain spatial regions in geo-spatial databases: Uncertain intersection and difference topological model. In *Proceedings of the 2010 Second International Conference on Advances in Databases, Knowledge, and Data Applications*, Menuires, April 11–16, 2010, 7–15.
47. Dube, M.; Egenhofer, M. An ordering of convex topological relations. In *Proceedings of the GIScience*, Columbus, OH, September 18–21, 2012, 72–86.
48. Egenhofer, M. A reference system for topological relations between compound spatial objects. In *Advances in Conceptual Modeling—Challenging Perspectives*, Vol. 5833; Heuser, C.; Pernul, G.; Eds.; Springer: Berlin, 2009, 307–316.
49. Behr, T.; Schneider, M. Topological relationships of complex points and complex regions. In *Proceedings of the International Conference on Conceptual Modeling*, 2001, 56–69.
50. Clementini, E.; Di Felice, P. A model for representing topological relationships between complex geometric features in spatial databases. Inf. Syst. **1996**, *90* (1–4), 121–136.
51. Clementini, E.; Di Felice, P.; Califano, G. Composite regions in topological queries. Inf. Syst. **1995**, *20* (7), 579–594.
52. Schneider, M. Implementing topological predicates for complex regions. In *Proceedings of the International Symposium on Spatial Data Handling*, 2002, 313–328.
53. Schneider, M.; Thomas, B. Topological relationships between complex spatial objects. ACM Trans. Database Syst. **2006**, *31* (1), 39–81.
54. Strand, E.J. Shapefiles shape GIS data transfer standards. GIS World **1998**, *11* (5), 28.
55. Claire, E.; Muky, H. Requirements for topology in 3D GIS. Trans. GIS **2006**, *10* (2), 157–175.
56. Molenaar, M. A topology for 3-D vector maps. ITC J. **1992**, (1), 25–33.
57. Molenaar, M. *An Introduction to the Theory of Spatial Object Modelling for GIS*; Taylor & Francis: New York, 1998.
58. Pigot, S.; Hazelton, B. The fundamentals of a topological model for a four-dimensional GIS. In *Proceedings of tile 5th International Symposium on Spatial Data Handling*, Charleston, SC, 1992, 580–591.
59. Burrough, P.A.; Frank, A.U. *Geographic Objects with Indeterminate Boundaries*; Taylor and Francis: Oxford, 1996.
60. Goodchild, M.F.; Gopal, S. *Accuracy of Spatial Databases*; Taylor & Francis: London, 1989, 107–113.
61. Guptill, S.C.; Morrison, J.L. *Elements of Spatial Data Quality*; Elsevier Scientific: Oxford, 1995.
62. Li, D.R.; Peng, M.Y.; Zhang, J.Q. Modeling positional uncertainty of line primitives in GIS. J. WTUSM **1995**, *21* (4), 283–288.
63. Li, Z.; Li, Y.; Chen, Y.Q. Basic topological models for spatial entities in 3-dimensional space. GeoInformatica **2000**, *4* (4), 419–433.
64. Liu, W.B.; Dai, H.L.; Xu, P.L. Models of positional error donut for planar polygon in GIS. Acta Geogdaetica et Cartographica Sinica **1998**, *27* (4), 338–344.

65. Gotts, N.; Gooday, J.; Cohn, A. A connection based approach to common-sense topological description and reasoning. Monist **1996**, *79* (1), 51–75.
66. Nedas, K.; Egenhofer, M.; Wilmsen, D. Metric details of topological line–line relations. Int. J. Geogr. Inf. Sci. **2007**, *21* (1), 21–48.
67. Randell, D.; Cohn, A. Modelling topological and metrical properties of physical processes. In *Proceedings of the International Conference on Principles of Knowledge Representation and Reasoning*, Toronto, ON, May 15–18, 1989, 357–368.
68. Egenhofer, M.; Dube, M. Topological relations from metric refinements. In *Proceedings of the 17th ACM Sigspatial International Symposium on Advances in Geographic Information Systems*, Seattle, WA, November 4–6, 2009, 158–167.
69. Gerevini, A. Combining topological and size constraints for spatial reasoning. Artif. Intell. **2002**, *137* (1–2), 1–42.
70. Sharma, J. Integrated topological and directional reasoning in geographic information systems. In *Geographic Information Research: Trans-Atlantic Perspectives*; Craglia, M.; Onsrud, H.; Taylor & Francis: London, 1999, 435–448.
71. Binbin, L.; Martin, C.; Stwart, A.F. Geographically weighted regression using a non-euclidean distance metric with a study on London house price data. Procedia Environ. Sci. **2011**, *7* (1), 92–97.
72. Worboys, M. Nearness relations in environmental space. Int. J. Geogr. Inf. Sci. **2001**, *15* (7), 633–651.
73. Varzi, A.C. On the boundary between mereology and topology. In *Philosophy and the Cognitive Sciences*; Casati, R.; Smith, B.; White, G.; Eds.; Holder-Pichler-Tempsky: Vienna, 1994, 423–442.
74. Sridhar, M.; Cohn, A.; Hogg, D. From video to RCC8: Exploiting a distance based semantics to stabilise the interpretation of mereotopological relations. In *Spatial Information Theory*; Springer: Heidelberg, 2011, 110–125.
75. Egenhofer, M.; Shariff, A. Metric details for natural-language spatial relations. ACM Trans. Inf. Syst. **1998**, *16* (4), 321–349.
76. Shariff, A.; Egenhofer, M.; Mark, D. Natural-language spatial relations between linear and areal objects: The topology and metric of English-language terms. Int. J. Geogr. Inf. Sci. **1998**, *12* (3), 215–246.
77. Brisson, E. Representing geometric structures in d dimensions: Topology and order. Discrete Comput. Geom. **1993**, *9* (1), 387–426.
78. Pigot, S. Topological models for 3D spatial information systems. In *Proceedings of the AutoCarto Conference*, Baltimore, MD, March 25–28, 1991, 368–392.
79. Ramos, F. A multi-level approach for 3D modeling geographical information sytems. In *Proceedings of the Symposium on Geospatial Theory, Processing and Applications*, Ottawa, ON, July 9–12, 2002, 1–5.
80. Tempfli, K. 3D topologic mapping for urban GIS. ITC Journal **1998**, *3* (4), 181–190.
81. Tse, R.O.C.; Gold, C. A proposed connectivity-based model for a 3-D cadastre. Comput. Environ. Urban Syst. **2003**, *27* (4), 427–445.
82. van Oosterom, P.; Stoter, J.; Quak, W.; Zlatanova, S. The balance between geometry and topology. In *10th International Symposium on Spatial Data Handling*; Richardson, D.; van Oosterom, P.; Eds.; Springer-Verlag: Berlin, 2002, 209–224.
83. Wei, G.; Ping, Z.; Jun, C. Topological data modelling for 3D GIS. In *Proceedings of ISPRS Commission IV*, Stuttgart, 1998, 657–661.
84. Lee, J.; Kwan, M. A combinatorial data model for representing topological relations among 3D geographical features in micro-spatial environments. Int. J. Geogr. Inf. Sci. **2005**, *19* (10), 1039–1056.
85. Molenaar, M. A formal data structure for 3D vector maps. In *Proceedings of the EGIS'90*, Amsterdam, September 10–14, 1990, 770–781.
86. Lee, J. A spatial access-oriented implementation of a 3-D GIS topological data model for Urban entities. GeoInformatica **2004**, *8* (3), 237–264.
87. Birkin, M.; Clarke, G.; Clarke, M.; Wilson, A. *Intelligent GIS: Location Decisions and Strategic Planning*; Geoinfomation International, John Wiley & Sons Inc.: New York, 1996.
88. Lee, J. *A 3-D Data Model for Representing Topological Relationships between Spatial Entities in Built-Environments*; PhD dissertation; The Ohio State University, 2001.
89. Raper, J. *Multidimensional Geographic Information Science*; Taylor & Francis: New York, 2000.
90. Rikkers, R.; Molenaar, M.; Stuiver, J. A query oriented implementation of a topologic data structure for 3-dimensional vector maps. Int. J. Geogr. Inf. Sci. **1994**, *8* (3), 243–260.
91. Pigot, S. A topological model for a 3-D spatial information system. In *Proceedings of the 5th International Symposium on Spatial Data Handling*, Charleston, SC, 1992, 344–359.
92. Pilouk, M. *Integrated Modeling for 3-D GIS*; PhD dissertation; ITC, The Netherlands, 1996.
93. Zlatanova, S.; Rahman, A.; Pilouk, M. Trends in 3-D GIS development. J. Geospatial Eng. **2002**, *4* (2), 1–10.
94. Zlatanova, S.; Rahman, A.; Shi, W. Topology for 3-D spatial objects. *International Symposium and Exhibition on Geoinformation* Oct 22–24, Kuala Lumpur, CDROM, 2002.
95. Cooke, D.F. Topology and TIGER: The Census Bureau's contribution. In *The History of Geographic Information Systems: Perspectives from the Pioneer*; Foresman, T.W.; Ed.; Prentice Hall: Upper Saddle River, NJ, 1998, 47–57.
96. Prasanna, K.N. A short note on the theory of perspective topology in GIS. Ann. GIS **2013**, *19* (2), 123–128.
97. Prasanna, K.N. On the topological situations in geographic spaces. Ann. GIS **2014**, *20* (2), 131–137.
98. Shihong, D.; Chen-Chieh, F.; Qiao, W.; Luo, G. Multi-scale qualitative location: A topology-based model. Trans. GIS **2014**, *18* (4), 604–631.
99. Yuan, L.; Yu, Z.; Luo, W.; Yi, L.; Lv, G. Multidimensional-unified topological relations computation: A hierarchical geometric algebra-based approach. Int. J. Geogr. Inf. Sci. **2014**, *28* (12), 2435–2455.
100. Dube, M.; Barrett, J.; Egenhofer, M. From metric to topology: Determining relations in discrete space. *In Proceedings of the 12th International Conference, COSIT 2015*, Santa Fe, NM, October 12–16, 2015, 151–171.
101. Dube, M.; Egenhofer, M.; Lewis, J.; Stephen, S.; Plummer, M. Swiss canton regions: A model for complex objects in geographic partitions. *In Proceedings of the COSIT 2015*, Santa Fe, NM, October 12–16, 2015, 309–330.

102. Yu, Z.; Luo, W.; Yuan, L.; Hu, Y.; Zhu, A.; Lv, G. Geometric algebra model for geometry-oriented topological relation computation. Trans. GIS **2016**, *20* (2), 259–279.
103. Lin, L.; Feng, L.; Haihong, Z.; Shen, Y.; Zhigang, Z. A two-level topological model for 3D features in City GML. Comput. Environ. Urban Syst. **2016**, *59*, 11–24.
104. Rolf, G.; Ross, S.P.; Lukas, W. Evaluating topological queries in linked data using DBpedia and GeoNames in Switzerland and Scotland. Trans. GIS **2017**, *21* (1), 114–133.
105. Leng, L.; Yang, G.; Chen, S. A combinatorial reasoning mechanism with topological and metric relations for change detection in river planforms: An application to globeland30's water bodies. ISPRS Int. J. Geo-Inf. **2017**, *6* (1), 1–13.
106. Jingwei, S.; Tinggang, Z.; Chen, M. A 27-Intersection model for representing detailed topological relations between spatial objects in two-dimensional space. ISPRS Int. J. Geo-Inf. **2017**, *37* (6), 1–16.

Traffic Analysis

Gabriel Ambrósio Archanjo
Marvin Project, Campinas, Brazil

Barry McCullagh
Dublin, Ireland

Abstract

The transportation system is a vital element of any city and consequently one of the center points of human society. Since motor vehicles are the primary mode of transportation for most, this system is mainly designed for them. Traffic solutions are used to optimize the use of road network, support driver decisions, and detect and react in abnormal situations among other applications. In order to collect the necessary information to be processed in these applications, sensors are necessary. This work describes the use of video cameras as the main sensor in traffic applications and image processing as an approach to interpret the information collected by these kinds of sensors. Concepts of image processing techniques such as background removal, environment modeling, and pedestrian and vehicle detection are discussed. The usage of these concepts is presented in traffic applications such as street intersection monitoring, congestion and accident detection, and driver assistance and self-driving systems. Results and challenges are discussed as well as future trends are outlined.

Keywords: Feature extraction; Image processing; Intelligent systems; Machine learning; Pattern recognition; Smart cities; Traffic analysis.

INTRODUCTION

Motor vehicles are the primary mode of transportation for most. Consequently, they have the most impact in the transportation system, affecting the lives of almost every citizen in cities around the world. The city size, population density, road network characteristics, traffic laws, traffic monitoring, and many other elements influence the traffic flow and consequently are objects of attention. Moreover, being the most widely used means of transportation implies some consequences like the significant number of deaths caused by motor vehicle accidents. The two most populated countries, China and India, together account for 500,000 deaths per year. Considering figures from all countries combined, the number of deaths exceeds one million per year. This data suggest that any effort to improve traffic safety is worthwhile.

With regard to more basic traffic applications like street intersection monitoring and congestion estimation, the use of physical sensors in the road is the most predominant approach to detect vehicles. Usually, inductive loop detectors (ILDs) are installed in the pavement for detecting objects with significant metal mass crossing them. The use of image processing for these purposes has gained attention since (i) ILD demands installation and maintenance interrupting the use of the lanes for some periods and (ii) it is necessary to install an ILD in each lane of each road intersection, whereas video cameras are capable of monitoring multiple lanes at the same time. Moreover, cameras perform more complex analysis and can be used for other purposes such as surveillance.

Improvements in real-time image and video analysis, the emergence of smart cities and self-guided vehicle applications are among other factors contributing to the increased research and applications in the field. Vehicle traffic analysis through image processing is a complex task which involves (i) understanding the environment by detecting its features such as roads, lane regions, transit signs, and traffic lights; (ii) detection and tracking of objects of interest such as vehicles and pedestrians; and (iii) the analysis of these elements in order to extract meaningful information and support decisions. Since the scene is usually outdoor, the applications have to face a variety of challenges including weather conditions which affect color and texture of the objects in the scene; shadows caused by buildings, trees, and other elements in the scene; object occlusion due to camera angle; and changes in the brightness throughout the day.

Before discussing the applications in traffic analysis, some image processing concepts are discussed. Section "Stationary Camera" addresses background modeling and subtraction, an important technique used for segmenting the foreground objects. Then, feature extraction and pattern recognition are introduced to readers not familiar with the concept.

Finally, in the following sections, techniques for traffic analysis and applications are presented. In Section "Conclusion," concluding remarks and future prospects are outlined.

BACKGROUND REMOVAL

An important image processing technique for many applications that need to detect, segment, and classify objects is background removal. In most cases, there are objects in the scene which are of interest and need to be identified and analyzed and others which are irrelevant and should not be considered in the analysis. Irrespective of their relative distances, the objects of interest are labeled as foreground or background objects. In most cases, the ability to separate the foreground elements from the background is fundamental for effectively segmenting and classifying scene objects.

Since the approaches are quite different for stationary and moving cameras, they are addressed separately in Sections "Stationary Camera" and "Moving Camera."

Stationary Camera

The first step to analyze foreground elements like vehicles and persons is to separate them from the rest of the scene. A common approach to detect such elements is the premise that these scene elements are usually in motion and their motion relative to the background objects allows them to be separated from the scene background. In basic terms, any stationary element in the scene should be considered part of the background and any moving element considered part of the foreground. A simple approach to segment the foreground elements is to store an image of the scene without foreground elements as background, denoted as B. Then, for each pixel at video frame $V(t)$ at time t, subtract the intensity of the background pixel at the same position and determine whether it is pixel from a foreground element or not as follows:

$$F(x,y) = \begin{cases} \text{foreground,} & \text{if } |V(x,y,t) - B(x,y)| > T \\ \text{background,} & \text{otherwise} \end{cases}$$

where $B(x, y)$ is the background pixel intensity at (x, y), $V(x, y, t)$ is the pixel intensity at (x, y) of the video frame at time t, and T is a tolerance threshold for comparing background and video frame pixel intensities. It is important to note that the subtraction operation is inherent to image color model.

However, the previous approach does not handle scene changes due to the passing of time such as brightness variance or changes in the background configuration for any other reason. A more sophisticated approach proposed by Wren et al.[1] analyzes each pixel (x, y) independently, fitting a Gaussian probability density function (pdf) based on the last n pixel values for each pixel coordinate. In other words, given the last N video frames, what are the most probable colors for each pixel position? The pdf is defined by the parameters mean μ and variance σ^2. The initial condition may assume the pixel's intensity of the first frame as the mean and some default value for the variance. For each new frame at time t, the mean and variance are updated as follows:

$$\mu_t = \rho I_t + (1-\rho)\mu_{t-1}$$

$$\mu_t = \rho I_t + (1-\rho)\mu_{t-1}$$

$$d = \left\|(I_t - \mu_t)\right\|$$

where d is the Euclidian distance between the value of the pixel and mean, and ρ is the temporal window which determines the impact on the pdf by each new video frame update. For $\rho=1$, the pdf mean and variance are determined by only the current frame, and therefore, every new frame is the background. The smaller the value of ρ, the larger the number of frames used to compute the pdf. A threshold k is used to determine whether a pixel value lies within the confidence interval of the background pixel intensities distribution. Pixels are classified as background or foreground accordingly:

$$\frac{\left\|(I_t - \mu_t)\right\|}{\sigma_t} > k \rightarrow \text{Foreground}$$

$$\frac{\left\|(I_t - \mu_t)\right\|}{\sigma_t} < k \rightarrow \text{Background}$$

This approach has the advantage of updating the background model in real time, adapting to changes in the scene such as illumination and the presence of non-static background objects. Figure 1 demonstrates the outcome of employing this approach for modeling the background of a street scene.

The simple approach presented previously is ideal to explain the concept. Nevertheless, there are more sophisticated approaches for background modeling which can achieve a better performance, as discussed by Piccardi.[2] Having created the background model, it is possible to subtract it from the current scene frame in order to perform the first step in foreground object segmentation. In the case of the pixel-based approach explained previously, for each pixel in the image, the pixel is disregarded if its value is considered to be background in that position, otherwise it is kept as a foreground element for further analysis.

Moving Camera

When isolating foreground objects with a moving camera, the movement of these objects must be identified and

Fig. 1 Background modeling using a stationary camera: (a) a scene frame with many pedestrians; (b) the modeled background after analyzing frames for a few seconds

distinguished from the apparent movement of the background due to the change in position of the camera. Separation of these movements can be performed by estimating the motion of the camera and building up a model of the background using probability distributions similar to those employed in the static camera scenario.

The approach taken by Szolgay et al.[3] to estimating camera motion is similar to that performed during the MPEG-2 encoding process. The current frame is divided into a series of blocks and these blocks are searched for in a reference frame. The displacement of each block between the current and reference frames is known as a displacement vector or a motion vector. The motion vectors calculated can be used to create an estimate of the reference frame in the current frame. Assume that $F(t-1)$ and $F(t)$ are the reference and current frames and D contains the motion vectors for the center pixels in each block, then an estimate of the current frame can be made using the reference frame and the displacements: $F'(t)=(F(t-1)+D)$. The difference between the two frames, E, is calculated using $E=|F'(t)-F(t)|$. Szolgay et al. create what they call a modified error image, MEI, using E and a threshold. If the value at a pixel of E is greater than a threshold, then $\text{MEI}(x, y)=I(x, y)$, else $\text{MEI}(x, y)=0$. An MEI is calculated for each frame.

The background model is created using the previous n frames ($n=15$ in the work of Szolgay et al.[3]). When a new frame is acquired the oldest is removed. The displacements for each of the stored frames are updated so that they are relative to the newest frame. A probability density function is created which is used to identify when pixels in the current MEI belong to the background or an independently moving object which can be passed to an identification stage.

FEATURE EXTRACTION AND PATTERN RECOGNITION

Section "Background Removal" discussed background removal which is one of the possible techniques to separate objects of interest from the scene static elements. This section discusses methods to recognize the scene objects in order to analyze and use them to support decisions. Concepts and techniques described in this section are embodied in the approaches and applications presented in the following sections. The field of feature extraction and pattern recognition is a very wide one; therefore, this section just presents an introduction for readers who are not familiar with the field.

Feature Extraction

When dealing with raw data, much of the information may be of little value for discrimination of the different patterns in the data. Feature extraction is the process of transforming or representing the raw data in a more meaningful way considering the target application. In the case of image processing, this is even more noticeable since the raw data of an image is just a sequence of pixel intensity values. For instance, consider an image with a black square. In the perspective of shape analysis, the pixel data itself does not provide any meaningful information that describes such shape. However, using edge and corner detection algorithms, the number of sides and their lengths, corners, and angle values could be extracted from the image, as shown in Fig. 2.

These new derived attributes, called features, represent the characteristics of the shape in the image much better than an array of pixel intensity values. Pattern recognition methods can use these new attributes to discriminate different shapes as discussed in Section "Pattern Recognition."

Pattern Recognition

Section "Feature Extraction" demonstrated how to extract meaningful information from images targeting a specific application. However, the extraction of that information does not solve any problem. It is just a premise for any approach that uses this new derived information for some purpose. Regarding the example shown in the previous section, having that features extracted from images with shape figures, simple *if-then* rules can be used to discriminate different shapes, as demonstrated in the three rules:

$$if\ sides = 3\ then\ shape = TRIANGLE \qquad (1)$$

$$if\ sides = 5\ then\ shape = PENTAGON \qquad (2)$$

$$\begin{aligned}&if\ sides = 4\ AND\ (sideLength[0] = sideLength[1]\\&\quad AND\ (sideLength[1] = sideLength[2]\\&\quad AND\ (sideLength[2] = sideLength[3]\\&\quad then\ shape = SQUARE\end{aligned} \qquad (3)$$

Fig. 2 (a) Raw information: pixel intensity values; (b) image visual representation; (c) features detected using edge and corner detection algorithms. Having the edges it is possible to estimate the angles. This new set of derived information is not available in the raw data and could not be used by algorithms before the feature extraction step

However, in many cases, it is not common to have high discriminant attributes like in the previous example. The higher is the complexity to distinguish samples from distinct classes, the harder is to create rules by hand as shown above. In these cases, machine learning algorithms can be used to induce models that map input instances to output labels from a set of examples. When the procedure used to generate the model uses a dataset with samples with their classes labeled correctly by hand, it is called supervised learning. There are many different approaches to induce models with this technique, but in general they employ an iterative process that minimizes a function that measures the distance between the current output of the model and the desired one. The backpropagation training method used in artificial neural networks, for instance, adjust de model parameters (neuron interconnection weights) iteratively computing the error between the network output and the values labeled by hand, as shown in Algorithm 1. This adjustment is performed iteratively until a maximum number of iteration max_it is achieved or the sum of errors $total\_error$ (considering N samples) is less than a predefined threshold. The error is a measure of the difference between of the network output and the desired one.

Algorithm 1. Pseudocode of Backpropagation Algorithm Used to Train an Artificial Neural Network

```
network = initialize_network_weights()
  t = 0
  while t < max_it and total_error < threshold do
    for i to N do
      sample = training_dataset[i]
      prediction = network_output(network, sample)
      desired = labeled_output(sample)
      error = compute_error(prediction, desired)
      delta = compute_weights_delta(error, network)
      update_weights(delta, network)
      total_error += (1/N * error)
    end
    t = t + 1
  end
```

In the didactic example presented previously, the features extracted from shape figures are the input instances. The classes of each figure are the labels or classes. Using a training dataset with multiple examples of shape figure features and the shape class specified by hand, a model that maps features to classes could be generated using a machine learning approach like the one presented earlier. Having the model, it can be used to classify new samples which are not in the training dataset. Along with artificial neural networks, there are other approaches to induce models for pattern recognition such as support vector machine (SVM), naive Bayes, and decision tree learning. In this work, some of these approaches are used for classification, and further reading is recommended for those who wish to understand the details behind it. For those more concerned with applications, models induced by machine learning approaches can be seen as black boxes which receive features as input and outputs the recognized class.

ENVIRONMENT MODELING

Traffic standardization predates the automobile industry. In the 19th century, horse-drawn vehicles were the most used mode of transport. The first traffic light, for instance, was installed in 1868 in London as pointed out by Taale et al.[4] Nevertheless, it was only in the beginning of the 20th century that started the long transition from horse-drawn vehicles to automobiles. From that period to the present many traffic rules have being created to standardize the use of such vehicles and improve safety. Among these standards are the signs in the street that specify where the vehicles must travel, plates indicating speed and rules for that road, and traffic light systems controlling the flow. Therefore, the detection and analysis of these elements is fundamental to understand the traffic situation and for making decisions in the many possible applications.

The existence of the traffic standards facilitate the understanding of traffic conditions by humans and therefore by computer automated systems. Just like humans, computer systems may be able to recognize these signs and the intrinsic information. This section describes the concepts and some approaches for such a challenging task.

Lane Detection

Street lanes are a traffic standard that specifies where the vehicle must travel and other possible lanes in the same or the opposite direction. There are two major issues related to detect lane markings: they are not always clearly visible due to print quality and natural wear, and the geometry of the markings cannot be used as a discriminating factor as there is no governing standard in this aspect as pointed out by Lai and Yung.[5]

Lai and Yung[5] proposed an approach to detect the lanes in the road using a stationary camera. Basically, the idea behind the proposed approach is to detect the lines in the scene using Sobel edge detection, then horizontal lines and vertical lines that do not match the characteristics of road lanes are discarded. The remaining lines are the road lanes. First, the background is estimated using a sequence of frames. The background is expected to have all visible lanes since there are no vehicles in the image. Then, Sobel edge detection is used to detect lines. In the next step, a transformation is performed, based on camera parameters such as height, tilt angle, and focal length, in order to compensate for distortions caused by perspective, resulting in a flat image. Finally, the lines detected in the image are clustered and features are extracted such as orientation and length. Rules created using these features determine whether a line is related to road or not.

The previous approach is only suitable for applications using stationary cameras since the background estimation and the following operations suppose the scene is static, except for the moving vehicles. However, lane detection is fundamental in many on-road applications such as driver assistance systems and autonomous vehicles. Wang et al.[6] modeled lanes using cubic b-splines in order to support curved road models. Images of the road collected by a camera placed in the car are divided horizontally into sections. The Hough transform is used to detect the vertical straight lines of the lane markings in each section. By connecting the straight lines of each section, it is possible to determine the curve of the road. Assuming the road boundaries are parallel, projecting the intersection of the boundaries of each side, it is possible to determine the vanishing points of the scene, a necessary parameter to handle perspective. Combining the curve and perspective parameters, splines are defined and used to model the boundaries and the midline of the lane.

Traffic Sign Detection and Recognition

Section "Lane Detection" discussed the importance of lane detection in traffic analysis. This section discusses the detection, analysis, and recognition of another group of important elements in the scene: transit signs. Driver assistance systems can detect and recognize signs and notify the driver in the case of doing a prohibited maneuver. Moreover, in the case of autonomous vehicle applications, the detection, recognition, and interpretation of traffic signs are mandatory in the absence of a priori knowledge of the environment.

Solutions to recognize traffic signs face similar challenges compared to the recognition of other traffic elements in the environment. Weather and hour of the day affect the brightness of the scene and consequently scene objects' color and texture. Like road lanes, traffic sign appearance is degraded by weathering. Shadow and occlusions may be caused by other objects in the scene such as buildings, trees, posts, and wires.

Fortunately, traffic sign design respects restrictive rules. Usually, the signs use simple geometric shapes such as triangles, circles, diamonds, and octagons, as well as different colors to distinguish from each other. The sign colors, mostly red and yellow, were selected in such a way that they are easily noticeable in a natural environment by humans. These definitions also facilitate detection and recognition by computers.

Zadeh et al.[7] proposed an approach that employs a supervised learning method, which uses a dataset composed by manually labeled samples to train an algorithm, to detect and recognize traffic signs. The first step filters image pixel removing non-sign pixels based on their colors. The algorithm was trained using samples of the traffic signs in different light conditions. Using the color distribution in the signs in these different conditions, it is possible to remove the pixels with values outside the distribution. In the next step, edge detection is performed and straight lines and curves are detected. Then, these connected lines are approximated to known classes of polygons and shape is identified. Finally, analyzing the shape and color ratios in subregions of the image, a traffic sign is assigned to the given candidate.

VEHICLE DETECTION

As expected, when analyzing traffic for any reason, vehicles are the components that deserve the most attention. Independent of the application, analyzing vehicles through image processing can be separated into two steps: vehicle detection and vehicle tracking. The first step is responsible for the first assignment of identification to a vehicle in the scene. Usually, an identification number (ID) and the initial coordinates are assigned to that vehicle. Detected vehicles are tracked until they leave the scene in the second step. When determining the current coordinate of a given vehicle, the past coordinates and the motion pattern might be used to improve the accuracy and performance.

Traffic is usually analyzed by the city infrastructure management, extracting useful information about the traffic condition in order to support decisions, detect incidents, congestions, and other undesired situations that demand rapid reaction, or by onboard computer systems in vehicles in order to support the driver or make autonomous decisions. In the former, fixed cameras are placed in strategic points of the city to monitor roads of interest. In the latter, the cameras are placed in the vehicle for analyzing the current road condition. The traffic scene is analyzed under considerably different perspectives in these two situations;

therefore, the approaches are considerably different too and are described separately in this work.

Stationary Camera

Considering scenes taken by stationary cameras, from an image processing point of view, vehicles might be assigned as foreground elements since they are usually in motion in relation to the scene's background elements. Thus, background modeling and subtraction techniques can be used for segmenting foreground elements, including the vehicles. Having separated the foreground elements from the scene background, the next step is the classification of them based on visual or motion features. After classification, these elements can be analyzed in order to provide useful information for the target applications.

Color and texture are important perceptual descriptors to describe objects, so they are commonly used as one of the first discriminative elements in order to determine segments of interest in images. Since roads are in shades of gray, in some scenarios just the color is sufficient to detect and segment vehicles with distinctive colors. Nevertheless, the majority of vehicles are in shades of gray too. Therefore other features such as edges and shapes must be considered. However, in outdoor environments like streets and roads, the color and texture of a given object may appear quite different under different weather and brightness conditions and these have to be taken into account. Rojas and Crisman[8] proposed an approach that classifies pixels as road or non-road color based on the color distribution of roads estimated using a set of images. Then, non-road pixels are grouped into regions and filter rules are applied to filter out non-vehicles. Back projection is used to find the region coordinates in the ground plane and finally, in the last step, regions that collectively form vehicles are joined.

Moving Camera

The problem of detecting objects using a moving camera is more challenging than in a fixed camera situation since there is no clear separation between foreground and background objects. Vehicles ahead at the same speed of the vehicle hosting the camera are seen as stationary objects. Therefore, their detection must be based on visual features that discriminate them from objects of other classes in the scene. Continuously changing landscapes along the road and vehicles that suddenly enter the scene with very different speeds, sizes, and appearances make the task challenging. The approaches presented in this section are fundamental for driver assistance and autonomous vehicle applications.

As discussed by Betke et al.,[9] the detection of passing and distant cars is addressed employing different methods. When other cars pass the camera-assisted car, they usually cover large portions of the image frame, changing the brightness in those regions. This brightness change in a given region is used to hypothesize the presence of a car on it. On the other hand, distant vehicles are detected as rectangular objects. A feature-based approach creates edge maps for vertical and horizontal edges, then aspect ratio and correlation are used to determine whether such an object is a car or not. Bucher et al.[10] focused on the detection of approaching vehicles for collision avoidance. It scans each road lane, starting from the bottom of the video frames, seeking strong horizontal segments. Using a priori knowledge of the range of car length in road lane positions, the approach can infer whether that horizontal line belongs to a vehicle or not.

When it is hard to separate foreground elements from the background with the precision necessary for the target application, stereovision approaches might be employed. Using a stereo camera or two cameras placed in slightly different positions it is possible to use the same principle of human binocular vision to estimate the distance of the objects in the field of view. The disparity map is computed using the difference in the coordinates of pixels in the left and right images corresponding to the same scene point. This coordinate difference is known as disparity and when combined with information about the cameras obtained through calibration (distance between centers of the optical systems, focal lengths of lenses, etc.) can be used to map disparity values to depth. The inverse mapping can be used in traffic analysis systems as discussed by Sun et al.[11] The set of disparity values corresponding to the range of crucial depths can be calculated. When the disparity map for the current scene has been computed, a histogram of disparity values is calculated. The number of peaks in the histogram within the critical range indicates the number of objects which must be identified and processed.

PEDESTRIAN DETECTION

Section "Moving Camera" described the detection of the most common object on roads: vehicles. However, pedestrians are as important as vehicles since any real-world application must respect safety constraints. In the case of infrastructure cameras, detecting pedestrians crossing roads or streets is useful in order to take action in the case they are in a prohibited area or even to notify drivers in some way. The latter depends on a nonexistent infrastructure to create a communication channel between smart city systems and vehicles, but it is already subject of study. On the other hand, for on road applications, like driver assistance systems, in which the camera is placed in the car, pedestrian detection is fundamental. In the case of autonomous vehicle applications, the ability to detect pedestrians is mandatory in order to make reliable decisions, ensuring the safety of passengers and pedestrians.

The detection, recognition, and tracking of objects in real time in outdoor scenes is a challenging task in image processing applications. In the case of pedestrian detection, as pointed out by Geronimo et al.,[12] the major challenges are as follows: (i) the appearance of pedestrians vary since they can change pose, wear different clothes, carry different objects; (ii) they must be detected in outdoor urban scenarios with cluttered background, a wide range of illumination and weather conditions; (iii) pedestrians can be partially occluded by other pedestrians and urban elements; and (iv) they must be identified in motion and different view angles. Moreover, for most applications, the solution must be able to react in real time to be practical in real-world applications.

Stationary Camera

Regarding segmenting pedestrians in scenes captured by stationary cameras, approaches for background modeling and foreground object segmentation—similar to that presented in the previous section—might be used. The fundamental difference between detection vehicles and pedestrian is on in the classification step. In the work presented by Viola and Snow,[13] for instance, a cascading classifier is employed. In this approach multiple classifiers are combined in order to estimate the final classification as shown in Fig. 3. Usually, each classifier uses a small set of features, and with a lower false positive rate, they are combined in order to obtain a more powerful classifier. In this case, classifiers use motion pattern and appearance features. An object is classified as a pedestrian in the case all classifiers classify it positively.

Moreover, pedestrians detected at intersections or crossing the street is useful information for drivers that could be sent to vehicle in order to avoid accidents. Schaack et al.[14] presented a solution for detecting and estimating pedestrians' position and send the information to a screen in vehicle through a car-to-infrastructure communication. For detecting pedestrians, a region of interest is manually defined in the scene as polygon. Comparing sequential frames, objects in motion, or foreground objects, are separated from the background. Since vehicles and pedestrians have a distinct pattern of movement—regarding path, direction, and velocity—they can be discriminated by analyzing their motion patterns. In the latter work, the classification is performed employing a SVM using motion features as parameters. In this specific case, the advantage of not using visual features for classification is the ability to detect not only pedestrians in the scene, but any moving object that can cause an accident such as cyclists or animals.

Moving Camera

The detection of pedestrian using a moving camera faces the same challenges of detecting vehicles due to the dynamic background and existence of a clear separation between foreground and background objects. Vehicles have a rigid body, considerable size, and must travel in specific regions. On the other hand, pedestrian appear in different poses, increasing the challenge for shape algorithms. They are also smaller than vehicles, in many cases presented in low resolution considering the distance to the camera. In addition, unlike vehicles, people can walk in most places of the scene as sidewalks, town squares, shoulders, and roads. Consequently, pedestrian can be occluded by many different objects in the scene and may appear in the scene from any direction. In combination, these factors make the pedestrian detection problem usually more complex than vehicle detection.

Alonso et al.[15] proposed an approach for person detection that uses stereovision for segmentation, with feature detection algorithms to create an input vector for an SVM classifier. First, points of interest are selected in the left and right images and their distance to the camera is computed. A clustering technique is employed grouping these points in order to have candidate regions of interest (ROIs) that are possibly related to a person. For each ROI, a series of features are extracted such as Canny-based features, Haar wavelets, gradient magnitude and orientation, co-occurrence matrix, histogram of intensity differences, and number of texture units. For each ROI, the features are combined to obtain a feature vector. A database is manually created, selecting positive and negative samples, and used to train a SVM classifier.

APPLICATIONS

The previous sections described approaches for environment modeling and vehicle and pedestrian detection and tracking. The simple detection and tracking of these elements does not solve any problem. They are the means, not the end. Nevertheless, the analysis of those elements can result in useful information that can be used into a wide range of traffic applications. The solutions described in this section were chosen based on its popularity or its potential for solving problems in future applications.

Street Intersection Monitoring

Street intersections are worthy of attention in traffic analysis since streets crossing each other have vehicles traveling in different directions. At these intersections, vehicles

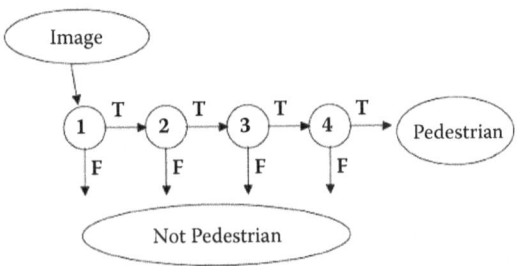

Fig. 3 An example of a cascading classifier for classifying image element as pedestrian

may stay in the same street or change their route going to another one. When there are traffic lights controlling the flow, the use of lanes is optimized allowing high speeds at intersections, but increase the gravity of accidents when vehicles violate the red light. Therefore, monitoring the intersection is fundamental for punishing drivers violating traffic laws and for detecting accidents in order to rapidly dispatch emergency services to minimize the consequences. Rapid medical assistance is crucial for people involved in the accident. The removal of vehicles involved in the accident is fundamental to avoid a traffic jam that may create bottleneck in the traffic network.

Regarding vehicle detection at intersection, ILDs are currently most widely used approach. Basically, it is a physical sensor buried in the traffic lane. This sensor has a wire loop that resonates at a constant frequency when there is no vehicle over the loop. This frequency is monitored by a detection device. When a large metal object, such as a vehicle, moves over the loop, the resonated frequency increases and the vehicle is detected. Since objects are detected by their metal mass, the sensor works in the same way in different weather and lighting conditions, which are among the facts that most affect image processing approaches. On the other hand, it is necessary to install a sensor in each lane intersection. A single vehicle might activate two sensors when passing between them, trucks may be detected as multiple vehicles in some situations, and maintenance issues are among some drawbacks of this approach. The possibility of using a single camera to monitor multiple lanes and the ability to perform more complex analysis, considering visual aspects of the scene, promotes the research for a camera-based solution for this issue.

Kamijo et al.[16] presented an approach to tackle intersection monitoring with the use of image processing. In the first step, the background model of the intersection is constructed by accumulating and averaging the pixel values of each position using an initial 20 min of sequential image frames. Then, virtual sensors are set perpendicular to the incoming roads. Analyzing the intensity variance in these virtual sensors (a rectangle area), it is possible to detect incoming vehicles. When a vehicle is detected by a virtual sensor, the background is subtracted and the vehicle is outlined. The intersection at each time is represented by a data structure that stores the state for each 8×8 pixel block of the scene. After the vehicle detection and segmentation, each block that is occupied by a part of the vehicle is assigned with the vehicle's ID. By analyzing the state of the each block in a sequence of time, its motion vector can be determined. By analyzing the motion of each block in the region of a given vehicle, the motion is estimated and the vehicle is tracked. This approach is employed because simply segmenting the vehicle at each time frame cannot handle occlusions due to the camera angle or vehicles very close to each other. The solution was able to detect vehicles at intersections with occlusion and clutter effects at the success rate of 93%–96%.

Congestion Detection

Traffic congestion is a problem faced by many cities with high population density or limited infrastructure. Among the various approaches, traffic control system optimization might contribute to reducing traffic congestion in some way, saving fuel, reducing emission of $CO(2)$, and consequently improving the lives of millions of citizens. Congestion can be defined as a situation in which the throughput of the road is significantly lower than expected and consequently the same for vehicles' speed. Detecting or predicting congestions can be used to take action in order to mitigate it or reduce its consequences. Dynamic lane reversal, for instance, is an approach that increases the capacity of the road in one direction by inverting the direction of some lanes in the other direction, as illustrated in Fig. 4. In experiments conducted by Hausknecht et al.,[17] the network efficiency was increased by up to 72% in some situations. The sooner an action is taken, the better the result may be. Thus, rapid reaction by the use of automatic detection systems monitoring multiple roads can become a tool to tackle this problem more efficiently.

In order to estimate the traffic condition in a given road, each lane can be monitored, analyzing the speed of the cars in some manner. For instance, Li et al.[18] presented a solution in which virtual line sensors are placed perpendicular to each lane. Using the pixel's values in that virtual line in a sequence of camera frames, a time-spatial image S is constructed where each row in the image is the pixel values in that line at a different time. When there is no vehicle, S is just a gray-shaded rectangle. When a vehicle crosses the line, a slice of the vehicle from each frame is copied to S and the result is a shape. The lower the height of the shape in S, higher the speed of the vehicle. On the other hand, the higher the height of the shape in S, the lower the speed of

Fig. 4 Dynamic lane reversal. In the right is shown a situation that three lanes are used in one direction and just one in another in order to increase road capacity in a direction of interest

the vehicle. Mapping the relationship between shape sizes and traffic conditions, it is possible to estimate traffic and detect congestions in order to take some action.

Instead of using a virtual sensor, Yang et al.[19] analyzed the corners of the vehicles. In each camera frame, a corner detector is applied and the corners are grouped based on the distance between them. Analyzing the motion of the corners, it is possible to discriminate the corners of vehicles from the corners in the background. Each corner group is assigned to a vehicle and by analyzing the coordinates of the corners in a sequence of frames, it is possible to estimate the speed in some sense, in that case, corner pixels moved per frame. Manual setup can map low-speed situations in pixels per frame in order to notify probable congestions. Another approach grouping corners to detect and track vehicles for congestion detection is presented by Chintalacheruvu and Muthukumar.[20] But, in this case, in order to assess the speed in miles per hour, a more intuitive unit for understanding the situation, two virtual detection zones are manually placed in the scene. The distance of the two zones in scene is also a parameter of the system. Analyzing the difference in time between the detection of the same vehicle in each zone, it is possible to assess the speed in distance per time unit.

In these works, the results analyze the capacity of the proposed solutions to detect congestion scenarios effectively. Manual setups, inability to work during the night or in heavy raining conditions are some barriers for broader adoption of these systems. The existence of other solutions using physical sensors, RFID, or GPS data, which work independently of illumination and weather conditions, contributes to this phenomenon either. Nevertheless, the possibility to use a single camera sensor for surveillance and traffic analysis for multiple purposes promote the research and application in the field.

Accident Detection

With much technological advancement in driver safety systems and road quality, the number of traffic fatalities per distance traveled has been decreasing in recent years. Figure 5 illustrates this aspect for the case of the United States. The decreasing line is the number of annual deaths per billion miles traveled. The increasing line is the number of vehicle miles traveled. As can be seen, people are statistically less involved in traffic accidents per distance traveled. However, the population and number of cars increased and, consequently, the absolute number of deaths per year still increases in many countries. In the case of the United States, the number of deaths increased to 33,561 in 2012, which is 1,082 more fatalities than in 2011. In the case of India and China, the two most populated countries, the scenario is more dramatic. In 2011, the numbers of China and India in together accounts more than 500,000 deaths.

Considering the data presented above, any effort to reduce these rates are worthwhile. An aspect that affects the number of fatal accidents is the lag time between an accident and the arrival of medical assistance, as mentioned by Evanco.[21] Automatic accident detection systems might be employed to decrease this lag time.

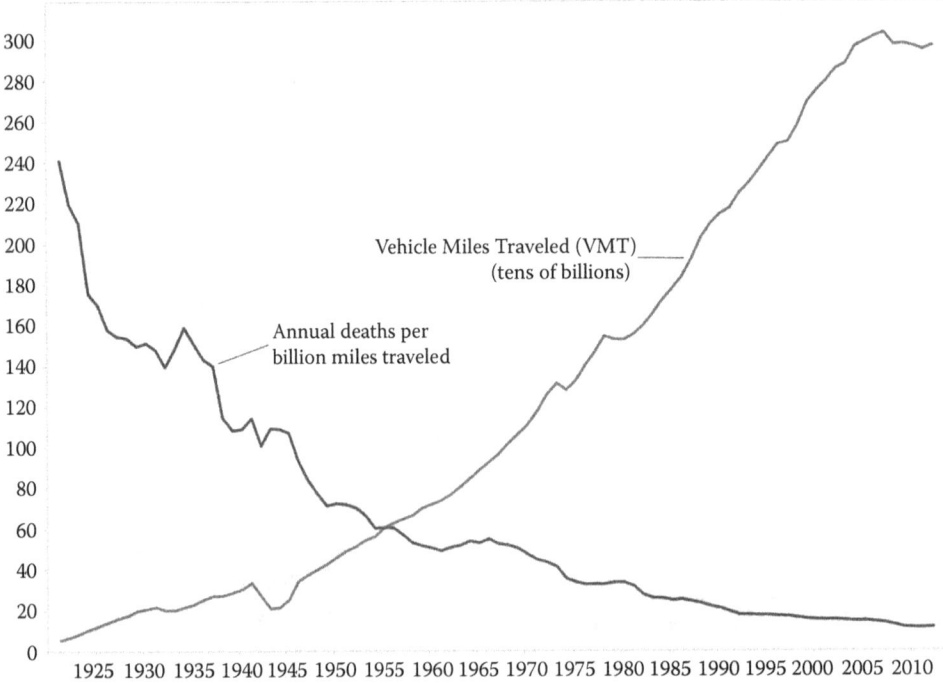

Fig. 5 Annual US traffic fatalities per billion vehicle miles traveled (decreasing line) and miles traveled (increasing line) from 1922 to 2012

In order to detect traffic accidents in real time, Sadeky et al.[22] demonstrates an approach that analyze a camera stream, estimating motion by computing optical flow for a sequence of video frames. The center of gravity is located for each motion pattern using a histogram of flow gradient and the distance between them are calculated. Analyzing the distance between patterns during a time period, a probabilistic classifier is obtained for predicting the probability of an accident in the scene. The approach was tested in a small dataset, depicting a total of 250 real scenes of traffic accidents or abnormal vehicle events captured by traffic surveillance cameras, and presented an expressive detection rate of 99.6% of accuracy.

Driver Assistance Systems

The first motor vehicles date from the end of the 20th century. Automatic transmission systems date from the end of the 1940s. Cruise control systems became popular in some countries in the 1970s and onboard computer systems in the 1980s. As can be seen, motor vehicle development is heading toward reducing manual operations of driving. The last advancements in this field are mainly in intelligent driver assistance systems. Basically, a computer system, monitoring the internal and external environment using sensors, is able to take simple actions and notify the driver of any important information. In order to get information about the vehicle operation, driver situation and road condition as reliably as possible, multiple types of sensors such as radar, laser, and cameras are employed. It is important to note that sensors are only responsible for getting raw information. This information must be processed and interpreted by a computer to be useful. In the case of the data provided by the cameras, image processing techniques are responsible for processing and extracting information.

In order to illustrate the use of image processing in driver assistance systems, two applications were chosen and discussed in the next subsections.

Driver Fatigue Detection

As can be seen in several statistics, driver fatigue accounts for a significant number of road accidents. Truck drivers and others who drive certain amount of hours a day are the most common victim of fatigue, risking their lives and the others. Therefore, the development and adoption of systems capable of detecting fatigue and notifying the driver in some manner could improve safety and save lives.

The most serious consequence of fatigue is the driver falling asleep. A firm closure of the eyes usually precedes the onset of sleep and it can be detected in order to send some signal to the driver. Devi and Bajaj[23] proposed an approach to tackle this issue by placing a camera in front of the driver. Applying an approach to detect skin-colored pixels, the face region is defined. Analyzing the average intensity of each image row, the eyes are detected since they have a distinctive pattern compared to the other face elements. Having identified the eye region, analyzing the color distribution and focusing on white, it is possible to determine whether the eye is open or closed. In the case of the eye being closed for a number of video frames, superior or a given threshold, fatigue is detected. The approach was tested in a controlled scenario with good image quality regarding brightness and sharpness. For real-world applications, improvements are necessary to deal with low-light conditions and a method to notify the driver has to be defined.

Collision Avoidance

When driving accompanied by family members or friends, it is a difficult task to remain focused on the road all the time. It is usual to get into conversations when driving. Consequently, many rear-end collisions are the consequence of drivers distraction. Abrupt braking maneuvers by a vehicle ahead are another cause of this type of collision. In this situation, any delay in making a decision is decisive. Therefore, collision avoidance systems which can notify the driver when an object is approaching the vehicle can improve safety and reduce accidents.

Laser- and radar-based solutions are the most common ones because of reliability. Nevertheless, the potential of using a video camera, a sensor that provides more rich information, for multiple applications and for dealing with more complex scenarios promotes research in image processing and computer vision targeting future applications.

Aiming to detect braking maneuvers of vehicles ahead, Chachuli et al.[24] proposed an approach to detect the beak lights being turned on by the use of a hybrid color model and morphological operations. The hybrid color model combines the red color from the RGB color model and saturation and intensity values from the HSI color model. With knowledge of the brake light position, it is possible to estimate the vehicle's position. Even though this approach is dependent on the presence of the brake light, it can be combined with other approaches since in the scenario in which there is a brake light, it could provide a better response.

Also using a camera placed in the vehicle, Bucher et al.[10] proposed a method that scans each road lane, starting from the bottom, applying image processing techniques in order to detect possible approaching vehicles. After the vehicle is detected its vertical position in the video frame is used to estimate the distance in some manner. The advantage of this approach is that it is able to handle approaching vehicles in different situations, including those stopped on the road.

Self-Driving Automobile

The previous section described the applications of driver assistance systems to support drivers, increasing safety and comfort. Since the 1980s, approaches to create autonomous motor vehicles, completely removing the driver in some situations, have been developed. In 2004, the first DARPA Grand Challenge—a prize competition for American autonomous vehicles—took place in order to promote the development in the field. The entry of a large company like Google into the field brought up the topic for the mainstream media, creating a lot of expectations. Consequently, the research efforts in the field have been increasing in the last few years.

The potential of future commercial applications will probably motivate even more investments in the next years. More reliable decisions in less controlled environments, for instance, in difficult weather conditions, in situations in which other drivers made unexpected or even illegal maneuvers, and in cases that objects suddenly appear in the scene such as distracted pedestrians, are essential to get people approval and political support.

Different to driver assistance systems which just support the driver with information and simple actions, self-driving capability implies total control of the vehicle functions. The system must be able to not only drive the car but also use the proper signals to inform other drivers about the next maneuver, and most importantly, it must be prepared for unexpected behavior from other drivers and people in the street. In order to accomplish that, the self-driving automobile combines a series of sensors and systems. As expected given its rich information, video cameras are among the most important sensors and image processing techniques at the heart of the solutions.

Image processing solutions have an important role in the development of autonomous driving solutions. Levinson et al.[25] and Teichman et al.[26] described an autonomous driving system consisting of two spherical cameras and two stereovision cameras that demands image processing to interpret their data. Moreover, the system also has a rotating LIDAR which uses a rotating laser system to produce an image of the scene. Other supporting systems such as radars and GPS are also used. However, the most significant data to the system are the images; therefore, image processing is on its core. For instance, in order to recognize objects, it is combined object shape and motion description classifiers. The system is able to recognize pedestrian, cyclists, and cars in real time. Many techniques described in this work such as pedestrian, vehicle and lane recognition, brake light detection, and traffic sign recognition might be combined to compose self-driving solutions. Many improvements and new approaches are expected to be released in this field of application and image processing is probable to be involved in many of them.

CONCLUSION

This entry presented an overview of image processing approaches and applications to analyze the traffic scene images in order to extract meaningful information to support decisions in a wide range of applications. The basic concepts of image processing techniques such as background removal, environment modeling, and pedestrian and vehicle detections are explained and references are given for details of each specific approach. Applications using these basic concepts and others for traffic analysis (e.g., street intersection monitoring, congestion and accident detection, collision avoidance systems, driver fatigue detection, self-driving automobiles) are discussed in order to give a panorama of the field. It is important to note that this work does not cover all existing traffic applications that employ image processing techniques, but it rather considers representative systems that demonstrate the most common concepts and highlight the major trends in the area.

Although this work addresses several sophisticated solutions, their adoption seems likely to happen gradually. In the case of driver assistance systems, a new solution must be well developed and reliable in many different conditions to go to the market. There is also an economic barrier for many of these solutions to get into mainstream market. Automobile industry is a very competitive one; therefore, the benefits of a new solution must justify the increasing in the vehicles prices. This phenomenon can be observed in the case of driver assistance systems and other more sophisticated solutions only available in high-end and more expensive vehicles. Economic indices of the markets also affect the configuration of the vehicles and the adoption of such solutions. For instance, seatbelt is mandatory in every country, whereas airbag is not mandatory in many developing countries. In the case of applications that take autonomous actions, in which for safety reasons must have high reliability, such as autonomous vehicles, the adoption will depend of the regulation and political efforts in each market.

In the forthcoming decades, much of the attention will be focused on smart cities applications and autonomous vehicles. Real-time automatic traffic analysis integrated with other city systems has applications in public safety, and traffic and emergency management, so it is expected to become more popular in next years. In the case of autonomous vehicles, a lot of expectations were created due to significant technological improvements and efforts from companies like Google that brought the subject into mainstream media. Traditional automakers like Volvo has announced that they will release autonomous vehicles for test with real drivers before 2020. Nevertheless, even after regulation approval and releasing of the first commercial solutions, the adoption is expected to happen gradually due to many reasons. Most notably, just the sensors used in current autonomous vehicle solutions are several hundred thousand US dollars' worth. The entrance of new players in the competition targeting this new and very valuable market may contribute for development of solutions and equipment and continuous drop in price in the next decades.

REFERENCES

1. Wren, C.R.; Azarbayejani, A.; Darrell, T.; Pentland, A.P. Pfinder: Real-time tracking of the human body. IEEE Trans. Pattern Anal. Mach. Intell. **1997**, *19*, 780–785.
2. Piccardi, M. Background subtraction techniques: A review. In *IEEE International Conference on Systems, Man and Cybernetics*, Vol. 4; 2004, 3099–3104.
3. Szolgay, D.; Benois-Pineau, J.; Mégret, R.; Gaëstel, Y.; Dartigues, J.F. Detection of moving foreground objects in videos with strong camera motion. Pattern Anal. Appl. **2011**, *14*, 311–328.
4. Taale, H.; Hoogendoorn, S.; van den Berg, M.; De Schutter, B. Anticiperende netwerkregelingen. NM Magazine **2006**, *1*, 22–27.
5. Lai, A.H.; Yung, N.H. Lane detection by orientation and length discrimination. IEEE Trans. Syst. Man Cybern. B Cybern. **2000**, *30*, 539–548.
6. Wang, Y.; Teoh, E.K.; Shen, D. Lane detection and tracking using B-Snake. Image Vis. Compu. **2004**, *22*, 269–280.
7. Zadeh, M.M.; Kasvand, T.; Suen, C.Y. Localization and recognition of traffic signs for automated vehicle control systems. In *Intelligent Systems & Advanced Manufacturing*, 1998, 272–282.
8. Rojas, J.C.; Crisman, J.D. Vehicle detection in color images. In *IEEE Conference on Intelligent Transportation System*, 1997.
9. Betke, M.; Haritaoglu, E.; Davis, L.S. Real-time multiple vehicle detection and tracking from a moving vehicle. Mach. Vis. Appl. **2000**, *12*, 69–83.
10. Bucher, T.; Curio, C.; Edelbrunner, J.; Igel, C.; Kastrup, D.; Leefken, I.; von Seelen, W. Image processing and behavior planning for intelligent vehicles. IEEE Trans. Ind. Electron. **2003**, *50*, 62–75.
11. Sun, Z.; Bebis, G.; Miller, R. On-road vehicle detection: A review. IEEE Trans. Pattern Anal. Mach. Intell. **2006**, *28*, 694–711.
12. Geronimo, D.; Lopez, A.M.; Sappa, A.D.; Graf, T. Survey of pedestrian detection for advanced driver assistance systems. IEEE Trans. Pattern Anal. Mach. Intell. **2010**, *32*, 1239–1258.
13. Viola, P.; Jones, M.J.; Snow, D. Detecting pedestrians using patterns of motion and appearance. In *Ninth IEEE International Conference on Computer Vision*, 2003, 734–741.
14. Schaack, S.; Mauthofer, A.; Brunsmann, U. Stationary video-based pedestrian recognition for driver assistance systems. In *Proceedings of 21st International Technical Conference on the Enhanced Safety of Vehicles*, 2009.
15. Alonso, I.P.; Llorca, D.F.; Sotelo, M.Á.; Bergasa, L.M.; Revenga de Toro, P.; Nuevo, J.; Garrido, M.G. Combination of feature extraction methods for SVM pedestrian detection. IEEE Trans. Intell. Transp. Syst. **2007**, *8*, 292–307.
16. Kamijo, S.; Matsushita, Y.; Ikeuchi, K.; Sakauchi, M. Traffic monitoring and accident detection at intersections. IEEE Trans. Intell. Transp. Syst. **2000**, *1*, 108–118.
17. Hausknecht, M.; Au, T.C.; Stone, P.; Fajardo, D.; Waller, T. Dynamic lane reversal in traffic management. In *IEEE Conference on Intelligent Transportation Systems (ITSC)*, 2011, 1929–1934.
18. Li, L.; Chen, L.; Huang, X.; Huang, J. A traffic congestion estimation approach from video using time-spatial imagery. In *Intelligent Networks and Intelligent Systems*, 2008, 465–469.
19. Yang, Z.; Meng, H.; Wei, Y.; Zhang, H.; Wang, X. Tracking ground vehicles in heavy-traffic video by grouping tracks of vehicle corners. In *Proceedings of the IEEE ITSC*, 2007, 396–399.
20. Chintalacheruvu, N.; Muthukumar, V. Video based vehicle detection and its application in intelligent transportation systems. J. Transp. Technol. **2012**, *2*.
21. Evanco, W.M. The impact of rapid incident detection on freeway accident fatalities. Mitretek 1996.
22. Sadeky, S.; Al-Hamadiy, A.; Michaelisy, B.; Sayed, U. Real-time automatic traffic accident recognition using HFG. In *Pattern Recognition (ICPR), 2010 20th International Conference on*, 2010, 3348–3351.
23. Devi, M.S.; Bajaj, P.R. Driver fatigue detection based on eye tracking. In *First International Conference on Emerging Trends in Engineering and Technology*, 2008, 649–652.
24. Chachuli, S.A.M.; Ishak, K.A.; Yusop, N. Vehicle brake light detection using hybrid color model. Appl. Mech. Mat. **2014**.
25. Levinson, J.; Askeland, J.; Becker, J.; Dolson, J.; Held, D.; Kammel, S.; Thrun, S. Towards fully autonomous driving: Systems and algorithms. In *Intelligent Vehicles Symposium (IV)*, 2011, 163–168.
26. Teichman, A.; Levinson, J.; Thrun, S. Towards 3D object recognition via classification of arbitrary object tracks. In *2011 IEEE International Conference on Robotics and Automation (ICRA)*, 2011, 4034–4041.

UAS Imaging

Yuxia (Lucy) Huang
Computing Sciences, Texas A&M University – Corpus Christi, Corpus Christi, Texas, U.S.A.

Lihong Su
Harte Research Institute for Gulf of Mexico Studies, Texas A&M University – Corpus Christi, Corpus Christi, Texas, U.S.A.

Abstract
Unmanned aerial systems (UASs) have been used as a robust tool for agricultural and environmental applications in recent years. Geo-referencing, convolution, texture calculation, and co-registration are the four critical components in UAS image processing. These components are compute-intensive due to the large amount of data and the complexity of the processes. On the other hand, the modern GPU promises to be significant for image processing. This entry introduces a novel GPU-based approach for great speed increases in these critical components in processing UAS imagery. The proposed approach takes advantage of the massive parallel computing in the modern GPU and demonstrates the capacities of UAS image processing.

Keywords: Cross correlation coefficient; GPU kernel design; GPU memory usage; Gray level co-occurrence matrix; Image processing; Parallel computing; UAS remote sensing.

INTRODUCTION

Unmanned aerial systems (UASs) are experiencing the greatest near-term growth in civil and commercial operations due to their versatility and relatively low initial cost and operating expenses.[1] In recent years, small UASs have been used as a robust tool for remote sensing in agricultural and environmental applications. The images acquired from small UAS usually have a small footprint, hyperspatial resolution (such as sub-decimeter), and high overlapping ratio due to its low flight altitude and high definition cameras. For a regular study area, such as the crop fields in a farm or the coastal shoreline of an island, image acquisition from small UAS commonly results in hundreds or thousands of large small-footprint images.[2–4] For example, the RS-16 UAS produced over 200 GB imagery by 1 h data acquisition. The numerous large UAS images cause a significant computing load and further analysis and processes. Furthermore, in near-real time agricultural and environmental applications, such as crop health and precision pest management, and flood relief and emergency management, the investigators usually need to see the classification maps derived from those large UAS images in a very short time, such as by the next day.

Big data volume and time-limit demands are the two main challenges that UAS remote sensing faces [Su]. Geo-referencing, convolution, texture calculation, and co-registration are the four critical components in UAS image processing. These processes typically are computationally intensive due to the large amount of data and the complexity of the process. The traditional sequential computing approach is difficult to meet these kinds of data intensive real-time applications. More efficient processes are necessary so that final products from UAS remote sensing can be effectively obtained in a reasonable time. Fortunately, with the fascinating evolution of the graphics processing unit (GPU), the modern GPU is promising to be a significant part in addressing these issues due to its massive parallel computation capability.[5–6] Some GPU methods have been proposed for remote sensing image processing, for example, geo-correction and orthorectification,[7–9] and segmentation and classification.[10] However, more comprehensive studies on GPU-based UAS image processing need to be investigated in order to provide tremendous speed increases over sequential computing approaches.

This entry explores a GPU-based approach for UAS image processing. The remainder of this entry is organized as follows. Section "Background" reviews remote sensing image processing and introduces CUDA (compute unified device architecture) architecture and programming model. CUDA, namely, compute unified device architecture, is a parallel computing platform and application programming interface model for general-purpose GPU computing. Section "The GPU-Based Approach" proposes a GPU-based approach and its implementation to address the aforementioned four critical components in UAS image processing. In this section, the algorithm and the GPU implementation for these components are presented in the first four subsections, one for each subsection, followed

Encyclopedia of Image Processing, First Edition
DOI: 10.1201/9781351110273-140000144
Copyright © 2018 by Taylor & Francis. All rights reserved.

by the discussions on the GPU memory usage and kernel design in the last subsection. Our conclusions are given in the final section.

BACKGROUND

Remote Sensing Image Processing

Procedures for conventional remote sensing image processing are often grouped into three board categories: image rectification and restoration, image enhancement, and image classification.[11]

According to this categorization, the four image processing methods investigated in this entry fall into the first two categories. On the other hand, from the respective of neighbor pixels used in the manipulations, those procedures also can be categorized into another three broad groups: pixelwise manipulations, spatial manipulations, and multi-image manipulations (Fig. 1). First, pixelwise manipulations can further be divided into brightness manipulations and position manipulations, which are based on independently modifying the brightness value or the position value of each pixel in an image. Color space transformations, such as from red–green–blue space to intensity–hue–saturation space, are examples of brightness manipulations while geo-reference is an example of position manipulation. The second group, spatial manipulations consist of local operations and global operations. Local operations modify the value of each pixel only based on neighboring brightness values, while global operations use the information from the entire image. Spatial filtering, texture calculation, and convolution are examples of the local operations, while Fourier analysis and principle components analysis are examples of the global operations. Geo-referencing can be considered as either a pixelwise manipulation or a local spatial manipulation depending on whether the bilinear interpolation is used or the nearest neighbor sampling is used. The last group, multi-image manipulations use two or more images in the same manipulation. The examples include image co-registration, three-dimensional (3D) point clouds creation, and digital surface model generation. In this entry, we adopt this categorization, which is from the respective of neighbor pixels used in the manipulations. This is because the GPU memory usage, a central concern in the GPU programming, is heavily affected by the neighbor pixels used in manipulations. In this regard, geo-referencing, convolution, texture calculation, and co-registration fall into pixelwise manipulations, spatial manipulations, and multi-image manipulations, respectively.

CUDA Architecture and Programming Model

CUDA is a general-purpose parallel computing architecture invented by NVIDIA, a world leader in visual computing technologies.[12] The CUDA platform is designed to work with programming languages such as C and C++ for an

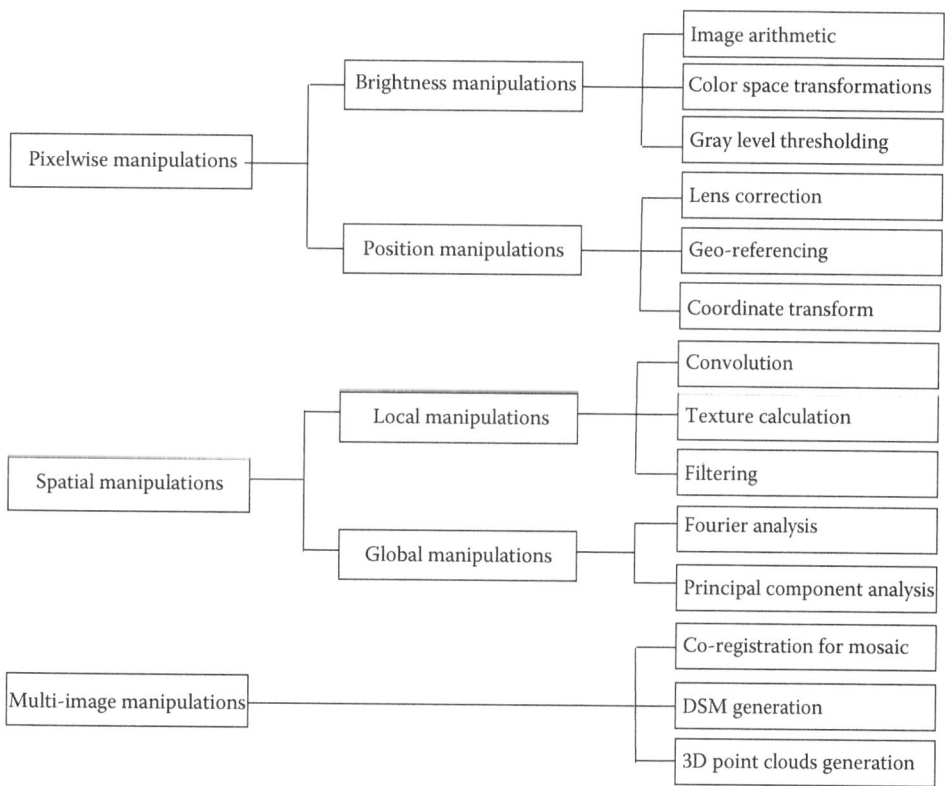

Fig. 1 The categorization of remote sensing image processing procedures: from the perspective of neighbor pixels used in the manipulations

easy use of the GPU resources. The CUDA computing system consists of a host and one or more devices, where the host is a traditional central processing unit (CPU) and the devices are GPU with massively parallel processors.

A CUDA program often consists of one or more phases that are executed either on the CPU or on the GPU, depending on specific tasks. The phases that exhibit little or no data parallelism are implemented by the CPU code, which is the straight ANSI C code. The phases that exhibit a rich amount of data parallelism are implemented by the GPU code, written by the so-called kernels. A CUDA kernel is executed by a group of threads, where a thread is the minimum execution unit (Fig. 2). The threads are executed in parallel and an array of the threads (typically set to a multiple of 16) form a block, and further an array of blocks form a grid. Normally the total number of threads in a block does not exceed 1,024. Moreover, the blocks are arranged in a grid of 3D array, and each dimensional value can be ranged from 1 to 65,535. All threads in the same grid execute the same kernel. A kernel function must be called with an execution configuration, which defines the array of blocks and the array of threads. The same kernel function can be executed with different configurations, namely, different grids. Therefore, a CUDA program can have multiple grids and they are executed sequentially.

At any time, threads belong to the same CUDA kernel execute the same instruction, but they operate on different data. Two variables provided by CUDA, blockIdx (the index of a block in a grid) and threadIdx (the index of a thread in a block), are used to localize the data needed to process and to distinguish threads from each other.

In CUDA, the host and the devices have separate memory spaces. The GPU has five types of memory spaces: global, constant, shared, local, and registers (Table 1). The threads can access to multiple memory spaces, and each thread has a private memory space namely registers and local memory. The automatic variables declared in a CUDA kernel are placed into the registers. If the variable is an array, the compiler would place it in the local memory. The local memory is a part of the global device memory. A local memory variable, however, is accessible only by the thread that declares the variable.

Each block has a shared memory and each thread in this block can access to the shared memory. All threads in the same block can share data efficiently through the shared memory, as well as synchronize their execution in order to coordinate the access to the shared memory. Threads in different blocks, however, cannot cooperate with each other. All blocks can run in parallel with each other. Each thread can access to both the global memory and the constant memory, which are used for data transfer between the host and the devices. Constant variables are often used for variables that provide input values to kernel functions because all threads in a grid see the same version of a constant variable during the entire application execution. Constant variables are stored in the global memory but they are cached for access in an efficient manner. The constant memory and global memory can be accessed by different kernel functions with different execution configurations.

The shared memory latency can be roughly 100 times lower than the global memory latency.[12] The local memory latency is the same as the global memory latency because the local memory is a part of the global device memory. The constant memory has the same efficiency in terms of access as the shared memory; however, the constant memory is read-only. The profitable strategy for performing computation on the GPUs include first divide data

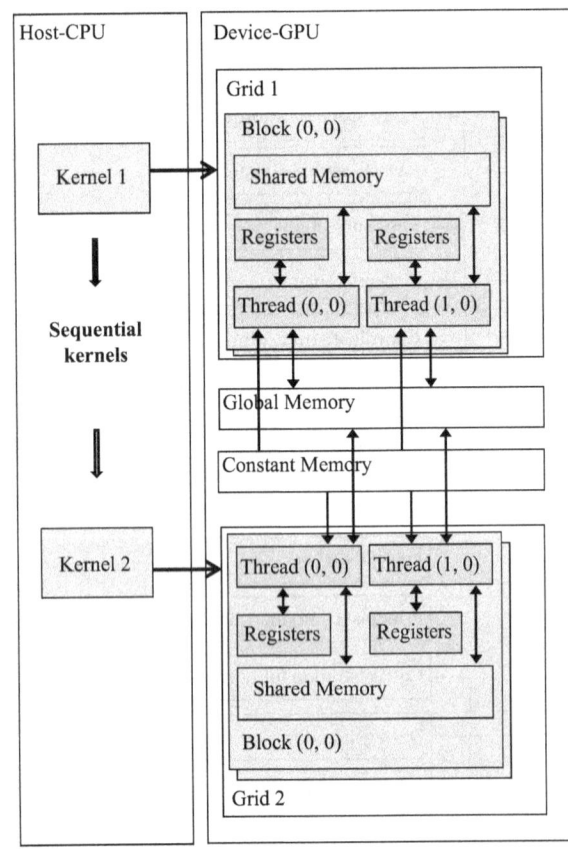

Fig. 2 CUDA memory model

Table 1 CUDA memory types and their characteristics

Variable declaration	Memory	Scope	Access	Penalty
int localVar	Register	One thread	Read/Write	1×
int localArray	Local	One thread	Read/Write	100×
__share__ int shareVar	Shared	All threads in a block	Read/Write	1×
__device__ int globalVar	Global	All threads+host	Read/Write	100×
__constant__ int constantVar	Constant	All threads+host	Read	1×

into subsets, then copy them from the global memory to the shared memory, achieve shared memory locally in threads, and finally copy the results back to the global memory. In addition, when we design a GPU application, we should balance the number of kernels to be employed and their complexities. A good design consists of a group of small kernels and each executes simple work. In this way, a large number of threads can be executed concurrently and at the same time the overall performance can be improved.

Different GPU hardware may have different configurations. The CUDA environment, however, takes care of the differences. The transparent scalability lets users focus on their applications. According to a CUDA programming model,[13] the basic flow of a standard GPU application is as follows:

a. Declare and implement kernel functions, for example, __global__ void KernelFunc(input parameters) { /*A series of commands */ }
b. Declare input and output variables in CPU memory.
c. Declare corresponding variables in the GPU global memory using cudaMalloc() function.
d. Transfer input data from GPU global memory to CPU memory using cudaMemcpy() function.
e. Run a kernel function by <<<dimGrid, dimBlock>>> instruction.
f. Transfer result data from GPU global memory to CPU memory using cudaMemcpy() function.
g. Free variables in GPU memory using cudaFree() function such as cudaFree(dev_a).

THE GPU-BASED APPROACH

GPU-Based Lens Distortion Correction and Geo-referencing

Lens Distortion Correction

The lens distortion correction uses the internal geometry of a camera that exists at the time of data capture to transform the photo pixel coordinate system to the imaging plane coordinate system.[13] The internal geometry includes the focal length and lens distortion coefficients. Two types of lens distortion might exist and they are radial lens distortion Δr and tangential lens distortion Δt. Since the tangential lens distortion is usually much smaller in magnitude than the radial lens distortion, it is considered negligible. For a digital camera, the radial lens distortion usually is an inward displacement of a given image point from its ideal location. The inward displacement should be eliminated by adding matching displacement on each pixel. The effects of radial lens distortion throughout an image can be approximated using a polynomial.[13] A commonly used forward correction model of nonmetric digital camera lens distortion is given in the following equations:[14]

$$\Delta x = (x - x_0)(k_0 r + k_1 r^3 + k_2 r^5) \qquad (1)$$

$$\Delta y = (y - y_0)(k_0 r + k_1 r^3 + k_2 r^5) \qquad (2)$$

where x_0 and y_0 are the principal point offsets from the image center; x and y are the coordinates of the image point; Δx and Δy are the corrections of the image point coordinates; k_0, k_1, and k_2 are the radial lens distortions; and $r = \sqrt{(x - x_0)^2 + (y - y_0)^2}$.

Geo-referencing

Geo-referencing establishes the location of an image in terms of ground coordinate systems. In other words, geo-referencing generates a destination image that has geographic coordinates from a photo (namely, an original image) with no geographic coordinates. As we know this is a key process to make UAS imagery useful for mapping and further analysis. A ground coordinate (X, Y, Z) is typically defined as a 3D Cartesian coordinate system, in which (X, Y) uses a known map projection such as the Universal Transverse Mercator (UTM), and the Z value is the elevation above the mean sea level by a given vertical datum. Geo-referencing can be empirically and manually completed through affine transformation between distorted photo coordinates (column, row) and map coordinates (X, Y) with well-distributed ground control points (GCPs).[11] Here lens distortion correction is included in the affine transformation. However, determining sufficient GCPs is a significant human resource component in an image processing system. To overcome this problem, geo-referencing is often conducted through a photogrammetric approach, which consists of four main components: boresight calibration, camera lens distortion correction, obtaining camera position and attitude, and orthoimage production. Typically, the camera, the global positioning system (GPS) receiver, and the inertial measurement unit (IMU) are installed individually on an airplane frame. The GPS position is required to shift to the camera position, namely, the perspective center. The boresight calibration adjusts the offset angles between IMU and camera reference frame axes.[15] The perspective center (X, Y, Z) and the three rotation angles (ω, ϕ, κ) are acquired by GPS and IMU measurement, and they are associated with the ground coordinate system. The camera position and the angles at the time of imaging can be obtained by interpolating the GPS and IMU data through dedicated software. According to the camera optical principle, namely, collinearity, a straight line can be extended from the perspective center of a camera to a pixel on the photo and further to the object of the pixel on the ground.[13] An orthoimage can be produced based on the collinearity, where the lens distortion correction is logically integrated into the collinearity.

A GPU Implementation of the Lens Distortion Correction and Geo-referencing

Figure 3 provides pseudocode of a GPU-based index array for UAS image geo-referencing.[3] The four photo corners are first projected onto the ground coordinate, such as UTM, so that the footprint of the destination image falls into the minimal bounding rectangle (MBR) formed by these four corners. This step is implemented on the CPU because little data parallelism is involved in the calculation. After that, given a ground pixel size, the ground coordinates of each pixel in the MBR can be calculated. The pixel's coordinates on the imaging plane are obtained through collinearity, and then the lens distortion correction computes the displacement of the location to obtain its position on the original photo. The calculation of each pixel of the destination image is implemented on the GPU due to the fact that an image commonly consists of millions pixels and each pixel can be calculated independently.

Several methods can be used to obtain the gray value for each pixel of the destination image based on its corresponding position and neighbors on the original photo. A commonly used method, the nearest neighbor resampling is adopted in this entry due to its simplicity and the capability of preserving original radiometric values. The preservation of the original gray value is very important because low radiometric resolution off-the-shelf cameras are often used in UAS remote sensing. Interpolation methods of producing the pixel gray values bring the effects of neighboring pixel gray value into the interpolated pixel. This mixing process, however, reduces the further radiometric resolution. Although the nearest neighbor resampling may produce some position errors, especially along linear features, the position errors of several pixels on UAS images do not produce large displacements due to its sub-decimeter pixel size.[3]

Another advantage of using the nearest neighbor resampling method is that it is not necessary to access the gray value of the original photo during geo-referencing. Pixels of the destination image store the location of its nearest neighbor pixel on the original photo, namely, row and column, instead of the gray value. The destination image can be produced later by assigning its pixel gray value with the index array. Remote sensing images usually are multispectral or hyperspectral images, which have spatially co-registered several spectral bands or hundreds of spectral bands, respectively. This index array reduces huge data transfer between the CPU and the GPU memory spaces.

The input parameters of geo-referencing are declared as constant variables because each thread uses the same set of parameters and coefficients. Both the original photo and the destination image are variables in the CPU memory without transferring between the CPU and the GPU. Because the index array is fully generated on GPU, the index array is only transferred from the GPU global memory space to the CPU memory space. Due to the independence of the index array elements, the threads do not need to share data among them. The GPU memory usage is simple, but the processing algorithm is complex that leads to a complex kernel. Complex kernels are those long kernels that may contain complex arithmetic or logic operations. The more complex a kernel is, the slower the kernel is. This is because launching each kernel takes an overhead time that could reduce its performance. The proposed GPU-based geo-referencing implementation uses the constant memory

```
constant__ struct-of-input-parameters GeoRef-parameters; /*declare input parameters as constant variables*/
    /*declare and implement the geo-referencing kernel function*/
    __global__ void GeoRef()
{
        Task 1 - calculate the imaging plane coordinates of each pixel in MBR by colinearity.
        Task 2 - calculate the row/col on photo pixel coordinates for the pixel by lens distortion correction.
        If the row/col is valid, the pixel is set row*Width-of-Photo + col,else set -1.
}
/*CUDA main program*/
int main()
{
        CPU task 1 - compute the UTM coordinates of four corners of an image
        CPU task 2 - obtain the minimal bounding rectangle (MBR) that the four corners form.
        cudaMemcpyToSymbol(GeoRef-parameters, oneCase-parameter, sizeof(oneCase-parameter));
        /*move the parameters for a photo into the GPU constant memory variable*/
        dim3 dimGrid(100, 50); /*for example, this grid has 100*50 blocks*/
        dim3 dimBlock(4, 8, 8); /* for example, 4*8*8 threads per block*/
        GeoRef <<< dimGrid, dimBlock >>>();
        CPU task 3 - read the original image
        CPU task 4 - assign grey value to the destination image pixel from the original image by the index array
}
```

Fig. 3 Pseudocode of the lens distortion correction and geo-referencing

to transfer input parameters from the CPU to the GPU. In this way, no parameters need to transfer when the kernel is lunched. For more details about this method, including the implementation, see Su et al.[3]

GPU-BASED CONVOLUTION

Convolution

Convolution, an important operation in image processing, usually involves the following two procedures:[16]

- A moving window that consists of an array of coefficients or weighting factors. These arrays are referred to as kernels, and they are normally an odd number of pixels in size (e.g., 3×3, 7×7).
- The kernel is moved throughout the original image, and the value at the center of the kernel is assigned to the corresponding pixel of the output image. This value at the center is derived from the brightness of a set of the surrounding pixels. Spatial interdependence of the pixel values leads to variations in geometrical details of the convoluted image.

For the discrete convolution, given an M by N pixel-sized kernel, the response for the image pixel (i, j) is calculated using the following equation.

$$r(i,j) = \sum_{\substack{0 \le m \le M \\ 0 \le n \le N}} \varphi\left(i+m-\frac{M}{2}, j+n-\frac{N}{2}\right) k(m,n) \quad (3)$$

where $\varphi(i, j)$ is the pixel brightness, addressed according to the kernel position and $k(m, n)$ is the kernel coefficient

$$\begin{bmatrix} -1 & 0 & +1 \\ -1 & 0 & +1 \\ -1 & 0 & +1 \end{bmatrix} \cdots \begin{bmatrix} -1 & -1 & -1 \\ 0 & 0 & 0 \\ +1 & +1 & +1 \end{bmatrix} \cdots \begin{bmatrix} 0 & +1 & +1 \\ -1 & 0 & +1 \\ -1 & -1 & 0 \end{bmatrix} \cdots \begin{bmatrix} +1 & +1 & 0 \\ +1 & 0 & -1 \\ 0 & -1 & -1 \end{bmatrix}$$

vertical → horizontal → diagonal

4. *Gaussian filtering*: Gaussian kernel is a low-pass filter, attenuating high-frequency components of the image. The 2D Gaussian function is the product of two 1D Gaussian functions.

$$G(x,y) = \frac{1}{2\pi\sigma^2} e^{-\frac{x^2+y^2}{2\sigma^2}} = \frac{1}{\sqrt{2\pi\sigma^2}} e^{-\frac{x^2}{2\sigma^2}} \cdot \frac{1}{\sqrt{2\pi\sigma^2}} e^{-\frac{y^2}{2\sigma^2}} = G(x) \cdot G(y) \quad (5)$$

2D discrete Gaussian kernel matrix is the product of two 1D Gaussian kernel vectors:

at this location. M and N are normally odd numbers, where $M/2$ and $N/2$ are the integer divisions. A 2D convolution kernel is separable if the convolution matrix can be expressed as the outer product of two 1D convolution kernels: one is in the horizontal and the other in the vertical direction. Only a small fraction of the possible 2D convolution kernels are separable. Mean value filtering, median filtering, linear edge detecting kernels, and Gaussian filtering are the commonly used filtering methods. Each one is briefly explained as follows:

1. *Mean value filtering*: $k(m,n) = 1/MN$ for all m, n, for example, $m = 3$, $n = 3$

$$\begin{bmatrix} \frac{1}{9} & \frac{1}{9} & \frac{1}{9} \\ \frac{1}{9} & \frac{1}{9} & \frac{1}{9} \\ \frac{1}{9} & \frac{1}{9} & \frac{1}{9} \end{bmatrix} = \begin{bmatrix} \frac{1}{3} & \frac{1}{3} & \frac{1}{3} \end{bmatrix} \begin{bmatrix} \frac{1}{3} \\ \frac{1}{3} \\ \frac{1}{3} \end{bmatrix} \quad (4)$$

The pixel at the center of the kernel is represented by the average brightness level in a neighborhood defined by the kernel dimensions. The mean kernel is separable.

2. *Median filtering*: The pixel at the center of the kernel is given by the median brightness value of all the pixels covered by the kernel. The median filter is non-separable.
3. *Linear edge detecting kernels*: They are non-separable including vertical, horizontal, and diagonal kernels.

$$\frac{1}{16}\begin{bmatrix} 1 & 2 & 1 \\ 2 & 4 & 2 \\ 1 & 2 & 1 \end{bmatrix} = \begin{bmatrix} \frac{1}{4} & \frac{2}{4} & \frac{1}{4} \end{bmatrix}\begin{bmatrix} \frac{1}{4} \\ \frac{2}{4} \\ \frac{1}{4} \end{bmatrix} \quad (6)$$

A GPU Implementation of the Convolution

Convolution is computationally expensive due to its complexity. For a given $M \times M$ kernel matrix, a total number of M^2WH multiplications and additions are executed to convolve an image of size $W \times H$. When a 2D convolution kernel is separable, we can apply these two kernels sequentially to the same image and the same results can be made with only a total of $2MWH$ multiplications and additions.[17]

Global memory access is a major concern in the GPU-based convolution implementation. Convolution typically leads to redundant access to the global memory because each pixel on the output image is derived from a set of the surrounding pixels of its corresponding pixel on the original image. In other words, each pixel in the original image is accessed repeatedly in order to support the kernel calculation on its neighboring pixels when the kernel window is moved through the original image. As we know that the shared memory is much faster than the global memory and all threads in a block can access to variables in the shared memory. We can use the shared memory to cache data in order to reduce redundant access to the global memory. In addition to the data elements of a block, processing of the block depends on pixels outside its boundary. Therefore, the block needs to load a halo of the kernel radius. The following is an example of 1D convolution.

__shared__ int sharedmem[SizeofBlock+2*Radius]

Radius defines the number of elements on either side of a point. Therefore, Radius = $M/2$, M = the size of the 1D convolution kernel. As illustrated in Fig. 4, each block needs a halo of Radius elements at its both left and right ends.

The kernel weights are stored in the constant memory. The input image is partitioned into small tiles. A tile corresponds to a block of an execution configuration. Each of the tiles is processed by a single block. Threads in a block copy the pixels of the tile and halo elements into the shared memory (red dots are loaded by the same thread and blue crosses are loaded by the other thread) (Fig. 4). Each thread can quickly access to neighboring pixels loaded by other threads. The use of the shared memory leads to magnitudes of speedup, especially for large kernels, due to M^2WH cost of convolution. Although an element in 1D convolution has only two neighboring elements, namely, left and right neighboring elements, a pixel on 2D convolution has eight neighboring pixels. Thereafter, in addition to its computational efficiency, a separable kernel also improves the memory bandwidth requirements.

GPU-Based Gray-Level Co-occurrence Matrix Texture Calculation

GPU-Based Gray-Level Co-occurrence Matrix Textures

Similar ground objects such as trees may have similar gray-level variation in an image, which is referred to as texture. Among numerous texture measures, the gray-level co-occurrence matrix (GLCM)[18] is the most used texture measure in remote sensing classification.[11,19] The GLCM method is a way to extract the second-order statistical texture features. The first-order texture measures are statistics calculated from the original image values, like variance, and they do not consider pixel neighbor relationships. The second-order texture measures, however, consider the relationships between groups of two (usually neighboring) pixels in the original image. The GLCM for a region, defined by a user-specified window, is the matrix of those measurements over all gray level pairs. In a common way, textures derived at different window sizes are separately added as additional ancillary bands for further classifications. The contribution of textures to classification accuracy depends on both the texture and ground objects scales. The highest accuracy could be achieved by inclusion of the optimal scale where ground objects represent the highest between-class variation and the lowest within-class variation.[19]

The GLCM describes the number of occurrences of the relationship between a pixel and its specified neighbor. Let $g(\rho_1, \rho_2, h, \theta)$ be the relative occurrence of pixels with gray levels ρ_1 and ρ_2 spaced h pixels apart in direction θ. Relative occurrence is the number of times that each gray level pair is counted divided by the total possible number of gray level pairs. The GLCM is a $G \times G$ matrix, where G is the total possible number of the gray level values. If G has a quite large value, the brightness value can be binned. The displacement h is commonly assigned to one pixel due to the highly correlated spatial relationships between each pixel and its neighbor. The GLCM computed for various values of θ is to detect whether the texture is orientation dependent or not. The common directions are horizontal, vertical, and diagonal. Figure 5 shows an example of a base window and a shift window, the derived co-occurrence matrix, and the probabilities of co-occurrences. The co-occurrence matrix is produced using each pixel and its horizontal neighbor (shift values of $X = 1$, $Y = 0$) for a 3×3 window. The pixels in the 3×3 base window and the pixels in a 3×3 window that is shifted by one pixel are used to create the co-occurrence matrix.

A number of statistical measures can be extracted from the GLCM including mean, variance, contrast, local homogeneity, sum average, and correlation.[18,20] The GLCM

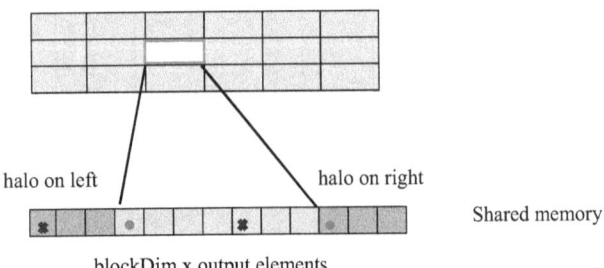

Fig. 4 The horizontal filter, as processed by a block. Red dots are loaded by the same thread, and blue crosses are loaded by the other thread

Base window	Shift window	Co-occurrence matrix					$P(i,j;\Delta x=1, \Delta y=0)$				
		Gray level	$j=3$	4	5	6	Gray level	$j=3$	4	5	6
4 3 5	3 5 6	$i=3$	0	0	2	1	$i=3$	0	0	2/9	1/9
3 5 6	5 6 3	4	2	0	0	0	4	2/9	0	0	0
6 4 3	4 3 6	5	0	0	0	2	5	0	0	0	2/9
		6	1	1	0	0	6	1/9	1/9	0	0

Fig. 5 A model of a base window and a shift window, the derived co-occurrence matrix, and the probabilities of the co-occurrences

variance is a measure of the dispersion of the values around the GLCM mean. The range of the local homogeneity is [0, 1]. If the image has a little variation, the homogeneity is high while if there is no variation, the homogeneity is equal to 1. If the neighboring pixels are very similar in their gray level values, the contrast of the image region is very low.

Given an $M \times N$ window of an input image containing G gray levels from 0 to $G-1$, let $f(m, n)$ be the intensity at sample m, line n of the window.

$$P(i,j \mid \Delta x, \Delta y) = W \cdot Q(i,j \mid \Delta x, \Delta y)$$

where

$$W = \frac{1}{(M-\Delta x)(N-\Delta y)}$$

$$Q(i,j \mid \Delta x, \Delta y) = \sum_{n=1}^{N-\Delta y} \sum_{m=1}^{M-\Delta x} A$$

$$\text{And } A = \begin{cases} 1 & \text{if } f(m,n) = i \text{ and } f(m+\Delta x, n+\Delta y) = j \\ 0 & \text{else where} \end{cases}$$

A number of texture features may be extracted from the GLCM. Here, the following notation is applied:

G is the number of gray levels used.

μ is the mean value of P.

$\mu_x, \mu_y, \sigma_x,$ and σ_y are the means and standard deviations of P_x and P_y. $P_x(i)$ is the ith entry in the marginal probability matrix obtained by summing the rows of $P_x(i,j)$:

$$P_x(i) = \sum_{j=0}^{G-1} P(i,j)$$

$$P_y(j) = \sum_{i=0}^{G-1} P(i,j)$$

$$\mu_x = \sum_{i=0}^{G-1} i \sum_{j=0}^{G-1} P(i,j) = \sum_{i=0}^{G-1} i \cdot P_x(i)$$

$$\mu_y = \sum_{i=0}^{G-1} \sum_{j=0}^{G-1} j \cdot P(i,j) = \sum_{j=0}^{G-1} j \cdot P_y(j)$$

$$\sigma_x^2 = \sum_{i=0}^{G-1} (i-\mu_x)^2 \sum_{j=0}^{G-1} P(i,j) = \sum_{i=0}^{G-1} (P_x(i) - \mu_x(i))^2$$

$$\sigma_y^2 = \sum_{j=0}^{G-1} (j-\mu_y)^2 \sum_{i=0}^{G-1} P(i,j) = \sum_{j=0}^{G-1} (P_y(j) - \mu_y(j))^2$$

and

$$P_{x+y}(k) = \sum_{i=0}^{G-1} \sum_{j=0}^{G-1} P(i,j) \quad i+j=k \quad \text{for } k=0,1,\ldots 2(G-1)$$

The following features are used:

Contrast: $\sum_{n=0}^{G-1} n^2 \left\{ \sum_{i=0}^{G-1} \sum_{j=0}^{G-1} P(i,j) \right\}, |i-j| = n$

This measure of contrast or local intensity variation will favor contributions from $P(i, j)$ away from the diagonal, that is, $i \neq j$.

Local homogeneity: $\sum_{i=0}^{G-1} \sum_{j=0}^{G-1} \frac{1}{1+(i+j)^2} P(i,j)$

Variance: $\sum_{i=0}^{G-1} \sum_{j=0}^{G-1} (i-\mu)^2 P(i,j)$

Sum average: $\sum_{i=0}^{2G-2} i \cdot P_{x+y}(i)$

Correlation: $\sum_{i=0}^{G-1} \sum_{j=0}^{G-1} \frac{\{i \times j\} \times P(i,j) - \{\mu_x \times \mu_y\}}{\sigma_x \times \sigma_y}$

Instead of providing all the statistical measures or features, only five features are presented here in order to show the natural dependencies among the GLCM textures. In other words, the contrast μ_x and P_{x+y} can be calculated directly from GLCM. Calculation of the variance, sum average, and σ_x is dependent on μ_x and P_{x+y}. Calculation of the correlation needs σ_x. Based on the dependencies, these features are grouped into three phases of calculation.

Figure 6 provides the pseudocode of a straightforward sequential CPU implementation of the GLCM texture calculation. The pseudocode does not include all texture features. Instead, the purpose is to explain time complexity and space complexity as well as the optimizations. Besides the size of the original image, the size of the window and

```
For i=1 to W*H /*number of pixels for an image by width W and height H*/
    Find min and max
End for
For i=1 to W*H
    Bin pixel brightness to G /*G is the grey-level, could be one of 64,128,256,512 or more*/
End for
For i=1 to W*H
    For m=1 to M*M /*for M*M window of the GLCM*/
        Generate GLCM
    End for
    For l=1 to G*G /*number of elements in GLCM, gray-level G*/
        Compute Energy, Contras, Homogeneity
    End for
    For l=1 to G*G
        Compute Variance, Sum Average
    End for
    For l=1 to G*G
        Compute Correlation
    End for
End for
```

Fig. 6 Pseudocode of the CPU sequential implementation of the GLCM textures

the number of the gray levels are another two key factors in computation efficiency. The first and second "for" loops are simple. For each pixel of the original image, building a co-occurrence matrix requires $O(M^2)$ data access to traverse the GLCM window on the base image and the shift image, and $O(G^2)$ space to hold the co-occurrence matrix. The calculation of the texture features are conducted with three $O(G^2)$ time by the three phases. With a gray level of 256, each co-occurrence matrix needs 256 KB (256 × 256 × 4 bytes) of storage capacity and $O(256^2)$ time for calculating the textures. However, actually the co-occurrence matrix is a sparse matrix where many elements are zero. These zero elements are not used to calculate texture features. When the loops span all elements, the most computation time is lost when reading these zero ones. In this regard, an optimization is needed to store only nonzero elements of the co-occurrence matrix in a linked list or an improved structure,[21] in which each node at least contains two co-occurring gray levels (i, j) and their probability of co-occurrence $P(i, j)$. The search for an existing (i, j) pair normally requires $O(N)$ time (N is the number of the nonzero elements). When N is less than G, searching all $P(i, j)$ is always shorter than looping the whole GLCM that includes all elements. Fortunately, N is typically less than G. For example, given a 7 × 7 window and 64 gray levels, there are only a total of 49 possible pairs of co-occurring gray levels (i, j). Since multiple pairs may have the same co-occurring gray levels (i, j), the number of the nonzero elements should be much smaller than 49. Given a small gray level and a large window, for example, 32 gray levels and a window of 256 × 256 pixels, the optimization of nonzero storage does not have any advantage. However, normal cases often have large gray levels but a relative small window, for example, gray levels may go up to 4,094 and the window size is smaller than 128 × 128 pixels. These optimizations are standard implementation provided in widely used image processing software, such as ENVI.

A GPU Implementation of the GLCM Texture Calculation

There are two main tasks in a GPU implementation of the GLCM textures:[22,23] building a co-occurrence matrix and calculating texture features. Building the co-occurrence matrix is the main concern. This is because as the number of gray levels increases the size of co-occurrence matrix grows quickly, and the entire matrix may not be able to fit into the shared memory over a certain gray level. The shared memory typically has 64 kB. Therefore, it cannot hold a co-occurrence matrix for a gray level of 256, which needs 128 kB (256 × 256 × 2 bytes) of storage capacity.

We use five kernels to complete the calculation of the GPU-based GLCM textures, and they are referred to as shift kernel, nonzero kernel, phase 1 kernel, phase 2 kernel, and phase 3 kernel. The shift kernel generates the shift image from the base image. The nonzero kernel generates an array that contains all nonzero co-occurring gray level pairs $P(i, j)$ and the probability of co-occurrence $P(i, j)$. Corresponding to the three phases involved in the texture calculation, phase 1 kernel computes phase 1 features, such as energy, contrast, and the intermediate variables, such as μ_x and P_{x+y}; phase 2 kernel computes phase 2 features, such as variance and sum average, and the intermediate variables, such as σ_x and σ_y; and phase 3 kernel that computes phase 3 features, such as correlation.

Figure 7 provides the pseudocode of a GPU implementation of the GLCM texture calculation. Similar to the GPU convolution, the GLCM window is moved throughout the entire input images. To reduce the redundant global memory access, the input images are partitioned into tiles. The nonzero kernel uses the shared memory to hold all pixels of a tile and its eight neighboring tiles. For all pixels of the tile, the nonzero array of each pixel is generated by a thread. In this kernel each thread allocates a local array to hold the entire co-occurrence matrix. The nonzero

```
/*define a kernel function to generate shift image by shifting base image*/
__global__ void shiftimage (int* baseImage, int*shiftImage)
{
    int sample= threadIdx.x + blockIdx.x * blockDim.x;
    int line = threadIdx.y + blockIdx.y * blockDim.y;
    /*locate data to be processed by the thread with block index and thread index*/
    shiftImage[line * Width + sample] = baseImage[(line + deltay) * Width + (sample + deltay)];
}
/*define a kernel function to build the co-occurrence matrix*/
__global__ void buildingCoMatrix(input: baseimage, shiftimage; output: array4nonzeros)
{
    __shared__ unsigned int cachedimage[N]; /*size of shared memory variable, N=9*TileSize*/
    unsigned int co-matrix[G*G];/*declare the co-occurrence matrix, G is number of grey levels*/
    Task 1 – load halo elements from the base image and the shift image into cachedimage.
    __syncthreads(); /*synchronize all threads in a block*/
    Task 2 – build the co-occurrence matrix, co-matrix, from cachedimage.
    Task 3 – generate output by filtering out all non-zero co-occurring grey level pairs
}
/*define a kernel function to phase 1 textures and intermediate features*/
__global__ void phase1(input: array4nonzeros; output: textures1 and intermediates1)
{
    /*declare shared memory for co-occurrence grey levels and probabilities for non-zero grey-level pairs*/
    __shared__ unsigned float coprob[M];/*M is size of the array generated in buildingCoMatrix */
    __shared__ unsigned short coi[M], coj[M];
    __shared__ unsigned float textures[M], intermediates[M];
    Task 1 –load halo elements and compute corresponding items of textures1 and intermediates1;
    __syncthreads(); /*synchronize all threads in a block*/
    Task 2 – sum items of textures1 and intermediates1 by reduction algorithms.
    Task 3 – write out textures1 and intermediates1.
}
/*define a kernel function to phase 2 textures and intermediate features*/
__global__ void phase2(input: array4nonzeros, intermediates1; output: textures2 and intermediates2)
{
    /*almost same to phase 1*/
}
/*definea kernel functionto phase 3 textures and intermediate features*/
__global__ void phase3(input: array4nonzeros, intermediates2; output: textures3)
{
    /*almost same to phase 1*/
}
/*CUDA main program*/
int main()
{
    Task 1 – execution of shiftimage kernel.
    Task 2 – execution of building CoMatrix kernel.
    Task 3 – execution of phase1 kernel.
    Task 4 – execution of phase2 kernel.
    Task 5 – execution of phase3 kernel.
}
```

Fig. 7 Pseudocode of the GPU implementation of the GLCM textures

array of (i, j) and $P(i, j)$ are then produced by traversing the co-occurrence matrix. This method can process the co-occurrence matrix with a large number of gray levels. The local array actually is in the global memory. Consequently, the local array can be large with the cost of slow access.

Similar to the previous CPU implementation, the textures also are calculated by the three phases. The reduction optimization steps described in the work of Harris[24] are implemented and applied to these parallel operations. Empirically, this GPU implementation shows 34× speedup compared with the function provided in ENVI, and the reduction algorithms contribute 14× speedup among the overall speedup.

GPU-Based Co-registration

Similarity Measures of Two Images

All individual orthoimages need to be co-registered so that they can be mosaicked in a correct way. The transformation for co-registration can be derived from the tie points generated by feature matching algorithms, such as the scale-invariant feature transform (SIFT) algorithm. The SIFT algorithm has been widely used in UAS remote sensing applications,[25] however, there are some problems in rangeland and agricultural areas.[26,27] On the other hand, the area-based image similarity could find correct locations in some failure cases that are performed by the SIFT algorithm.[3]

This entry adopts global similarity measures with spatial alignment. These similarity measures monotonically increase with decreasing spatial misalignment.[28] Specifically, the zero-mean cross correlation coefficient is used here, as defined in the following equation:

$$\rho = \frac{\sum_k (a_k - \bar{A})(b_k - \bar{B})}{\sqrt{\sum_k (a_k - \bar{A})^2 \sum_k (b_k - \bar{B})^2}} \quad (7)$$

where a_k and b_k are, respectively, the gray levels of the k pixel in images, or image patches, A and B. $\bar{A}$ and $\bar{B}$, respectively, are the mean gray levels of A and B. The cross correlation coefficient can be directly computed by using the raw intensity values of the match and base images due to almost the same image acquisition condition. The seagrass experiments in the work of Su et al.[3] showed that the coefficient is effective for the UAS images that consist of all water.

Two questions needed to be addressed before the similarity is applied to the co-registration: (i) what is the suitable size of the image patches? and (ii) how big should the search range be? To get reliable co-registration, the similarity calculation should use a large array of pixels on the UAS images. For example, neighboring cotton areas of 1 m by 1 m, even 10 m by 10 m, usually are similar due to the same agricultural crop management. Another example, in most coastal areas in Texas, the United States, natural stable beaches are usually more than 30 m wide.[29] The sand beach actually looks similar everywhere. Empirically, it is desirable for an image patch to be sufficiently large, such as 100 m by 100 m for Texas coastal beaches. An image patch of this size usually can encompass several land cover types, such as foredune vegetation, beach sand, and seawater. This magnitude means an extent of 1000 by 1000 pixels on an image of 0.1 m pixel size. The optimal match location can be found by traversing the image patch over its neighboring image. Hereafter, the image patch is referred to as the match image, and the neighboring image is referred to as the base image.

The search range is determined by both GPS (X, Y, Z) errors and IMU (ω, ϕ, κ) errors. The low cost IMU typically has an angular resolution of 0.05°. The 0.05° errors result in positional error up to roughly 10 pixels. Combining with GPS errors, a suitable search range roughly extends 20 m for raw GPS and IMU data, which is a search range of 200 pixels of 0.1 m pixel size.[3] The more inaccurate the GPS and IMU is, the larger the search range could be. The search range for the spatial alignment is quite small compared with the sizes of both the match and base images. If the higher accurate GPS and IMU are used and the postprocessing is well carried out, the search range could be reduced greatly.

A GPU Implementation of the Co-registration

This entry presents a dual tiled similarity method for the image co-registration in order to take advantage of the GPU capability. Figure 8 provides pseudocode of this method. After loading the match and base images into the global memory, this algorithm uses dual partition strategies, namely, partitions on both the match and base images. The first partition divides the match image into small pieces so that each piece can fit into the constant memories. The second partition organizes the search range on the base image into tiles. A tile and its eight neighboring tiles will be loaded into the shared memory by the threads of a block. A piece of match image traverses the search range over the base image by once the kernel execution. It is worth noting that the partitions should be carefully chosen in order to not exceed the capacity of these memories, because the size of both the constant and the shared memories can vary from device to device.

As shown in Eq. 7, in the cross correlation coefficient, there are three different sums of pixel values, namely, $\sum_k (a_k - \bar{A})^2$, $\sum_k (b_k - \bar{B})^2$, and $\sum_k (a_k - \bar{A})(b_k - \bar{B})$, over the overlapping region of the match image and the base image. This allows the three components to be calculated by regions. After the match image is partitioned into pieces, the three components of the entire search range can be produced by summing the corresponding components of all pieces of the match image. Each piece of the match image is iteratively loaded into the constant memory in order to calculate its three cross-correlation components. The search range on the base image has been divided into tiles. For simplicity's sake, the width of the match image piece is set equal to double width of the tile plus one, namely $P_{width} = 2 \times T_{width} + 1$. The three quantities calculated for a tile needs elements from its eight neighbor tiles, namely, halo elements in the neighbor tiles. Although computing the three quantities of all elements in a tile needs to load nine tiles into the shared memory, this tile method greatly eliminates redundancy in the global memory access. Time complexity and space complexity of the dual-tile algorithm have been discussed in the work of Su et al.[3]

Discussions

GPU memory usage and kernel organization are the two main concerns in any GPU implementation. Different types of the GPU memory have different access speed and size, which plays a key role is an efficient GPU implementation. The module organization is important because if a GPU function contains lengthy complex arithmetic and/or logic operations, it may have low performance.

As for the proposed GPU-based approach for georeferencing, the usage of GPU memory is quite simple where only constant and global memory spaces are used.

```
/*declare and implement the means kernel function*/
__global__ void Mean(input params)
{
    __shared__ unsigned char N_ds[3*TileWidth][3* TileWidth]; /*declare shared memory matrix*/
    /*locate data to be processed by the thread with block index and thread index*/
    int line = blockIdx.y * TileWidth + threadIdx.y;
    int sample = blockIdx.x * TileWidth+ threadIdx.x;
    Task 1 /*load halo elements from its neighboring tiles into the shared memory matrix*/
    __syncthreads();/*synchronize all threads in a block*/
    Task 2–Compute means
}
/*declare and implement the cross-correlation kernel function*/
__global__ void CrossCorrelation(input params)
{
            /*all preceding commands are same to Mean kernel except Task 2*/
    Task 2 – Compute cross correlation coefficient by Eq.3
}
/*CUDA main program*/
int main()
{
    int patchWidth= 1 + 2*TileWidth; /*set width of the patch of match image*/
    Task 1 -partition the match image into pieces by patchWidth* patchWidth.
    For i=1 to N /*N is number of the patch of match image*/
            /*iteratively launch the Mean and CrossCorrelation kernels for each piece*/
            int NrOfTiles_Row = ceil(RowsOfSearchRrange/TileWidth);
            int NrOfTiles_Col= ceil(ColsOfSearchRrange/TileWidth);
            /*compute number of tiles on row and column of the search range*/
            dim3 dimGrid(NrOfTiles_Row, NrOfTiles_Col);/*configure the search range into tiles*/
            dim3 dimBlock(TileWidth, TileWidth);
            Mean<<< dimGrid, dimBlock >>>(params);
            CrossCorrelation <<< dimGrid, dimBlock >>>(params);
    End for
            Task 2 - find the max coefficient.
}
```

Fig. 8 Pseudocode of the dual tiled image match algorithm

The kernel, however, is quite complicated due to the complexity of the algorithm. The convolution calculation shows an efficient use of the GPU shared memory as well as the constant and global memories. The number of kernel could be one for non-separable convolution and two for separable convolution. The proposed GPU approach for GLCM texture calculation illustrates the use of four GPU memory spaces (namely, local, shared, constant, and global) and of five small kernels in order to obtain a good overall performance. The dual tiled algorithm for co-registration efficiently uses the constant memory, the shared memory, and the global memory all together, and two small kernels are needed to implement the algorithm. The GPU memory usage and kernel organization are summarized in Table 2.

CONCLUSION

In conclusion, this entry presents a GPU-based approach for geo-referencing, convolution calculation, texture calculation, and co-registration, the four critical components in processing UAS images. The GPU provides a feasible solution to address big data volume and time-limit demands, the two main challenges that UAS remote sensing faces.

The GPU-based approach presented in this entry also demonstrates the capability of the GPU on UAS images processing. To take advantage of the GPU parallel computation, a good design of a GPU application should be composed of small kernels, and its kernels are capable to sufficiently use the fast memory spaces such as the shared memory and constant memory. In this way, a large number of the GPU threads can be executed concurrently. As a result, a good overall performance could be obtained.

Table 2 The memory usage and kernel design for the four GPU-based implementations

Component	Memory space	Kernel
Geo-referencing	Constant, global	One, complex
Convolution	Shared, constant, global	One or two, simple
GLCM texture	Local, shared, constant, global	Five, simple
Co-registration	Shared, constant, global	Two, simple

ACKNOWLEDGMENTS

Sections "CUDA Architecture and Programming Model", "GPU-based Lens Distortion Correction and Georeferencing," and " GPU-based Co-registration" are based in part on our previous paper "Su, L., Huang, Y., Gibeaut J., Li, L. The index array approach and the dual tiled similarity algorithm for UAS hyper-spatial image processing. *Geoinformatica* 2016, 20:859; doi:10.1007/s10707-016-0253-2."

REFERENCES

1. FAA. *Unmanned Aircraft Systems (UAS)—Regulations & Policies*, 2011. Available at http://www.faa.gov/about/initiatives/uas/reg/.
2. Laliberte, A.S.; Goforth, M.A.; Steele, C.M.; Rango, A. Multispectral remote sensing from unmanned aircraft: Image processing workflows and applications for rangeland environments. Remote Sens. **2011**, *3*, 2529–2551.
3. Su, L.; Huang, Y.; Gibeaut J.; Li, L. The index array approach and the dual tiled similarity algorithm for UAS hyper-spatial image processing. Geoinformatica **2016**, *20*, 859.
4. Su, L.; Gibeaut, J. Using UAS hyperspatial RGB imagery for identifying beach zones along the South Texas Coast. Remote Sens. **2017**, *9*, 159.
5. Lee, C.A.; Gasster, S.D.; Plaza, A.; Chang, C.; Huang, B. Recent developments in high performance computing for remote sensing: A review. IEEE J. Sel. Top. Appl. Earth Obs. Remote Sens. **2011**, *4* (3), 508–527.
6. Yang, H.; Du, Q.; Chen, G. Unsupervised hyperspectral band selection using graphics processing units. IEEE J. Sel. Top. Appl. Earth Obs. Remote Sens. **2011**, *4* (3), 660–668.
7. Lemoine, G.; Giovalli, M. Geo-correction of high-resolution imagery using fast template matching on a GPU in emergency mapping contexts. Remote Sens. **2013**, *5*, 4488–4502.
8. Fang, L.; Wang, M.; Li, D.; Pan, J. CPU/GPU near real-time preprocessing for ZY-3 satellite images: Relative radiometric correction, MTF compensation, and geocorrection. ISPRS J. Photogramm. Remote Sens. **2014**, *87*, 229–240.
9. Lei, Z.; Wang, M.; Li, D.; Lei, T.L. Stream model-based orthorectification in a GPU cluster environment. IEEE Geosci. Remote Sens. Lett. **2014**, *11* (12), 2115–2119.
10. Bernabe, S.; Plaza, A.; Marpu, P.R.; Benediktsson, J.A. A new parallel tool for classification of remotely sensed imagery. Comput. Geosci. **2012**, *46*, 208–218.
11. Lillesand, T.; Kiefer, R.W.; Chipman, J. *Remote Sensing and Image Interpretation*, 6th Ed.; John Wiley & Sons: Hoboken, NJ, 2007.
12. Kirk, D.B.; Hwu, W.W. *Programming Massively Parallel Processors, a Hands-On Approach*; Morgan Kaufmann Publisher: Burlington, MA, 2010.
13. Wolf, P.; DeWitt, B. *Elements of Photogrammetry with Applications in GIS*, 3rd Ed.; McGraw-Hill: New York, 2000.
14. Intergraph, ERDAS Field Guide, 2013.
15. Maune, D.F. *Digital Elevation Model Technologies and Applications: The DEM Users Manual*; The American Society for Photogrammetry and Remote Sensing: Bethesda, MD, 2007.
16. Richards, J.A; Jia, X. *Remote Sensing Digital Image Analysis: An Introduction*; Springer-Verlag: Berlin, Heidelberg, 2006.
17. Podlozhnyuk, V. Image convolution with CUDA. NVIDIA Corporation white paper, June 2007, Vol. 2097, No. 3.
18. Haralick, R.M.; Shanmugam, K.; Dinstein, I. Textural features for image classification. IEEE Trans. Syst. Man Cybern. **1973**, *3* (6), 610–621.
19. Laliberte, A.S.; Herrick, J.E.; Rango, A.; Winters, C. Acquisition, orthorectification, and object-based classification of Unmanned Aerial Vehicle (UAV) imagery for rangeland monitoring. Photogramm. Eng. Remote Sens. **2010**, *76* (6), 661–672.
20. Albregtsen, F. Statistical texture measures computed from gray level cooccurrence matrices. Technical Note; Department of Informatics, University of Oslo: Oslo, 2008.
21. Clausi, D.A.; Jernigan, M. A fast method to determine co-occurrence texture features. IEEE Trans. Geosci. Remote Sens. **1998**, *36* (1), 298–300.
22. Dixon, J.; Ding, J. An empirical study of parallel solutions for GLCM calculation of diffraction images. In *38th Annual International Conference of the IEEE Engineering in Medicine and Biology Society (EMBC)*, Orlando, FL, August 16–20, 2016.
23. Saladin, A.M.; Jiao, L.; Zhang, X. GPU-accelerated computation for texture features using OpenCL framework. In *Electrical Engineering/Electronics, Computer, Telecommunications and Information Technology (ECTI-CON), 2014 11th International Conference on 14–17 May 2014*, Nakhon Ratchasima, 2014.
24. Harris, M. Optimizing parallel reduction in CUDA. NVIDIA, November 2007.
25. Turner, D.; Lucieer, A.; Malenovsky, Z.; King, D.H.; Robinson, S.A. Spatial co-registration of ultra-high resolution visible, multispectral and thermal images acquired with a micro-UAV over Antarctic moss beds. Remote Sens. **2014**, *6* (5), 4003–4024.
26. Laliberte, A.S.; Herrick, J.E.; Rango, A.; Winters, C. Acquisition, orthorectification, and object-based classification of Unmanned Aerial Vehicle (UAV) imagery for rangeland monitoring. Photogramm. Eng. Remote Sens. **2010**, *76* (6), 661–672.
27. Du, Q.; Raksuntorn, N.; Orduyilmaz, A.; Bruce, L.M. Automatic registration and mosaicking for airborne multispectral image sequences. Photogramm. Eng. Remote Sens. **2008**, *74* (2), 169–181.
28. Mitchell, H.B. *Image Fusion Theories, Techniques and Applications*; Springer: Heidelberg, 2010.
29. Morton, R.A.; Peterson, R.L. *South Texas Coastal Classification Maps—Mansfield Channel to the Rio Grande. USGS Open File Report 2006-1133*. Available at http://pubs.usgs.gov/of/2006/1133/mapintro.html, (accessed in April 15, 2017).

UAV Imaging for Disaster Management

Caixia Wang
Geomatics, University of Alaska Anchorage, Anchorage, Alaska, U.S.A.

Qunying Huang
Geography, University of Wisconsin Madison, Madison, Wisconsin, U.S.A.

Abstract
In recent years, the use of unmanned aerial vehicle (UAV) images is offering increasing opportunities in disaster-related situations. This is driven, in part, by the ease of operation, deployment agility, quick turnaround time for high-resolution data, and low data acquisition costs. In this entry, we will review and discuss the development of image processing and analysis techniques for UAV-based imagery with an emphasis on the emergency management. Case studies will cover the wide spectrum of different applications of UAVs for disaster management. Open issues and future trends will also be discussed in this entry.

Keywords: Damage; Disaster management; Drone; Feature extraction; Hazards; Image processing and analysis; Monitoring; Unmanned aerial vehicles.

INTRODUCTION

Previously, methodologies such as phone calls, direct observations, or personal interviews are commonly used by disaster responders and damage evaluators to gain situational awareness (SA) and investigate impacted population during a disaster.[1] However, these data collection methods are time-consuming and laborious in processing the data. Unmanned aerial vehicles (UAVs) or drones equipped with sensor payloads can provide images and videos with very high spatial resolution (VHR) and appropriate imagery metadata about the impacted regions, therefore offer numerous opportunities in disaster-related situations.[2]

Specially, these images and videos can be processed to generate various products, such as hazard maps, dense surface models, detailed building renderings, comprehensive elevation models, and other disaster area characteristics, which in turn can be further analyzed to coordinate rescue efforts, detect building failures, investigate access issues, and verify experimental disaster modeling.[2] As a result, it is widely acknowledged that UAV systems could be leveraged to assist in a variety of disaster and emergency situations, such as severe storms, tornados, hurricanes, wild fires, tsunamis, floods, earthquakes, avalanches, civil disturbances, oil or chemical spills, and urban disasters.[3] For example, Quaritsch et al.[4] have tested and demonstrated a collaborative UAV system on a large fire service drill. The system provided high-resolution images of the large test site with regular updates, and proved that it is easy to handle, fast to deploy, and useful for the firefighters.

Although VHR images introduce great opportunities to characterize disaster situations, grand challenges also arise when dealing with such newly emerging data type.[2,5–8] With the image resolution becoming finer, smaller objects become visible revealing higher level of details.[9] Apparently, mixed-pixel effect can be greatly reduced in comparison to low-to-median resolution images.[10] However, since the small object can be represented by a group of pixels which vary in spectral values, such detailed information also leads to higher spectral variability for the object, which in turn reduces the spectral discrepancies between object classes.[11] Additionally, in terms of spatial resolution, images from UAVs can compete with traditional mapping solutions. Note that these traditional solutions require high-precision altitude parameters and ground control point (GCP) data,[12] which are harsh demands for UAVs to meet. One obstacle comes from their low flying altitude regulated for civil applications[13] that directly impacts the flight quality of their remote sensing systems.[14] In the use of UAVs for applications, such as disaster monitoring and management, this becomes more difficult as required GCP data might be damaged or destroyed by natural disasters.

Even though many techniques have been developed to tackle the problems of VHR imagery, UAV image processing and analysis remains challenging due to aforementioned characteristics. This entry provides an overview of the recent developments of UAV image processing and analysis techniques, with a particular focus on the area of disaster situations. Specifically, our discussion centers on techniques that can derive timely and accurate information

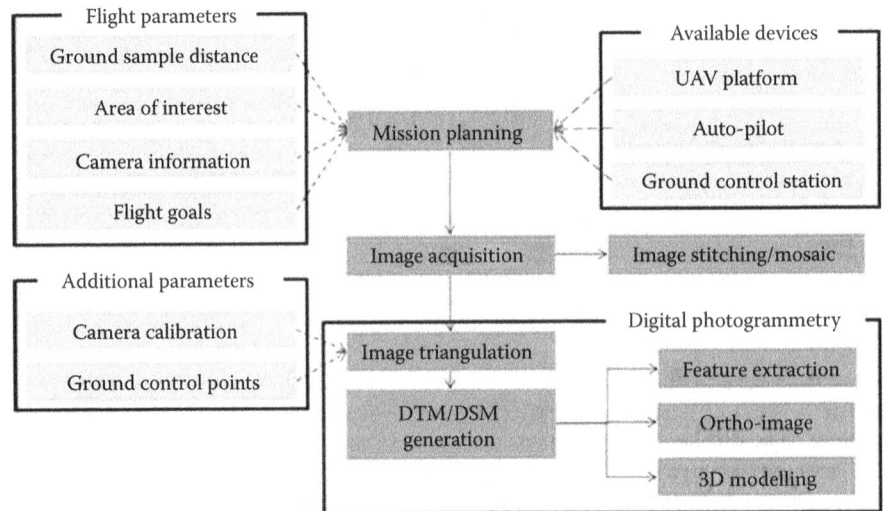

Fig. 1 An example of a photogrammetric workflow for UAV image acquisition and processing[6]

from UAV imagery collected during or after natural disasters as such information is critical to disaster responders. The followed part is structured as a grand array of application fields dealing with the analysis of UAV-based images for disaster management. This entry concludes with final remarks and an outlook for future trends.

RELATED WORK

Mapping affected areas with UAVs in disaster management offers a quick and well-founded overview of the situation. The mission is essentially a typical image-based aerial surveying, followed by sophisticated image analysis algorithms for object or damage detections. Many efforts have developed to use traditional photogrammetric methods on UAV image processing for stringent metric reconstruction.[12] Figure 1 demonstrates a general workflow in this group. In the case of disaster applications, the critical need is to fast obtain (quasi) orthorectified images of the impacted region for disaster alert, relief, or assessment. The absolute positioning accuracy of UAV images, on the other hand, is not as important, and the accuracy of direct navigation observations from onboard GNSS (global navigation satellite system)/INS (inertial navigation system) devices can be sufficient for direct geo-referencing.[15,16] This section presents recent advancements of image processing and analysis techniques driven by such needs.

Image Rectification

Without GCP data, direct geo-referencing is used to fast rectify the captured images with low metric quality requirements for disaster situations. It is time consuming, however, to process single imagery one by one since one flight mission acquires a large number of data. In addition, most UAV missions use nonmetric digital cameras, and the flight attitude is not stable, which results in complex projecting relations between single image and ground object space. Li et al.[17] proposed to automatically mosaic UAV images collected after earthquake disaster using SIFT (scale-invariant feature transform) algorithm which is based on local invariant image features.[17–19] The blocks of imagery are then rectified by using auxiliary data recorded by onboard GPS (global positioning system). Some studies investigate the great potential of sensor fusion techniques. For example, Xiang and Tian[12] integrated the navigation data provided by low-cost IMU (inertial measurement unit) sensors and GPS with a camera lens distortion model to automatically geo-reference the collected imagery.

Object Detection

In the case of natural hazards, for example, flooding, wildfire, and earthquake, it is often crucial to identify and rescue victims as soon as possible. Real-time people and vehicle detection from UAV imagery has become an important aspect in these missions. Due to the rapid motion of UAVs, varied image orientation and cluttered environment, such analysis of images from UAVs is difficult. Most recent work tackled the problem on moving objects using video feeds,[20–22] thermal images,[23] or the combination of color and thermal images.[24]

The problem becomes more difficult when addressing people detection in a UAV context due to small target size, low contrast of the target with background, or the presence of clutter.[25] A wide range of digital image processing and analysis techniques have been investigated. For example, Gaszaczak et al.[26] proposed a cascaded Haar classification technique that integrates multivariate Gaussian shape matching for people detection (Fig. 2). They achieved about 70% detection rate for people in thermal

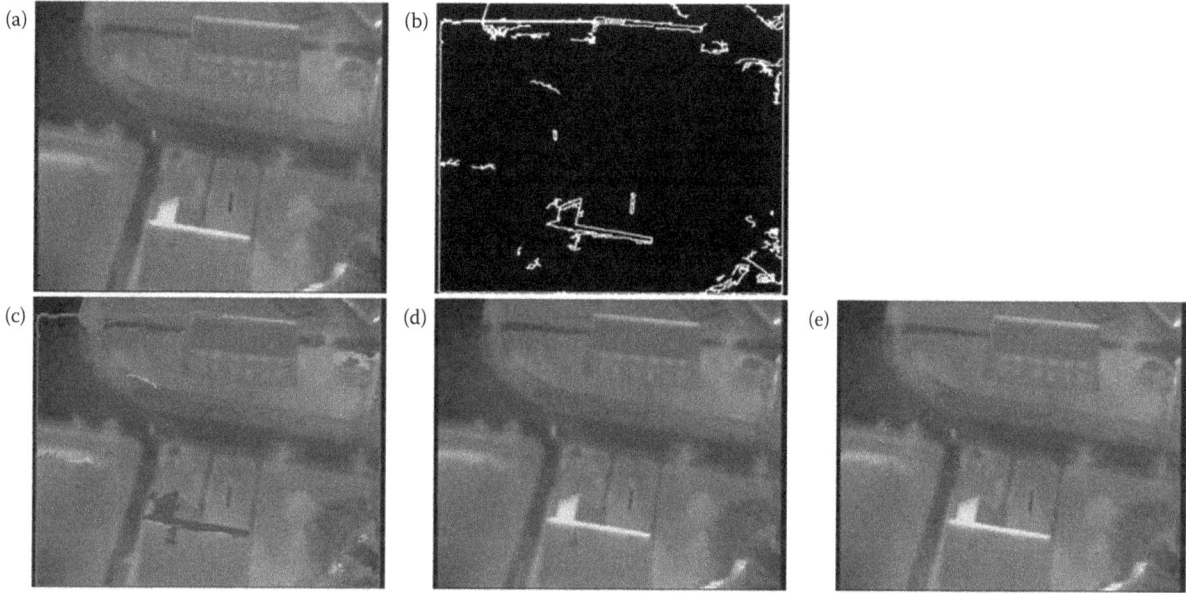

Fig. 2 An analysis technique for people detection: (a) mean-shift segmented thermal image; (b) edge image; (c) extracted shape contours; (d) highlighted contour matching the model; (e) Haar classifier detection[26]

images. The histogram of oriented gradients (HOG) introduced by Dalal and Triggs[27] has been extended in the works of Oreifej et al.[28] and Blondel et al.[29] for human detection, showing promising results in uncluttered environments. Some detectors are based on human detection methods, which were originally developed to deal with datasets containing imagery taken from the ground. Bose and Grimson[30] developed a scene-invariant classification system with identified discriminative object features for classifying vehicles and pedestrians. Reilly et al.[31] used people's shadows as geometric constraints to detect and localize people. Impressive results have been demonstrated, but strong assumptions such as weather conditions to capture shadows or temporal information have to be met.

Damage Detection

Detection of severe damages on man-made infrastructures (such as buildings, ports, and dams) after destructive disaster events is critical particularly in emergency response actions, rescue work, and post-disaster reconstruction. Structurally damaged regions of these infrastructures often reveal nonuniform radiometric distribution compared to undamaged areas.[8] Based on such characteristic distribution patterns (Fig. 3) using gradient orientation, many texture features such as HOG and Gabor features have been proposed as effective descriptors which are used to train classifiers (such as support vector machine or SVM) for building damage detection.[32–35] On the other hand, textures of man-made structures in the real world are complex, which likely present similar patterns from other factors such as aging or presence of dead leaves. Appearance-based image features from deep learning approaches such as convolutional neural networks (CNNs), a dominant method for visual recognition, have become highly desirable in many works for generalized supervised models.[36–39] Researchers, Vetrivel et al.[8] pioneered their work to integrate CNN features with 3D point cloud features and develop a supervised learning model for building damage detection based on multiple kernel learning. The authors claimed their model highly beneficial to the first responders giving its encouraging transferability.

In addition to damages expressed on building roofs, damages on building facades such as cracks or inclined walls are also important. Multi-view oblique images from UAVs capture both roof and facades in a VHR image, and thus overcome the perspective constraints of vertical imagery. They have received increasing interests for damage detection, though the multi-angle imagery also adds a new dimension of complexity.[40–42] In such detailed damage detection, object-based image analysis (OBIA) has been widely applied to extract damage indicators as it adds a cognitive dimension that is found useful in detailed classification from VHR imagery.[11,43] The research by Galarreta et al.[44] exploited OBIA to assess damages at a detailed level on both facades and roofs, with which the entire building damage was successfully identified using 3D point cloud data.

APPLICATIONS OF UAVs FOR DISASTER MANAGEMENT

Although a wide variety of system types and all sizes with varying capabilities already exist with even more under development, small-scale UAVs have recently attracted

Fig. 3 Histogram of gradient orientation pattern (shown in black boxes) for undamaged (annotated as A, B, C, and D) and damaged (no annotation) image samples[8]

various interests for applications such as surveillance, environmental monitoring, and emergency response operations.[4] In fact, the advancement of control engineering and material science makes it relatively easy to develop small-scale UAVs, which enables us to obtain a bird's eye view of the environment to build an "overview" of the environment and assess the current situation.[45] Correspondingly, such systems have the potential to improve the effectiveness of disaster relief and emergency response efforts by enhancing first responder capabilities with aerial photography of buildings and surveillance of crowds, and by providing advanced predictive capabilities and early warning.[3]

This section therefore introduces how small-scale UAVs can be used in SA and assist in disaster management in various scenarios.

Real-Time Data Collection

The UAVs are typically equipped with different sensors and cameras, and thus can collect real-time data about the impacted area of a disaster and provide information important for first responders. For example, Quaritsch et al.[45] deployed an "aerial sensor network" system with a set of small-scale, battery-powered, and wirelessly connected UAVs. Within this system, UAVs build a flight formation, fly over a disaster area, and deliver high-quality sensor data (e.g., images and videos). Using the wireless communication channels, the UAVs are able to communicate with each other and transmit the data to the ground stations. The data are then fused, analyzed in real time, and finally delivered to the disaster managers and first responders to gain SA.

Real-Time Event Tracking and Monitoring

As a result of the rapid or even immediate availability of information in UAV data, UAVs are widely applied for the tracking and monitoring of fast-evolving events, such as tornados and severe thunderstorm. Using tornadoes as an example, DeBusk[3] demonstrated how UAV systems can be used to obtain advanced warning and damage assessments when disaster situation changes rapidly. Specifically, UAVs are designed to fly ahead of storm fronts so weather data can be recorded to predict areas of heavy rain, hail, and high winds for surveillance purpose. We can also place vehicles inside the storms to forecast and detect tornados for advanced tornado warning.

Real-Time Map Generation

Much effort has been made to use UAVs for an automatic image acquisition and to automatically generate various associated data products, such as digital surface models (DSMs) and orthophotos, which are then used for map creation. For example, Marenchino[46] first developed a procedure for collecting UAV images automatically. The radiometric content of the images was then improved to increase the dynamic range of the images and optimized using Wallis filter[47] for subsequent interest point extraction and image matching. The final product of the image processing is the Solid True Orthophoto (STOP).

A STOP provides not only the geometric and radiometric information of a true orthophoto but also the altimetry data provided by the DSM.[46]

Additionally, advances in UAV instrumentation and associated data processing techniques have even enabled semiautomated and even fully automated map creation, a valuable tool for rapid response.[2] For example, Suzuki et al.[48] developed a small prototype UAV and onboard software using a GPS navigation system and proposed a unique disaster mitigation system that generates digital real-time hazard maps. To generate the hazard maps, the authors first generated the mosaic images based on the collected moving images. These images were then automatically transformed and projected onto the map by using the position and attitude of the UAVs. Next, the coordinates of important places, such as those of the location of disaster victims, and the images of hazardous area were integrated with the geographic information system (GIS) database for map generation.

Damage Assessment

The knowledge of the physical infrastructure conditions, and the evaluation of the damage in infrastructures, which could potentially increase the hazard (e.g., dams, dikes, and bridges), is paramount for evacuations and disaster relief. VHR images and videos collected from UAVs become a fundamental tool for the assessment of damage impacts on infrastructures and strategic areas.[46] After a disaster, UAVs could overfly a damaged area and assign it a "destruction rating" based on the classification of the destruction levels of the infrastructures that have been impacted. Any building collapses could also be determined to assist emergency personnel in concentrating their efforts where they would be most beneficial. To further aid responding vehicles, UAVs could be used to detect if any roads are blocked or damaged and route emergency vehicles around any trouble spots. This function is very similar as modern driving GPS units with "alternate route" and "traffic alert" capabilities but is tailored specifically to the disaster situation for real-time traffic monitoring and routing with an on-station aerial vehicle.[3]

OPEN DISCUSSION

Indeed, the ease and capability of UAVs to acquire VHR images and corresponding image processing technologies constitute a new disaster management context (Section "Applications of UAVs for Disaster Management"). Despite the enormous efforts spent, there are challenges and limitations while leveraging UAVs to obtain aerial images and assist in disaster management.[3,45] For example, UAVs cannot easily obtain images of large disaster areas. The images captured by the UAV are limited to small areas because of its low-altitude flight, and it is also difficult to find the coordinates of important places in the moving images transmitted from the UAV.[48] DeBusk[3] introduced the risk, cost, need, and design trade-offs for UAV systems. Optimal design of aerial systems will lead UAVs to provide maximum potentiality for relief and emergency response while accounting for public safety concerns and regulatory requirements.

In disaster situations, it is extremely important to process and analyze UAV images in a timely fashion, ideally on the fly, to coordinate rescue efforts, responses to the disaster, and automatic damage assessment. With the advances in control engineering and sensor technologies, a vision for the future is to develop smart sensors with computing capability. This type of sensors should be relatively light to be equipped on UAVs. The smart sensors would capture, process, and analyze data at real time during the mission of flight. Extracted information such as locations of detected victims or building damages is communicated real time to the ground and presented to the user. Another emerging trend, in our perspective, would be similar to wireless sensor networks, which have been widely used in environmental monitoring or smart environments. An aerial sensor network[45] formed by small-scale UAVs has great potential to be developed and deployed in disaster situations. These UAVs coordinates themselves during flight and operates as autonomous as possible. While impressive work have been developed, majority of them are still preliminary.[49–51] Additionally, much more research work is in need to fully automatically process and analyze UAV images for damage feature extraction, and combine the extracted information with engineering understanding for risk monitoring and prediction of the infrastructures.

REFERENCES

1. Huang, Q.; Xiao, Y. Geographic situational awareness: Mining tweets for disaster preparedness, emergency response, impact, and recovery. ISPRS Int. J. Geo-Inf. **2015**, *4*, 1549–1568.
2. Adams, S.M.; Friedland, C.J. A survey of unmanned aerial vehicle (UAV) usage for imagery collection in disaster research and management. In *Proceedings of the 9th International Workshop on Remote Sensing for Disaster Response*, Stanford University, Stanford, CA, September 14–16, 2011.
3. DeBusk, W.M. Unmanned aerial vehicle systems for disaster relief: Tornado Alley. In *Proceedings of the AIAA Infotech@Aerospace*, Atlanta, GA, 2010.
4. Quaritsch, M.; Kuschnig, R.; Hellwagner, H.; Rinner B. Fast aerial image acquisition and mosaicking for emergency response operations by collaborative UAVs. In *Proceedings of the 8th International Conference on Information Systems for Crisis Response and Management (ISCRAM 2011)*, Lisbon, May 8–11, 2011.
5. Watts, A.C.; Ambrosia, V.G.; Hinkley, E.A. Unmanned aircraft systems in remote sensing and scientific research: Classification and considerations of use. Remote Sens. **2012**, *4* (6), 1671–1692.

6. Nex, F.; Remondino, F. UAV for 3D mapping applications: A review. Applied Geomatics. **2014**, *6* (1), 1–15.
7. Henrickson, J.V.; Rogers, C.; Lu, H.H. Infrastructure assessment with small unmanned aircraft systems. In *Proceedings of the 2016 International Conference on Unmanned Aircraft Systems (ICUAS)*, Arlington, VA, 2016, 933–942.
8. Vetrivel, A.; Gerke, M.; Kerle, N.; Nex, F.; Vosselman, G. Disaster damage detection through synergistic use of deep learning and 3D point cloud features derived from very high resolution oblique aerial images, and multiple-kernel-learning. ISPRS J. Photogramm. Remote Sens. **2017** *140* (June), 45–59.
9. Lillesand, T.; Kiefer, R.W.; Chipman, J. *Remote Sensing and Image Interpretation*, 6th Ed.; John Wiley & Sons: New York, 2007.
10. Foody, G.M. Relating the land-cover composition of mixed pixels to artificial neural network classification output. Photogramm. Eng. Remote Sens. **1996**, *62* (5), 491–499.
11. Wang, C.; Pavlowsky, R.T.; Huang, Q.; Chang, C. Channel bar feature extraction for a mining-contaminated river using high-spatial multispectral remote-sensing imagery. GIScience Remote Sensing **2016**, *53* (3), 283–302.
12. Xiang, H.; Tian, L. Method for automatic georeferencing aerial remote sensing images from an unmanned aerial vehicle (UAV) platform. Biosystems Eng. **2011**, *108* (2), 104–113.
13. FAA. Available at https://www.faa.gov/uas/resources/uas_regulations_policy/, (accessed in June 2017).
14. Eisenbeiss, H. *UAV Photogrammetry*. Dissertation. ETH Zurich, 2009.
15. Zhou, G. Near real-time orthorectification and mosaic of small. IEEE Trans. Geosci. Remote Sens. **2009**, *47* (3), 739–747.
16. Pfeifer, N.; Glira, P.; Briese, C. Direct georeferencing with on board navigation components of light weight UAV platforms. In *Proceedings of the XXII ISPRS Congress*, Melbourne, VIC, August 25–September 1, 2012.
17. Li, C.; Ma, L. A new framework for feature descriptor based on SIFT. Pattern Recognit. Lett. **2009**, *30* (5), 544–557.
18. Lowe, D.G. Distinctive image feature from scale-invariant keypoints. Int. J. Comput. Vision **2004**, *60* (2), 91–110.
19. Zeng, L.; Zhou, D.; Liang, J.; Zhang, K. Polar scale-invariant feature transform for synthetic aperture radar image registration. IEEE Geosci. Remote Sens. Lett. **2017**, *14* (7), 1101–1105.
20. Miller, A.; Babenko, P.; Hu, M.; Shah, M. Person tracking in UAV video. In *Multimodal Technologies for Perception of Humans*; Stiefelhagen, R.; Bowers, R.; Fiscus, J.; Eds.; Springer: Berlin and Heidelberg, 2008, 215–220.
21. Xiao, J.; Yang, C.; Han, F.; Cheng, H. Vehicle and person tracking in aerial videos. In *Multimodal Technologies for Perception of Humans*; Stiefelhagen, R.; Bowers, R.; Fiscus, J.; Eds.; Springer: Berlin and Heidelberg, 2008, 203–214.
22. Huang, C.; Wu, Y.; Kao, J.; Shih, M.; Chou, C. A hybrid moving object detection method for aerial images. In *Advances in Multimedia Information Processing*; Qiu, G.; Lam, K.M.; Kiya, H.; Xue, X.-Y.; Kuo, C.-C.J.; Lew, M.S.; Eds.; Springer: Berlin, 2010, 357–268.
23. Hinz, S.; Stilla, U. Car detection in aerial thermal images by local and global evidence accumulation. Pattern Recognit. Lett. **2006**, *27* (4), 308–315.
24. Rudol, P.; Doherty, P. Human body detection and geolocalization for UAV search and rescue missions using color and thermal imagery. In *Proceedings of the IEEE Aerospace Conference*, Big Sky, MT, 2008, 1–8.
25. Andriluka, M.; Schnitaspan, P.; Meyer, J.; Kohlbrecher, S.; Petersen, K.; Stryk, O.V.; Roth, S.; Shiele, B. Vison based victim detection from unmanned aerial vehicles. In *Proceedings of the IEEE/RSJ International Conference on Intelligent Robots and Systems (IROS)*, Taipei, 2010.
26. Gąszczak, A.; Breckon, T.P.; Han, J. Real-time people and vehicle detection from UAV imagery. In *Proceedings of SPIE 7878: Intelligent Robots and Computer Vision XXVIII: Algorithms and Techniques*, San Francisco, CA, 2011, 78780B-1-13.
27. Dalal, N.; Triggs, B. Histograms of oriented gradients for human detection. In *Proceedings of the IEEE Computer Society Conference on Computer Vision and Pattern Recognition*, San Diego, CA, 2005, 886–893.
28. Oreifej, O.; Mehran, R.; Shah, M. Human identity recognition in aerial images. In *Proceedings of the IEEE Computer Society Conference on Computer Vision and Pattern Recognition*, San Fransisco, CA, 2010, 709–716.
29. Blondel, P.; Potelle, A.; Pégard, C.; Lozano, R. Human detection in uncluttered environments: from ground to UAV view. In *Proceedings of the International Conference on Control Automation Robotics and Vision (ICARCV 2014)*, Singapore, 2014, 76–81.
30. Bose, B.; Grimson, E. Improving object classification in farfield video. In *Proceedings of the IEEE Computer Society Conference on Computer Vision and Pattern Recognition*, Washington, DC, 2004.
31. Reilly, V.; Solmaz, M.; Shah, M. Geometric constraints for human detection in aerial imagery. In *Computer Vision—ECCV 2010*; Daniilidis, K.; Magagos, P.; Paragios, N.; Eds.; Lecture Notes in Computer Science. Springer: Berlin and Heidelberg, Vol. 6316.
32. Turker, M.; Cetinkaya, B. Automatic detection of earthquake-damaged buildings using DEMs created from pre- and post-earthquake stereo aerial photographs. Int. J. Remote Sens. **2005**, *26* (4), 823–832.
33. Sui, H.; Tu, J.; Song, Z.; Li, Q. A novel 3D building damage detection method using multiple overlapping UAV images. ISPRS Int. Arch. Photogramm. Remote Sens. Spatial Inf. Sci. **2014**, *40* (7), 173–179.
34. Vetrivel, A.; Gerke, M.; Kerle, N.; Vosselman, G. Identification of damage in buildings based on gaps in 3D point clouds from very high resolution oblique airborne images. ISPRS J. Photogramm. Remote Sens. **2015**, *105*, 61–78.
35. Tu, J.; Sui, H.; Feng, W.; Song, Z. Automatic building damage detection method using high-resolution remote sensing images and 3D GIS model. ISPRS Ann. Photogramm. Remote Sens. Spatial Inf. Sci. **2016**, *III-8*, 43–50.
36. Long, J.; Shelhamer, E.; Darrell, T. Fully convolutional networks for semantic segmentation. In *Proceedings of the IEEE Computer Society Conference on Computer Vision and Pattern Recognition*, Boston, MA, 2015, 3431–3440.

37. Szegedy, C.; Liu, W.; Jia, Y.; Sermanet, P.; Reed, S.; Anguelov, D.; Erhan, D.; Vanhoucke, V.; Rabinovich, A. Going deeper with convolutions. In *Proceedings of the IEEE Computer Society Conference on Computer Vision Pattern Recognition*, Boston, MA, 2015, 1–9.
38. Zhou, P.; Cheng, G.; Liu, Z.; Bu, S.; Hu, X. Weakly supervised target detection in remote sensing images based on transferred deep features and negative bootstrapping. Multidimension. Syst. Signal Process. **2015**, *27* (4), 925–944.
39. Sherrah, J. Fully Convolutional Networks for Dense Semantic Labelling of High Resolution Aerial Imagery. Available at https://arxiv.org/abs/1606.02585, (accessed in June 2017).
40. Mitomi, H.; Saita, J.; Matsuoka, M.; Yamazaki, F. Automated damage detection of buildings from aerial television images of the 2001 Gujarat, India earthquake. In *Proceedings of the IEEE 2001 International Geoscience and Remote Sensing Symposium*, Sydney, NSW, 2001, 147–149.
41. Gerke, M.; Kerle, N. Automatic structural seismic damage assessment with airborne oblique pictometry© imagery. Photogramm. Eng. Remote Sens. **2011**, *77* (9), 885–898.
42. Dell'Acqua, F.; Gamba, P. Remote sensing and earthquake damage assessment: experiences, limits, and perspectives. Proc. IEEE **2012**, *100* (10), 2876–2890.
43. Johnson, B.; Xie, Z. Unsupervised image segmentation evaluation and refinement using a multi-scale approach. ISPRS J. Photogramm. Remote Sens. **2011**, *66* (4), 473–483.
44. Galarreta, J.F.; Kerle, N.; Gerke, M. UAV-based urban structural damage assessment using object-based image analysis and semantic reasoning. Nat. Hazard. Earth Syst. Sci. **2015**, *15*, 1087–1101.
45. Quaritsch, M.; Kruggl, K.; Wischounig-Strucl, D.; Bhattacharya, S.; Shah, M.; Rinner, B. Networked UAVs as aerial sensor network for disaster management applications. Elektrotech. Informationstechnik (E&I) **2010**, *127* (3), 56–63.
46. Marenchino, D. *Low-Cost UAV for the Environmental Emergency Management. Photogrammetric Procedures for Rapid Mapping Activities*; PhD thesis; Politecnico di Torino, 2009.
47. Wallis, R. An approach to the space variant restoration and enhancement of images. In *Proceedings of the Symposium on Current Mathematical Problems in Image Science*, Monterey, CA, 1976.
48. Suzuki, T.; Miyoshi, D.; Meguro, J.; Amano, Y.; Hashizume, T.; Sato, K.; Takiguchi, J.I. Real-time hazard map generation using small unmanned aerial vehicle. In *Proceedings of the SICE Annual Conference*, Tokyo, 2008.
49. Murphy, R.R.; Steimle, E.; Griffin, C.; Cullins, C.; Hall, M.; Pratt, K. Cooperative use of unmanned sea surface and micro aerial vehicles at Hurricane Wilma. J. Field Rob. **2008**, *25* (3), 164–180.
50. Nardi, D. Intelligent systems for emergency response. In *Invited Talk in the Fourth International Workshop on Synthetic Simulation and Robotics to Mitigate Earthquake Disaster (SRMED)*, 2009.
51. Tuna, G.; Nefzi, B.; Conte, G. Unmanned aerial vehicle-aided communications system for disaster recovery. J. Network Comput. Appl. **2014**, *41*, 27–36.

Unusual Trajectory Detection

Cigdem Beyan
Pattern Analysis and Computer Vision (PAVIS) Department, Istituto Italiano di Tecnologia (IIT), Genoa, Italy

Abstract
There has been a growing interest in unusual behavior detection in computer vision and image processing because of its wide range of applications such as traffic surveillance, human/animal behavior understanding, and elderly surveillance. The most traditional methods for unusual behavior detection are trajectory based, which rely on different trajectory representations, feature extraction, and learning methods. In this work, a comprehensive review including different trajectory representations and learning methods for unusual trajectory detection is presented. Additionally, a comparative analysis using different computational methods applied to real-world datasets such as fish and pedestrian trajectory was performed. To the best of our knowledge for the first time in this work, active learning with feature selection is applied for unusual trajectory detection which presents sufficiently good results even with much less training data.

Keywords: Abnormal classification; Active learning trajectory; Trajectory analysis; Unusual trajectory detection.

INTRODUCTION

Behavior understanding research in computer vision and image processing fields has commonly presented research on human behavior analysis,[1-4] traffic surveillance,[5] animal behavior understanding,[6-9] nursing home surveillance,[10] etc. By definition, behavior includes activities and events in a specific context such that an activity is defined as a complex sequence of actions that last longer time and many include more than one agent,[1] an event is an occurrence of an activity in a specific place and time,[11] and action is characterized by simple motion patterns which happen in a short time by a single object.[1]

A fundamental research topic in behavior understanding is automatic detection of unusual events. This problem has deeply investigated in many works not only in academia but also in industry in the recent years.[12-17] A common approach to detect anomaly is to learn a model which describes normal activities in a video and then discover unusual events by detecting the patterns that are significantly different than the model learnt. However, the complexity of the scenes and the variety of the normal and abnormal behaviors make anomaly detection still a challenging task.[18]

Among previous works, many anomaly event detection approaches are based on trajectories, which are defined as the displacement of an object that is typically considered as positions in two or three dimensions over time. In those approaches, anomaly detection was performed by analyzing individual moving objects in the scene. Generally, the first step is the tracking of objects. Then, trajectory extraction is carried out to perform trajectory clustering as in the works of Hu et al.[5] and Fu et al.[19] or modeling typical activities as applied in Ref. [20] to detect anomalies. Since tracking accuracy decreases in the presence of occlusions, trajectory-based anomaly detection is not suitable for analyzing very complex and crowded scenes. However, trajectories have the advantage of representing object's motion for a long time, which is not possible for other features such as optical flow.[21] Additionally, trajectories represent the motions in a larger space compared to local features. For instance, the spatial–temporal patches used to represent the activities are limited within the space of a single patch defined. On the other hand, trajectories can move across a set of different patches.[21] These benefits make using trajectories desirable for unusual event detection, especially when the object detection and tracking algorithms used are robust.

The aim of this work is to present a comparative study which uses different passive and active learning methods (see the "Methods" section for the definition) for unusual trajectory detection. Here, it is worth mentioning that the definition of unusual is a bit ambiguous in the computer vision literature. The words such as abnormal, rare, outlier, suspicious, and anomaly have been used interchangeably with the word unusual. In this work, unusual refers to "not common," "not frequently observed," "rare," and "outlier," whereas normal means "frequently observed" and "common." Additionally, although there are slight differences in the literature, trajectory has been used interchangeably with the terms action, activity, event, and behavior. Similarly, in this work, even though all analyses were performed by using trajectories, the words behavior and event were used to refer them as well.

The contributions of this work can be listed as follows: (i) a comprehensive review regarding unusual trajectory detection studies, including the trajectory representation methods applied and the learning methods used, is presented; (ii) experimental analyses using different passive learning methods (with/without feature selection) including deep learning are performed; and (iii) to the best of our knowledge, for the first time, an unusual trajectory detection method which combines active learning with feature selection is presented in this work. The experimental analyses are realized using two real-world datasets: (i) fish trajectory dataset[7,9] and (ii) indoor pedestrian trajectory dataset.[22]

The rest of this work is organized as follows: The previous studies related to unusual trajectory detection are discussed in the "Related Work" section. Then, the definition of trajectory is given; the proposed and existing unusual trajectory detection methods used are defined. Later, the datasets used are introduced briefly. Following that, the experimental design and the results are presented. Finally, the paper is concluded with a discussion and future work.

RELATED WORK

In this section, unusual trajectory detection studies are reviewed in terms of the trajectory representation methods that they used and the learning methods that they used.

In general, the trajectory representation methods can be summarized as follows[8]:

(i) Using raw trajectory points and/or reproducing trajectory points such as by applying polynomial fitting and multi-resolution decomposition.
(ii) Extracting multiple features from trajectories, which represent trajectories' shape and temporal characteristics.
(iii) Combination of the above.

Meanwhile, the learning methods used can be categorized as follows[8]:

(i) Unsupervised.
(ii) Supervised.
(iii) Semi-supervised.

Trajectory Representation Methods

The trajectory representation methods were categorized in Ref. [11] which is a survey about trajectory learning that includes trajectory clustering, path modeling, activity analysis, activity recognition, and unusual trajectory detection. Since we are interested in unusual trajectory detection, in the following cases, we discussed only the corresponding representation methods.

i. *Polynomial fitting*: This includes least square polynomials, Chebyshev polynomials, and cubic B-spline curves. In general, the aim is to fit a 2D curve to trajectories. For unusual trajectory detection, the most common method is cubic B-spline curves. For example, Sillito and Fisher[23] proposed a semi-supervised anomalous trajectory detection method using cubic B-spline fitting. Li et al.[24] adapted the B-spline approach proposed in Ref. [23] for trajectory representation and applied sparse reconstruction for abnormal behavior detection. Similarly, to extract common pathways from pedestrians' trajectories, Makris and Ellis[25] also applied spline fitting. The advantage of spline fitting is that it does not need any parameter learning although the accuracy of it depends on the chosen number of control points. In case an incorrect number of points is used, some trajectory dynamics might be ignored. Such a case might result in losing sharp changes of a trajectory and result in low accuracy for the prediction of trajectories as normal or unusual.

ii. *Multi-resolution decomposition*: Haar and the discrete Fourier transform (DFT) are the most frequently used multi-resolution techniques for unusual trajectory detection. For instance, Naftel and Khalid[26] used the DFT to convert the time series trajectories to frequency domain. Then, using self-organizing map (SOM), trajectories were clustered. The anomalous trajectories were found in terms of the distance such that the trajectories sufficiently distant from all identified trajectory clusters were identified as anomalous. Recently, in Ref. [27], trajectories were represented with a feature vector that was composed of DFT coefficients of low-frequency components. K-means clustering was applied to discretize the features into a "bag-of-words" form, and different classifiers were applied for abnormal event detection. Although the DFT representation is simple, it has a disadvantage such that not being able to represent temporal occurrences. Additionally, it is not successful to represent complex trajectories. It only captures the global properties of the trajectories, which results in a loss in the local properties of them.[21,26]

iii. *Hidden Markov model (HMM)*: It is a frequently used technique to represent and model the trajectories.[28-30] For instance, in Ref. [2], the suspicious human activities were identified using HMM after detecting and tracking people. All possible normal activities were represented using HMMs such that each activity was modeled by one HMM.

During testing, any activity with low likelihoods from all the HMMs was determined as suspicious. The HMM is successful if all trajectories have the same length. However, usually this is not the case. Therefore, to use HMM, trajectory interpolation might be needed. Furthermore, HMM is based on learning, and thus requires training data to define the state and transition matrices.

iv. *Subspace methods*: Principal component analysis (PCA) is the most popular subspace method for trajectory representation. It uses eigenvectors to project data into a lower dimension space. In Ref. [30], segmented trajectories were represented using PCA. The trajectory segmentation was applied by perceptual discontinuities, which were defined using velocity and acceleration. A single data matrix composed of similar trajectories was formed, and the principal components of this matrix were used to find a more compact representation. Lastly, trajectories in terms of their compact representations were classified using an HMM. During testing, a trajectory was classified as normal or abnormal by using a likelihood function. PCA in general is useful since it provides a compact representation, but on the other hand, the number of component should be determined carefully not to lose any trajectory information.

Sillito and Fisher[31] compared trajectory representation methods: Haar wavelet coefficients, DFT, Chebyshev polynomial coefficients and cubic B-spline to pedestrian trajectories, vehicle trajectories, hand trajectories, and pen trajectories using the class separability metric which is a metric that can be used to evaluate the performance of unsupervised unusual trajectory detection methods as well. In the study by Sillito and Fisher,[31] it was shown that the Haar representation is better than the DFT, whereas Chebyshev or B-spline representations are the best.

The trajectory representation methods are summarized in Table 1 with their advantages, shortcomings, and the references cited therein.

An alternative to applying a trajectory representation method is to extract multiple features from the trajectories as applied in Refs. [3,10,32–39]. For example, in an earlier work, Zhong et al.[32] used color and texture histograms to classify the behavior patterns as normal and unusual using the co-occurrences of these features. Porikli and Haga[33] proposed using object-based (the histogram of aspect ratio, orientation, speed, color, size of the object, the HMM trajectory representation, duration, length, displacement, and global direction of the trajectory) and frame-based features (histogram of orientations, location, speed, and size of objects) together to detect the abnormal events.

The remaining studies which used multiple features extracted from raw trajectories were discussed in the "Learning Methods" section where the learning methods are more emphasized.

Table 1 Summary of popular trajectory representation methods used for unusual trajectory detection

Method	Advantages	Disadvantages	Example references
Cubic B-spline	No learning/training	The correct number of control points should be chosen for high accuracy	[23–25]
HMM	Performance with high accuracy	Needs training data to define the states and transition matrix. The lengths of trajectories should be the same	[2,28,29]
PCA	Compact representation	The correct number of PCA components should be chosen for high accuracy	[30]
Haar	No learning/training	Not able to represent complex trajectories	[31]
DFT	Simple	Not able to represent the temporal occurrences of trajectories	[26]

Learning Methods

To detect unusual trajectories, most of the works are unsupervised and usually based on clustering. However, there is not any single clustering method that was particularly successful. Using the similarity between the trajectories and the known clusters with a predefined threshold is one of the most common ways to detect unusual trajectories. On the other hand, first applying clustering and then modeling each cluster with an HMM was also frequently applied. In that case, the unusual trajectories were detected as the trajectories having low likelihood.

Unsupervised Learning

As mentioned, the majority of the unusual trajectory detection works have applied unsupervised learning. One common methodology is using clustering with an assumption which defines a predefined distance threshold such that the trajectories that are not similar (closer) to any clusters are defined as unusual or such that the trajectories belong to clusters having few trajectories are defined as unusual.[6] For instance, in Ref. [40], SOMs were used to cluster the trajectories which were represented in terms of time-smoothed positions and instantaneous velocity. Following the former assumption, the unusual trajectories were detected using Euclidean distance between trajectories and clusters with a predefined distance threshold such that having a distance larger than the defined threshold implies an unusual trajectory. Hu et al.,[34] on the other hand, defined a two-level trajectory clustering method for detecting abnormal trajectories and predicting behaviors. Features such as position, velocity, and size of the object were used. At the first level of the hierarchy, the trajectories were clustered using the spatial information only, and at the second level, the clustered trajectories were grouped using the temporal information. Finally, following the second assumption, abnormal trajectories were detected as the trajectories that belong to clusters having few samples.

Differently, Izo and Grimson[41] clustered the normal and unusual trajectories (a feature vector composed of the area of the object's bounding box, the speed, the direction of motion, and the position) individually using the normalized cuts spectral clustering method. A new trajectory was projected into the spectral embedding space of the obtained clusters and matched with normal and unusual clusters to determine the class of the new trajectory. The features of 4D histogram composed of trajectory position and instantaneous velocity and the direction of motion extracted from raw trajectories were used for unusual event detection.[42] In the paper by Jung et al.,[42] trajectories were first clustered using Gaussian mixture models (GMMs) with an outlier removal using the direction of motion. The number of clusters was determined by the finite mixture models[43] and the outlier removal which includes a split-merge procedure based on Bhattacharyya distance. Later, 4D histograms were analyzed to examine the local characteristics of the trajectory. A new trajectory was classified as normal or unusual by comparing its features with all the histograms that were obtained from training data and the thresholds which are specific to each cluster. In Ref. [44], three-stage hierarchical unsupervised trajectory and activity learning method was presented. The raw trajectory points and the velocity were used as features. In the first stage, GMM was applied to learn the interesting nodes. In the second stage, each route represents a trajectory cluster were extracted using the longest common subsequence distance and spectral clustering. In the last stage, using HMM, the dynamics of the activities was encoded. The abnormal trajectories were determined by a threshold using the trajectory's log-likelihood. In Ref. [45], multiple features such as velocity, directional distance, target trajectory mean, initial target position, speed, acceleration, PCA-transformed trajectory points, and trajectory turns were used to cluster the trajectories and then to detect anomalies. The mean-shift algorithm was applied to normalized trajectory features to obtain trajectory clusters. The abnormal trajectories were defined as outliers to the clusters that are different from the other trajectories in the same cluster or the trajectories that belong to a cluster which has few samples. In Ref. [5], trajectories were examined in three levels: (a) spatial level, (b) directional level, and (c) object type level (vehicles and pedestrians) for unusual traffic behavior detection. In each level, different clustering methods were applied. For the spatial level, spectral clustering was performed using a similarity matrix in terms of Hausdorff distance. For the directional level, GMM was applied using the starting and ending points of trajectories. For the object type level, k-means clustering was used. The output of each level was combined as multilevel motion patterns. Abnormalities were detected as the trajectories that do not fit to any multilevel motion patterns that were learnt during training. As a different approach, in Ref. [46], each trajectory was represented as a sequence of symbols (which are obtained by partitioning the scene into a fixed number of adaptive zones) that are based on some features such as shape and speed in each zone. Unlike many works, in the study,[46] the similarity between trajectories was defined in terms of string kernel which allowed defining a kernel-based clustering algorithm.

An alternative unsupervised learning approach is topic models. In Ref. [47], probabilistic latent semantic analysis (pLSA) was used to determine the co-occurrence of motion paths to find the unusual paths such that an unusual path was determined if the pLSA predicted it as rare or if its log-likelihood was below a threshold. Similarly, in Ref. [48], pLSA was applied using the location, the direction, and the shape of the trajectory. Abnormality detection was performed by different metrics such as log-likelihood, Kullback–Leibler divergence, and Bhattacharyya distance. For instance, when the log-likelihood is used, it was assumed that normal trajectories have a high log-likelihood, whereas abnormal trajectories do not fit any learned topic. Latent Dirichlet allocation (LDA) is another topic model which was applied in an unsupervised way for unusual trajectory detection in Refs. [4,49]. In those studies[4,49] trajectories were grouped by LDA after being represented using HMMs.

Supervised and Semi-Supervised Learning

Support vector machines (SVMs),[50,51] HMM,[3,10,35] and dynamic Bayesian networks (DBN)[36–39] are the popular classifiers which were applied in a supervised (using the trajectories fully labeled as normal and unusual) or a semi-supervised (using only the trajectories labeled as normal) way for unusual trajectory detection.

Ivanov et al.[51] used velocity and acceleration as features to detect unusual activities such as running or careless driving. SVM was applied using normal and unusual trajectories, and the learnt model was used to detect new unusual activities. Piciarelli et al.[50] proposed an SVM-based unusual event detection method as well. The trajectories were clustered with a single-class SVM, and the outliers in the training samples were detected as unusual events.

Xiang and Gong[36] proposed a time accumulative reliability measure to detect abnormalities such that once a sufficient number of trajectories that belong to the same behavior class is observed, the normal trajectories were determined on the fly. A trajectory was represented in terms of the center of bounding box of detection, the width and the height of the bounding box, and its shape and the first-order moment. Using the Bayesian information criterion, the number of behavior classes and patterns was found automatically. These behavior patterns were used to find the groupings, and each group was represented by a DBN with multi-observation hidden Markov model (MOHMM) topology. The log-likelihood of a new trajectory was determined by the MOHMM model. All log-likelihoods were used to determine the abnormality of the trajectory by comparing the reliability measure which is based on a threshold.

Following the work presented in Refs. [36,37], the performance of the behavior model trained using an unlabeled dataset and the performance of the behavior model trained using the same, but labeled dataset, were compared. The results have shown that the trained model using an unlabeled dataset was better than the trained model using the same but labeled dataset. The methods proposed in Refs. [36,37] were adapted in Ref. [38] to detect anomalies in video but using incremental learning which is based on expectation maximization. In the study,[38] likelihood ratio test was used to detect anomalies unlike using reliability measure as applied in Refs. [36,37].

In Ref. [39], using a two-stage cascade dynamic Bayesian network and the assumption such that a normal behavior should follow a typical order of atomic actions with certain durations while a deviation in temporal order or temporal duration should cause an anomaly, abnormal behaviors were detected and discriminated. In Ref. [39], features such as the blob center, the width, the height of the bounding box, the occupancy, the ratio of the dimension, the mean optical flow of the bounding box, and the scaled optical flow were used. In detail, three anomalies were considered: (i) patterns which are visually different from the patterns observed during training, (ii) ambiguous patterns since they are rare, and (iii) patterns having weak visual evidence. Different DBN models such as a first-order HMM to model the temporal order in the first stage and an MOHMM to model the temporal duration of a behavior in the second stage were used. In each stage, a threshold was defined which was used to detect and discriminate different classes of anomalies by comparing it with the normalized log-likelihood of a new behavior.

Beyan and Fisher[52] presented a rule-based trajectory filtering method which aims to detect normal trajectories while rejecting abnormal trajectories. Rules were defined in terms of direction of the trajectory points (straight, cross movements) and/or being stationary. In total, 21 rules were defined and applied as a cascade classifier. During training, the useful rules (the one which was successful to detect a normal trajectory while filtering out an abnormal trajectory) and the parameters which correspond to the selected rules were found. A new trajectory was classified as normal or abnormal by applying the selected rules with the learnt parameters only. For detecting unusual fish trajectories in another work,[6] multiple features were extracted from fish trajectories. The method used in Ref. [6] applied affinity propagation for clustering trajectories and outlier detection to find the unusual trajectories. During training, feature selection was applied to learn the best features. A new trajectory was classified using the outlier detection parameters learnt per cluster using the features selected during training.

Sillito and Fisher[23] proposed a semi-supervised trajectory detection method which was based on GMM that models normal and unusual trajectories. When a new trajectory was classified as unusual by the existing model, the human operator checked if that trajectory was really unusual or not. Based on the operator's decision, the model was incrementally updated. For the trajectories classified as normal, the human operator did not take any action. By the model, the trajectory was classified as unusual if its Mahalanobis distance to the closest component of the GMM exceeds a predefined threshold. The advantage of this system is the capability of classifying a new trajectory at any time during the training. Luhr et al.[3] and Duong et al.[10] presented semi-supervised methods for nursing home and smart home systems, respectively. They used variations of HMM, that is, a fully connected explicit state duration HMM and switching semi-HMM. Both of these studies detected anomalies based on the order of the activities and the durations of the activities. Differently, in Ref. [53], a weakly supervised joint topic model which is based on LDA was used to find rare and subtle behaviors which were defined as sparse and not frequent enough to be modeled precisely. The advantage of the work in Ref. [53] is being able to detect rare behaviors even with a few training data and being able to detect the anomalies which have very small spatiotemporal deviations.

Recently, deep learning has demonstrated great success in many topics of image processing and computer vision. Also for unusual event detection, many methods were presented. However, the studies that use trajectories for unusual event detection are very few. Generally, instead of using raw trajectories or multiple features extracted from trajectories, studies used the spatio/temporal activity patterns (as 2D image cells or 3D video volumes) such as Refs. [54,55]. Alternatively, as applied in Refs. [18,56], some learnt discriminative feature representations such as from appearance and motion patterns. In Ref. [57], restricted Boltzmann machines (RBMs) with Replicated Softmax (RS) model[58] was applied for trajectory analysis. Trajectories were represented using "bag-of-words" method. The video scene was divided into grid regions, and in each region, the motion directions were grouped into different bins which were used to build the dictionary. A trajectory was projected into dictionary, and its occupations rate was used as the feature representation which was used to train the RS model. For the detection of unusual trajectories, sparse reconstruction analysis, RS model with different

Table 2 Learning methods used for unusual trajectory detection

Method	Example references
Clustering	[5,13,19,23,29,33,34,40–42,44,45,67,69–81]
HMM and its variants	[3,10,16,35,44,82–86]
DBN	[36–39]
SVM	[50,51,87–89]
Topic models	[4,47–49,53,81,90,91]
Rule-based learning	[52]
Outlier detection	[6,7,20]
Restricted Boltzmann machines	[57]

number of hidden units, bag of words with SVM, and RS model with SVM were compared. As a result, it was shown that RS model and RS model with SVM performed better than other methods.

The learning methods with more references are summarized in Table 2.

METHODOLOGY

In this section, the methods that were used to detect the unusual trajectories are described, and the definition of a trajectory is given. SVM (since it was one of the most frequently used methods for this task) and hierarchical decomposition (HD)[7,9] (since it demonstrated the best performances in Refs. [7,9] for the datasets used) were used as supervised unusual trajectory detection methods. SVM was also combined with sequential forward feature selection (SFFS)[59] algorithm which has shown improved performance compared to SVM only. As an alternative to SFFS, similar to Ref. [57], RBM was used to learn potentially better feature representations, and the resulting features are used for the classification of trajectories using SVM. Lastly, different than passive learning (such as SVM, HD, and RBM), motivated from the studies which applied semi-supervised learning,[3,10,23,53] active learning with and without feature selection (which potentially can perform as good as passive learning even with less training data) was also applied. To the best of our knowledge, this work is the first attempt to use active learning with feature selection for unusual trajectory detection.

Methods

SVM: It was applied using a radial basis kernel function (RBF) with varying kernel parameters, while the hyperplanes were separated by sequential minimal optimization. It was combined with SFFS algorithm[59] as well (referred to as SVM-SFFS).

HD[7,9]: This method is based on clustering, outlier detection, and feature selection which are combined to decompose a hierarchy during training. In detail, at each level of the hierarchy, data is first clustered using the best feature subset which is found by SFFS. After clustering, outlier detection is applied to each cluster, and outliers (unusual trajectories) for the current level of the hierarchy are found. Then, using the ground-truth data, each cluster is examined to find if they include any misclassified normal or unusual trajectories. The trajectories belong to a cluster which do not contain any misclassified trajectory are not used for the construction of the rest of the hierarchy. On the other hand, the clusters having at least one misclassified trajectory are used for construction of the rest of the hierarchy which is repeated in the same way. By repeating clustering, outlier detection, and feature selection, the hierarchy construction continues until it reaches to a hierarchy level which has all clusters having at least one misclassified trajectory or all trajectories are perfectly classified. Shortly, at each level of the hierarchy, different trajectories are used and different feature subsets are selected. During testing, a rule-based heuristic which is based on finding the closest cluster at each level of the hierarchy for a given new trajectory is applied. In detail, based on the closest cluster and the new trajectory's position in the closest cluster, the class decision for the new trajectory is made as (i) unusual trajectory, (ii) candidate normal trajectory, or (iii) no effect on the decision in the single level of the hierarchy. In case, the decision is "candidate normal trajectory" or "no effect on the decision," the next level is visited to find the closest cluster and make a decision. Otherwise, the final class of the new trajectory is determined as unusual. All the parameters (outlier detection parameter) used were taken as applied in Refs. [7–9]. For more information, interested readers can refer to Refs. [7–9].

Restricted Boltzmann machines (Bernoulli–Bernoulli RBM [BB-RBM])[60,61]: As shown in the experimental work section, the performance of a method was improved when it was combined with SFFS (for the fish dataset only). This shows that some of the features are either irrelevant or redundant and because of that they misguide the classification which results in poor performance when all features are used together. Another way to handle this problem can be applying RBMs[60] to the features extracted from trajectories (after preprocessing) to learn better feature representations which are in terms of the hidden unit activations. Later, the obtained feature representations are used as an input to SVM to detect the unusual trajectories.

In this study, BB-RBM which has binary visible and hidden units was applied as described in Ref. [61] (other RBM types such as Gaussian-Bernoulli RBM and RBM with RS model[58] can be applied as well). Given that we have a continuous value data, the feature space was first binarized by applying thresholding to each feature vector individually. For each feature vector in the feature space, a threshold using (i) the mean, (ii) median, and (iii) sum of mean and standard deviation of the feature vector was found. These thresholds were applied such that any feature values greater than or equal to the estimated threshold were taken as one, and any feature values less than the estimated threshold were taken as zero. By using arbitrary chosen number of hidden units, visible units, and batch sizes, new features (i.e., the hidden unit activations) were learnt. Once the new features were learnt, SVM was applied as described earlier to detect unusual trajectories.

Active learning: The goal of active learning is to achieve sufficiently better learning performance (the performance as good as passive learning) with fewer training instances.[62,63] Especially, when unlabeled data is abundant, labeled data is noisy and limited, and labeling is expensive, active learning is desirable.

Basically, active learning seeks to choose the most informative unlabeled training instances with a query strategy. This requires labeling only selected instances which decreases the labeling cost in contrast to passive learning where the labels of all training examples are required (as all the other methods used in this work, that is, SVM, HD, and RBM, apply).

Active learning has been examined in different domains, and several studies have addressed this problem. The majority of the studies in this field focused on imbalanced dataset classification. Given that unusual trajectory detection is an imbalanced dataset classification problem, in this work, we propose to use active learning.

Active learning can be described as follows:

– Train a model using the existing *labeled training data*.
– Select the *informative samples* from the unlabeled training data using the model learnt and a *query strategy* (information density in our case).
– Label the selected *informative samples*.
– Combine the new labeled data with the existing labeled training data and repeat all these steps until reaching a *stopping criterion* (such as a decrease in classification performance, having a certain amount of labeled training data).

This mechanism is illustrated in Fig. 1.

Even though different studies proposed different *query strategies*, there are very popular strategies such as uncertainty,[64] information density,[65] and maximum probability.[63] Those popular query strategies are successful although no one query strategy is the best for all datasets,[62,63] and recent strategies proposed are indeed based on them.

In this work, we used information density[65] for our analysis. Information density assumes that the informative samples are those which are uncertain (having big entropy, Eq. 1) and which are more similar to the other unlabeled samples. To calculate the similarity, we applied inverse of the Euclidean distance (Eq. 2), while the information density is defined as follows:

$$E(x) = -\sum_{i=1}^{c} p_i(x) \log_2 p_i(x) \quad (1)$$

$$\text{sim}(x_1, x_2) = \frac{1}{1+d(x_1, x_2)} \quad (2)$$

$$\text{ID}(x) = E(x) \left[\frac{1}{U} \sum_{i=1, x_i \neq x}^{U} \text{sim}(x_1, x_2) \right] \quad (3)$$

where U is the number of the unlabeled instances, x and x_i refer to unlabeled training samples, d is the Euclidean distance, and $E(x)$ is the entropy of unlabeled instance x, while c is the number of classes (which is two in our case: normal and unusual), and p refers to posterior probabilities.

In this study, active learning was applied when all features were used and also it is combined with SFFS[59] motivating the successful performance of SFFS when it was combined with passive learning (such as SVM).

In the literature, there is a large amount of research on active learning and feature selection individually. However, there is not much work which combines these two methodologies together. To the best of our knowledge, previous studies about active learning with feature selection all belong to the natural language processing field (especially text classification), but this has never applied for image processing and computer vision especially for unusual trajectory detection. Hence, for the first time in this study, active learning with feature selection is applied to detect unusual trajectories.

Our claim is that a proper feature selection criterion which forces the classifier to pay attention to the

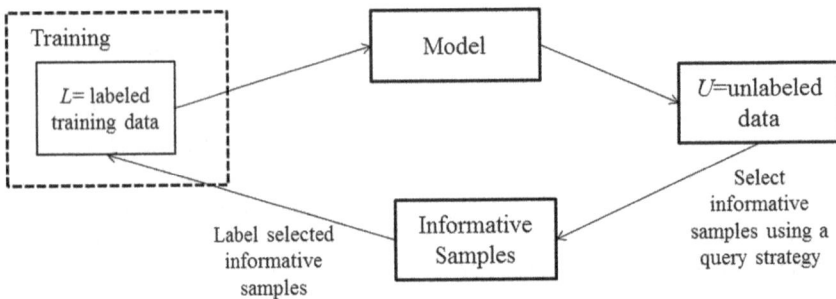

Fig. 1 Active learning[62]

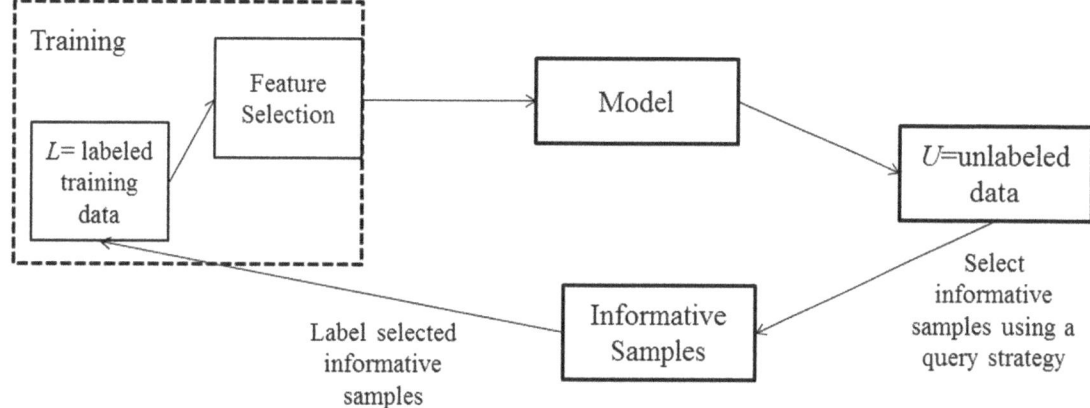

Fig. 2 Active learning with feature selection

classification of the unusual trajectories or both classes equally can give sufficient classification performance (the performance as good as passive learning) by using less training data and better performance compared to active learning without feature selection.

Feature selection is integrated with active learning as follows:

- During training using the *labeled training samples*, apply SFFS starting with an empty set of features.
- Once the *feature selection* is ended (the best features are determined), train a model using the *best features* which are specific for the current iteration of active learning.
- Determine the most *informative samples* using the learnt model and the *query strategy* (information density in our case).
- Label the *informative samples* and combine them with the existing training data.
- Repeat these steps until you reach a *stopping criterion*.

Active learning with feature selection mechanism is illustrated in Fig. 2.

For all active learning experiments, we continued iterations until all training data was labeled. As a learning method, SVM was used (referred to as SVM-active learning and SVM-active learning-SFFS for active learning and active learning with feature selection, respectively). As the kernel function, an RBF kernel was applied. Using the MATLAB *fitcsvm function*,[66] the scale values of the kernel functions were found automatically. Then, the data was standardized before applying SVM. MATLAB *fitPosterior function*[66] allowed us to find the posterior probabilities which were used by the query strategy to find the informative samples.

Trajectory Description

A trajectory is defined as

$$T = \left\{ \left(x_{f_1}, y_{f_1}\right), \left(x_{f_2}, y_{f_2}\right), \ldots, \left(x_{f_{n-1}}, y_{f_{n-1}}\right), \left(x_{f_n}, y_{f_n}\right) \right\} \quad (4)$$

where (x, y) refers to the center of the object's bounding box such that the object is tracked through n frames, and f_i is the frame number in a video.

DATASETS

In total, two different datasets were used for the experimental analysis: (i) fish trajectory and (ii) indoor pedestrian trajectory.

Fish Trajectory Dataset

This dataset[7,9] includes 3,102 fish trajectories such that 3,043 of them are normal and 59 of them are unusual. All trajectories belong to *Dascyllus reticulatus* which is a fish species living in the Taiwanese coral reef. The collected data is from 93 different underwater videos, all captured in the morning. Each video has 320×240 resolutions, and the frame rate is five frames per second. The normal and unusual trajectories were annotated manually and examined by marine biologist.[7,9]

Examples of normal fish behavior are given as follows: (i) fish hovering over the coral and (ii) freely swimming fish in the open area. Examples of unusual fish behavior are given as follows: (i) fish suddenly changing the direction for predator avoidance, (ii) fish biting at coral, (iii) fish diving quickly between the coral branches since it was frightened or to hide from the predator, and (iv) aggressive fish which were moving. In this dataset, a trajectory having unusual and normal segments was labeled as unusual.

As applied in Refs. [7–9], feature categories (i) curvature scale space based, (ii) moment based, (iii) velocity based, (iv) acceleration based, (v) turn based, (vi) centered distance function based, (vii) vicinity based, (viii) loop based, (ix) fish pass by based, (x) displacement based, and (xi) using the size of bounding box were extracted from each trajectory (for definitions of features, see Refs. [7,9]). In total, 776 features were obtained, but this decreased to 140 by applying PCA to each feature group individually as

Table 3 Details of features extracted from the datasets used

Feature group: details of features	Abbreviation	Datasets
Curvature scale space-based features: Mean and variance of length of the curves in curvature scale space image (CSS image), number of zero crossings in CSS image, total number of curves, mean and variance of peak points in CSS image, mean and variance of starting points of each curve	CSS	FT, PT
Moment descriptor-based features: The affine moment invariants, moments, central moments, and translation–scale invariant moments (see Refs. [7–9] for definitions) while the moment order was taken up to 4	MD	FT, PT
Velocity- and acceleration-based features: Mean, standard deviation, minimum, maximum, number of zero crossings, the total number of local minima, the total number of local maxima in the velocity and acceleration vector	VA	FT, PT
Turn-based features: Mean, standard deviation, minimum, maximum, the total number of zero crossings, the total number of local minima and maxima in the turn vector	T	FT, PT
Centered distance function-based features: Mean, maximum, minimum, standard deviation, number of mean crossings, number of local minima and maxima, skewness and kurtosis	CDF	FT, PT
Vicinity-based features: Using the aspect of vicinity, vicinity curliness, and vicinity linearity (see Refs. [7–9] for definitions), the statistics such as mean, standard deviation, skewness, kurtosis, the total number of mean crossings, the total number of local minima and maxima, and maximum, minimum, and median were calculated	V	FT, PT
Loop-based features: The total number of loops, maximum, minimum, and median number of points in a loop	L	FT, PT
Fish pass by-based features: The percentage of time that a fish was in a specific location (i.e., open sea, under coral and above coral) and the percentage of time that fish crosses from one specific location to another	FP	FT
Features based on DL: Mean, maximum, minimum, standard deviation, and median of average displacement in a specific location of the scene	DL	FT
Features based on the normalized size of bounding box: The total number of one crossings, the total number of local minima, and the total number of local maxima extracted from the normalized size of bounding box vector (see Refs. [7–9] for definitions)	NBB	FT
Trajectory points after cubic B-spline fitting	TPB	PT
The deviation between the reconstructed trajectory and the original trajectory: Mean, standard deviation, minimum, maximum, median, the total number of local minima and maxima, skewness and kurtosis of the deviation vector	D	PT

Note: See Refs. [7–9] for the details of feature extraction. FT, fish trajectory dataset; PT, pedestrian trajectory dataset.

suggested in Refs. [7–9]. The details of features for each feature category are listed in Table 3.

Forum Pedestrian Database

Forum pedestrian database[22] includes various pedestrian trajectory datasets belong to different dates. For our analysis, we chose the dataset belong to September 1, 2009, which is one of the largest datasets and also used in Ref. [8]. This dataset includes 2,342 trajectories. In total, 1,624 trajectories are normal and 718 are unusual, and they all belong to different people. The normal and unusual trajectories were labeled using Ref. [22], and the results were also manually inspected. Basically, the field of the camera was divided into regions such as the main entrance to the building, lifts, access to the atrium, access to the hall, staircase, reception desk, and four exits. A trajectory was annotated as normal if (i) it represents a clear goal such as going from one exit to another, and (ii) the goal was achieved in an efficient way which means with a trajectory close to a straight line. Otherwise, it was labeled as unusual.

For our analysis, as applied in Ref. [8], (i) velocity-based, (ii) acceleration-based, (iii) vicinity-based, (iv) curvature scale space-based, (v) center distance function in 2D based, (vi) loop-based, (vii) moment-based, and (viii) turn-based features were extracted (for definition of features, see Refs. [7,9]). Furthermore, (ix) trajectory points after cubic B-spline fitting and (x) the statistical features (mean, standard deviation, minimum, maximum, median, the number of local minima, the number of local maxima, skewness, and kurtosis) which were extracted from the deviation between the reconstructed trajectory and the original trajectory were also extracted.[8] To extract features (ix) and (x), each trajectory was approximated with a cubic spline with six control points using the implementation in Ref. [23]. In total, 758 features were obtained. To prevent possible overtraining or curse of dimensionality, PCA was applied to each group of features individually except the trajectory points that were obtained after cubic B-spline fitting (as applied in Refs. [7–9]). As a result, 57 features were obtained and used for the following analysis. The details of features for each feature category are listed in Table 3.

EXPERIMENTAL WORK

For all experiments presented in this section, ninefold cross-validation was performed. Training, validation, and test sets were constituted randomly, and the normal and unusual trajectories were distributed equally in each set. For the methods combined with feature selection, that is, SFFS,[59] validation sets were used to pick the best feature set for each method individually. For the methods which were not combined with SFFS, validation sets were not used. The training and testing sets were kept the same for all methods.

For active learning, at each cross-validation fold, one unusual trajectory and one normal trajectory were randomly chosen as the initial labeled training set. Information density was then used to pick trajectories from the unlabeled training data. At each iteration of active learning, 25 trajectories were chosen to be labeled. No early stopping criterion was applied; hence, active learning iterations continued until all training samples were labeled.

The performance evaluations were performed in terms of (i) *true positive rate* (*TPrate*) which represents the unusual trajectory detection given that positive class represents the unusual trajectories and is defined as given in Eq. 5, (ii) *true negative rate* (*TNrate*) which represents normal trajectory detection given that the negative class represents the normal trajectories and is defined as given in Eq. 6, and (iii) *geometric mean of TPrate and TNrate* (*GeoMean*) which represents the overall performance and is defined as given in Eq. 7. *GeoMean* is used since it does not ignore the importance of the classification of unusual trajectories (such as accuracy does) because unusual trajectories are underrepresented given that the number of normal trajectories is much more than unusual trajectories. Additionally, in Ref. [68], *GeoMean* was suggested as a proper metric for the evaluation of performance of classifiers when imbalanced datasets are analyzed.

$$TPrate = TP / (TP + FN) \quad (5)$$

$$TNrate = TN / (TN + FP) \quad (6)$$

$$GeoMean = \sqrt{TPrate \times TNrate} \quad (7)$$

In Eqs. 5–7, *TP* represents the number of correctly classified unusual trajectories, *TN* represents the number of correctly classified normal trajectories, *FN* represents the number of misclassified unusual trajectories, and *FP* represents the number of misclassified normal trajectories.

The best results of the methods using fish trajectory dataset are given in terms of the *GeoMean* with the corresponding *TPrate* and *TNrate* in Table 4. For active learning, the number of training data that was used to obtain the best performance is given in the parentheses. The results (Table 4) show that the best-performing (*GeoMean*) passive learning methods were HD and BB-RBM. HD was better at detecting unusual trajectories (since *TPrate* is higher) compared to BB-RBM. The performance of SVM was improved by applying feature selection, which shows that for this dataset not every feature is equally important, and some features actually misguide the unusual trajectory detection task. On the other hand, by applying active learning (without feature selection) even with a training set having 825 trajectories (the training set of passive learning has 2,452 trajectories), better performance compared to SVM (without feature selection) was obtained. This shows that information density is good at determining informative trajectories. Similarly, when active learning was combined with feature selection, using the training set composed of 1,177 trajectories, the best performance (0.98±0.01 *GeoMean*) was obtained. This result can be seen as an outlier since there were many fluctuations in the testing performance (see Fig. 3). However, given that the average *GeoMean* was 0.79±0.08 (from a training set size of 1,177–2,452) and the obtained *GeoMean* was above 0.86 (the performance of SVM-SFFS) in 10 iterations out of 52 (from a training set size of 1,177–2,452), this method

Table 4 The best results of each method in terms of *GeoMean* with the corresponding *TPrate* and *TNrate* using fish trajectory dataset

Methods	TPrate	TNrate	GeoMean
SVM	0.21±0.07	0.99±0.01	0.45±0.07
SVM-SFFS	0.81±0.16	0.93±0.03	0.86±0.09
HD	0.94±0.10	0.88±0.02	0.91±0.05
BB-RBM	0.88±0.13	0.95±0.01	0.91±0.07
SVM-active learning	0.61±0.19	0.72±0.02	0.66±0.14 (852)
SVM-active learning-SFFS	**0.97±0.03**	**0.99±0.01**	**0.98±0.01** (1177)

Note: The standard deviations (considering cross-validation folds) of the methods are given after ± sign. The best results are emphasized in boldface. For active learning, the number of training data that was used to obtain the best result is given in parentheses after the value of *GeoMean*.

can still be a good alternative to SVM-SFFS, HD, and BB-RBM, as it is capable of performing sufficiently good with less labeled data.

The validation set (the set used to test the learnt model in the training phase) and testing set (the set used to test the learnt model in the testing phase) performances (as the average of cross-validation folds) at each iteration of active learning with/without feature selection (SFFS) for fish trajectory dataset are given in Fig. 3. In these analyses, the evaluation metric *GeoMean* was used. As seen in Fig. 3, the performance of active learning with SFFS was much better than without SFFS during training and also during testing.

During training (Fig. 3a and c), only in 10 out of 99 iterations, active learning with SFFS performed worse than pure active learning. Here, it is worth to mention that the labeled training data used in a specific iteration (except the first and the last iterations) can be different for active learning and active learning with SFFS because the training set samples (the labeled data) are automatically selected by calculating the information density function.

During testing (Fig. 3b and d), especially for active learning with feature selection, the performance reached the performance of passive learning (especially SVM) very quickly (means using much less training data). This shows that information density combined with feature selection is a better query strategy to pick the most informative samples compared to information density only.

To better understand how SFFS works during training, in Table 5, the *GeoMean* for each cross-validation fold while features were adding one by one including the value that ends the SFFS is given. These results correspond to the last iteration of SVM-active learning-SFFS which is in other words SVMSFFS using the fish trajectory dataset. Although different features were selected in each cross-validation fold, in six out of nine folds, features from "the features based on displacement on the location (DL)" category were selected as the first feature. As seen, in total, not many features were selected, and even one to two features were enough to obtain good training performances.

The best results of the methods using pedestrian trajectory dataset are given in terms of the *GeoMean* with the corresponding *TPrate* and *TNrate* in Table 6. For active learning, the number of training data that was used to obtain the best performance was given in the parentheses. As seen in Table 6, BB-RBM performed better than any method including active learning while its *TNrate* and *TPrate* were the best as well. On the other hand, active learning with feature selection performed as good as HD and SVM-SFFS even with less data (after 15 iterations such that there are 352 trajectories in the training set, while in total 1667 trajectories were used for passive learning which is the 68th iteration of active learning). Between iterations 15 and 68 (having 352 trajectories to 1,667 trajectories in the training set), the average *GeoMean* was

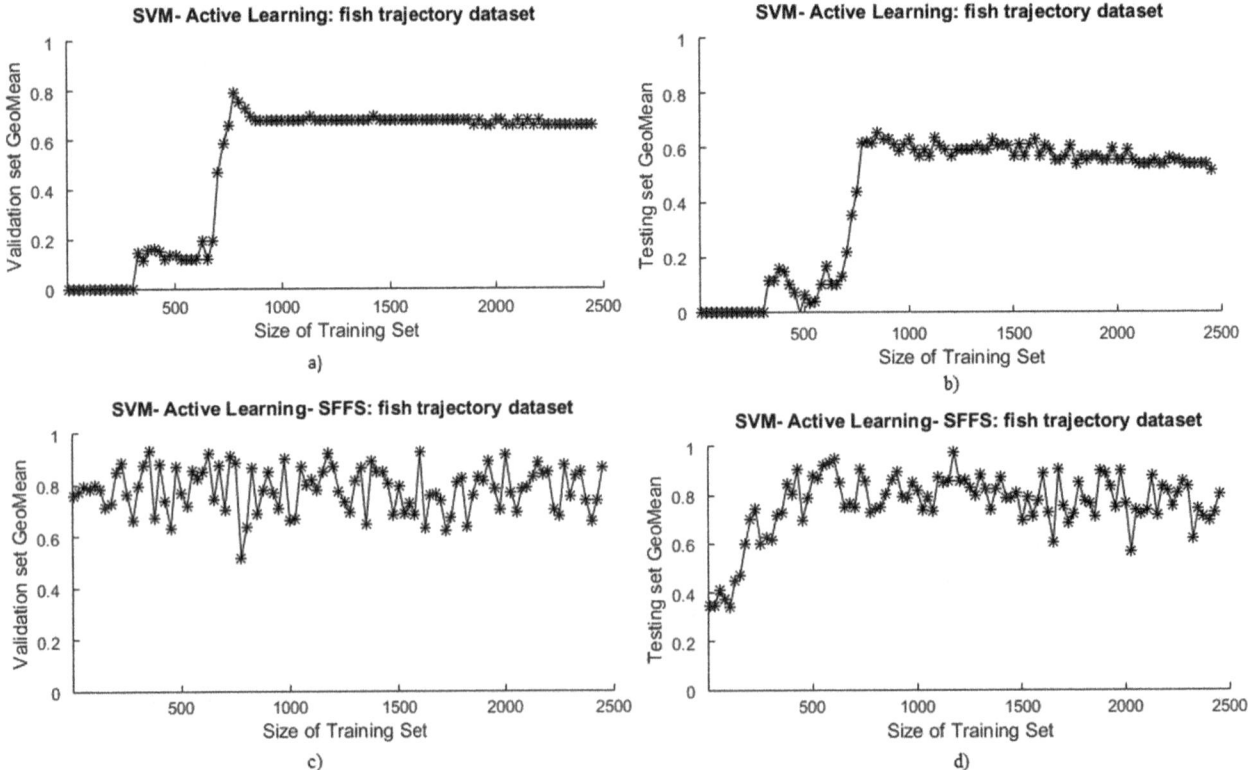

Fig. 3 The performances (as the average of cross-validation folds) in terms of *GeoMean* for the fish trajectory dataset: (a) active learning without feature selection (SVM-active learning) for validation set; (b) SVM-active learning for testing set; (c) active learning with feature selection (SVM-active learning-SFFS) for validation set; (d) SVM-active learning-SFFS for testing set

Unusual Trajectory Detection

Table 5 The selected features at each iteration of SFFS with the corresponding training performance (in terms of *GeoMean*) when all training set is used (the last iteration of SVM-active learning-SFFS/SVM-SFFS) for fish trajectory dataset

Cross-validation fold	*GeoMean*-selected features		
	SFFS iteration 1	SFFS iteration 2	SFFS iteration 3
1.	**0.91-DL**	0.65-CSS	
2.	0.78-DL	**0.79-DL**	0.68-CSS
3.	**0.84-DL**	0.81-DL	
4.	**0.85-FP**	0.76-DL	
5.	**0.92-DL**	0.82-DL	
6.	**0.96-DL**	0.72-DL	
7.	**0.94-V**	0.91-V	
8.	**0.93-V**	0.88-L	
9.	**0.97-DL**	0.72-DL	

Note: Best training *GeoMean* at each cross-validation fold is shown in bold.

Table 6 The best results of each method in terms of *GeoMean* with the corresponding *TPrate* and *TNrate* using pedestrian trajectory dataset

Method	*TPrate*	*TNrate*	*GeoMean*
SVM	0.78±0.04	0.81±0.02	0.79±0.02
SVM-SFFS	0.83±0.03	0.79±0.04	0.81±0.01
HD	0.87±0.06	0.86±0.05	0.86±0.02
BB-RBM	0.89±0.03	**0.91±0.03**	**0.90±0.02**
SVM-active learning	0.79±0.07	0.83±0.09	0.81±0.03 (827)
SVM-active learning-SFFS	**0.91±0.02**	0.79±0.12	0.85±0.02 (352)

Note: The standard deviations (considering cross-validation folds) of the methods are given after ± sign. The best results are emphasized in boldface. For active learning, the number of training data that was used to obtain the best result is given in parentheses after the value of *GeoMean*.

0.84±0.01 which is as good as HD and better than SVM and SVM-SFFS.

In Fig. 4, the validation set (training phase) and testing set (testing phase) performances (as the average of cross-validation folds) at each iteration of active learning with/without feature selection for pedestrian trajectory dataset are given. In these analyses, *GeoMean* was used as the evaluation metric. As seen in Fig. 4, similar to the results obtained using fish trajectory dataset, the performance of active learning with feature selection was much better than pure active learning during training and also during testing.

In the training phase (Fig. 4a and c), only in 4 out of 68 iterations, active learning with SFFS performed worse than pure active learning.

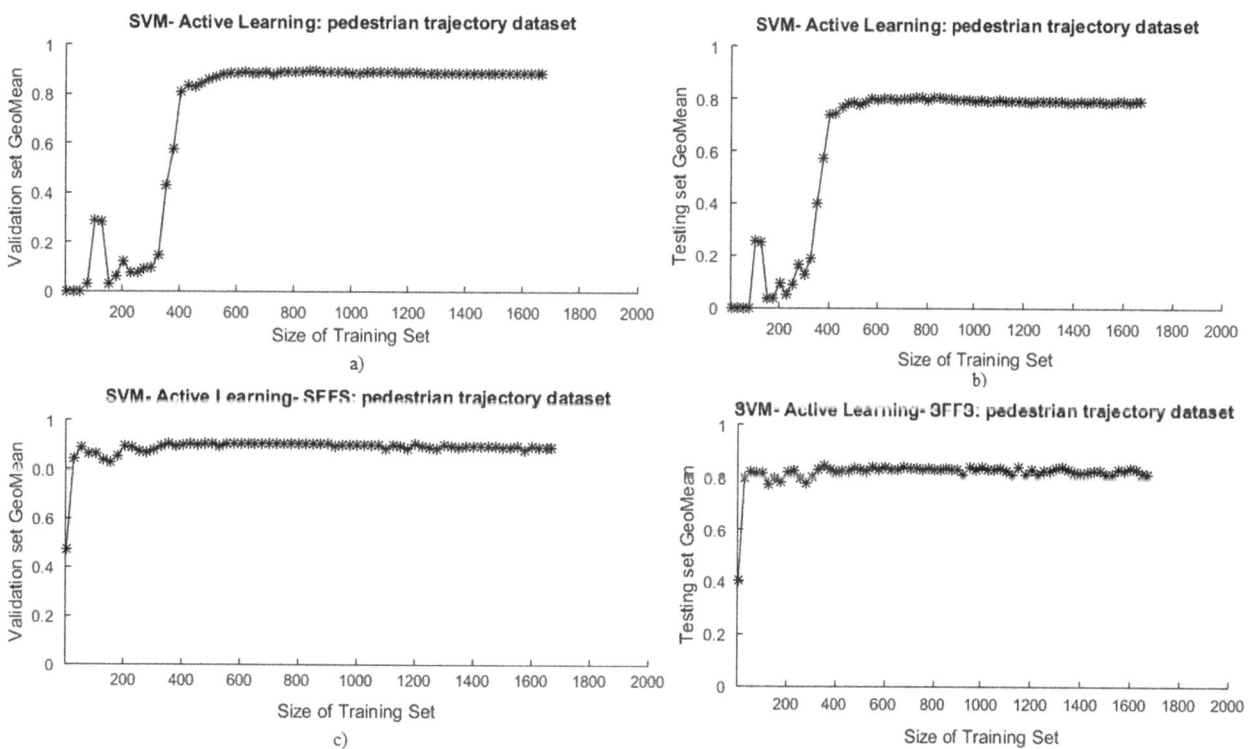

Fig. 4 The performances (as the average of cross-validation folds) in terms of *GeoMean* for the pedestrian trajectory dataset: (a) active learning without feature selection (SVM-active learning) for validation set; (b) SVM-active learning for testing set; (c) active learning with feature selection (SVM-active learning-SFFS) for validation set; (d) SVM-active learning-SFFS for testing set

Table 7 The selected features at each iteration of SFFS with the corresponding training performance (in terms of *GeoMean*) when all training set is used (the last iteration of SVM-active learning-SFFS/SVM-SFFS) for pedestrian trajectory dataset

Cross-validation fold	*GeoMean*-selected features						
	SFFS iteration 1	SFFS iteration 2	SFFS iteration 3	SFFS iteration 4	SFFS iteration 5	SFFS iteration 6	SFFS iteration 7
1.	0.84-CSS	0.87-MD	0.89-VA	0.90-CDF	0.91-T	**0.92-TPB**	0.91-V
2.	0.85-CSS	0.88-MD	0.91-D	**0.91-D**	0.90-D		
3.	0.84-CSS	0.88-MD	0.90-V	**0.91-CDF**	0.90-TPB		
4.	0.75-CSS	0.86-CSS	0.89-TPB	**0.90-TPB**	0.89-D		
5.	0.79-CSS	0.86-TPB	0.88-TPB	**0.89-D**	0.87-D		
6.	0.84-CSS	0.88-V	**0.89-TPB**	0.88-D			
7.	0.80-CSS	0.86-TPB	**0.87-D**	0.86-D			
8.	0.81-CSS	0.85-D	**0.87-D**	0.86-D			
9.	0.85-CSS	0.88-MD	0.89-VA	**0.91-TPB**	0.90-D		

Note: Best training *GeoMean* at each cross-validation fold is shown in bold.

In the testing phase (Fig. 4b and d), the performance of active learning with SFFS quickly reached the sufficient performance (the performance as good as SVM, SVM-SFFS, and HD). Even with 352 trajectories (as given in Table 6), the performance of active learning with feature selection was better than SVM (passive learning).

Overall, when active learning with/without feature selection is compared, it can be seen that (Figs. 3c and d and 4c and d) the fluctuation in the validation/testing performances when feature selection was applied was more compared to pure active learning. This might be because of the change in feature space at each iteration of active learning. However, the sufficient performance (the performance as good as passive learning) was reached very quickly when feature selection was applied. This suggests that training can be stopped rather quickly, but it is unclear how to define where to stop precisely. Similar to the analysis given in Table 5, the *GeoMean* for each cross-validation fold at each iteration of SFFS is given in Table 7. These results correspond to the last iteration of SVM-active learning-SFFS, which is also SVM-SFFS using the pedestrian trajectory dataset. Although various features were selected in each cross-validation fold, for all folds, the first feature was always from curvature scale space (CSS) category, and in four out of nine folds, the second feature was from moment descriptors (MDs), i.e., moment-based features category. Compared to fish trajectory dataset, more features were selected for pedestrian trajectories, and the selected features were more diverse, which belong to different feature categories.

CONCLUSIONS

A fundamental topic in video surveillance is detecting abnormal events in a video stream. Video anomaly detection is a challenging task, and different approaches have been proposed depending on how the complexity of the scene is. One way to detect video anomaly is utilizing trajectories of objects (i.e., unusual trajectory detection), which usually requires robust tracking algorithms but allows long-term and space-unlimited motion analysis. Many trajectory representation, feature extraction, and machine learning techniques have been proposed regarding unusual trajectory detection.

In this work, existing unusual trajectory detection methods which used unsupervised learning (such as clustering, topic models, and RBM), supervised learning, and semi-supervised learning (such as SVM, HMM, and DBN) were reviewed. Then, a comparative analysis was carried out using two real-world datasets. To detect unusual trajectories, different methodologies were used: supervised learning with/without feature selection, hierarchical decomposition based on clustering and outlier detection, and deep learning were used. Unlike the mentioned passive learning methods, active learning (with/without feature selection) was also adapted for this application to show that with a proper query strategy combined with feature selection, it is possible to perform as good as passive learning with much less training data.

As future work, the methods used here should be evaluated on more datasets. Moreover, active learning which presented a good potential for unusual trajectory detection should be better investigated by combining it with different learning algorithms, query strategies, and a criterion which decides where to stop training.

REFERENCES

1. Turaga, P.; Chellappa, R.; Subrahmanian, V.; Udrea, S.O. Machine recognition of human activities: A survey. IEEE Trans. Circuits Syst. Video Technol. **2008**, *18* (11), 1473–1488.

2. Nair, V.; Clark, J.J. Automated visual surveillance using Hidden Markov Models. In *Proceedings of 15th Vision Interface Conference*, Calgary, AB, 2002, 88–92.
3. Luhr, S.; Venkatesh, S.; West, G.; Bui, H.H. *Duration Abnormality Detection in Sequence of Human Activity*; Technical Report TR-2004/02, Curtin University of Technology, 2004, 1–5.
4. Zhu, X.; Liu, Z.; Zhang, J. Human activity clustering for online anomaly detection. J. Comput. **2011**, *6* (6), 1071–1079.
5. Hu, H.-Y.; Qu, Z.-W.; Li, Z.-H. Multi-level trajectory learning for traffic behavior detection and analysis. J. Chin. Inst. Eng. **2014**, *37* (8), 1–12.
6. Beyan, C.; Fisher, R.B. Detecting abnormal fish trajectories using clustered and labeled data. In *Proceedings of 21th International Conference on Image Processing (ICIP)*, Melbourne, 2013, 1476–1480.
7. Beyan, C.; Fisher, R.B. Detection of abnormal fish trajectories using a clustering based hierarchical classifier. In *Proceedings of British Machine Vision Conference (BMVC)*, Bristol, 2013, 1–11.
8. Beyan, C.; Fisher, R.B. Hierarchal decomposition for unusual fish trajectory detection. In *Computer Vision and Pattern Recognition in Environmental Informatics*; Zhou, J.; Bai, X.; Caelli, T.; Eds.; IGI Global: Hershey, PA, 2016, 1–21.
9. Beyan, C. Fish behavior analysis. In *Fish4Knowledge: Collecting and Analyzing Massive Coral Reef Fish Video Data*; Fisher, R.B.; Chen-Burger, Y.H.; Giordano, D.; Hardman, L.; Lin, F.-P.; Eds.; Springer: New York, 2016, 161–179.
10. Duong, T.V.; Bui, H.H.; Phung, D.Q.; Venkatesh, S. Activity recognition and abnormality detection with the switching Hidden Semi-Markov Model. In *Proceedings of Conference on Computer Vision and Pattern Recognition (CVPR)*, San Diego, CA, 2005, 838–845.
11. Morris, B.T.; Trivedi, M.M. A survey of vision-based trajectory learning and analysis for surveillance. IEEE Trans. Circuits Syst. Video Technol. **2008**, *18* (8), 1114–1127.
12. Cong, Y.; Yuan, J.; Liu, J. Sparse reconstruction cost for abnormal event detection. In *Proceedings of Conference on Computer Vision and Pattern Recognition (CVPR)*, Colorado Springs, CO, 2011, 3449–3456.
13. Wang, X.; Tieu, K.; Grimson, E. Learning semantic scene models by trajectory analysis. In *Proceedings of European Conference on Computer Vision (ECCV), ECCV*, Graz, 2006, 110–123.
14. Lu, C.; Shi, J.; Jia, J. Abnormal event detection at 150 fps in Matlab. In *Proceedings of International Conference on Computer Vision (ICCV)*, Sydney, NSW, 2013, 2720–2727.
15. Mahadevan, V.; Li, W.; Bhalodia, V.; Vasconcelos, N. Anomaly detection in crowded scenes. In *Proceedings of Conference on Computer Vision and Pattern Recognition (CVPR)*, San Francisco, CA, 2010, 1975–1981.
16. Vaswani, N.; Roy-Chowdhury, A.; Chellappa, R. Shape activity: A continuousstate hmm for moving/deforming shapes with application to abnormal activity detection. IEEE Trans. Image Process. **2005**, *14* (10), 1603–1616.
17. Saligrama, V.; Chen, Z. Video anomaly detection based on local statistical aggregates. In *Proceedings of Conference on Computer Vision and Pattern Recognition (CVPR)*, Providence, RI, 2012, 2112–2119.
18. Xu, D.; Ricci, E.; Yan, Y.; Song, J.; Sebe, N. Learning deep representations of appearance and motion for anomalous event detection. In *Proceedings of British Machine Vision Conference (BMVC)*, Swansea, 2015, 8.1–8.12.
19. Fu, Z.; Hu, W.; Tan, T. Similarity based vehicle trajectory clustering and anomaly detection. In *Proceedings of the International Conference on Image Processing (ICIP)*, Genova, 2005, 2.
20. Stauffer, C.; Grimson, W.E.L. Learning patterns of activity using real-time tracking. IEEE Trans. PAMI **2000**, *22* (8), 747–757.
21. Xu, J.; Fookes, C.; Sridharan, S. *Automatic Event Detection for Signal-Based Surveillance*; arXiv: 1612.01611 [cs. CV], 2016.
22. Majecka, B. *Statistical Models of Pedestrian Behavior in the Forum*; Master Thesis; School of Informatics, University of Edinburgh: Edinburgh, 2009.
23. Sillito, R.R.; Fisher, R.B. Semi-supervised learning for anomalous trajectory detection. In *Proceedings of British Machine Vision Conference (BMVC)*, Leed, 2008, 227–238.
24. Li, C.; Han, Z.; Ye, Q.; Jiao, J. Abnormal behavior detection via sparse reconstruction analysis of trajectory. In *Proceedings of International Conference on Image and Graphics (ICGIP)*, Cairo, 2011, 807–810.
25. Makris, D.; Ellis, T.J. Spatial and probabilistic modelling of pedestrian behaviour. In *Proceedings of British Machine Vision Conference (BMVC)*, Cardiff, vol. 2, 2002, 557–566.
26. Naftel, A.; Khalid, S. Classifying spatiotemporal object trajectories using unsupervised learning in the coefficient feature space. Multimedia Syst. **2006**, *12*, 227–238.
27. Xu, J.; Denman, S.; Sridharan, S.; Fookes, C. Activity analysis in complicated scenes using {DFT} coefficients of particle trajectories. In *Proceedings of Advanced Video and Signal-Based Surveillance (AVSS)*, Beijing, 2012, 82–87.
28. Brand, M.; Kettnaker, V. Discovery and segmentation of activities in video. IEEE Trans. Pattern Anal. Mach. Intell. **2000**, *22* (8), 844–851.
29. Porikli, F. Learning object trajectory patterns by spectral clustering. In *Proceedings of IEEE Conference Multimedia Expo (ICME)*, Taipei, vol. 2, 2004, 1171–1174.
30. Bashir, F.; Wu, Q.; Khokhar, A.; Schonfeld, D. Hmm-based motion recognition system using segmented PCA. In *Proceedings of IEEE International Conference on Image Processing (ICIP)*, Genova, 2005, 2286–2289.
31. Sillito, R.R.; Fisher, R.B. Parametric trajectory representations for behavior classification. In *British Machine Vision Conference (BMVC)*, London, 2009.
32. Zhong, H.; Shi, J.; Visontai, M. Detecting unusual activity in video. In *Proceedings of Computer Vision and Pattern Recognition (CVPR)*, Washington, DC, vol. 2, 2004, 819–826.
33. Porikli, F.; Haga, T. Event detection by eigenvector decomposition using object and frame features. In *Proceedings of Conference on Computer Vision and Pattern Recognition (CVPR)*, Washington, DC, 2004, 114–122.
34. Hu, W.; Xiao, X.; Fu, Z.; Xie, D.; Tan, T.; Maybank, S. A system for learning statistical motion patterns. IEEE Trans. Pattern Anal. Mach. Intell. **2006**, *28* (9), 450–1464.

35. Zhang, D.; Gatica-Prez, D.; Bengio, S.; McCowan, I. Semi-supervised adapted HMMs for unusual event detection. In *Proceedings of IEEE Computer Vision Pattern Recognition (CVPR)*, San Diego, CA, vol. 1, 2005, 611–618.
36. Xiang, T.; Gong, S. Video behaviour abnormality detection using reliability measure. In *British Machine Vision Conference (BMVC)*, Oxford, 2005.
37. Xiang, T.; Gong, S. Video behaviour profiling and abnormality detection without manual labelling. In *Proceedings of IEEE International Conference on Computer Vision (ICCV)*, Beijing, vol. 2, 2005, 1238–1245.
38. Xiang, T.; Gong, S. Incremental visual behaviour modelling. In *IEEE Visual Surveillance Workshop*, Graz, 2006, 65–72.
39. Loy, C.C.; Xiang, T.; Gong, S. Detecting and discriminating behavioural anomalies. Pattern Recognit. **2011**, *44*, 117–132.
40. Owens, J.; Hunter, A. Application of the self-organizing map to trajectory classification. In *Proceedings of IEEE International Workshop on Visual Surveillance*, Graz, 2000, 77–83.
41. Izo, T.; Grimson, W.E.L. Unsupervised modeling of object tracks for fast anomaly detection. In *Proceedings of IEEE International Conference on Image Processing (ICIP)*, San Antonio, TX, 2007, 529–532.
42. Jung, C.R.; Hennemann, L.; Musse, S.R. Event detection using trajectory clustering and 4-D histograms. IEEE Trans. Circuit Syst. Video Technol. **2008**, *18* (11), 1565–1575.
43. Figueiredo, M.T.A.; Jain, A.K. Unsupervised learning of finite mixture models. IEEE Trans. Pattern Anal. Mach. Intell. **2002**, *24* (3), 381–396.
44. Morris, B.T.; Trivedi, M.M. Trajectory learning for activity understanding: Unsupervised, multilevel and long-term adaptive approach. IEEE Trans. Pattern Anal. Mach. Intell. **2011**, *33* (11), 2287–2301.
45. Anjum, N.; Cavallaro, A. Multifeature object trajectory clustering for video analysis. IEEE Trans. Circuits Syst. Video Technol. **2008**, *18* (11), 1555–1564.
46. Brun, L.; Saggese, A.; Vento, M. Learning and classification of car trajectories in road video by string kernels. In *International Conference on Computer Vision Theory and Applications*, Barcelona, 2013, 709–714.
47. Choudhary, A.; Pal, M.; Banerjee, S.; Chaudhury, S. Unusual activity analysis using video epitomes and pLSA. In *Proceedings of 6th Indian Conference on Computer Vision, Graphics and Image Processing*, Bhubaneswar, 2008, 390–397.
48. Varadarajan, J.; Odobez, J.M. Topic models for scene analysis and abnormality detection. In *Proceedings of International Conference on Computer Vision Workshop*, Lisboa, 2009, 1338–1345.
49. Jeong, H.; Chang, H.J.; Choi, J.Y. Modelling of moving object trajectories by spatio-temporal learning for abnormal behaviour detection. In *Proceedings of IEEE International Conference on Advanced Video and Signal Based Surveillance (AVSS)*, Klagenfurt, 2011, 119–123.
50. Piciarelli, C.; Micheloni, C.; Foresti, G.L. Trajectory-based anomalous event detection. IEEE Trans. Circuits Syst. Video Technol. **2008**, *18* (11), 1544–1554.
51. Ivanov, I.; Dufaux, F.; Ha, T.M.; Ebrahimi, T. Towards generic detection of unusual events in video surveillance. In *Proceedings of Advanced Video and Signal Based Surveillance (AVSS)*, Genova, 2009, 61–66.
52. Beyan, C.; Fisher, R.B. A filtering mechanism for normal fish trajectories. In *International Conference on Pattern Recognition (ICPR)*, Tsukuba Science City, 2012, 2286–2289.
53. Hospedales, T.M.; Li, J.; Gong, S.; Xiang, T. Identifying rare and subtle behaviours: A weakly supervised joint topic model. IEEE Trans. Pattern Anal. Mach. Intell. **2011**, *33* (12), 2451–2464.
54. Hasan, M.; Choi, J.; Amit, J.N.; Roy-Chowdhury, K.; Davis, L.S. Learning temporal regularity in video sequences. In *Proceedings of IEEE Computer Vision Pattern Recognition (CVPR)*, Las vegas, NV, 2016, 796–803.
55. Sabokrou, M.; Fathy, M.; Hoseini, M. Video anomaly detection and localisation based on the sparsity and reconstruction error of auto-encoder. IET Electron. Lett. **2016**, *52* (13), 1122–1124.
56. Xua, D.; Yand, Y.; Ricci, E.; Sebea, N. Detecting anomalous events in videos by learning deep representations of appearance and motion. Comput. Vis. Image Underst. **2016**, *5* (42), 1–11.
57. Chen, X.; Ye, Q.; Zou, J.; Li, C.; Cui, Y.; Jiao, J. Visual trajectory analysis via replicated softmax-based models. Sig. Image Video Process **2014**, *8* (1), 183–190.
58. Hinton, G.E.; Salakhutdinov, R. Replicated Softmax: An undirected topic model. In *Proceedings of the Advances in Neural Information Processing Systems*, 2009, 1607–1614.
59. McDonagh, S.; Beyan, C.; Huang, P.X.; Fisher, R.B. Applying semi-synchronised task farming to large-scale computer vision problems. Int. J. High Perform. Comput. Appl. **2015**, *29* (4), 437–460.
60. Hinton, G.E. Training products of experts by minimizing contrastive divergence. Neural Comput. **2002**, *14* (8), 1771–1800.
61. Yamashita, T.; Tanaka, M.; Yoshida, E.; Yamauchi, Y.; Fujiyoshi, H. To be Bernoulli or to be Gaussian, for a restricted Boltzmann machine. In *Proceedings of International Conference on Pattern Recognition (ICPR)*, Stockholm, 2014, 1520–1525.
62. Settles, B. *Active Learning Literature Survey*; Computer Sciences Technical Report 1648, University of Wisconsin-Madison, 2009.
63. Uguroglu, S. *Robust Learning with Highly Skewed Category Distributions*; PhD thesis; Carnegie Mellon University, School of Computer Science, 2013.
64. Lewis, D.; Gale, W. A sequential algorithm for training text classifiers. In *Proceedings of SIGIR Conference on Research and Development in Information Retrieval*; ACM/Springer: New York, 1994, 3–12.
65. Settles, B.; Craven, M. An analysis of active learning strategies for sequence labeling tasks. In *Proceedings of Conference on Empirical Methods in Natural Language Processing (EMNLP)*, Edinburgh, 2008, 1069–1078.
66. MATLAB version 8.3.0.532. *Statistics and Machine Learning Toolbox* (version 9.0); The MathWorks Inc., 2014.
67. Junejo, I.N.; Javed, O.; Shah, M. Multi feature path modeling for video surveillance. In *Proceedings of IEEE International Conference on Pattern Recognition (ICPR)*, Cambridge, 2004, 716–719.

68. Kubat, M.; Holte, R.; Matwin, S. Machine learning for the detection of oil spills in satellite radar images. Mach. Learn. **1998**, *30*, 195–215.
69. Khalid, S.; Naftel, A. Classifying spatiotemporal object trajectories using unsupervised learning of basis function coefficients. In *Proceedings of ACM International Workshop on Video Surveillance and Sensor Networks*, New York, 2005, 45–52.
70. Makris, D.; Ellis, T. Learning semantic scene models from observing activity in visual surveillance. IEEE Trans. Syst. Man Cybern. **2005**, *35* (3), 397–408.
71. Dahlbom, A.; Niklasson, L. Trajectory clustering for coastal surveillance. In *Proceedings of IEEE International Conference on Information Fusion*, Val-Bélair, QC, 2007, 1–8.
72. Basharat, A.; Gritai, A.; Shah, M. Learning object motion patterns for anomaly detection and improved object detection. In *Proceedings of IEEE Computer Vision and Pattern Recognition (CVPR)*, Anchorage, AK, 2008, 1–8.
73. Wiliem, A.; Madasu, V.; Boles, W.; Yarlagadda, P. Detecting uncommon trajectories. In *Proceedings of Digital Image Computing: Techniques and Applications (DICTA)*, 2008, 398–404.
74. Bouttefroy, P.L.M.; Bouzerdoum, A.; Phung, S.L.; Beghdadi, A. Abnormal behavior detection using a multi-modal stochastic learning approach. In *Proceedings of Intelligent Sensors, Sensor Networks and Information Processing*, 2008, 121–126.
75. Zelniker, E.E.; Gong, S.; Xiang, T. Global abnormal behaviour detection using a network of CCTV cameras. In *Eighth International Workshop on Visual Surveillance*, Marseille, 2008.
76. Wiliem, A.; Madasu, V.; Boles, W.; Yarlagadda, P. A context-based approach for detecting suspicious behaviours. In *Proceedings of Digital Image Computing: Techniques and Applications (DICTA)*, 2009, 146–153.
77. Jiang, F.; Wu, Y.; Katsaggelos, A.K. A dynamic hierarchical clustering method for trajectory based unusual video event detection. IEEE Trans. Image Process. **2009**, *18* (4), 907–913.
78. Espinosa-Isidron, D.L.; Garcia-Reyes, E.B. A new dissimilarity measure for trajectories with applications in anomaly detection. Lecture Notes in Computer Science Progress in Pattern Recognition, Image Analysis, Computer Vision and Applications, vol. 6419, 2010, 193–201.
79. Al-Khateeb, H.; Petrou, M. An extended fuzzy SOM for anomalous behavior detection. In *Proceedings of IEEE Computer Vision and Pattern Recognition Workshops (CVPRW)*, Colorado Springs, CO, 2011, 31–36.
80. Shi, F.; Zhou, Z.; Xiao, J.; Wu, W. Robust trajectory clustering for motion segmentation. In *Proceedings of IEEE International Conference on Computer Vision (ICCV)*, Sydney, NSW, 2013, 3088–3095.
81. Morris, B.T.; Trivedi, M.M. Understanding vehicular traffic behavior from video: A survey of unsupervised approaches. J. Electron. Imaging **2013**, *22* (4), 1–15.
82. Dickinson, P.; Hunter, A. Using inactivity to detect unusual behaviour. In *Proceedings of IEEE Workshop on Motion and Video Computing*, Copper Mountain, CO, 2008, 1–6.
83. Yin, J.; Meng, Y. Abnormal behavior recognition using self-adaptive Hidden Markov Models. In *Proceedings of International Conference on Image Analysis and Recognition*, Halifax, NS, 2009, 337–346.
84. Nishio, S.; Okamoto, H.; Babaguchi, N. Hierarchical anomality detection based on situation. In *Proceedings of IEEE International Conference on Pattern Recognition (ICPR)*, Istanbul, 2010, 1108–1111.
85. Akoz, O.; Karsligil, M.E. Video-based traffic accident analysis at intersections using partial vehicle trajectories. In *Proceedings of IEEE International Conference on Image Processing (ICIP)*, Hong Kong, 2010, 4693–4696.
86. Wang, Y.; Wang, D.; Chen, F. Abnormal behavior detection using trajectory analysis in camera sensor networks. Int. J. Distrib. Sens. Netw. **2014**, *10* (1), 1–9.
87. Piciarelli, C.; Foresti, G.L. Anomalous trajectory detection using support vector machines. In *Proceedings of Advanced Video and Signal Based Surveillance*, London, 2007, 153–158.
88. Ma, Y.; Li, M. Detection for abnormal event based on trajectory analysis and FSVM. In *Proceedings of International Conference on Intelligent Computing*, Qingdao, 2007, 1112–1120.
89. Lui, C.; Wang, G.; Ning, W.; Lin, X.; Li, L.; Liu, Z. Anomaly detection in surveillance video using motion direction statistics. In *Proceedings of IEEE International Conference on Image Processing (ICIP)*, Hong Kong, 2010, 717–720.
90. Li, J.; Hospedales, T.M.; Gong, S.; Xiang, T. Learning rare behaviours. In *Proceedings of Asian Conference on Computer Vision (ACCV)*, Queenstown, 2010, 292–307.
91. Hendel, A.; Weinshall, D.; Peleg, S. Identifying surprising events in videos using Bayesian topic models. In *Proceedings of Asian Conference on Computer Vision (ACCV), Part 3, LNCS*, Queenstown, Vol. 6494, 2010, 448–459.

BIBLIOGRAPHY

1. Beyan, C. *Detection of Unusual Fish Trajectories from Underwater Videos*. PhD Thesis; School of Informatics, University of Edinburgh, Edinburgh, Scotland, 2014.

Video Analytics in the Compressed Domain

Edgar A. Bernal and Robert P. Loce
Palo Alto Research Center, Xerox Corporation, Webster, New York, U.S.A.

Abstract

Digital cameras are being deployed to capture video data used in applications such as surveillance, traffic law enforcement, and environmental monitoring. As cameras become interconnected, they make up large-scale, distributed, and heterogeneous camera networks. While the characteristics of the cameras that make up these camera networks may vary widely (e.g., pixel resolution, viewing angle, color capabilities, and frame rate, among others), all existing networks must cope with limitations common to their supporting infrastructures, namely, storage and transmission bandwidth limitations. Consequently, modern video compression technologies are as pervasive as digital cameras. While most ongoing research on video compression is aimed at improving computational complexity and compression efficiency, a subset of recent work on video compression and video analytics has focused on exploiting the feature-rich nature of video data in the compressed domain in order to more effectively support the aforementioned video analytics applications.

Keywords: Computer vision; Motion analysis; Motion detection; Motion vector; Object tracking; Video analytics; Video compression.

INTRODUCTION

Video compression or video encoding is the process by which the amount of data required to represent a digital video signal is reduced. Video compression is essential in applications where high quality video transmission and/or archival is required. Camera networks are the most prevalent example of extensive video compression use. Camera networks comprise multiple digital video cameras that relay video data to a central processing and/or archival facility. The communication network used to transport the video stream between the cameras and the central facility, typically Internet Protocol- (IP) compliant, usually has bandwidth constraints. Storage resources at data repositories are also limited in applications when video is acquired for archival and/or evidentiary purposes.

In addition to exploiting statistical redundancy via entropy or arithmetic encoding, which is a common element to most digital data compressors, video compression is achieved by exploiting two other types of redundancies within the video stream: spatial redundancies among neighboring pixels within a frame and temporal redundancies between adjacent frames. Spatial redundancies are exploited via the use of traditional transform-based techniques such as those relying on the discrete cosine transform (DCT) or the discrete wavelet transform (DWT), as well as with intra-frame prediction techniques, which exploit similarities among blocks in a given frame. Temporal redundancies, on the other hand, are exploited via the estimation of motion vectors that describe the apparent motion of pixels or blocks thereof across frames in the video. Together, the transform coefficients and the motion vectors comprise the compressed domain representation of the original video stream.

Most video analytics tasks, both automated (i.e., performed by computers) and manual (i.e., performed by humans), are performed in the uncompressed domain. This approach can be computationally inefficient because, as previously stated, digital video is generally compressed right after acquisition and prior to transmission or storage, which means that analysis tasks typically require a decompression stage. A subset of recent work in automated video analytics has focused on the development of algorithms that exploit the feature-rich and computationally efficient nature of the compressed video domain by relying on compressed domain features.

In this entry, we introduce a suite of video analytics technologies that operate in the compressed domain. Section "Background on Video Compression" introduces the main concepts of video compression. Section "Task-Oriented Video Compression" illustrates techniques aimed at tuning video compression method to specific video analytics tasks. Section "Motion Detection and Analysis in the Compressed Domain" describes techniques designed to perform motion analysis, one of the most fundamental tasks in computer vision. Section "Example Video

Analytics and Computer Vision Applications in the Compressed Domain" introduces algorithms for higher level computer vision tasks in the compressed domain. We conclude and provide future trends and challenges in Section "Conclusion."

BACKGROUND ON VIDEO COMPRESSION[1]

Video compression converts uncompressed digital video to a format more suitable for transmission and storage, more specifically, a representation requiring a smaller number of bits. Figure 1 illustrates a video transmission process involving a video encoder, which performs compression and produces a compressed video stream, and a video decoder, which converts the compressed stream into a representation of the original video data. Encoders perform one of two types of compression: lossless, which enables exact reconstruction of the compressed data upon decompression, and lossy, where the decompressed data is not identical to the original data. Due to their higher compression efficiency, lossy compression methods are typically preferred as they can better address the limitations of today's communication and archival infrastructures.

Lossy video compression techniques achieve data reduction by exploiting spatial and temporal redundancies in the video frames via predictive stages that estimate the pixel values of the frame or frame portion to be compressed, and encode the prediction residual error in a lossy manner. Spatial redundancies refer to the phenomenon whereby pixels in a given video frame tend to have values that are highly correlated with spatially proximal pixels. Temporal redundancies relate to the high correlation between values of a given pixel or pixels in a neighborhood across temporally proximal frames. Figure 2 shows a high-level block diagram describing the operation of a video encoder, which predicts the appearance of a target block in a frame being compressed from previously encoded data and forms a compressed stream by entropy encoding the quantized version of the prediction error (i.e., the difference between the actual and the predicted data), often in a transformed domain. The previously encoded data can be obtained either from previously encoded blocks in the current frame or in other frames. Reference frames or I-frames (Intra-frame encoded) are frames encoded in a standalone manner, that is, using prediction information from data contained in the frame itself. Non-reference frames, P-frames or B-frames are encoded using information contained in previously encoded frames. While P-frames (predicted frames) are encoded using information from frames in the past, B-frames (bidirectional predicted frames) can be encoded using information from either past or future frames, or a combination thereof. The prediction error or residual data is then transformed into a domain that enables high-energy compaction using DCT, DWT, or similar transformation techniques. The resulting transformed coefficients are quantized and entropy encoded, and the encoded bit stream transmitted or stored. The video decoder reverses the steps performed by the encoder as shown in Fig. 2.

When prediction is applied to reference frames, the values of image samples being processed are estimated from previously coded samples in the same frame in a process termed "intra-frame" prediction. In contrast, prediction across time for non-reference frames involves predicting the appearance of a current frame from the values in a previously encoded frame. Figure 3 illustrates this process, which is often termed "inter-frame prediction." Figure 3a shows the source image for an already encoded frame, or reference frame. Figure 3b shows the frame to be compressed, or target frame. The pixel values of the frame in Fig. 3b are predicted via motion estimation techniques, where the apparent motion of pixels or blocks thereof between the frames is estimated. To this end, the target frame is segmented into rectangular blocks of $m \times n$ pixels, or target blocks. For each target block, a search is performed in the previously encoded reference frame for the block that most resembles the target block being processed. Since exhaustive searches can be computationally expensive, a search neighborhood of a predetermined size is defined, usually centered about the location of the target

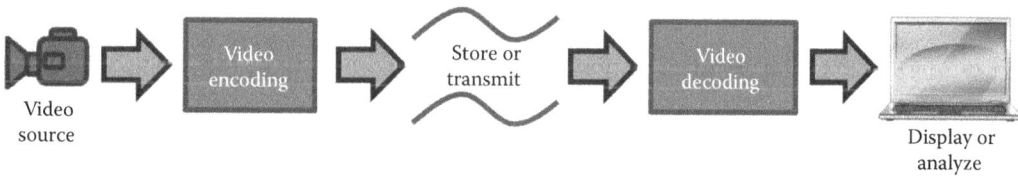

Fig. 1 The video compression and decompression process

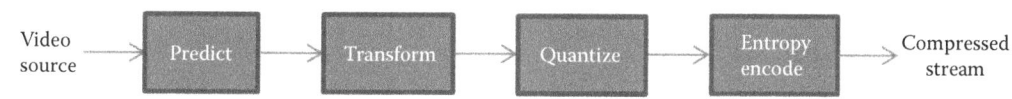

Fig. 2 High-level view of a video encoder

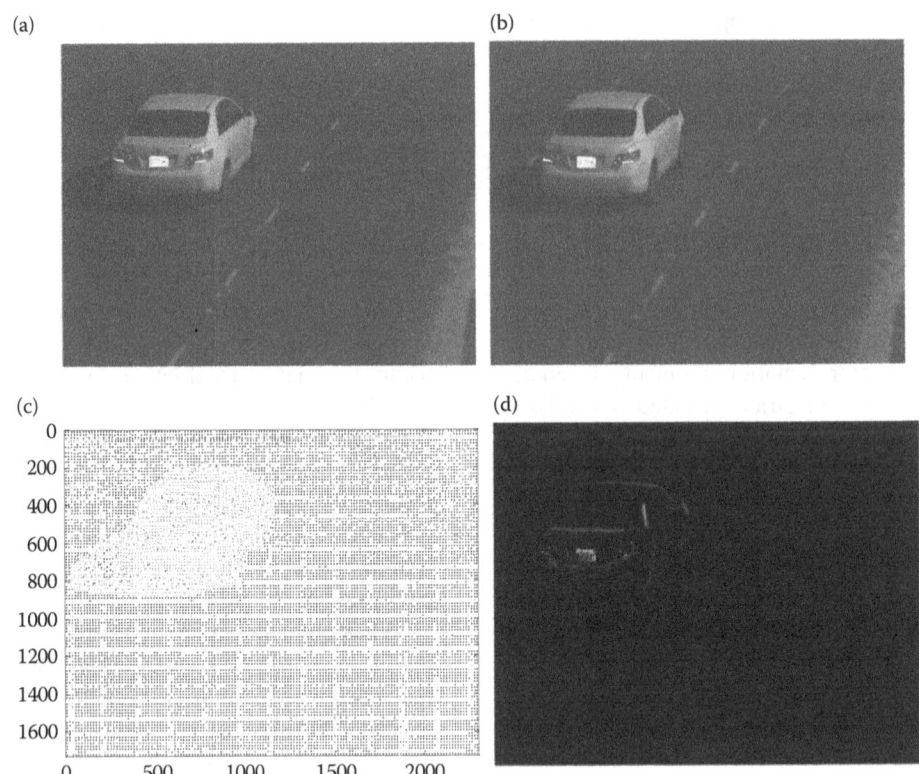

Fig. 3 Illustration of the motion compensation process: (a) reference frame; (b) target frame-; (c) estimated motion vector field; (d) resulting prediction error

block, as illustrated in Fig. 4. A similarity or dissimilarity metric is computed between every reference block in the search neighborhood and the target block. Example dissimilarity criteria are the mean squared error (MSE) and the mean absolute difference (MAD), which are calculated as follows:

$$\text{MSE}(d_1, d_2) = \frac{\sum_{k=1}^{m}\sum_{l=1}^{n}\left(B(k,l,j) - B(k+d_1, l+d_2, j-1)\right)^2}{(mn)}, \quad (1)$$

$$\text{MAD}(d_1, d_2) = \frac{\sum_{k=1}^{m}\sum_{l=1}^{n}\left|B(k, l, j) - B(k+d_1, l+d_2, j-1)\right|}{(mn)}. \quad (2)$$

Here, $B(k, l, j)$ denotes the pixel value of the pixel located on the kth row and the lth column of the $m \times n$ block of pixels in the jth frame. The vector (d_1, d_2) describing the relative displacement between the target block and the best matching reference block is termed "motion vector" (see Fig. 4). In this case, the $(j-1)$th frame is the reference frame, and the jth frame is the target frame. The reference block whose appearance best matches the appearance of the target block (because its corresponding MSE or MAD are the smallest out of the whole set of candidate blocks) is used as a predictor of the appearance of the target block. Figure 3c illustrates the motion vector field resulting from the application of motion estimation techniques to the target frame from Fig. 3b using the frame from Fig. 3a as a reference. Figure 3d illustrates the reconstruction error resulting from the approximation of the target frame from Fig. 3b with blocks from the reference frame from Fig. 3a and the computed motion vector field from Fig. 3c.

Rate-Distortion Optimization

Video data recovered from the decoding of lossy compressed video encoders is not identical to the video data that was originally compressed. Distortion is a metric that quantifies the errors that result from the use of lossy compression methods in terms of how much the recovered video stream differs from the original one. Example distortion models include the MSE and the peak signal-to-noise ratio (PSNR). Let $F(p)$ and $G(p)$ denote the pixel values of the original and the recovered video, respectively. Then, MSE is defined as follows:

$$\text{MSE}(F, G) = \frac{\sum (F(p) - G(p))^2}{|P|}, \quad (3)$$

where the sum is performed across all $p \in P$, that is to say, across all pixels in all frames of the videos, and the operator $|P|$ denotes cardinality of set P. The smaller the MSE, the more faithfully the compressed video represents the original video. The PSNR is defined as follows:

$$\text{PSNR}(F,G) = 10\log_{10}\frac{(2^n-1)^2}{\text{MSE}(F,G)}, \quad (4)$$

where n is the number of bits used to encode each pixel. The larger the PSNR, the more faithfully the compressed video represents the original video. The different compression modes offered by traditional video compression standards (e.g., intra-prediction, inter-prediction, and transform-based) result in varying degrees of compression error for different numbers of bits required to represent the compressed stream. Similarly, for a given compression mode, the compression error may vary relative to different parameter values (e.g., block size and quantization step size). The relationship between the distortion metric and the bit rate achieved by the different compression modes can be visualized in rate-distortion curves, as illustrated in Fig. 5.

Rate-distortion optimization refers to the selection of the most efficient coded representation for each image region in every frame in the rate-distortion sense.[2] For example, using intra-frame prediction after a scene change may result in more efficient encoding than using inter-frame prediction, given the lack of similarity between the target frame and previously encoded frames. The compression modes used to compress the different image regions have different rate-distortion characteristics; the goal of an encoder is to optimize overall fidelity. More specifically, this goal can be framed as "minimize a distortion metric D subject to an upper limit b_T on the number of bits used b." Mathematically, it can be posed as the following optimization task:

$$\min\{D\} \text{ subject to } b \leq b_T. \quad (5)$$

TASK-ORIENTED VIDEO COMPRESSION

Video compression standards are aimed at supporting a wide range of video data types, and, as such, may be suboptimal in terms of compression ratios and quality of the compressed video data relative to specific analytics

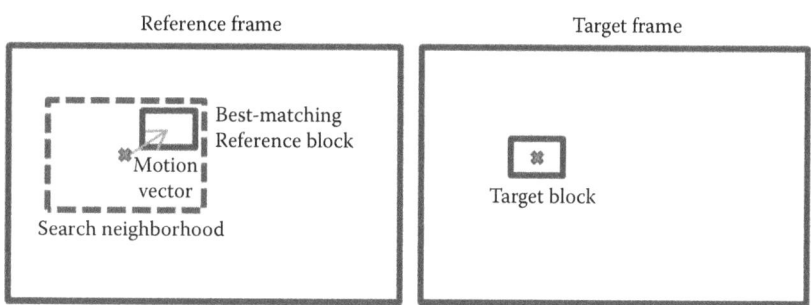

Fig. 4 Search neighborhood and motion vector in a motion estimation algorithm

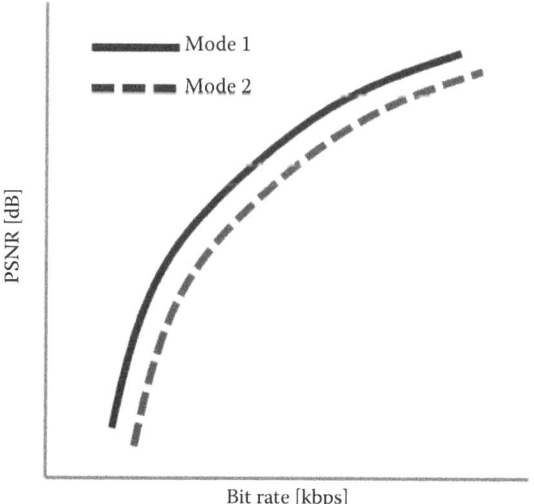

Fig. 5 Sample rate-distortion curves

tasks. Task-oriented video compression approaches optimizing the video compression process relative to specific types of data and associated post-processing tasks. This optimization is usually achieved by judiciously allocating resources associated with encoding of different portions of the content of the video stream.

Visual Attention-Based Video Compression

As illustrated earlier, video compression is achieved by removing statistical, spatial, and temporal redundancies in the video stream. Subjective quality-based video coding methods have been developed that achieve even higher compression ratios than traditional methods. The underlying assumption of these compression methods is that the video being acquired and compressed is intended for human consumption. Under these conditions, inherent physiological characteristics of the human visual system as well as psychological, behavioral, and cognitive processes that dictate human attention can be exploited to further eliminate redundancies.

In the more general implementation of attention-based video compression, saliency models that predict the level of conspicuity of every spatial and temporal location in the incoming video frames are used to determine the bit allocation strategy: Larger numbers of bits are allocated to more conspicuous video segments and vice versa.[3] Attention models usually take into account multiple dimensions of features or feature channels in order to determine conspicuity. Color (e.g., red–green and blue–yellow contrast), temporal intensity flicker, spatial orientation (e.g., 0°, 45°, 90°, and 135°) and motion orientation, and intensity are typically considered. Other features such as texture, shape, and object type (e.g., text and face) are sometimes considered. Figure 6 shows a sample video frame and its corresponding pseudo-colored saliency map.

Assuming that the saliency map of a video F has values ranging from 0 to S_N, the video can be segmented into N regions of interest $R_1, R_2, \ldots, R_N$ corresponding to portions of the video with saliency values in the intervals $[0, S_1)$, $[S_1, S_2), \ldots, [S_{N-1}, S_N]$. The total distortion can be written as follows:

$$D = \frac{\sum_{i=1}^{N} |R_i| D_i(R_i)}{\left| \bigcup_{i=1}^{N} R_i \right|}, \tag{6}$$

where $|R_i|$ denotes the cardinality of the spatiotemporal support of R_i, and $D_i(R_i)$ denotes the distortion of region of interest R_i. It is not uncommon to assume parametric distortion models such as[4]

$$D_i(R_i) = w_i \sigma_i^2 e^{-\gamma B_i}, \tag{7}$$

where w_i is the relative weight assigned to R_i, σ_i^2 is the variance of the pixel values in R_i, γ is a constant, and B_i is the number of bits required to represent R_i. Incorporating Eq. 7 into Eq. 6, the expression for the total distortion becomes:

$$D = \frac{\sum_{i=1}^{N} |R_i| w_i \sigma_i^2 e^{-\gamma B_i}}{\left| \bigcup_{i=1}^{N} R_i \right|}. \tag{8}$$

The number of bits required to represent each region in the video can be found by finding the B_i that minimize Eq. 8 via the Lagrange multiplier method to yield:

$$B_i = B + \frac{2 \sum_{j \neq i} |R_j| \ln\left(\frac{\sigma_i}{\sigma_j}\right)}{\gamma N \left| \bigcup_{i=1}^{N} R_i \right|} + \frac{\sum_{j \neq i} |R_j| \ln\left(\frac{w_i}{w_j}\right)}{\gamma N \left| \bigcup_{i=1}^{N} R_i \right|}, \tag{9}$$

for $i = 1, 2, \ldots, N$.

The bit allocation scheme described by Eq. 9 is carried out by using quantizers with different step sizes for the different regions of interest: since larger amounts of bits are allocated to more salient portions of video streams, finer quantization (with larger numbers of quantization levels) is used at more salient regions and vice versa.

 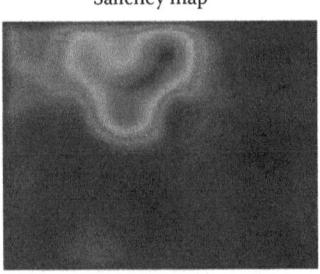

Original image   Saliency map

Fig. 6 Sample video frame and corresponding saliency map

Video Search- and Retrieval-Oriented Video Compression

Most search- and retrieval-oriented compression techniques take advantage of the fact that reference frames or I-frames are encoded in a standalone manner and, as such, their decoding does not require decompression of any additional frame data. Although less compressible than non-reference frames, reference frames can be decompressed in a more computationally efficient manner. This makes these frames ideal random access points for video decoders to initiate a decoding process.

Judicious selection of reference frames among frames in a video being compressed can facilitate search activities, particularly when these are performed on large video datasets. Important examples include Amber Alert, which is issued in the United States when a minor is abducted, and Silver Alert which is issued when a senior citizen or cognitively impaired individual goes missing. Amber Alert and Silver Alert typically trigger area-wide searches for a vehicle with certain known attributes, which could include color, make, model, and license plate. By compressing frames where vehicles are at an ideal position for license plate reading or vehicle type, make or model identification as reference frames, search tasks on surveillance video can be performed by decompressing such frames only, which can result in significant savings relative to time and computational complexity,[5] which is critical because 75% of abducted children are murdered in the first 3 hours. Figure 7 illustrates the video encoding process, as well as an outline of a possible search algorithm.

Note that, as illustrated in Fig. 7, a video compression method that automatically selects reference frame labels can be enabled with little modification to the traditional implementations of the compression standards. In one instance, a virtual trip wire can be defined to traverse the road or highway being monitored in a direction perpendicular to the traffic flow. Every frame in the incoming video is initially treated as a non-reference frame, so motion vectors corresponding to that frame can be computed. Intersection between the virtual trip wire and a cluster of motion vectors consistent with a vehicle in motion across the trip wire can provide information to determine if a vehicle is traversing the scene, as illustrated in Fig. 8. When that determination is made, the computed motion vectors can be discarded, and the frame can be encoded as a reference or I-frame.

In an alternative implementation, the selective insertion of reference frames can be performed as part of a transcoding task.[6] While it will be more difficult to implement such an approach inline into the compression process, the main advantage of an offline approach is that it may rely on the richer uncompressed video data, which in theory should enable more accurate performance and higher level abstraction. For example, motion detection can be performed in the uncompressed domain via traditional frame differencing techniques whereby the absolute pixel-wise difference between the gray-level representations of the current frame $f_i(x, y)$ and the previous frame $f_{i-1}(x, y)$ is thresholded with a threshold T to create a binary motion mask $M_i(x, y)$ as follows:

$$M_i(x,y) = \begin{cases} 1 & \text{if abs}\left(f_i(x,y) - f_{i-1}(x,y)\right) > T \\ 0 & \text{otherwise} \end{cases}. \quad (10)$$

Morphological operations can be applied to the binary mask resulting from the thresholding operation to eliminate the effects of spatial noise. The size of the resulting motion mask in pixels can be computed as follows:

$$m_i = \sum_{x=1}^{W} \sum_{y=1}^{H} M_i(x, y), \quad (11)$$

where W and H are the number of columns and rows of the video frames, respectively, and the summation is

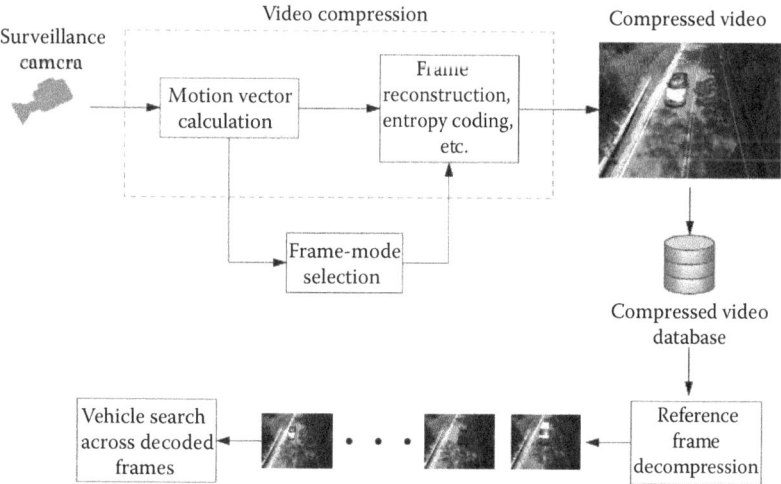

Fig. 7 Compression of traffic surveillance videos that enables efficient vehicle searches

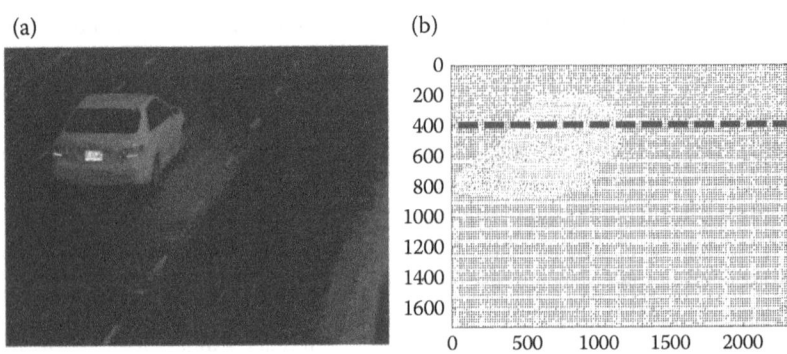

Fig. 8 (a) Incoming video frame and (b) corresponding motion vector field, where a motion vector cluster is determined to cross a virtual trip wire (dashed red line)

performed across all pixels in the two-dimensional mask. In order to counteract the effects of spurious sources of motion such as sudden changes in lighting and camera shake, an averaging filter can be applied to the resulting motion blob size counts to produce a count filtered over time according to

$$\bar{m}_i = \frac{\sum_{i=-N}^{N} m_i}{2N+1}, \qquad (12)$$

for some constant integer N. The start and end index frames, I_s and I_e, respectively, where a moving object traversed the scene can then be calculated based on the resulting sequence of filtered counts. Specifically, as the filtered average of number of motion pixels exceeds (falls below) a predetermined threshold, an object entering (exiting) the scene is detected. These heuristics are implemented as follows:

$$\begin{aligned} I_s &= i \text{ if } \bar{m}_{i-1} < T \text{ and } \bar{m}_i > T \\ I_e &= i \text{ if } \bar{m}_{i-1} > T \text{ and } \bar{m}_i < T \end{aligned}, \qquad (13)$$

for some threshold T. Judicious reference frame selection can be performed based on the values of I_s and I_e.

Video Compression for Traffic Surveillance Tracking Applications

Video cameras are commonly used in traffic monitoring and surveillance given the flexibility they provide to support a wide range of applications including anomaly detection, traffic pattern analysis, law enforcement, congestion prevention, roadway planning, and driver assistance. Core video analytics tasks required to support these applications include vehicle tracking, the performance of which is affected by video compression related parameters such as resolution, frame rate, and image quality. Needless to say, it is critical to maintain robust object tracking accuracy in the presence of compression.

As previously discussed, rate-distortion optimization is often performed with specific post-processing tasks in mind. In the case of video intended for automatic analysis via motion detection and tracking algorithms, distortion can be modeled in terms of task-dependent accuracy. Under these assumptions, rate-distortion optimization would result in the most efficient use of bit rate relative to the performance of the designated video analytics algorithms. A metric for tracking performance has to be defined to carry out this optimization task. Two commonly used performance metrics for tracking algorithms include precision and recall.[7] Given known ground truth about the location of the object(s) being tracked in a given video frame, recall is the fraction of relevant (ground truth) pixels identified by the tracker. Let GT_i and ε_i denote the set of pixels in video frame i that correspond to the object being tracked as indicated by the process of ground truthing and by the tracker, respectively. Recall is defined as follows:

$$\rho_i = \frac{|\varepsilon_i \cap GT_i|}{|GT_i|}. \qquad (14)$$

Precision measures the fraction of pixels identified by the tracker that are relevant (in the ground truth), and it is defined as follows:

$$v_i = \frac{|\varepsilon_i \cap GT_i|}{|\varepsilon_i|}. \qquad (15)$$

Precision and recall take values between 0 (no overlap) and 1 (complete overlap).

Noise in the video can be considered to be introduced by two sources: noise introduced by the video acquisition device, and noise or distortion introduced by the compression process. The compression method can be optimized relative to the tracking application at hand by minimizing distortion as seen by the tracking algorithm. This can be achieved by adjusting encoder parameters, measuring resulting tracking performance on the encoded video, and iterating to effectively implement a gradient search procedure that maximizes tracking performance as a function of encoder parameters,[8] all of which aim at limiting the effect of noise relative to the tracking needs. The resulting compressed video will best preserve features of the original video that are conducive to robust tracking.

In one particular implementation of the described optimization process, the quantization scheme of the transform coefficients of either the residual or the reference frames is adjusted so as to invest the fewest number of bits possible on transform features that are the least useful to tracking and vice versa. The gradient search is performed by generating a set of operating points describing a resulting tracking accuracy A, where the accuracy is a combination of the computed precision and recall, for a given bit rate R. Out of the resulting set of operating points, a subset S_{opt} is selected comprising those considered superior in terms of performance. Specifically, S_{opt} at iteration i is defined as follows:

$$S_{opt}^i = \left\{ (A, R) : A_k^i < A_{k+n}^i \,\middle|\, R_k^i < R_{k+n}^i, \forall n > 0, k \geq 0, k + n \leq N \right\}, \quad (16)$$

where A_k^i and R_k^i denote the tracking accuracy and the bit rate for the kth compression mode (out of a total of N compression modes) at iteration i, respectively. S_{opt} comprises a strictly increasing set of rate–accuracy pairs which include the lowest bit rate in the set, as illustrated in Fig. 9, and $S_{opt}^0 = \operatorname{argmin}\{R_k\}, \forall k$.

The next iteration is performed based on operating points that result from modifying the parameters corresponding to the selected subset. The process converges when the set of iterations does not yield significant improvement. At the end of the process, a rate–accuracy curve that describes tracking accuracy as a function of bit rate is obtained. Based on this curve, optimal quantization tables can be determined according to bitrate requirements input by a user.

MOTION DETECTION AND ANALYSIS IN THE COMPRESSED DOMAIN

Inspection of the motion vector field from Fig. 3c reveals that the motion vectors used for video compression tasks can be highly descriptive of the apparent motion in the scene captured by the reference and target frames. Consequently, accurate motion detection analysis may be enabled by extracting and processing motion vectors in a compressed video stream without the need to rely on full video stream decompression. Strictly speaking, compression motion vectors are derived with a goal of minimizing prediction error, and, as such, do not always represent perceived or acquired motion. Additionally, since motion vectors are assigned on a pixel-block basis, a motion vector field associated with a frame is inherently lower dimensional or sparser than the frame itself. In light of the above limitations, occlusion handling and segmentation of multiple objects based on their motion characteristics have been areas of significant study. Note that the relationship between compression motion vectors and the motion in the scene varies depending on the characteristics of the camera: for stationary cameras, motion vectors capture the apparent motion in the scene. For cameras in motion (e.g., zooming, panning, and tilting), on the other hand, motion vectors contain information about both camera and scene motion, which needs to be disambiguated. This phenomenon is illustrated in Fig. 10 which shows a motion vectors in a video that contains a vehicle moving from left to right in the scene and the camera panning with it. Figure 10a and b show the reference and target frames, respectively, and Fig. 10c shows the resulting motion vector field.

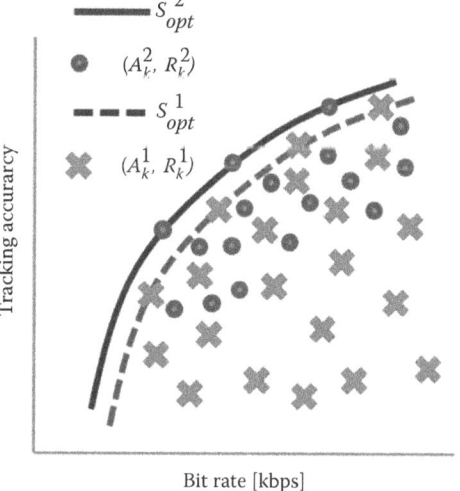

Fig. 9 Rate-distortion optimization relative to tracking accuracy

Note that motion vectors are only available for non-reference frames, namely P- or B-frames, as illustrated in the previous section. Reference frames can be skipped with little penalty to motion analysis when the frame rate is high enough relative to the speeds of motion being captured, which is typically the case for robust automated analytics algorithms. Alternatively, motion vectors corresponding to the reference frame can be estimated via weighted averages of the motion vectors of neighboring, non-reference frames. In the remainder of this document, it is assumed that motion vectors are available for every frame of interest.

Stationary Cameras

Conventional motion detection algorithms for stationary cameras that operate in the compressed domain can be classified into clustering- and filtering-based methods.[9]

Clustering-Based Methods

Clustering-based methods typically perform motion detection in two stages. In the first stage, blocks are classified as either in motion or stationary; in the second stage, blocks are clustered into foreground and background according to the homogeneity of their corresponding motion vectors. In one embodiment,[10] motion vectors that correspond to actual objects in motion can be identified based on their length. Specifically, let $\mathbf{MV}^i = \left(MV_x^i, MV_y^i\right)$ denote the motion vector corresponding to block i. Then, a length threshold T can be defined so that vectors can be identified as representative of an object in motion:

$$\|\mathbf{MV}^i\| < T, \quad (17)$$

where $\|\bullet\|$ denotes the norm operator. Once blocks containing motion are identified, they can be clustered based on the coherence of their orientation. Specifically, the orientation of motion vector $\mathbf{MV}^i$ can be computed as follows:

$$\theta_i = \frac{180}{\pi} \arctan\frac{MV_y^i}{MV_x^i}. \quad (18)$$

Once the orientation of every vector determined to correspond to motion is computed, clustering of motion vectors with coherent orientation is performed by defining a threshold $\Delta\theta$ and iteratively performing region growing segmentation. The segmentation process uses a motion vector with orientation θ_0 yet to be assigned to a cluster as seed and enforces the clustering criteria

$$|\theta_i - \theta_0| \leq \Delta\theta, \quad (19)$$

to decide if $\mathbf{MV}^i$ belongs to the cluster being processed. A new seed corresponding to an unassigned motion block

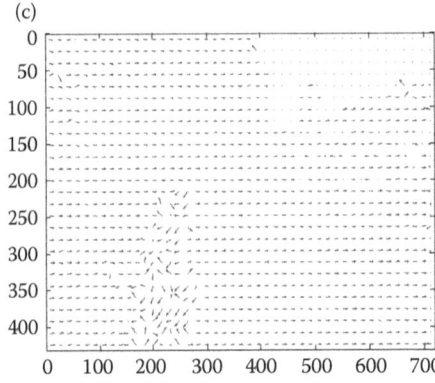

Fig. 10 Motion vectors for video containing camera motion: (a) reference frame; (b) target frame; (c) resulting motion vector field. (http://ls.wim.uni-mannheim.de/de/pi4/research/projects/retargeting/test-sequences/)

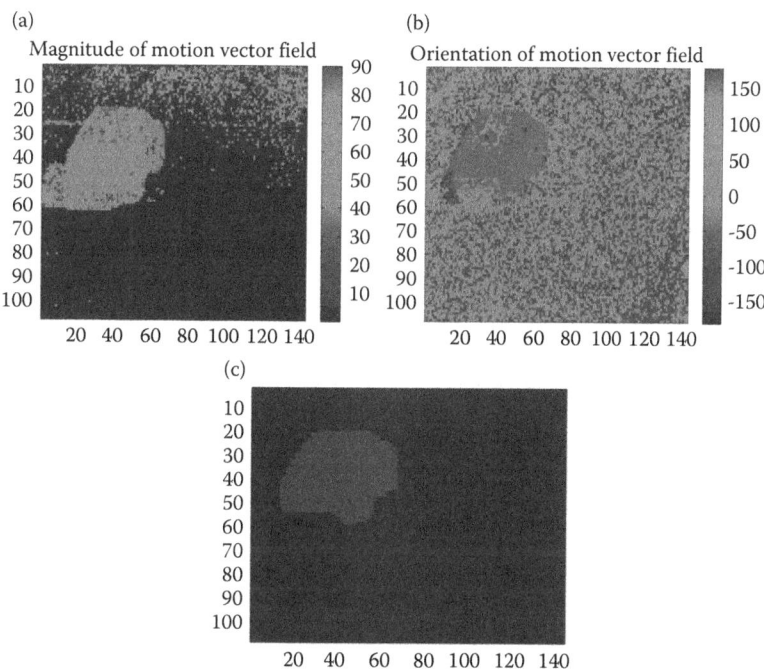

Fig. 11 Clustering of motion vectors from Fig. 3c: (a) magnitude; (b) orientation of motion vector field; (c) resulting motion vector clusters

is selected at the start of the next iteration, and iterations are performed until all motion vectors are assigned to respective motion clusters. Figure 11 illustrates the clustering procedure when applied to the motion vectors in the motion vector field from Fig. 3c. Figure 11a and b illustrates pseudo-colored versions of the magnitude and orientation of the motion vector field from Fig. 3c. Figure 11c illustrates an image containing the identified clusters, after morphological opening to eliminate clusters smaller than 5×5 motion blocks and morphological closing to fill in holes smaller than 5×5 motion blocks.

Filtering-Based Methods

Filtering-based methods alter the value of motion vectors based on the characteristics of spatially and temporally neighboring or adjacent motion vectors. This is in contrast with clustering-based methods, which do not modify the motion vector field. The goal is for the resulting modified vector field to be more semantically representative of perceived motion. While filtering-based methods typically outperform clustering-based methods when it refers to analysis of motion in the compressed domain, the increase in performance is achieved at the cost of higher computational complexity.

In one instance, filtering is performed by quantifying the degree of coherence of a motion vector with its surroundings, both in the temporal and in the spatial domain.[11] To this end, a spatial neighborhood Θ of motion blocks around motion vector MV^i is defined, typically according to 4- or 8-connectivity criteria. The magnitude $M_{conf}^{i,s}$ and orientation $O_{conf}^{i,s}$ spatial confidence measures of MV^i are defined as follows:

$$M_{conf}^{i,s} = \min\left\{1, \max\left\{0, 1 - \frac{\left|\left(\|MV^i\| - \sum_{j \in \Theta}\|MV^j\|\right)\big/|\Theta|\right|}{\sum_{j \in \Theta}\|MV^j\|\big/|\Theta|}\right\}\right\}, \quad (20)$$

$$O_{conf}^{i,s} = \min\left\{1, \max\left\{0, 1 - \frac{\left|\left(MV^i - \sum_{j \in \Theta} MV^j\right)\big/|\Theta|\right|}{\sum_{j \in \Theta} MV^j\big/|\Theta|}\right\}\right\}, \quad (21)$$

where $\|MV^i\|$ and $\measuredangle MV^i$ denote the length and orientation of motion vector MV^i, respectively. For a given motion vector, the spatial confidence metrics in Eqs. 20 and 21 indicate how coherent it is relative to its spatial neighborhood: higher confidence metrics imply higher coherence and vice versa. Similar confidence metrics for temporal coherence can be defined. Let $MV^{i,t_0-N},\ldots,MV^{i,t_0},\ldots,MV^{i,t_0+N}$ be motion vectors corresponding to the ith block in consecutive frames with temporal indices $t_0 - N$, through $t_0 + N$, respectively, for some positive integer $N \geq 1$. Temporal confidence metrics of MV^{i,t_0} can be computed as follows:

$$M_{conf}^{i,t_0} = \min\left\{1, \max\left\{0, 1 - \left|\frac{\left(\|MV^{i,t_0}\| - \left[\sum_{j=1}^{N}\|MV^{i,t_0-j}\| + \|MV^{i,t_0-j}\|\right]\right)/(2N)}{\left[\sum_{j=1}^{N}\|MV^{i,t_0-j}\| + \|MV^{i,t_0-j}\|\right]/(2N)}\right|\right\}\right\}, \quad (22)$$

$$O_{conf}^{i,t_0} = \min\left\{1, \max\left\{0, 1 - \left|\frac{\left(\angle MV^{i,t_0} - \left[\sum_{j=1}^{N}\angle MV^{i,t_0-j} + \angle MV^{i,t_0+j}\right]\right)/(2N)}{\left[\sum_{j=1}^{N}\angle MV^{i,t_0-j} + \angle MV^{i,t_0+j}\right]/(2N)}\right|\right\}\right\}. \quad (23)$$

Once spatial and temporal confidence metrics are computed, an overall confidence metric S^{i,t_0} for motion vector MV^{i,t_0} can be estimated based on a functional combination of the individual metrics. The simplest example is a weighted average of all four confidence metrics.

Once an overall confidence metric is computed, the filtered value of MV^{i,t_0} can be computed as follows:

$$MV^{i,t_0} = \frac{\sum_{j=-N}^{N} S^{i,t_0+j} MV^{i,t_0+j}}{\sum_{j=-N}^{N} S^{i,t_0+j}} \quad (24)$$

The preceding equation implies that a filtered motion vector can be obtained by computing a weighted average of temporally neighboring motion vectors at the same spatial location; alternative implementations can include terms where spatial neighbors of the motion vector to be filtered are considered in the filtering procedure. Figure 12b shows a portion of the motion vector field illustrated in Fig. 12a. Figure 12c shows the resulting motion vector field after application of the confidence-based filtering techniques described earlier. This is a particularly challenging scenario for any linear filtering approach, since the highlighted portion includes edge regions of abrupt transitions in the motion vector field. Consequently, some softening of the boundaries between the clusters of motion vectors is evident.

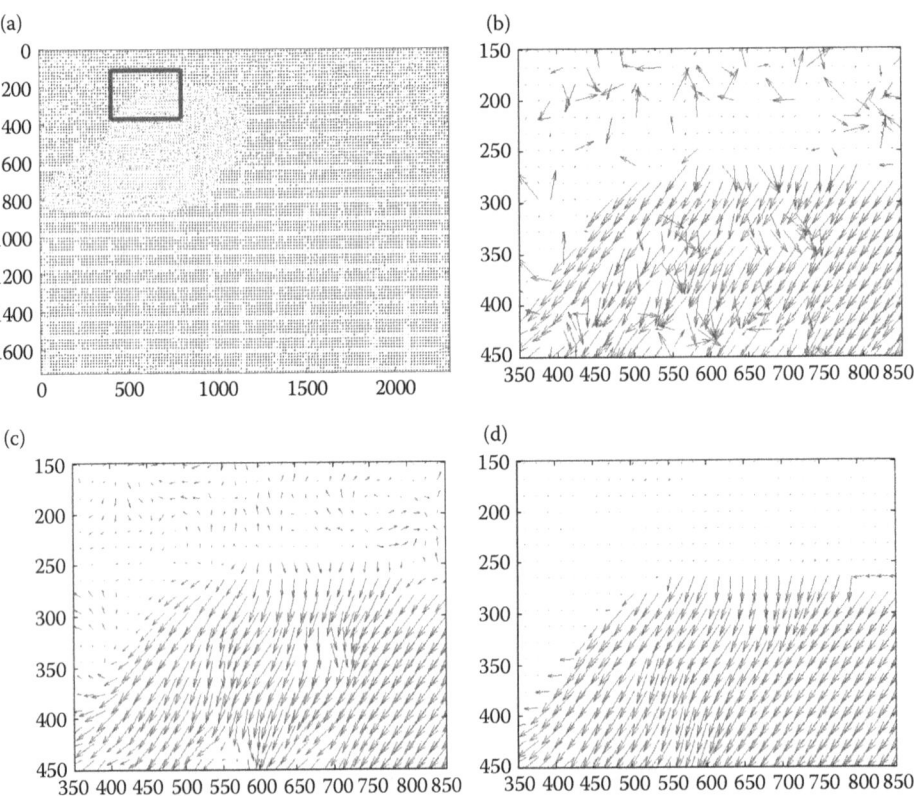

Fig. 12 (a) Original motion vector field; (b) magnified portion of the highlighted motion field of Fig. 11a; magnified portion after filtering with (c) confidence-based filter described earlier and (d) median filter

Additionally, residuals of the outlier motion vectors located in the stationary region of the vector field remain after filtering, also a characteristic of any linear filtering procedure. In contrast, Fig. 12d shows the motion vector field resulting from nonlinear median filtering of the horizontal and vertical vector fields individually. It can be seen that the nonlinear method handles outliers and discontinuities better.

Cameras in Motion

Let **X**, **Y**, and **Z** denote the axes of the camera coordinate system, as illustrated in Fig. 13. The (**x**, **y**) image plane is perpendicular to the optical axis, which in turn is parallel to the **Z** axis. The coordinates of the point where the optical axis intersects the image plane are $(0, 0, f)$, where f is the focal length of the camera. Let **x** and **y** denote the axes of the image coordinate system. A point in 3D space located at coordinates $P = (X, Y, Z)$ is projected onto the point located at image plane coordinates $p = (x, y)$, where $x = fX/Z$ and $y = fY/Z$.

The apparent motion that a camera captures between two frames can be modeled by five parameters[12]: (1) change in the focal length f, typically referred to as zooming; rotation around the (2) **X**-axis, β (3) **Y**-axis, α; and (4) **Z**-axis, γ, and (5) translation. As the camera moves, the coordinates of the stationary point P become $P' = (X', Y', Z')$. The relationship between P and P' is given by the following:

$$P' = RP + t, \qquad (25)$$

where **R** represents the rotation of the camera around an arbitrary axis through the origin of the camera coordinates, and $t = [t_x, t_y, t_z]^T$ represents the camera translation. Note that a three-dimensional rotation can be represented by successive rotations around the axes of the camera coordinate systems $\mathbf{R}_\alpha$, $\mathbf{R}_\beta$, and $\mathbf{R}_\gamma$. Specifically,

$$\mathbf{R} = \begin{bmatrix} r_{11} & r_{12} & r_{13} \\ r_{21} & r_{22} & r_{23} \\ r_{31} & r_{32} & r_{33} \end{bmatrix} = \mathbf{R}_\alpha \mathbf{R}_\beta \mathbf{R}_\gamma$$

$$= \begin{bmatrix} \cos\alpha & 0 & -\sin\alpha \\ 0 & 1 & 0 \\ \sin\alpha & 0 & \cos\alpha \end{bmatrix} \begin{bmatrix} 1 & 0 & 0 \\ 0 & \cos\beta & -\sin\beta \\ 0 & \sin\beta & \cos\beta \end{bmatrix}$$

$$\begin{bmatrix} \cos\gamma & -\sin\gamma & 0 \\ \sin\gamma & \cos\gamma & 0 \\ 0 & 0 & 1 \end{bmatrix} \qquad (26)$$

The relationship between the 3D point and image plane coordinates, P' and $p' = (x', y')$, after camera translation and change in focal length are given by $x' = f'X'/Z'$ and $y' = f'Y'/Z'$. Inserting into Eq. 25 and expanding,

$$X' = t_x + Z\left(r_{11}\frac{x}{f} + r_{12}\frac{y}{f} + r_{13}\right),$$

$$Y' = t_y + Z\left(r_{21}\frac{x}{f} + r_{22}\frac{y}{f} + r_{23}\right), \qquad (27)$$

$$Z' = t_z + Z\left(r_{31}\frac{x}{f} + r_{32}\frac{y}{f} + r_{33}\right).$$

Consequently,

$$x' = \frac{f'}{Z'}X' = f' \frac{t_x + Z\left(r_{11}\frac{x}{f} + r_{12}\frac{y}{f} + r_{13}\right)}{t_z + Z\left(r_{31}\frac{x}{f} + r_{32}\frac{y}{f} + r_{33}\right)}$$

$$= f' \frac{t_x/Z + r_{11}\frac{x}{f} + r_{12}\frac{y}{f} + r_{13}}{t_z/Z + r_{31}\frac{x}{f} + r_{32}\frac{y}{f} + r_{33}},$$

$$y' = \frac{f'}{Z'}Y' = f' \frac{t_y + Z\left(r_{21}\frac{x}{f} + r_{22}\frac{y}{f} + r_{23}\right)}{t_z + Z\left(r_{31}\frac{x}{f} + r_{32}\frac{y}{f} + r_{33}\right)} \qquad (28)$$

$$= f' \frac{t_y/Z + r_{21}\frac{x}{f} + r_{22}\frac{y}{f} + r_{23}}{t_z/Z + r_{31}\frac{x}{f} + r_{32}\frac{y}{f} + r_{33}}.$$

In some cases, it is reasonable to assume that the magnitude of the scalar components of the camera translation vector is significantly smaller than the distance between the point being imaged and the camera along the optical axis. Under these conditions, Eq. 28 can be simplified to

$$x' = f' \frac{r_{11}\frac{x}{f} + r_{12}\frac{y}{f} + r_{13}}{r_{31}\frac{x}{f} + r_{32}\frac{y}{f} + r_{33}},$$

$$y' = f' \frac{r_{21}\frac{x}{f} + r_{22}\frac{y}{f} + r_{23}}{r_{31}\frac{x}{f} + r_{32}\frac{y}{f} + r_{33}}. \qquad (29)$$

Note that this assumption holds, for example, when the frame rate of the video is high enough for camera translations to be negligible between frames, particularly relative to the distance between the point and the camera as measured along the optical axis. There are six unknown motion parameters to be estimated in Eq. 29, including the change in focal length.

In other cases, it may be reasonable to assume that the scene is a planar surface, so that points on the scene satisfy $k_1 X + k_2 Y + k_3 Z = 1$ for some constants k_1, k_2, and k_3. Under these conditions, Eq. 29 can be simplified into[13] the following:

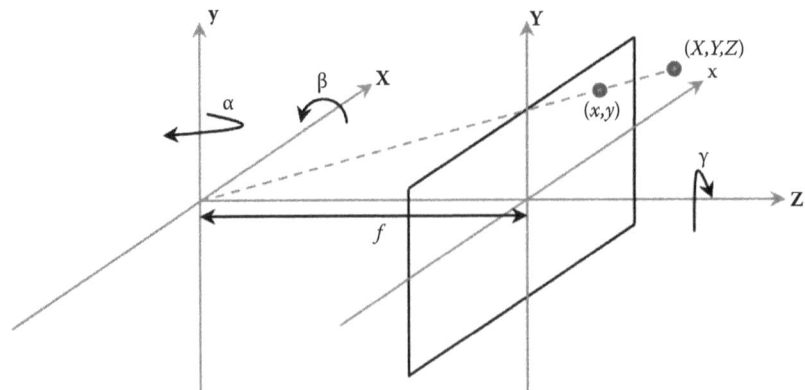

Fig. 13 The geometry of perspective imaging

$$x' = f' \frac{\left(\frac{r_{11}}{f}+t_x k_1\right)x + \left(\frac{r_{12}}{f}+t_x k_2\right)y + r_{13}+t_x k_3}{\left(\frac{r_{31}}{f}+t_z k_1\right)x + \left(\frac{r_{32}}{f}+t_z k_2\right)y + r_{33}+t_z k_3},$$

$$y' = f' \frac{\left(\frac{r_{21}}{f}+t_y k_1\right)x + \left(\frac{r_{22}}{f}+t_y k_2\right)y + r_{23}+t_y k_3}{\left(\frac{r_{31}}{f}+t_z k_1\right)x + \left(\frac{r_{32}}{f}+t_z k_2\right)y + r_{33}+t_z k_3}.$$

(30)

There are 11 unknown motion parameters in Eq. 30. Note that in both Eqs. 29 and 30, the relationship between the image coordinates before and after camera motion is expressed in terms of a projective transformation; consequently, estimating motion parameters is equivalent to estimating the parameters describing the transformation. Due to the nonlinear nature of Eqs. 29 and 30, absolute focal length cannot be estimated. It is common practice to solve for the unknown variables in Eqs. 29 and 30 by tracking sets of points whose locations in the image plane can be identified across frames. In other words, the corresponding image plane coordinates for image points $\{(x_1, y_1), (x_2, y_2), \ldots, (x_n, y_n)\}$ and $\{(x_1', y_1'), (x_2', y_2'), \ldots, (x_n', y_n')\}$ are measured and the resulting system of equations of the form described by Eq. 27 solved. In the absence of noise, the unknowns in Eq. 29 can be solved for by establishing at least three correspondences. Similarly, in the absence of noise, the unknowns in Eq. 30 can be solved for by establishing at least six correspondences. Since image plane coordinate measurements and correspondences can be noisy, enough measurements for the resulting system of equations to be highly over determined are usually taken, and the resulting over determined system is solved via least squares methods.

In the compressed domain, motion vectors can be interpreted as indicators of correspondence information for a given motion block for a pair of video frames. Specifically, if (x_i', y_i') denotes the pixel coordinates of the ith motion block in frame with temporal index t_0, then the corresponding motion block in frame with temporal index $t_0 - 1$ is located at coordinates $(x_i, y_i) = (x_i', y_i') + MV^{i, t_0}$. Figure 14 illustrates the estimation of the motion parameters of a camera in motion from compression motion vectors. Figure 14a shows a reference frame, Fig. 14b shows a target frame, and Fig. 14c shows the resulting motion vector field. Figure 14d illustrates the image obtained by applying the geometric transformation estimated from the correspondences described by the motion field in Fig. 14c to the frame in Fig. 14a. Note that the more accurate the geometric transformation estimate is, the closer the image in Fig. 14d will resemble the image in Fig. 14b. Figure 14e illustrates the pixel-wise absolute difference between the images in Fig. 14b and d.

EXAMPLE VIDEO ANALYTICS AND COMPUTER VISION APPLICATIONS IN THE COMPRESSED DOMAIN

Given the complexity of algorithms designed to perform high-level video analytics and computer vision tasks, the algorithms rarely perform in real time on high frame rate, high-resolution video data (e.g., 30 fps and higher, and 1,280 × 720 pixels and above). Algorithms that operate in the compressed domain have been proposed to address these shortcomings.

Action Recognition

The majority of algorithms that perform action recognition in the compressed domain operate by extracting motion-vector-based descriptors of a known set of actions from training video sequences, and then compute the same descriptors from the test video stream. Similarity metrics between the known actions and the unknown actions in the test video stream are computed to recognize the type of action contained in the incoming streams.[14]

In order to make the descriptors extracted from motion vector fields more representative of the captured motion, the motion vectors usually undergo a preprocessing filtering

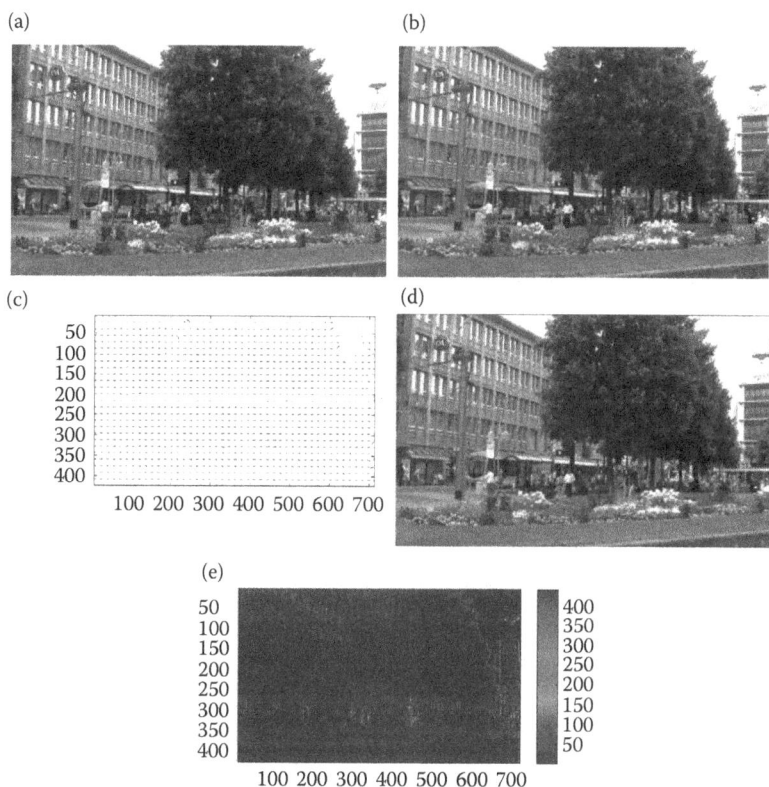

Fig. 14 Estimation of camera motion parameters from compression motion vectors: (a) reference frame; (b) target frame; (c) corresponding motion vector field; (d) image obtained by applying the geometric transformation estimated from motion vector in (c) to image in (a); (e) resulting pixel-wise transformation error

stage similar to the filtering procedure described in Section "Motion Detection and Analysis in the Compressed Domain." In addition, since the temporal gap between reference and target or predicted frames is variable, motion vectors are normalized by the actual gap between the frames used to estimate the vectors, so that their length is relative to a single frame interval.

Let $\{MV_i^{1,test}, MV_i^{2,test}, \ldots, MV_i^{K,test}\}$ and $\{MV_j^{1,train}, MV_j^{1,train}, \ldots, MV_j^{K,train}\}$, where K is the total number of motion vectors per frame, be the filtered motion vector fields corresponding to the ith and jth frames in the test and training sequences, respectively, where the action contained in the training sequence is known. The similarity between both motion vector fields can be computed via normalized correlation as follows:

$$S(i,j) = \frac{\langle u_{i,test}, u_{j,train} \rangle}{\|u_{i,test}\| \|u_{j,train}\|}, \quad (31)$$

where $u_{i,test}$ and $u_{j,train}$ are the vectorized representations of the filtered motion vector fields of the ith test and jth training frames, respectively, and $\langle u_1, u_2 \rangle$ denotes the inner product between vectors u_1 and u_2. One limitation of the similarity metric in Eq. 31 is that motion vectors with zero magnitude in *either* the test or the training sequence

contribute to the correlation as much as motion vectors with zero magnitude in *both* sequences. In practice, vector fields with common zero entries should be considered to be similar. An improved measure of similarity can then be defined as follows:

$$\hat{S}(i,j) = \frac{\sum_{k=1}^{K} d\left(MV_i^{k,test}, MV_j^{k,train}\right)}{Z\left(MV_i^{k,test}, MV_j^{k,train}\right)}, \quad (32)$$

where for two vectors u_1 and u_2,

$$d(u_1, u_2) = \begin{cases} \dfrac{\langle u_1, u_2 \rangle}{\max\left\{\|u_1\|^2, \|u_2\|^2\right\}}, & \text{if } \|u_1\|\|u_2\| \neq 0 \\ 0, & \text{otherwise} \end{cases}, \quad (33)$$

and $Z(MV^{test}, MV^{train}) = \sum_{k=1}^{K} \mathbf{I}(\|MV^{k,test}\| > 0 \text{ or } \|MV^{k,train}\| > 0)$

is a normalizing factor that counts the number of instances when a motion vector in either field has nonzero length. In this case, $\mathbf{I}(b)$ is the indicator function which equals 1 if the Boolean variable b is true and equals 0 otherwise.

Computing the similarity metric from Eq. 32 for every pair of motion vector fields in the training and test sequence will result in a similarity matrix of dimensionality $M \times N$, where M is the number of frames in the test sequence and N is the number of frames in the training sequence. Finding the optimal temporal alignment between the sequences is equivalent to finding the path with the largest total similarity within the similarity matrix. Since the length and/or the speed of the action in the sequences may vary, finding the optimal alignment between the sequences may require performing temporal warping of the sequences. Figure 15 illustrates the temporal alignment process between a 100 frame test sequence and a 200 frame training sequence. Figure 15a shows a pseudo-colored version of the 100×200 entry similarity matrix. Figure 15b highlights the optimal alignment path (in white) between both sequences, obtained via the dynamic time warping technique.[15] The aggregated similarity between the test and training sequences can be computed by integrating the local similarity measures along the optimal alignment path.

Once the optimal alignment between a test sequence and all the known training sequences is computed, the test sequence is assigned the label of the training sequence that is most similar to it among all the available training sequences.

Object Tracking

Object tracking is the process of locating an object as it moves across a scene being monitored with a video camera, usually on a frame-by-frame basis. Before an object is tracked, it is identified as an object of interest via motion analysis algorithms such as the ones described in the previous section. Once an initial location of the object of interest is determined, the tracking algorithm estimates its location across subsequent video frames. Traditional tracking algorithms that operate in the uncompressed domain treat tracking as a correspondence problem. Specifically, a set of appearance descriptive features of the object is extracted in the initial instance. Such features can include contour, shape, color, or interest/salient point descriptors. As the video progresses, features of areas neighboring the last known location of the object in the incoming frames are extracted, and the location of the region in the frame that best matches the object description is selected as the new location of the object.

The simplest embodiment of object tracking algorithms that operate in the compressed domain exploit the low-resolution appearance model of objects in motion provided by the motion vector field. For instance, an image obtained by finding clusters within a motion vector field based on magnitude and/or orientation (see Fig. 11c) can be descriptive of the outline and shape of a moving object.[16] To illustrate this point, consider the images in Fig. 16a and b, which correspond to motion vector clusters of two temporally adjacent frames. The clusters were obtained by applying median filtering on the original motion vector fields, followed by thresholding based on motion vector magnitude. Assuming that the initial detection of the vehicle was performed on the frame corresponding to Fig. 16a and that the feature to be tracked corresponds to the window highlighted in green, Fig. 16c shows the similarity map between the feature to be tracked and the motion vector cluster in Fig. 16b. The location of the tracked feature as determined by the tracking algorithm would correspond to the location where the similarity between the incoming motion vector cluster and the tracked feature is largest, as indicated by the dark red area in Fig. 16c. It can be seen that the largest similarity is achieved at the approximate new location of the feature being tracked. In this case, the similarity metric is computed in the binary domain by sliding the binary template of the tracked feature across the new motion vector cluster, at each location performing a pixel-wise XNOR operation between the tracked feature and the motion vector cluster, and subtracting the number of resulting "0"s from the number of resulting "1"s. Since this operation is performed in the binary domain, it is extremely computationally efficient.

More elaborate similarity metrics can be implemented, which, while more computationally expensive, can provide more robust feature matching, particularly in noisy video data. For example, Fig. 14d illustrates the similarity map obtained by measuring similarities between the tracked feature and the incoming motion vector field according to Eq. 33, which takes into account the magnitude and orientation of the motion vectors in both the cluster and the

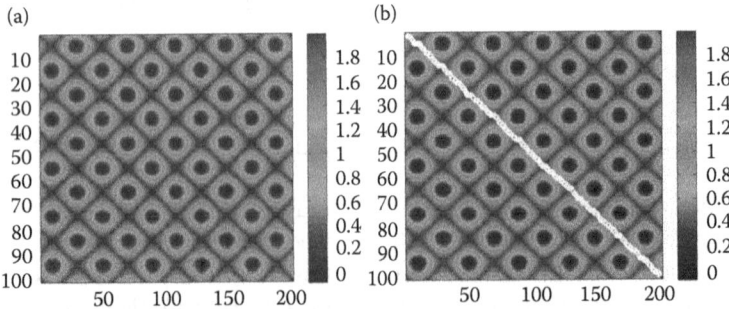

Fig. 15 Temporal alignment between training and test sequences: (a) pseudo-colored view of similarity matrix $S(i, j)$; (b) the optimal alignment path

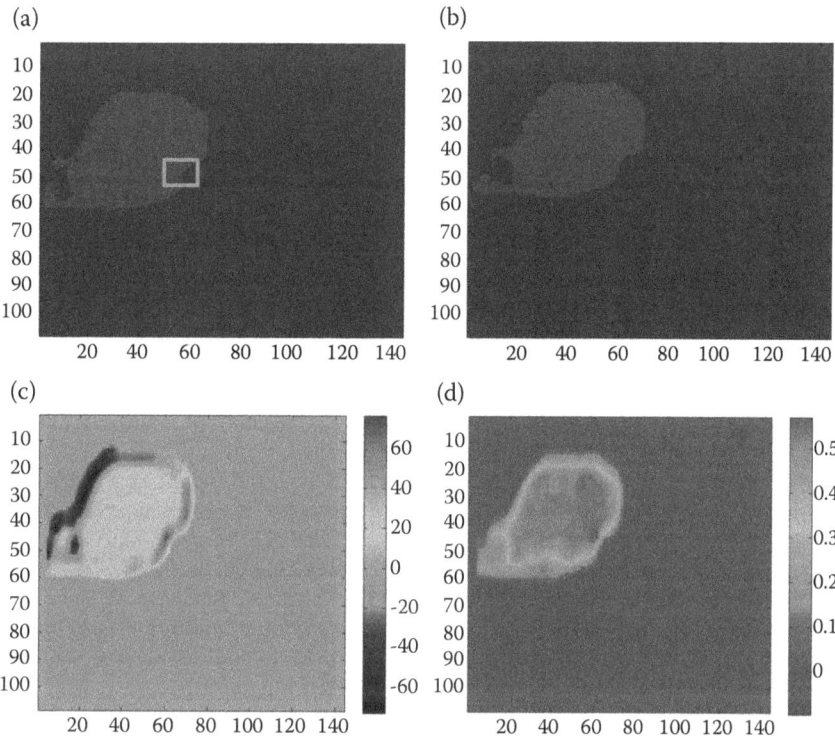

Fig. 16 Illustration of a tracking algorithm operating in the compressed domain: (a) initial instance and (b) subsequent appearance of object of interest; similarity maps between feature being tracked and motion vector cluster from (b) according to (c) sliding XNOR operation, and (d) similarity metric from Eq. 33

tracked feature. Once again, the largest measured similarity is located approximately at the location of the tracked feature. While we illustrate the tracking of a single feature, it should be clear that multiple-feature tracking can be implemented for added robustness.

CONCLUSION

Video compression is a ubiquitous technique that is commonly employed across a wide variety of camera networks so as to maximize their storage and transmission capabilities. In order to maximize compression efficiency, state-of-the-art video compression algorithms implement motion estimation and compensation, which involves computation of motion vectors describing block-wise displacements between frames in the video stream. The resulting motion vector fields are highly descriptive of captured apparent motion, both due to motion of objects in the scene and motion of the camera itself. Motion analysis from the processing of the motion vector fields used in the compression process can support a plethora of computer vision applications in the compressed domain which, while in theory less accurate than their uncompressed domain counterparts, are significantly more computationally efficient. As shown, one of the main limitations of the motion descriptive features is that they are not aimed at representing motion *per se*, but rather at minimizing the distortion that accompanies lossy compression procedures. Consequently, one of the main difficulties in processing video streams compressed generically lies in the preprocessing of the motion vectors so that they better represent actual motion. Alternatively, the semantic relevance of the motion vectors relative to a specific analytics task can be enforced at the time of compression via task-oriented compression algorithms. As video cameras become more specialized, we will see built-in compression algorithms provide the option to perform task-oriented compression for a variety of tasks, including surveillance, access control, transportation, and law enforcement. These task-oriented compression algorithms will in turn enable a variety of embedded processes operating on the compressed domain.

Potential applications of said compressed domain processes include smart camera networks, wherein the video processing is performed in the embedded hardware of the individual cameras, which is usually cost and complexity bound. As processing moves farther from a central computing unit into the distributed devices of the network, significant bandwidth and storage savings can be achieved by transmitting and storing only portions of videos that are deemed relevant by the camera, or, in more extreme cases, only metadata resulting from the embedded processing of the acquired video feed. This *modus operandi* will give rise to a multilayered analytics architecture wherein an initial layer of processing is performed in a distributed manner on each device, and a secondary layer of processing takes place in a centralized processor. Applications of the secondary level processing could include verification of the initial results for accuracy, or further processing aimed at extracting additional information.

REFERENCES

1. Richardson, I. *The H264 Advanced Video Compression Standard*, 2nd Ed.; John Wiley & Sons: Hoboken, NJ, 2010.
2. Sullivan, G.; Wiegand, T. Rate-distortion optimization for video compression. IEEE Signal Process Mag. **1998**, *15* (6), 74–90.
3. Li, Z.; Qin, S.; Itti, L. Visual attention guided bit allocation in video compression. Image Vision Comput. **2011**, *29*, 1–14.
4. Lai, W.; Gu, X.; Wang, R.; Ma, W.; Zhang, H. A content-based bit allocation model for video streaming. In *Proceedings of the IEEE International Conference on Multimedia and Expo*, Taipei, 2004.
5. Bulan, O.; Bernal, E.; Loce, R. Efficient processing of transportation surveillance videos in the compressed domain. J. Electron. Imaging **2013**, *22* (4), 041116.
6. Zhao, S.; You, Z.; Lan, S.; Zhou, X. An improved video compression algorithm for lane surveillance. In *Proceedings of the Fourth International Conference on Image and Graphics*, Sichuan, 2007.
7. Smith, K.; Gatica-Perez, D.; Odobez, J.; Ba, S. Evaluating multi-object tracking. In *Proceedings of the IEEE Computer Society Conference on Computer Vision and Pattern Recognition*, San Diego, CA, 2005.
8. Soyak, E.; Tsaftaris, S.; Katsaggelos, A. Quantization optimized H.264 encoding for traffic video tracking applications. In *Proceedings of the IEEE International Conference on Image Processing*, Hong Kong, 2010.
9. You, W.; Sabirin, M.; Kim, M. Real-time detection and tracking of multiple objects with partial decoding in H.264|AVC bitstream domain. In *Proceedings of the SPIE Real-Time Image and Video Processing Conference*, San Jose, CA, 2009.
10. Zen, H.; Hasegawa, T.; Ozawa, S. Moving object detection from MPEG coded picture. In *Proceedings of the IEEE International Conference on Image Processing*, Kobe, Japan, 1999.
11. Wung, R.; Zhang, H.; Zhang, Y. A confidence-measure-based moving object extraction system built for the compressed domain. In *Proceedings of the IEEE International Symposium on Circuits and Systems*, Geneva, Switzerland, 2000.
12. Tan, Y.; Kulkarni, S.; Ramadge, P. A new method for camera motion parameter estimation. In *Proceedings of the IEEE International Conference of Image Processing*, Washington, DC, 1995.
13. Radke, R.; Ramadge, P.; Echigo, T. Efficiently estimating projective transformations. In *Proceedings of the IEEE International Conference on Image Processing*, Vancouver, BC, 2000.
14. Yeo, C.; Ahammad, P.; Ramchandran, K.; Sastry, S. High-speed action recognition and localization in compressed domain videos. IEEE Trans. Circuits Syst. Video Technol. **2008**, *18* (8), 1006–1015.
15. Muller, M. *Information Retrieval for Music and Motion*; Springer: Berlin/Heidelberg, 2007.
16. Bernal, E.; Wu, W.; Bulan, O.; Loce, R. Monocular vision-based vehicular speed estimation from compressed video streams. In *Proceedings of the IEEE International Conference on Intelligent Transportation Systems*, The Hague, 2013.

Virtual Worlds

Ann Latham Cudworth
Ann Cudworth Projects/Alchemy Sims, New York University, New York, U.S.A.

Abstract
Designers can speak for the people by representing their needs and interests in the virtual world. The designer's job is about being human, thinking about what humans need, solving their problems, helping them communicate, and doing some good in the world, virtual or otherwise.

Keywords: Virtual worlds; History; Visual theory; Gaming; Professional use; Prebuilt grids; Hosted grids.

WELCOME TO THE INFINITE VISUALIZATION TOOL, A VIRTUAL WORLD

Like the universe with its glittering galaxies floating over our heads on a summer night, cyberspace continues to expand, full of people like you creating worlds for exploration, entertainment, and learning. Within this three-dimensional (3-D) manifestation of our collective imagination, you will find new ways of understanding time and space. Terrestrial and temporal identifiers become insignificant as you work with people from around the world. Unlike any visualization tool that precedes it, a virtual world in cyberspace provides you with a place where your creative concepts can be shared as a 3-D form with the world, in any scale, at any time. Let your mind unfold to the possibilities of how a virtual platform works, and you will be rewarded with a new understanding of design and the human perception of it. Almost 2.5 billion people are in cyberspace worldwide,[1] and according to KZero, the number of registered virtual world accounts has broken 1 billion.[2] Obviously, virtual worlds are here to stay, and they need people who will design and create content for them. If you are interested in becoming a virtual world designer, the virtual worlds that run on user-generated content (UGC), like the open grids created with OpenSim (OpenSimulator; http://opensimulator.org/wiki/Grid_List) or private membership grids like Second Life (http://secondlife.com/), are good places to start. Mesh model-based content created for those worlds can also be used on game development platforms like Unity.

A SHORT HISTORY OF VIRTUAL WORLDS

Figure 1 is an illustrated timeline showing an overview of virtual world concepts and how the evolution of presentational devices has created the possibility of immersive virtual environments. This process started long ago, in our ancient world.

Visual Theory and Creation of the First Illusions

Let us jump into an imaginary time machine and look at how historical concepts in philosophy and observations on perception can inform us about virtual reality and the virtual worlds it contains. What is it about perception and illusion that fascinates us? Perhaps when early humans noticed the effects of a flickering campfire on the painted animals that decorated their cave walls, they began to see a story in their minds. This imaginary story was brought into being by their primitive projection technology: firelight. As civilization developed, perceptions of reality and the attempt to describe it gave rise to philosophy, which gave rise to theories, experiments, and debate. Early concepts about the nature of reality and virtual reality may have started with Plato (approximately 424–348 B.C.) and Aristotle (384–322 B.C.). In his "Allegory of the Cave," Plato constructs a model for reality and how it is perceived.[3] Within the allegorical environment of a dark, deep cave, he describes four kinds of individuals: prisoners, puppeteers, the released prisoners, and observers, all experiencing a different reality. The prisoners are chained to a bench and forced to watch a shadow play performed by the puppeteers, carrying shapes and objects back and forth in front of torchlights. The prisoners think that the moving shadows they see and hear are reality. The third group, the released prisoners, has been unchained from the bench and, as they make their way out of the cave and into the sunlight, they are beginning the process of acknowledging that the shadows on the cave wall are not reality. The observers are standing outside the cave and learning about the sun and how it lights the world.

Plato disagreed with Aristotle regarding how humans perceived reality; he believed that the true "Forms" of

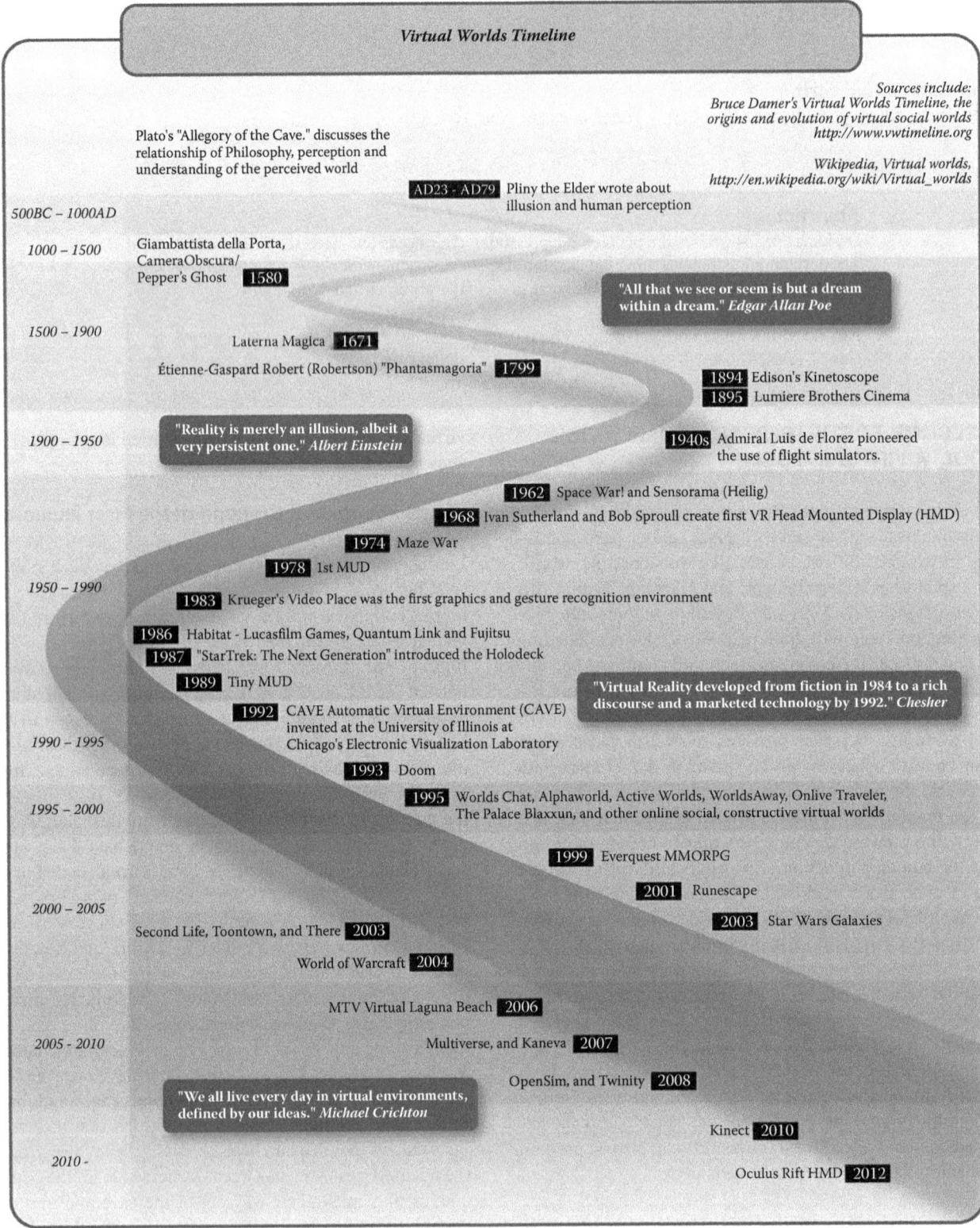

Fig. 1 Timeline showing some of the key ideas, devices, companies, games, and projects that have contributed to the development of virtual environments as we know them today

natural things or concepts were imperfectly understood by humans, whereas Aristotle believed that systematic observation and analysis could lead to the human understanding of the "Forms" of natural things or concepts.[4]

Plato's belief that our experience was only a shadow of the real and unknowable "Forms," represents an interesting philosophical juxtaposition to the virtual reality we create today. In today's world, we have another kind of cave automatic virtual environment (CAVE) for virtual reality. Invented in 1992, this CAVE is a virtual reality environment made from projected images on walls surrounding a person wearing a head-mounted display (HMD); it is not a holodeck yet, but it is approaching it.

Almost 400 years after Aristotle, Pliny the Elder wrote about the origin of painting, sculpture, illusion, and human perception in his book, *Natural History*, ca. A.D. 77–79. You can almost imagine Pliny, relaxing on the patio of his villa in Pompeii, telling the story of how Butades of Corinth, seeing a line drawing his daughter had traced on the wall from her lover's shadow, filled it in with clay and fired the relief to make the world's first portrait.[5] Maybe later in the evening, Pliny would tell of the painting contest between Zeuxis and Parrhasius, two great painters of the fifth century B.C. Smiling at the memory, Pliny tells you of how Zeuxis created a still life containing a bunch of grapes painted so realistically that the birds flew down from a nearby tree to eat them. Seeing that, Parrhasius invited Zeuxis to remove the curtain from his painting to reveal the image. When Zeuxis tried to do that, he discovered that the painting of the curtain was so realistic, that he was fooled into thinking it could be drawn aside. Pliny concludes with the account of Zeuxis ceding victory to Parrhasius, saying: "I have deceived the birds, but Parrhasius has deceived Zeuxis."[6] As Pliny undoubtedly noted, the creation of realistic images fascinates us, and each successive development in the visual arts has been influenced by that fascination.

Trompe L'oeil, Photorealism, and the Projected Image

In the centuries from A.D. 1000 to 1700, great painters and sculptors discovered more ways to make illusions. Trompe l'oeil was invented, and images that could create 3-D spaces in our mind's eye filled churches and mansions. The tools to create those illusions involved the scientific analysis of perception and optics by Alhazen Ibn al-Haytham (965–1039), the observations of Leonardo Da Vinci (1452–1519), and others. In 1580, Giambattista della Porta perfected the camera obscura and another device that eventually became known as Pepper's ghost.[7] Pepper's ghost is named after John Henry Pepper, who popularized it in 1862 from a device developed by Henry Dircks. This ancient device finds everyday use in the television studio as a teleprompter and occasionally makes an appearance in a stage show or fashion video when they want to include the animated image of someone in the performance space.

Projected reality began with the magic lantern (mid-seventeenth century); its invention is credited to both Athanasius Kircher and Christiaan Huygens. Étienne-Gaspard Robert (Robertson) and his "phantasmagoria" (ca. 1799) used the magic lantern to great theatrical effect with complex shows involving moving projectors, live voices, and elaborate arrangements of curtain masking and projection screens. In one long-running show, staged in the crypt of an abandoned Parisian monastery, he succeeded in creating the virtual reality of a supernatural world in the minds of the audience. As an eyewitness describes: "In fact, many people were so convinced of the reality of his shows that police temporarily halted the proceedings, believing that Robertson had the power to bring Louis XVI back to life."[8] Once the lens and a reliable source of illumination were worked out, moving images and the cinema were soon to follow.

The Birth of Cinema, Electronic Screens, and the Start of Immersive 3-D Design

On December 28, 1895, the Lumière brothers did something that changed our perception of reality again. In the first public screening of commercially produced cinema, they showed 10 short films at Salon Indien du Grand Café in Paris.[9] Later that year, one film in particular captured the public's imagination: *L'Arrivée d'un Train en Gare de la Ciotat* ("The Arrival of a Train at Ciotat Station"). By setting the camera intentionally close to the tracks, they captured a dramatic image of the train as it progressed diagonally across the screen, from long shot into close-up shot. There were many other creators of motion picture devices at the time, including Thomas Edison with his kinetoscope (ca. 1891), but the Lumière brothers are credited with being the first to see the potential for cinema and modern filmmaking. They went on to develop and establish many of the filmmaking techniques and cinematographic methodologies that are still used today.

Many of the modern imaging devices have long histories. The ancient Romans, in their time, created wonderful mosaics. They also created a conceptual model for the functioning of a computer screen—the concept of producing an image from many small colored dots, tiles to them, pixels to us.

At some time at the end of the nineteenth century, photographic manipulation began to appear; the Maison Bonfils Company connected four aerial photographs to create a panorama of the city of Beirut, Lebanon. Another step toward illusionary immersion was made and is now shown in the 360-degree panoramic stereographic projections stitched together from dozens of images and seen all over the World Wide Web today.[10]

Computer-Created 3-D Space and Early Virtual Worlds

The war years gave virtual reality and the means to create it a big boost. Admiral Luis de Florez (1889–1962), who fought in both World War I and II, pioneered the use of flight simulators to save pilots' lives. Military usage of virtual reality and training simulations continues to this day and now includes the use of virtual worlds built on Open-Sim platforms and others.[11] In 1962, Morton Heilig built the Sensorama device. It was described by a witness this way: "The Sensorama was able to display stereoscopic 3-D images in a wide-angle view, provide body tilting, supply stereo sound, and also had tracks for wind and aromas to be triggered during the film."[12] Shortly afterward, Ivan Sutherland, working with Bob Sproull, developed the first HMD and called it the "Sword of Damocles" because of the great elongated cable and arm hanging above the head of the wearer. With this device, they opened the door to full-immersion virtual reality.[13]

Meanwhile, haptic devices were being developed at the University of North Carolina's Haptics Research Department; in the late 1960s through the early 1980s, devices like Grope I, II, and III and the Sarcos Arms were created there. At the AT&T labs, Knowlton's virtual push-button device was built. It projected a virtual graphic of symbols on a half-silvered screen above the hands of an operator using a keyboard, effectively combining the virtual with the real.[14] More developments in virtual reality physical feedback (haptic) interfaces started to happen in the 1980s. The Sayre Data Glove (developed at the University of Illinois with a National Endowment for the Arts grant) led to the Mattel Power Glove. Thomas Zimmerman, Jaron Lanier, and Scott Fisher met at Atari and later worked on the VPL glove.[15]

At the same time, virtual worlds were being created in computers and in the early versions of the Internet. In 1974, Maze War was created, an early ancestor of the first-person shooter game; this included the first appearance of avatars, game space maps, and a first-person 3-D perspective within the game space.[16] By 1978, the first MUD (Multiuser Dungeon) arrived. Known as the "Essex MUD" and played on the Essex University (UK) network, it ran until late 1987.[17] The Essex MUD was a text-based game, creating a "constructivist" approach to virtual reality by allowing the players' imagination to construct the virtual world as they role-play with others online. Also notable was Krueger's Videoplace, created in 1983. It was the first graphics and gesture recognition environment.[18]

GAMING AND VIRTUAL WORLDS

By 1986, Lucasfilm Games, Quantum Link, and Fujitsu had opened "Habitat."[19] This was a significant step toward creating online gaming communities in virtual worlds. The imagination of the public and the appetite for immersive virtual worlds was stimulated by the appearance of the holodeck in *Star Trek, The Next Generation* (1987).[20] MUDs were reinvented with the appearance of TinyMUD in 1989. This codebase, which created a socially oriented MUD, was based on player cooperation rather than competition and opened the door for socially based virtual worlds.[21]

The early 1990s saw the invention and construction of the first CAVE at the University of Illinois in Chicago (1992). In the CAVE, all the technologies that had come before it were combined into one powerful device, creating intense immersive experiences. Still active today, the CAVE has video images projected in stereoscopic 3-D. When they are inside it, visitors wear an HMD containing stereoscopic liquid crystal digital (LCD) shutter glasses to view the environment. Sensors collect information about the location and body position of the visitor and adjust the projection fields accordingly.[22] In 1993, Doom started the craze for gamers' first-person shooter games, creating the foundation of a gamer subculture, and was played by over 10 million within the first 2 years of its appearance. Full of graphic violent imagery, Doom has been named one of the 10 most controversial games of all time by Yahoo Games.[23]

The mid-1990s ushered in a wave of online, socially based, constructive virtual worlds; among the most popular were Worlds Chat, Active Worlds, and WorldsAway.[24] Although it seems impossible these days, these worlds functioned on a dial-up connection. In 1995, the ban on commercial usage of the Internet was lifted, and a home-based connection to more sophisticated games became possible. Eventually, broadband cable and Internet connections became available, paving the way for increased popularity of online gaming and virtual worlds.[25] EverQuest and Runescape were early members of the online virtual world massively multiplayer online role-playing game (MMORPG) category.[26] Soon, large, established games and entertainment franchises like World of Warcraft and Star Wars created their own virtual worlds.[27] Also in the early 2000s, the virtual worlds of Second Life and There combined social connection with UGC that could be bought and sold in the virtual world market.[28,29] As the midpoint of the second decade of this century approaches, increasing interactivity and immersion is being interwoven into the online home-based experience of virtual worlds. Many game makers and virtual world developers are striving toward creating an open game, one without levels or barriers that creates a compelling story through the emergent play of its visitors.[30] In 2007, OpenSimulator (or OpenSim, the abbreviation used in this book) arrived and started the creation of a system of virtual world grids, the foundation of a 3-D Internet. This software, based on the Second Life protocols, does not seek to be a copy of Second Life; it seeks to expand the virtual worlds' Metaverse and provide connectivity among them all.[31] Kinect for Xbox 360 has been hacked to capture real-time motion tracking, and the Oculus Rift HMD holds promise for eager customers looking for immersive visual feedback in their virtual worlds.[32] There is undoubtedly much more to come, and for you, a designer of virtual environments, great challenges await.

HOW DO THEY WORK?

The best way to understand a virtual world is to visit one. There are hundreds of public virtual worlds online that

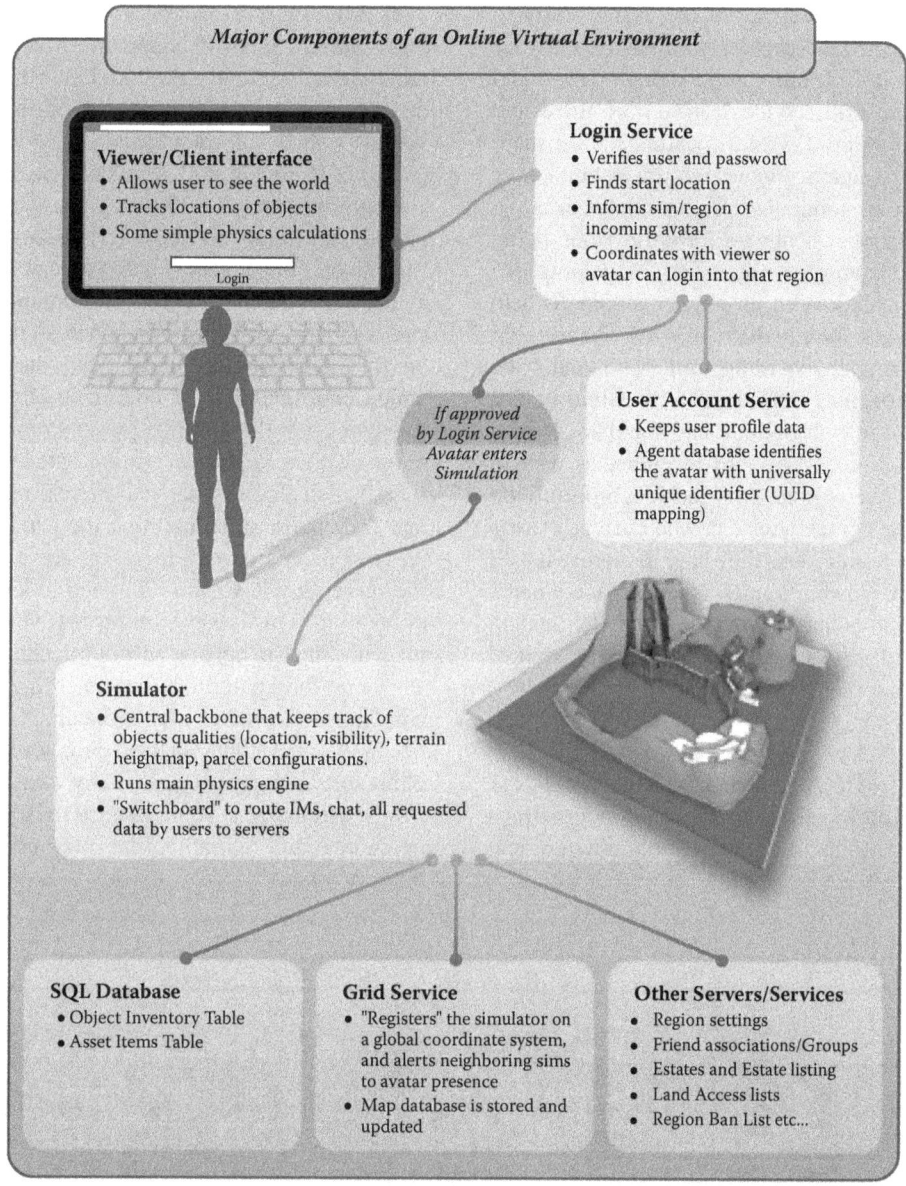

Fig. 2 Diagram showing the major components of an online virtual environment, and how these components interact to bring the virtual environment experience to the user

allow access free of charge or through a subscription. A list of the most popular virtual 3-D worlds include Active Worlds, Minecraft, OpenSim-based worlds, Second Life, and The Sims Online.

Essentially, virtual worlds are persistent 3-D spaces defined inside of a computer program running on a server. When you log out of one, the place still remains, it is "persistent," cycling through its daily settings and hosting other visitors to its location. Figure 2 shows the generic structure shared by most user content-generated virtual worlds such as OpenSim and Second Life. As you can see, even this bare-bones description is fairly complex. When you want to enter this world, you start with the viewer (or client) interface. This is a program that you download and run on your computer. For the purposes of this book, the Firestorm Viewer was selected since it is the most popular one and provides a customized setup for use in OpenSim as well as Second Life.

Assuming that you have set up an avatar account previously with whatever virtual world you would like to visit, let us look at the process of entering a virtual world.

After you have launched the viewer and logged in to your avatar account, your user name and password are verified by the login service, and if the account is valid, the avatar's last location or home location coordinates (x, y, z) are found on the land map of the world. The login service coordinates with the viewer so that the avatar can view the region it is entering and tells the simulator (sim) to expect an incoming avatar. Once the avatar starts to appear, you can see the landscape on your screen and your avatar standing on it. You are now in the simulator, and it is busy telling the grid service that you are there and that all the simulators around your location should share their data with your avatar and viewer so you can look over at them. There is a lot going on in the simulator; it is the central backbone and a switchboard for information being sent to and received from your avatar. The simulator is also running lots of calculations in its physics engine to help objects and your avatar have realistic physics, like keeping your feet on the ground, providing solid walls and open doorways, letting coconuts from the nearby palm tree fall with a soft thump and roll around. The simulator is also keeping track of the objects in your inventory as well as the other assets in the region. This includes information about who made the content you see, who currently owns it, and what kind of permissions they have to modify it, sell it, or give it away. As you build stuff, especially large items like buildings or mountains, the simulator talks to the map database in the grid service, and that updates the look of your sim on the land map, which is shown in the viewer. The simulator also relays information to other servers and services that keep track of who owns land, especially if it is divided into parcels. It relays information about the lists of your friends and groups, what your social/privacy settings are, and other sorts of details regarding your environment and affiliations.

In 2011, a group of fearless explorers called the Alchemy Sims Builders set off across the Metaverse in search of more space and land they could call their own. They settled in on a server maintained by SimHost (http://www.simhost.com) and established the Alchemy Sims Grid (ASG), a 16-region grid accessible via the hypergrid (grid.alchemysims.com:8002:triton). Most of the content and pictures taken for this book were made on this grid, which runs OpenSim 0.7.5. In Fig. 3, you can see what the "back end" of the system looks like on a computer screen. As you can see, there are 17 windows open all the time; 16 of them are for the various regions on ASG, and 1 is for ROBUST, which coordinates the whole grid. Having a system like this gives you the god-like power to make quick and dramatic changes to your terrains, create avatar accounts, and such. It also requires that you exercise responsible behavior and dedication to maintaining the persistence of the grid. At least once a month, there are server updates that shut the server down, so you will need to know how to restart everything. This is especially true if you share the simulator with other people or have it attached to the hypergrid. They may have plans to build something that day, and if the simulator is down because of the updates or upgrades, that becomes frustrating. Furthermore, never forget that this is alpha software right now, so many weird and unpredictable things will happen. You will need to be diligent in backing up content and reminding the other users on your grid to do so as well.

Fig. 3 A screen grab of Alchemy Sims Grid server shown via a Remote Desktop Connection login. On the screen are 17 command line windows, one each for the 16 regions of the grid, and 1 for the ROBUST (Redesigned OpenSimulator Basic Universal Server Technology) program that manages them. Alchemy Sims Grid can be reached via the Hypergrid address of grid.alchemysims.com:8002:triton

So Many Worlds, So Little Time

There are many kinds of virtual worlds and the grids they contain; as a designer, you should visit as many as possible to see what the grid "culture" is and how things are organized. Another excellent way to keep up with virtual worlds is to follow blogs like New World Notes (http://www.nwn.blogs.com) and Hypergrid Business (http://www.hypergridbusiness.com) and set up some Google alerts about virtual worlds and virtual design so you receive links to all the latest news in your e-mail.

WHO USES VIRTUAL WORLDS AND HOW THEY USE THEM

The topic of virtual worlds and who uses them is always under exploration by researchers. Rutgers University conducted an online survey about worlds with UGC and posted some of the initial results (http://player-authors.rutgers.edu/). If you are new to virtual worlds, you would probably like to know how other people use them. An alphabetical set of thumbnail reports about how they are used in various sectors and some specific instances of each is listed below.

Architects/Landscaping Designers

Architects and environmental designers like Jon Brouchoud (http://jonbrouchoud.com/) and David Denton (http://www.daviddenton.com/) utilize virtual worlds like Second Life and OpenSim and the gaming platform Unity. In these worlds, they work with prototyping building designs, building site planning, and developing 3-D models of both real and virtual worlds.

Artists/Painters, Sculptors, Dancers, Actors

The number of artists/painters, sculptors, dancers, and actors working in virtual worlds is huge. There are many performing arts groups in Second Life and OpenSim, and the creation of spectacular scenery and performance spaces is standard fare. Linden Endowment for the Arts (http://lindenarts.blogspot.com/) provides grants for full-region builds to artists and creators on a regular basis. In general, the folks in this category use virtual worlds for design planning, performance, environmental art, sculpture, game design, and gallery displays.

Engineers/Medical Professionals

Ever since their beginnings, Second Life, OpenSim, and Unity have displayed virtual versions of engineering problems, chemical modeling, medical models of the human body, and medical care facilities. Pam Broviak is a licensed civil engineer (state of Illinois) who utilizes virtual worlds for engineering training. Currently, she is working on developing virtual builds to simulate the layout and performance of civil engineering designs and plans, educational environments to teach civil engineering concepts, and a virtual learning and reference environment for local government. Other members of this category use virtual worlds for physical engine studies and structural 3-D design.

Designers, Set Designers, Interior Designers

Richard Finkelstein (http://www.rfdesigns.org/), a professional set designer and set design teacher, uses virtual worlds for prototyping real scenery, teaching set design principles, and designing scenery for virtual performances. Other set and interior designers use virtual worlds for space planning and color studies.

Scientists and Mathematicians

Scientists like Andrew Lang (http://journal.chemistrycentral.com/content/3/1/14) have used virtual worlds like Second Life to demonstrate the structure of molecules for chemical models. J. Gregory Moxness (http://theoryofeverything.org/MyToE/) uses virtual worlds to demonstrate models of E8 math, making the concept visible and beautiful.

Teachers in Primary, Secondary, and Graduate Schools

Teachers in primary, secondary, and graduate schools quickly recognized the value of virtual worlds. Jokay Wollongong (http://www.jokaydiagrid.com/about/) started with Second Life and now has branches of her educational virtual worlds in OpenSim and Minecraft. Jokaydia, her grid in OpenSim, is made for teachers and their classes in grades K–20 and shares resources across the grid. Since 2009, Kenneth Y. T. Lim (http://voyager.blogs.com/about.html) has developed the Six Learnings curricular design framework, which has been used to inform the conceptualization, design, and development of a series of lessons that leverage the affordances for learning with the use of immersive environments in the Singapore school system. Teachers use this "Six Learnings" program in Singapore with virtual worlds to help students deepen their knowledge of and to examine their intuitions about local environments, as well as to prototype their creative works and to plan the productions and staging of dramatic performances.

Trainers and Therapists

Virtual worlds are invaluable for training, simulations, and therapy. Silicon Valley Media Group's First Responders Simulator utilizes them to create training and practice scenarios for disaster workers and emergency response teams. SVMG, headed by Cynthia Stagner, also develops simulation environments for hazardous occupations, in fields such as utilities, law enforcement, and industrial manufacturing. Other trainers use them to teach languages, practice acculturation for military deployment, and teach military strategy. The therapists working in virtual worlds find that the results are comparable to or better than a real-life meeting with their clients, as they counsel them on motherhood, post-traumatic stress disorders (PTSDs), and even weight loss. Patti Abshier works in Second Life to assist counseling professionals who provide counseling using virtual worlds. Other uses are role-playing for cognitive therapy, technical systems training, and medical training.

VIRTUAL ENVIRONMENTS FROM A DESIGNER'S POINT OF VIEW

What are the challenges to a designer on a virtual environment project? In some ways, they are exactly the same challenges of a designer in the physical world.

The primary challenge is to keep the client happy by helping define and present his or her message and supporting those efforts within a positive working environment including you and your team. A happy client is a flexible client, and a flexible client will give you the design latitude you need to be creative.

The secondary challenge is to be professional in your demeanor and work ethic. This means that you have taken the time to learn your craft and the working rules involved with your community, and that you strive to maintain the highest professional standards possible.

The tertiary challenge is to understand the needs of people, how that affects the accessibility of a virtual world, and the principles of Design for All.

Defining the Job of a Virtual Environment Designer

At present, there is no union or guild of virtual environment designers to codify the job description or set the working practices, but you will probably want to have those defined in any letter of agreement you have with a new client. A good definition to include in your paperwork would be the following:

> The virtual environment designer shall be responsible for the creation of the following: 1) creative concept representations, including sketches, models, and descriptions that pertain to the client's stated list of requirements for the environment; 2) subsequent rough preliminary 3-D models and iterative progress representations as the design process progresses; and 3) the final environment in a virtual space, including all of the following items: terrain, landscaping, buildings, scripted objects, lighting, and sound elements.

You may also want to add in avatars and other kinds of special objects like vehicles if that is what you want to design and the client has requested it.

Being a Designer "In The Know"

As a designer of virtual environments, you should be well versed on the differences and similarities of various virtual platforms. One of the first questions to your client should be: "How many virtual worlds do you want to see this environment on?" If you have been designing for a while, especially if you started before early 2008, you are probably aware of the options provided by OpenSim and Unity. Each platform has created a paradigm shift in design thinking for virtual environments, and as a "metaversal" designer, you need to be aware of that. Figure 4 is a schematic drawing displaying some of the differences between Second Life, OpenSim, and Unity in terms of terrains, inventory backup, and content creation. A designer who is "in the know" plans for these differences so that if the client decides to change platforms, the designer is ready to go with it.

DESIGNING IN A PREEXISTING VIRTUAL WORLD OR MAKING ONE YOURSELF

The decision about where to build your first design for yourself or a client should not be taken lightly. There are many options, so you will need to do your homework regarding what these can provide and match them to your project's needs. Essentially, there are two basic categories: 1) prebuilt grids (or hosted grids) that allow you to have land regions, modify terrain, and create content that is hosted on someone else's server and 2) "do-it-yourself" virtual worlds that you install, run, and build on your own server or computer. Let us look at these and some examples of each.

Prebuilt Grids/Hosted Grids

The category of prebuilt and hosted grids includes the "full-service" grids like Second Life (http://www.secondlife.com) and over 200 OpenSim-based grids, such as 3rd Rock Grid (http://3rdrockgrid.com/), Avination Grid (https://www.avination.com), or InWorldz (http://inworldz.com/faq.php). Maria Korolov keeps an updated list on her site, Hypergrid Business (http://www.hypergridbusiness.com/opensim-hosting-providers/). You can also start your own grid and have companies like SimHost (http://www.simhost.com/) and Dreamland Metaverse (http://www.dreamlandmetaverse.com/) host it on their servers. Each hosting company will offer different levels of access to the server on which your grid is hosted, so check their hosting packages carefully so you obtain what you want. If the image in Fig. 3 makes you nervous, you probably want a hosting service that has created an interface for the server, which simplifies the process of uploading new terrain or saving an OAR (OpenSimulator archive) file for your built regions. If you want to dive into the system and access all those command windows, look for a host service that will allow that.

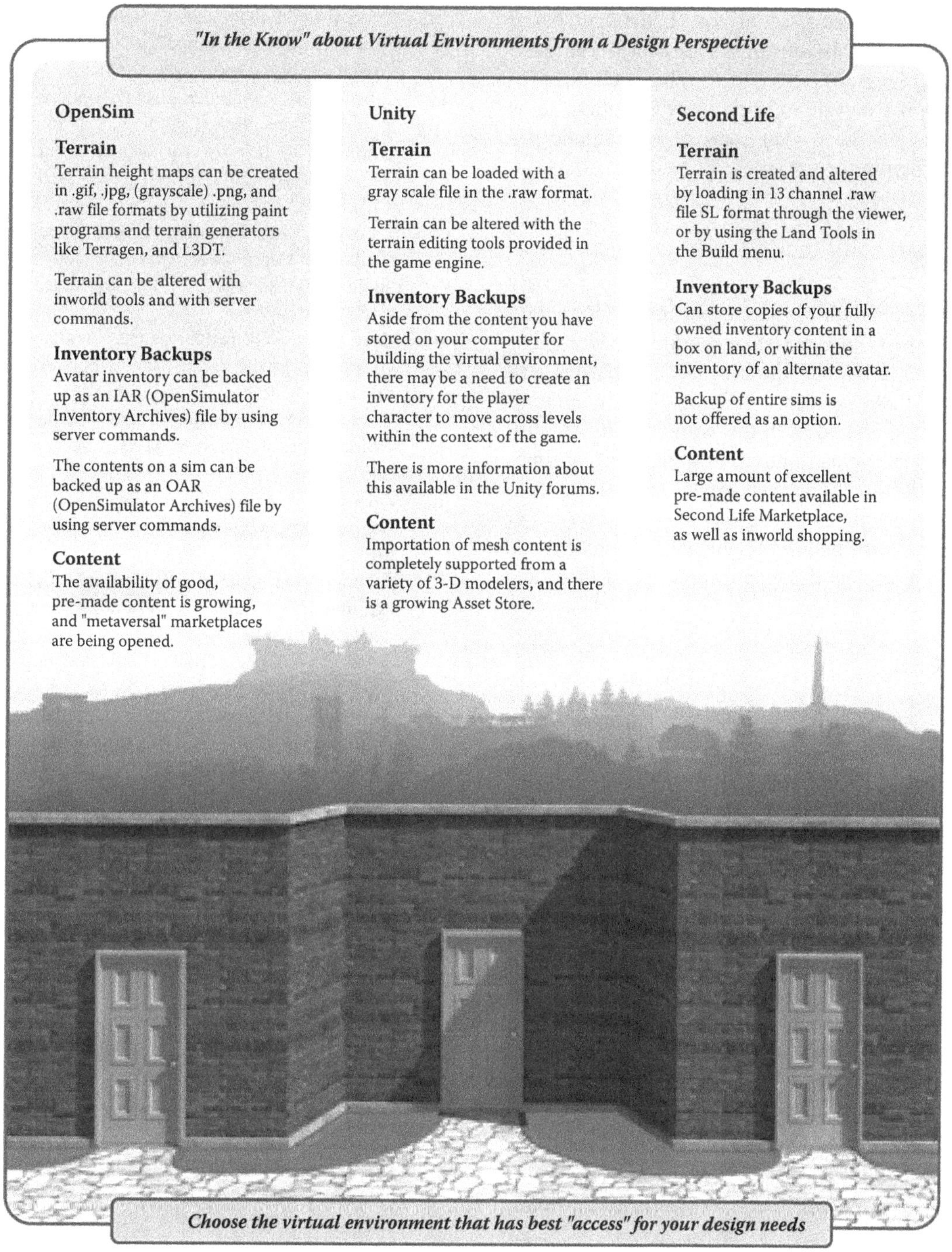

"In the Know" about Virtual Environments from a Design Perspective

OpenSim

Terrain

Terrain height maps can be created in .gif, .jpg, (grayscale) .png, and .raw file formats by utilizing paint programs and terrain generators like Terragen, and L3DT.

Terrain can be altered with inworld tools and with server commands.

Inventory Backups

Avatar inventory can be backed up as an IAR (OpenSimulator Inventory Archives) file by using server commands.

The contents on a sim can be backed up as an OAR (OpenSimulator Archives) file by using server commands.

Content

The availability of good, pre-made content is growing, and "metaversal" marketplaces are being opened.

Unity

Terrain

Terrain can be loaded with a gray scale file in the .raw format.

Terrain can be altered with the terrain editing tools provided in the game engine.

Inventory Backups

Aside from the content you have stored on your computer for building the virtual environment, there may be a need to create an inventory for the player character to move across levels within the context of the game.

There is more information about this available in the Unity forums.

Content

Importation of mesh content is completely supported from a variety of 3-D modelers, and there is a growing Asset Store.

Second Life

Terrain

Terrain is created and altered by loading in 13 channel .raw file SL format through the viewer, or by using the Land Tools in the Build menu.

Inventory Backups

Can store copies of your fully owned inventory content in a box on land, or within the inventory of an alternate avatar.

Backup of entire sims is not offered as an option.

Content

Large amount of excellent pre-made content available in Second Life Marketplace, as well as inworld shopping.

Choose the virtual environment that has best "access" for your design needs

Fig. 4 Diagram comparing some of the key features of three popular online virtual environments available for a designer to utilize. Two of them, OpenSim and Unity, are free to download and use. Second Life charges fees for membership and land usage

Do It Yourself

If you have the desire to create your own grid and host it on your own equipment, and words like configuration, network, firewall, and database settings do not make your head hurt, there are instructions for downloading your very own version of OpenSim and setting it up (http://opensimulator.org/wiki/User_Documentation).

If you want to do the same but use a self-installing version of OpenSim, you should look at New World Studio (http://newworldstudio.net/). While still in its early stages of development, this version of OpenSim holds promise for providing you with a relatively easy way to install OpenSim and get it up and running quickly.

The final option in this category is the simple and elegant Sim on a Stick (SoaS) (http://simonastick.com/).

As a single-user, universal serial bus (USB) stick-based version of OpenSim, this software is available in multiple configurations of regions from 1 to 16. This is a good program to obtain if you want to learn about OpenSim, show it to other folks, and keep some spare regions around just for design prototyping.

CONCLUSION

Designers can "speak for the people" by representing their needs and interests in the virtual world. The designer's job is about being human, thinking about what humans need, solving their problems, helping them communicate, and doing some good in the world, virtual or otherwise. With those goals in mind, this entry will at times take the "30,000-foot" perspective on virtual environments and at other times swoop down to obtain a closer look and deal with the details of a virtual environment and how to utilize it for your design purposes. Enjoy the ride.

REFERENCES

1. Internet World Stats. Available at http://www.internetworldstats.com/ stats.htm, (accessed in August 1, 2013).
2. KZero Virtual World Registered Accounts Statistics. Available at http://www.kzero.co.uk/blog/virtual-world-registered-accountsbreakthrough-1bn/, (accessed in August 1, 2013).
3. Allegory of the Cave, Wikipedia article. Available at http://en.wikipedia. org/wiki/Allegory_of_the_Cave, (accessed in August 4, 2013).
4. Aristotle, Wikipedia article. Available at http://en.wikipedia.org/wiki/Aristotle#Metaphysics, (accessed in August 4, 2013).
5. Butades, Wikipedia article. Available at http://en.wikipedia.org/wiki/Butades, (accessed in August 4, 2013).
6. Zeuxis and Parrhasius, Wikipedia article. Available at http://en.wikipedia. org/wiki/Zeuxis_and_Parrhasius, (accessed in August 3, 2013).
7. Pepper's Ghost, Wikipedia article. Available at http://en.wikipedia.org/wiki/Pepper%27s_ghost, (accessed in August 4, 2013).
8. Phantasmagorica, Wikipedia article. Available at http://en.wikipedia.org/wiki/Phantasmagoria, (accessed in August 4, 2013).
9. Auguste and Louis Lumière, Wikipedia article. Available at http://en.wikipedia.org/wiki/Auguste_and_Louis_Lumi%C3%A8re, (accessed in August 4, 2013).
10. Panorama, Wikipedia article. Available at http://en.wikipedia.org/wiki/Panorama, (accessed in August 4, 2013).
11. Luis de Florez, Wikipedia article. Available at http://en.wikipedia.org/wiki/Luis_de_Florez, (accessed in August 4, 2013).
12. Sensorama, Wikipedia article. Available at http://en.wikipedia.org/wiki/Sensorama, (accessed in August 2, 2013).
13. The Sword of Damocles (Virtual Reality). Available at http://en.wikipedia. org/wiki/The_Sword_of_Damocles_(virtual_reality), (accessed in August 4, 2013).
14. Knowlton, K. Computer displays optically superimposed on input devices. Bell Syst. Tech. J. 1976, 56 (3), 367–383. Available at http://www3.alcatel-lucent.com/bstj/vol56-1977/articles/bstj56-3-367.pdf, (accessed in August 3, 3013).
15. History of Virtual Reality, A Slide Show by Greg Welch. Available at http://www.cs.jhu.edu/~cohen/VirtualWorlds/media/pdf/Historical.color.pdf, (accessed in August 2, 2013).
16. Maze War, Wikipedia article. Available at http://en.wikipedia.org/wiki/Maze_war, (accessed in August 3, 2013).
17. Essex MUD, Wikipedia article. Available at http://en.wikipedia.org/wiki/Essex_MUD, (accessed in August 3, 2013).
18. Videoplace, Wikipedia article. Available at http://en.wikipedia.org/wiki/Videoplace, (accessed in August 4, 2013).
19. Habitat (Video Game), Wikipedia article. Available at http://en.wikipedia. org/wiki/Habitat_(video_game), (accessed in August 4, 2013).
20. Holodeck, Wikipedia article. Available at http://en.wikipedia.org/wiki/Holodeck, (accessed in August 4, 2013).
21. TinyMUD, Wikipedia article. Available at http://en.wikipedia.org/wiki/TinyMUD, (accessed in August 4, 2013).
22. CAVE Automatic Virtual Environment, Wikipedia article. Available at http://en.wikipedia.org/wiki/Cave_automatic_virtual_environment, (accessed in August 4, 2013).
23. Doom, Wikipedia article. Available at http://en.wikipedia.org/wiki/Doom_(video_game), (accessed in August 3, 2013).
24. Active Worlds, Wikipedia article. Available at http://en.wikipedia.org/wiki/Activeworlds, (accessed in August 3, 2013).
25. Broadband, Wikipedia article. Available at http://en.wikipedia.org/wiki/Broadband, (accessed in August 3, 2013).
26. EverQuest, Wikipedia article. Available at http://en.wikipedia.org/wiki/EverQuest, (accessed in August 2, 2013).
27. World of Warcraft, Wikipedia article. Available at http://en.wikipedia. org/wiki/World_of_Warcraft, (accessed in August 3, 2013).
28. Second Life, Wikipedia article. Available at http://en.wikipedia.org/wiki/Second_Life, (accessed in August 4, 2013).
29. There, Wikipedia article. Available at http://en.wikipedia.org/wiki/There_(virtual_world), (accessed in August 4, 2013).
30. Open World, Wikipedia article. Available at http://en.wikipedia.org/wiki/Open_world, (accessed in August 4, 2013).
31. OpenSimulator. Available at http://opensimulator.org/wiki/Main_Page, (accessed in August 4, 2013).
32. Oculus Rift, Wikipedia article. Available at http://en.wikipedia.org/wiki/Oculus_Rift, (accessed in August 4, 2013).

Visual Cryptography Applications

Bernd Borchert and Klaus Reinhardt
Universität Tübingen, Tübingen, Germany

Abstract
Visual cryptography allows visual information to be encrypted in such a way that decryption can be done via sight reading. Introduced by Naor and Shamir, in 1994, a version of the technique from Naor and Pinkas is introduced. This version can be used in several applications, for example, to protect online transactions against manipulation and to verify the correctness of the outcome of an election.

Keywords: Cryptography; Secret image; Security; Visual cryptography.

INTRODUCTION

Naor and Pinkas in their seminal paper[13] suggested to use visual cryptography in a transparency-on-screen version. Their main purpose was authentication, in the sense that an online server is able to authenticate itself to a user sitting in front of the screen. Implicitly, this already suggests the following application of visual cryptography to the problem of manipulation of online transactions, like online money transfers, by trojans:

Main Method. In order to secure online transactions, like online money transfers, the user gets a numbered set of transparencies, each with a visual cryptography pattern printed on it, from the transaction server. Now the user is able to command online transactions in a secure way, see Fig. 1 as follows. He fills out an online form containing the data for the intended transaction; in the case of a money transfer this would be the account number and bank number of the destination bank account and the amount of the money. This transaction data is submitted via Internet to the server. The server does not execute the transaction immediately because in that case it would be an easy task for a man-in-the-middle to manipulate the transaction: the man-in-the-middle would just send his manipulated transaction to the server. In order to prevent such a manipulation, the server sends a visual message containing the transaction data to the user's screen — but of course this image is not sent openly but instead it is encoded via visual cryptography: if the user puts the transparency with a certain number on top of the encoded image on the screen he can see the message contained within the image, i.e., the transaction data. Note that the image on the screen is random to a man-in-the-Middle, as this a guaranteed by visual cryptography. The number of the transparency requested to be used is shown by the server on the user's screen together with the secret image. In order to finally confirm the transaction the user types a transaction number (TAN), which is additionally shown on the secret image message from the server, into a form on the screen and submits this TAN to the server. When the server receives the right TAN it executes the transaction, otherwise not.

Why does this method protect the transaction from being manipulated by a man-in-the-middle (which may be, for example, a trojan sitting on the user's PC)? Because a man-in-the-middle does not know the transparencies the user got from the server, the man-in-the-middle is not able to manipulate the image message sent from the server to the user. In other words, the user will see the transaction that is planned to be executed by the server and will only confirm such a transaction—a clandestine manipulation of the transaction by the trojan is impossible.

In the original purpose of secret sharing in Ref. [14], the order of the slides was not relevant. Later work[15] showed that a better contrast can be achieved with colors if the first slide can have non-transparent colors. Furthermore, practical applications will work in the way that the (first) slide is sent first from Alice to Bob over a secure channel (i.e., by surface mail) and used later as a key to decrypt an image received over an insecure channel.

We regard visual cryptography as a special case of the Cardano grille, which works on pixels instead of letters. In both cases we can describe the slide (grille) as a 2-dimensional array over $\{0, 1\}$, where 0 stands for "transparent" and 1 for "black." We describe the encrypted image as a 2-dimensional array over $\Sigma = \{0, 1\}$ or $\Sigma = \{0, 1, red, green,...\}$ or $\Sigma = \{0, a, b, c, d,...\}$, where 0 stands for "white" and 1 for "black." Colors are used for pixel-oriented applications and in case the areas are big enough, any alphabet of symbols can be used, which the receiver of the image can read through a transparent area.

A compromise between pixel- and symbol-orientation is the segment-based method described in Ref. [2], which

Encyclopedia of Image Processing, First Edition
DOI: 10.1201/9781351110273-140000469
Copyright © 2018 by Taylor & Francis. All rights reserved.

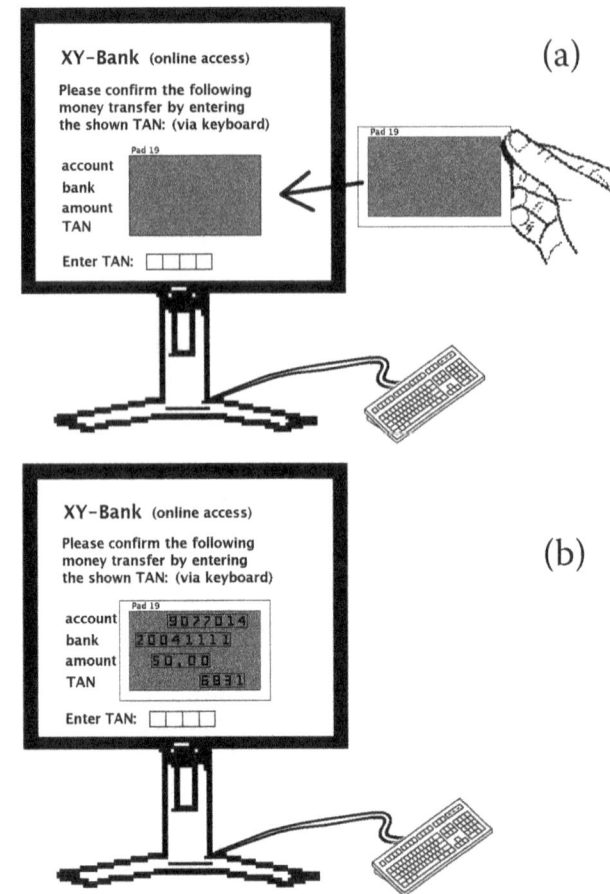

Fig. 1 (a) The bank sends the information to be confirmed in an encrypted image to the user's computer and (b) the user is able read this information using the transparency he got from the bank

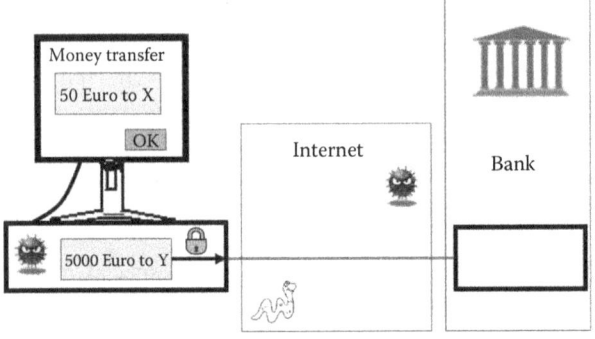

Fig. 2 A man-in-the-middle manipulation attack by a trojan on an online money transfer

works as demonstrated in Fig. 5. It is applicable whenever the message consists of symbols that can be represented by a segment code, for example the 10 digits by the well-known 7-segment code. The encryption method is basically the same: Instead of pixels, longish and larger segments are encoded via two possible parallel positions.

Outline: Based on the above idea in Ref. [13], we describe techniques in Section 12.2 where the user can confirm a transaction as shown in Fig. 1. Section 12.3 describes similar techniques that allow the user who received a slide from an account provider to securely enter a PIN or confirm transactions. The multiple use of a slide would be an economical and ecological asset and improve the convenience of the user who could, for example, leave the slide adjusted to the screen, but leads to security problems addressed in Ref. [13] and in Section 12.4. A further generalization concerning the slide leads to refractional (optical) cryptography, which is described in Section 12.5. Technical problems are discussed in Section 12.6. Section 12.7 describes Chaum's application of visual cryptography in elections. It verifies that a ballot was counted without giving the voter the possibility to show others what she voted for.

TROJAN-SECURE CONFIRMATION OF TRANSACTIONS

Naor and Pinkas state the application to online transactions implicitly in their conference paper.[13] Explicitly it is stated in Appendix A of their full paper, which can be found on their homepages. Klein, in 2005, describes this Naor/Pinkas "transparency onto screen" idea as a main application of visual cryptography in Ref. [11]. Hogl independently re-invents visual cryptography and the Naor/Pinkas idea in the patent application.[9] Greveler refines some of the aspects of the Naor/Pinkas idea in Ref. [8]. Borchert and Reinhardt[3] discuss variants of the Naor/Pinkas idea.

We assume the computer can be infected with a trojan, which is able to eavesdrop and manipulate all input- and output information. Even after a secure login, a trojan (Malice) can manipulate a transaction, which is confirmed with the TAN or iTAN method in the following way as in Fig. 2: Bob wants to instruct his banker Alice to transfer 50 dollars to X, but Malice chances this to "transfer 5000 dollars to Y." When Alice requests a confirmation by sending the message "To transfer 5000 dollars to Y enter the TAN No. 37," Malice changes it to "To transfer 50 dollars to X enter the TAN No. 37," and Bob will cluelessly enter the TAN No. 37.

To prevent this kind of attack, the authors proposed in Ref. [3] methods as in Figs. 1 and 4, with the idea that Eve is not able to produce a forged encrypted image of the original transaction. Here again, the image of message is shifted by a random offset in the x and y-direction to prevent Eve from concluding back to the slide.

However, the method in Fig. 1 has the disadvantage that the user still needs TANs and that Eve might place the image of the original transaction in an unencrypted way on the screen, which will have the same appearance with the slide as if it would be the encrypted image of the original transaction. Thus, Bob might get fooled, if he did not check that there should be a "gray" pattern without any information before he places the slide.

This can be improved using the method in Fig. 4; it makes sure that the user is able to see the black balls, which is only possible if (at least most of) the encrypted image from Alice was sent unchanged to Bob. Since Malice does not know the position of the parts of the transactions on the picture, any attempt to alter a part of the transaction would most likely lead to an incorrect image, which can easily be detected by Bob. A similar version using the segment-based method in Ref. [2] is shown in Fig. 6 and another similar version for mobile phones is shown in Fig. 3. Two versions using Cardano Cryptography, where the user has to verify the transaction consiting of an account number by following the blue path, are shown in Fig. 7 and implemented in Ref. [1]. In the 1-factor confirmation case, the user has to confirm entering the numbers along the red path; in the 2-factor confirmation case, the user has to confirm by typing his PIN on the keyboard below according to the permutation of the digits shown within the red-edged holes.

TROJAN-SECURE AUTHENTICATION USING A PIN

The purpose of this section is to apply visual cryptography in a way such that the user can enter a password to the server in a way such that the trojan is not able to get the password.

As shown in Fig. 8, the trojan on the computer is not able to see the permutation chosen by Alice on the keys, which was randomly chosen by the server. Thus, the mouse-clicks of the user cannot be interpreted by the trojan. This method can be generalized to any alphabet and allows Bob to send short messages to Alice in a secure way.

Note here that this method becomes insecure if the message contains multiple occurrences of the same symbol, a PIN should thus be chosen without repetitions. In case Bob wants to send messages of length l that may contain repetitions, this could still be accomplished in a secure way by extending the alphabet to $\Sigma \cup \{r_1, r_2, ..., r_{l-1}\}$, where r_i indicates the repetition of the symbol at position i. In this way, for example, the message "*messages*" could be submitted as "$mesr_3 agr_2 r_4$" containing no repetitions in the extended alphabet.

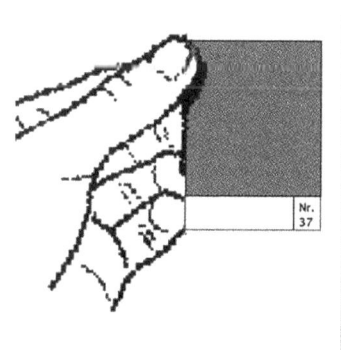

Fig. 3 The main method is also applicable to mobile banking

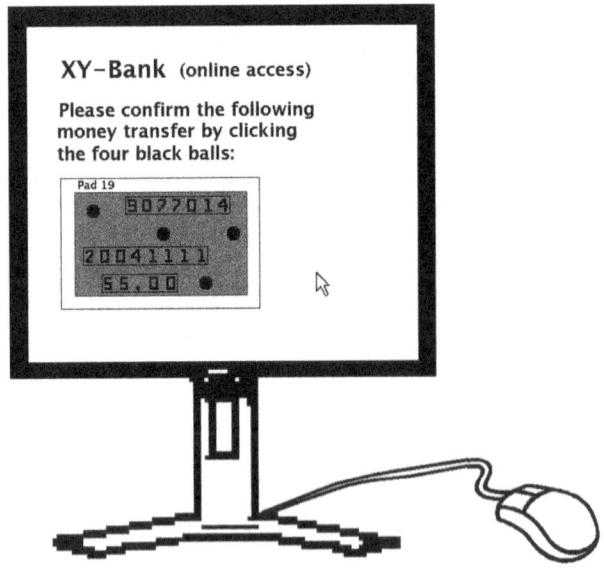

Fig. 4 For confirmation, the user has to click the black balls placed between parts of the transaction data

In order to achieve 2-factor security for transactions, we combine the PIN method with the confirmation method of Section 12.2 as described in Fig. 9.

Furthermore, to prevent the attack using the original transaction in an unencrypted way on the screen as in Section 12.2, we use the refined method of Fig. 9.

SECURITY VERSUS MULTIPLE USE

To achieve information theoretic security for a single use, we can divide the array into clusters of c pixels (resp. areas), where c is the size of the alphabet of the encrypted image. Only one pixel in each cluster has a 0 on the slide. To encrypt a pixel $p \in \Sigma$, place p at the position of the cluster with the 0 on the slide and fill the rest of the cluster with a random permutation of $\Sigma \setminus \{p\}$. Since each pixel-value in Σ occurs in each cluster of the encrypted image, each image is possible from the viewpoint of an a evesdropper.

In the model of a known plaintext attack, we assume that the a evesdropper Eve may receive the secret image later, then she can find out which position in each cluster has the o on the slide and thus the slide cannot be used securely a second time.

Known plaintext is relevant for authentication as considered in Ref. [13] as well as for confirmation as considered in Section 12.2; in both cases Bob has to be convinced that the message was sent by Alice. The problem of multiple usability is solved in Ref. [13] by dividing the slide in distinct areas, where each has to be big enough to contain the complete message; here we use an approach with a different distribution. To achieve information theoretic security use a slide n times, we propose the following two possibilities:

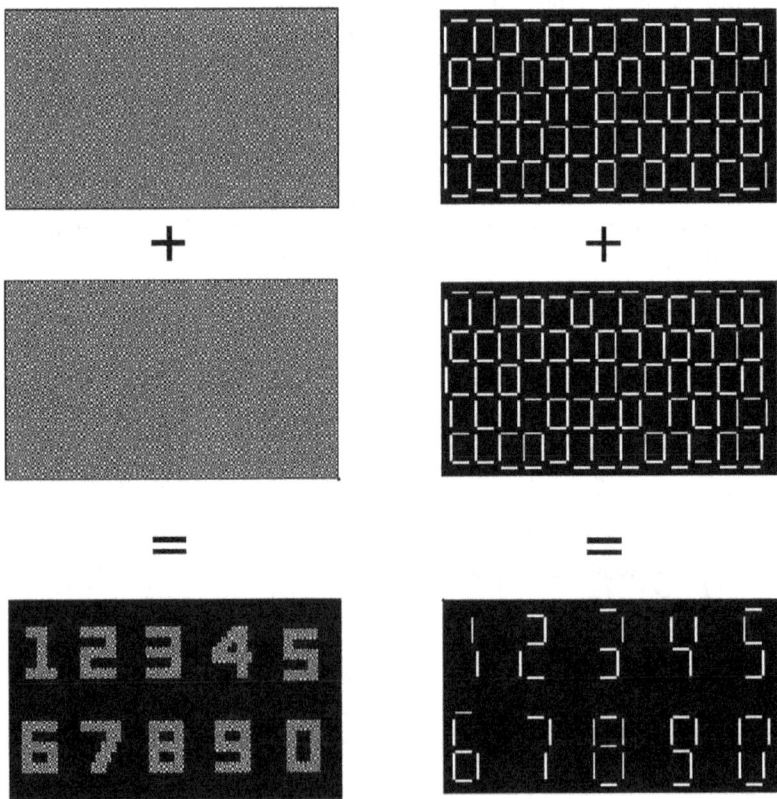

Fig. 5 Pixel-based (left) versus segment-based (right) visual cryptography

Visual Cryptography Applications

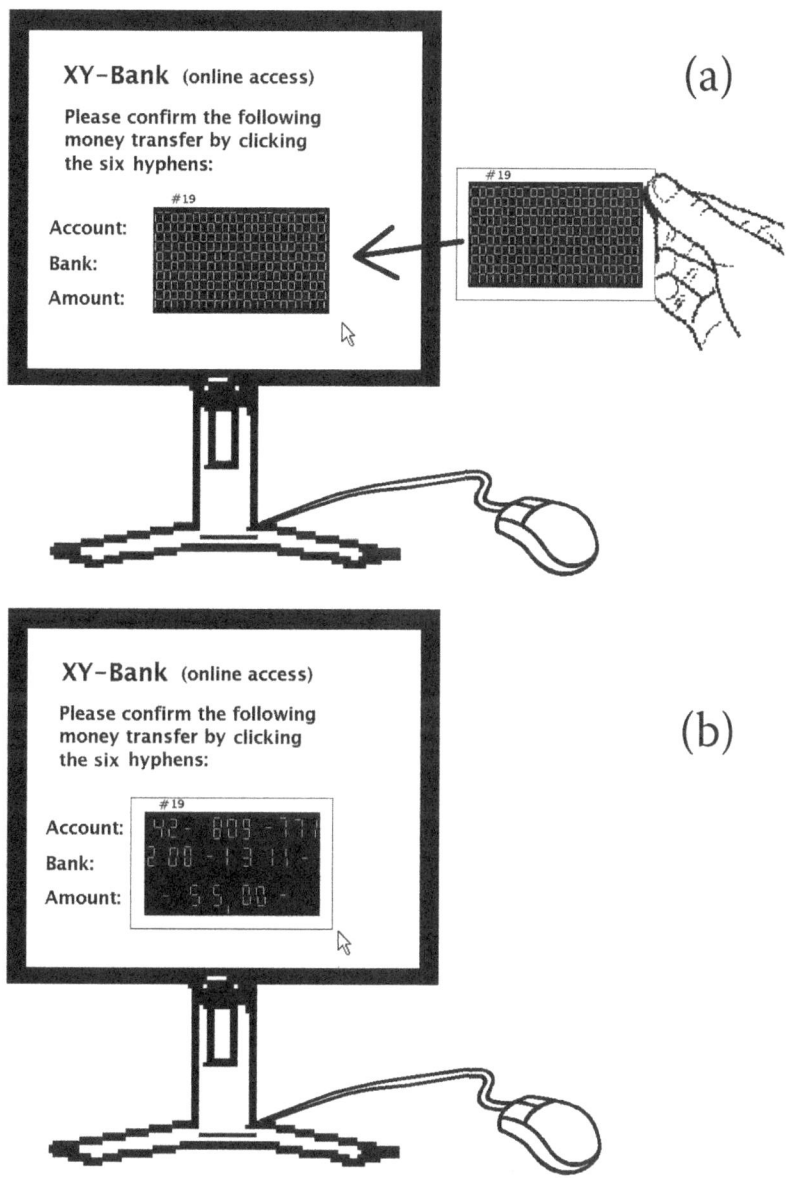

Fig. 6 The main method using segment-based visual cryptography in (a) and (b)

1. For one pixel use a cluster of nc pixels divided into n subclusters of c pixels in which each subcluster has a 0 on the slide. To encrypt the i-th image, we use only the i-th subcluster as above and fill the rest with 1. This leads to a contrast of $\frac{1}{nc}$.
2. For one pixel use a cluster of c^n pixels, which we address as an n-dimensional array (but arrange 2-dimensionally). Only one pixel in each cluster has a 0 on the slide. To encrypt a pixel $p \in \Sigma$ in the i-th image, place p at each position of the cluster, which has the same i-th coordinate as the 0 on the slide then choose a random permutation of $\Sigma \setminus \{p\}$ and fill the rest of the cluster in a way such that pixels with the same i-th coordinate get the same color. This leads to a contrast of $\frac{1}{c^n}$.

In both cases each combination of images is consistent with any combination of encrypted images. In the first case, the contrast can be improved only to $\frac{1}{n}$ using refraction, whereas it can be improved to 1 using refraction in the second case.

In the case of an unknown plaintext attack, Eve is still able to obtain the secret images if the slide was used too often and if she can anticipate patterns in the picture. Let us consider the simple case $\Sigma = \{0, 1\}$, $c = 2$ and the slide was used twice. Then Eve can XOR both encrypted images resulting in an image that is 0 at pixels where both original images coincide and 1 at pixels where both original images differ, thus she can see the difference of the original pictures as shown in Ref. [12] page 35 and Fig. 11. Now assume the original images depict messages using a Font

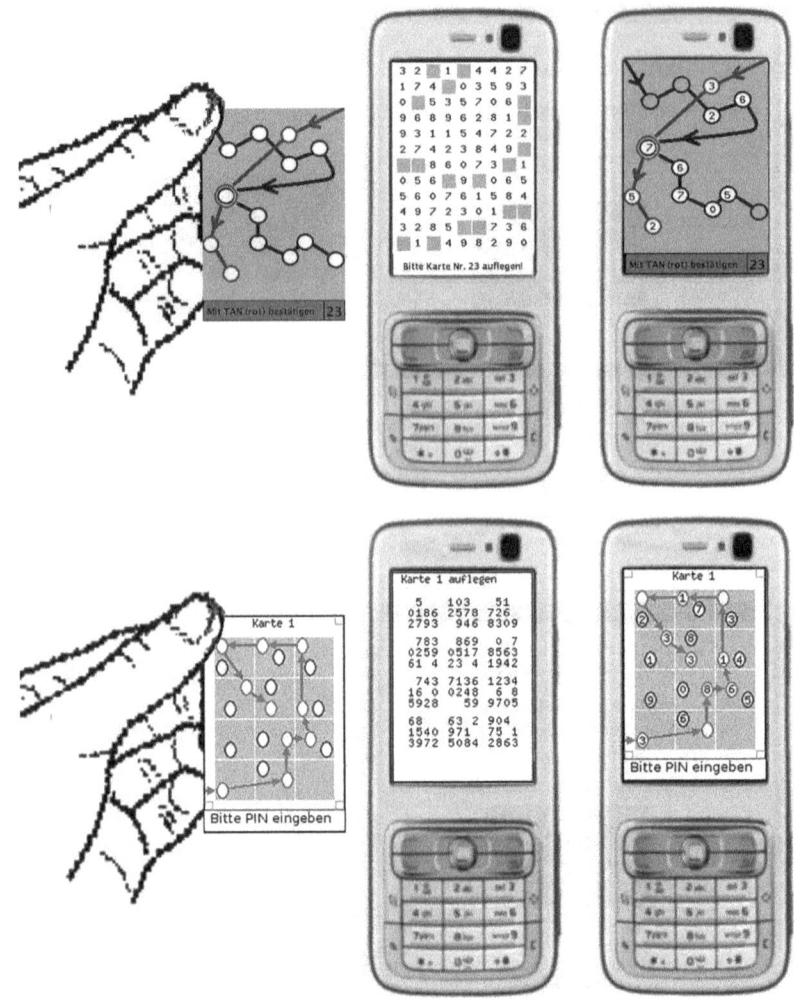

Fig. 7 Cardano cryptography: above a 1-factor confirmation (user types 3752), below a 2-factor confirmation (for example, in case his PIN is 1234, the user types in 4136)

F consisting of $f = |F|$ small symbols (symbol-pictures) having q pixels.

Assume furthermore the symbols are placed on fixed positions, then Eve can identify the pairs of symbols on corresponding positions as long as $f^2/2 \ll 2^q$. Then Eve can use, for example, the redundancy of natural languages to decipher the text. One measure to complicate this attack is to shift the picture by a random number of pixels to the left or the right. Then Eve will have to try out the position of the first symbol and consider $x \cdot y \cdot f^2/2$ combinations, where x and y are the differences of the shifts but smaller than the width and the height of the symbols. Further complications can be caused by filling the space around the text by partially random patterns, which Bob can easily distinguish from the symbols, but Eve will have to start analyzing parts in the middle of the image. Here she can only assume that about a quarter of the surface of the letters overlap, which means this method of attacking can be expected to be successfull if $x \cdot y \cdot f^2/2 \ll 2^{q/4}$.

Furthermore, the partially random patterns can compensate the statistical imbalance of the correlation of neighboring points. For example, Eve could look at pairs of pixels, where one is in some small distance above the other. If both are on a position having a 0 on the slide, then, given many encrypted images of texts, Eve could detect that they have the same color with a higher probability than other pairs. But the partially random patterns are made in a way such that this probability decreases in the overall image.

Let us now turn to the slides on the previous page: If the slide was used more than n times, then Eve can try the following attack: She chooses an area of $q = x \cdot y$ pixels somewhere in the image, then she tries each combination of positions of the 0 in each of the corresponding q clusters on the slide ($(2^n)^q$ possibilities) and checks if it is consistent with each encrypted image in the sense that there are 4 symbols in F overlapping the area. This takes $4 \cdot x \cdot y \cdot f$ steps. And thus $2^{n \cdot q} \cdot 4 \cdot x \cdot y \cdot f$ steps in total.

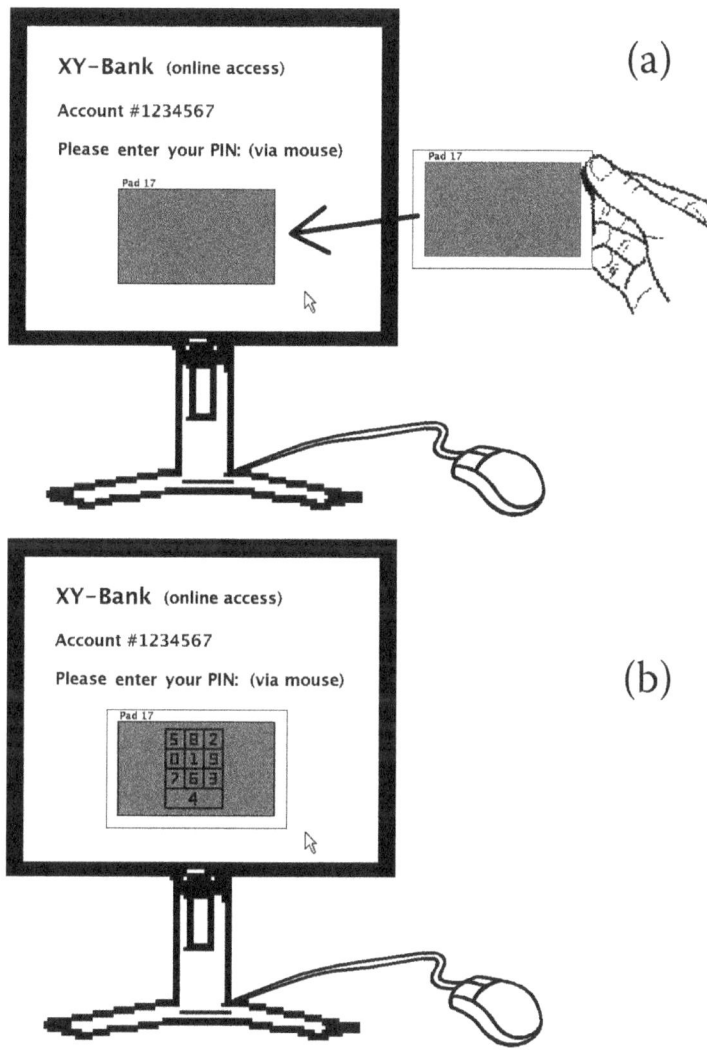

Fig. 8 (a) To log in, the server sends an encrypted image of a permutated keyboard, which the user can only read after placing the slide over it. (b) The user enters the PIN by clicking at the positions according to their order in the PIN

The number of possible subimages of a possible original image, where four symbols overlap, is approximately $(x \cdot y \cdot f)^4$. This means each observed encrypted image can help Eve to exclude a sufficient number of possible choices if $2^q \gg (x \cdot y \cdot f)^4$. We therefore estimate the number of steps for Eve as $\gg (x \cdot y \cdot f)^{4 \cdot n} \cdot 4 \cdot x \cdot y \cdot f = 4 \cdot (x \cdot y \cdot f)^{4 \cdot n + 1}$. Now assume we use $x = y = 10$ and a huge font Γ with many possibilities to depict a symbol that leads to $f = 1000$ and roughly $10^{20 \cdot n}$ steps for Eve. Considering many obvious and also less obvious improvements of the algorithm for Eve, we believe that the attack is still too expensive for Eve for $n = 3$.

USING REFRACTION

In Ref. [4] we generalize the slide from a 2-dimensional array over $\Sigma = \{0, 1\}$ to a 2-dimensional array over $\Sigma \subset \{1\} \cup \mathbb{R} \times \mathbb{R}$ with the idea that each pixel on the slide can either be black or contains a prism (x, y) that refracts the light from a region on the encrypted image or, from the perspective of the user as shown in Fig. 12, refracts the view to a region that is shifted by (x, y) from the pixel, which is directly behind the pixel on the slide. For example $(0, 0)$ would correspond to the 0 in the case of usual Visual Cryptography just showing the pixel directly behind. One possible application would be to use clusters of $2 \cdot 2 = 4$ pixels for each pixel of the original and randomly choose one of the 4 pixels to be visible. For example would direct the view to the upper right pixel on the encrypted image. This corresponds to construction 2 in Section 12.4. Using the slide two times is information theoretically secure and the contrast is 1.

$$(1, 0) \quad (0, 0)$$
$$(1, -1) \quad (0, -1)$$

If we use lenses or fragments of lenses instead of prisms, the view can be focused on a point inside a pixel. This has the

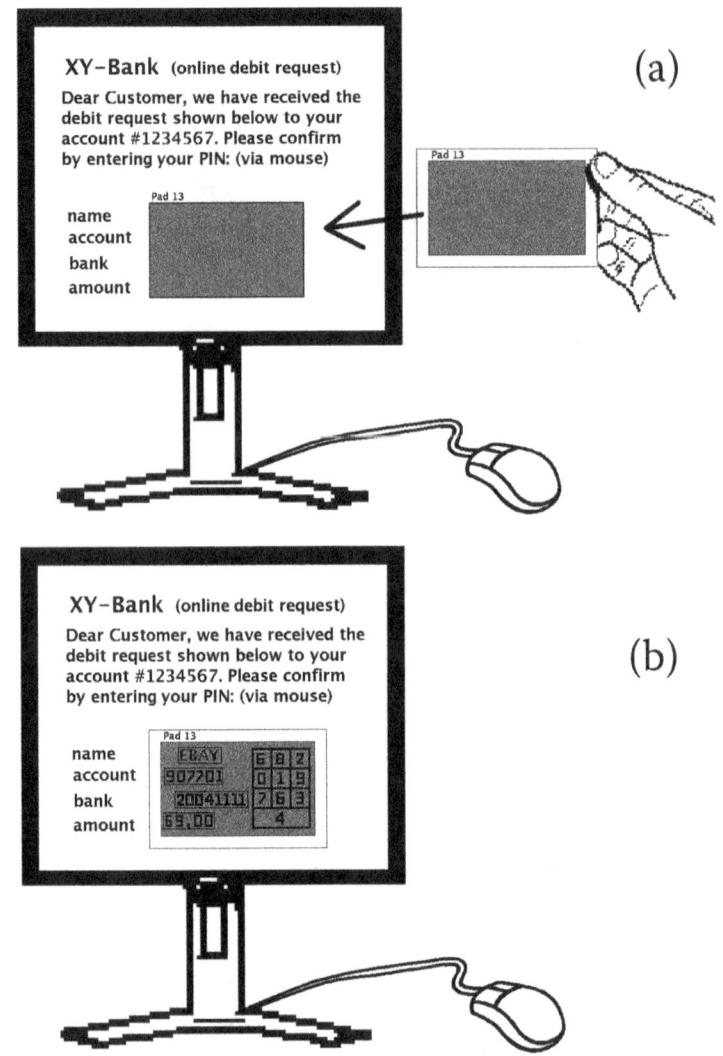

Fig. 9 For confirmation, the user has to click his PIN using a permutation of digits on the right side in (a) and (b)

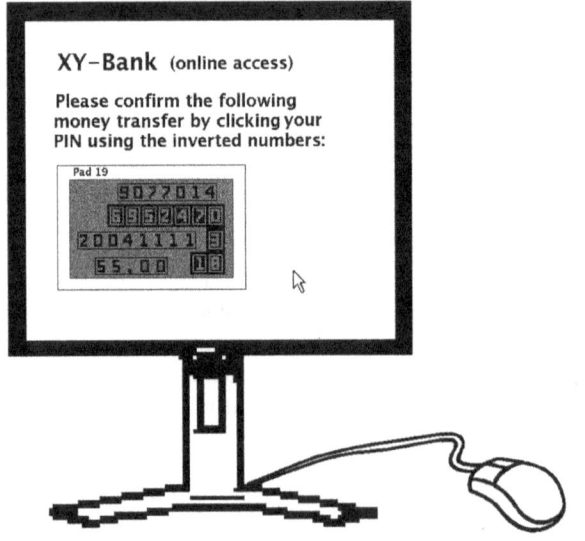

Fig. 10 For confirmation, the user has to click his PIN using inverted numbers

advantage that the positioning of the slide allows an error of up to half of a pixel. An example is shown in Fig. 13. While producing fragments of lenses and prisms on the slide might require expensive special machines, it will be much cheaper to produce complete lenses. Since the lenses do not need to have a perfect shape, it would be sufficient to place drops of a transparent liquid that becomes hard on the side. This can be done by either using a modified ink-printer or by spraying the liquid using physical randomness (and Alice can scan it before sending it to Bob). The use of such a slide is shown in Figs. 14 and 15 shows an optical generalization of Cardano cryptography. The advantage to Cardano cryptography is that the letters are magnified and can be read in a more natural ordering. The advantage to pixel-based visual cryptography is that much less precision is needed to position the slide. The disadvantage is that, in order to achieve a sufficient level of security, more than four choices (like in Fig. 15) would be required. A repeated use would be too insecure as the slide contains only little information. Furthermore, a bigger distance of slide and screen is required.

Visual Cryptography Applications

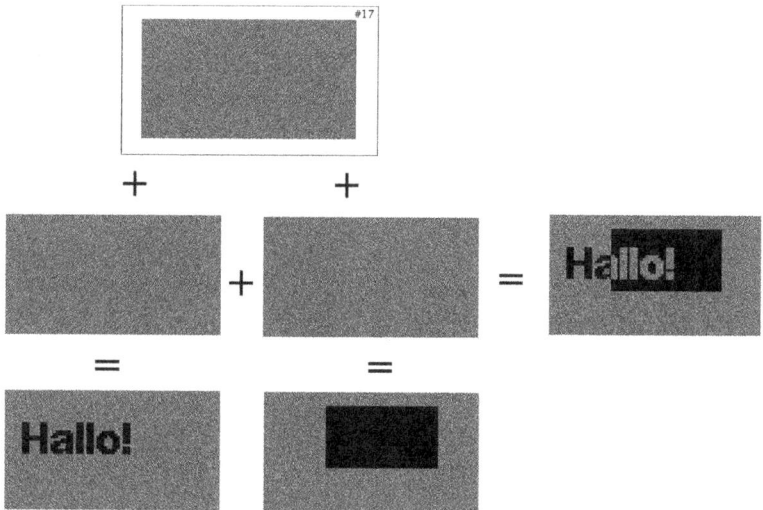

Fig. 11 Superimposing two encrypted images for the same key-slide shows the difference of the original picture

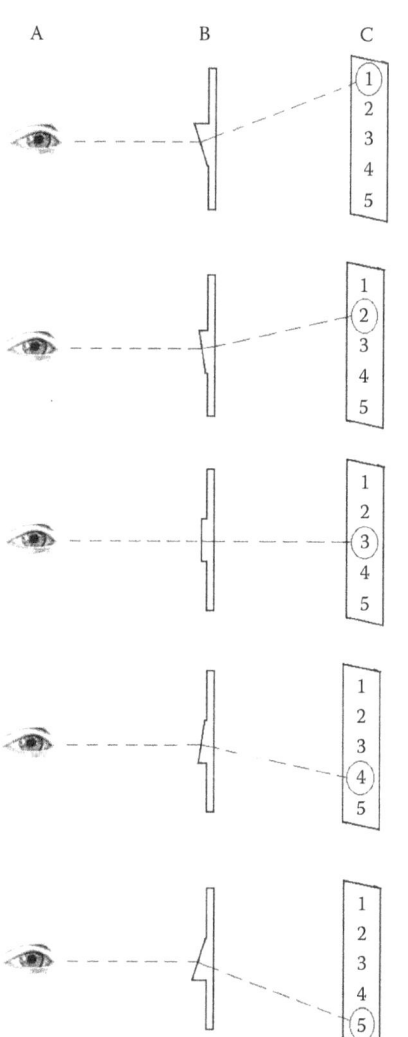

Fig. 12 This example shows how the view from the observer (A) through prisms (B) is directed to areas 1,2,... or 5 on the encrypted image (C); the deviation depends on the slope of the prism

TECHNICAL PROBLEMS CONCERNING ADJUSTMENT AND PARALLAXES

A disadvantage of pixel-based visual cryptography to Cardano cryptography is that the slide has to be placed at an exact position. We propose to position the slide at the (for example, left lower) corner of the screen making use of its frame. Then use the mouse to position and stretch the encrypted image on the screen accordingsly. The parallaxes is the shift of the pixel on the screen that can be seen through one point of the slide, which is caused by looking at it from a certain angle that differs from the right angle because of central projection and because the viewer has two eyes at different positions. This is estimated in Ref. [10] to be 0.25 mm in the case of a usual TFT screen, less for displays of new mobile phones, but might be more if we require a higher distance of slide and screen so that the use of refraction can take effect (depending on the size of the lenses). Slight misplacements of the slide cause effects like a bad contrast or even inverting the picture as described in Ref. [10] where the author also proposes methods to use interference effects at the frame of the slide as an aid for adjustment.

VOTING WITH A RECEIPT BASED ON VISUAL CRYPTOGRAPHY

The purpose is to give a voter a receipt, which allows her or anyone else to verify that her vote was counted in the final result. The difficulty comes from the requirement that the voter should not be able to prove to anyone else, what her vote was since this would make abuses such as vote selling possible.

The main procedural method in Ref. [6] lets the voter enter her ballot on a touch screen and then the voting machine produces two slides laminated together that show the ballot

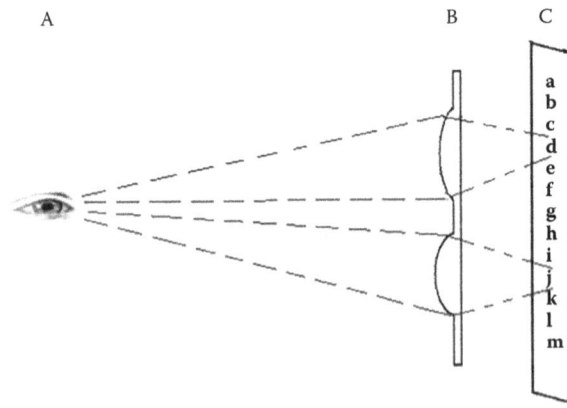

Fig. 13 Some parts of the encrypted image (C) is magnified for the observer (A), while other parts are hidden. For example, *d* and *j* are in the focus while *b, c, e, f, h, i, k*, and *l* are hidden

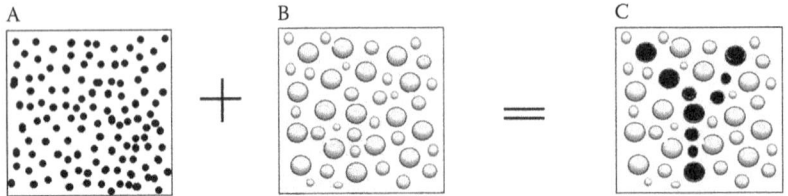

Fig. 14 Lenses are placed randomly on the slide (b). This can be done by spraying a transparent liquid that becomes hard on the side. The area in the focus of the lenses in the encrypted image (a) is colored in the color of the original image at this region. The rest of (a) is filled such that colors in (a) are equally distributed so that the original image can not be obtained from (a) alone but only together with the slide (b)

Fig. 15 Each area of the slide (b) has fragments of lenses, which direct the view (c) in a magnifying manor to one of the symbols on the encrypted text (a)

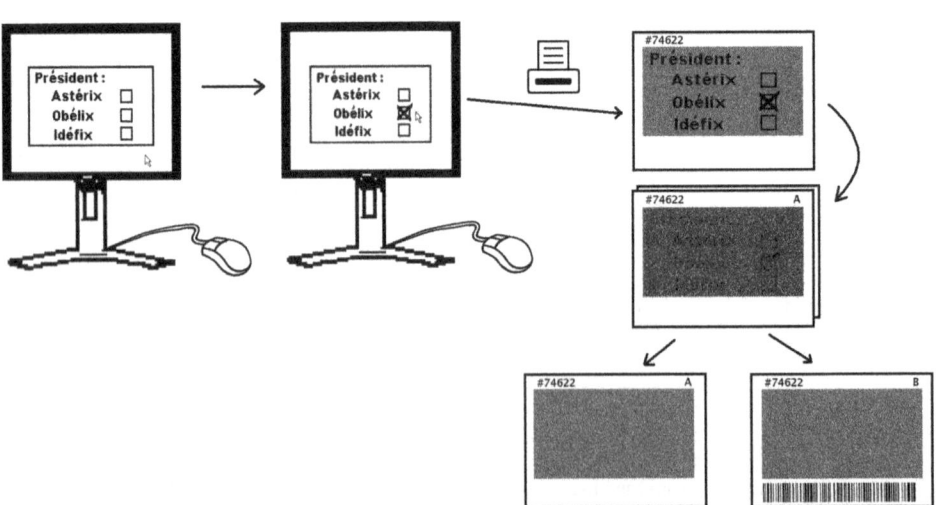

Fig. 16 The voter enters the vote, verifies the image, and separates the slides

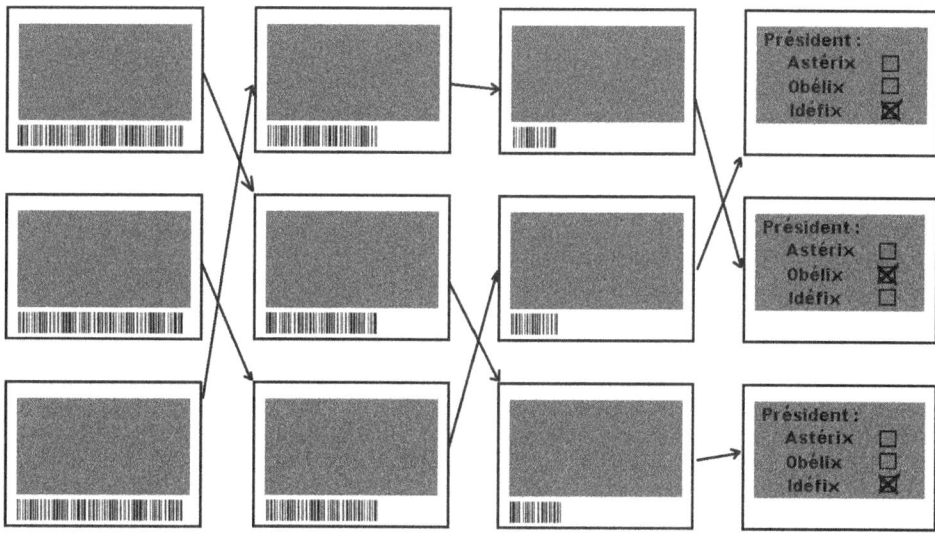

Fig. 17 Each trustee strips one layer of the doll (represented by the barcode) and uses it to modify the image. The order is randomly permutated

image with visual cryptography to the voter only as shown in Fig. 16. Then the voter separates the two slides and goes to the poll worker. Here one of the slides is destroyed and the other one is scanned and uploaded to the official election website. Furthermore, the voter keeps her slide as her receipt.

For confidentiality, the voter has to trust that the voting machine has no memory. But how can the results be computed only from the scanned slides? To make this possible, the slide cannot be completely random (as usual in visual cryptography) but is produced by a pseudorandom generator from the cryptographic version of a nested doll that contains the necessary information to reproduce the ballot image inside. This means that confidentiality is not information theoretically secure but only computationally secure under the usual cryptographic assumptions.

The cryptographic version of a nested doll is produced by the voting machine from the serial number of the ballot by successively encrypting with the public keys of a sequence of trustees and printed on the slide as a barcode. Only all trustees together would be able to compute the ballot image using their secret keys. The result of the election is computed by a sequence of mix-operations as described in Ref. [5]: The first trustee gets a batch of scanned slides as input. For each vote, he removes the first layer of the doll and modifies the encrypted image using the removed layer of the doll. Then he uploads a batch containing a random permutation of the results to the official election website. All other trustees do the same with the batch from their predecessor (see Fig. 17). In the batch produced by the last trustee, the dolls were used up and the image became the original ballot image, which can be seen (and thus counted) by anyone.

To verify that the trustees worked properly, a public random choice is used to audit the trustees, which have to reveal half of the connections to the following batch by publishing the removed layer of the doll, which again allows anyone to verify these connections. But the choice was made in a way such that there will be no complete path visible leading from a scanned slide to its ballot image. In Ref. [7], a refinement of the choice is described, which ensures that nothing can be learned about the ballots of even groups of voters.

But how can the voter be sure that she was not cheated by the voting machine printing a false second (and later destroyed) slide letting her see her ballot image but producing a different one after passing all the mixes? The idea is that after the voting machine printed the visual cryptography part on the laminated slides, the voter has still the choice to take the upper or lower slide as receipt and tell the choice to the voting machine before it continues by printing the barcode of the doll to it; in this way the voting machine would for each voter only have a 50% chance of cheating without being caught.

CONCLUSION

Visual cryptography is a fascinating technique and very intuitive to the user. However, it is surprising that within the last 15 years since its invention by Naor and Shamir only a few suggestions have been made to apply it to practical problems. In this paper we presented Naor and Pinkas technique to use visual cryptography in order to protect online transactions against manipulation and Chaum's idea to apply it to verify the correctness of the outcome of an election. Because of difficulties like adjustment, multiple use, and costs of special equipment, these suggestions did not lead to applications that are used for serious purposes. But in the future, further developments of the ideas presented in this paper, as well as new ideas, could spread practical applications of visual cryptography.

BIBLIOGRAPHY

1. Beschke, S. *Implementierung der Cardano-TAN (cTAN)*, Studienarbeit, 2009.
2. Borchert, B. Segment-based visual cryptography. Technical Report WSI–2007–04; Universität Tübingen (Germany), Wilhelm-Schickard-Institut für Informatik, 2007.
3. Borchert, B.; Reinhardt, K. *Abhör- und manipulationssichere Verschlüsselung für Online Accounts mittels Visueller Krytographie an der Bildschirmoberfläche*. Patent application DE-10–2007–018802.3 (approved 2008), 2007.
4. Borchert, B.; Reinhardt, K. *Lichtbrechungs-Kryptographie*. Patent application DE-10–2010–031 960.0, 2010.
5. Chaum, D. Untraceable electronic mail, return addresses, and digital pseudonyms. Commun. ACM **1981**, *24* (2), 84–88.
6. Chaum, D. Secret-ballot receipts: True voter-verifiable elections. IEEE Secur. Priv. **2004**, *2* (1), 38–47.
7. Gomulkiewicz, M.; Klonowski, M.; Kutylowski, M. Rapid mixing and security of chaum's visual electronic voting. In Snekkenes, E.; Gollmann, D.; Eds.; *ESORICS*, volume 2808 of Lecture Notes in Computer Science; Springer, 2003, 132–145.
8. Greveler, U. VTANs Eine Anwendung visueller Kryptographie in der Online-sicherheit. In *2. Workshop "Kryptologie in Theorie und Praxis," Bremen*, Lecture Notes in Informatics (LNT), 2007, 210–214.
9. Hogl, C. *Verfahren und System zum bertragen von Daten*. Patent application DE-10–2010–031 960.0, 2005.
10. Hunszinger, F. *Implementierung und Untersuchung eines Verfahrens zur visuellen Kryptographie*, Diplomathesis, 2010.
11. Klein, A. Eine Einführung in die visuelle Kryptographie. In *DMV Mitteilungen 1/2005*, 2005, 54–57.
12. Klein, A. *Visuelle Kryptographie*; Springer: Berlin, Heidelberg and New York, 2007.
13. Naor, M.; Pinkas, B. Visual authentication and identification. In *Lecture Notes in Computer Science*; Springer-Verlag: Berlin, 1997, 322–336.
14. Naor, M.; Shamir, A. Visual cryptography. In *EUROCRYPT*, 1994, 1–12.
15. Naor, M.; Shamir, A. Visual cryptography ii: Improving the contrast via the cover base. In *Proceedings of the International Workshop on Security Protocols*; Springer-Verlag: London, 1997, 197–202.

Wavelet Analysis

Tania Pouli
Technicolor, Rennes, France

Erik Reinhard
Technicolor Research and Innovation, Rennes, France

Douglas W. Cunningham
Brandenburg University of Technology, Cottbus, Germany

Abstract
This entry discusses methods that are used in the analysis of local structure in images—in particular, wavelets. That is, when an image's statistical structure does not vary by image location. Wavelets are introduced and its various properties are covered. Wavelets can be used to analyze structure in images and also in image compression, and denoising.

Keywords: Denoising; Image compression; Stationary image; Wavelets.

The Fourier series for periodic signals is used in many applications, including the analysis of natural images. It is the first example of a signal expansion,[1] and as shown before, it uses sines and cosines as its basis functions. A Fourier expansion provides insight into which frequencies exist in an image by means of the amplitude spectrum, as well as where they exist through the phase spectrum.

However, a Fourier expansion does not allow an analysis of local structure. This means that it is not possible to use a Fourier expansion to understand which frequencies and locations exist in a local area around a given pixel of interest. In other words, Fourier basis functions are localized in frequency, but not in space. A further disadvantage is that discontinuities in images require a large number of basis functions to be adequately represented. Figure 6.8, for instance, shows that summing 500 sinusoids is only very roughly beginning to approximate a step edge. Representations that require only a small number of basis functions to represent the image are useful in, for instance, data compression. Wavelets, which are discussed next, are one such example.

Fourier series are intended for periodic signals. Boundary effects can be minimized by applying a window to the center of the image. This strategy can be extended to create local Fourier bases, effectively centering a set of windows on a grid, and applying a Fourier transform for each window. This is known as the Gabor transform, short-time Fourier transform (STFT), or short-term Fourier transform.[2] Such an approach gives information about each grid location of the image and could therefore be used to assess local information.

Given that images are typically nonstationary, i.e., their statistical structure can vary by image location, an assessment of local features has revealed additional patterns unique to natural images and has even revealed a correlation with saliency.[3] In this entry we discuss methods that have been used in the analysis of local structure in images—in particular, wavelets.

Wavelets are an alternative decomposition of a function into sets of basis functions. As opposed to the Fourier transform, it can localize the analysis in both space and frequency, often allowing for a sparser representation.[4–6] This means that fewer basis functions are needed to represent the original function, which is, for instance, directly exploited in image compression.[7] Wavelets are amenable to statistical analysis, and due to their spatially local form they are able to reveal statistical regularities of local structure. In this section, we first outline the mathematical background to wavelets, followed by a presentation of the discoveries made by analyzing wavelet decompositions of natural images.

WAVELET TRANSFORM

An alternative to the Gabor transform is the wavelet transform, which can compute a linear expansion of a signal or image by using different scalings and shifts of a prototype wavelet. As the scale factors tend to be powers of 2—i.e., each subsequent scale is a factor of two larger—the frequencies represented at each scale halve. This means that the frequency axis is logarithmic rather than linear, as is used in Fourier and Gabor transforms.

The aforementioned wavelet itself is a basis function, which is analogous to the sines and cosines that form the bases in Fourier series. There are, however, many different choices of basis functions, each with their own trade-offs. In all cases, though, the basis function needs to be well localized and integrate to zero in order to qualify as a potential wavelet.

Wavelets can be defined in terms of a single function, if suitably parameterized. This function is normally called the *mother wavelet* and in one dimension is denoted by $\psi(x)$. The parameter a indicates scaling (dilation), whereas b indicates translation:

$$\psi_{a,b}(x) = \frac{1}{\sqrt{|a|}} \psi\left(\frac{x-b}{a}\right) \qquad (1)$$

where $(a,b) \in R^+ \times R$. Examples of specific wavelets ψ are given in later sections. We can define the Fourier transform of this wavelet as:

$$\Psi(\omega) = \int_{\mathbb{R}} \psi(x) e^{-ix\omega} dx \qquad (2)$$

Certain functions $f(x)$ can be represented by applying a wavelet transform to them. A requirement of such functions is that they are square-integrable, i.e.:

$$\int_{-\infty}^{\infty} |f(x)|^2 dx < \infty \qquad (3)$$

Square-integrable functions are said to be in $\mathbb{L}^2$. Pairs of such functions, $f(x)$ and $g(x)$, will have an inner product $\langle f,g \rangle$ defined as:

$$\langle f,g \rangle = \int_{-\infty}^{\infty} f(x) \overline{g(x)} dx \qquad (4)$$

where we note that the inner product of a function with itself is less than infinite due to the requirement of square integrability $\langle f,f \rangle < \infty$. The wavelet transform of a function $f(x) \in \mathbb{L}^2$ is given by $W_f(a,b)$:

$$W_f(a,b) = \langle f, \psi_{a,b} \rangle = \int_{-\infty}^{\infty} f(x) \overline{\psi_{a,b}(x)} dx \qquad (5)$$

showing that the wavelet itself needs to be square-integrable. Using Calderóns reproducing identity, it is possible to reconstruct the original function from the wavelet transform:

$$f(x) = C_\psi^{-1} \int_0^\infty \int_{-\infty}^\infty \frac{W_f(a,b)}{a^2} \psi_{a,b}(x) db\, da \qquad (6)$$

where

$$C_\psi = \int_0^\infty \frac{|\Psi(\omega)|^2}{\omega} d\omega \qquad (7)$$

This inverse transform requires that the admissibility condition be satisfied, which means that $C_\psi < \infty$. This implies that wavelets integrate to 0, i.e., $\int \psi(x) dx = 0$.

The wavelet transform has several interesting properties, which include shifting and scaling properties, as well as localization. Here, we assume that functions $f(x)$ and $g(x)$ have wavelet transforms $W_f(a, b)$ and $W_g(a, b)$. The shifting property states that if $g(x)=f(x-k)$, we have $W_g(a,b) = W_f(a, b-k)$. The scaling property states that if $g(x) = f(x/k)/\sqrt{k}$, then $W_g(a,b) = W_f(a/k, b/k)$. If $f(x)$ is zero everywhere, except at a single point in space x_0, i.e., $f(x) = \delta(x - x_0)$, then $W_f(a,b) = \psi((x_0 - b)/a)/\sqrt{a}$.

The wavelet transform is redundant in that a function of one variable x is transformed into a function of two variables (a, b). It is possible to select a subset of values for a and b and still have an invertible transformation. The coarsest sampling that can be chosen for a and b is called the *critical sampling*, and is given by:

$$a = 2^{-j} \qquad (8)$$

$$b = k2^{-j} \qquad (9)$$

where $k, j \in \mathbb{Z}$ are suitably chosen integers. Any coarser sampling will not allow the original function $f(x)$ to be uniquely recovered. Any finer sampling will cause redundancy to remain. Note that a can be thought of as the reciprocal of frequency. It can be shown that under certain conditions this sampling produces an orthonormal basis:

$$\psi_{j,k}(x) = 2^{j/2} \psi(2^j x - k) \quad j,k \in \mathbb{Z} \qquad (10)$$

where dilation is governed by parameter j and translation is given by k.

To simplify the subsequent discussion, we will now give an example of a wavelet, which is also the oldest and arguably the simplest wavelet transform. It is known as the *Haar wavelet*:[8]

$$\psi(x) = \begin{cases} 1 & x \in [0, 1/2) \\ -1 & x \in [1/2, 1) \\ 0 & \text{otherwise} \end{cases} \qquad (11)$$

We see that this function is 0 everywhere, except between 0 and 1, i.e., it has local support. The set of functions $\{\psi_{j,k} \mid j, k \in \mathbb{Z}\}$ forms an orthogonal basis in $\mathbb{L}^2$, i.e., the space of all square-integrable functions.

MULTIRESOLUTION ANALYSIS

For the choice of a and b outlined above, wavelet transformations can be analyzed as follows.[9,10] We would like to efficiently represent a function $f(x)$ in some hierarchical fashion. The space of functions that we are interested in is the set of square-integrable functions, i.e., $f(x) \in \mathbb{L}^2$. We could impose a hierarchy on this space, creating nested subspaces:

$$V_0 \subset V_1 \subset V_2 \subset \ldots \subset \mathbb{L}^2 \qquad (12)$$

Each of these spaces contain functions, starting with the smallest subspace V_0, the *reference space*, that contains only one function family. Each subsequent subspace then extends this space. The function in V_0 is known as the generating function or *father wavelet*, and is denoted by $\phi(x)$. Having a generating function allows the number of nested subspaces to be finite. In the case of the Haar wavelet, the generating function is given by:

$$\phi(x) = \begin{cases} 1 & x \in [0,1) \\ 0 & \text{otherwise} \end{cases} \quad (13)$$

This is the box function, defined to be 1 between $x=0$ and $x=1$. It is also known as the *characteristic function* or the *indicator function*. It is possible to construct an orthonormal basis for V_0 by introducing translation:

$$\phi_{0,k}(x) = \phi(x-k) \quad k \in \mathbb{Z} \quad (14)$$

Here we have introduced the subscript 0 to indicate that this family of functions forms an orthonormal basis for the reference space V_0. With this family of functions we can approximate any function $f(x)$ with step functions which are defined at unit intervals by appropriately summing these basis functions. In particular, we can compute scaling coefficients $c(n)$ from the characteristic function, and wavelet coefficients from $d(j,n)$ from the wavelet function $\psi(x)$ for a signal $f(x)$:

$$c(n) = \int_{-infty}^{\infty} f(x)\phi(x-n)dt \quad (15)$$

$$d(j,n) = 2^{j/2} \int_{-\infty}^{\infty} f(x)\psi(2^j t - n)dt \quad (16)$$

An example for $c(n)$ is shown in Fig. 1.

We cannot approximate any functions that step at places other than at integer positions within reference space V_0. However, we could augment V_0 with some additional functions that create a new subspace V_1. For instance, we could introduce a new subspace W_0 such that:

$$V_0 \oplus W_0 = V_1 \quad (17)$$

This means that W_0 is the complement of V_0 to V_1. To extend V_0 in a meaningful way, we could use translated versions of $\psi(x)$ as defined in (10) to form subspace W_0. Note that this function also has local support in that it is nonzero between 0 and 1. However, it makes a step from +1 to −1 at $x=1/2$. It is therefore not a function that is already in V_0 and is orthogonal to any of the functions $\phi_{0,k}(x)$ in V_0. The translated versions of $\psi(x)$ are given by:

$$\psi_{0,k}(x) = \psi(x-k) \quad k \in \mathbb{Z} \quad (18)$$

The functions $f(x)$ that can be represented in V_1 are now all step functions that step at half-intervals. By going from V_0 to V_1 we have therefore refined our representation, admitting more detailed functions. An example of a function that

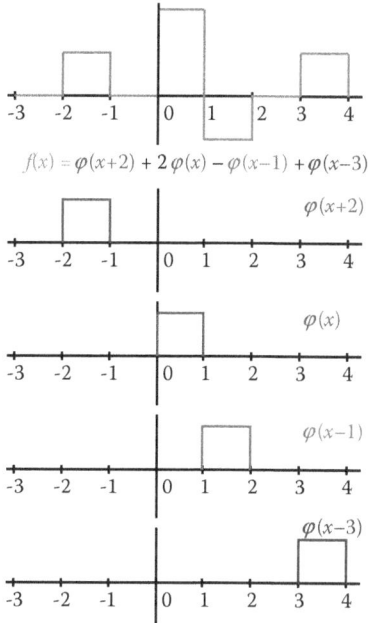

Fig. 1 A stepped function $f(x)$ with steps at integer locations can be approximated by summing instances of basis functions $\phi_{0,k}$ in reference space V_0

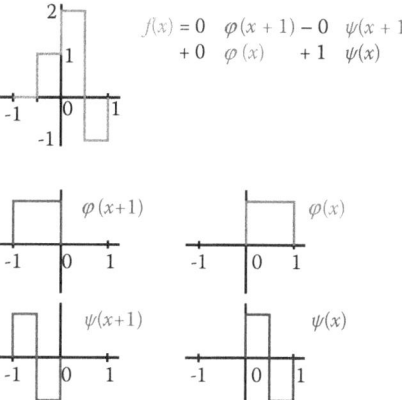

Fig. 2 A stepped function $f(x)$ with steps at half-integer locations can be approximated by summing instances of basis functions $\phi_{1,k}$ in reference space V_1

can be represented by the basis functions in $V_1 = V_0 \oplus W_0$ is given in Fig. 2.

We note that the principle of extending a subspace V_0 by W_0 to obtain subspace V_1 holds for all other subspaces V_j, i.e., in general we have $V_{j+1} = V_j \oplus W_j$. This also implies that by substitution V_j can be written as $V_0 \oplus W_0 \oplus W_1 \oplus \ldots \oplus W_j$. As a result, to represent a signal we only need the father wavelets $\phi_{0,k}(x)$ from reference space V_0 in addition to wavelet functions $\psi_{j,k}(x)$ from all subspaces V_j, although it is certainly possible (and often done) to construct a wavelet decomposition by means of a sequence of scaling functions $\phi_{j,k}(x)$ combined with a sequence of wavelet functions $\psi_{j,k}(x)$.

To extend the subspace V_1 to V_2 and beyond, the translation of functions by k is not sufficient. We also need to scale our basis functions, an operation known as *dilation*, to be able to represent all functions that step at intervals spaced by 2^{-j}. The amount by which each function is scaled depends on which subspace they are defined in. So, for subspace V_j and W_j, our basis functions $\phi_{j,k}(x)$ and $\psi_{j,k}(x)$ are written:

$$\phi_{j,k}(x) = 2^{j/2}\phi(2^j x - k) \quad j,k \in \mathbb{Z} \tag{19}$$

$$\psi_{j,k}(x) = 2^{j/2}\psi(2^j x - k) \quad j,k \in \mathbb{Z} \tag{20}$$

Note that only in the reference space V_0 and W_0 no dilations occur, as $2^0 = 1$. Assuming Haar wavelets, one way to think of this decomposition is that $\phi_{0,k}(x)$ represents averages, while $\psi_{j,k}(x)$ encodes deviations from the average at some scale. Alternatively, the scaling function acts as a low-pass filter, whereas the wavelet function acts as a band-pass filter.

Dilating by 2^j means that in each subsequent level the range of frequencies considered is halved/doubled. In other words, the band-pass filters with octave bandwidths and center frequencies that are one octave apart.

If enough scales are considered, in the limit continuous functions $f(x)$ could be approximated. In other words, the union of all subspaces is *dense*. However, as seen above, Haar wavelets do this by representing the function $f(x)$ as a sequence of piecewise constant segments. As the decomposition is into discontinuous functions, Haar wavelets are good at representing edges but not so good at representing smooth functions.

SIGNAL PROCESSING

Wavelets can be explained (and implemented) within a signal processing framework. Because V_0 is a subset of all functions in V_1 it should be possible to write the scaling functions in V_0 as a linear combination of the basis functions in V_1:

$$\phi(x) = \sum_{k \in \mathbb{Z}} h_\phi(k)\sqrt{2}\,\phi(2x - k) \tag{21}$$

where $\sqrt{2}\phi(2x-k)$ are the basis functions in V_1. The coefficients $h_\phi(k)$ are defined as follows:

$$\langle \phi(x), \sqrt{2}\,\phi(2x-k) \rangle \tag{22}$$

However, note that as $\phi(x)$ is nonzero only between 0 and 1, in the case of Haar wavelets the number of nonzero coefficients $h_\phi(k)$ is limited to 2. Similar to the scaling function in V_0, the wavelet function can be defined as:

$$\psi(x) = \sum_{k \in \mathbb{Z}} h_\psi(k)\sqrt{2}\,\phi(2x - k) \tag{23}$$

The coefficients $h_\psi(k)$ and $h_\phi(k)$ are then related as follows:

$$h_\psi(k) = (-1)^k h_\phi(1-k) \tag{24}$$

For Haar wavelets, it can be shown that:

$$\phi(x) = \phi(2x) + \phi(2x-1) \tag{25}$$

$$\psi(x) = \phi(2x) - \phi(2x-1) \tag{26}$$

As a result, the nonzero filter coefficients for $h_\phi(k)$ are $h_\phi(0) = 1/\sqrt{2}$ and $h_\phi(1) = 1/\sqrt{2}$. The nonzero coefficients for $h_\psi(k)$ are $h_\psi(0) = 1/\sqrt{2}$ and $h_\psi(1) = -1/\sqrt{2}$. It is therefore possible to compute coefficients for each of the (integer) elements of the input function $f(x)$ as follows:

$$c_\phi(x) = \frac{1}{\sqrt{2}} f(x) + \frac{1}{\sqrt{2}} f(x-1) \tag{27}$$

$$c_\psi(x) = \frac{1}{\sqrt{2}} f(x) - \frac{1}{\sqrt{2}} f(x-1) \tag{28}$$

where the coefficients $c_\phi(x)$ encode the average of $f(x)$ and $f(x-1)$ and the coefficients $c_\psi(x)$ represent deviations from this average for elements $f(x)$ and $f(x-1)$. To encode a signal of length n, the above coefficients would be computed for each pair of neighboring elements so that the total number of coefficients is $n/2$ for $c_\phi(x)$ as well as $n/2$ for $c_\psi(x)$. This could be written in matrix form as follows:

$$\begin{bmatrix} c_\phi^1(0) \\ c_\phi^1(1) \\ c_\phi^1(2) \\ \vdots \\ c_\phi^1(n/2) \\ \hline c_\psi^1(0) \\ c_\psi^1(1) \\ c_\psi^1(2) \\ \vdots \\ c_\psi^1(n/2) \end{bmatrix} = \begin{bmatrix} \frac{1}{\sqrt{2}} & \frac{1}{\sqrt{2}} & 0 & \cdots & \cdots & \cdots & \cdots & \cdots & \cdots & 0 \\ 0 & 0 & \frac{1}{\sqrt{2}} & \frac{1}{\sqrt{2}} & 0 & \cdots & \cdots & \cdots & \cdots & 0 \\ 0 & 0 & 0 & 0 & \frac{1}{\sqrt{2}} & \frac{1}{\sqrt{2}} & 0 & \cdots & \cdots & 0 \\ & & & & \ddots & \ddots & & & & \\ 0 & \cdots & \cdots & \cdots & \cdots & \cdots & \cdots & 0 & \frac{1}{\sqrt{2}} & \frac{1}{\sqrt{2}} \\ \hline \frac{1}{\sqrt{2}} & \frac{-1}{\sqrt{2}} & 0 & \cdots & \cdots & \cdots & \cdots & \cdots & \cdots & 0 \\ 0 & 0 & \frac{1}{\sqrt{2}} & \frac{-1}{\sqrt{2}} & 0 & \cdots & \cdots & \cdots & \cdots & 0 \\ 0 & 0 & 0 & 0 & \frac{1}{\sqrt{2}} & \frac{-1}{\sqrt{2}} & 0 & \cdots & \cdots & 0 \\ & & & & \ddots & \ddots & & & & \\ 0 & \cdots & \cdots & \cdots & \cdots & \cdots & \cdots & 0 & \frac{1}{\sqrt{2}} & \frac{-1}{\sqrt{2}} \end{bmatrix} \begin{bmatrix} f(0) \\ f(1) \\ f(2) \\ \vdots \\ f(n/2) \\ f(n/2+1) \\ f(n/2+2) \\ f(n/2+3) \\ \vdots \\ f(n) \end{bmatrix} \tag{29}$$

This matrix is set up such that the resulting vector contains all the averages in its first $n/2$ elements and all the differences in the remaining elements. An interesting observation is that on average we may expect that the elements $c_\phi^1(x)$ have values in the same range as the values in $f(x)$, whereas the elements $c_\psi^1(x)$ are on average much smaller.

$$\begin{bmatrix} c_\phi^2(0) \\ c_\phi^2(1) \\ c_\phi^2(2) \\ \vdots \\ c_\phi^2(n/4) \\ \hline c_\psi^2(0) \\ c_\psi^2(1) \\ c_\psi^2(2) \\ \vdots \\ c_\psi^2(n/4) \end{bmatrix} = \begin{bmatrix} \frac{1}{\sqrt{2}} & \frac{1}{\sqrt{2}} & 0 & \cdots & \cdots & \cdots & \cdots & \cdots & \cdots & 0 \\ 0 & 0 & \frac{1}{\sqrt{2}} & \frac{1}{\sqrt{2}} & 0 & \cdots & \cdots & \cdots & \cdots & 0 \\ 0 & 0 & 0 & 0 & \frac{1}{\sqrt{2}} & \frac{1}{\sqrt{2}} & 0 & \cdots & \cdots & 0 \\ & & & & & \ddots & \ddots & & & \\ 0 & \cdots & \cdots & \cdots & \cdots & \cdots & \cdots & 0 & \frac{1}{\sqrt{2}} & \frac{1}{\sqrt{2}} \\ \hline \frac{1}{\sqrt{2}} & \frac{-1}{\sqrt{2}} & 0 & \cdots & \cdots & \cdots & \cdots & \cdots & \cdots & 0 \\ 0 & 0 & \frac{1}{\sqrt{2}} & \frac{-1}{\sqrt{2}} & 0 & \cdots & \cdots & \cdots & \cdots & 0 \\ 0 & 0 & 0 & 0 & \frac{1}{\sqrt{2}} & \frac{-1}{\sqrt{2}} & 0 & \cdots & \cdots & 0 \\ & & & & & \ddots & \ddots & & & \\ 0 & \cdots & \cdots & \cdots & \cdots & \cdots & \cdots & 0 & \frac{1}{\sqrt{2}} & \frac{-1}{\sqrt{2}} \end{bmatrix} \begin{bmatrix} c_\phi^1(0) \\ c_\phi^1(1) \\ c_\phi^1(0) \\ \vdots \\ c_\phi^1(n/2) \end{bmatrix} \quad (30)$$

This has many implications for specific applications, which will be discussed later.

After this procedure, we have a signal that consists on average for the first half of large values and for the second half of small values. It is now possible to further encode the signal by repeating the procedure on $c_\phi^1(x)$:

As a result, by repeating this process until we have only one average of the entire signal followed by $n-1$ differences, we can create a vector that represents the final wavelet encoding:

$$\begin{bmatrix} c_\phi^m(0) & c_\psi^m(0) & c_\psi^{m-1}(0) & c_\psi^{m-1}(1) & c_\psi^{m-2}(0) \\ c_\psi^{m-2}(1) & c_\psi^{m-2}(2) & c_\psi^{m-2}(3) & \cdots & \end{bmatrix} \quad (31)$$

where $m = \log_2(n)$. Note that it is possible to exactly reconstruct the original signal $f(x)$ from this encoding. The first step of the inverse is given by:

$$\begin{bmatrix} c_\phi^{m-1}(0) \\ c_\phi^{m-1}(1) \end{bmatrix} = \begin{bmatrix} \frac{1}{\sqrt{2}} & \frac{-1}{\sqrt{2}} \\ \frac{1}{\sqrt{2}} & \frac{1}{\sqrt{2}} \end{bmatrix} \begin{bmatrix} c_\phi^m(0) \\ c_\psi^m(0) \end{bmatrix} \quad (32)$$

Here, the two averages are reconstructed from the top-level average and difference. Repeating this process gives us four averages:

$$\begin{bmatrix} c_\phi^{m-2}(0) \\ c_\phi^{m-2}(1) \\ c_\phi^{m-2}(2) \\ c_\phi^{m-2}(3) \end{bmatrix} = \begin{bmatrix} \frac{1}{\sqrt{2}} & 0 & \frac{-1}{\sqrt{2}} & 0 \\ \frac{1}{\sqrt{2}} & 0 & \frac{1}{\sqrt{2}} & 0 \\ 0 & \frac{1}{\sqrt{2}} & 0 & \frac{-1}{\sqrt{2}} \\ 0 & \frac{1}{\sqrt{2}} & 0 & \frac{1}{\sqrt{2}} \end{bmatrix} \begin{bmatrix} c_\phi^{m-1}(0) \\ c_\phi^{m-1}(1) \\ c_\psi^{m-1}(0) \\ c_\psi^{m-1}(1) \end{bmatrix} \quad (33)$$

This process can then be repeated a total of $m = \log_2(n)$ to reconstruct the original function $f(x)$. The final step in this reconstruction is:

$$\begin{bmatrix} f(0) \\ f(1) \\ \vdots \\ f(n) \end{bmatrix} = \mathbf{R} \begin{bmatrix} c_\phi^1(0) \\ c_\phi^1(1) \\ \vdots \\ c_\phi^1(n/2) \\ c_\psi^1(0) \\ c_\psi^1(1) \\ \vdots \\ c_\psi^1(n/2) \end{bmatrix} \quad (34)$$

where $\mathbf{R}$ is given by:

$$\mathbf{R} = \begin{bmatrix} \frac{1}{\sqrt{2}} & 0 & \cdots & 0 & \frac{-1}{\sqrt{2}} & 0 & \cdots & 0 \\ \frac{1}{\sqrt{2}} & 0 & \cdots & 0 & \frac{1}{\sqrt{2}} & 0 & \cdots & 0 \\ 0 & \frac{1}{\sqrt{2}} & 0 & \cdots & 0 & 0 & \frac{-1}{\sqrt{2}} & 0 & \cdots & 0 \\ 0 & \frac{1}{\sqrt{2}} & 0 & \cdots & 0 & 0 & \frac{1}{\sqrt{2}} & 0 & \cdots & 0 \\ 0 & 0 & \frac{1}{\sqrt{2}} & 0 & \cdots & 0 & 0 & 0 & \frac{-1}{\sqrt{2}} & 0 & \cdots & 0 \\ 0 & 0 & \frac{1}{\sqrt{2}} & 0 & \cdots & 0 & 0 & 0 & \frac{1}{\sqrt{2}} & 0 & \cdots & 0 \\ & & \ddots & & & & & \ddots & & \\ 0 & \cdots & \cdots & 0 & \frac{1}{\sqrt{2}} & 0 & \cdots & \cdots & 0 & \frac{-1}{\sqrt{2}} \\ 0 & \cdots & \cdots & 0 & \frac{1}{\sqrt{2}} & 0 & \cdots & \cdots & 0 & \frac{1}{\sqrt{2}} \end{bmatrix} \quad (35)$$

The wavelet-encoded signal consists of differences for all but the first element. As argued above, this means that most values in the encoded signal will be smaller than the signal's original values. For instance, the Haar wavelet-encoded signal of a sequence of eight numbers is given in Table 1. Although the average of the signal itself is 6.5 (first column on the $m=3$ row), the average of the magnitude of the remaining coefficients is just 1.5. Another example is shown in Fig. 3, where the signal in the left panel is encoded using Haar wavelets. The wavelet coefficients are then sorted by magnitude and plotted in the right panel. Note that most coefficients are very small.

As a result, it should be possible to encode each element with fewer bits than the original signal, leading to a simple form of data compression. If the signal is encoded as integers, many of these small differences would be the same value. The signal could therefore be encoded very efficiently with Huffman coding.

Table 1 The input signal $f(x)$ is successively decomposed into wavelet coefficients, leading to an average of 6.5 followed by a set of small coefficients. For clarity we have removed the factor of $1/\sqrt{2}$ so that $h_\phi = [1\ 1]$ and $h_\psi = [1\ -1]$

$f(x)$	8	10	3	7	6	2	7	9
$m=1$	9	5	4	8	1	2	−2	1
$m=2$	7	6	−2	2	1	2	−2	1
$m=3$	6.5	−0.5	−2	2	1	2	−2	1

Moreover, it would be possible to set all elements that are smaller than a given threshold to zero, effectively compressing the data in a lossy manner. An example for Haar wavelets is given in Fig. 4. Here, the signal was

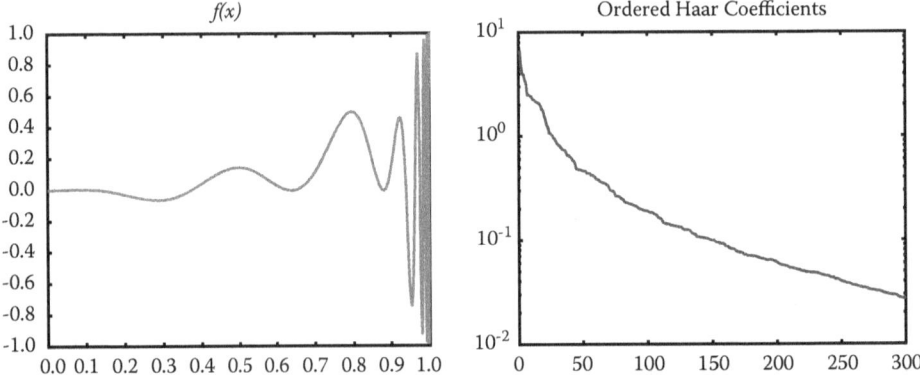

Fig. 3 The signal on the left is encoded using Haar wavelets, after which the resulting coefficients are sorted by magnitude and plotted on the right. Note how the magnitude of the coefficients is small relative to the magnitude of the signal

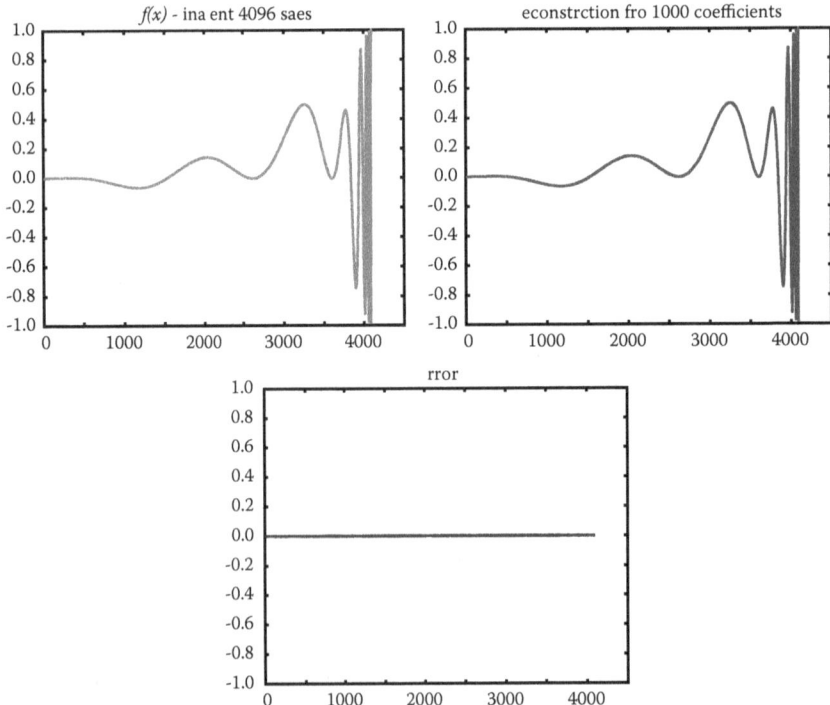

Fig. 4 The signal on the top left is encoded using Haar wavelets, after which the signal is reconstructed using 1,000 coefficients. The original function contained 4,096 samples, so that lossy reconstruction is achieved with low error and a compression of around 4:1

compressed by a ratio of around 4:1 as the signal $f(x)$ contained 4,096 samples, while the reconstruction used only 1,000 wavelets.

OTHER BASES

While the Haar wavelet basis is useful in many cases, in part due to its simplicity, and in part because it can represent sharp edges in the signal well, many coefficients are required to fully reconstruct a smooth signal, as seen in Fig. 4. Thus, in addition to local support, smoothness is often a second requirement for effective wavelet decomposition. The machinery required for all wavelets is similar: the generating function $\phi(x)$ from Eq. 13 will represent (weighted) averages, while the wavelet function $\psi(x)$ from Eq. 11 will encode differences. To efficiently encode smoother functions, the generating and wavelet function pairs can be replaced with new basis functions. This has given rise to many alternatives to Haar wavelets, including Daubechies,[11] Coiflet, and Symlet wavelet, as well as many others.

While the Haar wavelet can be implemented as a convolution with a filter that has two nonzero elements (or *taps*) per coefficient (see Eq. 30), all other wavelets require more taps. Daubechies, for instance, has developed a family of wavelets which have an increasing number of taps,[11,12] as shown in Fig. 5. Daubechies weights at each tap cannot be generated algorithmically, but are typically produced numerically by the *cascade algorithm*.[13] Daubechies wavelets have a number of *vanishing moments* equal to half the number of taps. A Daubechies wavelet with four taps (D4) therefore has two vanishing moments.

The importance of vanishing moments lies in the fact that its value determines the smoothness of the functions that can be encoded. With n vanishing moments, polynomials of degree $n-1$ can be encoded. This means that a D4 wavelet can encode constant and linear polynomials. An example demonstrating the implications for reconstruction is given in Fig. 6. As the input function is smooth, Daubechies D20 with ten vanishing moments can represent polynomials with degree 9 and is therefore able to reconstruct the function with only 50 wavelet coefficients. The Haar function can only represent piecewise constant functions and so does not perform well with only 50 coefficients. The Daubechies D20 wavelets have allowed a compression of around 82:1, which cannot be matched by the Haar wavelets.

The symlet wavelets improve symmetry properties relative to the Daubechies family of wavelets. The Coiflet wavelets[4,14] have vanishing moments in both wavelet function as well as the scaling/generating function,[15] while also being near-symmetric.

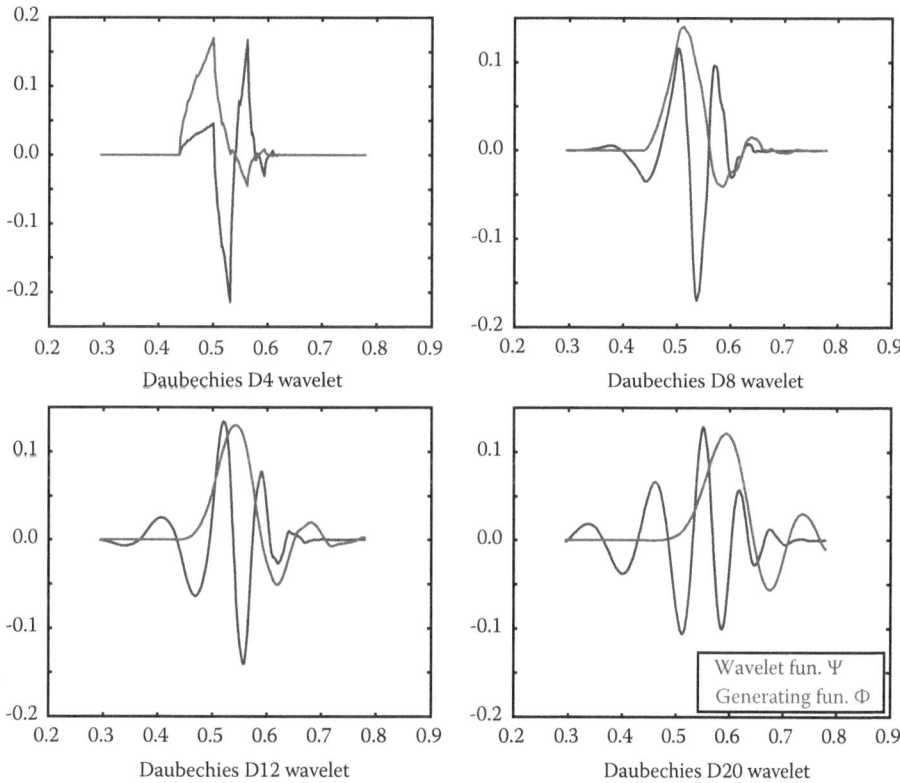

Fig. 5 Daubechies wavelets are orthonormal and can be constructed with a different number of taps. The more taps, the smoother the wavelet. Shown here are the D4, D8, D12, and D20 wavelets

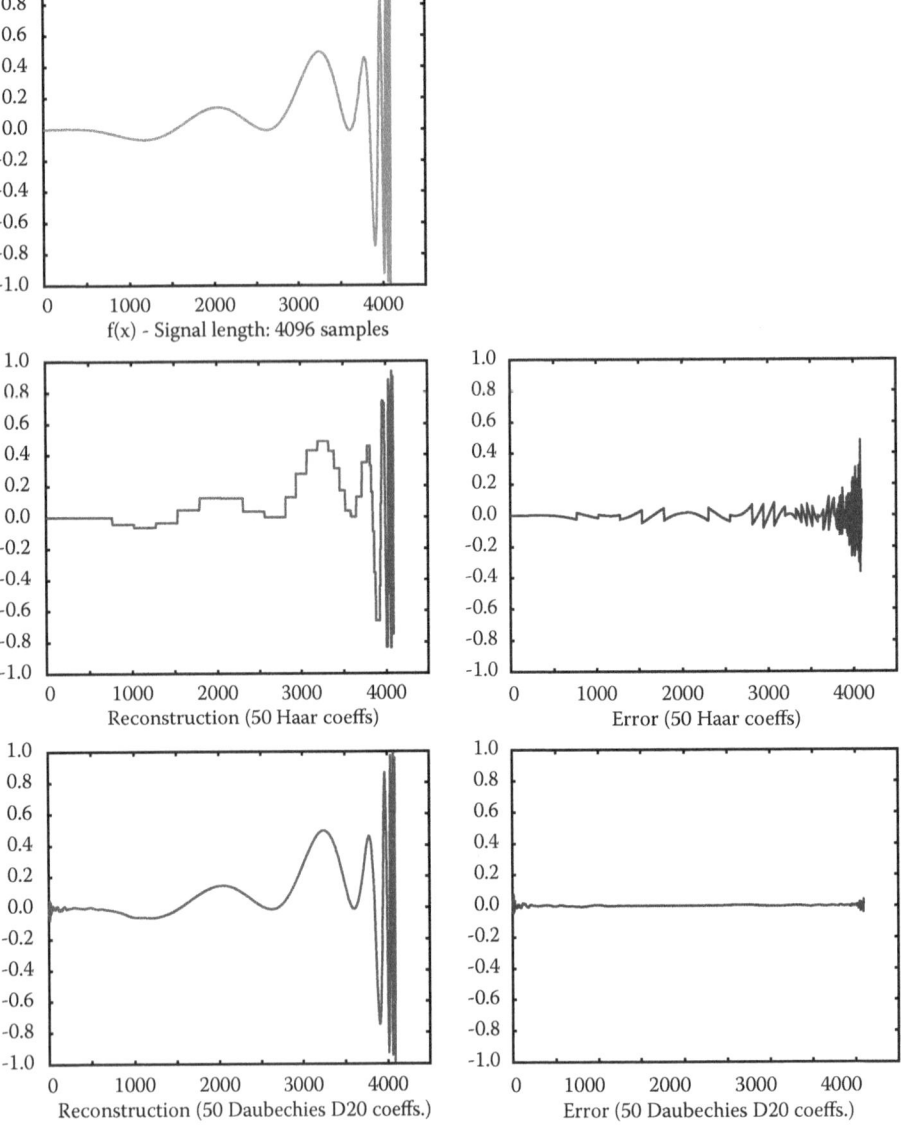

Fig. 6 Reconstruction comparison between Haar and Daubechies D20 wavelets. The signal $f(x)$ has 4,096 samples, which are here reconstructed from 50 wavelet coefficients

2D WAVELETS

So far we have considered functions of one variable. Images, however, are two-dimensional, and we therefore have to extend wavelet representations to two dimensions. The Haar wavelet is separable, which means that an image can be decomposed in one dimension first, followed by decomposition in the second dimension. In essence, the scaling function becomes:

$$\phi(x,y) = \phi(x)\phi(y) \tag{36}$$

The wavelet function is split into three wavelet functions, one each for the horizontal, vertical, and diagonal directions:

$$\psi^H(x,y) = \psi(x)\phi(y) \tag{37}$$

$$\psi^V(x,y) = \phi(x)\psi(y) \tag{38}$$

$$\psi^D(x,y) = \psi(x)\psi(y) \tag{39}$$

The only difference between the three functions $\psi^H(x, y)$, $\psi^V(x, y)$, and $\psi^D(x, y)$ lies, therefore, in which of the scaling and wavelet functions is applied and in which dimension. Otherwise, the computations proceed as before. A single level decomposition can be created by applying the 1D Haar decomposition to all the rows of the image. As averages will be moved toward the left half of the image and differences to the right half, each row will consist of the elements of $\phi(x)$ followed by the elements of $\psi(x)$.

If the procedure is repeated, but now on the columns of the image, the resulting output will consist of four quadrants, as shown in the top-right panel of Fig. 7. The top-left

Fig. 7 One, two, and three levels of Haar decomposition. Note that we have passed the resulting images through a gamma curve (exponent 0.4) to help visualize the coefficients

Fig. 8 A full Haar decomposition. Note that we have passed the resulting images through a gamma curve (exponent 0.2) to help visualize the coefficients. (Grand Canyon, Arizona, 2012)

quadrant will contain the averages $\phi(x, y)$ which is a downsampled version of the input image. The top-right and bottom-left quadrants contain horizontal and vertical differences. Finally, the bottom-right quadrant contains the vertical differences of previously computed horizontal differences. These therefore represent diagonal differences. This decomposition is said to be separable, as horizontal and vertical calculations are carried out consecutively.

As before, these procedures can be applied recursively on the averages, leading to a hierarchical wavelet decomposition. Two- and three-level Haar decompositions are shown at the bottom of Fig. 7. Of course, this procedure can be repeated until the top-left pixel contains the image average and all other pixels store differences. An example of a full decomposition is shown in Fig. 8. Here, most coefficients in the image have become very small so that the encoded image appears mostly black.

Although Haar wavelets are used extensively in image processing, they are by no means the only wavelets amenable to image encoding. Daubechies wavelets, for instance,

Fig. 9 Two levels of Daubechies decomposition for D4, D8, D12, and D20 wavelets. Note that we have passed the resulting images through a gamma curve (exponent 0.4) to help visualize the coefficients. (New Delhi, India, 2008)

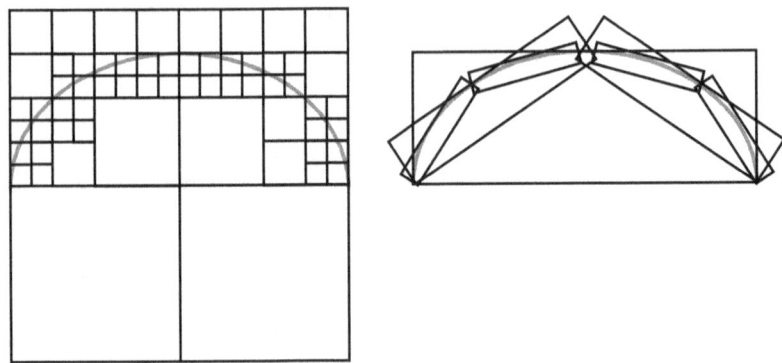

Fig. 10 An illustration of a wavelet decomposition (left) and a contourlet decomposition (right)

can be used to encode images as shown in Fig. 9, although note that as these wavelets use more than two taps, boundary effects can be more of a problem.

CONTOURLETS, CURVELETS, AND RIDGELETS

Further, wavelets have been extended to contourlets, which are more adaptable to the images being encoded,[16] as well as curvelets[17] and ridgelets,[18,19] which are especially adapted to edges. Consider a curved edge in an image. In a conventional wavelet decomposition, the image would be partitioned as illustrated in the left part of Fig. 10. By introducing more orientations, and by allowing the basis functions to follow image features more closely, a sparser decomposition can be achieved, as shown in the right part of Fig. 10.

COEFFICIENT HISTOGRAMS

So far, we have shown that wavelet coefficients tend to lead to smaller numbers than the values found in the original

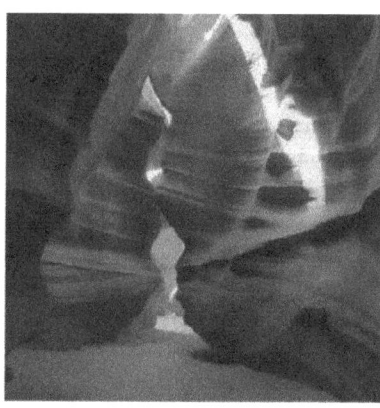
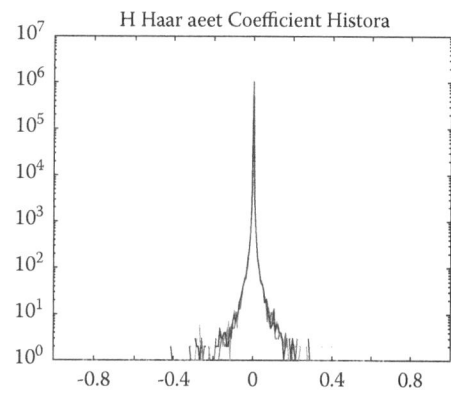

Fig. 11 The Haar wavelet coefficients are collected into a histogram for each of the red, green, and blue channels of the high dynamic range image depicted on the left. The image is tonemapped for display,[24] but coefficients are computed on the high dynamic range original. Note the highly kurtotic shape of the distribution. (Antelope Canyon, Arizona, 2012)

signal/image. It has been shown that wavelet coefficients follow a highly kurtotic distribution.[9,20] An example for a single image is shown in Fig. 11, where we have subjected a high dynamic range image to the Haar wavelet transform. The kurtosis observed in each channel is larger than 2,000, which is indeed pointing at very long tails. We have plotted the vertical axis on a log scale for the purpose of visualization.

The highly non-Gaussian histogram distribution of wavelet coefficients can be modeled with the *generalized Laplacian function*[9,21,22,23] (also known as the *exponential power distribution* or the *generalized error distribution*), as given by:

$$P(x) = \frac{\beta}{2\alpha \Gamma\left(\frac{1}{\beta}\right)} e^{-|(x-\mu)/\alpha|^{\beta}} \qquad (40)$$

where Γ is the gamma function, as defined by:

$$\Gamma(x) = \int_0^\infty e^{-t} t^{z-1} dt \qquad (41)$$

The parameters α, β, and μ can be used to fit the model to a specific image or image ensemble. For sparse distributions as seen with wavelet coefficients, the shape parameter β is less than 1. The parameter α scales the distribution, while μ is the mean. Note that to model the distribution of wavelet coefficients, the mean is zero. The variance of the distribution, as modeled with Eq. 40, is given by:

$$\sigma^2 = \frac{\alpha^2 \Gamma(3/\beta)}{\Gamma(1/\beta)} \qquad (42)$$

As the distribution is symmetric, the skewness is necessarily 0. The kurtosis of this distribution is:

$$\frac{\Gamma(5/\beta)\Gamma(1/\beta)}{\Gamma(3/\beta)^2} - 3 \qquad (43)$$

For instance, if the function fit yields a value for β of 0.5, then the kurtosis of the distribution was $\Gamma(10)\Gamma(2)/\Gamma(6) - 3 > 22$. This would be a highly peaked function. For the distribution shown in Fig. 11, the value of β is around 0.15, given that the pixel values in this image have a kurtosis of around 20,000. Also note that the kurtosis of the original pixel values of this image is around 135.

Wavelet encodings tend to be sparse. However, there will be cases where the encoding fails to be sparse. It has been shown that in such cases Markov random fields can be used to represent the local non-sparseness.[25]

High values for the kurtosis of the distribution of wavelet coefficients are common in wavelet decompositions. As expected, they have also been found in contourlets[26] as well as curvelets and ridgelets.[27] In fact, it appears that high kurtosis appears whenever localized zero-mean linear kernels are used as bases.[28,29] Sparseness of the representation, indicated by high kurtosis in wavelet coefficients, accounts for much of the success of wavelets in various applications. As an example, by explicitly estimating the parameters α and β, it is possible to construct an algorithm to denoise images[21] or to efficiently encode images.[22] Quality assessment algorithms have also been designed using the non-Gaussian distribution of wavelet coefficients.[30,31] Wavelets have also been used in the authentication of art,[32,33] steganalysis,[34,35] and in the differentiation between photographs and photorealistic renderings.[36]

SCALE INVARIANCE

One could analyze the distribution of wavelet coefficients at each scale of the wavelet tree separately. For natural images, the result would be that each distribution would have roughly the same, highly kurtotic shape.[37] This is the same as saying that the distribution of wavelet coefficients is invariant under scaling. If the histogram of a distribution remains invariant under scaling, then so are its associated moments.[37]

It can be argued that image formation is driven by objects (which are possibly textured[38]). The analysis of amplitude spectra of natural image ensembles (Section 6.4.2) shows scale invariance as a result. Here, we have a different way of expressing the same scale invariance, albeit that we could go one step further and assess scale invariance locally in small image regions. Scale invariance returns in the analysis of the dead leaves model in Section 11.1, where scenes are assumed to be composed of objects that can (partially) occlude each other. For instance, the dead leaves model gives clues as to the size distribution of objects required to match the observed scale invariant statistics.[39]

CORRELATIONS BETWEEN COEFFICIENTS

A second observation that can be made about wavelet coefficients is that corresponding coefficients in either scale, space, or orientation tend to be highly correlated.[40] For instance, when taking neighboring coefficients at a particular scale and orientation, their values tend to be similar. This can be seen by computing a joint histogram of coefficients and their neighbors at some distance. For instance, Fig. 12 shows the conditional histogram of the vertical coefficients for pixels separated by 1, 2, and 4 pixels in the vertical direction. The bow-tie shape of the conditional histogram shows that especially small values of one coefficient are good predictors for the value of coefficients some distance away. This holds for horizontal and diagonal coefficients as well, as shown in Figs. 13 and 14, and also applies to contourlets.[26]

This bow-tie shape suggests that there are still dependencies between the coefficients.[41] The variance scales with the absolute value on the abscissa. These dependencies cannot be further reduced with any linear transform. It is thought that occlusion of objects is one of the main processes giving rise to image formation[42] which is nonlinear. This tends to give rise to relatively flat areas in images with sharp discontinuities. Such edge-like structures tend to have substantial power across scales, orientations, and local position. This is reflected in the conditional coefficient histograms.

It is interesting to note that these dependencies can be reduced by introducing a divisive normalization akin to

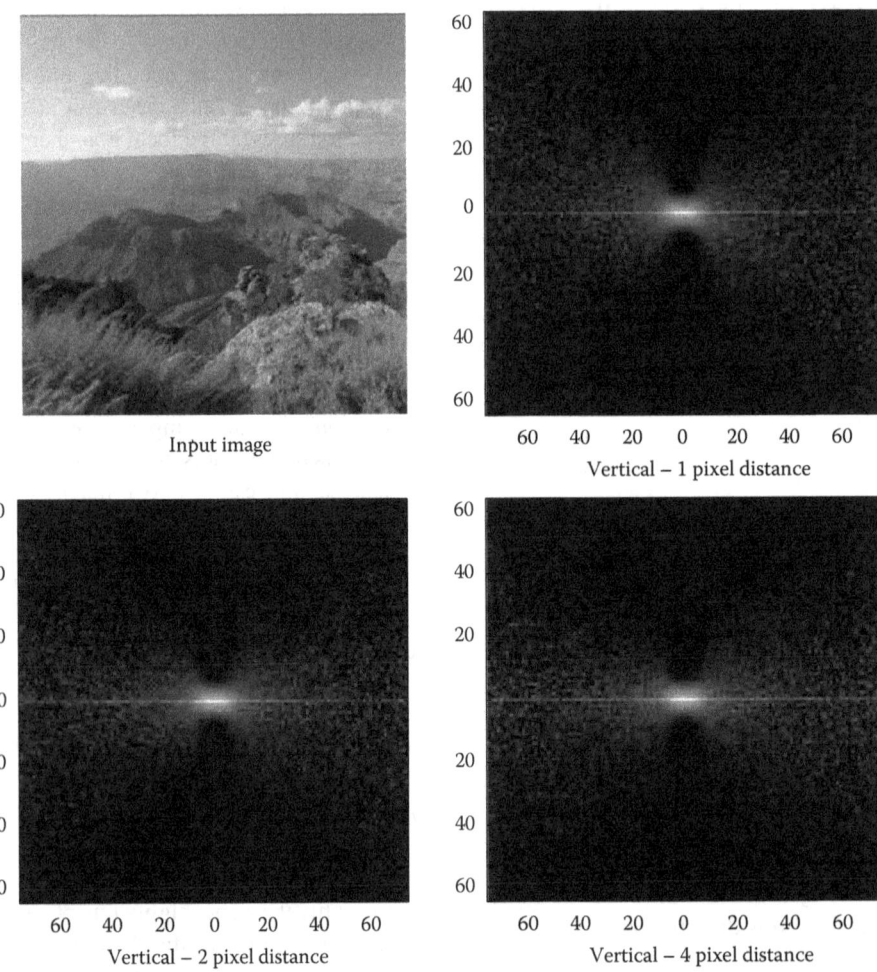

Fig. 12 Conditional histograms for vertical wavelet coefficients. The distance between the pixels was 1, 2, and 4 pixels. (Grand Canyon, Arizona, 2012)

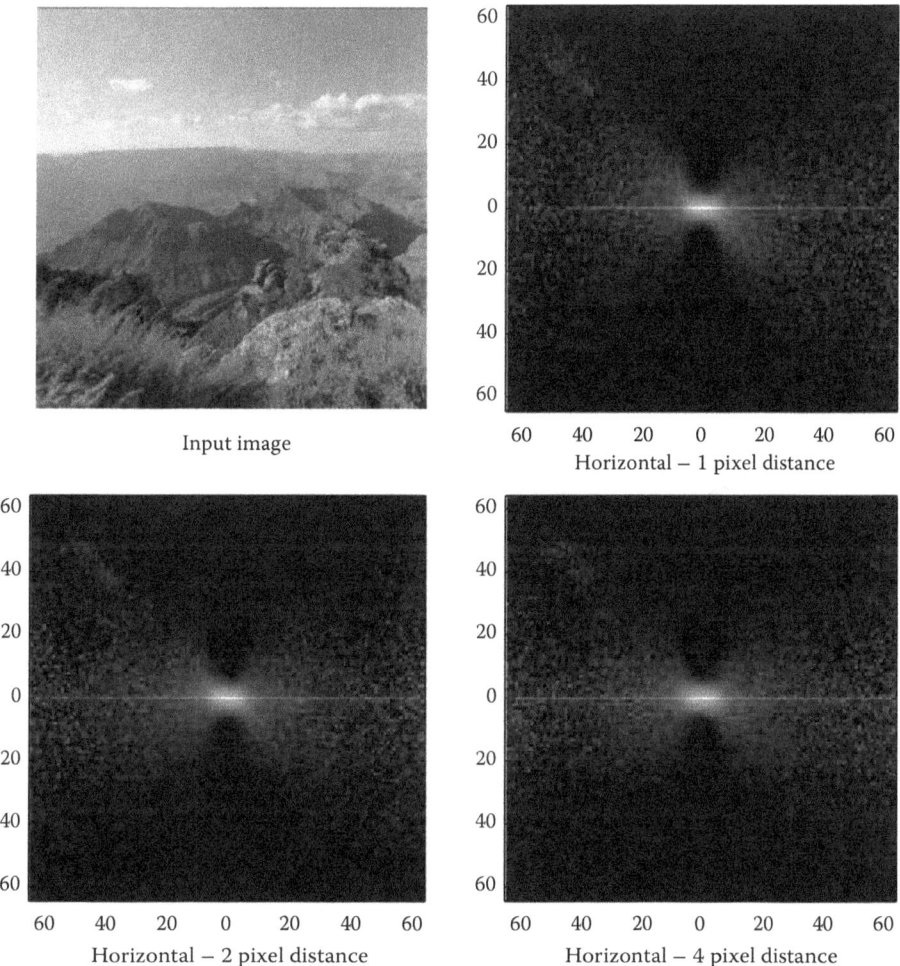

Fig. 13 Conditional histograms for horizontal wavelet coefficients. The distance between the pixels was 1, 2, and 4 pixels. (Grand Canyon, Arizona, 2012)

the Naka-Rushton equation, which models photoreceptors (see Section 2.3.2):[41,43,44]

$$V_j = \frac{I_j^2}{\sum_k w_{jk} I_k^2 + \sigma_j^2} \qquad (44)$$

where each coefficient I_j is replaced by V_j. Neighboring coefficients in scale, space, and orientation are indicated with I_k. The parameters σ_j and w_{jk} can be determined from the analysis of natural images. Note that σ_j represents the variance that cannot be accounted for by the neighboring coefficients.

Finally, we note that correlations between coefficients in a wavelet decompositions have been used in the field of steganography,[35] image restoration,[45] and image compression.[46]

COMPLEX WAVELETS

The wavelets described so far produce one real-valued coefficient per pixel. In analogy to Fourier transforms, it is possible to design wavelet pairs that produce complex-valued coefficients, which can be reinterpreted as representing amplitude and phase. However, contrary to a Fourier decomposition, the amplitude and phase representations of complex wavelets are localized in space.

Real-valued wavelets as discussed so far have advantages in that they can have nonredundant orthonormal bases, allow for multiresolution decomposition and have fast linear time algorithms to decompose and reconstruct signals and images. However, they also have some disadvantages in image analysis and reconstruction tasks:[47,48]

Sensitivity to shifts. If the input is shifted by small amounts, this can lead to major changes in the distribution and values of the coefficients. In many image processing tasks, this is an undesirable feature.

Poor directional selectivity. Separable and real wavelets tend to have poor selectivity for orientations other than horizontal and vertical.

No phase information. The discrete wavelet transform does not encode phase information and also tends to have poor frequency resolution.

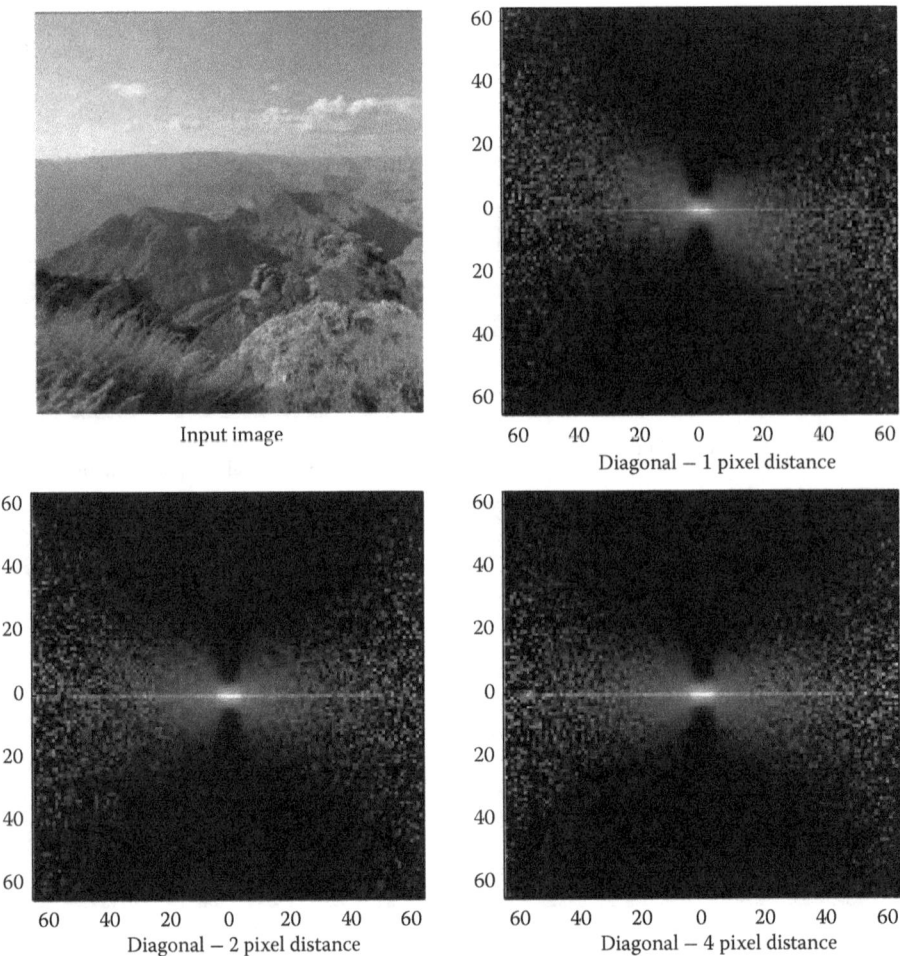

Fig. 14 Conditional histograms for diagonal wavelet coefficients. The distance between the pixels was 1, 2 and, 4 pixels. (Grand Canyon, Arizona, 2012)

Oscillations. Ideally wavelet coefficients are only large near edges. However, this is not necessarily the case with real-valued wavelets, which may make certain image processing tasks such as finding edges difficult.[49]

These disadvantages can be overcome with complex wavelets. These in essence introduce some redundancy, as every real-valued sample in the input results in a complex-valued coefficient. Complex wavelets solve many of these problems akin to Fourier transforms. In particular, the amplitude of a Fourier transform is fully shift invariant. Further, in two dimensions complex wavelets are highly directionally sensitive.

As argued in Section 6.2, the Fourier transform decomposes a signal into sine and cosine basis functions:

$$e^{i\omega x} = \cos(\omega x) + i\sin(\omega x) \qquad (45)$$

As the cosine and sine components are 90 degrees out of phase with each other, they form a Hilbert transform pair. The background to complex wavelet transforms relates to the Hilbert transform in a similar fashion. A real signal $f(x)$ can be extended into the complex domain using:[50]

$$h(x) = f(x) + ig(x) \qquad (46)$$

where $g(x)$ is the Hilbert transform of $f(x)$, denoted as:

$$g(x) = H\{f(x)\} \qquad (47)$$

As mentioned above, the Hilbert transform is defined as a 90 degree rotation in the complex plane of the input signal $f(x)$. This makes $g(x)$ orthogonal to $f(x)$. Being out of phase by a quarter period of $f(x)$ and $g(x)$ is also known as being in *quadrature*. The Hilbert transform is given by:[51]

$$g(x) = \frac{1}{\pi} \int_{-\infty}^{\infty} \frac{1}{x - \tau} f(x) d\tau \qquad (48)$$

$$= f(x) \otimes \frac{1}{\pi x} \qquad (49)$$

The *magnitude* $|h(x)|$ and *phase* $\angle h(x)$ of the complex signal $h(x)$ are then given by:

$$|h(x)| = \sqrt{f(x)^2 + g(x)^2} \qquad (50)$$

$$\angle h(x) = \tan^{-1}(g(x)/f(x)) \tag{51}$$

Note that the magnitude is also known as the *envelope* of the combined response of $f(x)$ and $g(x)$.

As any signal can be extended into the complex domain in this manner, it is certainly also possible to create a complex scaling function (where the subscripts c, r, and i denote complex, real, and imaginary components):[48]

$$\psi_c(x) = \psi_r(x) + i\psi_i(x) \tag{52}$$

Once again, the real and imaginary parts $\psi_i(x)$ and $\psi_r(x)$ should form a Hilbert transform pair. A 2D example is shown in Fig. 15, where the top row shows the real part of six oriented wavelets, and the second row shows the imaginary part of the same six oriented wavelets. Note that the real part of this wavelet is an even-symmetric filter, whereas the imaginary part is odd-symmetric. The third row in this figure plots the envelope or magnitude of these six wavelets, which has an approximately Gaussian shape. It is then possible to compute wavelet coefficients for a specific signal $f(x)$ using complex wavelets:[48]

$$d(j,n) = 2^{j/2} \int_{-\infty}^{\infty} f(x) \psi_c(2^j t - n) dt \tag{53}$$

The magnitude and (phase) angle can be computed from these coefficients as before:

$$|d_c(j,n)| = \sqrt{d_r(x)^2 + d_i(x)^2} \tag{54}$$

$$\angle d_c(x) = \tan^{-1}(d_i(x)/d_r(x)) \tag{55}$$

On average, the magnitude will be small for most coefficients but may become large for pixels that are on or near a contrast feature in a signal or image. The value of the phase then indicates the position of the feature within the support of the wavelet.

As with real-valued wavelet decompositions, complex wavelet decompositions can be either redundant or nonredundant. In the latter case, it is possible to construct

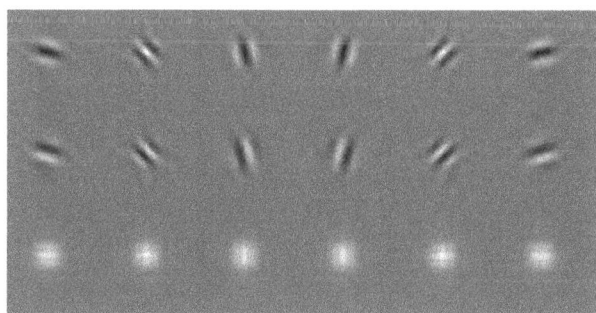

Fig. 15 Complex wavelets. The top row shows the real part, the second row shows the imaginary part, and the bottom row shows the magnitude for each of six orientations

complex wavelets that have orthonormal or biorthogonal bases.[52–57] It can be argued, however, that a redundant representation would be better able to overcome the shortcomings of real-valued wavelets,[48] as outlined on page 197. The resulting wavelet decomposition would be 2× redundant in one dimension,[11] as the real-valued samples in the original signal are replaced by complex-valued wavelet coefficients. Such redundant wavelet decompositions can be efficiently constructed by means of the dual-tree complex wavelet transform.[47,48,58,59,60,61,62] They can also be naturally extended to two dimensions, allowing images to be transformed into the complex wavelet domain. This, however, incurs a 4× redundancy.

Complex wavelets have advantages in practical applications as they are nearly shift and rotation invariant. In particular, they have been shown to be more effective than real-valued wavelets in image and video denoising.[49,63,64,65,66] Other applications in which complex wavelets have proven useful include image segmentation,[67] feature extraction,[68,69] texture analysis, and synthesis,[70–72] image classification,[73] deconvolution,[74,75] watermarking,[76,77] image sharpening,[78] motion estimation,[79] and image and video coding.[80–82] In addition, complex wavelets are studied in the context of natural image statistics, having revealed one regularity that is not observed in the analysis of real-valued wavelets. This is discussed in the following section.

CORRELATIONS BETWEEN SCALES

Given that edges tend to persist across scales, it is to be expected that if we were to compute correlations between scales, the structure in scenes should cause such correlations to be high. In other words, we might expect there to be some redundancy between neighboring scales in a wavelet decomposition.

However, as the filters applied in subsequent scales of a wavelet transform are normally orthogonal, measuring the correlation of wavelet coefficients in subsequent scales proves problematic. This can be seen as follows. Assume that at scales i and $i+1$ we have wavelet filters $f_i(x)$ and $f_{i+1}(x)$, which are orthogonal by definition, so that:

$$\int f_i(x) f_{i+1}(x - x_0) dx = 0 \quad \forall x_0 \tag{56}$$

This means that the convolution of two such filters with a given signal will also be orthogonal. As a result, the correlation between wavelet coefficients will be zero if the two filters are truly orthogonal.[83] An example of the correlations measured with a real-valued wavelet decomposition is shown in Fig. 16.

Thus, a different measure is required to assess the correlations that may occur between scales. In particular, complex wavelets offer a unique opportunity to understand how edges produce correlations (or redundancy) between

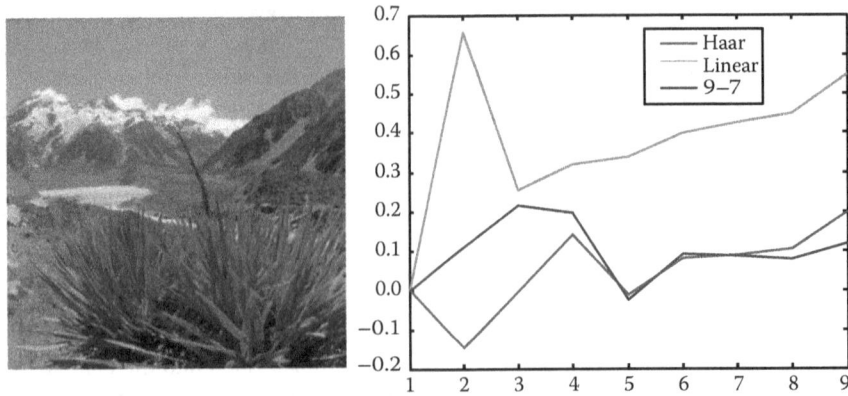

Fig. 16 The magnitude of real wavelets shows comparatively weak correlations across scale. For each of three different types of real-valued wavelets, the correlation between two neighboring scales are shown. The smaller scales represent detail coefficients, whereas larger scales represent coarser features. (Mt. Cook, New Zealand, 2012)

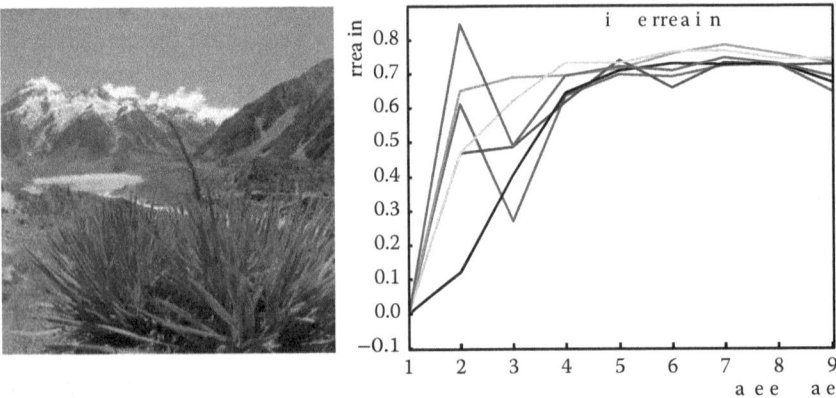

Fig. 17 The magnitude of complex wavelets shows strong correlations across scale. For each of six orientations, the correlation between two neighboring scales are shown. The smaller scales represent detail coefficients, whereas larger scales represent coarser features. (Mt. Cook, New Zealand, 2012)

scales. Although the real and complex wavelets by themselves show the same problems as outlined above, i.e., they are to a large extent orthogonal between scales, the complex wavelet decomposition can be transformed from a Euclidian coordinate space to a polar space, whereby real and complex values transform into magnitudes and angles.

It turns out that the amplitudes of complex wavelet coefficients tend to be correlated across scales.[83] In fact, if the phases of image features are aligned at similar scales in two frequency bands, the correlation between the magnitude of complex wavelet coefficients will be high. An example of the correlations found across scale in a single image are shown in Fig. 17, allowing a comparison with the real-valued wavelet composition of Fig. 16.

We can show that these correlations are largely due to information encoded in the phase spectrum (whereby we mean the phase spectrum as available in Fourier space). To this end, we have computed the Fourier transform of an image (the one shown in Figs. 16 and 17), permuted the phase spectrum, and converted back to the spatial domain. We then computed the complex wavelet transform as well as the correlations across scales of its amplitude spectrum. The results are shown in Fig. 18. Note that by removing structure from the Fourier domain phase spectrum, correlations across scale vanish for the most part. This means that local structure that persists across scale in a complex wavelet decomposition is due to an image's phase, rather than its amplitude.[83]

APPLICATION: IMAGE DENOISING

Wavelets can be used for image denoising. This can, for instance, be achieved by applying soft-thresholding on the coefficients $c_{j,k} = \langle f, \psi_{j,k} \rangle$, leading to new coefficients $c'_{j,k}$.[84,85]

$$c'_{j,k} = \begin{cases} c_{j,k} - t_j & c_{j,k} \geq t_j \\ 0 & |c_{j,k}| \leq t_j \\ c_{j,k} + t_j & c_{j,k} \leq -t_j \end{cases} \qquad (57)$$

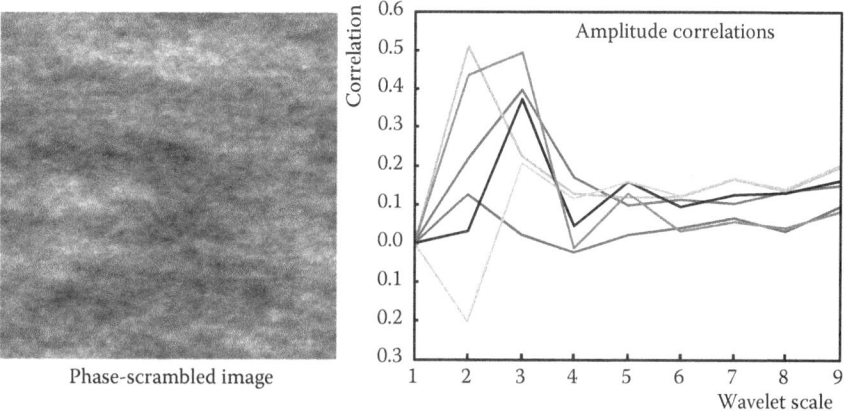

Fig. 18 If the phase spectrum of an image is permuted (left), then the amplitude of a complex wavelet decomposition does not show strong correlations across scale (right)

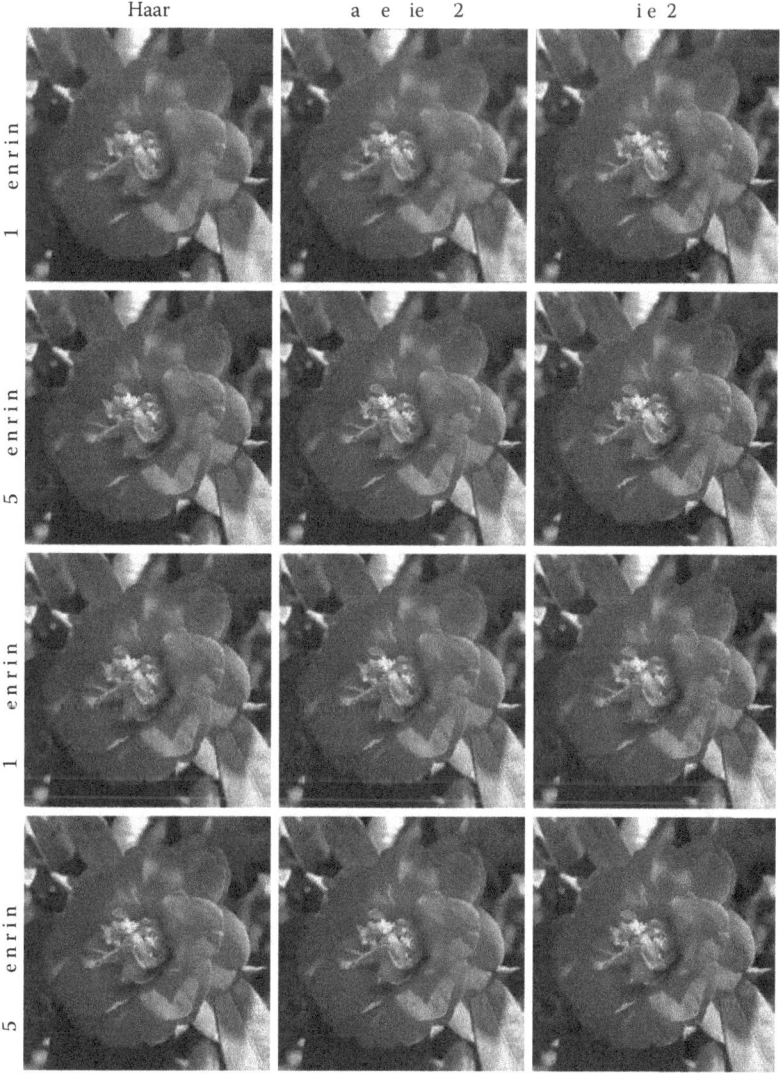

Fig. 19 Wavelets allow images to be progressively reconstructed, which can be used in transmission. Here, we use the best 1%, 5%, 10%, and 50% of the wavelet coefficients (top to bottom), using three different wavelets (left to right): Haar, Daubechies D20, and Coiflet (2)

This method is essentially the wavelet shrinkage denoising method.[85–88] It can be extended by replacing coefficients according to some function, i.e., $c'_{j,k} = g(c_{j,k})$. Assuming that the coefficients $c_{j,k}$ are polluted by additive noise:

$$c_{j,k} = c^s_{j,k} + n(i,k) \qquad (58)$$

Where $c^s_{j,k}$ is the noise-free coefficient and $n(i, k)$ is an independent noise component. It is then possible to replace the observed coefficients $c_{j,k}$ by an optimal linear estimate:[89]

$$c'_{j,k} = c_{j,k} \frac{E(c^2_{j,k})^2}{E(c^s_{j,k} + n(j,k))^2} \qquad (59)$$

APPLICATION: PROGRESSIVE RECONSTRUCTION

Wavelets can be configured to allow progressive reconstruction. Here, the wavelet coefficients can be sorted according to magnitude. If such sorted wavelet coefficients are then transmitted, the receiver can start reconstructing a crude image with few coefficients and refine the image as more coefficients are received.[12,90,91,92] For Haar, Coiflet, and Daubechies wavelets, an example is shown in Fig. 19.

APPLICATION: TEXTURE SYNTHESIS

This book is predominantly about natural image statistics. We see textures as a specific form of images, namely those that have stationary statistics. In other words, each local region of a texture has the same statistics, whereas this is not necessarily true for natural images. As an example, a natural image depicting a forest with a sky above it would have very different gradient statistics in the sky than in the forest. This means that textures are in some sense more amenable to statistical modeling than arbitrary natural images.

Textures have been described in terms of nth-order joint empirical densities of image pixels[93] and subsequently as statistical interactions in local neighborhoods by means

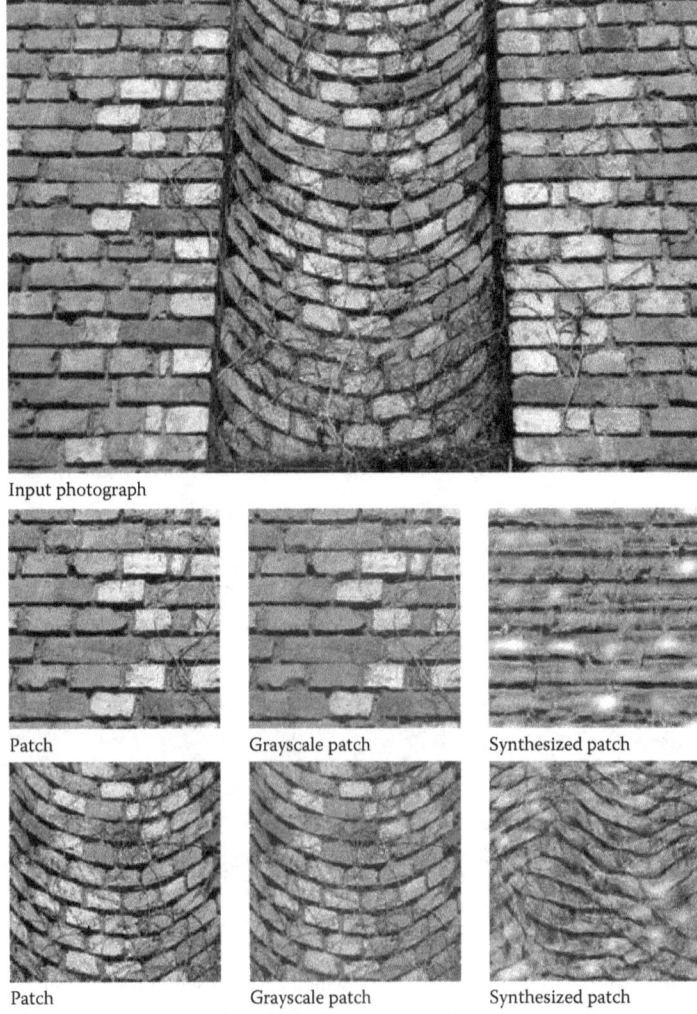

Fig. 20 Texture synthesis examples, using analysis/synthesis on complex wavelets.[107] (Stockholm, Sweden, 2009)

of Markov random fields[94–97]. Inspired by knowledge of early visual processing in human and mammalian vision, textures can also be analyzed and represented at multiple spatial scales with oriented linear kernels.[98,99] These, in turn, have given rise to the use of wavelet representations for the purpose of texture synthesis.[100–106]

Statistics of complex wavelet decompositions have also proven successful in the area of texture synthesis. Here, correlations between pairs of coefficients across space, scale, and orientation in an example texture form constraints, which are augmented with a selection of additional constraints, including the expected product of the magnitudes of coefficient pairs, marginal statistics of image pixels, and low-pass coefficients.[107] An efficient sampling algorithm then creates a new texture subject to these constraints. Thus, the new texture will have the same correlations as those measured in the wavelet decomposition of the example texture.

A texture can be represented as a two-dimensional homogeneous random field defined on integer positions.[a] Its statistical properties can be connected to human visual perception by making the basic assumption that a set of functions can be defined such that if we were to draw samples from two random fields X and Y that are equal in expectation over these functions, they would be visually indistinguishable.[93] This is known as the Julesz conjecture. Assuming we have such a set of functions ϕ_k, then perceptual equivalence would be obtained if:

$$E(\phi_k(X)) = E(\phi_k(Y)) \quad \forall k \tag{60}$$

In a synthesis-by-analysis approach, one would analyze an example texture to find that the expected value for each of these functions is a specific value c_k. The idea is then to generate a new random field such that it has the same expected values for each of the constraint functions:

$$E(\phi_k(X)) = c_k \quad \forall k \tag{61}$$

Portilla and Simoncelli have developed an efficient sampling algorithm that can generate such textures.[107] Rather than aim to satisfy all k constraints simultaneously, they treat each constraint sequentially.

Of interest, then, is the set of constraint functions that would lead to perceptual equivalence. It can be argued that to achieve this goal, constraint functions that emulate the early stages of human vision can be chosen. Portilla and Simoncelli have demonstrated that a viable route would be to linearly decompose an image into a multiscale representation such that the basis functions are localized, oriented, and about one-octave in bandwidth.[107] Their steerable pyramid[108,109] includes complex filters to allow local phase information to be incorporated in the analysis.

The statistical constraints are then chosen on this basis using a technique akin to greedy algorithms: one by one a new constraint is added that captures the visual quality most noticeably missing while verifying that previously added constraints are still relevant. The resulting constraint set contains marginal statistics on texture pixels, coefficient correlations, magnitude correlation, and cross-scale phase statistics. An example of texture synthesis achieved with this method is shown in Fig. 20.

REFERENCES

1. Vetterli, M.; Kovačević, J. *Wavelets and Subband Coding*; Prentice Hall: Englewood Cliffs, NJ, 2005.
2. Kovačević, J.; Goyal, V.K.; Vetterli, M. *Signal Processing Fourier and Wavelet Representations*, 2012, fourierandwavelets.org.
3. Tkacik, G.; Prentice, J.S.; Victor, J.D.; Balasubramanian, V. Local statistics in natural scenes predict the saliency of synthetic textures. Proc. Nat. Acad. Sci. **2010**, *107* (42), 18149–18154.
4. Daubechies, I. *Ten Lectures on Wavelets*; Society for Industrial and Applied Mathematics: Philadelphia, PA, 1992.
5. Mallat, S.G. *A Wavelet Tour of Signal Processing*, 3rd Ed.; Elsevier: Amsterdam, 2009.
6. Strang, G.; Nguyen, T. *Wavelets and Filter Banks*, 2nd Ed.; Wellesley-Cambridge Press: Wellesley, MA, 1996.
7. Shapiro, J. Embedded image coding using zerotrees of wavelet coefficients. IEEE Trans. Signal Process. **1993**, *41* (12), 3445–3462.
8. Haar, A. Zur Theorie der orthogonalen Funktionensysteme. Mathematische Annalen **1910**, *69* (3), 331–371.
9. Mallat, S.G. A theory for multiresolution signal decomposition: The wavelet representation. IEEE Trans. Pattern Anal. Mach. Intell. **1989**, *11* (7), 674–693.
10. Mallat, S.G. Multiresolution approximations and wavelet orthonormal bases of L 2 (R). Trans. Am. Math. Soc. **1989**, *315* (1), 69–87.
11. Daubechies, I. The wavelet transform, time-frequency localization and signal analysis. IEEE Trans. Inf. Theory **1990**, *36* (5), 961–1005.
12. Antonini, M.; Barlaud, M.; Mathieu, P.; Daubechies, I. Image coding using the wavelet transform. IEEE Trans. Image Process. **1992**, *1* (2), 205–220.
13. Burrus, C.S.; Gopinath, R.A.; Guo, H. *Introduction to Wavelets and Wavelet Transforms: A Primer*; Prentice-Hall: Upper Saddle River, NJ, 1998.
14. Beylkin, G.; Coifman, R.; Rokhlin, V. Fast wavelet transforms and numerical algorithms I. Commun. Pure. Appl. Math. **1991**, *44* (2), 141–183.
15. Huang, S.-J.; Hsieh, C.-T. Coiflet wavelet transform applied to inspect power system disturbance-generated signals. IEEE Trans. Aerosp. Electron. Syst. **2002**, *38* (1), 204–210.
16. Do, M.N.; Vetterli, M. The contourlet transform: An efficient directional multiresolution image representation. IEEE Trans. Image Process. **2005**, *14* (12), 2091–2106.

---

[a] Note that an image is not necessarily homogeneous, while a texture is assumed to be statistically homogeneous.

17. Candés, E.; Donoho, D.L. Curvelets: A surprisingly effective nonadaptive representation of objects with edges. In *Curve and Surface Fitting: Saint Malo 1999*, Cohen, A.; Rabut, C.; Schumaker, L.L.; Eds.; Vanderbilt University Press: Nashville, TN, 2000.
18. Candés, E. Ridgelets: Theory and applications. PhD Thesis, Department of Statistics, Stanford University, 1998.
19. Candés, E.; Donoho, D.L. Ridgelets: The key to high-dimensional intermittency. Philos. Trans. R. Soc. A **1999**, *357* (1760), 2495–2509.
20. Field, D.J. Relations between the statistics of natural images and the response properties of cortical cells. J. Opt. Soc. Am. A **1987**, *4* (12), 2379–2394.
21. Simoncelli, E.P.; Adelson, E.H. Noise removal via Bayesian wavelet coring. In *Proceedings of the 3rd IEEE International Conference on Image Processing*, Vol. I, 1996, 379–392.
22. Buccigrossi, R.W.; Simoncelli, E.P. Progressive wavelet image coding based on a conditional probability model. In *Proceedings of the IEEE International Conference on Acoustics, Speech, and Signal Processing*, Vol. 4, Munich, 1997, 2957–2960.
23. Simoncelli, E.P. Modeling the joint statistics of images in the wavelet domain. In *Proceedings of the SPIE 44th Annual Meeting*, Vol. 3813, 1999, 188–195.
24. Reinhard, E.; Stark, M.; Shirley, P.; Ferwerda, J. Photographic tone reproduction for digital images. ACM Trans. Graphics **2002**, *21* (3), 267–276.
25. Wu, Y.N.; Guo, C.E.; Zhu, S.C. From information scaling of natural images to regimes of statistical models. Q. Appl. Math. **2008**, *66* (1), 81–122.
26. Po, D.-Y.; Do, M.N. Directional multiscale modeling of images using the contourlet transform. IEEE Trans. Image Process. **2006**, *15* (6), 1610–1620.
27. Donoho, D.L.; Flesia, A. Can recent innovations in harmonic analysis "explain" key findings in natural image statistics? Network: Comput. Neural Syst. **2001**, *12* (3), 371–393.
28. Zetzsche, C.; Krieger, G. Exploitation of natural scene statistics by orientation selectivity and cortical gain control. Percept. ECVP **1998**, Supplement, 27, 154.
29. Srivastava, A.; Lee, A.B.; Simoncelli, E.P.; Zhu, S.-C. On advances in statistical modeling of natural images. J. Math. Imaging Vision **2003**, *18* (1), 17–33.
30. Sheikh, H.R.; Bovik, A.C.; De Veciana, G. An information fidelity criterion for image quality assessment using natural scene statistics. IEEE Trans. Image Process. **2005**, *14* (12), 2117–2128.
31. Sheikh, H.R.; Bovik, A.C.; Cormack, L. No-reference quality assessment using natural scene statistics: JPEG2000. IEEE Trans. Image Process. **2005**, *14* (11), 1918–1927.
32. Graham, D.J.; Friedenberg, J.D.; Rockmore, D.N.; Field, D.J. Mapping the similarity space of paintings: Image statistics and visual perception. Visual Cognition **2010**, *18* (4), 559–573.
33. Lyu, S.; Rockmore, D.; Farid, H. A digital technique for art authentication. Proc. Nat. Acad. Sci. USA **2004**, *101* (49), 17006–17010.
34. Lyu, S.; Farid, H. Detecting hidden messages using higher-order statistics and support vector machines. In *Information Hiding*, 2003, 340–354.
35. Lyu, S.; Farid, H. Steganalysis using higher-order image statistics. IEEE Trans. Inf. Forensics Secur. **2006**, *1* (1), 111–119.
36. Lyu, S.; Farid, H. How realistic is photorealistic? IEEE Trans. Signal Process. **2005**, *53* (2), 845–850.
37. Mumford, D.; Gidas, B. Stochastic models for generic images. Q. Appl. Math. **2001**, *59* (1), 85–112.
38. Zhu, Z.; Wu, S.; Rahardja, S.; Fränti, P. Real-time ghost removal for composing high dynamic range images. In *Proceedings of the 5th IEEE International Conference on Industrial Electronics and Applications*, 2010, 1627–1631.
39. Alvarez, L.; Gousseau, Y.; Morel, J.-M. The size of objects in natural and artificial images. Adv. Imaging Electron. Phys. **1999**, *111*, 167–242.
40. Liu, J.; Moulin, P. Information-theoretic analysis of interscale and intrascale dependencies between image wavelet coefficients. IEEE Trans. Image Process. **2001**, *10* (11), 1647–1658. 194.
41. Wainwright, M.J.; Schwartz, O.; Simoncelli, E.P. Natural image statistics and divisive normalization: Modeling non-linearities and adaptation in cortical neurons. In *Statistical Theories of the Brain*, Rao, R.; Olshausen, B.; Lewicki, M.; Eds.; MIT Press: Cambridge, MA, 2002, 203.
42. Grenander, U.; Srivastava, A. Probability models for clutter in natural images. IEEE Trans. Pattern Anal. Mach. Intell. **2001**, *23* (4), 424–429.
43. Teo, P.C.; Heeger, D.J. Perceptual image distortion. In *Proceedings of the IEEE International Conference on Image Processing*, Vol. 2, 1994, 982–986.
44. Schwartz, O.; Simoncelli, E.P. Natural signal statistics and sensory gain control. Nat. Neurosci. **2001**, *4* (8), 819–825.
45. Simoncelli, E.P. Statistical models for images: Compression, restoration and synthesis. In *Proceedings of the 31st Asilomar Conference on Signals, Systems and Computers*, 1997, 673–678.
46. Buccigrossi, R.W.; Simoncelli, E.P. Image compression via joint statistical characterization in the wavelet domain. IEEE Trans. Image Process. **1999**, *8* (12), 1688–1701.
47. Kingsbury, N.G. Image processing with complex wavelets. Philos. Trans. R. Soc. A **1999**, *357* (1760), 2543–2560.
48. XXX–XXX.
49. Choi, H.; Romberg, J.; Baraniuk, R.G.; Kingsbury, N. Hidden Markov tree modeling of complex wavelet transforms. In *Proceedings of the IEEE International Conference on Acoustics, Speech, and Signal Processing*, Vol. 1, 2000, 133–136.
50. Gabor, D. Theory of communication. J. Inst. Electr. Eng. (JIEE) **1946**, *93* (III), 429–457. 198.
51. Hahn, S. *Hilbert Transforms in Signal Processing*; Artech House: Boston, MA, 1996. 198.
52. Ates, H.F.; Orchard, M.T. A nonlinear image representation in wavelet domain using complex signals with single quadrant spectrum. In *Proceedings of the Asilomar Conference on Signals, Systems and Computers*, Vol. 2, 2003, 1966–1970.
53. Belzer, B.; Lina, J.M.; Villasenor, J. Complex, linear-phase filters for efficient image coding. IEEE Trans. Signal Process. **1995**, *40* (4), 2425–2427.
54. Fernandez, F.; Wakin, M.; Baraniuk, R.G. Non-redundant, linear-phase, semiorthogonal, directional complex wavelets. In *Proceedings of the IEEE International Conference*

on *Acoustics, Speech, and Signal Processing*, Vol. 2, 2004, 953–956.
55. Lina, J.M.; Mayrand, M. Complex Daubechies wavelets. Appl. Comput. Harmon. Anal. **1995**, *2* (3), 219–229. 199.
56. van Spaendonck, R.; Blu, T.; Baraniuk, R.; Vetterli, M. Orthogonal Hilbert transform filter banks and wavelets. In *Proceedings of the IEEE International Conference on Acoustics, Speech and Signal Processing*, Vol. 6, 2003, 505–508.
57. Wakin, M.; Orchard, M.; Baraniuk, R.G.; Chandrasekaran, V. Phase and magnitude perceptual sensitivities in nonredundant complex wavelet representations. In *Proceedings of the Asilomar Conference on Signals, Systems and Computers*, vol. 2, 2003, 1413–1417.
58. Kingsbury, N.G. Shift invariant properties of the dual-tree complex wavelet transform. In *Proceedings of the IEEE International Conference on Acoustics, Speech and Signal Processing*, vol. 3, 1999, 1221–1224.
59. Kingsbury, N.G. Complex wavelets for shift invariant analysis and filtering of signals. J. Appl. Comput. Harmon. Anal. **2001**, *10* (3), 234–253.
60. Kingsbury, N.G. Design of Q-shift complex wavelets for image processing using frequency domain energy minimization. In *Proceedings of the IEEE International Conference on Image Processing*, Vol. 1, 2003, 1013–1016.
61. XXX-XXX.
62. Selesnick, I.W. The design of Hilbert transform pairs of wavelet bases via the flat delay filter. In *Proceedings of the IEEE International Conference on Acoustics, Speech, and Signal Processing*, vol. 6, 3.
63. Romberg, J.K.; Choi, H.; Baraniuk, R.G. Multiscale edge grammars for complex wavelet transforms. In *Proceedings of the IEEE International Conference on Image Processing*, vol. 1, 2003, 614–617.
64. Ye, Z.; Lu, C.-C. A complex wavelet domain Markov model for image denoising. In *Proceedings of the IEEE International Conference on Image Processing*, vol. 3, 2003, 365–368.
65. XXX-XXX.
66. Shi, F.; Selesnick, I.W. Video denoising using oriented complex wavelet transforms. In *Proceedings of the IEEE International Conference on Acoustics, Speech and Signal Processing*, Vol. 2, 2004, 949–952.
67. XXX-XXX.
68. Kokare, M.; Biswas, P.K.; Chatterji, B.N. Rotation invariant features using rotated complex wavelet for content based image retrieval. In *Proceedings of the IEEE International Conference on Image Processing*, vol. 1, 2004, 393–396.
69. Lo, E.; Pickering, M.; Frater, M.; Arnold, J. Scale and rotation invariant texture features from the dual-tree complex wavelet transform. In *Proceedings of the IEEE International Conference on Image Processing*, vol. 1, 2004, 227–230.
70. de Rivaz, P.F.C.; Kingsbury, N.G. Complex wavelet features for fast texture image retrieval. In *Proceedings of the IEEE International Conference on Image Processing*, vol. 1, 1999, 109–113.
71. Hatipoglu, S.; Mitra, S.K.; Kingsbury, N.G. Image texture description using complex wavelet transform. In *Proceedings of the IEEE International Conference on Image Processing*, vol. 2, 2000, 530–533.
72. Hill, P.R.; Bull, D.R.; Canagarajah, C.N. Rotationally invariant texture features using the dual-tree complex wavelet transform. In *Proceedings of the IEEE International Conference on Image Processing*, vol. 3, 2000, 901–904.
73. Romberg, J.K.; Choi, H.; Baraniuk, R.G.; Kingsbury, N.G. Multiscale classification using complex wavelets and hidden Markov tree models. In *Proceedings of the IEEE International Conference on Image Processing*, vol. 2, 2000, 371–374.
74. de Rivaz, P.F.C.; Kingsbury, N.G. Bayesian image deconvolution and denoising using complex wavelets. In *Proceedings of the IEEE International Conference on Image Processing*, vol. 2, 2001, 273–276.
75. Jalobeanu, A.; Kingsbury, N.G.; Zerubia, J. Image deconvolution using hidden Markov tree modeling of complex wavelet packets. In *Proceedings of the IEEE International Conference on Image Processing*, vol. 1, 2001, 201–204.
76. Earl, J.W.; Kingsbury, N.G. Spread transform watermarking for video sources. In *Proceedings of the IEEE International Conference on Image Processing*, vol. 2, 2001, 491–494.
77. Loo, P.; Kingsbury, N.G. Digital watermarking using complex wavelets. In *Proceedings of the IEEE International Conference on Image Processing*, vol. 3, 2000, 29–32.
78. Shi, F.; Selesnick, I.W.; Cai, S. Image sharpening via image denoising in the complex wavelet domain. In *Proceedings of Wavelet Applications and Signal Processing X*, 2003, 467–474.
79. Margarey, J.F.A.; Kingsbury, N.G. Motion estimation using a complex-valued wavelet transform. IEEE Trans. Signal Process. **1998**, *46* (4), 1069–1084.
80. Reeves, T.H.; Kingsbury, N.G. Overcomplete image coding using iterative projection-based noise shaping. In *Proceedings of the IEEE International Conference on Image Processing*, vol. 3, 2002, 597–600.
81. Sivaramakrishnan, K.; Nguyen, T. A uniform transform domain video codec based on dual tree complex wavelet transform. In *Proceedings of the IEEE International Conference on Acoustics, Speech and Signal Processing*, vol. 3, 2001, 1821–1824.
82. Wang, B.; Wang, Y.; Selesnick, I.W.; Vetro, A. An investigation of 3D dualtree wavelet transforms for video coding. In *Proceedings of the IEEE International Conference on Image Processing*, vol. 2, 2004, 1317–1320.
83. Field, D.J. Scale-invariance and self-similar "wavelet" transforms: An analysis of natural scenes and mammalian visual systems. In *Wavelets, Fractals and Fourier Transforms*, Farge, M.; Hunt, J.C.R.; Vassilicos, J.C.; Eds.; Clarendon Press: Oxford, 1993, 151–193.
84. Weaver, J.B.; Yansun, X.; Healy Jr., D.M.; Cromwell, L.D. Filtering noise from images with wavelet transforms. Magn. Reson. Med. **1991**, *21* (2), 288–295.
85. Unser, M. Wavelets, statistics, and biomedical applications. In *Proceedings of the 8th IEEE Signal Processing Workshop on Statistical Signal and Array Processing*, 1996, 244–249.
86. Donoho, D.L.; Johnstone, I.M. Ideal spatial adaptation via wavelet shrinkage. Biometrika **1994**, *81* (3), 425–455.
87. Donoho, D.L. De-noising by soft-thresholding. IEEE Trans. Inf. Theory **1995**, *41* (3), 613–627.

88. DeVore, R.A.; Lucier, B.J. Fast wavelet techniques for near-optimal image processing. In *Proceedings of the IEEE Military Communications Conference*, New York, 1992, 48.3.1–48.3.7.
89. Bertrand, O.; Bohorquez, J.; Pernier, J. Time-frequency digital filtering based on an invertible wavelet transform: An application to evoked potentials. IEEE Trans. Biomed. Eng. **1994**, *41* (1), 77–88.
90. Berman, D.F.; Bartell, J.T.; Salesin, D.H. Multiresolution painting and compositing. In *SIGGRAPH '94: Proceedings of the 21st Annual Conference on Computer Graphics and Interactive Systems*, Orlando, FL, 1994, 85–90.
91. Certain, A.; Popović, J.; DeRose, T.; Duchamp, T.; Salesin, D.; Stuetzle, W. Interactive multiresolution surface viewing. In *SIGGRAPH '96: Proceedings of the 23rd Annual Conference on Computer Graphics and Interactive Techniques*, New Orleans, 1996, 91–98.
92. Rempel, A.G. Fast progressive transmission of images using wavelets with sorted coefficients. Master's Thesis, University of Saskatchewan, 1993.
93. Julesz, B. Visual pattern discrimination. IRE Trans. Inf. Theory **1962**, *8* (2), 84–92.
94. Hassner, M.; Sklansky, J. The use of Markov random fields as models of texture. Comput. Graphics Image Process. **1980**, *12* (4), 357–370.
95. Cross, G.R.; Jain, A.K. Markov random field texture models. IEEE Trans. Pattern Anal. Mach. Intell. **1983**, *5* (1), 25–39.
96. Geman, S.; Geman, D. Stochastic relaxation, Gibbs distributions, and the bayesian restoration of images. IEEE Trans. Pattern Anal. Mach. Intell. **1984**, *6* (6), 721–741.
97. Derin, H.; Elliott, H. Modeling and segmentation of noisy and textured images using Gibbs random fields. IEEE Trans. Pattern Anal. Mach. Intell. **1987**, *9* (1), 39–55.
98. Turner, M.R. Texture discrimination by Gabor functions. Biol. Cybern. **1986**, *55* (2), 71–82.
99. Malik, J.; Perona, P. Preattentive texture discrimination with early vision mechanisms. J. Opt. Soc. Am. A **1990**, *7* (5), 923–932.
100. Cano, D.; Minh, T.H. Texture synthesis using hierarchical linear transforms. Signal Process. **1988**, *15* (2), 131–148.
101. Porat, M.; Zeevi, Y.Y. Localized texture processing in vision: Analysis and synthesis in the Gaborian space. IEEE Trans. Biomed. Eng. **1989**, *36* (1), 115–129.
102. Popat, K.; Picard, R.W. Novel cluster-based probability model for texture synthesis, classification, and compression. In *Visual Communications '93*; International Society for Optics and Photonics, 1993, 756–768.
103. Heeger, D.J.; Bergen, J.R. Pyramid-based texture analysis/synthesis. In *SIGGRAPH '95: Proceedings of the 22nd Annual Conference on Computer Graphics and Interactive Techniques*, ACM: New York, 1995, 229–238.
104. Portilla, J.; Navarro, R.; Nestares, O.; Tabernero, A. Texture synthesis-by-analysis method based on a multiscale early-vision model. Opt. Eng. **1996**, *35* (8), 2403–2417.
105. Zhu, S.C.; Wu, Y.; Mumford, D. Filters, random fields and maximum entropy (FRAME): Towards a unified theory for texture modeling. Int. J. Comput. Vision **1998**, *27* (2), 107–126.
106. De Bonet, J.S.; Viola, P.A. A non-parametric multi-scale statistical model for natural images. In *NIPS-11: Proceedings of the 1998 Conference on Advances in Neural Information Processing Systems*, 1998, 773–779.
107. Portilla, J.; Simoncelli, E.P. Texture modeling and synthesis using joint statistics of complex wavelet coefficients. In *Proceedings of the IEEE Workshop on Statistical and Computational Theories of Vision*, 1999.
108. Simoncelli, E.P.; Freeman, W.T.; Adelson, E.H.; Heeger, D.J. Shiftable multiscale transforms. IEEE Trans. Inf. Theory **1992**, *38* (2), 587–607.
109. Simoncelli, E.P.; Freeman, W.T. The steerable pyramid: A flexible architecture for multi-scale derivative computation. In *Proceedings of the 2nd IEEE International Conference on Image Processing*, 1995, 444–447.

Index

A

Abelian group and group laws, 452–453
Absolute difference method, 575
Absolute emissivity, 694
Accident detection, 743–744
Action recognition, 795–797
Active learning, 773
Active remote sensing, 590
Adaptation, 675
Adaptive perceptual image, and video compression, 545–546
Adder modeling, 469–471
AddRoundKey, 448
Advanced encryption standard, 445–446
AEB (Automatic Exposure Bracketing), 550
Agriculture, and remote sensing image fusion, 595
Alchemy Sims Grid (ASG), 805
Al-Haytham, Alhazen Ibn, 802
Amacrine and ganglion cells, 678
Amber Alert, 788
Angle perception, 117
Angular dependence, 205–206
Anisotropic diffusion, 295–296
Anticipation, 161
Appeal, 163
Archive interfaces, 650
Arcs, 162
Arithmetic coding, 602
Arrayed waveguide gratings (AWGs), 513–515
Art statistics, 210–212
ASG, *see* Alchemy Sims Grid (ASG)
Associative mechanisms, fuzzy, 220
  clustering for documents, 230–231
  fuzzy thesauri, 229–230
  performance measures, 231
Asymptotic consistency, 669
Autocorrelation function, 195–197
Automatic Exposure Bracketing (AEB), 550
Automatic thresholding, 327–328
Auto-models, 415
Avatars, 157–158
Average distributions, 169–171
AWGs, *see* arrayed waveguide gratings (AWGs)
Axiomatization, 43–44

B

BABC, *see* binary artificial bee colony (bABC)
Background removal, 736–737
Background subtractions, 575–576
Backscattering, 635–636
Backward difference method, 236
Back water/fluctuating lake levels, 178
Basis functions, 150
Basis pursuit approach, 665
BB-RBM, *see* Bernoulli–Bernoulli RBM (BB-RBM)
Believability flip, 159
Bernoulli–Bernoulli RBM (BB-RBM), 772–773
BGA, *see* binary genetic algorithm (BGA)

Bicubic interpolation, 310–311
Bidirectional reflectance distribution function (BRDF), 425–426
Big data, 408–409
Bilinear interpolation, 309–310
Binary artificial bee colony (bABC), 127
Binary GCD processor, 466–467
Binary genetic algorithm (BGA), 126
Binary particle swarm optimization (BPSO), 126–127
Binary robust independent elementary features (BRIEF), 573–574
Biological motion and people detection, 718
Bipolar cells, 677–678
Bitstream-based visual quality assessment, 528
Block ciphers
  AddRoundKey, 448
  advanced encryption standard, 445–446
  inner structures of, 444–445
  key-scheduling, 448
  MixColumns, 447–448
  overview, 443–444
  ShiftRows, 447
  SubBytes, 446–447
Blood-oxygenation-level-dependent (BOLD), 666
BOLD, *see* blood-oxygenation-level-dependent (BOLD)
Boolean algebras
  application, 44–45
  axiomatization, 43–44
  description, 40–41
  examples, 41–43
  overview, 40
BOW-based method, 190
BPSO, *see* binary particle swarm optimization (BPSO)
BRDF, *see* bidirectional reflectance distribution function (BRDF)
BRIEF, *see* Binary robust independent elementary features (BRIEF)
Broviak, Pam, 806
B-spline approach, 768

C

Cameras
  in motion detection and analysis, 794–795
  stationary, 791–794
Cantor sets, 46–47
Cardano cryptography, 817
Cascade algorithm, 828
CAVE, *see* cave automatic virtual environment (CAVE)
Cave automatic virtual environment (CAVE), 801–802
CCSDS, *see* Consultative Committee for Space Data Systems (CCSDS)
Central difference method, 236
Central processing unit (CPU), 749
Change detection, 595
Characteristic 2, 453–456
Characteristic function, 824

Chiral admittance, 634–635
Chiral/chiro-ferrite-coated twisted PEC cylinder
  backscattering, 635–636
  chiral admittance, 634–635
  description, 632–634
  gyrotropy effect, 635
Chiral-coated twisted cylindrical mediums
  circular dielectric cylinder, 630–631
  elliptical dielectric cylinder, 628–629
Chiral cylinder, 616–617
Chiral nihility medium-based cylinder, 617–618
CIE Lab systems, 552–553
Circular convolution *versus* linear convolution, 151–152
Circular dielectric cylinder, 630–631
CityGML, 729
City/landscape classification, 658
Cliques, 413–414
Clustering-based methods, 791–792
CNNs, *see* convolutional neural networks (CNNs)
Coastal floodplains, 178
Coefficient histograms, 831–832
Coiflet wavelets, 828
Collision avoidance, 744
Color
  opponent processing, 55–58
  overview, 54
  space statistics
    constancy and white balancing, 65–67
    correlation analysis, 63–65
    Minkowski norm, 67–68
    normalization, 63
    white-balance algorithm selection, 68
  as 3D space, 55
  transfer
    as 3D problem, 61–62
    histogram features, 60–61
    through higher-order manipulation, 59–60
    through simple moments, 58–59
  trichromacy and metamerism, 55
Color adjustment, 316–317
Color classification, 656
Color descriptors, 340
Color difference representation, 75–76
Color distance, 552–553
Color features, 340
Color image denoising
  experimental results and discussions
    mean absolute error (MAE), 80
    normalized color distance (NCD), 80–92
    peak signal-to-noise ratio (PSNR), 80
    structural similarity index (SSIM), 80
  overview, 72–73
  quaternions
    application of, 76–78
    common properties of, 73–75
    representation of color differences, 75–76
    representation of color pixel, 75
    switching vector median filter, 78–79
Color image enhancement, 290–291

Color image segmentation metrics
 discussion and suggestion, 104–108
 hierarchical structure in, 96–99
 overview, 95–96
  quantitative metrics
   distance-based metrics, 103–104
   region/volume based, 99–103
  system design, 96
Color matching, 340
Color opponent theory, 682
Color pixel representation, 75
Color processing, 681–682
Color quality assessment, 528
Color restoration, 539
Color space transformations, 748
Combinatorial group testing, 665
Commission Internationale de l'Eclairage (CIE) Luv, 552–553
Complex models and patch-based regularities
 fields of experts, 417
 products of experts, 417
Complex wavelets, 834–836
Component substitution (CS)
 Gram-Schmidt fusion, 592–593
 intensity hue saturation fusion, 591–592
 principal component substitution, 592
Composite objects, 339
Compressed sensing, 667–668
Computer-created 3-D space, 802–803
Computer network diagnosis, 665–666
Congestion detection, 742–743
Constancy and white balancing, 65–67
Consultative Committee for Space Data Systems (CCSDS), 602
Content-based image search and retrieval, 656
Context of content analysis, 654
Contourlets, 831
Contour- or silhouette-based object detection, 574
Contrast, 681
Contrast enhancement
 color restoration, 539
 histogram equalization approach using GMM, 540–541
 image fusion, 541–542
 tone mapping and HDR, 542–543
Contrast/texture masking, 533–535
Convolution, 752–753
Convolutional neural networks (CNNs), 762
Copy forgery detection techniques, 184
Co-registration, 756–757
Correlation analysis, 63–65
Correlation-based motion detection, 717
Correlations
 between coefficients, 833–834
 between scales, 836–837
Cost function, 125–126
CPU, *see* central processing unit (CPU)
Craik-O'Brien-Cornsweet illusion, 683
Cross-correlation, 325–327
Crossover effect, 390–391
Cryptography
 block ciphers
  AddRoundKey, 448
  advanced encryption standard, 445–446
  inner structures of, 444–445
  key-scheduling, 448

MixColumns, 447–448
 overview, 443–444
 ShiftRows, 447
 SubBytes, 446–447
elliptic curves
 abelian group and group laws, 452–453
 with characteristic 2, 453–456
 projective coordinate representation, 456
 singularity of curves, 452
 Weierstraß equation, 451–452
modern hardware design practices
 adder modeling, 469–471
 binary GCD processor, 466–467
 components of, 465–466
 delay estimates, 471–472
 enhancing performance of, 467–469
 field programmable gate arrays (FPGAs), 460–465
 LUT utilization, 471
 multiplexer modeling, 471
 overview, 460
 total LUT estimate, 471
Montgomery's algorithm
 faster multiplication on EC, 458
 Montgomery's ladder, 457–458
 projective co-ordinates, 458–459
overview, 439–440
Rijndael in composite field
 expressing an element, 448–449
 inversion of element, 449
 round of AES in, 449–450
 scalar multiplications, 456–457
 technical details, 440–443
Cryptosystems
 key binding technique, 481
 key generation technique, 481–482
CS, *see* component substitution (CS)
Cubic trivariate lagrange polynomial, 698–703
CUDA architecture and programming model, 748–750
Current density distributions, 624–628
Curvelets, 831
Curve singularity, 452

D

Damage assessment, 764
Damage detection, 762
Dark-is-deep paradigm, 173–174
Dascyllus reticulatus, 774
Data analysis, 332–334
Databases, image and video quality
 image content in, 526
 representative, 526
 test methodologies, 527–528
 video content in, 526–527
Data compression, 336–337
Data hiding scheme, 480
Data models, 650–651
Datasets
 fish trajectory, 774–775
 forum pedestrian, 775–776
DBN, *see* dynamic Bayesian networks (DBN)
DCT, *see* discrete cosine transform (DCT)
"Dead leaves" model, 115–116
Deblurring, 248–249, 536–537
Decision-level fusion, 479
Deconvolution, 324–325

Defining edges, 322–323
De Florez, Luis, 802
Delay estimation, 471–472
Della Porta, Giambattista, 802
Denoising, 536
Dense optical flow technique (Gunner Farneback's optical flow), 577
Depth completion results, 436–437
Depth image completion, 436
Depth reconstruction, 119–120
Depth statistics
 "dead leaves" model, 115–116
 depth reconstruction, 119–120
 overview, 114–115
 scene geometry perception
  description, 116
  length perception, 117
  orientation and angle perception, 117
  2D and range statistics, 117–119
Description graphs, 661
Design/measurement matrix, 668
DFT, *see* discrete Fourier transform (DFT)
Dictionary learning, 668
Differential absorption, 694
Diffusion-based image inpainting, 296–297
Digital aerial photogrammetry, 179
Digital elevation model (DEM), 178–179
 remotely sensed, 179
Digital halftoning
 binary artificial bee colony (bABC), 127
 binary genetic algorithm (BGA), 126
 binary particle swarm optimization (BPSO), 126–127
 objective function formulation
  cost function, 125–126
  human vision system in, 125
  overview, 123–125
 p-LUT implementation, 127–129
Digital image
 description, 271–272
 visualization, 272–273
Digital libraries, 346–349
 medical image databases, 347–348
 for oil industry, 348–349
 remotely sensed image databases, 348–349
Digital photographic image similarity index, 554
Dilation, 825
Dimensionality reduction, 667
 Gaussian mixture models, 144–146
 independent component analysis (ICA)
  description, 141–142
  on natural images, 142–144
 overview, 135
 principal component analysis (PCA)
  description, 135–137
  eigenfaces, 139–141
  on images, 139
  on patches, 138–139
  on pixels, 138
  whitening, 137–138
Dircks, Henry, 802
Disaster management, UAVs for
 damage assessment, 764
 real-time data collection, 763–764
 real-time event tracking and monitoring, 763
 real-time map generation, 763–764

Index

Discrete cosine transform (DCT), 602, 783
Discrete Fourier transform (DFT), 602, 768
   basis functions, 150
   description, 148–149
   linear convolution *versus* circular convolution, 151–152
   in matrix form, 149–150
   properties of, 150–151
      m-point, 151
   RADIX-2 algorithm for FFT, 152–153
   zero padding, 151
Discrete wavelet transform (DWT), 602, 783
Disney's principles of animation
   anticipation, 161
   appeal, 163
   arcs, 162
   ease-in and ease-out, 161–162
   exaggeration, 162
   follow-through and overlapping action, 162
   secondary motion, 162
   squash and stretch, 161
   staging, 161
Distance-based metrics, 103–104
Distortion measurements, 609–610
Document indexing, 219
   vector space, probabilistic, and generalized Boolean, 220–224
Document quality, 491–492
Double opponency, 678
Driver assistance systems, 744
Driver fatigue detection, 744
Drop holes
   description, 559–561
   field comparisons, 561–564
2D wavelets, 829–831
DWT, *see* Discrete wavelet transform (DWT)
Dynamic Bayesian networks (DBN), 770
Dynamic occlusion, 718–719
Dynamic range, 203–205, 552

E

Ease-in and ease-out, 161–162
Edge detection
   image, 275
   processes, 239–241
Edge detection, information sets for
   edge strength factor
      effect of crossover on, 390–391
      effect of threshold, 391–392
      effect of parameters on FoM, 392–393
   fractional derivatives, 379
      application, 382–383
      FOM measure comparison using, 387–390
      qualitative comparison using, 386
      simulations on noisy images, 386
   gamma function
      fuzzification of fractional gradient, 381–382
      Grunwald–Letnikov fraction derivative, 380
      Sobel fractional derivative masks, 380–381
   overview, 378–379
   proposed edge detector using fractional derivative mask, 384–385
   quantitative measures of analysis
      edge strength factor, 383–384
      optimal crossover, 384
      Pratt's figure of merit, 383
      proposed algorithm using USAN, 384
      structural similarity (SSIM) index, 383
   SUSAN principle
      application, 382
      qualitative comparison using, 386
      quantitative comparison using, 386–387
      simulations on noisy images, 386
      univalue segment assimilating nucleus (USAN), 379
   use of, 382
Edges
   definition, 239
   statistics, 241–242
Edge strength factor
   effect of crossover on, 390–391
   effect of threshold, 391–392
   quantitative analysis measures, 383–384
Edison, Thomas, 802
Eigenfaces, 139–141
Electronic screens, 802
Elliptical dielectric cylinder, 628–629
Elliptical track feature (ETF)-based method, 189–190
Elliptic curves
   abelian group and group laws, 452–453
   with characteristic 2, 453–456
   projective coordinate representation, 456
   singularity of curves, 452
   Weierstraß equation, 451–452
Embedded methods, 667
Emissivity
   absolute, 694
   and Kirchhoff's law, 689
   relative, 694
Environment modeling
   lane detection, 738–739
   traffic sign detection and recognition, 739
ESA, *see* European Space Agency (ESA)
Essex MUD, 803
ETF, *see* elliptical track feature (ETF)-based method
European Space Agency (ESA), 589
Exaggeration, 162
Exemplar-based image inpainting, 300–302
Exner, Sigmund, 717
Exponential power distribution, *see* generalized Laplacian function
Eyes
   adaptation, 675
   amacrine and ganglion cells, 678
   horizontal and bipolar cells, 677–678
   light sensitivity, 675–677
   optical media and retina, 674
   overview, 673–674
   photoreceptor(s), 674–677
   photoreceptor mosaic, 677

F

Facial animation
   application areas
      games industry, 156
      medicine, 156–157
      social agents and avatars, 157–158
      social robots, 158–159
      video teleconferencing, 157
   believability flip and uncanny valley, 159
   description, 154–155
   Disney's principles of animation
      anticipation, 161
      appeal, 163
      arcs, 162
      ease-in and ease-out, 161–162
      exaggeration, 162
      follow-through and overlapping action, 162
      secondary motion, 162
      squash and stretch, 161
      staging, 161
   historical sketch of, 155–156
   overview, 154
   turing test for, 159–160
False color, 290
Fast directional chamfer matching (FDCM), 574
Father wavelet, 824
FBGs, *see* Fiber Bragg gratings (FBGs)
FDCM, *see* fast directional chamfer matching (FDCM)
Feature-based object detection
   features from accelerated segment test (FAST), 569–570
   Haar-like features, 573
   Harris corner detection, 569
   scale-invariant feature transform, 570–571
   speeded-up robust features (SURF), 571–573
Feature brightness, size, and location, 331–332
Feature combination, 187
Feature detectors, and N-Jet, 244–245
Feature extraction, 737
   description, 184–185
   feature combination, 187
   global features, 185
   learning-based features, 187–188
   local features, 185–186
Feature indexing, 188
Feature-level fusion, 477
Feature matching, 574
Feature selection, 774
Features from accelerated segment test (FAST), 569–570
Feature shape, 332
Feature transformation
   invertible, 480
   non-invertible, 480
Feature vectors, 273
Feedback (haptic) interfaces, 803
Fiber Bragg gratings (FBGs), 512–513
Field programmable gate arrays (FPGAs)
   architecture, 460–462
   design flow, 462–465
Fields of experts, 417
Filtering-based methods, 792–794
Finite-dimensional signal recovery, 669
Firestorm Viewer, 804
First-order statistics
   dark-is-deep paradigm, 173–174
   histograms and moments
      description, 165–166
      histogram adjustments, 168
      image moments and moment invariants, 166–168

First-order statistics (*Cont.*)
  material properties, 171–172
  moment statistics and average distributions, 169–171
  nonlinear compression, 172–173
  overview, 165
Fisher, Scott, 803
Fish trajectory datasets, 774–775
Floodplain mapping
  back water/fluctuating lake levels, 178
  challenges in, 178
  coastal, 178
  computation of slope and aspect, 180
  description, 177–178, 180–181
  digital aerial photogrammetry, 179
  digital elevation model (DEM), 178–179
    remotely sensed, 179
  light detection and ranging, 179–180
  optical and microwave remote sensors, 179
  overview, 177
  riverine, 178
  urban, 178
FMRI, *see* functional magnetic resonance imaging (fMRI)
Follow-through and overlapping action, 162
FOM measure comparison, 387–390
Foreign language OCR, 492–493
Forensics and intellectual property rights, 554
Forgery detection
  BOW-based method, 190
  copy techniques, 184
  elliptical track feature (ETF)-based method, 189–190
  feature extraction
    description, 184–185
    feature combination, 187
    global features, 185
    learning-based features, 187–188
    local features, 185–186
  geometric coding (GC)-based method, 190
  image matching
    feature indexing, 188
    image similarity measurement, 188–189
  LF-based method, 190–191
  ordinal measure method, 189
  overview, 183–184
  RANSAC-based method, 190
Forum pedestrian datasets, 775–776
Forward difference method, 235–236
Fourier analysis
  art statistics, 210–212
  autocorrelation function, 195–197
  Fourier transform, 197–199
  fractal forgeries, 208
  human perception, 207–208
  image processing and categorization, 208–209
  overview, 194–195
  phase spectra, 207
  power spectra
    angular dependence, 205–206
    dynamic range, 203–205
    1/f failures, 206–207
    image representation, 205
    slope computation, 199–201
    spectral slope analysis, 201–203
    temporal dependence, 206
  terrain synthesis, 210
  texture descriptors, 209–210
  Wiener-Khintchine theorem, 199
Fourier transform, 197–199
FPGAs, *see* field programmable gate arrays (FPGAs)
Fractal forgeries, 208
Fractional derivatives, 379
  application, 382–383
  FOM measure comparison using, 387–390
  qualitative comparison using, 386
  simulations on noisy images, 386
Fractional gradient fuzzification, 381–382
Frequency domain
  image enhancement in, 288–290
Frequency domain SR approach, 366–368
Full-reference visual quality assessment, 523–524
Functional magnetic resonance imaging (fMRI), 666
Fusion schemes
  decision-level, 479
  feature-level, 477
  score-level, 477–479
  sensor-level, 476–477
Fuzzy sets
  in image analysis, 50
  near, 50
  notion of, 50
Fuzzy set theory/fuzzy retrieval models
  associative mechanisms, 220
    clustering for documents, 230–231
    fuzzy thesauri, 229–230
    performance measures, 231
  document indexing, 219
    vector space, probabilistic, and generalized Boolean, 220–224
  information retrieval (IR)
    current trends in, 217
    imprecision/vagueness/uncertainty/inconsistency in, 217–219
    key issues in, 217
  overview, 216–217
  query weights
    evaluation mechanism, 224–225
    ideal, 226
    implicit, 226
    linguistic, 227–228
    linguistic quantifiers, 228
    query languages, 220
    relative importance, 226
    semantics, 226–227
    threshold, 226
Fuzzy taxonomic relations, 659–660
Fuzzy thesauri, 229–230

G

Games industry, 156
Gaming, and virtual worlds, 803
Gamma function
  fuzzification of fractional gradient, 381–382
  Grunwald–Letnikov fraction derivative, 380
  Sobel fractional derivative masks, 380–381
Ganglion cells, 678
Gaussian mixture models, 144–146
Gaussian mixture models (GMMs), 770
GC, *see* geometric coding (GC)-based method
GCP, *see* ground control point (GCP)
GCPs, *see* ground control points (GCPs)
Generalized error distribution, *see* generalized Laplacian function
Generalized Laplacian function, 832
General Linear Models (GLMs), 667
GEOBIA, *see* geographic object-based image analysis (GEOBIA)
Geographic information system (GIS), 764
  topology in
    description, 727
    metric-induced modeling, 728
    overview, 726–727
    qualitative description, 728
    recent advances, 729
    spatial relationships and topological maps, 727
    3D data handling, 728–729
Geographic object-based image analysis (GEOBIA), 612
Geology, and remote sensing image fusion, 595
GeoMean, 776
Geometric coding (GC)-based method, 190
Geometric distortion, 319
Geo-referencing, 750
Gibbs distributions, 414
GIS, *see* geographic information system (GIS)
GLCM, *see* gray-level co-occurrence matrix (GLCM) textures
GLMs, *see* General Linear Models (GLMs)
Global features, 185
Global navigation satellite system (GNSS), 761
Global positioning system (GPS), 761
GMMs, *see* Gaussian mixture models (GMMs)
GNSS, *see* global navigation satellite system (GNSS)
GPS, *see* global positioning system (GPS)
Gradients
  backward difference method, 236
  -based motion detection, 717
  central difference method, 236
  forward difference method, 235–236
  grayscale morphological image enhancement, 255
  second derivative methods, 237–238
  Söbel operator, 236–237
  statistics, 238
    single-image, 238–239
Gram-Schmidt fusion, 592–593
Graphics processing unit (GPU)-based approach
  convolution, 752–753
  co-registration, 756–757
  geo-referencing, 750
  gray-level co-occurrence matrix (GLCM) textures, 753–756
  implementation, 751–752
  lens distortion correction, 750
Graphs
  cliques, 413–414
  neighborhood systems, 413
Gray-level co-occurrence matrix (GLCM) textures, 753–756
Grayscale morphological image enhancement
  gradients, 255
  human vision perception, 260–261

morphological contrast, 256
  measuring, 266–267
morphological slope filters, 256–257
  sequential, 257
  weighted, 257–258
overview, 254–255
in poor lighting, 263–266
rational multiscale contrast, 261–263
top hat, 255
transformations by reconstruction, 255–256
using connected top hats, 258–260
Grid/cloud computing, 407–408
Ground control point (GCP), 750, 760
Ground-truth model, 669
Grunwald-Letnikov fraction derivative, 380
Gunner Farneback's optical flow (dense optical flow technique), 577
Gyrotropy effect, 635

H

Haar-like features, 573
Haar wavelets, 825
Handwriting character recognition, 490–491
Hardware-based OCR, 489
Hardware implementation challenges, 578–580
Harris corner detection, 569
HD, see hierarchical decomposition (HD)
HDR, see high dynamic range (HDR)
  and tone mapping, 542–543
Head-mounted display (HMD), 802
Heat Capacity Mapping Mission, 689
Heilig, Morton, 802
Helmholtz reciprocity, 426
Herschel, Frederick William, 688
Heterogeneous computing, 408
Hidden Markov model (HMM), 768–769
Hierarchical decomposition (HD), 772
High dynamic range (HDR), 550
Higher-order manipulation, 59–60
Hilbert transform, 835
Histogram adjustments, 319–321
Histogram equalization approach, 540–541
Histogram features
  transfer, 60–61
Histogram of oriented gradients (HOG), 762
Histograms, 554–555
  and moments
    description, 165–166
    histogram adjustments, 168
    image moments and moment invariants, 166–168
HMD, see head-mounted display (HMD)
HMM, see Hidden Markov model (HMM)
HOG, see histogram of oriented gradients (HOG)
Horizontal and bipolar cells, 677–678
Hosted grids/prebuilt grids, 807–808
HPC technologies
  and large-scale remote sensing (RS) image processing, 405
Huffman coding, 827
Human perception, 207–208, 245
Human vision, 672–673
Human vision perception, 260–261
Human vision system (HVS)
  in objective function formulation, 125

Human visual system (HVS)
  color processing, 681–682
  contrast, 681
  eyes
    adaptation, 675
    amacrine and ganglion cells, 678
    horizontal and bipolar cells, 677–678
    light sensitivity, 675–677
    optical media and retina, 674
    overview, 673–674
    photoreceptor mosaic, 677
    photoreceptors, 674–677
  human vision, 672–673
  lateral geniculate nucleus (LGN), 678–680
  overview, 672
  and perceptual image and video quality, 518–519
  and perceptual image enhancement
    contrast/texture masking, 533–535
    luminance masking, 535
    modeling statistical regularities of natural images, 535
    multiscale and multiorientation decomposition, 533
    spatial and temporal contrast sensitivity function, 532–533
    temporal masking and silencing, 535
  radiometric and photometric terms, 672
  temporal resolution, 681
  visual acuity, 680–681
  visual illusions, 682–684
Huygens, Christiaan, 802
HVS, see human visual system (HVS)
Hybrid image fusion, 594
Hybrid methods, 304
Hybrid objective–subjective evaluation, 99
Hyperacuity, 680

I

ICF, see intensity-curvature functional (ICF)
ICMAs, see three-dimensional intensity-curvature measurement approaches (ICMAs)
ICs, see independent components (ICs)
ICTAI, see intensity-curvature term after interpolation (ICTAI)
ICTBI, see intensity-curvature term before interpolation (ICTBI)
Ideal query weights, 226
I-frames/reference frames, 784
Image and scene analysis
  digital image
    description, 271–272
    visualization, 272–273
  distance between feature vectors, 273
  image clustering, 274–275
  image edge detection, 275
  neighborhoods of image pixels, 273–274
  overview, 270
  scene analysis, 275–276
  skeletonization, 276
  Voronoï diagram, 277–278
Image and video quality databases
  image content in, 526
  representative, 526
  test methodologies, 527–528
  video content in, 526–527

Image clustering, 274–275
Image combinations, 324
Image compression
  OnBoard compression, 602–603
  overview, 601–602
  quality evaluation
    distortion measurements, 609–610
    metrics, 609
    object-based classification, 612
    pixel-by-pixel classification, 612
    quality assessment, 611–612
    spatial pattern alterations, 610–611
  remotely sensed images
    metadata for, 606
    missing values, 606–609
    spatial dimension, 604
    spectral dimension, 604–606
    time dimension, 606
  software, 609
  standards in, 603
  user-side compression, 602
Image content, in databases, 526
Image contrast, 246–248
Image deblurring, 248–249
Image defogging, and scene visibility, 543–544
Image denoising, 837–839
Image edge detection, 275
Image enhancement
  application of, 291
  color, 290–291
  false color, true color, and pseudo color, 290
  in frequency domain, 288–290
  multispectral, 290
  overview, 281
  in spatial domain, 281–288
Image formats, and data compression, 336–337
Image fusion, 541–542
Image inpainting
  applications of, 304–305
  exemplar-based image inpainting, 300–302
  general problem, 294–295
  hybrid methods, 304
  overview, 293–294
  PDE-based diffusion
    anisotropic diffusion, 295–296
    diffusion-based image inpainting, 296–297
    total variational image inpainting, 298
  sparsity-based image inpainting, 302–304
Image interpolation
  bicubic interpolation, 310–311
  bilinear interpolation, 309–310
  historical review of, 309
  NEDI and ICBI, 311
  overview, 308–309
  SAI and NARM, 311–312
  SR
    example-based, 312
    local self-example-based upsampling, 313
    via neighborhood embedding, 312
    via sparse coding, 313
Image interpretation, 412
Image matching
  feature indexing, 188
  image similarity measurement, 188–189

Image modalities, 529
Image moments, 166–168
Image processing
    and categorization, 208–209
Image processing and measurement
    color adjustment, 316–317
    cross-correlation, 325–327
    data analysis, 332–334
    deconvolution, 324–325
    defining edges, 322–323
    feature brightness, size, and location, 331–332
    feature shape, 332
    geometric distortion, 319
    histogram adjustments, 319–321
    image combinations, 324
    morphological processing, 328–329
    noise reduction, 317–318
    non-uniform illumination, 318–319
    overview, 316
    photogrammetry, 329–330
    principal components, 323–324
    revealing texture, 323
    sharpening detail, 321–322
    stereology, 330–331
    thresholding
        automatic, 327–328
        interactive, 328
Image repositories, 346–349
    medical image databases, 347–348
    for oil industry, 348–349
    remotely sensed image databases, 348–349
    searching, 337–338
Image representation, 205
Image restoration, 416
Image retrieval
    color descriptors and color matching, 340
    color features, 340
    features, 340
    image formats and data compression, 336–337
    image repositories and digital libraries, 346–349
        medical image databases, 347–348
        for oil industry, 348–349
        remotely sensed image databases, 348–349
    metadata, 345
    overview, 335–336
    progressive pixel-level retrieval, 339–340
    progressive search, 345–346
    progressive semantic retrieval, 345
    progressive texture retrieval, 342
    query specification
        composite objects, 339
        simple objects and attributes, 338–339
    searching image repositories, 337–338
    searching images at raw data level, 339
    semantic content characterization, 343
    semantic content extraction, 343–344
    shape features, 342–343
    texture descriptors, 341–342
    texture features, 341–342
    texture similarity and texture-based retrieval, 342
Images, PCA on, 139
Image secret sharing
    based on moving lines, 360–361
    experiment and evaluation, 362–363
    improved algorithm, 361–362
    and Lagrange interpolation, 358–359
    overview, 354–355
    Shamir's secret sharing scheme, 358
    state of the art, 355–358
Image similarity measurement, 188–189
Image space consequences, 235
Image super resolution
    frequency domain SR approach, 366–368
    imaging model, 365–366
    LR image acquisition, 366
    overview, 365
    spatial domain SR approach
        interpolation-based approach, 369–370
        iterative back projection-based approach, 371
        learning-based singe-frame, 372–374
        POCS approach, 371–372
        reconstruction approach, 374–375
        stochastic approach, 370–371
Imaging model, 365–366
Immersive 3-D design, 802
Implicit query weights, 226
Imprecision/vagueness/uncertainty/inconsistency, 217–219
IMU, see inertial measurement unit (IMU)
Inconsistency/imprecision/vagueness/uncertainty, 217–219
Independent component analysis (ICA)
    description, 141–142
    on natural images, 142–144
Independent components (ICs), 716
Indicator function, 824
Indoor/outdoor classification, 658
Inertial measurement unit (IMU), 750, 761
Inertial navigation system (INS), 761
Information retrieval (IR)
    current trends in, 217
    imprecision/vagueness/uncertainty/inconsistency in, 217–219
    key issues in, 217
Information sets for edge detection
    edge strength factor
        effect of crossover on, 390–391
        effect of threshold, 391–392
        effect of parameters on FoM, 392–393
    fractional derivatives, 379
        application, 382–383
        FOM measure comparison using, 387–390
        qualitative comparison using, 386
        simulations on noisy images, 386
    gamma function
        fuzzification of fractional gradient, 381–382
        Grunwald–Letnikov fraction derivative, 380
        Sobel fractional derivative masks, 380–381
    overview, 378–379
    proposed edge detector using fractional derivative mask, 384–385
    quantitative measures of analysis
        edge strength factor, 383–384
        optimal crossover, 384
        Pratt's figure of merit, 383
        proposed algorithm using USAN, 384
        structural similarity (SSIM) index, 383
    SUSAN principle
        application, 382
        qualitative comparison using, 386
        quantitative comparison using, 386–387
        simulations on noisy images, 386
        univalue segment assimilating nucleus (USAN), 379
    use of, 382
Inner structures of, 444–445
Inpainting, 249–250
INS, see inertial navigation system (INS)
Intellectual property rights, 554
Intelligent character recognition, 490–491
Intelligent content vision, 649
Intensity-curvature functional (ICF), 698
Intensity-curvature term after interpolation (ICTAI), 698
Intensity-curvature term before interpolation (ICTBI), 698
Intensity hue saturation fusion, 591–592
Interactive thresholding, 328
Internet Protocol (IP), 783
Interpolation-based approach, 369–370
Invertible feature transformation, 480
IP, see Internet Protocol (IP)
IR, see information retrieval (IR)
Iterative back projection-based approach, 371
Ivan Sutherland, 803

J

Joss–Waldvogel disdrometer (JWD), 558
JPEG (Joint Photographic Expert Group), 548
    symmetric exponential quantization, 395–399
    YCgCb color space *vs.* YCbCr, 400–403
JWD, see Joss–Waldvogel disdrometer (JWD)

K

Karhunen–Loeve transform (KLT), 602
Key binding technique, 481
Key generation technique, 481–482
Key-scheduling, 448
Kircher, Athanasius, 802
Kirchhoff's law, 689
KLT, see Karhunen–Loeve transform (KLT)
Knowledge acquisition and modeling, 649
Knowledge-assisted content analysis, 654
Knowledge representation
    fuzzy taxonomic relations, 659–660
    notion of context, 651–653
    taxonomic context model, 660–661
    visual context
        identification, 654–655
        in image analysis, 655–659
Koniocellular layer, 678
Krueger's Videoplace, 803

L

Lagrange interpolation, 358–359
Land surface emissivity (LSE), 688
Land surface temperature (LST)
    estimation
        from multispectral TIR data, 694–695
        from single-channel algorithms, 694
        from two-channel/dual-angle techniques, 694
Land Use/Land Cover (LULC), 611
Land use/land cover mapping, 595

Lane detection, 738–739
Lanier, Jaron, 803
Laplacian of Gaussian (LoG), 570
Large-scale remote sensing (RS) image processing
  in big data, 408–409
  description, 404–405
  developments in, 405
  with grid/cloud computing, 407–408
  with heterogeneous computing, 408
  overview, 404
  with parallel cluster computing, 406–407
  using HPC technologies, 405
Lateral geniculate nucleus (LGN), 678–680
Layer assignment method, 434–436
LBP, see local binary pattern (LBP)
LCD, see liquid crystal digital (LCD)
Learning-based features, 187–188
Learning-based singe-frame, 372–374
Learning methods
  supervised and semi-supervised, 770–772
  unsupervised, 769–770
Lee's model
  applications, 586
  model, 585–586
  overview, 585
Lempel Ziv (LZ), 602
Length perception, 117
Lens distortion correction, 750
LES, see local edge segment (LES)
LF-based method, 190–191
LGN, see lateral geniculate nucleus (LGN)
LiDAR, see light detection and ranging (LiDAR)
Light detection and ranging (LiDAR), 179–180, 589
Light sensitivity, 675–677
Linear convolution *versus* circular convolution, 151–152
Linear edge detecting kernels, 752
Linear scale space
  contrast in images, 246–248
  description, 242–244
  human perception, 245
  N-Jet and feature detectors, 244–245
  scale-space statistics, 245–246
Linguistic quantifiers, 228
Linguistic query weights, 227–228
Liquid crystal digital (LCD), 803
LMT, see Lorenz–Mie theory (LMT)
Local binary pattern (LBP)
  binary robust independent elementary features, 573–574
  feature matching, 574
Local edge segment (LES), 719
Local features, 185–186
Lorenz–Mie theory (LMT), 615
Low-rank matrix completion, 670
LR image acquisition, 366
LSE, see land surface emissivity (LSE)
LST, see land surface temperature (LST)
LULC, see Land Use/Land Cover (LULC)
Luminance masking, 535
LUT utilization, 471
Lyzenga's model
  applications, 584
  example, 584–585
  model, 584
  overview, 583–584

M

MAD, see mean absolute difference (MAD)
MAE, see mean absolute error (MAE)
Magic lantern, 802
Magnocellular layer, 678
Maison Bonfils Company, 802
MAP-MRF, 415–416
Markov Random fields
  complex models and patch-based regularities
    fields of experts, 417
    products of experts, 417
  graphs
    cliques, 413–414
    neighborhood systems, 413
  image interpretation, 412
  image restoration, 416
  MAP-MRF, 415–416
  object segmentation, 416
  overview, 411–412
  probabilities and
    auto-models, 415
    Gibbs distributions, 414
  statistical analysis, 417–419
Markov random fields (MRFs), 718
Massively multiplayer online role-playing game (MMORPG), 803
Material properties, 171–172
Mathematical formula recognition, 490
Matrix form, DFT in, 149–150
Mattel Power Glove, 803
Maximum intensity projection (MIP), 704
Maximum likelihood approach, 664
Maze War, 803
MBR, see minimal bounding rectangle (MBR)
Mean absolute difference (MAD), 785
Mean absolute error (MAE), 80
Meanshift and camshift method, 577–578
Mean squared error (MSE), 785
Measurement/design matrix, 668
Median filtering, 752
Mediator systems, 649
Medical image databases, 347–348
Medicine, facial animation in, 156–157
Metadata, 345
Metamerism, 55
Metamers, 675
Metric-based models, 520–523
Metric-induced modeling, 728
Metrics, quality evaluation, 609
Microwave remote sensors, 179
Minimal bounding rectangle (MBR), 751
Minimum mean square error (MMSE), 593
Minkowski norm, 67–68
MIP, see maximum intensity projection (MIP)
Missing depth data in-painting
  depth completion results, 436–437
  depth image completion, 436
  existing methods, 433–434
  layer assignment method, 434–436
  problem definition, 431–433
MixColumns, 447–448
Mixture of Gaussians (MoG), 575
MMORPG, see massively multiplayer online role-playing game (MMORPG)
MMSE, see minimum mean square error (MMSE)
Model efficiency, 669
Modeling statistical regularities, of natural images, 535
Model selection consistency, 669
Model-selection error, 669
Moderate Resolution Imaging Spectroradiometer (MODIS), 606
Modern hardware design practices, for cryptography
  adder modeling, 469–471
  binary GCD processor, 466–467
  components of, 465–466
  delay estimates, 471–472
  enhancing performance of, 467–469
  field programmable gate arrays (FPGAs), 460–465
  LUT utilization, 471
  multiplexer modeling, 471
  overview, 460
  total LUT estimate, 471
MODIS, see Moderate Resolution Imaging Spectroradiometer (MODIS)
MODTRAN simulations, 691–693
Modulation transfer function (MTF), 593
MoG, see mixture of Gaussians (MoG)
MOHMM, see multi-observation hidden Markov model (MOHMM)
Moment invariants, 166–168
Moment statistics, and average distributions, 169–171
Montgomery's algorithm
  faster multiplication on EC, 458
  Montgomery's ladder, 457–458
  projective co-ordinates, 458–459
Montgomery's ladder, 457–458
Morphological contrast, 256
  measuring, 266–267
Morphological processing, 328–329
Morphological slope filters, 256–257
  sequential, 257
  weighted, 257–258
Motion and time
  correlation-based motion detection, 717
  gradient-based motion detection, 717
  Markov random fields (MRFs), 718
  motion blur, 718
  optical flow
    probabilistic, 720–721
    real-world motion and retinal flow, 721
    statistics of, 721–722
  overview, 714
  people detection and biological motion, 718
  shape from dynamic occlusion, 718–719
  specular highlights, 717–718
  statistics of, 714–716
Motion blur, 718
Motion detection and analysis
  action recognition, 795–797
  cameras in, 794–795
  clustering-based methods, 791–792
  description, 790–791
  filtering-based methods, 792–794

Motion detection and analysis (*Cont.*)
   object tracking, 797–798
   stationary cameras, 791–794
Motion vector, 785
Moving camera
   background removal, 736–737
   pedestrian detection, 741
   vehicle detection, 740
Moving lines, 360–361
Moxness, J. Gregory, 806
M-point properties, 151
MRFs, *see* Markov random fields (MRFs)
MSE, *see* mean squared error (MSE)
MTF, *see* modulation transfer function (MTF)
MUD, *see* Multiuser Dungeon (MUD)
Multibiometric fusion and template protection
   cryptosystems
      key binding technique, 481
      key generation technique, 481–482
   data hiding scheme, 480
   feature transformation
      invertible, 480
      non-invertible, 480
   and fusion schemes
      decision-level, 479
      feature-level, 477
      score-level, 477–479
      sensor-level, 476–477
   issues and challenges, 474–475
   open research problems, 475
   overview, 473–474
   standard encryption algorithms, 480
Multicode keying and shifted-code keying, 506–508
Multi-observation hidden Markov model (MOHMM), 771
Multiorientation decomposition, 533
Multiplexer modeling, 471
Multirate and multiple-QOS coding, 505–506
Multiresolution analysis, 823–825
Multi-resolution approaches, 594
Multi-resolution decomposition, 768
Multiscale and multiorientation decomposition, 533
Multispectral image enhancement, 290
Multiuser Dungeon (MUD), 803
Music-based OCR, 490

N
Naka-Rushton equation, 834
Natural hazards and disasters, 595
Natural images: ICA on, 142–144; repair of, 537
Natural scene statistics, 519–520
NCD, *see* normalized color distance (NCD)
NCDD, *see* normalized current density distribution (NCDD)
Near sets
   basic approach, 47–48
   in image analysis, 49–50
   psychophysics and Merleau-Ponty, 48–49
   and rough sets, 47
   of similar images, 49
   tolerance, 49
   visual acuity tolerance, 49
NEDI and ICBI, 311
Negative index metamaterials (NIMs), 628

Neighborhood embedding, 312
Neighborhood systems, 413
Network tomography, 666
Neuroimaging analysis, 666–667
NIMs, *see* negative index metamaterials (NIMs)
N-Jet and feature detectors, 244–245
NODATA regions, 607
Noise reduction, 317–318
Non-invertible feature transformation, 480
Nonlinear compression, 172–173
Nonlinear unsharp masking, for mammogram enhancement, 544–545
Non-uniform illumination, 318–319
No-reference (NR) visual quality assessment, 524–525
Normalization, 63
Normalized color distance (NCD), 80–92
Normalized current density distribution (NCDD), 624
North American Space Association (NASA), 589
Nuclear norm, 670
Numerical fusion techniques, 593

O
OAR, *see* OpenSimulator archive (OAR)
OBIA, *see* object-based image analysis (OBIA)
Object-based classification, 612
Object-based image analysis (OBIA), 762
Object detection
   applications of, 566–567
   challenges in, 567–568
   contour- or silhouette-based, 574
   feature-based
      features from accelerated segment test (FAST), 569–570
      Haar-like features, 573
      Harris corner detection, 569
      scale-invariant feature transform, 570–571
      speeded-up robust features (SURF), 571–573
   hardware implementation challenges, 578–580
   local binary pattern (LBP)
      binary robust independent elementary features, 573–574
      feature matching, 574
   overview, 566
Object detection/recognition, 655, 658
Objective evaluation, 98
Objective function formulation
   cost function, 125–126
   human vision system in, 125
Object segmentation, 416
Object tracking, 797–798
   absolute difference method, 575
   applications of, 566–567
   background subtractions, 575–576
   challenges in, 567–568
   dense optical flow technique (Gunner Farneback's optical flow), 577
   hardware implementation challenges, 578–580
   meanshift and camshift method, 577–578

   optical flow technique, 576–577
   overview, 574–575
Occam's razor/Ockham's razor, 664
Oculus Rift HMD, 803
Oil industry, image repositories for, 348–349
OnBoard compression, 602–603
1-D spectral amplitude coding, 500–503
1-D spectral phase coding, 499–500
1-D temporal phase coding, 499
1/f failures, 206–207
Open research problems, 475
OpenSim, 804
OpenSimulator archive (OAR), 807
Opponent processing, 55–58
Optical and microwave remote sensors, 179
Optical character recognition (OCR)
   detection, 487–488
   document quality, 491–492
   foreign language, 492–493
   handwriting and intelligent character recognition, 490–491
   hardware-based, 489
   history of, 486–487
   output, 488
   overview, 486
   recognition, 488
   software-based
      mathematical formula recognition, 490
      music, 490
Optical coding
   applications, 494–495
   arrayed waveguide gratings (AWGs), 513–515
   Fiber Bragg gratings (FBGs), 512–513
   multicode keying and shifted-code keying, 506–508
   multirate and multiple-QOS coding, 505–506
   1-D spectral amplitude coding, 500–503
   1-D spectral phase coding, 499–500
   1-D temporal phase coding, 499
   overview, 496–499
   three-dimensional coding, 504–505
   2-D spatial-temporal amplitude coding, 503–504
   2-D spectral-temporal amplitude coding, 504
   wavelength-aware hard-limiting detector, 512
Optical flow
   probabilistic, 720–721
   real-world motion and retinal flow, 721
   statistics of, 721–722
Optical flow technique, 576–577
Optical media and retina, 674
Optical remote sensing, 589–590
Optic nerve bundle, 674
Optimal crossover, 384
Ordinal measure method, 189
Orientation and angle perception, 117
Overlapping action, 162

P
Pansharpening, 588
Parallel cluster computing, 406–407
Parallel polarization (TM), 619–621
Parameter estimation consistency, 669
Parameter estimation error, 669
Parvocellular layer, 678

Patches, PCA on, 138–139
Pattern recognition, 737–738
PCA, *see* principal component analysis (PCA)
PCS, *see* principal component substitution (PCS)
PDE-based diffusion
 anisotropic diffusion, 295–296
 diffusion-based image inpainting, 296–297
 total variational image inpainting, 298
Peak signal-to-noise ratio (PSNR), 80, 785
Pedestrian detection, 741
People detection and biological motion, 718
Pepper, John Henry, 802
Pepper's ghost, 802
Perceptual consequences, 234
Perceptual image and video quality
 databases
  image content in, 526
  representative, 526
  test methodologies, 527–528
  video content in, 526–527
 and human visual system, 518–519
 versus metric-based models, 520–523
 and natural scene statistics, 519–520
 overview, 517
 perceptually optimized image acquisition, 529
 visual quality assessment, 518
  bitstream-based, 528
  color quality assessment, 528
  in different image modalities, 529
  full-reference, 523–524
  no-reference (NR), 524–525
  reduced reference (RR), 524
  stereoscopic (3D image and video), 525–526
  temporal visual masking, 525
  time-varying, 529
Perceptual image enhancement
 adaptive perceptual image and video compression, 545–546
 contrast enhancement
  color restoration, 539
  histogram equalization approach using GMM, 540–541
  image fusion, 541–542
  tone mapping and HDR, 542–543
 deblurring, 536–537
 denoising, 536
 HVS
  contrast/texture masking, 533–535
  luminance masking, 535
  modeling statistical regularities of natural images, 535
  multiscale and multiorientation decomposition, 533
  spatial and temporal contrast sensitivity function, 532–533
  temporal masking and silencing, 535
 image defogging and scene visibility, 543–544
 nonlinear unsharp masking for mammogram enhancement, 544–545
 overview, 531–532
 repair of natural images, 537
 robust face recognition, 543
Perceptually optimized image acquisition, 529

Perpendicular polarization (TE), 621
PET, *see* positron emission tomography (PET)
Phantasmagoria, 802
Phase spectra, 207
Photogrammetry, 329–330
Photographic image characteristics
 CIE Lab systems, 552–553
 color distance, 552–553
 Commission Internationale de l'Eclairage (CIE) Luv, 552–553
 digital photographic image similarity index, 554
 dynamic range, 552
 forensics and intellectual property rights, 554
 histograms, 554–555
 overview, 548–552
 resolution and sharpness assessment, 553
 visual technical assessment, 553–554
Photopic conditions, 674
Photopic luminous efficiency, 674
Photorealism, 802
Photoreceptor mosaic, 677
Photoreceptors, 674–677
Pixel-by-pixel classification, 612
Pixels, PCA on, 138
P-LUT implementation, 127–129
POCS approach, 371–372
Point spread function (PSF), 680
Polynomial fitting, 768
Positron emission tomography (PET), 708
Post-traumatic stress disorders (PTSDs), 807
Power spectra
 angular dependence, 205–206
 dynamic range, 203–205
 1/f failures, 206–207
 image representation, 205
 slope computation, 199–201
 spectral slope analysis, 201–203
 temporal dependence, 206
Pratt's merit figure, 383
Prebuilt grids/hosted grids, 807–808
Prediction error, 669
Principal component analysis (PCA), 670, 769
 description, 135–137
 eigenfaces, 139–141
 on images, 139
 on patches, 138–139
 on pixels, 138
 whitening, 137–138
Principal components, 323–324
Principal component substitution (PCS), 592
Probabilistic optical flow, 720–721
Probabilities, Markov Random fields and
 auto-models, 415
 Gibbs distributions, 414
Products of experts, 417
Progressive pixel-level retrieval, 339–340
Progressive reconstruction, 839
Progressive search, 345–346
Progressive semantic retrieval, 345
Progressive texture retrieval, 342
Projected image, 802
Projective coordinate representation, 456
Projective co-ordinates, 458–459
Pseudo color, 290
PSF, *see* point spread function (PSF)

PSNR, *see* peak signal-to-noise ratio (PSNR)
PTSDs, *see* post-traumatic stress disorders (PTSDs)

Q

QNR, *see* Quality with No Reference (QNR)
Quadrature, 835
Qualitative comparison, 386
Quality assessment, 611–612
 remote sensing image fusion, 596–598
Quality evaluation
 distortion measurements, 609–610
 metrics, 609
 object-based classification, 612
 pixel-by-pixel classification, 612
 quality assessment, 611–612
 spatial pattern alterations, 610–611
Quality with No Reference (QNR), 597
Quantitative analysis measures
 edge strength factor, 383–384
 optimal crossover, 384
 Pratt's figure of merit, 383
 proposed algorithm using USAN, 384
 structural similarity (SSIM) index, 383
Quantitative comparison, 386–387
Quantitative metrics
 distance-based metrics, 103–104
 region/volume based, 99–103
Quaternions
 application of, 76–78
 common properties of, 73–75
 representation of color differences, 75–76
 representation of color pixel, 75
Query languages, 220
Query specification
 composite objects, 339
 simple objects and attributes, 338–339
Query weights
 evaluation mechanism, 224–225
 ideal, 226
 implicit, 226
 linguistic, 227–228
 quantifiers, 228
 query languages, 220
 relative importance, 226
 semantics, 226–227
 threshold, 226

R

Radar cross section (RCS), 615
Radiative transfer codes and spectral libraries, 690–691
Radiative transfer equation (RTE), 689–690
Radiometric and photometric terms, 672
Radiometric magnitudes, 689
RADIX-2 algorithm for FFT, 152–153
Raindrop imaging
 drop holes
  description, 559–561
  field comparisons, 561–564
 overview, 557–559
RANSAC-based method, 190
Rate-distortion optimization, 785–786
Rational multiscale contrast, 261–263
RBMs, *see* restricted Boltzmann machines (RBMs)
RCS, *see* radar cross section (RCS)

Real-time data collection, 763–764
Real-time event tracking and monitoring, 763
Real-time map generation, 763–764
Real-valued wavelets, 834
Real-world motion and retinal flow, 721
Reconstruction approach, 374–375
Reduced reference (RR) visual quality assessment, 524
Reference frames/I-frames, 784
Reflectance modeling
   Lee's model
      applications, 586
      model, 585–586
      overview, 585
   Lyzenga's model
      applications, 584
      example, 584–585
      model, 584
      overview, 583–584
   overview, 583
   propagation of visible light in shallow water, 583
Refraction, visual cryptography using, 816–818
Region-based segmentation, 655
Region/volume based quantitative metrics, 99–103
Reichardt, Werner, 717
Reichardt detector, 715
Relative emissivity, 694
Remotely sensed digital elevation model, 179
Remotely sensed image compression
   metadata for, 606
   missing values, 606–609
   spatial dimension, 604
   spectral dimension, 604–606
   time dimension, 606
Remotely sensed image databases, 348–349
Remote sensing image fusion
   active remote sensing, 590
   and agriculture, 595
   change detection, 595
   component substitution
      Gram-Schmidt fusion, 592–593
      intensity hue saturation fusion, 591–592
      principal component substitution, 592
   description, 590–591
   geology, 595
   hybrid image fusion, 594
   issues in, 596
   land use/land cover mapping, 595
   multi-resolution approaches, 594
   and natural hazards and disasters, 595
   numerical fusion techniques, 593
   optical remote sensing, 589–590
   overview, 588–589
   quality assessment, 596–598
   spatiotemporal fusion algorithms, 594–595
   statistical fusion methods, 593–594
   urban applications, 595
Remote sensing image processing, 748
Rendering algorithm, 422–425
Rendering equation, 429–430
Replicated Softmax (RS) model, 771
Representation
   of color differences, 75–76
   of color pixel, 75

Representative databases, 526
Resolution, and sharpness assessment, 553
Restricted Boltzmann machines (RBMs), 771
Restricted isometry property (RIP), 669
Ridgelets, 831
Rijndael in composite field
   expressing an element, 448–449
   inversion of element, 449
   round of AES in, 449–450
RIP, *see* restricted isometry property (RIP)
Riverine floodplains, 178
River planform changes (RPCs), 729
Robert, Étienne-Gaspard, 802
Robust face recognition, 543
Rough sets, 50–51
Routing matrix, 666
RPCs, *see* river planform changes (RPCs)
RS, *see* Replicated Softmax (RS) model
RTE, *see* radiative transfer equation (RTE)
Run-length encoding (RLE), 602

S
SAI and NARM, 311–312
SAR, *see* synthetic aperture radar (SAR)
Sarcos Arms, 803
Sayre Data Glove, 803
SBF, *see* Spatiotemporal Boundary Formation (SBF)
Scalar multiplications, 456–457
Scale invariance, 832–833
Scale-invariant feature transform (SIFT), 570–571, 756
Scale-space statistics, 245–246
Scattering features of complex mediums
   applications in invisibility, 618–619
   case of chiral cylinder, 616–617
   chiral/chiro-ferrite-coated twisted PEC cylinder
      backscattering, 635–636
      chiral admittance, 634–635
      description, 632–634
      gyrotropy effect, 635
   chiral-coated twisted cylindrical mediums
      circular dielectric cylinder, 630–631
      elliptical dielectric cylinder, 628–629
   chiral nihility medium-based cylinder, 617–618
   overview, 615–616
   sheath helix and l-nihilitybased silver metal cylinder
      current density distributions, 624–628
      parallel polarization (TM), 619–621
      perpendicular polarization (TE), 621
      structure analysis, 621–624
   uniaxial anisotropic chiral mediums (UACMs), 636–640
Scene classification, 655, 657
Scene context, 657
Scene geometry perception
   description, 116
   length perception, 117
   orientation and angle perception, 117
Scene visibility, 543–544
Score-level fusion, 477–479
SDIs, *see* spatial data infrastructures (SDIs)
Secondary motion, 162
Second derivative methods, 237–238

Second Life, 804
Security *versus* multiple use, of visual cryptography, 813–816
Self-driving automobile, 744–745
Self-organizing map (SOM), 768
Semantic annotation interpretation, 661
Semantic content characterization, 343
Semantic content extraction, 343–344
Semantic gap, 649
Semantic index/indexing, 651, 661
Semantic processing
   architecture, 650
   data models, 650–651
   knowledge model, 650–651
   knowledge representation
      fuzzy taxonomic relations, 659–660
      notion of context, 651–653
      taxonomic context model, 660–661
      visual context identification, 654–655
      visual context in image analysis, 655–659
   overview, 649–650
   semantic annotation interpretation, 661
   semantic index/indexing, 651, 661
Semantics
   query weights, 226–227
Semi-supervised learning methods, 770–772
Sensor-level fusion, 476–477
Sensory gaps, 652
Sequential forward feature selection (SFFS), 772
Sequential morphological slope filters, 257
Set partitioning in hierarchical trees (SPIHT), 602
SFFS, *see* sequential forward feature selection (SFFS)
Shamir's secret sharing scheme, 358
Shape features, 342–343
Sharpening detail, 321–322
Sharpness assessment, and resolution, 553
Sheath helix and l-nihilitybased silver metal cylinder
   current density distributions, 624–628
   parallel polarization (TM), 619–621
   perpendicular polarization (TE), 621
   structure analysis, 621–624
ShiftRows, 447
Short-time Fourier transform (STFT), 822
Short-wave infrared (SWIR), 589
SIFT, *see* scale-invariant feature transform (SIFT)
Sigmoid, 677
Signal processing, 825–828
Signal-to-noise ratio (SNR), 607
Silver Alert, 788
SimHost, 805
Sim on a Stick (SoaS), 808
Simple objects and attributes, 338–339
Simultaneous-binocular viewing technique, 553
Simultaneous-haploscopic viewing technique, 554
Single-image gradient statistics, 238–239
Skeletonization, 276
Slope and aspect, 180
Slope computation, 199–201
SNR, *see* signal-to-noise ratio (SNR)
SoaS, *see* Sim on a Stick (SoaS)

Sobel fractional derivative masks, 380–381
Söbel operator, 236–237
Social agents, 157–158
Social robots, 158–159
Software, for image compression, 609
Software-based OCR
  mathematical formula recognition, 490
  music, 490
SOM, see self-organizing map (SOM)
Space statistics
  constancy and white balancing, 65–67
  correlation analysis, 63–65
  Minkowski norm, 67–68
  normalization, 63
  white-balance algorithm selection, 68
Sparse Bayesian learning, 670
Sparse coding, 313
Sparse modeling
  compressed sensing, 667–668
  computer network diagnosis, 665–666
  neuroimaging analysis, 666–667
  overview, 664–665
  recovery, 668–669
  statistical learning versus compressed sensing, 669–670
Sparse optimization methods, 668
Sparse regression, 667
Sparsistency, 669
Sparsity-based image inpainting, 302–304
Spatial and temporal adaptive reflectance fusion model (STARFM), 594
Spatial context, 656
Spatial contrast sensitivity function, 532–533
Spatial data infrastructures (SDIs), 603
Spatial dimension, 604
Spatial domain, image enhancement in, 281–288
Spatial domain SR approach
  interpolation-based approach, 369–370
  iterative back projection-based approach, 371
  learning-based singe-frame, 372–374
  POCS approach, 371–372
  reconstruction approach, 374–375
  stochastic approach, 370–371
Spatial pattern alterations, 610–611
Spatiotemporal Boundary Formation (SBF), 719
Spatiotemporal fusion algorithms, 594–595
Spectral dimension, 604–606
Spectral slope analysis, 201–203
Specular highlights, 717–718
Speeded-up robust features (SURF), 571–573
SPIHT, see Set partitioning in hierarchical trees (SPIHT)
Sproull, Bob, 803
Squash and stretch, 161
SR image interpolation
  example-based, 312
  local self-example-based upsampling, 313
  via neighborhood embedding, 312
  via sparse coding, 313
SSIM, see structural similarity (SSIM) index
Staging, 161
Standard encryption algorithms, 480
STARFM, see spatial and temporal adaptive reflectance fusion model (STARFM)

State of the art image secret sharing, 355–358
Stationary camera
  background removal, 736
  pedestrian detection, 741
  vehicle detection, 740
Stationary cameras, 791–794
Statistical analysis, 417–419
Statistical fusion methods, 593–594
Statistical learning versus compressed sensing, 669–670
Stereology, 330–331
Stereoscopic/3D image and video quality assessment, 525–526
STFT, see short-time Fourier transform (STFT)
Stochastic approach, 370–371
Street intersection monitoring, 741–742
Structural similarity (SSIM) index, 80, 383
SubBytes, 446–447
Subjective evaluation, 98
Subspace methods, 769
Successive binocular viewing technique, 553
Successive-Ganzfeld-haploscopic viewing technique, 554
Superresolution, 249
Supervised learning methods, 770–772
Support recovery, 669
Support vector machine (SVM), 575, 770
Support vector regression (SVR), 575
SURF, see speeded-up robust features (SURF)
SUSAN principle
  application, 382
  qualitative comparison using, 386
  quantitative comparison using, 386–387
  simulations on noisy images, 386
SVM, see support vector machine (SVM)
SVR, see support vector regression (SVR)
SWIR, see short-wave infrared (SWIR)
Switching vector median filter, 78–79
"Sword of Damocles," 803
Symmetric exponential quantization, 395–399
Synthetic aperture radar (SAR), 588

T
TAN, see transaction number (TAN)
Task-oriented video compression, 786–790
Taxonomic context model, 660–661
Temperature and emissivity separation (TES), 694
Temporal context, 656
Temporal contrast sensitivity function, 532–533
Temporal dependence, 206
Temporal masking and silencing, 535
Temporal resolution, 681
Temporal visual masking, 525
Terrain synthesis, 210
TES, see temperature and emissivity separation (TES)
Texture-based retrieval, 342
Texture descriptors, 209–210, 341–342
Texture features, 341–342
Texture similarity, 342
Texture synthesis, 839–840
Thermal infrared remote sensing
  absolute emissivity, 694
  description, 688–689

  emissivity and Kirchhoff's law, 689
  laws of radiation, 689
  LST estimation
    from multispectral TIR data, 694–695
    from single-channel algorithms, 694
    from two-channel/dual-angle techniques, 694
  MODTRAN simulations, 691–693
  overview, 688
  radiative transfer codes and spectral libraries, 690–691
  radiative transfer equation (RTE), 689–690
  radiometric magnitudes, 689
  relative emissivity, 694
3D data handling, 728–729
3-D graphics
  bidirectional reflectance distribution function (BRDF), 425–426
  colorimetry and radiometry, 426–427
  Helmholtz reciprocity, 426
  overview, 421
  rendering algorithm, 422–425
  rendering equation, 429–430
Three-dimensional coding, 504–505
Three-dimensional intensity-curvature measurement approaches (ICMAs)
  of cubic trivariate lagrange polynomial, 698–703
  description, 697–698
  of MR images, 704–708
  MRI of human brain, 708–710
  overview, 697
  preliminary study, 704
  results, 704
3D problem, color transfer as, 61–62
3D space, color as, 55
Threshold
  query weights, 226
Threshold effect, 391–392
Thresholding
  automatic, 327–328
  interactive, 328
TIFF (Tagged Image File Format), 548
Time and motion
  correlation-based motion detection, 717
  gradient-based motion detection, 717
  Markov random fields (MRFs), 718
  motion blur, 718
  optical flow
    probabilistic, 720–721
    real-world motion and retinal flow, 721
    statistics of, 721–722
  overview, 714
  people detection and biological motion, 718
  shape from dynamic occlusion, 718–719
  specular highlights, 717–718
  statistics of, 714–716
Time dimension, 606
Time repetitions (TRs), 666
Time-varying visual quality assessment, 529
TinyMUD, 803
Tolerance near sets, 49
Tone mapping and HDR, 542–543
Topology, in GIS
  description, 727
  metric-induced modelling, 728
  overview, 726–727

Topology, in GIS (*Cont.*)
    qualitative description, 728
    recent advances, 729
    spatial relationships and topological maps, 727
    3D data handling, 728–729
Total LUT estimation, 471
Total variational image inpainting, 298
Trace norm, 670
Traffic analysis
    accident detection, 743–744
    collision avoidance, 744
    congestion detection, 742–743
    driver assistance systems, 744
    driver fatigue detection, 744
    environment modelling
        lane detection, 738–739
        traffic sign detection and recognition, 739
    feature extraction, 737
    moving camera
        background removal, 736–737
        pedestrian detection, 741
        vehicle detection, 740
    overview, 735–736
    pattern recognition, 737–738
    self-driving automobile, 744–745
    stationary camera
        background removal, 736
        pedestrian detection, 741
        vehicle detection, 740
    street intersection monitoring, 741–742
Traffic sign detection and recognition, 739
Traffic surveillance tracking, 789–790
Transaction number (TAN), 810
Transfer, color
    histogram features, 60–61
    as 3D problem, 61–62
    through higher-order manipulation, 59–60
    through simple moments, 58–59
Trichromacy and metamerism, 55
Trojan-secure authentication, 812–813
Trojan-secure confirmation, 811–812
Trompe l'oeil, 802
TRs, *see* time repetitions (TRs)
True color, 290
True negative rate (TNrate), 776
True positive rate (TPrate), 776
Turing test, 159–160
2-D spatial-temporal amplitude coding, 503–504
2-D spectral-temporal amplitude coding, 504

U

UACMs, *see* uniaxial anisotropic chiral mediums (UACMs)
UGC, *see* user-generated content (UGC)
UIQI, *see* universal image quality index (UIQI)
Uncanny valley, 159
Uncertainty/inconsistency/imprecision/vagueness, 217–219
Uniaxial anisotropic chiral mediums (UACMs), 636–640
United States Geological Survey (USGS), 589
Univalue segment assimilating nucleus (USAN), 379
Universal image quality index (UIQI), 597
Universal serial bus (USB), 809
Unmanned aerial systems (UASs)
    CUDA architecture and programming model, 748–750
    GPU-based approach
        convolution, 752–753
        co-registration, 756–757
        geo-referencing, 750
        gray-level co-occurrence matrix (GLCM) textures, 753–756
        implementation, 751–752
        lens distortion correction, 750
    overview, 747–748
    remote sensing image processing, 748
Unmanned aerial vehicles (UAVs) imaging
    damage detection, 762
    for disaster management
        damage assessment, 764
        real-time data collection, 763–764
        real-time event tracking and monitoring, 763
        real-time map generation, 763–764
    image rectification, 761
    object detection, 761–762
    overview, 760–761
Unsupervised learning methods, 769–770
Unusual trajectory detection
    datasets
        fish trajectory, 774–775
        forum pedestrian, 775–776
    description, 774
    experimental work, 776–779
    learning methods
        supervised and semi-supervised, 770–772
        unsupervised, 769–770
    methodology, 772–774
    overview, 767–768
    trajectory representation methods, 768–769
Urban floodplain mapping, 178
USAN, *see* univalue segment assimilating nucleus (USAN)
USB, *see* universal serial bus (USB)
User-generated content (UGC), 800
User-side compression, 602
USGS, *see* United States Geological Survey (USGS)

V

Vagueness/uncertainty/inconsistency/imprecision, 217–219
Vanishing moments, 828
Variable selection, 667
VBM, *see* voxel-based morphometry (VBM)
Vehicle detection, 740
Very high spatial resolution (VHR), 760
VHR, *see* very high spatial resolution (VHR)
Video analytics
    motion detection and analysis
        action recognition, 795–797
        cameras in, 794–795
        clustering-based methods, 791–792
        description, 790–791
        filtering-based methods, 792–794
        object tracking, 797–798
        stationary cameras, 791–794
    overview, 783–784
    video compression
        description, 784–785
        rate-distortion optimization, 785–786
        task-oriented, 786–790
        for traffic surveillance tracking, 789–790
        video search- and retrieval-oriented, 788–789
        visual attention-based, 787
Video compression
    and adaptive perceptual image, 545–546
    description, 784–785
    rate-distortion optimization, 785–786
    task-oriented, 786–790
    for traffic surveillance tracking, 789–790
    video search- and retrieval-oriented, 788–789
    visual attention-based, 787
Video content, in databases, 526–527
Video search- and retrieval-oriented video compression, 788–789
Video teleconferencing, 157
Virtual worlds
    applications of, 806–807
    cinema, electronic screens, and immersive 3-D design, 802
    components of, 803–806
    computer-created 3-D space, 802–803
    designer's point of view, 807
    gaming and, 803
    overview, 800
    prebuilt grids/hosted grids, 807–808
    trompe l'oeil, photorealism, and projected image, 802
    visual theory and creation of first illusions, 800–802
Visible light propagation, in shallow water, 583
Visual acuity, 680–681
Visual acuity tolerance, 49
Visual attention-based video compression, 787
Visual context
    identification, 654–655
    in image analysis, 655–659
Visual cryptography
    overview, 810–811
    security *versus* multiple use, 813–816
    trojan-secure authentication, 812–813
    trojan-secure confirmation, 811–812
    using refraction, 816–818
    voting with receipt based on, 818–820
Visual illusions, 682–684
Visualization
    digital image, 272–273
Visual quality assessment, 518
    bitstream-based, 528
    color quality assessment, 528
    in different image modalities, 529
    full-reference, 523–524
    no-reference (NR), 524–525
    reduced reference (RR), 524
    stereoscopic (3D image and video), 525–526
    temporal visual masking, 525
    time-varying, 529
Visual technical assessment, 553–554
Voronoï diagram, 277–278
Voxel-based morphometry (VBM), 710

Index

W

Wavelength-aware hard-limiting detector, 512
Wavelet analysis
   coefficient histograms, 831–832
   complex wavelets, 834–836
   contourlets, curvelets, and ridgelets, 831
   correlations
      between coefficients, 833–834
      between scales, 836–837
   2D wavelets, 829–831
   image denoising, 837–839
   multiresolution analysis, 823–825
   other bases, 828–829
   overview, 822
   progressive reconstruction, 839
   scale invariance, 832–833
   signal processing, 825–828
   texture synthesis, 839–840
   wavelet transform, 822–823
Wavelet transform, 822–823
Weierstraß equation, 451–452
Weighted morphological slope filters, 257–258
White-balance algorithm selection, 68
Whitening, 137–138
Wiener-Khintchine theorem, 199
World Wide Web, 802

Y

YCgCb color space *vs.* YCbCr, 400–403

Z

Zero padding, 151
Zimmerman, Thomas, 803